616.8   SCH

DATE DUE			

NEUROPSYCHIATRY

SECOND EDITION

NEUROPSYCHIATRY

SECOND EDITION

Editors

RANDOLPH B. SCHIFFER, M.D.

The Haggarton Chair and Professor
Neuropsychiatry and Behavioral Science
Texas Tech University Health Sciences Center
Lubbock, Texas

STEPHEN M. RAO, PH.D.

Director, Functional Imaging Research Center
Professor of Neurology
Medical College of Wisconsin
Milwaukee, Wisconsin

BARRY S. FOGEL, M.D.

Clinical Professor
Brigham Behavioral Neurology Group
Harvard Medical School
Boston, Massachusetts

LIPPINCOTT WILLIAMS & WILKINS
A **Wolters Kluwer** Company

Philadelphia · Baltimore · New York · London
Buenos Aires · Hong Kong · Sydney · Tokyo

Acquisitions Editor: Charles W. Mitchell
Developmental Editor: Raymond E. Reter
Production Editor: Frank Aversa
Manufacturing Manager: Ben Rivera
Cover Designer: Christine Jenny
Compositor: TechBooks
Printer: Maple Press

Library of Congress Cataloging-in-Publication Data

Neuropsychiatry /editors, Randolph B. Schiffer, Stephen M. Rao, Barry S. Fogel.—2nd ed.
    p. ; cm.
  Includes bibliographical references and index.
  ISBN 0-7817-2655-7
  1. Neuropsychiatry.   I. Schiffer, Randolph B.   II. Rao, Stephen M.
III. Fogel, Barry S.
  [DNLM: 1. Neuropsychology.   2. Neurophysiology.   WL 103.5 N493 2003]
RC341 .N4354   2003
616.8—dc21

                                       2002043075

CONTENTS

CONTRIBUTING AUTHORS

Kurt Ackerman, M.D., Ph.D. Assistant Professor, Department of Psychiatry, Western Psychiatric Institute and Clinic, Pittsburgh, Pennsylvania

Michael J. Aminoff, M.D., D.Sc., F.R.C.P. Professor, Department of Neurology, University of California, School of Medicine; Attending Physician, Department of Neurology, University of California Medical Center, San Francisco, California

Luis R. Arce, M.D. Clinical Fellow, Department of Endocrinology and Metabolism, Brown Medical School, Rhode Island Hospital, Providence, Rhode Island

Catherine M. Arrington, Ph.D. Postdoctoral Fellow, Department of Psychology, Vanderbilt University, Nashville, Tennessee

Deanna M. Barch, Ph.D. Assistant Professor, Departments of Psychology and Psychiatry, Washington University, St. Louis, Missouri

Denise L. Bellinger, Ph.D. Associate Professor, Department of Pathology and Human Anatomy, Loma Linda University, School of Medicine, Loma Linda, California

Nicholas M. Bogod, M.Sc. Doctoral Candidate, Department of Psychology, University of Victoria, Victoria, British Columbia, Canada

Omer Bonne, M.D. Mood and Anxiety Disorder Program, National Institute of Mental Health, Bethesda, Maryland

Kara M. Brooklier, M.A. Graduate Student, Department of Psychology, Wayne State University, Detroit, Michigan

Randy L. Buckner, Ph.D. Howard Hughes Medical Institute; Departments of Radiology, Anatomy, Neurobiology, and Psychology, Washington University, St. Louis, Missouri

Amer Burhan, M.D. University of Rochester Medical Center, Rochester, New York

David Caplan, M.D., Ph.D. Professor, Department of Neurology, Harvard Medical School; Neurologist, Department of Neurology, Massachusetts General Hospital, Boston, Massachusetts

Thomas H. Carr, M.D. Professor, Department of Psychology, Michigan State University, East Lansing, Michigan

Dennis S. Charney, M.D. Mood and Anxiety Disorders Program, National Institute of Mental Health, National Institutes of Health, Bethesda, Maryland

David W. Collins, M.A. Department of Psychology, University of Windsor, Windsor, Ontario, Canada

Peggy Compton, R.N., Ph.D. Assistant Clinical Professor of Acute Care Nursing, University of California, Los Angeles; Research Scientist, Los Angeles Addiction Treatment Research Center, Los Angeles, California

Yeates Conwell, M.D. Professor, Department of Psychiatry, University of Rochester School of Medicine, Rochester, New York

Jody Corey-Bloom, M.D., Ph.D. Professor, Department of Neurosciences, University of California, San Diego; Professor, Department of Neurosciences, University of California, San Diego Medical Center, San Diego, California

Richard J. Davidson, Ph.D. William James and Vilas Research Professor, Department of Psychology and Psychiatry, University of Wisconsin-Madison, Madison, Wisconsin

Mark D'Esposito, M.D. Professor of Neuroscience and Psychology, Director, Henry H. Wheeler Brain Imaging Center, University of California, Berkeley, Berkeley, California

Maria L. De León, M.D. De León Neurological Clinic, Nacogdoches, Texas

Beatriz M. DeMoranville, M.D. Staff Endocrinologist, Department of Medicine, South County Hospital, Wakefield, Rhode Island

Christian R. Dolder, Pharm.D. Assistant Clinical Professor, Department of Psychiatry, University of California, San Diego, La Jolla, California; Clinical Pharmacist, Veterans Affairs San Diego Healthcare System, San Diego, California

Valsamma Eapen, M.D.   Associate Professor, Department of Psychiatry, United Arab Emirates University, Al Ain, United Arab Emirates; Honorary Lecturer, Department of Psychiatry, University of London Medical School, London, United Kingdom

Deborah Fein, Ph.D.   Professor, Department of Psychology, University of Connecticut, Storrs, Connecticut

Robert G. Feldman, M.D.   Chairman Emeritus, Professor of Neurology, Pharmacology, and Public Health, Boston University; Boston Medical Center, Boston, Massachusetts

Francisco Fernandez, M.D.   Professor and Chairperson, Department of Psychiatry and Behavioral Medicine, University of South Florida, Tampa, Florida

Howard L. Fields, M.D., Ph.D.   Professor, Departments of Neurology, Physiology, and Psychiatry, University of California, San Francisco, San Francisco, California

Christopher M. Filley, M.D.   Professor of Neurology and Psychiatry, Behavioral Neurology Section, University of Colorado School of Medicine; Attending Physician, Department of Neurology, Denver Veterans Affairs Medical Center, Denver, Colorado

Mariellen Fischer, Ph.D.   Department of Neurology, Medical College of Wisconsin, Milwaukee, Wisconsin

David A. Fishbain, M.D., F.A.P.A.   Professor, Department of Neurological Surgery and Anesthesiology, University of Miami School of Medicine; Psychiatrist, University of Miami Pain Center, South Shore Hospital, Miami Beach, Florida

Barry S. Fogel, M.D.   Clinical Professor of Psychiatry, Brigham Behavioral Neurology Group, Harvard Medical School, Boston, Massachusetts

Norman L. Foster, M.D.   Professor, Department of Neurology, Senior Research Scientist, Institute of Gerontology, University of Michigan, Ann Arbor, Michigan

Thomas A. Gennarelli, M.D.   Department of Neurosurgery, Medical College of Wisconsin, Milwaukee, Wisconsin

Steven J. Grant, Ph.D.   Cognitive Neuroscience of Addiction Program, Clinical Neurobiology Unit, Division of Treatment Research and Development, National Institute on Drug Abuse, National Institutes of Health, Bethesda, Maryland

LeeAnne Green, Ph.D.   Adjunct Assistant Professor, Department of Psychiatry and Behavioral Neuroscience, Wayne State University Medical School; Director, Autism Assessment Clinic, Department of Psychiatry/Psychology, Children's Hospital of Michigan, Detroit, Michigan

Thomas A. Hammeke, Ph.D.   Professor, Department of Neurology (Neuropsychology), Medical College of Wisconsin; Co-Director, Department of Neuropsychology, Froedtert Memorial Lutheran Hospital, Milwaukee, Wisconsin

Allan T. Hanretta, M.D., Ph.D.   Assistant Professor, Department of Neuropsychiatry, Texas Tech University Health Sciences Center; Medical Director, Behavioral Health Services, Covenant Health System, Lubbock, Texas

Brent A. Hayman-Abello, M.Sc.   Department of Psychology, University of Windsor, Windsor, Ontario, Canada

Nkangineme Ike, M.D.   Department of Psychiatry, Metrohealth Medical Center, Case Western Reserve University, Cleveland, Ohio; Nova Behavioral Health, Inc., Non-profit Cooperation, Canton, Ohio

Sharon S. Ishikawa, Ph.D.   Department of Psychology, University of Southern California, Los Angeles, California

Ivor M. D. Jackson, M.D.   Professor, Department of Medicine, Division of Endocrinology, Brown University; Physician, Department of Medicine, Division of Endocrinology, Rhode Island Hospital, Providence, Rhode Island

Joseph Jankovic, M.D.   Professor, Department of Neurology, Director, Parkinson's Disease Center and Movement Disorders Clinic, Baylor College of Medicine, Houston, Texas

Terry L. Jernigan, Ph.D.   Clinical Research Psychologist, Associate Chief, Department of Psychology Service, San Diego Veterans Administration Healthcare System; Full Professor, Department of Psychiatry and Radiology, University of California, San Diego, La Jolla, California

Dilip V. Jeste, M.D.   Estelle and Edgar Levi Chair in Aging, Professor, Department of Psychiatry, University of California, San Diego, La Jolla, California; Psychiatrist, Department of Psychiatry Service, Veterans Affairs San Diego Healthcare System, San Diego, California

Stephen P. Joy, Ph.D.   Assistant Professor, Department of Psychology, Albertus Magnus College, New Haven, Connecticut

Johanna Sara Kaplan, B.A.   Research Assistant, Mood and Anxiety Disorders Program, National Institutes of Mental Health, Bethesda, Maryland

Gina R. Kuperberg, M.B.B.S., M.R.C.Psych., Ph.D.   Instructor, Harvard Medical School; Clinical Assistant, Department of Psychiatry, Massachusetts General Hospital, Charlestown, Massachusetts

Walter A. Lajara-Nanson, M.D. Assistant Professor, Department of Neuropsychiatry, Texas Tech University; Attending Physician, Neuropsychiatry Consultation Service, University Medical Centers, Lubbock, Texas

Michelle V. Lambert, M.D. Neurobehavioral Service, Collingwood Clinic, Newcastle, United Kingdom

Walter Ling, M.D. Professor, Department of Psychiatry and Behavioral Sciences, David Geffen School of Medicine at UCLA, Los Angeles, California

Edythe D. London, Ph.D. Department of Psychiatry and Behavioral Science, Department of Molecular and Medical Pharmacology, University of California, Los Angeles, Los Angeles, California

Jessica Stewart Lord, B.A. Doctoral Student in Clinical Psychology, Department of Psychology, University of Connecticut, Storrs, Connecticut

Tina K. Machu, Ph.D. Associate Professor, Department of Pharmacology, Texas Tech University Health Sciences Center, Lubbock, Texas

Kelley S. Madden, Ph.D. Research Assistant Professor, Department of Psychiatry, University of Rochester School of Medicine and Dentistry, Rochester, New York

Jorge Luis Maldonado, M.D. Psychiatrist, Department of Psychiatry, Alamo Mental Health Group; Department of Psychiatry, Methodist Healthcare System, San Antonio, Texas

Catherine A. Mateer, Ph.D., A.B.P.P./C.N. Professor, Department of Psychology, University of Victoria, Victoria, British Columbia, Canada

R. Dayne Mayfield, Ph.D. Research Associate Professor, Department of Neurobiology, Waggoner Center for Alcohol and Addiction Research, University of Texas at Austin, Austin, Texas

Robert W. McCarley, M.D. Professor and Chair, Director, Neuroscience Laboratory, Department of Psychiatry, Harvard Medical School; Deputy Chief of Staff, Mental Health Services, Veterans Affairs Boston Healthcare, Brockton Campus, Brockton, Massachusetts

Robert J. Morecraft, Ph.D. Associate Professor, Division of Biomedical Sciences, University of South Dakota School of Medicine, Vermillion, South Dakota

Michael John Morgan, Ph.D. Senior Lecturer, Department of Experimental Psychology, University of Sussex, Falmer, Brighton, United Kingdom

William S. Musser, M.D. Clinical Assistant Professor, Department of Psychiatry, University of Pittsburgh; Attending Psychiatrist/Neurologist, Department of Psychiatry, Thomas Detre Hall of the Western Psychiatric Institute and Clinic, Pittsburgh, Pennsylvania

Alexander Neumeister, M.D. Mood and Anxiety Disorders Program, National Institute of Mental Health, Bethesda, Maryland

Tomáš Paus, M.D., Ph.D. Associate Professor, Department of Neurology, Neurosurgery, and Psychology, Montreal Neurological Institute, McGill University, Montreal, Quebec, Canada

John B. Penney, Jr., M.D. (Deceased) Professor of Neurology, Neurology Service, Massachusetts General Hospital, Boston, Massachusetts

R. Lisa Popp, Ph.D. Assistant Professor, Department of Pharmacology, Texas Tech University Health Sciences Center, Lubbock, Texas

Adrian Raine, D.Phil. Robert G. Wright Professor of Psychology, Department of Psychology, University of Southern California, Los Angeles, California

Stephen M. Rao, Ph.D. Professor, Department of Neurology, Director, Functional Imaging Research Center, Medical College of Wisconsin, Milwaukee, Wisconsin

Marcia H. Ratner, B.A. Associate Director, Environmental and Occupational Neurology Program, Department of Neurology, Boston University School of Medicine, Boston, Massachusetts

Richard Rawson, Ph.D. Executive Director, Matrix Center, Deputy Director, Alcoholism and Addiction Medicine Services, University of California, Los Angeles, Los Angeles, California

Alya Reeve, M.D. Associate Professor, Department of Psychiatry, University of New Mexico; Director, MI-DD Clinic, University Psychiatric Center, University Hospital, Albuquerque, New Mexico

Howard A. Ring, M.D. Reader in Neuropsychiatry, Barts and the London School of Medicine; Consultant Psychiatrist, Barts and the London NHS Trust, London, United Kingdom

Lynn C. Robertson, Ph.D. Veterans Affairs Medical Center, Martinez, California; Department of Psychology, University of California, Berkeley, California

Mary May Robertson, M.B.Ch.B., M.D., D.P.M., M.R.C.P.C.H., F.R.C.Psych. Professor, Department of Psychiatry and Behavioral Sciences, University College, London; Consultant Neuropsychiatrist, Department of Neuropsychiatry, National Hospital for Neurology and Neurosurgery, London, United Kingdom

Diana L. Robins, Ph.D. Postdoctoral Fellow, Child Study Center, Yale University, New Haven, Connecticut

Robert G. Robinson, M.D.   Paul W. Penningroth Professor, Department of Psychiatry, University of Iowa; Head, Department of Psychiatry, University of Iowa Hospitals and Clinics, Iowa City, Iowa

Byron P. Rourke, Ph.D., F.R.S.C.   Department of Psychology, University of Windsor, Windsor, Ontario, Canada; Child Study Center, Yale University, New Haven, Connecticut

Ruth Salo, Ph.D.   Assistant Research Psychologist, Department of Psychiatry and Behavioral Sciences, University of California, Davis, Sacramento, California

Randolph B. Schiffer, M.D.   Vernon and Elizabeth Haggerton Chair in Neurology, Chair and Professor, Department of Neuropsychiatry and Behavioral Science, Texas Tech University Health Sciences Center, Lubbock, Texas

E. Bettina Schmitz, M.D., Ph.D.   Head, Epilepsy Research Group, Department of Neurology, Humboldt University Berlin, Berlin, Germany

Robert Taylor Segraves, M.D., Ph.D.   Professor, Department of Psychiatry, Case Western Reserve; Chairperson, Department of Psychiatry, Metro Health, Cleveland, Ohio

Stephen D. Silberstein, M.D.   Professor of Neurology, Jefferson Medical College, Thomas Jefferson University; Director, Jefferson Headache Center, Thomas Jefferson University Hospital, Philadelphia, Pennsylvania

Christopher M. Sinton, Ph.D.   Lecturer in Psychiatry, Department of Psychiatry, Harvard Medical School, Brockton Veterans Affairs Medical Center, Brockton, Massachusetts

Hua Chiang Siow, M.D.   Clinical Instructor, Department of Neurology, Thomas Jefferson University; Attending Neurologist, Department of Neurology, Thomas Jefferson University Hospital, Philadelphia, Pennsylvania

David G. Standaert, M.D., Ph.D.   Associate Professor, Department of Neurology, Harvard Medical School, Boston, Massachusetts; Associate Neurologist, Department of Neurology, Massachusetts General Hospital, Charlestown, Massachusetts

Sara J. Swanson, Ph.D., A.B.P.P.   Associate Professor, Department of Neurology, Division of Neuropsychology, Medical College of Wisconsin, Milwaukee, Wisconsin

Michael Trimble, M.D.   Professor, Institute of Neurology, London, United Kingdom

Elizabeth W. Twamley, Ph.D.   Postdoctoral Fellow, Department of Psychiatry, University of California, San Diego, La Jolla, California

Gary W. Van Hoesen, Ph.D.   Professor, Department of Anatomy and Cell Biology and Neurology, University of Iowa, Iowa City, Iowa

Jasmin Vassileva, Ph.D.   Postdoctoral Research Associate, Department of Psychiatry, University of Illinois at Chicago; Neuropsychology Postdoctoral Fellow, Department of Psychiatry, University of Illinois Medical Center, Chicago, Illinois

Meena Vythilingam, M.D.   Medical Director, Anxiety Disorders Research Clinic, Mood and Anxiety Disorders Program, National Institute of Mental Health, Bethesda, Maryland

Lynn H. Waterhouse, Ph.D.   Director, Child Behavior Studies, College of New Jersey, Ewing, New Jersey

Donald R. Wesson, M.D.   Medical and Scientific Director, MPI Treatment Services, Summit Medical Center, Oakland, California

Jessica W. Yakeley, M.D.   Specialist Registrar, Psychotherapy Unit, Maudsley Hospital, London, United Kingdom

Stephen R. Zukin, M.D.   Professor of Psychiatry and Neuroscience, Albert Einstein College of Medicine of Yeshiva University, Bronx, New York; Director, Division of Clinical and Services Research, National Institute on Drug Abuse, National Institutes of Health, Rockville, Maryland

FOREWORD

The current degree of separation between psychiatry and neurology makes little sense either clinically or scientifically. After all, both neurologists and psychiatrists treat individuals with diseases of the nervous system, frequently the same diseases. Moreover, the basic science underlying both psychiatry and neurology is the same—neuroscience in its full range from molecular neurobiology to behavior. It is not the case that neurology is more closely related to molecular neuroscience and psychiatry to behavioral science. An understanding of the diseases treated by both specialties requires the whole gamut of neuroscience. The accelerating progress of clinical neuroscience that is documented in the second edition of *Neuropsychiatry* only serves to underscore what is shared between these two clinical disciplines—not what is different.

In reflecting on the training of physicians, it is instructive to compare the state of psychiatry and neurology with the various subspecialties that are joined within the umbrella of internal medicine. It is hard to argue that neurology and psychiatry are more different from each other than, say, interventional cardiology is from endocrinology or rheumatology. Indeed, such diverse subspecialties of internal medicine clearly benefit from shared foundations in clinical training and shared scientific understandings, which can then be followed by subspecialty training. Unfortunately, most residency training programs in psychiatry and neurology intersect very little. True, specialty board requirements in the United States stipulate that psychiatry residents must spend a few months in neurology training and vice-versa, but most of the didactic training that is provided to residents in these two fields, and essentially all of the remaining clinical training, is excessively compartmentalized. In his foreword to the first edition of this textbook, the neurologist Fred Plum traced some of the historical factors that drove psychiatry and neurology apart. If that separation ever served patients, it certainly does not do so today. One could generate a long list of problems that result from this separation, but to give some limited examples, one could argue that neurologists often spend too little time on important affective components of the diseases that they treat, resulting in excess distress and disability, while psychiatrists too often fail to perform physical examinations of the nervous system, risking a failure to reach appropriate diagnoses in some cases. Fortunately, concerns are beginning to be heard for bringing the initial training of psychiatrists and neurologists more closely together again (Martin, 2002)*. *Neuropsychiatry,* 2nd edition is a superb text for such a residency, but in our current world of largely separated training this volume is even more important. It could serve well both the fundamental understandings and the practice of psychiatrists and neurologists.

To my mind, one of the great strengths of *Neuropsychiatry* is that it puts the brain appropriately at the center of psychiatry. The practicing neurologist never doubts the organ system on which he or she works. Indeed, every neurology resident learns that the first step in understanding a patient's symptoms and signs is to localize the lesion within the nervous system, and only then to address the etiology of the disorder. Psychiatry, in contrast, may still be practiced without much reference to the brain. In fairness, given the early state of our knowledge of most psychiatric disorders, it has been difficult to usefully relate neurobiology to the clinical situation of psychiatric patients. Why, then, does it seem imperative for clinical psychiatry to become more closely connected to understandings of the brain at this time? The answer is quite clear—we are living in an extraordinary period of progress for many areas of biology ranging from genomics and genetics to systems-level neuroscience and cognitive science. Without engagement by clinical psychiatry, modern scientific tools may not be brought to bear with the urgency and effectiveness that psychiatric patients deserve. Additionally, without a change in approaches to training, the practicing psychiatrists of the next generation will not be equipped to incorporate new and useful scientific knowledge as it becomes available. This could leave psychiatry even more divorced from mainstream medicine than it now is. A failure to close this gap would not only be harmful to patients with psychiatric disorders, it would also hamper the recruitment of the brightest possible medical students and graduate students into clinical psychiatry and psychiatry research. A healthy future for psychiatry lies in putting the brain at the center of its understanding. Those who believe that focusing on stronger scientific foundations for psychiatry entails a weakening of the doctor-patient relationship or a sacrifice of caring and humanism, are simply wrong. This present volume is a fine place to begin relating clinical psychiatry to both neurology and neuroscience.

Steven E. Hyman, M.D.
Provost, Harvard University
Cambridge, Massachusetts
October 1, 2002

---

*Martin JB. The integration of neurology, psychiatry, and neuroscience in the 21st century. *Am J Psychiatry* 2002;159(5):695—704.

FOREWORD TO THE FIRST EDITION

A hundred years ago, the disciplinary activities and interests of neurology, psychiatry, and neuropathology could hardly be distinguished, either by the clinical pursuits of their practitioners or the way those who chose one or the other field diagnosed and attempted to treat patients. Few persons nowadays know that Alzheimer, who first described the morphology of the disease that bears his name, was a psychiatrist as well as a neuropathologist. Freud himself was initially a child neurologist and only later committed his life to the study of the subconscious-preconscious mind and how it might influence human behavior. Along with Jung and Adler, Freud and his followers provided psychological theories and therapeutics that led psychiatry forward into a psychological revolution, and away from the neurobiological explanation of behavior. Neurology marched deliberately forward, superimposing technological advances on meticulous clinical observations in order to identify and classify the signs, symptoms, and neuropathology of structural disorders affecting the central and peripheral nervous systems. There were struggles for theoretical and clinical territory, and for status. As Pearce Bailey, the first director of the National Institute of Neurological Diseases and Blindness put it (1951), "At the end of the nineteenth century, a psychiatrist almost had to be a neurologist to be respectable."

Gradually, however, a gulf of mutual disinterest arose between the anatomically grounded basis of neurological thinking and the psychological preoccupation of many psychiatrists. This shift peaked just after World War II, over 50 years ago. The separation of interests largely reflected the huge demand for psychiatric counseling for the many persons who became psychologically devastated by the disastrous political and material damage created by the war. Psychiatrists were learning how to talk with their patients and how to soothe their suffering. Meanwhile, neurology, with its demanding emphasis on physical causes and provable treatments at that time, was viewed by a large proportion of both the public and the medical profession as a rather arcane discipline. Although meticulously accurate in clinical diagnosis, its leaders seemed bereft of specific treatment for any neurological disease except epilepsy. Its practitioners were seen as, by-and-large, not very interested in patient care.

During the late 1940s, several things halted the interdisciplinary drift and began to set the pattern by which psychiatry and neurology would eventually close ranks in the medical battle against brain disease. Neurology and psychiatry had been militarily unified into the category of neuropsychiatry during World War I, and the designation was continued during World War II. After the war ended, however, the large number of neurologically damaged veterans not only forced an immediate need for neurologists, but emphasized the principle that for neurology to survive it must become a therapeutic specialty as well as a diagnostic one. By the late 1940s, military needs had greatly improved neurophysiological instrumentation and had stimulated the advent of antibiotics capable of treating neurological infections. The development of the Kety-Schmidt cerebral blood flow technique, first introduced in 1948, whetted the appetite of neurologists and psychiatrists alike to know about brain mechanisms in neurological and psychiatric illness. Probably most important of all the stimuli bringing psychiatry and neurology together, however, was the rapid discovery of new psychoactive and soma-active drugs. Once and for all, the effects of these agents demonstrated not only that a large proportion of psychiatric illnesses have their biologically specific origins in abnormal brain mechanisms, but also that many of the aches and pains that neurologists attempt to treat are due to psychophysiological disturbances that can be explained in biological terms. Always bonded by their common efforts to understand the brain but sometimes treating each other as estranged relatives, the disciplines of neurology and psychiatry now find themselves addressing similar scientific questions, using similar approaches to treatment and, even more importantly, applying similar scientific techniques to the analysis of their problems. If not representing a revolution in the way that neurologists and psychiatrists face the problems of brain disease, this certainly represents an enormous change from the barriers that divided the two specialties only a few years ago.

Neuropsychiatry helps to strengthen the common features of the disciplines. A review of its contents illustrates the remarkable breadth of the topic. Similarly, a review of its authors and their backgrounds illustrates how greatly the fields of neurology, psychiatry, and neuropsychology already have joined forces to define, solve, and treat disturbances of the mind and behavior. In this textbook, we hear from prominent experts in the neurosciences, as well as from published neuropsychiatric researchers. Many of the chapters suggest great progress in dealing with their topics. Others reflect how readily one can obtain clinical and scientific material about certain behavioral problems, but also how difficult it is to link the results of science into a testable, mechanistically governed, and treatable process. Still other aspects of the mind-brain mystery remain almost entirely clinically descriptive, offering invitations to all readers to explore the basic pathogenesis and develop rational treatments of the still poorly understood disorders.

Having had an opportunity to review *Neuropsychiatry* I regard it as an original and indispensable single-resource volume for every practicing neurologist and psychiatrist, which will be equally valuable for many academicians as well. Of one thing I am sure: this volume will catalyze much future thinking on this subject. Undoubtedly, many subsequent editions of this text will follow.

Fred Plum, M.D.
New York, 1995

PREFACE

Brain-based theories of behavior have led several epochal periods of change in medical practice since the early nineteenth century. Benjamin Rush, in his 1812 textbook, *Medical Inquiries and Observations upon the Diseases of the Mind,* clearly described an interactionist view of mental illness. He envisioned behavior as a transitional state between general medicine and psychology, at the crossroads of causation between subjectivity and physical forces. Dr. Rush also demonstrated a broad sensitivity to the importance of general medicine in psychiatry. His example of a psychiatrist who was clearly a physician as well is much needed in the modern world. Later in the nineteenth century, the newly minted medical specialty of neurology, in the person of George M. Beard, gave us our first truly American behavioral disorder, in neurasthenia. Physicians in practice might quip that we need him now to take it away.

Psychoanalytic and psychodynamic ideas dominated the first half of the twentieth century; especially in the Western Hemisphere, where these ideas found fertile intellectual grounding. Even in the halcyon days of psychodynamics there were other voices about the importance of the brain in determining behavior. Early in the twentieth century, discerning psychiatric observers described the psychotic and neurotic behavioral features of patients in the great viral encephalitis epidemics. Here, indeed, was a model of neuropsychiatric behavioral theory; if only it could have been understood at the time. By the mid-twentieth century, the psychopharmacologic revolutions had begun, and the laboratory neurosciences had been born, setting the substructures for the neuropsychiatric revolution in which we presently labor.

This neuropsychiatric revolution has changed the way we think about behavior, mental illness, and neurology. We now know that medications can change the way we think, feel, and act. We have learned that the major psychiatric disorders have signature neurochemistries, neurophysiologies, and neurogenetics. Conversely, almost all of the major neurologic disorders are now understood as producing fundamental changes in behavior. Behavior may become the medical frontier of the twenty-first century.

We are pleased that the 1st edition of *Neuropsychiatry,* published in 1996, made its contribution to the scholarly library of evolving changes in clinical medicine. We are even more pleased to bring to press the 2nd edition of *Neuropsychiatry,* which brings to date the impact of many recent changes in clinical and basic neuropsychiatry that have occurred since the mid-1990s. The book is designed for clinicians of many backgrounds, as well as for clinical scientists. We hope that you will find it enjoyable reading, and a useful reference as we collectively move forward the frontiers of knowledge in the twenty-first century.

Randolph B. Schiffer, M.D.
Stephen M. Rao, Ph.D.

ACKNOWLEDGMENTS

We recognize the steady support and guidance that we have received from the publishing leadership at Lippincott Williams & Wilkins. Mr. Ray Reter has steadfastly supported the development and editorial management of the manuscript. Mr. Charley Mitchell has helped with the strategic vision that allowed us to construct a 2nd edition of *Neuropsychiatry* upon the foundations of the first. Ms. Dorothy Williams, at Texas Tech University, has provided indefatigable manuscript assistance.

Above all, we recognize those individuals who have been our mentors and inspirators over the years: my father, A. Brenton Schiffer, my first teacher in medicine; and to my wife, Rebecca Winner, my children, Jess and Julia, and parents, John and Anne Rao, for their love, support, and encouragement.

Randolph B. Schiffer, M.D.
Stephen M. Rao, Ph.D.

SECTION

I

ASSESSMENT AND TREATMENT

NEUROPSYCHIATRIC EXAMINATION

RANDOLPH B. SCHIFFER
WALTER A. LAJARA-NANSON

The neuropsychiatric examination is a specialized clinical evaluation that supports clinical diagnosis and management in behavioral disease (1). The elements of the neuropsychiatric examination include the adult psychiatric evaluation, as described in the American Psychiatric Association Practice Guidelines for the Psychiatric Evaluation of Adults (2); the neurologic examination, as described comprehensively by DeJong (3); and the neurobehavioral mental status examination, as described by Strub and Black (4), Trzepacz and Baker (5), and Weintraub and Mesulam (6). The judicious use of neuroimaging and neurophysiologic testing usually is included in this examination, as described in Chapters 4 and 5 of this book. This chapter reviews the clinical components of the neuropsychiatric examination. A comprehensive discussion of examination techniques is beyond the scope of this chapter. Readers unfamiliar with the references cited are encouraged to consult them.

DISTINCTIVE FEATURES OF THE NEUROPSYCHIATRIC APPROACH

The neuropsychiatric examination differs from conventional psychiatric evaluation by its inclusion of cognitive assessment, the neurologic examination, and attention to supportive neuroimaging or neurophysiologic test data. In addition, the neuropsychiatric examination aims to produce an explicit neurobehavioral formulation of behavioral disorders, which identifies neuroanatomy, neurophysiology, and neurochemistry of relevance to the behavioral syndrome at issue. The management plans that evolve from neuropsychiatric examinations typically include a broader range of therapeutic options than those of general psychiatry, often including the range of available neurotropic drugs in addition to the psychotropic agents. Figure 1.1 provides an overview of the neuropsychiatric examination elements.

There are other differences of emphasis in the conduct of a neuropsychiatric examination compared with general psychiatric work.

The clinical data in neuropsychiatry are viewed by the examiner as possibly demonstrating various forms of underlying brain dysfunction. In addition to general psychiatric history taking and interview skills, the neuropsychiatrist also must consider core neurologic symptomatology, such as alterations of consciousness, or alterations in sensorimotor function. Life development events capable of affecting neuropsychiatric function are reviewed, such as trauma, or significant metabolic disease. Corroborative medical records describing such events may need to be obtained.

When mental symptoms are described, precise description of the core psychopathology takes precedence over organization of the patient's symptoms into standard psychiatric syndromes and disorders. The possibility is kept in mind that the patient may have a recognized psychiatric syndrome whose symptomatic expression is modified by brain dysfunction or a recognized neurobehavioral syndrome whose symptomatic expression is modified by psychological defense mechanisms or sociocultural influences.

Specific neurobehavioral syndromes are considered throughout the neuropsychiatric examination, which may not routinely be recognized in general psychiatry or be listed in general psychiatric diagnostic manuals. These include clinically categorized syndromes such as frontal system syndromes, temporal lobe syndromes, basal ganglia syndromes, the dementias, and various forms of aphasia, alexia, agraphia, agnosia, apraxia, and amusia.

Neurologic and mental status examinations must follow up clues identified by the clinical history. These examinations must include an array of specific cognitive assessments, flexibly suited to the specific problems of individual patients. The neurologic examination must be adapted to the symptomatology at hand, sensitized to confirming or disconfirming hypotheses of regional brain dysfunction suggested by the clinical history. The neurologic examination always is conducted in the light of the patient's clinical history so that diagnostic possibilities are amplified or decreased in probability in a systematic way.

Based on the history and examination, hypotheses are developed that specifically include multidimensional

Assessment Skills
Interview; Cognitive assessment; Neurologic examination

Laboratory and Imaging
EEG; CT; MRI

Therapeutics

FIGURE 1.1. Neuropsychiatric examination elements.

interactions involving regional brain dysfunction, psychological development, genetics, sociocultural factors, current environment, and individual psychology. Proximate causes of behavioral changes are identified, as are potentially remediable factors that contribute to the behavioral disorder.

We now present a survey of the functional elements of the neuropsychiatric examination.

THE CLINICAL HISTORY

The proper taking of the clinical history is above all else a listening skill. The listening must be active, scrutinizing silently the entire neurobehavioral production of the patient, for clues about strengths and weaknesses of his or her neuropsychiatric function.

The opening interchange should include a salutation signal from the clinician, words from everyday life that serve to greet and reassure. Each clinician must individualize the specifics of such interactions, but greetings such as "Welcome to our clinic," "I am sorry you have had to come to the emergency room," and similar statements can put the patient at ease.

The second interchange from the physician should include an open-ended question that invites the patient to elaborate about the clinical concern that has brought him or her to the health care facility, whether this be an inpatient setting, emergency room, or ambulatory clinic. Appropriate examples include such statements as "Tell me what happened," "Why did you come to see me today," or "What caused your doctor to admit you to the hospital?" This initial invitation for the patient to speak is key to the interview. It should be terse and neither too vacuous ("How are you?") nor too specific ("When did the headache start?"). The clinician signals to the patient that he or she is focused upon the patient's most acute problem but that the patient will be the teacher about that problem.

The second and third history-eliciting questions should serve only to elaborate upon the patient's narrative text. These questions should be individualized to the narrative of each patient and should encourage the enrichment of that narrative. These interventions always should be open-ended, even in acute clinical situations. Examples include such statements as "Tell me more" and "What happened before that?" Questions or statements from the interviewer

that are too specific serve to inhibit the narrative production from the patient or to distort that production so that the clinical history becomes a "virtual" one, removed from the actual experience of the patient. This is important, because if the history is wrong the diagnosis in neuropsychiatry cannot be correct.

Only after the patient's illness narrative has been completed to the best of that person's ability should the clinician contemplate asking some direct questions. These direct questions are aimed at key historical issues of diagnostic importance. Such questions can be used to clarify important time order issues, such as "Did the headache come before the rash?" These questions should be asked in short sentences, not in compound or complex sentence structures, which confuse the patient about what is important. Only a limited number of these questions can be asked for each patient before ritualized or invalid information begins to emerge. The direct questions should be selected carefully. There is no "review of systems" in the neuropsychiatric evaluation.

Important dimensions of the neuropsychiatric examination are conducted during the clinical interview. Key observations are made of the patient's level of arousal, language function, thought process, mood and affect, and psychomotor function. The goal of the interview is to obtain enough visual and historical information to focus the examination upon a few salient clinical issues. In diagnostic terms, the interview presents a formulation with two or three diagnostic possibilities, and the examination allows the clinician to choose among the possibilities.

Only after the outline of the patient's principal illness narrative is complete should a few questions be asked about past medical, family, and personal/social issues. These questions, too, should be selected by the interviewer to help clarify the diagnostic formulation that emerges during the clinical history. Questions should never have a rote or ritualized feel for the patient, lest the same quality of information be returned.

Notes should never be taken during the interview, nor should the patient's medical record preoccupy the clinician. The clinician should focus his or her entire attention upon the patient, and the patient should feel "heard."

The manner of the interviewer should connote patience, receptivity, and acceptance. Reserve and interest without distance should invest the "feeling" of the interaction. It is not the actual time of the interview, which counts in establishing this relationship, but the quality of use of the time that is available.

At the conclusion of the clinical history, whether it last 2 minutes in an emergency room cubicle or 45 minutes in a psychiatric office, the examiner should have an initial formulation of the patient's presenting problem. By "formulation" we mean a brief, three-sentence summary of the issues that captures the major time course and subjective quality of the patient's principal problem. This formulation should be crafted in words that allow testable hypotheses about diagnosis to be tested during the subsequent neuropsychiatric examination. Is it an aphasia or a frontal lobe syndrome? Is the time course 3 months or 3 years? Has motor function been affected, or has just cognition been affected? The neuropsychiatric examination will permit the selection of the best diagnostic hypothesis consistent with the formulation developed during the clinical history.

MENTAL STATUS EXAMINATION

Initial Observations

Appearance

Initial observations of the patient should include notes about dress, demeanor, eye contact, level of motor activity, and general interpersonal style. Defensive, hostile, seductive, playful, perplexed, attentive, and evasive are commonly used terms. The examiner should note the patient's height, body type, posture, hygiene, clothing, and gross evidence of health or illness. Are there stigmata of alcoholism, such as palmar erythema, facial flushing, spider angiomata, or jaundice? Does the appearance suggest an endocrine disorder, such as Cushing disease, or the use of exogenous steroids (moon facies and "buffalo hump")? Facial asymmetry, pupillary asymmetry, or discrepancy in the size of a patient's hands may be observed before the formal neurologic examination begins. Are abnormal movements suggestive of seizures present?

Level of Arousal

This portion of the evaluation assesses the patient's spontaneous behavior and reactivity to the surrounding environment. *Arousal* refers to the individual's state of wakefulness and alertness. When there is an impairment of arousal, descriptive terms such as "clouding of consciousness," "obtundation," "stupor," and "coma" are used. One should make note of whether the patient's eyes are open or closed and whether the patient focuses attention on the examiner. Response to auditory stimuli, such as calling the patient's name or the need to use loud auditory stimuli such as clapping to arouse the patient, should be noted. The same should be applied to the use of mild physical stimuli, such as touching the patient or shaking him or her, or the use of painful stimuli in order to arouse the patient. The neural substrates of arousal reside in the brainstem tegmental nuclei of the reticular activating system, which project diffusely to forebrain and diencephalic structures. Factors that may affect level of arousal include use of drugs (prescription or other), sleep schedule, and other medical illnesses. If arousal is affected by these variables, the behavioral assessment may be better scheduled for a more appropriate time (i.e., when drugs have been reduced or adequate sleep obtained).

Assessment Test: Glasgow Coma Scale

The Glasgow Coma Scale (GCS) (Table 1.1) is a simple, easy-to-administer technique that is universally used to rate the severity of coma via the patient's ability to open his or her eyes, move, and speak. A patient is assigned a number in each of three categories: eye opening, motor response, and verbal response.

The GCS can be used to quantitate and follow clinical syndromes of decreased arousal (7). The GCS quantifies reactivity to verbal and physical stimulation in terms of a 15-point scale, subdivided into eye-opening (score: 1–4), motor responses (score: 1–6), and verbal responses (score: 1–5). On this scale, a score of 3 represents deep coma, whereas a score of 15 represents alert wakefulness. The GCS is of less value in quantitating the severity of mild or very mild brain injury. Patients with such injury score 13 to 15 on presentation, and the vast majority of patients who are conscious score 15.

Orientation

It is customary in neuropsychiatry to assess the patient's awareness of name, place (state, town, county, hospital floor), and time (month, date, day of week, year, season, or time of day). Such information is relatively diffuse and nonspecific with regard to its neurodiagnostic usefulness. Personal identity and orientation often are preserved in the face off diffuse brain disease and sometimes disturbed in relatively minor neurotic mental illness. Metabolic brain disease perhaps is the characteristic disorder in which disturbances of orientation are seen and can be followed clinically.

TABLE 1.1. GLASGOW COMA SCALE

Eye opening	
Spontaneous	4
To speech	3
To pain	2
None	1
Best motor response	
Obeys commands	6
Moves within the general locale	5
Withdraws	4
Abnormal muscle bending and flexing	3
Involuntary muscle straightening and extending	2
None	1
Verbal response	
Is orientated	5
Confused conversation	4
Inappropriate words	3
Incomprehensible sounds	2
None	1

Minimum score 3; maximum score 15.
From Teasdale C, Jennett B. Assessment of coma and impaired consciousness. A practical scale. *Lancet* 1974;2:81–84, with permission.

Assessment Tests
Record the patient's answers to these questions:

Who are you?
Where are we?
What is the date?

Attention and Concentration

Attention refers to the ability to focus awareness over a period of time. The focus may be internal or external, or upon an idea or an object, but the act of attention highlights the target and sets it apart from other competing stimuli. *Concentration* is one aspect of attention, connoting a more prolonged or intense attending to some stimulus or problem. The neural support for these functions comes from prefrontal cortical systems, along with connecting circuitry from basal ganglia and reticular systems.

Disorders of attention are the hallmark of confusional states (either quiet confusional states or agitated confusional states, sometimes referred to as delirium). Confusional states most commonly occur in the context of toxic, infectious, and metabolic encephalopathies. Structural hemispheric brain disease, including stroke and dementia, may adversely affect attention and concentration. Emotional distress, pain, fatigue, and drug intoxication are common confounds in patients with attentional impairments. Hearing loss and aphasia must be considered as contributors to inattention. Identification and assessment of attention and concentration deficits are important early in the mental status examination, because all subsequent cognitive evaluation is affected by the presence of a confusional state.

The mental status tests that are most sensitive to the functions of attention and concentration are vigilance and persistence tasks, which require a response to the presentation of a designated stimulus. For example, a patient may be asked to push a button every time he or she sees an "x" appear on a monitor or every time she hears the word "dog" in a continuous list of words. More complex forms of attention include *divided attention* (switching attention back and forth between two ongoing stimuli) and *selective attention* (abstracting a designated stimulus for attention from all ongoing stimuli). These complex forms of attention are representative of the real life behavioral tasks, which our nervous systems face in daily functioning.

Because some of the attentional functions described can be difficult to test at the bedside without computers and other apparatus, other tests are more commonly used for clinical purposes. Digit span is part of many standardized batteries, including the Wechsler Memory Scale-Revised (8) and the Wechsler Adult Intelligence Scale-Revised (9). Because these tests also require the mental manipulation of information held in short-term memory, they provide partial measures of working memory as well.

Tests that measure other dimensions of attentional capacities include (a) the digit-symbol test; (b) cancellation tasks (a page covered with letters or digits is presented to the patient, who is asked to strike out a particular digit or letter whenever it occurs); (c) trails, or the Trail-Making Test (the patient's task is to convert in sequence a set of numbers or a set of letters alternating with numbers); (d) the Stroop test (the patient begins by naming several different colors and then is asked to identify the color of the ink in which each of a list of color names is printed and then to name the color of the ink in which the names of different colors are printed where the names do not correspond to the color of the ink); and (e) the Corsi Block Tapping Test (a series of numbered blocks arrayed on a board are tapped in a sequence that the patient is asked to imitate). These supplementary tests of attention almost certainly commit other cortical systems for their successful performance, including frontal and parietal systems.

Assessment Tests
Digit span forward requires the patient to listen to a string of digits (e.g., "2–4") and then repeat them immediately after presentation. The digits should be spoken at 1-second intervals, without verbal cues about groupings. For digit span backward, patients listen to a string of numbers (e.g., "2–4") and are required to repeat the digits in reverse order (i.e., "4–2"). For both portions, the number of digits in each series is increased consecutively, and testing is terminated once the patient's threshold has been realized. The longest repeatable span is recorded, both forward and backward. Normal performance varies from five to seven digits forward, depending upon age, education, and general medical status. Reverse digit span should be three to five digits.

Another frequently used task is subtracting serial 7s. The patient is asked to subtract 7 from 100 and continue until asked to stop. Some patients may have difficulty with this and may perform better with serial 3's, where the patient is asked to subtract 3 from 20 and continue until asked to stop. Normal performance is one error or less.

Cognitive Functions

The cognitive functions are a family of neuropsychiatric behaviors that have more to do with knowing than with feeling or acting. The field of neuropsychology has emerged as a specialized discipline committed to the assessment, understanding, and treatment of the cognitive disorders (see Chapter 2). Performance on cognitive tests of all types is sensitive to modulating factors, including fatigue, motivation, and use of prescribed drugs or drugs of abuse. Premorbid educational achievement is another powerful factor that determines cognitive performance in the face of neuropsychiatric disease. There may be some age-related cognitive deterioration that is part of "normal aging," but more likely cognitive deterioration in late life is a manifestation of neuropsychiatric disease (10). Cognitive functions in clinical neuropsychiatry generally are assessed in a hierarchical order, according to the following outline.

Speech and Language

Speech is not the same behavioral function as language, yet they often are assessed together. Speech is essentially a brainstem, cerebellar, and cranial nerve function, whereas language is a distributed higher cortical function. Spontaneous speech can be characterized in terms of its flow, volume, pressure, rhythm, and intonation. The prosodic dimension of speech refers to the emotional intonation speech. Alteration in prosody produces a mechanical or engineered speech pattern. Disturbances of speech related to brainstem dysfunction are called *dysarthrias.* Dysarthria can be spastic, flaccid, ataxic, hypokinetic, or hyperkinetic. Stammering and stuttering are types of speech disorders that can be either developmental or acquired.

Speech abnormalities occur in a wide array of neuropsychiatric conditions, including mania, Parkinson disease, stroke, white matter diseases, and neuromuscular diseases. Localization of the pathologic process usually depends upon data from other dimensions of the cognitive and neurologic examinations (see later).

In contrast to disorders of speech such as mutism and dysarthria, in which writing is preserved, disorders of language represent central problems in the reception and manipulation of linguistic symbols. Disturbances of writing usually parallel those of spoken language in aphasic conditions. Figure 1.2 enables clinicians to rapidly classify an aphasia by assessing fluency, comprehension, and repetition. Adequate assessment of language entails consideration of (a) fluency, (b) comprehension, (c) repetition, (d) naming, (e) reading, and (f) writing.

Fluency can be assessed by careful listening during the interview phase of the neuropsychiatric examination. Fluency refers to the rate, rhythm, and degree of effort demonstrated in producing spontaneous speech. More formal assessments of this function can be made by asking the patient to perform a verbal elaboration task, such as describing the Cookie Theft Picture from the Boston Diagnostic Aphasia Examination [see Lezak (11) for an overview of formal cognitive tests]. In describing spontaneous speech, the examiner records word-finding pauses, agrammatisms, and paraphasic errors (substitution of inappropriate syllables or entire words). Most neurologists and psychiatrists describe speech in all-or-nothing terms as either fluent or nonfluent.

Comprehension of spoken language refers to the patient's ability to hear and process spoken or written communications. Proper comprehension function requires intact superior temporal and inferior parietal lobe systems, usually in the left hemisphere for right-handed people. Problems with hearing and comprehension initially may be unappreciated in patients with gross cognitive deficits. Such unappreciated disturbances in comprehension may lead caregivers to conclude that a patient is uncooperative, negativistic, or inattentive, or that the patient's memory is poor. Even subtle comprehension disturbance can significantly constrain the patient's capacity to interact with other people. Any tendency on the part of the patient to consistently echo or repeat what others say should raise the question of a comprehension disturbance. Such individuals may be thought of as attempting to

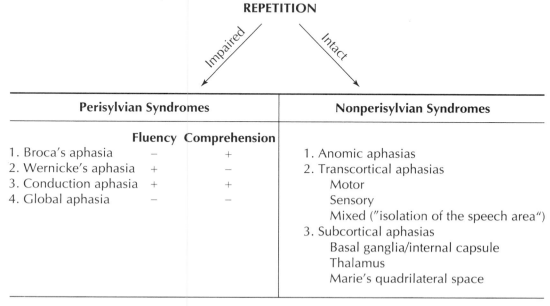

FIGURE 1.2. Classification of aphasic syndromes. (From Mueller J, Flynn FG, Fields HL. Brain and behavior. In: Goldman HH, ed. *Review of general psychiatry,* 4th ed. East Norwalk, CT: Appleton & Lange, 1995:58, with permission.)

"run the tape by" one more time in order to extract as much information as possible.

Repetition is impaired by all four of the major aphasic disorders that result from damage to the perisylvian speech areas of the dominant hemisphere: Broca area, Wernicke area, conduction aphasia, and global aphasia (Fig. 1.2). Any patient who can accurately repeat phrases of significant length either has no language disorder or has one of the nonperisylvian aphasias: anomic, subcortical, or transcortical. Repetition can be assessed for single words, short phrases, and lengthy sentences. Patients with receptive aphasias, particularly those with damage to the Wernicke area, often make paraphasic errors when attempting to repeat longer phrases. Paraphasias may involve substitution of wholly inappropriate words (verbal paraphasias) or of individual syllables (phonemic or literal paraphasias). Demonstration of paraphasic errors on a repetition task provides convincing evidence that a patient suffers from a language disturbance.

Naming refers to the language functions of word finding and word selection. This discrete cognitive function does not localize as well as other language subfunctions in the brain, although it is generally associated with left hemisphere structures in right-handed people. Most patients with aphasic disorders of all types have at least some word-finding difficulty or *dysnomia*. If the word-finding difficulty is more severe, it is termed *anomia*.

Word-finding difficulty can be caused by other conditions. Word-finding difficulties may occur in metabolic encephalopathies or as a side effect of psychotropic medication. Physical exhaustion, sleep deprivation, extreme anxiety, and depression may produce erroneous word selections. For such reasons, dysnomia as a finding must be considered in the context of performance in the other language subfunctions.

Reading is impaired to some extent in most individuals with aphasic disorders (12). The presence of such an alexia syndrome indicates damage to the more posterior language processing cortex, with associations to the visual cortex. Localized damage to the left inferior parietal lobule produces a syndrome known as *alexia with agraphia,* whereas damage to the left occipital cortex produces a syndrome of pure alexia *(alexia without agraphia)*. In the assessment of reading ability it is important to realize that an intact visual system is required, and that reading aloud is different from reading for comprehension. The fact that a patient can read a passage aloud does not indicate that he or she is able to understand or remember it. Previous educational attainment is to be remembered as well, because someone who could not read before onset of a neuropsychiatric illness will not be able to read after!

Writing and spelling usually are assessed last in the bedside aphasia assessment. The motor abilities involved in writing draw upon more anterior language system centers than those involved in reading. Occasionally these functions can be differentially affected by neuropsychiatric disease, as noted earlier. Writing samples are invaluable for documenting para-

phasic errors. Serial writing samples can be used to document the resolution of aphasias that are due to treatable neuropsychiatric disease. Spelling errors are common after insults that produce aphasia, but it is often difficult to separate acquired "dysorthographia" from the baseline poor spelling in the United States today.

Assessment Tests

Fluency (Spontaneous Speech)
The patient can be asked to generate words beginning with a particular letter over 60 seconds, or three letters such as "F," "A," and "S" (FAS test). Word production is counted and should exceed 15 words per minute for each letter.

Comprehension
A common method for assessing comprehension is to have the patient perform one-, two-, or three-step tasks. Individuals who correctly perform three-step tasks may be considered to have intact comprehension. Patients who are able to correctly perform one- or two-step tasks, but not three-step tasks, are likely to have difficulties in interactions with hospital staff and others. A typical sequence of such questions is as follows:

> "Close your eyes!"
> "Show me your right hand!"
> "Put your right hand on your left ear!"
> "Is it true that your foot is not on the bed?"

With patients who have motor system impairments that preclude limb movements, other variations of yes-or-no type questions are used. Eyeblink movement is always retained, even in the most severely motor impaired individual. Correct answers to five consecutive yes-or-no questions could occur by chance alone in only 3% of cases.

Repetition
One example of a series of phrases and sentences of increasing difficulty to repeat is as follows:

Please repeat:
"Ball"
"Methodist"
"Methodist Episcopal"
"No *if*'s *and*'s, or *but*'s!"
"Around the rugged rock the ragged rascal ran!"

Naming
Initial testing of word-finding ability also uses body parts. The patient is asked to name simple, discrete body parts first, such as "nose" or "ear." To escalate test difficulty, the five digits of the hand can be held up sequentially for naming by the patient. As an alternative, simple objects, such as objects in the room, elements of a pen, or parts of a watch, may be used to interrogate the patient's naming ability. For greater

depth in this assessment, one can use the Boston Naming Test 15-item version, which requires the patient to name pictured items. The threshold for judgment of an anomic aphasia must be individualized for each patient, considering age, education, and metabolic stressors. In general, no errors should be made on the simple object-naming tasks and not more than one on the 15-item version of the Boston Naming Test.

Reading

Patients should be asked to read words or sentences of escalating degrees of complexity. "Close your eyes!" can be hand written at the bedside as an initial test. Newspaper text is readily available in most hospital rooms and can be used at the next degree of difficulty. Newspaper articles are generally written at the eighth grade level of language expertise. For a more in-depth assessment, the patient can be asked to read the Cowboy Story. The patient is asked to read the story aloud while the examiner listens for nonfluency, paraphasic errors, and errors of articulation. The utility of the test can be expanded to include a measure of context-based verbal learning and memory by asking the patient to recall details of the story.

Cowboy Story Bill Rogers, a cowboy from Arizona, went to Texas with Roy, his German Shepherd, whom he left at a friend's when he went to buy a new suit of clothes. While he was away, his friend fed the dog four bones and tied him to a tree. Two days later, dressed in his new suit, Bill returned, whistled to his dog, snapped his fingers, and called out his dog's name. The dog sniffed his pants and began to growl. The cowboy went and changed his clothes. Now, when Roy saw his master, he jumped for joy.

Writing and Spelling

As in all language assessments, the writing tasks should be performed in escalating order of complexity. The most simple initial task is to ask the patient to write his or her name. Next, he or she should be asked to write a sentence of his or her choosing. The open-ended nature of such a task allows some initial projective themes to emerge, which sometimes helps the physician in understanding the patient. The next task can consist of asking the patient to write dictated sentences or a paragraph about a scene or theme.

Verbal Memory

Almost all neuropsychiatric diseases are capable of interfering with one or more steps in the process of acquiring, storing, and retrieving information through verbal mechanisms (see Chapter 18). The neuroanatomic substrates of verbal memory involve bilateral hippocampal systems of the medial temporal lobes, as well as multiple supporting connections through cortical and limbic structures. Reviews of memory functional systems and neuroanatomic supports have been provided by Squire (13,14). Verbal memory is somewhat more of a left hemisphere function in right-handed people compared with nonverbal memory, which localizes somewhat more to the right hemisphere.

It is considered that "memory" as a cognitive function can be broken down into functional subsystems, which may be differentially affected in some neuropsychiatric diseases. The goal in the assessment of verbal learning abilities is to derive useful objective information about three subfunctions of memory:

Ability to learn *(acquisition)*
Ability to hold onto information *(retention)*
Ability to use stored information *(retrieval)*

A large number of tests for verbal memory have been developed, including the Logical Memory Paragraphs from the Wechsler Memory Scale (8). In this test, patients are read two paragraphs, each containing 25 idea units, and are asked to repeat as much of the two stories as possible. After a 30- or 60-minute interval, patients are again asked to repeat whatever they recall from each of the two stories. This allows a measure of decay of memory over time. The Rey Auditory Verbal Learning Test (RAVLT) provides a measure of supraspan word list learning (11). The RAVLT consists of a list of 15 words that are read to the patient five times. After each reading, the patient is asked to recall as many words as possible. This test allows the examiner to look at the patient's learning curve when confronted with a "supraspan" list of words. After the five trials, a second list of 15 words is read and the patient is again asked to recall as many words as possible. After this "interference" trial, the patient is asked to recall as many words as possible from the original list. Finally, the patient is read a list of 30 words, half of which were on the first list. He or she is asked to say "yes" after each word that appeared on the initial list. Similar word list learning tests include the Selective Reminding Task of Buschke (15) and the California Verbal Learning Test (16). The Hopkins Verbal Learning Test (12 words) allows for assessment of encoding strategies (17).

Assessment of verbal memory can occur more informally by asking the patient autobiographical questions referring to different periods in his or her life (to compare recent vs. remote memory), by asking the patient to recall information presented previously during the interview, and providing cues when necessary. The clinician should carefully observe the patient throughout the evaluation to gauge verbal memory in action in responses to task instructions. Remote memory also can be assessed by asking the patient to recall past events from his or her personal life, historic events, or famous people. Such information has limited clinical utility, however, because remote memory is generally preserved even in patients with advanced dementia.

Assessment Tests

Registration, storage, and delayed recall of verbal information is assessed initially by asking a patient to register a list of five words:

> Elastic
> Dog
> Polish
> Red
> Careful

The short list has three lower-frequency words and two more common words to afford the examiner an opportunity to observe subtle thresholds in the patient's memory functions.

After distraction for 1 minute, the patient is asked to recall the list *(active recall)*.

If there are errors on active recall, the patient is given a list of ten words, which includes matched distractors to the original five words:

> Elastic (old)
> Brown
> Plastic (old)
> Cat
> Dog (old)
> Polish (old)
> Red (old)
> Burnish
> Careful (old)
> Callous

The patient is asked which are old words and which are new as a measure of recognition recall.

Visual Memory

A second type of memory testing is that which assesses visuospatial recall. Visuospatial functioning is supported in the brain by a widely distributed neural network involving parietal, temporal, or occipital anatomic systems. Visuospatial memory functions are more strongly dependent upon intact right hemisphere structures in right-handed persons than are verbal memory functions, but one should never localize neurologic deficits on the basis of cognitive performance alone. Additional complexity exists in the subfunctions of visuospatial cognition, because visuospatial copying ability must be assessed. There is not a precise analogy to this function in the verbal learning domain, and it almost certainly calls upon neural systems that are more widely distributed than purely memory functions.

The basic neuropsychiatric assessment strategy for visuospatial functions is to ask the patient to copy a drawn or printed figure and then reproduce it from memory shortly afterward. Cognitive performance on such tasks must be interpreted from the perspective of supporting neuropsychiatric functions that are required for successful performance of the task. Visual systems must be intact, as well as praxis capabilities and frontal-executive functions.

Assessment Tests

A useful screening test for visuospatial skills involves asking the patient to draw a clock and to place the hands at "ten after eleven." The manner in which he or she carries out this task should be observed and recorded, and the drawing can be kept in the record for comparison with future performance.

The Rey-Osterrieth Complex Figure (Fig. 1.3) is a line drawing patients may be asked to copy (18).

Other Visuospatial Skills

Visuospatial skills initially are assessed during the visual memory testing described earlier. Further tests may enhance understanding of related, nonmemory visuospatial cognition. Patients may be asked to draw circles, triangles, and three-dimensional cubes, a person, a bicycle, a house, or the face of a clock with the hands placed at a particular time. The clock drawing in particular has established value as a screening test for dementia. However, highly intelligent patients can use left hemisphere strategies to plan and draw an acceptable clock face. Line-cancellation and line-direction tests are more sensitive and more specific tests of visuospatial skills.

In addition to drawing tasks, patients may be asked to manipulate either tokens or three-dimensional blocks to make a series of designs of increasing complexity. The examiner records the time required to complete each design.

Visuospatial skills can be disrupted because of perceptual problems, including visual field cuts due to damage to the retina, optic nerve, optic chiasm, optic tract, lateral geniculate body, optic radiations, or calcarine cortex. It is important that limitations of visual acuity be detected and that sensory/perceptual deficits be ruled out before the examiner ascribes any visuospatial deficit to "higher cortical" dysfunction. Cerebral visuospatial problems may also derive from damage to the visual association cortex in the parietal and temporal lobes.

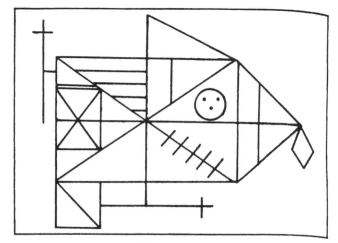

FIGURE 1.3. Rey-Osterrieth Complex Figure.

Impairment of visuospatial ability is associated with parietal lobe damage, particularly that of the right hemisphere. The drawings of these individuals often omit major elements of the figure being copied, with particular difficulty manifested with the left side of the drawing. On the other hand, damage to the dominant parietal lobe may result in some difficulty reproducing some aspects of figures, but the deficits tend to be less coarse and may be confined to difficulty with accurate depiction of the inner details of complex figures.

It should be noted that most techniques for assessing visuospatial skills rely on motor activity of the patient. An alternative approach is to ask an individual to identify a particular geometric figure among a series of figures oriented in different planes. Such tasks require mental manipulation of figures without physical effort; therefore, they may be useful in examining individuals who have difficulty using their upper extremities.

Calculation

Difficulties with calculation may be seen in *Gerstmann syndrome* (acalculia, agraphia, right–left confusion, and finger agnosia) and other disorders that affect left hemisphere posterior parietal lobe functions. Some degree of receptive language difficulty typically accompanies damage to this area. Assessment of this and other mathematical cognitive functions presume a certain premorbid educational achievement, which should be ascertained. Calculation tasks can be set in clinical situations using examples that are at hand in the real world. Hypothetical situations of relevance to daily life, such as the purchase of items at a store, may be used. Comparison of oral and written modes of presentation and response can provide more extensive information about calculation abilities.

Frontal, Executive, and Higher-Order Cognition

The executive functions are cognitive processes that orchestrate complex, goal-directed activities (19). They include functions that we label variously as planning, sequencing, persistence, concept formation, organization, and prioritization. For successful implementation, these higher-order functions also require a certain baseline of initiative, drive, and energy. Sometimes these capacities are referred to as *frontal lobe* functions, because persons with frontal lobe brain disease often show deficit patterns in executive performance. Discrete neural localization of such executive functions should not be assumed, however, because they are dependent upon widespread neural networks in forebrain and hindbrain areas.

Assessment Tests
Royall and colleagues at the University of Texas Health Sciences Center at San Antonio have developed the EXIT25 tool for assessment of executive functions and developed data on its reliability and validity. Copies of this instrument should

be requested directly from Dr. Royall at the Department of Psychiatry, The University of Texas Health Science Center at San Antonio, 7703 Floyd Curl Drive, San Antonio, TX 78284-7792.

Many other bedside cognitive assessments can be performed that provide some insight into dimensions of these executive functions.

Verbal Fluency Task: Category Based. A score sheet for concept-based category retrieval task is shown in Table 1.2.

This is also a test of language, but comparing performance on a "category" trial with performance on a "letter" trial can provide insight into the ability to organize information and use strategies (i.e., category retrieval should be better because the patient has been provided with a strategy for retrieving). The examiner asks the patient to generate a list of all animals that he or she can think of in a 60-second period. Proper names are not allowed. The rate of production of animal types is recorded on the score sheet, according to each 15-second quartile. Expectable "normal" performance is 15 to 20 types of animals for persons with a high-school education. The test draws on a variety of cognitive skills, including conceptualization abilities, persistence, and creativity.

Letter Generation Test. As in the category fluency test, in the FAS test the patient is asked to produce all of the words he or she can think of starting with the letter "F", "A", and "S". One minute is allowed for each letter. This test also invokes language and verbal fluency function.

Trail-Making Test. The "Trails Test" is a more visuospatially oriented task that draws upon a variety of higher-order cognitive skills, including following directions, understanding concepts, changing psychological sets, and persistence. The B version of the Trails Test is shown in Figure 1.4.

In part A, the task is for the subject to connect the eight digits in numerical sequence. In part B, the task is to alternately connect letter A to number 1, letter B number 2, and so on until the task is completed without removing the pencil from the paper. The examiner determines how long it takes to complete (5-minute maximum). The examiner should point out mistakes during testing (by drawing a slash through incorrect lines) but should not indicate what the correct response should be; this is left for the patient to figure out. The test draws upon cognitive abilities of concentration and attention, and the ability to alternate cognitive sets.

Behavior, Action, and Style. Patients act, move, and interact in characteristic patterns that can be seen even in the clinical evaluation. They demonstrate patterns of psychomotor activity, which can be observed and described. These psychomotor patterns include a wide range of features, including hyperkinetic activity, restlessness, pacing, hand wringing, agitation, tics, automatisms, and mannerisms. Hypokinetic activity, such as psychomotor retardation or slow body movements, can be noted. Involuntary purposeless activity should

TABLE 1.2. CATEGORY RETRIEVAL: ANIMALS IN SIXTY SECONDS

Time Interval	Animals			
0–15 seconds				
16–30 seconds				
31–45 seconds				
46–60 seconds				

Word count: 0–15 seconds _____
16–30 seconds _____
31–45 seconds _____
46–60 seconds _____

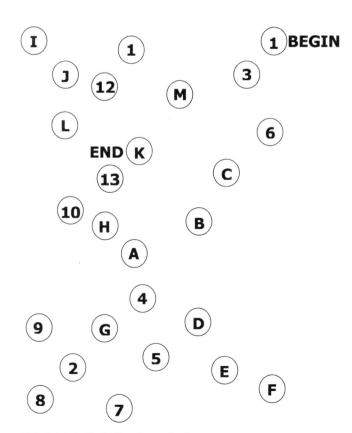

FIGURE 1.4. Trails Test (B version).

be described. Repetitive movements should be noted and described so that determination of neuroanatomic origin can be made, whether neocortical, limbic, basal ganglia, or other.

Action and style are additional dimensions of behavior that must be assessed during the neuropsychiatric examination. Each of us has an enduring set of predispositions with which we face life's challenges. These predispositions are for the most part *not* context dependent, nor are they easily changed from one point in time to another. We call them *traits,* as opposed to moods, thoughts, or symptoms. Dysfunctional patterns of coping with the world are called *character disorders.* There are many character qualities, such as honesty, timeliness, reliability, aggressiveness, and submissiveness.

Clusters of dysfunctional traits recur as patterns or archetypes in various individuals. When this happens, we speak of the *character disorders* or *personality disorders.* A number of these are listed with tables of qualifying features in the American Psychiatric Association's *Diagnostic and Statistical Manual of Mental Disorders,* 4th edition (DSM-IV) (20). An overview of these disorders deriving from DSM-IV is given in Table 1.3.

In DSM-IV, these disorders are classified as "axis II" disorders as opposed to the "axis I" classification of the more overt, state-dependent symptom complexes such as depression or anxiety.

Affect, Emotion, and Mood. *Affect* refers to the outward display of emotion as manifested by the patient's facial expression, speech, behavioral demeanor, and range of expressive behavior. Patients are generally categorized as normal,

TABLE 1.3. CHARACTER DISORDERS

Type	Features
Paranoid personality	Pattern of distrust and suspiciousness
Schizoid personality	Pattern of detachment from social relationships, with a restricted range of emotional expression
Schizotypal personality	Pattern of eccentricities in behavior and cognitive distortions; acute discomfort in close relationships
Antisocial personality	Pattern of disregard for rights of others; defect in the experience of compunction or remorse for harming others
Borderline personality	Pattern of instability in interpersonal relationships, self-image, and affective regulation
Histrionic personality	Pattern of emotional overreactivity, theatrical behaviors, and seductiveness
Narcissistic personality	Pattern of persisting grandiosity, need for admiration, and lack of empathy for others
Avoidant personality	Pattern of social inhibition, feelings of inadequacy, and hypersensivity to negative evaluation
Dependent personality	Pattern of submission and clinging behaviors
Obsessive-compulsive personality	Pattern of rigid, detail-oriented behaviors, often associated with compulsions
Personality disorders not otherwise specified	Disorders that do not fit into the above categories

euphoric, labile, constricted, restricted, blunted, or flat in affect. A patient's affect can be mood congruent or incongruent, which refers to the association between the patient's internal emotional state and present affect. For instance, a constricted or blunted affect associated with depressed mood would be congruent as opposed to a patient with a depressed mood and a euphoric affect. *Mood* can be defined as the individual's internal or subjective emotional experience. The examiner's access to subjective emotional experience is limited to the self-reports of the individual patient. Adjectives such as depressed, anxious, angry, irritable, guilty, labile, and euphoric may be used to describe this reported, internal emotional state. Mood disturbances are seen in a variety of neuropsychiatric disorders affecting various neural systems, including the frontal lobes or their connections, the right hemisphere, and a variety of subcortical and temporal lobe systems.

A number of unusual disorders of emotional function occur in neuropsychiatric disease, many of which have no precise counterpart in general psychiatry. An example is pathologic laughing and weeping, in which upper motor neuron regulation of the brainstem nuclei subserving emotional expression is disrupted. Such patients may complain of excessive and sometimes even explosive affective display in response to trivial emotional stimuli. They may describe "emotional incontinence," either sobbing uncontrollably when they feel only somewhat sad, or laughing in a grotesquely loud and disinhibited fashion in response to something that may strike them as only mildly amusing. Occasionally such patients become tearful when they are confronted with humorous events. At other times they may laugh when they experience sadness. Patients with schizophrenia or bilateral

frontal lobe disease may demonstrate strange affective disorders, variously demonstrating unusual detachment from serious life events or sometimes disinhibition from social norms. Apathy and anergy syndromes can appear in these and related disorders. In disorders involving the right hemisphere in right-handed people, disorders of emotional expression can occur, which are called the *aprosodias.*

Assessment Tests. Clinical descriptions of affective display and subjective affective experience usually are recorded in clinical work with neuropsychiatric patients. The use of more objective rating scales that have evolved from general psychiatry is limited in neuropsychiatric disease by the problem of generalizability. The Neuropsychiatric Inventory (NPI) is a relatively new scale that is adapted to several dimensions of behavioral function in neuropsychiatric patients (21).

Thought Process. *Thought process* refers to the connectedness of the patient's flow of ideas or to the tightness of his or her responsiveness to questions from the examiner. Presumably this observed quality of the mental status provides a picture of the internal connectedness of the *patient's thinking.* Thought process is best observed during the open-ended, initial portion of the interview, when the patient is encouraged to speak freely about whatever is on his or her mind. The examiner notes whether the patient's flow of thoughts is focused, organized, and goal directed, and what themes are followed. There is no discrete neuroanatomy or neural distributed network that uniquely subserves this quality of the mental state. Abnormalities of thought process are common in patients with nonfluent aphasias, the schizophrenias, mania, depressive psychosis, delirium, and other forms of encephalopathy.

The capacity of the patient to demonstrate abstraction and conceptual reasoning should be considered during the assessment of thought process. Patients may be asked to define similarities between objects such as a ball and orange or to interpret proverbs. One must be careful with placing too much importance on certain proverbs, because educational background and cultural issues may have a significant impact on their interpretation.

Abnormalities of thought process may be described, depending upon the severity, as flight of ideas, circumstantiality, tangentiality, loosening, and similar adjectives. *Thought blocking* refers to a total interruption of thought process such that the patient seems to experience "thought arrests." *Flight of ideas* refers to a pattern in which the patient leaps from one idea to another on the basis of idiosyncratic associations. Some manic individuals may make "clang" associations, in which the sound of a word, or its rhyming with another word, serves as the basis for shifting from one topic to another. Patients with psychomotor retardation may appear to have a poverty of thought and to lack bridging associations from one idea to the next. Similar slowing may be seen in brain diseases affecting the frontal lobe or basal ganglia, as well as in major depressive disorder. The term *Witzelsucht* refers to a facetious punning style exhibited by some patients with frontal lobe disinhibition. This jocularity resembles the playful quality of some hypomanic patients.

Thought Content. *Thought content* refers to the ideas, themes, and subjects that occur during the interview with the patient. If patients are allowed to speak in an open way, they usually bring out the themes and topics that are of importance to them. Traumatized patients may have recurring intrusive preoccupations with sounds, images, or feelings associated with the trauma. Psychotic individuals may be preoccupied with recurring hallucinations. Other themes of a patient's thought may be anger, guilt, diminished self-esteem, desire for closeness, or fear of intimacy. Obsessive thoughts, ruminations, fears, and concerns all are part of the thought content portion of the examination. Clinical assessment of suicidal and homicidal thoughts also belong in this portion of the examination. The examiner should appreciate the presence of delusions, grandiosity, ideas of reference (receiving commands of information from the television or radio), paranoid ideas, and magical or bizarre thoughts, as well as delusions of guilt, hopelessness, and worthlessness during the assessment of thought content.

Perception. Primary abnormalities of auditory, visual, olfactory, gustatory, or somesthetic function should be suspected initially during the interview portion of the examination. Specification of such deficits is accomplished later, during the neurologic examination itself. Higher-order abnormalities of perception can be focused during the interview. Hallucina-

tions and illusions represent such higher-order abnormalities of perceptual functioning. Patients with panic attacks and temporal lobe epilepsy may describe strange or bizarre abnormalities of perception, such as derealization or depersonalization. Experiences of déjà vu and jamais vous also may be described.

Fund of Knowledge. Patients can be asked where the location of the nearest major city is, what is the name of the president, or about current news events. Patients suffering from attentional and memory disturbances, as well as those with a limited level of education, may perform poorly.

Remote memory, which refers to a patient's recall of past events, may be difficult unless the examiner is familiar with the patient's biography. An alternate way of exploring remote memory is to ask the patient about famous people or historical events. Patients with dementia or with amnesia due to acquired brain disease may accurately recall events from the distant past but not from the more immediate past. Absence of such a temporal gradient should always raise the possibility of a psychogenic memory disturbance. Patients with major depression, for instance, may find all memory tasks equally difficult. Delirious patients may find it difficult to recall either remote or recent information.

Judgment and Insight. Over the course of the evaluation the examiner should have a reasonable idea regarding the patient's judgment and level of insight. A patient's *judgment* can be assessed by determining the level of self-understanding regarding behavior and potential consequences. Patients can be asked hypothetical questions concerning complex situations, which permits the examiner to assess the ability of the individual to elaborate and to weigh competing conceptual issues. *Insight* refers to the patient's level of self-awareness, whether of his or her illness, or of other self-referential issues.

Disturbances in insight can be seen in a variety of psychiatric disorders, but they also can be seen in neurologic conditions such as Wernicke's aphasia, dementia, or nondominant parietal lesions. Patients with borderline and narcissistic personality disorders, as well as patients with frontal lobe injuries, may be aware of certain laws, rules, or regulations but then behave in a contrary manner.

Reliability. If applicable, one should make mention of the accuracy or reliability of the information provided during the evaluation. Patients with dementia, amnestic disorders, confusional states, or major mental illnesses should be accompanied by a family member, neighbor, friend, or other individual who can help with the taking of the history.

Insight. Patients vary tremendously with regard to their capacity for self-awareness. With lesions involving the parietal lobe of the right hemisphere, patients may be unaware of, deny, or minimize problems involving the contralateral hemibody. Babinski coined the term *anosognosia* to describe the striking lack of awareness that he observed in patients whose right hemisphere strokes were associated with left hemiplegia. For other patients this quality seems to be much more characterologic and psychodevelopmental in origin.

If a patient shows no awareness of an apparent problem, the patient may be asked directly whether the apparent deficit has created any problems for the patient in relation to his or her family or employer. Awareness of a problem does not imply that the patient will be able to modify behavior based on that awareness. For instance, antisocial patients and individuals with frontal lobe syndromes may speak clearly about their awareness of society's regulations or the feelings of their spouse but then behave in a fashion that seems to ignore both.

Interpretation of the Mental Status Examination

In the end it is clinical judgment that synthesizes the mental status findings in a meaningful way and discerns the normal from the abnormal. Experience and patience improve these clinical judgments greatly. There are several issues to be considered.

The mental status examination looks at a patient's mental "state" at a *particular time*. Many factors, such as anxiety, insomnia, exhaustion, and medications, can affect patients' mental states and, therefore, their test performance. To determine whether impaired performance at one time reflects an enduring deficit, the examiner may need to repeat portions of the mental status examination at different times after the initial evaluation. Ideally, the clinician seeks to determine the *range* (both the zenith and the nadir) of a patient's cognitive capacity. For patients who are suspected of having a progressive dementing illness, assessment of cognition at different times over the course of several months or years may be required. Such serial testing becomes affected by practice effects, and alternate forms of formal tests should be used when feasible.

The levels of premorbid intellectual and psychological functioning of each patient must be considered. Patients with superior premorbid intelligence may have significant decrements in intellectual functioning but still perform within the normal range on cognitive screening tests designed to detect major impairments. Many of the briefer bedside instruments for cognitive assessment, such as the Mini-Mental State Examination (MMSE) (22) and the Cognitive Capacity Screening Examination (CCSE) (23), approach cognition as univariate (i.e., global) and have such low cutoff scores for the diagnosis of impairment.

Low premorbid educational and psychological achievements also limit the interpretation of mental status findings. Difficulties with spatial orientation, sentence structure, or arithmetic, for instance, commonly reflect learning disabilities that may not have been previously identified.

The absence of formal education makes it especially difficult to judge cognitive impairment that has been acquired due to neuropsychiatric disease.

Medications taken at the time of cognitive assessment or even a few days earlier can affect all mental status functioning. Medication-related cognitive deficits usually involve attentional mechanisms. Impaired performance in one or more areas of "higher cortical function" in a medicated patient should be interpreted conservatively if attentional tests such as digit span or letter cancellation show deficits. Whenever a patient is less than fully alert for whatever reason and particularly when the level of consciousness fluctuates (secondary to physical exhaustion, sleep deprivation, narcolepsy, and postictal states), cognitive test results must be interpreted conservatively.

Any impairment of language, such as word-finding difficulty or comprehension, may impair performance on other parts of the mental status examination, such as verbal memory, judgment, or thought process assessment, which are evaluated through verbal communication.

The time taken to complete cognitive tests provides information complementary to the accuracy with which they are completed. Patients may be able to successfully complete cognitive tasks given sufficient time, but they may be unable to do them in the time usually allocated to the task. Patients with bradyphrenia due to basal ganglia diseases are an example. Affective disorders and frontal lobe disorders also can affect speed of mental processing. The effects of medication on cognitive function can show themselves as psychomotor slowing.

Intercurrent systemic illness may produce changes in mental status or behavior (24). Early childhood illnesses may cause enduring effects on the brain or other major organ systems. Major medical illnesses in a child, the child's sibling, or other members of the family may be of psychodevelopmental relevance, affecting resultant attachment capacity or self-esteem. Surgical procedures, regardless of the age at which they are performed, may have special importance during development as events during which control of the body is relinquished to medical personnel.

General Physical Examination

It has been said that the general medical examination is that part of the neuropsychiatric examination that has to do with the review of systems. Some portions of this general medical examination must always be performed during a neuropsychiatric examination, at least the observational elements. Discerning observation of the patient's general appearance gives important clues about cardiorespiratory, endocrine, dermatologic, and musculoskeletal disease. Vital signs should be recorded either routinely or when suspicion arises during observational portions of the examination, depending upon the practice pattern of the clinician.

NEUROLOGIC EXAMINATION

The neurologic examination complements the clinical history and mental status examinations that preceded it (3). In the neurologic examination the examiner confirms and disconfirms impressions that arise during the interview. No examination is ever "comprehensive." Measures are emphasized from the neurologic examination that amplify and corroborate the clinician's developing impression of neuropsychiatric disease, individualized in each case. The examination has powerful operator characteristics related to the experience of the clinician, which are difficult to measure. The basic outlines of the major components of the neurologic examination follow.

Cranial Nerves

Cranial Nerve I

Smell should be tested separately in each nostril using a distinctive stimulus such as cloves. The substance should be nonirritating. "Scratch-and-sniff" cards are available for systematic, validated olfactory screening using standardized stimuli. Disturbance of the sense of smell may occur in a wide variety of neurologic disorders. Shearing of the olfactory filaments as they traverse the cribriform plate occurs frequently with closed head injury and can take place without loss of consciousness. Smoking probably is the most common cause of impaired smell in Western societies. Complete anosmia should be distinguished from partial loss of olfactory capacity (hyposmia). Some patients are unaware of their olfactory loss and may not appear concerned even when a clear olfactory deficit is documented, whereas other individuals complain bitterly of their loss of olfactory perception.

Cranial Nerve II

Examination of the optic nerve entails examination of the optic fundus (retina), visual fields, visual acuity, and pupil. Funduscopic examination allows consideration of the retinal vascularity and the optic disk. Visual acuity can be assessed with a pocket screener or wall chart. It is important to indicate whether acuity is assessed with patients wearing corrective lenses or glasses. In the absence of corrective lenses, a pinhole can be used to correct for refractive errors. Visual fields can be tested with small red-and-white objects on the tip of a pointer. Cotton swabs on long sticks will serve the purpose at the bedside if standard test objects are not at hand. Each eye is examined independently. Pupillary reactivity to direct and consensual light is recorded, as is the near reflex (accommodation and convergence).

Cranial Nerves III, IV, and VI

Extraocular motility is examined by observing the patient's upward, downward, lateral, and angular gaze. The examiner looks closely for nystagmus, diplopia, or limitation of gaze in any direction.

Cranial Nerve V

Facial sensation over each of the three branches of the trigeminal nerve is tested with light touch and pinprick, and each side compared to the other. The corneal reflex (afferent limb via cranial nerve V, efferent limb via cranial nerve VII) should be tested with a wisp of cotton. Its absence can be the sole evidence of trigeminal nerve dysfunction; its presence is evidence against facial anesthesia.

Cranial Nerve VII

Evidence of facial asymmetry is noted both at rest and during facial expression. In peripheral facial nerve lesions, the entire hemiface is involved, whereas in a central lesion the forehead is spared. Central lesions involving the right face may be associated with aphasia.

Cranial Nerve VIII

Sensorineural hearing loss should be distinguished from conductive hearing loss, if possible. High-frequency hearing loss (presbycusis) is common in the elderly. Hearing may be assessed at the bedside with whispering, tuning forks, or ideally an audioscope. The examiner also performs the Weber test in which a 512-Hz tuning fork is placed on the vertex and the patient is asked whether the sound is heard equally well in both ears. In conductive hearing loss, the sound will be louder on the defective side. In sensorineural hearing loss, the sound will be greater on the less impaired side. Finally, bone conduction is compared with air conduction by placing the base of the vibrating tuning first on the mastoid process behind the ear and asking the patient to compare the strength of that sound with the sound of the tuning fork in the air. In conductive hearing loss, bone conduction is greater than air conduction.

The vestibular component of cranial nerve VIII can be tested with the Nylen-Báarány maneuver, which stimulates the vestibular apparatus. The patient initially is seated and then moved rapidly into a supine posture with head tilted 45 degrees backward below the level of the table and 45 degrees to the left. The patient resumes the seated posture and the maneuver is repeated, moving the head backward and 45 degrees to the right. The examiner notes whether nystagmus or symptoms of vertigo are produced by this maneuver and compares the intensity and duration of nystagmus with turning of the head toward the left and the right sides.

Cranial Nerves IX and X

The afferent limb of the gag reflex is via cranial nerve IX and the efferent limb via cranial nerve X. Patients with upper motor neuron palsy may have an overly brisk reflex with excessive coughing, whereas patients with lower motor neuron palsy may have a diminished or absent gag reflex leading to increased risk of aspiration. Swallowing is a complex act that can be impaired even if the gag reflex is intact. Timed swallowing of liquid is a sensitive bedside test for dysphagia. A rate less than 10 mL/s is strongly associated with objective demonstration of dysphagia on cine esophagram.

Cranial Nerve XI

Function of the accessory nerve is tested by palpation of the sternocleidomastoid muscle as the patient turns his or her head to the left and right alternately while being opposed by the hand of the examiner. Weakness of the *left* sternocleidomastoid impairs head turning to the *right*, and vice versa. This point occasionally is useful in identifying weakness due to (hysterical) conversion. Shrugging or elevation of the shoulders is accomplished through activation of the trapezius muscle. The examiner requests the patient to elevate his or her shoulders while the examiner palpates the involved muscle.

Cranial Nerve XII

The patient is asked to protrude the tongue and move it from side to side. In unilateral paralysis of cranial nerve XII, the tongue will be observed to deviate toward the side of the weakness.

Motor Examination

Muscle mass is observed for any asymmetry of bulk and for any focal atrophy due to peripheral lesions or disuse.

Muscle Tone

Muscle tone is tested by passively moving the patient's limb. The examiner compares tone in both extremities and notes the presence or absence of increased tone. Increased tone may be characterized as having a "jackknife" quality typical of upper motor neuron disease: paratonic (a ratchetlike intermittent tone seen in frontal lobe disease), cog wheeling (parkinsonian), or negativistic (the rigid resistance of catatonia). Absence or flaccidity of muscle tone is indicative of lower motor neuron disease, and decreased tone is associated with cerebellar lesions. Tone varies considerably among normal individuals.

Muscle Strength

Muscle strength is graded from 0 to 5, where 0 is no movement, 1 is trace movement, 2 is movement with gravity in a horizontal plane, 3 is movement against gravity, 4 is movement against gravity and against applied force, and 5

is normal strength. The examiner compares left versus right, proximal versus distal, and upper versus lower extremity strength.

Reflexes

Reflexes are graded from 0 to 4, with 2 being average. Briskness of reflexes varies in healthy normal individuals, and areflexia that is not asymmetric and not associated with weakness or other motor or sensory problems is not necessarily pathologic. In individuals who are anxious, consuming benzodiazepines, or withdrawing from alcohol, reflexes may be diffusely increased. Asymmetry between reflexes and discrepancy between reflexes in upper and lower extremities can have localizing significance.

Pathologic reflexes: In a normal adult, stimulation of the plantar surface of the foot produces either no response or plantar flexion (a downgoing toe). In individuals with upper motor neuron pathology (i.e., a pyramidal tract lesion), the great toe may dorsiflex (i.e., extend) and the toes fan out. An upgoing toe with or without fanning of the other toes is referred to as a *Babinski sign.* The Babinski sign in an adult usually implies structural disease of the central nervous system or a toxic metabolic encephalopathy.

There are a number of *"primitive" reflexes* that may be seen in infants during the first year of life. Reappearance of these reflexes (i.e., grasp, suck, snout, root) suggests frontal lobe disorders. The occurrence of primitive reflexes in elderly persons is of relatively low diagnostic specificity.

Coordination

Coordination (cerebellar) testing consists of finger-nose-finger, heel-knee-shin, and rapid alternating movements. Weakness can interfere with coordination testing. Spasticity can cause clumsiness but not the dysmetria typical of cerebellar lesions.

Station

Is the patient able to stand comfortably with his or her feet together? Is there evidence of swaying when the patient is asked to close his or her eyes (positive Romberg sign)? Can the patient stand on one foot?

Gait

The examiner should observe several components in a patient's gait. How wide is the stride? How broad based is the gait? Are movements made in a shuffling tentative fashion? Is there a footdrop with a "steppage" gait? Is turning done smoothly or "en bloc"? Is there asymmetry of arm swing? The time taken to walk a fixed distance can be used as a quantitative measure of a patient's gait.

Abnormal Movements

Is there a paucity or an excess of movements (hypokinesia vs. hyperkinesia)? Are there spontaneous dyskinesias (such as those of Huntington disease or tardive dyskinesia)? Is there a tremor visible at rest (parkinsonian tremor), postural tremor (familial or essential tremor), or tremor that appears only with intention (cerebellar tremor)?

Myoclonic jerks may involve a single limb or the entire body. Fasciculations (movements of small muscle groups) can be seen in amyotrophic lateral sclerosis (ALS) and in nerve-root lesions.

Sensation

The examiner tests primary sensory perception by examining the patient's awareness of pinprick, light touch, position, and vibration. Parietal sensory testing consists of examining a patient's capacity to recognize numbers or letters written on the palm or on the sole of the foot while the patient is not looking (graphesthesia); the ability to recognize an object from manipulating or palpating it (stereognosis); and the ability to detect sensory stimuli applied at the same time in different anatomic regions (double-simultaneous stimulation). It is meaningless to test for parietal deficits in a region of significant primary sensory loss.

Soft Signs

The neurologic and psychiatric literature has been inconsistent as to whether the traditional neurologic examination is of assistance in the evaluation of learning difficulties. Several pediatric neurologists have attempted to refine the neurologic examination by asking patients to repetitively perform certain tasks, such as jumping on one foot, and noting the nature of difficulties seen over specific periods of time.

In contrast to "subtle" signs of neurologic abnormalities, such as reflex asymmetries or equivocal Babinski signs, "soft" neurologic signs are unlikely to imply lateralized pathology and more often reflect developmental immaturity of the central nervous system.

Soft neurologic signs include the following:

Fine motor incoordination: Awkwardness at tasks such as handwriting, finger-nose-finger, and finger pursuit in the absence of frank cerebellar dysmetria

Dysrhythmia: Lack of smooth transitions between different motor tasks

Mirror movements: Contralateral overflow of motor activity in homologous muscle groups

Synkinesis: Ipsilateral overflow of extraneous associated movements when the patient is asked to perform an activity involving a specific set of muscle groups

Both mirror movements and synkinesis can be elicited by asking the patient to perform activities that involve discrete muscle groups: (a) tap one foot; (b) alternately tap heel and toe; (c) pat one's thigh; (d) flip-flop one hand on one's thigh; (e) tap one's thumb and index finger repeatedly; and (f) successively touch one's index, middle, ring, and little fingers to one's thumb.

Mirror movements and synkinesis may be seen as a failure to inhibit or suppress movements in larger or distant muscle groups while attempting to perform a discrete task.

The reliability of rating of soft signs can be enhanced by having a standard routine for eliciting the signs, such as having the patient performing movements *a* through *f* ten times in succession on the right and then on the left. Timing the sets of ten with a stopwatch can give further useful information about lateralized differences in speed of performance.

Soft signs do not correlate sharply with neuroanatomic systems, as do the preceding components of the neurologic examination. They are useful occasionally as indices of more global cerebral dysfunction of various etiologies and should be used to confirm impressions of such global impairment.

CONCLUSION

The neuropsychiatric examination allows the clinician to diagnose neuropsychiatric disturbance, to use laboratory and imaging ancillary testing in an informed way, to set therapeutic plans, and to monitor the success of the therapies. The examination may clarify the functional consequences of various neuropsychiatric diseases and the unique meaning of those functional consequences for each individual. Only prolonged experience and self-correction by the clinician can bring these examination skills to maturity.

REFERENCES

1. Lajara-Nanson WA, Schiffer RB. Neuropsychiatry. In: *Baker's clinical neurology,* cd-rom. Philadelphia: Lippincott Williams & Wilkins, 2001.
2. American Psychiatric Association Work Group on Practice Guidelines for the Psychiatric Evaluation of Adults. Practice guidelines for the psychiatric evaluation of adults. *Am J Psychiatry* 1995;152[Suppl]:65–80.
3. DeJong R. *The neurologic examination,* 4th ed. New York: Harper & Row, 1979.
4. Strub RL, Black FW. *The mental status examination in neurology,* 3rd ed. Philadelphia: FA Davis, 1993.
5. Trzepacz PT, Baker RW. *The psychiatric mental status examination.* New York: Oxford University Press, 1993.
6. Weintraub S, Mesulam M-M. Mental state assessment of young and elderly adults in behavioral neurology. In: Mesulam M-M, ed. *Principles of behavioral neurology.* Philadelphia: FA Davis, 1985:71–124.
7. Jennett B, Bond M. Assessment of outcome after severe brain damage. A practical scale. *Lancet* 1975;I:480–481.
8. Wechsler D. *Wechsler Memory Scale-Revised manual.* San Antonio, TX: The Psychological Corporation, 1987.
9. Wechsler D. *WAIS-R manual.* San Antonio, TX: The Psychological Corporation, 1981.
10. Wilson RS, Beckett LA, Bennett DA, et al. Change in cognitive function in older persons from a community population. *Arch Neurol* 1999;56:1274–1279.
11. Lezak M. *Neuropsychological assessment,* 3rd ed. New York: Oxford University Press, 1995:523–558.

12. Benson DF. *Aphasia, alexia, and agraphia.* New York: Churchill Livingstone, 1979.
13. Squire LA. Declarative and nondeclarative memory: multiple brain systems supporting learning and memory. *J Cogn Neurosci* 1992:4;232–243.
14. Squire LA. Memory and the hippocampus: a synthesis from findings with rats and humans. *Psychol Rev* 1992;99:195–231.
15. Buschke H, Fuld PA. Evaluation of storage, retention, and retrieval in disordered memory and learning. *Neurology* 1974;11:1019–1025.
16. Delis DC, Kramer JH, Kaplan E, et al. *California Verbal Learning Test: Adult Version.* San Antonio, TX: The Psychological Corporation, 1987.
17. Brandt J. The Hopkins Verbal Learning Test: development of a new verbal memory test with six equivalent forms. *Clin Neuropsychol* 1991;5:125–142.
18. Rey A. *L'examen clinique en psychologie.* Paris: Presses Universitaires de France, 1964.
19. Royall DR. Executive cognitive impairment: a novel perspective on dementia. *Neuroepidemiology* 2000;19: 293–299.
20. American Psychiatric Association. *Diagnostic and Statistical Manual of Mental Disorders,* 4th ed. Washington, DC: American Psychiatric Association, 1994.
21. Wood S, Cummings JL, Hsu M-A, et al. The use of the neuropsychiatric inventory in nursing home residents: characterization and measure. *Am J Geriatr Psychiatry* 2000;8: 75–83.
22. Folstein M, Folstein S, McHugh PR. Mini-Mental State: a practical method for grading the cognitive state of the patient for the clinician. *J Psychiatr Res* 1975;12:189–198.
23. Jacobs JW, Bernard MR, Delgado A, et al. Screening for organic mental syndromes in the medically ill. *Ann Intern Med* 1977;86:40–47.
24. Schiffer RB, Klein RF, Sider RC. *The medical evaluation of psychiatric patients.* New York: Plenum Press, 1988.

2

NEUROPSYCHOLOGICAL ASSESSMENT

STEPHEN M. RAO
SARA J. SWANSON

This chapter describes the clinical neuropsychological examination and its role in the management of individuals with presumed or verified brain dysfunction. Whereas the study of brain-behavior relationships has a rich history extending from the nineteenth-century revelations of Broca and Wernicke, the field of human neuropsychology has only recently evolved (1). The application of psychological tests to study brain dysfunction took hold in the decades following World War II with the pioneering experimental studies of focal brain damage by Hans-Lukas Teuber, Brenda Milner, Alexander Luria, Arthur Benton, Henry Hécaen, and Ward Halstead. These psychological investigations, based on accidents of nature (strokes and tumors), warfare (penetrating head wounds), and surgery (cortical excision of epileptic foci), provided important new insights regarding cerebral localization and lateralization. These early studies formed the basis for the scientific field of human neuropsychology.

As a scientific endeavor, human neuropsychology has grown exponentially, along with all other areas of neuroscience research during the past 3 decades. The emphasis in human neuropsychology has broadened to include more diffusely brain-damaged individuals (closed head injury, Alzheimer disease, acquired immunodeficiency syndrome, demyelinating disease). The field also has turned its attention to traditional psychiatric disorders, such as schizophrenia and affective disorders. In recent years increasing emphasis has been placed on symptom validity testing, improving the empirical basis of normative standards for detecting impairment as well as statistical methods for determining the existence of meaningful cognitive change and development of briefer test batteries targeted to specific diseases or clinical questions.

At the same time, validation of the lesion models of localization can be achieved by comparison with an unprecedented variety of sophisticated brain imaging technologies used to extract structural and functional information from the intact human brain. These technologies include computed tomography (CT), structural and functional magnetic resonance imaging, positron emission tomography, single photon emission computed tomography, magnetoencephalography, event-related potentials, transcranial magnetic stimulation, magnetic resonance spectroscopy, diffusion- and perfusion-weighted imaging, magnetization transfer imaging, fluid-attenuated inversion recovery imaging, and diffusion tensor imaging (2–6). Not only are they providing more sensitive means for detecting neuropathology and concurrent validation of existing models, but they also are generating new ideas about brain functional organization (7,8). Functional neuroimaging studies allow exploration of the parts of specific brain systems that are active during cognitive tasks. In addition, human neuropsychological models and techniques have become increasingly more sophisticated with the emergence of cognitive neuropsychology as a scientific discipline (9). These new paradigms are theory driven and parcel components of cognitive systems, such as attention, memory, and language, in ways that more closely match the distributed neural networks of the human brain. Further theoretical advances in cognitive neuroscience have influenced clinical neuropsychology through the development of cognitive tests and interpretive strategies that are based on conceptual understanding of cognitive models (e.g., working memory).

Clinical neuropsychology as a professional specialty is a relatively recent development, with its roots originating in the 1970's. Whereas human neuropsychology is multidisciplinary involving specialists in experimental, cognitive, and clinical psychology, neurology, psychiatry, linguistics, speech pathology, and neuroscience, clinical neuropsychology typically is practiced by clinical psychologists, who have completed specialized doctoral, internship, and residency education and training in clinical neuropsychology. Clinical neuropsychologists are charged with applying the scientific knowledge derived from human neuropsychological research to the evaluation and treatment of individuals suspected of brain dysfunction. (At the end of this chapter we briefly discuss the training, qualifications, and credentialing of clinical neuropsychologists.) The purpose of this chapter is to provide an overview of the field of clinical neuropsychological practice. For more detailed discussions of the points raised in this chapter, the reader is referred to textbooks pertaining to clinical neuropsychology (10) or to the scientific field of human neuropsychology (11).

Because of space limitations, this chapter focuses on the assessment of adults. The tests and assessment issues involved in pediatric clinical neuropsychology are sufficiently different that this area could not be covered in detail (for a review of pediatric neuropsychology, see references 12–14). Likewise, this chapter emphasizes the assessment of acquired rather than developmental brain disorders (attention deficit/hyperactivity disorder, learning disabilities) (see Chapters 23 and 24). Finally, the emphasis of the chapter is on the assessment rather than the treatment of neurobehavioral disorders. Several comprehensive textbooks have been published that address issues associated with cognitive and behavioral interventions (15–22) (see Chapter 7).

REASONS FOR CONDUCTING A NEUROPSYCHOLOGICAL ASSESSMENT

The reasons for performing a clinical neuropsychological assessment have broadened over the years. During the 1950's and 1960's, evaluations frequently were performed to address a single question: Is the patient experiencing symptoms of "organicity"? To address this question, clinical psychologists relied on conventional testing instruments, such as the Wechsler intelligence scales, the Bender-Gestalt figure copying test, or personality tests such as the Minnesota Multiphasic Personality Inventory (MMPI) and the Rorschach Inkblot Test. With the proliferation of more specialized neuropsychological procedures, such as the Halstead-Reitan Battery in the 1960's, clinicians began to broaden their role and started to compete with the neurologic examination and diagnostic procedures of the period (e.g., electroencephalogram, pneumoencephalogram) for localizing lesions within the cerebral hemispheres. With the advent of more sophisticated brain imaging techniques, such as CT in the late 1970's and magnetic resonance imaging in the mid-1980's, localization of lesions with neuropsychological testing became less important. Some neuropsychologists began to wonder whether these technologic advancements would eliminate the need for psychological testing of organic brain syndromes.

Subsequent studies showed the limitations of structural brain imaging in predicting the type and extent of behavioral change in dementing and traumatic brain disorders (23). For example, neuropsychological testing often is more sensitive than structural neuroimaging and neurologic examination for assessing neurobehavioral changes after closed head trauma, detecting cognitive decline in the early stages of dementia, and distinguishing dementia from depression. Neuroimaging findings alone have limited predictive ability with regard to the behavioral effects of lesions, whereas neuropsychological examination can define the neurobehavioral syndrome associated with specific brain abnormalities (24). Thus, the emphasis of neuropsychological assessment

shifted from lesion localization to differential diagnosis of neurobehavioral disorders and a more comprehensive characterization of the patient's cognitive and emotional status. The more frequently asked referral question is no longer "Where is the lesion?" but rather "What is the neurobehavioral diagnosis in this patient with a right frontal glioma?" (e.g., hemispatial neglect) and "What are the nature and severity of the cognitive and emotional symptoms?" (e.g., tactile allesthesia, perseveration, depression).

There are three primary ways that neuropsychological testing is used relative to diagnosis. First, documentation of neuropsychological deficits is necessary for diagnosis of certain neurologic conditions, particularly dementia. The clinical criteria for probable Alzheimer disease and mild cognitive impairment require confirmatory neuropsychological test data showing evidence of cognitive deficits. Impairment must be documented in two or more cognitive domains, including memory, language, perceptual skills, attention, praxis, orientation, problem solving, and functional abilities based on the diagnostic criteria of Alzheimer disease from the National Institute of Neurological and Communicative Disorders and Stroke and the Alzheimer's Disease and Related Disorders Association (NINCDS-ADRDA) (25). Neuropsychological testing is critical for differentiating mild cognitive impairment from normal age-related memory changes, and depression or pseudodementia where a depressed individual presents with complaints suggestive of dementia. Early detection of dementia leads to intervention strategies such as the use of memory-enhancing medications at a time when such medications are thought to be most useful for potentially altering the rate of progression from mild cognitive impairment or presymptomatic Alzheimer disease to dementia (26,27). In a 10-year prospective follow-up study of 603 community-based nondemented elderly, measures of executive function and delayed recall of word lists were the most effective at discriminating between those who would develop dementia and those who would remain nondemented 1.5 years later (28). Thus neuropsychological testing plays a role in the early detection and initial diagnosis of some neurologic disorders.

Second, analysis of a profile of neuropsychological test scores is used for differential diagnosis of the major neurobehavioral syndromes, such as dementia, confusion (29), amnesia, attentional disorders, aphasia, nonaphasic focal deficit syndromes of the dominant hemisphere (e.g., Gerstmann syndrome, alexia without agraphia), focal deficit syndromes of the nondominant hemisphere, frontal lobe syndrome, and developmental disorders (for a review of neurobehavioral syndromes, see reference 30). The pattern of performance across measures of immediate, recent, and remote memory is particularly useful for differentiating dementia, confusion, dementia with confusion, and amnesia (Table 2.1). Table 2.1 shows that dementia is characterized by impairment in recent and remote memory in the context of intact immediate or span memory. The memory profile of an individual with

TABLE 2.1. MEMORY FUNCTIONING ASSOCIATED WITH THE MAJOR NEUROBEHAVIORAL SYNDROMES

	Memory Functions		
	Immediate	Recent	Remote
Confusional states (delirium)	−	−	+
Dementia	+	−	−
Dementia with confusion	−	−	−
Primary amnesia	+	−	+
Attentional disorder (aprosexia)	−	+	+

Adapted from Hamsher KD. Specialized neuropsychological assessment methods. In: Goldsein G, Hersen M, eds. *Handbook of psychological assessment,* 2nd ed. New York: Pergamon Press, 1990.

amnesia shows impaired recent memory. Amnestic individuals recall no information after a delay but have intact remote and immediate memory.

Third, cognitive testing can be used to differentiate between subtypes within neurobehavioral syndromes, such as classifying the type of aphasia or dementia or elucidating psychological factors that are contributing to the symptoms. For example, specific cognitive profiles are associated with cortical (Alzheimer disease) versus subcortical dementias (31,32), such as vascular dementias associated with small-vessel ischemic disease, Huntington disease, progressive supranuclear palsy, Parkinson disease (33), human immunodeficiency virus infection (34), and multiple sclerosis (35). Whereas patients with cortical dementias typically show impaired encoding affecting both recall and recognition memory, impaired naming, and reduced semantic fluency, patients with subcortical dementias often show impaired recall but normal recognition indicating intact encoding, psychomotor slowing, attentional difficulties, executive deficits, and motor dysfunction. The lobar variants of dementia, such as frontotemporal degeneration (36), primary progressive aphasia (37), or the visual spatial variant of Alzheimer disease (38,39), present with focal or lateralized cognitive impairment that can be detected with neuropsychological testing. In addition, distinct cognitive profiles are associated with specific dementias such as dementia with Lewy bodies, where disproportionate impairment is seen in working memory and in executive and visuospatial domains in the context of relatively preserved memory (40,41).

Integration of the results of cognitive and personality testing is useful for differentiating organic from psychiatric conditions such as depression or somatoform disorders that present with neurologic symptoms. Patients who present with memory complaints secondary to depression have a generally normal profile on psychometric testing or may show psychomotor slowing, reduced working memory/attention, and poorer performance on effortful memory measures (42–45). They show evidence of depression on personality testing or mood scales. Patients with somatoform disorders pre-

senting with neurologic symptoms and cognitive complaints have generally normal cognitive profiles or display a functional presentation on cognitive tests in the context of a profile suggesting conversion or somatization on the MMPI-2 [elevations on scales 1 (hypochondriasis) and 3 (hysteria)].

Although differential diagnosis of neurobehavioral disorders remains one of the primary purposes of neuropsychological testing, the applications for neuropsychological testing have widened considerably in recent years. Neuropsychological testing has been used to establish baseline measures to monitor changes in time associated with progressive cerebral diseases (neoplasms, demyelinating, and dementing conditions) or recovery from acute brain disorders (traumatic head injury, stroke). Neuropsychological testing has been used as an outcome measure in clinical efficacy trials involving surgical, pharmacologic, and behavioral interventions. Examples include medically oriented treatments (e.g., drug trials, revascularization, resective surgery of tumors or seizure foci, ventriculoperitoneal shunts) designed to treat specific symptoms or monitor side effects, and psychological treatments (e.g., language or cognitive retraining) oriented to improving various aspects of neurobehavioral functioning. Although practical problems arise in exposing subjects to repeated neuropsychological testing (see discussion of practice effects later), such testing can be a useful adjunct to the neurologic examination, brain imaging, and other biologically oriented outcome measures. Neuropsychological testing has the advantage of providing information that may be useful in predicting changes in quality of life, for example, employment and social adjustment (46). Furthermore, improvements in brain structure or physiology may be viewed with some skepticism if there is no measurable behavioral change, particularly in the context of a managed health care environment.

In the forensic area, neuropsychological testing has become an important component of civil and criminal cases involving individuals with suspected or acquired brain damage. The legal system is the third most prevalent source of referrals for neuropsychologists (47,48). In personal injury lawsuits, the most common of which arise from motor vehicle accidents, neuropsychological testing is frequently the only objective method for describing the neurobehavioral sequelae associated with mild closed head injuries, whereas CT and magnetic resonance scans may show little if any observable signs of brain damage (23). Neuropsychological testing is uniquely well suited for establishing whether a person is competent to make judgments regarding his or her medical management and personal finances and whether the person is in need of protective placement. In criminal cases, neuropsychological assessments have provided assistance to determine if an individual is competent to stand trial (49,50), is capable of assisting in his or her own counsel, or is not guilty by reason of insanity secondary to acute or chronic brain dysfunction. Neuropsychological testing is used in criminal cases to determine if an individual has cognitive deficits associated with a brain injury or neurologic disorder that may be used by the defense to mitigate the sentence.

Finally, neuropsychological testing can be applied to determine the cognitive and affective status of the patient for formulation and design of rehabilitation and remedial interventions. By defining a patient's strengths and weaknesses, neuropsychological testing also can be useful for educational and vocational planning or for determining appropriate placement (51). Psychometric testing can be used to make recommendations about driving or the need for a formal driving evaluation in patients with hemispatial neglect, visual perceptual disturbance, impaired attention, mental flexibility, or psychomotor speed. Family members and caretakers can experience considerable emotional turmoil in trying to adjust to the neurobehavioral changes of the patient (10) and may require education or psychotherapeutic intervention. Neuropsychological assessment can provide valuable information to family members who are trying to cope with the cognitive and emotional changes associated with brain dysfunction. Such an assessment can determine the patient's level of insight and awareness and define his or her readiness and appropriateness for psychotherapy. Based on a recent review of 125 metaanalytic studies (52), compelling evidence was found to support the validity of neuropsychological assessment for the purposes of differential diagnosis, prediction of functional capacity, and determination of treatment needs. Further, the validity of psychological tests was found to be comparable to medical test validity, particularly when multimethod assessments were conducted.

NEUROPSYCHOLOGICAL ASSESSMENT APPROACHES

In light of the diverse origins of human neuropsychology and the ever-expanding scientific literature in this field, it should come as no surprise that there are different philosophies regarding how to conduct a clinical neuropsychological examination. In the developing years of clinical neuropsychology, many clinicians favored a fixed or standard battery of neuropsychological tests, such as the Halstead-Reitan Battery (53) or the Luria-Nebraska Battery (54). The standard battery typically takes 3 to 8 hours to administer and provides a comprehensive assessment of cognitive, perceptual, linguistic, and sensorimotor skills. Normative standards have been generated in healthy individuals, and a large body of research comparing the performance of various patient groups on these batteries has been published (54,55). Traditionally, the clinical assessment was noninteractive in that the clinician rarely changed the assessment procedures in light of the referring questions or the patient's clinical status. Clinicians who adopt a battery approach typically integrate test data with results from a comprehensive psychosocial interview of the patient and family members and from pertinent medical, school, and legal records.

Historically, the field has debated the merits of fixed versus flexible batteries (56,57), but changes in the health care environment have necessitated assessments that are briefer and targeted to the referral question. Fixed batteries such as the Halstead-Reitan Battery and now the Expanded Halstead-Reitan Battery provide excellent sources for normative data that are corrected for demographic variables such as age, education, and gender (55).

Currently, the most common approach to assessment, advocated initially by Luria (58) and described by Lezak (10), calls for greater flexibility in the selection of neuropsychological tests. According to a recent survey, this flexible battery approach is preferred by 85% of neuropsychologists (47). The test battery is specifically tailored to the referral question and is influenced by the interview and prior medical and psychosocial records. Thus, an obviously aphasic patient with a left hemisphere stroke verified by CT scan may receive a more comprehensive language assessment than a patient with no obvious signs of aphasia in spontaneous conversation. The testing approach is hypothesis driven and highly interactive. The clinical neuropsychologist may include additional tests based on the performance of the patient on earlier testing to evaluate a specific hypothesis regarding the nature of a cognitive deficit. This testing philosophy also emphasizes a qualitative analysis of test performance, as best exemplified by the Boston process approach (59). In point of fact, most clinical neuropsychologists who advocate a flexible approach give their patients a core battery of tests that provide a brief screen of a wide range of neuropsychological functions, which then is supplemented by further testing depending on the outcome of the screening battery.

Clinical research involving group studies of various patient populations can be hindered by the flexible battery approach because the same testing procedures may not be used across patient groups. As a result, several scientific committees have advocated the use of specific test batteries tailored to a particular disease.

BRIEF TEST BATTERIES

Brief batteries have been developed to assess the cognitive deficits common to particular neurologic diseases. A review of all the brief or tailored batteries is beyond the scope of this chapter, but an overview of some brief or tailored assessment batteries is provided. Several brief batteries for the assessment of multiple sclerosis have been developed (60–62). Our multiple sclerosis screening battery includes measures of sustained attention, verbal fluency, word list learning, spatial learning, and mental processing speed (60). These tests are sensitive to the cognitive changes associated with white matter pathology, are free of significant motor demands, and have multiple alternate forms that allow for repeat testing over time and thus assessment of disease progression. A similar battery has been developed by Beatty et al. (61).

Various test batteries have been developed to assess cognitive functioning in presurgical epilepsy patients. For example, the Bozeman Epilepsy Consortium has adopted a set of neuropsychological tests for assessing patients across several epilepsy surgical centers, a practice that allows for research using much larger patient samples when data are pooled. Tests included in epilepsy batteries are sensitive to temporal lobe pathology and thus emphasize verbal and nonverbal memory [e.g., Wechsler Memory Scale-Revised (WMS-R) or WMS-III, California Verbal Learning Test-II (CVLT-II)] and language (Boston Naming Test or Multilingual Aphasia Examination Visual Naming), but they also assess intelligence and the presence of lateralized brain dysfunction associated with a seizure-onset focus. Material-specific memory deficits, primarily reduced verbal memory in association with complex partial seizures originating in the dominant temporal lobe, are found in patients with mesial temporal lobe sclerosis (63). Both neuropsychological testing and asymmetric performance on memory testing during the intracarotid sodium amobarbital (Amytal) test are used to predict side of seizure focus, risk for postsurgical amnesia or verbal memory decline, and seizure outcome after temporal lobectomy (64–70).

Brief assessment batteries covering several cognitive domains have been developed to streamline the assessment of dementia and are particularly well suited for testing inpatient or impaired populations. The Repeatable Battery for the Assessment of Neuropsychological Status (RBANS) includes measures of language, immediate memory, delayed memory, attention, and visuospatial abilities. This battery can be administered in 30 minutes to assess dementia or screen younger patients (71). Alternatively, the Dementia Rating Scale-2 (DRS-2) is a measure that was designed to track cognitive changes in degenerative dementia patients ranging in age from 55 to 89 years and includes age- and education-corrected normative data. This test provides a global measure of dementia based on subtests measuring attention, perseveration, initiation, construction, abstract reasoning, and memory (72,73). The DRS-2 and RBANS are brief, not limited by floor effects, and provide more information about performance within specific cognitive domains than, for example, a Mini-Mental State Examination (MMSE).

Specific batteries have been recommended for the assessment of patients with human immunodeficiency virus infection (74), Parkinson disease (75), pallidotomy candidates (76), Huntington disease (77), and Alzheimer disease [the Consortium to Establish a Registry for Alzheimer's Disease (CERAD) Battery (78,79)] and for evaluation of acute concussion in athletes (80). Assessment of specific patient populations can be streamlined when measures sensitive to particular pathologies are selected. Such brief core batteries can be altered flexibly by the examiner based on the patient's performance. Use of published selected batteries

also facilitates large multicenter studies when test results are pooled.

METHODS OF INTERPRETATION

Neuropsychological assessment typically involves an integration of interview data, medical and school records when available, and neuropsychological test data. Determination of the presence, type, and degree of brain dysfunction is based on several inferential methods. The most common methods include measurement of the patient's level of performance, appearance of pathognomonic signs for brain dysfunction, testing for lateralized brain dysfunction, and profile analysis that is consistent with known pathology.

Level of Performance

The most common method for inferring brain impairment from neuropsychological testing is based on a deficit model (10). This model assumes that cognitive impairment has occurred when a discrepancy is observed between the patient's level of test performance and his or her estimated ability level before the onset of brain dysfunction. This method assumes that the patient's premorbid ability level can be accurately predicted. Rarely, however, does the clinical neuropsychologist have available information regarding the patient's premorbid cognitive ability level at the time of assessment. School records, when available, may yield pertinent data, such as group-administered intellectual and academic achievement test scores. Such information may indicate an individual's level of functioning relative to a local or national normative standard before the onset of brain dysfunction. However, group-administered tests may not provide a measure of optimal ability. Tests of intelligence also may be negatively affected by the presence of a learning disability.

More commonly, the patient's highest level of functioning is inferred indirectly through demographic data or from the current testing session. Several investigators have generated regression formulas from the standardization sample for the Wechsler intelligence scales for calculating premorbid intelligence from demographic variables, such as gender, age, race, education, occupational status, and region of the country (81–83). Because these predictor sets have less than perfect correlations with intelligence, the estimated IQs generated from these regression equations tend to underestimate deviation from the mean that is most apparent in cases at the extremes of the normal distribution.

An alternative method is to use the patient's current level of performance on tests that are less sensitive to brain dysfunction. Lezak (10) advocated a method whereby the patient's highest performance score on a test battery is used as an estimate of premorbid ability. More recently, clinicians have advocated the use of specialized tests of premorbid intellectual ability. Two such tests are the National Adult

Reading Test (NART) (84,85) and the Wide Range Achievement Test-Revised (WRAT-R)/III reading scores (86,87). These tests assume that the ability to read words aloud is retained in the presence of cognitive decline. Studies have shown, however, that the NART may provide a reasonably accurate estimate of premorbid ability in mild dementing conditions, but NART scores drop significantly with more severe forms of dementia (88). Furthermore, Wiens et al. (89) found that the correlation between the NART-estimated IQ and the Wechsler full-scale IQ was relatively low ($r = 0.46$), resulting in lower premorbid estimates in the high-IQ group and higher estimates in the low-IQ group. The NART was developed with the specific goal of IQ estimation, whereas the WRAT-R reading subtest was not developed for this purpose but is commonly used as a "hold" measure. WRAT-R reading scores have similar correlations with IQ in healthy adults ($r = 0.60$) (90), but less data are available about the ability of WRAT-R/III scores to predict premorbid IQ in brain-injured populations. When WRAT-R/III scores were examined longitudinally in stable, declining, or improving patient groups, reading scores were stable in the stable and declining patient groups but showed improvement in patients recovering from brain injuries. The patients who showed intellectual improvement after brain injury also showed significant WRAT-R/III improvements over time, which suggests caution in using reading to estimate premorbid IQ after brain injury (91).

One is left with the impression that only in circumstances where the neuropathologic process does not affect the neural circuitry critical to the "hold" function might this strategy of estimation of premorbid function be useful or appropriate.

In addition to the inherent limitations of inferring decline from a level of functioning that is imperfectly predicted, this approach also assumes that the patient's premorbid level of intellectual functioning will be uniform across a wide range of cognitive domains. Intercorrelations between measures of intelligence, attention, memory, and visuospatial abilities are less than perfect in healthy adults. Dodrill (92) found that intelligence is significantly correlated with neuropsychological test performance when IQ is below average (66–89), but this same correlation was not found for individuals with average (90–109) or above average (110–131) IQ. Further, individuals with a developmental learning disability will be difficult to detect using such premorbid estimation models.

Pathognomonic Signs

An additional method of inferring cognitive decline is based on detection of qualitative changes that are characteristic of specific types of brain dysfunction, particularly in the acute state. Such changes have an extremely low base rate in the normal population. Examples include the appearance of left-sided neglect, motor perseverations, confabulation, and paraphasic speech. A patient with a midline frontal lobe lesion

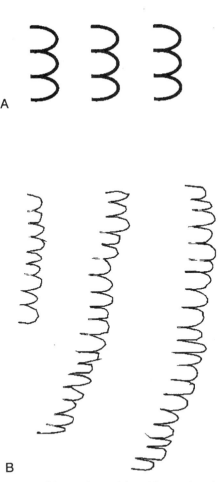

FIGURE 2.1. Repetitive series writing. The patient is asked to copy the line drawings shown in **(A)**. The patient's copy **(B)** demonstrates motor perseveration, an inability to suppress a motor program once started.

with a deep extension involving the basal ganglia was asked to copy a design consisting of three connected half-circles (Fig. 2.1A). The reproduction (Fig. 2.1B) demonstrates the inability of the patient to voluntarily stop a motor program after it has begun. A spontaneous drawing of a flower (Fig. 2.2A) and a copy of a butterfly (Figs. 2.2B and C) reveal hemispatial neglect in a patient with a right parietal stroke.

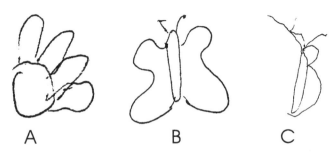

FIGURE 2.2. The patient was asked to draw a flower **(A)** and then copy the examiner's butterfly **(B)**. When the patient was asked if her butterfly looked like the examiner's, she said "no," and added a third antenna on the left side of her drawing **(C)**.

Laterality

The third inferential method involves testing of abilities that imply dysfunction in one cerebral hemisphere relative to the other. Most neuropsychological examinations include measurements of upper extremity motor strength, speed, and coordination as a means of comparing right and left body sides. Assessments of visual, auditory, and tactile perceptual asymmetries are performed for similar purposes. Clinicians also test more complex perceptual and memory processes with verbal and visual test stimuli as a means of inferring lateralized dysfunction. It is not uncommon to compare a patient's recall of a story (as derived from the logical memory subtest of the WMS-R/III) to their recall of a complex geometric design (from the visual reproduction subtest of the WMS-R/III) to determine if there may be selective involvement of the left or right medial temporal lobe, respectively (93).

Profile Analysis

This interpretive approach posits that a neurobehavioral syndrome will demonstrate a fairly specific pattern or profile of deficits on a comprehensive neuropsychological test battery (10). It is assumed that there is consistency in the expression of cognitive functions within an individual. When test scores are converted to a similar metric (standard scores, T-scores, percentiles) and plotted, one can readily observe a patient's strengths and weaknesses. A good example of this approach is used in the Boston Diagnostic Aphasia Examination-Revised (94), in which different test profiles are believed to be indicative of specific aphasic syndromes (e.g., Broca vs. Wernicke vs. anomic aphasia).

ISSUES AFFECTING THE INTERPRETATION OF NEUROPSYCHOLOGICAL TESTING

Numerous factors can influence or modify the interpretation of neuropsychological testing. In this section, some of the more relevant factors are discussed, including issues pertaining to test reliability and validity, adequacy of normative standards, motivational factors, influence of affective disorders and medication effects, validity of specific neuropsychological tests in patients with sensorimotor dysfunction, problems of interpretation created by practice effects with repeated testing, and fatigue effects with extended neuropsychological testing.

Test Reliability and Validity

An important issue in psychometric testing is the ability of a test to demonstrate adequate levels of reliability. Reliability can be assessed by examining a test's consistency over time or its internal consistency (whether all test items assess the same psychological construct). Whereas tests with high levels of reliability may be insensitive in detecting brain dysfunction, the converse is not true: unreliable measures cannot be valid measures of brain dysfunction.

An important question for all neuropsychological instruments is whether the test is capable of discriminating between healthy individuals and patients with presumed brain dysfunction (test validity). An increasingly important question pertains to whether a test also can discriminate between various brain-damaged groups (e.g., frontal lobe degeneration vs. Alzheimer disease). Whereas a test may demonstrate a statistically significant group difference based on means scores, one also may ask whether the instrument has *clinical* utility, that is, whether the test can be used in making discriminations on a case-by-case basis.

Some neuropsychological tests are claimed to be sensitive to brain disruption in a specific brain region (e.g., the frontal lobes). The double dissociation method is a common approach for validating such claims. This validation method typically involves two focal lesion groups and two test instruments. According to this procedure, if patients with a lesion in brain region X are impaired on test A but not on test B, and patients with a lesion in region Y are impaired on test B but not on test A, it is assumed that the two tests are selectively sensitive to disruption of each of the separate brain regions. As an example, Heindel et al. (95) showed that patients exhibiting a subcortical dementia from Huntington disease were impaired on implicit memory measures involving perceptuomotor learning, but they performed normally on priming tasks. In contrast, the cortical dementia associated with Alzheimer disease was found to disrupt priming but not perceptuomotor learning.

Finally, test validity can involve an evaluation to determine if the neuropsychological test measures the psychological construct (e.g., recent memory) it is purported to evaluate. Factor analysis studies performed on tests of the same construct (e.g., concept formation) can indicate whether they are redundant (i.e., load on a single factor) or tap different constructs. One method for assessing construct validity is to use the multitrait-multimethod approach developed by Campbell and Fiske (96). In this approach, three or more cognitive domains (e.g., recent memory, attention, conceptual reasoning) would be assessed by at least three neuropsychological tests per domain. Tests that use different methods to measure the same cognitive domain should correlate at a higher level than tests that use the same method but test different cognitive domains. This method can be used to establish both convergent and discriminant validity. Moreover, using a multimethod assessment battery improves the validity of clinical judgments made about individual patients (52).

Normative Standards

Estimates of cognitive decline require the use of normative data to determine the ranking of the patient's test performance relative to an appropriate peer group.

Neuropsychological tests are frequently influenced by age, education, race, socioeconomic status, and, less frequently, gender (97). The effects of these demographic factors can be complex and nonuniform either across or within cognitive domains. For example, numerous studies have demonstrated a reliable age decrement in memory tests that require spontaneous or cued recall of previously presented material (98). In contrast, age differences are much less pronounced when memory is assessed using recognition tasks, which require the subject to discriminate between previously presented information and new material. Presumably, recall tasks require more self-initiated and effortful processing than recognition tasks (99). Furthermore, recognition tests have a large degree of environmental support (i.e., cues) that is absent in recall tests.

The ideal neuropsychological test, therefore, has normative data derived from a large (preferably nationwide), stratified, and randomized standardization sample. For example, the Wechsler Adult Intelligence Scale-III (WAIS-III) and WMS-III standardization samples included 2,450 and 1,250 adults, respectively, ranging in age from 16 to 89 years. The sample matched the 1995 census population proportions on gender, race/ethnicity, educational level, and geographic region (100). Few neuropsychological tests approach this ideal, and in some cases the "norms" are based on a small control sample (n < 30) reported in a clinical investigation. Whereas some neuropsychological tests have well-documented manuals available from the test publishers, other tests in the public domain have norms that are available on a less formal basis. Several textbooks provide norms for the latter tests (10,55,101,102).

Motivational Factors

The outcome of neuropsychological testing can be influenced by the degree of cooperation and effort put forth by the patient. The validity of neuropsychological test performance has been called into question in patients seeking financial compensation for injuries (103,104). Evaluation of the veridicality of the patient's responses has become increasingly important in forensic cases where patients are seeking financial compensation after mild closed head injury, for example. A complex set of nonneurologic factors, such as symptom magnification, poor effort, psychological need to maintain a sick or disabled role (e.g., factitious disorder), somatoform disorder (e.g., somatization, conversion), and malingering can result in poor performance on cognitive tests (105). Precise prevalence estimates of malingering on neuropsychological testing are unknown, although data from personal injury and worker's compensation cases suggest rates of 33% to 64% (104,106). To compound this issue, several investigations have found that neuropsychologists were unable to discriminate between neuropsychological test protocols of patients without financial incentives and subjects asked by the investigators to simulate cognitive deficits (107,108). It

should be noted that the neuropsychologists in these studies were blinded to interview data and medical records.

Assessment of malingering has become more sophisticated as worker's compensation and forensic referrals increase and neuropsychological data is relied upon for determination of financial settlements. Standards have been proposed outlining the diagnostic criteria for malingered cognitive impairment (109,110). The criteria for definite malingering of neurocognitive dysfunction include (a) the presence of a substantial external incentive such as material gain and avoiding duty or financial responsibility; (b) a negative response bias as indicated by a below chance performance on one or more forced-choice measures (see later) or discrepancy between the test data and known patterns of brain functioning, observed behavior, or collateral reports; and (c) the factors listed in (b) that are not better accounted for by psychiatric, developmental, or neurologic factors.

In recent years, the increased interest in the detection of malingering has resulted in an increased effort to develop measures sensitive to malingering. One of the first widely used measures of malingering is the Rey (111) 15-Item Memory Test. For this test, the patient is presented with 15 items on a sheet of paper and asked to reproduce the items from memory after a 10-second exposure interval (Fig. 2.3A). The instructions emphasize the large number of items to be recalled. The actual amount of information required to correctly recall items in this array, however, is small and within span memory. The majority of brain-impaired patients are capable of recalling at least two of the rows in the correct order. Figure 2.3B shows the performance of a suspected malingerer. Although this test has been criticized for having low sensitivity (4%–22%) (104), Arnett et al. (112), using a cutoff score of less than two correct rows, yielded acceptable sensitivity and specificity values in the ranges from 47% to 64% and 96% to 97% in discriminating neurologically impaired patients from subjects instructed to feign memory impairment.

Although the Rey 15-item test is best suited for detecting floor effects, an alternative method for assessing malingering, referred to as symptom validity testing (113), uses a probabilistic analysis of performance on a forced two-choice testing format where the chance of obtaining a correct response typically is 50% (114). Malingerers are assumed to perform below chance levels as determined from the binomial probability for wrong responses; thus, a subject who obtains only 10% correct is believed to be feigning impairment. One such technique, the Portland Digit Recognition Test (115), consists of a forced-choice measure of recognition memory. Numerous studies have examined the sensitivity and specificity of this test. Using the below chance criterion, low sensitivity rates (less than 35% of simulators) have been achieved (104). Cutoff scores that are less restrictive (90% correct), however, have produced higher sensitivity rates for detection of feigned memory loss without lowering specificity rates (116).

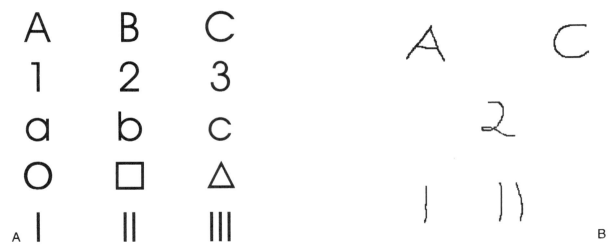

FIGURE 2.3. The 15-item test. The patient is shown the design in **(A)** for 10 seconds. The design is removed, and the patient is given a blank piece of paper and asked to draw the design from memory. The drawing in **(B)** is from a patient suspected of malingering.

The Computerized Assessment of Response Bias (CARB) is another well-validated, forced-choice measure of response bias (117). This measure is sensitive to effort in head trauma patients. In a study in which litigating patients who sustained head traumas ranging from trivial to severe were tested on the CARB, patients with mild closed head injuries were found to perform more poorly than patients with severe head traumas (105). Alternatively, the Validity Indicator Profile (VIP) (118) combines the forced-choice paradigm of symptom validity testing with the comparison of the patient's performance on hard and easy test items and provides a more general measure of response bias (119). This test includes normative data from normal subjects, compliant brain-injured patients, normal subjects instructed to feign impairment, suspected malingerers, random responders, and mentally retarded individuals. Even with the varied groups included in the normative sample, differentiation between careless responding due to low effort and a malingered response style with high motivation to fail was difficult (119). However, the VIP is more sensitive and sophisticated than older simpler methods for assessing malingering, and the test was designed to assess both effort and motivation.

The 21-Item Memory Test is another forced-choice measure developed for assessment of malingered memory deficits. A 21-item word list is presented, followed by free recall and then a forced-choice recognition trial. One hundred percent of the normal and memory-impaired subjects and 65% of the experimental malingerers were correctly classified on this task using a cutting score of 9 of a possible 21 on the recognition trial. Using a cutting score of 13, no normal controls and only one memory-impaired subject were falsely classified, resulting in an overall accuracy rate of 98% (120).

One of the more widely used measures of effort, the Test of Memory Malingering (TOMM) (121), involves presenta-

tion of line drawings of common objects, followed by forced-choice recognition testing with corrective feedback for two trials, followed by a retention trial 15 minutes later. Patients with documented brain dysfunction perform well on this measure, with more than 90% correct on trial 2 and during the retention trial. This test has been well validated, has a high degree of specificity, and is not affected by age or educational level (122,123).

Affective Disorders

Patients with major affective disorders may exhibit potentially reversible cognitive decline on neuropsychological testing. Weingartner (124) proposed a central motivational hypothesis to account for the memory disorders observed on neuropsychological testing in depressed patients. According to this hypothesis, depressed patients will perform poorly on tests requiring high degrees of mental effort, such as free recall tests, but perform normally on memory tests that are more automatic and requiring less effort, as in recognition testing formats or incidental learning tasks. Using an incidental memory task as a modification of the digit symbol subtest of the WAIS-R, Hart et al. (125) were able to successfully differentiate patients with Alzheimer disease from those diagnosed with pseudodementia.

Some investigators view the effects of major depression as not simply a complication in interpreting neuropsychological testing. Functional neuroimaging differences have been found in prefrontal cortex, anterior cingulate, and amygdala in depressed individuals relative to controls in resting state activation, response to emotional stimuli, and correlations between negative affect and amygdalar activation (126–129). Evidence has accumulated that some patients with major affective disorders may be experiencing a form of subcortical dementia, which may have correlates with changes on structural and functional neuroimaging (130). There also appears to be an interaction between age and depressive mood state and cognitive decline (131). Finally, there is some evidence that the cognitive decline is not reversed with treatment of depression (132). Thus, it often is difficult to determine whether depression is a normal psychological reaction to chronic physical illness or part of a specific neurobehavioral syndrome associated with brain dysfunction (133).

Medication Effects

A variety of prescription medications can have an impact on cognitive functioning and thereby alter the interpretation of neuropsychological testing (for a review, see references 134–136). Examples of such drug classes include antidepressants with significant anticholinergic side effects (e.g., amitriptyline), anticonvulsants (particularly phenobarbital), anxiolytics (particularly the benzodiazepines), and the antipsychotic agents. In addition, various drugs of abuse (alcohol, stimulants, cocaine, heroine, cannabinol, and nicotine) can have short- and long-term effects on cognitive test performance. Typically, these effects are observed most often on measures of attention, memory, information processing speed, and fine motor dexterity.

Sensorimotor Dysfunction

Many neuropsychological tests that assess higher cognitive functions make the assumption that patients have adequate primary sensory and motor functions to perform the task. Many neurologic and nonneurologic conditions produce sensorimotor impairment from disease or trauma that have its primary effects outside the cerebral hemispheres. For example, patients with severe rheumatoid arthritis will perform worse on the performance subtests of the WAIS-R because of an inability to manipulate the test stimuli in a speeded fashion. Patients with multiple sclerosis will experience visual and tactile sensory loss or motor impairment from white matter lesions affecting the optic nerve, brainstem, spinal cord, or cerebellum. Such lesions should have little or no effect on cognitive processes, but they can reduce performance scores on cognitive tests, particularly those that demand motor speed/dexterity or fine visual acuity (137). Despite these obvious limitations, it is not unusual for inexperienced neuropsychological clinicians to administer an extensive test battery to patients with sensorimotor deficits and attempt to infer cognitive impairment.

Several methods are available to address these interpretive problems. Clinicians knowledgeable of the nature of the primary sensorimotor deficit will attempt to select neuropsychological tests that provide unambiguous test findings. As an example, paper-and-pencil tests can be avoided in favor of tests that require a vocal response. When information regarding the speed of cognitive processing is required, tests that use the identical motor response across a range of conditions varying in cognitive complexity can provide useful information. For example, subtracting timed scores on Trails A from Trails B, tests of visual search and sequencing of different levels of cognitive difficulty, may be a more valid approach than comparing each test score separately to normative standards.

Practice Effects

As noted earlier, neuropsychological testing has been used to monitor changes in a patient's status over time or as an outcome variable to evaluate various forms of treatment interventions. Several studies have demonstrated that repeated neuropsychological testing results in improvements in test scores in both normal and brain-damaged populations (138–144). *Practice effects*, as they have come to be known, can present challenges in test interpretation. For example, a patient may obtain an identical score on a test administered 6 months apart. From these data, it is unclear whether the patient's status is unchanged or a deterioration occurred that was canceled out by an improvement associated with practice.

It is unclear why practice effects occur. Some clinicians have assumed that practice effects result from the explicit and conscious recollection of the identical test stimuli. If this is the case, it might be assumed that practice effects will be minimized after longer retest intervals (>1 year) due to normal forgetting from long-term memory. Furthermore, practice effects could be eliminated over shorter intervals by constructing alternate test forms of equivalent difficulty. Unfortunately, most of the studies cited used brief retest intervals and did not use alternate, equivalent test forms. In an unpublished study, we showed that both assumptions may be incorrect. Practice effects were demonstrated after a 3-year retest period in the majority of test scores derived from a lengthy neuropsychological test battery. Furthermore, alternate equivalent test forms, administered in counterbalanced fashion, did not appear to eliminate the effect. These data suggest that practice effects do not result entirely from the explicit recall of identical test stimuli, rather they occur as a function of learning how to take a test. This type of procedural or implicit memory may be more resistant to decay over time than explicit memory.

To address the problem of practice effects, manuals for neuropsychological tests might include retest norms for various follow-up periods. In group treatment studies that use neuropsychological outcome measures, it is highly recommended that a yoked control group be incorporated into the experimental design. Empirical methods for detecting cognitive change after temporal lobectomy have been developed that control for practice effects, test–retest reliability, and regression to the mean (145–148). Reliable change indexes or regression-based change norms have been statistically derived by testing a control group of unoperated epilepsy patients twice using an identical battery of neuropsychological tests at the same time interval as the epilepsy group undergoing surgery. Using the regression-based models, each patient's baseline score on a measure (raw score on an object naming test, for example) is entered into a regression equation that controls for practice effects and test–retest unreliability, and includes any demographic variables that predict scores on that measure at time 2 in the control group. The predicted score is compared to the observed score, divided by the standard error of estimate from the regression equation, and converted to a standard score. Thus, normative data exist for a number of neuropsychological tests that clinicians can use to determine the significance of change from a particular baseline score. Eventually such norms will be developed in other patient populations. In this way, change norms can be used that are specific to diagnostic groups because different patient populations may be more or less susceptible to practice effects.

Fatigue Effects and Duration of Testing

A common complaint about neuropsychological testing is the length of testing, which for fixed batteries like the Halstead-Reitan Battery can extend over a 6- to 8-hour period. Patients frequently experience fatigue and reduced motivation at the end of a lengthy testing session. These effects may influence the validity of such tests. To counteract this problem, most neuropsychologists administer tests that are most likely to be influenced by fatigue (e.g., attention and memory) at the beginning of the test session.

A related question pertains to the need for extended testing to perform a reliable, valid, and clinically meaningful assessment. This issue is becoming of greater importance in an era when cost containment of medical and psychological procedures is being carefully scrutinized. Clinical neuropsychologists who adapt a flexible approach to testing tend to tailor the amount of testing to answer the referral questions while taking into consideration the capacity of the patient to tolerate an extended examination. Thus, it may not be reasonable to subject an elderly patient in the middle to late stages of Alzheimer disease to an extended examination when the referral question is a relatively simple one, that is, to determine if the patient has deteriorated cognitively from the previous evaluation 6 months ago.

On the other hand, the same patient presenting at an earlier stage of the illness may display symptoms of mild memory loss in the context of a clinical depression. Evaluation at this stage of the disease is more complicated because the cognitive deficits may be subtle or confined to a single domain, such as memory as in mild cognitive impairment (26). Furthermore, more complicated questions arise regarding the patient's capacity to work, drive, and make financial decisions. The outcome of this neuropsychological evaluation may assist in determining whether the patient has a potentially treatable disorder (depression) or a progressive dementing illness. Thus, the length of neuropsychological testing likely will increase with the complexity of the referral question and the associated risk of making errors in interpretation.

Several investigators have developed screening examinations for quickly assessing cognitive functions at the bedside (149). The most commonly used screening instrument, the MMSE (150), is brief (5–10 minutes), can be administered by a health care professional without psychological training, and yields a single score that minimizes interpretative skills. Unfortunately, the MMSE is insensitive to most of the milder forms of cognitive dysfunction or to dementing disorders affecting primarily subcortical structures (60). The Mattis Dementia Rating Scale (151) addresses many of the criticisms of the MMSE, but it takes longer to administer and requires greater psychological expertise to administer and interpret.

SPECIFIC NEUROPSYCHOLOGICAL TESTING PROCEDURES

This section describes the neurobehavioral areas typically assessed in a neuropsychological examination. The emphasis is on describing those standardized tests most commonly used by neuropsychologists in clinical practice. In addition, newly developed standardized tests are described, along with occasional experimental measures that hold promise as clinical assessment tools but have not yet been standardized.

Intelligence

Neuropsychological assessments typically begin with a measurement of intellectual functioning. The most common test used in clinical practice is the WAIS-III (100), which consists of six verbal and five performance subtests. Three intelligence quotients, a full-scale, verbal, and performance IQ, are derived. A major advantage of this test is its large, stratified, normative database, which corrects for age differences. Comparisons of tested IQ values with premorbid IQ estimates (from demographic variables or tests such as the NART, as noted earlier) can be achieved to obtain a gross measure of the degree of cognitive deterioration. A major disadvantage of this test is that it was designed to provide a measure of an individual's general ability levels. Consequently, performance on each subtest is influenced by multiple cognitive operations. A low score on a subtest, such as block design, may result from deficits in focused or lateralized attention, general planning and organizational skills, visuospatial perception, constructional abilities, psychomotor slowing, or primary sensorimotor impairment. Thus, although the intelligence measures may be sensitive, they lack specificity for understanding the reasons why an individual may perform in an impaired fashion. Recent attempts to modify the examination process to extract the qualitative features of the patient's test performance may address this limitation to some degree, although most clinicians will continue to supplement the WAIS-III with more specific neuropsychological tests (152).

In addition to comparing estimated versus actual IQ test scores, clinicians frequently will examine various patterns of test performance. One such method is to compare the verbal comprehension index and perceptual organization index scores from the WAIS-III as a measure of lateralized brain dysfunction. Whereas low verbal abilities relative to perceptual organizational abilities are suggestive of dominant hemisphere damage, the opposite pattern (a low perceptual organization index score relative to verbal comprehension index score) is thought to be indicative of nondominant hemisphere damage. Likewise, lower scores on WMS-III auditory memory relative to visual memory tasks may be seen in patients with dominant hemisphere dysfunction. Group studies of patients with focal epilepsy showed lower verbal comprehension index scores and lower verbal memory scores on the WAIS-III and WMS-III (WAIS-III WMS-III Manual). Older group studies of patients with lateralized lesions generally support this test pattern (153), although the effect is attenuated in females (154) and in patients with more chronic lesions (153). Still, it is easy to find clinical examples of patients who are misdiagnosed by this index (e.g., patients with subcortical pathology, nondominant parietal lobe lesions). It should be noted that verbal performance IQ differences by as many as 15 points can occur in 20% of college-educated, healthy individuals (155). The WAIS-III Manual includes data on the frequency of differences between IQ and index scores by full-scale IQ. In the normative sample, 11% of the individuals with an IQ of 79 or below had a verbal comprehension/perceptual organization difference of 15 or more, whereas 29% of the sample with an IQ ranging from 110 to 119 had verbal comprehension/perceptual organization differences of 15 or more points. Thus, caution should be exercised in interpreting this index unless there is a large verbal performance IQ discrepancy (>21 points).

Another method of interpretation involves analysis of intersubtest scatter. Individual subtests of the WAIS-III are standardized with a mean of 10, standard deviation of 3, and range from 1 to 19. Scatter analysis is achieved by comparing the highest and lowest scores of the 11 subtests. An implicit assumption is that the highest scores reflect premorbid skill levels, whereas the lowest scores are indicative of brain dysfunction. Matarazzo et al. (156) showed that scatter is common in the WAIS-R standardization sample. Typically, a scatter of at least 9 to 12 points, depending on the full-scale IQ level, is necessary to be clinically significant (defined as less than the fifth percentile). The WAIS-III Manual includes intercorrelations of subtest scaled scores for each age group, subtest score differences that are significant at either a 0.05 or 0.15 level, and percentages of the normative sample that obtained various subtest score differences.

A final method of interpretation involves an examination of intrasubtest scatter. Most of the subtests of the WAIS-R and WAIS-III are composed of items that become progressively more difficult. It is assumed that healthy individuals will begin to fail items abruptly as they approach the upper end of their ability level. In contrast, brain-damaged patients, particularly those of high premorbid ability, may pass difficult items at the end of the subtest but fail easy items at the beginning. Several investigators have developed various indices of intrasubtest scatter and shown that they may be useful in discriminating some types of brain dysfunction (157,158).

As noted earlier, results from the WAIS-III performance subtests may be of questionable validity in some patient groups because of the emphasis on upper extremity motor speed. Patients with cervical spinal cord injuries or rheumatoid arthritis, for example, may achieve lower than expected IQ scores because they were unable to obtain "bonus" points for rapidly manipulating the test stimuli. However, a new subtest added to the WAIS-III, matrix reasoning, is useful for making intersubtest comparisons because it is an untimed measure of spatial reasoning, is free of motor demands, and has a similar format to the Raven Standard Progressive Matrices (159).

Conceptual Reasoning and Executive Functions

Neuropsychological assessments traditionally have included measurement of nonverbal abstract thinking or concept formation. The two most common tasks used in this regard are the Wisconsin Card Sorting Test (WCST) (160,161) and the Category Test (162). The WCST, in particular, is used by more than 70% of practicing neuropsychologists (163). Both measures are sensitive to cerebral dysfunction in neurologic (164–169) and psychiatric (165,170–172) disorders. The WCST is viewed as a measure particularly sensitive to focal frontal lobe dysfunction. Several studies have found a greater number of WCST perseverative errors in patients with frontal lobe lesions than those with nonfrontal lesions (166,173,174), although this relationship has not been uniformly observed (175,176).

Error scores on the WCST and the Category Test are correlated on a statistical basis, but they share only 12% to 30% of the common test variance. A study by Perrine (177) suggests that the two tests may emphasize different concept formation operations. Specifically, the WCST may be related to the identification of stimulus attributes, whereas the Category Test involves a greater degree of rule learning. Not surprisingly, therefore, the Category Test has a stronger correlation with the Wechsler full-scale IQ than the WCST (167) and is more likely to be influenced by premorbid intellectual ability.

Conceptual reasoning may be subsumed under a broader category of cognitive operations, commonly referred to as *executive functions*. Executive functions reflect the "ability to spontaneously generate efficient strategies when relying on self-directed and task-specific planning" (178).

According to Shallice and Burgess (179), executive or supervisory functions involve planning or decision making that requires error correction or troubleshooting, resulting in the performance of novel sequences of actions that overcome strong habitual responses. The frontal lobes are assumed to be the critical structures involved in the performance of executive functions. Clinically, patients with executive disorders exhibit perseveration (an inability to stop a sequence of actions once begun), loss of initiative or intention to act, inability to generate plans, tendency to act impulsively, and problems incorporating feedback in modifying their behavior.

A wide variety of tasks have been used to tap this diverse cluster of cognitive abilities. Some, like the Tower of Hanoi or the Tower of London (180,181), require patients to rearrange blocks from an initial position into a goal position in a minimum number of moves. Verbal (182) and design (183) fluency tasks assess the ability of the patient to spontaneously generate items consistent with established rules. The Porteus Maze Test (184) assesses the patients' ability to trace the correct path through a maze. Copying repetitive line drawings (multiple loops; Fig. 2.1) or recurrent series writing ("*mnomnomno*") can be used to assess motor perseveration (185). Finally, executive functions can involve the ability to plan, follow, arrange, or recall the temporal order or sequence of events. Frontal lobe patients, for example, may experience problems in judging which of two stimuli was presented more recently, but they have little or no impairment on recognition testing in which the patient must discriminate whether or not a stimulus was presented previously (186). Tasks that require patients to generate a script (187) that describes the sequence of events they would perform, for example, in getting ready for work in the morning or selecting and attending a restaurant may prove useful in understanding the executive deficits of patients and may provide useful information for developing meaningful rehabilitation strategies.

Executive functions are associated with the patients' ability to evaluate the accuracy of their own performance. Some brain-damaged patients, particularly those with severe impairment, show no or limited awareness of deficits (i.e., anosognosia) despite profound cognitive deficits (188). Patients can be particularly impaired in rating their memory ability. This ability, referred to as *metamemory*, can be assessed by asking patients to rate their confidence in recalling or recognizing previously presented material. Such testing can be performed in conjunction with standardized memory testing.

Attention

Attention frequently is impaired in neurologic disorders and may be the primary area of cognitive dysfunction in psychiatric disorders, such as schizophrenia (189). Attention can be segregated into focal, sustained, and divided processes (190). *Focused attention* refers to the process of search for, and localization of, target stimuli. *Sustained attention*, sometimes used interchangeably with the term *vigilance*, refers to the monitoring of target stimuli over an extended duration of time. *Divided attention* refers to the ability to perform two tasks simultaneously. Most neuropsychological tests of attention used in clinical practice assess one or more of these attentional components.

Some attentional measures emphasize speed of information processing. Slowing frequently occurs in patients with symptoms of a subcortical dementia (191). One of the commonly used tests in clinical practice, the Trail-Making Test (53), requires the patient to "connect the dots" using a paper-and-pencil format. Two forms are administered: Trails A involves a simple numerical series, whereas Trails B alternates between numerical and alphabetical series. While norms are available for time to completion for each form, a pure measure of mental processing speed with increasing task complexity can be achieved by subtracting Trails A from Trails B, because the same motor response is used for both conditions. The Stroop test (192) also has multiple forms of increasing complexity. Unlike the Trails test, which examines primarily focused and divided attention (alternation of two series), the Stroop test assesses the patient's vulnerability to interference effects (focused attention). In the simpler (control) conditions of this test, patients are timed while they are either reading a list of color word names printed in black or reading the color name to a series of identical nonverbal stimuli printed in various colors. In the interference condition, a series of color words is printed in colors different from what the word represents (e.g., the word "blue" is printed in the color green). The patient is asked to name the color the word is printed in rather than read the word. Healthy subjects typically take longer to complete the interference task than the control tasks. Norms are available for assessing abnormally long interference effects in patient populations (101). Another sensitive test of mental processing speed with excellent norms that allows either an oral or written test format is the Symbol Digit Modalities Test (193).

Although sometimes lengthy to administer, some clinicians prefer reaction time tasks to tease apart mental from motor processing speed. One method involves comparing reaction times from simple (one-choice) versus complex (multi-choice) stimulus presentations where the same motor response is required for all conditions, enabling a subtraction of the simple from complex reaction times. Similarly, the Sternberg paradigm (194) has been used to demonstrate slowing of mental processing in patients with substantial motor impairment, such as Parkinson disease (195) and multiple sclerosis (196). For this task, the patient must hold a digit sequence of varying length (one to five digits) in working memory and decide whether a single probe digit presented on a computer screen matches one of the numbers in memory. Most individuals achieve better than 95% correct on this task; however, reaction times increase linearly with the number of digits held in memory. Abnormal slowing is demonstrated by a steeper slope than in healthy individuals.

Other attentional measures emphasize accuracy of performance, such as the Paced Auditory Serial Addition Test (197). This demanding test requires the patient to add consecutive single-digit numbers presented at rates ranging from about one every second to one every 3 seconds. The test does not require a cumulative tally but rather addition of the current digit with the immediately preceding digit. This test has been particularly sensitive to the subtle attentional problems observed in patients with mild head injury (197) and multiple sclerosis (60). Finally, some attentional tasks have been designed to assess unilateral spatial neglect. Two of the more common include the letter cancellation (198) and line bisection (199) tasks. For the cancellation tasks, the patient must cross out with a pencil specified letters (e.g., *C*'s and *E*'s) from an array of letters presented on a piece of paper. The number of omissions and commissions, as well as the total time to completion, are recorded for this task. The line bisection task requires the patient to mark the middle of 20 horizontally arranged lines that are centered in the left, right, or center of the page. Deviation from the true center of the lines is recorded. Patients with left-sided hemineglect typically will make more omission errors on the left side of the stimulus array than on the right on the letter cancellation task or will mark the center of the line to the right of midline on the line bisection task.

Memory

The theoretical and empirical underpinnings of assessing memory impairment are described in detail in Chapter 18. Briefly, one view of memory posits that information flows "linearly" from sensory input to a limited-capacity storage, referred to as *primary (short-term) memory* (Fig. 2.4) (200). *Primary memory* is defined as the information processing system dedicated to the temporary storage of information. Information held in primary memory is lost to conscious

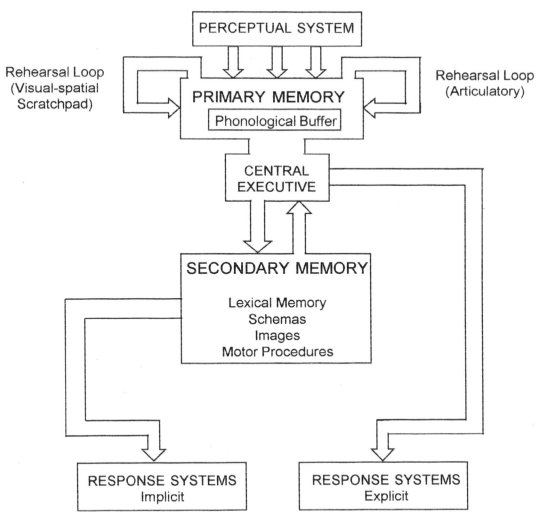

FIGURE 2.4. Model of multiple interacting memory systems (see text for explanation).

awareness if it is not immediately rehearsed. This system is hypothesized to include postperceptual storage of information, rehearsal mechanisms required to reactivate information for greater processing, and consolidation processes. Each part of this memory system has capacity and speed of processing constraints. Clinically, this system can be assessed by using tasks that measure the amount of information that can be briefly held in short-term storage, such as the digit span subtest of the WAIS-III or the rate of forgetting from short-term storage using the Brown-Peterson task (201,202). More experimental procedures, developed by Baddeley (203), have examined the effects of a parallel task load upon temporary storage and rehearsal mechanisms in short-term storage (referred to as *working memory*), although such tasks have not as yet been applied routinely to clinical assessment.

Secondary (long-term) memory represents a larger capacity, more permanent storage of newly acquired information that has been consolidated from primary memory. Both recent and remote personal information and historical events are considered part of secondary memory. The preponderance of published memory studies typically evaluate secondary memory impairment. Tasks that assess secondary memory typically ask the patient to recall or recognize units of information that exceed the capacity of primary memory (i.e., at least eight or nine units). The importance of the distinction between span and supraspan stimulus presentations is illustrated by the "memory cliff" associated with increasing the number of digits beyond the patient's maximal digit capacity (204). The most commonly used secondary memory tasks used in clinical practice involve the immediate and delayed (typically 30–60 minutes) recall of a paragraph-length story, such as the logical memory subtest of the WMS-III. The delayed recall score enables the examiner to assess the rate of forgetting from secondary memory.

Secondary memory also can be evaluated by examining word list learning on multitrial stimulus presentations. Capacity for learning with repeated stimulus presentations can be assessed with such free recall tasks. In addition, delayed free recall, cued recall, and recognition memory typically are assessed. Recognition memory tests are useful for examining the relative contributions of encoding/storage versus retrieval failure to secondary memory impairment. It is assumed that if a patient is unable to spontaneously recall an item of information but can successfully discriminate this item in a list combining old and new information, then the item must have entered into long-term storage. Hence, the patient's deficits occur primarily in gaining access to information in long-term storage. The most common list-learning tests used in clinical practice include the Rey Auditory Verbal Learning Test (111), the CVLT (205), the CVLT-II (206), and the Buschke Selective Reminding Test (207). The Buschke Selective Reminding Test has the advantage of differentiating words recalled from primary versus secondary memory (208). The CVLT also assesses the patient's ability to use semantic clustering to aid encoding, examines encoding versus retrieval dichotomies, primacy/recency recall, learning slope, and proactive and retroactive interference. This test includes an extensive normative sample that is stratified by age from 16 to 89 years and by gender. The CVLT and CVLT-II include an option for a forced-choice recognition trial and a critical item analysis for detection of dissimulation or poor effort. Exaggeration of memory impairment was detected using the CVLT with a cutting score of less than 87% accuracy on the forced-choice recognition, with 80% sensitivity and 97% specificity (207). The CVLT-II includes a parametric measure of positive and negative response bias.

Since the early discovery by Milner (93) that right temporal lobectomy produces a selective *visuospatial* memory impairment, whereas a left temporal lobectomy results in a memory loss for *verbal* material, most clinical neuropsychologists assess nonverbal as well as verbal memory. The most commonly used tests of nonverbal memory are the Benton Visual Retention Test (209), the visual reproduction subtests of the WMS-III, and the Rey-Osterrieth Complex Figure Test (111). These tests require the patient to copy from memory a previously presented geometric design and have excellent norms, but they may be invalid in patients with a constructional apraxia or in patients with primary motor impairment, such as ataxia or hemiplegia. As an alternative, we have adapted Barbizet and Cany's 7/24 Spatial Recall Test (210) for clinical practice (111,211). This task requires the patient to reproduce a design generated with seven checkers presented on a 6 × 4 checkerboard. The checkerboard array reduces the perceptual demands of the test. In addition, this test can be administered to motor-impaired patients who have the ability to point to the appropriate squares and having the examiner place the checkers. A more demanding variation of the 7/24 test, called the Visual Spatial Learning Test, has been developed (212).

All of the secondary memory tasks used in the studies described were designed to measure the patient's conscious and explicit recollection of factual material. Studies have suggested that there is an alternate memory system in which learning is expressed implicitly (213). On implicit memory tasks, such as motor skill acquisition (e.g., pursuit rotor task) or priming (e.g., stem completion task), patients with global amnesia have been found to perform normally (214,215). As noted previously, an interesting double dissociation has been noted in two dementing conditions: patients with Huntington disease exhibit deficits in motor skill learning (216) but have normal priming (215), whereas patients with Alzheimer disease have the opposite pattern of deficits, abnormal priming (217) and intact motor skill learning (216). Although of considerable theoretical interest, experimental tests of implicit memory have not been routinely administered in clinical assessments.

Language

Aphasia is a common symptom of focal dominant hemisphere lesions and diffuse cerebral dysfunction. The assessment of aphasia typically involves evaluation of spontaneous speech, repetition, comprehension, naming, reading, and writing. Several comprehensive aphasia batteries have been developed over the years, including the Boston Diagnostic Aphasia Examination-3 (94), Western Aphasia Battery (218), Multilingual Aphasia Examination (219), Illinois Test of Psycholinguistic Ability (220), and Porch Index of Communicative Ability (221). The Reitan Aphasia Screening Test (222) provides a less thorough language assessment, but it can be administered rapidly. More specific language tasks include the token test (223) to assess verbal comprehension of commands of increasing difficulty, the Boston Naming Test (224) to assess the ability to name pictured objects, and the Peabody Picture Vocabulary Test-Revised (225) to assess auditory comprehension of picture names. As mentioned in a previous section, the Controlled Oral Word Association Test (182) (also known as the word fluency, FAS, or letter fluency test), which assesses spontaneous generation of words beginning with a given letter within a limited amount of time, is impaired in nonaphasic patients with left frontal lobe lesions (226). Animal naming or category fluency is assessed by having the patient generate a list of animals, items found in a grocery store, or clothing items in 60 seconds. Studies of dementia patients show a greater vulnerability to decline on semantic (category) than phonemic (letter) fluency tasks (227).

Perception

Assessment of higher perceptual processes is commonly included in a comprehensive neuropsychological evaluation (see Chapter 21 for a comprehensive discussion of perceptual disorders). *Agnosia* classically is defined as a failure to recognize a percept that cannot be accounted for by defects in elementary sensory function. The correct interpretation of perceptual tasks, therefore, cannot be made unless more primary sensory data are collected using screening tests of visual, auditory, tactile, and olfactory function either during the neuropsychological examination (e.g., visual field assessment through bedside confrontation, two-point tactile stimulation) or from specialized diagnostic procedures performed by other specialties (e.g., formal perimetric visual field testing, audiometric testing). Double simultaneous stimulation frequently can identify patients experiencing a mild, residual neglect that may not be observed during unilateral presentations.

The most common tests for assessing visuospatial perception include the Benton Facial Recognition Test (182), Benton Line Orientation Test (182), hidden figures test (228), and Hooper Visual Organization Test (229). The first three of these tests use a recognition format and the fourth examines visual organization processes by having patients name fragmented objects. The Halstead-Reitan Battery (53) contains several tests of tactile perception, including the finger localization test, tactile form perception (stereognosis), fingertip number-writing perception, and tactual performance

test. Two tests from the Halstead-Reitan Battery, the seashore rhythm test and the speech-sounds perception test, also are used to assess auditory perception, although the diagnostic validity of these tests has been called into question (230) because such tests place a large demand on attentional systems. The dichotic listening task is used routinely in several medical centers. In this task, different speech or nonverbal auditory stimuli are presented simultaneously to both ears (231). Unilateral brain damage (232) or lesions involving the corpus callosum (233) may be inferred if a large deviation from the normal right ear accuracy advantage for speech sounds is observed. Smell perception also can be assessed using bedside techniques (234) or standardized measures of olfactory perception, such as the University of Pennsylvania Smell Identification Test (235).

Praxis and Motor Dexterity

Apraxia is a disorder of skilled movements that cannot be accounted for by primary motor dysfunction (diminished strength, speed, and coordination), sensory loss, impaired language comprehension, or inattention to commands. Lesion (236) and functional imaging (237) studies point to the motor association cortex (premotor and supplementary motor areas) and the parietal cortex, particularly of the left hemisphere, as critical brain structures in performing skilled movements. Testing for apraxia involves assessment of the motor system at the highest level of programming. Specialized screening batteries have been developed for assessing limb apraxia (238,239); these batteries generally require the production of gestures to command and/or imitation.

Constructional apraxia frequently is associated with parietal lobe lesions, although it also can occur with frontal lesions. Clinical assessments consist of asking patients to copy designs from a two- or three-dimensional model or to spontaneously generate common objects, such as a house, clock, or daisy. More commonly used tests include the Bender-Gestalt Test (240), the copy forms of the Rey-Osterrieth Complex Figure Test and the Benton Visual Retention Test, the block design subtest of the Wechsler intelligence tests, and the Benton Test of Three-Dimensional Constructional Praxis (182).

To determine the integrity of *primary* motor functions, neuropsychological assessments include tests of strength, speed, and coordination involving the upper extremities. Tests derived from the Halstead-Reitan Battery (53) can be used to assess motor strength (hand dynamometer test), speed (finger tapping test), and coordination (grooved pegboard test). The Purdue Pegboard Test (10) also is used to measure motor coordination. In addition to these three tests, the Wisconsin Motor Battery (102) also includes measures of maze coordination and static steadiness. Such tests can provide useful information regarding the absolute level of performance and the degree of lateralized impairment.

Academic Achievement

A common complaint of patients is impairment in their ability to read, spell, or calculate as well as they did before onset of brain damage in the dominant hemisphere. Academic achievement tests have been developed specifically for assessing developmental learning disorders, but they also can be useful for evaluating and understanding acquired cognitive deficits in the academic skill areas. The most commonly used tests in neuropsychological assessments include the Woodcock-Johnson Psycho-Educational Battery-Revised (241) and the Wide Range Achievement Test-3 (242).

Personality and Socioadaptive Functions

Personality change is one of the most common and debilitating symptoms associated with acquired brain disorders (243–245). Such changes can occur as a direct result of the brain damage or disease, a psychological reaction to experiencing a chronic injury/disease, premorbid personality characteristics, or a combination of these factors. Common interpersonal problems resulting directly from brain dysfunction include loss of impulse control, insensitivity, emotional lability, irritability, loss of self-awareness, lack of initiative, euphoria, and inability to profit from experience. Brain-damaged patients can experience emotional reactions to their disability, including anger, anxiety, denial, dependency, repression, and depression. The development of psychometric instruments for assessing personality change in neurologic disorders has lagged behind the development of instruments for assessing cognition.

One of the most common techniques for assessing personality involves patient self-ratings. The Minnesota Multiphasic Personality Inventory (MMPI-2), which was restandardized in 1989 (246), is frequently used for evaluation of personality and emotional status in neurologic patients. More specialized tests for depression, such as the Beck Depression Inventory (247) and the Zung Depression Scale (248), and for anxiety, such as the State-Trait Anxiety Inventory (249), also are used. Several problems can occur, however, when interpreting these scales in brain-damaged populations, because these tests were standardized on psychiatric populations. For example, numerous items on these scales (particularly the MMPI-2) assess valid symptoms of neurologic or systemic diseases (250,251), thereby giving inflated estimates of psychopathology in neurologic patients. Self-rating scales also may lack validity in patients who experience gross changes in personal insight and self-awareness (15).

An alternative method of assessing personality change is to use relative ratings, which can be compared to self-reported symptoms. One such instrument, the Katz Adjustment Scale (252), has been used with success to assess interpersonal problems of neurologic patients from the perspective of a close relative or friend. Such information can be invaluable for developing a psychological intervention program. Finally, standardized instruments developed from structured psychiatric interviews of the patient can provide useful information about the patient's socioadaptive capabilities and can be applied to treatment planning.

NEUROPSYCHOLOGICAL ASSESSMENT OF NEUROLOGIC AND PSYCHIATRIC DISORDERS

The past 20 years has witnessed an explosion of neuropsychological studies relating to a wide range of neurologic and neuropsychiatric disorders. A detailed discussion of the applications of neuropsychological assessment to specific disorders is beyond the scope of this chapter. The reader is referred to integrative reviews of the neuropsychological research in cerebrovascular disease (253), head injury (Chapter 39) (254,255), Alzheimer disease (256), Parkinson disease (257), Huntington disease (256) and other basal ganglia disorders (Chapter 33), multiple sclerosis (258) and other white matter disorders (Chapter 35) (259), progressive supranuclear palsy (260), amnesic disorders (Chapter 18) (261), neoplasms (256), neurotoxic disorders (Chapter 41) (136), alcohol and drug abuse (Chapter 31), epilepsy (Chapter 38) (262), human immunodeficiency virus infection (263), schizophrenia (Chapter 29) (264), affective disorders (Chapter 27) (133), and anxiety disorders (Chapter 28).

QUALIFICATIONS TO PRACTICE CLINICAL NEUROPSYCHOLOGY

The practice of clinical neuropsychology is relatively new in the health care delivery system. Before 1980, training in neuropsychology was limited to a few doctoral programs and clinical internships. It was not uncommon for clinical psychologists in clinical practice to take brief workshops on neuropsychological assessment and, without a comprehensive knowledge of the behavioral neurosciences or clinical neurology, begin assessing patients with neuropsychological test procedures. Recognizing the need to identify qualified neuropsychological practitioners, in 1983 the American Board of Professional Psychology, in conjunction with the newly formed American Board of Clinical Neuropsychology (ABCN) (265), identified clinical neuropsychology as a specialty area and developed an examination process to evaluate the training, knowledge, and skills of psychologists specializing in neuropsychology. As of late 2001, there were more than 375 board-certified clinical neuropsychologists in North America.

The training of clinical neuropsychologists has expanded rapidly in recent years. As of late 2001, there were 39 doctoral programs, 56 internship sites, and 87 postdoctoral residency programs with specialty training in clinical neuropsychology (see list of training programs on the American Psychological Association (APA) Division 40 Web site, *http://www.div40.org*). In 1987, a task force sponsored by the International Neuropsychological Society (INS) and the neuropsychology division of the APA developed guidelines for education and training at the doctoral, internship, and postdoctoral level (266). To evaluate and accredit postdoctoral training programs and to determine if these programs are meeting the INS/APA guidelines, the Association of Postdoctoral Programs in Clinical Neuropsychology was formed (267). In 1995, clinical neuropsychology was recognized as a clinical practice specialty by the APA (268). In 1997, the Houston Conference on Specialty Education and Training in Clinical Neuropsychology convened with the goal of developing an integrated model for training in clinical neuropsychology. This conference developed a policy statement outlining (a) a definition of a clinical neuropsychologist, (b) the core knowledge and skills expected of a neuropsychologist, and (c) an integrated model of education and training in the science and professional practice of clinical neuropsychology, extending from doctoral program to internship to residency education and training (269). This conference also recommended 2 years of residency training after completion of a doctoral degree in clinical psychology or a 1-year residency after completion of a doctoral program in clinical psychology with a specialty track in neuropsychology. The committee developed the following definition of a clinical neuropsychologist: "A clinical neuropsychologist is a professional psychologist trained in the science of brain-behavior relationships. The clinical neuropsychologist specializes in the application of assessment and intervention principles based on the scientific study of human behavior across the lifespan as it relates to normal and abnormal functioning of the central nervous system" (269, p. 161).

The work of such committees is necessary to ensure that the clinical neuropsychological assessment is conducted by highly qualified, trained, and competent professionals. The ABCN web site (*http://www.med.umich.edu/abcn/diplomates.html*) provides a listing of board-certified clinical neuropsychologists. In this way, the consumers, that is, the patient, family members, referring physicians, mental health professionals, teachers, and attorneys, can be assured of the highest quality of neuropsychological services.

ACKNOWLEDGMENT

The authors appreciate the critical suggestions to this chapter provided by T.A. Hammeke.

REFERENCES

1. Benton A. Evolution of a clinical specialty. *Clin Neuropsychologist* 1987;1:5–8.

2. Simos PG, Papanicolaou AC, Breier JI, et al. Insights into brain function and neural plasticity using magnetic source imaging. *J Clin Neurophysiol* 2000;17:143–162.

3. Walsh V, Cowey A. Transcranial magnetic stimulation and cognitive neuroscience. *Nat Rev Neurosci* 2000;1:73–79.

4. Schaefer PW, Grant PE, Gonzalez RG. Diffusion-weighted MR imaging of the brain. *Radiology* 2000;217:331–345.

5. Neumann-Haefelin T, Moseley ME, Albers GW. New magnetic resonance imaging methods for cerebrovascular disease: emerging clinical applications. *Ann Neurol* 2000;47:559–570.

6. Pomper MG, Port JD. New techniques in MR imaging of brain tumors. *Magn Reson Imaging Clin N Am* 2000;8:691–713.

7. Raichle ME. Functional neuroimaging: a historical and physiological perspective. In: Cabeza R, Kingstone A, eds. *The handbook of functional neuroimaging of cognition.* Cambridge, MA: MIT Press, 2001:3–26.

8. Mazziotta JC. Imaging: window on the brain. *Arch Neurol* 2000;57:1413–1421.

9. Posner MI, DiGirolamo GJ. Cognitive neuroscience: origins and promise. *Psychol Bull* 2000;126:873–889.

10. Lezak MD. *Neuropsychological assessment,* 3rd ed. New York: Oxford University Press, 1995.

11. Kolb B, Whishaw IQ. *Fundamentals of human neuropsychology,* 3rd ed. New York: WH Freeman & Co., 1990.

12. Yeates OK, Ris DM, Taylor HG. *Pediatric neuropsychology.* New York: Guilford Press, 2000.

13. Spreen O, Risser AT, Edgell D. *Developmental neuropsychology.* New York: Oxford University Press, 1995.

14. Baron IS, Fennell EB, Voeller KKS. *Pediatric neuropsychology in the medical setting.* New York: Oxford University Press, 1995.

15. Prigatano GP, Schacter DL. *Awareness of deficits after brain injury: clinical and theoretical issues.* New York: Oxford University Press, 1992.

16. Meier MJ, Benton AL, Diller LD. *Neuropsychological rehabilitation.* New York: Guilford Press, 1987.

17. Wilson BA. *Rehabilitation of memory.* New York: Guilford Press, 1987.

18. Sohlberg MM, Mateer CA. *Introduction to cognitive rehabilitation: theory and practice.* New York: Guilford Publications, 1989.

19. Prigatano GP. *Principles of neuropsychological rehabilitation.* New York: Oxford University Press, 1999.

20. Sohlberg MM, Mateer CA. *Introduction to cognitive rehabilitation: theory and practice.* New York: Guilford Press, 1989.

21. Raskin SA, Mateer CA. *Neuropsychological management of mild traumatic brain injury.* New York: Oxford University Press, 2000.

22. Wilson BA. *Case studies in neuropsychological rehabilitation.* New York: Oxford University Press, 1999.

23. Charletta DA, Bennett DA, Wilson RS. Computed tomography and magnetic resonance imaging. In: Parks RW, Zec RF, Wilson RS, eds. *Neuropsychology of Alzheimer's disease and other dementias.* New York: Oxford University Press, 1993:534–561.

24. Bigler E. Neuropsychological testing defines the neurobehavioral significance of neuroimaging-identified abnormalities. *Arch Clin Neuropsychol* 2001;16:227–236.

25. McKhann G, Drachmann D, Folstein M, et al. Clinical diagnosis of Alzheimer's disease: report of the NINCDS-ADRDA work group under the auspices of the Department of Health and Human Services Task Force on Alzheimer's disease. *Neurology* 1984;34:939–944.

26. Petersen RC, Stevens JC, Ganguli M, et al. Practice parameter: early detection of dementia: mild cognitive impairment (an evidence-based review). Report of the Quality Standards Subcommittee of the American Academy of Neurology. *Neurology* 2001;56:1133–1142.

27. Pasquier F. Early diagnosis of dementia: neuropsychology. *J Neurol* 1999;246:6–15.

28. Chen P, Ratcliff G, Belle SH, et al. Cognitive tests that best discriminate between presymptomatic AD and those who remain nondemented. *Neurology* 2000;55:1847–1853.

29. Lee GP, Hamsher KD. Neuropsychological findings in toxicometabolic confusional states. *J Clin Exp Neuropsychol* 1988;10:769–778.

30. Heilman KM, Valenstein E. *Clinical neuropsychology,* 3rd ed. New York: Oxford University Press, 1993.

31. Cummings JL. Subcortical dementia. Neuropsychology, neuropsychiatry, and pathophysiology. *Br J Psychiatry* 1986;149:682–697.

32. Duke LM, Kaszniak AW. Executive control functions in degenerative dementias: a comparative review. *Neuropsychol Rev* 2000;10:75–99.

33. Savage CR. Neuropsychology of subcortical dementias. *Psychiatr Clin North Am* 1997;20:911–931.

34. Heaton RK, Grant I, Butters N, et al. The HNRC 500—neuropsychology of HIV infection at different disease stages. HIV Neurobehavioral Research Center. *J Int Neuropsychol Soc* 1995;1:231–251.

35. Rao SM. Neuropsychology of multiple sclerosis: a critical review. *J Clin Exp Neuropsychol* 1986;8:503–542.

36. Neary D, Snowden JS, Gustafson L, et al. Frontotemporal lobar degeneration: a consensus on clinical diagnostic criteria. *Neurology* 1998;51:1546–1554.

37. Mesulam MM. Primary progressive aphasia. *Ann Neurol* 2001;49:425–432.

38. Caine D, Hodges JR. Heterogeneity of semantic and visuospatial deficits in early Alzheimer's disease. *Neuropsychology* 2001;15:155–164.

39. Levine DN, Lee JM, Fisher CM. The visual variant of Alzheimer's disease: a clinicopathologic case study. *Neurology* 1993;43:305–313.

40. McKeith IG, Galasko D, Kosaka K, et al. Consensus guidelines for the clinical and pathologic diagnosis of dementia with Lewy bodies (DLB): report of the consortium on DLB international workshop. *Neurology* 1996;47:1113–1124.

41. Simard M, van Reekum R, Cohen T. A review of the cognitive and behavioral symptoms in dementia with Lewy bodies. *J Neuropsychiatry Clin Neurosci* 2000;12:425–450.

42. Cullum CM, Heaton RK, Nemiroff B. Neuropsychology of late-life psychoses. *Psychiatr Clin North Am* 1988;11:47–59.

43. Gainotti G, Marra C. Progress and controversies in neuropsychology of memory. *Acta Neurol (Napoli)* 1992;14:561–577.

44. Mialet JP, Pope HG, Yurgelun-Todd D. Impaired attention in depressive states: a non-specific deficit? *Psychol Med* 1996;26:1009–1020.

45. Rosenstein LD. Differential diagnosis of the major progressive dementias and depression in middle and late adulthood: a summary of the literature of the early 1990s. *Neuropsychol Rev* 1998;8:109–167.

46. Perrine K, Hermann BP, Meador KJ, et al. The relationship of neuropsychological functioning to quality of life in epilepsy. *Arch Neurol* 1995;52:997–1003.

47. Sweet JJ, Moberg PJ, Westergaard CK. Five-year follow-up survey of practices and beliefs of clinical neuropsychologists. *Clin Neuropsychol* 1996;10:202–221.

48. Essig SM, Mittenberg W, Petersen RS, et al. Practices in forensic neuropsychology: perspectives of neuropsychologists and trial attorneys. *Arch Clin Neuropsychol* 2001;16:271–291.

49. Borum R, Grisso T. Psychologist test use in criminal forensic evaluations. *Prof Psychol Res Pract* 1995;26:465–473.

50. Cruise KR, Rogers R. An analysis of competency to stand trial: an integration of case law and clinical knowledge. *Behav Sci Law* 1998;16:35–50.

51. Bennett TL. Neuropsychological evaluation in rehabilitation planning and evaluation of functional skills. *Arch Clin Neuropsychol* 2001;16:237–253.

52. Meyer GJ, Finn SE, Eyde LD, et al. Psychological testing and psychological assessment. A review of evidence and issues. *Am Psychol* 2001;56:128–165.

53. Reitan RM, Davison LA. *Clinical neuropsychology: current status and applications.* New York: Wiley, 1974.

54. Golden CJ, Hammeke TA, Purisch AD. *Luria-Nebraska Neuropsychological Battery.* Los Angeles, CA: Western Psychological Services, 1976.

55. Heaton RK, Grant I, Matthews CG. *Comprehensive norms for an expanded Halstead-Reitan Battery: demographic corrections, research findings, and clinical applications.* Odessa, FL: Psychological Assessment Resources, 1991.

56. Delis DC, Kaplan E. Hazards of a standardized neuropsychological test with low content validity: comment on the Luria-Nebraska Neuropsychological Battery. *J Consult Clin Psychol* 1983;51:396–398.

57. Delis DC, Kaplan E. The assessment of aphasia with the Luria-Nebraska Neuropsychological Battery: a case critique. *J Consult Clin Psychol* 1982;50:32–39.

58. Luria AR. *Higher cortical functions in man.* New York: Basic Books, 1966.

59. Milberg WP, Hebben N, Kaplan E. The Boston process approach to neuropsychological assessment. In: Grant I, Adams KM, eds. *Neuropsychological assessment of neuropsychiatric disorders,* 2nd ed. New York: Oxford University Press, 1996:58–80.

60. Rao SM, Leo GJ, Bernardin L, et al. Cognitive dysfunction in multiple sclerosis: I. Frequency, patterns, and prediction. *Neurology* 1991;41:685–691.

61. Beatty WW, Paul RH, Wilbanks SL, et al. Identifying multiple sclerosis patients with mild or global cognitive impairment using the Screening Examination for Cognitive Impairment (SEFCI). *Neurology* 1995;45:718–723.

62. Basso MR, Beason-Hazen S, Lynn J, et al. Screening for cognitive dysfunction in multiple sclerosis. *Arch Neurol* 1996;53:980–984.

63. Hermann BP, Seidenberg M, Schoenfeld J, et al. Neuropsychological characteristics of the syndrome of mesial temporal lobe epilepsy. *Arch Neurol* 1997;54:369–376.

64. Chelune GJ. Using neuropsychological data to forecast postsurgical cognitive outcome. In: Luders H, ed. *Epilepsy surgery.* New York: Raven Press, 1992:477–485.

65. Chelune GJ, Naugle RI, Hermann BP, et al. Does presurgical IQ predict seizure outcome after temporal lobectomy? Evidence from the Bozeman Epilepsy Consortium. *Epilepsia* 1998;39:314–318.

66. Loring DW, Meador KJ, Lee GP, et al. Stimulus timing effects on Wada memory testing. *Arch Neurol* 1994;51:806–810.

67. Loring DW, Meador KJ, Lee GP, et al. Wada memory asymmetries predict verbal memory decline after anterior temporal lobectomy. *Neurology* 1995;45:1329–1333.

68. Loring DW, Meador KJ, Lee GP, et al. Wada memory performance predicts seizure outcome following anterior temporal lobectomy. *Neurology* 1994;44:2322–2324.

69. Jones-Gotman M. Neuropsychological techniques in the identification of epileptic foci. *Epilepsy Res Suppl* 1992;5:87–94.

70. Perrine K, Westerveld M, Sass KJ, et al. Wada memory disparities predict seizure laterality and postoperative seizure control. *Epilepsia* 1995;36:851–856.

71. Randolph C, Tierney MC, Mohr E, et al. The Repeatable Battery for the Assessment of Neuropsychological Status

(RBANS): preliminary clinical validity. *J Clin Exp Neuropsychol* 1998;20:310–319.

72. Mattis S. *Dementia Rating Scale: professional manual.* Odessa, FL: Psychological Assessment Resources, 1988.

73. Vangel SJ Jr, Lichtenberg PA. Mattis Dementia Rating Scale: clinical utility and relationship with demographic variables. *Clin Neuropsychol* 1995;9:209–213.

74. Butters N, Grant I, Haxby J, et al. Assessment of AIDS-related cognitive changes: recommendations of the NIMH Workshop on neuropsychological assessment approaches. *J Clin Exp Neuropsychol* 1990;12:963–978.

75. Peavy GM, Salmon D, Bear PI, et al. Detection of mild cognitive deficits in Parkinson's disease patients with the WAIS-R NI. *J Int Neuropsychol Soc* 2001;7:535–543.

76. Alterman RL, Kelly PJ. Pallidotomy technique and results: the New York University experience. *Neurosurg Clin N Am* 1998;9:337–343.

77. Snowden J, Craufurd D, Griffiths H, et al. Longitudinal evaluation of cognitive disorder in Huntington's disease. *J Int Neuropsychol Soc* 2001;7:33–44.

78. Morris JC, Heyman A, Mohs RC, et al. The Consortium to Establish a Registry for Alzheimer's Disease (CERAD). Part I. Clinical and neuropsychological assessment of Alzheimer's disease. *Neurology* 1989;39:1159–1165.

79. Fillenbaum GG, Beekly D, Edland SD, et al. Consortium to establish a registry for Alzheimer's disease: development, database structure, and selected findings. *Top Health Inf Manage* 1997;18:47–58.

80. McCrea M, Kelly JP, Randolph C, et al. Standardized assessment of concussion (SAC): on-site mental status evaluation of the athlete. *J Head Trauma Rehabil* 1998;13:27–35.

81. Wilson RS, Rosenbaum G, Brown G, et al. An index of premorbid intelligence. *J Consult Clin Psychol* 1978;46:1554–1555.

82. Barona A, Reynolds CR, Chastain R. A demographically based index of premorbid intelligence for the WAIS-R. *J Consult Clin Psychol* 1984;52:885–887.

83. Barona A, Chastain R. An improved estimate of premorbid IQ for blacks and whites on the WAIS-R. *Int J Clin Neuropsychol* 1986;8:169–172.

84. Nelson HE. *National Adult Reading Test (NART): test manual.* Windsor, UK: NFER Nelson, 1982.

85. Nelson HE, O'Connell A. Dementia: the estimation of premorbid intelligence levels using the new adult reading test. *Cortex* 1978;14:234–244.

86. Jastak S, Wilkinson GS. *Wide Range Achievement Test-Revised administration manual.* Wilmington, DE: Jastak Associates, 1984.

87. Wilkinson GS. *WRAT-3: Wide Range Achievement Test administration manual,* 3rd ed. Wilmington, DE: Wide Range, Inc., 1993.

88. Stebbins GT, Wilson RS, Gilley DW, et al. Use of the National Adult Reading Test to estimate premorbid IQ in dementia. *Clin Neuropsychol* 1990;4:18–24.

89. Wiens AN, Bryan JE, Crossen JR. Estimating WAIS-R FSIQ from the National Adult Reading Test-Revised in normal subjects. *Clin Neuropsychol* 1993;7:70–84.

90. Kareken DA, Gur RC, Saykin AJ. Reading on the Wide Range Achievement Test-Revised and parental education as predictors of IQ: comparison with the Barona Formula. *Arch Clin Neuropsychol* 1995;10:147–157.

91. Johnstone B, Wilhelm KL. The longitudinal stability of the WRAT-R Reading subtest: is it an appropriate estimate of premorbid intelligence? *J Int Neuropsychol Soc* 1996;2:282–285.

92. Dodrill CB. Myths of neuropsychology: further considerations. *Clin Neuropsychol* 1999;13:562–572.

93. Delaney RC, Rosen AJ, Mattson RH, Novelly RA. Memory function in focal epilepsy: a comparison of non-surgical unilateral temporal lobe and frontal lobe samples. *Cortex* 1980;16:103–107.

94. Goodglass H, Kaplan E, Barresi B. *The assessment of aphasia and related disorders,* 3rd ed. Philadelphia: Lippincott Williams & Wilkins, 2000.

95. Heindel WC, Salmon DP, Shults CW, et al. Neuropsychological evidence for multiple implicit memory systems: a comparison of Alzheimer's, Huntington's, and Parkinson's disease patients. *J Neurosci* 1989;9:582–587.

96. Campbell DT, Fiske DW. Convergent and discriminant validation by the multitrait-multimethod matrix. *Psychol Bull* 1959;56:81–105.

97. Heaton RK, Grant I, Matthews CG. Differences in neuropsychological test performance associated with age, education, and sex. In: Grant I, Adams KM, eds. *Neuropsychological assessment of neuropsychiatric disorders.* New York: Oxford University Press, 1986:100–120.

98. Craik FIM, McDowd JM. Age differences in recall and recognition. *J Exp Psychol Learn Mem Cogn* 1987;13:474–479.

99. Hasher L, Zacks RT. Automatic and effortful processes in memory. *J Exp Psychol Gen* 1979;108:356–388.

100. The Psychological Corporation. *WAIS-III WMS-III technical manual.* San Antonio, TX: Harcourt Brace & Company, 1997.

101. Spreen O, Strauss E. *A compendium of neuropsychological tests: administration, norms, and commentary.* New York: Oxford University Press, 1991.

102. Beardsley JV, Matthews CG, Cleeland CS, et al. *Experimental T-score norms on the Wisconsin Neuropsychological Test Battery.* Madison, WI: University of Wisconsin Center for Health Science, 1978.

103. Dorward J, Posthuma A. Validity limits of forensic neuropsychological testing. *Am J Forensic Psychol* 1993;11:17–26.

104. Rogers R, Harrell EH, Liff CD. Feigning neuropsychological impairment: a critical review of methodological and clinical considerations. *Clin Psychol Rev* 1993;13:255–274.

105. Iverson GL, Binder LM. Detecting exaggeration and malingering in neuropsychological assessment. *J Head Trauma Rehabil* 2000;15:829–858.

106. Mittenberg W, Fichera ST, Zielinski RE, et al. Identification of malingered head injury on the Wechsler Adult Intelligence Scale-Revised. *Prof Psychol Res Pract* 1995;1995:491–498.

107. Heaton RK, Smith HH, Lehman RA, et al. Prospects for faking believable deficits on neuropsychological testing. *J Consult Clin Psychol* 1978;46:892–900.

108. Faust D, Hart K, Guilmette TJ, et al. Pediatric malingering: the capacity of children to fake believable deficits on neuropsychological testing. *J Consult Clin Psychol* 1988;56:578–582.

109. Nies KJ, Sweet JL. Neuropsychological assessment and malingering: a critical review of past and present strategies. *Arch Clin Neuropsychol* 1994;9:501–552.

110. Slick DJ, Sherman EM, Iverson GL. Diagnostic criteria for malingered neurocognitive dysfunction: proposed standards for clinical practice and research. *Clin Neuropsychol* 1999;13:545–561.

111. Rey A. *L'examen clinique en psychologie.* Paris: Presses Universitaires de France, 1964.

112. Arnett PA, Hammeke TA, Schwartz L. Quantitative and qualitative performance on Rey's 15-Item Test in neurological patients and dissimulators. *Clin Neuropsychol* 1995;9:17–26.

113. Pankratz L. Symptom validity testing and symptom retraining:

procedures for the assessment and treatment of functional sensory deficits. *J Consult Clin Psychol* 1979;47:409–410.

114. Hiscock M, Hiscock CK. Refining the forced-choice method for the detection of malingering. *J Clin Exp Neuropsychol* 1989; 11:967–974.

115. Binder LM. Assessment of malingering after mild head trauma with the Portland Digit Recognition Test. *J Clin Exp Neuropsychol* 1993;15:170–182.

116. Guilmette TJ, Hart KJ, Giuliano AJ. Malingering detection: the use of a forced-choice method in identifying organic versus simulated memory impairment. *Clin Neuropsychol* 1993;7:59–69.

117. Allen LM, Conder RL, Green P, et al. *Manual for the computerized assessment of response bias.* Durham, NC: Cognisyst, 1997.

118. Frederick RI. *Validity Indicator Profile manual.* Minneapolis, MN: National Computer Systems, 1997.

119. Ross SR, Adams KM. One more test of malingering? *Clin Neuropsychol* 1999;13:112–116.

120. Iverson GL, Franzen MD, McCracken LM. Evaluation of an objective assessment technique for the detection of malingered memory deficits. *Law Hum Behav* 1991;15:667–676.

121. Tombaugh TN. *Test of Memory Malingering.* Toronto, CA: Multi-Health Systems, 1996.

122. Rees LM, Tombaugh TN, Dansler DA, et al. Five validation experiments of the test of memory malingering (TOMM). *Psychol Assess* 1998;10:10–20.

123. Tombaugh TN. The Test of Memory Malingering (TOMM): normative data from cognitively intact and cognitively impaired individuals. *Psychol Assess* 1997;9:260–268.

124. Weingartner H. Automatic and effort-demanding cognitive processes in depression. In: Poon LW, ed. *Handbook for clinical memory assessment of older adults.* Washington, DC: American Psychological Association, 1986:218–225.

125. Hart RP, Kwentus JA, Wade JB, et al. Digit symbol performance in mild dementia and depression. *J Consult Clin Psychol* 1987;55:236–238.

126. Davidson RJ, Abercrombie H, Nitschke JB, et al. Regional brain function, emotion and disorders of emotion. *Curr Opin Neurobiol* 1999;9:228–234.

127. Drevets WC. Functional anatomical abnormalities in limbic and prefrontal cortical structures in major depression. *Prog Brain Res* 2000;126:413–431.

128. Abercrombie HC, Schaefer SM, Larson CL, et al. Metabolic rate in the right amygdala predicts negative affect in depressed patients. *Neuroreport* 1998;9:3301–3307.

129. Liotti M, Mayberg HS. The role of functional neuroimaging in the neuropsychology of depression. *J Clin Exp Neuropsychol* 2001;23:121–136.

130. Massman PJ, Delis DC, Butters N, et al. The subcortical dysfunction hypothesis of memory deficits in depression: neuropsychological validation in a subgroup of patients. *J Clin Exp Neuropsychol* 1992;14:687–706.

131. King DA, Caine ED, Cox C. Influence of depression and age on selected cognitive functions. *Clin Neuropsychol* 1993;7:443–453.

132. Sackeim HA, Freeman J, McElhiney M, et al. Effects of major depression on estimates of intelligence. *J Clin Exp Neuropsychol* 1992;14:268–288.

133. Starkstein SE, Robinson RG. *Depression in neurologic disease.* Baltimore: The Johns Hopkins University Press, 1993.

134. Wolkowitz OM, Tinklenberg JR, Weingartner H. A psychopharmacological perspective of cognitive functions. II. Specific pharmacologic agents. *Neuropsychobiology* 1985;14:133–156.

135. Wolkowitz OM, Tinklenberg JR, Weingartner H. A psychopharmacological perspective of cognitive functions. I. Theoretical overview and methodological considerations. *Neuropsychobiology* 1985;14:88–96.

136. Hartman DE. *Neuropsychological toxicology: identification and assessment of human neurotoxic syndromes,* 2 ed. New York: Plenum Press, 1995.

137. Kempen JH, Kritchevsky M, Feldman ST. Effect of visual impairment on neuropsychological test performance. *J Clin Exp Neuropsychol* 1994;16:223–231.

138. Casey JE, Ferguson GG, Kimura D, et al. Neuropsychological improvement versus practice effect following unilateral carotid endarterectomy in patients without stroke. *J Clin Exp Neuropsychol* 1989;11:461–470.

139. Dodrill CB, Troupin AS. Effects of repeated administrations of a comprehensive neuropsychological battery among chronic epileptics. *J Nerv Ment Dis* 1975;161:185–190.

140. Duke R, Bloor B, Nugent R, et al. Changes in performance on WAIS, Trail Making Test and Finger Tapping Test associated with carotid artery surgery. *Percept Mot Skills* 1968;26:399–404.

141. McCaffrey RJ, Ortega A, Orsillo SM, et al. Practice effects in repeated neuropsychological assessments. *Clin Neuropsychol* 1992;6:32–42.

142. Putnam SH, Adams KM, Schneider AM. One-day test-retest reliability of neuropsychological tests in a personal injury case. *Psychol Assess* 1992;4:312–316.

143. Ryan JJ, Paolo AM, Brungardt TM. WAIS-R test-retest stability in normal persons 75 years and older. *Clin Neuropsychol* 1992;6:3–8.

144. Ryan JJ, Georgemiller RJ, Geisser ME, et al. Test-retest stability of the WAIS-R in a clinical sample. *J Clin Psychol* 1985;41:552–556.

145. McSweeny AJ, Naugle RI, Chelune GJ, et al. "T scores for change": An illustration of a regression approach to depicting change in clinical neuropsychology. *Clin Neuropsychol* 1993;7:300–312.

146. Sawrie SM, Chelune GJ, Naugle RI, et al. Empirical methods for assessing meaningful neuropsychological change following epilepsy surgery. *J Int Neuropsychol Soc* 1996;2:556–564.

147. Seidenberg M, Hermann B, Wyler AR, et al. Neuropsychological outcome following anterior temporal lobectomy in patients with and without the syndrome of mesial temporal lobe epilepsy. *Neuropsychology* 1998;12:303–316.

148. Temkin NR, Heaton RK, Grant I, et al. Detecting significant change in neuropsychological test performance: a comparison of four models. *J Int Neuropsychol Soc* 1999;5:357–369.

149. Nelson A, Fogel BS, Faust D. Bedside cognitive screening instruments. *J Nerv Ment Dis* 1986;174:73–83.

150. Folstein MF, Folstein SE, McHugh PR. Mini-Mental State: a practical method for grading the cognitive state of patients for the clinician. *J Psychiatr Res* 1975;12:189–198.

151. Mattis S. *Dementia rating scale: professional manual.* Odessa, FL: Psychological Assessment Resources, 1988.

152. Kaplan E, Fein D, Morris R, et al. *WAIS-R as a neuropsychological instrument.* New York: The Psychological Corporation, 1991.

153. Matarazzo JD. *Wechsler's measurement and appraisal of adult intelligence,* 5th ed. New York: Oxford University Press, 1972.

154. Inglis J, Lawson JS. Sex differences in the effects of unilateral brain damage on intelligence. *Science* 1981;212:693–695.

155. Bornstein RA, Suga L, Prifitera A. Incidence of verbal IQ-performance IQ discrepancies at various levels of education. *J Clin Psychol* 1987;43:387–389.

156. Matarazzo JD, Daniel MH, Prifitera A, et al. Inter-subtest scatter in the WAIS-R standardization sample. *J Clin Psychol* 1988;44:940–949.

157. Mittenberg W, Hammeke TA, Rao SM. Intrasubtest scatter

on the WAIS-R as a pathognomonic sign of brain injury. *Psychol Assess* 1989;1:273–276.

158. Hallenbeck CE, Fink SL, Grossman JS. Measurement of intellectual inefficiency. *Psychol Rep* 1965;17:339–349.

159. Raven JC. *Guide to the Standard Progressive Matrices.* London: HK Lewis, 1960.

160. Grant DA, Berg EA. A behavioral analysis of degree of reinforcement and ease of shifting to new responses in a Weigl-type card-sorting problem. *J Exp Psychol* 1948;38:404–411.

161. Heaton RK, Chelune GJ, Talley JL, et al. *Wisconsin Card Sorting Test manual: revised and expanded.* Odessa, FL: Psychological Assessment Resources, 1993.

162. Halstead WC. *Brain and intelligence: a quantitative study of the frontal lobes.* Chicago: University of Chicago Press, 1947.

163. Butler M, Retzlaff P, Vanderploeg R. Neuropsychological test usage. *Prof Psychol Res Pract* 1991;22:510–512.

164. Corrigan JD, Agresti AA, Hinkeldey NS. Psychometric characteristics of the Category Test: replication and extension. *J Clin Psychol* 1987;43:368–376.

165. Crockett D, Bilsker D, Hurwitz T, et al. Clinical utility of three measures of frontal lobe dysfunction in neuropsychiatric samples. *Int J Neurosci* 1986;30:241–248.

166. Milner B. Effects of different brain lesions on card sorting. *Arch Neurol* 1963;9:90–100.

167. Pendleton MG, Heaton RK. A comparison of the Wisconsin Card Sorting Test and the Category Test. *J Clin Psychol* 1982;38:392–396.

168. Rao SM, Hammeke TA, Speech TJ. Wisconsin Card Sorting Test performance in relapsing-remitting and chronic-progressive multiple sclerosis. *J Consult Clin Psychol* 1987;55:263–265.

169. Teuber HL, Battersby WS, Bender MB. Performance of complex visual tasks after cerebral lesions. *J Nerv Ment Dis* 1951;114:413–429.

170. Fey ET. The performance of young schizophrenics and young normals on the Wisconsin Card Sorting Test. *J Consult Psychol* 1951;15:311–319.

171. Goldberg TE, Kelsoe JR, Weinberger DR, et al. Performance of schizophrenic patients on putative neuropsychological tests of frontal lobe function. *Int J Neurosci* 1988;42:51–58.

172. Stuss DT, Benson DF, Kaplan EF, et al. The involvement of the orbitofrontal cerebrum in cognitive tasks. *Neuropsychologia* 1983;21:235–248.

173. Arnett PA, Rao SM, Bernardin L, et al. Relationship between frontal lobe lesions and Wisconsin Card Sorting Test performance in patients with multiple sclerosis. *Neurology* 1994;44:420–425.

174. Robinson AL, Heaton RK, Lehman RAW, et al. The utility of the Wisconsin Card Sorting Test in detecting and localizing frontal lobe lesions. *J Consult Clin Psychol* 1980;48:605–614.

175. Anderson SW, Damasio H, Jones RD, et al. Wisconsin Card Sorting Test performance as a measure of frontal lobe damage. *J Clin Exp Neuropsychol* 1991;13:909–922.

176. Mountain MA, Snow WG. Wisconsin Card Sorting Test as a measure of frontal pathology: a review. *Clin Neuropsychol* 1993;7:108–118.

177. Perrine K. Differential aspects of conceptual processing in the Category Test and Wisconsin Card Sorting Test. *J Clin Exp Neuropsychol* 1993;15:461–473.

178. Bondi MW, Kaszniak AW, Bayles KA, et al. Contributions of frontal system dysfunction to memory and perceptual abilities in Parkinson's disease. *Neuropsychology* 1993;7:89–102.

179. Shallice T, Burgess P. Higher-order cognitive impairments and frontal lobe lesions in man. In: Levin HS, Eisenberg HM, Benton AL, eds. *Frontal lobe function and dysfunction.* New York: Oxford University Press, 1991:125–138.

180. Lezak MD. The problem of assessing executive functions. *Int J Psychol* 1982;17:281–297.

181. Lezak MD. Newer contributions to the neuropsychological assessment of executive functions. *J Head Trauma Rehabil* 1993;8:24–31.

182. Benton AL, Hamsher KD, Varney NR, et al. *Contributions to neuropsychological assessment: a clinical manual.* New York: Oxford University Press, 1983.

183. Jones-Gottman M, Milner B. Design fluency: the invention of nonsense drawings after focal cortical lesions. *Neuropsychologia* 1977;15:653–674.

184. Porteus SD. *Porteus Maze Test: fifty years' application.* Palo Alto, CA: Pacific Books, 1965.

185. Luria AR. *Higher cortical functions in man. Revised and expanded,* 2 ed. New York: Basic Books, 1980.

186. Shimamura AP, Janowsky JS, Squire LR. Memory for the temporal order of events in patients with frontal lobe lesions and patients with amnesia. *Neuropsychologia* 1990;28:803–813.

187. Grafman J, Thompson K, Weingartner H, et al. Script generation as an indicator of knowledge representation in patients with Alzheimer's disease. *Brain Lang* 1991;40:344–358.

188. McGlynn SM, Schacter DL. Unawareness of deficits in neuropsychological syndromes. *J Clin Exp Neuropsychol* 1989;11:143–205.

189. Weinberger DR. Schizophrenia and the frontal lobes. *Trends Neurosci* 1988;11:367–370.

190. Posner MI, Boies SJ. Components of attention. *Psychol Rev* 1971;78:391–408.

191. Cummings JL. *Subcortical dementia.* New York: Oxford University Press, 1990.

192. Stroop JR. Studies of interference in serial verbal reactions. *J Exp Psychol* 1935;18:643–662.

193. Smith A. *Symbol Digit Modalities Test manual.* Los Angeles, CA: Western Psychological Services, 1973.

194. Sternberg S. Memory scanning: mental processes revealed by reaction-time experiments. *Am Sci* 1969;57:421–457.

195. Wilson RS, Kaszniak AW, Klawans HL, et al. High speed memory scanning in parkinsonism. *Cortex* 1980;16:67–72.

196. Rao SM, St. Aubin-Faubert P, Leo GJ. Information processing speed in patients with multiple sclerosis. *J Clin Exp Neuropsychol* 1989;11:471–477.

197. Gronwall DMA. Paced auditory serial-addition task: a measure of recovery from concussion. *Percept Mot Skills* 1977;44:367–373.

198. Diller L, Ben-Yishay Y, Gertsman LJ, et al. *Studies in cognition and rehabilitation in hemiplegia.* New York: New York University Medical Center Institute of Rehabilitation Medicine, 1974.

199. Schenkenberg T, Bradford DC, Ajax ET. Line bisection and unilateral visual neglect in patients with neurologic impairment. *Neurology* 1980;30:509–517.

200. Waugh NC, Norman DA. Primary memory. *Psychol Rev* 1965;72:89–104.

201. Brown J. Some tests of the decay theory of immediate memory. *Q J Exp Psychol* 1958;10:12–21.

202. Peterson LR, Peterson MJ. Short-term retention of individual verbal items. *J Exp Psychol* 1959;58:193–198.

203. Baddeley A. *Working memory.* Oxford: Clarendon Press, 1986.

204. Drachman DA, Zaks MS. The "memory cliff" beyond span in immediate recall. *Psychol Rep* 1967;21:105–112.

205. Delis DC, Kramer JH, Kaplan E, et al. *California Verbal Learning Test, research edition.* Manual. New York: Psychological Corporation, 1987.

206. Delis DC, Kramer JH, Kaplan E, et al. *California Verbal Learning Test manual,* 2nd ed. San Antonio, TX: The Psychological Corporation, 2000.

207. Buschke H. Selective reminding for analysis of memory and learning. *J Verb Learn Verb Behav* 1973;12:543–550.

208. Buschke H, Fuld PA. Evaluating storage, retention, and retrieval in disordered memory and learning. *Neurology* 1974;24:1019–1025.

209. Benton AL. *Revised Visual Retention Test,* 4th ed. New York: The Psychological Corporation, 1974.

210. Vowels LM. Memory impairment in multiple sclerosis. In: Molloy M, Stanley GV, Walsh KW, eds. *Brain impairment: proceedings of the 1978 Brain Impairment Workshop.* Melbourne: University of Melbourne, 1979:10–22.

211. Rao SM, Hammeke TA, McQuillen MP, et al. Memory disturbance in chronic progressive multiple sclerosis. *Arch Neurol* 1984;41:625–631.

212. Malec JF, Ivnik RJ, Hinkeldey NS. Visual Spatial Learning Test. *Psychol Assess* 1991;3:82–88.

213. Schacter DL. Priming and multiple memory systems: perceptual mechanisms of implicit memory. *J Cogn Neurosci* 1992;4:244–256.

214. Corkin S. Acquisition of a motor skill after bilateral medial temporal-lobe excision. *Neuropsychologia* 1968;6:255–264.

215. Shimamura AP, Salmon DP, Squire LR, et al. Memory dysfunction and word priming in dementia and amnesia. *Behav Neurosci* 1987;101:347–351.

216. Heindel WC, Butters N, Salmon DP. Impaired learning of a motor skill in patients with Huntington's disease. *Behav Neurosci* 1988;102:141–147.

217. Salmon DP, Shimamura AP, Butters N, et al. Lexical and semantic priming deficits in patients with Alzheimer's disease. *J Clin Exp Neuropsychol* 1988;10:477–494.

218. Kertesz A. *Western Aphasia Battery.* London, Ontario: University of Western Ontario, 1980.

219. Benton AL, Hamsher KD. *Multilingual Aphasia Examination.* Iowa City: University of Iowa, 1976.

220. Kirk SA, McCarthy J, Kirk W. *The Illinois Test of Psycholinguistic Ability,* rev. ed. Urbana, IL: Illinois University Press, 1968.

221. Porch B. *The Porch Index of Communicative Ability, vol. 2: administration and scoring.* Palo Alto, CA: Consulting Psychologists Press, 1971.

222. Reitan RM. *Aphasia and sensory-perceptual deficits in adults.* Tucson, AZ: Reitan Neuropsychology Laboratory, 1984.

223. De Renzi E, Vignolo L. The Token Test: a sensitive test to detect receptive disturbances in aphasics. *Brain* 1962;85:665–678.

224. Kaplan EF, Goodglass H, Weintraub S. *The Boston Naming Test,* 2nd ed. Philadelphia: Lea & Febiger, 1983.

225. Dunn LM, Dunn LN. *Peabody Picture Vocabulary Test—revised manual.* Circle Pines, MN: American Guidance Service, 1981.

226. Benton AL. Differential behavioral effects in frontal lobe disease. *Neuropsychologia* 1968;6:53–60.

227. Barr A, Brandt J. Word-list generation deficits in dementia. *J Clin Exp Neuropsychol* 1996;18:810–822.

228. Talland GA. *Deranged memory.* New York: Academic Press, 1965.

229. Hooper HE. *The Hooper Visual Organization Test manual.* Los Angeles, CA: Western Psychological Services, 1958.

230. Sherer M, Parsons OA, Nixon SJ, et al. Clinical validity of the Speech-Sounds Perception Test and the Seashore Rhythm Test. *J Clin Exp Neuropsychol* 1991;13:741–751.

231. Kimura D. Functional asymmetry of the brain in dichotic listening. *Cortex* 1967;3:163–178.

232. Berlin CI, Lowe-Bell SS, Jannetta PJ, et al. Central auditory deficits after temporal lobectomy. *Arch Otolaryngol* 1972;96:4–10.

233. Rao SM, Bernardin L, Leo GJ, et al. Cerebral disconnection in multiple sclerosis: relationship to atrophy of the corpus callosum. *Arch Neurol* 1989;46:918–920.

234. Doty RL. *Smell Identification Test administration manual.* Haddonfield, NJ: Sensonics, Inc., 1983.

235. Doty RL. *The Smell Identification Test administration manual,* 3rd ed. Haddon Heights, NJ: Sensonics, Inc., 1995.

236. Haaland KY, Yeo RA. Neuropsychological function and neuroanatomic aspects of complex motor control. In: Bigler ED, Yeo RA, Turkheimer E, eds. *Neuropsychological function and brain imaging.* New York: Plenum Publishing Corp., 1989:219–243.

237. Rao SM, Binder JR, Bandettini PA, et al. Functional magnetic resonance imaging of complex human movements. *Neurology* 1993;43:2311–2318.

238. Haaland KY, Flaherty D. The different types of limb apraxia errors made by patients with left vs. right hemisphere damage. *Brain Cogn* 1984;3:370–384.

239. Poizner H, Mack L, Verfallie M, et al. Three-dimensional computer graphic analysis of apraxia. *Brain* 1990;113:85–101.

240. Hutt ML. *The Hutt adaptation of the Bender-Gestalt test,* 3rd ed. New York: Grune & Stratton, 1977.

241. Woodcock RW, Mather N. *Woodcock-Johnson Psycho-Educational Battery—revised.* Allen, TX: DLM Teaching Resources, 1989.

242. Wilkinson GS. *WRAT3: administration manual.* Wilmington, DE: Wide Range, Inc., 1993.

243. Crosson B. Treatment of interpersonal deficits for head-trauma patients in inpatient rehabilitation settings. *Clin Neuropsychol* 1987;1:335–352.

244. Lezak MD. Living with the characterologically altered brain injured patient. *J Clin Psychiatry* 1978;39:592–598.

245. Levin HS, Benton AL, Grossman RG. *Neurobehavioral consequences of closed head injury.* New York: Oxford University Press, 1982.

246. Butcher JN, Dahlstrom WG, Graham JR, et al. *MMPI-2: manual for administration and scoring.* Minneapolis, MN: University of Minnesota Press, 1989.

247. Beck AT, Ward CH, Mendelson M, et al. An inventory for measuring depression. *Arch Gen Psychiatry* 1961;41:561–571.

248. Zung WK. A self-rating depression scale. *Arch Gen Psychiatry* 1965;12:63–70.

249. Spielberger CD, Gorsuch RL, Lushene RE. *STAI manual for the State-Trait Anxiety Inventory.* Palo Alto, CA: Consulting Psychologists Press Inc., 1970.

250. Alfano DP, Finlayson MAJ, Stearns GM, et al. The MMPI and neurologic dysfunction: profile configuration and analysis. *Clin Neuropsychol* 1990;4:69–79.

251. Peyser JM, Rao SM, LaRocca NG, et al. Guidelines for neuropsychological research in multiple sclerosis. *Arch Neurol* 1990;47:94–97.

252. Katz MM, Lyerly SB. Methods of measuring adjustment and behavior in the community: I. Rationale, description, discriminative validity, and scale development. *Psychol Rep* 1963;13:503–535.

253. Bornstein RA, Brown GG. *Neurobehavioral aspects of cerebrovascular disease.* New York: Oxford University Press, 1991.

254. Levin HS, Grafman J, Eisenberg HM. *Neurobehavioral recovery from head injury.* New York: Oxford University Press, 1987.

255. Levin HS, Eisenberg HM, Benton AL. *Mild head injury.* New York: Oxford University Press, 1989.

256. Parks RW, Zec RF, Wilson RS. *Neuropsychology of Alzheimer's disease and other dementias.* New York: Oxford University Press, 1993.

257. Huber SJ, Cummings JL. *Parkinson's disease: neurobehavioral aspects.* New York: Oxford University Press, 1992.

258. Rao SM. *Neurobehavioral aspects of multiple sclerosis.* New York: Oxford University Press, 1990.

259. Rao SM. White matter dementias. In: Parks RW, Zec RF, Wilson

RS, eds. *Neuropsychology of Alzheimer's disease and other dementias.* New York: Oxford University Press, 1993:438–456.

260. Litvan I, Agid Y. *Progressive supranuclear palsy: clinical and research approaches.* New York: Oxford University Press, 1992.

261. Squire LR, Butters N. *Neuropsychology of memory,* 2nd ed. New York: Guilford Press, 1992.

262. Hermann BP, Whitman S. Behavioral and personality correlates of epilepsy: a review, methodological critique, and conceptual model. *Psychol Bull* 1984;95:451–497.

263. Grant I, Martin A. *Neuropsychology of HIV infection.* New York: Oxford University Press, 1994.

264. Weinberger DR, Berman KF, Daniel DG. Prefrontal cortex dysfunction in schizophrenia. In: Levin HS, Eisenberg HM, Benton AL, eds. *Frontal lobe function and dysfunction.* New York: Oxford University Press, 1991:275–287.

265. Bieliauskas LA, Matthews CG. American Board of Clinical Neuropsychology: policies and procedures. *Clin Neuropsychol* 1987;1:21–28.

266. Reports of the INS-Division 40 Task Force on Education, Accreditation, and Credentialing. *Clin Neuropsychol* 1987;1:29–34.

267. Hammeke TA. The Association of Postdoctoral Programs in Clinical Neuropsychology (APPCN). *Clin Neuropsychol* 1993; 7:197–204.

268. Meier MJ, Crosson BA, Eubanks D. *Division 40 petition for the recognition of a specialty in professional psychology.* Washington, DC: American Psychological Association, 1995.

269. Hannay HJ, Bieliauskas LA, Crosson BA, et al. The Houston conference on specialty education and training in neuropsychology. *Arch Clin Neuropsychol* 1998;13:157–250.

3

ANATOMIC NEUROIMAGING

TERRY L. JERNIGAN

The availability of *in vivo* brain imaging techniques has led to dramatic advances in neuroanatomic investigation of psychiatric disorders. Until about 3 decades ago, evidence for brain abnormalities in psychiatric disorders came almost entirely from autopsy studies. Although neuropathologic examination was, and remains, a powerful investigative tool, the limitations of the approach are particularly severe for the study of neuropsychiatric disorders because of their early onset and decades-long duration. *In vivo* brain imaging provides the opportunity to study living patients at various points in the course of the illness, as well as before and after treatment. Several neuropsychiatric disorders, such as schizophrenia spectrum and bipolar disorders, emerge during either adolescence or early adulthood. Others, such as the dementing disorders, occur most frequently in the elderly. Increasingly, models of these disorders invoke critical interactions between disease-specific factors and the processes of ongoing brain maturation and brain aging, respectively. Noninvasive brain imaging studies offer the potential to chart the diverging courses of normal and pathologic development in human subjects. Although x-ray computed tomography (CT) provided the first images of soft tissues of the head in living subjects, this method has been superseded for neuroanatomic studies by magnetic resonance imaging (MRI). The latter method is the focus of this chapter.

STRUCTURAL MAGNETIC RESONANCE IMAGING METHODS

Physical Basis for Structural Magnetic Resonance Imaging

MRI of the type described here is possible because the abundant protons in our body tissues, when placed in a magnetic field, behave like small magnets. MRI of the brain requires that a strong static magnetic field is created into which the subject is placed for imaging. Protons tend to align parallel to this applied magnetic field, much as small bar magnets would. In reality, the protons behave like spinning tops precessing around the axis of the static field (Fig. 3.1) (1). They are more likely to align parallel to, rather than in the oppo-

site orientation to, the field, reflecting the net magnetization in the tissues along the long (z) axis of the static magnetic field. The rate of precession of the protons is determined by the field strength; thus, all of the protons precessing in a homogenous magnetic field will precess at the same rate. This precession results in a rotating magnetic field at right angles (orthogonal) to the long axis of the applied static field. In other words, for each proton there is a rotating magnetic moment in the x–y plane. However, the spins of the protons are at random phases, so at equilibrium there is no net magnetization in the x–y plane of the field because the rotating components cancel each other out. The MR signal is created by applying a radiofrequency (RF) pulse orthogonal to the long (z) axis of the static field and at the precession frequency of the protons. This pulse creates a rotating magnetic field in the x–y plane by inducing greater phase coherence among the individual rotating fields of the protons. In other words, the RF pulse creates net magnetization in the x–y plane where there was no net magnetization before the application of the RF pulse. The resulting magnetization produces a detectable electromagnetic signal.

Much of the useful information in the MR signal can be extracted by modeling the decay (or relaxation) of the signal. Two important parameters in the model are the exponential time constants that describe what is referred to as *longitudinal relaxation rate* (T1) and *transverse relaxation rate* (T2). The former relates to the rate at which equilibrium magnetization occurs when the protons are placed in the static magnetic field. The latter refers to the loss of signal (i.e., magnetization in the x–y plane), associated with loss of phase coherence among the rotating fields of the protons, after the RF pulse is discontinued. This transverse relaxation rate (T2) is influenced by several different factors, including interactions among the protons, various other sources of inhomogeneity in the magnetic field, and diffusion (or motion). Different pulse sequences result in differing contributions of these different factors to the MR signal.

Thus far the process by which an MR signal is produced within an ensemble of protons has been described. However, MR images would be of little interest if they provided only information about the nature of the composite MR signal

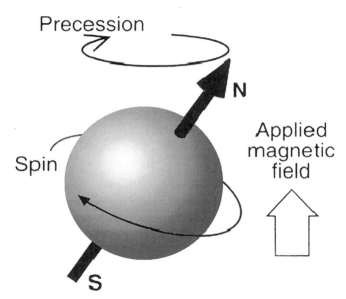

FIGURE 3.1. Illustration from Cohen and Bookheimer (1) showing the behavior of protons in a static magnetic field. (From Cohen MS, Bookheimer SY. Localization of brain function using magnetic resonance imaging. *Trends Neurosci* 1994;17:268–277, with permission.)

produced by the body's protons. To reveal the internal structure of the body, the MR signals from different locations in the body must be isolated from each other. This is possible because, as mentioned previously, the precession frequency of protons is determined by field strength. To isolate MR signals from different spatial locations in the body, magnetic field gradients are introduced across the field of view. This results in variability in the precession frequency across the field of view, due to the variation in field strength, and the signal from different points along the gradient can therefore be isolated by frequency analysis.

The basis for an MR image is a two-dimensional matrix of numbers. The information in the matrix is visualized by assigning gray levels or colors to the values in the matrix and displaying the information as an image. Whether a particular kind of anatomic information is detectable visually in the image depends upon both the degree to which that information is present within the numbers themselves and the degree to which it is preserved by the gray scaling or color coding used to display the image data. As described earlier, the MR signal is complex, and different pulse sequences produce MR signals more or less strongly influenced by factors such as T1 and T2. Because different kinds of biologic information are represented in different components of the MR signal and because the different components can be differentially weighted by varying the pulse sequence, many different views of the underlying anatomy are possible with MRI. Generally speaking, T1 weighting of images (such as those in the top panels of Fig. 3.2) results in good contrast between the gray and white matter of the brain. Strong T2 weighting (bottom panels of Fig. 3.2) yields strong contrast between brain and

cerebrospinal fluid (CSF) structures and better sensitivity to the subtle increases in tissue water content that can occur in association with ischemia, gliosis, edema, demyelination, and other tissue abnormalities.

The level of anatomic detail possible with MRI has increased rapidly over the last decade. It now is routine to acquire full-brain image volumes with 1–mm resolution in all three axes, and specialized procedures permit even higher spatial resolution in more focused fields of view. Many structural MRI (sMRI) procedures also afford excellent tissue contrast (Fig. 3.2). However, in spite of the remarkable visual quality of these images, accurate quantitation and interpretation of the information they provide remain difficult problems.

Interpretation and Quantitation of Imaging Signals: Magnetic Resonance Morphometry

In the early years after the introduction of MRI, much clinical research was needed to describe the features and anomalies that could be detected visually with this new technique. Modifications to the acquisition protocols were devised to improve tissue and lesion contrast and thereby enhance the detectability of such features. This important work continues apace, and new methods for extracting novel, biologically informative signals from the brain are constantly emerging. However, investigators soon became dissatisfied with the limitations of purely visual, and necessarily subjective, interpretations of the imaging data. This was nowhere a more significant problem than in the investigation of neuropsychiatric disorders in which the abnormalities, when they are present, are rarely focal and often very subtle. It was soon surmised that the inherently numerical nature of the imaging data could be further exploited by developing and applying more sensitive quantitative methods.

Although the merits of objective quantitative methods for characterizing brain structure with MRI are beyond dispute, there are a number of significant challenges associated with their application. MRI produces large matrices of signal values in which each value represents the estimated signal from a small volume *(voxel)* within the imaged object *(field of view)*. As described earlier, visualization of the structure is possible by coding the signal values with a gray scale or a color scale. However, although an expert observer can readily identify the underlying structure, the images contain, in addition to the information of interest, considerable signal fluctuation that is artifactual or irrelevant. For example, signals from structures of the brain itself are accompanied in the images by signals from cranial and extracranial structures (such as skull, eyes, and facial tissues). These structures sometimes have signals numerically similar to brain signals, which therefore can contaminate measurements of brain structures. Furthermore, due to imperfect magnetic field gradients, motion, or magnetic properties of the tissues themselves, the signals

Spiral Fast Spin Echo Images

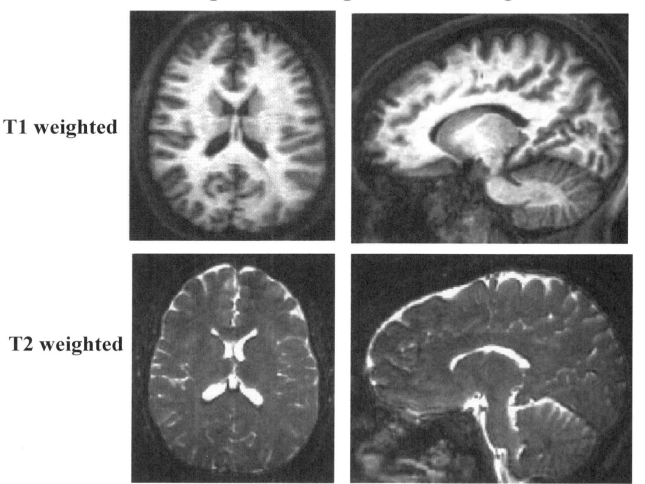

T1 weighted

T2 weighted

FIGURE 3.2. **Top:** High-resolution T1-weighted spiral fast spin-echo images exhibiting excellent tissue contrast. **Bottom:** High-resolution T2-weighted spiral fast spin-echo images. (From Wong EC, Luh W-M, Buxton RB, et al. Single slab high resolution 3D whole brain imaging using spiral FSE. *Proc Int Soc Magn Reson Med* 2000;8:683.117, with permission.)

can be distorted, leading to subtle artifacts that nonetheless significantly degrade measurement validity. For example, most MRI data sets (three-dimensional matrices of signal values) exhibit low-frequency signal drift across the field of view that results in slightly different values for the same tissue in different parts of the brain. When attempting to detect subtle tissue changes or abnormalities, this fluctuation can introduce an unacceptable level of noise. Methods for addressing these and other problems have been developed and will be outlined later.

Quantitative analysis of MRI data for the study of brain structure is referred to as *MR morphometry*. These techniques are used in neuroanatomic studies of neuropsychiatric disorders to determine whether abnormalities are present in specific brain structures. The impetus for many of these studies came from early observations with x-ray CT that revealed increases in the amount of CSF in the cerebral ventricles and subarachnoid spaces surrounding the brain. When observed, this was taken as evidence that brain shrinkage had occurred either as a result of earlier damage or in association with ongoing neurodegenerative processes. Unfortunately, CT offered limited soft tissue contrast and little opportunity to determine the anatomic distribution of the effects on specific brain structures. Later MRI studies exploited the improved tissue contrast and spatial resolution to extend these early CT studies, by estimating the volumes of different brain tissues and individual brain structures (Fig. 3.3). In the following section, the basic methodologic requirements for such studies are outlined.

Specific MR morphometric methods vary across laboratories, but most involve a set of important steps as follows.

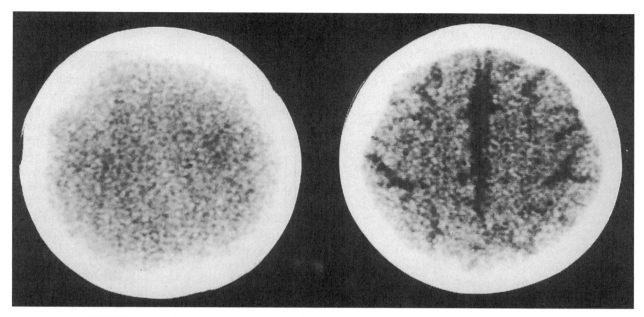

FIGURE 3.3. Sections from x-ray computed tomography examinations of two individuals. The sections are from near the vertex (through the cerebral cortex). The section on the **left** is from a normal control and shows little cerebrospinal fluid in cortical sulci. The section on the **right** is from a chronic alcoholic of similar age and shows the increase in sulcal cerebrospinal fluid often reported in these patients.

Isolation of Brain from Nonbrain Areas Within the Images

A surprisingly challenging problem is the separation of the brain from adjacent structures within the images (sometimes referred to as *stripping*). Although trained anatomists can generally distinguish the boundaries of the brain fairly reliably, the process of "tracing" these boundaries within each imaged brain section can be tedious and time consuming. Also, within sections at the edges of the brain the problem is complicated by the presence of overlapping structures with signal values similar to brain. Most laboratories have adopted substantially automated methods that exploit the change in signal values at the interfaces between the cerebral cortex, the adjacent subarachnoid space, and the cranium. Sometimes models or templates of the brain's modal shape aid in the process. However, even using these algorithms, in most cases some human supervision (or editing) of the process is necessary to achieve results acceptable for volumetric studies.

Filtering or Other Methods for Correcting Signal Drift in Image Values

As described earlier, the signal values for a given tissue sometimes differ in different parts of the image due primarily to magnetic field inhomogeneity. A variety of bias-correction methods have been employed (2) to reduce these effects, most of which produce similar results. In most cases the methods attempt, by various means, to isolate and reduce low spatial frequency signal modulation present in the images. Although recent improvements in instrumentation have mitigated the problems leading to this signal bias to some extent, most investigators still find that better tissue segmentation results are obtained after processing the image volumes to further reduce drift artifacts.

Identification of Voxels Within Different Tissues (Usually White Matter, Gray Matter, and Cerebrospinal Fluid)

Because the ultimate goal of these methods is to estimate the size of specific structures and because a structure usually is defined in part by its tissue type, it is important to operationalize the definition of the different tissues. For example, when attempting to estimate the volume of the cerebral cortex, one generally intends to include the gray matter of the cortical ribbon while excluding the underlying white matter. Automated, or semiautomated, tissue classification is an important step in the process of extracting a usable three-dimensional model of the brain's structure from a raw MR image volume. The utility of efficient methods for tissue segmentation of MR images has been widely acknowledged (3–6), and several approaches have emerged (for reviews, see references 4,7–19). However, the basic rationale underlying tissue segmentation merits some scrutiny. Most schemes attempt either (a) to classify each voxel as gray matter, white matter, or CSF (sometimes additional categories, such as lesion or tumor are included); or (b) to assign some portion of each voxel to each of these categories. The latter sometimes are referred to as *partial-volume* methods and are inherently appealing because it is known that many voxels span the boundaries of gray, white, and CSF structures and thus contain a mix of tissues. Although a more accurate volume estimate of a given tissue is obtained from a partial-volume method than a classification method, in some cases the simpler structure of the segmented image volume resulting from a classification method is of greater value to the investigator than the increase in sensitivity gained by a partial-volume method. If a classification method assigns each voxel to the tissue category that represents the largest percentage of its volume, an equally accurate partial-volume method affords little increase in sensitivity when the voxel size is small relative to the sizes of the target structures. However, when the target structure has an expected volume that is a small multiple of the voxel size, the advantages of partial-volume methods can be considerable. Because most morphometric studies of neuropsychiatric disorders examine brain structures hundreds to thousands of times larger than the voxels, it is not surprising that most of these studies have used voxel classification methods.

From a practical point of view, a much more significant challenge to tissue segmentation occurs in the form of signal inhomogeneity within tissue classes. As mentioned earlier, nonbiologic sources of signal nonuniformity, such as RF inhomogeneity, contribute substantially to this variability. However, biologic sources also exist. For example, even to the naked eye, the MR signal is not uniform across regions deemed by anatomic nomenclature to be "white matter" regions. This is because the white matter is, to some extent, biochemically heterogeneous. Similar heterogeneity is present within the cerebral gray matter. With most MR acquisition sequences, this results in some fully volumed voxels within the "white matter" regions having signal values identical to those of some fully volumed voxels within the "gray matter" regions. When this occurs because of variability within a tissue class in the biochemical composition of the voxels, it is not always clear what is meant by "white matter" and "gray matter" (i.e., is it defined by location or by MR signal value?). There also exists variation across individual brains in the signal characteristics of the tissues and in the tissue contrast, and this variation is by no means limited to "pathologic" cases. These factors considerably complicate the already formidable task of validating tissue classification methods, and readers of the MR morphometry literature should consider them when interpreting the results.

Most methods identify tissues by first defining characteristic signal values for each of the tissue types to be distinguished. Supervised methods generally require that an operator select regions within the tissues of interest and then these values are used to "seed" the segmentation algorithm. In other words, an operator designates certain voxels to be sampled for an estimate of the characteristic value of each tissue. A few methods "seed" the algorithm with voxel val-

ues selected using an automated method. Other methods, not requiring supervision, estimate the characteristic tissue values by assuming that the histogram of signal values from the matrix consists of overlapping distributions from a given number of tissues with distinct values. By making assumptions about the underlying distributions, the best fit to the model can be determined statistically and the characteristic value of each tissue can be estimated from the histogram. With either approach, after the values for the different tissues have been defined, each voxel can be classified based on the similarity of its signal value to these tissue values, or each voxel's tissue composition can be estimated using a prediction model.

Most of these tissue segmentation methods produce reasonable and similar results when applied to human MRI data. They also produce similar results when tested on phantom data (images of models constructed with simulated "tissues"), and they produce highly reproducible results (particularly the more highly automated ones, of course) when applied repeatedly to the same data set. Unfortunately, few have been validated on serial MRI examinations in the same individuals. As mentioned earlier, the assumption on which these tissue segmentation methods are based, namely the assumption that the brain consists of a discrete number of tissue types with uniform values within each type, is not strictly true. Therefore, the results of all of the methods contain errors. Nevertheless, the results of different methods applied to MRI from young healthy volunteers are remarkably similar and have good face validity (i.e., the segmentations look right). Unfortunately, there is ample evidence that the signal values of brain tissues are affected by factors associated with aging, and by toxicity and disease. This raises concern about the effects of such factors on results of tissue segmentation in older individuals and patients, because violation of the assumptions of the segmentation models may be more severe in these cases. An association of segmentation bias with diagnosis or with other clinical variables can lead to significant confounding of the experimental effects.

Delineation of the Boundaries of Different Brain Structures of Interest

Because most MR morphometric studies of neuropsychiatric disorders have focused on structure volumes, the accurate delineation of the boundaries of target structures has been of utmost importance. If one is attempting to measure the volume of the amygdala, for example, in an attempt to detect disease-related change in the volume in that structure, it is necessary to define the contour that separates the amygdala from adjacent entorhinal cortex, basal forebrain, hippocampus, and white matter. Any inconsistency in setting these boundaries can result in considerable unreliability in the estimated amygdalar volumes. Because of the difficulty involved, most investigators have relied upon manual delineation of boundaries by trained anatomists. Nevertheless, for many brain structures of interest there are few gross morphologic features visible on MRI that can be used to define their boundaries. For this reason, precise criteria must be developed, sometimes involving the positions of nearby anatomic landmarks, to guide operators in setting reliable boundaries. Often, it is not possible to define boundaries for a brain structure following cytoarchitectural conventions. In these cases, there are two goals: first, to define in a highly consistent way the boundaries of a region that includes the structure; and second, to ensure that the structure of interest contributes a sufficiently large percentage of the total volume of that region that any variability in the volume of the structure is likely to make up the majority of the variability in the volume of the total region. When reporting the results of volumetric studies, investigators usually report both the specific criteria used to define boundaries between structures and estimates of the reliability with which human operators can reproduce the boundaries using those criteria. The reader must judge whether it is likely that variability in the size of the target structure is likely to be measured faithfully when the reported criteria are applied.

Computation of Volume Estimates (and Correction for Overall Cranial Volume)

Once the boundaries have been defined, structural volumes can be estimated by summing the voxels, or voxel tissue proportions, within the boundaries, usually across the imaged brain sections containing the structures. Most investigators are well aware that these estimates of raw volume show wide variability across individuals. For example, raw volumes of the caudate nuclei of two adult individuals can easily vary by a factor of nearly two! Much of this variation appears simply to be related to body size. Larger people tend to have larger heads (and brains) than do smaller people. Consistent with this, on average, women have smaller heads than do men. Furthermore, there appears to be additional individual difference variability in head size, just as there is in nose size. This large degree of individual difference variability in head size, which also is present in volumes of individual brain structures, is a problem when one is attempting to infer volume *loss* or volume increase (associated with hyperplasia) from volume per se. When there is good reason to believe that the brains of the individuals studied developed normally (i.e., that any tissue loss occurred after the brain reached its full size), it often is helpful to correct the raw volumes for overall cranial volume. The logic here is that the cranial volume indexes the premorbid brain size and that any reduction in the volume of a brain structure relative to the size of the cranial volume is likely to reflect volume loss in that structure. Frequently the volumes are expressed as proportions of cranial volume. Most results of volumetric studies in neuropsychiatric patients and their controls are reported in this way.

Obviously, however, the same logic cannot be applied to estimate the degree of brain hypoplasia. In this case the cranial volume is itself reduced as a result of the pathologic process. This is a particularly thorny issue when studying a disorder, such as schizophrenia, for which there are hypothesized neurodevelopmental, as well as possible neurodegenerative, factors contributing to the etiology.

It should be noted that MR morphometry sometimes is used to define anomalies of brain shape, as might result from aberrant brain development, when no evidence for atrophy (or loss of brain tissue) is present. The goal in this case is to examine the relative sizes of brain structures. Again, the practice of expressing structural volumes as a function of cranial volume often is used in an attempt to reveal shape anomalies while controlling for individual (or group) differences in overall brain size.

Effects of Neuropsychiatric Disorders Occur in the Context of Developmental Effects on Brain Structure

It is known that there are brain structural changes that accompany normal aging (for a summary from the author's laboratory, see reference 20). These changes in the volumes of brain structures can be substantial and must be considered carefully in the interpretation of morphometric results in neuropsychiatric populations. Adjusting brain volumes for known effects of aging can increase the likelihood of detecting differences between diagnostic groups when the subjects vary considerably in age, and can help to differentiate the effects of factors such as disease duration or treatment duration from aging effects. For the study of neuropsychiatric disorders in which brain damage or neurodegeneration is suspected, the loss of brain tissue should be measured relative to the losses that are typical for normal individuals of similar age. This is particularly important for the study of such disorders as alcohol and drug abuse, or the dementing disorders. Some investigators have incorporated age adjustments based on larger normative samples into the morphometric methods themselves by computing volume z-scores (21,22).

The effects of age on structural volumes has most often been considered in studies of adult-onset disorders. However, many neuropsychiatric disorders have their onset in childhood, adolescence, or early adulthood. MR morphometric studies have provided strong evidence that the brain's structure changes substantially during normal development and that the changes continue well into adulthood (23–26). It is possible that some neuropsychiatric disorders involve interactions between disease factors and the ongoing processes of brain maturation. Structural MRI studies of these disorders should incorporate, in the same way that studies of adult-onset disorders do, data regarding age effects on brain morphology. In other words, it may be important to consider not only whether anatomic measures in young patients are in the normal range but also the extent to which they diverge from those in normal age-peers.

Voxel-Based Analyses of Brain Dysmorphology

In recent years, methods have emerged that attempt to characterize brain morphologic effects on MRI on a voxel-by-voxel basis (27,28). These methods have the potential to provide a more detailed description of the regional distribution of structural anomalies produced by hypoplasia or degeneration in the brain than the conventional volumetric techniques described earlier. Although voxel-based approaches sometimes require the identification of anchoring landmarks within the brain, they obviate the time-consuming delineation of the complete outside boundaries of structures using conventional anatomic criteria. The desired results of these methods are statistical maps of the brain revealing the pattern of structural anomaly present in a group of patients (relative to a control group) with high spatial resolution. Because the precise anatomic plane of section varies in MRI volumes and because there is considerable normal subject-to-subject variability in brain morphology, these methods require that the data sets across individuals be registered to each other accurately. Then, at each voxel location, signal differences between the groups must be compared statistically with signal variability within the groups.

As in the case of the more conventional morphometric techniques described earlier, a significant challenge to these methods is the large degree of variability in head size. One approach to this problem is to "size normalize" the data in the MR images before examining the distribution of shape anomalies. This results in considerably more precision in the maps. Normalizing the maps using a strictly linear method leads to a reasonably straightforward distribution for subsequent shape analysis, but there still is considerable variability in the MR signals at a given (standardized) voxel location due to individual differences in brain shape. Voxel-based approaches to describing the abnormalities resulting from degenerative processes in patients with adult-onset disorders generally have used such linear methods for size normalization. Here an assumption that the overall size of the brain (or other structure to be examined) is unrelated to clinical status seems reasonable; in fact, this assumption can be examined empirically by comparing the groups first on these measures before performing voxel-based shape analysis. Under these circumstances, a voxel-wise map of the variability associated with clinical group (taking into account the within-group variability) should reveal where in the brain or structure the greatest morphologic anomalies are present.

It should be noted that the effect of spatial normalization must be considered carefully when using voxel-based approaches to compare morphology in patient groups. Imagine for example, that a brain structure is smaller in a clinical group than in controls and further that this size reduction is entirely due to a reduction of the more anterior parts of the structure. Assuming that the structures are properly aligned in a standard anatomic space, voxel-wise statistical maps (comparing the groups) would be expected to reveal anomalies in the anterior, but not the posterior, parts. However, if within-group size variability is great, the resulting noise at each voxel that is due to this factor may reduce power. On the other hand, if a size normalization procedure were used to render each structure equal in size, the statistical maps would be expected to show anomalies throughout the extent of the structure, with anterior portions exhibiting anomalies in one direction and posterior portions exhibiting anomalies in the opposite direction. In other words, the maps would show that when size is held constant, the smaller anterior portions are *relatively* smaller but the posterior portions are also *relatively* larger in the clinical group. This is strictly true, but in this case it probably should not be interpreted to mean that the posterior as well as the anterior portions are *abnormal.* This is just one example of the interpretive difficulties that arise when examining morphology using voxel-based maps in subjects in whom systematic differences in size and shape of the structures are present. Note that similar problems could occur in mapping age effects on tissue composition in samples within which brain size varies systematically with age. Given these caveats, it is clear that mapping of morphologic anomalies, particularly in groups with known or suspected brain hypoplasia, using voxel-based methods must be approached cautiously. However, such methods have considerable potential to reveal differences not apparent in more conventional analyses, and their application is substantially more efficient.

Many of the voxel-based image processing methods were developed for use with functional imaging. The original goal was to devise methods for intersubject averaging that would best bring functionally homologous brain structures into spatial alignment so that modal brain maps for particular functions could be defined. The so-called *affine* methods were superseded by new methods that involve nonlinear warping of the brain image data to minimize the structural discrepancies in size and shape among individuals. This is a reasonable approach if one can make the assumption that individual differences in brain size and shape are relatively independent of functional characteristics of the structures. The use of voxel-based analysis for mapping structural differences has in some cases been "bootstrapped" from these methods.

The proper interpretation of maps of sMRI signal anomaly that compare groups that may differ substantially in brain size and disease related dysmorphology, particularly when intensive spatial normalization has occurred, is a complex subject. There is a danger, as described briefly earlier, that interactions between the effects of diagnosis and the effects of spatial normalization may produce apparent abnormalities in the maps that are remote in location from the sites of the actual structural abnormalities present in the clinical group. Another example of the problem relates to the tendency, using most warping techniques, for warping of unaffected structures near large affected structures to be influenced by their proximity to these structures. Consider the case of a comparison of a group of patients with ventricular enlargement (due perhaps to white matter volume loss) to a group without such changes. Warping techniques might accomplish alignment of the enlarged ventricles with the smaller ventricles at the expense of distortions in the adjacent subcortical structures. The resulting map of morphologic anomalies may display apparent group differences in these structures that do not exist.

It is likely that these methods will have different levels of sensitivity to changes in different brain structures so that even large differences in structures that vary widely in position or shape may not be as easily detected in voxel-based maps. The problem of unequal measurement sensitivity across different structures examined is equally important in the application of more conventional morphometric techniques, but for these reliability and validity data for each structure can be provided, compared, and considered in the interpretation. Given the lack of this information regarding anatomic sensitivity variability for voxel-based methods, they may in their currently implemented form be most useful as exploratory methods. At any rate, apparent discrepancies between anatomic findings that emerge from studies using conventional morphometrics and voxel-based mapping should be analyzed carefully with consideration of the relative strengths and weaknesses of the different approaches. Bookstein (29) was among the first to propose mathematical (voxel-based) approaches to characterizing dysmorphology. He and his associates have attempted to define patterns of callosal shape abnormality in neuropsychiatric patients on midsagittal MR images using a thin-plate spline method. Bookstein (30) also has cautioned against the use of voxel-based methods for group comparisons of brain morphology, citing problems resulting from the effects of the spatial normalization.

APPLICATIONS OF STRUCTURAL MAGNETIC RESONANCE IMAGING IN NEUROPSYCHIATRIC RESEARCH

A review of the wide range of studies examining the many different neuropsychiatric disorders with anatomic imaging is beyond the scope of this chapter. What follows is a selective review of studies with a bias toward studies of those neuropsychiatric disorders for which a number of independent investigations have emerged and for which the imaging findings seem clearly to have extended our understanding of the illness. It is hoped that the brief review of clinical studies that follows will provide examples of the kinds of questions that may be addressed with anatomic imaging studies. The review focuses on two lines of research: one aimed at eluci-

dating the effects of heavy alcohol exposure on the brain and the second on defining anomalies of brain morphology that characterize schizophrenia.

Studies of Alcohol Effects on Brain Structure

Brain Structural Anomalies in Chronic Alcoholics

Early CT studies suggested that many chronic alcoholic patients had suffered brain volume loss as indexed by increased CSF. MRI was the first method with which the volumes of individual brain structures could be estimated in living individuals. This made it possible to survey the regional pattern of brain volume loss in chronic alcoholic patients to determine whether some structures appeared to be more vulnerable than others to the effects of alcohol abuse. In the author's laboratory, morphometric techniques of the type described in the previous section were developed for application to MRI. The first study examined the effects of normal aging, uncomplicated by substance abuse or other significant neurologic or neuropsychiatric conditions (31). This study revealed that over the range from 30 to 79 years, age-related volume loss occurred in both cortical and subcortical gray matter of the cerebrum and that the effects were best measured by correcting the raw volumes for the volume of the cranial vault. This early MR morphometric study of normal individuals produced data that were used to derive measures of brain structural volumes that were corrected both for cranial volume and for the effects of "normal aging." These measures were used to examine the volumes of subcortical brain structures and cortical regions in a group of 28 chronic alcoholic patients and matched controls (32). Regarding the CSF-filled structures, the results confirmed those of several earlier CT studies (33) showing modest increases in ventricular size but striking increases in subarachnoid spaces. Because the subarachnoid spaces are adjacent to the cerebral cortex whereas the ventricles lie deep in the center of the brain, this pattern had been interpreted by many to indicate that the cerebral cortex might be particularly vulnerable to alcohol-related damage, whereas the effects on subcortical structures might be less severe. The study by Jernigan et al. (32) revealed that the cerebral cortex was reduced in volume in chronic alcoholic patients; however, there were losses in subcortical gray matter structures (e.g., basal ganglia and diencephalon) that were at least as striking as those in cortex. Furthermore, although the losses in gray matter structures throughout the brain were substantial, they did not seem to explain entirely the dramatic increases in subarachnoid CSF. The authors pointed out that some of the volume loss probably was in white matter, and their study revealed significant abnormality of the cerebral white matter in the form of increased volume of white matter with abnormally high signal on MRI (32,34). Whatever the sources of brain volume loss in the patients of this study, there seemed to be a moderate correlation between the degree of loss (as reflected in increased CSF) and the severity of their performance decrements on tests of memory, attention, and speed of information processing.

Reports by Pfefferbaum et al. (22,35) substantially extended the findings, revealing in a larger sample of 49 recently detoxified patients significantly decreased volumes of white matter, as well as gray matter, in cortical regions. In addition, an interaction between the effects of age and those of alcoholism clearly seemed to be observable in this study. The older patients showed considerably more volume loss relative to their controls than did the younger patients. Surprisingly, within this sample there was little association between severity of alcohol history and age; therefore, the authors interpreted their results as suggesting that the older brain was more vulnerable to the effects of alcohol. Although in the study by Pfefferbaum et al. and the earlier study by Jernigan et al. (32), there was some minor variability in the amount of loss observed in different cortical regions, the evidence for greater vulnerability of specific cortical regions was not strong. However, it should be noted that in both studies the anatomic boundaries between cortical regions were set using stereotactic criteria rather than conventional sulcal landmarks, and they did not specifically delineate the cortical lobules or gyri.

Several subsequent morphometric studies have focused on structures of the temporal lobe (and in particular the hippocampus) in chronic alcoholic patients. Sullivan et al. (36,37) reported on specific measures of the volume of the anterior and posterior segments of the hippocampus. Measures of gray and white matter volumes of the temporal lobe (excluding hippocampus) were available. They observed that the anterior hippocampus, like temporal lobe cortex, was reduced in the alcoholic patients and that the losses in both areas were greater in the alcoholics than would be expected from the ages of the subjects. Attempts to correlate these losses with severity of alcohol history suggested little or no association, providing further evidence for an age-related increase in brain vulnerability to alcohol. Of particular interest were the comparisons by this research group of patients with and without histories of alcohol withdrawal seizures. Although patients with no history of seizures showed comparable hippocampal and temporal lobe gray matter losses to those with seizure histories, the latter had significantly less temporal lobe white matter. This suggests that subsequent investigation of the mechanisms by which alcohol withdrawal induces seizures should consider a possible role for temporal lobe white matter damage.

A recent MR morphometric study also focused on the hippocampus. Beresford et al. (38) examined the volumes of the hippocampus and the pituitary in alcohol-dependent subjects seeking evidence that the hippocampal damage is mediated via the hypothalamic–pituitary–adrenal axis. They reasoned that if the hippocampal damage were so mediated, the pituitary might be enlarged in association with hypercortisolemia. Consistent with this hypothesis, their study showed that the ratio of hippocampal volume to pituitary volume was significantly reduced in a small group of alcohol-dependent individuals relative to controls.

Two studies by Pfefferbaum, Sullivan, and their associates examined in greater detail the regional pattern of gray and white matter losses in chronic alcoholics (39,40). In a comparison of younger and older alcoholics, they found evidence that the older patients exhibited disproportionate damage to gray and white matter of the frontal lobes, whereas in younger patients the cortical losses were more uniform (39). In a separate study, the regional pattern of volume loss in chronic alcoholics was compared to that in schizophrenic patients (40). The latter had somewhat greater loss of cortical gray matter overall but disproportionate loss in more anterior areas. In contrast, the cortical gray matter losses in the alcoholics were more uniform across the cortex. White matter losses, which were more severe overall in alcoholics than in schizophrenics, were greatest in frontal and posterior temporal areas.

A final MR morphometric study of alcoholics examined the cerebellar vermis and hemispheres (41). Alcoholic patients with and without Korsakoff syndrome (alcoholic amnesia) were studied. Gray matter abnormalities of the cerebellar hemispheres were observed in both groups of alcoholics, and volume loss in both gray and white matter tissue compartments was present in the Korsakoff patients. There was evidence that the degree of volume loss in anterior vermal white matter was related to degree of ataxia within the alcoholic patients.

An early study applying voxel-based, rather than morphometric, methods was conducted by Pfefferbaum et al. (42), who compared corpus callosum shapes in older alcoholics and older controls. Their analysis was performed on the averaged midsagittal sections through the corpus callosum and revealed significant reductions in the midbody area of the callosum in the alcoholics, with relative sparing of genu and splenium.

In summary, results of MR morphometric studies of chronic alcoholic patients suggest that there are volume losses in the cerebral cortex, the underlying cerebral white matter, the cerebellum, and the midline basal ganglia, limbic, and diencephalic structures. In addition, there appear to be some MR signal alterations in the remaining cerebral white matter. Older patients appear to have disproportionate, more frontally distributed losses, suggesting that the older brain, and particularly the older frontal lobes, may be more vulnerable to alcohol effects. These studies contribute to hypotheses about the pathogenesis of alcohol-related neurodegeneration, and they demonstrate that it is feasible to monitor these changes in brain structure during life. Interestingly they suggest that the early CT findings of marked increases in subarachnoid spaces were not so much due to disproportionate damage to the adjacent cortex but rather to widespread white matter damage. Nevertheless, there is much work left to be done to reveal the relevant neurodegenerative mechanisms associated with these phenomena and to define their clinical implications.

Studies of Effects of Abstinence on Brain Structural Abnormalities

Given the early CT findings of increased CSF in alcoholics, the presumption that such increases reflected loss of brain parenchyma, and the prevalent view at that time that little brain regeneration occurred, it was surprising to many that subsequent CT studies of recovering patients suggested that the CSF spaces were smaller after successful treatment (43,44). The reversibility of CSF changes with abstinence subsequently was confirmed in MRI studies (45,46). An MR morphometric study comparing abstaining to relapsing alcoholics 3 months after baseline revealed significant increases in cerebral white matter volumes in the abstaining group (47). Pfefferbaum et al. (48) examined changes on MRI over a period of about 3 weeks in 58 patients and over a longer period in a smaller group. This study was particularly informative because normal controls also were followed with MRI. The findings suggested that the short-term reductions in CSF spaces observed in abstaining patients were associated with increases in gray matter and that in the longer term follow-up white matter volume was observed to decrease in relapsers. Taken together, these studies seem to confirm that there is alcohol-related shrinkage of brain tissue in many chronic alcoholics that is to some substantial degree reversible with abstinence. Some have suggested that the changes might reflect rehydration of the brain associated with abstinence. Mann et al. (49) tested this hypothesis directly by examining CT values of tissue in recovering alcoholics. Rehydration of the brain in association with abstinence would be expected to decrease CT values; however, these authors reported slightly increased values in the abstaining patients in association with the expected reductions in CSF volumes. Given these findings, it seems likely that, rather than rehydration, neuroplastic changes, perhaps trophic increases in cell size, remyelination, or growth of neuronal processes, lead to the observed tissue volume increases.

Studies of Brain Structure in Alcoholic Amnesics

A rare syndrome of anterograde and retrograde amnesia sometimes occurs in chronic alcoholics in the context of the Wernicke-Korsakoff syndrome. Autopsy studies have generally implicated diencephalic structures (particularly mammillary bodies and medial thalamus) in the pathogenesis of the memory deficits of these patients (50–52); however, before the advent of MRI, it was not possible to examine these structures directly in living patients. Because the mammillary bodies in particular often are involved in Wernicke-Korsakoff cases and because these structures are well visualized in midsagittal MR images, they were the object of scrutiny in several early MRI studies. Most of these studies used subjective evaluations of the degree of mammillary body damage. Numerous reports confirmed the high incidence of mammillary body damage in alcoholic amnesics (53–57), but they also documented an increase in the incidence of similar damage in nonamnesic alcoholics (55–57). Furthermore, there were some reports of an absence of any visually apparent mammillary body damage in some alcoholic amnesics (57). In a later study, a high-resolution MRI protocol and morphometric techniques were used for estimating the volume of these small midline structures (58). Mammillary body volumes were compared in demented alcoholics, amnesic alcoholics, nonamnesic alcoholics, and matched controls. The demented and amnesic patient groups were small, but the results suggest that mammillary body volume loss, although present in nonamnesic alcoholics, nevertheless is more severe in amnesic patients and even more severe in demented alcoholic patients. Overall, the results of this study suggested that both degree of mammillary body damage and cognitive impairment within nonamnesic alcoholics are on a continuum with that seen in Wernicke-Korsakoff syndrome. These authors also examined the relationship between a history of Wernicke encephalopathy and the degree of mammillary body damage. They found an association of both with declarative memory impairment across the groups of alcoholics and interpreted this as evidence for a distinct form of neuropathology related to Wernicke encephalopathy.

Jernigan et al. (59) did not examine mammillary bodies but did estimate the volumes of a number of other brain structures in their study of alcoholic amnesia. Eight alcoholic amnesics were compared to matched groups of nonamnesic alcoholics and normal controls. The volumes of CSF structures, subcortical gray matter structures, and regions of the cerebral cortex were examined. Moderate increases in subarachnoid CSF volumes, roughly comparable in degree, were present in both alcoholic groups. In contrast, the increases in ventricular CSF were only modest in the nonamnesic patients but were significantly more severe in the amnesics. Circumscribed gray matter losses were present in the nonamnesic patients in the diencephalon and ventral cortex. In contrast, the amnesic patients had considerably more extensive gray matter loss. In particular, their losses in the anterior diencephalon (which did not include mammillary bodies), the limbic structures on the mesial surface of the temporal lobes, and orbitofrontal cortex were significantly greater than those of the nonamnesic patients. This suggests that damage in structures other than the mammillary bodies and midline thalamus may play a role in the amnesia of Wernicke-Korsakoff syndrome. Some investigators (60), but not all (54), also observed mesial temporal lobe damage on MRI in alcoholic amnesics.

Studies of the Effects of Fetal Alcohol Exposure on Brain Structure

Jones et al. (61) were the first to describe a syndrome of facial dysmorphology, growth retardation, mental retardation, and other neurologic symptoms in children who prenatally were exposed to severe levels of alcohol. A later autopsy study revealed a small abnormal brain lacking the corpus callosum (the largest interhemispheric fiber tract in the brain) in a case with this syndrome (62). Most children with severe fetal alcohol exposure have a relatively normal life expectancy and, fortunately, do not come to autopsy during their development. However, such children can and have been examined with MRI in attempts to define the prevalence and pattern of brain dysmorphology associated with fetal alcohol exposure.

In an early case study, two young patients with severe fetal alcohol syndrome (FAS) were examined with MR morphometric techniques (63). The brains of these two patients were compared to those of a group of age-matched controls and with a small group of children with Down syndrome. On visual inspection, it was apparent that the corpus callosum was absent in one of the FAS patients and was moderately hypoplastic in the other. Volumetric analyses revealed severe overall brain size reduction, affecting both cerebrum and cerebellum, and suggested disproportionate reduction in the volume of subcortical gray matter structures, such as basal ganglia and diencephalon. Data for two additional individuals were added in a second preliminary report (64). These two young subjects had histories of severe prenatal alcohol exposure, mental retardation, and behavioral problems, but they lacked the facial features required for the diagnosis of FAS. These youngsters also had significantly reduced cerebral and cerebellar volumes, and again there was evidence for disproportionate hypoplasia of the basal ganglia. In another study by this group, measurements of the cerebellar vermis in children with severe prenatal alcohol exposure (FAS and non-FAS) were compared to those in normal controls (65). The children with fetal alcohol exposure had reduced anterior vermal size but no significant reductions in posterior vermal areas.

A subsequent report described morphometric results in an additional 6 children with FAS and 7 control children, none of whom had agenesis of the corpus callosum, though the callosum may have been thin (66). The study focused on overall brain hypoplasia and on the volumes of cortical and subcortical structures. Again, there was evidence for cerebral and cerebellar hypoplasia. Volumes of subcortical structures were expressed as proportions of the total cranial volume to determine whether the reductions in these structures were commensurate with or disproportionate to overall cerebral hypoplasia. The basal ganglia proportion was significantly reduced in the FAS subjects. In a secondary analysis, volumes of the caudate nuclei and the lenticular nuclei (components of the basal ganglia) were examined separately. The basal ganglia effects appeared to be particularly severe in caudate nuclei, and there was some indication that the caudate nuclei might be more severely hypoplastic than the lenticular nuclei (although the region by group interaction failed to reach significance).

The data in the latter report were generally consistent with those from another MRI report that appeared about the same time (67), in which visually identifiable anomalies on MRI were noted in a group of nine cases of FAS. These authors stressed the prevalence in the disorder of midline abnormalities, that is, in corpus callosum, cavum septi pellucidum, and brainstem. Hippocampal and amygdalar volumetry was performed using MRI in a separate study of 11 FAS individuals (68). There was some evidence for disturbed asymmetry of the hippocampus (with relatively smaller left hippocampi) in FAS subjects relative to controls. Surprisingly, in the latter report the authors described findings they interpreted as indicative of atrophy, that is, loss of brain tissue.

A more comprehensive MR morphometric study compared 14 FAS individuals, 8 subjects with severe prenatal alcohol exposure (but without facial dysmorphology), and 41 controls (69). No subjects with agenesis of the corpus callosum were included. Detailed anatomic analyses of different tissues and structures within cerebrum, cerebellum, and brainstem were performed (Fig. 3.4). The results in the FAS subjects were generally consistent with, and substantially extended, previous observations; however, they provided no evidence for atrophy (as indexed by CSF increases). Instead there was evidence for a specific pattern of hypoplasia affecting the cerebrum and brainstem but perhaps affecting the cerebellum even more dramatically. The results also indicated that within the cerebrum, white matter was more severely hypoplastic than was gray matter and that the parietal lobe was disproportionately affected overall. There again was definite evidence for disproportionately reduced size of the caudate nuclei in FAS subjects, but in contrast the hippocampus appeared to be significantly spared. In general, brain morphology in the alcohol-exposed subjects without facial dysmorphology was more similar to that in controls than to FAS. However, *post hoc* analyses focusing on those regions that were most abnormal in the FAS subjects, namely caudate nuclei and parietal lobes, produced some evidence for modest volume reductions in this less affected group.

An example of an interesting study of fetal alcohol effects using a voxel-based approach is that of Sowell et al. (70). The subjects in this study included many of the fetal alcohol-exposed subjects and controls examined in the study by Archibald et al. (69). A particularly interesting analysis attempted to map the pattern of hypoplasia as represented at the cortical surface. For this analysis the authors defined the surface of the cerebral cortex for each subject and aligned the brain volumes so that they were registered based on the positions of principal cortical sulci. They then determined the distance from each of 65,536 points on the cortical surface to a point deep in the brain on the midline (at the decussation of the anterior commissure). They refer to these values as *differences from center* (DFC). Figure 3.5 shows the statistical map comparing the alcohol-exposed subjects to the control subjects in terms of these DFC values. Note that no size normalization was applied, because the purpose was to map the pattern of brain shape anomaly that results from the hypoplasia. There appear to be large reductions in DFC in parietal areas bilaterally, which could be expected given the disproportionate hypoplasia in parietal lobe observed by

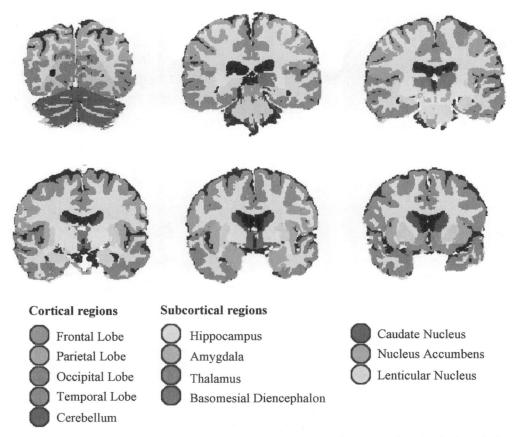

Cortical regions

- Frontal Lobe
- Parietal Lobe
- Occipital Lobe
- Temporal Lobe
- Cerebellum

Subcortical regions

- Hippocampus
- Amygdala
- Thalamus
- Basomesial Diencephalon

- Caudate Nucleus
- Nucleus Accumbens
- Lenticular Nucleus

FIGURE 3.4. A set of "processed" magnetic resonance images from a single individual studied in a magnetic resonance morphometric study performed in the author's laboratory. The different regions and structures examined are color coded to show the structural boundaries. (See Color Figure 3.4 following page 526.)

Archibald et al. (69). However, the pattern shown within this area may provide clues about the specific gyri most affected. Furthermore, there is an unexpected area of reduced DFC visible in the ventral part of the frontal lobe, in the left hemisphere. The only measures of frontal lobe available in the earlier MR morphometric study included the entire lobe. A focal decrease in volume in the area implicated by the DFC statistical map may easily have been masked in this study using large regions of interest. Thus, the voxel-based approach in this case has raised an important question about the effects of prenatal alcohol exposure on ventral structures within frontal lobe.

Anatomic Studies of Schizophrenia

Anatomic imaging studies of schizophrenia with MRI were stimulated by early CT studies suggesting ventricular enlargement in this population (71,72). Initial reactions to these observations linked them putatively to a chronic progressive course characterized in the later stages by prominent negative symptoms (73,74). Subsequent studies with MR addressed the prevalence and temporal stability of cerebral

dysmorphology in schizophrenic patients and identified specific anatomic anomalies that characterize brain morphology in this population.

Brain Structural Anomalies in Chronic Schizophrenics

Since the mid-1980's more than 200 studies using MRI to examine the brains of chronic schizophrenics have been reported in the literature. Unfortunately, a summary of these findings is complicated by a wide range in quality of the basic imaging data obtained in the studies and even greater diversity in the types of neuroanatomic methods applied. Fortunately, the findings have been summarized well in several reviews (75–77). The initial CT findings of ventricular enlargement have been confirmed in the majority of studies with MRI. The implication of this finding is that either there are anomalies of midline brain development that lead to larger ventricles (as occurs in some known disorders of brain development) or the ventricles are enlarged as a result of brain tissue loss. Brain dysmorphology attributable to neurodevelopmental factors might be expected to result in

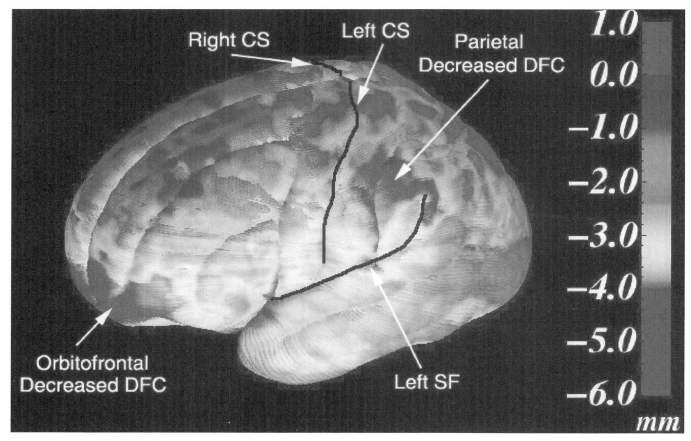

FIGURE 3.5. Differences from center (DFC) group difference maps (70) showing DFC (in millimeters) between the alcohol-exposed subjects and control subjects according to the color bar on the right. Note the negative effects in nearly all regions, most prominent in parietal and orbital frontal regions, that is, DFC is greater in the controls than in the alcohol-exposed patients, with regional patterns of DFC reduction up to 8 mm. Although only the left hemisphere is fully visualized here, the DFC reduction in both parietal and orbital frontal regions is bilateral. The left and right central sulcus (CS) and the left hemisphere sylvian fissure (SF) are highlighted in black. (See Color Figure 3.5 following page 526.)

a change in the overall cranial volume. Findings of reduced overall brain volume (as reflected in lower volume of the cranial cavity) have been reported (78,79) but were absent in a large number of studies. Failures to observe significant differences should not, in general, be given the same inferential status as positive results (because they may be due to lack of power or measurement sensitivity). However, it seems likely that any tendency for the brains of schizophrenics to be hypoplastic is either subtle or confined to a subgroup. Relevant to the latter possibility, Jernigan et al. (21) observed reduced cranial volume only in a subgroup of patients diagnosed as "disorganized type."

As noted earlier, developmental explanations for ventricular dysmorphology might predict accompanying dysmorphology of midline brain structures. There have been a number of reports of shape anomalies of the corpus callosum (80–82) and septum pellucidum (83,84), which although not confirmatory are at least consistent with a developmental explanation. Other reported anomalies suggestive

of a neurodevelopmental process are increases in the size of some basal ganglia structures, including caudate, putamen, and globus pallidus (21,85). Some authors have linked these increases to neuroleptic treatment; however, it is possible that larger basal ganglia structures could result from altered brain maturation because it is known that these structures normally undergo volume reductions during childhood (21,23).

Many studies reporting ventricular enlargement have noted increases in the size of the subarachnoid spaces, which is further evidence for brain volume loss. Whether the anomalies arise entirely during development or represent neurodegenerative changes (or some combination of developmental and degenerative factors) is a matter of continuing debate. However, it is instructive to review what has been observed regarding the regional pattern of the anomalies present in groups of chronic schizophrenic patients. The studies examining gray and white matter tissue compartments suggest that gray matter reductions are more prominent than white matter reductions (40), although there is some evidence for signal abnormality in the white matter (21). The distribution of the gray matter reductions has been examined in a number of studies. Most studies reveal reduced volume of the cerebral cortex (21,40,86–89), and the cortex has been the focus of many studies.

Of the cerebral lobes, the frontal and temporal lobes have been examined most frequently. Findings of reduced volumes of temporal lobe structures are perhaps the most consistent across studies, particularly in medial temporal lobe structures but also including neocortical gyri. Superior temporal cortex has been the object of much scrutiny, not only because of the consistency with which volume reductions are reported (87,90–92) but also because these have been linked to the severity of symptoms (87,90). Interestingly, reductions of superior temporal lobe volume have been reported in first-episode patients (93), and there is evidence that these may be more severe in the left than in the right hemisphere.

The frontal lobes have been examined in a large number of anatomic studies. Findings of decreased volumes of frontal lobe structures often have been reported (21,40,85,94,95); however, significant differences on frontal lobe measures were not observed in a number of studies. More detailed anatomic studies of the principal gyri of the frontal lobe in schizophrenic patients have been reported only recently. Gur et al. (96) reported reductions in dorsolateral frontal cortex, and additional gender-specific reductions in dorsomedial and orbitofrontal regions, and these reductions were related to symptom severity and cognitive function. Convit et al. (97) reported reductions in superior frontal and orbitofrontal gyri but not other frontal lobe gyri. Given this, it seems prudent to conclude that there are anomalies in the frontal lobe in a substantial number of schizophrenic patients, but these may be relatively confined to specific frontal lobe regions that continue to be defined.

Although there have been studies examining other regions of the cerebral cortex, they have been few in number and have produced inconsistent results. There is continuing controversy regarding the regional extent and anatomic specificity of the cortical volume reductions in schizophrenia. In part, this is because no credible studies surveying the entire cerebral cortex in detail, with sulcal-based regions of interest, have appeared. Another source of confusion is the willingness of some investigators to "interpret" the contrast between a group difference that reaches significance and those that do not when such comparisons usually are not meaningful.

The subcortical gray matter structures have been examined in several studies. Reductions in the size of the thalamus have been reported by some investigators (78,98,99). The basal ganglia also have been examined. Jernigan et al. (21) found significant volume *increases* in the lenticular nucleus (putamen and globus pallidus), but no difference could be detected in the volume of the caudate nucleus. In this study, due to pallidal signal loss on T2–weighted images, the lenticular measure primarily measured

putamen volume. Interestingly the lenticular increase was present in all clinical subgroups but was more pronounced in patients with an earlier onset of the illness. Subsequent studies produced evidence for caudate enlargement, as well as enlargement of lenticular structures (85). Interpretation of these findings of increased volume of basal ganglia has focused on the role of neuroleptic exposure. Studies of neuroleptic effects on basal ganglia volumes are discussed later.

Stability of Structural Anomalies Over Time

A great advantage of sMRI over other neuroanatomic methods is its repeatability. Investigators have exploited this feature to address two important questions in schizophrenia research: Do the brain structural changes progress over time, and does neuroleptic treatment itself alter brain structure?

The results of early anatomic imaging studies provided little evidence for progressive worsening of structural abnormalities, because in these cross-sectional studies the abnormalities were rarely found to be related to either age or estimates of illness duration within the patient samples. However, a recent focus on the findings in first-episode patients, and the appearance of longitudinal studies, has yielded increasing evidence for change over time in the severity of anatomic abnormalities. Several controlled studies of patients over follow-up intervals from 1 to 5 years revealed accelerated increases in ventricular size (100–105) and decreases in brain volume, particularly in frontal and temporal lobe structures (101,103,105–109) in schizophrenic patients. These findings suggest that whether or not there are important neurodevelopmental factors that contribute to the pathogenesis of schizophrenia, there are clearly also neuropathologic events that occur in the period immediately after the onset of symptoms. The findings add urgency to the search for an understanding of the basic mechanisms that give rise to these changes, because effective intervention could prevent neurodegenerative changes in recent-onset patients.

Several studies have examined the effects of neuroleptic treatment on brain structure. These studies were stimulated by observations (reviewed earlier) of enlarged basal ganglia structures in chronic patients. Chakos et al. (110) reported that caudate nucleus volumes increase after the initiation of typical neuroleptic treatment, and Corson et al. (111) observed such changes in the lenticular nucleus and in the caudate. Subsequent studies provided evidence that the replacement of typical with atypical antipsychotics reverses these increases in basal ganglia volumes (112–114).

In summary, MR morphometric studies of the brains of schizophrenic patients have helped to define the range of brain morphologic abnormalities that attend the disorder. They also have added significantly to our understanding of the neurologic course of the illness and of the effects of major treatments.

Voxel-Based Approaches to Mapping Structural Abnormalities in Schizophrenia

Several early attempts to map brain structural abnormalities with voxel-based approaches were applications in studies of schizophrenia. The early "image averaging" report by Andreasen et al. (98) presented a map obtained by examining the voxel differences observed after size normalizing the brains in patient and control groups with a "bounding box" method. This map suggested that the largest differences occurred in the thalamus and adjacent white matter areas. A similar study performed statistical parametric mapping of gray matter, white matter, and CSF, and compared groups of childhood-onset schizophrenics and matched controls (115). This study also suggested abnormalities in the periventricular areas, including the basal ganglia, thalamus, and cingulate cortex. DeQuardo et al. (82) compared the morphology in the midsagittal plane from MRI in patients and controls by computing and presenting a deformation map that related the positions of specific anatomic landmarks in the two groups. This map suggested that the abnormalities involved primarily the posterior corpus callosum, upper brainstem, and quadrigeminal cistern.

Although these studies provide important information that is complementary to the region-of-interest information, their interpretation is complicated by the issues described earlier. For example, the absence of cortical findings in these studies may have been due to lack of sensitivity in these parts of the map associated with large individual difference variability, and some of the findings that implicated midline gray matter structures may have been influenced by the presence of ventricular enlargement in the schizophrenic patients. Still, these methods represent early examples of what is certain to be the future of morphologic analysis of sMRI data.

INTERPRETATION OF STRUCTURAL MAGNETIC RESONANCE IMAGING RESULTS AND IMPLICATIONS FOR FUNCTIONAL IMAGING STUDIES

It is clear that MRI and MR morphometry have provided much important information about schizophrenia and the effects of alcohol neurotoxicity. There remains little doubt that substantial dysmorphology exists in a significant proportion of individuals affected by these neuropsychiatric disorders. Conventional morphometric methods in these studies and in studies of many other psychiatric populations have focused on sizes of brain structures. The consistency of the results suggest that most investigators are measuring brain volumes with reasonable accuracy. However, as described earlier, most volumetric techniques make the assumption that the different tissues are reasonably homogenous in signal value and that the tissue values themselves are relatively independent of diagnosis. This may not be true. For example,

it is possible that effects of some diseases on white matter biochemistry could lead to widespread alteration of the MR signal of white matter in patients afflicted with these diseases. Such signal changes could result in segmentation biases resulting in apparent tissue volume changes that were the result of white matter signal alteration. Further study of tissue characteristics in neuropsychiatric populations could help to resolve such uncertainty.

Functional imaging modalities have the potential to add critical information about the functional implications of the anatomic changes described in this chapter, as well as to reveal brain functional aspects of neuropsychiatric disorders that are independent of measurable anatomic changes. However, it should be noted that in virtually all functional imaging, structural imaging is used in the image analysis path to localize the effects. Functional imaging modalities, such as functional MRI (fMRI), usually involve averaging across individuals, and they produce maps of the effects observed, plotted in anatomically standardized coordinates. Producing these maps requires several complex postprocessing steps. Structural MR images sometimes are processed with automated algorithms to define the different tissues. Size normalization almost invariably is performed. To further reduce noise associated with individual morphologic variability, nonlinear warping methods are applied in an attempt to bring homologous structures in different individuals into better spatial registration. These processes usually are applied before group differences are tested. Most of the methods for analyzing functional imaging data in common use today were designed making the assumption that morphologic variability is independent of any of the experimental variables. The implications of systematic morphologic effects on the outcome of these analyses are poorly understood, but there is no doubt that the results are affected by them to some extent. The results of an fMRI study that compared groups of patients with cerebral atrophy (or cerebral hypoplasia) with normal controls (a number of common analysis packages) would, by themselves, be ambiguous. It would be unclear whether any group differences observed in brain activation were due entirely to morphologic differences, were caused by the same processes that caused the anatomic changes, or were independent of any morphologic changes. Furthermore, it would be difficult to conclude that the effects were in the structures to which they would be assigned on the basis of the standardized coordinates, because it would not be certain that the spatial normalization algorithms would affect subjects in the two groups similarly. These are important problems that must be addressed as functional imaging modalities are applied more widely to investigate neuropsychiatric disorders.

FUTURE DIRECTIONS

The spatial resolution and anatomic precision of MRI will continue to improve. This will provide more detailed information about the neuroanatomic correlates of neuropsychiatric disorders. One of the major shortcomings of the anatomic work done to date is the paucity of information provided about specific effects on cerebral white matter. The importance of this limitation is underscored by the evidence that in many neuropsychiatric disorders, such as multiple sclerosis, chronic alcoholism, and human immunodeficiency virus-related neurobehavioral disorders, effects on white matter are among the most pronounced. Even in schizophrenia, some of the few studies to examine white matter have detected signal abnormalities (21). Diffusion tensor imaging (DTI) promises to add vital information about the location of individual fiber tracts and about their structural integrity in these and other populations. DTI measures the motion associated with diffusion of water molecules in tissues. Diffusion is influenced by the tissue microstructure. For example, within the CSF the protons in water molecules move randomly in all directions. This random motion of the protons is referred to as *diffusion isotropy*. When the microstructure of the tissue constrains the diffusion of water molecules, the diffusion is said to be *anisotropic*. Diffusion of water protons has the greatest degree of anisotropy in the white matter of the brain, presumably because the molecules diffuse along the long axes of the axons. With DTI, the degree of diffusion anisotropy can be mapped and the tensor data can be further analyzed to distinguish specific fiber tracts and to detect alterations of the normal pattern of anisotropy associated with damage to myelin or axons. Methods for modeling DTI data to "trace tracts" are being developed. If these methods are successful, it may be possible in the future to define the pattern of connectivity in individual brains. This would add an important dimension to *in vivo* characterization of brain structure. This and other new imaging modalities, such as MR spectroscopy and MR perfusion studies, will provide much needed additional information about the structure, function, and biochemistry of brain tissues. Already an early study using DTI to examine schizophrenic patients has yielded interesting findings of reduced anisotropy in the white matter of these patients (116).

One of the greatest challenges facing investigators in the future is that of integrating the vast amounts of information obtained with brain imaging. New methods for analyzing these huge data sets are needed that will allow us to extract the information that is most relevant to the pathogenesis and resulting dysfunction afflicting neuropsychiatric patients. These methods must yield an accurate characterization of brain structural anomaly in individual subjects; they must be efficient; and they must provide the means for defining the relationship between anatomic factors and the other effects under investigation. This work represents a branch of a new field referred to as *neuroinformatics*. To succeed in developing these new modeling techniques, investigators must form teams that include, in addition to neuroscientists and behavioral scientists, mathematicians and information scientists who bring knowledge of relevant recent developments in those fields. The urgency of this problem cannot be overemphasized. Although there are numerous sophisticated informatics tools for accomplishing some or all of the processes involved in basic brain mapping, they frequently are inadequate for use in investigations of clinical problems. There is a serious need for appropriate data reduction and data analysis schemes with sufficient complexity to model even the major known factors contributing to the results of functional imaging in patients with brain dysfunction. This probably is the most significant problem reducing the yield of neuroimaging investigations today.

ACKNOWLEDGMENTS

The author acknowledges the support of the Medical Research Service of the Department of Veterans Affairs and the National Institutes of Health. Substantial portions of the chapter overlap significantly with similar sections in the chapter entitled "Structural magnetic resonance imaging (MRI) of the human brain" in Liu Y, Lovinger DM, eds. *Methods for Alcohol-Related Neuroscience Research.* Boca Raton, FL: CRC Press LLC, 2002.

REFERENCES

1. Cohen MS, Bookheimer SY. Localization of brain function using magnetic resonance imaging. *Trends Neurosci* 1994;17:268–277.
2. Arnold JB, Liow JS, Schaper KA, et al. Qualitative and quantitative evaluation of six bias-correction algorithms for correcting intensity nonuniformity effects. *Neuroimage* 2001;135:931–943.
3. Jernigan TL, Ostergaard AL. Word priming and recognition memory are both affected by mesial temporal lobe damage. *Neuropsychology* 1993;71:14–26.
4. Collins DL, Holmes CJ, Peters TM, et al. Automatic 3-D model-based neuroanatomical segmentation. *Human Brain Mapping* 1995;3:190–208.
5. Caviness VS Jr, Kennedy DN, Makris N, et al. Advanced application of magnetic resonance imaging in human brain science. *Brain Dev* 1995;17:399–408.
6. Andreasen NC, Rajarethinam R, Cizadlo T, et al. Automatic atlas-based volume estimation of human brain regions from MR images. *J Comput Assist Tomogr* 1996;201:98–106.
7. Cagnoni S, Coppini G, Rucci M, et al. Neural network segmentation of magnetic resonance spin echo images of the brain. *J Biomed Eng* 1993;15:355–362.
8. Clarke LP, Velthuizen RP, Phuphanich S, et al. MRI: stability of three supervised segmentation techniques. *Magn Reson Imaging* 1993;11:95–106.
9. Bonar DC, Schaper KA, Anderson JR, et al. Graphical analysis of MR feature space for measurement of CSF, gray-matter and white-matter volumes. *J Comput Assist Tomogr* 1993;17:461–470.
10. Fletcher LM, Barsotti JB, Hornak JP. A multispectral analysis of brain tissues. *Magn Reson Med* 1993;29:623–630.
11. Jackson EF, Narayana PA, Falconer JC. Reproducibility of nonparametric feature map segmentation for determination of normal human intracranial volumes with MR imaging data. *J Magn Reson Imaging* 1994;45:692–700.
12. Kao YH, Sorenson JA, Bahn MM, et al. Dual-echo MRI segmentation using vector decomposition and probability

techniques: a two-tissue model. *Magn Reson Med* 1994;32:342–357.

13. Kennedy DN, Meyer JW, Filipek PA, et al., *MRI-based topographic segmentation in MRI-based topographic segmentation.* San Diego, CA: Academic Press, 1994:201–208.

14. Simmons A, Arridge SR, Barker GJ, et al. Simulation of MRI cluster plots and application to neurological segmentation. *Magn Reson Imaging* 1996;14:73–92.

15. Vaidyanathan M, Clarke LP, Heidtman C, et al. Normal brain volume measurements using multispectral MRI segmentation. *Magn Reson Imaging* 1997;151:87–97.

16. Vinitski S, Gonzalez C, Mohamed F, et al. Improved intracranial lesion characterization by tissue segmentation based on a 3D feature map. *Magn Reson Med* 1997;373:457–469.

17. Bezdek JC, Hall LO, Clarke LP. Review of MR image segmentation techniques using pattern recognition. *Med Phys* 1993;20:1033–1048.

18. Zijdenbos AP, Dawant BM. Brain segmentation and white matter lesion detection in MR images. *Crit Rev Biomed Eng* 1994;22:401–465.

19. Clarke LP, Velthuizen RP, Camacho MA, et al. MRI segmentation: methods and applications. *Magn Reson Imaging* 1994;13:343–368.

20. Jernigan TL, Archibald SL, Fennema-Notestine C, et al. Effects of age on tissues and regions of the cerebrum and cerebellum. *Neurobiol Aging* 2001;224:581–594.

21. Jernigan TL, Zisook S, Heaton RK, et al. Magnetic resonance imaging abnormalities in lenticular nuclei and cerebral cortex in schizophrenia. *Arch Gen Psychiatry* 1991;48:881–890.

22. Pfefferbaum A, Lim KO, Zipursky RB, et al. Brain gray and white matter volume loss accelerates with aging in chronic alcoholics: a quantitative MRI study. *Alcoholism Clin Exp Res* 1992;166:1078–1089.

23. Jernigan TL, Trauner DA, Hesselink JR, et al. Maturation of human cerebrum observed in vivo during adolescence. *Brain* 1991;114:2037–2049.

24. Sowell ER, Jernigan TL. Further MRI evidence of late brain maturation: limbic volume increases and changing asymmetries during childhood and adolescence. *Dev Neuropsychol* 1998;144:599–617.

25. Sowell ER, Thompson PM, Holmes CJ, et al. Localizing age-related changes in brain structure between childhood and adolescence using statistical parametric mapping. *Neuroimage* 1999;96[Pt 1]:587–597.

26. Sowell ER, Thompson PM, Holmes CJ, et al. In vivo evidence for post-adolescent brain maturation in frontal and striatal regions [Letter]. *Nat Neurosci* 1999;210:859–861.

27. Thompson PM, Moussai J, Zohoori S, et al. Cortical variability and asymmetry in normal aging and Alzheimer's disease. *Cereb Cortex* 1998;86:492–509.

28. Ashburner J, Friston KJ. Voxel-based morphometry—the methods. *Neuroimage* 2000;11:805–821.

29. Bookstein FL. Endophrenology: new statistical techniques for studies of brain form. Life on the hyphen in neuro-informatics. *Neuroimage* 1996;43[Pt 2]:S36–S38.

30. Bookstein FL. "Voxel-based morphometry" should not be used with imperfectly registered images. *Neuroimage* 2001;146:1454–1462.

31. Jernigan TL, Archibald SL, Berhow MT, et al. Cerebral structure on MRI, part I: localization of age-related changes. *Biol Psychiatry* 1991;291:55–67.

32. Jernigan TL, Butters N, DiTraglia G, et al. Reduced cerebral grey matter observed in alcoholics using magnetic resonance imaging. *Alcoholism Clin Exp Res* 1991;15:418–427.

33. Jernigan TL, Zatz LM, Ahumada AJ Jr, et al. CT measures of cerebrospinal fluid volume in alcoholics and normal volunteers. *Psychiatry Res* 1982;7:9–17.

34. Jernigan TL, Butters N, Cermak LS. Studies of brain structure in chronic alcoholism using magnetic resonance imaging. In: Zakhari S, Witt E, eds. *National Institute on Alcohol Abuse and Alcoholism Research monograph 21: imaging in alcohol research.* Rockville, MD: US Department of Health and Human Services. 1991:14.

35. Pfefferbaum A, Sullivan EV, Rosenbloom MJ, et al. Increase in brain cerebrospinal fluid volume is greater in older than in younger alcoholic patients: a replication study and CT/MRI comparison. *Psychiatry Res* 1993;504:257–274.

36. Sullivan EV, Marsh L, Mathalon DH, et al. Anterior hippocampal volume deficits in nonamnesic, aging chronic alcoholics. *Alcoholism Clin Exp Res* 1995;191:110–122.

37. Sullivan EV, Marsh L, Mathalon DH, et al. Relationship between alcohol withdrawal seizures and temporal lobe white matter volume deficits. *Alcohol Clin Exp Res* 1996;202:348–354.

38. Beresford T, Arciniegas D, Rojas D, et al. Hippocampal to pituitary volume ratio: a specific measure of reciprocal neuroendocrine alterations in alcohol dependence. *J Stud Alcohol* 1999;605:586–588.

39. Pfefferbaum A, Sullivan EV, Mathalon DH, et al. Frontal lobe volume loss observed with magnetic resonance imaging in older chronic alcoholics. *Alcohol Clin Exp Res* 1997;213:521–529.

40. Sullivan EV, Mathalon DH, Lim KO, et al. Patterns of regional cortical dysmorphology distinguishing schizophrenia and chronic alcoholism. *Biol Psychiatry* 1998;432:118–131.

41. Sullivan EV, Deshmukh A, Desmond JE, et al. Cerebellar volume decline in normal aging, alcoholism, and Korsakoff's syndrome: relation to ataxia. *Neuropsychology* 2000;143:341–352.

42. Pfefferbaum A, Lim KO, Desmond JE, et al. Thinning of the corpus callosum in older alcoholic men: a magnetic resonance imaging study. *Alcohol Clin Exp Res* 1996;204:752–757.

43. Carlen PL, Wortzman G, Holgate RC, et al. Reversible cerebral atrophy in recently abstinent chronic alcoholics measured by computed tomography scans. *Science* 1978;200:1076–1078.

44. Artmann H, Gall MV, Hacker H, et al. Reversible enlargement of cerebral spinal fluid spaces in chronic alcoholics. *AJNR Am J Neuroradiol* 1981;2:23–27.

45. Schroth G, Naegele T, Klose U, et al. Reversible brain shrinkage in abstinent alcoholics, measured by MRI. *Neuroradiology* 1988;305:385–389.

46. Zipursky RB, Lim KC, Pfefferbaum A. MRI study of brain changes with short-term abstinence from alcohol. *Alcohol Clin Exp Res* 1989;135:664–666.

47. Shear PK, Jernigan TL, Butters N. Volumetric magnetic resonance imaging quantification of longitudinal brain changes in abstinent alcoholics [published erratum appears in *Alcohol Clin Exp Res* 1994;18:766]. *Alcoholism Clin Exp Res* 1994;181:172–176.

48. Pfefferbaum A, Sullivan EV, Mathalon DH, et al. Longitudinal changes in magnetic resonance imaging brain volumes in abstinent and relapsed alcoholics. *Alcohol Clin Exp Res* 1995;195:1177–1191.

49. Mann K, Mundle G, Langle G, et al. The reversibility of alcoholic brain damage is not due to rehydration: a CT study. *Addiction* 1993;88:649–653.

50. Victor M, Adams RD, Collins GH. The Wernicke-Korsakoff syndrome. A clinical and pathological study of 245 patients, 82 with post-mortem examinations. *Contemp Neurol Ser* 1971;7:1–206.

51. Mair WG, Warrington EK, Weiskrantz L. Memory disorder in Korsakoff's psychosis: a neuropathological and neuropsychological investigation of two cases. *Brain* 1979;102:749–783.

52. Mayes A, Meudell P, Pickering A, et al. Locations of lesions in Korsakoff's syndrome: neuropsychological and neuropathological data on two patients. *Cortex* 1988;243:367–388.

53. Charness ME, DeLaPaz RL. Mamillary body atrophy in Wernicke's encephalopathy: antemortem identification using magnetic resonance imaging. *Ann Neurol* 1987;22:595–600.

54. Squire LR, Amaral DG, Press GA. Magnetic resonance imaging of the hippocampal formation and mammillary nuclei distinguish medial temporal lobe and diencephalic amnesia. *J Neurosci* 1990;10:3106–3317.

55. Davila MD, Shear PK, Lane B, et al. Mammillary body and cerebellar shrinkage in chronic alcoholics: an MRI and neuropsychological study. *Neuropsychology* 1994;83:433–444.

56. Blansjaar BA, Vielvoye GJ, van Dijk JG, et al. Similar brain lesions in alcoholics and Korsakoff patients: MRI, psychometric and clinical findings. *Clin Neurol Neurosurg* 1992;943:197–203.

57. Shear PK, Sullivan EV, Lane B, et al. Mammillary body and cerebellar shrinkage in chronic alcoholics with and without amnesia. *Alcohol Clin Exp Res* 1996;208:1489–1495.

58. Sullivan EV, Lane B, Deshmukh A, et al. In vivo mammillary body volume deficits in amnesic and nonamnesic alcoholics. *Alcohol Clin Exp Res* 1999;2310:1629–1636.

59. Jernigan TL, Schafer K, Butters N, et al. Magnetic resonance imaging of alcoholic Korsakoff patients. *Neuropsychopharmacology* 1991;43:175–186.

60. Sullivan EV, Marsh L, Shear PK, et al. Hippocampal but not cortical volumes distinguish amnesic and nonamnesic alcoholics. *Soc Neurosci Abstr* 1996.

61. Jones KL, Smith DW, Ulleland CN, et al. Pattern of malformation in offspring of chronic alcoholic mothers. *Lancet* 1973;1:1267–1271.

62. Jones KL, Smith DW. The fetal alcohol syndrome. *Teratology* 1975;12:1–10.

63. Mattson SN, Riley EP, Jernigan TL, et al. Fetal alcohol syndrome: a case report of neuropsychological, MRI, and EEG assessment of two children. *Alcohol Clin Exp Res* 1992;16:1001–1003.

64. Mattson SN, Riley EP, Jernigan TL, et al. A decrease in the size of the basal ganglia following prenatal alcohol exposure: a preliminary report. *Neurotoxicol Teratol* 1994;163:283–289.

65. Sowell ER, Jernigan TL, Mattson SN, et al. Abnormal development of the cerebellar vermis in children prenatally exposed to alcohol: size reduction in lobules I–V. *Alcoholism Clin Exp Res* 1996;201:31–34.

66. Mattson SN, Riley EP, Sowell ER, et al. A decrease in the size of the basal ganglia in children with fetal alcohol syndrome. *Alcoholism Clin Exp Res* 1996;206:1088–1093.

67. Johnson VP, Swayze VW, Sato Y, et al. Fetal alcohol syndrome: craniofacial and central nervous system manifestations. *Am J Med Genet* 1996;61:329–339.

68. Riikonen R, Salonen I, Partanen K, et al. Brain perfusion SPECT and MRI in foetal alcohol syndrome. *Dev Med Child Neurol* 1999;4110:652–659.

69. Archibald SL, Fennema-Notestine C, Gamst A, et al. Brain dysmorphology in individuals with severe prenatal alcohol exposure. *Dev Med Child Neurol* 2001;433:148–154.

70. Sowell ER, Thompson PM, Mattson SN, et al. Regional brain shape abnormalities persist into adolescence after heavy prenatal alcohol exposure. *Cerebral Cortex* 2002;8:856–865.

71. Johnstone EC, Crow TJ, Frith CD, et al. Cerebral ventricular size and cognitive impairment in chronic schizophrenia. *Lancet* 1976;2:924–996.

72. Weinberger DR, Torrey EF, Neophytides AN, et al. Lateral cerebral ventricular enlargement in chronic schizophrenia. *Arch Gen Psychiatry* 1979;36:735–779.

73. Crow TJ. Is schizophrenia an infectious disease? *Lancet* 1983;1:173–115.

74. Crow TJ. Schizophrenia as an infection [Letter]. *Lancet* 1983;1:819–820.

75. McCarley RW, Niznikiewicz MA, Salisbury DF, et al. Cognitive dysfunction in schizophrenia: unifying basic research and clinical aspects. *Eur Arch Psychiatry Clin Neurosci* 1999;249[Suppl 4]:69–82.

76. Henn FA, Braus DF. Structural neuroimaging in schizophrenia. An integrative view of neuromorphology. *Eur Arch Psychiatry Clin Neurosci* 1999;249[Suppl 4]:48–56.

77. Shenton ME, Dickey CC, Frumin M, et al. A review of MRI findings in schizophrenia. *Schizophr Res* 2001;491:1–52.

78. Andreasen NC, Ehrhardt JC, Swayze VW, et al. Magnetic resonance imaging of the brain in schizophrenia. The pathophysiologic significance of structural abnormalities. *Arch Gen Psychiatry* 1990;47:35–44.

79. Andreasen NC, Flashman L, Flaum M, et al. Regional brain abnormalities in schizophrenia measured with magnetic resonance imaging. *JAMA* 1994;272:1763–1769.

80. Casanova MF, Zito M, Goldberg T, et al. Shape distortion of the corpus callosum of monozygotic twins discordant for schizophrenia. *Schizophr Res* 1990;32:155–156.

81. Raine A, Harrison GN, Reynolds GP, et al. Structural and functional characteristics of the corpus callosum in schizophrenics, psychiatric controls, and normal controls. A magnetic resonance imaging and neuropsychological evaluation. *Arch Gen Psychiatry* 1990;4711:1060–1064.

82. DeQuardo JR, Bookstein FL, Green WD, et al. Spatial relationships of neuroanatomic landmarks in schizophrenia. *Psychiatry Res* 1996;671:81–95.

83. Degreef G, Ashtari M, Bogerts B, et al. Volumes of ventricular system subdivisions measured from magnetic resonance images in first-episode schizophrenic patients. *Arch Gen Psychiatry* 1992;49:531–557.

84. Degreef G, Bogerts B, Falkai P, et al. Increased prevalence of the cavum septum pellucidum in magnetic resonance scans and post-mortem brains of schizophrenic patients. *Psychiatry Res* 1992;451:1–13.

85. Breier A, Buchanan RW, Elkashef A, et al. Brain morphology and schizophrenia: a magnetic resonance imaging study of limbic, prefrontal cortex, and caudate structures. *Arch Gen Psychiatry* 1992;49:921–926.

86. Zipursky RB, Lim KO, Sullivan EV, et al. Widespread cerebral gray matter volume deficits in schizophrenia. *Arch Gen Psychiatry* 1992;493:195–205.

87. Shenton ME, Kikinis R, Jolesz FA, et al. Abnormalities of the left temporal lobe and thought disorder in schizophrenia: a quantitative magnetic resonance imaging study. *N Engl J Med* 1992;327:604–612.

88. Harvey I, Ron MA, Du Boulay G, et al. Reduction of cortical volume in schizophrenia on magnetic resonance imaging. *Psychol Med* 1993;233:591–604.

89. Zipursky RB, Seeman MV, Bury A, et al. Deficits in gray matter volume are present in schizophrenia but not bipolar disorder. *Schizophr Res* 1997;262:85–92.

90. Barta PE, Pearlson GD, Powers RE, et al. Auditory hallucinations and smaller superior temporal gyral volume in schizophrenia. *Am J Psychiatry* 1990;147:1457–1162.

91. Schlaepfer TE, Harris GJ, Tien AY, et al. Decreased regional cortical gray matter volume in schizophrenia. *Am J Psychiatry* 1994;151:842–848.

92. Menon RR, Barta PE, Aylward EH, et al. Posterior superior

temporal gyrus in schizophrenia: grey matter changes and clinical correlates. *Schizophr Res* 1995;162:127–135.

93. Hirayasu Y, Shenton ME, Salisbury DF, et al. Lower left temporal lobe MRI volumes in patients with first-episode schizophrenia compared with psychotic patients with first-episode affective disorder and normal subjects. *Am J Psychiatry* 1998;155:1384–1391.

94. Raine A, Lencz T, Reynolds GP, et al. An evaluation of structural and functional prefrontal deficits in schizophrenia: MRI and neuropsychological measures. *Psychiatry Res* 1992;452:123–137.

95. Andreasen NC, Flashman L, Flaum M, et al. Regional brain abnormalities in schizophrenia measured with magnetic resonance imaging. *JAMA* 1994;272:1763–1769.

96. Gur RE, Cowell PE, Latshaw A, et al. Reduced dorsal and orbital prefrontal gray matter volumes in schizophrenia. *Arch Gen Psychiatry* 2000;578:761–768.

97. Convit A, Wolf OT, de Leon MJ, et al. Volumetric analysis of the pre-frontal regions: findings in aging and schizophrenia. *Psychiatry Res* 2001;1072:61–73.

98. Andreasen NC, Arndt S, Swayze V 2nd, et al. Thalamic abnormalities in schizophrenia visualized through magnetic resonance image averaging. *Science* 1994;266:294–298.

99. Buchsbaum MS, Someya T, Teng CY, et al. PET and MRI of the thalamus in never-medicated patients with schizophrenia. *Am J Psychiatry* 1996;153:191–199.

100. DeLisi LE, Tew W, Xie S, et al. A prospective follow-up study of brain morphology and cognition in first-episode schizophrenic patients: preliminary findings. *Biol Psychiatry* 1995;386:349–360.

101. DeLisi LE, Sakuma M, Tew W, et al. Schizophrenia as a chronic active brain process: a study of progressive brain structural change subsequent to the onset of schizophrenia. *Psychiatry Res* 1997;743:129–140.

102. Nair TR, Christensen JD, Kingsbury SJ, et al. Progression of cerebroventricular enlargement and the subtyping of schizophrenia. *Psychiatry Res* 1997;743:141–150.

103. Rapoport JL, Giedd J, Kumra S, et al. Childhood-onset schizophrenia. Progressive ventricular change during adolescence. *Arch Gen Psychiatry* 1997;5410:897–903.

104. Lieberman J, Chakos M, Wu H, et al. Longitudinal study of brain morphology in first episode schizophrenia. *Biol Psychiatry* 2001;496:487–499.

105. Mathalon DH, Sullivan EV, Lim KO, et al. Progressive brain volume changes and the clinical course of schizophrenia

in men: a longitudinal magnetic resonance imaging study. *Arch Gen Psychiatry* 2001;582:148–157.

106. Gur RE, Cowell P, Turetsky BI, et al. A follow-up magnetic resonance imaging study of schizophrenia. Relationship of neuroanatomical changes to clinical and neurobehavioral measures. *Arch Gen Psychiatry* 1998;552:145–152.

107. Jacobsen LK, Giedd JN, Castellanos FX, et al. Progressive reduction of temporal lobe structures in childhood-onset schizophrenia. *Am J Psychiatry* 1998;155:678–685.

108. Keshavan MS, Haas GL, Kahn CE, et al. Superior temporal gyrus and the course of early schizophrenia: progressive, static, or reversible? *J Psychiatr Res* 1998;32:161–167.

109. Rapoport JL, Giedd JN, Blumenthal J, et al. Progressive cortical change during adolescence in childhood-onset schizophrenia. A longitudinal magnetic resonance imaging study. *Arch Gen Psychiatry* 1999;567:649–654.

110. Chakos MH, Lieberman JA, Bilder RM, et al. Increase in caudate nuclei volumes of first-episode schizophrenic patients taking antipsychotic drugs. *Am J Psychiatry* 1994;151:1430–1436.

111. Corson PW, Nopoulos P, Miller DD, et al. Change in basal ganglia volume over 2 years in patients with schizophrenia: typical versus atypical neuroleptics. *Am J Psychiatry* 1999;156:1200–1204.

112. Chakos MH, Lieberman JA, Alvir J, et al. Caudate nuclei volumes in schizophrenic patients treated with typical antipsychotics or clozapine. *Lancet* 1995;345:456–457.

113. Corson PW, Nopoulos P, Andreasen NC, et al. Caudate size in first-episode neuroleptic-naive schizophrenic patients measured using an artificial neural network. *Biol Psychiatry* 1999;465:712–720.

114. Scheepers FE, de Wied CC, Pol HE, et al. The effect of clozapine on caudate nucleus volume in schizophrenic patients previously treated with typical antipsychotics. *Neuropsychopharmacology* 2001;241:47–54.

115. Sowell ER, Levitt J, Thompson PM, et al. Brain abnormalities in early-onset schizophrenia spectrum disorder observed with statistical parametric mapping of structural magnetic resonance images. *Am J Psychiatry* 2000;157:1475–1484.

116. Lim KO, Hedehus M, Moseley M, et al. Compromised white matter tract integrity in schizophrenia inferred from diffusion tensor imaging. *Arch Gen Psychiatry* 1999;564:367–374.

117. Wong EC, Luh W-M, Buxton RB, et al. Single slab high resolution 3D whole brain imaging using spiral FSE. *Proc Int Soc Magn Reson Med* 2000;8:683.

PRINCIPLES OF FUNCTIONAL NEUROIMAGING

TOMÁŠ PAUS

This chapter provides an introduction to the basic concepts and techniques used in studies of brain–behavior relationships in human subjects. The chief motivation for such research is identification—in time and space—of neural circuits associated with particular sensory, motor, and cognitive functions. Although the initial research focused on the healthy brain, functional neuroimaging is beginning to make inroads into psychiatry. Discovery of symptom-specific abnormalities in particular neural circuits represents a first step toward understanding the pathogenesis of psychiatric disorders.

A wide range of brain mapping techniques is available today, including positron emission tomography (PET), functional magnetic resonance imaging (fMRI), electroencephalography (EEG), magnetoencephalography (MEG), and transcranial magnetic stimulation (TMS). These methods allow the researcher to measure (PET, fMRI, EEG, MEG) and manipulate (TMS) neural activity in the human brain. They differ with regard to spatial and temporal resolution, nature of the measured signal, and extent of brain coverage.

In the following three sections, an overview of the neurophysiologic underpinnings of the measured signal and the basic principles of data acquisition is provided, followed by a brief discussion of issues pertinent to the design of brain mapping studies. The last section outlines the rationale for combining brain stimulation (TMS) with brain imaging (fMRI, PET, and EEG) and illustrates this approach with several examples from our laboratory.

NEUROPHYSIOLOGIC UNDERPINNINGS OF THE SIGNAL

Positron Emission Tomography and Functional Magnetic Resonance Imaging

The most common parameters measured with PET and fMRI in studies of brain–behavior relationships are cerebral blood flow and blood oxygenation level-dependent (BOLD)

signals, respectively. Cerebral blood flow (CBF) typically is measured with PET by tracking the regional distribution in the brain of ^{15}O-water, a positron-emitting tracer with a short half-life. BOLD signal, measured on $T_2{}^*$-weighted MR images, is based on endogenous contrast properties of the paramagnetic deoxyhemoglobin and the fact that neural activation is accompanied by local CBF increases that are disproportionate relative to changes in oxygen consumption.

In the case of both CBF and BOLD, regional changes in brain activity are inferred from the measured changes in local hemodynamics. Although there is no dispute about the existence of a relationship between local hemodynamics and brain activity, there is little agreement about the role of various neural events in driving the hemodynamic signal. At least two issues need to be considered when interpreting the functional significance of CBF/BOLD responses to a behavioral challenge: (a) the relative importance of neuronal firing versus synaptic activity occurring in the sampled tissue; and (b) the relative contributions of excitatory and inhibitory neurotransmission.

Intuitively, a brain region requires more energy and, hence, more blood flow when neurons located in that region increase their firing rate. This notion is conceptually attractive; if true, one would be able to relate directly findings obtained with functional imaging of the human brain to those acquired with single-unit recordings in nonhuman primates. In the latter, firing rate and firing pattern are viewed as the best available proxy of the information code. Several recent experiments suggest, however, that firing rate is not the best predictor of local changes in hemodynamics. Using simultaneous recordings of single-unit activity, field potentials, and CBF in the rat cerebellar cortex, Mathiesen et al. (1) demonstrated that electrical stimulation of parallel fibers inhibited spontaneous firing of Purkinje cells located in the sampled cortex while it *increased* CBF and field potentials in the same tissue sample. A strong correlation ($r = 0.985$) was observed between postsynaptic activity, indicated by the summed field potentials (for definition see section on Electroencephalography and Magnetoencephalography) and activity-dependent increase in CBF. Similar findings were obtained in the monkey visual cortex, where the local field potentials provided a better estimate of BOLD responses to visual stimulation than did multiunit activity. Multiunit activity increased only briefly at the onset of stimulation, whereas the field-potential increase was sustained throughout the presence of the stimulus (2). In addition, Moore et al. (3,4) showed that even subthreshold synaptic activity in the rat primary somatosensory cortex is accompanied by significant variations in hemodynamic signal.

Although it is likely that the exact relationship among neuronal firing, input synaptic activity, and CBF might vary across different brain regions, it is safe to assume that postsynaptic activity is the primary driving force of the signal measured with PET and fMRI. The study by Mathiesen et al. also addressed the second issue of interest, namely the relative contribution of excitatory and inhibitory neurotransmission to changes in local hemodynamics. In their study, blockage of γ-aminobutyric acid (GABA)-ergic transmission during electrical stimulation of parallel fibers did not attenuate stimulation-induced increase in local blood flow, suggesting that the increase was primarily due to the excitatory component of the postsynaptic input (1). As suggested previously (5,6), a direct link between excitatory neurotransmission and local blood flow may be related to the role of nitric oxide (NO) in coupling blood flow to synaptic activity. NO is one of the signals leading to dilation of

small vessels in the vicinity of "active" synapses (7,8). It is known that glutamate activates NO synthase through an increase in the intracellular level of calcium and that, under physiologic conditions, entry of calcium into a cell is almost exclusively linked to excitatory neurotransmission (Fig. 4.1).

Overall, it is likely that changes in local hemodynamics reflect a sum of excitatory postsynaptic inputs in the sample of scanned tissue. The firing rate of "output" neurons may be related to local blood flow (9,10) but only inasmuch as it is linked in linear fashion to excitatory postsynaptic input. Inhibitory neurotransmission may lead to decreases in CBF indirectly through its presynaptic effects on postsynaptic excitation.

Electroencephalography and Magnetoencephalography

Compared with PET and fMRI, more direct and real-time measurements of electrical and magnetic signals generated by brain tissue are recorded by EEG and MEG, respectively (11,12).

The main sources of these signals are intracellular and extracellular currents, or field potentials, elicited by the activation of excitatory and inhibitory synapses. Spatial and temporal summation is necessary, however, to generate signals strong enough to be detected from outside the head

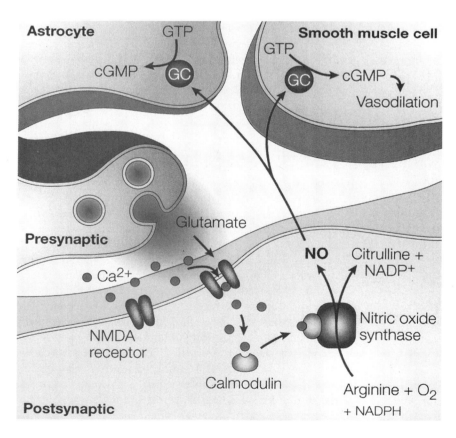

FIGURE 4.1. Synaptic activity and its coupling to regional cerebral blood flow. Nitric-oxide (NO) based model of coupling between local blood flow response and changes in excitatory neurotransmission. The cascade leading to the production of NO begins with a release of the excitatory neurotransmitter glutamate. Through its action on *N*-acetyl-D-aspartate (NMDA) receptors, glutamate increases the intracellular level of calcium that, in turn, binds to calmodulin and activates NO synthase (130). NO diffuses freely into the surrounding tissue and activates cyclic guanylate cyclase, producing cyclic guanosine monophosphate (cGMP) and eventually relaxing smooth muscle cells in the blood vessel. (Drawing by Simon Fenwick, art editor of *Nature Reviews Neuroscience*, based on the original design by Carolina Echevaria.) (See Color Figure 4.1 following page 526.)

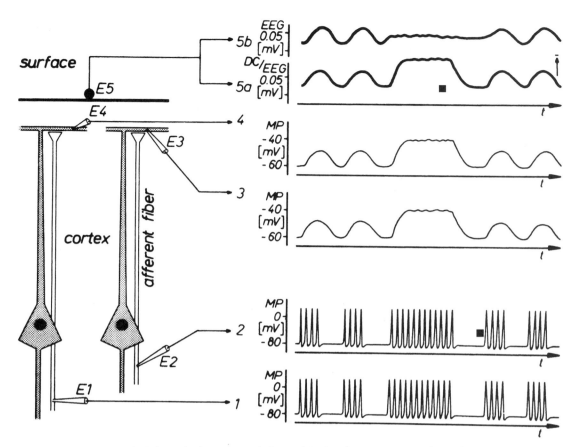

FIGURE 4.2. Principles of electroencephalographic (EEG) wave generation. The excitatory synapses of two afferent fibers contact the superficial dendritic arborization of two longitudinal neuronal elements. The afferent fiber activity is recorded by the intracellular electrodes E_1 and E_2, and the membrane potentials (MP) of the dendritic elements are recorded by electrodes E_3 and E_4. The field potential at the surface of the neuronal structure (cortex) is led by electrode E_5. Synchronized groups of action potentials in the afferent fibers (E_1, E_2) generate wavelike excitatory postsynaptic potentials (EPSP) in the dendritic area (E_3, E_4) and corresponding field potentials in the EEG and direct-current/EEG recording (E_5). Tonic activity in the afferent fibers results in a long-lasting EPSP with small fluctuations. During this period the EEG (5b) shows only a reduction in amplitude, whereas the DC/EEG recording (5a) reflects the depolarization of the neuronal elements as well. (From Speckmann EJ, Elger CE. Introduction to the neurophysiological basis of the EEG and DC potentials. In: Niedermeyer E, Lopes da Silva F, eds. *Electroencephalography: basic principles, clinical applications, and related fields,* 4th ed. Baltimore: Williams & Wilkins, 1999:15–27, with permission.)

(Fig. 4.2). Such a summation occurs most often during simultaneous excitatory inputs onto apical dendrites of pyramidal cells. The apical dendrites are for the most part oriented in parallel with each other and are perpendicular to the cortical surface. Release of excitatory neurotransmitter causes a local change in membrane potential, namely excitatory postsynaptic potential (EPSP). Each EPSP generates a potential gradient along the membrane that, in turn, gives rise to the field potential.

EEG detects field potentials regardless of their orientation relative to the skull. In contrast, MEG can measure only magnetic fields perpendicular to the skull. Such fields are generated by tangential current dipoles and, due to the orientation of pyramidal cells and their apical dendrites in the cortex, reflect primarily postsynaptic activity occurring in the cerebral sulci.

Transcranial Magnetic Stimulation

TMS allows investigators to induce, in a noninvasive fashion, electrical currents in a spatially restricted region of the cerebral cortex. In brain mapping studies, TMS is used in three different ways: (a) to elicit a clear physiologic response, such as a muscle twitch or a phosphene, which in turn serves as a dependent variable; (b) to interfere with ongoing brain activity, thus creating a temporary "virtual lesion"; and (c) to induce short-term (minutes) changes in cortical excitability. Here, some of the relevant neurophysiology underlying these three phenomena is reviewed briefly; technical aspects of TMS are discussed in the section on Data Acquisition and Analysis: Transcranial Magnetic Stimulation.

Single-pulse TMS applied over the primary motor cortex elicits motor evoked potentials (MEPs) in contralateral muscles. TMS exerts its influence on corticospinal neurons primarily through activation of their afferents; direct excitation of corticospinal axons is small (13,14). Transcranial electrical stimulation (TES), on the other hand, elicits a muscle response largely through direct activation of the corticospinal axons. This TMS-TES difference illustrates an important principle: TMS-induced effects depend on the geometry of the coil relative to that of the stimulated neural elements. Cortical currents induced by a coil positioned tangentially to the skull would be maximal in dendrites and axons located in the plane parallel to the plane of the coil and oriented along the axis passing through the virtual anode (+) and cathode (−). TES and TMS induce most of the currents perpendicular and tangential to the skull/brain surface, respectively. This leads to predominant activation of vertical (TES) and horizontal (TMS) fibers and, therefore, direct (D wave) and indirect (I wave) activation of corticospinal neurons.

TMS appears to activate both excitatory and inhibitory interneurons. This is best illustrated in studies of the silent period and those of intracortical inhibition and excitation (for a mini-review see reference 15). A single TMS pulse applied during a tonic contraction of the muscle suppresses ongoing muscle activity. This suppression, or silent period, lasts for about 200 milliseconds when induced by TMS but only about 100 milliseconds when elicited by TES (16). It is believed that the late part of a TMS-induced silent period is of cortical origin (17–19). Intracortical inhibition and excitation are assessed using the paired-pulse paradigm whereby two TMS pulses are applied in rapid succession (1–30 milliseconds between-pulse interval). The first pulse is of subthreshold intensity (conditioning stimulus); the second pulse is suprathreshold (test stimulus). Paired-pulse TMS with short (1–5 milliseconds) and long (8–30 milliseconds) between-pulse intervals elicits suppression and facilitation of the muscle (MEP) response, respectively (20–22). Early suppression and later facilitation suggest somewhat different dynamics of inhibitory and excitatory processes activated by TMS in the motor cortex under these conditions. Pharmacologic (23,24) and *in vitro* (25) studies suggest that the short- (paired-pulse) and long- (silent period) lasting cortical inhibition is mediated by GABA-A and GABA-B receptors, respectively. It remains to be seen whether similar dynamics of TMS-induced excitation and inhibition exist in other cortical regions.

Neurophysiologic mechanisms underlying TMS-induced interference with brain activity ("virtual lesion") are not well understood. Pascual-Leone et al. (26) proposed two theoretical explanations of this phenomenon: (a) TMS simply adds "noise" to the system; and (b) TMS induces a burst of synchronous activity followed by a brief period of neuronal

inhibition. Based on the studies of the motor system and on empirical data obtained with single-pulse TMS applied over other cortical regions (27,28), it is assumed that neural effects elicited by a single TMS pulse last for tens of milliseconds. Application of a train of pulses (repetitive TMS) would simply extend such interfering effects over the train duration. It should be pointed out, however, that single-pulse TMS can lead to facilitation of behavior (29,30) and that repetitive TMS can both decrease and increase cortical excitability beyond the time of stimulation (discussed later).

Long-lasting effects of repetitive TMS (rTMS) were first described in studies of mood (31,32). Repetitive TMS-induced effects on mood and plasma levels of thyroid stimulating hormone lasted up to 3 hours (31). The neural mechanisms of such complex behavioral effects are unclear, but rTMS-induced changes in motor excitability can provide useful information regarding stimulation parameters and duration of rTMS-induced changes in cortical excitability. In the majority of studies, decreases and increases in MEP amplitude have been observed immediately after low (1 Hz) and high (15–20 Hz) rTMS applied for 4 to 15 minutes at different intensities of stimulation, respectively (33–35). Intermediate frequency of rTMS (5 Hz) has been shown to reduce intracortical inhibition when tested with the paired-pulse paradigm (36,37). These effects appear to last up to 30 minutes after the cessation of rTMS (33,36,38,39). Using the combined TMS-PET method, we have shown that similar effects can be induced in other cortical regions (40) (see section on Combination of Brain Stimulation and Imaging: Transcranial Magnetic Stimulation During Positron Emission Tomography).

DATA ACQUISITION AND ANALYSIS
Positron Emission Tomography

In a typical blood-flow activation study, a small amount of the positron emitting ^{15}O-water is injected into a vein, and the distribution of the tracer is measured over a period of 60 seconds by PET detectors. As it decays, the tracer emits positrons that annihilate by interacting with electrons. In the process, two γ-rays are emitted in the opposite direction. Using a coincidence circuit, a pair of detectors records the γ-rays emitted along the line passing through the site of positron–electron interaction (Fig. 4.3). A two-dimensional (2D) tomographic measurement of the number and location of positron–electron interactions is carried out for the entire brain, allowing us to map the pattern of changes in regional CBF. With the Siemens Exact HR+ scanner, for example, one can acquire sixty-three 2.4-mm-thick axial slices covering a 15.2-cm field of view. This is achieved with 18,432 detectors arranged in 32 rings. Sensitivity of the scanner can be increased by retracting septa separating the individual rings and collecting the counts in a three-dimensional (3D) mode (for details see reference 41). One of the main benefits

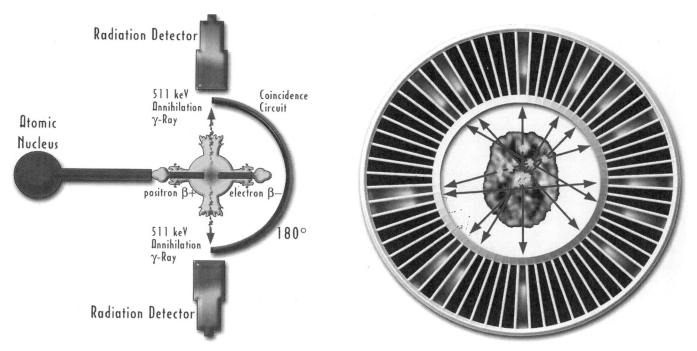

FIGURE 4.3. Principles of positron emission tomography. **Left:** Two γ-rays are emitted along the line passing through the site of positron–electron interaction. A fast timing circuit detects a temporal coincidence of the two γ-rays reaching a particular pair of radiation detectors. **Right:** An array of radiation detectors provides tomographic measurements of the tissue concentration of positron-labeled compounds in two-dimensional space. (Drawing by Helmut Bernhard, Neurophotography, Montreal Neurological Institute.) (See Color Figure 4.3 following page 526.)

of the 3D mode of acquisition is the reduction in radiation exposure to the subject; about 10 mCi of ^{15}O-water per scan is sufficient to obtain an image.

The 2D record of γ-rays (sinogram) is reconstructed into a 3D image. A series of postprocessing steps is required before statistical analysis can be performed. First, PET-PET registration is performed. PET images acquired during one session in a given subject are registered, typically to the first scan, to correct for possible between-scan movements of the head. Second, PET-MRI registration is done. The first PET image is registered with the structural MR image of the subject's brain to facilitate mapping of functional data onto the structural ones. In our Brain Imaging Centre (BIC), both PET-PET and PET-MRI registration is achieved with a modified version of the Automatic Image Registration (AIR) software developed originally by Woods et al. (42,43). Third, the images are normalized with respect to global number of counts. This step is necessary because of the absence of absolute measurements of CBF, which are not carried out in most studies to avoid arterial blood sampling. At the BIC, proportional normalization is used to remove such between-scan and between-subject differences in the total number of counts in each 3D image. The final step involves transformation of all images into standardized stereotaxic space (MRI_{native}–$MRI_{Talairach}$). It is carried out with an automatic image registration method (44) based on multi-

scale 3D cross-correlation with an average (n = 305) MR image aligned with the "Talairach" stereotaxic space (45). This transformation is linear, yielding three scaling factors for the width (X-axis), length (Y-axis), and height (Z-axis) of the brain and effectively removing interindividual differences in brain size. Transformation of PET and MRI images into standardized stereotaxic space serves a dual purpose: (a) it allows for a voxel-based statistical analysis of the data (discussed later); and (b) it facilitates comparison of brain mapping results across laboratories by identifying the 3D location of a significant "peak" with a set of standard X, Y, and Z coordinates.

Intensity and spatial normalization of images allows one to compare scans acquired across different conditions in many subjects. To identify brain regions that differ in task-related activity, PET images are compared with each other using either subtraction or regression analysis. In subtraction analysis, count values obtained in scans A and B are simply subtracted from each other for each 3D element of the image (voxel). In regression analysis, voxel count values are regressed either against those of an experimental variable, such as the number of movements executed during a scan, or against count values measured in a particular brain region. The latter analysis can tell us about a coordinated blood flow response within a putative neural "network", that is, it infers functional connectivity between different brain regions (46).

The statistical significance of the differences (or correlations) is evaluated by calculating t statistics for each voxel. The resulting t-statistic maps test whether, at a given voxel, the difference (subtraction) or the slope of the regression (regression) is significantly different from zero. An adjustment for multiple comparisons involved in assessing statistical significance in each of the voxels constituting the entire brain volume is based on the number of resolution elements (resels). The resel size reflects the final spatial resolution of the image (47). For example, a t value of 4.5 is deemed statistically significant at $p < 0.00001$ (two-tailed, uncorrected) and, correcting for multiple comparisons, yields a false-positive rate of 0.06 in 600 resels constituting the scanned volume of the whole brain (spatial smoothing at $1.4 \times 1.4 \times 1.4$ cm; full-width-at-half-maximum; resel size $= 2.74$ cm^3; scanned volume $= 1,644.00$ cm^3). In other words, if the experiment were repeated 100 times, on only 6 occasions would any false positive peaks appear above the critical t-value of 4.5.

Functional Magnetic Resonance Imaging

In the majority of fMRI studies, task-related changes in local hemodynamics are inferred from changes in the BOLD signal. As its name suggests, BOLD signal is based on the ability to detect, with MRI, changes in blood concentration of deoxyhemoglobin. Deoxyhemoglobin is paramagnetic and, as such, acts as an endogenous contrast agent that creates microscopic magnetic field inhomogeneities in the sampled tissue. These local inhomogeneities increase the rate of decay of transverse magnetization, that is, they shorten T_2 and T_2* relaxation times. Upon neural activation, the disproportionate increase in local blood flow "dilutes" the amount of deoxyhemoglobin and, in turn, reduces the amount of local inhomogeneities. This leads to a positive signal on T_2*-weighted images.

Before proceeding to the specifics of fMRI acquisition, the general principles of MRI are reviewed briefly (48). Nuclei that have an odd number of nucleons (protons and neutrons) possess both a magnetic moment and angular momentum *(spin)*. Hydrogen atoms contain only a single proton and therefore precess when exposed to a magnetic field, emitting electromagnetic energy in the process. The MR signal is generated and measured in the following way. First, the subject is exposed to a large static magnetic field (B_0) that preferentially aligns hydrogen nuclei along the direction of the applied field. Second, a pulse of electromagnetic energy is applied at a specific radiofrequency (RF) with an RF coil placed around the head. The frequency is selected to be the same as the frequency of precession of the imaged nuclei at a given strength of B_0; for hydrogen, this "resonant" frequency is about 64 MHz at 1.5 T. The RF pulse rotates (or "flips") the precessing nuclei away from their axes, thus allowing one to measure, with a receiver coil, the time it takes for the nuclei to "relax" back to their original position aligned with

B_0. The spatial origin of the signal is determined using subtle position-related changes in B_0 induced by gradient coils (these generate the knocking noise heard during scanning). Two relaxation times characterize the return to equilibrium: (a) longitudinal relaxation time (T_1); and (b) transverse relaxation time (T_2). T_1 is the time constant of the exponential recovery of the magnetization time. Due to local magnetic field inhomogeneities, individual hydrogen nuclei precess at slightly different rates, leading to their magnetic moments eventually pointing in different directions, with a concomitant decay of the total MR signal. T_2 is the time constant of the local rate of such "dephasing" within a population of nuclei. The presence of static field inhomogeneities enhances the rate of dephasing, with the resulting overall relaxation time of T_2*; $1/T_2^* = 1/T_2 + 1/T'_2$, where T'_2 is the decay time constant due just to field inhomogeneities (Fig. 4.4). The effect of local inhomogeneities can be minimized using spin-echo (SE) imaging with two successive RF excitations, yielding T_2-weighted images. In the case of BOLD-based fMRI, however, one is interested in maximizing the detection of local field inhomogeneities. Therefore, gradient-echo (GE) imaging is used whereby a single RF excitation is followed, after a time delay of echo time (TE), by the measurement of the amount of transverse relaxation. TE is adjusted so as to provide T_2*-weighted images.

In a typical BOLD-based fMRI study, a set of T_2*-weighted images is acquired. Each 2D image has a given thickness (e.g., 5 mm) and size (e.g., 64×64 pixels). If the field of view is a 320×320-mm square, the resulting in-plane pixel size would be 5×5 mm. To cover most of the cerebrum, about 25 contiguous 5-mm thick slices must be acquired. The exact number of slices depends on the brain size and, to some extent, on slice orientation. Unlike PET, which acquires data from the entire scanned volume simultaneously, fMRI measurements usually are serial, that is, obtained one 2D slice at a time. To collect many slices as quickly as possible, the use of fast acquisition techniques is required. These techniques allow one to acquire data in a "single shot," that is, after a single RF excitation. The most common acquisition technique of this type is *echo-planar imaging* (EPI). Using EPI, a slice (or plane) can be acquired in about 50 milliseconds that elapses from the time of RF excitation to the time of signal measurement (i.e., TE of 50 milliseconds). Such a rapid acquisition has two major advantages. First, it allows one to collect a large number of slices within a relatively short interval. For example, 25 slices would be acquired in about 3.75 seconds using TE = 50 milliseconds and repetition time (TR) = 100 milliseconds (TR is the time interval between two successive acquisitions). Second, it virtually eliminates the possible effect of intrascan head motion on the data. The slices are acquired in a particular order, for example, from the bottom slice upward or in an interleaved sequence. The latter approach minimizes possible interference between successively excited adjacent

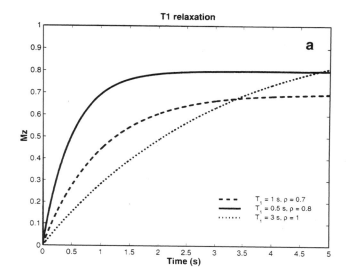

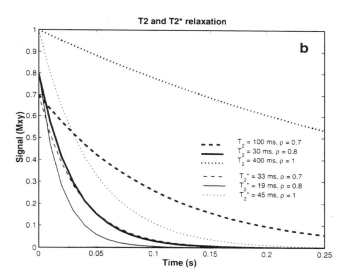

FIGURE 4.4. Magnetic resonance relaxation time constants. Spin-lattice **(A)** and spin-spin **(B)** relaxation curves for three fictitious materials. Spin-lattice (or longitudinal) relaxation is the process by which the magnetization (M_z) forms along the direction of the external magnetic field (B_0) and is characterized by the time constant T_1. Spin-spin (or transverse) relaxation describes the decay of transverse magnetization (signal) following an excitation and is characterized by the time constants T_2 *(thick lines)* and T_2^* *(thin lines)*, which exclude and include magnetic field inhomogeneities, respectively. (From Pike GB, Hoge RD. Functional magnetic resonance imaging: technical aspects. In: Casey KL, Bushnell MC, eds. *Pain imaging. Progress in pain research and management.* Seattle, WA: IASP Press, 2000:157–194, with permission.)

slices. Once the full set of slices has been acquired, the whole process is repeated many times to collect the entire time series of fMRI data. Depending on the experimental design (see section on Study Design: Functional Magnetic Resonance Imaging), the time interval between the acquisitions of each

set of slices ("frame") can range between a couple of seconds (block design or rapid event-related fMRI) to tens of seconds (slow event-related fMRI).

Analysis of fMRI data involves several steps that are similar to those described for PET (see section on Data Acquisition and Analysis: Positron Emission Tomography). Correction for head motion that may occur during scanning often is necessary. This correction can be achieved with a variety of algorithms (42,43) and, in principle, corresponds to the PET-PET registration of images described earlier. The next steps involve spatial and temporal filtering of the data. Spatial low-pass filtering *(blurring)* of the images reduces noise and residual misregistration artifacts. It also decreases the number of resels and, in turn, the number of multiple comparisons involved in the statistical analysis of the data (see section on Data Acquisition and Analysis: Positron Emission Tomography). Temporal filtering is carried out to remove high-frequency noise and low-frequency fluctuations (49). The latter may be due to cardiac and respiratory cycles or to linear drifts in MR signal. In block designs, it is worth considering the cycle frequency such that it does not match the size of the "temporal" filters. At the BIC, statistical analysis of such preprocessed images is carried out using the General Linear Model, a parametric method that assesses, at each voxel, the relationship between the BOLD signal and a given input function (available at *http://euclid.math.mcgill.ca/keith/fmristat/*). In the simplest case, two input functions are convolved: one that characterizes the temporal relationship between the fMRI time series and the presence/absence of the stimulus (e.g., boxcar function) and another that provides information about the assumed hemodynamic response [so-called *hemodynamic response function* (HRF)]. The use of an optimal HRF is critical for proper characterization of the delay between the onset of the stimulus/neural activation and the hemodynamic response. The delay typically varies between 4 and 6 seconds (50). The statistical significance of the relationship between the stimulus input function and the signal is evaluated by calculating t statistics for each voxel (51). Corrections for multiple comparisons involved in searching through the entire scanned volume take into account the number of resels (see section on Data Acquisition and Analysis: Positron Emission Tomography) and the number of degrees of freedom. In addition to this "single-voxel"–based t threshold, the size of a cluster of voxels ("cluster size") can be used as a criterion of statistical significance (52). All these steps apply to a single individual and session. Multiple individuals and sessions can be pooled together for group analyses as described in the section on Data Acquisition and Analysis: Positron Emission Tomography. In the case of both single- and group-analyses, localization of statistically significant task-related changes in BOLD signal is best carried out by first registering the functional MR images to the structural image acquired in the same subject and second by transferring *t*-maps and structural images to standardized stereotaxic space.

Electroencephalography

EEG is the oldest of all brain mapping techniques. It has been used to measure activity of the human brain for more than 70 years. The wealth of knowledge accumulated over this period is contained in many handbooks that can be consulted for detailed explanations of issues briefly discussed here (53–55).

The magnitude of brain electrical signals recorded from the scalp is on the order of microvolts. Three elements are essential for signal acquisition: electrodes, amplifiers, and a computer. The silver/silver chloride (Ag/AgCl) electrode is the most common type used for EEG recordings from the scalp because of its low resistance for direct-current and low-frequency potentials. To ensure a low-resistance contact between the electrode and the skin, an electrode jelly with a high concentration of sodium chloride (NaCl, 5%–10%) is applied on the skin from which the most superficial horny layer is scraped off. The resulting electrode impedance should be less than 3,000 ω; impedance greater than 5000 ω should not be accepted as adequate (56). Electrodes can be attached to the scalp in various ways. The most secure method is gluing electrodes to the head with collodion and then applying electrode jelly through a small hole in the center of a concave electrode. But the use of multiple (32, 64, or 128) electrodes makes this method impractical, and different types of electrode caps are used. It should be pointed out that proper electrode attachment and low electrode impedance throughout the entire recording session are critical for acquisition of high-quality signal; motion of an electrode in relation to the scalp can cause significant EEG artifacts. The scalp positions of the electrodes often are based on the International 10-20 System for 21 electrodes (57); the letters refer to an anatomic area (e.g., *F* for frontal) and the numbers indicate the relative distance from the midline (odd and even numbers are used for the left and right hemisphere, respectively). Although other schemes have been developed more recently to accommodate higher number of electrode sites, the best way to relate electrode positions to the brain is through their registration with structural MRI. This can be achieved, for example, with frameless stereotaxy (see section on Data Acquisition and Analysis: Transcranial Magnetic Stimulation).

The EEG signal picked up by the electrodes is fed to a set of EEG amplifiers. The band width of the amplifiers must be such as to allow undistorted measurement of the expected signal frequencies. In most studies, the adequate range is between 0.16 and 100 Hz and can be changed using high-pass and low-pass filters, respectively (58). Signal from the scalp electrodes is measured either with a bipolar montage or relative to a reference electrode. The most common reference electrodes used during the recording are the (linked) earlobes and the (linked) mastoids. At the postprocessing stage, one also can use any single scalp electrode or the average of all scalp electrodes as a reference. Finally, the continuous analog outputs of the amplifiers are sampled with analog-to-digital converters and processed on line (display) and off line (analysis) by a computer. The sampling rate is important because of the possible signal distortions it can introduce. The highest frequency that can be represented in the digitized signal is half the sampling rate; this is the so-called *Nyquist frequency* or *limit* (59). The most common sampling rate sufficient for the majority of brain mapping applications is 256 samples per second. In addition to EEG signal, timing of different events, such as the onset of stimuli and subject's responses, is recorded in the same file.

Analysis of EEG data for brain mapping purposes focuses typically on two domains: (a) event-related potentials, and (b) event-related synchronization/desynchronization of brain activity.

Event-related potentials (ERPs) are stimulus-locked, small (2–20 μV) potential changes extracted from the ongoing "background" EEG activity by averaging tens of brief EEG epochs surrounding the stimulus. According to their latency and functional attributes, they can be classified into early "exogenous" and late "endogenous" potentials (Fig. 4.5). The latency and amplitude of individual ERP components can be quantified in the individual averages. This usually is done at an electrode yielding a prototypical waveform. Localization of the brain sources of a given ERP component, or any other feature of the EEG signal, from the scalp-recorded potentials is a complex undertaking (for an overview of different methods see reference 60). Model-dependent methods approach the so-called *inverse problem,* that is, identification of the intracranial source that generated a particular distribution of scalp potentials, by making *a priori* assumptions about the number, type, and localization of dipole. The "goodness of fit" of the model to the recorded data is evaluated by the sum of squares of the differences between the computed and measured potentials at each electrode and each time point (61). Model-independent methods do not require any assumptions be made about the number and configuration of sources in the brain. A simple topographic display of the scalp potentials provides spatial information about the possible sources of the signal with a minimal manipulation of the raw data. Interpolation of potentials to intermediate points between the scalp electrodes is the only computation required (for different interpolation algorithms see reference 60). Due to the "blurring" properties of the tissue located between the assumed (cortical) source and the recording electrodes (i.e., the skull and scalp), maps of scalp potentials are rather broad even when recorded with a relatively high number (e.g., 32) of scalp electrodes. Spatial resolution of the topographic maps can be increased with a variety of methods (reviewed in reference 60), including the calculation of the Laplacian of the scalp potential (62), minimum norm estimates (63,64) and the use of "deblurring"

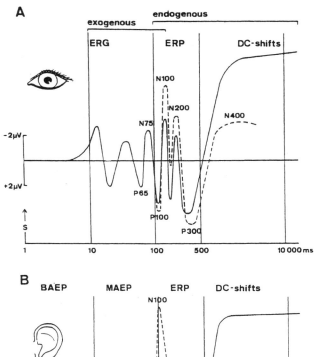

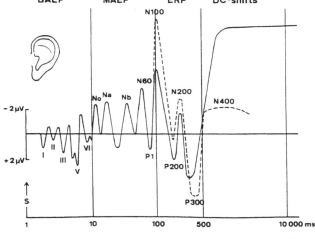

FIGURE 4.5. Averaged event-related responses to visual **(A)** and acoustic **(B)** stimuli. Schematic potential traces on a logarithmic time scale. **A:** Exogenous components comprise the electroretinogram (ERG) and the P65 and N75. Components with latencies longer than 100-millisecond latency are considered endogenous components. The P100 and N100 components can be modified by orienting and selective attention *(dashed lines)*, the N200 by stimulus evaluation, and the P300 by context updating. The N400 is related to semantic expectancy. Large direct-current shifts occur when complex cognitive tasks have to be solved. **B:** In the acoustic modality, exogenous components comprise the acoustic brainstem auditory evoked potentials (BAEP) and the mid-latency auditory evoked potentials components (MAEP). Endogenous components can be modified analogous to the visual modality but have a tendency toward shorter latencies. Whereas exogenous event-related potentials exhibit modality-specific potential traces, endogenous components are similar in both modalities. (From Altenmüller EO, Gerloff C. Psychophysiology and EEG. In: Niedermeyer E, Lopes da Silva F, eds. *Electroencephalography: basic principles, clinical applications, and related fields,* 4th ed. Baltimore: Williams & Wilkins, 1999:637–655, with permission.)

algorithms (65). Acquisition of structural MR image of the subject's brain together with the recording of the scalp position of the electrodes greatly facilitates localization of the neural generators.

Event-related synchronization (ERS) and event-related desynchronization (ERD) allow study of changes in spontaneous rhythmic activity of the brain elicited by different events (for overview see reference 66). Unlike ERPs, ERDs and ERSs typically are not phase locked to the event onset (but see section on Combination of Brain Stimulation and Imaging: Transcranial Magnetic Stimulation During Electroencephalography and reference 67). The temporal relationship between an event and ERD/ERS is less sharp, developing over a period of tens of milliseconds. The basic steps involved in the analysis of ERD/ERS are illustrated in Figure 4.6 and, for each frequency band, include the following: (a) band-pass filtering of raw EEG obtained in each trial epoch (e.g., 4 seconds before and 3 seconds after the event onset); (b) squaring (rectifying) the amplitude of the filtered signal to obtain power measurements; (c) averaging over all trials; and (d) averaging over a small number of consecutive power samples to reduce the variance (66). Decreases and increases in power relative to a reference interval are referred to as ERDs and ERSs, respectively.

Magnetoencephalography

Magnetic fields generated by the brain are on the order of tens of femtotesla (fT, where $1\ fT = 10^{-15}T$). For comparison, the Earth's magnetic field is 10^{-3} T and magnetic fields generated by electrical equipment (e.g., elevators) are of the order of 10^{-9} T (see Table 60.1 in reference 11). To record such a weak signal, sensitive measuring instruments are operated in magnetically shielded rooms. The basic element of all commercially available MEG systems is the superconducting quantum interference device (SQUID) sensor. The SQUID-based magnetometer measures the brain's magnetic field using a superconducting flux transformer containing a single pickup loop. To minimize the effect of distal environmental noise, another coil is added either above (axial first-order gradiometer) or within (planar gradiometer) the plane of the pickup coil. Thus, gradiometers and magnetometers allow investigators to measure relative and absolute magnitudes of the brain's magnetic fields, respectively. Current MEG systems contain between 122 and 306 channels that cover the entire convexity of the two cerebral hemispheres (Fig. 4.7).

As explained in the section on Neurophysiologic Underpinnings of the Signal: Electroencephalography and Magnetoencephalography, the neurophysiologic underpinnings of EEG and MEG signals are the same. Unlike EEG, however, MEG detects primarily current dipoles generated in the cerebral sulci. The main advantage of the MEG over EEG lies in (a) virtual transparency of the skull and other extracerebral tissue to magnetic fields and (b) reference-free

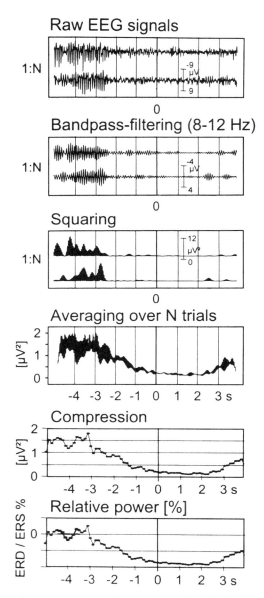

Raw EEG signals

1:N

-9
μV
9

0

Bandpass-filtering (8-12 Hz)

1:N

-4
μV
4

0

Squaring

1:N

12
μV²
0

0

Averaging over N trials

[μV²]

-4 -3 -2 -1 0 1 2 3 s

Compression

[μV²]

-4 -3 -2 -1 0 1 2 3 s

Relative power [%]

ERD / ERS %

-4 -3 -2 -1 0 1 2 3 s

FIGURE 4.6. Event-related desynchronization (ERD). Schema for ERD processing showing an example with movement-related μ-desynchronization. The raw electroencephalographic signals (0.5–50 Hz) are first band-pass filtered (8–12 Hz), then all samples are squared and averaged over N trials. After averaging over consecutive power samples (compression) and specification of a reference interval (e.g., −3.5 to −4.5), the relative band power values are displayed. A power decrease corresponds to ERD. (From Pfurtscheller G. EEG event-related desynchronization (ERD) and event-related synchronization (ERS). In: Niedermeyer E, Lopes da Silva F, eds. *Electroencephalography: basic principles, clinical applications, and related fields,* 4th ed. Baltimore: Williams & Wilkins, 1999:958–967, with permission.)

recording of local magnetic fields. The two features facilitate localization of brain sources of the signal recorded outside the skull. Analysis of MEG signals uses virtually the same set of techniques described earlier (see section on Data Acquisition and Analysis: Electroencephalography). Both ERPs

and ERDs/ERSs are of main interest in brain mapping studies carried out with MEG (for an overview see reference 11).

Transcranial Magnetic Stimulation

TMS is based on Faraday's law of induction, which describes a quantitative relationship between a changing magnetic field and the electric field created by the change. The law was developed on the basis of experimental observations made in 1831 by the English scientist Michael Faraday. Figure 4.8 illustrates the key events resulting in the induction of stimulating current in the brain by TMS: (a) a high current (~5,000 A) is discharged from a capacitor through the coil, generating a strong magnetic field (~2 T); (b) the magnetic field is rapidly switched on and off (~200 microseconds), leading to a high rate of field change (~30 kT/s); and (c) this time-varying magnetic field induces electrical current in the tissue under the coil.

Commercially available stimulators allow investigators to deliver monophasic or biphasic pulses with different between-pulse intervals. Single-pulse stimulators (e.g., Magstim model 200, Cadwell MES-10) can deliver pulses not faster than every 2 to 3 seconds. Rapid-rate stimulators (e.g., Magstim Standard Rapid, Cadwell High Speed Magnetic Stimulator) can deliver trains of rTMS with frequency of pulses up to 30 Hz. Paired-pulse stimulation used for assessment of intracortical facilitation and inhibition requires very short between-pulse intervals (1–20 milliseconds) and independent setting of stimulation intensities for each of the two pulses. This can be achieved by combining two single-pulse stimulators through a connecting module (Magstim BiStim Module).

The size and shape of the stimulating coils both determine the geometry of the induced magnetic field and, in turn, that of electrical current induced in the brain. Figure 4.9 shows six different coils and the distribution of the modeled electric field calculated from the coil geometry (68). The highest focality is achieved with a figure-of-eight coil in which the current flows in opposite directions in the two wings, therefore adding up in the central junction of the coil. For any coil, the magnitude of the induced electric field falls off quickly with distance; for example, the modeled electrical field is reduced by 38% over a 10-mm distance from the cortical surface (68). However, such theoretical calculations provide only the first approximation of the spatial resolution of TMS. Complex interactions between the magnetic field and all conductive neural elements in the brain tissue under the coil (see section on Neurophysiologic Underpinnings of the Signal: Transcranial Magnetic Stimulation) determine the magnitude of the actual physiologic effects. For example, small changes in coil position and/or orientation relative to the central sulcus lead to significant variations in the MEP amplitudes (Fig. 4.10). The steep falloff of magnetic field with increasing distance from the coil

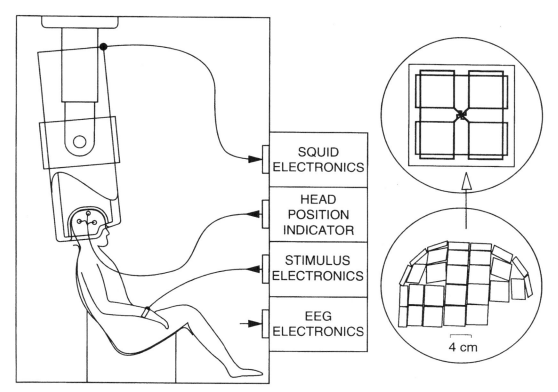

FIGURE 4.7. Schematic illustration of a magnetoencephalographic system. **Left:** The subject in a magnetically shielded room during the measurement. **Right:** Schematic illustration of the device's flux transformer array. Each of the 61 sensor units has two figure-of-eight loops to detect the orthogonal tangential derivatives $\delta B_r/\delta x$ and $\delta B_r/\delta y$ of the radial field component B_r. (From Hari R. Magnetoencephalography as a tool of clinical neurophysiology. In: Niedermeyer E, Lopes da Silva F, eds. *Electroencephalography: basic principles, clinical applications, and related fields,* 4th ed. Baltimore: Williams &Wilkins, 1999:1107–1134, with permission.)

prevents the use of TMS for direct stimulation of subcortical structures. Furthermore, cortical regions located at some distance from the skull convexity cannot be stimulated either. Recent studies combining TMS with PET suggest that neural activity in such distal structures can be influenced through existing connections with regions accessible to TMS (see section on Combination of Brain Stimulation and Imaging: Transcranial Magnetic Stimulation During Positron Emission Tomography).

Precise localization of the TMS coil relative to the brain is critical for interpretation of brain mapping studies. This is best achieved by acquiring a structural MR image of the subject's brain and using the image to guide positioning of the coil in real time. Thus, the first step in planning a TMS study is acquisition of a high-resolution T_1-weighted image of the subject's brain. In the TMS laboratory, the next step involves coregistration of the subject's MRI with the actual position of his or her head. The MRI-to-head registration can be carried out either with a fiducial frame attached to the subject's head (69) or without a frame, that is, with frameless stereotaxy (70). Frameless stereotaxy uses a set of anatomic landmarks, such as the bridge of the nose and the tragus of the ear, that are visible on both the subject's MRI and on his or

her head. The 3D location of the landmark is measured with a position sensor (Fig. 4.11). The accuracy of the frameless stereotaxy is slightly inferior to that based on a fiducial frame and varies between 4 and 8 mm (71). A frameless stereotaxy system consists of two major components: a position sensor and tracking software. A commercial TMS-dedicated system has been developed that combines hardware and software necessary for coil positioning and tracking in real time (BrainSight Frameless, Rogue Research Inc., Montreal, Canada).

STUDY DESIGN

Positron Emission Tomography

Use of a radioactive tracer imposes significant constraints on the design of PET studies. In blood-flow activation studies, about 10 mCi of O^{15}-water is sufficient for acquisition of one scan with the scanner operating in the 3D mode. Depending on the local radiation safety regulations, several scans can be obtained in a single volunteer. At the BIC, the current annual limit of radiation exposure allows the use of up to 120 mCi, thus allowing acquisition of up to 12 O^{15}-water scans per

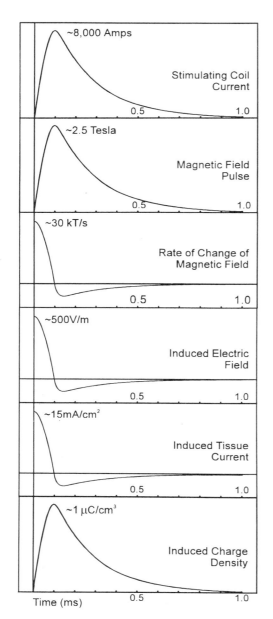

FIGURE 4.8. Induction of stimulating current in the brain by transcranial magnetic stimulation. **Top:** A high current (~5,000 A) is discharged from a capacitor through the coil, generating a strong magnetic field (~2 T). **Middle:** The magnetic field is rapidly switched on and off (~200 μs), leading to a high rate of field change (~30 kT/s). **Bottom:** The time-varying magnetic field induces electrical current in the tissue under the coil. (From Jalinous, *Guide to magnetic stimulation,* with permission.)

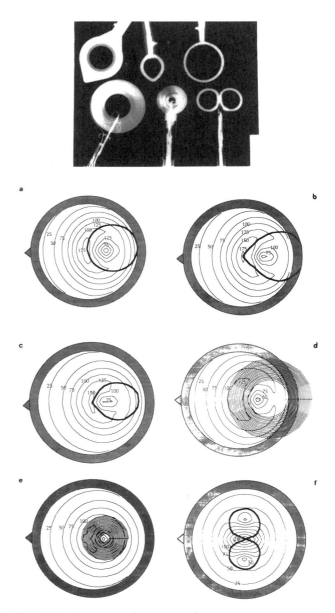

FIGURE 4.9. Geometry of transcranial magnetic stimulation coils and associated electric fields. **Top:** X-ray photographs of six coils. *Clockwise from upper left:* coil b, coil c, coil a, coil f, coil e, coil d. **Bottom:** *a–f.* Magnitude of the modeled electric field 3 mm below the surface of the brain for the six coils. (From Roth BJ, Saypol JM, Hallett M, et al. A theoretical calculation of the electric field induced in the cortex during magnetic stimulation. *Electroencephalogr Clin Neurophysiol* 1991;81:47–56, with permission.)

subject per year. Because of the 2-minute half-life of O^{15}, the minimal between-scan interval needed for radiation to return to the background levels is about 10 minutes. Taken together, twelve 1-minute emission (i.e., O^{15}-water) scans and one 10-minute transmission scan can be acquired in about 2.5 hours. The transmission scan is necessary to calculate attenuation and scatter corrections, and it is typically acquired before the emission scans. A session of this duration is generally well tolerated by healthy volunteers even when head restraint devices are used.

Each scan provides a single measure of regional CBF representing the sum of brain activity occurring over the entire scanning period. No within-scan comparisons of CBF related to different events are possible. Therefore, PET studies used the so-called *block* design, where a block of similar events is administered during one scan and a block of

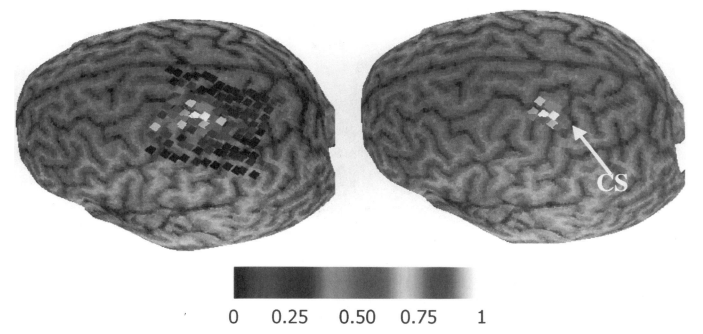

FIGURE 4.10. Motor evoked potentials (MEP) and coil position. Normalized amplitude of MEPs acquired with the Magstim figure-of-eight coil and plotted over the cortical surface. The coil locations were sampled with frameless stereotaxy on a 5-mm grid at the time of each transcranial magnetic stimulation pulse. Each point indicates the location of the presumed virtual "maximum" of the coil, estimated to be 10 mm in front of its physical center. The *arrow* indicates the central sulcus at the level of the presumed hand representation ("the knob"). (See Color Figure 4.10 following page 526.)

different events is used in another scan. Differences in the type of events can be qualitative or quantitative. For example, one can contrast a scan involving eye movements with the scan during which the subject simply fixates the screen center. Subtraction of the two scans would reveal brain regions involved in oculomotor control (i.e., a qualitative difference). A similar study can be designed in a parametric manner in which a different number of eye movements is made during each scan. Regression of blood flow against the number of eye movements would reveal regions responding to the behavioral manipulation in a "dose-dependent" manner. Depending on the question of interest, the researcher may decide to include several qualitatively different conditions, vary the number of operations of interest across scans, or combine the two. Furthermore, the same condition may be repeated more than once in a given subject in order to increase statistical power. The only limit is the total number of scans per session.

The successive nature of scanning over a long period requires careful counterbalancing of the order of conditions. This should minimize time-in-the-scanner effects, such as reduction of the initial anxiety or increasing discomfort (e.g., back pain or need to urinate). The performance of subjects during the scans should be documented carefully. The open scanning environment makes it relatively easy to collect a variety of measures, including the latency and accuracy of button presses, verbal responses, eye movements, and auto-

nomic (e.g., galvanic skin response) or cortical (EEG) indicators of the level of arousal. These measures are important not only for the documentation of the expected behavioral task-related differences, but they also can be used in various regression analyses of the blood-flow data.

PET-based blood-flow activation studies are being increasingly replaced by fMRI, mostly because of its greater accessibility to researchers and the absence of radiation exposure. Use of PET for brain mapping studies continues in certain clinical populations and in task conditions that require a relatively opened scanning environment, easy access to the subject, and limited head restraint. More importantly, brain mapping studies are gradually moving into the next stage in which the use of ligands other than O^{15}-water allow investigators to link behavior with particular neurochemical systems (72,73).

Functional Magnetic Resonance Imaging

By collecting a set of images over a relatively brief period of time (seconds), fMRI affords a great deal of flexibility in designing brain mapping studies. Two types of fMRI designs are in use: (a) block-design fMRI, and (b) event-related fMRI.

The fMRI block design is similar to that described for PET, namely a number of similar events are presented over a block of time and compared with different events

REGISTRATION

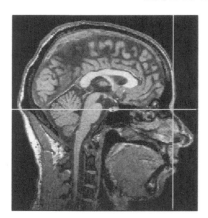
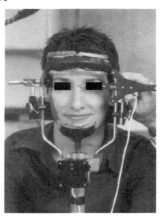

POSITIONING OF THE COIL

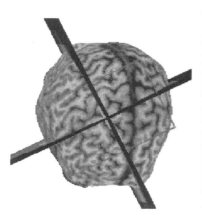
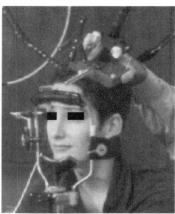

FIGURE 4.11. Frameless stereotaxy. **Top:** Registration of the subject's head with the corresponding magnetic resonance (MR) image. A computer-linked probe is touching the bridge of the nose **(right)**. The matching location is highlighted by a *cross-hair* on the MR image **(left)**. **Bottom:** A location targeted by transcranial magnetic stimulation **(left)**, and the probe–coil interface used to position the coil over this location **(right)**.

administered during a different block. For example, the subject may be moving a finger during one 30-second block and resting during another block. Typically the two conditions would alternate for several cycles. The images are acquired in a rapid succession throughout the session so that, for example, a total of 120 "frames" (e.g., 1 frame = one set of twenty-five 5-mm-thick axial slices) would be collected in 6 minutes. The frames collected during the two respective conditions are pooled together (i.e., 60 frames per condition) and analyzed without the use of information about the exact temporal relationship between an event (i.e., onset of finger movement) and the frame acquisition.

Event-related design, on the other hand, takes advantage of the discrete sampling of brain activity made possible with EPI. Thus, the acquisition of a frame is time locked to the onset of an event of interest. Due to the delay of the hemodynamic response relative to the onset of event-related neural activity, the acquisition typically is delayed by 4 to 6 seconds. In the "slow" event-related fMRI, successive events are spaced by 10 to 12 seconds to allow for return of the signal to the baseline before the next acquisition. This arrangement also makes it possible to present stimuli (e.g., tones) (74) and record verbal output (Fig. 4.12) in a quiet environment. This

also minimizes the effect of speech-related head movements on image acquisition. In the "rapid" event-related fMRI, the events are occurring quickly (<2 seconds) one after another, resulting in the overlapping hemodynamic responses generated by the successive events. This overlap is deconvolved at the processing stage (75,76). In both types of event-related fMRI, the relative timing of the events and acquisitions can be systematically varied ("jittered") to allow for a detailed temporal analysis of the hemodynamic response function. Although the latter approach virtually guarantees that the "peak" response would not be missed due to suboptimal event-acquisition timing, it also reduces the proportion of samples with high-enough signal. The event-related design is particularly useful when various types of events are interleaved (e.g., verbs and nouns in a sentence) or when the events are classified *post hoc* according to the subject's response during scanning (e.g., errors, onset of hallucinations).

Electroencephalography and Magnetoencephalography

Design of EEG and MEG studies is conceptually similar to that of event-related fMRI. A variable number of events/trials

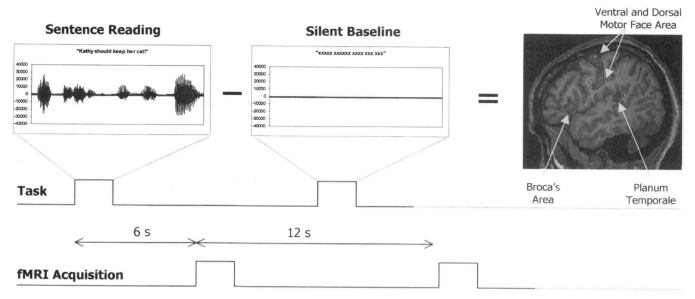

FIGURE 4.12. Design and results of a single-subject functional magnetic resonance imaging (MRI) study of sentence reading. Functional MR acquisitions are separated by 12 seconds and start 6 seconds after the onset of stimulus presentation. Forty frames were acquired after reading of a sentence and 20 frames were acquired after a silent baseline. **Top left:** Speech output recorded in the scanner. Colored areas superimposed on the subject's MRI scan indicate regions of significantly ($t > 5.0$, $p < 0.05$ corrected) increased blood oxygen level-dependent signal during the sentence reading compared to the silent baseline. This sagittal slice shows a set of cortical areas related to speaking and hearing oneself speak. (From Barrett J, Paus T, Pike B. Affect and speech: an fMRI investigation of brain correlates underlying variation in range of pitch. *Neuroimage* 2002, with permission.) (See Color Figure 4.12 following page 526.)

is presented during a session, while brain electric or magnetic signals are continuously recorded together with the stimulus and/or response onset. Depending on the task, the number of trials necessary to obtain a clear EEG/MEG response to a given event varies widely (40–200 trials per event). The length of an intertrial interval (ITI) depends on the expected latency of a given ERP component of interest and can vary between 0.2 and 5 seconds. ITIs are generally longer when studying ERD/ERS. If two types of events were studied, the recording session would last for about 20 minutes (2 events × 100 trials × 5-second ITI). In the case of multichannel EEG, a significant amount of time is required for electrode placement and impedance check (about 30 minutes) and for sampling of the exact electrode positions with frameless stereotaxy (15 minutes). The main advantage of multichannel EEG is the minimal demand on the subject vis-à-vis head restraint and low (if any) discomfort during the actual recording session. This is why EEG is still the most popular technique used in developmental studies (77–79).

Transcranial Magnetic Stimulation

As pointed out in the section on Neurophysiologic Underpinnings of the Signal: Transcranial Magnetic Stimulation, TMS can be used in brain mapping studies in the following

ways: (a) to elicit a physiologic response that, in turn, serves as a dependent variable; (b) to interfere with ongoing brain activity; and (c) to induce changes in cortical excitability.

TMS of the primary motor cortex (M1) combined with electromyography allows researchers to assess changes in motor physiology brought about by experimental manipulations, as well as those associated with various neurologic and psychiatric disorders. Several parameters can be measured, including resting and active motor thresholds, MEP amplitudes, duration of the silent period, and interhemispheric interactions [for a mini-review see Chen (15)]. Resting motor threshold typically refers to the minimum stimulation intensity necessary to elicit MEPs, in a relaxed muscle, greater than 50 μV in at least five of ten successive trials. Because of the ease of electromyographic recording, the first dorsal interosseus muscle is used in most studies. Active motor threshold is obtained in a similar way by applying TMS during a tonic contraction of the target muscle. The subject is asked to maintain muscle contraction at about 30% of his or her maximum while he or she is receiving auditory or visual feedback. The amplitude of MEPs elicited by suprathreshold single-pulse TMS provides an index of corticospinal excitability. Although reflecting changes at the cortical level, this approach does not allow the researcher to exclude the possibility of the observed changes occurring at the spinal level. This also is true for the duration of the silent period elicited by suprathreshold single-pulse TMS in a contracting muscle. The contribution of the α motoneurons of the spinal cord can be assessed with a concurrent measurement of H-reflex (19). The best way to evaluate changes in cortical excitability, however, is the use of paired-pulse TMS paradigm whereby two TMS pulses are applied in rapid succession (between-pulse interval of 1–30 milliseconds). The first pulse is of subthreshold intensity, i.e., conditioning stimulus, whereas the second pulse is suprathreshold, i.e., test stimulus (80). Paired-pulse TMS with short (1–5 milliseconds) and long (8–30 milliseconds) between-pulse intervals elicits, respectively, suppression and facilitation of the muscle (MEP) response mediated by cortical mechanisms (20–22,24,81).

These techniques have been used successfully in developmental studies of the motor system (82–84) and in experiments aimed at elucidating brain mechanisms of motor imagery (85–87) and action observation (88–92). Using the motor cortex as a "sample" cortical region, they also provide a window into the state of cortical excitability in various psychiatric disorders (93,94) and its changes in response to different treatments (95).

The second and perhaps most valuable mode of using TMS in cognitive neuroscience is its use to interfere with brain activity, thus inducing focal and temporary "virtual lesions" (96,97). The importance of this approach lies in the need to establish whether or not a cortical region, identified for example in a neuroimaging study (see the section on Combination of Brain Stimulation and Imaging: Imaging-Based Planning of Transcranial Magnetic Stimulation Studies), is critical for a given cognitive process. Experimentally changing brain activity and evaluating behavioral consequences of this change can best assess the presence of such a critical link between structure and function. Both single-pulse and repetitive TMS are being used for this purpose. As a general rule, single-pulse TMS is particularly useful when a relatively invariable temporal relationship exists between an external event and a given cognitive process. For example, it can be predicted that visual information about a visual stimulus would be processed in the primary visual cortex about 50 to 100 milliseconds after the stimulus onset. The presence of such a relationship also allows the researcher to establish the critical period during which a given process is interrupted by TMS and, by inference, implemented by the stimulated region. This is simply achieved by applying TMS pulses at several different intervals after the stimulus onset. On the other hand, single-pulse TMS may not be the best approach in studies of cognitive processes that are characterized by relatively large interindividual and intertrial variabilities in response latencies. In such experiments, trains of repetitive TMS often are used to compensate for the lack of an invariable temporal relationship between the external event and cognitive process of interest. Frequency of stimulation typically varies between 5 and 10 Hz and the duration of each train typically varies between 0.4 and 1.0 second. Safety guidelines regarding the frequency

intensity and duration of TMS trains must be followed to minimize the possibility of serious adverse effects (98).

Three important methodologic issues need to be considered in the "virtual lesions" studies: (a) choice of dependent variable, (b) choice of control condition, and (c) localization of the stimulation site. With few exceptions, the most common effects of single-pulse and repetitive TMS on behavior are changes in response latency but not in the rate of errors. This may simply be a "dose" issue. To comply with safety guidelines and to make the experiment as comfortable as possible for the subject, relatively low frequency and intensity of stimulation is used. Therefore, information processing in the stimulated region may not be fully interrupted for the entire stimulus-to-response interval. Furthermore, one should keep in mind that the majority of cognitive functions are represented in both hemispheres; therefore, unilateral stimulation interferes with only part of the system. The second important issue is the choice of a control condition. Most of the "sham" procedures, such as tilting the coil away from the skull, are not adequate to control for peripheral effects of TMS and related distractions. This is particularly the case when the target region lies in the lower parts of the brain, such as the ventrolateral prefrontal cortex or inferior temporal or occipital cortex. In these cases, TMS often stimulates underlying muscles and/or nerves causing, in turn, significant peripheral effects. For these reasons, the best approach is to choose a control region in which TMS, applied with the same parameters as those used over the "experimental" region, induces comparable peripheral effects without affecting cortical tissue known to be involved in the cognitive process of interest. When significant hemispheric asymmetries are expected (e.g., picture naming, discussed later), this can be best achieved by comparing the left and right stimulation sites. The third issue is that of site localization. Interpretation of "virtual lesion" experiments depends critically on the availability of detailed information about the coil position relative to the subject's brain. As explained in the section Data Acquisition and Analysis, this can be best achieved with the use of frameless stereotaxy. Other approaches, such as the relative distance of the coil from the primary motor cortex or the International 10-20 System, are suboptimal in this respect.

The third mode of TMS use in behavioral experiments is that of studying the consequences of long-term (minutes) changes in cortical excitability induced by repetitive TMS on a subject's performance. This approach is a variation of the "virtual lesion" paradigm described earlier. The only difference is that TMS is not used during the actual task performance but is applied before the task. As explained in the section on Neurophysiologic Underpinnings of the Signal: Transcranial Magnetic Stimulation, the rationale for this approach is based on the observed long-term (up to 30 minutes) reduction in cortical excitability following low-frequency (1 Hz) stimulation of the primary motor cortex. It is assumed that similar changes take place in other cortical regions and that reduction of excitability leads to less efficient processing of information by the stimulated region. In a few studies using this approach, 10 to 15 minutes of 1-Hz TMS was applied at motor-threshold intensity and the cognitive function of interest was tested both before and immediately after the stimulation (99). The main advantages of this approach are the absence of any potentially distracting effects of TMS and no need for head restraint during the actual task performance. Compared with intermittent brief trains of rTMS, however, such continuous stimulation may be more likely to affect other regions connected to the stimulation site. If this is true, this may limit interpretations vis-à-vis the exclusivity of the stimulated region's involvement in mediating the observed effects.

COMBINATION OF BRAIN STIMULATION AND IMAGING

Imaging-Based Planning of Transcranial Magnetic Stimulation Studies

The complementarity of TMS and functional neuroimaging is best exemplified by the following "two-stage" approach to studies of brain–behavior relationships. In stage 1, PET or fMRI is used to identify a cortical region with statistically significant task-related changes in local hemodynamics. In stage 2, TMS is used to interfere with neural activity in this "target" region during the performance of the same task. In this way, TMS is used to test whether or not a region in which change in neural activity is *associated* with a given task also is *necessary* for the performance of this task.

Identification of the target region can be based on a single-subject imaging study or on group averages of PET/fMRI data. In the former case, the same individual participates in both the imaging and TMS stages. For example, in one of our recent studies (100) we used fMRI to identify, in each individual, a region located in the inferior temporal cortex (ITC) in which the BOLD signal increases with the difficulty of a picture-naming task (Fig. 4.13A). Subsequently, each subject participated in an rTMS experiment where the coil was positioned, with frameless stereotaxy, over the "activated" region and a brief train of rTMS was applied during the picture naming (Fig. 4.13B). Repetitive TMS was applied to the left and right ITC, as well as to a control region in the left superior parietal cortex. The results showed a significant increase in naming latency during stimulation of the left ITC. No changes were observed when rTMS was applied over the right ITC or the parietal cortex (Fig. 4.13C). Although both the left and right ITC showed task-specific changes in BOLD signal, the rTMS results suggest that only the left ITC is necessary for picture naming.

Considering a relatively high consistency in the location of task-related "activations" across individuals, single-subject imaging studies may not always be required to identify the target region. An alternative approach takes advantage of standardized stereotaxic space and uses X, Y, and Z coordinates of "activation" peaks identified in previous group-

based PET or fMRI studies (101). For example, we have used a probabilistic location of the frontal eye field in a TMS study of visuospatial attention (29). The frontal eye field was identified through a metaanalysis of previous oculomotor blood-flow activation studies (102). Its X, Y and Z coordinates were transformed from the standardized stereotaxic space (45) to the subject's brain coordinate ("native") space, and the coil was positioned over this location with frameless stereotaxy. The transformation from the "Talairach" to the "native" space simply uses the inverse version of the $MRI_{native}–MRI_{Talairach}$ transformation matrix described in the section on Data Acquisition and Analysis: Positron Emission Tomography. As shown in Figure 4.14, single-pulse TMS applied 50 milliseconds before the expected onset of a saccadic eye movement increased latency of contralateral saccades in half of the subjects included in the study. The lack of a similar effect in the other subjects most likely was due to the low intensity of stimulation and high temporal uncertainty in predicting the saccade onset. To validate further this "probabilistic" approach, we used average coordinates of the hand and face representations in the primary motor cortex and positioned a figure-of-eight coil over these locations. Figure 4.15 provides a comparison of the MEP values obtained when stimulating over such probabilistic locations (Fig. 4.15A) with those obtained in the same subjects by sampling many locations over the putative primary motor cortex until the strongest response was elicited (Fig. 4.15B). No statistically significant differences were found between the two approaches, thus confirming the usefulness of probabilistic locations for targeting cortical regions of interest.

Transcranial Magnetic Stimulation During Positron Emission Tomography

The main motivation behind the development of the combined TMS/PET approach has been its potential for *in vivo* assessment of cortical connectivity (101). Although we know a great deal about neural connectivity in nonhuman primates, little information is available in human primates because of the limited repertoire of suitable anatomic techniques (103). Functional techniques based on the correlational analysis of EEG (104–106) or CBF (46,107–110) are indirect and may not reveal actual neural connectivity. The correlational studies suffer a major limitation in that the engagement of a subject in performing a task confounds the data being acquired. The observed "coactivations" may reflect relationships between different components of behavior rather than connectivity. The TMS/PET approach allows direct assessment of neural connectivity, without requiring the subject to engage in any specific behavior. The key principle of this approach is that of measuring the effect of a focal TMS-induced electrical stimulation of one region on activity, indexed by changes in CBF, elsewhere in the brain.

Several technical issues need to be considered in TMS/PET experiments. First, the strong, albeit brief,

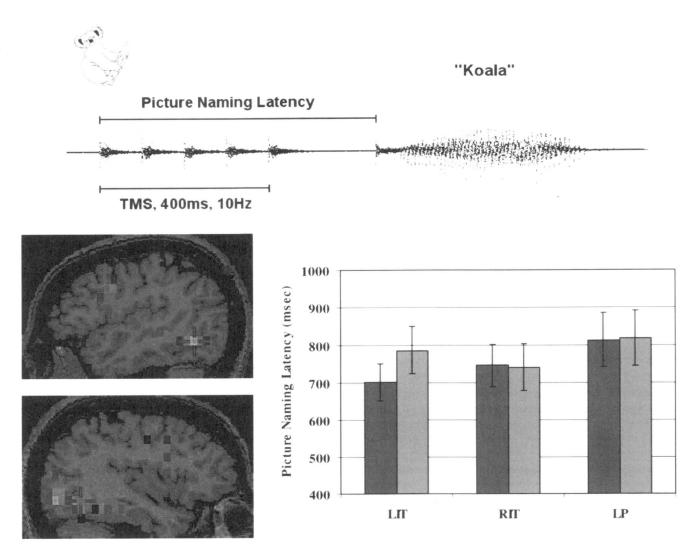

FIGURE 4.13. Two-phase brain mapping experiment. In phase I, functional magnetic resonance imaging was used to identify, in each individual, a region located in the inferior temporal cortex (ITC) in which the blood oxygen level-dependent signal increased with the difficulty of a picture-naming task **(bottom left)**. In phase II, such "activated" regions were stimulated with a brief train of repetitive transcranial magnetic stimulation (TMS) during picture naming **(top)**. The naming latency is shown for the left and right ITC, and a control region in the left parietal (LP) lobe **(bottom right)**. TMS increased significantly the latency when applied over the left IC only [from Fung TD, et al. (100)]. (See Color Figure 4.13 following page 526.)

coil-generated magnetic field can affect photomultipliers and the related electronic circuits housed about 20 cm away from the subject's head in the gantry of the PET scanner. Even though the field falls off quickly with distance, photomultipliers are sensitive to the interfering effects of magnetic fields as small as 10^{-4} T. Using a single-detector assembly and a figure-of-eight coil positioned 19 cm from the photomultipliers, we examined such possible effects and showed that operating the coil at even 40% of the maximum output of the stimulator causes serious distortions in the crystal identification matrix (101,111). We were able to prevent these distortions by placing four sheets of well-grounded μ-metal between the coil and the PET detector; however, the μ-metal causes attenuation of γ-rays and, in turn, a decrease in the

number of detected coincidence counts. The second important issue is that of possible movement of the coil relative to the subject's head and the effect of such a misalignment on attenuation corrections. As explained in the section on Study Design: Positron Emission Tomography, each PET session begins with a transmission scan that provides information about the location and density of various objects located between the brain, that is, the source of γ-rays in subsequent emission scans, and the PET detectors. In the combined TMS/PET studies, the coil is the single most attenuating object. Any movement of the coil or the head after the transmission scan will result in an incorrect application of the attenuation corrections when calculating the distribution of counts measured in emission scans (112). Thus,

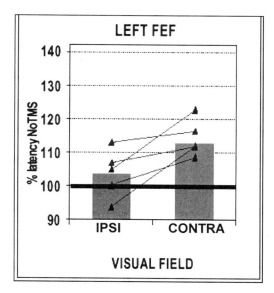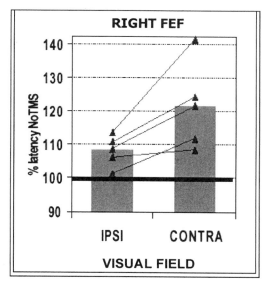

FIGURE 4.14. Single-pulse transcranial magnetic stimulation (TMS) of the probabilistic frontal eye field. The saccadic latencies in trials with TMS are expressed as a percentage of the median latency in trials without TMS. The left part of each graph displays the latencies for saccades directed toward the hemifield ipsilateral to the site of stimulation; the right part displays latencies for saccades directed towards the contralateral hemifield. *Lines* represent individual data, and *vertical bars* represent the average across five subjects. (From Grosbras MH, Paus T. Transcranial magnetic stimulation of the human frontal eye-field: effects on visual perception and attention. *J Cogn Neurosci* 2002, with permission.)

false-positive areas of significant differences in rCBF could emerge in cases where a movement occurred between the TMS-on and TMS-off scans. Lee et al. (113) used a phantom to evaluate the effect of coil movement that occurred between the acquisition of transmission and emission scans, respectively, and found that a 2-cm movement of the coil in the tangential plane caused up to 50% change in the number of counts recorded from beneath the coil. To minimize the possibility of such movement-related artifacts, we use a bite bar in all TMS/PET studies. The bite bar is placed in the subject's mouth during coil positioning, transmission scan, and all emission scans, but it is retracted during each 10-minute break between successive emission scans. Return of the bite bar to the same position is achieved with a special lock-in mechanism.

PET-based measurements of local and distal effects of TMS on blood flow allow assessment of cortical excitability and connectivity, respectively, and the changes brought about by various interventions. As an example, the design and some results of a TMS/PET study aimed at studying cortico-cortical connectivity of the mid dorsolateral frontal cortex (MDL-FC) and its modulation by brief periods of repetitive TMS are described (40). In this study, the target region was chosen based on a probabilistic location of the left MDL-FC (X = −40, Y = 32, Z = 30), as revealed by a previous PET study of verbal working memory (Fig. 4.16B). Using the inverse MRI_{native}–$MRI_{Talairach}$ transformation matrix, the X, Y, and Z coordinates of this location were calculated for each individual's "native" brain space and a figure-of-eight coil was positioned over this location using

frameless stereotaxy. During the study, the coil was held with a rigid arm mounted at the back of the scanner's gantry. The transmission scan was carried out and used not only for the attenuation corrections but also for verification of coil positioning (Fig. 4.16C) (114). Six 60-second water-bolus emission scans were acquired afterward: two baseline scans with no TMS applied and four TMS scans during which 30 pairs of pulses were administered with intensity at the individual's motor threshold (Fig. 4.16A). White noise (90 dB) was played over insert earphones during all scans to attenuate coil-generated clicks. The double-pulse TMS applied during the scans provided a measure of cortical excitability and connectivity of the left MDL-FC. In addition, to examine the putative modulatory effect of repetitive TMS on MDL-FC excitability/connectivity, two series of rTMS were applied between the first and last TMS scans. The following TMS parameters were used for each series: fifteen 1-second trains; ten pulses in each train (i.e., 10 Hz); 10-second between-train intervals; and intensity at motor threshold (Fig. 4.16A). The stimulation site was identical for the double-pulse and repetitive TMS. In response to the double-pulse TMS applied before rTMS, CBF decreased both at the stimulation site and in several distal regions presumably connected to the site, including the anterior cingulate cortex (Fig. 4.16D–E). Following the two series of rTMS, this "suppression" response was reverted, resulting in double-pulse–induced increases in CBF that were maximal during the last TMS scan (Fig. 4.16F). Using correlational analysis, a network of cortical regions was revealed in which blood-flow response to the double-pulse TMS covaried with

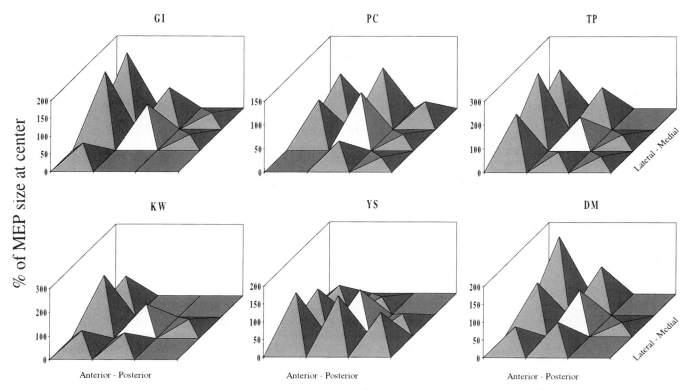

FIGURE 4.15. Motor evoked potentials (MEP) obtained at the probabilistic location of the left sensorimotor cortex and its vicinity. Each plot shows the amplitude of MEPs obtained in six subjects by stimulating at the probabilistic location of the left primary sensorimotor cortex (X = −31, Y = −22, Z = 52) (131) and at eight other sites on a 10-mm grid around it. Average amplitudes were calculated from MEPs obtained in response to five single pulses applied at each location and expressed as a percentage of the average obtained at the probabilistic location (center of graph, *stippled pyramid*). The stimulation was carried out using the Magstim figure-of-eight coil, the virtual "maximum" of which (see legend to Fig. 4.10) was positioned over each of the sites using frameless stereotaxy. Note that in most subjects the largest MEP amplitude was obtained anterior to the probabilistic location of the primary *sensorimotor* cortex, which was located on the central sulcus in all subjects.

that at the stimulation site, including the contralateral MDL-FC and the anterior cingulate cortex (Fig. 4.16G). Overall, this study demonstrated that only 30 trains (300 pulses) of 10-Hz rTMS are capable of inducing subtle changes in cortical excitability and connectivity of the stimulated region.

As explained in the section on Neurophysiologic Underpinnings of the Signal: Positron Emission Tomography and Functional Magnetic Resonance Imaging, the blood flow technique provides a robust but only indirect index of synaptic activity. Assessment of stimulation-induced release of neurotransmitters allows one to begin dissecting specific neurochemical pathways involved in mediating some TMS-induced effects. In a recent study, we used ^{11}C-raclopride to measure release of dopamine in the human striatum in response to rTMS of the left MDL-FC; the left occipital cortex was used as a control site (115). On two successive days, three series of fifteen 10-Hz trains of rTMS were applied with a circular coil 10 minutes apart at either of the two sites. The ^{11}C-raclopride was injected immediately after

the end of the last rTMS series, and the tracer uptake in the brain was measured over the next 60 minutes. Binding potential of ^{11}C-raclopride was calculated voxel-wise using a simplified reference tissue method (116,117) to generate statistical parametric images of change in binding potential (118). This analysis revealed a significant decrease in binding potential in the left caudate nucleus following rTMS of the left MDL-FC compared with rTMS of the occipital cortex (Fig. 4.17A–B). Such a reduction in ^{11}C-raclopride binding potential is indicative of an increase in extracellular dopamine concentration (119,120). It is likely that this TMS-induced focal release of dopamine in the ipsilateral caudate nucleus is mediated by excitatory cortico-striatal projections known to originate at high density in the primate prefrontal cortex and to synapse at the vicinity of nigrostriatal dopaminergic nerve terminals. Overall, this study confirms the feasibility of using PET to investigate TMS-induced changes in specific neurotransmitter systems of the human brain and opens up new avenues for studies of the pathophysiology of neurologic and psychiatric disorders.

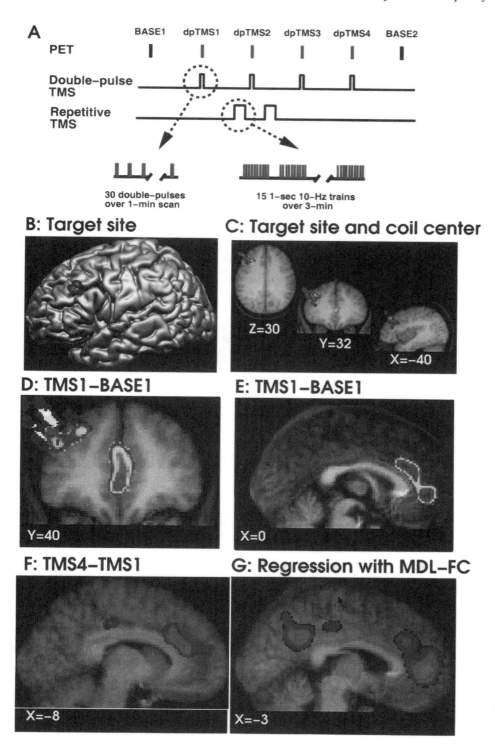

FIGURE 4.16. Modulation of cortico-cortical connectivity by repetitive transcranial magnetic stimulation (TMS). **A:** Flowchart indicates sequence of events during the TMS/positron emission tomographic (PET) study. The PET scans were repeated every 10 minutes. Base, no-TMS applied; dpTMS, double-pulse TMS. **B:** The target site within the left mid dorsolateral frontal cortex was selected from a previous blood flow activation study by Petrides et al. (132). The "peak" is located just above the left inferior frontal sulcus. The proximity of the target site (*cross-hair*) and the coil center derived from the transmission scans in the eight subjects (*color bars*) demonstrate the successful positioning of the coil with frameless stereotaxy (**C**). The results of subtraction (**D–F**) and regression (**G**) analyses of blood flow data. The images depict the exact locations that showed statistically significant decreases (**D,E**) and increases (**F**) in blood flow, and significant positive correlation with blood flow at the stimulation site (**G**). The thresholded maps of *t*-statistic values (t > 3.0 or t < −3.0) are superimposed on coronal (**D**) and sagittal (**E–G**) sections through the average magnetic resonance image of the eight subjects. All images are aligned within the standardized stereotaxic space. Reprinted from Paus et al. (104), with permission. (See Color Figure 4.16 following page 526.)

Transcranial Magnetic Stimulation During Electroencephalography

The high temporal resolution of EEG affords unique insights into the dynamics of TMS-induced changes in cortical excitability and connectivity. Cracco, Amassian, and colleagues were the first to combine TMS with EEG in their stud-ies of transcallosal and frontocerebellar responses (121,122). Ilmoniemi et al. (123) perfected this technique by combining TMS with a 60-channel EEG system. They observed clear transcallosal EEG responses to magnetic stimulation of the primary motor and visual cortex. To prevent saturation of EEG amplifiers by the magnetic pulse, these authors designed a sample-and-hold circuit that pins the amplifier

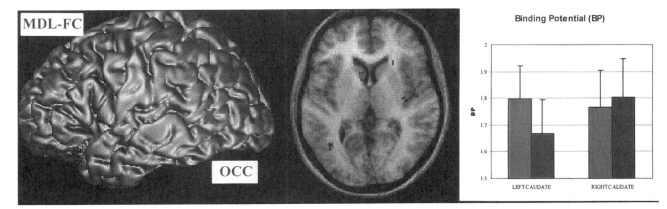

FIGURE 4.17. Dopamine release induced by repetitive transcranial magnetic stimulation (TMS) of the prefrontal cortex. **Left:** Location (*red markers*) of the two stimulation sites, the left mid-dorsolateral prefrontal cortex (MDL-FC), and the left occipital (OCC) cortex, on the magnetic resonance image of one subject in stereotaxic space. **Middle:** Transverse (Z = 6) section of the statistical parametric map of the change in ^{11}C-raclopride binding potential overlaid upon the average magnetic resonance image of all subjects in stereotaxic space. The *peak* in the left caudate nucleus shows the location where ^{11}C-raclopride binding potential changed significantly after repetitive TMS of the left mid dorsolateral prefrontal cortex. **Right:** Mean ± SEM values of the binding potential in the left and right caudate nucleus in response to repetitive TMS of the left MDL-FC (*purple*) and the left occipital cortex (*blue*). (From Strafella A, Paus T. Cerebral blood-flow changes induced by paired-pulse transcranial magnetic stimulation of the primary motor cortex. *J Neurophysiol* 2001;85:2624–2629, with permission.) (See Color Figure 4.17 following page 526.)

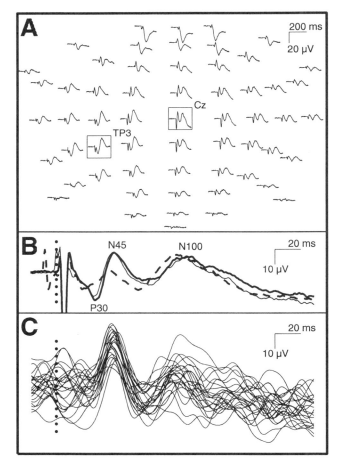

FIGURE 4.18. Electroencephalographic (EEG) potentials elicited by single- and paired-pulse transcranial magnetic stimulation (TMS) applied over the left motor cortex. **A:** Grand average (seven subjects, 100–120 trials per subject) of the EEG response to single-pulse TMS at all scalp locations. **B:** Grand average of the EEG response recorded at vertex (Cz) in single-pulse (*thick solid line*), 3-millisecond paired-pulse (*thin solid line*), and 12-millisecond paired-pulse (*dashed line*) trials. **C:** Twenty single traces of EEG randomly selected from 120 traces recorded during single-pulse TMS in one subject (P.S.). The *dotted line* indicates TMS onset. (From Paus T, Sipila PK, Strafella AP. Synchronization of neuronal activity in the human sensori-motor cortex by transcranial magnetic stimulation: a combined TMS/EEG study. *J Neurophysiol* 2001;86:1983–1990, with permission.)

output to a constant level during the pulse. The amplifiers recover within 3 milliseconds after the pulse; if there were no other sources of artifacts (discussed later), such a fast recovering amplification system would allow investigators to measure immediate TMS-induced changes in the EEG signal. Overheating of the electrodes placed close to the coil (124) is prevented by making the electrodes from low-conductivity material (purified silver with Ag/AgCl coating) (125) and by cutting a slit in the electrode to interrupt eddy currents. As pointed out in the section on Study Design: Transcranial Magnetic Stimulation, TMS often induces peripheral effects, including activation of scalp muscles. In our experience, local muscle artifacts, as well as widespread artifacts, due to the relative movement of the scalp against the electrode cap persist for up to 20 milliseconds after the pulse. This effectively prevents evaluation of evoked potentials elicited by TMS within this time window and, hence, measurement of the speed of neural transmission between and within the hemispheres. Until satisfactory technical solutions to this problem

are found, we have focused our TMS/EEG studies on the assessment of TMS-induced cortical oscillations.

As explained in the section on Data Acquisition and Analysis: Electroencephalography, ERS and ERD allow study of changes in spontaneous rhythmic activity of the brain elicited by different events. In the case of a TMS/EEG study, an application of a TMS pulse becomes the event of interest. Instead of "natural" activation of a given cortical region by sensory, motor, or cognitive events, the region is stimulated directly. In the first study of this type, we applied single suprathreshold TMS pulses to the left primary motor cortex and recorded EEG with the TMS-compatible 60-channel system (126). The waveform elicited by single-pulse TMS consisted of a positive peak (30 milliseconds) followed by two negative peaks (45 and 100 milliseconds) (Fig. 4.18A–B). The potential maps (Fig. 4.18C) revealed that the P30 component was distributed centrally, the N45 component formed a dipole centered over the stimulation site, and the N100 component had a wide distribution with

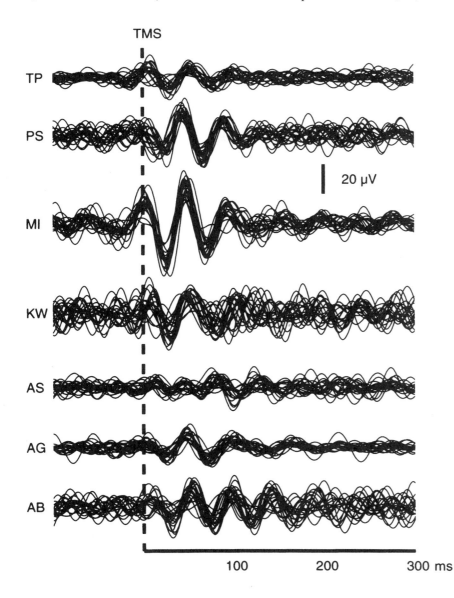

FIGURE 4.19. Cortical oscillations elicited by single-pulse transcranial magnetic stimulation (TMS) applied over the left motor cortex. Temporal evolution of electroencephalographic activity filtered in the β range (15–30 Hz) using single traces recorded during single-pulse TMS in the seven subjects. Twenty traces are superimposed in each subject. Note that the onset of the oscillation appears to precede the TMS pulse, but this time difference is a technical one, resulting from the filtering. (From Paus T, Sipila PK, Strafella AP. Synchronization of neuronal activity in the human sensori-motor cortex by transcranial magnetic stimulation: a combined TMS/EEG study. *J Neurophysiol* 2001;86:1983–1990, with permission.)

slight predominance over the left central region. Using the ERS analytic approach described in Figure 4.6, we observed a striking burst of EEG synchronization in the β frequency range (Fig. 4.19). This synchronization was time locked to the onset of the TMS pulse. Several control experiments were performed to rule out the possibility that the observed changes were related to different TMS-induced peripheral effects, including somatosensory stimulation of the scalp, peripheral feedback from the activated muscle, and auditory stimulation due to the coil-generated click. Altogether, our findings are consistent with the possibility that TMS applied to the primary motor cortex induces transient synchronization of spontaneous activity of cortical neurons within the frequency range from 15 to 30 Hz. As such, the results corroborate previous studies of cortical oscillations in the motor cortex and point to the potential of the combined TMS/EEG approach for further investigations of cortical rhythms in the human brain.

CONCLUSION

This review of the most commonly used brain mapping techniques makes it clear that availability of experimental tools is not a limiting factor in our quest for knowledge about functional organization of the human brain in health and disease. In addition to the proper design and analysis of brain mapping experiments, the main challenge lies in insightful integration of the previously known facts about the structural and functional organization of the primate cerebral cortex with the new data obtained in a given experiment. One should not ignore, for example, the wealth of knowledge about brain–behavior relationships obtained from studies of patients with circumscribed brain lesions. In the same vein, the detailed maps of anatomic connectivity obtained in track tracing studies in nonhuman primates provide valuable clues *vis-à-vis* interpretation of interregional interactions observed in functional imaging data. Finally, the reader is referred to the original sources for brain mapping methods not discussed in this chapter, such as near infrared optical imaging (127,128) and TMS during fMRI (129).

ACKNOWLEDGMENTS

I thank Drs. Bruce Pike, Antonio Strafella, Kate Watkins, and Gabriel Leonard for comments on the manuscript. The author's research is supported by the Canadian Institutes of Health Research, Canadian Foundation for Innovation, and National Science and Engineering Research Council of Canada. Cadwell Laboratories Inc. and Rogue Research Inc. donated equipment and services to the TMS laboratory.

REFERENCES

1. Mathiesen C, Caesar K, Akgoren N, et al. Modification of activity-dependent increases of cerebral blood flow by excitatory synaptic activity and spikes in rat cerebellar cortex. *J Physiol* 1998;512:555–566.
2. Logothetis NK, Pauls J, Augath M, et al. Neurophysiological investigation of the basis of the fMRI signal. *Nature* 2001412:150–157.
3. Moore CI, Nelson SB, Sur M. Dynamics of neuronal integration in rat somatosensory cortex. *Trends Neurosci* 1999;22:513–520.
4. Moore CI, Sheth B, Basu A, et al. What is the neural correlate of the optical imaging signal? Intracellular receptive field maps and optical imaging in rat barrel cortex. *Soc Neurosci Abst* 1996:22.
5. Akgoren N, Dalgaard P, Lauritzen M. Cerebral blood flow increases evoked by electrical stimulation of rat cerebellar cortex: relation to excitatory synaptic activity and nitric oxide synthesis. *Brain Res* 1996;710:204–214.
6. Paus T, Marrett S, Worsley KJ, et al. Extra-retinal modulation of cerebral blood-flow in the human visual cortex: implications for saccadic suppression. *J Neurophysiol* 1995;74:2179–2183.
7. Iadecola C. Regulation of the cerebral microcirculation during neural activity: is nitric oxide the missing link? *Trends Neurosci* 1993;16:206–214.
8. Northington FJ, Matherne GP, Berne RM. Competitive inhibition of nitric oxide synthase prevents the cortical hyperaemia associated with peripheral nerve stimulation. *Proc Natl Acad Sci USA* 1992; 89:6649–6652.
9. Heeger DJ, Huk AC, Geisler WS, et al. Spikes versus BOLD: what does neuroimaging tell us about neuronal activity? *Nat Neurosci* 2000;3:631–633.
10. Rees G, Friston K, Koch C. A direct quantitative relationship between the functional properties of human and macaque V5. *Nat Neurosci* 2000;3:716–723.
11. Hari R. Magnetoencephalography as a tool of clinical neurophysiology. In: Niedermeyer E, Lopes da Silva F, eds. *Electroencephalography: basic principles, clinical applications, and related fields,* 4th ed. Baltimore: Williams & Wilkins, 1999:1107–1134.
12. Speckmann EJ, Elger CE.Introduction to the neurophysiological basis of the EEG and DC potentials. In: Niedermeyer E, Lopes da Silva F, eds. *Electroencephalography: basic principles, clinical applications, and related fields,* 4th ed. Baltimore: Williams & Wilkins, 1999:15–27.
13. Nakamura H, Kitagawa H, Kawaguchi Y, Tsuji H. Direct and indirect activation of human corticospinal neurons by transcranial magnetic and electrical stimulation. *Neurosci Lett* 1996;210:45–48.
14. Rothwell JC. Techniques and mechanisms of action of transcranial stimulation of the human motor cortex. *J Neurosci Methods* 1997;74:113–122.
15. Chen R. Studies of human motor physiology with transcranial magnetic stimulation. *Muscle Nerve* 2000;9:S26–S32.
16. Inghilleri M, Berardelli A, Cruccu G, et al. Silent period evoked by transcranial stimulation of the human cortex and cervicomedullary junction. *J Physiol* 1993;466:521–534.
17. Chen R, Lozano AM, Ashby P. Mechanism of the silent period following transcranial magnetic stimulation. Evidence from epidural recordings. *Exp Brain Res* 1999;128:539–542.
18. Fuhr P, Agostino R, Hallett M. Spinal motor neuron excitability during the silent period after cortical stimulation. *Electroencephalogr Clin Neurophysiol* 1991;81:257–262.
19. Hallett M. Transcranial magnetic stimulation. Negative effects. *Adv Neurol* 1995;67:107–113.
20. Kujirai T, Caramia MD, Rothwell JC, et al. Corticocortical inhibition in human motor cortex. *J Physiol (Lond)* 1993;471:501–519.
21. Nakamura H, Kitagawa H, Kawaguchi Y, et al. Intracortical facilitation and inhibition after transcranial magnetic stimulation in conscious humans. *J Physiol (Lond)* 1997;498:871–823.

22. Ziemann U, Rothwell JC, Ridding MC. Interaction between intracortical inhibition and facilitation in human motor cortex. *J Physiol (Lond)* 1996;496:873–881.
23. Werhahn KJ, Kunesch E, Noachtar S, et al. Differential effects on motor cortical inhibition induced by blockade of GABA uptake in humans. *J Physiol* 1999;517:591–597.
24. Ziemann U, Lonnecker S, Steinhoff BJ, et al. Effects of antiepileptic drugs on motor cortex excitability in humans: a transcranial magnetic stimulation study. *Ann Neurol* 1996;40:367–378.
25. Avoli M, Hwa G, Louvel J, et al. Functional and pharmacological properties of GABA-mediated inhibition in the human neocortex. *Can J Physiol Pharmacol* 1997;75:526–534.
26. Pascual-Leone A, Walsh V, Rothwell J. Transcranial magnetic stimulation in cognitive neuroscience—virtual lesion, chronometry, and functional connectivity. *Curr Opin Neurobiol* 2000;10:232–237.
27. Amassian VE, Cracco RQ, Maccabee PJ, et al. Transcranial magnetic stimulation in study of the visual pathway. *J Clin Neurophysiol* 1998;15:288–304.
28. Ashbridge E, Walsh V, Cowey A. Temporal aspects of visual search studied by transcranial magnetic stimulation. *Neuropsychologia* 1997;35:1121–1131.
29. Grosbras MH, Paus T. Transcranial magnetic stimulation of the human frontal eye-field: effects on visual perception and attention. *J Cog Neurosci* 2002;14:1109–1120.
30. Topper R, Mottaghy FM, Brugmann M, et al. Facilitation of picture naming by focal transcranial magnetic stimulation of Wernicke's area. *Exp Brain Res* 1998;121:371–378.
31. George MS, Wassermann EM, Williams WA, et al. Changes in mood and hormone levels after rTMS of the prefrontal cortex. *J Neuropsychiatry Clin Neurosci* 1996;8:172.
32. Pascual-Leone A, Catala MD, Pascual-Leone Pascal A. Lateralized effect of rapid rate transcranial magnetic stimulation of the prefrontal cortex on mood. *Neurology* 1996;46:499.
33. Chen R, Classen J, Gerloff C, et al. Depression of motor cortex excitability by low-frequency transcranial magnetic stimulation. *Neurology* 1997;48:1398–1403.
34. Maeda F, Keenan JP, Tormos JM, et al. Modulation of corticospinal excitability by repetitive transcranial magnetic stimulation. *Clin Neurophysiol* 2000;111:800–805.
35. Maeda F, Keenan JP, Tormos JM, et al. Interindividual variability of the modulatory effects of repetitive transcranial magnetic stimulation on cortical excitability. *Exp Brain Res* 2000;133:425–430.
36. Peinemann A, Lehner C, Mentschel C, et al. Subthreshold 5-Hz repetitive transcranial magnetic stimulation of the human primary motor cortex reduces intracortical paired-pulse inhibition. *Neurosci Lett* 2000;296:21–24.
37. Wu T, Sommer M, Tergau F, et al. Lasting influence of repetitive transcranial magnetic stimulation on intracortical excitability in human subjects. *Neurosci Lett* 2000;287:37–40.
38. Gerschlager W, Siebner HR, Rothwell JC. Decreased corticospinal excitability after subthreshold 1 Hz rTMS over lateral premotor cortex. *Neurology* 2001;57:449–455.
39. Muellbacher W, Ziemann U, Boroojerdi B, et al. Effects of low-frequency transcranial magnetic stimulation on motor excitability and basic motor behavior. *Clin Neurophysiol* 2000;111:1002–1007.
40. Paus T, Castro-Alamancos M, Petrides M. Cortico-cortical connectivity of the human mid-dorsolateral frontal cortex and its modulation by repetitive transcranial magnetic stimulation. *Eur J Neurosci* 2001;14:1405–1411.
41. Cherry SR, Phelps ME. Imaging brain function with positron emission tomography. In: Toga AW, Mazziotta JC, eds. *Brain mapping: the methods.* San Diego, CA: Academic Press, 1996:191–222.
42. Woods RP, Grafton ST, Holmes CJ, et al. Automated image registration: I. General methods and intrasubject, intramodality validation. *J Comput Assist Tomogr* 1998;22:139–152.
43. Woods RP, Grafton ST, Watson JD, et al. Automated image registration: II. Intersubject validation of linear and nonlinear models. *J Comput Assist Tomogr* 1998;22:153–165.
44. Collins DL, Neelin P, Peters TM, et al. Automatic 3D intersubject registration of MR volumetric data in standardized Talairach space. *J Comput Assist Tomogr* 1994;18:192.
45. Talairach J, Tournoux P. *Co-planar stereotaxic atlas of the human brain.* New York: Thieme Medical Publishers, 1988.
46. Friston KJ, Frith CD, Fletcher P, et al. Functional topography: multidimensional scaling and functional connectivity in the brain. *Cereb Cortex* 1996;6,156–164.
47. Worsley KJ, Evans AC, Marrett S, et al. Determining the number of statistically significant areas of activation in subtracted activation studies from PET. *J Cereb Blood Flow Metab* 1992;12:900.
48. Pike GB, Hoge RD. Functional magnetic resonance imaging: technical aspects. In: Casey KL, Bushnell MC, eds. *Pain imaging. Progress in pain research and management.* Seattle, WA: IASP Press, 2000:157–194.
49. Biswal B, DeYoe AE, Hyde JS. Reduction of physiological fluctuations in fMRI using digital filters. *Magn Reson Med* 1996;35:107–113.
50. Glover GH. Deconvolution of impulse response in event-related BOLD fMRI. *Neuroimage* 1999;9:416–429.
51. Worsley KJ, Poline JB, Friston KJ, et al. Characterizing the response of PET and fMRI data using multivariate linear models. *Neuroimage* 1997;6:305–319.
52. Worsley KJ, Liao CH, Aston J, et al. A general statistical analysis for fMRI data. *Neuro Image* 2002;15:1–15.
53. Daly DD, Pedley TA, eds. *Current practice of clinical electroencephalography,* 2nd ed. New York: Raven Press, 1990.
54. Niedermeyer E, Lopes da Silva F, eds. *Electroencephalography: basic principles, clinical applications, and related fields,* 4th ed. Baltimore: Williams & Wilkins, 1999.
55. Rugg MD, Coles MG. *Electrophysiology of mind.* Oxford: Oxford University Press, 1995.
56. American EEG Society guidelines (1). *Electroencephalogr Clin Neurophysiol* 1994;11:2–5.
57. Jasper H. Report of committee on methods of clinical exam in EEG. *Electroencephalogr Clin Neurophysiol* 1958;10:370–375.
58. Kamp A, Lopes da Silva F. Technological basis of EEG recording. In: Niedermeyer E, Lopes da Silva F, eds. *Electroencephalography: basic principles, clinical applications, and related fields,* 4th ed. Baltimore: Williams & Wilkins, 1999:110–121.
59. Krauss GL, Webber WRS. Digital EEG. In: Niedermeyer E, Lopes da Silva F, eds. *Electroencephalography: basic principles, clinical applications, and related fields,* 4th ed. Baltimore: Williams & Wilkins, 1999:781–796.
60. Lagerlund TD. EEG source localization (model-dependent and model-independent methods). In: Niedermeyer E, Lopes da Silva F, eds. *Electroencephalography: basic principles, clinical applications, and related fields,* 4th ed. Baltimore: Williams & Wilkins, 1999:809–822.
61. Scherg M. Fundamentals of dipole source potential analysis. In: Grandori F, Hoke M, Romani GL, eds. *Auditory evoked magnetic fields and evoked potentials. Advances in audiology, vol. 6.* Basel: Karger, 1990:40–69.
62. Nunez PL. Estimation of large scale neocortical source activity with EEG surface Laplacians. *Brain Topogr* 1989;2:141–154.
63. Hamalainen MS, Ilmoniemi RJ. Interpreting magnetic fields of the brain: minimum norm estimates. *Med Biol Eng Comput* 1994;32:35–42.
64. Srebro R. Iterative refinement of the minimum norm solution of the bioelectric inverse problem. *IEEE Trans Biomed Eng* 1996;43:547–552.

65. Gevins A, Le J, Brickett P, et al. Seeing through the skull: advanced EEGs use MRIs to accurately measure cortical activity from the scalp. *Brain Topogr* 1991;4:125–131.

66. Pfurtscheller G. EEG event-related desynchronization (ERD) and event-related synchronization (ERS). In: Niedermeyer E, Lopes da Silva F, eds. *Electroencephalography: basic principles, clinical applications, and related fields,* 4th ed. Baltimore: Williams & Wilkins, 1999:958–967.

67. Basar E, Bullock TH, eds. *Brain dynamics: progress and perspectives.* Boston: Birkhauser 1992.

68. Roth BJ, Saypol JM, Hallett M, et al. A theoretical calculation of the electric field induced in the cortex during magnetic stimulation. *Electroencephalogr Clin Neurophysiol* 1991;81:47–56.

69. Singh KD, Hamdy S, Aziz Q, et al. Topographic mapping of trans-cranial magnetic stimulation data on surface rendered MR images of the brain. *Electroencephalogr Clin Neurophysiol* 1997;105:345–351.

70. Paus T. Imaging the brain before, during, and after transcranial magnetic stimulation. *Neuropsychologia* 1999;37:219–224.

71. Zinreich SJ, Tebo S, Long DM, et al. Frameless stereotactic integration of CT imaging data. *Radiology* 1993;188:735–742.

72. de La Fuente-Fernandez R, Ruth TJ, Sossi V, et al. Expectation and dopamine release: mechanism of the placebo effect in Parkinson's disease. *Science* 2001;293:1164–1166.

73. Koepp MJ, Gunn RN, Lawrence AD, et al. Evidence for striatal dopamine release during a video game. *Nature* 1998;393:266–268.

74. Belin P, Zatorre RJ, Hoge R, et al. Event-related fMRI of the auditory cortex. *Neuroimage* 1999;10:417–429.

75. Burock MA, Buckner RL, Woldorff MG, et al. Randomized event-related experimental designs allow for extremely rapid presentation rates using functional MRI. *Neuroreport* 1998;9:3735–3739.

76. Clark VP, Maisog JM, Haxby JV. fMRI study of face perception and memory using random stimulus sequences. *J Neurophysiol* 1998;79:3257–3265.

77. Mochizuki Y, Go T, Ohkubo H, et al. Developmental changes of brainstem auditory evoked potentials (BAEPs) in normal human subjects from infants to young adults. *Brain Dev* 1982;4:127–136.

78. Nelson CA, Collins PF. Neural and behavioral correlates of visual recognition memory in 4- and 8-month-old infants. *Brain Cogn* 1992;19:105–121.

79. Thatcher RW. Cyclic cortical reorganization during early childhood. *Brain Cogn* 1992;20:24–50.

80. Rothwell JC. Paired-pulse investigations of short-latency intracortical facilitation using TMS in humans. *Electroencephalogr Clin Neurophysiol Suppl* 1999;51:113–119.

81. Strafella A, Paus T, Barrett J, et al. Repetitive transcranial magnetic stimulation of the human prefrontal cortex induces dopamine release in the caudate nucleus. *J Neurosci* 2001;21:RC157.

82. Eyre JA, Miller S, Ramesh V. Constancy of central conduction delays during development in man: investigation of motor and somatosensory pathways. *J Physiol* 1991;434:441–452.

83. Muller K, Homberg V. Development of speed of repetitive movements in children is determined by structural changes in corticospinal efferents. *Neurosci Lett* 1992;144:57–60.

84. Nezu A, Kimura S, Uehara S, et al. Magnetic stimulation of motor cortex in children: maturity of corticospinal pathway and problem of clinical application. *Brain Dev* 1997;19:176–180.

85. Fadiga L, Buccino G, Craighero L, et al. Corticospinal excitability is specifically modulated by motor imagery: a magnetic stimulation study. *Neuropsychologia* 1999;37:147–158.

86. Hashimoto R, Rothwell JC. Dynamic changes in corticospinal excitability during motor imagery. *Exp Brain Res* 1999;125:75–81.

87. Kasai T, Kawai S, Kawanishi M, et al. Evidence for facilitation of motor evoked potentials (MEPs) induced by motor imagery. *Brain Res* 1997;744:147–150.

88. Fadiga L, Fogassi L, Pavesi G, et al. Motor facilitation during action observation: a magnetic stimulation study. *J Neurophysiol* 1995;73:2608–2611.

89. Fadiga L, Craighero L, Buccino G, et al. Speech listening specifically modulates the excitability of tongue muscles: a TMS study. *Eur J Neurosci* 2002;15:399–402.

90. Gangitano M, Mottaghy FM, Pascual-Leone A. Phase-specific modulation of cortical motor output during movement observation. *Neuroreport* 2001;12:1489–1492.

91. Watkins KE, Strafella AP, Paus T. Seeing and hearing speech excites the motor system involved in speech production. *Neuropsychologia* (in press).

92. Strafella A, Paus T. Modulation of cortical excitability during action observation: a transcranial magnetic stimulation study. *Neuroreport* 2000;11:2289–2292.

93. Boroojerdi B, Topper R, Foltys H, et al. Transcallosal inhibition and motor conduction studies in patients with schizophrenia using transcranial magnetic stimulation. *Br J Psychiatry* 1999;175:375–379.

94. Shajahan PM, Glabus MF, Gooding PA, et al. Reduced cortical excitability in depression. Impaired post-exercise motor facilitation with transcranial magnetic stimulation. *Br J Psychiatry* 1999;174:449–454.

95. Davey NJ, Puri BK, Lewis HS, et al. Effects of antipsychotic medication on electromyographic responses to transcranial magnetic stimulation of the motor cortex in schizophrenia. *J Neurol Neurosurg Psychiatry* 1997;63:468–473.

96. Jahanshahi M, Rothwell J. Transcranial magnetic stimulation studies of cognition: an emerging field. *Exp Brain Res* 2000;131:1–9.

97. Walsh V, Cowey A. Transcranial magnetic stimulation and cognitive neuroscience. *Nat Rev Neurosci* 2000;1:73–79.

98. Wassermann EM. Risk and safety of repetitive transcranial magnetic stimulation: report and suggested guidelines from the International Workshop on the Safety of Repetitive Transcranial Magnetic Stimulation, June 5–7, 1996. *Electroencephalogr Clin Neurophysiol* 1998;108:1–16.

99. Kosslyn SM, Pascual-Leone A, Felician O, et al. The role of area 17 in visual imagery: convergent evidence from PET and rTMS. *Science* 1999;284:167–170.

100. Fung TD, Chertkow H, Paus T, et al. Transcranial magnetic stimulation of the left inferior temporal cortex slows down picture naming. *J Int Neuropsychol Soc* 2002;8:206 (abst).

101. Paus T, Jech R, Thompson CJ, et al. Transcranial magnetic stimulation during positron emission tomography: a new method for studying connectivity of the human cerebral cortex. *J Neurosci* 1997;17:3178–3184.

102. Paus T. Location and function of the human frontal eye-field: a selective review. *Neuropsychologia* 1996;34:475–483.

103. Mesulam MM. Tracing neural connections of human brain with selective silver impregnation. Observations on geniculocalcarine, spinothalamic, and entorhinal pathways. *Arch Neurol* 1979;36:814–818.

104. Gevins A, Doyle JC, Cuttilo BA, et al. Electrical potentials in human brain during cognition: new method reveals dynamic patterns of correlation. *Science* 1981; 213:918–922.

105. Thatcher RW, Krause PJ, Hrybyk M Cortico-cortical associations and EEG coherence: a two-compartmental model. *Electroencephalogr Clin Neurophysiol* 1986;64:123–143.

106. Tucker DM, Roth DL, Bair TB. Functional connections

among cortical regions: topography of EEG coherence. *Electroencephalogr Clin Neurophysiol* 1986;63:242–250.

107. Friston KJ. Functional and effective connectivity in neuroimaging: a synthesis. *Hum Brain Mapping* 1994;2:56–78.
108. McIntosh AR, Gonzalez-Lima F. Structural equation modelling and its application to network analysis in functional brain imaging. *Hum Brain Mapping* 1994;2:2–22.
109. Paus T, Marrett S, Worsley K, et al. Imaging motor-to-sensory discharges in the human brain: an experimental tool for the assessment of functional connectivity. *Neuroimage* 1996;4:78–86.
110. Strother SC, Kanno I, Rottenberg DA. Principal component analysis, variance partitioning, and "functional connectivity". *J Cereb Blood Flow Metab* 1995;15:353–360.
111. Thompson CJ, Paus T, Clancy R. Magnetic shielding requirements for PET detectors during transcranial magnetic stimulation. *IEEE Trans Nucl Sci* 1998;45:1303–1307.
112. Turkington TG, Coleman RE, Schubert SF, et al. Evaluation of post-injection transmission measurement in PET. *IEEE Trans Nucl Sci* 1994;41:1538–154.
113. Lee JS, Kim KM, Lee DS, et al. Significant effect of subject movement during brain PET imaging with transcranial magnetic stimulation. *J Nucl Med* 2000;41:845.
114. Paus T, Wolforth M. Transcranial magnetic stimulation during PET: reaching and verifying the target site. *Hum Brain Mapping* 1998;6:399–402.
115. Strafella A, Paus T. Cerebral blood-flow changes induced by paired-pulse transcranial magnetic stimulation of the primary motor cortex. *J Neurophysiol* 2001;85:2624–2629.
116. Gunn RN, Lammertsma A, Hume SP, et al. Parametric imaging of ligand-receptor binding in PET using a simplified reference region model. *Neuroimage* 1997;6:279–287.
117. Lammertsma AA, Hume SP. Simplified reference tissue model for PET receptor studies. *Neuroimage* 1996;4:153–158.
118. Aston JA, Gunn RN, Worsley KJ, et al. A statistical method for the analysis of positron emission tomography neuroreceptor ligand data. *Neuroimage* 2000;12:245–256.
119. Endres CJ, Kolachana BS, Saunders RC, et al. Kinetic modelling of [^{11}C] raclopride: combined PET-microdialysis studies. *J Cereb Blood Flow Metab* 1997;9:932–942.
120. Laruelle M, Iyer RN, Al-Tikriti MS, et al. Microdialysis and SPECT measurements of amphetamine-induced dopamine release in non human primates. *Synapse* 1997;25:1–14.
121. Amassian VE, Cracco RQ, Maccabee PJ, et al. Cerebello-frontal cortical projections in humans studied with the magnetic coil. *Electroencephalogr Clin Neurophysiol* 1992;85:265–272.
122. Cracco RQ, Amassian VE, Maccabee PJ, et al. Comparison of human transcallosal responses evoked by magnetic coil and electrical stimulation. *Electroencephalogr Clin Neurophysiol* 1989;74:417–424.
123. Ilmoniemi RJ, Virtanen J, Ruohonen J, et al. Neuronal responses to magnetic stimulation reveal cortical reactivity and connectivity. *Neuroreport* 1997;8:3537–3540.
124. Roth BJ, Pascual-Leone A, Cohen LG, et al. The heating of metal electrodes during rapid-rate magnetic stimulation: a possible safety hazard. *Electroencephalogr Clin Neurophysiol* 1992;85:116–123.
125. Virtanen J, Rinne T, Ilmoniemi RJ, et al. MEG-compatible multichannel EEG electrode array. *Electroencephalogr Clin Neurophysiol* 1996;99:568–570.
126. Paus T, Sipila PK, Strafella AP. Synchronization of neuronal activity in the human sensori-motor cortex by transcranial magnetic stimulation: a combined TMS/EEG study. *J Neurophysiol* 2001;86:1983–1990.
127. Gratton G, Fabiani M. Shedding light on brain function: the event-related optical signal. *Trends Cogn Sci* 2001;5:357–363.
128. Kleinschmidt A, Obrig H, Requardt M, et al. Simulta-

neous recording of cerebral blood oxygenation changes during human brain activation by magnetic resonance imaging and near-infrared spectroscopy. *J Cereb Blood Flow Metab* 1996;16:817–826.
129. Bohning DE, Shastri A, Wassermann EM, et al. BOLD-f MRI response to single-pulse transcranial magnetic stimulation (TMS). *J Magn Reson Imaging* 2000;11:569–574.
130. Knowles RG, Palacios M, Palmer RMJ, et al. Formation of nitric oxide from L-arginine in the central nervous system: a transduction mechanism for stimulation of the soluble guanylate cyclase. *Proc Natl Acad Sci USA* 1989;86:5159–5162.
131. Paus T, Jech R, Thompson CJ, et al. Dose-dependent reduction of cerebral blood flow during rapid-rate transcranial magnetic stimulation of the human sensorimotor cortex. *J Neurophysiol* 1998;79:1102–1107.
132. Petrides M, Alivisatos B, Meyer E, et al. Functional activation of the human frontal cortex during the performance of verbal working memory tasks. *Proc Natl Acad Sci USA* 1993;90:878.

SUGGESTED READING

Ashbridge E, Walsh V, Cowey A. Temporal aspects of visual search studied by transcranial magnetic stimulation. *Neuropsychologia* 1997;35:1121–1131.

Basar E, Bullock TH. *Brain dynamics: progress and perspectives.* Birkhauser Boston, 1992.

Bohning DE, Shastri A, Wassermann EM, et al. BOLD-f MRI response to single-pulse transcranial magnetic stimulation (TMS). *J Magn Reson Imaging* 2000;11:569–574.

Chen R, Classen J, Gerloff C, et al. Depression of motor cortex excitability by low-frequency transcranial magnetic stimulation. *Neurology* 1997;48:1398–1403.

Clark VP, Maisog JM, Haxby JV. fMRI study of face perception and memory using random stimulus sequences. *J Neurophysiol* 1998;79:3257–3265.

de Haan M, Nelson CA. Brain activity differentiates face and object processing in 6-month-old infants. *Dev Psychol* 1999;35:1113–1121.

Fadiga L, Fogassi L, Pavesi G, Rizzolatti G. Motor facilitation during action observation: a magnetic stimulation study. *J Neurophysiol* 1995;73:2608–2611.

George MS, Wassermann EM, Williams WA, et al. Changes in mood and hormone levels after rTMS of the prefrontal cortex. *Journal of Neuropsychiatry and Clinical Neuroscience* 1996;8:172.

Gratton G, Fabiani M. Shedding light on brain function: the event-related optical signal. *Trends Cogn Sci* 2001;5:357–363.

Grosbras MH, Paus T. Transcranial magnetic stimulation of the frontal eye-field: effects on visual perception and attention. *Journal of Cognitive Neuroscience* 2002;14:1109–1120.

Hashimoto R, Rothwell JC. Dynamic changes in corticospinal excitability during motor imagery. *Exp Brain Res* 1999;125:75–81.

Ilmoniemi RJ, Virtanen J, Ruohonen J, et al. Neuronal responses to magnetic stimulation reveal cortical reactivity and connectivity. *Neuroreport* 1997;8:3537–3540.

Kleinschmidt A, Obrig H, Requardt M, et al. Simultaneous recording of cerebral blood oxygenation changes during human brain activation by magnetic resonance imaging and near-infrared spectroscopy. *J Cereb Blood Flow Metab* 1996;16:817–826.

Koepp MJ, Gunn RN, Lawrence AD, et al. Evidence for striatal dopamine release during a video game. *Nature* 1998;393:266–268.

Kosslyn SM, Pascual-Leone A, Felician O, et al. The role of area 17 in visual imagery: convergent evidence from PET and rTMS. *Science* 1999;284:167–170.

Laruelle M, Iyer RN, A1-Tikriti MS, et al. Microdialysis and SPECT measurements of amphetamine-induced dopamine release in non human primates. *Synapse* 1997;25:1–14.

Logothetis NK, Pauls J, Augath M, et al. Neurophysiological investigation of the basis of the fMRI signal. *Nature* 2001;412:150–157.

Mathiesen C, Caesar K, Akgoren N, Lauritzen M. Modification of activity-dependent increases of cerebral blood flow by excitatory synaptic activity and spikes in rat cerebellar cortex. *J Physiol* 1998;512:555–566.

Muellbacher W, Ziemann U, Boroojerdi B, Hallett M. Effects of low-frequency transcranial magnetic stimulation on motor excitability and basic motor behavior. *Clin Neurophysiol* 2000;111:1002–1007.

Muller K, Homberg V. Development of speed of repetitive movements in children is determined by structural changes in corticospinal efferents. *Neurosci Lett* 1992;144:57–60.

Niedermeyer E, Lopes da Silva F. *Electroencephalography: basic principles, clinical applications, and related fields,* 4th ed. Baltimore: Williams & Wilkins, 1999.

Pascual-Leone A, Catala MD. Lateralized effect of rapid rate transcranial magnetic stimulation of the prefrontal cortex on mood. *Neurology* 1996;46:499.

Paus T, Castro-Alamancos M, Petrides M. Cortico-cortical connectivity of the human mid-dorsolateral frontal cortex and its modulation by repetitive transcranial magnetic stimulation. *European Journal of Neuroscience* 2001;14:1405–1411.

Paus T, Jech R, Thompson CJ, et al. Transcranial magnetic stimulation during positron emission tomography: a new method for studying connectivity of the human cerebral cortex. *Journal of Neuroscience* 1997;17:3178–3184.

Paus T, Sipila PK, Strafella AP. Synchronization of neuronal activity in the human sensori-motor cortex by transcranial magnetic stimulation: a combined TMS/EEG study. *Journal of Neurophysiology* 2001;86:1983–1990.

Rothwell JC. Techniques and mechanisms of action of transcranial stimulation of the human motor cortex. *J Neurosci Methods* 1997;74:113–122.

Rugg MD, Coles MG. *Electrophysiology of mind.* Oxford: Oxford University Press, 1995.

Strafella A, Paus T, Barrett J, Dagher A. Repetitive transcranial magnetic stimulation of the human prefrontal cortex induces dopamine release in the caudate nucleus. *Journal of Neuroscience* 2001;21:RC157(1–4).

Thatcher RW, Krause PJ, Hrybyk M. Cortico-cortical associations and EEG coherence: a two-compartmental model. *EEG Clin Neurophysiol* 1986;64;123–143.

Toga AW, Mazziotta JC. *Brain mapping: the methods,* 2nd ed. San Diego: Academic Press, 2002.

Wagner AD, Schacter DL, Rotte M, et al. Building memories: remembering and forgetting of verbal experiences as predicted by brain activity. *Science* 1998;281:1188–1191.

Walsh V, Cowey A. Transcranial magnetic stimulation and cognitive neuroscience. *Nat Rev Neurosci* 2000;1:73–79.

Wassermann EM. Risk and safety of repetitive transcranial magnetic stimulation: report and suggested guidelines from the international workshop on the safety of repetitive transcranial magnetic stimulation, June 5–7, 1996. *Electroencephalogr Clin Neurophysiol* 1998;108:1–16.

Ziemann U, Rothwell JC, Ridding MC. Interaction between intracortical inhibition and facilitation in human motor cortex. *J Physiol (Lond)* 1996;496:873–881.

5

NEUROPHYSIOLOGIC STUDY OF NEUROPSYCHIATRIC DISORDERS

ALYA REEVE

Study of brain and behavior using objective tools furthers our understanding of the mechanisms underlying complex activities. The tools that operate within the time frame of neuronal events have the greatest usefulness for this purpose because the brain is the source and repository for sensory and intellectual experience, attribution of meaning, and integration of memory.

The common technique at our disposal involves recording electrical events. This is done by electroencephalography (EEG) and magnetoencephalography (MEG). The discussion that follows reviews some basic aspects of these technologies and their application to neuropsychiatry. EEG and MEG are particularly meaningful techniques because they collect information in the millisecond time range that can be linked to brain function (or dysfunction).

Since the first edition of this textbook, 5 to 10 years of changes in technology and mathematics have resulted in progress in data collection, disease specificity, and generating new implications for the coming decade. Historically, EEG findings are notorious for their frequent nonspecificity. Part of the reason for an apparent lack of specificity in findings lies in the organization of the brain (with replication of function, multiple representations, plasticity of neurons and/or function). Disappointment over initial hopes of diagnostic precision from EEG may have led to an underestimation of the information provided by the technique. Neuropsychiatric applications of electrophysiology are receiving greater attention than ever before. Unlike many brain imaging tools, EEG and MEG are noninvasive. EEG has the advantage of becoming a smaller, more portable technology and increasingly less expensive.

NEUROPHYSIOLOGIC MEASUREMENTS

Electroencephalography

EEG is a standardized, seemingly straightforward method of recording the electrical activity of the brain at the scalp. The field was founded by Hans Berger in the 1920s. As discussed more fully in later sections of this chapter, the clinical EEG is both a useful and, at times, misleading source of information about the brain. The following sections are a brief description of the basic technology. A comprehensive discussion can be found in several sources (1,2,3).

Definition

The EEG measures a summation of electrical activity, as recorded as a difference in electrical potential between two active recording electrodes. It is not a direct measure of brain electrical activity. Rather, brain activity is measured indirectly as a difference between electrical potentials received at two recording electrodes. The choice of placement of electrodes and the properties of material between the source (the brain) and the electrodes affect the type, intensity, and specificity of the activity measured.

Factors that Influence the Electroencephalographic Record

Sources that can alter the electrical signal include the skull, scalp, sweat, metal, and poor physical contact of the electrode with the scalp. Electrode paste has properties that increase conductance of electrical signal through it. Because of this, it is important that the area of paste not be significantly larger than the recording electrode, otherwise there will be some loss of regional specificity around that electrode. These different factors all are contributors to resistance that must be minimized at the recording electrode. Of course, sources of resistance within the scalp (such as thickness of the skull; scar tissue in the meninges; and blood, fresh or coagulated) are factors that usually cannot be controlled. Every effort has to be made in the laboratory setting to minimize the confounding factors that *can* be controlled, by assuring clean electrodes, little to minimal oil on the scalp surface (by cleaning with an abrasive agent), a discrete area of conductive gel, and minimal to no movement of the electrode from its contact with the scalp.

The electrical signal of the neurons of the brain is relatively weak; therefore, the signal coming from the recording electrode must be amplified. The electromechanical properties of amplifiers vary greatly. The configuration of the amplifier may be designed to maximize the type of recording one is trying to obtain. Amplifiers are used to increase the voltage difference between the input to that of the output. They are characterized in terms of sensitivity and gain.

Sensitivity

Sensitivity is reflected as the ratio of the input voltage to the output pen deflections in the standard EEG recording. A typical ratio is 7 μV/mm. A higher numerical value for sensitivity reflects less amplification, that is, a sensitivity of 10 indicates 10 μV/mm. EEG machines have an array of amplifiers set up to allow individual sensitivity adjustment for each channel or general adjustment for all channels.

Gain

Gain is the ratio of the voltage obtained at the output of the amplifier to the voltage applied at the input. Often described in terms of volts (V) and microvolts (μV), gain sometimes is expressed in terms of decibels (dB) or 20 times the logarithm in base 10 of the gain. For example, a gain of 10 equals 20 dB; a gain of 1,000,000 equals 120 dB. In clinical EEG, sensitivity usually is used to describe the amplification because the gain is not directly measured from the output of the amplifier.

Filters

Electrical signals coming through a recording channel have a characteristic frequency. The bandwidth of the signal is defined by the frequency range within which the signal is contained. Some information will be lost if the frequency range of the recording channels is narrower than the frequency range of the EEG signal. Likewise, extraneous information will be included if the frequency range of the recording channels is wider than the bandwidth of the EEG signal. Different types of filters are used to maximize the EEG signal and minimize noise and signals of noncerebral origin. In practice, EEG recording channels can have high-pass, low-pass, or band-pass filters. High-pass filters are designed to let higher frequencies pass through and block lower frequencies, such as those below 1 Hz. Low-pass filters are designed to let low-frequency activity through and to filter high-frequency activity. Band-pass filters are designed around a range of frequencies to filter out both the high and low ends of the frequency spectrum. Most filters are designed to have the signal drop off gradually, to minimize the artifacts that can be seen with an abrupt attenuation of signal outside the chosen range [for further introductory discussion see Speckmann et al. (4) and Epstein (5)].

Montage and Reference Electrodes

Recordings over the head usually are done following the International 10-20 System. This localizes the electrodes proportionately (10% and 20% spacing) between certain landmarks—inion, nasion, and preauricular points (Fig. 5.1). By convention, the electrodes over the left side of the head are numbered odd and those on the right side of the head are even. Z stands for zero or midline. The letters of the electrode placement reflect relative position over the head: in other words, Fp (frontopolar), F (frontal), C (central), P (parietal), T (temporal), O (occipital), A (auricular), and M (mastoid).

The number of electrodes that are applied varies depending on the montage (series of patterns in which the electrodes are connected to each other), the population being studied, the type of activity considered important to detect, and the number of channels/amplifiers available. Standard clinical laboratories may have from eight- to 32-channel EEG machines with which to make recordings. Research laboratories are routinely using four to 164 channels of electrodes, depending upon the question they are addressing!

The montages used in EEG are of two general types. The first is referential; the second is bipolar. A common form of referential montage is to have a series of electrodes attached to the same common reference, often the earlobe or mastoid. The printout then can be alternated between left and right, which aids in the detection of hemispheric asymmetry. To maximize the ability to detect subtle differences, the left temporal electrodes can be referenced to the right ear or mastoid and vice versa for the right side. The earlobe tends to be easily "contaminated" by electrical activity of the heart (extremely rhythmic and of higher amplitude than brain signals). One technique to avoid this interference involves putting one electrode over the right sternoclavicular junction and another on the spine of the first thoracic vertebrae. These two electrodes are connected together through a 20,000-Ω variable resistor that can change its resistance such that the electrocardiogram is canceled or essentially canceled (6). A different type of reference is a common average reference created by a computerized analysis system after the recording is completed. These various attempts to make a "neutral reference" must always be carefully checked for artifacts. The details by which the reference was chosen and recorded must be carefully documented and remain with the record.

Bipolar montages are particularly useful to compare the polarity of signals and to detect the so-called *phase reversal* that occurs around an area of high amplitude signal change. To be useful, bipolar montages should cover relatively large

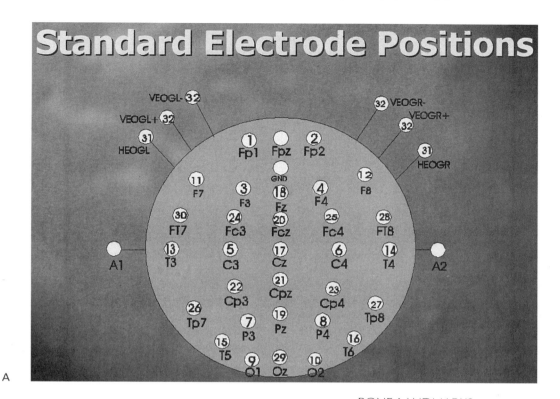

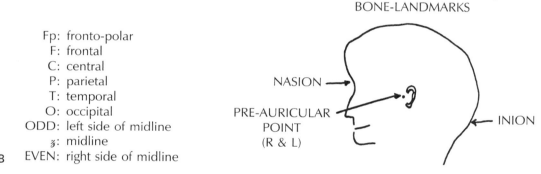

Fp: fronto-polar
F: frontal
C: central
P: parietal
T: temporal
O: occipital
ODD: left side of midline
z: midline
EVEN: right side of midline

FIGURE 5.1. International 10-20 System of electrode placement. This system is a proportionate placement of electrodes (at 10% and 20% increments) between anterior-posterior/right-left landmarks.

areas of the head so that interpretation of the direction (or polarity) can be made in a reasonable manner. There is a general rule that higher voltages are seen as the interelectrode distance increases. Switching from bipolar montage to referential montage often will accomplish this. The voltage may increase so much after the switch that the sensitivity of the record needs to be decreased.

The American EEG Society recommends that the full 21 electrodes of the 10-20 system be used and that a record be obtained in bipolar and referential montages. They recommend that electrode connections for each channel be indicated at the beginning of each montage and that over the course of the record, bipolar recordings be made in both longitudinal and transverse directions (i.e., front to back

and left to right). The number of montages recorded will depend upon the number of channels available to make simultaneous recordings.

The clinical record (Table 5.1) on the actual paper tracing, or in the active computer recording, should document that the mechanical instrumentation is intact. In other words, the calibration signal should show up in proper alignment across all channels, the voltage of that signal should be consistent across all channels, and the tracing should show little or no variation in deflections when not connected to a biologic substrate. This is necessary so that changes in waveforms can be attributed to the brain and the conditions under which the recordings were made rather than the electrical/mechanical properties of the instrumentation! The speed of recording

TABLE 5.1. CLINICAL RECORD: VARIABLES IN ELEC-TROENCEPHALOGRAPHIC DATA

Artifacts
- Muscle tension and movement
- Mechanical
- Electrical
- Drug effects

Reactivity
- Awake/sleep
- Drug effects
- Level of consciousness

Interpretation
- State of subject
- Human versus computer
- Clinical factors
- Subject age

should be noted, along with the specific montage used and the activity the patient/subject is undertaking at the time of the recording. The duration of record in each montage can be deduced by knowing the speed at which the record was made. There should be several montages, and there should be a period of certain kinds of activation such as hyperventilation (for 3 minutes at least) and photostimulation. Other types of stimulation can be done depending on the condition. For example, a patient who experiences seizurelike symptoms during reading should be asked to read; conversely, a person who exhibits specific behaviors upon hearing music, or another activity, should be exposed, whenever possible, to the provoking stimulus. The technician should change the montages to maximize views of any abnormalities seen in the EEG. Standard frequency waveforms are listed in Table 5.2. Disorders and brain states that elicit a particular frequency are not exclusive but rather characteristic for a frequency type. (The discussion later in this chapter highlights some relevant findings.)

Alternate Electrode Placements

Electrode placement other than that of the 10-20 system can make the information provided by the EEG record much more useful for specific studies. The closer the spacing of electrodes, the more specific the information can be obtained regarding a region of interest. However, the signal must be large enough for it to be resolved or focal enough so that the electrodes can differentiate the activity over the region. In some laboratories, 128- and 164-channel recordings are not uncommon (7, 8). The resolution that can be obtained by very close spacing of electrodes and careful analysis of the record may begin to provide the kind of specificity of information that clinicians working with neuropsychiatric disorders would like to have. There may be some loss of information as more electrodes are placed over a specific region because the amplitude of the signal is lower. If the signal-to-noise ratio is too low the desired improvement in spatial resolution cannot be attained.

Telemetry

The type of disorder one is trying to study will influence the recording strategy. Intermittent infrequent activity is the most difficult to catch. At times, correlating behaviors of the patients with changes in EEG activity is essential. Simultaneous video recording with EEG recording is one way to differentiate clinical and electrical events that are occurring in relation to one another from events that are occurring completely independent from each other. The camera can capture whole body movements or be focused to zoom in on particular areas. Portable systems allow the equipment to be set up in any hospital room and can be used with scalp electrodes or implanted electrodes, and with cabled or radio transmitted connections to amplifiers. This technique often is used to demonstrate nonphysiologic seizure activity. Nonepileptic seizures may involve so much muscle artifact in the EEG that an occasional true seizure may be obscured.

Other forms of telemetry are useful for long-term monitoring of EEG activity during normal daily routines for a patient. Here the video aspect is not so important. The purpose of these recordings is to obtain a record of infrequently occurring activity (such as sharp waves or spikes). Patients are asked to keep a diary of their activity, especially noting any sensory changes that occur and the time of day and duration of the symptoms. Patients who are candidates for surgical removal of an epileptic focus need to be carefully assessed for the stability of the focus and the degree of the disability it is producing.

EEG that has been collected over hours to days can be analyzed by computer to yield a summary of frequency and amplitude patterns over small chunks of time (e.g., 2, 4, or 10 seconds). If there is an area of interest, one can return to the original (expanded) data set to read the activity more carefully. During certain procedures, such as carotid endarterectomy, on-line analysis of EEG in 2-second bits can provide immediate and clearly visible change in brain activity to the surgical team. Immediate feedback about decrements in brain activity can guide the vigor of the procedure and help to improve outcomes for the patient.

Sleep Studies

Sleep monitoring is of two types. The first, obtained frequently during routine EEG, is brief, often drug induced, and may not require any special adaptations of technique. The second is to monitor through all phases and types of sleep, usually over a period longer than 8 hours. Again, maintaining adequate electrode-to-head contact is essential. This may be done by an electrode cap, needle electrodes, or implanted (cortical or depth) electrodes, and some form of cable-to-transmitter box arrangement, rather than direct connection to amplifiers. A lot of movement occurs during sleep, and one does not want the patient to become wrapped up in wires or to disconnect the electrodes. Evaluation for

TABLE 5.2. ELECTROENCEPHALOGRAPHIC FREQUENCY BAND AND ITS CHARACTERISTICS

Name	Frequency (Hz)	Location	Factors
β	13–25	Frontal	Increased by benzodiazepines and barbiturates
α	8–12	Occipital	Awake; resting state with eyes closed (meditative); does not slow appreciably with increased age
θ	5–8	Central	Awake, task oriented; may be generalized in drowsiness
δ	1–4	Frontal and central	Generalized in deep sleep; seen in encephalopathy, coma, and toxic states

EEG FREQUENCY EXAMPLES

A.

B.

C.

D.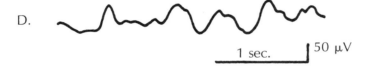

1 sec.   50 µV

A. Beta: 20-25 Hz.;   B. Alpha: 9-10 Hz.;
C. Theta: 5-6 Hz.;   D. Delta: 2-2.5 Hz.

sleep apnea and other respiratory compromise requires monitoring oxygenation, heart rate, and peripheral muscle tone, in addition to brain activity. Disorders of excessive daytime drowsiness are important to evaluate because of the associated psychological malfunction they produce. For example, average sleep latency as measured during the Multiple Sleep Latency Test of less than 5 minutes is a sign of pathologic drowsiness and correlates with poor work performance because of the intrusion of sleep episodes into the wakeful state (9).

Standards for Interpretation

Over the decades, the paper presentation of the EEG record has become more standardized. Usually, this tracing is read by the naked eye of a certified specialist (encephalographer). This technique emphasizes recognition of familiar patterns and deviation from those patterns. It is difficult for an encephalographer to quantitate the distribution of power in adjacent frequency bands, to differentiate minor interhemispheric and intrahemispheric differences, and to subtract eye movement artifacts from frontal lead potentials. Notwithstanding, the paper record, interpreted by an experienced EEG reader, permits expert comparison of the patient's EEG with the hundreds of others seen previously, enabling rejection of subtle artifacts recognizable to the trained eye.

The advent of computerized topographic mapping of the EEG has made analysis of the record much easier in many ways. It has tempted people to rely on computer-based algorithms for determination of abnormal and normal activity. Algorithms for distinguishing eye movement potentials from cerebral activity and eliminating it from the record by subtraction are one example of a useful automated feature. Group comparisons of specific features or general results must always be put in the proper statistical context—the representation of the disorder against the normative background (by sampling appropriate numbers and distribution of subjects). A major advantage of computerized quantitative EEG (QEEG) is that the display can be rearranged easily, after the recording session is over. Distributions of frequency, lateralization patterns, coherence between electrical signals, intermittent activity, and reactivity to stimuli—all are examples of analysis of the EEG record. This *post hoc* data analysis remains dependent on the data being collected initially under reliable conditions of instrumentation and accurate recording of the patient's state and activity.

Understanding the Sources of the Electroencephalogram

The electrical potential at any specific point in the brain is the sum of potentials generated by underlying cellular activity. Biochemical activity at the cellular level generates ionic currents that create the electrical potentials and currents. The electrical and magnetic fields associated with these potentials and currents follow well-known physical laws. For EEG, the field potential of a group or population of neurons reflects the sum of the contributions of a collection of individual neurons. Asynchronous and/or irregular firing of neurons cannot be detected with accuracy by EEG electrodes distant from the source; therefore, the scalp EEG reflects activity from regularly arranged neuronal sources, firing in more or less synchronous patterns. The contribution from radially and tangentially oriented currents is equally appreciated at the EEG electrode. The pyramidal neurons of layers IV and V of the cortex often are activated synchronously, producing brain activity measured at the scalp.

An electrical event recorded at the cortex will seem to be of shorter duration than the same event recorded at the scalp. This is because a moving dipole observed from a greater distance can be recorded for a longer period of time than when seen from nearby. Another factor that influences recordings obtained at the scalp is the conduction of electrical activity by the different layers between the neuronal population and the scalp. One of the difficulties with solving for the source of electrical activity, as it is recorded in the EEG, is to correctly account for the (conductive) inhomogeneities of the different layers. Mathematical models exist that account for concentric spheres of different conductive properties. In addition, research studies are comparing the mathematical solutions to actual recordings of skull, scalp, meninges, and fluid in the same subject (usually animal). A difficulty inherent in finding the sole solution to the question of which group of cells or what sequence of activities originated the electrical recording obtained in an EEG is that there may be no unique solution to this problem (named the *inverse problem*). Many different combinations of activity can result in a similar or even the same EEG pattern. One way to reduce the number of potential sources is to solve the problem for

an equivalent current dipole (ECD) of activity in the brain. Thus, when we speak of solving for dipole localization, we are actually talking about an abstract representation of brain activity occurring within a defined region.

There are a number of programs written that deal with this problem of dipole localization. By providing information about multiple sources of activity in several different montages, the EEG may permit deductions about the underlying source of activity. Large contiguous areas of activity as opposed to many discrete areas that are near to each other are difficult to distinguish from each other unless a great number of electrodes are placed over the scalp (providing sufficient discrimination). Another factor that is accounted for only with difficulty in dipole localization algorithms is that of activity having different time constants. In other words, closely superimposed sources with different rates of onset and offset of electrical signal can be confusing to distinguish from each other. They tend to blur into one source. Different algorithms for dipole localization use different approaches. Some make assumptions regarding what activity is "background" at all electrodes and what other activity represents "true" activity of interest. Others make assumptions about the timing or orientation of the EEG signal of interest.

Magnetoencephalography

Definition

The electrical activity of the brain, as all electrical activity, generates magnetic fields. Compared with the earth's magnetic field, the brain's magnetic field is roughly one billion times weaker. Unlike the EEG, which is affected by the attenuation of electrical potential by the boundaries between brain and scalp, as well as the inherent properties within those boundaries, the MEG records a magnetic field of brain and surrounding tissues in a region of constant magnetic permeability. Specific dipole activity within the brain produces measurable perturbations in the magnetic field. The magnetic field is affected by the strength, direction, and location of the electrical source. To begin to deduce the equivalent source for the recorded magnetic field, some properties of biomagnetism must be kept in mind.

Biomagnetic Properties

Every electrical current generates an associated magnetic field, following what is commonly called the *right-hand rule*. If the thumb of a person's right hand indicates the direction of electrical current flow, the curl of the fingers indicates the direction of flux of the magnetic field around that current (Fig. 5.2). All electrical currents in the brain can be modeled as the sum of their tangential and radial components. The interaction of cerebral magnetic fields within certain geometric configurations causes the radial components of the bioelectrical sources to be virtually invisible in the MEG.

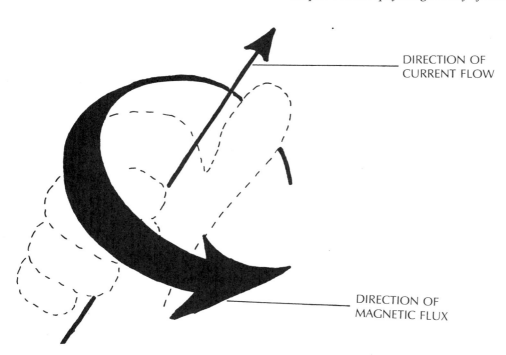

DIRECTION OF
CURRENT FLOW

DIRECTION OF
MAGNETIC FLUX

FIGURE 5.2. The "right-hand rule" of magnetic flux around electrical current.

Thus, the fields recorded in MEG are those from the tangential components of electrical signals.

Instrumentation

It is possible to detect the weak magnetic field of brain activity because of the advent of superconducting quantum interference devices (SQUID). As the name implies, the SQUID must be bathed in a supercooled environment, usually helium. The SQUID is a component of a gradiometer, a device intended to differentiate nearby magnetic fields from those that are further away and therefore more invariant. Gradiometers can be arranged with different numbers of coils and with different shapes of coils. Each configuration enhances the recording of a specific phenomenon: weak magnetic fields, activities that are located close together, or activities that occur in a plane versus a curved surface. New configurations are continually being made. Physical limitations on how small the components can be made and the biophysical properties of the instrumentation also place constraints on the spacing of SQUIDs from each other and their geometric array.

The earliest human MEG was collected with single-channel instruments. Full-head systems are now in use, as well as many other models. Historically, field maps were generated by putting together information collected sequentially. When pieces of recorded information that are not recorded simultaneously are put together for a composite map as if representing simultaneous activity, assumptions are suggested about the temporal relationship of all the data. The time that has passed between successive recordings is reflected in changes in the state of the brain. To truly compare field strength arising simultaneously at different channels, composite maps should not be made. Under the repetitive conditions of evoked responses, it is commonly assumed that the stimulus strength overrides underlying brain state changes. One can never remove the variable influence that progression of time possibly had on the recorded signal and on the state of the individual during the recording session.

Artifact and Noise Reduction

In addition to the magnetometer design, placing the instrument inside a specially designed room assists in reducing background "noise." Two sheets of μ-metal arranged approximately 8 inches apart provide a major source of deflection of nearby strong magnetic fields by dispersing them within the metal sheets. A shielded room should be degaussed about every 2 to 3 years. A shielded room, usually located within a laboratory environment, is the chamber in which the recordings are made. The patient/subject is supported in a bed or chair so that muscles can be relaxed and not obscure brain activity. It is customary that there be some form of communication (e.g., video monitoring and radio link) between the room and the operators outside the shielded room. This is both to reassure the subject and to provide a means of communication.

There are many sources of magnetic artifact. The most common include metal objects on the subject. Undergarments with metal portions in them, shoes with metal shanks, belt buckles, watches, and anything else with metal content must be removed. Dental fillings cannot be removed and may be preclude a valid recording. It often is helpful to degauss the subject. Extraneous sources of electromagnetic

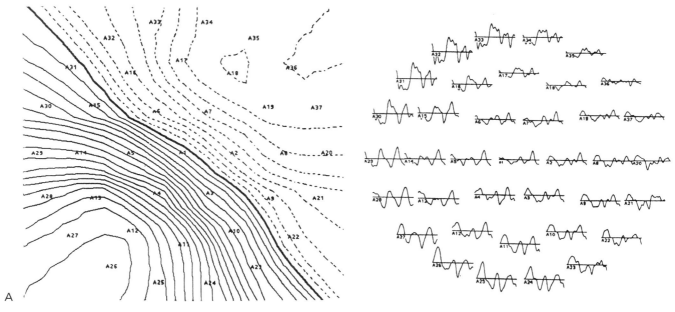

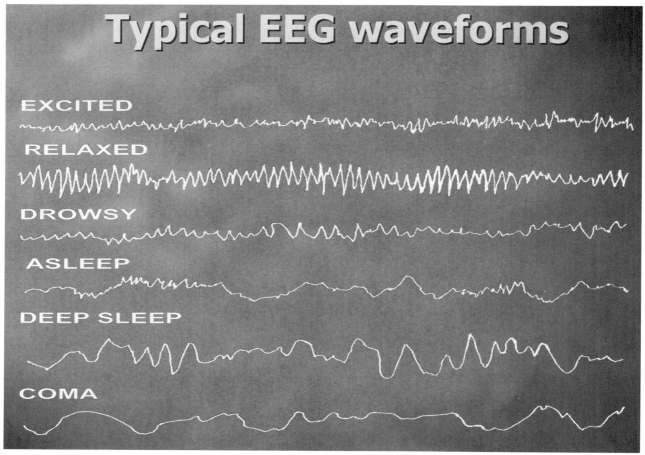

FIGURE 5.3. Exemplars. **A:** Magnetic field map and waveforms. **B:** Electroencephalogram.

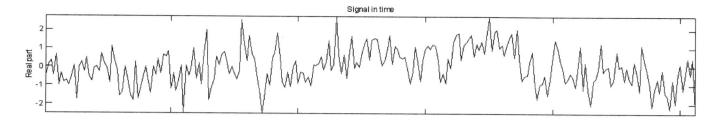

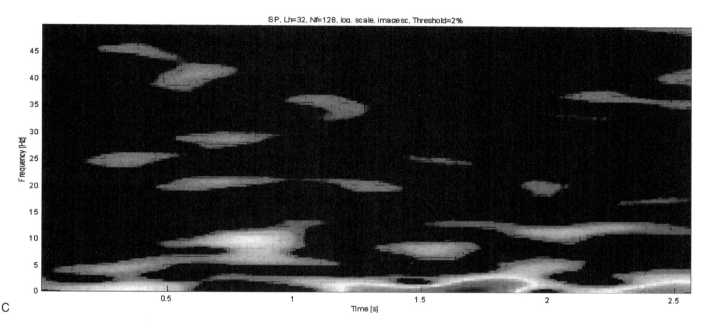

C

FIGURE 5.3. (*continued*) **C:** EEG displayed as individual frequencies (1 Hz increments) over time (seconds of recording).

fields include equipment brought into the room, such as a bad electrode, or metal fragments inside the subject's head. The cardiac pulse can disturb both MEG and EEG when it is of an extreme polarity, or the subject is tall and thin. If the subject's respiratory effort or cardiac pulse disturb the recording because of rhythmic bodily movement, repositioning the person may correct the artifact. As with EEG, all attempts to remove sources of artifact, including mechanical problems (e.g., EEG equipment, bad MEG channel, too low a level of helium) must be made before the recording is in progress.

Clinical Application

At present, there are no uniform standards of practice or predetermined clinical indications for obtaining an MEG. Magnetic field data are presented as waveforms, contour maps, and dynamic color montages (Fig. 5.3). Even though they resemble EEG data, it is important to remember that these are renditions of magnetic flux, not representations of electrical current. Magnetic field data require selective filtering in a manner similar to EEG data. Often spontaneous activity or evoked responses are of weak amplitude. Appropriately chosen filters can improve the signal-to-noise ratio.

The clinical report of MEG resembles a research report in that recording conditions and equipment must be specified. It is assumed that magnetic resonance imaging (MRI) is performed with the same set of reference landmarks. At least three, if not five, external landmark points should be identified that are common to the MRI and MEG. Data analysis, even using the most automated programs, takes at least as long as the recording session or up to five times the duration of a recording session, representing a significant investment of computer and personnel time.

The Inverse Problem

Solving the inverse problem is a particular challenge in MEG because it is virtually impossible to prove the source for the magnetic field. Any validating technique uses other technology. The simplest algorithms solve for single ECDs. Multiple ECDs have the advantage that when a field map is constructed based upon the solution, the resulting theoretical magnetic field map looks more like the empirically recorded evoked field map. Multiple ECDs make sense when one considers the associative functions that must occur in the brain. Depending on where these dipoles are located (i.e., *in* a gyrus, *on* a gyrus, or *in* a sulcus) the effect of their activity, and the time course, the associated magnetic field will vary and contribute differentially to the evoked field.

Interpolation

In MEG, as in EEG, there is a need to extrapolate or interpolate between points of data collection. The spatial separation of individual channels has less of a direct effect upon the magnetic field recording by each channel than differences in spatial relationships affect the potentials recorded at different located electrodes. One iterative program type, called *minimum norm localization,* relies on determining local changes in magnetic field strength or direction. There is a risk that a local norm may be identified that does not represent the largest change in field polarity for the data being analyzed. This is more likely to occur if the magnetic field has multiple areas of different directions of flux. It is least likely to happen when a single dipolar flux pattern is present. Other techniques of extrapolation are based on comparing the coherence of channels to each other (10). Processing difficulties or abnormal connections of circuits of neurons can be detected using spatial filtering techniques that assess sequences of changes in activity.

Electroencephalography and Magnetoencephalography: Comparison

These two neurophysiologic imaging techniques for brain activity are very different from each other. The MEG and EEG record activity from the same substrate, yet they are primarily recording different aspects or components of that activity (Table 5.3). EEG records all electrical activity at the scalp. MEG records the absolute magnetic field at one location in space. MEG records magnetic flux of cortical sources with specific orientation, whereas EEG records electrical current information from sources throughout the brain. Both techniques require extensive analysis of large amounts of data.

EEG has an advantage over MEG in that, using at least the International 10-20 System of electrodes, recording is made over the entire scalp simultaneously. Most MEG systems do not cover the entire head. To obtain full-field maps

around each extremum of the magnetic field, the magnetometer must be repositioned several times. The state of the subject, drug effects, or other variables may change over time and contribute to variability in the record.

Limitations

A challenge for both techniques is to differentiate multiple superficial sources of activity that are occurring simultaneously or nearly simultaneously from those that are a combination of superficial and deeper sources with close temporal relationships to each other. Close approximations are achieved by calculating multiple forward iterative solutions for the varying contributions to the magnetic field. For the EEG, it is difficult to be certain that the apparent contribution to the recorded potential, or difference in potential, over several electrodes is coming from sources close to the recording electrodes and not from the reference electrode. In some ways this is mathematically easier with MEG because one is analyzing an absolute field rather than a difference between two electrodes.

The cost of establishing and maintaining an MEG laboratory is high. EEG is relatively less expensive and is becoming a more portable modality. There is disagreement in the general field of neurophysiology on the clinical need for these tools and their relevance to the study and treatment of neuropsychiatric disorders (11–13).

Applications of the Technologies

These technologies collect data from neuronal activity in real time. The technologic advances accompanying MEG have greatly benefited the entire field of EEG. Three major challenges face any application of these technologies: group or individual comparisons, the nature of an illness or disease, and variability of the subject's state of mind.

Among the challenges when studying disease and diseased states is the choice of appropriate controls. Comparing small groups of an identified illness with normal controls does not control for variation within the normal population. Large numbers of people need to be studied before conclusions about a disorder can be drawn. The assumptions about the distribution of symptoms and signs that are the subject of

TABLE 5.3. COMPARISON OF ELECTROENCEPHALOGRAPHY AND MAGNETO-ENCEPHALOGRAPHY

Variable	Electroencephalogram	Magnetoencephalogram
Time domain	Milliseconds	Milliseconds
Physical sources of artifact	Scalp, bone, fluids; poor electrode contact	Nonphysiologic magnetic fields and electrical currents
Records directly	Electrical activity	Magnetic fields
Direction of recorded dipoles	All that come to electrode	Tangential
Cost	++	++++++
Established clinical standards	Yes	No

the study need to meet the expectation of distribution of the statistical measures used to determine differences between findings. When conditions have a high degree of individual variation, it may be more appropriate to perform repeated measures on the same individual over time (in different relative state of homeostasis) rather than compare the individual's findings to a group finding.

Most subjects who join research studies on an outpatient basis represent less severe forms of a given illness. They are competent to participate voluntarily and can understand and cooperate with the protocols. This skews some data toward less severe pathology and, at least theoretically, increases the likelihood of overlap with normal findings.

There are many variables that need to be carefully controlled and accounted for in all neurophysiologic studies. Recordings will be affected by medications and drugs, chronicity of illness, age, time of day, and level of arousal, as well as by specific diseases. Pharmacologic effects of medications are different in the normal person than in the person who has demonstrated a clinical need for the effects of the medication. Claims about drug effects on the EEG should be considered with regard to the clinical state of the individual receiving the drugs.

EEG and MEG frequently are combined for clinical research studies. This method controls for the effects of variables on the results at the same time. Changes in state of arousal, drug effects, disease effects, and time will affect the EEG and MEG equally. It permits direct comparison and cross-validation of results from both EEG and MEG. The development of improved modeling of sources and solutions to the inverse problem are strengthened by simultaneous recording. To date, these technologies are providing improved complementary information (particularly for the presurgical localization of epileptic foci) but have not demonstrated a compelling improvement in sensitivity and specificity beyond that of EEG alone.

Evoked Potentials and Fields

Evoked Responses

Evoked responses are brain-specific activity, linked to a specific stimulus, distinguished from background activity. Multiple presentations of the stimulus are given and the activity is summed and averaged. Better "noise" reduction can be obtained with increased repetitions of the stimulus, but the identified response signal can be lost as the brain habituates to a repeated stimulus.

The summation (averaging) period in an evoked response study usually is selected to start just before the stimulus and to continue for some time longer than the response being studied (Fig. 5.4). The number of repetitions for cortical evoked responses usually is between 50 and 200 repetitions; for subcortical responses between 1,000 and 4000 repetitions.

Sensory Modality

Traditional evoked responses have used three sensory modalities: visual, somatosensory, and auditory. There are standardized norms for early, middle, and late responses. In the visual domain, several standardized types of stimuli are used for eliciting these responses (e.g., checkerboard, flash, central and peripheral visual field). Somatosensory evoked potentials (SSEP) are elicited by a painful stimulus, generally an electric shock just above threshold and sufficient to cause a muscle twitch (e.g., a thumb twitch). Auditory stimuli include broad-band clicks presented in rapid succession for brainstem evoked responses or series of tone pips or more complicated sound presentations for later cortical responses. Cross-modal tasks, across sensory modalities and including variations within each modality, are constructed to evaluate mental manipulation. For example, word stimuli describing colors (blue, red, green) that are presented in written form in the true or false color of the name (within modality) and are interspersed with simultaneously spoken words of the same name or different/nonsense words (cross-modality) challenge attentional, perceptual accuracy, and discrimination capabilities of the subject. These tasks that require mental manipulation, such as retention in short-term memory or processing of two types of information simultaneously, help to differentiate processes occurring in connected brain regions.

Timing

The rate of stimulus presentation in each domain is critical for producing a maximal response. For example, when auditory tones are presented at rates faster than one per second, the amplitude of the response is significantly decreased. If the interstimulus interval is closer to 4 to 6 seconds, various components associated with the early cortical response can begin to be identified. There are major differences in uses of evoked potential studies between a clinical EEG laboratory running evoked responses for neurologic diagnostic testing and a research laboratory geared toward understanding what areas of brain are being activated by different stimuli. When a study is performed to determine the contribution to an abnormal finding or abnormal sensation for a given individual, the stimulus should be designed in all aspects of presentation to most nearly replicate that which the subject indicates produces the abnormal response.

For this reason, the quality of the stimulus, the instrumentation used to generate it, and the specific timing characteristics all must be documented in any publication so that repeat studies can be done. Subtle differences between different types of equipment can produce minor differences in the stimulus, particularly in visual and auditory systems, such that the response may be either enhanced or reduced and therefore not replicate previous findings. When in doubt, it is best to call the investigator who first reported the finding that one is trying to replicate.

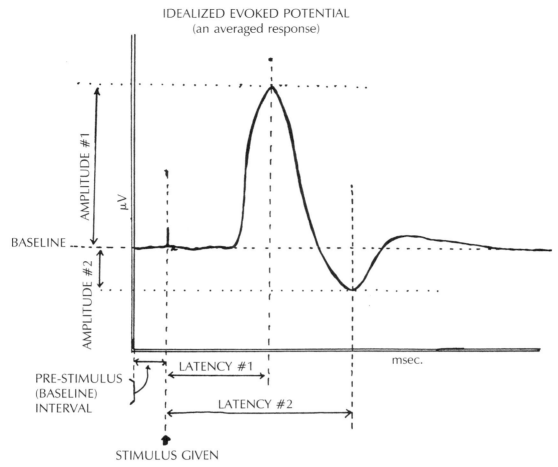

IDEALIZED EVOKED POTENTIAL
(an averaged response)

FIGURE 5.4. Idealized evoked potential illustrating general points of reference.

Brainstem auditory evoked potentials (BAEP) are a sensitive test of central nervous system (CNS) myelination and of acquired demyelination. The interpeak latency from peak I-V (for further information see reference 1) is a sensitive measure of the degree of myelination. For example, in newborns it is considerably longer and progressively shortens with increasing age, reaching a plateau at about age 3 years. However, Brivio et al. (14) reported somewhat longer interpeak latencies in males at age 3 years than in females.

Findings of Questionable Significance

A number of findings are seen on EEG and MEG that are of questionable significance, that is, these patterns have low specificity or uncertain sensitivity. Incidental sharp waves, spikes, slow waves, exaggerated transitional phenomena (such as excessive hyperventilation high-voltage activity), and low-grade dysrhythmia all can occur without diagnostic specificity. Patients with neuropsychiatric disorders appear to have a greater incidence of these types of findings in settings where there are no clearly distinct accompanying behavioral changes.

Although no encephalographer would be willing to ascribe specific meaning to such findings, they should not be categorically dismissed. First, ascertain whether the pattern in the patient is caused by medication. Second, when the patient's behavior changes markedly or the class of medication is changed, repeat the EEG and compare it to the initial EEG. Third, note whether there is a temporal relationship between the EEG findings and physiologic symptoms.

The caveat to this discussion is that these same incidental findings occasionally can be seen in normal people who are asymptomatic. The challenge is to determine whether there is any meaning to a disruption in the well-modulated EEG. If one is inclined to ignore this finding in a normal person, how much can be made of the same finding in someone who has a neuropsychiatric illness without a clearly defined behavioral abnormality?

APPLICATIONS IN NEUROPSYCHIATRY

Epilepsy

The EEG is widely used to document the existence of seizures or the presence of focal electrical abnormality that is causing behavioral disturbances. EEG/MEG has limited sensitivity in accurate diagnosis of partial seizures. The epileptic (ictal) electrical activity may not be generated in or near the cortical surface and will reflect normal brain patterns interictally. A clinical history of alterations in behavior, perception, and/or cognition is required to make the diagnosis when the EEG is normal.

Comprehensive reviews of the applications of neurophysiology in diagnostic evaluation of the epilepsies are provided in major textbooks (1–3). New adaptations include computer-assisted ambulatory EEG for detection of seizures on an outpatient basis (15). The recordings analyzed showed 28% of partial seizures were not recognized by the patients during a 24-hour recording. Relying upon the patient's reported awareness of seizure activity may not provide the necessary information to fully treat partial seizures.

EEG and MEG are combined in many studies, demonstrating that the technologies are complementary rather than duplicative (16,17). Surgical excision of epileptic brain tissue and other causes of holes in the skull cause large errors in localization in EEG that often have negligible effects on the MEG (18). However, the sensitivity distributions of EEG and planar MEG measurements have been shown to be very similar and to have smaller sensitivity than axial gradiometer (MEG) recordings (19). When both technologies are readily available, it is standard practice to complete presurgical evaluation using both techniques. Addition of other brain imaging technologies, such as functional MRI, can aid in the accurate spatial resolution (20,21). Noninvasive physiologic measures have a moderate degree of inaccuracy in localizing electrical sources, as can be demonstrated with comparison of interictal and intraictal scalp EEG and subdural electrocorticogram recordings (22). In seven of 12 patients in this study, no interictal sharp waves were detected in the EEG but significant interictal seizure activity was seen on the subdural electrode recordings.

Nonepileptic Seizure

One frequently asked clinical question is how to differentiate seizure from "pseudoseizure" or nonepileptic seizure. Not surprisingly, many patients with an established diagnosis of epilepsy have the perception of seizures when no epileptiform activity can be detected during this perception (23). Often there is little benefit to treating nonepileptic seizures with anticonvulsant medications. Clinicians must include the possibility of nonepileptic seizure in the evaluation of all patients, rather than confront the patient with proof that their seizures are not epileptic. In a small number of patients who appear only to have psychologically induced seizures, a definite neurologic disorder may be diagnosed over the succeeding 5 to 10 years (such as aneurysm, tumor, or inflammation).

Behaviors of dyscontrol, such as nonepileptic seizures, are indicators and expressions of psychological difficulty that the person has found no other way to express or understand. Illness behavior(s) is learned over time and serves as a form of adaptation to an alteration in functioning. Clinicians must express directly to the patient that illness behavior can be understood and changed, but it often is not responsive to medications. This discussion should occur from the outset of an evaluation for seizures. The least helpful for the epileptologist or the patient is to leave these "psychiatric spasms" as a discard diagnosis at the end of a long and grueling evaluation. The anticonvulsant medications are effective for treatment of impulsivity, pain, and less severe depression; therefore, there may be more than one clinical effect from the treatment effect intended.

Collaboration between neurologic and psychiatric and psychological disciplines is required for ongoing treatment and management of persons with epileptic and nonepileptic seizures. Repeated evaluation is indicated when patients have changes in their symptoms, do not respond as expected to medical intervention, or have recurrence of their presenting symptoms after a period of apparent relief. People with recurrent seizures have a chronic condition that waxes and wanes in severity and frequency, both with epileptic and nonepileptic seizures. The clinical team must be prepared to support the individual through these periods over the decades of their life.

Schizophrenia

During the first decades of EEG technology, it was hoped that this "new" measurement of brain function would demonstrate the etiology of the severe and disturbing symptoms of schizophrenia. However, the paper EEG record was not sensitive enough to characterize the differences in brain function underlying the behaviors observed. Mild generalized disturbances, such as "nonspecific dysrhythmia" and "slow waves" without lateralization patterns, were frequently described. There has been widespread consensus among psychiatrists and neurologists that the EEG was essentially useless for detection of clinically relevant abnormalities in patients with schizophrenia. The quantification inherent in evoked responses, with the requirement to repeat stimuli until a significant signal to noise ratio is achieved, and the development of computerized recording systems slowly began to change both what was being measured and the questions that were being asked.

A skeptical view of the utility of neurophysiology for understanding the pathophysiology of schizophrenia is at once realistic and unrealistic. It is true that there are no immediate definitive answers from these technologies to the search for the etiology of schizophrenia. For example, Canive et al. (24) found no gross MEG abnormalities in patients with schizophrenia who were in the medicated state, although they reported lower peak frequency and overall power than in normal controls. On the other hand, no other techniques can as reliably and repeatedly measure brain function in real time.

Other investigators have found marked slowing widespread over cortical areas that may be drug induced rather than based upon the pathophysiology of schizophrenia itself (25).

Transient hallucinations affect the auditory evoked fields (MEG), allowing clear differentiation of hallucinating and nonhallucinating states in two individuals (26). During the hallucinating period, the N100m was narrower in distribution and the response was dampened in several channels, with no effect on the amplitude of response. The source localizations of the equivalent dipoles were 1 cm apart in the supratemporal auditory cortex, consistent with previous findings for normal subjects.

This study had several important design features (26). It used two cooperative subjects who could accurately report the onset and offset of symptoms and who could tolerate repeated studies. The subjects were used as their own controls. During nonhallucinating and hallucinating periods, the latency values for their N100m were within a range distribution for normal subjects. The data from the subjects were compared to data from normal subjects in whom binaural tones were administered with and without unilateral masking noise (an auditory interference paradigm). Similar effects on the N100m were found, suggesting that the functioning neuronal substrate could be manipulated to produce abnormalities in the evoked response. The authors concluded that "a distorted interplay between auditory associative areas and the limbic structures, rather than with a malfunction of the primary auditory cortex itself" (26, p. 257) could produce alterations in measurements of auditory cortex functioning (i.e., the evoked response). Rightly, they did not draw conclusions about the etiology of the hallucinations (only one of the two subjects met criteria for schizophrenia).

Uses of evoked responses have generally been applied to groups, such as patients versus normal subjects, or selected treatment or symptom-based groups. Auditory evoked responses using EEG and MEG have shown lowered amplitude and more latency variability in patients with schizophrenia compared with normal subjects (27,28). Somatosensory responses tend to have higher overall amplitude before 100 milliseconds but lower amplitude after 100 milliseconds. The distributions of P30 and N60 responses were found to be more posterior in patients with schizophrenia than for normal subjects, depressed individuals, or nonpsychotic subjects (29,30). The P300, in both auditory and visual domains, tends to have low amplitude and be less responsive to uncertainty in the stimulus presentation. A wide range of latencies in individual trials contributes to the longer averaged response seen in groups of patients with schizophrenia compared to groups of normal controls.

The variation in averaged responses does not generally result as a direct effect of the antipsychotic medications. Faux et al. (31) compared P300 auditory event-related potentials in 14 neuroleptic-withdrawn male subjects with schizophrenia and 14 healthy male controls. In the normal group, the P300 mean amplitude was largest in the midline with symmetrically reduced amplitudes and lateral positions. The group with schizophrenia showed amplitude reduction overall, with a greater right-sided P300 component amplitude than left-sided amplitude. The intent of this study was to show that medication did not exert the effect in the patient group. A serious caveat is that a large range of drug-free duration was included in this study. It is possible that the duration of the drug-free state was insufficient to remove actual drug effects. The authors related the decreased amplitude of left P300 to left temporal lobe abnormalities in schizophrenia. Their findings supported the conclusion that neuroleptic medication alone does not influence the measure of the P300 asymmetry found in schizophrenia.

An important hypothesis to explain the symptoms of schizophrenia is that of impaired filtering of sensory information, or altered sensory gating. Using paired auditory clicks presented 1 second apart, with 5 to 8 seconds between pairs [interstimulus interval (ISI)], Adler et al. (32) and Waldo et al. (33) demonstrated that normal subjects and their relatives have a reduced amplitude of response (P50) to the second click. Patients with schizophrenia have no reduction in their response to the second click; their first-degree relatives are between the patients and normal subjects in size of response to the second click. This finding has been replicated by independent laboratories (34). The idea is that schizophrenia impairs the ability of the individual to dampen an automatic processing response (at 40–50 milliseconds, signals are thought to be arriving at the primary auditory cortex) when there is no new information contained in the signal.

When considering explanations for the symptoms of schizophrenia (impaired attention, auditory and visual hallucinations, impaired reality testing, disinhibited behaviors, disorganization of thought and action, poor ability to change mode of operation with new information), one tends to look at the functioning of primary and association cortex in frontal and temporal lobes. Reduced contingent negative variation responses appear to be a stable marker of schizophrenia (35). In the same study, larger contingent negative variation evoked responses contralateral to the responding hand were found in patients with schizophrenia, similar to patients with Parkinson disease (PD), suggesting that a medication side effect may be contributing to the finding. Attention to oddball tones did not show improved localization over time or enhanced strength of response of N100m in male patients with paranoid schizophrenia (36), suggesting abnormal processing in superior temporal gyrus compared with normal subjects. Reite's group further reported loss of normal asymmetry of equivalent dipoles for N100m in male patients with schizophrenia compared to male normal subjects. Subsequent studies suggest that the N100m is a compound evoked response and that the second subcomponent may be the major contributor to these laterality differences (37). Time constants for 100-millisecond responses are longer in women than in men (38) and may represent an underrecognized contributing factor to group-averaged responses. In previous studies, Rojas et al. (39) and Reite et al. (40) found greater left-to-right asymmetries in N100m responses in women with schizophrenia compared to men with schizophrenia.

Sleep recordings reveal persistent abnormality of sleep structure in treated and untreated patients with schizophrenia (41). Findings of delayed sleep onset, more arousals, and more early morning awakening are not specific for schizophrenia. They most likely contribute to impaired arousal, excessive daytime sleepiness, and the tendency to be disorganized. Sleep disturbance reflects the altered balance of several neurotransmitters. Cholinergic, dopaminergic, serotonergic, and noradrenergic systems all have been implicated in sleep disorders. Significantly reduced amount of δ sleep activity has been recorded in several studies (42). This reduction is independent of medication treatment. Shortened rapid eye movement (REM) latency is associated with severity of negative symptoms, consistent with a hypothesis of increased cholinergic activity contributing to both schizophrenia and sleep disorder (43). Auditory and visual evoked potentials were measured during sleep to assess transfer properties in brain function (44). The patients with schizophrenia demonstrated different patterns from both patients with depression and normal controls and exhibited increased θ activity during REM sleep.

The issue of medication effects on patients with schizophrenia is complicated. Antipsychotic medications have been shown to cause intermittent slow waves of large amplitude over central and frontal regions. These patterns may be described as low-grade dysrhythmia. On the other hand, patients who have not been taking antipsychotic medications or have never been exposed to antipsychotic medications also have been found to have nonspecific changes in their waveform patterns. Some measures of EEG reactivity are associated with responsiveness to medication, such as responders having higher amplitudes in α spectrum than nonresponders (45). It is difficult to assess the direct effect of antipsychotics because in normal controls this type of medication tends to rapidly induce sedation, and at very low dose it does not cause gross changes in EEG patterns.

The challenges of assessing antipsychotic medication effects on sensory processing were demonstrated in a series of experiments conducted by Malaspina et al. (46). Amphetamine (0.3 mg/kg) was given to normal subjects who subsequently were given 2 mg haloperidol intramuscularly and then studied with a standard continuous performance task (CPT) and auditory conditioning testing to elicit paired P50 (C-T). Amphetamine improved performance on CPT, and haloperidol caused a larger decrement in performance than when given alone. Sensory gating on C-T was increased by amphetamine (second P50 was even smaller or absent than in the drug-free condition); this effect was eliminated by haloperidol. Haloperidol alone reduced both first and second responses (P50) to the paired stimuli. In contrast, in a patient with schizophrenia, neuroleptic treatment can improve performance on CPT and increase sensory gating in C-T paradigms. These results suggested that there may be very different effects of haloperidol on EEG and evoked potentials (EP) in acute and chronic treatment; that the effects of haloperidol in normal subjects are quite different from those in patients with psychotic illnesses, and that the state of the individual (e.g., amphetamine, drug free, resting, or task oriented) may have important influences on the effects of an antipsychotic medication.

Cognitive processing is associated with synchronous activity of neurons in the γ frequency (40 Hz). Normal controls showed significant suppression of the transient 40 Hz after receiving 2 mg haloperidol during auditory testing (47). No effect was seen to unattended stimuli in either EEG or MEG records, suggesting that dopamine (D_2) receptors modulate selective attention.

Benzodiazepines are known to increase fast (β) activity and may contribute to identification of fast frequency activity in studies of schizophrenia. Increased left-sided β activity was reported in patients with schizophrenia who were not taking benzodiazepines (48). Benzodiazepines in either low or high doses were not found to affect mismatch negativity responses in patients with schizophrenia or even in normal controls (49).

Future studies will need to continue to concentrate on combining region-specific tasks during EEG and MEG measurements. EEG and MEG in combination will be most useful as tools to develop understanding of sequential processing of information during periods of exacerbation and remission. Pharmaco-EEG and pharmaco-MEG will continue to become more sophisticated, provide meaningful information about rapidity of response to different classes of drugs, and predict individuals who will respond to specific classes of drugs.

Parkinson Disease

Idiopathic PD is identified through clinical diagnosis, including history, physical examination, and medication trials. Typically, the EEG remains normal in early stages of the disease. During symptomatic stages of the illness, tremor-correlated cortical activity over contralateral motor cortex areas was recorded by Hellwig et al. (50). This underscored the fact that motor cortex is involved in the neural network generating the tremor. Tremor related to cerebellar disease uses different pathways, although a fair amount of coherence between a specific thalamic nucleus (nucleus ventralis intermedius) and sensorimotor cortex was noted during a study of several causes of tremor (51).

Attempts to understand the postural instability and bradykinesia found in patients with PD have effectively utilized evoked responses. For example, audiospinal facilitation is significantly reduced in amplitude during the 75- to 150-millisecond period after conditioning stimulation using the soleus H-reflex test. It is improved by treatment with lev-odopa (L-dopa) but not by anticholinergic agents. This represents an example of dysfunction of the reticular nuclei in PD (52).

Beckley et al. (53) examined the interaction between motor response and cognitive set using EMG to look at long-latency postural reflexes in patients with PD compared with normal controls and young normal subjects. Patients with PD had postural long-latency reflexes of lower amplitude and slower onset, especially under unpredictable testing conditions. Young normal subjects and older healthy subjects were able to scale long-latency responses under both predictable and unpredictable conditions to anticipate the degree of perturbation encountered. Patients with PD were unable to anticipate the need to modify the effort in the long-latency response. This defective modulation may be affected by abnormal frontal influence on the basal ganglia through the caudate. These results correlate with studies showing that PD often causes difficulty with complex motor tasks involving shifts in cognitive set. The response latencies were not affected by whether there was a predictable or unpredictable stimulus, consistent with previous findings that latencies are programmed from minimal sensory input. Response amplitudes were sensitive to predictability.

To distinguish further between the attention-related components and the cognitive challenge of switching sets, patients with PD were compared with age-matched normal controls (54). Force of response and contingent negative variation were slowed in the patient group, indicating difficulty of activation in movement preparation. There was no difference between groups on lateralization responses, indicating unimpaired response selection. The difficulty in activation was more pronounced during simple monotonous task. The delay of responses during invalidly cued signals supported a cognitive difficulty during a more complex task.

The role of the reticular formation in motor control in PD was evaluated by pairing sound stimulation with the electrically elicited blink reflex (55). The blink reflex, measured by EMG, was conditioned by sound stimuli that were not loud enough to produce blink responses by themselves. The time interval between the conditioning sound and the electrical stimulus to the supraorbital nerve was varied. In control subjects, sound stimulation facilitated the first component of the blink reflex. In PD subjects, rigidity in the neck positively correlated with the degree of suppression of the second component, when there was a 500-millisecond interval between the conditioning sound and electrical stimulation. The durations and amplitudes of the first and second components of the electrically elicited blink reflex were the same in both normal and PD subjects, suggesting that the reflex pathway components were normal in the PD patients. The authors' test paradigm differentiated between the functioning of the reflex arc itself and the function of facilitatory and inhibitory interneurons, as modulated by the basal ganglia in parkinsonism.

The rate of recovery in motor cortex after movement appears to be impaired in PD. In a study of 17 patients, markedly reduced event-related synchronization in the β frequency range (12–24 Hz) was found in PD patients compared with normal controls (56). Another measure of increased motor cortex excitability is the redundant signals effect. However, patients with PD and healthy controls had similar latencies of movement-related potentials (57). In addition, the size of force signals in both groups was not different as the subjects performed simple manual responses to visual, auditory, and auditory-visual signals. The increased motor cortex excitability found in PD does not appear to directly affect the mechanism underlying the redundant signals effect. To evaluate whether medications, such as L-dopa, play a direct or indirect role affecting the rhythms of the motor cortex, α and β frequency rhythms were studied in 12 patients with PD while they moved a wrist (58). Patients were evaluated during a 24-hour drug-free period and 1 hour after L-dopa was reinstituted. Attenuation of α and β rhythms correlated with improvement in size and speed of movement affected by levodopa. This finding was the result of a direct effect by L-dopa on the recorded frequency that correlated with improvements of bradykinesia. Premovement desynchronization also improved in conjunction with reduction of bradykinesia over the contralateral sensorimotor and supplementary motor area (59). Basal ganglia appear to liberate the frontal cortex from idling rhythms, but it (the degree of "liberation" of frontal cortical neurons) changes with the demands of the tasks.

In a study by Chen et al. (60), patients with parkinsonism and dementia showed slowing of background α rates during clinical EEG. They documented the earliest changes occurred in α frequency that dropped below 7 Hz. In the last 2 1/2 years of life, three fourths of the patients studied had diffuse bilateral synchronous bursts of 5- to 7-Hz waves of 30- to 50-μV amplitude. The EEG rhythm slowed to 4 to 5 Hz, with interspersed asynchronous irregular waves at 3 to 4 Hz, with progressive diminution of amplitude and reactivity. These irregular waves occasionally were located over temporal lobes. There were no clear epileptiform discharges; sleep records were generally normal. Tremors had no obvious relationship with the EEG. There was no correlation with extrapyramidal signs, for example, patients who had marked rigidity and akinesia but no dementia had only moderate changes on the EEG that were not specific. In a separate study, nondemented patients with PD had increased amplitude of the event-related potential component P3 (61). As dementia increased, amplitude and power decreased and the latency to P3 largely increased.

Vascular parkinsonism differs from idiopathic PD and has significantly less slowing on the resting EEG (62). The results further indicated that vascular parkinsonism is not simply PD with subcortical vascular lesions. In later stages of disease, comparison between amounts of slower activity will show similar patterns between the two groups. All patients with severe dementia have abnormal EEGs early in the illness course.

Depression

The literature on EEG in depression is confusing to evaluate because a broad range of symptoms and severity have been lumped under the label *depression*. EEG sleep studies have been the most consistent with one another (63). Biologic markers of abnormalities of sleep are found in people prone to this disorder compared with control subjects. Krieg et al. (64) found that subjects who are at high risk for depressive disorder but did not yet meet the Diagnostic and Statistical Manual of Mental Disorders (DSM) criteria had decreased sleep efficacy and decreased amount of slow wave sleep during the second non-REM period. The high-risk group also had twice the number of awakenings during the night than the control subjects. Of 20 subjects at high risk for developing a depressive disorder, the EEG in four was found to have alterations characteristic of major depression (generally considered to be decreased sleep efficiency, prolonged sleep-onset latency, frequent spontaneous awakenings, early-morning awakening, reduced amounts of slow wave sleep with a slight increase in REM sleep, and, often, shortened latency to the first REM period). The group at high risk also showed greater overall variability in sleep parameters compared with the normal controls. Effective treatment of depression has been documented to restore altered sleep patterns, at least in the short term (65). There may be separate effects on sleep patterns due to race. African-American subjects with depression were reported to have less total sleep and less slow wave sleep and less REM sleep than Euro-American subjects with equivalent severity of depression (66).

Studies of depression with EEG have used various data reduction techniques. Dierks (67) used fast Fourier transformation within a narrow bandwidth (1–30 Hz), followed by equivalency dipole source localization for various frequencies. The five depressed subjects in his study showed more superficial and more anterior localization of the ECD than age-matched medication-free patients with schizophrenia or normal controls. The ECD localizations were intriguing because there were differences among the three diagnostic groups, raising the question of whether there are different actual generators for EEG in each group. With such a low number of subjects, it is difficult to generalize to all patients with depressive symptoms. In that study, the α-equivalent sources were more superficial and anterior in the people with depression. As a person moves from wakefulness to light sleep, the α generators appear to move more anterior and temporally (68). A more recent study found that variability of asymmetric distribution of α across frontal leads was characteristic in persons with depression compared with normal controls (69). Lindgren et al. (70) hypothesized that abnormalities in thalamocortical circuitry attributed to depression may contribute to the variations in

α power observed in depression. Increased thalamic metabolic activity was associated with decreased α power in normal controls and was not correlated in the patient group with depression.

One wonders if, in depression, a person is more ready to make this stage shift or if this localization reflects an abnormality in sleep regulation as part of the dysfunction of depression. Using a combination of MEG/EEG, Lu et al. (71) demonstrated the various contributions from EEG and MEG in determining equivalents for localizations of various components in the sleep EEG. The temporal localization of α tends to be more resistant to changes or decreases in vigilance than the occipital generators (72,73). What contribution depressive disorder and its attendant neurotransmitter defects may have to the abnormalities reflected in electrophysiologic measures is difficult to separate from the effects caused by disruption of sleep and wakefulness patterns overall. In other words, if the increase in amount of α activity noted during depressive phases reflects a change in arousal that was not noted clinically, then previously reported results need be revised to identify changes in state (arousal–somnolence) rather than mood. In one study, sleep EEG β-δ coherence changes were evident in men and women at high risk for developing depression (74).

Many studies have shown power and coherence changes in many frequency bands over both hemispheres under different stimulation conditions, not always in the same direction, not always consistently, but differing from normal controls (75–78). In a study attempting to separate patients with depression, schizophrenia, and epilepsy from each other using EEG, Nagakubo et al. (79) found that the three groups had approximately the same rate of irregular β patterns (about 11%–14%) compared with 4% in healthy controls. The prevalence of spikes and sharp waves in the epileptic groups was significantly higher (35%) compared with either of the psychiatric disorders (4–5%).

One intriguing question is the relationship between specific frequencies seen in the EEG and the level of activity of the individual. In rats, θ activity is correlated with motor activity and can be correlated in non-REM slow wave sleep during periods of isolation (80). In humans, θ is not directly correlated with movement. However, in typical unipolar depression, subjects are reporting that it is difficult to get themselves motivated, to complete tasks, and to be mobile. The increased amount of slower-frequency activity in the EEG may be related to this depressed state. To understand how various abnormalities may be reflected or represented in the EEG, structural underpinnings of cortical activity need to be studied. Severity of depression and decreased size of caudate nuclei on MRI in patients with depression (81) were speculated to parallel the prevalence of depressive symptoms in Huntington disease, which is known to have decreased caudate nucleus size. The authors further speculated that this anatomic finding may be related to reported difficulties in negotiating complex social behavior, decreased initiation of activities, and decreased ability to solve problems when the caudate nuclei are known to be damaged.

Prefrontal, parietal, and occipital cortices have been studied by evoked potential paradigms because of their role in maintaining sustained attention. Using the selective attention paradigm of Hilliard (slightly adapted), Burkhart and Thomas (82) studied subjects who had moderate depression, did not meet the criteria for major depressive disorder, and were not taking any antidepressant medication. They studied the negative difference wave, which is the difference in the evoked potential response between the attended and the ignored stimuli. Overall there seemed to be a larger response in the control group compared with the depressed group, regardless of attention condition. There were no significant differences between groups in terms of accuracy of response or type of response.

The attention changes that can be seen with mild-to-moderate levels of depression are more likely to be evident with affectively toned stimuli and less likely to be present when there are neutral stimuli (such as auditory tones). A behavioral study showed that depressed subjects could recall essentially the same *amount* of information that normal controls could, but the *nature* of the information was different (83). They selectively recalled negatively toned material better. This suggests that a simple "objective measure of selective attention" may not be specific enough to differentiate depressive disorder and/or major depression from other neuropsychiatric disorders or from normal functioning, because the design is not specific enough for the characteristics being investigated.

Treatment issues have been studied from several angles, such as differential diagnosis of depression and aging effects on brain function, normalization of findings, predictors of responsiveness to medications, and electrophysiologic signs of medication. A frequent clinical challenge is to differentiate depression from dementia in elderly patients. Confounding this problem is the fact that structural changes (seen on MRI) are greater in depression with onset in later life. Depression and EEG findings were separately studied in elderly patients with late-onset and early-onset depression (84). Increased slow wave activity and slower reaction times accompanied global cognitive impairment, which persisted after clinical recovery in late-onset depression. Nortriptyline improved clinical symptoms in elderly patients with depression and decreased REM sleep time that reverted to baseline levels after discontinuation of medication (85). Subjects with bereavement-related depression in the same study had persistent abnormalities in sleep structure after clinical remission. Alternative therapies also have measurable effects on the EEG. Massage and music therapy reduced right frontal activation on resting EEG in adolescents with depression (86).

QEEG cordance, a measure reflecting cerebral energy utilization (87,88), was a sensitive measure of response to fluoxetine in patients with major depression (89). In a related study, changes in cordance measures were seen in patients as early as 48 hours after initiation of treatment (90). These changes presaged clinical response to the therapy. The authors compared both serotonin reuptake inhibitors and tricyclic medications and found robust ability to predict responsiveness to treatments (91). Placebo responders produce increased cordance measures, medication responders produce decreased cordance, and nonresponders in both groups showed no significant changes during a 9-week comparison study (92). The overall QEEG power in all groups was not different. This suggests that although similar clinical outcomes were achieved in the responder groups, placebo and medication treatments each have distinctive ways of effecting their improvements.

Aging and Dementia

One of the difficulties in identifying characteristic abnormalities or patterns of both normal and abnormal aging processes is that the clinical diagnoses for both these conditions have not been well defined for many years. Preliminary results from the Kunzs-Holmen project in Stockholm, Sweden, revealed a "disturbed behavior" factor helped to distinguish Alzheimer-type dementia (AD) cases from other dementias and nondemented subjects (93). Knowing the premorbid cognitive capacities and educational background of individuals is essential to adequately evaluate a dementing process (94). Previous studies indicated that spatial abilities were most heritable, followed in order by vocabulary, word fluency, speed in arithmetic, and reasoning (95). QEEG measures (cordance and coherence) were used to assess synaptic function in patients with AD and elderly controls in studies comparable to evaluation of depression and other disruptions of neurophysiology due to neuropsychiatric disorders (96).

AD of mild-to-moderate severity is reported to produce increased slowing in the EEG. Increased slow activity was reported over frontal and midline recording locations, especially in δ and θ frequencies (97,98). This is consistent with the cortical damage theory of AD. Significantly, in this study there was no slowing in the α frequency for the control group (age-matched, nondemented subjects) (97). A study of EEG changes in persons with mild, moderate, or severe dementia revealed a positive correlation between EEG changes (abnormalities) and worsening performance on psychometric scales (99). EEG changes included presence of δ and θ waves, change in background activity, focal abnormalities with lateralization, presence of spike and sharp waves, and increased synchronization. Increased amount of δ wave activity did not independently correlate with progressive dementia (99–101). Bennys et al. (102) found that use of indices calculated from EEG power discriminated dementia from matched controls, with increased slow activity and decreased fast activity in the dementia subjects. This was corroborated by Ihl et al. (103), who also found consistently increased

δ amplitude, and in similar studies using MEG (104). Using a δ plus θ percentage score, Brenner et al. (105,106) identified 17 of 35 demented patients, with no misclassification. Eleven of the 17 demented patients had moderately generalized abnormal EEG; all of them had lower Folstein Mini-Mental Status Examination (MMSE) score than those who were incorrectly classified.

Because there is no *a priori* reason for one area of brain to suffer degeneration before another, the EEG may be a useful tool for determining clinical variations. Matousek et al. (107) found that parietal lobe syndromes of dementia were particularly correlated with slow activity on the EEG. An earlier study demonstrated temporospatial EEG patterns of reduced reactivity between eyes open/closed states and reduced microstate duration in patients with early cognitive loss (108,109). The progressive increase in δ activity that was associated with parietal lobe symptoms was hypothesized to support parietal decline as a major contributing factor to functional decline in AD (110). However, Duffy et al. (111) and Prichep et al. (112) reported that maximal group differences between controls and patients were over frontal and temporal regions. EEG and regional positron emission tomography performed on ten patients with AD showed strong correlation between slowing in EEG and parietotemporal hypometabolism (113).

A separate study found a 75% classification rate using QEEG on a group of elderly subjects with probable AD compared with matched elderly controls (114). Comparison of the complexity in EEG recordings in AD versus vascular dementia and elderly controls also supports decreased complexity in this form of dementia (115) and decreased coherence (116–119). In patients with mild cognitive symptoms of dementia, progression of illness may be anticipated by EEG changes of increased left hemispheric δ power and uneven power of δ and α over the right hemisphere (120). Simple evoked potentials, such as P300 (auditory), are not appreciably altered in AD (121).

Normal aging effects in the sensory systems must be accounted for when ascribing particular characteristics to premature decline in function. Visual evoked responses to checkerboard pattern-reversal stimulation show a curvilinear relationship between P100 latency and age over the life span. The shortest latency occurs during middle adulthood, with increasing latencies found at ages over 60 years. Tobimatsu et al. (122) reported the differential age effect on the P100 latency caused by contrast changes also depends on the size of check used as stimulus. In general, as the pattern contrast decreased, P100 latency was prolonged and the amplitude decreased. Other groups found that the evoked potential latency of elderly subjects only became comparable to younger subjects when the stimulus contrast was appropriately increased to accommodate for physical changes in the lens and eyeball that accompany increasing age. Similarly, the response time was fastest in these elderly subjects when the stimulus was longest, that is, 320-millisecond cues elicited

faster responses than 2-millisecond cues. When there was a delay between the stimulus cue and the response, patients who had a history of alcoholism performed much worse than normal elderly subjects. There was a significant increase in extraneous responses when there was a 5-second delay rather than no delay, whether or not a correct response was given. Korsakoff syndrome patients were the most severely affected by short visual exposure durations and had the most clearly progressive decline in response, with prolonged delay between stimulus cue and response in both auditory and visual dimensions. Visual cues were much worse than auditory. Patients who had a significant history of alcoholism without Korsakoff syndrome were more impaired than normal aged controls. These data provide further support for a disorder of orbitofrontal cortex, particularly in Korsakoff-type dementia.

Cerebral metabolic changes were documented to correlate with EEG changes in patients with vascular dementia (123). δ Power was negatively correlated to glucose metabolism in selected regions over both hemispheres. Thalamic measures of glucose metabolism were correlated with frontal θ activity, whereas α activity and occipital glucose metabolism were correlated. β Power was not correlated with any specific region. More severe dementia correlated with decreased ratio of α activity between eyes closed and eyes open, but not with the amplitudes of other frequencies in patients with moderate-to-severe vascular dementia (124).

Progression of AD may include psychotic symptoms. Patients with dementia and psychosis are noted to have more abnormalities on EEG (125), but these abnormalities do not predict survival rates. EEG abnormalities in conjunction with MRI abnormalities and other physical abnormalities often are predictive of progressive cognitive decline (126,127) but are insufficiently specific to make accurate prognostication.

Persons with Down syndrome are considered to be at risk for early development of AD. A prospective study documented progressive slowing of α activity and eventual disappearance of α activity correlated with cognitive decline (128). Nondemented Down syndrome demonstrated increased amount of slowed α and decreased responsiveness to photic stimulation, similar to AD (129).

Although a normal EEG may be found in persons with dementing illness, the presence of specific abnormalities in EEG are not consistent with depression or pseudodementia within dissociative disorder (130). Increased slow activity, slowing of α frequency, and cordance measures all are indicators of the cognitive decline of dementia.

Pharmaco-Electroencephalography and Pharmaco-Magnetoencephalography

EEG and MEG can be useful tools for measurement of drug effects on brain function (131–133). What is most clear from all research to date is that there is a lack of drug-specific

TABLE 5.4. EFFECT OF PSYCHOACTIVE SUBSTANCES ON ELECTROENCEPHALO-GRAPHIC FREQUENCIES

	δ	θ	α	β
Benzodiazepines	↓			↑
Antipsychotics	↑	↑		
Tricyclics	↑			(↑)
Opioids	↑		↑	
Phenobarbital	↑		↓	(↑)
Methylphenidate				↓
Nicotine			↓	
Alcohol			↑	

Adapted from Sannita WG, Quantitative EEG in human neuropharmacology. Rationale, history, and recent developments. *Acta Neurol* 1990;12:389–409.

characteristic findings (EEG/MEG) for all individuals under all conditions.

Sannita (134) reviewed the history and use of QEEG in human neuropharmacology. His article reviews only studies in humans (Table 5.4), whereas much pharmaco-EEG has been performed in animals. The purposes of performing pharmaco-EEG are summarized as follows:

1. Study drugs of abuse
2. Classify psychoactive compounds
3. Attempt to predict the therapeutic indications for, or responses to, certain classes of drugs
4. Develop bioavailability and pharmacodynamic correlations
5. Attempt to classify subgroups of psychiatric patients
6. Predict responders to certain types of treatment.

Other factors also affect drug kinetics and bioavailability, including gastrointestinal and hormonal effects of shift work (135,136). Direct and indirect hormonal changes also may alter EEG patterns and influence responsiveness to medication (137).

One must be cautious in extrapolating pharmaco-EEG/pharmaco-MEG findings based on small doses of a compound or medicine given to normal controls. The results, although characteristic for the drug in normal subjects, will not necessarily indicate its mechanism of effect in the illness condition. For example, 1 to 5 mg haloperidol in a drug-free normal human control produces a different response than it would in a person with chronic schizophrenia who had been taking 40 mg haloperidol daily. Notwithstanding, certain categories of drug produce predictable responses; for example, benzodiazepines increase fast (β) activity. The best approach is to combine resting state and some form of activation condition specific to both the disease state being studied and to the presumed mechanism of action for the class of drug being studied.

Testing a medication for its efficacy to reduce pain is an example of an appropriate state-dependent application. Pain experienced acutely, however, is different from unremitting

chronic pain. Rate of onset of action and attributes of brain changes in response to different dosage of pain medication can be readily studied in healthy controls using the appropriate pain stimulus (138).

Within diseases, there are pharmacologic variables to study as well. Anticonvulsant medications have differing bioavailability, actions on CNS, and side effect profiles (139). Some of these drugs, such as valproic acid, have distinct acute and chronic effects on power spectrum and distribution of activity over the skull (140,141).

As mentioned in other sections of this chapter, EEG/MEG is a useful tool in the assessment of the different effects of similar classes of drug for treatment of neuropsychiatric disorders, such as schizophrenia or depression (45,91,92). Developing accurate predictive measures of response to treatment will be of real clinical utility in the near future.

Multiple Sclerosis

Multiple sclerosis (MS) affects the myelination of the axon, slowing their conduction. EEG and MEG show diffuse or focal patterns of slowing, depending on the area affected and severity of disease. Early in the course of illness, abnormal findings will return to normative values during remission. Impaired mental function can be associated with generalized slowing and, more rarely, epileptiform discharges (142). Periodic lateralized epileptiform discharges have been reported (143,144) but are neither specific for, nor characteristic of, MS.

BAEPs frequently are abnormal. Prolongation of the I-V peak interval indicates brainstem dysfunction. In contrast to patients with metabolic encephalopathy, patients with MS typically have a normal wave I. BAEP is most useful in patients who have a history consistent with MS and do not have clinically evident brainstem lesion(s). Middle latency auditory evoked potentials were found to detect abnormalities in the auditory system, without false-positive identification of cases (145). Many evoked potentials, however, are not sensitive enough to capture milder subclinical lesions (146).

Not infrequently, patients with MS develop seizure disorders due to lesions that have inflammatory processes and myelin breakdown products within them (147). If the lesion is small and transient, the EEG abnormalities resolve in the same time course as the flare. Unresolving lesions tend to be associated with continuing seizure(s). The prevalence of seizure in 2,353 cases of MS was 1.7% (148). In a study of 100 patients with MS with cortical atrophy, paroxysmal events in the EEG (25%) were observed more frequently than seizure (5%) (149). Sudden neurologic impairment may include nonconvulsive status and may respond to steroid and antiepileptic medication therapy (150).

Abnormal SSEPs are reported in most patients with MS (151). The most frequently ordered, confirmatory test for MS is the visual evoked potential (VEP) during an initial episode of acute onset of loss of visual acuity. Waveforms are prolonged overall and at specific latencies corresponding to sites of demyelinating plaques. Brau and Ulrich (152) reported improved visuomotor tracking performance and increased absolute α power during periods of remission compared to periods of exacerbation in 23 patients. Repeated SSEPs do not correlate with the course of the disease (153,154). Higher rates of VEP abnormalities were found with equiluminant chromatic stimuli compared with achromatic stimuli in 30 patients with MS (155). This supported the general vulnerability of color-opponent visual pathways in MS, with red-green and blue-yellow being about equally affected.

Pattern of cortical activation for voluntary movement in patients with fatigue were compared to patients without fatigue and normal controls (156). Event-related desynchronization was more widespread anteriorly in the fatigue group. Postmovement contralateral sensorimotor event-related synchronization (indicating cortical inhibition) was lower in the fatigue group. Other cortical measures affected in MS include decreased θ and α frequency coherence in patients with subcortical lesions on MRI and cognitive involvement (157).

Rudkowska et al. (158) reviewed patients who had clinical findings consistent with brainstem lesions that could be responsible for paroxysmal EEG activity. VEP N100 latency was slightly slower and the amplitude of the P100/N120 complex was slightly larger in these 12 patients compared with matched MS controls without paroxysmal activity on the EEG. Subcortical lesions were found to prolong P200 latencies, and P300 latency prolongation correlated with duration of illness in 101 patients with varying severity of MS (159). P300 prolongation and decreased amplitude are associated with cognitive decline in MS (160,161) but not necessarily with degree of disablement (162).

Complicated cases of MS reveal both the usefulness and limitations of electrophysiology to localize disturbances within the brain. Periodic synchronous discharges were reported in a 12-year-old girl with MS (163). This EEG pattern more frequently is identified with subacute sclerosing

panencephalitis. An example of the limits of EEG specificity is a case report of nonspecific diffuse θ activity in a 53-year-old woman with a 6-year history of MS (164). Cerebrospinal fluid (CSF), MRI, and other imaging techniques were used to confirm the diagnosis and evaluate her state of health during a complicated course with multiple pneumonias. Autopsy revealed continuous cystic lesions along the lateral ventricles from the frontal tips of the anterior horns to posterior and temporal horns of the lateral ventricles. There also were more classic demyelinated lesions in the white matter, brainstem, cerebellum, and spinal cord.

Attention Deficit/Hyperactivity Disorder

Attention deficit disorder (ADD) is drawing more clinical attention as the prevalence of this diagnosis is increasing. As might be imagined, EEG shows both increased and decreased activity in frontal-striatal systems (165). Increased frontal δ and θ activity is reported in most groups with ADD (166). Asymmetry of evoked responses (seen in hyperactivity) or of resting EEG (found in straight ADD) are not confirmed diagnostic group characteristics (167). Excess frontal θ activity was found in a separate study to correlate with good response to stimulant medications (168). Increased θ or α power and frontal θ/α coherence were correlated with good behavioral response to stimulant medications (78% correspondence) (169). Linden et al. (170) reported increased attentional functioning in children who received biofeedback to enhance β activity and suppress θ activity. Biofeedback was combined with teaching strategies for academic tasks, when the feedback indicated the patient was focused, in a study of people with ADD ages 5 to 63 years (171). Successful outcome was achieved as the individual achieved a decrease in the ratio of θ/β activity. Widespread recommendations for biofeedback, although popular in some regions of the country, are awaiting results of controlled, double-blinded trials.

Stroke

Physiologic measures can assist in the assessment of recovery from stroke and in quantifying the degree of alteration in normal function resulting from the stroke, such as motor function, sleep patterns, and alerting or sensory perceptual changes. Clinical neurologic examination combined with topographic EEG compares favorably to computed tomographic (CT) scan in accurate localization of lesions in the same patients (172).

To better understand the mechanisms of damage during stroke, several experimental models have been used, including carotid artery clamping. Patient candidates for endarterectomy have been studied preoperatively and intraoperatively with EEG to assess the risk of developing intraoperative or postoperative strokes. EEG changes indicating poorer neuronal function along with transcranial Doppler ultrasonography give immediate and accurate feedback to the surgeon about the state of perfusion of the brain, allowing for a reduced rate of intraoperative stroke (173,174). Preoperative reduction of sensory evoked potentials (SEPs), specifically P25 latency and amplitude, determined when patients should receive shunting during endarterectomy (175). Another study (176) found that risk of intraoperative stroke was correlated with major EEG changes but not with the amount of carotid occlusion or other risk factors, and shunting was not a useful technique to alter the risk of intraoperative stroke.

Topographic EEG provided more information on lateralization of deficit after stroke than routine EEG, including patients with negative CT findings (177). High correlation between QEEG focal abnormalities and lacunar infarct location by high CT was reported by Kappelle et al. (178).

Milandre et al. (179) reported that the incidence of stroke-induced epilepsy was greater with hemorrhagic stroke. Follow-up EEG after several weeks increased the specificity of EEG abnormalities in 63% of records, which initially showed only focal or diffuse slowing. Gras et al. (180) reported that periodic lateralized epileptiform discharges correlated with ischemic stroke in older persons.

As might be expected, sleep function is altered in stroke. As acute cerebral changes resolve, there is a shorter time to onset of sleep and to the first REM period. Domzal et al. (181) studied patients 4 weeks after stoke and found a higher proportion of stage 1 non-REM sleep in patients with right-sided middle cerebral artery stroke. Hachinski et al. (182) studied one patient, a 70-year-old woman, in whom they were able to obtain preinfarct and postinfarct sleep EEG records. In the immediate infarct state, slow activity increased on the side of the infarct, REM was not recorded at all until day 3, and sleep spindles almost disappeared. One year later, there remained persistent δ activity over the unaffected hemisphere, with other aspects of the sleep recording having normalized over both hemispheres.

EEG has been recorded in some cases of transient ischemic attacks, usually after resolution of the episode, because of the short duration of symptoms in most patients. Rapid dysrhythmia and diffuse temporal (right or left) θ activity were found by Constantinovici et al. (183) in their retrospective study of transient global amnesia. In a single case study, MEG and neuropsychological testing (Wechsler Memory Scale-Revised) were administered simultaneously 5 days and 1 month after transient global ischemia (184). Although there were no significant changes seen in routine EEG, MRI, or single photon emission computed tomographic (SPECT) studies, the MEG patterns of activation paralleled those of the memory performance, showing deficits in the immediate acute phase that resolved by 1 month.

A common problem that particularly affects elderly persons is chronic subdural hematoma. In a careful study of ten patients, Tanaka et al. (185) found that changes in brain wave activity correlated with reduced thalamic blood flow on the side of the hematoma. Over the hemisphere contralateral to the hematoma, EEG correlated with overall hemisphere blood flow.

One risk factor for stroke is arteriovenous malformation, which too frequently is discovered only after the stroke has occurred. Five patients with arteriovenous malformation, who presented with epilepsy and no intracranial hemorrhage, were studied using MEG, EEG, MRI, SPECT, and intraoperative electrocorticography and laser Doppler flow (186). MEG demonstrated high-frequency magnetic activity with an equivalent dipole around the nidus of epileptic activity confirmed by intraoperative electrocorticography. The same areas showed hypoperfusion areas on SPECT that were confirmed with intraoperative laser Doppler.

These studies demonstrate that EEG and MEG can be of assistance in assessing how much brain will be at risk for injury during an ischemic episode. These noninvasive techniques will be particularly useful in studies of neuroprotective strategies to prevent and reduce the size of strokes. They currently have demonstrated utility in clinical stroke cases when there are discrepant findings on anatomic brain imaging studies.

Migraine

Etiology of migraine headache is thought to differ from that of tension headache. Prevalent theories include an electrophysiologic spreading depression, and vascular contraction and dilation as causal factors in both the production of aura and subsequent painful headache. There are likely to be several etiologies for this clinical syndrome (187). When tension headache is present simultaneously with migraine, muscle artifact often obscures any other EEG patterns of activity.

Basilar migraine is characterized by symptoms referable to dysfunction of the brainstem. EEG studies show increased generalized δ activity that resolves after symptoms have subsided (188–190). Transient occipital spike wave complexes during basilar migraine attacks have been reported by several groups (191,192). Follow-up of children who outgrew basilar migraines showed normal EEGs (192). Childhood cyclic vomiting has been attributed to varying etiologies, including epilepsy, migraine, and somatoform disorder. In one case of cyclic vomiting, QEEG showed δ activity during the episodes that resolved with termination of vomiting (193). Transient alteration of the EEG can assist in the diagnosis of basilar migraine and confirmation of more than one illness process occurring simultaneously, as in one patient in whom basilar migraine was noted during coma (194). Fuller et al. (195) recorded symmetric frontal δ activity during migraine in an elderly man with time-limited psychosis, whose symptoms included formed visual hallucinations, delusions, and reduplicative paramnesia.

Some EEG patterns are seen with regularity in migraineurs. Corona et al. (196) postulate that EEG abnormalities indicate lowered ischemic threshold. They reported a higher incidence of abnormalities during headache in migraineurs with aura (93%) compared to those without aura (38%). α Power is reported to be decreased both during migraine attacks and during auras. Differentiating different types of headache still is not an easy task. Children with migraine with aura have an increased θ/α ratio in posterior temporal and occipital areas compared to those without aura, those with tension headache, or those serving as normal age-matched controls (197). Migraine more frequently is associated with a lateralized decrease in α power during an attack than tension headache (198). Routine EEGs may show minor and nonspecific slowing in 8% to 15% of those studied. Topographic EEG of young adults with common migraine, those with classic migraine, and those without headache (normal controls) showed lower peak α power, lower α reactivity, and slightly faster peak α frequency in the eyes closed state (199). Adults with migraine show increased asymmetry of α power (42%) and peak frequency (55%) over posterior regions and increased θ and δ activity over temporal regions bilaterally (65%) compared with normal matched controls (200). An experimental model of headache induction with nitroglycerin was treated with sumatriptan injection (201). During the induced common migraine attack, increased rhythmic θ and δ activity and reduced α and β activity were recorded. The EEG normalized within 30 minutes of the sumatriptan injection.

Topographic EEG was recorded during the course of migraines in children (202). α Power was decreased over the occipital area contralateral to the hemifield affected by the visual aura, followed by increased δ power over the frontal electrodes, and then, during the headache, increased δ power in posterior temporal and occipital areas. All abnormalities resolved with resolution of the headache.

In children, EEG is useful in establishing a correct diagnosis, particularly when the symptoms are vague. Unilateral hemispheric reduction in EEG activity during abdominal migraine (203) and episodic continuous diffuse β activity not due to a drug-induced confusional state (204) are examples of EEG as an important diagnostic measure.

Migraine sometimes is associated with epilepsy. Increased background slow activity is particularly seen in patients with epilepsy and classic migraine (205). Marks and Ehrenberg (206) studied 395 patients with seizures, 13 of whom had migraine aura immediately preceding their seizures. Distinctive EEG changes occurred during the aura in two patients and could be distinguished from partial seizure activity. Five patients had lateralized epileptiform discharges in close temporal relation to migraine attacks. Six patients who did not achieve good seizure control with antiepileptic drugs had improved seizure control with a combination of treatments directed at migraine and seizure. A case report of a 23-year-old man with visual aura and photosensitive seizures linked his epilepsy and migraine attacks (207). His seizures responded to valproate, although he continued to have rare visual auras.

De Romanis et al. (208) recorded occipital spike wave complexes that occurred only during migraine attacks in 13 children who also had seizures. Wilder-Smith and Nirkko (209) were able to distinguish between focal epilepsy and migraine by the combination of EEG and Doppler sonography. They reported one patient with episodes of visual symptoms who had nonspecific generalized changes on the EEG and showed increased blood flow velocity with a latency of a few seconds only in the posterior cerebral arteries. The pattern was consistent with autoregulatory hyperperfusion in response to increased neuronal activity.

The causation of both the aura and painful headache remain controversial. MEG studies have shown slow current shifts associated with auras (personal communications) that could support both a primary neuronal and a primary vascular/ischemic etiology. de Tommaso et al. (210) reported slowing and asymmetry of α activity during steady-state VEPs (27-Hz stimulation) during migraine attacks. The evoked response waveform was reduced in amplitude as well, but there was a marked increase in reactivity over the entire scalp during the intercritical period. These findings were consistent with the theory of spreading depression occurring in a situation of increased neuronal excitability. Others would argue that abnormalities shown in blood flow, using SPECT, support a vascular dysfunction for aura and headache (211,212). The brevity of most auras makes it difficult to reliably duplicate the clinical syndrome in the laboratory setting.

An innovative treatment was objectively recorded using EEG (213). Twenty children with clinically diagnosed migraine wore either rose-tinted or blue-tinted glasses for 4 months. The group that wore rose-tinted glasses had a reduction in mean migraine frequency from 6.2 to 1.6 episodes per month and showed a concomitant reduction in visually evoked β activity. The group wearing blue-tinted glasses had a transient improvement in headache at 1 month, which was not sustained, and no change in EEG recording.

As intriguing as many of these findings are, clinical judgment will continue to be the most reliable and cost-efficient method for routine evaluation of chronic headache.

Mental Retardation/Intellectual Disability

This heterogeneous group of disorders has a wide clinical spectrum, ranging from very mild to severe intellectual deficits. EEG and MEG do not diagnose the etiology for a particular type of intellectual disability or mental retardation. Often EEG shows a generalized dysrhythmia with preservation of α rhythm. The more severe the level of neurologic impairment, the more likely it is that the EEG will show disorganization, large slow waves, and occasional spike activity.

Diagnostic considerations are complicated by the fact that in more severe cases of retardation there often are concomitant disorders, such as epilepsy, drug intoxication, and movement disorders. EEG abnormalities typical of dementia were seen in patients older than 35 years who had epilepsy and

Down syndrome. Patients younger than 35 years, with or without epilepsy, showed no EEG abnormalities. The prevalence of epilepsy in adults with Down syndrome was 9.4% (214) and increased with age (46% in patients older than 50 years). Later age of onset was associated with poorer outcome and clinical signs and symptoms of dementia.

One attempt to integrate neuropsychological findings, clinical observations, and evoked potential models is the model proposed by Elliott and Weeks (215). They hypothesized that people with Down syndrome may have trouble performing movements on the basis of verbal information, because right hemisphere speech perception systems appear to be functionally isolated from left hemisphere systems of movement, planning, and execution. This was tested using a verbal-motor impairment index and dichotic listening tests. Persons with Down syndrome had a slight tendency for left ear advantage for speech sounds. Under verbal testing and instruction, patients with Down syndrome had more difficulty preserving the order of responses, and they produced more irrelevant responses. It seemed that the tendency for the right hemisphere to develop receptive language functions interferes with the more typical right hemisphere visual and spatial functions for these individuals. A different study found impairment in α activity as measured by the ratio between eyes closed/open states in 32 persons with Down syndrome compared to normal controls. This was an independent factor from visual activation impairment and did not correlate in a linear fashion with neuropsychological impairment.

The differential diagnosis of causes of mental retardation is complex and challenging. Evoked potentials were used to differentiate between children with Down syndrome and fetal alcohol syndrome and normal controls. Measures of latency and amplitude of P300 auditory (oddball-plus-noise) evoked responses effectively identified each group from the others (217). Early and later visual evoked potentials were reported to be delayed in a group of 7- to 10-year-old boys with mental retardation compared with age-matched normal controls (218). Cognitive evoked potentials, P300, and resting EEG were compared in a group of persons with mild mental retardation with equivalent verbal and performance IQ measures and in a smaller cohort whose performance IQ was ten points or greater than their verbal IQ (219). Decreased amplitude of P300 was observed in the first group and the α frequency was higher in the second group, with P300 closer to normal values. A different study compared a small group of adults with mental retardation in the moderate-to-profound range to a matched sample of nonretarded adults (220). Auditory oddball paradigm under inattentive conditions revealed no significant differences in P3a responses between groups but a markedly smaller amplitude and increased latency for the mismatch negativity response. This does not represent an idiosyncratic finding for this group of patients, but as with schizophrenia and other more circumscribed neuropsychiatric disorders it reflects difficulties in automatic detection of changes in sensory input.

More often, generalized EEG abnormalities are reported and concomitant epileptic syndromes are identified (221–225). Whenever episodic behavioral or cognitive changes are blindly attributed to mental retardation, nonconvulsive status epilepticus should be given serious consideration. Nonconvulsive status epilepticus can be identified using EEG/MEG (226) and will respond to treatment with anticonvulsant medications. In an isolated report, three patients with ring chromosome 20 were reported to have long bursts of rhythmic, sharply contoured θ waves, and spike waves associated with seizures of the frontotemporal origin (227).

The breadth of etiologies for mental retardation renders electrophysiologic techniques most useful as descriptors of particular states or as focused neurophysiologic examinations of a person's brain responding to specific stimuli.

Pain

Theory of pain mechanisms and perception has changed dramatically over the past few decades to include the type of stimulus and its energy, the individual's physiologic responses and perceptions, and the complexity of circuits in the CNS influencing continuing amplification or dampening of this altered perception (228). Memory for pain may produce significant alterations in EEG that are statistically indistinguishable from response to hypertonic saline (229). Sleep is reported to be interrupted in patients with chronic pain. A recent study suggests that the electrophysiologic measures of insomnia are the same in chronic pain as in primary insomnia (230). People with and without actual nighttime pain attributed their sleep difficulty to having pain; however, treating the pain did not improve the insomnia. Isolating components of brain activity that are involved in pain perception requires using complementary technologies. In one study comparing MEG and EEG measures of skin stimulus, the MEG response at 164 milliseconds was not affected by distraction; however, the 240- to 340-millisecond responses on the EEG were affected. Healthy subjects who underwent three different types of painful stimuli (muscle, joint, and cutaneous) during slow wave sleep showed different responses in a nine-channel EEG (231). Muscle pain induced more of an arousal response. Joint pain induced global changes of less δ and more α and β activity. No background changes were observed with cutaneous stimulation. It is difficult to know whether such acute changes apply to chronic states of pain.

There are several models for pain, ranging from diffuse perception such as in headache, to the acute pain experienced in physical trauma and the pain from habitually stimulated fibers as in phantom limb pain. Migraine is a form of headache that is poorly characterized. EEG, EMG, and EPs are used to better characterize effects of medications and mechanisms of pain induction and resolution (232). One to four days before a migraine attack, Siniatchkin et al. (233) found marked increases in δ power, β power, and early component of contingent negative variation in 20 female patients with migraine compared to 12 control subjects without migraine. After the migraine attack had passed, there were no differences between the EEGs of the two groups. Another approach used was to study the sleep architecture of patients during a nocturnal attack and during a pain-free interval (234). There was a loss of dimensional complexity during the first two non-REM sleep states during the migraine night, providing evidence of a global change of decreased activity during an attack.

Tonic muscle pain induced by a small injection of hypertonic saline was compared to vibratory stimulation of the same muscle group (brachioradialis muscle) (235). The δ and α frequency powers were significantly increased over contralateral parietal areas. Vibratory stimulus increased β power in the contralateral frontal area. Pain and somatotopic representation of painful input are known to change as a result of changes in input to the somatosensory cortex. Chronic pain associated with back or phantom limb pain is associated with changes in somatosensory evoked responses measured by EEG (236). EEG was combined with transcranial magnetic stimulation to document the extent of reorganization and plasticity in motor and sensory cortex of amputees with phantom limb pain (237). Maps of motor-evoked responses had larger amplitude over the hemisphere corresponding to the amputated side than for the intact side, documenting the enhanced cortical area involved in the response.

Perception of pain can be attenuated by distraction techniques, which are documented to affect EEG recordings more than MEG recordings to a painful CO_2 stimulus (238). Ice-cold pressor testing leads to decreased EEG amplitude and increased coherence over central recording areas (239). The most pronounced amplitude decrease occurred in the α frequency. Thus, peripheral stimuli produced identifiable central EEG changes. A similar study of 15 young right-handed adults showed α desynchronization on the side contralateral to stimulation (240). δ Activity increased bilaterally over frontal areas, particularly with immersion of the left hand.

Nonsteroidal antiinflammatory drugs are used commonly for control of pain. Bromm et al. (241) measured the effects of acetaminophen and antipyrine on pain perception, on evoked potentials, and on resting EEG in healthy volunteers in a placebo-controlled, double-blinded crossover study, using 20-millisecond electrical pulses as the noxious stimulus. Both drugs reduced pain perception by 6%, reduced stimulus-induced δ power in the EEG by 21%, and did not alter auditory evoked potentials (AEP) or reaction times. On the resting EEG, acetaminophen enhanced power in the θ range and antipyrine depressed α frequency. It would be worthwhile to determine if people who experience little to no relief of pain from these drugs have the same physiologic responses as the subjects of this study.

EEG is a useful tool in animal models of pain and the effect of pain eradication. Van Praag et al. (242) demonstrated

in a developing rat that the excitatory effects of opiates (injections of morphine into neonatal rats) do not occur until the third week of life. EEG spikes can be produced by increasing doses of morphine during the second week of life, but these spikes are not reversible by opiate antagonists. It is not until the third week that the effects of morphine begin to mimic the effects in adults. Human subjects also are used to assess response to analgesic drugs. A small number of epochs are required to detect somatosensory evoked responses to tooth pulp electrical stimulation (243). This model was used in a study of efficacy of codeine. Clinical efficacy was documented by visual analogue scale and a decrease in evoked potential amplitudes with codeine.

EEG and MEG are only in the preliminary stages of assisting the understanding of pain. Working on sensory mechanisms, pain perception, and tolerance for types of pain, electrophysiologic techniques are adding to data obtained from pharmacologic and behavioral studies.

Traumatic Brain Injury

Physical examination and anatomic imaging studies (e.g., CT) are the most relevant sources of information in the immediate injury period. When the traumatic brain injury (TBI) is mild, subtle and persistent deficits in cognitive function can be difficult to evaluate. The routine EEG rarely shows focal or global changes (244). The challenge to the patient and the clinician is to differentiate among psychiatric illness, embellishment or malingering, and integrative difficulties (sensory, cognitive, motor). Sleep EEG and neuroendocrine dysregulation were studied in young men who were not depressed after severe TBI (245). The patterns were similar to persons with depression in remission, giving support to the importance of monitoring for possible reversible conditions.

Evoked potentials have been used as tools to assess potential for recovery from severe TBI, before cognitive assessment was clinically possible. SSEPs have been reported to be useful predictors of good recovery in patients who were in coma due to TBI (246). Eight of 29 patients with bilaterally absent N20 responses died, whereas 14 of 20 with normal SSEPs had good outcomes. In a prospective study of 60 patients who were comatose due to severe head injury, BAEPs and EEG were correlated with 6-month and 1-year outcomes (247). β Activity in frontocentral and centrotemporal regions of the left hemisphere correlated positively with outcomes at 6 months and 1 year. Interhemispheric coherence did not help to predict the severity of diffuse axonal injury. BAEPs were not correlated with outcome measures. A separate study of patients with documented intracerebral lesions also found utility in SSEPs of median and tibial nerves (248). All patients with bilateral loss of median nerve SSEP had poor outcomes (Glasgow Outcome Score = 2). The volume of brainstem lesions was strongly correlated with SSEP abnormalities. From these initial studies the reader can infer that

EEG and EP, as independent measures of brain activity, are useful tools in developing prognostications for recovery from coma.

EEG is used to monitor the stability of adult and pediatric patients in the intensive care unit during the acute injury phase of illness. Because of the limitations of space and demands for other equipment in the same physical space, EEG often has been limited to single-channel recordings. Even a single channel can provide documentation of paroxysmal events, general arousal, change in level of arousal, and drug effects (249). Limitations of single-channel recordings include the lack of regional information and relatively small chances of noting changes distant from the electrode site. Different types of EEG abnormalities are associated with different modes of brain injury in the preterm infant. Watanabe et al. (250) reported that serial EEGs were useful in demonstrating the origin (antenatal vs. postnatal) of cerebral palsies and periventricular leukomalacias. An example of the utility of EEG in evaluating drug efficacy is a study of the addition of ketamine during propofol sedation (251). At three doses of ketamine, intracranial pressure was objectively reduced, and the EEG showed a low-amplitude fast activity with burst suppression pattern in the eight patients studied. There was no reduction in cerebral perfusion pressure, oxygen saturation, or middle cerebral artery blood flow.

Physiologic measures of cognitive and integrative functions after brain injury are becoming more standard and more readily available. Mild TBI may result in generalized increases in low-frequency magnetic field activity (252). EEG coherence between pairs of closely spaced electrodes and between pairs of distantly spaced electrodes may reflect the effectiveness of neural synchronization after TBI (253). The severity of TBI has more recently been modeled by interaction between the resting QEEG and three-dimensional T2 MRI relaxation times (254). Thatcher et al. (255) proposed a severity index of TBI using QEEG spectral analysis. This work has yet to be independently replicated. However, QEEG (256) and cognitive ERPs (257) provide more sensitive and objective measures of deficits of cognitive function than traditional EEG or EPs. Even in the absence of detectable lesions by neuroimaging (CT, MRI), P300 event-related potentials had increased latency in patients with mild TBI with traumatic headache (258). Difficulties in sensory integration may interfere with cognitive performance as well. One measure of sensory gating is elicited by paired clicks to the ears and measuring the auditory evoked potentials at 50 milliseconds. The auditory P50 ratio was significantly greater in patients with symptomatic TBI compared with control subjects (259).

Seizures are frequent sequelae of TBI, resulting from neuronal lesions or the secondary scarring. Evaluation of the source of the seizures, as in idiopathic epilepsy, may help to direct treatment and improve outcomes (260). Patients with TBI may develop nonepileptic seizures as well. The risk of treating nonepileptic seizures with antiepileptic medications is greater in these patients because of the likelihood of impairing cognitive function and reinforcing their inability to function. Self-control relaxation paradigms can be effective treatments for nonepileptic seizures and for some types of partial seizures. Conder and Zasler (261) reported that an aphasic brain-injured patient responded well to individualized self-control treatment with improved responsiveness to antiepileptic medication, as well as a decrease in nonepileptic seizures. In this situation, EEG was used to make the differential diagnosis between the two types of seizures.

Sports often cause brain trauma. Boxing is the sport in which the head is one of the targets. Dementia pugilistica is a traumatic encephalopathy (262). Critchley (263) found increased frequency of EEG abnormalities in boxers who had this encephalopathy, although there was no correlation between the degree of encephalopathy and abnormalities of the EEG. Repeated fights in short periods of time do not give the brain sufficient time to recover from the trauma. Immediately (15–30 minutes) after a fight, the EEG shows diminished amplitude and slowed irregular θ activity (264). In older fighters these differences due to the fight are not seen as readily. In all likelihood, the effects of aging and repeated injury have damaged the brain to a degree that acute change is not as likely to stand out against the chronically slower background activity. The reader is reminded that the presence of a normal EEG does not exclude either the presence of encephalopathy or its imminent development.

Over the past decade, novel uses and exploratory research into applications of QEEG are helping us understand specific conditions and how to derive useful information from the EEG. QEEG was applied in follow-up study of individuals who survived the Chernobyl accident in 1986 (265). Three to five years after radiation exposure, paroxysmal activity had shifted to the left frontotemporal region. Five to eight years later, excess fast and slow activity with reduced amounts of α and θ activity were recognizable patterns. The authors noted that dose–effect relationships were identified in the EEG patterns of activity and affected cortical-limbic systems preferentially.

Use of QEEG as a tool in the rehabilitative process is being actively pursued, although it is difficult to document that the biofeedback has a direct effect (266,267). Subjective improvements in rates of depression, fatigue, and occupational and social functioning may not be dependent upon the focus on changing a specific brain wave pattern (the task given to the subject) but may reflect the benefit of repeated low-stress focused attention.

Autism

Disorders of developmental onset, including autism, pervasive developmental disorder, and Asperger syndrome, do not have pathognomonic EEG/MEG findings. Minor generalized EEG abnormalities are not uncommon but are made more compelling by their association with a family history significant for affective disorder (268). Epileptic seizures are reported frequently in children with autism (4%–32%) (269). Rossi et al. (269) noted that the higher incidence of epilepsy and paroxysmal activity on EEG in a sample of 106 patients with autism was related to genetic factors responsible for autism and for epilepsy. Epileptiform activity was noted in 82% of a sample of 50 children with autism or pervasive developmental disorder (270) studied during slow wave sleep. The multifocal epileptiform pattern was similar to that seen in children with Landau-Kleffner syndrome (acquired language disorder caused by epileptiform activity) in this study sample. Nass et al. (271) reported on seven children with autism who had occipital spikes and cognitive difficulties, suggesting a causal relation. The rate of seizure disorder may be higher in females than in males with autism (272), although the clinical severity of autistic symptoms does not correlate with severity of seizure disorder. Rarely do the symptoms of autism improve with the treatment for concomitant epilepsy (273,274). At least one case has been reported where the autistic syndrome appeared to be due to temporal lobe epilepsy (275).

Dysfunctional maturation of monoaminergic neurons may contribute to the poorly differentiated sleep pattern seen in persons with autism (276,277). Abnormalities of the phasic components of sleep and REM sleep component intrusion into non-REM sleep periods were seen in patients with autism and in patients with Down syndrome (276). In the Down syndrome group lower IQ was correlated with reduced REM sleep, whereas there was no correlation with IQ among patients with autism and reduced REM sleep.

Attention processing deficits are implicated in autism and corroborated in electrophysiologic studies. Ciesielski et al. (278) demonstrated lack of normal modulation of the slow negative wave (P3b) during final decision making about target selection in high-functioning autistic adults compared to matched controls. Visuospatial processing was demonstrated to be abnormal in patients with autism with normal visual processing in a study by Townsend et al (279). The authors suspected that the spatial attention deficits reflected cerebellar influence on frontal and parietal spatial attention functions. Mid latency auditory evoked responses also were found to be abnormal in adult subjects with autism (280). P1 responses did not show the normal decrement as the rate of stimulation was increased.

Other studies implicate disordered functioning in frontal and parietal areas in autism. A 25-year-old man with autism was given neuropsychological tests and found to have right cerebral dysfunction, with QEEG relative activation over the right frontal region (281). Frontal paroxysmal activity was identified in a separate series of EEGs from children with autism as they moved into adolescence (282). Of 86 patients with autism studied during sleep, 37 had epileptic discharges, with a majority over the mid frontal region (283). A separate study of EEG in children with autism versus normal controls revealed reduced EEG power in frontal and temporal regions, but no differences in parietal regions (284). Passive autistic children also had reduced α power in the frontal region. An earlier review of pineal gland function and maturation of the α rhythm suggested that disorders of cerebral maturation, such as autism, may be related to disturbances of pineal melatonin function in early life (285). This hypothesis implies a differential influence of the pineal on maturation of the left hemisphere, perhaps accounting for left hemisphere vulnerability to cerebral insult.

Acquired Immunodeficiency Syndrome/Human Immunodeficiency Virus

As treatment for human immunodeficiency virus (HIV) infection has prolonged the duration that people can survive viral load without development of full-blown illness, electrophysiology has become a useful tool to monitor cognitive function and evaluate the progressive effect of the virus on brain functioning. Anatomic imaging techniques are generally not sensitive enough to follow disease progression. In the late stages of disease, when dementia is the predominant neuropsychiatric clinical feature, EEG and MEG show patterns consistent with neuronal loss and the disorganization of cognitive processing.

In studies of asymptomatic HIV-positive individuals, lateralization of EEG spectra correlated with immune competence and compromise 2 to 3 years later (286). Those with superior performance on left hemisphere functions (word fluency, semantic processing, and word/face processing) and more optimistic mood did not convert their immunologic status over the 30-month prospective study. P300 latency increases with increased viral load and decline of cognitive capability. Repeated auditory evoked responses were compared between HIV-infected men and unaffected young controls to search for task-induced fatigue (287). HIV patients had greater prestimulus δ power over frontal areas and less prestimulus δ, θ, and α power over central and parietal areas compared with the control group.

Serial EEGs demonstrate progressive gradual slowing of background activity over 1 to 2 years (288). The percentage of spontaneous dysrhythmias increased from 31% to 42%; foci did not change in frequency; α rhythm decreased from 10.7 to 10.0 Hz and increased in amplitude from 60.9 to 69.5 μV. QEEG changes closely correlated with psychiatric symptoms (using the Present State Examination) in symptomatic, but not asymptomatic, HIV infection (289). Several studies report that an increase in amplitude of α precedes cognitive and neurologic impairment in the symptomatic stage of infection (290–292). Antiretroviral medication suppressed this α elevation (290). Although unspecific, the changes in α rhythm appear to be the earliest sign of HIV brain involvement. This is consistent with the hypothesis that the pattern of brain infection by HIV is subcortical to cortical.

Event-related potentials (auditory) are sensitive markers for neuropathologic processes. N100 responses were delayed in seropositive stage IV individuals and N200 responses were delayed in HIV-positive individuals, irrespective of clinical symptoms (293). P300 responses to identification of "odd-ball" target tones showed abnormal morphology to the rapid dichotic stimulus presentations in asymptomatic persons with HIV infection for 3 months to 8 years (294). Comparisons between groups infected with HIV consistently reveal worsening of multimodal or complex evoked potential responses that correlate with worsening immunologic status or worsening of clinical status (295–301). However, abnormalities in event-related potentials do not predict imminent dementia (302). There may be significant benefit from instituting or continuing relevant treatment. Antiviral treatment (zidovudine) has been associated with lack of progressive change in δ and θ amplitude and amount in a drug-versus-placebo comparison (303). Effectiveness of treatment also was documented using EEG in children with severe acquired immunodeficiency virus encephalopathy (304).

Dementia is accompanied by progressive neuropsychological decline, including slowed verbal responses and mutism in end-stage illness. Generalized low-amplitude activity with increased amounts of slow wave activity was found in subjects with HIV dementia (305). Severe prolongation of N200 and P300 latencies have been reported (306). At the terminal stages of illness, electrophysiology is not likely to be used for evaluating disease progression or response to interventions.

Other Neuropsychiatric Disorders

There are many appropriate uses of electrophysiology to better comprehend how the brain is functioning under unusual or novel conditions. For example, children and adolescents with mitochondrial encephalomyopathies have slowing of α rhythm and epileptiform discharges (307). Depending upon the interval until repeat study, these findings may be fairly constant. Combining MEG/EEG with positron emission tomography and MRI has led to progressive understanding of the topographic organization of sensorimotor cortex (308). Evoked potentials not only are used to demonstrate intact sensory function, but also to document alterations in response to disease processes.

Somatosensory evoked potentials/fields (SSEP/SSEF) may be a useful tool when there is a focal lesion. The amplitude of response will generally be larger over the side, or site, of the lesion. For example, Furlong et al. (309) described the case of a 75-year-old man with a highly cellular, poorly differentiated metastatic tumor located in the area of the right central sulcus and slightly displacing the prefrontal gyrus anteriorly. There was a significant interhemispheric difference in SSEPs. The patient had presented with a history of paresthesia and occasional involuntary movements of the left thumb and index finger. On examination there was some loss of joint position sense and two-point dis-

crimination in the left hand, with diminished stereognosis. During recording, no involuntary movement or EEG epileptiform activity was recorded. SSEPs were obtained for both the right and left wrist stimulation. This resulted in augmentation of the right hemisphere P22 component with relative attenuation of all other right hemisphere components. The P22/M31 complex represented a "giant" potential relative to a comparison control group, associated with an interhemispheric amplitude asymmetry greater than 50% and maxima located over C3/C4 electrodes for the respective sides. Other latencies were not grossly different between the hemispheres or from standard controls. This suggests that the P22 and M31 complexes arise from central and prefrontal regions.

Evoked potential studies and resting and activated-state studies using EEG in myasthenia gravis were reviewed by Keesey (310) and do not support a direct effect of myasthenia gravis upon the CNS. It was noted that inadequate oxygenation and inadequate sleep may produce secondary impairments in CNS functioning.

QEEG was used to document an association between gut and brain (311) in irritable bowel syndrome. Twenty-four patients with irritable bowel syndrome and 24 matched normal controls were studied with EEG, at rest and under mental tasks, without and with neostigmine. Patients with irritable bowel syndrome had a significantly greater percentage of β power. This CNS activity correlated with measures of colonic motility only in the patient group.

Working memory and attention have different patterns of cortical activation in high-ability individuals (more parietal activation) than in low-ability individuals (more frontal activation) (312). α Activity was relatively attenuated during spatial tasks, especially over the posterior right hemisphere (313), whereas θ activity increased with both increased practice and more effortful attention. Frontal θ increased and α activity decreased during progressively more difficult working memory load (314). There is much fluctuation in levels of attention and concentration during any task. For this reason, the robustness of EEG characteristics and of the tasks was tested at 1-hour and 7-day intervals (315). Resting EEG showed high reliability ($r = 0.7$), and task-related EEG was extremely stable during working memory tasks ($r = 0.9$) and psychomotor vigilance ($r = 0.8$). Clinical changes in cognitive status may be reliably assessed using these tools. Similar results were obtained in assessing people during a controlled alcohol ingestion setting (316). Practice-related changes in patterns of cortical activity, particularly θ and α amplitude and distribution, are associated with task processing (317) and may be usefully monitored with EEG.

Activation patterns and responsiveness to feedback were examined with EEG in people with obsessive-compulsive disorder (318). Patients with obsessive-compulsive disorder showed delayed onset of μ event-related desynchronization with movement preparation (moving the thumb spontaneously) and less movement β synchronization compared with normal subjects. Delayed event-related desynchronization in obsessive-compulsive disorder is consistent with involvement of structures related to motor programming, such as basal ganglia.

CONCLUSIONS

The fields of EEG and MEG are evolving rapidly. There is a long-standing realization among users of these physiologic techniques that individualized applications are relevant (319) and further our understanding of the expression of disorders in the brain (320,321).

The greatest importance of EEG and MEG are as techniques for recording brain *function.* The particular value of EEG/MEG to the neuropsychiatrist is that it remains directly tied to neuronal function, normal and abnormal, without reliance on uptake of labeled substances such as required by metabolic or blood flow measures.

Focused application to clinical and research questions will be the most rewarding approaches to the use of EEG or MEG. Most disorders that catch the attention and interest of the neuropsychiatrist wax and wane, vary in their response to pharmacologic treatment(s), and are affected by the state of the individual (stressed/nonstressed). Careful study of physiologic changes that fluctuate in meaningful patterns from minute to minute, even from millisecond to millisecond, are needed to develop understanding of CNS function in neuropsychiatric disorders.

REFERENCES

1. Niedermeyer E, Lopes da Silva F, eds. *Electroencephalography: basic principles, clinical applications, and related fields,* 4th ed. Philadelphia: Williams & Wilkins, 1999.
2. Hughes JR, Wilson WP, eds. *EEG and evoked potentials in psychiatry and behavioral neurology.* Boston: Butterworths, 1983.
3. Wyllie E. *The treatment of epilepsy: principles and practice,* 3rd ed. Philadelphia: Lippincott Williams & Wilkins, 2001.
4. Speckmann E-J, Elger CE, Altrup U. Neurophysiologic basis of the electroencephalogram. In: Wyllie E, ed. *The treatment of epilepsy: principles and practice,* 3rd ed. Philadelphia: Williams & Wilkins, 2001:149–164.
5. Epstein CM. Technical aspects of EEG: an overview. In: Wyllie E, *The treatment of epilepsy: principles and practice,* 2nd ed. Baltimore: Williams & Wilkins, 1997:218–227.
6. Stephenson WA, Goetz FA. Balanced non-cephalic reference electrode. *Electroencephalogr Clin Neurophysiol* 1951;3:237–240.
7. Gevins A. High resolution EEG. *Brain Topogr* 1993;5:321–325.
8. Gevins A, Le J, Martin NK, et al. High resolution EEG: 124-channel recording, spatial deblurring and MRI integration methods. *Electroencephalogr Clin Neurophysiol* 1994;90:337–358.
9. Mitler MM. The multiple sleep latency test as an evaluation for excessive somnolence. In: Guilleminault C, ed. *Sleeping and waking disorders: indications and techniques.* Menlo Park, CA: Addison-Wesley, 1982:145–153.
10. Robinson SE. Theory and properties of lead field synthesis analysis. In: Williamson SJ, Hoke M, Stroink G, et al., eds. *Advances in biomagnetism.* New York: Plenum Press, 1989:599–602.
11. Nuwer, M. Assessment of digital EEG, quantitative EEG, and EEG brain mapping: report of the American Academy of Neurology and the American Clinical Neurophysiology Society. *Neurology* 1997;49:277–292.
12. Hoffman DA, Lubar JF, Thatcher RW, et al. Limitations of the American Academy of Neurology and American Clinical Neurophysiology Society paper on QEEG. *J Neuropsychiatry Clin Neurosci* 1999;11:401–407.
13. Thatcher RW, Moore N, John ER, et al. QEEG and traumatic brain injury: rebuttal of the American Academy of Neurology 1997 report by the EEG and Clinical Neuroscience Society. *Clin Electroencephalogy* 1999;30:94–98.
14. Brivio L, Grasso R, Slavaggio A, et al. Brain stem auditory evoked potentials (BAEPs): maturation of interpeak latency I-V (IPL I-V) in the first years of life. *Electroencephalogr Clin Neurophysiol* 1993;88:28–31.
15. Tatum WO, Winters L, Gieron M, et al. Outpatient seizure identification. Results of 502 patients using computer-assisted ambulatory EEG. *J Clin Neurophysiol* 2001;18:14–19.
16. Ebersole JS. New applications of EEG/MEG in epilepsy evaluation. *Epilepsy Res Suppl* 1996;11:227–237.
17. Huotilainen M, Winkler I, Alho K, et al. Combined mapping of human auditory EEG and MEG responses. *Electroencephalogr Clin Neurophysiol* 1998;108:370–379.
18. van den Broek SP, Reinders F, Donderwindel M, et al. Volume conduction effects in EEG and MEG. *Electroencephalogr Clin Neurophysiol* 1998;106:522–534.
19. Malmivuo J, Suihko V, Eskola H. Sensitivity distributions of EEG and MEG measurements. *IEEE Trans Biomed Eng* 1997;44:196–208.
20. Dale AM, Halgren E. Spatiotemporal mapping of brain activity by integration of multiple imaging modalities. *Curr Opin Neurobiol* 2001;11:202–208.
21. Lantz G, Spinelli L, Menendez RG, et al. Localization of distributed sources and comparison with functional MRI. *Epileptic Disord Special Issue* 2001;July:45–58.
22. Salanova V, Morris HH, Van Ness PC, et al. Comparison of scalp electroencephalogram with subdural electrocortigram recordings and functional mapping on frontal lobe epilepsy. *Arch Neurol* 1993;50:294–299.
23. Devinsky O. Patients with refractory seizures. *N Engl J Med* 1999;340:1565–1570.
24. Canive JM, Lewine JD, Edgar JC, et al. Magnetoencephalographic assessment of spontaneous brain activity in schizophrenia. *Psychopharmacol Bull* 1996;32:741–750.
25. Joutsiniemi SL, Gross A, Appelberg B. Marked clozapine-induced slowing of EEG background over frontal, central, and parietal scalp areas in schizophrenic patients. *J Clin Neurophysiol* 2001;18:9–13.
26. Tiihonen J, Hari R, Naukkarinen H, et al. Modified activity of the human auditory cortex during auditory hallucinations. *Am J Psychiatry* 1992;149:255–257.
27. Ortiz T, Fernandez A, Maestu F, et al. Applications of the magnetoencephalography in the study of schizophrenia. *Actas Esp Psiquiatr* 1999;27:259–263.
28. Reite M, Teale P, Goldstein L, et al. Late auditory magnetic sources may differ in the left hemisphere of schizophrenic patients. *Arch Gen Psychiatry* 1989;46:565–572.
29. Shagass C, Roemer RA, Straumanis JJ, et al. Temporal variability of somatosensory, visual and auditory evoked potentials in schizophrenia. *Arch Gen Psychiatry* 1979;36:1341–1351.
30. Shagass C, Roemer RA, Straumanis JJ, et al. Topography of sensory evoked potentials in depressive disorders. *Biol Psychiatry* 1980;15:183–207.
31. Faux SF, McCarley RW, Nestor PG, et al. P300 topographic asymmetries are present in unmedicated schizophrenics. *Electroencephalogr Clin Neurophysiol* 1993;88:32–41.

32. Adler LE, Waldo MC, Freedman R. Neurophysiologic studies of sensory gating in schizophrenia: comparison of auditory and visual responses. *Biol Psychiatry* 1985;20:1284–1296.

33. Waldo MC, Adler LE, Freedman R. Defects in auditory sensory gating and their apparent compensation in relatives of schizophrenics. *Schizophr Res* 1988;1:19–24.

34. Braff DL, Grillon C, Geyer MA. Gating and habituation of the startle reflex in schizophrenic patients. *Arch Gen Psychiatry* 1992;49:206–215.

35. Verleger R, Wascher E, Arolt V, et al. Slow EEG potentials contingent negative variation and post-imperative negative variation in schizophrenia: their association to the present state and to Parkinsonian medication effects. *Clin Neurophysiol* 1999;110:1175–1192.

36. Linnville S, Teal P, Scheunemann D, et al. Schizophrenia may alter neuromagnetic representations of attention. *J Neuropsychiatry Clin Neurosci* 1995;7:92–95.

37. Teale P, Reite M, Rojas, et al. Fine structure of the auditory M100 in schizophrenia and schizoaffective disorder. *Biol Psychiatry* 2000;48:1109–1112.

38. Rojas DC, Teale P, Sheeder J, et al. Sex differences in the refractory period of the 100 ms auditory evoked magnetic field. *Neuroreport* 1999;10:3321–3325.

39. Rojas DC, Teale P, Sheeder J, et al. Sex-specific expression of Heschl's gyrus functional and structural abnormalities in paranoid schizophrenia. *Am J Psychiatry* 1997;154:1655–1662.

40. Reite M, Sheeder J, Teale P, et al. Magnetic source imaging evidence of sex differences in cerebral lateralization in schizophrenia. *Arch Gen Psychiatry* 1997;54:433–440.

41. Tandon R, Shipley JE, Taylor S, et al. Electroencephalographic sleep abnormalities in schizophrenia. Relationship to positive/negative symptoms and prior neuroleptic treatment. *Arch Gen Psychiatry* 1992;49:185–194.

42. Keshavean MS, Reynolds CF, Miewald JM, et al. Delta sleep deficits in schizophrenia: evidence from automated analyses of sleep data *Arch Gen Psychiatry* 1998;55:443–448.

43. Tandon R, Shipley JE, Taylor S, et al. Sleep abnormalities in schizophrenia: cholinergic contribution. *Clin Neuropharmacol* 1992;15:294A–295A.

44. Roschke J, Wagner P, Mann K, et al. An analysis of the brain's transfer properties in schizophrenia: amplitude frequency characteristics and evoked potentials during sleep. *Biol Psychiatry* 1998;43:503–510.

45. Small JG, Milstein VM, Small IF, et al. Computerized EEG profiles of haloperidol, chlorpromazine, clozapine and placebo in treatment resistant schizophrenia. *Clin Electroencephalogr* 1987;18:124–135.

46. Malaspina D, Maclin E, Comblatt B, et al. Stimulant/Haldol study of SPEM, CPT, and P50 in normals. APA Annual Meeting, San Francisco, California, 1989.

47. Ahveninen J, Kahkonen S, Tiitinen H, et al. Suppression of transient 40-Hz auditory response by haloperidol suggests modulation of human selective attention by dopamine D2. *Neurosci Lett* 2000;292:29–32.

48. Karson CN, Coppola R, Morihisa JM, et al. Computed electroencephalographic activity mapping in schizophrenia: the resting state reconsidered. *Arch Gen Psychiatry* 1987;44:514–417

49. Kasai K, Yamada H, Kamio S, et al. Do high or low doses of anxiolytics and hypnotics affect mismatch negativity in schizophrenic subjects? An EEG and MEG study. *Clin Neurophysiol* 2002;113:141–150.

50. Hellwig B, Haussler S, Lauk M, et al. Tremor-correlated cortical activity detected by electroencephalography. *Clin Neurophysiol* 2000;111:806–809.

51. Marsden JF, Ashby P, Limousin-Dowsey P, et al. Coherence between cerebellar thalamus, cortex and muscle in man: cerebellar thalamus interactions. *Brain* 2000;123[Pt 1]:1459–1470.

52. Delwaide PJ, Pepih JL, Maertens de Noordhout A. The audiospinal reaction in parkinsonian patients reflects functional changes in reticular nuclei. *Ann Neurol* 1993;33:63–69.

53. Beckley DJ, Bloem BR, Remler MP. Impaired scaling of long latency postural reflexes in patients with Parkinson's disease. *Electroencephogr Clin Neurophysiol* 1993;89:22–28.

54. Wascher E, Verleger R, Vieregge P, et al. Responses to cued signals in Parkinson's disease. Distinguishing between disorders of cognition and of activation. *Brain* 1997;120[Pt 8]:1355–1375.

55. Nakashima K, Shimoyama R, Yokoyama Y, et al. Auditory effects on the electrically elicited blink reflex in patients with Parkinson's disease. *Electroencephalogr Clin Neurophysiol* 1993;89:108–112.

56. Pfurtscheller G, Pichler Zalaudek K, Ortmayr B, et al. Postmovement beta synchronization in patients with Parkinson's disease. *J Clin Neurophysiol* 1998;15:243–250.

57. Plat FM, Praamstra P, Horstink MW. Redundant-signals effects on reaction time, response force, and movement-related potentials in Parkinson's disease. *Exp Brain Res* 2000;130:533–539.

58. Brown P, Marsden CD. Bradykinesia and impairment of EEG desynchronization in Parkinson's disease. *Mov Disord* 1999;14:423–429.

59. Wang HC, Lees AJ, Brown P. Impairment of EEG desynchronization before and during movement and its relation to bradykinesia in Parkinson's disease. *J Neurol Neurosurg Psychiatry* 1999;66:442–446.

60. Chen K, Abrams BM, Brody JA. Serial EEGs of patients with Parkinsonism-dementia syndrome of Guam. *Electroencephogr Clin Neurophysiol* 1968;25:380–385.

61. Tanaka H, Keonig T, Pascual-Marqui RD, et al. Event-related potential and EEG measures in Parkinson's disease without and with dementia. *Dement Geriatr Cogn Disord* 2000;11:39–45.

62. Zijlmans JC, Pasman JW, Horstink MW, et al. EEG findings in patients with vascular parkinsonism. *Acta Neurol Scand* 1998;98:243–247.

63. Buysse DJ, Hall M, Tu XM, et al. Latent structure of EEG sleep variables in depressed and control subjects: descriptions and clinical correlates. *Psychiatry Res* 1998;79:105–122.

64. Krieg JC, Lauer CJ, Hermle L, et al. Psychometric, polysomnographic, and neuroendocrine measures in subjects at high risk for psychiatric disorders: preliminary results. *Neuropsychobiology* 1990;23:57–67.

65. Buysse DJ, Kupfrer DJ, Cherry C, et al. Effects of prior fluoxetine treatment on EEG sleep in women with recurrent depression. *Neuropsychopharmacology* 1999;21:258–267.

66. Giles DE, Perlis ML, Reynolds CF 3rd, et al. EEG sleep in African-American patients with major depression: a historical case control study. *Depress Anxiety* 1998;8:58–64.

67. Dierks T. Equivalent EEG sources determined by FFT approximation in healthy subjects, schizophrenia and depressive patients. *Brain Topogr* 1992;4:207–213.

68. Zeitlhofer J, Anderer P, Obergottsberger S, et al. Topographic mapping of EEG during sleep. *Brain Topogr* 1993;6:123–129.

69. Debener S, Beauducel A, Nessler D, et al. Is resting anterior EEG alpha asymmetry a trait marker for depression? Findings for healthy adults and clinically depressed patients. *Neuropsychobiology* 2000;41:31–37.

70. Lindgren KA, Larson CL, Schaefer SM, et al. Thalamic metabolic rate predicts EEG alpha power in health control subjects but not in depressed patients. *Biol Psychiatry* 1999;45:943–952.

71. Lu ST, Kajola M, Joutsiniemi SL, et al. Generator sites of spontaneous MEG activity during sleep. *Electromyogr Clin Neurophysiol* 1992;82:182–196.

72. Wang G, Takigawa M. A non-linear method for estimating the alpha generators from an EEG over the scalp. *Frontiers Med Biol Eng* 1992;4:169–179.

73. Tiihonen J, Hari R, Kajola M, et al. Magnetoencephalographic 10-Hz rhythm from the human auditory cortex. *Neurosci Lett* 1991;129:303–305.

74. Fulton MK, Armitage R, Rush AJ. Sleep electroencephalographic coherence abnormalities in individuals at high risk for depression: a pilot study. *Biol Psychiatry* 2000;47:618–625.

75. Schatzberg AF, et al. Topographic mapping in depressed patients. In: Duffy FH, ed. *Topographic mapping of brain electrical activity.* Boston: Butterworths, 1986.

76. Williamson PC, Kaye H. EEG mapping applications in psychiatric disorders. *Can J Psychiatry* 1989;34:680–686.

77. Nofzinger EA, Price JC, Meltzer CC, et al. Towards a neurobiology of dysfunctional arousal in depression: the relationship between beta EEG power and regional cerebral glucose metabolism during NREM sleep. *Psychiatry Res* 2000;98:71–91.

78. Armitage R, Hoffmann RF, Rush AJ. Biological rhythm disturbance in depression: temporal coherence of ultradian sleep EEG rhythms. *Psychol Med* 1999;29:1435–1448.

79. Nagakubo S, Kumagai N, Kameyama T, et al. Diagnostic reliability and significance of irregular beta patterns. *Jpn J Psychiatry Neurol* 1991;45:631–640.

80. Ehlers CL, Kaneko WM, Owens MJ, et al. Effects of gender and social isolation on electroencephalogram and neuroendocrine parameters in rats. *Biol Psychiatry* 1993;33:358–366.

81. Krishnan KRR, McDonald WM, Escalona R, et al. Magnetic resonance imaging of the caudate nuclei in depression. *Arch Gen Psychiatry* 1992;49:553–557.

82. Burkhart MA, Thomas DG. Event-related potential measures of attention in moderately depressed subjects. *Electromyogr Clin Neurophysiol* 1993;88:42–50.

83. Goldstein GK, NcCue M, Rogers J, et al. Diagnostic differences in memory test based predictions of functional capacity in the elderly. *Neuropsychol Rehabil* 1992;2:307–317.

84. Dahabra S, Ashton CH, Bahrainian M, et al. Structural and functional abnormalities in elderly patients clinically recovered from early- and late-onset depression. *Biol Psychiatry* 1998;44:34–46.

85. Taylor MP, Reynolds CF 3rd, Frank E, et al. EEG sleep measures in later-life bereavement depression. A randomized, double-blind, placebo-controlled evaluation of nortriptyline. *Am J Geriatr Psychiatry* 1999;7:41–47.

86. Jones NA, Field T. Massage and music therapies attenuate frontal EEG asymmetry in depressed adolescents. *Adolescence* 1999;34:529–534.

87. Leuchter AF, Uijtdehaage SH, Cook IA, et al. Relationship between brain electrical activity and cortical perfusion in normal subjects. *Psychiatry Res* 1999;90:125–140.

88. Cook IA, Leuchter AF, Uijtdehaage SH, et al. Altered cerebral energy utilization in late life depression. *J Affect Disord* 1998;49:89–99.

89. Cook IA, Leuchter AF, Witte E, et al. Neurophysiologic predictors of treatment response to fluoxetine in major depression. *Psychiatry Res* 1999;85:263–273.

90. Leuchter AF, Cook IA, Uijtdehaage SH, et al. Brain structure and function and the outcomes of treatment for depression. *J Clin Psychiatry* 1997;59[Suppl 16]:22–31.

91. Cook IA, Leuchter AF. Prefrontal changes and treatment response prediction in depression. *Semin Clin Neuropsychiatry* 2001;6:113–120.

92. Leuchter AF, Cook IA, Witte EA, et al. Changes in brain function of depressed subjects during treatment with placebo. *Am J Psychiatry* 2002;159:122–129.

93. Jorm AF, Fratiglioni L, Winblad B. Differential diagnosis in dementia. Principal components analysis of clinical data from a population survey. *Arch Neurol* 1993;50:73–77.

94. Kurita A, Blass JP, Nolan KA, et al. Relationship between cognitive status and behavioral symptoms in Alzheimer's disease and mixed dementia. *J Am Geriatr Soc* 1993;41:732–736.

95. DeFries JC, Vandenberg SG, McClearn GE. Genetics of specific cognitive abilities. *Annu Rev Genet* 1976;10:179–207.

96. Cook IA, Leuchter AF. Synaptic dysfunction in Alzheimer's disease: clinical assessment using quantitative EEG. *Behav Brain Res* 1996;78:15–23.

97. Martin-Loeches M, Gil P, Rubia FJ. Two-Hz wide EEG bands in Alzheimer's disease. *Biol Psychiatry* 1993;33:153–159.

98. Adler G, Bramesfeld A, Jajcevic A. Mild cognitive impairment in old-age depression is associated with increased EEG slow-wave power. *Neuropsychobiology* 1999;40:218–222.

99. Kowalski JW, Gawel M, Pfeffer A, et al. The diagnostic value of EEG in Alzheimer disease: correlation with the severity of mental impairment. *J Clin Neurophysiol* 2001;18:570–575.

100. Lehtovirta M, Partanen J, Konenen M, et al. A longitudinal quantitative EEG study of Alzheimer's disease: relation to apolipoprotein E polymorphism. *Dement Geriatr Cogn Disord* 2000;11:29–35.

101. Adler G, Bramesdeld A, Jajcevic A. Mild cognitive impairment in old-age depression is associated with increased EEG slow-wave power. *Neuropsychobiology* 1999;40:218–222.

102. Bennys K, Kandoum G, Vergmes C, et al. Diagnostic value of quantitative EEG in Alzheimer's disease. *Neurophysiol Clin* 2001;31:153–160.

103. Ihl R, Brinkmeyer J, Janner M, et al. A comparison of ADAS and EEG in the discrimination of patients with dementia of the Alzheimer type from healthy controls. *Neuropsychobiology* 2000;41:102–107.

104. Berendse HW, Verbunt JP, Scheltens P, et al. Magnetoencephalographic analysis of cortical activity in Alzheimer's disease: a pilot study. *Clin Neurophysiol* 2000;111:604–612.

105. Brenner RP, Ulrich RF, Spiker DG, et al. Computerized EEG spectral analysis in elderly normal, demented and depressed subjects. *Electroencephalogr Clin Neurophysiol* 1986;64:483–492.

106. Brenner RP, Reynolds CF III, Ulrich RF. Diagnostic efficacy of computerized spectral versus visual EEG analysis in elderly normal, demented and depressed subjects. *Electroencephalogr Clin Neurophysiol* 1988;69:110–117.

107. Matousek M, Brunovsky M, Edman A, et al. EEG abnormalities in dementia reflect the parietal lobe syndrome. *Clin Neurophysiol* 2001;112:1001–1005.

108. Stevens A, Kircher T. Cognitive decline unlike normal aging is associated with alterations of EEG temporo-spatial characteristics. *Eur Arch Psychiatry Clin Neurosci* 1998;248:259–266.

109. Strik WK, Chiaramonti R, Muscas GC, et al. Decreased EEG microstate duration and anteriorisation of the brain electrical fields in mild and moderate dementia of the Alzheimer type. *Psychiatry Res* 1997;75:183–191.

110. Edman A, Matousek M, Sjogren M, et al. Longitudinal EEG findings in dementia related to the parietal brain syndrome and the degree of dementia. *Dement Geriatr Cogn Disord* 1998;9:199–204.

111. Duffy FH, Albert MS, McAnulty G. Brain electrical activity in patients with presenile and senile dementia of the Alzheimer type. *Ann Neurol* 1984;16:439–448.

112. Prichep LS, John ER, Ferris SH, et al. Quantitative EEG correlates of cognitive deterioration in the elderly. *Neurobiol Aging* 1994;15:85–90.

113. Buchan RJ, Nagata K, Yokoyama E, et al. Regional correlations between the EEG and oxygen metabolism in dementia of Alzheimer's type. *Electroencephalogr Clin Neurophysiol* 1997;103:409–417.

114. Knott V, Mohr L, Mahoney C, Hrvitsky V. Quantitative electroencephalography in Alzheimer's disease: comparison with a control group, population norms and mental status. *J Psychiatry Neurosci* 2001;26:106–116.

115. Jeong J, Chao J, Kuhn BT, et al. Nonlinear dynamic analysis of the EEG in patients with Alzheimer's disease and vascular dementia. *J Clin Neurophysiol* 2001;18:58–67.

116. Cantero JL, Alienza M, Salas RM. Clinical value of EEG coherence as electrophysiological index of cortico-cortical connections during sleep. [Spanish]. *Rev Neurol* 2000;31:442–454.

117. Anghinah R, Kanda PA, Jorge MS, et al. Alpha band coherence analysis of EEG in healthy adult's and Alzheimer's type dementia patients. [Portuguese] *Arq Neuro-Psiquiatr* 2000;58(2A):272–275.

118. Wada Y, Nanbu Y, Koshino Y, et al. Reduced interhemispheric EEG coherence in Alzheimer disease: analysis during rest and photic stimulation. *Alzheimer Dis Assoc Disord* 1998;12:175–181.

119. Locatehl T, Cursi M, Liberati D, et al. EEG coherence in Alzheimer's disease. *Electroencephalogr Clin Neurophysiol* 1998;106:229–237.

120. Nobill T, Copello T, Vitali P, et al. Timing of disease progression by quantitative EEG in Alzheimer's patients. *J Clin Neurophysiol* 1999;16:566–573.

121. Knorr V, Mohr E, Hache NK, et al. EEG and the passive P300 in dementia of the Alzheimer type. *Clin Electroencephalogr* 1999;30:64–72.

122. Tobimatsu S, Kurita-Tashima S, Nakayama-Hiromatsu M, et al. Age-related changes in pattern visual evoked potentials and differential effects of luminance, contrast and check size. *Electroencephalogr Clin Neurophysiol* 1993;88:12–19.

123. Szelies B, Mielke R, Kessler J, et al. EEG power changes are related to regional cerebral glucose metabolism in vascular dementia. *Clin Neurophysiol* 1999;11:615–620.

124. Partanen J, Soinenen II, Ilelkula EL, et al. Relationship between EEG reactivity and neuropsychological tests in vascular dementia. *J Neural Transm* 1997;104:905–912.

125. Lopez OL, Brenner RP, Becker JT, et al. EEG spectral abnormalities and psychosis as predictors of cognitive and functional decline in probable Alzheimer's disease. *Neurology* 1997;48:1521–1525.

126. Strijers RL, Scheltens P, Jonkman EJ, et al. Diagnosing Alzheimer's disease in community-dwelling elderly: a comparison of EEG and MRI. *Dement Geriatr Cogn Disord* 1997;8:198–202.

127. Elmstahl S, Rosen I. Postural hypotension and EEG variables predict cognitive decline: results from a 5-year follow-up of healthy elderly women. *Dement Geriatr Cogn Disord* 1997;8:180–187.

128. Visser FE, Kuilman M, Oosting J, et al. Use of electroencephalography to detect Alzheimer's disease in Down's syndrome. *Acta Neurol Scand* 1996;94:97–103.

129. Politoff AL, Stadter RP, Monson N, et al. Cognition-related EEG abnormalities in nondemented Down syndrome subjects. *Dementia* 1996;7:69–75.

130. Hegerl U, Moller HJ. Electroencephalography as a diagnostic instrument in Alzheimer's disease: reviews and perspectives. *Int Psychogeriatr* 1997;9[Suppl 1]:237–246 and discussion 247–252.

131. Fink M. EEG and psychopharmacology. *Electroencephalogr Clin Neurophysiol* 1978;34[Suppl]:41–56.

132. Bente D. Differential and general effects of psychotropic pharmaceuticals on the human EEG. [German] *Psychiatr Neurol Med Psychol Beihefte* 1975;20 21:123–130.

133. Scott DF, Schwartz MS, Farrant GW, et al. The importance of methodology in drug studies using EEG. *Electroencephalogr Clin Neurophysiol* 1974;37:670–672.

134. Sannita WG. Quantitative EEG in human neuropharmacology. Rationale, history, and recent developments. *Acta Neurol* 1990;12:389–409.

135. Gabarino S, Beelke M, Costa G, et al. Brain function and effects of shift work: implications for clinical neuropharmacology. *Neuropsychobiology* 2002;45:50–56.

136. Sannita WG. Shift work and clinical (neuro)pharmacology. *Neuropsychobiology* 2002;45:49.

137. Sannita WG, Loizzo A, Garbarino S, et al. Adrenocorticotropin-related modulation of the human EEG and individual variability. *Neurosci Lett* 1999;262:147–150.

138. Sannita WG, Crimi E, Riela S, et al. Cutaneous antihistaminic action of cetirizine and dose-related EEG concomitants of sedation in man. *Eur J Pharmacol* 1996;300:33–41.

139. Sannita WG. Dose-related and treatment-dependent quantitative EEG effects of antiepileptic compounds in humans. *Clin Neuropharmacol* 1992;15[Suppl 1 Pt A]:423A–424A.

140. Sannita WG, Balestra V, DiBon G, et al. Ammonia-independent modifications of the background EEG signal and paradoxical enhancement of epileptic abnormalities in EEG after acute administration of valproate to epileptic patients. *Neuropharmacology* 1993;32:919–928.

141. Sannita WG, Gervasio L, Zagnoni P. Quantitative EEG effects and plasma concentration of sodium valproate: acute and long-term administration to epileptic patients. *Neuropsychobiology* 1989;22:231–235.

142. Levic ZM. Electroencephalographic studies in multiple sclerosis. Specific changes in benign multiple sclerosis. *Electroencephalogr Clin Neurophysiol* 1978;44:471–478.

143. Awerbuch GI, Verma NP. Periodic lateralized epileptiform discharges from a patient with definite multiple sclerosis. *Clin Electroencephalogr* 1987;18:38–40.

144. Lawn ND, Westmoreland BF, Sharbrough RW. Multifocal periodic lateralized epileptiform discharges (PLEDs): EEG features and clinical correlations. *Clin Neurophysiol* 2000;111:2125–2129.

145. Versino M, Bergamaschi R, Romani A, et al. Middle latency auditory evoked potentials improve the detection of abnormalities along auditory pathways in multiple sclerosis patients. *Electroencephalogr Clin Neurophysiol* 1992;84:296–299.

146. Leocani L, Medaglini S, Comi G. Evoked potentials in monitoring multiple sclerosis. *Neurol Sci* 2000;21[4 Suppl 2]:S889–S891.

147. Thompson AJ, Kermode AG, Moseley IF, et al. Seizures due to multiple sclerosis: seven patients with MRI correlations. *J Neurol Neurosurg Psychiatry* 1993;56:1317–1320.

148. Ghezzi A, Montanini R, Basso PF, et al. Epilepsy in multiple sclerosis. *Eur Neurol* 1990;30:218–223.

149. Korwin-Piotrowska T, Nocon D. Is there any correlation between cerebral atrophy, EEG abnormalities and epileptic attacks in patients with multiple sclerosis? [Polish] *Neurol Neurochir Polska* 1999;33:1305–1310.

150. Maingueneua F, Honnorat J, Isnard J, et al. Partial nonconvulsive status epilepsy in multiple sclerosis. [French] *Neurophysiol Clin* 1999;29:463–472.

151. Chiappa KH. Pattern shift visual brainstem auditory and short-latency somatosensory evoked potentials in multiple sclerosis. *Neurology* 1980;30:110–123.

152. Brau H, Ulrich G. Electroencephalographic vigilance dynamics in multiple sclerosis during an acute episode and after remission. *Eur Arch Psychiatry Clin Neurosci* 1990;239:320–324.

153. Aminoff MJ, Davis SL, Panitch HS. Serial evoked potential studies in patients with definite multiple sclerosis. *Arch Neurol* 1984;41:1197–1202.

154. Matthews WB, Small JG. Serial recordings of visual and somatosensory evoked potentials in multiple sclerosis. *J Neurol Sci* 1979;40:11–21.

155. Sartucci F, Murri L, Orsini C, et al. Equiluminant red-green and blue-yellow VEPs in multiple sclerosis. *J Clin Neurophysiol* 2001;18:583–591.

156. Leocani L, Colombo B, Magnani G, et al. Fatigue in multiple sclerosis is associated with abnormal cortical activation to voluntary movement—EEG evidence. *Neuroimage* 2001;13[6 Pt 1]:1186–1192.

157. Leocani L, Locatelli T, Martinelli V, et al. Electroencephalographic coherence analysis in multiple sclerosis: correlation with clinical, neuropsychological, and MRI findings. *J Neurology Neurosurg Psychiatry* 2000;69:192–198.

158. Rudkowska A, Gruska E, Serwacka B, et al. Paroxysmal EEG changes in patients with multiple sclerosis. [Polish] *Neurol Neurochir Polska* 1992;26:466–472.

159. Gil R, Zai L, Neau JP, et al. Event-related auditory evoked potentials and multiple sclerosis. *Electroencephalogr Clin Neurophysiol* 1993;88:182–187.

160. Polich J, Romine JS, Sipe JC, et al. P300 in multiple sclerosis: a preliminary report. *Int J Psychophysiol* 1992;12:155–163.

161. Triantafyllou NI, Voumvourakis K, Zalonis I, et al. Cognition in relapsing-remitting multiple sclerosis: a multichannel event-related potential (P300) study. *Acta Neurol Scand* 1992;85:10–13.

162. Honig LS, Ramsay RE, Sheremata WA. Event-related potential P300 in multiple sclerosis. Relation to magnetic resonance imaging and cognitive impairment. *Arch Neurol* 1992;49:44–50.

163. Tsuda M, Miyazaki M, Tanaka Y, et al. A case report of childhood multiple sclerosis with periodic synchronous discharge on EEG. [Japanese] *No To Hattatsu* 1991;23:612–616.

164. Miura H, Mukoyama M, Kamei N. An autopsy case of multiple sclerosis with bilateral continuous cystic lesions along lateral ventricles and caudate-callosal angles (Wetterwinkel). [Japanese] *No To Shinkei Brain Nerve* 1991;43:1087–1091.

165. Chabot RJ, Serfontein G. Quantitative electroencephalographic profiles of children with attention deficit disorder. *Biol Psychiatry* 1996;40:951–963.

166. Matsuura M, Okubo Y, Toru M, et al. A cross-national EEG study of children with emotional and behavioral problems: a WHO collaborative study in the Western Pacific Region. *Biol Psychiatry* 1993;34:59–65.

167. Kuperman S, Johnson B, Arndt S, et al. Quantitative EEG differences in a nonclinical sample of children with ADHD and undifferentiated ADD. *J Am Acad Child Adolesc Psychiatry* 1996;35:1009–1017.

168. Suffin SC, Emory WH. Neurometric subgroups in attentional and affective disorders and their association with pharmacotherapeutic outcome. *Clin Electroencephalogr* 1995;26:76–83.

169. Chabot RJ, Orgill AA, Crawford G, et al. Behavioral and electrophysiologic predictors of treatment response to stimulants in children with attention disorders. *J Child Neurol* 1999;14:343–351.

170. Linden M, Habib T, Radojevic V. A controlled study of the effects of EEG biofeedback on cognition and behavior of children with attention deficit disorder and learning disabilities. *Biofeedback Self-Regul* 1996;21:35–49.

171. Thompson L, Thompson M. Neurofeedback combined with training in metacognitive strategies: effectiveness in students with ADD. *Appl Psychophysiol Biofeedback* 1998;23:243–246.

172. Luckacher GI, Strelets VB, Marsakova GD, et al. A comparison of the results of topographic EEG mapping with the data from a neurological examination and computed tomography of the brain. [Russian] *Zhurnal Nevropatologii I Psikhiatrii Imeni S-S-Korsakova* 1994;94:26–30.

173. Jansen C, Moll FL, Vermeulen FE, et al. Continuous transcranial Doppler ultrasonography and electroencephalography during carotid endarterectomy: a multimodal monitoring system to detect intraoperative ischemia. *Ann Vasc Surg* 1993;7:95–101.

174. Facco E, Deriu GP, Cona B, et al. EEG monitoring of carotid endarterectomy with routine patch graft angioplasty: an experience in a large series. *Neurophysiol Clin* 1992;6:437–446.

175. Fava, E, Bortolani E, Ducati A, et al. Role of SEP in identifying patients requiring temporary shunt during carotid endarterectomy. *Electroencephalogr Clin Neurophysiol* 1992;84:426–432.

176. Redekop G, Ferguson G. Correlation of contralateral stenosis and intraoperative electroencephalogram change with risk of stroke during carotid endarterectomy. *Neurosurgery* 1992;30:191–194.

177. Logar C. The place of EEG mapping in cerebral ischemia. [German] *EEG-EMG Zeitschr Elektroenzephalogr Elektromyogr Verwandte Geviete* 1990;21:161–162.

178. Kappelle LJ, van Huffelen AC, van Gijn J. Is the EEG really normal in lacunar stroke? *J Neurol Neurosurg Psychiatry* 1990;53:53–66.

179. Milandre L, Broca P, Sambuc R, et al. Epileptic crisis during and after cerebrovascular diseases. A clinical analysis of 78 cases. [French] *Rev Neurol* 1992;148:767–772.

180. Gras P, Grosmaire E, Soichot P, et al. EEG periodic lateralized activities associated with ischemic cerebro-vascular strokes. [French] *Neurophysiol Clin* 1991;21:293–299.

181. Domzal T, Malowidzka-Serwinska M, Mroz K. Electrophysiological pattern of sleep after stroke. [Polish] *Neurol Neurochir Polska* 1994;28:27–34.

182. Hachinski VC, Mamelak M, Norris JW. Clinical recovery and sleep architecture degradation. *Can J Neurol Sci* 1990;17:332–335.

183. Constantinovici A, Radutoiu E, Osanu M, et al. Transient global amnesia (a study of 30 cases). [Rumanian] *Revista de Medicina-Interna, Neurolgie, Psihiatrie, Neurochirurgie, Dermato-Venerologie-Neurologie* 1990;35:61–68.

184. Mizuno-Matsumoto Y, Ishijima M, Shinosaki K, et al. Transient global amnesia (TGA) in an MEG study. *Brain Topogr* 2001;13:269–274.

185. Tanaka A, Kimur M, Yoshinaga S, et al. Quantitative electroencephalographic correlates of cerebral blood flow in patients with chronic subdural hematomas. *Surg Neurol* 1998;50:235–240.

186. Morioka T, Nishio S, Hisada K, et al. Neuromagnetic assessment of epileptogenicity in cerebral arteriovenous malformation. *Neurosurg Rev* 2000;23:206–212.

187. Sand T. EEG in migraine: a review of the literature. *Funct Neurol* 1991;6:7–22.

188. Muellbacher W, Mamoli B. Prolonged impaired consciousness in basilar artery migraine. *Headache* 1994;34:282–285.

189. Passier PE, Vredeveld JW, de Krom MC. Basilar migraine with severe EEG abnormalities. *Headache* 1994;34:56–58.

190. Ganji S, Hellman S, Stagg S, et al. Episodic coma due to acute basilar artery migraine: correlation of EEG and brainstem auditory evoked potential patterns. *Clin Electroencephalogr* 1993;24:44–48.

191. Morimoto Y, Nakajima S, Nishioka R, et al. Basilar artery migraine with transient MRI and EEG abnormalities. [Japanese] *Rinsho Shinkeigaku [Clin Neurol]* 1993;33:61–67.

192. De Romanis F, Buzzi MG, Assenza S, et al. Basilar migraine with electroencephalographic findings of occipital

spike-wave complexes: a long-term study in seven children. *Cephalalgia* 1993;13:192–196.

193. Jernigan SA, Ware LM. Reversible quantitative EEG changes in a case of cyclic vomiting: evidence for migraine equivalent. *Dev Med Child Neurol* 1991;33:80–85.

194. Frequin ST, Linssen WH, Pasman JW, et al. Recurrent prolonged coma due to basilar artery migraine. A case report. *Headache* 1991;31:75–81.

195. Fuller GN, Marshall A, Flint J, et al. Migraine madness: recurrent psychosis after migraine. *J Neurol Neurosurg Psychiatry* 1993;56:416–418.

196. Corona T, Otero-Siliceo E, Teyes Baez B, et al. Electroencephalographic alterations in patients with migraine with and without aura. [Spanish] *Neurologia* 1994;9:81–84.

197. Valdizan JR, Andreu C, Almarcegui C, et al. Quantitative EEG in children with headache. *Headache* 1994;34:53–55.

198. Pothmann R. Topographic EEG mapping in childhood headaches. *Cephalalgia* 1993;13:57–58.

199. Neufeld MY, Treves TA, Korczyn AD. EEG and topographic frequency analysis in common and classic migraine. *Headache* 1991;31:232–236.

200. Facchetti D, Marsile C, Faggi L, et al. Cerebral mapping in subjects suffering from migraine with aura. *Cephalalgia* 1990;10:279–284.

201. Thomaides T, Tagaris G, Karageorgiou C. EEG and topographic frequency analysis in migraine attack before and after sumatriptan infusion. *Headache* 1996;36:111–114.

202. Seri S, Cerquiglini A, Guidette V. Computerized EEG topography in childhood migraine between and during attacks. *Cephalalgia* 1993;13:53–56.

203. Sangermani R, Priovano S, Vaccari R, et al. Abdominal migraine simulating acute abdomen. [Italian] *Pediatr Med Chirurg* 1992;14:163–165.

204. Soriani S, Scarpa P, Faggioli R, et al. Uncommon EEG pattern in an 8-year-old boy with recurrent migraine aura without headache. *Headache* 1993;33:509–511.

205. Farkas V, Kohlheb O, Benninger C, et al. Comparison of the EEG background activity of epileptic children and children with migraine. *Epilepsy Res* 1992;6:199–205.

206. Marks DA, Ehrenberg BL. Migraine-related seizures in adults with epilepsy, with EEG correlation. *Neurology* 1993;43:2476–2483.

207. Donnet A, Bartolomei F. Migraine with visual aura and photosensitive epileptic seizures. *Epilepsia* 1997;38:1032–1034.

208. De Romanis F, Buzzi MG, Cerbo R, et al. Migraine and epilepsy with infantile onset and electroencephalographic findings of occipital spike-wave complexes. *Headache* 1991;31:378–383.

209. Wilder-Smith E, Nirkko AC. Contribution of concurrent Doppler and EEG in differentiating occipital epileptic discharges from migraine. *Neurology* 1991;41:2005–2007.

210. de Tommaso M, Sciruicchio V, Guido M, et al. EEG spectral analysis in migraine without aura attacks. *Cephalalgia* 1998;18:324–328.

211. Olsen TS, Friberg L, Lassen NA. Migraine aura—vascular or neuronal disease? [Danish] *Ugeskrift Laeger* 1990;152:1507–1509.

212. Tomaiolo S, Stiglich F, Bonomo F, et al. SPECT with 99m-TC HMPAO in the study of classical hemicrania. [Italian] *Radiol Med (Torino)* 1991;81:537–541.

213. Good PA, Taylor RH, Mortimer MJ. The use of tinted glasses in childhood migraine. *Headache* 1991;31:533–536.

214. McVicker RW, Shanks OE, McClelland RJ. Prevalence and associated feature of epilepsy in adults with Down syndrome. *Br J Psychiatry* 1994;163:117.

215. Elliot D, Weeks DJ. Cerebral specialization for speech perception and movement organization in adults with Down's syndrome. *Br J Psychiatry* 1993;29:103–113.

216. Partanen J, Soininen H, Kononen M, et al. EEG reactivity correlates with neuropsychological test scores in Down's syndrome. *Acta Neurol Scand* 1996;94:242–246.

217. Kaneko WN, Ehlers CL, Philips EL, et al. Auditory event-related potentials in fetal alcohol syndrome and Down's syndrome children. *Alcohol Clin Exp Res* 1996;20:35–42.

218. Puchinskaia LM, Katargina TA, Kryzhanovskaia IL. The temporal correlations of the VEP of the visual and motor cortices during the perception and mental reproduction of an image in normal children and in those with intellectual disorders. [Russian] *Zhurnal Vysshei Nervnoi Deiatelnosti Imeni I. P. Pavlova* 1997;47:49–57.

219. Munoz-Ruata J, Gomez-Jarabo G, Marin-Loeches M, et al. Neurophysiological and neuropsychological differences related to performance and verbal abilities in subjects with mild intellectual disabilities. *J Intellect Disabil Res* 2000;44[Part 5]:567–578.

220. Ikeda K, Okuzumi H, Hayashi A, et al. Automatic auditory processing and event-related brain potentials in persons with mental retardation. *Percept Motor Skills* 2000;91[3 Part 2]:1145–1150.

221. Watanabe K, Hayakawa F, Okumura A. Neonatal EEG: a powerful tool in the assessment of brain damage in preterm infants. *Brain Dev* 1999;21:361–372.

222. Sheth RD. Electroencephalogram in developmental delay: specific electroclinical syndromes. *Semin Pediatr Neurol* 1998;5:45–51.

223. Buoni S, Grosso S, Pucci L, et al. Diagnosis of Angelman syndrome: clinical and EEG criteria. *Brain Dev* 1999;21:296–302.

224. Casara GL, Vecchi M, Boniver C, et al. Electroclinical diagnosis of Angelman syndrome: a study of 7 cases. *Brain Dev* 1995;17:64–68.

225. Deb S. Electrophysiological correlates of psychopathology in individuals with mental retardation and epilepsy. *J Intellect Disabil Res* 1995;39[Part 2]:129–135.

226. Kaplan PW. Assessing the outcomes in patients with nonconvulsive status epilepticus: nonconvulsive status epilepticus is underdiagnosed, potentially overtreated, and confounded by comorbidity. *J Clin Neurophysiol* 1999;16:341–352.

227. Canevini MP, Sgro V, Zuffardi O, et al. Chromosome 20 ring: a chromosomal disorder associated with a particular electroclinical pattern. *Epilepsia* 1998;39:943–951.

228. Chen AC. New perspectives in EEG/MEG brain mapping and PET/fMRI neuroimaging of human pain. *Int J Psychophysiol* 2001;42:147–159.

229. Veerasarn P, Stohler CS. The effect of experimental muscle pain on the background electrical brain activity. *Pain* 1992;49:349–360.

230. Schneider-Helmert D, Whitehouse I, Kumar A, et al. Insomnia and alpha sleep in chronic non-organic pain as compared to primary insomnia. *Neuropsychobiology* 2001;43:54–58.

231. Drewes AM, Nielsen KD, Arendt-Nielsen L, et al. The effect of cutaneous and deep pain on the electroencephalogram during sleep—an experimental study. *Sleep* 1997;20:632–640.

232. Lozza A, Proiettei Cecchini A, Afra J, et al. Neurophysiological approach to primary headache pathophysiology. *Cephalalgia* 1998;18[Suppl 21]:12–16.

233. Siniatchkin M, Gerber WD, Kropp P, Vein A. How the brain anticipates an attack: a study of neurophysiological periodicity in migraine. *Funct Neurol* 1999;14:69–77.

234. Strenge H, Fritzer G, Goder R, et al. Non-linear electroencephalogram dynamics in patients with spontaneous nocturnal migraine attacks. *Neurosci Lett* 2001;309:105–108.

235. Le Pera D, Svensson P, Valeriani M, et al. Long-lasting effect evoked by tonic muscle pain on parietal EEG activity in humans. *Clin Neurophysiol* 2000;111:2130–2137.

236. Wiech K, Preissl H, Birbaumer N. Neuroimaging of chronic pain: phantom limb and musculoskeletal pain. *Scand J Rhematol Suppl* 2000;113:13–18.

237. Karl A, Birbaumer N, Lutzenberger W, et al. Reorganization of motor and somatosensory cortex in upper extremity amputees with phantom limb pain. *J Neurosci* 2001;21:3609–3618.

238. Yamasaki H, Kakigi R, Watanabe S, et al. Effects of distraction on pain perception: magneto- and electro-encephalographic studies. *Cogn Brain Res* 1999;8:73–76.

239. Chen AC, Rappelsberger P. Brain and human pain: topographic EEG amplitude and coherence mapping. *Brain Topogr* 1994;7:129–140.

240. Ferracuti S, Seri S, Mattia D, et al. Quantitative EEG modifications during the Cold Water Pressor Test: hemispheric and hand differences. *Int J Psychophysiol* 1994;17:261–268.

241. Bromm B, Forth W, Richter E, et al. Effects of acetaminophen and antipyrine on non-inflammatory pain, and EEG activity. *Pain* 1992;50:213–221.

242. Van Praag H, Falcon M, Guendelman D, et al. The development of analgesic pro- and anti-convulsant opiate effects in the rat. *Annali Dell Lurtituto Superiore di Sanita* 1993;29:419–429.

243. Suri A, Kaltenbach ML, Grundy BL, et al. Pharmacodynamic evaluation of codeine using tooth pulp evoked potentials. *J Clin Pharmacol* 1996;36:1126–1131.

244. Voller B, Benke T, Benedetto K, et al. Neuropsychological, MRI and EEG findings after very mild traumatic brain injury. *Brain Inj* 1999;13:821–827.

245. Frieboes RM, Muller U, Murck H, et al. Nocturnal hormone secretion and the sleep EEG in patients several months after traumatic brain injury. *J Neuropsychiatry Clin Neurosci* 1999;11:354–60.

246. Hume AL, Cant BR. Central somatosensory conduction after head injury. *Ann Neurol* 1981;10:411–419.

247. Kane NM, Moss TH, Curry SH, et al. Quantitative electroencephalographic evaluation of non-fatal and fatal traumatic coma. *Electroencephalogr Clin Neurophysiol* 1998;106:244–250.

248. Soldner F, Holper BM, Chone L, et al. Evoked potentials in acute head injured patients with MRI-detected intracerebral lesions. *Acta Neurochirurg* 2001;143:873–883.

249. Tasker RC, Boyd SG, Harden A, et al. The cerebral function analyzing monitor in pediatric medical intensive care: applications and limitations. *Intens Care Med* 1990;16:60–68.

250. Watanabe K, Hayakawa F, Okumura A. Neonatal EEG: a powerful tool in the assessment of brain damage in preterm infants. *Brain Dev* 1999;21:361–372.

251. Albanese J, Arnaud S, Rey M, et al. Ketamine decreases intracranial pressure and electroencephalographic activity in traumatic brain injury patients during propofol sedation. *Anesthesiology* 1997;87:1328–1334.

252. Lewine JD, Davis JT, Sloan JH, et al. Neuromagnetic assessment of pathophysiologic brain activity induced by minor head trauma. *AJNR Am J Neuroradiol* 1999;20:857–866.

253. Thatcher RW, Biver C, McAlaster R, et al. Biophysical linkage between MRI and EEG coherence in closed head injury. *Neuroimage* 1998;8:307–326.

254. Thatcher RW, Biver C, Gomez JF, et al. Estimation of the EEG power spectrum using MRI T2 relaxation time in traumatic brain injury. *Clin Neurophysiol* 2001;112:1729–1745.

255. Thatcher RW, North DM, Curtin RT, et al. An EEG severity index of traumatic brain injury. *J Neuropsychiatry Clin Neurosci* 2001;13:77–87.

256. Wallace BE, Wagner AK, Wagner EP, et al. A history and review of quantitative electroencephalography in traumatic brain injury. *J Head Trauma Rehabil* 2001;16:165–190.

257. Gaetz M, Bernstein DM. The current status of electrophysiologic procedures for the assessment of mild traumatic brain injury. *J Head Trauma Rehabil* 2001;16:386–405.

258. Alberti A, Sarchielli P, Mazzotta G, et al. Event-related potentials in posttraumatic headache. *Headache* 2001;41:579–585.

259. Arciniegas D, Olincy A, Topkoff J, et al. Impaired auditory gating and P50 nonsuppression following traumatic brain injury. *J Neuropsychiatry Clin Neurosci* 2000;12:77–85.

260. Diaz-Arrastia R, Agostini MA, Frol AB, et al. Neurophysiologic and neuroradiologic features of intractable epilepsy after traumatic brain injury in adults. *Arch Neurol* 2000;57:1611–1616.

261. Conder RL, Zasler ND. Psychogenic seizures in brain injury: diagnosis, treatment and case study. *Brain Inj* 1990;4:391–397.

262. Millspaugh JA. Dementia pugilistica. *US Naval Military Bull* 1937;35:297–303.

263. Critchley M. Medical aspects of boxing, particularly from a neurological standpoint. *Br Med J* 1957;5015:357–362.

264. Pampus F, Grote W. Electroencephalographic and clinical findings in boxers and their significance in the pathophysiology of traumatic brain disorders. [German] *Arch Psychiatr* 1956;194:152–178.

265. Loganovsky KN, Yuryev KL. EEG patterns in persons exposed to ionizing radiation as a result of the Chernobyl accident: part 1: conventional EEG analysis. *J Neuropsychiatry Clin Neurosci* 2001;13:441–458.

266. Schoenberger NE, Shif SC, Esty ML, et al. Flexyx Neurotherapy System in the treatment of traumatic brain injury: an initial evaluation. *J Head Trauma Rehabil* 2001;16:260–274.

267. Thatcher RW. EEG operant conditioning (biofeedback) and traumatic brain injury. *Clin Electroencephalogr* 2000;31:38–44.

268. DeLong R, Nohria C. Psychiatric family history and neurological disease in autistic spectrum disorders. *Dev Med Child Neurol* 1994;36:441–448.

269. Rossi PG, Parmeggiani A, Bach V, et al. EEG features and epilepsy in patients with autism. *Brain Dev* 1995;17:169–174.

270. Lewine JD, Andrews R, Chez M, et al. Magnetoencephalographic patterns of epileptiform activity in children with regressive autism spectrum disorders. *Pediatrics* 1999;104[3 Part 1]:405–418.

271. Nass R, Gross A, Devinsky O. Autism and autistic epileptiform regression with occipital spikes. *Dev Med Child Neurol* 1998;40:453–458.

272. Elia M, Musumeci SA, Ferri R, et al. Clinical and neurophysiological aspects of epilepsy in subjects with autism and mental retardation. *Am J Ment Retard* 1995;100:6–16.

273. Tuchman RF. Epilepsy, language, and behavior: clinical models in childhood. *J Child Neurol* 1994;9:95–102.

274. Plioplys AV. Autism: electroencephalogram abnormalities and clinical improvement with valproic acid. *Arch Pediatr Adolesc Med* 1994;148:220–222.

275. Carracedo A, Martin Murcia F, Garcia Penas JJ, et al. Autistic syndrome associated with refractory temporal epilepsy. [Spanish] *Rev Neurol* 1995;23:1239–1241.

276. Diomedi M, Curatolo P, Scalise A, et al. Sleep abnormalities in mentally retarded autistic subjects: Down's syndrome with mental retardation and normal subjects. *Brain Dev* 1999;21:548–553.

277. Richdale AL. Sleep problems in autism: prevalence, cause, and intervention. *Dev Med Child Neurol* 1999;41:60–66.

278. Ciesielski KT, Knight JE, Prince RJ, et al. Event-related potentials in cross-modal divided attention in autism. *Neuropsychologia* 1995;33:225–246.

279. Townsend J, Westerfield M, Leaver E, et al. Event-related brain response abnormalities in autism: evidence for impaired cerebello-frontal spatial attention networks. *Cogn Brain Res* 2001;11:127–145.

280. Buchwald JS, Erwin R, Van Lancker D, et al. Midlatency auditory evoked responses: P1 abnormalities in adult autistic subjects. *Electroencephalogr Clin Neurophysiol* 1992;84:164–171.

281. Harrison DW, Demaree HA, Shenal RV, et al. QEEG assisted neuropsychological evaluation of autism. *Int J Neurosci* 1998;93:133–140.

282. Kawasaki Y, Yokota K, Shinomiaya M, et al. Brief report: electroencephalographic paroxysmal activities in the frontal area emerged in middle childhood and during adolescence in a follow-up study of autism. *J Autism Dev Disord* 1997;27:605–620.

283. Hashimoto T, Sasaki M, Sugai K, et al. Paroxysmal discharges on EEG in young autistic patients are frequent in frontal regions. *J Med Investig* 2001;48:175–180.

284. Dawson G, Klinger LGK, Panagiotides H, et al. Subgroups of autistic children based on social behavior display distinct patterns of brain activity. *J Abnorm Child Psychol* 1995;23:569–583.

285. Sandyk R. Alpha rhythm and the pineal gland. *Int J Neurosci* 1992;63:221–227.

286. Gruzelier J, Burgess A, Baldeweg T, et al. Prospective associations between lateralised brain function and immune status in HIV infection: analysis of EEG, cognition and mood over 30 months. *Int J Psychophysiol* 1996;23:215–224.

287. Polich J, Ilan A, Poceta JS, et al. Neuroelectric assessment of HIV: EEG, ERP, and viral load. *Int J Psychophysiol* 2000;38:97–108.

288. Diehl B, Evers S, Sylvester E, et al. Routine electroencephalogram in follow-up of patients with HIV infections of different stages. A long-term study. [German] *Nervenarzt* 1998;69:485–489.

289. Baldeweg T, Catalan J, Pugh K, et al. Neurophysiological changes associated with psychiatric symptoms in HIV-infected individuals without AIDS. *Biol Psychiatry* 1997;41:474–487.

290. Baldeweg T, Gruzelier JH. Alpha EEG activity and subcortical pathology in HIV infection. *Int J Psychophysiol* 1997;26:431–442.

291. Newton TF, Leuchter AF, Miller EN, et al. Quantitative EEG in patients with AIDS and asymptomatic HIV infection. *Clin Electroencephalogr* 1994;25:18–25.

292. Fletcher DJ, Raz J, Fein G. Intra-hemispheric alpha coherence decreases with increasing cognitive impairment in HIV patients. *Electroencephalogr Clin Neurophysiol* 1997;102:286–294.

293. Schroeder MM, Handelsman L, Torres L, et al. Consistency of repeated event-related potentials in clinically stable HIV-1–infected drug users. *J Neuropsychiatry Clin Neurosci* 1996;8:305–310.

294. Linnville SE, Elliott FS, Larson GE. Event-related potentials as indices of subclinical neurological differences in HIV patients during rapid decision making. *J Neuropsychiatry Clin Neurosci* 1996;8:293–304.

295. Pierelli F, Garrubba C, Tilia G, et al. Multimodal evoked potentials in HIV-1–seropositive patients: relationship between the immune impairment and the neurophysiological function. *Acta Neurol Scand* 1996;93:266–271.

296. Linnville SE. Brain-emitted event-related potentials recorded during rapid decision-making may be useful in monitoring the neuropathogenesis of the human immunodeficiency virus (HIV). *Percept Motor Skills* 1995;81:209–210.

297. Vigliano P, Rigardetto R, Capizzi G, et al. EEG diagnostic and predictive value on HIV infection in childhood. *Neurophysiol Clin* 1994;24:367–379.

298. Iragui VJ, Kalmijn J, Thal LJ, Grant I. Neurological dysfunction in asymptomatic HIV-1 infected men: evidence from evoked potentials. *Electroencephalogr Clin Neurophysiol* 1994;92:1–10.

299. Egan VG, Chiswick A, Bretle RP, et al. The Edinburgh cohort of HIV-positive drug users: the relationship between auditory P3 latency, cognitive function and self-rated mood. *Psychol Med* 1993;23:613–622.

300. Baldeweg T, Gruzelier JH, Catalan J, et al. Auditory and visual event-related potentials in a controlled investigation of HIV infection. *Electroencephalogr Clin Neurophysiol* 1993;88:356–368.

301. Jabbari B, Coats M, Salazar A, et al. Longitudinal study of EEG and evoked potentials in neurologically asymptomatic HIV infected subjects. *Electroencephalogr Clin Neurophysiol* 1993;86:145–151.

302. Birdsall HH, Ozluogluy, Lew HL, et al. Auditory P300 abnormalities and leukocyte activation in HIVB infection. *Otolaryngol Head Neck Surg* 1994;110:53–59.

303. Baldeweg T, Riccio M, Gruzelier J, et al. Neurophysiological evaluation of zidovudine in asymptomatic HIV-1 infection: a longitudinal placebo-controlled study. *J Neurol Sci* 1995;132:162–169.

304. Schmitt B, Seeger J, Jacobi G. EEG and evoked potentials in HIV-infected children. *Clin Electroencephalogr* 1992;23:111–117.

305. Harden CL, Daras M, Tuchman AJ, et al. Low amplitude EEGs in demented AIDS patients. *Electroencephalogr Clin Neurophysiol* 1993;87:54–56.

306. Arendt G, Hefter H, Jablonowski H. Acoustically evoked event-related potentials in HIV-associated dementia. *Electroencephalogr Clin Neurophysiol* 1993;86:152–160.

307. Tulinius MH, Hagne I. EEG findings in children and adolescents with mitochondrial encephalomyopathies: a study of 25 cases. *Brain Dev* 1991;13:167–173.

308. Walter H, Kristeva R, Knorr U. Individual somatotopy of primary sensorimotor cortex revealed by intermodal matching of MEG, PET, MRI. *Brain Topogr* 1992;5:83–87.

309. Furlong PL, Wimalaratna S, Harding GFA. Augmented P22–N31 SEP component in a patient with a unilateral space occupying lesion. *Electromyogr Clin Neurophysiol* 1993;88:91–123.

310. Keesey JC. Does myasthenia gravis affect the brain? *J Neurol Sci* 1999;170:77–89.

311. Nomura T, Fukudo S, Matsuoka H, et al. Abnormal electroencephalogram in irritable bowel syndrome. *Scand J Gastroenterol* 1999;34:478–484.

312. Gevins A, Smith ME. Neurophysiological measures of working memory and individual differences in cognitive ability and cognitive style. *Cereb Cortex* 2000;10:829–839.

313. Gevins A, Smith ME, McEvoy LK, et al. High-resolution EEG mapping of cortical activation related to working memory: effects of task difficulty, type of processing, and practice. *Cereb Cortex* 1997;7:374–385.

314. Gevins A, Smith ME, Leong H, et al. Monitoring working memory load during computer-based tasks with EEG pattern recognition methods. *Human Factors* 1998;40:79–91.

315. McEvoy LK, Smith ME, Gevins A. Test-retest reliability of cognitive EEG. *Clin Neurophysiol* 2000;111:457–463.

316. Gevins A, Smith ME. Detecting transient cognitive impairment with EEG pattern recognition methods. *Aviat Space Environ Med* 1999;70:1018–1024.

317. Smith ME, McEvoy LK, Gevins A. Neurophysiological indices of strategy development and skill acquisition. *Cogn Brain Res* 1999;7:389–404.

318. Leocani L, Locatelli M, Bellodi L, et al. Abnormal pattern of cortical activation associated with voluntary movement in obsessive-compulsive disorder: an EEG study. *Am J Psychiatry* 2001;158:140–142.

319. Myslobodsky MS, Coppola R, Weinberger DR. EEG laterality in the era of structural brain imaging. *Brain Topogr* 1991;3:381–390.

320. Reite M. Advances in the study of mental illness. *Adv Neurol* 1990;54:207–222.

321. Reite M, Teale P, Sheeder J, et al. Neuropsychiatric applications of MEG. *Electroencephalogr Clin Neurophysiol Suppl* 1996;47:363–382.

6

DRUG THERAPY IN NEUROPSYCHIATRY

ALLAN T. HANRETTA
BARRY S. FOGEL

Although various psychotherapeutic, psychoeducational, environmental, and behavioral interventions may be useful in patients with neuropsychiatric disorders, the use of drugs acting on the central nervous system (CNS) is the cornerstone of therapy for most patients seen in practice. This chapter discusses the ways in which drug therapy in neuropsychiatry is distinguished from general psychopharmacologic practice. Advice is offered concerning the selection and monitoring of drug therapy for patients with neuropsychiatric disorders, including the decision to use drug treatment, choice of target symptoms and syndromes, methods of monitoring treatment response and adverse effects, use of drug combinations, and concerns regarding side effects and drug interactions. Specific guidance is also provided for the use of antidepressants, neuroleptics, antianxiety drugs, stimulants, anti-Parkinson drugs, antiepileptic drugs (AED), autonomic agents, and opiate antagonists. Discussion of clinical trials of drug therapy for specific conditions, however, is limited to the illustration of general principles; further details are found in the chapters on each of the specific neuropsychiatric conditions.

DISTINCTIVE FEATURES OF DRUG THERAPY IN NEUROPSYCHIATRY

Limits of Diagnosis and the Need to Extrapolate from the Literature

In general psychopharmacology, the indications for drug treatment usually are (primary) mental disorders as defined in the American Psychiatric Association *Diagnostic and Statistical Manual* (1) or in the World Health Organization's *International Classification of Diseases* (2). The efficacy of the treatment of these conditions is established by randomized, controlled clinical trials (RCT). For many standard psychopharmacologic agents, such as the tricyclic antidepressants (TCA), literally hundreds of RCT support the efficacy of their principal indication, in this case, major depression (3). In contrast, the syndromes of mood and behavior owing to gross brain diseases, although they have a special place in

systems of diagnostic classification, may differ in important ways from the primary mental disorders that have been so intensively studied. Even when patients with a specific neurologic disease have a syndrome that meets formal criteria for a common mental disorder such as major depression, RCT specifically addressing drug efficacy for that disorder in that specific population are few, and the aggregate number of patients studied is small. Thus, for example, only five RCT address the treatment of major depression after stroke (4–7): one studied nortriptyline, one studied trazodone, one studied citalopram, one studied fluoxetine, and one compared nortriptyline and fluoxetine with placebo.

The limitations of the RCT literature may restrict the choice of treatment options, or place the clinician in the position of choosing therapy based on extrapolation from studies in which either the patients did not have exactly the same syndrome or they did not have the neurologic disease that afflicts the current patient. When a clinician undertakes the treatment of poststroke major depression using a selective serotonin reuptake inhibitor (SSRI), the empiric basis consists of studies of the efficacy of the SSRI for major depression in patients without strokes, plus only three RCT of SSRI in poststroke depression involving only two SSRI, citalopram and fluoxetine (5–7).

Atypical and Subsyndromal Conditions

In general, patients with syndromes that do not meet the criteria for established mental disorders do not necessarily have milder or less functionally significant problems. Criteria sets are chosen by consensus, with the aim of maximizing accuracy, not eliminating all false-negative classifications (8). The presence of brain disease adds another issue to the interpretation of syndromes that do not meet conventional criteria: the modification of the expression of psychopathology by brain dysfunction. Some ways in which regional brain dysfunction modifies the symptomatic expression of mental disorders are by (a) producing indifference or shallowness of affect, as in frontal lobe or right-hemisphere dysfunction; (b) affecting the ability to organize a coherent account of symptoms, as with diffuse dysfunction or frontal system dysfunction; (c) producing apathy regarding one's condition, as with subcortical and basal ganglia dysfunction; and (d) affecting the expression of problems in spoken language, as when left-hemisphere dysfunction produces aphasia.

In patients with mental syndromes that do not meet the consensus criteria for a major mental disorder, the clinical decision in favor of drug therapy is based on the severity, persistence, tolerability, and functional consequences of the symptoms. To this, an additional bias in favor of treatment is added if there is an obvious neurologic deficit that would be expected to prevent expression of the full syndrome if it were present.

Major mental disorders in patients with neurologic disease may represent either coincident diagnoses or true secondary effects of the neurologic disease itself. A history of a chronic, recurrent, or relapsing/remitting mental disorder similar to the current problem and preceding the onset of the neurologic disease favors coincident diagnoses, as does a family history of a similar problem in a neurologically healthy first-degree relative. When a patient with a neurologic disease has a coincident psychiatric diagnosis, a choice of therapy based on RCT for the mental disorder in question involves less extrapolation and can be made with greater confidence. A true secondary mental disorder, even if it comprises the same syndrome as a primary mental disorder, may have different biology and a different pattern of drug response. For example, secondary major depression in patients with significant medical illness may respond to lower doses of antidepressants than are usually necessary to treat primary major depression (9).

Behavioral Syndromes of Gross Brain Dysfunction

In addition to psychiatric syndromes and partial versions of such syndromes, neuropsychiatrists also apply drug therapy to symptoms and syndromes that are primarily found in populations with gross brain dysfunction. Such symptoms and syndromes, when they occur in connection with primary mental disorders, usually are seen as dimensions of illness rather than as the objects of drug treatment in their own right. Such symptoms and syndromes include

1. Affective instability or emotional lability,
2. Apathy or abulia,
3. Aggressive behavior, either spontaneous or reactive to provocation,
4. Irritability without overt aggression,
5. Impaired attention,
6. Disordered sleep,
7. Pain,
8. Involuntary movements,
9. Paroxysmal phenomena
10. Recurrent self-injurious behavior (SIB).

Obviously, patients may have these phenomena either in isolation or in connection with disturbances of cognition and mood. As a rule, if a symptom (e.g., apathy) is part of a major disorder (e.g., depression), then the disorder will be treated. However, the symptom may become the focus of subsequent or concurrent treatment if it does not resolve with treatment of the accompanying polysymptomatic disorder. Thus, drug treatment is given for apathy that persists after successful treatment of depression, for aggressiveness that persists after successful treatment of a delusional disorder, and for sleep disturbance that persists after optimal remediation of the cognitive deficits of dementia.

Increased Sensitivity to Central Nervous System Side Effects of Drugs

Patients with gross brain disease are known to be more sensitive to the CNS side effects of psychotropic drugs. Appropriate cautions are given to start with a low dose of drugs, slowly increase the dose, and, if possible, avoid drugs known to have an especially high rate of CNS toxicity. When a drug is chosen, its specific CNS toxicities should be considered in relation to the patient's specific CNS vulnerabilities. Thus, when giving a benzodiazepine for anxiety, the major concern in a young epileptic patient would be the effects on cognition; in an older person with cerebellar degeneration, the major concern would be the potential for falls. Medications given for incidental medical conditions that have high rates of CNS side effects should be eliminated or replaced whenever possible. CNS toxicity of drugs usually is greater if the patient is depressed, sedated, or otherwise underaroused. Successful treatment of depression facilitates the treatment of other neuropsychiatric conditions. Elimination of sedating drugs may improve the tolerability of other drugs. Stimulants may be added to a drug regimen if the patient is underaroused or sedated and can lead to a generalized improvement in CNS tolerability of the entire regimen.

Potential for Personality Change

Gross changes in brain structure or function have the potential to change a person's enduring patterns of social interaction. Such personality changes interact with the patient's premorbid personality. Premorbid traits may either disappear or become more intense. An example of the latter case is when a demanding, irritable person becomes physically agitated and violent in the early stages of a dementing illness. When premorbid traits are intensified by gross brain disease, drug therapy can be directed at the premorbid trait (10). In the given example, an SSRI might be administered to reduce the patient's constitutional irritability. If the behavioral manifestations of the trait had not been intensified by the brain disease, it is unlikely that it would be treated with psychotropic drugs.

Frequency of Comorbid Medical Conditions

In addition to brain diseases and mental disorders, neuropsychiatric patients often have general medical diseases that are related to their neurologic conditions as causes or consequences. Patients with cerebrovascular disease usually have hypertension; patients with debilitating conditions frequently develop infectious complications. Treatment of these general medical diseases must be compatible with the treatment of the patient's neurologic and psychiatric problems. In addition to screening patients for the well-known pharmacokinetic interactions that now are summarized in electronic databases, this requires consideration of the CNS actions of the drugs given for the medical condition. For example, theophylline generally should not be used to treat bronchospasm in a person with epilepsy and an associated anxiety disorder because of its proconvulsant and anxiogenic effects (11). CNS actions of general medical drugs can on occasion be exploited for neuropsychiatric benefit. Thus, patients with recurrent agitation who also have hypertension can be treated with agents such as propranolol that both lower blood pressure and may decrease the frequency and severity of agitation (12,13).

Comorbid endocrine disorders are of particular importance in neuropsychiatric patients because of the effect of hormones on central neurotransmission. When there is a relatively mild disorder of endocrine function, such as hypothalamic hypogonadism in a man with limbic epilepsy (14) or subclinical hypothyroidism with a normal thyroxine level and an elevated thyroid-stimulating hormone level (15), the neuropsychiatric bias is to offer treatment for the endocrine disorder. Experience suggests that interventions to normalize endocrine status facilitate the effectiveness of psychotropic drug treatment, even when they do not in themselves eliminate the patient's mental and behavioral symptoms.

A Broader Pharmacopeia

Neuropsychiatrists use the full range of traditional psychotropic drugs. In addition, they frequently use "neurologic" drugs to modify behavior. Specific categories of neurologic drugs widely used by neuropsychiatrists as psychotropics include AED, anti-Parkinson drugs, agents that affect adrenergic function, and opiate agonists and antagonists. In contrast to general psychiatrists, neuropsychiatrists appear more likely to use these agents for behavioral indications before their efficacy has been established by RCT. Thus, neuropsychiatrists widely used carbamazepine for mood sta-

bilization for years before RCT in bipolar disorder brought it into the general psychiatric mainstream.

Hypothesis-Driven Prescribing Versus Indication-Driven Prescribing

Indication-driven prescribing matches drugs with disorders for which RCT have demonstrated drug efficacy. Hypothesis-driven prescribing begins with a hypothesis about the biochemical or physiologic basis of the patient's symptoms or syndrome and selects a drug with a mechanism of action relevant to the hypothesis. Because of the shortcomings of both classification systems and the literature on neuropsychiatric RCT, neuropsychiatrists often must resort to hypothesis-driven prescribing. Prescription of drugs on this basis amounts to conducting an "N of 1" drug study (a clinical trial with a single subject). A poor response to the drug suggests, but does not prove, that the hypothesis is flawed.

As an example of hypothesis-based prescribing, consider the initial use of the opiate receptor blocker naltrexone in the treatment of SIB in a patient with severe mental retardation. The neuropsychiatric hypothesis is that self-inflicted pain releases endogenous opiates, which in turn reinforce the behavior (16). Initial successes with naltrexone led to RCT (17,18), which showed effectiveness for some autistic patients and an apparent adverse effect in patients with Rett syndrome. The neuropsychiatrist who now prescribes naltrexone in this situation combines an indication-based approach with a hypothesis-based approach. If a particular patient does not respond, the neuropsychiatrist suspects that opiate-related mechanisms may not be the essential driver of the SIB in the individual case.

Need for Combining Drug Therapy with Nonpharmacologic Therapy

Like general psychiatric patients, neuropsychiatric patients often need a combination of drug therapy with nonpharmacologic therapy to get a satisfactory outcome. In neuropsychiatry, there are important situations in which appropriate drug therapy will have little benefit unless it is combined with environmental or behavioral intervention. These situations involve patients with major impairments in executive cognitive function who are extremely responsive to immediate environmental contingencies because of their deficits in working memory and self-regulation (19). Drug therapy for such patients, beyond treating established psychiatric syndromes, aims to optimize patients' attention and arousal so that they can make use of whatever residual executive functions remain. However, if the optimal level of executive function attainable with drug therapy remains low, well-planned external contingencies will still be essential for behavioral management.

Presenting Neuropsychiatric Drug Therapy to Patients and Caregivers

Drug therapy in neuropsychiatry requires the consent and cooperation of the patient, or, if the patient is incompetent, the responsible caregiver or guardian. Obtaining consent and cooperation is based on explaining to the patient or caregiver the hypotheses on which the drug trial is based, the background from relevant RCT or case series, the desired result of treatment, the expected risks, and how therapy will be monitored. When there is a reasonable choice of therapies or several therapies to be tried in sequence until one is effective, the patient and/or caregiver should participate in the decision of which to try first.

The process of obtaining informed consent and eliciting cooperation has the same general features in neuropsychiatry as in other specialties, but the presentation tends to emphasize more of what is *not* known about the field and the physician's reasons for recommending treatment that is not thoroughly established by RCT. The patient or the caregiver sometimes fears that the patient will be treated as an experimental animal; an emphasis on the individualization of treatment, close monitoring, and the human situation that compels a trial of drug treatment will help to alleviate these fears.

DECISION TO USE DRUG THERAPY

The decision to use drug therapy in a neuropsychiatric patient is based on the traditional balancing of risks and benefits while taking into account the wishes of the patient, caregiver, or guardian. With regard to expected benefits, considerations include the results of RCT addressing the syndrome that the patient has or those similar to it, the results described in case studies or case series in which the patients reported most closely resemble the patient being treated, the condition of the patient without drug treatment, and the likely natural history of the patient's symptoms or syndrome, drawn both from the literature and the clinician's experience. The impact of the mental and behavioral symptoms or syndrome on the patient's everyday function is considered, and the likely functional consequences of improvement are estimated. Risk is estimated from the literature on the drug to be used, emphasizing literature in which the drug was used in a population sharing important clinical and demographic features with the patient to be treated. Risks of permanent injury are carefully distinguished from risks of temporary discomfort.

Before initiating drug therapy, the relevance and implementation of nonpharmacologic therapies should be carefully reviewed, and general medical or environmental factors influencing the patient's mental or behavioral symptoms should be addressed. If the need for symptom control is urgent, drug therapy can be proposed in combination with efforts to modify environmental or general medical factors.

Patients and caregivers should be aware that a holistic approach will be taken and that drug therapy will not substitute for attention to general medical and environmental issues or for psychotherapeutic and psychoeducational efforts.

CHOICE OF TARGET SYMPTOMS

Neuropsychiatric patients may present with well-recognized mental syndromes and disorders, such as major depression. When they do, drug therapy is directed at the disorder rather than at a specific symptom. When a specific symptom, such as aggressiveness, is the impetus for drug therapy, it is first determined whether the aggressiveness is directly related to a mental disorder such as hypomania or a delusional disorder. If it is, the disorder is the focus of treatment. If it is not, the aggressiveness as such becomes the focus of treatment.

Even when a recognized mental disorder is the basis for selecting treatment, the symptoms should be identified that are most closely connected to the patient's distress and disability. These key symptoms may direct the choice of drug among equally reasonable alternatives and form the basis of ancillary measures of treatment response.

MONITORING TREATMENT RESPONSE AND ADVERSE EFFECTS

Monitoring the drug therapy in neuropsychiatry is based on the principle that optimal therapy should improve the patient's function and well-being as well as reducing or eliminating symptoms and signs that cause distress or concern to others. Thus, monitoring emphasizes a multidimensional approach that incorporates measures of function and measures of symptoms that are most distressing to the patient.

When patients are treated for recognized mental disorders, such as major depression or obsessive-compulsive disorder (OCD), a standard rating scale is used to monitor treatment response. The choice of rating scale takes into account any limitations that the patient may have in self-reporting symptoms. Thus, when monitoring treatment of major depression in dementia, an observer-rated scale such as the Cornell Scale for Depression in Dementia (20) would be used because a self-rated instrument like the Geriatric Depression Scale (21) is known to underestimate depressive symptoms in this situation (22).

Whether or not a disorder-specific scale is used, a determination is made regarding the symptom or syndrome causing the patient the greatest distress or disability. A scale is used to periodically measure the response of this symptom or syndrome to the drug treatment. For some syndromes, such as for aggression (23) and apathy (24), well-validated multi-item scales are available. When no such scale is relevant, an alternative is to use a visual analogue scale (25) or have the patient or caregivers rate the symptom on a scale of 1 to 10, with mutually agreed-on anchor points. This latter approach has shown considerable value in monitoring the treatment of chronic pain (26) (see also Chapter 31 on chronic pain management).

In addition to measures of disorders, symptoms, and syndromes, patients should be periodically assessed for functional performance. In patients with more severe neurologic impairment, this can be done with standard scales of physical activities of daily living and instrumental activities of daily living, such as those used regularly by occupational therapists (27). For less severely impaired outpatients, general-purpose self-rating questionnaires are available that measure the impact of illness on everyday physical and social activities (28).

Monitoring for adverse reactions is based on two lists of potential adverse effects: those that are uncommon but serious and require immediate action and those that are common but not immediately dangerous. The patient and caregivers are informed of both kinds of potential problems and are instructed to notify the physician immediately about problems of the first type. When specific monitoring of blood tests or other examinations are needed to evaluate the more serious type of adverse reaction, clear responsibility for these must be taken by one individual. The patient and caregivers are encouraged to tell the physician about less serious but common adverse reactions, and the more common ones are asked about systematically on each follow-up visit with the patient. When the patient is an outpatient who is seen infrequently, telephone follow-up should include a systematic review of common or potentially serious adverse reactions.

USE OF DRUG COMBINATIONS

There is a long tradition in general psychiatry of avoiding drug combinations, which has gradually yielded to scientific evidence on additive and synergistic therapeutic effects. Neurologists have come to accept the necessity of drug combinations for some cases of epilepsy (29) and some cases of Parkinson disease (30). Notwithstanding, polypharmacy continues to be viewed with suspicion by both consumers and providers, in part because of the greater complexity of monitoring multiple drug therapy and the much greater incidence of drug interactions.

In a limited number of situations in neuropsychiatry, drug combinations are supported by well-designed RCT of combined therapy. Studies of add-on therapy for poorly controlled epilepsy are an outstanding example (31,32). In most cases, however, drug combinations are justified either by the presence of multiple coincident diseases or a hypothesis of multiple mechanisms contributing to the patient's mental or behavioral syndrome. In the former case, the justification for multiple-drug therapy is clear, and the major precaution to be taken is checking a recent text or database for po-

tential drug interactions. In the latter case, clarity regarding the hypothesis of multiple causation is the best way to prevent inappropriate prescribing. As an example, consider a patient with recent traumatic brain injury (TBI), a premorbid irritable personality, and a current syndrome of affective instability, emotional lability, and irritable outbursts. The working hypothesis is that the affective instability and emotional lability are consequences of diffuse cerebral injury but that the irritability represents a premorbid personality trait. Initial treatment of the affective instability with a valproate reduces the patient's mood swings, but he or she remains irritable. An SSRI is added, directed at the hypothesis of an underlying irritable temperament. A third type of rational polypharmacy is the application of two treatments that work on the same neurotransmitter system in different ways, as when levodopa-carbidopa and a direct dopamine agonist are combined in the treatment of advanced Parkinson disease.

SIDE EFFECTS AND INTERACTIONS

Comprehensive treatment of drug side effects is beyond the scope of this chapter, as is a comprehensive review of all potentially relevant drug interactions. Neuropsychiatric patients often have general medical illnesses and take multiple medications. The implications of these factors for choosing and monitoring psychotropic drugs are discussed in detail in a series of analytic reviews (33–35). This chapter focuses specifically on neurologic side effects of drugs and drug interactions that are either very common or illustrate an important principle of drug interaction.

ANTIDEPRESSANT DRUGS
Efficacy

The efficacy of antidepressant drugs is best established for the treatment of primary major depression, but a relatively small number of RCT support their efficacy for major depression secondary to common neurologic diseases. Successful trials have been reported for nortriptyline (6,36), citalopram (5), fluoxetine (7), and trazodone (37) in the treatment of poststroke depression. Robinson et al. (6) reported that nortriptyline was superior to fluoxetine for the treatment of poststroke depression. Nortriptyline (38), amitriptyline (39), citalopram (40), and sertraline (41) have been shown to help pathologic crying after stroke. Reifler et al. (42) failed to find imipramine superior to placebo for the treatment of major depression associated with Alzheimer disease, although numerous case series and uncontrolled studies suggest that usual antidepressant drug treatments are efficacious. Schiffer and Wineman (43) found desipramine superior to placebo for the treatment of major depression associated with multiple sclerosis. Four RCT support the efficacy of TCA for major depression associated with Parkinson disease (44–47).

The benefit of selegiline for depressed patients as well as for parkinsonism per se is supported by a well-designed study by Allain et al. (48). An open trial supported a combined antidepressant–anti-Parkinson effect of bupropion (49). To our knowledge, RCT of antidepressants for major depression after TBI and depression accompanying multi-infarct dementia have not been carried out.

In contrast to the paucity of RCT, there are numerous case reports, case series, and open trials suggesting that antidepressants of various classes will relieve symptoms of major depression in patients with common brain diseases. This uncontrolled literature does not support the choice of one antidepressant type over another in any specific neurologic disease. Peyser and Folstein (50) observe in their review of depression in Huntington disease that even though there is a known genetic lesion on chromosome 4 that underlies all cases, not all patients respond to the same antidepressant medication.

With regard to efficacy in primary nonmelancholic, nondelusional major depression, there is reasonable consensus that all currently available antidepressants are of equivalent efficacy (51,52). In cases of melancholia, TCA and venlafaxine may be superior to the SSRI, although the issue remains unresolved because there have been few direct comparative studies with severely ill patients (52–55). Delusional depression requires treatment with an antidepressant–neuroleptic combination or electroconvulsive therapy.

Apart from major depression, the best established indication for antidepressants is the treatment of OCD. Here, clomipramine, fluoxetine, paroxetine, sertraline, and fluvoxamine have proven efficacious by RCT sufficient to satisfy the U.S. Food and Drug Administration (FDA).

Neuropsychiatric Applications Other Than Major Depression and Obsessive-Compulsive Disorder

In the past several years, several reports have appeared regarding the application of antidepressant drugs to neuropsychiatric problems other than major depression. These applications have included the following:

1. *Dysthymia in patients with mental retardation.* Jancar and Gunaratne (56) reported two cases of successful treatment with SSRI and psychotherapy.
2. *Adult attention deficit disorders.* Wilens et al. (57) reviewed several studies supporting the effectiveness of TCA in some cases.
3. *SIB in developmentally disabled persons.* Case reports, summarized by Aman (58), suggest that SSRI may inhibit this behavior. The evidence for their efficacy, however, lags behind the evidence supporting the efficacy of thioridazine, lithium, and naltrexone.
4. *Chronic fatigue syndrome and fibromyalgia.* Twelve double-blind studies, reviewed by Goodnick and Sandoval (59), support the efficacy of serotonergic antidepressants for pain relief in these conditions and the efficacy of the catecholaminergic agents maprotiline and bupropion for depressive symptoms as such.
5. *Chronic nonmalignant pain.* A metaanalysis of 39 controlled studies of antidepressants suggested that the average chronic pain patient who received an antidepressant drug had less pain than 74% of comparison patients who received a placebo (60). The study suggested that antidepressants have analgesic effects. More recent articles continue to support the role of antidepressants as analgesics (61) (see Chapter 31).
6. *Miscellaneous disorders of impulse control.* Pathologic gambling and certain sexual compulsions and paraphilias have been conceptualized as part of a continuum of obsessive-compulsive spectrum disorders that also include OCD itself and body dysmorphic disorder. These conditions may respond to treatment with SSRI (62,63).
7. *Emotional lability, including pathologic laughing and crying, in patients with brain injury.* Fluoxetine and TCA have both been shown effective for this symptom in case series (64,65).
8. *Posttraumatic stress disorder.* Sutherland and Davidson (66) treated patients with posttraumatic stress disorder with high-dose SSRI for 8 weeks. They found improvement, especially in psychic numbing and avoidant behavior.
9. *Tourette syndrome with attention deficits.* Sweeney and Henry (67) reported the efficacy of imipramine.

Neurotoxicity of Antidepressant Drugs

In choosing an antidepressant drug for a patient with gross brain disease, CNS toxicity of antidepressants will be a primary concern. Other aspects of antidepressant toxicity, such as the cardiac toxicity of TCA, are of course important but are not discussed here. The reader is referred to the comprehensive reviews of Stoudemire et al. (33–35,68) for detailed discussion of general medical issues in the choice of antidepressants.

The SSRI, which have emerged as the antidepressants of first choice for most physicians (52,69), rarely cause delirium or gross cognitive impairment. However, at high doses, they can produce an apathetic frontal lobe syndrome in some patients (70,71). SSRI also increase neuromuscular excitability and can produce or aggravate tremor or myoclonus. Overt extrapyramidal disorders, such as akathisia and parkinsonism, occur more rarely but have been seen in patients not exposed to neuroleptics, either concurrently or in the past (72). Although the mechanism is not known for certain, SSRI may produce extrapyramidal effects by indirect down-regulation of dopamine turnover by increased synaptic serotonin (73). Among the SSRI, fluoxetine is most likely to cause nervousness or restlessness. Although there is some question about whether this represents akathisia in all cases, it frequently requires dose reduction or switching to a different SSRI. Finally, all the SSRI can cause headaches.

TCA, all of which have potent anticholinergic properties, can cause hallucinations or delirium, particularly in elderly patients and those with dementia. Like the SSRI, they can cause or aggravate tremor or myoclonus, but, with the exception of the antidepressant-neuroleptic amoxapine, they are rarely associated with extrapyramidal reactions such as akathisia and acute dyskinesia (74). TCA have been associated with speech arrest, particularly at high doses. They can have discrete effects on memory in the absence of delirium (75).

The monoamine oxidase inhibitors (MAOI) increase neuromuscular excitability. In usual doses, they do not cause delirium, cognitive impairment, or extrapyramidal reactions.

Bupropion has the greatest propensity to induce seizures of all the currently marketed antidepressants. In patients without neurologic risk factors for seizures who are taking a full therapeutic dose, the risk is approximately 0.4%. Bupropion is contraindicated in patients with seizure disorders and in patients with current or prior diagnosis of bulimia or anorexia nervosa because they have a higher risk of seizures. Accordingly, patients with risk factors for seizures should be on a prophylactic AED if bupropion is to be prescribed. Bupropion is a stimulating antidepressant and can produce agitation or a delirious psychosis in occasional vulnerable individuals (76).

Nefazodone has a relatively benign neurologic side effect profile. Sedation and ataxia are the most common limiting neurologic side effects. However, nefazodone is a fairly weak antidepressant and must be pushed to the upper end of its therapeutic range (i.e., 400–600 mg per day) to have efficacy equivalent to that of TCA or conventional SSRI (77). At this higher dose, sedation and/or ataxia can pose significant problems in patients with preexisting brain disease. In one animal model, nefazodone significantly potentiates opioid analgesia (78). It is not known whether this is an applicable effect in humans at the usual clinical dose.

Venlafaxine has a neurologic side effect profile resembling that of the SSRI. In addition, it is associated with hypertension: the rates are 3% to 5% for doses of less than 200 mg per day, 7% for doses of 201 to 300 mg per day, and 13% for doses of more than 300 mg per day (79). Paroxysmal sweats, sometimes drenching the patient's bedding or clothing, can also be seen. Cognitive side effects appear minimal.

Mirtazapine resembles venlafaxine in its neurologic action to enhance both noradrenergic and serotonergic neurotransmission. It is well tolerated in depressed patients, and its efficacy is comparable with that of amitriptyline, SSRI, and venlafaxine (80,81). Adverse effects most frequently associated with mirtazapine include, in decreasing frequency, weight gain, dry mouth, headache, and somnolence (80,81). A rare risk of severe neutropenia (incidence of 1.1 per 1,000

patients) has been reported in three patients, with the development of agranulocytosis in two of these patients (82). All three patients recovered when mirtazapine was discontinued. Mirtazapine should be discontinued in patients who develop signs of infection with a low white cell count.

Drug Interactions

Concern about drug interactions is greatest with the SSRI and nefazodone, which are potent inhibitors of the hepatic cytochrome P-450 system. The inhibition of the cytochromes leads to clinically significant increases in drug levels when a drug is coadministered that is primarily metabolized by the isoenzyme that is inhibited. The SSRI inhibit the isoenzyme CYP2D6. This isoenzyme metabolizes the TCA (83). It is well established that TCA levels are significantly higher in patients taking SSRI concomitantly (83,84). Levels of clozapine, the atypical neuroleptic, are raised by as much as an order of magnitude when the SSRI fluvoxamine is given concomitantly (85). Fluvoxamine inhibits several cytochrome isoenzymes; it has clinically significant effects to slow the metabolism and increase the blood level of TCA, alprazolam, diazepam, theophylline, propranolol, warfarin, methadone, and carbamazepine (86). The size and clinical importance of pharmacokinetic interactions of other drugs with SSRI depend in part on the genetic polymorphism of CYP2D6; different alleles corresponding to different levels of metabolic activity are distributed in the general population (87). Citalopram, the most recently approved SSRI for use in the United States, is as equally effective and well tolerated as other SSRI (88). Unlike other SSRI, citalopram produces only a mild to negligible inhibition of CYP2D6 and other cytochromes and, thus, is less likely than other SSRI to have pharmacokinetic interactions with other drugs (89,90). Nefazodone is a potent inhibitor of CYP3A4, the enzyme that metabolizes terfenadine and astemizole as well as carbamazepine. Because of the risk of ventricular arrhythmia at high serum levels of terfenadine or astemizole, these drugs should not be used together with nefazodone and are no longer available for use. When given together with nefazodone, the carbamazepine dose should be reduced and blood levels monitored frequently during dose adjustments.

A particularly neuropsychiatric concern is the interaction of antidepressant drugs with AED. The concurrent administration of an SSRI and carbamazepine may raise carbamazepine levels. The concurrent administration of carbamazepine, phenytoin, or primidone together with a TCA may lower tricyclic levels (91,92). In general, when an antidepressant and an AED are given together, the level of the AED should be monitored frequently until a stable and therapeutic blood level is confirmed. TCA blood levels as well as carbamazepine levels should be monitored when combining the two drugs. The choice of desipramine or nortriptyline, each of which has meaningful blood levels and no significant active metabolites, facilitates this task.

Pharmacodynamic drug interactions involving antidepressant drugs usually are related to the additive effects of the antidepressant and other drugs on neurotransmitter receptors, specifically α-adrenergic, serotonergic, and cholinergic receptors. Thus, hypotension can occur when the TCA trazodone or mirtazapine is combined with antihypertensive α-adrenergic blockers. The serotonin syndrome, a state of delirium, tremulous rigidity, and autonomic instability, can occur when an SSRI is combined with an MAOI or clomipramine. An anticholinergic delirium can occur when TCA are combined with other drugs that have central anticholinergic effects, such as first-generation antihistamines.

Choice of Antidepressant in the Neuropsychiatric Patient

Table 6.1 summarizes some of the issues relevant to choosing and initiating antidepressant therapy in a neuropsychiatric patient.

The process of choosing an antidepressant begins by ruling out any that are contraindicated. In practical terms, this means rejecting TCA for patients with heart block and rejecting drugs to which the patient has shown a significant hypersensitivity reaction in the past.

The second step is to consider whether the patient has significant melancholic features, especially substantial weight loss and early morning awakening. Such patients may do better on a TCA or venlafaxine.

Next, the patient should be placed on a continuum between apathetic-retarded and agitated-anxious. Patients with significant agitation and anxiety are more likely to tolerate sertraline, paroxetine, citalopram, or nefazodone than bupropion or fluoxetine, both of which are stimulating. Patients with apathy and retardation may benefit from the activating effects of bupropion or fluoxetine.

Among anxious patients, those with obsessions, compulsions, or a premorbid obsessive-compulsive personality should receive SSRI, with fluvoxamine as an additional consideration. Those with predominant anxiety and insomnia might be treated with nefazodone. Patients with compulsive behaviors such as paraphilias and compulsive gambling should be treated with SSRI.

MAOI generally should be regarded as second-line agents. Consideration of an MAOI as first-line therapy would be given for patients with severe phobic symptomatology or with an active migraine problem in need of prophylaxis. Regarding the latter, phenelzine is a well-established agent for migraine prophylaxis.

Patients with a history of poor response to antidepressant drugs, with rapid cycling of mood induced by antidepressants, or with a mixture of depressive symptoms and paroxysmal symptoms typical of partial seizures should not be initially treated with antidepressants. Instead, the first-line treatment would be a mood-stabilizing AED such as carbamazepine (93).

Titration of Antidepressant Therapy

Starting doses of antidepressants in patients with gross brain disease should be low enough to test the patient for unusual sensitivity to the drug. The recommended starting doses are therefore below those recommended by the manufacturer. They are:

1. TCA (excluding protriptyline): 10 mg q.h.s. (every bedtime),
2. Fluoxetine and paroxetine: 5 mg q.d. (every day),
3. Sertraline: 25 mg q.d.,
4. Nefazodone: 50 mg q.h.s.,
5. Bupropion: 75 mg q.d.,
6. MAOI: tranylcypromine 10 mg every morning and noon; phenelzine 15 mg b.i.d. (twice daily),
7. Protriptyline: 5 mg q.d.

If the patient tolerates the test dose of the drug, the daily dose may be increased every 2 to 3 days until limiting side effects develop, a conventional therapeutic dose is reached, or, in the case of the secondary amine tricyclics, a therapeutic blood level is attained. Along the way, if the patient experiences significant improvement in symptoms, the dose is held at that level for at least a week. If the patient is continuing to improve, the dose is not increased further. If a plateau has been reached and there are persistent symptoms, dose increases are resumed. The increment of dose increase is approximately the same as the initial dose, with the proviso that bupropion and the MAOI are given on a t.i.d. (three times daily) schedule and venlafaxine and nefazodone are given on a b.i.d. schedule.

Generally, a slow start is emphasized because of the greater vulnerability of neuropsychiatric patients to CNS side effects. However, there will be some patients who, because of some combination of fast metabolism and pharmacodynamic insensitivity, will require doses at the top of the recommended range for physically healthy patients or even slightly beyond it. If a patient does not experience significant side effects from an SSRI, venlafaxine, nefazodone, or an MAOI, one may continue dose increases to the following limits:

1. Fluoxetine and paroxetine: 80 mg per day,
2. Sertraline: 300 mg per day,
3. Nefazodone: 600 mg per day,
4. MAOI: tranylcypromine 60 mg per day; phenelzine 90 mg per day.

Bupropion is not increased beyond 450 mg per day (150 mg t.i.d.) because of an unacceptable risk of seizures at higher doses. TCA may on occasion be increased beyond 300 mg per day for desipramine or imipramine or 150 mg per day for nortriptyline. When this is done, a tricyclic serum level

TABLE 6.1. ISSUES RELEVANT TO CHOOSING AND INSTITUTING AN ANTIDEPRESSANT FOR NEUROPSYCHIATRIC PATIENTS

Drug	Benefits	Precautions	Relevant Controlled Trials	Initial Daily Dose
Amitriptyline	Sedation, analgesia	Strongly anticholinergic; quinidinelike cardiac effects; orthostatic hypotension	Pathologic laughing and crying; poststroke depression; pain; headache	10–25 mg, usually given at bedtime
Nortriptyline	Meaningful blood levels; less hypotension, sedation, and anticholinergic effect than amitriptyline	Quinidinelike cardiac effects; orthostatic hypotension; anticholinergic effects	Pathologic laughing and crying; poststroke depression	10–25 mg, usually given at bedtime (target blood level 50–150 ng/mL)
Imipramine	Antipanic and antiphobic actions; less sedating than amitriptyline	Quinidinelike cardiac effects; orthostatic hypotension; anticholinergic effects	Depression in Parkinson disease	10–25 mg
Desipramine	Least anticholinergic of tricyclic antidepressants; can be stimulating	Quinidinelike cardiac effects; orthostatic hypotension; anticholinergic effects	Depression in multiple sclerosis; depression in Parkinson disease	10–25 mg (target blood level >125 ng/mL)
Clomipramine	Potent antiobsessional effects	As for amitriptyline, plus myoclonus	None	12.5–25 mg
Protriptyline	Potent stimulant; alternative to direct stimulants	As for amitriptyline	None; case reports suggest utility as a stimulant in traumatic brain injury patients	5 mg
Fluoxetine	Helpful for dysphoria and irritability; can be activating; effective for OCD at higher doses	Akathisia, agitation, anxiety may appear on treatment initiation; apathy at very high doses; drug interactions due to P-450 inhibition	None; numerous case series suggest efficacy for depression owing to general medical diseases, including neurologic diseases	5–10 mg
Fluvoxamine	Specific for OCD	Potent enzyme inhibitor with significant drug interactions; sedation, myoclonus, GI side effects from hyperserotonergic state all possible	None	25–50 mg
Paroxetine	Less activating than fluoxetine; may cause less sedation and fewer GI symptoms than sertraline; chief benefits are relief of dysphoria and irritability	Drug interactions from P-450 inhibition; nausea; potential for sedation	None	5–10 mg
Sertraline	Helpful for dysphoria and irritability; may enhance executive cognitive function	Sedation; fewer GI side effects; drug interactions from P-450 inhibition	None	12.5–25 mg
Citalopram	Fewer adverse drug interactions relative to other SSRI	Somnolence, insomnia, nausea, sweating, asthenia/tiredness in elderly	Poststroke depression; emotional disturbances in dementia	10 mg
Trazodone	Sedating and anxiolytic; may help agitation in dementia	Orthostatic hypotension; excessive sedation; priapism risk in men	Poststroke depression (improved rehabilitation outcome)	25 mg, usually at bedtime
Nefazodone	Anxiolytic, less sedating than trazodone; promotes sleep without disrupting sleep architecture	Sedation (orthostatic hypotension rare and priapism not seen); drug interactions owing to inhibition of P-450 3A4	None	50 mg at bedtime, or 50 mg b.i.d.
Bupropion	Stimulating, antiapathy, mildly antiparkinson; no cardiotoxicity	Risk of seizures, especially with single doses >150 mg or total daily dose >450 mg; mild anticholinergic effects; insomnia, weight loss possible	Depression in Parkinson disease (levodopa requirement reduced)	75 b.i.d.
Venlafaxine	Anxiolytic, less orthostatic hypotension, no significant weight changes; no significant inhibition of cytochrome P-450 isoenzymes	Nausea, dizziness, somnolence, insomnia, and sweating; potential for sustained hypertension	None	25–37.5 mg
Mirtazapine	Sedating, low frequency of sexual dysfunction; not a potent inhibitor of cytochrome P-450 isoenzymes	Weight gain, dry mouth, somnolence, constipation, dizziness; rare severe neutropenia	None	7.5 mg
Tranylcypromine	Stimulating, antiapathy; no cardiotoxicity; broad spectrum of antidepressant efficacy	Orthostatic hypotension, insomnia, risk of drug interactions or hypertensive crisis from tyramine-rich foods	None; case series suggest utility in subcortical vascular dementia with apathetic depression	10 mg morning and noon
Phenelzine	Antianxiety and antiphobic effects; analgesic and antimigraine effects	As for tranylcypromine; more likely to be sedating; may cause major weight gain	Migraine	15 mg morning and noon
Selegiline	Well tolerated; stimulating and antiparkinson effects	Risk of drug interactions or drug–food interactions with dose >10 mg/d; can cause nausea	In early Parkinson disease helps motor symptoms; mood, cognition	5 mg b.i.d.

OCD, obsessive-compulsive disorder; GI, gastrointestinal; SSRI, selective serotonin reuptake inhibitors; b.i.d., twice daily.

and an electrocardiogram should be done to establish the safety of going to a higher dose.

Duration of Therapy

When antidepressants are used to treat primary major depression, maximal treatment response takes 4 to 8 weeks. Therapy, if successful, is continued for at least 6 months at the dose used to induce a remission of depression. It is not known whether these durations apply equally to major depression secondary to gross brain diseases. However, in the absence of other data, one should allow at least 4 weeks at full dose (or maximal tolerated dose) before determining that an antidepressant trial is unsuccessful. If a drug is helpful, treatment should be continued for several months at the remission-inducing dose before attempting to taper the dose of antidepressant.

ANTIPSYCHOTIC DRUGS

Efficacy

Antipsychotic drugs, or neuroleptics, have extremely well-established efficacy for the treatment of schizophrenia and mood disorders with delusions. They have also been tested in numerous RCT for psychotic symptoms and behavioral disturbances in dementia and mental retardation. In patients with dementia, they have their most clear-cut benefit for delusions and hallucinations (94–96) but also benefit agitation and aggressiveness (97). When used to treat agitation without overt psychotic symptoms, they are superior to placebo, but fewer than half of patients treated will benefit (98). No particular antipsychotic drug has shown consistent superiority to any other in the treatment of psychotic complications of dementia. In mental retardation, RCT have not shown consistent benefit of neuroleptics over placebo for behavioral disturbances in general, although low-dose neuroleptics have been shown to alleviate stereotypies in some small controlled studies (96,98). Risperidone was shown to be superior to placebo for aberrant behavior in a group of 30 mentally retarded individuals (99), for behavioral symptoms of autism in adults (100), and for psychosis and behavioral disturbances associated with dementia (101).

The efficacy of haloperidol and pimozide for reducing tics in Tourette syndrome has been established by numerous clinical trials (102), with more than three-fourths of patients showing some benefit (103).

Haloperidol has been tested for efficacy in children with autism (pervasive developmental disorder) and has been shown to reduce hyperactivity, stereotypies, and behavioral problems (104,105). However, the drug is associated with a high rate of dyskinesias, both acute dyskinesia and withdrawal dyskinesia, in this population (84). Risperidone has demonstrated a similar benefit in adults with autism and is well tolerated without acute extrapyramidal effects (100).

Other Neuropsychiatric Indications

Substantial published experience, including several RCT, support the use of neuroleptics for several common neuropsychiatric problems:

1. *The chorea of Huntington disease.* Haloperidol, perphenazine, and sulpiride have all been reported to be effective (106).
2. Agitated behavior, hallucinations, and paranoia in patients with delirium. Lipowski (107), in his comprehensive synthesis of the literature on delirium, concluded that haloperidol was the drug of first choice for delirious agitation.
3. *Some cases of late-life delusional disorder.* However, there is general agreement that neuroleptics are less effective for delusional disorder than for schizophrenia (108).
4. *Psychosis in Parkinson disease.* Clozapine is the only neuroleptic consistently reported to alleviate hallucinations and delusions in patients with treated Parkinson disease, without aggravating patients' movement disorder (109). Two recent RCT report beneficial clozapine treatment of psychosis and tremor without worsening of motor symptoms (110–112). Doses as low as 6.25 mg per day may provide substantial improvement. Risperidone, despite its lower rate of parkinsonian side effects in the schizophrenic population, nonetheless aggravates rigidity when given to patients with Parkinson disease (109). Trials with olanzapine suggest improvement of psychosis but provide conflicting reports on tolerability, including worsening of motor symptoms (110,113). Quetiapine appears better tolerated in Parkinson disease with fewer adverse effects than risperidone and olanzapine and with beneficial antipsychotic action (110,113).
5. *Obsessive-compulsive symptoms poorly responsive to SSRI alone and accompanied by tics or psychotic features* (114,115). In a controlled clinical trial, McDougle et al. (116) showed that patients with OCD and tics usually improved when haloperidol was added to fluvoxamine, but those without tics did so only occasionally.

In summary, the neuroleptics have their most consistent beneficial effects on the positive symptoms of psychosis and on motor tics, chorea, and stereotypies. In addition, they appear to help a poorly defined subset of persons with agitation owing to gross brain disease or cerebral metabolic dysfunction.

Neurotoxicity of Neuroleptics

Significant neurologic toxicity is a feature of all the neuroleptics, and their propensity to cause neurologic side effects parallels their antipsychotic potency. The atypical neuroleptics, which include clozapine, risperidone, olanzapine, quetiapine, and ziprasidone, have fewer neurologic side effects than the typical agents for an equivalent degree of antipsychotic efficacy.

Neurologic toxicity of neuroleptics is divided into two types: *acute* toxicity, seen within days to weeks of drug administration, and *tardive* toxicity, which develops after more prolonged exposure and may persist, even permanently, after drug discontinuation. Advanced age is a risk factor for two of the most common neuroleptic side effects: drug-induced parkinsonism and tardive dyskinesia.

Acute neurologic toxicity includes parkinsonism, akathisia, dystonia (usually an axial dystonia), dyskinesia of the face or limbs, akinesia, and apathy. More rarely, patients can show extreme rigidity exceeding that typical of ordinary parkinsonism. One of the rigid conditions, neuroleptic-induced catatonia, may be prolonged but does not necessarily lead to dire systemic complications. The other, neuroleptic malignant syndrome, is a combination of tremulous rigidity, fever, delirium, autonomic instability, and rhabdomyolysis that has the potential for death owing to renal failure, pneumonia, and shock.

Chronic neurologic toxicity includes tardive dyskinesia, tardive dystonia, tardive akathisia, tardive dysmentia, and dopamine supersensitivity psychosis. Tardive dyskinesia characteristically involves the buccal, oral, and lingual muscles but may also include choreoathetosis of the upper extremities or dyskinesia of the larynx and pharynx, leading to dysphagia and dysarthria. Tardive dystonia, like acute dystonia, is usually axial. Tardive akathisia resembles acute akathisia. Tardive dysmentia is a chronic disturbance of attention, concentration, and executive cognitive function that usually accompanies a tardive movement disorder. Dopamine supersensitivity psychosis is a state of agitation, confusion, and affective instability that emerges after neuroleptics are withdrawn after long-term use and differs from the psychosis that originally led to neuroleptic use (117).

All the neuroleptic drugs also can cause or provoke seizures. However, neuroleptic-induced seizures are most common with the atypical agent clozapine, with which the incidence rate reaches 10% at doses of 900 mg per day or higher. Among the typical agents, chlorpromazine is the most likely to cause seizures. The agent least likely to cause seizures is not definitively established, although molindone and fluphenazine were least likely to produce seizure discharges in one *in vitro* model (118).

An extensive literature on risk factors for tardive dyskinesia, which was surveyed by Kane (119), intermittently suggests that gross brain disease is a risk factor, but only advanced age and female sex have been established as risk factors beyond all doubt. Tardive dyskinesia in older patients is more likely to be irreversible. Regarding acute side effects, age is a risk factor for drug-induced parkinsonism.

Drug Interactions

The neuroleptics are metabolized by oxidative enzymes in the liver; specific cytochromes are responsible for the metabolism of each specific drug. There is substantial interindividual variability in drug metabolism (102–122). Most of the neuroleptic drugs have many active metabolites, complicating the interpretation of blood levels. Haloperidol is notable for not having significant active metabolites.

Pharmacokinetic interactions with neuroleptics are primarily related to the effects of hepatic enzyme induction or inhibition on oxidative metabolism. Carbamazepine, a potent enzyme inducer, lowers the blood levels of haloperidol. Conversely, phenothiazine and thioxanthene neuroleptics can raise carbamazepine levels by inhibiting carbamazepine metabolism (122). As noted above, fluvoxamine raises the blood level of clozapine when the two drugs are given together; other SSRI may have similar but weaker effects. It is not clear whether the enzyme inhibition associated with valproate reduces neuroleptic metabolism to a clinically significant degree.

However, in practice, pharmacodynamic interactions with the neuroleptics are far more common and clinically relevant. The principle is that neuroleptics, which block dopamine receptors, will have additive or synergistic effects when combined with drugs that reduce dopamine release or affect dopamine's interaction with second messengers. These additive or synergistic effects may be therapeutic, toxic, or both. When they are toxic, one sees the usual neuroleptic toxicity, or, in more severe cases, sedation or delirium. Drugs that interact in this way with neuroleptics include carbamazepine, lithium, and calcium channel blockers.

A second set of pharmacodynamic interactions, which may also be either beneficial or problematic, occurs between the neuroleptics and drugs that potentiate inhibition by the γ-aminobutyric acid system, such as benzodiazepines and barbiturates. On the therapeutic side, tranquilizing effects may be enhanced, with less likelihood of tremor or akathisia. Conversely, sedation, lethargy, quiet delirium, or apathy and akinesia may be more common.

Choice of Neuroleptic

Considerations relevant to the choice and initiation of neuroleptic therapy are presented in Table 6.2.

Before the availability of atypical agents, choosing a neuroleptic was primarily an issue of deciding among a high-, mid-, or low-potency agent, taking into account past experiences that a patient encountered with any specific drug. The high-potency agents, of which haloperidol is the prototype, have less sedating, anticholinergic, and hypotensive effects but frequently cause extrapyramidal reactions, whereas the low-potency agents, of which chlorpromazine is the prototype, are associated with a lower incidence of extrapyramidal side effects (i.e., acute neurotoxicity). The midpotency agents represent a compromise. Among midpotency agents, molindone has the distinguishing characteristic of being the only neuroleptic that does not cause weight gain and actually causes a modest weight loss; it is also the neuroleptic with the

TABLE 6.2. ISSUES RELEVANT TO CHOOSING AND INITIATING A NEUROLEPTIC IN NEUROPSYCHIATRIC PATIENTS

Drug	Initial Dose (mg)	Usual Maximal Dose (mg)	Motor	Anticholinergic	Hypotensive	Sedative	Special Benefits	Special Concerns
Haloperidol	0.5–1.0	5–10	+++	+	±	+	Can give intravenously; depot form; cardiac safety	Frequent extrapyramidal effects, some malignant
Fluphenazine	0.5–1.0	5–10	+++	+	±	+	Depot form; least effect on seizures	Same as haloperidol
Thiothixene	1.0–2.0	10	+++	+	±	+	Somewhat less motor effect than haloperidol	
Perphenazine	2–4	16–24	++	++	±	++	Antianxiety effect; combines well with TCA for psychotic depression	Drug interaction; raises TCA blood levels
Molindone	5–10	50–100	++	0	0	+	Lack of systemic side effects; does not cause weight gain	No IM preparation available; not sedative; poor choice for behavioral crises
Chlorpromazine	10–25	100	+	+++	+++	+++	Marked sedation may be useful in crises; good antiemetic	Systemic side effects risky in frail elderly; causes seizures
Thioridazine	10–25	100–150	+	+++	++	++	Lowest motor effect of "typical" antipsychotics	QTc prolongation; lack of IM preparation
Mesoridazine	10–25	100	+	+++	++	++	Like thioridazine but available IM	Like thioridazine
Risperidone	0.25–0.5	4–6	+	±	+	++	Rare motor side effects at low doses; available as an oral solution	High cost; hypotension, especially if dose increased quickly; no IM preparation
Olanzapine	1.25–2.5	5–15	+	++	++	+++	Weight gain may be useful in anorexic states; sedation may be helpful in insomnia; available in an orally disintegrating tablet	Somnolence; weight gain and risk of diabetes; anticholinergic side effects
Quetiapine	25	150–400	±	+	+	++	Minimal weight gain	Dizziness, postural hypotension, dry mouth; requires routine eye examinations
Ziprasidone	20	80–160	±	±	+	+	Low incidence of weight gain	QTc prolongation; somnolence, nausea, dizziness, respiratory disorders
Clozapine	12.5–25	200	0	+++	+++	+++	Virtually no motor side effects; drug of choice in patients with Parkinson and psychosis	Weekly blood counts required; many systemic side effects; may cause seizures

Side effect intensity and/or frequency: 0, none; ±, minimal; +, mild; ++, moderate; +++, high. TCA, tricyclic antidepressants; IM, intramuscular.

least effect on nondopaminergic systems. Perphenazine may have an anxiolytic effect disproportionate to its antipsychotic effect.

Warnings were issued in 2000 for thioridazine and mesoridazine because of the potential for these drugs to cause prolongation of the QTc interval of the electrocardiogram and an increased risk of the development of torsade de pointes, a malignant ventricular arrhythmia, and sudden death in some cases (123). Therefore, these drugs have restrictions to minimize their potential adverse cardiac effects and are reserved for use in treatment-resistant schizophrenia. The most recent atypical antipsychotic approved for use in the United States, ziprasidone, also prolongs the QTc interval. QTc prolongation associated with pimozide has resulted in restricting its use to patients with Tourette syndrome whose symptoms are severe and who have not responded to, or cannot tolerate, treatment with haloperidol (124).

Marked patient preferences for particular antipsychotic drugs may reflect patients' subjective experiences of pharmacokinetic differences among drugs or the relevance for them of a particular drug's effects on nondopamine systems. Such preferences are associated with better therapeutic outcomes and better adherence to treatment (125). Drugs of equal potency should not be presumed therapeutically equivalent for all patients.

There are currently five atypical neuroleptics available in the United States: clozapine, risperidone, olanzapine, quetiapine, and ziprasidone. Atypical neuroleptics are distinguished from older or conventional neuroleptics by their relatively low potential to cause extrapyramidal side effects, tardive dyskinesia, and elevation of serum prolactin levels. In addition, they are effective in treating both the positive and negative signs and symptoms of schizophrenia. Atypical neuroleptics have relative affinities and potencies for antagonism at both the dopamine D_2 and serotonin $5HT_2$ receptors and may confer on them the clinical characteristics of atypicality (126). The cost and lack of parenteral preparations for atypical agents may preclude their use in some cases, but their overall higher degree of tolerability with fewer neurologic side effects compared with the conventional neuroleptics has resulted in their increased use as first-line drugs in schizophrenia.

The atypical agent risperidone is a high-potency D_2 receptor blocker that also has potent $5HT_2$ antagonist effects. Risperidone, particularly at doses of 6 mg per day or less, causes substantially fewer extrapyramidal side effects than haloperidol. The most commonly encountered side effects are weight gain, sedation, orthostatic hypotension, and hyperprolactinemia. Compared with typical agents, risperidone appears to have more effect on so-called negative symptoms (e.g., apathy, anhedonia, social disengagement) (127) and may also relieve symptoms in patients unresponsive to typical agents. Risperidone is also available as an oral liquid preparation.

Risperidone's relative lack of extrapyramidal side effects at lower doses, combined with its high potency, make it an

appealing antipsychotic drug for neuropsychiatric patients. Early, uncontrolled reports of its use in patients with dementia and psychotic features suggest that it may be better tolerated than haloperidol, with at least equal efficacy.

Olanzapine is effective at doses ranging from 5 to 15 mg per day and is usually given as a single dose at bedtime. Common treatment-emergent adverse events include somnolence, dry mouth, constipation, and significant weight gain.

Effective doses of quetiapine range from 150 to 400 mg per day in divided doses. Common side effects associated with quetiapine include dizziness, postural hypotension, dry mouth, and dyspepsia. Quetiapine requires eye examinations every 6 months to monitor for lenticular changes to detect cataract formation.

The potential for QTc prolongation associated with ziprasidone generally precludes its use as a first-line drug. Ziprasidone should be avoided in combination with other drugs known to prolong the QTc interval and in patients who have multiple risk factors for QTc prolongation and torsade de pointes. In short-term, controlled trials in schizophrenia and schizoaffective disorder, the most common adverse events were somnolence, nausea, constipation, akathisia, dyspepsia, dizziness, and respiratory disorders (128). Ziprasidone is initiated at 20 mg per day to 20 mg twice daily with slow upward titration to therapeutic doses of 40 to 80 mg twice daily.

As a group, the atypical neuroleptics have a more favorable extrapyramidal side effect profile than conventional neuroleptics. Among the atypical neuroleptics, the risk of extrapyramidal or motor symptoms is least likely with clozapine, somewhat more likely with quetiapine and ziprasidone, and probably greatest with risperidone and olanzapine.

Weight gain is an important consideration in the use of neuroleptic drugs because it affects compliance and may increase the risks of diabetes mellitus and hypertriglyceridemia (129–131). The atypical neuroleptics have been associated with significant weight gain. Among the atypical agents, the relative potential for weight gain appears to be greatest with clozapine and olanzapine, less with risperidone and quetiapine, and least with ziprasidone (129–131).

Clozapine, the first atypical neuroleptic available in the United States, is a low-potency agent that is a relatively weak blocker of D_2 receptors, a significant blocker of D_3 and D_4 receptors, and a potent blocker of $5HT_2$ receptors. It has an extremely low rate of extrapyramidal side effects. When a movement disorder occurs, it usually consists of bradykinesia or mild akathisia rather than rigidity or tremor (132). Clozapine can effectively treat psychosis in patients with severe Parkinson disease without aggravating their movement disorder and may actually ameliorate tardive dyskinesia and tardive dystonia, not merely mask them (133,134). Clozapine always alters the electroencephalogram (EEG) and causes seizures at a rate that increases with the dose: the annual incidence is 1% at doses of less than 300 mg per day, 2.7% at doses of 300 to 599 mg per day, and 4.4% at doses between

600 and 900 mg per day. [Olanzapine is also associated with a risk of EEG abnormalities but has a low reported seizure risk of 0.88% (135).] It has many systemic side effects, including fever, hypotension, anticholinergic effects, weight gain, hypersalivation, and suppression of bone marrow. Agranulocytosis occurs in approximately 1% of patients, requiring weekly monitoring of blood counts and immediate discontinuation of the drug if the white blood count falls below 3,000/mm^3. If there are no abnormal blood events during the first 6 months of clozapine treatment, then blood counts are monitored every other week thereafter.

In the treatment of schizophrenia, clozapine is effective in more than one-third of patients who are unresponsive to typical neuroleptics. Also, clozapine is more effective than typical agents for relieving negative symptoms and may alleviate some of the cognitive impairments associated with chronic schizophrenia. At present, however, because of the risk of agranulocytosis, clozapine is reserved for patients with treatment-resistant schizophrenia. It has not been established whether it is beneficial to routinely attempt to switch clozapine responders to a different atypical neuroleptic with less potential systemic toxicity. A related issue of specifically neuropsychiatric interest is whether a person with schizophrenia who has good control of positive symptoms on a typical agent but has negative symptoms and cognitive impairment should have a trial of an atypical neuroleptic in the hope of improving motivation and cognition (136). Neuropsychiatrists, with a tradition of attending to cognitive and motivational deficits in patients with brain diseases, may favor an aggressive position toward trials of atypical neuroleptics in this situation, particularly if the cognitive and motivational symptoms are causing distress to the patient or producing functional disability.

When choosing a neuroleptic for neuropsychiatric indications, the following guidelines may be useful:

1. In acute behavioral emergencies in which a neuroleptic is to be used for short-term stabilization or "chemical restraint," haloperidol is the best established agent. When it cannot be given orally, it may be given by either the intramuscular or intravenous route; the intravenous route, although not FDA approved, offers the advantage of not causing muscle damage that might later obscure the interpretation of a creatine phosphokinase level. Dose is titrated in increments every hour, with increments of 0.5 to 2 mg for fragile patients and 5 to 10 mg for patients who are physically robust. A combination of haloperidol with lorazepam (0.5–1 mg for fragile patients, 2 mg for robust patients) gives a more rapid behavioral response and may prevent some acute extrapyramidal symptoms.

2. When the patient's primary problem appears to be overwhelming anxiety with some psychotic features such as mild paranoia or disorganization, the patient may be treated with a sedative neuroleptic such as chlorpromazine or perphenazine. Typical starting doses of chlorpromazine would be 10 mg t.i.d. for fragile patients and 25 mg t.i.d.

for more robust patients; the dose is increased as tolerated, with sedation, hypotension, and anticholinergic effects usually the limiting factors. Perphenazine can be used similarly; this midrange agent presents fewer problems with sedation, hypotension, and anticholinergic effects but is more likely to cause acute extrapyramidal effects. Starting dose would be 2 mg b.i.d. at the lower end and 4 mg t.i.d. at the higher end.

3. For psychotic phenomena such as delusions and hallucinations in persons with gross brain disease who do not present a behavioral emergency, risperidone is an appealing first choice because of its low rate of motor side effects. However, it can cause substantial sedation at first, so the dose should be built up slowly. Starting doses range from 0.5 mg at bedtime to 1.0 mg b.i.d. in a more severely disturbed yet physically robust person. The dose is increased if necessary and as tolerated to 3 mg b.i.d. Above this level, risperidone's distinctive lack of extrapyramidal side effects is less evident, and the use of a high- or midpotency typical agent should be considered because of the lower cost and lesser sedation.

Clozapine is probably the best currently available drug for treating psychosis in patients with system degenerations with parkinsonian features (137) but has the potential for agranulocytosis and requires blood count monitoring. Therefore, quetiapine has been recommended as a first-line drug for psychosis in Parkinson disease (110). It is well tolerated, seems to have fewer adverse motor effects than risperidone and olanzapine, and does not have the potential for severe adverse systemic toxicity that is associated with clozapine. Quetiapine is initiated at 12.5 mg at bedtime, then titrated upward by 12.5 mg every 4 to 7 days to a final dose of 50 mg per day. If quetiapine is ineffective or poorly tolerated, then a trial of olanzapine or clozapine may be considered as second- and third-line drugs. Because olanzapine has been associated with confusion and delirium in Parkinson disease, it is recommended that it be initiated at a low dose with slow upward titration (111). Olanzapine is initiated at 2.5 mg at bedtime and increased by 2.5 mg weekly. Clozapine is initiated at 6.25 mg at bedtime and increased by 6.25 mg every 4 to 7 days (110). Doses of all medications are increased until psychosis remits or side effects occur (110). Risperidone has not proved to be a good alternative because it aggravates the movement disorder even at a low dose. Data for the use of ziprasidone in psychosis associated with Parkinson disease are not yet available.

Titration of Antipsychotic Drug Therapy

Titration of neuroleptic treatment in neuropsychiatric patients comprises several issues: dose increase in emergent and nonemergent situations, dose reduction when behavior or symptoms have stabilized, the use of adjuncts to minimize neuroleptic dose, and the use of anti-Parkinson drugs. Particularly in the neuropsychiatric population, there are essentially no comparative studies of dose titration strategies, so what follows must be regarded only as informed opinion. However, the dose strategy should take into account that neuroleptics alter behavior first, affect second, and thought and perception last and that the full effect of a given dose on thought and perception may take several weeks.

In the emergency situation, a neuroleptic (typically haloperidol) is given every hour until the patient's behavior is manageable. Then, the total dose used to stabilize the patient is repeated every 24 hours, usually in divided doses. Doses are reduced or withheld if the patient appears excessively sedated. Once the neurologic, medical, or psychiatric condition underlying the behavioral emergency is diagnosed and treatment is under way, neuroleptics should be tapered. If the patient has a psychotic disorder and will need longer term neuroleptic therapy, the taper should aim for a typical long-term maintenance dose within 2 weeks. If the patient has a transient psychosis owing to an underlying medical condition that is resolved, the neuroleptic should be tapered and discontinued over 1 week. If problematic behavioral symptoms recur, the dose should be raised to the lowest dose that controlled the symptoms, and withdrawal should be reattempted in another week.

When a neuroleptic is to be used for a nonemergent indication, the therapeutic dose is approached from below rather than from above. The dose is increased weekly until behavioral symptoms are clearly improved; it is then kept at that level for another month to assess its effect on thought disorder before considering further dose increases.

Frequently, neuroleptics will suppress specifically psychotic symptoms such as hallucinations and delusions, while leaving the patient with residual symptoms such as affective instability, irritability, anxiety, or intermittent explosive behavior. Because neuroleptics are the most neurotoxic of the psychotropic classes, it is generally preferable to use adjunctive drugs to treat these symptoms rather than to increase the neuroleptic dose in an effort to control them. Preferred options include mood-stabilizing AED or lithium for affective instability; SSRI or buspirone for irritability; SSRI, buspirone, or benzodiazepines for anxiety; and mood-stabilizing AED, propranolol, or clonidine for intermittent explosive behavior. Recent studies have supported the use of atypical neuroleptics, particularly olanzapine and ziprasidone, in the treatment of mania and are further discussed in the section on antiepileptic and mood-stabilizing drugs.

Neuropsychiatric patients may be more likely than general psychiatric patients to develop extrapyramidal reactions to neuroleptics and may be more likely to have significant functional impairment as a consequence of those reactions. For example, an aged stroke patient is at increased risk of parkinsonism and tardive dyskinesia because of age, and the effects of a movement disorder on mobility and activities of daily living performance will add to the effects of the stroke itself. For this reason, special efforts are warranted to identify extrapyramidal reactions early and to treat them effectively.

When extrapyramidal reactions occur, the first consideration is whether a dose reduction or a switch to an atypical neuroleptic is reasonable in a given patient. If these measures are not feasible, or have been taken, drug treatment of the reaction comes next (138). First-line drugs for treatment of extrapyramidal reactions are benztropine or trihexyphenidyl for tremor, acute dystonia, or acute dyskinesia; amantadine or bromocriptine for rigidity and akinesia; and propranolol for akathisia. There is substantial interindividual variability in response to the anti-Parkinson drugs for pharmacokinetic reasons alone. Typical starting doses are: benztropine 2 mg b.i.d. to q.i.d. (four times daily); trihexyphenidyl 5 mg b.i.d. to q.i.d.; amantadine 100 mg b.i.d. to t.i.d.; bromocriptine 2.5 mg q.d. to b.i.d., and propranolol 20 mg t.i.d. Fragile patients should begin at approximately half of the usual dose. Rapid upward dose titration should be carried out until symptoms are relieved. Usual maximal doses are benztropine 10 mg per day, trihexyphenidyl 20 mg per day, amantadine 300 mg per day, bromocriptine 30 mg per day, and propranolol 360 mg per day. If the maximal dose is reached with persistent extrapyramidal symptoms, either the neuroleptic or its dose should be changed or a different anti-Parkinson agent should be used.

Duration of Therapy

The duration of neuroleptic therapy depends on the indication for which the drug is used. Strategy for dose reduction in schizophrenia is beyond the scope of this chapter. When neuroleptics are used for psychotic symptoms or for severe agitation in patients with dementia or gross brain disease, periodic efforts at drug discontinuation are warranted because changes in the brain and the patient's situation over time may have removed the necessity for neuroleptic use. A reasonable approach is to attempt a taper after the patient has had stable behavior or well-controlled psychotic symptoms for 3 months. The drug should then be tapered over the next 3 months, with an interruption of the taper if clinically significant symptoms recur.

Patients with chronic neuropsychiatric disorders may present for neuropsychiatric attention on neuroleptic therapy of uncertain duration, for which the original indications are unclear. In this situation, vigorous detective work is warranted to determine the preneuroleptic mental status and whether the indications for the drug were sound. If it cannot be determined that the patient is receiving neuroleptics for an appropriate indication, a gradual taper should be attempted over no less than 3 months. The value of withdrawing unnecessary neuroleptics is illustrated by a recent study of Thapa et al. (139) in elderly nursing home residents in which withdrawal of neuroleptics prescribed for questionable indications did not increase behavioral problems but led to a highly significant improvement in patients' affect. Excessively rapid tapering of neuroleptics can cause somatic withdrawal symptoms, especially anorexia and weight loss. Also, patients may show a mental disorder characterized by confusion, agitation, and affective instability without the thought disorder typical of schizophrenia. This dopamine supersensitivity psychosis can be treated with mood-stabilizing AED. The rate of tapering of the neuroleptic should be slowed, but the effort to discontinue the neuroleptic should not be abandoned unless a convincing indication for its use can be established.

ANTIANXIETY DRUGS

The category of antianxiety drugs reviewed in this section comprises the benzodiazepines and the $5HT_{1A}$ partial agonist drug buspirone. However, many other classes of drugs have been used to treat specific anxiety disorders. The SSRI are the drugs of choice for OCD, and numerous antidepressants have been effective for treating recurrent panic attacks. The more sedating neuroleptics may be the drugs of choice for anxiety associated with paranoid phenomena and disorganized thinking. Anxiety owing to hypomania is appropriately treated with lithium or a mood-stabilizing AED. The unique role of antianxiety drugs is the acute, short-term treatment of anxiety, panic, and insomnia. Antianxiety drugs are also used for the treatment of generalized anxiety disorder and chronic anxious traits. Antidepressants are an acceptable alternative therapy for many of these patients.

Efficacy

The efficacy of benzodiazepines for symptoms of anxiety and insomnia has been established by hundreds of RCT conducted in patients without significant neurologic disease (140,141). The benzodiazepines are especially effective for autonomic symptoms of anxiety and for hypervigilance. Buspirone is more effective than placebo for symptoms of generalized anxiety but is less consistently effective than the benzodiazepines (142). It may be more effective than benzodiazepines for cognitive symptoms and less for hypervigilance and autonomic arousal. It does not block panic attacks and does not have a direct hypnotic effect, although it can sometimes improve sleep through relieving anxiety.

Although there are no RCT specifically confirming the antianxiety effects of benzodiazepines or buspirone in neurologically ill populations, there is general consensus among clinicians that benzodiazepines are therapeutically effective for anxiety in patients with gross brain disease. Concerns about the use of benzodiazepines in neuropsychiatric populations center on side effects rather than efficacy.

Other Neuropsychiatric Indications

Clonazepam is an established AED and is discussed in more detail in the section on AED. Benzodiazepines have been used effectively for the following additional indications:

1. *Myoclonus, including nocturnal myoclonus.* Clonazepam is the benzodiazepine of choice (143–145). It also can be used to suppress myoclonus associated with antidepressants.
2. *Parasomnia associated with slow-wave sleep.* This can be alleviated by any of the benzodiazepines, which suppress stage 4 sleep.
3. *Rapid behavioral stabilization of acutely agitated patients.* The combination of a short-acting benzodiazepine, such as lorazepam, with a neuroleptic, such as haloperidol, is more effective and safer than single-drug neuroleptic therapy. The combination of midazolam and droperidol has been used at some institutions.
4. *Catatonia.* Both lorazepam and diazepam have been reported to reverse the motor manifestations of catatonia without affecting the underlying disorder of thinking or mood.

Benzodiazepines generally are not used in the treatment of aggressive behavior or SIB because they may disinhibit behavior, and such disinhibition may occur unpredictably. Buspirone has been efficacious for such problems. Neuropsychiatric indications for buspirone supported by recent uncontrolled experience include the following:

1. *SIB in some adults with mental retardation.* A modest reduction was seen with an average of 30 mg per day of buspirone (146).
2. *Agitated and impulsive behavior in patients with "organic" mental syndromes.* Partial response of this behavior has been reported (147,148).
3. *Tardive dyskinesia.* Relatively high doses, as high as 180 mg per day, partially relieved dyskinesia in five of seven patients in an open study (149).
4. *Hostility and irritable "type A" behavior.* Buspirone treatment was associated with decreased hostility, impatience, and anxiety in an 8-week open trial of therapy in ten men with coronary artery disease and no psychiatric diagnosis (150).

Neurologic Toxicity

The neurologic side effects of the benzodiazepines are directly related to their effect on the benzodiazepine receptor to facilitate γ-aminobutyric acid–related inhibitory neurotransmission. They include cognitive impairment, memory impairment, decreased alertness, and impaired coordination. The more severe forms of these side effects include confusion, amnesia, lethargy, apathy or somnolence, ataxia, and falls. These neurologic side effects are more likely to occur in elderly patients (151).

The neurologic side effects of buspirone are less predictable. Both insomnia and sedation can occur, but most patients do not have measurable cognitive impairment at usual doses. Akathisia and dyskinesia have been reported as rare occurrences; the mechanism of these movement disorders is not known.

Drug Interactions

The benzodiazepines subdivide into two groups according to their metabolism: some are metabolized by hepatic oxidation; others are conjugated with glucuronide and are eliminated by the kidney. The benzodiazepines that are oxidized are more often involved in pharmacokinetic interactions because any drug that inhibits the relevant hepatic oxidative enzymes can raise benzodiazepine levels and prolong the drug's half-life. Likewise, agents that induce hepatic oxidative enzymes can reduce blood levels of these benzodiazepines and shorten their effective duration of action. Diazepam, chlordiazepoxide, alprazolam, and clonazepam all undergo oxidative metabolism; oxazepam and lorazepam do not. Common drugs that induce benzodiazepine metabolism include theophylline and carbamazepine, with clinically significant effects of the latter drug on the metabolism of alprazolam (152,153) and clonazepam (154). Common drugs that inhibit benzodiazepine metabolism include valproate, SSRI, nefazodone, erythromycin, and cimetidine.

Choice of Antianxiety Drug

After determining that a patient should be treated with an antianxiety drug rather than a drug of another class, the first decision is whether to use buspirone or a benzodiazepine. Buspirone is most likely to be helpful when the indication is either generalized anxiety disorder, chronic anxiety traits and symptoms falling short of diagnostic criteria for generalized anxiety disorder, or chronic tension and irritability. It tends to work best in patients who either never have used benzodiazepines or do not like benzodiazepines. Patients rarely experience buspirone as a satisfactory substitute for chronic benzodiazepine therapy if they found such therapy helpful and well tolerated.

Benzodiazepines are the drugs of first choice in situations of acute anxiety or panic and in situations in which a specific benzodiazepine effect, such as an antiepileptic action, is desired. For some indications (e.g., myoclonic epilepsy), only one benzodiazepine is known to be efficacious. For others, such as panic attacks or generalized anxiety, there is a choice of many agents. The choice of drug in the latter case can be guided by these principles:

1. Patients on a complex, multiple-drug regimen usually should be given a benzodiazepine that is eliminated by the kidney, does not require extensive oxidative metabolism, and does not have active metabolites. Oxazepam is the paradigm for such a drug.
2. Patients who have panic attacks are more likely to respond to alprazolam or clonazepam than to other benzo-

diazepines. Alprazolam, with a relatively short duration of action, is more likely to cause problems with rebound anxiety or interdose anxiety than clonazepam. Therefore, clonazepam is usually preferable. However, clonazepam does cause more sedation and ataxia, so alprazolam would be preferable for patients with unsteady gait or apathy at baseline.

3. When benzodiazepines are to be used for treating insomnia, the important distinction is between use for a day or 2 and use for a week or more. For very short-term use, long-acting drugs such as flurazepam or quazepam offer the advantage of minimal rebound insomnia after the drug is stopped, and their propensity to accumulate is not relevant. When a drug is to be used for a week or more, accumulation is a problem for the longer acting drugs, and a drug with a medium half-life, such as temazepam, would be a better choice. The ultrashort-acting drug triazolam is not advisable in the neuropsychiatric population. Although the issue remains controversial, it may have slightly more behavioral toxicity than the other hypnotic agents.

Dose Titration

Dose titration for buspirone must deal with the wide interindividual variability in first-pass metabolism for this drug. For some patients, 5 mg t.i.d. is adequate; for others, 20 mg t.i.d. will have little effect. Excessively high blood levels of buspirone give patients unpleasant side effects such as nausea, dizziness, and tinnitus. Accordingly, patients should be started on 2.5 mg b.i.d. to t.i.d. if they are fragile and 5 mg t.i.d. if they are not, and then, if there are neither therapeutic effects nor side effects, the dose should be increased every 2 to 3 days until the patient gets relief, develops a limiting side effect, or a dose of 20 mg t.i.d. is reached. Increments are 2.5 to 5 mg at a time in fragile patients and 5 to 10 mg at a time in healthier patients. Once limiting side effects have been reached, the dose should be reduced to the highest level that does not cause unpleasant side effects and kept there for 3 to 4 weeks. If a patient reaches 20 mg t.i.d. without side effects but without benefit, the dose should be maintained for 2 weeks. If there is still no effect, the dose may be increased to 30 mg t.i.d. for an additional 2 weeks.

When short-acting benzodiazepines are used, patients are started on one-half of the manufacturer's recommended dose, given as divided doses according to the manufacturer's recommendation. Every other day, the dose can be increased until the patient gets relief or limiting side effects develop. When long-acting agents are used, 2 weeks should be allowed between dose increases. If necessary, as-needed doses of a shorter acting agent can be used to alleviate severe symptoms during the titration period.

During upward dose titration of benzodiazepines, the patient should be checked periodically for the effects of the drug on gait and coordination, alertness, and memory. Patients without insight or with a history of drug misuse should be tested directly for side effects; the clinician should not rely only on the patient's self-reports.

When patients on long-term benzodiazepine therapy develop brain diseases, they may become intolerant to their usual benzodiazepine dose and show side effects, most often decreased alertness, memory disturbance, or impaired gait or coordination. Under such circumstances, a slow withdrawal of benzodiazepines is indicated. The dose should be reduced by no more than 10% per week unless the clinical situation is urgent. If the patient is known to have an underlying anxiety disorder, alternative therapy for that anxiety disorder should be initiated during the benzodiazepine taper, before severe symptoms break through.

Duration of Therapy

Anxiety disorders can have either a relapsing/remitting or chronic course. Once it is established that a patient has a chronic course, he or she can be kept on antianxiety drugs indefinitely, as long as there are few or no side effects. If the patient's course is unknown or the patient has never had a trial off medication since it was started, gradual withdrawal of medication may be attempted 3 to 6 months after the patient is free of major symptoms of the anxiety disorder. Buspirone should be tapered over approximately 1 month and benzodiazepines over 2 to 3 months. Benzodiazepine tapering usually is easier with longer acting agents, and crossover from short- to long-acting benzodiazepines should be considered if the patient is on a short-acting agent. If anxiety symptoms emerge during the taper, nonpharmacologic treatment should be provided. Ideally, such treatment would precede an effort to taper a patient off benzodiazepine therapy. If an unacceptable level of symptoms recurs, the dose should be restored to the minimum that controls the symptoms, and a taper should be reattempted in another 3 to 6 months. Nonpharmacologic therapy should be offered in the interim. A patient who fails three well-implemented attempts to taper antianxiety drugs probably has a chronic condition. Alternate drug therapy, such as antidepressants, should be considered if there is a problem with the use of benzodiazepines in the individual patient.

LITHIUM

Lithium, a simple monovalent cation, has an impressive range of effects on the neurochemical processes in the brain. Among these are enhancement of serotonergic neurotransmission, at least at some subtypes of serotonergic synapses; prevention of dopamine receptor supersensitivity induced by receptor blockade; facilitation of norepinephrine release; reduction of the adenylate cyclase response to β-adrenergic stimulation; stimulation of acetylcholine synthesis and re-

lease; and prevention of up-regulation of muscarinic receptors in response to chronic blockade. Lithium also affects neurochemical signal transduction through effects on phosphoinositide turnover, adenylate cyclase activity, G proteins, and protein kinase C (155,156). This broad range of pharmacodynamic actions is the basis of lithium's applicability to several different neuropsychiatric disorders and may be the basis of pharmacodynamic interactions between lithium and other CNS drugs.

Efficacy

Lithium has been shown by RCT to be an effective agent for acute mania in 70% to 80% of manic patients. Well-designed, double-blind studies have shown it to prevent relapses in approximately two-thirds of bipolar patients (157). RCT have not been extended to the population of patients with mania secondary to gross brain diseases (e.g., right-hemisphere stroke).

Substantial literature, albeit with fewer RCT, supports the use of lithium for augmentation of antidepressants in major depression and combined use with neuroleptics in the treatment of schizoaffective disorder (158).

Other Neuropsychiatric Indications

Treatment of Impulsive Aggression

For more than 20 years, reports with varying levels of rigor have supported an antiaggressive effect of lithium at blood levels similar to those used to treat bipolar disorder (158,159). Studies and case reports have been conducted mainly with institutionalized populations, including people with mental retardation (159,160), hospitalized children with conduct disorder (161), prison inmates (162), and a patient recovering from a severe TBI (163).

Self-Injurious Behavior

SIB in people with mental retardation has been treated successfully with lithium (164). A chart review study examining 3 months of care before and after institution of lithium therapy in 11 mentally retarded patients showed a significant reduction in episodes of SIB in the treatment period.

Huntington Disease

Irritability and aggression in patients with Huntington disease have also been reduced by lithium. Reports suggest that lithium may improve these behavioral features without affecting the movement disorder (165,166).

Prevention of Cluster Headache

This application has been established by RCT (167).

Neurotoxicity of Lithium

Lithium is associated with numerous neurologic side effects, all of which are dose related. The occurrence of neurologic side effects increases very rapidly with lithium levels greater than 1.5 mEq/mL (168). At lithium levels in the therapeutic range, some patients have virtually no neurologic side effects apart from a barely detectable action tremor. For others, including many patients with gross brain disease or dementia, neurologic side effects at levels in the usual therapeutic range are limiting factors in the use of lithium.

Neurologic side effects of lithium seen in the usual therapeutic range include impaired concentration or memory, apathy, restlessness, myoclonus, and tremor. The tremor typically seen at therapeutic doses is a fine action tremor. With lithium levels in the toxic range, the tremor is more coarse and proximal and may be accompanied by asterixis. Other side effects seen with toxic levels of lithium are confusion, hallucinations, and rigidity resembling that produced by neuroleptics. At extreme levels of lithium (e.g., more than 2.5 mEq/L), patients can develop seizures and may develop muscular weakness owing to anterior horn cell damage and/or interference with the function of the neuromuscular junction. Elderly patients and patients with Parkinson disease and related disorders can show rigidity and coarse tremor with lithium levels in the usual therapeutic range.

Neurologic side effects of lithium can also be produced indirectly through lithium's effect on the endocrine system (169). Lithium-induced diabetes insipidus can lead to hypernatremia and thereby cause a typical metabolic encephalopathy. Lithium-induced hyperparathyroidism can cause the typical neurologic syndromes of hypercalcemia. The most common lithium-induced endocrinopathy, hypothyroidism, causes depression, ataxia, and slowed cognition; it also increases the patient's sensitivity to adverse effects of other drugs on alertness, cognition, or gait and coordination. Accordingly, when a patient on lithium therapy develops new neurologic side effects, the diagnostic assessment should include not only a lithium level but also measures of electrolytes, calcium, renal function, and thyroid function.

Lithium Versus Alternatives

When lithium is considered as a mood-stabilizing agent, the major alternatives are the mood-stabilizing AED, of which carbamazepine and valproate are best established. Both of the latter drugs have been shown to be as efficacious as lithium in the treatment of mania and appear to work as prophylaxis of bipolar disorder. Although individual patients may respond to lithium and not to AED, to AED but not to lithium, or to either one, there is no well-established way to know in advance which drug will be best for a given individual. However, the relatively poor response of patients with dysphoric mania or rapid cycling to lithium has led some authors to recommend that valproate or carbamazepine be the first choice

for such patients (170). Also, the presence of "organic" features in patients with bipolar disorder has been shown to favor a good response to valproate therapy (171). Lithium is the standard first-line mood stabilizer used in patients with typical bipolar disorder, and AED are the first-line mood stabilizers in all other situations, including rapid cycling, mixed manic and depressive symptoms, dysphoric mania, and mood instability accompanying gross brain disease or epilepsy. Atypical neuroleptic drugs are increasingly being employed either as monotherapy or adjunctively in bipolar mania. The clinical experience of their use in neuropsychiatric practice is limited, but atypical neuroleptic drugs that have a propensity toward seizures as side effects (i.e., clozapine) should be avoided in patients who are at risk of seizures. When mania results from a specific brain lesion, such as a right-hemisphere stroke, AED are preferred because the embolic strokes that usually cause secondary mania are potential seizure foci. The use of an AED both addresses the mania and minimizes the risk of seizures. If AED are the mood stabilizers of choice except for classic primary bipolar disorder, an EEG is not needed as an aid to choosing treatment. It may, however, be useful for diagnostic purposes.

When lithium is considered as an antiaggressive agent, its competition includes the AED, adrenergic agents such as propranolol and clonidine, and serotonergic agents such as the SSRI antidepressants and buspirone. Various authors have suggested sequences of drug trials in this situation, with different rationales for the order in which drugs are tried (170–173). Unless there is a contraindication, lithium should be tried in any seriously aggressive but nonpsychotic patient before accepting long-term neuroleptic therapy. Lithium is tried first, or early on, in patients with (a) a family history of bipolar disorder or another lithium-responsive mental disorder; (b) long periods of elevated mood, energy, or rate of function alternating with periods of normal or decreased function; (c) grandiosity or expansiveness; or (d) recurrent depression with clear-cut discrete episodes.

Drug Interactions

Like all electrolytes, lithium is eliminated by the kidney. It is processed by the proximal tubule in parallel with sodium; drugs and clinical situations that increase sodium retention also increase lithium retention. Most significant pharmacokinetic interactions involving lithium arise in situations in which a drug increases proximal tubular reabsorption of sodium, thereby increasing lithium levels for a given oral dose. Drugs that can raise lithium levels in this way include thiazide diuretics and nonsteroidal antiinflammatory drugs. When starting lithium in a patient taking a diuretic or a nonsteroidal antiinflammatory drug, the dose should be decreased by approximately half. When adding a diuretic or nonsteroidal antiinflammatory drug to the regimen of a patient already taking lithium, a similar adjustment should be made, and lithium levels should be checked frequently (e.g., every other day) until they are stabilized in the desired therapeutic range.

Pharmacodynamic interactions with lithium are many, and most can be understood by considering lithium's manifold effects on neurotransmitters and second messengers. Coadministration of lithium with a neuroleptic increases the risk of an acute extrapyramidal reaction over the risk associated with the neuroleptic alone. Coadministration of lithium with an antidepressant increases the risk of tremor, myoclonus, or cognitive side effects. Coadministration of lithium with an AED increases the risk of confusion, ataxia, sedation, or tremor. However, all these combinations have greater therapeutic effectiveness for mania, depression, and bipolar disorder, respectively, than either drug given alone.

Titration of Lithium Dose

In patients not deemed to be at increased risk for lithium neurotoxicity, the target blood level for dose titration is 1.2 to 1.4 mEq/L. In elderly, demented, or parkinsonian patients, in whom the risk is higher, the target blood level is 0.8 to 1.0 mEq/L. The oral dose needed to attain the target blood level will depend on renal function and the size of the patient (i.e., the volume of distribution). A small, fragile patient with normal or perhaps slightly decreased renal function might be started on 300 mg b.i.d. A large, robust patient with known normal renal function might be started on 600 mg t.i.d.

During aggressive dose titration, as would be carried out for acute mania, blood levels are determined every other day. The patient is also examined for neurologic symptoms, which are correlated with the blood level obtained at the time of the examination. The dose is adjusted as needed to rapidly attain the desired blood level; if unacceptable neurologic symptoms occur at a level lower than the original target, the target is lowered.

Gradual dose titration, as might be carried out for an outpatient with intermittent explosiveness and no behavioral emergency, would be based on weekly blood level determinations. Incremental changes in dose would be conservative to minimize the risk of overshooting the target blood level.

Duration of Therapy

The usual psychiatric indication for lithium (prophylaxis of bipolar disorder) implies very long-term treatment. When lithium is used for mood disorders or aggressiveness associated with gross brain diseases, there are no rigorous studies of continuation therapy to supplant clinical judgment as the criterion for continued therapy. A reasonable recommendation is to determine first whether the patient has bipolar disorder that is simply altered in its presentation because of brain disease (e.g., mental retardation or dementia). If so, prolonged treatment would be carried out, in keeping with the hypothesis of an underlying case of primary bipolar disorder. If the current syndrome is the consequence of an acute brain insult, such as stroke or TBI, it would be reasonable to attempt a taper of lithium after the patient had more time to recover from the insult. For example, if the patient's mental and behavioral symptoms were well controlled for 3 months and neurologic recovery is proceeding well, an attempt would be made to taper the lithium. The dose would be tapered for approximately 2 weeks, then the drug is discontinued. If symptoms recurred, the drug would be reinstated for another 3 months, and another attempt might be made at that time.

If the patient's behavioral syndrome is related to a progressive or degenerative brain disease, a reasonable time to attempt tapering of lithium is when clinically significant progression of the disease has been noted. Under those circumstances, it is possible that lithium is no longer necessary or the relative risk:benefit ratio has changed.

If a patient has a stable, lifelong neurologic basis for mood instability or aggressiveness, long-term therapy with lithium may be justified if lithium has substantially improved safety, enhanced function, relieved distress, or permitted greater freedom in the patient's life. Under these circumstances, an annual reassessment of the patient's symptoms and drug treatment is a valuable precaution.

ANTIEPILEPTIC AND MOOD-STABILIZING DRUGS

The use of AED for mood and behavior and for seizure disorders is one of the distinguishing features of the neuropsychiatric practitioner. Neuropsychiatrists used AED for a wide range of indications long before they entered the mainstream of general psychiatry. As of the mid-1990's, clonazepam has been broadly accepted by general psychiatrists as a treatment for anxiety and panic, and both carbamazepine and valproate have been accepted as efficacious treatments for mania and the prophylaxis of bipolar disorder. Issues relevant to choosing and initiating lithium or a mood-stabilizing AED are presented in Table 6.3.

Efficacy

The efficacy of the AED for seizures has been established by multiple RCT. Among AED approved in the United States in the past 30 years, valproate, carbamazepine, clonazepam, felbamate, and oxcarbazepine have been shown to be effective as monotherapy, whereas lamotrigine, gabapentin, and topiramate have been shown to be effective as add-on therapy. However, lamotrigine in particular has been used effectively as monotherapy in Europe.

The efficacy of carbamazepine for the treatment of acute mania and its comparability with lithium for this purpose have been demonstrated by 11 RCT (174). Five studies, only one of which was placebo controlled, demonstrate the efficacy of carbamazepine in the prophylaxis of mania in patients

TABLE 6.3. ISSUES IN CHOOSING AND INITIATING LITHIUM OR ANTIEPILEPTIC DRUGS

Drug	Benefits	Controlled Trials	Precautions	Initial Dose
Lithium	Low cost; very well-known side effects and established monitoring schedule	None; case series support efficacy for mania owing to gross brain disease, impulsive aggression, and self-injurious behavior	Low therapeutic index; may induce hypothyroidism; causes tremor and may aggravate parkinsonism; synergistic toxicity with neuroleptics	300 mg b.i.d.; usual target blood level approximately 1.0 mEq/mL
Carbamazepine	Efficacy for rapid cycling; antiepileptic effects	None; case series support efficacy for mania owing to gross brain disease and impulsive aggression	Quinidinelike cardiac effects; anticholinergic effects; drug interactions owing to enzyme induction; rare hematologic side effects require patient warning but not routine monitoring of complete blood cell count	100 mg b.i.d.; slow titration to target blood level of 8 to 12 $\mu g/dL$ when the drug is used as sole therapy
Valproate	Efficacy for rapid cycling, antianxiety effects; antiepileptic effects	None; case series support efficacy for mania owing to gross brain disease and for mood instability and aggression in mentally retarded persons and dementia patients	Rare hepatic toxicity requires warning but not routine monitoring of enzymes; can cause pancreatitis or hyperammonemia; weight gain common; may cause tremor	250 mg b.i.d. (125 mg if unusually sensitive to side effects); gradual titration to blood level of 50–100 $\mu g/mL$; may load rapidly in case of behavioral emergency owing to mania
Clonazepam	Strong antianxiety and antipanic effects; sedative; treats myoclonus	None; case series support efficacy for anxiety, panic, and agitation owing to gross brain disease	Long half-life predisposes to accumulation and risk of falling; ataxia and sedation the main side effects that limit therapy; some risk of disinhibition	0.25–0.5 mg, usually at bedtime
Gabapentin	Antianxiety and mood-stabilizing actions suggested by patients' experiences in epilepsy trials; no significant drug interactions	No systemic trials or psychotropic activity	Additive sedation and ataxia with other drugs	100 mg t.i.d.; may increase gradually to maximum of 900 mg t.i.d., but mechanism of action implies diminishing effect of increases
Lamotrigine	Subjective well-being improved in patients enrolled in epilepsy clinical trials	None; anecdotes support efficacy for rapid-cycling mood disorders	Interactions with valproate; can cause severe skin rashes, especially if dose increased rapidly	50 mg at bedtime or b.i.d.; 25 mg every other day if patient is on valproate because of increased risk of rash
Topiramate	No significant drug–drug interactions	None	Slow dose titration is essential; cognitive changes; weight loss; renal stones	25 mg at bedtime or b.i.d.
Oxcarbazepine	Fewer pharmacokinetic interactions than carbamazepine with less frequent and less severe side effects; may be titrated rapidly	None	Sedation, dizziness, ataxia, nausea, rash, hyponatremia	300 mg at bedtime

b.i.d., twice daily; t.i.d., three times daily.

with bipolar disorder. The efficacy of valproate for acute mania is supported by six RCT. Its efficacy for prophylaxis of mania is supported by open studies only (174). None of the controlled studies of carbamazepine or valproate addressed the use of those agents in patients with mood disorders secondary to gross brain disease, epilepsy, or dementia.

The evidence for psychiatric indications of the newer AED felbamate, lamotrigine, gabapentin, topiramate, and oxcarbazepine is not as well established as that for carbamazepine and valproate. Anecdotally, several of the newer AED are in use by mood disorder specialists as alternatives or adjuncts to carbamazepine and valproate for the treatment of bipolar disorder, particularly atypical or treatment-refractory cases. Lamotrigine monotherapy has been reported in one RCT to have significant antidepressant efficacy in treating bipolar I depression (175). Another RCT reported lamotrigine to be superior to both placebo and gabapentin in patients with mania and depression, whereas gabapentin did not differ from placebo (176). The efficacy of gabapentin monotherapy in bipolar disorder is not supported by research, but many pilot open-label trials suggest that it may be beneficial as an adjunctive therapy in mania or bipolar depression (177,178). Nevertheless, a recent RCT reported that adjunctive therapy of mania with gabapentin did not differ from placebo (179). Several trials and open-label studies suggest that patients with bipolar disorder may benefit from topiramate adjunctive therapy (177).

Neuroleptics have typically been used in combination with mood-stabilizing agents for bipolar disorder associated with mania and psychotic symptoms. More recently, studies evaluating the efficacy of atypical antipsychotic monotherapy in bipolar disorder have demonstrated the superiority of olanzapine and ziprasidone to placebo in patients with acute mania (176,177). Olanzapine has been approved by the FDA for treatment of acute mania associated with bipolar disorder.

Other Neuropsychiatric Indications

The AED have been used empirically in neuropsychiatric conditions in which paroxysmal brain activity is part of a hypothesized mechanism of symptom production. Numerous case reports, case series, and open studies have been reported. Some of the potential indications for AED include the following:

1. *Posttraumatic stress disorder.* Both carbamazepine and valproate (180–182) have been found to reduce symptoms in patients with posttraumatic stress disorder, especially the symptom of episodic hyperarousal.
2. *Neuropathic pain.* Although carbamazepine for trigeminal neuralgia is recognized as indicated by the FDA, clinical experience and numerous reports, some based on controlled studies, suggest that carbamazepine, valproate, lamotrigine, gabapentin, topiramate, and clonazepam are

effective for a wide range of neuropathic pain syndromes (183–185) (see also Chapter 31).
3. *Aggressive behavior.* Carbamazepine has been used to suppress aggressive outbursts in patients with and without epilepsy in clinical populations with dementia (186,187) as well as in other "organic" populations (188–190). Similar positive reports of clinical experience and open trials have appeared for valproate in the treatment of aggressive behavior (191–194).
4. *Treatment-resistant depression with partial seizurelike symptoms.* Carbamazepine led to improvement in 11 of 13 tricyclic-unresponsive patients in an open-trial study (195).
5. *SIB in mentally retarded persons.* Open trials have supported the efficacy of carbamazepine for some patients with this condition, although it has not been shown to be generally useful (196).
6. *Withdrawal from long-term benzodiazepine therapy.* Carbamazepine during benzodiazepine withdrawal increased the success of the withdrawal attempt in a sample of 40 patients with a history of difficulty withdrawing from benzodiazepines. The trial was conducted under placebo-controlled, double-blind conditions, with a 25% per week reduction in the benzodiazepine dose (197).
7. *Affective symptoms in people with mental retardation.* Kastner et al. (198) gave valproate to a cohort of 18 mentally retarded patients with three of four of the following symptoms: behavioral cycling, aggressive behavior or SIB, sleep disturbance, and irritability. Fourteen of the 18 improved and continued to improve on valproate over 2 years of follow-up.

Neurologic Toxicity

The AED, with the exception of felbamate, all share the tendency to cause sedation, ataxia, nystagmus, and other signs of CNS depression. These signs can be seen if the serum drug concentration is too high, the dose is increased too rapidly, or the patient is unusually sensitive to the depressant effects of the drug. In addition, several of the AED have characteristic neurologic side effects.

Carbamazepine toxicity frequently manifests with diplopia. Carbamazepine also is anticholinergic and can cause typical signs of anticholinergic toxicity in susceptible patients, usually elderly persons or those with Alzheimer disease.

Valproate often causes an action tremor; at times, this is severe enough to be a limiting side effect. Valproate raises the serum ammonia level; signs of hepatic encephalopathy can develop in patients with occult liver disease treated with this drug. Phenytoin usually produces nystagmus and ataxia at toxic levels. Prolonged exposure to phenytoin at excessive levels can cause permanent cerebellar damage.

Clonazepam toxicity typically presents with sedation and ataxia. An apathetic confusional state can also occur. Lamotrigine and gabapentin tend to have the "generic" AED side effects of sedation, dizziness, ataxia, or impaired coordination. Remarkably, patients on lamotrigine frequently describe a sense of increased well-being (199). Felbamate can cause headache, insomnia, or agitation; more severe toxicity can include an agitated confusional state. Felbamate is the only AED that commonly produces side effects related to CNS stimulation.

Topiramate is slowly titrated upward by 25 to 50 mg every week to minimize the potential for dose-related neurotoxic adverse effects on alertness, concentration, and cognitive function (200,201). Topiramate is also associated with fatigue, poor appetite and weight loss, dizziness, ataxia, and nephrolithiasis (202,203).

In contrast to topiramate, oxcarbazepine can be titrated rapidly to maintenance dose levels. Oxcarbazepine has adverse effects similar to those of carbamazepine but are less frequent and less severe (201). The most common adverse effects include sedation, dizziness, headache, vertigo, ataxia, diplopia, nausea, and an allergic skin rash. A cross-allergy between oxcarbazepine and carbamazepine occurs in 25% to 31% of patients (203). Oxcarbazepine has been associated with hyponatremia, most often in elderly patients (200,204).

Phenobarbital, primidone, phenytoin, and ethosuximide are prescribed as AED but are very rarely used as psychotropics. Phenobarbital and primidone, in particular, are generally regarded as psychotoxic, with a propensity to cause depression, cognitive impairment, or behavioral disturbances. Common neuropsychiatric practice is to replace these medications by other AED if patients develop significant cognitive, behavioral, or emotional problems while being treated for epilepsy with one of them.

Drug Interactions

Pharmacokinetic interactions involving AED usually result from the induction or inhibition of oxidative enzymes. In these interactions, the AED may either be the cause of altered metabolism or the drug affected by it. Some typical and clinically relevant interactions are summarized here.

Carbamazepine levels are increased by drugs that inhibit the enzymes that metabolize it or compete with carbamazepine at the active sites of those enzymes. Calcium channel blockers (verapamil and diltiazem), macrolide antibiotics (erythromycin and clarithromycin), and cimetidine all can raise carbamazepine levels to a clinically significant degree. Carbamazepine itself can induce enzymes that metabolize other drugs as well as carbamazepine itself. Four important examples are warfarin, oral contraceptives, opiate analgesics, and TCA. In each of these cases, the other drug may lack the expected efficacy if given at its usual dose.

Valproate inhibits both oxidative metabolism and glucuronide conjugation of a number of drugs. Its effect on glucuronide conjugation is relevant to the combination of valproate with lorazepam; with coadministration, lorazepam has a higher level and a longer elimination half-life. Impairment of oxidation by valproate affects the metabolism of carbamazepine-10,11-epoxide. When valproate and carbamazepine are given together, carbamazepine-10,11-epoxide can accumulate, leading to typical symptoms of carbamazepine toxicity at an apparently therapeutic level of the parent compound, carbamazepine. Valproate also prolongs the metabolism of lamotrigine, increasing its half-life by 15% to 25% (205). This interaction necessitates slower dose titration of lamotrigine when it is added to valproate therapy. Valproate is highly bound to serum albumin. Aspirin displaces valproate from serum albumin, raising the free drug level, sometimes to the toxic range.

Topiramate may reduce the elimination of phenytoin but otherwise does not have a significant pharmacokinetic interaction with other AED. However, enzyme-inducing drugs may increase the metabolism of topiramate by 50% when given together (203). Topiramate levels are decreased by 15% when coadministered with valproic acid (206).

Although oxcarbazepine induces isoenzymes of the CYP3A group, it does so at a significantly lower level than carbamazepine and has only minor pharmacokinetic interactions. However, lamotrigine levels are reduced by 29% when oxcarbazepine and lamotrigine are given together (203). Levels of the active metabolite of oxcarbazepine, the 10-monohydroxy derivative, are decreased with the coadministration of carbamazepine, phenobarbital, or phenytoin (203). In contrast to carbamazepine, oxcarbazepine does not reduce the anticoagulant effect of warfarin (204,206). Although both oxcarbazepine and topiramate are weak inducers of CYP enzymes, their use may reduce the level and efficacy of oral contraceptives (203,206).

Felbamate is an inhibitor of oxidative enzymes, particularly CYP2C19. This isoenzyme metabolizes omeprazole, imipramine, and diazepam, any of which can have increased levels for a given dose if given concurrently with felbamate. It raises levels of valproate by a mechanism not fully elucidated. Conversely, concurrent felbamate lowers carbamazepine levels by 20% to 25% while raising levels of carbamazepine-10,11-epoxide (207). Lamotrigine does not induce or inhibit oxidative enzymes, so it does not influence the metabolism of other drugs. Gabapentin, which is excreted by the kidney unchanged, is essentially free of pharmacokinetic drug interactions (208).

Pharmacodynamic interactions involving the AED generally consist of additive CNS depression. However, carbamazepine also decreases dopamine turnover and can intensify extrapyramidal side effects when coadministered with neuroleptics. This interaction may be offset by a pharmacokinetic interaction that lowers the blood level of the neuroleptic drug.

A common and clinically important interaction occurs between theophylline and AED. Theophylline, a drug still prescribed for bronchospasm despite its being superseded as a drug of choice by inhaled bronchodilators and inhaled corticosteroids, is a proconvulsant agent. It antagonizes the antiepileptic effect of a number of AED, including carbamazepine and diazepam.

Choice of Antiepileptic Drug

Neuropsychiatrists are called on to choose AED for epilepsy, mood stabilization, or one of the other neuropsychiatric indications mentioned previously. In the treatment of epilepsy, guidance is provided by standard references on AED therapy (209,210). At this point, because of their behavioral toxicity, barbiturates and primidone are no longer first-line agents for the treatment of epilepsy.

When choosing an AED for treatment of bipolar disorder, the fact the valproate has an FDA indication for the treatment of mania is a point in its favor, although the evidence for the therapeutic efficacy of carbamazepine is just as good. When choosing an AED for mood stabilization in a patient with gross brain disease, the FDA indication is not relevant and the drugs can be viewed as equivalent. In some cases, however, safety favors one drug over the other. For example, carbamazepine would be preferred to valproate in a patient with a history of pancreatitis or the possibility of occult liver cirrhosis because of the tendency of valproate to increase amylase and cause hyperammonemia. Valproate would be preferred to carbamazepine in a patient with heart block that might be aggravated by carbamazepine's quinidinelike effects. Oxcarbazepine is increasingly being substituted for carbamazepine because it has fewer severe adverse effects and fewer pharmacokinetic interactions compared with carbamazepine. Although its efficacy as a mood stabilizer has not yet been established, topiramate is the only AED used as a mood stabilizer that is associated with weight loss and is preferred when weight gain is a liability or, adjunctively, to lessen weight gain caused by other psychotropic drugs.

When an AED is used as an antidepressant, a case can be made for preferring carbamazepine to valproate because there is slightly more uncontrolled evidence of its antidepressant efficacy.

When carbamazepine and valproate are ineffective or not tolerated, the neuropsychiatric picture might still suggest that an AED might be helpful. For example, a patient with a right-hemisphere lesion may have a rapidly cycling mood disorder, or a patient with a history of TBI may have sudden rages associated with a paroxysmal EEG. In these situations, a reasonable preference is to employ lamotrigine. Other neuropsychiatrists use gabapentin in this situation. Both drugs are considered as promising neuropsychiatric drugs because of reports of positive effects on mood when the drugs underwent clinical testing as antiepileptic agents (211,212).

When AED are ineffective or intolerable or pharmacokinetic interactions are a concern, a trial of an atypical neuroleptic may be considered. As discussed earlier, olanzapine and ziprasidone have shown therapeutic efficacy for acute mania. The preference of either atypical neuroleptic would depend on the combined consideration of the drug's potential adverse effects and the medical and neuropsychiatric status of a particular patient.

An extensive earlier literature described the psychotropic actions of phenytoin. Because these were not confirmed by subsequent controlled clinical trials and the robust psychotropic actions of valproate and carbamazepine became evident, phenytoin fell out of use as a psychotropic. It is actually unknown whether phenytoin might be useful as a psychotropic drug for carefully selected patients and indications.

Dose Titration

In the absence of data suggesting that an alternative would be better, the doses and blood levels appropriate for the treatment of epilepsy are followed when using AED as psychotropics. Therapeutic doses and levels usually are approached gradually. When AED are discontinued for reasons other than serious adverse reactions, discontinuation is by gradual taper. A typical, conservative dose strategy might be as follows:

1. *Carbamazepine.* Start with 100 mg q.h.s. If tolerated, increase every 2 to 3 days by 100 mg, building up to an evenly divided t.i.d. schedule (thus, 100 mg b.i.d., 100 mg t.i.d., 100–100–200 on a t.i.d. schedule, etc.) A predose blood (trough) level would be checked after reaching 200 mg t.i.d. The target level would be 8 to 12 μg/mL. Dose increases would continue until the target range is reached; the dose would then be held constant for another 2 weeks. The blood level would be checked again and further dose adjustment made if necessary.

2. *Valproate.* Start with 250 mg q.h.s. If tolerated, increase every 2 to 3 days by 250 mg, building up to a t.i.d. schedule (thus, 250 b.i.d., 250 t.i.d., 250–250–500, etc.) A trough blood level would be checked at 1,000 mg per day and weekly thereafter. Dose increases would continue until the target range of 75 to 100 μg/mL was reached.

3. *Lamotrigine.* Start with 50 mg q.h.s. If tolerated, increase by 50 mg every week on a b.i.d. schedule, eventually reaching 200 mg b.i.d.

4. *Gabapentin.* Start with 100 mg q.h.s. Build up in 100-mg increments every 2 to 3 days on a t.i.d. schedule, eventually reaching 600 mg t.i.d. After reaching 300 mg t.i.d., increments may be 300 mg.

5. *Topiramate.* Start low and go slow to avoid dose-related adverse effects. Start with 25 to 50 mg per day and increase the dose every 1 to 2 weeks by 25 to 50 mg on a b.i.d. schedule with a target range of 200 to 400 mg per day, given in divided doses, achieved at 8 to 12 weeks.

6. *Oxcarbazepine.* Start with 300 mg q.h.s. and increase the daily dose by 300 mg every third day, given on a b.i.d. or

t.i.d. schedule, to a target dose range of 900 to 2,400 mg per day. Patients and elderly with renal impairment will require lower doses.

In all cases, dose increases are stopped if the desired therapeutic effect is attained or limiting side effects develop.

An exception to gradual dose titration is in the treatment of acute mania, in which the need to rapidly attain a therapeutic blood level outweighs the greater incidence of side effects with a high initial dose. When treating acute mania, the drug dose likely to produce a therapeutic level in steady state (e.g., valproate 500 mg t.i.d., carbamazepine 200 t.i.d.) is estimated and initiated at that dose.

Duration of Therapy

Considerations regarding the duration of therapy with AED as mood stabilizers are identical with those presented earlier for lithium therapy. When AED are used for epilepsy, contemporary guidelines for AED discontinuation are followed if the patient has been seizure free for more than 1 year.

STIMULANTS

Stimulants (dextroamphetamine, methylphenidate, and pemoline) are the well-established drug treatment for attention deficit/hyperactivity disorder (ADHD). In sleep medicine, they are the treatment of choice for narcolepsy. In consultation psychiatry, they are used to rapidly mobilize hospitalized medically ill patients with apathy and depressed mood (213). In neuropsychiatric practice, additional applications draw on the capacity of stimulants to enhance arousal and motivation and ability to sustain attention.

Efficacy

The efficacy of the stimulants for ADHD has been established by more than 100 RCT (214). However, as many as 30% of ADHD patients treated with any given stimulant will not improve. Some will respond to a different stimulant and others to an alternate therapy such as a TCA.

The efficacy of stimulants for patients with narcolepsy has been established by sleep laboratory studies using the Multiple Sleep Latency Test or the Maintenance of Wakefulness Tests as measures. These studies have shown that pemoline is inferior to methylphenidate and dextroamphetamine. Also, they suggest that patients with narcolepsy are still sleepier than control subjects, even after therapy with relatively high doses of stimulants (215).

Methylphenidate has been shown to produce significant improvement in depressive symptoms within 1 week in a double-blind, randomized, placebo-controlled crossover trial in 16 older medically ill patients (216). This recent RCT is in accord with a large anecdotal literature on the effectiveness of stimulants for depression in the medically ill.

Other Neuropsychiatric Indications

Poststroke Depression

Methylphenidate may be able to produce a remission of depressive symptoms after stroke in approximately half of patients treated. Lazarus et al. (217) retrospectively reviewed the treatment of 58 elderly stroke patients for poststroke major depression, 28 of whom received methylphenidate and 30 of whom received nortriptyline. Fifteen (53%) of the methylphenidate group had a complete remission of depression, whereas 13 (43%) of the nortriptyline group had a complete remission. The difference was not statistically significant. However, the average response time was 2.4 days in the methylphenidate group and 27 days in the nortriptyline group ($p < .001$).

Attention Deficit/Hyperactivity Disorder Among Patients with Developmental Disabilities

ADHD symptomatology in patients with mental retardation and other developmental disabilities has improved with the use of stimulants. The improvement rate of ADHD in blind methylphenidate trials does not differ significantly between children with ADHD alone and those with ADHD plus other neurodevelopmental disorders (218). However, the comparison did not include severely and profoundly retarded children.

Apathy

Methylphenidate or dextroamphetamine can reduce apathy in patients of various causes, including depression, TBI, side effects of AED, antidepressants, or neuroleptics (219–221). The use of stimulants for the treatment of apathy has not been subjected to RCT.

Neurologic Toxicity

In usual therapeutic doses, neurologic side effects of the stimulants include insomnia, irritability, nervousness, headache, and tremor. Acute overdose can produce an agitated confusional state with hyperreflexia and seizures; a maniclike psychosis is also possible. Prolonged use at unusually high doses can produce a paranoid psychosis with features of acute schizophrenia. Withdrawal from long-term use can be associated with headache, apathy, fatigue, and hypersomnia.

Between 5% and 10% of children with ADHD treated with stimulant medications will develop tics or dyskinesias (222). Most are transient, with fewer than 1% of treated children developing Tourette syndrome. Although risk factors have not been fully evaluated, a personal or family history of tics may increase the risk of developing persistent tics on stimulant therapy (222).

Drug Interactions

Methylphenidate inhibits oxidative enzymes that metabolize TCA, warfarin, and AED, including phenobarbital, phenytoin, and primidone. All these drugs may require downward dose adjustments to avoid toxicity.

All the stimulants have pharmacodynamic interactions with psychotropic drugs. They enhance antidepressant effects. Their stimulating and anorexiant effects are antagonized by neuroleptics and lithium.

Stimulants Versus Other Therapies

For the treatment of ADHD, stimulants are first-line therapy, and other therapies would be used only if the patient objected to stimulants, did not tolerate them, did not respond to them, had a personal or family history of tics or Tourette syndrome, or was deemed to be at an unacceptably high risk of abusing or diverting stimulant drugs. For these patients, therapeutic alternatives include most of the nonsedating antidepressants (desipramine, nortriptyline, imipramine, phenelzine, tranylcypromine, bupropion, sertraline, and fluoxetine), and the adrenergic agents clonidine and propranolol (223).

For depression, stimulants are preferred to antidepressants when apathy is a prominent feature of the depression and a prompt response to treatment is of great importance to the patient. For example, a patient with a recent stroke who is depressed and too apathetic to participate in physical therapy would be an ideal candidate for stimulant therapy. Patients with anxious or delusional depressions should not be treated with stimulants, nor should stimulants be the primary treatment for a physically healthy person with a primary unipolar depression.

Dopamine agonist drugs are the current alternative therapy for apathy. Stimulants are preferred when an apathetic patient is sleepy or lethargic and maximal efforts had already been made to remove sedating drugs from the patient's regimen. Dopamine agonists are preferred when the patient has rigidity, bradykinesia, or other motor signs of parkinsonism. For apathy with neither sleepiness nor parkinsonian features, either a stimulant or a dopamine agonist is a reasonable first choice.

Choice of Stimulant

Issues related to the choice of stimulant and to initiating therapy are outlined in Table 6.4 along with similar details for the dopamine agonist drugs.

Patients may respond to dextroamphetamine or methylphenidate but not to pemoline; the reverse situation can occur, but it is uncommon. Therefore, one of the former two drugs usually should be tried first. During dose titration,

the drug is given two or three times per day. Once a dose has been established, the use of long-acting preparations may allow q.d. or b.i.d. dosing. Pemoline can be given on a q.d. schedule at the outset because it is long acting.

Dose Titration

The initial dose of dextroamphetamine or methylphenidate is 2.5 mg in the morning and at noon in a fragile or small person and 5 mg in the morning and at noon in a more robust and larger person. Therapeutic effects, if any, are immediate. If there are neither side effects nor an apparent therapeutic response, the dose is increased the next day to 5 mg in the morning and at noon or 10 mg in the morning and at noon, respectively. The dose may be increased further if needed, in units of 2.5 mg for the small or fragile patient and 5 mg for the larger and healthier patient. When the patient feels some change for the better in attention, mood, or motivation or the clinician sees an apparent change, the dose is maintained at the same level for a week, after which the response is reevaluated and the dose adjusted if necessary to maximize response or to minimize side effects.

Pemoline is started at 18.75 mg once daily. The dose is raised by 18.75 mg once per week until the patient improves, limiting side effects develop, or a dose of 112.5 mg per day is reached.

Duration of Therapy

Treatment for ADHD and other attention disorders is continual and long term. However, in children, drug-free intervals are given when feasible to allow growth unimpeded by the anorexiant and potential growth-suppressing effects of stimulants. Weekends and school vacations are natural opportunities for children whose behavior at home without medication is tolerable. For adults with attention disorders, growth is not an issue, but tolerance or tachyphylaxis to stimulants may be. If an adult on stimulant therapy for ADHD finds that the benefits of stimulants are wearing off, a drug holiday should be considered. Vacations or periods of relatively lower work demand may be suitable times.

When stimulants are used to treat illness-associated depressions, the drug should be withdrawn once the acute illness has resolved or stabilized. If the patient's depression persists after withdrawal of the stimulant, a standard antidepressant should be initiated. For the patient's comfort, the stimulant could be restarted until there was time for the antidepressant to take effect. If stimulants are given for a week or more, they should be withdrawn gradually, ideally by no more than 25% every 2 to 3 days.

When stimulants are used to treat apathy, therapy, if effective, may continue until the underlying cause of the apathy has resolved. If the cause of the apathy is permanent, stimulant therapy may continue unless tachyphylaxis or tolerance develops. In that case, a planned drug holiday might restore

TABLE 6.4. ISSUES RELEVANT TO CHOOSING AND INITIATING A DOPAMINE AGONIST OR STIMULANT

Drug	Benefits	Precautions	Controlled Trials	Initial Daily Dose
Bromocriptine	Antidepressant, antiparkinson, antiapathetic effects	Orthostatic hypotension; nausea; confusion; depression or mania; ergot hypersensitivity (pulmonary or retroperitoneal fibrosis)	Many for Parkinson disease and several for primary major depression; case series support for apathy and abulia from gross brain disease and drug-induced parkinsonism	2.5–5 mg; drug is usually given on a b.i.d. schedule
Pergolide	Can be given once daily because of long half-life; may be less expensive because of its high potency; otherwise like bromocriptine	As for bromocriptine; however, some patients do better on one than the other	Many for Parkinson disease; case series support for depression; case reports for apathy and abulia	0.05–0.10 mg q.d.
Amantadine	Anti-Parkinson, antiapathetic; rapid onset of action	Orthostatic hypotension, confusion, anticholinergic effects	In drug-induced parkinsonism has fewer cognitive side effects than benztropine	50–100 mg b.i.d.
Dextroamphetamine	Improves alertness, attention, motivation	May increase anxiety to induce mania; subject to abuse because of euphoriant effects	For narcolepsy; case series support benefit for apathy and depression in the medically ill	2.5–5 mg morning and noon
Methylphenidate	Improves alertness, attention, motivation	May increase anxiety to induce mania; subject to abuse because of euphoriant effects	Many for ADHD; one for depression in the medically ill; case series support benefit for apathy	2.5–5 mg morning and noon
Pemoline	Improves alertness; attention, motivation; less subject to abuse than dextroamphetamine and less tightly controlled; longer duration of action permits once daily dosing	Onset of action may take weeks; dysphoric reactions more common than with dextroamphetamine; overall efficacy probably lower; may induce mania or anxiety	Many for ADHD; no studies in other neuropsychiatric conditions	18.75 mg q.d.

b.i.d., twice daily; q.d., every day; ADHD; attention deficit/hyperactivity disorder.

sensitivity to the stimulant. Alternatively, the specific stimulant might be rotated to a different member of the class.

DOPAMINE AGONISTS

The dopamine agonist, anti-Parkinson drugs selegiline, amantadine, bromocriptine, and pergolide have recently been used for neuropsychiatric indications other than Parkinson disease, drug-induced parkinsonism, and other movement disorders. The best established neuropsychiatric indication is bromocriptine for treatment-resistant depression.

Efficacy

Bromocriptine has been tested as an antidepressant in three RCT with a total of 125 patients (224). An open trial of pergolide as an adjunct to standard antidepressants in 20 patients with treatment-resistant depression converted 11 cases to remissions (225).

Other Neuropsychiatric Indications

Dopamine agonists have other neuropsychiatric indications. These include the following:

1. *Neuroleptic-induced apathy and abulia.* Amantadine has a stimulating effect on patients with neuroleptic-treated schizophrenia (226).
2. *Agitation and assaultiveness during recovery from traumatic coma.* Amantadine was reported to suppress agitation in two patients with traumatic frontotemporal lesions (227) and was then tested by Gualtieri et al. (228) in a group of patients recovering from coma who were agitated and emotionally labile. More than 50% responded in this open trial; doses of 100 to 400 mg per day were given in this study. The response was all or nothing—either dramatic improvement or no benefit.
3. *Akinetic mutism.* Ross and Stewart (229) reported a case in which bromocriptine relieved stimulated spontaneous speech and movement in a man with akinetic mutism from hypothalamic damage.
4. *Hemineglect.* Fleet et al. (230) reported alleviation of neglect when bromocriptine was administered to two patients with right-hemisphere strokes.
5. *Amnesia owing to mediobasal forebrain injury.* In a single surgical case of damage to the area of the septum and nucleus accumbens, bromocriptine partially alleviated anterograde amnesia (231).
6. *Transcortical motor aphasia.* Albert et al. (232) reported recovery of spontaneous speech when bromocriptine was given, which relapsed when bromocriptine was discontinued.

In summary, there is evidence for an antidepressant, psychomotor activating effect of dopamine agonist drugs but insufficient empiric study to determine the appropriate indications for dopamine agonist therapy.

Neurologic Toxicity

The most serious neurologic side effects of the dopamine agonist drugs are changes in mental status. The dopamine agonists can cause confusion, agitation, insomnia, hallucinations, paranoid ideation, disorganized thinking, incoherent speech, emotional lability, hypersexuality, hypomanic and manic phenomena, depression, or anxiety. They also can cause involuntary movements, including choreoathetosis, dyskinesia, and tremor. Although the list of potential side effects is long, many patients take those drugs with no side effects at all.

Drug Interactions

Drug interactions with selegiline may become problematic if the dose of selegiline is increased to more than 10 mg per day. At higher doses, selegiline, a selective MAO-B inhibitor, loses its selectivity. When MAO is nonselectively inhibited, patients can develop serious hypertension from indirect adrenergic agents or manifest a serotonin syndrome if they take an SSRI.

Combination therapy, using multiple dopamine agonists, typically the case in the treatment of Parkinson disease, increases both the therapeutic effect and risk of side effects, particularly the adverse effects on mental status discussed previously.

Choice of Dopamine Agonist

The choice of dopamine agonist drugs for Parkinson disease is beyond the scope of this chapter. When a dopamine agonist drug is to be used for neuroleptic-induced parkinsonism, amantadine usually is selected first because of ease of administration and extensive experience with its use for this indication. If amantadine at a full dose does not relieve the patient's symptoms, bromocriptine is used next because the combination of bromocriptine and neuroleptics in schizophrenia treatment is documented, and bromocriptine rarely appears to cause aggravation of psychosis during neuroleptic treatment.

When a dopamine agonist drug is to be used as a primary or adjunctive treatment for depression resistant to usual treatments, bromocriptine should be considered as a first choice, in view of the support for its use from RCT. If bromocriptine is not tolerated, then pergolide would be substituted.

When a dopamine agonist is to be used to treat apathy or to attempt pharmacologic remediation of a focal cortical deficit (as in the reported cases of hemineglect and transcortical motor aphasia), there is no strong reason to prefer bromocriptine over pergolide. Because successful cases have been reported using bromocriptine, some clinicians might prefer to begin with that agent.

Dose Titration

The starting dose of amantadine is 50 mg b.i.d. in a small or fragile person and 100 mg b.i.d. in a larger and more robust person. Dose increases should be in 50-mg increments in the former case and 100-mg increments in the latter case. The dose usually should not exceed 100 mg t.i.d. and must be reduced in renal insufficiency.

The starting dose of selegiline is 5 mg q.d. This can be increased to 5 mg b.i.d. after 2 days, if the initial dose is tolerated.

The starting dose of bromocriptine is 1.25 mg b.i.d. in a smaller or more fragile patient and 2.5 mg b.i.d. in a larger and more robust person. Dose increases are in 1.25-mg increments in the former type of patient and 2.5 mg in the latter type. There is no absolute maximal dose for neuropsychiatric indications if the drug is well tolerated; however, the dose should not be increased to more than 10 mg t.i.d. unless there is clear evidence of benefit.

The starting dose of pergolide is 0.05 mg q.d. in a small or fragile person and 0.1 mg q.d. in a larger and more robust person. Dose increments begin at 0.05 to 0.1 mg and gradually rise to 0.1 to 0.25 mg per day. Dosage would not ordinarily be increased to more than 5 mg per day unless partial benefit is observed.

Duration of Therapy

Duration of therapy for major depression is 6 to 12 months, followed by an attempt to taper the medication. When dopamine agonists are used for drug-induced parkinsonism, efforts should be made periodically (e.g., at 2 and 4 months) to see whether they are still needed because some patients become more tolerant of neuroleptics. If extrapyramidal effects recur promptly with two or three attempts to taper dopamine agonists, they probably will be needed long term unless the patient is switched to an atypical neuroleptic.

When dopamine agonists are used to partially reverse cognitive deficits caused by acquired lesions such as strokes and contusions, the clinical hope is that further recovery of function in the brain will make pharmacologic palliation of deficits unnecessary or that the drug-induced improvement will be relevant to everyday functioning. Attempts to taper the drugs should be made every 3 months or so during the recovery process. Deficits persisting at 3 years after acquisition of the lesion are likely to be permanent. At that point, if the drug makes a meaningful difference to the patient's function and well-being, it may be continued indefinitely.

ADRENERGIC AGENTS

The adrenergic agents most often used in neuropsychiatry are the β-adrenergic blockers, especially propranolol, and clonidine, a mixed α_1-α_2 agonist. In addition to their original indication of hypertension, these agents have well-established efficacy for several neuropsychiatric indications and have been tried for numerous others. The appeal of these agents, which either block adrenergic receptors or indirectly decrease adrenergic activity, is their ability to diminish the autonomic component of emotional response. Through feedback to the limbic system, decreased autonomic response can lead to decreased cognitive and affective responses.

Efficacy

The efficacy of propranolol has been established by RCT for migraine prophylaxis (233) and the treatment of essential tremor (234). The efficacy of clonidine has been established by RCT for reducing tics in Tourette syndrome (235) and decreasing the intensity of withdrawal symptoms after discontinuation of opiates (236).

Other Neuropsychiatric Indications

Both propranolol and clonidine have been used in a number of conditions in which anxiety or symptoms of anxiety are present or inferred. These include the following:

1. *ADHD.* Clonidine (237,238) has shown efficacy and is a primary therapy for ADHD. The dose, reached by gradual titration, is 3 to 10 μg/kg per day. Propranolol has been used as an adjunct to stimulants to reduce impulsive behavior at a dose of 2 to 8 mg/kg per day (239). Guanfacine, a more selective α_2 agonist than clonidine, was shown in an open study of 13 outpatients with ADHD to substantially improve hyperactivity, inattention, and immaturity with less sedation than would be expected with an equivalent dose of clonidine (240).
2. *Agitation in dementia.* There have been several double-blind studies showing propranolol to be superior to placebo for assaultiveness and agitation; doses have varied greatly, from 10 to 600 mg per day (241,242).
3. *Rage attacks in children, adolescents, and adults with "organic" brain dysfunction.* Williams et al. (242) and Yudofsky et al. (243) have summarized a number of early studies supporting an antiaggressive action of propranolol.
4. *Akathisia due to neuroleptics.* Propranolol has been found to be superior to benztropine and placebo in the treatment of akathisia at doses as high as 120 mg per day (244,245).

Neurologic Toxicity

The most common neurologic side effects of propranolol and clonidine are fatigue, sedation, and decreased libido. The fatigue and sedation occasionally combine with other symptoms to form a depressive syndrome, although it is unclear whether the drugs actually can cause a major depression in a person with no psychiatric vulnerability. Rarely, and especially in patients with preexisting cognitive impairment, either drug can cause a delirious or agitated state, with visual or auditory hallucinations.

Drug Interactions

Pharmacokinetic interactions with propranolol are related to its oxidative metabolism. Propranolol levels are raised by coadministration of chlorpromazine or cimetidine; they are lowered by coadministration of phenytoin or phenobarbital. Propranolol inhibits the metabolism of theophylline, leading to increased serum drug concentrations at a fixed dose.

The major reported interaction of clonidine is with TCA; coadministration decreases the hypotensive effect of clonidine. Pharmacodynamic interactions are mainly with other antihypertensive drugs (additive or synergistic antihypertensive effect) and with other drugs with sedative or CNS depressant effects (increased lethargy, fatigue, and depressive symptoms).

Decision to Use an Adrenergic Agent

Issues related to the choice and initiation of an adrenergic agent are summarized in Table 6.5.

Propranolol is a first-choice agent for migraine and for essential tremor; clonidine is a second-line agent for Tourette syndrome. For akathisia, propranolol is preferred and is almost always more effective than either benzodiazepines or anticholinergics. For aggressive behavior and rage attacks, there is no general way to choose between a mood-stabilizing drug, an antidepressant, and propranolol as a first treatment. In a specific patient, affective instability or cycling of behavior suggests trying a mood stabilizer first, chronic irritability suggests trying an SSRI, and marked autonomic arousal at the time of behavioral display suggests propranolol. If more than one picture fits, such considerations as vulnerability to side effects and past experiences of the patient or the physician will affect the decision. At present, no laboratory test, including the EEG, supersedes clinical considerations in drug choice.

Clonidine should be regarded as a potential adjunct in the treatment of ADHD, Tourette syndrome, or the combination of the two to address attention impairment, impulsivity, and autonomic arousal, if these persist after first-line treatments are instituted. It is possible that guanfacine, with its lesser sedation and possibly greater benefit than clonidine, may become more of a routine agent in the treatment of ADHD and related conditions.

OPIATE RECEPTOR BLOCKERS

Oral opiate receptor blockers, of which naltrexone is the only one currently available in the United States, were introduced as an aid in the treatment of opiate abuse and subsequently in the treatment of alcoholism (246). However, neuropsychiatrists have perennially speculated on the role of endogenous opiate mechanisms in sustaining other forms of habitual behaviors, particularly those that involve

TABLE 6.5. ISSUES RELEVANT TO CHOOSING AND INITIATING AN ADRENERGIC AGENT

Drug	Benefits	Precautions	Controlled Trials	Initial Daily Dose
Propranolol	Antianxiety; helpful for migraine, tremor, irritability, impulsive aggression	Sedation; hypotension; can aggravate asthma or heart failure; various drug interactions related to its oxidative metabolism	Migraine, essential tremor; case series support utility for agitation and aggression and neuroleptic-induced akathisia	20–40 mg t.i.d.
Nadolol	Less fat soluble and more β_1 selective; therefore less sedation and aggravation of asthma; may help impulsive aggression; long half-life so may be taken once daily	Same as propranolol	None; open trials support benefit for aggression in mental retardation	40 mg
Clonidine	Major anxiolytic effects, including blocking of anxiety from drug withdrawal; decreases firing of locus ceruleus	Sedation, hypotension, confusion	ADHD; Tourette syndrome, opiate or nicotine withdrawal; case series support use for impulsive aggression and memory loss in Korsakoff syndrome	0.1 mg q.d. to t.i.d., depending on indication
Guanfacine	Decreases firing of locus ceruleus; stimulates frontal lobe α_2 receptors to improve executive function; long duration of action permits once daily dosing	As with clonidine, but less sedation; can cause dyspepsia or nausea	ADHD; anecdotal support for use in impulsive aggression	1 mg q.d.

t.i.d., three times daily; ADHD, attention deficit/hyperactivity disorder; q.d., every day.

some self-inflicted pain or discomfort. Theoretically, patients with self-inflicted injuries may be releasing substantial amounts of endogenous opiates when they perform self-injurious acts.

The evidence for the efficacy of naltrexone in addiction treatment is provided in Chapter 30. Other neuropsychiatric indications for naltrexone include the following:

1. *SIB in mentally retarded adults.* Sandman (247), reviewed studies of naloxone and naltrexone in SIB and concluded that sometimes they worked and sometimes they did not. Gillberg (248) reached similar conclusions. Herman et al. (249) reported that patients with the severe SIB of head and face hitting improved with naltrexone. Possibly, more severe forms of SIB are related to opiate mechanisms, and less severe forms represent a form of stereotypic behavior with a different pattern of pharmacologic response (250). In a study of SIB in eight mentally retarded adults, naltrexone decreased the number of days with frequent SIB and increased the number of days with infrequent SIB (251).

2. *Autism.* Because of the role of opiates in systems mediating attachment, there has been speculation that opiate antagonists might restore more normal social behavior. In an RCT of naltrexone in autistic children, Campbell et al. (252) found that the naltrexone-treated patients communicated more and withdrew less. Kolman (253) found improved social behavior in eight of 13 autistic children treated with naltrexone. Lensing et al. (254) reported a case of a 5-year-old boy in whom treatment with naltrexone increased smiling, crying, and playing. However, Zingarelli et al. (255) failed to find benefit in a trial of naltrexone in eight autistic young adults. Apparently, naltrexone affects social behavior, but the response may be dependent on the measures used, the individual's level of function before treatment, and the social environment.

3. *Bulimia.* A study comparing low- and high-dose naltrexone in the treatment of bulimia found a lower rate of binge eating and purging in the high-dose group (256). Subsequently, a single case appeared reporting that naltrexone reduced a bulimic patient's subjective urge to binge eat (257).

4. *Tourette syndrome.* Kurlan et al. (246) found that naltrexone reduced tics and also improved patients' concentration.

Neurologic Toxicity

As might be expected from a drug used exclusively in populations with mental illness or substance abuse, patients on naltrexone have reported a wide range of mental symptoms, including restlessness, insomnia, nightmares, confusion, hallucinations, paranoia, and fatigue. The relation of these symptoms to drug effects is not clear.

Drug Interactions

There is only one drug interaction of consequence—the precipitation of an acute and severe withdrawal state if the drug is given to a person still taking narcotics on a regular basis.

When to Consider Using Naltrexone

In the neuropsychiatric context, naltrexone should be considered as an option for the treatment of recurrent SIB. It might be expanded for other repetitive behaviors that are hypothesized to be caused by opiate-related mechanisms. The use of naltrexone in the treatment of alcohol and drug abuse is discussed fully by Ling et al. in Chapter 30.

Dose Titration

No titration is needed. A fixed dose of 50 mg per day, 100 mg every other day, or 150 mg every third day should provide adequate opiate receptor blockade.

Duration of Therapy

After blocking opiate receptors, there theoretically can be a brief rebound of the behavior that was previously reinforced by endogenous opiate release. Such a rebound was actually observed in a case reported by Benjamin and Buot-Smith (258). An adequate therapeutic trial should permit this rebound period to pass. After that, behavioral and environmental therapies to extinguish the unwanted behavior should be instituted, taking advantage of the change in internal reinforcement. An attempt to taper and discontinue naltrexone could be made once the patient had a stable period without significant self-injury. If the SIB recurred, therapy with naltrexone would be reinstated.

CONCLUSION

RCT focusing specifically on neuropsychiatric populations are limited. Still, the combination of RCT in general psychiatric populations and a rich literature of case series and open trials should provide the neuropsychiatrist with an empiric basis for planning drug therapy in a neuropsychiatric practice. Neuropsychiatrists working with well-defined and homogeneous populations would greatly advance the field by conducting methodologically rigorous tests of the drug therapies that they find clinically useful in their populations of interest.

REFERENCES

1. American Psychiatric Association. *Diagnostic and statistical manual of mental disorders*, 4th ed. Washington, DC: American Psychiatric Association, 1994.

2. World Health Organization. *International classification of diseases*, 10th ed. Geneva: World Health Organization, 1994.

3. Burke MJ, Preskorn SH. Short-term treatment of mood disorders with standard antidepressants. In: Bloom FE, Kupfer DJ, eds. *Psychopharmacology: the fourth generation of progress*. New York: Raven Press, 1995:1053–1066.

4. Starkstein SE, Robinson RG. Depression in cerebrovascular disease. In: Starkstein SE, Robinson RG, eds. *Depression in neurologic disease*. Baltimore: Johns Hopkins University Press, 1993:28–49.

5. Andersen G, Vestergaard K, Lauritzen L. Effective treatment of post-stroke depression with the selective serotonin reuptake inhibitor citalopram. *Stroke* 1994;25:1099–1104.

6. Robinson RG, Shultz SK, Castillo C, et al. Nortriptyline versus fluoxetine in the treatment of depression and in short-term recovery after stroke: a placebo-controlled, double-blind study. *Am J Psychiatry* 2000;157:351–359.

7. Wiart L, Petit H, Joseph PA, et al. Fluoxetine in early poststroke depression. *Stroke* 2000;31:1829–1832.

8. Berrios GE. History of descriptive psychopathology. In: Mezzich JE, Jorge MR, Salloum IM, eds. *Psychiatric epidemiology: assessment concepts and methods*. Baltimore: Johns Hopkins University Press, 1994:47–68.

9. Lakshamanan M, Mion LC, Frengley JD. Effective low dose tricyclic antidepressant treatment for depressed geriatric rehabilitation patients. A double-blind study. *J Am Geriatr Soc* 1986;34:421–426.

10. Fogel BS, Ratey JJ. A neuropsychiatric approach to personality and behavior. In: Ratey JJ, ed. *Neuropsychiatry of personality disorders*. Cambridge: Blackwell Science, 1995:1–16.

11. Wlaz P, Rolinski Z, Kleinrok Z, et al. Influence of chronic aminophylline on antielectroshock activity of diazepam and aminophylline induced convulsions in mice. *Pharmacol Biochem Behav* 1994;49:609–613.

12. Yudofsky S, Williams D, Gorman J. Propranolol in the treatment of rage and violent behavior in patients with chronic brain syndromes. *Am J Psychiatry* 1981;138:218–220.

13. Silver JM, Yudofsky SC. Aggressive behavior in patients with neuropsychiatric disorders. *Psychiatr Ann* 1987;17:367–370.

14. Herzog AG. Reproductive endocrine considerations and hormonal therapy for men with epilepsy. *Epilepsia* 1991;32[Suppl]:S34–S37.

15. Haggerty JJ, Prange AJ. Borderline hyperthyroidism and depression. In: Coggins CH, Hancock EW, Levitt LJ, eds. *Annual review of medicine: selected topics in the clinical sciences*. Palo Alto: Annual Reviews, Inc., 1995:37–46.

16. Sandman CA, Datta PC, Barrum J, et al. Naloxone attenuates self abusive behavior in developmentally disabled clients. *Appl Res Ment Retard* 1983;4:5–11.

17. Campbell M, Anderson LT, Small AM, et al. Naltrexone in autistic children: behavioral symptoms and attentional learning. *J Am Acad Child Adolesc Psychiatry* 1993;32:1283–1291.

18. Percy AK, Glaze DG, Schultz RJ, et al. Rett syndrome: controlled study of an oral opiate antagonist, naltrexone. *Ann Neurol* 1994;35:464–470.

19. Royall DR, Mahurin RK, True JE, et al. Executive impairment among the functionally dependent: comparisons between schizophrenic and elderly subjects. *Am J Psychiatry* 1993;150:1813.

20. Alexopoulous GS, Abrams RC, Young RC, et al. Cornell Scale for Depression in Dementia. *Biol Psychiatry* 1988;23:271–284.

21. Yesavage JA, Brink TL, Rose TL, et al. Development and validation of a geriatric depression screening scale: a preliminary report. *Psychiatry Res* 1983;17:37–39.

22. Ott B, Fogel BS. Measurement of depression in dementia: self vs. clinician rating. *Int J Geriatr Psychiatry* 1993;7:899–904.

23. Yudofsky SC, Silver JM, Jackson W, et al. The overt aggression scale for the objective rating of verbal and physical aggression. *Am J Psychiatry* 1986;143:35–39.

24. Marin RS, Biedrzyck RC, Firinciogullari S. Reliability and validity of the apathy evaluation scale. *Psychiatry Res* 1991;38:143–162.

25. Huskisson EC. Visual analogue scales. In: Melzack R, ed. *Pain measurement and assessment*. New York: Raven Press, 1983:33–37.

26. Melzack R, ed. *Pain measurement and assessment*. New York: Raven Press, 1983.

27. Siu AL, Reuben DB, Hays RD. Hierarchical reaction of physical function in ambulatory geriatrics. *J Am Geriatr Soc* 1990;38:1113–1119.

28. Stewart AL, Hays RD, Ware JE. The MOS short form health survey. *Med Care* 1988;26:724–732.

29. Jenner P. The rationale for the use of dopamine agonists in Parkinson's disease. *Neurology* 1995;45[Suppl 3]:S6–S12.

30. Koller WC, Silver DE, Lieberman A. An algorithm for the management of Parkinson's disease. *Neurology* 1994;44[Suppl 10]:S1–S52.

31. Leach JP, Brodie MJ. Lamotrigine–clinical use. In: Levy RH, Mattson RH, Meldrum BS, eds. *Antiepileptic drugs*, 4th ed. New York: Raven Press, 1995:889–896.

32. Chadwick D. Gabapentin–clinical use. In: Levy RH, Mattson RH, Meldrum BS, eds. *Antiepileptic drugs*, 4th ed. New York: Raven Press, 1995:851–856.

33. Stoudemire A, Fogel BS. Psychopharmacology update. In: Stoudemire A, Fogel BS, eds. *Medical psychiatric practice, vol. 1*. Washington, DC: American Psychiatric Press, 1991:29–98.

34. Stoudemire A, Fogel BS. New psychotropics in medically ill patients. In: Stoudemire A, Fogel BS, eds. *Medical psychiatric practice, vol. 2*. Washington, DC: American Psychiatric Press, 1993:69–112.

35. Stoudermire A, Fogel BS. Psychopharmacology in medical patients: an update. In: Stoudemire A, Fogel BS, eds. *Medical psychiatric practice, vol. 3*. Washington, DC: American Psychiatric Press, 1995:79–150.

36. Lipsey JR, Robinson, RG, Pearlson FD, et al. Nortriptyline treatment of post-stroke depression: a double-blind treatment trial. *Lancet* 1985;1:297–323.

37. Reding MJ, Orto LA, Winter SW, et al. Antidepressant therapy after stroke: a double-blind trial. *Arch Neurol* 1986;47:785–789.

38. Robinson RG, Parikh RM, Lipsey JR, et al. Pathologic laughing and crying following stroke: validation of a measurement scale and a double-blind treatment study. *Am J Psychiatry* 193;150:286–293.

39. Schiffer RB, Herndon RM, Rudick RA. Treatment of pathological laughing and weeping with amitriptyline. *N Engl J Med* 1985;312:1480–1482.

40. Anderson G, Vestergaard K, Riis JO. Citalopram for post-stroke pathological crying. *Lancet* 1993;342:837–839.

41. Burns A, Russell E, Stratton-Powell H, et al. Sertraline in stroke-associated lability of mood. *Int J Geriatr Psychiatry* 1999;14:681–685.

42. Reifler BV, Teri L, Rasking M, et al. Double blind trial of imipramine in Alzheimer's disease patients with and without depression. *Am J Psychiatry* 1989;146:45–49.

43. Schiffer RB, Wineman NM. Antidepressant pharmacotherapy of depression associated with multiple sclerosis. *Am J Psychiatry* 1990;147:1493–1497.

44. Denmark JC, Powell JD, McComb SG. Imipramine hydrochloride in parkinsonism. *Br J Clin Pract* 1961;15:523–524.

45. Laitinena L. Desipramine in treatment of Parkinson's disease. *Acta Neurol Scand* 1969;45:109–113.

46. Strang RR. Imipramine in the treatment of parkinsonism: a double blind placebo study. *BMJ* 1965;2:33–34.

47. Andersen J, Aabro E, Gulmann N, et al. Antidepressive treatment in Parkinson's disease. *Acta Neurol Scand* 1980;62:210–219.

48. Allain H, Cougnard, Neukirch HC. Selegiline in *de novo* parkinsonian patients: the French Multicenter Trial (FSMT). *Acta Neurol Scand Suppl* 1991;136:73–78.

49. Goetz CG, Tanner CM, Klawans HL. Bupropion in Parkinson's disease. *Neurology* 1984;34:1092–1094.

50. Peyser CE, Folstein SE. Huntington's disease as a model for mood disorders: clues from neuropathology and neurochemistry. *Mol Chem Neuropathol* 1990;12:99–119.

51. Moller HJ, Fuger J, Kasper S. Efficacy of new generation antidepressants: meta-analysis of imipramine-controlled studies. *Pharmacopsychiatry* 1994;27:215–223.

52. Montgomery SA, Henry J, McDonald G, et al. Selective serotonin reuptake inhibitors: meta analysis of discontinuation rates. *Int Clin Psychopharmacol* 1994;9:47–53.

53. Andrews JM, Nemeroff CB. Contemporary management of depression. *Am J Med* 1994;97:24S–32S.

54. Clerc GE, Ruimy P, Verdeau Palles J. A double-blind comparison of venlafaxine and fluoxetine in patients hospitalized for major depression and melancholia. The Venlafaxine French Inpatient Study Group. *Int Clin Psychopharmacol* 1994;9:139–143.

55. Nierenberg AA, Feighner JP, Rudolph R, et al. Venlafaxine for treatment resistant unipolar depression. *J Clin Psychopharmacol* 1944;14:419–423.

56. Jancar J, Gunaratne JJ. Dysthymia and mental handicap. *Br J Psychiatry* 1994;164:691–693.

57. Wilens TE, Biederman J, Mick E, et al. A systematic assessment of tricyclic antidepressants in the treatment of adult attention-deficit hyperactivity disorder. *J Nerv Ment Dis* 1995;183:48–50.

58. Aman MG. Efficacy of psychotropic drugs for reducing self-injurious behavior in the developmental disabilities. *Ann Clin Psychiatry* 1993;5:171–188.

59. Goodnick PJ, Sandoval R. Psychotropic treatment of chronic fatigue syndrome and related disorders. *J Clin Psychiatry* 1993;54:13–20.

60. Onghena P, Van Houdenhove B. Antidepressant induced analgesia in chronic non-malignant pain: a meta-analysis of 39 placebo controlled studies. *Pain* 1992;49:205–219.

61. Philipp M, Fickinger M. Psychotropic drugs in the management of chronic pain syndromes. *Pharmacopsychiatry* 1993;26:221–234.

62. Hollander E, Wong CM. Body dysmorphic disorder, pathological gambling, and sexual compulsions. *J Clin Psychiatry* 1995;56[Suppl 4]:7–12.

63. Kafka MP. Sertraline pharmacotherapy for paraphilias and paraphilia-related disorders: an open trial. *Ann Clin Psychiatry* 1994;6:189–195.

64. Sloan RL, Brown KW, Pentland B. Fluoxetine as a treatment of emotional lability after brain injury. *Brain Inj* 1992;6:315–319.

65. Panzer MJ, Mellow AM. Antidepressant treatment of pathologic laughing or crying in elderly stroke patients. *J Geriatr Psychiatry Neurol* 1992;5:195–199.

66. Sutherland SM, Davidson JR. Pharmacotherapy for post-traumatic stress disorder. *Psychiatr Clin North Am* 1994;17:409–423.

67. Sweeney S, Henry A. The use of imipramine in Tourette's syndrome and attention deficit disorder. *J Clin Psychiatry* 1994;46:348.

68. Stoudemire A, Fogel BS, Gulley L, et al. Psychopharmacology in the medical patient. In: Stoudemire A, Fogel S, eds. *Psychiatric care of the medical patient.* New York: Oxford University Press, 1993:155–206.

69. Cassano GB, Musetti L, Soriani A, et al. The pharmacologic treatment of depression: drug selection criteria. *Pharmacopsychiatry* 1995;26[Suppl 1]:17–23.

70. George MS, Trimble MR. A fluvoxamine-induced frontal lobe syndrome in a patient with comorbid Gilles de la Tourette's syndrome and obsessive compulsive disorder. *J Clin Psychiatry* 1992;53:379–380.

71. Hoehn SR, Harris GJ, Pearlson GD, et al. A fluoxetine induced frontal lobe syndrome in an obsessive compulsive patient. *J Clin Psychiatry* 1991;52:131–133.

72. Coulter DM, Pillans PI. Fluoxetine and extrapyramidal side effects. *Am J Psychiatry* 1995;152:122–125.

73. Kahn RS, David KL. New developments in dopamine and schizophrenia. In: Floom FE, Kupfer DJ, eds. *Psychopharmacology: the fourth generation of progress.* New York: Raven Press, 1995:1193–1204.

74. Lejoyeux M, Rouillon F, Ades J, et al. Neural symptoms induced by tricyclic antidepressants: phenomenology and pathophysiology. *Acta Psychiatr Scand* 1992;85:249–256.

75. Knegtering H, Eijck M, Huijsman A. Effects of antidepressants on cognitive functioning of elderly patients. A review. *Drugs Aging* 1994;5:192–199.

76. Ames D, Wirshing WC, Szuba MP. Organic mental disorders associated with bupropion in three patients. *J Clin Psychiatry* 1992;53:53–55.

77. Ansseau M, Darimont P, Lecoq A, et al. Controlled comparison of nefazodone and amitriptyline in major depressive inpatients. *Psychopharmacology (Berl)* 1994;115:254–260.

78. Pick CG, Paul D, Eison MS, et al. Potentiation of opioid analgesia by the antidepressant nefazodone. *Eur J Pharmacol* 1992;211:375–381.

79. Feighner JP. The role of venlafaxine in rational antidepressant therapy. *J Clin Psychiatry* 1994;55[Suppl A]:62–68.

80. Guelfi JD, Ansseau M, Timmerman L, et al., and the Mirtazapine-Venlafaxine Study Group. Mirtazapine versus venlafaxine in hospitalized severely depressed patients with melancholic features. *J Clin Psychopharmacol* 2000;21:425–431.

81. Anderson IM. Meta-analytical studies on new antidepressants. *Br Med Bull* 2001;57:161–178.

82. Kent JM. SNaRIs, NaSSAs, and NaRIs: new agents for the treatment of depression. *Lancet* 2000;355:911–918.

83. Jerling M, Bertilsson L, Sjoqvist F. The issue of therapeutic drug monitoring data to document kinetic drug interactions: an example with amitriptyline and nortriptyline. *Ther Drug Monit* 1994;16:1–12.

84. Von Moltke LL, Greenblatt DJ, Cotreau Bibbo MM, et al. Inhibition of desipramine hydroxylation in vitro by serotonin reuptake inhibitor antidepressants and by quinidine and ketoconazole: a model system to predict drug interactions in vivo. *J Pharmcol Exp Ther* 1994;268:1278–1283.

85. Jerling M, Lindstrom L, Bondesson U, et al. Fluvoxamine inhibition and carbamazepine induction of the metabolism of clozapine: evidence from a therapeutic drug monitoring service. *Ther Drug Monit* 1994;16:368–374.

86. Perucca E, Gatti G, Spina E. Clinical pharmacokinetics of fluvoxamine. *Clin Pharmcokinet* 1994;27:175–190.

87. DeVane CL, Ware MR, Lydiard RB. Pharmacokinetics, pharmacodynamics, and treatment issues of benzodiazepines: alprazolam, adinazolam, and clonazepam. *Psychopharmacol Bull* 1991;27:463–473.

88. Thomas KV, Holimon TD. Citalopram versus other selective serotonin-reuptake inhibitors. *Am J Health Syst Pharm* 1999;56:2242–2244.

89. Keller MB. Citalopram therapy for depression: a review of 10 years of European experience and data from U.S. clinical trials. *J Clin Psychiatry* 2000;61:896–908.

90. Parker NG, Brown CS. Citalopram in the treatment of depression. *Ann Pharmacother* 2000;34:761–771.

91. Brathwaite RA, Flanagan RA, Richens A. Steady state plasma nortriptyline concentrations in epileptic patients. *Br J Clin Pharmacol* 1975;2:469–471.

92. Brosen K, Kragh Sorensen P. Concomitant intake of nortriptyline and carbamazepine. *Ther Drug Monit* 1993;15:258–260.

93. Akiskal HS. Dysthymic and cyclothymic depressions: therapeutic considerations. *J Clin Psychiatry* 1994;55[Suppl]:46–52.

94. Wragg RE, Jeste DV. Neuroleptics and alternative treatments: management of behavioral symptoms and psychosis in Alzheimer's disease and related conditions. *Psychiatr Clin North Am* 1988;11:195–213.

95. Schneider LS, Pollock VE, Lyness SA. A metaanalysis of controlled trials of neuroleptic treatment in dementia. *J Am Geriatric Soc* 1990;38:553–563.

96. Feinstein C, Leroy D. Pharmacotherapy of severe psychiatric disorders in mentally retarded individuals. In: Stoudemire A, Fogel BS, eds. *Medical psychiatric practice, vol. 1.* Washington, DC: American Psychiatric Press, 1991:501–537.

97. Daniel DG. Antipsychotic treatment of psychosis and agitation in the elderly. *J Clin Psychiatry* 2000;61[Suppl 14]:49–52.

98. Aman MG, Singh NN. A critical appraisal of recent drug research in mental retardation: the Coldwater studies. *J Ment Defic Res* 1988;30:203–216.

99. Vanden Borre R, Vermote R, Buttiens M, et al. Risperidone as add-on therapy in behavioral disturbances in mental retardation: a double-blind placebo controlled cross over study. *Acta Psychiatr Scand* 1993;87:167–171.

100. McDougle CJ, Holmes JP, Carlson DC, et al. A double-blind, placebo-controlled study of risperidone in adults with autistic disorder and other pervasive developmental disorders. *Arch Gen Psychiatry* 1998;55:633–641.

101. Katz IR, Jeste DV, Mintzes JE, et al. Comparison of risperidone and placebo for psychosis and behavioral disturbances associated with dementia: a randomized, double-blind trial. *J Clin Psychiatry* 1999;60:107–115.

102. Shapiro AK, Shapiro E. Neuroleptic drugs. In: Kurlan R, ed. *Handbook of Tourette's syndrome and related tic and behavioral disorders.* New York: Marcel Dekker, 1993:347–377.

103. Cohen DJ. The pathology of the self in primary childhood autism and Gilles de la Tourette syndrome. *Psychiatr Clin North Am* 1980;3:383–402.

104. Locascio JJ, Malone RP, Small AM, et al. Factors related to haloperidol response and dyskinesias in autistic children. *Psychopharmacol Bull* 1991;27:119–126.

105. Campbell M, Adams P, Perry R, et al. Tardive and withdrawal dyskinesia in autistic children: a prospective study. *Psychopharmacol Bull* 1988;24:251–255.

106. Morris M, Tyler A. Management and therapy. In: Harper PS, ed. *Huntington's disease.* Philadelphia: WB Saunders, 1991:205–250.

107. Lipowski ZJ. *Delirium: acute confusional states.* New York: Oxford University Press, 1990.

108. Mori E, Yamadori A. Acute confusional state and acute agitated delirium. Occurrence after infection in the right middle cerebral artery. *Arch Neurol* 1987;44:1139–1143.

109. Rich SS, Friedman JH. Treatment of psychosis in Parkinson's disease. In: Stoudemire A, Fogel BS, eds. *Medical psychiatric practice.* Washington, DC: American Psychiatric Press, 1995:151–182.

110. Friedman JH, Factor SA. Atypical antipsychotics in the treatment of drug-induced psychosis in Parkinson's disease. *Mov Disord* 2000;15:201–211.

111. The Parkinson Study Group. Low-dose clozapine for the treatment of drug-induced psychosis in Parkinson's disease. *N Engl J Med* 1999;340:757–763.

112. The French Clozapine Parkinson Study Group. Clozapine in drug-induced psychosis in Parkinson's disease. *Lancet* 1999;353:2041–2042.

113. Juncos JL. Management of psychotic aspects of Parkinson's disease. *J Clin Psychiatry* 1999;60[Suppl 8]:42–53.

114. McDougle CJ, Price LH, Goodman WK. Fluvoxamine treatment of coincident autistic disorder and obsessive-compulsive disorder: a case report. *J Autism Dev Disord* 1990;20:537–543.

115. McDougle CJ, Goodman WK, Price LH, et al. Neuroleptic addition in fluvoxamine-refractory obsessive-compulsive disorder. *Am J Psychiatry* 1990;147:652–654.

116. McDougle CJ, Goodman WK, Leckman JF, et al. Haloperidol addition to fluvoxamine-refractory obsessive-compulsive disorder: a double-blind, placebo-controlled study in patients with and without tics. *Arch Gen Psychiatry* 1994;51:302–308.

117. Chouinard G, Sultan S. Treatment of supersensitivity psychosis with antiepileptic drugs: report of a series of 43 cases. *Psychopharmacol Bull* 1990;26:337–341.

118. Luchins DJ, Oliver AP, Wyatt RJ. Seizures with antidepressants: an in vitro technique to assess relative risk. *Epilepsia* 1984;25:25–32.

119. Kane JM. Tardive dyskinesia: epidemiological and clinical presentation. In: Bloom FE, Kupfer DJ, eds. *Psychopharmacology: the fourth generation of progress.* New York: Raven Press, 1995:1485–1495.

120. Balant-Gorgia AE, Balant LP, Andreoli A. Pharmacokinetic optimisation of the treatment of psychosis. *Clin Pharmacokinet* 1993;25:217–236.

121. Javaid JI. Clinical pharmacokinetics of antipsychotics. *J Clin Pharmacol* 1994;34:286–295.

122. Daniel W, Janczar L, Danek L, et al. Pharmacokinetic interaction between carbamazepine and neuroleptics after combined prolonged treatment in rats. *Naunyn Schmiedebergs Arch Pharmacol* 1992;345:598–605.

123. Glassman AH, Bigger JT. Antipsychotic drugs: prolonged QTc interval, torsade de pointes, and sudden death. *Am J Psychiatry* 2001;158:1774–1782.

124. *Physicians' Desk Reference.* Orap (pimozide), 56th ed. Montvale, NJ: Medical Economics, 2002:1407–1409.

125. Awad AG. Subjective response to neuroleptics in schizophrenia. *Schizophr Bull* 1993;19:609–618.

126. Richelson E. Receptor pharmacology of neuroleptics: relation to clinical effects. *J Clin Psychiatry* 1999;60[Suppl 10]:5–14.

127. Schooler NR. Negative symptoms in schizophrenia: assessment of the effect of risperidone. *J Clin Psychiatry* 1994;55[Suppl]:22–28.

128. Carnahan RM, Lund BC, Perry PJ. Ziprasidone, a new atypical antipsychotic drug. *Pharmacotherapy* 2001;21:717–730.

129. McIntyre RS, McCann SM, Kennedy SH. Antipsychotic metabolic effects: weight gain, diabetes mellitus, and lipid abnormalities. *Can J Psychiatry* 2001;46:273–281.

130. Sussman N. Review of atypical antipsychotics and weight gain. *J Clin Psychiatry* 2001;62[Suppl 23]:5–12.

131. Allison DB, Mentore JL, Heo M, et al. Antipsychotic-induced weight gain: a comprehensive research synthesis. *Am J Psychiatry* 1999;156:1686–1696.

132. Gerlach J, Peacock L. Motor and mental side effects of clozapine. *J Clin Psychiatry* 1994;55[Suppl B]:107–109.

133. Chengappa KN, Shelton MD, Baker RW, et al. The prevalence of akathisia in patients receiving stable doses of clozapine. *J Clin Psychiatry* 1994;55:142–145.

134. Friedman JH. Clozapine treatment of psychosis in patients with tardive dystonia: report of three cases. *Mov Disord* 1994;9:321–324.

135. Centorrino F, Price BH, Tuttle M, et al. EEG abnormalities during treatment with typical and atypical antipsychotics. *Am J Psychiatry* 2002;159:109–115.

136. Lindstrom LH. Long-term clinical and social outcome studies in schizophrenia in relation to the cognitive and emotional side effects of antipsychotic drugs. *Acta Psychiatr Scand* 1994;380:74–76.

137. Safferman AZ, Kane JM, Aronowitz JS, et al. The use of clozapine in neurologic disorders. *J Clin Psychiatry* 1994;55[Suppl B]:98–101.

138. Bezchlibnyk-Butler KZ, Remington GJ. Antiparkinsonian drugs in the treatment of neuroleptic-induced extrapyramidal symptoms. *Can J Psychiatry* 1994;39:74–84.

139. Thapa PB, Meador KG, Dieon P, et al. Effects of antipsychotic withdrawal in elderly nursing home residents. *J Am Geriatr Soc* 1994;42:280–286.

140. Shader RI, Greenblatt DJ. Use of benzodiazepines in anxiety disorders. *N Engl J Med* 1993;328:1398–1405.

141. Woods JH, Katz JL, Winger G. Benzodiazepine use, abuse, and consequences. *Pharmacol Rev* 1992;44:151–347.

142. Jann MW, Kurtz NM. Treatment of panic and phobic disorders. *Clin Pharmacol* 1987;6:947–962.

143. Tarsy D. Restless legs syndrome. In: Joseph AB, Young RR, eds. *Movement disorders in neurology and neuropsychiatry.* Boston: Blackwell Scientific Publications, 1992:397–400.

144. Ronthal M. Myoclonus and asterixis. In: Joseph AB, Young RR, eds. *Movement disorders in neurology and neuropsychiatry.* Boston: Blackwell Scientific Publications, 1992:479–486.

145. Joseph AB. Catatonia. In: Joseph AB, Young RR, eds. *Movement disorders in neurology and neuropsychiatry.* Boston: Blackwell Scientific Publications, 1992:335–342.

146. Ricketts RW, Goza AB, Ellis CR, et al. Clinical effects of buspirone on intractable self-injury in adults with mental retardation. *J Am Acad Child Adolesc Psychiatry* 1994;33:270–276.

147. Stanislav SW, Fabre T, Crismon ML, et al. Buspirone's efficacy in organic induced aggression. *J Clin Psychopharmacol* 1994;14:126–130.

148. Ratey J, Sovner R, Parks A, et al. Buspirone treatment of aggression and anxiety in mentally retarded patients: a multiple-baseline, placebo lead in study. *J Clin Psychiatry* 1991;52:159–162.

149. Moss LE, Neppe M, Drevets WC. Buspirone in the treatment of tardive dyskinesia. *J Clin Psychopharmacol* 1993;13:204–209.

150. Littman AB, Fava M, McKool K, et al. Buspirone therapy for type A behavior, hostility, and perceived stress in cardiac patients. *Psychother Psychosom* 1993;59:107–110.

151. Salzman C. Treatment of anxiety. In: Salzman C, ed. *Clinical geriatric psychopharmacology,* 2nd ed. Baltimore: Williams & Wilkins, 1992:189–212.

152. Arana GW, Epstein S, Molloy M, et al. Carbamazepine-induced reduction of plasma alprazolam concentrations: a clinical case report. *J Clin Psychiatry* 1988;49:448–449.

153. Tuncok Y, Akpina R, Guven H, et al. The effects of theophylline on serum alprazolam levels. *Int J Clin Pharmacol Ther* 1994;32:642–645.

154. Lai AA, Levy RH, Cutler RE. Time course of interaction between carbamazepine and clonazepam in normal man. *Clin Pharmacol Ther* 1978;24:316–323.

155. Manji HK, Chen G, Hsiao JK, et al. Regulation of signal trans-duction pathways by mood-stabilizing agents. In: Manji HK, Bowden CL, Belmaker RH, eds. *Bipolar medications: mechanisms of action.* Washington, DC: American Psychiatric Press, 2000:129–177.

156. Lenox RH, Manji HK. Lithium. In: Schatzberg AF, Nemeroff CB, eds. *Textbook of psychopharmacology.* Washington, DC: American Psychiatric Press, 1995:303–358.

157. Goodwin FK, Jamison KR. *Manic depressive illness.* New York: Oxford University Press, 1990.

158. Schou M. Use in other psychiatric conditions. In: Johnson FN, ed. *Depression and mania: modern lithium therapy.* Oxford: IRL Press, 1987:44–50.

159. Dale PG. Lithium therapy in aggressive mentally subnormal patients. *Br J Psychiatry* 1980;137:469–474.

160. Smith DA, Perry PJ. Nonneuroleptic treatment of disruptive behavior in organic mental syndromes. *Ann Pharmacother* 1992;26:1400–1408.

161. Campbell M, Small AM, Green WH, et al. Behavioral efficacy of haloperidol and lithium carbonate–a comparison in hospitalized aggressive children with conduct disorder. *Arch Gen Psychiatry* 1984;41:650–656.

162. Sheard MH, Marini JL, Bridges C, et al. The effects of lithium in impulsive aggressive behavior in man. *Am J Psychiatry* 1976;133:1409–1413.

163. Haas JF, Cope N. Neuropharmacologic management of behavior sequelae in head injury: a case report. *Arch Phys Med Rehabil* 1985;66:472–474.

164. Luchins DF, Dojka D. Lithium and propranolol in aggression and self-injurious behavior in the mentally retarded. *Psychopharmacol Bull* 1989;25:372–375.

165. Leonard DP, Kidson MA, Shannon PJ, et al. Double-blind trial of lithium carbonate and haloperidol in Huntington's chorea. *Lancet* 1974;2:1208–1209.

166. Morris M, Tyler A. Management and therapy. In: Harper PS, ed. *Huntington's disease.* Philadelphia: WB Saunders, 1991:205–250.

167. Bussone G, Leone M, Peccarisi C, et al. Double-blind comparison of lithium and verapamil in cluster headache prophylaxis. *Headache* 1990;30:411–417.

168. Delgado PL, Gelenberg AJ. Antidepressant and antimanic medication. In: Gabbard GO, ed. *Treatments of psychiatric disorders.* Washington DC: American Psychiatric Press, 1995:1132–1168.

169. Lazarus JH. *Endocrine and metabolic effects of lithium.* New York: Plenum, 1986.

170. McElroy SL, Keck PE. Antiepileptic drugs. In: Schatzberg AF, Nemeroff CB, eds. *Textbook of psychopharmacology.* Washington, DC: American Psychiatric Press, 1995:351–376.

171. Stoll AL, Banov M, Kolbrener M, et al. Neurologic factors predict a favorable valproate response in bipolar and schizoaffective disorders. *J Clin Psychopharmacol* 1994;14:311–313.

172. Fogel BS, Duffy J. Elderly patients. In: Silver JM, Yudofsky SC, Hales RE, eds. *Neuropsychiatry of traumatic brain injury.* Washington, DC: American Psychiatric Press, 1994:413–442.

173. Silver JM, Yudofsky SC. Aggressive disorders. In: Silver JM, Yudofsky SC, Hales RE, eds. *Neuropsychiatry of traumatic brain injury.* Washington, DC: American Psychiatric Press, 1994:313–356.

174. Gualtieri CT. *Neuropsychiatry and behavioral pharmacology.* New York: Springer-Verlag, 1991.

175. Calabrese JR, Bowden CL, Sachs GS, et al., for the Lamictal 602 Study Group. A double-blind placebo-controlled study of lamotrigine monotherapy in outpatients with bipolar I depression. *J Clin Psychiatry* 1999;60:79–88.

176. Frye MA, Ketter TA, Kimbrell TA, et al. A placebo controlled

study of lamotrigine and gabapentin monotherapy in refractory mood disorders. *J Clin Psychopharmacol* 2000;20:607–614.

177. Nemeroff CB. An ever-increasing pharmacopoeia for the management of patients with bipolar disorder. *J Clin Psychiatry* 2000;61[Suppl 13]:19–25.

178. Bowden CL. Advances in the treatment of bipolar disorder. *Essential Psychopharmacol* 2001;4:215–231.

179. Pande AC, Crockatt JG, Janney CA, et al., Gabapentin Bipolar Disorder Study Group. Gabapentin in bipolar disorder: a placebo-controlled trial of adjunctive therapy. *Bipolar Disord* 2000;2:249–255.

180. Keck PE, McElroy SL, Friedman LM. Valproate and carbamazepine in the treatment of panic and posttraumatic stress disorders, withdrawal states, and behavioral dyscontrol syndromes. *J Clin Psychopharmacol* 1992;12:36S–41S.

181. Fesler FA. Valproate in combat-related posttraumatic stress disorder. *J Clin Psychiatry* 1991;52:361–364.

182. Silver JM, Sandberg DP, Hales RE. New approaches in the pharmacotherapy of posttraumatic stress disorder. *J Clin Psychiatry* 1990;51[Suppl]:33–38.

183. McQuay H, Carroll D, Jadad AR, et al. Anticonvulsant drugs for management of pain: a systematic review. *BMJ* 1995;311:1047–1052.

184. Guiec R, Mesdjian E, Rochat H, et al. Central analgesic effect of valproate in patients with epilepsy. *Seizure* 1993;2:147–150.

185. Tremont-Lukats IW, Megeff C, Backonja MM. Anticonvulsants for neuropathic pain syndromes: mechanisms of action and place in therapy. *Drugs* 2000;60:1029–1052.

186. Lemke MR. Effect of carbamazepine on agitation in Alzheimer's in patients refractory to neuroleptics. *J Clin Psychiatry* 1995;56:354–357.

187. Tariot PN, Erb R, Leibovici A, et al. Carbamazepine treatment of agitation in nursing home patients with dementia: a preliminary study. *J Am Geriatr Soc* 1994;42:1160–1166.

188. Young JL, Hillbrand M. Carbamazepine lowers aggression: a review. *Bull Am Acad Psychiatry Law* 1994;22:53–61.

189. Barratt ES. The use of anticonvulsants in aggression and violence. *Psychopharmacol Bull* 1993;29:75–81.

190. Lewin J, Sumners D. Successful treatment of episodic dyscontrol with carbamazepine. *Br J Psychiatry* 1992;161:261–262.

191. Lott AD, McElroy SL, Keys MA. Valproate in the treatment of behavioral agitation in elderly patients with dementia. *J Neuropsychiatry Clin Neurosci* 1995;7:314–319.

192. Geracioti TD. Valproic acid treatment of episodic explosiveness related to brain injury. *J Clin Psychiatry* 1994;55:416–417.

193. Mazure CM, Druss BG, Cellar JS. Valproate treatment of older psychotic patients with organic mental syndromes and behavioral dyscontrol. *J Am Geriatr Soc* 1992;40:914–916.

194. Mellow AM, Solano Lopez C, Davis S. Sodium valproate in the treatment of behavioral disturbance in dementia. *J Geriatr Psychiatry Neurol* 1993;6:205–209.

195. Varney NR, Garvey MJ, Cook BL, et al. Identification of treatment resistant depressives who respond favorably to carbamazepine. *Ann Clin Psychiatry* 1993;5:117–122.

196. Winchel RM, Stanley M. Self-injurious behavior: a review of the behavior and biology of self-mutilation. *Am J Psychiatry* 1991;148:306–317.

197. Schweizer E, Rickels K, Case WG, et al. Carbamazepine treatment in patients discontinuing long-term benzodiazepine therapy. Effects on withdrawal severity and outcome. *Arch Gen Psychiatry* 1991;48:448–452.

198. Kastner T, Finesmith R, Walsh K. Long-term administration of valproic acid in the treatment of affective symptoms in people with mental retardation. *J Clin Psychopharmacol* 1993;13:448–451.

199. Pellock JM. The clinical efficacy of lamotrigine as an antiepileptic drug. *Neurology* 1994;44:S29–S35.

200. Wallace SJ. Newer antiepileptic drugs: advantages and disadvantages. *Brain Dev* 2001;23:277–283.

201. Ferrendelli JA. Concerns with antiepileptic drug initiation: safety, tolerability, and efficacy. *Epilepsia* 2001;42[Suppl 4]:28–30.

202. Glauser TA. Topiramate. *Epilepsia* 1999;40[Suppl 5]:S71–S80.

203. Sabers A, Gram L. Newer anticonvulsants. *Drugs* 2000;60:23–33.

204. Tecoma ES. Oxcarbazepine. *Epilepsia* 1999;40[Suppl 5]:S37–S46.

205. Yau MK, Wargin WA, Wolf KB, et al. Effect of valproate on the pharmacokinetics of lamotrigine at steady state. *Epilepsia* 1992;33[Suppl 3]:82.

206. Perucca E. The clinical pharmacokinetics of the newer antiepileptic drugs. *Epilepsia* 1999;40[Suppl 9]:S7–S13.

207. Howard JR, Dix RK, Shumaker RC, et al. Effect of felbamate on carbamazepine pharmacokinetics. *Epilepsia* 1992;33[Suppl 3]:84–85.

208. McLean MJ. Gabapentin: chemistry, absorption, distribution, and excretion. In: Levy RH, Mattson RH, Meldrum BS, eds. *Antiepileptic drugs*, 4th ed. New York: Raven Press, 1995:843–849.

209. Levy RH, Mattson RH, Meldrum BS, eds. *Antiepileptic drugs*, 4th ed. New York: Raven Press, 1995.

210. Laidlaw J, Richens A, Chadwick D, eds. *A textbook of epilepsy*, 4th ed. Edinburgh: Churchill Livingstone, 1993.

211. Saletu B, Grunberger J, Linzmayer L. Evaluation of encephalotropic and psychotropic properties of gabapentin in man by pharmaco-EEG and psychometry. *Int J Clin Pharmacol Ther Toxicol* 1986;24:364–373.

212. Smith D, Baker G, Davies G, et al. Outcomes of add-on treatment with lamotrigine in partial epilepsy. *Epilepsia* 1993;34:312–322.

213. Kaplitz SE. Withdrawn, apathetic geriatric patients responsive to methylphenidate. *J Am Geriatric Soc* 1975;23:271–276.

214. Wilens TE, Biederman J. The stimulants. *Psychiatr Clin North Am* 1992;15:191–222.

215. Mitler MM. Evaluation of treatment with stimulants in narcolepsy. *Sleep* 1993;17[Suppl 8]:S103–S106.

216. Wallace AE, Kofoed LL, West AN. Double blind, placebo controlled trial of methylphenidate in older, depressed, medically ill patients. *Am J Psychiatry* 1995;152:929–931.

217. Lazarus LW, Moberg PJ, Langsley PR, et al. Methylphenidate and nortriptyline in the treatment of poststroke depression: a retrospective comparison. *Arch Phys Med Rehabil* 1994;75:403–406.

218. Mayes SD, Crites DL, Bixler EO, et al. Methylphenidate and ADHD: influence of age, IQ and neurodevelopmental status. *Dev Med Child Neurol* 1994;36:1099–1107.

219. Marin R, Fogel BS, Hawkins J, et al. Apathy: a treatable disorder. *J Neuropsychiatr Clin Neurosci* 1995;7:23–30.

220. Pritchard JG, Mykyta LJ. Use of a combination of methylphenidate and oxprenolol in the management of physically disabled, apathetic, elderly patients: a pilot study. *Curr Med Res Opin* 1975;3:26–29.

221. Chiarello RJ, Cole JO. The use of psychostimulants in general psychiatry. *Arch Gen Psychiatry* 1987;44:286–295.

222. Lipkin PH, Goldstein IJ, Adesman AR. Tics and dyskinesias associated with stimulant treatment in attention deficit hyperactivity disorder. *Arch Pediatr Adolesc Med* 1994;148:859–861.

223. Biederman J, Spencer T, Wilens TE, et al. Attention deficit hyperactivity disorder: pharmacotherapy. In: Gabbard G, ed. *Treatment*

of psychiatric disorders. Washington, DC: American Psychiatric Press, 1995.

224. Wells BG, Marken PA. Bromocriptine in the treatment of depression DICP. *Ann Pharmacother* 1989;23:600–601.

225. Bouckoms A, Manigini L. Pergolide: an antidepressant adjuvant for mood disorders? *Psychopharmacol Bull* 1993;29:207–211.

226. Borison RI. Amantadine in the management of extrapyramidal side effects. *Clin Neuropharmacol* 1983;6:557–563.

227. Chandler MC, Barnhill JB, Gualtieri CT. Amantadine for the agitated head injury patient. *Brain Inj* 1988;2:309–311.

228. Gualtieri CT, Chandler M, Coons T, et al. Amantadine: a new clinical profile for traumatic brain injury. *Clin Neuropharmacol* 1989;12:258–270.

229. Ross ED, Stewart RM. Akinetic mutism from hypothalamic damage: successful treatment with dopamine agonists. *Neurology* 1981;31:1435–1439.

230. Fleet WS, Valenstein E, Watson RT, et al. Dopamine agonist therapy for neglect in humans. *Neurology* 1987;37:1765–1770.

231. Dobkin BH, Hanlon R. Dopamine agonist treatment of anterograde amnesia from a mediobasal forebrain injury. *Ann Neurol* 1993;33:313–316.

232. Albert ML, Bachman DL, Morgan A, et al. Pharmacotherapy for aphasia. *Neurology* 1988;38:877–879.

233. Ziegler DK, Hurwitz A, Hassanein RS, et al. Migraine prophylaxis: a comparison of propranolol and amitriptyline. *Arch Neurol* 1987;44:486–489.

234. Larsen TA, Teravainen H. Beta I versus nonselective blockade in therapy of essential tremor. *Adv Neurol* 1983;37:247–251.

235. Leckman JF, Knorr AM, Rasmusson AM, et al. Basal ganglia research and Tourette's syndrome. *Trends Neurosci* 1991;14:94.

236. Jasinski DR, Johnson RE, Kocher TR. Clonidine in morphine withdrawal: differential effects on signs and symptoms. *Arch Gen Psychiatry* 1985;42:1063–1065.

237. Hunt RD. Treatment effects of oral and transdermal clonidine in relation to methylphenidate: an open pilot study in ADD-H. *Psychopharmacol Bull* 1987;23:111–114.

238. Steingard RJ, Biederman J, Spencer T, et al. Comparison of clonidine response in the treatment of attention deficit hyperactivity disorder with and without comorbid tic disorders. *J Am Acad Child Adolesc Psychiatry* 1993;32:350–353.

239. Ratey JJ, Greenberg MS, Lindem DJ. Combination of treatments for attention deficit hyperactivity disorder in adults. *J Nerv Ment Dis* 1991;179:699–701.

240. Hunt RD, Arnsten AF, Asbell MD. An open trial of guanfacine in the treatment of attention deficit hyperactivity disorder. *J Am Acad Child Adolesc Psychiatry* 1995;34:50–54.

241. Schneider LS, Sobin PB. Non-neuroleptic treatment of behavioral symptoms and agitation in Alzheimer's disease and other dementias. *Psychopharmacol Bull* 1992;28:71–79.

242. Williams DT, Mehl R, Yudofsky S, et al. The effect of propranolol on uncontrolled rage outbursts in children and adolescents with organic brain dysfunction. *J Am Acad Child Psychiatry* 1982;21:129–135.

243. Yudofsky SC, Silver JM, Schneider SE. Pharmacologic treatment of aggression. *Psychiatr Ann* 1987;17:397–407.

244. Adler LA, Angrist B, Reiter S, et al. Neuroleptic induced akathisia: a review. *Psychopharmacology* 1989;97:1–11.

245. Fleischhacker W, Roth SD, Kane JM. The pharmacologic treatment of neuroleptic-induced akathisia. *J Clin Psychopharmacol* 1990;10:12–21.

246. Kurlan R, Majumdar L, Deeley C, et al. A controlled trial of propoxyphene and naltrexone in patients with Tourette's syndrome. *Ann Neurol* 1991;30:19–23.

247. Sandman CA. The opiate hypothesis in autism and self injury. *J Child Adolesc Psychopharmacol* 1991;1:237–248.

248. Gillberg C. Endogenous opioids and opiate antagonists in autism: brief review of empirical findings and implications for clinicians. *Dev Med Child Neurol* 1995;37:239–245.

249. Herman BH, Hammock MK, Egan J, et al. Role for opioid peptides in self-injurious behavior: dissociation from autonomic nervous system functioning. *Dev Pharmacol Ther* 1989;12:81–89.

250. Gillberg C. Endogenous opioids and opiate antagonists in autism: brief review of empirical findings and implications for clinicians. *Dev Med Child Neurol* 1995;37:239–245.

251. Thompson T, Hackenberg T, Cerutti D, et al. Opioid antagonist effects on self-injury in adults with mental retardation: response form and location as determinants of medication effects. *Am J Ment Retard* 1994;99:85–102.

252. Campbell M, Anderson LT, Small AM, et al. Naltrexone in autistic children: a double-blind and placebo-controlled study. *Psychopharmacol Bull* 1990;26:130–135.

253. Kolman D. The use of opiate antagonists in treatment of bulimia: a study of low-dose versus high-dose naltrexone. *Psychiatry Res* 1988;24:195–199.

254. Lensing P, Klingler D, Lampl C, et al. Naltrexone open trial with a 5 year old boy. A social rebound reaction. *Acta Paedopsychiatr* 1992;55:169–173.

255. Zingarelli G, Ellman G, Hom A, et al. Clinical effects of naltrexone on autistic behavior. *Am J Ment Retard* 1992;97:57–63.

256. Jonas JM, Gold MS. The use of opiate antagonists in treating bulimia: a study of low-dose versus high-dose naltrexone. *Psychiatry Res* 1988;24:195–199.

257. Chatoor I, Herman BH, Hartzler J. Effects of the opiate antagonist, naltrexone, on binging antecedents and plasma beta-endorphin concentrations. *J Am Acad Child Adolesc Psychiatry* 1994;33:748–752.

258. Benjamin E, Buot-Smith T. Naltrexone and fluoxetine in Prader-Willi syndrome. *J Am Acad Child Adolesc Psychiatry* 1993;32:870–873.

7

COGNITIVE, BEHAVIORAL, AND SELECTED PHARMACOLOGIC INTERVENTIONS IN REHABILITATION AFTER ACQUIRED BRAIN INJURY

CATHERINE A. MATEER
NICHOLAS M. BOGOD

Brain injury and disease can result in a wide range of physical, sensory, cognitive, and behavioral impairments. Cognitive and behavioral impairments are among the most common and disruptive sequelae of brain injury, particularly with respect to their impact on adaptive functioning and are commonly the focus of rehabilitative efforts. Impairments in orientation, attention, memory, communication, visuospatial processing, reasoning, insight, and judgment are common cognitive sequelae of many forms of brain insult. Problems with emotional regulation, stability of mood, and behavioral dyscontrol are other common neuropsychiatric sequelae of brain damage and disease. Diminished self-awareness and appreciation of the nature and impact of deficits is a frequent problem (1). The goals of cognitive, behavioral, and pharmacologic approaches to rehabilitation are to guide and enhance the recovery of a person's capacity to process and interpret information and to improve his or her ability to self-regulate and adapt emotionally to residual changes in function. Outcomes focus on increasing adaptive functioning of the individual in all aspects of home, family, and community life.

The history of rehabilitation for individuals with brain injury began in Europe when pioneers of behavioral neurology such as Alexander Luria in Russia and Kurt Goldstein in Germany began systematic programs of treatment and research with soldiers who were injured in the First World War. Many of our ideas about important brain functions, such as those of the frontal lobes, grew out of the observations these neurologists made in their work with men who had sustained war-related traumatic brain injury (TBI). Similarly, in North America, physical therapists, occupational therapists, and speech language pathologists actively developed neurorehabilitation services within specialized physical medicine and rehabilitation units. During peacetime, rehabilitation efforts tended to be targeted mainly toward indi-

viduals with cerebrovascular accidents for whom services focused on recovery of motor functions, speech and language, and perceptual abilities. Cognitive rehabilitation gained momentum during the 1970's with advances in emergency and trauma care, and the consequent increase in survival from even severe TBI. Due to advances in neuropsychology and behavioral neurology, there was increasing recognition of the nature of cognitive and behavioral problems associated with various forms of brain injury. Impairments in attention, memory, executive functions, and self-regulation of behavior and mood often posed marked impediments when individuals returned to home and community, even when physical recovery appeared to be good. Cognitive rehabilitation was developed within existing services of neuropsychology, speech language pathology, and occupational therapy programs, or as stand-alone but integrated day-treatment community integration programs. At the present time, 95% of rehabilitation facilities serving persons with brain injury provide some form of cognitive rehabilitation, involving individual, group, and/or community-based therapies. As with any emerging discipline, controversies about the nature of the service, its method of delivery, its goals, and its outcomes have been inevitable. However, there has been a growing body of published research, review, and opinion that continues to shape, define, and examine best practices and to define and evaluate the efficacy and functional impact of cognitively oriented interventions.

Cognition includes the ability to pay attention to relevant information; to understand, acquire, and retain new information; and to organize one's behavior to meet functional objectives and goals. *Cognitive rehabilitation* is defined as a systematic, functionally oriented set of therapeutic activities that are designed to facilitate, improve, and/or compensate for cognitive impairments so as to improve everyday adaptive abilities and functioning. The following sections discuss mechanisms of recovery, challenges to rehabilitation research, intervention models, and provide a brief review of approaches to working with impairments in several commonly impaired cognitive and behavioral domains.

PHARMACOLOGIC MANAGEMENT OF ACUTE BRAIN INJURY

Treatment of acute brain injury can be divided into two broad categories: attempts to minimize the damage and disability resulting from the initial insult; and attempts to encourage awakening and arousal in comatose patients.

Damage Reduction

Acute treatment represents an area of significant interest because agents that reduce the ultimate quantity of affected brain tissue operate within a unique temporal window to moderate the potential degree of disability that will result from the insult/injury. Acute treatment of significant brain injury generally demands ventilation, control of perfusion pressures, and prophylactic anticonvulsant treatment. Sodium concentrations in the blood are kept above a certain level, and hypoglycemia and hyperglycemia are avoided. Beyond these standard interventions, little has been established thus far in terms of the efficacy of pharmacologic interventions in reducing damage after brain injury (2,3). Observations of dramatic recovery in brain anoxia involving cold-water drowning have prompted scientists to examine the neuroprotective effects of moderate hypothermia in brain injury recovery, whereas others have focused on corticosteroids/cerebral edema, apoptotic cell death, excitatory amino acids (e.g., glutamate), and the effects of free radicals (2–8).

Corticosteroids have been used for almost 30 years to aid in the reduction of intracranial pressure. However, in a systematic review of all relevant trials of any type of steroid treatment from 1966 through 1995, Alderson and Roberts (9) found inconsistent and inconclusive benefits. Although the use of steroids remains in question, Yates and Roberts (10) point out that they continue to be used routinely in as many as 50% of patients with severe brain injury. In 1999 Kmietowicz (11) reported that a multicenter randomized trial of 20,000 patients was being initiated to demonstrate the use of steroid treatment in acute TBI; the results are not yet available.

Recent investigations into the mechanisms of cell death have suggested that cells die from one of two mechanisms: necrosis or apoptosis (6). Cells that suffer fatal structural damage due to trauma and/or severe hypoxia die within minutes to hours of necrotic cell death; cells that suffer only moderate hypoxia experience increased cellular toxicity that activates cell death cysteine proteases (caspases) that result in cell death in hours to days (5,6). Thus, damage in less injured tissues need not result in cell death but generally does over

the ensuing days. Preliminary research suggests that timely reperfusion may prevent apoptosis, but also suggests that caspase inhibition may prove to be an important neuroprotective therapy (6,12). Caspase inhibition has promise in both acute brain damage and progressive neurologic conditions such as Huntington, Alzheimer, and Parkinson disease (6). An interesting finding is that treatment strategies aimed at preventing necrotic cell death in acute TBI appear to increase apoptotic cell death, suggesting that combination therapies selected to both reduce necrotic death and block apoptosis may be necessary to achieve full benefit in this area (5).

The amino acid glutamate is the primary excitatory neurotransmitter discovered to date in the human brain. Excess extracellular glutamate resulting from cell damage and/or death is a primary factor in the cascade of excitotoxic death experienced by cells adjacent to damaged areas, and its excitotoxic effects have been linked to activation of excitatory amino acid receptors for many years (13). More recently, research has demonstrated that dysfunction in glutamate transport by neurons and astroglia may contribute to cell death by glutamate excitotoxicity (7). Maragakis and Rothstein (7, p. 368) report that under ischemic conditions, the ion pump that maintains glutamate in the cell reverses and "swamps the extracellular environment with large amounts of intracellular glutamate." In both animal and human tissue, increased glutamate transporting astroglia are observed in the areas of ischemic penumbra, suggesting an attempt to increase removal of excess glutamate. This also provides a possible explanation for the vulnerability of hippocampal tissue to anoxia/ischemia, as few astrocyte glutamate transporters typically are present in this region and hence available for activation should excess glutamate be present (7). Present neuroprotective therapies acting on glutamate include riluzole (used in amyotrophic lateral sclerosis (ALS) to prevent glutamate release) and topiramate (an antiepileptic antagonist of the AMPA/kainate glutamate receptor), although Faden (5) argues that glutamate antagonists thus far have not proved beneficial in human trials.

Free radicals produce oxidative damage that accentuates neuronal injury and may lead to or hasten the progression of neurologic disease (4). Delanty and Dichter (4, p. 1266) report numerous clinical trials of various antioxidant compounds with "positive, negative, marginal and conflicting results." Animal studies strongly implicate oxidative mechanisms in neurologic disorder (14), but human clinical trials are inconclusive. Delanty and Dichter (4) suggest that suboptimal dosing and poor matching of antioxidant with disorder/disease have contributed to poor results. They suggest that better measurement of oxidative damage and antioxidant effectiveness will allow for better and more favorable outcome studies, which ultimately should provide better guidelines for which compounds, if any, can meaningfully improve outcome. In summary, modern neuropharmacology attempts to duplicate, prevent, enhance, and/or attenuate the effects of neurotransmitters thought to play a role in the condition or disorder of interest (15). However, as this brief review of neuroprotective agents indicates, the field is in its infancy, and few treatments have been clearly established with regard to efficacy. Nevertheless, research in this area, particularly around neuropharmacologic agents, receptor sites, and specific actions of receptor subtypes, is paving the way for increasingly tailored treatments. As our understanding of the mechanisms of secondary damage after brain trauma increases, it seems inevitable that newer and more effective treatments will emerge. In the meantime, refinement of clinical trials in terms of sample size, better identified target populations (e.g., severe brain injury), and better stratification of these samples will permit more fruitful research both with existing agents and with agents yet to come (5,15).

Coma Duration

There are few studies examining pharmacologic interventions in the comatose patient after brain injury, and no medication to date has been reliably demonstrated to stimulate consciousness/emergence from coma (16). However, a few studies scattered throughout the recent literature suggest possible applications for a number of existing compounds in the comatose or minimally responsive patient. Of the reported studies, the most common intervention is with methylphenidate (Ritalin), presumably due to its established involvement in attention, alertness, and wakefulness via the reticular formation. Anecdotal reports suggest significant improvements in comatose or minimally aroused patients with administration of stimulants (17,18). Clearly there is a need for further research in this area, although preliminary studies appear to hold some promise. As Worzniak et al. (18) suggest, methylphenidate represents an inexpensive, low-risk intervention for treatment of patients who remain in a comatose or semicomatose state. Further research is needed with more rigorous designs and larger samples to demonstrate whether stimulants will serve a more general role in patients who remain in states of underarousal or coma.

Other research has examined whether selective serotonin reuptake inhibitors (SSRI) can assist with recovery from coma, because serotonin acts as both an excitatory and inhibitory neurotransmitter and is implicated in many aspects of cognitive and emotional functioning. A prospective placebo-controlled randomized trial of the SSRI antidepressant sertraline in 11 patients with severe brain injury (Glasgow Coma Scale ≤ 8) failed to find a significant difference between control and experimental conditions on measures of orientation, agitation, and memory. Both the control group and the sertraline (200 mg/day) group demonstrated equal improvement over a 2-week period (19). The authors suggested that a longer study and larger sample size would be needed to confirm the findings, but preliminary indications were equivocal.

Other interventions have examined the use of dopaminergic agonists commonly used to improve cognitive speed and motor movement in Parkinson disease. A single-case investigation of amantadine found substantial improvement in a patient who was in a "minimally conscious state" 5 months after injury (20). An investigation of bromocriptine combined with neuropsychological testing, sensory stimulation, physical therapy, occupational therapy, and speech therapy in five patients in a "vegetative state" found significant improvement compared to reviews of literature reports on 33 vegetative state and 37 "minimally conscious" patients (21). Large, randomized controlled trials of dopamine agonists such as amantadine and bromocriptine are required to extend these preliminary findings.

Finally, Showalter and Kimmel (16) examined the effect of lamotrigine in 14 individuals in a minimally conscious state after observing a surprising recovery in a 53-year-old man administered lamotrigine after experiencing a seizure (and an allergic reaction to phenytoin) 8 months after injury. The authors report that 11 of the 14 patients returned home; the remaining three patients were transferred to skilled nursing facilities, a result they considered unusual based on their experience with patients who had sustained this severity of injury. They recommend further investigation of lamotrigine.

In summary, exciting preliminary research is starting to emerge in the literature suggesting the possibility of hastened recovery and emergence from coma in individuals who typically would have been considered likely to be transferred to long-term care without meaningful recovery. The International Working Party on the Management of the Vegetative State's summary report found that there was insufficient evidence to suggest whether drug therapy could influence the recovery of vegetative patients in a meaningful way (22). The report noted the need for further research and suggested that research should focus only on those drugs that have the least cerebral inhibitory effect (i.e., avoid using antiepileptic and antispasticity medications wherever possible). It appears that most of the emerging studies in the last 5 years have avoided drugs that are known to dampen cortical activity, the exception being the use of lamotrigine, which the authors discovered incidentally while treating an emerging seizure disorder. However, their findings suggest that understanding the role of pharmacology in recovery after brain injury will be more complex than simply knowing whether a drug's expected main effect is to reduce or increase cortical activity across a certain neurotransmitter substrate. As knowledge of the sites and mechanisms of drug action improves, better and more informed choices can be made about which drugs are likely to hasten rather than hinder the recovery process and in which individuals.

RECOVERY OF FUNCTION

Neuroscience and observation of human behavior leaves little doubt that the brain is fundamentally altered by experience.

The ability to learn and to adapt one's behavior to changing needs and circumstances must depend, at some level, on changes in nervous system activity and organization. The notion of *neuroplasticity,* the brain's capacity to change and alter its structure and function, is particularly relevant to rehabilitation and an understanding of both natural and induced recovery processes. It is widely acknowledged that after nonprogressive brain damage most individuals demonstrate some recovery of cognitive and behavioral functions and that many make significant recovery (23).

It is possible to predict recovery with some degree of accuracy. Although there is a wide range of variability in recovery from neurologic insult, group studies suggest that certain prognostic factors, including demographic variables, injury-related variables, and patterns of cognitive loss, can lead to reasonable predictions about the likely nature and extent of recovery. Age at injury, severity of injury, and a variety of premorbid risk and protective factors related to intelligence, achievement, history of substance use, and psychosocial adjustment, can be used to estimate likely outcomes (24). Older age, more severe injury, more anterior/frontal lobe involvement, less formal education, and preinjury or postinjury alcohol or drug abuse all contribute to a poorer prognosis for recovery. Some cognitive impairments, in attention for example, seem to respond well to interventions and often show substantial recovery, whereas others, such as severe episodic memory impairment secondary to prolonged anoxia, may be devastatingly and permanently impaired. This may be due to the fact that some functions are more widely distributed than others. It also is increasingly recognized that psychosocial adjustment, emotional state, and social support make critical contributions to recovery.

Some cognitive functions appear to be capable of being fully restored and some partially recovered, whereas others may be permanently lost or substantially degraded and must be compensated for. Thus, recovery is believed to occur through a combination of two fundamental processes: *restoration* and *compensation.* The basic mechanism underlying both restoration and compensation is *reorganization,* which can occur at neural, cognitive, and/or behavioral levels. There are different forms of compensation that can be characterized in terms of how conscious the process of compensation is and the degree to which the compensation is internally implemented by the patient or externally implemented and supported by someone else. Automatic compensations occur within a neural or cognitive system to compensate for damage to the system, and they emerge from the system's attempt to fulfill its automated function. Such compensation is not accessible to the individual. This unconscious form contrasts with compensation where an individual consciously uses a compensatory strategy to alleviate problems caused by the cognitive deficit. For example, someone may use a mnemonic strategy, such as mental imagery, to assist in remembering some piece of new information. Intermediate to these forms of compensation are the use of cues, prompts, and other techniques implemented by a caregiver or therapist in order to provide needed information or guidance for the patient. Most rehabilitation approaches involve a combination of restorative and compensatory approaches, depending on the nature and degree of the patient's problems and course of recovery.

New Perspectives on Neuroplasticity

For most of the past century there was a fairly nihilistic view of the potential for recovery from neurologic insult beyond an early stage of apparently spontaneous, but perhaps only partial, resolution of difficulties. The prominent view of neural recovery in adults was that restoration only occurred through neural sparing. Von Monakow (25) introduced the term *diaschisis,* which he used to describe depression of activity in remote, nondamaged brain sites that are functionally connected to lesioned areas. This deprivation of synaptic input and output alters the functioning of undamaged areas and results in a large area of functional lesion. As these nonlesioned areas become reconnected with other areas or adjust to their lack of input from the lesioned areas, they begin to resume functioning. A significant amount of early recovery from what initially appears to be very severe functional impairments was hypothesized to be due to recovery from diaschisis. As late as the mid-1970's, Le Vere (26, p.351) stated "Recovery following damage is dependent upon what survives and what continues to function in the normal manner." The adult nervous system was believed to be incapable of any substantial reorganization.

Exciting research on neuroplasticity in the last decade has led to a more positive and promising perspective with respect to potential neurologic recovery of adults with brain injury. Work primarily in nonhuman animal research has shown that neural reorganization can take place following damage to the adult brain (27). New "growth," which is believed to enhance neural connectivity, has been shown to occur in both normal and damaged brains of adult animals. Mechanisms, which enable the establishment of new connections between neurons, occur primarily at the synapse and include long-term potentiation, which increases the efficiency of neural transmission at a given synapse, dendritic branching, and axonal sprouting. There are, however, limits on plasticity, and reorganization of neural connections can have negative effects (e.g., spasticity) as well as positive effects (28–30).

Of critical importance to models of rehabilitation, research with both animals and humans has shown that environmental manipulation has a marked influence on behavioral recovery (29,31,32), in other words, change in the underlying neural mechanisms that support recovery are experience dependent. The goal for rehabilitation specialists is to understand how best to provide experience that facilitates the positive aspects of recovery and reduces potentially negative consequences of neuroplasticity.

Hebb (33) proposed that *networks,* or *cell assemblies,* constituted the functional units underlying particular motor, cognitive, and behavioral functions. According to his theory, neurons strengthen their interconnections if they frequently are activated simultaneously, a notion popularized by the adage that "neurons that fire together wire together." This simultaneous neural activation serves to strengthen the local network, thereby constituting learning, which is seen as the establishment of such integrated connections. Particularly strongly connected assemblies are believed to underlie highly automatic behaviors for which only a small percentage of cells may be needed to stimulate the entire assembly firing (34). Relative sparing of even portions of these cell assemblies may account for the relatively stronger recovery of highly automatic, well-learned behaviors and behavioral sequences. As such, small lesions might not disrupt utilization of local neural networks to the same degree as a large lesion. Large lesions may damage too much or too critical a part of the underlying tissue needed for a particular function. After large lesions occur, there is some evidence that neural compensation takes place but may use intact neural circuits not originally concerned with the lost function, perhaps at some distance from the damaged circuits, and even in the opposite hemisphere (29,35). After damage to language systems in the left hemisphere, for example, there is evidence that tissue immediately adjacent to the lesion is recruited during recovery, but also that distant tissue in other areas, including homologous areas of the right hemisphere, becomes active during language processing activities (36). These may reflect two different forms of recovery or compensation and may take place at different intervals or time periods after injury. As yet, little data are available with respect to if or how speech/language interventions affect these neuroplastic processes.

Another consideration in understanding neuroplasticity in the context of recovery is the somewhat competitive relationship that exists between the hemispheres (37). This relationship becomes important in that, after damage to one hemisphere, there is some evidence that its natural inhibitory influence on the other hemisphere may be lost or reduced. It has been proposed that damaged circuits in the brain can suffer further loss of function because of inhibitory competition from undamaged circuits. There is evidence that the undamaged hemisphere of the brain in a group of patients who had suffered unilateral strokes showed higher levels of regional blood flow than did that hemisphere of the brain in control subjects (38). Although such increased activity may contribute to some of the compensatory gains, it also may be exerting an inhibitory effect on the damaged hemisphere that reduces the potential recovery of its damaged circuits. Paradoxical improvements in performance may occur after a second lesion on the opposite side of the brain. One of our clients had marked left neglect after a right hemisphere stroke but demonstrated resolution of the neglect after a left hemisphere stroke several months later. Perhaps this occurred be-

cause the new lesion reduced the activity of left-sided neural networks that were excessively inhibiting the originally damaged right hemisphere circuitry. There also is new evidence that restricting the movement of the intact limb enhances recovery of function in the hemiparetic limb. A therapeutic approach that embodies this belief, called *constraint-induced* therapy, holds that functional loss in a hemiparetic limb may come about during the period of initial injury due to the effects of learned nonuse. Constraining the unaffected limb, together with forced use of the affected limb, may facilitate motor recovery by encouraging limb movement that is simply not produced spontaneously and/or by reducing some of the inhibitory influence from the intact hemisphere (39).

Despite substantial evidence for experience-dependent neural change after brain damage, it is important to recognize that "recovered" behavior is not always performed in the same way as it was before the injury. Some researchers have argued that careful analysis of apparently recovered behaviors always reveals subtle differences in how the behavior is actually carried out. In addition, neural reorganization can produce maladaptive compensation at sensorimotor, cognitive, and behavioral levels (40). Spasticity, increased sensory sensitivity (e.g., photosensitivity), and behavioral irritability may be examples of such maladaptive compensation.

In order to understand *recovery of cognitive function* it is important to have a clear conceptualization of the functional architecture of cognitive processes and the nature of the cognitive impairments demonstrated by brain-injured individuals (41). The disciplines of neuropsychology and neurolinguistics have made significant advances in specifying the nature of functional and cognitive impairments. Research has increasingly specified the nature and structure of cognitive processes underlying attention, language, visuospatial ability, memory, and executive functions. There also is a large body of research specifying the nature of neurologically determined and functionally based aspects of psychosocial and emotional functioning. Rehabilitation research is beginning to identify which processes may be restored, which partially restored, and which compensated for in individual patients.

CHALLENGES IN REHABILITATION RESEARCH

There are significant challenges for research in cognitive rehabilitation. One is the nature of the appropriate subjects or subject groups to include in research studies. Another is a limited understanding of the natural course of recovery and how to best influence it. Yet another is identifying the appropriate level of measurement to quantify and characterize change.

Individual differences in normal functioning and individual responses to brain damage are such that information derived from heterogenous group research often is difficult to interpret or generalize. No two brain injuries are alike, and factors such as age, size and location of injury, time since injury, and a wide range of preinjury and postinjury variables can dramatically influence recovery. It is well documented that group data in neuropsychology and rehabilitation mask potentially important and meaningful individual data. As such, theory-driven approaches to the investigation of cognitive rehabilitation in individual patients or with replication in small groups of patients often is preferable to attempts to compare heterogenous groups of patients categorized according to etiology. In addition, most therapists argue that interventions need to be individually tailored to suit the particular patient and meet his or her specific needs. This approach, however, is somewhat at odds with the large-scale, randomized, treatment-control studies that have been the hallmark of popular forms of treatment efficacy research, as exemplified in pharmacologic and medical research. Although some larger-scale research studies have been reported, careful single-case designs with replications in small numbers of patients have led to important insights about cognitive deficits and the efficacy of particular interventions. Concerns remain about the wide-scale applicability of the small-sample approaches and the sensitivity and meaningfulness of some of the larger studies.

Several ambitious undertakings have begun to provide the evidenced-based reviews of practice in cognitive rehabilitation that can give rehabilitation specialists greater direction and support for their activities. The first was undertaken by a panel of experts that reviewed 10 years of cognitive rehabilitation literature (42). Their findings and recommendations were reported by the National Institutes of Health as a consensus statement on the effectiveness of various approaches to cognitive rehabilitation (43). The second was a meta-analysis of cognitive rehabilitation research reported by Carney et al. (44). The third was a review undertaken by members of a subcommittee of the Brain Injury–Interdisciplinary Special Interest Group of the American Congress of Rehabilitation Medicine (45). This last article also describes results of a meta-analysis of cognitive rehabilitation research based on an initial bibliography of 655 published papers. Based on their findings, the authors provide best practice guidelines for treatments within the domains of attention, visuospatial functioning, speech and language, memory, executive functions, and problem solving.

Most individuals show some, and many substantial, recovery of cognitive and functional capacities, particularly in the first few weeks and months after injury. Even after severe injury, changes in functioning often can be seen even years after injury. It is extremely difficult, especially further after the injury, to distinguish changes that may have recovered spontaneously from changes that are a direct result of interventions. Ethical considerations do not allow for studies in which some patients are not provided with standard treatment protocols, even though the efficacy of these treatments may not be fully established. For this reason, many studies of efficacy involving cognitive rehabilitation have used patients who are believed to be beyond the period of spontaneous recovery and are relatively stable. Nevertheless, there is some evidence that it is best to capitalize on early neuroplasticity and to begin treatments early. The issue of how and when to most effectively implement interventions remains largely unresolved.

Outcome measurement has become a major area of interest and investigation in rehabilitation research. Advances in neuropsychological assessment have allowed greater specification of cognitive impairments. However, evidence of cognitive impairment on psychometric tests does not necessarily translate directly to functional impairments in everyday life. Meaningful recovery of cognitive abilities entails more than improvement on psychometric tests, and progress in, or benefit from, therapy is only genuine to the extent that there are functional changes in the patient's everyday life and psychosocial experience. In addition, psychometric testing, by nature, typically assesses cognitive abilities without the opportunity for extra cueing, support, or use of compensation strategies. These may be used effectively in more functional settings yet not be evident in neuropsychological test findings.

Increasingly, attention is being paid to the psychosocial impact of brain injury and its interaction with rehabilitation. The ultimate goals of rehabilitation include better adaptive functioning, psychological adjustment to changes in function or lifestyle, and effective social and community integration of the individual. There is increasing evidence that cognitive and psychosocial processes interact freely in normal life (46), and there is a high correlation between response to rehabilitation and the emotional and psychosocial state of patients (24,47). Cognitive functioning can be significantly affected by mood. Depression, for example, can cause cognitive slowing, reduced motivation, psychological inertia, reduced motor activity, and impaired memory, attention, and concentration. As such, psychological adjustment typically makes a significant contribution to recovery, although it frequently is not taken into account in predicting recovery.

Emotional reactions after brain injury can arise from both direct/primary reactions and indirect/secondary effects. Direct behavioral and emotional reactions, such as adynamia, irritability, agitation, quickness to anger, and unawareness, are believed to be due to direct damage or disturbance to the nervous system and particularly to frontal and frontally connected structures. Indirect reactions, which may include depression, anxiety, and social withdrawal, are believed to be due, at least in part, to a natural reaction to disability, loss, and handicap.

Recovery has different meanings for different people. Studies that have asked people to rank and weigh the relevance of different psychosocial factors have shown that individual perceptions about recovery often do not coincide (47). Different professional groups also have different perspectives on what aspects of recovery are most salient (48), and the amount of experience in rehabilitation impacts on a professional's view of what is important (49). Most brain-injured individuals and their families are concerned with the disabling and handicapping effects of brain injury and are little focused on the specific cognitive impairments that usually are the target of assessment. Effective rehabilitation involves partnering with patients and families and working toward mutually developed and agreed upon goals.

MAJOR APPROACHES TO COGNITIVE REHABILITATION

Problems in adaptive functioning arise in relation to cognitive impairments when there is either (a) a decrease in a given skill without an accompanying decrease in the environmental demands placed on the person or (b) an increase in environmental demands without an accompanying increase in the skills required for successful performance (50). Consistent with this view, most rehabilitative approaches for management of cognitive impairments take one of four forms: (a) environmental modifications, (b) use of compensatory behaviors, (c) direct retraining in the areas of cognitive impairment, and (d) pharmacologic interventions.

Environmental Modifications

This category includes approaches to rehabilitation that alter factors within the client's environment with minimal or no expectation of underlying change in the individual's capacities. Included in this category are manipulations that decrease the demands on the individual, including simplifying tasks, eliminating the need to do certain tasks, or allowing longer time frames to complete activities. Other manipulations consistent with this approach include the provision of external support in the form of oral or written cue systems, checklists to follow, or alteration of environmental parameters such as reducing noise or other potential distractions. One of the simplest ways to assist persons with a memory disorder, for example, is to structure the environment so that they can rely less on memory. Labeling of closets and drawers and the posting of checklists and other written reminders are examples of contextual supports for memory. Although the person using the reminder might need to be trained to refer to it, self-initiated compensation is not necessarily anticipated. There is an inherent assumption that such external manipulations may need to remain in place if the improvement in functioning is to continue, although it is possible that the behaviors or the required knowledge might become routinized or that skill levels may recover at some point.

Compensatory Approaches

In contrast to approaches that focus on environmental manipulations or modification of factors external to the patient, there are approaches that have as their goal a primary change in the behavior or actions of the patient himself or herself, such that he or she is performing a task in another way. This approach would include the training of compensatory behaviors or skills that require concerted effort and commitment on the part of the impaired person, such as the ability to independently record in, and refer to, a memory system or organizer. Approaches that attempt to increase self-awareness or teach self-regulatory or metacognitive strategies can be included in this category. Compensation might include a new behavior or substitute skill (such as using a memory notebook) and/or an increase in time and/or effort used in completing a task (such as in studying). The patient also may adapt to the new situation by changing self-expectations, selecting new tasks, or relaxing his or her own criteria for success. Whether the person is taught to use the compensation or develops it on his or her own, the patient is an active participant in its application and continued use.

To some extent, severity of impairment is believed to affect the extent to which compensation is spontaneously adopted. Interestingly, moderately impaired individuals are most likely to compensate, whereas mildly impaired individuals may be unaware of the need to compensate, and severely impaired individuals may lack the skill and insight to implement compensatory behavior without substantial training and support. It is important to recognize that the use of a particular compensation may have a negative trade-off. Compensatory behaviors should optimize, not hinder, utilization of available resources, including the residual capacities of the injured system. Compensatory behaviors should balance the consequences for the individual against the individual's environment.

Restorative Interventions

Restorative interventions involve the use of procedures that have as their goal improving or restoring some underlying ability or cognitive capacity. This includes a myriad of approaches for improving underlying cognitive skills, such as attention, memory, and problem solving. Sohlberg and Mateer (41) applied what they termed a *process oriented approach* to the rehabilitation of cognitive impairments. The basic tenets of this approach included a solid understanding of the specific cognitive area involved and a detailed analysis of the nature of impairments in that area. This was followed by hierarchical training exercises designed to stimulate and rebuild the impaired skills. Repetition was believed critical to the reautomatization of skills. Only some areas of cognitive ability, such as attention, were purported to be modified by this type of intervention, whereas other areas, such as episodic and semantic memory, were not targeted using this form of treatment. The argument for training broad-based cognitive skills was that they were seen to underlie many functional impairments. Impairments in sustained or divided attention, for example, could affect reading, listening, meal preparation, driving, and many work-related skills. The problem often is not that specific behaviors or tasks cannot be done, but that accuracy and efficiency are hampered by distractibility and poor regulation of attention resources. The belief is that if attention can be improved, then there will be a positive impact on many other functions, including memory, organization, and communication. The process of training attention also helps to increase understanding of the nature and impact of attention deficits for the individual and actively promotes better self-management of the internal and external environments.

Pharmacologic Interventions

In today's restrictive financial climate, funding for inpatient rehabilitation services and outpatient day programs has dwindled, and the luxury of time that rehabilitation professionals once enjoyed is a thing of the past (51). The pressure and necessity to produce rapid results, combined with an increased understanding of the effects of medications on brain systems and behavior after brain injury (15), has resulted in an increasing focus on pharmacologic intervention in recovery and rehabilitation (52). Pharmacologic interventions in cognitive rehabilitation are best viewed as adjunctive to traditional cognitive rehabilitation techniques. Postacute treatments address three main areas of disability: attempts to restore and/or compensate for deficits in cognitive and physical functioning; attempts to address behaviors that are a barrier to treatment or less restrictive living environments; and interventions targeting psychiatric disorders emerging or exacerbated by the brain injury.

GUIDELINES FOR SELECTING INTERVENTION STRATEGIES

In general, patients who demonstrate little behavioral initiative or flexibility, who are environmentally dependent with apparently minimal response to internal cues, and/or who are minimally aware of their deficits tend to respond better and more consistently to external manipulations. For these patients, environmental manipulations, behavioral strategies, and external cueing systems often are effective in increasing function. Patients who demonstrate greater behavioral initiative and flexibility, who initiate and direct their own behavior to some degree, and who are somewhat aware of the change in their abilities resulting from their injury are more likely to demonstrate improvements with cognitive-behavioral interventions, process-oriented cognitive training, training in the

use of compensatory devices, and training in the use of self-instructional and metacognitive strategies. When working with cognitively impaired patients it is important to match the profile of the patient with the intervention approach (24,53).

STRATEGIES FOR PROMOTING MAINTENANCE AND GENERALIZATION

A major and continuing concern with regard to cognitive rehabilitation is whether abilities or skills targeted in treatment will maintain and generalize and thus lead to sustained improvement in targeted aspects of everyday function. Generalization can be measured at multiple levels, from other similar but untrained treatment activities, to psychometric measures of the process or function addressed, to other abilities that are presumably related to or subserve the process, to structured functional activities, and to spontaneous functional activities. As an example, successful training on a high-level working memory task, such as alphabetizing words in sentences, might be expected to result in better performance on other high-level working memory exercises (e.g., number sequencing), better performance on psychometric measures that require working memory such as the Paced Auditory Serial Addition Test, and better performance on a functional task such as preparing a shopping list for a dinner. Therapists should not "expect" generalization, rather they should "program" for generalization.

It has become abundantly clear that spontaneous generalization of skills is difficult, if not impossible, for many clients with brain injury; however, steps can be taken to facilitate and ensure generalization. Some of the principles to keep in mind with respect to increasing the likelihood of generalization include:

- Be explicit in training, but train a variety of target skills and practice beyond criteria (over-learn)
- Train general strategies and practice in a variety of natural settings
- Change the environment to support new skills and behaviors
- Enlist help and involvement from caregivers or significant others
- Promote internal attributions of change
- Identify barriers to maintenance and plan for high-risk situations
- Plan for recovery from setbacks and schedule booster sessions and long-term maintenance plans

In addition to these approaches, it has become increasingly evident that many individuals need assistance in dealing with the emotional and behavioral consequences of acquired cognitive impairment. It often is difficult to adjust to changes in one's thinking ability, and fear, frustration, and feelings of loss are common. The importance of providing assistance in dealing with the emotional responses to these changes in functioning, on the part of both the affected individual and their family, cannot be underestimated. Educating the family and significant others on how to respond to a person with cognitive impairments is important. An appreciation for the organic or nonvolitional nature of the behavior often is helpful in alleviating the fears and misconceptions of family members and caregivers.

Repetition is a key factor in rehabilitation. No matter what kind of intervention is used, whether it is restorative or compensatory, multiple opportunities for practice must be incorporated into the treatment program. Behavioral programs need to be well planned and consistently carried out. Finally, it is vital that clinicians actively train for generalization. One should not expect generalization but rather provide systematic opportunities during which skills and behaviors can be trained and stabilized.

A comprehensive evaluation of each patient's cognitive profile is a critical first step in developing the rehabilitation plan. Although beyond the scope of this chapter, detailed cognitive, communicative, behavioral, emotional, and psychosocial evaluations should be conducted with each client. After evaluation of the cognitive/behavioral profile of the individual and determination of the observed or anticipated real world impact of the cognitive deficits, it is necessary to establish specific mutually agreed upon rehabilitation goals given the client's current and future circumstances (54). Mutual goal setting can boost motivation and result in increased persistence. Other principles include encouraging choice in therapy, working collaboratively with clients and their families, creating a supportive environment, reinforcing clients for their efforts, asking clients to assess their progress, and providing clients with an effective and meaningful method for accomplishing tasks. The intervention plan should be monitored constantly for efficacy, in concert with the goals and plan for generalization. In the following sections specific considerations for assessment and intervention are provided for three of the most common cognitive sequelae of TBI: attention, memory, and executive function disorders.

COGNITIVE REHABILITATION IN SPECIFIC DOMAINS

Management Strategies for Disorders of Attention

Impairments in attention, concentration, and distractibility are among the most commonly reported cognitive problems following TBI (55,56). Persons who have suffered a TBI or stroke often describe taking more time to complete tasks, having difficulty concentrating on tasks in noisy or busy environments, experiencing problems doing more than one thing at once, and forgetting what he or she was about to say or do. Although a comprehensive understanding of the na-

ture of attention problems in TBI remains to be developed, contributors to attention deficits have been experimentally investigated. These include an overall slowing of information processing, problems with distractibility, difficulty with divided attention tasks, and problems with attention control for both sustained and alternating attention.

Memory difficulties are commonly recognized sequelae of TBI. Because attention paid to information is a crucial factor in its later recall, it has been suggested that some of the memory impairment in individuals with TBI is related to limitations in attention. Mateer et al. (81) reported results of a self-report questionnaire on the frequency of different kinds of forgetting failures. Subjects who had sustained TBI reported that the forgetting experiences, which occurred most frequently, appeared to be those related to attention, mental control/working memory, and prospective memory. It has been hypothesized that for a portion of the TBI population, memory may be secondarily impaired due to a disorder of attention (58–60).

Psychological research does not support a unitary concept of attention. Many experimental studies have shown that at least four rather independent aspects of attention can be separated (60). *Phasic alertness* is defined as the ability to enhance the activation level after a warning stimulus. *Selective attention* is the ability to focus on certain aspects of a situation while ignoring irrelevant ones. *Divided attention* requires monitoring of at least two stimulus sources at the same time and having to react to relevant stimuli that appear in one or the other or in both sources simultaneously. *Vigilance* or *sustained attention* requires maintaining attention on a task over an extended time period. Neuropsychological studies have shown that these different aspects of attention can be impaired selectively by focal brain damage.

A rehabilitation-oriented evaluation for an individual who may have experienced a TBI should include assessment of the different components of the attention system. This can be done by conducting a careful interview and observing attention behaviors, using a balanced set of psychometric measures sensitive to attention, and using rating scales or questionnaires that focus on everyday attention functioning. Chan (61) notes that attention difficulties after brain injury are likely to represent a mixed pattern with varying degrees of impairment in different attention subtypes.

Five approaches to addressing difficulties with attention are described here. Most clinicians implement a combination of approaches, either simultaneously or at different times in the recovery process.

- *Attention training* involves the use of cognitive exercises that are designed to remediate and improve attention functioning.
- *Use of strategies and environmental supports* includes both self-management strategies and modifications to the environment to help a client compensate for attention

problems. Self-management strategies also may work to remediate attention problems.

- *Psychosocial support* addresses the emotional and social factors that can result from and/or exacerbate an attention deficit.
- *Pharmacologic interventions* for attention difficulties after brain injury have focused primarily on the use of stimulant medications, with some additional attention given to dopamine agonists.

Attention Training

Many brain-injured individuals, including those with significant frontal lobe impairment, have been shown to benefit substantially from exercise and training of attention skills (62–67). Sohlberg and Mateer (41,68) developed and evaluated the efficacy of a package of attention training materials [Attention Process Training (APT)] that was based on a hierarchical model with five levels of attention: focused, sustained, selective, alternating, and divided attention. A large set of both auditory and visual tasks designed to exercise and challenge different aspects of attention was used in treatment sessions over periods of 6 to 8 weeks. The efficacy of this training in improving attention capacities has been supported in a series of single-case designs and in-group pretreatment and posttreatment comparisons (68,69). It also was demonstrated that improved attention function was associated with improved anterograde memory function in individuals who received attention but not memory training (58,59). In a larger-scale study of more than 20 patients who received training with the APT materials, gains were seen in the number of correct consecutive responses on the PASAT over baseline performance (70). These researchers hypothesized that attention training improves working memory capacity. Finally, on a neurophysiologic index using evoked potential data, subjects with TBI showed a delayed P3 before training and reduced latency after training of attention skills (11).

Sturm et al. (40) addressed the issue of specificity in attention training in a group of patients with lateralized focal vascular lesions. The domains of alertness, vigilance, selective attention, and divided attention were evaluated, and subjects received consecutive training in the two most impaired of the four attention domains. There were significant training effects for both alertness and vigilance; in addition, subjects demonstrated shorter response times on selective attention tasks and reduced error rates on divided attention tasks. Interestingly, the study not only revealed a high degree of specific training effects but a substantial number of deteriorations in performance after inadequate or nonspecific training. The authors suggested a negative effect of training when the treatment focused on complex aspects of attention (e.g., selectivity training) when basic aspects of attention (e.g., alertness and/or vigilance) had not been trained. These results supported the hypothesis of a hierarchical organiza-

tion of attention functions and the need to incorporate this information into treatment paradigms.

There is growing agreement that certain aspects of attention, particularly sustained attention and working memory, can be improved with targeted training. However, questions remain about the nature and underlying source of the change. Although some of the electrophysiologic data cited suggests underlying neurophysiologic changes, others have argued that attention training can serve as a forum for gaining insight regarding attention failures. Developments in attention training programs have emphasized not only practice on attention tasks but also activities that require the client to monitor and evaluate attention failure and successes. They also focus on assisting the individual to become more knowledgeable and active in managing situations such as enhancing his or her capacity to attend (APT II) (72). This reflects a shift to techniques that increase self-awareness and enhance a sense of self-control and mastery over cognitive weaknesses.

Self-management Strategies

Self-management strategies for dealing with attention difficulties encompass self-instructional routines that help an individual focus attention on a task. Examples of self-management strategies include orienting and pacing procedures. Orienting procedures may be helpful for clients who have difficulty sustaining attention or screening out distractions. Clients are taught to monitor their activities consciously, thereby avoiding attention lapses. For example, a client may be taught to ask three orienting questions each time his or her watch beeps on the hour: (1) "What am I currently doing?" (2) "What was I doing before this?" (3) "What am I supposed to do next?" The goal is to assist the client in focusing attention. When clients experience difficulty with fatigue, use of pacing strategies can be helpful. The client is assisted to develop realistic expectations for productivity, to build in breaks, and to schedule demanding activities during the part of the day when he or she is most alert and rested.

Environmental supports for attention include developing task management strategies. A list of "difficult" and "helpful" environments is generated. Typically loud, busy, or otherwise distracting environments are most difficult for clients with attention impairments. Choosing to shop in a smaller store or at a quieter time of day may be helpful. Other strategies to reduce distraction include turning off the television or stereo, shutting a door or curtains, using earplugs, and turning off the ringer for the phone answering machine. Other environmental modifications include organizing the client's living space to reduce the load on attention, memory, and organizational abilities. Setting up filing systems, message centers, and bill payment systems often are useful in assisting individuals with attention and other cognitive problems.

Psychosocial Support

Psychosocial support is critical for working with many individuals who suffer attention impairments. As indicated earlier, suggestions for dealing with attention problems focus on avoiding distraction, taking breaks, and reducing "cognitive overload." Such suggestions can be useful in working with individuals with TBI in several respects. First, by reducing the effects of overload and fatigue, they often result in increased performance and decreased frustration. Second, they serve as an acknowledgment to the patient of his or her subjective experience. Recognition of the person's feelings of frustration and perception of changed mental abilities often increases a sense of trust and mutual understanding. The suggestions assist the individual who is feeling overwhelmed and "victimized" by his or her cognitive difficulties in gaining a sense of control and self-efficacy. Use of such techniques can improve the individual's ability to manage both his or her external environment and internal emotional state.

The APT generalization program (72) makes use of two attention "logs": the attention lapse log and the attention success log. The purpose of these protocols is to provide an organized way of recording both breakdowns and successes in attention in naturalistic settings throughout the person's day. This increases the person's awareness of the situations and times in which attention is difficult and helps in focusing treatment. Not only the nature of the lapse or the success is recorded but also how the individual responded to the lapse or failure. This often yields insights about ways in which the individual responds emotionally and behaviorally to variations in his or her performance. Reasons for successes and failures become more apparent, and an assessment of the frequency and impact of the attention problems may become clearer. Use of a success log provides a focus for discussing how the patient can take a more active role in managing and controlling situations that demand attention and their own response to cognitive successes or failures. It can become a focus for discussing and evaluating the effectiveness of stress management techniques. The individual can be assisted to feel a greater sense of self-control and control over the environment. This "empowerment" phase often leads to a redefinition of self in relation to cognitive inefficiency.

Pharmacologic Approaches

Treatment of attention problems has been heavily influenced by findings from the treatment of deficits of arousal, attention, concentration, and memory in developmental attention deficit disorder (73). As a result, stimulants have been the most common treatment for attention problems after TBI (52). Stimulant medications such as methylphenidate (Ritalin) and dextroamphetamine (Dexedrine) work by increasing synaptic availability of dopamine and norepinephrine, methylphenidate by primarily inhibiting reuptake and dextroamphetamine by primarily promoting the release of intracellular dopamine and norepinephrine. Primary areas of effect for these stimulants are found in the mesolimbic dopaminergic pathway from the ventral tegmental area via the medial forebrain bundle to the nucleus accumbens and prefrontal cortex (74) and the dorsal noradrenergic ascending bundle from the locus ceruleus diffusely to the cortex (75).

In terms of research, studies of stimulant effects in individuals who sustained TBI have produced mixed results. Studies have demonstrated improvement in attention and related systems (76), improvement in attention with a dose–response gradient (77), subjective and measurable improvement in attention (78), better performance on cognitive tests with improved response latency and consistency (79), significant improvements in attention compared to natural recovery (80), and improvements in performance on working memory tasks (81). Other studies, however, have failed to find improvement in attention, learning, or processing speed (82), found initial improvements that were no longer demonstrable after 3 months of treatment (83), or found improvements only in speed of mental processing without significant improvement in measures of orienting to distraction, sustained attention, or motor speed (84).

Overall, uncertainty remains as to the utility of stimulant medications in improving attention abilities after brain injury, although there appears to be a positive trend in terms of enhancing slowed processing speed. Large, randomized, double-blinded, placebo-controlled studies are needed to establish whether stimulant medications can treat attention difficulties in adult brain injury. In addition, the heterogeneity of attention processes and their presumed underpinnings, and of brain injury in general, contributes to the lack of consistency in pharmacologic intervention research. As argued by Arciniegas et al. (85), psychostimulants appear to modestly improve arousal and speed of information processing and to reduce distractibility. The effectiveness of stimulants for the treatment of other aspects of attention remains unproved.

Although the attention system appears to be vulnerable to disruption, this system also appears to be one that can be modified with targeted intervention. Among those areas of cognitive processing that have been addressed in the cognitive rehabilitation literature, perhaps the most compelling findings are those that pertain to the improvement and management of attention impairments. A concerted and well-founded focus on these areas often can result in significant improvements in satisfaction on the part of the patient and in demonstrable functional improvements. Interventions for attention should include training in different modalities, levels of complexity, and response demands. Therapists should be involved in monitoring subjects' performance, providing feedback, teaching strategies, and assisting the clients in understanding and managing emotional response to attention failure. Stimulant medications may be a useful adjunct to these approaches, particularly if speed of information processing is poor.

Management of Memory Impairments

Memory impairments can affect many aspects of everyday functioning. Individuals with memory impairment may have difficulty remembering names, dates, routes, faces, appointments, routines, and many other kinds of information. Approaches to working with memory-impaired individuals must be tailored to the nature and severity of the deficit, the particular types of everyday memory failures experienced by the person, and the nature and degree of other spared or impaired cognitive abilities. The major categories of memory interventions include the following:

- *Memory enhancing mnemonic strategies.* These techniques, sometimes called *mnemonic strategies,* are used in a person with memory impairment to enhance encoding, storage, and retrieval of new information.
- *Use of external aids.* External aids provide a mechanism to record and store new information to alleviate the functional impact of memory impairments.
- *Specialized teaching techniques.* Individuals with memory impairment often show spared domains of learning and can learn with sufficient repetition and/or when particular kinds of teaching strategies are used.
- *Pharmacologic interventions.* Augmentation of acetylcholine availability, which effectively slows memory loss in Alzheimer disease, is being explored as a treatment for memory impairment after brain injury.

Memory Enhancing Mnemonic Strategies

In some individuals, memory and new learning are inefficient and unreliable secondary to attention or information processing deficits. Learning can take place, but it is slow and effortful. For these individuals, treatment may take the form of assisting them to better attend to new information, to more fully process new information by reframing it or by using multimodal exposure. Clients also might be taught compensatory approaches such as overt or covert rehearsal strategies, organizational strategies, visual imagery, and other mnemonics. Studies comparing memory remediation with alternative treatment conditions have generally shown beneficial effects of memory remediation on neuropsychological indices of memory functioning or reductions in subjective reports of everyday memory failures (86,87).

Use of External Aids

In other individuals, the neural substrate for storing and/or retrieving new information may be so compromised that it is extremely difficult to acquire new information. Individuals with significant amnestic deficits tend not to benefit from mnemonic techniques or strategies. This may be because the neural substrates necessary for storage of new semantic memory simply are not available and information is quickly lost from memory, or it may be that the mnemonic techniques require levels of insight, initiation, and self-monitoring that are too great for individuals with significant memory impair-

ment and other cognitive losses. For many individuals with memory impairment, but particularly for individuals with severe memory deficits, use of compensatory aids is the most (and often the only) helpful strategy. Use of some kind of external memory aid(s) is the most frequently reported strategy by individuals with persistent memory impairment. There is a substantial literature supporting the use of compensatory memory aids (88–91).

Compensatory aids include memory notebooks, paper-based or electronic organizers, paging systems, computers, calendars, posted lists, and other reminding devices. Together with the memory impaired individual and their caregivers, therapists determine what needs for memory aids exist, and what cognitive and financial resources the individual has with respect to purchasing and using an external memory system. In order to use a memory notebook, for example, the individual generally needs to be able to read and write. Once a memory system is designed and developed, the individual needs to be trained to understand and use the system. Sohlberg and Mateer (92) described a three-stage approach to memory book training that involves learning about the system and its parts. This is done through use of teaching techniques that provide repetition and stabilization of the new information. The client also needs to learn to record systematically in the system and to refer to the system at appropriate intervals. This is done through intensive behavioral training. The system must be applied in a wide variety of functionally based settings to provide sufficient practice and application of the system. This is done through system use in role play and real-life activities. Brain-injured clients often are not aware of the extent or impact of memory deficits and initially may resist use of a system. In addition, family members may be concerned that using an external system will limit potential recovery of memory. This has not been shown to be the case. Research has suggested that memory remediation is most effective when subjects are fairly independent in daily function, are actively involved in identifying the memory problem to be treated, and are capable and motivated to continue active, independent strategy use. Thus, information and support of system use and involvement of the individual and caregivers in system development are extremely important. Efforts need to be taken to ensure that patients continue to use compensations through appropriate preparation and follow-up.

Specialized Teaching Techniques

Other interventions for memory have been directed at facilitating the acquisition of specific skills and domain-specific knowledge rather than on improving or compensating for memory functioning *per se.* It is known, for example, that procedural memory often is relatively spared relative to declarative memory. The person gradually changes behavior in response to a particular setting or in response to a particular set of stimuli, but he or she has no conscious sense of "knowing" or remembering. Even severely amnestic patients often can be taught new behavioral routines without conscious awareness (93). Teaching techniques such as the "method of vanishing cues" (94) or "errorless learning" (95,96) capitalize on this ability to learn implicitly. With repetitive direct training, which minimizes the potential to make errors or mistakes, amnestic individuals have been shown to learn new face–name associations, routes, routines for entering data into a computer, and orientation information. These techniques do not attempt to change or alter memory functioning *per se,* but they use information about how different memory systems work to facilitate acquisition of new knowledge or behaviors.

Pharmacologic Interventions

New research suggests the promise of pharmacologic treatment for some forms of memory impairment after brain injury. Acetylcholine deficiency in the basal forebrain pathways to the frontal and temporal cortices and hippocampus has been identified as one of the mechanisms of memory loss in Alzheimer disease (97). Cholinergic agonist drugs such as physostigmine, tacrine, and donepezil have been used successfully to slow the progression of memory loss in Alzheimer disease (52). Amnestic disorders after brain injury often result from damage to the hippocampal portion of this cholinergic pathway, because the hippocampi are especially vulnerable to even brief hypoxic episodes (98). Thus, researchers have attempted to augment acetylcholine availability in amnestic brain-injured patients and evaluated the evidence for consequent memory improvement. Donepezil has been used for this purpose because its effects are more centrally cholinergic, which minimizes the prohibitive peripheral side effects seen with both physostigmine and tacrine (99). Four preliminary trials using 5 to 10 mg/day donepezil found improvement on memory tests and staff and family observations in two patients (100); improvement of five or more points on the Mini-Mental Status Examination in nine patients who were 6 months to 16 years after cerebral injury (99); improvement in full-scale IQ score and on clinician and family ratings of daily functioning in 53 patients (101); and, improvement on the Rey Auditory Verbal Learning Test and Complex Figure Test, with positive trends on the Rivermead Behavioral Memory Test and Neuropsychiatric Inventory in four patients (102). Overall, the efficacy in the preliminary studies to date appears to provide converging lines of evidence for the use of the acetylcholinesterase inhibitor donepezil for memory impairment and possibly other cognitive sequelae of TBI. As in other areas of pharmacology for brain injury rehabilitation, large-scale trials are needed to firmly establish the efficacy of this class of treatments.

Remediation and Management of Executive Function Impairments

Individuals with frontal lobe damage and associated executive function compromise can be among the most baffling and challenging patients with whom rehabilitation specialists work. These patients may demonstrate average or even above average intellectual abilities, seemingly adequate recall of information, and apparent knowledge when a verbal response is asked for, yet they may demonstrate organizational, self-regulatory, and behavioral problems such that they fail to accomplish goals or complete even the simplest of tasks in the absence of cueing or structure.

The exact nature of the functions of the frontal lobe have been somewhat elusive, even though frontal lobe dysfunction can seriously impair many aspects of cognitive ability and independent functioning. The frontal lobes do not have a unitary function; rather different regions of the frontal system are important in different aspects of cognition and behavior (103,104). Portions of the frontal lobes play an important role in the initiation of action and behavior. Disruption of this system can result in aspontaneity or behavioral inertia (105). Individuals with such disorders often are capable of speaking or behaving quite normally, but without substantial external cueing they fail to do so. The frontal lobes are important in identifying goals, planning and organizing behavior to accomplish relevant goals, and monitoring goal-directed activity (106–108). Disruption in this system may result in behavior that seems random, disorganized, or without purpose.

The frontal lobes are important in many aspects of attention (for review see reference 60). *Working memory,* the capacity to hold information in a mental store during active information processing and to relate what has gone on just previously with what is going on now, appears dependent on frontal systems. The ability to switch attention between tasks and the capacity for divided attention also appears, at least in part, dependent on frontal systems. In novel situations where overlearned skills, knowledge, or responses cannot be used successfully, the frontal lobes are important in monitoring the effectiveness of behavior (108). As a situation becomes more familiar or a pattern of response more routine, there is a decreased need for frontal involvement.

Comprehensive evaluation of executive functioning should include the use of psychometric tests shown to be sensitive to frontal involvement; behavioral observations in different settings and under conditions of varying executive demand; and comprehensive interviews and rating scales with family and caregivers. Determinations should be made about the individual's capacities in the areas of initiation, goal setting, planning, organizing, self-awareness, and cognitive and behavioral flexibility.

A variety of approaches have been discussed with reference to the treatment of individuals with frontal lobe impairment (109). In general, they involve moving from simple structured activities with significant external cuing and support to more complex, multistep activities in which external support is gradually reduced and internal support or self-direction is required. Although a variety of articles have suggested success in using these techniques, there are as yet a very small number of cases and a limited number of studies in which such techniques have been experimentally evaluated. Approaches to working with executive function impairments include the following:

- Behavior modification techniques
- Pharmacologic management of behavior
- Strategies to improve initiation and drive
- Teaching task-specific routines
- Training problem-solving strategies
- Training metacognitve strategies

Behavior Modification

Loss of executive skills can impair the ability of the individual to initiate use of specific preserved abilities, monitor performance, and use feedback effectively to regulate behavior. Disruptive and aggressive behaviors often are a major impediment to rehabilitation. Alderman and Brown (110) and Alderman and Burgess (111) demonstrated effective use of several behavior modification techniques in assisting individuals to gain greater inhibitory control over their behavior. There are a wide variety of behavioral interventions that have been shown to be useful in work with brain-injured patients. The goal of such techniques is to either develop and stabilize certain adaptive behaviors or to reduce or eliminate disruptive or potentially injurious behaviors. The techniques involve doing a careful analysis of factors that trigger and/or maintain the behavior. Task analysis, behavioral charting, and behavioral planning are well-established techniques. Careful observation often provides insights into disruptive behaviors that arise out of confusion, frustration, fatigue, or inability to accurately process information. Modifying the setting and the task demands often results in a reduction of behavioral difficulties.

Pharmacologic Management of Behavior

Individuals who demonstrate persistent aggression that is not quickly amenable to behavioral intervention alone should be initiated on adjunctive drug therapy with continued non-pharmacologic intervention. Pharmacologic interventions for aggression that are represented in the literature include β-adrenergic blockers (112), carbamazepine (113), clozapine (114), lithium (115), SSRIs (116), and valproic acid (117). One or more compounds may be necessary for long-term management of aggressive behavior. Pabis and Stanislav (118) recommend lithium or propranolol as first-line anti-aggressive agents in patients without comorbid psychiatric disorders and suggest a 6- to 8-week trial of any compound to establish its efficacy. For aggressive and disinhibited behavior of a sexual nature, pharmacologic intervention may be used to target and reduce the libidinal urges of the individual. Three primary classes of medications have been used to manage sexual urges: SSRIs; antiandrogens and hormonal agents; and luteinizing hormone releasing hormone (LHRH) agonists (119). In a previous study, unacceptable sexual be-

havior was eliminated or reduced using a combination of antiandrogen medication and outpatient counseling in eight males who demonstrated sexually aggressive behavior after TBI (120). In a similar study, Britton (121) used medroxyprogesterone in a weekly 300-mg intramuscular injection to successfully control hypersexual behavior in a 36-year-old man who sustained a severe TBI.

Saver et al. (122) present a model of different types of aggression and suggest pharmacologic strategies based on the type of observed aggressive behavior and its purported neurocognitive substrate. They recommend that impulsive aggressive acts presenting after diencephalic injury or congenital brain damage respond best to β-adrenergic blockers (e.g., propranolol) or SSRIs (e.g., fluoxetine, sertraline, paroxetine) and possibly anticholinergic agents. For reflexive aggression seen after prefrontal TBI associated with social disinhibition, they recommend similar medications to the diencephalic aggression group but suggest avoiding the use of medications that may increase disinhibition (e.g., barbiturates, benzodiazepines). This is interesting because it appears to run against the historical approach of heavily sedating aggressive patients, and the ongoing prevalence of using benzodiazepines. For acute violence they recommend neuroleptics and benzodiazepines, but they caution the clinician to be aware that benzodiazepines can cause paradoxical rage (and, in the authors' experience with benzodiazepines, rebound aggression as the drug wears off).

Use of "serenic" agents specifically targeted to attenuate aggressive behavior is being investigated (123). Initial studies examined the role of serotonin (5-HT) in aggression (124) and suggested that 5-HT_1 agonists such as eltoprazine and fluprazine may be beneficial (125,126). Kravitz and Fawcette (127) argue that a limited number of animal studies and human studies on selected populations have demonstrated some use for serenic agents in aggression control. However, at least one animal study found increased emotional reactivity in mice with both eltoprazine and fluprazine (128). A human trial using eltoprazine to treat aggressive behavior in four participants with epilepsy and five with Tourette syndrome found no utility in aggression reduction, and the trial was terminated prematurely because of psychotic reactions in two participants (129). Further, a double-blind, placebo-controlled, and baseline-controlled study of eltoprazine for aggressive behavior in 160 developmentally disabled adults found no improvement on the Overt Aggression Scale, Global Aggression Score, or Social Dysfunction and Aggression Scale, or on clinical evaluation (130). Although the relationship between serotonin and aggression has been established (131), the use of antiaggressive drugs designed to act selectively on 5-HT requires further research to demonstrate use in humans. In terms of alternate serenic agents, Hurwitz (132) achieved ongoing moderate benefit in two brain-injured males using the synthetic cannabinoid nabilone. One patient displayed reduction in aggressive and socially inappropriate outbursts; the other displayed a reduction in both intensity and frequency of aggressive behaviors. Hurwitz (132) notes that because of the lack of empirical data supporting the use of cannabinoids for treatment of aggression and the expense of treatment, they should be considered only after all other reasonable therapies have failed. These preliminary findings are encouraging, but further research is necessary to identify whether nabilone and/or other synthetic cannabinoids [e.g., dronabinol (Marinol)] will prove to be a useful therapy for aggressive behaviors after acquired brain injury.

In general, pharmacologic agents should be thought of as a second-line treatment to permit a better opportunity to build trust, implement behavioral control strategies, and identify environmental triggers. Once the behavior is well controlled with a combination of medication and behavioral interventions, gradual withdrawal of the medication is recommended with maintenance or increased intensity of the behavioral strategy(s).

Strategies to Improve Initiation and Drive

The frontal lobes are essential for effective initiation and sequencing of action. Damage to this system can manifest in apparent disinterest or inactivity. There may be a significant disassociation between verbal output (i.e., what someone says they will do) and what he or she actually does. Often, the problem is not that the patient with executive function deficits cannot perform an action, but that it may not occur to them to perform the action at the appropriate time or place, or that they begin an activity but fail to maintain it. Generally, such patients respond well to external cues or prompts to initiate activity, and their behavior can be modified through behavioral techniques (133).

Teaching Task-Specific Routines

Executive functions involve the capacity to organize sequences of behavior and to carry out both familiar and novel activities. If the patient is very acute in the rehabilitation process and/or is demonstrating very severe executive function disturbance, it may be profitable to initially focus on *teaching task-specific routines*. The assumption here is that the patient will not be capable of a wide variety of different action plans in different settings because of stimulus boundedness, perseveration, severe related cognitive disorders of attention or memory, or extremely limited insight and awareness. In such individuals, it may not be reasonable to facilitate flexible individually determined sequences, but training of particular sequences for familiar, highly repetitious functional activities may be possible (134). Included here are a variety of grooming procedures, such as showering and dressing. Other common task-specific routines include preparation of simple meals or going to the cafeteria for lunch. Systematic use of checklists to guide behavior combined with self-instructional techniques have been shown to be effective (135).

Training Problem-Solving Strategies

Von Cramon et al. (136) described positive results in a series of patients with frontal lobe dysfunction using a training procedure designed to build problem-solving skills. The procedure focused on enabling patients to reduce the complexity of multistage problems by breaking them down into more manageable units. Therapy tasks involved working on hypothetical problems, such as making travel arrangements, planning a party, and identifying clues to solve a mystery. The most useful therapeutic activities for working on problem-solving skills involve applications and examples from everyday situations and functional activities.

Metacognitive Training Approaches

Insight, self-awareness, and self-regulatory capacity are believed to reflect the highest level of frontal lobe activity and executive function (137). Verbal self-regulation strategies are based on the observation that it is possible to regulate one's own behavior through self-talk. Stuss et al. (138) used a verbal self-regulation approach in an individual with motor impersistence who could not maintain a simple motor movement over time. The patient learned to alter his behavior, although he continued to need to be cued to initiate and maintain the self-regulation strategy. Additional work by Cicerone and Wood (139) and Cicerone and Giacino (140) reported successful use of self-instructional strategies in brain-injured patients. The goal of these cognitive interventions is to promote internalization of self-regulation strategies through the use of verbal self-instruction, self-questioning, and self-monitoring (45). Cognitive rehabilitation, in general, should include efforts to assess patients' awareness of their deficits and to improve the accuracy of patients' self-appraisal of their performance.

Remediation of Visuospatial Deficits

Most of the work on rehabilitation of visual spatial deficits has addressed problems related to visual neglect. A number of studies have shown that practice in visual scanning improves compensation for visual neglect after right hemisphere stroke (45,141). Training on more complex visuospatial tasks appears to enhance the benefits of scanning treatment and facilitates generalization to other visuospatial, academic, and everyday activities that require visual scanning. Treatment effects have been shown to enhance reading, driving, and other activities of daily living. In order to be effective, training needs to be relatively intense (i.e., daily) and involve apparatus that is large enough to engage peripheral vision. Training of scanning only via computer screen is less likely to be helpful. The positive effects translate into shorter lengths of stay in acute rehabilitation, and the effects appear to be maintained. Other techniques to decrease neglect include movement of the limb on the same side as the neglect, which is believed to activate the damaged hemisphere, as well as a variety of cuing and prompting devices (e.g., peripheral lights, tactile vibrators) that also draw attention to the affected side (142,143). Beyond work on neglect, there is no good evidence that training in other aspects of visuospatial analysis or constructional/drawing abilities are aided by rehabilitation, although there have been positive anecdotal reports of improvement in individual cases.

Remediation of Speech and Language Deficits

Language deficits after TBI and stroke can include specific speech and language disorders (i.e., aphasia, oral/verbal apraxia, agraphia, and alexia), as well as impairments in the pragmatics of communication. There also is a dynamic interaction between language and cognition such that linguistic processes are involved in the mediation of cognitive processes, and cognitive impairments result in related communication impairments. Good evidence exists to support the effectiveness of cognitive linguistic therapies beyond the period of spontaneous recovery in patients with language deficits from left hemisphere stroke (45). Many aspects of language and communication, including naming, oral expression, language comprehension, reading, writing, and gesturing, have been the focus of intervention. Both linguistic and pragmatic aspects of language show improvement but, from a functional perspective, there should be particular emphasis on improving pragmatic communication and conversational skills.

Principles of Cognitive Rehabilitation

Cognitive, speech language, and behavioral interventions take place in a wide variety of settings, including inpatient settings, acute rehabilitation, outpatient and day treatment programs, school settings, and private offices and clinics. Regardless of the setting, the following set of principles for implementing effective rehabilitation with individuals who demonstrate cognitive, behavioral, emotional, and psychosocial difficulties after acquired brain injury should be adhered to.

- Cognitive rehabilitation should be informed by medical and neuropsychological diagnosis, but it is based on an ever-evolving formulation of the individual client's needs and his or her problems and strengths in physical, cognitive, emotional, and social domains.
- Cognitive rehabilitation requires a sound therapeutic alliance among the therapist, the client, and the family or caregiver, and emphasizes collaboration, mutual goal setting, and active participation.
- Cognitive rehabilitation is goal oriented and focused on everyday problems and functional limitations resulting from the impairments. It builds on cognitive and behavioral strengths while addressing areas of deficit or concern.
- Cognitive rehabilitation is built on a theoretical framework with respect to brain and behavior relationships and

incorporates information from clinical psychology, behavioral psychology/learning theory, neuropsychology, and behavioral neurology.
- Cognitive rehabilitation involves a combination of techniques involving environmental modifications that allow more adaptive functioning, the identification and implementation of external and internal compensatory strategies, and the application of procedures to improve cognitive functioning.
- Cognitive rehabilitation is eclectic and uses a variety of techniques and strategies to improve abilities, teach new and compensatory skills, facilitate regulation of behavior, and modify negative or disruptive thoughts, feelings, and emotions.
- Cognitive rehabilitation assists clients in achieving a more accurate understanding of their strengths and limitations and in adjusting to injury-related changes in functioning and life circumstances.
- Cognitive rehabilitation emphasizes empowerment, self-regulation, and self-sufficiency.
- Cognitive rehabilitation professionals recognize and respond to the need to objectively evaluate the effectiveness of interventions. The field needs to be responsive to new evidence as it emerges and to changing theories and technologies. This is reflected in the more frequent use of adjunctive pharmacologic interventions as options and efficacy studies increase.

SUMMARY AND CONCLUSIONS

This chapter covered the major directions, findings, trends, and challenges facing rehabilitation therapists and researchers in their work with cognitively impaired individuals. There have been many exciting new developments in cognitive theory, knowledge about the effects of brain injury, support for neuroplasticity, and both acute and postacute pharmacologic interventions. The literature provides substantial support for a variety of rehabilitation practices that target attention, memory compensation, executive functions, visual scanning, and language. There is growing appreciation of the impact of premorbid and postmorbid variables that affect outcome and expanding efforts to minimize resultant challenges through neuroprotectant treatments in the acute phase of injury. Rehabilitation must improve everyday functioning and needs to take into account the individual client's needs and goals. Psychosocial and emotional functioning are critical variables that must be addressed within the rehabilitation plan. Additional outcome and efficacy research in neurorehabilitation are needed, but the field has come of age, with guidelines for current clinical practice and evidence-based interventions well articulated.

REFERENCES

1. Prigatano GP. Disturbances of self-awareness of deficit after traumatic brain injury. In: Prigatano GP, Schacter DL, eds. *Awareness*

of deficit after brain injury: clinical and theoretical perspectives. New York: Oxford University Press, 1991:115–135.

2. Doppenberg EMR, Choi SC, Bullock R. Clinical trials in traumatic brain injury. What can we learn from previous studies? In: Slikker W Jr, Trembly B, eds. *Neuroprotective agents. Annals of the New York Academy of Science, Vol. 825.* New York: New York Academy of Sciences, 1997:305–322.

3. Grant IS, Andrews PJD. ABC of intensive care: neurological support. *Br Med J* 1999;319;110–113.

4. Delanty N, Dichter MA. Antioxidant therapy in neurologic disease. *Arch Neurol* 2000;57:1265–1270.

5. Faden AI. Neuroprotection and traumatic brain injury: the search continues. *Arch Neurol* 2001;58:1553–1555.

6. Friedlander RM. Role of caspase 1 in neurologic disease. *Arch Neurol* 2000;57:1273–1276.

7. Maragakis NJ, Rothstein JD. Glutamate transporters in neurologic disease. *Arch Neurol* 2001;58,365–370.

8. Novack TA, Dillon MC, Jackson WT. Neurochemical mechanisms in brain injury and treatment: a review. *J Clin Exp Neuropsychol* 1996;18:685–706.

9. Alderson P, Roberts I. Corticosteroids in acute traumatic brain injury: systematic review of randomized controlled trials. *BMJ* 1997;314:1855–1863.

10. Yates D, Roberts I. Corticosteroids in head injury: it's time for a large sample randomized trial. *BMJ* 2000;321:128–129.

11. Kmictowicz Z. Trial of steroids for treating head injury begins. *BMJ* 1999;318:1441.

12. Fink KB, Andrews LJ, Butler WE, et al. Reduction of post-traumatic brain injury and free radical production by inhibition of the caspase cascade. *Neuroscience* 1999;94:1213–1218.

13. Olney JW. The toxic effects of glutamate and related compounds in the retina and the brain. *Retina* 1982;2:341–359.

14. Nakao N, Grasbon-Frodl EM, Widner H, et al. Antioxidant treatment protects striatal neurons against excitotoxic insults. *Neuroscience* 1996;73:185–200.

15. Greenfield S. Brain drugs of the future. *BMJ* 1998;317:1698–1701.

16. Showalter PEC, Kimmel DN. Stimulating consciousness and cognition following severe brain injury: a new potential use of lamotrigine. *Brain Inj* 2000;14:997–1001.

17. Piguet O, King AC, Harrison DP. Assessment of minimally responsive patients: clinical difficulties of single-case design. *Brain Inj* 1999;13:829–837

18. Worzniak M, Fetters MD, Comfort M. Methylphenidate in the treatment of coma. *J Fam Pract* 1997;44:495–498.

19. Meythaler JM, DePalma L, DeVivo MJ, et al. Sertraline to improve arousal and alertness in severe traumatic brain injury secondary to motor vehicle crashes. *Brain Inj* 2001;15:321–331.

20. Zafonte RD, Watanabe T, Mann NR. Amantadine: a potential treatment for the minimally conscious state. *Brain Inj* 1998;12:617–621.

21. Passler MA, Riggs RV. Positive outcomes in traumatic brain injury-vegetative state: patients treated with bromocriptine. *Arch Phys Med Rehabil* 2001;82:311–315.

22. Andrews K. International working party on the management of the vegetative state: summary report. *Brain Inj* 1996;10:797–806.

23. Blomert L. Recovery from language disorders: interactions between brain and rehabilitation. In: Stemmer B, ed. *Handbook of neurolinguistics.* New York: Academic Press, 1998:535–596.

24. Sohlberg MM, Mateer CA. *Cognitive rehabilitation: an integrated neuropsychological approach.* New York: Guilford Press, 2001.

25. Von Monakow C. *Localization in the cerebrum and the degeneration of functions through cortical sources.* Wiesbaden: JF Bergmann, 1914.

26. Le Vere TE. Neural stability, sparing, and behavioural recovery following brain damage. *Psychol Rev* 1975;82:344–358.

27. Kolb B, Gibb R. Neuroplasticity and recovery of function after brain injuries. In: Stuss DT, Winocur G, Robertson IH, eds. *Cognitive neurorehabilitation.* Cambridge, England: Cambridge University Press, 1999:9–25.

28. Keefe KA. Applying basic neuroscience to aphasia therapy: what are the animals telling us. *Am J Speech Pathol* 1995;4:88–93.

29. Kolb B. *Brain plasticity and behavior.* Hillsdale, NJ: Erlbaum, 1996.

30. Robertson IH, Murre JMJ. Rehabilitation of brain damage: brain plasticity and principles of guided recovery. *Psychol Bull* 1999;125:544–575.

31. Jones TA, Schallert T. Use-dependent growth of pyramidal neurons after neocortical damage. *J Neurosci* 1994;14:2140–2152.

32. Turner AM, Greenough WT. Differential rearing effects on rat visual cortex synapses. I. synaptic and neuronal densities and synapses per neuron. *Brain Res* 1985;329:195–203.

33. Hebb DO. *The organization of behavior: a neuropsychological theory.* New York: Wiley, 1949.

34. Pulvermuller F. Words in the brain's language. *Behav Brain Sci* 1999;22:253–336.

35. Le Vere TE. Recovery of function after brain damage: a theory of the behavioural deficit. *Physiol Psychol* 1980;8:297–308.

36. Gainotti G. The riddle of the right hemisphere's contribution to the recovery of language. *Eur J Disord Commun* 1993;28:227–246.

37. Kinsbourne M. Orientation bias model of unilateral neglect: evidence from attentional gradients within hemispace. In: Robertson IH, Marshall JC, eds. *Unilateral neglect: clinical and experimental studies.* Hillsdale, NJ: Erlbaum, 1993:63–86.

38. Weiller C, Chollet F, Friston KJ, et al. Functional reorganization of the brain in recovery from striato-capsular infarction in man. *Ann Neurol* 1992;31:463–472.

39. Taub E, Miller NE, Novack TA, et al. Technique to improve chronic motor deficit after stroke. *Arch Phys Med Rehabil* 1993;74:347–353.

40. Sturm W, Willmes K, Orgass B, et al. Do specific attention deficits need specific training? *Neuropsychol Rehabil* 1997;7:81–103.

41. Sohlberg MM, Mateer CA. *Introduction to cognitive rehabilitation: theory and practice.* New York: Guilford Press, 1989.

42. Chestnut RM, Carney N, Maynard H, et al. Agency for health care policy and research evidence-based practice report. NIH consensus development conference on rehabilitation of persons with traumatic brain injury. Programs and abstracts, 1998:73–82.

43. NIH Consensus Development Panel on Rehabilitation of Persons with Traumatic Brain Injury. Rehabilitation of persons with traumatic brain injury. *JAMA* 1999;282:974–983.

44. Carney N, Chetnut RM, Maynard H, et al. Effect of cognitive rehabilitation on outcomes for persons with traumatic brain injury: a systematic review. *J Head Trauma Rehabil* 1999;14:277–307.

45. Cicerone KD, Dahlberg C, Kalmar K, et al. Evidence-based cognitive rehabilitation: recommendations for clinical practice. *Arch Phys Med Rehabil* 2000;81:1596–1615.

46. Power M, Dalgleish T. *Cognition and emotion: from order to disorder.* Hove, UK: Psychology Press, 1997.

47. Helmsley G, Code C. Interactions between recovery in aphasia, emotional and psychosocial factors in subjects with aphasia, their significant others and speech pathologists. *Disabil Rehabil* 1996;18:567–584.

48. Herrmann M, Wallesch CW. Expectations of psychosocial adjustment in aphasia: a MAUT study with the Code-Muller protocols. *Aphasiology* 1990;3:513–526.

49. Herrman M, Code C. Weighting of items of the Code-Muller protocols: the effects of clinical experience of aphasia therapy. *Disabil Rehabil* 1996;18:509–514.

50. Bäckman L, Dixon RA. Psychological compensation: a theoretical framework. *Psychol Bull* 1992;112:259–283.
51. Banja J. Patient advocacy at risk: ethical, legal and political dimensions of adverse reimbursement practices in brain injury rehabilitation in the US. *Brain Inj* 1999;13:745–758.
52. Perna RB, Bordini EJ, Newman SA. Pharmacological treatments: considerations in brain injury. *J Cogn Rehabil* 2001;Spring:4–7.
53. Sohlberg MM, Mateer CA, Stuss DT. Contemporary approaches to the management of executive control dysfunction. *J Head Trauma Rehabil* 1993;8:45–58.
54. Webb PM, Glueckauf RL. The effects of direct involvement in goal setting on rehabilitation outcome for persons with traumatic brain injuries. *Rehabil Psychol* 1994;39:179–188.
55. Lezak M. Newer contributions to the neuropsychological assessment of executive functions. *J Head Trauma Rehabil* 1993;8:24–31.
56. Mateer CA, Mapou R. Assessment and management of attentional disorders following closed head injury. *J Head Trauma Rehabil* 1996;11:1–16.
57. Mateer CA, Sohlberg MM, Crinean J. Perceptions of memory functions in individuals with closed head injury. *J Head Trauma Rehabil* 1987;2:79–84.
58. Mateer CA. Systems of care for post-concussive syndrome. In: Horn LJ, Zasler ND, eds. *Rehabilitation of post-concussive disorders.* Philadelphia: Henley Belfus, 1992:143–155.
59. Mateer CA, Sohlberg MM. A paradigm shift in memory rehabilitation. In: Whitaker H, ed. *Neuropsychological studies of nonfocal brain injury: dementia and closed head injury.* New York: Springer-Verlag, 1988:202–225.
60. Van Zomeren AH, Brouwer WH, Deelman BG. Attention deficits: the riddle of selectivity speed and alertness. In: Brooks N, ed. *Closed head injury: psychological, social, and family consequences.* Oxford: Oxford University Press, 1984:74–107.
61. Chan RC. Attentional deficits in patients with post-concussion symptoms: a componential perspective. *Brain Inj* 2001;15:71–94.
62. Ben-Yishay Y, Piasetsky EB, Rattock J. A systematic method for ameliorating disorders in basic attention. In: Meyer MJ, Benton AL, Diller L, eds. *Neuropsychological rehabilitation.* Edinburgh: Churchill Livingstone, 1987:165–181.
63. Ethier M, Baribeau JMC, Braun CMJ. Computer-dispensed cognitive-perceptual training of closed head injury patients after spontaneous recovery. Study 2: non-speeded tasks. *Can J Rehabil* 1989;3:7–16.
64. Gray JM, Robertson I. Remediation of attentional difficulties following brain injury: three experimental single case designs. *Brain Inj* 1989;3:163–170.
65. Gray JM, Robertson I, Pentland B, et al. Microcomputer-based attentional retraining after brain damage: a randomised group controlled trial. *Neuropsychol Rehabil* 1992;2:97–115.
66. Kewman DG, Seigerman C, Kinter H, et al. Stimulation and training of psychomotor skills: teaching the brain-injured to drive. *Rehabil Psychol* 1985;30:11–27.
67. Niemann H, Ruff RM, Baser CA. Computer assisted attention training in head injured individuals: a controlled efficacy study of an outpatient program. *J Clin Consult Psychol* 1990;58:811–817.
68. Sohlberg MM, Mateer CA. Effectiveness of an attention training program. *J Clin Exp Neuropsychol* 1987;19:117–130.
69. Mateer CA, Sohlberg MM, Youngman P. The management of acquired attentional and memory disorders following mild closed head injury. In: Wood R, ed. *Cognitive rehabilitation in perspective.* London: Taylor and Francis, 1990:68–95.
70. Park N, Proulx GB, Towers WM. Evaluation of the attention process training programme. *Neuropsychol Rehabil* 1999;9:135–154.
71. Baribeau J, Ethier M, Braun C. A neurophysiological assessment of attention before and after cognitive remediation in patients with severe closed head injury. *J Neurol Rehabil* 1989;3:71–92.
72. Sohlberg MM, Johnson L, Paule L, et al. *Attention Process Training-II: A program to address attentional deficits for persons with mild cognitive dysfunction.* Wake Forest, NC: Lash Publishing, 1993.
73. Karli DC, Burke DT, Kim HJ, et al. Effects of dopaminergic combination therapy for frontal lobe dysfunction in traumatic brain injury rehabilitation. *Brain Inj* 1999;13:63–68.
74. Challman TD, Lipsky JJ. Methylphenidate: its pharmacology and uses. *Mayo Clinic Proc* 2000;75:711–721.
75. Cope DN. The pharmacology of attention and memory. *J Head Trauma Rehabil* 1986;1:34–42.
76. Lipper S, Tuchman MM. Treatment of chronic post-traumatic organic brain syndrome and dextroamphetamine: first reported case. *J Nerv Ment Dis* 1976;162:366–371.
77. Evans RW, Gualtieri CT, Paterson D. Treatment of chronic closed head injury with psychostimulant drugs. A controlled case study and an appropriate evaluation procedure. *J Nerv Ment Dis* 1987;175.
78. Gualtieri CT, Evans RW. Stimulant treatment for the neurobehavioural sequelae of traumatic brain injury. *Brain Inj* 1988;2:273–290.
79. Bleiburg J, Garmoe W, Cederquest J, et al. Effects of Dexedrine on performance consistency following brain injury. *Neuropsychiatry Neuropsychol Behav Neurol* 1993;6:245–248.
80. Kaelin DL, Cifu DX, Matthies B. Methylphenidate effect on attention deficit in the acutely brain-injured adult. *Arch Phys Med Rehabil* 1996;77:6–9.
81. Mehta MA, Owen AM, Sahakian BJ, et al. Methylphenidate enhances working memory by modulating discrete frontal and parietal lobe regions in the human brain. *J Neurosci* 2000;20RC65:1–6.
82. Speech TJ, Rao SM, Osmon DC, et al. A double-blind controlled study of methylphenidate treatment in closed head injury. *Brain Inj* 1993;7:333–338.
83. Plenger PM, Dixon CE, Castillo RM, et al. Subacute methylphenidate treatment for moderate to moderately severe traumatic brain injury: a preliminary double-blind placebo-controlled study. *Arch Phys Med Rehabil* 1996;77:536–540.
84. Whyte J, Hart T, Schuster K, et al. Effects of methylphenidate on attentional function after traumatic brain injury. *Am J Phys Med Rehabil* 1997;76:440–450.
85. Arciniegas DB, Held K, Wagner P. Cognitive impairment following traumatic brain injury. *Curr Treat Options Neurol* 2002;4:43–57.
86. Berg I, Konnig-Haaanstra M, Deelman B. Long term effects of memory rehabilitation. A controlled study. *Neuropsychol Rehabil* 1991;1:97–111.
87. Freeman MR, Mittenberg W, DiCowden M, et al. Executive and compensatory memory retraining in traumatic brain injury. *Brain Inj* 1992;6:65–70.
88. Burke J, Danick J, Bemis B, et al. A process approach to memory book training for neurological patients. *Brain Inj* 1994;8:71–81.
89. Schmitter-Edgecombe M, Fahy J, Whelan J, et al. Memory remediation after severe closed head injury. Notebook training versus supportive therapy. *J Consult Clin Psychol* 1995;63:484–489.
90. Squires EJ, Hunkin NM, Parkin AJ. Memory notebook training in a case of severe amnesia: generalization from paired

associate learning to real life. *Neuropsychol Rehabil* 1996;6:55–65.

91. Wilson BA, Evans JJ, Emslie H, et al. Evaluation of NeuroPage: a new memory aid. *J Neurol Neurosurg Psychiatry* 1997;63:113–115.

92. Sohlberg MM, Mateer CA. Training use of compensatory memory books: a three stage behavioral approach. *J Clin Exp Neuropsychol* 1989;11:871–891.

93. Kime S, Lamb D. Wilson G. Use of a comprehensive programme of external cueing to enhance procedural memory in a patient with dense amnesia. *Brain Inj* 1996;10:17–25.

94. Glisky EL. Acquisition and transfer of word processing skills by an amnesic patient. *Neuropsychol Rehabil* 1995;5:299–318.

95. Evans JJ, Wilson BA, Schuri U, et al. A comparison of "errorless" and "trial-and-error" learning methods for teaching individuals with acquired memory deficits. *Neuropsychol Rehabil* 2000;10:67–101.

96. Wilson BA, Baddeley AD, Evans E, et al. Errorless learning in the rehabilitation of memory impaired people. *Neuropsychol Rehabil* 1994;4:307–326.

97. Gottwald MD, Rozanski RI. Rivastigmine, a brain-region selective acetylcholinesterase inhibitor for treating Alzheimer's disease: review and current status. *Expert Opin Investig Drugs* 1999;8:1673–1682.

98. Aggleton JP, Brown MW. Episodic memory, amnesia, and the hippocampal–anterior thalamic axis. *Behav Brain Sci* 1999;22:425–489.

99. Whitlock JA Jr. Brain injury, cognitive impairment and donepezil. *J Head Trauma Rehabil* 1999;14:424–427.

100. Taverni JP, Seliger G, Lichtman SW. Donepezil mediated memory improvement in traumatic brain injury during post acute rehabilitation. *Brain Inj* 1998;12:77–80.

101. Whelan FJ, Walker MS, Schultz SK. Donepezil in the treatment of cognitive dysfunction associated with traumatic brain injury. *Ann Clin Psychiatry* 2000;12:131–135.

102. Masanic CA, Bayley MT, Van Reekum R, et al. Open-label study of donepezil in traumatic brain injury. *Arch Phys Med Rehabil* 2001;82:896–901.

103. Luria AR. *Higher cortical functions in man.* London: Tavistock, 1966.

104. Stuss DT, Benson DF. *The frontal lobes.* New York: Raven Press, 1986.

105. Fuster JM. *The prefrontal cortex: anatomy, physiology, and neuropsychology of the frontal lobe,* 2nd ed. New York: Raven Press, 1989.

106. Shallice T. Specific impairments of planning. *Phil Trans R Soc Lond* 1982;298:199–209.

107. Shallice T, Burgess PW. Higher-order cognitive impairments and frontal-lobe lesions in man. In: Levin H, Eisenberg HM, Benton AL, eds. *Frontal lobe function and injury.* Oxford, England: Oxford University Press, 1991:125–138.

108. Shallice T, Burgess PW. Deficits in strategy application following frontal lobe damage in man. *Brain* 1991;114:727–741.

109. Mateer C. Rehabilitation of individuals with frontal lobe impairment. In: Leon-Carrion J, ed. *Neuropsychological rehabilitation: fundamentals, innovations and directions.* Delray Beach, FL: GR/St. Lucie Press, 1997:285–300.

110. Alderman N, Burgess PW. Integrating cognition and behaviour: a pragmatic approach to brain injury rehabilitation. In: Wood RL, Fussey I, eds. *Cognitive rehabilitation in perspective.* London: Taylor Francis Ltd., 1990.

111. Alderman N, Ward A. Behavioural treatment of the dysexecutive syndrome: reduction of repetitive speech using response cost and cognitive overlearning. *Neuropsychol Rehabil* 1991;1:65–80.

112. Silver JM, Yudofsky SC, Slater JA, et al. Propranolol treatment of chronically hospitalized aggressive patients. *J Neuropsychiatry Clin Neurosci* 1999;11:328–335.

113. Azouvi P, Jokic C, Attal N, et al. Carbamazepine in agitation and aggressive behavior following severe closed-head injury: results of an open trial. *Brain Inj* 1999;13:797–804.

114. Stanislav SW. Cognitive effects of antipsychotic agents in persons with traumatic brain injury. *Brain Inj* 1997;11:335–341.

115. Bellus SB, Stewart D, Vergo JG, et al. The use of lithium in the treatment of aggressive behaviors with two brain-injured individuals in a state psychiatric hospital. *Brain Inj* 1996;10:849–860.

116. Lavine R. Psychopharmacological treatment of aggression and violence in the substance using population. *J Psychoact Drugs* 1997;29:321–329.

117. Wrobelski BA, Joseph AB, Kupfer J, et al. Effectiveness of valproic acid on destructive and aggressive behaviors in patients with acquired brain injury. *Brain Inj* 1997;11:37–47.

118. Pabis DJ, Stanislav SW. Pharmacotherapy of aggressive behavior. *Ann Pharmacother* 1996;30:278–87.

119. Bradford JM. The neurobiology, neuropharmacology, and the pharmacological treatment of the paraphilias and compulsive sexual behavior. *Can J Psychiatry* 2000;46:26–34.

120. Emory LE, Cole CM, Meyer WJ. Use of Depo Provera to control sexual aggression in persons with traumatic brain injury. *J Head Trauma Rehabil* 1995;10:47–58.

121. Britton KR. Medroxyprogesterone in the treatment of aggressive hypersexual behavior in traumatic brain injury. *Brain Inj* 1998;12:703–707.

122. Saver JL, Salloway SP, Devinsky O, et al. Neuropsychiatry of aggression. In: Fogel R, Schiffer R, Rao S, eds. *Neuropsychiatry.* Philadelphia: Williams & Wilkins, 1996:331–349.

123. Olivier B, VanDalen D, Hartog J. A new class of psychoactive drugs, serenics. *Drugs Future* 1986;11:473–499.

124. Coccaro EF. Central serotonin and impulsive aggression. *Br J Psychiatry* 1989;155:52–62.

125. Flannelly KJ, Muraoka MY, Blanchard DC, et al. Specific anti-aggressive effects of fluprazine hydrochloride. *Psychopharmacology* 1985;87:86–89.

126. Meek LR, Kemble ED. Effects of eltoprazine hydrochloride on predatory behavior in rats. *Psychol Rec* 1991;41:143–146.

127. Kravitz HM, Fawcett J. Serenics for aggressive behaviors. *Psychiatr Ann* 1994;24:453–459.

128. Griebel G, Saffroy-Spittler M, Misslin R, et al. Serenics fluprazine (DU 27716) and eltoprazine (DU 28853) enhance neophobic and emotional behaviour in mice. *Psychopharmacology* 1990;102:498–502.

129. Moriarty J, Schmitz B, Trimble MR, et al. A trial of eltoprazine in the treatment of aggressive behaviours in two populations: patients with epilepsy or Gilles de la Tourette's syndrome. *Hum Psychopharmacol Clin Exp* 1994;9:253–258.

130. de-Koning P, Mak M, de Vries MH, et al. Eltoprazine in aggressive mentally handicapped patients: a double-blind, placebo- and baseline-controlled multi-centre study. *Int Clin Psychopharmacol* 1994;9:187–194.

131. Lane SD, Cherek DR. Biological and behavioral investigation of aggression and impulsivity. In: Fishbein DH, ed. *The science, treatment, and prevention of antisocial behaviors: application to the criminal justice system.* Kingston, NJ: Civic Research Institute, 2000:5-1–5-21.

132. Hurwitz T. The synthetic cannabinoid nabilone in the treatment of aggressive behaviour in 2 males with acquired brain injuries (unpublished data). 2001.

133. Sohlberg MM, Sprunk H, Metzelaar K. Efficacy of an external

cuing system in an individual with severe frontal lobe damage. *Cogn Rehabil* 1988;4:36–40.

134. Craine SF. The retraining of frontal lobe dysfunction. In: Trexler LE, ed. *Cognitive rehabilitation: conceptualization and intervention.* New York: Plenum, 1982:239–262.

135. Burke WH, Zenicus AH, Wesolowski MD, et al. Improving executive function disorders in brain-injured clients. *Brain Inj* 1991;5:25–28.

136. Von Cramon DY, Matthes-von Cramon G, Mai N. Problem solving deficits in brain injured patients: a therapeutic approach. *Neuropsychol Rehabil* 1991;1:45–64.

137. Stuss DT. Disturbances of self-awareness after frontal lobe damage. In: Prigatano GP, Schacter DL, eds. *Awareness of deficit after brain injury: clinical and theoretical issues.* New York: Oxford University Press, 1991:63–83.

138. Stuss DT, Delgado M, Guzman DA. Verbal regulation in the control of motor impersistence. *J Neurol Rehabil* 1987;1:19–24.

139. Cicerone KD, Wood JC. Planning disorder after closed head injury: a case study. *Arch Phys Med Rehabil* 1987;68:111–115.

140. Cicerone KD, Giacino JT. Remediation of executive function deficits after traumatic brain injury. *Neuropsychol Rehabil* 1992;2:12–22.

141. Pizzamiglio L, Antonucci G, Judica A, et al. Cognitive rehabilitation of the hemineglect disorder in chronic patients with unilateral right brain damage. *J Clin Exp Neuropsychol* 1992;14:901–923.

142. Robertson IH, North N, Geggie C. Spatio-motor cueing in unilateral neglect: three single case studies of its therapeutic effectiveness. *J Neurol Neurosurg Psychiatry* 1992;55:799–805.

143. Robertson IH, Tegner R, Tham K, et al. Sustained attention training for unilateral neglect: theoretical and rehabilitation implications. *J Clin Exp Neuropsychol* 1995;17:416–430.

SECTION II

FUNCTIONAL BRAIN SYSTEMS

8

NEUROCHEMICAL NEUROANATOMY

DAVID G. STANDAERT
JOHN B. PENNEY, JR.

The human brain is the most complex organized structure known to exist. In addition to the billions of nerve cells with their trillions of synapses, there are dozens of neurotransmitters and neuromodulators. These transmitters, along with their receptors and second messenger systems, make the brain's chemistry unique. This chapter first reviews the synaptic organization that makes neurotransmission possible. It continues with a review of the anatomic distribution of these transmitters, modulators, and their receptors. Finally, the distributions of second messenger systems are briefly described. The numerous other compounds unique to the brain are omitted from this chapter. These include some compounds with distinct regional distribution such as growth factors and others such as neurofilament proteins, ion channel proteins in axons, and specific glial markers that are without special regional localization. This review is based on secondary sources, which are cited at the end of the chapter.

SYNAPTIC ORGANIZATION AND NOMENCLATURE

Synapses are sites of close apposition of neuronal processes that serve to convey information from one nerve cell to another. Information is carried across the synapse by one or more chemical neurotransmitter. Except for the gases, whose mechanisms of action as neuronal messengers are entirely different, all neurotransmitters require a similar set of specialized structures at the synapse to convey a message from one neuron to another (Fig. 8.1). Most synapses in the brain are unidirectional, with a presynaptic terminal that releases a signal and a postsynaptic dendrite or spine that is the recipient. Some examples of reciprocal synapses are known to exist that have both kinds of components on each side of the point of contact.

The Presynaptic Terminal

The presynaptic terminal is responsible for the release of neurotransmitters. Most neurotransmitters are produced lo-

cally by enzymes within the presynaptic terminal. After synthesis, they are stored in small, spherical, membrane-bound structures, the presynaptic vesicles, through the action of a class of proteins known as vesicular transporters. These transporters are found on the surface of presynaptic vesicles and use energy from adenosine 5'-triphosphate to translocate the transmitters from the cytoplasm of the terminal into the vesicles. Each vesicular transporter protein stores a different type of neurotransmitter. A different process is used for the peptides, which are synthesized on ribosomes and packaged into vesicles in the neuronal cell body, and the vesicles are then transported down the axon to the terminal. Transmitters stored in presynaptic vesicles are released through a process of exocytosis, which requires that the membranes of the vesicles fuse with the membrane of the presynaptic terminal and release the vesicular contents into the synaptic cleft. This fusion and release occur along the presynaptic active zone, which contains a number of specialized molecules required for the process. In most cases, the release of vesicles is triggered by the entry of calcium ion into the presynaptic terminal. Calcium triggers a molecular cascade resulting in activation of a molecular machinery, the SNAP and SNARE proteins, responsible for the movement of presynaptic vesicles to the active zone and release of their contents.

Postsynaptic Receptors

The postsynaptic membrane is studded with a high density of neurotransmitter receptors. Binding of a neurotransmitter to its receptor starts the postsynaptic cell's response to the transmitter. Receptors are either linked to ion channels that open to initiate the cell's response or are linked to guanosine triphosphate binding proteins (G proteins), which, in turn, activate second messenger systems within the postsynaptic cell. Numerous types of receptor may be located at a single synapse.

The effect that activation of the receptor has on the postsynaptic neuron is highly dependent on the synapse's location on the postsynaptic neuron. Ion channel–linked receptors located on the neuronal soma near the axon hillock (where

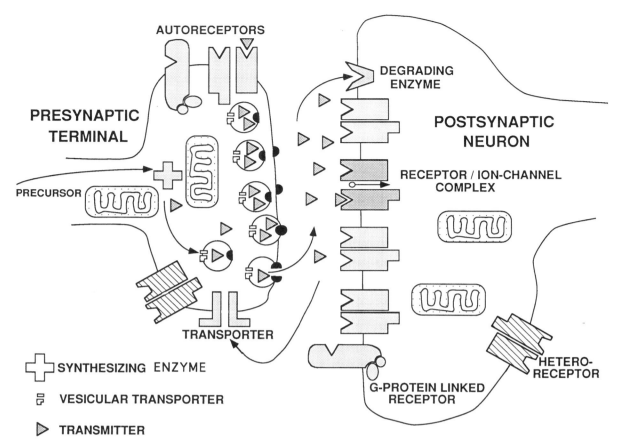

FIGURE 8.1. Schematic drawing of the fate of neurotransmitters in synaptic transmission. In the presynaptic terminal, the transmitter is synthesized from a precursor, transported into storage vesicles by a vesicular transporter, and released into the synaptic cleft by depolarization of the presynaptic terminal. Once released, the transmitter diffuses into the synaptic cleft, where it can interact with receptors that have sites to which it can bind. Receptors can be either on the post- or presynaptic cell. Autoreceptors on the presynaptic terminal regulate transmitter synthesis and release. Receptors on the postsynaptic membrane initiate changes in the postsynaptic cell that are the ultimate result of transmission at the synapse. Heteroreceptors (receptors for other neurotransmitters) may also regulate synaptic activity. Receptors may be either of the ligand gated ion channel or the G protein–coupled type. Ligand gated ion channels open on binding the transmitter to permit passage of ions through the cell's plasma membrane. (One channel is shown being opened and passing an ion on the postsynaptic cell.) G protein–coupled receptors initiate second messenger cascades within the cells on which they reside. Regardless of whether the transmitter binds to a receptor, it eventually is removed from the synapse either by being metabolized by a degrading enzyme so that it is no longer active or by being transported back into the presynaptic terminal by a high-affinity transporter.

the cell's action potential is generated) will strongly influence the hillock's membrane potential and thus whether an action potential will be generated. Activation of ion channel–linked receptors at synapses on spines of distant dendrites will have only a very small effect on the neuron membrane's potential, and simultaneous activation of many such distant synapses will be required to alter a neuron's activity. The effects of G protein–linked receptors are not as limited by their location because they activate second messenger cascades.

The ion channel receptors, known as ligand gated ion channels, are each assembled from several protein subunits,

usually four or five, that are coded by different genes. Each subunit has several membrane-spanning regions with the amino terminal end of the protein being extracellular (Fig. 8.2). The assembled subunits form a pore through the membrane, which, when the receptor is activated, allows the flux of ions driven by their concentration gradients. For most of the receptor families, the second membrane-spanning portion of each subunit forms the lining of this pore, and specific amino acids in this region regulate which ions may permeate.

G protein–linked receptors typically have seven membrane-spanning regions (Fig. 8.2). The amino terminal end is extracellular. These receptors do not form channels; the binding of transmitters to the extracellular part of the protein triggers conformational changes that, in turn, alter the interaction of the intracellular portion of the receptors with G proteins. In their resting state, before activation by a transmitter, G proteins are bound to the intracellular part of the receptors, and the guanine nucleotide binding site of the G protein is occupied by guanosine 5′-diphosphate (GDP). Neurotransmitter binding produces conformational changes that result in the G protein assuming a guanosine 5′-triphosphate (GTP)–preferring form. Once this occurs, GTP displaces GDP and the G protein dissociates from the receptor, leaving it free to diffuse through the cytoplasm of the cell. These GTP-associated G proteins can trigger a variety of second messenger systems, for example, by activating adenylate cyclase, the enzyme that forms cyclic adenosine 3′,5′-monophosphate (cyclic AMP), or phospholipase C, the enzyme that initiates the phosphatidylinositol cycle. The effects of the G proteins are terminated when GTP is hydrolyzed to GDP, returning it to a state with a preference for binding to a G protein–linked receptor.

G proteins are themselves composed of three different subunits, the most important of which are the α subunits. The α subunits determine which receptors a G protein complex will interact with and what kind of effector system will be modified. More than a dozen different α subunits are known to be present in the brain, although many are expressed only in specific regions. Each of these has a distinct role. For example, $G\alpha_s$ subunits activate adenylate cyclase (and thus cyclic AMP) and calcium channels. $G\alpha_i$ proteins inhibit the activity of adenylate cyclase and activate phospholipase C, phospholipase A_2, and potassium channels. $G\alpha_o$ subunits appear to be involved in modulating the activity of voltage-sensitive calcium channels and phospholipase C. $G\alpha_{olf}$ subunits are located in the neurons in the nasal mucosa, where they are responsible for transducing signals received from the olfactory receptors.

Controlling Synaptic Activation

Turning a synaptic signal off at the proper time is just as important as turning it on. The brain has several mechanisms for limiting the duration of the synaptic signal. One important mechanism is reuptake, accomplished by a class of high-affinity transporters. Most of these transporters are found on presynaptic terminals, although some are also present on glial processes that surround the synapse. Uptake of transmitters by the transporters serves two purposes: it terminates the action of the neurotransmitters by removing them from the synaptic cleft and it spares the presynaptic cell the work of resynthesizing the neurotransmitter by providing a premade source of the chemical. These cell surface neuronal transporters are different proteins from the vesicular transporters. Most of the known transporters are in a single family of proteins. The family includes the transporters for

γ-aminobutyric acid (GABA), norepinephrine, dopamine, serotonin, glycine, adenosine, and proline. Members of this family transport neurotransmitters across the cell membrane by transporting a sodium and, frequently, a chloride ion along with each molecule of transmitter. These transporters all share a similar structure having 12 membrane-spanning regions (Fig. 8.2). However, the transporters for the excitatory messenger glutamate are different. Instead of transporting a glutamate molecule with a sodium and a chloride ion, the glutamate is transported along with two sodium ions while one potassium ion and one hydroxyl ion are counter-transported during the process. The glutamate transporters have a different structure as well with either eight (or possibly ten) transmembrane regions rather than the 12 transmembrane regions found in the main family of neurotransmitter transporters.

A second important mechanism for limiting the duration of synaptic signaling is the presence of enzymes that degrade neurotransmitters after they have been released. Enzymatic degradation is particularly important for terminating the actions of acetylcholine, which is broken down by acetylcholinesterase, which is present in the synaptic cleft. Enzymatic degradation is also the main mechanism for limiting the action of many of the peptides and makes an important contribution to regulating the action of dopamine, epinephrine, and norepinephrine.

A third regulatory mechanism is the presence of presynaptic receptors located on the presynaptic terminal's membrane. Binding of the transmitter to these sites stimulates the receptor to modulate the activity of the terminal through one of two mechanisms. Presynaptic receptors can influence neurotransmitter release by being coupled to ion channels that change the presynaptic terminal's membrane potential. Because neurotransmitter release is governed by the amount of membrane depolarization that occurs when the impulse that has been conducted along the axon reaches the terminal, depolarizing the presynaptic terminal increases neurotransmitter release, whereas hyperpolarizing it decreases release. Alternatively, presynaptic receptors may be coupled via second messengers to either change the rate of neurotransmitter synthesis or to modulate it. Autoreceptors bind the neurotransmitter that the presynaptic terminal is releasing. Heteroreceptors bind other neurotransmitters.

A final means of limiting the action of neurotransmitters is compensatory changes in the receptor system. This adaptation of the receptors can occur through several mechanisms and over different time scales. Ligand gated ion channels often exhibit desensitization, a loss of sensitivity to a transmitter after continuous stimulation. Desensitization is often owing to conformational changes in the receptor and occurs in a fraction of a second. Intermediate duration changes can occur through covalent modifications of receptor subunits, such as phosphorylation, or internalization of the receptor, which removes it from the synapse and makes it inaccessible to the transmitter. Very long duration changes can occur

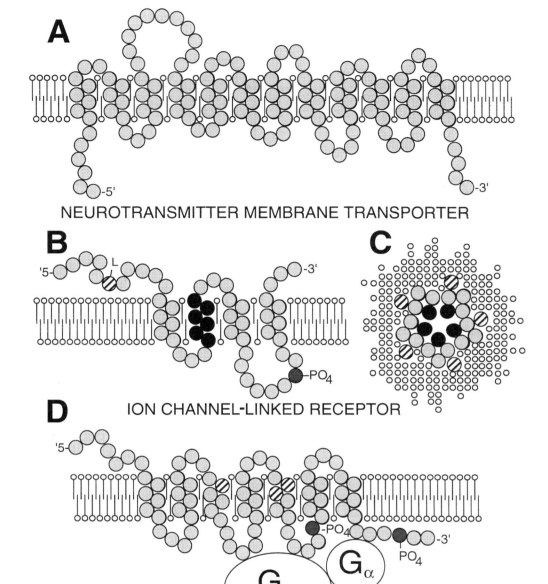

A NEUROTRANSMITTER MEMBRANE TRANSPORTER

B ION CHANNEL-LINKED RECEPTOR

C

D G-PROTEIN-LINKED RECEPTOR

FIGURE 8.2. Schematic drawing of the presumed shapes of the proteins that bind to neurotransmitters and are inserted in the synaptic membrane. The proteins are shown as a chain of circles that are a simplified representation of the protein's amino acid that makes up the sequence running from the amino (5′) to the carboxy (3′) terminal. In reality, all the proteins depicted have several times as many amino acids as are shown in these schematics. **A:** Cross-section of a member of the family of high-affinity transporters of neurotransmitters that includes the γ-aminobutyric acid, norepinephrine, dopamine, and serotonin transporters. These proteins remove transmitters from the synaptic cleft by transporting them into the presynaptic terminal along with sodium and chloride ions. These transporters are thought to have 12 membrane-spanning regions with long 5′ and 3′ intracytoplasmic ends. The second extracellular loop is also thought to be longer than the other loops. **B:** Cross-section of the presumed structure of one of the peptides that make up the ligand gated ion channel type of neurotransmitter. Each peptide probably has four membrane-spanning regions, with the 5′ end of the peptide on the extracellular side of the membrane. The portion of the peptide chain that binds the transmitter *(striped)* is in the 5′ extracellular end. Regulatory phosphorylation sites *(stippled)* are typically on the second intracytoplasmic loop. The second transmembrane segment is shown in black because this is the region believed to line the ion channel itself. **C:** Depiction of a face-on view of how five peptide chains are thought to fit

through changes in gene expression, altering the synthesis of neurotransmitter receptors, or synaptic plasticity, altering the number and structure of synapses.

NEUROTRANSMITTERS AND MODULATORS

The neurotransmitters and neuromodulators fall into several categories. The first category is the amino acids. These compounds are responsible for rapid synaptic activity at the vast majority (~90%) of synapses in the brain. Amino acid pathways tend to go from one specific spot in the nervous system to another. The second category comprises the amines, acetylcholine, and adenosine. These compounds largely act through second messenger systems. The axon terminals of neurons that use this class of transmitter tend to be more widely distributed than those of the amino acids. The third major class is the peptides. Like the amines, these neuromodulators act through second messenger systems, but the axon terminals of peptide neurons tend to have a more limited distribution than those of the amines. The fourth, and newly discovered class of neurotransmitters, is the gases. The gases seem to be used in local circuit neurons rather than in long projection pathways.

Many neurons use more than one type of molecule as a neurotransmitter or modulator. Peptides are often used by cells that also use an amino acid, amine, or gas. This enables neurons to give signals of differing durations. Whether all the neurotransmitters/modulators contained in a neuron are released at every synapse remains unknown. Thus, it is possible that neurons can release one neurotransmitter at some of its terminals and another at other terminals. Alternatively, although every neurotransmitter that a neuron possesses may be released at every synapse, the postsynaptic cell may not be sensitive to all the transmitters released. This concept is consistent with the finding that the distribution of receptors, particularly peptide receptors, often does not match the distribution of presynaptic terminals containing that transmitter. It has been hypothesized that the transmitters released at a synapse where there are no receptors for them may diffuse away to distant places, contributing to a general neuropep-

tide milieu reflecting the total ongoing peptidergic activity in the brain. This hypothesis remains highly controversial. All that is known with certainty is that the distribution of neurotransmitters and that of their receptors do not always match.

The remainder of this section describes the distribution of neurotransmitter pathways and their receptors. The distributions are presented in a series of schematic drawings showing the neurotransmitter-specific pathways and the densities of their receptors in a variety of brain regions. The names of these regions are shown in Fig. 8.3. In the subsequent figures, the neurotransmitter specific pathways and densities of receptors are shown. The densities shown are based on available human and animal reports. Where the density of a receptor in a region has not been reported, an estimate of the density has been made based on the known distribution in similar regions.

Amino Acid Neurotransmitters

Amino acids are the neurotransmitters used by the majority of neurons for rapid (millisecond timescale) intercellular communication. More than 90% of neurons use an amino acid for neurotransmission. Frequently, cells that use an amino acid will also use an amine or peptide. The amino acids used for excitatory neurotransmission are glutamic acid and, possibly, aspartic acid. The inhibitory amino acids are glycine, GABA, and possibly taurine and proline. Aspartic acid, taurine, and proline are not discussed further.

Glutamic Acid

Glutamic acid (glutamate) is by far the most important excitatory neurotransmitter in vertebrate brains. This important role was recognized only relatively recently because glutamate is also an essential amino acid and a key component of intracellular metabolism. Consequently, glutamate is found throughout the brain. Its actions are so ubiquitous that, at first, investigators thought that glutamate's ability to excite neurons must be a nonspecific artifact rather than a true neurotransmitter action because all neurons tested responded to glutamate's excitatory effects.

FIGURE 8.2. together to make a functional ion channel. The five membrane-spanning regions form the walls of the ion channel, with the ligand-binding regions on the outside of the membrane. Usually two molecules of neurotransmitter are required to bind to the receptor to cause the allosteric change that opens the ion channel. **D:** Typical G protein–coupled receptor. These receptors have seven membrane-spanning regions, with the transmitter-binding region located on the third, fifth, and possibly more membrane-spanning regions. Phosphorylation sites are located on the third intracytoplasmic loop and the 3′ end of the peptide. The G protein complex is thought to bind to the second and third intracytoplasmic loops and the 3′ end of the receptor. Somehow binding of neurotransmitter to the transmembrane regions of the receptor induces an allosteric change in the G protein's α subunit so that it binds guanosine 5′-triphosphate (GTP) with higher affinity than guanosine 5′-diphosphate. Once GTP binds to the α subunit, the subunit dissociates from the β and γ subunits and diffuses into the cell, where it can stimulate or inhibit specific second messenger systems. The ultimate action induced in the cell is determined largely by the α subunit.

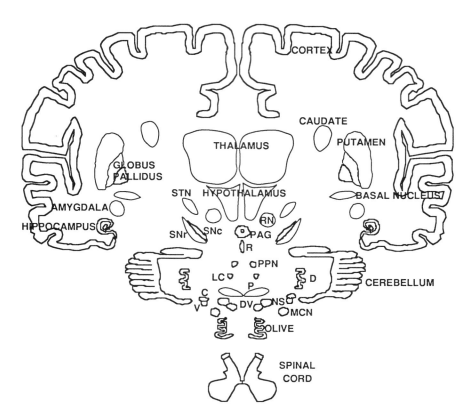

FIGURE 8.3. Key to the subsequent neurotransmitter and receptor distribution figures. The names of various structures in the brain are provided here. The subsequent figures are all schematic drawings that show the current concept of the distribution of a neurotransmitter and/or its receptors. The loci of transmitter-specific neurons are shown as filled circles with their pathways shown as lines projecting from the neurons. The density of receptors is shown as cross-hatching of the various regions. Where the density of a receptor in human brain is unclear, the density has been estimated from reports of other species and from measurements from similar structures. C, cochlear nucleus; D, dentate nucleus of cerebellum; DV, dorsal motor nucleus of the vagus nerve; LC, locus ceruleus; MCN, somatic motor cranial nerve nuclei; NS, nucleus of the solitary tract; P, pontine nuclei; PAG, periaqueductal gray; PPN, pedunculopontine nucleus; R, raphe nuclei; RN, red nucleus; SNc, substantia nigra pars compacta; SNr, substantia nigra pars reticulata; STN, subthalamic nucleus; V, vestibular nuclei.

It is now clear that glutamate is a transmitter that is employed by specific neurons in the brain. Only approximately 25% of the glutamate in the central nervous system is in the neurotransmitter pool. Although much of the neurotransmitter glutamate may come from the ordinary metabolic pool, there are also enzymes present at some excitatory synapses that can synthesize glutamate from other precursors, including aspartic acid and glutamine. Glutamate is stored within presynaptic vesicles by a vesicular transporter. Glutamate release occurs through a typical process of calcium ion–dependent exocytosis. Glutamate released into the synaptic cleft can be taken back up by transporters found on both the presynaptic glutamatergic terminals and glia.

Glutamate Pathways

Most of the excitatory pathways in the brain are known to use glutamate as a neurotransmitter (Fig. 8.4). These include the cortical output pathways to subcortical structures, which include the basal ganglia, thalamus, pons, brainstem, and spinal cord. The long intracortical pathways are glutamatergic as are the pathways within the hippocampal formation, including the perforant pathway from the entorhinal cortex to the molecular layers of the dentate gyrus and CA1, the mossy fiber pathway from the dentate granule cells to the pyramidal cells of CA3, and the Schaeffer collateral pathway from CA3 to CA1 as well as the hippocampal commissural pathway of which the Schaeffer pathway fibers are collaterals. Subthalamic nucleus neurons use glutamate. The main

inputs to the cerebellum, the climbing fiber pathway from the inferior olive and the mossy fibers from the pontine nuclei, use glutamate. The only excitatory neurons in the cerebellar cortex, the cerebellar granule cells, are glutamatergic. Many, if not all, of the primary sensory fibers use glutamate as (one of) their transmitters.

A number of the other major anatomic pathways in the brain are probably glutamatergic, although the glutamatergic nature of these pathways has not been proven. These include the projections from the thalamus to the cortex and the ascending sensory pathways. Some brainstem pathways such as the median longitudinal fasciculus probably also use glutamate.

Glutamate Pharmacology and Receptors

Four pharmacologic types of glutamate receptor are known to be present in the mammalian brain. Three of the types are ligand gated ion channels. These have each been named for a prototypic drug that acts specifically at that receptor, although in the brain, all of them are activated by glutamate. The fourth type is the G protein–linked metabotropic glutamate receptors.

α-Amino-3-hydroxy-5-methylisoxazole-4-propionic Acid Receptors

Receptors of the α-amino-3-hydroxy-5-methylisoxazole-4-propionic acid (AMPA) class mediate most of the fast,

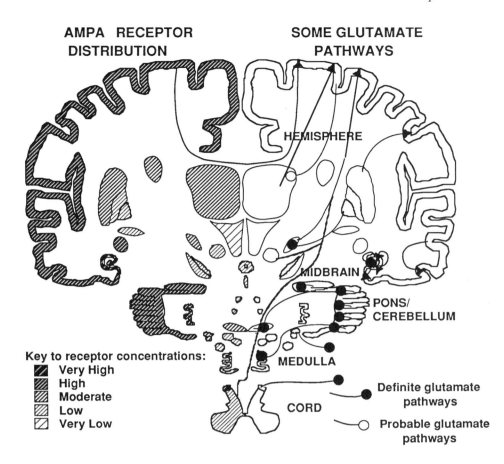

AMPA RECEPTOR DISTRIBUTION

SOME GLUTAMATE PATHWAYS

HEMISPHERE

MIDBRAIN

PONS/ CEREBELLUM

MEDULLA

CORD

Key to receptor concentrations:
- Very High
- High
- Moderate
- Low
- Very Low

● Definite glutamate pathways

○ Probable glutamate pathways

FIGURE 8.4. Some glutamate pathways and the distribution of the α-amino-3-hydroxy-5-methylisoxazole-4-propionic acid subtype of glutamate receptors.

excitatory neurotransmission in the brain. Most of these channels act by regulating the flux of sodium ions across the membrane and do not allow the passage of divalent cations such as calcium. Structural variants capable of allowing entry of both calcium and sodium ions do exist; these are scarce in the normal adult brain but may be increased during neural development or in pathologic states. AMPA receptors desensitize rapidly so that the stimulation of the postsynaptic cell by activation of an AMPA receptor is brief. This makes it likely that AMPA receptors are used to convey highly time- and location-specific information along the excitatory pathways.

AMPA receptors are present throughout the brain. Their distribution is shown in Figure 8.4. They are densest in the hippocampus; dense in neocortex, on the dendrites of Purkinje cells in the cerebellum, and in the dorsal horn of the spinal cord; moderate in the caudate, putamen, thalamus, and subthalamus; and of lower density in the rest of the brain.

A family of genes that code for the AMPA receptor has been cloned. Four genes (*GluR1, GluR2, GluR3,* and *GluR4*) are known whose sequences are approximately 70% identical. The functional receptors that are present on neuronal membranes probably consist of a complex of four of these proteins. Each complex seems to be made up of at least one GluR2 molecule and three other molecules. The exact composition may vary from cell to cell and from region to region within a cell. The diversity of these channels is further increased by variations introduced in the process of DNA transcription. In the transcription of any mammalian gene, the noncoding introns that are present as inserts within the DNA coding sequence are removed from the primary RNA transcript and the exons that code for the amino acid sequence of the protein are spliced together. During transcription of the AMPA subunits, one of the exons may be spliced in two different ways. Thus, the resultant mRNA can have two different splice variants, which have been called "flip" and "flop." In addition, the RNA sequence that codes for the second transmembrane domain of *GluR2* is "edited" so that a codon that carries the instructions for a glutamine is changed into a codon that codes for an arginine. This unusual biochemical process has a crucial function because it occurs in the second transmembrane domain that lines the ion channel, and it is this arginine that usually prevents calcium ions from flowing through the channel.

N-Methyl-D-Aspartate Receptors

The *N*-methyl-D-aspartate (NMDA) type of glutamate receptor has been studied more extensively than all the other classes. NMDA receptors are also ligand gated ion channels, but, unlike the AMPA receptors, when activated, NMDA receptors are highly permeable to calcium and sodium ions. The NMDA receptors have an important regulatory feature, which is voltage-dependent blockade by magnesium ions. At normal resting membrane potentials, extracellular magnesium ions will block the pore of NMDA channels so that activation of the receptor by glutamate does not produce any transmembrane ion flux. Depolarization of the postsynaptic membrane relieves the blockade by magnesium ions and allows glutamate to induce a flux of sodium and calcium ions. This property allows the NMDA receptors to function as "coincidence detectors." Calcium ion flux takes place only when two different events occur simultaneously: (a) glutamate is bound to the NMDA receptor and (b) the cell membrane is depolarized through some process independent of NMDA receptors, for example, by the action of glutamate at nearby AMPA receptors. In addition, the NMDA receptor has two other regulator sites, one that binds the amino acid glycine (although recent evidence indicates that the endogenous ligand for this site may be the unusual amino acid D-serine) and another that binds polyamines.

NMDA receptors have a crucial role in the normal functioning of the brain, but excessive activation can lead to neuronal injury or death, a process termed excitotoxicity. Under physiologic conditions, the calcium ions that enter through activated NMDA receptors can induce a variety of cellular changes that contribute to learning and the formation and storage of memories. Blockade of NMDA receptors by drugs such as the dissociative anesthetic ketamine or the drug of abuse phencyclidine can produce amnesia and hallucinosis. Conversely, excessive activation of NMDA receptors, especially under conditions such as ischemia in which cell membranes are already depolarized, can lead to massive calcium influx and trigger irreversible processes leading to cell death.

The distribution of NMDA receptors is shown in Figure 8.5. NMDA receptors are present on practically all neurons. They are densest on the pyramidal neurons of the hippocampus. They are very dense throughout the neocortex, less dense in subcortical structures, and quite low in number in the brainstem. In the cerebellum, they are located on granule cell neurons, with very few on the Purkinje cells. There is a moderate number of NMDA receptors in the dorsal horn of the spinal cord, where they may be involved in pain transmission pathways.

Structurally, the NMDA receptors are similar to other types of ligand gated ion channels in that they are composed of several (probably four) different subunits assembled to form a pore through the membrane. Two families of genes

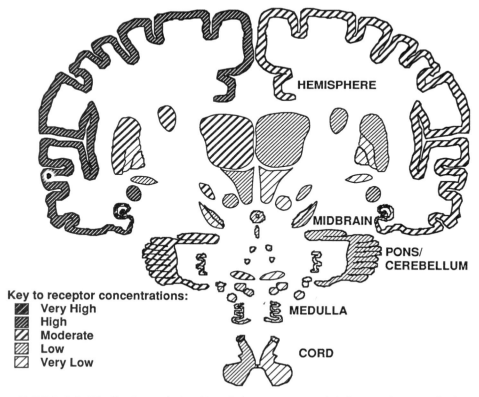

NMDA RECEPTOR DISTRIBUTION   **KAINATE RECEPTOR DISTRIBUTION**

HEMISPHERE

MIDBRAIN

PONS/
CEREBELLUM

MEDULLA

CORD

Key to receptor concentrations:
Very High
High
Moderate
Low
Very Low

FIGURE 8.5. Distributions of the *N*-methyl-D-aspartate and kainate subtypes of glutamate receptors.

have been cloned that encode NMDA receptor subunits. One gene codes for the subunit NMDAR1 and is expressed in virtually all neurons. Like the AMPA receptors, the mRNA for this single gene is edited into a number of different splice variants before it is translated into a protein. These splice variants differ in how sensitive they are to modulation by glycine and how many phosphorylation sites they possess. Neurons differ in which splice variant they express, thus controlling their sensitivity to glycine and phosphorylation. The NMDAR2 subunit family consists of at least four different genes that are highly (approximately 70%) homologous to one another but share only a 25% homology with the *NMDAR1* gene. NMDAR2A is expressed throughout the brain, NMDAR2B in the forebrain, and NMDAR2C in the cerebellum. NMDAR2D is expressed in the diencephalon, midbrain, and brainstem. Although the exact composition of brain NMDA receptors has not yet been determined, it is thought that all active receptors contain at least one NMDAR1 subunit and one or more NMDAR2 subunits. There are of course many different possible combinations that can be assembled under these rules. It is hoped that medicinal chemists will be able to exploit these properties to produce NMDA receptor antagonists that are relatively selective for particular types of NMDA receptors, but at present no such selective agents are available for clinical use.

Kainate Receptors

The kainate family of glutamate receptors shares structural and functional properties with the AMPA receptors. Like the AMPA receptors, when kainate receptors are activated, they allow the flux of sodium ions. They are also assembled from a set of receptor subunit proteins that are similar to the AMPA receptor subunits. Five kainate receptor subunits have been cloned and are named GluR5, GluR6, GluR7, KA1, and KA2. Kainate receptors do not desensitize as rapidly as AMPA receptors, a property that contributes to a form of neurotoxicity caused by eating contaminated shellfish. Microorganisms found in the shellfish may produce either kainate itself or the more potent domoic acid. These toxins cause long-lasting activation of kainate receptors, allowing sodium ions to flood into the neuron and kill it through an osmotic overload. Much less is known about the function of kainate receptors under physiologic conditions, largely because of a lack of experimentally useful selective antagonists. Recently, a variety of genetically engineered mice with deletions of particular kainate receptor subunits has been produced, and these animals have begun to reveal the role of kainate receptors in synaptic plasticity.

The distribution of kainate receptors is shown in Figure 8.5. The receptors are extremely dense on postsynaptic neurons in the stratum lucidum of the CA3 region of the hippocampus. This is the area where the mossy fiber pathway from the dentate gyrus to CA3 terminates. CA3 neurons are, by far, the most susceptible to the toxic effects of kainate and

domoate. There is a moderate number of kainate receptors in the deep layers of the cortex and few to very few of them elsewhere in the brain.

Metabotropic Glutamate Receptors

In addition to the three types of ion channel–linked excitatory amino acid receptors, there are also G protein–linked receptors, the metabotropic glutamate receptors (mGluR), that are activated by glutamate. In contrast to the ion channel glutamate receptors, each mGluR is made up of only a single protein. Eight different mGluR (mGluR1–mGluR8) proteins have been identified, and each of these has seven membrane-spanning domains, as do all other G protein–linked receptors. The eight different mGluRs can be subdivided into three groups based on their structure and pharmacologic properties. Group 1 mGluRs are coded by the genes *mGluR1* and *mGluR5*. They are linked to phosphatidylinositol metabolism and mGluR1 is also linked to weak stimulation of cyclic AMP. The group 2 mGluRs (coded by the genes *mGluR2* and *mGluR3*) are linked to cyclic AMP inhibition. The Group 3 receptors, coded by the *mGluR4*, *mGluR6*, and *mGluR7* genes are linked to cyclic AMP stimulation. The distribution of the mGluRs is shown in Figure 8.6. Group 1 receptors are extremely dense in the Purkinje cells of the cerebellum, with moderate amounts in the thalamus, and smaller amounts in the basal ganglia, hippocampus, and cortex. Group 2 receptors are relatively dense in the cortex, hippocampus, basal ganglia, and cerebellum. Group 2 receptors are also present in glial cells, coded by the *mGluR3* gene. The mGluR4 and mGluR7 receptors have complementary distributions in the brain, with mGluR4 receptors being almost exclusively located in cerebellar granule cells, with a few in the thalamus, caudate nucleus, putamen, and the CA2 region of the hippocampus. mGluR7 receptors are located in most other neurons. mGluR6 receptors are found only in the "on" bipolar cells of the retina. mGluR8 is also very scarce in the adult brain.

Glycine

Glycine, like glutamate, is an essential amino acid that is present in all cells. Because the alkaloid poison strychnine acts by blocking glycine receptors, glycine was shown to be a neurotransmitter before glutamate, even though there are many more glutamatergic than glycinergic neurons. The principal action of glycine, besides its role in intermediary metabolism, is as an inhibitory neurotransmitter.

All known glycinergic neurons are inhibitory interneurons. These cells release glycine when stimulated and have high-affinity transporters for removing glycine from the synaptic cleft. The best known example is the inhibitory interneurons of the ventral horn of the spinal cord. These include the Renshaw cell that mediates recurrent inhibition of the α motor neurons and the inhibitory Ia afferent-coupled

PHOSPHATIDYLINOSITOL-LINKED
METABOTROPIC RECEPTORS

CYCLIC AMP-LINKED
METABOTROPIC RECEPTORS

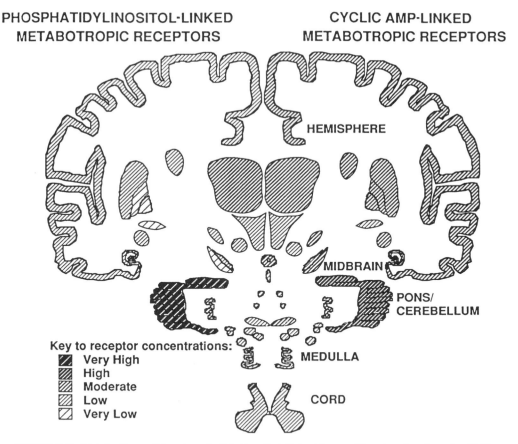

FIGURE 8.6. Distributions of the phosphatidyl inositol–linked (group 1) and cyclic adenosine 3',5'-monophosphate–linked (groups 2 and 3) subtypes of metabotropic glutamate receptors.

interneurons that mediate crossed inhibition of the spinal reflexes. Similar interneurons are present in the motor nuclei of the brainstem. The distribution of glycinergic neurons is shown in Figure 8.7.

Postsynaptic glycine receptors are typical ligand gated ion channels consisting of several, probably five, interacting subunits. The distribution of these receptors is shown in Figure 8.7. The channel is permeable to chloride and, when open, causes the neuron to become hyperpolarized.

Dysfunction of the synapses formed by glycinergic neurons causes a marked increase in the reflexes with the spread of innervation from one spinal cord segment to another. The rare inherited syndrome hyperekplexia, characterized by excessive startle reactions, is caused by minor mutations in the α subunit of the glycine receptor. Total failure of these synapses causes the sensory input from one contracting muscle to stimulate other muscles to contract. Shortly thereafter, all muscles are maximally contracting owing to positive feedback. This leads to opisthotonus, paralysis, and respiratory failure. These symptoms are observed after strychnine poisoning because of blockade of glycine receptors and as a result of exposure to tetanus toxin, which blocks the release of glycine from the presynaptic terminal.

γ-Aminobutyric Acid

GABA is the predominant fast-acting inhibitory neurotransmitter in the nervous system. GABA is synthesized from glutamate by the specific enzyme glutamic acid decarboxylase. This enzyme occurs in two isoforms, long and short, that are under separate metabolic regulation. Some GABA neurons have only one form of the enzyme, and some have both. Glutamic acid decarboxylase is found only in neurons that utilize GABA as a neurotransmitter and serves as a useful immunohistochemical marker for these neurons. GABA is ordinarily removed from the synapse by a high-affinity neuronal transport system and restored in synaptic terminals. GABA may also be transported into glia, where it is converted to succinic semialdehyde, a Krebs cycle intermediate, by the enzyme GABA transaminase. There are several classes of drugs whose mechanism of action is to enhance GABA's actions. These drugs function as sedatives, anticonvulsants, and anxiolytics.

γ-Aminobutyric Acid Pathways

GABA neurons serve as the local circuit inhibitory neurons throughout the central nervous system except in the

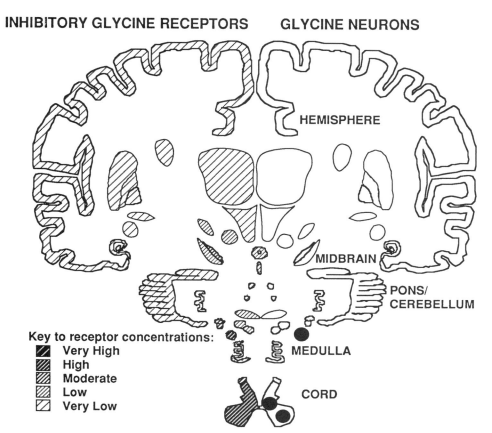

INHIBITORY GLYCINE RECEPTORS   GLYCINE NEURONS

HEMISPHERE

MIDBRAIN

PONS/
CEREBELLUM

Key to receptor concentrations:
- Very High
- High
- Moderate
- Low
- Very Low

MEDULLA

CORD

FIGURE 8.7. Location of some glycine neurons and the distribution of inhibitory glycine receptors.

spinal and brainstem motor nuclei, where glycine serves this purpose. There are also several long projection pathways that utilize GABA. One is the Purkinje cell pathway from cerebellar cortex to the deep cerebellar nuclei. The Purkinje cell is the only output cell of the cerebellar cortex, and it functions to inhibit cells in the deep cerebellar nuclei. Similarly, the only output neurons of the caudate nucleus, putamen, and nucleus accumbens (striatum) are GABA-ergic, medium-sized neurons whose dendrites are covered with synaptic spines (medium spiny neurons). These GABA-ergic cells also contain a neuromodulatory peptide, either enkephalin or substance P. These neurons project to the globus pallidus and substantia nigra. The neurons of the globus pallidus and pars reticulata of the substantia nigra are large, tonically firing neurons that also use GABA. These neurons project to the subthalamic nucleus and thalamus. Activation of the striatal spiny neurons thus disinhibits (inhibits the inhibitor of) the thalamus. The distribution of GABA neurons and their pathways are shown in Figure 8.8.

γ-Aminobutyric Acid Receptors: Benzodiazepines and Barbiturates

Many clinically useful drugs, the benzodiazepines, the barbiturates, some anticonvulsants, and baclofen (Lioresal), act by modulating the activity of one of the two types of postsynaptic GABA receptors.

The first type of GABA receptor to be discovered, the GABA$_A$ receptor, is a typical ligand gated ion channel. When open, these receptors provide a pathway for chloride ions to enter the cell and hyperpolarize it. GABA$_A$ receptors are quite dense throughout the brain, wherever there are GABA synapses. Both benzodiazepines and barbiturates modulate the activity of these receptors through binding sites that are different from the one to which GABA binds. The benzodiazepines bind to sites on the extracellular part of the receptor complex and cannot open the channel unless GABA is present. The barbiturates bind to a site in the chloride channel and can open the channel without GABA being present. This difference in the actions of the two classes of drugs explains the greater toxicity of the barbiturates. The two classes of drugs also affect the opening of the channel in different ways. Benzodiazepines increase the frequency with which the channel opens when GABA is present but does not affect the duration of opening or the channel conductance. In contrast, barbiturates do not alter the frequency of channel opening but prolong the duration of the open state.

At least 15 different genes code for subunits of the GABA$_A$ receptor. These can be grouped into families of genes designated by Greek alphabetical names: α, β, γ, δ, and ρ. Each of the functional receptor–ion channel complexes contains at

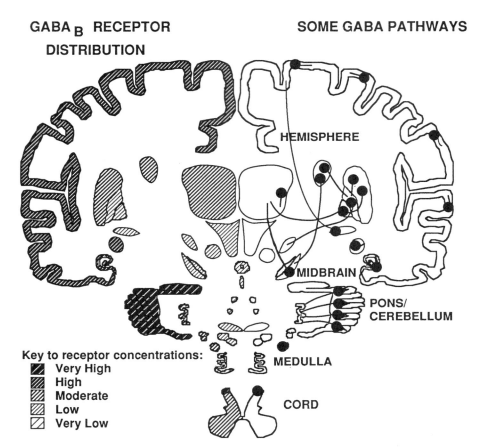

FIGURE 8.8. Some γ-aminobutyric acid (GABA) pathways and the distribution of GABA$_B$ receptors.

least one α and one β subunit. A γ or δ subunit may or may not be included with the α and β subunits. The α and β subunits have binding sites for GABA. All GABA$_A$ receptors are sensitive to barbiturates, but sensitivity to benzodiazepines requires coassembly of α and γ subunits. Recent studies suggest the site of benzodiazepine binding is near the point of contact of these two subunits. The β subunits contain a number of phosphorylation sites through which the receptor can be regulated by intracellular events. The α subunit is commonly colocalized with the β_2 subunit in the globus pallidus, substantia nigra, and cerebellum. The α_3 subunit is commonly colocalized with the β_1 subunit in the cerebral cortex. An unusual α subunit, the α_6 unit, is located on cerebellar granule cells and is not sensitive to benzodiazepines. The ρ subunit is located only in the retina. The frequent colocalization of α_1 with β_2 and α_3 with β_1 suggests that these subunits join together to make the functional receptors and that their gene expression is under a single control system.

The diversity of GABA$_A$ receptor subunits gives rise to the complex pharmacology of the benzodiazepines. Two broad classes of benzodiazepine sites can be distinguished in the brain. Type 1 benzodiazepine receptors are more associated with the anxiolytic properties of the drugs, whereas type 2 receptors are more associated with benzodiazepine's sedative

and anticonvulsant properties. The α_1 subunit is associated with benzodiazepine type 1 pharmacology, whereas the α_2, α_3, and α_4 subunits are associated with benzodiazepine type 2 pharmacology. The distribution of these two types of central benzodiazepine receptors is shown in Figure 8.9. There is also a third, peripheral benzodiazepine binding site on a mitochondrial transporter that has nothing to do with brain GABA pharmacology.

The other major type of GABA receptor is the GABA$_B$ receptor. GABA$_B$ receptors are G protein linked. Activation of these receptors results in inhibition of cyclic AMP, an increase in the permeability of potassium channels, and a decrease in the permeability of sodium channels. Physiologic studies have indicated that many GABA$_B$ receptors are located on presynaptic terminals of non-GABA neurons, where they function as heteroreceptors, regulating the release of other neurotransmitters. Conversely, ligand-binding studies indicate that most GABA$_B$ receptors are located postsynaptically. Whether presynaptic or postsynaptic GABA$_B$ receptors are functionally more important remains to be resolved. The distribution of GABA$_B$ receptors as revealed by ligand-binding studies is shown in Figure 8.8. Binding is very high in the cerebellum and dorsal horn of the spinal cord. High binding is present in the cortex. The very high binding in dorsal horn or spinal cord likely represents receptors that regulate the

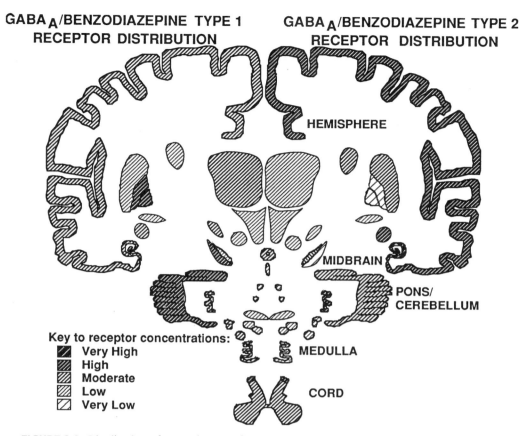

GABA_A/BENZODIAZEPINE TYPE 1 RECEPTOR DISTRIBUTION

GABA_A/BENZODIAZEPINE TYPE 2 RECEPTOR DISTRIBUTION

HEMISPHERE

MIDBRAIN

PONS/CEREBELLUM

MEDULLA

CORD

Key to receptor concentrations:
- Very High
- High
- Moderate
- Low
- Very Low

FIGURE 8.9. Distribution of two subtypes of γ-aminobutyric acid_A/benzodiazepine receptors.

release of neurotransmitter from primary afferents through presynaptic inhibition.

Acetylcholine

Acetylcholine was the first neurotransmitter to be clearly described. It is the major transmitter of the peripheral motor system, the presynaptic autonomic nervous system, the postsynaptic innervation of sweat glands, and the entire postsynaptic parasympathetic system. It also serves as a neurotransmitter in a number of central nervous system pathways. Acetylcholine is released from presynaptic terminals and interacts with postsynaptic receptors in a manner very similar to that of amino acids. However, its fate in the synapse differs from that of the amino acids. There is no high-affinity transport system for acetylcholine. Instead, the transmitter is broken down within the synaptic cleft by an enzyme, acetylcholinesterase, forming choline and acetate. The choline is taken back up into the presynaptic nerve terminal by a transporter and used for synthesis of new acetylcholine. Several drugs that are useful for treating myasthenia gravis act by blocking the activity of acetylcholinesterase at the neuromuscular junction. Several centrally active inhibitors of acetylcholinesterase have been developed as treatments for Alzheimer disease, which tends to cause particularly severe

degeneration of cholinergic neurons in the brain. However, excessive inhibition can produce severe consequences, as illustrated by the rapidly lethal effects of nerve gas chemical warfare agents, which also act through inhibition of acetylcholinesterase.

Acetylcholine Pathways

There are six major types of acetylcholine neurons within the central nervous system. The distribution of these neurons is shown in Figure 8.10. The most numerous are the motor neurons of the ventral horn of the spinal cord and the motor nuclei of the brainstem. These neurons send their axons to innervate the somatic muscles. Recurrent collaterals of these neurons innervate the Renshaw cells. The second major type of neuron is the presynaptic sympathetic neurons, whose cell bodies are in the intermediolateral cell column of the spinal cord in the thoracic and lumbar regions. The first-order parasympathetic neurons in cranial nerves III, VII, IX, X, and XI and the sacral spinal cord are also cholinergic.

Within the central nervous system there are two major cholinergic projections. One arises in the pedunculopontine nuclei of the midbrain and projects to the medulla, substantia nigra, thalamus, and globus pallidus. These neurons are part of the ascending reticular activating system. These

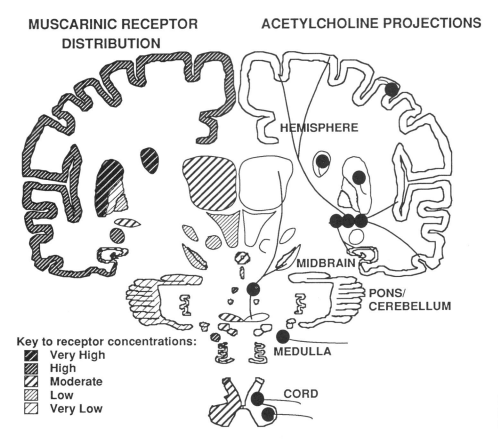

MUSCARINIC RECEPTOR DISTRIBUTION

ACETYLCHOLINE PROJECTIONS

HEMISPHERE

MIDBRAIN

PONS/ CEREBELLUM

MEDULLA

CORD

Key to receptor concentrations:
- Very High
- High
- Moderate
- Low
- Very Low

FIGURE 8.10. Some acetylcholine pathways and the distribution of muscarinic acetylcholine receptors.

neurons may play an important role in regulating the sleep/wake cycles. A second major central nervous system projection for acetylcholine is provided by a complex of nuclei, including the medial septal nucleus, the diagonal band of Broca, and the basal nucleus of Meynert. These neurons innervate all the cortical structures including the neocortex, the hippocampus, and the amygdala. The acetylcholine provided by these neurons seems to be vital for normal learning and memory function. The degeneration of these neurons is one of the major hallmarks of Alzheimer disease.

There are a few acetylcholine neurons within the neocortex that function as local circuit neurons. However, the major region where acetylcholine neurons serve as local circuit neurons is within the caudate nucleus and putamen. Here, approximately 5% of the neurons are large cholinergic neurons that have no long projections. These neurons participate in the cholinergic/dopaminergic balance that characterizes the pharmacology of the movement disorders.

Acetylcholine Receptors

There are two major classes of acetylcholine receptors named after chemical compounds that can activate each class. Nicotinic receptors are sensitive to nicotine. They are ligand gated ion channels that allow the flux of sodium and calcium

ions. Muscarinic receptors are sensitive to the compound muscarine, and these are receptors with seven transmembrane domains that are linked through G proteins to effector systems.

The acetylcholine receptors present at the neuromuscular junction are nicotinic, and their structure has been studied extensively. They are formed from pentamers of protein subunits, usually two α and two β subunits with one γ, δ, or ϵ subunit. Activation of these receptors leads to depolarization of the muscle fiber and muscular contraction. Drugs that can block the nicotinic receptors found on muscle, such as curare, are used to produce paralysis in preparation for surgery.

Nicotinic receptors are also found in the central nervous system and sympathetic ganglia, but these receptors are composed of different subunits and are pharmacologically different from those present at the neuromuscular junction. Many of these receptors are located on presynaptic terminals, where they may govern neurotransmitter release. They are also particularly dense on the postsynaptic neurons of the interpeduncular nucleus and superior colliculus.

The other major type of acetylcholine receptor is the muscarinic receptor. Like the nicotinic receptors, muscarinic receptors are found both in the central and peripheral nervous systems. Muscarinic receptors are present at the synapses where the postsynaptic parasympathetic neurons terminate, such as those in the heart, intestine, and sweat glands. Atropine is a compound that can block muscarinic receptors, producing tachycardia, constipation, and decreased sweating.

At least five different muscarinic receptor proteins are present in the brain, which can be divided into two pharmacologic subclasses: M_1 receptors, which are susceptible to pirenzepine and M_2 receptors, which are not. The M_1 receptors are located postsynaptically and are often linked to cyclic AMP and/or phosphatidylinositol metabolism. The M_2 receptors are located on the presynaptic terminals of acetylcholine neurons and function as autoreceptors governing acetylcholine release. The distribution of the M_1 receptors is shown in Figure 8.10. Blockade of central muscarinic receptors produces sedation and can profoundly affect learning and memory. Many commonly used drugs have some central antimuscarinic effects. These include not only anticholinergics, such as trihexyphenidyl, but also tricyclic antidepressants, antihistamines, and many of the typical and atypical neuroleptics.

AMINE NEUROTRANSMITTERS

Dopamine

The discovery that the catecholamine dopamine was a neurotransmitter was largely the result of the work of Dr. Arvid Carrlson, for which he was awarded the Nobel Prize in 2000. This neurotransmitter plays major roles in governing both motor activity and behavior and is the target of a large number of neurologic and psychiatric pharmaceuticals.

Dopamine is synthesized from the amino acid tyrosine, derived from dietary protein. The rate-limiting step is the conversion of tyrosine to L-dihydroxyphenylalanine (L-dopa) by tyrosine hydroxylase. L-Dopa is in turn converted into dopamine by the enzyme aromatic acid decarboxylase. Tyrosine hydroxylase is found only in catecholaminergic neurons. Aromatic acid decarboxylase is much more widely distributed in the brain and is found in many neurons and in glia. In neurons using the other catecholamine neurotransmitters, additional enzymes are present: dopamine β-hydroxylase, which converts dopamine into norepinephrine, and phenylethanolamine N-methyltransferase, which converts norepinephrine into epinephrine.

Several mechanisms are present to limit the actions of dopamine in the brain. Dopamine is metabolized by two different enzymes, monoamine oxidase and catechol-O-methyltransferase, producing the biologically inactive product homovanillic acid. The actions of dopamine at dopamine synapses may also be terminated by a specific high-affinity transporter that translocates dopamine from the synaptic cleft back into the presynaptic terminal. The stimulant drugs of abuse, cocaine and the amphetamines, act by blocking the reuptake of dopamine via this transporter. In addition, amphetamines can stimulate the release of dopamine.

Dopamine Pathways

Dopamine cell bodies are located in a number of distinct nuclei in the brainstem and hypothalamus. The location of these nuclei and their pathways are shown in Figure 8.11. There are dopamine neurons in the area postrema and dorsal motor nucleus of the vagus nerve that may be involved in the generation of emesis. There is a descending dopaminergic pathway from the zona incerta that projects to the dorsal horn of the spinal cord and regulates some aspects of pain sensation. There are also dopamine neurons located in the median eminence of the hypothalamus. These neurons govern the release of pituitary hormones, particularly prolactin release, which dopamine inhibits. Thus, prolactin-secreting pituitary tumors can be treated with dopamine agonists. The major source of dopamine for the brain, however, is the chain of cells running from the substantia nigra pars compacta through the ventral tegmental area. The nigral neurons project to the caudate nucleus and putamen, where they govern motor activity. Dysfunction of these neurons produces the symptoms of parkinsonism. The ventral tegmental area neurons project to the nucleus accumbens and the entire cortex, particularly the frontal lobe. Dysfunction of these neurons is thought to contribute to the symptoms of Tourette syndrome and schizophrenia.

Dopamine Receptors

All dopamine receptors are G protein–linked receptors. There are two main pharmacologic classes of such receptors. D_1 receptors are linked to the stimulation of cyclic AMP, whereas D_2 receptors are linked to cyclic AMP inhibition. The distribution of these receptors is shown in Figure 8.12. Both receptor subtypes are extremely abundant in the caudate nucleus, putamen, and nucleus accumbens. D_1 receptors are located on postsynaptic striatal neurons. They are also located on the terminals of these neurons in the medial global pallidus and substantia nigra pars reticulata. Most striatal D_2 receptors are also on intrinsic striatal neurons, although some exist as autoreceptors on dopamine neuron terminals, where they govern dopamine release. The D_2 receptors are blocked by the classic neuroleptics, such as haloperidol, and the antiemetics, such as metoclopramide. Blockade of these receptors produces parkinsonism as a prominent side effect.

Five genes that code for dopamine receptors have been cloned. The d_1 and d_5 proteins have D_1 pharmacology and stimulate adenylate cyclase. The other three proteins, the d_2, d_3, and d_4 proteins, all have a D_2 pharmacology. The d_1 and d_5 proteins are primarily located in the striatum, with a few cortical cells expressing the d_1 receptor protein. The d_2 receptor gene is expressed in the striatum, hypothalamus, and substantia nigra. These nigral d_2 receptors presumably represent autoreceptors. d_3 Receptors are expressed in the nucleus accumbens and parts of the caudate and may be responsible

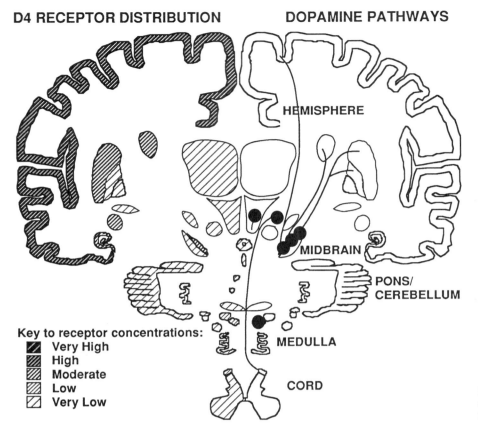

FIGURE 8.11. Some dopamine pathways and the distribution of the dopamine D₄ receptor.

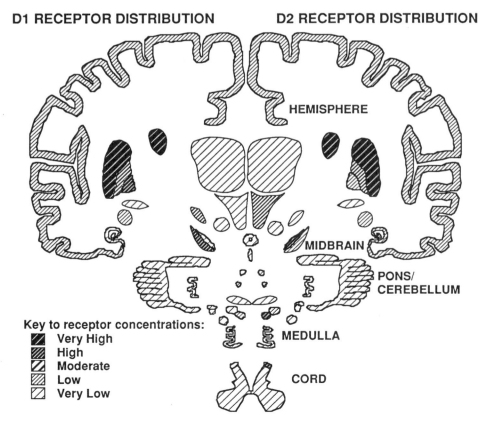

FIGURE 8.12. Distribution of the dopamine D₁ and D₂ receptors.

for mediating some of the cognitive effects of dopamine. However, most of the cognitive effects of dopamine are probably mediated by the d_4 receptor protein, which is expressed mainly in the cortex. The d_4 receptor protein is sensitive to the atypical neuroleptics, such as clozapine, and the traditional neuroleptics.

There is abundant evidence for interaction between the D_1 and D_2 receptor subtypes, particularly in the caudate and putamen. Both are required for normal motor activity. However, it is not clear whether the two receptor subtypes are localized on the same cells. Some studies have shown a complete segregation of receptors, with D_1 receptors being located on striatal output cells that project to the medial globus pallidus and substantia nigra and contain substance P. D_2 receptors are located on cells that project to the lateral globus pallidus and that contain enkephalin. Other studies have suggested an overlap of the receptors types, with as many as 25% of cells expressing both types of receptor. These disparate results may reflect a dynamic situation in which the expression of the dopamine receptors is modulated by external stimuli.

Norepinephrine

Norepinephrine is synthesized from dopamine by the enzyme dopamine β-hydroxylase. The presence of this enzyme in cells distinguishes norepinephrine from dopamine neurons. Like dopamine, norepinephrine is metabolized by monoamine oxidase and catechol-O-methyltransferase. Again, like dopamine, the major route of inactivation of norepinephrine at the synapse is via a high-affinity transport system into the presynaptic terminal. Paroxetine is a specific inhibitor of the norepinephrine transport system. Tricyclic antidepressants, particularly desipramine, also block norepinephrine reuptake. This may be part of their mechanism of action.

Norepinephrine Pathways

In the peripheral nervous system, norepinephrine is the major transmitter of the postsynaptic sympathetic nervous system. These neurons have their cell bodies in the chain of sympathetic ganglia that lie bilaterally in the paravertebral gutter from the lower cervical to lumbar regions. The axons of these neurons are distributed to the blood vessels throughout the body, irides, salivary glands, heart, lungs, intestines, and bladder. These neurons play important roles in the regulation of blood pressure, pupillary dilation, cardiac strength and rhythm, pulmonary airway dilation, and intestinal and vesicle mobility.

There are several major norepinephrine-containing cell groups in the mammalian brain. The distribution of these neurons is shown in Figure 8.13. One group is distributed in the reticular formation of the lateral medulla. Descending axons of these neurons project into the spinal cord, where

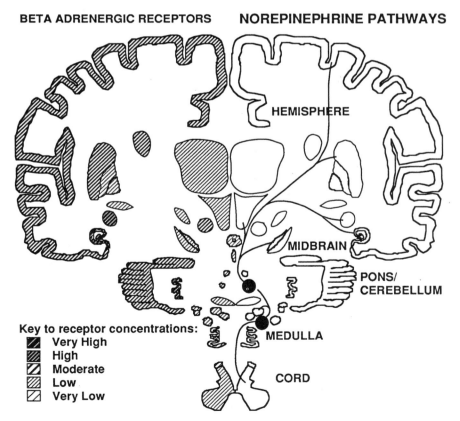

BETA ADRENERGIC RECEPTORS   NOREPINEPHRINE PATHWAYS

HEMISPHERE

MIDBRAIN

PONS/
CEREBELLUM

MEDULLA

CORD

Key to receptor concentrations:
Very High
High
Moderate
Low
Very Low

FIGURE 8.13. Some norepinephrine pathways and the distribution of β-adrenergic receptors (receptors for norepinephrine and epinephrine).

they have an inhibitory effect on muscle tone and spinal cutaneous pain reflexes. Ascending projections of these neurons go to the locus ceruleus. The locus ceruleus itself serves as the major source of norepinephrine for the forebrain. This dense cluster of pigmented neurons is located in the dorsolateral pons. A few cells project from there to the cerebellum, but most project rostrally. The axons of the ceruleus neurons project in the median forebrain bundle, where they distribute to the hypothalamus, thalamus, basal ganglia, amygdala, hippocampus, and the entire neocortex. Norepinephrine terminals are distributed most densely in cortical layers 2, 5, and 6. These projections from the locus ceruleus are thought to play an important role in arousal, memory (being responsible for the hippocampal theta rhythm), and affect.

Norepinephrine Receptors

Norepinephrine receptors are all the G protein–coupled type. There are two main pharmacologic subtypes of these receptors, α- and β-receptors. α-Receptors are linked to the inositol phosphate system and the inhibition of cyclic AMP, whereas β-receptors are linked to stimulation of cyclic AMP. Both receptors are found not only in the brain but also in the periphery. Each major pharmacologic class has several subtypes as well.

The distribution of α-receptors in the brain is shown in Figure 8.14. α_1-Receptors are located postsynaptically

throughout the brain and periphery. There are many in the thalamus and hippocampus, a moderate amount in the basal ganglia, cerebral, and cerebellar cortices, and some in brainstem nuclei, whereas there are few elsewhere. α_2-Receptors, conversely, are located postsynaptically and as autoreceptors on the synaptic terminals of norepinephrine neurons, where they regulate norepinephrine release. They are particularly abundant in the cerebral cortex and locus ceruleus.

β-Receptors are also distributed throughout the brain and spinal cord. The distribution of these receptors is shown in Figure 8.13. They are particularly concentrated in the basal ganglia structures, caudate nucleus, globus pallidus, and subthalamic nucleus as well as in the deep nuclei of the cerebellum. β-Receptors have been found to be increased in some cases of drug-free suicide, and there is evidence that β-receptors down-regulate with antidepressant treatment. Perhaps this plays a role in the therapeutic response to tricyclics.

At least 11 distinct adrenergic receptor genes are known to exist. There are four different genes that code for α_1 proteins. All these proteins are linked to phosphatidylinositol metabolism. Three other genes code for slightly different pharmacologic types of α_2-receptors. All are linked to cyclic AMP inhibition. One of these, α_{2B}, is found largely in the kidney, whereas the other two, α_{2A} and α_{2C}, are found throughout the brain. α_{2A} Protein is located both pre- and postsynaptically on neurons. It is also found in the

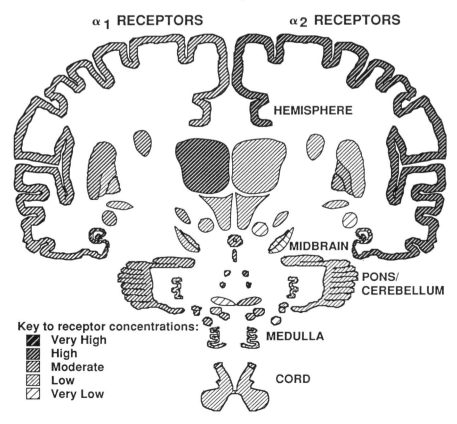

FIGURE 8.14. Distributions of the α_1 and α_2 subtypes of adrenergic receptors.

pituitary, blood, lungs, spleen, and muscles, where it mediates postsynaptic α effects. Postsynaptically, these α_2-receptors modulate smooth muscle tone in the arteries and play roles in sodium excretion by the kidney in platelet aggregation and fat and carbohydrate metabolism. There are three genes that code for β-receptor proteins. All are linked to cyclic AMP stimulation.

Serotonin

Although serotonin has a small role in regulating the gut, most of its actions are in the central nervous system. Serotonin is synthesized from tryptophan in a manner entirely analogous to that by which norepinephrine is synthesized from tyrosine. 5-Hydroxytryptophan is the immediate precursor for serotonin and has had some success as a drug in the treatment of myoclonus. Serotonin is metabolized to 5-hydroxyindole acetic acid. Like norepinephrine, serotonin is thought to play a major role in the arousal, sleep, and affective systems within the brain.

Serotonin is mostly inactivated at the synapse by a high-affinity transport system. The specific serotonin transport inhibitors fluoxetine and sertraline increase synaptic concentrations of serotonin and are useful antidepressants. Tricyclic antidepressants are also potent serotonin reuptake inhibitors. The usefulness of these drugs has led to the hypothesis that serotonin plays a major role in depression.

Serotonin Pathways

All serotonin pathways have their origin in a series of nuclei that are located on the midline of the brainstem, the raphe nuclei. Serotonin pathways are shown in Figure 8.15. Some serotonin axons descend to the spinal cord, where they modulate pain transmission. Within the brainstem, serotonin neurons play a role in arousal and sleep. Ascending pathways from the raphe nuclei distribute to the substantia nigra, the rest of the basal ganglia, the thalamus, hypothalamus, cortex, amygdala, and hippocampi. Within the cortex, the serotonin terminals are concentrated in the superficial layers.

Serotonin Receptors

The pharmacology of serotonin receptors is complex, and their nomenclature has been under continual revision for nearly 50 years. The major subtypes of serotonin receptor were identified and named long before the proteins responsible were cloned. Originally, the serotonin receptors were divided into three main types, $5HT_1$, $5HT_2$, and $5HT_3$. Subsequently, work has revealed a total of at least 15 different serotonin receptor proteins, which are now grouped into a total of seven families. The receptors that are members of the $5HT_1$ and $5HT_2$ families are proteins with seven transmembrane domains and are linked to effector systems through the action of G proteins. The $5HT_3$ receptors are ligand gated

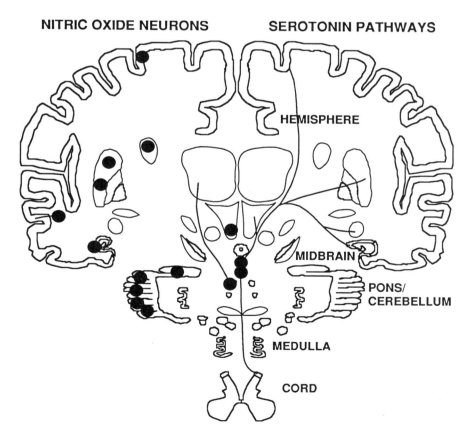

NITRIC OXIDE NEURONS   **SEROTONIN PATHWAYS**

HEMISPHERE

MIDBRAIN

PONS/
CEREBELLUM

MEDULLA

CORD

FIGURE 8.15. Some serotonin pathways and the distribution of nitric oxide–producing neurons.

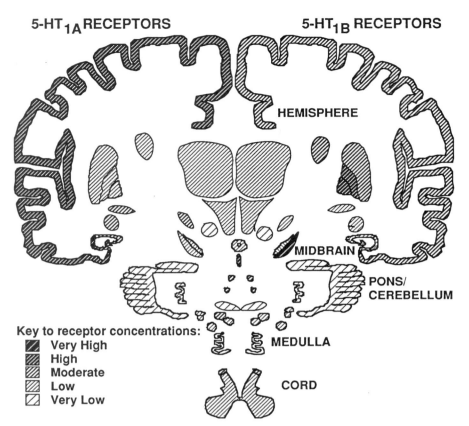

5-HT₁ₐ RECEPTORS   5-HT₁ʙ RECEPTORS

HEMISPHERE

MIDBRAIN

PONS/
CEREBELLUM

MEDULLA

CORD

Key to receptor concentrations:
- Very High
- High
- Moderate
- Low
- Very Low

FIGURE 8.16. Distribution of the 5HT₁ₐ and 5HT₁ʙ subtypes of serotonin receptors.

ion channels, similar to the nicotinic acetylcholine receptors. When activated, 5HT₃ receptors allow the passage of both sodium and potassium ions. The distributions of some of the serotonin receptor families are illustrated in Figures 8.16 and 8.17.

The 5HT₁ family contains two important subtypes, 5HT₁ₐ and 5HT₁ʙ. Both of these subtypes are coupled by G proteins to inhibit adenylate cyclase activity. 5HT₁ₐ Receptors are particularly prominent in the limbic system, including the hippocampus, septum, and thalamus. 5HT₁ₐ Receptors also appear to be present as presynaptic autoreceptors on raphe nuclei dendrites. Drugs with actions at 5HT₁ₐ receptors have antidepressant and anxiolytic actions. Buspirone, a clinically useful anxiolytic, is a partial agonist of 5HT₁ₐ receptors. In human brain, 5HT₁ʙ receptors are located in the substantia nigra and globus pallidus preferentially, where they are present on the terminals on striatal output neurons.

The two major subtypes of 5HT₂ receptors are 5HT₂ₐ and 5HT₂c. These receptors are closely related to one another and are both linked to stimulation of the phosphatidylinositol pathway. The 5HT₂ₐ receptor is found in high abundance only in the cerebral cortex. This provides a substrate for the actions of the serotonergic agonist lysergic acid diethylamide. It is present in much lower amounts in the hypothalamus, hippocampus, spinal cord, platelets, and muscle cells. The 5HT₂c receptor was originally described

in the choroid plexus. These receptors are also present in the pyramidal cells of the hippocampus and to a lesser extent in the cerebral cortex as a whole. They are also present in sensory nuclei of the thalamus, the dopaminergic neurons of the pars compacta of the substantia nigra, and the raphe nuclei themselves. In addition, they are found in pain transmission areas, such as the periaqueductal gray, descending raphe serotonin neurons, and the spinal thalamic tract, suggesting that these receptors play a role in the pain system.

Adenosine

The purine adenosine, which is involved in many biosynthetic processes, is also a neuromodulator in the central nervous system. Adenosine has a well-established role as a precursor to both DNA and RNA. It is also a precursor to ATP, adenosine 5′-diphosphate, and cyclic AMP. Thus, enzymes for its synthesis and metabolism are ubiquitously distributed. There is also a high-affinity transport system for adenosine. Adenosine is found in the extracellular space in the brain, where it is available to bind to receptors. However, specific neurons that use adenosine as their neurotransmitter have not been identified. It may be that the release of adenosine is a more distributed phenomenon, occurring as a consequence of the metabolic activity of neurons and glia.

Adenosine interacts with two types of postsynaptic receptors in the brain. The pharmacology of these receptors

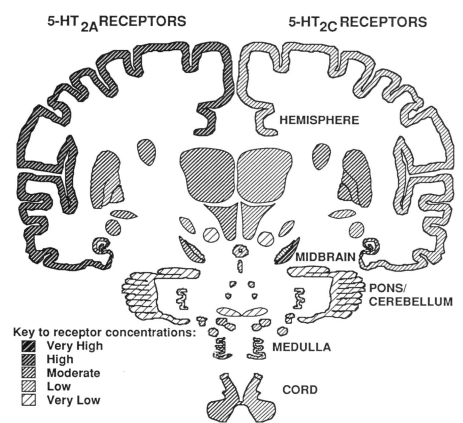

5-HT$_{2A}$ RECEPTORS   **5-HT$_{2C}$ RECEPTORS**

HEMISPHERE

MIDBRAIN

PONS/ CEREBELLUM

Key to receptor concentrations:
- Very High
- High
- Moderate
- Low
- Very Low

MEDULLA

CORD

FIGURE 8.17. Distribution of the 5HT$_{2A}$ and 5HT$_{2C}$ subtypes of serotonin receptors.

is important because the xanthines, such as caffeine, act as antagonists at adenosine receptors. In fact, adenosine antagonism appears to be responsible for the major pharmacologic and behavioral effects of caffeine. The distribution of these receptors is shown in Figure 8.18. Adenosine A1 receptors have a wide distribution with high binding in the cortex (particularly the outer layers), basal ganglia, hippocampus, amygdala, and thalamus. Lower amounts are found in the hypothalamus brainstem structures and spinal cord. There are two subtypes of adenosine A2 receptors. The A2$_A$ receptor is found primarily in the striatum, and the A2$_B$ receptor is found only in the periphery and is not thought to play a role in central neurotransmission.

PEPTIDE NEUROTRANSMITTERS

A bewildering variety of peptides has been shown to act as neuromodulators within the central nervous system. Most of these peptides were originally isolated in another bodily system and then shown to be present in the brain either by purification or demonstration of physiologic action or (most commonly) their presence has been inferred by immunohistochemical staining. In every case that has been studied, neuropeptides are found to be cotransmitters with some other neuroactive compound such as GABA or serotonin, being

released at the same synapse. The best studied of these neuromodulators are the endogenous opiates and substance P. These compounds are reviewed in some detail here.

Endorphins

The endorphins were all isolated as endogenous compounds that have activity at opiate receptors. The endorphins are all synthesized by cleavage of much larger peptide precursor molecules. The three known precursors are proopiomelanocortin, proenkephalin, and prodynorphin. Methionine enkephalin is derived from both proopiomelanocortin and proenkephalin. Leucine enkephalin is derived from proenkephalin and prodynorphin. In addition, the two dynorphins, dynorphin A and dynorphin B, are derived from prodynorphin.

Endorphin Pathways

The distribution of enkephalin neurons is shown in Figure 8.19. Enkephalin neurons serve as local interneurons in the primary sensory receiving areas of the dorsal horn of the spinal cord and the spinal tract of the trigeminal nucleus. It is thought that this is a primary site where the enkephalins suppress pain sensation. There is light to moderate enkephalinergic innervation of the solitary nucleus and other dorsal

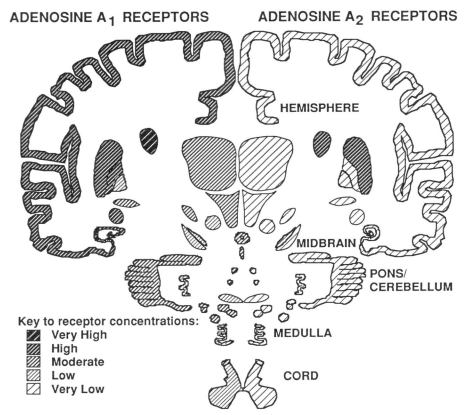

FIGURE 8.18. Distribution of two sub-types of adenosine receptors.

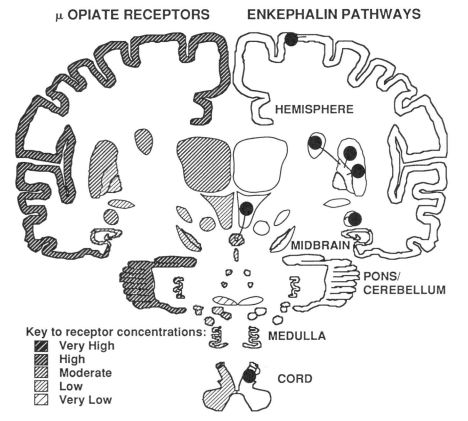

FIGURE 8.19. Some enkephalin pathways and the distribution of μ-opiate receptors.

tegmental nuclei of the medulla. There is also dense enkephalinergic innervation of the central gray and periaqueductal gray from the floor of the fourth ventricle through the entire midbrain. In addition, the interpeduncular nucleus has dense enkephalinergic innervation. There are enkephalinergic neurons in the parabrachial nucleus that appear to be important for enkephalinergic innervation of the substantia nigra. Within the forebrain, there are some enkephalinergic neurons in the cortex and hippocampus, although no cell contains a high concentration of this neuromodulator. There is relatively dense enkephalinergic innervation of the central nucleus of the amygdala. In the bed nucleus of the stria terminalis, there are large numbers of enkephalinergic neurons. These serve as a relay nucleus for the amygdala's projection to the hypothalamus and nucleus accumbens. However, the major source of enkephalin in the forebrain is in the striatum. Approximately half of the spiny neurons of the caudate nucleus and putamen contain enkephalin as a coneurotransmitter. These neurons send their projections almost exclusively to the lateral segment of the globus pallidus, with recurrent collaterals innervating the striatum itself. The concentration of enkephalin in the globus pallidus is by far the highest of any place in the brain. The precise role of enkephalin in this structure remains unknown. It appears to have nothing to do with the pain system.

Dynorphin neurons are distributed very similarly to the enkephalin neurons, with concentrations in the pain-receiving areas of the spinal cord and spinal tract of the fifth nerve being particularly prominent. A major difference between enkephalin and dynorphin systems occurs, however, in the basal ganglia. Here, dynorphin is colocalized with substance P in striatal neurons that project to the medial segment of the globus pallidus and substantia nigra rather than in the striatal cells that project to the lateral globus pallidus.

Opiate Receptors

Several different binding sites for opiates have been described using ligand-binding techniques. The three that are probably most specific for opiate actions are the μ-, δ-, and κ-receptors. μ-Receptors seem to bind the morphinelike opiate pain medications most specifically. δ-Receptors are most specific for the enkephalins, whereas the κ-receptors are most specific for the dynorphins. The paradox of these binding sites and binding sites for ligands that bind to many of the other peptide transmitters is that the distribution of ligand-binding sites is radically different from the distribution of peptide-containing nerve terminals. Whether the receptors are present in places to which the peptides can diffuse over long distances, whether the peptides are somehow paradoxically present in great concentrations at places where they have no functional role, or whether the presence of high concentrations of peptide causes down-regulation of receptor numbers remains to be determined.

The distribution of μ-opiate receptors is shown in Figure 8.19. These receptors are found selectively throughout the brain, with significant numbers in the dorsal horn of the spinal cord and spinal trigeminal nucleus. There are also receptors present in the cerebellum, where there is no enkephalinergic innervation, or in the central gray of the brainstem. In the forebrain, they are found in the cortex, particularly in association cortices, being most concentrated in layers 1, 2, and 4. There are numerous μ binding sites in the amygdala, whereas there are virtually no binding sites in the hippocampus or claustrum. The basal ganglia contain numerous sites in the caudate nucleus and putamen but practically no binding sites in the globus pallidus, particularly in the lateral segment. μ-Receptors are also in high concentration in the medial dorsal nucleus of the thalamus.

The distribution of δ-receptors is shown in Figure 8.20. This distribution closely resembles that of μ-receptors being present in the dorsal horn or spinal cord and trigeminal nucleus, in layers 1 and 2 of the cortex, caudate nucleus, and putamen, medial dorsal nucleus of thalamus, and amygdala with low binding in the globus pallidus. The δ sites are different from the μ sites in that there are significant numbers of δ-receptors in the hippocampal formation, particularly in the dentate gyrus, whereas there are very few δ-receptors in the cerebellum.

κ-Receptor distribution is also shown in Figure 8.20. κ-Receptors are concentrated in the deep layers of the cortex, particularly layers 5 and 6. It has been suggested that this deep layer binding is responsible for the sedative properties of κ-agonist drugs. There is relatively little binding in the caudate nucleus and putamen except in the striosomes. There is slightly more binding in globus pallidus for κ-receptors than there is for either μ- or κ-receptors. There is also a significant concentration of κ-receptors in the basal lateral nucleus of the amygdala. κ-Receptor sites are also found in the granule cell layer of the cerebellum, whereas the μ-receptors seem to be more concentrated in the molecular cell layer.

Tachykinins

Three main tachykinins have been described in the mammalian brain: substance P, neurokinin A, and neurokinin B. By far, the best described of these is substance P. The tachykinins all share the same carboxy terminal amino acid sequence, phenylalanine–X-glycine-leucine–methionine–NH_2, where X may be any amino acid. All the tachykinins are produced as larger proteins that are then cleaved by specific enzymes. These larger precursors are coded by two different genes, preprotachykinin A and B. After transcription, the preprotachykinin A gene is further processed by differential splicing into either α-preprotachykinin A mRNA (the resultant protein is then cleaved to give substance P) or into β- and γ-preprotachykinin A mRNA, which contains the coding sequences for both substance P and neurokinin A.

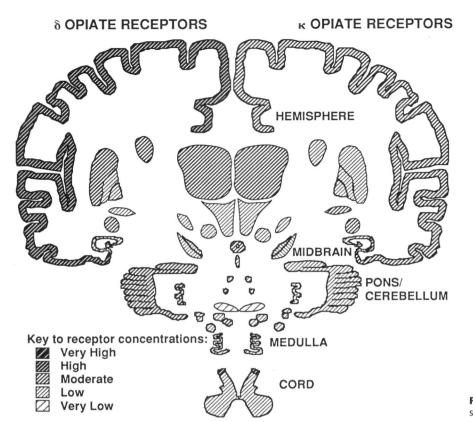

δ **OPIATE RECEPTORS**   κ **OPIATE RECEPTORS**

HEMISPHERE

MIDBRAIN

PONS/
CEREBELLUM

Key to receptor concentrations:
Very High
High
Moderate
Low
Very Low

MEDULLA

CORD

FIGURE 8.20. Distribution of the δ and κ subtypes of opiate receptors.

Neurokinin B is produced from the preprotachykinin B gene product, which is present in both the central nervous system and periphery. Preprotachykinin A and, therefore, substance P and neurokinin A are mainly expressed in the trigeminal ganglion, dorsal root ganglia, and striatum, whereas preprotachykinin B and, thus, neurokinin B are primarily synthesized in the hypothalamus and intestines.

SUBSTANCE P DISTRIBUTION

The distribution of substance P neurons is shown in Figure 8.21. Substance P is a major neurotransmitter of primary afferent nerve terminals. Its primary roles seem to be in pain transmission via the small, unmyelinated, primary afferent nerve fibers. There is dense innervation of substantia gelatinosa of the spinal cord and spinal tract of the trigeminal nerve with substance P fibers. Stimulation of substance P fibers produces burning pain. Capsaicin, the active ingredient in chili peppers, stimulates the release of substance P both where it is applied locally to a nerve, such as the skin, or the mucous membranes of the mouth and centrally. Capsaicin applications produce a temporary sensation of burning, followed by relative anesthesia as the substance P in the affected nerve is first released and then depleted.

Within the central nervous system (Figure 8.21), substance P neurons are densely concentrated within the nucleus of the solitary tract, moderately around the cranial nerve nuclei, and moderately in the other medullary tegmental nuclei. There is also a dense cluster of substance P neurons in the parabrachial nucleus. These neurons innervate other pontine nuclei as well as the caudal parts of the substantia nigra.

In the forebrain, there are a few substance P neurons in the cortex. These are mainly located in layer 6 and in deep parts of layer 5. They seem to project to the upper three layers. The dentate gyrus of the hippocampus contains some substance P innervation, whereas the central nucleus of the amygdala contains more. As with enkephalin, approximately half of the spiny neurons of the striatum contain substance P. This is a different population from that of the enkephalin neurons. These neurons project to the medial segment of the globus pallidus and substantia nigra pars reticulata rather than to the lateral segment of the globus pallidus.

Tachykinin Receptors

The tachykinin receptors are coupled by G proteins to the activation of the phosphatidylinositol system. There are three receptors named the NK1, NK2, and NK3. These three receptors share approximately 60% homology in the seven transmembrane regions, which makes them a family as closely related as the adrenergic or muscarinic receptors. The NK1 receptor is most sensitive to substance P, the NK2 is

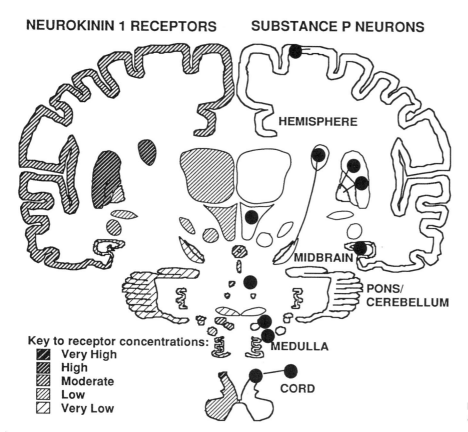

NEUROKININ 1 RECEPTORS   SUBSTANCE P NEURONS

Key to receptor concentrations:
- Very High
- High
- Moderate
- Low
- Very Low

FIGURE 8.21. Some substance P pathways and the distribution of NK1 receptors.

most sensitive to neurokinin A, and the NK3 receptor gene has highest affinity for neurokinin B.

In the human brain, the NK1 receptor is widely distributed (Figure 8.21). It is highly expressed in the hypothalamus, dorsal horn in the spinal cord, olfactory bulb, and striatum within the central nervous system. In the periphery, it is expressed in the bladder, salivary glands, and intestines. The NK2 receptor is restricted to peripheral tissues, particularly the bladder, stomach, large intestine, and adrenal gland. The NK3 receptor gene, similar to the NK1 receptor gene, is more highly expressed in the central nervous system than in the periphery. The highest densities of expression are in the cerebellum, hypothalamus, and cortex.

Similar to the opiates, there is a large divergence between the location of the neuropeptides and the location of their receptors. For example, neurokinin A is present in the brain, but the receptor is not located in either the brain or spinal cord. Thus, the actual role that these neuromodulators play in central nervous system function remains in doubt because of the lack of correlation between presence of the neuromodulator and presence of its effector.

Somatostatin

Somatostatin is a peptide that was first discovered to play a role in the control of growth hormone secretion. It was also found to be present in the intestinal system as well as in scattered interneurons throughout the brain. These interneurons are particularly prominent in layer 4 of the neocortex and in the hippocampus and striatum. Five subtypes of somatostatin receptors have been described. In contrast to the opiate and neurokinin receptors, these receptors seem to be located in regions close to the location of somatostatinergic presynaptic terminals.

Other Peptides

A large number of other peptides that were first discovered because they had a role in bodily functions outside the brain have subsequently been identified within the brain. For the most part, the distribution of these neuropeptides is much more restricted than that of the endorphins, somatostatin, or tachykinins. The three main locations of these other neuropeptides are the primary afferent neurons, medulla, and hypothalamus. Thus, these peptides are positioned to influence sensation through the primary afferents and vegetative functions through the medulla and hypothalamus.

Among these peptides are two that were originally discovered to play hormonal roles in intestinal functions. These are cholecystokinin and bombesin. Both are present in primary afferents. In addition, cholecystokinin is found in the dopamine neurons that project from the ventral tegmental area to the nucleus accumbens in the rat but not in higher primates. Cholecystokinin is found in the caudate nucleus of humans, but the source of this projection is unknown. Both peptides are thought to play roles in the sensation of satiety. Several regulators of vascular tone (angiotensin, atrial natriuretic peptide, bradykinin, and vasoactive intestinal peptide) are present in hypothalamus and medulla. The peptides that were originally described as hormones (calcitonin gene related peptide, melatonin, oxytocin, and vasopressin) have projecting axons within the hypothalamus, outside the pituitary stalk. Finally, recently several of the interleukins have been shown to be used as neuromodulators by cells of the hypothalamus. Interleukins are peptides used as intercellular signaling messengers by the cells of the immune system and are involved in the inflammatory response to foreign antigens. In the brain, interleukins serve as neurotransmitters in the hypothalamic pathways mediating the febrile response to infection.

Gases

Within the past decade, it has been discovered that at least two diffusible gases are used as intercellular messengers within the central nervous system and the periphery. Several years ago, such a role was described for nitric oxide (NO). More recently, it has been proposed that carbon monoxide is also a neurotransmitter. The mechanism of action of the gases is completely different from that of the other neurotransmitters and neuromodulators because they are so diffusible. They are not stored in vesicles. There is no release mechanism at the synaptic terminal and no receptor on the surface of the postsynaptic cell for these gases. These gases are synthesized by enzymes present in the presynaptic terminals on an as-needed basis. They then diffuse freely moving through lipid membranes to reach target neurons, where they interact directly with intracellular proteins connected to signaling pathways.

Nitric Oxide

The first gas to be discovered to play a role as a neurotransmitter was NO. It had long been known that in addition to the acetylcholine and norepinephrine systems, there was at least one more transmitter involved in the autonomic nervous system. This noncholinergic/nonadrenergic neurotransmitter was responsible for the relaxation of arterioles and was called endothelial relaxing factor. Eventually, it was realized that NO is the endothelial relaxing factor. In blood vessels, NO is synthesized by a specific enzyme, endothelial NO synthase, and it diffuses from there into the smooth muscle cells that surround the arterioles. Inside the muscle cells, it stimulates guanylate cyclase, the synthetic enzyme for the second messenger molecule, cyclic guanosine $3',5'$-monophosphate (GMP). It is the cyclic GMP that produces relaxation of the arteriolar smooth muscle. This NO/cyclic GMP–mediated system is vital to the arteriolar relaxation that allows blood to engorge the penis. Thus, NO is the neurotransmitter responsible for penile erections.

The mechanism of NO action in the brain is similar. It is produced by neuronal NO synthase (NOS) and diffuses across neuronal membranes to interact with guanylate cyclase. In the brain, only a relatively small number of neurons contain neuronal NOS. These are easy to detect using a histochemical technique, nicotinamide adenine dinucleotide phosphate diaphorase staining. The distribution of NO neurons in the brain is shown in Figure 8.15. NOS is located densely in the granule cells of the cerebellum. Other cells containing NOS are scattered throughout the brain, including the cortex, striatum, and hippocampus. In these locations, the NOS-containing neurons are interneurons with large numbers of local axon collaterals.

The actions of NO in the brain are complex and incompletely understood. NO may play a very important role in the stimulation of long-term potentiation, a model of memory formation in the hippocampus. Like glutamate, NO may also contribute to neuronal death in some disease states. This is because NO is a free radical and can combine with oxygen to form even more toxic free radicals. This same mechanism is used by neutrophils to kill invading bacteria. Interestingly, the NO-containing neurons themselves are quite resistant to neurotoxic processes such as ischemia and Huntington disease. It has been proposed that as producers of NO, these neurons may somehow have extra protective mechanisms against its neurotoxic actions.

Carbon Monoxide

Carbon monoxide can be produced in the brain by the enzyme hemeoxygenase-2. Like NO, carbon monoxide is a freely diffusible compound. Once synthesized, it appears to be able to stimulate guanylate cyclase, just as NO can. Thus, carbon monoxide and NO both stimulate the same enzyme, and carbon monoxide can greatly potentiate the response to NO. Hemeoxygenase-2 is enriched in olfactory structures, pyramidal cells of the hippocampus, and Purkinje cells of the cerebellum, but it is present throughout the brain.

CONCLUDING REMARKS

In the past two decades, enormous progress has been made in identifying brain neurotransmitters, discovering their receptors and signaling systems, and defining the anatomic localizations of these molecules. This information is the basis for many of the current hypotheses about brain function. Much important work remains to be done, not only in further elucidating the mechanisms of neurotransmitter action but also in determining how these systems are modified in disease states.

ACKNOWLEDGMENTS

This work was supported by USPHS grant NS34361.

SUGGESTED READINGS
General

1998 Receptor and ion channel nomenclature supplement. *Trends Pharmacol Sci* 1998;1–98.
Cooper JR, Bloom FE, Roth RH. *The biochemical basis of neuropharmacology*, 6th ed. New York: Oxford University Press, 1991.
Mendelsohn FAO, Paxinos G, eds. *Receptors in the human nervous system*. San Diego: Academic Press, 1991.
Paxinos G, ed. *The human nervous system*. San Diego: Academic Press, 1990.

Transporters

Iversen L. Neurotransmitter transporters: fruitful targets for CNS drug discovery. *Mol Psychiatry* 2000;5:357–362.
Masson J, Sagn C, Hamon M, et al. Neurotransmitter transporters in the central nervous system. *Pharmacol Rev* 1999;51:439–464.
Sims KD, Robinson MB. Expression patterns and regulation of glutamate transporters in the developing and adult nervous system. *Crit Rev Neurobiol* 1999;13:169–197.

Glutamate Receptors

Nakanishi S. Molecular diversity of glutamate receptors and implications for brain function. *Science* 1992;258:597–603.
Hollmann M, Heinemann S. Cloned glutamate receptors. *Annu Rev Neurosci* 1994;17:31–108.

γ-Aminobutyric Acid Receptors

Barnard EA, Skolnick P, Olsen RW, et al. International Union of Pharmacology. XV. Subtypes of gamma-aminobutyric acid$_A$ receptors: classification on the basis of subunit structure and receptor function. *Pharmacol Rev* 1998;50:291–313.
Burt DR, Kamatchi GL. GABA$_B$ receptor subtypes: from pharmacology to molecular biology. *FASEB J* 1991;5:2916–2923.

Acetylcholine Receptors

Brann MR, Ellis J, Jorgensen H, et al. Muscarinic acetylcholine receptor subtypes: localization and structure/function. *Prog Brain Res* 1993;98:121–127.
Caulfield MP, Birdsall NJ. International Union of Pharmacology. XVII. Classification of muscarinic acetylcholine receptors. *Pharmacol Rev* 1998;50:279–290.
Paterson D, Nordberg A. Neuronal nicotinic receptors in the human brain. *Prog Neurobiol* 2000;61:75–111.

Dopamine Receptors

O'Dowd BF. Structures of dopamine receptors. *J Neurochem* 1993;60:804–816.
Sibley DR, Monsma FJ. Molecular biology of dopamine receptors. *Trend Pharmacol Sci* 1992;13:61–69.

Norepinephrine Receptors

Guarino RD, Perez DM, Piascik MT. Recent advances in the molecular pharmacology of the alpha 1-adrenergic receptors. *Cell Signal* 1996;8:323–333.
Kobilka B. Adrenergic receptors as models for G protein-coupled receptors. *Annu Rev Neurosci* 1992;15:87–114.

Serotonin Receptors

Glennon RA, Dukat M. Serotonin receptor subtypes. In: Bloom FE, Kupfer DJ, eds. *Psychopharmacology: the fourth generation of progress*. New York: Raven Press, 1995:415–429.
Julius D. Molecular biology of serotonin receptors. *Annu Rev Neurosci* 1991;14:335–360.

Adenosine Receptors

Augood SJ, Emson PC, Standaert DG. Localization of adenosine receptors in brain and periphery. In: Kase H, Richerdson PJ, Jenner JP, eds. *Adenosine receptors and Parkinson's disease*. London: Academic Press, 2000:17–30.

Peptide Receptors

Herkenham M. Mismatches between neurotransmitter and receptor localizations in brain: observations and implications. *Neuroscience* 1987;23:1–38.
Hoyer D, Bell GI, Berelowitz M, et al. Classification and nomenclature of somatostatin receptors. *Trends Pharmacol Sci* 1995;16:86–88.
Nakanishi S. Mammalian tachykinin receptors. *Annu Rev Neurosci* 1991;14:123–136.

Gases

Baranano DE, Ferris CD, Snyder SH. Atypical neural messengers. *Trends Neurosci* 2001;24:99–106.
Snyder SH, Jaffrey SR, Zakhary R. Nitric oxide and carbon monoxide: parallel roles as neural messengers. *Brain Res Rev* 1998;26:167–175.

9

PSYCHONEUROENDOCRINOLOGY

LUIS R. ARCE
BEATRIZ M. DEMORANVILLE
IVOR M. D. JACKSON

The central nervous system (CNS) has a fundamental role in the regulation of the endocrine system. The hypothalamus, part of the diencephalon, secretes specific neuroregulatory hormones or releasing factors that can activate or inhibit anterior pituitary function. Additionally, within the CNS itself, there are peptide hormone–producing neurons (neuroendocrine cells) responsible for interneuronal communication and neuromodulation but not directly involved in pituitary regulation (1–3).

Hypothalamic peptidergic neurons are regulated by monoamine neurotransmitters: dopamine (DA), norepinephrine (NE), and serotonin (5-HT). Peptidergic neurons act as neuroendocrine transducers, converting neural information from the brain into chemical (hormonal) information (4,5). The hypothalamic releasing hormones are part of a family of neural peptides with a widespread distribution throughout the CNS.

Alterations in monoamine neurotransmitters play a major role in the pathophysiology of psychiatric and neurologic disorders. For this reason, the hypothalamic–pituitary axis has been extensively investigated in neuropsychiatric disorders as a window on CNS pathophysiology.

PSYCHONEUROENDOCRINOLOGY: DEFINITION

Psychoneuroendocrinology is the field of experimental and clinical neurosciences in which psychiatry, neurology, and endocrinology intersect. Study of this discipline enhances knowledge about the relationships between the neurosecretory systems and behavior and leads to a better understanding of the pathogenesis and management of psychiatric disorders (Fig. 9.1).

Objectives of Psychoneuroendocrinology

The field of psychoneuroendocrinology involves the following objectives (1):

1. To define the hormonal, neurotransmitter, or neuromodulatory defect in a psychiatric or neurologic disorder;
2. To utilize this information to elucidate the differential diagnosis of psychiatric disorders;
3. To predict the response to treatment and its prognosis by attempting to correct underlying neuroendocrine abnormalities;
4. To utilize this approach to characterize the mechanisms of action of the neuroactive compounds used for therapy of psychiatric conditions.

HYPOTHALAMIC–PITUITARY SYSTEM: ANATOMIC CONSIDERATIONS

This section reviews the anatomy of the hypothalamic–pituitary system. Familiarity with it is important for the understanding of the role of the neuroendocrine system in psychiatric disorders.

Hypothalamus

The hypothalamus is part of the diencephalon (6). Anteriorly, it is limited by the optic chiasm and lamina terminalis, and it is continuous with the preoptic area, substantia innominata, and septal region. Posteriorly, it is limited by an imaginary plane defined by the posterior mammillary bodies ventrally and the posterior commissure dorsally. Caudally, the hypothalamus merges with the midbrain periaqueductal gray and tegmental reticular formation. The dorsal limit of the hypothalamus is determined by the horizontal level of the hypothalamic sulcus on the medial wall of the third ventricle, at the horizontal level of the anterior commissure. Here the hypothalamus is continuous with the subthalamus and the zona incerta. Laterally, the hypothalamus is limited by the internal capsule and the basis of the cerebral peduncles (7).

The hypothalamic area involved with the regulation of the anterior pituitary has been named the hypophysiotropic

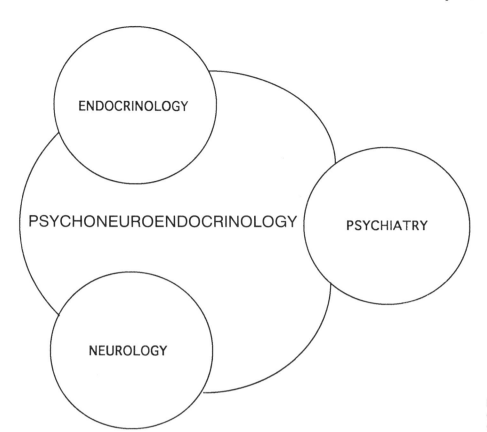

FIGURE 9.1. Diagram of psychoneuroendocrinology and its interactions with endocrinology, psychiatry, and neurology.

area (6). Neuroendocrine cells in this area form the following nuclei: the supraoptic (SON), paraventricular (PVN), periventricular, medial preoptic, and arcuate nuclei of the hypothalamus (3,6). These nuclei project to the median eminence (6) and secrete the hypophysiotropic factors that regulate pituitary function.

Median Eminence

The *median eminence* or *infundibulum* gives rise to the pituitary stalk at the base of the hypothalamus, in the floor of the third ventricle. The median eminence is the site where (a) the hypothalamic neuroendocrine cells release their secretions to the primary plexus of the hypophysial portal system for regulation of the adenohypophysis (i.e., the anterior pituitary); (b) hypothalamic neural fibers that end in the neurohypophysis and intermediate lobe (the hypothalamic-neurohypophysial tract) pass through; and (c) the portal venous system, which provides the only significant blood flow to the anterior pituitary, originates (6,8).

The Pituitary Gland or Hypophysis

The *pituitary gland* or *hypophysis* lies close to the medial basal hypothalamus, to which it is connected by the pituitary stalk (7). In most vertebrates, it is divided into three lobes: the anterior lobe or adenohypophysis, the posterior lobe or neurohypophysis, and the intermediate lobe (6,8). In the adult human, there are only rudimentary vestiges of the intermediate lobe (6,8). During fetal life and pregnancy, however, an intermediate lobe is evident (8).

The Anterior Pituitary or Adenohypophysis

The *anterior pituitary* or *adenohypophysis* contains cells that secrete the following hormones: adrenocorticotropic hormone (ACTH; corticotropin), thyroid-stimulating hormone (TSH), luteinizing hormone (LH), follicle-stimulating hormone (FSH), growth hormone (GH), and prolactin (PRL) (Fig. 9.2).

Five cell types have been recognized in the adenohypophysis and are responsible for the synthesis of the classic anterior pituitary hormones: somatotrophs, mammotrophs, corticotrophs, thyrotrophs, and gonadotrophs. The somatotrophs are the GH-producing cells, which account for 50% of the cells in the adenohypophysis. The mammotrophs are the PRL-producing cells, which account for 15% to 25% of the cells. They increase in number and size with pregnancy, lactation, and estrogen therapy. The corticotrophs are ACTH-producing cells, which constitute approximately 20% of the anterior pituitary cells. Thyrotrophs produce TSH and constitute approximately 5% of the cells in the anterior pituitary. Gonadotrophs, the least numerous

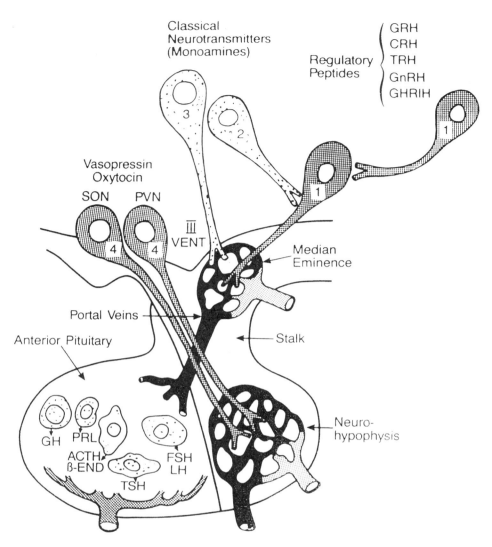

Classical
Neurotransmitters
(Monoamines)

Regulatory
Peptides
⎰ GRH
⎱ CRH
  TRH
  GnRH
  GHRIH

Vasopressin
Oxytocin

SON   PVN

III VENT

Median
Eminence

Portal Veins

Stalk

Anterior Pituitary

GH   PRL

ACTH
β-END

TSH

FSH
LH

Neuro-
hypophysis

FIGURE 9.2. Diagram of the hypothalamic–pituitary axis. Neuron 4 represents the magnocellular peptidergic neurons of the hypothalamo–neurohypophysial tract with cell bodies in the supraoptic (SON) and paraventricular (PVN) nuclei and terminals in the neurohypophysis. Neurons 2 and 3 are monoaminergic neurons. Neuron 2 represents a neuron contacting the cell body of a peptidergic neuron, whereas neuron 3 represents a dopaminergic neuron projecting to the median eminence where release of dopamine occurs. The neurons 1 are the peptidergic neurons that secrete regulatory peptides into the pituitary portal plexus. The regulatory peptides and dopamine are involved in the control of the secretion of different anterior pituitary hormones (β-END, β-endorphin). (Reproduced from Pelletier G. Anatomy of the hypothalamic-pituitary axis. *Methods Achiev Exp Pathol* 1992;14:2, with permission.)

hormone-secreting cells in the anterior pituitary, produce LH and FSH. Each of these cell types can also be identified histologically by immunostaining techniques using specific antisera against the specific hormone produced (9) or by *in situ* hybridization. Additionally, there are other cell types within the anterior pituitary, including the folliculostellate cells shown to contain interleukin-6. Furthermore, there are numerous neural peptides produced within the anterior pituitary, including vasoactive intestinal polypeptide (VIP), which regulates PRL and substance P (10). It is likely that these peptides modulate anterior pituitary hormone secretion, especially under conditions of stress (11,12). VIP may have a role in regulating the pulsatile secretion of LH (13) and in determining circadian rhythms (14–16), along with substance P (17,18).

Neurohypophysis

The *neurohypophysis* includes the neural stalk, the neural lobe or posterior pituitary, and the specialized neurons at the base of the hypothalamus. The major nerve tracts of

the neurohypophysis arise from the accessory magnocellular, the SON, the PVN, and cells scattered in the perifornical and lateral hypothalamic areas and the bed nucleus of the stria terminalis. The SON is located above the optic tract, and the PVN is located on each side of the third ventricle. Most of their unmyelinated fibers descend through the infundibulum and neural stalk within the zona interna to end in the neural lobe. Most of these fibers contain arginine vasopressin (also known as antidiuretic hormone) and oxytocin. Some vasopressin or oxytocin fibers derived from the parvocellular division descend in the zona externa of the median eminence, where they are involved in the regulation of the anterior pituitary, specifically, ACTH release during stress (19). Other neuropeptides, including thyrotropin releasing hormone (TRH), corticotropin releasing hormone (CRH), VIP, and neurotensin, have been found to be secreted from smaller cells or parvicellular neurons (8). In addition, enkephalins, dynorphins, galanin, cholecystokinin, and angiotensin II have been found in the fibers of the supraopticohypophysial tract and in neuronal terminals in the neurohypophysis (3).

Hypothalamic Hypophysiotropic Factors

It is known that the hypothalamic neurons produce neuropeptides that regulate the function of the adenohypophysis. Two regulators are inhibitory: DA, a monoamine tonic inhibitory factor for PRL, and somatostatin (also known as somatotropin inhibitory factor), which inhibits the production of GH. The other known factors stimulate the release of anterior pituitary hormones and are therefore referred to as releasing factors.

The hypophysiotropic factors or hormones are synthesized by neurons in the hypothalamus, transported to nerve endings in the stalk–median eminence, released into the interstitial space in contiguity with the primary portal capillary plexus, and distributed to the anterior pituitary by means of the portal circulation (Fig. 9.2) (5).

The arcuate nucleus of the hypothalamus contains GH releasing hormone (GHRH), the PVN nucleus contains mainly TRH and CRH, and the preoptic nucleus contains mainly gonadotropin releasing hormone [GnRH or LH releasing hormone (LHRH)]. Somatotropin inhibitory factor is found in the PVN. These nuclei also contain many other peptides, including proopiomelanocortin (POMC) and derived peptides ACTH, β-endorphin, and α-melanocyte stimulating hormone, neuropeptide Y, galanin, substance P, enkephalins, atrial natriuretic peptide, angiotensin II, cholecystokinin, and dynorphins (3,20).

Although the actions of each of the hypophysiotropic factors are not limited to a single pituitary hormone, these factors have been named in accordance with the anterior pituitary hormone first recognized to be regulated. Releasing factors are synthesized in neuronal cell bodies, then migrate in dense core vesicles to the median eminence where they are stored in axonal endings, ready for their release into the pituitary portal circulation on stimulation. The hypophysiotropic neurons are regulated by neurotransmitters and neuropeptide modulators and by feedback effects. These are exerted both by the hormones produced by the end organs and pituitary hormones—"short-loop" feedback (Fig. 9.3).

Hypophysiotropic factors of the hypothalamus in relationship to the corresponding pituitary–end organ axis are shown schematically in Figure 9.4.

Gonadotropin Releasing Hormone or Luteinizing Hormone Releasing Hormone

This decapeptide was the first hypophysiotropic factor to be localized by immunohistochemistry. In humans, the GnRH neurons are located in the highest concentrations in the medial basal hypothalamus (infundibular and mammillary nuclei) and the preoptic area (6). GnRH stimulates the release of LH and FSH by the pituitary gland. In females, it controls the menstrual cycle. In males, it controls testosterone secretion and spermatogenesis. It has been shown to stimulate sexual activity in rats of both sexes, independently of its hypophysiotropic action, suggesting a direct influence on

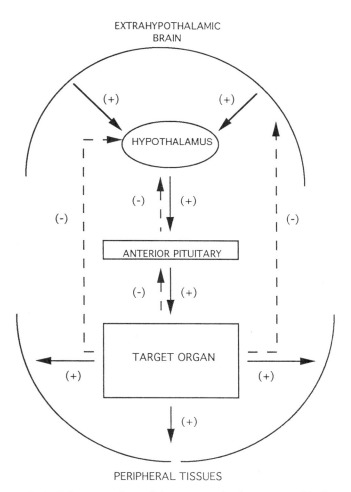

FIGURE 9.3. Interactions of the neuroendocrine system with the extrahypothalamic brain and peripheral tissues. Lines indicate regulatory interactions.

brain functions relevant to sexual drive. For example, LHRH enhances lordosis behavior and sexual receptivity in ovariectomized, estradiol-benzoate–treated rats (21–23).

Corticotropin Releasing Hormone
This 41-peptide is found in neuronal cell bodies of the PVN of the human hypothalamus (24,25). CRH is responsible for stimulating the secretion of ACTH and other POMC-related peptides from the anterior pituitary. ACTH regulates the secretion of cortisol from the adrenal cortex. The hypothalamic–pituitary–adrenocortical (HPA) axis is essential to the neuroendocrine response to physical or mental stress. Additionally, cerebral CRH is involved in the stress response independently of its pituitary–adrenal effects (26,27).

Thyrotropin Releasing Hormone
In the rat brain, cell bodies containing immunoreactive TRH and its precursor pro-TRH have been demonstrated in the preoptic nucleus, parvocellular subdivision of the PVN, perifornical region, dorsomedial nucleus, and basolateral

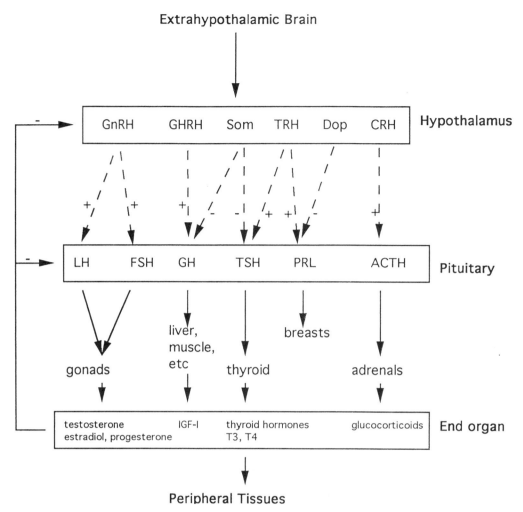

FIGURE 9.4. Diagram representing the interactions of the different hypothalamic–pituitary–end organ axes.

hypothalamus (6). The TRH-containing cells of the PVN are most important to the regulation of the pituitary–thyroid axis. TRH regulates the thyroid axis by stimulating the release of TSH from the anterior pituitary, which acts directly on receptors on the secretory cells of the thyroid gland. TRH is also a potent stimulatory factor for PRL secretion, but its physiologic role in the regulation of PRL is uncertain at this time. TRH neurons are also present in extrahypothalamic tissues, particularly the raphe nucleus, where there is colocalization with 5-HT. The relevance of this colocalization is not yet known. TRH may be a neurotransmitter or neuromodulator in view of its localization in nerve endings and the presence of TRH receptors in brain tissue.

Growth Hormone Releasing Hormone

In primates, neuronal bodies containing GHRH are found in high concentrations in the arcuate nucleus (6). GHRH stimulates GH secretion. GH is required for normal growth and development.

Somatostatin

Bodies of the neurons containing this tetradecapeptide are mainly found in the anterior portion of the arcuate nucleus close to the infundibular recess in humans (28). Somatostatin inhibits GH release. It is widely distributed throughout the nervous system and in extraneuronal tissues, including the gastrointestinal tract and endocrine pancreas. Somatostatin inhibits the secretion of TSH and, under some circumstances, PRL and ACTH. In addition, it has inhibitory effects on endocrine and exocrine secretions of the pancreas, gallbladder, and gut.

Prolactin Regulatory Factors: Dopamine

The most important is the inhibitory factor DA. DA acts on the lactotroph to inhibit the biosynthesis and release of PRL.

PSYCHONEUROENDOCRINE APPROACH OF THE HYPOTHALAMIC–PITUITARY AXIS

In the following section, we review the different hypothalamic–pituitary systems in the context of neuropsychiatric conditions.

The Hypothalamic–Pituitary–Adrenocortical Axis

The HPA system has been the subject of much of the literature and research in the field of psychoneuroendocrinology since it was recognized that hypercortisolism is a consequence of stress and a frequent accompaniment of depression.

Regulation of the Hypothalamic–Pituitary–Adrenocortical Axis

At the highest level, a variety of neurotransmitter pathways (including serotonergic and cholinergic excitatory pathways and adrenergic and possibly GABA (γ-aminobutyric acid)-ergic inhibitory pathways) influences the release of CRH and cosecretagogues, including arginine vasopressin and oxytocin, into the portal circulation. In response to CRH, POMC, a large precursor molecule, is synthesized in the anterior pituitary and is cleaved into ACTH and other neuropeptides, including β-endorphin. ACTH is then secreted into the systemic circulation. At the adrenal cortex, ACTH stimulates the release of cortisol (19,29,30). The major HPA neuropeptides CRH and ACTH not only coordinate the neuroendocrine response to stress but also facilitate behavioral adaptation. They play a key role as regulators of cell development, homeostatic maintenance, and adaptation to environmental challenges.

Regulation of the HPA axis is dependent on three major mechanisms (30,31):

1. Pulsatile CRH release linked to the endogenous circadian rhythm of the CNS. This mechanism involves an oscillator located in the suprachiasmatic nucleus (SCN) and is mediated by serotonergic pathways.
2. Physical and psychological stresses that affect the input from the limbic and reticular activating systems to CRH-secreting neurons. Vasopressin secretion, which is increased by stress, enhances the CRH-induced release of ACTH from corticotrophs.
3. Feedback loops, by which circulating glucocorticoids feed back to the pituitary, hypothalamus, and some extrahypothalamic regions to inhibit HPA activity (29).

The hippocampus appears to have an important role in this inhibitory action of glucocorticoids. There are numerous glucocorticoid receptors in the hippocampus. A reduction in number of these receptors, as by prenatal dexamethasone treatment of experimental monkeys (32) or by hippocam-

pal lesions such as those of Alzheimer disease, is associated with hypersecretion of glucocorticoids, increased hypothalamic levels of CRH and reduced suppression of ACTH by exogenous glucocorticoid administration (29,33–35). Chronic but variable stress down-regulates glucocorticoid receptors in both the hippocampus and paraventricular nucleus of the hypothalamus, thereby reducing the suppression of CRH by glucocorticoids and leading to tonic hypersecretion of cortisol (36).

The Hypothalamic–Pituitary–Adrenocortical Axis in Stress and Depression

Stress can be defined as any threat, real or perceived, to homeostasis or survival (37). Stimuli in this category include injury, hemorrhage, significant illness, fear, and bereavement. In recent years, immune activation has been found to be a stimulus for HPA activation as well through the role of interleukins. Interleukin-1, for example, has been shown to directly stimulate the release of both CRH and ACTH in experimental animals (37–42); it also down-regulates PRL secretion (31). Interleukin effects have been previously reviewed by Rivier (37) and Abraham (41).

In 1936, Hans Seyle (43) was the first to describe the stress syndrome produced by noxious agents, which included three stages: alarm reaction, resistance, and exhaustion. There is a major role for glucocorticoids during the alarm reaction. The initial hormonal response to stress occurs within seconds; this involves liberation of epinephrine and NE from the adrenal medulla followed by release of CRH that enhances the secretion of ACTH by the adenohypophysis. Almost simultaneously, the secretion of GnRH decreases, owing to inhibition by CRH mediated by an opioid peptidergic pathway, and the liberation of pituitary gonadotropins is suppressed. Arginine vasopressin and renin secretion are included in this wave of responses, allowing the body to defend against potential fluid or blood loss. A second phase of response to stress entails the augmented production of glucocorticoids and the diminished secretion of sex hormones. The above alterations in homeostasis result in modulations of the immune system, energy mobilization, increased cerebral blood flow and glucose utilization, loss of sexual behaviors, increased cardiovascular tone, water retention, and vasoconstriction (44). Glucocorticoids are highly catabolic and induce lipolysis, glycogenolysis, and proteolysis, thus increasing the levels of free fatty acids, glucose, and amino acids as readily available fuels for the body (29,45). The ultimate effect is to prepare the organism for strenuous activity by increasing the availability of energy substrates. In addition, glucocorticoids suppress immunologic function and decrease the inflammatory response (45), involving alterations in both T and B cell function and interleukin release (41). The suppression of immunity during stress may protect the animal at a time when maximal mobility may be important (29) but also may explain the increased incidence of infection after injury (41). Other physiologic changes induced by prolonged stress include suppression of thyroid and GH activity and sexual and reproductive behavior (45–47). Growth and reproduction are subordinated to the need to make energy readily available to confront an immediate danger.

Psychological stressors are less potent stimuli of the HPA axis than physical stressors, but they can cause changes in ACTH and cortisol production. Acute psychological stressors augment memory, a phenomenon mediated by catecholamine secretion, which increases cerebral perfusion and delivery of glucose to the CNS (48). At the initiation of the stress, basal levels of glucocorticoids help to activate the cognitive response. The effect of glucocorticoids on memory is focused on the hippocampus, which possesses large quantities of mineralocorticoid and glucocorticoid receptors. Glucocorticoids, in the basal state, by way of the mineralocorticoid receptors, amplify hippocampal excitability and synaptic plasticity (49). In adrenalectomized animals, the memory process is disrupted and function is restored after glucocorticoid administration, primarily via occupation of the mineralocorticoid receptors (50). Once an increase in glucocorticoids occurs in response to stress, the cognitive response is suppressed. Over a short period of time, the effects described above on the hippocampus are blunted and disrupted, but, in this instance, the effect is mediated by the glucocorticoid receptors instead of the mineralocorticoid receptors. Long-term exposure to stress levels of glucocorticoids produces atrophy of the hippocampal neurons and ultimately neuron loss. In addition, glucocorticoids at stress levels (>100 nM) inhibit local cerebral glucose utilization and inhibit glucose transport in neurons, glia, and possibly endothelial cells, causing disruption of the memory process (51–54).

Patients with uncomplicated anxiety or panic disorder generally do not demonstrate excess urinary free cortisol levels. Conversely, when panic disorder is coupled with depression, the urinary free cortisol levels are elevated (55). However, in recent studies of HPA axis activity in panic disorder, wherein salivary and plasma cortisol levels were measured during the attack and 24 hours after, it was found that both were significantly elevated in patients presenting with severe symptoms compared with control subjects (56–58).

In depression, there are both increased secretion of CRH and a neurally mediated hyperresponsivity of the adrenal gland to ACTH (59). A group of ten patients with melancholic depression underwent measurements of plasma and cerebrospinal fluid (CSF) levels of cortisol, ACTH, CRH, and NE for 30 hours. The results obtained showed that the CSF NE and plasma cortisol levels were significantly increased around the clock when compared with controls, although diurnal variations correlated between groups. Considering the hypercortisolism state, when plasma ACTH and CSF CRH were obtained, these values were inappropriately high for that degree of cortisol. Therefore, this suggests a presence of hypernoradrenergic stimulation of CRH and increased adrenocortical secretion (60).

Measures of Basal Hypothalamic–Pituitary–Adrenocortical Activity

Several measures of HPA activity have been employed in neuroendocrine studies of depression.

Basal plasma cortisol. This increases with increased activity at any level of the HPA axis. It shows a good correlation with levels of urinary free cortisol.

Urinary free cortisol. This is an integrated measure of HPA activity over time and directly measures the unbound portion of total plasma cortisol. It reflects the cortisol secretion rate. It is less sensitive to time of collection.

Cortisol in saliva and CSF. These correlate well with unbound plasma cortisol. The former provides a potential means for frequent noninvasive sampling of ambient circulating levels, whereas the latter provides a direct indication of the levels of glucocorticoids to which CNS is exposed.

Dynamic Measures of Hypothalamic–Pituitary–Adrenocortical Axis Activity

The Dexamethasone Suppression Test

The lack of inhibition of the HPA axis by exogenous glucocorticoid was first observed by Carrol et al. (61) in the presence of severe depression. It is found in children and adolescents as well as in adults (62). A "psychiatric" dexamethasone suppression test (DST) is normal if there is suppression of the plasma cortisol level to less than 5 μg/dL 8 to 24 hours after an oral dose of 1 mg dexamethasone given at 11 p.m. (63,64). An abnormal DST is found in approximately 45% of patients with depression. The test has shown a specificity as high as 80% in some studies, but the specificity depends on the population tested. Abnormal DST are seen in other psychiatric disorders, including anorexia nervosa (AN) (65), obsessive-compulsive disorder (66), degenerative dementia (67), mania (68), schizophrenia (69), alcoholism (70), psychosexual dysfunction (71), and schizoaffective psychosis (72) as well as gross brain diseases, including Parkinson disease (73) and stroke. Other variables, such as weight loss, acute hospitalization, and drug and alcohol withdrawal, can affect the test significantly, as can many commonly prescribed medications (74,75). The DST is a more sensitive test for major depression in older persons and children and adolescents (76). It is abnormal more frequently in patients with psychotic affective disorders and melancholia ($>50\%$) than in those with minor depression (23%) or grief reactions (10%) (63,77). In a metaanalysis reported by Nelson and Davis (78) that reviewed 14 studies of DST in psychotic and nonpsychotic patients, there was a greater rate of nonsuppression in those patients with psychotic depression (64%). Patients with melancholia had a lower rate of nonsuppression (36%); when rates were corrected for inpatient/outpatient status, melancholic depression was not significantly associated with nonsuppression.

Several hypotheses have been proposed to explain the nonsuppression of the DST in major depression (79), including increased metabolism of dexamethasone, decreased sensitivity of the pituitary glucocorticoid receptors to dexamethasone, hyperresponsivity of the adrenal gland to ACTH stimulation, and the increased central drive of the pituitary from hypothalamic/limbic structures that overrides the action of the dexamethasone. The latter seems to be the most coherent and is supported by the blunted ACTH response to exogenous CRH administration (80–82) and insulin-induced hypoglycemia (83), which reflects the down-regulation of corticotropin receptors (84) in the pituitary from prolonged hypothalamic CRH hypersecretion and the consequent feedback by the elevated cortisol levels. In the hippocampus of chronically stressed rats, which plays an important role in the regulation of the feedback system, the binding of [^{3}H]dexamethasone was decreased, which suggests that chronic stress induces a hyposuppressive state for cortisol caused by down-regulation of the glucocorticoid receptors (85).

The DST has not evolved into a routine diagnostic test in clinical psychiatry because of its lack of specificity. There is, however, some evidence that an abnormal DST identifies depressed patients who respond more slowly to treatment and that nonsuppression persisting after treatment may predict earlier relapse (77,86). In studies of patients with bipolar disease who underwent DST both in the manic phase and after remission, cortisol levels were higher and dexamethasone levels lower during the manic phase (87). This suggests that impaired suppression during mania might reflect, at least in part, more rapid metabolism of the administered dexamethasone.

The DST also has been applied to patients with neurologic diseases, with and without a concomitant major depression. In Parkinson disease, one study showed 74% nonsuppression in patients with concurrent major depression and 25% suppression in those with Parkinson disease alone (88); another study showed 14% nonsuppression among depressed patients and no nonsuppression among those with Parkinson disease alone (89). Alzheimer disease yields nonsuppression rates comparable with those seen in primary melancholia (90). Stroke is associated with an abnormal DST in 25% or more of patients without clinical depression (91) but with nonsuppression in more than half of patients with concomitant major depression (91–94).

Adrenocorticotropic Hormone Stimulation Test

Adrenocortical hyperresponsiveness to exogenous ACTH is associated with the hypercortisolism of Cushing syndrome. Depressed patients also may show an increased response to ACTH, which correlates with adrenal enlargement found

by computed tomography (95) and in postmortem studies of adrenals of subjects who committed suicide (96). The enhanced response to ACTH stimulation may represent a persistent and prolonged state of endogenous ACTH elevation, leading to adrenal hyperplasia.

Corticotropin Releasing Hormone Stimulation Test

Patients with melancholia show blunted ACTH and cortisol response to exogenous CRH administration (80–82). Corticotrophs in the anterior pituitary are down-regulated by hypercortisolism and persistent CRH release. This decreased response to exogenous CRH distinguishes depressed patients from Cushing disease patients, who have both an ACTH and a cortisol hyperresponse to exogenous CRH. In Cushing disease, endogenous CRH secretion is decreased; in depression, it is increased. The use of the combined dexamethasone/CRH (DEX/CRH) test has been explored as a means to predict medium-term (i.e., 6 months) outcome after treatment of major depression. Established protocol for this test includes giving 1.5 mg dexamethasone at 11:00 p.m. followed by intravenous administration of 1 μg/kg CRH at 8:00 a.m. the following day. Plasma cortisol is determined before the CRH administration and repeated at 30 and 60 minutes after the bolus. In normal subjects, the cortisol response should be less than 1.5 μg/dL. In some patients with depression, cortisol levels fail to suppress. A study of 74 patients with remission of depressive symptoms after inpatient treatment suggests that an elevated cortisol response after DEX/CRH testing, performed at the time of discharge, correlates with a four- to six-fold higher risk of relapse (97). A modification in the DEX/CRH applied by some is using salivary instead of plasma cortisol. In a recent study, the levels of both salivary and plasma cortisol showed a comparable stimulation pattern and significant correlation before and after antidepressant therapy in patients with major depression (98).

Role of Corticotropin Releasing Hormone in Stress and Depression

Since its identification and sequencing in 1981, CRH, a 41-amino-acid peptide, has been recognized as the principal organizer of the neuroendocrine stress response (99) and an activator of the sympathoadrenal and sympathetic systems. CRH plays a major role in coordinating the endocrine, autonomic, behavioral, and immune responses to stress through actions in the brain and peripheral tissues. In autoradiographic studies, CRH receptors are localized in highest densities in anterior and intermediate lobes of the pituitary, olfactory bulb, cerebral cortex, amygdala and cerebellum, and spleen (100). In addition to activation of the HPA axis, CRH causes behavioral arousal, sympathetic stimulation, and a decrease in appetite (99,101). In depression, CRH is hypersecreted not only from the hypothalamus but also from extrahypothalamic sites in the brain, leading to hyperactivity of the HPA axis and behavioral changes. In such instances, the concentrations of CRH within the CSF are elevated. These alterations in the HPA axis mediate some of the symptoms of depression (i.e., sleep, psychomotor, libido, appetite alterations) (102). The overactivity normalizes with treatment; evidence of this is the suppression of cortisol after a DEX/CRH test while in remission. The alterations of the HPA axis in patients with chronic posttraumatic stress disorder (PTSD) differ from those with depression. In PTSD, levels of NE and reactivity of α_2-adrenergic receptors are increased. Stress leads to activation of both the HPA axis and the catecholaminergic systems (which include the sympathetic nervous systems and CNS catecholamines) (103). CRH stimulates noradrenergic neurons in the locus ceruleus (LC) (104), suggesting that brain CRH coordinates the behavioral and autonomic nervous system responses in stress with the neuroendocrine response (105). There is evidence of a reciprocal effect of NE on CRH; α_1-adrenergic agonists raise the levels of CRH within the CNS (104). Therefore, in PTSD, we find increased concentrations of CRH in the CSF, but, interestingly, the cortisol levels tend to be low or inappropriately normal (106). Studies looking at combat veterans with PTSD found an elevation of CSF CRH concentrations and normal 24-hour urinary free cortisol excretion, the latter correlated negatively and significantly with symptoms of PTSD (107). The sensitivity of the negative feedback response in the hypothalamus and hypophysis is increased, reflected by exaggerated suppression of cortisol by dexamethasone. An alternative mechanism that may account for low cortisol concentrations is a decrease in adrenocortical responsiveness; a recent study exploring this option examined the HPA axis of PTSD subjects after metyrapone followed by a cortisol infusion. Metyrapone blocks the conversion of 11-deoxycortisol to cortisol by 11-β-hydroxylase, thus cortisol synthesis and secretion fall markedly, although usually not to hazardously low levels; 11-deoxycortisol accumulates in serum but is essentially devoid of glucocorticoid bioactivity and does not inhibit ACTH secretion. Therefore, in normal subjects, the decrease in serum cortisol concentrations leads sequentially to increases in ACTH secretion, adrenal steroidogenesis, and the release of cortisol precursors, especially 11-deoxycortisol. This study showed the basal cortisol and 11-deoxycortisol levels after metyrapone were significantly lower in PTSD patients when compared with controls; interestingly, the ACTH response to the cortisol infusion after metyrapone was not different between the groups (108).

The current belief is that the development of PTSD is related to an inadequate biologic response to that overwhelming, stressful situation, thus representing an abnormal process of adaptation. Typically, cortisol concentrations increase in response to a stressful event, whereas in subjects who develop PTSD, the cortisol increase is attenuated (Fig. 9.5). Lower cortisol levels at the time of trauma may prolong the availability of NE to synapses. In animals, learning is facilitated by increased adrenergic activation in the face of low cortisol levels. It is thought that this combination of abnor-malities promotes the consolidation of the memory of the traumatic event. If these events do occur in humans who develop PTSD, it might explain why the experience is embedded in the memory and associated with feelings of distress and danger (109). After a motor vehicle accident, a lower NE/cortisol ratio has been found in amnesic victims, contrary to the nonamnesic subjects in which the level of NE was higher and the cortisol level was lower. On further evaluation, a month after the accident, the incidence of PTSD was higher in the nonamnesic group, thus suggesting that amnesia from a traumatic event may serve as buffer in the risk of developing PTSD (110). Other studies of motor vehicle accident victims have looked at the urinary excretion of cortisol collected on arrival at the hospital; a month after the event, they were evaluated for the presence of PTSD. The results showed that those who met the criteria for PTSD had lower concentrations of urinary cortisol on the day of admission to the hospital, and this correlated with a significant percentage of the variance in intrusive and avoidant thoughts at the time of evaluation (111). How much of this cerebral CRH emanates from hypothalamic PVN neuronal outflow or arises *in situ* from other limbic system neuronal sources such as the amygdala is unknown, but it is likely that most of the CRH in the brain is derived from extrahypothalamic neurons. Recent studies using positron emission tomography (PET) and functional magnetic resonance imaging in patients with PTSD have shown that the reactivity of the amygdala and anterior paralimbic region to trauma-related stimuli is increased and the reactivity of the anterior cingulate and orbitofrontal areas is decreased. These regions are involved in the response to fear. At the level of the hippocampus, the functional differences noted suggest a neuroanatomic substrate for the recollections of the events.

Intracerebroventricular administration of CRH in rats produces behavioral activation (112). When the CRH antagonist α-helical CRH9-41 (αhCRH) was given to rats before experimental stress, it attenuated expected increases in NE in several brain regions, including the LC region, cerebral cortex, hippocampus, amygdala, and hypothalamus. Plasma corticosterone levels were not significantly decreased. These findings suggest that CRH is necessary for stress-induced NE release in these regions (113). A study by Matzusaki et al. (112) showed that intracerebroventricular administration of CRH in rats increased DA utilization in the frontal cortex, striatum, hippocampus, and amygdala.

CRH exerts its effects through two receptors, CRH-R1 and CRH-R2. Within the pituitary there is a predominance of CRH-R1. Both are found in the extrahypothalamic brain; however, CRH-R2 occurs more extensively in peripheral organs (114).

The antidepressive potential of CRH-R1 antagonists has been investigated, and currently there is a promising role for the therapeutic use of CRH-R1 antagonists in conditions such as major depression and anxiety disorder. The compound R-121919, a water-soluble pyrrolopyrimidine that

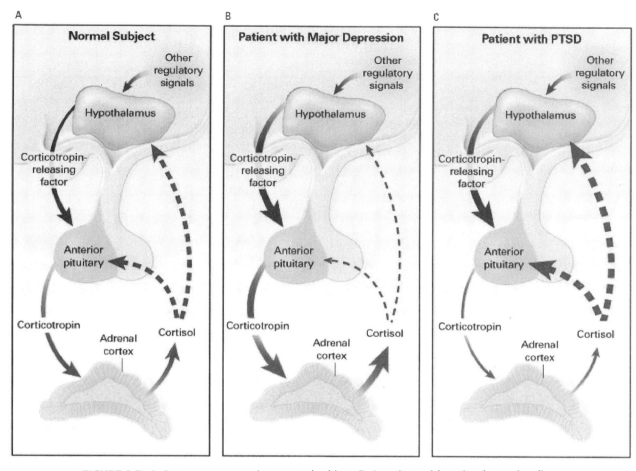

FIGURE 9.5. **A:** Response to stress in a normal subject. **B:** A patient with major depressive disorder. **C:** A patient with posttraumatic stress disorder (PTSD). In normal subjects **(A)** and in patients with major depression **(B)**, brief or sustained periods of stress are typically associated with increased levels of both cortisol and corticotropin releasing factor. In each panel, the thickness of the interconnecting arrows denotes the magnitude of the biologic response. Corticotropin releasing factor stimulates the production of corticotropin, which in turn stimulates the production of cortisol. Cortisol inhibits the release of corticotropin from the pituitary and the release of corticotropin releasing factor from the hypothalamus. It is also responsible for the containment of many stress-activated biologic reactions. In patients with PTSD **(C)**, levels of cortisol are low and levels of corticotropin releasing factor are high. In addition, the sensitivity of the negative feedback system of the hypothalamic–pituitary–adrenal axis is increased in patients with PTSD rather than decreased, as often occurs in patients with major depression. (Reprinted from Yehuda R. Posttraumatic stress disorder. *N Engl J Med* 2002;346:108–114, with permission.) (See Color Figure 9.5 following page 526.)

binds to CRH-R1, seems to be tolerated well and apparently does not impair ACTH or cortisol secretory activity in humans. After its administration, the depression and anxiety scores were significantly lower in patients with major depression (115).

There is little doubt that CRH receptors are significant during stress, but the role of the CRH-R2 is controversial. The CRH-R2 knockout mouse is extremely sensitive to stress and displays anxiety–like behavior (116). However, a more recent study found that mice presented a reduction in stress-induced freezing after blocking the CRH-R2 located in the lateral septum (117). These studies suggest that the CRH-R2 is implicated in modulating the behavioral responses to

stress; however, the mechanism and importance of this process remain to be determined.

CRH release from the hypothalamus is subject to stimulatory serotonergic control. Owens et al. (118) showed that acute administration of 5-HT agonists to rats activates the HPA axis, with increased plasma ACTH and corticosterone levels. Chronic administration of 5-HT agonists does not increase CRH levels in the median eminence or CRH receptor number or receptor affinity in the anterior pituitary. It does, however, increase concentrations of CRH in the piriform cortex and hippocampus (118).

In depression, neuroendocrine and catecholamine dysfunction may be linked to CRH effects on LC neurons via persistent elevated levels of LC discharge and diminished responses to phasic sensory stimuli (27,119). Antidepressant drugs appear to decrease LC sensory-evoked discharge after acute administration. Desipramine and mianserin appear to attenuate LC activation by suppressing endogenous CRH release (27,119).

Some forms of depression are associated with hyperactivity of the noradrenergic system (120), which, by direct or indirect connections, leads to enhanced CRH production in the hypothalamic PVN. This causes activation of the HPA axis and increased sympathetic outflow. Evidence in favor of a major role for CRH in depression is shown by studies of Nemeroff et al. (121) documenting reduced CRH binding sites in the frontal cortex of suicide victims and the increased levels of CRH in the CSF of depressed patients (121–123). However, the latter finding was not confirmed in a subsequent study (124).

Suicide

Postmortem studies of brains of suicide victims support an association with altered 5-HT levels and HPA axis activation. Levels of both 5-HT and its metabolite 5-hydroxyindoleacetic acid are lower in the brains of suicide victims than in control brains (125–127). A positive DST has been reported in 80% of violent suicide attempters (128,129); elevated plasma and urine corticosteroid levels have also been reported (130,131). Brain CRH levels are increased (121–123), as discussed earlier. However, reports from other studies of suicide attempters (129,132–134) and studies of depressed people with suicidal ideation showed no difference in DST response from those without suicidal ideation (135). In a recent study of 57 medication-free schizophrenics, those with a history of suicide attempts had higher levels of post–dexamethasone cortisol control than those who did not (136). The validity of the DST was compared with demographic and historical risk factors for suicide in 78 patients with depression. After a 1-mg overnight DST, 32 patients had an abnormal result. Survival analysis showed that the estimated risk of eventual suicide in the nonsuppressor group was 26.8% compared with 2.9% among the suppressors. Of the demographic and historical risk factors, none provided distinction between those who did or did not commit suicide. It is possible that the DST might be a tool with more power to predict risk of suicide than the clinical predictors in use (137).

The Hypothalamic–Pituitary–Thyroid Axis

Primary thyroid disorders may present with mental and behavioral symptoms, and both physical and psychological stress can affect hypothalamic–pituitary–thyroid (HPT) axis function. Knowledge of these phenomena aids in the management of mood disorders and early recognition of thyroid disease.

Regulation of the Hypothalamic–Pituitary–Thyroid Axis

The synthesis and secretion of the thyroid hormones, thyroxine (T_4) and 3,5,3′-triiodothyronine (T_3), by the thyroid gland is regulated primarily by TSH or thyrotropin, a glycoprotein synthesized by the thyrotropic cells of the anterior pituitary. TSH is secreted from the pituitary in a circadian rhythm determined by rhythmic TRH secretion by the SCN of the hypothalamus. The highest serum TSH concentrations occur between 9:00 p.m. and 5:30 a.m. The TSH nadir occurs between 4:00 and 7:00 p.m. TSH is liberated in the circulation and stimulates the thyrocytes to produce thyroid hormones via the thyrotropin receptor. Once T_4 and T_3 are liberated, they exert an inhibitory effect on TSH and TRH secretion (Fig. 9.6). Although negative feedback of T_4

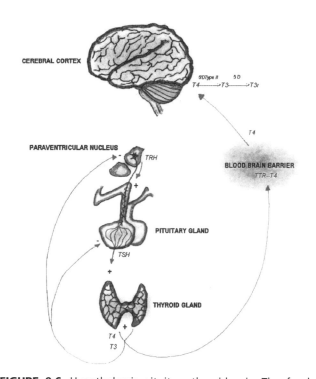

FIGURE 9.6. Hypothalamic–pituitary–thyroid axis. The fundamental action of the hypothalamic–pituitary–thyroid axis is based on stimulation of the pituitary thyrotrophs by thyrotropin releasing hormone (TRH). In response to this signal, thyrotropin (TSH) is liberated into the circulation and reaches the thyroid gland, which secretes thyroxine (T_4) and 3,5,3′-triiodothyronine (T_3), the former being the most abundant hormone produced by the gland. Through a negative feedback response, both TRH and TSH secretion are inhibited by the thyroid hormones. Other substances can also inhibit TSH secretion to a certain degree. Once T_4 is in the circulation, it crosses the blood–brain barrier with the aid of transthyretin (TTR). Within the cerebral cortex, T_4 is converted into T_3 by 5′-deiodinase type II (5′D Type II). Finally, T_3 is inactivated by 5-deiodinase (5D) into reverse T_3 (rT3). T_3 is the active thyroid hormone at the cellular level. (Reprinted from Jackson I, Arce LR. Thyroid function and major depression. In: Martini L, ed. *Encyclopedia of endocrinology and endocrine diseases.* San Diego: Academic Press, with permission.) (See Color Figure 9.6 following page 526.)

and T_3 on the pituitary is the most potent factor affecting TSH secretion, the hypothalamus plays an important role through releasing and inhibiting factors. TRH is a tripeptide synthesized in the thyrotropic area of the hypothalamus, principally the parvocellular division of the PVN (138). It is the dominant hypophysiotropic factor modulating the secretion of TSH through a tonic stimulating action. TRH interacts with high-affinity receptors on the pituitary thyrotrophs. Two other hypothalamic factors, DA (139) and somatostatin (140), have inhibitory effects on TSH secretion at the level of the thyrotroph and may function as physiologic thyrotropin-inhibitory factors; they reduce the degree of TSH release by TRH. Antiserum to somatostatin increases basal TSH secretion and potentiates the TSH response to exogenous TRH (141). Drugs that block pituitary DA receptors, such as metoclopramide or neuroleptics, have a similar effect (7,142). The levels of cortisol (138), sex steroids, and GH in the peripheral circulation (138) also modulate TSH secretion and its responsivity to TRH. Conversely, α-adrenergic agonists stimulate the thyrotrophs (144).

The major product of thyroid gland is T_4. The monodeiodination of T_4 in extrathyroidal tissues gives rise to T_3. While the brain contributes little to the circulating levels of T_3, local deiodination of T_4 in the brain is of great importance because it permits maintenance of an optimal T_3 content in the CNS. Type II 5′-deiodinase and 5-deiodinase are the major mediators of this process. The first is responsible for the conversion of T_4 to T_3 and is found in the cerebral cortex, pituitary, and hypothalamus (145). Propylthiouracil, which inhibits type I 5′-deiodinase in the periphery, does not affect type II 5′-deiodinase. The latter is also found in the cerebral cortex but not in the anterior pituitary. It results in degradation of T_3 and inactivates T_4, yielding reverse T_3 (rT_3) (146). 5-Deiodinase activity is predominant in the euthyroid state, and the rT_3 produced plays a role in the regulation of local deiodination. The distribution of these enzymes within the different regions of the nervous system varies. The cortex has the highest level of activity, followed by the midbrain, pons, hypothalamus, and brainstem.

The hypothalamic neurons that secrete TRH are regulated by monoamine neurotransmitters. 5-HT, NE, and histamine can all stimulate the release of TRH (147,148).

Stress inhibits TSH release. In rats, this effect may be owing to the stress-induced release of somatostatin (141). In humans subjected to physical stress, such as in the euthyroid sick syndrome, TSH levels do not compensate for the low T_3 and T_4 levels found in these situations (149). Patients with this syndrome have decreased levels of TSH secretion as well as a diminished circadian periodicity of TSH release (150).

Extrahypothalamic Thyroid Releasing Hormone

TRH is widely distributed throughout the mammalian brain and is also found in the pancreas, gastrointestinal tract, and reproductive system (104). Its widespread presence in the brain may suggest that TRH functions as a neurotransmitter or neuromodulator in addition to its hypophysiotropic actions. Despite the fact that TRH crosses the brain–blood barrier poorly, effects on the CNS occur after systematic administration. In the rat, TRH reverses the CNS depression induced by barbiturate and ethanol administration and causes increased motor activity in hypophysectomized mice pretreated with pargyline and then given L-dopa (151). In normal human subjects, TRH administration produces mild euphoria. In subjects undergoing alcohol withdrawal, TRH increases subjective well-being. The antidepressant actions of TRH are controversial. Intravenous administration of TRH in normal subjects often produces a transient increase or decrease in blood pressure as well as a sensation of urinary urgency (152,153). These responses reflect the effects of TRH on the autonomic nervous system.

Thyroid Axis and Psychiatric Disorders

Approximately one-fourth of patients with unipolar major depression have a blunted TSH response to TRH administration (153–156). Although increased cortisol levels reduce the thyrotropin response to TRH in patients with endocrine disorders and normal individuals (143,157), investigations have not confirmed a significant correlation between excess endogenous cortisol and the blunted TRH responses seen in depression (154). Because an abnormality in brain catecholamine metabolism may underlie some depressive disorders (156), it is conceivable that a DA excess, for example, may account for this blunted TSH response to TRH seen in depression, but this notion has not been confirmed (138,155). Additionally, neurotensin and somatostatin, which may inhibit TSH secretion, have also been suggested as possible factors in the blunted TSH responses seen in depressed patients. Other hypotheses include an altered pituitary receptor sensitivity (158) and the possibility that the blunted TRH response might reflect enhanced endogenous TRH secretion with down-regulation of TRH receptors.

People who abuse amphetamines and patients with paranoid schizophrenia may develop hyperthyroxinemia (increased T_4). Interestingly, amphetamine is structurally similar to phenylethylamine, an endogenous brain amine that may be elevated in some cases of paranoid schizophrenia (159,160). Activation of the HPT axis by endogenous biogenic amines may be responsible for the hyperthyroxinemia found in many patients who are hospitalized for acute psychosis.

Thyroid Axis in Depression

Major depression may affect as many as 15% of the general population. This syndrome entails alterations in mood manifested as a feeling of sadness, irritability, dejection, despair, or loss of interest or pleasure. Associated neurovegetative or biologic signs of depression include impairment in sleep, appetite, energy level, libido, and psychomotor activity. The diagnosis can be made in those who present with a depressed or irritable mood and have a lack of interest or pleasure for much of the day nearly every day for at least 2 weeks, in combination with the criteria illustrated in *Diagnostic and Statistical Manual of Mental Disorders IV.* The clinical syndrome is associated with significant psychological distress or impairment in psychosocial or work functioning.

Depression most frequently occurs during the fourth or fifth decade of life but may occur at any age. The duration of an episode can vary greatly, from a few months to 1 or more years. Occasionally, it may persist as chronic depression. A family history of mood disorders is common. Recurrence, which can be seen in half of all patients, generally occurs within 2 years of the first episode. Depression and thyroid disease occur more often in women than in men (161). The peak incidence occurs in women between the ages of 35 and 45, but currently the diagnosis is being made more in older women than previously (162).

The prevalence of autoimmune thyroid disease is highest among women older than 40 years of age (163). Depression and thyroid dysfunction, particularly hypothyroidism, share many clinical features. Most subjects with depression do not have biochemical evidence of thyroid dysfunction. When a patient presents with depression and findings that suggest thyroid dysfunction, it may be difficult to distinguish which entity is the cause of the symptoms from a clinical standpoint. Several studies have found either high normal or borderline low thyroid hormone levels in a substantial minority of depressed subjects. rT_3, the inactive analogue of T_3, is increased in patients with unipolar depression. In general, thyroid hormone levels are not correlated with severity of depression (155), although some measures of HPT axis function are altered in patients with major depression.

In a fetal rat hypothalamic culture system (164), there is an increase in TRH gene expression on glucocorticoid exposure (165). This effect is notable because human depression is characterized by hypercortisolemia (166), which leads to activation of the TRH neuron (167) and, consequently, thyroid function. The hypercortisolemia of depression probably results from the loss of inhibitory pathways emanating from the hippocampus, a locus for glucocorticoid negative feedback for the HPA axis. Although glucocorticoids generally inhibit the thyroid axis in humans (168) and rats (169) *in vivo,* a "functional" disconnection of the hypothalamus from the rest of the brain (as is postulated to occur in depression) would remove this inhibitory influence from the hypothalamus. Indeed, a fornical lesion that severs the hypothalamus from hippocampal regulation will increase thyroid function (170). In many ways, the hypothalamus in culture is analogous to a deafferentated hypothalamus *in vivo.* It is likely that the direct stimulation of the TRH neuron by glucocorticoids seen *in vitro* is overridden *in vivo* by an inhibitory influence emanating from the hippocam-

pus in the normal human or rat but not in some persons with clinical depression. In these individuals, activation of the TRH neuron could lead to increased hypothalamic TRH secretion with down-regulation of the TRH receptor. This could result in a blunted TSH response to exogenous TRH and spillover of TRH into the CSF, increased levels of which have been reported (171).

Thyrotropin Releasing Hormone Test

This test consists of the measurement of serum TSH before and after administration of exogenous TRH. A standardized dose of 200 to 500 μg TRH is injected intravenously after an overnight fast, with the subject recumbent throughout the procedure. Blood samples for serum TSH are drawn at baseline and at 30-minute intervals for 3 hours. A blunted response is defined as a peak response that is less than 5 μU/mL above baseline. With this definition, approximately 30% of patients with primary depression will have a blunted response (154). Hyperthyroid patients will have suppressed TSH and a flat response to TRH (<1 μU/L). Hypothyroid patients will show a high normal or elevated TSH and an exaggerated response because of a lack of feedback inhibition (>30 μU/L) (160). Several factors may influence the thyrotropin response; therefore, the clinical value of the TRH test is controversial. Age, especially in men older than 60 years of age, acute starvation, chronic renal failure, Klinefelter syndrome, and numerous medications also can reduce the response of TSH to TRH to as little as 2 μg/mL (138). There appears to be no association of the response with severity of depression or previous intake of antidepressant drugs (except for long-term lithium use). The test does not seem to aid in the distinction between primary or secondary depression or between unipolar and bipolar subgroups. However, blunted TSH responses are more common in patients with chronic depression. In addition, there is a negative association between the TSH peak value and a history of violent suicidal behavior. In a study by Linkowski et al. (172), 12 of 51 female patients with a history of suicidal behavior had an absent thyrotropin response to TRH, and of these, seven had a history of violent suicidal behavior. In contrast, only four of 39 patients with a TSH response greater than 1 μU/mL had a history of violent suicidal behavior. Also, during a 5-year follow-up study, there was a higher frequency of suicides in the group with absent TSH response to TRH stimulation.

Many studies have evaluated the TRH test as a marker of depression and a tool of prognostic value regarding outcome of treatment, but their results have been discrepant. In some patients, a blunted TSH response to TRH stimulation is a "state" marker, abnormal during a state of depression and normal on recovery. In others, it persists even after remission, suggesting that it is a biologic "trait" that identifies the patient as at risk of depression (155,173).

Hypotheses advanced to explain this blunted response include suppression of TSH secretion by increased levels of

DA, somatostatin, and/or neurotensin secretion. Another suggested mechanism is down-regulation of TRH receptors in the anterior pituitary, possibly owing to depression-associated hypersecretion of TRH (158). Previous studies show that TRH is elevated in the spinal fluid of depressed patients (174) and that chronic administration of TRH to healthy subjects causes blunted TSH response to TRH (175) as well as loss of the normal diurnal rhythm (176).

Other investigators have suggested a role for inhibition of the pituitary by an increased T_3/T_4 ratio or hypercortisolism (143,177,178).

Other Alterations of the Hypothalamic–Pituitary–Thyroid Axis in Depression

Patients with depression can have many abnormalities of the HPT axis besides blunting of the TSH response after TRH administration; these include the following (Fig. 9.7):

1. An increase, within the conventional normal range, in total and/or free T_4 levels, which regress after successful treatment of depression (179). Some patients admitted with acute psychosis, including depression, manifest a transient elevation above normal of T_4 and/or free T_4 and occasionally TSH (180). These findings usually resolve spontaneously within 2 weeks. There seems to be a direct effect of antidepressants on the TRH neuron,

and this may explain the reversal of hyperthyroxinemia with successful treatment of clinical depression. The results show that the selective serotonin reuptake inhibitors (SSRI) and the tricyclic antidepressants (TCA) inhibit TRH secretion (181). These studies indicate that the fall in circulating T_4 levels seen with antidepressant medication may reflect a direct effect on the TRH neuron and consequently a reduction of the activation of the thyroid axis. Antidepressants may also be clinically efficacious by enhancing T_4 to T_3 conversion in the CNS (182).

2. Depression leads to inhibition of type II deiodinase, possibly owing to the elevated cortisol levels. Therefore, T_4 is converted to rT_3 by 5-deiodinase. Levels of rT_3 are elevated levels in the CSF of subjects with unipolar depression (183).

3. Transthyretin (TTR), a thyroid hormone transport protein, is synthesized by the choroid plexus and accounts for as much as 25% of the protein in the CSF (184); it has a higher binding affinity for T_4 (185). TTR has an insignificant role in the transport of T_3 across the blood–brain barrier. In patients with refractory depression, CSF levels of TTR were much lower when compared with patients with neurologic disease but without depression. It has been proposed that low levels of TTR could give rise to "brain hypothyroidism" with normal peripheral thyroid hormone concentrations in depression (186).

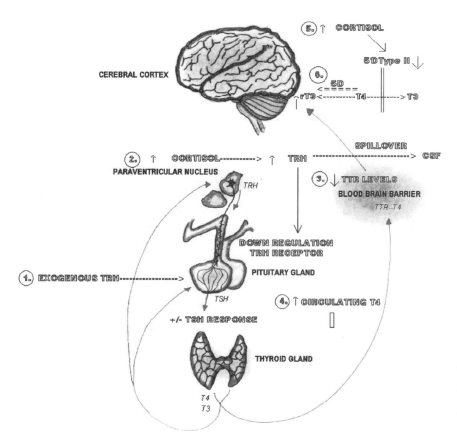

FIGURE 9.7. Alterations of the hypothalamic–pituitary–thyroid axis in depression. In depression, various alterations of the hypothalamic–pituitary–thyroid axis have been described: (1) blunted (+/−) thyrotropin (TSH) response to exogenous thyrotropin releasing hormone (TRH) administration; (2) elevated cortisol levels activate the TRH neuron, which can lead to downregulation of the TRH receptor. Thus, a spillover of TRH into the cerebrospinal fluid (CSF) can occur; (3) decreased levels of transthyretin (TTR), a thyroxine (T_4) transport protein, have been found; (4) circulating T_4 levels have been found to be elevated; (5) inhibition of 5'-deiodinase type II (5'D Type II) occurs, possibly owing to elevation of cortisol, resulting in decreased 3,5,3'-triiodothyronine (T_3); (6) T_4 is converted to reverse T_3 (T_3r), resulting in higher T_3r levels in the CSF. (See Color Figure 9.7 following page 526.)

4. Autoimmune thyroiditis is found in at least 15% of depressed patients (187) and is associated with an exaggerated response to TRH stimulation. In rapid cycling bipolar disease (four or more episodes of manic depression per year), the prevalence of autoimmune thyroid disease is even higher, reaching 50% in one series (188).

5. Alterations in the circadian rhythm of TSH have been described, including the absence of the normal nocturnal TSH surge, which may result in an overall diminution of thyroid hormone secretion. The loss of the nocturnal TSH surge is a more sensitive indicator of HPT axis alterations in depressive illness than the TRH test. Bartalena et al. (189) found that the normal nocturnal TSH surge was abolished in 14 of 15 depressed patients. Six of them had low morning TSH levels and a blunted response to the TRH test: all of them lacked the normal nocturnal TSH surge. The nine depressed patients with normal morning TSH values all had a normal response to the TRH test, but the TSH surge was lost in all but one. The SCN regulates the circadian rhythm in the body including that of TRH, which may be suppressed by the elevated circulating glucocorticoids (160), although a direct effect on the nucleus through its neural input is also possible. This suggests that there may be a degree of central hypothyroidism in some patients with depression. A similar loss of the nocturnal TSH surge has been reported in Cushing syndrome (190) and during the rigors of military boot camp (191). Furthermore, sleep deprivation, which has an antidepressant effect, leads to restoration of the nocturnal TSH rise and an elevation in the levels of T_4 and T_3 (192).

Thyroid Hormone Supplementation in Depression

In the evaluation of thyroid hormone as adjuvant therapy for depression, there is a need to exclude patients with borderline TSH elevation as well as those with detectable antithyroid antibodies. This population may respond favorably to thyroid hormone supplementation simply because of underlying hypothyroidism (193). Additionally, when parameters of thyroid function are evaluated, the antithyroid peroxidase antibody should be included. It is conceivable that patients with evidence of autoimmune thyroid disease might benefit from treatment with levothyroxine because the treatment corrects a cryptic underlying decrease in thyroid function. Patients with rapid cycling bipolar disease (three to four episodes per year) may benefit from pharmacologic doses of levothyroxine. Such cases may reflect an aberrant expression of hypothyroidism. In studies involving thyroid hormone as adjuvant therapy in depression, T_3 appears to be superior to T_4 and thus is used much more often. Cooke et al. (194) reported that T_3 augmented the response to antidepressant therapy in T_4-replaced hypothyroid patients in a randomized, controlled trial during a 3-week pe-

riod. Joffe (195) found T_3 to be significantly more effective than T_4 in patients with depression who did not respond to TCA. However, because T_4 equilibrates in tissues more slowly than T_3, 6 to 8 weeks of T_4 therapy may be required for adequate comparison of its efficacy with the more rapidly acting T_3.

Use of Thyroid Hormone to Hasten Response to Antidepressants

The use of liothyronine has been proposed to hasten antidepressant effect. The therapeutic response of TCA is delayed as long as 1 month; Prange et al. (196) studied 20 patients with retarded depression. Ten subjects received 25 μg liothyronine sodium per day in addition to 150 mg imipramine daily. The remaining group was given imipramine, at equivalent doses, as well as placebo. After 10 days of treatment, they reported a significant improvement in the Hamilton Rating Scale (HRS) for depression. The scores improved by 50%, when compared with the initial value, in those taking the combination of imipramine and liothyronine. The patients treated only with imipramine reached the same reduction in the HRS but not until after the third week of treatment. The benefit of the combination was limited to women. These workers reported a similar benefit from adjuvant liothyronine in 20 individuals with nonretarded depression.

A double-blind comparison of amitriptyline alone and in combination with two doses of liothyronine demonstrated that there was a steady and significant improvement in depression scores after receiving the higher dose of liothyronine. When adjustments were made for sex, there was a significant difference in females compared with males, which confirmed previous findings. There was no significant difference in response once results were adjusted to age and anxiety score (197).

Some studies have not found any benefit from the use of thyroid hormone as an accelerator for patients with primary depression treated with imipramine alone or in combination with liothyronine (198).

As Joffe et al. (199) pointed out, these and other studies performed more than two decades ago had the major limitations of a small sample size and what are currently judged to be inadequate doses of antidepressants. At present, therefore, liothyronine has not been established as a means of accelerating the onset of response to TCA (200).

Use of Thyroid Hormone to Convert Nonresponders to Antidepressants into Responders

Approximately 25% of patients with depression fail to improve with TCA alone. As a consequence, there has been much interest in the possibility that augmentation therapy with liothyronine could convert a nonresponder to a responder (201). Additionally, a recent study looked at the relationship between basal thyroid hormone levels and the life course of depressive illness in 75 outpatients with unipolar major depressive disorder. The significant positive predictors of recurrence were the presence of comorbid anxiety, higher numbers of previous episodes of depression, prolonged course of episodes, and T_3 level. Increases in T_3 were associated with a 22% decrease in risk of recurrence (202).

Goodwin et al. (203) reported a significant improvement in the HRS that was sufficient to allow discharge from the hospital. One of the largest studies published to date took 51 nonresponders to TCA and added liothyronine, lithium, or placebo to the antidepressant. There was a significant response in half of those included in the liothyronine and lithium groups when compared with placebo. No superiority was seen between the uses of either medication for augmentation (204).

However, Thase et al. (205) evaluated the efficacy of augmentation with liothyronine in a series of patients with recurrent unipolar depression who had failed an extended and closely monitored course (>12 weeks) of imipramine and interpersonal psychotherapy. In their open clinical trial of adjunctive liothyronine (25 μg daily) for 4 weeks, the overall response did not differ significantly from a matched historical comparison group managed with imipramine alone. Similar findings were noted by Steiner et al. (206) who compared in a double-blind study the addition of liothyronine to imipramine with imipramine and placebo and electroconvulsive therapy. At the end of the 5 weeks, three of the four women with recurrent unipolar endogenous depression in each group responded to the specific regimen administered.

Currently, the most frequently prescribed antidepressants are the SSRI. There are no controlled trials of liothyronine augmentation with these agents. However, there are case reports of beneficial effects from T_3 after failure to respond to them (207).

Although liothyronine adjuvant therapy may help 25% of TCA nonresponders, as some researchers claim, observers must be cautious about this conclusion for many reasons (208). The first is that we do not know with certainty which clinical and/or biochemical parameters determine the subset of depressed patients who will benefit from adjuvant liothyronine. Second, the clarification of the thyroid status was not universally reported in the studies that are available and unrecognized "subclinical" hypothyroidism or autoimmune thyroid disease could have led researchers to overestimate the therapeutic response to T_3. Third, the relationship between T_3 dose and clinical response is unclear. Fourth, since the longest duration of liothyronine administration in studies has been only for a few weeks, the extent of treatment remains imprecise, although it has been recommended by some investigators that thyroid augmentation be discontinued 8 to 12 weeks after a response and then reinstituted if symptoms recur. Fifth, studies regarding long-term side effects from liothyronine in depressed patients are not available, and when instituting this form of therapy, the potential effect of thyroid hormone overreplacement on the heart and bones must be strongly taken into account. Sixth, longer and larger randomized, double-blind, placebo-controlled trials are needed, especially with SSRI because these are the contemporary antidepressants.

Use of Thyrotropin Releasing Hormone in Refractory Depression

The therapeutic use of TRH in mood disorders has been proposed on the basis that this hormone has direct effects on the CNS independent of pituitary and thyroid stimulation, that high-affinity receptors are found throughout the brain, especially in the amygdala and hippocampus, and that this hormone modulates the effects of 5-HT and DA (209). Additionally, in some studies, the concentrations of TRH in the CSF of untreated individuals with mood disorders have been found to be elevated (210). It remains unclear whether this corresponds to a natural compensatory mechanism or a pathologic occurrence. This elevation could render an explanation for the blunted TSH response after exogenous administration of TRH because if the relatively high endogenous levels of TRH are causing receptor downregulation, the pituitary might be less responsive to exogenous stimulation. Carbamazepine, an anticonvulsant that has been used to treat mood disorders, has been reported to increase TRH levels in the CSF, suggesting that its beneficial clinical effect is mediated through enhanced CNS TRH production (211). However, in other studies, carbamazepine given as an adjunct to lithium in the treatment of bipolar disorder was found to lower levels of T_4 and free T_4 without a compensatory increase in TSH (212). It is thus possible that the effect of carbamazepine on TRH may differ at the hypothalamic PVN compared with extrahypothalamic locations.

Marangell and Callahan (213) administered 500 μg TRH intrathecally to eight drug-free patients with refractory depression. An identical sham spinal puncture was performed a week apart. The study was carried out in a double-blind fashion. Five patients had a clinically significant, rapid and robust but short-lived improvement in mood after TRH infusion but not after the sham puncture. Systemic thyroid function was unaltered. None of the subjects showed a worsening of their depression after infusing TRH; therefore, this finding suggests that the elevation of TRH in the CSF is more a physiologic response. In conclusion, intrathecal TRH might be a positive modulator of mood and its administration in refractory depression might be a reasonable strategy in its management (213), but further studies are needed.

Lithium and Thyroid Function

Drugs frequently used in psychiatry can affect thyroid function. Lithium at therapeutic levels inhibits thyroid hormone release by decreasing endocytosis of T_3- and T_4-laden thyroglobulin on the luminal side of the thyroid follicle. At higher levels, it can inhibit iodine uptake and organification. Clinical hypothyroidism develops in approximately 7% of patients taking lithium, over and above the rate expected for patients of the same age and sex not on lithium. Women and patients with positive antithyroid antibodies are especially vulnerable. In some cases, lithium unmasks a subclinical case of autoimmune thyroiditis. Lithium inhibits the conversion of T_4 to T_3 in the rodent brain and pituitary, suggesting that part of the therapeutic effect of lithium in patients with bipolar disorder may be achieved by reducing brain T_3 levels (160), the converse of the effect of the administration of T_3 (along with TCA) in unipolar depression.

Reproductive Axis

In this section, several issues of clinical relevance are discussed. These are stress and reproductive function, neuroendocrine accompaniments of eating disorders, and reproductive endocrine aspects of epilepsy.

Sexual Dimorphism of the Central Nervous System: Sex Hormones and Central Nervous System Function

Sexual brain organization is dependent on sex hormone and neurotransmitter levels occurring during critical developmental periods, including peri- and neonatal life (214–219). In males, studies have suggested that somatic components of the testes may contribute to a male type of differentiation of germ cells from the beginning of sexual differentiation (220). The testes actively produce androgens *in utero*, whereas physiologically important steroid hormone production by the ovary does not start until puberty (221). Androgens can produce a male phenotype, regardless of genotypic gender. Their action is mediated by the androgen receptor (222). Higher androgen levels during brain organization cause a characteristically male pattern of brain development.

The levels of androgen necessary at the time of differentiation in the second trimester of pregnancy are only three times greater in male fetuses than female fetuses during the second trimester compared with 20 times higher levels of free testosterone in the adult male than in the adult female. Androgen excess in females can be caused by genetic conditions, such as the 21-hydroxylase deficiency or exogenous androgenic drugs. Studies in rats showed a significant sex-related difference in the hypothalamic somatostatin and GHRH mRNA levels after full sexual maturation was attained. In addition, ovariectomized rats that received dihydrotestosterone had a GH secretory pattern that was indistinguishable from that of intact male rats, with the higher somatostatin and GHRH mRNA levels typical of male rats (223). These effects on the genetic expression of GH were therefore mediated by androgen receptors. A functional androgen receptor is an absolute requirement for male sexual differentiation (222).

Estrogens are also important during brain differentiation. Estrogen exposure during fetal development is necessary for the sex-specific organization of gonadotropin secretion, including positive estrogen feedback (215) and the establishment of the cyclic LH surge mechanism. In the rat, fetal estrogen exposure is necessary for the expression of lordosis behavior in the adult (217).

Many neuroanatomic sex differences have been identified in both animals and humans that may underlie sex-specific behavior in both reproductive and nonreproductive functions. For instance, in the adult pig, the cell count and volume of the vasopressin- and oxytocin-containing nucleus is 260% larger in the male than in the female. Hoffman and Swaab (224) reviewed observations on the human hypothalamus. They observed that the SCN, a structure involved in the regulation of circadian rhythms and reproductive cycles, is ellipsoidal in females and more spherical in males. In addition, an extremely large SCN was observed in the brains of homosexual men who died of acquired immunodeficiency syndrome. Both the volume of the SCN and the number of vasopressin neurons were approximately twice as large as in the male reference group. The volume and cell count of the so-called sexually dimorphic nucleus of the preoptic area show a marked sexual dimorphism. The mean volume of the sexually dimorphic nucleus of the preoptic area is 2.2 times larger in males than in females and contains approximately twice as many cells. This area might be involved in the control of male sexual behavior. Asymmetries in temporoparietal regions of the brain also differ between men and women (225). However, a correlation between these findings and gender-specific behaviors has not yet been established.

Studies performed by LeVay (226) on autopsy material from 41 subjects (19 homosexual men who died of acquired immunodeficiency syndrome, 16 heterosexual men, and six heterosexual women) showed that the third interstitial nuclei of the anterior hypothalamus were twice as large in heterosexual men as in the women or homosexual men. Studies by Allen and Gorski (227) on postmortem tissue have shown that there are neuroanatomic differences in the area of both the anterior commissure and massa intermedia, both being areas larger in females than in males.

Play behavior in children seems to be affected by the history of their mothers of exposure to progestational agents. In this case, the history of progesterone exposure in the prenatal period was associated with hypomasculinization effects (218). Progesterone exposure also reduces play-fighting in male rats (228). Perinatal androgen exposure masculinizes social play in children, but after this period of critical neuronal differentiation, androgen levels apparently have no effect on the expression of social play. This effect may involve androgen receptors in the amygdala (228).

There are also large sex differences in the incidence or age of onset of many mental disorders. For example, depression is more prevalent in women, and antisocial personality is more prevalent in men. Also, schizophrenia tends to have a later onset in women. These differences are likely to involve both social and biologic factors (229,230).

Stress and Reproductive Function

Stress-related hormones, i.e., CRH, the POMC derivatives ACTH and β-endorphin, and adrenal corticosteroids, can influence reproductive function at three different levels of the hypothalamic–pituitary–gonadal (HPG) axis: in the brain by inhibiting GnRH secretion, in the pituitary by interfering with LH release, and in the gonads by interfering with the stimulatory effect of gonadotropin on sex steroid secretion (231).

The mechanisms through which CRH suppresses GnRH and LH in stress are not fully understood. In rats, direct anatomic connections have been demonstrated between CRH axon terminals and dendrites of GnRH-secreting neurons (232). But, again, there are CRH-immunoreactive cells and fibers in many telencephalic, diencephalic, and brainstem structures (233). Studies suggest that the effect of CRH on GnRH may involve the activation of several pathways, including those dependent on endogenous opiates (234) and catecholamines (235). There is substantial evidence that the stress-induced suppression of gonadotropins results from the effect of CRH on the hypothalamus. *In vitro* studies have shown that CRH inhibits hypothalamic GnRH release (236). In ovariectomized rats, infusion of CRH into both sides of the medial preoptic area significantly decrease GnRH release and plasma LH levels (231). In women, CRH infusions inhibit LH secretion (237,238). Conversely, in rats, the inhibition of LH secretion by experimental stress can be prevented by prior administration of a CRH antagonist (α-helical CRH9-41) (239). In addition, stress levels of glucocorticoids disrupt reproductive physiology by decreasing hypothalamic GnRH liberation and basal pituitary LH secretion as well as after stimulation with GnRH (44).

In male and female CRH-deficient mice, LH is suppressed after restraint and food withdrawal; thus, CRH is not the sole inhibitor of gonadotropin release during stress (240). Endogenous opiates (POMC derivatives) and morphine decrease GnRH concentrations (241,242) in rats, whereas naloxone increases GnRH release from hypothalamic fragments (243). CRH stimulates the release of POMC-derived peptides, including β-endorphin. Both CRH and POMC-derived peptides may be directly released onto GnRH neurons of the preoptic area and act synergistically to reduce their activity (231,244,245).

Functional hypothalamic amenorrhea (FHA) is a common and potentially reversible disorder of the HPG axis. This disorder is characterized by psychophysiologic and behavioral responses to life events, which cause dysregulation of central neuronal activity and thus disruption of GnRH secretion from the mediobasal hypothalamus. Women with FHA are reported to have more depressive symptoms, dysfunc-

tional attitudes, and disordered eating than eumenorrheic controls (246). Initially, patients may deny abnormal eating patterns, but after careful evaluation, they seem to have a significantly higher incidence of AN or bulimia (247). As noted above, GnRH is affected by stress through CRH. In patients with FHA, basal cortisol levels are significantly higher than eumenorrheic or non-FHA amenorrheic women (248), and the responses of ACTH and cortisol to CRH injection are blunted when compared with controls (249–251). The latter findings suggest hypothalamic overactivity with down-regulation of the CRH receptors in the pituitary, although some studies have been unable to show elevated CRH levels in the CSF (252). Long-term follow-up studies of patients with FHA suggest there may be a correlation between the baseline body mass index, cortisol, and androstenedione levels with recovery of menses. A lower basal cortisol level and higher androstenedione value and body mass index, as well as an increase of the latter, were positive prognostic factors for the resolution of FHA (253).

Eating Disorders: Anorexia Nervosa and Bulimia

Eating disorders, especially AN, are associated with alterations in the reproductive endocrine system. Other neuroendocrine systems are altered simultaneously, i.e., the HPA, GH, and HPT axes.

Anorexia Nervosa

AN is a psychiatric disorder characterized by voluntary emaciation because of an intense fear of weight gain and a disturbance in body image. The criteria include a weight below 15% of expected body weight and at least a 3-month history of amenorrhea (254). Multiple endocrine disturbances are present in this condition that are not explained by weight loss alone. Some of them may persist despite weight recovery.

The amenorrhea of AN is a form of hypothalamic dysfunction in which LH and FSH secretion reverts to a prepubertal pattern, that is, LH and FSH secretion are significantly decreased when compared with control women. LH shows a greater depression than FSH. The number of their secretory spikes is also reduced (255,256). These abnormalities may be mild or severe, ranging from a luteal phase defect to the complete picture of hypogonadotropic hypogonadism, with loss of the LH and FSH response to a clomiphene challenge (256). The response of LH to GnRH is blunted, whereas the FSH response may be increased (255,256). The characteristic immature pattern of 24-hour LH secretion has been reproduced in experimental starvation of healthy individuals. However, the onset of amenorrhea and return to menses in AN is poorly correlated with body weight (255). Resumption of menses has been correlated with a return to at least 90% of the patient's standard body weight (257).

The circadian rhythm of other hormones has also been found to be altered. Leptin is a peptide that has the capacity to cause weight loss and diminish adiposity when injected in mice. It is encoded by the *ob* gene and expressed mainly in adipocytes. Adipose tissue and plasma leptin concentrations are concordant with amount of stored energy. Obese subjects have higher levels of plasma leptin. Leptin also responds to insulin because the latter increases the synthesis of leptin after infused *in vivo*. It follows a circadian rhythm; the levels of leptin increase at night and reach a nadir in the morning. Leptin can be considered as a signal of nutritional status and exerts an effect on the hypothalamic–pituitary axis. Some studies seem to suggest that the body fat mass and eating behavior have a significant correlation with the levels of plasma leptin in AN (258). Behavior is not the sole determinant of leptin levels because in overweight women with a binge-eating disorder, plasma leptin concentrations are elevated despite an aberrant eating pattern (259). In women with eating disorders, leptin levels are extremely low, but when these patients regain weight, the plasma concentrations increase (260). LH and FSH levels appear to track increments of leptin during the initial stages of therapy for AN; very low basal levels of leptin seem to correlate with a slow increase in gonadotropins (261).

GH is significantly higher in many women with AN, and the elevation is inversely proportional to the deviation of body weight from the ideal. Basal GH levels are elevated in 29% of women with AN, and 22% of AN patients do not exhibit a significant GH nocturnal peak (262). An abnormal GH increase after TRH stimulation has been found in approximately 50% of patients (263). Insulin–like growth factor-I (IGF-I) is decreased. These abnormalities are most likely tied to starvation. Elevated GH levels are found in protein or caloric malnutrition states, and the levels revert to normal after refeeding (264).

Ghrelin is a GH-releasing peptide secreted by the stomach (265) in response to food intake, which stimulates GH release from the anterior pituitary. Studies suggest that ghrelin responds to changes in energy homeostasis; in subjects with AN, plasma ghrelin concentrations are higher than in healthy subjects with normal body weight, whereas it is decreased in nondiabetic obese individuals (266). It has been observed that ghrelin concentrations decrease and return to normal levels in anorectics during refeeding and after weight gain (267). The significance of these findings still needs to be determined with further studies.

Cortisol levels are also significantly elevated in AN patients and are inversely correlated with body mass index and directly correlated with the percentage of decrease from ideal body weight and HRS scores for depression (262). Endogenous hypothalamic CRH levels are also likely elevated, and the ACTH/cortisol response to CRH is decreased, as in depression (255,268a). These findings indicate a significant relationship between the HPA axis abnormalities and the severity of weight loss and depressive symptoms in AN.

A delayed TSH response is found in 86% of anorectics, and it may normalize with recovery of weight. This abnormal TSH response is not correlated with depressive symptoms (263). PRL levels and the PRL response to TRH stimulation usually are normal (263) but occasionally are decreased.

Bulimia

Bulimia is an eating disorder characterized by recurrent binge eating with fear or inability to stop eating voluntarily, alternating with behavior directed toward losing weight or avoiding weight gain, such as self-induced vomiting, laxative or diuretic abuse, dieting, and exercising to lose weight, accompanied by depressed mood after binge eating (268b). Approximately one-third of patients with bulimia nervosa have a history of AN.

Studies performed in bulimic patients by Fichter et al. (269) showed that there was a tendency toward nonsuppression of cortisol in the DST of patients with bulimia. This was more pronounced in patients with evidence of restricted caloric intake, but there was no association with depressive symptoms. However, plasma dexamethasone levels after a standard dose were lower in bulimics than in controls. TRH stimulation produced blunted responses in TSH levels in 80% of patients with both bulimia and AN and in 22% of those with bulimia alone (263). PRL response was blunted in one study (269) but normal in another (263). LH and FSH were reduced in bulimics with evidence of reduced caloric intake (269). Increased basal GH levels were found in 33% of bulimic patients, regardless of food restriction. A paradoxical increase of GH levels on TRH stimulation was found in 15% to 20% of patients with bulimia (263). All these results indicate that multiple neuroendocrine disturbances occur in bulimia but to a lesser degree than in AN. This also suggests that caloric intake is an important factor, but not the only one, in the maintenance of a normal neuroendocrine function.

Gonadotropin levels to GnRH stimulation were found to be abnormal in bulimic patients (220). LH response to GnRH and the number of LH peaks in 24 hours tend to be very heterogeneous. FSH response, as in AN, is sometimes greater than that of LH.

Reproductive Endocrine Aspects of Epilepsy

Epilepsy is associated with reduced fertility in women, impotence in men, and hyposexuality in both men and women. These changes are often owing to hormonal phenomena that may be affected by pharmacologic treatment.

In one recent study of 37 adult male epileptic patients who were seizure free on monotherapy, eight were impotent, and two of those lacked desire. Patients with epilepsy had significantly lower levels of free testosterone and dihydrotestosterone than healthy controls, and those with impotence had higher levels of estradiol than those with normal sexual function (270). Other authors have observed that both carbamazepine and phenytoin increase serum levels of sex hormone binding globulin, which, without an increase in total testosterone, reduces the free fraction present (271). Measurement of free testosterone and estradiol would thus be expected to aid in understanding impotence or hyposexuality in a treated epileptic man.

The decreased fertility of women with epilepsy is owing to impairment of ovulation (272). More than one-third of 17 women with temporal lobe epilepsy had anovulatory cycles, in contrast to none of seven women with primary generalized epilepsy studied recently by Cummings et al. (272). Women with epilepsy and anovulation may have either hypothalamic amenorrhea or the polycystic ovary syndrome (273). In both cases, the pulsatile secretion of LH is altered; Drislane et al. (274) found it to be both lower and more variable among patients with temporal lobe epilepsy than among healthy controls. In contrast, Meo et al. (275) studied 11 drug-free epileptic women with normal menstrual cycles and found them to have increased pulsatility of LH secretion, with increased basal secretion during increases in interictal activity.

Herzog (276) presented impressive evidence for laterality effects of temporal lobe foci. In a study of 30 women with unilateral temporal lobe foci, complex partial seizures, and reproductive endocrine abnormality, 15 of the 16 with polycystic ovary syndrome had left-sided electroencephalogram discharges, and 12 of 14 with hypothalamic hypogonadism had right-sided electroencephalogram discharges.

The specific choice of drug therapy for epilepsy in women apparently affects its reproductive endocrine consequences. Menstrual disturbances, polycystic ovaries by ultrasound, and elevated serum testosterone all are more common in women taking valproate than those taking carbamazepine or other medications (277). In fact, the majority of women taking valproate for epilepsy in this study had either elevated testosterone, polycystic ovaries, or both. Women who had started valproate before age 20 had the highest rate of reproductive endocrine problems (80%).

Serum PRL is known to increase after generalized seizures and complex partial seizures with bilateral medial temporal involvement (278,279). PRL peaks at two to three times normal in 15 to 20 minutes and returns to baseline within 3 hours. Unilateral temporal lobe activity, in contrast, does not raise PRL levels (280). Failure of PRL levels to rise after an apparent generalized seizure suggests pseudoseizures.

In summary, specific reproductive endocrine abnormalities can be demonstrated in most patients with epilepsy and altered sexual or menstrual function. Although patients with such abnormalities should be offered endocrine treatment if they are troubled by their symptoms, the efficacy of endocrine interventions in fully restoring sexual function in epileptic patients is not known.

Hypothalamic–Pituitary–Somatotropic System

GH release is regulated by the noncompetitive antagonism of GHRH and somatostatin (158). The effects of a number of other hormones, such as opioid peptides, neurotensin, and neuropeptide Y, appear to be mediated via alterations in the secretion of both GHRH and somatostatin. However, ghrelin, which is a GH releasing peptide that initially was characterized in the stomach and is a potent stimulator of GH release by the pituitary, is probably also synthesized and secreted in the hypothalamus. Receptors for ghrelin have been identified on the somatotrophs. Its precise relationship to GHRH and somatostatin in the regulation of GH secretion is not yet fully understood.

It has been observed that somatostatin concentration is decreased in the CSF of patients with depression (281), and its levels are inversely related to the maximal post–dexamethasone plasma cortisol concentration (282,283).

Somatostatin-containing neuronal cell bodies, terminals, and receptors are distributed throughout the brain where they likely play a role in neuronal function unrelated to regulation of pituitary GH secretion. There is a widespread reduction of somatostatin immunoreactivity in the cerebral cortex of patients with Alzheimer disease caused by degeneration of intrinsic somatostatin cortical neurons (284,285). The somatostatin reduction correlates with impaired cognitive function and brain hypometabolism on PET scanning (286). It is not known whether the somatostatin changes play an etiologic role in the dementia or are a secondary response to the neuronal degeneration.

Prolactin

Conflicting results on basal PRL levels and response to TRH have been reported in depression. Basal PRL has been found to be low, normal, and elevated. Similarly, the response to TRH has been variable (172). The underlying cause of the disturbances in PRL regulation in affective disorders is still not clear. Increased activity of the hypothalamic tuberoinfundibular dopaminergic neurons or deficiency of central 5-HT pathways could account for the suppressed TRH-induced PRL response. DA is the principal physiologic regulator of PRL secretion; however, a number of neurotransmitters, including GABA, 5-HT, histamine, and several neural peptides (VIP, β-endorphin, neurotensin, substance P) affect PRL secretion. This effect may occur either at the pituitary or hypothalamic level (157), and a disturbance in any of these peptides may account for the abnormalities found in psychiatric disorders.

Endocrine Diseases Presenting as Psychiatric Disorders

Hypothalamic Lesions

Experiments of nature demonstrate *in vivo* the role of the CNS, particularly, the hypothalamus in human behavior. Anatomic lesions of the hypothalamus can present as mental or behavioral disorders, such as explosive disorder (287),

TABLE 9.1. HYPOTHALAMIC LESIONS PRESENTING AS PSYCHIATRIC DISORDERS

Hypothalamic Lesions	Psychiatric Manifestations
Granulomatous disorders	Explosive disorder
Sarcoidosis	Polyphagia
Eosinophilic granuloma	Hypodipsia
Histiocytosis X	Polydipsia
Congenital defects	Hypersomnolence
Hemochromatosis	Dementia
Pseudotumor cerebri	
Whole brain irradiation	

polyphagia and morbid obesity (288), hypodipsia (289,290), primary polydipsia (291), hypersomnolence (292), and dementia (293,294).

The causes cited are varied and include not only parasellar tumors but also granulomatous disorders such as sarcoidosis (288,291), eosinophilic granuloma (295), histiocytosis X (293,296), congenital defects (289), pseudotumor cerebri (290), hemochromatosis (297), and whole brain irradiation (Table 9.1) (294).

Hypo- and Hyperthyroidism

Disturbances of thyroid function can cause mental and behavioral symptoms. Thus, appropriate treatment directed at the thyroid gland is a necessary and often sufficient condition for their total reversal. In addition, patients with subtle defects in thyroid function may be more susceptible to mental disorders because the prevalence of thyroid autoantibodies is higher in psychiatric patients than in the general population (Table 9.2) (298).

Hypothyroidism indicates any degree of thyroid hormone deficiency. The Whickham survey found an incidence of 2% among 2,800 individuals. The mean age at diagnosis was 57 years. Female predominance was noted, although in debilitated geriatric populations, the prevalence does not have a gender difference (299). Subclinical hypothyroidism is char-

TABLE 9.2. PSYCHIATRIC MANIFESTATIONS OF THYROID DISEASE

Hyperthyroidism	Hypothyroidism
Anxiety	Depression
Dysphoria	Inattention
Irritability	Poor memory
Insomnia	Sleepiness
Difficulty concentrating	Difficulty concentrating
Restlessness	Slow motor function
Manic symptoms	Paucity of speech
Decreased energy	Hallucinations
Delusions	Bizarre behavior
Paranoia	Lethargy
Apathy	Drowsiness
Lethargy	Stupor
Depression	Coma

acterized by a mild elevation of the serum TSH, normal circulating levels of thyroid hormones, and the absence of the typical symptoms of overt hypothyroidism (300). The prevalence of subclinical hypothyroidism is 2% to 8%. It is higher in women than in men (7.5% vs. 2.8%, respectively), particularly if the individual is older than 60 years of age. In that instance, the prevalence can be as high as 16%. Progression to overt hypothyroidism occurs in 20% to 50% of individuals within a period of 4 to 8 years. Those patients with elevated titers of antithyroid antibodies have a higher incidence of progression (close to 80% within 4 years) (301). As noted before, T_3 in the brain is derived from deiodination of T_4 by type II deiodinase. The activity of this enzyme is increased in hypothyroidism. Intracerebral generation of T_3 increases as serum concentrations of T_4 decline. It seems that intracellular T_3 concentrations will remain fairly stable until serum T_4 has been depleted (302).

Depression has been described in all grades of hypothyroidism. The occurrence of depression is believed to be higher in those who have a history of a first-degree relative with depression. As many as 10% of patients admitted for treatment of depression are found to have subclinical or frank hypothyroidism. Clinical manifestations are indistinguishable from nonthyroid-related depression. The neuropsychiatric features of adult-onset hypothyroidism are not specific and consist of ill-defined complaints and disturbances in cognition and/or mood. There is lack of attention, concentration, and poor memory. Symptoms may resemble those of depression, with apathy, social withdrawal, paucity of speech, and slow motor function. In severe hypothyroidism, delusions and hallucinations may occur, giving rise to bizarre behavior and paranoid ideas. This may be followed by lethargy, drowsiness, and, finally, stupor and coma (303). Response is usually refractory to antidepressants alone; on some occasions, it has also been resistant to thyroid hormone replacement (304). The pathophysiology of these symptoms and signs probably involves both direct effects of T_3 on neurons and indirect effects via other neurotransmitters. T_3 nuclear receptors are present in neurons, with high concentrations in the amygdala and hippocampus and the lowest concentrations in the brainstem and cerebellum (305). Studies in adult patients with hypothyroidism have demonstrated decreased cerebral blood flow and consumption of oxygen and glucose (27% below normal). Results from PET studies indicate that TSH levels correlate inversely with global and regional blood flow as well as cerebral glucose metabolism (306). Additionally, when PET studies were performed in individuals with severe hypothyroidism of short duration, the brain activity was globally reduced without the regional modifications observed in primary hypothyroidism (307).

Hyperthyroidism can be seen in as many as 3% of subjects older than 60 years of age. Typically, Graves disease is still the most common cause of thyrotoxicosis in the elderly, but toxic multinodular goiter or adenomas are more frequent than in young people. Patients with hyperthyroidism may present with anxiety and dysphoria, emotional liability, insomnia, difficulty concentrating, irritability, restlessness, and tremulousness. This may resemble mania because motor activity is increased, but, unlike mania, energy is usually decreased (303). In some cases, severe thyrotoxicosis may present as a psychotic illness, with delusional and paranoid thoughts. Elderly individuals, however, may present with apathy, lethargy, and depression (apathetic thyrotoxicosis). The clinical diagnosis of apathetic thyrotoxicosis is often delayed, with the patient developing tachyarrhythmia or congestive heart failure before thyroid disease is appreciated. It is probable that the interactions between the thyroid hormones and the catecholaminergic system play an important role in the development of symptoms and signs of hyperthyroidism. Catecholamines and thyroid hormones share the amino acid tyrosine as a precursor and have synergistic cellular actions. In hyperthyroidism, the turnover of catecholamines is decreased, but β-adrenergic receptor numbers are increased (305). These mechanisms may explain the role of β-adrenergic antagonists in the reduction of thyrotoxic symptoms. Successful treatment of thyrotoxicosis with antithyroid agents, thyroidectomy, or radioactive iodine usually leads to the resolution of the neuropsychiatric syndrome. The laboratory data may show a suppressed TSH with or without frank elevation of T_4 and T_3. In studies of patients with thyrotoxicosis, depression has been documented in as many as 30% to 60% of the cases.

Hypo- and Hypercortisolism

Patients with severe or chronic adrenal insufficiency may present with mild to moderate cognitive impairment (5%–20%), apathetic depression (20%–40%), or psychosis (20%–40%). Symptoms may include irritability, social withdrawal, poor judgment, hallucinations, paranoid delusions, bizarre thinking, and even catatonic posturing (308).

Hypercortisolism, such as in Cushing syndrome or exogenous glucocorticoid excess, presents with mental or behavioral symptoms in as many as 50% of cases (Table 9.3). Symptoms and signs include emotional lability, depression, loss of energy and poor libido, irritability, anxiety, panic attacks, and paranoia. Occasional patients present with a manic psychosis. A study performed by Martignoni et al. (309) of

TABLE 9.3. PSYCHIATRIC MANIFESTATIONS OF HYPERCORTISOLISM

Emotional lability
Depression
Lack of energy
Loss of libido
Irritability
Anxiety
Panic attacks
Paranoia
Increased appetite

24 patients with Cushing disease (hypercortisolism induced by an ACTH–pituitary tumor) matched with same number of control subjects showed significant difficulty with everyday tasks and impairment in verbal and nonverbal episodic memory, which did not correlate with plasma ACTH levels, urinary free cortisol, or DST response. Memory impairment was worse in older patients. The behavioral abnormalities were significantly reversed after surgical treatment of the pituitary tumor.

The brain represents one of the principal targets for steroid hormones. Steroids secreted by the adrenal cortex, such as glucocorticoids, mineralocorticoids, and stress hormones, easily cross the blood–brain barrier and exert their actions in the CNS at classic steroid receptors (309). Glucocorticoids have prominent actions on the hippocampus, which is the area of the brain with the greatest number of glucocorticoid receptors. Either a lack or an excess of glucocorticoids can produce significant functional impairment or structural change in CA3 cells in the hippocampus. Studies in monkeys subjected to stress showed degenerative changes at the level of the hippocampus (310). Furthermore, in humans subjected to severe stress, such as concentration camp survivors (311) and war sailors (312), cognitive and psychological disturbances have been documented. Brain computed tomography scans of political prisoners who were subjected to physical torture showed cerebral atrophy (313).

Exogenous glucocorticoids have been shown to produce impairment of higher brain functions in humans. The term *steroid psychosis* has been used to describe a psychotic syndrome in patients who are given high doses of corticosteroids for medical illnesses that have not directly affected the CNS (314). A cognitive impairment syndrome was described by Varney et al. (315) in such patients, consisting of decreased attention, concentration, retention, and mental speed. Administration of exogenous glucocorticoids in healthy subjects has also been associated with cognitive impairment and poor performance on memory tasks (316). It has been postulated that corticosteroids may impair selective attention (316) and may suppress the activity of the hippocampus, where the stimuli may be initially filtered (317).

Endocrine dysfunction is a common accompaniment of neuropsychiatric disorders, and neuropsychiatric symptoms frequently are part of the preservation of primary endocrine disorders. Clinical assessment and, where appropriate, laboratory investigation of endocrine function are an integral part of comprehensive neuropsychiatric evaluation. Endocrine interventions are likely to find an increasing role in neuropsychiatric therapy.

REFERENCES

1. Müller EE. The neuroendocrine approach to psychiatric disorders: a critical appraisal. *J Neural Transm* 1990;81:1–15.
2. Palkovits M. Neuropeptides in the median eminence. *Neurochem Int* 1986;9:131–137.
3. Palkovits M. Peptidergic neurotransmitters in the endocrine

hypothalamus. In: *Functional anatomy of the neuroendocrine hypothalamus (Ciba Foundation Symposium 168)*. Chichester: John Wiley & Sons, 1992:3–5.

4. Wurtman RJ, Hefti F, Melamed E. Precursor control of neurotransmitter synthesis. *Pharmacol Rev* 1980;32:315–335.

5. Jackson IMD. Hypothalamic releasing hormones: mechanisms underlying neuroendocrine dysfunction in affective disorders. In: Brown GM, et al., eds. *Neuroendocrinology of psychiatric disorders*. New York: Raven Press, 1984:255–266.

6. Pelletier G. Anatomy of the hypothalamic-pituitary axis. *Methods Achiev Exp Pathol* 1991;14:1–22.

7. Martin J, Reichlin S. *Clinical neuroendocrinology*. Philadelphia: FA Davis, 1987:11–44.

8. Reichlin S. Neuroendocrinology. In: Wilson JD, Foster DW, eds. *Williams textbook of endocrinology*. Philadelphia: WB Saunders, 1992:135–219.

9. Kovacs K, Horvath E, Ezrin C. Anatomy and histology of the normal and abnormal pituitary gland. In: DeGroot L, ed. *Endocrinology*. Philadelphia: WB Saunders, 1989:264–274.

10. Denef C, Schramme C, Baes M. Stimulation of growth hormone release by vasoactive intestinal peptide and peptide PHI in rat anterior pituitary reaggregates. Permissive action of a glucocorticoid and inhibition by thyrotropin releasing hormone. *Neuroendocrinology* 1985;40:88–91.

11. Nowak M, Markowska A, Nussdorfer GG, et al. Evidence that endogenous vasoactive intestinal peptide (VIP) is involved in the regulation of rat pituitary-adrenocortical function: *in vivo* studies with a VIP antagonist. *Neuropeptides* 1994;27:297–303.

12. Youngren OM, Silsby JL, Rozenboim I, et al. Active immunization with vasoactive intestinal peptide prevents the secretion of prolactin induced by electrical stimulation of the turkey hypothalamus. *Gen Comp Endocrinol* 1994;95:330–336.

13. Lafuente A, Marco J, Esquifino AI. Possible changes in the regulatory mechanisms of pulsatile luteinizing hormone secretion in adult pituitary-grafted female rats. *Proc Soc Exp Biol Med* 1995;209:163–169.

14. Shibata S, Ono M, Tominaga K, et al. Involvement of vasoactive intestinal polypeptide in NMDA-induced phase delay of firing activity rhythm in the suprachiasmatic nucleus *in vitro*. *Neurosci Biobehav Rev* 1994;18:591–595.

15. Aguilar-Roblero R, Morin LP, Moore RY. Morphological correlates of circadian rhythm restoration induced by transplantation of the suprachiasmatic nucleus in hamsters. *Exp Neurol* 1994;130:250–260.

16. Peters RV, Zoeller RT, Hennessey AC, et al. The control of circadian rhythms and the levels of vasoactive intestinal peptide mRNA in the suprachiasmatic nucleus are altered in spontaneously hypertensive rats. *Brain Res* 1994;639:217–227.

17. Mick G, Shigemoto R, Kitahama K. Localization of substance P receptors in central neural structures controlling daily rhythms in nocturnal rodents. *C R Acad Sci* 1995;318:209–217.

18. Mick G, Maeno H, Kiyama H, et al. Marginal topography of neurons expressing the substance P receptor in the rat suprachiasmatic nucleus. *Brain Res Mol Brain Res* 1994;21:157–161.

19. Makara GB. The relative importance of hypothalamic neurons containing corticotropin-releasing factor or vasopressin in the regulation of adrenocorticotropic hormone secretion. In: *Functional anatomy of the neuroendocrine hypothalamus (Ciba Foundation Symposium)*. Chichester: John Wiley & Sons, 1992:43–53.

20. Hokfelt T, Meister B, Everitt B, et al. Chemical neuroanatomy of the hypothalamo-pituitary axis: focus on multimessenger systems. In: McCann SM, Weiner RI, eds. *Integrative neuroendocrinology: molecular, cellular and clinical aspects*. Basel: Karger, 1987:1–34.

21. Dudley CA, Moss RL. LHRH and mating behavior: sexual receptivity versus sexual preference. *Pharmacol Biochem Behav* 1985;22:967–972.

22. Dudley CA, Moss RL. Facilitation of lordosis in female rats by CNS-site specific infusions of an LHRH fragment, Ac-LHRH-(5-10). *Brain Res* 1988;44:161–167.

23. Moss RL, Dudley CA. Differential effects of a luteinizing-hormone (LHRH) antagonist analogue on lordosis behavior induced by LHRH and the LHTH fragment Ac-LHRH 5-10. *Neuroendocrinology* 1990;52:138–142.

24. Vale W, Spiess J, Rivier C, et al. Characterization of a 41 residue ovine hypothalamic peptide that stimulates secretion of corticotropin and β-endorphin. *Science* 1981;213:1394–1397.

25. Pelletier G, Desy L, Cote J, et al. Immunocytochemical localization of corticotropin-releasing factor-like immunoreactivity in the human hypothalamus. *Neurosci Lett* 1983;41:259–263.

26. Dunn AJ, Berridge CW. Physiological and behavioral responses to corticotropin-releasing factor administration: is CRF a mediator of anxiety or stress responses? *Brain Res Brain Res Rev* 1990;15:71–100.

27. Valentino RJ, Page ME, Curtis AL. Activation of noradrenergic locus coeruleus neurons by hemodynamic stress is due to local release of corticotropin-releasing factor. *Brain Res* 1991;555:25–34.

28. Desy L, Pelletier G. Immunohistochemical localization of somatostatin in the human hypothalamus. *Cell Tissue Res* 1977;184:491–497.

Hypothalamic–Pituitary–Adrenal Axis

29. Meaney JM, Viau V, Bhatnagar S, et al. Cellular mechanisms underlying the development and expression of individual differences in the hypothalamic-pituitary-adrenal stress response. *J Steroid Biochem Mol Biol* 1991;39:265–274.

30. Stokes PE, Sikes CR. The hypothalamic-pituitary-adrenocortical axis in major depression. *Endocrinol Metab Clin North Am* 1988;17:1–17.

31. Stokes PE, Sikes CR. Hypothalamic-pituitary-adrenal axis in psychiatric disorders. *Annu Rev Med* 1991;42:519–531.

32. Uno H, Eisele S, Sakai A, et al. Neurotoxicity of glucocorticoids in the primate brain. *Horm Behav* 1994;28:336–348.

33. Sapolsky RM, Krey LC, McEwen BS. Glucocorticoid sensitive hippocampal neurons are involved in terminating the adrenocortical stress response. *Proc Natl Acad Sci U S A* 1984;81:6147–6177.

34. Herman JP, Schafter MK, Young EA, et al. Evidence for the hippocampal regulation of neuroendocrine neurons of the hypothalamo-pituitary-adrenocortical axis. *J Neurosci* 1989;9:3072–3082.

35. McEwen BS, De Kloet ER, Rostene WH. Adrenal steroid receptors in the nervous system. *Physiol Rev* 1986;66:1121–1150.

36. Herman JP, Adams D, Prewitt C. Regulatory changes in neuroendocrine stress-integrative circuitry produced by a variable stress paradigm. *Neuroendocrinology* 1995;61:180–190.

37. Rivier C. Role of interleukins in the stress response. *Methods Achiev Exp Pathol* 1991;14:63–79.

38. Woloski BM, Smith EM, Meyer WI, et al. Corticotropin-releasing activities of monokines. *Science* 1985;230:1035–2037.

39. Ovadia H, Abramsky O, Barak V, et al. Effect of interleukin-1 on adrenocortical activity in intact and hypothalamic deafferentated rats. *Exp Brain Res* 1989;76:246–249.

40. Harbuz MS, Stephanou A, Sarlis N, et al. The effects of recombinant human interleukin (IL)-1 alpha, IL-1 beta or IL-6 on hypothalamo-pituitary-adrenal axis aviation. *J Endocrinol* 1992;133:349–355.

41. Abraham E. Effects of stress on cytokine production. In: Jasmin G, Cantin M, eds. Stress revisited. 1. Neuroendocrinology of stress. *Methods Achiev Exp Pathol* 1991;14:45–62.

42. Jorgensen C, Sany J. Modulation of the immune response by the neuro-endocrine axis in rheumatoid arthritis. *Clin Exp Rheumatol* 1994;12:435–441.

43. Seyle H. General adaption syndrome and diseases adaption. *J Clin Endocrinol Metab* 1946;6:117–230.

44. Sapolsky RM, Romero LM, Munck AU. How do glucocorticoids influence stress responses? Integrating permissive, suppressive, stimulatory and preparative actions. *Endocrinol Rev* 2000;21:55–89.

45. Van der Kar LD, Richardson-Morton KD, et al. Stress: neuroendocrine and pharmacological mechanisms. In: Jasmin G, Cantin M, eds. Stress revisited. I. Neuroendocrinology of stress. *Methods Achiev Exp Pathol* 1991;14:133–173.

46. Opstad PK, Aakvaag A. Decreased serum levels of oestradiol, testosterone and prolactin during prolonged physical strain and sleep deprivation. *Eur J Appl Physiol* 1981;49:343–348.

47. Opstad PK. Androgenic hormones during prolonged physical stress, sleep and energy deficiency. *J Clin Endocrinol Metab* 1992;74:1174–1183.

48. McEwwn BS, Sapolsky RM. Stress and cognitive function. *Curr Opin Neurobiol* 1995;5:205–216.

49. Beck SG, List TJ, Choi KC. Long and short term administration of corticosterone alters CA1 hippocampal neuronal properties. *Neuroendocrinology* 1994;656:71–78.

50. Vaher PR, Luine VN, Gould E, et al. Effects of adrenalectomy on spatial memory performance and dentate gyrus morphology. *Brain Res* 1994;60:261–272.

51. Sapolsky R. Why stress is bad for your brain. *Science* 1996;273:749–750.

52. Keenan PA, Jacobson MW, Soleymani RM, et al. The effect on memory of chronic prednisone treatment in patients with systemic disease. *Neurology* 1996;47:1396–1402.

53. Doyle P, Guillaume-Gentile C, Rohner-Jeanrenaud F, et al. Effects of corticosterone administration on local cerebral glucose utilization of rats. *Brain Res* 1994;645:225–230.

54. Horner HC, Packan DR, Sapolsky RM. Glucocorticoids inhibit glucose transport in cultured hippocampal neurons and glia. *Neuroendocrinology* 1990;52:57–63.

55. Kathol RG, Noyes R, Lopez A. Similarities in hypothalamic adrenal axis activity between patients with panic disorder and those experiencing external stress. *Psychiatr Clin North Am* 1988;11:335–348.

56. Wedekind D, Bandelow B, Broocks A, et al. Salivary, total plasma and plasma free cortisol in panic disorder. *J Neural Transm* 2000;107:831–837.

57. Bandelow B, Wedekind D, Sandvoss V, et al. Diurnal variation of cortisol in panic disorder. *Psychiatry Res* 2000;95:245–250.

58. Bandelow B, Wedekind D, Pauls J, et al. Salivary cortisol in panic attacks. *Am J Psychiatry* 2000;157:454–456.

59. Charlton BG, Ferrier IN. Hypothalamo-pituitary-adrenal axis abnormalities in depression: a review and a model. *Psychol Med* 1989;19:331–336.

60. Wong ML, Kling MA, Munson PJ, et al. Pronounced and sustained central hypernoradrenergic function in major depression with melancholic features: relation to hypercortisolism and corticotropin-releasing hormone. *Proc Natl Acad Sci U S A* 2000;7:325–330.

61. Carrol BJ, Martin FI, Davis BM. Resistance to suppression by dexamethasone of plasma 11(OH)CS levels in severe depressive illness. *BMJ* 1968;3:285–287.

62. Weller EB, Weller RA. Neuroendocrine changes in affectively ill children and adolescents. *Neurol Clin* 1988;6:41–54.

63. Arana GW, Mossman D. The dexamethasone suppression test and depression. *Endocrinol Metab Clin North Am* 1988;17:21–39.

64. Carrol BJ. The dexamethasone suppression test for melancholia. *Br J Psychiatry* 1982;140:293–304.

65. Gerner R, Gwirtsman HE. Abnormalities of dexamethasone suppression and urinary MPHG in anorexia nervosa. *Am J Psychiatry* 1981;138:650–653.

66. Insel TR, Kalin HH, Guttmacher LB, et al. The dexamethasone suppression test in patients with primary obsessive-compulsive disorder. *Psychiatry Res* 1982;6:153–160.

67. Spar JE, Gerner R. Does the dexamethasone suppression test distinguish dementia from depression? *Am J Psychiatry* 1982;139:238–240.

68. Graham PM, Booth H, Boranga G, et al. The dexamethasone suppression test in mania. *J Affect Disord* 1982;4:201–211.

69. Dewan MJ, Pandurangi AK, Boucher ML, et al. Abnormal dexamethasone suppression test results in chronic schizophrenic patients. *Am J Psychiatry* 1982;131:1501–1503.

70. Kroll P, Palmer C, Greden JF. The dexamethasone suppression test in patients with alcoholism. *Biol Psychiatry* 1983;14:441–450.

71. Rupprecht R, Noder M, Jecht E, et al. Pre- and post dexamethasone cortisol and prolactin levels in sexual dysfunction and normal controls. *Biol Psychiatry* 1988;23:527–530.

72. Greden JF, Kronfol Z, Gardner R, et al. Neuroendocrine evaluation of schizoaffectives with the dexamethasone suppression test. *Biol Psychiatry* 1984;145:372–382.

73. Frochtengarten ML, Villares JCB, Maluf E, et al. Depressive symptoms and the dexamethasone suppression test in parkinsonian patients. *Biol Psychiatry* 1987;22:386–389.

74. Berger M, Pirke KM, Doerr P. The limited utility of the dexamethasone suppression test for the diagnostic process in psychiatry. *Br J Psychiatry* 1984;145:372–382.

75. Berger M, Krieg C, Bossert S, et al. Past and present strategies of research on the HPA-axis in psychiatry. *Acta Psychiatr Scand Suppl* 1988;341:112–125.

76. Weller E, Weller R. Neuroendocrine changes in affectively ill children and adolescents. *Endocrinol Metab Clin North Am* 1988;17:41–54.

77. Arana GW, Baldessarini RJ, Ornsteen M. The dexamethasone suppression test for diagnoses and prognosis in psychiatry: commentary and review. *Arch Gen Psychiatry* 1985;42:1193–1204.

78. Nelson JC, Davis JM. DST studies in psychotic depression: a meta-analysis. *Am J Psychiatry* 1997;154:1497–1503.

79. Kathol RG, Jaeckle RS, Lopez JF, et al. Pathophysiology of HPA axis abnormalities in patients with major depression: an update. *Am J Psychiatry* 1989;146:311–317.

80. Gold PW, Chrousos GP. Clinical studies with corticotropin releasing factor: implications for the diagnosis and pathophysiology of depression, Cushing disease and adrenal insufficiency. *Psychoneuroendocrinology* 1985;10:401–419.

81. Gold PW, Loriaux DL, Roy A, et al. Responses to corticotropin-releasing hormone in the hypercortisolism of

depression and Cushing's disease: pathophysiologic and diagnostic implications. *N Engl J Med* 1986;314:1329–1335.

82. Amsterdam JD, Maislin G, Winokur A, et al. Pituitary and adrenocortical responses to the ovine corticotropin releasing hormone in depressed patients and healthy volunteers. *Arch Gen Psychiatry* 1987;44:775–781.

83. Lopez JF, Kathol RG, Jaeckle RS, et al. The HPA axis response to insulin hypoglycemia in depression. *Biol Psychiatry* 1987;22:153–166.

84. Reisine T, Hoffman AA. Desensitization of corticotropin-releasing factor receptors. *Biochem Biophys Res Commun* 1983;3:919–925.

85. Mixoguchi K, Yuzurihara M, Ishige A, et al. Chronic stress differentially regulates glucocorticoid negative feedback response in rats. *Psychoneuroendocrinology* 2001;26:443–459.

86. Brown GM. Psychoneuroendocrinology of depression. *Psychiatr J University Ottawa* 1989;14:344–348.

87. Cassidy F, Ritchie JC, Carroll BJ. Plasma dexamethasone concentration and cortisol response during manic episodes. *Biol Psychiatry* 1998;43:747–754.

88. Kostic VS, Sternic NC, Bumbasirevic LB, et al. Dexamethasone suppression test in patients with Parkinson's disease. *Mov Disord* 1990;5:23–26.

89. Frochtengarten ML, Villares JCB, Maluf E, et al. Depressive symptoms and the dexamethasone suppression test in Parkinsonian patients. *Biol Psychiatry* 1987;22:386–389.

90. Skare S, Pew B, Dysken M. The dexamethasone suppression test in dementia: a review of the literature. *J Geriatr Psychiatry Neurol* 1990;3:124–138.

91. Lipsey JR, Robinson RG, Pearlson GD, et al. Dexamethasone suppression test and mood following strokes. *Am J Psychiatry* 1985;142:318–323.

92. Finklestein S, Benowitz LI, Baldessarini RJ, et al. Mood, vegetative disturbance, and dexamethasone suppression test after stroke. *Ann Neurol* 1981;12:463–468.

93. Bauer M, Gans JS, Harley JP, et al. Dexamethasone suppression test and depression in a rehabilitation setting. *Arch Phys Med Rehabil* 1983;64:421–422.

94. Olsson T, Astrom M, Eriksson S, et al. Hypercorticolism revealed by the dexamethasone suppression test with acute ischemic stroke. *Stroke* 1989;20:1685–1690.

95. Amsterdam JD, Winokur A, Abelman E, et al. Cosyntropin stimulation test in depressed patients and healthy controls. *Am J Psychiatry* 1983;140:907–909.

96. Zis AP, Dorovini-Zis K. Increased adrenal weight in victims of violent suicides. *Am J Psychiatry* 1987;144:1214–1215.

97. Zobel AW, Nickel T, Sonntag A, et al. Cortisol response in the combined dexamethasone/CRH test as predictor of relapse in patients with remitted depression, a prospective study. *J Psychiatr Res* 2001;35:83–94.

98. Baghai TC, Schule C, Zwanzger P, et al. Evaluation of a salivary based combined dexamethasone/CRH test in patients with major depression. *Psychoneuroendocrinology* 2002;27:385–399.

99. Ur E, Grossman A. Corticotropin releasing hormone in health and disease: an update. *Acta Endocrinol (Copenh)* 1992;127:193–199.

100. Grigoriadis DE, Heroux JA, De Souza EB. Characterization and regulation of corticotropin releasing factor receptors in the central nervous, endocrine and immune systems. *Ciba Found Symp* 1993;172:85–101.

101. Sawchenko PE, Imaki T, Potter E, et al. The functional neuroanatomy of corticotropin releasing factor. *Ciba Found Symp* 1993;172:5–29.

102. Arborelius L, Owens MJ, Plotsky PM, et al. The role of corticotropin-releasing factor in depression and anxiety disorders. *J Endocrinol* 1999;160:1–12.

103. Axelrod J, Reisine TD. Stress hormones: their interaction and regulation. *Science* 1984;224:452–459.

104. Dunn AJ, Berridge CW. Physiological and behavioral responses to corticotropin-releasing factor administration: is CRF a mediator of anxiety or stress responses? *Brain Res Brain Res Rev* 1990;15:71–100.

105. Koob GF, Bloom FE. Corticotropin-releasing factor and behavior. *Fed Proc* 1985;44:259–263.

106. Anisman H, Griffiths J, Matheson K, et al. Posttraumatic stress symptoms and salivary cortisol levels. *Am J Psychiatry* 2001;158:1509–1511.

107. Baker DG, West SA, Nicholson WE, et al. Serial CSF corticotropin-releasing hormone levels and adrenocortical activity in combat veterans with posttraumatic stress disorder. *Am J Psychiatry* 1999;156:585–588.

108. Kanter ED, Wilkinson CW, Radant AD, et al. Glucocorticoid feedback sensitivity and adrenocortical responsiveness in posttraumatic stress disorder. *Biol Psychiatry* 2001;50:238–245.

109. Yehuda R. Post-traumatic stress disorder. *N Engl J Med* 2002;346:108–114.

110. Flesher MR, Delahanty DL, Raimonde AJ, et al. Amnesia, neuroendocrine levels and PTSD in motor vehicle accident victims. *Brain Inj* 2001;15:879–889.

111. Delahanty DL, Raimonde AJ, Spoonster E. Initial posttraumatic urinary cortisol levels predict subsequent PTSD symptoms in motor vehicle accident victims. *Biol Psychiatry* 2000;48:940–947.

112. Matsuzaki I, Takamatsu Y, Moroji T. The effects of intracerebroventricularly injected corticotropin-releasing factor (CRF) on the central nervous system: behavioral and biochemical studies. *Neuropeptides* 1989;13:147–155.

113. Emoto H, Koga C, Ishii H, et al. The effect of CRF antagonist on immobilization stress induced increases in noradrenaline release in rat brain regions. *Yakubutsu Seishin Kodo* 1993;13:81–87.

114. Hiroi N, Wong ML, Licinio J, et al. Expression of corticotropin releasing hormone receptors type I and type II mRNA in suicide victims and controls. *Mol Psychiatry* 2001;6:540–546.

115. Zobel AW, Nickel T, Kunzel HE, et al. Effects of the high-affinity corticotropin-releasing hormone receptor 1 antagonist R121919 in major depression: the first 20 patients treated. *J Psychiatry Res* 2000;34:171–181.

116. Bale TL, Contarino A, Smith GW, et al. Mice deficient for corticotropin-releasing hormone receptor-2 display anxiety-like behavior and are hypersensitive to stress. *Nat Genet* 2000;24:410–414.

117. Bakshi VP, Smith-Roe S, Newman SM, et al. Reduction of stress-induced behavior by antagonism of corticotropin-releasing hormone 2 (CRH_2) receptors in lateral septum or CRH_1 receptors in amygdala. *J Neurosci* 2002;22:2926–2935.

118. Owens MJ, Edwards E, Nemeroff CB. Effects of 5-HT1A receptor agonists on hypothalamo-pituitary adrenal axis activity and corticotropin-releasing factor containing neurons in the rat brain. *Eur J Pharmacol* 1990;190:113–122.

119. Valentino RJ, Curtis AL. Pharmacology of locus coeruleus spontaneous and sensory evoked activity. *Prog Brain Res* 1991;88:249–256.

120. Butler PD, Weiss JM, Stout JC, et al. Corticotropin releasing factor produces fear-enhancing and behavioral activating effects following infusion into the locus coeruleus. *J Neurosci* 1990;10:176–183.

121. Nemeroff CB, Owens MJ, Bissette G, et al. Reduced

corticotropin releasing factor binding sites in the frontal cortex of suicide victims. *Arch Gen Psychiatry* 1988;45:577–579.

122. Banki CM, Bissette G, Arato M, et al. CSF corticotropin-releasing factor-like immunoreactivity in depression and schizophrenia. *Am J Psychiatry* 1987;144:873–877.

123. Arato M, Banki CM, Bissette G, et al. Elevated CSF CRF in suicide victims. *Biol Psychiatry* 1989;25:355–359.

124. Geracioti TD, Orth DN, Ekhator NN, et al. Serial cerebrospinal fluid corticotropin releasing hormone concentrations in healthy and depressed humans. *J Clin Endocrinol Metab* 1992;74:1325–1330.

Suicide

125. Stahl SM. Neuroendocrine markers of serotonin responsivity in depression. *Prog Neuropsychopharmacol Biol Psychiatry* 1992;16:655–659.

126. Beskow J, Gottfries EF, Ross BE. Determination of monoamine metabolites in the human brain: postmortem studies in a group of suicides and in a control group. *Acta Psychiatr Scand* 1976;53:7–20.

127. Lloyd KG, Farley IJ, Deck JHN, et al. Serotonin and 5-hydroxyindoleacetic acid in discrete areas of the brain stem of suicide victims and control patients. *Adv Biochem* 1974;11:387–398.

128. Banki CM, Arato M, Papp Z, et al. Biochemical markers in suicidal patients. Investigations with cerebrospinal fluid amine metabolites and neuroendocrine test. *J Affect Disord* 1984;6:341–350.

129. Roy A. Hypothalamic-pituitary adrenal function and suicidal behavior in depression. *Biol Psychiatry* 1992;32:812–816.

130. Bunney WE, Fawcett JA, Davis JM, et al. Further evaluation of urinary hydroxycorticosteroid in suicidal patients. *Arch Gen Psychiatry* 1969;21:138–150.

131. Krieger J. The plasma level of cortisol as a predictor of suicide. *J Dis Nerv Sys* 1974;35:273–240.

132. De Leo D, Pellegrini C, Serraiotto L, et al. Assessment of severity of suicide attempts. A trial with the dexamethasone suppression test and 2 rating scales. *Psychopathology* 1986;19:186–191.

133. Modestin J, Ruef C. Dexamethasone suppression test (DST) in relation to depressive somatic and suicidal manifestations. *Acta Psychiatr Scand* 1987;75:491–494.

134. Wilmotte J, Van Wettere JP, Depauw Y, et al. Dexamethasone suppression test repeated after a suicide attempt. *Acta Psychiatr Belg* 1986;86:242–248.

135. Maes M, Vandewoude M, Schotte C, et al. Hypothalamic-pituitary-adrenal and thyroid axis dysfunctions and decrements in the availability of L-tryptophan as biological markers of suicidal ideation in major depressed females. *Acta Psychiatr Scand* 1989;80:13–17.

136. Jones JS, Stein DJ, Stanley B, et al. Negative and depressive symptoms in suicidal schizophrenics. *Acta Psychiatr Scand* 1994;89:81–87.

137. Coryell W, Schlesser M. The dexamethasone suppression test and suicide prediction. *Am J Psychiatry* 2001;158:748–753.

Hypothalamic–Pituitary–Thyroid Axis

138. Jackson IMD. Thyrotropin releasing hormone. *N Engl J Med* 1983;306:145–155.

139. Scanlon MF, Weightman DR, Shale DJ, et al. Dopamine is a physiological regulator of thyrotropin (TSH) secretion in normal man. *Clin Endocrinol* 1979;10:7–15.

140. Tanjasiri P, Kozbur X, Floersheim WH. Somatostatin in the physiologic feedback control of thyrotropin secretion. *Life Sci* 1976;19:657–660.

141. Arimura A, Schally AV. Increases in basal and thyrotropin releasing hormone (TRH)-stimulated secretion of thyrotropin (TSH) by passive immunization with antiserum to somatostatin in rats. *Endocrinology* 1976;98:1069–1072.

142. Birkhaeuser MH, Staubb JC, Crani R, et al. Dopaminergic control of TSH response to TRH in depressive patients. Abstracts of the XIth International Congress of the International Society of Psychoneuroendocrinology, Florence, Italy, 1980:61.

143. Nicoloff JT, Fisher DA, Appleman MD. The role of glucocorticoids in the regulation of thyroid function in man. *J Clin Invest* 1979;49:1922–1929.

144. Scanlon MF, Toft AD. Regulation of thyrotropin secretion. In: Braverman LE, Utiger RD, eds. *Werner and Ingbar's the thyroid: a fundamental and clinical text*, 8th ed. Philadelphia: Lippincott Williams & Wilkins, 2000:234–253.

145. Larsen PR, Silva JE, Kaplan MM. Relationships between circulating and intracellular thyroid hormones: physiological and clinical implications. *Endocr Rev* 1981;2:87–102.

146. Leonard JL, Koehrle J. Intracellular pathways of iodothyronine metabolism. In: Braverman LE, Utiger RD, eds. *Werner and Ingbar's the thyroid: a fundamental and clinical text,* 8th ed. Philadelphia: Lippincott Williams & Wilkins, 2000:136–173.

147. Weiner RL, Ganong WF. Role of brain monoamines and histamine in regulation of anterior pituitary secretion. *Physiol Rev* 1978;58:905–976.

148. Burger HG, Patel YC. TSH and TRH: their physiological regulation and the clinical applications of TRH. In: Martini L, Besser GM, eds. *Clinical neuroendocrinology*. New York: Academic Press, 1977:67–131.

149. Wartofsky L, Burman KD. Alterations in thyroid function in patients with systemic illness: the "euthyroid sick syndrome." *Endocr Rev* 1982;3:164–217.

150. Custro N, Scafidi V, Gallo S, et al. Deficient pulsatile thyrotropin secretion in the low-thyroid-hormone state of severe non-thyroidal illness. *Eur J Endocrinol* 1994;130:132–136.

151. Yarbrough GG. On the neuropharmacology of thyrotropin releasing hormone (TRH). *Prog Neurobiol* 1979;12:291–312.

152. Hershman JM. Clinical application of thyrotropin-releasing hormone. *N Engl J Med* 1974;290:886–890.

153. Prange AJ, Numeroff CB, Loosen PT, et al. Behavioral effects of thyrotropin releasing hormone in animals and man: a review. In: Collu R, Brabeau A, Ducharme JR, et al., eds. *Central nervous system effects of the hypothalamic hormones and other peptides.* New York: Raven Press, 1979:75–96.

154. Loosen PT, Prange AJ. Serum thyrotropin response to thyrotropin-releasing hormone in psychiatric patients: a review. *Am J Psychiatry* 1982;139:405–416.

155. Loosen PT. Thyroid function in affective disorders and alcoholism. *Endocrinol Metab Clin North Am* 1988;17:55–81.

156. Terry LC. Catecholamine regulation of growth hormone and thyrotropin in mood disorders. In: Brown GM, ed. *Neuroendocrinology and psychiatric disorders.* New York: Raven Press, 1984:237–254.

157. Re RB, Kourides IA, Ridgway EC, et al. The effect of glucocorticoid administration on human pituitary secretion of thyrotropin and prolactin. *J Clin Endocrinol Metab* 1976;43:338–346.

158. Lesch KP, Rupprecht R. Psychoneuroendocrine research in depression. *J Neural Transm* 1989;75:179–194.

159. Morley JE, Shafer RB, Elson MK, et al. Amphetamine induced hyperthyroxinemia. *Ann Intern Med* 1980;93:707–709.

160. Hein MD, Jackson IMD. Review: thyroid function in psychiatric illness. *Gen Hosp Psychiatry* 1990;12:232–244.

161. Tucker GJ. Psychiatric disorders in medical practice. In: Bennett JC, Plum F, eds. *Cecil textbook of medicine*, 20th ed. Philadelphia: WB Saunders, 1996:2006–2010.

162. Cohen GD. Neuropsychiatric aspects of aging. In: Bennett JC, Plum F, eds. *Cecil textbook of medicine*, 20th ed. Philadelphia: WB Saunders, 1996:17–21.

163. Rosenthal MJ, Hint WC, Garry PJ, et al. Thyroid failure in the elderly: microsomal antibodies as discriminant for therapy. *JAMA* 1987;258:209–213.

164. Bruhn TO, Rondeel JMM, Bolduc TG, et al. Induction of thyrotropin-releasing hormone gene expression in cultured fetal diencephalic neurons by differentiating agents. *Endocrinology* 1996;137:572–579.

165. Luo G-L, Bruhn TO, Jackson IMD. Glucocorticoids stimulate thyrotropin-releasing hormone gene expression in cultured hypothalamic neurons. *Endocrinology* 1995;136:4945–4950.

166. Axelson DA, Doraiswamy PM, McDonald WM, et al. Hypercortisolemia and hippocampal changes in depression. *Psychiatry Res* 1996;47:163–173.

167. Jackson IMD. Thyrotropin-releasing hormone and corticotropin-releasing hormone—what's the message? [Editorial]. *Endocrinology* 1995;136:2793–2794.

168. Brabant A, Brabant G, Schuermeyer T, et al. The role of glucocorticoids in the regulation of thyrotropin. *Acta Endocrinol (Copenh)* 1989;121:95–100.

169. Kakucska I, Qi Y, Lechan RM. Changes in adrenal status affect hypothalamic thyrotropin-releasing hormone gene expression in parallel with corticotropin-releasing hormone. *Endocrinology* 1995;136:2795–2802.

170. Shi Z-X, Levy A, Lightman SL. Hippocampal input to the hypothalamus inhibits thyrotropin and thyrotropin-releasing hormone gene expression. *Neuroendocrinology* 1993;57:576–580.

171. Kirkegaard C, Faber J, Hummer L, et al. Increased levels of TRH in cerebrospinal fluid from patients with endogenous depression. *Psychoneuroendocrinology* 1979;4:227–233.

172. Linkowski P, VanWettere JP, Kerkhofs M, et al. Thyrotropin response to thyreostimulin in affectively ill women: relationships to suicidal behavior. *Br J Psychiatry* 1983;143:401–405.

173. Prange AJ, Loosen PT. Findings in affective disorders relevant to the thyroid axis, menotropin, oxytocin and vasopressin. In: Brown GM, ed. *Neuroendocrinology and psychiatric disorder.* New York: Raven Press, 1984:191–200.

174. Kirkegard C, Faber J, Hummer L. Increased levels of TRH in cerebrospinal fluids from patients with endogenous depression. *Psychoneuroendocrinology* 1979;4:227–235.

175. Snyder PJ, Utiger RD. Repetitive administration of thyrotropin-releasing hormone results in small elevations of serum thyroid hormones and marked inhibition of thyrotropin response. *J Clin Invest* 1973;52:2305–2312.

176. Spencer CA, Greenstadt MA, Wheeler WS, et al. The influence of long-term low dose thyrotropin-hormone infusions on serum thyrotropin and prolactin concentrations in man. *J Clin Endcrinol Metab* 1980;51:771–775.

177. Otsuki M, Dakoda M, Baba S. Influence of glucocorticoids in TRF induced TSH response in man. *J Clin Endocrinol Metab* 1973;36:95–102.

178. Wilber JF, Utiger RD. The effects of glucocorticoids on thyrotropin secretion. *J Clin Invest* 1969;48:2096–2103.

179. Bauer MS, Whybrow PC. Thyroid hormones and the central nervous system in affective illness: interactions that may have clinical significance. *Integr Psychiatry* 1988;6:75–100.

180. Chopra IJ, Solomon DH, Huang T-S. Serum thyrotropin in hospitalized psychiatric patients: Evidence for hyperthyrotropinemia as measured by an ultrasensitive thyrotropin assay. *Metabolism* 1990;39:538–543.

181. Jackson IMD, Luo L-G. Antidepressants inhibit the glucocorticoid stimulation of TRH expression in cultured hypothalamic neurons. *J Investig Med* 1998;46:470–474.

182. Campos-Barros A, Meinhold H, Stula M, et al. The influence of desipramine on thyroid hormone metabolism in rat brain. *J Pharmacol Exp Ther* 1994;268:1143–1152.

183. Linnoila M, Crowley R, Lamberg B-A, et al. CSF triiodothyronine (rT$_3$) levels in patients with affective disorders. *Biol Psychiatry* 1983;18:1489–1492.

184. Schreiber G, Aldred AR, Jaworowski A, et al. Thyroxine transport from blood to brain via transthyretin synthesis in choroid plexus. *Am J Physiol* 1990;258:R338–R345.

185. Cody V. Thyroid hormone interactions: molecular conformation, protein binding, and hormone action. *Endocr Rev* 1980;1:140–166.

186. Hatterer JA, Herbert J, Hidaka C, et al. CSF transthyretin in patients with depression. *Am J Psychiatry* 1993;150:813–815.

187. Nemeroff CB, Simon JS, Haggerty JJ, et al. Antithyroid antibodies in depressed patients. *Am J Psychiatry* 1983;142:840–843.

188. Bauer MS, Whybrow PC, Wynokour A. Rapid cycling bipolar affective disorder. 1. Association with grade 1 hypothyroidism. *Arch Gen Psychiatry* 1990;47:427–432.

189. Bartalena L, Placidi GF, Martin, et al. Nocturnal serum thyrotropin (TSH) surge and TSH response to TSH-releasing hormone: dissociated behavior in untreated depressives. *J Clin Endocrinol Metab* 1990;71:650–655.

190. Adriaanse R, Brabant G, Endert E, et al. Pulsatile thyrotropin secretion in patients with Cushing's syndrome. *Metabolism* 1994;43:782–786.

191. Opstad K. Circadian rhythm of hormones is extinguished during prolonged physical stress, sleep and energy deficiency in young men. *Eur J Endocrinol* 1994;131:56–66.

192. Baumgartner A, Graf K-J, Kurten I, et al. Thyrotropin (TSH) and thyroid hormone concentrations during partial sleep deprivation in patients with major depressive disorders. *J Psychiatr Res* 1990;24:281–292.

193. Targum SD, Greenburg RD, Harmon RH, et al. Thyroid hormone and the TRH stimulation test in refractory depression. *J Clin Psychiatry* 1984;45:345–346.

194. Cooke RG, Joffe RT, Levett AJ. T$_3$ augmentation of antidepressant treatment in T$_4$-replaced thyroid patients. *J Clin Psychiatry* 1992;53:16–18.

195. Joffe RT. A perspective on the thyroid and depression. *Can J Psychiatry* 1990;35:754–758.

196. Prange AJ Jr, Wilson IC, Rabon AM. Enhancement of imipramine and antidepressant activity by thyroid hormone. *Am J Psychiatry* 1969;126:457–469.

197. Wheatley D. Potentiation of amitriptyline by thyroid hormone. *Arch Gen Psychiatry* 1972;26:229–233.

198. Feighner JP, King LJ, Schuckit MA, et al. Hormonal potentiation of imipramine and ECT in primary depression. *Am J Psychiatry* 1972;128:1230–1238.

199. Joffe RT, Sokolov STH, Singer W. Thyroid hormone treatment of depression. *Thyroid* 1995;5:235–239.

200. Nierenberg AA. Depression: recent developments and innovative treatments. Antidepressant augmentation and combinations. *Psychiatr Clin North Am* 2000;23:743–755.

201. Nelson JC. Augmentation strategies in depression 2000. *J Clin Psychiatry* 2000;61:13–19.
202. Joffe RT, Marriott M. Thyroid hormone levels and recurrence of major depression. *Am J Psychiatry* 2000;157:1689–1691.
203. Goodwin FK, Prange AJ Jr, Post RM, et al. Potentiation of antidepressant effects by L-triiodothyronine in tricyclic nonresponders. *Am J Psychiatry* 1982;139:34–38.
204. Joffe RT, Singer W, Levitt AJ, et al. A placebo-controlled comparison of lithium and triiodothyronine augmentation of tricyclic antidepressants in unipolar refractory depression. *Arch Gen Psychiatry* 1993;50:387–393.
205. Thase ME, Kupfer DJ, Jarrett DB. Treatment of imipramine-resistant recurrent depression: I. An open clinical trial of adjunctive L-triiodothyronine. *J Clin Psychiatry* 1989;50:385–388.
206. Steiner M, Radwan M, Elizur A, et al. Failure of L-triiodothyronine (T₃) to potentiate tricyclic antidepressant response. *Curr Ther Res* 1978;23:655–659.
207. Gupta S, Masand P, Tanquary JF. Thyroid hormone supplementation of fluoxetine in the treatment of major depression. *Br J Psychiatry* 1991;159:866–867.
208. Joffe RT. The use of thyroid supplements to augment antidepressant medication. *J Clin Psychiatry* 1998;59[Suppl 5]:26–29.
209. Marangell LB, George MS, Callahan AM, et al. Effects of intrathecal thyrotropin-releasing hormone (Protirelin) in refractory depressed patients. *Arch Gen Psychiatry* 1997;54:214–222.
210. Banki CM, Bissette G, Arato M, et al. Elevation of immunoreactive CSF TRH in depressed patients. *Am J Psychiatry* 1988;145:1526–1531.
211. Marangell LB, George MS, Bissette G, et al. Carbamazepine increases cerebrospinal fluid thyrotropin-releasing hormone levels in affectively ill patients. *Arch Gen Psychiatry* 1994;51:625–628.
212. Isojarvi JIT, Pakarinen AJ, Ylipalosaari, et al. Serum hormones in male epileptic patients receiving anticonvulsant medication. *Arch Neurol* 1990;47:670–676.
213. Marangell LB, Callahan AM. Mood disorders and the thyroid axis. *Curr Opin Psychiatry* 1998;11:67–70.
214. Dörner G, Poppe I, Stahl F, et al. Gene- and environment-dependent neuroendocrine etiogenesis of homosexuality and transsexualism. *Exp Clin Endocrinol* 1991;98:141–150.
215. Dörner G. Neuroendocrine response to estrogen and brain differentiation in heterosexuals, homosexuals and transsexuals. *Arch Sex Behav* 1988;17:57–75.
216. Dörner G. Hormone-dependent brain development and neuroendocrine prophylaxis. *Exp Clin Endocrinol* 1989;94:4–22.
217. Dohler KD. The pre- and post-natal influence of hormones and neurotransmitters on sexual differentiation of the mammalian hypothalamus. *Int Rev Cytol* 1991;131:1–57.
218. Meyer-Bahlburg HF, Feldman JF, Cohen P, et al. Perinatal factors in the development of gender-related play behavior: sex hormones versus pregnancy complications. *Psychiatry* 1988;51:260–271.
219. Vanderticele H, Eechaute W, Lacroix E, et al. The effects of neonatal androgenization of male rats on testosterone metabolism by the hypothalamus-pituitary-gonadal axis. *J Steroid Biochem Mol Biol* 1987;26:493–497.
220. Francavilla S, Cordeschi G, Properzi G, et al. Ultrastructure of fetal human gonad before sexual differentiation and during early testicular and ovarian development. *J Submicrosc Cytol Pathol* 1990;22:389–400.
221. Huhtaniemi I. Endocrine function and regulation of the fetal and neonatal testis. *Int J Dev Biol* 1989;33:117–123.
222. Mowszowics I, Stamatiadis D, Wright F, et al. Androgen receptor in sexual differentiation. *J Steroid Biochem Mol Biol* 1989;32:157–162.
223. Hasegawa O, Sugihara H, Minami S, et al. Masculinization of growth hormone (GH) secretory pattern by dihydrotestosterone is associated with augmentation of hypothalamic somatostatin and GH-releasing hormone mRNA levels in ovariectomized adult rats. *Peptides* 1992;13:475–481.
224. Hoffman MA, Swaab DF. Sexual dimorphism of the human brain: myth and reality. *Exp Clin Endocrinol* 1991;98:161–170.
225. Witelson SF. Neural sexual mosaicism: sexual differentiation of the human temporoparietal region for functional asymmetry. *Psychoneuroendocrinology* 1991;16:131–153.
226. LeVay SA. A difference in hypothalamic structure between heterosexual and homosexual men. *Science* 1991;253:1034–1037.
227. Allen LS, Gorski RA. Sexual dimorphism of the anterior commissure and massa intermedia of the human brain. *J Comp Neurol* 1991;312:97–104.
228. Meaney MJ. The sexual differentiation of social play. *Psychiatr Dev* 1989;7:247–261.
229. Seeman MV. Prenatal gonadal hormones and schizophrenia in men and women. *Psychiatr J University Ottawa* 1989;14:473–475.
230. Sikich L, Todd RD. Are the neurodevelopmental effects of gonadal hormones related to sex differences in psychiatric illnesses? *Psychiatr Dev* 1988;6:277–309.

Stress and Reproductive Function

231. Rivier C, Rivest S. Review: effect of stress on the activity of the hypothalamic-pituitary-gonadal axis: peripheral and central mechanisms. *Biol Reprod* 1991;45:523–532.
232. MacLuskey NJ, Naftolin F, Leranth C. Immunocytochemical evidence for direct synaptic connections between corticotropin releasing factor (CRF) and gonadotropin releasing hormone (GnRH)-containing neurons in the preoptic area of the rat. *Brain Res* 1988;439:391–395.
233. Sawchenko PE, Swanson LW. Organization of CRF immunoreactive cells and fibers in the rat brain: immunohistochemical studies. In: DeSouza EB, Numeroff CB, eds. *Corticotropin-releasing factor: basic and clinical studies of a neuropeptide.* Boca Raton, FL: CRC Press, 1990:29–51.
234. Almeida OFX, Nikolarakis KE, Herz A. Evidence for the involvement of endogenous opioids in the inhibition of luteinizing hormone by corticotropin releasing factor. *Endocrinology* 1988;122:1034–1041.
235. Butler PD, Weiss JM, Stout JC, et al. Corticotropin-releasing factor produces fear-enhancing and behavioral activating effects following infusion into the locus coeruleus. *J Neurosci* 1990;10:176–183.
236. Gambacciani M, Yen S, Rasmussen D. GnRH release from the mediobasal hypothalamus: *in vitro* inhibition by CRF. *Neuroendocrinology* 1986;43:533–536.
237. Barbarino A, De Marinis L, Fillo G, et al. Corticotropin-releasing hormone inhibition of gonadotropin secretion during the menstrual cycle. *Metabolism* 1989;38:504–506.
238. Barbarino A, De Marinis L, Tofani A, et al. Corticotropin-releasing hormone inhibition of gonadotropin release and the effect of opioid blockade. *J Clin Endocrinol Metab* 1989;68:523–528.
239. Tazi A, Dantzer R, Le Moal M, et al. Corticotropin releasing-factor antagonist blocks stress induced fighting in rats. *Regul Pept* 1987;18:37–42.
240. Jeong KH, Jacobson L, Widmaier EP, et al. Normal

suppression of the reproductive axis following stress in corticotropin-releasing hormone-deficient mice. *Endocrinology* 1999;140:1702–1708.

241. Sarkar D, Yen S. Hyperprolactinemia decreases the luteinizing hormone-releasing hormone concentration in pituitary portal plasma: a possible role for β-endorphin as a mediator. *Endocrinology* 1985;116:2080–2084.

242. Ching M. Morphine suppresses the proestrous surge of GnRH in pituitary portal plasma of rats. *Endocrinology* 1983;112:2209–2211.

243. Leadem CA, Crowley WR, Simpkins JW, et al. Effects of naloxone on catecholamine and LHRH release from the perifused hypothalamus of the steroid primed rat. *Neuroendocrinology* 1985;40:497–500.

244. Sirinathsinghji DJS. Regulation of lordosis behavior in the female rat by corticotropin releasing factor, β-endorphin/corticotropin and luteinizing hormone releasing-hormone neuronal systems in the medial preoptic area. *Brain Res* 1986;375:49–56.

245. Gopalan C, Gilmore DP, Brown CH. Effects of different opiates on hypothalamic monoamine turnover and on plasma LH levels in pro-estrous rats. *J Neurol Sci* 1989;94:211–219.

246. Marcus MD, Loucks TL, Berga SL. Psychological correlates of functional hypothalamic amenorrhea. *Fertil Steril* 2001;76:310–316.

247. Warren MP, Voussoughian F, Geer EB, et al. Functional hypothalamic amenorrhea: hypoleptinemia and disordered eating. *J Clin Endocrinol Metab* 1999;84:873–877.

248. Berga SL, Daniels TL, Giles DE. Women with functional hypothalamic amenorrhea but not other forms of anovulation display amplified cortisol concentrations. *Fertil Steril* 1997;67:1024–1030.

249. Kondoh Y, Uemura T, Murase M, et al. A longitudinal study of the hypothalamic-pituitary-adrenal axis in women with progestin-negative functional hypothalamic amenorrhea. *Fertil Steril* 2001;76:748–752.

250. Meczekalski B, Tonetti A, Monteleone P, et al. Hypothalamic amenorrhea with normal body weight: ACTH, allopregnanolone and cortisol responses to corticotropin-releasing hormone test. *Eur J Endocrinol* 2000;142:280–285.

251. Genazzani AD, Bersi C, Luisi S, et al. Increased adrenal steroid secretion in response to CRF in women with hypothalamic amenorrhea. *J Steroid Biochem Mol Biol* 2001;78:247–252.

252. Berga SL, Loucks-Daniels TL, Adler LJ, et al. Cerebrospinal fluid levels of corticotropin-releasing hormone in women with functional hypothalamic amenorrhea. *Am J Obstet Gynecol* 2000;182:776–781.

253. Falsetti L, Gambera A, Barbetti L, et al. Long-term follow-up of functional hypothalamic amenorrhea and prognostic factors. *J Clin Endocrinol Metab* 2002;87:500–505.

Anorexia and Bulimia

254. Newman MM, Halmi KA. The endocrinology of anorexia nervosa and bulimia nervosa. *Endocrinol Metab Clin North Am* 1988;17:195–211.

255. Devlin M, Walsh T, Katz J, et al. Hypothalamic-pituitary gonadal function in anorexia nervosa and bulimia. *Psychiatry Res* 1988;28:11–24.

256. Sherman BM. Hypothalamic control of the menstrual cycle: implications for the study of anorexia nervosa. In: Brown GM, ed. *Neuroendocrinology and psychiatric disorder*. New York: Raven Press, 1984:315–324.

257. Golden NH, Jacobson MS, Schebendach J, et al. Resumption of menses in anorexia nervosa. *Arch Pediatr Adolesc Med* 1997;151:16–21.

258. Nakai Y, Hamagaki S, Kato S, et al. Leptin in women with eating disorders. *Metabolism* 1999;48:217–220.

259. Monteleone P, Di Lieto A, Tortorella A, et al. Circulating leptin in patients with anorexia nervosa, bulimia nervosa or binge-eating disorder: relationship to body weight, eating patterns, psychopathology and endocrine changes. *Psychiatry Res* 2000;94:121–129.

260. Herpertz S, Wagner R, Albers N, et al. Circadian plasma leptin levels in patients with anorexia nervosa: relation to insulin and cortisol. *Horm Res* 1998;50:197–204.

261. Ballauf A, Ziegler A, Emons G, et al. Serum leptin and gonadotropin levels in patients with anorexia nervosa during weight gain. *Mol Psychiatry* 1999;4:71–75.

262. Ferrari E, Fraschini F, Brambilla F. Hormonal circadian rhythms in eating disorders. *Biol Psychiatry* 1990;27:1007–1020.

263. Kiriike N, Nishiwaki S, Izumiya Y, et al. Thyrotropin, prolactin, and growth hormone responses to thyrotropin-releasing hormone in anorexia nervosa and bulimia. *Biol Psychiatry* 1987;22:167–176.

264. Garfinkel PE. Anorexia nervosa: an overview of hypothalamic-pituitary function. In: Brown GM, ed. *Neuroendocrinology and psychiatric disorder*. New York: Raven Press, 1984:301–314.

265. Ariyasu H, Takaya K, Tagami T, et al. Stomach is a major source of circulating ghrelin, and feeding state determines plasma ghrelin-like immunoreactivity levels in humans. *J Clin Endocrinol Metab* 2001;86:4753–4758.

266. Shiiya T, Nakazato M, Mizuta M, et al. Plasma ghrelin levels in lean and obese humans and the effect of glucose on ghrelin secretion. *J Clin Endocrinol Mol* 2002;87:240–244.

267. Otto B, Cuntz U, Fruehauf E, et al. Weight gain decreases elevated plasma ghrelin concentrations of patients with anorexia nervosa. *Eur J Endocrinol* 2001;145:669–673.

268a. Duclos M, Corcuff JB, Roger P, et al. The dexamethasone-suppressed corticotrophin-releasing hormone stimulation test in anorexia nervosa. *Clin Endocrinol (Oxf)* 1999;51:725–731.

268b. American Psychiatric Association. *DSM-III-R: diagnostic and statistical manual of mental disorders,* 3rd ed. revised. Washington, DC: American Psychiatric Association, 1987.

269. Fichter MM, Pirke KM, Pöllinger J, et al. Disturbances in the hypothalamo-pituitary-adrenal and other neuroendocrine axes in bulimia. *Biol Psychiatry* 1990;27:1021–1037.

270. Murialdo G, Galimberti CA, Fonzi S, et al. Sex hormones and pituitary function in male epileptic patients with altered or normal sexuality. *Epilepsia* 1995;36:360–365.

271. Isojarvi JI, Repo M, Pakarinen AJ, et al. Carbamazepine, phenytoin, sex hormones, and sexual function in men with epilepsy. *Epilepsia* 1995;36:366–370.

272. Cummings LN, Giudice L, Morrell MJ. Ovulatory function in epilepsy. *Epilepsia* 1995;36:355–359.

273. Nappi C, Meo R, Di Carlo C, et al. Reduced fertility and neuroendocrine dysfunction in women with epilepsy. *Gynecol Endocrinol* 1994;8:133–145.

274. Drislane FW, Coleman AE, Schomer DL, et al. Altered pulsatile secretion of luteinizing hormone in women with epilepsy. *Neurology* 1994;44:306–310.

275. Meo R, Bilo L, Nappi C, et al. Derangement of the hypothalamic GnRH pulse generator in women with epilepsy. *Seizure* 1993;2:241–252.

276. Herzog AG. A relationship between particular reproductive endocrine disorders and the laterality of epileptiform discharges in women with epilepsy. *Neurology* 1993;43:1907–1910.

277. Isojarvi JI, Laatikainen TJ, Pakarinen AJ, et al. Polycystic ovaries and hyperandrogenism in women taking valproate for epilepsy. *N Engl J Med* 1993;329:1383–1388.
278. Sperling MR, Pritchard PB III, Engel J Jr, et al. Prolactin in partial epilepsy: an indicator of limbic seizures. *Ann Neurol* 1986;20:716–722.
279. Yerby MS, vanBelle G, Friel PN, et al. Serum prolactins in the diagnosis of epilepsy. Sensitivity, specificity, and predictive value. *Neurology* 1987;37:1224–1226.
280. Matthew E, Woods JF. Growth hormone and prolactin in temporal lobe epilepsy. *Epilepsy Res* 1993;16:215–222.

Somatostatin

281. Rubinow DR, Gold PW, Post RM, et al. CSF somatostatin in affective illness. *Arch Gen Psychiatry* 1983;40:409–412.
282. Doran AR, Rubinow DR, Roy A, et al. CSF somatostatin and abnormal response to dexamethasone administration in schizophrenic and depressed patients. *Arch Gen Psychiatry* 1986;43:365–369.
283. Serby M, Richardson SB, Rypma B, et al. Somatostatin regulation of the CRF-ACTH-cortisol axis. *Biol Psychiatry* 1986;21:971–974.
284. Beal MF, Kowall NW, Mazurek MF. Neuropeptides in Alzheimer's disease. *J Neural Transm Suppl* 1987;24:163–174.
285. Beal MF, Mazurek MF, Svendsen CN, et al. Widespread reduction of somatostatin-like immunoreactivity in the cerebral cortex in Alzheimer's disease. *Ann Neurol* 1986;20:489–495.
286. Tamminga CA, Foster NL, Fedio P. Alzheimer's disease: low cerebral somatostatin levels correlate with impaired cognitive function and cortical metabolism. *Neurology* 1987;37:161–165.

Hypothalamic Lesions

287. Tonkonogy JM, Geller JL. Hypothalamic lesions and intermittent explosive disorder. *J Neuropsychiatry Clin Neurosci* 1992;4:45–50.
288. Vesely DL. Hypothalamic sarcoidosis: a new cause of morbid obesity. *South Med J* 1989;82:758–761.
289. Ben-Amitai D, Rachmel A, Levy Y, et al. Hypodypsic hypernatremia and hypertriglyceridemia associated with cleft lip and cleft palate: a new hypothalamic dysfunction syndrome? *Am J Med Genet* 1990;36:275–278.
290. Verdin E, Smitz S, Thibaut A, et al. Adipsic hypernatremia in a patient with pseudotumor cerebri and the primary empty sella syndrome. *J Endocrinol Invest* 1985;8:369–372.
291. Chiang R, Marshall MC, Rosman PM, et al. Empty sella turcica in intracranial sarcoidosis. Pituitary insufficiency, primary polydipsia, and changing neuroradiologic findings. *Arch Neurol* 1984;41:662–665.
292. Gurewitz R, Blum I, Lavie P, et al. Recurrent hypothermia, hypersomnolence, central sleep apnea, hypodipsia, hypernatremia, hypothyroidism, hyperprolactinemia and growth hormone deficiency in a boy—treatment with clomipramine. *Acta Endocrinol Suppl* 1986;279:468–472.
293. Yoshikawa M, Yamamoto M, Ohba S, et al. Hypothalamic histiocytosis X with diabetes insipidus and Korsakoff's syndrome: a case report. *Neurol Med Chir* 1991;31:529–534.
294. Mechanick JI, Hochberg FH, LaRocque A. Hypothalamic dysfunction following whole-brain irradiation. *J Neurosurg* 1986;65:490–494.
295. Moore JB, Kulkarni R, Crutcher DC, et al. MRI in multifocal eosinophilic granuloma: staging disease and monitoring response to therapy. *Am J Pediatr Hematol Oncol* 1989;11:174–177.
296. Ober KP, Alexander E, Challa VR, et al. Histiocytosis X of the hypothalamus. *Neurosurgery* 1989;24:93–95.
297. Williams TC, Frohman LA. Hypothalamic dysfunction associated with hemochromatosis. *Ann Intern Med* 1985;103:550–551.

Thyroid Disease

298. Nemeroff CB, Simon JS, Haggerty JJ, et al. Antithyroid antibodies in depressed patients. *Am J Psychiatry* 1983;142:840–843.
299. Tumbridge WM, Evered DC, Hall R, et al. Lipid profiles and cardiovascular disease in the Whickham area with particular reference to thyroid failure. *Clin Endocrinol (Oxf)* 1977;7:481.
300. Ross DS. Subclinical hypothyroidism. In: Braverman LE, Utiger RD, eds. *Werner and Ingbar's the thyroid: a fundamental and clinical text,* 8th ed. Philadelphia: Lippincott Williams & Wilkins, 2000:1001–1006.
301. Surks MI, Ocampo E. Subclinical thyroid disease. *Am J Med* 1996;100:217–223.
302. Dratman MB, Crutchfield FL, Gordon JT, et al. Iodothyronine homeostasis in rat brain during hypo- and hyperthyroidism. *Am J Physiol* 1983;245:E185–E193.
303. Whybrow PC. Behavioral and psychiatric aspects of hypothyroidism. In: Braverman L, Utiger R, eds. *The thyroid.* Philadelphia: JB Lippincott, 1991:1078–1083.
304. Leigh H. Cerebral effects of endocrine disease. In: Becker KL, ed. *Principles and practice of endocrinology and metabolism,* 3rd ed. Philadelphia: Lippincott Williams & Wilkins, 2001:1834–1837.
305. Ruel J, Faure R, Dussault JH. Regional distribution of nuclear T4 receptors in rat brain and evidence for preferential localization in neurons. *J Endocrinol Invest* 1985;8:343–348.
306. Marangell LB, Ketter TA, George MS, et al. Inverse relationship of peripheral thyrotropin-stimulating hormone levels to brain activity in mood disorders. *Am J Psychiatry* 1997;154:224–230.
307. Constant EL, De Volder AG, Ivanoiu A, et al. Cerebral blood flow and glucose metabolism in hypothyroidism: a positron emission tomography study. *J Clin Endocrinol Metab* 2001;86:3864–3870.

Hypo- and Hypercortisolism

308. Orth DN, Kovacs WJ, DeBold CR. The adrenal cortex. In: Wilson JD, Foster DW, eds. *Williams textbook of endocrinology.* Philadelphia: WB Saunders, 1992:489.
309. Martignoni E, Costa A, Sinforani E, et al. The brain as a target for adrenocortical steroids: cognitive implications. *Psychoneuroendocrinology* 1992;17:343–354.
310. Uno H, Tarara R, Else J, et al. Hippocampal damage associated with prolonged and fatal stress in primates. *J Neurosci* 1988;9:1705–1711.
311. Thuggesen P, Herman K, Willanger R. Concentration camp survivors in Denmark: persecution, disease, disability, compensation. *Dan Med Bull* 1970;17:65–70.
312. Sjaastad O. The war sailor and KZ syndromes. *Funct Neurol* 1986;1:5–19.

313. Jensen T, Genefke I, Hyldebrandt N. Cerebral atrophy in young torture victims. *N Engl J Med* 1982;307:1341.

314. Ling M, Perry P, Tsaung M. Side effects of corticosteroid therapy. *Arch Gen Psychiatry* 1981;38:471–477.

315. Varney NR, Alexander B, MacIndoe JH. Reversible steroid dementia in patients without steroid psychosis. *Am J Psychiatry* 1984;141:369–372.

316. Wolkowitz OM, Reus VI, Weingartner H, et al. Cognitive effects of corticosteroids. *Am J Psychiatry* 1990;147:1297–1303.

317. McEwen BS. Glucocorticoids and hippocampus: receptors in search of a function. In: Ganten D, Pfaff D, eds. *Adrenal actions on brain*. New York: Springer-Verlag 1982: 1–22.

10

PSYCHONEUROIMMUNOLOGY: INTERACTIONS BETWEEN THE BRAIN AND THE IMMUNE SYSTEM

KELLEY S. MADDEN
DENISE L. BELLINGER
KURT ACKERMAN

The immune system comprises lymphocytes, macrophages, and other cells that work together to eliminate disease-causing viruses, bacteria, and parasites. During innate immunity, neutrophils, macrophages, and natural killer (NK) cells are rapidly mobilized to the site of infection. These cells can directly kill pathogens and also recruit T and B lymphocytes. In the next phase of the response, called acquired immunity, pathogen-specific T and B lymphocytes are activated and proliferate in response to antigen and major histocompatibility molecules expressed on the surface of antigen presenting cells (dendritic cells, macrophages, B cells). Activation elicits protein cytokines that are necessary to direct proliferation, maturation, differentiation, and termination of the response. One type of T lymphocyte, the $CD4^+$ T helper (Th) cell, is characterized by the pattern of cytokines produced after receptor activation and the class of acquired immunity that these cytokines elicit. Th1 cells produce interleukin (IL)-2 and interferon (IFN)-γ and elicit cell-mediated responses, such as delayed-type hypersensitivity and generation of $CD8^+$ cytotoxic T lymphocytes. Th2 responses are characterized by the production of cytokines IL-4, IL-10, and IL-13, among others, and elicit humoral (antibody) responses. Thus, acquired immunity consists of the generation of effector functions that ensure the complete elimination of the pathogen. During this phase, memory T and B lymphocytes are also produced that protect against subsequent exposure to the pathogen. The potential for autoimmune destruction of normal tissue is limited by immune and nonimmune mechanisms, such as the production of antiinflammatory cytokines and neuroendocrine hormones.

Over the past 25 years, a large body of evidence has accumulated demonstrating that external stimuli can alter immune reactivity and influence susceptibility to infectious disease, autoimmune disease, and cancer. This chapter reviews the evidence for central nervous system (CNS) regulation of the immune system via its two outflow pathways,

the hypothalamic–pituitary–adrenal (HPA) axis and the autonomic nervous system, through activation of the sympathetic nervous system (SNS). Also, activation of the immune system in the periphery is communicated to the CNS, suggesting that the immune system can serve as a sensory organ. The clinical relevance of CNS–immune system interactions to disease processes is discussed briefly. For more detailed reviews of many of the topics covered here, see reference 1.

EVIDENCE FOR CENTRAL NERVOUS SYSTEM–IMMUNE SYSTEM INTERACTIONS

Immune reactivity can be altered by psychological, emotional, and physical stressors and classic behavioral conditioning in humans and animals. This evidence serves as the basis for the contention that the CNS can influence health and disease through altered immune reactivity.

Effects of Stress

Stressors tend to exacerbate infectious and autoimmune diseases, suggesting that the immune system is sensitive to changes in the external stimuli (reviewed in reference 2). Many studies provide direct evidence for stress-related immunosuppression. In humans, chronic stress is associated with reduced responses to vaccination (3) and decreased wound healing (4). High levels of stress may also produce undesirable reductions in immune function in some disease states. For example, Andersen et al. (5) showed an inverse correlation between stress levels and NK and T cell function in breast cancer patients, suggesting that stressor exposure may contribute to a diminished ability to eliminate tumor cells and prevent metastasis. In animals, immune reactivity to influenza virus and herpes simplex virus is reduced after exposure to chronic restraint stress (6–8). Stressor-induced immunosuppression may reduce viral or bacterial clearance (6,9) and reactivate latent viruses, such as herpes simplex virus (10). The neuroendocrine mechanism(s) underlying stress-induced immunosuppression includes glucocorticoid (GC) elevation and sympathetic activation, and both systems can work together to regulate a virus-specific response. For example, after restraint stress in herpes simplex virus– or influenza-infected animals, increased GC were responsible for stress-induced alterations in lymphocyte migration, but SNS activation contributed to an impairment of the cytotoxic T lymphocyte response (11).

Based simply on the notion that stress is immunosuppressive, one might predict that autoimmune inflammatory processes would be reduced after chronic exposure to stressors. In rats, chronic stress reduced clinical signs and histologic manifestations of experimental allergic encephalomyelitis (EAE), an animal model of multiple sclerosis (MS) (12), but in humans, exposure to stressors tends to exacerbate autoimmune pathologies (13,14). Furthermore, stressor-induced immunoenhancement has been reported in animals (15–18). For example, exposure of BALB/c mice to odors produced by a foot-shocked donor mouse elicits an opiate-induced anesthesia in recipient mice in conjunction with increased antibody production and increased IL-4 production *in vitro* (15). These data demonstrate that stressors are not solely immunosuppressive and suggest that multiple mechanisms may be responsible for stress-induced alterations in immune function.

Experimental evidence from laboratory animals indicates that the immune-modulating effects of stress are dependent on genetic background (7), the route of infection (16), and, perhaps most important, the duration and severity of the stressor (reviewed in references 19,20). The importance of stressor duration and intensity as factors in stress-induced immunomodulation is probably a reflection of differences in the neuroendocrine effector mechanisms elicited. However, even in the same animal, disparate effects of stress can be observed, depending on the lymphoid tissue and immune parameter examined. For example, in influenza-infected mice, restraint stress suppressed IgM and IgG antibody secreting cells in the lung, mediastinal lymph nodes, and spleen but enhanced the antibody response in the superficial cervical lymph nodes (8). These tissue-specific responses after stressor exposure may be related to a variety of factors that have yet to be elucidated but most likely include alterations in lymphocyte migration patterns and differences in the neuroendocrine milieu at each site.

The apparent complexity of stress-induced immunomodulation led Dhabhar and McEwen (17) to postulate that exposure to stress, at a level termed *eustress*, may promote immune reactivity (19,21). When a stressor becomes severe or chronic in nature (distress), maladaptive immunosuppression occurs, with increased susceptibility to infection. As discussed above, maladaptive, stress-induced immunosuppression has been demonstrated, but evidence

for an adaptive role for stressor exposure is not as abundant. Dhabhar and McEwen (17,22) showed that when mice were exposed to an acute stressor before antigen challenge, delayed-type hypersensitivity was increased in association with elevated GC levels and increased leukocytes entering the skin and draining lymphoid organs. In another study using rats, exposure to an acute stressor immediately before immunization produced an enhancement of macrophage nitric oxide production (23). A stress-induced increase in nitric oxide production accompanied by a stress-induced increase in acute phase reactants (24) may promote bacterial clearance, although this proposition has yet to be thoroughly tested. These results constitute evidence that alterations in immune responses after exposure to an acute stressor can be adaptive, i.e., promote rapid responses to a foreign agent, but more work is required to determine whether an adaptive effect of acute stressors is generally present under normal conditions and in disease states.

Aging and Stress-Induced Alterations in Immunity

In aged rats, stressors elicit increased or prolonged GC and catecholamine production relative to young animals (25,26). In animals and humans, aging is also associated with impaired immune responsiveness, particularly T cell reactivity, suggesting that stress-induced changes in immunity may be more detrimental to the aged individual. In mice, chronic restraint stress exacerbated the age-related impairment in influenza virus–specific immune responsiveness in association with increased mortality after infection (27). Similarly, an age-related reduction in splenic NK cell activity and mitogen-induced T cell proliferation was potentiated after acute or chronic stressor exposure in rats (28,29). In very old rats, however, when the age-related suppression of T cell proliferation is the greatest, chronic stressor exposure had no effect (29). The inability to respond to the stressor may be owing to an inherent "floor effect" in which T cell proliferation cannot be inhibited further. Alternatively, deficits in signaling pathways from the CNS to the periphery may render the very old individual insensitive to stressor-induced immunomodulation. Collectively, these reports indicate that stressor exposure may be an important factor dictating the rate of immunologic decline and may contribute to increased morbidity and mortality in aging hosts.

Behavioral Conditioning Alters Immune Reactivity

Conditioned Immunosuppression

Numerous investigators have demonstrated modulation of immune reactivity in animals that were behaviorally conditioned by pairing a novel, immunologically neutral stimulus [the conditioned stimulus (CS)], such as ingestion of saccharin, with an immunosuppressive unconditioned stimulus (UCS), such as cyclophosphamide or a physical stressor (reviewed in reference 30). Subsequent exposure of conditioned animals to the CS alone suppressed antibody and cell-mediated immune responses compared with animals that were not reexposed to the CS or compared with a nonconditioned group in which CS exposure is temporally separate from the UCS at conditioning. Conditioned immunosuppression is not restricted to a limited set of CS, UCS, or immune responses. Conditioned immune suppression is not dependent on a noxious UCS or the development of conditioned taste aversion. For example, morphine, a euphoria-inducing immunosuppressive drug, has been successfully used as a UCS (31). Roudebush and Bryant (32) tested both dexamethasone and cyclophosphamide as immunosuppressive UCS that elicit conditioned taste aversion. Using the same conditioning procedures, they reported conditioned immunosuppression only when cyclophosphamide was the UCS. These results demonstrate that conditioned immunosuppression can be elicited under a variety of conditions. Although not as thoroughly investigated, conditioned immunoenhancement using behavioral conditioning techniques can also be achieved (see below).

Conditioned Immunoenhancement

Antigen, the most biologically relevant immune activator, can be paired with a variety of CS to elicit conditioned immunoenhancement. In two allergic animal models, when antigen as the UCS was paired with olfactory or audiovisual stimuli, CS reexposure elicited the release of histamine or rat mast cell protease I (33,34). Cell-mediated and antibody responses are also conditionable with antigen as the UCS (35–38). In Ader et al. (33), mice underwent three to five conditioning trials consisting of chocolate milk ingestion followed immediately by intraperitoneal antigen challenge (36). Reexposure to the CS, chocolate milk ingestion, and injection with a nonimmunogenic dose of antigen induced an elevated antibody response. By contrast, Alvarez-Borda et al. (37) used a single conditioning trial, pairing saccharin ingestion with an intraperitoneal injection of a protein antigen in rats (37). When conditioned animals were reexposed to saccharin, in the absence of exogenously administered antigen, an enhanced IgG antibody response to the immunizing antigen was observed. Our laboratory was able to duplicate this work (38), although the magnitude of the conditioned response in our hands was much lower than that reported by Alvarez-Borda et al. Nevertheless, the finding that antibody levels were elevated in the conditioned group without requiring exogenous administration of antigen is potentially significant. Antibody responses are not easily elicited in the absence of an antigenic or mitogenic signal; even memory T and B cell activation, although less dependent on antigenic and costimulatory molecule interactions, requires a minimal level of antigenic stimulation (39). One possibility

is that reexposure of conditioned animals to the CS evoked release of antigen from an endogenous source, such as follicular dendritic cells or B cells (40). Alternatively, alterations in neuroendocrine hormones may directly increase antibody production, although there is little evidence to support this possibility. The ability to elicit a conditioned immune response in the absence of exogenous antigen implies that the CNS may help to maintain serum antibody levels and memory responses without requiring reexposure to antigen. Also, successful behavioral conditioning using antigen as UCS indicates that the CNS is made aware of peripheral exposure to antigen. The pathways used to communicate immune system activation with the CNS are discussed in the section on immune system communication with the nervous system.

Mechanisms Underlying Conditioned Immune Responses

Lesions in the insular cortex and amygdala before conditioning prevented conditioned immunosuppression and immunoenhancement but did not prevent conditioned taste aversion elicited by cyclophosphamide (41–43). Bilateral excitotoxic lesioning of the insular cortex, but not the amygdala, before reexposure of conditioned animals to the CS also prevented conditioned immunosuppression. Importantly, amygdala or insular cortex lesions in normal animals did not alter immune function. The neurochemical pathways used in conditioned immune responses are not well understood, although central μ-opioid pathways were demonstrated to be involved in conditioned immunosuppression (44).

The efferent mechanisms that elicit conditioned immunosuppression have been evaluated in a few reports. Conditioned immunosuppression occurred in adrenalectomized animals (45), and plasma corticosteroid levels were not elevated after CS reexposure (32,46), suggesting that the expression of conditioned immunosuppression is not dependent on adrenal GC production. Conversely, splenic denervation before pairing the immunosuppressant cyclosporine with saccharin ingestion prevented conditioned immunosuppression but not conditioned taste aversion (47). This suggests that the afferent pathways remained intact in the denervated animals and that loss of sympathetic noradrenergic innervation of the spleen prevented elicitation of conditioned immunosuppression. Also, several effector mechanisms may interact to mediate conditioned immunosuppression. For example, in rats conditioned by exposure to an aversive stimulus paired with an audiovisual cue, treatment with β-antagonists before CS reexposure prevented the conditioned suppression of splenic T cell proliferation but did not block the suppression of splenic B cell proliferation, NK cell activity, or peripheral blood lymphocyte mitogenesis (48). These results demonstrate that the neuroendocrine mechanisms underlying conditioned immunosuppression are dependent on the immune cell type involved and their anatomic location.

Potential for Clinical and Therapeutic Uses

The demonstration of conditioned immune responses in laboratory animals suggests that behavioral conditioning may be successful in treating a number of immune-related diseases, especially if the beneficial conditioned immune effects can be separated from undesirable side effects (e.g., taste aversion). Conditioned immunosuppression may be used to reduce the dose of drug required to suppress immune function, as was elegantly demonstrated by Bovbjerg et al. (49). Transplant rejection is mediated by a cell-mediated graft-versus-host response and is suppressed by the administration of three low doses of cyclophosphamide; a single low dose is only minimally effective. Rats were conditioned by multiple pairings of the UCS (cyclophosphamide) with the CS (ingestion of saccharin). Seven weeks later, reexposure to the CS in combination with low-dose cyclophosphamide, followed by two more exposures to the CS alone, resulted in immunosuppression that was equivalent to the nonconditioned group that had received three injections of low-dose cyclophosphamide. In another example, heart allograft rejection was delayed in animals conditioned by multiple pairings of cyclosporine with saccharin ingestion and reexposed to saccharin alone compared with animals that were not conditioned (47). The conditioned response was similar in magnitude to those animals that received a therapeutic dose of cyclosporine at each test trial. Finally, evidence from animal models suggests that conditioned immunosuppression may have clinical application in the treatment of autoimmune disease (50–52). Conditioned immune responses reported in human subjects have been difficult to reproduce, perhaps because of the impact of cognitive processes on conditioning (46,53,54). Further work is required to determine whether behavioral conditioning can be used therapeutically in the treatment of cancer and autoimmune disease in humans.

Brain Lesions and Intracerebroventricular Infusions Alter Peripheral Immunity

Discrete lesions in the brain, particularly in specific nuclei of the hypothalamus and limbic system, resulted in structural and functional changes in the immune system, demonstrating that specific brain regions can exert control over immune activity in the periphery (reviewed in reference 55). Furthermore, intracerebroventricular administration of agents that alter CNS communication with the periphery can modulate immune function. For example, injection of gp120, the envelope protein of human immunodeficiency virus (HIV), or IL-1 into rat lateral ventricles elevated plasma corticosteroid levels and reduced immune responses in the periphery (56,57). Central administration of corticotropin releasing hormone (CRH), a key hypothalamic neurotransmitter, or the cytokine IFN-α reduced splenic NK cell activity via activation of the SNS (58,59). Thus, chemical stimulation of specific brain regions can alter CNS output to influence

immune function in the periphery using the HPA axis and autonomic nervous system.

The complexity of CNS communication with the immune system demonstrated by the mixed effects of stress and conditioning on immune reactivity is mirrored in the interactions between the immune system and the effector molecules of the CNS. The following sections discuss the influence of neuroendocrine hormones, neuropeptides, and catecholamines on immune reactivity.

ENDOCRINE-IMMUNE SYSTEM INTERACTIONS

Early studies revealed that neuroendocrine hormones produced by the anterior pituitary influenced the development of immune organs and the ability of the host to respond to foreign antigen. Removal of the pituitary from normal animals (hypophysectomy) produced small lymphoid organs and reduced immune reactivity (60). These effects are similar to those reported in dwarf animals that are unable to produce growth hormone (GH), prolactin (PRL), and thyroid stimulating hormone (61,62). In hypophysectomized rats infected with *Salmonella typhimurium,* mortality was significantly increased in association with reduced macrophage lysis of *S. typhimurium* (63). Reconstitution with GH, PRL, or thyroxine (T$_4$) alone or in combination partially or fully restored many of the immunodeficiencies in these animals. This discussion incorporates the most recent advancements in our knowledge of the role of these pituitary hormones and others in the regulation of the immune system.

The Hypothalamic–Pituitary–Adrenal Axis

GC production by the adrenal gland is often considered the primary mediator of immunosuppression after HPA axis activation. In fact, all components of this pathway can influence immune or inflammatory processes through endocrine or autocrine interactions with cells of the immune system.

Corticotropin Releasing Hormone

CRH, a 41-amino-acid neuropeptide, plays a major role in coordinating and modulating endocrine, autonomic, behavioral, and immune responses to stressful stimuli, including inflammatory stimuli (64). One of the first examples of immune system communication with the CNS was the capacity of peripheral IL-1 to activate the HPA axis by inducing hypothalamic CRH release (65,66), a response that is important in muting an inflammatory response (67). Interestingly, mice deficient in CRH receptor 1, the receptor type that mediates CRH-induced adrenocorticotropic hormone (ACTH) secretion, cannot produce GC in response to acute stressors but can still mount a pituitary–adrenal response to turpentine-induced inflammation, suggesting that redundant mechanisms exist to elicit inflammation-induced GC production (68). In addition to HPA axis activation, CRH- or stress-induced activation of CRH-containing descending autonomic also activates the SNS, leading to immunosuppression (58,69).

Genetic and protein evidence suggests that macrophages and lymphocytes can produce CRH (70–72). CRH can increase or decrease immune function *in vitro*, depending on the concentration or type of activational signal, probably reflecting direct and indirect mechanisms of CRH (73–75). High-affinity receptors for CRH are present on monocytes/macrophages, neutrophils, and some activated lymphocytes (76,77). CRH can also induce ACTH or β-endorphin production by lymphocytes or macrophages (78,79), and these molecules may serve as intermediate molecules for CRH-induced immunomodulation. Localized CRH may modulate the perception of pain in inflamed tissue by eliciting immune cell production of β-endorphin and other opioids (79,80). This suggests that CRH present in inflamed joints of rheumatoid arthritis (RA) patients and in animals with experimental arthritis plays a complex role in the pathogenesis of arthritis (81,82).

Adrenocorticotropic Hormone and Melanocyte Stimulating Hormone

In the anterior pituitary, CRH induces transcription of the proopiomelanocortin (POMC) gene that is translated into ACTH, α-melanocyte stimulating hormone (α-MSH), β-endorphin, and other peptides. Unstimulated macrophages also produce a full-length POMC mRNA, as do T lymphocytes after mitogen stimulation (83). At the protein level, immunoreactive ACTH species have been identified in spleen cell extracts identical to that of anterior pituitary extracts. Furthermore, ACTH and its peptide precursors were increased by mitogen-induced lymphocyte activation or HIV infection of T lymphoma cells (84,85). Early studies demonstrated that ACTH potently suppressed *in vitro* antibody production (86), but ACTH can increase concanavalin A–induced proliferation at low doses (<1 nM) through induction of IL-2 (87). ACTH had no effect on spontaneous NK cell activity by human peripheral blood mononuclear cells, but enhanced IL-2 induced lymphokine-activated killer cells (88). ACTH (and β-endorphin) blunted the suppressive effect of 10^{-6} M cortisol (88), indicating that these neuropeptides can counteract the immunosuppressive effects of GC. Radioligand-binding and genetic studies indicate that ACTH receptors, specifically the melanocortin-5 receptor, are present on cells of the immune system (89,90).

Some of the immunomodulatory properties of ACTH may by mediated indirectly by enzymatic shortening of ACTH to α-MSH (ACTH 1–13). For example, the ACTH-induced reduction in spontaneous neutrophil activity was prevented by a peptidase inhibitor that prevented ACTH conversion to α-MSH (91). α-MSH exhibits potent antiin-flammatory properties directed toward mast cells, including reduced histamine release and inhibition of mast cell inflammatory cytokine synthesis (92). T-lymphocyte function is also suppressed by α-MSH. When applied to the skin, α-MSH inhibited T cell reactivity to a contact-sensitizing agent. Furthermore, intravenous administration of α-MSH before sensitization produced a long-lasting antigen-specific tolerance through induction of the monocyte-derived anti-inflammatory cytokine IL-10 (93). α-MSH may also downregulate the activity of the cytokine network in the brain. For example, peripheral immunosuppression induced by intracerebroventricular administration of IL-1 was prevented by concurrent central administration of α-MSH (94).

Glucocorticoids

Pharmacologic doses of GC or their synthetic analogues suppress many lymphocyte and macrophage functions, including cellular proliferation, cytokine and chemokine production, adhesion molecule expression, and migration into inflammatory sites. These immunosuppressive/antiinflammatory effects are the basis for their usefulness in the treatment of autoimmune diseases, allergic atopy, and leukemias/lymphomas (95). Endogenous GC production (cortisol in humans, corticosterone in rodents) is induced by endotoxin, viral and bacterial infections, and the proinflammatory cytokines induced by these agents (reviewed in reference 96). A higher rate of mortality is reported in infected animals when the increase in endogenous GC production is blocked by adrenalectomy or when GC responsiveness by GC receptor blockade is prevented (97,98). Unlike many neurohormones that bind to cell surface receptors, GC diffuse into cells and bind to cytoplasmic receptors present in all cells of the immune system (99,100). Multiple molecular mechanisms underlie GC-induced immunosuppression, including reduced transcription and nuclear binding capacity of the transcription factors induced by proinflammatory cytokines (AP-1 and nuclear factor-κB), reduced cytokine gene transcription, and inhibition of cytokine-induced signal transduction pathways (95,101).

GC are not universally inhibitory, however, especially at low concentrations. For example, corticosterone increased T cell CD4 expression and anti–T cell receptor-induced proliferation *in vitro* (102). An elevation in endogenous GC late in the response to a protein antigen may promote IgG antibody production (103). Hydrocortisone-induced elevation in IgE production by IL-4–stimulated human B cells was mediated by induction of the costimulatory molecule CD40L (104). GC may skew toward a Th2-type response in conjunction with an inhibition of Th1-type cytokine production, although this finding was not replicated (105,106). GC enhanced the antiinflammatory Th2 cytokine IL-10 but reduced CD8$^+$ T cell production of IL-4 and IL-5, two Th2-type cytokines; however, IL-4 and IL-5 were reduced only in the presence of antigen presenting cells (107). These results suggest that other cell types may influence the effects of GC on Th2-type cytokine production.

Immature thymocytes are exquisitely sensitive to apoptosis by stress- and cytokine-induced endogenous GC (108,109), suggesting that GC participate in negative selection, the deletion of immature CD4+8+ thymocytes that do not respond appropriately to self-antigen. Somewhat surprisingly, however, GC were able to protect thymocytes from anti–CD3-induced apoptosis *in vitro* (110). To determine the role of GC in T cell development *in vivo*, King et al. (111) generated mice that produce an antisense GC receptor transgene only in the thymus, rendering thymocytes hyporesponsive to exogenous GC (111). T cells that normally mature in the thymus of wild-type mice were absent in the transgenic animals, suggesting that T cells were deleted in the thymus in the absence of normal GC signaling (112). Similarly, in MRL-*lpr/lpr* mice [a model for systemic lupus erythematosus (SLE)] expressing the antisense GC receptor transgene, the loss of thymocyte GC responsiveness resulted in the deletion of the T cells that produce the disease pathologies (113). The transgenic MLR mice had lower autoantibody production and milder pathologic symptoms of autoimmune disease and lived longer than the MRL-*lpr/lpr* wild-type controls. All precursors and enzymes required for corticosteroid synthesis have been detected in the thymus, although the specific cell types(s) that produce GC have yet to be elucidated (114,115). Collectively, these results imply that locally produced GC regulate the process of thymocyte selection by promoting maturation of some T cells that undergo cell death in the absence of sufficient GC signaling.

Other groups have reported that GC mediate thymocyte deletion or negative selection. In mice treated with the GC receptor blocker RU-486, antigen-specific thymocyte and anti–CD3-induced programmed cell death (apoptosis) was reduced *in vivo* (116). Moreover, Purton et al. (117) recently reported that GC receptor −/− transgenic mice that are completely unresponsive to GC exhibit fetal thymocyte development that is phenotypically indistinguishable from wild-type controls. It should be noted that the GC receptor −/− fetuses do not live past embryonic day 18, so these data do not address the issue of GC modulation of thymocyte development postnatally; however, the results do differ from results obtained in the antisense GC receptor mice at the equivalent developmental time points. Further studies are required to determine whether the differences in thymocyte development between the GC receptor transgenic strains are related to the differences in GC responsiveness. These issues are important to resolve; if endogenous GC regulate thymocyte development, they may play an important role in the initiation of autoimmune processes.

Susceptibility to particular autoimmune diseases has been correlated with the ability to produce GC. For example, susceptibility to streptococcal cell wall–induced arthritis in young, female LEW/N rats is associated with reduced corticosterone responses to stress and other pharmacologic agents (118). Administration of cortisone increased the resistance of these animals to SCW streptococcal cell wall–induced arthritis. In contrast, the major histocompatibility complex–identical F344/N strain is relatively resistant to streptococcal cell wall–induced arthritis, and removal of the adrenal gland produced greater susceptibility. Susceptibility to and recovery from EAE, an animal model of MS, are also influenced by endogenous GC (119). A deficit in cortisol production may be a factor in RA disease severity. Circadian rhythmicity is absent in RA patients compared with patients with chronic osteomyelitis, and postsurgical plasma cortisol levels were reduced in RA patients compared with patients with chronic osteomyelitis and patients with noninflammatory osteoarthritis (120). In MS patients, GC responsiveness may be impaired. Antigen-specific T cells from MS patients with progressive disease are less responsive to GC-induced apoptosis compared with T cells from patients with the less severe relapsing/remitting form of MS (121). These results suggest that endogenous GCs dampen inflammation and reduce the severity of autoimmune disease symptoms, but more work is required to determine how HPA axis activation influences initiation of autoimmune disease.

Opioid Peptides

Several opioid peptides are produced through differential splicing of the intermediate POMC product β-lipotropin into smaller peptides, such as β-, α-, and γ-endorphin. These peptides and the related opioids leu- and met-enkephalin are produced in the pituitary and CNS and by cells of the immune system. For example, β-endorphin is produced by lymphocytes in response to CRH and IL-1 (78,80) and may regulate its own synthesis through induction of IL-1 (122). β-Endorphin promotes a variety of immune responses. For example, it enhances spontaneous NK- and IL-2–induced lymphokine-activated killer cell activity, increased neutrophil activity, and enhanced mitogen-induced T cell proliferation (88,123–125). α- and γ-Endorphin and leu- and met-enkephalin suppressed *in vitro* antibody responses (86). Expression of leu- and met-enkephalin mRNA in T and B lymphocytes has been reported (126,127), and the enkephalins and β-endorphin are potent human T cell chemoattractants (128). As discussed previously, immune cell–derived opioids may also help to reduce pain when produced at inflammatory sites (80).

Although lymphocyte μ-, κ-, and δ-opioidlike receptors have been reported (reviewed in references 129,130), their presence (especially μ) on cells of the immune system remains somewhat controversial. Opioid-induced immune effects are not always blocked by standard opioid antagonists. Also, the immunosuppressive effects of peripherally administered morphine, a selective μ-opioid agonist, appear to be mediated centrally, not peripherally, in rodents (131,132). Recent reports suggest that morphine can up-regulate μ-opioid receptors on human lymphocytes (133) and can activate the same intracellular signaling pathways as T cell mitogens (134). These distinctions are particularly important in light of recent findings that morphine induces a chemokine receptor that is used by HIV-1 to gain entry to T lymphocytes (135), suggesting a means by which abuse of morphinelike drugs can increase susceptibility to HIV-1 infection. This finding, in combination with the evidence for opioid mediation of stress-induced alterations in immune function (136), suggests that both central and peripheral endogenous opioid production can be deployed to modulate immune function.

Growth Hormone and Insulinlike Growth Factor

GH and insulinlike growth factor (IGF)-1 and their receptors contain similarities in gene organization and spatial structure that place them in the cytokine/hemopoietin supergene family with PRL and the growth-promoting cytokine IL-2, among others (137). Early studies implicated GH as a promoter of immune reactivity and lymphocyte development (62,138). More recently, similar immunoenhancing activity of IGF-1, whose production is initiated by GH, has been demonstrated.

Growth Hormone/Insulinlike Growth Factor-1 Production/Receptor Expression

GH is best known for its ability to induce longitudinal growth. It is predominantly produced by the anterior pituitary and placenta, and its production declines with age. Many of its physiologic effects are linked to its ability to stimulate IGF-1 in the liver. GH and its mRNA have been detected *in vitro* in minor populations of lymphocytes, but *in vivo,* GH mRNA expression in rat bone marrow and thymus was detectable only during the neonatal period; no splenic GH mRNA was detectable in the neonate or adult (139,140). IGF-1 mRNA is barely detectable in lymphocytes and monocytes but is strongly expressed in activated macrophage and differentiated bone marrow macrophages (141). Proinflammatory mediators, such as tumor necrosis factor (TNF)-α and prostaglandin E$_2$ can stimulate IGF-1 synthesis in macrophages (142). T and B lymphocytes, thymocytes, macrophages, and bone marrow cells possess varying levels of GH and IGF-1 receptors (143–145). This heterogeneity in receptor expression indicates that some lymphocytes subsets may be more susceptible to GH or IGF-1 regulation than others.

Animal Studies

B and T Lymphocyte Development

In dwarf animals, GH and IGF-1 administration increased the number of B and T lymphocytes in primary and secondary lymphoid organs (62), but the growth-promoting effects in primary lymphoid organs are not associated with enhanced hematopoiesis. For example, in Snell-Bagg dwarf (*dw/dw*) mice, a deficit in the frequency of bone marrow stem cells at the pre–B cell stage could not be overcome by GH, IGF-1, or PRL (alone or in combination with GH) administration, although the number of B lineage cells in the bone marrow increased (146,147). Similarly, IGF-1 administration doubled thymus cellularity in young rats in the absence of changes in the proportion of CD4/CD8-expressing subpopulations (148). In DW/J Ames dwarf mice, GH increased thymus cellularity and the frequency of CD4/CD8 double-positive thymocytes, suggesting an influence on T cell maturation, but GH produced no change in peripheral immune reactivity to antigen (146,149). In genetically GH/IGF-1 deficient mice, thymus and bone marrow cellularity was reduced, but T and B cell development, as assessed by precursor frequency, was not significantly altered, suggesting that the primary role of GH and IGF-1 *in vivo* is to expand primary lymphoid compartments without promoting hematopoiesis (150).

Immune Reactivity

In vivo and *in vitro* immune reactivity is augmented in the presence of GH or IGF-1. The primary and secondary antibody response to DNP-ovalbumin was increased after IGF-1 administration to mice (151). GH and IGF-1 potentiated concanavalin A– or anti–CD3-induced proliferation of mouse spleen cells in part through increased IL-2 production; resting lymphocytes did not respond to GH (152,153). Recently, it has been determined that IGF-1 can also increase proliferation by inhibiting apoptosis in lymphocyte populations (144). Both GH and IGF-1 increased mitogen-induced B cell proliferation and differentiation with increases in IgG4 and IgE (154). GH can directly prime macrophages and neutrophils to produce reactive oxygen metabolites necessary to kill pathogens (155,156) and provide an antiinflammatory signal by reducing TNF-α production and blocking nuclear translocation of nuclear factor-κB (157). The biologic significance of these effects was demonstrated by Edwards et al. (63), who showed that the increased rate of death in hypophysectomized rats infected with *S. typhimurium* was reduced by GH administration. These results suggest that GH and IGF-1 may be useful in the therapeutic treatment of infectious disease and septic shock.

Clinical Relevance

Despite the evidence that GH and IGF-1 can expand lymphoid compartments and promote reactivity to foreign agents *in vivo,* GH and IGF-1 knockout mice display few deficiencies in T and B cell number and function, suggesting that these neuroendocrine hormones do not play an obligatory role in immune regulation (reviewed in reference 158). Similarly, immune function in Snell-Bagg hypopituitary mice is normal if animals are weaned later than

21 days postnatally (159). In humans, GH deficiency does not contribute to poor immune reactivity or increased risks of infection, and GH treatment has few effects on peripheral blood lymphocyte number or function (160). These findings suggest that other hormonal systems overlap the physiologic roles of GH and IGF-1 in augmenting immune function.

It is possible that GH and IGF-1 have a more profound impact under specific conditions, such as in aging, during which neuroendocrine and immune functions decline. Implantation of a GH- and PRL-secreting pituitary adenoma reversed age-related thymic involution and partially restored splenic mitogen–induced T cell proliferation of aged rats (161). GH treatment, in combination with IFN-γ, reversed the decline in neutrophil bactericidal activity (162). In aged animals, the infusion of bone marrow cells from young animals followed by IGF-1 treatment increased thymus cellularity to levels significantly above that of each treatment by itself (148). Treatment with IGF-1 also helped to restore bone marrow and thymic cellularity in young animals after irradiation and treatment with cyclosporine or dexamethasone (163,164). Thus, IGF-1 or GH may be most useful therapeutically when restoring homeostasis to the immune system under some pathologic conditions.

Prolactin

Prolactin Production/Receptor Expression

The lactogenic hormone PRL is produced by the anterior pituitary and released into the bloodstream to regulate reproduction and growth and function of mammary glands and ovaries. Almost all hematopoietic cells express PRL receptors (165). These receptors induce phosphorylation of the Janus kinases and signal transducer and activator of transcription proteins, pathways used by IL-2, GH, and other cytokines in the cytokine/hematopoietin supergene family (137,165). T cells have been demonstrated to produce PRL mRNA and a corresponding PRL-like protein with or without activation (166,167), suggesting that locally produced PRL can serve as an autocrine regulator of lymphocyte function.

Animal Studies

In laboratory animals, pituitary-derived PRL can influence susceptibility to infectious disease. Inhibition of PRL release from the pituitary by bromocriptine treatment increased the rate of death in mice infected with the bacterial pathogens *Listeria monocytogenes* and *Mycobacterium bovis* (168). This increased mortality was associated with markedly diminished macrophage activation, reduced production of IFN-γ, and reduced lymphocyte proliferation; treatment with PRL restored these responses to those of untreated infected animals. Daily administration of PRL for 2 weeks before *S. typhimurium* infection reduced mortality and increased bacterial killing by peritoneal macrophages in association

with elevated nitric oxide production (169). PRL may also promote resistance to pathogens by acting on NK cells, important early responders to infection. NK cell activity is increased in the presence of PRL, in part through increased IFN-γ production (170).

Most *in vitro* studies have demonstrated that PRL can stimulate T and B lymphocyte function in conjunction with lymphocyte activation by antigen or mitogen. B cell proliferation and differentiation into antibody producing cells were potentiated by PRL, especially in the presence of suboptimal doses of mitogen (171,172). IL-2–induced activation of NK cells was also promoted by PRL (172). In one study, PRL was able to directly induce T cell proliferation and IL-2 production but only in spleen cells from ovariectomized rats, suggesting that other neuroendocrine hormones, in this case, estrogen, may provide inhibitory signals to prevent an unregulated T cell response to PRL (173). This is important because T cell PRL receptor stimulation phosphorylates proteins activated by T cell receptor signaling, such as ZAP-70 and T cell receptor–associated molecule CD3ϵ (174,175). *In vitro*, neutralizing anti-PRL antibodies inhibited mitogen-induced proliferation of T lymphocytes in the absence of exogenous PRL, suggesting that PRL may be a cell cycle progression factor in T cells (166,176). These results confirm the *in vivo* evidence for PRL-induced immunoenhancement and indicate that PRL acts by augmenting activational signals received by T cells, B cells, macrophages, and NK cells.

Despite the experimental evidence for the immunomodulating properties of PRL in normal animals, recent studies using genetically modified mice have raised questions regarding the relevance of PRL interactions with the immune system *in vivo*. In female mice unable to synthesize PRL, reproduction and mammary gland development were defective, but B and T lymphocyte hematopoiesis was normal (177). Similarly, in PRL receptor knockout mice, antigen- and mitogen-induced lymphocyte proliferation and resistance to bacterial infection were all normal (178). One possible interpretation of these results is that PRL is not obligatory for immune reactivity; other mechanisms can compensate for the deficiencies in PRL signaling. For example, in the PRL knockout mice, GH production or responsiveness may be increased. In the PRL receptor knockout strain, PRL may bind to other hormone receptors on cells of the immune system (e.g., placental lactogen receptor). It should be noted that PRL signaling in the knockout animals is absent throughout development, resulting in an animal that may compensate for that deficiency very differently compared with an adult animal with an acute loss of PRL signaling. Resolution of these important issues will result in a greater understanding of the conditions under which PRL influences immune physiology.

Clinical Relevance

The evidence for PRL-induced immunoenhancement suggests that PRL may have the capacity to contribute to autoimmune disease. In New Zealand black autoimmune mice, a naturally occurring model of SLE, lymphocyte PRL receptor expression increased with disease progression (179). PRL-producing lymphocytes have also been detected in the synovium of patients with RA (180). It is not known whether the presence of PRL or its receptors under these conditions contributes to disease induction or progression; PRL or up-regulation of its receptors may simply serve as markers of lymphocyte activation. PRL may be useful therapeutically to enhance immune function in immunocompromised patients, but interactions with other hormones or cytokines must be considered. For example, PRL (and GH) can block the immunosuppressive effects of GC *in vitro* (181,182). Clearly, the clinical use of these neurohormones must be approached with caution, and potential autoimmune reactivity must be carefully assessed in human subjects.

The Thyroid Hormones

Animal Studies

In Snell-Bagg dwarf *(dw/dw)* mice, a pituitary defect in the synthesis of TSH leads to deficiencies in production of the thyroid hormones 3,5,3'-triiodothyronine (T_3) and thyroxine (T_4). Recent evidence suggests that this defect, and not the defect in GH or PRL production, is responsible for the defects in the B cell compartment in *dw/dw* mice. The reduced frequency of bone marrow CD45R$^+$/surface IgM$^-$ pre–B cells in *dw/dw* mice was restored to that of the heterozygous control by T_4 (converted to T_3 *in vivo*) and not by GH, although GH did increase bone marrow cellularity (146,147). Hypothyroid *(hyt/hyt)* mice that cannot produce T_3 and T_4 and knockout mice unable to express the receptor for T_3 also exhibited reduced frequencies of immature and mature B cells (150,183). Conversely, T cell development and myelopoiesis appeared normal in the Snell-Bagg dwarf mice, *hyt/hyt* mice, and GH- and IGF-1–defective strains, although lymphoid compartment cellularity was reduced in all these strains (150). An increase in thymus cellularity after T_4 treatment of *dw/dw* and *hyt/hyt* mice was not accompanied by changes in frequencies in thymocyte subpopulations, suggesting that thyroid hormones may be important in expanding thymocyte populations (150).

Clinical Relevance and Aging Effects

Age-related deficits in antibody production, mitogen-induced T cell proliferation, and NK cell activity were improved by T_4 treatment (184,185). However, the association of thyroid hormone treatment with augmented B cell development in laboratory animals may also exist in patients who overproduce these hormones. In Graves disease, autoimmune hyperthyroidism is associated with an increase in the proportion of peripheral blood B lymphocytes and, more specifically, in the proportion of CD5$^+$ B cells, a B cell subset associated with autoimmune reactions (186). The number of CD5$^+$ B cells correlated with serum levels of free T_4 and total T_3 and was reduced with restoration of thyroid function. Therefore, caution should be used when contemplating the use of thyroid hormone treatment in immunocompromised hosts.

Gonadal Steroids

Several observations have implicated sex steroids as important regulatory hormones of the immune system (reviewed in references 187,188). Females have higher levels of serum IgG, IgG1, IgM, and IgA than males in several species, and antibody responses are greater in magnitude and more prolonged in females than in males (189). The hormonal changes associated with pregnancy alter immune function, and such alterations may contribute to a successful pregnancy (190–192). The incidence of autoimmune diseases such as MS, SLE, and RA is higher in females than in males (193). Removal of sex organs accompanied by androgen or estrogen replacement in animals confirmed that immune differences associated with gender were related to the presence of sex steroids (194). T and B lymphocyte expression of classic nuclear androgen and estrogen receptors has been reported (195,196). Recently, nonreceptor, plasma membrane–associated estrogen and testosterone binding sites that deliver a signal for elevation of intracellular calcium have been reported on CD4$^+$ and CD8$^+$ T cells (197,198); the function of these binding sites has yet to be clarified.

Recently, several groups have demonstrated that B cell hematopoiesis in the bone marrow is dramatically reduced by gonadal steroids. Castration of male mice produced a selective increase in B cell number in the periphery, driven primarily by an increase in B cell progenitors in the bone marrow (199). Genetically hypogonadal *(hpg)* female mice displayed increased B cell progenitors in the bone marrow, and estrogen replacement reduced this effect (200). Similarly, in animals unable to produce the androgen receptor *(Tfm* mice) and pregnant normal mice, the presence of sex steroids in females reduced the ability of immature B cells to produce mature B cells that migrate into the periphery (191,200,201). Bone marrow stromal cells may be indirectly responsible for reduced B cell development in the presence of estrogen; genetic analysis suggests that these supporting cells may produce both the α and β form of the estrogen receptor (201).

The higher incidence of autoimmune diseases in women has led to the view that estrogen may promote autoimmune immunoreactivity (reviewed in references 202,203). For example, estradiol treatment in mice activated T cell differentiation in the liver, detected by an increase in "forbidden" T cell clones (Vβ3+) (204). Also, estrogen elicited autoantibody production by CD5$^+$ B cells *in vitro* (205). Conversely, elevated estrogen levels can reduce autoimmune disease progression, especially in diseases considered to be Th1 predominant (206). For example, reduced clinical symptoms are reported in patients with RA and MS in the last trimester of pregnancy when estrogen and other hormones are dramatically elevated (203). Some investigators have hypothesized that low doses of estrogen may promote autoimmune development, whereas high doses, such as those produced during pregnancy, are protective. Others have suggested that estrogen may shift the cytokine profile from disease-inducing Th1-like to Th2-like. EAE induction in mice was impaired by pretreatment with doses of estrogen ranging from very low (equivalent to diestrus) to very high (associated with pregnancy) (207). T cells obtained from the estrogen-treated mice were less able to transfer the disease compared with T cells from control mice, but no dramatic shift from Th1-like to Th2-like cytokine production was observed. Another group demonstrated that estrogen improved the efficacy of an anti–T cell vaccine that delays the progression of EAE in female mice in association with a significant shift toward IL-10 and TGF-β production by antigen-specific T cells (208). In these latter two studies, estrogen was administered before disease initiation, limiting its therapeutic potential. However, a recent study demonstrated that when estriol, the form of estrogen produced by the placenta, was administered after disease initiation, development of EAE in mice was impaired in association with increased IL-10 production by myelin basic protein–specific T cells and increased IgG1 anti–myelin basic protein antibody production (209). This effect is consistent with the shift in Th2-like cytokine production that is observed in pregnant mice (192) and suggests that estrogen treatment may have therapeutic benefit during disease progression.

Testosterone may also inhibit Th1-mediated autoimmune disease processes. For example, administration of 5α-dihydrotestosterone to nonobese diabetic female mice after the onset of insulitis prevented the islet cell destruction and spontaneous development of diabetes (210). Androgens appear able to skew a Th1-like response to a Th2-like response when present at the time of the initial exposure to antigen, an effect that contributed to slower EAE progression (211). More studies are required to explain the apparently contradictory effects of estrogen on immune reactivity and the development of autoimmune disease. Manipulation of gonadal steroids may be a useful component in therapy of some autoimmune diseases, but further work is necessary to ensure the safety and efficacy of potential treatment protocols.

NERVOUS SYSTEM–IMMUNE SYSTEM INTERACTIONS

Behaviorally conditioned or stressor-induced alterations in immune function can be prevented by removing sympathetic neurotransmission (47,212). Cells of the immune system may encounter the catecholamine norepinephrine (NE) and

other neuropeptides released from sympathetic noradrenergic nerve fibers. NE and epinephrine (EPI) are also released from the adrenal medulla on exposure to stressors. Evidence has also been accumulating that catecholamines detected within cells of the immune system can regulate immune reactivity in an autocrine fashion.

Catecholamines

Intracellular Catecholamines

NE, EPI, dopamine (DA), and their metabolites have been detected in lymphocytes, macrophages, and neutrophils (213–215). Several lines of evidence indicate that lymphocytes may synthesize catecholamines. Incubation of lymphocytes with α-methylparatyrosine to inhibit tyrosine hydroxylase, the rate-limiting enzyme in catecholamine biosynthesis, reduced intracellular and extracellular NE and DA (213,214). L-Tyrosine, the amino acid precursor in the catecholamine synthetic pathway, increased lymphocyte NE and L-dopa, an intermediate product (215). It is also possible that cells of the immune system take up and store catecholamines from extracellular sources. In neutrophils, the concentration of intracellular catecholamines was positively correlated with plasma catecholamine levels (216), and blocking catecholamine uptake with desipramine increased extracellular NE, DA, and EPI (214,216).

The functional impact of endogenously derived catecholamines on immune reactivity is not well understood. Spengler et al. (217) showed that endogenous catecholamines influenced macrophage cytokine production. First, lipopolysaccharide (LPS) reduced intracellular levels of NE and EPI in macrophages, suggesting that activation elicited catecholamine release. Second, LPS-induced TNF-α production by macrophages was increased by β-adrenergic receptor (AR) blockade and decreased by α-AR blockade in the absence of exogenous adrenergic agonists. Recently, Miller et al. (218) showed that synovial macrophages from RA patients contain tyrosine hydroxylase and secrete NE, suggesting that macrophages may contribute to disease progression by local production of NE. Further work is necessary to elucidate the significance of autocrine regulation of lymphocyte function by catecholamines.

Sympathetic Innervation of Lymphoid Organs

In the spleen, lymph nodes, thymus, and bone marrow, sites that serve as residence for developing and mature lymphocytes, sympathetic noradrenergic nerve fibers distribute with blood vessels and branch into the parenchymal areas containing lymphocytes and macrophages of the immune system (reviewed in reference 219). A few noradrenergic nerve terminals directly contact lymphocytes by forming a direct synapse with these cells (220). Noradrenergic nerves contain NE and other colocalized effector molecules such as neuropeptide

Y and adenosine triphosphate (221,222). These molecules are coreleased with NE, depending on the strength of the stimuli, and potentiate NE-induced effects (222,223). The noradrenergic neuron–selective toxin 6-hydroxydopamine (6-OHDA) depletes NE in the spleen and lymph nodes as much as 95% (224), suggesting that the primary source of NE in these tissues is sympathetic noradrenergic nerve terminals. In contrast, there is little evidence that the parasympathetic nervous system is present in the spleen and lymph nodes. The functional impact of sympathetic innervation of lymphoid organs is discussed below.

Alterations in Catecholamine Availability with Aging

Changes in sympathetic noradrenergic innervation and plasma catecholamines may contribute to age-related alterations in immune function. In Fisher 344 (F344) rats, noradrenergic innervation and NE were markedly reduced in the spleen and lymph nodes with age (225). Ablation of noradrenergic nerve fibers by 6-OHDA treatment altered immune reactivity in aged F344 rats, suggesting that despite the dramatic reduction in splenic noradrenergic innervation, the remaining noradrenergic sympathetic nerves are functional (226,227). Chronic treatment with L-deprenyl, a monoamine oxidase B inhibitor used in the treatment of Parkinson disease, partially restored sympathetic noradrenergic nerves in the spleen of aged F344 rats (228). The restoration of noradrenergic nerves was associated with increased splenic NK cell activity and mitogen-induced IL-2 and IFN-γ production, suggesting that noradrenergic innervation can positively modulate immune function (229). Interestingly, L-deprenyl can also significantly reduce the number and size of spontaneously and chemically induced mammary tumors in young and old female rats, although the precise antitumor mechanisms are not yet known (230). Besides restoring noradrenergic innervation, L-deprenyl can act both centrally and peripherally to alter neuroendocrine pathways and increase superoxide dismutase activity. Nevertheless, the antitumor effects of L-deprenyl administration demonstrate that interactions between neuroendocrine pathways and the immune system can modulate disease outcome in both young and old hosts.

α- and β-Adrenergic Receptor Expression by Cells of the Immune System

NE and EPI can signal α-AR and β-AR with different affinities; therefore, regulation of lymphocyte function will depend on NE concentration and the AR subtype available. Based on receptor ligand-binding studies, α-AR are not present on resting lymphocytes but may be up-regulated after macrophage activation (231). One of the more striking changes in AR expression is the induction of lymphocyte α_1-AR in children with a severe form of juvenile RA; lymphocyte α_1-AR were not detected in healthy subjects or children with less severe forms of the disease (232). By contrast, β-AR are found on T and B lymphocytes, macrophages, and other cells of the immune system, and the density varies with lymphocyte subset and activational status (reviewed in reference 233). β-AR expression may be differentially regulated within the Th population. Th1 clones express a density and affinity of β-AR similar to normal T cells, but β-AR are not detectable on Th2 clones (234). Lymphocyte β-AR density and signaling can be altered by disease states, including RA (235) and MS (236,237). The supporting cells of lymphoid organs may also express AR. For example, the thymic epithelium appears to express β-AR (238); NE binding to these stromal cells may influence thymocyte development at a developmental time point when thymocytes do not express the receptor themselves (239).

In Vitro Immune Regulation by Catecholamines

The functional significance of β-AR heterogeneity has been investigated by Sanders and Powell-Oliver (24) using an elegant *in vitro* system. To determine which lymphocyte populations were sensitive to β-AR stimulation, purified antigen-specific B cells, which serve as both the antigen presenting cell and the antibody secreting cell, were incubated with either Th1 or Th2 antigen–specific clones and the antigen. When the β_2-selective agonist terbutaline was added to cultures containing Th1 or Th2 clones and B cells, the number of antibody-forming cells and the amount of secreted antibody increased. Conversely, incubation of Th1 clones with terbutaline before their interaction with B cells inhibited antigen-induced IFN-γ production and the IFN-γ–induced immunoglobulin isotype IgG2a (234). Preincubation of Th2 clones with terbutaline did not affect either IL-2 production or IgG1 (Th2-driven) antibody production by B cells, as would be predicted by the deficiency in Th2 β_2-AR. These results suggest that temporal factors can determine whether NE augments or inhibits a T-dependent antibody response. This example also demonstrates β-AR–induced inhibition when maturation to a Th1-type response has already occurred. By contrast, the maturation of naïve Th0 cells into Th1 cells is enhanced in the presence of NE. When naïve CD4$^+$ T cells were activated with either anti-CD3/anti-CD28 or antigen-pulsed antigen presenting cells, the presence of NE (10^{-6} M) increased IFN-γ production on a per-cell basis (240). These results demonstrate the complexity of NE interactions with T and B cells; this complexity is also apparent *in vivo*.

In Vivo Immune Regulation by Catecholamines

One approach used to examine SNS regulation of immune function *in vivo* is chemical sympathectomy with the noradrenergic-selective neurotoxin 6-OHDA. In adults, 6-OHDA destroys noradrenergic nerves in the periphery without crossing the blood–brain barrier and significantly depletes NE in lymphoid organs (227,241). Sympathectomy can enhance or suppress antibody production and can inhibit cell-mediated responses, such as delayed-type hypersensitivity (reviewed in references 239,242). The reason for these mixed effects of sympathectomy is probably related to the type of antigen used (T dependent versus T independent), the timing of sympathectomy relative to immunization, and animal strain. For example, sympathectomy increased antibody production in C57Bl/6 mice (Th1 dominant) in conjunction with elevated IFN-γ. Sympathectomy did not significantly alter antibody production in BALB/c mice (Th2 dominant) that cannot produce detectable levels of IFN-γ (243,244). These results suggest that Th1, and not Th2, responses are inhibited by intact noradrenergic nerves. However, the Th2-like cytokine IL-4 was elevated in both strains, suggesting that the SNS does not selectively drive Th1 or Th2 responses. Recently, Kohm and Sanders (245) used severe combined immunodeficiency (SCID) mice that are unable to produce lymphocytes to demonstrate that Th2-mediated responses *in vivo* could be promoted by NE. SCID mice were sympathectomized and injected with antigen-specific donor B cells and Th2 clones and antigen. The primary and secondary responses to antigen were inhibited in sympathectomized mice compared with intact vehicle controls, an effect that was partially reversed by NE. β-Blockade with nadolol also inhibited the antibody response in noradrenergic nerve–intact SCID mice. These results suggest that loss of β-AR signaling was responsible for the reduced antibody response. Because the B cells in this model were the only donor cells that express β-AR, the effect appeared to be mediated through antigen-specific B cells, an interpretation that is consistent with NE-induced enhancement of antibody production *in vitro* (246). Together, these results suggest that sympathetic regulation of immune reactivity can be inhibited or enhanced by SNS neurotransmission, depending on the cell types involved in the response and their AR expression.

Lymphocyte Redistribution and Trafficking

The SNS is an important regulator of lymphocyte trafficking under a variety of conditions. NK cells, lymphocytes, and monocytes enter the blood after physical exercise or exposure to stressors. These effects are blocked by β-AR blockers and can be mimicked by administering the β-AR agonists EPI or isoproterenol (ISO), in part by reducing adhesion to endothelial cells (reviewed in reference 247). Early studies demonstrated that direct infusion of catecholamines into the spleen increased leukocyte output in the absence of changes in blood flow (248). More recently, it has been demonstrated that LPS-induced leukocyte output is regulated by sympathetic innervation of the spleen (249). Together, these results suggest that the SNS and NE can regulate cell entry, retention, and release from lymphoid organs. Understanding the

mechanisms underlying catecholamine-induced changes in lymphocyte trafficking may help to provide the means by which cells can be mobilized or immobilized during local immune and inflammatory responses.

Sympathetic Activation–Induced Immunomodulation

In contrast to the evidence for a complex role of the SNS under resting conditions, sympathetic activation consistently elicits immunosuppression. For example, peripherally administered morphine or central administration of IFN-α or CRH reduced T cell proliferation and NK cell activity in the spleen; these effects were blocked by peripheral β-AR blockade or ganglionic blockade (58,250,251). Recently, progress has been made in the area of LPS-induced sympathetic activation and catecholamine regulation of proinflammatory and antiinflammatory mediators of endotoxic shock. This research has demonstrated the complex role of the SNS in maintaining homeostasis in endotoxic shock through regulation of cytokine production.

LPS administration in laboratory animals rapidly elevates plasma EPI and NE and increases splenic nerve activity and NE turnover (252,253). Sympathetic activation may be responsible for muting the LPS-induced inflammatory response by reducing the proinflammatory cytokine response and augmenting the antiinflammatory cytokine response (reviewed in reference 222). In laboratory animals and humans, *in vivo* EPI and ISO administration reduced plasma levels of TNF-α, nitric oxide, and IL-12, IFN-γ but enhanced IL-6 and IL-10 levels (254–257). Conversely, β-AR blockade before LPS treatment in mice increased plasma TNF-α and reduced plasma IL-10 (258). These *in vivo* results are consistent with β-AR–mediated inhibition of TNF-α and elevation of IL-10 production by macrophages *in vitro* (259–261). Furthermore, ISO- and EPI-induced inhibition of TNF-α production *in vitro* was partially blocked by neutralizing anti-IL-10 antibodies, suggesting that IL-10 can itself reduce TNF-α (254,258).

α_2-AR also participate in SNS modulation of the response to LPS, although their role has yet to be fully elucidated. Fessler et al. (262) reported that mortality was increased in rats treated with the α_2-agonists xylazine and UK 14304 before LPS, whereas pretreatment with the α_2-antagonist rauwolscine protected rats from an LD100 dose of LPS. In these animals, α_2-blockade reduced plasma TNF-α levels, indicating that α_2-AR stimulation elevates TNF-α production. This notion is consistent with evidence for α_2-AR-induced elevation of macrophage TNF-α production *in vitro* (217). In another report, however, either α_2-AR stimulation or α_2-AR blockade before LPS reduced plasma TNF-α *in vivo* (257). In this case, the effect of α_2-AR blockade was prevented by ganglionic blockade of sympathetic activation or by blocking β_1-AR but not β_2-AR. This suggests that the inhibitory effect of α_2-AR blockade may be produced by blocking inhibitory presynaptic α_2-AR receptors and promoting NE release through excitatory presynaptic β_1-AR. Thus, SNS activation of α_2-AR may have multiple functions in regulating the endotoxic response.

Together, these data suggest that SNS activation after LPS administration serves several purposes. First, increased catecholamine release inhibits the inflammatory response by reducing proinflammatory cytokine production and augmenting antiinflammatory cytokine production. At the same time, the increase in sympathetic outflow may be dampened by catecholamine stimulation of inhibitory presynaptic α_2-AR. β-AR stimulation may also inhibit Th1 responses by inhibiting IL-12 production (255,263). SNS activation may be a general response to inflammation. For example, high levels of catecholamines induced by severe stressors (e.g., brain injury, myocardial infarction, major surgery) have been demonstrated to inhibit TNF-α and elevate IL-10 production by monocytes (264,265). Basic studies clarifying the AR subtypes and cell types responsible for the catecholamine-induced changes in LPS-induced cytokine production *in vivo* may lead to clinical studies that maximize the benefits and minimize the unwanted side effects of stimulation of AR in endotoxic shock and inflammatory processes contributing to disease entities such as RA and MS (see below).

Sympathetic Regulation of Autoimmune Disease

In animal models of human autoimmune disease, alterations in sympathetic innervation, NE concentration, and lymphocyte AR expression have been demonstrated. For example, in MRL-*lpr/lpr* mice, a naturally occurring lupus model, reduced splenic noradrenergic innervation and decreased splenic NE concentration were apparent before the onset of disease symptoms (266). In myelin basic protein–induced EAE, an MS-like disease, a reduction in splenic NE concentration was reported at the time of maximal antigen-induced lymphocyte proliferation and was accompanied by an increase in the density of splenic lymphocyte β-AR (267). In chronic/relapsing EAE (CREAE) induced in rats, splenocyte β-AR density correlated positively with the severity of CREAE (268). Removal of noradrenergic innervation by chemical sympathectomy with 6-OHDA enhanced the severity of symptoms in EAE (269), and administration of ISO or the β_2-selective agonist terbutaline before disease induction reduced the severity of symptoms in EAE and CREAE and reduced the number of relapses in CREAE (268,270). When donor T cells from sympathectomized animals with EAE were transferred to normal hosts, EAE was more severe compared with that in animals that received cells from nonsympathectomized animals (271). This result demonstrated that sympathectomy influenced the disease-inducing T cells and suggests that alterations in catecholamine availability, β-AR signaling, or both are an integral component of autoimmune disease progression.

The SNS may elicit different, and often opposing, functions at different anatomic sites. Localized denervation of draining lymph nodes, with sparing of the nerves innervating the joint, exacerbated joint pathology (272), but systemic removal of sympathetic input, either by β-AR blockade or chemical sympathectomy with 6-OHDA or guanethidine, reduced arthritic symptoms (273). These results suggest that noradrenergic innervation of draining lymph nodes inhibits the generation of antigen-specific T cells but promotes inflammation in the joints. A similar complexity was demonstrated after β-agonist administration. Administration of a high dose of EPI reduced the severity of experimental arthritis by an α_2-AR–mediated mechanism, but a low dose of EPI exacerbated joint injury (274). These results indicate that care must be used in manipulating the SNS therapeutically in complex diseases.

The SNS may also influence autoimmune processes in humans. Alterations in autonomic activity have been reported in RA and MS, but it is not known whether these changes are induced in response to disease or whether alterations in the SNS play a role in initiating the disease. In children with juvenile RA, increased sympathetic activity and autonomic hyporesponsiveness were associated with disease exacerbation (275). Adults with RA also exhibit autonomic dysfunction, depending on the severity of the disease (276,277), and regional removal of sympathetic input by guanethidine treatment improved arthritic symptoms, probably by removing a proinflammatory component in the joint itself (278). Increased and decreased β-AR density has been reported in peripheral blood mononuclear cells from RA patients (235,279). Increased β-AR expression and signaling capacity in CD8$^+$ T cells have been reported in MS patients with active disease, in conjunction with reduced sympathetic tonus (236,279,280).

Zoukos et al. (279) proposed that the functional significance of increased lymphocyte β-AR sensitivity with RA and MS disease progression may be induced by, and thus be an indicator of, underlying inflammation. They demonstrated that increased lymphocyte β-AR correlated with increased high-affinity IL-2 receptors on T cells and with worsening of disease, as assessed by clinical observations and magnetic resonance imaging activity. IL-1 and cortisol, which may be present during inflammation, increased β-AR in normal cells *in vitro* (237). β-Agonist stimulation of normal lymphocytes *in vitro* reduced IL-2 receptor expression and suppressed mitogen-induced proliferation, suggesting that increased β-AR expression may be an attempt to mute the proliferative response of T cells by increasing sensitivity to endogenous catecholamines (237). As we have seen after endotoxin administration, β-AR stimulation may be important in reducing proinflammatory cytokine production (especially TNF-α) and enhancing antiinflammatory cytokine production (IL-10) in autoimmune phenomena. These cytokines are key mediators of RA and MS (281–283). Therefore, effective therapies may be devised incorporating our knowledge of the SNS and AR to reduce the progression of the disease and minimize pathology in these diseases and others associated with dysregulation of the immune system and inflammation (284).

Sympathetic Regulation of Tumor Growth and Metastases

Circumstantial evidence suggests that the SNS may regulate tumor growth through modulation of antitumor immunity. For example, ThyagaRajan et al. (230) demonstrated that treatment with L-deprenyl, a monoamine oxidase B inhibitor, reduced spontaneous mammary and pituitary tumor growth in old rats in association with partial restoration of splenic noradrenergic innervation and improved immune function in the spleen. Tumor metastases may be influenced by sympathetic activation through modulation of NK cells, whose main function is to prevent metastasis. Stress-induced inhibition of NK cell activity in rats was associated with increased numbers of the NK cell–sensitive MADB106 adenocarcinoma cell line colonizing the lung (285). This effect was prevented by adrenal demedullation, β-AR antagonist administration, and ganglionic blockade with chlorisondamine. Administration of metaproterenol (MET), a β-AR nonselective agonist, before injection of MADB106 tumor cells, reduced peripheral blood NK cell activity and elevated tumor metastases (286). Peripheral β-AR blockade with nadolol or propranolol prevented the MET-induced effects. The MET-induced increase in lung tumor retention *in vivo* was not observed after depletion of NK cells by treatment of rats with anti-NKR-P1 antibody, suggesting that MET acted by directly stimulating NK cell β-AR. These results indicate that catecholamine release can inhibit NK cell activity and thus promote tumor cell metastases.

Neuropeptides

Many neuropeptides are found in nerve fibers innervating T and B cell–containing compartments of primary and secondary lymphoid organs. One such peptide, neuropeptide Y, is colocalized in noradrenergic nerve terminals in rat spleen and thymus (221,287), whereas others, such as vasoactive intestinal peptide, substance P (SP), and calcitonin gene related peptide (CGRP), appear to be associated with nonnoradrenergic nerve fibers in the spleen (287–289). Analysis of the regulatory role of these neuropeptides in inflammatory and immune processes has revealed that they are important modulators of lymphocyte and accessory cell activity (reviewed in references 290,291).

SP, an 11-amino-acid peptide, and CGRP, a 37-amino-acid peptide, are elevated in chronically inflamed tissue, where they act on the vasculature to promote vascular permeability and vasodilation (292). SP potentiates the activity and release of vasoactive amines, such as histamine, from mast cells, and enhances phagocytosis by neutrophils (293).

The proinflammatory cytokines IL-1, TNF-α, and IL-6 can be induced by SP in the absence of endotoxin or other stimulators, suggesting that SP is an early inducer of local and systemic host defense responses to inflammation and injury (291,294). SP also can regulate cell-mediated and humoral immune responses via interactions with SP receptors on T and B lymphocytes (295). SP-enhanced antibody production *in vivo* and *in vitro*, particularly IL-4-induced switching to IgE (296–298), and increased T cell responses *in vitro* (299,300). SP may influence thymocyte differentiation by preventing GC-induced thymocyte apoptosis (301).

In contrast to SP, CGRP suppresses T cell function and reduces inflammation. CGRP reduced T cell proliferation and IL-2 production, inhibited thymocytes apoptosis, and reduced B cell hematopoiesis (302–304). CGRP interactions with epithelial Langerhans cells, the specialized antigen presenting cells of the skin, decreased antigen presenting capacity in association with reduced expression of the costimulatory molecule B7-2 (CD86) and increased IL-10 production (305,306). Functionally, local administration of CGRP inhibited a cutaneous delayed-type hypersensitivity response (305). In nonobese diabetic mice engineered to produce CGRP in beta cells of the pancreas, no male mice developed the disease, and the incidence in female mice was reduced by 63% compared with their diabetic littermates (307). These results suggest that CGRP may be useful therapeutically as an immunosuppressive agent in the skin and other sites. Vasoactive intestinal peptide also inhibited lymphocyte proliferation to T cell mitogens but not B cell mitogens (308,309), and it can influence T lymphocyte trafficking and localization, depending on the subtype of receptor expressed (310,311). Further investigation is required to further elucidate the conditions under which these and other neuropeptides contribute to immune/inflammatory processes.

Immune System Communication with the Nervous System

Viral or bacterial infection induces "sickness behavior" characterized by fever, hyperalgesia, reduced intake of food, increased sleep, and other behavioral changes (reviewed in reference 312). Experimentally, peripheral administration of LPS is most often used to induce sickness behavior, but other bacterial products can elicit CNS-mediated changes in behavior. It is noteworthy that staphylococcal enterotoxin B (SEB) did not induce the same sickness behaviors as LPS administration (313). The difference in the behavioral response to LPS versus SEB may reflect differences in immune cell cytokine production: SEB elicits T cell activation, whereas LPS activates primarily monocytes/macrophages. LPS-induced sickness behavior is mediated by the proinflammatory cytokines IL-1, IL-6, and TNF-α because peripheral or central blockade of these cytokines prevents various components of LPS-induced sickness behavior (312). Thus, cytokine interactions with the brain constitute an integral part of the host response to peripheral and central infection and inflammation.

Another manifestation of infection, LPS, or SEB treatment is the activation of the HPA axis and the SNS (64,252,314,315). The resulting elevation in plasma GC and catecholamines may contribute to restoration of homeostasis by down-regulating the host inflammatory response. In LPS-treated or virally infected animals, when sympathetic or HPA axis activation is impaired, morbidity and mortality are increased in association with dysregulation of proinflammatory cytokine production, particularly elevated TNF-α (97,262). Defects in sympathetic and HPA axis activation also promote autoimmune disease in laboratory animals and possibly in humans (118,269,275,316). Thus, communication from the periphery to the CNS is necessary to resolve immunologic and inflammatory responses elicited by autoimmune or infectious disease processes. The initial report that peripheral administration of cytokine-containing T cell supernatants could activate specific brain regions (317) has led to extensive examination of the mechanisms by which cytokines convey information from the periphery to the CNS.

Pathways of Cytokine Communication with the Central Nervous System

Peripheral IL-1β administration elicits the full range of sickness-induced behaviors, activates the HPA axis, and mimics changes in brain neurochemistry induced by viral infection or LPS (reviewed in reference 318). The large literature examining IL-1 interactions with the CNS is the primary focus of this discussion of cytokine-to-brain pathways. However, not all aspects of LPS- or virus-induced, CNS-mediated responses are blocked by neutralizing the actions of IL-1. For example, the fever response to LPS is not completely blocked in IL-1β knockout mice (319), suggesting that other cytokines can elicit sickness behaviors. IL-6 is responsible for activation of the HPA axis after infection with murine cytomegalovirus (320), and peripheral administration of TNF-α and IL-6 can act independent of, or synergize with, IL-1 to activate the HPA axis (64). Therefore, other cytokines induced by infection or LPS can also effectively communicate with the CNS, perhaps by similar mechanisms.

Initially, peripheral administration of IL-1 was shown to activate CRH-containing neurons of the paraventricular nucleus, leading to increased plasma ACTH and GC levels (65,66). Subsequent work demonstrated that other brain nuclei essential for neuroendocrine and autonomic regulation also participate in the CNS response to peripheral LPS or IL-1β, including the nucleus of the solitary tract, central amygdala, and ventral medial preoptic area (321–323). Therefore, any mechanism underlying IL-1 signaling of the brain had to explain how IL-1, a large, lipophobic molecule that cannot cross the blood–brain barrier, activates anatomic regions deep within the brain in regions devoid of IL-1 receptor protein or mRNA expression (324). It is now apparent that several mechanisms may account for IL-1 signaling of these brain regions (reviewed in reference 318). First, blood-borne IL-1 can access the brain by active transport across the blood–brain barrier, but this mechanism appears to be physiologically relevant only at high levels of plasma IL-1. Second, IL-1 can induce the synthesis of small, diffusable secondary molecules, such as prostaglandins in the vasculature and neurons of the brain. Finally, rapid CNS signaling by peripheral IL-1 may occur via afferent sensory neurons contained within the vagus nerve.

Prostaglandins

Prostaglandins are small, lipophilic molecules that can easily enter the brain parenchyma and have been implicated as indirect mediators of IL-1–to–CNS communication. Prostaglandin E_2 and cyclooxygenase-2, an important prostaglandin synthetic enzyme, are induced in the brain after peripheral LPS administration (325,326). Prostaglandins administered centrally can induce sickness behavior and patterns of c-fos activation similar to IL-1 (327,328). Finally, IL-1–induced sickness behavior and c-fos activation in the brain and LPS-induced sympathetic activation in the spleen are inhibited by central or peripheral administration of indomethacin, a prostaglandin synthesis inhibitor (252,328,329). Ericsson et al. (328) suggest that the differences in the density of prostaglandin receptors within specific brain regions may help to produce the brain-region selective effects of peripheral immune activation.

Afferent Sensory Fibers of the Vagus Nerve

Two brainstem regions, the nucleus of the solitary tract and the area postrema, are activated within 30 minutes of LPS or IL-1 administration. The rapidity of this response suggests a neuronal mechanism of communication with the CNS that does not require lengthy processes such as protein synthesis. Activation of vagal afferents that terminate in the nucleus of the solitary tract is evident after intraperitoneal administration of LPS or IL-1 (330). Surgical disruption of the vagus nerve (vagotomy) prevented LPS- or cytokine-induced monophasic fever, neuroendocrine activation, and cytokine synthesis in the brain (reviewed in reference 331). LPS-induced IL-1 production by dendritic cells and macrophages within connective tissue surrounding the abdominal vagus nerve suggests that local production of IL-1 can signal afferent sensory neurons to the brain (332). Despite this evidence, the effectiveness of vagotomy is somewhat controversial (328,333), and the role of the vagus nerve may be limited to particular doses and/or route of administration of LPS or IL-1. For example, fever induced by low-dose, but not high-dose, IL-1β is prevented in vagotomized in rats, suggesting that the neural route of cytokine–brain communication is elicited only when very low levels of cytokine are induced locally, not systemically (334). Wan et al. (32) demonstrated that c-fos activation in the brain was prevented by subdiaphragmatic vagotomy after intraperitoneal, but not intravenous, administration of LPS. Further work is required to determine the conditions under which each pathway to the CNS predominates. Nonetheless, the multiple mechanisms by which IL-1 can signal the brain are indicative of the importance of CNS communication after peripheral immune activation for maintenance of homeostasis after exposure to a pathogen, tissue injury, or autoimmune disease.

CLINICAL IMPLICATIONS OF PSYCHONEUROIMMUNOLOGY

Given the large body of evidence for CNS modulation of the immune system, it is important to understand the impact of altered CNS function on diseases mediated by the immune system. The primary difficulty lies in the extraordinary breadth of factors that can disturb CNS function and the number of disorders influenced by immune system activity. Changes in CNS signaling may stem from normal or aberrant development, behavioral conditioning, stressful life events, depression and other psychiatric conditions, drug and alcohol usage, aging, and CNS structural lesions such as tumors or strokes. The range of clinical immunology is similarly broad. The immune system plays a pivotal role in host defense, allergies and asthma, cancer surveillance, wound healing, and the prevention or development of autoimmune disorders. Furthermore, many other diseases such as atherosclerotic heart disease, peptic ulcer disease, and psychiatric disorders may be directly or indirectly influenced by the immune system.

A general model of clinical psychoneuroimmunology is shown in Figure 10.1. Previous sections of this chapter deal with individual elements of the model such as stressor effects on immune responses and the influence of autonomic and neuroendocrine signaling on immune function. This section

focuses on human research in which disease activity is the primary outcome and in which a linkage has been established between psychosocial and behavioral factors and alterations in clinical outcome.

Susceptibility to Viral Infection

Like many areas of clinical psychoneuroimmunology, the belief that stress increases susceptibility to infection is widespread and has predated rigorous scientific study. Epidemiologic studies have suggested that psychosocial stress, such as increased family conflict, is associated with the development of upper respiratory infections (335,336). Although this relationship can be attributed in part to increased social contacts during periods of distress, carefully controlled viral challenge studies by Cohen et al. (337) have confirmed that emotional factors alter susceptibility to upper respiratory infections (reviewed in reference 2). In these studies, individuals with no prior exposure to a particular pathogen were isolated for several days and inoculated with a standard dose of intranasal virus. Investigators found that stressful life events and negative affect predicted the development of infection with five different upper respiratory infection viruses, independent of health practices. Furthermore, individuals with smaller social networks and more chronic stressors were at greater risk.

Herpes viruses are somewhat unique in that they frequently remain present in a latent state and can be reactivated during periods of immune suppression. This characteristic provides an excellent opportunity for studying the impact of stress-related immune changes in a longitudinal fashion. Several studies have suggested that stress and negative affect increase oral and genital herpes outbreaks (337–340). Furthermore, psychosocial stress has also been shown to contribute to reactivation of other herpes viruses including Epstein-Barr virus and cytomegalovirus (341,342). Recent work has also focused on identifying the role of psychosocial factors in HIV infection (reviewed in references 343,344).

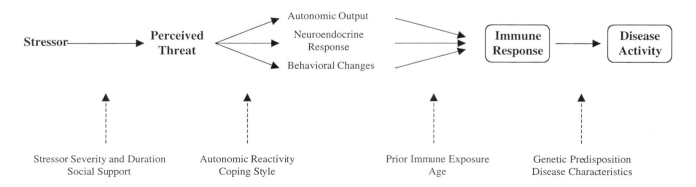

Biological and Psychosocial Moderating Factors

FIGURE 10.1. A model of clinical psychoneuroimmunology.

At least 50 prospective, naturalistic studies have been conducted, yielding somewhat conflicting results; however, several psychosocial factors appear relevant to disease progression. These factors include the presence or absence of social support, coping style for stressful life events (particularly HIV-related events such as seroconversion), and depression related to death of significant others. Treatment studies focused on increasing support and active coping or prevention of depression have shown some promise in altering immune indices of HIV infection; however, the effects of these interventions on the development and progression of AIDS have yet to be established.

Wound Healing

Although wound healing is not classically considered to be an immunologic process, the immune system plays a critical role in each stage of wound restoration from initial inflammation to epithelialization, wound contraction and reorganization, and prevention of infection. One of the earliest suggestions that stress can influence wound formation and healing came from the observation in the 1950's that peptic ulcer disease was more prevalent after stressful life experiences (345). Subsequent studies (4,346,347) extended this work to demonstrate that psychological or physical stress can have a strong impact on recovery from perioperative, mucosal, and cutaneous wounds.

The most compelling of these studies was conducted by Kiecolt-Glaser et al. (4). They initially examined how quickly caregivers of Alzheimer disease patients recover from a standardized punch biopsy. This group of chronically stressed individuals took an average of 9 days longer than controls to close a 3.5-mm wound. Subsequent studies of dental students revealed a 40% delay in wound healing during examinations compared with a control period (346). Peripheral blood mononuclear cells from the stressed subjects in these studies produced less IL-1, a cytokine necessary for wound healing (4,346). Glaser et al. (348) also demonstrated that lower levels of IL-1 and IL-8 were produced within the wound site in individuals with higher perceived stress. It is important to acknowledge that stress-induced immunosuppression is likely only one of several pathways by which stress can alter wound healing. Stress may also lead to altered blood flow, deceased nutritional intake, sleep disruption, and behavioral changes ranging from decreased compliance with wound care to increased smoking and alcohol consumption. Each of these factors may contribute, alone or in concert, to produce an inadequate tissue response.

Cancer

Early studies of tumor biology suggested that most tumors are relatively autonomous in their development and progression. Although genetic and other intrinsic factors clearly play a large role in cancer development, increasing attention has been focused on how external factors alter tumor immunity. Animal studies have shown that environmental stressors such as handling and separation can influence tumor incidence and progression (reviewed in references 349,350); however, the literature on effects of stress on tumor development in humans has been less consistent. Anecdotal reports of increased cancer susceptibility after grief and depression date back to 200 AD (reviewed in reference 351). Several prospective 10- to 20-year follow-up studies confirmed that higher baseline depression and hopelessness increased the risk of cancer incidence and death (352–354), although other studies failed to reproduce their finding (355,356). This inconsistency may be owing in part to wide variability between cancers in their responsiveness to external stimuli and the fluctuating nature of depression and other mood states.

Studies that examine cancer progression and recurrence within specific cancer populations are beginning to yield more consistent information on the potential impact of emotions on cancer development. For example, in breast cancer patients, a "fighting spirit" was associated with a better prognosis, but avoidance, denial, and hopeless/helpless responses were associated with a poorer cancer prognosis (reviewed in reference 357). Perhaps the most compelling research in the area of psychooncology involved psychosocial intervention studies in patients with early stage cancers. Fawzy et al. (358,359) administered a 6-week structured group psychotherapy intervention to patients with stage I or II melanoma. They demonstrated reduced short-term distress in treated patients compared with controls, as well as an increase in NK cell number and cytotoxicity. At 6-year follow-up, the intervention group had significantly lower mortality rate and a trend toward decreased recurrence. Similarly, Spiegel et al. (360,361) conducted three randomized clinical trials of supportive–expressive group therapy in breast cancer patients and found an increase in survival time of 18 months at 10-year follow-up. A similar impact on survival has been reported for lymphoma and leukemia patients after psychosocial intervention (362). There have also been several negative studies of psychosocial treatments and cancer (363,364); however, these studies had inadequate "dosing" of therapy or other methodologic issues.

Autoimmune Disorders

Although the strength of evidence varies for each disease, psychosocial factors have been described as a potent trigger of disease activity in RA, insulin-dependent diabetes mellitus, SLE, psoriasis, Graves disease, Crohn disease, and MS (reviewed in references 13,14,365–369). This suggests that a person's emotional state cannot only contribute to suppression of immune system activity but may exacerbate hyperimmune states as well.

The strongest evidence comes from studies of MS. In the first description of MS in 1868, Charcot noted that grief and worry might precipitate the onset of the disease. Since that time, there have been more than 40 epidemiologic and experimental studies of stress and MS. Although the results have been somewhat mixed, recent studies employing better controls and more accurate assessments of stressful life events have shown a consistent relationship between stress and MS disease activity. Warren et al. (370) interviewed 100 MS subjects along with rheumatologic and neurologic controls and found that MS subjects had more stressful life events in the 2 years before the onset of their disease. In addition, Grant et al. (371) found that 77% of MS patients reported severely stressful life events in the 6 months before the onset of MS versus a 35% incidence of stressful events in a comparable period for controls. MS subjects experienced significantly more life events before exacerbations compared with remissions (372–374), even when rigorous magnetic resonance imaging criteria were used to define exacerbations. Interestingly, the life events that appear to precipitate MS exacerbations were not the most severely threatening but rather more modest stressors such as housing, financial stress, and disruptions in routine. In fact, in a prospective study of Israelis during the Persian Gulf War, MS patients reported a decrease in the number of exacerbations (375).

The contrasting effects of moderate and severe stressors may provide some insight into the discrepancy between human studies of MS and animal studies of EAE. A biphasic relationship between stress and MS disease activity may exist such that moderate stressors are associated with an increase in the production of inflammatory cytokines and subsequent disease activity, whereas extreme stressors produce high levels of GC activity that may override other systems and suppress immune activity. This variation of the eustress/distress hypothesis by Dhabhar and McEwen (19; see section on evidence for CNS–immune system interactions) is supported by recent studies of EAE, in which severe standard stressors such as electric shock and restraint stress generally delayed or diminished symptoms of the disease, but milder stressors increased disease severity (reviewed in reference 369).

CONCLUSIONS

There is substantial scientific and clinical evidence that psychological factors can influence the course of a variety of immune-mediated disorders including viral infections, wound healing, and autoimmune disorders. Furthermore, although the evidence for a role of psychoneuroimmunology in cancer is somewhat less established, recent clinical studies of cognitive therapy and relapse prevention appear compelling. This research is hampered by the possibility that changes in disease activity may stem from alterations in behavior or changes in physiologic systems other than the immune system. For example, in the cancer studies, reductions in cancer recurrence may be related to improved compliance, decreased smoking, and other positive health behaviors. Further research is needed to obtain direct evidence

that psychosocial factors influence the course of clinical diseases through changes in CNS outflow and immune system activity.

Research in psychoneuroimmunology has revealed that an immune reaction does not occur in a vacuum; external factors such as emotional and physical stressors can modulate immune reactivity and influence health and disease. Investigators have established the range of effects that the neuroendocrine environment has on immune reactivity and susceptibility to pathogens, cancer, and autoimmune disease. However, important basic questions have yet to be answered. What role does the external environment play in the initiation of human diseases such as cancer and autoimmunity? How does stress influence the progression or severity of these diseases, once established? How does the neuroendocrine environment influence the response to vaccination, especially in very young and very old populations? Finally, how can the interactions between the nervous system and immune system be used in the therapeutic treatment of human diseases? This multidisciplinary research will require collaborations between psychologists, neuroscientists, and immunologists and the removal of barriers of language and scientific bias to answer these important questions.

ACKNOWLEDGMENT

This research was supported by funds to K.D.A. from NIH grants K01-MH01468, HL65112, MHCRC- MH30915, and NIH/NCRR/GCRC grant M01-RR00056.

REFERENCES

1. Ader R, Felten DL, Cohen N. *Psychoneuroimmunology,* 3rd ed. San Diego: Academic Press, 2001.
2. Cohen S, Herbert TB. Health psychology: psychological factors and physical disease from the perspective of human psychoneuroimmunology. *Annu Rev Psychol* 1996;47:113–142.
3. Cohen S, Miller GE, Rabin BS. Psychological stress and antibody response to immunization: a critical review of the human literature. *Psychosom Med* 2001;63:7–18.
4. Kiecolt-Glaser JK, Marucha PT, Malarkey WB, et al. Slowing of wound healing by psychological stress. *Lancet* 1995;346:1194–1196.
5. Andersen BL, Farrar WB, Golden-Kreutz D, et al. Stress and immune responses after surgical treatment for regional breast cancer. *J Natl Cancer Inst* 1998;90:30–36.
6. Kusnecov AV, Grota LJ, Schmidt SG, et al. Decreased herpes simplex viral immunity and enhanced pathogenesis following stressor administration in mice. *J Neuroimmunol* 1992;38:129–138.
7. Hermann G, Tovar CA, Beck FM, et al. Restraint stress differentially affects the pathogenesis of an experimental influenza viral infection in three inbred strains of mice. *J Neuroimmunol* 1993;47:83–94.
8. Sheridan JF, Dobbs C, Jung J, et al. Stress-induced neuroendocrine modulation of viral pathogenesis and immunity. *Ann N Y Acad Sci* 1998;840:803–808.

9. Rojas I-G, Padgett DA, Sheridan JF, et al. Stress-induced susceptibility to bacterial infection during cutaneous wound healing. *Brain Behav Immun* 2002;16:74–84.

10. Padgett DA, Sheridan JF, Dorne J, et al. Social stress and the reactivation of latent herpes simplex virus type 1. *Proc Natl Acad Sci U S A* 1998;95:7231–7235.

11. Dobbs CM, Vasquez M, Glaser R, et al. Mechanisms of stress-induced modulation of viral pathogenesis and immunity. *J Neuroimmunol* 1993;48:151–160.

12. Griffin AC, Lo WD, Wolny AC, et al. Suppression of experimental autoimmune encephalomyelitis by restraint stress: sex differences. *J Neuroimmunol* 1993;44:103–116.

13. Herrmann M, Scholmerich J, Straub RH. Stress and rheumatic diseases. *Rheum Dis Clin North Am* 2000;26:737–763.

14. Rogers MP, Fozdar M. Psychoneuroimmunology of autoimmune disorders. *Adv Neuroimmunol* 1996;6:169–177.

15. Cocke R, Moynihan JA, Cohen N, et al. Exposure to conspecific alarm chemosignals alters immune responses in BALB/c mice. *Brain Behav Immun* 1993;7:36–46.

16. Brenner GJ, Moynihan JA. Stressor-induced alterations in immune response and viral clearance following infection with herpes simplex virus-type 1 in BALB/c and C57Bl/6 mice. *Brain Behav Immun* 1997;11:9–23.

17. Dhabhar FS, McEwen BS. Stress-induced enhancement of antigen-specific cell-mediated immunity. *J Immunol* 1996;156:2608–2615.

18. Persoons JH, Berkenbosch F, Schornagel K, et al. Increased specific IgE production in lungs after the induction of acute stress in rats. *J Allergy Clin Immunol* 1995;95:765–770.

19. Dhabhar FS, McEwen BS. Bidirectional effects of stress and glucocorticoid hormones on immune function: possible explanations for paradoxical observations. In: Ader R, Felten DL, Cohen N, eds. *Psychoneuroimmunology, volume 2,* 3rd ed. San Diego: Academic Press, 2001:301–338.

20. Kusnecov AW, Sved A, Rabin BS. Immunologic effects of acute versus chronic stress in animals. In: Ader R, Felten DL, Cohen N, eds. *Psychoneuroimmunology, volume 2,* 3rd ed. San Diego: Academic Press, 2001:265–278.

21. Dhabhar FS, McEwen BS. Acute stress enhances while chronic stress suppresses immune function in vivo: a potential role for leukocyte trafficking. *Brain Behav Immun* 1997;11:286–306.

22. Dhabhar FS, McEwen BS. Enhancing versus suppressive effects of stress hormones on skin immune function. *Proc Natl Acad Sci U S A* 1999;96:1059–1064.

23. Fleshner M, Nguyen KT, Cotter CS, et al. Acute stressor exposure both suppresses acquired immunity and potentiates innate immunity. *Am J Physiol* 1998;275:R870–R878.

24. Deak T, Meriwether JL, Fleshner M, et al. Evidence that brief stress may induce the acute phase response in rats. *Am J Physiol* 1997;273:R1998–R2004.

25. Mabry TR, Gold PE, McCarty R. Stress, aging, and memory involvement of peripheral catecholamines. *Ann N Y Acad Sci* 1995;771:512–522.

26. Sapolsky RM, Krey LC, McEwen BS. The adrenocortical stress-response in the aged male rat: impairment of recovery from stress. *Exp Gerontol* 1983;18:55–64.

27. Padgett DA, MacCallum RC, Sheridan JF. Stress exacerbates age-related decrements in the immune response to an experimental influenza viral infection. *J Gerontol A Biol Sci Med Sci* 1998;53:347–353.

28. Lorens SA, Hata N, Handa RJ, et al. Neurochemical, endocrine and immunological responses to stress in young and old Fischer 344 male rats. *Neurobiol Aging* 1990;11:139–150.

29. Odio M, Brodish A, Ricardo MJ Jr. Effects on immune responses by chronic stress are modulated by aging. *Brain Behav Immun* 1987;1:204–215.

30. Ader R, Cohen N. Conditioning and immunity. In: Ader R, Felten DL, Cohen N, eds. *Psychoneuroimmunology, volume 2,* 3rd ed. San Diego: Academic Press, 2001:3–34.

31. Lysle DT, Coussons ME, Watts VJ, et al. Morphine-induced alterations of immune status: dose dependency, compartment specificity and antagonism by naltrexone. *J Pharmacol Exp Ther* 1993;265:1071–1078.

32. Roudebush RE, Bryant HU. Conditioned immunosuppression of a murine delayed type hypersensitivity response: dissociation from corticosterone elevation. *Brain Behav Immun* 1991;5:308–317.

33. Russell M, Dark KA, Cummins RW, et al. Learned histamine release. *Science* 1984;225:733–734.

34. MacQueen G, Marshall J, Perdue M, et al. Pavlovian conditioning of rat mucosal mast cells to secrete rat mast cell protease II. *Science* 1989;243:83–85.

35. Gorczynski RM, Macrae S, Kennedy M. Conditioned immune response associated with allogenic skin grafts in mice. *J Immunol* 1982;129:704–709.

36. Ader R, Kelly K, Moynihan JA, et al. Conditioned enhancement of antibody production using antigen as the unconditioned stimulus. *Brain Behav Immun* 1993;7:334–343.

37. Alvarez-Borda B, Ramírez-Amaya V, Pérez-Montfort R, et al. Enhancement of antibody production by a learning paradigm. *Neurobiol Learn Mem* 1995;64:103–105.

38. Madden KS, Boehm GW, Lee SC, et al. One-trial conditioning of the antibody response to hen egg lysozyme in rats. *J Neuroimmunol* 2001;113:236–239.

39. Dutton RW, Bradley LM, Swain SL. T cell memory. *Annu Rev Immunol* 1998;16:201–223.

40. Matzinger P. Memories are made of this? *Nature* 1994;369:605–606.

41. Ramírez-Amaya V, Alvarez-Borda B, Bermúdez-Rattoni F. Differential effects of NMDA-induced lesions into the insular cortex and amygdala on the acquisition and evocation of conditioned immunosuppression. *Brain Behav Immun* 1998;12:149–160.

42. Ramírez-Amaya V, Bermúdez-Rattoni F. Conditioned enhancement of antibody production is disrupted by insular cortex and amygdala but not hippocampal lesions. *Brain Behav Immun* 1999;13:46–60.

43. Ramírez-Amaya V, Alvarez-Borda B, Ormsby CE, et al. Insular cortex lesions impair the acquisition of conditioned immunosuppression. *Brain Behav Immun* 1996;10:103–114.

44. Perez L, Lysle DT. Conditioned immunomodulation: investigations of the role of endogenous activity at μ, κ, and δ opioid receptor subtypes. *J Neuroimmunol* 1997;79:101–112.

45. Ader R, Cohen N, Grota LJ. Adrenal involvement in conditioned immunosuppression. *Int J Pharm* 1979;1:141–145.

46. Longo DL, Duffey PL, Kopp WC, et al. Conditioned immune response to interferon-γ in humans. *Clin Immunol* 1999;90:173–181.

47. Exton MS, von Hörsten S, Schult M, et al. Behaviorally conditioned immunosuppression using cyclosporine A: central nervous system reduces IL-2 production via splenic innervation. *J Neuroimmunol* 1998;88:182–191.

48. Leuken LJ, Lysle DT. Evidence for the involvement of β-adrenergic receptors in conditioned immunomodulation. *J Neuroimmunol* 1992;38:209–220.

49. Bovbjerg DH, Ader R, Cohen N. Behaviorally conditioned suppression of a graft-verus-host response. *Proc Natl Acad Sci U S A* 1982;79:583–585.

50. Ader R, Cohen N. Behaviorally conditioned immunosuppression and murine systemic lupus erythematosus. *Science* 1982;215:1534–1536.

51. Klosterhalfen S, Klosterhalfen W. Conditioned cyclosporine effects but not conditioned taste aversion in immunized rats. *Behav Neurosci* 1990;104:716–724.
52. Lysle DT, Luecken LJ, Maslonek KA. Suppression of the development of adjuvant arthritis by a conditioned aversive stimulus. *Brain Behav Immun* 1992;6:64–73.
53. Buske-Kirschbaum A, Kirschbaum C, Stierle H, et al. Conditioned increase of natural killer cell activity (NKCA) in humans. *Psychosom Med* 1992;54:123–132.
54. Kirschbaum C, Jabaij L, Buske-Kirschbaum A, et al. Conditioning of drug-induced immunomodulation in human volunteers: a European collaborative study. *Br J Clin Psychol* 1992;31:459–472.
55. Felten DL, Cohen N, Ader R, et al. Central neural circuits involved in neural-immune interactions. In: Ader R, Felten DL, Cohen N, eds. *Psychoneuroimmunology*, 2nd ed. San Diego: Academic Press, 1991:3–25.
56. Sundar SK, Cierpial MA, Kamaraju LS, et al. Human immunodeficiency virus glycoprotein (gp120) infused into rat brain induces interleukin 1 to elevate pituitary-adrenal activity and decrease peripheral cellular immune responses. *Proc Natl Acad Sci U S A* 1991;88:11246–11250.
57. Brown R, Li Z, Vriend C, et al. Suppression of splenic macrophage interleukin-1 secretion following intracerebroventricular injection of interleukin-1β: evidence for pituitary-adrenal and sympathetic control. *Cell Immunol* 1991;132:84–93.
58. Irwin M, Hauger RL, Brown M, et al. CRF activates autonomic nervous system and reduces natural killer cytotoxicity. *Am J Physiol* 1988;255:R744–R747.
59. Take S, Mori T, Katafuchi T, et al. Central interferon-α inhibits natural killer cytotoxicity through sympathetic innervation. *Am J Physiol* 1993;265:R453–R459.
60. Nagy E, Berczi I, Friesen HG. Regulation of immunity in rats by lactogenic and growth hormones. *Acta Endocrinol* 1983;102:351–357.
61. Pierpaoli W, Baroni C, Fabris N, et al. Hormones and immunological capacity. II. Reconstitution of antibody production in hormonally deficient mice by somatotropic hormone, thyrotropic hormone and thyroxin. *Immunology* 1969;16:217–230.
62. Fabris N, Pierpaoli W, Sorkin E. Hormones and the immunological capacity. IV. Restorative effects of developmental hormones or of lymphocytes on the immunodeficiency syndrome of the dwarf mouse. *Clin Exp Immunol* 1971;9:227–240.
63. Edwards CK III, Yunger LM, Lorence RM, et al. The pituitary gland is required for protection against lethal effects of *Salmonella typhimurium*. *Proc Natl Acad Sci U S A* 1991;88:2274–2277.
64. Turnbull AV, Rivier CL. Regulation of the hypothalamic-pituitary-adrenal axis by cytokines: actions and mechanisms of action. *Physiol Rev* 1999;79:1–71.
65. Berkenbosch F, van Oers J, del Rey A, et al. Corticotropin-releasing factor-producing neurons in the rat activated by interleukin-1. *Science* 1987;238:524–526.
66. Sapolsky R, Rivier C, Yamamoto G, et al. Interleukin-1 stimulates the secretion of hypothalamic corticotropin-releasing factor. *Science* 1987;238:522–524.
67. Casadevall M, Saperas E, Panés J, et al. Mechanisms underlying the anti-inflammatory actions of central corticotropin-releasing factor. *Am J Physiol* 1999;276:G1016–G1026.
68. Turnbull AV, Smith GW, Lee S, et al. CRF type I receptor deficient mice exhibit a pronounced pituitary-adrenal response to local inflammation. *Endocrinology* 1999;140:1013–1017.
69. Irwin M, Hauger R, Brown M. Central corticotropin-releasing hormone activates the sympathetic nervous system and reduces immune function: increased responsivity of the aged rat. *Endocrinology* 1992;131:1047–1053.
70. Stephanou A, Jessop DS, Knight RA, et al. Corticotrophin-releasing factor-like immunoreactivity and mRNA in human leukocytes. *Brain Behav Immun* 1990;4:67–73.
71. Aird F, Clevenger CV, Prystowsky MB, et al. Corticotropin-releasing factor mRNA in rat thymus and spleen. *Proc Natl Acad Sci U S A* 1993;90:7104–7108.
72. Brouxhon SM, Prasad AV, Joseph SA, et al. Localization of corticotropin-releasing factor in primary and secondary lymphoid organs of the rat. *Brain Behav Immun* 1998;12:107–122.
73. McGillis JP, Park A, Rubin-Fletter P, et al. Stimulation of rat B-lymphocyte proliferation by corticotropin-releasing factor. *J Neurosci Res* 1989;23:346–352.
74. Paez Pereda M, Sauer J, Perez Castro C, et al. Corticotropin-releasing hormone differentially modulates the interleukin-1 system according to the level of monocyte activation by endotoxin. *Endocrinology* 1995;136:5504–5510.
75. Leu SJ, Singh VK. Suppression of in vitro antibody production by corticotropin-releasing factor neurohormone. *J Neuroimmunol* 1993;45:23–29.
76. Webster EL, Tracey DE, Jutila MA, et al. Corticotropin-releasing factor receptors in mouse spleen: identification of receptor-bearing cells as resident macrophages. *Endocrinology* 1990;127:440–452.
77. Radulovic M, Dautzenberg FM, Sydow S, et al. Corticotropin-releasing factor receptor 1 in mouse spleen: expression after immune stimulation and identification of receptor-bearing cells. *J Immunol* 1999;162:3013–3021.
78. Kavelaars A, Ballieux RE, Heijnen CJ. The role of IL-1 in the corticotropin-releasing factor and arginine-vasopressin-induced secretion of immunoreactive β-endorphin by human peripheral blood mononuclear cells. *J Immunol* 1989;142:2338–2342.
79. Cabot PJ, Carter L, Gaiddon C, et al. Immune cell-derived β-endorphin production, release, and control of inflammatory pain in rats. *J Clin Invest* 1997;100:142–148.
80. Schäfer M, Carter L, Stein C. Interleukin-1β and corticotropin-releasing factor inhibit pain by releasing opioids from immune cells in inflamed tissue. *Proc Natl Acad Sci U S A* 1994;91:4219–4223.
81. Crofford LJ, Sano H, Karalis K, et al. Local secretion of corticotropin-releasing hormone in the joints of Lewis rats with inflammatory arthritis. *J Clin Invest* 1992;90:2555–2564.
82. Crofford LJ, Sano H, Karalis K, et al. Corticotropin-releasing hormone in synovial fluids and tissues of patients with rheumatoid arthritis and osteoarthritis. *J Immunol* 1993;151:1587–1596.
83. Lyons PD, Blalock JE. Pro-opiomelanocortin gene expression and protein processing in rat mononuclear leukocytes. *J Neuroimmunol* 1997;78:47–56.
84. Hashemi FB, Hughes TK, Smith EM. Human immunodeficiency virus induction of corticotropin in lymphoid cells. *J Clin Endocrinol Metab* 1998;83:4363–4381.
85. Lyons PD, Blalock JE. The kinetics of ACTH expression in rat leukocyte subpopulations. *J Neuroimmunol* 1995;63:103–112.
86. Johnson HM, Smith EM, Torres BA, et al. Regulation of the in vitro antibody response by neuroendocrine hormones. *Proc Natl Acad Sci U S A* 1982;79:4171–4174.
87. Wermerskirchen AS, LaTocha DH, Clarke BL. Adrenocorticotrophic hormone controls concanavalin A activation of rat lymphocytes by modulating IL-2 production. *Life Sci* 2000;67:2177–2187.
88. Gatti G, Masera RG, Pallavicini L, et al. Interplay in vitro between ACTH, beta-endorphin, and glucocorticoids in the modulation of spontaneous and lymphokine-inducible human natural killer (NK) cell activity. *Brain Behav Immun* 1993;7:16–28.

89. Clarke BL, Bost KL. Differential expression of functional adrenocorticotropic hormone receptors by subpopulations of lymphocytes. *J Immunol* 1989;143:464–469.

90. Akbulut S, Byersdorfer CA, Larsen CP, et al. Expression of the melanocortin 5 receptor on rat lymphocytes. *Biochem Biophys Res Commun* 2001;281:1086–1092.

91. Smith EM, Hughes TK Jr, Hashemi F, et al. Immunosuppressive effects of corticotropin and melanotropin and their possible significance in human immunodeficiency virus infection. *Proc Natl Acad Sci U S A* 1992;89:782–786.

92. Adachi A, Nakano T, Vliagoftis H, et al. Receptor-mediated modulation of murine mast cell function by alpha-melanocyte stimulating hormone. *J Immunol* 1999;163:3363–3368.

93. Grabbe S, Bhardwaj RS, Mahnke K, et al. Alpha-melanocyte-stimulating hormone induces hapten-specific tolerance in mice. *J Immunol* 1996;156:473–478.

94. Sundar SK, Becker KJ, Cierpial MA, et al. Intracerebroventricular infusion of interleukin 1 rapidly decreases peripheral cellular immune responses. *Proc Natl Acad Sci U S A* 1989;86:6398–6402.

95. Rook GA. Glucocorticoids and immune function. *Ballieres Best Pract Res Clin Endocrinol Metab* 1999;13:567–581.

96. Wilckens T, De Rijk R. Glucocorticoids and immune function: unknown dimensions and new frontiers. *Immunol Today* 1997;18:418–423.

97. Ruzek MC, Pearce BD, Miller AH, et al. Endogenous glucocorticoids protect against cytokine-mediated lethality during viral infection. *J Immunol* 1999;162:3527–3533.

98. Gelin JL, Moldawer LL, Iresjö BM, et al. The role of the adrenals in the acute phase response to interleukin-1 and tumor necrosis factor-alpha. *J Surg Res* 1993;54:70–78.

99. Miller AH, Spencer RL, Pearce BD, et al. Glucocorticoid receptors are differentially expressed in the cells and tissues of the immune system. *Cell Immunol* 1998;186:45–54.

100. Homo-Delarche F. Glucocorticoid receptors and steroid sensitivity in normal and neoplastic human lymphoid tissues: a review. *Cancer Res* 1984;44:431–437.

101. Bianchi M, Meng C, Ivashkiv LB. Inhibition of IL-2-induced Jak-STAT signaling by glucocorticoids. *Proc Natl Acad Sci U S A* 2000;97:9573–9578.

102. Wiegers GJ, Stec IEM, Klinkert WEF, et al. Glucocorticoids regulate TCR-induced elevation of CD4: functional implications. *J Immunol* 2000;164:6213–6220.

103. Fleshner M, Deak T, Nguyen KT, et al. Endogenous glucocorticoids play a positive regulatory role in the anti-keyhole limpet hemocyanin in vivo antibody response. *J Immunol* 2001;166:3813–3819.

104. Jabara HH, Brodeur SR, Geha RS. Glucocorticoids up-regulate CD40 ligand expression and induce CD40L-dependent immunoglobulin isotype switching. *J Clin Invest* 2001;107:371–378.

105. Daynes RA, Araneo BA. Contrasting effects of glucocorticoids on the capacity of T cells to produce the growth factors interleukin 2 and interleukin 4. *Eur J Immunol* 1989;19:2319–2325.

106. Moynihan JA, Callahan TA, Kelley SP, et al. Adrenal hormone modulation of type 1 and type 2 cytokine production by spleen cells: dexamethasone and dehydroepiandrosterone suppress interleukin 2, interleukin-4, and interferon-γ production in vitro. *Cell Immunol* 1998;184:58–64.

107. Richards DF, Fernandez M, Caulfield J, et al. Glucocorticoids drive human CD8+ T cell differentiation towards a phenotype with high IL-10 and reduced IL-4, IL-5 and IL-13 production. *Eur J Immunol* 2000;30:2344–2354.

108. Tarcic N, Ovadia H, Weiss DW, et al. Restraint stress-induced thymic involution and cell apoptosis are dependent on endogenous glucocorticoids. *J Neuroimmunol* 1998;82:40–46.

109. Gruber J, Sgone R, Hu YH, et al. Thymocyte apoptosis induced by elevated endogenous corticosterone levels. *Eur J Immunol* 1994;24:1115–1121.

110. Iwata M, Hanaoka S, Sato K. Rescue of thymocytes and T cell hybridomas from glucocorticoid-induced apoptosis by stimulation via the T cell receptor/CD3 complex: a possible in vitro model for positive selection of the T cell repertoire. *Eur J Immunol* 1991;21:643–648.

111. King LB, Vacchio MS, Dixon K, et al. A targeted glucocorticoid receptor anti-sense transgene increases thymocyte apoptosis and alters thymocyte development. *Immunity* 1995;3:647–656.

112. Lu FW, Yasutomo K, Goodman GB, et al. Thymocyte resistance to glucocorticoids leads to antigen specific unresponsiveness due to "holes" in the T cell repertoire. *Immunity* 2000;12:183–192.

113. Tolosa E, King LB, Ashwell JD. Thymocyte glucocorticoid resistance alters positive selection and inhibits autoimmunity and lymphoproliferative disease in MRL-*lpr/lpr* mice. *Immunity* 1998;8:67–76.

114. Vacchio MS, Papadopoulos V, Ashwell JD. Steroid production in the thymus: implications for thymocyte selection. *J Exp Med* 1994;179:1835–1846.

115. Lechner O, Wiegers GJ, Oliveira-Dos-Santos AJ, et al. Glucocorticoid production in the murine thymus. *Eur J Immunol* 2000;30:337–346.

116. Xue Y, Murdjeva M, Okret S, et al. Inhibition of I-Ad-, but not Kb-restricted peptide-induced thymic apoptosis by glucocorticoid receptor antagonist RU486 in T cell receptor transgenic mice. *Eur J Immunol* 1996;26:428–434.

117. Purton JF, Boyd RL, Cole TJ, et al. Intrathymic T cell development and selection proceeds normally in the absence of glucocorticoid receptor signaling. *Immunity* 2000;13:179–186.

118. Sternberg EM, Hill JM, Chrousos GP, et al. Inflammatory mediator-induced hypothalamic-pituitary-adrenal axis activation is defective in streptococcal cell wall arthritis-susceptible Lewis rats. *Proc Natl Acad Sci U S A* 1989;86:2374–2378.

119. Mason D. Genetic variation in the stress response: susceptibility to experimental allergic encephalomyelitis and implications for human inflammatory disease. *Immunol Today* 1991;12:57–60.

120. Chikanza IC, Petrou P, Kingsley G, et al. Defective hypothalamic response to immune and inflammatory stimuli in patients with rheumatoid arthritis. *Arthritis Rheum* 1992;35:1281–1288.

121. Correale J, Gilmore W, Li S, et al. Resistance to glucocorticoid-induced apoptosis in PLP peptide-specific T cell clones from patients with progressive MS. *J Neuroimmunol* 2000;109:197–210.

122. Apte RN, Oppenheim JJ, Durum SK. β-Endorphin regulates interleukin 1 production and release by murine bone marrow macrophages. *Int Immunol* 1989;1:465–470.

123. Sharp BM, Tsukayama DT, Gekker G, et al. β-Endorphin stimulates human polymorphonuclear leukocyte superoxide production via a stereoselective opiate receptor. *J Pharmacol Exp Ther* 1987;242:579–582.

124. Van Den Bergh P, Rozing J, Nagelkerken L. Two opposing modes of action of β-endorphin on lymphocyte function. *Immunology* 1991;72:537–543.

125. Hemmick LM, Bidlack JM. β-Endorphin stimulates rat T lymphocyte proliferation. *J Neuroimmunol* 1990;29:239–248.

126. Zurawski G, Benedik M, Kamb BJ, et al. Activation of mouse T-helper cells induces abundant preproenkephalin mRNA synthesis. *Science* 1986;232:772–775.

127. Behar OZ, Ovadia H, Polakiewicz RD, et al. Regulation of

proenkephalin A messenger ribonucleic acid levels in normal B lymphocytes: specific inhibition by glucocorticoid hormones and superinduction by cycloheximide. *Endocrinology* 1991;129:649–655.

128. Heagy W, Laurance M, Cohen E, et al. Neurohormones regulate T cell function. *J Exp Med* 1990;171:1625–1633.

129. Roy S, Loh HH. Effects of opioids on the immune system. *Neurochem Res* 1996;21:1375–1386.

130. Eisenstein TK, Hilburger ME. Opioid modulation of immune responses: effects on phagocyte and lymphoid cell populations. *J Neuroimmunol* 1998;83:36–44.

131. Fecho K, Maslonek KA, Dykstra LA, et al. Assessment of the involvement of central nervous system and peripheral opioid receptors in the immunomodulatory effects of acute morphine treatment in rats. *J Pharmacol Exp Ther* 1996;276:626–636.

132. Hall DM, Suo J-L, Weber RJ. Opioid mediated effects on the immune system: sympathetic nervous system involvement. *J Neuroimmunol* 1998;83:29–35.

133. Suzuki S, Miyagi T, Chuang TK, et al. Morphine upregulates mu opioid receptors of human and monkey lymphocytes. *Biochem Biophys Res Commun* 2000;279:621–628.

134. Chuang LF, Killam KF Jr, Chuang RY. Induction and activation of mitogen-activated protein kinases of human lymphocytes as one of the signaling pathways of the immunomodulatory effects of morphine sulfate. *J Biol Chem* 1997;272:26815–26817.

135. Miyagi T, Chuang LF, Doi RH, et al. Morphine induces gene expression of CCR5 in human CEMx174 lymphocytes. *J Biol Chem* 2000;275:31305–31310.

136. Moynihan JA, Karp JD, Cohen N, et al. Immune deviation following stress odor exposure: role of endogenous opioids. *J Neuroimmunol* 2000;102:145–153.

137. Kooijman R, Hooghe-Peters EL, Hooghe R. Prolactin, growth hormone, and insulin-like growth factor-1 in the immune system. *Adv Immunol* 1996;63:377–454.

138. Saxena QB, Saxena RK, Adler WH. Regulation of natural killer activity in vivo. III. Effect of hypophysectomy and growth hormone treatment on the natural killer activity of the mouse spleen cell population. *Int Arch Allergy Appl Immunol* 1982;67:169–174.

139. Binder G, Revskoy S, Gupta D. In vivo growth hormone gene expression in neonatal rat thymus and bone marrow. *J Endocrinol* 1994;140:137–143.

140. Weigent DA, Blalock JE, LeBoeuf RD. An antisense oligodeoxynucleotide to growth hormone messenger ribonucleic acid inhibits lymphocyte proliferation. *Endocrinology* 1991;128:2053–2057.

141. Arkins S, Rebeiz N, Biragyn A, et al. Murine macrophages express abundant insulin-like growth factor-I class I Eα and Eβ transcripts. *Endocrinology* 1993;133:2334–2343.

142. Fournier T, Riches DW, Winston BW, et al. Divergence in macrophage insulin-like growth factor-I (IGF-I) synthesis induced by TNF-alpha and prostaglandin E2. *J Immunol* 1995;155:2123–2133.

143. Gagnerault M-C, Postel-Vinay M-C, Dardenne M. Expression of growth hormone receptors in murine lymphoid cells analyzed by flow cytofluorometry. *Endocrinology* 1996;137:1719–1726.

144. Walsh PT, O'Connor R. The insulin-like growth factor-I receptor is regulated by CD28 and protects activated T cells from apoptosis. *Eur J Immunol* 2000;30:1010–1018.

145. Kooijman RK, Scholtens LE, Rijkers GT, et al. Differential expression of type I insulin-like growth factor receptors in different stages of human T cells. *Eur J Immunol* 1995;25:931–935.

146. Murphy WJ, Durum SK, Longo DL. Role of neuroendocrine hormones in murine T cell development growth hormone exerts thymopoietic effects in vivo. *J Immunol* 1992;49:3851–3857.

147. Montecino-Rodrquez E, Clark R, Johnson A, et al. Defective B cell development in Snell dwarf (dw/dw) mice can be corrected by thyroxine treatment. *J Immunol* 1996;157:3334–3340.

148. Montecino-Rodriguez E, Clark R, Dorshkind K. Effects of insulin-like growth factor administration and bone marrow transplantation on thymopoiesis in aged mice. *Endocrinology* 1998;139:4120–4126.

149. Murphy WJ, Durum SK, Longo DL. Differential effects of growth hormone and prolactin on murine T cell development and function. *J Exp Med* 1993;178:231–236.

150. Montecino-Rodriguez E, Clark RG, Powell-Braxton L, et al. Bone marrow B cell development is impaired in mice with defects of the pituitary/thyroid axis. *J Immunol* 1997;159:2712–2719.

151. Robbins K, McCabe S, Scheiner T, et al. Immunological effects of insulin-like growth factor-I—enhancement of immunoglobulin synthesis. *Clin Exp Immunol* 1994;95:337–342.

152. Postel-Vinay M-C, de Mello Coelho V, Gagnerault M-C, et al. Growth hormone stimulates the proliferation of activated mouse T lymphocytes. *Endocrinology* 1997;138:1816–1820.

153. Kooijman R, Rijkers GT, Zegers BJ. IGF-1 potentiates interleukin-2 production in human peripheral T cells. *J Endocrinol* 1996;149:351–356.

154. Yoshida A, Ishioka C, Kimata H, et al. Recombinant human growth hormone stimulates B cell immunoglobulin synthesis and proliferation in serum-free medium. *Acta Endocrinol* 1992;126:524–529.

155. Edwards CK, Ghiasuddin SM, Schepper JM, et al. A newly defined property of somatotropin: priming of macrophages for production of superoxide anion. *Science* 1988;239:769–771.

156. Warwick-Davies J, Lowrie DB, Cole PJ. Growth hormone is a human macrophage activating factor. Priming of human monocytes for enhanced release of H2O2. *J Immunol* 1995;154:1909–1918.

157. Haeffner A, Thieblemont N, Déas O, et al. Inhibitory effect of growth hormone on TNF-α secretion and nuclear factor-β translocation in lipopolysaccharide-stimulated human monocytes. *J Immunol* 1997;158:1310–1314.

158. Dorshkind K, Horseman ND. The roles of prolactin, growth hormone, insulin-like growth factor-I, and thyroid hormones in lymphocyte development and function: insights from genetic models of hormone and hormone receptor deficiency. *Endocr Rev* 2000;21:292–312.

159. Cross RJ, Bryson JS, Roszman TL. Immunologic disparity in the hypopituitary dwarf mouse. *J Immunol* 1992;148:1347–1352.

160. Petersen BH, Rapaport R, Henry DP, et al. Effect of treatment with biosynthetic human growth hormone (GH) on peripheral blood lymphocyte populations in growth hormone-deficient children. *J Clin Endocrinol Metab* 1990;70:1756–1760.

161. Kelley KW, Brief S, Westly HJ, et al. GH3 pituitary adenoma cells can reverse thymic ageing in rats. *Proc Natl Acad Sci U S A* 1986;83:5663–5667.

162. Fu Y-K, Arkins S, Li YM, et al. Reduction in superoxide anion secretion and bactericidal activity of neutrophils from aged rats: reversal by the combination of gamma interferon and growth hormone. *Infect Immun* 1994;62:1–8.

163. Jardieu P, Clark R, Mortensen D, et al. In vivo administration of insulin-like growth factor-I stimulates primary B lymphopoiesis and enhances lymphocyte recovery after bone marrow transplantation. *J Immunol* 1994;152:4320–4327.

164. Hinton PS, Peterson CA, Dahly EM, et al. IGF-I alters lymphocyte survival and regeneration in thymus and spleen after dexamethasone treatment. *Am J Physiol* 1998;274:R912–R920.

165. Yu-Lee L-Y. Molecular actions of prolactin in the immune system. *Proc Soc Exp Biol Med* 1997;215:35–52.

166. Sabharwal P, Glaser R, Lafuse W, et al. Prolactin synthesized and secreted by human peripheral blood mononuclear cells: an

autocrine growth factor for lymphoproliferation. *Proc Natl Acad Sci U S A* 1992;89:7713–7716.

167. Pellegrini I, Lebrun J-J, Ali S, et al. Expression of prolactin and its receptor in human lymphoid cells. *Mol Endocrinol* 1992;6:1023–1031.

168. Bernton EW, Meltzer MS, Holaday JW. Suppression of macrophage activation and T-lymphocyte function in hypoprolactinemic mice. *Science* 1988;239:401–404.

169. Meli R, Raso GM, Bentivoglio C, et al. Recombinant human prolactin induces protection against *Salmonella typhimurium* infection in the mouse: role of nitric oxide. *Immunopharmacology* 1996;34:1–7.

170. Matera L, Contarini M, Bellone G, et al. Up-modulation of interferon-γ mediates the enhancement of spontaneous cytotoxicity in prolactin-activated natural killer cells. *Immunology* 1999;98:386–392.

171. Lahat N, Miller A, Shtiller R, et al. Differential effects of prolactin upon activation and differentiation of human B lymphocytes. *J Neuroimmunol* 1993;47:35–40.

172. Matera L, Cesano A, Bellone G, et al. Modulatory effect of prolactin on the resting and mitogen-induced activity of T, B, and NK lymphocytes. *Brain Behav Immun* 1992;6:409–417.

173. Viselli SM, Stanek EM, Mukherjee P, et al. Prolactin-induced mitogenesis of lymphocytes from ovariectomized rats. *Endocrinology* 1991;129:983–990.

174. Montgomery DW, Krumenacker JS, Buckley AR. Prolactin stimulates phosphorylation of the human T-cell antigen receptor complex and ZAP-70 tyrosine kinase: a potential mechanism for its immunomodulation. *Endocrinology* 1998;139:811–814.

175. Dusanter-Fourt I, Muller O, Ziemiecki A, et al. Identification of JAK protein tyrosine kinases as signaling molecules for prolactin. Functional analysis of prolactin receptor and prolactin-erythropoietin receptor chimera expressed in lymphoid cells. *EMBO J* 1994;13:2583–2591.

176. Hartmann DP, Holaday JW, Bernton EW. Inhibition of lymphocyte proliferation by antibodies to prolactin. *FASEB J* 1989;3:2194–2202.

177. Horseman ND, Zhao W, Montecino-Rodriguez E, et al. Defective mammopoiesis, but normal hematopoiesis, in mice with a targeted disruption of the prolactin gene. *EMBO J* 1997;16:6926–6935.

178. Bouchard B, Ormandy CJ, Di Santo JP, et al. Immune system development and function in prolactin receptor-deficient mice. *J Immunol* 1999;163:576–582.

179. Gagnerault M-C, Touraine P, Savino W, et al. Expression of prolactin receptors in murine lymphoid cells in normal and autoimmune situations. *J Immunol* 1993;150:5673–5681.

180. Nagafuchi H, Suzuki N, Kaneko A, et al. Prolactin locally produced by synovium infiltrating T lymphocytes induces excessive synovial cell functions in patients with rheumatoid arthritis. *J Rheumatol* 1999;26:1890–1900.

181. Sandi C, Cambronero JC, Borrell J, et al. Mutually antagonistic effects of corticosterone and prolactin on rat lymphocyte proliferation. *Neuroendocrinology* 1992;56:574–581.

182. Dobashi H, Sato M, Tanaka T, et al. Growth hormone restores glucocorticoid-induced T cell suppression. *FASEB J* 2001;15:1861–1863.

183. Arpin CA, Pihlgren M, Fraichard A, et al. Effects of T3Rα1 and T3Rα2 gene deletion on T and B lymphocyte development. *J Immunol* 2000;164:152–160.

184. Provinciali M, Muzzioli M, DiStefano G, et al. Recovery of spleen cell natural killer activity by thyroid hormone treatment in old mice. *Nat Immun Cell Growth Regul* 1991;10:226–236.

185. Fabris N, Muzzioli M, Mocchegiani E. Recovery of age-dependent immunological deterioration in Balb/c mice by short-term treatment with L-thyroxine. *Mech Ageing Dev* 1982;18:327–338.

186. Corrales JJ, Orfao A, Lopez A, et al. CD5+ B cells in Graves' disease: correlation with disease activity. *Horm Metab Res* 1996;28:280–285.

187. Lahita RG. Sex hormones and the immune system—part 1. Human data. *Baillieres Clin Rheumatol* 1990;4:1–12.

188. Ansar Ahmed S, Talal N. Sex hormones and the immune system—part 2. Animal data. *Baillieres Clin Rheumatol* 1990;4:13–31.

189. Grossman CJ. Interactions between the gonadal steroids and the immune system. *Science* 1985;227:257–261.

190. Hunt JS, Robertson SA. Uterine macrophages and environmental programming for pregnancy success. *J Reprod Immunol* 1996;32:1–25.

191. Medina KL, Smithson G, Kincade PW. Suppression of B lymphopoiesis during normal pregnancy. *J Exp Med* 1993;178:1507–1515.

192. Wegmann TG, Lin H, Guilbert L, et al. Bidirectional cytokine interactions in the maternal-fetal relationship: is successful pregnancy a TH2 phenomenon? *Immunol Today* 1993;14:353–356.

193. Grossman CJ, Roselle GA, Mendenhall CL. Sex steroid regulation of autoimmunity. *J Steroid Biochem Mol Biol* 1991;40:649–659.

194. Pearce P, Khalid BAK, Funder JW. Androgens and the thymus. *Endocrinology* 1981;109:1073–1077.

195. Danel L, Souweine G, Monier JC, et al. Specific estrogen binding sites in human lymphoid cells and thymic cells. *J Steroid Biochem* 1983;18:559–563.

196. Stimson WH. Oestrogen and human T lymphocytes: presence of specific receptors in the T-suppressor/cytotoxic subset. *Scand J Immunol* 1988;28:345–350.

197. Benten WP, Lieberherr M, Giese G, et al. Estradiol binding to cell surface raises cytosolic free calcium in T cells. *FEBS Lett* 1998;422:349–353.

198. Benten WP, Lieberherr M, Giese G, et al. Functional testosterone receptors in plasma membranes of T cells. *FASEB J* 1999;13:123–133.

199. Ellis TM, Moser MT, Le PT, et al. Alterations in peripheral B cells and B cell progenitors following androgen ablation in mice. *Int Immunol* 2001;13:553–558.

200. Smithson G, Beamer WG, Shultz KL, et al. Increased B lymphopoiesis in genetically sex steroid-deficient hypogonadal (hpg) mice. *J Exp Med* 1994;180:717–720.

201. Smithson G, Couse JF, Lubahn DB, et al. The role of estrogen receptors and androgen receptors in sex steroid regulation of B lymphopoiesis. *J Immunol* 1998;161:27–34.

202. Ansar Ahmed S, Penhale WJ, Talal N. Sex hormones, immune responses, and autoimmune diseases. *Am J Pathol* 1985;121:531–551.

203. Cutolo M. Sex hormone adjuvant therapy in rheumatoid arthritis. *Rheum Dis Clin North Am* 2000;26:881–895.

204. Okuyama R, Abo T, Seki S, et al. Estrogen administration activates extrathymic T cell differentiation in the liver. *J Exp Med* 1992;175:661–669.

205. Ansar Ahmed S, Dauphinee MJ, Montoya AI, et al. Estrogen induces normal murine CD5+ B cells to produce autoantibodies. *J Immunol* 1989;142:2647–2653.

206. Kanik KS, Wilder RL. Hormonal alterations in rheumatoid arthritis, including the effects of pregnancy. *Rheum Dis Clin North Am* 2000;26:805–823.

207. Bebo BF Jr, Fyfe-Johnson A, Adlard K, et al. Low dose estrogen therapy ameliorates experimental autoimmune encephalomyelitis in two different inbred mouse strains. *J Immunol* 2001;166:2080–2089.

208. Offner H, Adlard K, Zamora A, et al. Estrogen potentiates treatment with T-cell receptor protein of female mice with experimental encephalomyelitis. *J Clin Invest* 2000;105:1465–1472.

209. Kim S, Liva SM, Dalal MA, et al. Estriol ameliorates autoimmune demyelinating disease: implications for multiple sclerosis. *Neurology* 1999;52:1230–1238.

210. Fox HS. Androgen treatment prevents diabetes in nonobese diabetic mice. *J Exp Med* 1992;175:1409–1412.

211. Bebo BF Jr, Schuster JC, Vandenbark AA, et al. Androgens alter the cytokine profile and reduce encephalitogenicity of myelin-reactive T cells. *J Immunol* 1999;162:35–40.

212. Cunnick JE, Lysle DT, Kucinski BJ, et al. Evidence that shock-induced immune suppression is mediated by adrenal hormones and peripheral β-adrenergic receptors. *Pharmacol Biochem Behav* 1990;36:645–651.

213. Bergquist J, Tarkowski A, Ekman R, et al. Discovery of endogenous catecholamine in lymphocytes and evidence for catecholamine regulation of lymphocyte function via an autocrine loop. *Proc Natl Acad Sci U S A* 1994;91:12912–12916.

214. Marino F, Cosentino M, Bombelli R, et al. Endogenous catecholamine synthesis, metabolism, storage, and uptake in human peripheral blood mononuclear cells. *Exp Hematol* 1999;27:489–495.

215. Musso NR, Brenci S, Indiveri F, et al. L-tyrosine and nicotine induce synthesis of L-Dopa and norepinephrine in human lymphocytes. *J Neuroimmunol* 1997;74:117–120.

216. Cosentino M, Marino F, Bombelli R, et al. Endogenous catecholamine synthesis, metabolism, storage and uptake in human neutrophils. *Life Sci* 1999;64:975–981.

217. Spengler RN, Chensue SW, Giacherio DA, et al. Endogenous norepinephrine regulates tumor necrosis factor-α production from macrophages in vitro. *J Immunol* 1994;152:3024–3031.

218. Miller LE, Hans-Peter J, Schölmerich J, et al. The loss of sympathetic nerve fibers in the synovial tissue of patients with rheumatoid arthritis is accompanied by increased norepinephrine release from synovial macrophages. *FASEB J* 2000;14:2097–2107.

219. Bellinger DL, Lorton D, Lubahn C, et al. Innervation of lymphoid organs—Association of nerves with cells of the immune system and their implications in disease. In: Ader R, Felten DL, Cohen N, eds. *Psychoneuroimmunology, volume 1,* 3rd ed. San Diego: Academic Press, 2001:55–111.

220. Felten SY, Olschowka JA. Noradrenergic sympathetic innervation of the spleen: II. Tyrosine hydroxylase (TH)-positive nerve terminals form synaptic-like contacts on lymphocytes in the splenic white pulp. *J Neurosci Res* 1987;18:37–48.

221. Romano TA, Felten SY, Felten DL, et al. Neuropeptide-Y innervation of the rat spleen: another potential immunomodulatory neuropeptide. *Brain Behav Immun* 1991;5:116–131.

222. Haskó G, Szabó C. Regulation of cytokine and chemokine production by transmitters and co-transmitters of the autonomic nervous system. *Biochem Pharmacol* 1998;56:1079–1087.

223. Straub RH, Schaller T, Miller LE, et al. Neuropeptide Y cotransmission with norepinephrine in the sympathetic nerve-macrophage interplay. *J Neurochem* 2000;75:2464–2471.

224. Lorton D, Hewitt D, Bellinger DL, et al. Noradrenergic reinnervation of the rat spleen following chemical sympathectomy with 6-hydroxydopamine: pattern and time course of reinnervation. *Brain Behav Immun* 1990;4:198–222.

225. Bellinger DL, Ackerman KD, Felten SY, et al. A longitudinal study of age-related loss of noradrenergic nerves and lymphoid cells in the rat spleen. *Exp Neurol* 1992;116:295–311.

226. Madden KS, Felten SY, Felten DL, et al. Sympathetic nervous system-immune system interactions in young and old Fischer 344 rats. *Ann N Y Acad Sci* 1995;771:523–534.

227. Madden KS, Stevens SY, Felten DL, et al. Alterations in T lymphocyte activity following chemical sympathectomy in young and old Fischer 344 rats. *J Neuroimmunol* 2000;103:131–145.

228. ThyagaRajan S, Felten SY, Felten DL. Restoration of sympathetic noradrenergic nerve fibers in the spleen by low doses of L-deprenyl treatment in young sympathectomized and old Fischer 344 rats. *J Neuroimmunol* 1998;81:144–157.

229. ThyagaRajan S, Madden KS, Kalvass JC, et al. L-deprenyl-induced increase in IL-2 and NK cell activity accompanies restoration of noradrenergic nerve fibers in the spleens of old F344 rats. *J Neuroimmunol* 1998;92:9–21.

230. ThyagaRajan S, Madden KS, Stevens SY, et al. Anti-tumor effect of L-deprenyl is associated with enhanced central and peripheral neurotransmission and immune reactivity in rats with carcinogen-induced mammary tumors. *J Neuroimmunol* 2000;109:95–104.

231. Spengler RN, Allen RM, Remick DG, et al. Stimulation of alpha-adrenergic receptor augments the production of macrophage-derived tumor necrosis factor. *J Immunol* 1990;145:1430–1434.

232. Heijnen CJ, Rouppe van der Voort C, Wulffraat N, et al. Functional α1-adrenergic receptors on leukocytes of patients with polyarticular juvenile rheumatoid arthritis. *J Neuroimmunol* 1996;71:223–226.

233. Landmann R. Beta-adrenergic receptors in human leukocyte subpopulations. *Eur J Clin Invest* 1992;22:30–36.

234. Sanders VM, Baker RA, Ramer-Quinn DS, et al. Differential expression of the β2-adrenergic receptor by Th1 and Th2 clones. *J Immunol* 1997;158:4200–4210.

235. Baerwald C, Graefe C, von Wichert P, et al. Decreased density of β-adrenergic receptors on peripheral blood mononuclear cells in patients with rheumatoid arthritis. *J Rheumatol* 1992;19:204–210.

236. Karaszewski JW, Reder AT, Anlar B, et al. Increased high affinity beta-adrenergic receptor densities and cyclic AMP responses of CD8 cells in multiple sclerosis. *J Neuroimmunol* 1993;43:1–8.

237. Zoukos Y, Kidd D, Woodroofe MN, et al. Increased expression of high affinity IL-2 receptors and β-adrenoceptors on peripheral blood mononuclear cells is associated with clinical and MRI activity in multiple sclerosis. *Brain* 1994;117:307–315.

238. Kurz B, Feindt J, von Gaudecker B, et al. β-Adrenoceptor-mediated effects in rat cultured thymic epithelial cells. *Br J Pharmacol* 1997;120:1401–1408.

239. Madden KS. Catecholamines, sympathetic nerves, and immunity. In: Ader R, Felten DL, Cohen N, eds. *Psychoneuroimmunology, volume 1,* 3rd ed. San Diego: Academic Press, 2001:197–216.

240. Swanson MA, Lee WT, Sanders VM. IFN-γ production by Th1 cells generated from naive CD4+ T cells exposed to norepinephrine. *J Immunol* 2001;166:232–240.

241. Madden KS, Felten SY, Felten DL, et al. Sympathetic nervous system modulation of the immune system II. Induction of lymphocyte proliferation and migration in vivo by chemical sympathectomy. *J Neuroimmunol* 1994;49:67–75.

242. Madden KS, Sanders VM, Felten DL. Catecholamine influences and sympathetic neural modulation of immune responsiveness. *Annu Rev Pharmacol Toxicol* 1995;35:417–448.

243. Kruszewska B, Felten SY, Moynihan JA. Alterations in cytokine and antibody production following chemical sympathectomy in two strains of mice. *J Immunol* 1995;155:4613–4620.

244. Kruszewska B, Felten DL, Stevens SY, et al. Sympathectomy-induced immune changes are not abrogated by the glucocorticoid receptor blocker RU-486. *Brain Behav Immun* 1998;12:181–200.

245. Kohm A, Sanders VM. Suppression of antigen-specific Th2 cell-dependent IgM and IgG1 production following norepinephrine depletion in vivo. *J Immunol* 1999;162:5299–5308.

246. Sanders VM, Powell-Oliver FE. β2-Adrenoceptor stimulation increases the number of antigen-specific precursor B lymphocytes that differentiate into IgM-secreting cells without affecting burst size. *J Immunol* 1992;148:1822–1828.

247. Benschop RJ, Rodriquez-Feuerhahn M, Schedlowski M. Catecholamine-induced leukocytosis. Early observations, current research, and future directions. *Brain Behav Immun* 1996;10:77–91.

248. Ernström U, Sandberg G. Effects of alpha- and beta-receptor stimulation on the release of lymphocytes and granulocytes from the spleen. *Scand J Haematol* 1973;11:275–286.

249. Rogausch H, del Rey A, Oertel J, et al. Norepinephrine stimulates lymphoid cell mobilization from the perfused rat spleen via β-adrenergic receptors. *Am J Physiol* 1999;276:R724–R730.

250. Fecho K, Dykstra LA, Lysle DT. Evidence for beta adrenergic receptor involvement in the immunomodulatory effects of morphine. *J Pharmacol Exp Ther* 1993;265:1079–1087.

251. Katafuchi T, Take S, Hori T. Roles of sympathetic nervous system in the suppression of cytotoxicity of splenic natural killer cells in the rat. *J Physiol* 1993;465:343–357.

252. MacNeil BJ, Jansen AH, Janz LJ, et al. Peripheral endotoxin increases splenic sympathetic nerve activity via central prostaglandin synthesis. *Am J Physiol* 1997;273:R609–R614.

253. Qi M, Zhou Z, Wurster RD, et al. Mechanisms involved in the rapid dissipation of plasma epinephrine response to bacterial endotoxin in conscious rats. *Am J Physiol* 1991;261:R1431–R1437.

254. van der Poll T, Coyle SM, Barbosa K, et al. Epinephrine inhibits tumor necrosis factor-α and potentiates interleukin 10 production during human endotoxemia. *J Clin Invest* 1996;97:713–719.

255. Haskó G, Szabó C, Németh ZH, et al. Stimulation of β-adrenoceptors inhibits endotoxin-induced IL-12 production in normal and IL-10 deficient mice. *J Neuroimmunol* 1998;88:57–61.

256. Szabó C, Haskó G, Zingarelli B, et al. Isoproterenol regulates tumour necrosis factor, interleukin-10, interleukin-6 and nitric oxide production and protects against the development of vascular hyporeactivity in endotoxaemia. *Immunology* 1997;90:95–100.

257. Elenkov IJ, Haskó G, Kovács KJ, et al. Modulation of lipopolysaccharide-induced tumor necrosis factor-α production by selective α- and β-adrenergic drugs in mice. *J Neuroimmunol* 1995;61:123–131.

258. Suberville S, Bellocq A, Fouqueray B, et al. Regulation of interleukin-10 production by β-adrenergic agonists. *Eur J Immunol* 1996;26:2601–2605.

259. Severn A, Rapson NT, Hunter CA, et al. Regulation of tumor necrosis factor production by adrenaline and β-adrenergic agonists. *J Immunol* 1992;148:3441–3445.

260. Chou RC, Stinson MW, Noble BK, et al. β-Adrenergic receptor regulation of macrophage-derived tumor necrosis factor-a production from rats with experimental arthritis. *J Neuroimmunol* 1996;67:7–16.

261. van der Poll T, Jansen J, Endert E, et al. Noradrenaline inhibits lipopolysaccharide-induced tumor necrosis factor and interleukin 6 production in human whole blood. *Infect Immun* 1994;62:2046–2050.

262. Fessler HE, Otterbein L, Chung HS, et al. Alpha-2 adrenoceptor blockade protects rats against lipopolysaccharide. *Am J Respir Crit Care Med* 1996;154:1689–1693.

263. Panina-Bordignon P, Mazzeo D, Di Lucia P, et al. β2-Agonists prevent Th1 development by selective inhibition of IL-12. *J Clin Invest* 1997;100:1513–1519.

264. Platzer C, Döcke W-D, Volk H-D, et al. Catecholamines trigger IL-10 release in acute systemic stress reaction by direct stimulation of its promoter/enhancer activity in monocytic cells. *J Neuroimmunol* 2000;105:31–38.

265. Woiciechowsky C, Asadullah K, Nestler D, et al. Sympathetic activation triggers systemic IL-10 release in immunodepression induced by brain injury. *Nat Med* 1998;4:808–813.

266. Breneman SM, Moynihan JA, Grota LJ, et al. Splenic norepinephrine is decreased in MRL-lpr/lpr mice. *Brain Behav Immun* 1993;7:135–143.

267. Mackenzie FJ, Leonard JP, Cuzner ML. Changes in lymphocyte β-adrenergic receptor density and noradrenaline content of the spleen are early indicators of immune reactivity in acute experimental allergic encephalomyelitis in the Lewis rat. *J Neuroimmunol* 1989;23:93–100.

268. Wiegmann K, Muthyala S, Kim DH, et al. β-Adrenergic agonists suppress chronic/relapsing experimental allergic encephalomyelitis (CREAE) in Lewis rats. *J Neuroimmunol* 1995;56:201–206.

269. Chelmicka-Schorr E, Checinski M, Arnason BGW. Chemical sympathectomy augments the severity of experimental allergic encephalomyelitis. *J Neuroimmunol* 1988;17:347–350.

270. Chelmicka-Schorr E, Kwasniewski MN, Thomas BE, et al. The β-adrenergic agonist isoproterenol suppresses experimental allergic encephalomyelitis in Lewis rats. *J Neuroimmunol* 1989;25:203–207.

271. Chelmicka-Schorr E, Kwasniewski MN, Wollmann RL. Sympathectomy augments adoptively transferred experimental allergic encephalomyelitis. *J Neuroimmunol* 1992;37:99–103.

272. Lorton D, Bellinger DL, Duclos M, et al. Application of 6-hydroxydopamine into the fatpads surrounding the draining lymph nodes exacerbates adjuvant-induced arthritis. *J Neuroimmunol* 1996;64:103–113.

273. Levine JD, Coderre TJ, Helms C, et al. β-2 adrenergic mechanisms in experimental arthritis. *Proc Natl Acad Sci U S A* 1988;85:4553–4556.

274. Coderre TJ, Basbaum AI, Helms C, et al. High-dose epinephrine acts at a2-adrenoceptors to suppress experimental arthritis. *Brain Res* 1991;544:325–328.

275. Kuis W, de Jong-de Vos van Steenwuk CCE, et al. The autonomic nervous system and the immune system in juvenile rheumatoid arthritis. *Brain Behav Immun* 1996;10:387–398.

276. Leden I, Eriksson A, Lilja B, et al. Autonomic nerve function in rheumatoid arthritis of varying severity. *Scand J Rheumatol* 1983;12:166–170.

277. Perry F, Heller PH, Kamiya J, et al. Altered autonomic function in patients with arthritis or with chronic myofascial pain. *Pain* 1989;39:77–84.

278. Levine JD, Fye K, Heller P, et al. Clinical response to regional intravenous guanethidine in patients with rheumatoid arthritis. *J Rheumatol* 1986;13:1040–1043.

279. Zoukos Y, Leonard JP, Thomaides T, et al. β-Adrenergic receptor density and function of peripheral blood mononuclear cells are increased in multiple sclerosis: a regulatory role for cortisol and interleukin-1. *Ann Neurol* 1992;31:657–662.

280. Karaszewski JW, Reder AT, Maselli R, et al. Sympathetic skin responses are decreased and lymphocyte beta-adrenergic receptors are increased in progressive multiple sclerosis. *Ann Neurol* 1990;27:366–372.

281. Saxne T, Palladino MA Jr, Heinegård D, et al. Detection of tumor necrosis factor a but not tumor necrosis factor b in rheumatoid arthritis synovial fluid and serum. *Arthritis Rheum* 1988;31:1041–1045.

282. Balashov KE, Smith DR, Khoury SJ, et al. Increased interleukin 12 production in progressive multiple sclerosis: induction by activated CD4+ T cells via CD40 ligand. *Proc Natl Acad Sci U S A* 1997;94:599–603.

283. van Boxel-Dezaire AH, Hoff SC, van Oosten BW, et al. Decreased interleukin-10 and increased interleukin-12p40 mRNA are associated with disease activity and characterize different disease stages in multiple sclerosis. *Ann Neurol* 1999;45:695–703.

284. Chou RC, Dong XL, Noble BK, et al. Adrenergic regulation of macrophage-derived tumor necrosis factor—a generation during a chronic polyarthritis pain model. *J Neuroimmunol* 1998;82:140–148.

285. Ben-Eliyahu S, Shakhar G, Page GG, et al. Suppression of NK cell activity and of resistance to metastasis by stress: a role for adrenal catecholamines and beta-adrenoceptors. *Neuroimmunomodulation* 2000;8:154–164.

286. Shakhar G, Ben-Eliyahu S. In vivo β-adrenergic stimulation suppresses natural killer activity and compromises resistance to tumor metastasis in rats. *J Immunol* 1998;160:3251–3258.

287. Weihe E, Müller S, Fink T, et al. Tachykinins, calcitonin gene-related peptide and neuropeptide Y in nerves of the mammalian thymus: interactions with mast cells in autonomic and sensory neuroimmunomodulation. *Neurosci Lett* 1989;100:77–82.

288. Lundberg JM, Anggard A, Pernow J, et al. Neuropeptide Y-, substance P- and VIP-immunoreactive nerves in cat spleen in relation to autonomic vascular and volume control. *Cell Tissue Res* 1985;239:9–18.

289. Lorton D, Bellinger DL, Felten SY, et al. Substance P innervation of spleen in rats: nerve fibers associate with lymphocytes and macrophages in specific compartments of the spleen. *Brain Behav Immun* 1991;5:29–40.

290. McGillis JP, Fernandez S, Knopf MA. Regulation of immune and inflammatory reactions in local microenvironments by sensory neuropeptides. In: Ader R, Felten DL, Cohen N, eds. *Psychoneuroimmunology, volume 1,* 3rd ed. San Diego: Academic Press, 2001:217–229.

291. Maggi CA. The effects of tachykinins on inflammatory and immune cells. *Regul Pept* 1997;70:75–90.

292. Payan DG. The role of neuropeptides in inflammation. In: Gallin JI, Goldstein IM, Snyderman R, eds. *Inflammation: basic principles and clinical correlates,* 2nd ed. New York: Raven Press, 1992:177–192.

293. McGillis JP, Organist ML, Payan DG. Substance P and immunoregulation. *Fed Proc* 1987;46:196–199.

294. Lotz M, Vaughan JH, Carson DA. Effect of neuropeptides on production of inflammatory cytokines by human monocytes. *Science* 1988;241:1218–1221.

295. Stanisz AM, Scicchitano R, Dazin P, et al. Distribution of substance P receptors on murine spleen and Peyer's patch T and B cells. *J Immunol* 1987;139:749–754.

296. Helme RD, Eglezos A, Dandie GW, et al. The effect of substance P on the regional lymph node antibody response to antigenic stimulation in capsaicin-pretreated rats. *J Immunol* 1987;139:3470–3473.

297. Stanisz AM, Befus D, Bienenstock J. Differential effects of vasoactive intestinal peptide, substance P, and somatostatin on immunoglobulin synthesis and proliferations by lymphocytes from Peyer's patches, mesenteric lymph nodes, and spleen. *J Immunol* 1986;136:152–156.

298. Aebischer I, Stampfli MR, Miescher S, et al. Neuropeptides accentuate interleukin-4 induced human immunoglobulin E synthesis in vitro. *Exp Dermatol* 1996;5:38–44.

299. Calvo C-F, Chavanel G, Senik A. Substance P enhances IL-2 expression in activated human T cells. *J Immunol* 1992;148:3498–3504.

300. Levite M. Neuropeptides, by direct interaction with T cells, induce cytokine secretion and break the commitment to a distinct T helper phenotype. *Proc Natl Acad Sci U S A* 1998;95:12544–12549.

301. Dimri R, Sharabi Y, Shoham J. Specific inhibition of glucocorticoid-induced thymocyte apoptosis by substance P. *J Immunol* 2000;164:2479–2486.

302. Wang F, Millet I, Bottomly K, et al. Calcitonin gene-related peptide inhibits interleukin 2 production by murine T lymphocytes. *J Biol Chem* 1992;267:21052–21057.

303. Millet I, Phillips RJ, Sherwin RS, et al. Inhibition of NF-kappaB activity and enhancement of apoptosis by the neuropeptide calcitonin gene-related peptide. *J Biol Chem* 2000;275:15114–151121.

304. Fernandez S, Knopf MA, McGillis JP. Calcitonin-gene related peptide (CGRP) inhibits interleukin-7-induced pre-B cell colony formation. *J Leukocyte Biol* 2000;67:669–676.

305. Asahina A, Moro O, Hosoi J, et al. Specific induction of cAMP in Langerhans cells by calcitonin gene-related peptide: relevance to functional effects. *Proc Natl Acad Sci U S A* 1995;92:8323–8327.

306. Fox FE, Kubin M, Cassin M, et al. Calcitonin gene-related peptide inhibits proliferation and antigen presentation by human peripheral blood mononuclear cells: effects on B7, interleukin 10, and interleukin 12. *J Invest Dermatol* 1997;108:43–48.

307. Khachatryan A, Guerder S, Palluault F, et al. Targeted expression of the neuropeptide calcitonin gene-related peptide to beta cells prevents diabetes in NOD mice. *J Immunol* 1997;158:1409–1416.

308. Ottaway CA, Greenberg GR. Interaction of vasoactive intestinal peptide with mouse lymphocytes: Specific binding and the modulation of mitogen responses. *J Immunol* 1984;132:417–423.

309. O'Dorisio MS, Shannon BT, Fleshman DJ, et al. Identification of high affinity receptors for vasoactive intestinal peptide on human lymphocytes of B cell lineage. *J Immunol* 1989;142:3533–3536.

310. Ottaway CA. In vitro alteration of receptors for vasoactive intestinal polypeptide changes the in vivo localization of mouse T cells. *J Exp Med* 1984;160:1054–1069.

311. Xia M, Gaufo GO, Wang Q, et al. Transduction of specific inhibition of HuT 78 human T cell chemotaxis by type I vasoactive intestinal peptide receptors. *J Immunol* 1996;157:1132–1138.

312. Dantzer R, Bluthé R-M, Castanon N, et al. Cytokine effects on behavior. In: Ader R, Felten DL, Cohen N, eds. *Psychoneuroimmunology, volume 2,* 3rd ed. San Diego: Academic Press, 2001:703–727.

313. Kusnecov AW, Liang R, Shurin G. T-lymphocyte activation increases hypothalamic and amygdaloid expression of CRH mRNA and emotional reactivity to novelty. *J Neurosci* 1999;19:4533–4543.

314. Dunn A, Vickers S. Neurochemical and neuroendocrine responses of Newcastle disease virus administration in mice. *Brain Res* 1994;645:103–112.

315. Shurin G, Shanks N, Nelson L, et al. Hypothalamic-pituitary-adrenal activation by the bacterial superantigen staphylococcal enterotoxin B: role of macrophages and T cells. *Neuroendocrinology* 1997;65:18–28.

316. Besedovsky HO, Da Prada M, del Rey AE, et al. Immunoregulation by sympathetic nervous system. *Trends Pharmacol Sci* 1981;2:236–238.

317. Besedovsky HO, del Rey AE, Sorkin E, et al. The immune response evokes changes in brain noradrenergic neurons. *Science* 1983;221:564–565.

318. Maier SF, Watkins LR, Nance DM. Multiple routes of action of interleukin-1 on the nervous system. In: Ader R, Felten DL, Cohen N, eds. *Psychoneuroimmunology, volume 1,* 3rd ed. San Diego: Academic Press, 2001:563–583.

319. Kozak W, Zheng H, Conn CA, et al. Thermal and behavioral effects of lipopolysaccharide and influenza in interleukin-1β-deficient mice. *Am J Physiol* 1995;269:R969–R977.

320. Ruzek MC, Miller AH, Opal SM, et al. Characterization of early cytokine responses and an interleukin (IL)-6-dependent pathway of endogenous glucocorticoid induction during murine cytomegalovirus infection. *J Exp Med* 1997;185:1185–1192.

321. Konsman JP, Kelley K, Dantzer R. Temporal and spatial relationships between lipopolysaccharide-induced expression of fos, interleukin-1β and inducible nitric oxide synthase in rat brain. *Neuroscience* 1999;89:535–548.

322. Elmquist JK, Scammell TE, Jacobson CD, et al. Distribution of fos-like immunoreactivity in the rat brain following intravenous lipopolysaccharide administration. *J Comp Neurol* 1996;371:85–103.

323. Ericsson A, Kovacs KJ, Sawchenko PE. A functional anatomical analysis of central pathways subserving the effects of interleukin-1 on stress-related neuroendocrine neurons. *J Neurosci* 1994;14:897–913.

324. Dunn AJ. Effects of cytokines and infections on brain neurochemistry. In: Ader R, Felten DL, Cohen N, eds. *Psychoneuroimmunology, volume 2,* 3rd ed. San Diego: Academic Press, 2001:649–666.

325. Van Dam A-M, Brouns M, Man-A-Hing W, et al. Immunocytochemical detection of prostaglandin E2 in microvasculature and in neurons of rat brain after administration of bacterial endotoxin. *Brain Res* 1993;613:331–336.

326. Cao C, Matsumura K, Yamagata K, et al. Involvement of cyclooxygenase-2 in LPS-induced fever and regulation of its mRNA in the rat brain by LPS. *Am J Physiol* 1997;272:R1712–R1725.

327. Milligan ED, McGorry MM, Fleshner M, et al. Subdiaphragmatic vagotomy does not prevent fever following intracerebroventricular prostaglandin E2: further evidence for the importance of vagal afferents in immune-to-brain communication. *Brain Res* 1997;766:240–243.

328. Ericsson A, Arias C, Sawchenko PE. Evidence for an intramedullary prostaglandin-dependent mechanism in the activation of stress-related neuroendocrine circuitry by intravenous interleukin-1. *J Neurosci* 1997;17:7166–7179.

329. Wan W, Wetmore L, Sorensen CM, et al. Neural and biochemical mediators of endotoxin and stress-induced c-fos expression in the rat brain. *Brain Res Bull* 1994;34:7–14.

330. Goehler LE, Gaykema RP, Hammack SE, et al. Interleukin-1 induces c-Fos immunoreactivity in primary afferent neurons of the vagus nerve. *Brain Res* 1998;804:306–310.

331. Maier SF, Goehler LE, Fleshner M, et al. The role of the vagus nerve in cytokine-to-brain communication. *Ann N Y Acad Sci* 1998;289:289–300.

332. Goehler LE, Gaykema RPA, Nguyen KT, et al. Interleukin-1β in immune cells of the abdominal vagus nerve: a link between the immune and nervous systems? *J Neurosci* 1999;19:2799–2806.

333. Turrin NP, Plata-Salamán CR. Cytokine-cytokine interactions and the brain. *Brain Res Bull* 2000;51:3–9.

334. Hansen MK, O'Connor KA, Goehler LE, et al. The contribution of the vagus nerve in interleukin-1beta-induced fever is dependent on dose. *Am J Physiol* 2001;280:R929–R934.

335. Graham NM, Douglas RM, Ryan P. Stress and acute respiratory infection. *Am J Epidemiol* 1986;124:389–401.

336. Clover RD, Abell T, Becker LA, et al. Family functioning and stress as predictors of influenza B infection. *J Fam Pract* 1989;28:535–539.

337. Cohen F, Kemeny ME, Kearney KA, et al. Persistent stress as a predictor of genital herpes recurrence. *Arch Intern Med* 1999;159:2430–2436.

338. Glaser R, Kiecolt-Glaser JK. Chronic stress modulates the virus-specific immune response to latent herpes simplex virus type 1. *Ann Behav Med* 1997;19:78–82.

339. McLarnon LD, Kaloupek DG. Psychological investigation of genital herpes recurrence: prospective assessment and cognitive-behavioral intervention for a chronic physical disorder. *Health Psychol* 1988;7:231–249.

340. Luborsky L, Mintz J, Brightman VJ, et al. Herpes simplex virus and moods: a longitudinal study. *J Psychosom Res* 1976;20:543–548.

341. Glaser R, Pearson GR, Jones JF, et al. Stress-related activation of Epstein-Barr virus. *Brain Behav Immun* 1991;5:219–232.

342. Glaser R, Kiecolt-Glaser JK, Speicher CE, et al. Stress, loneliness, and changes in herpes virus latency. *J Behav Med* 1985;8:249–260.

343. Cole SW, Kemeny ME. Psychosocial influence on the progression of HIV infection. In: Ader R, Felten DL, Cohen N, eds. *Psychoneuroimmunology, volume 2,* 3rd ed. San Diego: Academic Press, 2001:583–612.

344. Solomon GF, Ironson GH, Balbin EG. Psychoneuroimmunology and HIV/AIDS. *Ann N Y Acad Sci* 2000;917:500–504.

345. Mirsky A. Physiologic, psychological, and social determinants in the etiology of duodenal ulcer. *Am J Dig Dis* 1958;3:285–313.

346. Marucha PT, Kiecolt-Glaser JK, Favagehi M. Mucosal wound healing is impaired by examination stress. *Psychosom Med* 1998;60:362–365.

347. Mathews A, Ridgeway V. Personality and surgical recovery: a review. *Br J Clin Psychol* 1981;20:243–260.

348. Glaser R, Kiecolt-Glaser JK, Marucha PT, et al. Stress-related changes in proinflammatory cytokine production in wounds. *Arch Gen Psychiatry* 1999;56:450–456.

349. Conti A. Oncology in neuroimmunomodulation. What progress has been made? *Ann N Y Acad Sci* 2000;917:68–83.

350. Kiecolt-Glaser JK, Glaser R. Psychoneuroimmunology and cancer: fact or fiction? *Eur J Cancer* 1999;35:1603–1607.

351. Butow PN, Hiller JE, Price MA, et al. Epidemiological evidence for a relationship between life events, coping style, and personality factors in the development of breast cancer. *J Psychosom Res* 2000;49:169–181.

352. Persky VW, Kempthorne-Rawson J, Shekelle RB. Personality and risk of cancer: 20-year follow-up of the Western Electric Study. *Psychosom Med* 1987;49:435–449.

353. Shekelle RB, Raynor WJ, Ostfeld AM, et al. Psychological depression and 17-year risk of death from cancer. *Psychosom Med* 1981;43:117–125.

354. Grossarth-Maticek R, Bastiaans J, Kanazir DT. Psychosocial factors as strong predictors of mortality from cancer, ischaemic heart disease and stroke: the Yugoslav prospective study. *J Psychosom Res* 1985;29:167–176.

355. Zonderman AB, Costa PT, McCrae RR. Depression as a risk for cancer morbidity and mortality in a nationally representative sample. *JAMA* 1989;262:1191–1195.

356. Hahn RC, Petitti DB. Minnesota Multiphasic Personality Inventory-rated depression and the incidence of breast cancer. *Cancer* 1988;61:845–848.

357. Turner-Cobb JM, Sephton SE, Spiegel D. Psychosocial effects on immune function and disease progression in cancer: human studies. In: Ader R, Felten DL, Cohen N, eds. *Psychoneuroimmunology, volume 2,* 3rd ed. San Diego: Academic Press, 2001:565–582.

358. Fawzy FI, Fawzy NW. A structured psychoeducational intervention for cancer patients. *Gen Hosp Psychiatry* 1994;16:149–192.

359. Fawzy FI, Fawzy NW, Hyun CS, et al. Malignant melanoma. Effects of an early structured psychiatric intervention, coping, and affective state on recurrence and survival 6 years later. *Arch Gen Psychiatry* 1993;50:681–689.

360. Spiegel D, Bloom JR, Kraemer HC, et al. Effect of psychosocial treatment on survival of patients with metastatic breast cancer. *Lancet* 1989;2:888–891.

361. Spiegel D, Sephton SE, Terr AI, et al. Effects of psychosocial treatment in prolonging cancer survival may be mediated by neuroimmune pathways. *Ann N Y Acad Sci* 1998;840:674–683.

362. Richardson JL, Zarnegar Z, Bisno B, et al. Psychosocial status at initiation of cancer treatment and survival. *J Psychosom Res* 1990;34:189–201.

363. Gellert GA, Maxwell RM, Siegel BS. Survival of breast cancer patients receiving adjunctive psychosocial support therapy: a 10-year follow-up study. *J Clin Oncol* 1993;11:66–69.

364. Cunningham AJ, Edmonds CV, Jenkins GP, et al. A randomized controlled trial of the effects of group psychological therapy on survival in women with metastatic breast cancer. *Psychooncology* 1998;7:508–517.

365. Griffiths CEM. Psychological influences in psoriasis. *Clin Exp Dermatol* 2001;26:338–342.

366. Ringel Y, Drossman DA. Psychosocial aspects of Crohn's disease. *Surg Clin North Am* 2001;81:231–252.

367. Robinson N, Lloyd CE, Fuller JH, et al. Psychosocial factors and the onset of type 1 diabetes. *Diabet Med* 1989;6:53–58.

368. Radosavljevic VR, Jankovic SM, Marinkovic JM. Stressful life events in the pathogenesis of Graves' disease. *Eur J Endocrinol* 1996;134:699–701.

369. Ackerman KD, Heyman R, Rabin BS, et al. Stress and its relationship to disease activity in multiple sclerosis. *Int J Mult Scler* 2000;7:20–29.

370. Warren S, Greenhill S, Warren KG. Emotional stress and the development of multiple sclerosis: case-control evidence of a relationship. *J Chronic Dis* 1982;35:821–831.

371. Grant I, Brown GW, Harris T, et al. Severely threatening events and marked life difficulties preceding onset or exacerbation of multiple sclerosis. *J Neurol Neurosurg Psychiatry* 1989;52:8–13.

372. Franklin GM, Nelson LM, Heaton RK. Stress and its relationship to acute exacerbations in multiple sclerosis. *J Neurol Rehab* 1988;2:7–11.

373. Stip E, Truelle JL. [Organic personality syndrome in multiple sclerosis and effect of stress on recurrent attacks] [French]. *Can J Psychiatry* 1994;39:27–33.

374. Mohr DC, Goodkin DE, Bacchetti P, et al. Psychological stress and the subsequent appearance of new brain MRI lesions in MS. *Neurology* 2000;55:55–61.

375. Nisipeanu P, Korczyn AD. Psychological stress as risk factor for exacerbations in multiple sclerosis. *Neurology* 1993;43:1311–1312.

11

AUTONOMIC NERVOUS SYSTEM

MICHAEL J. AMINOFF

The autonomic nervous system is concerned primarily with the innervation of various internal organs, the maintenance of the internal environment, and the regulation of processes that are not usually considered to be under voluntary control. It is traditionally divided into the sympathetic and parasympathetic systems, which have seemingly opposing activities, ensuring the harmonious integration of different functions. The autonomic nervous system is also instrumental in generating the physical responses to emotional stimuli that characterize some aspects of behavior. Thus, the cardiovascular changes accompanying anger (i.e., an increase in heart rate and blood pressure) or embarrassment (i.e., a cutaneous vasodilation or "blush") depend on autonomic activity. Similarly, the dry mouth of excitement, the excessive sweating that occurs with anxiety, the urinary incontinence associated with intense fear, and the wide-eyed appearance of surprise all depend on autonomic activity. Such autonomic responses have to be integrated with any somatic responses necessitated by the emotional stimulus, such as offensive or predatory maneuvers occasioned by anger or defensive behavior elicited by fear.

ANATOMY OF THE AUTONOMIC NERVOUS SYSTEM

Autonomic Efferent Pathway

The efferent autonomic pathways consist of two neurons, one of which has its nerve cell body in the central nervous system (CNS), the preganglionic neuron, and the other with its perikaryon situated peripherally, the postganglionic neuron. The preganglionic neurons of the parasympathetic system are located intracranially and sacrally (Fig. 11.1). The neurons of the cranial division are located in the brainstem adjacent to the somatic efferent nuclei of the third, seventh, ninth, and tenth cranial nerves. More specifically, parasympathetic fibers destined for the third nerve arise from the Edinger-Westphal and anteromedian nuclei, those for the seventh nerve from the superior salivatory nucleus, those for the ninth nerve from the inferior salivatory nucleus, and those for the tenth nerve from the dorsal motor nucleus

of the vagus and the nucleus ambiguus. In the sacral cord, preganglionic parasympathetic neurons are grouped laterally and medially in the intermediate gray matter of the cord between the anterior and posterior horns.

Axons of these neurons extend as far as, or close to, the organs to be innervated before synapsing with postganglionic cells in distinct ganglia or more diffuse intramural plexuses in the target organs. For example, parasympathetic fibers pass in the third (oculomotor) nerve to the ciliary ganglion in the orbit, from which postganglionic fibers pass in the short ciliary nerves to the ciliary muscle and sphincter of the iris. The course of parasympathetic fibers traveling in the facial nerve (to the lacrimal and some salivary glands and the mucosal glands in the nose) and glossopharyngeal nerve (to the parotid gland) is documented in standard texts, but its importance mandates further comment concerning the vagus nerve. Preganglionic parasympathetic fibers arise from the dorsal motor nucleus of the vagus and nucleus ambiguus and pass toward their target organs, where they end in ganglia or plexuses related to the heart, pharynx, esophagus, and abdominal viscera. Parasympathetic fibers to the heart travel through the superior, middle, and inferior cardiac nerves to terminate on cells in the cardiac plexus, especially around the origin of the aorta and pulmonary artery and along the coronary arteries. The vagus nerves pass through the diaphragm to supply the stomach and contribute to the celiac plexus from which fibers pass (with sympathetic fibers in the periarterial plexuses) to the abdominal viscera.

The preganglionic parasympathetic sacral fibers pass with the anterior sacral roots (S2–S4) and eventually form the nervi erigentes or pelvic nerves on each side of the rectum, synapsing with postganglionic neurons in the walls of the bladder, rectum, and genitalia.

In the sympathetic system, the small, preganglionic efferent neurons are located in the thoracic and upper two lumbar segments of the spinal cord. Their axons leave the cord in the anterior roots and pass via the white rami communicantes (Fig. 11.2) to the sympathetic trunk, a chain of paravertebral ganglia situated on either side of the vertebral column, and interconnected by longitudinally arranged nerve fibers. In addition, there are transverse connections below the level of

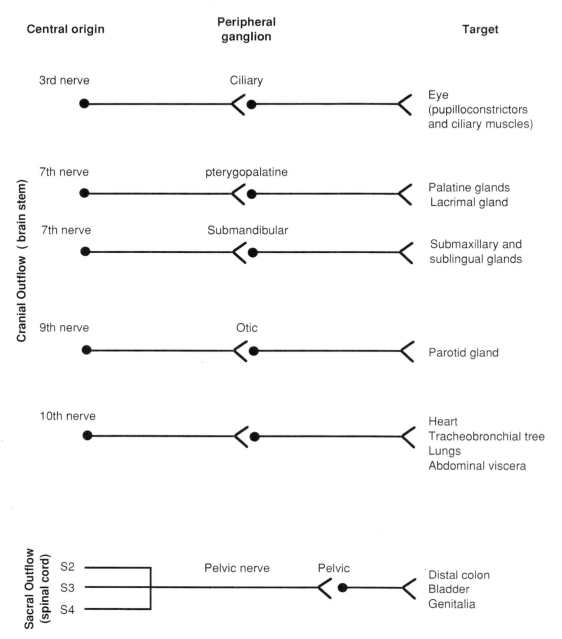

FIGURE 11.1. Diagram of the parasympathetic nervous system.

the fifth lumbar vertebra. Postganglionic sympathetic neurons are found in the paravertebral ganglia and also in the prevertebral ganglia that are situated more peripherally (Fig. 11.3). The precise anatomy of the sympathetic trunk and its contained ganglia varies in different individuals, but some generalizations can be made.

There are three cervical ganglia: the superior (at the level of the upper two cervical vertebrae), middle (at the sixth cervical vertebra), and inferior (behind the subclavian artery). The inferior cervical ganglion is often fused with the first thoracic ganglion to form the stellate ganglion. An intermediate ganglion may be present between the middle

and inferior ganglia. The ansa subclavia goes around the subclavian artery to join the intermediate with the inferior ganglion. The superior cervical ganglion sends postganglionic fibers to the upper four cervical spinal nerves and branches to the vagal and glossopharyngeal ganglia, hypoglossal nerve, carotid body, jugular bulb, cardiac plexus, periarterial plexus around the carotid arteries, and some salivary and facial sweat glands. The middle cervical ganglion has a major cardiac branch and also supplies the fifth and sixth cervical nerves, thyroid and parathyroid glands, esophagus, and trachea. The stellate ganglion supplies the lower cervical and first thoracic spinal nerves and contributes

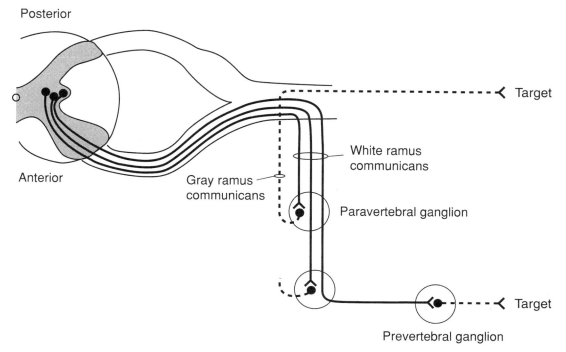

FIGURE 11.2. Sympathetic pathways from the cord.

to the cardiac plexus and various perivascular plexuses, especially about the vertebral artery. Preganglionic fibers destined for the upper limb arise in the upper thoracic cord and ascend in the sympathetic trunk to synapse in the stellate ganglion, whence postganglionic fibers pass to the brachial plexus.

In the thoracic and lumbar regions, there is generally one ganglion for each segment, and in the sacral region, there is usually a total of four or five ganglia; in addition, a single (unpaired) coccygeal ganglion is frequently present. Sympathetic ganglion cells are typically present along the gray rami communicantes (discussed below) as well as in the ganglia themselves.

The unmyelinated axons of the postganglionic sympathetic neurons in the paravertebral and prevertebral ganglia pass to the skin, muscles, and organs that they innervate. They either return to the spinal nerves as gray rami communicantes or pass along the major arteries as perivascular plexuses of nerve fibers. These perivascular plexuses are exemplified by the fibers that accompany the internal carotid artery and its branches to supply various cranial structures or join particular cranial nerves. Fibers from the stellate ganglion may pass with the subclavian artery to the upper limb or in the gray rami communicantes to the lower cervical nerves. The superior, middle, and inferior cardiac nerves arise from postganglionic neurons in the cervical ganglia and form the cardiac plexus, which also contains postganglionic sympathetic fibers from the upper thoracic ganglia and parasympathetic fibers from the vagus nerve. Nerves to the coronary vessels are given off from this plexus.

Postganglionic sympathetic fibers pass from various prevertebral abdominal ganglia (including the celiac) with aortic branches to form autonomic plexuses that also contain parasympathetic fibers related to the viscera and with the gonadal and iliac arteries to supply eventually the pelvic organs. Although the intermediolateral cell columns of the spinal cord are restricted to the thoracic and upper two lumbar segments, all regions of the body are supplied by sympathetic fibers. This is because some preganglionic sympathetic fibers pass without interruption through the nearest sympathetic ganglia to ascend or descend to ganglia at other levels in the sympathetic trunk before synapsing or exit without synapsing to relay in the prevertebral (preaortic) ganglia, i.e., the celiac, aorticorenal, superior mesenteric, and inferior mesenteric ganglia.

Several different types of postganglionic sympathetic fibers can be distinguished, depending on their target organ, and these, in turn, may have different physiologic characteristics. The postganglionic sympathetic fibers to the skin, for example, consist of vasomotor and sudomotor fibers with different conduction velocities (1).

Afferent Fibers

Afferent fibers are important in permitting autonomic reflex activity, but their precise pathways have been defined less clearly than those taken by efferent fibers.

The afferent pathway subserving the pupillary reflex response to light originates in the retina, whence information concerning luminance passes to the pretectum

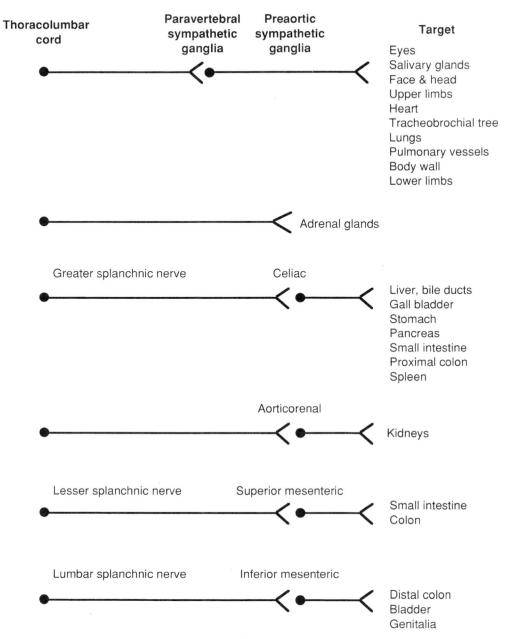

Thoracolumbar cord

Paravertebral sympathetic ganglia

Preaortic sympathetic ganglia

Target

Eyes
Salivary glands
Face & head
Upper limbs
Heart
Tracheobrochial tree
Lungs
Pulmonary vessels
Body wall
Lower limbs

Adrenal glands

Greater splanchnic nerve

Celiac

Liver, bile ducts
Gall bladder
Stomach
Pancreas
Small intestine
Proximal colon
Spleen

Aorticorenal

Kidneys

Lesser splanchnic nerve

Superior mesenteric

Small intestine
Colon

Lumbar splanchnic nerve

Inferior mesenteric

Distal colon
Bladder
Genitalia

FIGURE 11.3. Diagram of the efferent sympathetic nervous system.

via fibers in the optic tract. From there, neurons project to the pupilloconstrictor neurons in the Edinger-Westphal nucleus. Afferent fibers from the cornea and nasal and oropharyngeal mucosa pass in the trigeminal nerve to the trigeminal nuclei and nucleus tractus solitarius in the brainstem; stimulation of these afferent fibers elicits lacrimal, nasal, and oral secretions by brainstem reflexes.

Afferent fibers from baroreceptors in the carotid sinus and aortic arch are carried in branches of the ninth (glossopharyngeal) and tenth (vagus) nerves, respectively. Cell bodies for the fibers in the ninth nerve lie in its inferior ganglion

and for the tenth nerve in the nodose ganglion. Cardiac afferent fibers transmit impulses from mechanoreceptors or chemoreceptors about the openings of the great veins, in the walls of the atria and interatrial septum, and in the walls of the ventricles, atrioventricular valves, and aorta. They pass in the vagus and sympathetic nerves. The vagus nerve also contains afferent fibers from the tracheobronchial tree and abdominal viscera that pass, via the nodose ganglion, to the nucleus tractus solitarius.

Visceral afferent fibers arise in the S2 through S4 dorsal root ganglia and are involved in anorectal and urogenital reflex activity.

Sensory neurons associated with the sympathetic system are in the dorsal root ganglia. Afferent information is derived from both the sympathetic ganglia and the peripheral target organs and is transmitted to the dorsal horn of the spinal cord (2).

Central Autonomic Structures

The autonomic nervous system is represented at many different levels of the CNS, including the *cortex* of the superior frontal gyrus and areas 4 and 6 of the cerebral cortex (3). Cerebral pathology may lead to disturbances of cardiovascular, pilomotor, sudomotor, or gastric function. For more than a century, it has been known that cortical lesions or stimulation may influence the heart, respiratory rate, and blood pressure. The anterior cingulate cortex is involved in the control of bladder and bowels, and loss of sphincter function accompanies bilateral cingulate lesions. The temporal lobe and amygdala also have autonomic functions. This probably accounts for the autonomic accompaniments of some complex partial seizures (4). There are profuse connections between those parts of the cerebral cortex involved in autonomic activity and other CNS regions having autonomic functions, but the specific pathways mediating this activity are unknown.

The *hypothalamus* seems to be a major (direct or indirect) relay station for autonomic pathways from the spinal cord, brainstem, and hippocampus and is connected with the premotor frontal cortex. There are also rich efferent connections (direct or indirect) with autonomic neurons in the spinal cord and a close association between the hypothalamus and hypophysis. The anterior hypothalamic or preoptic area contains temperature-sensitive neurons that respond to either heat or cold. This area has a major role in integrating thermal inputs from different sources. Hypothalamic stimulation influences cardiovascular, pilomotor, and thermoregulatory function. For example, with electrical stimulation of the hypothalamus in cats anesthetized with chloralose, there is an increase in heart rate and mean arterial blood pressure and inhibition of cardiac and vasomotor components of baroreceptor reflexes (5,6). The hypothalamic region also influences feeding behavior; hypothalamic pathology leads either to hyperphagia and obesity or aphagia and weight loss, depending on the precise site of the lesion.

The *cerebellum* has profound influences on autonomic function. Thus, Bradley et al. (7) noted that electrical or chemical activation of a localized region of the posterior vermis may markedly influence vagal and sympathetic activity, leading in turn to changes in heart rate, arterial blood pressure, regional blood flow, and renal sympathetic nerve discharge. Other studies have revealed that the posterior vermis of the cerebellum (including the uvula) is essential for the acquisition of classically conditioned bradycardias; vermian lesions lead to severe attenuation of bradycardic responses to simple conditioning situations without altering resting heart rate and unconditioned heart rate orienting responses to a tone stimulus (8). The cerebellum also influences respiratory activity, but it is not clear whether it influences gastrointestinal function and sphincter control.

There are major, often reciprocal, connections between some *pontine nuclei* and lower brainstem (medullary), forebrain, and hypothalamic structures that are also concerned with cardiovascular and respiratory control (9). The pontine regions influencing cardiovascular control connect with the *nucleus tractus solitarius*, which is the site of termination of afferent fibers from the arterial baroreceptors and chemoreceptors (Fig. 11.4), the heart and lungs, and the gastrointestinal system (10). Bilateral lesions of the nucleus tractus solitarius at the level of the obex in rats abolishes baroreceptor reflexes and results in an elevation in blood pressure without a change in heart rate (11). The nucleus tractus solitarius is also influenced by, and influences, neocortical regions and forebrain, diencephalic, and rostral brainstem nuclei. It is situated in the dorsomedial portion of the medulla and connects with the dorsal nucleus of the vagus and neurons in the lateral region of the reticular formation that project via the bulbospinal pathway to the spinal cord, thereby influencing the cardiovascular system (Fig. 11.4). Through connections with the dorsal motor nucleus of the vagus, the nucleus tractus solitarius also influences gastrointestinal motility and secretions (12).

The brainstem has important influences on ventilation. Neuronal activity related to respiration occurs in discrete regions of the upper pons, and electrical stimulation of these regions leads to changes in the ventilatory phase (13). Brainstem transection below the level of the so-called pontine pneumotaxic center in vagotomized animals produces apneustic ventilation, with prolonged end-inspiratory pauses (14). In the medulla, a dorsal respiratory group of neurons has been identified in part of the nucleus tractus solitarius; these are active during inspiration and receive vagal pulmonary afferents. A ventral respiratory group of neurons has also been identified within the nucleus ambiguus and nucleus retroambigualis and contains cells that are active in inspiration or expiration (15). Fibers from these respiratory neurons in the medulla pass to the contralateral motor neurons, whose axons constitute the phrenic and intercostal nerves.

Descending pathways conduct impulses from the brainstem to the preganglionic sympathetic neurons in the intermediolateral cell columns in the thoracic and upper lumbar regions of the cord. These cells have the staining characteristics of motor neurons. Their number diminishes with age at a rate of approximately 8% per decade (16). The axons of these cells exit with the anterior nerve roots as preganglionic sympathetic fibers passing to adjacent ganglia. Descending spinal pathways also connect with the parasympathetic outflow in the cranial and sacral regions. The spinal autonomic

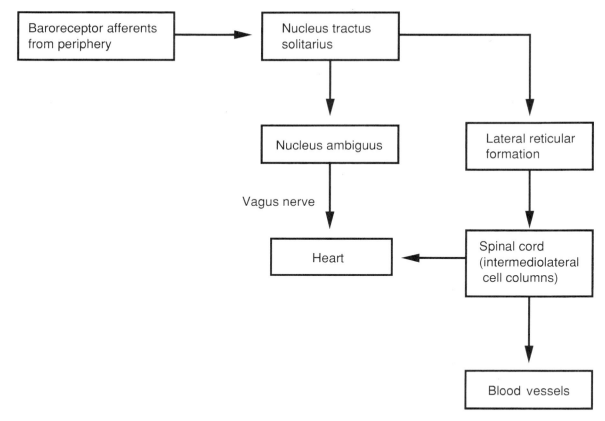

FIGURE 11.4. Baroreceptor reflex regulation of the circulation.

fibers are small in diameter and are probably most profuse in the lateral funiculi, but their precise pathway is unknown.

NEUROTRANSMITTER AND NEUROMODULATORS

The major postganglionic neurotransmitter of the sympathetic system is norepinephrine, which is transported along postganglionic fibers to their terminals, where it is stored in dense-core vesicles until released. There are two types of adrenergic receptors, the so-called α- and β-receptors, and subclasses within these two broad categories. Plasma norepinephrine levels normally increase markedly with change from recumbency to an upright posture, reflecting an increase in sympathetic activity. Epinephrine is released from the adrenal medulla and has important effects on cardiovascular function as well as other responses to stress. In a rare disorder characterized by deficiency of dopamine β-hydroxylase, sympathetic dysfunction results from the inability to synthesize norepinephrine (and epinephrine). In this disorder, dopamine, not norepinephrine, is released from adrenergic nerve terminals.

Acetylcholine is the main peripheral neurotransmitter in the parasympathetic system, but it is also released from preganglionic sympathetic fibers and at postganglionic sympa-

thetic nerve terminals to the sweat glands and some blood vessels in the skeletal muscles as well as from fibers supplying the adrenal medulla. Numerous central pathways are cholinergic. Cholinergic receptors are divided into muscarinic receptors (e.g., in CNS neurons, skeletal postganglionic sympathetic neurons, smooth and cardiac muscle) and nicotinic receptors (e.g., autonomic ganglia, skeletal neuromuscular junctions, spinal cord). There are at least three different muscarinic and nicotinic receptors. Serotonin is probably important also as a neurotransmitter in central autonomic pathways.

Neuropeptides are present in autonomic neurons and may function as cotransmitters or modulators of neurotransmission, but their precise role has yet to be clarified. Neuropeptide Y is present in a variety of sympathetic, parasympathetic, and enteric neurons; it is contained, for example, in many nonadrenergic vasomotor neurons in the sympathetic ganglia. Opioid (and other) peptides are similarly widespread in peripheral autonomic neurons and may coexist with neuropeptide Y in the same nerve cells (17). Various neuropeptides also seem to function as cotransmitters at the nerve terminals of sudomotor fibers (18). The neurons in the intermediate cell columns are apparently influenced by several different neuropeptides. For example, immunoreactivity to substance P exists in several of the projections to these columns, immunoreactivity to somatostatin occurs in the intermediolateral cell columns and sacral preganglionic parasympathetic neurons, and vasoactive intestinal polypeptide occurs in a number of different autonomic (and somatic) neurons including cells in the cord, hypothalamus, and sympathetic ganglia (19). Numerous peptides are involved in the regulation of cardiovascular function in concert with the autonomic nervous system. Thus, the release of atrial natriuretic factor, a peptide with diuretic and vasodilating properties, depends on atrial distention but seems to be modulated by the sympathetic system (20).

During development, the environment into which the axon of a sympathetic neuron grows affects the phenotype of the cell. Thus, cholinergic sympathetic sudomotor neurons in rats initially are noradrenergic and acquire their cholinergic properties only after contact with their target sweat glands in the postnatal period by a mechanism that is unclear (21). The neuropeptide phenotype of sympathetic neurons is also influenced by environmental factors (17). Studies of vasomotor neurons projecting to different levels of the vascular bed in the skin of the ears and paws of guinea pigs have shown differences in immunoreactivity to various peptides, implying remarkable specificity among nerve cells supplying targets of similar type (17). Thus, neuropeptide Y is present in sympathetic noradrenergic neurons innervating large distributing arteries, whereas neurons supplying the smallest cutaneous arterioles contain prodynorphin-derived peptides but not neuropeptide Y; neurons innervating the rest of the vasculature contain prodynorphin-derived peptides and neuropeptide Y (17).

Nerve growth factor and other neurotrophic factors may suppress built-in self-destruction programs as well as stimulating growth (22–24). Some rats are prone to hereditary hypertension, and their superior cervical ganglia contain fewer neurons and many more axonal terminals containing substance P than control animals (25,26). These changes are prevented by neonatal treatment with nerve growth factor (27).

AUTONOMIC INNERVATION AND REGULATION OF VARIOUS STRUCTURES

In this section, further details of the autonomic innervation of specific structures are provided. In addition, some clinical deficits resulting from disturbed autonomic regulation are considered.

Heart and Blood Vessels

Vasomotor tone and peripheral resistance are determined primarily by the sympathetic nervous system. The sympathetic outflow to different regions and structures (such as the skin and muscle) is regulated separately. Postganglionic sympathetic fibers pass mainly with the somatic nerves to the distant peripheral blood vessels and from the celiac and other abdominal ganglia to the splanchnic vascular bed.

The main action of the sympathetic nervous system on blood vessels is constrictive and is induced by norepinephrine. The existence and importance of sympathetic vasodilator fibers are less clear, and vasodilation in limb vessels probably relates to axon reflexes, antidromic conduction, or, most important, reduced sympathetic activity. Peripheral nerve lesions may disrupt vasoconstrictor fibers, thereby leading to erythema and an increased cutaneous temperature in an appropriate distribution in the limbs. Microelectrode studies in humans have shown that baroreceptor activity has a major influence on impulse traffic in sympathetic fibers to muscle, whereas brief mental stress has very little effect (1). By contrast, sympathetic (vasoconstrictor) activity in human cutaneous nerves is increased by emotional and mental stress. Some of the blood vessels in muscle are probably also supplied by sympathetic cholinergic vasodilator fibers. Vasodilator cholinergic fibers are present in the parasympathetic supply to the cranial, pelvic, and visceral blood vessels.

Tachycardia results from sympathetic stimulation of fibers (derived from T1 to T4) to the heart, and bradycardia follows parasympathetic (vagal) stimulation.

The baroreceptor reflexes are important in the control of the heart and blood pressure (Fig. 11.4). Arterial high-pressure baroreceptors are located primarily in the aortic arch (and send their afferent fibers in branches of the vagus nerve) and carotid sinus (branches of the glossopharyngeal nerve). The wall of the carotid sinus is characterized by more elastic tissue and less muscle than the wall of the adjacent vessel. The baroreceptor endings are located in the adventitia and are stimulated by mechanical deformation of the vessel wall. Their sensitivity may be influenced by the sympathetic nerves that terminate in the carotid sinus.

Direct recordings from sympathetic efferent vasomotor fibers to blood vessels in the skin or muscles reveal that bursts of impulses occur rhythmically, time-locked to the pulse and often to the respiratory cycle. This rhythmic sympathetic activity is virtually eliminated below the level of complete cord transection, indicating its dependence on supraspinal mechanisms. Sympathetic efferent activity can be inhibited by an increase in blood pressure in the carotid sinus and aortic arch. The baroreceptor reflexes elicited by this increased pressure also cause slowing of the heart, which is mediated primarily through the vagus nerve. In contrast, hypotension leads to tachycardia by inhibition of vagal activity as well as to peripheral vasoconstriction from increased sympathetic activity.

Change in posture from a recumbent to an erect position leads to pooling of blood in the legs and lower abdomen. This may lead to a slight fall in systolic blood pressure, with little or no fall in the diastolic pressure. The reduced systolic pressure leads to activation of the baroreceptor reflexes so that vasoconstriction occurs in the periphery, and there is an increase in the rate and contractile force of the heart. Intraneural recordings in humans have shown that the response to postural head-up tilt and the Valsalva maneuver consists of an increase in sympathetic activity in vasomotor nerves to muscle but not skin.

Assumption of the upright posture also leads to a reduction in hepatic blood flow, and there is increasing evidence that regulation of the splanchnic vasculature is important in maintaining the blood pressure. Other regulatory mechanisms to permit hemodynamic adjustments during postural changes include constriction of the capacitance (venous) beds and activation of the renin–angiotensin system.

The initiation of spontaneous physical activity is accompanied by a central command that arises above the level of the pons and influences cardiovascular and respiratory centers in the brainstem or elsewhere, leading to an immediate increase in cardiac and respiratory rates and blood pressure (28,29). The motor cortex is not essential for this purpose, and feedback mechanisms are not required for the genesis of respiratory and circulatory changes during exercise (28). There may also be cholinergically mediated vasodilation in some muscles (29–31).

The CNS plays an important integrative role in regulating cardiovascular function. Inputs to various brainstem "centers" arise not only in the periphery but also from other CNS structures. Neurogenic hypertension may result from lesions at different sites within the CNS, such as the hypothalamus. With brainstem lesions, such as bilateral lesions of the nucleus tractus solitarius, hypertension results from an increase in vasomotor tone and total peripheral resistance (32), but other factors, such as elevated plasma vasopressin levels, probably contribute as well (33). Posterior fossa tumors may cause hypertension, and compression or ischemia of the dorsal medullary reticular formation from either intrinsic pathology or increased intracranial pressure (Cushing response) may lead to hypertension, bradycardia, and apnea. By contrast, lesions of the ventrolateral medulla or rostral transection of the spinal cord lead to profound resting or postural hypotension in animals. Humans with brainstem tumors (34), syringobulbia (35), or spinal transection (36) also exhibit major disturbances of blood pressure regulation.

Ventilation and Bronchial Function

The importance of some medullary and pontine centers in the regulation of ventilation was noted earlier. The medullary respiratory neurons are influenced by peripheral (carotid sinus and aortic body) and central (medullary) chemoreceptors and by more rostral regions. Spinal transection in the upper cervical region interrupts the descending influence from the brainstem and abolishes all rhythmic ventilatory movements. Integration of various inputs to respiratory motor neurons, e.g., from the forebrain, also occurs at a spinal (segmental) level. The forebrain's influence on ventilation is evident from the changes that can be exerted volitionally, as during breath holding, and from the apnea that sometimes occurs during seizures (37). In animals, electrical stimulation of the uncus,

fornix, and amygdala and some cortical areas (anterior cingulate, anterior insular, inferior medial temporal, and posterior lateral frontal cortex) has an inhibitory effect on ventilation (38).

Sleep apnea, the impairment of automatic but not volitional ventilatory movements, may be central, obstructive, or mixed in type. Central sleep apnea is usually idiopathic but may result from unilateral or bilateral medullary infarction (39) or other brainstem pathology (40). Central pathology, such as syringobulbia (41) and olivopontocerebellar atrophy (40), may also cause obstructive sleep apnea. Sleep apnea sometimes occurs in patients with the dysautonomia of multisystem atrophy, who may also have hypopnea and laryngeal stridor. Iatrogenic sleep apnea has followed bilateral cervical tractotomy for pain (42).

Neuromuscular disorders can impair ventilation, necessitating supportive measures, as exemplified by some cases of poliomyelitis, amyotrophic lateral sclerosis, Guillain-Barré syndrome, and myasthenia gravis.

Bronchodilation results from sympathetic stimulation, whereas vagal stimulation leads to constriction of the bronchioles and may increase bronchial secretions.

Eyes, Pupils, and Lacrimal Glands

The sympathetic control of the pupil depends on an uncrossed three-neuron pathway. Axons descend from hypothalamic neurons to the intermediolateral cell column of the cord at the level of the first thoracic segment. Preganglionic sympathetic neurons project from there to the superior cervical ganglion. The adrenergic sympathetic postganglionic fibers to the pupil arise from the superior cervical ganglion, proceed adjacent to the internal carotid artery to the cavernous sinus, and then pass to the orbit to supply the dilator muscle of the pupil. Parasympathetic fibers to the pupil travel with the third nerve from the mesencephalic Edinger-Westphal nucleus to the ciliary ganglion in the orbit, from which cholinergic postganglionic fibers pass in the short ciliary nerves to the pupillary constrictor and ciliary muscle (accommodation).

Stimulation of the parasympathetic fibers (or application of pilocarpine) leads to pupillary constriction and contraction of the ciliary muscle; mydriasis occurs in response to atropine or with complete third nerve palsy. Degeneration of the ciliary ganglion or a postganglionic parasympathetic lesion is responsible for the tonic pupil of Adie; the pupil is large and reacts only sluggishly to light and accommodation. The abnormality is usually unilateral and may be associated with absent tendon reflexes or segmental anhidrosis (Ross syndrome). As a result of denervation supersensitivity, the pupil constricts in response to instillation of 0.125% pilocarpine or 2.5% methacholine, which has no effect on normal pupils.

Argyll Robertson pupils have come to be regarded as a hallmark of neurosyphilis, but they also occur in other disorders such as encephalitis, multiple sclerosis, and diabetes. They are small, irregular, unequal pupils that are unresponsive to light but reactive to convergence and accommodation. The responsible pathology probably involves the rostral midbrain.

Stimulation of sympathetic fibers causes pupillary dilation, whereas sympathetic lesions cause a small pupil, often associated with ptosis, enophthalmos, and anhidrosis (Horner syndrome). This may result from preganglionic or postganglionic (superior cervical ganglion) lesions and thus occurs, for example, with lateral medullary infarcts, cervical cord lesions, carotid artery thrombosis, pulmonary apical or mediastinal tumors, and injuries to the neck.

The lacrimal glands produce tears. The main gland receives parasympathetic innervation that arises in the pontine tegmentum and passes with the nervus intermedius to join the seventh cranial nerve as it enters the internal auditory meatus. The parasympathetic fibers eventually emerge as the greater superficial petrosal nerve and pass to the sphenopalatine ganglion before reaching the lacrimal glands, where they stimulate the production of tears.

Bladder

Bladder function is regulated by several different parts of the CNS, including the frontal cortex, basal ganglia, and lower brainstem. Lesions of the anterior frontal lobes are notorious for causing disturbances of micturition (43,44). There may be urinary incontinence, voluntary micturition in inappropriate circumstances, or both. Less frequently, urinary hesitancy or retention occurs with frontal lesions (43). In patients with hydrocephalus, the enlarged ventricles may stretch or distort corticobulbar fibers, leading to incontinence. It thus appears that descending fibers from the frontal lobe have a regulatory effect on micturition. Frequency, urgency, and urgency incontinence are common in patients with basal ganglia dysfunction resulting from Parkinson disease. Hesitancy and difficulty of micturition occur less often. Detrusor hyperreflexia and abnormal cystometric findings are common in parkinsonism (45), regardless of whether there are any symptoms of bladder dysfunction. Experimental studies in animals suggest that various pontomedullary regions (46) are also involved in regulating the bladder, but evidence in humans is lacking.

The cerebral influences on micturition (and defecation) are exerted through pathways traversing the spinal cord. With lesions interrupting afferent pathways in the cord, the normal sensation of bladder fullness is lost; the bladder becomes overdistended and overflow incontinence occurs. In addition, bladder emptying will occur reflexly if the influence of the supraspinal control mechanisms is lost, and detrusor-sphincter dyssynergia occurs. The pattern of voiding after complete traumatic myelopathy depends on the site of the lesion. With spinal lesions above the conus, urinary retention occurs in the acute phase and is then followed by reflex

voiding that is initiated by various stimuli and is incomplete. With lesions of the conus medullaris (or cauda equina), by contrast, there is urinary retention. In patients with incomplete cord lesions above the conus, such as cervical spondylotic myelopathy, urinary frequency, urgency, and urgency incontinence are common, and cystometry reveals detrusor hyperreflexia.

The bladder is supplied by both sympathetic and parasympathetic fibers, whereas the external sphincter receives a somatic innervation. Parasympathetic preganglionic fibers originate from the S2 through S4 segments of the cord, passing with the anterior roots and then in the pelvic nerves to form a diffuse network over the bladder, where they synapse with postganglionic cells innervating the bladder and urethra. The external urinary sphincter and penis/clitoris are supplied by pudendal (somatic) fibers arising from the sacral plexus. Afferent fibers responsible for reflex bladder contraction pass with the parasympathetic nerves to enter the spinal cord through the posterior S2 through S4 roots. Sympathetic preganglionic efferent fibers arise from the intermediolateral column of the spinal cord and pass through the sympathetic ganglia and splanchnic nerves to the hypogastric plexus. Postganglionic fibers arise from cells in this plexus or the vesical plexus and supply the bladder muscle. The role of the sympathetic efferent system in regulating micturition is unclear.

Urinary incontinence results from disruption of somatic afferent and efferent fibers to the external urinary sphincter, as by tumors, trauma, spinal stenosis, and diabetic polyradiculopathy. Selective damage to the neurons supplying the external sphincter occurs in Shy-Drager syndrome or multisystem atrophy (47).

Many elderly patients have urinary incontinence without evidence of any underlying neurologic disorder. In some instances, the incontinence occurs only at times of stress to the control system, as with coughing. Recent studies relate this to damage that has previously occurred to the striated pelvic floor sphincter muscles or their nerve supply, e.g., during childbirth (47). In other instances, incontinence occurs during confusional states or under circumstances in which voluntary control of sphincter function is reduced.

Gastrointestinal Tract

Parasympathetic fibers to the organs associated with digestion are secretory; they also cause increased gastrointestinal peristalsis and relaxation of sphincters. Conversely, sympathetic stimulation reduces peristalsis and secretion and increases sphincter tone. The anteromedial frontal lobes are involved in the regulation of defecation. Lesions in this region may lead to fecal incontinence (43).

Intramural plexuses (Auerbach plexus in the external muscular coat, Meissner submucosal plexus, and other ill-defined neuronal networks) are important for the regulation of gastrointestinal motility and secretions (48,49). This so-called enteric nervous system is regulated, in turn, by the sympathetic and parasympathetic systems, the former primarily via the celiac and mesenteric ganglia and the latter via the vagus and sacral (S2–S4) spinal nerves. Functionally complete circuits within the enteric nervous systems permit integrated motor and reflex responses to occur regardless of parasympathetic input. Such responses can, however, be modulated by vagal and sympathetic activity.

Dysphagia may occur for many different reasons, such as structural abnormalities or neuromuscular disorders (e.g., progressive bulbar palsy, brainstem stroke, or bulbar poliomyelitis). In some patients, however, it relates to a failure of the esophageal sphincter to relax (as in achalasia) or impaired esophageal peristalsis. Gastroparesis leads to a sense of postprandial fullness, discomfort, and vomiting and is a common feature of some dysautonomias, especially diabetic autonomic neuropathy. Somewhat similar symptoms may result from intestinal pseudoobstruction, such as sometimes occurs in amyloidosis. Constipation is a very common and rather nonspecific symptom, and its cause is frequently multifactorial. In Parkinson disease, for example, it may relate to antiparkinsonian medication, bradykinesia, reduced gastrointestinal motility (perhaps related to involvement of the dorsal motor nucleus of the vagus), or pelvic floor dysfunction; primary involvement of the enteric nervous system may also have a role (50). Traumatic myelopathy leads to paralytic ileus and fecal incontinence in the acute phase; after this period, the defecatory reflexes recover, but there is loss of voluntary control of defecation and an inability to strain.

There are many causes of fecal incontinence, but, in general, the neurologic ones are similar to those discussed earlier as causing disturbances of bladder function. Both pure autonomic failure and Shy-Drager syndrome may lead to constipation, fecal incontinence, and impaired gastrointestinal motility.

Sexual Function

The cerebral cortex and thalamus receive sensory inputs from the genitalia. The genitalia are represented in the parasagittal area of the primary sensory cortex. Frontal lobe lesions, especially those affecting the basomedial area, may affect social control and sexual behavior. The hypothalamus and limbic regions have an important role in sexual arousal. Electrical stimulation of various hypothalamic areas causes erection, whereas lesions in these regions, and especially of the medial preoptic–anterior hypothalamic area, suppress sexual behavior (51). Bilateral temporal lobectomy in monkeys (52) and humans (53) leads to an increase in sexual activity as well as other behavioral changes, and a similar clinical picture may occur with other encephalopathies (54). In a recent study of eight men during visually evoked sexual arousal, positron emission tomography revealed bilateral activation of the inferior temporal cortex (a visual association area) as well as activation of the right insula and inferior frontal cortex and left cingulate cortex (55).

Descending pathways traverse the midbrain, lower brainstem, and lateral columns of the cord to the thoracolumbar and sacral regions. Spinal cord lesions may have profound effects on sexual function, depending on the extent and level of the lesion. Many patients with complete cervical lesions have reflex erections, but the pleasurable experience of orgasm is abolished. With more caudal lesions, erectile and ejaculatory failure are common.

The thoracolumbar sympathetic supply to the sexual organs has a complex and variable course. Some preganglionic sympathetic fibers synapse with postganglionic fibers in the sympathetic chain, whereas others pass through to reach the inferior mesenteric and superior hypogastric plexuses where they synapse with postganglionic fibers that reach the pelvic organs via the hypogastric nerves, pelvic plexus, and cavernous nerves. Other sympathetic fibers reach the pelvic plexus with the parasympathetic nerves, and some travel in the pudendal nerves. Preganglionic parasympathetic (sacral) fibers form the pelvic nerves, which join the pelvic plexus. Both sympathetic and parasympathetic pathways are probably involved in arousal and orgasm (55). Lesions of the sacral nerve roots and pelvic nerves lead to erectile failure in men and failure of arousal in women.

The sympathetic and parasympathetic fibers innervate the blood vessels of various pelvic structures, erectile tissues in the penis and clitoris, and smooth muscle of the vagina, uterus, prostate, and seminal vesicles. The mechanisms involved in erection are complex and poorly understood. They are initiated by the parasympathetic system. Cavernosal tissue becomes engorged because of increased blood flow resulting from dilation of cavernosal and helicine arteries. This engorgement leads to compression of emissary veins, and venous drainage is therefore reduced. Penile or clitoral tumescence is maintained as intracorporeal pressure stabilizes at approximately the level of the systolic blood pressure. Detumescence results from contraction of the helicine arteries and the trabecular walls of the lacunar spaces of cavernosal tissue. Blood inflow is thus reduced, and decompression of veins leads to increased drainage. The sympathetic nervous system is primarily responsible for emission. The smooth muscle of the epididymis, vas deferens, seminal vesicles, and prostate contracts, as does the sphincter at the bladder neck to prevent retrograde ejaculation. Ejaculation (i.e., rhythmic seminal transport along, and expulsion from, the urethra) and orgasm are mediated primarily by pudendal somatic efferent and afferent fibers, respectively. Sympathetic outflow is also important; moreover, the secretion of fluids contributing to the seminal fluid is under sympathetic control. The neuropharmacologic basis of these events is unclear, but erection probably involves nitric oxide, vasoactive intestinal polypeptide, and endothelium-derived releasing factor, which are activated by the parasympathetic system (51). Neurologic mechanisms of sexual function and their clinical assessment are discussed in detail by Ike and Seagraves in Chapter 14 of this volume.

Thermoregulation and Sweating

Descending fibers from the hypothalamic preoptic area pass through the ipsilateral brainstem to the cells in the intermediolateral columns of the spinal cord. Preganglionic axons from these neurons pass through the white rami communicantes to the sympathetic ganglia, from which postganglionic cholinergic fibers emerge to join the peripheral nerves and ultimately to innervate the eccrine sweat glands. Several neuropeptides are found in the nerve terminals related to these glands and may function as cotransmitters (18). Thermoregulatory sweating occurs after a rise in central temperature that activates the preoptic hypothalamic area, but thermal receptors in other regions of the CNS and the skin are probably involved as well. Sweating may occur in response to emotion in some regions, such as the axillae, from apocrine glands. Palmar and pedal sweating also occurs in the absence of heat and is enhanced by emotional stress.

Excessive sweating, or hyperhidrosis, may occur as a generalized phenomenon with exercise, infections, or some metabolic disorders such as thyrotoxicosis, pheochromocytoma, or carcinoid syndrome. It also occurs with brainstem (pontine) ischemia. The most common cause, however, is probably anxiety. In essential hyperhidrosis, no specific cause can be found. Localized or regional hyperhidrosis may occur in association with lesions of the cord or peripheral nerves or as a result of sweat gland abnormalities.

Anhidrosis or hypohidrosis occurs distally in some patients with polyneuropathy and in a more restricted territory with discrete peripheral nerve or root involvement. Segmental anhidrosis occurring in association with Adie tonic pupils is labeled Ross syndrome. Anhidrosis sometimes occurs in dermatologic disorders or with a widespread distribution as a consequence of neuroleptic, anticholinergic, or tricyclic antidepressant drugs. Generalized or patchy anhidrosis is a feature of pure autonomic failure and central dysautonomias, such as Shy-Drager syndrome, and may occur with hypothalamic or posterior fossa tumors, syringomyelia or syringobulbia, multiple sclerosis, and other diffuse brainstem lesions. In Wallenberg syndrome and other unilateral brainstem lesions, there is ipsilateral hypohidrosis rather than anhidrosis. Chronic idiopathic anhidrosis is characterized by a disturbance of thermoregulatory sweating without evidence of more extensive autonomic involvement except for occasional pupillary abnormalities (56). Patients with this uncommon disorder are intolerant of heat and become hot, flushed, weak, and dyspneic, but do not sweat with activity or when the ambient temperature is high.

CLINICAL EVALUATION OF AUTONOMIC FUNCTION

A variety of noninvasive tests of autonomic function has been developed to test the integrity of parasympathetic (va-gal) fibers to the heart and sympathetic vasomotor and sudomotor fibers. Such tests can be used to evaluate the autonomic nervous system in patients with symptoms suggestive of dysautonomia. The aim is to determine the presence and severity of any autonomic disturbance and the site of the underlying lesion. These studies can also be used to determine the functional integrity of the unmyelinated and small myelinated fibers that comprise the peripheral component of the autonomic nervous system in patients with small-fiber peripheral neuropathies.

Heart Rate and Blood Pressure Responses

Measurement of the heart rate response to deep breathing is a noninvasive, sensitive, quantitative test that is simple to perform and provides a reliable index of the afferent and efferent parasympathetic (vagal) innervation of the heart. A tachycardia normally occurs during inspiration because of a reduction in cardiac vagal activity (Fig. 11.5). For clinical purposes, an electrocardiogram is used to record the R-R intervals continuously over 60 seconds in the recumbent patient. After a 5-minute rest period, the patient takes six deep breaths over 1 minute. The difference between the longest and shortest R-R intervals is noted. The variation of heart rate with respiration is age dependent (57) but is normally at least 15 beats per minute; values of less than 10 are clearly abnormal (58). In addition, an expiratory:inspiratory ratio is calculated from the ratio of the mean of the maximal R-R intervals during expiration to the mean of the minimal R-R intervals during inspiration. This ratio also declines with advancing age.

The heart rate response to standing is another useful, simple, noninvasive test of vagal function. On standing from the recumbent position, there is initial tachycardia followed by bradycardia that begins after approximately 20 seconds and stabilizes after approximately 30 heartbeats. For clinical purposes, the R-R interval of the electrocardiogram can be measured during the performance of the maneuver, and the ratio of the R-R interval at the thirtieth beat to the fifteenth beat is determined (59). This 30:15 ratio depends on vagal function and is age dependent (57), but in young adults, it normally exceeds 1.04.

The heart rate responses to passive head-up tilt can also be measured after the patient has remained supine for 10 minutes. There is normally an increase of between 10 and 30 beats per minute, reflecting a change in both sympathetic and parasympathetic activity, with 60-degree head-up tilt. The systolic and diastolic blood pressures usually fall slightly, but this does not exceed 20 and 10 mm Hg, respectively, in normal subjects. Recent studies have emphasized the value of maintaining the head-up tilt for as long as 60 minutes, particularly in patients in whom vasodepressor syncope is suspected (60).

The cardiovascular responses to sustained handgrip (30% of maximum for 5 minutes) have also been used as a means

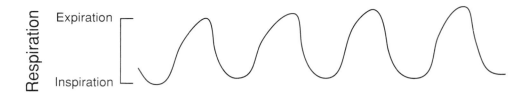

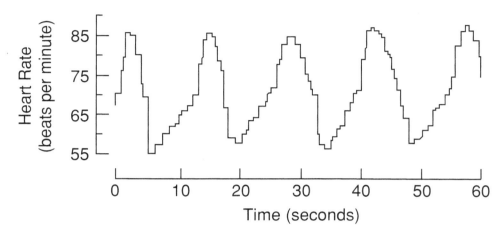

FIGURE 11.5. Variation in heart rate with deep breathing in a normal subject.

of assessing adrenergic function. The sustained muscle contraction produces a rise in heart rate and blood pressure. The diastolic pressure normally rises by at least 15 mm Hg, when the last value recorded before release of the handgrip is compared with the mean value obtained in the 3 minutes before commencing the maneuver. The test, however, is of limited sensitivity and specificity.

The Valsalva maneuver, in which the subject performs a forced expiration against a closed glottis, provides important information concerning the integrity of cardiovascular innervation. For clinical purposes, an electrocardiography machine or heart rate monitor can be used to record the heart rate responses (Fig. 11.6). The recumbent subject makes a sustained expiration that is sufficient to maintain a column of mercury at 40 mm for 15 seconds by blowing into a mouthpiece with a calibrated air leak. A Valsalva ratio can then be determined from the shortest interbeat interval (or fastest heart rate) generated during the forced expiration divided into the longest interbeat interval (slowest rate) that occurs after it. This reflects both sympathetic and vagal function. Ewing et al. (61) arbitrarily defined a value of 1.1 or less as abnormal and 1.21 or greater as normal, but the ratio is age dependent; in subjects younger than 40 years, the Valsalva ratio normally exceeds 1.4.

The Valsalva response was originally studied using an intraarterial cannula to record the blood pressure and heart rate (Fig. 11.7). This can now be determined noninvasively using a photoplethysmographic recording device (Finapres) that also permits beat-to-beat variation in blood pressure during the maneuver to be determined. The response to the Valsalva maneuver is traditionally divided into four phases. In phase I, there is a transient increase in blood pressure as the increase in intrathoracic pressure is transmitted to the great vessels. In phase II, the blood pressure falls owing to a reduced cardiac output (resulting from the decline in venous return). This is associated with tachycardia caused by diminished vagal activity. The decline in blood pressure is arrested within approximately 5 seconds and then reversed as increased sympathetic activity increases the peripheral resistance. With release of the forced expiratory maneuver, there is a transient decline in blood pressure owing to pooling of blood and expansion of the pulmonary vascular bed with the reduced intrathoracic pressure (phase III), after which the blood pressure overshoots its original baseline value as cardiac output returns to normal while the tachycardia and peripheral vasoconstriction persist (phase IV). This overshoot of the blood pressure leads in turn to compensatory bradycardia. Dysautonomic patients typically show a continuous and excessive decline in blood pressure in phase II and no overshoot in phase IV (Fig. 11.8).

Startle (as by a sudden loud noise) or mental stress (induced, for example, by attempting to perform the serial-sevens test despite constant distraction) normally leads to an increase in heart rate and blood pressure. In the cold

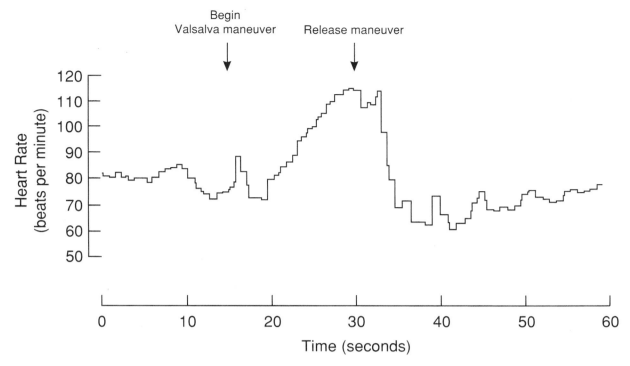

FIGURE 11.6. Response to the Valsalva maneuver, recorded with a heart rate monitor in a normal subject. There is tachycardia during the forced expiratory maneuver and compensatory bradycardia when the maneuver is released.

pressor test, immersion of one hand in cold (4°C) water normally produces an increase in systolic pressure of 15 mm Hg within 1 minute. Abnormalities of these responses imply a lesion centrally or in sympathetic efferent pathways. Normal responses in patients with an abnormal response to the Valsalva maneuver suggest an afferent baroreceptor lesion.

Tests of Cutaneous Vasomotor Function

Cutaneous blood flow can be studied by plethysmography or a laser Doppler flow meter. The vasomotor responses to various stimuli are evaluated as a measure of adrenergic function. For convenience, digital blood flow is recorded for clinical purposes.

Valsalva Maneuver

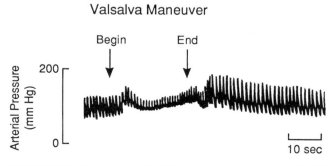

FIGURE 11.7. Response to the Valsalva maneuver, recorded with an intraarterial needle in a normal subject.

A sudden inspiratory gasp normally produces a digital vasoconstriction as a spinal reflex (Fig. 11.9). The response is lost or impaired in patients with a cord lesion or a disturbance of sympathetic efferent pathways, such as in peripheral neuropathies (62,63). A cold stimulus (ice cold water at 4°C) to the opposite hand also produces a reflex vasoconstriction, as does mental stress.

Sweat Tests

The thermoregulatory sweat test is a highly sensitive test in which the body temperature is raised by 1°C by exposure of the subject to radiant heat from a heat cradle. The presence and distribution of sweating are determined by a change in color of an indicator powder placed on the skin. Abnormalities may reflect pre- or postganglionic lesions.

The quantitative sudomotor axon reflex test is a test of postganglionic sudomotor sympathetic function (64). Axon terminals are activated by iontophoresed acetylcholine; the resulting impulses travel antidromically to a branching point and then orthodromically down another branch of the axon to its terminals, where they release acetylcholine, which generates a sweat response. The sudomotor responses elicited are measured quantitatively. The test is highly sensitive and specific and yields reproducible results. Normal findings are obtained in preganglionic disorders.

The sympathetic skin response is another means of evaluating sudomotor function. It depends on the electrical activity arising from sweat glands and adjacent tissues either

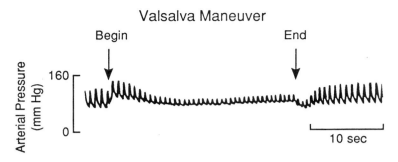

FIGURE 11.8. Abnormal, intraarterially recorded response to the Valsalva maneuver in a patient with Shy-Drager syndrome.

spontaneously or in response to particular stimuli. For clinical purposes, electrical stimulation of a mixed or cutaneous nerve is used to elicit a response from the palms of the hands and soles of the feet, and the resulting change in voltage is measured from the skin surface (65). Although easy to perform, the sympathetic skin response is of uncertain specificity, is not quantitative, and habituates. Accordingly, it can reliably be considered as abnormal only when it is absent.

Plasma Catecholamine Levels and Norepinephrine Infusion

The resting plasma norepinephrine level provides an index of sympathetic activity. It is diminished in disorders with postganglionic (as opposed to preganglionic) pathology. The plasma norepinephrine level normally increases with a change in posture from recumbent to standing, and this postural response may be markedly attenuated or absent with pre- or postganglionic lesions.

Another approach is to measure the blood pressure changes that occur with intravenous infusion of norepinephrine at different dose rates up to 20 μg per minute (66). To increase the systolic pressure to 40 mm Hg above baseline, normal subjects require infusion at a rate of 15 to 20 μg per minute, whereas patients with Shy-Drager syndrome require 5 to 10 μg per minute and those with primary autonomic failure less than 2.5 μg per minute (66).

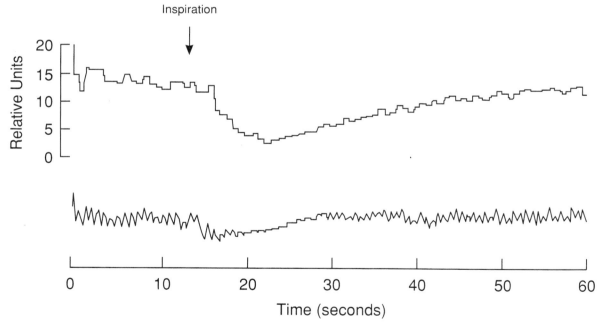

FIGURE 11.9. Normal variation in blood volume after a deep inspiration recorded photoplethysmographically by an infrared emitter and detector on the pad of the index finger. The bottom trace represents the sensor output after amplification by the photoplethysmographic module of a computerized autonomic testing system and is a function of the absolute blood volume in the finger. Each peak represents a heartbeat, and the amplitude of each wave reflects blood volume in the area around the sensor. The apparent shift of the DC signal component results from the long time constant required to prevent loss of signal information. The upper trace shows the relative voltage, representing the amplitude of each pulse. After the deep inspiration, there is a reduction in digital blood flow (i.e., reduced amplitude of the waveforms in the lower trace and a corresponding decline in the upper trace).

Bladder and Gastrointestinal Function

Bladder function is assessed by several methods. These include determination of the volume of residual urine after attempted voiding, which is normally less than 100 mL. Cystometry is an important investigative approach in which the bladder is filled with water or radiologic contrast medium or inflated with carbon dioxide via a self-retaining catheter, and the relationship between intravesical pressure and volume is determined. Bladder filling is normally first appreciated at volumes of 100 to 200 mL, and the desire to void increases as filling continues. Impaired sensation may reflect peripheral or central pathology. In the normal subject, vesical pressure does not increase by more than 10 cm H_2O until maximal capacity (generally between 300 and 600 mL) is reached. Detrusor hyperreflexia is found in patients with central (cerebral or spinal) disturbances of micturition, and detrusor areflexia (i.e., an absence of detrusor reflex contractions during cystometry) occurs in patients with peripheral (root or nerve) pathology. It is also helpful to measure urinary flow and, in particular, the mean and maximal flow rates, duration and pattern of flow, and volume voided. The technique is helpful in distinguishing between flow disturbances from obstruction and abnormal detrusor activity.

Radiologic studies are important in evaluating gastrointestinal function and especially in excluding mechanical causes of symptoms such as dysphagia, delayed gastric emptying, intestinal pseudoobstruction, or intractable constipation. Fluoroscopy and studies of colonic transit time may also be helpful. Catheter probes connected to a pressure transducer can be used to measure anorectal pressure. In the region of the anal sphincter, resting pressure reflects internal anal sphincter tone. Activity in the external anal sphincter muscle is reflected by the pressure recorded on voluntary contraction. Rectal distention with a balloon causes reflex relaxation of the internal anal sphincter. This rectoanal reflex depends on intramural nerves (67) and is absent in Hirschsprung disease.

Sphincter Electromyography and Evoked Potential Studies

Denervation of the urethral and anal sphincters may result from cell loss in the Onuf nucleus in the sacral cord or from pathology situated more distally. Needle electromyography permits recognition of abnormal motor unit action potentials in the sphincteric muscles in patients with dysautonomia and thereby helps to localize the lesion. With pyramidal lesions, volitional control of sphincter function is impaired, and the urethral striated sphincter muscle fails to relax during detrusor contraction (detrusor–sphincter dyssynergia).

The cerebral potentials elicited by electrical stimulation in the anorectum, at the neck or trigone of the bladder, or of the pudendal nerve (dorsal nerve of the penis or clitoris) and pelvic urethral nerve are easy to record. The utility of such evoked potential studies, which are somewhat uncomfortable, in the investigation of patients with impotence or a disturbance of voiding or defecation is unclear.

Pharmacologic Evaluation of Pupillary Reactivity

Anticholinergic (parasympatholytic) agents applied topically cause pupillary dilation, as do sympathomimetic agents. Conversely, parasympathomimetic or sympatholytic agents cause pupillary constriction.

Pilocarpine and methylcholine are cholinergic (parasympathomimetic) agents. In patients with parasympathetic denervation, very weak solutions of pilocarpine (e.g., 0.125%) will cause pupillary constriction because of denervation supersensitivity. A dilated pupil that is unresponsive to instilled cholinergic agents probably reflects sympathetic overactivity.

Sympathomimetic agents cause dilation of denervated pupils at concentrations that are ineffective on normal pupils. For example, 1% phenylephrine hydrochloride produces dilation of the pupil in Horner syndrome because of denervation supersensitivity. Cocaine hydrochloride (4%) can also be instilled into the conjunctival sac and normally causes pupillary dilation by sympathetic activation. In Horner syndrome resulting from peripheral sympathetic denervation, this effect of cocaine is lost.

CLINICAL FEATURES OF DYSAUTONOMIA

The clinical features of dysautonomia vary in different patients and depend in part on the nature of the underlying disorder, as do the existence and nature of any somatic neurologic manifestations. For example, dysautonomia occurring in diabetics may be associated with symptoms and signs suggestive of a peripheral neuropathy, whereas parkinsonism, cerebellar dysfunction, a pyramidal disturbance, or some combination of these abnormalities occurs in patients with central disorders of autonomic function, such as Shy-Drager syndrome.

Impotence and loss of libido are common presenting features in men with dysautonomia and are often mistakenly attributed to psychogenic factors until other autonomic disturbances develop. Ejaculation may also be impaired. When retrograde ejaculation occurs, patients feel as if they have ejaculated but little is produced; the subsequently voided urine is found to be discolored and to contain spermatozoa. Abnormalities of micturition are another common early symptom and in men may be attributed to prostatic hypertrophy.

The most disabling symptom of autonomic dysfunction is usually postural hypotension. A feeling of impending loss of consciousness occurs on standing or walking, and the patient may ultimately fall to the ground unconscious unless symptoms are aborted by sitting or lying down. Unlike a typical syncopal episode, there is no preceding sweating. Symptomatic postural hypotension is more likely to occur in the early part of the day, postprandially, with activity, or in hot weather because of the cardiovascular and hemodynamic alterations that occur under these circumstances.

An impairment or loss of thermoregulatory sweating is common and may be life threatening in hot climates. The distribution of any anhidrosis may suggest the site of pathology. For example, impaired sweating distally in the extremities suggests an underlying peripheral neuropathy, whereas a more generalized disturbance of sweating, with involvement of the trunk, implies a central lesion.

Disturbances of bladder or bowel regulation are especially distressing. Urinary urgency, frequency, and nocturia may occur, and there may be urgency incontinence. In other instances, an atonic bladder leads to overflow incontinence. Cystometry clarifies the nature of the bladder disturbance and thereby indicates the appropriate therapeutic approach. Intractable constipation, intermittent diarrhea, or an alternation between constipation and diarrhea also occurs. Rectal incontinence during episodes of diarrhea is particularly troublesome. Nausea, postprandial fullness, and severe vomiting may reflect gastroparesis.

Visual blurring and dryness of the eyes are sometimes troublesome, and there may be a variety of pupillary abnormalities in patients with dysautonomia. The effect of lesions affecting the parasympathetic innervation (i.e., the third cranial nerve) or sympathetic system (causing Horner syndrome) have already been described.

Disturbances of breathing may consist of inspiratory gasps or cluster breathing in patients with the Shy-Drager syndrome of autonomic failure. Occasional patients experience central sleep apnea. Laryngeal stridor may also occur.

SELECTED CAUSES OF DYSAUTONOMIA
Acute Peripheral Pathology

There are many causes of autonomic dysfunction. Neuropathies are among the most important causes. Acute involvement of peripheral autonomic fibers may occur in several contexts. First, a pure autonomic neuropathy may occur acutely or subacutely, either as an idiopathic or postviral disorder or a paraneoplastic syndrome, with diffuse, generalized sympathetic and parasympathetic dysfunction (68–70). Examination typically reveals significant postural hypotension, a fixed heart rate, anhidrosis, xerostomia, dry eyes, a distended bladder, and dilated unresponsive pupils. There is often no motor or sensory deficit, but minor sensory abnormalities are sometimes evident, especially an impairment of those sensory modalities mediated by small fibers (pain and temperature appreciation). Recovery occurs gradually and may be incomplete. The dysautonomia is sometimes confined to the cholinergic system (71), in which case postural hypotension does not occur.

Autonomic dysfunction is common in patients with a peripheral neuropathy but is usually overshadowed by the coexisting motor and sensory deficits. In Guillain-Barré syndrome, however, autonomic disturbances may be life threatening (72–74). There may be severe hypotension, paroxysmal hypertension, or extreme fluctuations in blood pressure; disturbances of cardiac rhythm, regional anhidrosis or hyperhidrosis, pupillary abnormalities, gastroparesis, constipation, urinary retention, and urinary or fecal incontinence also occur. The cause of the paroxysmal hypertension is unclear, but denervation supersensitivity to circulating catecholamines or denervation of baroreceptors may be responsible.

Autonomic dysfunction, resulting from a disturbance of cholinergic mechanisms, is a feature of botulism. Constipation, urinary retention, dryness of the eyes, anhidrosis, and xerostomia present the most common evidence of such dysfunction. Disorders of autonomic, especially cholinergic, function are well described in Lambert-Eaton myasthenic syndrome (75).

A number of iatrogenic neuropathies may be associated with autonomic involvement, such as those related to amiodarone, perhexiline maleate, cis-platinum, or vincristine. Dysautonomia also follows exposure to acrylamide, organic solvents, and Vacor (a rodenticide).

Chronic Peripheral Pathology

The most common cause of dysautonomia in developed countries is probably diabetes. Diabetic autonomic neuropathy may occur in isolation or in association with any of the neuromuscular complications of diabetes, especially a symmetric sensory or sensorimotor polyneuropathy, entrapment neuropathy, mononeuropathy multiplex, polyradiculopathy, or plexopathy. Postural hypotension, abnormal cardiovagal function, and impaired distal thermoregulatory sweating are typically found; impotence, gastroparesis, constipation, fecal incontinence (often in association with diarrhea), and bladder disturbances are not uncommon (76,77).

Autonomic disturbances also occur in patients with the neuropathy of chronic renal failure (78), leprosy (79), vitamin B_{12} deficiency (80), and various connective tissue diseases (81). Indeed, a chronic autonomic disturbance may occur with any polyneuropathy but especially a neuropathy involving small fibers. Axonal neuropathies are more likely to be associated with autonomic disturbances than are demyelinating neuropathies.

Chagas disease occurs mainly in patients from South or Central America, results from infection with the protozoan *Trypanosoma cruzi*, and is especially likely to cause chronic dysautonomia. Cholinergic dysfunction predominates, with involvement of the heart and gastrointestinal system. Cardiomegaly and cardiac conduction defects are frequent, and there may be distention of parts of the gastrointestinal tract, especially the esophagus, duodenum, and colon. Postural hypotension, abnormal cardiovascular reflexes

(82–84), and reduced plasma norepinephrine levels (85) may be found.

Autonomic dysfunction is an early cause of symptoms in either sporadic systemic amyloid neuropathy or familial amyloidosis. Postural hypotension is common and disabling (86,87). Abnormalities of gastrointestinal motility reflect infiltration of the enteric plexuses. Anhidrosis is common, as is impotence in men. Somatic neurologic deficits typically consist of impaired pain and temperature appreciation resulting from an underlying axonal neuropathy, with deposition of amyloid. Examination of autonomic function reveals marked postural hypotension, abnormal heart rate responses to deep inspiration and the Valsalva maneuver, widely impaired thermoregulatory sweating, and impaired peripheral vasomotor responses to such maneuvers as deep inspiration.

Various hereditary disorders are associated with autonomic failure. Familial dysautonomia (Riley-Day syndrome) is a recessively inherited neuropathy that begins in infancy with disturbances of thermoregulation, lacrimation, blood pressure regulation, and gastrointestinal function. Repeated episodes of pneumonia are typical. Somatic involvement is manifest by impaired pain and temperature appreciation, weakness, poor sucking, depressed tendon reflexes, and arthropathy. Several other types of hereditary sensory and autonomic neuropathy have been described (88).

Autonomic dysfunction sometimes occurs in hereditary motor and sensory neuropathy type I or II and is manifest especially by pupillary changes, impaired vasomotor regulation of the distal blood vessels, and occasionally by impaired cardiovascular reflexes (89,90). Minor abnormalities of dubious clinical relevance have also been reported in myotonic dystrophy (89).

Dopamine β-hydroxylase deficiency is a rare disorder affecting central and peripheral adrenergic neurons that was described earlier in this chapter. It leads to postural hypotension and a variety of other symptoms. Improvement occurs with administration of dihydroxyphenylserine, which is converted endogenously to norepinephrine (91).

Central Dysautonomias

Autonomic manifestations are a feature of many different disorders of the CNS. Seizures, for example, may have autonomic accompaniments, especially when they arise from limbic and paralimbic structures. The most common clinical evidence of autonomic involvement during seizures is a sinus tachycardia, but a variety of other cardiac manifestations may occur ictally, including atypical anginal pain, bradycardias, conduction defects, sinus arrest, tachyarrhythmias, and atrial fibrillation (92). Another common, possibly dysautonomic, feature of seizures, especially complex partial attacks arising from the temporal lobe, is a curious feeling of discomfort that ascends from the epigastrium, accompanied sometimes by nausea. Pupillary dilation, respiratory arrest, other changes in respiratory rate or pattern, blood pressure changes, cu-

taneous vasodilation or vasoconstriction, piloerection (93), and vomiting (94) are other autonomic accompaniments of seizures, especially those arising from the temporal lobe. When vomiting occurs as a seizure phenomenon, patients are typically unaware of vomiting, and other ictal phenomena are associated (94).

Lesions of the mesial frontal lobe, especially the cingulate gyrus, may produce disturbances of sphincter function, with urinary incontinence, fecal incontinence, or both. Hydrocephalus sometimes leads to a similar effect, presumably because of distortion and stretch of corticobulbar fibers.

Hypothalamic pathology may affect temperature regulation, mainly leading to hypothermia. This may be the basis of the hypothermia that is sometimes associated with episodic hyperhidrosis and agenesis of the corpus callosum (95,96). In other instances, episodic hyperthermia occurs after head injury or with hypothalamic pathology or acute hydrocephalus (97) and may be associated with tachycardia, hyperpnea, transient hypertension, cutaneous vasoconstriction or vasodilation, pupillary changes, and increased muscle tone. Hypothalamic pathology such as infarction occasionally leads to ipsilateral Horner syndrome (98). Pathologic involvement of the anterior hypothalamus may affect circadian rhythms (99).

Diseases of the basal ganglia are also accompanied by autonomic dysfunction. In patients with Parkinson disease, many symptoms, such as disturbances of sweating, postural dizziness, and sphincter dysfunction, suggest autonomic involvement. However, the cardiovascular reflexes are generally normal, although there may be increased sensitivity to infused catecholamines. This finding may suggest a subtle disturbance of autonomic function owing to a central rather than a peripheral lesion (100,101). In some patients, there seems to be a disturbance in "set" of the baroreceptor reflexes without disruption of their integrity (100,102). Autonomic symptoms also occur in other extrapyramidal disorders. Some patients with progressive supranuclear palsy have postural hypotension, but this does not usually exceed the postural change sometimes encountered in normal subjects (103,104). In Huntington disease, there may be sphincter disturbances, abnormalities of swallowing and respiration, hyperhidrosis, sialorrhea, polyuria, polydipsia, and hypogenitalism (101). Pathologic changes in this disorder involve the caudate nuclei, and these structures have been implicated in the regulation of blood pressure during change in posture. Aminoff and Gross (105) investigated 11 patients with Huntington disease and found no abnormality of either resting blood pressure or baroreceptor reflex responses; there was nevertheless a significantly greater fall in blood pressure on a 60-degree head-up tilt compared with controls, implying that suprabulbar structures including the caudate nuclei may indeed influence postural vasoregulatory mechanisms without affecting baroreceptor reflexes.

Autonomic hyperactivity (with tachycardia, hypertension, hyperhidrosis, and hyperthermia) is a major feature of fatal familial insomnia, in which it is conjoined with intractable insomnia and motor abnormalities including myoclonus, ataxia, and pyramidal deficits. The disorder is associated with pathology involving the anterior ventral and dorsomedial nuclei of the thalamus (106).

Tumors, ischemia, or degenerative disorders affecting the brainstem or cerebellum may lead to postural hypotension (34,107) or, less commonly, to hypertension (33). In some cases, the blood pressure abnormality precedes development of other neurologic deficits, and its cause may go unrecognized unless the CNS is imaged. Blood pressure regulation may be impaired in Wernicke encephalopathy owing to central or peripheral pathology (108,109).

Shy-Drager syndrome (or multisystem atrophy) is a condition in which marked autonomic dysfunction occurs in association with a somatic neurologic deficit characterized primarily by parkinsonian features but also by cerebellar signs, pyramidal deficits, and sometimes lower motor neuron involvement (110). It can therefore be distinguished by its neurologic accompaniments from the syndrome of primary (or pure) autonomic failure, in which the autonomic dysfunction occurs in isolation. Shy-Drager syndrome may simulate classic Parkinson disease, but the existence of more widespread neurologic deficits and an impairment of baroreceptor reflexes distinguish it from the latter disorder. The disorder tends to have a progressive course, and many patients die within 5 years of the diagnosis being established. The dysautonomia results primarily from a loss of sympathetic cells in the intermediolateral cell columns of the spinal cord, and there may also be pathologic changes in the dorsal nucleus of the vagus, nucleus tractus solitarius, and locus ceruleus in the brainstem. Dysautonomia may occur in patients with olivopontocerebellar atrophy or striatonigral degeneration; these disorders probably reflect different manifestations of multisystem atrophy (111).

Complete lesions of the spinal cord (e.g., from trauma) have major effects on autonomic function. After transection of the cervical cord, reflex function returns to the isolated spinal segment after a variable period (usually a few weeks), but cerebral regulation of autonomic activity is lost. With lesions above T6, the resting blood pressure is reduced, and there is marked orthostatic hypotension, with an overshoot of the blood pressure on resumption of a recumbent posture (36). Changes in cardiac rate still occur with change in posture, so that on head-up tilt, the heart rate increases; this is because the influence of the vagus nerve on the heart is preserved. There may be disturbances of temperature regulations because of an inability to sweat or alter vasomotor function below the level of the lesion. Bladder, bowel, and sexual function are markedly impaired, as discussed earlier. Autonomic dysreflexia is sometimes a major management problem in these patients. Visceral, muscle, or cutaneous stimulation below the level of the lesion leads to reflex sympathetic and parasympathetic excitation, with consequent activity in a number of organs supplied by the autonomic

nervous system. A marked and rapid elevation of the blood pressure may lead to intracranial hemorrhage. Stroke volume and cardiac output also increase. Cutaneous vasodilation and sweating sometimes occur above the level of the lesion, but the mechanism of this is unclear.

Cardiac arrhythmias, neurogenic hypertension, and acute pulmonary edema may occur in patients with acute intracranial pathology, such as subarachnoid hemorrhage or increased intracranial pressure, and sometimes lead to sudden death. They probably relate to excessive sympathetic activity.

Old Age

Many subjects older than 70 years of age have a postural drop in systolic pressure of 20 mm Hg or more on standing (112). There may be alterations in baroreceptor sensitivity with age, and reduction in the elasticity of blood vessels and of adrenoreceptor sensitivity may also account for the postural drop in blood pressure (113). Syncope is a common problem in the elderly and often occurs for unclear reasons (114).

TREATMENT OF DYSAUTONOMIA

The treatment of the various manifestations of autonomic dysfunction is beyond the scope of this chapter. For further information, the reader is referred to other sources (115,116).

REFERENCES

1. Wallin BG, Elam M. Microneurography and autonomic dysfunction. In: Low PA, ed. *Clinical autonomic disorders*, 2nd ed. Philadelphia: Lippincott–Raven, 1997:233–243.
2. Loewy AD. Anatomy of the autonomic nervous system: an overview. In: Loewy AD, Spyer KM, eds. *Central regulation of autonomic functions.* New York: Oxford University Press, 1990:3–16.
3. Cechetto DF, Saper CB. Role of the cerebral cortex in autonomic function. In: Loewy AD, Spyer KM, eds. *Central regulation of autonomic functions.* New York: Oxford University Press, 1990:208–223.
4. Wannamaker BB. Autonomic nervous system and epilepsy. *Epilepsia* 1985;26[Suppl 1]:S31–S39.
5. Hilton SM. Inhibition of baroreceptor reflexes on hypothalamic stimulation. *J Physiol* 1963;165:56P–57P.
6. McAllen RM. Inhibition of the baroreceptor input to the medulla by stimulation of the hypothalamic defence area. *J Physiol* 1976;257:45P–46P.
7. Bradley DJ, Ghelarducci B, Paton JFR, et al. The cardiovascular responses elicited from the posterior cerebellar cortex in the anaesthetized and decerebrate rabbit. *J Physiol* 1987;383:537–550.
8. Supple WF, Leaton RN. Cerebellar vermis: essential for classically conditioned bradycardia in the rat. *Brain Res* 1990;509:17–23.
9. Spyer KM. Physiology of the autonomic nervous system: CNS control of the cardiovascular system. *Curr Opin Neurol Neurosurg* 1991;4:528–532.

10. Loewy AD. Central autonomic pathways. In: Loewy AD, Spyer KM, eds. *Central regulation of autonomic functions.* New York: Oxford University Press, 1990:88–103.
11. Doba N, Reis DJ. Acute fulminating neurogenic hypertension produced by brainstem lesions in the rat. *Circ Res* 1973;32:584–593.
12. Barron KD, Chokroverty S. Anatomy of the autonomic nervous system: brain and brainstem. In: Low PA, ed. *Clinical autonomic disorders.* Boston: Little, Brown, 1993:3–15.
13. Mitchell RA, Berger AJ. Neural regulation of respiration. In: Hornbein TF, ed. *Regulation of breathing, part I.* New York: Marcel Dekker, 1981:541–620.
14. Lumsden T. Observations on the respiratory centres in the cat. *J Physiol* 1923;57:153–160.
15. Berger AJ, Mitchell RA, Severinghaus JW. Regulation of respiration. *N Engl J Med* 1977;297:92–97, 138–143, 194–201.
16. Low PA, Okazaki H, Dyck PJ. Splanchnic preganglionic neurons in man: I. Morphometry of preganglionic cytons. *Acta Neuropathol (Berl)* 1977;40:55–61.
17. Gibbins IL, Morris JL. Sympathetic noradrenergic neurons containing dynorphin but not neuropeptide Y innervate small cutaneous blood vessels of guinea-pigs. *J Auton Nerv Syst* 1990;29:137–149.
18. Tainio H, Vaalasti A, Rechardt L. The distribution of substance P-, CGRP-, galanin-, and ANP-like immunoreactive nerves in human sweat glands. *Histochem J* 1987;19:375–380.
19. Harati Y. Anatomy of the spinal and peripheral autonomic nervous system. In: Low PA, ed. *Clinical autonomic disorders.* Boston: Little, Brown, 1993:17–37.
20. Pettersson A, Ricksten S-E, Towle AC, et al. Effect of blood volume expansion and sympathetic denervation on plasma levels of atrial natriuretic factor (ANF) in the rat. *Acta Physiol Scand* 1985;124:309–311.
21. Schotzinger RJ, Landis SC. Acquisition of cholinergic and peptidergic properties by sympathetic innervation of rat sweat glands requires interaction with normal target. *Neuron* 1990;5:91–100.
22. Bell C. Anatomical aspects of growth and ageing in the autonomic nervous system. *Curr Opin Neurol Neurosurg* 1991;4:524–527.
23. Scott SA, Davies AM. Inhibition of protein synthesis prevents cell death in sensory and parasympathetic neurons deprived of neurotrophic factors in vitro. *J Neurobiol* 1990;21:630–638.
24. Chang JY, Martin DP, Johnson EM. Interferon suppresses sympathetic neuronal cell death caused by nerve growth factor deprivation. *J Neurochem* 1990;55:436–445.
25. Gurusinghe CJ, Bell C. Substance P immunoreactivity in the superior cervical ganglia of normotensive and genetically hypertensive rats. *J Auton Nerv Syst* 1989;27:249–256.
26. Gurusinghe CJ, Harris PJ, Abbott DF, et al. Neuropeptide Y in rat sympathetic neurons is altered by genetic hypertension and by age. *Hypertension* 1990;16:63–71.
27. Messina A, Bell C. Are genetically hypertensive rats deficient in nerve growth factor? *Neuroreport* 1991;2:45–48.
28. Eldridge FL, Millhorn DE, Kiley JP, et al. Stimulation by central command of locomotion, respiration and circulation during exercise. *Respir Physiol* 1985;59:313–337.
29. Joyner MJ, Shepherd JT. Autonomic regulation of circulation. In: Low PA, ed. *Clinical autonomic disorders,* 2nd ed. Philadelphia: Lippincott–Raven, 1997:61–71.
30. Sanders JS, Mark AL, Ferguson DW. Evidence for cholinergically mediated vasodilation at the beginning of isometric exercise in humans. *Circulation* 1989;79:815–824.
31. Dietz NM, Rivera JM, Eggener SE, et al. Nitric oxide contributes to the rise in forearm blood flow during mental stress in humans. *J Physiol* 1994;480:361–368.
32. Doba N, Reis DJ. Role of central and peripheral adrenergic mechanisms in neurogenic hypertension produced by brainstem lesions in rat. *Circ Res* 1974;34:293–301.
33. Sved AF, Imaizumi T, Talman WT, et al. Vasopressin contributes to hypertension caused by nucleus tractus solitarius lesions. *Hypertension* 1985;7:262–267.
34. Hsu CY, Hogan EL, Wingfield W, et al. Orthostatic hypotension with brain stem tumors. *Neurology* 1984;34:1137–1143.
35. Aminoff MJ, Wilcox CS. Autonomic dysfunction in syringomyelia. *Postgrad Med J* 1972;48:113–115.
36. Mathias CJ, Frankel HL. Cardiovascular control in spinal man. *Annu Rev Physiol* 1988;50:577–592.
37. Plum F. Neurological integration of behavioural and metabolic control of breathing. In: Porter R, ed. *Breathing: Hering-Breuer centenary symposium.* London: Churchill, 1970:159–175.
38. Kaada BR. Somato-motor, autonomic and electrocorticographic responses to electrical stimulation of "rhinencephalic" and other structures in primates, cat and dog. *Acta Physiol Scand Suppl* 1951;24:83.
39. Levin BE, Margolis G. Acute failure of automatic respirations secondary to a unilateral brainstem infarct. *Ann Neurol* 1977;1:583–586.
40. Adelman S, Dinner DS, Goren H, et al. Obstructive sleep apnea in association with posterior fossa neurologic disease. *Arch Neurol* 1984;41:509–510.
41. Haponik EF, Givens D, Angelo J. Syringobulbia-myelia with obstructive sleep apnea. *Neurology* 1983;33:1046–1049.
42. Tranmer BI, Tucker WS, Bilbao JM. Sleep apnea following percutaneous cervical cordotomy. *Can J Neurol Sci* 1987;14:262–267.
43. Andrew J, Nathan PW. Lesions of the anterior frontal lobes and disturbances of micturition and defaecation. *Brain* 1964;87:233–262.
44. Maurice-Williams RS. Micturition symptoms in frontal tumours. *J Neurol Neurosurg Psychiatry* 1974;37:431–436.
45. Murnaghan GF. Neurogenic disorders of the bladder in parkinsonism. *Br J Urol* 1961;33:403–409.
46. Holstege G, Tan J. Supraspinal control of motoneurons innervating the striated muscles of the pelvic floor including urethral and anal sphincters in the cat. *Brain* 1987;110:1323–1344.
47. Swash M. Sphincter disorders and the nervous system. In: Aminoff MJ, ed. *Neurology and general medicine,* 3rd ed. Philadelphia: Churchill Livingstone, 2001:537–555.
48. Wood JD. Enteric neurophysiology. *Am J Physiol* 1984;247:G585–G598.
49. Furness JB, Bornstein JC, Smith TK. The normal structure of gastrointestinal innervation. *J Gastroenterol Hepatol* 1990;1:1–9.
50. Camilleri M. Disturbances of gastrointestinal motility and the nervous system. In: Aminoff MJ ed. *Neurology and general medicine,* 3rd ed. Philadelphia: Churchill Livingstone, 2001:261–276.
51. Steward JD. Autonomic regulation of sexual function. In: Low PA, ed. *Clinical autonomic disorders,* 2nd ed. Philadelphia: Lippincott–Raven, 1997:129–134.
52. Klüver H, Bucy PC. "Psychic blindness" and other symptoms following bilateral temporal lobectomy in rhesus monkeys. *Am J Physiol* 1937;119:352–353.
53. Terzian H, Ore GD. Syndrome of Klüduver and Bucy reproduced in man by bilateral removal of the temporal lobes. *Neurology* 1955;5:373–380.
54. Lilly R, Cummings JL, Benson DF, et al. The human Klüduver-Bucy syndrome. *Neurology* 1983;33:1141–1145.
55. Stuleru S, Gregoire MC, Gerard D, et al. Neuroanatomical correlates of visually evoked sexual arousal in human males. *Arch Sex Behav* 1999;28:1–21.

56. Low PA, Fealey RD, Sheps SG, et al. Chronic idiopathic anhidrosis. *Ann Neurol* 1985;18:344–348.

57. Vita G, Princi P, Calabro R, et al. Cardiovascular reflex tests: assessment of age-adjusted normal range. *J Neurol Sci* 1986;75:263–274.

58. Watkins PJ, MacKay JD. Cardiac denervation in diabetic neuropathy. *Ann Intern Med* 1980;92:304–307.

59. Ewing DJ, Campbell IW, Clarke BF. Assessment of cardiovascular effects in diabetic autonomic neuropathy and prognostic implications. *Ann Intern Med* 1980;92:308–311.

60. Low PA. Laboratory evaluation of autonomic failure. In: Low PA, ed. *Clinical autonomic disorders*, 2nd ed. Philadelphia: Lippincott–Raven, 1997:179–208.

61. Ewing DJ, Campbell IW, Burt AA, et al. Vascular reflexes in diabetic autonomic neuropathy. *Lancet* 1973;2:1354–1356.

62. Aminoff MJ. Involvement of peripheral vasomotor fibres in carpal tunnel syndrome. *J Neurol Neurosurg Psychiatry* 1979;42:649–655.

63. Aminoff MJ. Peripheral sympathetic function in patients with a polyneuropathy. *J Neurol Sci* 1980;44:213–219.

64. Low PA, Caskey PE, Tuck RR, et al. Quantitative sudomotor axon reflex test in normal and neuropathic subjects. *Ann Neurol* 1983;14:573–580.

65. Shahani BT, Halperin JJ, Boulu P, et al. Sympathetic skin response—a method of assessing unmyelinated axon dysfunction in peripheral neuropathies. *J Neurol Neurosurg Psychiatry* 1984;47:536–542.

66. Polinsky RJ. Multiple system atrophy: clinical aspects, pathophysiology, and treatment. *Neurol Clin* 1984;2:487–498.

67. Lubowski DZ, Nicholls RJ, Swash M, et al. Neural control of internal anal sphincter function. *Br J Surg* 1987;74:668–670.

68. Chiappa KH, Young RR. A case of paracarcinomatous pandysautonomia. *Neurology* 1973;23:423.

69. Young RR, Asbury AK, Corbett JL, et al. Pure pandysautonomia with recovery. Description and discussion of diagnostic criteria. *Brain* 1975;98:613–636.

70. Low PA, Dyck PJ, Lambert EH, et al. Acute panautonomic neuropathy. *Ann Neurol* 1983;13:412–417.

71. Hopkins A, Neville B, Bannister R. Autonomic neuropathy of acute onset. *Lancet* 1974;1:769–771.

72. Birchfield RI, Shaw CM. Postural hypotension in the Guillain-Barré syndrome. *Arch Neurol* 1964;10:149–157.

73. Lichtenfeld P. Autonomic dysfunction in the Guillain-Barré syndrome. *Am J Med* 1971;50:772–780.

74. Tuck RR, McLeod JG. Autonomic dysfunction in Guillain-Barré syndrome. *J Neurol Neurosurg Psychiatry* 1981;44:983–990.

75. Khurana RK, Koski CL, Mayer RF. Autonomic dysfunction in Lambert-Eaton myasthenic syndrome. *J Neurol Sci* 1988;85:77–86.

76. Clarke BF, Ewing DJ, Campbell IW. Diabetic autonomic neuropathy. *Diabetologia* 1979;17:195–212.

77. Ewing DJ, Campbell IW, Clarke BF. The natural history of diabetic autonomic neuropathy. *Q J Med* 1980;49:95–108.

78. Naik RB, Mathias CJ, Wilson CA, et al. Cardiovascular and autonomic reflexes in haemodialysis patients. *Clin Sci* 1981;60:165–170.

79. Kyriakidis MK, Noutsis CG, Robinson-Kyriakidis CA, et al. Autonomic neuropathy in leprosy. *Int J Lepr* 1983;51:331–335.

80. McCombe PA, McLeod JG. The peripheral neuropathy of vitamin B$_{12}$ deficiency. *J Neurol Sci* 1984;66:117–126.

81. McLeod JG. Autonomic dysfunction in peripheral nerve disease. *J Clin Neurophysiol* 1993;10:51–60.

82. Manco JC, Gallo L Jr, Godoy RA, et al. Degeneration of the cardiac nerves in Chagas' disease. Further studies. *Circulation* 1969;40:879–885.

83. Gallo L Jr, Marin Neto JA, Manücco JC, et al. Abnormal heart rate responses during exercise in patients with Chagas' disease. *Cardiology* 1975;60:147–162.

84. Sousa ACS, Marin Neto JA, Maciel BC, et al. Cardiac parasympathetic impairment in gastrointestinal Chagas' disease. *Lancet* 1987;1:985.

85. Iosa D, DeQuattro V, Lee DD-P, et al. Plasma norepinephrine in Chagas' cardioneuromyopathy: a marker of progressive dysautonomia. *Am Heart J* 1989;117:882–887.

86. Kyle RA, Greipp PR. Amyloidosis: clinical and laboratory features in 229 cases. *Mayo Clin Proc* 1983;58:665–683.

87. Kyle RA, Kottke BA, Schirger A. Orthostatic hypotension as a clue to primary systemic amyloidosis. *Circulation* 1966;34:883–888.

88. Dyck PJ. Neuronal atrophy and degeneration predominantly affecting peripheral sensory and autonomic neurons. In: Dyck PJ, Thomas PK, Griffin JW, et al., eds. *Peripheral neuropathy*, 3rd ed. Philadelphia: WB Saunders, 1993:1065–1093.

89. Bird TD, Reenan AM, Pfeifer M. Autonomic nervous system function in genetic neuromuscular disorders. *Arch Neurol* 1984;41:43–46.

90. Brooks AP. Abnormal vascular reflexes in Charcot-Marie-Tooth disease. *J Neurol Neurosurg Psychiatry* 1980;43:348–350.

91. Biaggioni I, Goldstein DS, Atkinson T, et al. Dopamine-β-hydroxylase deficiency in humans. *Neurology* 1990;40:370–373.

92. Devinsky O, Price BH, Cohen SI. Cardiac manifestations of complex partial seizures. *Am J Med* 1986;80:195–202.

93. Green JB. Pilomotor seizures. *Neurology* 1984;34:837–839.

94. Kramer RE, Lüduders H, Goldstick LP, et al. Ictus emeticus: an electroclinical analysis. *Neurology* 1988;38:1048–1052.

95. LeWitt PA, Newman RP, Greenberg HS, et al. Episodic hyperhidrosis, hypothermia, and agenesis of corpus callosum. *Neurology* 1983;33:1122–1129.

96. LeWitt P. Hyperhidrosis and hypothermia responsive to oxybutynin. *Neurology* 1988;38:506–507.

97. Erickson TC. Neurogenic hyperthermia (A clinical syndrome and its treatment). *Brain* 1939;62:172–190.

98. Stone WM, de Toledo J, Romanul FCA. Horner's syndrome due to hypothalamic infarction. *Arch Neurol* 1985;43:199–200.

99. Schwartz WJ, Busis NA, Hedley-Whyte ET. A discrete lesion of ventral hypothalamus and optic chiasm that disturbed the daily temperature rhythm. *J Neurol* 1986;233:1–4.

100. Aminoff MJ, Wilcox CS. Assessment of autonomic function in patients with a parkinsonian syndrome. *BMJ* 1971;4:80–84.

101. Aminoff MJ. Other extrapyramidal disorders. In: Low PA, ed. *Clinical autonomic disorders*, 2nd ed. Philadelphia: Lippincott–Raven, 1997:577–584.

102. Gross M, Bannister R, Godwin-Austen R. Orthostatic hypotension in Parkinson's disease. *Lancet* 1972;1:174–176.

103. Sandroni P, Ahlskog JE, Fealey RD, et al. Autonomic involvement in extrapyramidal and cerebellar disorders. *Clin Auton Res* 1991;1:147–155.

104. Gutrecht JA. Autonomic cardiovascular reflexes in progressive supranuclear palsy. *J Auton Nerv Syst* 1992;39:29–36.

105. Aminoff MJ, Gross M. Vasoregulatory activity in patients with Huntington's chorea. *J Neurol Sci* 1974;21:33–38.

106. Manetto V, Medori R, Cortelli P, et al. Fatal familial insomnia: clinical and pathologic study of five new cases. *Neurology* 1992;42:312–319.

107. Thomas JE, Schirger A, Love JG, et al. Orthostatic hypotension as the presenting sign in craniopharyngioma. *Neurology* 1961;11:418–423.

108. Birchfield RI. Postural hypotension in Wernicke's disease. *Am J Med* 1964;36:404–414.

109. Cravioto H, Korein J, Silberman J. Wernicke's encephalopathy. *Arch Neurol* 1961;4:510–519.

110. Shy GM, Drager GA. A neurological syndrome associated with orthostatic hypotension. *Arch Neurol* 1960;3:511–527.

111. Oppenheimer D. Neuropathology and neurochemistry of autonomic failure. In: Bannister R, ed. *Autonomic failure*, 2nd ed. Oxford: Oxford University Press, 1988:451–463.

112. Johnson RH, Smith AC, Spalding JMK, et al. Effect of posture on blood-pressure in elderly patients. *Lancet* 1965;1:731–733.

113. Aminoff MJ. Postural hypotension. In: Aminoff MJ, ed. *Neurology and general medicine*, 3rd ed. Philadelphia: Churchill Livingstone, 2001:131–150.

114. Lipsitz LA. Syncope in the elderly. *Ann Intern Med* 1983;99:92–105.

115. McLeod JG, Tuck RR. Disorders of the autonomic nervous system. Part 2: investigation and treatment. *Ann Neurol* 1987;21:519–529.

116. Low PA, ed. *Clinical autonomic disorders*, 2nd ed. Philadelphia: Lippincott–Raven, 1997.

12

FUNCTIONAL NEUROANATOMY OF LIMBIC STRUCTURES AND SOME RELATIONSHIPS WITH PREFRONTAL CORTEX

ROBERT J. MORECRAFT
GARY W. VAN HOESEN

INTRODUCTION

The limbic system concept was coined by MacLean (1) near the midpoint of the previous century to reintroduce and reemphasize the seminal thinking of Papez (2) on the neuroanatomic correlates of emotion and to integrate his deductions with advances from both the clinic and the laboratory. This led to a larger "circuit for emotion" in neuroanatomic terms and a somewhat more multifaceted one in functional terms. A scant 5 years later memory function was added to the functional correlates (3) because acquiring new information and learning relies critically on an intact Papez's circuit. MacLean's efforts along with Brodal's influential review (4) a few years earlier laid to rest the "smell brain" or rhinencephalon concept of limbic lobe function and relegated this sensation to its appropriate place as one of only many sensory inputs to the limbic system. Elaboration of the latter has since been a major showpiece of modern experimental neuroanatomy.

Although obscured somewhat by anatomic detail, the backbone of Papez's circuit deals with the rather simple anatomic notion of how the cerebral cortex (and, by definition, all sensory systems) influences the hypothalamus (the building up of the "emotive process") and how the latter acts back on the cerebral cortex ("psychic coloring"). He was successful in elaborating one facet of these relationships, but the full panorama was not appreciated until the past 4 decades with the advent of newer and more powerful experimental methodology and a major effort by neuroanatomists, which led to an expanded conception of the limbic system in both neuroanatomic and functional terms. Although confusing because the term *system* implies unity of function, the core of the concept is the limbic system as the mediator of two-way communication between the cerebral cortex and the external world, and the hypothalamus and the internal world of the organism. Behavior in general, whether it is the consequence of autonomic, endocrine, or somatic effectors, is governed by this interplay of the external and internal worlds. Thus, a multifunctional limbic system should be expected where many behaviors are involved (5).

The term *prefrontal cortex* is attributed to Sir Richard Owen, the great nineteenth-century English pioneer of comparative anatomy (6). It refers to that cortex anterior to the electrically excitable motor cortex of the frontal lobe. This is a large area of cortex in the human and nonhuman primate brain that includes dorsolateral, orbital, and medial sectors. The medial sector typically includes the anterior cortices of the cingulate gyrus, which wrap around the genu of the corpus callosum as it follows the rostrum ventrally and posteriorly.

There have been many conceptualizations of prefrontal cortical function, ranging from the most lofty domains of behavior to those dealing with more fundamental matters such as emotion, motivation, social behavior, and inhibition. Recent studies have linked it to working memory and decision making (7–10). *Working memory* refers to the ordering of behavior or the manner in which time is bridged neurally to complete a sequence of intended acts or tasks. Spatial abilities and motor acts are essential to place one in favorable positions to carry out the intended behavior. *Decision-making behavior* complements working memory in critical ways and involves an assessment of somatic markers (11) that impart personal reality to choices of behavior. Some patients with frontal lobe damage have preserved intellect, analytic abilities, social consciousness, and sensory awareness, but they lack the ability to make accurate predictions about the outcome of decisions on their own well-being. The apparent overlap and interactions between functional correlates of the prefrontal cortices and limbic system is of great interest and a matter of active inquiry in psychiatric research (12). Our goal in this chapter is to examine recent neuroanatomic findings regarding the major parts of the limbic system and the prefrontal cortices and new concepts that emerge from them.

LIMBIC SYSTEM

Anatomic structures included under the term *limbic system* have diverse locations in the cerebral hemisphere and occupy parts of the telencephalon, diencephalon, and mesencephalon. Comparative neuroanatomists have labeled them the so-called *conservative* parts of the brain—those found in a wide variety of mammals and to some extent in vertebrates in general. The cortical parts comprise what are frequently termed the *older* parts of the cerebral cortex and are characterized as simple in structure and common to many species that form the edge, or *limbus,* of the cerebral cortex (13). Care has to be taken in the use of terms such as *older* and *conservative.* They are best reserved for discussions relating to phylogeny, because limbic system structures have evolved like all other parts of the cerebral hemisphere and have assumed functional roles specific to the adaptations of a given extant species. Although a general core of behaviors may unite diverse species, species-specific functions may be more impressive. However, these may be mediated neurally by structures that resemble each other across species.

A useful way of dealing with the complex anatomy of the limbic system is shown in Figure 12.1, which divides the term into four conceptual units: landmarks, cortical components, subcortical structures, and interconnecting pathways. The cortical structures of the limbic system are well demarcated on the medial surface of the hemisphere by the cingulate sulcus dorsally and the collateral sulcus ventrally (Fig. 12.1A). Bridging areas such as the subcallosal gyrus, posterior orbital, anterior insular, temporal polar, and perirhinal cortices connect the cingulate and parahippocampal gyri rostrally, whereas the retrosplenial and retrocalcarine cortices provide a bridge caudally (Fig. 12.1B). Secondary sulci may be present, but major sulci never invade or interrupt the continuity of the limbic lobe. Altogether, they form the classic *limbus* or edge of the hemisphere, in the sense of Broca's original intent. The areas differ widely in cytoarchitecture and include Brodmann areas 23, 24, 25, 27, 28, 29, 35, 36, and 38, but none are true isocortical areas. Instead they fall under the categories of periallocortex and proisocortex of Sanides' terminology or, when combined, the mesocortices of Filimonoff, which are intermediate in structure between the allocortices and the isocortices (5).

The subcortical structures included in the limbic system (Fig. 12.1C) vary somewhat among authors and are scattered throughout many parts of the hemisphere. However, it seems appropriate to include the amygdala, septum, nucleus basalis of Meynert, anterior thalamus, habenula complex,

interpeduncular nucleus, and some additional limbic midbrain areas. A structural criterion common to this list relates to the fact that all are connected among themselves, as well as with the hypothalamus. Additionally, as will be discussed later, many of these nuclei receive direct cortical projections from one or more parts of the limbic lobe.

The final units of the limbic system are the interconnecting pathways (Fig. 12.1D). These include pathways within the limbic lobe, those that connect the limbic lobe with subcortical limbic structures, those that connect subcortical limbic structures to each other, and those that connect elements of the limbic system to the hypothalamus. Specifically, these include pathways such as the cingulum, uncinate fasciculus, fimbria-fornix, perforant pathway, stria terminalis, ventroamygdalofugal pathway, mammillothalamic tract, mammillotegmental tract, stria medullaris, and habenulointerpeduncular pathway. These compact bundles or tracts are well-known features of classic neuroanatomy. However, a lesson from more modern experimental neuroanatomy is that innumerable diffuse systems exist, uniquely or in parallel, and the limbic system neural relationships cannot be understood without consideration of them. For example, the amygdaloid nuclei receive their largest sources of input from diffuse corticoamygdaloid systems, yet no single corticoamygdaloid bundle or tract carries such input.

All aspects of this extensive topic of neuroanatomy are not reviewed here. Instead, we deal with it selectively, with an emphasis on new concepts and the relationships between the limbic system and frontal lobe. We begin with the limbic lobe.

Cingulate Gyrus

The cingulate gyrus is the major dorsal part of the limbic lobe on the medial wall of the cerebral hemisphere. It forms the upper half of the cortical ring of gray matter (Fig. 12.1B). Its main portion is located dorsal to the corpus callosum and callosal sulcus and ventral to the cingulate sulcus, although it extends into the depths and fundus of the latter. Its anterior part coincides with Brodmann area 24 and its posterior part with area 23 (Fig. 12.2) (14–19). The cingulate gyrus continues anteriorly and arches ventrally, around the genu of the corpus callosum. This ventral extension of the subgenual cingulate cortex forms area 25 and lies adjacent to the rostrum of the corpus callosum. Posteriorly, the cingulate gyrus arches ventrally, around the splenium of the corpus callosum. Most of this ventral extension is designated as area 23, except for cortex buried in the callosal sulcus, which is referred to as *retrosplenial cortex* (areas 29 and 30). The most ventral part of this posterior continuation is called the *isthmus* of the cingulate gyrus and ends in the upper bank of the anterior portion of the calcarine sulcus. Buried within the cingulate gyrus is a prominent interconnecting white matter pathway known as the *cingulum bundle.* The cingulum bundle follows the curvature of the cingulate gyrus and

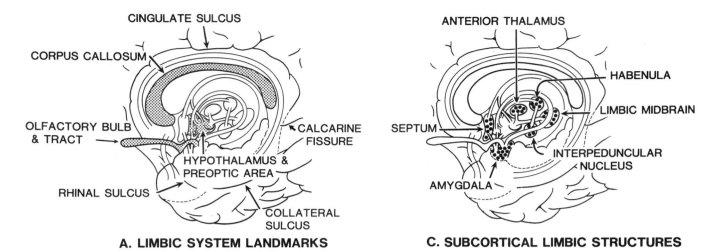

FIGURE 12.1. Four schematic representations of the medial surface of the cerebral hemisphere depicting the anatomic components of the limbic system. **A:** Relevant landmarks. **B:** Limbic lobe or cortical components. **C:** Subcortical limbic structures (nucleus basalis of Meynert and diagonal bands of Broca, both of which contain cholinergic neurons, are not shown). **D:** Interconnecting limbic system pathways. (From Damasio AR, Van Hoesen GW. Emotional disturbances associated with focal lesions of the limbic frontal lobe. In: Heilman KM, Satz P, eds. *Neuropsychology of human emotion.* New York: Guilford Press, 1983:85–108, with permission.)

some of its axons form the longest corticocortical association connections in the cerebral hemisphere. Along its length, axons enter and exit to interconnect the cingulate cortex with many cortical and subcortical targets, such as the multimodal association cortices, basal ganglia, and thalamus. Projections to precuneus or medial parietal lobule are especially prominent.

The various subsectors of the cingulate cortex, particularly areas 24 and 23, are coupled anatomically by a vast set of intracingulate connections (20–22). From a cortical perspective, the cingulate gyrus also is connected to the frontal lobe, parietal lobe, occipital lobe, temporal lobe, other parts of the limbic lobe, and the insula (20,23–35). In general, both

the anterior and posterior divisions of the cingulate gyrus are connected with the prefrontal, orbitofrontal, posterior parietal, posterior parahippocampal, perirhinal, entorhinal, and lateral temporal cortices, as well as the presubicular and subicular/CA1 parts of the parahippocampal gyrus and hippocampal formation, respectively. Topographic differences in these connections occur. For instance, more anterior levels of the orbitofrontal cortex are linked primarily to area 23, whereas more posterior levels are linked to area 24. The superior and inferior parietal lobules both are connected with area 23, whereas the parietal connection with area 24 is comparatively less intense and involves primarily the inferior parietal lobule. With respect to the parahippocampal region, medial

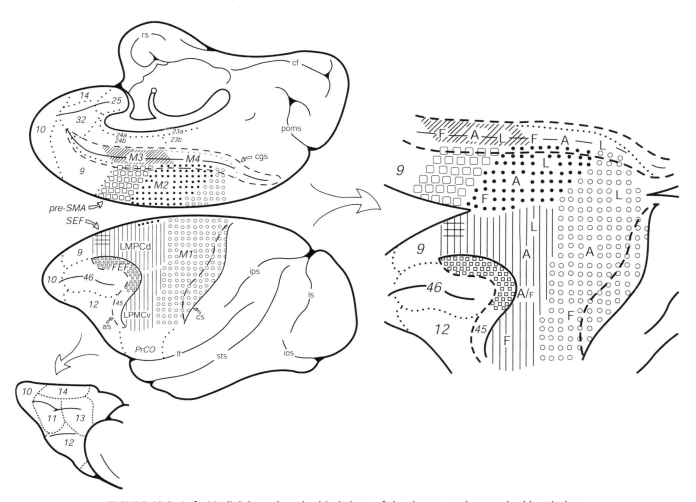

FIGURE 12.2. Left: Medial, lateral, and orbital views of the rhesus monkey cerebral hemisphere depicting the basic organization of the frontal lobe and anterior cingulate region. **Right:** Enlarged representation of the motor fields emphasizing the general somatotopic organization. Numbers denote Brodmann and Walker cytoarchitectonic areas. A, arm; as, arcuate sulcus; cf, calcarine fissure; cgs, cingulate sulcus; cs, central sulcus; F, face; ios, inferior occipital sulcus; ips, intraparietal sulcus; L, leg; lf, lateral fissure; LPMCd, dorsal lateral premotor cortex; LPMCv, ventral lateral premotor cortex; ls, lunate sulcus; M1, primary motor cortex; M2, supplementary motor cortex; M3, rostral cingulate motor cortex; M4, caudal cingulate motor cortex; poms, medial parietooccipital sulcus; pre-SMA, presupplementary motor cortex; rs, rhinal sulcus; SEF, supplementary eye field; sts, superior temporal sulcus. (Adapted from Morecraft RJ, Louie JL, Herrick JL, et al. Cortical innervation of the facial nucleus in the non-human primate. A new interpretation of the effects of stroke and related subtotal brain trauma on the muscles of facial expression. *Brain* 2001;124:176–208, with permission.)

levels are linked primarily to area 24, whereas lateral levels are linked to area 23.

Other connections are known to target only the anterior cingulate region, which underscores the cytoarchitectural distinction of this more anterior "agranular" part of the cingulate gyrus (areas 24, 25, and 32) from the more posterior "granular" part of the cingulate gyrus (area 23). For example, frontal area 32, temporal polar area 38, and the amygdala are connected with only the anterior cingulate gyrus. This unique set of projections suggests that neural events taking place in the anterior cingulate gyrus are different, in part, than those occurring posteriorly. However, the

fact that areas 24 and 23 are strongly interconnected suggests that projections from the anterior cingulate gyrus are likely to heavily influence the posterior cingulate gyrus and vice versa.

Like the different patterns of connectivity that distinguish the anterior from posterior parts of the cingulate gyrus, the dorsal part of the cingulate gyrus (subdivision c) differs on a connectional basis from the more ventral parts of the cingulate gyrus (subdivisions a and b). For example, input from the prefrontal and limbic cortices more heavily target cingulate cortex located on the exposed surface of the gyrus (areas 24a, 24b, 23a, and 23b) than cingulate cortex lining the lower bank and fundus of the cingulate sulcus (areas 24c and 23c). The subicular/CA1 sector of the hippocampal formation projects only to the more ventrally located cingulate areas (areas 24a, 23a, and 25). Areas 24c and 23c, which form the dorsal part of the cingulate gyrus, are located in the depths of the cingulate sulcus and are the only parts of the cingulate gyrus connected with the primary motor cortex (M1) (Fig. 12.3). Similarly areas 24c (M3) and 23c (M4) are strongly interconnected with the supplementary motor cortex (M2), whereas fewer neurons in area 24b project to M2. Area 24c is the only part of the cingulate gyrus connected with the ventral part of the lateral premotor cortex and area prostriata of the anterior calcarine region. Likewise, area 23c appears to be the only part of the cingulate gyrus connected with the adjacent primary somatosensory cortex (S1 or areas 1 and 2). Complementing the motor-related corticocortical interconnections are direct subcortical projections from areas 24c (M3) and 23c (M4) to the facial nucleus (Fig. 12.4) (36,37) and spinal cord (38,39).

Several conclusion can be made based upon these anatomic observations. First, the vast network of intracingulate connections, particularly those linking area 24 with area 23, provide multifaceted avenues for information exchange between anterior and posterior parts of the cingulate gyrus. Second, widespread parts of the cingulate gyrus are linked to multimodal association and other limbic cortices. Multimodal sources include prefrontal, rostral orbitofrontal, posterior parietal, and lateral temporal cortices. Limbic sources include posterior orbitofrontal, temporopolar, posterior parahippocampal, perirhinal, and entorhinal cortices (Fig. 12.5). Thus, highly processed and abstract information from neocortical association areas, and emotionally and motivationally relevant information from limbic sources, can influence a wide variety of cingulate subsectors. Third, general cingulocortical connections appear to be made throughout much of the cingulate gyrus. However, direct motor cortex interactions appear to occur through cingulate cortex lining the depths of the cingulate sulcus (areas 24c and 23c) and direct somatosensory cortex interactions occur through area 23c. Likewise, spinal cord projections and facial nucleus projections are known to arise only from areas 24c and 23c.

Summary and Functional Considerations

Although the cingulate gyrus does not contain an area that specifically subserves a primary modality, selected parts of it may be involved in regulating various complex behaviors as indicated by the widespread and diverse connections affiliated with this cortex. For example, lesions of the posterior cingulate gyrus alter an animal's ability to navigate appropriately in its environment despite receiving extended training to accomplish a learned task (40). Therefore, posterior cingulate cortex may be an essential component of spatially guided orientation. The anterior cingulate gyrus appears to

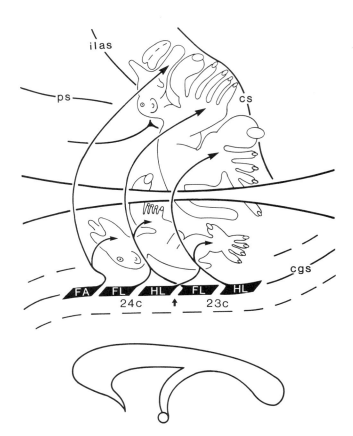

FIGURE 12.3. Somatic topography of the primary and supplementary motor cortices shown in a partial flattened view of the monkey hemisphere. The opposing convex lines in the center of the illustration represent the true dorsal convexity of the superior frontal lobule such that the lateral surface extends inverted toward the top of the page and the medial surface of the hemisphere extends upright toward the bottom of the page. The projections of cingulate motor areas 24c (or M3) and 23c (or M4) with the somatic topography of the supplementary and primary motor cortices are shown. Note that the face area (FA), forelimb (FL), and hind limb (HL) of area 24c projects to the corresponding representations of the other motor maps. Not shown is the recent finding demonstrating that area 23c also projects to the face areas of the primary and supplementary motor areas. The origin of this projection arises anterior to the area 23c arm region and posterior to the area 24c leg region (see Fig. 12.2), thus forming a complete body representation in the caudal cingulate motor area (e.g., M4 or area 23c). Both areas 24c and 23c contribute descending axons to the facial nucleus and spinal cord. The corticofacial projection arises from the face areas, whereas the corticospinal projections to the cervical and lumbosacral regions originate from the arm and leg areas, respectively. cgs, cingulate sulcus; ilas, inferior limb of arcuate sulcus; ps, principal sulcus. (From Morecraft RJ, Van Hoesen GW. Cingulate input to the primary and supplementary motor cortices in the rhesus monkey: evidence for somatotopy in areas 24c and 23c. *J Comp Neurol* 1992;322:471–489, with permission.)

participate in cognitive brain operations related to evaluating environmental stimuli. For example, it is involved in regulating attention (41) and may guide the environmental selection process that triggers a specific goal-directed action (42). More recently, an emerging theory suggests that the anterior cingulate plays a role in detecting when an error

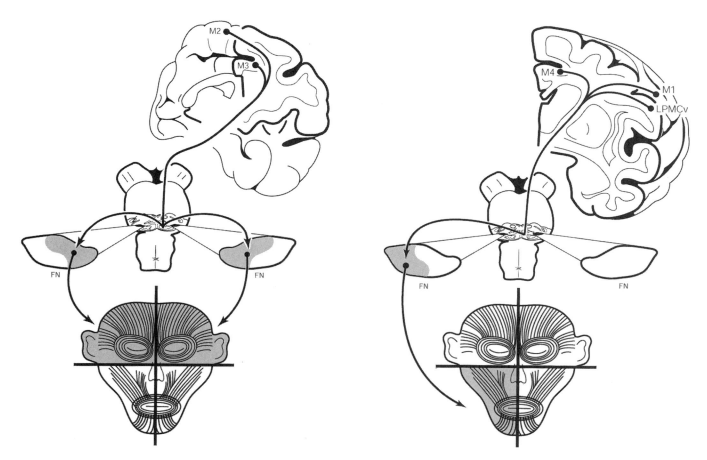

FIGURE 12.4. Summary diagram illustrating cortical face areas giving rise to bilateral innervation of the upper face **(left)** and contralateral innervation of the lower face **(right)**. Sparing of the upper facial muscles after middle cerebral artery infarction may be a result of sparing of the rostral cingulate motor cortex (M3) and supplementary motor cortex (M2) projection to the facial nucleus. The origin of these projections resides within the territory of the anterior cerebral artery. The fact that the rostral cingulate motor cortex receives widespread limbic and prefrontal inputs suggests that M3 mediates emotionally relevant upper facial expressions. Because the caudal cingulate motor cortex (M4) also is located medially and is the recipient of widespread limbic and prefrontal inputs, sparing of M4 after MCA infarction may underlie the preserved reflex smiling in the lower facial muscles contralateral to the cortical lesion. Similarly, anterior cerebral artery occlusion affecting M3 and M4 may contribute to emotional facial paralysis. FN, facial nucleus; LPMCv, ventral lateral premotor cortex; M1, primary motor area; M2, supplementary motor area; M3, rostral cingulate motor cortex; M4, caudal cingulate motor cortex. (From Morecraft RJ, Louie JL, Herrick JL, et al. Cortical innervation of the facial nucleus in the non-human primate. A new interpretation of the effects of stroke and related subtotal brain trauma on the muscles of facial expression. *Brain* 2001;124:176–208, with permission.)

in choice is likely to occur and when resolving conflict between competing response representations (e.g., conflict resolution) (43,44). These functions fit well with the extensive connections formed between the cingulate cortex and the multimodal limbic, prefrontal, and posterior parietal cortices.

Functional imaging and recording studies, as well as classic observations from patients with naturally occurring or surgically guided lesions, demonstrate that the anterior cingulate cortex plays a vital role in various behavioral responses related to painful stimuli (45–51). For example, the anterior cingulate is involved in generating one's self-perception

of pain. The anterior cingulate gyrus also is active during phases preceding our interaction with, or potential interaction with, painful stimuli. Related is the suggestion that the anterior cingulate gyrus is involved in generating movements affiliated with painful stimuli (52). For example, units here respond selectively to a variety of noxious stimuli. In line with this observation, it has been demonstrated that the anterior cingulate cortex may mediate vocal expressions that reflect the internal state of the animal. Stimulation of the anterior cingulate gyrus evokes simple vocalizations, whereas lesions placed rostral to the periaqueductal gray matter (PAG), which apparently disrupt the cingulo-periaqueductal

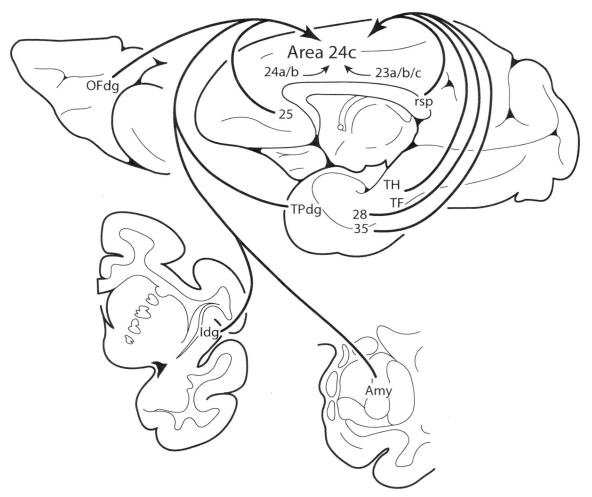

FIGURE 12.5. Schematic representation of the convergence of limbic system inputs to the dorsal region of the anterior cingulate cortex (area 24c or M3). Note the extensive input from other parts of the limbic lobe as well as the amygdala (Amy). Idg, insular dysgranular cortex; Ofdg, orbitofrontal dysgranular cortex; rsp, retrosplenial cortex; Tpdg, temporopolar dysgranular cortex. (Adapted from Morecraft RJ and Van Hoesen GW. Convergence of limbic input to the cingulate motor cortex in rhesus monkey. *Brain Res Bull* 1998;45:209–232, with permission.)

projection, abolish the response. It is well known that the PAG has a role in brainstem pain mechanisms and vocalization.

The anterior cingulate cortex has been implicated in the production of other forms of vocal expression that are linked to emotional expression. The separation cry, which is elicited by primates to maintain contact with a distant group of individuals or is induced by setting apart a mother from her offspring, is adversely affected after ablation of the anterior subcallosal region of the cingulate gyrus (53). A different form of vocalization, namely that of laughter, also may be mediated by the anterior cingulate gyrus (54). Based upon observations of patients with epileptic seizures, Arroyo et al. (54) suggested that the anterior cingulate gyrus is involved in generating the motor act of laughter and basal temporal cortex with the development of laughter's emotional content. It

seems that regardless of the emotional phenomena expressed, the anterior cingulate gyrus plays an important role in developing the associated motor response. It also appears from the common engagement of the head and neck region in these responses that brainstem centers mediating the operation of structures such as the larynx, tongue, and muscles of facial expression are heavily influenced by ongoing activity in the anterior cingulate gyrus.

The anterior cingulate gyrus may regulate autonomic responses, given that stimulation evokes pupillary dilation, piloerection, altered heart rate, and changes in blood pressure (55–58). Specifically, these responses are elicited from the cortex located below the genu of the corpus callosum or Brodmann area 25. These physiologic findings, coupled with the underlying neuroanatomic circuitry described for this part of the cingulate gyrus, has led to the suggestion that this part of the cingulate gyrus serves as a "visceral motor cortex" (59).

In addition to autonomic responses, physiologic studies demonstrate that the anterior cingulate gyrus in humans (60,61) and nonhuman primates (55,56,62–67) also gives rise to complex forms of somatic movement. Although surface stimulation studies conducted more than 50 years ago noted the potential influence of the primate cingulate gyrus on somatomotor mechanisms, more recent physiologic (62–67) and anatomic (15,23,32,33,36,67–70) studies have been able to pinpoint a cingulate motor region and identify within this region an organized somatotopy (Figs. 12.2 and 12.6). The cingulate motor cortex lies in the depths of the cingulate sulcus and corresponds cytoarchitectonically to areas 24c and 23c. In both areas in the monkey, the face representation is anterior to the forelimb representation, which in turn is anterior to the hind limb representation. Using neuroimaging information, it appears that a similar somatotopic organization in the human anterior cingulate cortex is consistent with the reported somatotopy in the monkey anterior cingulate cortex (71,72). The observations of Paus et al. (71) also indicate that the anterior cingulate gyrus mediates the execution of appropriate motor responses and suppresses inappropriate responses. Based upon the suggestive physiologic and anatomic data collected in nonhuman primates, as well as the suggestion that area 24 may be the developmental progenitor of the isocortical motor fields (73), areas 24c and 23c may be conceptualized as an M3 and M4, respectively, in the scheme of cortical motor representation (Fig. 12.6) (33,36). As in the more traditional isocortical motor areas (M1 and M2), modern tract tracing techniques used in monkeys show that M3 and M4 give rise to a host of projections that target various motor centers positioned at all parts of the neuraxis. For example, M3 and area 23c give rise to somatotopically organized projections that target the primary and supplementary motor cortices (Fig. 12.3). Therefore, M1, M2, M3, and M4 are highly interconnected at the cortical level. From a subcortical standpoint, the same motor centers targeted by M1 and M2 are targeted by M3. For example, area M3 projects heavily to the ventrolateral part of the putamen, medial and dorsal parts of the parvicellular red nucleus, ventromedial part of the pontine nuclei, facial nucleus, and intermediate zone of the cervical enlargement of the spinal cord (68,70,74–77). Judging from detailed comparisons of the descending projections arising from M1 and M2, those from M3 target different as well as similar parts of the basal ganglia, red nucleus, pontine gray matter, and spinal cord. Therefore, it can be stated that the highly interconnected motor cortices seem to be characterized by overlap in some of their corticofugal projection targets. As noted earlier, the cortical inputs to the cingulate motor cortex are unique compared to those targeting M2 and M1 in that the cingulate motor cortex receives strong and widespread inputs from the prefrontal cortex and limbic lobes (34,78–80). From

the perspective of motor cortex organization, the powerful and direct input from prefrontal cortex, limbic lobe, and amygdala serve, in part, as a distinguishing element of M3 (Fig. 12.5).

Hippocampal Formation and Parahippocampal Gyrus

The hippocampal formation is composed of three allocortical areas: (a) the pyramids that form the CA zones of the hippocampus (CA1, CA2, CA3); (b) the dentate gyrus, including the CA4 polymorph neurons found in its hilum; and (c) the various subicular cortices (81–85). The latter includes the subiculum proper, a true allocortical zone, and two periallocortical zones, the presubiculum and parasubiculum. The latter are multilayered and associated closely with the hippocampal formation. They have continuity with the subiculum in their deep layers. Because they are multilayered periallocortical areas also related to the entorhinal cortex of the parahippocampal gyrus, it is appropriate to include them with this part of the ventromedial temporal lobe. In short, they form the medial boundary of the parahippocampal gyrus, intervening between the entorhinal cortex and the subiculum proper of the hippocampal formation.

Cajal (86,87) reported many seminal observations on the cytoarchitecture, fiber architecture, and connections of the hippocampal formation using descriptive methods such as the Golgi technique. His work led to a conceptualization of the hippocampal formation that persisted for many decades. It is only in recent years that these findings have been embellished using newer and more sensitive experimental methods (Fig. 12.7 A–D) (82). The major input to the hippocampal formation was thought to arrive via two conspicuous white matter pathways, the fimbria-fornix system and what Cajal termed the temporo-ammonic or perforant pathway. Axons from these pathways were observed to end on the pyramidal cells of the hippocampus and the granule cells of the dentate gyrus. A two-part sequential series of intrinsic connections also was demonstrated. The first part is a large projection from the dentate gyrus granule cells to the proximal part of the apical dendrites of the CA3 pyramidal cells. This system sends no axons outside of the hippocampal formation and is known as the *mossy fiber system*. The second part is known as the *Schaffer collateral system*. It arises from the axons of CA3 pyramids and terminates on the basal and apical dendrites of the adjacent CA1 pyramids.

The view that emerged from these studies of hippocampal connectivity held that input would arrive via the fimbria-fornix system and perforant pathway and activate the pyramidal neurons directly, or indirectly, via the dentate gyrus and intrinsic pathways. Pyramidal axons entered the alveus and the fimbria-fornix system and conveyed hippocampal output to subcortical structures. There are no errors in this anatomy; nearly all aspects of Cajal's careful research

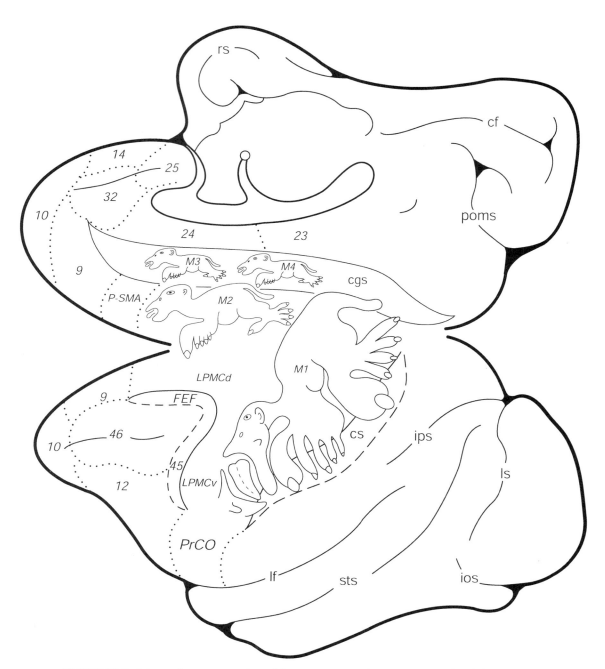

FIGURE 12.6. Schematic representation of four motor areas (M1, M2, M3, and M4) found in the frontal (M1 and M2) and cingulate (M3 and M4) cortices. (Adapted from Woolsey CN, Settlage PH, Meyer DR, et al. Patterns of localization in precentral and "supplementary" motor areas and their relation to the concept of a premotor area. *Res Publ Assoc Res Nerv Ment Dis* 1952;30:238–264.) As carefully noted by Woolsey et al. in their classic paper, these caricatures show only simple relations and do not illustrate the complex patterns of multiple movement representations that occur within a given somatotopic region (e.g., within the forelimb area of M1). Specifically, Woolsey et al. state "It must be emphasized, however, that this diagram is an inadequate representation of the localization pattern, since in a line drawing one cannot indicate the successive overlap which is so characteristic a feature of cortical representation, not only in the motor but also in the sensory areas. If this warning is heeded, Figure 131 may serve to outline in general, the localization pattern of the precentral motor field and to show its relation to the supplementary motor area." For simplicity, somatotopic representation in the lateral premotor region is omitted in this diagram because a portion of M1 as originally depicted by Woolsey is located more anterior than currently recognized. As discussed in the text, the cingulate motor areas (M3 and M4) represent major cortical entry points for both limbic and prefrontal input to motor cortex, and both motor areas give rise to corticofacial and corticospinal projections. as, arcuate sulcus; cf, calcarine fissure; cgs, cingulate sulcus; cs, central sulcus; FEF, frontal eye field; ips, intraparietal sulcus; lf, lateral fissure; LPMCd, dorsal lateral premotor cortex; LPMCv, ventral lateral premotor cortex; M1, primary motor cortex; M2, supplementary motor cortex; M3, rostral cingulate motor cortex; M4, caudal cingulate motor cortex; ls, lunate sulcus; poms, medial parietooccipital sulcus; ps, principal sulcus; P-SMA; presupplementary motor area; ros, rostral sulcus; rs, rhinal sulcus; sts, superior temporal sulcus.

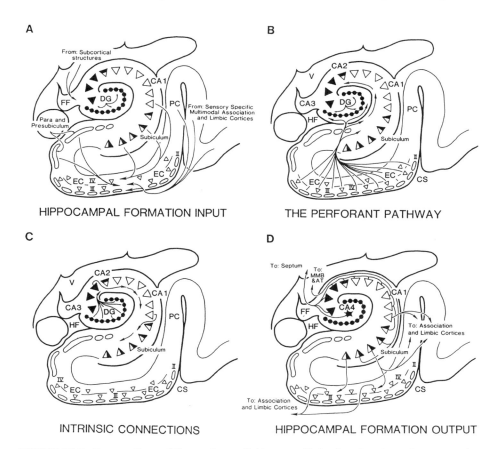

FIGURE 12.7. Cross sections of the ventromedial temporal lobe showing the major connection of the entorhinal (EC) and perirhinal cortices (PC) and the hippocampal formation. **A:** Entorhinal input from neighboring cortical areas, the subiculum, and from the sensory-specific and multimodal association areas and the limbic lobe. Subcortical input directly to the hippocampal formation arrives via the fimbria-fornix (FF). **B:** Origin and course of the major output pathway of the EC, the perforant pathway. Note its strong distribution to the pyramidal neurons of the subiculum, hippocampus (CA1–CA3), and dentate gyrus (DG). **C:** Major intrinsic connections of the hippocampal formation. Note in particular the DG projections to the CA3 pyramids, CA3 pyramid projections to CA1 pyramids, and CA1 pyramid projections to the subiculum. **D:** Major output projections of the hippocampal formation and the EC. Note that all hippocampal formation pyramidal neurons project to the anterior thalamus (AT) and mamillary bodies (MMB). However, these neurons give rise to output projections to the association and limbic cortices. Direct projection to the deep layers of the EC and their projections to the same areas provide the anatomic basis for a powerful hippocampal influence on other cortical areas. cs, collateral sulcus; hf, hippocampal fissure; v, inferior horn of the lateral ventricle.

observations have been verified. However, both he and his student Lorente de Nó failed to fully document an important extension of the intrinsic circuitry of the hippocampal formation, namely, CA1 projections to the adjacent subicular cortices (88,89). It now is known that the neurons forming the subiculum, and not the hippocampal pyramids *per se,* are responsible for a large amount of hippocampal formation output and nearly all of its diversity with regard to influencing other brain areas (90–92). For example, the axons of CA3 pyramids project mainly to the septum, although they also give rise to major intrinsic and commissural interhippocampal and intrahippocampal projections. With the exception of commissural projections, this is somewhat true for CA1 pyramidal neurons as well. However, the neurons

that form the subicular cortices, and to a lesser extent those of the CA1 zone, have extensive extrinsic projections that divide hippocampal output into major components, one to a variety of cortical areas and another to a variety of subcortical structures such as the basal forebrain, amygdala, thalamus and hypothalamus (90,92–94). Thus, hippocampal output is disseminated much more widely than previously thought and, importantly, projects not only to subcortical areas but to cortical ones as well. The latter are thought to be the neural basis of whether information is preserved in less labile forms in the association areas (82).

Another feature of hippocampal anatomy not described by early anatomists concerns the issue of afferent input, particularly input to the entorhinal cortex. Early investigators

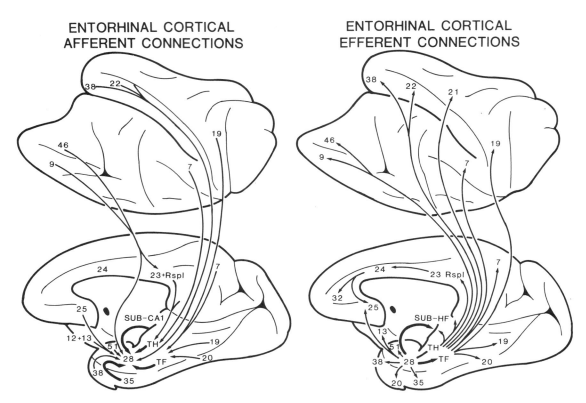

FIGURE 12.8. Major afferent (or input) and efferent (or output) connections of the entorhinal (Brodmann area 28) and posterior parahippocampal (area TF and FH) cortices are summarized on lateral (inverted) and medial (upright) views of the rhesus monkey hemisphere. Note the convergence of afferents to this part of ventromedial temporal lobe and the divergence of efferents from it back to the association and limbic cortices. These link the hippocampal formation to the association and limbic cortices in a bi-directional manner. (From Van Hoesen GW. The parahippocampal gyrus: new observations regarding its cortical connections in the monkey. *Trends Neurosci* 1982;5:345–350, with permission.)

saw axons entering the entorhinal cortex, but use of the Golgi method precluded ascertaining their origin. This meant that the input to the major source of afferents to the hippocampal formation was left uncharacterized. It was shown in later studies that the entorhinal cortex receives powerful projections from many cortical areas (Figs. 12.8 and 12.9), particularly in the temporal lobe (95–103), and from subcortical structures, such as the amygdala and midline thalamus (104–106). It is important to note that amygdaloid input to the entorhinal cortex is derived from amygdaloid nuclei (e.g., the lateral basal complex) that receive both limbic cortical input as well as input from association cortices located in both the frontal and temporal lobes (107–111). Additionally, these same nuclei receive direct or indirect hypothalamic (111) and basal forebrain input (112,113). Thus, input to the entorhinal cortex is extensive. It arises largely from the cortices that form the limbic lobe, amygdala, and midline thalamus. In terms of limbic cortex, proisocortical areas, such as the perirhinal, posterior parahippocampal, cingulate, temporal polar, and posterior orbitofrontal cortices, and periallocortical areas, such as the retrosplenial, presubicular, and parasubicular cortices, are the major contributors.

These transitional cortical areas, which are interposed between the allocortex and neocortex, receive major projections from both sensory-specific and multimodal association cortices. Neocortical projections are largely indirect and arise from lateral prefrontal, lateral, inferior temporal, and lateral and medial parietal sources (114). Thus, the entorhinal cortices receive a digest or abstract of the sensory output generated by subcortical and cortical areas, and this includes both interoceptive and exteroceptive information. Like a censor and archivist, it seems to determine whether this is relevant for the organism, is worthy of storage, or is to be left to decay.

Summary and Function Considerations

The hippocampal formation is the focal point for major forebrain neural systems that are interconnected with the sensory-specific association cortices and the multimodal association cortices. These are widespread systems that involve much of the cortical mantle. As mentioned, the cortices that form the limbic lobe, in general, and the amygdala and posterior parahippocampal area, in particular, receive input from

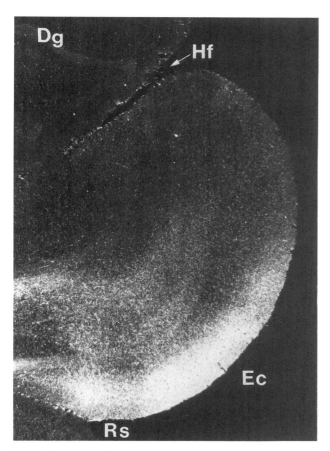

FIGURE 12.9. Dark-field photomicrograph of the entorhinal cortex (EC) in a rhesus monkey experiment in which tritiated amino acids were injected into the posterior parahippocampal cortex to label axons and their terminals. White areas indicate the location of terminal labeling. Note the dense band of labeling in the superficial layer of the EC, indicating that the posterior parahippocampal area projects powerfully to the EC. Dg, dentate gyrus; HF, hippocampal fissure; Rs, rhinal sulcus.

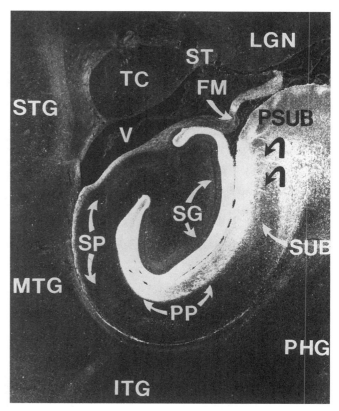

FIGURE 12.10. Dark-field photomicrograph of the hippocampal formation in a rhesus monkey experiment in which tritiated amino acids were injected into the entorhinal cortex to label perforant pathway axons and their terminals. White areas indicate the location of terminal labeling. Note the dense terminal labeling in the subiculum (Sub) CA1, CA3, and dentate gyrus (Dg) molecular layers, indicating a powerful linkage between the entorhinal cortex and hippocampal formation. The perforant pathway is one of the largest projections of the cerebral cortex and the major avenue by which the limbic and association cortices influence the hippocampal formation. HP, hippocampus; ITG, inferior temporal gyrus; LGN, lateral geniculate nucleus; MTG, middle temporal gyrus; PHG, parahippocampal gyrus; PP, perforant pathway; STG, superior temporal gyrus; Tc, tail of caudate nucleus; V, ventricle.

the various association cortices and either project directly to the hippocampal formation or first to the entorhinal cortex, which then projects to the hippocampal formation. The most compact part of this latter system, the perforant pathway, is a major input system from the entorhinal cortex (Fig. 12.10). It mediates a powerful excitatory input to the hippocampal formation that culminates in extrinsic output to the septum via the fimbria-fornix or intrinsic output to the subicular cortices. These latter areas then project to several basal forebrain areas, including the amygdala, various diencephalic nuclei, and many parts of the limbic lobe and association cortices. Thus, hippocampal output is disseminated widely by the subicular and CA1 pyramids of the hippocampal formation (Figs. 12.8, 12.11, and 12.12). The interplay of these neural systems forms a critical foundation in processes pertaining to sensory analysis and integration of ongoing environmental events and the eventual preservation of this information in the association cortices. Many of these systems are compromised or destroyed entirely at

end-stage Alzheimer disease, underscoring their critical role in memory systems (Figs. 12.13–12.15) (115).

Amygdaloid Complex

As mentioned earlier, several amygdaloid nuclei receive inputs from the cerebral cortex (Fig. 12.16) and from a host of subcortical structures of both diencephalic and mesencephalic origin. The latter includes structures such as the midline thalamic nuclei, hypothalamus, supramamillary nucleus, PAG, peripeduncular nucleus, and ventral tegmental area (111,112,116).

Unlike the hippocampal formation whose input is derived largely from limbic lobe areas that receive input from the association cortices, many of the latter project directly to the amygdala without relays in the limbic lobe.

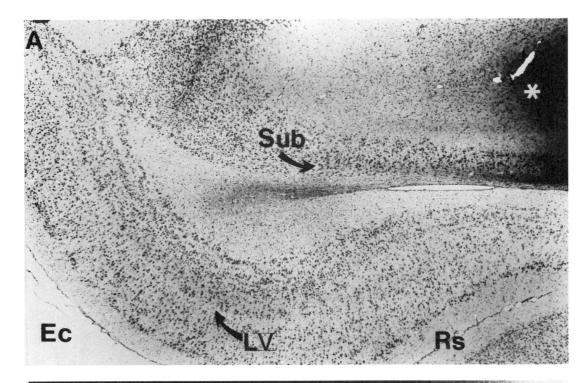

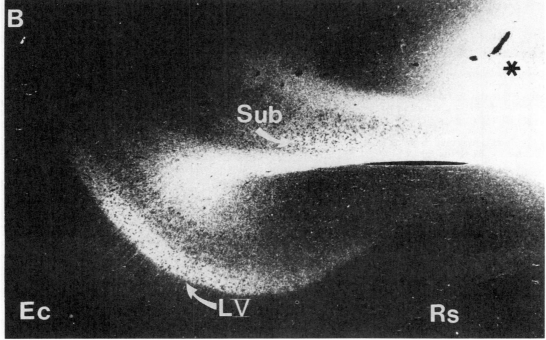

FIGURE 12.11. **A:** Nissl-stained bright-field photomicrograph of the anterior subiculum (Sub) and entorhinal cortex (EC) in a rhesus monkey. The dark area (marked with an *asterisk*) shows where tritiated amino acids are injected into the subiculum (Sub). **B:** Same microscopic field shown in A, but with dark-field viewing conditions. Note the dense subicular projection to layer V of the EC. This direct projection from the hippocampal formation to the cortex is one of many cortical projections that arise from the subiculum and CA1 parts of the hippocampal formation. Lv, lateral ventricle; Rs, rhinal sulcus.

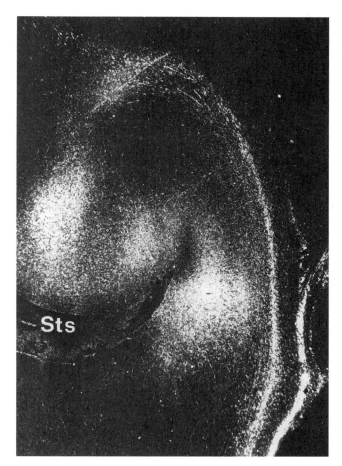

FIGURE 12.12. Dark-field photomicrograph showing terminal labeling in the depths of the superior temporal sulcus (Sts) in a rhesus monkey experiment in which labeled amino acids were injected into the cortex of the posterior parahippocampal gyrus. The latter receives a strong hippocampal output and projects in turn to association areas in all other lobes.

tices and the periallocortices. The accessory basal nucleus, for example, receives strong projections from the temporal polar cortex, insular cortex, medial frontal cortex, and, to some extent, the orbitofrontal cortex. The mediobasal nucleus is not well characterized in terms of input, but it receives projections from the perirhinal and subicular cortices (92). The laterobasal nucleus receives input from many of the cortical areas listed, but it also is characterized by having input from the anterior cingulate cortex. The subiculum and entorhinal cortices also project to part of the basal amygdaloid complex. These nuclei collectively form the basal complex, the largest mass of the amygdala; thus, limbic lobe input must be regarded as the major source of cortical input.

The central amygdaloid nucleus is unusual in the sense that it receives input derived from all types of cortex. For example, it receives input from the lateral temporal isocortex; temporal polar, orbitofrontal, and insular proisocortices; entorhinal periallocortex; and periamygdaloid and primary olfactory allocortices (110). The superficial nuclei of the amygdala, such as the medial nucleus, and the various cortical nuclei receive input largely from allocortical origin, such as the subicular and periamygdaloid cortices and olfactory piriform cortex.

It was believed until recently that the major input and output relationships of the amygdala were with the hypothalamus. Such connections are strong (111), but the diversity of amygdaloid output is far more extensive than previously appreciated. For example, the lateral nucleus and some components of the basal complex project strongly to the entorhinal cortex (120), several association areas, and even the primary visual cortex (121). Additionally, the basal amygdaloid complex has strong reciprocal interconnections with the subiculum, which is the major source of hippocampal output (92,120,122,123). From these studies it is clear that the amygdala is highly interrelated with the hippocampal formation in anatomic terms and that these two temporal neighbors undoubtedly influence each other to a great degree.

Nonhypothalamic subcortical projections arise from several amygdaloid nuclei and link the structure with many parts of the neuraxis. Of special interest are powerful projections to parts of the basal forebrain, including the nucleus basalis of Meynert (124). Additional nonhypothalamic subcortical projections course to the dorsomedial thalamic nucleus (125–127) and to autonomic centers in the brainstem (124,128).

Among the more surprising aspects of amygdaloid anatomy described recently is that this structure has strong projections to many parts of the temporal association, insular, and frontal cortices (120,123,125,126,129–131). In the frontal lobe, these projections end on parts of the isocortices that form the frontal granular cortex, frontal agranular cortex, and cingulate, medial frontal, and posterior orbitofrontal proisocortices. Powerful projections from the basal complex of the amygdala to the neostriatum and ventral striatum also have been described (132).

For example, the visual association cortices of the lateral temporal neocortex send direct projections to the lateral amygdaloid nucleus and to the dorsal part of the laterobasal amygdaloid nucleus (107–109,117,118). Some investigators have shown that the auditory association cortex of the superior temporal gyrus also projects directly to the lateral amygdaloid nucleus. Input related to somatic sensation also converges on this nucleus from the insular cortex (119).

Although neocortical input from association cortices constitutes a major source of input to the lateral amygdaloid nucleus, it is erroneous to characterize all corticoamygdaloid input as derived from the neocortex. The proisocortices, periallocortices, and allocortices make a large contribution (110). For example, the medial half of the lateral nucleus receives input from the insular, temporal polar, and orbitofrontal cortices, which all are parts of the limbic lobe (Fig. 12.16). The basal complex of the amygdala, which consists of the laterobasal, accessory basal, and mediobasal nuclei, receives cortical projections derived almost exclusively from the proisocor-

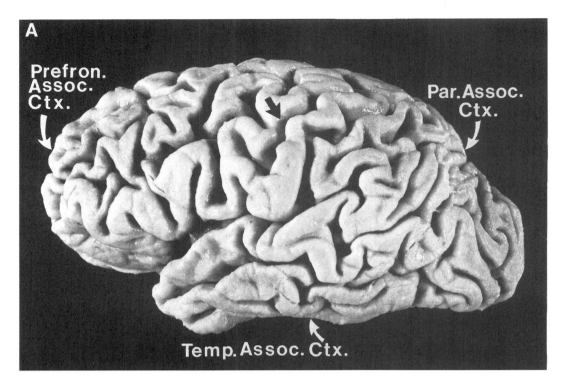

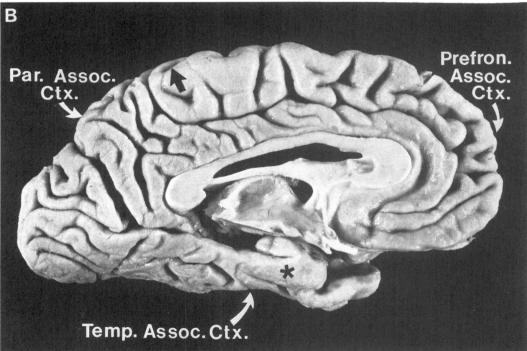

FIGURE 12.13. Lateral **(A)** and medial **(B)** views of the cerebral hemisphere at end-stage Alzheimer disease after a long duration of illness (13 years) in an 83-year-old woman. Note the pronounced atrophic changes in the prefrontal, parietal, and temporal association cortices but the relative preservation of the precentral and postcentral gyri on either side of the central sulcus (the latter is marked by the *arrow* in both photographs). The *asterisk* in panel **B** marks the location of the entorhinal cortex.

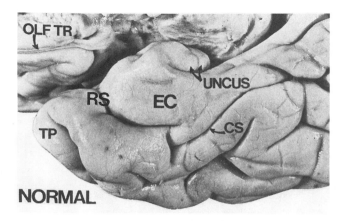

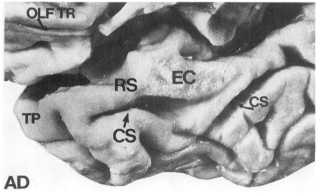

FIGURE 12.14. Two views of the ventromedial temporal area focusing on the entorhinal cortex (EC). **Top:** Normal nondemented elderly donor (age 76 years). **Bottom:** Pathologically confirmed Alzheimer's disease donor (age 71 years). Note the atrophic pitted appearance of the EC. cs, collateral sulcus; OLF TR, olfactory tracts; TP, temporal pole.

Summary and Functional Considerations

From a neural systems viewpoint, the amygdala must be viewed from a broader perspective than its classically described interrelationships with the hypothalamus and olfactory system. For example, the amygdala has powerful direct interconnections with much of the anterior cortex of the limbic lobe and with neocortical areas of the frontal, temporal, and even occipital lobes. Additional smaller projections have been reported to terminate in the premotor cortices and anterior cingulate area 24, both of which project directly to the supplementary motor cortex and primary motor cortex. Moreover, projections to the rostral cingulate motor area (M3) provide a means for emotionally relevant input to influence corticofacial and corticospinal axons (Fig. 12.17). Additional projections connect the amygdala with the neostriatum and ventral striatum, involving it in basal ganglia circuitry. Certain amygdaloid nuclei also project strongly to the nucleus basalis of Meynert, whose axons provide powerful cholinergic input to the cerebral cortex. Descending amygdaloid projections from the central amygdaloid nucleus also provide input to important autonomic centers in the brainstem. Although much of the input to the amygdala,

particularly from subcortical areas, cannot be characterized well in functional terms, this is not the case for amygdaloid output. Overall it can be concluded that amygdaloid output is directed toward the origin of what may be termed *effector systems* that influence motor, endocrine, and autonomic areas along much of the full extent of the supraspinal neuraxis. Thus, it is highly unlike the hippocampal formation, whose output to such areas is either less strong or more indirect, and instead is shifted more toward the association cortices. A persuasive argument could be made that the amygdala more greatly influences overt behavior whereas the hippocampus more greatly influences more covert aspects of behavior such as cognition.

Nucleus Basalis of Meynert

The neurons that form the nucleus basalis of Meynert have attracted substantial attention because they project to the cerebral cortex (133–137) and are affected frequently in Alzheimer disease (138). Such projections had been suggested in earlier ablation experiments because of retrograde degeneration, but the magnitude of this projection was not fully appreciated. It now is clear that probably all of the nucleus basalis of Meynert projects to the cortex and most parts of it. These findings are intriguing, but they assume an added significance with the demonstration that the majority of these neurons contain cholinergic enzymes and, in fact, provide the major source of cholinergic input to the cerebral cortex (136,139). Thus, they mirror the projections of their counterparts in the diagonal band nuclei and medial septum that project to the allocortices of the hippocampal formation.

The neurons that form the nucleus basalis are large, hyperchromatic, multipolar, and fusiform-shaped cells that lie among the ascending and descending limbic, hypothalamic, and brainstem pathways that course through the basal forebrain. Part of the nucleus is found within the substantia innominata, but cholinergic neurons span the anteroposterior expanse of the ventral surface of the hemisphere all the way from the septum anteriorly to the midbrain posteriorly. They also have a lateral extension that follows the course of the anterior commissure into the temporal lobe (135,139). Scattered acetylcholinesterase- and choline acetyltransferase-positive neurons are found within the internal and external medullary lamina of the globus pallidus, in the lateral hypothalamus, and in the dorsal parts of various amygdaloid nuclei. Given this topography, calling the nucleus basalis a "nucleus" is somewhat generous, although the common cholinergic nature of its neurons somewhat salvages the term.

The output of the nucleus basalis to the cortex has been well characterized in the rat and monkey. These studies reveal that a rigid topography exists with regard to where in the cortical mantle they end (137,139–141). Additionally, it has been demonstrated that the nucleus basalis projects to the basal complex of the amygdala (141,142). Importantly, it is known that this nucleus projects to the reticular nucleus of

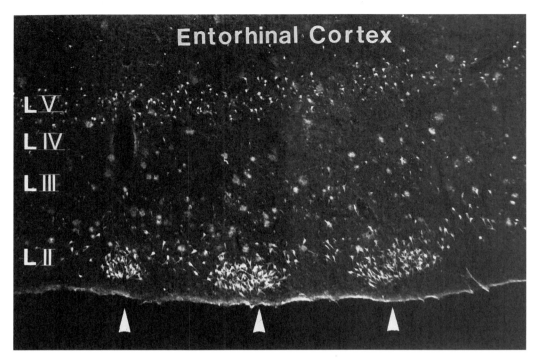

FIGURE 12.15. Cross section through the entorhinal cortex in Alzheimer disease stained with the fluorochrome thioflavin S to reveal neurofibrillary tangles. Note that the neurons of layers II and V are heavily invested with this form of pathology. The former link the hippocampal formation to the cortex, whereas the latter receive hippocampal output and project to the cerebral cortex. The disease-related pathology of Alzheimer disease thereby disconnects the hippocampal formation from the cerebral cortex.

the thalamus (142–145). This places the nucleus basalis in a position to influence the cortex directly as well as indirectly, because the reticular nucleus governs thalamic transmission via intrinsic thalamic connections (146). However, beyond these observations, little else is known about the efferent connections of the nucleus basalis of Meynert, and there remains a fundamental need for further study of this in experimental neuroanatomy. Suggestive evidence was provided in early studies that nucleus basalis axons project at least as far causally as the midbrain (134).

The input to the nucleus basalis of Meynert with respect to topography is better understood. In terms of cortex, it has been shown that it receives projections from only a small percentage of the cortical areas to which it sends axons (147). These include areas such as the olfactory, orbitofrontal, anterior insular, temporal polar, entorhinal, and medial temporal cortices—all components of the limbic lobe. Subcortical projections to the nucleus basalis arise from the septum, nucleus accumbens, hypothalamus, amygdala, preoptic nucleus, and peripeduncular nucleus of the midbrain (Fig. 12.18) (135,148–150).

Summary and Functional Considerations

From a neural systems viewpoint, it is reasonable to believe that the nucleus basalis of Meynert is a key structure. For example, the widespread projections to the cortex and the fact

that acetylcholine serves as the neurotransmitter for these projections are of fundamental importance to normal brain functions. These neurons are much like serotonergic neurons in the raphe complex, noradrenergic neurons in the locus caeruleus, and dopaminergic neurons in the ventral tegmental area with similar cortical projections. Like these neurons, many of their projections are not reciprocated by projections from the cortex their axons innervate. The input to nucleus basalis neurons seems topographically organized and rather specific. At least two investigations have reported afferent input that seemingly "picks out" the clusters of nucleus basalis neurons. Some of these originate in the amygdala and may provide a highly specific, albeit indirect, manner for this structure to exert its influence on widespread parts of the cortical mantle (Fig. 12.18). These projections arise from amygdaloid nuclei that receive intrinsic amygdaloid projections, suggesting that, at least with regard to the amygdala, its output to the nucleus basalis reflects output derived from much of the structure. In this context, it should not be overlooked that the subiculum of the hippocampal formation projects both to the basal complex of the amygdala and to other basal forebrain areas that project to the nucleus basalis. Thus, a highly synthesized output from the hippocampal formation seems plausible. On these grounds and on the basis of its limbic cortical input, it can be concluded that the major input to the nucleus basalis is from nearly the entire limbic system. Interestingly, the nucleus basalis of Meynert is not

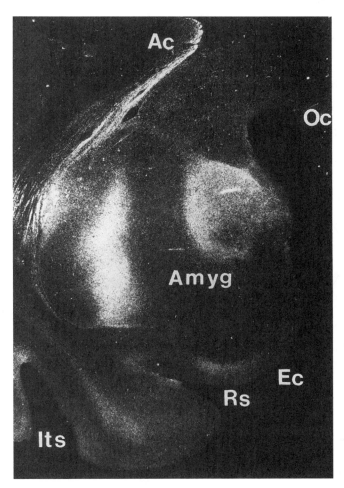

FIGURE 12.16. Dark-field photomicrograph of the amygdala (Amyg) and entorhinal cortex (Ec) showing terminal labeling in a rhesus monkey experiment in which labeled amino acids were injected into the cortex of the temporal pole. The amygdaloid terminal labeling is primary over the medial part of the lateral nucleus and over the accessory basal nucleus. Ac, anterior commissure; Its, inferior temporal sulcus; Oc, optic chiasm; Rs, rhinal sulcus.

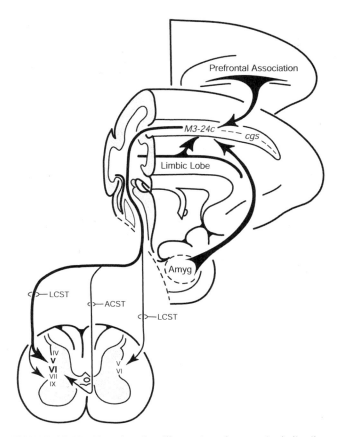

FIGURE 12.17. Line drawing illustrating the terminal distribution of the corticospinal projection from the rostral cingulate motor area (M3 or area 24c). Also illustrated are some important inputs that characterize the origin of the M3 projection and likely provide unique higher-order and emotionally relevant influences on this corticospinal projection. ACST, anterior corticospinal tract; Amyg, amygdala; cgs, cingulate sulcus; I–X, Rexed laminae; LCST, lateral corticospinal tract. (From Morecraft RJ, Louie JL, Schroeder CM, et al. Segregated parallel inputs to the brachial spinal cord from the cingulate motor cortex in the monkey. *Neuroreport* 1997;8:3933–3938, with permission.)

influenced directly by the major part of the cortex to which it projects. It is only influenced indirectly, after the outflow of corticocortical connections reach their end stations in limbic cortical areas, many synapses removed from primary motor cortex.

Finally, the nucleus basalis of Meynert receives projections from the hypothalamus. These require further study but provide a structural basis by which the internal state of the organism can indirectly influence both the motor and the sensory cortices and, ultimately, the manner by which the organism interacts with its environment. Many well-documented behavioral observations suggesting such influences have not had strong anatomic backing in the past. In this regard, however, it should be noted that several limbic structures, notably the amygdala and hippocampal formation, receive hypothalamic projections and project to the cortex. Thus, the anatomic interface maintained by the nu-

cleus basalis of Meynert between the hypothalamus and cerebral cortex is not unique in this sense but instead is part of many widespread limbic cortical neural systems.

Dorsomedial and Midline Thalamic Nuclei

The dorsomedial thalamic nucleus is known to play a role in many behaviors in humans, including visuospatial processing, attention, and memory. Contributing roles have been argued for aphasia, dementia, and temporal disorientation when the nucleus is diseased or damaged. Some authors attribute damage to this nucleus as the pathologic basis for the debilitating cognitive changes that occur in the alcoholic Korsakoff syndrome (151). Evidence from penetrating wounds (152) and thalamic infarcts (153) support this contention despite the radically different nature of the lesion. The literature regarding prefrontal lobotomy also applies here, because one would think that this surgery would cause

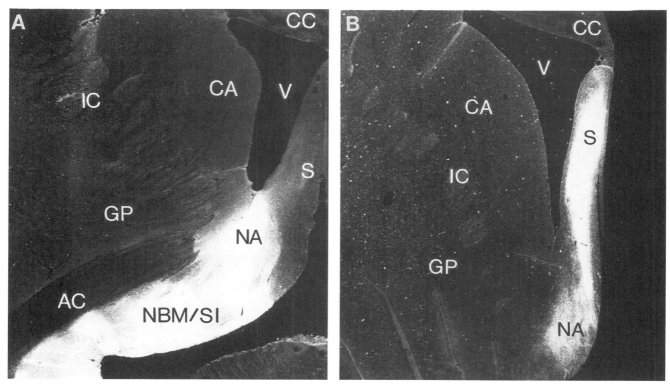

FIGURE 12.18. A: Dark-field photomicrograph of a coronal section showing terminal labeling (white) over the nucleus basalis of Meynert (NBM/SI) and nucleus accumbens (NA) after injection of tritiated amino acids into the amygdala. The labeled axon terminals are known to contact nucleus basalis of Meynert cholinergic neurons. **B:** Dark-field photomicrograph of a coronal section showing terminal labeling (white) over the septum/diagonal band (S) and the NA after injection of tritiated amino acids into the CA1/subiculum sectors of the hippocampal formation. Many of these axon terminals contact cholinergic neurons in the medial septal nucleus and vertical limb of the diagonal band of Broca. Ac, anterior commissure; CA, caudate nucleus; CC, corpus callosum; GP, globus pallidus; IC, internal capsule; S, septum; V, lateral ventricle.

extensive retrograde cell changes in the dorsomedial thalamic nucleus. However, only a subset of the behavioral changes listed earlier was reported in individuals who underwent this surgical procedure, and purportedly these changes were confined largely to the realm of personality.

The dorsomedial thalamic nucleus is a large midline association nucleus that has powerful interconnections with the prefrontal granular association cortex (Fig. 12.19) (154–156). From a cytoarchitectural viewpoint, it is a complex composed of several subdivisions that form partially concentric areas around the third ventricle. In general terms, they have topographically organized reciprocal connections with the prefrontal cortex in the monkey (154,155–158). For example, the most medial subdivision of the dorsomedial nucleus projects to and receives projections from the posterior orbital, anterior cingulate, and medial frontal cortex. A more lateral subdivision projects to and receives projections from the prefrontal association cortex dorsal and ventral to the principal sulcus and the anterior most parts of the orbitofrontal cortex. The lateral-most subdivision of the dorsomedial nucleus sends projections to and receives

projections from the periarcuate cortex in the anterior bank of the arcuate sulcus.

With the exception of its prefrontal cortex connections, the neural systems of the dorsomedial nucleus are poorly understood. In fact, the known input and output relationships of the mediodorsal nucleus with other structures are decidedly sparse in comparison to other nuclei of the thalamus and even other association nuclei, such as the pulvinar.

Some evidence is accumulating that enables at least partial characterization of this structure. For example, early ablation degeneration experiments identified another cortical projection to this large nucleus from the lateral temporal cortex (107). It also has been shown that the inferior parietal lobule and cortex lining the upper bank of the superior temporal sulcus receive projections from the lateral sector of the mediodorsal nucleus (159). Additional evidence suggests that certain cortical areas of the limbic lobe, such as the anterior cingulate cortex, have connections with the dorsomedial nucleus (160). This is of some interest because the dorsal part of the cingulate cortex (area 24c or M3) contributes descending axons to the facial nucleus and spinal

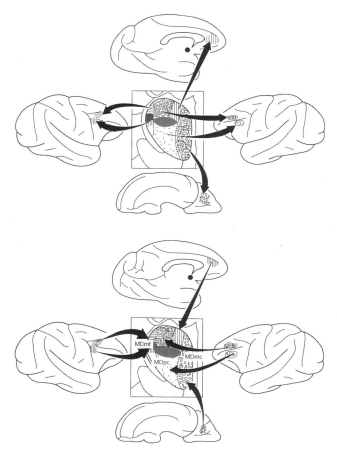

FIGURE 12.19. Summary diagram illustrating the topographic organization of thalamocortical **(top)** and corticothalamic **(bottom)** connections between the prefrontal cortex and the mediodorsal (MD) nucleus. MDmc; magnocellular division of the medial dorsal nucleus; MDmf, multiform division of the medial dorsal nucleus; MDpc, parvicellular division of the medial dorsal nucleus. (From Siwek DF, Pandya DN. Prefrontal projections to the mediodorsal nucleus of the thalamus in the rhesus monkey. *J Comp Neurol* 1991;312:509–524, with permission.)

cord, and it sends corticocortical projections directly to the supplementary and primary motor cortices.

Direct input from the amygdala to the dorsomedial nucleus has been known for years, and later tracing experiments buttressed and extended these findings (104,126,127). These axons arise from the basal complex of the amygdala and terminate in the more medial parts of the dorsomedial nucleus. The mediobasal nucleus seems to be the primary focus for this projection, although other basal nuclei (laterobasal and accessory basal) also appear to contribute. These axons course largely via the ventroamygdalofugal pathway and the inferior thalamic peduncle. Curiously, the amygdala projection to the mediodorsal nucleus is not reciprocated by thalamoamygdaloid projections. Additional input to the mediodorsal nucleus from the ventral pallidum, substantia nigra, septum, superior colliculus, and hypothalamus have been reported.

Summary and Functional Considerations

Overall, the full complement of neural systems associated with the dorsomedial thalamic nucleus are not well known. The position of this nucleus ventral to two large fiber systems, the fimbria-fornix and corpus callosum, its encasement within the internal medullary lamina of the thalamus, and the fact that the mammillothalamic tract traverses its ventral parts discouraged early experimental study. Therefore, and with regard to memory mechanisms, most attention has been focused squarely on the amygdalothalamic and temporothalamic projections because they clearly link the dorsomedial nucleus with temporal structures known to play a role in memory. For example, the inferotemporal cortices have been characterized as playing a mnemonic role in certain perceptual learning tasks. In addition, the mediobasal amygdaloid nucleus, which contributes strongly to the amygdalothalamic projection, is a direct recipient of subicular output from the hippocampal formation. Finally, anatomic findings reveal that the frontal association cortices that receive powerful input from the dorsomedial nucleus themselves receive projections from the hippocampal formation and project to the cortex around the rhinal sulcus, which, in turn, projects directly to the subiculum.

These findings themselves may be sufficient to implicate the dorsomedial thalamic nucleus in some aspects of memory and many other related behaviors. However, the anatomic proximity of multiple thalamic nuclei, as well as the multitude of adjacent white matter pathways, suggests that extreme caution must be exercised when making specific clinicopathologic inferences involving this nucleus. For example, several adjacent midline nuclei of the thalamus project directly to the hippocampal formation, entorhinal cortices, and amygdala. These include the nucleus reuniens, paracentral nucleus, and paraventricular nucleus. These are likely involved in hemorrhagic and nonhemorrhagic infarcts to the midline thalamus and in penetrating wounds that reach the thalamus. Their involvement in the alcoholic Korsakoff patient seems likely as well. At present, an association of midline thalamic damage with memory impairments and behavioral changes is known, but the role of specific thalamic nuclei remains to be evaluated.

LIMBIC PREFRONTAL INTERACTIONS

The frontal lobe lies anterior to the central sulcus and can be divided into two major parts, a caudal part containing the electrophysiologically "excitable" motor cortices and a rostral part containing the prefrontal association cortex (Fig. 12.2). The motor cortices include the primary (M1 or area 4), supplementary (M2 or area 6m), and lateral premotor cortices (LPMC or areas 6D and 6V) (161–165). All are characterized as agranular cortex, attesting to the fact that their internal granular layer, or layer IV, is not conspicuous. It is well known that M1 plays a critical role in activating and facilitating independent body movements. On the other hand, M2 seems more involved with movement and premovement activity possibly related to the internal generation and sequencing of movements (166–168). Significant modulation of neuronal activity in the lateral premotor cortex, particularly the ventral part of lateral area 6, has been shown to be coupled with stimulus triggered (visual and somatosensory cued) motor responses and during the execution of purposeful movements such as grasping and/or bringing hand to mouth and object manipulation (168).

Also included as part of the motor cortices are the frontal eye field (FEF; area 8), supplementary eye field (SEF; area F6) and presupplementary motor area (pre-SMA) (Fig. 12.2). The FEF and SEF are located on the lateral surface of the hemisphere (169–171). The FEF is located anterior to the midportion of lateral premotor cortex and is dysgranular in cytoarchitecture. This refers to the fact that FEF is characterized by a poorly defined or incipient, internal granular layer IV. The rostral part of area 6D of the lateral premotor cortex also is dysgranular (172). Both the FEF and SEF regulate contralateral saccadic eye movements. The pre-SMA is located on the medial wall of the hemisphere (173), rostral to M2, and would correspond to Walker area 8B (174). Neurons in this field modulate their activity before and during movement, and imaging studies in humans suggest the pre-SMA is involved in the movement decision-making process (166,175).

The prefrontal cortex lies rostral to the motor-related cortices and extends to the frontal pole (Fig. 12.2). On the medial surface, the prefrontal cortex lies anterior to the medial component of motor cortex and the anterior part of cingulate gyrus. The primate prefrontal cortex is subdivided commonly on broad anatomic grounds. Its major partitions include ventrolateral, dorsolateral, medial, and orbitofrontal regions. In the monkey, the principal sulcus is located on the lateral surface of the hemisphere, and its depths form the boundary between the ventrolateral and dorsolateral regions of prefrontal cortex. As named, the medial region of the prefrontal cortex is located on the medial surface of the hemisphere. Finally, the orbitofrontal region of the prefrontal cortex lies in the anterior cranial fossa above the bony orbit and forms the basal, or ventral, surface of the frontal lobe.

A large portion of the prefrontal cortex is classified cytoarchitecturally as granular cortex, which has six well-differentiated layers including a prominent external granular layer II and internal granular layer IV (176,177). However, differences in layers II and IV, as well as the other cortical laminae, serve as a basis for partitioning the frontal granular cortex into several subfields. They are designated numerically, according to Brodmann and others. The prefrontal cortex includes areas 45, 12, 46, 10, and 9 laterally; 9, 10, and 32 medially; and 12, 13, 11, 10, and 14 ventrally (Fig. 12.2). Many of these areas have been redefined and subdivided even further.

Cortex on the orbitofrontal surface lobe can be subdivided based on cytoarchitectural criteria into a caudal agranular sector, an anterior granular sector, and a transitional dysgranular sector between them (176,178–180). From this perspective, general trends and distinguishing features of major orbitofrontal organization are clearly recognizable. For example, agranular and dysgranular components are located caudally on the orbitofrontal surface and are strongly connected with the limbic cortices, amygdala, and midline nuclei of the thalamus (Fig. 12.20). In contrast, the granular component is situated rostrally on the orbitofrontal surface. This cortex is linked strongly to isocortical association areas and association nuclei of the thalamus. As expected, these unique patterns of neural interconnections would differentially influence activity and events processed in rostral versus caudal parts of the orbitofrontal cortex. Lesions rarely affect small parts of the orbitofrontal surface and thus are correlated with more global behavioral changes and not specific impairments (Fig. 12.21). For example, posterior lesions disrupt limbic and medial temporal connections and lead to changes in emotional and social behavior, as well as decision making and autonomic regulation. More anterior lesions that disrupt association inputs affect more complex behaviors.

Cortical Association Connections of Prefrontal Cortex

In addition to its obvious role in motor behavior, frontal lobe function has long been associated with a variety of higher-order behaviors and cognitive processes. Some of the more notable ones include working memory, motor planning, developing behavioral strategies, decision making and problem solving (181–194). When considering the higher-order functions mediated by the prefrontal cortex, the finding that prefrontal cortex is linked directly to a constellation of cortical association areas should not be surprising. The prefrontal cortex is well known for its widespread corticocortical connections that involve distal parts of the cerebral cortex and, specifically, primary association and multimodal association cortices.

Primary association cortex is committed functionally to the early processing of sensory data conveyed by the neurons of an adjacent primary sensory area. Thus, primary association cortex is thought to operate in a more integrative fashion than primary sensory area. In contrast, multimodal association cortex is not committed to processing information related to one modality but rather integrates highly transformed information whose source can be traced back to a number of, or multiple, sensory modalities. Although the traditional dogma suggests that multimodal association cortex represents the highest or end stages of cortical processing, it is becoming more clear that information flowing in the reverse direction, that is, directed from multimodal association back to the primary association areas, may initiate and synchronize neural elements that are responsible for forming a

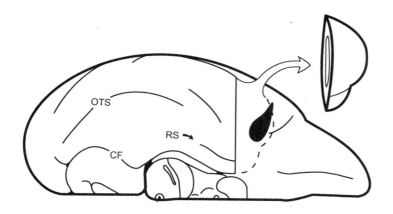

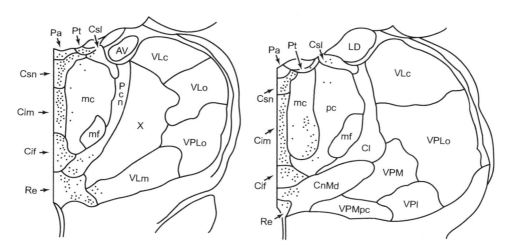

FIGURE 12.20. Line drawings illustrating a horseradish peroxidase (HRP) injection site in the posterior orbitofrontal cortex and the locations of retrogradely labeled cells at rostral (bottom left) and middle (bottom right) levels of the thalamus. Note the distinctive and heavy labeling in the central subnuclei underscoring the significant midline thalamic input to the posterior orbitofrontal cortex. AV, anteroventral nucleus; CF, calcarine fissure; Cif, central inferior nucleus; Cim, central intermediate nucleus; Cl, central lateral nucleus; CnMd, centromedian nucleus; Csl, central superior lateral nucleus; Csn, central superior nucleus; LD, lateral dorsal nucleus; mc, magnocellular division of the medial dorsal nucleus; mf, multiform division of the medial dorsal nucleus; OTS, occipitotemporal sulcus; Pa, paraventricular nucleus; pc, parvicellular division of the medial dorsal nucleus; Pcn, paracentral nucleus; Pt, parataenialis; Re, reunions; RS, rhinal sulcus; VLc, caudal division of the ventral lateral nucleus; VLm, medial division of the ventral lateral nucleus; VLo, oral division of the ventral lateral nucleus; VPl, ventral posterior inferior nucleus; VPLo, oral division of the ventral posterior lateral nucleus; VPMpc, parvicellular division of the ventral posterior medial nucleus; VPM, ventral posterior medial nucleus; X, area X. (Adapted from Morecraft RJ, Geula C, Mesulam M-M. Cytoarchitecture and neural afferents of orbitofrontal cortex in the brain of the monkey. *J Comp Neurol* 1992;323:341–358, with permission.)

selective preception by reactivation of multiple earlier processes (189,195).

The prefrontal cortex is linked to sensory association and multimodal association cortices of the parietal lobe (areas 7m, 7a, and 7b) and temporal lobe (areas V4t, MT, and MST), as well as sensory association cortices of the anterior part of the occipital lobe (area V3) (27,29,31,82,117,178,196,197). In terms of topography, it

has been shown in the monkey that the posterior part of the inferior parietal lobule (area 7a), anterior part of the occipital lobe, and medial parietal lobule (area 7m) are reciprocally connected with the dorsolateral and sulcal principalis regions of the prefrontal cortex. Physiologic studies indicate that the projections between the dorsolateral prefrontal region and posterior parietal cortex engage in a symmetric exchange of neuronal signaling where the output of each

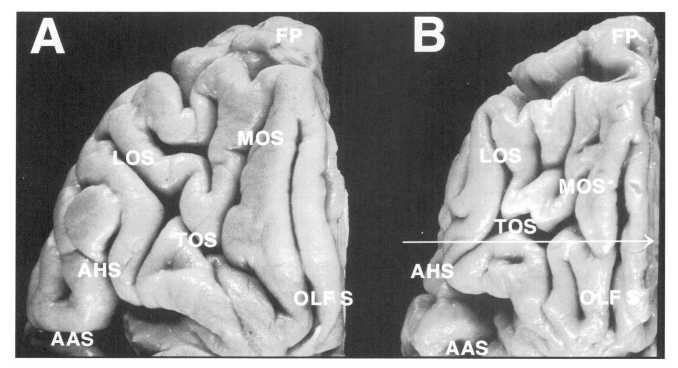

FIGURE 12.21. A: Human orbitofrontal surface of an elderly control brain. **B:** Orbitofrontal surface of an Alzheimer disease (AD) donor. Major orbital sulci are identified in both panels. Note the atrophic appearance of the orbitofrontal surface in panel *B* and the widening sulci. The widespread orbitofrontal damage in AD affecting the projection neurons suggests that this pathology may contribute heavily to the many non–memory-related behavioral changes observed in this disorder. AHS, anterior horizontal branch of the sylvian fissure; ASS, anterior ascending branch of the sylvian fissure; FP, frontal pole; LOS, lateral orbital sulcus; MOS, medial orbital sulcus; OLF S, olfactory sulcus; TOS, transverse orbitofrontal sulcus. (From Van Hoesen GW, Parvizi J, Chu C-C. Orbitofrontal cortex in Alzheimer's disease. *Cereb Cortex* 2000;10:243–251, with permission.)

cortical area produces a mixture of excitatory and inhibitory influences within its target (198). Anatomic and behavioral investigations conducted over the past 3 decades have led to the conclusion that long association pathways, reciprocally linking posterior parietal cortex and prefrontal cortex, are particularly important for appropriate execution of visually guided occulomotor and arm movements. Presumably, somatosensory and visual inputs converge on posterior part of the inferior parietal lobule, and information related to spatial orientation and motion analysis is advanced from this stage to the dorsolateral prefrontal cortex (199,200). Therefore, the dorsolateral part of the prefrontal cortex is thought to process information concerned with understanding *where* an object is in space. It also is known that the more rostral part of the inferior parietal lobule (area 7b) and ventrolateral part of the temporal lobe project to the ventrolateral part of the prefrontal cortex. Specifically, the projection from the ventrolateral part of temporal lobe is thought to carry information dealing with form and object recognition. Therefore, it has been suggested that the ventral pathway may constitute a processing stream that addresses *what* an object represents in the extrapersonal environment. The prefrontal cortex also is influenced by other parts of the temporal lobe through a subcomponent of the ventral pathway whose origin arises

from the rostral part of the superior temporal gyrus and the temporal pole. This projection probably represents an important source of auditory input to the ventrolateral part of the prefrontal cortex.

In addition to the long association pathways linking the prefrontal cortex with the parietal, occipital, and temporal lobes, short association pathways interlock the various parts of the prefrontal cortex with one another in an organized fashion (176). The less differentiated (in terms of cytoarchitectonic lamination) agranular and dysgranular cortices, located posteriorly on the basal and medial surface of the prefrontal cortex, give rise to widespread *intrinsic* prefrontal connections. In contrast, the more differentiated isocortical granular areas, which are situated anteriorly and laterally, are characterized by more limited *intrinsic* connections; instead they account for a large component of the widespread *extrinsic* prefrontal connections.

Cortical Limbic Connections of the Prefrontal Cortex

The strong structural relation between prefrontal cortex and association cortex has played a dominant role in shaping our views on prefrontal organization and function.

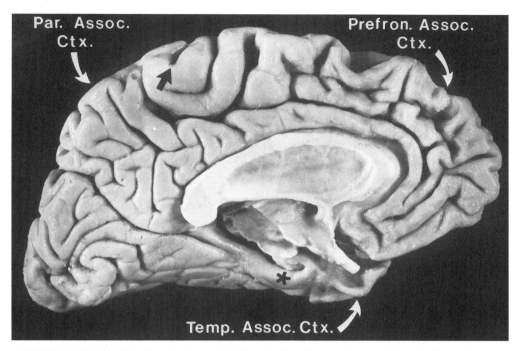

FIGURE 12.22. Medial view of the cerebral hemisphere from a donor with pathologically confirmed Pick disease. The *arrow* marks the medial tip of the central sulcus, and the *asterisk* marks the location of the entorhinal cortex. Note the marked atrophy of the prefrontal association and the temporal association cortices compared with preservation of the parietal association cortices. Neuron loss, gliosis, and occasional Pick bodies were observed in the former areas.

However, it is important not to neglect the structural interaction between the prefrontal cortex and limbic lobe, which also is very strong and functionally influential. For example, it is of great interest that these heavily interconnected areas, such as the prefrontal cortex and temporal pole, are severely atrophied in many cases of Pick disease and/or frontotemporal dementia (Fig. 12.22). There are many direct interconnections between the limbic lobe and the prefrontal cortex. Specifically, limbic projections to the prefrontal cortex arise from diverse and widespread parts of the limbic lobe, including the cingulate, orbitofrontal, temporopolar, perirhinal, entorhinal, posterior parahippocampal, and insular cortices (25,26,31,35,96,100,102,201,202). Although the lateral prefrontal cortex is a target for some of these connections, the bulk of this anatomic interrelationship is established with the posterior orbitofrontal and medial prefrontal regions.

As outlined previously, the prefrontal cortex maintains a highly organized anatomic affiliation with the cingulate gyrus. The more anterior levels of the dorsolateral prefrontal cortex are connected with the posterior cingulate cortex (area 23), whereas the posterior levels of the dorsolateral prefrontal cortex seem to be more strongly linked to the anterior cingulate gyrus. Similarly, anterior levels of the orbitofrontal cortex are related to area 23 and posterior levels with area 24.

Early studies that relied on the Marchi technique to trace neural connections failed to recognize a strong connection between the frontal and temporal lobes. However, the use of more sensitive tracing techniques enabled investigators to demonstrate that fibers forming a subcomponent of the uncinate fasciculus, as well as the extreme capsule, served to interconnect the frontal and temporal lobes in a strong and highly specific fashion. The strongest links with the limbic portion of temporal lobe involve the posterior orbitofrontal cortex and medial prefrontal cortex, followed by the lateral prefrontal cortex.

Although a precise topography has yet to be determined, a number of other areas in the temporal portion of the limbic lobe are connected to the prefrontal cortex. They include the temporal pole (area 38), perirhinal (area 35), entorhinal (area 28), posterior parahippocampal (areas TH and TF), presubicular, and subicular cortices. All but the subicular connection have been shown to be reciprocal. The subiculum of the hippocampal formation projects to the posterior part of the orbitofrontal cortex and sends some afferents to the dorsolateral part of the prefrontal cortex. A particularly heavy component of this projection terminates in the posterior part of the gyrus rectus, on the orbitofrontal surface. Hippocampal output is known to be mediated heavily by the subiculum and to some extent the CA1 sector of the hippocampal formation. Thus, subicular/CA1 output represents a direct hippocampal influence on the prefrontal cortex.

All parts of the prefrontal cortex are reciprocally connected with the insula, and a distinct anatomic relationship between cortex forming the orbitofrontal surface and the insula has been demonstrated. The agranular part of the orbitofrontal is preferentially linked to the agranular part of the insula. In terms of topography, this translates into distinct connections between the posterior orbitofrontal cortex and the anterior insula. Likewise, the granular orbitofrontal cortex, which is located anteriorly, is preferentially linked to the granular insula, which is located posteriorly. Interestingly, anterior parts of the insula receive direct input from the gustatory and olfactory cortex and the posterior parts of primary somatosensory and auditory areas. This suggests that the insula may be a common cite of *direct* convergence of all somatic sensory afferents. This is remarkable because integration of sensory information arising from primary areas more commonly occurs *after* a polysynaptic relay through sensory association and multimodal cortices. All parts of the insular appear to receive input from sensory association and multimodal association cortices. Therefore, the diversity of projections to the insula suggests that prefrontal input arising from the insula may be as little as one synapse away from a primary sensory area or conversely highly processed at the cortical level.

Amygdala projections to prefrontal cortex arise primarily from the basolateral and accessory basal nuclei and to a lesser extent from cortical and lateral nuclei (121,178,203). The strongest amygdalo/frontal projection ends in the posterior part (agranular and dysgranular sectors) of the orbitofrontal cortex. Another strong projection terminates in the medial prefrontal cortices and adjacent anterior cingulate cortex (areas 14, 25, 32, and 24). Projections to isocortical areas on the lateral convexity (areas 45, 46, 9, and 10) are less strong. As mentioned previously, the amygdala is the recipient of a wide variety of cortical inputs from allocortical, periallocortical, proisocortical, and isocortical association areas. The latter includes converging input from both auditory and visual association areas, as well as multimodal association cortices. Therefore, there is reason to believe that amygdala output directed toward the prefrontal cortices is influenced by a variety of neural systems related to the interplay of both the internal and external environments of the organism.

Motor Cortex Connections with Prefrontal Cortex

The available data suggest that prefrontal corticofugal axons are not directed to cranial nerve nuclei or the spinal cord. However, since the latter part of the nineteenth century, it has been appreciated that prefrontal cortex plays a special and important role in guiding the outcome of voluntary motor behavior. How this is accomplished automatically has been an enigma for many years. The view that emerged from classic studies suggested that corticospinal neurons in M1 were influenced by prefrontal input through an indirect series of connections passing through the greater extent of lateral area 6 or premotor cortex (117,196). However, the discovery that

cortex located outside the primary motor cortex, including the lateral premotor cortex, supplementary motor cortex, and cingulate motor cortex, contained corticospinal and corticobulbar neurons suggested otherwise (37–39,70,204).

As summarized in Figure 12.23, more recent efforts have shown clearly that the prefrontal cortex projects directly to parts of the motor cortices giving rise to the corticospinal and corticofacial axons (34,78,79). In the lateral premotor cortex, only the very rostral parts of areas 6D and 6V, which contain corticospinal and corticofacial neurons, receive prefrontal input, with area 6V being the primary link. Prefrontal input to M2 seems to converge on the rostral part of M2, which contains the face area that projects to the facial nucleus, and less so on corticospinal output zones that subserve the arm (Fig. 12.23B). Therefore, the more rostral parts of the lateral premotor cortex and supplementary motor cortex receive prefrontal input. However, the rostral cingulate motor cortex (M3 or area 24c) and caudal cingulate motor cortex (M4 or area 23c) has also been shown to receive strong prefrontal input that converges on parts of these cortices, giving rise to the corticofacial and corticospinal pathways (Figs. 12.17 and 12.23A). Thus, recent work implies that several anatomically distinct sources of corticofacial and corticospinal axons are directly influenced by prefrontal output. Moreover, the ventral part of M1 receives input from the caudal part of the ventrolateral operculum of the prefrontal cortex (Fig. 12.18C). The target of this projection resides outside the M1 corticospinal projection zone and may correspond to the face representation. In many nonhuman primate models, the FEF has been shown to give rise to projections that innervate midbrain centers regulating ocular motor behavior. Although no corticospinal projections arise from the FEF, it does receive significant prefrontal inputs.

Subcortical Connections of Prefrontal Cortex

The corticostriate projection from the prefrontal cortex is powerful and directed toward targets in the caudate nucleus and to a much lesser extent the putamen (29,205–207). Because a portion of the outflow of the basal ganglia is directed to the thalamus and eventually back to all parts of the frontal lobe, the corticostriate projection represents initial stages of a sequential pathway by which prefrontal cortex can influence a wide variety of neural systems. Specifically, the prefrontal corticostriatal projection is nonreciprocal and has a distinctive mediolateral topography that is related to the origin of the projection. Medial and orbital prefrontal regions project mainly to the medial and intermediate portions of the head and body of the caudate nucleus. This includes the nucleus accumbens of the ventral striatum (208), which may play a role in the genesis of drug addiction, major depression, and possibly psychosis in schizophrenia. The lateral prefrontal cortex is related preferentially to the intermediate portion of the head and body of the caudate nucleus, whereas the

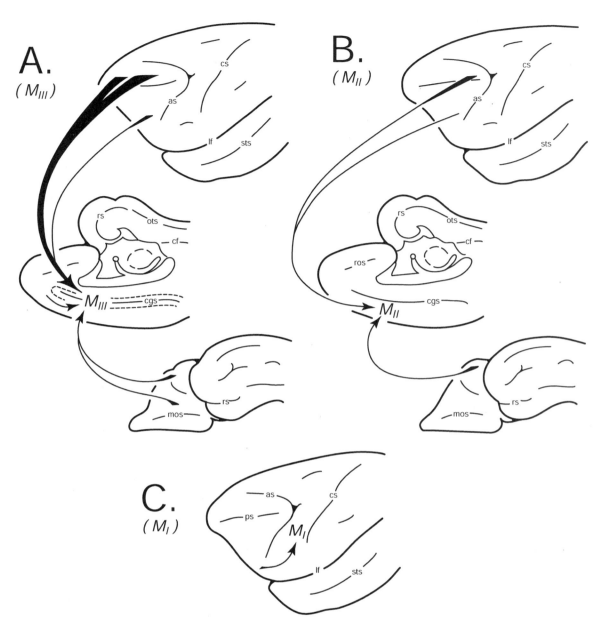

FIGURE 12.23. Line diagram illustrating the diversity and strength of prefrontal input to the rostral cingulate motor cortex (MIII) **(A),** supplementary motor cortex (MII) **(B),** and primary motor cortex (MI) **(C)** in the rhesus monkey. Note that the rostral cingulate motor cortex receives dorsolateral, medial, and orbitofrontal projections. The supplementary motor cortex receives some direct prefrontal input, which is primarily directed to its rostral part (e.g., face representation). The primary motor cortex receives little direct prefrontal cortex input. These afferents arise from the agranular/granular transition region located below the arcuate sulcus and end in the inferior region (e.g., face representation) of the primary motor cortex. as, arcuate sulcus; cf, calcarine sulcus; cgs, cingulate sulcus; cs, central sulcus; lf, lateral fissure; mos, medial orbital sulcus; ots, occipitotemporal sulcus; ros, rostral sulcus; rs, rhinal sulcus; sts, superior temporal sulcus. (From Morecraft RJ, Van Hoesen GW. Frontal granular cortex input to the cingulate (M3), supplementary (M2) and primary (M1) motor cortices in the rhesus monkey. *J Comp Neurol* 1993;337:669–689, with permission.)

caudal prefrontal region (area 8) projects mainly to the lateral and intermediate portion of the nucleus. A dorsoventral topography is evident as well. Dorsal prefrontal areas project primarily to the dorsal portion of the head and body of the caudate, whereas ventral prefrontal areas are connected preferentially with the ventral portion. It is important to note that a certain degree of overlap occurs in these projections within the central sector of the head and body of the caudate. The prefrontal striatal projection to the tail of the caudate is less distinct. The medial portion of the tail receives input from the dorsal and ventral proisocortical prefrontal regions, as well as from the dorsomedial (dorsal area 10) and dorsolateral (dorsal areas 9 and 46) portions of the prefrontal cortices. In comparison to prefrontal caudate projections, those to the putamen are less extensive. In most circumstances these appear to be continuous with connections to the caudate nucleus and have a dorsoventral topography. Dorsal trend areas (areas 10, 9, 46, and 8) project mainly to dorsal and central parts of the putamen, with the exception of ventral area 8, which projects to the dorsal part of the putamen.

The corticothalamic projection is one of the most studied corticofugal pathways from the prefrontal cortex (155,156,209–212). The seminal work of Akert (154) and Nauta (209) demonstrated that the mediodorsal thalamic nucleus is connected to all parts of the prefrontal cortex in the primate, including the granular sectors located anteriorly and laterally and the dysgranular and agranular sectors located medially and ventrally (the posterior orbitofrontal cortex). The medial part (magnocellular division) of the mediodorsal nucleus was found to be connected primarily with the orbital surface of the prefrontal cortex and the lateral part (parvocellular division) with the lateral surface of the prefrontal cortex. Anatomic findings and functional observations since have been combined to suggest that the prefrontal cortex can be viewed as having at least four major subdivisions. They include dorsolateral, ventrolateral, medial, and orbitofrontal divisions, each having its own unique thalamic projection pattern. Although the strong reciprocal anatomic relationship between the mediodorsal nucleus and prefrontal cortex often is emphasized, like all other cortical areas, numerous thalamic nuclei are linked to this brain region. They include the midline, ventral anterior, intralaminar, anterior medial, and pulvinar nuclei (Figs. 12.19 and 12.20). The hypothalamus is another important subcortical diencephalic structure interconnected with the prefrontal cortex (213). The hypothalamus is well known for mediating core autonomic functions, including feeding, fluid regulation, heart rate, respiratory rate, reproduction, and self-defense. The agranular and dysgranular portions of the prefrontal cortex are reciprocally interconnected with the hypothalamus. Specifically, medial areas 25 and 32 are connected with the anterior and ventromedial hypothalamic regions, whereas the posterior orbitofrontal cortex is interconnected primarily with the posterior hypothalamic region. There also is a

mediolateral topography suggesting that lateral parts of the hypothalamus are connected with the posterior orbitofrontal cortices, whereas the medial hypothalamus is interconnected with the medial prefrontal cortices. Lateral prefrontal connections with the hypothalamus are weaker than those established with the orbitofrontal and medial prefrontal cortices and target primarily the posterior hypothalamic region. The patterns of descending projections indicate that prefrontal activity may directly initiate autonomic responses accompanying stressful and important decision-making processes. Similarly, direct ascending hypothalamic projections to prefrontal cortex indicate that basic and powerful physiologic drives likely affect the outcome of higher-order behavior. These connections may be involved in recruiting a prefrontal contribution to appropriately address the autonomic needs of the body (e.g., maintenance of physiologic homeostasis) or to solve a potential life-threatening situation. Also relevant is the possibility that these connections serve as "somatic markers" (11,190), that is, development of a "body state" evoked by the experience of a reward, or punishment, can influence the potential outcome of a mental process in a manner that guides the selection of a behavior that is advantageous to the organism. In the absence of such markers, decision making is faulty.

Several important brainstem projections from prefrontal cortex have been identified that may play a role in motor control (29,214,215). Corticotectal projections have been demonstrated to arise from the dorsolateral principalis region in the monkey and terminate in the intermediate and deep layers of the superior colliculus of the midbrain. Such projections are likely candidates for influencing behaviors linked to eye movement, particularly from the standpoint of ocular motor tasks requiring on-line or working memory, to order and issue an appropriate set of behavioral commands. Additional prefrontal projections terminate in the substantia nigra, PAG, cuneiform nucleus, median raphe nucleus, and midbrain reticular formation (216). Medial, lateral, and orbitofrontal projections to the PAG are highly organized and terminate in distinct columns. The prefrontal projection to the PAG is heaviest from the medial and posterior orbitofrontal regions and terminates in the dorsolateral and ventrolateral columns, respectively. In addition, projections from areas 9 and 24b end mainly in the lateral column. The prefrontal PAG projection from areas 9, 46, and 8 is weak and directed mainly to the dorsolateral column. Finally, the rostral cingulate motor cortex (M3) projects to the lateral and ventrolateral sectors of the PAG.

Projections from the prefrontal cortex to the pontine reticular formation have been described. This projection appears to be distributed over the paramedian portion reticular formation, an area corresponding in location to the central superior nucleus. Several prefrontal regions also give rise to a corticopontine projection, which may contribute to executing cognitive-based tasks supported by cerebellar activity (217–219). Lateral prefrontal areas 8, 9, 46d, and 10 have been found to give rise to a strong corticopontine projection ending primarily in the paramedian nucleus and in the medial sectors of the peripeduncular nucleus. Interestingly, ventrolateral prefrontal and orbitofrontal cortex do not appear to contribute strongly to the corticopontine system. System wise, the corticopontine projection represents a major component of the corticopontocerebellar circuit. Transneuronal labeling techniques have shown that cerebellar as well as pallidal neurons are labeled after injections of herpes simplex virus type 1 into the dorsolateral prefrontal cortex (218,220). This indicates that thalamocortical input to the prefrontal cortex is influenced by both basal ganglia (e.g., corticocaudate projections) as well as cerebellar circuits (e.g., corticopontine projections).

The prefrontal cortex is connected to several small but important brainstem nuclei that synthesize and transmit selective neurotransmitters to widespread parts of the prefrontal cortex. It has been suggested that these projections play an important role in regulating global as well as discrete behavioral states (221,222). For example, as discussed earlier, the nucleus basalis of Meynert gives rise to cholinergic projections that innervate all parts of the prefrontal cortex. The ventral tegmental area, dorsal raphe nucleus, and locus caeruleus also belong to pharmacologically distinct classes of subcortical nuclei and. like the nucleus basalis, project to all parts of the prefrontal cortex. The ventral tegmental area is a dopaminergic mesencephalic nucleus located ventral and posterior to the red nucleus. The dorsal raphe nucleus is situated in the midbrain and pons, immediately ventral to the PAG, and consists of serotonergic projection neurons. Finally, neurons of the locus caeruleus reside in the pons, located near the periventricular gray of the upper fourth ventricle, give rise to norepinephrine projections. There are several notable and interesting exceptions regarding the issue of reciprocity when considering the pharmacologically specific subcortical nuclei. For example, the nucleus basalis appears to project to all the cerebral cortex, including the prefrontal cortex, but it receives input from only the limbic lobe. From the standpoint of the frontal lobe, this includes the posterior parts of the orbitofrontal cortex and medial prefrontal cortex. The dorsolateral and medial parts of the prefrontal cortex have been found to send a reciprocal subcortical efferent projection back to the dorsal raphe nucleus and locus caeruleus (223). Therefore, the selective reciprocity of these connections gives distinct parts of the prefrontal cortex feedback control over its own monoaminergic and cholinergic innervation, as well innervation that distributes to widespread parts of the cerebral cortex.

Summary and Conclusions

The widespread extent of corticocortical and subcortical connections underlies the complex functions subserved by prefrontal cortex. The diverse set of association connections converging on prefrontal cortex indicate that inte-

grated information, associated with multiple sensory modalities, shape the outcome of prefrontal-guided behaviors. The anatomic interaction between prefrontal and limbic cortices may affect the motivational state and emotional tone or temperament of prefrontal behaviors, in addition to involving memory features, such as the storage and retrieval of information.

The prefrontal cortex gives rise to a host of descending projections that contribute to the corticostriate, corticothalamic, corticotectal, corticoreticular, and corticopontine pathways. Collectively, prefrontal mesencephalic projections are thought to play a role in influencing vocalization, autonomic mechanisms, and pain-related responses. Information directed away from the prefrontal cortex through nonreciprocating projections, such as the corticostriate and corticopontine projections, eventually converges back on the prefrontal cortex only after coursing through a sequential and parallel set of subcortical circuits or "loops." Other projections, such as corticothalamic and corticohypothalamic projections, being reciprocal in nature, allow for direct interaction between prefrontal cortex and discrete subcortical diencephalic nuclei. Finally, chemically specific subcortical projections from cholinergic and monoaminergic sources to all parts of the prefrontal cortex may globally affect the operation of cortical states associated with arousal, attention, motivation, and learning. Likewise, projections from selected parts of the prefrontal cortex to monoaminergic and cholinergic subcortical centers suggest that prefrontal cortex may modulate its own afferent neurochemical innervation. The uniqueness of frontal lobe connections is not matched in any other lobe.

COMMENT AND CONCLUSIONS

The neural components of the limbic system and frontal lobes have been studied extensively for more than a century. The results attest to a continually evolving experimental neuroanatomy in which progressively better methods have supplanted earlier and less informative ones or methods that had very restrictive application, such as the Golgi method. Despite the neuroanatomic complexity of these brain areas, the skeleton, or the common wisdom, that has survived is largely accurate. For example, Papez's original circuit has been found to have few flaws; Cajal's early descriptions of hippocampal and limbic lobe circuitry remain intact; and the descriptions of extensive interconnections of the frontal lobe with other association areas and the limbic system are truer than ever. It seems that what has changed most is an appreciation for the degree of interaction between structures. These findings have provided fresh insights. In the case of more recent research, the strength and topography of connections between the limbic system and the association cortices has been one of the major payoffs of better methodology. Structures such as the amygdala, cingulate cortex, and hippocampal formation now are known to have extensive and widespread connections with the association cortices. These occur directly, without having to execute the cumbersome pathways of Papez's circuit. For example, both the amygdala and hippocampal formation project strongly to parts of the prefrontal cortex, and the amygdala has output to the occipital lobe and cingulate cortex neurons that give rise to corticofacial and corticospinal axons. The entorhinal cortex conveys cortical association input to the hippocampal formation and, after receiving hippocampal output, projects back strongly to association areas. Among the truly new findings, the nucleus basalis of Meynert projections to literally all of the cerebral cortex link the hypothalamus to the cortex and provide a manner for the basal forebrain to have widespread cholinergic influence on the cortical mantle. Its projections to the reticular nucleus of the thalamus also establish a morphologic basis for its influence on thalamic output before it relays to the cortex. Direct amygdaloid projections to the dorsomedial thalamic nucleus provide a manner for this temporal limbic structure to influence not only the orbital and medial prefrontal cortices but also the extensive dorsolateral prefrontal cortices. Direct prefrontal–hypothalamic interconnections link basic physiologic drives with cognitive operations. Finally, and of critical importance, direct projections from the rostral (M3) and caudal (M4) cingulate motor areas to the facial nucleus and spinal cord provide an unambiguous manner for the limbic system to influence somatic effectors and behavior, and the extensive prefrontal cortex output to this area gives this cortex more direct influence on lower motor neurons than previously appreciated. This circuitry possibly takes part in the expression of psychogenic movement disorders.

The refined and augmented older findings, and truly new neuroanatomic findings, ensure the generation of fruitful hypotheses regarding neural networks that might explain higher-order behavior. Furthermore, they provide a richer description of normality against which neurologic and psychiatric diseases affecting the limbic system and frontal cortex can be assessed. This already has been of benefit in the case of Alzheimer disease (224) and likely will contribute to a better understanding of frontotemporal dementia.

ACKNOWLEDGMENT

This work was supported by National Institutes of Health Grants NS 33003 and RR15567 (R.J.M.) and PO 19632 and NS 14944 (G.V.H.).

REFERENCES

1. MacLean P. Some psychiatric implications of physiological studies on frontotemporal portion of limbic system (visceral brain). *Electroencephalogr Clin Neurophysiol* 1952;4:407–418.
2. Papez JW. A proposed mechanism of emotion. *Arch Neurol Psychiatry* 1937;38:725–743.
3. Scoville WB, Milner B. Loss of recent memory after bilateral hippocampal lesions. *J Neurol Neurosurg Psychiatry* 1957;20:11–21.
4. Brodal A. The hippocampus and the sense of smell. A review. *Brain* 1947;70:179–222.
5. Damasio AR, Van Hoesen GW. Emotional disturbances associated with focal lesions of the limbic frontal lobe. In: Heilman KM, Satz P, eds. *Neuropsychology of human emotion.* New York: Guilford Press, 1983:85–108.
6. Finger S. *Origins of neuroscience: a history of explorations into brain function.* New York: Oxford University Press, 1994.
7. Goldman-Rakic PS. Working memory dysfunction in schizophrenia. *J Neuropsychiatry* 1994;6:348–357.
8. Eslinger PJ, Damasio AR. Severe disturbance of higher cognition after bilateral frontal lobe ablation: patient EVR. *Neurology* 1985;35:1731–1741.
9. Duncan J, Seitz RJ, Kolodny J, et al. A neural basis for general intelligence. *Science* 2000;28:457–460.
10. Rowe JB, Toni I, Josephs O, et al. The prefrontal cortex: response selection or maintenance within working memory? *Science* 2000;288:1656–1660.
11. Damasio AR. *Descartes' error.* New York: Grosset and Putnam Publishers, 1994.
12. Weinberger DR, Aloia MS, Goldberg TE, et al. The frontal lobes and schizophrenia. *J Neuropsychiatry* 1994;6:419–427.
13. Van Hoesen GW, Alheid GF, Heimer L. Major brain structures. In: Kaplan HI, Sadock BJ, eds. *Comprehensive textbook of psychiatry,* 5th ed. Baltimore: Williams & Wilkins, 1989:5–26.
14. Brodmann K. Beitrage zur histologischen localisation der grosshirnrinde. III. Mitteilung: die rindenfelder der niederen. *J Psych Neurol* 1905;4:177–266.
15. Braak H. A primitive gigantopyramidal field buried in the depth of the cingulate sulcus of the human brain. *Brain Res* 1976;109:219–233.
16. Vogt BA, Pandya DN, Rosene DL. Cingulate cortex of the rhesus monkey: I. Cytoarchitecture and thalamic afferents. *J Comp Neurol* 1987;262:256–270.
17. Vogt BA, Nimchinsky EA, Vogt LJ, et al. Human cingulate cortex: surface features, flat maps, and cytoarchitecture. *J Comp Neurol* 1995;359:490–506.
18. Nimchinsky EA, Vogt BA, Morrison JH, et al. Neurofilament and calcium-binding proteins in the human cingulate cortex. *J Comp Neurol* 1997;384:597–620.
19. Morris R, Paxinos G, Petrides M. Architectonic analysis of the human retrosplenial cortex. *J Comp Neurol* 2000;421:14–28.
20. Baleydier C, Mauguiere F. The duality of the cingulate gyrus in monkey. Neuroanatomical study and functional hypotheses. *Brain* 1980;103:525–554.
21. Vogt BA, Pandya DN. Cingulate cortex of rhesus monkey: II. Cortical afferents. *J Comp Neurol* 1987;262:271–289.
22. Van Hoesen GW, Morecraft RJ, Vogt BA. Connections of the monkey cingulate cortex. In: Vogt BA, Gabriel M, eds. *Neurobiology of the cingulate cortex and limbic thalamus: a comprehensive handbook.* Boston: Birkhauser, 1993:249–284.
23. Muakkassa KF, Strick PL. Frontal lobe inputs to primate motor cortex: evidence for four somatotopically organized "premotor" areas. *Brain Res* 1979;177:176–182.
24. Pandya DN, Van Hoesen GW, Mesulam M-M. Efferent connections of the cingulate gyrus in the rhesus monkey. *Exp Brain Res* 1981;42:319–330.
25. Mesulam M-M, Mufson EJ. Insula of the old world monkey. III. Efferent cortical output and comments on function. *J Comp Neurol* 1982;212:38–52.
26. Mufson EJ, Mesulam M-M. Insula of the old world monkey. II: afferent cortical input and comments on the claustrum. *J Comp Neurol* 1982;212:23–37.
27. Petrides M, Pandya DN. Projections to the frontal cortex from

the posterior parietal region in the rhesus monkey. *J Comp Neurol* 1984;228:105–116.

28. Vogt BA, Pandya DN. Cingulate cortex of rhesus monkey: II. Cortical afferents. *J Comp Neurol* 1987;262:271–289.
29. Selemon LD, Goldman-Rakic PS. Common cortical and subcortical targets of the dorsolateral prefrontal and posterior parietal cortices in the rhesus monkey: evidence for a distributed neural network subserving spatially guided behavior. *J Neurosci* 1988;8:4049–4068.
30. Vogt BA, Barbas H. Structure and connections of the cingulate vocalization region in the rhesus monkey. In: Newmann JD, ed. *The physiological control of mammalian vocalization.* New York: Plenum Press, 1988:203–225.
31. Cavada CC, Goldman-Rakic PS. Posterior parietal cortex in rhesus monkey. I: parcellation of areas based on distinctive limbic and sensory corticocortical connections. *J Comp Neurol* 1989;287:393–421.
32. Morecraft RJ. The cortical and subcortical efferent and afferent connections of a proposed cingulate motor cortex and its topographical relationship to the primary and supplementary motor cortices of the rhesus monkey. Thesis, The University of Iowa, Iowa City, IA, 1989:1–297.
33. Morecraft RJ, Van Hoesen GW. Cingulate input to the primary and supplementary motor cortices in the rhesus monkey: evidence for somatotopy in areas 24c and 23c. *J Comp Neurol* 1992;322:471–489.
34. Morecraft RJ, Van Hoesen GW. Frontal granular cortex input to the cingulate (M3), supplementary (M2) and primary (M1) motor cortices in the rhesus monkey. *J Comp Neurol* 1993;337:669–688.
35. Morecraft RJ, Geula C, Mesulam M-M. Architecture of connectivity within a cingulo-fronto-parietal neurocognitive network for directed attention. *Arch Neurol* 1993;50:279–284.
36. Morecraft RJ, Schroeder CM, Keifer J. Organization of face representation in the cingulate cortex of the rhesus monkey. *Neuroreport* 1996;7:1343–1348.
37. Morecraft RJ, Louie JL, Herrick JL, et al. Cortical innervation of the facial nucleus in the non-human primate. A new interpretation of the effects of stroke and related subtotal brain trauma on the muscles of facial expression. *Brain* 2001;124:176–208.
38. Biber MP, Kneisley LW, LaVail JH. Cortical neurons projecting to the cervical and lumbar enlargements of the spinal cord in young adult rhesus monkeys. *Exp Neurol* 1978;59:492–508.
39. Murray EA, Coulter JD. Organization of corticospinal neurons in the monkey. *J Comp Neurol* 1981;195:339–365.
40. Sutherland RJ, Whishaw IQ, Kolb B. Contributions of cingulate cortex to two forms of spatial learning and memory. *J Neurosci* 1988;8:1863–1872.
41. D'Esposito M, Detre JA, Alsop DC, et al. The neural basis of the central executive system of working memory. *Nature* 16;378:279–281.
42. Frith CD, Friston K, Liddle PF, et al. Willed action and the prefrontal cortex in man: a study with PET. *Proc R Soc Lond B Biol Sci* 1991;244:241–246.
43. Carter CS, Braver TS, Barch DM, et al. Anterior cingulate cortex, error detection, and the online monitoring of performance. *Science* 1998;280:747–749.
44. Botvinick M, Nystrom LE, Fissell K, et al. Conflict monitoring versus selection-for-action in anterior cingulate cortex. *Nature* 1999;402:179–181.
45. Foltz EL, White LE. Pain "relief" by frontal cingulumotomy. *J Neurosurg* 1962;19:89–100.
46. Davis KD, Taylor SJ, Crawley AP, et al. Functional MRI of pain- and attention related activations in the human cingulate cortex. *J Neurophysiol* 1997;77:3370–3380.
47. Rainville P, Duncan GH, Price D, et al. Pain affect encoded in human anterior cingulate but not somatosensory cortex. *Science* 1997;277:968–971.
48. Derbyshire SWG, Vogt BA, Jones AKP. Pain and Stroop interference tasks activate separate processing modules in anterior cingulate cortex. *Exp Brain Res* 1998;118:52–60.
49. Koyyama T, Tanaka YZ, Mikami A. Nociceptive neurons in the macaque anterior cingulate activate during anticipation of pain. *Neuroreport* 1998;9:2663–2667.
50. Lenz FA, Rios M, Zirh A, et al. Painful stimuli evoke potentials recorded over the human anterior cingulate gyrus. *J Neurophysiol* 1998;79:2231–2234.
51. Ploghaus A, Tracey I, Gati JS, et al. Dissociating pain from its anticipation in the human brain. *Science* 1999;284:1978–1981.
52. Sikes RW, Vogt BA. Nociceptive neurons in area 24 of the rabbit cingulate cortex. *J Neurophysiol* 1992;68:1720–1732.
53. MacLean PD, Newman JD. Role of midline frontolimbic cortex in the production of the isolation call of squirrel monkeys. *Brain Res* 1988;45:111–123.
54. Arroyo A, Lesser RA, Gordon B, et al. Mirth, laughter and gelastic seizures. *Brain* 1993;116:757–780.
55. Smith WK. The functional significance of the rostral cingular cortex as revealed by its responses to electrical excitation. *J Neurophysiol* 1945;8:241–259.
56. Ward AA. The cingular gyrus: area 24. *J Neurophysiol* 1948;11:13–23.
57. Kaada BR, Pribram KH, Epstein JA. Respiratory and vascular response in monkeys from temporal pole, insula, orbital surface and cingulate gyrus. *J Neurophysiol* 1949;12:348–356.
58. Neafsey EJ. Prefrontal autonomic control in the rat: anatomical and electrophysiological observations. *Prog Brain Res* 1990;85:147–166.
59. Neafsey EJ, Terreberry RR, Hurley KM, et al. Anterior cingulate cortex in rodents: connections, visceral control functions, and implications for emotion. In: Vogt BA, Gabriel M, eds. *Neurobiology of the cingulate cortex and limbic thalamus: a comprehensive handbook.* Boston: Birkhauser, 1993:206–223.
60. Talairach J, Bancaud J, Geier S, et al. The cingulate gyrus and human behavior. *Electroencephagr Clin Neurophysiol* 1973;34:45–52.
61. Kremer S, Chassagnon S, Hoffmann D, et al. The cingulate hidden hand. *J Neurol Neurosurg Psychiatry* 2001;70:264–265.
62. Mitz AR, Wise SP. The somatotopic organization of the supplementary motor area: intracortical microstimulation mapping. *J Neurosci* 1987;7:1010–1021.
63. Shima K, Aya K, Mushiake H, et al. Two movement-related foci in the primate cingulate cortex observed in signal-triggered and self-paced forelimb movements. *J Neurophysiol* 1991;65:188–202.
64. Luppino G, Matelli M, Camarda RM, et al. Multiple representations of body movements in mesial area 6 and adjacent cingulate cortex: an intracortical microstimulation study in the macaque monkey. *J Comp Neurol* 1991;311:463–482.
65. Shima K, Tanji J. Role for cingulate motor area cells in voluntary movement selection based on reward. *Science* 1998;282:1335–1338.
66. Wu CW, Bichot NP, Kaas JH. Converging evidence from microstimulation, architecture, and connections for multiple motor areas in the frontal and cingulate cortex of prosimian primates. *J Comp Neurol* 2000;423:140–177.
67. Wang Yan Y, Shima K, Sawamura H, et al. Spatial distribution of cingulate cells projecting to the primary, supplementary, and pre-supplementary motor areas: a retrograde multiple labeling study in the macaque monkey. *Neurosci Res* 2001;39:39–49.
68. Hutchins KD, Martino AM, Strick PL. Corticospinal projections from the medial wall of the hemisphere. *Exp Brain Res* 1988;71:667–672.

69. Morecraft RJ, Van Hoesen GW. Somatotopical organization of cingulate projections to the primary and supplementary motor cortices in the old-world monkey. *Soc Neurosci Abstr* 1988;14:820.

70. Dum RP, Strick PL. The origin of corticospinal projections from the premotor areas in the frontal lobe. *J Neurosci* 1991;11:667–689.

71. Paus T, Petrides M, Evans AC, et al. Role of the human anterior cingulate cortex in the control of oculomotor, manual, and speech responses: a positron emission tomography study. *J Neurophysiol* 1993;70:453–469.

72. Picard N, Strick PL. Motor areas of the medial wall: a review or their location and functional activation. *Cereb Cortex* 1996;6:342–353.

73. Sanides F. Comparative architectonics of the neocortex of mammals and their evolutionary interpretation. *Ann N Y Acad Sci* 1969;167:404–423.

74. Humphrey DR, Gold R, Reed DJ. Sizes, laminar and topographic origins of cortical projections to the major divisions of the red nucleus in the monk. *J Comp Neurol* 1984;225:75–94.

75. Morecraft RJ, Van Hoesen GW. Descending projections to the basal ganglia, red nucleus and pontine nuclei from the cingulate motor cortex (M3 or area 24c) in the rhesus monkey. *Soc Neurosci Abstr* 1994;20:986.

76. Morecraft RJ, Keifer J, Saoi DJ. The corticospinal projection from the cingulate motor cortex (M3 or area 24c) to the cervical enlargement in rhesus monkey. *Soc Neurosci Abstr* 1995;21:411.

77. Morecraft RJ, Louie JL, Schroeder CM, et al. Segregated parallel inputs to the spinal cord from the cingulate motor cortex in the monkey. *Neuroreport* 1997;8:3933–3938.

78. Bates JF, Goldman-Rakic PS. Prefrontal connections of medial motor areas in the rhesus monkey. *J Comp Neurol* 1993;336:211–228.

79. Lu MT, Preston JB, Strick PL. Interconnections between the prefrontal cortex and the premotor areas in the frontal lobe. *J Comp Neurol* 1994;341:375–392.

80. Morecraft RJ, Van Hoesen GW. Convergence of limbic input to the cingulate motor cortex in rhesus monkey. *Brain Res Bull* 1998;45:209–232.

81. Blackstad TW. Commissural connections of the hippocampal region in the rat, with special reference to their mode of termination. *J Comp Neurol* 1956;105:417–537.

82. Van Hoesen GW. The parahippocampal gyrus: new observations regarding its cortical connections in the monkey. *Trends Neurosci* 1982;5:345–350.

83. Rosene DL, Van Hoesen GW. The hippocampal formation of the primate brain: a review of some comparative aspects of cytoarchitecture and connections. In: Jones EG, Peters A, ed. *Cerebral cortex. Further aspects of cortical function, including hippocampus, vol. 6.* New York: Plenum Press, 1987:345–456.

84. Duvernoy HM. *The human hippocampus. An atlas of applied anatomy.* München: Bergmann, 1988:1–66.

85. Amaral DG, Insausti R. The hippocampal formation. In: Paxinos G, ed. *The human nervous system.* New York: Academic Press, 1990.

86. Cajal Ramón y S. *Studies on the cerebral cortex.* Chicago: Year Book, 1955:1–174.

87. Cajal Ramón y S. *The structure of Ammon's horn.* Kraft LM, trans. Springfield: Charles C Thomas, 1968:78.

88. Hjorth-Simonsen A. Some intrinsic connections of the hippocampus in the rat: an experimental analysis. *J Comp Neurol* 1973;147:145–162.

89. Andersen P, Bland BH, Dudar JD. Organization of the hippocampal output. *Exp Brain Res* 1973;17:152–168.

90. Swanson LW, Cowan WM. Hippocampo-hypothalamic connec-

tions: origin in subicular cortex, not Ammon's horn. *Science* 1975;189:303–304.

91. Meibach RC, Siegel A. The origin of fornix fibers which project to the mammillary bodies of the rat: a horseradish peroxidase study. *Brain Res* 1975;88:518–522.

92. Rosene DL, Van Hoesen GW. Hippocampal efferents reach widespread areas of the cerebral cortex and amygdala in the rhesus monkey. *Science* 1977;198:315–317.

93. Swanson LW, Wyss JM, Cowan WM. An autoradiographic study of the organization of intrahippocampal association pathways in the rat. *J Comp Neurol* 1978;181:681–716.

94. Sorenson KE, Shipley MT. Projections from the subiculum to the deep layers of the ipsilateral presubicular and entorhinal cortices in the guinea pig. *J Comp Neurol* 1979;188:313–334.

95. Van Hoesen GW, Pandya DN, Butters N. Cortical afferents to the entorhinal cortex of the rhesus monkey. *Science* 1972;175:1471–1473.

96. Van Hoesen GW, Pandya DN. Some connections of the entorhinal (area 28) and perirhinal (area 35) cortices of the rhesus monkey. II. Efferent connections. *Brain Res* 1975;95:39–59.

97. Shipley MT. The topographic and laminar organization of the presubiculum's projection to the ipsi- and contralateral entorhinal cortex in the guinea pig. *J Comp Neurol* 1975;160:127–146.

98. Van Hoesen GW, Rosene DL, Mesulam M-M. Subicular input from temporal cortex in the rhesus monkey. *Science* 1979;205:608–610.

99. Amaral DG, Insausti R, Cowan WM. Evidence for a direct projection from the superior temporal gyrus to the entorhinal cortex in the monkey. *Brain Res* 1983;275:263–277.

100. Insausti R, Amaral DG, Cowen WM. The entorhinal cortex of the monkey: II. Cortical afferents. *J Comp Neurol* 1987;264:356–395.

101. Witter MP. Organization of the entorhinal–hippocampal system: a review of current anatomical data. *Hippocampus* 1993;3:33–44.

102. Suzuki WA, Amaral DG. Topographic organization of the reciprocal connections between the monkey entorhinal cortex and the perirhinal and parahippocampal cortices. *J Neurosci Methods* 1994;14:1856–1877.

103. Yukie M. Connections between the medial temporal cortex and the CA1 subfield of the hippocampal formation in the Japanese monkey. *J Comp Neurol* 2000;423:282–298.

104. Krettek JE, Price JL. Projections from the amygdala to the perirhinal and entorhinal cortices and the subiculum. *Brain Res* 1974;71:150–154.

105. Herkenham M. The connections of the nucleus reuniens thalami: evidence for a direct thalamo-hippocampal pathway in the rat. *J Comp Neurol* 1978;177:589–610.

106. Amaral DG, Cowan WM. Subcortical afferents to the hippocampal formation in the monkey. *J Comp Neurol* 1980;189:573–591.

107. Whitlock DG, Nauta WJH. Subcortical projections from the temporal neocortex in Macaca mulatta. *J Comp Neurol* 1956;106:183–212.

108. Herzog AG, Van Hoesen GW. Temporal neocortical afferent connections to the amygdala in the rhesus monkey. *Brain Res* 1976;115:57–69.

109. Turner BH, Mishkin M, Knapp M. Organization of the amygdalopetal projections from modality-specific cortical association areas in the monkey. *J Comp Neurol* 1980;191:515–543.

110. Van Hoesen GW. The different distribution, diversity and sprouting of cortical projections to the amygdala in the rhesus monkey. In: Ben Ari Y, ed. *The amygdaloid complex.* New York: Elsevier, 1981:77–90.

111. Amaral DG, Veazey RB, Cowan WM. Some observations on hypothalamo-amygdaloid connections in the monkey. *Brain Res* 1982;252:13–27.

112. Aggleton JP, Burton MJ, Passingham RE. Cortical and subcortical afferents to the amygdala of the rhesus monkey (Macaca mulatta). *Brain Res* 1980;190:347–368.

113. Woolf NJ, Butcher LL. Cholinergic projections to the basolateral amygdala: a combined Evans blue and acetylcholinesterase analysis. *Brain Res Bull* 1982;8:751–763.

114. Ding SL, Van Hoesen GV, Rockland KS. Inferior parietal lobule projections to the presubiculum and neighboring ventromedial temporal cortical areas. *J Comp Neurol* 2000;425:510–530.

115. Van Hoesen GW, Augustinack JC, Dierking J, et al. The parahippocampal gyrus in Alzheimer's disease. *Ann N Y Acad Sci* 2000;911;254–274.

116. Mehler WR. Subcortical afferent connections of the amygdala in the monkey. *J Comp Neurol* 1980;190:733–762.

117. Jones EG, Powell TPS. An anatomical study of converging sensory pathways within the cerebral cortex of the monkey. *Brain* 1970;93:793–820.

118. Klinger J, Gloor P. The connections of the amygdala and of the anterior temporal cortex in the human brain. *J Comp Neurol* 1960;115:333–369.

119. Mufson EJ, Mesulam M-M, Pandya DN. Insular interconnections with the amygdala in the rhesus monkey. *Neuroscience* 1981;6:1231–1248.

120. Krettek JE, Price JL. Projections from the amygdaloid complex and adjacent olfactory structures to the entorhinal cortex and to the subiculum in the rat and cat. *J Comp Neurol* 1977;172:723–752.

121. Amaral DG, Price JL. Amygdalo-cortical projections in the monkey (Macaca fascicularis). *J Comp Neurol* 1984;230:465–496.

122. Saunders RC, Rosene DL. A comparison of the efferents of the amygdala and the hippocampal formation in the rhesus monkey: I. Convergence in the entorhinal, prorhinal and perirhinal cortices. *J Comp Neurol* 1988;271:153–184.

123. Saunders RC, Rosene DL, Van Hoesen GW. Comparison of the afferents of the amygdala and hippocampal formation in the rhesus monkey: II. Reciprocal and non-reciprocal connections. *J Comp Neurol* 1988;271:185–207.

124. Price JL, Amaral DG. An autoradiographic study of the projections of the central nucleus of the monkey amygdala. *J Neurosci* 1982;1:1242–1259.

125. Nauta WJH. Fibre degeneration following lesions of the amygdaloid complex in the monkey. *J Anat* 1961;95:515–531.

126. Porrino LJ, Crane AM, Goldman-Rakic PS. Direct and indirect pathways from the amygdala to the frontal lobe in rhesus monkeys. *J Comp Neurol* 1981;198:121–136.

127. Aggleton JP, Mishkin M. Projections of the amygdala to the thalamus in the cynomolgus monkey. *J Comp Neurol* 1984;222:56–68.

128. Hopkins DA, Holstege G. Amygdaloid projections to the mesencephalon, pons and medulla oblongata in the cat. *Exp Brain Res* 1978;32:529–547.

129. Jacobson S, Trojanowski JQ. Amygdaloid projections to prefrontal granular cortex in rhesus monkey demonstrated with horseradish peroxidase. *Brain Res* 1975;100:132–139.

130. Krettek JE, Price JL. Projections from the amygdaloid complex and adjacent olfactory structures to the entorhinal cortex and to the subiculum in the rat and cat. *J Comp Neurol* 1977;172:723–752.

131. Avendano C, Price JL, Amaral DG. Evidence for an amygdaloid projection to premotor cortex but not to motor cortex in the monkey. *Brain Res* 1983;264:111–117.

132. Kelley AE, Domesick VB, Nauta WJH. The amygdalostriatal projection in the rat—an anatomical study by anterograde and retrograde tracing methods. *Neuroscience* 1982;7:615–630.

133. Kievit J, Kuypers HGJM. Basal forebrain and hypothalamic connections to the frontal and parietal cortex in the rhesus monkey. *Science* 1975;187:660–662.

134. Divac I. Magnocellular nuclei of the basal forebrain project to neocortex, brain stem and olfactory bulb: review of some functional correlates. *Brain Res* 1975;93:385–398.

135. Jones EG, Burton H, Saper CB, et al. Midbrain, diencephalic and cortical relationships of the basal nucleus of Meynert and associated structures in primates. *J Comp Neurol* 1976;167:385–420.

136. Mesulam M-M, Van Hoesen GW. Acetylcholinesterase containing basal forebrain neurons in the rhesus monkey project to neocortex. *Brain Res* 1976;109:152–157.

137. Pearson RCA, Gather KC, Bridal P, et al. The projection of the basal nucleus of Meynert upon the neocortex in the monkey. *Brain Res* 1983;259:132–136.

138. Whitehouse PJ, Price DL, Clark AW, et al. Alzheimer disease: evidence for selective loss of cholinergic neurons in the nucleus basalis. *Ann Neurol* 1981;10:122–126.

139. Mesulam M-M, Mufson EJ, Levey AI, et al. Cholinergic innervation of cortex by the basal forebrain: cytochemistry and cortical connections of the septal area, diagonal band nuclei, nucleus basalis (substantia innominata), and hypothalamus in the rhesus monkey. *J Comp Neurol* 1983;214:170–197.

140. Wenk H, Bigl V, Meyer V. Cholinergic projections from magnocellular nuclei of the basal forebrain to cortical areas in rats. *Brain Res Rev* 1980;2:295–316.

141. Fibiger HC. The organization and some projections of cholinergic neurons of the mammalian forebrain. *Brain Res Rev* 1982;4:327–388.

142. Levey AI, Hallanger AE, Wainer BH. Cholinergic nucleus basalis neurons may influence the cortex via the thalamus. *Neurosci Lett* 1987;74:7–13.

143. Buzsaki G, Bickford RG, Ponomareff G, et al. Nucleus basalis and thalamic control of neocortical activity in the freely moving rat. *J Neurosci* 1988;8:4007–4026.

144. Asanuma C. Axonal arborizations of a magnocellular basal nucleus input and their relation to the neurons in the thalamic reticular nucleus of rats. *Proc Natl Acad Sci USA* 1989;86:4746–4750.

145. Tourtellotte WG, Van Hoesen GW, Hyman BT, et al. Alz-50 immunoreactivity in the thalamic reticular nucleus in Alzheimer's disease. *Brain Res* 1989;515:227–234.

146. Jones EG. Some aspects of the organization of the thalamic reticular complex. *J Comp Neurol* 1975;162:285–308.

147. Mesulam M-M, Mufson EJ. Neural inputs into the nucleus basalis of the substantia innominata in the rhesus monkey. *Brain Res* 1984;107:253–274.

148. Nauta WJH, Haymaker W. Hypothalamic nuclear and fiber connections. In: Haymaker W, Anderson E, Nauta WJH, eds. *The hypothalamus.* Springfield, IL: Charles C Thomas 1969:136–209.

149. Saper CB, Swanson LW, Cowan WM. Some afferent connections of the rostral hypothalamus in the squirrel monkey (saimiri sciureus) cat. *J Comp Neurol* 1979;184:205–242.

150. Price JL, Amaral DG. An autoradiographic study of the projections of the central nucleus of the monkey amygdala. *J Neurosci* 1981;1:1242–1259.

151. Victor M, Adams RD, Collins GH. *The Wernicke-Korsakoff syndrome.* Philadelphia: FA Davis, 1971:1–206.

152. Squire LR, Moore RY. Dorsal thalamic lesion in a noted case of chronic memory dysfunction. *Ann Neurol* 1979;6:505–506.

153. Graff-Radford NR, Eslinger PJ, Damasio AR, et al. Nonhemorrhagic infarction of the thalamus: behavioral, anatomical and physiological correlates. *Neurology* 1984;34:14–23.

154. Akert K. Comparative anatomy of the frontal cortex and

thalamocortical connections. In: Warren JM, Akert K, eds. *The frontal granular cortex and behavior.* New York: McGraw-Hill, 1964:372–396.

155. Goldman-Rakic PS, Porrino LJ. The primate medial dorsal (MD) nucleus and its projection to the frontal lobe. *J Comp Neurol* 1985;242:535–560.

156. Ray JP, Price JL. The organization of projections from the mediodorsal nucleus of the thalamus to orbital and medial prefrontal cortex in macaque monkeys. *J Comp Neurol* 1993;337:1–31.

157. Tobias TJ. Afferents to prefrontal cortex from the thalamic mediodorsal nucleus in the rhesus monkey. *Brain Res* 1975;83:191–212.

158. Tanaka D. Thalamic projections of the dorso-medial prefrontal cortex in the rhesus monkey (Macaca mulatta). *Brain Res* 1976;110:21–38.

159. Cavada C, Company T, Hernandez-Gonzalez A, et al. Acetylcholinesterase histochemistry in the macaque thalamus reveals territories selectively connected to frontal, parietal and temporal association cortices. *J Chem Neuroanat* 1995;8:245–257.

160. Arikuni T, Sakai M, Kubota K. Columnar aggregation of prefrontal and anterior cingulate cortical cells projecting to the thalamic mediodorsal nucleus in the monkey. *J Comp Neurol* 1983;220:116–125.

161. Penfield W, Welch K. The supplementary motor area of the cerebral cortex: a clinical and experimental study. *Arch Neurol Psychiatry* 1951;66:289–317.

162. Woolsey CN, Settlage PH, Meyer DR, et al. Patterns of localization in precentral and "supplementary" motor areas and their relation to the concept of a premotor area. *Res Publ Assoc Res Nerv Ment Dis* 1952;30:238–264.

163. Barbas H, Pandya DN. Architecture and frontal cortical connections of the premotor cortex (area 6) in the rhesus monkey. *J Comp Neurol* 1987;256:211–228.

164. Mitz AR, Wise SP. The somatotopic organization of the supplementary motor area: intracortical microstimulation mapping. *J Neurosci* 1987;7:1010–1021.

165. Rizzolatti G, Camarda R, Fogassi L, et al. Functional organization of inferior area 6 in the macaque monkey II. Area F5 and the control of distal movements. *Exp Brain Res* 1988;71:491.

166. Humberstone M, Sawle GV, Clare S, et al. Functional magnetic resonance imaging of single motor events reveals human presupplementary motor area. *Ann Neurol* 1997;42:632–637.

167. Ikeda A, Yazawa S, Kunieda T, et al. Cognitive motor control in human pre-supplementary motor area studied by subdural recording of discrimination/selection-related potentials. *Brain* 1999;122:915–931.

168. Luppino G, Rizzolatti G. The organization of the frontal motor cortex. *News Physiol Sci* 2000;15:219–224.

169. Wurtz RH, Mohler CW. Enhancement of visual responses in the monkey striate cortex and frontal eye fields. *J Neurophysiol* 1976;39:766–772.

170. Schlage J, Schlag-Rey M. Evidence for a supplementary eye field. *J Neurophysiol* 1987;57:179–200.

171. Huerta MF, Krubitzer LA, Kaas JH. Frontal eye field as defined by intracortical microstimulation in squirrel monkeys, owl monkeys and macaque monkeys: II. Cortical connections. *J Comp Neurol* 1987;265:332–361.

172. Di Pellegrino G, Wise SP. A neurophysiological comparison of three distinct regions of the primate frontal lobe. *Brain* 1991;114:951–978.

173. Luppino G, Matelli M, Camarda RM, et al. Corticocortical connections of area F3 (SMA-proper) and area F6 (pre-SMA) in the macaque monkey. *J Comp Neurol* 1993;338:114–140.

174. Walker AE. A cytoarchitectural study of the prefrontal area in the macaque monkey. *J Comp Neurol* 1940;262:256–270.

175. Matsuzaka Y, Aizawa H, Tanji J. A motor area rostral to the supplementary motor area (presupplementary motor area) in the monkey: neuronal activity during a learned motor task. *J Neurophysiol* 1992;68:653–662.

176. Barbas H, Pandya DN. Architecture and intrinsic connections of the prefrontal cortex in the rhesus monkey. *J Comp Neurol* 1989;286:353–375.

177. Preuss TM, Goldman-Rakic PS. Myelo- and cytoarchitecture of the granular frontal cortex and surrounding regions in the strepsirhine primate Galago and the anthropoid primate Macaca. *J Comp Neurol* 1991;310:429–474.

178. Morecraft RJ, Geula C, Mesulam M-M. Cytoarchitecture and neural afferents of orbitofrontal cortex in the brain of the monkey. *J Comp Neurol* 1992;323:341–358.

179. Carmichael ST, Price JL. Architectonic subdivision of the orbital and medial prefrontal cortex in the Macaque monkey. *J Comp Neurol* 1994;346:366–402.

180. Cavada C, Company T, Tejedor J, et al. The anatomical connections of the macaque monkey Orbitofrontal cortex. A review. *Cereb Cortex* 2000;10:220–242.

181. Harlow JM. Recovery from the passage of an iron bar through the head. *Mass Med Soc Pub* 1868;2:327–346.

182. Geschwind N. Disconnexion syndromes in animals and man. I. *Brain* 1965;88:237–294.

183. Roland PE, Friberg L. Localization of cortical areas activated by thinking. *J Neurophysiol* 1985;53:1219–1243.

184. Goldman-Rakic PS. Circuitry of prefrontal cortex and regulation of behavior by representational memory. In: Mountcastle VB, Plum F, eds. *Handbook of physiology. The nervous system,* vol. 5. Bethesda, MD: American Physiological Society, 1987:373–417.

185. Fuster JM. *The prefrontal cortex. Anatomy, physiology and neuropsychiatry of the frontal lobe,* 2nd ed. New York: Raven Press, 1989.

186. Freund H-J. Abnormalities of motor behavior after cortical lesions in humans. In: Mountcastle VB, Plum F, eds. *Handbook of physiology. The nervous system, vol. 5.* Bethesda, MD: American Physiological Society 1990:763–810.

187. Mesulam M-M. Large-scale neurocognitive networks and distributed processing for attention, language, and memory. *Ann Neurol* 1990;28:597–613.

188. Damasio H, Grabowski T, Frank R, et al. The return of Phineas Gage: clues about the brain from the skull of a famous patient. *Science* 1994;264:1102–1105.

189. Damasio AR. The time-locked multiregional retroactivation: a systems level proposal for the neural substrates of recall and recognition. *Cognition* 1989;33:25.

190. Bechara A, Tranel D, Damasio H. Characterization of the decision-making deficit of patients with ventromedial prefrontal cortex lesions. *Brain* 2000;123:2189–2202.

191. Daffner KR, Mesulam MM, Scinto LFM, et al. The central role of the prefrontal cortex in directing attention to novel events. *Brain* 2000;123:927–939.

192. Savage CR, Deckersbach T, Heckers S, et al. Prefrontal regions supporting spontaneous and directed application of verbal learning strategies. Evidence from PET. *Brain* 2001;124:219–231.

193. Franowicz MN, Goldman-Rakic PS. The sensory nature of mnemonic representation in the primate prefrontal cortex. *Nat Neurosci* 2001;4:311–316

194. Morecraft RJ, Yeterian EH. Prefrontal cortex. In: Ramachrandran VS, ed. *Encyclopedia of the human brain,* vol. 4. San Diego: Academic Press, 2002:11–26.

195. Adolphs R, Damasio H, Cooper TD, et al. A role for somatosensory cortices in the visual recognition of emotion as revealed by

three-dimensional lesion mapping. *J Neurosci* 2000;20:2683–2690.

196. Pandya DN, Kuypers HGJM. Cortico-cortical connections in the rhesus monkey. *Brain Res* 1969;13:13–36.

197. Felleman DJ, Van Essen DC. Distributed hierarchical processing in the primate cerebral cortex. *Cereb Cortex* 1991;1:1–47.

198. Chafee MV, Goldman-Rakic PS. Inactivation of parietal and prefrontal cortex reveals interdependence of neural activity during memory-guided saccades. *J Neurophysiol* 2000;83:1550–1566.

199. Ungerleider LG, Mishkin M. Two cortical visual systems: In: Ingle DG, Goodale MA, Mansfield RJQ, eds. *Analysis of visual behavior.* Cambridge, MA: MIT Press, 1992:549–586.

200. Wilson FAW, Scalaidhe SPO, Goldman-Rakic PS. Dissociation of object and spatial processing domains in primate prefrontal cortex. *Science* 1993;260:1955.

201. Goldman-Rakic, PS, Selemon LD, Schwartz ML. Dual pathways connecting the dorsolateral prefrontal cortex with the hippocampal formation and parahippocampal cortex in the rhesus monkey. *Neuroscience* 1984;12:719–743.

202. Moran MA, Mufson EJ, Mesulam M-M. Neural input into the temporopolar cortex of the rhesus monkey. *J Comp Neurol* 1987;256:88–103.

203. Barbas H, De Olmos J. Projections from the amygdala to basoventral and mediodorsal prefrontal regions in the rhesus monkey. *J Comp Neurol* 1990;300:549–571.

204. He SQ, Dum RP, Strick PL. Topographic organization of corticospinal projections from the frontal lobe: motor areas on the lateral surface of the hemisphere. *J Neurosci* 1993;13:952–980.

205. Goldman PS, Nauta WJH. An intricately patterned prefronto-caudate projection in the rhesus monkey. *J Comp Neurol* 1977;171:369–386.

206. Yeterian EH, Van Hoesen GW. Cortico-striate projections in the rhesus monkey: the organization of certain cortico-caudate connections. *Brain Res* 1978;139:43–63.

207. Yeterian EH, Pandya DN. Prefrontostriatal connections in relation to cortical architectonic organization in rhesus monkeys. *J Comp Neurol* 1991;312:43–67.

208. Haber SN, McFarland NR. The concept of the ventral striatum in nonhuman primates. *Ann N Y Acad Sci* 2000;911:33–48.

209. Nauta WJH. Neural associations of the frontal cortex. *Acta Neurobiol Exp* 1972;32:125–140.

210. Kievit J, Kuypers HGJM. Organization of the thalamo-cortical connexions to the frontal lobe in the rhesus monkey. *Exp Brain Res* 1977;29:299–322.

211. Barbas H, Haswell-Henion TH, Dermon CR. Diverse thalamic projections to the prefrontal cortex in the rhesus monkey. *J Comp Neurol* 1991;313:65–94.

212. Yeterian EH, Pandya DN. Corticothalamic connections of paralimbic regions in the rhesus monkey. *J Comp Neurol* 1988;269:130–146.

213. Ongur D, An X, Price JL. Prefrontal cortical projections to the hypothalamus in macaque monkeys. *J Comp Neurol* 1998;401:480–505.

214. Goldman PS, Nauta WJH. Autoradiographic demonstration of a projection from prefrontal association cortex to the superior colliculus in the rhesus monkey. *Brain Res* 1976;116:145–149.

215. Leichnetz GR. The prefrontal cortico-oculomotor trajectories in the monkey. *J Neurol Sci* 1981;49:387–396.

216. An X, Bandler R, Ongur D, et al. Prefrontal cortical projections to longitudinal columns in the midbrain periaqueductal gray in macaque monkeys. *J Comp Neurol* 1998;401:455–479.

217. Brodal P. The corticopontine projection in the rhesus monkey: origin and principles of organization. *Brain* 1978;101:251–283.

218. Middleton FA, Strick PL. Anatomical evidence for cerebellar and basal ganglia involvement in higher cognitive function. *Science* 1994;266:458–461.

219. Schmahmann JD, Pandya DN. Anatomic organization of the basilar pontine projections from prefrontal cortices in rhesus monkey. *J Neurosci* 1997;17:438–458.

220. Middleton FA, Strick PL. Cerebellar projections to prefrontal cortex of the primate. *J Neurosci* 2001;21:700–712.

221. Aston-Jones G, Ennis M, Pieribone VA, et al. The brain nucleus locus caeruleus: restricted afferent control of a broad efferent network. *Science* 1986;234:734–737.

222. Mesulam M-M. Asymmetry of neural feedback in the organization of behavioral states [Letter]. *Science* 1987;237:537–538.

223. Arnsten AFT, Goldman-Rakic PS. Selective prefrontal cortical projections to the region of the locus caeruleus and raphe nuclei in the rhesus monkey. *Brain Res* 1984;306:9–18.

224. Van Hoesen GW, Augustinack JC, Redman SJ. Ventromedial temporal lobe pathology in dementia, brain trauma, and schizophrenia. *Ann N Y Acad Sci* 2000;877:575–594.

13

EXECUTIVE FUNCTION AND FRONTAL SYSTEMS

MARK D'ESPOSITO

The clinical neuropsychological literature includes under the rubric of *executive function* a wide range of cognitive processes, such as focused and sustained attention, fluency, and flexibility of thought, for generating solutions to novel problems and in planning and regulating adaptive and goal-directed behavior (1–3). As evident by the wide scope of these processes, executive function has been used to capture the highest order of cognitive abilities. Such abilities sometimes are not only difficult to operationally define but difficult to measure, which has led to the development of a large number of clinical and experimental neuropsychological tests in an attempt to tap this range of abilities (2,4).

What are the neural mechanisms underlying executive function? Studies addressing this question often are biased toward the assumption that "executive function" is synonymous with "frontal lobe function" and that "the frontal lobes" are an undifferentiated cortical mass. It is likely that not all abilities thought to represent executive function are subserved by the frontal lobes, that the frontal lobes are not the only brain region contributing to executive function, and that particular portions of the frontal lobes are specialized to contribute to specific aspects of executive function as well as other higher cognitive processes.

This chapter begins by describing the cognitive deficits observed in patients with frontal lobe damage, which has resulted in the concept of a *dysexecutive syndrome.* Next, the definition of the concept of executive function in a theoretically motivated manner is described. Such a definition hopefully can lead to the direct investigation of these cognitive processes at the neurophysiologic level. Finally, the evidence for the neural basis of this executive function and the specific role of prefrontal cortex are reviewed.

DYSEXECUTIVE SYNDROME

It is widely recognized that extensive frontal lobe damage may have little impact on the abilities measured by standardized intelligence tests or other neuropsychological tests, but these findings are in marked contrast to the way that these patients perform in unintelligent ways in real life (5). Based on this observation, it is obvious that neuropsychological tests designed for the laboratory do not always capture the abilities that are necessary for success in real life. For example, "real-life behavior requires heavy time processing demands and a core system of values based on both inherited (drives, instincts) and acquired (education, socialization) information that is probably not necessary for most artificial problems posed by neuropsychological tasks" (6). However, tests designed to tap executive function, which are difficult to administer at the bedside, do seem to capture the type of abilities that typically are impaired after damage to frontal lobes. The types of deficits observed in patients with frontal lobe damage are briefly reviewed here.

Inability to Initiate, Stop, and Modify Behavior in Response to Changing Stimuli

Impairment of this type is seen when patients with frontal lobe damage perform the Wisconsin Card Sorting Test, which is a test thought to be sensitive to frontal function (7). In this test, a deck of cards is presented one at time to a patient who must sort each one according to various stimulus dimensions (color, form, or number). Each card from the deck contains from one to four identical figures (stars, triangles, crosses, or circles) in one of four colors. The patient is told after each response whether the response is correct and must infer from this information only (the sorting principle is not given by the examiner) what the next response should be. After ten correct sorts, the sorting principle is changed without warning. During this test, frontal patients usually understand and can repeat the rules of the test, but they are unable to follow them or use knowledge of incorrect performance based on feedback to alter their behavior (7,8). Patients with frontal lobe lesions make both random errors and perseverative errors (9). Perseverative errors traditionally are viewed as a failure in inhibition of a previous response pattern and on the Wisconsin Card Sorting Test is due to failure to shift set to a new sorting criteria. A random error occurs when a patient is sorting correctly and switches to a new incorrect sorting category without any prompt from the examiner; this can be viewed as a transient failure in maintaining the goal at hand.

Inability to Handle Sequential Behavior Necessary for Organization, Planning, and Problem Solving

Patients with frontal lobe lesions often have no difficulty with the basic operations of a given task, but nevertheless they perform poorly. For example, when performing complex mathematical problems requiring multiple steps, the patient initially may respond impulsively to an early step and will be unable to string together and execute the component steps required for solving the problem (10). However, the ability to perform each of the mathematical operations in isolation (i.e., adding and subtracting) required to complete the complex task might be intact. The following problem, "The price of canned peas is two cans for 31 cents. What is the price of one dozen cans?," is almost impossible for patients with frontal lobe damage, even though these patients perform the direct arithmetical task of multiplying 6 times 31 with ease (10). Similar errors are observed in routine everyday tasks that require a series of simple steps such as wrapping a present or making a sandwich (11). Careful observation of patients with frontal lobe lesions will reveal that even though they are capable of each individual step of the task if the steps are presented to them in isolation, they will fail to fully complete the entire task.

Inability to Inhibit Responses

One way in which this impairment can be demonstrated is with a measure called the Stroop paradigm (12). It is based on the observation that it takes longer to name the color of a series of conflicting color words (e.g., "red" printed in blue ink) than to name the color of a series of color blocks. This phenomenon is exaggerated in patients with frontal lobe lesions (13). A related phenomena is that frontal patients may display a remarkable tendency to imitate the examiner's gestures and behaviors even when no instruction has been given to do so, and even when this imitation entails considerable personal embarrassment. The mere sight of an object may elicit the compulsion to use it, even though the patient has not been asked to do so and the context is inappropriate, for example, a patient sees a pair of glasses and puts them on, even though he or she is already wearing his or her own pair. These symptoms have been called the *environmental dependency syndrome*. It has been postulated that the frontal lobes promote distance from the environment and the parietal lobes foster approach toward one's environment. Therefore, loss of frontal inhibition may result in overactivity of the parietal lobes. Without the frontal lobes our autonomy

from our environment would not be possible. A given stimulus would automatically call up a predetermined response regardless of context (14,15).

Perseveration

Perseveration is defined as an abnormal repetition of a specific behavior. It can be observed in patients after frontal lobe damage in a wide range of tasks, including motor acts, verbalizations, sorting tests, drawing, and writing. Several different types of perseverative behavior have been described in patients with brain damage, such as (a) *recurrent perseveration,* which is the recurrence of a previous response to a subsequent stimulus within the context of an established set; (b) *stuck-in-set perseveration,* which is the inappropriate maintenance of a category or framework of activity; and (c) *continuous perseveration,* which is abnormal prolongation or continuation without cessation of a current behavior (16).

Together, the range of deficits described that are observed in patients with frontal lobe lesions captures the essence of the dysexecutive syndrome. However, the dysexecutive syndrome cannot be considered unitary given the diverse nature of these deficits. Moreover, any single patient with a frontal lobe lesion may exhibit some of these behavioral deficits and not others. Nevertheless, careful observation of the type of deficits observed in patients with frontal lesions has allowed for the development of cognitive models of executive function.

COGNITIVE MODELS OF EXECUTIVE FUNCTION

How can we operationally define *executive function* in a way that captures the types of deficits seen in patients with frontal lobe damage? Such a definition of executive function begins with a definition of the concept of working memory. *Working memory* is defined as the set of processes that allow for the temporary storage and manipulation of information that allows one to flexibly guide future behavior. This ability is critical for almost all other higher cognitive abilities, such as reasoning, language comprehension, and visuospatial processing. From a psychological point of view, working memory is conceptualized as comprising multiple components that support executive control processes and basic mnemonic processes. For example, to maintain and manipulate information when that information is not accessible in the environment, the brain needs storage processes, rehearsal processes that can prevent the contents of the storage system from decaying, and executive control processes to act on the mnemonic representations being stored and rehearsed (17).

Based on behavioral studies of normal subjects, Baddeley (18,19) proposed that working memory includes a central executive system that actively regulates the distribution of limited attentional resources and coordinates information within limited capacity verbal and spatial slave memory buffers. The concept of the central executive system was based on the analogous supervisory attentional system introduced by Norman and Shallice (20) and is proposed to take control over cognitive processing when novel tasks are involved and when existing habits must be overridden. Thus, a major component of the cognitive architecture of working memory, in addition to storage and rehearsal processes, are executive control processes that allow for the contents being rehearsed and stored to be acted upon.

The central executive system, like other models of executive control, postulates that it is a system with a limited capacity, that is, each additional cognitive operation that a subject performs at one time places increasing demands on this system. Performance of two tasks simultaneously (i.e., a dual task) can be used to directly test the capacity of executive control processes. For example, two tasks that are performed sequentially will make minimal demands on executive control processes because these tasks can be performed successfully by using separate processing systems. However, two tasks performed concurrently will lead to a decrement in performance, compared to performance on either task alone, because dual tasking requires similar processing systems and will make greater demands on executive control processes. This finding from the experimental psychology literature parallels our experience in everyday life—there is clearly a limit to how many tasks one can perform at any one time before performance suffers. Just imagine your ability to fully comprehend and remember what is being told to you while you are talking on your mobile telephone and driving your car.

Two broad cognitive models of executive control exist: those that propose that there is single executive control system that directs and monitors the activities of lower-level systems in order to guide behavior (such as proposed by Baddeley and Shallice), and those that posit that there is not a dedicated "controller" in the brain but rather executive control emerges from the maintenance of task rules and goals (for review see reference 21). Regardless of the exact nature of the psychological constructs of models of executive control, proponents of both types of models seem to have reached a consensus that these types of processes likely are implemented by the frontal lobes. Thus, any theory of frontal lobe function is essentially a theory that helps explain the neural basis of executive control.

For example, Fuster (22,23) has proposed that the frontal cortex is critically important in tasks that require the temporal integration of information. In proposing his model, Fuster argues explicitly against the interpretation of the homoncular view of executive control, writing that "the prefrontal cortex would not superimpose a steering or directing function on the remainder of the nervous system, but rather, by expanding the temporal perspectives of the system, it would allow it to integrate longer, newer, and more complex structures of behavior."

Likewise, Cohen and Servan-Schreiber (24) have proposed that frontal lobe damage results in "a degradation in the ability to construct and maintain an internal representation of context, (by which) we mean information held in mind in such a form that it can be used to mediate an appropriate behavioral response." In their model, disordered performance in executive function is seen as a consequence of a change to a single low-level parameter. In this way, two behaviors that appear outwardly different as indexed by poor performance on seemingly different tasks (such as the Stroop paradigm and Wisconsin Card Sorting Test) may have their roots in similar fundamental processes (for a similar model see reference 25).

NEURAL BASIS OF EXECUTIVE CONTROL AND WORKING MEMORY

As mentioned, what is common to each of these cognitive models of executive control is the notion that the frontal lobes are critical for implementing processes necessary for the control of behavior, and frontal lobe damage results in loss of executive control. Direct insight into the role of the frontal lobes in executive control in humans has come from neuropsychological, electrophysiologic, and functional neuroimaging studies. Several physiologic questions can be derived from these cognitive models. For example, what specifically is the role of the frontal lobes in executive control? Are the different behavioral components of executive control subserved by different regions within the frontal lobes? Are there specific neurotransmitter systems that modulate executive function? The following sections address these questions.

Role of the Prefrontal Cortex in Executive Control

A complex cognitive process such as executive control is likely to be subserved by a distributed network of distinct brain regions (26). However, evidence from neuropsychological, electrophysiologic, and functional neuroimaging research supports a role of the frontal lobes as a critical node in the network supporting in executive control of goal-directed behavior (23). The frontal lobes make up over one third of the human cerebral cortex and can be divided into three major subdivisions: motor/premotor cortex, paralimbic cortex (which includes the anterior cingulate gyrus), and prefrontal cortex. It is the prefrontal cortex that has been specifically linked to executive control.

The extensive reciprocal connections from prefrontal cortex to virtually all cortical and subcortical structures place the prefrontal cortex in a unique neuroanatomic position to monitor and manipulate diverse cognitive processes. For example, there are at least two major neural networks that interact with the prefrontal cortex (27). The first network involves reciprocal cortical–cortical connections between the prefrontal cortex and the posterior parietal cortex, as well as connections with the anterior and posterior cingulate, and medial temporal lobe regions including the entorhinal and parahippocampal cortex (28). The second network involves cortical–subcortical connections between the prefrontal cortex and the striatum, globus pallidus, substantia nigra, and mediodorsal nucleus of the thalamus (29). Each of these networks likely subserves different aspects of executive control, such as maintenance of goals (cortical network), response selection, and motor control (subcortical network).

Neuropsychological studies investigating patients with focal lesions of the prefrontal cortex have been reported infrequently (30,31). This is primarily due to the clinical observation that few patients have selective lesions confined to the prefrontal cortex but rather have lesions that extend to involve other brain structures. Etiologic factors in patients with restricted lesions of the prefrontal cortex typically include strokes within the middle or anterior cerebral artery territory, focal cerebral trauma, tumor resection, or in epileptic patients after frontal lobe excisions to treat the epilepsy. Patients with prefrontal cortical lesions, compared to patients with lesions in other areas or normal subjects, have shown deficits on a wide range of measures that tap aspects of executive control. For example, studies have shown that patients with frontal lesions appear to have a specific deficit on the Tower of London task (31), which is a test that requires planning; on self-ordered memory tasks that assess the ability to monitor information held in working memory (30); and on dual tasks (32).

Functional neuroimaging studies of healthy adults have provided converging evidence from a different perspective concerning the role of the prefrontal cortex in executive control. For example, the dual-task paradigm, which has been used as an effective behavioral tool for probing executive control processes, has been studied using functional neuroimaging. In one such functional magnetic resonance imaging (MRI) study, healthy young subjects performed concurrently a spatial task (mental rotation of visual stimuli) and a verbal task (semantic judgments of auditory stimuli), which are cognitive tasks previously reported to activate predominantly posterior brain regions (i.e., not the prefrontal cortex) (33). It was reasoned that any activation within the prefrontal cortex would be due to recruitment of cognitive processes involved in the concurrent performance of two tasks simultaneously and not cognitive processes engaged by the individual tasks per se. This study demonstrated lateral prefrontal cortex activation only during the dual-task condition and not during either single-task condition. This finding suggested that the prefrontal cortex likely is involved in this aspect of executive control.

Another study using positron emission tomographic scanning also explored the neural basis of executive control with a dual-task paradigm (34). Normal subjects were scanned while they performed two cognitive tasks (the Wisconsin Card Sorting test and an auditory verbal shadowing task),

both individually and simultaneously. As expected, when both tasks were performed simultaneously there were significant decrements in performance compared with the individual task performance scores. Different than the previous dual-task study discussed (33), one of the cognitive tasks (the card sorting test) activated the prefrontal cortex when performed individually. In this study, there was less prefrontal cortex activation under the dual-task condition compared to when the card sorting task was performed individually. This result suggests that, under circumstances in which the capacity of executive control is exceeded, cortical activity in the prefrontal cortex may be attenuated. Thus, under dual-task conditions, prefrontal cortical activity may increase to meet the processing demands until some level of asymptote before it attenuates.

In another dual-task positron emission tomographic study, subjects were scanned while they were performing an auditory working memory task, a visual working memory task, both auditory and working memory tasks, and a control task (35). Unlike the study by D'Esposito et al. (33) but similar to the study by Goldberg et al. (34), each of the single tasks, compared to the control task, activated the prefrontal cortex. During the dual-task condition, there was no distinct region within the prefrontal cortex (or any other cortical region) that was activated only when both tasks were performed concurrently. It was proposed that these results were consistent with the hypothesis that tasks performed concurrently interfere with each other if they require recruitment of the same cortical regions. This conclusion is supported by an earlier study by this group in which it was found that the larger the extent of overlap of the cortical regions that were activated when each task was performed individually, the greater the decrement in behavioral performance when these two tasks were performed concurrently (36).

There now seems to be a critical mass of functional neuroimaging studies, such as those presented, that have demonstrated activation of the lateral prefrontal cortex during tasks that tap executive control processes (for review see reference 37). Thus, the diverse spectrum of cognitive deficits in patients with damage to the prefrontal cortex may be considered to arise from difficulties in executive control. Moreover, executive control likely comprises both inhibitory and excitatory modulation of a distributed neural network. The following sections briefly review the role of the prefrontal cortex in excitatory and inhibitory control.

Inhibitory Control

Research in animals first provided evidence for a prefrontal-thalamic inhibitory system that regulates flow of sensory information from the external world to the cerebral cortex. For example, it was observed that cooling of the prefrontal cortex in cat increased the amplitudes of evoked electrophysiologic responses recorded from primary cortex in all sensory modalities (38). Conversely, stimulation of specific regions of the thalamus (i.e., nucleus reticularis thalami) that surround the sensory relay thalamic nuclei resulted in modality-specific suppression of activity in primary sensory cortex (39). These findings are evidence for an inhibitory pathway from the prefrontal cortex that regulates the flow of sensory information through thalamic relay nuclei. This prefrontal-thalamic inhibitory system provides a mechanism for modality-specific suppression of irrelevant inputs at an early stage of sensory processing.

There is evidence in humans that the prefrontal cortex exhibits inhibitory control on other cortical and subcortical regions. For example, event-related potential (ERP) studies in patients with focal prefrontal damage have shown that primary auditory and somatosensory evoked responses are enhanced, suggesting disinhibition of sensory flow to these regions. In a series of ERP experiments, task irrelevant auditory and somatosensory stimuli (monaural clicks or brief electric shocks to the median nerve) were presented to patients with comparably sized lesions in the lateral prefrontal cortex, temporal-parietal junction, or lateral parietal cortex. Evoked responses from primary auditory (40) and somatosensory (41) cortices were recorded. Unlike damage in temporal and parietal regions, lateral prefrontal cortex damage resulted in enhanced amplitudes of both primary auditory and somatosensory evoked responses (42,43). Spinal cord and brainstem potentials were not affected by lateral prefrontal cortex damage, suggesting that the amplitude enhancements were due to abnormalities in either a prefrontal-thalamic or a prefrontal-sensory cortex mechanism.

As mentioned previously, patients with frontal lobe damage exhibit behavior that is consistent with a deficit in inhibitory control. Moreover, several functional neuroimaging studies in healthy young subjects have demonstrated a link between lateral prefrontal cortical function and inhibitory control (44–46). For example, inhibitory control has been investigated with the go/no go task. In this task, during go trials a green square is presented and subjects have to respond by promptly pushing a button. In no go trials a red square is presented and subjects are instructed not to respond. During this type of task, patients with frontal lesions typically respond to no go trials despite being told repeatedly not to do so. During functional MRI scanning, normal healthy subjects exhibit greater activity within prefrontal cortex during no go trials than during go trials, suggesting a role for the prefrontal cortex inhibiting inherent response tendencies. Thus, functional neuroimaging studies, combined with electrophysiologic studies in patients with prefrontal lesions, provide powerful evidence that the prefrontal cortex in humans provides a net inhibitory regulation of early sensory transmission.

Excitatory Control

To voluntarily guide our behavior based on events in the outside world, not only must we implement inhibitory control

to suppress irrelevant information but we also must excite and sustain neural activity in distributed brain regions in order to maintain necessary information in an active state. It has been observed that when an individual selectively attends to one ear, a region of the visual field, or a particular finger, there are increases in the amplitude of sensory evoked potentials to all stimuli delivered to that sensory channel (47). This provides evidence that overt attention processes can reliably modulate neural activity at early sensory cortices. Single-cell recordings in monkeys (48), lesion (49), functional neuroimaging (50), and ERP studies in humans have implicated the prefrontal cortex as the site of such top-down modulation of processing within early visual pathways. For example, patients with unilateral prefrontal cortex lesions (centered within the middle frontal gyrus) exhibit a reduction in the amplitude of evoked potentials arising from extrastriate cortex (49).

It is likely that all tasks require parallel inhibitory and excitatory control of neural activity, which forms the basic neural architecture underlying executive control as implemented by the prefrontal cortex. However, the findings supporting a role of the prefrontal cortex in both excitatory and inhibitory control presented thus far raise a critical conceptual issue. Why does lateral prefrontal cortex implement parallel inhibition and excitation to control distributed cognitive systems?

It is well documented that the nervous system uses an interleaved inhibition and excitation mechanism throughout the neuraxis. Examples include spinal reflexes, cerebellar outputs, and basal ganglia movement control networks. Thus, it is not surprising that prefrontal systems also use inhibitory and excitatory mechanisms to control cognitive processing. It is likely that such parallel excitatory-inhibitory control entails large-scale neural control, as might be involved in a prefrontal cortex-thalamic gating network as well as direct excitatory prefrontal cortical input to a specific cortical region. Local cortical attention tuning through inhibition might entail long excitatory prefrontal cortical projections that then activate local inhibitory neurons. There is evidence in rodents that long distance excitatory prefrontal cortical projections terminate on γ-aminobutyric acid–immunoreactive neurons, providing a potential neuronal architecture for prefrontal dependent inhibitory modulation (51).

Functional Subdivisions of the Prefrontal Cortex

The prefrontal cortex appears to be critical for some aspects of executive control, but it is unclear whether there are functional subdivisions within the prefrontal cortex that are specialized for different types of executive control processes. Based on clinical observations, there are two major behavioral/cognitive syndromes (52) that occur after damage to different regions of the prefrontal cortex (dorsolateral vs. orbitofrontal). These syndromes reflect separable circuits of connections of the prefrontal cortex with subcortical structures.

Only damage to the dorsolateral prefrontal cortex causes the most severe impairments in executive dysfunction, as described previously. In contrast, damage to the orbitofrontal cortex, which is intimately connected to the limbic system, spares many cognitive skills but dramatically affects all spheres of social and emotional behavior (53,54). The orbitofrontal patient frequently is impulsive, hyperactive, and labile, and lacks proper social skills despite showing reasonable performance on cognitive tasks that typically are impaired in patients with damage to the lateral prefrontal cortex. There may be further subdivisions of the orbitofrontal cortex. The ventromedial portion has been associated with use of internal autonomic states in the guidance of goal-directed behavior and involved in inhibitory processing of emotional stimuli. The ventrolateral portion of the orbitofrontal cortex has been implicated in the rapid establishment of reward–punishment associations (for a review of these current ideas see reference 55).

Given that the lateral prefrontal cortex seems to be the most critical region for executive control, the question that arises is whether functional subdivisions exist within lateral prefrontal cortex. Petrides has proposed one possible functional organization of the lateral prefrontal cortex (56). According to this "two-stage model," there are two executive control systems, one dorsal and the other ventral, within lateral prefrontal cortex. It is proposed that the ventrolateral prefrontal cortex, which comprises the inferior frontal gyrus (Brodmann areas 44, 45, and 47), is the site where information initially is received from posterior cortical association areas and where active comparisons of information held in working memory are made. In contrast, the dorsolateral prefrontal cortex, which comprises the middle frontal gyrus (Brodmann areas 9 and 46), is recruited only when monitoring and manipulation of information held within working memory is required. In this way, it is proposed that there is hierarchical processing of information from ventrolateral to dorsolateral prefrontal regions.

Results of several functional neuroimaging studies in healthy young subjects support this type of functional organization of lateral prefrontal cortex (37,56,57). For example, in a functional MRI study, healthy young subjects were asked to either maintain or manipulate information being held in working memory (58). Thus, in this task, subjects were presented with two types of trials in which they were required either to (a) *maintain* a randomly ordered sequence of five letters across a delay period or (b) *manipulate* a comparable sequence of letters by arranging them into alphabetical order during the delay period. The *maintain* condition simply required the temporary retention of letters in the same format as presented at the beginning of the trial. In the *manipulate* condition, subjects were required to transpose the order of the five items presented at the beginning of the trial during the delay period. In each subject, activity during the delay period was found in both the dorsolateral and ventrolateral prefrontal cortex in both types of trials. Additionally, in each subject, dorsolateral prefrontal activity was significantly greater in trials during which information held in working memory was manipulated. Thus, these results are consistent with a two-stage processing model of the organization of lateral prefrontal cortex. It is likely that future studies of the prefrontal cortex will reveal further subdivisions supporting different aspects of executive control.

Role of Dopamine in Executive Control

The function of the cerebral cortex is clearly influenced by the diffuse inputs from brainstem neuromodulatory systems mediated by neurotransmitters such as dopamine, acetylcholine, and serotonin, yet little is known about the relationship between neurotransmitter function and cognition. A key to understanding the function of the prefrontal cortex will arise from an understanding of its functional relationship with other brain regions and how it is modulated by such brainstem projections. Fuster (23) writes, "certain neural structures and pathways have been grouped and characterized as forming separate functional systems because they share a prevalent neurotransmitter. . .none of them can have more functional specialization than the system of interconnected brain structures in which it operates." Thus, knowledge about the role of a specific neurotransmitter to cognition can be derived by understanding the function of the brain regions that receive input from neurons producing that neurotransmitter.

Based on the anatomic distribution of brainstem dopaminergic projections, there is a logical basis for proposing a role for dopamine in higher-level cognitive abilities and specifically in prefrontal cortical function (for review, see reference 59). Dopaminergic neurons in the human brain are organized into several major subsystems (mesocortical, mesolimbic, and nigrostriatal). The mesocortical and mesolimbic dopaminergic systems originate in the ventral tegmental area of the midbrain and project to prefrontal cortex; anterior cingulate cortex; anterior temporal structures such as the amygdala, hippocampus, and entorhinal cortex; and basal forebrain (60). There is an anteroposterior gradient in the brain for concentration of dopamine where it is highest in the prefrontal cortex (61). Thus, the anatomic distribution of the mesocortical dopaminergic system suggests that it will have a greater influence on anterior than posterior brain structures.

The functional importance of dopamine to prefrontal function has been demonstrated in several ways. First, in monkeys depletion of dopamine in the prefrontal cortex or pharmacologic blockade of dopamine receptors induces impairment in working memory tasks (62,63). This working memory impairment is as severe as the impairment in monkeys with lesions of the prefrontal cortex and is not observed in monkeys in which other neurotransmitters, such as serotonin or norepinephrine, are depleted. Furthermore, dopaminergic agonists administered to these same monkeys reverse their working memory impairments (62,64). These findings provide evidence that impairments in prefrontal function can be caused by dopamine deficiency and replacement of dopamine can lead to improvement.

One method of assessing dopamine's influence on cognitive function in humans is by testing patients with Parkinson disease who are taking and who are not taking their dopaminergic replacement medication. Although the motor symptoms of Parkinson disease are caused by degeneration of the nigrostriatal dopaminergic system, a large proportion of patients with Parkinson disease also have degeneration of dopaminergic neurons in the ventrotegmental area comprising the mesocortical system (65). Thus, Parkinson disease provides an excellent model for investigating the role of dopamine in prefrontal function. Also, because the half-life of some of the dopaminergic replacement drugs (i.e., levodopa) administered to these patients is short, central nervous system levels of dopamine can be manipulated over short periods of time and monitored by observing a deterioration in the patient's motor status.

Several studies have reported findings using this method in which patients with Parkinson disease were tested on tasks thought to be sensitive to frontal lobe dysfunction (66–69). In each of these studies, patients were impaired on these tasks when they were not taking their dopaminergic medications. For example, in one study the tasks that were performed poorly by Parkinson disease patients not taking their medications (the Tower of London, a spatial working memory task, and a test of attention set shifting) also have been shown to be specifically impaired in patients with frontal lobe lesions (70). This evidence for a specific role of dopamine in prefrontal function is strengthened by the concurrent findings that patients with Parkinson disease perform similarly on long-term memory tasks thought to be sensitive to medial temporal lobe function. Taken together, these studies provide evidence that prefrontal function is influenced by the dopaminergic neurotransmitter system.

Administration of dopamine receptor agonists, which stimulate dopamine receptors in the same manner as dopamine, to healthy young subjects provides a viable method for examining the role of dopaminergic systems in higher cognitive functions in humans. Most dopamine receptor agonists are relatively selective for a particular receptor subtype, the two most common being the D_1 and the D_2. Two such drugs approved for human use are bromocriptine, which is relatively selective for the D_2 receptor subtype, and pergolide, which affects D_1 and D_2 receptor subtypes. Because both drugs are relatively safe to administer to normal human subjects and have well-understood agonist properties, they offer a useful probe for investigating the relationship between dopamine and prefrontal cortical function.

Healthy young human subjects given bromocriptine (71,72) or pergolide (73) perform better on working memory tasks compared to when they are given a placebo. In these studies, the dopaminergic medication had a highly specific effect on working memory but had no effect on other cognitive abilities such as attention or sensorimotor function. Converging on these findings, normal subjects who were given sulpiride, a D_2 receptor antagonist, were impaired on several tasks sensitive to frontal lobe function. Importantly, the impairments could not be accounted for by generalized sedative or motoric influences of the medication (74).

Interestingly, in another study the effects of bromocriptine on prefrontal function were not the same for all subjects but interacted with the subject's working memory capacity (75). Subjects with lower baseline working memory abilities when they were not taking the drug tended to demonstrate cognitive improvement on the drug, whereas those with higher baseline working memory abilities worsened. A similar relationship between dopamine and prefrontal function has been observed in monkeys given dopaminergic agonist and antagonists. Specifically, a U-shaped dose–response curve is observed, demonstrating that a specific dosage produces optimal performance on working memory tasks (59). This observation suggests that "more" is not "better" but rather there is an optimal level of dopamine concentration that is necessary for optimal function of the prefrontal cortex.

The cognitive effect of the administration of dopamine agonists on patients with frontal lesions provides insight into the relationship between dopamine and executive control. In one such study, patients who suffered prefrontal damage from traumatic brain injury were assessed with and without bromocriptine therapy while performing several clinical experimental measures of executive function (e.g., Stroop task, Wisconsin Card Sorting Test, Trail-Making Test, dual task) (76). Significant improvement in performance on all tasks requiring executive control processes was observed in traumatic brain injury patients with bromocriptine compared to placebo. In contrast, bromocriptine did not improve performance on measures with minimal executive control demands, even if they were cognitively demanding, or other simpler tasks requiring basic attentional, mnemonic, or sensorimotor processes. This pattern of findings provides evidence that the dopaminergic system may specifically modulate executive control processes and may not be critical for basic mnemonic processes.

As mentioned earlier, dopamine receptors are found in large concentrations in the prefrontal cortex. D_2 dopamine receptors are present in much lower concentrations in the cortex than D_1 receptors and are located mostly within the striatum (77). However, D_2 receptors are at their highest concentrations in layer V of the prefrontal cortex, which makes them especially well placed to interact with behavior (78). An important area for future investigation will be to determine the relative contribution of each of these dopamine receptors to specific aspects of cognitive processing. It also will be critical to determine the synergistic interaction between these two dopamine receptor types (79).

Knowledge of the functioning of the dopaminergic system has shed light on the role of the prefrontal cortex in behavior and may shed light on the mechanisms underlying cognitive deficits in a wide range of disorders, such as attention deficit disorder, drug addiction, and schizophrenia; in states commonly encountered by normal healthy individuals, such as stress and sleep deprivation; and in the normal aging process. As Arnsten (59) aptly states, "the PFC [prefrontal cortex] is extraordinarily sensitive to its neurochemical environment, and small changes in levels of dopamine...may contribute substantially to altered prefrontal cortical cognitive function."

CONCLUSIONS

Executive function is a concept meant to capture the highest of cognitive abilities. The type of cognitive operations thought to be "executive" in nature allow us to act and control the enormous number of internal and external representations available to us that are necessary to guide our behavior in real time, either moment by moment or year by year. Voluntary guidance of behavior requires not only implement inhibitory control to suppress irrelevant information but the ability to excite and sustain neural activity in distributed brain regions in order to maintain necessary information in an active state. The neural bases of these executive control processes are beginning to be mapped out, both on the neuroanatomic and neurochemical level, using sophisticated cognitive neuroscience methodologies such as functional MRI. Improved understanding of the physiologic basis of executive control will lead to a narrower and more useful view of prefrontal cortical function that hopefully will allow the development of new therapies in patients with specific cognitive difficulties caused by damage to this critical region of the brain.

ACKNOWLEDGMENTS

Supported by the American Federation for Aging Research and the National Institutes of Health.

REFERENCES

1. Hecaen H, Albert ML. *Human neuropsychology.* New York: John Wiley and Sons, 1978.
2. Lezak M. *Neuropsychological assessment,* 3rd ed. New York: Oxford University Press, 1995.
3. Luria AR. *Higher cortical functions in man.* New York: Basic Books, 1966.
4. Spreen O, Strauss E. *A compendium of neuropsychological tests: administration, norms, and commentary.* New York: Oxford University Press, 1991.
5. Shallice T, Burgess PW. Deficits in strategy application following frontal lobe damage in man. *Brain* 1991;114:727–741.
6. Damasio AR, Anderson SW. The frontal lobes. In: Heilman K, Valenstein E, eds. *Clinical neuropsychology.* New York: Oxford University Press, 1993:409–460.
7. Milner B. Effects of different brain regions on card sorting. *Arch Neurol* 1963;9:90–100.
8. Eslinger PJ, Damasio AR. Severe disturbance of higher cognition following bilateral frontal lobe ablation: patient EVR. *Neurology* 1985;35:1731–1741.
9. Barcelo F, Knight RT. Both random and perseverative errors underlie WCST deficits in prefrontal patients. *Neuropsychologia* 2002;40:349–356.
10. Stuss DT, Benson DF. Neuropsychological studies of the frontal lobes. *Psychol Bull* 1984;95:3–28.
11. Schwartz MF, Montgomery MW, Buxbaum LJ, et al. Naturalistic action impairment in closed head injury. *Neuropsychology* 1998;12:13–28.
12. Stroop JR. Studies of interference in serial verbal reactions. *J Exp Psychol* 1935;18:643–662.
13. Perret E. The left frontal lobe of man and the suppression of habitual responses in verbal categorical behaviour. *Neuropsychologia* 1974;12:323–330.
14. Lhermitte F. Human autonomy and the frontal lobes. Part II: patient behavior in complex and social situations: the "environmental dependency syndrome." *Ann Neurol* 1986;19:335–343.
15. Lhermitte F, Pillon B, Serdaru M. Human autonomy and the frontal lobes. Part I: imitation and utilization behavior: a neuropsychological study of 75 patients. *Ann Neurol* 1986;19:326–334.
16. Sandson J, Albert ML. Perseveration in behavioral neurology. *Neurology* 1987;37:1736–1741.
17. Smith EE, Jonides J. Neuroimaging analyses of human working memory. *Proc Natl Acad Sci U S A* 1998;95:12061–12068.
18. Baddeley A. Working memory. *Science* 1992;255:556–559.
19. Baddeley A. *Working memory.* New York: Oxford University Press, 1986.
20. Shallice T. *From neuropsychology to mental structure.* Cambridge, UK: Cambridge University Press, 1988.
21. Kimberg DK, D'Esposito M, Farah MJ. Cognitive functions in the prefrontal cortex: working memory and executive control. *Curr Dir Psychol Sci* 1998;6:185–192.
22. Fuster JM, ed. *The prefrontal cortex and temporal integration, vol. 4.* New York: Plenum Press, 1985.
23. Fuster J. *The prefrontal cortex: anatomy, physiology, and neuropsychology of the frontal lobes,* 3rd ed. New York: Raven Press, 1997.
24. Cohen JD, Servan-Schreiber D. Context, cortex, and dopamine: a connectionist approach to behavior and biology in schizophrenia. *Psychol Rev* 1992;99:45–77.
25. Kimberg D, Farah M. A unified account of cognitive impairments following frontal damage: the role of working memory in complex, organized behavior. *J Exp Psychol Learn Mem Cogn* 1993;122:411–428.
26. Mesulam M-M. Large-scale neurocognitive networks and distributed processing in attention, language and memory. *Ann Neurol* 1990;28:597–613.
27. Goldman-Rakic PS, Friedman HR. The circuitry of working memory revealed by anatomy and metabolic imaging. In: Levin H, Eisenberg H, Benton A, eds. *Frontal lobe function and dysfunction.* New York: Oxford University Press, 1991:72–91.
28. Selemon LD, Goldman-Rakic PS. Common cortical and subcortical targets of the dorsolateral prefrontal and parietal cortices in the rhesus monkey: evidence for a distributed neural network subserving spatially guided behavior. *J Neurosci* 1988;8:4049–4068.
29. Ilinisky IA, Jouandet ML, Goldman-Rakic PS. Organization of the nigrothalamocortical system in the rhesus monkey. *J Comp Neurol* 1985;236:315–330.

30. Petrides M, Milner B. Deficits on subject-ordered tasks after frontal- and temporal-lobe excisions in man. *Neuropsyhcologia* 1982;23:601–614.
31. Owen AM, Downes JJ, Sahakian BJ, et al. Planning and spatial working memory following frontal lobe lesions in man. *Neuropsychologia* 1990;28:1021–1034.
32. McDowell S, Whyte J, D'Esposito M. Working memory impairments in traumatic brain injury: evidence from a dual-task paradigm. *Neuropsychologia* 1997;35:1341–1353.
33. D'Esposito M, Detre JA, Alsop DC, et al. The neural basis of the central executive system of working memory. *Nature* 1995;378:279–281.
34. Goldberg TE, Berman KF, Fleming K, et al. Uncoupling cognitive workload and prefrontal cortical physiology: a PET rCBF study. *Neuroimage* 1998;7:296–303.
35. Klingberg T. Concurrent performance of two working memory tasks: potential mechanisms of interference. *Cereb Cortex* 1998;8:593–601.
36. Klingberg T, Roland PE. Interference between two concurrent tasks is associated with activation of overlapping fields in the cortex. *Cogn Brain Res* 1997;6:1–8.
37. D'Esposito M, Aguirre GK, Zarahn E, et al. Functional MRI studies of spatial and non-spatial working memory. *Cogn Brain Res* 1998;7:1–13.
38. Skinner JE, Yingling CD. Central gating mechanisms that regulate event-related potentials and behavior. In: Desmedt JE, ed. *Progress in clinical neurophysiology, vol. 1.* Basel: S. Karger, 1977;30–69.
39. Yingling CD, Skinner JE. Gating of thalamic input to cerebral cortex by nucleus reticularis thalami. In: Desmedt JE, ed. *Progress in clinical neurophysiology, vol. 1.* Basel: S. Karger, 1977:70–96.
40. Kraus N, Ozdamar O, Stein L. Auditory middle latency responses (MLRs) in patients with cortical lesions. *Electroencephalogr Clin Neurophysiol* 1982;54:275–287.
41. Leuders H, Leser RP, Harn J, et al. Cortical somatosensory evoked potentials in response to hand stimulation. *J Neurosurg* 1983;58:885–894.
42. Yamaguchi S, Knight RT. Gating of somatosensory inputs by human prefrontal cortex. *Brain Res* 1990;521:281–288.
43. Knight RT, Scabini D, Woods DL. Prefrontal cortex gating of auditory transmission in humans. *Brain Res* 1989;504:338–342.
44. D'Esposito M, Postle BR, Jonides J, et al. The neural substrate and temporal dynamics of interference effects in working memory as revealed by event-related functional MRI. *Proc Natl Acad Sci U S A* 1999;96:7514–7519.
45. Garavan H, Ross TJ, Stein EA. Right hemispheric dominance of inhibitory control: an event-related functional MRI study. *Proc Natl Acad Sci U S A* 1999;96:8301–8306.
46. Konishi S, Nakajima K, Uchida I, et al. No-go dominant brain activity in human inferior prefrontal cortex revealed by functional magnetic resonance imaging. *Eur J Neurosci* 1998;10:1209–1213.
47. Hillyard SA, Hink RF, Schwent UL, et al. Electrical signs of selective attention in the human brain. *Science* 1973;182:177–180.
48. Rainer G, Asaad WF, Miller EK. Selective representation of relevant information by neurons in the primate prefrontal cortex. *Nature* 1998;393:577–579.
49. Barcelo F, Suwazono S, Knight RT. Prefrontal modulation of visual processing in humans. *Nat Neurosci* 2000;3:399–403.
50. Hopfinger JB, Buonocore MH, Mangun GR. The neural mechanisms of top-down attentional control. *Nat Neurosci* 2000;3:284–291.
51. Carr DB, Sesack SR. Callosal terminals in the rat prefrontal cortex: synaptic targets and association with GABA-immunoreactive structures. *Synapse* 1998;29:193–205.
52. Cummings JL. Frontal-subcortical circuits and human behavior. *Arch Neurol* 1993;50:873–880.
53. Bechara A, Damasio H, Tranel D, et al. Dissociation of working memory from decision making within the human prefrontal cortex. *J Neurosci* 1998;18:428–437.
54. Stone VE, Baron-Cohen S, Knight RT. Frontal lobe contributions to theory of mind. *J Cogn Neurosci* 1998;10:640–656.
55. Elliot R, Dolan RJ, Frith CD. Dissociable functions in the medial and lateral orbitofrontal cortex: Evidence from human neuroimaging studies. *Cereb Cortex* 2000;10:308–317.
56. Owen AM, Evans AC, Petrides M. Evidence for a two-stage model of spatial working memory processing within the lateral frontal cortex: a positron emission tomography study. *Cereb Cortex* 1996;6:31–38.
57. Owen AM, Herrod NJ, Menon DK, et al. Redefining the functional organization of working memory processes within human lateral prefrontal cortex. *Eur J Neurosci* 1999;11:567–574.
58. D'Esposito M, Postle BR, Ballard D, et al. Maintenance versus manipulation of information held in working memory: an event-related fMRI study. *Brain Cogn* 1999;41:66–86.
59. Arnsten A. Catecholamine regulation of the prefrontal cortex. *J Psychopharmacol* 1997;11:151–162.
60. Bannon MJ, Roth RH. Pharmacology of mesocortical dopamine neurons. *Pharmacol Rev* 1983;35:53–68.
61. Brown RM, Crane AM, Goldman PS. Regional distribution of monoamines in the cerebral cortex and subcortical structures of the rhesus monkey: concentrations and in vitro synthesis rates. *Brain Res* 1979;168:133–150.
62. Brozoski TJ, Brown RM, Rosvold HE, et al. Cognitive deficit caused by regional depletion of dopamine in prefrontal cortex of rhesus monkey. *Science* 1979;205:929–932.
63. Sawaguchi T, Goldman-Rakic PS. D1 dopamine receptors in prefrontal cortex: involvement in working memory. *Science* 1991;251:947–950.
64. Arnsten KT, Cai JX, Murphy BL, et al. Dopamine D1 receptor mechanisms in the cognitive performance of young adult and aged monkeys. *Psychopharmacology* 1994;116:143–151.
65. Javoy-Agid F, Agid Y. Is the mesocortical dopaminergic system involved in Parkinson disease? *Neurology* 1980;30:1326–1330.
66. Gotham AM, Brown RG, Marsden CD. "Frontal" cognitive function in patients with Parkinson's disease "on" and "off" levodopa. *Brain* 1988;111:299–321.
67. Lange KW, Paul GM, Naumann M, et al. Dopaminergic effects of cognitive performance in patients with Parkinson's disease. *J Neural Transm Suppl* 1995;46:423–432.
68. Cooper JA, Sagar HJ, Doherty M, et al. Different effects of dopaminergic and anticholinergic therapies on cognitive and motor function in Parkinson's disease. *Brain* 1992;115:1701–1725.
69. Fournet N, Moreaud O, Roulin JL, et al. Working memory in medicated patients with Parkinson's disease: the central executive seems to work. *J Neurol Neurosurg Psychiatry* 1996;60:313–317.
70. Lange KW, Robbins TW, Marsden CD, et al. L-Dopa withdrawal in Parkinson's disease selectively impairs cognitive performance in tests sensitive to frontal lobe dysfunction. *Psychopharmacology* 1992;107:394–404.
71. Luciana M, Depue RA, Arbisi P, et al. Facilitation of working memory in humans by a D2 dopamine receptor agonist. *J Cogn Neurosci* 1992;4:58–68.
72. Luciana M, Collins PF. Dopaminergic modulation of working memory for spatial but not object cues in normal humans. *J Cogn Neurosci* 1997;9:330–347.
73. Müller U, Pollmann S, von Cramon DY. D1 versus D2-receptor modulation of visuospatial working memory in humans. *J Neurosci* 1998;18:2720–2728.

74. Mehta MA, Sahakian BJ, McKenna PJ, et al. Systemic sulpiride in young adult volunteers simulates the profile of cognitive deficits in Parkinson's disease. *Psychopharmacology (Berl)* 1999;146:162–174.

75. Kimberg DY, D'Esposito M, Farah M. Effects of bromocriptine on human subjects depend on working memory capacity. *Neuroreport* 1997;8:3581–3585.

76. McDowell S, Whyte J, D'Esposito M. Differential effect of a dopaminergic agonist on prefrontal function in traumatic brain injury patients. *Brain* 1998;121[Pt 6]:1155–1164.

77. Camps M, Cortés R, Gueye B, et al. Dopamine receptors in human brain: autoradiographic distribution of D1 sites. *Neuroscience* 1989;28:275–290.

78. Goldman-Rakic PS, Lidow MS, Gallager DW. Overlap of dopaminergic, adrenergic, and serotoninergic receptors and complementarity of their subtypes in primate prefrontal cortex. *J Neurosci* 1990;10:2125–2138.

79. Arnsten AFT, Cai JX, Steere JC, et al. Dopamine D2 receptor mechanisms contribute to age-related cognitive decline: the effects of quinpirole on memory and motor performance in monkeys. *J Neurosci* 1995;15:3429–3439.

14

NEUROPSYCHIATRIC ASPECTS
OF SEXUAL DYSFUNCTION

NKANGINEME IKE
ROBERT TAYLOR SEGRAVES

Sexual function is an integral part of most of our patients' lives; therefore, some assessment of sexual function should be part of a comprehensive medical evaluation. Unfortunately, this component often is omitted from psychiatric, neurologic, and *general medical* evaluations. This omission occurs despite the high prevalence of sexual problems in the general population (1,2), medical outpatient populations (3–6), neurologic patients (7), and certain psychiatric patients (8). Clearly, neurologists and neuropsychiatrists should be able to assess sexual function routinely.

Because sexual behavior is multidimensional, assessing sexual function must consider biologic as well as psychological variables. In the past, sexual dysfunctions were largely thought to be caused by psychological and social factors, and much less by biologic factors. The current thinking, which evolved over the past decade, is that biologic factors are involved in the etiology of most of the cases of sexual dysfunction.

Within the field of psychiatry, there has been an emphasis on the interpersonal and intrapsychic determinants of sexual behavior. Except for a small number of investigators, few biologic psychiatrists have shown a major interest in sexual disorders. A similar situation exists within the field of neurology.

This chapter reviews a new and somewhat neglected area of neuropsychiatry: the neuropsychiatry of sexual dysfunction. Basic neurophysiology of sexual behavior is reviewed briefly, followed by a discussion of drug effects on sexual behavior, and by a review of sexual dysfunction occurring in association with neuropsychiatric disorders. The chapter concludes with a brief review of evaluation and treatment of sexual disorders. Neuropsychiatric aspects of gender identity and sexual preference are summarized.

NEUROPHYSIOLOGY OF THE HUMAN
SEXUAL RESPONSE

Sexual behavior can be subdivided into four overlapping phases: desire, arousal, orgasm, and resolution. Dysfunction can occur in any of these phases either singly or in various combinations. We also must recognize that human sexuality has three main roots: biologic, motivational-affective-relational, and cognitive.

Sexual physiology of the male has been more widely studied than that of the female. Accordingly, our knowledge of male sexual neurophysiology is more advanced than our knowledge of female physiology.

Neurophysiology of Male Erection
and Female Vaginal Congestion

Reflex vasodilation of the genital vasculature in response to sexual stimuli is responsible for both male penile erection and female lubrication (9,10). Decreased vascular resistance in the penile corpora appears to be the major factor causing penile corpora vascular engorgement and penile erection (11). The relaxation of these smooth muscles in males allows the penile arteries to dilate and cause the inflow of blood resulting in penile erection. Relaxation of these cavernosal smooth muscle is thought to be aided by an endothelial derived factor, nitric oxide, which activates the enzyme guanylate cyclase, which then leads to increased levels of cyclic guanosine monophosphate (cGMP), which causes cavernosal smooth muscle relaxation. Sildenafil (Viagra) is thought to enhance the effect of nitric oxide by inhibiting phosphodiesterase type 5, which is responsible for the degradation of cGMP in the corpus cavernosum. The smooth muscle of the corpora cavernosa is predominately innervated by adrenergic fibers, although cholinergic fibers also are present (12).

α-Adrenergic impulses appear to maintain the penis in a nontumescent state (13). It is unclear whether a parallel innervation exists in the human female.

Neuroanatomic studies have demonstrated a dual innervation of the genitals in both sexes: sympathetic innervation from the T12 to L4 *(thoracolumber)* segments of the spinal cord and parasympathetic innervation from the S2 to S4 *(via the pelvic splanchnic nerves)* cord segments (14,15). Stimulation of the sacral parasympathetic fibers has been shown to elicit penile erection in many species, and ablation of these nerves interferes with reflexogenic erections. These fibers also are thought to mediate the lubrication response in females. The postganglionic neurotransmitter in these parasympathetic fibers is unclear but does not appear to be acetylcholine.

The sympathetic outflow from the T12 to L4 area contains both vasodilator and vasoconstrictor fibers. The exact nature of the contribution of these fibers to erection and lubrication is unclear. *It seems a synergy exists between the sympathetic and parasympathetic system.* Some investigators have proposed that the sympathetic fibers mediate erections produced by erotic imagery and thoughts (e.g., psychogenic erections) as opposed to erections produced by tactile stimulation (e.g., reflexogenic reactions) (16), which are thought to be parasympathetic. A parallel mechanism presumably exists in human females.

From the foregoing it is clear that because the autonomic nervous system is outside voluntary control, and because it has such a pervading influence on major aspects of sexuality; such external influences as stress and drugs and internal influences such as emotions, cortical events, and hypothalamic events, would impact on sexual function.

Sexuality and Higher Central Nervous System Functions

Stimulation and ablation experiments in laboratory animals, including mammals, have identified cerebral areas mediating erection. The major areas that elicit penile erection upon stimulation include the medial septopreoptic region and the medial part of the medial dorsal nucleus of the thalamus. Other areas involved in penile erection include septal projections of the hippocampus, the anterior cingulate gyrus, the mamillothalamic tract, and the mamillary bodies (17–19). The cerebral representation of vaginal lubrication is unknown.

Orgasm

Orgasm can be conceptualized as the sensory experience of a series of spinal cord reflexes. These reflexes are triggered when a series of sensory stimuli reach threshold values (20,21). In the male, sensory impulses eliciting the ejaculatory reflex travel in the pudendal nerve to the sacral cord. Once a threshold value is reached, contractions of the vas deferens, seminal vesicles, and prostatic smooth muscle occur, resulting in delivery of the ejaculate into the pelvic urethra. Stimulation of the urethral bulb by the inflowing ejaculate elicits reflex closure of the bladder neck, preventing retrograde ejaculation, and rhythmic contractions of the perineal muscles and urethral bulb, resulting in expulsion of the ejaculate (22).

Efferent fibers mediating ejaculation arise from the thoracolumbar cord, travel in the hypogastric nerve, and synapse with short adrenergic fibers that innervate the organs involved in orgasm. These fibers appear to be mainly α-adrenergic fibers, although these organs also are innervated by cholinergic fibers (23). Sensory impulses from the pudendal nerve to the sacral cord presumably are relayed cranially to the thalamus and sensory cortex, resulting in the experience of orgasm.

Female orgasm also appears to be a genital reflex. Sensory impulses travel to the sacral cord in the pudendal nerve, and efferent fibers innervate the ovary, fallopian tubes, vaginal musculature, and uterus. Rhythmic contractions of these structures appear to be mediated by α-adrenergic fibers, although the female sexual organs also have a cholinergic innervation.

Animal research has identified both subcortical and cortical structures associated with the ejaculatory reflex (24). In the monkey, ejaculation can be elicited by stimulation of the anterior thalamus and preoptic area (17). In both the human male and female, direct stimulation of the septal region of the brain has been reported to produce sexual orgasm (25,26). Using deep recording electrodes, researchers have demonstrated that sexual orgasm in the human is accompanied by electrical discharges in the septal area (27).

Sensory representation of genital sensations appears to be localized in the paracentral lobule. Penfield and Rasmussen (28) reported that stimulation of the posterior part of the postcentral gyrus produced genital sensations, and epileptics with lesions in the paracentral lobule reported genital sensations as part of the aura (29).

Current evidence suggests that central dopamine and serotonin pathways are involved in sexual behavior. Animal studies have demonstrated that drugs that increase brain dopaminergic activity lower thresholds for ejaculatory and erectile reflexes (30). Infusions of dopamine agonists into the medial preoptic region and the lumbar cord augment male sexual behavior (31,32). Male patterns of sexual behavior are *said to be* increased by drugs that lower brain serotonin levels (33,34). Other studies have demonstrated that destruction of central nervous system serotonergic fibers in the medial forebrain bundle facilitates ejaculation in laboratory animals (35).

From the preceding review of the neurophysiology of sexual behavior, it is clear that various neurologic lesions could interfere with sexual behavior. These include lesions in the brain, cord, and peripheral nerves. It also is clear that many commonly prescribed drugs, especially psychiatric and hypotensive drugs that affect monoamine neuromodulators, might alter both central nervous system function and sexual behavior.

Age and Sexuality

With advancing age, greater physical stimulation is required to attain and maintain erections in men, and orgasms are less intense. In women, menopause terminates fertility and produces changes stemming from estrogen deficiency. The extent to which aging affects sexual function depends largely on psychological, pharmacologic, and illness-related factors. Although it is important that older men and women not fall into the psychosocial trap of expecting (or worse, trying to force) the kind and degree of sexual response characteristic of their youth, it is equally as important that they not fall prey to the negative folklore according to which decreased physical intimacy is an inevitable consequence of the passage of time (36).

ENDOCRINE FUNCTION AND SEXUAL BEHAVIOR

Clinicians should be aware of possible sexual problems resulting from changes in circulating sex hormone, binding globulin, and free testosterone in men and women, which may be due to endogenous or exogenous hormonal changes (37).

Hormones are necessary, but not sufficient, factors to maintain a satisfying human libido. We must recognize that human sexuality has three main roots: biologic, motivational-affective-relational, and cognitive. In women, estrogens prime the central nervous system, acting as neurotrophic and psychotropic factors throughout life. They also prime the sensory organs, including the skin with its sebaceous and sweat glands, which are the key receptors for external sexual stimuli. Estrogens also are the "permitting factors" for the action of vasoactive intestinal peptide, the key neurotransmitter involved in the endothelial and vascular changes leading to vaginal lubrication. Other factors, such as medication, alcohol, and other health problems, can modify the biologic impact of hormones on libido. Well-tailored hormone replacement therapy, including androgens in selected cases, may reduce the biologic causes of loss of libido. Comprehensive treatment requires a balanced evaluation between biologic and psychological factors (38).

Testosterone: Testosterone is believed to affect libido in both men and women. Interestingly, stress is inversely correlated with blood testosterone levels in men. Factors such as sleep, mood, and lifestyle affect circulating testosterone levels. Testosterone is released in a pulsatile and diurnal manner; the highest levels occur in the morning and the lowest levels in the evening. By age 50 years, decreased levels of testosterone become apparent and are said to decrease at a rate of 100 ng/dL per decade. This does not necessarily correlate with impending hypogonadism; however, there have been suggestions of diminished androgen sensitivity with advancing age (39).

Prolactin: Prolactin is described in subsequent paragraphs.

Oxytocin: Concentrations of oxytocin increase in both men and women during orgasm. It has been suggested that oxytocin enhances sexual behavior (39).

Cortisol: Abnormally high levels of cortisol decrease sexual drive in men and women, possibly due to increased corticotropin releasing hormone (40).

Histamine: Case studies suggest histamine facilitates erection in men with erectile failure. One comparable case study has been reported in women (40).

Pheromones: These are substances secreted from glands at the anus, urinary outlet, breasts, and mouth. Minimal research suggests they may facilitate sexual attractiveness in men in the presence of women, given that this change is not seen during masturbatory activity in men (40).

Opioids: Long-term opioid use impairs erection in men, possibly via suppression of circulating hormones such as testosterone. Case reports suggest that opioid antagonists may restore erectile dysfunction in men. Limited studies suggest similar effects may exist in women (40).

Ephedrine: Twenty sexually functional women participated in two experimental conditions in which subjective (self-report) and physiologic (vaginal photoplethysmography) sexual responses to erotic stimuli were measured after administration of either ephedrine sulfate (50 mg), an α- and β-adrenergic agonist, or placebo in a randomized, double-blinded, crossover protocol. The study concluded that ephedrine can significantly facilitate the initial stages of physiologic sexual arousal in women. These findings have implications for deriving new pharmacologic approaches to the management of sexual dysfunction in women (41).

In humans, normal endocrine function is necessary for reproduction. In subprimate mammals, sexual behavior is clearly hormonally dependent. In humans, the precise relationship between sexual behavior and endocrine function has not been established; the relationship is less well understood for women. The tremendous influence of social learning on human sexuality contributes to the difficulty in determining endocrine effects on human sexual behavior.

In the human male, evidence on the relationship between endocrine function and sexual behavior can be obtained from studies of men who have been surgically castrated, men prescribed antiandrogens, and hypogonadal men receiving androgen replacement therapy. Bilateral orchidectomy is used to treat sexual offenders in some countries and as a palliative treatment for some neoplasms. Studies of patients after bilateral orchidectomy reveal a dramatic loss of libido, usually followed by an inability to ejaculate in most patients (42–44). However, a small number of patients remain sexually active for years after castration (45). The effects of castration are generally reversed by the administration of exogenous androgen. Some castrated males subsequently develop erectile problems. It has been hypothesized that these problems are secondary to attempted coitus in the presence of low libido and do not reflect a direct effect of androgen deficiency on erectile function (6). Studies of the effects of the estrogenic compounds medroxyprogesterone and cyproterone acetate have shown that these drugs markedly diminish libido without interfering with erectile capacity (42). In men with a disease state of hypogonadism, it has been shown that androgen therapy restores libido and seminal emission (46). Thus, evidence from three different sources of information indicates that androgen levels are closely linked to seminal emission and sexual drive in the human male. Current evidence suggests that a certain minimal level of androgen is necessary for sexual function but that excess androgen above these levels has minimal or no effects.

In many subprimates, female sexual activity is restricted to estrus and is clearly related to estrogen or progesterone levels. In primates, the relationship between sexual activity and endocrine variables is more obscure. Current evidence suggests that estrogen levels are essential to vaginal epithelial integrity and lubrication, whereas androgen levels may be related to libido. Most of the evidence on endocrine effects on female sexual behavior consists of studies of sexual behavior across the menstrual cycle, with the use of androgen therapy, and during hormonal replacement therapy.

A number of investigators have studied sexual activity across the menstrual cycle and in relationship to cyclic changes of progesterone and estradiol. A variety of measures, including coital frequency, self-rating of libido, and arousability to sexual stimuli as measured by vaginal plethysmography, have been used (42). To date, there are no consistent data relating cyclic changes in estrogens or progesterone during the menstrual cycle and sexual activity. In a study of 115 women who presented with complaints of premenstrual symptoms, Clayton et al. (47) investigated the relationship between the menstrual cycle and different aspects of sexual functioning, using the Changes in Sexual Functioning Questionnaire. Women who were assessed at the screening visit during the late luteal phase of their menstrual cycle reported less desire to engage in sexual activity and less frequent sexual activity than women who were assessed during other phases of the menstrual cycle. The 24 women who returned for the second and third visits reported less frequent orgasms and less satisfaction from their orgasms premenstrually than during midcycle.

Slob et al. (48) found that the relationship between menstrual cycle phase and sexual arousal was further substantiated by the reported greater increase in sexual desire after the first erotic video in follicular women than in luteal women. However, they cautioned that this increased desire, as well as more erotic fantasies, persisted during the next 24 hours and concluded that studies on the effects of menstrual cycle phase on sexual arousability in the laboratory should seriously consider the possible learning and conditioning effects as suggested by the present investigation. Harvey (49) also concluded that human sexual behavior may be influenced by hormonal fluctuations and cognitive factors associated with the menstrual cycle. However, Meuwissen and Over (50) noted that three experiments assessing sexual arousal to erotic stimulation at several phases within the menstrual cycle indicated that the subjective sexual arousal elicited during fantasy depicting specific themes was stable across the menstrual cycle. They concluded that the experiments provide no support for the claim that female sexual response varies systematically across the menstrual cycle. Another study also

noted that only one of 12 women tested during their luteal phase indicated an increase in desire to make love, whereas six of 12 women tested during their follicular phase indicated an increase (51). Apparently the menstrual cycle phase during the first test determines the immediate response and indirectly affects the response during the second test approximately 10 days later. In a random block design the difference in sexual response between the follicular and luteal phase of the menstrual cycle disappeared.

There is some evidence that sexual libido and arousability are related to serum androgen levels (52–56). Increased libido was noted as a side effect of androgen therapy used to treat a number of medical conditions (42). Some clinicians have reported using androgen therapy to treat hypoactive sexual desire disorders; others have reported that antiandrogens lower libido in females. The strongest evidence relating androgens to female libido was reported by the McGill University research group. In a number of controlled studies they showed that estrogen-androgen preparations are superior to estrogen and placebo in restoring sexual function to patients with surgical menopause (57,58).

Hyperprolactinemia has been reported to be associated with decreased libido in both sexes (42). Male patients may complain of erectile problems. In many cases, bromocriptine therapy restores normal function.

DRUGS AND SEXUAL BEHAVIOR

Within the last decade, physicians have become increasingly aware that many commonly prescribed drugs adversely affect sexual functioning. Although the list of drugs with adverse sexual side effects is extensive, the worst offenders are psychiatric drugs and drugs used to treat arterial hypertensive disease (59).

The presence of sexual dysfunction may predate commencement of psychotropic or other medications. We must encourage the acquisition of baseline sexual function from patients before giving them these medications. The *Diagnostic and Statistical Manual of Mental Disorders, Fourth Edition (DSM-IV)* recognizes seven major categories of sexual dysfunction: (a) sexual desire disorder, (b) sexual arousal disorder, (c) orgasmic disorder, (d) sexual pain disorder, (e) sexual disorder due to general medical condition, (f) substance-induced sexual dysfunction, and (g) sexual disorder not otherwise specified (60).

Some patients with schizophrenia may discontinue taking their antipsychotic medications because of sexual side effects resulting from either the antipsychotics themselves or the anticholinergic medications used to treat EPS (61). Antipsychotic drugs have been reported to cause disturbances in libido, ejaculatory impairment, female anorgasmia, and erectile failure (62,63).

Current evidence suggests that interference with orgasm is secondary to α-adrenergic blockade and that interference with libido and erectile capacity probably is secondary to central dopamine blockade (63). Case reports and clinical series have documented diminished libido with chlorpromazine, thioridazine, thiothixene, fluphenazine, and haloperidol (64). Erectile failure has been reported with chlorpromazine (65), pimozide (66), thiothixene (67), thioridazine (68–70), sulpiride (71), haloperidol (72), and fluphenazine (73). Many of these same neuroleptics have been reported to interfere with ejaculatory function (14). One controlled, double-blinded study of the effect of neuroleptics on sexual function in humans by Tennett et al. (74) found that low doses of benperidol (1.25 mg) and chlorpromazine (125 mg) had no effect on erectile function. When a patient complains of sexual dysfunction due to an antipsychotic drug, changing to a different neuroleptic such as loxapine or molindone (with lesser α-adrenergic effects) may resolve the sexual problem.

Most of the atypical antipsychotics are reported to cause lower prolactin elevation than conventional neuroleptics, so sexual dysfunction might be supposed to be less common. However, as of 1999 there are no adequate data comparing actual sexual side effects of atypical antipsychotics with placebo or with conventional neuroleptics. Limited data available at that time suggested that adverse effects in this regard are no more common with atypical antipsychotics than with conventional drugs, except for the unexplained reduction in ejaculatory volume seen with sertindole, which is reversible (75).

A Collaborative Working Group on Clinical Trial Evaluations (76) in 1998 noted the following: Clozapine has been reported to increase libido and restore menstrual function in some patients, resulting in a fair number of pregnancies; thus, patients starting clozapine treatment should be alerted about potential pregnancy. The increased libido may be related to improvement in negative symptoms and improved fertility. Ejaculatory dysfunction has not been reported for clozapine, risperidone or olanzapine, and it has not been specifically evaluated with quetiapine. Risperidone is unique among the atypicals in that it produces dose-related increases in prolactin levels comparable or perhaps higher than those of the typicals. Only a few patients discontinue treatment due to breast tenderness, menstrual irregularities, or galactorrhea. In women risperidone dose and endpoint prolactin levels was not correlated with adverse events, but in men the number of adverse events within a dose range from 4 to 10 mg/day was not significantly higher than placebo, whereas the endpoint prolactin level was not correlated with adverse events.

A double-blinded comparison of olanzapine versus risperidone in 1997 showed that there was no increase in the prevalence of amenorrhea, galactorrhea, or gynecomastia, even though risperidone produced higher prolactin levels than olanzapine (77).

Although benzodiazepines probably do not have adverse effects on erectile function (14), there is evidence that these drugs cause ejaculatory delay and may be used to treat premature ejaculation (78–83). They have direct effects on cognition, memory, and motor control. However, benzodiazepines may improve sexual function by its anxiolytic properties if anxiety is the root cause of sexual difficulties. They decrease plasma concentration of epinephrine and reduce anxiety through its γ-aminobutyric acid receptor action.

If ejaculatory delay is a serious difficulty, buspirone may be substituted for the offending agent. Buspirone is reported to reverse the sexual side effects of some selective serotonin reuptake inhibitor (SSRI) antidepressants, probably by suppressing SSRI-induced elevation of prolactin and by its 5-HT_a agonist action.

There have been case studies showing that lithium carbonate may cause erectile impairment (84). This was confirmed in a double-blinded, placebo-controlled trial using therapeutic doses in men with affective disorder (85). *This may be due to the dopamine antagonist property of lithium.* Valproate may be considered an alternative to lithium in bipolar patients who experience erectile failure while taking lithium.

Antidepressants have been reported to have a variety of sexual side effects. Heterocyclics, monoamine oxidase inhibitors, and SSRIs including imipramine, desipramine, nortriptyline, amitriptyline, doxepin, protriptyline, amoxapine, trazodone, maprotiline, tranylcypromine, phenelzine, bupropion, and fluoxetine (86) reportedly have caused diminished libido. The only controlled, double-blinded study of antidepressant effects on libido found that both phenelzine and imipramine decrease libido (87). Of the SSRIs the most frequent sexual adverse effects are seen with paroxetine, next with fluoxetine, and the least with sertraline (Zoloft). Similar adverse sexual symptoms were seen with fluvoxamine and citalopram (88).

Such sexual side effects affect quality of life and may result in noncompliance with medication and the associated risk of recurrence of depression. Depression also may be associated with sexual disturbances, especially reduced libido. It is important to unravel the origin of sexual problems during depression and determine whether they were present before depression started, whether they are associated with the depression, or whether they are an effect of medication. Baseline measurements, objective measures, and accurate instruments all are essential for scientific research into the sexual side effects of antidepressants. Various human factors that may influence measurements of sexual behavior must be considered (89).

Hindmarch (90) in the United Kingdom pointed out that many antidepressants have a direct pharmacologic action on the central nervous system and disrupt cognitive function, thus increasing anhedonia and impairing sexual function. Drug actions on cognitive structures, which in turn increase anhedonia and reduce sexual libido, are over and above any direct pharmacologic effects on the more overt behavioral activities associated with sex, including orgasm, erectile function, potency, and ejaculation. The article reports the following: (a) The tricyclic antidepressants destroy the cognitive structures that are vital to maintaining normal libido as well as disturbing overt sexual behaviors. (b) Some SSRIs (paroxetine and sertraline) are associated with behavioral activation that is responsible for impairment of sexual function, but there are clear differences among the SSRIs. (c) Fluvoxamine (relative to the other SSRIs) has little effect on objective measures of cognition or on cerebral and behavioral components of sexual function (90).

Treatment approaches for antidepressant-induced low libido include coadministration of 7.5 to 15 mg of neostigmine before coitus (91), coadministration of yohimbine with fluoxetine (92), or substitution of bupropion because of this drugs' very low incidence of sexual side effects (93). The mechanism by which antidepressant drugs influence libido is unclear. A similarly large group of antidepressants has been reported to cause erectile problems. Only one controlled study has documented an adverse effect of antidepressant drugs on erectile function (86). In a double-blinded, placebo-controlled study, Kowalski et al. (94) found that both amitriptyline and mianserin decreased erectile capacity as measured by nocturnal penile tumescence. It appears that substitution of bupropion may be an effective intervention for antidepressant-induced erectile impairment (93). Anorgasmia and delayed ejaculation have been reported with phenelzine (95), amoxapine (96), amitriptyline (97), clomipramine (98), imipramine (99), fluoxetine (100), trazodone (101), and sertraline (102). Double-blinded studies have confirmed the effects of phenelzine, imipramine, and clomipramine on orgasm (103); however, clomipramine has been reported to increase sex drive in some persons.

Increased serotonergic activity may be a mechanism by which antidepressant drugs inhibit orgasm (103). It is thought that activation of 5-HT$_2$ receptors impairs sexual function, whereas activation of 5-HT$_1$ facilitates sexual function. This may explain why cyproheptadine may reverse antidepressant-induced sexual dysfunction by acting postsynaptically to reduce activity at the 5-HT$_2$ receptor. Nefazodone may have a similar mode of action; however, it also is a serotonin reuptake inhibitor and may have less risk of causing a recurrence of depression compared to cyproheptadine (40). Bupropion (Wellbutrin), on the other hand, is a norepinephrine-dopamine reuptake inhibitor and is not usually associated with any sexual side effects. Mirtazapine (Remeron) is rarely associated with any sexual side effects.

Antidepressant-induced anorgasmia has been reported to be reversed by the use of bethanechol (104), cyproheptadine (105), and yohimbine (106). Drug substitution with desipramine or bupropion has been reported to be effective (86).

It is noteworthy that spontaneous orgasm has been reported with clomipramine (107) and fluoxetine (108). There also has been a case report of successful use of trazodone for treatment of male erectile disorder (109). Psychiatric drugs

TABLE 14.1. PSYCHIATRIC DRUGS REPORTED TO CAUSE ERECTILE DYSFUNCTION

Imipramine	Amoxapine
Desipramine	Trazodone
Nortriptyline	Maprotiline
Amitriptyline	Tranylcypromine
Doxepin	Phenelzine
Protriptyline	Isocarboxazid
Clomipramine	Lithium carbonate
Chlorpromazine	Thioridazine
Pimozide	Sulpiride
Thiothixene	Haloperidol
Fluphenazine	

that have been reported to cause erectile and ejaculatory problems are listed in Tables 14.1 and 14.2.

The ejaculation delaying effects of the SSRIs (fluvoxamine, fluoxetine, paroxetine, and sertraline) were investigated in a double-blinded, placebo-controlled study in men with rapid ejaculation. The SSRIs were given at their recommended daily dosages for 6 weeks, and the men measured their intravaginal ejaculation latency time at home using a stopwatch. The results showed a clear difference among the SSRIs, with fluvoxamine having by far the least disturbing effect on ejaculation (89).

Hypotensive Agents

Antihypertensive drugs are another class of drugs that have been found to frequently interfere with sexual function. Current evidence indicates that any clinician encountering a patient with sexual dysfunction should immediately inquire about drug usage and suspect hypotensive agents as a possible etiologic agent. These drugs have been reported to interfere with erection, ejaculation, orgasm, and libido (for more extensive reviews, see references 5, 14, 62, and 63).

Diuretics are frequently used for management of mild hypertension and previously were thought to be devoid

TABLE 14.2. PSYCHIATRIC DRUGS REPORTED TO CAUSE ORGASM DISTURBANCES

Fluoxetine	Amitryptyline
Sertraline	Doxepin
Paroxetine	Protriptyline
Imipramine	Clomipramine
Desipramine	Amoxapine
Nortriptyline	Trazodone
Tranylcypromine	Maprotiline
Phenylzine	Alprazolam
Isocarboxazid	Chlordiazepoxide
Lorazepam	Thioridazine
Chlorpromazine	Mesoridazine
Chlorprothixene	Fluphenazine
Perphenazine	Thiothixene
Trifluoperazine	Haloperidol

of sexual side effects. Current evidence suggests that hydrochlorothiazide, chlorthalidone, and spironolactone may decrease libido and cause erectile problems (110–113). Spironolactone's antilibidinal effects may be related to its antiandrogenic effect (114). Other antiandrogens that may cause sexual problems include ketoconazole and finasteride. The mechanism by which other diuretics influence sexual behavior is unclear. Antihypertensive drugs with central antiadrenergic effects, such as methyldopa and reserpine, have been reported to cause diminished libido, erectile failure, and ejaculatory impairment (62,63). Current evidence also suggests that β-blockers may be associated with erectile failure (115,116). The available evidence suggests that β-blockers that are more lipophilic are more likely to cause sexual dysfunction than those that are more hydrophilic, although impotence has been reported with atenolol, a hydrophilic β-blocker. Sexual problems appear to be less frequent with pindolol, metoprolol, and nadolol. Impotence has been reported as a side effect of timolol eyedrops, which are systemically absorbed (117). Ejaculatory problems appear to be common with guanethidine, bethanidine, labetalol, and nifedipine (59). Among antihypertensive drugs, the angiotensin-converting-enzyme inhibitors, such as captopril, enalapril, and lisinopril, appear to lack sexual side effects.

Other Drugs

Controlled studies indicate that cimetidine can cause decreased libido and erectile failure (118). Other H_2-receptor antagonists, such as ranitidine, appear to have this side effect less often. The mechanism for this phenomenon is unclear, because cimetidine has both antiandrogenic effects and ganglion-blocking effects. Controlled studies have found decreased libido and erectile problems associated with long-term use of digoxin (119,120). Antiepileptic drugs have been reported to decrease libido. Phenobarbital and primidone have been reported to have more adverse effects than phenytoin and carbamazepine (121). It is difficult to evaluate these reports because epilepsy may affect libido independent of drug effects. The reader is referred elsewhere for more extensive information on the effects of drugs on sexual function (59). Neuroendocrine issues in epilepsy are discussed in Chapter 12.

NEUROLOGIC DISEASE AND SEXUAL DYSFUNCTION

A number of neurologic diseases have been reported to be associated with sexual dysfunction and should be considered in the differential diagnosis of a patient with sexual difficulties. In diseases causing peripheral neuropathies, the mechanism usually is clear. In diseases affecting the brain, the specific mechanisms usually are unclear.

Multiple Sclerosis

A number of different investigators have established that multiple sclerosis can cause decreased libido, arousal, sexual sensations, and orgasmic capacity in both sexes, and erectile response in male patients (122–126). Estimates of the frequency of sexual problems in such patients have ranged widely, from 26% to 90% (7,122). Sexual life in such patients can be altered by various mechanisms, including neuropathologic lesions in the brain and spinal cord. In this regard, hypersexuality has been reported in a multiple sclerosis patient with frontal and temporal lesions (127). Psychosocial factors may contribute to altered sexual function.

Diabetes Mellitus

Diabetes mellitus is known to be associated with erectile failure and to a lesser extent with ejaculatory disturbance (128–134). Decreased erectile function in diabetic males has been confirmed by nocturnal penile tumescence studies (135,136) and by laboratory-based erotic stimulation studies (137). These sexual difficulties are suspected to be secondary to peripheral neuropathy of the autonomic nervous system (138,139), although one investigator has suggested that cerebral dysregulation of autonomic nervous system activity may be responsible (140). Kolodny (141) reported that diabetic women had an increased frequency of acquired orgasmic dysfunction; however, numerous other investigators failed to replicate this finding (142). Decreased vaginal lubrication in insulin-dependent diabetics has been consistently reported (143–145). One exception is a study by Schreiner-Engel et al. at the Mount Sinai School of Medicine (New York, New York) (146). In this study, type I (insulin-dependent) diabetics had minimal evidence of sexual impairment, whereas type II diabetics suffered both sexual and relationship discord. This finding is discrepant from many other studies and may reflect the interaction of psychosocial variables with the disease process. For example, it is postulated that some of the sexual impairment in type II diabetes may be secondary to the emotional consequences of a late-onset disease.

Spinal Cord Injury

One of the most common causes of neurogenic sexual difficulty in males is spinal cord injury. Although certain general statements can be made about the site of the lesion and the resulting sexual impairment, there is some inconsistency in the literature. In most series, the location and completeness of the series are deduced by clinical examination, thus allowing for incorrect assignment of the level of lesion and the presence of partial communication between the brain and the partially severed cord. Complete cervical and high thoracic cord lesions are extremely destructive to ejaculatory function (147). Depending on the level of the cord lesion, reflexogenic erections to local stimuli or psychogenic erections to psychic stimuli may be intact. Reflexogenic erections occur most often in patients with complete upper motor lesions, especially in men with cervical lesions. Psychogenic erections are intact most often with lower motor neuron lesions below T9.

There is relatively little information on sexual behavior in female patients with cord lesions. Comarr and Vigue (148) could not correlate sexual function with any given level of cord lesion, although Berard (149), in an investigation of 15 cord-injured women, concluded that orgasm was impossible if the lesion was below T12, because of decreased sensation.

Damage to the Autonomic Nervous System

A number of surgical procedures interfere with autonomic nervous system innervation of the pelvic organs and result in sexual impairment. These procedures include sympathectomies (150), retroperitoneal lymphadenectomy (151), abdominoperitoneal resection (152), anterior resection of the rectum (153), aortoiliac surgery (154,155), and radical retropubic and transvesical prostectomy (156). Most of these same procedures have been reported to cause orgasm disorders in female patients (157).

Seizure Disorder

Both ictal and intraictal sexual abnormalities have been noted in patients with epilepsy (158). Ictal sexual manifestations, such as sexual emotions, genital sensations, and sexual automatisms, are rare and appear to be more common during partial complex seizures (159). Sexual auras are reported to have a temporal lobe origin (160–162), although the origin of other ictal sexual behavior is less clear.

Interictal sexual abnormalities are more common than ictal sexual abnormalities. Although hypersexuality (163) and paraphilia (164–166) have been reported, hyposexuality is the more common finding (167–173). Sexual abnormalities appear to be more common in patients with partial complex seizures (158). Partial complex seizures of temporal lobe origin are more commonly associated with sexual abnormalities than seizures from extratemporal foci (173), although there are examples of altered sexuality from extratemporal foci (172). Several different factors might explain altered sexual behavior in patients with epilepsy, "[t]hese include social and psychological factors, disruption of normal limbic function by epileptic discharges, altered pituitary and gonadal hormones, and alteration of behavior and hormones by AEDs [antiepileptic drugs]" (158, p. 541).

Judging the importance of these factors in a given case can be difficult. In most cases the degree of alteration of pituitary and gonadal dysfunction associated with the epilepsy itself is insufficient to explain the degree of hyposexuality observed (121,174). Some of the hyposexuality observed in some epileptics might be explained by the effects of antiepileptic drugs on endocrine function (175,176). Testosterone exists in three forms in the serum: a free form, which is biologi-

cally active; a form loosely bound to albumin; and an inactive form, which is tightly bound to sex hormone binding globulin. A number of antiepileptic drugs, notably carbamazepine, decrease the amount of free testosterone. Low levels of free testosterone appear to be related to decreased libido. Unfortunately, androgen replacement therapy has been only moderately successful in improving libido in men with epilepsy.

Other Neurologic Diseases

A number of clinicians have noted a high frequency of sexual problems in patients with Parkinson disease (177,178). However, Lipe et al. (179) reported that the degree of sexual impairment in patients with Parkinson disease is no higher than that in patients with arthritis. This suggests that the impairment may be due to chronic illness in general and not specific to the disease process of Parkinson disease. *Hyper*sexuality has repeatedly been reported in parkinsonian patients treated with dopaminergic drugs (180).

Another syndrome frequently associated with a decreased frequency of sexual activity is cerebral vascular accident (181–183). Although some investigators report that most men who were sexually active before stroke resume sexual activity, other investigators have found more devastating effects on sexual behavior (184–187). There are several potential mechanisms for the decrease in sexual activity, including hypotensive medications, partner reaction, immobility, and deformity. There is minimal evidence regarding specific brain regions infarcted and the likelihood of resulting sexual impairment. An important exception is the report of Monga and colleagues *of hypersexuality* after stroke involving the temporal lobe close to the amygdaloid nucleus (188).

Similarly, most studies of brain trauma patients have documented hyposexuality (189,190); however, hypersexuality has been noted after injury to the temporal lobes and dorsal septal region (191,192). Damage to the hypothalamo-pituitary region of the brain by tumor results in decreased libido, which appears to be correlated with the degree of hypogonadism (193). Erectile failure has been reported to be common in patients with Alzheimer disease (194). One case of hypersexuality in Huntington chorea has been reported (195).

PSYCHIATRIC DISORDERS AND SEXUAL DYSFUNCTION

The prevalence of sexual disorders in various psychiatric diseases has been reviewed (196). A number of investigators documented loss of libido and decreased sexual function in the presence of depressive disorder (197–199). Loss of libido is a common symptom in both dysthymia and major depression (200). After age 70 years, loss of libido is less indicative of depression, primarily because loss of libido is common among nondepressed elderly (201,202). In addition to loss of libido, major depression may be associated with decreased ability to attain and maintain penile erections (203,204). Interestingly, Schreiner-Engel and Schiavi (205) reported that nondepressed patients with low sexual desire had an elevated lifetime prevalence rate of affective disorders. They hypothesized that there may be a common etiology to hypoactive sexual desire disorder and depressive disorders.

In an ongoing Zurich Cohort Study, 591 males and females from the general population of Zurich were interviewed five times over the 15-year period between the ages of 20 and 35 years. Emotional problems or sexual dysfunction alone are rare and almost always are associated with each other or with changes in libido. However, changes in libido without associated emotional problems or sexual dysfunction are relatively common. Libido therefore appears to be the core problem with which other sexual problems overlap. Overall, sexual problems of some type were found in 26% of normal subjects, 45% of nontreated depressed patients, and 63% of treated depressed patients. This increase in sexual problems in treated depressed patients is mainly due to an increase in sexual dysfunction and emotional problems; the level of libido appears not to be affected by treatment. There was no difference in the prevalence of sexual problems of any kind between patients treated with medication and those treated only with psychotherapy (206). Depression may cause a progressive decline in interest in sexual behavior leading to low libido, difficulty in sexual arousal, secondary anorgasmia, and/or frank sexual aversion. Increasing attention of physicians to the sexual problems of women will dramatically improve female quality of life, especially during difficult periods of transition (207).

Arousal disorders, dyspareunia, orgasmic difficulties, and dissatisfaction, both physical and emotional, may contribute to a secondary loss of libido. Depression, anxiety, and chronic stress may interfere with central and peripheral pathways of the sexual response and reduce the quality of sexual function mostly in its motivational root. Relational conflicts and/or delusions and partner-specific problems, erectile deficit first, may contribute to fading of sexual drive in the postmenopausal years (38).

During manic episodes, an increase in sexual thoughts, conduct, and number of sexual partners may occur. Some of the hypersexuality seen in manic patients can be a reflection of premorbid personality (196).

As Offit (196, p. 2255) concluded in the summary of evidence on sexuality and schizophrenia, "The literature is quite inconsistent about changes in sexual behavior among schizophrenics." Part of this inconsistency may be related to changing diagnostic criteria and alternative assessments of sexual behavior. Several well-conducted studies using good methodology have reported decreased sexual interest and activity among patients with schizophrenia (208,209). Raboch (210) looked at the sexual development of 51 female schizophrenics, their attitudes toward sex, and their sexual activity and arousability. The control group consisted of 101 gynecologic patients. The sexual development of the female schizophrenics was significantly retarded, and their sexual activity and arousability were significantly lowered in adult life compared with the control group. The hypothesis was advanced that the retarded development may be caused by some biologic factor, whereas sexuality in adult life is negatively affected primarily by social isolation (210). Friedman and Harrison (211) found that 60% of women with schizophrenia had never experienced an orgasm. Aizenberg et al. (212) reported that libido was diminished in untreated patients with schizophrenia but returned with treatment, although erection and orgasm problems developed with treatment.

Information on sexual behavior in other psychiatric disorders is scarce. Women with anorexia nervosa appear to have markedly diminished sexual activity and interest (213). There is little evidence linking any specific personality disorder with sexual difficulties, apart from the high frequency of sexual identity issues in young adults with borderline personality disorder. Patients with sensitivity to rejection (214) and a predisposition to anxiety and depression may be more likely to develop sexual problems (215).

NEUROPSYCHIATRIC EVALUATION OF SEXUAL DYSFUNCTION

A critical part of the evaluation of any sexual complaint is a careful history. The clinician needs to document the precise difficulty, the associated symptoms, and its onset and course, and consider if the difficulty is secondary to another psychiatric disorder, such as affective disorder. One needs to ascertain gender identity and sexual preference for partners, as well as assess whether the problem is secondary to relationship discord or life stress. One also needs to carefully delineate whether the disorder is generalized or partner specific. It is unusual for biogenic sexual problems to be partner specific. Thus, a history of a partner-specific difficulty is presumptive evidence of a psychogenic difficulty.

In cases of hypoactive sexual desire disorders, one needs to inquire closely regarding the frequency of romantic daydreams and sexual thoughts. A high frequency of sexual daydreams suggests that the problem is not biogenic. One also needs to inquire about masturbatory frequency and the preferred masturbatory fantasy. Obviously, a high frequency of masturbatory behavior or an aberrant sexual fantasy life suggests that the problem is not biogenic. If one determines that the decreased desire is present in all contexts, one would have a greater suspicion that the problem is biogenic and would evaluate androgen and prolactin levels.

Deficient sexual arousal as evidenced by decreased vaginal lubrication as an isolated complaint almost always is caused by atrophic vulvovaginitis, related to estrogen deficiency. Psychogenic arousal disorder in the female almost always occurs in conjunction with hypoactive sexual desire disorder (216).

Diagnostic evaluation of male erectile disorder can be complex. A careful sexual history should be a major part of the evaluation. The presence of full erections upon awakening is highly suggestive of a psychogenic erectile problem (217). To date there is minimal evidence that psychometric assessment aids in the differential diagnosis of biogenic from psychogenic erectile disorder (218). A global screening procedure to measure erectile capacity is to monitor nocturnal erections with a nocturnal penile tumescence study (219,220). The clinician needs to be aware of a number of possibly confounding influences, including affective disorder, aging, hypoandrogenic states, and hypoactive desire states. Barring these limitations, a large amount of normative data allows the clinician to judge deviancy from these values given normal sleep parameters. Alternative nonlaboratory approaches include snap gauges and portable nocturnal erection monitors (e.g., RigiScan). The lack of sleep monitoring limits the diagnostic accuracy of these approaches. Another general screening procedure is the visual sexual stimulation method. In this procedure, the erectile response to erotica is monitored (221). The reliability, validity, sensitivity, and specificity of this approach have not been established.

A variety of procedures can be used to test the integrity of the vascular system. Penile blood pressure, strain gauge plethysmography, and pulse wave assessment all have been used; however, all of these procedures only assess hemodynamics in the flaccid penis. Intracorporal injection of papaverine/phentolamine or prostaglandin E_1 can be used to assess functional integrity of the penile vasculature. Unfortunately, it is not clear how much vascular disease must be present to cause diminution of pharmacologically induced erection. A decreased response provides suggestive but not definitive evidence of vasculogenic impotence (219). More invasive procedures include cavernosometry and cavernosography (222,223). To date, lack of standardization limits the usefulness of these procedures.

Most of the procedures currently available to assess integrity of the neurologic components of erection primarily assess the pudendal nerve. Bulbocavernosus reflex latency testing consists of recording the electromyographic response of the bulbocavernosus muscle to electrical stimulation of the dorsal nerve of the penis. Absent or prolonged reflex latency indicates pathology within the reflex arc (224). A variation of this procedure is to stimulate the prostatic urethra and test for autonomic neuropathy. Other procedures include measuring dorsal nerve conduction velocity, the threshold for perception of vibratory sensations in the penis, and somatosensory evoked potentials evoked by penile stimulation. All of the procedures can detect neurologic abnormalities. However, the absence of adequate normative data limits the interpretation of these findings (219). In other words, it is unclear what values are clearly indicative of neurogenic impotence.

Premature ejaculation rarely has an organic cause, with the possible exception of acquired premature ejaculation as an early manifestation of multiple sclerosis. If the patient complains of retarded ejaculation, one can distinguish between retrograde ejaculation and an ejaculatory orgasm by examination of the postorgasm spun urine sample. Anything that interrupts the orgasmic reflex arc can interfere with both male and female orgasmic function. Acquired orgasm disorder in the absence of marital discord mandates a search for treatable biogenic causes. Numerous drugs can cause orgasmic dysfunction. Numerous neuropathies (e.g., diabetes mellitus, multiple sclerosis, alcoholic neuropathy) and surgical procedures (e.g., aortoiliac surgery, lumbar sympathectomy, retroperitoneal lymphadenectomy) can cause anorgasmia. Severe hypogonadism in the male can result in anorgasmia. The sensory part of the orgasmic reflex arc theoretically could be measured by somatosensory evoked potentials or by the threshold for vibratory sensation. To our knowledge, these procedures have not been used for that purpose.

PSYCHOLOGICAL TREATMENT OF SEXUAL DISORDERS

Psychological treatment of sexual dysfunction is based on techniques introduced by behavioral psychologists and then elaborated by Masters and Johnson (225). To briefly summarize, these techniques focus on attitude change and anxiety reduction. Sexual activity is structured such that there is the progressive experience of emotional and sexual intimacy at a pace tolerable to both partners. Preferably, the couple is seen together on a once-a-week basis. Either a dual-sex therapy team or a solo therapist of either sex may be used. More detailed descriptions of this treatment approach can be found elsewhere (196,225,226).

PHARMACOLOGIC TREATMENT OF SEXUAL DISORDERS

Currently, there is evidence that both premature ejaculation and iatrogenic erectile disorder may respond to pharmacologic interventions. Clinicians have reported successful treatment of premature ejaculation using low doses of thioridazine (227,228), monoamine oxidase inhibitors (229), lorazepam (180), and clomipramine (230). One double-blinded, placebo-controlled study demonstrated that 25 to 50 mg clomipramine taken 6 hours before coitus extends the time to ejaculation by 6 to 8 minutes in men with premature ejaculation (231).

Current evidence suggests that yohimbine is effective in reversing certain cases of idiopathic erectile dysfunction (232). In a 10-week, double-blinded, partial crossover study of yohimbine for treatment of psychogenic erectile problems, Reid et al. (233) found yohimbine was clearly superior to placebo. Two other studies found yohimbine was effective for treatment of idiopathic erectile dysfunction (234,235). Current evidence suggests that yohimbine has its mechanism of action in the central nervous system. Direct injection of yohimbine into the corpora cavernosa does not induce erection, whereas intracerebral injection increases various parameters of sexual behavior in laboratory animals (232).

There is evidence suggesting that dopaminergic drugs and/or opioid receptor blockers may prove useful in the treatment of human sexual disorders. Anecdotal evidence suggests that dopaminergic drugs such as levodopa and pergolide may increase libido (14). A number of independent investigators demonstrated in controlled studies that apomorphine administered subcutaneously in the arm elicits penile erections (236–238). In an open trial, Lal (239) reported that oral bromocriptine is effective in men with impotence who previously demonstrated an erectile response to injected apomorphine. Unfortunately, side effects from apomorphine render it unsuitable as a treatment for erectile problems. Two controlled studies of the efficacy of levodopa for treatment of erectile problems reported contradictory results (240,241). Although it is well known that opiate abuse may diminish libido, there are relatively few studies on the effects of opioid antagonists on sexual function. Charney and Heninger (242) reported that infusion of 1 mg/kg solution of naloxone resulted in partial penile erections in normal volunteers. In a placebo-controlled study of the efficacy of naltrexone in the treatment of erectile dysfunction, Fabbri et al. (243) reported that 50 mg naltrexone significantly improved erectile function.

Moore and Rothschild (244) noted that sexual dysfunction is a common side effect seen in patients taking SSRIs. Spontaneous resolution may occur in some patients, but therapy can be modified to eliminate undesirable side effects. Modifications include reducing drug dosages, altering timing of drug dosages, taking drug holidays, adding an adjunctive drug, and switching to alternative antidepressants. Reported incidences of SSRI-induced sexual side effects have ranged widely, from 1.9% to 75%. A prospective multicenter study of 344 patients by Montejo-Gonzalez et al. (245) found that drug-related sexual dysfunction occurred in 54% of patients taking fluoxetine, 56% of patients taking sertraline, and 65% of patients taking paroxetine. Another prospective study showed that difficulty or inabil-

ity to achieve orgasm is the principal sexual side effect associated with SSRIs (246). One controlled study showed that 70% to 80% of depressed patients have diminished libido (246). Anorgasmia, however, is infrequently due to depressive illness, making this symptom helpful in distinguishing between depression-related and treatment-related dysfunction.

Treatment options for SSRI-induced sexual dysfunction include waiting 4 to 6 weeks for sexual side effects to resolve, decreasing dosage, altering timing of daily dose, planning a 2-day drug holiday (for short-acting antidepressants such as sertraline and paroxetine), coadministering an antidote, or substituting another antidepressant (preferably nefazodone or bupropion) (Table 14.3). Like antidepressant-induced weight gain, however, sexual side effects are likely to persist. Spontaneous resolution of other side effects, such as nausea and dry mouth, are far more common. When using adjunctive drugs, the patient should be instructed to take the antidote 1 to 2 hours before anticipated sexual activity (except in the case of amantadine, which may require 2 days of dosing before it is effective).

Treatment Caveats

In some patients, bupropion had its own side effects, most commonly anxiety and tremor. Cyproheptadine may relieve sexual side effects of SSRIs, but the improvement induced by this drug may be transitory and sedation is a common side effect. Cyproheptadine should be used only on an as-needed basis because daily dosing may interfere with the antidepressant action of the SSRI. Yohimbine may be more effective than either cyproheptadine or amantadine. A retrospective case review of 45 patients by Ashton et al. (247) compared yohimbine, cyproheptadine, and amantadine as antidotes for SSRI-induced sexual dysfunction. Of the 21 patients taking yohimbine, 17 (81%) showed improvement in sexual functioning compared to only 48% of those taking cyproheptadine and 42% of those taking amantadine. The antidotes had a low rate of adverse effects. Agitation occurred in three patients taking yohimbine; sedation and weight gain in two patients taking cyproheptadine; and depressive

TABLE 14.3. ADJUNCTIVE DRUG THERAPY FOR SSRI-INDUCED SEXUAL DYSFUNCTION

Drug	Symptom	Dosage as Needed	Daily
Amantadine	Anorgasmia, decreased libido, erectile dysfunction	100–400 mg (for 2 days before coitus)	75–100 mg b.i.d. or t.i.d.
Bupropion	Anorgasmia	75–150 mg	75 mg b.i.d. or t.i.d.
Buspirone	Anorgasmia, decreased libido, erectile dysfunction	15–60 mg	5–15 mg b.i.d.
Cyproheptadine	Anorgasmia, decreased libido, erectile dysfunction		4–12 mg
Dextro-amphetamine	Anorgasmia	5–20 mg	2.5–5 mg b.i.d. or t.i.d.
Pemoline	Anorgasmia		18.75 mg
Yohimbine	Anorgasmia, decreased libido, erectile dysfunction	5.4–10.8 mg	5.4 mg t.i.d.

SSRI, selective serotonin reuptake inhibitor.
Adapted from references 273–277.

symptoms in one patient taking amantadine. Switching to nefazodone, bupropion, or possibly mirtazapine, after a washout period of 1 to 2 weeks, might be a reasonable choice. It should be noted, however, that the alternative medications may have their own side effects, notably sedation and dizziness with nefazodone; dry mouth, drowsiness, and weight gain with mirtazapine; and agitation with bupropion. Cyproheptadine, which has antiserotonergic properties, may exert its beneficial effect because it preferentially antagonizes the 5-HT$_{2A}$ receptor. Coordination of other neurotransmitter pathways also is essential to normal sexual functioning. Animal studies suggest that central adrenergic activation increases sexual behavior, which would account for the ameliorative effects of dextroamphetamine and pemoline (sympathomimetics) and yohimbine (an α_2-adrenergic receptor antagonist) (246). Increased central dopaminergic activity also appears to increase sexual behavior and functioning. Accordingly, bupropion's lower rate of sexual dysfunction side effects may stem from its combined norepinephrine and dopamine reuptake inhibition. Finally, cholinergic receptors are thought to modulate adrenergic activity for optimal sexual functioning but do not seem to act directly.

NEUROPSYCHIATRIC ASPECTS OF GENDER IDENTITY AND SEXUAL PREFERENCE

The origins of gender identity and sexual orientation have been the subject of intense debate within the scientific community (248). Early sexologists such as Kraft-Ebing (249) argued that homosexuality must be innate, and numerous investigations have attempted to find evidence to support this viewpoint (250–252). In the twentieth century, the work by Money et al. (253,254) has been very influential in the United States. From studies of psychosexual development in children with ambiguous sexual development, it was concluded that gender identity is predominately determined by sex of rearing before age 2.5 years. In contrast, Donner (255) suggested that a neuroendocrine predisposition for homosexuality might be based on the effects of prenatal androgen deficiency on the developing hypothalamus. This theory was largely ignored by many American sexologists. A major challenge to the concept of the predominant influence of rearing on sexual identity and orientation came from the work of Imperato-McGinley et al. (256) on the psychosexual development of children with 5α-reductase deficiency. Genetic male children with this disorder are raised as females and then assume masculine roles at puberty. Although other investigators have questioned whether these children are raised unambiguously as females early in life (257), the work by this group rekindled interest in the search for biologic etiologies to gender identity and sexual orientation. A fascinating series of reports suggests that sexual orientation may have biologic correlates; however, most of these findings to date have not been verified by other investigations.

Biologic correlates of gender identity have received somewhat less study.

Before considering the biologic factors that influence gender identity and sexual orientation, it is important to review basic definitions and to emphasize the complexity of this subject. For humans, gender identity can be defined as the persistent belief that one is male, female, or ambivalent. Current DSM-IV classification divides gender identity disorder into four subtypes: (a) attracted to males; (b) attracted to females; (c) attracted to males and females; and (d) attracted to neither males nor females. As many as five boys for each girl are referred for gender identity disorder. An overall rough estimate of prevalence of gender identity disorder is 3% for boys and 1% for girls. The male-to-female ratio is roughly 4:1 in adults. Referrals for children usually are made in early grade school, whereas adults usually present in their mid 20's (258).

Current evidence suggests that sex assignment by parents play a major role in determining gender identity. Gender role may be defined as behavior that society designates as masculine or feminine. Current evidence suggests that gender role behavior is influenced by gonadal hormones during development (259). Sexual orientation refers to the erotic responsiveness of one individual to others of the same or opposite sex. The complexity of the area can be comprehended if one considers the situation where a genetic male is raised by his parents as a male, but who has a behavior pattern that society defines as extremely feminine or effeminate. Such a male raised in a subculture that rigidly defines sex-specific behavior patterns may have difficulty being accepted as masculine by his peers or as an attractive love object by the opposite sex. A similar male raised in a different subculture might have considerably less difficulty. One can further complicate the situation by assuming that one parent encouraged role-discrepant behavior and gender identity whereas the other did not. Clearly the interaction between psychological and biologic factors could be exceedingly complex to unravel because the manifestation of these early influences may not be evident until adult life.

Three basic strategies have been used in the investigation of biologic correlates of gender identity and sexual orientation: heritability, neuroendocrine, and neuroanatomic studies. The neuroendocrine and neuroanatomic studies are based on animal models of the effects of prenatal hormones on sexual behavior and sexual differentiation of the brain.

Heritability

A number of twin studies have suggested approximately a 50% concordance rate for homosexuality in monozygotic twins (248). Most of these studies are deficient in terms of sample size or lack systematic sampling of a well-specified population. Bailey and Pillard (260) and Bailey et al. (261) reported their work on concordance for homosexuality in co-twins. In a study of male homosexuality, probands were solicited in a homophile publication by advertising for homosexual or bisexual men with either co-twins or adoptive brothers. Probands were asked to rate the sexual orientation of the co-twin or adoptive brother. Fifty-two percent of the 56 monozygotic co-twins were either homosexual or bisexual compared with 22% of the 54 dizygotic co-twins and 11% of the 57 adoptive brothers. A similar methodology was used in a study of heritability of sexual orientation in women. In this study, 48% of monozygotic co-twins were homosexual or bisexual compared with 16% of dizygotic co-twins and 6% of adoptive sisters. Similar findings were reported by Whitam et al. (262) in 1993. All of these studies can be criticized for not having a representative sample of the homosexual population. Hamer et al. (263) later reported a pedigree and linkage analysis on 114 families of homosexual men. Index subjects, recruited from a human immunodeficiency virus (HIV) clinic and local homophile organizations, were asked to rate the sexuality of their fathers, sons, brothers, uncles, and male cousins. The highest rates of homosexuality were found in brothers, maternal uncles, and sons of maternal aunts, suggesting that some instances of homosexuality might be male limited and maternally inherited. To test this, they recruited families in which there were two homosexual brothers, no more than one lesbian relative, and no indication of direct father-to-son transmission of homosexuality. The sample for the linkage analysis consisted of 40 pairs of homosexual brothers together with mothers or siblings, if available. This study found a significant correlation between sexual orientation and the inheritance of genetic markers on chromosomal region Xq 28. The authors state that confirmation of their results is essential.

Two gene linkage studies add further weight to a possible genetic basis of male homosexuality showing a marker shared on the X chromosome inherited from the mother shared more often between two homosexual brothers and less often shared between heterosexual and homosexual brothers (264).

The available evidence is strongly suggestive of a genetic contribution to sexual orientation. The twin studies, despite possible sampling bias, are consistent in their findings. Pedigree studies must be confirmed. The need for more studies on the genetic basis of "gender identity" cannot be overemphasized.

Neuroendocrine Studies

Most neuroendocrine studies are based on the assumption that one can extrapolate from animal studies to sexual behavior in man. The underlying model differentiates the organizational aspects (enduring effects on the developing brain) from the activating (reversible) aspects of hormones. The basic assumption is that male heterosexuality, female homosexuality, and female-to-male transsexualism all result from prenatal exposure to high levels of testicular hormones. Male homosexuality, female heterosexuality, and male-to-

female transsexualism are postulated to result from lower levels of prenatal testicular hormones, thus retaining a female pattern of brain organization. Studies of gender identity and sexual orientation in syndromes involving prenatal androgen insensitivity or deficiency in males and studies of females with syndromes involving androgen excess have, by and large, not produced evidence consistent with the prenatal hormonal hypothesis (248). The other major approach has been to study hormonal feedback mechanisms. The underlying assumption in these studies is that androgens have an organizing effect on the developing brain, abolishing the ability of estrogen to exert positive feedback on luteinizing hormone (LH) release. Thus, if male homosexuality or male-to-female transsexualism is associated with a deficiency of prenatal androgenization, then these groups should demonstrate more positive feedback than heterosexuals. Several studies found that male homosexuals had LH responses to estrogen stimulation that were intermediate between those of women and those of heterosexual men (265,266). Similar findings were reported in male-to-female transsexuals. Subsequent studies have failed to find differences in the neuroendocrine regulation of LH secretion between transsexuals, homosexuals, and heterosexuals (267).

Neuroanatomic Studies

LeVay (268) reported that heterosexual men have greater volume of the third interstitial nucleus of the anterior hypothalamus compared with heterosexual females and homosexual men. Byne and Parsons (248) criticized this study on several grounds, the most important of which is that all of the brain tissue in homosexuals were from men who died of complications of acquired immunodeficiency syndrome. The other major issues of note are the small sample size (19 homosexual men, 16 heterosexual men, and 8 women) and the absence of replication. Byne and Parsons (248) also pointed out that there is no evidence that this area of the hypothalamus plays a critical role in generation of male-typical sexual behavior.

There has been a report that the suprachiasmatic nucleus of the hypothalamus is larger in homosexual than heterosexual men (269). The meaning of this report is unclear because the size of this nucleus does not vary between men and women, and there is no evidence that this nucleus regulates sexual behavior (261). This study has not been confirmed by an independent laboratory.

Allen and Gorski (270) reported that the anterior commissure is larger in its midsagittal area in women and homosexual men than in heterosexual men. This study also has not been replicated. This is especially important because the only other study of sex differences in the anterior commissure found a tendency for it to be larger in men than in women (261).

Other Etiologic Theories of Sexual Behavior

Psychosocial Theories

In an early study, boys with extensive mother–son symbiosis in their early years appeared later to manifest significant feminine behavior (271). Mothers who appeared detached or did not support their daughter's femininity were thought to predispose to female gender dysphoria. Traumatic loss events in early years have been associated with gender identity disorder.

Social Learning Theories

Social learning theories emphasize the importance of differential reinforcement in the normal or abnormal development of gender identity. In this construct the role of imitative behavior, vicarious learning to live up to a perceived role, and cognitive recognition of oneself as male or female before finding the behaviors associated with that cognitive picture all are recognized as part of the complex etiologic picture.

Psychodynamic Theory

The psychodynamic theory as advanced by Ethel Person and Lionel Ovesey posits that transsexualism in males originates from unresolved separation anxiety during the separation-individuation phase of infantile development. The child in this construct reverts to a reparative fantasy of symbiotic fusion with the missing parent to cope with this anxiety. Robert Stoller propagated another theory, which traced the origin of true transsexuals to the maternal grandmother who, being emotionally nonnurturant, imbued the mother of the male transsexual with "penis envy" (272). In the absence of a strong father figure this penis envy manifests itself in a strong mother–son symbiosis, with the son wanting to have his penis surgically removed in order to achieve true happiness.

Stress

The role of stress during pregnancy has been examined. Stressing pregnant rats was associated with feminized behavior in their male offspring. A possible explanation was that stress led to a mistiming of androgen secretion at critical times *in utero*. Another explanation was that there was competition between the adrenal stress steroids and the testicular androgens. It was noted that there was higher than average rate of homosexuality in male children born in Germany during the stressful years of World War II, between 1941 and 1946; however an environmental explanation can be inferred because the fathers of these children were away during these periods (264).

Immunologic Theory

An immunologic theory has been advanced from the observation that homosexual males (including males with gender dysphoria who are attracted to males only) are more likely to have more older brothers than nonhomosexual males. The assumption would be that the earlier nonhomosexual males successively primed antigenic reactions in the mother to testosterone, which then crossed the placenta in subsequent pregnancies to render the fetal testosterone less effective at the critical time; however, this is however because steroid hormones are not generally antigenic. Another more plausible explanation is the possible action on a Y-linked minor histocompatibility antigen (called H-Y antigen) that is primed with each subsequent pregnancy. In support of this theory is the finding that male mice whose mothers were immunized to H-Y antigen before pregnancy were less likely to mate with receptive females (264).

CONCLUSION

Our understanding of the neuropsychiatry of sexual dysfunction is in its infancy but has grown considerably in the last decade. One factor hindering the development of a coherent model of the biologic basis of sexual behavior is that the requisite knowledge base is spread across numerous disciplines. The adverse effects of both hypotensive and psychiatric drugs on sexual behavior, as well as the effects of specific neurologic lesions on sexual behavior, can be seen as an opportunity to deduce probable neurochemical and neuroanatomic substrates for such behavior. The shortage of literature concerning biologic treatment of sexual dysfunction attests to the need for further exploration. Fortunately, there is growing awareness that events and structures within the central nervous system are intimately associated with human sexual behavior. It is our hope that the field of neuropsychiatry will incorporate the study of human sexual behavior within its purview and outside the clear possibility of making it a distinct subspecialty.

REFERENCES

1. Frank E, Anderson C, Rubinstein D. Frequency of sexual dysfunction in normal couples. *N Engl J Med* 1978;299:111–115.
2. Nettelbladt P, Uddenberg N. Sexual dysfunction and sexual satisfaction in 58 married Swedish men. *Psychom Res* 1979;23:141–147.
3. Slag MF, Mortem JE, Elson MK. Impotence in medical clinical outpatients. *JAMA* 1983;249:1736–1740.
4. Segraves RT, Schoenberg NW. Diagnosis and treatment of erectile problems: current status. In: RT Segraves, NW Schoenberg, eds. *Diagnosis and treatment of erectile disturbances: a guide for clinicians.* New York: Plenum, 1985:1–22.
5. Papadopoulos C. *Sexual aspects of cardiovascular disease.* New York: Praeger, 1989.
6. Bancroft J. *Human sexuality and its problems.* London: Churchill Livingstone, 1983.

7. Lundberg PO, Brattberg A, Hulter, B. Sexual dysfunction in patients with neurological and neuroendocrine disorders. In: Bezemer W, Cohen-Kettenis P, Slob K, et al., eds. *Sex matters.* Amsterdam: Excerpta Medica, 1992:281–284.

8. Hawton K. Sexual dysfunctions and psychiatric disorders. In: Bezemer W, Cohen-Kettenis P, Slob K, et al., eds. *Sex matters.* Amsterdam: Excerpta Medica, 1992:79–84.

9. Kaufman SA.The gynecologic evaluation of female excitement disorders. In: Kaplan HS, ed. *The evaluation of sexual disorders: psychological and medical aspects.* New York: Brunner/Mazel, 1983:122–127.

10. Rivard DJ.Anatomy, physiology, and neurophysiology of male sexual function. In: Bennett AH, ed. *Management of male impotence.* Baltimore: Williams & Wilkins, 1982:1–25.

11. Melman A.Neural and vascular control of education. In: Rosen RC, Leiblum SR, eds. *Erectile disorders: assessment and treatment.* New York: Guilford, 1992:55–71.

12. Benson GS, McConnell J, Lipshultz LI, et al. Neuromorphology and neuropharmacology of the human penis. *J Clin Invest* 1980;65:605–612.

13. Sjostrand NO, Klinge E. Principal mechanisms controlling penile retraction and protrusion in rabbits. *Acta Physiol Scand* 1979; 106:199–214.

14. Segraves RT. Effects of psychotropic drugs on human erection and ejaculation. *Arch Gen Psychiatry* 1989;46:275–284.

15. Kaufman SA. The gynecologic evaluation of female orgasm disorders. In: Kaplan HS, ed. *The evaluation of sexual disorders: psychological and medical aspects.* New York: Brunner/Mazel, 1983:117–121.

16. Bors E, Comarr AE. Neurological disturbances of sexual function with special reference to 529 patients with spinal cord injury. *Urol Surv* 1960;10:191–222.

17. MacLean PD.Brain mechanisms of primal sexual functions and related behavior. In: Sandler M, Gessa GL, eds. *Sexual behavior: pharmacology and biochemistry.* New York: Raven Press, 1975:1–12.

18. Dua S, MacLean PD. Localization for penile erection in medial frontal lobe. *Am J Physiol* 1964;207:1425–1434.

19. MacLean PD, Ploog DW. Cerebral representation of penile erection. *J Neurophysiol* 1962;25:29–55.

20. Bell C. Autonomic control of reproduction circulatory and other factors. *Pharmacol Rev* 1972;24:657–736.

21. Reckler JM. The urologic evaluation of ejaculatory disorders (male orgasm disorders, RE and PE). In: Kaplan HS, ed. *The evaluation of sexual disorders: psychological and medical aspects.* New York: Brunner/Mazel, 1983:139–149.

22. Kleeman FJ. The physiology of the internal urinary sphincter. *J Urol* 1970;104:549–554.

23. Robinson PM. A cholinergic component in the innervation of the longitudinal smooth muscle of the guinea pig vas deferens. *J Cell Biol* 1969;41:462–476.

24. Blumer D, Walker EA. The neural basis of sexual behavior. In: Besson DF, Blumer D, eds. *Psychiatric aspects of neurologic disease.* New York: Grune & Stratton, 1975:199–217.

25. Heath RG. Electrical self-stimulation of the brain in man. *Am J Psychiatry* 1963;120:571–577.

26. Heath RG. Brain function and behavior. *J Nerv Ment Dis* 1975; 160:159–175.

27. Heath RG. Pleasure and brain activity in man: deep and surface electroencephalograms during orgasm. *J Nerv Ment Dis* 1972;154:3–18.

28. Penfield W, Rasmussen T. The cerebral cortex of man. New York: MacMillan, 1950.

29. Smith BH, Khatri AM. Cortical localization of sexual feeling. *Psychosomatics* 1979;20:771–776.

30. Gessa GL, Tagliamonte A. Role of brain monoamines in male sexual behavior. *Life Sci* 1974;14:425–436.

31. Foreman MM, Hall JL. Effects of D_2-dopaminergic receptor stimulation on male rat sexual behavior. *J Neural Transm* 1987;68: 153–170.

32. Hasen S. Spiral control of sexual behavior: effects of intrathecal administration of lisuride. *Neurosci Lett* 1982;33:329–332.

33. Tucker TC, Ale SE.Serotonin and sexual behavior. In: Wheatley D, ed. *Psychopharmacology and sexual disorders.* New York: Oxford University Press, 1983:22–49.

34. Whalen RE, Luttge WG. Para-chlorophenylalanine methylester: an aphrodisiac. *Science* 1970;169:1000–1001.

35. Rodriquez M, Castro R, Hernandez G, et al. Different roles of catecholaminergic and serotoninergic neurons of the medial forebrain bundle on male rat sexual behavior. *Physiol Behav* 1984;33:5–11.

36. Meston CM. Aging and sexuality. *Arch Sex Behav* 1984;13:341–349.

37. van Lunsen RH, Laan E. Hormones and sexuality in postmenopausal women. *J Psychosom* 1998;13(Suppl 6):515–522.

38. Graziottin A. Libido: the biologic scenario. *Int Clin Psychopharmacol* 1998l;13[Suppl 6]:S1–S4.

39. Sadock BJ, Sadock VA. Normal human sexuality and sexual and gender identity disorders. In: *Kaplan and Sadock's comprehensive textbook of psychiatry,* vol. 1, 7th ed. Philadelphia: Lippincott Williams & Wilkins, 2000.

40. Meston CM, Frohlich PF. Neurobiology of sexual dysfunction. *Arch Gen Psychiatry* 2000;57:1012–1030.

41. Meston CM. Ephedrine-activated physiological sexual arousal in women. *Arch Gen Psychiatry* 1998;55:7652–7656.

42. Segraves RT.Hormones and libido. In: Leiblum SR, Rosen RC, eds. *Sexual desire disorders.* London: Guilford, 1988:271–312.

43. Luttge WB. The role of gonadal hormones in the sexual behavior of the rhesus monkey and human: a literature survey. *Arch Sex Behav* 1971;1:61–68.

44. Davidson JM, Rosen RC.Hormonal determinants of erectile function. In: Rosen RC, Leiblum SR, eds. *Erectile disorders: assessment and treatment.* New York: Guilford, 1992:72–95.

45. Bremer J.*Asexualization: a follow-up study of 244 cases.* New York: MacMillan, 1959.

46. Luisi M, Franchi F. Double-blind comparison study of testosterone undecanoate and mesterolone in hypogonadal male patients. *J Endocrinol Invest* 1980;3:305–308.

47. Clayton AH, Clavet GJ, McGarvey EL, et al. Assessment of sexual functioning during menstrual cycle. *Sex Marital Ther* 1999; 25:281–291.

48. Slob AK, Bax CM, Hop WC, et al. Sexual arousability and menstrual cycle. *J Psychoneuroendocrinol* 1996;21:545–558.

49. Harvey SM. Female sexual behavior: fluctuations during menstrual cycle. *Psychosom Res* 1987;31:101–110.

50. Meuwissen I, Over R. Sexual arousal across phases of the human menstrual cycle. *Arch Sex Behav* 1992;21:101–119.

51. Slob AK, Ernste M, van der Werff ten Bosch JJ. Menstrual cycle phase and sexual arousability in women. *Arch Sex Behav* 1991;20:567–577.

52. Schreiner-Engel P, Schiavi RC, Smith H, et al. Sexual arousability and the menstrual cycle. *Psychosom Med* 1981;43:199–214.

53. Backstrom T, Sanders D, Leask R, et al. Mood, sexuality, hormones, and the menstrual cycle. II. Hormone levels and their relationship to the premenstrual syndrome. *Psychosom Med* 1983;45:503–507.

54. Bancroft J, Sanders D, Davidson D, et al. Mood, sexuality, hormones and the menstrual cycle. III. Sexuality and the role of androgens. *Psychosom Med* 1983;45:509–516.

55. Sanders D, Warner P, Backstrom T, et al. Mood, sexuality,

hormones and the menstrual cycle. I. Changes in mood and physical state: description of subjects and method. *Psychosom Med* 1983;15:487–501.

56. Persky H, Lief HI, Strauss D, et al. Plasma testosterone level and sexual behavior of couples. *Arch Sex Behav* 1978;7:157–173.

57. Sherwin BB, Gelfand MM. Differential symptom response to parental estrogen and/or androgen administration in surgical menopause. *Am J Obstet Gynecol* 1985;151:153–160.

58. Sherwin BB, Gelfand MM, Brender W. Androgen enhances sexual masturbation in females: a prospective crossover study of sex steroid administration in the surgical menopause. *Psychosom Med* 1985;47:339–351.

59. Segraves RT, Madsen R, Carter CS, et al.Erectile dysfunction with pharmacological agents. In: Segraves RT, Schoenberg HW, eds. *Diagnosis and treatment of erectile disturbances: a guide for clinicians.* New York: Plenum Press, 1985:23–64.

60. Sadock BJ, Sadock VA. Normal human sexuality and sexual and gender identity disorders. In: *Kaplan and Sadock's comprehensive textbook of psychiatry,* vol. 1, 7th ed. Philadelphia: Lippincott Williams & Wilkins, 2000.

61. Collaborative Working Group on Clinical Trial Evaluations. Adverse effects of the atypical antipsychotics. *J Clin Psychiatry* 1998;59[Suppl 12]:17–22.

62. Segraves RT. Sexual side-effects of psychiatric drugs. *Int J Psychiatry Med* 1988;18:243–252.

63. Segraves RT, Segraves KB.Aging and drug effects on male sexuality. In: Rosen SR, Leiblum SR, eds. *Erectile disorders: assessment and treatment.* New York: Guilford Press, 1992:96–138.

64. Segraves RT.Drugs and desire. In: Leiblum SR, Rosen RC, eds. *Sexual desire disorders.* New York: Guilford Press, 1988:313–347.

65. Greenberg HR. Inhibition of ejaculation by chlorpromazine. *J Nerv Ment Dis* 1971;152:364–366.

66. Ananth J. Impotence associated with pimozide. *Am J Psychiatry* 1982;139:1374.

67. Charalampous KD, Freemesser GF, Maleu J, et al. Loxapine succinate: a controlled double-blind study in schizophrenia. *Curr Ther Res* 1974;16:829–837.

68. Sandison RA, Whitelaw E, Currie JDC. Clinical trials with Mellaril (TPZI) in the treatment of schizophrenia. *J Ment Sci* 1960;106:732–741.

69. Haider I. Thioridazine and sexual dysfunctions. *Int J Neuropsychiatry* 1966;2:255–257.

70. Witton K. Sexual dysfunction secondary to Mellaril. *Dis Nerv Syst* 1962;23:175.

71. Weizman A, Maoz B, Treves I, et al. Sulpiride-induced hyperprolactinemia and impotence in male psychiatric outpatients. *Prog Neuropsychopharmacol Biol Psychiatry* 1985;9:193–198.

72. Meco G, Falachi P, Casacchia M, et al. Neuroendocrine effects of haloperidol decanoate in patients with chronic schizophrenia. In: Kemal D, Ragagni G, eds. *Chronic treatments in neuropsychiatry.* New York: Raven Press, 1985:88–93.

73. Bartholomew AA. A long-acting phenothiazine as a possible agent to control deviant sexual behavior. *Am J Psychiatry* 1968; 124:917–922.

74. Tennett G, Bancroft J, Cass J. The control of deviant sexual behavior by drugs: a double-blind controlled study of benperidol, chlorpromazine, and placebo. *Arch Sex Behav* 1974;3:261–271.

75. Barnes TR, McPhillips MA. Critical analysis and comparison of the side-effect and safety profiles of the new antipsychotics. *Br J Psychiatry Suppl* 1999;38:34–43.

76. Collaborative Working Group on Clinical Trial Evaluations. Adverse effects of the atypical antipsychotics. *J Clin Psychiatry* 1998;59[Suppl 12]:17–22.

77. Tran PV, Hamilton SH, Kuntz AJ, et al. Double-blind comparison of olanzapine versus risperidone in the treatment of schizophrenia and other psychotic disorders. *J Clin Psychopharmacol* 1997;17:407–418.

78. Riley AJ, Riley EJ. The effect of single dose diazepam on female sexual response induced by masturbation. *Sex Marital Ther* 1986;1:49–53.

79. Hughes JM. Failure to ejaculate with chlordiazepoxide. *Am J Psychiatry* 1964;121:610–611.

80. Segraves RT. Treatment of premature ejaculation with lorazepam. *Am J Psychiatry* 1987;144:1240.

81. Uhde TW, Tancer ME, Shea CA. Sexual dysfunction related to alprazolam treatment of social phobia. *Am J Psychiatry* 1988;145:531–532.

82. Sangal R. Inhibited female orgasm as a side-effect of alprazolam. *Am J Psychiatry* 1985;142:1223–1224.

83. Munjack DJ, Crocker B. Alprazolam-induced ejaculatory inhibition. *J Clin Psychopharmacol* 1986;6:57–58.

84. Blay SL, Ferraz MPT, Calil HM. Lithium-induced male sexual impairment: two case reports. *J Clin Psychiatry* 1982;43:497–498.

85. Vinarova E, Uhlir O, Stika L. Side-effects of lithium administration. *Activ Nerv Superior (Praha)* 1972;14:105–107.

86. Segraves RT. Sexual dysfunction complicating the treatment of depression. *J Clin Psychiatry (Monogr Ser)* 1992;1(vol. 10):75–79.

87. Harrison WM, Rabkin JG, Ehrhardt AA, et al. Effects on antidepressant medication on sexual function: a controlled study. *J Clin Psychopharmacol* 1986;6:144–149.

88. Sadock BJ, Sadock VA. Normal human sexuality and sexual and gender identity disorders. In: *Kaplan and Sadock's comprehensive textbook of psychiatry,* vol. 1, 7th ed. Philadelphia: Lippincott Williams & Wilkins, 2000.

89. Waldinger MD, Olivier B. Selective serotonin reuptake inhibitor-induced sexual dysfunction: clinical and research considerations. *West J Med* 1997;167:285–290.

90. Hindmarch I. The behavioral toxicity of antidepressants: effects on cognition and sexual function [published erratum appears in *Qual Life Res* 1997;6:606]. *Qual Life Res* 1996;5:81–90.

91. Kraupl-Taylor F. Loss of libido in depression [Letter]. *Br Med J* 1972;1:305.

92. Jacobsen FM. Fluoxetine-induced sexual dysfunction and an open trial of yohimbine. *J Clin Psychiatry* 1992;53:119–122.

93. Gardner EA, Johnston JA. Bupropion: an antidepressant without sexual pathophysiological action. *J Clin Psychopharmacol* 1985;5:24–29.

94. Kowalski A, Stanley RO, Dennerstein L, et al. The sexual side-effects of antidepressant medication: a double-blind comparison of two antidepressants in a non-psychiatric population. *Br J Psychiatry* 1985;147:413–418.

95. Hollander MH, Ban TA. Ejaculation retardation due to phenelzine. *Psychiatr J Univ Ottawa* 1980;4:233–234.

96. Schwarcz G. Case report of inhibition of ejaculation and retrograde ejaculation as side effects of amoxapine. *Am J Psychiatry* 1982;139:233–234.

97. Nininger JE. Inhibition of ejaculation by amitriptyline. *Am J Psychiatry* 1978;135:750–751.

98. Monteiro WO, Noshirvani HF, Marks IM, et al. Anorgasmia from clomipramine in obsessive-compulsive disorder: a controlled trial. *Br J Psychiatry* 1987;151:107–112.

99. Glass RM. Ejaculatory impairment from both phenelzine and imipramine, with tinnitus from phenelzine. *J Clin Psychopharmacol* 1981;3:152–154.

100. Goldbloom DS, Kennedy SH. Adverse interaction of fluoxetine and cyproheptadine in two patients with bulimia nervosa. *J Clin Psychiatry* 1991;52:261–262.

101. Jones SD. Ejaculatory inhibition with trazodone. *J Clin Psychopharmacol* 1984;4:279–281.

102. Reimherr FW, Chouinard G, Cohn CK. Antidepressant efficacy of sertraline: a double-blind, placebo and amitriptyline-controlled multicenter comparison study in outpatients with major depression. *J Clin Psychiatry* 1990;51[Suppl B]:18–27.

103. Zajecka J, Fawcett J, Schaff M, et al. The role of serotonin in sexual dysfunction: fluoxetine-associated orgasm dysfunction. *J Clin Psychiatry* 1991;52:66–68.

104. Segraves RT. Reversal by bethanechol of imipramine-induced ejaculatory dysfunction [Letter]. *Am J Psychiatry* 1987; 144:1243–1244.

105. Steele TE, Howell EF. Cyproheptadine for imipramine-induced anorgasmia [Letter]. *J Clin Psychopharmacol* 1986;6:326–327.

106. Price J, Grunhaus LJ. Treatment of clomipramine-induced anorgasmia with yohimbine: a case report. *J Clin Psychiatry* 1990;51:32–33.

107. McLean JD, Forsythe RG, Kapkin IA. Unusual side effects of clomipramine associated with yawning. *Can J Psychiatry* 1983;28:569–570.

108. Modell JG. Repeated observations of yawning, clitoral engorgement, and orgasm associated with fluoxetine administration. *J Clin Psychopharmacol* 1989;9:63–65.

109. Lal S, Rios O, Thavundayil JX. Treatment of impotence with trazodone: a case report. *J Urol* 1990;143:819–820.

110. Fletcher A, Bulpitt C. Antihypertensive medication and sexual dysfunction. In: Bezemer W, Cohen-Kettenis P, Slob K, et al., eds. *Sex matters.* Amsterdam: Excerpta Medica, 1992:265–273.

111. Yendt RE, Gray GF, Garcia DA. The use of thiazides in the prevention of renal calculi. *Can Med Assoc J* 1970;102:614–620.

112. Curd JD, Borhani NO, Blaszkowski TP, et al. Long-term surveillance for adverse effects of antihypertensive drugs. *JAMA* 1985;253:3263–3268.

113. Spark RF, Melby JC. Aldosteronism in hypertension. *Ann Intern Med* 1968;69:685–691.

114. Loriaux DL, Menard R, Taylor A, et al. Spironolactone and endocrine dysfunction. *Ann Intern Med* 1976;85:630–636.

115. Kostis JB, Rosen RC, Holzer BC, et al. CNS side effects of centrally-active antihypertensive agents: a prospective, placebo-controlled study of sleep, mood state, cognitive and sexual function in hypertensive males. *Psychopharmacology* 1990;102:163–170.

116. Croog SH, Levine S, Sudilovsky A, et al. Sexual symptoms in hypertensive patients: a clinical trial of antihypertensive medications. *Arch Intern Med* 1988;148:788–794.

117. Fraunfelder FT, Meyer SM. Sexual dysfunction secondary to topical ophthalmic timolol. *JAMA* 1985;253:3092–3093.

118. Jensen RT, Collen MJ, Pandol SJ, et al. Cimetidine-induced impotence and breast changes in patients with gastric hypersecretory states. *N Engl J Med* 1983;308:883–887.

119. Neri A, Aygen M, Zuckerman Z, et al. Subjective assessment of sexual dysfunction of patients on long-term administration of digoxin. *Arch Sex Behav* 1980;9:343–347.

120. Neri A, Zuckerman Z, Aygen M, et al. The effect of long-term administration of digoxin or plasma androgens and sexual dysfunction. *J Sex Marital Ther* 1987;13:58–63.

121. Mattson RH, Cramer JA. Epilepsy, sex hormones, and antiepileptic drugs. *Epilepsia* 1985;26[Suppl 1]:540–551.

122. Ivers R, Goldstein N. Multiple sclerosis: a current appraisal of signs and symptoms. *Proc Mayo Clin* 1963;38:457–466.

123. Vas C. Sexual impotence and some autonomic disturbances in men with multiple sclerosis. *Acta Neurol Scand* 1969;45:166–182.

124. Lilius H, Valtoren E, Wilkstrom J. Sexual problems in patients suffering from multiple sclerosis. *J Chronic Dis* 1976;29:643–647.

125. Szasz G, Paty D, Maurice WL. Sexual dysfunctions in multiple sclerosis. *Ann N Y Acad Sci* 1984;436:443–452.

126. Valleroy ML, Kraft GH. Sexual dysfunction in multiple sclerosis. *Arch Phys Med Rehabil* 1984;65:125–128.

127. Huns R, Shubsachs AP, Taylor PJ. Hypersexuality, fetishism and multiple sclerosis. *Br J Psychiatry* 1991;158:280–281.

128. Jensen SB. Diabetic sexual dysfunction: a comparative study of 160 insulin treated diabetic men and women and an age-matched control group. *Arch Sex Behav* 1981;10:493–504.

129. Nathan D, Singer DE, Godine JE. Insulin dependent diabetes in older patients. *Am J Med* 1986;81:837–842.

130. McCulloch DK, Campbell IW, Wu FC, et al. The prevalence of diabetic impotence. *Diabetiologia* 1980;18:279–283.

131. Fairburn CG, Wu FCW, McCulloch DK, et al. The clinical features of diabetic impotence: a preliminary study. *Br J Psychiatry* 1982;140:447–452.

132. Kolodny RC, Kahn CB, Goldstein HH, et al. Sexual dysfunction in diabetic men. *Diabetes* 1973;23:306–309.

133. Fairburn CG, McCulloch DK, Wu FC. The effects of diabetes on male sexual function. *Clin Endocrinol Metab* 1982;11:749–767.

134. McCulloch DK, Young RJ, Prescott RJ, et al. The natural history of impotence in diabetic men. *Diabetologia* 1984;26:437–440.

135. Schiavi RC, Fisher C, Quadland M, et al. Nocturnal penile tumescence evaluation of erectile function in insulin-dependent diabetic men. *Diabetologia* 1985;28:90–94.

136. House WC, Pendleton L. Sexual functioning in male diabetics with impotence problems. *Sex Marital Ther* 1988;3:205–212.

137. Bancroft J, Bell C, Ewing DJ, et al. Assessment of erectile function in diabetic and non-diabetic impotence by simultaneous recording of penile diameter and penile arterial pulse. *J Psychosom Res* 1985;29:315–334.

138. Karacan I. Diagnosis of erectile impotence in diabetes mellitus. *Ann Intern Med* 1980;92[Pt 2]:334–337.

139. Quadri R, Veglio M, Flecchia D, et al. Autonomic neuropathy and sexual impotence in diabetic patients. *Andrologia* 1989;21:346–352.

140. Nofzinger EA, Schmidt HS. An exploration of central dysregulation of erectile function as a contributing cause of diabetic impotence. *J Nerv Ment Dis* 1990;178:90–95.

141. Kolodny RC. Sexual dysfunction in diabetic females. *Diabetes* 1971;20:557–559.

142. Prather RC. Sexual dysfunction in the diabetic female: a review. *Arch Sex Behav* 1988;17:277–284.

143. Jensen SB. Sexual dysfunction in younger insulin-treated diabetic females. *Diabetes Metab* 1985;11:278–282.

144. Newman AS, Bertelson AD. Sexual dysfunction in diabetic women. *J Behav Med* 1986;9:261–270.

145. Bancroft J. Sexuality of diabetic women. *Clin Endocrinol Metab* 1982;11:785–789.

146. Schreiner-Engel P, Schiavi RC, Vietorisz D, et al. The differential impact of diabetes type on female sexuality. *J Psychosom Res* 1987;31:23–33.

147. Yalla SV. Sexual dysfunction in the paraplegic and quadriplegic. In: Bennett AH, ed. *Management of male impotence.* Baltimore: Williams & Wilkins, 1994:181–191.

148. Comarr AE, Vigue M. Sexual counseling among male and female patients with spinal cord and/or cauda equina injury. *Am J Phys Med* 1978;57:215–227.

149. Berard EJJ. The sexuality of spinal cord injured women: physiology and pathophysiology: a review. *Paraplegia* 1989;27:99–112.

150. Rose SS. An investigation into sterility after lumbar ganglionectomy. *Br Med J* 1953;1:247–250.

151. Schover LR, Von Eschbach AC. Sexual and marital relationships

after treatment for nonseminomatous testicular cancer. *Urology* 1985;25:251–255.

152. Aso R, Yasutom M. Urinary and sexual disturbances following radical surgery for rectal cancer and pudendal nerve block as a countermeasure for urinary disturbance. *Am J Proctol* 1974;25:60–74.
153. Hellstrom P. Urinary and sexual dysfunction after rectosigmoid surgery. *Ann Chir Gynaecol* 1988;77:51–56.
154. May AG, DeWeese J, Rob CG. Changes in sexual function following operation of the abdominal aorta. *Surgery* 1969;65:41–47.
155. Ohshiro T, Kosaki G. Sexual function after aorto-iliac vascular reconstruction. *J Cardiovasc Surg* 1984;25:47–50.
156. Hargreave TB, Stephenson TP. Potency and prostatectomy. *Br J Urol* 1977;49:683–688.
157. Segraves RT, Segraves KB. Medical aspects of orgasm disorders. In: O'Donohue W, Geer JH, eds. *Handbook of sexual dysfunctions: assessment and treatment.* Boston: Allyn & Bacon, 1993:225–252.
158. Morrell MJ. Sexual dysfunction in epilepsy. *Epilepsia* 1991;32 [Suppl 6]:538–545.
159. Inthaler S, Donati F, Pavlincova E, et al. Partial complex epileptic seizures with ictal urogenital manifestations in a child. *Neurology* 1991;31:212–215.
160. Remillard GM, Andermann F, Testa GF, et al. Sexual ictal manifestations predominate in women with temporal lobe epilepsy: a finding suggesting sexual dimorphism in the human brain. *Neurology* 1983;33:323–330.
161. Backman DS, Rossel CW. Orgasmic epilepsy. *Neurology* 1984;34:559–560.
162. Warneke LB. A case of temporal lobe epilepsy with an orgasmic component. *Can Psychiatr Assoc J* 1976;21:319–324.
163. Blumer D. Hypersexual episodes in temporal lobe epilepsy. *Am J Psychiatry* 1970;126:1099–1106.
164. Kolarsky A, Freund K, Machek J, et al. Male sexual deviation: association with early temporal lobe damage. *Arch Gen Psychiatry* 1967;17:735–743.
165. Mitchell W, Falconer MA, Hill D. Epilepsy with fetishism relieved by temporal lobectomy. *Lancet* 1954;2:626–630.
166. Davies BM, Morgenstern FS. A case of cysticercosis, temporal lobe epilepsy, and transvestism. *J Neurol Neurosurg Psychiatry* 1960;23:247–249.
167. Taylor DC. Appetitive inadequacy in the sex behavior of temporal lobe epileptics. *J Neurovisc Rel* 1971;[Suppl 10]:486–490.
168. Hierons R. Impotence in temporal lobe lesions. *J Neurovisc Rel* 1971;[Suppl 10]:477–481.
169. Hierons R, Saunders M. Impotence in patients with temporal-lobe lesions. *Lancet* 1966;2:761–764.
170. Shukla GD, Srivastava ON, Katiyar BC. Sexual disturbances in temporal lobe epilepsy: a controlled study. *Br J Psychiatry* 1979;134:288–292.
171. Saunders M, Rawson M. Sexuality in male epileptics. *J Neurol Sci* 1970;10:577–583.
172. Toone BK, Edem J, Nanjee MN, et al. Hyposexuality and epilepsy: a community survey of hormonal and behavioral changes in male epileptics. *Psychol Med* 1989;19:937–943.
173. Demerdash A, Shaalan M, Midani A, et al. Sexual behavior of a sample of females with epilepsy. *Epilepsia* 1991;32:82–85.
174. Mancall EL, Alonso RJ, Marlowe WB. Sexual dysfunction in neurological disease. In: Segraves RT, Schoenberg HW, eds. *Diagnosis and treatment of erectile disturbances.* New York: Plenum, 1985:65–86.
175. Herzog AG. Reproductive endocrine considerations and hormonal therapy for men with epilepsy. *Epilepsia* 1991;32[Suppl 6]:534–537.
176. Ramsay RE, Slater JD. Effects of antiepileptic drugs on hormones. *Epilepsia* 1991;32[Suppl 6]:560–567.
177. Brown RG, Jahan Shahi M, Quinn N, et al. Sexual function in patients with Parkinson's disease and their partners. *J Neurol Neurosurg Psychiatry* 1990;53:480–486.
178. Esibill N. Impact of Parkinson's disease on sexuality. *Sex Disabil* 1983;6:120–125.
179. Lipe H, Longstreth WT, Bird TD, et al. Sexual function in married men with Parkinson's disease compared to married men with arthritis. *Neurology* 1990;40:1347–1349.
180. Uiti R, Tanner CM, Rajput AH, et al. Hypersexuality with antiparkinsonian therapy. *Neuropharmacology* 1989;12:375–383.
181. Kalliomaki JL, Markkanen TK, Mustonen VA. Sexual behavior after cerebral vascular accident. *Fertil Steril* 1961;12:156–158.
182. Ford AB, Orfirer AP. Sexual behavior and the chronically ill patient. *Med Aspects Hum Sex* 1967;1:51–61.
183. Goddess ED, Wagner NN, Silverman DR. Post-stroke sexual activity of CVA patients. *Med Aspects Hum Sex* 1979;13:16–30.
184. Hawton K. Sexual adjustment of men who have had strokes. *J Psychosom Res* 1984;28:243–249.
185. Monga TN, Lawson JS, Inglis J. Sexual dysfunction in stroke patients. *Arch Phys Med Rehabil* 1986;67:19–22.
186. Bray GP, DeFrank RS, Wolfe TL. Sexual functioning in stroke patients. *Arch Phys Med Rehabil* 1981;62:286–288.
187. Sjogren K, Fugl-Meyer AR. Adjustment to life after stroke with special reference to sexual intercourse and leisure. *J Psychosom Res* 1982;26:409–417.
188. Monga TN, Monga M, Raina MS, et al. Hypersexuality in stroke. *Arch Phys Med Rehabil* 1986;67:415–417.
189. Kreutzer JS, Zasler ND. Psychosexual consequences of traumatic brain injury: methodology and preliminary findings. *Brain Inj* 1989;3:177–186.
190. Lusk MD, Kott JA. Effects of head injury on libido. *Med Aspects Hum Sex* 1982;16:22–30.
191. Isern RD. Family violence and the Kluver-Bucy syndrome. *South Med J* 1987;80:373–377.
192. Gorman DG. Hypersexuality following septal injury. *Arch Neurol* 1992;49:308–310.
193. Lundberg PO. Sexual dysfunction in patients with neurological disorders. In: Gemme R, Wheeler CC, eds. *Progress in sexuality.* New York: Plenum, 1977:129–139.
194. Zeiss AM, Davies HD, Wood M, et al. The incidence and correlates of erectile problems in patients with Alzheimer's disease. *Arch Sex Behav* 1990;19:325–331.
195. Janati A. Kluver-Bucy syndrome in Huntington's chorea. *J Nerv Ment Dis* 1985;173:632–635.
196. Offit AK. Psychiatric disorders and sexual functioning. In: Karasu TB, ed. *Treatments of psychiatric disorders, vol. 3.* Washington, DC: American Psychiatric Association, 1989:2253–2263.
197. Mathew RJ, Weinman ML. Sexual dysfunctions in depression. *Arch Sex Behav* 1982;11:323–328.
198. Casper RC, Redmond E, Katz MM, et al. Somatic symptoms in primary effective disorder. *Arch Gen Psychiatry* 1985;42:1098–1104.
199. Tamburello A, Seppecher MF. The effects of depression on sexual behavior: preliminary results of research. In: Gemme R, Wheeler CC, eds. *Progress in sexology.* New York: Plenum, 1977:107–128.
200. Kivela SL, Pahkala K, Eronen A. Depressive symptoms and signs that differentiate major and atypical depression from dysthymic disorder in elderly Finns. *Int J Geriatr Psychiatry* 1989;4:79–85.
201. Kivela SL, Pahkala K. Symptoms of depression in old people in Finland. *Z Gerontol* 1988;21:257–263.
202. Kivela SL, Pahkala K. Clinician-rated symptoms and signs of depression in aged Finns. *Int J Soc Psychiatry* 1988;34:274–284.

203. Roose SP, Glassman AH, Walsh BT, et al. Reversible loss of nocturnal penile tumescence during depression: a preliminary report. *Neuropsychobiology* 1982;8:284–288.
204. Thase ME, Reynolds CF, Jennings JR, et al. Nocturnal penile tumescence diminished in depressed men. *Biol Psychiatry* 1988;24:33–46.
205. Schreiner-Engel P, Schiavi RC. Lifetime psychopathology in individuals with low sexual desire. *J Nerv Ment Dis* 1986;174:646–651.
206. Angst J. Sexual problems in healthy and depressed persons. *Eur J Contracept Reprod Health Care* 1997;2:247–51.
207. Graziottin A. The biological basis of female sexuality. *Int J Clin Psychopharmacol* 1998;13[Suppl 6]:S15–S22.
208. Friedman S, Harrison G. Sexual histories, attitudes, and behavior of schizophrenic and normal women. *Arch Sex Behav* 1984;13:555–567.
209. Raboch J. The sexual development and life of female schizophrenic patients. *Arch Sex Behav* 1984;13:341–349.
210. Raboch J. The sexual development and life of female schizophrenic patients. *Int J Clin Psychopharmacol* 1998;13[Suppl 6]:S5–S8.
211. Friedman S, Harrison G. Sexual history, attitudes, and behavior of schizophrenic and "normal" women. *Arch Sex Behav* 1984;14:555–567.
212. Aizenberg D, Zemishlany Z, Dorfman-Etrog P, et al. Sexual dysfunction in male schizophrenic patients. *J Clin Psychiatry* 1995;56:137–141.
213. Raboch J, Faltus F. Sexuality of women with anorexia nervosa. *Acta Psychiatr Scand* 1991;84:9–11.
214. Rosenheim E, Neumann M. Personality characteristics of sexually dysfunctional males and their wives. *J Sex Res* 1981;17:124–138.
215. Schiavi RC. Psychological determinants of erectile disorders. *Sex Disabil* 1981;4:86–92.
216. Segraves KB, Segraves RT. Diagnosis of female arousal disorder. *Sex Marital Ther* 1991;6:9–13.
217. Segraves KAB, Segraves RT. Differentiation of biogenic and psychogenic impotence by sexual symptomatology. *Arch Sex Behav* 1987;16:125–137.
218. Segraves RT. Discrimination of psychogenic from organogenic impotence with psychometric instruments. *Sex Disabil* 1987;8:138–142.
219. Schiavi RC. Laboratory methods for evaluating erectile dysfunction. In: Rosen RC, Leiblum SR, eds. *Erectile disorders: assessment and treatment.* New York: Guilford, 1992:141–170.
220. Moore C. Evaluation of sexual disorders. In: Karasu TB, ed. *Treatment of psychiatric disorders, vol. 3.* Washington, DC: APA Press, 1989:2238–2247.
221. Melman A, Kaplan D, Redfield J. Evaluation of the first 70 patients in the Center for Male Sexual Dysfunction of Beth Israel Medical Center. *J Urol* 1984;131:53–55.
222. Buvat JJ, Lemaire A, Dehaene JL, et al. Venous incompetence: critical study of the organic basis of high maintenance flow rates during artificial erection test. *J Urol* 1986;135:926–928.
223. Lue TF, Hricak H, Marich KW, et al. Evaluation of vasculogenic impotence with high-resolution ultrasonography and pulse Doppler spectrum analysis. *Radiology* 1985;155:777–781.
224. Ertekin C, Reel F. Bulbocavernosus reflex in normal men and in patients with neurogenic bladder and/or impotence. *J Neurol Sci* 1976;28:1–15.
225. Segraves RT. Individual and couple therapy. In: Karasu TB, ed. *Treatments of psychiatric disorders, vol. 3.* Washington, DC: APA Press, 1989:2334–2342.
226. Levine SB. Hypoactive sexual desire and other problems of sexual desire. In: Karasu TB, ed. *Treatments of psychiatric disorders, vol. 3.* Washington, DC: APA Press, 1989:2264–2278.
227. Mellgren A. Treatment of ejaculation praecox with thioridazine. *Psychother Psychosom* 1967;15:454–460.
228. Singh H. Therapeutic use of thioridazine in premature ejaculation. *Am J Psychiatry* 1963;119:891.
229. Simpson GM, Blair JH, Amvsso D. Effect of antidepressants on genital urinary function. *Dis Nerv Syst* 1965;26:787–789.
230. Eaton H. Clomipramine (Anafranil) in the treatment of premature ejaculation. *J Int Med Res* 1973;1:432–434.
231. Segraves RT, Saran A, Segraves K, et al. Clomipramine vs. placebo in the treatment of premature ejaculation. Poster presented at New Clinical Drug Evaluation Unit Conference, Boca Raton, Florida, May 26–29, 1992.
232. Segraves RT. Pharmacological enhancement of human sexual behavior. *J Sex Educ Ther* 1991;17:283–289.
233. Reid K, Morales A, Harris C. Double-blind trial of yohimbine in treatment of psychogenic impotence. *Lancet* 1987;2:421–423.
234. Riley AJ, Goodman RE, Kellet JM, et al. Double-blind trial of yohimbine hydrochloride in the treatment of erection inadequacy. *Sex Marital Ther* 1989;4:17–26.
235. Susset JG, Tessier CD, Wincze J, et al. Effect of yohimbine hydrochloride on erectile impotence: a double-blind study. *J Urol* 1989;141:1360–1363.
236. Lal S, Ackman D, Thavundayil JX, et al. Effect of apomorphine, a dopamine receptor agonist, on penile tumescence in normal subjects. *Prog Neuropsychopharmacol Biol Psychiatry* 1984;8:695–699.
237. Danjov P, Alexandre L, Warot D, et al. Assessment of erectogenic properties of apomorphine and yohimbine in man. *Br J Clin Pharmacol* 1988;26:733–739.
238. Segraves RT, Bari M, Segraves K, et al. Effect of apomorphine on penile tumescence in men with psychogenic impotence. *J Urol* 1991;145:1174–1175.
239. Lal S. Apomorphine in the evaluation of dopaminergic function in man. *Prog Neuropsychopharmacol Biol Psychiatry* 1988;12:117–164.
240. Benkert O, Crombach G, Kockott G. Effect of l-dopa on sexually impotent patients. *Psychopharmacology* 1972;23:91–95.
241. Pierini AA, Nusimovich B. Male diabetic sexual impotence: effect of dopaminergic agents. *Arch Androl* 1981;6:347–350.
242. Charney DS, Heninger GR. Alpha$_2$ adrenergic and opiate receptor blockage. *Arch Gen Psychiatry* 1986;43:1037–1041.
243. Fabbri A, Jannini EA, Gnessi L, et al. Endorphins in male impotence: evidence for naltrexone stimulation of erectile activity in patient therapy. *Psychoneuroendocrinology* 1989;14:103–111.
244. Moore BE, Rothschild AJ. *www.hosppract.com/issues/1999/01/moore.htm.*
245. Montejo-Gonzalez AL, et al. SSRI-induced sexual dysfunction: fluoxetine, paroxetine, sertraline, and fluvoxamine in a prospective, multicenter, and descriptive clinical study of 344 patients. *J Sex Martial Ther* 1997;23:176.
246. Moore BE, Rothschild AJ. Treatment of antidepressant-induced sexual dysfunction. Hospital Practice 1999, available at: *www.hosppract.com/issues/1999/01/moore.htm.*
247. Ashton AK, Hamer R, Rosen RC. Serotonin reuptake inhibitor-induced sexual dysfunction and its treatment: a large-scale retrospective study of 596 psychiatric outpatients. *J Sex Marital Ther* 1997;23:165.
248. Byne W, Parsons B. Human sexual orientation: the biological theories reappraised. *Arch Gen Psychiatry* 1993;50:228–239.
249. Kraft-Ebing VR. *Psychopathia sexualis,* 12th rev. ed. New York: Physicians and Surgeons, 1925.
250. McCauley E, Urquiza AJ. Endocrine influences on human sexual behavior. In: Sitsen JMA, ed. *Handbook of sexology, vol. 6. The pharmacology and endocrinology of sexual function.* Amsterdam: Elsevier, 1988:352–387.

251. Gladve BA. Hormones in relationship to homosexual/bisexual/heterosexual gender identify. In: Sitsen JMA, ed. *Handbook of sexology, vol. 6. The pharmacology and endocrinology of sexual function.* Amsterdam: Elsevier, 1988:388–409.

252. Gooren LJG. An appraisal of endocrine theories of homosexuality and gender dysphoria. In: Sitsen JMA, ed. *Handbook of sexology, vol. 6. The pharmacology and endocrinology of sexual function.* Amsterdam: Elsevier, 1988:410–424.

253. Money J, Dalery J. Iatrogenic homosexuality: gender identity in seven 46XX chromosomal females with hyperadrenocortical hermaphrodism born with a penis, three reared as boys, four reared as girls. *J Homosex* 1976;1:357–371.

254. Money J, Ehrhardt AA. *Man and woman, boy and girl.* Baltimore: Johns Hopkins University Press, 1972.

255. Donner G. Hormones and sexual differentiation of the brain. In: Porter R, Whelan J, eds. *Sex, hormones, and behavior. Ciba Foundation Symposium 62 (new series).* Amsterdam: Excerpta Medica, 1979:81–112.

256. Imperato-McGinley J, Peterson RE, Gautier T, et al. Androgens and the evolution of male gender identity among male pseudohermaphrodites with 5 alpha-reductase deficiency. *N Engl J Med* 1979;300:1233–1237.

257. Meyer-Bahlburg HFL. Psychoendocrine research on sexual orientation. Current status and future option. *Prog Brain Res* 1984; 61:375–398.

258. Sadock BJ, Sadock VA. Normal human sexuality and sexual and gender identity disorders. In: *Kaplan and Sadock's comprehensive textbook of psychiatry,* vol. 1, 7th ed. Philadelphia: Lippincott Williams & Wilkins, 2000.

259. Donovan BT. *Humors, hormones and the mind. An approach to the understanding of behavior.* New York: Stockton Press, 1988.

260. Bailey JM, Pillard RC. A genetic study of male sexual orientation. *Arch Gen Psychiatry* 1991;48:1089–1096.

261. Bailey JM, Pillard RC, Neale MC, et al. Heritable factors influence sexual orientation in women. *Arch Gen Psychiatry* 1993; 50:217–223.

262. Whitam FL, Diamond M, Martin J. Homosexual orientation in twins: a report on GI pairs and three triplet sets. *Arch Sex Behav* 1993;22:187–206.

263. Hamer DH, Hu S, Magnvson VL, et al. A linkage between DNA markers on the X chromosome and male sexual orientation. *Science* 1993;261:321–327.

264. Sadock BJ, Sadock VA. Normal human sexuality and sexual and gender identity disorders. In: *Kaplan and Sadock's comprehensive textbook of psychiatry,* vol. 1, 7th ed. Philadelphia: Lippincott Williams & Wilkins, 2000.

265. Gladue BA, Green R, Hellman RE. Neuroendocrine response to estrogen and sexual orientation. *Science* 1984;225:1496–1499.

266. Dorner G, Docke F, Gotz F, et al. Sexual differentiation of gonadotropin secretion, sexual orientation and gender role behavior. *J Steroid Biochem Mol Biol* 1987;27:1081–1087.

267. Gooren L, Fliers E, Courtney K. Biological determinants of sexual orientation. *Annu Rev Sex Res* 1990;1:175–196.

268. Levay S. A difference in hypothalamic structure between heterosexual and homosexual men. *Science* 1991;253:1034–1037.

269. Swaab DF, Hoffman MA. An enlarged suprachiasmatic nucleus in homosexual men. *Brain Res* 1990;537:141–148.

270. Allen LS, Gorski RA. Sexual orientation and the size of the anterior commissure in the human brain. *Proc Natl Acad Sci U S A* 1992;89:7199–7202.

271. Sadock BJ, Sadock VA. Normal human sexuality and sexual and gender identity disorders. In: *Kaplan and Sadock's comprehensive textbook of psychiatry,* vol. 1, 7th ed. Philadelphia: Lippincott Williams & Wilkins, 2000.

272. Sadock BJ, Sadock VA. Normal human sexuality and sexual and gender identity disorders. In: *Kaplan and Sadock's comprehensive textbook of psychiatry,* vol. 1, 7th ed. Philadelphia: Lippincott Williams & Wilkins, 2000.

273. Labbate LA, et al. Sexual dysfunction induced by serotonin reuptake antidepressants. *J Sex Marital Ther* 1998;24:3.

274. Shrivastava RK, et al. Amantadine in the treatment of sexual dysfunction associated with selective serotonin reuptake inhibitors. *J Clin Psychopharmacol* 1995;15:83.

275. Ashton AK, Rosen RC. Bupropion as an antidote for serotonin reuptake inhibitor-induced sexual function. *J Clin Psychiatry* 1998;59:112.

276. Bartlik B, Kaplan P, Kaplan HS. Psychostimulants apparently reverse sexual dysfunction secondary to selective serotonin reuptake inhibitors. *J Sex Marital Ther* 1995;21:264.

277. Gitlin MJ. Treatment of sexual side effects with dopaminergic agents [Letter]. *J Clin Psychiatry* 1995;56:124.

15

NEUROPHYSIOLOGY AND NEUROPSYCHIATRY OF SLEEP AND SLEEP DISORDERS

CHRISTOPHER M. SINTON
ROBERT W. MCCARLEY

This is an exciting time for sleep medicine and sleep research as we begin to understand some of the mechanisms that control the changes in consciousness associated with sleep and wakefulness and the bases of some sleep disorders. Witness the recent discoveries of an involvement of the neuropeptide orexin in narcolepsy, the identification of hypothalamic structures promoting sleep, the genetics of the circadian pacemaker, and the mounting evidence that adenosine is an endogenous sleep factor. In this chapter we review these and other recent developments that aid in our understanding of the neuroanatomic and neurophysiologic basis of sleep disorders. Our emphasis is on the behavioral, cognitive, and affective issues related to sleep disorders (for a detailed discussion of specific sleep disorders and related clinical issues, see references 1 and 2). Overviews also are available covering rapid eye movement (REM) and non-REM sleep physiology (3,4), the role of humoral factors in sleep (5), and the relationship between the immune system and sleep (6). Where appropriate, we draw directly on material from our earlier summaries of work in the field (3). We begin with a brief review of the organization of sleep and wakefulness and the recording techniques that are used in sleep disorder centers. This provides a background for discussion of the anatomy and neurophysiology of the neural control of different vigilance states. We then review selected sleep disorders, grouped according to their presentation and symptoms (for the standard classification schemes for these disorders, see references 1 and 7). Differential diagnosis and treatment options are noted where appropriate, although this chapter should not be considered an exhaustive guide to available treatments.

SLEEP ARCHITECTURE AND POLYSOMNOGRAPHY

The electroencephalogram (EEG) is a relatively crude measure of brain activity that uses sensors (i.e., electrodes) placed on the scalp to detect small (i.e., microvolt) signals produced by the synchronous activity of cortical neurons. The amplified and filtered EEG signal has characteristic and recognizable patterns that are primarily due to the regular laminar structure of the cerebral cortex, which allows the electrical signals of individual neurons to summate. The EEG can be understood as a record of the spontaneous voltage fluctuations produced by brain activity. However, the exact source of the recorded fluctuations often is difficult to pinpoint, with the important exception of large-amplitude changes due to pathologic synchronization of neural elements, as in the case of spikes from seizure discharge. This has made the EEG an important tool for detecting the presence of seizure activity, but the EEG also is extensively used to describe changes in alertness and sleep stages, the topic of this chapter.

By describing the recorded voltage fluctuations in terms of the amplitude of the resulting waves and their frequency, the EEG plays a significant role in the study of the basic structure of sleep in humans. EEG frequencies are grouped into bands, which range from the low frequencies (δ 0.5–4 Hz) through θ (4–8 Hz) and α (8–14 Hz) to the fast frequencies (β 14–30 Hz and γ 30–50 Hz, centered on 40 Hz). As a general rule, the δ EEG frequencies are associated with states of consciousness involving less complex processing, such as non-REM sleep, and higher frequencies are associated with states that involve more complex processing, such as wakefulness and REM sleep.

Sleep is Organized into a Definite Structure

One third of our lives is spent in sleep. No other single behavior occupies so much of our time, yet few other behaviors have been so difficult to understand. We now are beginning to unravel at least some of the mysteries of sleep, but much work remains. The most basic description of sleep divides it into two principal states: *REM sleep,* typically associated with a higher level of brain neuronal activity; and *non-REM sleep,* typically associated with a lower level of neuronal activity. A study of human sleep includes records of the EEG, of eye movements with the electrooculogram (EOG), and of muscle tone with the electromyogram (EMG). Other measures often are recorded concurrently during sleep, and these depend on the patient being evaluated in the sleep clinic. They might include records of nasobuccal airflow with thermistors, respiratory effort with abdominal and thoracic strain gauges, oxygen saturation with an oximeter, penile tumescence monitoring with a strain gauge, and leg muscle EMG. This ensemble of records is known as a *polysomnogram,* and the recording process is called *polysomnography.* The standard sleep montage (EEG, EOG, EMG) enables us to determine the main stages of sleep, which are described in the first part of this chapter. Other records are important for the diagnosis of sleep disorders, which are described later. Recent developments in electronics have enabled manufacture of small, portable polysomnographic recorders. These include units capable of complete polysomnographic recordings, but they require careful positioning and maintenance of the electrodes while the subject is outside the sleep center environment. A simpler ambulatory solution, when only periods of sleep and wakefulness need to be distinguished, is for the subject to wear a small watchlike device, known as an *actigraph.* The actigraph records any movement of the subject.

As sleep onset approaches, the low-voltage fast (LVF) EEG of alert wakefulness (Fig. 15.1A), often with α waves, yields to stage 1 sleep, a brief transitional phase between wakefulness and "true" sleep. This stage often is called *descending stage 1* because it is the precursor to deeper sleep stages and is characterized by a low-amplitude (i.e., voltage), relatively fast-frequency EEG pattern, associated with loss of the α rhythm and slow, rolling eye movements. The α rhythm has a frequency range from 8 to 13 Hz and is best recorded over the occipital scalp region. As noted, this frequency pattern occurs during wakefulness, often appearing upon eye closure but disappearing with eye opening (8). Depth recordings in animals indicate that α rhythm frequencies also may be present in nuclei of the visual thalamus, including the lateral geniculate body (9). The cortical component appears to be generated in relatively small cortical areas that act as epicenters.

Stage 1 is followed by stage 2 sleep, when EEG frequencies slow still further. During stage 2, episodic bursts of rhythmic, 14- to 16-Hz waveforms occur in the EEG. These bursts, known as *sleep spindles* (10), are interspersed with occasional short-duration, high-amplitude *K complexes,* so named because of their morphologic resemblance to the letter *K* (11,12). Spindles have been noted during the light slow wave sleep phase of animals, as well as during stage 2 human sleep. The amplitude of the burst of waves that comprises the spindle waxes and then wanes over a duration of 1 to 2 seconds. Wave frequency varies among species and is higher in primates. Spindles are relatively well understood at the cel-

lular level. Studies by Steriade and collaborators (reviewed in 13) indicate spindle waves arise from discharge interactions between spindle pacemaker γ-aminobutyric acid (GABA)-ergic thalamic nucleus reticularis (RE) neurons and thalamocortical neurons, as the level of polarization of RE neurons changes as sleep begins.

Stage 2 sleep is followed by stages 3 and 4, which are defined by the presence of high-amplitude, slow (0.5–4 Hz) δ waves. Stages 3 and 4 often are grouped together using the term *slow wave sleep* (SWS). The cellular basis of cortical δ waves has been shown to depend on thalamocortical neurons (for reviews see references 14–17), and δ waves during sleep represent thalamocortical oscillations occurring in the absence of activating or arousing inputs. The basic concept is that a hyperpolarized membrane potential in these cells permits the occurrence of δ waves in thalamocortical circuits. Thus, any factors that depolarize the membrane will block δ waves. During wakefulness, input from the cholinergic forebrain nucleus basalis is important for suppression of slow wave activity, as shown by preclinical lesion studies (18) (see section on Neurophysiologic Basis for Electroencephalographic Synchronization and Desynchronization). In addition, brainstem noradrenergic and serotonergic projections help maintain a depolarized state in waking, although these neurons are quiescent during REM sleep, when brainstem cholinergic input is the principal depolarizing factor. There is a gradual reduction in depolarizing input to the thalamus as non-REM sleep deepens through stage 4; hence, δ activity increases. Thus, the relative intensity of cortical synchronization, including δ waves, correlates well with the relative absence of cholinergic activity.

The low-voltage, fast-frequency EEG pattern of REM sleep is in marked contrast to δ sleep, and in REM sleep the EEG pattern resembles that of active wakefulness and stage 1 descending (Fig. 15.1A). REM sleep is further characterized by the presence of bursts of rapid eye movements (hence the name) and by loss of muscle tone in most major muscle groups of the limbs, trunk, and neck.

There is a typical and predictable progression from one sleep state to another during a night's sleep (Fig. 15.1B). As the night begins, there is a stepwise descent from wakefulness to stage 1 through stage 4 sleep, followed by a more abrupt ascent back toward stage 1. However, in place of stage 1, the first REM sleep episode usually occurs at this transition point, about 70 to 90 minutes after sleep onset. The first REM sleep episode in humans is short. After the first REM sleep episode, the sleep cycle repeats with the appearance of non-REM sleep and then, about 90 minutes after the start of the first REM period, another REM sleep episode occurs. This rhythmic cycling persists throughout the night. The REM sleep cycle length is 90 minutes in humans, and the duration of each REM sleep episode after the first is approximately 30 minutes. Over the course of the night, δ wave activity tends to diminish. Thus, SWS tends to occur primarily during the first part of the night, and, as the night

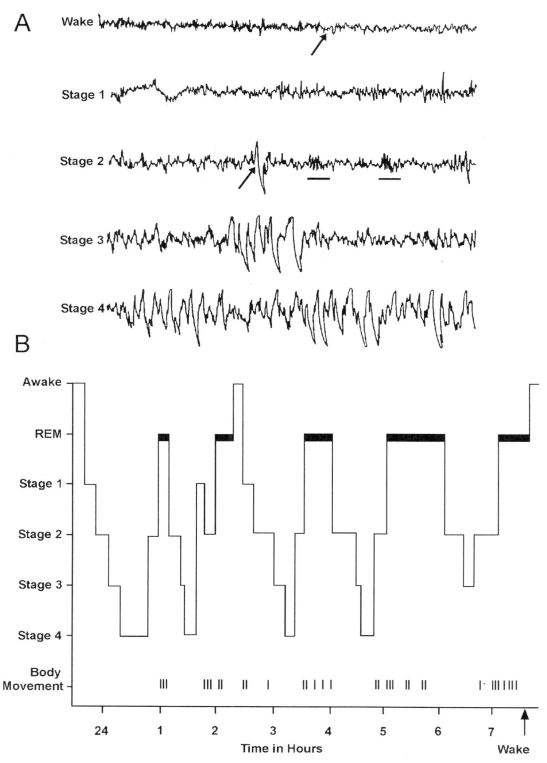

FIGURE 15.1. **A:** Electroencephalographic (EEG) patterns associated with wakefulness and stages of sleep. **B:** Time course of sleep stages during a night's sleep in a healthy young adult shown as a hypnogram. During wakefulness, a low-voltage fast EEG pattern typically is present, often with α waves, as shown here. The α rhythm (8–13 Hz) occurs during relaxed wakefulness, often appearing on eye closure and disappearing on eye opening. At the *arrow*, there is a transition to stage 1 sleep, with loss of the α rhythm but continued presence of a low-voltage fast EEG. As sleep deepens, the EEG frequency slows. Stage 2 is characterized by the presence of K complexes *(arrow)* and sleep spindles *(underlined)*. During stage 3, δ waves (0.5–4 Hz) appear, and in stage 4

progresses, non-REM sleep is comprised of waves of higher frequencies and lower amplitude. Brief, phasic body movements during sleep tend to cluster just before and during REM sleep (Fig. 15.1B). In general the ease of arousal from sleep parallels the ordering of the sleep stages, with REM and stage 1 being the easiest for arousal and stage 4 the most difficult.

Ontogeny of Sleep

Although the full characteristics of adult sleep stages are not fully recognized in infants, their sleep can be divided into two principal stages. One, *active sleep,* is considered the precursor for REM sleep and the other, *quiet sleep,* the precursor of non-REM sleep (19). In the human infant, active sleep is first apparent by 30 to 35 weeks of gestational age. *In utero,* all mammals spend a large percentage of time in active sleep, ranging from 50% to 80% of a 24-hour day. At birth, animals born with immature nervous systems have a much higher percentage of active sleep than do adults of the same species. For example, sleep in the human newborn occupies about 20 hours of the day, with active sleep occupying half of the total sleep time, or about one third of the entire 24-hour period. The characteristics of REM and non-REM sleep stages can first be differentiated between age 3 and 6 months; subsequently the percentage of REM sleep declines rapidly to about 25% of total sleep time as early as 9 months (20). By approximately age 10 years the adult percentage of REM sleep is reached, 20% of total sleep time. The predominance of REM sleep in the young suggests an important function for this sleep stage in promoting nervous system growth and development.

δ Activity is minimally present in the newborn, but it can be easily identified by 9 months. It then increases over the first years of life, reaches a maximum at about age 10 years, and declines thereafter. Feinberg et al. (21) examined the time course of δ wave intensity over the first 3 decades of life. They found that this time course of δ wave intensity can be modeled by a γ probability distribution and that, to a good approximation, the same time course is observed for synaptic density and positron emission tomographic measurements of metabolic rate in human frontal cortex. They speculate that the correlated reduction in these three variables reflects the synaptic change that is a key factor in cognitive maturation.

Sleep Deprivation

Physicians frequently are asked how much sleep is needed. As noted earlier, the answer first depends on the age of the individual, but a good general rule is that enough sleep is needed to prevent daytime drowsiness. Each individual seems to have a particular "set point" of need. Both long sleepers and short sleepers have been identified, although it is important to note that the amount of time individuals spend in SWS (i.e., stages 3 and 4) apparently is independent of total sleep time (22). In adults, the modal value of sleep need appears close to the traditional 8 hours, but there is considerable individual variation. If someone functions and feels well on less sleep, there is little need for concern.

The main consequence of sleep deprivation in normal people seems to be the presence of *microsleeps,* which are very brief episodes of sleep during which sensory input from the outside is diminished and cognitive function is markedly altered (23). Nevertheless, evaluation of the effects of sleep deprivation in normal people is important in the context of this chapter because of the disruption of sleep that accompanies psychiatric disorders (for detailed discussions of the effects of sleep deprivation, see references 24 and 25). Here we comment briefly on some of the consequences of the buildup of a "sleep debt."

Studies report that sleep deprivation invariably leads not only to poor performance and attentional deficit, but also to mood changes and irritability (26). Furthermore, long-term sleep debt may be associated with perceptual abnormalities and even hallucinations. However, the latter are rarely so extreme as the perceptual distortions reported by the disc jockey Peter Tripp after a marathon public sleep deprivation test carried out in New York City in 1959 (27). This type of study was driven by a hope at that time that the hallucinations of psychosis may represent a breakthrough of dreams into wakefulness (see section on Psychoses), but that expectation was not realized in later work. It is apparent that there are long-lasting effects of sleep deprivation on performance and physiology, and these effects persist even after a night of recovery sleep. Preclinical studies are continuing to elucidate the neurophysiologic and molecular changes associated with sleep deprivation in an attempt to understand these long-lasting deficits. For example, deprivation-induced changes in protein transcription may underlie the presence of sleep debt (see discussion in the context of adenosine as a sleep

they are present more than 50% of the time. During rapid eye movement (REM) sleep (*dark bars* in **B**), the EEG pattern returns to a low-voltage fast pattern. The percentage of time spent in REM sleep increases with successive sleep cycles, whereas the percentages of stages 3 and 4 decrease. These EEG segments were recorded over the parietal lobe (C3), except in waking, when occipital lobe (O2) recording was used to show the α rhythm most clearly. (Adapted from Carskadon MA, Dement WC. Normal human sleep: an overview. In: Kryger MH, Roth T, Dement WC, eds. *Principles and practice of sleep medicine,* 2nd ed. Philadelphia: WB Saunders, 1994:16–25, with permission.)

factor in section on Humoral Factors). Paradoxically, sleep deprivation has been found to be an effective but short-lived treatment for depression (discussed in greater detail in the section on Sleep Deprivation as a Therapy for Depression).

SLEEP PHYSIOLOGY AND RELEVANT ANATOMY

Here we review briefly our current understanding of brain physiology and anatomy related to the mechanisms of sleep, with particular emphasis on the aspects that have implications for sleep disorders. In this section, we refer frequently to changes in neuronal discharge patterns that result from altered polarization levels of neurons, typically at thalamic and cortical levels of the neuraxis. The reader is referred to other reviews of this important concept (28,29), although the following synopsis is sufficient in the context of this chapter. The polarization level of a neuron describes the difference in voltage between the inside and outside of the cell. A potential can be measured across the cell membrane, and at rest this potential is negative, typically on the order of -60 mV. This potential difference is maintained by ion channels, which allow the passage of charged ions across the membrane. Summarized briefly, neurotransmitters acting at receptor sites alter ion channel opening and so change neuronal polarization levels. After a reduction in the level of polarization, an action potential is generated, the neuron is said to discharge, and the action potential propagates down the axon. Membrane potential can be changed gradually by neurotransmitter input and local influences. Against such a different background level of polarization, altered discharge patterns can result. Thalamic neurons are a significant example of a type of cell in which this occurs. During neuronal discharge, specific ion channels that are voltage sensitive open and close rapidly as the action potential is propagated. Patterns of discharge in thalamic neurons follow different patterns, depending on whether the neuron is hyperpolarized (i.e., lower potential, such as -70 mV) or depolarized (i.e., higher potential, such as -50 mV) from its normal resting membrane potential at the time of initiation of the action potential.

Humoral Factors

The search for a humoral sleep factor has a long history and a large literature (5). Summarized briefly, a series of studies by Pappenheimer, Karnovsky, Krueger, and co-workers demonstrated that certain factors, later identified as muramyl peptides, are concentrated in the cerebrospinal fluid, brain, and urine of sleep-deprived animals (30–32). These muramyl peptides have the capability to reliably induce non-REM sleep when they are injected into the lateral ventricles or basal forebrain. The compounds, derived from bacterial

cell walls, also induce hyperthermia. Another sleep factor is interleukin-1, a cytokine, which is produced in response to infections, but also by injections of components of bacterial cell walls, which, as noted, include muramyl peptides. Interleukin-1 increases non-REM sleep and it produces hyperthermia (33). Hyperthermia itself may enhance sleep, but blocking the hyperthermic effects of interleukin-1 does not block the sleep-inducing effects (5). The argument that interleukin-1 is important in the hypersomnia associated with infections is strong. There also is evidence supporting a role for interleukin-1, tumor necrosis factor, and growth hormone releasing hormone as part of the humoral mechanisms regulating physiologic sleep (34). Injection of these substances enhances non-REM sleep, whereas their inhibition reduces spontaneous sleep and sleep rebound after sleep deprivation. Changes in the levels of messenger ribonucleic acid (mRNA) and of the proteins themselves in the brain are consistent with their proposed role in sleep regulation, as are results from transgenic and mutant animals. However, these factors appear to be involved in regulation of the propensity to sleep over longer time periods than the actions of the purine nucleoside adenosine, which we discuss later.

Hayaishi et al. (6,35) reported that injections of prostaglandin D_2 (PGD_2) into the third ventricle reliably produces non-REM sleep, and they proposed that this compound is a natural sleep regulatory factor. Interestingly, Hayaishi et al. found that at least some of the sleep-inducing effects of PGD_2 could be mediated by changes in extracellular levels of adenosine. Krueger and Fang (36) suggested a model in which some of the effects of interleukin-1 also might be mediated by adenosine. The possibility exists that adenosine might be a "final common factor" for other sleep-inducing substances. Reviews by Hayaishi (37) and Krueger (38) should be consulted for further details of this developing field.

There is a growing body of evidence supporting the role of adenosine as a mediator of the sleepiness that follows prolonged wakefulness, a role in which its inhibitory action on the basal forebrain wakefulness-promoting neurons may be especially important (for reviews see references 39 and 40). Figure 15.2A shows a schematic of this action of adenosine on basal forebrain neurons. Common-sense evidence for a sleep-enhancing effect of adenosine comes from the ubiquitous use of coffee and tea to increase alertness, because these beverages contain caffeine, which is an adenosine receptor antagonist (41). McCarley et al. (39) and Strecker et al. (40) hypothesized that, during prolonged wakefulness, adenosine accumulates selectively in the basal forebrain and promotes the transition from wakefulness to sleep by inhibiting, via the adenosine A1 receptor, cholinergic and noncholinergic wakefulness-promoting basal forebrain neurons. Direct evidence was obtained for adenosinergic inhibition of neurons in the cholinergic basal forebrain, as well as in the mesopontine

A

Adenosine (Ad) - mediated inhibition of basal forebrain wake - active neurons

B

Adenosine metabolism

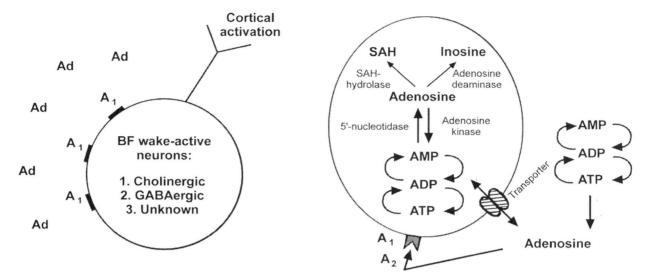

FIGURE 15.2. A: Schematic of the inhibitory effects of adenosine on wake-active neurons in the basal forebrain. Extracellular adenosine (AD) acts on the A_1 adenosine receptor subtype to inhibit neurons that promote electroencephalographic activation and wakefulness. **B:** Simplified schematic of the main intracellular and extracellular metabolic pathways of adenosine. The stages of the intracellular pathway from adenosine 5'-triphosphate (ATP) to adenosine diphosphate (ADP) to adenosine monophosphate (AMP) to adenosine are regulated by the enzymes ATP-ase, ADP-ase and 5'-nucleotidase, respectively. Extracellularly, the same pathway is regulated by the respective ectoenzymes. Adenosine kinase converts adenosine to AMP, whereas adenosine deaminase converts adenosine to inosine. The third enzyme to metabolize adenosine is S-adenosylhomocysteine (SAH) hydrolase, which converts adenosine to SAH. The adenosine concentration between the intracellular and extracellular compartments is equilibrated by nucleoside transporters. The effects of adenosine on the cell are mediated through specific adenosine receptor subtypes A_1 and A_2. (Adapted from Porkka-Heiskanen T, Strecker RE, McCarley RW, et al. Brain site-specificity of extracellular adenosine concentration changes during sleep deprivation and spontaneous sleep: an in vivo microdialysis study. *Neuroscience* 2000;99:507–517, with permission.)

nuclei of the laterodorsal tegmental (LDT) and pedunculopontine tegmental (PPT) nuclei (42). Both cholinergic nucleus basalis and LDT/PPT neurons are implicated in alertness and EEG activation (see section on Neurophysiologic Basis for Electroencephalographic Synchronization and Desynchronization). Regulation of the levels of extracellular adenosine, which inhibits these neurons that are important in the promotion of wakefulness, depends primarily on metabolic rate. Increased metabolism leads to reduced high-energy phosphate stores and increased adenosine, which, via an equilibrative transporter, leads to increased extracellular adenosine (Fig. 15.2B). Extracellular adenosine also may be increased by the release of adenosine triphosphate (ATP) as a cotransmitter and its breakdown, by the relevant ectoenzymes, to adenosine. Support for a putative link between adenosine and metabolism comes from the fact that EEG activation is known to diminish as a function of the duration

of prior wakefulness and with brain hyperthermia, both of which are associated with an increase in brain metabolic rate.

Active Non–Rapid Eye Movement Sleep-Promoting Mechanisms

Electrophysiologic recordings of basal forebrain/anterior hypothalamic neurons have indicated that some of these neurons selectively discharge during non-REM sleep, and this might represent an active sleep-promoting mechanism (for review see reference 43). Studies by Sherin et al. (44) used the immediate early gene protein product c-Fos to detect neurons in the ventrolateral preoptic (VLPO) area that were selectively active during non-REM sleep. Immunohistochemical studies subsequently suggested that these neurons were GABA-ergic, and anatomic work revealed their projections to wakefulness-promoting histaminergic neurons in the

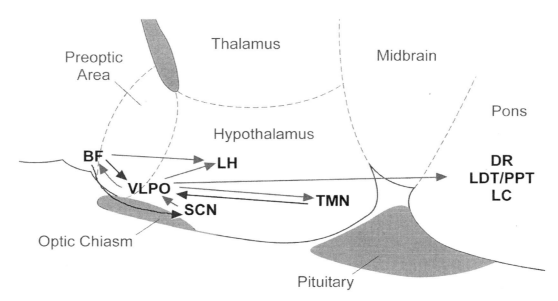

FIGURE 15.3. Schematic of a sagittal section of the basal forebrain and hypothalamic areas in the mammalian brain showing important centers in this region involved in the regulation of sleep and wakefulness and some of their major local projections. Excluded from this figure are the widespread projections from the lateral hypothalamus (LH) and the efferents from the basal forebrain/hypothalamic region to the thalamus and cortex (cf. Figs. 15.4 and 15.5 for these projections). Excitatory projections (e.g., using acetylcholine as the neurotransmitter) are in *red,* and inhibitory projections [e.g., using γ-aminobutyric acid (GABA)] are in *blue.* Ascending histaminergic wake-active neurons of the tuberomamillary nucleus (TMN) innervate widespread regions, including the cortex and hypothalamus. Excitatory cholinergic neurons of the basal forebrain (BF) preferentially discharge during wakefulness and rapid eye movement (REM) sleep and send projections to cortical and thalamic areas. Multisynaptic pathways from the suprachiasmatic nucleus (SCN) and the retina innervate, *inter alia,* the sleep-active neurons of the ventrolateral preoptic (VLPO) area. These projections allow modulation of the discharge rate of the VLPO neurons according to light levels and circadian phase. Reciprocal projections between the BF and VLPO regions may be a substrate for the integration of circadian and homeostatic sleep drives. GABA-ergic VLPO neurons, and closely associated GABA-ergic neurons in the BF, project to the TMN, the orexin-containing neurons of the LH, and brainstem nuclei involved in EEG activation and REM sleep cycle control. The latter include the dorsal raphe (DR), laterodorsal tegmental and pedunculopontine tegmental (LDT/PPT), and locus caeruleus (LC). These inhibitory GABA-ergic VLPO/BF projections may be important for the coordinated decrease in discharge activity of EEG-activating neurons as non-REM sleep begins and deepens, and then for the changes in the discharge activity of these neurons as the REM sleep cycle is initiated. The LH orexin-containing neurons, implicated in the control of the sleep/wake cycle and the transition between vigilance states, send widespread projections throughout the neuraxis, as shown in Figure 15.5. Shown here are inhibitory efferents from the VLPO and GABA-ergic neurons in the BF region to the LH orexin-containing neurons. (See Color Figure 15.3 following page 526.)

posterior hypothalamus and to brainstem nuclei important in EEG arousal (45). Figure 15.3 shows the location and some of the interactions of these important neuronal groups in the basal forebrain and hypothalamus. Current studies continue to investigate the interaction of these neurons with other systems that are important in sleep.

Neurophysiologic Basis for Electroencephalographic Synchronization and Desynchronization

The high-voltage, slow wave activity in the cortex during most of non-REM sleep, called *EEG synchronization,* contrasts sharply with the LVF pattern, called *desynchronized* or *activated EEG,* which is a characteristic of waking and

REM sleep. As noted earlier, slow wave activity is primarily in the δ (0.5–4 Hz) range, but desynchronized EEG consists mainly of frequencies in the β (14–30 Hz) and γ (30–50 Hz) ranges. We also reviewed how the polarization levels of thalamic RE and thalamocortical neurons are critical for the progression of changes seen on the EEG as sleep begins and then deepens. Thus, as sleep progresses past stage 2, the spindles characteristic of this stage are blocked primarily by brainstem cholinergic neurons that synapse in the thalamus and change the polarization level of RE neurons. The basal forebrain also provides depolarizing cholinergic and hyperpolarizing GABA-ergic input to the RE that assist brainstem input in disrupting spindling activity.

One of the major advances in the past few years has been the establishment of the importance of a cholinergic activating system in EEG desynchronization. This is probably the major component of the *ascending reticular activating system* (ARAS), a concept that arose from the work of Moruzzi and Magoun (46) before methods were available to label neurons that use specific neurotransmitters. This advance in our understanding of the ARAS can be dated to studies that labeled specific activating systems. Steriade et al. (47) identified thalamically projecting cells located near the pons–midbrain junction that increased their discharge rate about 60 seconds before the first change to a desynchronized state that was noted on the EEG. Although these neurons were localized to an area that was not precisely coincident with the areas identified in the earlier experiments as being critical for inducing arousal, the change in their discharge rate was clearly the first indication of arousal. Subsequent work identified these neurons as containing acetylcholine as their neurotransmitter and being localized to the LDT/PPT region. Figure 15.4 shows the location of these cholinergic LDT/PPT neurons. They have high discharge rates in REM

sleep and waking and low discharge rates in non-REM sleep. Extensive anatomic evidence obtained subsequently showed that these cholinergic neurons project to thalamic nuclei important for the processes of EEG desynchronization and synchronization. Neurophysiologic studies have indicated that the target neurons in the thalamus respond to cholinergic agonists in a manner consistent with EEG activation.

Cholinergic systems are not the exclusive substrate of EEG desynchronization, and evidence that multiple systems are involved in EEG desynchronization comes from the inability of lesions of any single one of these systems to disrupt EEG desynchronization on a permanent basis. Thus, other brain stem reticular neuronal projections to thalamus, probably using excitatory amino acid neurotransmission, as well as noradrenergic and serotonergic projections from locus caeruleus and raphe nuclei, respectively (in wakefulness, because these neurons are quiescent in REM sleep), also play important roles (16). In addition to brainstem nuclei, cholinergic input to the thalamus and cortex from the basal forebrain nuclei

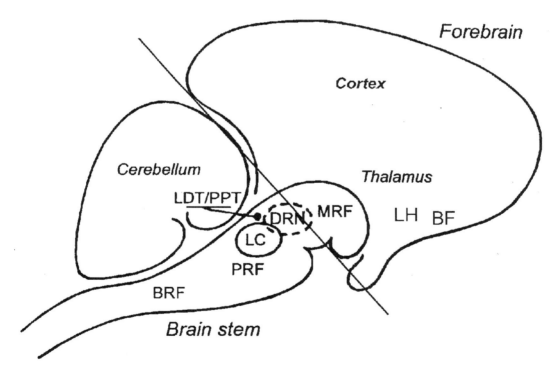

FIGURE 15.4. Schematic of a sagittal section of a mammalian brain showing the plane of transection that preserves rapid eye movement (REM) sleep signs caudal to the transection but abolishes them rostral to the transection. The nuclei important for the control of sleep are annotated as follows: BRF, bulbar reticular formation; LC, locus caeruleus (where most norepinephrine-containing neurons are located); LDT/PPT, laterodorsal tegmental/pedunculopontine tegmental nuclei (principal sites of the acetylcholine-containing neurons important for REM sleep and electroencephalographic desynchronization); LH = lateral hypothalamus (location of all the orexin-containing neurons); MRF, mesencephalic reticular formation; PRF, pontine reticular formation; RN, dorsal raphe nucleus (site of many serotonin-containing neurons). The basal forebrain (BF), in the ventral part of the forebrain, lies rostral to the hypothalamus and its boundaries are defined by the magnocellular cholinergic neurons. See text for relevant projections from these nuclei to thalamic and cortical centers that are important for control of sleep and wakefulness. (Adapted from McCarley RW. Dreams and the biology of sleep. In: Kryger MH, Roth T, Dement WC, eds. *Principles and practice of sleep medicine,* 2nd ed. Philadelphia: WB Saunders, 1994:373–383, with permission.)

(the nucleus basalis of Meynert in humans) play a role in EEG desynchronization (Fig. 15.4). As we begin to understand the neuronal mechanisms that underlie EEG synchronization and desynchronization, we can see how deficits in the control of these states can be manifested as insomnia or hypersomnia, which primarily results from deficiencies in components of the ARAS. Thus, brainstem tumors and other neurologic lesions have provided confirmation in humans of critical areas such as the LDT/PPT and the nucleus basalis in the maintenance of wakefulness. Insomnia is the opposite; it can be induced by psychiatric medications that activate ascending brainstem arousal projections, particularly noradrenergic (see section on Insomnia and Hypersomnia). Identification of desynchronizing processes in sleep with ascending brainstem cholinergic and reticular activation also means that the increasing intensity of EEG desynchronization preceding REM sleep is related to the increasing level of activity of REM-related cholinergic and reticular cell groups that precedes this state (see section on Rapid Eye Movement Sleep).

At a cellular level, extracellular recordings by McCarley et al. (48; reviewed in reference 13) demonstrated that thalamocortical relay neurons show a burst discharge pattern during non-REM sleep but not during waking or REM sleep. Subsequent *in vivo* studies by Steriade et al. (14,15,49), and *in vitro* investigations by McCormick et al. (16,17,50) indicated that this bursting mode in thalamocortical neurons occurs when the membrane is hyperpolarized. This hyperpolarization removes the inactivation of specific Ca^{2+} channels and enables the production of a "calcium spike" (i.e., a rapid entry of depolarizing calcium ions into the cell) when a small depolarization occurs. This depolarizing calcium spike is termed a *low threshold spike* (LTS) to distinguish it from other calcium currents with different triggering thresholds. The LTS depolarizes the neuron sufficiently to reach the threshold for fast sodium action potentials, and a burst of these rides on the LTS. However, the production of an LTS limits the frequency at which thalamocortical relay neurons can follow input discharges and hence blocks the rapid information transmission that is characteristic of waking and REM sleep.

Rapid Eye Movement Sleep

The brain physiology and neurotransmitters important for generation of REM sleep are becoming known, although many important questions about REM sleep mechanisms and especially the function of REM sleep remain unanswered. This chapter focuses on the aspects of REM sleep mechanisms that are important for clinical sleep disorders (for a detailed assessment of REM sleep mechanisms, see reference 3). Current findings demonstrate that brainstem cholinergic neurons act as promoters of REM sleep phenomena via their excitatory projections to several populations of cells in the pontine reticular formation (PRF). These PRF cells control the different features of REM sleep (e.g., atonia, eye movements). In contrast, the nearby serotonergic and noradrenergic neurons act to suppress most components of REM sleep via inhibitory inputs to the cholinergic neurons. Other neurotransmitters, such as GABA, are being investigated for their role in REM sleep.

As early as the 1960's, transection studies localized the neural circuitry required for generation of REM sleep to the brainstem. These lesion studies by Jouvet et al. (reviewed in reference 51) showed that a transection just above the junction of the pons and midbrain produces a state in which periodic occurrence of REM sleep phenomena could be found in recordings made in the isolated brainstem, whereas recordings in the isolated forebrain showed no sign of REM sleep (Fig. 15.4). Having established the importance of the brainstem in REM sleep, subsequent work has provided a detailed picture of the brainstem circuits controlling the REM sleep rhythm. As in humans, the cardinal signs of REM sleep in all species in which the state occurs are muscle atonia, EEG desynchronization (LVF pattern), and REMs. PGO waves, which are spiky EEG waves that arise in the *p*ons and are transmitted to the thalamic lateral *g*eniculate nucleus (a visual system nucleus) and to the visual *o*ccipital cortex, have been noted in recordings from deep brain structures in many animals and are an important component of REM sleep (3,13). Evidence suggests that PGO waves also are present in humans, but depth recordings would be necessary to establish their existence. PGO waves represent an important mode of brainstem activation of the forebrain during REM sleep; they also are present in nonvisual thalamic nuclei. It is worth noting that PGO waves can be recorded from nonvisual thalamic nuclei, although their timing is linked to eye movements, with the first wave of the usual burst of three to five waves occurring just before an eye movement.

A series of preclinical investigations over several decades has shown that REM sleep can be dissociated into its different components, including muscle atonia, EEG desynchronization, PGO waves, and REMs. Each of these components is under the control of different mechanisms and different anatomic loci, with important groups of neurons especially concentrated in the PRF. The cell groups that control the various cardinal signs of REM sleep have been termed *effector* neurons. REM sleep effector neurons are neurons that are directly in the neural pathways and that lead to the production of the features that characterize REM sleep. The reader familiar with the pathology associated with human REM sleep will find this concept easy to understand, because much of this pathology consists of inappropriate expression or suppression of individual components of REM sleep.

Because most of the physiologic events of REM sleep have effector neurons located in the PRF, neuronal recordings in this area are of particular interest to obtain information on mechanisms that produce these events. Intracellular recordings of PRF neurons have shown that effector neurons have relatively hyperpolarized membrane potentials and generate almost no action potentials during non-REM sleep. However, they begin to depolarize even before the occurrence of the first EEG sign of the approach of REM sleep, the PGO waves that occur 30 to 60 seconds before the onset of the remaining signs of REM sleep. As PRF neuronal depolarization proceeds and the threshold for action potential production is reached, these neurons start discharging. Their discharge rate increases as REM sleep is approached, and this high level of discharge is maintained throughout REM sleep because of continuing membrane depolarization. Several groups of PRF neurons are important for the various components of REM sleep, including those responsible for REMs (the generator for these saccades is in the PRF); PGO waves (a different group of PRF neurons); and a group of dorsolateral PRF neurons controlling the muscle atonia of REM sleep (these neurons become active just before the onset of muscle atonia). Neurons in the midbrain reticular formation are especially important for EEG desynchronization, the LVF pattern that is another characteristic of REM sleep. As noted previously (see section on Neurophysiologic Basis for Electroencephalographic Synchronization and Desynchronization), these neurons originally were described as comprising the ARAS, but subsequent work expanded on this concept to highlight the contribution of LDT/PPT and nucleus basalis cholinergic neurons to EEG desynchronization.

Several early studies indicated that activation of cholinergic neurons in the mesopontine tegmentum was involved in the generation of REM sleep via projections to the PRF effector neurons (for detailed reviews of this literature see references 13, 52, 53). Summarized, the data indicate that cholinergic neurons may directly mediate PGO wave transmission from brainstem to forebrain, but these cholinergic influences act primarily by increasing the excitability of PRF effector neurons. Evidence for the latter concept was established gradually over many years in a series of studies. It had been known since the mid-1960's that cholinergic agonists, injected into the PRF, produce a state that closely mimics natural REM sleep. Muscarinic receptors appear to be especially critical; nicotinic receptors are of lesser importance. The latency to onset and duration of the REM sleep episode are dose dependent. Within the PRF, most workers reported the shortest latencies with injections in dorsorostral pontine reticular sites. Anatomic studies subsequently confirmed that these mesopontine tegmental cholinergic neurons were located, as noted earlier, in two nuclei at the pons–midbrain junction, the LDT and PPT (54), and that these cholinergic neurons have projections to those areas in the PRF (54) in which local injections of cholinergic agonists are most effective in producing a REM-like state. It also was found that lesions (55) and electrical stimulation (56) of the LDT/PPT produce a decrease and an increase in REM sleep, respectively. Subsequent electrophysiologic studies of LDT/PPT neurons revealed

subpopulations that discharge preferentially just before and during REM sleep. Furthermore, as predicted, direct *in vivo* measurements of acetylcholine demonstrated that pontine levels of this neurotransmitter were higher during REM sleep (57,58).

In contrast to mesopontine cholinergic neurons that preferentially discharge in REM sleep, monoaminergic neurons in the noradrenergic locus caeruleus and serotonergic dorsal raphe nucleus exhibit a pattern of discharge activity that is nearly opposite. Their discharge rate is greatest during waking, declines during non-REM sleep, and virtually ceases before and during REM sleep. This inverse correlation with REM sleep led to suggestions that norepinephrine (59) and serotonin (60) might suppress REM sleep, and this concept has formed the basis of a structural and mathematical model of REM sleep control, the reciprocal interaction model (61). One of the postulates of this model is that these "REM-off" monoaminergic neurons inhibit the REM-promoting, "REM-on" neurons (53,61). For many years this was regarded as a highly controversial postulate, although work during the last few years has supported the hypothesis. Thus, *in vitro* studies demonstrated that serotonin directly inhibits LDT/PPT cholinergic neurons (62–64) by hyperpolarizing the membrane of these cells. Significantly, subsequent *in vivo* studies demonstrated that serotonergic inhibitory control of the LDT/PPT REM-generating region is sufficiently strong to influence the expression of REM sleep (65–68). Williams and Reiner (69) obtained similar inhibitory data with norepinephrine in an *in vitro* preparation. They found that 92% of histologically identified cholinergic LDT/PPT neurons in the rat were hyperpolarized in response to norepinephrine, whereas noncholinergic neurons exhibited mixed responses. In this regard, we should note that histamine-containing neurons, located in the tuberomamillary nucleus of the posterior hypothalamus (Fig. 15.3) also show a REM-off discharge pattern (70). These neurons are conceptualized as one of the wakefulness-promoting systems, in agreement with drowsiness as a common side effect of antihistamine drugs. Transection studies indicate, however, that the histaminergic neurons are not essential for REM sleep cyclicity.

Monoaminergic neurons have been found to play a role in the initiation of REM sleep. When the discharge activity of these neurons decreases in drowsiness and non-REM sleep, less monoamine neurotransmitter is released onto the inhibitory receptors found on LDT/PPT cholinergic neurons, causing these neurons to be disinhibited. This in turn leads to increased discharge of the LDT/PPT neurons, which themselves initiate the process of REM sleep, primarily via their projections to the REM sleep effector neurons in the PRF.

Rapid Eye Movement Sleep Muscle Atonia

We noted that one of the cardinal signs of REM sleep is muscle atonia, which is controlled by a group of effector neurons in the PRF (see section on Rapid Eye Movement Sleep). Muscle atonia is an important REM feature from a clinical point of view because disorders of this system are evident in many patients who present to sleep disorders clinicians. The target of the muscle inhibition effector system is the α motoneuron. Chase et al. investigated the effects on motoneurons that correlate with REM sleep at the cellular level both in trigeminal motoneurons and spinal α motoneurons. Morales and Chase (71,72) recorded antidromically identified lumbar motoneurons in naturally sleeping cats using a chronic intracellular recording technique that they developed. Their data demonstrated that there was a slight hyperpolarization in these neurons during the transition from active waking to non-REM sleep, but on passage to REM sleep a marked hyperpolarization was noted, temporally coincident with the loss of nuchal EMG activity. Chase et al. (73) later presented evidence suggesting that glycine is the principal neurotransmitter mediating this hyperpolarization in lumbar motoneurons. Additional results now indicate that the projections providing this inhibitory input ultimately arise from the bulbar reticular formation, with a probable intermediate synapse in spinal cord (for an overview of reticular areas important in muscle atonia and evidence about important neurotransmitters involved in this function, see reference 74).

Orexin and Control of Sleep and Wakefulness

Orexin (also known as hypocretin) is a neuropeptide contained in cell bodies situated uniquely in the anterior lateral hypothalamus (75,76). Orexin was linked to narcolepsy in two unrelated publications (77; see section on Narcolepsy). An abnormality in the gene for the orexin type-2 receptor (OX_2R) was found to be the basis of canine inherited narcolepsy (78). Orexin$^{-/-}$ "knockout" mice have increased REM sleep, sleep-onset REM periods, and cataplexy-like episodes entered directly from states of active movement, primarily during the dark, or active, phase (Fig. 15.5B and 15.5C) (79). These unexpected exciting findings will be the focus of intense research as the mechanisms involved in the action of orexin are elucidated. Although orexin-containing neurons project widely to all levels of the neuraxis, there is a particularly significant orexin innervation of some major nuclei that have been implicated in sleep regulation (Fig. 15.5A) (80–84). These include the VLPO sleep-promoting cell group, as well as basal forebrain cholinergic neurons and other components of the diffuse generalized activating system situated in the brainstem. The latter include the noradrenergic neurons of the locus caeruleus, serotonergic neurons of the dorsal raphe, and LDT/PPT cholinergic neurons.

Orexin/hypocretin had been identified by two independent groups. De Lecea et al. (75) identified two related peptides, hypocretin-1 and hypocretin-2, using a direct tag polymerase chain reaction subtraction technique to isolate mRNA from hypothalamic tissue. Shortly thereafter, while searching for endogenous ligands that would bind to orphan G-protein–coupled receptors, Sakurai et al. (76) identified the same two peptides, which they termed orexin-A (i.e., hypocretin-1) and orexin-B (i.e., hypocretin-2). The first data indicated that orexin neuropeptides may function as neurotransmitters because they were localized in synaptic vesicles and had neuroexcitatory effects on hypothalamic neurons (75). Orexin-A and orexin-B are 33- and 28-amino-acid neuropeptides, respectively, derived from a single precursor protein. Two orexin receptors have been identified (76). Orexin-A is a high-affinity ligand for the orexin type-1 receptor (OX_1R), which has an affinity for orexin-B that is one to two orders of magnitude lower. OX_2R exhibits equally high affinity for both peptides. Currently there are no ligands sufficiently specific for OX_1R and OX_2R to define their distribution. However, *in situ* hybridization studies of orexin receptor mRNA have shown a diffuse pattern, consistent with the widespread nature of orexin projections (79,85,86). Of the regions involved in vigilance state control, only the dorsal raphe and the locus caeruleus appear to show a predominance of OX_1R mRNA.

Orexin-A has been shown to excite noradrenergic neurons of the locus caeruleus, providing at least one mechanism by which orexin can promote wakefulness (83,87) and suppress REM sleep. However, orexin is not always excitatory, and the work by van den Pol et al. (88) reveals a variety and complexity of orexinergic effects at the cellular level. In the absence of an effective antagonist, antisense oligodeoxynucleotides against the mRNA for OX_2R were applied by microdialysis perfusion in the PRF of rats (89). This treatment increased REM sleep time during both the light and dark phases and, significantly, produced increases in behavioral cataplexy. These findings indicate that the control exerted by orexin on REM sleep may be mediated via the PRF brainstem nuclei that modulate the expression of REM sleep signs.

The importance of orexin in humans was confirmed by Nishino et al. (90), who reported that a group of narcoleptic patients had undetectable levels of orexin-A in cerebrospinal fluid. Furthermore, postmortem studies on brain tissue from narcoleptic patients revealed a loss of orexin-containing cells in the perifornical hypothalamus (91,92). In addition to being involved in the control of wakefulness and sleep, orexins may have a neuromodulatory role in several neuroendocrine/homeostatic functions, such as food intake, body temperature regulation, and blood pressure regulation (75,79,81,88; for review see reference 93).

CIRCADIAN FACTORS IN SLEEP

We noted that sleepiness depends on the extent of prior wakefulness. A factor such as increasing levels of adenosine

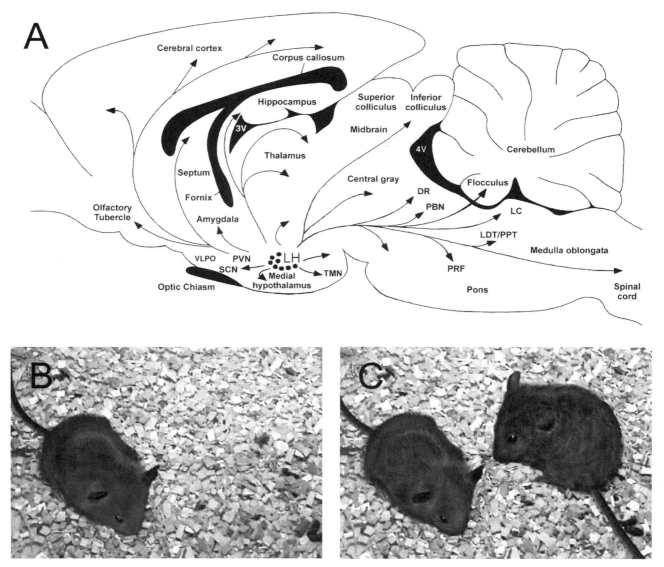

FIGURE 15.5. **A:** Schematic of a sagittal section showing the location of orexin-containing neurons [*dots*] in the lateral hypothalamus (LH) and their widely distributed projection pathways throughout the neuraxis. PBN, parabrachial nucleus. Other abbreviations as in Figure 15.4. **B, C:** Behavioral arrest in an orexin knockout mouse at age 4 weeks. The mouse in **(B)** was ambulating and burrowing and then abruptly collapsed onto his ventral surface during a direct transition from wakefulness to rapid eye movement sleep. This abnormal vigilance state transition, a characteristic of narcolepsy, was confirmed in other orexin$^{-/-}$ mice using electroencephalographic/electromyographic recording. **C:** A littermate approaches the mouse and, despite contact, does not awaken the mouse for several seconds. On awakening, the mouse abruptly continues the interrupted behavior without pausing. (Photographs courtesy of Jon Willie). With access to the *Cell* journal web site, videos of the orexin knockout mice [from Chemelli et al. (79)], view at *www.cell.com/cgi/content/full/98/4/437/DC1/.*) (See Color Figure 15.5 following page 526.)

in the basal forebrain (see section on Humoral Factors) may be a candidate for the underlying sleep factor(s) that creates this homeostatic drive for sleep. However, sleepiness also is dependent on a second major influence, circadian phase. Figure 15.6 shows on the same plot both the homeostatic and circadian drives for sleepiness so that their relative influences at different times of the day can be appreciated. *Circadian* means about a day (*circa* = about and *dies* = day), more specifically a rhythm with a period between 20 and 28 hours. The circadian rhythm of wakefulness in humans can be simplified as a sine wave function with a minimum that occurs between 4:00 and 7:00 a.m. in subjects with a normal daytime activity schedule. It is not surprising that accidents are most frequent around this

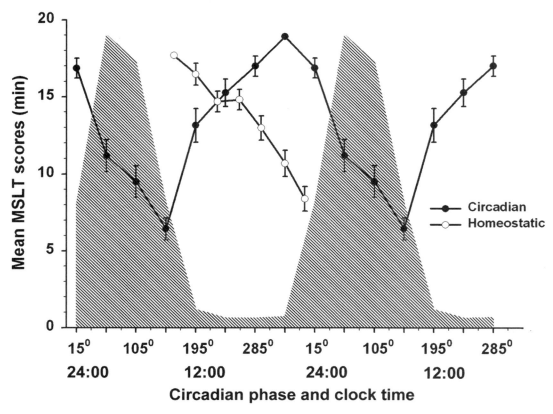

FIGURE 15.6. Circadian and homeostatic influences on sleepiness, displayed as multiple sleep latency test (MSLT) latencies (mean ± SEM). The ordinate is the mean latency score on the MSLT in minutes, that is, the time from when lights were turned out to the first polysomnographic indication of sleep, with a cutoff of 20 minutes. Higher scores on the MSLT indicate less tendency for sleepiness. Data were derived from a forced desynchrony protocol in which adolescent subjects were constrained to a schedule that lies outside the range of entrainment of the circadian pacemaker so that they "free run"; hence sleepiness could be assessed by examining MSLT results both as a function of circadian time, regardless of the time since waking, and as a function of the time since waking, regardless of the circadian phase. The abscissa displays the circadian phase, where 0° is the time of onset of melatonin secretion. In a healthy adolescent on a "normal" schedule, this occurs, on average, about 1 hour before sleep onset, or 22:00 hours. On this basis, indicative clock times are labeled on the abscissa. The *shaded area* was plotted, on an arbitrary scale, to show the variation in melatonin secretion, which is maximal during the circadian "night." The circadian drive for sleepiness *(solid circles)* and melatonin secretion were double plotted to show their diurnal rhythm, and the homeostatic drive for sleepiness *(open circles)* was plotted as if obtained during a "normal" day that falls between the two circadian "nights." The first point for the homeostatic drive is plotted at a time that would correspond to about 10:00 hours. Note the apparent paradoxical phase relationship between the circadian and homeostatic influences on sleepiness. (Adapted from unpublished data, cf. Carskadon MA. Consequences of the interaction between circadian and sleep systems for adolescents. *Physiologist* 1999;42:A16.)

circadian time of maximal sleepiness. There is a secondary peak of sleepiness that occurs about 3 p.m., corresponding to a favored time for naps. Human newborns do not have a strong circadian modulation of sleep, and some species, such as the cat, do not have much circadian variation even as adults.

Humans, like many other species, continue to show regular, circadian sleep/wake cycles and other physiologic and hormonal rhythms in the absence of a 24-hour light/dark cycle. These rhythms must depend on an internal "clock" or pacemaker that is self-sustaining in the absence of external

time cues and can be reset by changes in the environment. The mechanisms of this internal clock have been subject to research over many years, and significant progress has been made using genetic data obtained from a wide variety of species, including the bread mold *Neurospora,* the fruit fly *Drosophila,* and the mouse (94). The diversity of the species from which these results have been obtained underline a remarkable conservation of function in timekeeping in biologic systems during evolution.

The fundamental molecular events that act as the basis for the circadian clock depend on *clock genes,* which are defined as genes that can maintain a circadian rhythm of transcription in constant darkness and can be entrained to a new light/dark cycle in response to light exposure. Several such clock genes have been identified, including *clock, timeless (tim),* and *period (per)*. The details of the molecular events that lead to cyclical changes in the expression of these genes are beyond the scope of the present chapter (for comprehensive reviews see references 95–97). Summarized briefly, the clock genes are transcribed to encode proteins that subsequently dimerize. The resulting heterodimers are transported to the cell nucleus, where they inhibit the transcription of their own genes. The net result is a cyclical process in which proteins, once encoded, induce the feedback inhibition of their transcription. Then, as these protein products are subsequently degraded, gene transcription is again initiated and the cycle recommences on a circadian time scale. To this simplified picture must be added light-induced degradation of certain protein cofactors that play a role in the dimerization and nuclear translocation. Furthermore, the cells that express the cycling of protein encoding have output pathways whereby these cellular events are converted into transmissible circadian signals.

The emphasis in this chapter will be to review current results in the area of circadian influences on the sleep/wake cycle, especially where they are likely to impact sleep and psychiatric disorders. We will maintain a systems approach, although it must be emphasized that, as progress continues in understanding biologic rhythm-generating mechanisms at a molecular level, some conclusions may be modified. For example, the distributed pattern of expression of clock and clock-related genes outside the anatomic center of the circadian pacemaker in the mammalian central nervous system suggests that rhythm generation may be a more distributed and complex process than is currently understood, with additional pacemakers that are subordinate to a primary clock (but see references 98 and 99).

Entrainment of the Endogenous Oscillator

In principle, a circadian system depends on an endogenous pacemaker(s) that has efferent projections to couple activity to the variables that exhibit circadian function. Such a system must be able to respond to external conditions to mediate entrainment of the cycle to the environment. Under most circumstances, the light/dark cycle is a powerful stimulus for entrainment, and photic information couples the internal pacemaker to the environment. The hypothalamic suprachiasmatic nuclei (SCN), which is located in the anterior hypothalamus immediately above the optic chiasm (100,101), were recognized as important pacemakers during early studies of circadian timekeeping. The existence of direct retinohypothalamic projections terminating in the SCN (102,103) had suggested this function of the SCN before lesions of the nuclei were found to abolish circadian rhythms (104,105). Subsequent studies showed that the SCN could

exhibit rhythms when maintained *in vivo* while isolated from the rest of the brain by transections (106,107) and *in vitro* (108,109). These results confirmed that loss of circadian rhythmicity after SCN lesions was not due to an interruption of effector mechanisms for the expression of circadian function. If fetal anterior hypothalamus containing the SCN was transplanted into the rostral third ventricle in adult animals rendered arrhythmic by SCN lesions, rhythmicity could be restored (110,111).

In the rat, the SCN have been found to exhibit a prenatal rhythm independent of maternal rhythms (112,113). When prenatal oscillations first develop, the SCN are poorly differentiated (114), suggesting that the pacemaker function of the SCN is a network property of individual neuronal oscillators that are coupled before they eventually project to effector functional groups as the central nervous system matures. The SCN can be thought of as the principal locations of neurons that express clock genes in mammals. It is apparent that our current understanding of the molecular basis of cyclical molecular changes in these neurons is consistent with these earlier studies indicating that the SCN are composed of individual neuronal oscillators. However, more recent work has challenged the concept that the SCN are the only or even the dominant pacemaker. For example, it now is known that shorter (i.e., ultradian) rhythms become apparent after SCN lesions (115), and some circadian rhythms even survive SCN ablation. An example of the latter is the anticipatory activity associated with food availability under a restricted food schedule in the rodent (98).

Under normal circumstances, the light/dark cycle is the dominant entraining stimulus, and photic information sets the phase of the internal pacemaker. The SCN have been shown to receive two sources of photic information that participate in its entrainment. The primary source, the retinohypothalamic tract, projects from retinal photoreceptors to the anterior hypothalamus, including the SCN, the paraventricular nucleus, and the lateral hypothalamus (116). A secondary source of photic information passes from the retina to the thalamic intergeniculate leaflet of the lateral geniculate nucleus (117), which in turn projects to the SCN. Lesion and stimulation studies have shown that the intergeniculate leaflet is also a source of nonphotic information for the SCN (118–120), and additional nonphotic information to the SCN is likely to be provided by inputs from the median raphe, ventral subiculum, and infralimbic cortex (121). The importance of nonphotic phase setting of the pacemaker only recently has become known, as careful studies were necessary before it could be confirmed that the exogenous influences being examined were not simply imposing a rhythm without setting an underlying oscillator (122). In rodents, activity in anticipation of food availability (123) and arousal induced by varied stimuli, including wheel running (124) or sleep deprivation (125), have been established as nonphotic synchronizers, or zeitgebers. These stimuli interact in complex ways to affect the pacemaker (122). Thus, the circadian oscillator in these species is entrained by arousal and by the light/dark cycle.

Circadian Rhythms and the Sleep Period

A fundamental circadian pattern that reflects a basic adaptation of the organism to its environment is expressed as sleep/wake behavior. The daily appearance of a major sleep period, typically during the night, is one of the most characteristic features of human sleep. The appearance of the sleep phase is a circadian function that is known to be dependent on the SCN, as SCN lesions in rodents result in a random distribution of sleep across the nycthemeron (126). The initial studies with rodents suggested that total sleep remained the same after SCN lesions, that is, the SCN did not promote either wakefulness or sleep, but only gated the appearance of the homeostatic sleep process. A later study with squirrel monkeys clearly demonstrated that this was not the case in this primate, in which total sleep time was increased after SCN lesions (127). This important result led to the concept that the circadian tendency to sleep facilitated the initiation and maintenance of wakefulness by opposing the homeostatic sleep tendency. Hence, the circadian influence helps to consolidate sleep in diurnal primates, a concept that is addressed in greater detail later (see section on Combination of Circadian and Homeostatic Drives for Sleep). Subsequently, the concept of an interaction of circadian and homeostatic influences was extended to sleep in rodents. This is a small effect in rodents and had not been noted earlier because of the lack of a consolidated sleep phase.

Projections from the SCN to the adjacent anterior hypothalamus, retrochiasmatic area, lateral hypothalamus, basal forebrain, and paraventricular thalamus (128) are important in mediating the effect of the SCN on the expression of the sleep/wake rhythm. Glutamatergic innervation from the SCN to orexin cells in the lateral hypothalamus (see section on Orexin and the Control of Sleep and Wakefulness) has been described in the rat, with synapses on the cell bodies and proximal dendrites of the orexin cells (129). We noted that widespread projections of orexin cells to the PRF, locus caeruleus, dorsal raphe, and LDT/PPT areas are critical in vigilance state control. The anterior and retrochiasmatic hypothalamic areas also innervate these structures.

In addition to the sleep/wake cycle, variations in several physiologic and hormonal parameters, such as core body temperature, melatonin release, and plasma levels of glucocorticoid hormones, follow a circadian oscillatory rhythm. Studies over a number of years have been directed at accurate determination of the periodicity and phase of the circadian pacemaker and how this daily pacemaker drives observable physiologic functions. Under normal circumstances, the timing of the sleep period shows a consistent phase relationship with other circadian rhythms. For example, core body temperature typically peaks in the evening, begins to decrease before sleep onset, reaches its nadir in the last part of the night, then starts to increase before awakening, increasing progres-

sively during the day (130,131). However, the characteristic pattern of these variations results from a complex interaction among the circadian pacemaker itself, the sleep/wake cycle, and exogenous stimuli that modulate these drives. Hence, isolating the contribution from the pacemaker *per se* was not a trivial task in early studies. The development of the "free running" protocol, which is conducted in an environment totally lacking in time cues, was a significant milestone in this work, because such an environment should reveal the period of the endogenous pacemaker independent of environmental influence. Using this protocol and measuring the rest-activity cycle provided data on the remarkable stability and lack of interindividual variation in the endogenous rhythm in many species (122).

With human subjects under completely "free-running" conditions, it was believed that the circadian pacemaker and the rest-activity cycle could drift apart, a condition known as *spontaneous internal desynchrony* (132). In some subjects, the circadian pacemaker of core body temperature and the rest-activity cycle became uncoupled. It now is believed that this effect likely was an artifactual consequence of the conditions imposed on subjects in these isolation studies, in particular, they were not allowed afternoon naps (133). Nevertheless, these earlier studies that adopted the free-running protocol provided important data. For example, under these conditions the period of the core body temperature pacemaker occasionally was found to be shorter than 24 hours when the rest-activity cycle desynchronized, with an average period shorter than the period of the intrinsic pacemaker. Hence, even in the majority of subjects in whom the circadian pacemaker and rest-activity cycle remained in synchrony, the intrinsic period of the endogenous pacemaker probably was being modulated by rest-activity rhythms. In continuing efforts to isolate the pacemaker, constant routine conditions were adopted (134,135). In this protocol the subject was kept awake with a constant posture and a constant level of activity for extended periods. These studies provided information on physiologic parameters driven by the circadian pacemaker, but their highly restrictive nature limited their usefulness and the time over which they could be continued. Thus, determination of the period of the endogenous oscillator remained imprecise with this protocol.

For these reasons, more accurate estimation of the endogenous oscillator can be obtained only during desynchrony but when the period of the circadian pacemaker cannot be entrained to the rest-activity cycle. Furthermore, the study needs to be continued for a sufficiently long time for a mean estimate of the pacemaker period to be made. These considerations led to the adoption of the forced desynchrony protocol, originally developed by Kleitman (136) in 1938. In this type of study, subjects are kept on a rest-activity cycle that is sufficiently long (e.g., 28 hours) to prevent entrainment of the circadian pacemaker to this rhythm, thus exposing the natural rhythm of the oscillator. Although this protocol had been extensively adopted in studies conducted over many years to determine the period of the endogenous circadian pacemaker, results were variable and suggested that humans, unlike other species, did not show a precise circadian period. A modification to this protocol apparently has revealed the precision and stability of the circadian oscillator in humans for the first time (137). A critical aspect of this study was exposure of the subjects to very dim light (about 10–15 lux) throughout, because even ordinary levels of room lighting can impact the circadian system in humans. In this way, Czeisler et al. (137) were able to determine that the period of the human circadian pacemaker is 24.18 hours. They also reported that healthy older subjects had the same periodicity, stability, and precision as younger subjects. This latter result has important implications for circadian rhythm disorders and sleep disturbances that are more common in older patients, a topic that is reviewed later (see section on Circadian Rhythm Sleep Disorders). However, can we be sure that an accurate estimate of the endogenous circadian period in humans has now been determined? Not necessarily, because subjects enter an experimental paradigm with a circadian rhythm "inertia," an aftereffect of entrainment that can last for months (138,139). This important factor, in addition to any remaining influence of exogenous conditions on the period of the pacemaker, may make it impossible to determine the intrinsic period with greater precision.

Rest-Activity Cycles

Results from the free-running studies in humans show that rest-activity rhythm can influence the circadian period. This effect essentially can be considered a regular resetting of circadian phase. We also noted earlier (see section on Entrainment of the Endogenous Oscillator) that arousal and activity can be a zeitgeber in the rodent. Taken together, this work suggests that changes in arousal, whether endogenously generated or enforced, can act as input to the circadian system.

In rodents, ultradian rhythms are evident after lesions of the SCN (115) or of specific output pathways of the SCN, including the dorsomedial hypothalamus (140). Ultradian rhythms also are evident in humans. For example, nocturnal sleep in humans typically is expressed as an alternation of non-REM and REM sleep (see section on Sleep is Organized into a Definite Structure) with a periodicity of about 90 to 120 minutes in adults (141). Kleitman (142) proposed that this cyclical organization of sleep reflected the existence of a basic rest-activity cycle (BRAC) that continues throughout the 24-hour period. At night the BRAC would be expressed as the non-REM/REM cycle and during the day as periodic oscillations of drowsiness and alertness. EEG (143–146) and vigilance performance (147) studies confirmed this daytime periodicity, even while subjects were trying to maintain vigilance during prolonged performance testing. The reciprocal interaction model of McCarley and Massaquoi (61) (see section on Rapid Eye Movement Sleep) described a non-REM/REM cyclicity based on interactive inhibition of neuronal populations, although data are insufficient to extend the model to oscillations of vigilance during wakefulness.

However, in narcoleptics the timing of nocturnal REM periods can be predicted from daytime episodes (148), suggesting that daytime oscillations of vigilance in normal subjects also may be in phase with the nocturnal sleep cycle. This would be an essential prediction of the BRAC hypothesis and allow extension of the reciprocal interaction model to changes in vigilance throughout the nyctohemeron. Studies that unmask and determine the mechanisms of ultradian rhythms in rodents may provide important data for such an extension to the model.

REM propensity follows an ultradian rhythm of about 90 min, but it also demonstrates circadian variation. Under a forced desynchrony protocol, the peak in REM propensity occurs coincident with the peak of sleepiness (149). Interestingly, if subjects under such a protocol elect to initiate their sleep phase close to this peak of REM sleep propensity, they frequently exhibit sleep-onset REM periods (149), a rare phenomenon usually associated with narcolepsy. A study in rodents demonstrated that changes in cerebrospinal fluid levels of orexin follow a diurnal pattern, with a minimum that occurs toward the end of the light phase (150). If these data can be tentatively extrapolated, they suggest that orexin levels also may follow a diurnal pattern in humans, with a minimum corresponding to the time of maximum sleepiness and maximal REM propensity. The normal physiologic role of such an oscillator coupled to orexin remains unknown, but it may involve the way that feeding and food availability can influence the circadian rhythm or act as a subordinate oscillator. Current preliminary experimental data examining these possibilities are conflicting. It is apparent, however, that any disturbances in such an oscillator, linked to rest-activity and modulating circadian phase, may have important implications for understanding some sleep and rhythm pathophysiologies.

Perturbations in the period of endogenous oscillators, both circadian and ultradian, especially those that lie outside the range of entrainment of the light/dark cycle, are likely to create sleep disturbances and altered phase relationships among physiologic functions that are under circadian control.

Combination of Circadian and Homeostatic Drives for Sleep

We noted that the distribution and depth of the sleep period depend on both the circadian phase and the homeostatic drive for sleep, which is determined by the extent of prior wakefulness (Fig. 15.6). The homeostatic process reflects the propensity for sleep, which builds up during wakefulness and dissipates during non-REM sleep. As noted in the section on Circadian Rhythms and the Sleep Period, the circadian influence on sleep in a normal subject acts in concert with this homeostatic drive to maintain sleep by consolidating the sleep period. The circadian pacemaker achieves this consolidation by a mechanism that, at first sight, seems to be in a paradoxical phase relationship to the normal timing of the sleep period. This follows from the fact that the circadian drive for wakefulness is strongest in the evening hours, just before the normal time of sleep onset. Conversely, the circadian drive for sleepiness is strongest in the morning hours, just before the usual waking time. This process helps to consolidate the sleep phase despite the homeostatic drive for sleepiness in the evening and homeostatic drive for wakefulness in the morning (151).

Combining the characteristics of the endogenous sleep/wake rhythm with those of a circadian oscillator has led to testable mathematical models of sleep propensity. The most significant were developed by Kronauer et al. (152), who emphasize circadian control, and by Borbély (153), who emphasizes the extent of prior wakefulness. These workers did not specify the nature of the underlying sleep factor(s), but candidates were discussed earlier in the section on Humoral Factors. A two-process single oscillator model was proposed by Borbély (153). In this model, sleep is seen as the net result of two processes. The first process, S (or sleep propensity), builds during wakefulness and declines exponentially during sleep and is indexed by the δ power of EEG slow wave activity, as measured by power spectral analysis. The second process, C, is an endogenous circadian oscillator that closely parallels core body temperature. Another model based on two oscillators was proposed by Kronauer et al. (152). In their model, a strong X-oscillator that governs core body temperature and REM sleep is modulated by a weak Y-oscillator that governs the sleep/wake state. Both state and amplitude are defined in this model and differential equations are used to solve the model and make predictions. Despite excluding the homeostatic properties of sleep, this model has been found to make useful heuristic predictions.

We should note that circadian sleep/wake and physiologic rhythms typically are treated as sinusoidal variables in these models; these assumptions are not supported by actual data. The sleep/wake state is essentially a binary process, and the actual shape of the variation in physiologic variables across the nyctohemeron is asymmetric and under normal conditions is modulated by changes related, for example, to activity and sleep onset. Recording of core body temperature under several different conditions, including normal expression of the sleep/wake cycle, sleep deprivation with constant activity over 24 hours, and continuous bed rest with minimal activity but normal sleep/wake behavior, might provide more accurate modeling data after appropriate subtractive manipulation (154).

SLEEP MEDICINE

Although treatment of sleep disorders in the past was the principal domain of the primary care physician, sleep medicine now is recognized as an important specialty in its own right, and sleep disorder centers are found worldwide. The sleep medicine specialist plays a vital role in the treatment of dyssomnias and parasomnias and in the prevention of the psychiatric, psychological, neurologic, forensic, cardiovascular, and respiratory complications of sleep disorders. Our emphasis is on neuropsychiatric issues of sleep medicine, with particular regard to the presentation, differential diagnosis, and treatment of significant sleep disorders (Fig. 15.7).

The beginning of sleep medicine can be traced back at least 100 years, although it required the development of EEG recording (Berger in 1930) and the measurement of respiration (Gujer in 1928) and cardiac function (Klewitz in 1919) while the subject was sleeping to create the discipline as it is known today. The abnormalities and disorders of sleep were first described by the French physician Roger in 1932, a work that can be considered a precursor to the *International Classification of Sleep Disorders* (ICSD), first published by the American Academy of Sleep Medicine (previously ASDA) in 1977 (1). The 40 years prior to publication of the ICSD are noted for many of the discoveries about sleep that are the basis for the current practice of sleep medicine. Aserinsky, working with Kleitman at the Division of Psychiatry of the University of Chicago Clinics, discovered REMs during sleep. A few years later Dement and Kleitman were the first to recognize the sleep cycle. In Britain, Oswald at Edinburgh University began clinical sleep studies in the 1950's, and in France, Jouvet initiated seminal basic research studies and significant clinical work. Apnea was first identified by French (Gastault et al.) and German (Jung and Kuhlo) researchers independently in 1965. Dement had established a sleep medicine center at Stanford and began the study of narcolepsy in 1963. Guilleminault joined Dement from France in 1972 and added respiratory and cardiac measures to routine clinical sleep recording, enabling sleep apnea to be discovered at Stanford in 1975. Increasingly wide interest in sleep medicine can be dated to the 1970's, when the American Board of Sleep Medicine was founded.

SLEEP DISORDERS ASSOCIATED WITH NON–RAPID EYE MOVEMENT SLEEP

Insomnia and Hypersomnia

Insomnia is defined as insufficient, unsatisfying, or nonrecuperative sleep despite an adequate opportunity to sleep (155,156). Insomnia can be a presenting complaint, and the perception of the insomniac is important in diagnosis. Here we consider insomnia to be a symptom of disturbed sleep, which might be manifested as difficulty in getting to sleep or in maintaining sleep because of multiple nocturnal or early morning awakenings. Clearly, such sleep disturbances can and should be confirmed by polysomnographic recording. In a Gallup survey, 9% of American adults reported chronic insomnia and an additional 27% reported transient insomnia during the preceding 12 months (157). Thus, a remarkably high 36% of the surveyed population exhibited

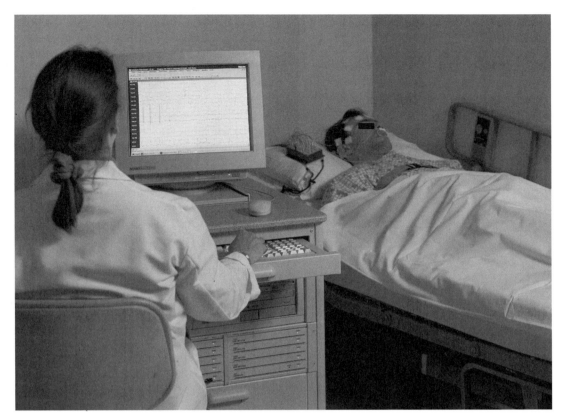

FIGURE 15.7. The polysomnographic technician prepares to record sleep data from a patient in a sleep clinic. The electroencephalographic electrodes are positioned at standard locations over the scalp. Digital polysomnography systems, such as the Grass Instruments system being used here, allow data to be recorded directly to the computer, thus avoiding the need for paper records. The data are easier to archive, analyze, and review. (Photograph courtesy of Astro-Med, Grass Instruments Division.)

some form of insomnia over this period. Only about 30% of all insomniacs actually seek help for the condition, however, which makes accurate unbiased data on the prevalence of insomnia difficult to obtain. Twenty percent of insomniacs have taken some form of prescription medication for insomnia, and 40% have self-medicated with over-the-counter compounds.

The diagnosis of insomnia depends initially on interviews with the patient, the sleeping partner, and the family. Whether the insomnia is related to difficulties in getting to sleep, multiple nocturnal or early morning awakenings must first be determined and, if necessary, confirmed in the sleep clinic. Associated problems, such as excessive daytime sleepiness, difficulties in maintaining concentration, mood disorder, or irritability, can be ascertained during an interview. Sleeping conditions, the normal need for sleep (i.e., short sleeper, long sleeper; see section on Ontogeny of Sleep), and constraints on sleeping related to lifestyle, sleep irregularity, or shift work can impact insomnia. It is important to evaluate the personality of the patient and the psychological context of the insomnia. The possibility of *night terror, somnambulism* (see section on Arousal Disorders), or *nocturnal epilepsy* (particularly focal) should be considered. If sleep is fragmented or disturbed it can produce excessive daytime sleepiness, so the patient should be questioned about sporadic medical conditions, such as asthma or heart disease; psychiatric conditions; or use of medications that might affect sleep. A detailed interview with the patient will alert the clinician to problems concerning sleep hygiene, such as excessive caffeine intake or irregular bedtime hours. A sleep log, in which the patient is required to note sleep and waking times over several weeks, is a useful adjunct during diagnosis. Any possibility of a medical condition will require polysomnographic evaluation, accompanied by a *multiple sleep latency test* (MSLT) if necessary (158). The latter evaluates the tendency for daytime sleepiness at several different times during the day by measuring the latency to fall sleep in a situation that is conducive to sleep.

Classification and diagnosis of insomnia vary with the system used (159), and there is no international consensus. Although none of the current classification systems follows an etiologically based scheme, such a classification can be useful for appreciating the multiple causes of insomnia. In considering insomnia in this way, we can see how the etiology of insomnia is related to disordered mechanisms of sleep onset and sleepiness:

1. Environment-related insomnia is caused by factors such as stress, high altitude, or rapid travel across time zones. This type of insomnia usually is transient.
2. Drug-related insomnia is provoked by use of any medication that affects sleep.
3. Psychiatric-related insomnia is secondary to a psychiatric condition. This type of insomnia is the most chronic and by far the most prevalent.
4. Insomnia caused by a medical condition.
5. Insomnia related to a sleep disorder, such as sleep apnea syndrome, restless legs syndrome (RLS), or narcolepsy.

A few types of insomnia, presented by a small percentage of patients, do not fit easily into such a classification because of lack of knowledge of the etiology. Of these, we consider two (sleep state misperception and psychophysiological insomnia) because of their importance from a psychiatric perspective. *Sleep state misperception* is the complaint of insomnia associated with an inability to appreciate the true amount of sleep (160). About 5% of patients who complain of insomnia have normal sleep by polysomnographic criteria, including normal sleep efficiency index (i.e., a measure of total sleep time expressed as a percentage of time spent in bed), total sleep time, and MSLT results. Despite these apparently normal objective sleep measures, these patients present with convincing and honest complaints of insomnia and excessive daytime sleepiness, which they believe is due to their poor sleep. In extreme instances they show sleep agnosia, a total inability to perceive subjective sleep, and they affirm that they have not slept for months or even years (161). Nocturnal polysomnographic recording usually is not sufficient to convince these patients of the reality of their sleep. Poor sleep perception has been proposed as a variant of psychophysiological insomnia, in this case characterized by imperceptibly small objective changes to sleep, so this type of insomnia may not be a true clinical entity. It also is feasible that criteria of poor sleep not currently identifiable on a standard polysomnogram eventually may be discovered.

Psychophysiological insomnia, or learned insomnia (162), is similar to the sleep seen in patients with generalized anxiety disorder (see section on Sleep and Anxiety Disorders). This type of insomnia typically is chronic and usually is the result of synergism between somatized tension/anxiety and a negative affect toward sleep, exemplified by excessive worry about not being able to sleep. Despite, or perhaps because of, intense effort to sleep, the patient is unable to sleep in the normal sleep environment. In contrast, sleep occurs easily when the patient is not attempting to sleep, for example, while reading or watching television. Although not specific to the disorder, the most frequently observed polysomnographic characteristics of sleep in these patients is prolonged sleep onset latency and increased nocturnal awakenings, combined

with a decrease in sleep efficiency. The organization of REM sleep is little affected, although the percentage of REM sleep can be reduced if sleep fragmentation is extreme. Treatment of chronic psychophysiological insomnia begins with review and application of the principles of sleep hygiene, including retiring to bed when drowsy, rising at the same time each day, avoiding naps, using the bed to sleep and not to eat or watch television, and refraining from alcohol and caffeine consumption close to bedtime. Concurrent cognitive or behavioral therapy also is effective.

Fatal familial insomnia, a rare condition, is significant because it impacts our understanding of the involvement of the thalamus in sleep mechanisms. A genetically linked, prion protein disorder (163–165), fatal familial insomnia is characterized by progressive and selective neurodegenerative atrophy of thalamic nuclei. The onset of the disorder, frequently at about age 50 years, is signaled by difficulty in going to sleep, and this typically evolves within a few months to total insomnia that is refractory to treatment. Polysomnographic monitoring of these patients reveals a progressive decline in the duration of non-REM and REM sleep, culminating in total insomnia that progresses to stupor and then coma and death, usually within 15 months.

In contrast to insomnia, *hypersomnia* is marked by profound, prolonged nocturnal sleep, difficult morning awakenings, and excessive daytime sleepiness, but less irresistible sleep onset than in narcolepsy (166). There is evidence that *idiopathic hypersomnia* appears in families, indicating a possible genetic linkage, which also is suggested by early onset: this type of hypersomnia usually occurs before age 25 years. Polysomnographic evaluation reveals shortened sleep onset latency and an increase in total sleep time, but no significant change in sleep architecture. *Posttraumatic hypersomnia* is defined as occurring within 12 months of a head injury that is otherwise benign and produces no other symptoms (167). Hypersomnolence in this disorder is typically at its most severe during the immediate posttraumatic period, before the patient gradually returns to normal sleep over a period that can vary between several weeks and several months. It should be noted that some of these patients show a gradually worsening hypersomnia as the posttraumatic interval increases. Radiologic examination or computed tomographic scans of these patients can reveal posterior hypothalamic or pontomesencephalic involvement, so it is probable that brainstem arousal mechanisms (see section on Neurophysiologic Basis for Electroencephalographic Synchronization and Desynchronization) are compromised in this form of hypersomnia.

Recurrent hypersomnia, another type of hypersomnia, is characterized by repetitive episodes of excessive sleepiness that last between several days and several weeks and recur approximately every month. Like posttraumatic hypersomnia, recurrent hypersomnia (168,169) can be caused by neuropathology, particularly intraventricular tumors, encephalitis, head trauma, or stroke. The *Kleine-Levin syndrome* is a significant but rare type of recurrent hypersomnia associated with hyperphagia, hypersexuality, irritability, aggression, and hallucinations. Males in late adolescence or their early twenties are principally affected, and onset can be either abrupt or gradual. The initial episode frequently occurs after fatigue or stress or the appearance of flulike symptoms. The patient sleeps about 20 hours per day for 1 to 3 weeks, and episodes recur at intervals ranging from several weeks to several months. This benign disorder with unknown etiology usually remits spontaneously after a gradual decline in the duration and frequency of episodes over several years (170). After the initial episode, differential diagnosis against depressive disorder, schizophrenia, encephalitis, and other neurologic disorders is necessary. Other diagnoses can be eliminated because the characteristic hypersomnia is accompanied by abnormal behaviors but no neurologic deficit.

Arousal Disorders

Sleepwalking, confusional arousals, and night terrors form a symptomatic cluster because they show several similarities, including age of onset. Typically they resolve spontaneously during adolescence, but in adults they commonly occur together in the same individual. They occur at about the same time of the night during stage 4 SWS, when the first REM sleep period would be expected, and their episodes are similar in length. They are considered disorders of arousal in that they all reflect impaired mechanisms of arousal from sleep. Importantly, there is evidence that sleepwalking, confusional arousals, and night terrors show similar familial patterns of incidence, typically in families of deep sleepers, and this implies genetic linkage (171). For example, one case in the literature describes a family of sleepwalkers awakening to find themselves all sitting around the dining table without any memory of how they got there. Arousal disorders were first described clinically as a related cluster by Broughton (172), but this early description was incomplete in the case of confusional arousals. When the intensity and/or frequency of an arousal disorder necessitate treatment in the adult, medications that suppress SWS such as benzodiazepines, including diazepam, or tricyclic antidepressants usually are effective when administered in the evening before bedtime. Children with arousal disorders are medicated only if behaviors are so severe that they are likely to cause injury.

Sleepwalking or somnambulism (173) is characterized by automatic involuntary ambulation that occurs during non-REM sleep. Onset typically is between ages 3 and 6 years, which, as noted earlier (see section on Ontogeny of Sleep), corresponds to the age at which δ sleep increases markedly. Sleepwalking usually remits spontaneously with puberty (174). Its prevalence has been estimated at about 20% in children up to age 15 years, although isolated episodes occur in almost all children. In adults, the prevalence is about 2.5%. In all age groups, the incidence in males is slightly higher than in females. In addition to maturational changes that clearly play a role in the appearance of sleepwalking, several studies have demonstrated the importance of innate factors. Incidence increases from about 20% when neither parent is affected to about 60% when both parents walked in their sleep and/or had night terrors at some time during their lives (171). Concordance is higher in monozygotic twins. The somnambulist shows no evidence of epilepsy or any form of seizure activity. Like other arousal disorders, sleepwalking typically occurs during stage 4 non-REM sleep, about 1 to 3 hours after sleep onset and at a time when the first period of REM sleep would be expected. During an episode of sleepwalking, the EEG often is dominated by slow 1- to 4-Hz δ waves, but their amplitude is less than during the preceding SWS.

At the onset of an episode, the child will get up in the middle of sleep and begin undirected walking while staring vacantly ahead with open eyes, usually for 3 to 5 minutes. More than one incident in the same night is rare, and the child will not remember the episode by the next day. Fever or stressful life events can trigger sleepwalking. In adults, automatic aggressive behavior might occur during an episode of somnambulism and it is this fact, as well as the way that sleepwalkers can injure themselves, that make this a significant sleep disorder requiring treatment. Because of the timing of sleepwalking during the night's sleep, it is possible that sleepwalking reflects an immaturity in the control mechanisms that gate forebrain arousal from slow wave δ sleep into the beginning of a REM sleep episode.

Night terrors (also called *pavor nocturnus* or *sleep terrors*), like sleepwalking, first appear at about age 4 years. They usually remit spontaneously by puberty (174), although they can occur in adults. The prevalence of night terrors has been estimated at about 6% during preschool years but is less than 1% in adults. A typical episode lasts for 1 to 15 minutes; begins between 1 and 3 hours after sleep onset; and only rarely occurs more than once per night. The child typically sits up, cries with terror, screams, and shows intense anxiety accompanied by autonomic arousal, including polypnea, tachycardia, sweating, mydriasis, and muscular hypertonia. By the next day, amnesia for the episode is complete. Night terrors occur during stage 4 non-REM sleep, usually at the beginning of the night when the first episode of REM sleep might be expected. The similarity of timing of these episodes to sleepwalking indicates that night terrors also might be the result of a disorder in the mechanism that initiates REM sleep at the end of stage 4 δ sleep. Night terrors should be distinguished from *nightmares,* which occur primarily during REM sleep in the latter part of the night. Nightmares also are accompanied by less autonomic activation, such as tachycardia and sweating, than night terrors. Furthermore, after awakening from a nightmare, the child is lucid and frequently has a precise memory of the dream scenario. This contrasts with the confusion and amnesia that follows a night terror.

A mild form of night terror is *confusional arousal,* which is an awakening from stage 4 non-REM sleep usually during the first part of the night, accompanied by confusion, disorientation, and amnesia of varying degree (174). This sleep disorder occurs particularly in the younger child, less than 5 years old. The arousal may be accompanied by vocalization and thrashing movements. An episode may last up to 40 minutes, although 5 to 15 minutes is typical. The manifestation of confusional arousal suggests that sleep is being maintained in the presence of brain activation. This hypothesis is supported by the presence of θ or nonreactive diffuse α on the EEG during episodes. Isolated instances can be caused by a child being overtired, or when the drive for sleep is excessive because of fever or illness. Confusional arousals are infrequent in older children or adolescents, in whom they may be caused by psychological factors. Recurrent confusional arousals and night terrors, especially in the adult, show a significant familial pattern, implying genetic linkage (171). Nocturnal confusional arousal in children should be distinguished from a confused state of waking that occurs in the morning in about 50% of patients presenting with hypersomnia. The term *nocturnal sleep drunkenness* has been used to describe this state, which can continue for some time with signs of microsleep on the EEG (175). Another type of confused awakening with hallucinations occurs in the elderly during a REM sleep episode.

CIRCADIAN RHYTHM SLEEP DISORDERS

Patients with sleep disorders that involve the circadian control of sleep and wakefulness show a rhythm of sleepiness that is markedly abnormal, appearing later or earlier in the day. For comparison, Figure 15.6 shows the normal temporal relationship between the circadian rhythm of sleepiness and the daytime hours, as defined by the increasing homeostatic sleep drive. Delayed sleep phase syndrome is the more common complaint; advanced sleep phase syndrome is rare and found only in the elderly. The existence of these disorders show us how circadian pacekeeping mechanisms are critical to normal functioning and how these mechanisms are controlled by exogenous synchronizers, such as the light/dark cycle and social activity. Profound changes to the external environment using enforced schedules or bright lights currently are the most effective treatments for these disorders. The discovery that exposure to bright light could shift the phase of endogenous rhythms, independent of the sleep/wake cycle, led to the concept of a pacemaker entrained by environmental cues and that this pacemaker, in turn, entrains other oscillators. In 1983, Wever (132) showed that bright light could increase the range of entrainment of the core body temperature rhythm in humans. Subsequent work demonstrated that endogenous rhythms could be shifted by altering the bright light/dark cycle, that is, bright light shifts the phase of the oscillator, and this effect is independent of the sleep/wake cycle. This led to the demonstration that evening exposure to bright light phase delays circadian rhythms (i.e., shifts to a later time) and bright light in the morning advances them

(i.e., shifts to an earlier time). These discoveries introduced the treatments described in this section.

The typical source of light used in therapy is a battery of fluorescent tubes behind an ultraviolet filter, providing a light intensity in the range from 2,500 to 10,000 lux at the eye (176). Side effects of this treatment usually appear at the beginning of treatment and can include anxiety, weariness, irritability, headache, nausea, eye fatigue, or redness and irritation of the eye. Migraine attacks also can be triggered by bright light exposure. Retinopathies, including macular degeneration and retinitis pigmentosa aggravated by sun exposure, aphakia, albinism, congenital iris abnormalities, cutaneous porphyria, and photoallergies, are the principal contraindications. Psychiatrists should note that any patient who has shown signs of aggression or has a history of psychotic hallucinations should be monitored carefully during light therapy.

Delayed sleep phase syndrome (DSPS) was first described by Weitzman et al. (177). It is characterized by a circadian sleep/wake rhythm that is persistently unaffected by normal external synchronizers and in which the major sleep period is delayed with respect to clock time. DSPS can affect individuals of all ages, but the most frequent age of onset is during adolescence or young adulthood. The prevalence of the syndrome is probably in the range from one to four per 1,000; men are twice as likely to be affected as women (178). A patient with DSPS usually shows sleep-onset insomnia and difficulty in morning awakening. When a strict sleep schedule need not be maintained, complaints about insomnia diminish and the patient sleeps normally but on a delayed phase. Diagnosis of the syndrome depends on an accurately kept sleep log covering at least a 2-week period and/or monitoring by actigraphy to quantify the phase shift of the sleep period. Nocturnal polysomnography reveals no major abnormality other than very prolonged sleep onset latency when recording is initiated at normal bedtime.

There are two standard ways of treating DSPS; both methods rely on changing the sleep phase with respect to clock time. The first, chronotherapy, was developed by Czeisler and colleagues (177,179). During chronotherapy, bedtime is delayed by 3 hours each day, which requires the patient to live a 27-hour day. Normal daytime activities, particularly the hours of light exposure and meals, follow this change in sleep time. The amount of sleep allowed during treatment is determined from sleep log data and set so that a small sleep debt is maintained over the 27-hour period to facilitate sleep onset. Once the new sleep phase is established and the patient returns home, bedtime and awakening hours must be maintained for at least 3 weeks so that the new sleep rhythm can become established. The second treatment option depends on exposure to a bright light, a method that was pioneered by Lewy et al. (180,181). When undergoing this treatment, the DSPS patient is exposed to bright light for 2 to 3 hours each morning. Restricting light exposure after 4 p.m. with sunglasses has been found to be a useful adjunct to morning light therapy. The sleep phase can be improved after about 2 weeks of this treatment, but relapse frequently occurs once treatment ends.

Advanced sleep phase syndrome (ASPS) is similarly a disorder of the timing of the sleep phase, but in this case the major sleep period is advanced with respect to clock time (182). The ASPS patient shows an irresistible need for sleep before normal bedtime, followed by early morning awakening when return to sleep is difficult. The syndrome is rare and essentially is observed only in the elderly, in whom the presenting symptoms frequently can be confused with depression. As noted by Czeisler et al. (137) (see section on Circadian Factors in Sleep), the early entrainment of the sleep cycle and early morning awakening of the healthy older subjects in their study could not be attributed to shortening of the circadian period as a normal consequence of aging. Hence, age-related changes in the homeostatic drive for sleep, or age-related changes in the entrainment of the circadian pacemaker to the light/dark cycle, must be considered possible alternative etiologies for ASPS. These possibilities are in addition to the pathologically shortened circadian period that previously was described in some elderly ASPS patients (183). It should be noted that an autosomal dominant familial ASPS, which mapped to a mutation within the sequence for a human homologue of the *per* gene, has been reported (184). These data may have profound implications for understanding the underlying pathophysiology of ASPS and other circadian rhythm disorders (see section on Circadian Factors in Sleep) (185).

Chronotherapy is effective for treatment of ASPS: bedtime is systematically advanced by about 3 hours each day until the desired time of sleep onset is achieved (177,179). Once sleep is fixed at the desired time of day, bedtime and awakening hours are maintained for several weeks so that the new sleep period can be established. Light therapy is an alternative approach, or it can be used as an adjunct to chronotherapy (180,181). For light treatment to be effective, the patient is exposed to 2 to 3 hours of bright light in the early evening during a period that corresponds as closely as possible to the hours preceding the minimum in core body temperature.

Hypernyctohemeral syndrome, or non–24-hour sleep/wake syndrome, is a disorder in which the sleep/wake cycle is no longer synchronized to a 24-hour period and drifts with respect to both clock time and core body temperature rhythm by 1 to 2 hours per day (186). In these patients, the capacity to phase advance is insufficient to compensate for the difference between the endogenous circadian cycle and the 24-hour environmental cycle, thus producing a rhythm that is reminiscent of that observed in normal subjects living under free-running conditions. Each day the patient goes to bed later and awakens later, but the total sleep time remains approximately constant if social constraints allow. Sleep time periodically returns to normal "in-phase" hours at which time there is no sleep complaint, but as sleep drifts out of phase the patient complains of difficulty getting to sleep combined with daytime somnolence. The vast majority of patients with this disorder are blind. In sighted patients, age of onset is variable, but the disorder frequently begins in adolescence. With these patients, an extensive sleep log is essential for diagnosis. Once diagnosis is confirmed, possible etiologic factors, including neurologic deficits (particularly SCN tumors), can be evaluated. Mental retardation, schizophrenia, or drug addiction can be associated with this disorder. Treatment of idiopathic hypernyctohemeral syndrome depends on finding synchronizers that will stabilize the sleep/wake cycle. Chronotherapy can be effective by fixing the hours for awakening and bedtime, combined with regular daytime activities, and then maintaining the schedule for several weeks (177,179). Rhythmic melatonin treatment in blind patients is an effective synchronizer (187). Among sighted patients, light therapy also is effective (180,181).

Everyone who has flown across several time zones by transmeridian flight is aware of the physiologic symptoms of *jet lag* (188). Unable to resynchronize rapidly to a sudden shift in external clock time, the circadian pacemaker of the traveler attempts to maintain the phase set to the time zone of departure. Jet lag is characterized by fatigue, mood disturbance, disordered sleep, anxiety, difficulty in concentrating, and, at times, gastrointestinal disorder. Of these, chronic fatigue and difficulties in getting to sleep and in awakening are the most significant. Biologic rhythms adapt at different rates. For example, sleep time resynchronizes rapidly with the rest-activity cycle, whereas hormonal rhythms, including plasma cortisol, and core body temperature rhythms resynchronize more slowly with the pacemaker (see section on Circadian Factors in Sleep). During the first few nights after the journey, short half-life hypnotic medications are an effective way of reducing sleep debt without inducing daytime sleepiness. Several studies on the effectiveness of melatonin as a way of manipulating the circadian pacemaker have been conducted, and there are some indications that low doses may be capable of resynchronizing biologic rhythms after jet lag (189).

NARCOLEPSY

One of the first descriptions of narcolepsy was made in 1877 by Westphal, a German physician, based on his observation of a patient with sudden sleep attacks associated with symptoms of motor incapacity and aphasia. He considered it a form of epilepsy. Westphal also noted the genetic character of the disorder, given that the mother of his patient had been affected by the same symptoms as her son. Three years later, the French physician Gélineau described the disorder for the first time as a distinct clinical entity rather than as a symptom of another condition, and proposed the term *narcolepsy,* meaning *sleep seizure*. Gélineau's description in his 1880 paper is of a patient who, from age 36 years, began to fall asleep suddenly during the day and had sudden falls, or *astasia* (*cataplexy* in the current terminology), that occurred whenever he became emotional.

Narcolepsy is characterized by two principal symptoms (excessive daytime sleepiness and cataplexy) and three supplementary symptoms (hypnagogic or hypnopompic hallucinations, sleep paralysis, and fragmented nocturnal sleep) (190).

■ *Excessive daytime sleepiness* is an irresistible need for sleep during the day, associated with a chronically low level of alertness and lapses of memory. Sudden sleep attacks, which last from several minutes to more than 1 hour, might occur while the narcoleptic patient is actively engaged in another task, although usually the patient will wake up refreshed and remain alert for some time after such attacks.

■ *Cataplexy* is an abrupt and reversible decrease or loss in muscle tone without alteration in the level of consciousness (191). This is a defining symptom of narcolepsy, but it should be noted that a precise definition of cataplexy remains uncertain because it can best be described as a transition between, or mixed state of, REM sleep and wakefulness (192). Patients who collapse in a position that allows sleep typically will progress from a cataplectic episode into stage REM sleep.

■ About 25% to 30% of narcoleptic patients report *hypnagogic* or *hypnopompic hallucinations,* which can be auditory, visual, or occasionally somatesthetic. These per-

ceptual experiences occur at sleep onset (i.e., hypnagogic) or at the moment of awakening (i.e., hypnopompic), and are vivid hallucinations, appearing as if the surroundings are being integrated into a dream. Narcoleptic patients report that these hallucinations usually appear when they try to avoid sleep and often are accompanied by feelings of fear or anxiety. It should be noted that normal subjects also can report these hallucinations after they have been deprived of sleep.

■ *Sleep paralysis* is a terrifying experience for the narcoleptic patient at sleep onset or on awakening from sleep or a nap. Patients find that a total loss of muscle tone, which can last for more than 1 minute, prevents all movement, speech, or even the ability to breathe normally.

■ *Nocturnal sleep fragmentation* and occasional complaints of insomnia are paradoxical symptoms of narcolepsy. Although the narcoleptic patient sleeps easily, the sleep is continuously interrupted by awakenings during the night. Figure 15.8 shows a nocturnal hypnogram recorded from a narcoleptic patient. Sleep fragmentation can be seen clearly on this record.

Narcolepsy usually begins before age 25 years, but age of onset is highly variable and the disorder can first occur at any time between ages 5 and 55 years (193). Males and females are equally affected. Prevalence of the disorder is estimated

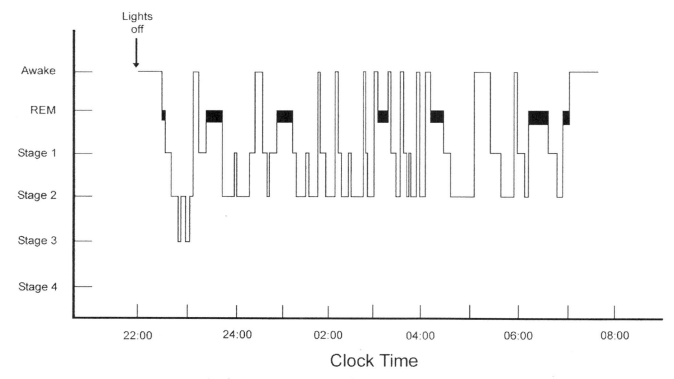

FIGURE 15.8. Example of a hypnogram recorded from a narcoleptic patient (cf. Fig. 15.1B). Nocturnal sleep is fragmented by multiple awakenings and the number of rapid eye movement (REM) sleep episodes is increased, which can lead to an increased duration of REM sleep. Sleep disruption also limits the time spent in slow wave sleep (stages 3–4). Note a sleep-onset REM period (REM sleep occurring within 15 minutes of sleep onset), which is a characteristic of the disorder.

to range from 0.02% and 0.18%. The risk to first-degree relatives is higher, probably about 1% to 2% (194), indicating genetic transmissibility, although concordance rates in monozygotic twins are only about 30%. In about half of the patients, the onset of narcolepsy is associated with an exogenous factor, such as head injury or severe stress, indicating that the disorder probably requires both genetic predisposition and a triggering environmental event. Because most cases of narcolepsy occur sporadically, the disorder has been considered multigenetic and environmentally influenced.

A genetic linkage for the disorder was found in early studies showing that narcolepsy was associated with two major histocompatibility human leukocyte antigen (HLA) class II alleles, called DR2 and DQw1. Since these initial observations, the nomenclature for HLA typing has been extended and refined and data from diverse populations included, but this initial finding still remains true: human narcoleptics frequently carry certain HLA haplotypes, specifically DQB1*0602, DQA1*0102, and DRB1*1501 (194,195). Although exceptions to these associations are rare, the existence of such exceptions, combined with the fact that the same alleles are found in the general population, means that factors other than HLA alleles must be implicated in the etiology of narcolepsy. Orexin pathophysiology (see section on Orexin and Control of Sleep and Wakefulness) is clearly one such factor, although we know that the disorder does not involve highly penetrant, orexin gene mutations (91). An exceptionally detailed study of HLA class II alleles in narcoleptic patients confirmed the influence of HLA haplotypes on predisposition to the disorder and reported significant effects from interactions between specific HLA-DR and HLA-DQ alleles to susceptibility (196). Taken in conjunction with the orexin findings, these data imply that an immunologically mediated degeneration of orexin-containing cells is occurring in human narcolepsy. It is evident that the possibility that narcolepsy involves an autoimmune-caused loss of a specific group of brain cells may have implications for other neurodegenerative disorders. Although the exact mechanisms by which HLA alleles influence susceptibility to narcolepsy remain to be determined, HLA typing remains of practical use in the evaluation of patients with excessive daytime sleepiness. If the presence of these histocompatibility antigens cannot be verified in cases of hypersomnolence, narcolepsy usually can be eliminated as a diagnosis.

To evaluate the severity of narcolepsy, continuous ambulatory or laboratory polysomnography provides a measure of total sleep time, together with information about the timing of sleep attacks, their duration, and the sleep stage in which they occur. Analysis of nocturnal sleep of the narcoleptic patient typically reveals that sleep begins quickly after lights are turned off, within about 10 minutes, but that sleep is interrupted by multiple awakenings throughout the night. A significant characteristic of narcolepsy, however, is sleep-onset REM during the day or at the beginning of the night (Fig. 15.8). The MSLT is an effective method for demonstrating sleep-onset REM during the day. With this test, sleep-onset REM is defined as sleep with a REM episode occurring within 15 minutes of sleep onset (158).

Narcolepsy currently is treated with medications directed at each of the symptoms. Daytime sleepiness and sleep attacks have been treated with amphetamine-type drugs and their derivatives (197). These include methylphenidate and amphetamine, although treatment with these drugs is complicated by the risk of abuse and development of tolerance. They also induce irritability, anxiety disorders, convulsions, and dependence as side effects. An alternative drug, modafinil, is a central nervous system stimulant with a high efficacy against sleep attacks (198). It appears to be well tolerated and shows little dependency with chronic use. Cataplexy is treated with tricyclic antidepressants, including clomipramine and imipramine, but if treatment is withdrawn for any reason after several months, reappearance of cataplexy is likely (191). Under these circumstances, selective serotonin uptake inhibitors, including fluoxetine and fluvoxamine, are alternatives. Sleep paralysis and hallucinations usually are ameliorated by treatment of excessive daytime somnolence and cataplexy. Typical hypnotic medications are effective against insomnia, and they improve nocturnal sleep fragmentation.

Narcolepsy should be differentiated in diagnosis from both hypersomnias and disorders in which daytime sleepiness is symptomatic, such as sleep apnea, mood disorders, hysteria, chronic use of tranquilizers, and focal epilepsy. However, it should be noted that it is not unusual to find narcolepsy associated with these disorders, particularly sleep apnea. Narcolepsy is an exceptionally disabling, chronic disorder (199). Although symptoms decrease in intensity with age, the disorder shows no tendency for spontaneous remission. Patients can fail scholastically, be involved in accidents, perform poorly at work, and be socially isolated. Subintrant cataplectic attacks, which confine the patient to bed for periods as long as several weeks, have been observed after withdrawal from treatment.

DISORDERS OF MOVEMENT ASSOCIATED WITH SLEEP

Narcolepsy is so incapacitating because muscle atonia, normally associated only with REM sleep, intrudes into waking as a cataplectic or narcoleptic attack. Muscle atonia is an important REM feature from a clinical point of view because dysfunctions of this system are present in other patients who present with sleep disorders (see section on Rapid Eye Movement Sleep Muscle Atonia). An example of sleep-related movement disorder resulting from REM sleep without atonia is *REM sleep behavior disorder* (RBD). The typical age of onset of RBD is around 50 years. It occurs predominantly in men, and the disorder is characterized by the intermittent loss of normal muscle atonia during REM sleep. This results in the patient acting out elaborate behavior patterns that are associated with dream mentation. RBD was first recognized as a clinical entity by Schenck et al. (200), although Sastre and Jouvet (201) previously reported that bilateral lesions of areas medial to the locus subcaeruleus region in the dorsal PRF caused a persistent absence of normal REM sleep atonia and the appearance of REM sleep behaviors.

Sometimes the behavior associated with RBD puts the patients or their partners at risk. One patient was dreaming he was a football player and was running against an opposing lineman, when he awoke to find himself out of bed charging against a bureau. Another found himself enacting the leaps that were occurring in his dream. Sometimes dreams of aggressive behavior might endanger the sleeping partner as the mentation is acted out. If awoken during an episode the patient will report a dream or nightmare consistent with the observed behavior. RBD episodes have been reported in association with a wide variety of neurologic pathology, including dementia, olivopontocerebellar degeneration, and Guillain-Barré syndrome (202,203). Polysomnographic recording, including EMG of the anterior tibial muscles, in conjunction with video monitoring is necessary to differentiate RBD from somnambulism or temporal lobe epilepsy. Recordings reveal abnormal behavior associated with abrupt REMs accompanying myoclonic EMG potentials. Diagnosis is confirmed when these behaviors are observed during REM sleep with incomplete muscle atonia. Clonazepam has been found to be effective for RBD, inhibiting nightmares and reducing the number and intensity of REM sleep behaviors while also inhibiting the phasic EMG bursts during REM sleep (202). Carbamazepine is also reported to be useful in RBD (204). Neither medication, however, reestablishes muscle atonia.

RBD must be distinguished both in theoretical understanding and in the clinic from another movement disorder that occurs during sleep, *periodic limb movement disorder.* Periodic limb movements (PLM) are characterized by episodic stereotyped movements, primarily of the lower limbs. The disorder does not occur during REM sleep. The prevalence of PLM increases with age: 30% of the population older than 50 years are affected, and 44% of those older than 65 years are affected (205). During initial diagnosis, PLM must be differentiated from *sleep starts* (also known as *hypnagogic jerks*), which occur upon falling asleep and are characterized by benign brief, sudden movements of the limbs or neck.

PLM are the primary cause of about 13% of the complaints of insomnia and 7% of the complaints of excessive daytime sleepiness. Clinically, the disorder presents as extension of the big toe combined with flexion of the ankle and knee, sometimes associated with partial flexion of the hip. Movements last 0.5 to 5 seconds and recur with fixed periodicity every 5 to 90 seconds. Episodes occur primarily at the beginning of the night, during stage 1 to 2 non-REM sleep, and frequently are associated with arousal or awakening, although many patients with PLM show no concomi-

tant awakening. Diagnosis of the disorder depends primarily on polysomnographic evaluation. Bilateral EMG activity from the anterior tibialis is diagnostic, and movements can be quantified from these records. Several etiologies are associated with PLM, which, like RLS, can occur in association with spinal lesion, peripheral neuropathy, anemia, chronic uremia, and Parkinson disease. Movements can be aggravated by tricyclic antidepressants, abrupt withdrawal from anticonvulsants, or withdrawal from benzodiazepines. Dopamimetics, including L-dopa for short-term use, and carbamazepine have been found to be effective for treatment of PLM (206).

As a nosologic entity, PLM are a state-dependent motor response and entirely separable from Restless Leg Syndrome (RLS), which is a state-dependent sensory stimulus. However, there is considerable overlap between the two conditions. Eighty percent of patients who present with RLS also demonstrate PLM. RLS is characterized by an irresistible need to move the legs accompanied by diffuse, unpleasant, creeping, and irritating sensations and pain deep in the limbs (205). Symptoms can appear at bedtime and interfere with sleep onset or during the day, particularly toward the end of the afternoon or in the early evening if the patient remains immobile for any length of time. Relief from the unpleasant sensations can be obtained only by walking or moving the legs. A family history of the syndrome has been frequently noted and autosomal dominant transmission is indicated. Although the syndrome can occur at any age, it is more common after age 60 years.

The prevalence of RLS has been estimated to be as high as 10% (207). Symptoms vary in intensity among patients and can change over time within an individual. RLS can be aggravated by excessive caffeine consumption, prolonged exposure to cold or heat, and pregnancy. About 50% of patients presenting with RLS indicate that symptoms began or worsened during pregnancy. RLS occasionally is related to hypothyroidism, rheumatoid arthritis, diabetes, and hyperparathyroidism. Tricyclic antidepressants may induce the disorder, as does withdrawal from benzodiazepines. In rare instances, RLS signals the onset of peripheral neuropathy. The most important differential diagnosis is with akathisia as a side effect of neuroleptic treatment. Like PLM, dopamimetics, including L-dopa, and carbamazepine have been found to be effective for treatment of RLS after elimination of any aggravating factors (206,208).

Another movement disorder associated primarily with the transition between waking and sleeping and observed essentially only in children is *rhythmic movement disorder* (RMD) (209). In adults, RMD is associated with autism (210). The disorder is characterized by a variety of stereotyped movements of major muscle groups. It rarely begins after age 2 years and declines with age, with a prevalence of about 5% by 5 years. The etiology of RMD is unknown, although the disorder presents more violently and is observed more frequently in association with neurologic dysfunction.

Movements can cover a broad spectrum, including antero-posterior rocking and banging of the head (head banging), side-to-side stereotyped movements of the head with the subject in a supine position (head rolling), anteroposterior thrusting of the whole body, supporting the body on hands and knees (body rocking), and side-to-side movements of the whole body in a supine position (i.e., body rolling). During polysomnographic evaluation, RMD is observed in waking and all sleep stages, but very infrequently in SWS or REM sleep. Movements are not associated with seizure activity. Episodes typically last about 15 minutes, but occasionally they can last several hours.

Most infants and small children require no treatment; RMD typically remits spontaneously before age 4 years. Family members need reassurance of the benign nature of the disorder and the absence of any underlying neuropathology. Violent behavior may result in injury, and protection usually is required if the child is mentally retarded or if there is evidence of psychiatric disorder. Behavioral therapy may be useful in older children and adults, in whom clonazepam has been reported as effective (210).

DISORDERS OF BREATHING DURING SLEEP

The discovery of sleep apnea was a defining moment for clinical sleep research and to this day remains the principal presenting complaint at sleep disorder centers. Sleep apnea is an age-related disorder that affects as many as 2% to 4% of the population older than 40 years (211). In a study of polysomnographic diagnoses made in sleep disorder centers, more than 40% of the patients referred with daytime somnolence had sleep apnea (212).

Sleep apnea (absence of airflow at the nose or mouth during sleep) is defined as an interruption in respiration lasting more than 10 seconds in both the adult and child (1, pp. 198–205). In the neonate, however, sleep apnea is defined as a respiratory pause of longer than 20 seconds, unless accompanied by cyanosis or bradycardia (213). Hypopnea is a partial absence of respiration during sleep, lasting for more than 10 seconds, and accompanied by at least a 50% reduction in thoracoabdominal movement. Polysomnography is the only technique that can confirm diagnosis of sleep apnea syndrome, which is characterized by more than five apneas per hour of sleep or more than ten apneas plus hypopneas per hour of sleep (214). A polysomnographic evaluation also enables identification of the type of apnea, whether obstructive (intermittent upper airway obstruction), central (impairment in central respiratory drive), or mixed (combination of the two). Figure 15.9 is a schematic display of the variation in respiratory movements and other parameters that are recorded during evaluation of apnea in the sleep laboratory and shows how the various types of apnea can be differentiated. The polysomnographic technique provides information about the severity of the disorder by means of arterial

oxygen desaturation measures, associated heart rhythm disturbances, and abnormalities in sleep architecture. Multiple nocturnal awakenings caused by apneic episodes markedly reduce total sleep time and frequently prevent the occurrence of SWS. The longest apneas and those associated with the greatest arterial oxygen desaturation tend to occur during REM sleep. Daytime sleepiness can be quantified with the MSLT (158), which typically reveals a mean sleep onset latency of less than 10 minutes in apneic patients.

Obstructive sleep apnea (OSA) is characterized by a collapse or partial collapse of the upper airways, resulting in repeated interruption of nasobuccal airflow during sleep. It usually is accompanied by a reduction in blood oxygen saturation. The principal clinical signs of OSA in the adult are excessive daytime sleepiness, loud snoring, and respiratory pauses reported by a sleeping partner (215). The disorder may be associated with other nocturnal symptoms, including polyuria, enuresis, night sweats, agitation, and dry mouth. Daytime symptoms include morning headache, mood disorder, and impaired memory and cognition. Although OSA patients occasionally complain of insomnia, daytime somnolence and unrestful naps as a result of disturbed nocturnal sleep are more typical. OSA in the child is associated with snoring in about 75% of patients, together with, or replaced by, a stridor or cough. Nocturnal breathing through the mouth, sweating, and enuresis are typical symptoms in the child. Excessive daytime sleepiness in the young child is characterized by prolonged naps that can alternate with aggressive episodes and hyperactivity. Delayed growth is not unusual. In addition to symptoms described in children, OSA in the neonate can be accompanied by thoracic deformation during inspiratory effort (213).

Several different therapeutic approaches are available for OSA, and the treatment of choice depends on the severity of the disorder and its etiology, which can include oronasomaxillofacial anatomy, hyperthyroidism, or acromegaly. Sleep hygiene and diet should be monitored and excessive fatigue avoided. Alcohol and, most importantly, sedative-hypnotic medications should be avoided. Weight reduction, which can reduce the incidence of apnea without eliminating it entirely, can be recommended if the patient is obese. The patient should be advised regarding sleep position training, particularly to avoid the supine position because the occurrence of apnea can depend on the position adopted during sleep (216). Surgical interventions are available to increase the lumen size of the oropharynx, including uvulopalatopharyngoplasty and maxillomandibular advancement. These interventions can be effective in the treatment of both snoring and apnea (217). However, mechanical approaches, which include tongue-restraining devices, orthodontic appliances, and continuous positive airway pressure (CPAP), currently are the treatment of choice for OSA. Nocturnal CPAP using a nasal mask to aid inspiratory effort and to maintain patency of the oropharynx during respiration with a pneumatic splint frequently is adopted for moderate-to-severe cases of apnea

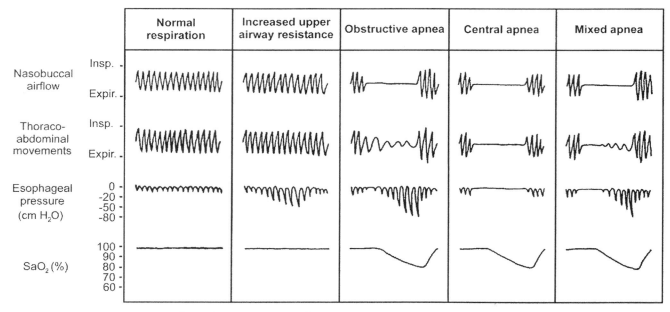

		Normal respiration	Increased upper airway resistance	Obstructive apnea	Central apnea	Mixed apnea

FIGURE 15.9. Schematic representation of changes in respiratory parameters during sleep-related breathing disorders. During normal respiration, esophageal pressure varies between 0 and −10 cm H$_2$O, and arterial oxygen saturation (S$_a$O$_2$) remains constant at about 98%. Increased upper airway resistance (UAR) is characterized by a reduction in esophageal pressure to −20 or −30 cm H$_2$O, without apneic pauses or significant oxygen desaturation. Note that the increase in inspiratory effort resulting from increased UAR is not recorded as a change in thoracoabdominal movements. Obstructive apnea is defined as an interruption of nasobuccal airflow lasting more than 10 seconds, despite the persistence of thoracoabdominal movements. Esophageal pressure can reach values as low as −50 to −80 cm H$_2$O, and S$_a$O$_2$ is reduced. Central apnea corresponds to an interruption of respiratory commands lasting more than 10 seconds, as revealed by the absence of thoracoabdominal movements. Mixed apnea typically is characterized by a central apnea followed by an obstructive apnea with respiratory interruption of at least 10 seconds. Hypoxia increases in severity as a function of apnea duration.

(218). CPAP can be very effective, although nightly compliance is required for optimal benefit (219).

Central sleep apnea (CSA) associates apneic episodes during sleep with the absence of ventilatory effort (220). Typically these patients also show obstructive and mixed apneas. Unlike OSA, CSA patients frequently complain of insomnia and depression; daytime hypersomnolence is rare. Idiopathic central apnea is characterized by a tendency to hyperventilation so that the associated hypocapnic alkalosis inhibits subsequent respiration. This tendency is increased during sleep by recurrent arousals. The resulting effect on respiration is comparable to high-altitude hyperventilation, which is induced by hypoxia and is the cause of altitude insomnia. The fact that hypocapnic alkalosis is an important factor in CSA suggests that P$_a$CO$_2$ is a significant stimulus to ventilation during sleep, and loss of this drive, as occurs in hypocapnia, can produce irregular breathing.

CSA is frequently seen in patients with congestive heart failure, neuromuscular disease, and some neurologic disorders. The latter, particularly associated with brainstem pathologies, is manifested as central alveolar hypoventilation (221). CSA in these patients is an exaggeration during sleep of a hypoventilation disorder, and the sleep disturbance is a secondary feature of recurrent respiratory failure. This type of respiratory disorder is likely to be caused by altered chemoreceptor function resulting in defective ventilatory control. A congenital form of hypoventilation disorder in the newborn must be distinguished from a usually acquired form in the adult. Although the condition is unusual in the newborn, correct diagnosis of the condition is rarely made. Guilleminault et al. (222) suggested that mortality during the first year of life can be due to an occult congenital hypoventilation disorder.

Treatment of CSA rarely is satisfactory, and management of respiratory insufficiency should primarily be directed at the underlying disorder. CPAP can be effective if CSA is severe. Few options are available for the patient with a limited number of central apneas and the complaint of insomnia. Ventilatory stimulants are rarely effective, although acetazolamide is useful for reducing CSA during short-term use. Several studies report that oxygen administration can reduce CSA. Tracheotomy often has been performed in patients with central alveolar hypoventilation, but the technique now is becoming increasingly replaced by assisted ventilation. With infants, CPAP usually is reserved for the most severe forms of the congenital hypoventilation syndrome.

Sleep apnea is a serious medical condition. In the adult, OSA is associated with significant cardiovascular morbidity and mortality. About 50% of patients presenting with the disorder are hypertensive; conversely, about 30% of patients with essential hypertension present with sleep apnea (211). The disorder is associated with an increased risk of myocardial infarct, coronary heart disease, and stroke. The risk of mortality is higher if a patient shows more than 20 apneas per hour of sleep. Even if the idiopathic form of CSA is relatively benign and is not associated with cardiorespiratory failure, these patients have other risk factors. Daytime sleepiness increases the chance of automobile accidents by a factor of seven in sleep apnea patients compared with normal subjects. The impact of hypoxia and sleep fragmentation on cognitive measures is considerable: successful treatment can improve cognitive performance on most, but not all, indices, suggesting that permanent anoxic brain injury might occur in these patients (223). In the infant, sleep apnea may be one of the causes of sudden infant death syndrome.

SLEEP ASSOCIATED WITH MENTAL DISORDERS

Mood Disorders

Episodes of affective illness often are accompanied by marked changes in sleep. As many as 90% of depressed patients show disordered sleep, and there is evidence that sleep is most affected in the most severe cases of depression (224). Insomnia frequently is associated with mania and hypersomnia or insomnia with depression. Sleep and mood disorders are related through exogenous manipulations of sleep time: sleep deprivation has been found to improve depression but relapse is rapid after a recovery sleep period (225). From these observations in depressed patients, sleep appears to have a depressant effect and wakefulness a mood-elevating effect. Some patients, especially those who are older, agitated, or psychotic, frequently have insomnia as a result of early morning awakening (226). Another group of patients (about 10%–15% of depressives) sleeps excessively, especially patients who are younger, lethargic, and bipolar. Recurrent seasonal affective disorder frequently is associated with hypersomnia (227).

Nocturnal polysomnographic recording confirms disturbed sleep continuity as a primary characteristic of depressive disorder (228,229). An increase in sleep onset latency and in the number of awakenings, combined with a reduction in total sleep time and sleep efficiency index, are frequently observed. These sleep disturbances are absent or minimal in hypersomniac depressed patients, but they are observed in many hyposomniac patients. Changes in sleep architecture that are highly indicative of depression include a reduction in SWS, an increase in the density of phasic events in REM sleep, and a greater amount of REM sleep at the beginning of the night, usually accompanied by a reduction in the latency to the first REM period. Although these

changes in sleep parameters can be considered to characterize depression, none is pathognomonic of the disorder. For example, reduced REM sleep latency has been described in other psychiatric disorders; conversely, normal REM latency has been noted in several studies of depression (230).

Abnormalities in the period of the circadian pacemaker have been implicated in the pathogenesis of some affective disorders. Early morning wakefulness, diurnal variation in mood, morning sadness followed by slight evening improvement, and the seasonality and cyclicity of depression have led to interest in the relationship between mood disorder and biologic rhythms. This led to a phase advance hypothesis of depression, based on Kronauer's two-oscillator model of the regulation of the sleep/wake cycle (152). As noted in the section on Combination of Circadian and Homeostatic Drives for Sleep, this model depends on an X-oscillator that entrains the circadian rhythm for core body temperature, cortisol secretion, and REM sleep, and a weaker Y-oscillator that entrains the sleep/wake rhythm, non-REM sleep, and growth hormone secretion. The early appearance of REM sleep and the increase in this sleep stage at the beginning of the night in depression is hypothesized to result from a phase advance of the X-oscillator with respect to the Y-oscillator (231). The circadian rhythms of core body temperature and cortisol release are phase advanced and decreased in amplitude in depressed patients. Treatment of depression has been based on the concept of a disturbed circadian rhythm in depression (232). During phase advance treatment, the bedtime of patients is progressively advanced to increasingly earlier hours. Over a 2-week period, bedtime, awakening, meals, and social activities are advanced by at least 5 hours. Conversely, in a phase delay treatment, the patient is sleep deprived during habitual abnormal sleep hours and a progressively later bedtime is imposed. The patient is required to delay bedtime and time of awakening by about 2 to 3 hours each day. After about 10 to 15 days, normal sleep hours can be achieved, although stabilization of the new sleep pattern is necessary for a minimum of 3 weeks.

An hypothesis on the pathophysiology of depression has been based on Borbély's two-process, single-oscillator model of sleep regulation (153; see section on Combination of Circadian and Homeostatic Drives for Sleep). In this model, the propensity and maintenance of sleep depend on the interaction of two processes: a circadian process C and a propensity for non-REM sleep, process S. Process S inhibits REM sleep. In depression, an hypothesized deficit in process S is proposed to lead to a disinhibition of REM sleep and hence a reduction in REM latency (233). A deficient process S also is associated with a decrease in SWS. Another model-based explanation of depression has been developed from the reciprocal interaction model (61). As noted in the section on Rapid Eye Movement Sleep, this model describes a periodic cycling between REM and non-REM sleep episodes that depends on the equilibrium between a cholinergic REM promoting system and a monoaminergic REM inhibiting system. In this hypothesis, the early appearance of REM sleep and increase in REM density (i.e., a measure of the frequency of REMs) in depression is caused by a relative increase in cholinergic activity with respect to monoaminergic activity (234). In support of this hypothesis, it should be noted that cholinergic agonists have been shown to induce REM sleep more rapidly in depressed patients than in normal subjects (235). Conversely, several antidepressants with anticholinergic activity inhibit REM sleep and delay its appearance.

Sleep Deprivation as a Therapy for Depression

Inhibition of REM sleep is considered a possible explanation for the antidepressant effect of mood-elevating medications. This led to hypotheses that the mood-elevating effects of sleep deprivation also are the result of REM sleep deprivation (234). The antidepressant properties of sleep deprivation therapy were first recognized by Schulte in 1966, when he reported a depressed patient who felt better after he slept poorly. Since that discovery, the technique has been modified and adapted to take several different forms so as to change the relative durations of sleep and wakefulness by total, partial, or selective suppression of sleep (236). Sleep deprivation can involve phase shifting the time of sleep from habitual abnormal hours. Total sleep deprivation consists of depriving the patient of all sleep for continuous periods of from 1 to 3 days, interrupted by 2-day intervals during which the patient is allowed to sleep at night (237). Total sleep deprivation can produce brief improvement and occasionally spectacular improvement in the symptoms of depression; however relapse to a depressed state on the day after the night of sleep recovery occurs frequently. The clinical response does not seem to be affected by the age or gender of the patient, the polarity of the disorder, the number of previous episodes, or the season during which treatment occurs. Total sleep deprivation can be a useful adjunct to antidepressants by accelerating the onset of improvement after medication begins.

Partial sleep deprivation has the advantage of conserving some sleep and was developed because total insomnia was frequently perceived by the patient as being too stressful. Although partial sleep deprivation in the first half of the night has little effect on depression, awakening the patient after about 2 a.m. to deprive sleep in the second half of the night frequently induces a significant but transitory improvement in mood (238). Partial sleep deprivation can be used to complement antidepressant medication. An alternative to partial sleep deprivation is to deprive the depressed patient specifically of REM sleep (234). During polysomnographic monitoring, the patient is woken as soon as each period of REM sleep begins until about 20 to 30 awakenings have been achieved during the night, at which point sleep recovery is allowed.

Currently, sleep deprivation is rarely used alone in therapy except in an experimental setting. However, the technique can be a useful adjunct to accelerate the onset of therapeutic action of antidepressants, particularly when there is danger of suicide. Side effects of sleep deprivation are rare but include increased anxiety, occasional worsening of depression, daytime confusion, mood lability, and triggering of manic episodes in bipolar patients. Sleep deprivation therapy is contraindicated in elderly patients or those who demonstrate agitation or anxiety.

Psychoses

When Aserinsky and Kleitman (239) discovered REM sleep, they noted that heart and respiratory rates were typically increased during periods of REMs. Combined with the activated EEG pattern, they conjectured that this phase of sleep might be associated with emotional disturbance, such as might be caused by dreaming. To test this hypothesis, sleepers were aroused and interrogated during or shortly after periods of REM sleep, and they almost invariably reported that they had dreamed. Subsequent studies confirmed that REM sleep is closely associated with dreaming, with about 80% of awakenings during the REM state producing a dream report. Dreams have a long history of interest in psychiatry. The British neurologist Hughlings Jackson (1835–1911) had remarked: "find out all about dreams, and you will find out all about insanity." Today, the neural systems responsible for REM sleep, and the resulting activated EEG, seem to be an accurate and simple explanation for the instigation of the subjective dream state that is linked to the cyclic appearance of REM sleep. However, in the first years after discovery of REM sleep, researchers believed they had acquired a tool for investigation of Hughlings Jackson's premise. They speculated that schizophrenia might represent the intrusion of the dream state into wakefulness, and, therefore, there was the prospect of being able to link dream cognition to psychosis through REM sleep (240).

The first measures of REM sleep and dream recall in the schizophrenic were reported soon after the discovery of REM sleep (241). Many studies followed; by the 1960's about half of all sleep studies in psychiatric patients were with schizophrenics (242). No link between REM sleep and psychosis was found, but most studies reported that schizophrenia was associated with disturbed sleep. Many schizophrenic patients complain of disrupted sleep and prolonged insomnia. Lengthy periods of total sleeplessness are not uncommon. Insomnia and/or particularly intense or frequent nightmares and profoundly disturbing hypnagogic hallucinations typically precede acute schizophrenic episodes or worsening of the disorder. Some psychotic patients return to normal sleep patterns during periods of remission.

The polysomnographic data obtained from schizophrenic patients can be summarized as showing that their sleep, compared with that of normal subjects, is characterized by increased sleep onset latency combined with decreased total sleep time, sleep efficiency index, SWS duration, and REM sleep latency (243). However, REM density during the first REM period and the duration of REM sleep are not significantly different from those found in normal subjects. None of these sleep disturbances is specific to schizophrenia and not all schizophrenics show disturbed sleep, although reduced SWS duration and δ power is one of the most consistently reported abnormalities in the sleep of schizophrenics (244,245). REM sleep latency has been significantly correlated with negative symptoms; medication-induced tardive dyskinesia and/or drug withdrawal has an impact on this parameter. A high prevalence of both PLM and sleep apnea has been noted in schizophrenic patients.

Antipsychotic medication usually is sedative, but the schizophrenic patient will benefit from medication and treatment specifically directed against sleep disorders (246). Sleep hygiene counseling can be effective, because many schizophrenic patients have developed poor sleep habits, including sleep reversals and polyphasic sleep patterns, as a way of avoiding social interaction.

Sleep Associated with Eating Disorders

Anorexia nervosa and *bulimia nervosa* are eating disorders that result from a voluntarily and deliberately restricted food intake related to a fear of lack of control of body weight (247). Both disorders typically begin in adolescence and most likely affect girls. All patients have a problem with the perception of body image, seeing themselves as consistently overweight.

Spontaneous complaints about sleep are rarely made by anorectics, and they seldom abuse hypnotics. Polysomnographic evaluation shows, however, that total sleep time is reduced and the number of nocturnal awakenings is increased in these patients. A reduced percentage of SWS has been noted in some anorectics. These sleep disturbances usually are related to the severity of weight loss. Any weight gain typically is accompanied by an improvement in sleep quality and a rebound in SWS. In the absence of depression, no other specific sleep abnormalities are associated with anorexia nervosa.

Bulimia nervosa is characterized by diurnal or nocturnal periods of binge eating (periods of excessive and uncontrollable overeating). Nocturnal bulimia occasionally can cause chronic insomnia. Inappropriate and repetitive compensatory behaviors, such as self-induced vomiting and abuse of diuretics, frequently are adopted by these patients in an attempt to prevent weight gain. In the absence of any associated mood disorder, the sleep of normal weight bulimics cannot be distinguished from that of normal subjects.

A syndrome associated with multiple nocturnal awakenings, accompanied by uncontrollable hyperphagia, was first described in a group of adult obese subjects (248). It has been termed the *nocturnal eating (drinking) syndrome*. Abnormal compulsive eating and drinking during the night lead to chronic insomnia and a lack of appetite in the morning. Prevalence of the disorder is highest in overweight women who show no evidence of a daytime feeding disorder. Most of these patients are able to control their caloric intake during the day, but they are incapable of regulating their feeding behavior at night. Anxiety, mood disorders, and stressful life events are evident in a high proportion of these patients. Comorbidity of nocturnal eating syndrome has been noted with somnambulism, PLM, and OSA (249). Treatment typically is with a selective serotonin uptake inhibitor, such as fluoxetine, whether or not the syndrome is associated with depression or obsessive-compulsive disorder. Clonazepam has been shown to give satisfactory results if PLM are present. Neither daytime restrictions on caloric intake nor treatment with hypnotics is effective.

Sleep and Anxiety Disorders

Generalized anxiety disorder (GAD) frequently gives rise to severe and prolonged sleep disturbances. It is associated with difficulties in falling asleep and, less typically, with multiple nocturnal awakenings (250). In contrast to other insomnias, anxiety-related insomnia has few daytime sequelae. Anxious patients usually do not show diurnal symptoms of sleep deprivation, such as excessive daytime sleepiness or memory disturbance. Polysomnographic evaluation reveals that sleep onset latency is increased in these patients, who also have difficulty in maintaining sleep with sleep fragmentation and a lowered sleep efficiency index. However, REM sleep duration is normal and any reduction in REM sleep latency is unusual, thus differentiating these patients from depressives (251). Treatment primarily depends on benzodiazepine hypnotics, sometimes combined with psychotherapy. Reduction or elimination of the symptoms of anxiety usually is accompanied by a subjective improvement in sleep quality.

Most benzodiazepine hypnotics and anxiolytics modify sleep architecture by increasing the duration of stage 1 to 2 non-REM sleep at the expense of SWS and REM sleep. Although these medications do not assure normal sleep, they reduce sleep onset latency and increase total sleep time. Thus, they can be said to provide a "good night's rest" without multiple nocturnal awakenings that interrupt sleep continuity and make going back to sleep difficult for the patient with generalized anxiety. The clinician should be aware that prolonged use of these medications, especially if they are abruptly withdrawn from use, can lead to rebound anxiety and insomnia, together with increased REM sleep rebound and nightmares. These effects are present in all patients, but their consequences more likely will be severe when generalized anxiety disorder is present.

Panic disorder is characterized by an acute episode of intense anxiety *(panic attack)* without apparent cause and associated with a wide range of somatic symptoms. *Nocturnal panic attacks* occur at night during sleep (252,253), can have a major impact on sleep, and can be recurrent, making return to sleep difficult. During a nocturnal panic attack, the patient usually awakens with palpitations, sweating, the sensation of being strangled, and even with the sensation of imminent death. Patients may develop secondary phobias and avoidance behaviors if left untreated. Compared with sleep in normal subjects, sleep in patients with nocturnal panic attacks is characterized by reduced sleep efficiency index resulting from increased sleep onset latency and reduced total sleep time. Differential diagnosis is necessary against other awakenings when similar expressions of acute anxiety are present, but nocturnal panic attacks tend to occur during non-REM sleep, particularly at the time of transition from stage 2 to stage 3. This characteristic enables them to be differentiated polysomnographically from night terrors, which appear in stage 4, and nightmares and RBD, which occur during REM sleep. Nocturnal anxiety associated with postapneic awakening can be distinguished by associated arterial oxygen desaturation (254). Patients with nocturnal panic attacks show similar sleep stage percentages to normal subjects, although the durations of the first two sleep cycles can be shortened. Benzodiazepine anxiolytics are effective for treatment during periods of attack, and high-dose tricyclic and serotoninergic antidepressants have been used successfully for maintenance therapy.

The sleep of patients with *posttraumatic stress disorder* is disturbed (255). They frequently report difficulties in going to sleep, sleep interrupted by multiple awakenings, and nightmares centered on the traumatizing event, reliving or reexperiencing a highly stressful incident, such as rape, hostage kidnapping, or battle. A holocaust survivor returns to Auschwitz because her nightmares are so vivid that she is convinced that the reality has begun again (256). Posttraumatic stress disorder was first recognized in war veterans. Polysomnographic evaluation of posttraumatic stress disorder patients frequently reveals reduced total sleep time and sleep efficiency index, although REM sleep-related abnormalities appear to be the most consistently reported across different studies. For example, Mellman et al. (257) observed reduced REM sleep latency, reduced REM sleep duration, and increased REM phasic activity, similar to that seen in depression. Depression frequently is associated with posttraumatic stress disorder. Several studies indicate that nightmares are likely to be uniquely related to the exposure to traumatic stress (258,259). Like nocturnal panic attacks and night terrors, anxiety arousals or "traumatic dreams" can occur early in the sleep cycle in non-REM sleep in the posttraumatic stress disorder patient. Anxiolytics, antidepressants, and hypnotics have been shown to be effective in treatment, especially if they are given in conjunction with psychotherapy to aid the patient in coping with intrusive thoughts and memories.

CONCLUSION

With the many rapid advances taking place in the field of sleep medicine, it is evident that the psychiatrist can only become more involved in the treatment of sleep disorders and the work of the sleep disorders center in the future. As we noted, not only are psychiatric complaints frequently associated with disordered sleep, but also polysomnographic evaluation of patients is a vital tool in differential diagnosis for the psychiatrist. To this picture, we must add basic research findings in sleep neurophysiology and neuroanatomy and the impact that these findings are having on our understanding of brain function and psychiatric disorder. Psychiatrists will continue to make major contributions to the field of sleep medicine, particularly in the study of the consequences and causes of sleep disruption.

ACKNOWLEDGMENTS

Supported in part by grants from the Department of Veterans Affairs, Medical Research Service, and NIMH (R37 MH 39,683 and R01 MH 40,799), and an MSTP NIH award (MH18825) to C.M.S.

REFERENCES

1. Diagnostic Classification Steering Committee, Thorpy MJ, Chairman. *International classification of sleep disorders: diagnostic and coding manual.* Rochester, MN: American Sleep Disorders Association, 1990.
2. Poceta JS, Mitler MM, eds. *Sleep disorders: diagnosis and treatment.* Totowa, NJ: Humana Press, 1998.
3. McCarley RW. Sleep neurophysiology: basic mechanisms underlying control of wakefulness and sleep. In: Chokroverty S, ed. *Sleep disorders medicine: basic science, technical considerations, and clinical aspects,* 2nd ed. Woburn, MA, Butterworth-Heinemann, 1999;21–50.
4. McCarley RW. Human electrophysiology and basic sleep mechanisms. In: Yudofsky SC, Hales RE, eds. *The american psychiatric publishing textbook of neuropsychiatry and clinical neurosciences,* 4th ed. Washington, DC: American Psychiatric Press, 2002;43–70.
5. Krueger JM. Somnogenic activity of immune response modifiers. *Trends Pharm Sci* 1990;11:122–126.
6. Hayaishi O, Inoue S, eds. *Sleep and sleep disorders: from molecule to behavior.* Tokyo: Academic Press/Harcourt Brace Japan, 1997.
7. Task Force on DSM-IV, Frances A, Chairman. *Diagnostic and statistical manual of mental disorders,* 4th ed. Washington, DC: American Psychiatric Association, 1994.
8. Chatrian GE, Lairy GC. The EEG of the waking adult. In: Rémond A, ed. *Handbook of electroencephography and clinical neurophysiology, vol. 6A.* Amsterdam: Elsevier, 1976.
9. Chatilla M, Milleret C, Rougeul A, et al. Alpha rhythms in the cat thalamus. *C R Biol* 1993;316:51–58.
10. Jankel WR, Niedermeyer E. Sleep spindles. *J Clin Neurophysiol* 1985;2:1–35.
11. Niedermeyer E. Sleep and EEG. In: Niedermeyer E, Lopes da Silva F, eds. *Electroencephalography. Basic principles, clinical applications, and related fields,* 3rd ed. Baltimore: Williams & Wilkins, 1993:153–166.
12. Amzica F, Steriade M. Cellular substrates and laminar profile of sleep K-complex. *Neuroscience* 1998;82:671–686.
13. Steriade M, McCarley RW. *Brainstem control of wakefulness and sleep.* New York: Plenum Press, 1990.
14. Steriade M, Curró Dossi R, Nuñez A. Network modulation

of a slow intrinsic oscillation of cat thalamocortical neurons implicated in sleep delta waves: cortically induced synchronization and brainstem cholinergic suppression. *J Neurosci* 1991;11:3200–3217.

15. Steriade M, Amzica F. Coalescence of sleep rhythms and their chronology in corticothalamic networks. *Sleep Res Online* 1998; 1:1–10.

16. McCormick DA. Cholinergic and noradrenergic modulation of thalamocortical processing. *Trends Neurosci* 1989;12:215–221.

17. von Krosigk M, Bal T, McCormick DA. Cellular mechanisms of a synchronized oscillation in the thalamus. *Science* 1993;261:361–364.

18. Buzsaki G, Bickford RG, Ponomareff G, et al. Nucleus basalis and thalamic control of neocortical activity in the freely moving rat. *J Neurosci* 1988;8:4007–4026.

19. Anders T, Emde R, Parmelee A, eds. *A manual of standardized terminology, techniques and criteria for scoring of states of sleep and wakefulness in newborn infants.* Los Angeles: UCLA Brain Information Services, 1971.

20. Dreyfus-Brisac C, Curzi-Dascalova, L. The EEG during the first year of life (section II). In: Rémond A, ed. *Handbook of electroencephography and clinical neurophysiology, vol. 6B.* Amsterdam: Elsevier, 1975:24–30.

21. Feinberg I, Thode HC, Chugani HT, et al. Gamma function describes maturational curves for delta wave amplitude, cortical metabolic rate and synaptic density. *J Theor Biol* 1990;142:149–161.

22. Borbély A. Sleep: a theme with variations. In: *Secrets of sleep.* New York: Basic Books, 1986:31–47.

23. Dinges DF, Pack F, Williams K, et al. Cumulative sleepiness, mood disturbance and psychomotor vigilance performance decrements during a week of sleep restricted to 4–5 hours per night. *Sleep* 1997;20:267–277.

24. Horne JA. Sleep function, with particular reference to sleep deprivation. *Ann Clin Res* 1985;17:199–208.

25. Mendelson WB. *Human sleep research and clinical care.* New York: Plenum, 1987.

26. Leung L, Becker CE. Sleep deprivation and house staff performance: update 1984–1991. *J Occup Med* 1992;34:1153–1160.

27. Luce GG, Segal J. *Sleep.* New York: Coward, McCann & Geoghegan, 1966:90–93.

28. Nicholls JG, Martin AR, Wallace BG, et al. *From neuron to brain,* 4th ed. Sunderland, MA: Sinauer Associates, 2000.

29. Kandel ER, Schwartz JH, Jessell TM. *Principles of neural science,* 4th ed. New York: McGraw-Hill, 2000.

30. Krueger JM, Pappenheimer JR, Karnovsky ML. The composition of sleep-promoting factor isolated from human urine. *J Biol Chem* 1982;257:1664–1669.

31. Krueger JM, Pappenheimer JR, Karnovsky ML. Sleep promoting effects of muramyl peptides. *Proc Natl Acad Sci USA* 1982;79:6102–6106.

32. Pappenheimer JR, Miller TB, Goodrich CA. Sleep-promoting effects of cerebrospinal fluid from sleep deprived goats. *Proc Natl Acad Sci U S A* 1967;58:513–517.

33. Krueger JM, Majde JA. Sleep as a host defense: its regulation by microbial products and cytokines. *Clin Immunol Immunopathol* 1990;57:188–199.

34. Krueger JM, Obal F, Fang J. Humoral regulation of physiological sleep: cytokines and GHRH. *J Sleep Res* 1999;8[Suppl 1]:53–59.

35. Hayaishi O. Sleep-wake regulation by prostaglandins D_2 and E_2. *J Biol Chem* 1988;263:14593–14596.

36. Krueger JM, Fang J. Cytokines in sleep regulation. In: Hayaishi O, Inoue S, eds. *Sleep and sleep disorders: from molecule to behavior.* Tokyo: Academic Press/Harcourt Brace Japan, 1997.

37. Hayaishi O. Regulation of sleep by prostaglandin D2 and adenosine. In: Borbély AA, Hayaishi O, Sejnowski TJ, et al., eds. *The regulation of sleep.* Strasbourg: Human Frontier Science Program, 2000:97–102.

38. Krueger JM. Cytokines and growth factors in sleep regulation. In: Borbély AA, Hayaishi O, Sejnowski TJ, et al., eds. *The regulation of sleep.* Strasbourg: Human Frontier Science Program, 2000:122–130.

39. McCarley RW, Strecker RE, Thakkar MM, et al. Adenosine and 5-HT as regulators of behavioural state. In: Borbély AA, Hayaishi O, Sejnowski TJ, et al., eds. *The regulation of sleep.* Strasbourg: Human Frontier Science Program, 2000:103–112.

40. Strecker RE, Moriarty S, Thakkar M, et al. Adenosinergic modulation of basal forebrain and preoptic/anterior hypothalamic neuronal activity in the control of behavioral state. *Behav Brain Res* 2000;115:183–204.

41. Fredholm BB. Adenosine receptors in the central nervous system. *News Physiol Sci* 1995;10:122–128.

42. Rainnie DG, Grunze HC, McCarley RW, et al. Adenosine inhibition of mesopontine cholinergic neurons: implications for EEG arousal. *Science* 1994;263:689–692.

43. Szymusiak R. Magnocellular nuclei of the basal forebrain: substrates of sleep and arousal regulation. *Sleep* 1995;18:478–500.

44. Sherin JE, Shiromani PJ, McCarley RW, et al. Activation of ventrolateral preoptic neurons during sleep. *Science* 1996;271:216–219.

45. Sherin JE, Elmquist JK, Torrealba F, et al. Innervation of histaminergic tuberomamillary neurons by GABAergic and galaninergic neurons in the ventrolateral preoptic nucleus of the rat. *J Neurosci* 1998;18:4705–4721.

46. Moruzzi G, Magoun HW. Brain stem reticular formation and activation of the EEG. *Electroenceph Clin Neurophysiol* 1949;1:455–473.

47. Steriade M, Datta S, Paré D, et al. Neuronal activities in brainstem cholinergic nuclei related to tonic activation processes in thalamocortical systems. *J Neurosci* 1990;10:2541–2559.

48. McCarley RW, Benoit O, Barrionuevo G. Lateral geniculate nucleus unitary discharge in sleep and waking: state and rate specific aspects. *J Neurophysiol* 1983;50:798–818.

49. Amzica F, Steriade M. Electrophysiological correlates of sleep delta waves. *Electroenceph Clin Neurophysiol* 1998;107:69–83.

50. McCormick DA, Bal T. Sleep and arousal: thalamocortical mechanisms. *Annu Rev Neurosci* 1997;20:185–215.

51. Jouvet M. What does a cat dream about? *Trends Neurosci* 1979;2:15–16.

52. Greene RW, McCarley RW. Cholinergic neurotransmission in the brainstem: implications for behavioral state control. In: Steriade M, Biesold D, eds. *Brain cholinergic systems.* New York: Oxford University Press, 1990:224–235.

53. McCarley RW, Greene RW, Rainnie D, et al. Brain stem neuromodulation and REM sleep. *Sem Neurosci* 1995;7:341–354.

54. Mitani A, Ito K, Hallanger AH, et al. Cholinergic projections from the laterodorsal and pedunculopontine tegmental nuclei to the pontine gigantocellular tegmental field in the cat. *Brain Res* 1988;451:397–402.

55. Webster HH, Jones BE. Neurotoxic lesions of the dorsolateral pontomesencephalic tegmentum-cholinergic area in the cat. II. Effects upon sleep-waking states. *Brain Res* 1988;458:285–302.

56. Thakkar M, Portas CM, McCarley RW. Chronic low amplitude electrical stimulation of the laterodorsal tegmental nucleus of freely moving cats increases REM sleep. *Brain Res* 1996;723:223–227.

57. Kodama T, Lai YY, Siegel JM. Enhancement of acetylcholine release during REM sleep in the caudomedial medulla as measured by in vivo microdialysis. *Brain Res* 1992;580:348–350.

58. Leonard TO, Lydic R. Pontine nitric oxide modulates acetylcholine release, rapid eye movement sleep generation, and respiratory rate. *J Neuroscience* 1997;17:774–785.

59. McCarley RW, Hobson JA. Neuronal excitability modulation over the sleep cycle: a structural and mathematical model. *Science* 1975;189:58–60.

60. McGinty DJ, Harper RM. Dorsal raphe neurons: depression of firing during sleep in cats. *Brain Res* 1976;101:569–575.

61. McCarley RW, Massaquoi SG. Neurobiological structure of the revised limit cycle reciprocal interaction model of REM sleep cycle control. *J Sleep Res* 1992;1:132–137.

62. Mühlethaler M, Khateb A, Serafin M. Effects of monoamines and opiates on pedunculopontine neurones. In: Mancia M, Marini G, eds. *The diencephalon and sleep.* New York: Raven Press, 1990.

63. Luebke JI, Greene RW, Semba K, et al. Serotonin hyperpolarizes cholinergic low-threshold burst neurons in the rat laterodorsal tegmental nucleus in vitro. *Proc Natl Acad Sci U S A* 1992;89:743–747.

64. Leonard CS, Llinás R. Serotonergic and cholinergic inhibition of mesopontine cholinergic neurons controlling REM sleep: an in vitro electrophysiological study. *Neuroscience* 1994;59:309–330.

65. Cespuglio R, Gomez ME, Walker E, et al. Effect du refroidissement et de la stimulation des noyaux du systeme du raphe sur les etats de vigilance chez le chat. *Electroenceph Clin Neurophysiol* 1979;47:289–308.

66. Portas CM, Thakkar M, Rainnie D, et al. Microdialysis perfusion of 8-hydroxy-2-(di-n-propylamino)tetralin (8-OH-DPAT) in the dorsal raphe nucleus decreases serotonin release and increases rapid eye movement sleep in the freely moving cat. *J Neurosci* 1996;16:2820–2828.

67. Sanford LD, Ross RJ, Seggos AE, et al. Central administration of two 5-HT receptor agonists: effect on REM sleep initiation and PGO waves. *Pharmacol Biochem Behav* 1994;49:93–100.

68. Horner RL, Sanford LD, Annis D, et al. Serotonin at the laterodorsal tegmental nucleus suppresses rapid-eye-movement sleep in freely behaving rats. *J Neurosci* 1997;17:7541–7552.

69. Williams JA, Reiner PB. Noradrenaline hyperpolarizes identified rat mesopontine cholinergic neurons *in vitro. J Neurosci* 1993;13:3878–3883.

70. Monti JM. Involvement of histamine in the control of the waking state. *Life Sci* 1993;53:1331–1338.

71. Morales FR, Chase MH. Intracellular recording of lumbar motoneuron membrane potential during sleep and wakefulness. *Exp Neurol* 1978;62:821–827.

72. Morales FR, Chase MH. Postsynaptic control of lumbar motoneuron excitability during active sleep in the chronic cat. *Brain Res* 1981;225:279–295.

73. Chase MH, Soja PJ, Morales FR. Evidence that glycine mediates the postsynaptic potentials that inhibit lumbar motoneuron during the atonia of active sleep. *J Neurosci* 1989;9:743–751.

74. Siegel JM. Brainstem mechanisms generating REM sleep. In: Kryger MH, Roth T, Dement WC, eds. *Principles and practice of sleep medicine,* 3rd ed. Philadelphia: WB Saunders, 2000:112–133.

75. de Lecea L, Kilduff TS, Peyron C, et al. The hypocretins: hypothalamus-specific peptides with neuroexcitatory activity. *Proc Natl Acad Sci USA* 1998;95:322–327.

76. Sakurai T, Amemiya A, Ishii M, et al. Orexins and orexin receptors: a family of hypothalamic neuropeptides and G protein-coupled receptors that regulate feeding behavior. *Cell* 1998;92:573–585.

77. Siegel JM. Narcolepsy: a key role for hypocretins. *Cell* 1999;98: 409–412.

78. Lin L, Faraco J, Li R, et al. The sleep disorder canine narcolepsy is caused by a mutation in the hypocretin (orexin) receptor 2 gene. *Cell* 1999;98:365–376.

79. Chemelli RM, Willie JT, Sinton CM, et al. Narcolepsy in orexin knockout mice: molecular genetics of sleep regulation. *Cell* 1999;98:437–451.

80. Elias CF, Saper CB, Maratos-Flier E, et al. Chemically defined projections linking the mediobasal hypothalamus and the lateral hypothalamic area. *J Comp Neurol* 1998;402:442–459.

81. Peyron C, Tighe D, van den Pol A, et al. Neurons containing hypocretin (orexin) project to multiple neuronal systems. *J Neurosci* 1998;18:9996–10015.

82. Date Y, Ueta Y, Yamashita H, et al. Orexins, orexigenic hypothalamic peptides, interact with autonomic, neuroendocrine and neuroregulatory systems. *Proc Natl Acad Sci U S A* 1999;96:748–753.

83. Horvath TL, Peyron C, Diano S, et al. Hypocretin (orexin) activation and synaptic innervation of the locus caeruleus noradrenergic system. *J Comp Neurol* 1999;415:145–159.

84. Nambu T, Sakurai T, Mizukami K, et al. Distribution of orexin neurons in the adult rat brain. *Brain Res* 1999;827:243–260.

85. Trivedi P, Yu H, MacNeil DJ, et al. Distribution of orexin receptor mRNA in the rat brain. *FEBS Lett* 1998;438:71–75.

86. Marcus JN, Aschkenasi CJ, Lee CE, et al. Differential expression of orexin receptors 1 and 2 in the rat brain. *J Comp Neurol* 2001;435:6–25.

87. Hagan JJ, Leslie RA, Patel S, et al. Orexin A activates locus caeruleus cell firing and increases arousal in the rat. *Proc Natl Acad Sci U S A* 1999;96:10911–10916.

88. van den Pol AN, Gao XB, Obrietan K, et al. Presynaptic and postsynaptic actions and modulation of neuroendocrine neurons by a new hypothalamic peptide, hypocretin/orexin. *J Neurosci* 1998;18:7962–7971.

89. Thakkar MM, Ramesh V, Cape EG, et al. REM sleep enhancement and behavioral cataplexy following orexin (hypocretin) II receptor antisense perfusion in the pontine reticular formation. *Sleep Res Online* 1999;2:113–120.

90. Nishino S, Ripley B, Overeem S, et al. Hypocretin (orexin) deficiency in human narcolepsy. *Lancet* 2000;355:39–40.

91. Peyron C, Faraco J, Rogers W, et al. A mutation in a case of early onset narcolepsy and a generalized absence of hypocretin peptides in human narcoleptic brains. *Nat Med* 2000;6:991–997.

92. Thannickal TC, Moore RY, Nienhuis R, et al. Reduced number of hypocretin neurons in human narcolepsy. *Neuron* 2000;27:469–474.

93. Willie JT, Chemelli RM, Sinton CM, et al. To eat or to sleep? Orexin in the regulation of feeding and wakefulness. *Annu Rev Neurosci* 2001;24:429–458.

94. Lakin-Thomas PL. Circadian rhythms: new functions for old clock genes? *Trends Genet* 2000;16:135–142.

95. Edery I. Circadian rhythms in a nutshell. *Physiol Genomics* 2000;3:59–74.

96. Loudin ASI, Semikhodskii AG, Crosthwaite SK. A brief history of circadian time. *Trends Genet* 2000;16:477–481.

97. Young MW. Life's 24-hour clock: molecular control of circadian rhythms in animal cells. *Trends Biol Sci* 2000;25:601–606.

98. Mistlberger RE. Circadian food-anticipatory activity: formal models and physiological mechanisms. *Neurosci Biobehav Rev* 1994;18:171–195.

99. Stephan FK. Interactions between light- and feeding-entrainable circadian rhythms in the rat. *Physiol Behav* 1986;38:127–133.

100. Cassone VN, Speh JC, Card JP, et al. Comparative anatomy of the mammalian hypothalamic suprachiasmatic nucleus. *J Biol Rhythms* 1988;3:71–92.

101. Moore RY. Retinohypothalamic projections in mammals: a comparative study. *Brain Res* 1973;49:403–409.

102. Hendrickson AE, Wagoner N, Cowan WM. Autoradiographic

and electron microscopic study of retinohypothalamic projections. *Z Zellforsch Mikrosk Anat* 1972;125:1–26.

103. Moore RY, Lenn NJ. A retinohypothalamic projection in the rat. *J Comp Neurol* 1972;142:1–14.

104. Moore RY, Eichler VB. Loss of a circadian adrenal corticosterone rhythm following suprachiasmatic lesions in the rat. *Brain Res* 1972;42:201–206.

105. Stephan FK, Zucker I. Circadian rhythms in drinking behavior and locomotor activity of rats are eliminated by hypothalamic lesions. *Proc Natl Acad Sci U S A* 1972;69:1583–1586.

106. Inouye ST, Kawamura H. Persistence of circadian rhythmicity in a mammalian hypothalamic "island" containing the suprachiasmatic nucleus. *Proc Natl Acad Sci U S A* 1979;76:5962–5966.

107. Inouye ST, Kawamura H. Characteristics of a circadian pacemaker in the suprachiasmatic nucleus. *J Comp Physiol [A]* 1982;146:143–160.

108. Newman GC, Hospod FE. Rhythm of suprachiasmatic nucleus 2-deoxyglucose uptake in vitro. *Brain Res* 1986;381:345–350.

109. Shibata S, Moore RY. Electrical and metabolic activity of suprachiasmatic nucleus neurons in hamster hypothalamic slices. *Brain Res* 1988;438:374–378.

110. Aguiler-Roblero R, Garcia-Hernandez F, Fernandez-Cancino F, et al. Fetal suprachiasmatic nucleus transplants: diurnal rhythm recovery of lesioned rats. *Brain Res* 1984;311:353–357.

111. Sawaki Y, Nihommatsu I, Kawamura H. Transplantation of the neonatal suprachiasmatic nuclei into rats with complete bilateral suprachiasmatic lesions. *Neurosci Res* 1984;1:67–72.

112. Reppert SM, Schwartz WJ. Maternal coordination of the fetal biological clock in utero. *Science* 1983;220:969–971.

113. Shibata S, Moore RY. Development of a fetal circadian rhythm after disruption of the maternal circadian system. *Dev Brain Res* 1988;41:313–317.

114. Moore RY, Bernstein ME. Synaptogenesis in the rat suprachiasmatic nucleus demonstrated by electron microscopy and synapsin I immunoreactivity. *J Neurosci* 1989;9:2151–2162.

115. Rusak B. The role of suprachiasmatic nuclei in the generation of circadian rhythms in the golden hamster, *Mesocricetus auratus*. *J Comp Physiol* 1977;118:145–146.

116. Johnson RF, Morin LP, Moore RY. Retinohypothalamic projections in the hamster and rat demonstrated using cholera toxin. *Brain Res* 1988;462:301–312.

117. Card JP, Moore RY. Ventral lateral geniculate nucleus efferents to the rat suprachiasmatic nucleus exhibit avian pancreatic polypeptide-like immunoreactivity. *J Comp Neurol* 1982;206:390–396.

118. Challet E, Pevet P, Malan A. Intergeniculate leaflet lesion and daily rhythms in food-restricted rats fed during daytime. *Neurosci Lett* 1996;216:214–218.

119. Maywood ES, Smith E, Hall SJ, et al. A thalamic contribution to arousal-induced, non-photic entrainment of the circadian clock in the Syrian hamster. *Eur J Neurosci* 1997;9:1739–1747.

120. Harrington ME. The ventral lateral geniculate nucleus and the intergeniculate leaflet: interrelated structures in the visual and circadian systems. *Neurosci Biobehav Rev* 1997;21:705–727.

121. Moga MM, Moore RY. Organization of neural inputs to the suprachiasmatic nucleus in the rat. *J Comp Neurol* 1997;389:508–534.

122. Mistleberger RE, Rusak B. Circadian rhythms in mammals: formal properties and environmental influences. In: Kryger MH, Roth T, Dement WC, eds. *Principles and practice of sleep medicine*, 3rd ed. Philadelphia: WB Saunders, 2000:321–333.

123. Mistleberger RE. Circadian food anticipatory activity: formal models and physiological mechanisms. *Neurosci Biobehav Rev* 1994;18:1–25.

124. Reebs SG, Mrosovsky N. Effects of induced wheel running on the circadian activity rhythms of Syrian hamsters: entrainment and phase-response curve. *J Biol Rhythms* 1989;4:39–48.

125. Antle MC, Mistleberger RE. Circadian clock resetting by sleep deprivation without exercise in the Syrian hamster. *J Neurosci* 2000;20:9326–9332.

126. Ibuka N, Inouye S-I, Kawamura H. Analysis of sleep-wakefulness rhythms in male rats after suprachiasmatic nucleus lesions and ocular enucleation. *Brain Res* 1977;122:33–47.

127. Edgar DM, Dement WC, Fuller CA. Effect of SCN lesions on sleep in squirrel monkeys: evidence for opponent processes in sleep-wake regulation. *J Neurosci* 1993;13:1065–1079.

128. Watts AG, Swanson LW, Sanchez-Watts G. Efferent projections of the suprachiasmatic nucleus. I. Studies using anterograde transport of Phaseolus vulgaris leukoagglutinin in the rat. *J Comp Neurol* 1987;258:204–229.

129. Moore RY. Circadian system influences on sleep. Presented at Associated Professional Sleep Societies Meeting, June 2001.

130. Colquhon WP, Blake MJF, Edwards RS. Experimental studies of shift work: a comparison of "rotating" and "stabilized" 4-hour shift systems. *Ergonomics* 1968;11:437–453.

131. Kleitman N, Jackson DP. Body temperature and performance under different routines. *Am J Appl Physiol* 1950;3:309–328.

132. Wever RA. *The circadian system of man.* New York: Springer-Verlag, 1979.

133. Zulley J, Campbell SS. Napping behavior during "spontaneous internal desynchronization": sleep remains in synchrony with body temperature. *Hum Neurobiol* 1985;4:123–126.

134. Mills JN, Minors DS, Waterhouse JM. Adaptation to abrupt time shifts of the oscillator(s) controlling human circadian rhythms. *J Physiol (Lond)* 1978;285:455–470.

135. Duffy JF. Constant routine. In: Carskadon MA, ed. *Encyclopedia of sleep and dreaming.* New York: MacMillan, 1993:134–136.

136. Kleitman N. *Sleep and wakefulness.* Chicago: University of Chicago Press, 1963:172–184.

137. Czeisler CA, Duffy JF, Shanahan TL, et al. Stability, precision, and near-24-hour period of the human circadian pacemaker. *Science* 1999;284:2177–2181.

138. Campbell S. Is there an intrinsic period of the circadian clock? *Science* 2000;288:1174–1175.

139. Czeisler CA, Dijk D-J, Kronauer RE, et al. Response. *Science* 2000;288:1174–1175.

140. Saper CB. Circadian system influences on sleep. Presented at Associated Professional Sleep Societies Meeting, June 2001.

141. Dement WC, Kleitman N. Cyclic variations in EEG during sleep and their relation to eye movements, body motility and dreaming. *Electroenceph Clin Neurophysiol* 1956;9:673–690.

142. Kleitman N. *Sleep and wakefulness.* Chicago: University of Chicago Press, 1963:364–365.

143. Kripke DF. An ultradian biological rhythm associated with perceptual deprivation and REM sleep. *Psychosom Med* 1972;34:221–234.

144. Manseau C, Broughton RJ. Bilaterally synchronous ultradian EEG rhythms in adult humans. *Psychophysiology* 1984;21:265–273.

145. Okawa M, Matousek M, Petersen I. Spontaneous vigilance fluctuations in the daytime. *Psychophysiology* 1984;21:207–211.

146. Gronfier C, Simon C, Piquard F, et al. Neuroendocrine processes underlying ultradian sleep regulation in man. *J Clin Endocrinol Metab* 1999;84:2686–2690.

147. Orr WC, Hoffman HG, Hegge FW. Ultradian rhythms in extended performance. *Aerospace Med* 1974;45:995–1000.

148. De Koninck J, Quera Salva M, Besset A, et al. Are REM cycle narcoleptic patients governed by an ultradian rhythm? *Sleep* 1986;9:162–166.

149. Czeisler CA, Zimmerman JC, Ronda JM, et al. Timing of REM

sleep is coupled to the circadian rhythm of body temperature in man. *Sleep* 1980;2:329–346.

150. Fujiki N, Yoshida Y, Ripley B, et al. Changes in CSF hypocretin-1 (orexin A) levels in rats across 24 hours and in response to food deprivation. *Neuroreport* 2001;12:993–997.

151. Dijk D-J, Duffy JF. A circadian perspective on human sleep-wake regulation and ageing. In: Borbély AA, Hayaishi O, Sejnowski TJ, et al., eds. *The regulation of sleep.* Strasbourg: Human Frontier Science Program, 2000:212–222.

152. Kronauer RE, Czeisler CA, Pilato SF, et al. Mathematical model of the human circadian system with two interacting oscillators. *Am J Physiol Regul Integr Comp Physiol* 1982;242:R3–R17.

153. Borbély AA. A two process model of sleep regulation. *Hum Neurobiol* 1982;1:195–204.

154. Folkard S. The pragmatic approach to masking. *Chronobiol Int* 1989;6:55–64.

155. Bearpark HM. Insomnia: causes, effects and treatment. In: Cooper R, ed. *Sleep.* London: Chapman & Hall Medical, 1994.

156. Hartmann E. Insomnia: diagnosis and treatment. In: Williams RL, Karacan I, Moore CA, eds. *Sleep disorders: diagnosis and treatment.* New York: Wiley, 1988.

157. Ancoli-Israel S, Roth T. Characteristics of insomnia in the United States: results of the 1991 National Sleep Foundation Survey. I. *Sleep* 1999;22[Suppl 2]:S347–S353.

158. Carskadon MA, Dement WC, Merrill MM, et al. Guidelines for the Multiple Sleep Latency Test (MSLT): a standard measure of sleepiness. *Sleep* 1986;9:519–524.

159. Kryger M, Lavie P, Rosen R. Recognition and diagnosis of insomnia. *Sleep* 1999;22[Suppl 3]:S421–S426.

160. Hauri PJ. Primary insomnia. In: Kryger MH, Roth T, Dement WC, eds. *Principles and practice of sleep medicine,* 3rd ed. Philadelphia: WB Saunders, 2000:633–639.

161. McCall WV, Edinger JD. Subjective total insomnia: an example of sleep state misperception. *Sleep* 1992;15:71–73.

162. Hauri PJ, Fischer J. Persistent psychophysiological (learned) insomnia. *Sleep* 1986;9:38–53.

163. Lugaresi E, Medori R, Montagna P, et al. Fatal familial insomnia and dysautonomia with selective degeneration of thalamic nuclei. *N Engl J Med* 1986;315:997–1003.

164. Medori R, Tritschler HJ, LeBlanc A, et al. Fatal familial insomnia: a prion disease with a mutation at codon 178 of the prion protein gene. *N Engl J Med* 1992;326:444–449.

165. Parchi P, Petersen RB, Chen SG, et al. Molecular pathology of fatal familial insomnia. *Brain Pathol* 1998;8:539–548.

166. Guilleminault C. Disorders of excessive daytime sleepiness. *Annu Clin Res* 1985;17:209–219.

167. Guilleminault C, Faull KM, Miles L, et al. Posttraumatic excessive daytime sleepiness: a review of 20 patients. *Neurology* 1980;33:1584–1589.

168. Takahashi Y. Clinical studies of periodic somnolence. Analysis of 28 personal cases. *Psych Neurol (Jpn)* 1965;853–889.

169. Tanabe E, Yara K, Mastuura M, et al. Prolonged polysomnography in a case with recurrent hypersomnia. *Psychiatry Clin Neurosci* 1998;52:204–205.

170. Critchley M. Periodic hypersomnia and megaphagia in adolescent males. *Brain* 1962;59:494–515.

171. Kales A, Soldatos CR, Bixler EO, et al. Hereditary factors in sleepwalking and night terrors. *Br J Psychiatry* 1980;137:111–118.

172. Broughton RJ. Sleep disorders. Disorders of arousal? *Science* 1968;159:1070–1078.

173. Broughton RJ. NREM arousal phenomena In: Kryger MH, Roth T, Dement WC, eds. *Principles and practice of sleep medicine,* 3rd ed. Philadelphia: WB Saunders, 2000:693–706.

174. Rosen G, Mahowald MW, Ferber R. Sleepwalking, confusional arousals and sleep terrors in the child. In: Ferber R, Kryger M, eds. *Principles and practice of sleep medicine in the child.* Philadelphia: WB Saunders, 1995.

175. Roth B, Nevsimolova S, Rechtschaffen A. Hypersomnia with "sleep drunkenness." *Arch Gen Psychiatry* 1972;26:456–462.

176. Terman M, Terman JS. Light therapy. In: Kryger MH, Roth T, Dement WC, eds. *Principles and practice of sleep medicine,* 3rd ed. Philadelphia: WB Saunders, 2000:1258–1274.

177. Weitzman ED, Czeisler CA, Coleman RM, et al. Delayed sleep phase syndrome. A chronobiological disorder with sleep-onset insomnia. *Arch Gen Psychiatry* 1981;38:737–746.

178. Regestein QR, Monk TH. Delayed sleep phase syndrome: a review of its clinical aspects. *Am J Psychiatry* 1995;152:602–608.

179. Czeisler CA, Richardson GS, Coleman RM, et al. Chronotherapy: resetting the circadian clocks of patients with delayed sleep phase insomnia. *Sleep* 1981;4:1–21.

180. Lewy AJ, Sack RL, Singer CM. Treating phase typed chronobiologic sleep and mood disorders using appropriately timed bright artificial light. *Psychopharm Bull* 1985;21:368–72.

181. Lewy A, Sack RL, Singer CM. Melatonin, light and chronobiological disorders. *Ciba Found Symposium* 1985;117:231–52.

182. Baker SK, Zee PC. Circadian disorders of the sleep-wake cycle. In: Kryger MH, Roth T, Dement WC, eds. *Principles and practice of sleep medicine,* 3rd ed. Philadelphia: WB Saunders, 2000:606–614.

183. Czeisler CA, Allan JS, Strogatz SH, et al. Bright light resets the human circadian pacemaker independent of the timing of the sleep-wake cycle. *Science* 1986;233:667–671.

184. Toh KL, Jones CR, He Y, et al. An hPer2 phosphorylation site mutation in familial advanced phase syndrome. *Science* 2001;291:1040–1043.

185. Chicurel M. Mutant gene speeds up the human clock. *Science* 2001;291:226–227.

186. Weber AL, Cary MS, Connor N, et al. Human non-24-hour sleep-wake cycles in an everyday environment. *Sleep* 1980;2:347–354.

187. Arendt J, Aldhous M, Wright J. Synchronisation of a disturbed sleep-wake cycle in a blind man by melatonin treatment [Letter]. *Lancet,* 1988;1:772–773.

188. Arendt J, Stone B, Skene D. Jet lag and sleep disruption. In: Kryger MH, Roth T, Dement WC, eds. *Principles and practice of sleep medicine,* 3rd ed. Philadelphia: WB Saunders, 2000:591–599.

189. Claustrat B, Brun J, David M, et al. Melatonin and jet lag: confirmatory result using a simplified protocol. *Biol Psychiatry* 1992;32:705–711.

190. Choo KL, Guilleminault C. Narcolepsy and idiopathic hypersomnolence. *Clin Chest Med* 1998;19:169–181.

191. Guilleminault C, Gelb M. Clinical aspects and features of cataplexy. *Adv Neurol* 1995;67:65–77.

192. Hishikawa Y, Shimizu T. Physiology of REM sleep, cataplexy, and sleep paralysis. In: Fahn S, Hallett M, Lüders HO, et al., eds. *Negative motor phenomena.* Philadelphia: Lippincott-Raven, 1995:245–271.

193. Green P, Stillman MJ. Narcolepsy. Signs, symptoms, differential diagnosis, and management. *Arch Family Med* 1998;7:472–478.

194. Mignot E. Genetic and familial aspects of narcolepsy. *Neurology* 1998;50[Suppl 1]:S16–S22.

195. Honda Y, Takahashi Y, Honda M, et al. Genetic aspects of narcolepsy. In: Hayaishi O, Inoue S, eds. *Sleep and sleep disorders: from molecules to behavior.* New York: Academic Press, 1997:341–358.

196. Mignot E, Lin L, Rogers W, et al. Complex HLA-DR and -DQ

interactions confer risk of narcolepsy-cataplexy in three ethnic groups. *Am J Hum Genet* 2001;68:686–699.

197. Mitler MM, Hajkukovic R. Relative efficacy of drugs for the treatment of sleepiness in narcolepsy. *Sleep* 1991;14:218–220.

198. Broughton RJ, Fleming JA, George CF, et al. Randomized, double-blind, placebo-controlled crossover trial of modafinil in the treatment of excessive daytime sleepiness in narcolepsy. *Neurology* 1997;49:444–451.

199. Broughton WA, Broughton RJ. Psychosocial impact of narcolepsy. *Sleep* 1994;17[Suppl 8]:S45–S49.

200. Schenck CH, Bundlie SR, Ettinger MG, et al. Chronic behavioral disorders of human sleep: a new category of parasomnia. *Sleep* 1986;9:293–308.

201. Sastre JP, Jouvet M. Oneiric behavior in cats. *Physiol Behav* 1979;22:979–989.

202. Mahowald MW, Schenck CH. REM sleep parasomnias. In: Kryger MH, Roth T, Dement WC, eds. *Principles and practice of sleep medicine,* 3rd ed. Philadelphia: WB Saunders, 2000:724–741.

203. Nofzinger EA, Reynolds CF III. REM sleep behavior disorder. *JAMA* 1994;271:820.

204. Bamford CR. Carbamazepine in REM sleep behavior disorder. *Sleep* 1993;16:33–34.

205. Montplaisir J, Nicolas A, Godbout R, et al. Restless legs syndrome and periodic limb movement disorders. In: Kryger MH, Roth T, Dement WC, eds. *Principles and practice of sleep medicine,* 3rd ed. Philadelphia: WB Saunders, 2000:742–752.

206. Hening W, Allen R, Earley C, et al. The treatment of restless legs syndrome and periodic limb movement disorder. An American Academy of Sleep Medicine review. *Sleep* 1999;22:970–999.

207. Phillips B, Young T, Finn L, et al. Epidemiology of restless legs syndrome in adults. *Arch Intern Med* 2000;160:2137–2141.

208. Tan EK, Ondo W. Restless legs syndrome: clinical features and treatment. *Am J Med Sci* 2000;319:397–403.

209. Mahowald MW, Thorpy MJ. Nonarousal parasomnias in the child. In: Ferber R, Kryger M, eds. *Principles and practice of sleep medicine in the child.* Philadelphia: WB Saunders, 1995:115–123.

210. Chisholm T, Morehouse RL. Adult headbanging: sleep studies and treatment. *Sleep* 1996;19:343–346.

211. Douglas NJ. The sleep apnoea/hypopnoea syndrome. In: Cooper R, ed. *Sleep.* London: Chapman & Hall Medical, 1994:272–292.

212. Coleman RM, Roffwarg HP, Kennedy SJ, et al. Sleep-wake disorders based on a polysomnographic diagnosis. A national cooperative study. *JAMA* 1982;247:997–1003.

213. Henderson-Smart DJ. Regulation of breathing in the fetus and the newborn. In: Saunders NA, Sullivan CE, eds. *Sleep and breathing.* New York: Marcel Dekker, 1994:605–647.

214. Guilleminault C, Tilkian AG, Dement WC. The sleep apnea syndromes. *Annu Rev Med* 1976;27:465–484.

215. Guilleminault C. Sleep and breathing. In: Guilleminault C, ed. *Sleep and waking.* Menlo Park, CA: Addison-Wesley, 1982:155–182.

216. Cartwright R, Ristanovic R, Diaz F, et al. A comparative study of treatments for positional sleep apnea. *Sleep* 1991;14:546–552.

217. Fujita S. Pharyngeal surgery for obstructive sleep apnea and snoring. In: Fairbanks DNF, Fujita S, Ikematsu T, et al., eds. *Snoring and obstructive sleep apnea.* New York: Raven Press, 1987:101–128.

218. Sullivan CE, Issa FG, Berthon-Jones M, et al. Reversal of obstructive sleep apnea by continuous positive airway pressure applied through the nares. *Lancet* 1981;1:862–865.

219. Kribbs NB, Pack AI, Kline LR, et al. Objective measurement of patterns of nasal CPAP use by patients with obstructive sleep apnea. *Am Rev Respir Dis* 1993;147:887–895.

220. Bradley TD, Phillipson EA. Central sleep apnea. *Clin Chest Med* 1992;13:493–505.

221. Severinghaus JW, Mitchell RA. Ondine's curse—failure of respiratory center automaticity while awake. *Clin Res* 1962;10:122.

222. Guilleminault C, McQuitty J, Ariagno RL, et al. Congenital central alveolar hypoventilation syndrome in six infants. *Pediatrics* 1982;70:684–694.

223. Bedard MA, Montplaisir J, Malo J, et al. Persistent neuropsychological deficits and vigilance impairment in sleep apnea syndrome after continuous positive airway pressure (CPAP). *J Clin Exp Neuropsychol* 1993;15:330–341.

224. Kupfer DJ, Reynolds CF. Sleep and affective disorders. In: Paykel ES, ed. *Handbook of affective disorders,* 2nd ed. New York: Guilford, 1992:311–323.

225. Wu JC, Bunney WE. The biological basis of an antidepressant response to sleep deprivation and relapse: review and hypothesis. *Am J Psychiatry* 1990;147:14–21.

226. Kvist J, Kirkegaard C. Effects of repeated sleep deprivation on clinical symptoms and the TRH test in endogenous depression. *Acta Psychiatr Scand* 1980;62:494–502.

227. Dalgleish T, Rosen K, Marks M. Rhythm and blues: the theory and treatment of seasonal affective disorder. *Br J Clin Psychol* 1996;35:163–182.

228. Benca RM. Mood disorders. In: Kryger MH, Roth T, Dement WC, eds. *Principles and practice of sleep medicine,* 3rd ed. Philadelphia: WB Saunders, 2000:1140–1158.

229. Buysse DJ, Kupfer DJ. Diagnostic and research applications of electroencephalographic sleep studies in depression: conceptual and methodological issues. *J Nerv Ment Dis* 1990;178:405–414.

230. Wehr T. Effects of wakefulness and sleep on depression and mania. In: Montplaisir J, Godbout R, eds. *Sleep and biological rhythms: basic mechanisms and applications to psychiatry.* New York: Oxford University Press, 1990:42–86.

231. Czeisler CA, Kronauer RE, Nooney JJ, et al. Biologic rhythm disorders, depression, and phototherapy. A new hypothesis. *Psychiatr Clin North Am* 1987;10:687–709.

232. Sack DA, Nurnburger J, Rosenthal NE, et al. The potentiation of antidepressant medications by phase-advance of the sleep-wake cycle. *Am J Psychiatry* 1985;142:606–608.

233. Borbély AA, Wirz-Justice A. Sleep, sleep deprivation, and depression. *Hum Neurobiol* 1982;1:205–210.

234. Vogel GW, Vogel F, McAbee RS, et al. Improvement of depression by REM sleep deprivation. New findings and a new hypothesis. *Arch Gen Psychiatry* 1980;37:247–253.

235. Gillin JC, Sutton L, Ruiz C, et al. The cholinergic rapid eye movement induction test with arecoline in depression. *Arch Gen Psychiatry* 1991;48:264–270.

236. Kuhs H, Tolle R. Sleep deprivation therapy. *Biol Psychiatry* 1991;29:1129–1148.

237. Wirz-Justice A, Pühringer W, Hole G. Response to sleep deprivation as a predictor of therapeutic results with antidepressant drugs. *Am J Psychiatry* 1979;136:1222–1223.

238. Sack DA, Duncan W, Rosenthal NE, et al. The timing and duration of sleep in partial sleep deprivation therapy of depression. *Acta Psychiatr Scand* 1988;77:219–224.

239. Aserinsky E, Kleitman N. Regularly occurring periods of eye motility and concomitant phenomena during sleep. *Science* 1953;118:273–274.

240. Zarcone VP. Sleep and schizophrenia. In: Williams RL, Karacan I, Moore CA, eds. *Sleep disorders: diagnosis and treatment,* 2nd ed. New York: Wiley, 1988:165–188.

241. Dement WC. Dream recall and eye movements during sleep

in schizophrenics and normals. *J Nerv Ment Dis* 1955;122:263–269.

242. Nofzinger EA, Buysse DJ, Reynolds CF III, et al. Sleep disorders related to another mental disorder (nonsubstance/primary): a DSM-IV literature review. *J Clin Psychiatry* 1993;54:244–255.

243. Tanden R, Shipley JE, Taylor S, et al. Electroencephalographic sleep abnormalities in schizophrenia. *Arch Gen Psychiatry* 1992;49:185–194.

244. Feinberg I, Hiatt JF. Sleep patterns in schizophrenia: a selective review. In: Williams RL, Karacan I, ed. *Sleep disorders: diagnosis and treatment.* New York: Wiley, 1978:205–231.

245. Keshaven MS, Reynolds CF, Miewald MJ, et al. Delta sleep deficits in schizophrenia: evidence from automated analyses of sleep data. *Arch Gen Psychiatry* 1998;55:443–448.

246. Maixner S, Tandon R, Eiser A, et al. Effects of antipsychotic treatment on polysomnographic measures in schizophrenia: a replication and extension. *Am J Psychiatry* 1998;155:1600–1602.

247. Benca RM, Casper RC. Eating disorders. In: Kryger MH, Roth T, Dement WC, eds. *Principles and practice of sleep medicine,* 3rd ed. Philadelphia: WB Saunders, 2000:1169–1175.

248. Stunkard AJ, Grace WJ, Wolff HG. The night-eating syndrome. A pattern of food intake among certain obese patients. *Am J Med* 1955;19:78–86.

249. Schenck CH, Hurwitz TD, O'Connor KA, et al. Additional categories of sleep-related eating disorders and current status of treatment. *Sleep* 1993;16:457–466.

250. Uhde TW. Anxiety disorders. In: Kryger MH, Roth T, Dement WC, eds. *Principles and practice of sleep medicine,* 3rd ed. Philadelphia: WB Saunders, 2000:1123–1139.

251. Reynolds CF III, Shaw DH, Newton TF, et al. EEG sleep in outpatients with generalized anxiety: a preliminary comparison with depressed outpatients. *Psychiatry Res* 1983;8:81–89.

252. Mellman TA, Uhde TW. Electroencephalographic sleep in panic disorder. *Arch Gen Psychiatry* 1989;46:178–184.

253. Hauri PJ, Friedman M, Ravaris CL. Sleep in patients with spontaneous panic attacks. *Sleep* 1989;12:323–337.

254. Edlund MJ, McNamara ME, Millman RP. Sleep apnea and panic attacks. *Comp Psychiatry* 1991;32:130–132.

255. Mellman TA. Psychobiology of sleep disturbances in posttraumatic stress disorder. *Ann N Y Acad Sci* 1997;821:142–149.

256. Kuch K, Cox BJ. Symptoms of PTSD in 124 survivors of the Holocaust. *Am J Psychiatry* 1992;149:337–340.

257. Mellman TA, Nolan B, Hebding J, et al. A polysomnographic comparison of veterans with combat-related PTSD, depressed men and non-ill controls. *Sleep* 1997;20:46–51.

258. Rosen J, Reynolds CF III, Yeager AL, et al. Sleep disturbances in survivors of the Nazi holocaust. *Am J Psychiatry* 1991;148:62–66.

259. True WR, Rice J, Eisen SA, et al. A twin study of genetic and environmental contributions to liability for posttraumatic stress symptoms. *Arch Gen Psychiatry* 1993;50:257–264.

16

EVALUATION AND TREATMENT OF NEUROPATHIC PAIN

HOWARD L. FIELDS

Clinically, persistent pain presents a major challenge. This is because pain is subjective, and the knowledge required for optimal evaluation and treatment crosses traditional disciplinary lines. Of particular relevance to the readers of this book, severe persistent pain can have a strikingly destructive impact on the psychological state of the patient. Furthermore, psychological problems that either predate or are induced by persistent pain can add to the intractability of the problem (see Chapter 22). Ultimately, the challenge for the clinician is to distinguish somatic from psychological contributions to the suffering of each patient. Thus, patients with chronic pain are ideal candidates for neuropsychiatric referrals.

Evaluation of the patient with neuropathic pain represents a potentially rewarding opportunity for the diagnostically oriented neuropsychiatrist. Several therapies can significantly reduce the somatically generated components of this type of pain; however, accurate diagnosis is essential. Among the useful skills required for optimal evaluation and treatment of patients with persistent pain is the ability to elicit symptoms and signs consistent with the diagnosis of neuropathic pain. Most important among the findings are the location and quality of the pain and the presence of specific types of sensory abnormalities. This chapter reviews the clinical features of patients with neuropathic pain, the current thinking about the pathophysiology of different syndromes, and a strategy for treatment of patients with this condition.

CLINICAL FINDINGS IN PATIENTS WITH PAIN DUE TO NEURAL INJURY OR DYSFUNCTION

It is important to understand that the clinical features of neuropathic pain in themselves usually are insufficient for diagnosis, and rarely are all features present in an individual. Another point of confusion is that pain itself, regardless of the precipitating cause, can produce secondary changes such as reflex muscle contraction, guarding with immobilization of the painful area, and increased sympathetic outflow. Each of these secondary processes can generate new (and potentially independent) sources of pain with their own characteristic features. It is not unusual for a patient to complain of several distinct pains with different anatomic locations and sensory qualities. It is essential to detail the time course and relative severity of each separate component.

The following is a list of the common features of neuropathic pain (1):

1. Association of the pain with evidence of neural damage, particularly sensory deficits. Although not conclusive, this feature is by far the most reliable indicator of neuropathic pain. Except in specific, easily identifiable syndromes such as trigeminal neuralgia or pain due to epilepsy, the absence of this feature should raise doubt about the diagnosis. Commonly, the reported location of the neuropathic pain is at least partially coextensive with a sensory disturbance. Usually this is a sensory loss, but sometimes pathologically enhanced sensitivity to sensory stimuli (see later) is present without a deficit.

2. The sensory quality of the pain is characteristically unique to the patient's experience. Burning and tingling are frequently used descriptors. Shooting, shocking, and/or electrical feelings are other common descriptors. Words such as crawling, squeezing, and tearing are suggestive of a neuropathic cause of pain. The point is that the sensation is unusual and difficult to ignore. Neurologists use words such as *dysesthesias* (unpleasant sensation) or *paresthesias* (abnormal sensations) to categorize sensations that are described in these terms.

3. There is often a significant delay between the causative insult and the onset of pain. For example, in patients with pain due to central nervous system (CNS) lesions, the pain often begins after a delay of months, usually after they have achieved partial recovery of motor and/or sensory function.

4. Hypersensitivity phenomena are present. It is not unusual for patients with painful injuries of any type to complain of tenderness of the affected part and to avoid threatened

contact with it. With neuropathic pain, such hypersensitivity phenomena are exaggerated, often to the point of contributing in a major way to the patient's disability. For example, many patients with posttraumatic neuralgias complain that any movements involving the affected nerve trigger severe pain. Many neuropathic pain patients report that light moving stimuli, such as a gust of wind or the brushing of their skin by clothing, induce rapid bursts of pain. This phenomenon, where severe pain is evoked by very light, moving tactile stimuli, is termed *allodynia* and, when prominent, is suggestive of neuropathic pain.

Other hypersensitivity phenomena include reduced threshold for heat pain, a striking buildup of reported pain with repeated stimuli that are near threshold (summation), the spatial spread of perceived pain from the site of the noxious stimulus, and a prolonged paroxysm of pain after a brief stimulus (after discharge). These hypersensitivity phenomena often are thought of as defining the hyperpathic state.

PATHOPHYSIOLOGY-BASED TREATMENT APPROACH TO NEUROPATHIC PAIN

Various mechanisms have been proposed to account for the paradoxical appearance of pain after injury or dysfunction of the peripheral nervous system or CNS. Unfortunately, little is known about the origin of any painful neuropathic condition. On the other hand, for each mechanism described here, there is a reasonable body of evidence that it contributes to some clinical condition. Where there is such evidence, it will be pointed out.

There is a major practical reason for describing a broad range of possible mechanisms in neuropathic pain, namely, that each potential mechanism suggests a particular treatment approach. In a given patient, several mechanisms may contribute to the generation of pain (e.g., sympathetic activity, muscle spasm). Importantly, ascertaining the etiology of the neural damage (e.g., trauma, herpes zoster) does not necessarily specify the underlying mechanism of the pain. Until better diagnostic tools that can determine the proximate cause of the pain are developed, the optimal strategy for treatment of an individual patient will remain a sequence of therapeutic trials targeted at specific pain-generating mechanisms. In this way, knowledge of multiple potential pain mechanisms, their expected clinical manifestations, and their sensitivity to treatments leads to a general algorithm for finding the optimal therapy in the shortest period of time for the largest number of patients.

This mechanism-based approach is only as good as the information upon which it is based. Currently, the relevant information is fraught with uncertainty, and many patients are not helped. On the other hand, the use of this approach

gives patients the maximum benefit of current knowledge about mechanism and available therapies. Furthermore, it is an approach that has evolved directly from clinical experience and a careful reading of the recent pertinent literature. At a minimum, this framework can help physicians organize their own clinical experience and evaluate the published work of others.

Deafferentation Hyperactivity

The loss of small-diameter primary afferents, such as occurs in some polyneuropathies, usually leads to loss of pain sensation. In the majority of cases, interruption of central pain transmission pathways results in impaired pain sensation without spontaneous pain. Anterolateral cordotomy, which interrupts the spinothalamic tract, is effective for treatment of pain in patients with terminal illness. Paradoxically, deafferentation can produce pain. Complete peripheral deafferentation is rare, but occasionally it is associated with severe pain. Lesions of the CNS, particularly of the spinal cord, can cause pain in some patients. Such central pain syndromes almost always are accompanied by impaired pain and temperature sensation (2,3).

The most persuasive clinical example of peripheral deafferentation pain results from avulsion of the brachial plexus. In this condition, backward hyperextension of the arm places severe traction on the brachial plexus, and some dorsal roots are anatomically separated from the spinal cord. Most of these patients have spontaneous pain referred to the anesthetic extremity (4). In animal studies, cutting the dorsal roots results in the development of high-frequency spontaneous activity in dorsal horn neurons, some of which may be pain transmission neurons (5). In support of the concept that spontaneous activity in dorsal horn neurons contributes to the pain of brachial plexus avulsion, Nashold and Ostdahl (6) showed that destructive lesions of the dorsal horn can give significant relief to patients with pain due to avulsion of the brachial plexus.

It is possible that some pain syndromes associated with lesions of the CNS are due to deafferentation hyperactivity of central pain transmission neurons. For example, thalamic pain could be caused by deafferentation of the cortical neurons to which they project. Unfortunately, there is no direct evidence supporting this proposed mechanism, nor is there any known therapy that works as well as dorsal horn lesions for the pain of brachial plexus avulsion.

Loss of Inhibition Produced by Myelinated Primary Afferents

Primary afferents are classified by their response to peripheral stimuli, their axonal diameter, and their conduction velocity. Primary afferent nociceptors have axons that are of small diameter, mostly unmyelinated, and therefore conduct at slow velocities. When large-diameter axons in a peripheral nerve are selectively blocked by pressure/ischemia, pain threshold is unaffected, whereas discriminative aspects of sensation (joint position sense, two-point discrimination, vibration sense) are lost. More importantly, experimental blocking of myelinated axons produces an exaggerated response to noxious stimuli (7). Stimuli that usually are not painful can produce severe burning pain when only unmyelinated primary afferents (C fibers) are functioning in a peripheral nerve. In parallel animal studies, dorsal horn neurons that respond to noxious stimuli show an exaggerated response to nociceptor activation when their predominantly inhibitory input from myelinated axons is blocked (8).

These observations lead to the conclusion that some central pain transmission neurons have inhibitory input from myelinated primary afferents (9). It is possible that pain results from damage to peripheral nerves when it is relatively selective for myelinated fibers. Traumatic mononeuropathies due to compression would be most likely to produce damage of this type because larger-diameter, myelinated axons are more susceptible to compression/ischemic damage. Consistent with this idea is the clinical evidence that selective electrical stimulation of large-diameter axons in a peripheral nerve can be dramatically effective in relieving pain caused by traumatic nerve injury (10).

Even if the large-diameter primary afferents are undamaged, their inhibitory effect would be reduced if the function of the dorsal horn interneurons that mediate their inhibitory effect were impaired. Bennett and Laird (11) presented evidence that certain types of peripheral nerve injury are associated with atrophy of dorsal horn neurons, some of which could be inhibitory interneurons.

Ectopic Impulse Generation

Although it is likely that deafferentation hyperactivity of central pain transmission neurons and selective damage to large-diameter primary afferents are major contributing factors in some patients with neuropathic pain, such patients are a minority. Other mechanisms must be postulated for most patients. For example, patients with common painful conditions such as postherpetic neuralgia and painful diabetic neuropathy often have minimal deafferentation, and small-diameter axons may show relatively greater damage than large-diameter axons. Thomas (12) has pointed out that some painful neuropathies are associated with specific damage to small-diameter fibers.

A major breakthrough in our understanding of neuropathic pain was the discovery that many primary afferents become spontaneously active when they are damaged (13–17). In addition to providing a possible explanation for the association of pain with damage to small-diameter primary afferents, the discovery of ectopic impulse generation may help us understand why certain patients with neuropathic pain obtain relief with membrane-stabilizing drugs. For example, the antiepileptic drugs (e.g., carbamazepine) and the

antiarrhythmics (e.g., lidocaine) are not known to have general analgesic efficacy, but they can be dramatically helpful to some patients with neuropathic pain (see following discussion). Presumably, such membrane-stabilizing drugs block the ectopic impulse generation in damaged primary afferents at concentrations that spare normal axonal conduction (16,18).

The spontaneous activity in primary afferents is prominent in partially damaged peripheral nerve and involves nociceptors that still are connected to their peripheral targets, for example, in the skin. Such spontaneous activity in "intact" primary afferent nociceptors is a likely source of pain in patients with nerve injury because topical agents that inactivate primary afferents can provide significant relief for some patients with neuropathic pain. For example, low-dose topical local anesthetics relieve the pain of postherpetic neuralgia (19). Another topical agent that is effective is capsaicin, the active ingredient of the hot chili pepper. Capsaicin acts specifically at the vanilloid receptor VR1, which is present only in primary afferent nociceptors (19,20). At increasing concentrations, capsaicin can activate, reversibly inactivate, peptide deplete, or destroy primary afferent nociceptors (21). In human subjects, 0.1% formulations of capsaicin produce reversible cutaneous analgesia (22). A 0.075% preparation provides some relief for patients with postherpetic neuralgia (23,24), postmastectomy syndrome (25), and diabetic neuropathy (26,27).

Sympathetic Nervous System Outflow

It is a well-established clinical fact that activity in the sympathetic nervous system can produce pain (28). Importantly, this does not occur in normal individuals. Thus, although electrical stimulation of the sympathetic chain normally is painless, a small percentage of patients with peripheral nerve injury develop a severe pain syndrome that is exacerbated by sympathetic activity and reversed by blockade of the sympathetic nervous system.

Mitchell (29) was the first to describe what is probably the most dramatic example of sympathetically maintained pain—causalgia. Causalgia is the syndrome of burning pain that occasionally follows peripheral nerve injury. In addition to the pain, which is the most prominent and disabling feature of the syndrome, patients with causalgia often are observed to have a cold, sweaty, and swollen extremity, especially distally in the limb. There may be focal arthritis and exquisite hypersensitivity to light moving touch. In many patients the pain is exacerbated by loud noises, movements, and cold.

A major breakthrough in understanding causalgia was the discovery by Leriche (30) that early sympathectomy could cure causalgia. Subsequently, experimental models of neuropathic pain provided a basis for understanding how the sympathetic nervous system can elicit and/or maintain pain. Sympathetic efferent activity sensitizes primary afferent nociceptors, but only in damaged peripheral nerve (31). Damaged nociceptive afferents that have regenerated into a neuroma can be excited by activating sympathetic efferents that have grown into the same neuroma (32). Partial peripheral nerve injury induces a state of increased sensitivity to adrenergic agonists in primary afferent nociceptors that have grown into the same neuroma (33). This ability of sympathetic outflow to evoke afferent activity can persist for months in rats (34).

The knowledge that sympathetic outflow can cause pain, particularly in patients with partial nerve injury, has important practical implications. First, the use of sympathetic blockade becomes a major tool in the evaluation of patients with neuropathic pain. If a patient responds to sympathetic blockade, most likely there is a major sympathetic component. Such patients should be treated with repeated sympathetic blocks, physical therapy, and antiinflammatory drugs (33,34).

There are a variety of ways to carry out a sympathetic block. The traditional method is by regional blockade of the sympathetic chain with local anesthetic (34). The use of intravenous phentolamine has become popular because it is less invasive and produces little discomfort. Furthermore, it does not produce a false-positive response due to local anesthetic block of nearby somatic sensory axons (35,36). Regional infusions with adrenergic antagonists, such as bretylium and guanethidine, have been used (37); however, the latter drug is not available in the United States. There is anecdotal evidence that oral sympatholytic agents such as phenoxybenzamine (Dibenzyline), guanethidine, and prazosin are helpful (38). On the other hand, dramatic responses are unusual, and there are no controlled clinical trials demonstrating their efficacy.

Prolonged Changes in Central Neurons Generated by Synaptic Activity

Independent of nerve injury, tissue-damaging stimuli can elicit prolonged changes in the excitability of central pain transmission neurons (39). Stimuli that are intense and involve deep somatic and visceral structures are capable of eliciting a prolonged hyperexcitable state in spinal cord pain transmission neurons (40). Depending on stimulus intensity and duration and the particular nerve stimulated, this hyperexcitable state may last for hours. There is evidence that a sensory barrage occurring at the time of a peripheral nerve injury can contribute to the hyperpathic syndrome that develops later (41).

These long-term changes in the CNS that occur after intense nociceptor activation make a convincing case for the theory that there is a persistent memory trace for pain. The clinical evidence supporting this idea is controversial (42); however, there are reports that patients having preoperative or intraoperative local anesthetic block to prevent the massive sensory input from surgical trauma have less severe postoperative pain. Similarly, preoperative anesthetic block seems to lower the incidence of postamputation phantom limb pain (43).

The presence of a prolonged central hyperexcitable state, although not unique to neuropathic pain, may help to explain the hyperpathia often observed in these patients. For example, in patients with causalgia or postherpetic neuralgia, severe pain usually can be elicited with light, moving, mechanical stimuli. This phenomenon, termed *allodynia,* is mediated by activity in large-diameter myelinated fibers, whose activity normally elicits light tactile sensations (44). Perhaps spinal pain transmission neurons can be sufficiently excited by large-diameter myelinated afferents to elicit pain sensations only when they are sensitized by prior noxious input.

Animal studies indicate that this state of hyperexcitability can be partially and selectively blocked by antagonists to the N-methyl-D-aspartate (NMDA)-type glutamate receptor. For example, ketamine and dextrorphan have NMDA receptor-blocking action and have been shown to reduce hyperalgesia in a rat model of neuropathic pain (45,46). This knowledge has provided encouraging leads. For example, results of a clinical trial suggest that NMDA antagonists can relieve neuropathic pain (47).

Nociceptive Nerve Pain

Pain may arise from activation of nociceptive primary afferents that innervate the connective tissue sheath of nerve trunks. This concept is supported by the clinical observation that inflammation of nerve trunks is characteristically painful (48). Inflamed peripheral nerves are characteristically tender and the pain has a local aching quality that is not distributed to the dermatomes innervated by the affected nerve. Peripheral nerve and nerve root sheaths are known to be innervated by nervi nervorum, including identified nociceptors (49,50). There is clear evidence that the vasa nervorum are innervated by axons containing peptides, including SP (51). Because SP is a marker for unmyelinated primary afferents, mostly nociceptors, it is likely that these axons render the nerve sensitive to noxious stimuli and inflammation.

Inflammation may be a common contributing factor to neuropathic pain. Acute demyelinating inflammatory polyneuropathy and leprosy are examples of peripheral neuropathies with a major inflammatory component. Animal studies indicate that minor trauma to a peripheral nerve causes an increase in inflammatory mediators, specifically cytokines such as interleukin-1 and tumor necrosis factor (52). These cytokines have been linked to increased nociceptor activity (53) and pain behaviors (54).

Thus, there are at least two distinct pathophysiologic types of nociceptor activation arising from diseased nerve: one due to increased activity in nociceptors whose axons in the nerve trunk pass through a site of injury or terminate in a neuroma, and the other due to "physiologic" activation of the nociceptors that innervate the nerve sheath. This latter type of pain can be thought of as nociceptive nerve pain. It may account for the deep aching pain of nerve root irritation, brachial neuritis, or acute inflammatory demyelinating polyneuropathy. One would predict that nociceptive nerve pain would respond well to opioids and antiinflammatory agents, but there are no controlled human clinical trials.

TREATMENT APPROACHES TO PATIENTS WITH NEUROPATHIC PAIN

Antiepileptic drugs and tricyclic antidepressants (TCA) are the mainstays for neurologists in the treatment of neuropathic pain. Antiepileptic drugs and TCAs can be effective by judiciously adjusting their dose and by combining them. However, even with their optimal use and the addition of other modalities, many patients continue to suffer. Hopefully, new and more effective drugs will be available in the near future.

Tricyclic Antidepressants

Tricyclics have a long history in the treatment of chronic pain. Tolerance and physical dependence are negligible, and they are effective for a broad range of problems, including migraine and muscle contraction headache, low back pain, cancer pain, and any problem accompanied by significant depression (55). Importantly, tricyclics are effective for a range of conditions different from those of the cyclooxygenase inhibitors (e.g. aspirin, nonsteroidal antiinflammatory drugs), which is the other major category of drug that has achieved wide acceptance for chronically painful conditions.

Amitriptyline (Elavil) is the most commonly prescribed tricyclic for treatment of pain patients. It is unclear which of its manifold pharmacologic actions is responsible for its analgesic action. It not only blocks the reuptake of serotonin and norepinephrine, it blocks α-adrenergic, muscarinic, cholinergic, and histamine receptors (56). This broad range of actions produces a significant number of dose-related side effects from the harmless but annoying dry mouth to the more serious orthostatic hypotension, urinary retention, cardiac conduction abnormalities, and memory disturbance. It also produces sedation, which is desirable in some patients and unwanted in others. So pervasive and unpleasant are the side effects of this drug that patient compliance requires starting amitriptyline at a very low dose (10 mg q.h.s.) and building slowly to therapeutic effect or to limiting side effects.

Other TCAs are effective for treatment of neuropathic pain. For example, desipramine is effective for postherpetic neuralgia and is almost as effective as amitriptyline for diabetic neuropathy (57). Desipramine is less potent as a histamine and acetylcholine antagonist; thus, it is less sedating and less likely to cause memory disturbance. As with amitriptyline, it is best to initiate therapy with a very low dose of this drug (10 to 25 mg per day) and raise it about every third day. In nondepressed patients, relatively low doses of a tricyclic may be sufficient for optimal pain control (e.g., 75 mg amitriptyline or desipramine per day).

The newer generation, non-TCA agents are less effective than TCAs for treatment of neuropathic pain (57). Fluoxetine is a good example; it is a highly selective serotonin [5-hydroxytryptamine (5-HT)] uptake inhibitor and is an excellent antidepressant. Because it has no cholinergic, adrenergic, or histamine blocking action, it is remarkably free of the troubling and persistent side effects that are pervasive with TCAs. Unfortunately, it has little or no effect in patients with neuropathic pain who are not depressed. These clinical observations suggest that 5-HT uptake blockade is insufficient to produce pain relief. In fact, there is evidence that antidepressants with relatively greater norepinephrine reuptake blockade have better analgesic efficacy than those that are more serotonin selective (58).

It should be kept in mind that the complaint of pain is a common symptom in depressive disorders (1), and it is not unusual for pain symptoms to either clear or become less bothersome when depression is controlled. For this reason, if the patient's pain does not respond to lower doses of a TCA, the dose should be increased into the antidepressant range. Obtaining blood TCA levels is mandatory if the dose of a TCA is to be raised above the maximum recommended level. If the patient is depressed, fluoxetine may be a valuable drug to use because of its relative lack of serious or uncomfortable side effects.

Phenothiazines

Some reports suggest that drugs of the phenothiazine class are useful for neuropathic pain, especially when they are used as adjuncts to tricyclics (59). However, controlled studies have failed to demonstrate phenothiazine efficacy for pain management (60). Furthermore, the side effect profile of these drugs should discourage their use. On the other hand, when more selective dopamine receptor subtype-specific ligands become available, their use should be evaluated in patients with chronic pain.

Anticonvulsants and Antiepileptic Agents

Compared with the TCAs, this group of drugs is helpful for a more restricted patient population. Carbamazepine and phenytoin have been used for a long time for treatment of neuralgic pains. They are effective for trigeminal neuralgia, and carbamazepine has been reported to be helpful for some patients with painful diabetic neuropathy (1). Beyond these specific conditions, these drugs are most likely to be helpful when there is a shooting, shocklike, lancinating pain. If carbamazepine and phenytoin are ineffective or are helpful but not well tolerated, alternative choices include baclofen (a second-line drug for trigeminal neuralgia), clonazepam, and valproic acid. For neuropathic pain, the most commonly used antiepileptic drug is gabapentin. This agent has proven efficacy in postherpetic neuralgia (61) and diabetic neuropathy (62). Patients need to be titrated to efficacy, which is generally in the range from 2,400 to 4,200 mg daily in divided doses. Sedation is often reported, but the side effect profile is otherwise benign. In many clinics, gabapentin is replacing TCAs as the first-line agent for neuropathic pain.

As discussed previously, *in vitro* studies have shown that relatively low concentrations of local anesthetics can block the ectopic impulses that arise in damaged small-diameter primary afferents (16,18). In clinical studies, lidocaine given by intravenous infusion provides immediate and often dramatic pain relief. A variety of neuropathic pains respond (63), including postherpetic neuralgia (64).

Because of the effectiveness of lidocaine-like drugs in the treatment of cardiac arrhythmias, several oral drugs of this class are available; tocainide and mexiletine are two examples. Tocainide has considerable toxicity, making mexiletine the preferred first-line drug of this class. Mexiletine has been shown to be effective for diabetic neuropathy (65) and other neuropathic pains (66).

Although mexiletine helps some patients and occasionally is dramatically effective, many patients are not helped even at the maximal allowable doses, or they tolerate the drug poorly because of gastrointestinal symptoms. A promising approach for some of these patients, particularly those with cutaneous hyperpathia, is the use of topical lidocaine preparations (19).

Opioids

Although the use of opioids in patients with chronic neuropathic pain has been the subject of significant controversy (67,68), there is a growing consensus that they are effective (69). Morphine, given acutely and in a high dose, provides significant relief in patients with postherpetic neuralgia (61). Several studies now indicate that opioids are effective when they are used for prolonged periods in patients with neuropathic pain (70–72). It should be pointed out that two of these studies were carried out using tramadol as the opioid medication. Tramadol is a weak opioid that has some norepinephrine and serotonin reuptake blocking action. This combined action may be critical to its efficacy for neuropathic pain.

Although many patients obtain significant relief with opioids, for many the relief is incomplete and for some the benefit is minimal to nonexistent. Unfortunately, in the absence of a therapeutic trial, it is difficult to predict the degree of relief that will be obtained by a given patient. When other options have been exhausted and the patient still has significant pain, opioids should be offered unless there are strong contraindications.

The major argument that has been raised against a trial of opioids is the possibility that the patient will become

an addict. However, available evidence indicates that the risk of addiction is extremely small for patients on short-term opioid therapy (73). On the other hand, if the drug is effective, the patient may need opioids on a long-term basis, which probably increases the likelihood of addiction, especially in patients with a history of drug abuse (74). This risk should be explained to the patient. Once an effective dose is established, the patient should not be permitted to increase it on his or her own initiative.

Once the decision is made and the patient agrees to chronic opioid therapy, my preference is to use long-acting or sustained release preparations. The pharmacokinetics of such compounds (levorphanol, sustained release morphine, or oxycodone) avoids the plasma level peaks (associated with increased side effects) and valleys (associated with breakthrough pain and mild abstinence). In the extreme, use of the transdermal fentanyl patch (Duragesic) provides nearly steady-state kinetics, and there is some clinical evidence for prolonged efficacy of this modality (75).

Some patients treated with opioids experience a fading of drug efficacy over time. This could represent a change in the underlying pain problem or the development of tolerance. Dose escalation in these patients should be undertaken cautiously and only after they are reevaluated both neurologically and psychologically and adjuvant drugs have been added (e.g., TCA, membrane-stabilizing drugs, or α_2-adrenergic agonists such as clonidine).

General Algorithm for Management of Patients with Neuropathic Pain

This algorithm is useful as a rough guideline to evaluate and treat patients with peripheral neuropathic pain. It is not necessary to use this approach if the diagnosis is obvious and there is an accepted treatment of choice. Examples of the latter situation include (a) patients with trigeminal neuralgia, who should be started immediately on carbamazepine; or (b) patients with a progressive compression-induced mononeuropathy, such as carpal tunnel syndrome, who should be considered for early surgical decompression. On the other hand, if the location, time course, or quality of the pain do not suggest a diagnosis for which there is a generally accepted and effective treatment, this algorithm can provide an organized approach to a diverse patient population.

Phase I: Initial Evaluation and Local Treatments

Figure 16.1 shows the initial treatment process. If there is any evidence for a sympathetically maintained pain, the initial step is a diagnostic sympathetic block. This should be considered in patients with pain in a single extremity and whose pain has a burning quality, is made worse by cold, and is associated with swelling and/or sweating abnormalities. If the patient has a good response, these blocks should be repeated when their pain-relieving effect subsides. They

Neuropathic Pain Management
Phase I (Local treatments)

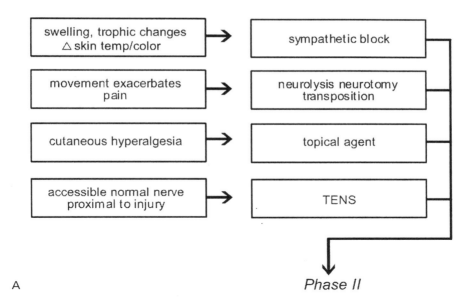

A

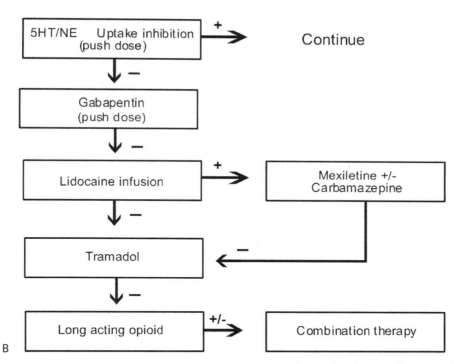

B

FIGURE 16.1. Algorithm for treatment of pain associated with peripheral nerve damage. Phase I outlines the steps to be taken before instituting long-term medical management. Phase II illustrates one systematic approach to long-term pharmacological treatment. (Adapted from Fields HL, Rowbotham MC. Multiple mechanisms of neuropathic pain: a clinical perspective. In: Gebhart GF, Hammond DL, Jensen TS, eds. *Progress in pain research and management, vol. 2.* Seattle, WA: IASP Press, 1994:437–454.) Transcutaneous electronic nerve stimulation (TENS): The way TENS is generally used, only local cutaneous stimulation is achieved. Because the sensation produced by TENS cannot be avoided, there will always be a cue that a treatment is being given and thus a placebo effect is difficult to avoid. In my experience, TENS is effective in patients with traumatic mononeuropathies when the stimulating electrodes can be placed on the trunk of the nerve proximal to the site of injury (10). Lidocaine infusion: We administer 5 mg per kilogram body weight via continuous intravenous infusion over 90 minutes. Electrocardiogram (ECG) and blood pressure monitoring are recommended (76).

should be continued as long as each successive block produces a longer-lasting pain-relieving effect.

If the patient has pain associated with a traumatic or compressive mononeuropathy and the pain is made worse by movement, it is possible that the pain is due to traction on a mechanically sensitive neuroma at the site of injury. Such patients often can be helped by either decompression or neurotomy and/or moving the nerve to reduce the traction on it (76).

Topical agents should be considered in patients who experience cutaneous discomfort. Such an approach is especially effective if the patients have allodynia or hyperalgesia to cutaneous stimulation. Topical lidocaine and capsaicin preparations are commercially available. Topical nonsteroidal anti-inflammatory drugs and local anesthetic preparations are not currently available; however, aspirin tablets can be crushed in chloroform (77,78).

Phase II: Oral Medications

If local approaches leave the patient with significant pain, we initiate therapy with gabapentin or a TCA (as outlined). Gabapentin is almost as effective as TCAs, and its side effect profile is more benign (61,62). If further relief is needed, we proceed to a trial of a membrane-stabilizing drug. We use a lidocaine infusion to evaluate the usefulness of this approach (79). If there is no relief with lidocaine, we have found that oral membrane-stabilizing drugs will not work, and we can save the patient the time it would take to build up to the dose required for a full evaluation of related oral medications. If there is immediate and dramatic relief with lidocaine infusion, we are willing to administer oral mexiletine to high levels (up to 1,200 mg per day, provided plasma levels are within the acceptable therapeutic range).

For many patients, TCAs, gabapentin, and membrane-stabilizing drugs either alone or in combination are inadequate. For many of these patients, opioids are clearly effective.

One can question the need for the prognostic infusions in this algorithm. Why not just proceed with the oral medications? The main argument for the prognostic infusions is to save the time required for dose titrations of drugs that may prove to be worthless for an individual. These patients are suffering and often disabled by their pain. Thus, we try to assess as many options as quickly as possible to reduce their pain to a bearable level. The infusions also help us to be more aggressive and patients to be more tolerant with a particular approach in the face of initial failure or bothersome side effects.

Finally, for patients with lower extremity pain, more invasive approaches may be appropriate. For those with a significant opioid responsive component, an intraspinal pump may provide increased efficacy with less sedation. Although intraspinal opioids are standard practice for control of operative and postoperative pain, their use for chronic pain is supported largely by anecdotal data (80). Another approach is to use electrical stimulation of the spinal cord, with the electrodes generally placed in the thoracic region. This modality is supported by extensive but uncontrolled clinical data (81).

SUMMARY

This chapter outlined a rational mechanism-based approach to the evaluation and treatment of patients with neuropathic pain. It is important to point out that this approach is evolving and that the algorithm will change as we learn more about the mechanisms of neuropathic pain and new treatment options become available. The therapeutic outcome for a given patient will be suboptimal unless the assessment and treatment of the psychosocial aspects of their problem are carried out concurrently with the somatically based algorithm outlined in this chapter.

REFERENCES

1. Fields HL. *Pain.* New York: McGraw-Hill, 1987:133–170.
2. Leijon G, Boivie J, Johansson I. Central post-stroke pain—neurological symptoms and pain characteristics. *Pain* 1989;36: 13–25.
3. Cassinari V, Pagni CA. *Central pain: a neurosurgical survey.* Cambridge, MA: Harvard University Press, 1969.
4. Wynn-Parry CB. Pain in avulsion lesions of the brachial plexus. *Pain* 1980;9:41–53.
5. Lombard MC, Larabi Y. Electrophysiological study of cervical dorsal horn cells in partially deafferented rats. In: Bonica JJ, et al., eds. *Advances in pain research and therapy.* New York: Raven Press, 1983.
6. Nashold BS Jr, Ostdahl RH. Dorsal root entry zone lesions for pain relief. *J Neurosurg* 1979;51:59–69.
7. Landau W, Bishop GG. Pain from dermal, periosteal and fascial endings and from inflammation. *Arch Neurol Psychiatry* 1953;69:490–504.
8. Price DD, Hayes RL, Ruda MA, et al. Spatial and temporal transformations of input to spinothalamic tract neurons and their relation to somatic sensations. *J Neurophysiol* 1978;41:933–947.
9. Melzack R, Wall PD. Pain mechanisms: a new theory. *Science* 1965;150:971–978.
10. Meyer GA, Fields HL. Causalgia treated by selective large fibre stimulation of peripheral nerve. *Brain* 1972;95:163–168.
11. Bennett GJ, Laird JM. Central changes contributing to neuropathic hyperalgesia. In: Willis W, ed. *Hyperalgesia and allodynia.* New York: Raven Press, 1992:305–310.
12. Thomas PK. The anatomical substratum of pain. *Can J Neurol Sci* 1974;1:92–97.
13. Wall PD, Gutnick M. Ongoing activity in peripheral nerves: the physiology and pharmacology of impulses originating from a neuroma. *Exp Neurol* 1974;43:580–593.
14. Scadding JW. Development of ongoing activity, mechanosensitivity, and adrenaline sensitivity in severed peripheral nerve axons. *Exp Neurol* 1981;73:345–364.
15. Welk E, Leah JD, Zimmerman M. Characteristics of A- and C-fibers ending in a sensory nerve neuroma in the rat. *J Neurophysiol* 1990;63:759–766.
16. Tanelian DL, MacIver MB. Analgesic concentrations of lidocaine suppress tonic A-delta and C-fiber discharges produced by acute injury. *Anesthesiology* 1991;74:934–936.

17. Ali Z, Ringkamp M, Hartke TV, et al. Uninjured C-fiber nociceptors develop spontaneous activity and alpha-adrenergic sensitivity following L6 spinal nerve ligation in monkey. *J Neurophysiol* 1999;81:455–466.
18. Fields HL, Rowbotham MC, Devor M. Excitability blockers: anticonvulsants and low concentration local anesthetics in the treatment of chronic pain. In: Dickenson A, Besson JM, eds. *Handbook of experimental pharmacology.* Berlin: Springer Verlag, 1997;93–116.
19. Rowbotham MC, Davies PS, Fields HL. Topical lidocaine gel relieves post-herpetic neuralgia. *Ann Neurol* 1995;37:246–253.
20. Holzer P. Capsaicin: cellular targets, mechanisms of action, and selectivity for thin sensory neurons. *Pharmacol Rev* 1991;43:143–201.
21. Caterina MJ, Julius D. The vanilloid receptor: a molecular gateway to the pain pathway. *Annu Rev Neurosci* 2001;24:487–517.
22. Bjerring P, Arendt-Nielsen L. Use of a new argon laser technique to evaluate changes in sensory and pain thresholds in human skin following topical capsaicin treatment. *Skin Pharmacol* 1989;2:161–167.
23. Watson CPN, Evans RJ, Watt VR. Post-herpetic neuralgia and topical capsaicin. *Pain* 1988;33:333–340.
24. Bernstein JE, Korman NJ, Bickers DR, et al. Topical capsaicin treatment of chronic postherpetic neuralgia. *J Am Acad Dermatol* 1989;21:265–270.
25. Watson CP, Evans RJ. The postmastectomy pain syndrome and topical capsaicin: a randomized trial. *Pain* 1992;51:375–379.
26. Scheffler NM, Sheitel PL, Lipton MN. Treatment of painful diabetic neuropathy with capsaicin 0.075%. *J Am Podiatr Med Assoc* 1991;81:288–293.
27. Tandan R, Lewis GA, Krusinski PB, et al. Topical capsaicin in painful diabetic neuropathy. Controlled study with long-term follow-up. *Diabetes Care* 1992;15:8–14.
28. Loh L, Nathan PW. Painful peripheral states and sympathetic blocks. *J Neurol Neurosurg Psychiatry* 1978;41:664–671.
29. Mitchell SW. *Injuries of nerves and their consequences.* New York: Dover Publications, 1965.
30. Leriche R, ed. *The surgery of pain.* London: Bailliere, Tindall & Cox, 1939.
31. Sato J, Perl ER. Adrenergic excitation of cutaneous pain receptors induced by peripheral nerve injury. *Science* 1991;251:1608–1610.
32. Devor M, Janig W. Activation of myelinated afferents ending in neuroma by stimulation of the sympathetic supply in the rat. *Neurosci Lett* 1981;24:43–47.
33. Kozin F, McCarty DJ, Sims J, et al. The reflex sympathetic dystrophy syndrome. 1. Clinical and histological studies: evidence for bilaterality, response to corticosteroids and articular involvement. *Am J Med* 1976;60:321–331.
34. Bonica JJ. Causalgia and other reflex sympathetic dystrophies. *Adv Pain Res Ther* 1979;3:141–161.
35. Dellemijn PL, Fields HL, Allen RR, et al. The interpretation of pain relief and sensory changes following sympathetic blockade. *Brain* 1994;117:1475–87.
36. Arner S. Intravenous phentolamine test: diagnostic and prognostic use in reflex sympathetic dystrophy. *Pain* 1991;46:17–22.
37. Hannington-Kiff JG. Intravenous regional sympathetic block with guanethidine. *Lancet* 1974;7865:1019–1020.
38. Ghostine SY, Comair YG, Turner DM, et al. Phenoxybenzamine in the treatment of causalgia: report of 40 cases. *J Neurosurg* 1984;60:1263–1268.
39. Willis WD. Central plastic responses to pain. In: Gebhart GF, Hammond DL, Jensen TS, eds. *Progress in pain research and management, vol. 2.* Seattle, WA: IASP Press, 1994:301–324.
40. Woolf CJ. Excitability changes in central neurons following peripheral damage: role of central sensitization in the pathogenesis of pain. In: Willis W, ed. *Hyperalgesia and allodynia.* New York: Raven Press, 1992:221–243.
41. Dougherty PM, Garrison CJ, Carlton SM. Differential influence of local anesthetic upon two models of experimentally induced peripheral mononeuropathy in the rat. *Brain Res* 1992;570:109–115.
42. Dahl JB, Kehlet H. The value of pre-emptive analgesia in the treatment of postoperative pain. *Br J Anaesth* 1993;70:434–439.
43. Bach S, Noreng MF, Tjellden NV. Phantom limb pain in amputees during the first 12 months following limb amputation, after reoperative epidural blockade. *Pain* 1988;33:297–301.
44. Ochoa JL, Yarnitsky D. Mechanical hyperalgesias in neuropathic pain patients: dynamic and static subtypes. *Ann Neurol* 1993;33:465–472.
45. Mao J, Price DD, Hayes RL, et al. Intrathecal treatment with dextrorphan or ketamine potently reduces pain-related behaviors in a rat model of peripheral mononeuropathy. *Brain Res* 1993;605:164–168.
46. Tal M, Bennett GJ. Dextrorphan relieves neuropathic heat-evoked hyperalgesia in the rat. *Neurosci Lett* 1993;151:107–110.
47. Eide PK, Jorum E, Stubhaug A, et al. Relief of postherpetic neuralgia with the N-methyl-D-aspartic acid receptor antagonist ketamine: a double-blind, cross-over comparison with morphine and placebo. *Pain* 1994;58:347–354.
48. Asbury AK, Fields HL. Pain due to peripheral nerve damage: an hypothesis. *Neurology* 1984;34:1587–1590.
49. Hromada J. On the nerve supply of the connective tissue of some peripheral nervous system components. *Acta Anat* 1963;55:343–351.
50. Bove GM, Light AR. Unmyelinated nociceptors of rat paraspinal tissues. *J Neurophysiol* 1995;73:1752–1762.
51. Appenzeller O, Dhital KK, Cowen T, et al. The nerves to blood vessels supplying blood to nerves: the innervation of vasa nervorum. *Brain Res* 1984;304:383–386.
52. Sommer, C. Cytokines and neuropathic pain. In: Hansson PT, Fields HL, Hill RG, et al., eds. *Neuropathic pain: pathophysiology and treatment. Progress in pain research and management, vol. 21.* Seattle, WA: IASP Press, 2001;37–62.
53. Sorkin LS, Doom CM. Epineurial application of TNF elicits an acute mechanical hyperalgesia in the awake rat. *J Periph Nerv Syst* 2000;5:96–100.
54. Sorkin LS, Xiao WH, Wagner R, et al. Tumour necrosis factor-alpha induces ectopic activity in nociceptive primary afferent fibres. *Neuroscience* 1997;81:255–262.
55. Onghena P, van Houdenhove B. Antidepressant-induced analgesia in chronic non-malignant pain: a meta analysis of 39 placebo controlled studies. *Pain* 1992;49:205–220.
56. Richelson E. Antidepressants and brain neurochemistry. *Mayo Clin Proc* 1990;65:1227–1236.
57. McQuay HJ, Tramer M, Nye BA, et al. A systematic review of antidepressants in neuropathic pain. *Pain* 1996;68:217–227.
58. Atkinson JH, Slater MA, Wahlgren DR, et al. Effects of noradrenergic and serotonergic antidepressants on chronic low back pain intensity. *Pain* 1999;83:137–145.
59. Taub A. Relief of postherpetic neuralgia with psychotropic drugs. *J Neurosurg* 1973;39:235–239.
60. McGee JL, Alexander MR. Phenothiazine analgesia–fact or fantasy? *Am J Hosp Pharm* 1979;1:39–49.
61. Rowbotham M, Harden N, Stacey B, et al. Gabapentin for the treatment of postherpetic neuralgia. *JAMA* 1998;280:1837–1842.
62. Backonja M, Beydoun A, Edwards KR, et al. Gabapentin for the symptomatic treatment of painful neuropathy in patients with diabetes mellitus: a randomized controlled trial. *JAMA* 1998;280:1831.

63. Glazer S, Portenoy RK. Systemic local anesthetics in pain control. *J Pain Symptom Manage* 1991;6:30–39.
64. Rowbotham MC, Reisner LM, Fields HL. Both intravenous lidocaine and morphine reduce the pain of post-herpetic neuralgia. *Neurology* 1991;41:1024–1028.
65. Dejgard A, Petersen P, Kastrup J. Mexiletine for the treatment of chronic painful diabetic neuropathy. *Lancet* 1988;1:9–11.
66. Chabal C, Jacobson L, Mariano A, et al. The use of oral mexiletine for the treatment of pain after peripheral nerve injury. *Anesthesiology* 1992;76:513–717.
67. Arner S, Meyerson B. Lack of analgesic effect of opioids on neuropathic and idiopathic forms of pain. *Pain* 1988;33:11–23.
68. Fields HL. Can opioids relieve neuropathic pain? *Pain* 1988;35:365
69. Rowbotham MC. Efficacy of opioids in neuropathic pain. In: Hansson PT, Fields HL, Hill RG, et al., eds. *Neuropathic pain: pathophysiology and treatment. Progress in pain research and management, vol. 21.* Seattle, WA: IASP Press, 2001;203–213.
70. Harati Y, Gooch C, Swenson M, et al. Double-blind randomized trial of tramadol for the treatment of the pain of diabetic neuropathy. *Neurology* 1998;50:1842–1846.
71. Sindrup SH, Andersen G, Madsen C, et al. Tramadol relieves pain and allodynia in polyneuropathy: a randomised, double-blind, controlled trial. *Pain* 1999;83:85–90.
72. Watson CPN, Babul N. Efficacy of oxycodone in neuropathic pain: a randomized trial in postherpetic neuralgia. *Neurology* 1998;50:1837–1841.
73. Porter J, Jick H. Addiction rare in patients treated with narcotics. *N Engl J Med* 1980;302:123.
74. Portenoy R, Foley K. Chronic use of opioid analgesics in nonmalignant pain: report of 38 cases. *Pain* 1986;25:171–186.
75. Allan L, Hays H, Jensen NH, et al. Randomised crossover trial of transdermal fentanyl and sustained release oral morphine for treating chronic non-cancer pain. *BMJ* 2001;322:1154–1158.
76. Dawson DM, Hallett M, Millender LH. *Entrapment neuropathies.* Boston: Little, Brown and Company, 1983.
77. King RB. Concerning the management of pain associated with herpes zoster and of postherpetic neuralgia. *Pain* 1988;33:73–78.
78. De Benedittis G, Besana F, Lorenzetti A. A new topical treatment for acute herpetic neuralgia and post-herpetic neuralgia: the aspirin/diethyl ether mixture. An open-label study plus a double-blind controlled clinical trial. *Pain* 1992;48:383–390.
79. Galer BS, Miller KV, Rowbotham MC. Response to intravenous lidocaine infusion differs based on clinical diagnosis and site of nervous system injury. *Neurology* 1993;43:1233–1235.
80. Anderson VC, Burchiel KJ. A prospective study of long-term intrathecal morphine in the management of chronic nonmalignant pain. *Neurosurgery* 1999;44:289–300.
81. North RB, Linderoth B. Spinal cord stimulation for chronic pain. In: Schmidek HH, ed. *Schmidek and Sweet's operative neurosurgical techniques,* 4th ed. Philadelphia: WB Saunders, 2000:2407–2422.

17

ATTENTION

CATHERINE M. ARRINGTON
THOMAS H. CARR

In a review article that influenced a decade of cognitive neuroscience research on the neural substrates of attention, Posner and Petersen (1) presented three principles thought to underlie the neural implementation of attention. First, they proposed that the neural systems executing attention are composed of anatomical regions that are separate from the regions of the brain involved in the processing of information: the sensory, motor, and memory regions. Second, these regions devoted specifically to attention consist of a network of structures rather than either a single attentional center in the brain or a diffuse attention function involving the entire brain. Finally, within this network of structures different regions govern specific aspects of the attentional process. These three principles guide our discussion of the functional brain systems of attention.

In this chapter we examine two primary processes of attention in detail and briefly consider a third. We begin with *selective attention,* the selection of input from the senses into the cognitive system, with emphasis on the orienting of visual attention. Next, we turn to *executive control,* the processes of attention involved in the supervision of task performance, including coordination of perceptual, memory, and response requirements of a task; switching of attentional resources among tasks; and monitoring of the success of task performance outcomes. Finally, we give brief coverage to general *arousal and vigilance processes.* For each topic area we consider in turn the functional and computational accounts of attentional processes, followed by the neural substrates underlying these processes, drawing on studies using a variety of cognitive neuroscience methodologies. We conclude with a discussion of attention deficits in clinical populations and the importance of considering attentional processes in therapeutic situations.

SELECTIVE ATTENTION

Environmental input to the sensory system at any given moment includes far more information than can be successfully processed. Limiting and selecting what input will receive further cognitive processing is one role of attention. The operations fulfilling this role are referred to as *selective attention.* In everyday conversation this notion of attention can be equated to uses such as, "The bright flash of the marquee lights caught my attention." Much of the attentional research over the preceding decades has focused on selective attention. Particular emphasis has been devoted to the processes of visuospatial selection, although selection in the auditory and tactile modalities also has been investigated. Key issues in the study of selective attention include (a) the point in processing at which selection occurs, (b) the impact of selection on the processing of selected stimuli compared to stimuli that are ignored, and (c) the locus of control over what gets selected.

Functional and Computational Approaches

Attention as Filter

Early studies of attention conceived of attention as a filter that limits the amount of sensory information that reaches consciousness. A critical debate in this approach to selective attention is the locus of selection. Does attention operate early in the stream of information processing, allowing only a few stimuli to be processed beyond simple physical characteristics? Or are all stimuli given a greater degree of perceptual processing before the attentional filter acts to limit stimuli only at the point of conscious awareness? Traditionally there have been two primary views on this issue. Early selection theories propose that only the most basic sensory and perceptual processing occurs in parallel for a large number of stimuli prior to the attentional filter, and that more sophisticated perceptual processing and all semantic processing occurs for only a small number of stimuli. In contrast, late selection theories place the locus of selection at a point just prior to conscious awareness, with a large number of stimuli receiving a considerable degree of perceptual and even semantic processing. Only after such processing has identified and attached meaning to the stimuli being processed in parallel are attentional limitations imposed. Evidence for both of these theories initially came from dichotic listening experiments in which participants listened through headphones to two simultaneous auditory messages presented one in each ear. Participants were instructed to attend to and shadow (repeat vocally) the message in one ear, while ignoring the message in the other ear. Within the dichotic listening paradigm, the locus of attentional selection was determined by examining the degree of processing of the unattended message. Early experimental evidence using this procedure supported early selection based on findings that the meaning of stimuli presented in the unattended message could not be reported even when presented repeatedly (2). Based on such findings, Broadbent (3) proposed one of the earliest and most influential models of attention. However, other evidence demonstrated a greater degree of semantic processing of stimuli in the unattended message, supporting late selection theories. In a classic experiment, Treisman (4) demonstrated processing of semantic information in the unattended ear by showing that participants would switch the ear that they were shadowing if the message originally presented in the attended ear switched midstream to the unattended ear. These results showed that participants must have been processing the meaning of the message in the unattended ear in order to "follow" the original message when it switched ears.

Debates about early versus late locus of selection have been informed by electrophysiological research measuring brain activity associated with attended versus unattended stimuli. In these studies, participants were instructed to attend to one portion of the environment, searching for targets either in one visual hemifield or in a message presented to one ear. Event-related potentials (ERP) were compared for targets in the attended versus the unattended stream of stimuli. In both the auditory and visual modalities, initial research demonstrated differences in neural activity associated with attended and unattended stimuli beginning approximately 100 milliseconds after stimulus onset, suggesting an early locus of selection (5). Late selection models also find support from ERP studies demonstrating that unattended stimuli are processed for word meaning (6). Items that are semantically inconsistent within a context elicit a larger response in the late N400 waveform. This effect was found even for unattended items that did not reach conscious awareness.

More recently, it has become generally accepted that attentional selection can occur both early and late in processing. Indeed, selection may occur at multiple points during processing. This being the case, the goal for researchers is to define the conditions under which selection will occur at the various stages of processing. A general condition for the locus of selective attention may be that selection will occur at any stage of processing in which interference is encountered (7). For example, one factor that appears to play a key role in determining the locus of selection is the amount of stimulation in the environment, or the perceptual load (8). Lavie (9) demonstrated the effect of perceptual load in a series of visual selective attention experiments. In these experiments participants had to discriminate a target letter in

a display that included either a small number (low load) or a large number (high load) of nontarget distractor items. In addition, one of the distractor items was either compatible with the target item (the same letter as the target that would thus elicit the same response) or incompatible with the target item (the other potential target letter that would elicit a competing response). Thus, if the distractor were identified, it might either facilitate or interfere with choosing the appropriate response to the target, depending on its compatibility. Under low perceptual load, the compatibility of the distractor item influenced the speed with which participants were able to make the discrimination judgment, suggesting a late locus of selection that allowed for full processing of the irrelevant distractor items; however, under high perceptual load, the compatibility of the distractor item had no effect on the discrimination judgment, suggesting an early locus of selection that limited processing of the irrelevant distractors. ERP data demonstrating changes in the differential neural responding to distractors under conditions of high and low perceptual load support the idea that perceptual load influences attentional selection (10). Developmental studies also demonstrate effects of perceptual load on selective attention. Both older adults (11) and children (12) show patterns of greater interference than do young adults. These patterns suggest developmental changes in selective attention involving interactions between interference control processes and selection mechanisms. The attentional mechanisms that implement early selection based on perceptual features mature earlier in the development of children and remain efficient longer in the face of aging than do the more fragile attentional processes that deal with interference and competition among stimuli allowed to reach the stage of late selection based on meaning and goal relevance.

Attention as Facilitator

Another extensively researched aspect of selective attention is attentional orienting. Not only must attention serve the function of limiting the amount of information proceeding through the cognitive system, attention also selects which information will be given priority in the procession. Within the visual modality this selection and orienting of attention most often occurs in concert with overt movements of the eyes such that the attended portion of the environment coincides with the region of space represented at the fovea. However, shifts of attention can occur without any corresponding overt movement. These covert shifts of attention have been widely studied both in terms of their relationship to overt orienting (13) and as a separate mechanism of attentional orienting in the absence of overt movement of the eyes (14).

The paradigm most often used to study attentional orienting is the spatial cueing task (14,15). In the modal version of the task shown in Figure 17.1, participants are engaged in a visual detection or discrimination task in which the target

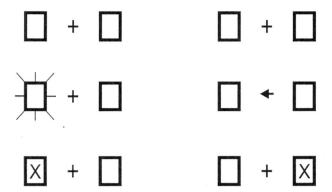

FIGURE 17.1. Diagram of trial lines for standard spatial cueing task paradigm. Exogenous peripheral cue and a valid cue target relationship **(left)**. Endogenous central cue and an invalid cue target relationship **(right)**.

may appear in one of multiple positions on a display. The appearance of the target stimulus is preceded by a cue that provides information about the most likely position for the upcoming target. The typical results of this manipulation are that on trials when the cue correctly indicates the position of the target (valid trials), participants are faster and more accurate in responding to the target. On trials when the cue incorrectly indicates the position of the target (invalid trials), participants are slower and less accurate in responding to the target. This pattern of results provides behavioral evidence suggesting that attention serves to facilitate the processing of stimuli at attended locations while inhibiting processing at unattended locations. The simplicity of this chronometric analysis of attentional orienting has resulted in the application of this methodology in an exceedingly wide range of experimental situations.

Numerous variations of the basic spatial cueing task have been used to study attentional orienting processes. Two types of cues are commonly used in the spatial cueing paradigm and are thought to initiate the operation of two distinct orienting mechanisms (14,16). A central arrow or other symbolic cue that predicts the target location with some certainty engages the endogenous orienting system that allows for intentional shifts of attention to the position indicated by the cue. A peripheral flash or sudden onset, whether or not predictive of target location, engages the exogenous orienting mechanism that draws attention to the location of the cue. These two orienting mechanisms have been distinguished on a number of different characteristics. Exogenous orienting is stimulus driven and automatic, occurring in response to the presence of a stimulus in the environment, regardless of intention and unhampered by concurrent tasks. Endogenous attention, on the other hand, is goal directed, with orienting of attention controlled by intention to use information in the environment to guide allocation of attention. As a controlled process, endogenous orienting can be reduced or eliminated by concurrent tasks or nonpredictive cues (16–18). The two orienting mechanisms differ in terms of the time course of

their facilitation of processing. Endogenous orienting has a slow rise in facilitation over a time period of approximately 200 milliseconds (the result presumably of having to interpret the cue and intentionally initiate a shift of attention) and then maintains the facilitatory effects over an extended period of several seconds. Exogenous orienting, on the other hand, shows a faster rise in the facilitatory component of orienting that occurs within as little as 50 milliseconds after cue presentation (19). However, this facilitation is brief and is followed by a period of slowed responding at the cued location known as the *inhibition of return* (20). Endogenous and exogenous orienting may differ in terms of the impact that attention has on the stimuli that are at the focus of attention, with exogenous orienting affecting processing of all stimuli at an attended location and endogenous orienting interacting with other factors such as task expectancies (21).

Models of attention based on research conducted using the spatial cueing paradigm have tended to conceive of attention metaphorically in terms of a spotlight or zoom lens (22). Within this framework, attention functions as a gain control mechanism that modulates activity in perceptual processing areas within the spotlight of attention. These models propose that selection occurs within a spatial medium with all stimuli falling within the attended region of environmental space receiving the benefits of attention. Spatial models of selective attention have been challenged in recent years by models of attention that suggest that objects rather than spatial locations are selected for attention. In an early study of object-based attention, Duncan (23) showed that viewers were more easily able to report two features associated with objects if they appeared on the same object than if they appeared on two separate objects, even when the objects overlapped in space. Further studies demonstrated that selection of portions of the visual field for processing can occur based on a variety of features, including shared motion, shared color, connectivity, and figure ground segmentation (24–26). Further, attentional selection may be influenced by factors such as illusory figures and completion of occluded objects (27). The mounting evidence that selection can occur based on properties other than spatial location must be accounted for in models of attentional selection and, as will be shown later, in models of the neural substrate of attentional control.

Attention as Integrator

Along with the spatial cueing task, the visual search paradigm has been used most often to study the processes of visual attention. In a visual search task, participants scan a visual array in order to locate a target stimulus among varying numbers of distractor items. The efficiency with which a particular search can be conducted is measured by the effect of increasing the number of distractor items presented in the display. Using this paradigm, Treisman and Gelade (28) demonstrated that when the target item can be distinguished from the distractors based on a single salient feature (e.g., a green *X* among red *X*s), search times do not increase substantially with increasing numbers of distractors. This type of "feature search" occurs without focal attention such that all of the items appear to be processed in parallel. In contrast, when the target differs from the distractors based on a conjunction of features (e.g., a green *X* among red *X*s and green *O*s), search times increase with increasing numbers of distractors, suggesting that a "conjunctive search" must occur in a serial fashion with focal attention directed to each stimulus in the display. Figure 17.2 shows typical visual search displays for both feature and conjunctive search tasks.

Based on findings from a variety of visual search tasks, Treisman and colleagues developed the feature integration theory (FIT) of selective attention (28,29). Within FIT, initial visual processing of feature information (e.g., color, orientation) occurs preattentively and in parallel across the visual field in separate feature maps. These feature maps may be thought of as cognitive representations that are loosely equivalent to the neural regions that have been shown to process different characteristics of visual stimuli (30,31). Just as regions of neural tissues that process visual information have been shown to contain retinotopic mapping of cells, feature maps also contain spatial information. The function of attention is to integrate the features that occupy the same

region in space across multiple feature maps. This process of binding together separate features results in the formation of object files, stable yet temporary mental representations of objects in the visual world (32). Thus, although recognizing and even locating the presence of a particular feature in the visual environment (e.g., a red square moving up and to the left) may be relatively undemanding of attentional resources, combining multiple features in a single representation of an object at a particular place is a limited capacity process that requires the allocation of visual attention.

Top-Down versus Bottom-Up Control of Selection

One issue that has been central to the study of attentional selection is the degree to which selective attention is under the control of top-down intentional processes versus bottom-up stimulus driven processes (33,34). Studies comparing endogenous and exogenous orienting demonstrate that both processes can occur, and a number of models have been aimed at accounting for the interaction between these two forces. Research in this area has been conducted examining the effects of task irrelevant singletons in visual search (e.g., a red stimulus within a display of green stimuli when the target is defined in terms of the shape of the stimulus and does not coincide with the red item.) Early studies emphasized the importance of abrupt onsets of stimuli in capturing attention (35,36). Other evidence suggests that attentional capture can occur as a result of a variety of singletons, such as color or shape. Folk et al. (37) provide integration across these apparently disparate experimental findings. Endogenously generated attentional control settings can influence what types of task irrelevant stimuli will capture attention in a given task environment. Thus, if the target is defined as a singleton (a particular shape that is unique among other distractor shapes), then the attentional system will be set up to respond to singleton stimuli, even if the dimension on which the item is a singleton (color) is not the dimension on which the target is defined (shape). A similar idea emphasizing the interaction between top-down and bottom-up control of attention is found in the work of Wolfe et al. (38) on "guided search." In guided search, visual perception initially occurs in a largely parallel preattentive manner, as in FIT. Following this preattentive processing, a combination of bottom-up influences (driven by points of high dissimilarity among neighboring items on a particular feature) and top-down influences (based on matches to features of the target in the search task that are already known and held in working memory) guides which items in the display will be processed attentively (for review see references 33 and 39).

Neural Substrates

The neural implementation of attentional selection has been examined using a wide range of cognitive neuroscience techniques, including electrophysiological, neuroimaging, and

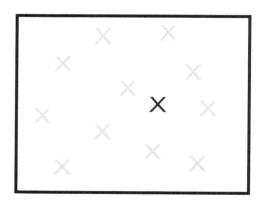

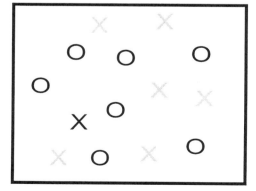

FIGURE 17.2. Example stimulus displays in feature search **(top)** and conjunction search **(bottom)** conditions of a visual search task. In each display the target stimulus is a black *X*.

patient studies. Regardless of the method of study, examination of the neural regions involved in attentional selection must consider the distinction between structures that act as the initiator or source of the attentional effect and structures that are impacted or modulated by attention.

Posterior Attention System

Early studies of the neural structures involved in directing processes of selective attentional indicated that a large and distributed network of cortical and subcortical structures is involved in attention (40). Investigations of patients with damage to a variety of neuroanatomical regions who showed deficits in attentional selection contributed to our understanding of the roles played by these neural structures. Based on studies examining patients with several types of focal brain damage, Posner and colleagues described a network of structures critical for attentional orienting, which they labeled the *posterior attention system* (1,41). Within the posterior attention system, three structures have been identified: the posterior parietal lobe, the superior colliculus, and the pulvinar nucleus of the thalamus. Each of these structures is thought to be involved in a particular aspect of orienting indicated by the specific pattern of behavioral impairment on tasks of attentional orienting. Patients with damage to the posterior parietal lobe show large deficits in responding on an orienting task when they must disengage attention from the "good" or ipsilesional visual field in order to respond to targets in the "bad" or contralesional visual field (42). These deficits are demonstrated by slowed response times in an attentional cueing task when the cue initially draws attention to the ipsilesional visual field and then the target appears in the contralesional visual field. Patients with damage to the midbrain in the area of the superior colliculus resulting from the degenerative disorder progressive supranuclear palsy show a profound slowing of attentional shifts in response to both cues and targets (43). This same slowing is seen with overt orienting involved in saccadic eye movements and suggests a

general deficit in moving attention from one location to another. Damage to the pulvinar nuclei of the thalamus results in dramatic increases in response time to targets appearing in the contralesional visual field even when the target location is correctly cued and the patient has ample time to shift attention to the target location (44). This deficit suggests a role of the pulvinar nuclei in engaging attention at a location in the visual field. Based on the findings from these three lines of investigation, a neural circuitry has been described in which specific neural regions underlie separate processes involved in attentional orienting: the parietal lobes disengage attention from its current focus; the superior colliculi shift attention; and the pulvinar nuclei engage attention in order to process stimulus information that may be present at the next focus (Fig. 17.3).

The posterior attention system described by Posner and colleagues specifies a limited and clearly defined network of structures involved in orienting attention. Although evidence from neuroimaging studies of intact subjects has generally supported the proposal of a posterior attention system, neuroimaging conducted during visual selection tasks points to a wider network of structures, including both parietal and frontal regions, involved in attentional selection (45–48). In a seminal neuroimaging study of attention, Corbetta et al. (49) used positron emission tomography (PET) to examine the neural activation associated with shifting attention. Participants engaged in a predictable and systematic series of attentional shifts to locations in the periphery during PET scanning. Regions involved in attentional orienting included superior parietal and superior frontal cortex. The superior parietal regions were active during both active viewing conditions (requiring responding) and passive viewing conditions (in which no response was required), suggesting a role for this region in shifts of attention regardless of whether the task involved overt action toward the attended stimuli. On the other hand, superior frontal regions were selectively involved in the active task conditions, perhaps reflecting a role for the frontal cortex in intentional control of visual

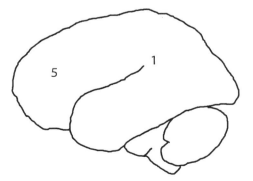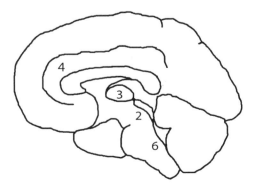

FIGURE 17.3. Regions involved in attention systems: posterior attention system includes (1) temporoparietal junction, (2) superior colliculus, and (3) thalamus; anterior attention systems includes (4) anterior cingulate cortex and (5) prefrontal cortex; and vigilance system includes (6) reticular formation in concert with thalamus, basal forebrain, and right prefrontal regions.

attention and coordinating of attended information with action. The distinction of parietal and other posterior cortical areas being involved in the specific disengage, shift, and engage processes of orienting while frontal areas are involved in control processes that underlie the intention to orient attention has been suggested in other experiments. Rosen et al. (47) conducted a functional magnetic resonance imaging (fMRI) study examining the differences in neural activity in centrally cued endogenous shifts of attention versus peripherally cued exogenous shifts. Although the neural regions associated with the two tasks were largely overlapping, superior frontal activation was found to be present only in endogenous orienting.

The role of frontal cortex in attentional control will be discussed more thoroughly in the following sections. The role of parietal cortex in selective attention has been the focus of continued investigations with increased spatial resolution of neuroanatomical imaging allowing for more refined analysis of the specific aspects of attention associated with regions of the parietal lobe. The original proposal of the posterior attention system specified the superior aspect of the posterior parietal lobe as being primarily involved in the disengage operation (41). However, a number of studies have suggested a more inferior locus of the disengage operation in the region of the temporoparietal junction and superior and middle temporal gyri (50). Event-related fMRI (ER-fMRI) studies that contrasted neural activation associated with valid cueing and invalid cueing trials in a visual cueing task indicated greater activity in the right temporoparietal junction on invalid trials when individuals must disengage attention from the cued location in order to respond to a target at an unexpected location (48). Further, activity in the region of the intraparietal sulcus may be associated with voluntary orienting of attention in response to the presentation of the cue, but not with disengaging of attention from a cued location (51). Hence, different regions of parietal cortex may respond to control signals from frontal cortex and to exogenously generated signals from salient stimuli. Evidence for the multifaceted nature of parietal cortex in selective attention and more broadly in the representation of space has received strong support from research involving single-cell recording in primates performing a variety of both overt and covert attention tasks. Multiple regions in parietal cortex demonstrate a retinotopic mapping of space and may be involved variously in spatial attention, spatial memory, motion perception, and guidance of movement (52).

Attentional Modulation of Visual Cortex

At a behavioral level, the effects of selective attention can be measured in terms of faster and more accurate responding to stimuli appearing at an attended location or containing an attended feature. A neural correlate of this behavioral effect can be seen in attentional modulation of neural activity in tissue devoted to processing visual information (53,54).

Attentional modulation has been demonstrated in both ERP and fMRI studies across a variety of experimental tasks and neural regions.

Experiments examining attentional modulation of sensory processing have been carried out in both the auditory and visual modalities (55,56). In these experiments attention is directed to one stream of stimuli while a second stream is unattended. Figure 17.4 shows an example of stimuli from a visual version of this experiment and the ERP waveforms elicited by stimuli presented to either the attended or unattended portions of the display (5). Increases in amplitude of the P1 and N1 components of the visual waveform are seen for attended stimuli; however, attention does not appear to modulate responses in the earlier C1 component of the waveform. These results in combination with PET data localize the neural source of these effects in extrastriate cortex rather than primary visual cortex (57). In addition to the

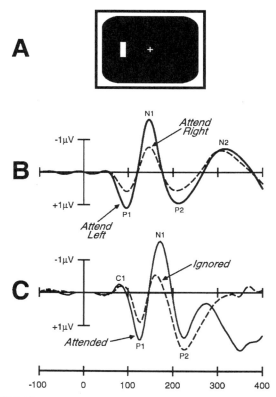

FIGURE 17.4. A: Example stimulus display for an event-related potential (ERP) study of selective attention. Participants are instructed to attend to one side of the display or the other. **B:** Idealized ERP waveforms expected from the two attention conditions in which the stimulus display was presented. P1 and N1 components show greater amplitude when attention is directed to the position at which the stimulus appears. **C:** Idealized ERP waveform demonstrating attention modulation in P1 and N1 components and the absence of attention modulations in the earlier C1 component. (From Luck SJ, Girelli M. Electrophysiological approaches to the study of selective attention in the human brain. In: Parasuraman R, ed. *The attentive brain.* Cambridge, MA: The MIT Press, 1998:71–94, with permission.)

modulations of the early P1 and N1 components found in experiments of spatial attention, selective attention based on other nonspatial features, such as color, motion, and shape, result in modulation of components of the visual waveform. However, these effects occur only for later components of the waveform. The relationship between these early and late components and the nature of the selection tasks that elicit them have led researchers to propose a hierarchy within attentional selection in which selection for location precedes selection on other feature dimensions (7). Behavioral evidence consistent with the primacy of location in visual selection comes from a variety of cueing and visual search tasks (58,59).

Further examination of the processes of attentional modulation of activity in visual cortical areas using PET and fMRI revealed several characteristics of this modulation. Neuroimaging studies showed increased or enhanced neural activity occurring for attended versus unattended stimuli. The greater spatial resolution afforded in neuroimaging techniques has allowed researchers to provide cortical mapping of attentional modulation in regions associated with processing of shape, color, and motion, as well as more complex characteristics such as face processing (60). Kastner and colleagues performed a series of fMRI experiments designed to examine activation of neural regions in the visual information processing stream under various conditions (61,62). The basic experimental design required participants to view a group of four visually complex images presented in one quadrant of the visual field either sequentially or simultaneously. The initial results under passive viewing conditions, when participants simply viewed the displays without performing any explicit task, indicated that activity in visual processing regions was suppressed during simultaneous presentation, suggesting a mutual inhibition of stimuli competing for neural representation. Focused visual attention appears to enhance the activity at the attended location through a mechanism that reduces the suppressive interactions with competing stimuli. Behavioral evidence consistent with their ideas comes from spatial cueing in Stroop color naming experiments (63). Further experimentation by Kastner and colleagues found that attentional modulation of activity in visual cortex occurs both in the presence and absence of visual stimulation. This finding suggests that attention serves to increase the baseline level of activity in cells that have receptive fields covering an attended region, as well as enhancing neural responses triggered by the appearance of a stimulus in the attended region.

The consensus of studies using both ERP and fMRI technologies appears to be that attentional modulation of stimulus processing occurs at a variety of stages of visual processing, but there is some disagreement about whether attentional modulation acts in primary visual cortex. Whereas ERP studies failed to find attentional modulation in the earliest components of the visual waveform associated with activity in the primary visual cortex (64), neuroimaging studies have localized some attentional modulation effects in V1 (65). These effects appear to be smaller in magnitude and inconsistent in occurrence relative to effects arising later in the stream of processing. One factor that may affect whether attentional modulation occurs in primary visual cortex is the complexity of the visual display. When displays are more complex, attentional modulation may involve primary visual cortex in order to resolve competing visual stimuli, suggesting a neural mechanism similar to the behavioral changes seen with increased perceptual load. Combined ERP and fMRI results indicate that the neuronal modulation in primary visual cortex may occur not in the initial stimulus processing but later in time via a feedback mechanism (60).

Hemispheric Asymmetries

Like many sensory and motor system functions, orienting of attention is controlled by neural regions contralateral to the direction of orienting (e.g., orienting to the left visual field is controlled by right hemisphere structures and vice versa). Early evidence for contralateral attentional control came from behavioral studies that looked for directional attentional biases following activation of one hemisphere or the other (66,67). This early behavioral work was supported by ERP and neuroimaging findings that attention to one visual field will be differentially associated with neural activity in the contralateral hemisphere (49,68). The relationship between the two hemispheres appears to be one of mutual inhibition generating hemispheric rivalry in attentional control.

In addition to the mechanisms of contralateral control of attention, evidence suggests that the right hemisphere is dominant in attentional functioning. The left hemisphere is involved in attending to the right side of space, whereas the right hemisphere appears to be involved in attending to both the left and right visual fields (69). The behavioral manifestations of the neglect syndrome support hemispheric asymmetries in attention. Neglect occurs more often and more severely after right hemisphere damage than left hemisphere damage (70). Along with the neuropsychological evidence, neuroimaging and electrophysiological studies have provided further support for the dominance of the right hemisphere in attentional orienting to one region of environmental space or another (46).

Although the left hemisphere has been generally considered to have less involvement in the processes of attention, a number of studies have suggested specific conditions under which the left hemisphere is more heavily engaged in attentional processes. Both patient studies and neuroimaging research have demonstrated left hemisphere involvement in the use of objects in guiding deployment of spatial attention. Damage in the region of the temporoparietal junction in the left hemisphere disrupted the effects of objects on the deployment of attention in a task where object boundaries normally appear to guide attention (71). An fMRI study has confirmed a left hemisphere role in object-based attentional processes (48). Comparisons of neural activation associated with location-based and object-based orienting tasks demonstrated a network of left hemisphere structures engaged when individuals attend to regions of space bounded by an object. This widespread network included classic attention regions such as temporal and parietal lobes and pulvinar nuclei, as well as areas associated with object processing (extrastriate and inferior temporal cortex) and executive control (dorsal lateral and medial prefrontal cortex).

Another attentional process that appears to engage the left hemisphere is strategic or guided visual search. Evidence was gathered in a study of a split-brain patient participating in a visual search task, where search could be strategically guided based on the characteristics of distractor items (72). Under conditions that allowed for guided search in each hemifield of the display (i.e., search for a target could be limited to a subset of items based on the color of those items), only the left hemisphere engaged in guided search and showed faster reaction times when search could be guided. The right hemisphere used guided search only under conditions when stimuli presented to the left hemisphere did not allow for guided search, suggesting a dominance of the left hemisphere in strategic control of attentional orienting that inhibits the right hemisphere's use of these mechanisms.

The previous two examples provided evidence for particular engagement of the left hemisphere in certain aspects of visual attention, including use of objects and strategic guidance of search, but other research suggests that the left hemisphere may take a lead role in attentional processes associated with other functions. One such study used PET to investigate neural regions associated with attention to upcoming motor responses. Results showed largely left lateralized activity in the parietal lobe in the region of the supramarginal gyrus associated with attention to motor activity (73). This finding of left hemisphere activation during motor attention may represent a specific manifestation of a tendency for the left hemisphere to orient attention in toward the body while the right hemisphere is involved in orienting attention away from the body (74).

EXECUTIVE ATTENTION AND EXECUTIVE CONTROL OF TASK PERFORMANCE

Whereas attentional selection deals primarily with the processing of information coming into the cognitive system, the second major area within the study of attention focuses on the role of attention in coordinating mental activities including memory, motor responses, and perception in order to carry through to completion a particular task or goal-directed activity. *Executive control* is the term used to specify the attentional functions that serve this role. Use of the word *attention* in everyday contexts, such as "Pay attention to what you are doing" or "Attend to business—you're off task again," are suggestive of the executive control functions of attention.

Historically, executive control processes have not received as much focus as selective attention, in part because the methods for studying executive control were not as well worked out, the behavioral impacts of executive control were less well defined, and the underlying mechanisms of executive control were likely to be more complex than those associated with attentional selection. However, there has been an increase in the number and analyticity of studies devoted to executive control processes both in cognitive psychology and neuroscience. These studies focus on two primary circumstances in which executive control may be necessary for cognitive tasks to be carried out appropriately and successfully: (a) multitask environments that demand maintenance and coordination of multiple stages of task performance, and (b) task environments that require inhibition of external stimuli or overlearned responses in order to correctly perform a task relevant to current goals.

Functional and Computational Approaches

Models of Executive Control

Executive control functions have been considered in a number of models of cognition, most notably Baddeley's model of working memory (75), which is shown schematically in Figure 17.5. Baddeley's model contains three components: two subordinate or slave systems, the visuospatial sketchpad and the phonologic loop, which serve as buffers for short-term storage of visual and auditory-linguistic information; and a central executive that allocates resources to and coordinates the functions of the two subordinate systems. Numerous criticisms have been leveled at Baddeley's model of the central executive, focusing primarily on the notion that the central executive was little more than a homunculus without descriptive or predictive value (76). Much of the initial experimental work examining Baddeley's model was devoted to characterizing the two subordinate systems, leaving the central executive largely undefined. However, more recently Baddeley and colleagues (77,78), as well as other researchers, have begun to conduct systematic investigations of the functions attributed to the central executive, including

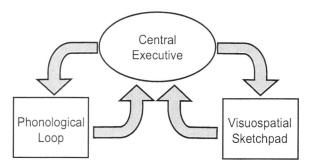

FIGURE 17.5. Diagram of working memory model proposed by Baddeley (75).

(a) coordinating the time allocated for processes associated with the slave systems of working memory, (b) switching retrieval plans in a dual-task environment, (c) temporary activation of information in long-term memory, and (d) top-down control of selective attention. In examining and proposing these functions, Baddeley drew heavily on an early theory of executive control introduced by Norman and Shallice (79).

At the core of Norman and Shallice's model was the idea that behavior is either controlled (willed) by executive processes or else it occurs automatically. They outlined five types of task situations that require executive control: (a) planning and decision making; (b) troubleshooting or error correction; (c) novel or unfamiliar action sequences; (d) dangerous or difficult tasks; and (e) overcoming habitual or prepotent responses. In contrast to these task situations are the automatic tasks that can be carried out in the absence of attention. Figure 17.6 depicts the architecture of Norman and Shallice's model. Actions are carried out under the guidance of schemas capable of executing each action. Behavior will follow the operations within a schema unless blocked by another action schema or by lack of resources. The process of "contention scheduling" coordinates competition between multiple schemas, with the activation level of a schema determining whether a schema will be selected. Activation of a schema is the result of both excitatory and inhibitory processes. Further, the activation of schemas may be initiated by either external stimulus conditions or internal control structures. For automatic tasks, action occurs simply as the result of a triggering mechanism associated with the presence of the appropriate stimuli in the environment. Under conditions when tasks cannot be performed automatically, the necessary executive control is carried out by the supervisory attention system (SAS). The SAS operates by increasing the activation values for currently preferred schemas while decreasing or inhibiting the values for other competing schemas, thus biasing the likelihood that a particular schema will reach threshold level and be initiated by the contention scheduling mechanism. The interplay of controlled and automatic, or top-down and bottom-up, processes put forth in Norman

and Shallice's model continues to be a question of theoretical interest to researchers of attentional control.

Unfortunately, empirical research on the properties of executive control lags behind the apparent coherence and generality of theoretical proposals such as those of Norman and Shallice or Baddeley. Furthermore, there is considerable tension between a unitary concept of a single executive control system responsible for the ordering of all cognitive functions and a drive toward fractionating executive control into individual elements serving separable and independent mechanisms of control specific to particular tasks or particular domains of task activity. Rigorous experimentation and theoretical integration across tasks and paradigms are needed to settle the question of unitary or multiple executive control functions. Nevertheless, progress has been made at both the computational and the neural levels in beginning to understand some of the specifics of executive control (76). We turn now to some of the available evidence.

Behavioral Studies of Executive Control

Executive control of attention has been studied within a number of experimental environments that reflect the circumstances theorized by Norman and Shallice (79) to lead to the necessity for executive control over behavior. These environments can be divided into two general types: paradigms that require the coordination of two or more tasks performed concurrently or sequentially, and paradigms that involve focused attention on a single task in conjunction with ignoring or inhibiting other task irrelevant stimuli and responses.

Dual-Task Environments

A variety of dual-task and complex-task paradigms have been developed to allow for investigation of different aspects of executive control. Key components in these tasks include the presence of multiple environmental stimuli presented either simultaneously or in rapid succession, task goals that must be maintained in working memory, and multiple stimulus–response mappings that may conflict based on changing task

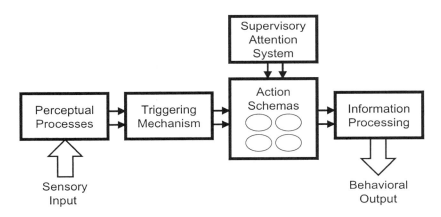

FIGURE 17.6. Diagram of supervisory attention system and executive control of behavior proposed by Norman and Shallice (79).

goals. Executive control processes serve to prepare the cognitive system for an upcoming task, update and maintain the task goals, and minimize cross-talk or interference from other task goals that are potentially but not currently relevant.

In this context, task switching paradigms have become extremely popular venues for studying executive control. In task switching experiments, when seemingly simple tasks such as identifying a digit or naming a color are interleaved in environments that ask for fast reaction, performance on these tasks is substantially slowed, especially on trials for which the task changed from the immediately preceding trial. This slowing is thought to reflect the executive control processes involved in reconfiguring the cognitive system to respond based on the goals of the current task set. A common variation of task switching involves alternating runs of trials (e.g., AABBAABB), which allows for the examination of switch and no switch transitions within the same block of trials. Rogers and Monsell (80) used an alternating runs paradigm along with a manipulation of the time between the trials to demonstrate that increased preparation time—a longer interval between one trial and the next—can significantly reduce switch costs. Rogers and Monsell introduced the concept of task set reconfiguration as an explanation of switch costs. When switching between tasks, the executive control mechanisms are engaged in a time-consuming set of prospective or anticipatory operations that activate or encode information about new processing components in preparation for the upcoming task. Support for task set reconfiguration as a responsibility of executive control in preparing for a switch between tasks comes from several studies using cueing of the upcoming task (81–83).

In contrast to this view of switch costs based on activating the processes that will be needed by the new task, other studies of task switching have emphasized processes associated with deactivation of the task being switched from. Mayr and Keele (84) proposed that inhibitory rather than facilitatory processes may underlie switch costs. In particular, the commonly used alternating runs procedure requires participants to switch to a task that recently has been switched from (e.g., AABBAA). They reasoned that if the initial switch from A to B involves actively inhibiting task A, then returning to task A at the next switch may require overcoming this active inhibition. They confirmed this hypothesis when they found that participants were slower to respond to the final trial in a task sequence such as ABA involving returning to a recently switched-from task when compared to a sequence of tasks such as CBA.

Both the reconfiguration and backward inhibition accounts of task switching suggest an active role of an executive control mechanism in accounting for task switch costs. Further, these accounts suggest that multiple mechanisms may be involved in the control process, some aiming to activate the upcoming task and others aiming to deactivate the preceding task. In addition, there may be exogenous processes involved in switching tasks—processes that can-

not take place until the stimulus appears on which the new task depends. Even when participants have ample time to prepare for an upcoming task, small residual switch costs remain (80). Thus, executing the goal-directed activity defined by a particular task is a complex mix of endogenously controlled and exogenously triggered processes. This mix is reminiscent of the interplay between endogenous and exogenous factors that influence selection of sensory information from the environment. It only makes sense that all functions of the attentional system would be tuned to the balancing act required to pursue internally held goals while remaining responsive to a dynamically changing environment.

Another dual-task paradigm often used to study executive control processes is the psychological refractory period paradigm, which involves separate speeded responses to two discrete tasks. In the standard version of the psychological refractory period paradigm, two stimuli (S1 and S2) affording two different responses (R1 and R2) are presented temporally close together, creating a potential overlap in the performance of the two tasks. Systematic manipulation of the time between presentation of S1 and S2 reveals a slowing in R2 but not in R1 when the time between the two stimuli is brief. This phenomenon is referred to as the *psychological refractory period* (PRP), from which the paradigm name is derived. Figure 17.7A shows an idealized example of data demonstrating the PRP effect. The PRP effect appears highly robust even under conditions in which the stimuli are presented in different stimulus modalities reducing any effect of sensory interference and when response modes differ, thereby removing motor interference from R1 and R2. Explanations of the PRP effect have focused primarily on the concept of a central bottleneck in processing, thus implicating executive control processes in dual-task slowing and providing a controlled environment for studying these processes. In an extended and systematic line of investigation, Pashler (85,86) examined the PRP effect and demonstrated that although some task components occur in parallel (such as perceptual processes), other seemingly computationally simple task components are subject to a bottleneck requiring serial processing (such as response selection and more generally effortful memory retrieval and/or effortful manipulation or transformation of perceived or retrieved information.) This central bottleneck account of the PRP effect is shown in the diagram of processing stages in Figure 17.7B.

The attentional blink paradigm represents another variation of the dual-task environment. In these tasks stimuli are presented in rapid serial visual presentation (RSVP) format, where briefly presented stimuli occur in quick succession. Embedded in the sequence are target stimuli for two separate detection tasks. For example, task 1 may be to report the identity of a letter printed in an oddball color and task 2 may be to report whether a particular target letter (an *X*) occurred in the display. If the appearance of the second target stimulus—the *X*—occurs between 200 and 600 milliseconds after the first target, then performance on task 2 is

A

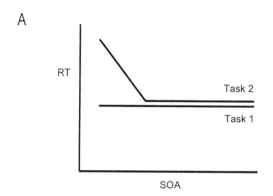

B

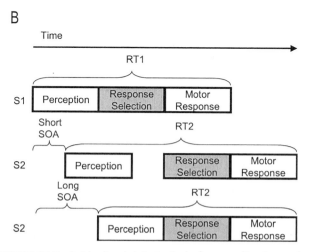

FIGURE 17.7. A: Idealized data typical of responding in psychological refractory period (PRP) tasks. RT for task 2 increases with decreasing SOA. **(B)** Diagram of processing stages in the PRP task, showing the central bottleneck at the stage of response selection.

diminished: the attentional blink. This bottleneck in processing stimuli presented in close temporal proximity initially was attributed to attentional limitations at perceptual processing stages (87), but later work demonstrated that the stimulus for task 2 receives perceptual processing but fails to enter working memory; therefore, performance is poor (88,89). The attentional blink is a late selection phenomenon in which perceptually processed stimuli are excluded from storage in memory because executive control processes are occupied with the processing of a stimulus that occurred earlier.

A final dual-task paradigm was introduced by Baddeley and colleagues to study executive control processes (77,78). In this task participants generate a sequence of random responses, either digits or finger movements. The pattern of responses made in the random generation task varies in a systematic fashion depending on the amount of attention paid to the task by the participant, with more stereotyped or patterned responses occurring when participants are allowed less

time to make each response or are responding while engaging in a primary task that requires executive attentional resources. That is, generation of a random sequence requires participants to continually bear in mind what a random sequence ought to look like and avoid introducing patterns. Therefore, random generation serves as a continuous measure of the degree to which a primary task requires executive control. Using this method, Baddeley and colleagues showed that tasks requiring little executive control (e.g., simple counting that is a highly practiced action) do not disrupt the generation of a random sequence, whereas tasks requiring executive control (e.g., a memory task that involves maintaining information in working memory) cause more patterned or stereotyped responding, resulting in deviation from randomness in the random generation task. The random generation task is relatively new but may prove to be a highly useful technique for studying executive control processes, particularly when the degree of executive control may vary over the course of a task.

We have reviewed a variety of dual-task paradigms from which progress has been made in specifying executive control processes. However, simply studying these tasks in isolation will not allow for an integrated account of the executive control processes that presumably are engaged across many task environments. Researchers have begun to examine the relationships between findings from various paradigms. Examples of such integrative work can be found in a number of contributions to the most recent publication in the *Attention and Performance* series. Jolicœur et al. (90) apply data from PRP and attentional blink paradigms to address questions of strategic control versus structural capacity limitation accounts of performance deficits in these tasks. Similar considerations are made by Pashler (86) focusing on PRP and task switching paradigms. Allport and Wylie (91) examined processes of negative priming (described later) in relation to task switching. Works such as these undoubtedly will continue to be critical in furthering our understanding of executive control processes across a variety of dual-task environments both in and out of the laboratory.

Inhibitory Control

Dual-task and multitask environments require executive control processes to manage the coordination of multiple stimulus and task requirements. Some circumstances require us to perform multiple tasks, whereas others necessitate ignoring certain environmental stimuli and inhibiting the responses they typically elicit in order to direct our attention to a particular task. Another function of executive control is this inhibition of unwanted thoughts and behaviors.

One of the tasks most commonly used to address the inhibition of automatic or overlearned responses is the Stroop task (63,92,93). In the Stroop test, participants are asked to name the color of ink in which a stimulus word is written. In the standard version of the Stroop task, the stimulus word can be the same color word as the ink in which it is presented (congruent condition), a color word different from the ink color (incongruent condition), or a non–color word or string of non–letter characters (neutral condition). Performance on congruent trials typically shows a benefit when compared to neutral trials, whereas performance on incongruent trials shows a cost. These effects are commonly considered to result from activation of the color spelled by the word despite its logical irrelevance to naming the color of the ink, because reading is an automatic response for literate persons presented with a printed word. In particular, the cost on incongruent trials reflects the need for the participant to inhibit the response that would be elicited by this automatic reading of the color word.

The Stroop task is just one example of a task that involves inhibitory control, in this case over an automatic or well-learned response. However, inhibition appears to be a part of the control mechanisms for a number of cognitive processes. A wide variety of tasks have been developed to assess inhibitory control mechanisms within a number of domains of cognitive processing. Inhibition in attentional selection has been seen in studies of negative priming. The negative priming paradigm presents participants with pairs of stimuli, one a target requiring a response and the other a distractor that must be ignored (e.g., see pairs of words, one printed in red and the other green, under instructions to read each word printed in red while ignoring each word printed in green). On trials where the target stimulus had served as the distractor stimulus in the previous trial, participants typically are slower to respond than if the target stimulus had not appeared at all in the previous trial, presumably because the stimulus remains inhibited based on the processes involved in ignoring its occurrence as a distractor on the previous trial (94–96). Other tasks have specifically examined inhibition of responding including go/no go tasks and stop signal tasks. In these paradigms participants again see a series of stimuli, for example *X*s and *O*s, to which a different response is required. In the case of the go/no go tasks, one stimulus that appears with high frequency requires participants to make a response (go task) and the second stimulus that appears with low frequency signals the participant to withhold responding (no go task). The stop signal paradigm presents each stimulus with equal frequency, but on a small number of trials a secondary stimulus (such as a tone) is presented that indicates that the participant should withhold responding on that trial. Within these paradigms the nature of inhibitory control can be examined through manipulations of variables such as the timing of the events and the difficulty of the individual tasks involved (97). Still other tasks have been developed to study inhibitory processes in memory. In directed forgetting tasks, participants are presented with lists of stimuli and then are directed to forget one of the lists. Later memory tests of both to-be-remembered and to-be-forgotten lists measure the degree to which the participant was able to inhibit memory for the items they were instructed to forget (98). Nigg (99) provides succinct summaries of these and other tasks measuring inhibitory control.

A theory of inhibitory control that has been influential in models of executive control within working memory was proposed by Hasher and Zacks (100,101). These researchers conducted a series of investigations aimed at examining the specific functions that inhibition fills in the manipulation of information in working memory. They propose three primary functions of inhibition in the control of working memory: access, deletion, and restraint. Inhibition serves to limit access into working memory, excluding task irrelevant information from entering working memory. Information that has entered working memory may be deleted or suppressed when it is no longer necessary. Finally, inhibition restrains prepotent responding. In combination, these functions allow for maintenance of goal-relevant information in working memory by reducing interference from task irrelevant information and limiting automatic or dominant responses. Hasher and Zacks' theoretical framework provides unification across the wide variety of tasks that have been used to study inhibitory processing.

Neural Substrates

Just as the functional and computational understanding of executive control has lagged behind research on selective attention, the research investigating the neural substrates of executive control also is just beginning to come into its own. There is a long history in clinical and experimental neuropsychology of studying patients who appear to show difficulties with executive control functions (such as keeping track of goals or switching from one task strategy to another) after brain damage, particularly to the frontal lobes. Only recently, however, has the study of patients been supplemented with application of neuroimaging techniques to goal maintenance, task switching, dual-task performance, and inhibition of prepotent responses by normal subjects.

Neuropsychological Approaches to Studying Executive Control

Within the neuropsychological tradition, impairments in a variety of tasks purporting to measure executive control have been linked to damage involving frontal cortex (102,103). Referred to generally as *dysexecutive disorder,* damage to diffuse regions of the frontal cortex, often due to closed head injury, results in the loss of planning and organization skills; deficits in problem solving, particularly when multiple steps or forethought are required; and perseveration, the repetition of a behavior even though the environment dictates change. These observable behavioral symptoms appear in multiple complex environments in the daily lives of patients; however, they are generally vague and poorly defined in terms of the underlying deficits in specific cognitive processes.

To better assess the specific processes that are deficient in dysexecutive disorder, a variety of tasks often are used in neuropsychological settings when frontal lobe damage is suspected. One of the more well known is the Wisconsin Card Sorting Task, which requires patients to place cards into groups based on features of the stimuli on the cards. The feature on which the cards are to be sorted is dynamic, changing without warning to the patient, and requires the patient to try to deduce the category based on feedback from the tester. Frontal lobe patients often have trouble switching away from a sorting category that is no longer relevant, demonstrating in this task the perseveration common in frontal lobe patients. Additional evidence for problems with inhibiting no longer appropriate responses is seen in verbal fluency tasks, where patients often repeat a small set of words when asked to generate as many words as possible beginning with a particular letter, and in go/no go tasks where patients respond at higher rates than controls on the no go trials. Other tasks are designed to detect deficits in planning or sequencing of behaviors. A well-studied problem-solving task that requires planning is the Tower of Hanoi. In this task (and in similar tasks such as the Tower of London), a series of moves must be made to reposition a set of rings on three posts. Specifics of the types of moves that are allowed vary for the different versions of the task. However, all versions require planning in order to complete the task in the fewest number of moves. Frontal lobe patients typically require more moves to reach the goal state of the task. This general deficit in planning and sequencing is seen in other tasks as well. We have mentioned only a few of the many tasks commonly used to assess dysexecutive disorder and the frontal lobe damage that typically underlies the disorder (more thorough reviews of these and other tasks can be found in references 102 and 103). However, the set we reviewed should be sufficient to demonstrate the complexity of the tasks used to assess the planning and sequencing deficits produced by frontal lobe damage. Each task requires many different kinds of processes, in addition to anything one might specifically call "planning," "sequencing," "maintenance of goals," or "inhibition." In order to more carefully define the neural substrate of the various components of executive control, more analytic methods of investigation will be necessary.

Anterior Attention System: Frontal and Anterior Cingulate Cortex

Theories of and behavioral data associated with executive control, including the data collected in neuropsychological assessment of frontal lobe patients, point to the fact that executive functions are involved in a wide variety of task situations that require effortful processing, such as those described by Norman and Shallice (79) as requiring the SAS for appropriate actions to be successfully carried out. Thus, when researchers began to use neuroimaging to examine the neural substrates of executive control, the strategy was to look for regions of the brain that are active in a wide array of tasks with different perceptual and motor demands. Common areas of activation across tasks that differ with respect to their

input and output demands but which all require effortful processing of the type attributed to SAS and working memory then become prime candidates for regions involved in some sort of executive control of processing. Based on such studies, researchers have suggested an anterior attention system (1,41). One critical region in this system is the anterior cingulate cortex (ACC), a midline structure comprising the inferior aspect of left and right medial frontal cortex posterior to orbitofrontal cortex and superior to the amygdala and corpus callosum. More generally a number of prefrontal regions, including dorsolateral prefrontal cortex (DLPFC), medial prefrontal cortex, and inferior frontal cortex, all appear to be involved in some aspect of executive control (Fig. 17.3). These regions are highly interconnected with areas of the brain involved in primary sensory and motor functions and with the posterior regions discussed above that make up the posterior attention system.

Neuroimaging studies using a number of experimental tasks have indicated ACC activity in task conditions putatively involving executive control functions. In a seminal investigation examining regions of brain activity associated with a controlled effortful task (104), participants were scanned using PET while reading words aloud (see *car,* say *car;* see *chair,* say *chair*) and while generating a verb that would be an appropriate action associated with the noun that was actually presented as the stimulus (see *car,* say *drive;* see *chair,* say *sit*). When the activity associated with the reading condition (which serves as a control or baseline measurement of the perceptual activity involved in recognizing the stimulus word and the motor activity involved in saying a word aloud) is subtracted from the generate-verb condition, the remaining activity was found in two left lateralized areas associated with classic language centers in frontal and temporal cortex and more importantly with a region of the ACC that the investigators interpreted as being involved in the executive control functions necessary for carrying out this unusual task. Such an interpretation is supported by further investigations in which participants were given extended practice with generating verbs for a single list of nouns. After extended practice at this task, after which responses were quite routine and executive control processes no longer seemed necessary to carry out what now looked like an automatic task, participants were again scanned. Results showed that the pattern of brain activity no longer involved ACC or the frontal language generation areas and looked very similar to activation seen when participants performed the presumably already highly automated word reading task. The generate-verbs task is an unfamiliar task environment that fits several of the conditions that Norman and Shallice (79) suggested result in engagement of the SAS, including novel situations and difficult tasks. However, when the task has been practiced enough to become overlearned, thus removing the conditions that are supposed to require control by SAS, then the pattern of brain activity associated with the task changes as well (105).

Another highly studied behavioral task that was used early in neuroimaging investigations of executive control is the Stroop task. As noted earlier, performing the Stroop task involves overcoming the prepotent response of reading the word in order to name the color in which the word is presented. Meeting this task demand presents more difficulties on incongruent trials when there is a mismatch between the word and the color. Succeeding at the Stroop task is thought to require the SAS. A number of neuroimaging experiments have found greater activation in the ACC associated with performance on the incongruent trials in the Stroop task. However, the relative degree and specific foci of activation vary based on the parameters of a given experimental design (106). Recently, fMRI methodology in combination with detailed analysis of task requirements introduced in specialized conditions within Strooplike experiments has been applied to questions of the specific executive control functions implemented in ACC. Based on this research, theories of ACC function have been proposed that emphasize processes of error detection and monitoring of response conflict as the special province of ACC (107–110).

In addition to ACC, neuroimaging studies of the Stroop task have implicated other cortical regions, both anterior and posterior, in processes of executive control. Banich et al. (111) examined the functions associated with these separate regions in an fMRI study that varied the degree of task relevant and task irrelevant information. Both anterior and posterior regions were sensitive to task relevant information; the posterior regions also were sensitive to task irrelevant information. Based on the specific pattern of results, these researchers proposed separate functions for the two networks. They attribute top-down control functions to the DLPFC that serve to maintain an attentional set for task relevant information that biases perceptual selection processes. In contrast, posterior regions are thought to carry out the function of selecting stimuli through downward modulation of activity associated with task irrelevant information. The specific findings in this Stroop experiment are closely related to a number of broader theories of the function and interrelation of areas in DLPFC and more posterior cortical regions.

The DLPFC has been implicated in numerous functions, and researchers continue to pursue fractionation of this large region of tissue and the separable cognitive functions that it supports. Two of these functions are of particular interest for the current purposes: top-down control of visual selective attention and working memory. As previously noted in the discussion on the neural substrates of selective attention, regions in the DLPFC are active in cases involving endogenously controlled attentional orienting (112). Supported by a large body of electrophysiological and neuroimaging research, the "biased competition" model presented by Desimone and Duncan (53) provides an account of the interactions between these frontal regions and the posterior sensory cortex in which modulations of activity are controlled through top-down biasing signals from DLPFC. This ac-

count of how DLPFC operates is similar in spirit to the description of SAS proposed by Norman and Shallice (79), providing congruence between the computational properties of SAS and the functional properties of its purported neural implementation. There also exists extensive evidence that DLPFC is involved in working memory functions. Again, a combination of single-cell recording techniques in primates and more recent event-related neuroimaging studies in humans has been used to uncover activity in this region that occurs during retention intervals (the time between stimulus presentation and response during which a short-term memory of task relevant stimulus characteristics must be maintained) in tasks involving working memory such as delayed matching tasks (113–116). Such sustained activity in the absence of external stimulation clearly suggests involvement of these areas in maintenance of information in memory.

As highly analytic behavioral and neuroimaging studies continue to gather information about particular cognitive functions and their associations with specific sites in frontal cortex, integrating the relevant data on memory and executive control functions becomes a daunting task. Increasing spatial and temporal resolution of neuroimaging techniques has allowed researchers to make finer grained distinctions between regions of tissue and the functions attributed to them (117,118). At the same time, strides are being made to understand the interconnections and commonalities among these separable tasks. One excellent example of a line of work that combines a variety of cognitive neuroscience methodologies to address such a question has been carried out by Awh and colleagues examining the relationship between mechanisms of spatial attention and working memory (119–121). Based on behavioral, ERP, and fMRI studies, they propose a functional overlap of attention and working memory processes (specific to spatial information, although they suggest similar relationships are likely to exist in other domains such as object processing). The relationship proposed is one in which attention serves as a mechanism for rehearsal in memory by maintaining activity in the perceptual structures that originally registered the stimulus information being remembered.

The inferior aspect of the frontal lobes, also called orbital prefrontal cortex due to the proximity to the orbits of the eyes, appears to play a key role in the inhibitory components of executive control. Several fMRI studies have examined inhibitory control using variations of the go/no go task procedure that requires participants to withhold responding on a small percentage of trials. These studies have shown activity in the right inferior frontal cortex on no go trials when inhibition occurs (122–124). Similar findings have been seen using ERP measures of neural activity. Using a stop signal paradigm, researchers investigating inhibition in children with attention deficit/hyperactivity disorder (ADHD) demonstrated a component of a negative waveform associated with stopping trials that was localized to the right inferior frontal gyrus. Interestingly this component of the waveform was larger in control participants than in those diagnosed with ADHD who showed behavioral deficits in the task (125). Further evidence of a role for inferior frontal cortex in inhibition of responding comes from studies of attentional cueing. On invalid trials when the target does not appear at the cued location, participants must refrain from responding while reorienting attention to the target location. Results from fMRI studies indicate greater activation on invalidly cued trials in the inferior frontal cortex (48,126).

VIGILANCE

We turn now to the third primary domain of attentional function, *vigilance.* As Posner and Raichle (41) state, "A major problem of attention is maintaining a sustained state of alertness." Variations in the overall level of arousal of the cortex exert dramatic effects on task performance. Many of these effects are captured by the inverted-U–shaped function of the Yerkes-Dodson Law (127), in which very low and very high levels of waking arousal, especially as induced by stressors of various kinds, produce poor performance—slow or inaccurate or both. Intermediate levels of arousal support the best performance possible given the complexity of the task and the individual's level of practice and skill at performing it. The more complex the task, the lower the level of arousal at which optimal performance is achieved, that is, more complex tasks reach their asymptotic levels of performance and begin to degrade at earlier points in the rise of arousal above quiescent baselines (128,129). The transition from waking levels of arousal to sleep introduces some additional and interesting complications. As indicated by studies in which participants are awakened from various stages of sleep and immediately asked to engage in task performance, many basic cortical functions, including even the nature of associative connections in semantic memory that bring one thought to mind in response to another, are altered by sleep (130). Finally, sustained on-task behavior involving the continued performance of a single task for an extended period of time places an additional burden on attention mechanisms. This burden is especially great when relevant task stimuli occur only infrequently, creating long periods of supposedly on-task behavior in which no imperative events are happening. In such situations, a "vigilance decrement" occurs: performance becomes increasingly worse as time passes (131).

It is generally agreed that interactions among the reticular formation, thalamus, basal forebrain, and regions of right prefrontal cortex regulate overall levels of arousal and hence many of the phenomena related to vigilance (132,133). The additional problem of vigilance decrement remains unresolved. One might suppose that vigilance decrement would result from fatigue or malfunction within the same network of structures that supports overall arousal, but this does not appear to be the case (132).

Obviously these considerations regarding vigilance and vigilance decrement have profound implications for the proper interpretation of attention deficits, especially in light of the complex organization of the other two primary attentional functions already discussed, selective attention and executive control. *Attention* is an extremely wide-ranging term, both in its ordinary usage and in its technical meanings. One might be tempted by application of the same term "attention" to suppose that all of these phenomena should be attributed to the same computational processes and hence to the same brain structures: missing an important stimulus, failing to perform up to standards in a distracting environment, failing at the most important task when several tasks are going on at once, and straying off task after 15 minutes of a half-hour duty interval. All of these in some sense are "failures of attention." However, as we have seen, the definition, computational demands, and neural substrates of "attention" vary substantially from one of these phenomena to the next.

ATTENTION DISORDERS

In the preceding sections, we presented the functional brain systems of attention, considering evidence gathered from cognitive psychology, neuropsychology, and cognitive neuroscience research. In the final section of the chapter, we turn briefly to a discussion of the implications of our understanding of these systems on the field of neuropsychiatry. The disciplines of cognitive psychology and human neuropsychology have traditionally had a substantial degree of interaction, with studies of patients with brain lesions serving as a testing ground for cognitive theory, as well as theories of cognitive functions being used to gain understanding of diagnosis and treatment of neuropsychological patients. However, a similar relationship has been slower in developing between cognitive psychology, cognitive neuroscience, and psychiatry. Studies of psychopathology have not had similar interactive relationships with cognitive theorizing. In the last decade, growing interest in reaching across these disciplines has developed. This blending of fields, which focuses on the application of models of cognitive and neural functioning to the understanding and treatment of psychiatric disorders, has been coined *cognitive neuropsychiatry* (134). The cognitive neuropsychiatry approach has reciprocal benefits. Not only does the understanding of psychopathology benefit, but also theories of normal cognitive function are advanced by insights drawn from studies conducted with various psychopathological and neuropsychological populations. Further advances in this direction should prove beneficial for each of the disciplines involved.

This final section is intended to provide only the briefest coverage of a sampling of disorders involving attentional components, in order to illustrate the integral relationship between the understanding of functional brain systems and the diagnosis and treatment of clinical disorders. The particular disorders covered are intended to provide examples of the range of disorders that include attentional components, to demonstrate the varying degree of knowledge about the functional and neural etiology of various disorders and to suggest potentially intriguing areas for future research. As these accounts are by necessity limited, references for more thorough reviews are provided for each topic.

Neglect

Neglect refers to a failure in orienting or responding to stimuli contralateral to a brain lesion that does not result from deficits in primary sensory or motor function. In the acute stages, neglect can result in severe cognitive and behavioral consequences, such as eating food from only one side of the plate, grooming only one side of the body, and even denying that limbs on the neglected side are part of one's body. Such severe effects often are present only in the more acute stages immediately after the causal brain damage. Recovery of function can be substantial, reaching the point where neglect can be measured only in carefully controlled environments involving simultaneous presentation of stimuli to both the good and bad visual fields. In such situations, attention will be captured by the stimulus ipsilateral to the brain lesion and the contralateral stimulus will go unnoticed. While generally referring to a failure to orient or respond to stimuli contralateral to the site of a lesion, fractionation of the neglect syndrome has been suggested based on specific characteristics associated with neglect. A primary division has been made between more perceptual aspects, which present difficulties with perceiving stimuli occurring in contralesional space, and more motor aspects, which involve a bias in motor movements to the contralesional side of the body (70).

The neuroanatomical basis for neglect has been studied extensively and is understood more completely than that of the other disorders discussed in this section (for review see reference 70). The most common presenting cause of neglect is damage to the posterior regions of cortex in the area of the temporoparietal junction. The resulting neglect typically is more frequent and more severe after right hemisphere lesions than in cases of homologous damage in the left hemisphere. Other lesions to cortex have been shown to result in neglect, including damage to frontal regions. Interestingly the frontal lesions appear to be more closely linked to the motor aspects of neglect, whereas posterior lesions tend to manifest in perceptual neglect. In addition, damage to a number of subcortical regions has been indicated in cases of neglect including thalamus and basal ganglia. As noted earlier, patient studies involving restricted lesions in these and other areas have played an integral role in the development of models of the neural substrates of selective attention. There are a number of excellent brief reviews of neglect (135,136), as well as more extensive consideration (137).

Attention Deficit/Hyperactivity Disorder

ADHD is one of the most widely diagnosed childhood disorders. The complex of symptoms includes levels of activity, inattention, and impulsiveness that are developmentally inappropriate, impairing, chronic, and likely to occur across multiple environments. These symptoms are not well defined and represent diagnostic criteria rather than clearly understood cognitive mechanisms for the disorder. The looseness of definition can be seen in the multiple and often extreme changes in the definition of the disorder in subsequent versions of the *Diagnostic and Statistical Manual of Mental Disorders* from the American Psychiatric Association (138), currently in its fourth edition (DSM-IV).

There is strong evidence that ADHD primarily involves a deficit in inhibitory control rather than inattention. The origins of this hypothesis arise from a number of areas, but the inhibitory deficit view is most closely associated with the work of Barkley (139). A review and reformulation of this proposal was presented by Nigg (99). Studies using a wide variety of tasks measuring inhibitory control have demonstrated such deficits. For example, ADHD children require a longer time in order to inhibit responding in the stop signal task. Interestingly, in comparison, those same children have response times that are equivalent to control groups on the go trials in this task, demonstrating that attention and responding to some aspects of the task are relatively normal in these children (140). This places the onus for the behavioral problems associated with ADHD squarely on difficulties with inhibiting prepotent responses that are activated with normal alacrity, enabling them to escape the control of debilitated inhibitory mechanisms. Although it is clear that inhibitory control plays a major role in the functional deficits associated with ADHD, further research is needed to examine the role of other executive attention processes in ADHD.

Exciting attempts are being made to advance an integrated understanding of ADHD in terms of the underlying cognitive and neural systems of attention that have been the focus of this review. Swanson et al. (141) propose that research should strive to "align" the clinical symptoms, cognitive functions, and neural substrates of ADHD. To this end, they present an effort to categorize the specific symptoms listed in the *DSM-IV* criteria for ADHD as resulting from underlying deficits in selective attention, executive control, and alerting (sustained attention, also called vigilance). In addition, they link these factors to neuroimaging studies designed to examine neural correlates of ADHD. Other work informed by considerations of attentional systems, such as that by Nigg (99), is pursuing an agenda consistent with Swanson's proposal by separating out problems due to inhibitory control from those that may be associated with other aspects of attending to and processing information from the environment and maintaining the goals or goal sequences required to sustain and advance on-task mental activity.

Attention and Affective Disorders

Cognitive theories of emotional psychopathology consider the role of attentional biasing in affective disorders (142). One of the most common tasks used to study the relationship between attention and affect is a variant of the standard

Stroop task known as the emotional Stroop task. In the emotional Stroop task, participants are instructed to name the color in which either neutral or emotional words are printed. Slowed color naming for emotional words as compared to neutral words serves as a measure of attentional bias toward emotional words. The emotional Stroop task has been used with a wide variety of clinical populations. A review of this literature found that the emotional Stroop task appears to be sensitive to different psychopathologies, with patients showing greater Stroop interference for emotional words that were relevant to their specific disorder. In addition, recovery from emotional disorder is linked with reduced emotional Stroop interference effects. Results from the emotional Stroop studies support the idea that attention to emotional stimuli is biased for individuals with affective disorders, thus resulting in more interference with other task activity when emotionally salient stimuli are present in the environment (143).

Studies of the neural basis of the emotional Stroop effect point to the importance of the ACC in attention to emotional stimuli. A recent metaanalysis of neuroimaging studies of cognitive Stroop and emotional Stroop tasks shows a distinct region of the ACC active in the emotional Stroop tasks that is rostral and ventral to the region active during the cognitive Stroop task (106). Interestingly these same tasks and regions show reciprocal inhibition. Emotional Stroop tasks show deactivation in regions that are activated in the cognitive Stroop task and cognitive Stroop tasks show deactivation in regions that are activated in the emotional Stroop task. Recent neuroimaging data suggest that in addition to ACC, more anterior portions of the medial frontal cortex may show activity related to an interaction between attention demanding cognitive tasks and emotional experience, particularly performance anxiety (144). These midline regions have a high degree of connectivity with structures in the limbic system known to be involved in processing of emotion.

Broader Implications in Clinical Settings: Attentional Demands of Psychiatric Treatment and Cognitive Rehabilitation

We close by discussing a final way in which attentional and an understanding of attentional systems are relevant to clinical treatment and rehabilitation, even if the presenting symptomatology for which treatment is being sought does not specifically involve an attentional deficit as the primary concern. Clinical interventions and rehabilitation training programs are learning environments in which a part of benefiting from treatment requires obtaining new knowledge, altering habitual ways of thinking and responding to difficult or troublesome situations, applying new strategies for dealing with such situations, and developing new cognitive and/or social skills. Learning, memory, strategy application, and skill development all draw heavily on attentional resources; therefore, attentional limitations may come to play a large role in treatment success, whether they arise from secondary attentional involvement in the

disability being treated or from naturally occurring individual differences in a normal but relatively low-capacity attentional system. The theoretical implications of limited-capacity attentional systems for learning new skills have been worked out with respect to mental retardation and severe learning disability (145,146). More recently the empirical consequences of individual differences in executive control and the higher-level problem-solving processes that make substantial demands on working memory capacity have begun to be investigated in aphasiology (147–149). A general principle emerging from this work is that treatment regimens that load working memory and require problem-solving capacity are highly susceptible to variation in attentional systems, whereas highly structured treatment environments that provide close supervision and routinized practice are much less so. Although such a simple principle may seem obvious, it is easy to overlook when an attention deficit is not the presenting symptomological reason for a patient being in treatment.

REFERENCES

1. Posner MI, Petersen SE. The attention system of the human brain. *Annu Rev Neurosci.* 1990;13:25–42.
2. Moray N. Attention in dichotic listening: affective cues and the influence of instructions. *Q J Exp Psychol* 1959;11:56–60.
3. Broadbent DE. *Perception and communication.* New York: Pergamon, 1958.
4. Treisman AM. Contextual cues in selective listening. *Q J Exp Psychol* 1960;12:242–248.
5. Luck SJ, Girelli M. Electrophysiological approaches to the study of selective attention in the human brain. In: Parasuraman R, ed. *The attentive brain.* Cambridge, MA: The MIT Press, 1998: 71–94.
6. Kutas M, Hillyard SA. An electrophysiological probe of incidental semantic association. *J Cogn Neurosci* 1989;1:38–49.
7. Luck SJ, Hillyard SA. The operation of selective attention at multiple stages of processing: evidence from human and monkey electrophysiology. In: Gazzaniga MS, ed. *The new cognitive neurosciences,* 2nd ed. Cambridge, MA: The MIT Press, 2000: 687–700.
8. Johnston WA, Dark VJ. Selective attention. *Annu Rev Psychol* 1986;37:43–75.
9. Lavie N. Perceptual load as a necessary condition for selective attention. *J Exp Psychol Hum Percept Perform* 1995;21:451–468.
10. Handy TC, Mangun GR. Attention and spatial selection: electrophysiological evidence for modulation by perceptual load. *Percept Psychophys* 2000;62:175–186.
11. Maylor E, Lavie N. The influence of perceptual load on age differences in selective attention. *Psychol Aging* 1998;13: 563–573.
12. Huang-Pollock CL, Carr TH, Nigg JT. Development of selective attention: perceptual load influences early versus late attentional selection in children and adults. *Dev Psychol* 2002;38:363–375.
13. Henderson JM. Visual attention and eye movement control during reading and picture viewing. In: Rayner K, ed. *Eye movement and visual cognition.* Berlin: Springer, 1992:260–283.
14. Posner MI. Orienting of attention. *Q J Exp Psychol* 1980;32: 3–25.
15. Posner MI, Snyder CRR, Davidson BJ. Attention and the detection of signals. *J Exp Psychol Gen* 1980;109:160–174.

16. Jonides J. Voluntary versus automatic control over the mind's eye's movement. In: Long JB, Baddeley AD, eds. *Attention and performance IX.* Hillsdale, NJ: Erlbaum, 1981:187–203.

17. Theeuwes J. Exogenous and endogenous control of attention: the effect of visual onsets and offsets. *Percept Psychophys* 1991;49:83–90.

18. Yantis S, Jonides J. Abrupt visual onsets and selective attention: voluntary versus automatic allocation. *J Exp Psychol Hum Percept Perform* 1990;16:121–134.

19. Müller HJ, Rabbitt PMA. Reflexive and voluntary orienting of visual attention: time course of activation and resistance to interruption. *J Exp Psychol Hum Percept Perform* 1989;15:315–330.

20. Klein RM. Inhibition of return. *Trends Cogn Sci* 2000;4:138–147.

21. Klein RM. Perceptual-motor expectancies interact with covert visual orienting under conditions of endogenous but not exogenous control. *Can J Exp Psychol* 1994;48:167–181.

22. Eriksen CW, St. James JD. Visual attention within and around the field of focal attention: a zoom lens model. *Percept Psychophys* 1986;40:225–240.

23. Duncan J. Selective attention and the organization of visual information. *J Exp Psychol Gen* 1984;113:501–517.

24. Driver J, Baylis GC. Movement and visual attention: the spotlight metaphor breaks down. *J Exp Psychol Hum Percept Perform* 1989;15:448–456.

25. Kramer AF, Jacobson A. Perceptual organization and focused attention: the role of objects and proximity in visual processing. *Percept Psychophys* 1991;50:267–284.

26. Baylis GC, Driver JS. Visual attention and objects: evidence for hierarchical coding of location. *J Exp Psychol Hum Percept Perform* 1993;19:451–470.

27. Moore CM, Yantis S, Vaughan B. Object-based visual selection: evidence from perceptual completion. *Psychol Sci* 1998;9:104–110.

28. Treisman A, Gelade G. A feature integration theory of attention. *Cogn Psychol* 1980;12:97–136.

29. Treisman A. Features and objects: the Fourteenth Bartlett Memorial Lecture. *Q J Exp Psychol A* 1988;40:201–237.

30. Hubel DH, Wiesel TN. Receptive fields and functional architecture of monkey striate cortex. *J Physiol (Lond)* 1968;195:215–243.

31. Palmer SE. *Vision science: photons to phenomenology.* Cambridge, MA: The MIT Press, 1999.

32. Kahneman D, Treisman A, Gibbs BJ. The reviewing of object files: object-specific integration of information. *Cogn Psychol* 1992;24:175–219.

33. Egeth HE, Yantis S. Visual attention: control, representation, and time course. *Annu Rev Psychol* 1997;48:269–297.

34. Theeuwes J, Atchley P, Kramer AF. On the time course of top-down and bottom-up control of visual attention. In: Monsell S, Driver J, eds. *Attention and performance XVIII.* Cambridge, MA: The MIT Press, 2000:105–124.

35. Jonides J, Yantis S. Uniqueness of abrupt visual onset in capturing attention. *Percept Psychophys* 1988;43:346–354.

36. Yantis S, Jonides J. Abrupt visual onsets and selective attention: evidence from visual search. *J Exp Psychol Hum Percept Perform* 1984;10:601–621.

37. Folk CL, Remington RW, Johnston JC. Involuntary covert orienting is contingent on control settings. *J Exp Psychol Hum Percept Perform* 1992;18:1030–1044.

38. Wolfe JM. Guided search 2.0: a revised model of visual search. *Psychon Bull Rev* 1994;1:202–238.

39. Yantis S. Goal-directed and stimulus-driven determinants of attentional control. In: Monsell S, Driver J, eds. *Attention and performance XVIII.* Cambridge, MA: The MIT Press, 2000:73–103.

40. Mesulam M-M. A cortical network for directed attention and unilateral neglect. *Ann Neurol* 1981;10:309–325.

41. Posner MI, Raichle ME. *Images of mind.* New York: Scientific American Library, 1994.

42. Posner MI, Walker JA, Friedrich FJ, et al. Effects of parietal lobe injury on covert orienting of visual attention. *J Neurosci* 1984;4:1863–1874.

43. Posner MI, Cohen Y, Rafal RD. Neural systems control of spatial orienting. *Philos Trans R Soc Lond B Biol Sci* 1982;298:187–198.

44. Rafal R, Posner MI. Deficits in human visual spatial attention following thalamic lesions. *Proc Natl Acad Sci U S A* 1987;84:7349–7353.

45. Nobre AC, Sebestyen GN, Gitelman DR, et al. Functional localization of the system for visuospatial attention using positron emission tomography. *Brain* 1997;120:515–533.

46. Gitelman DR, Nobre AC, Parrish TB, et al. A large-scale distributed network for covert spatial attention: further anatomical delineation based on stringent behavioural and cognitive controls. *Brain* 1999;122:1093–1106.

47. Rosen AC, Rao SM, Caffarra P, et al. Neural basis of endogenous and exogenous spatial orienting: a functional MRI study. *J Cogn Neurosci* 1999;11:135–152.

48. Arrington CM, Carr TH, Mayer AR, et al. Neural mechanisms of visual attention: object-based selection of a region in space. *J Cogn Neurosci* 2000;12[Suppl 2]:106–117.

49. Corbetta M, Miezin FM, Shulman GL, et al. A PET study of visuospatial attention. *J Neurosci* 1993;13:1202–1226.

50. Friedrich FJ, Egly R, Rafal RD, et al. Spatial attention deficits in humans: a comparison of superior parietal and temporal-parietal junction lesions. *Neuropsychology* 1998;12:193–207.

51. Corbetta M, Kincade JM, Ollinger JM, et al. Voluntary orienting is dissociated from target detection in human posterior parietal cortex. *Nat Neurosci* 2000;3:292–297.

52. Colby CL, Goldberg ME. Space and attention in parietal cortex. *Annu Rev Neurosci* 1999;22:319–349.

53. Desimone R, Duncan J. Neural mechanisms of selective visual attention. *Annu Rev Neurosci* 1995;18:193–222.

54. Hillyard SA, Vogel EK, Luck SJ. Sensory gain control (amplification) as a mechanism of selective attention: electrophysiological and neuroimaging evidence. *Philos Trans R Soc Lond B Biol Sci* 1998;353:1257–1270.

55. Woldorff MG, Hallen CC, Hampson SA, et al. Modulation of early sensory processing in human auditory cortex during auditory selective attention. *Proc Natl Acad Sci U S A* 1993;90:8722–8726.

56. Mangun GR, Hillyard SA. Modulations of sensory-evoked brain potentials indicate changes in perceptual processing during visual-spatial priming. *J Exp Psychol Hum Percept Perform* 1991;17:1057–1074.

57. Heinze H-J, Mangun GR, Burchert W, et al. Combined spatial and temporal imaging of brain activity during visual selective attention in humans. *Nature* 1994;372:543–546.

58. Tsal Y, Lavie N. Location dominance in attending to color and shape. *J Exp Psychol Hum Percept Perform* 1993;19:131–139.

59. Moore CM, Egeth H. How does feature-based attention affect visual processing? *J Exp Psychol Hum Percept Perform* 1998;24:1296–1310.

60. Handy TC, Hopfinger JB, Mangun GR. Functional neuroimaging of attention. In: Cabeza R, Kingstone A, eds. *Handbook of functional neuroimaging of cognition.* Cambridge, MA: The MIT Press, 2001:75–108.

61. Kastner S, De Weerd P, Desimone R, et al. Mechanisms of

directed attention in the human extrastriate cortex as revealed by functional MRI. *Science* 1998;282:108–111.

62. Kastner S, Ungerleider LG. Mechanisms of visual attention in the human cortex. *Annu Rev Neurosci* 2000;23:315–341.

63. Brown TL, Gore CL, Carr TH. Visual attention and word recognition in Stroop color-naming: is word recognition "automatic"? *J Exp Psychol Gen* 2002;131:220–240.

64. Clark VP, Hillyard SA. Spatial selective attention affects early extrastriate but not striate components of the visual evoked potential. *J Cogn Neurosci* 1996;8:387–402.

65. Gandhi SP, Heeger DJ, Boynton GM. Spatial attention affects brain activity in human primary visual cortex. *Proc Natl Acad Sci U S A* 1999;96:3314–3319.

66. Kinsbourne M. The cerebral basis of lateral asymmetries in attention. *Acta Psychol (Amst)* 1970;33:193–201.

67. Reuter-Lorenz PA, Kinsbourne M, Moscovitch M. Hemispheric control of spatial attention. *Brain Cogn* 1990;12:240–266.

68. Nobre AC, Sebestyen GN, Miniussi C. The dynamics of shifting visuospatial attention revealed by event-related potentials. *Neuropsychologia* 2000;38:964–974.

69. Heilman KM, Van Den Abell T. Right hemisphere dominance for attention: the mechanism underlying hemispheric asymmetries of inattention (neglect). *Neurology* 1980;30:327–330.

70. Vallar G. The anatomical basis of spatial hemineglect in humans. In: Robertson IH, Marshall JC, eds. *Unilateral neglect: clinical and experimental studies.* Hillsdale, NJ: Erlbaum, 1993:27–59.

71. Egly R, Driver J, Rafal RD. Shifting visual attention between objects and locations: evidence from normal and parietal lesion subjects. *J Exp Psychol Gen* 1994;123:161–177.

72. Kingstone A, Grabowecky M, Mangun GR, et al. Paying attention to the brain: the study of selective visual attention in cognitive neuroscience. In: Burack JA, Enns JT, eds. *Attention, development, and psychopathology.* New York: The Guilford Press, 1997:263–287.

73. Rushworth MFS, Krams M, Passingham RE. The attentional role of the left parietal cortex: the distinct lateralization and localization of motor attention in the human brain. *J Cogn Neurosci* 2001;13:698–710.

74. Heilman KM, Chatterjee A, Doty LC. Hemispheric asymmetries of near-far spatial attention. *Neuropsychology* 1995;9:58–61.

75. Baddeley A. *Working memory.* Oxford: Oxford University Press, 1986.

76. Monsell S, Driver J. Banishing the control homunculus. In: Monsell S, Driver J, eds. *Attention and performance XVIII.* Cambridge, MA: The MIT Press, 2000:3–32.

77. Baddeley A. Exploring the central executive. *Q J Exp Psychol A* 1996;49:5–28.

78. Baddeley A, Emslie H, Kolodny J, et al. Random generation and the executive control of working memory. *Q J Exp Psychol A* 1998;51:819–852.

79. Norman DA, Shallice T. Attention to action: willed and automatic control of behavior. In: Davidson RJ, Schwartz GE, Shapiro D, eds. *Consciousness and self-regulation: advances in research and theory, vol. 4.* New York: Plenum Press, 1986:1–18.

80. Rogers RD, Monsell S. Costs of a predictable switch between simple cognitive tasks. *J Exp Psychol Gen* 1995;124:207–231.

81. Sudevan P, Taylor DA. The cuing and priming of cognitive operations. *J Exp Psychol Hum Percept Perform* 1987;13:89–103.

82. Meiran N. Reconfiguration of processing mode prior to task performance. *J Exp Psychol Learn Mem Cogn* 1996;22:1423–1442.

83. Sohn M-H, Carlson RA. Effects of repetition and foreknowledge in task-set reconfiguration. *J Exp Psychol Learn Mem Cogn* 2000;26:1445–1460.

84. Mayr U, Keele SW. Changing internal constraints on action: the role of backward inhibition. *J Exp Psychol Gen* 2000;129:4–26.

85. Pashler H. Dual-task interference in simple tasks: data and theory. *Psychol Bull* 1994;116:220–244.

86. Pashler H. Task switching and multitask performance. In: Monsell S, Driver J, eds. *Attention and performance XVIII.* Cambridge, MA: The MIT Press, 2000:277–307.

87. Raymond JE, Shapiro KL, Arnell KM. Temporary suppression of visual processing in an RSVP task: an attentional blink? *J Exp Psychol Hum Percept Perform* 1992;18:849–860.

88. Chun MM, Potter MC. A two-stage model for multiple target detection in rapid serial visual presentation. *J Exp Psychol Hum Percept Perform* 1995;21:109–127.

89. Jolicœur P, Dell'Acqua R. The demonstration of short-term consolidation. *Cogn Psychol* 1998;36:138–202.

90. Jolicœur P, Dell'Acqua R, Crebolder J. Multitasking performance deficits: forging links between the attentional blink and the psychological refractory period. In: Monsell S, Driver J, eds. *Attention and performance XVIII.* Cambridge, MA: The MIT Press, 2000:309–330.

91. Allport A, Wylie G. Task switching, stimulus-response bindings, and negative priming. In: Monsell S, Driver J, eds. *Attention and performance XVIII.* Cambridge, MA: The MIT Press, 2000:35–70.

92. MacLeod CM. Half a century of research on the Stroop effect: an integrative review. *Psychol Bull* 1991;109:163–203.

93. Sugg MJ, McDonald JE. Time course of inhibition in color-response and word-response versions of the Stroop task. *J Exp Psychol Hum Percept Perform* 1994;20:647–675.

94. May CP, Kane MJ, Hasher L. Determinants of negative priming. *Psychol Bull* 1995;118:35–54.

95. Kane MJ, May CP, Hasher L, et al. Dual mechanisms of negative priming. *J Exp Psychol Hum Percept Perform* 1997;23:632–650.

96. Fox E. Negative priming from ignored distractors in visual selection: a review. *Psychonom Bull Rev* 1995;2:145–173.

97. Logan GD. On the ability to inhibit thought and action: a user's guide to the stop signal paradigm. In: Dagenbach D, Carr TH, eds. *Inhibitory processes in attention, memory, and language.* San Diego, CA: Academic Press, 1994:189–239.

98. Zacks RT, Hasher L. Directed ignoring: inhibitory regulation of working memory. In: Dagenbach D, Carr TH, eds. *Inhibitory processes in attention, memory, and language.* San Diego, CA: Academic Press, 1994:241–264.

99. Nigg JT. Is ADHD a disinhibitory disorder? *Psychol Bull* 2001;127:571–598.

100. Hasher L, Zacks RT. Working memory, comprehension, and aging: a review and a new view. In: Bower GH, ed. *The psychology of learning and motivation, vol. 22.* San Diego: Academic Press, 1988:193–225.

101. Hasher L, Zacks RT, May CP. Inhibitory control, circadian arousal, and age. In: Gopher D, Koriat A, eds. *Attention and performance XVII.* Cambridge, MA: The MIT Press, 1999:653–675.

102. Banich MT. *Neuropsychology: the neural bases of mental function.* Boston: Houghton Mifflin Company, 1997.

103. Kimberg DY, D'Esposito M, Farah MJ. Frontal lobes: cognitive neuropsychological aspects. In: Feinberg TE, Farah MJ, eds. *Behavioral neurology and neuropsychology.* New York: McGraw-Hill, 1997:409–418.

104. Petersen SE, Fox PT, Posner MI, et al. Positron emission tomographic studies of the cortical anatomy of single-word processing. *Nature* 1988;331:585–589.

105. Posner MI, DiGirolamo GJ. Executive attention: conflict, target

detection, and cognitive control. In: Parasuraman, ed. *The attentive brain.* Cambridge, MA: The MIT Press, 1998:401–423.

106. Bush G, Luu P, Posner MI. Cognitive and emotional influences in anterior cingulate cortex. *Trends Cogn Sci* 2000;4:215–222.

107. Braver TS, Barch DM, Gray JR, et al. Anterior cingulate cortex and response conflict: effects of frequency, inhibition and errors. *Cereb Cortex* 2001;11:825–836.

108. Barch DM, Braver TS, Akbudak E, et al. Anterior cingulate cortex and response conflict: effects of response modality and processing domain. *Cereb Cortex* 2001;11:837–848.

109. Carter CS, Macdonald AM, Botvinick M, et al. Parsing executive processes: strategic vs. evaluative functions of the anterior cingulate cortex. *Proc Natl Acad Sci U S A* 2000;97:1944–1948.

110. MacLeod CM, MacDonald PA. Interdimensional interference in the Stroop effect: uncovering the cognitive and neural anatomy of attention. *Trends Cogn Sci* 2000;4:383–391.

111. Banich MT, Milham MP, Atchley R, et al. fMRI studies of Stroop tasks reveal unique roles of anterior and posterior brain systems in attentional selection. *J Cogn Neurosci* 2000;12:988–1000.

112. Miller EK. The neural basis of top-down control of visual attention in the prefrontal cortex. In: Monsell S, Driver J, eds. *Attention and performance XVIII.* Cambridge, MA: The MIT Press, 2000:511–534.

113. Goldman-Rakic PS. Circuitry of primate prefrontal cortex and regulation of behaviour by representational knowledge. In: Mountcastle VB, ed. *Handbook of physiology: the nervous system, vol. 5.* Baltimore: Williams & Wilkins, 1987:373–417.

114. Cohen JD, Perlstein WM, Braver TS, et al. Temporal dynamics of brain activation during a working memory task. *Nature* 1997;386:604–607.

115. Courtney SM, Ungerleider LG, Keil K, et al. Transient and sustained activity in a distributed neural system for human working memory. *Nature* 1997;386:608–611.

116. Fletcher PC, Henson RNA. Frontal lobes and human memory: insights from functional neuroimaging. *Brain* 2001;124:849–881.

117. Petrides M. Middorsolateral and midventrolateral prefrontal cortex: two levels of executive control for the processing of mnemonic information. In: Monsell S, Driver J, eds. *Attention and performance XVIII.* Cambridge, MA: The MIT Press, 2000:535–548.

118. Postle BR, Berger JS, D'Esposito M. Functional neuroanatomical double dissociation of mnemonic and executive control processes contributing to working memory performance. *Proc Natl Acad Sci U S A* 1999;96:12959–12964.

119. Awh E, Jonides J, Reuter-Lorenz PA. Rehearsal in spatial working memory. *J Exp Psychol Hum Percept Perform* 1998;24:780–790.

120. Awh E, Anllo-Vento L, Hillyard SA. The role of spatial selective attention in working memory for locations: evidence from event-related potentials. *J Cogn Neurosci* 2000;12:840–847.

121. Awh E, Jonides J. Overlapping mechanisms of attention and spatial working memory. *Trends Cogn Sci* 2001;5:119–126.

122. Garavan H, Ross TJ, Stein EA. Right hemispheric dominance of inhibitory control: an event-related functional MRI study. *Proc Natl Acad Sci U S A* 1999;96:8301–8306.

123. Konishi S, Nakajima K, Uchida I, et al. Common inhibitory mechanism in human inferior prefrontal cortex revealed by event-related functional MRI. *Brain* 1999;122:981–991.

124. Casey BJ, Trainor RJ, Orendi JL, et al. A developmental functional MRI study of prefrontal activation during performance of a go-no-go task. *J Cogn Neurosci* 1997;9:835–847.

125. Pliszka SR, Liotti M, Woldorff MG. Inhibitory control in children with attention-deficit/hyperactivity disorder: event-related potentials identify the processing component and timing of an impaired right-frontal response-inhibition mechanism. *Biol Psychiatry* 2000;48:238–246.

126. Nobre AC, Coull JT, Frith CD, et al. Orbitofrontal cortex is activated during breaches of expectation in tasks of visual attention. *Nat Neurosci* 1999;2:11–12.

127. Yerkes RM, Dodson JD. The relation of strength of stimulus to rapidity of habit-formation. *J Comp Neurol Psychol* 1908;18:459–482.

128. Eysenck MW. *Attention and arousal: cognition and performance.* New York: Springer-Verlag, 1982.

129. Hockey GRJ. Changes in operator efficiency as a function of environmental stress, fatigue, and circadian rhythms. In: Boff KR, Kaufman L, Thomas JP, eds. *Handbook of perception and human performance, vol. 11.* New York: John Wiley, 1985:44:1–49.

130. Stickgold R, Scott L, Rittenhouse C, et al. Sleep-induced changes in associative memory. *J Cogn Neurosci* 1999;11:182–193.

131. Parasuraman R. Vigilance, monitoring, and search. In: Boff KR, Kaufman L, Thomas JP, eds. *Handbook of perception and human performance, vol. 11.* New York: John Wiley, 1985:43:1–39.

132. Parasuraman R, Warm JS, See JE. Brain systems of vigilance. In: Parasuraman R, ed. *The attentive brain.* Cambridge, MA: MIT Press, 1998:221–256.

133. Robbins TW. Arousal and attention: psychopharmacological and neuropsychological studies in experimental animals. In: Parasuraman R, ed. *The attentive brain.* Cambridge, MA: MIT Press, 1998:189–220.

134. Halligan PW, David AS. Cognitive neuropsychiatry: towards a scientific psychopathology. *Nat Rev Neurosci* 2001;2:209–215.

135. Rafal R, Henik A. The neurology of inhibition: integrating controlled and automatic processes. In: Dagenbach D, Carr TH, eds. *Inhibitory processes in attention, memory, and language.* New York: Academic Press, 1994:1–51.

136. Robertson LC, Rafal R. Disorders of visual attention. In: Gazzaniga MS, ed. *The new cognitive neurosciences,* 2nd ed. Cambridge, MA: The MIT Press, 2000:633–649.

137. Robertson I, Marshall JC, eds. *Unilateral neglect: clinical and experimental studies.* Hillsdale, NJ: Erlbaum, 1993.

138. American Psychiatric Association. *Diagnostic and statistical manual of mental disorders,* 4th ed. Washington, DC, American Psychiatric Association, 1994.

139. Barkley RA. Behavioral inhibition, sustained attention, and executive functions: constructing a unifying theory of ADHD. *Psychol Bull* 1997;121:65–94.

140. Logan GD, Schachar RJ, Tannock R. Executive control problems in childhood psychopathology: stop signal studies of attention deficit hyperactivity disorder. In: Monsell S, Driver J, eds. *Attention and performance XVIII.* Cambridge, MA: The MIT Press, 2000:653–677.

141. Swanson J, Posner MI, Cantwell D, et al. Attention-deficit/hyperactivity disorder: symptom domains, cognitive processes, and neural networks. In: Parasuraman R, ed. *The attentive brain.* Cambridge, MA: The MIT Press, 1998:445–460.

142. Williams JMG, Watts FN, MacLeod C, et al. *Cognitive psychology and emotional disorders.* New York: John Wiley & Sons, 1988.

143. Williams JMG, Mathews A, MacLeod C. The emotional Stroop task and psychopathology. *Psychol Bull* 1996;120:3–24.

144. Simpson JR, Snyder AZ, Gusnard DA, et al. Emotion-induced changes in human medial prefrontal cortex: I. During cognitive task performance. *Proc Natl Acad Sci U S A* 2001;98:683–687.

145. Brown AL, DeLoache JS. Skills, plans, and self-regulation. In:

Siegler RS, ed. *Children's thinking: what develops?* New York: John Wiley, 1978:3–35.

146. Carr TH. Attention, skill, and intelligence: some speculations on extreme individual differences in human performance. In: Brooks P, McCauley C, Sperber RD, eds. *Learning, cognition, and mental retardation.* Hillsdale, NJ: Lawrence Erlbaum, 1984.

147. Helm-Estabrooks N. A "cognitive approach" to treatment of an aphasic patient. In: Helm-Estabrooks N, Holland AL, eds. *Approaches to the treatment of aphasia.* San Diego, CA: Singular Publishing Group, 1998.

148. Hinckley JJ, Carr TH. *Differential contributions of cognitive abilities to success in skill-based versus context-based aphasia treatment.* Boulder, CO: Academy of Aphasia, 2001.

149. van Mouirik M, Verschaeve M, Boon P, et al. Cognition in global aphasia: indicators for therapy. *Aphasiology* 1992;6:491–499.

18

MEMORY

DEANNA M. BARCH
RANDY L. BUCKNER

The topic of this chapter is memory, broadly defined. Memory is a complex, multidimensional set of functions with numerous distinct forms that interact to provide the rich and diverse cognitive capacities that we possess. In this chapter, a broad overview is provided regarding the relationships between particular forms of memory and their associated neural substrates. In doing so, we use some of the currently accepted distinctions between forms of memory, such as the distinction between episodic memory and working memory. However, by doing so, we do not wish to imply that there is always a complete dissociation between the cognitive processes and neural systems involved across memory forms. Instead, we hope to convey the growing evidence that putatively different memory forms often share common processing elements and that many brain regions are important for multiple aspects of memory function. Such interactions are, quite frankly, just beginning to be understood. It should also become clear through the course of this chapter that damage to specific brain regions often leads to deficits in multiple memory domains rather than to isolated disturbances in specific forms of memory.

SUBDIVISIONS OF MEMORY

Researchers and theorists have long attempted to subdivide the construct of memory into multiple forms. Such attempts have included distinctions between episodic and semantic memory (186), declarative and procedural memory (43,203), working (short-term) memory and long-term memory (8), and implicit and explicit memory (169). In an attempt to summarize this literature, Schacter and Tulving suggested that there was evidence of the existence of at least five different major memory systems in humans, including the perceptual representation, working memory, semantic memory, episodic memory, and procedural memory systems (172). Several elements of these different memory forms are shown in Figure 18.1. One broad, first-level distinction particularly relevant to this chapter is the distinction between working memory and long-term memory (4).

Working memory has been defined as the ability to temporarily maintain and manipulate information over time (8). Long-term memory is used to refer to a diverse set of memory forms, some of which may be more related than others. In general, however, the term long-term memory is used to refer to memory forms that allow information to be available over extended periods of time (9).

Each of these broad divisions of memory has been hypothesized to include further important subdivisions and component processes. For example, within the domain of working memory, Baddeley's (8) influential theory distinguishes among three subcomponents, including short-term storage buffers for visual (visuospatial scratch pad) and verbal (phonologic loop) information, and a central executive component that guides the manipulation and transformation of information held within the storage buffers. To illustrate, holding a phone number in mind while dialing would require the phonologic loop, whereas remembering the locations of checkers on a board would entail using the visuospatial scratch pad.

Long-term memory has also been further subdivided, with distinctions commonly made between episodic memory (events), semantic memory (facts), and procedural memory (skills) (172). Episodic memory refers to the ability to learn and retrieve memories about personal experiences that are situated in a specific time and place (186). Semantic memory includes an individual's general knowledge about the world, including facts, the meanings of words and concepts, and the relationships among concepts and ideas. Semantic memory differs from episodic memory in that the information is not necessarily tied to any specific learning episode or context associated with that episode, as is the case for episodic memory (186). Procedural memory refers to the acquisition of new motor and cognitive skills (i.e., roller skating, reading), the contents of which may or may not be consciously available to the individual (151–153). For example, while learning to ride a bike, procedural memory would allow skills associated with riding to be acquired; semantic memory would support the knowledge of bike parts and rules of riding, and episodic memory would be required to remember the specific time

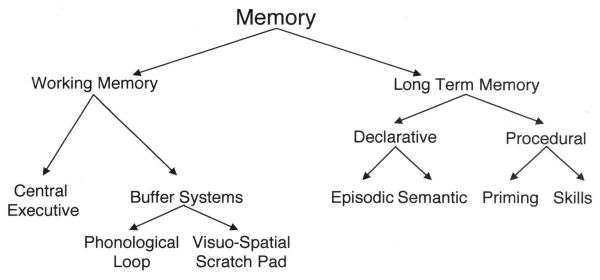

FIGURE 18.1. A schematic diagram represents a summary of different schemes for subdividing the construct of memory into dissociable subcomponents. (Developed in collaboration with Endel Tulving.)

and events associated with the first experience of trying to ride a bike.

The scheme for subdividing forms of memory outlined above is only one of several possible organizational systems suggested in the literature. In this review, the primary focus is on the functional neuroanatomy of working memory and two forms of long-term memory (semantic and episodic).

EPISODIC MEMORY

Episodic memory requires the coordination of a complex set of processes that involves the activity of multiple brain regions. Findings from both human neuropsychology research (i.e., lesion studies) and human functional neuroimaging research have identified multiple brain regions as being important for episodic memory. Historically, research has highlighted the contributions of the hippocampus and the surrounding medial temporal structures as being critical for the formation and retrieval of new episodic and semantic memories (44). One of the initial impetuses for the view that the hippocampal formation is critically involved in episodic memory is work with amnesic patients, such as H.M., who have had lesions to the hippocampus and/or surrounding medial temporal areas (173). After these lesions, such patients have profound deficits in the ability to learn and/or retrieve new episodic and semantic memories, despite relatively intact cognitive functioning in other domains (46,173,182). Nonhuman primate models of memory loss also demonstrate that lesions within the hippocampus and adjacent cortex (within the medial temporal lobes) result in an impaired ability to retrieve recently acquired information (118,203).

Several different theoretical models have been proposed to account for the role of the medial temporal cortex in memory formation. One theory is that the hippocampal formation is critical for the rapid binding of novel configurations of information and that this function is the basis of its role in episodic memory formation (44).

At the same time, a growing literature highlights the important contributions of prefrontal structures to episodic memory. Human neuropsychology research has shown that damage to the prefrontal cortex can also lead to episodic memory deficits, although episodic memory is typically not the only cognitive function impaired in these individuals. Instead, damage to the prefrontal cortex, depending on its location, can also lead to disturbances in language function and/or a variety of higher level cognitive processes, including planning, problem solving, inhibition, and working memory. Such results have led to the hypothesis that damage to prefrontal cortex impairs episodic memory by impairing strategic contributions to memory formation and retrieval (176), whereas hippocampal damage impairs the actual binding of information into new memories, a topic discussed below.

Research with individuals with Korsakoff syndrome has also brought to light the importance of diencephalic regions for memory function. Korsakoff syndrome is thought to result, at least in part, from a severe thiamine deficiency linked to the malnutrition that often occurs in chronic alcoholism. These individuals can display memory impairment that is as severe as those shown by individuals with bilateral medial temporal lesions (176,177). There are at least four major parts of the diencephalon, including the epithalamus, thalamus, hypothalamus, and subthalamus. Amnesia secondary to diencephalic damage (as is seen in Korsakoff syndrome) is most frequently associated with damage to the dorsomedial nucleus of the thalamus, the mamillary bodies of the hypothalamus, and the mamillothalamic white matter tract (54). Functional neuroimaging studies of episodic memory also often reveal activation in these regions, but to date their precise contributions to episodic memory are unclear.

The advantage of neuropsychological research with lesion patients is that it is particularly helpful in identifying brain regions that may be necessary to carry out particular cognitive functions. However, the fact that damage to a number of different brain regions can lead to impairment of episodic memory highlights the fact that the construct of episodic memory itself contains multiple subcomponents. One clear subdivision to be found within episodic memory is the distinction between encoding and retrieval. Unfortunately, it is more difficult to identify brain regions involved in encoding versus retrieval in research with lesion patients because such lesions are often diffuse, affecting multiple brain regions, and because behavioral tasks used to measure memory impairment simultaneously often depend on both encoding and retrieval processes. It is in this domain that human functional neuroimaging research has been particularly helpful. Although functional neuroimaging research is less able to determine which brain regions are *necessary* for cognitive task performance, it can help to determine whether activity in specific brain areas is present only during encoding or during retrieval or both.

Episodic Memory Encoding

Prefrontal Cortex Contributions

Human episodic encoding has been explored in many studies using a variety of materials, including both verbal and nonverbal stimuli. Many of these studies have targeted correlates of brain activity associated with the active attempt to remember words or sentences (commonly referred to as intentional encoding). In such studies, participants are explicitly told that a memory test will follow and are directly instructed to try to memorize the materials. Results obtained from both functional magnetic resonance imaging (fMRI) and positron emission tomography studies consistently demonstrate that specific regions within the frontal cortex are active when subjects intentionally memorize or encode stimuli such as words, sentences, and faces (65,99,111,122). The issue of domain specificity (i.e., verbal vs. nonverbal materials) within frontal regions is raised in a later section.

Although episodic encoding paradigms in the laboratory often involve participants being explicitly aware of the need to memorize stimuli, most instances of episodic memory formation in everyday life occur incidentally, without any specific intention to remember. A long history of research in cognitive psychology (48,87,146) has shown that episodic memories can form as a by-product of particular forms of information processing, independent of the individual's intent

to memorize the information. For example, words that are elaborated on in terms of their meaning and how they relate to other words and concepts in memory are better remembered than words processed in a shallow fashion in which only surface characteristics are examined—the well-known levels of processing effect (48,64) [but see also Fisher and Craik (64) for important caveats on this principle]. William James, in 1890, noted this observation elegantly by suggesting that "the more other facts a fact is associated with in the mind, the better possession of it our memory retains."

Functional neuroimaging research has demonstrated that prefrontal regions active during intentional memorization are also active during behavioral manipulations that incidentally encourage effective memory encoding through meaning-based elaboration, even when the participant is unaware of the need to encode. For example, when participants perform tasks that require meaning-based judgments of words (i.e., abstract/concrete, living/nonliving judgments), multiple regions within the left frontal cortex are activated. On subsequent surprise memory tests, the words on which these judgments were made are remembered, even though the participants made no direct attempt to memorize the words at the time of encoding (37,55,75,98,157,193). In contrast, when participants perform a task in which words are evaluated on surface level characteristics (i.e., are the letters uppercase or lowercase?), left frontal activity is reduced, and memory for the words is poor.

Further evidence of a link between left frontal activity and verbal encoding comes from neuroimaging studies that have examined neural activity at the time of encoding as a function of whether a particular item is subsequently remembered or forgotten. The idea for such investigations stemmed from early studies using electrical scalp recording techniques (60,130,158,192). These studies recorded event-related response potentials from subjects at the time of memorization and revealed differences for words that were later remembered compared with words that were later forgotten. Recent developments in fMRI methods [27,52,97,105,202; for a review, see Rosen (156)] have allowed similar phenomena to be examined with better spatial (anatomic) localization.

In many of the fMRI studies, participants performed tasks that encourage semantic processing of words without being told to expect a later memory test. A surprise recognition test was then administered, the results of which were used to identify the neural correlates during encoding of words that were remembered versus those that were forgotten. These studies have consistently demonstrated that regions of the frontal cortex are among those most strongly correlated with subsequent memory performance (2,10,22,23,37,102,125,194). For example, Figure 18.2 shows representative results from Wagner et al. (191), in which activity in the left prefrontal cortex along the ventral portion of the inferior frontal gyrus at the time of encoding was greater for words that were subsequently remembered than for those that were forgotten. Such findings provide compelling examples of a direct link

Sorting Based on Subsequent Memory Performance

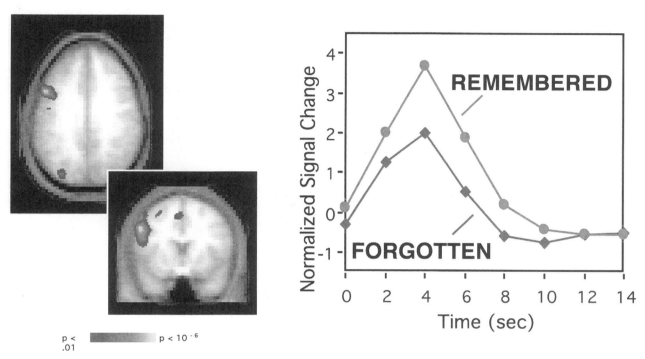

p <
.01   p < 10⁻⁶

FIGURE 18.2. A schematic diagram presents representative results of activity in the left frontal cortex as a function of whether a word was subsequently remembered or forgotten. The brain images display the location of activity in the left frontal cortex that was predictive of later memory performance. The time course (in seconds) shows the evolving functional magnetic resonance imaging signal over a 14-s period after the onset of the word during encoding. (Adapted from Buckner RL, Logan JM. Frontal contributions to episodic memory encoding in the young and elderly *(in press)*, with permission.)

between brain activity and behavior during episodic memory encoding. Of particular theoretical interest, the relationship between left frontal activity and subsequent memory for words has spanned a range of task contexts and even includes tasks involving episodic retrieval (37).

Domain Specificity of Prefrontal Cortex Contributions to Episodic Memory

As described above, activity within the frontal cortex is consistently associated with the encoding of verbal materials into episodic memory. However, the frontal cortex is not a homogeneous region. It contains numerous anatomically distinct areas as defined by changes in the distribution of cell types and density, interconnections to other brain areas, and physiologic properties (80). Multiple distinct areas have been identified in nonhuman primates and to a lesser extent in humans, based on such characteristics (13,24,40,140,147–149). Consistent with the hypothesis that the frontal cortex is a heterogeneous region, a growing body of research sug-

gests that the relationship between frontal cortex activity and episodic memory encoding is regionally specific. In particular, the existing research suggests at least two separate functional–anatomic dissociations in the frontal cortex that relate to episodic encoding. The first is dissociation between separate regions in the left frontal cortex and the second is dissociation between the left and right frontal cortex regions (35,135,136). Dissociation of left frontal cortex regions is discussed first.

Dissociation within Left Frontal Cortex Regions

In functional neuroimaging studies, the location of activity within the left frontal cortex associated with encoding has extended spatially from the dorsal extent of the inferior frontal gyrus near Brodmann areas (BA) 44 and 6 to more ventral and anterior regions, encompassing the classically defined Broca area and portions of the dorsolateral prefrontal cortex (near BA 45 and 47). Importantly, two distinct left frontal cortex regions have dissociated themselves functionally in a number of neuroimaging studies. The first region is located near the dorsal extent of the inferior frontal gyrus (BA 44/6), and the second is located more ventrally and anteriorly (near BA 44/45/47). We often refer to these two regions as the dorsal and ventral regions, respectively. However, these labels should not be taken to reflect specific anatomic distinctions (such as a relationship with the dorsolateral prefrontal cortex). Other researchers have labeled these same or similar regions as posterior and anterior (100,193). The topmost panel of Figure 18.3, which is adapted from Buckner and Logan (33), displays the approximate locations of these regions,

which have been dissociated by considering their divergent behavior across multiple task comparisons (35,36,143).

Several researchers have speculated on the possible functional roles of these two regions. One hypothesis is that the BA 44/6 region may provide access to lower-level, more generally utilized forms of representation, perhaps based on phonology or lexical access and perhaps interacting with parietal regions also participating in phonologic access. Support for this possibility comes from the observation that elaborate verbal processing tasks almost universally activate this region independent of whether the task demands require access to phonology or more elaborate meaning-based

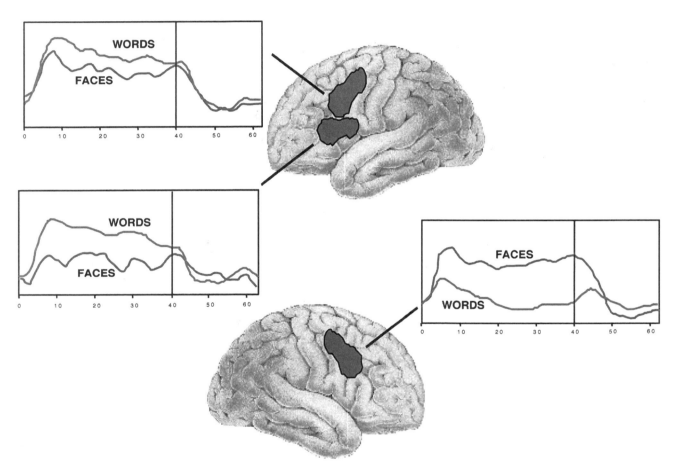

FIGURE 18.3. A schematic diagram heuristically represents the three frontal regions most consistently implicated in episodic memory encoding. Two separate left frontal regions, including Brodmann areas (BA) 44/6 and BA 44/6 regions, are plotted (top lateral view of brain). The third region is located in the right BA 44/6 (bottom lateral view of brain). Importantly, these regions dissociate functionally across encoding paradigms. One dissociation is shown by plotting the response over an encoding block for each region, separately for word and face encoding conditions. Each time course (in seconds) shows the evolving functional magnetic resonance imaging signal over a 40-s encoding epoch followed by a 24-s control period (encoding epoch ends and control period begins at the vertical line in each plot). Data were averaged over 272 separate encoding epochs. FACE and WORD time course data are plotted separately. Three important points are worth noting. First, there is increased involvement for WORD encoding in the left frontal regions and for FACE encoding in the right frontal regions. Second, within the left frontal cortex regions, the ventral (anterior) region is most selective for WORD encoding. Finally, for those regions showing little or no sustained responses (e.g., the word-encoding condition in the right frontal region), there is nonetheless a transient increase at the beginning and end of the epoch, perhaps reflecting an initial recruitment of the region. (See Color Figure 18.3 following page 526.)

processing (33). That is, to the extent that a word or wordlike representation is being extensively processed, the BA 44/6 region becomes active. Putatively nonverbal tasks have also activated (to a lesser degree) this BA 44/6 region, suggesting either an extremely general role of this frontal region in elaborate processing or the tendency of humans to incorporate verbal codes across almost all tasks. The BA 45/47 region may provide access to higher-level representations, perhaps based on meaning and related semantic associations, or selection among such representations (185). Evidence of this latter distinction comes from the finding that tasks requiring access to word meaning often activate BA 45/47 regions of the prefrontal cortex as well as the BA 44/6 regions, although it seems unlikely that the role of these more ventral regions will be exclusive to tasks tapping access to word meaning.

The relevance of the dissociation of these two regions to a discussion of episodic memory is that the BA 45/47 region appears to be more predictive of episodic encoding than the BA 44/6 region (33). In other words, in general, activity in the BA 45/47 region seems to be a better predictor of later memory performance. Considered from a functional perspective, this may directly relate to the possibility that the BA 45/47 region is required to access and/or manipulate representations associated with word meaning (36,100). As noted earlier, tasks requiring meaning-based elaboration are usually those most conducive to forming robust episodic memories.

Dissociation between the Left and Right Frontal Cortex Regions

The second prominent dissociation among frontal cortex regions relates to differences in encoding verbal and nonverbal materials. Cognitive theories have long suggested that memory formation relies on multiple kinds of information, with one important (albeit heuristic) distinction being between verbal and nonverbal codes. Behavioral studies have shown that a picture of an object, such as a lion, is more likely to be remembered than the presentation of the word "lion" (a finding known as the picture superiority effect). The implication is that pictures are associated with both nonverbal (image-based) and verbal codes, whereas words (particularly abstract words) are predominantly associated with just a verbal code (128,129). Moreover, patients with lateralized frontal lesions can show differences in memorization of different material types (115,155,199), suggesting code-specific regional specialization in the frontal cortex.

Several recent brain imaging studies have demonstrated that memorization of materials associated with different verbal and nonverbal codes can activate distinct regions of the left and right frontal cortex. As discussed above, the encoding of verbal materials such as words is associated with activation in specific left frontal cortex regions. By contrast, memorization of unfamiliar faces (101) and texture patterns (193), neither of which can be easily associated with a verbal label,

strongly activates the right frontal cortex regions near the homologue to the left BA 44/6 frontal region. The locations of the left and right frontal cortex regions showing dissociation between material types can be seen in Figure 18.3. Several other studies using both positron emission tomography and fMRI have also noted similar effects (17,107,110). However, it should be noted that the pattern of right versus left frontal cortex activity as a function of material type likely reflects material *specialization* rather than material *specificity*. In other words, although verbal materials tend to elicit greater left than right frontal cortex activity and nonverbal materials tend to elicit greater right than left frontal cortex activity, this dissociation is not absolute. For example, research often finds activity in the left frontal cortex during the encoding of nonverbal materials, even when such information is not easily verbalizable (17,101,111).

Medial Temporal Cortex Contributions to Episodic Memory Encoding

As noted earlier, neuropsychological research on memory has long focused on the role of the medial temporal cortex in semantic and episodic memory, with a particular focus on the hippocampus and surrounding parahippocampal gyrus. Human functional neuroimaging studies have sometimes yielded results consistent with the involvement of the medial temporal cortex regions in episodic encoding. However, many studies that would be expected to show a relationship between medial temporal activity and episodic encoding have not, making this a particularly perplexing area of neuroimaging memory research. In terms of positive results, many of the same studies that have identified regions of the frontal cortex as active during episodic encoding have also demonstrated activity in the hippocampal and parahippocampal gyrus regions during both intentional and incidental episodic encoding (101,102,110,125,194). Further, studies that have examined neural activity at the time of encoding as a function of whether information is subsequently remembered or forgotten have also highlighted the importance of medial temporal activity in mediating the acquisition of new information (23,102,125,194). For example, Kirchhoff et al. (102) found that activity in the hippocampus and parahippocampal gyrus at the time of encoding is greater for words subsequently remembered than for words subsequently forgotten. Recent work with depth electrodes implanted in humans undergoing surgery for epilepsy has provided further evidence of the specific involvement of the hippocampus in encoding, demonstrating that hippocampal activity at the time of encoding predicted subsequent memory for verbal stimuli (39).

As with work on the frontal cortex, research on medial temporal involvement in episodic memory encoding has provided some evidence of functional dissociation among medial temporal regions. One such proposed dissociation is similar to that proposed for the frontal cortex, namely, dissociation between the right and left medial temporal regions as a function of material type.

Dissociation between the Right and Left Medial Temporal Cortex

Early work with lesion patients suggested the possibility that the left medial temporal cortex was relatively more involved in the encoding of verbal materials and the right medial temporal cortex was relatively more involved in the encoding of nonverbal materials (114). Several subsequent studies with patients with unilateral lesions to either the left or right medial temporal cortex provide support for this hypothesis. These studies have demonstrated that patients with unilateral left medial temporal lobe lesions are relatively more impaired on tests of verbal learning and memory (134,168,179), whereas patients with unilateral right medial temporal lobe lesions are relatively more impaired on tests requiring memory for items that are more difficult to verbalize, such as visuospatial materials (47,63,94,95,120,141,168,179,180). However, it should be emphasized that, as with the frontal cortex, the links between left temporal lobe lesions and verbal memory deficits and right medial temporal lobe lesions and nonverbal memory deficits are often relative rather than absolute. In other words, research has demonstrated that right medial temporal lobe lesions can impair verbal episodic memory, although perhaps not to the same extent as left medial temporal lobe lesions (57,165). Similarly, some studies have found that left medial temporal lobe lesions can impair nonverbal episodic memory, again although perhaps not to the same extent as right medial temporal lobe lesions (106).

Episodic Memory Retrieval

Prefrontal Cortex Contributions

As with episodic memory encoding, human episodic memory retrieval has been explored in many studies using a variety of materials, again including both verbal and nonverbal stimuli. Similar to studies of episodic memory, retrieval studies consistently engage the activity of a number of different prefrontal brain regions. In particular, the same BA 44/6 and BA 45/47 frontal regions activated by episodic memory encoding are activated by episodic memory retrieval, with similar lateralization as a function of material type (i.e., verbal vs. nonverbal) (17,111). Such results suggest that the processes supported by these regions of the frontal cortex are not specific to encoding per se, but rather reflect engaging the functions necessary for multiple aspects of memory performance. However, studies of episodic memory retrieval often activate regions of the frontal cortex not typically activated by studies targeting encoding. In particular, the more anterior and/or superior regions of the dorsolateral prefrontal cortex (BA 46/9) and regions of the frontal polar cortex (BA 10, with a tendency

to be right lateralized) are often activated by episodic memory retrieval, but not by episodic memory encoding (3,17, 25,29,31,32,34,65,84,110,112,160,170,183,188,191).

Several different explanations have been put forth regarding the functional significance of activity in the frontal polar cortex during memory retrieval. For example, it has been suggested that frontal polar activation may represent the "set" or goal of attempting to retrieve past experiences, sometimes referred to as retrieval mode (30,113,121,188). Alternatively, it has also been suggested that the level of effort required during retrieval tasks modulate the level of activity in the frontal polar cortex (170). At a more specific level, others have suggested that activity in the frontal polar cortex may reflect successful recognition of items (112,159,160). Interestingly, the frontal polar cortex also appears to be reliably activated during planning, problem-solving, and reasoning tasks. For example, Baker et al. (11) observed frontal polar–prefrontal cortex activity in the Tower of London paradigm selectively under conditions that involved extensive planning. Such findings have led to the hypothesis that the frontal polar cortex may be more broadly involved in the monitoring of internally (vs. externally) generated information (42) or the maintenance of primary task goals while simultaneously allocating attention to subgoals (18,103). As should be clear by this discussion, despite the consistent findings of frontal polar activity in episodic memory retrieval, its precise role in human cognitive function is still unclear.

The findings of BA 46/9 activity during episodic memory retrieval has served to confirm findings from the neuropsychological literature highlighting the influence of damage in this region on episodic memory performance. However, individuals with circumscribed lesions of the frontal cortex are not grossly amnesic because they often score quite well on standardized measures of memory function (177). Nonetheless, their memory impairment can be detected on more sensitive tests of new learning. Further, damage to BA 46/9 does not selectively impair episodic memory but instead can impair a range of higher cognitive processes, including working memory, planning, and problem solving (176). In addition, the BA 46/9 regions activated in episodic memory retrieval tasks are often the same as those found in studies of working memory and planning (17,29). Taken together, such findings have led to the hypothesis that damage to BA 46/9 impairs episodic memory via an impact on the use of strategies that can enhance memory retrieval (176,177).

Consistent with this hypothesis, several studies have demonstrated that patients with damage to BA 46/9 are clearly impaired on episodic memory tasks requiring free recall of information (i.e., spontaneous generation of studied items) but are much less impaired (or even unimpaired) on tasks that simply require them to recognize whether an item was previously studied (177). In contrast, individuals with damage to the medial temporal cortex are often impaired on both recall and recognition measures (203). This difference in memory performance as a function of task requirements has been explained in terms of the difference between the operation of familiarity versus explicit recollective processes. Familiarity refers to the ability to evaluate the contextual memory strength of an item and does not necessarily involve specific access to the episode in which the item was learned (90,187). In contrast, explicit recollection refers to the ability to access a specific memory of the learning episode and is thought to be much more influenced by the use of strategies at the time of retrieval as well as at the time of encoding (90,91,93,187). Thus, individuals with damage to BA 46/9 may be impaired on tasks requiring explicit recollection, in part because they have difficulties spontaneously using strategies that aid memory formation and retrieval (59,184). Patients with damage to BA 46/9 are also impaired on tasks that specifically require them to encode and/or retrieve the source of information (93) or the order in which the information was presented (177), deficits that may result from, or contribute to, impairment of explicit recollective processes.

In line with the hypothesis that BA 46/9 damage impairs strategy use, Gershberg and Shimamura (77) demonstrated that patients with damage to BA 46/9 are impaired in the ability to spontaneously use subjective organizational strategies and semantic clustering strategies during episodic memory tasks. Further, patients with lesions to BA 46/9 benefit from instruction in the use of strategies, both at the time of encoding and at the time of retrieval (77,89). Individuals with other neurologic and psychiatric disorders (i.e., schizophrenia, Parkinson disease, obsessive-compulsive disorder) also thought to involve frontal lobe dysfunction have been found to show similar patterns of strategic memory impairment (88,104,142,166,174). Recent fMRI work pro-vides further support for the involvement of the dorsolateral prefrontal cortex in strategy use, demonstrating that activity in BA 46/9 can be elicited during encoding and retrieval if participants are engaging in strategies such as semantic clustering at the time of encoding (67,167).

Medial Temporal Cortex Contributions

As with the prefrontal cortex regions, studies of episodic memory retrieval often activate regions of the medial temporal cortex similar to those identified in studies of episodic memory encoding. However, a growing number of studies in this area has been devoted to trying to tease apart the specific contributions of different areas of the medial temporal lobe to memory formation and retrieval. As shown in Figure 18.4, the medial temporal cortex contains a number of anatomically separate regions that maintain hierarchical relationships with one another and receive convergent input from separate regions of the cortex. Lesions in many medial temporal lobe patients are relatively large, including both the hippocampus proper and surrounding entorhinal, perirhinal, and parahippocampal cortex, making it difficult to determine the specific roles that each of these regions plays in episodic memory. Nonetheless, some research has suggested that lesions restricted solely to the hippocampus in humans lead to moderate memory impairment, which primarily involves anterograde amnesia (inability to form new memories). More extensive damage that includes adjacent entorhinal and parahippocampal regions leads to much more severe impairment (203), which can include both anterograde and retrograde (loss of old memories) amnesia. Work with nonhuman primates provides a similar picture,

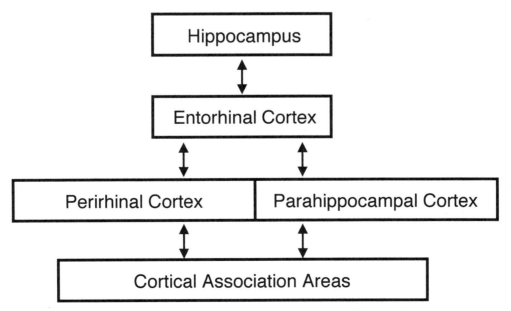

FIGURE 18.4. A schematic diagram outlines the major subdivisions of the medial temporal cortex and their connections with each other and with other brain regions.

demonstrating that lesions restricted to the hippocampus proper lead to moderate memory impairment, but lesions including the surrounding cortex lead to much more severe memory impairment (203). The anatomic organization of structures within the medial temporal cortex is such that one might expect such a specialized role for the hippocampus versus other medial temporal structures in memory function. At a simplified level, as shown in Figure 18.4, structures such as the parahippocampal and perirhinal cortex receive segregated inputs from distributed regions of the cortex and then send projections that converge on the hippocampus, resulting in a hierarchical organization of medial temporal structures.

As one means of making sense of the relationship between the extent of medial temporal cortex damage and the extent of memory impairment, Aggleton and Shaw (1) suggested that damage to the hippocampus proper leads to impairment of explicit recollection of episodic information but does not impair the use of more familiarity-based processes. In contrast, damage that includes the cortical structures surrounding the hippocampus (i.e., entorhinal cortex, parahippocampal gyrus) may impair both familiarity-based processing and conscious recollection. Aggleton and Shaw based this hypothesis on a review of 33 studies that examined recognition memory in patients with amnesia owing to a variety of different types of brain damage. Their review found that individuals whose lesions were restricted to the hippocampus had relatively intact recognition memory performance compared with recall performance. In contrast, individuals whose damage extended beyond the hippocampus tended to be impaired on recognition as well as recall performance. However, more recent work by Reed and Squire (154) calls this conclusion into question. They examined a somewhat larger group (six subjects compared with three in the study by Aggleton and Shaw) of individuals with lesions restricted to the hippocampus and found clear evidence of impairment of recognition and recall in these individuals. Further, research in the cognitive psychology domain has pointed out that one cannot necessarily equate performance of recall tasks with conscious recollection and performance of recognition tasks with familiarity, in that both recollection and familiarity can contribute to recognition memory task performance (90,201).

Despite this ongoing debate in the neuropsychology literature, recent event-related fMRI research provides evidence consistent with the hypothesis that the hippocampus may be particularly important for the explicit recollection. Eldridge et al. (58) examined neural activity at the time of memory retrieval as a function of whether the participants reported remembering or knowing that they had seen the item before. Participants were told to give a remember response (R) if their memory for the item was based on a distinct recollection of having seen the word at the time of encoding and to give a know response (K) if they had a feeling of familiarity that was not accompanied by explicit recollection of the learning episode (58). Activity in the left hippocampus was

significantly greater during retrieval of R responses compared with K responses, whereas activity to K responses did not differ from activity to correct rejections. A similar pattern was found in the right hippocampus, although the pattern was not as strong as in the left hippocampus. In contrast, regions of the parahippocampal gyrus were active in both R and K responses compared with correct rejections. These results provide support to the hypothesis that the hippocampus is more involved in explicit episodic memory recognition than in familiarity-based processes, but adjacent parahippocampal regions may also be important for familiarity-based processes. Clearly, however, further research is needed to tease apart the differential contributions of the different medial temporal regions to episodic memory function and its relationship with other forms of memory.

Relationship between Frontal and Medial Temporal Cortex Contributions to Episodic Memory

This review of the functional anatomy of episodic memory has attempted to summarize research demonstrating that multiple brain regions are critical for episodic memory function, including both the frontal and medial temporal cortex. Up to this point, however, we have not addressed the question of how the frontal cortex interacts with the medial temporal cortex to support episodic memory encoding and retrieval. We believe that the evidence to date suggests that the frontal cortex provides a source of information (an input or some form of modulatory influence) to medial temporal lobe structures (26,28,100,116) that serves to guide both memory formation and memory retrieval. This idea fits well with the hypothesis that medial temporal lobe structures (including the hippocampus and adjacent cortex) play a role in the integration and cohesion (binding) of incoming information to form memories (45,109,117,171). The frontal cortex may provide critical input to these medial temporal cortex structures, supplying the necessary "ingredients" that must be bound together to form an enduring episodic memory and providing important contextual cues that can help to facilitate memory retrieval. This idea is not new, and in fact other researchers using different methodologies have come to similar conclusions. For example, Squire (182) suggests that the "frontal cortex presumably performs its computations on many kinds of information, which are analyzed concurrently for other purposes by other regions of cortex. Frontal cortex allows information to be remembered in its appropriate context, that is, in the correct temporal coincident event. The medial temporal region then operates upon this information, allowing it to endure in the organized form it has achieved in neocortex." Moscovitch (117) suggests a related idea that the frontal lobes "are prototypical organization structures crucial for selecting and implementing encoding strategies that organize the input to the hippocampal component." Thus, both the frontal and medial temporal cortex regions are important to the formation of episodic memories, and both the frontal and medial temporal cortex regions are critical for the successful retrieval of episodic memories, although potentially not for the same reasons.

WORKING MEMORY

As described earlier in this chapter, working memory is typically defined as the ability to temporarily maintain and manipulate information over a short period of time. Over time, the construct of working memory has evolved to encompass earlier definitions of short-term memory and to describe the interactions between processes that support the maintenance of information and those that operate on the maintained information. As with episodic memory, the construct of working memory contains several different subcomponents, each of which may map onto the function of different neural systems. For example, Baddeley's (8) model of working memory contains at least three different subcomponents, including two content-specific buffer systems and a central executive system, defined as a limited-capacity attention system that coordinates the activities of the phonologic loop and visuospatial scratch pad and operates on the contents of these systems. The buffer systems are thought to include a phonologic loop system that subserves the rehearsal of verbal information (similar to earlier concepts of verbal short-term memory) and a visuospatial scratch pad that supports nonverbal maintenance. As with episodic memory, working memory is clearly dependent on the coordinated activity of multiple brain regions. However, research from multiple domains, including neuropsychology, nonhuman primates, and human functional neuroimaging have implicated the frontal cortex as an area of the brain important for working memory function, although the precise functions supported by different regions of the prefrontal cortex and how they specifically contribute to working memory are an area of ongoing debate.

In particular, at least two potential means of functional dissociating regions of the frontal cortex involved in working memory have been proposed. One means of dissociation is based on process (storage/maintenance processes versus executive/manipulation processes) and is analogous to the distinction between the central executive and buffer systems. Another means of dissociation is based on the content of the information to be maintained (i.e., verbal, spatial, object), which can be thought of as analogous to a distinction between the phonologic loop and the visuospatial scratch pad.

Working Memory Process Dissociations in the Prefrontal Cortex

The hypothesis that regions of the prefrontal cortex, including the dorsolateral prefrontal cortex (BA 46/9) may be involved in the maintenance of information in working memory stems, in part, from research using nonhu-

man primate models of working memory. This literature has shown that circumscribed lesions in the region of the principal sulcus (argued to be nonhuman primate homologous of BA 9/46 in humans) can impair delayed response task performance in monkeys (69,73). Further, single-unit recordings demonstrate sustained neural firing in neurons in the principal sulcus, arcuate sulcus, and lateral convexity during the delay period of such delayed-response tasks in monkeys (68,70,71,74,124,200). Based on such findings, it has been argued that prefrontal cortex supports working memory by actively holding information "on-line" through maintained neural activity and that delay-period activity in the prefrontal cortex serves as the "cellular basis of working memory" (79). This view of prefrontal cortex involvement in working memory suggests that regions such as the dorsolateral prefrontal cortex support the processes carried out by the buffer system component of working memory rather than the central executive system, at least according to Baddeley's formulation.

In contrast, the neuropsychological literature on the influence of BA 46/9 lesions on working memory and executive control has emphasized the role of BA 46/9 in the central executive components of working memory rather than the mnemonic processes supported by the buffer systems (7,82,137). For example, a recent review by D'Esposito and Postle (50) found minimal evidence that lesions to BA 46/9 impair performance on span tasks (i.e., forward digit span, spatial span). Such tasks are thought to rely primarily on the function of the phonologic loop or the visuospatial scratch pad and to require little or no involvement of the central executive system. In contrast, they found more evidence that BA 46/9 lesions impaired performance on delayed-response tasks, particularly when the delay period was filled with distracting information. The inclusion of distraction during the delay period is likely to elicit executive functions such as interference control (41), and thus impairment in such tasks is consistent with the hypothesis that BA 46/9 plays a role in the executive components of working memory.

However, delayed-response tasks without distraction seem more dependent on maintenance functions as opposed to executive functions. Thus, the fact that individuals with BA 46/9 lesions can show impairment in such tasks leaves open the possibility that BA 46/9 regions do play a role in maintenance functions as well as executive control functions. Interestingly, individuals with neurologic or psychiatric disorders also thought to involve BA 46/9 show a similar pattern of performance on different working memory tasks. For example, patients with schizophrenia show relatively little impairment on span tasks (especially at subspan lengths) (161) but do show impairment on delayed-response tasks (with and without distraction) (131,132) as well as more complex working memory tasks (15,78).

The debate regarding which regions of the prefrontal cortex support maintenance versus executive control functions extends to the functional neuroimaging and nonhuman primate literature as well as the neuropsychology literature.

For example, Petrides (138,139) suggests that ventral (BA 44/45/47) and dorsal (BA 46/9) regions of the prefrontal cortex may differ in their relative involvement in maintenance versus executive control processes. Specifically, Petrides postulates a two-stage hierarchical model of lateral prefrontal cortex function. In Petrides' model, ventral (BA 44/45/47) regions of the prefrontal cortex perform simple executive- and maintenance-related processes from input from the posterior cortical regions as well as help to select and retrieve information from short- and long-term memory. In contrast, Petrides argues that the dorsal (BA 46/9) prefrontal cortex regions operate on input from these ventral regions and other cortical regions and are able to perform more complex executive operations, such as monitoring and manipulating the contents of information maintained in ventral regions. Animal lesion studies have provided some support for Petrides' model (138), and the human functional neuroimaging literature has also produced evidence of a similar division between ventral (BA 44/45/47) and dorsal prefrontal (BA 46/9) cortex regions (49,51,126,127,144).

In the human neuroimaging literature, the distinction between the operation of the buffer storage systems (phonologic loop, scratch pad) and the central executive system has been characterized as a distinction between maintenance and manipulation. A review by D'Esposito et al. (49) of neuroimaging studies using a variety of working memory tasks provided support for the hypothesis that a distinction between maintenance and manipulation corresponded to a division between BA 46/9 and BA 44/45/47 prefrontal cortex activity during working memory tasks. In this review, tasks thought to primarily involve maintenance (i.e., delayed-response type tasks) were associated with activation in BA 44/45/47 but not BA 46/9, whereas those involving both maintenance and manipulation (i.e., self-ordered pointing, "N-back" type tasks) activated both BA 44/45/47 and BA 46/9 regions of the prefrontal cortex. Of interest, the BA 44/45/47 regions associated with maintenance tasks in the working memory literature are essentially the same prefrontal cortex regions identified in studies of episodic memory encoding (both incidental and intentional) and retrieval. As such, although these regions may be engaged by the need to maintain information across time, their involvement in incidental episodic memory encoding paradigms suggests that the maintenance of information is not necessary to activate these regions, which is also consistent with the single-unit literature on monkeys.

Several event-related fMRI studies have since been conducted to test explicitly hypotheses regarding functional distinctions between BA 44/45/47 and BA 46/9 regions of the prefrontal cortex in working memory (21,51,144,162,163). Most of these studies have demonstrated that activity in BA 44/45/47 is reliably modulated by maintenance demands but not by manipulation demands [for an alternative view, see Braver and Speer (21)]. However, these studies suggest that BA 46/9 activity may be modulated by both mainte-

nance and manipulation demands. Additional research suggests that the degree to which dorsolateral prefrontal cortex activity is elicited by maintenance demands may depend on factors such as the amount of information to be maintained, i.e., high memory loads may necessitate the use of chunking strategies (21,66,76), and the nature of the information to be maintained, e.g., memory for contextual information may be more dependent on BA 46/9 function than memory for stimulus identity (14,19,20).

Working Memory Content Dissociations in the Prefrontal Cortex

The hypothesis that there may be dissociable working memory subsystems for different types of information stems from multiple sources. One source is behavioral studies demonstrating that in dual-task paradigms, verbal secondary tasks are much more likely to interfere with primary working memory tasks that are verbal rather than nonverbal, whereas nonverbal secondary tasks show the opposite pattern (8,119). Such results contribute to the hypothesis that there may be separate verbal and nonverbal subsystems within working memory. Another source of support comes from animal neurophysiology research, suggesting regional specificity in delay-related neuronal activity in the prefrontal cortex as a function of stimulus content. Goldman-Rakic (79,80) has made the strongest arguments for this type of material-specific organization scheme in the prefrontal cortex based on data from single-cell recordings in nonhuman primates. For example, Goldman-Rakic and colleagues (200) showed that neurons in the ventral prefrontal cortex demonstrate greater sensitivity to object identity than to spatial location, whereas neurons in the dorsolateral prefrontal cortex show greater sensitivity to spatial location than to object identity (200). Further, even within the dorsolateral prefrontal cortex, Goldman-Rakic and colleagues (69) demonstrated that focal lesions can lead to "memory scotomas" causing selective impairment of delayed-response performance with specific spatial locations. Based on such data, Goldman-Rakic (81) postulates a multiple-domain model in which object and spatial dimensions of working memory are one possible functional subdivision found within the prefrontal cortex. This hypothesis has appeal in that it parallels a similar distinction made regarding dorsal and ventral processing streams in posterior brain regions, referred to as the "what" (ventral stream, object identity) and "where" (dorsal stream, spatial location) pathways (189).

Such hypotheses regarding material-specific subdivisions within working memory have inspired a large body of research examining such distinctions in humans. This research fairly consistently demonstrates evidence of material-type specialization related to working memory function in the ventrolateral (BA 44/45/47) regions of the prefrontal cortex. In particular, multiple studies have demonstrated greater activity in the left BA 44/45/47 prefrontal regions during verbal working memory tasks but greater activity in the right BA 44/45/47 prefrontal regions during visuospatial working memory tasks (17,178). This body of findings was recently reviewed by D'Esposito et al., who concluded that the research to date provides consistent support for material-specialized activity in the BA 44/45/47 regions of the prefrontal cortex during working memory performance (49). As hinted at above, these regions showing material-sensitive activity during the performance of working memory tasks are similar to those showing material sensitivity during the performance of episodic memory encoding and retrieval tasks (17), again suggesting that the cognitive functions supported by these regions are not selective to any one memory domain but rather may provide more general processing resources that are adapted to multiple memory forms.

Despite the evidence of material-sensitive activation patterns in the BA 44/45/47 regions of the prefrontal cortex, there has been relatively little evidence of differences in the location of BA 9/46 activity during working memory performance as a function of material type (49,123,145). Instead, numerous studies have suggested that similar regions of BA 46/9 are activated by verbal, spatial, and object working memory tasks, with activity typically either bilateral or right lateralized (17,38,49). Consistent with the results from neuroimaging studies, work in patients with lesions to BA 46/9 suggests that lesions to either the right or left hemisphere can impair performance of visuospatial working memory tasks (12). Recent work in nonhuman primates also calls into question the issue of material specificity in BA 46/9. Specifically, Rao et al. (150) showed that neuronal activity in the dorsolateral and ventrolateral prefrontal cortex can be sensitive to both spatial and object characteristics during the performance of working memory tasks, if successful performance of the task requires attention to both spatial and object characteristics of the stimulus. In many ways, a lack of material-selective processing in BA 46/9 is consistent with a role for this region in more executive or control components of working memory because such processes are likely to be engaged by many or all types of material (i.e., words, faces, objects, spatial locations).

In summary, the literature on prefrontal cortex involvement in working memory function suggests that BA 44/45/47 regions are involved in the maintenance of information in working memory and demonstrate patterns of material-sensitive activity that parallel those found during performance of episodic memory tasks. In contrast, BA 46/9 regions show relatively little evidence of material-sensitive patterns of activity and appear to be more involved in the executive components of working memory. The involvement of BA 46/9 regions in executive components of working memory is consistent with the evidence demonstrating that these same regions are also important for mediating strategic aspects of episodic memory processing. Nonetheless, further research is clearly needed to isolate the specific cognitive functions supported by BA 46/9 regions of the prefrontal cortex and to determine what principles guide functional organization in these regions if material type is not the determining factor.

Nonfrontal Cortex Contributions to Working Memory

The vast majority of research on the correlates of working memory function has focused on the role of the prefrontal cortex. However, as with episodic memory, neuropsychological research and functional neuroimaging studies have demonstrated that regions outside the prefrontal cortex also play important roles in working memory, including the basal ganglia, thalamus, and parietal cortex. For example, almost all neuroimaging studies of working memory find activation in the parietal cortex, both with verbal and nonverbal materials.

One hypothesis about the role of the parietal cortex in working memory is that the parietal cortex in the language-dominant hemisphere (typically the left hemisphere in the region of the supramarginal and angular gyri) is the anatomic locus of the phonologic storage component of the phonologic loop (16,96,133) as opposed to rehearsal components (associated more with the inferior frontal cortex) (5,6). As such, it seems possible that the frontal regions near BA 6 may interact with the parietal cortex to subserve Baddeley's phonologic loop. Further, it has also been argued that the right parietal cortex may play a role in visuospatial rehearsal processes via its involvement in visual-selective attention (6). Consistent with these hypotheses, neuropsychological research has shown that lesions of the left inferior parietal cortex can impair working memory tasks that tap phonologic storage and/or rehearsal (span performance, in particular) (164,175,190,196,197). In contrast, lesions of the right inferior parietal cortex can impair nonverbal working memory tasks that tap visuospatial storage and/or rehearsal (56,83).

SEMANTIC MEMORY

As described at the beginning of the chapter, semantic memory refers to an individual's general knowledge about the world, including facts, the meanings of words and concepts, and the relationships among concepts and ideas. Some theorists have argued that the critical difference between semantic memory and episodic memory is that the semantic knowledge is not necessarily tied to any specific learning experience, as it often is in episodic memory (186). As with episodic memory, semantic memory impairment is often associated with lesions of the medial temporal cortex, in that damage to the hippocampal formation can lead to impairment of the ability to learn new semantic as well as new episodic information. However, selective impairment of the retrieval of already formed semantic memories, compared with episodic memory, is typically *not* associated with selective damage to the hippocampus *per se* but rather to other regions of the temporal cortex. Warrington (195,198) presented the first research on individuals who demonstrate selective semantic memory impairment (i.e., confrontation naming, word-picture matching, tasks requiring individuals to access knowledge about the attributes and use of specific objects) in the context of relatively preserved functions in other areas of language and cognition. More recently, disorders of this type have been referred to as semantic dementia (181). A clear assessment of the intactness of episodic memory is often difficult in these individuals, in part because the semantic impairment makes it difficult for them to process words and pictures appropriately. However, some have suggested that there is evidence of preserved autobiographical memory in such individuals (86). Such selective impairment of semantic memory has been found in individuals who have recovered from herpes encephalitis (198) and in individuals with Pick disease (61). Individuals with late-stage Alzheimer disease can also demonstrate impairment of semantic memory, but these individuals also clearly show impairments of many other aspects of memory, including episodic and working memory.

Interestingly, several reports suggest the presence of category-specific semantic memory impairment in some individuals, such as more impairment in naming living things than nonliving things (85). Such findings have been hypothesized to reflect separate memory systems/mechanisms for the recognition of living versus nonliving things (62). Consistent with this hypothesis, functional neuroimaging studies have revealed differences in the anatomic location of brain activity during the naming of pictures of living things (e.g., animals) versus nonliving things (e.g., tools) (108). For example, naming pictures of animals was found to produce greater activation in the occipital cortex, whereas naming tools produced greater activation in the left premotor and left medial temporal cortex (108). Somewhat similar, although not identical, results have been found by Damasio et al. (53). Retrieval of other types of semantic information (i.e., color information, action information, face) has also been found to elicit activity in neuroanatomically distinct regions of the temporal cortex, often in regions close to the temporal cortex regions associated with the initial perception of such attributes (172).

The data on differential deficits in semantic knowledge about living versus nonliving objects, combined with the functional neuroimaging data suggesting differences across domains of semantic knowledge, have been used to support the hypothesis that semantic information about different categories of objects and events is stored in different cortical regions. However, this view has been criticized on the grounds that putative dissociations in the ability to name living versus nonliving objects may simply reflect different degrees of naming difficulty, with the stimulus materials often used to assess living objects being more difficult than those used to assess nonliving objects (i.e., lower frequency words, less familiar objects, more visually complex) (72). Farah et al. (62) argue against this view and provide support for the existence of a living/nonliving dissociation with stimulus materials of equal difficulty. Such findings, together with the functional neuroimaging research described above, suggest that there may be important differences in the location and/or functions of the brain regions supporting semantic memory for different types of information. Further research will be needed to resolve these debates.

SUMMARY

The explosion of functional neuroimaging research in humans over the past 20 years, combined with neuropsychological and nonhuman primate research, has vastly increased the amount of information that we have about the relationships between different aspects of memory function and their associated neurobiologic systems. As such, we have a growing understanding of the expected similarities and differences in the profiles of memory impairment demonstrated by individuals with lesions to different areas of the human brain. At the same time, research on the functional anatomy of memory continues to demonstrate that there are few, if any, one-to-one mappings between our current concepts of memory forms and the function of any particular brain region. Instead, it is clear that many putatively different subcomponents of memory share common processing elements and are supported by the overlapping brain regions. Nonetheless, the recognition that humans have access to multiple forms of memory has helped to make sense of the varying profiles of spared and intact cognitive function that can occur with diseases and lesions that can influence the human brain.

ACKNOWLEDGMENT

This work was supported by National Institute of Mental Health grants MH60887 and MH57506, National Institute of Aging grant AG05681, and a James S. McDonnell Foundation Program in Cognitive Neuroscience grant (99-63/9900003).

REFERENCES

1. Aggleton JP, Shaw C. Amnesia and recognition memory: a re-analysis of psychometric data. *Neuropsychologia* 1996;34:51–62.
2. Alkire MT, Haier RJ, Fallon JH, et al. Hippocampal, but not amygdala, activity at encoding correlates with long-term, free recall of nonemotional information. *Proc Natl Acad Sci U S A* 1998;95:14506–14510.
3. Andreasen NC, O'Leary DS, Arndt S, et al. Short-term and long-term verbal memory: a positron emission tomography study. *Proc Natl Acad Sci U S A* 1995;92:5111–5115.

4. Atkinson RC, Shiffrin RM. Human memory: a proposed system and its control processes. In: Spence KW, ed. *The psychology of learning and motivation: advances in research and theory.* New York: Academic Press, 1968:89–195.

5. Awh E, Jonides J, Smith EE, et al. Dissociation of storage and rehearsal in verbal working memory: evidence from PET. Unpublished paper, University of Michigan, Ann Arbor, Michigan, 1995.

6. Awh E, Smith EE, Jonides J. Human rehearsal processes and the frontal lobes: PET evidence. *Ann N Y Acad Sci* 1995;769:97–117.

7. Baddeley A, Della Sala S, Papagno C, et al. Dual-task performance in dysexecutive and nondysexecutive patients with frontal lesion. *Neuropsychology* 1997;11:187–194.

8. Baddeley AD. *Working memory.* New York: Oxford University Press, 1986.

9. Baddeley AD. *Essentials of human memory.* Hove, England: Psychology Press, 1999.

10. Baker JT, Sanders AL, Maccotta L, et al. Neural correlates of verbal memory encoding during semantic and structural processing tasks. *Neuroreport* 2001;12:1251–1256.

11. Baker SC, Rogers RD, Owen AM, et al. Neural systems engaged by planning: a PET study of the Tower of London Task. *Neuropsychologia* 1996;34:515–526.

12. Baldo JV, Shimamura AP. Spatial and color working memory in patients with lateral prefrontal cortex lesions. *Psychobiology* 2000;28:156–167.

13. Barbas H, Pandya DN. Architecture and intrinsic connections of the prefrontal cortex in the rhesus monkey. *J Comp Neurol* 1989;286:353–375.

14. Barch DM, Braver TS, Nystom LE, et al. Dissociating working memory from task difficulty in human prefrontal cortex. *Neuropsychologia* 1997;35:1373–1380.

15. Barch DM, Csernansky J, Conturo T, et al. Working and long-term memory deficits in schizophrenia. Is there a common underlying prefrontal mechanism? *J Abnorm Psych* 2002; 111(3):478–494.

16. Becker JT, MacAndrew DK, Fiez JA. A comment on the functional localization of the phonological storage subsystem of working memory. *Brain Cogn* 1999;41:27–38.

17. Braver TS, Barch DM, Kelley WM, et al. Direct comparison of prefrontal cortex regions engaged by working and long-term memory tasks. *Neuroimage* 2001;14:48–59.

18. Braver TS, Bongiolatti SR. The role of frontopolar prefrontal cortex in subgoal processing during working memory. *Neuroimage* 2002;15(3):523–536.

19. Braver TS, Cohen JD. On the control of control: the role of dopamine in regulating prefrontal function and working memory. In: Monsell S, Driver J, eds. *Attention and performance XVIII.* Cambridge, MA: MIT Press, 2000:713–738.

20. Braver TS, Cohen JD. Working memory, cognitive control, and the prefrontal cortex: computational and empirical studies. *Cogn Process* 2001;2:25–55.

21. Braver TS, Speer NK. Maintenance and manipulation in working memory: evidence against a dichotomy of function in prefrontal cortex, 2002 *(submitted).*

22. Brewer J, Zhao ZH, Gabrieli JDE. Parahippocampal and frontal responses to single events predict whether those events are remembered or forgotten. *Science* 1998;281: 1185–1187.

23. Brewer JB, Zhao Z, Glover GH, et al. Making memories: brain activity that predicts how well visual experience will be remembered. *Science* 1998;281:1185–1187.

24. Brodmann K. *Localisation in the cerebral cortex.* London: Smith-Gordon, 1909/1994. Garey LJ, translator.

25. Buckner RL. Beyond HERA: contributions of specific pre-frontal brain areas to long-term memory retrieval. *Psychon Bull Rev* 1996;3:149–158.

26. Buckner RL. Dual effect theory of episodic encoding. In: Tulving E, ed. *Memory, consciousness, and the brain.* Philadelphia: Psychology Press, 1999:278–292.

27. Buckner RL, Bandettini PA, O'Craven KM, et al. Detection of cortical activation during averaged single trials of a cognitive task using functional magnetic resonance imaging. *Proc Natl Acad Sci U S A* 1996;93:14878–14883.

28. Buckner RL, Kelley WM, Petersen SE. Frontal cortex contributes to human memory formation. *Nat Neurosci* 1999;2:311–314.

29. Buckner RL, Koutstaal W. Functional neuroimaging studies of encoding, priming and explicit memory retrieval. *Proc Natl Acad Sci U S A* 1998;95:891–898.

30. Buckner RL, Koutstaal W, Schacter DL, et al. Functional-anatomic study of episodic retrieval. II. Selective averaging of event-related fMRI trials to test the retrieval success hypothesis. *Neuroimage* 1998;7:163–175.

31. Buckner RL, Koutstaal W, Schacter DL, et al. Functional-anatomic study of episodic retrieval using fMRI. *Neuroimage* 1998;7:151–162.

32. Buckner RL, Koutstaal W, Schacter DL, et al. Functional-anatomic study of episodic retrieval using fMRI. I. Retrieval effort vs. retrieval success. *Neuroimage* 1998;7:151–162.

33. Buckner RL, Logan JM. Frontal contributions to episodic memory encoding in the young and elderly. In: Parker A, Wilding EL, Bussey TJ, eds. *The cognitive neuroscience of memory: encoding and retrieval.* New York: Psychology Press, in press (2003).

34. Buckner RL, Petersen SE, Ojemann JG, et al. Functional anatomical studies of explicit and implicit memory retrieval tasks. *J Neurosci* 1995;15:12–29.

35. Buckner RL, Raichle ME, Miezin FM, et al. Functional anatomic studies of memory retrieval for auditory words and visual pictures. *J Neurosci* 1996;16:6219–6235.

36. Buckner RL, Tulving E. Neuroimaging studies of memory: theory and recent PET results. In: Boller F, Grafman J, eds. *Handbook of neuropsychology, volume 10.* Amsterdam: Elsevier, 1995:439–466.

37. Buckner RL, Wheeler M, Sheridan M. Encoding processes during retrieval tasks. *J Cogn Neurosci* 2001;13:406–415.

38. Cabeza R, Nyberg L. Imaging cognition II: an empirical review of 275 PET and fMRI studies. *J Cogn Neurosci* 2000;12:1–47.

39. Cameron KA, Yashar S, Wilson CL, et al. Human hippocampal neurons predict how well word pairs will be remembered. *Neuron* 2001;30:289–298.

40. Carmichael ST, Price JL. Architectonic subdivision of the orbital and medial prefrontal cortex in the macaque monkey. *J Comp Neurol* 1994;4:249–259.

41. Chao LL, Knight RT. Human prefrontal lesions increase distractibility to irrelevant sensory inputs. *Neuroreport* 1995;6:1605–1610.

42. Christoff K, Gabrieli JDE. The frontopolar cortex and human cognition: evidence for a rostrocaudal hierarchical organization within the human prefrontal cortex. *Psychobiology* 2000;28:168–186.

43. Cohen NJ, Eichenbaum H. *Memory, amnesia, and the hippocampal system.* Cambridge, MA: MIT Press, 1993.

44. Cohen NJ, Eichenbaum H. *From conditioning to conscious recollection.* New York: Oxford University Press, 2001.

45. Cohen NJ, Eichenbaum HE. *Memory, amnesia, and the hippocampal system.* Cambridge, MA: MIT Press, 1993.

46. Corkin S. Lasting consequences of bilateral medial temporal lobectomy: clinical course and experimental findings in H. M. *Semin Neurol* 1984;4:249–259.

47. Corkin S. Tactually-guided maze learning in man: effects of

unilateral and bilateral hippocampal lesions. *Neuropsychologia* 1965;3:339–351.

48. Craik FIM, Lockhart RS. Levels of processing: a framework for memory research. *J Verbal Learning Verbal Behav* 1972;11:671–684.

49. D'Esposito M, Aguirre GK, Zarahn E, et al. Functional MRI studies of spatial and nonspatial working memory. *Cogn Brain Res* 1998;7:1–13.

50. D'Esposito M, Postle BR. The dependence of span and delayed response performance on prefrontal cortex. *Neuropsychologia* 1999;37:1303–1315.

51. D'Esposito M, Postle BR, Ballard D, et al. Maintenance versus manipulation of information held in working memory: an event-related fMRI study. *Brain Cogn* 1999;41:66–86.

52. Dale AM, Buckner RL. Selective averaging of rapidly presented individual trials using fMRI. *Hum Brain Mapp* 1997;5:329–340.

53. Damasio H, Grabowski T, Tranel D, et al. A neural basis for lexical retrieval. *Nature* 1995;380:499–505.

54. Delis DC, Lucas JA, Kopelman MD. Memory. In: Fogel BS, Schiffer RB, Rao SM, eds. *Synopsis of neuropsychiatry*. Philadelphia: Lippincott Williams & Wilkins, 2000.

55. Demb JB, Desmond JE, Wagner AD, et al. Semantic encoding and retrieval in the left inferior prefrontal cortex: a functional MRI study of task difficulty and process specificity. *J Neurosci* 1995;15:5870–5878.

56. DeRenzi E, Nichelli P. Verbal and nonverbal short-term memory impairment following hemispheric damage. *Cortex* 1975;11:341–354.

57. Dobbins IG, Kroll NEA, Tulving E, et al. Unilateral medial temporal lobe memory impairment: type deficit, function deficit, or both? *Neuropsychologia* 1998;36:115–127.

58. Eldridge LL, Knowlton BJ, Furmanski CS, et al. Remembering episodes: a selective role for the hippocampus during retrieval. *Nat Neurosci* 1999;3:1149–1152.

59. Eslinger PJ, Grattan M. Altered serial position learning after frontal lobe lesion. *Neuropsychologia* 1994;32:729–739.

60. Fabiani M, Karis M, Donchin E. P300 and recall in an incidental memory paradigm. *Psychophysiology* 1986;23:298–308.

61. Farah MJ, Grossman M. Semantic memory impairments. In: Farah MJ, Feinberg TE, eds. *Patient-based approaches to cognitive neuroscience*. Cambridge, MA: MIT Press, 2000:301–306.

62. Farah MJ, Meyer MM, McMullem PA. The living/nonliving dissociation is not an artifact: giving an a priori implausible hypothesis a strong test. *Cogn Neuropsychol* 1996;13:137–154.

63. Feigenbaum JD, Polkey CE, Morris RG. Deficits in spatial working memory after unilateral temporal lobectomy in man. *Neuropsychologia* 1996;34:163–176.

64. Fisher RP, Craik FIM. Interaction between encoding and retrieval operations in cued recall. *J Exp Psychol Hum Learn Mem* 1997;3:701–711.

65. Fletcher PC, Frith CD, Grasby PM, et al. Brain systems for encoding and retrieval of auditory-verbal memory: an in vivo study in humans. *Brain* 1995;118:401–416.

66. Fletcher PC, McKenna PJ, Frith CD, et al. Brain activation in schizophrenia during a graded memory task studied with functional neuroimaging. *Arch Gen Psychiatry* 1998;55:1001–1008.

67. Fletcher PC, Shallice T, Dolan RJ. The functional roles of prefrontal cortex in episodic memory. *Brain* 1998;121:1239–1248.

68. Funahashi S, Bruce CJ, Goldman-Rakic PS. Mnemonic coding of visual space in the monkey's dorsolateral prefrontal cortex. *J Neurophysiol* 1989;61:331–349.

69. Funahashi S, Bruce CJ, Goldman-Rakic PS. Dorsolateral prefrontal lesions and oculomotor delayed-response performance: evidence for mnemonic "scotomas." *J Neurosci* 1993;13:1479–1497.

70. Funahashi S, Chafee MV, Goldman-Rakic PS. Prefrontal neuronal activity in rhesus monkeys performing a delayed antisaccade task. *Nature* 1993;365:753–755.

71. Funahashi S, Inoue M, Kubota K. Delay-period activity in the primate prefrontal cortex encoding multiple spatial positions and their order of presentation. *Behav Brain Res* 1997;84:203–223.

72. Funnell E, Sheridan J. Categories of knowledge? Unfamiliar aspects of living and nonliving things. *Cogn Neuropsychol* 1992;9:135–153.

73. Fuster JM. *The prefrontal cortex*, 2nd ed. New York: Raven Press, 1989.

74. Fuster JM, Alexander GE. Neuron activity related to short-term memory. *Science* 1971;173:652–654.

75. Gabrieli JDE, Desmond JE, Demb JB, et al. Functional magnetic resonance imaging of semantic memory processes in the frontal lobes. *Psychol Sci* 1996;7:278–283.

76. Ganguli R, Carter CS, Mintun M, et al. PET brain mapping study of auditory verbal supraspan memory versus visual fixation in schizophrenia. *Biol Psychiatry* 1997;41:33–42.

77. Gershberg FB, Shimamura AP. Impaired use of organizational strategies in free recall following frontal lobe damage. *Neuropsychologia* 1995;13:1305–1333.

78. Gold JM, Carpenter C, Randolph C, et al. Auditory working memory and Wisconsin Card Sorting Test performance in schizophrenia. *Arch Gen Psychiatry* 1997;54:159–165.

79. Goldman-Rakic PS. Cellular basis of working memory. *Neuron* 1995;14:477–485.

80. Goldman-Rakic PS. Circuitry of primate prefrontal cortex and regulation of behavior by representational memory. In: Plum F, Mountcastle V, eds. *Handbook of physiology–the nervous system V*. Bethesda, MD: American Physiological Society, 1987:373–417

81. Goldman-Rakic PS. The prefrontal landscape: implications of functional architecture for understanding human mentation and the central executive. In: Roberts AC, Robbins TW, Weiskrantz L, eds. *The prefrontal cortex: executive and cognitive functions*. New York: Oxford University Press, 1998:87–102.

82. Grafman J, Sirigu A, Spector L, et al. Damage to the prefrontal cortex leads to decomposition of structured event complexes. *J Head Trauma Rehabil* 1993;8:73–87.

83. Hanley J, Young A, Pearson N. Impairment of the visuospatial scratchpad. *Q J Exp Psychol* 1991;43A:101–125.

84. Haxby JV, Ungerleider JV, Horwitz B, et al. Face encoding and recognition in the human brain. *Proc Natl Acad Sci U S A* 1996;93:922–927.

85. Hillis A, Caramazza A. Category specific naming and comprehension impairment: a double dissociation. *Brain* 1991;114:2081–2094.

86. Hodges JR, Salmon DP, Butters N. Semantic memory impairment in Alzheimer's disease: failure of access of degraded knowledge. *Neuropsychologia* 1992;30:301–314.

87. Hyde T, Jenkins JJ. Recall of words as a function of semantic, graphic, and syntactic orienting tasks. *J Verbal Learn Verbal Behav* 1973;12:471–480.

88. Iddon JL, McKenna PJ, Sahakian BJ, et al. Impaired generation and use of strategy in schizophrenia: evidence from visuospatial and verbal tasks. *Psychol Med* 1998;28:1049–1062.

89. Incisa della Rocchetta A, Milner B. Strategic search and retrieval inhibition: the role of the frontal lobes. *Neuropsychologia* 1993;31:503–524.

90. Jacoby LL. A process dissociation framework: separating automatic from intentional uses of memory. *J Mem Lang* 1991;30:513–541.

91. Jacoby LL, Kelley CM, McElree BD. The role of cognitive control: early selection versus late correction. In: Chaiken S, Trope E, eds. *Dual process theories in social psychology.* New York: Guilford, 1999:383–400.

92. Jacoby LL, Toth JP, Yonelinas AP. Separating conscious and unconscious influences of memory: measuring recollection. *J Exp Psychol Gen* 1993;122:139–154.

92a. James W. *Principles of Psychology.* New York: Holt, Rinchart, and Winston, 1890.

93. Janowsky J, Shimamura AP. Cognitive impairment following frontal lobe damage and its relevance to human amnesia. *Behav Neurosci* 1989;103:548–560.

94. Jones-Gotman M. Memory for designs: the hippocampal contribution. *Neuropsychologia* 1986;24:193–203.

95. Jones-Gotman M. Right hippocampal excision impairs learning and recall of a list of abstract designs. *Neuropsychologia* 1986;24:659–670.

96. Jonides J, Schumacher EH, Smith TE, et al. The role of parietal cortex in verbal working memory. *J Neurosci* 1998;18:5026–5034.

97. Josephs O, Turner R, Friston KJ. Event-related fMRI. *Hum Brain Mapp* 1997;5:243–248.

98. Kapur S, Craik FI, Tulving E, et al. Neuroanatomical correlates of encoding in episodic memory: levels of processing effect. *Proc Natl Acad Sci U S A* 1994;91:2008–2011.

99. Kapur S, Tulving E, Cabeza R, et al. The neural correlates of intentional learning of verbal materials: a PET study in humans. *Brain Res Cogn Brain Res* 1996;4:243–249.

100. Kapur S, Tulving E, Cabeza R, et al. The neural correlates of intentional learning of verbal materials: a PET study in humans. *Cogn Brain Res* 1996;4:243–249.

101. Kelley WM, Miezin FM, McDermott KB, et al. Hemispheric specialization in human dorsal frontal cortex and medial temporal lobe for verbal and non-verbal memory encoding. *Neuron* 1998;20:927–936.

102. Kirchhoff BA, Wagner AD, Maril A, et al. Prefrontal-temporal circuitry for episodic encoding and subsequent memory. *J Neurosci* 2000;20:6173–6180.

103. Koechlin E, Basso G, Pietrini P, et al. The role of the anterior prefrontal cortex in human cognition. *Nature* 1999;399:148–151.

104. Koh SD. Remembering of verbal material by schizophrenic young adults. In: Schwartz S, ed. *Language and cognitive in schizophrenia.* Hillsdale, NJ: Lawrence Erlbaum, 1978.

105. Konishi S, Wheeler ME, Donaldson DI, et al. Neural correlates of episodic retrieval success. *Neuroimage* 2000;12:276–286.

106. Kroll NEA, Knight RT, Metcalfe J, et al. Cohesion failures as a source of memory illusions. *J Mem Lang* 1996;35:176–196.

107. Lee ACH, Robbins TW, Pickard JD, et al. Asymmetric frontal activation during episodic memory: the effects of stimulus type on encoding and retrieval. *Neuropsychologia* 2000;38:677–692.

108. Martin A, Wiggs CL, Ungerleider LG, et al. Neural correlates of category-specific knowledge. *Nature* 1996;379:649–652.

109. McClelland JL, McNaughton BL, O'Reilly RC. Why there are complementary learning systems in the hippocampus and neocortex: insights from the successes and failures of connectionist models of learning and memory. *Psychol Rev* 1995;102:419–457.

110. McDermott KB, Buckner RL, Petersen SE, et al. Set- and code specific activation in the frontal cortex: an fMRI study of encoding and retrieval of faces and words. *J Cogn Neurosci* 1999;11:631–640.

111. McDermott KB, Buckner RL, Petersen SE, et al. Set- and code-specific activation in the frontal cortex: an fMRI study of encoding and retrieval of faces and words. *J Cogn Neurosci* 1999;11:631–640.

112. McDermott KB, Jones TC, Petersen SE, et al. Retrieval success is accompanied by enhanced activation in anterior prefrontal cortex during recognition memory: an event related MRI study. *J Cogn Neurosci* 2000;12:965–976.

113. McDermott KB, Ojermann JG, Petersen SE, et al. Direct comparison of episodic encoding and retrieval of words: an event-related fMRI study. *Memory* 1999;7:661–678.

114. Milner B. Interhemispheric differences in the localization of psychological processes in man. *Br Med Bull* 1971;27:272–277.

115. Milner B, Petrides M, Smith ML. Frontal lobes and the temporal organization of memory. *Hum Neurobiol* 1985;4:137–142.

116. Moscovitch M. Memory and working-with-memory: a component process model based on modules and central systems. *J Cogn Neurosci* 1992;4:257–267.

117. Moscovitch M. Memory and working with memory: evaluation of a component process model and comparisons with other models. In: Schacter DL, Tulving E, eds. *Memory stems.* Cambridge, MA: MIT Press, 1994:269–310

118. Murray EA. What have ablation studies told us about neural substrates of stimulus memory? *Semin Neurosci* 1996;8:13–22.

119. Myerson J, Hale S, Rhee SH, et al. Selective interference with verbal and spatial working memory in young and older adults. *J Gerontol B Psychol Sci Soc Sci* 1999;54:161–164.

120. Nunn JA, Polkey CE, Morris RG. Selective spatial memory impairment after right unilateral temporal lobectomy. *Neuropsychologia* 1998;36:837–848.

121. Nyberg L, Cabeza R, Tulving E. PET studies of encoding and retrieval: the HERA model. *Psychon Bull Rev* 1996;3:135–148.

122. Nyberg L, McIntosh AR, Cabeza R, et al. General and specific brain regions involved in encoding and retrieval of events: what, where and when. *Proc Natl Acad Sci U S A* 1996;93:11280–11285.

123. Nystrom LE, Braver TS, Sabb FW, et al. Working memory for letters, shapes, and locations: fMRI evidence against stimulus-based regional organization of human prefrontal cortex. *Neuroimage* 2000;11:424–446.

124. O Scalaidhe SP, Wilson FA, Goldman-Rakic PS. Areal segregation of face-processing neurons in prefrontal cortex. *Science* 1997;278:1135–1138.

125. Otten LJ, Henson RNA, Rugg MD. Depth of processing effects on neural correlates of memory encoding. *Brain* 2001;124:399–412.

126. Owen AM. The functional organization of working memory processes within human lateral frontal cortex: the contribution of functional neuroimaging. *Eur J Neurosci* 1997;9:1329–1339.

127. Owen AM, Morris RG, Sahakian BJ, et al. Double dissociations of memory and executive functions in working memory tasks following frontal lobe excisions, temporal lobe excisions or amygdalo-hippocampectomy in man. *Brain* 1996;119:1597–1615.

128. Paivio A. *Mental representations.* New York: Oxford University Press, 1986.

129. Paivio A, Csapo K. Picture superiority in free recall: imagery or dual coding? *Cogn Psychol* 1973;5:176–206.

130. Paller KA. Recall and stem-completion priming have different electrophysiological. *J Exp Psychol Learn Mem Cogn* 1990;16:1021–1032.

131. Park S, Holzman PS. Schizophrenics show spatial working memory deficits. *Arch Gen Psychiatry* 1992;49:975–982.
132. Park S, Holzman PS. Association of working memory deficit and eye tracking dysfunction in schizophrenia. *Schizophr Res* 1993;11:55–61.
133. Paulesu E, Frith CD, Frackowiak RSJ. The neural correlates of the verbal component of working memory. *Nature* 1993;362:342–345.
134. Petrides M. Deficits on conditional associative learning tasks after frontal- and temporal-lobe lesions in man. *Neuropsychologia* 1985;25:601–614.
135. Petrides M. Frontal lobes and working memory: evidence from investigations of the effects of cortical excisions in non-human primates. In: Boller F, Grafman J, eds. *Handbook of neuropsychology, volume 9.* Amsterdam: Elsevier, 1994:59–82.
136. Petrides M. Functional organization of the human frontal cortex for mnemonic processing: evidence from neuroimaging studies. In: Grafman J, Holyoak K, Boller F, eds. *Structure and functions of the human prefrontal cortex, volume 769.* New York: New York Academy of Sciences, 1995:85–97.
137. Petrides M. Impairments on nonspatial self-ordered and externally ordered working memory tasks after lesions of the mid-dorsal part of the lateral frontal cortex in the monkey. *J Neurosci* 1995;15:359–375.
138. Petrides M. Lateral frontal cortical contribution to memory. *Semin Neurosci* 1996;8:57–63.
139. Petrides M. Specialized systems for the processing of mnemonic information within the primate frontal cortex. *Philos Trans R Soc Lond B Biol Sci* 1996;351:1455–1462.
140. Petrides M, Pandya DN. Comparative architectonic analysis of the human and the macaque frontal cortex. In: Boller F, Grafman J, eds. *Handbook of neuropsychology, volume 9.* Elsevier: Amsterdam, 1994:17–58.
141. Piggott S, Milner B. Memory for different aspects of complex visual scenes after unilateral temporal or frontal resection. *Neuropsychologia* 1989;20:1–15.
142. Pillon B, Deweer B, Vidailhet M, et al. Is impaired memory for spatial location in Parkinson's disease domain specific or dependent on "strategic" processing? *Neuropsychologia* 1998;36:1–9.
143. Poldrack RA, Wagner AD, Prull MW, et al. Functional specialization for semantic and phonological processing in the left inferior prefrontal cortex. *Neuroimage* 1999;10:15–35.
144. Postle BR, Berger JS, D'Esposito M. Functional neuroanatomical double dissociation of mnemonic and executive control processes contributing to working memory performance. *Proc Natl Acad Sci U S A* 1999;96:12959–12964.
145. Postle BR, D'Esposito M. Evaluating models of the topographical organization of working memory function in frontal cortex with event-related fMRI. *Psychobiology* 2000;28:132–145.
146. Postman L. Short-term memory and incidental learning. In: Melton A, ed. *Categories of human learning.* New York: Academic Press, 1964:146–201.
147. Preuss TM, Goldman-Rakic PS. Myelo- and cytoarchitecture of the granular frontal cortex and surrounding regions in the strepsirhine Galago and the anthropoid primate Macaca. *J Comp Neurol* 1991;310:429–474.
148. Rajkowska G, Goldman-Rakic PS. Cytoarchitectonic definition of prefrontal areas in the normal human cortex: II. Variability in locations of areas 9 and 46 and relationship to the Talairach coordinate system. *Cereb Cortex* 1995;5:323–337.
149. Rajkowska G, Goldman-Rakic PS. Cytoarchitectonic definition of prefrontal areas in normal human cortex: I. Remapping of areas 9 and 46 using quantitative criteria. *Cereb Cortex* 1995;5:307–322.
150. Rao SC, Rainer G, Miller EK. Integration of what and where in the primate prefrontal cortex. *Science* 1997;276:821–824.
151. Reber AS. Implicit learning and tacit knowledge. *J Exp Psychol Gen* 1989;118:219–235.
152. Reber AS. How to differentiate implicit and explicit modes of acquisition. Paper presented at the 25th Carnegie Symposium: "Scientific Approaches to the Question of Consciousness." Pittsburgh, May 1992.
153. Reber PJ, Squire LR. Parallel brain systems for learning with and without awareness. *Learn Mem* 1994;1:217–229.
154. Reed JM, Squire LR. Impaired recognition memory in patients with lesions limited to the hippocampal formation. *Behav Neurosci* 1997;111:667–675.
155. Riege WH, Metter EJ, Hanson WR. Verbal and nonverbal recognition memory in aphasic and nonaphasic stroke patients. *Brain Lang* 1980;10:60–70.
156. Rosen BR, Buckner RL, Dale AM. Event-related functional MRI: past, present, and future. *Proc Natl Acad Sci U S A* 1998;95:773–780.
157. Rotte M, Koustaal W, Schacter DL, et al. Left prefrontal activation correlates with the levels of processing during verbal encoding: an event-related fMRI study. *Neuroimage* 1998;7:S813.
158. Rugg MD. ERP studies of memory. In: Rugg MD, Cols MGH, eds. *Electrophysiology of mind: event-related brain potentials and cognition.* Oxford: Oxford University Press, 1995:133–170.
159. Rugg MD, Fletcher PC, Allan K, et al. Neural correlates of memory retrieval during recognition memory and cued recall. *Neuroimage* 1998;8:262–273.
160. Rugg MD, Fletcher PC, Frith CD, et al. Differential activation of the prefrontal cortex in successful and unsuccessful memory retrieval. *Brain* 1996;119:2073–2083.
161. Rushe TM, Woodruff PWR, Murray RM, et al. Episodic memory and learning in patients with chronic schizophrenia. *Schizophr Res* 1998;35:85–96.
162. Rypma B, D'Esposito M. The roles of prefrontal brain regions in components of working memory: effects of memory load and individual differences. *Proc Natl Acad Sci U S A* 1999;96:6558–6563.
163. Rypma B, Prabhakaran V, Desmond JE, et al. Load-dependent roles of frontal brain regions in the maintenance of working memory. *Neuroimage* 1999;9:216–226.
164. Saffran EM, Marin SM. Immediate memory for word lists and sentences in a patient with deficient auditory short-term memory. *Brain Lang* 1975;2:420–433.
165. Saling MM, Berkovic SF, O'Shea MF, et al. Lateralization of verbal memory and unilateral hippocampal sclerosis: evidence of task-specific effects. *J Clin Exp Neuropsychol* 1993;15:608–618.
166. Savage CR, Baer L, Keuthen NJ, et al. Organizational strategies mediate nonverbal memory impairment in obsessive-compulsive disorder. *Biol Psychiatry* 1999;45:905–916.
167. Savage CR, Deckersbach T, Heckers S, et al. Prefrontal regions supporting spontaneous and directed application of verbal learning strategies: evidence from PET. *Brain* 2001;124:219–231.
168. Saykin AJ, Robinson LJ, Stafiniak P, et al. Neuropsychological changes after anterior temporal lobectomy: acute effects on memory, language and music. In: Bennett TL, ed. *The neuropsychology of epilepsy.* New York: Plenum Press, 1992.
169. Schacter DL. Memory, amnesia, and frontal lobe dysfunction. *Psychobiology* 1987;15:21–36.
170. Schacter DL, Alpert NM, Savage CR, et al. Conscious

recollection and the human hippocampal formation: evidence from positron emission tomography. *Proc Natl Acad Sci U S A* 1996;93:321–325.

171. Schacter DL, Normal DA, Koustaal W. The cognitive neuroscience of constructive memory. *Annu Rev Psychol* 1998;49:289–318.

172. Schacter DL, Wagner AD, Buckner RL. Memory systems of 1999. In: Tulving E, Craik FI, eds. *The Oxford handbook of memory*. New York: Oxford University Press, 2000.

173. Scoville WB, Milner B. Loss of recent memory after bilateral hippocampal lesions. *J Neurol Neurosurg Psychiatry* 1957;20:11–12.

174. Sengel RA, Lovallo WR. Effects of cueing on immediate and recent memory in schizophrenics. *J Nerv Ment Dis* 1983;171:426–430.

175. Shallice T, Vallar G. The impairment of auditory-verbal short-term storage. In: Vallar G, Shallice T, eds. *Neuropsychological impairments of short-term memory*. Cambridge, MA: Cambridge University Press, 1990.

176. Shimamura AP, Janowksy JS, Squire LR. What is the role of frontal lobe damage in memory disorders? In: Levin HS, Eisenberg HM, Benton AL, eds. *Frontal lobe function and dysfunction*. New York: Oxford University Press, 1991:173–195.

177. Shimamura AP, Janowsky JS, Squire LR. Memory for the temporal order of events in patients with frontal lobe lesions and amnesic patients. *Neuropsychologia* 1990;28:803–813.

178. Smith EE, Jonides J, Koeppe RA. Dissociating verbal and spatial working memory using PET. *Cereb Cortex* 1996;6:11–20.

179. Smith ML, Milner B. The role of the right hippocampus in the recall of spatial information. *Neuropsychologia* 1981;19:781–795.

180. Smith ML, Milner B. Right hippocampal impairment in the recall of location: encoding deficit or rapid forgetting. *Neuropsychologia* 1989;27:71–82.

181. Snowden JS, Goudling PJ, Neary D. Semantic dementia: a form of circumscribed cerebral atrophy. *Behav Neurol* 1989;2:167–182.

182. Squire LR. *Memory and brain*. New York: Oxford University Press, 1987.

183. Squire LR, Ojemann JG, Miezin FM, et al. Activation of the hippocampus in normal humans: a functional anatomical study of memory. *Proc Natl Acad Sci U S A* 1992;89:1837–1841.

184. Stuss DT, Alexander MP, Palumbo CL, et al. Organizational strategies of patients with unilateral or bilateral frontal lobe injury in word list learning tasks. *Neuropsychology* 1994;8:355–373.

185. Thompson-Schill SL, D'Esposito M, Aguire GK, et al. Role of left inferior prefrontal cortex in retrieval of semantic knowledge: a re-evaluation. *Proc Natl Acad Sci U S A* 1997;94:14792–14797.

186. Tulving E. Episodic and semantic memory. In: Tulving E, Donaldson W, eds. *Organization of memory*. New York: Academic Press, 1972:381–403.

187. Tulving E. Memory and consciousness. *Can Psychol* 1985;26:1–12.

188. Tulving E, Kapur S, Craik FI, et al. Hemispheric encoding/retrieval asymmetry in episodic memory: positron emission tomography findings. *Proc Natl Acad Sci U S A* 1994;91:2016–2020.

189. Ungerleider LG, Mishkin M. Two cortical visual systems. In: Ingle DJ, Goodale MA, Mansfield RJW, eds. *Analysis of visual behavior*. Cambridge, MA: MIT Press, 1982:549–587.

190. Vallar G, Papagno C. Neuropsychological impairments of short-term memory. In: Baddleley AD, Wilson BA, Watts FN, eds. *Handbook of memory disorders*. New York: John Wiley & Sons, 1995:135–165.

191. Wagner AD, Desmond JE, Glover GH, et al. Prefrontal cortex and recognition memory: functional MRI evidence for context-dependent retrieval processes. *Brain* 1998;121:1985–2002.

192. Wagner AD, Koutstaal W, Schacter DL. When encoding yields remembering: insights from event-related neuroimaging. *Philos Trans R Soc* 1999;354:1307–1324.

193. Wagner AD, Poldrack RA, Eldridge LL, et al. Material-specific lateralization of prefrontal activation during episodic encoding and retrieval. *Neuroreport* 1998;9:3711–3717.

194. Wagner AD, Schacter D, Rotte M, et al. Building memories: remembering and forgetting of verbal experiences as predicted by brain activity. *Science* 1998;281:1188–1191.

195. Warrington EK. The selective impairment of semantic memory. *Q J Exp Psychol* 1975;27:635–657.

196. Warrington EK, Logue V, Pratt RTC. The anatomical localization of selective impairment of auditory verbal short-term memory. *Neuropsychologia* 1971;9:377–387.

197. Warrington EK, Shallice T. The selective impairment of auditory verbal short-term memory. *Brain* 1969;92:885–896.

198. Warrington EK, Shallice T. Category specific naming impairments. *Brain* 1984;107:829–854.

199. Whitehouse PJ. Imagery and verbal encoding in left and right hemisphere damaged patients. *Brain Lang* 1981;14:315–332.

200. Wilson FAW, Scalaidhe SPO, Goldman-Rakic PS. Dissociation of object and spatial processing domains in primate prefrontal cortex. *Science* 1993;260:1955–1957.

201. Yonelinas A. Receiver-operating characteristics in recognition memory: evidence for a dual-process model. *J Exp Psychol Learn Mem Cogn* 1994;20:1341–1354.

202. Zarahn E, Aguirre G, D'Esposito M. A trial-based experimental design for fMRI. *Neuroimage* 1997;6:122–138.

203. Zola SM, Squire LR. The medial temporal lobe and the hippocampus. In: Tulving E, Craik FIM, eds. *The Oxford handbook of memory*. New York: Oxford University Press, 2000:485–500.

19

LANGUAGE DYSFUNCTION IN SCHIZOPHRENIA

GINA R. KUPERBERG
DAVID CAPLAN

A REVIEW OF LANGUAGE DYSFUNCTION IN SCHIZOPHRENIA

Abnormalities in language are central to psychosis, particularly the schizophrenic syndrome. This chapter first discusses one clinical manifestation of abnormal language in schizophrenia: 'thought disorder'. We then give a framework for understanding normal language structure and processing. Within this framework, we review studies of language processing in schizophrenia. Finally, we review some recent neuroimaging and electrophysiologic studies that have attempted to examine the neural correlates of language abnormalities in schizophrenia.

THOUGHT DISORDER

Perhaps the most extreme and obvious manifestation of a language disorder in schizophrenia is the abnormal speech produced by some patients. This disturbance is heterogeneous and has traditionally been termed 'thought disorder'. Positive thought disorder, with disorganized unintelligible speech, is a strong predictor of maladaptive social and vocational functioning (121,141,223). Yet its definition and study has challenged psychopathologists since schizophrenia was first described. In this section, we discuss some of the fundamental questions that have shaped our understanding of thought disorder. First, is it truly a disorder of thought or one of language? Second, is it characterized by problems in the content or the form of speech or both? Third, is it unique to schizophrenia? Fourth, how has it been classified and studied?

Thought Versus Language?

The traditional viewpoint of most psychopathologists has been to regard speech disturbances as reflective of an underlying disorder of thinking rather than a primary disorder of language. For example, Kraepelin (168) attributed abnormal speech of schizophrenic patients to derailments and incoherence in the "train of thought." Bleuler (28) believed that disorders of the association of thoughts were fundamental to schizophrenia but that these disorders did not "lie in the language itself." Indeed, he classified disorders of speech and writing as secondary "accessory symptoms" of schizophrenia. Despite these early attempts to distinguish between thought and language disturbances in schizophrenia, there has been considerable confusion in the terminology used to refer to these phenomena.

It has long been recognized that most disorders of thought can only be deduced from the speech of patients. Rochester and Martin (240) pointed out that, because thought cannot be accessed directly, attributing thought disorder to a speaker is tautologic, i.e., we infer thought disorder based on disordered speech. However, it has also long been acknowledged that there is no simple, one-to-one relationship between the language system, on the one hand, and abnormalities in accomplishing the goals of language use on the other. One patient could theoretically have major problems in his/her thought processes but choose to say nothing. Another patient's thought processes may be intact but he/she may find him/herself unable to use the tools of language to express him/herself. Thus, to some researchers, the term 'thought disorder' refers to subjective changes experienced and reported by a patient. Observed abnormalities of spoken or written language are referred to as speech or language disorders. Other researchers, however, refer to spoken and written language abnormalities as 'thought disorder'.

A second and related source of confusion has arisen because of differing assumptions about what phenomena and processes are classified as thought versus language. This debate has often been at a theoretical and philosophical level, perhaps because the very concept of thought is ill defined. Thought has been studied not just by examining speech, but by assessing logic, problem solving, various nonverbal analogy-making abilities, and so on. Nonetheless, we still know relatively little about the relationships between thought, knowledge, and language and its expression.

Content Versus Form of Thought

Traditional psychiatric teaching has distinguished between problems in the *content* and *form* of thought (253). Content has been defined simply as what the patient is talking about (291), whereas form has been defined as the way ideas, sentences, and words are put together.

Disorders of Content

In psychosis, the most extreme example of a disorder of content is the delusion. Indeed, some psychopathologists narrow their definition of content abnormalities to include only delusions (86). However, the content of speech produced by schizophrenic patients can be distinguished from that of control subjects in ways that do not meet the criteria for a delusion, particularly in terms of the conviction, consistency, and strength with which such beliefs are expressed. The content of schizophrenic speech has been described as deviant in the use of conventional social norms (121,196,241), the degree to which personal themes have an inappropriate impact (122,302), and in how subjects think about or judge events in the real world (33,68). For example, Brown (33) describes a patient who told him that "When I get out of here, I'm going to fly to Scotland where they are making a movie of *Fiddler on the Roof* because I'd really like to try out for the lead." Brown points out that this patient had incorrect knowledge about the nature of *Fiddler on the Roof,* Scotland was an unlikely venue for it, and the patient was the wrong age to play its leading role. Cutting and Murphy (68) give a similar example of a patient who claimed that "a thermometer made everyone of the age of 21 get either flu or pneumonia."

Disorders of content such as those described above form essential components of several commonly used rating scales of thought disorder (see below). For example, they are included, by definition, in the Assessment of Bizarre-Idiosyncratic Thinking (195,196). Similarly, criteria such as "queer" responses (expressions and imagery), fabulized combinations (impossible or bizarre), absurd responses, and autistic logic are included in the Thought Disorder Index (150,270).

Disorders of the Form of Thought

Disorders in the way ideas, sentences, and words are put together in psychotic speech were first codified in the late nineteenth and early twentieth centuries. For example, Schneider (254) described a number of phenomena including *Verschmezung* (fusion), *Faseln* (muddling), and *Entgleiten* (snapping off). These abnormalities ranged from the use of vague sentences that are difficult to follow, non sequitur re-

sponses to questions, through fragmented incomprehensible speech with neologisms, word approximations, and private word usage.

More recently, many of these phenomena have been included as essential components in the three most commonly used instruments to assess thought disorder: the Scale for Assessment of Thought, Language, and Communication (TLC) (12,15), the Thought Disorder Index (TDI) (150,270), and the Assessment of Bizarre-Idiosyncratic Thinking (195,196). Of these, the 20-item Thought, Language, and Communication scale (12) is notable because, unlike other scales that make assumptions about the nature of thought disorder, it simply describes abnormalities in speech during psychiatric interviews. For example, 'derailment' and 'loss of goal' would be scored highly on this instrument if a patient's utterances were either unrelated or only obliquely related to one another, and, as a result, the patient's discourse ended with something completely unrelated to how it started (e.g., "I always liked geography. My last teacher in that subject was Professor August A. He was a man with black eyes. I also like black eyes. There are also blue and grey eyes and other sorts, too. . ." (28).

Studies using this scale have reported that phenomena occurring at the level of sentences and discourse, such as derailment, loss of goal, perseveration, and tangentiality, are much more common in schizophrenia than phenomena that occur at the level of single words such as neologisms (11,80,200).

Distinction between Content and Form in Schizophrenia

The distinction between form and content, as defined above, is sometimes blurred. They often co-occur (26), and, in clinical practice, it is sometimes difficult to distinguish the two. For example, a tangential answer to a question (a disorder of form) might arise because a patient is preoccupied with a bizarre belief (a disorder of content). Moreover, in many patients, form and content interact. For example, recent studies have shown that speech becomes more disjointed when patients talk about negative emotional themes (75,117).

Is Thought Disorder Unique to Schizophrenia?

The question of whether thought disorder is unique to schizophrenia has stimulated extensive debate for two main reasons. First, for many years, it was believed that the identification and characterization of speech disturbances might aid in the differential diagnosis of schizophrenia from other psychoses. Second, many researchers believed that identifying similarities between the speech of patients with schizophrenia and that of patients with identifiable brain damage would give clues about the sites and origins of brain dysfunction in schizophrenia (161).

Speech in Schizophrenia Versus Other Psychoses

Because Bleuler (28) considered thought disorder to be a primary symptom and a core feature of the schizophrenic syndrome, it was once the norm for clinicians to assign a diagnosis of schizophrenia when any kind of thought disorder was present. However, with the development of more stringent diagnostic criteria, it became clear that thought disorder occurred in other psychoses, particularly in mania (10,11,120,150,271). Although several studies have documented overlap in the quantity and quality of thought disorder in schizophrenic and manic patients, some differences have also been highlighted. For example, compared with manic patients, the thought disorder of schizophrenic patients is often more disorganized and confused, with an increased use of peculiar words and phrases (271).

Speech in Schizophrenic Versus Brain-Damaged Patients

The speech of schizophrenic patients has been likened to that produced by two main groups of brain-damaged patients: those with aphasias and those with right-hemisphere lesions.

Aphasias

The speech of some schizophrenic patients appears, at least superficially, similar to Wernicke's aphasia, i.e., well formed syntactically but giving an overall impression of nonsensical jargon. Like Wernicke's aphasia, it can include paraphasic-like semantic substitutions of words and phrases and a tendency to string words together based on phonologic or semantic relationships rather than whole themes. Moreover, like some patients with Wernicke's aphasia, schizophrenic patients with thought disorder often show little awareness of their speech abnormalities.

These comparisons, however, have been largely based on clinical impression, and there have been few systematic comparisons of these two patient groups. One study reported that only one of five specialist raters was able to differentiate accurately between the speech produced by schizophrenic patients and that produced by aphasic patients (83). Another study, however, identified several differences between the two groups: the speech of eight patients with posterior aphasic syndromes (Wernicke's aphasia, transcortical sensory aphasia and conduction aphasia) tended to show paraphasic errors, whereas the speech of schizophrenic patients was characterized by one or two bizarre themes (100). Studies that have examined the performance of schizophrenic patients on neuropsychological language tasks traditionally used in aphasic patients, such as the Boston Naming Test (109), have identified both differences and similarities between the two groups.

The design and rationale of studies that have compared the speech of schizophrenic and aphasic patients may be inherently flawed: many of these studies failed to take into account the heterogeneity of schizophrenia: only some schizophrenic patients produce abnormal speech. Moreover, the aphasias themselves are syndromes that are relatively broadly defined: the criteria for inclusion of patients in different aphasic groups overlap. For example, phonemic paraphasias are observed in several categories, and anomia is said to occur in almost all categories.

Another reason why comparisons between groups of schizophrenic and brain-damaged patients may not yield useful information about the neural basis of schizophrenic language abnormalities is that the localization of brain lesions in the aphasic syndromes remains controversial (see later).

Right-Hemisphere Patients

The speech of schizophrenic patients has also been likened to that produced by patients with disorders of hemispheric imbalance, particularly those with isolated right-hemisphere lesions. Like schizophrenic patients, such patients are said to follow "associations that are tangential to the overall meaning of a discourse. . .and are often stuck with, or are satisfied with, a limited and piecemeal understanding, one based on personalization as well as on other inappropriate associations" (34). Right-hemisphere patients also show a range of other deficits including problems with nonliteral language, stories, jokes, and conversations. To our knowledge, there have been no systematic comparisons of the speech of right-hemisphere and schizophrenic patients.

Classifying Thought Disorder

There have been numerous attempts to classify the phenomena constituting thought disorder. Psychopathologists originally attempted to group these phenomena together on a conceptual basis. More recent studies have examined their co-occurrence in large numbers of patients, classifying patients on this more empirical basis. To the extent that what is usually called 'thought disorder' is manifest in language (see earlier discussion), a third approach is to describe and classify language itself within a framework used by linguists and psycholinguists to describe normal language structure and processing. We adopt this last approach in the following review of studies of language output and processing in schizophrenia.

Conceptual Classification

In the first half of this century, psychopathologists described a number of deficits that they believed to underlie the abnormal speech produced by many patients with schizophrenia.

These were essentially descriptions of what was inferred about patients' thinking—intrinsic disturbance in thinking or 'dyslogia' (13,14). They included concepts such as "looseness of association" (28), "overinclusive thinking," [a tendency of patients to use concepts beyond their usual boundaries (36,37)], concrete thinking [an inability to think abstractly (108)], and logical deficits (304).

Empirical Classification

With the development and use of scales that rated thought disorder more systematically came an attempt to group together its various phenomena based on the frequency of their co-occurrence. The first widely recognized classification of thought disorder was one that distinguished positive from negative thought disorders, paralleling the traditional distinction between positive and negative symptoms of schizophrenia (65). Positive thought disorder included tangentiality, derailment, neologisms, and several other phenomena that appeared to be highly correlated in patients (11,126,229). Negative thought disorder included phenomena such as "poverty of speech." A distinction between positive and negative thought disorders has been confirmed by several factor analytic studies of speech disturbances in schizophrenia (15,131,233,292). Positive thought disorder is generally associated with acute schizophrenia and often improves with neuroleptic treatment (105). However, it can sometimes persist after the acute phase and become chronic with a poor prognosis (123,200).

Positive thought disorder is now generally conceptualized as part of the disorganization subsyndrome of schizophrenia (79a). It is also termed disorganized speech.

Psycholinguistic Classification

It is clear from the discussions above that there are several problems with the traditional phenomenologic approach to studying thought disorder: there is confusion about terminology, phenomena are often imprecisely defined, and there is no obvious way to link clinical disturbances with dysfunction at the neurocognitive level. One way to overcome these problems is to approach the study of thought disorder in schizophrenia within a framework of psycholinguistics. Such an approach has the advantages of having a sound theoretical basis and of highlighting the close relationship between normal and abnormal processing. This, in turn, encourages the generation of specific hypotheses that can be tested experimentally.

A psycholinguistic approach involves the identification of disturbances in the major components of the language processing system that are present in patients with thought disorder. From a practical point of view, a very detailed taxonomy based on all possible deficits is unrealistic. One way

to approach a psycholinguistic classification is to identify language-processing deficits (e.g., semantic, syntactic) at the three basic levels of the language code: simple words (the lexical level), sentences (the sentential level), and discourse (the discourse level). It is also important to examine how psycholinguistic processes interact with other cognitive processes such as nonverbal semantic processing, attention, and working memory.

We first give an overview of the normal language system, highlighting its essential features and its interaction with other cognitive processes. Within this framework, we then review relevant studies in schizophrenia.

NORMAL LANGUAGE: STRUCTURE, REPRESENTATION, AND PROCESSING
Language Structure

The three basic levels of the language code are simple words (the lexical level), sentences (the sentential level), and discourse (the discourse level).

The Lexical Level

The lexical level of language consists of simple words. The basic form of a simple word (or lexical item) consists of a phonologic representation that specifies the segmental elements (phonemes) of the word and its organization into metrical structures such as syllables (118). The form of a word can also be represented orthographically (135). In addition, simple words are assigned to different syntactic categories, such as noun, verb, adjective, article, and position. Finally, words are represented at a lexico-semantic level. Each of these lexico-semantic representations is associated with concepts and categories in the nonlinguistic world. Simple words tend to designate concrete objects, abstract concepts, actions, properties, and logical connectives.

The Sentential Level

The sentential level of language consists of syntactic structures—hierarchical sets of syntactic categories, e.g., noun phrases, verb phrases (50–52), into which words are inserted. The meaning of a sentence, known as its propositional content, is determined by the way the meanings of words combine in syntactic structures. Propositions convey aspects of the structure of events and states in the world. These include information about who did what to whom (thematic roles), which adjectives go with which nouns (attribution of modification), and which words in a set of sentences refer to the same items or actions (the reference of pronouns and other anaphoric elements). For instance, in the sentence "The big boy told the little girl to wash herself," the agent of "told" is "the big boy" and its theme is "the little girl"; "big" is associated with "boy" and "little" with "girl," and "herself" refers to the same person as "girl." Sentences are a crucial level of the language code because the propositions that they express make assertions about the world. These assertions can be entered into logical systems and can be used to add to an individual's knowledge of the world.

The Discourse Level

The discourse level constitutes higher-order structures formed by the propositional meanings conveyed by sentences (113,301). Discourse includes information about the general topic under discussion, the focus of a speaker's attention, the novelty of the information in a given sentence, the temporal order of events, and causation. The structure and processing of discourse involve many nonlinguistic elements and operations, such as logical inferences, as well as more purely linguistic operations. For instance, consider the following set of sentences:

> John and Henry went to Peter's last night. They were very glad they did. They raved about the dessert all the next day.

The reader infers that John and Henry ate dinner at Peter's. This is an inference that is based on information that is outside the language system. On the other hand, the reader takes the word "They" in the second sentence to refer to John and Henry, not John and Peter, or Henry and Peter, or all three men. This assignment is probably based on the fact that "They" and "John and Henry" are both noun phrases in the subject positions of their sentences—a linguistic fact.

Information conveyed by the discourse level of language serves as a basis for updating an individual's knowledge of the world and for reasoning and planning action.

Representations: Linguistic and Other Types

Different linguistic representations (e.g., semantics and syntax) have different rules and are generally acknowledged to be independent of one another. It is clear from the above discussion that some linguistic representations cut across the three levels of the language code. For example, meaning—semantics—is represented at the word, sentence, and discourse levels. Similarly, individual words are assigned to specific syntactic categories, but the ways in which they are put together (the syntactic structure) are defined at the sentential level.

The boundary between linguistic representations and other types of representation is not always easy to draw. This is particularly the case in the study of semantics. One of the reasons for this blurring is inconsistent terminology. In the study of language, the word semantics is generally used as an umbrella term to refer to all aspects of meaning. For example, at the level of single words, it may refer to conceptual associations and groups, whereas at the level of sentences, it refers to propositions that are derived from a combination of word meanings and the syntactic structure. Another source of confusion is that the distinction between amodal semantic representations and lexico-semantic representations is not always specified.

There is also a blurred boundary between the study of semantics and pragmatics. Pragmatics is a widely used term that encompasses our social and real world knowledge—the way people use language in natural settings, particularly in the study of discourse. Some researchers but not others, explicitly exclude pragmatic inferences, discourse context, and knowledge of the world from the study of semantics.

Semantics has not only been studied from the perspective of language but also of memory. Semantic memory is traditionally conceptualized as the component of long-term memory that constitutes representations of objects, facts, concepts, word meanings, and their relationships. This is thought to be distinguishable from episodic memories that are temporally specific for personal events, i.e., the place or time of encoding (298).

Language Processing

The different forms of the language code are thought to be computed by a set of processors whose operations range from conversion of the acoustic signal to speech sounds, through visual word recognition through determination of sentence and discourse structure.

Information-processing models of language often depict a sequence of operations of the different components required to perform a language-related task. These models are based on the results of experimental psychological research in both normal subjects and in patient populations (39,135,181,259). They can become extremely detailed and complex when all the operations and components used in a task are specified. For our current purposes, it is adequate to identify the major components of the language-processing system as those processors that activate units at the lexical, sentential, and discourse levels of the language code.

Temporal and Spatial Distinctions

As noted above, different linguistic representations (e.g., semantic, syntactic) are thought to be independent of one another. A fundamental question is whether these representational distinctions are respected during language processing, both with respect to the temporal sequence of these processes and their spatial localization in the brain.

The question of relative *timing* of different linguistic processess addresses the extent to which the components of the system operate serially or in parallel. This question remains controversial. At one extreme, serial models hold that processing of one form of linguistic information awaits the completion of processing of another form of linguistic information. Parallel-processing models, in contrast, suggest that processing different types of information occurs at the same time. It is important to note that some parallel models are still consistent with different processes being temporally distinct. For example, cascade models hold that individual processing stages are arranged in a temporal sequence with one stage starting before another, but with information continually flowing in feed-forward fashion. Alternatively, the output of one type of processing may be delayed relative to another type of processing. Strict parallel models, however, assume that all processes are initiated at the same time and that the output of each process is available at the same time. Information about the timing of processes is given by on-line behavioral studies and by event-related potential (ERP) studies (see later discussion).

Information regarding whether different levels of language processing are mediated by distinct neural systems has traditionally come from studies of patients with brain lesions. So-called double dissociations in which one patient performs normally on one task and abnormally on a second task and a second patient shows the opposite pattern, provide evidence of the existence of separate processors, each involved in only one of the two tasks (259). These observations support localizationist theories of brain function.

Another somewhat orthogonal question is whether processing different types of linguistic information is autonomous or interactive. A purely modular view (91) holds that different processing components are each dedicated to activating particular elements of the language code, accepting only particular types of representations as input (encapsulation) and producing only specific types of representations as output (domain specificity). Pure interactionalist models hold that processing different types of linguistic information are dependent on each other (81,197,199).

Relationship with Other Cognitive Processes

Each component of the language-processing system interacts with and can be influenced by *outside* cognitive systems and processes, such as attention, working memory, executive function, and non-linguistic semantic memory.

Researchers have developed a number of theories describing such interactions (38,154,155,259). Some of these processes may be necessary for successful language comprehension. For example, consider the sentence "The daughter of the king's son shaved himself." It is relatively easy to understand its individual parts ("The daughter of the king" and "The king's son shaved himself"), but putting them all together to parse it as "[the daughter of the king]'s son," i.e., the king's grandson, may require additional processing resources.

Other cognitive processes may be under strategic control. For example, we exercise control over the entire language-processing system when we decide whether to use language to convey our intentions. We exercise control over the choice of vocabulary elements in our speech based on our estimation of the listeners' ability to understand different sets of words. We control the rate of our speech, the formality of the vocabulary, and the syntax that we choose.

OFF-LINE STUDIES OF LANGUAGE PROCESSING IN PSYCHOSIS

Neuropsychological profiles of patients with schizophrenia depict deficits across a broad range of cognitive functions. There is some evidence that, among cognitive domains, language processing and verbal memory are particularly susceptible to disruption. For example, in one study, patients performed comparably with controls on a tone serial position task but poorly on an auditory and visual verbal task (283).

In this section, we review studies of language output and processing in schizophrenia. First, we describe studies at the level of single words. These have been conducted mainly from the perspective of the structure and function of semantic memory. Second, we review studies that have examined the meaning of words in relation to their context within sentences and discourse. Third, we discuss studies that have specifically examined syntactic processing in schizophrenia. Finally, we review studies of discourse that have examined the relationships between sentences in schizophrenia.

At each of these levels of processing and/or representation, we consider studies of language production and language processing in turn. Our review of language-processing studies in schizophrenia in this section is limited to those that have used traditional off-line methods. These are methods that do not measure psycholinguistic operations at the time they occur. In most off-line studies, the task is untimed (immediate judgment or immediate/delayed recall or recognition), and the dependent variable is usually accuracy or error type. The interest of the study lies in what linguistic representations a patient can or cannot deal with. Studies that have made use of so-called on-line methods are reviewed in the next section.

Single Words

It has long been noted that the speech of some patients with schizophrenia is characterized by a "preoccupation with too many of the semantic features of words" (42). After Bleuler (28) who conceived of schizophrenia as a disorder of association, most studies in schizophrenia at the level of single words have focussed on the structure of and access to semantic memory.

As discussed below, patients with schizophrenia perform poorly on several different verbal semantic tasks including semantic fluency, naming, categorization and recall. Recent studies have suggested that poor performance on some semantic tasks is particularly impaired in patients with thought disorder (104).

Many of these studies of semantic memory in schizophrenia have followed the neuropsychological lesion literature in attempting to distinguish between storage and access/retrieval deficits in semantic memory (259,260). There are said to be five hallmarks of the loss of items in semantic memory:

1. Consistent production of semantic errors on particular items across different inputs (pictures, written words, spoken words),
2. Relative preservation of superordinate information as opposed to information about an item's specific semantic features,
3. Relative preservation of information about higher frequency items,
4. Improvement of performance by priming and cueing,
5. No effect of the rate at which a task is performed on performance.

Disorders of retrieval of items and information from semantic memory are said to be characterized by the opposite effects of these variables on performance. Patients have been described who are said to show characteristics indicating storage versus retrieval deficits and vice versa, but the interpretation of these data remains controversial (39,40). As discussed further below, in schizophrenia, the traditional empirical distinctions between storage and access deficits may be even more blurred because several researchers have proposed that in these patients the semantic memory store may be disorganized rather than degraded (8,231).

Production of Single Words

Word Association Tasks and Verbal Fluency

Early experiments carried out by Bleuler, Jung, and Kraeplin used word-association tasks to show that schizo-

phrenic patients produced more idiosyncratic associations than normal controls (153,273). These findings were confirmed by some later studies (149,157,208,214,217, 261,262).

Increased associative word production may be particularly characteristic of positive thought disorder, not only in schizophrenia but also in schizoaffective disorder and mania (182).

The word-association task has inherent limitations: there exist different sets of norms that have changed over the years, and deciding what constitutes a "rare" response is largely subjective. Another paradigm commonly used in schizophrenia research is verbal fluency, otherwise known as the controlled oral association test. The semantic or category fluency version of this task requires subjects to produce words within specific categories (e.g., animals, body parts, furniture) (296). Many researchers using the semantic fluency task have reported that schizophrenic patients generate fewer words in a specified period than controls (2,6,8,23,49,62,103,106,111,115,152,205,231). Such deficits are evident at an early stage of illness (49,231) and are particularly associated with negative symptoms (2,48). They do not increase with increased illness duration (49,103) and cannot be explained by intellectual deficits (62).

Temporal analyses of semantic fluency in schizophrenia suggests that, given enough time, patients do eventually produce the same total number of category exemplars as controls (5,6), suggesting that the deficit is not due to a degradation in knowledge but rather reflects an impairment in access to or retrieval from the semantic system. There have been two studies in which the experimenter provided cues to verbal production, but the results of these studies are difficult to interpret, partly because of ceiling effects in the control groups (103,152).

If the problem is indeed one of retrieval, one question is whether it is specific to retrieval from semantic memory or whether it reflects a general problem in the retrieval of verbal items. Some studies have reported that patients perform selectively poorly on semantic versus letter fluency tasks (84,104,111), suggesting a differential deficit in the semantic system. However, in two other studies, patients produced 60% to 70% of the number of words produced by the controls' on both letter and semantic fluency tasks (23, 152).

During semantic fluency, some patients produce words that are inappropriate for a given category. Multidimensional scaling and clustering techniques have been used to examine the relationships between words produced within given categories in more depth (2,8,231). These studies have suggested that patients are less likely than controls to group superordinate exemplars in clusters and are more likely to produce bizarre associations. Whether these patterns reflect a disorganization of the storage of items or specific deficits in the task-appropriate selection of items in semantic memory (see below) requires further investigation.

Naming and Repetition Tasks

Anomia is a characteristic feature of several types of aphasic syndromes and suggests problems in accessing lexical phonologic representations from semantic memory and/or planning speech production. Naming tasks in schizophrenia have usually been administered as part of neuropsychological batteries designed for use with the aphasia syndromes, e.g., the Boston Diagnostic Aphasia Examination (109). Schizophrenic patients perform less well than controls (82) and at times as poorly as fluent aphasic patients (176) and patients with Alzheimer disease (69) on such tasks. This deficit may not, however, be specifically associated with thought disorder (104).

Improved naming with semantic cuing (178,206) suggests problems with access to semantic memory rather than a storage loss (see above), although the questions of whether patients show variability in performance across trials or an effect of familiarity remain controversial (178,206).

Comprehension and Memory of Single Words

Semantic Categorization and Immediate Recall

Several investigators have reported that patients with schizophrenia are slower and less accurate in classifying words or word pairs as members of conceptual categories (47,53,116). Moreover, in a timed short-term memory recognition task, schizophrenic patients with positive thought disorder failed to use semantic information to (a) improve recall of items that were originally encoded among semantically related words and (b) elicit false recognitions of targets that were semantically related to the originally encoded words (210).

Semantic Categorization and Long-Term Recall

When normal subjects learn a list of words, their recall is better if the list can be organized into semantic categories than if it consists of a sequence of unrelated words. This is thought to reflect the tendency to organize words in semantic memory during encoding (61,160). An encoding strategy of semantic organization is also reflected by the organization of words at recall (282).

Patients with schizophrenia fail to spontaneously use semantic categorization strategies (201), often producing largely unorganized word lists at recall (32,35,102,146, 156,162,177,186,201,243). Nonetheless, most studies (146,163,164,177,243), although not all (102), have reported that, if material is preorganized or if patients are given enough time to organize material during encoding (see above), they have the capacity to use semantic information to improve recall.

One study examined the recall of words whose associative properties were already known. In healthy controls, recall of a particular word was dependent on both its associative strength and the number of associative links within its associative network. In schizophrenic patients, however, cued recall was impaired, particularly for words of low associative strength (221).

Summary

Most of the studies reviewed above suggest that there is no loss of semantic information in schizophrenia. The problem appears to be one of access/retrieval and of using semantic knowledge effectively. Access and retrieval of items in semantic memory can be subdivided into several subcomponents: successful retrieval (or recovery) involves both the activation and selection of target items in semantic memory (194). Either of these processes could be disturbed in schizophrenia. Moreover, some of the studies described above suggest that schizophrenia may be characterized by a disorganized semantic memory store. This might, in turn, lead to deficits in effective retrieval.

Another important question requiring further study is the degree to which semantic deficits occur specifically at the lexical level as opposed to at an amodal level of semantic representation. Some patients have been shown to perform poorly on nonverbal or cross-modal semantic tasks such as word-to-picture matching and picture classification (205,289), suggesting at least some dysfunction at an amodal level of semantic memory.

Words in Sentences and Discourse

The observation that some patients with schizophrenia produce "sentences according to the semantic features of previously uttered words, rather than according to a topic" (42) has led to several studies investigating the relationship between the meanings of individual words within whole sentences and discourse in schizophrenia.

Production of Words in Sentences

Word Associations in Sentences

In an elaboration of the free word-association task (discussed above), subjects are asked to place the words that they produced in the context of a sentence (147). In one study using this paradigm, at least 70% of responses by both patients and controls that were judged to be pathologic based on the word-association test alone became meaningful in the context of sentences (110). In a more recent study, however, patients did have difficulty in producing words in the context of sentences: some patients with negative symptoms were unable to put their idiosyncratic associations into meaningful sentences, and patients with positive symptoms were unable to place common associations in meaningful sentences (149).

This method, however, has the same drawbacks as the single word-association tasks in that the decision of whether or not a word is appropriate in the context of a sentence is largely subjective.

Predictability of Words in Sentences

A more systematic method of examining how well a word fits in with its surrounding context is to determine its predictability or redundancy using the technique of Cloze analysis. Speech is transcribed, and words in the resultant text are perodically omitted. Normal readers are then asked to determine the missing words. In one of the first studies using this technique, judges (healthy individuals) were less able to guess the words omitted from the first 100 words of a schizophrenic patient's transcript than those omitted from the transcript of a control subject (248). Judges performed even more poorly for the second 100 words of the schizophrenic transcript, although, for the control transcript, their judgment actually improved. Further experiments indicated that judges performed differentially poorly on schizophrenic transcripts when lengthy (14 words surrounding the omitted word) as opposed to more immediate contexts (four surrounding words) were provided (249,250). These findings were interpreted as supporting an 'immediacy hypothesis', which proposed that schizophrenic behavior (verbal and nonverbal) was primarily controlled by stimuli immediate to the environment. Later studies, however, failed to replicate these findings (244) and suggested that only patients with thought disorder produced unpredictable speech (124, 189).

Repetition of Words in Sentences

Another statistical measure of language is the type:token ratio. This is considered a measure of flexibility or variability in the use of words in discourse. It measures the number of different words (types) in relation to the total number of words used (tokens). Several studies have reported that the type:token ratio is generally lower in speech and writing produced by schizophrenic patients than that produced by healthy controls (119,192,232). This is particularly apparent in patients with thought disorder (3,190).

Comprehension and Memory of Words in Sentences

Cloze Technique

Schizophrenic patients do not only produce speech that is relatively hard to predict (as described above), but they also appear to be specifically impaired in their ability to make predictability judgments on normal speech. This has been demonstrated using a reverse Cloze technique in which patients are asked to judge whether a word is appropriate in the context of normal transcribed speech (27,144). Moreover, when levels of context are systematically varied (by deleting words with different periodicity), acute schizophrenic patients not only fail to improve with greater context, but their performance deteriorates (71). These deficits do not appear to be due to differences in verbal IQ (71,269).

Lexical Ambiguity

The effects of sentential context on the meaning of individual words have been examined in a series of experiments of lexical ambiguity (45). Participants in these studies were asked to use context to judge the meaning of homographs—words with multiple unrelated meanings (e.g., bridge: a structure across a river or a card game). For many homographs, some meaning(s) occur more frequently than others and are referred to as *dominant* and *subordinate*, respectively. To interpret the subordinate meaning of homographs correctly, the surrounding context plays a crucial role. When the homograph and the preceding alternative sentence contexts are presented in the form of a multiple-choice test, schizophrenic patients tend to misjudge the subordinate meanings of the homographs more frequently than controls (24,27,45,285). For example, patients tended to select "writing implement" when given a sentence "When the farmer bought a herd of cattle, he needed a new pen."

Recall of Words in and out of Context

Just as the recall of words of normal subjects improves if they can be semantically categorized, recall is also better when words are encoded in the context of whole sentences rather than as isolated strings. Indeed, when normal subjects are presented with speech samples varying in their degree of contextual constraint, recall of these samples improves with increasing constraint (211). Findings have been mixed in schizophrenic patients. Some studies have documented that patients are less able to benefit from increasing context than normal controls (179,183,192), whereas others report that only some patients are impaired: chronic but not acute patients (180); thought-disordered but not non–thought-disordered patients (187), and left-handed but not right-handed patients (193). Intriguingly, there appears to be an interaction between the effects of context and the serial position of the word to be recalled (192); schizophrenic patients were able to use contextual constraint to recall words that were recently presented but failed to do so when words were presented in primacy and middle positions.

In most of these studies, contexts were derived statistically (211). Another approach is to use stimuli in which linguistic rules are violated in different ways. For example, in one study, the recall of words encoded in the context of normal sentences was compared with those encoded in the context of semantically anomalous sentences, semantically related word strings, and random word strings (297). Recall of words encoded in the context of real sentences was selectively impaired in the schizophrenic group. This was interpreted as suggesting that schizophrenic patients were unable to use a combination of syntactic and semantic information to improve recall. This interpretation is discussed in more detail below.

Judgment of Words in and out of Context

Three studies have examined the ability of schizophrenic patients to explicitly judge the acceptability of sentences in which linguistic rules have been violated. In the first study, chronic schizophrenic patients performed normally in judging the acceptability of selection–restriction violations (e.g., animacy, concrete/abstract violations) in sentences (212). In the second and third more recent studies, acutely psychotic patients (9) and thought-disordered patients (169) were relatively impaired in judging the acceptability of semantically anomalous sentences.

Sentences: Syntax

The extent to which syntactic relationships within sentences break down in schizophrenia is somewhat controversial. Although several classes of syntactic errors have been identified in the speech of a single schizophrenic patient (42), most of these errors can occur in the utterances of normal speakers (97). There have been several formal analyses of the syntactic structure of speech produced by schizophrenic patients and a few studies that have examined speech comprehension by manipulating syntactic parameters. One of the difficulties in using off-line tasks (that do not measure psycholinguistic operations at the time that they occur) to determine whether or not schizophrenic patients are selectively impaired in their use of syntax, is that syntactic structure and the meaning of words are eventually combined. Such tasks may therefore not distinguish between specific deficits in syntactic processing *per se* and deficits in the combination of syntactic with lexico-semantic processes. For example, in the study described above that reported a selective impairment in the ability of patients to recall words encoded in the context of normal sentences versus word strings might suggest either a problem in using syntactic information or in using a combination of syntactic and semantic information to improve recall.

Production of Syntactic Structure

Formal analyses of the speech produced by schizophrenic patients show that it is more grammatically deviant (139) and less complex than that of controls, as reflected by a higher percentage of simple sentences and, in compound sentences, fewer dependent clauses that are not deeply embedded (95,215). The latter findings are particularly associated with chronicity (158,294), early onset of the illness (216), and negative symptoms (293). A study that used another measure of syntactic complexity (number of clauses and proportion of relative:total clauses), however, reported no differences between patients and controls (251).

Comprehension of Syntactic Structure

Early studies suggested that schizophrenic patients perform as poorly as many aphasic brain-damaged patients on measures of comprehension (82,236,265). More recent studies have confirmed comprehension deficits in schizophrenic patients (57–59). Although manipulating grammatical structure appears to have no effect on language comprehension (58), when patients are asked to explicitly identify syntactic errors in sentences, their performance is relatively poor (9). This deficit, however, is not as marked as for the identification of semantic errors. Moreover, the identification of syntactic errors, unlike semantic errors, correlates negatively with educational achievement (9).

Discourse: Relationships between Sentences

Phenomena that are most frequent in schizophrenic thought disorder—tangentiality and derailment—occur primarily at the level of whole discourse. It is not surprising, therefore, that several investigators have attempted to examine discourse structure and the relationships between sentences in the speech produced by schizophrenic patients.

Production of Discourse

Cohesion Analysis

One of the most systematic examinations of schizophrenic discourse is cohesion analysis, first applied by Rochester and Martin (240) who reported that thought-disordered schizophrenic patients used fewer cohesive ties than normal and non–thought-disordered patients. In this first study, three types of cohesive ties were examined: reference (e.g., "I've known Bill for years. He is a great guy."), conjunction ("First I went to school and then I came back."), and lexical cohesion ("My sister is pretty independent. Independence has always been one of her strengths."). The finding that schizophrenic patients used abnormally few reference ties has been replicated by several investigators (7,125,129,207,246). Earlier measures have been refined and expanded. For example, the Communication Disturbance Index classifies unclear communication into several subtypes (73). The use of unclear and ambiguous verbal references appears to be a stable trait of schizophrenia (72), although it remains unclear exactly how this trait is linked to the symptom of thought disorder. One study, for example, reported no differences in cohesive elements between segments of speech that did and did not meet criteria for thought disorder as rated by the Thought, Language and Communication scale (129).

Predictability of Discourse

The links between sentences produced by schizophrenic patients have also been examined using measures of predictability. In one experiment, discourse produced by normal controls and schizophrenic patients was transcribed, and judges were asked to arrange groups of randomly arrayed sentences of schizophrenic patients and normal subjects into their correct original sequential order. The correct arrangement of three or more sentences in their original order was achieved more often for normal than for schizophrenic discourse (245). This was also true of transcripts of the conversations of schizophrenic patients with others (246).

Structure of Discourse

Normal discourse exhibits a systematic hierarchical structure in which propositions branch out from a central proposition. This tree structure has been reported to be relatively unconnected in psychotic speech (138,140).

Thematic Content of Discourse

Another approach has been to analyze the thematic content of discourse. In one study, patients were asked to describe pictures, and speech transcripts were decomposed into ideas (individual sentences, semantic propositions, phrases, and words) and then rated according to whether they were thematically appropriate to the picture or inferential. Thought-disordered patients produced significantly fewer inferences than controls, but there was a trend toward an increase in the number of ideas classified as inappropriate (4).

Manic Versus Schizophrenic Discourse

Some of the discourse measures described above have been used to compare the speech produced by schizophrenic and manic patients. Studies that included patients irrespective of their level of thought disorder have failed to detect differences in language coherence between these two groups (72,125,129,130). However, studies that limited comparisons to thought-disordered speech suggest that the discourse of manic patients is more cohesive than that of schizophrenic patients (140,207,310).

Comprehension and Memory of Discourse

There have been relatively few studies examining discourse processing in schizophrenia. An early study suggested that schizophrenic listeners were able to identify referents in the speech of normal controls (54). Later studies have focused on the effects of discourse organization and of the number of propositions during encoding on later recall and recognition.

Recall of Sentences in Discourse

As described above, in healthy individuals the recall of single words is superior if items are organized by semantic categories or are presented within sentences during encoding. Similarly, whole sentences are better remembered when presented as part of coherent discourse than when presented in random order. Schizophrenic patients may fail to take advantage of the organizational structure of sentences during encoding. In one study, organization of material presented during encoding did not influence organization at recall (127). In this study, even when patients generated their own discourse passages, recall performance remained inferior to that of controls. Intriguingly, thought-disordered patients (schizophrenics and manics) remembered significantly *more* than controls when sentences were presented randomly than when they were presented within a coherent text (272).

A gist paradigm (31) has also been used to study the memory of discourse in schizophrenia. In this task, participants learn four complex ideas from sentences with different numbers of propositions and later complete a recognition test. Normal participants are usually more confident in their recognition of sentences that describe ideas with a greater number of propositions, even when asked to recognize sentences that were not presented during encoding. This response pattern is thought to reflect a tendency of comprehenders to integrate semantic information and to store the meaning of the whole message rather than its verbatim form. The results of two studies using this gist paradigm in schizophrenia have been contradictory: When sentences were presented verbally, a normal pattern of recognition confidence levels in patients was reported (114), but when sentences were presented visually, some patients showed an abnormal response pattern (165).

Relationships between Comprehension and Production of Discourse

Performance on a task that probes the ability to select information relevant to discourse topics has also been examined in schizophrenic patients (19). Participants were given story topics and asked to select five (of 20) pictures that best told the story and to put these pictures in sequential order. Interestingly, within the schizophrenic group, discourse planning performance deficits indexed by this task were selectively correlated with the number of incomplete references (see above) produced in speech.

ON-LINE STUDIES OF LANGUAGE PROCESSING AND PRODUCTION IN PSYCHOSIS

By the mid-1960's, psycholinguistic researchers had begun to use methods that required a subject to make responses to ongoing language stimuli and to respond to a stimulus in a way that did not require conscious consideration of the representation under investigation. As opposed to 'off-line' tasks that can be thought of as tapping into the final representations, these on-line methods probe implicit language processing and index the *intermediate representations* formed as language unfolds. One example of an on-line task is the semantic priming paradigm (with a lexical decision task) used to study the semantic relationships between single words (218,219). Most on-line methods, however, have been used to investigate sentence processing. They include the localization of extraneous noises (clicks) in sentences (99), monitoring for phonemes (94), and more complex tasks such as self-paced reading (pressing a key to call up subsequent words or passages) (159,305).

Researchers have been concerned with the "ecologic validity" of some of the more complex on-line tasks. Monitoring the loci and durations of eye fixations during reading and recording ERPs (see below) does not require subjects to make behavioral responses and may therefore be more naturalistic measures of on-line language processing.

Characteristics of the Normal Language-Processing System

Studies of on-line processing have established a number of features that characterize the normal language-processing system (39,181,300). First, most processors are obligatorily activated when their inputs are presented to them. For instance, if we attend to a sound that happens to be the word "elephant," we must hear and understand that word; we cannot hear this sound as just a noise (198). Second, language processors generally operate unconsciously. The unconscious nature of most of language processing can be appreciated by considering that when we listen to a lecture, converse with an interlocutor, read a novel, or engage in some other language-processing task, we usually have the subjective impression that we are extracting another person's meaning and producing linguistic forms appropriate to our intentions without being aware of the details of the sounds of words or sentence structure. Third, components of the system operate remarkably quickly and accurately. For instance, it has been estimated that spoken words are usually recognized less than 125 ms after their onset, i.e., while they are still being uttered (199,299). This speed is achieved because of the massively parallel functional architecture of the language-processing system, leading to many of its components being simultaneously active.

Normal word production in speech occurs at the rate of approximately three words per second, with an error rate of approximately one per 1,000 words (181). Words that are appropriate to our conceptual preparation must be retrieved from a mental word production dictionary of more than 30,000 items (181). Thus, the ability to use linguistic context on-line is essential in speech production. Higher-order semantic–lexical connections are thought to be reciprocally interconnected or shared by speech input and output systems (181,213).

On-line methods have begun to be applied to the study of psychiatric disorders. Indeed, some of the earliest on-line studies of language processing were in schizophrenia (41,238), but these were not followed up. The use of these techniques in patient populations is valuable because they can give quite different views of language processing abnormalities than those emerging from off-line studies.

On-line Studies of lexico-semantic Processing in Schizophrenia

Numerous behavioral studies have shown that a subject's processing of related targets (e.g., "doctor–nurse") is enhanced or facilitated compared with naming or making lexical decisions about unrelated targets (e.g., "window–nurse"). Similarly, when subjects are asked to name or make lexical decisions about words in sentences, they respond more quickly to words preceded by a related context than to words preceded by an unrelated context (85,255–257,279–281). This is also true of scripts and texts (46,263,264). This facilitation typically takes the form of faster reaction times and is termed semantic or sentential priming.

There have been several studies of semantic priming in schizophrenia. Some have demonstrated greater priming effects in schizophrenic subjects than in controls (137,177,191,274,277). Furthermore, Spitzer et al. (275) showed that schizophrenics, particularly those who were thought disordered, had greater *indirect* priming effects (i.e., priming when there was a mediating word between prime and target) than normal subjects. These findings are consistent with Maher's proposal (188) that thought-disordered schizophrenic patients have an activated or disinhibited semantic associative network. Conversely, some groups have shown that priming in schizophrenic subjects is no greater than in normal subjects (20,29,43,44,136,225,303) and may even be reduced (8,21,137,225,303). These contradictory results may be due to a variety of methodologic factors, including the failure to distinguish between thought-disordered and non–thought-disordered schizophrenic patients. However, as discussed above, it may also depend on the particular experimental conditions and the paradigm used (219). Thus, it has been argued that the decreased priming shown by schizophrenic subjects under particular experimental conditions reflects a deficit in conceptually mediated priming that involves controlled rather than automatic mechanisms (20,225,303).

On-line Studies of Sentence Processing in Schizophrenia

One of the first applications of online techniques to the study of schizophrenia was the use of the 'click' paradigm (92,98) to investigate syntactic processing in sentences. With this technique, a short burst of noise (the click) presented during speech is usually perceived as occurring at, or near, a clause boundary, when, in fact, the click might have occurred somewhere in the middle of the clause. This automatic perceptual displacement of clicks is thought to be due to the operation of syntactic constraints. Three studies in schizophrenia using this paradigm have suggested that patients perceive the click at or near clause boundaries to the same extent as matched controls (41,114,238). These findings suggest that at least some aspects of syntactic processing are intact in schizophrenia.

Online studies examining the use of different contextual constraints in schizophrenia have yielded contradictory findings. In one study, schizophrenic patients were presented with words that were masked to a greater or lesser extent by white noise. The word strings formed grammatical and meaningful sentences, grammatical and meaningless sentences, or were randomized word strings. Schizophrenic patients benefited from increases in sentence cohesion to the same degree as healthy controls (101). In another study that examined the influence of contextual constraints on word perception in both visual and auditory domains, schizophrenic patients were no less sensitive in their detection threshold of words in linguistically anomalous sentences than healthy controls (79).

Other online studies suggest that patients with schizophrenia may be impaired in using contextual constraints during language processing. In one study in which schizophrenic patients shadowed (i.e., listened and repeated each word) texts, their errors tended to be semantically irrelevant more often than the errors of patients with affective disorders and normal controls. This was interpreted as suggesting that schizophrenic patients were impaired in their ability to follow semantic context (235). Another study reported a negative correlation between positive thought disorder and the ability of schizophrenic patients to monitor the level of organization in passages that they were required to shadow under distraction conditions (306).

A third study (169) investigated the use of linguistic context in positively thought disordered schizophrenics by examining their performance on an online word-monitoring task. Controls and non-thought disordered patients took longer to recognize words preceded by linguistic anomalies compared with words in normal sentences. Compared with both other groups, thought disordered schizophrenics showed significantly smaller differences in reaction time, suggesting that they were relatively insensitive to linguistic violations within individual patients over time. This suggested

that the impairment in the use of linguistic context were related to the state, rather than the trait, of thought disorder (171).

On-line Language Production in Schizophrenia

The term on-line is generally reserved for studies of language processing. However, as discussed above, speech production occurs implicitly and extremely fast. A few investigators have begun to investigate implicit processes during speech production in schizophrenia. For example, Spitzer et al. (276) compared the implicit use of context by thought-disordered and non–thought-disordered schizophrenic subjects by examining the distribution of pauses in the speech spontaneously produced by these two patient groups. Whereas the proportion of pauses before words produced in context was smaller than the proportion of pauses before words produced out of context in normal controls and non–thought-disordered patients, no such pattern was observed in thought-disordered schizophrenic patients.

RELATIONSHIP BETWEEN LANGUAGE DEFICITS AND OTHER COGNITIVE DEFICITS

Several investigators have argued that many of the language-processing deficits in schizophrenia arise directly from cognitive deficits outside the language system. These include deficits in selective attention and pigeonholing (134,258), working memory, and updating and retrieval from short-term memory (107,239), using an "internal representation of context" to guide action (55), use of strategy (146), and other executive and frontal lobe functions (138,202). In this section, we review some of the studies cited to support such theories. We first discuss studies that have documented associations between clinical measures of thought disorder and performance of cognitive tasks of attention and working memory. We then review the few studies that have investigated relationships between these more general cognitive functions and some of the measures of language production and processing outlined above.

Clinical Measures of Thought Disorder and General Cognitive Deficits

There is evidence that severity of disorganized speech in schizophrenia correlates with distractibility (76,130), deficits in short-term verbal memory (221), selective attention as measured by the Stroop task (21), sustained attention as measured by the Continuous Performance Test (224,230,286), measures of executive dysfunction (221), and lower-level information processing deficits such as prepulse inhibition (70,234). The question of whether thought disorder is linked to working memory deficits remains unresolved. On the one hand, there is a modest correlation of thought disorder severity with deficits in verbal working memory (221). On the other hand, although clinical thought disorder generally improves with standard neuroleptic medication treatment and worsens with withdrawal of medication, working memory deficits are usually resistant to neuroleptic treatment (105).

Measures of Language Dysfunction and General Cognitive Deficits

Some of the language deficits described above can theoretically be attributed to cognitive deficits outside the language system. For example, a deficit in rehearsing information (a component of working memory) might prevent deep encoding of semantic information (32). A short-term or working memory deficit might account for the interaction of contextual constraint with word position (primacy or recency) in schizophrenia (192).

At the level of single words, reduced semantic priming does not correlate with impaired performance on the Stroop test (21). Evidence of an association between performance of working memory and language tasks in schizophrenia is strongest at the level of sentences and discourse. Verbal working memory deficits are correlated with language comprehension deficits (58), and referential communication disturbances are associated with poor performance on tasks of immediate auditory memory (76), distractibility (76,128, 145), and working memory and attention (74). In a study of written language in schizophrenic patients, syntactic errors were partly explained by deficiencies in working memory and attention, although significant differences between the groups remained (295).

There are several caveats to the interpretation of most of these studies. First, functions such as working memory and attention have been defined very broadly. There is growing recognition of the importance of "fractionating" cognitive systems conceptually and empirically. For example, studies of working memory involve tasks that range from spatial to linguistic domains, with different levels of analysis from behavioral to neuroanatomic. Second, even in healthy volunteers, the precise relationships and interactions between general cognitive processes such as working memory and different types of language processing are not fully understood.

THE NEURAL SUBSTRATES OF LANGUAGE PROCESSING IN SCHIZOPHRENIA

Three different techniques have been used to investigate the neural basis of language deficits in schizophrenia. First, structural imaging studies have reported abnormalities in regions that are known to play an important role in language. Second, functional neuroimaging studies have described abnormal patterns of activation during the presentation of linguistic stimuli (usually at the level of single words). Third, electrophysiologic studies have reported abnormalities of ERP components that are known to be sensitive to Levels of Language processing in schizophrenia, using both single-word and whole-sentence paradigms.

Structural Imaging Studies

Morphometric studies examining cortical gray matter volume in schizophrenia have traditionally focused on specific regions of interest (ROIs) that are usually selected on the basis of lobar neuroanatomy. Such studies have reported small reductions in the volume of several ROIs, particularly within the temporal and prefrontal cortices, in patients with schizophrenia (265a). In addition, there have been attempts to automate the measurement of gray matter volume or density throughout the brain (17,307,37a,295a). Taken together, such studies have confirmed subtle volumetric reductions in multiple anatomical regions within the prefrontal and temporal cortices (145a,265a,267,308). Interestingly, several studies suggest that these structural abnormalities may be more extensive on the left than the right. One study suggested that, within a group of schizophrenic patients, the degree of atrophy within the left temporal cortex was correlated with the severity of thought disorder (266). It is therefore possible that subtle temporal and frontal gray matter atrophy may contribute to some of the abnormalities in semantic and language function discussed in this review.

Functional Neuroimaging Studies

Functional neuroimaging studies in schizophrenia using a wide variety of cognitive paradigms have shown that perturbations of brain activity in schizophrenia are not localized to one brain region but to networks comprising multiple regions, particularly involving the frontal and temporal cortices and subcortical structures such as the cerebellum and thalamus (170).

Most functional neuroimaging studies in schizophrenia that have used linguistic stimuli have been at the level of individual words rather than whole sentences and discourse. Studies that have used tasks such as verbal fluency (66,88,96,311), semantic categorization (67,148), and the recognition of learned words (64,133) have reported abnormal patterns of activity within both frontal and temporal regions. Some of these studies have also reported relatively increased activity in parietal regions (64,66).

Another approach to investigating localized dysfunction is to examine the relationships between brain regions in schizophrenia. Structural equation modeling (path analysis) of a PET semantic processing study suggested differences between schizophrenic and control groups in interactions among frontal regions, between frontal and temporal regions, and between the lateral, frontal and anterior cingulate cortices (148). An effective connectivity analysis of PET data from a graded memory study suggested that the normal anterior cingulate modulation of frontal–temporal interaction is disrupted in schizophrenia (89,90).

Relative increases in activation of several other regions in patients relative to controls have been described during functional neuroimaging studies of language in schizophrenia. For example, greater activity in the parietal cortex in schizophrenic patients (relative to controls) has been reported in association with covert word production (66) as well as learning and recalling word lists (90). Increased activation of the inferior temporal cortex and/or fusiform cortex has been reported in schizophrenic patients (relative to controls) in association with cued stem recall of semantically encoded words (133), the recognition of novel words (64), the completion of sentences (160a) and during speech production (203,160b).

Event-Related Potential Studies

ERPs are voltage fluctuations derived from the ongoing electroencephalography that are time locked to specific sensory, motor, or cognitive events (56). Particular regions or temporal windows of the ERP waveform (components) have been differentiated and labeled according to their polarity (positive or negative), peak latency, and/or spatial position on the scalp. Several ERP components have been particularly useful in the study of on-line processes. These components can be measured without subjects having to make an overt response (e.g., pressing a button after each word or sentence), giving them an advantage over the on-line behavioral measures reviewed earlier in this chapter.

The best studied ERP component in schizophrenia is the P300 which is thought to index the process by which contextual information is updated within memory (78). The amplitude of the P300 is reduced in schizophrenic patients; indeed, this is one of the most robust and consistent biologic markers of schizophrenia, although it is not specific to this disorder. In addition, many (although not all) studies have reported an increased latency of the P300 in schizophrenia (93).

More recently, researchers have begun to investigate the use of semantic context in schizophrenia by examining the N400 ERP component. The N400 is a large, negative-going waveform, peaking at approximately 400 ms that was first described in association with contextually inappropriate words in sentences (172–174) and with unprimed words in word-pair semantic priming paradigms (25,142,143,242). The difference in N400 amplitude to primed and unprimed words was termed the N400 effect.

In many studies of the N400 in schizophrenia, patients showed a relatively intact N400 congruity effect, i.e., with larger amplitude N400 elicited to unprimed than primed words in semantic priming word-pair paradigms (30,112,166,167,228,278) and to words preceded by incongruent than congruent contexts in sentence paradigms (16,220,222,268). Nonetheless, other studies have reported an abnormally reduced N400 effect in both sentence (1,227,247) and word-pair paradigms (60,284). One reason for these contradictory findings may be heterogeneity in the schizophrenic patient samples used in different studies. Consistent with this idea is the finding of an inverse correlation between the N400 effect and severity of thought disorder within a group of schizophrenic patients (16). This supports the behavioral findings described above and suggests that an on-line deficit in using semantic context may be specifically associated with positive thought disorder.

Some of the above studies have reported an increase in the absolute amplitude of the N400 waveform to primed and unprimed words in word-pair paradigms (30) as well as to congruous and incongruous words in sentence paradigms (220,222,227). This has been argued to support the idea that schizophrenic patients have difficulty in processing the meaning of words, regardless of their context.

Modifications of both word and sentence paradigms give additional insights into the nature of on-line language-processing deficits in schizophrenia. In a mediated priming paradigm (18), an N400 congruity effect to target words that were preceded by indirectly related words (e.g., "lion—stripes"—both related to "tiger") was reported in schizophrenic patients but not in healthy participants (278). This is consistent with the idea that activity spreads abnormally far across interconnected representations in semantic memory in patients, supporting the online behavioral studies using the indirect semantic priming paradigm described earlier in this chapter. In a sentence paradigm, an N400 effect was elicited to words preceded by a semantically associated homonym when the surrounding context suggested the secondary meaning of the homonym in healthy volunteers but not in patients with schizophrenia (268). In other words, in patients, the context of the whole sentence failed to override the semantic associative effects of its individual words. This suggests that sentence and discourse deficits in schizophrenia may be, to some degree, driven by abnormalities in a lexico-semantic network.

Finally, probably the most robust abnormality described across studies is an increased N400 latency. This has been reported in both word-pair (30,112,167) and sentence (1,16,220,222,228,278) paradigms and suggests that the contextual integration of words may be delayed in schizophrenia.

CONCLUSIONS

In summary, schizophrenia is a complex disorder that is frequently manifest in language and related cognitive dysfunction. We have reviewed studies that have identified abnormalities in both language output and comprehension in schizophrenic patients. Abnormalities have been described at the level of single words (deficits in the structure and function of lexico-semantic memory), sentences (impaired use of different types of linguistic context) and whole discourse (abnormal relationships between sentences). The neurocognitive basis of language dysfunction in schizophrenia has been investigated using structural and functional neuroimaging as well as electrophysiological techniques. These techniques give complementary information. ERP studies suggest neurophysiological anormalities in the online use of semantic context, while neuroimaging studies suggest widespread structural and functional neuroanatomical abnormalities, particularly in the temporal and frontal cortices.

ACKNOWLEDGMENT

This work was supported by grants from the National Institute of Mental Health (K23MH02034) and National Institutes of Health (DC00942).

REFERENCES

1. Adams J, Faux SF, Nestor PG, et al. ERP abnormalities during semantic processing in schizophrenia. *Schizophr Res* 1993; 10:247–257.
2. Allen HA. Do positive symptom and negative symptom subtypes of schizophrenia show qualitative differences in language production? *Psychol Med* 1983;13:787–797.
3. Allen HA. Selective retrieval and free emission of category exemplars in schizophrenia. *Br J Psychol* 1983;74:481–490.
4. Allen HA. Positive and negative symptoms and the thematic organisation of schizophrenic speech. *Br J Psychiatry* 1984;144: 611–617.
5. Allen HA, Frith CD. Selective retrieval and free emission of category exemplars in schizophrenia. *Br J Psychol* 1983;74:481–490.
6. Allen HA, Liddle PF, Frith CD. Negative features, retrieval processes and verbal fluency in schizophrenia. *Br J Psychiatry* 1993;163:769–775.
7. Allen HA, Allen DS. Positive symptoms and the organization within and between ideas in schizophrenic speech. *Psychol Med* 1985;15:71–80.
8. Aloia MS, Gourovitch ML, Weinberger DR, et al. An investigation of semantic space in patients with schizophrenia. *J Int Neuropsychol Soc* 1996;2:267–273.
9. Anand A, Wales RJ, Jackson HJ, et al. Linguistic impairment in early psychosis. *J Nerv Ment Dis* 1994;182:488–493.
10. Andreasen NC, Powers PS. Overinclusive thinking in mania and schizophrenia. *Br J Psychiatry* 1974;125:452–456.
11. Andreasen NC. Thought, language and communication disorders: I. Clinical assessment, definition of terms, and evaluation of their reliability. *Arch Gen Psychiatry* 1979;36:1315–1321.
12. Andreasen NC. Thought, language and communication disorders. II. Diagnostic significance. *Arch Gen Psychiatry* 1979;36: 1325–1330.
13. Andreasen NC. Negative symptoms in schizophrenia: definition and reliability. *Arch Gen Psychiatry* 1982;39:784–788.
14. Andreasen NC. Should the term "thought disorder" be revised? *Compr Psychiatry* 1982;23:291–299.
15. Andreasen NC. The scale for assessment of thought, language and communication (TLC). *Schizophr Bull* 1986;12:473–482.
16. Andrews S, Shelley A, Ward PB, et al. Event-related potential indices of semantic processing in schizophrenia. *Biol Psychiatry* 1993;34:443–458.
17. Ashburner J, Friston KJ. Voxel-based morphometry—the methods. *Neuroimage* 2000;11:805–821.
18. Balota DA, Lorch RF Jr. *J Exp Psychol Learn Mem Cogn* 1986; 12:336–345.
19. Barch DM, Berenbaum H. Language production and thought disorder in schizophrenia. *J Abnorm Psychol* 1996;105:81–88.
20. Barch DM, Cohen JD, Servan-Schreiber D, et al. Semantic priming in schizophrenia: an examination of spreading activation using word pronunciation and multiple SOAs. *J Abnorm Psychol* 1996;105:592–601.
21. Barch DM, Carter CS, Perlstein W, et al. Increased Stroop facilitation effects in schizophrenia are not due to increased automatic spreading activation. *Schizophr Res* 1999;39:51–64.
22. [Reserved.]
23. Beatty WW, Jocic Z, Monson N, et al. Memory and frontal lobe dysfunction in schizophrenia and schizoaffective disorder. *J Nerv Ment Dis* 1993;181:448–453.
24. Benjamin TB, Watt NF. Psychopathology and semantic interpretation of ambiguous words. *J Abnorm Psychol* 1969;74:706–714.
25. Bentin S. Event-related potentials, semantic processes, and expectancy factors in word recognition. *Brain Lang* 1987;31:308–327.
26. Berenbaum H, Barch D. The categorization of thought disorder. *J Psycholinguist Res* 1995;24:349–376.
27. Blaney PH. Two studies on the language behavior of schizophrenics. *J Abnorm Psychol* 1974;83:23–31.
28. Bleuler E. *Dementia praecox, or the group of schizophrenias.* New York: International Universities Press, 1911/1950.
29. Blum NA, Freides D. Investigating thought disorder in schizophrenia with the lexical decision task. *Schizophr Res* 1995; 16:217–224.
30. Bobes MA, Lei ZX, Ibanez S, et al. Semantic matching of pictures in schizophrenia: a cross-cultural ERP study. *Biol Psychiatry* 1996;40:189–202.
31. Bransford JD, Franks JJ. The abstraction of linguistic ideas. *Cogn Psychol* 1971;2:331–350.
32. Brebion G, Amador X, Smith MJ, et al. Mechanisms

underlying memory impairment in schizophrenia. *Psychol Med* 1997;27:383–393.

33. Brown R. Schizophrenia, language and reality. *Am Psychol* 1973;28:395–403.

34. Brownell HH, Potter HH, Bihrle AM, et al. Inference deficits in right brain-damaged patients. *Brain Lang* 1986;29:310–321.

35. Calev A, Venables PH, Monk AF. Evidence for distinct verbal memory pathologies in severely and mildly disturbed schizophrenics. *Schizophr Bull* 1983;9:247–264.

36. Cameron N. Schizophrenic thinking in a problem-solving situation. *J Ment Sci* 1939;85:1012–1035.

37. Cameron N. Experimental analysis of schizophrenia thinking. In: Kasanin J, ed. *Language and thought in schizophrenia.* Berkeley: University of California Press, 1964.

37a. Cannon TD, Thompson PM, van Erp TG, et al. Cortex mapping reveals regionally specific patterns of genetic and disease-specific gray-matter deficits in twins discordant for schizophrenia. *Proc Natl Acad Sci USA* 2002;99(5):3228–3233.

38. Caplan D. *Neurolinguistics and linguistic aphasiology.* Cambridge, UK: Cambridge University Press, 1987.

39. Caplan D. *Language structure, processing and disorders.* Cambridge, MA: MIT Press, 1992.

40. Caramazza A, Hillis AE, et al. The multiple semantics hypothesis: multiple confusions? *Cogn Neuropsychol* 1990;7:161–189.

41. Carpenter MD. Sensitivity to syntactic structure: good versus poor premorbid schizophrenics. *J Abnorm Psychol* 1976;85:41–50.

42. Chaika E. A linguist looks at 'schizophrenic' language. *Brain Lang* 1974;1:257–276.

43. Chapin K, Vann LE, Lycaki H, et al. Investigation of the associative network in schizophrenia using the semantic priming paradigm. *Schizophr Res* 1989;2:355–360.

44. Chapin K, McCown J, Vann L, et al. Activation and facilitation in the lexicon of schizophrenics. *Schizophr Res* 1992;6:251–255.

45. Chapman LJ, Chapman JP, et al. A theory of verbal behaviour in schizophrenia. In: Maher BA, ed. *Progress in experimental personality research, volume 1.* San Diego: Academic Press, 1964:49–77.

46. Chawarsky MC, Sternberg JR. Negative priming in word recognition: a context effect. *J Exp Psychol* 1993;122:195–206.

47. Chen EYH, Wilkins AJ, McKenna PJ. Semantic memory is both impaired and anomalous in schizophrenia. *Psychol Med* 1994;24:193–202.

48. Chen EY, Lam LC, Chen RY, et al. Negative symptoms, neurological signs and neuropsychological impairments in 204 Hong Kong Chinese patients with schizophrenia. *Br J Psychiatry* 1996;168:227–233.

49. Chen EY, Lam LC, Chen RY, et al. Prefrontal neuropsychological impairment and illness duration in schizophrenia: a study of 204 patients in Hong Kong. *Acta Psychiatr Scand* 1996;93:144–150.

50. Chomsky N. *Aspects of the theory of syntax.* Cambridge, MA: MIT Press, 1965.

51. Chomsky N. *Lectures on government and binding.* Dordrecht: Foris, 1981.

52. Chomsky N. *Knowledge of language.* New York: Praeger, 1986.

53. Clare L, McKenna PJ, Mortimer AM, et al. Memory in schizophrenia: what is impaired and what is preserved? *Neuropsychologia* 1993;31:1225–1241.

54. [Reserved.]

55. Cohen JD, Servan-Schreiber D. Context, cortex, and dopamine: a connectionist approach to behaviour and biology in schizophrenia. *Psychol Rev* 1992;99:45–77.

56. Coles MGH, Rugg MD. In: Rugg MD, Coles MGH, eds. *Electrophysiology of mind.* Oxford: Oxford University Press, 1995.

57. Condray R, Steinhauer SR, Goldstein G, et al. Language comprehension in schizophrenics and their brothers. *Biol Psychiatry* 1992;32:790–802.

58. Condray R, Steinhauer SR, van Kammen DP, et al. Working memory capacity predicts language comprehension in schizophrenic patients. *Schizophr Res* 1996;20:1–13.

59. Condray R, van Kammen DP, Steinhauer SR, et al. Language comprehension in schizophrenia: trait or state indicator? *Biol Psychiatry* 1995;38:287–296.

60. Condray R, Steinhauer SR, Cohen JD, et al. Modulation of language processing in schizophrenia: effects of context and haloperidol on the event-related potential. *Biol Psychiatry* 1999;45:1336–1355.

61. Craik F, Lockhart R. Levels of processing: a framework for memory research. *J Verbal Learn Verbal Behav* 1971;11:671–684.

62. Crawford JR, Obonsawin MC, Bremner M. Frontal lobe impairment in schizophrenia: relationship to intellectual functioning. *Psychol Med* 1993;23:787–790.

63. [Reserved.]

64. Crespo-Facorro B, Wiser AK, Andreasen NC, et al. Neural basis of novel and well-learned recognition memory in schizophrenia: a positron emission tomography study. *Hum Brain Mapp* 2001;12:219–231.

65. Crow TJ. Molecular pathology of schizophrenia: more than one disease process. *BMJ* 1980;280:1–9.

66. Curtis VA, Bullmore ET, Brammer MJ, et al. Attenuated frontal activation during a verbal fluency task in patients with schizophrenia. *Am J Psychiatry* 1998;155:1056–1063.

67. Curtis VA, Bullmore ET, Morris RG, et al. Attenuated frontal activation in schizophrenia may be task dependent. *Schizophr Res* 1999;37:35–44.

68. Cutting J, Murphy D. Schizophrenic thought disorder. A psychological and organic interpretation. *Br J Psychiatry* 1988; 152:310–319.

69. Davidson M, Harvey P, Welsh KA, et al. Cognitive functioning in late-life schizophrenia: a comparison of elderly schizophrenic patients and patients with Alzheimer's disease. *Am J Psychiatry* 1996;153:1274–1279.

70. Dawson ME, Schell AM, Hazlett EA, et al. On the clinical and cognitive meaning of impaired sensorimotor gating in schizophrenia. *Psychiatry Res* 2000;96:187–197.

71. de Silva WP, Hemsley DR. The influence of context on language perception in schizophrenia. *Br J Soc Clin Psychol* 1977;16:337–345.

72. Docherty N, Schnur M, Harvey PD. Reference performance and positive and negative thought disorder: a follow-up study of manics and schizophrenics. *J Abnorm Psychol* 1988;97:437–442.

73. Docherty NM, DeRosa M, Andreason NC. Communication disturbances in schizophrenia and mania. *Arch Gen Psychiatry* 1996;53:358–364.

74. Docherty NM, Hawkins KA, Hoffman RE, et al. Working memory, attention, and communication disturbances in schizophrenia. *J Abnorm Psychol* 1996;105:212–219.

75. Docherty NM, Hall MJ, Gordinier SW. Affective reactivity of speech in schizophrenia patients and their nonschizophrenic relatives. *J Abnorm Psychol* 1998;107:461–467.

76. Docherty NM, Gordinier SW. Immediate memory, attention and communication disturbances in schizophrenia patients and their relatives. *Psychol Med* 1999;29:189–197.

77. [Reserved.]

78. Donchin E, Coles MGH. Is the P300 component a manifestation of context updating? *Behav Brain Sci* 1988;11:355–372.
79. Done DJ, Frith CD. The effect of context during word perception in schizophrenic patients. *Brain Lang* 1984;23:318–336.
79a. DSM-IV: Diagnostic and statistical manual of mental disorders, 4th ed. Washington, D.C.: American Psychiatric Press, 1990.
80. Earle-Boyer EA, Levinson JC, Grant R, et al. The consistency of thought disorder in mania and schizophrenia. II. An assessment at consecutive admissions. *J Nerv Ment Dis* 1986;174:443–447.
81. Elman JL, McClelland JL. Speech perception as a cognitive process: the interactive activation model. In: Lass N, ed. *Speech and language.* New York: Academic Press, 1984:10.
82. Faber R, Reichstein MB. Language dysfunction in schizophrenia. *Br J Psychiatry* 1981;139:519–522.
83. Faber R, Abrams R, Taylor MA, et al. Comparison of schizophrenic patients with formal thought disorder and neurologically impaired patients with aphasia. *Am J Psychiatry* 1983;140:1348–1351.
84. Feinstein A, Goldberg TE, Nowlin B, et al. Types and characteristics of remote memory impairment in schizophrenia. *Schizophr Res* 1998;30:155–163.
85. Fischler IS, Bloom PA. Automatic and attentional processes in the effects of sentence contexts on word recognition. *J Verbal Learn Verbal Behav* 1979;5:1–20.
86. Fish F. *An outline of psychiatry for students and practitioners.* Bristol: John Wright & Sons, 1964.
87. [Reserved.]
88. Fletcher PC, Shallice T, Frith CD, et al. Brain activity during memory retrieval. The influence of imagery and semantic cueing. *Brain* 1996;119:1587–1596.
89. Fletcher P, McKenna PJ, Friston KJ, et al. Abnormal cingulate modulation of fronto-temporal connectivity in schizophrenia. *Neuroimage* 1999;9:337–342.
90. Fletcher PC, McKenna PJ, Frith CD, et al. Brain activations in schizophrenia during a graded memory task studied with functional neuroimaging. *Arch Gen Psychiatry* 1998;55:1001–1008.
91. Fodor JA. *The modularity of the mind.* Cambridge, MA: MIT Press, 1983.
92. Fodor JA, Bever TG. The psychological reality of linguistic segments. *J Verbal Learn Verbal Behav* 1965;4:414–420.
93. Ford JM. Schizophrenia: the broken P300 and beyond. *Psychophysiology* 1999;36:667–682.
94. Foss DJ. Some effects of ambiguity upon sentence comprehension. *J Verbal Learn Verbal Behav* 1970;9:699–706.
95. Fraser WI, King KM, Thomas P, et al. The diagnosis of schizophrenia by language analysis. *Br J Psychiatry* 1986;148:275–278.
96. Frith CD, Friston KJ, Herold S, et al. Regional brain activity in chronic schizophrenic patients during the performance of a verbal fluency task. *Br J Psychiatry* 1995;167(3):343–349.
97. Fromkin VA. A linguist looks at 'schizophrenic' language. *Brain Lang* 1975;2:498–503.
98. Garrett M, Bever TG, et al. The active use of grammar in speech perception. *Percept Psycholing* 1966;1:30–32.
99. Garrett MF. The analysis of sentence production. In: Bower B, ed. *Psychology of learning and motivation: vol. 9.* New York: Academic Press, 1975:137–177.
100. Gerson S, Benson DF, Frazier SH. Diagnosis: schizophrenic versus posterior aphasia. *Am J Psychiatry* 1977;134:966–969.
101. Gerver D. Linguistic rules and the perception and recall of speech by schizophrenic patients. *J Soc Clin Psychol* 1967;6:204–211.
102. Gold JM, Randolph C, Carpenter CJ, et al. Forms of memory failure in schizophrenia. *J Abnorm Psychol* 1992;10:487–494.
103. Goldberg TE, Hyde TM, Kleinman JE, et al. Course of schizophrenia: neuropsychological evidence for a static encephalopathy. *Schizophr Bull* 1993;19:797–804.
104. Goldberg TE, Aloia MS, Gourovitch ML, et al. Cognitive substrates of thought disorder, I: the semantic system. *Am J Psychiatry* 1998;155:1671–1676.
105. Goldberg TE, Weinberger DR. Thought disorder, working memory and attention: interrelationships and the effects of neuroleptic medications. *Int Clin Psychopharmacol* 1995;10[Suppl 3]:99–104.
106. Goldberg TE, Torrey EF, Gold JM, et al. Learning and memory in monozygotic twins discordant for schizophrenia. *Psychol Med* 1993;23:71–85.
107. Goldman-Rakic PS. Working memory dysfunction in schizophrenia. *J Neuropsychiatry Clin Neurosci* 1994;6:348–357.
108. Goldstein K. Methodological approach to the study of schizophrenic thought disorder. In: Kasanin J, ed. *Language and thought in schizophrenia.* Berkeley: University of California Press, 1944.
109. Goodglass H, Kaplan E. *Boston diagnostic aphasia examination booklet.* Philadelphia: Lea & Febiger, 1972.
110. Gordon R, Silverstein ML, Harrow M. Associative thinking in schizophrenia: a contextualist approach. *J Clin Psychol* 1982;38:684–696.
111. Gourovitch ML, Goldberg TE, et al. Verbal fluency deficits in patients with schizophrenia: semantic fluency is differentially impaired as compared with phonologic fluency. *Neuropsychology* 1996;10:573–577.
112. Grillon C, Rezvan A, Glazer WM. N400 and semantic categorization in schizophrenia. *Biol Psychiatry* 1991;29:467–480.
113. Grosz BJ, Pollack ME, Sidner CL. In: Posner M, ed. *Foundations of cognitive science.* Cambridge, MA: MIT Press, 1989:437–468.
114. Grove WM, Andreasen NC. Language and thinking in psychosis: is there an input abnormality? *Arch Gen Psychiatry* 1985;42:26–32.
115. Gruzelier J, Seymour K, Wilson L, et al. Impairments on neuropsychologic tests of temporohippocampal and frontohippocampal functions and word fluency in remitting schizophrenia and affective disorders. *Arch Gen Psychiatry* 1988;45:623–629.
116. Gurd JM, Elvevaag B, et al. Semantic category word search impairment in schizophrenia. *Cogn Neuropsychiatry* 1997;2:291–302.
117. Haddock G, Wolfenden M, Lowens I, et al. Effect of emotional salience on thought disorder in patients with schizophrenia. *Br J Psychiatry* 1995;167:618–620.
118. Halle M, Vergnaud J-R. *An essay on stress.* Cambridge, MA: MIT Press, 1987.
119. Hammer M, Salzinger K. Some formal characteristics of schizophrenic speech as a measure of social deviance. *Ann N Y Acad Sci* 1965;105:861–869.
120. Harrow M, Quinlan D. Is disordered thinking unique to schizophrenia? *Arch Gen Psychiatry* 1977;34:15–21.
121. Harrow M, Quinlan DM. *Disordered thinking and schizophrenic psychopathology.* New York: Gardner, 1985.
122. Harrow M, Prosen M. Schizophrenic thought disorders: bizarre associations and intermingling. *Am J Psychiatry* 1979;136:293–296.
123. Harrow M, Marengo JT. Schizophrenic thought disorder at

follow-up: its persistence and prognostic significance. *Schizophr Bull* 1986;12:373–393.

124. Hart DS, Payne RW. Language structure and predictability in overinclusive patients. *Br J Psychiatry* 1973;123:643–652.

125. Harvey P. Speech competence in manic and schizophrenic psychoses: the association between clinically rated thought disorder and cohesion and reference performance. *J Abnorm Psychol* 1983;92:368–377.

126. Harvey PD, Earle-Boyer EA, Wielgus MS. The consistency of thought disorder in mania and schizophrenia. An assessment of acute psychotics. *J Nerv Ment Dis* 1984;172:458–463.

127. Harvey PD, Earle-Boyer EA, Wielgus MS, et al. Encoding, memory, and thought disorder in schizophrenia and mania. *Schizophr Bull* 1986;12:252–261.

128. Harvey PD, Earle-Boyer EA, Levinson JC. Distractibility and discourse failure. Their association in mania and schizophrenia. *J Nerv Ment Dis* 1986;174:274–279.

129. Harvey PD, Brault J. Speech performance in mania and schizophrenia: the association of positive and negative thought disorders and reference failures. *J Commun Disord* 1986;19:161–173.

130. Harvey PD, Serper MR. Linguistic and cognitive failures in schizophrenia. *J Nerv Mental Dis* 1990;178:487–319.

131. Harvey PD, Lenzenweger MF, Keefe RS, et al. Empirical assessment of the factorial structure of clinical symptoms in schizophrenic patients: formal thought disorder. *Psychiatry Res* 1992;44:141–151.

132. [Reserved.]

133. Heckers S, Rauch SL, Goff D, et al. Impaired recruitment of the hippocampus during conscious recollection in schizophrenia. *Nat Neurosci* 1998;1:318–323.

134. Hemsley DR. A two-stage model of attention in schizophrenia research. *Br J Soc Clin Psychol* 1975;14:81–89.

135. Henderson L. *Orthography and word recognition in reading.* London: Academic Press, 1982.

136. Henik A, Priel B, et al. Attention and automaticity in semantic processing of schizophrenic patients. *Neuropsychiatry Neuropsychol Behav Neurol* 1992;5:161–169.

137. Henik A, Nissimov E, Priel B, et al. Effects of cognitive load on semantic priming in patients with schizophrenia. *J Abnorm Psychol* 1995;104:576–584.

138. Hoffman RE, Kirstein L, Stopek S, et al. Apprehending schizophrenic discourse: a structural analysis of the listener's task. *Brain Lang* 1982;15:207–233.

139. Hoffman RE, Sledge W. An analysis of grammatical deviance occurring in spontaneous schizophrenic speech. *J Neuroling* 1988;3:89–101.

140. Hoffman RE, Stopek S, Andreason NC. A comparative study of manic vs schizophrenic speech disorganization. *Arch Gen Psychiatry* 1986;43:831–838.

141. Hoffmann H, Kuper Z. Relationships between social competence, psychopathology and work performance and their predictive value for vocational rehabilitation of schizophrenic outpatients. *Schizophr Res* 1997;23:69–79.

142. Holcomb PJ. Automatic and attentional processing: an event-related brain potential analysis of semantic priming. *Brain Lang* 1988;35:66–85.

143. Holcomb PJ, Ackerman PT, Dykman RA. Auditory event-related potentials in attention and reading disabled boys. *Int J Psychophysiol* 1986;3:263–273.

144. Honigfeld G. The ability of schizophrenics to understand normal, psychotic and pseudo-psychotic speech. *Dis Nerv Syst* 1963;24:692–694.

145. Hotchkiss AP, Harvey PD. Effect of distraction on commu-

nication failures in schizophrenic patients. *Am J Psychiatry* 1990;147:513–515.

145a. Hulshoff Pol HE, Schnack HG, Mandl RC, et al. Focal gray matter density changes in schizophrenia. *Arch Gen Psychiatry* 2001;58 (12):1118–1125.

146. Iddon JL, McKenna PJ, Sahakian BJ, et al. Impaired generation and use of strategy in schizophrenia: evidence from visuospatial and verbal tasks. *Psychol Med* 1998;28:1049–1062.

147. Jenkins JJ. Remember that old theory of memory? Well, forget it! *Am Psychol* 1974;29:785–795.

148. Jennings J, McIntosh AR, Kapur S, et al. Functional network differences in schizophrenia: a rCBF study of semantic processing. *Neuroreport* 1998;9:1697–1700.

149. Johnson DE, Shean GD. Word associations and schizophrenic symptoms. *J Psychiatr Res* 1993;27:69–77.

150. Johnston MH, Holtzman PS. *Assessing schizophrenic thinking.* San Francisco: Jossey-Bass, 1979.

151. Johnstone EC, Crow TJ, et al. Cerebral ventricle size and cognitive impairment in chronic schizophrenia. *Lancet* 1976;2:924–926.

152. Joyce EM, Collinson SL, Crichton P. Verbal fluency in schizophrenia: relationship with executive function, semantic memory and clinical alogia. *Psychol Med* 1996;26:39–49.

153. Jung CG. Reaction time ratio in the association experiment. In: Read H, Forman M, Alder G, eds. *The collected works of C. G. Jung.* Princeton: Princeton University Press, 1981:227–265.

154. Just MA, Carpenter PA. A capacity theory of comprehension: individual differences in working memory. *Psychol Rev* 1992;99:122–149.

155. Kahneman D. *Attention and effort.* Englewood Cliffs, NJ: Prentice Hall, 1973.

156. Kareken DA, Moberg PJ, et al. Proactive inhibition and semantic organization: relationship with verbal memory in patients with schizophrenia. *J Int Neuropsychol Soc* 1996;2:486–493.

157. Kent GH, Rosanoff AJ. *A study of association in insanity.* Baltimore: Lord Baltimore Press, 1910.

158. King K, Fraser WI, Thomas P, et al. Re-examination of the language of psychotic subjects. *Br J Psychiatry* 1991;156:211–215.

159. King J, Just MA. Individual differences in syntactic processing: the role of working memory. *J Mem Lang* 1991;30:580–602.

160. Kintsch W. Recognition and free recall of organised lists. *J Exp Psychol Gen* 1968;78:481–487.

160a. Kircher TT, Bulimore ET, Brammer MJ, et al. Differential activation of temporal cortex during sentence completion in schizophrenic patients with and without formal thought disorder. *Schizophr Res* 2001;50(1–2):27–40.

160b. Kircher TT, Liddle PF, Brammer MJ, et al. Neural correlates of formal thought disorder in schizophrenia: preliminary findings from a functional magnetic resonance imaging study. *Arch Gen Psychiatry* 2001;58 (8):769–774.

161. Kleist KK. Schizophrenic symptoms and cerebral pathology. *J Ment Sci* 1969;106:246–255.

162. Koh SD, Kayton L, Berry R. Mnemonic organization in young nonpsychotic schizophrenics. *J Abnorm Psychol* 1973;81:299–310.

163. Koh SD, Kayton L. Memorization of 'unrelated' word strings by young nonpsychotic schizophrenics. *J Abnorm Psychol* 1974;83:14–22.

164. Koh SD. Remembering of verbal materials by schizophrenic young adults. *Language and cognition in schizophrenia.* Hillsdale, NJ: Lawrence Erlbaum, 1978.

165. Knight RA, Sims-Knight JE. Integration of linguistic ideas in schizophrenics. *J Abnorm Psychol* 1979;88:191–202.
166. Koyama S, Hokama H, Miyatani M, et al. ERPs in schizophrenic patients during word recognition task and reaction times. *Electroencephalogr Clinical Neurophysiol* 1994;92:546–554.
167. Koyama S, Nageishi Y, Shimokochi M, et al. The N400 component of event-related potentials in schizophrenic patients: a preliminary study. *Electroencephalogr Clin Neurophysiol* 1991;78:124–132.
168. Kraepelin E.*Dementia praecox and paraphrenia.* New York: Krieger, 1971.
169. Kuperberg GR, McGuire PK, David A. Reduced sensitivity to linguistic context in schizophrenic thought disorder: evidence from online monitoring for words in linguistically-anomalous sentences. *J Abnorm Psychol* 1998;107:423–434.
170. Kuperberg G, Heckers S. Schizophrenia and cognitive function. *Curr Opin Neurobiol* 2000;10:205–210.
171. Kuperberg GR, McGuire PK, David AS. Sensitivity to linguistic anomalies in spoken sentences: a case study approach to understanding thought disorder in schizophrenia. *Psychol Med* 2000;30:345–357.
172. Kutas M, Hillyar SA. Event-related potentials to grammatical errors and semantic anomalies. *Mem Cogn* 1983;11:539–550.
173. Kutas M, Hillyard SA. Reading senseless sentences: brain potential reflect semantic incongruity. *Science* 1980;207:203–205.
174. Kutas M, Hillyard SA. Brain potentials during reading reflect word expectancy and semantic association. *Nature* 1984;307:161–163.
175. Kwapil TR, Hegley DC, Chapman LJ , et al. Facilitation of word-recognition by semantic priming in schizophrenia. *J Abnorm Psychol* 1990;99:215–221.
176. Landre NA, Taylor MA, et al. Language functioning in schizophrenic and aphasic patients. *Neuropsychiatry Neuropsychol Behav Neurol* 1992;5:7–14.
177. Larsen SF, Fromholt P. Mnemonic organization and free recall in schizophrenia. *J Abnorm Psychol* 1976;85:61–65.
178. Laws KR, McKenna PJ, Kondel TK, et al. On the distinction between access and store disorders in schizophrenia: a question of deficit severity? *Neuropsychologia* 1998;36:313–321.
179. Lawson JS, McGhie A, et al. Perception of speech in schizophrenia. *Br J Psychiatry* 1964;110:375–380.
180. Lewinsohn PM, Elwood DL. The role of contextual constraint in the learning of language samples in schizophrenia. *J Nerv Ment Dis* 1961;133:79–81.
181. Levelt WJM. *Speaking: from intention to articulation.* Cambridge, MA: MIT Press, 1989.
182. Levine J, Schild K, Kimhi R, et al. Word associative production in affective versus schizophrenic psychoses. *Psychopathology* 1996;29:7–13.
183. Levy R, Maxwell AE. The effect of verbal context on the recall of schizophrenics and other psychiatric patients. *Br J Psychiatry* 1968;114:311–316.
184. [Reserved.]
185. [Reserved.]
186. Lutz J, Marsh TK. The effect of a dual level word list on schizophrenic free recall. *Schizophr Bull* 1981;7:509–515.
187. Maher BA, Manschreck TC, Rucklos ME. Contextual constraint and the recall of verbal material in schizophrenia: the effect of thought disorder. *Br J Psychiatry* 1980;137:69–73.
188. Maher BA. A tentative theory of schizophrenic utterances. In: Maher BA, Maher WB, eds. *Progress in experimental personality research XI.* San Diego: Academic Press, 1983.
189. Manschreck TC, Maher BA, Rucklos ME, et al. The predictability of thought disordered speech in schizophrenic patients. *Br J Psychiatry* 1979;134:595–601.
190. Manschreck TC, Maher BA, Ader DN. Formal thought disorder, the type-token ratio and disturbed voluntary motor movement in schizophrenia. *Br J Psychiatry* 1981;139:7–15.
191. Manschreck TC, Maher BA, Milaretz JJ, et al. Semantic priming in thought disordered schizophrenic patients. *Schizophr Res* 1988;1:61–66.
192. Manschreck TC, Maher BA, Rosenthal E, et al. Reduced primacy and related features in schizophrenia. *Schizophr Res* 1991;5:35–41.
193. Manschreck TC, Maher BA, et al. Laterality, memory and thought disorder in schizophrenia. *Neuropsychiatry Neuropsychol Behav Neurol* 1996;9:1–7.
194. Marcel A J. Conscious and unconscious perception: an approach to the relations between phenomenal experience and perceptual processes. *Cognit Psychol* 1983;15:238–300.
195. Marengo JT, Harrow M, et al. A manual for assessing aspects of bizarre-idiosyncratic thinking. In: Harrow M, Quinlan D, eds. *Disordered thinking and schizophrenic psychopathology.* New York: Gardner Press, 1983:394–411.
196. Marengo JT, Harrow M, Lanin-Kettering I, et al. Evaluating bizarre-idiosyncratic thinking: a comprehensive index of positive thought disorder. *Schizophr Bull* 1986;12:497–509.
197. Marslen-Wilson WD, Tyler LK. The temporal structure of spoken language understanding. *Cognition* 1980;8:1–71.
198. Marslen-Wilson WD. Functional parallelism in spoken word-recognition. *Cognition* 1987;25:71–102.
199. Marslen-Wilson WD, Welsh A. Processing interactions during word-recognition in continuous speech. *Cognit Psychol* 1978;10:29–63.
200. Mazumdar PK, Chaturvedi SK, Gopinath PS. A comparative study of thought disorder in acute and chronic schizophrenia. *Psychopathology* 1995;28:185–189.
201. McClain L. Encoding and retrieval in schizophrenics' free recall. *J Nerv Ment Dis* 1983;171:471–479.
202. McGrath J. Ordering thoughts on thought disorder. *Br J Psychiatry* 1983;158:307–316.
203. McGuire PK, Quested DJ, Spence SA, et al. Pathophysiology of 'positive' thought disorder in schizophrenia. *Br J Psychiatry* 1998;173:231–235.
204. [Reserved.]
205. McKay P, McKenna PJ, Bentham P, et al. Semantic memory is impaired in schizophrenia. *Biol Psychiatry* 1996;39:929–937.
206. McKenna PK, Mortimer AM, et al. Semantic memory and schizophrenia. In: David AS, Cutting JD, eds. *The neuropsychology of schizophrenia.* LEA, 1994.
207. McPherson L, Harvey PD. Discourse connectedness in manic and schizophrenic patients: associations with derailment and other clinical thought disorders. *Cogn Neurospsychiatry* 1996;1:41–53.
208. Mefferd RB Jr. Word association: grammatical, semantic, and affective dimensions of associates to homonyms. *Psychol Rep* 1979;45:359–374.
209. [Reserved.]
210. Mesure G, Passerieux C, Besche C, et al. Impairment of semantic categorization processes among thought-disordered schizophrenic patients. *Can J Psychiatry* 1998;43:271–278.
211. Miller G, Selfridge J. Verbal context and the recall of meaningful material. *Am J Psychol* 1950;63:176–185.
212. Miller WK, Phelan JG. Comparison of adult schizophrenics with matched normal native speakers of English as to "acceptability" of English sentences. *J Psycholing Res* 1980;9:579–593.

213. Monsell S. On the relation between lexical input and output pathways for speech. In: Allport DA, Mackay DG, Prinz W, et al., eds. *Language perception and production: relationships among listening, speaking, reading and writing.* London: Academic Press, 1987:273–311.

214. Moran LJ, Mefferd RB, et al. Idiodynamic sets in word association. *Psychol Monogr Gen Appl* 1964;78:1–22.

215. Morice RD, Ingram JCL. Language analysis in schizophrenia: diagnostic implications. *Aust N Z J Psychiatry* 1982;16:11–21.

216. Morice RD, Ingram JCL. Language complexity and age of onset of schizophrenia. *Psychiatry Res* 1983;9:233–242.

217. Namyslowska I. Thought disorders in schizophrenia before and after pharmacological treatment. *Compr Psychiatry* 1975; 16:37–42.

218. Neely JH. Semantic priming and retrieval from lexical memory: roles of inhibitionless spreading activation and limited-capacity attention. *J Exp Psychol Gen* 1977;106:226–254.

219. Neely JH. Semantic priming effects in visual word recognition: a selective review of current findings and theories. In: Besner D, Humphreys GW, eds. *Basic processes in reading and visual word recognition.* Hillsdale, NJ: Lawrence Erlbaum, 1991:264–333.

220. Nestor PG, Kimble MO, O'Donnell BF, et al. Aberrant semantic activation in schizophrenia: a neurophysiological study. *Am J Psychiatry* 1997;154:640–646.

221. Nestor PG, Shenton ME, Wible C, et al. A neuropsychological analysis of schizophrenic thought disorder. *Schizophr Res* 1998;29:217–225.

222. Niznikiewicz MA, O'Donnell BF, Nestor PG, et al. ERP assessment of visual and auditory language processing in schizophrenia. *J Abnorm Psychol* 1997;106:85–94.

223. Norman RM, et al. Symptoms and cognition as predictors of community functioning: a prospective analysis. *Am J Psychiatry* 1999;156:400–405.

224. Nuecheterlein KH, Edell WS, Norris M, et al. Attentional vulnerability indicators, thought disorder, and negative symptoms. *Schizophr Bull* 1986;12:408–426.

225. Ober BA, Vinogradov S, Shenaut GK. Semantic priming of category relations in schizophrenia. *Neuropsychology* 1995;9:220–228.

226. Ogawa S, Lee TM, Kay AR, et al. Brain magnetic resonance imaging with contrast dependent blood oxygenation. *Proc Natl Acad Sci U S A* 1990;87:8868–8872.

227. Ohta K, Uchiyama M, Matsushima E, et al. An event-related potential study in schizophrenia using Japanese sentences. *Schizophr Res* 1999;40:159–170.

228. Olichney JM, Iragui VJ, Kutas M, et al. N400 abnormalities in late life schizophrenia and related psychoses. *Biol Psychiatry* 1997;42:13–23.

229. Oltmanns TF, Murphy Berenbaum H, et al. Rating verbal communication impairment in schizophrenia and affective disorders. *Schizophr Bull* 1985;11:292–299.

230. Pandurangi AK, Sax KW, Pelonero AL, et al. Sustained attention and positive formal thought disorder in schizophrenia. *Schizophr Res* 1994;13:109–116.

231. Paulsen JS, Romero R, Chan A, et al. Impairment of the semantic network in schizophrenia. *Psychiatry Res* 1996;63:109–121.

232. Pavy D, Grinspoon L, et al. Word frequency measures of verbal disorders in schizophrenia. *Dis Nerv Syst* 1969;32:3–25.

233. Peralta V, Cuesta MJ, de Leon J. Formal thought disorder in schizophrenia: a factor analytic study. *Compr Psychiatry* 1992;33:105–110.

234. Perry W, Braff DL. Information-processing deficits and thought disorder in schizophrenia. *Am J Psychiatry* 1994;151:363–367.

235. Pogue-Geile MF, Oltmanns TF. Sentence perception and dis-

236. Purisch AD, Golden CJ, Hammeke TA. Discrimination of schizophrenic and brain-injured patients by a standardized version of Luria's neuropsychological tests. *J Consult Clin Psychol* 1978;46:1266–1273.

237. Raichle ME, Mintun MA, Herscovitch P. Positron emission tomography with 15oxygen radiopharmaceuticals. *Res Publ Assoc Res Nerv Ment Dis* 1985;63:51–59.

238. Rochester SR, Harris J, Seeman MV. Sentence processing in schizophrenic listeners. *J Abnorm Psychol* 1973;82:350–356.

239. Rochester SR. Are language disorders in acute schizophrenia actually information processing problems. *J Psychiatr Res* 1978;14:275–283.

240. Rochester S, Martin JR. *Crazy talk: a study of the discourse of schizophrenic speakers.* New York: Plenum Press, 1979.

241. Rosenberg SD, Tucker GJ. Verbal behavior and schizophrenia. *Arch Gen Psychiatry* 1979;36:1331–1337.

242. Rugg MD. The effects of semantic priming and word repetition on event-related potentials. *Psychophysiology* 1985;22:642–647.

243. Russell PN, Beekhuis ME. Organization in memory: a comparison of psychotics and normals. *J Abnorm Psychol* 1976;85:527–534.

244. Rutter DJ, Wishner J, et al. The predictability of speech in schizophrenic patients. *Br J Psychiatry* 1978;132:228–232.

245. Rutter DR. The reconstruction of schizophrenic speech. *Br J Psychiatry* 1979;134:356–359.

246. Rutter DR. Language in schizophrenia. The structure of monologues and conversations. *Br J Psychiatry* 1985;146:399–404.

247. Salisbury DF, O'Donnell BF, McCarley RW, et al. Event-related potentials elicited during a context-free homograph task in normal versus schizophrenic subjects. *Psychophysiology* 2000;37:456–463.

248. Salzinger K, Portnoy S, et al. Verbal behavior of schizophrenic and normal subjects. *Ann N Y Acad Sci* 1964;105:845–860.

249. Salzinger K, Pisoni DB, Portnoy S, et al. The immediacy hypothesis and response-produced stimuli in schizophrenic speech. *J Abnorm Psychol* 1970;76:258–264.

250. Salzinger K, Portnoy S, Feldman RS. The predictability of speech in schizophrenic patients [Letter]. *Br J Psychiatry* 1979; 135:284–287.

251. Sanders LM, Adams J, et al. A comparison of clinical and linguistic indices of deviance in the verbal discourse of schizophrenics. *Appl Psycholing* 1995;16:325–338.

252. [Reserved.]

253. Schilder P. On the development of thoughts. In: Rapaport D, ed. *Organization and pathology of thought.* New York: Columbia University Press, 1920 (Trans.1951).

254. Schneider C. *Psychologie der Schizophrenie.* Leipzig, 1930.

255. Schuberth RE, Eimas PD. Effects of context on the classification of words and nonwords. *J Exp Psychol Hum Percept Perform* 1977;3:27–36.

256. Schuberth RE, Spoehr KT, et al. Effects of stimulus and contextual information on the lexical decision process. *Mem Cogn* 1981;9:68–77.

257. Schwanenflugel PJ, Shoben EJ. The influence of sentence constraint on the scope of facilitation for upcoming words. *J Mem Lang* 1985;24:232–252.

258. Schwartz S. Is there a schizophrenic language? *Behav Brain Sci* 1982;5:579–626.

259. Shallice T. *From neuropsychology to mental structure.* Cambridge: Cambridge University Press, 1988.

260. Shallice T. Specialisation within the semantic system. *Cognit Neuropsychol* 1988;5:133–142.

261. Shakow D. Kent-Rosanoff association and its implications for segmental set theory. *Schizophr Bull* 1980;6:676–685.

262. Shakow D, Jellinek EM. Composite index of the Kent-Rosanoff free association test. *J Abnorm Psychol* 1965;70:403–404.

263. Sharkey A JC, Sharkey NE. Weak contextual constraints in text and word priming. *J Mem Lang* 1992;31:543–572.

264. Sharkey NE, Mitchell DC. Word recognition in a functional context: the use of scripts in reading. *J Mem Lang* 1985;24:253–270.

265. Shelly C, Goldstein G. Discrimination of chronic schizophrenia and brain damage with the Luria-Nebraska Battery: a partially successful replication. *Clin Neuropsychol* 1983;5:82–85.

265a. Shenton ME, Dickey CC, Frumin M, McCarley RW. A review of MRI findings in schizophrenia. *Schizophr Res* 2001;49(1–2):1–52.

266. Shenton ME, Kikinis R, Jolesz FA, et al. Abnormalities of the left temporal lobe and thought disorder in schizophrenia. *N Engl J Med* 1992;327:604–612.

267. Sigmundsson T, Suckling J, Maier M, et al. Structural abnormalities in frontal, temporal, and limbic regions and interconnecting white matter tracts in schizophrenic patients with prominent negative symptoms. *Am J Psychiatry* 2001;158:234–243.

268. Sitnikova T, Salisbury DF, Kuperberg G, Holcomb PI. Electrophysiological insights into language processing in schizophrenia. *Psychophysiology* 2002;39(6):851–860.

269. Soares I, Collet L. Relationships between verbal intelligence, educational level and reconstitution of linguistic messages in schizophrenia. *Int J Neurosci* 1988;38:69–74.

270. Solovay MR, Shenton ME, Gasparetti C, et al. Scoring manual for the thought disorder index. *Schizophr Bull* 1986;12:483–496.

271. Solovay MR, Shenton ME, Holzman PS, et al. Comparative studies of thought disorders. *Arch Gen Psychiatry* 1987;44:13–20.

272. Speed M, Shugar G, et al. Thought disorder and verbal recall in acutely psychotic patients. *J Clin Psychol* 1991;47:735–744.

273. Spitzer M. Word associations in experimental psychology: a historical perspective. In: Spitzer M, Uehlein FA, Schwartz MA, eds. *Phenomenology, language and schizophrenia.* Berlin/Heidelberg/New York: Springer, 1992:160–196.

274. Spitzer M, Braun U, Hermle L, et al. Associative semantic network dysfunction in thought-disordered schizophrenic patients: direct evidence from indirect semantic priming. *Biol Psychiatry* 1993;34:864–877.

275. Spitzer M, Braun U, Maier S, et al. Indirect semantic priming in schizophrenic patients. *Schizophr Res* 1993;11:71–80.

276. Spitzer M, Beuckers J, Beyer S, et al. Contextual insensitivity in thought-disordered schizophrenic patients: evidence from pauses in spontaneous speech. *Lang Speech* 1994;37:171–185.

277. Spitzer M, Weisker I, Winter M, et al. Semantic and phonological priming in schizophrenia. *J Abnorm Psychol* 1994;103:485–494.

278. Spitzer M, Weisbrod M, Winkler S, et al. [Event-related potentials in semantic speech processing by schizophrenic patients.] *Nervenarzt* 1997;68:212–225.

279. Stanovich KE, West RF. Mechanisms of sentence context effects in reading: automatic activation and conscious attention. *Mem Cogn* 1979;7:77–85.

280. Stanovich KE, West RF. The effect of a sentence context on ongoing word recognition: tests of a two-process theory. *J Exp Psychol Hum Percept Perform* 1981;7:658–772.

281. Stanovich KE, West RF. On priming by a sentence context. *J Exp Psychol Gene* 1983;112:1–36.

282. Sternberg RJ, Tulving E. The measurement of subjective organization in free recall. *Psychol Bull* 1977;84:539–556.

283. Stevens AA, Donegan NH, Anderson M, et al. Verbal processing deficits in schizophrenia. *J Abnorm Psychol* 2000;109:461–471.

284. Strandburg RJ, Marsh JT, Brown WS, et al. Event-related potential correlates of linguistic information processing in schizophrenics. *Biol Psychiatry* 1997;42:596–608.

285. Strauss ME. Strong meaning-response bias in schizophrenia. *J Abnorm Psychol* 1975;84:295–298.

286. Strauss ME, Buchanan RW, Hale J. Relations between attentional deficits and clinical symptoms in schizophrenic outpatients. *Psychiatr Res* 1993;47:205–213.

287. [Reserved.]

288. [Reserved.]

289. Tamlyn D, McKenna PJ, Mortimer AM, et al. Memory impairment in schizophrenia—its extent, affiliations and neuropsychological character. *Psychol Med* 1992;22:101–115.

290. [Reserved.]

291. Taylor MA. *The neuropsychiatric mental state examination.* New York: Spectrum Publications, 1981.

292. Taylor MA, Reed R, Berenbaum S. Patterns of speech disorders in schizophrenia and mania. *J Nerv Ment Dis* 1994;182:319–326.

293. Thomas PF, King K, Fraser WI. Positive and negative symptoms of schizophrenia and linguistic performance. *Acta Psychiatr Scand* 1987;76:144–151.

294. Thomas P, King K, Fraser WI, et al. Linguistic performance in schizophrenia: a comparison of acute and chronic patients. *Br J Psychiatry* 1990;156:204–210.

295. Thomas P, Leudar I, Newby D, et al. Syntactic processing in the written language output of first episode psychotics. *J Commun Disord* 1993;26:209–230.

295a. Thompson PM, Vidal C, Giedd JN, et al. Mapping adolescent brain change reveals dynamic wave of accelerated gray matter loss in very early-onset schizophrenia. *Proc Natl Acad Sci USA* 2001;98(20):11650–11655.

296. Thurstone LL. *Tests used to study mental abilities.* Chicago: Chicago University Press. 1934.

297. Truscott IP. Contextual constraint and schizophrenic language. *J Consult Clin Psychol* 1970;35:189–194.

298. Tulving E. Episodic and semantic memory. In: Tulving E, Donaldson W, eds. *Organization of memory.* New York: Academic Press, 1972:381–403.

299. Tyler L, Wessels J. Quantifying contextual contributions to word-recognition processes. *Percept Psychophys* 1983;34:409–420.

300. Tyler LK. *Spoken language comprehension: an experimental approach to disordered and normal processing.* Cambridge, MA: MIT Press, 1992.

301. Van Dijk TA, Kintsch W. *Strategies of discourse comprehension.* New York: Academic Press, 1983.

302. Vigotsky L. Thought in schizophrenia. *Arch Neurol Psychiatry* 1934;31:1063.

303. Vinogradov S, Ober BA, Shenaut GK. Semantic priming of word pronunciation and lexical decision in schizophrenia. *Schizophr Res* 1992;8:171–181.

304. von Domarus E. The specific laws of logic in schizophrenia. In: Kasanin J, ed. *Language and thought in schizophrenia.* Berkeley: University of California Press, 1944.

305. Waters G, Caplan D, et al. Working memory and written sentence comprehension. In: Coltheart M, ed. *Attention and*

performance XII: the psychology of reading. Hillsdale, NJ: Lawrence Erlbaum, 1987:531–555.

306. Wielgus MS, Harvey PD. Dichotic listening and recall in schizophrenia and mania. *Schizophr Bull* 1988;14:689–700.

307. Wright IC, McGuire PK, Poline JB, et al. A voxel-based method for the statistical analysis of gray and white matter density applied to schizophrenia. *Neuroimage* 1995;2:244–252.

308. Wright IC, Ellison ZR, Sharma T, et al. Mapping of grey matter changes in schizophrenia. *Schizophr Res* 1999;35:1–14.

309. [Reserved.]

310. Wykes T, Leff J. Disordered speech: differences between manics and schizophrenics. *Brain Lang* 1982;15:117–124.

311. Yurgelun-Todd DA, Waternaux CM, Cohen BM, et al. Functional magnetic resonance imaging of schizophrenic patients and comparison subjects during word production. *Am J Psychiatry* 1996;153:200–205.

EMOTION AND DISORDERS OF EMOTION: PERSPECTIVES FROM AFFECTIVE NEUROSCIENCE

RICHARD J. DAVIDSON

Affective neuroscience is the subdiscipline of the biobehavioral sciences that examines the underlying neural bases of mood and emotion. The application of this body of theory and data to the understanding of individual differences in affective style, mood regulation, and mood disorders is helping to generate a new understanding of the brain circuitry underlying these phenomena. At a more general level, this approach is helping to bridge the wide chasm between the literature that has focused on normal emotion and the literature that has focused on the disorders of emotion. Historically, these research traditions have had little to do with one another and have emerged completely independently. However, affective neuroscience has helped to integrate these approaches into a more unified project that is focused on the understanding of normal and pathologic individual differences in affective style, its components, and their neural bases (28,29).

Affective neuroscience takes as its overall aim a project that is similar to that pursued by its cognate discipline, cognitive neuroscience, although focused instead on affective processes. The decomposition of cognitive processes into more elementary constituents that can then be studied in neural terms has been remarkably successful. We no longer query subjects about the contents of their cognitive processes because many of the processes so central to important aspects of cognitive function are opaque to consciousness. Instead, modern cognitive scientists and neuroscientists have developed laboratory tasks to interrogate and reveal more elementary cognitive function. These more elementary processes can then be studied using imaging methods in humans, lesion methods in animals, and the study of human patients with focal brain damage. Affective neuroscience approaches emotion using the same strategy. Global constructs of emotion are giving way to more specific and elementary constituents that can be examined with objective laboratory measures. For example, the time course of emotional responding and the mechanisms that are brought into play during the regulation of emotion can now be probed using objective laboratory measures. These constructs may be particularly important for understanding individual differences in affective style because one of the key characteristics of variations in mood among individuals is the extent to which negative affect persists instead of rapidly subsiding. Moreover, these ideas have significant import for understanding mood disorders. Some patients with mood disorders may have a particular problem with persistence of negative affect, whereas other patients may have a primary deficit in reactivity to positive incentives.

Previously, constructs such as emotion regulation have mostly been gleaned from self-report measures whose validity has been seriously questioned (63). Although the phenomenology of emotion provides critical information to the subject that helps to guide behavior, it may not be a particularly good source for making inferences about the processes and mechanisms that underlie emotion and its regulation. Although it is still tempting and often important to obtain measures of subjects' conscious experience of the contents of their emotional states and traits, these no longer constitute the sole source of information about emotion.

Because there are recent reviews of the basic literature on the circuitry underlying emotion and emotion regulation (27,29,31,112), these data are not systematically reviewed in this chapter. The emphasis is on studies that have been published in the past 3 years because two recent reviews cover much of the literature before this time (26,36). A major focus of this chapter is on individual differences in affective style and how such variability across individuals can be captured using objective laboratory probes rather than relying exclusively on self-report data.

There are two broad goals for this chapter:

1. To review the functional role of the prefrontal cortex (PFC), anterior cingulate cortex (ACC), hippocampus, and amygdala in affect and emotion regulation (Figure 20.1 depicts these structures and their locations);

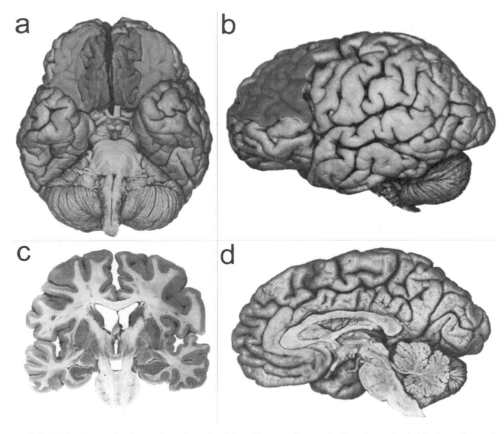

FIGURE 20.1. Key brain regions involved in affect and mood disorders. **A:** Orbital prefrontal cortex (green) and ventromedial prefrontal cortex (red). **B:** Dorsolateral prefrontal cortex (blue). **C:** Hippocampus (purple) and amygdala (orange). **D:** Anterior cingulate cortex (yellow). (See Color Figure 20.1 following page 526.)

2. To review the functional and structural variations in these regions that have been linked to affective style and affective disorders.

THE CIRCUITRY OF EMOTION

Prefrontal Cortex: Functional and Anatomic Considerations for Understanding Its Role in Affect

Although the PFC is often considered the province of higher cognitive control, it has also consistently been linked to various features of affective processing [for an early preview, see Nauta (90)]. Miller and Cohen (85) recently outlined a comprehensive theory of prefrontal function based on nonhuman primate anatomic and neurophysiologic studies, human neuroimaging findings, and computational modeling. The core feature of their model holds that the PFC maintains the representation of goals and the means to achieve them. Particularly in situations that are ambiguous, the PFC sends bias signals to other areas of the brain to facilitate the expression of task-appropriate responses in the face of competition with potentially stronger alternatives. In the affective domain, we often confront situations in which the arousal of emotion is inconsistent with other goals that have already been instantiated. For example, the availability of an immediate reward may provide a potent response alternative that may not be in the best interest of the overall goals of the person. In such a case, the PFC is required to produce a bias signal to other brain regions that guide behavior toward the acquisition of a more adaptive goal, which in this case would entail delay of gratification. Affect-guided planning and anticipation that involve the experience of emotion associated with an anticipated choice are the hallmark of adaptive, emotion-based decision making that has repeatedly been found to become impaired in patients with lesions of the ventromedial PFC (25). Affect-guided anticipation is most often accomplished in situations that are heavily laden with competition from potentially stronger alternatives. In such cases in particular, we would expect PFC activation to occur. Some disorders of emotional processing such as depression may be caused by abnormalities of affect-guided anticipation. For example, the failure to anticipate positive incentives and direct behavior toward the acquisition of appetitive goals are symptoms of depression that may arise from abnormalities in the circuitry that implements positive affect-guided anticipation.

Our laboratory has contributed extensively to the literature on asymmetries in PFC function associated with approach- and withdrawal-related emotion and mood (27,29). In this context, we suggest that left-sided PFC regions are particularly involved in approach-related, appetitive goals. The instantiation of such goals, particularly in the face of strong alternative responses, requires left-sided PFC activation, and hypoactivation in these circuits has been linked to depression. Right-sided PFC regions, alternatively, are hypothesized to be particularly important in the maintenance of goals that require behavioral inhibition and withdrawal in situations that involve strong alternative response options to approach. The prototype of such a process has recently been captured in several neuroimaging studies that involve variants of a go/no go task in which a dominant response set is established to respond quickly, except in those trials in which a cue to inhibit the response is presented. Two recent studies using event-related functional magnetic resonance imaging found a lateralized focus of activation in the right lateral PFC (inferior frontal sulcus) to cues that signaled response inhibition that were presented in the context of other stimuli toward which a strong approach set was established (47,69).

Depressed individuals with hypoactivation in some regions of the PFC may be deficient in the instantiation of goal-directed behavior and the overriding of more automatic responses that may involve the perseveration of negative affect and dysfunctional attitudes. Such deficits would be expected to be unmasked in situations in which decision making is ambiguous and the maintenance of goal-directed behavior is required in the face of potentially strong alternative responses. As argued below, when the strong alternative responses involve affect, which they often do, the ventromedial PFC is particularly implicated.

Recent neuroimaging and electrophysiologic studies suggest that the orbital and ventral frontal cortex, in particular, may be especially important for the representation of rewards and punishments, and different sectors within this cortex may emphasize reward versus punishment (65,94). In particular, a left-sided medial region of the orbitofrontal cortex (OFC) appears particularly responsive to rewards, whereas a lateral right-sided region appears particularly responsive to punishments (94). Kawasaki et al. (65) recorded from single units in the right ventral PFC of patients with implanted depth electrodes for presurgical planning. They found that these neurons in healthy tissue exhibited short-latency responses to aversive visual stimuli. Such studies provide important clues regarding the circuitry that might be most relevant to understanding differences among individuals in affective style. For example, there are individual differences in responsivity to rewards versus punishments that can be probed behaviorally using signal detection methods (51,52). Most normal individuals exhibit systematic modification of response bias to monetary reward, but some do not. Those who do not showed elevated depressed mood. We would also predict that the left medial OFC would be hyporespon-

sive to manipulations of reward in such individuals, whereas right lateral OFC responsivity to punishment would be either normal or perhaps accentuated.

Anterior Cingulate Cortex: Functional and Anatomic Considerations for Understanding Its Role in Affect

Several theories have proposed that the ACC acts as a bridge between attention and emotion (33,41,80,141). In their recent review, Thayer and Lane (137) describe the ACC as "a point of integration for visceral, attentional, and affective information that is critical for self-regulation and adaptability." In light of its anatomic connections (see below), the ACC appears well equipped for assessing and responding to the behavioral significance of external stimuli. Critical roles of the ACC in selective attention (i.e., prioritizing incoming information), affect, and specific characteristic mammalian social behaviors have been described (33,140). However, to fully understand the role of the ACC in psychopathology, affective states, and emotional processing, it is critical to recognize that the ACC is far from being a functionally homogeneous region, and at least two subdivisions can be discerned (33,140,141). The first, referred to as the affect subdivision, encompasses rostral and ventral areas of the ACC (areas 25, 32, 33, and rostral area 24). The second, referred to as the cognitive subdivision, involves dorsal regions of the ACC (caudal area 24' and 32', cingulate motor area). The affect subdivision possesses extensive connections with limbic and paralimbic regions, such as the amygdala, nucleus accumbens, OFC, periaqueductal gray, anterior insula, and autonomic brainstem motor nuclei, and is assumed to be involved in regulating visceral and autonomic responses to stressful behavioral and emotional events, emotional expression, and social behavior. Because of its strong connections with the lateral hypothalamus, the subgenual ACC [Brodmann area (BA) 25] is considered the most important region within the frontal cortex for regulating autonomic function (96).

Conversely, the cognitive subdivision is intimately connected with the dorsolateral PFC (BA 46/9), posterior cingulate, parietal cortex (BA 7), supplementary motor area, and spinal cord and plays an important role in response selection and processing of cognitively demanding information. In functional neuroimaging studies, evidence suggesting a functional differentiation between ventral (affective) and dorsal (cognitive) ACC subdivisions is emerging (15,16,44).

From a functional perspective, activation of the cognitive subdivision of the ACC has been reported during interference between competing information (99), visual attention (91), monitoring of cognitive (19,76) and reward-related (111) conflicts, task difficulty (101), and increased risk-associated outcome uncertainty (24), among other experimental manipulations. A common denominator among these experimental conditions is that they all required modulation of attention or executive functions and monitoring of competition (16). The role of the ACC in conflict monitoring has been especially emphasized by Cohen and colleagues (18,19,85). These authors propose that the ACC may serve an evaluative function, reflecting the degree of response conflict elicited by a given task. Conflict occurs when two or more possible task-related decisions compete or interfere with each other. According to the "competition monitoring hypothesis," the cognitive subdivision of the ACC monitors conflicts or cross talk between brain regions. If a signal of competition emerges, this output signals the need for controlled processing. The dorsolateral PFC (BA 9) is assumed to be critical for this form of controlled processing in that it represents and maintains task demands necessary for such control and inhibits (47) or increases neural activity in brain regions implicated in the competition. Thus, dorsal ACC activation leading to a call for further processing by other brain regions may represent a mechanism for effortful control.

From a functional perspective, activation of the affective subdivision of the ACC has been reported during various emotional states and manipulations (16,109,144). What could be a common denominator underlying activation of the rostral/ventral ACC in such disparate experimental conditions as pain, classical conditioning, transient mood, primal affect, Stroop task, and perceiving facial expressions, all of which have been reported in the literature? A possible answer to this question is that the affective subdivision of the ACC may be critical for assessing the presence of possible conflicts between the current functional state of the organism and incoming information with potentially relevant motivational and emotional consequences. This suggestion is based on the observation that the affective subdivision of the ACC is involved in behaviors characterized by monitoring and evaluation of performance, internal states, and presence of reward or punishment, which often require a change in behavior.

Extant evidence suggests that ACC activation may be present when effortful emotion regulation is required in situations in which behavior is failing to achieve a desired outcome or when affect is elicited in contexts that are not normative, which includes most laboratory situations (16,95). Similarly, it is not surprising that the ACC is one of the most consistently activated regions in patients with different anxiety disorders, such as obsessive-compulsive disorder (9,107), simple phobia (105), and posttraumatic stress disorder (106,128), in which conflicts between response tendencies and environments are prominent. Interestingly, psychosurgical lesions of the ACC have been used as a treatment for mood and anxiety disorders (4,6), possibly because of a reduction of conflict monitoring and uncertainty that otherwise characterize these psychiatric conditions.

The interplay between the affective and cognitive subdivision of the ACC is currently unknown. From a theoretical perspective, several authors have suggested that the affective subdivision of the ACC may integrate salient affective and cognitive information (such as that derived from environmental stimuli or task demands) and subsequently modulate attentional processes within the cognitive subdivision accordingly (80,81,83,103). In agreement with this hypothesis, dorsal anterior and posterior cingulate pathways devoted to attentional processes and amygdalar pathways devoted to affective processing converge within area BA 24 (83). These mechanisms may be especially important for understanding the now replicated finding in depressed patients that increased *pre* treatment activity in the rostral ACC is associated with an eventually better treatment response (40,83,103,145,146). In an influential paper, Mayberg (80) reported that unipolar depressed patients who responded to treatment after 6 weeks showed higher pretreatment glucose metabolism in a rostral region of the ACC (BA 24a/b) compared with both nonresponders and nonpsychiatric comparison subjects. Recently, we replicated this finding (103) with electroencephalographic source localization techniques and demonstrated that even among those patients who respond to treatment, the magnitude of treatment response was predicted by baseline levels of activation in the same region of the ACC as identified by Mayberg et al (81). In addition, we suggested that hyperactivation of the rostral ACC in depression might reflect an increased sensitivity to affective conflict such that the disparity between one's current mood and the responses expected in a particular context activates this region of the ACC, which in turn issues a call for further processing to help to resolve the conflict. This call for further processing is hypothesized to aid the treatment response. In other words, individuals exhibiting high levels of activation in the rostral ACC may be affectively resilient because these individuals would be motivated to resolve discrepancies between their current mood state and the behavior that is most appropriate for the situation at hand.

One of the major outputs from the ACC is a projection to the PFC. This pathway may be the route via which the ACC issues a call to the PFC for further processing to address a conflict that has been detected. Individual differences in PFC function that are relevant to affective style may arise as a consequence of variations in signals from the ACC or may be intrinsic to the PFC or both. There may be ACC-based variations in affective style that are reflected phenomenologically in the motivation or "will to change" particular habits or patterns of affective reactivity. Individuals with low levels of rostral ACC activation would not experience conflict between their current state and the demands of everyday life and would thus be unmotivated to alter their behavior. PFC-based variations in affective style may predominantly revolve around differences among individuals in the capacity to organize and guide behavior in a goal-directed fashion.

An important issue not considered above is the anatomic and functional connectivity between the different regions of the PFC and ACC. Future studies need to examine both structural and functional variations in these connections because it is likely that some individual differences in affective style are primarily associated with connectivity between the PFC, ACC, and amygdala rather than with activation differences in any single or even multiple regions. This issue is discussed in more detail below.

Hippocampus: Functional and Anatomic Considerations for Understanding Its Role in Affect

The hippocampus is critically involved in episodic, declarative, contextual, and spatial learning and memory (44,129). Additionally, it is also significantly involved in the regulation of adrenocorticotropic hormone secretion (61). With respect to conditioning, in recent years, rodent studies have convincingly shown that the hippocampus plays an essential role in the formation, storage, and consolidation of contextual fear conditioning (44). In this form of hippocampus-dependent, pavlovian conditioning, fear (e.g., expressed in increased freezing) is acquired in places or contexts (e.g., a specific cage) previously associated with aversive events (e.g., shock). This fact has important implications for our understanding of the abnormalities in affective function that may arise as a consequence of hippocampal dysfunction.

In functional neuroimaging studies, hippocampal/parahippocampal activation has been reported during perception of several negatively valenced stimuli and/or experiencing of negatively valenced affective states, such as trace conditioning (14), perception of aversive complex stimuli (70), threat-related words (59), increasing music dissonance (8), tinnituslike aversive auditory stimulation (86), vocal expressions of fear (102), aversive taste (148), anticipatory anxiety (62), procaine-induced affect (67,121), and monetary penalties (42). However, it seems that valence is not the critical variable for evoking hippocampal activation. Indeed, hippocampal activation has been also reported during experimental manipulation of positive affect, such as reevoking pleasant affective autobiographical memories (46), increases in winning in a gamelike task (149), and perception of the loved person (5). Hippocampal activation was also correlated with long-term recognition memory for pleasant films (50).

To reconcile these findings, we suggest that most of the experimental manipulations leading to hippocampal activation contain contextual cues. That is, we assume that they involve the consolidation of a memory for an integrated representation of a context similar to that associated with the presented stimulus (44). This is clearly the case during pavlovian and trace conditioning, for instance, but also during presentation of both positively and negatively valenced visual, olfactory, and auditory cues that may induce reevocation and consolidation of contextual information associated with a similar situation in the past (89).

Although in humans, the mechanisms underlying contextual conditioning are still unclear, it is possible that plasticity in functional connectivity between the hippocampus and regions crucially involved in decoding the behavioral

significance of incoming information, such as the amygdala and pulvinar, may critically contribute to contextual learning (87,88), even when the information is presented below the level of conscious awareness (88). As recently reviewed by Davis and Whalen (32), animal studies clearly suggest that the amygdala exerts a modulatory influence on hippocampus-dependent memory systems, possibly through direct projections from the basolateral nucleus of the amygdala. Consistent with this view, stimulation of the amygdala causes long-term potentiation induction in the dentate gyrus of the hippocampus (57). Conversely, lesions to (56) or local anesthetics within (58) the basolateral nucleus of the amygdala attenuate long-term potentiation in the dentate gyrus. Although drawing conclusions from these rodent studies to humans is speculative at this stage, it is intriguing that most of the human neuroimaging studies reporting hippocampal activation during aversive affective manipulations also found amygdalar activation (14,59,67,86,121,148). Future neuroimaging studies should directly test the interplay between the hippocampus and amygdala in these processes and in fear-related learning and memory, especially in light of recent animal data suggesting an interplay between these regions for modulating extinction of conditioned fear (22).

In their recent review, Davidson et al. (29) note that various forms of psychopathology involving disorders of affect could be characterized as disorders in context regulation of affect. That is, patients with mood and anxiety disorders often display normative affective responses but in *inappropriate* contexts. For example, fear that may be appropriate in response to an actual physical threat but persists after the removal of that threat or sadness that may be appropriate in the acute period after a loss but persists for a year after that loss are both examples of context-inappropriate emotional responding. In these examples, the intensity and form of the emotion would be perfectly appropriate in response to the acute challenges, but when they occur in the absence of those acute stresses, they can be viewed as context inappropriate.

In a series of studies with nonhuman primates, Kalin and Shelton (64) have used the human intruder paradigm to probe the context specificity of emotional responding. In this paradigm, the monkey is exposed to different contexts that elicit a specific normative pattern of affective responses. In response to the profile of a human intruder (no eye contact condition), the animal tends to freeze, whereas in response to the same human staring at the animal, agonistic, aggressive behavior is elicited. When the animal is alone, freezing and aggression decline and the animal vocalizes. These very well defined normative patterns can be used to identify responses that are context inappropriate. In a group of 100 monkeys, approximately five showed highly context-inappropriate responding. The most dramatic example of this is the small group of animals that freeze during the alone condition at levels that are comparable with what they exhibit during the no eye contact condition. These are the animals that we predict would have hippocampal dysfunction because they are not regulating their emotions in a context-appropriate fashion.

Given the preclinical and functional neuroimaging literature reviewed above, one may hypothesize that subjects displaying inappropriate context regulation of affect may be characterized by hippocampal dysfunction. Consistent with this conjecture, recent morphometric studies using magnetic resonance imaging indeed reported smaller hippocampal volumes in patients with major depression (3,13,84,123,124,126,131,138,142), bipolar disorder (93), posttraumatic stress disorder (10,11,132), and borderline personality disorder (38,116,127). Where hippocampal volume reductions in depression have been found, the magnitude of reduction ranges from 8% to 19%. Recently, functional hippocampal abnormalities in major depression have been also reported at baseline using positron emission tomography measures of glucose metabolism (117). Whether hippocampal dysfunction precedes or follows onset of depressive symptomatology is still unknown.

In depression, inconsistencies across studies may be explained by several methodologic considerations. First, as pointed out by Sheline (127), studies reporting positive findings generally used magnetic resonance imaging with higher spatial resolution (~0.5–2 mm) compared with those reporting negative findings (~3–10 mm). Second, it seems that age, severity of depression, and, most significantly, duration of recurrent depression may be important moderator variables. Indeed, studies reporting negative findings either studied younger cohorts [e.g.,Vakili et al. (138): 38 ± 10 years vs. Sheline et al. (124):69 ± 10 years; von Gunten et al. (142): 58 ± 9 years; Steffens et al. (131): 72 ± 8 years] or less severe and less chronic cohorts [Ashtari et al. (3) vs. Sheline et al. (124), Shah et al. (123), and Bremner et al. (13)]. In a recent study, Rusch et al. (113) also failed to find hippocampal atrophy in a relatively young subject sample (33.2 ± 9.5 years) with moderate depression severity. Notably, in normal early adulthood (18–42 years), decreased bilateral hippocampal volume has been reported with increasing age in healthy male but not healthy female subjects (104). Finally, in female subjects, initial evidence suggests that total lifetime duration of depression rather than age is associated with hippocampal atrophy (126), inviting the possibility that hippocampal atrophy may be a symptom rather than a cause of depression. Future studies should carefully assess the relative contribution of these possible modulatory variables in the hippocampal pathophysiology and examine hippocampal changes longitudinally in individuals at risk of mood disorders.

Structurally, the hippocampal changes may arise owing to neuronal loss through chronic hypercortisolemia, glial cell loss, stress-induced reduction in neurotrophic factors, or stress-induced reduction in neurogenesis, but the precise mechanisms are not completely known (127). In depression, the hypothesis of an association between sustained, stress-related elevations of cortisol and hippocampal damage has received considerable attention. This hypothesis is based on the observation that the pathophysiology of depression involves dysfunction in negative feedback of the hypothalamic–pituitary–adrenal axis (100), which results in increased levels of cortisol during depressive episodes (17). Higher levels of cortisol may, in turn, lead to neuronal damage in the hippocampus because this region possesses high levels of glucocorticoid receptors (110) and glucocorticoids are neurotoxic (114). Because the hippocampus is involved in negative feedback control of cortisol (61), hippocampal dysfunction may result in the reduction of the inhibitory regulation of the hypothalamic–pituitary–adrenal axis, which could then lead to hypercortisolemia. Consistent with this view, chronic exposure to increased glucocorticoid concentrations has been shown to lower the threshold for hippocampal neuronal degeneration in animals (48,82,115) and humans (75). At least in nonhuman primates, this association is qualified by the observation that chronically elevated cortisol concentrations in the absence of chronic psychosocial stress do not produce hippocampal neuronal loss (73). Conversely, naturalistic, chronic psychosocial stress has been shown to induce structural changes in hippocampal neurons of subordinate animals (78). In depression, hippocampal volume loss has been shown to be associated with lifetime duration of depression (126), consistent with the assumption that long-term exposure to high cortisol levels may lead to hippocampal atrophy. However, this conjecture has not been empirically verified in humans.

Although intriguing, these findings cannot inform us about the causality between hippocampal dysfunction, elevated levels of cortisol, and, most important, inappropriate context regulation of affect. Unfortunately, none of the structural neuroimaging studies in depression investigating hippocampal volume was prospective and took into account cortisol data in an effort to unravel the causal link between cortisol output and hippocampal dysfunction.

The possibility of plasticity in the hippocampus deserves particular comment. In rodents, recent studies have shown hippocampal neurogenesis as a consequence of antidepressant pharmacologic treatment (20,79), electroconvulsive shock (77), and, most intriguingly, positive handling, learning, and exposure to an enriched environment (49,66). In humans, neurogenesis in the adult human hippocampus has been also reported (43). Further, in patients with Cushing disease who are characterized by very high levels of cortisol, increases in hippocampal volume were significantly associated with the magnitude of cortisol decrease produced by microadrenomectomy (130). As a corpus, these animal and human data clearly suggest that plasticity in the human hippocampus is possible (39,49,60), a finding that suggests that structural and functional changes in the hippocampus of depressed patients may be reversible.

In summary, preclinical and clinical studies converge in suggesting an association between context modulation of affective responding and hippocampal function. Future studies should (a) assess whether hippocampal atrophy precedes or follows the onset of depression or other syndromes of affective dysregulation, (b) assess the causal relationship between hypercortisolemia and hippocampal volume reduction, (c) directly test a putative link between inappropriate context-dependent affective responding and hippocampal atrophy, and (d) assess putative treatment-mediated plastic changes in the hippocampus.

Amygdala: Functional and Anatomic Considerations for Understanding Its Role in Affect

Although a link between amygdala activity and negative affect has been a prevalent view in the literature, particularly when examined in response to exteroceptive aversive stimuli (72), recent findings from invasive animal studies and human lesion and functional neuroimaging studies are converging on a broader view that regards the amygdala's role in negative affect as a special case of its more general role in directing attention to affectively salient stimuli and issuing a call for further processing of stimuli that have major significance for the individual. Extant evidence is consistent with the argument that the amygdala is critical for recruiting and coordinating cortical arousal and vigilant attention for optimizing sensory and perceptual processing of stimuli associated with underdetermined contingencies, such as novel, "surprising," or "ambiguous" stimuli (32,54,143). Most stimuli in this class may be conceptualized as having an aversive valence because we tend to have a negativity bias in the face of uncertainty (134).

Both structural and functional differences in the amygdala have been reported in disorders of emotion, particularly depression. Structurally, several recent studies reported an association between enlargement of amygdala volume and depression. This association has been found in depressed patients with bipolar disorders (2,133) and temporal lobe epilepsy (135,136). In a recent study, Mervaala et al. (84) observed significant asymmetry in amygdalar volumes (right smaller than left) in patients with major depressive disorder (MDD) but not the controls. In temporal lobe epilepsy patients with dysthymia, left amygdala volume was positively correlated with depression severity, as assessed with the Beck Depression Inventory (135). Although these findings depict a relationship between increased amygdalar volume and depression, it is important to stress that (a) the causal relationships between the two entities are still unknown and (b) some inconsistencies among studies are present. Indeed, some studies reported either decreased bilateral volume in the amygdala core nuclei (125) or null findings (3,21,98). Although the reasons are still unclear, it is interesting to note that two null findings were found in geriatric depression (3,98).

Functionally, abnormal elevations of resting regional cerebral blood flow or glucose metabolism in the amygdala have been reported in depression during both wakefulness (34) and sleep (53,92). In an [^{18}F]fluorodeoxyglucose–positron emission tomography study, Ho et al. (53) reports increased absolute cerebral glucose metabolism in several brain regions, particularly the amygdala (+44%), in ten unmedicated men with unipolar depression during the nonrapid eye movement sleep period. Further, in his recent review, Drevets (37) reports data from five consecutive studies in which increased regional cerebral blood flow or glucose metabolism has been consistently replicated in depressives with familial MDD or melancholic features. In a postmortem study, $5HT_2$ receptor density was significantly increased in the amygdala of depressive patients committing suicide (55). Abnormally increased amygdalar activation has also been recently reported in bipolar depression (68) and anxiety disorders, which often show a high degree of comorbidity with depression (7,74,106,108,119,122,128). Further establishing a link between depression and amygdalar activation, two studies reported a positive correlation between amygdalar activation and depression severity or dispositional negative affect in patients with MDD (1,34). After pharmacologically induced remission from depression, amygdalar activation has been observed to decrease to normative values (37). In familial pure depressive disease, however, increased (left) amygdalar activation persists during the remittent phases (34), suggesting that, at least in some subtypes of depression, amygdalar dysfunction may be traitlike. Interestingly, remitted MDD patients demonstrating symptom relapse as a consequence of serotonin depletion showed increased amygdalar activation before the depletion compared with those who did not relapse (12). Finally, in one of the first functional magnetic resonance imaging studies using an activation paradigm, Yurgelun-Todd et al. (147) report higher left amygdalar activation for bipolar patients than controls in response to fearful faces.

In light of the pivotal role of the amygdala in recruiting and coordinating vigilant behavior toward stimuli with underdetermined contingencies, hyperactivation of the amygdala in major depression may bias the initial evaluation of and response to incoming information. Although still speculative, this mechanism may rely on norepinephrine, which (a) is often abnormally elevated in depression (139), (b) is involved in amygdala-mediated emotional learning (45), and (c) is affected by glucocorticoid secretion, which is often elevated in MDD (17). Thus, these findings may explain cognitive biases toward aversive or emotionally arousing information observed in depression.

Increased amygdalar activation in depression may also represent a possible biologic substrate for anxiety, which is often comorbid with depression. In this respect, elevated levels of glucocorticoid hormones, which characterize at least some subgroups of patients with depression, may be especially relevant because elevated glucocorticoid hormones have been shown to be associated with increased corticotropin releasing hormone (CRH) in the amygdala. Increased corticotropin releasing hormone availability may increase anxiety, fear, and expectation for adversity (120).

In light of evidence suggesting a link between amygdalar activation, on the one hand, and memory consolidation and acquisition of long-term declarative knowledge about emotionally salient information on the other, the observations of dysfunctionally increased amygdalar activation in major depression are intriguing. As recently pointed out by Drevets (37), tonically *increased* amygdalar activation during depressive episodes may favor the emergence of rumination based on increased availability of emotionally negative memories. Although still untested, it is possible that these aberrant processes may rely on dysfunctional interactions between the amygdala, PFC, and ACC. Notably, structural abnormalities have been reported in territories of the PFC intimately connected with the ACC (35,97). ACC dysfunction, in particular, may lead to a decreased capability of monitoring potential conflict between memory-based ruminative processes and sensory information coming from the environment.

SUMMARY AND CONCLUSIONS

This chapter reviewed circuitry that underlies the representation and regulation of emotion. This circuitry is responsible for many of the emotional variations among people and for governing vulnerability and resilience in the face of stressful events. Different territories of the PFC and ACC, hippocampus, and amygdala were considered. These structures are all interconnected in regionally specific ways and exhibit bidirectional feedback. Variations in the morphometry and functioning of each of these structures have been reported in disorders of emotion, and functional variations are associated with several parameters of affective style in normal individuals. The establishment of differences in brain function or structure in cross-sectional studies that involve only a single assessment have been informative. However, such studies cannot specify which variations may be more primary and that may be a consequence of more primary variations. For example, an individual may have a low threshold for activation in the amygdala that will predispose him or her to react with more intense and more prolonged negative affect in response to a stressful event. Territories of the PFC may display accentuated activation as part of a regulatory strategy to attenuate activation in the amygdala. In this instance, one might refer to the amygdala difference as primary and the PFC difference as secondary. In the absence of longitudinal research, however, it is difficult to tease apart.

In addition, a paucity of work has examined functional and/or structural connectivity among these regions. Some of the variations in affective style that have been identified may arise as a consequence of variations in connectivity, functional, structural, or both. Future research should include measures of both functional (23) and structural connectivity. The latter can be measured with diffusion tensor imaging (71).

Animal and human literature on basic processes in emotion and emotion regulation was drawn on to help interpret normal and pathologic variations in affective style and to

highlight the kinds of studies that have not yet been performed but are important to conduct. The findings on the basic processes in animals and normal humans provide the foundation for a model of the major components in affect representation and regulation. The input to affect representation can be either a sensory stimulus or a memory. Most sensory stimuli are relayed through the thalamus, and from there, they can take a short route to the amygdala (72) and/or go up to the cortex. From both association cortex and subcortical regions including the amygdala, information is relayed to different zones of the PFC. The PFC plays a crucial role in the representation of goals. In the presence of ambiguous situations, the PFC sends bias signals to other brain regions to facilitate the expression of task-appropriate responses in the face of competition with potentially stronger alternatives. We argue that in the affective domain, the PFC implements affect-guided anticipatory processes. Left-sided PFC regions are particularly involved in approach-related appetitive goals, whereas right-sided PFC regions are involved in the maintenance of goals that require behavioral inhibition. Abnormalities in PFC function would be expected to compromise goal instantiation in patients with depression. Left-sided hypoactivation would result in deficits specifically in pregoal attainment forms of positive affect, whereas right-sided hyperactivation would result in excessive behavioral inhibition and anticipatory anxiety. Hypoactivation in regions of the PFC with which the amygdala is interconnected may result in a decrease in the regulatory influence on the amygdala and a prolonged time course of amygdala activation in response to challenge. This might be expressed phenomenologically as perseveration of negative affect and rumination.

The ACC is critically involved in conflict monitoring and is activated whenever an individual is confronted with a challenge that involves conflict among two or more response options. According to an influential theory of ACC function (18), the ACC monitors the conflicts among brain regions. When such conflict is detected, the ACC issues a call for further processing to the PFC that then adjudicates among the various response options and guides behavior toward a goal. The ACC is very frequently activated in neuroimaging studies of human emotion (16), in part because when emotion is elicited in the laboratory, it produces response conflict. There is the general expectation to behave in an unemotional fashion because subjects are participating in a scientific experiment, yet there are the responses that are pulled for by the emotional challenge, such as particular patterns of facial expression. This is commonly reported by subjects and is associated with ACC activation. The ACC is also activated when an individual is exposed to a conflict among different channels of emotional communication. For example, when the face and voice each express inconsistent emotions simultaneously, a conflict in the viewer is created and the ACC is activated (150). Individuals with low levels of ACC activation would be expected to be less sensitive or less reactive to these inconsistent affective cues.

There is sometimes a conflict between an individual's mood state and the behavior that is expected of the individual in a particular social or role context. For example, among depressed individuals, their dispositional mood state may predispose them to set few goals and engage in little intentional action, yet the demands of their environments may include expectations to behave and act in specific ways. In an individual with normal levels of ACC activation, the signal from the ACC would issue a call to other brain regions, the PFC being the most important, to resolve the conflict and engage in the appropriate goal-directed behavior. However, in an individual with abnormally low levels of ACC activation, the conflict between his/her dispositional mood state and the expectations of his/her context would not be effectively monitored, and, thus, the usual call for further processing would not be issued. The data on ACC function in depression most consistently reveal a pattern of decreased activation in particular regions of the ACC. Interestingly, those depressed patients with greater activation in the ventral ACC before antidepressant treatment are the ones most likely to show the greatest treatment responses. In normal individuals, activation of the affective subdivision of the ACC may also be associated phenomenologically with the will to change.

The hippocampus appears to play an important role in encoding context. Lesions to the hippocampus in animals impair context conditioning. In addition, this structure has a high density of glucocorticoid receptors, and elevated levels of cortisol in animal models have been found to produce hippocampal cell death. In humans, various stress-related disorders, including depression, have been found to be associated with hippocampal volume reductions. Whether such hippocampal volume differences are a cause or consequence of the depression cannot be answered from the extant data. However, to the extent that hippocampal dysfunction is present, we would expect that such individuals would show abnormalities in the context-appropriate modulation of emotional behavior. This type of abnormality would be expressed as the display of normal emotion in inappropriate contexts. Thus, the persistence of sadness in situations that would ordinarily engender happiness could arise, in part, as a consequence of a hippocampus-dependent problem in the context modulation of emotional responses. We have shown such effects in rhesus monkeys (29), although they have not yet been studied systematically in humans. The extensive connections between the hippocampus and PFC would presumably provide the requisite anatomic substrate for conveying the contextual information to the PFC to regulate emotional behavior in a context-appropriate fashion. The connections between the hippocampus and PFC are another potential target of dysfunction in depression and other disorders of emotion. It is possible that a particular subtype of individual exists in whom contextual encoding is intact and the PFC-implemented, goal-directed behavior is intact, but the context fails to adequately guide and repri-

oritize goals. In such cases, the functional and/or anatomic connectivity between the hippocampus and PFC might be a prime candidate for dysfunction. The tools are now available to examine both types of connectivity using noninvasive measures.

The amygdala has long been viewed as a key site for both the perception of cues that signal threat and the production of behavioral and autonomic responses associated with aversive responding. As we have noted above, current evidence suggests that the amygdala's role in negative affect may be a special case of its more general role in directing attention and resources to affectively salient stimuli and issuing a call for further processing of stimuli that have potentially major significance for the individual. As with other parts of the circuitry that we have addressed, there are extensive connections between the amygdala and each of the other structures that we have considered. The amygdala receives input from a wide range of cortical zones and has even more extensive projections back to the cortex, enabling the biasing of cortical processing as a function of the early evaluation of a stimulus as affectively salient. Also like the other components of the circuitry described, there are individual differences in amygdala activation both at baseline (118) and in response to challenge (27). Moreover, it is likely that regions of the PFC play an important role in modulating activation in the amygdala and thus influencing the time course of amygdala-driven negative affective responding. In light of the associations that have been reported between individual differences in amygdala activation and affect measures, it is likely that when it occurs, hyperactivation of the amygdala in depression is associated more with the fearlike and anxiety components of the symptoms than with the sad mood and anhedonia. In our own work, we have found that amygdala activation predicts dispositional negative affect in depressed patients but is unrelated to variations in positive affect (1). Excessive activation of the amygdala in depressed patients may also be associated with hypervigilance, particularly toward threat-related cues, which further exacerbates some of the symptoms of depression.

There are several types of studies that critically need to be performed in light of the extant evidence reviewed in this chapter. First, studies that relate specific variations in activation in particular brain regions to objective laboratory tasks that are neurally inspired and designed to capture the particular kinds of processing that are hypothesized to be implemented in those brain regions are needed. Relatively few studies of that kind have been conducted. Most studies that examine relationships between individual differences in neural activity and affective style, either normal or abnormal, usually relate such neural variation to either self-report or interview-based indices. In the future, it will be important to complement the phenomenologic description with laboratory measures that are explicitly designed to highlight the processes implemented in different parts of the circuitry described here.

Such future studies should include measures of both functional and structural connectivity to complement the activation measures. It is clear that interactions among the various components of the circuitry described are likely to play a crucial role in determining behavioral output. Moreover, it is possible that connectional abnormalities may exist in the absence of abnormalities in specific structures.

Longitudinal studies of at-risk samples with the types of imaging measures that are featured in this review are crucial. We do not know whether any of the variations discussed, of both a structural and functional variety, precede the onset of a disorder, co-occur with the onset of a disorder, or follow by some time the expression of a disorder. It is likely that the timing of the abnormalities in relation to the clinical course of the disorder varies for different parts of the circuitry. For example, data showing a relationship between the number of cumulative days depressed over the course of the lifetime and hippocampal volume (124,126) suggest that this abnormality may follow the expression of the disorder and represent a consequence rather than a primary cause of the disorder. However, before such a conclusion is accepted, it is important to conduct the requisite longitudinal studies to begin to disentangle these complex causal factors.

Finally, we regard the evidence presented in this review as offering very strong support for the view that specific constituents of emotion regulation and affective style will be identified that have not been directly uncovered with self-report methods. For example, the rapidity of recovery from a stressful stimulus and variations in context sensitivity of emotional responding are separable processes that will influence self-reports of emotion, yet such reports will be unrevealing with respect to the constituents that led to these influences. Thus, two individuals, each reporting high levels of dispositional negative affect, may be doing so because of variations in different parts of the circuitry reviewed. It is also very likely that some of the important variation in affective style, such as individual differences in the rapidity of recovery from a negative event, may not map precisely onto extant personality or self-report descriptors. A major challenge for the future will be to build a more neurobiologically plausible scheme for parsing the heterogeneity of emotion, emotion regulation, disorders of emotion, and affective style, based on the location and nature of the abnormality in the featured circuitry. We believe that this ambitious effort will lead to considerably more consistent findings at the biologic level and enable us to more rigorously characterize different endophenotypes that could then be exploited for genetic studies.

ACKNOWLEDGMENT

The author thanks Alexander J. Shackman, William Irwin, Diego Pizzagalli, and Jack Nitschke for invaluable comments and Jenna Topolovich for skilled and dedicated assistance in the preparation of the manuscript. This work was supported by National Institute of Mental Health grants (MH40747, P50-MH52354, MH43454, P50-MH61083) and a National Institute of Mental Health Research Scientist Award (K05-MH00875). Portions of this chapter appear in reference 30.

REFERENCES

1. Abercrombie HC, Schaefer SM, Larson CL, et al. Metabolic rate in the right amygdala predicts negative affect in depressed patients. *Neuroreport* 1998;9:3301–3307.
2. Altshuler LL, Bartzokis G, Grieder T, et al. Amygdala enlargement in bipolar disorder and hippocampal reduction in schizophrenia: an MRI study demonstrating neuroanatomic specificity. *Arch Gen Psychiatry* 1998;55:663–664.
3. Ashtari M, Greenwald BS, Kramer-Ginsberg E, et al. Hippocampal/amygdala volumes in geriatric depression. *Psychol Med* 1999;29:629–638.
4. Baer L, Rauch SL, Ballantine HTJ, et al. Cingulotomy for intractable obsessive-compulsive disorder. Prospective long-term follow-up of 18 patients. *Arch Gen Psychiatry* 1995;52:384–392.
5. Bartels A, Zeki S. The neural basis of romantic love. *Neuroreport* 2000;11:3829–3834.
6. Binder DK, Iskandar BJ. Modern neurosurgery for psychiatric disorders. *Neurosurgery* 2000;47:9–21.
7. Birbaumer N, Grodd W, Diedrich O, et al. fMRI reveals amygdala activation to human faces in social phobics. *Neuroreport* 1998;9:1223–1226.
8. Blood AJ, Zatorre RJ, Bermudez P, et al. Emotional responses to pleasant and unpleasant music correlate with activity in paralimbic brain regions. *Nat Neurosci* 1999;2:382–387.
9. Breiter HC, Rauch SL, Kwong KK, et al. Functional magnetic resonance imaging of symptom provocation in obsessive-compulsive disorder. *Arch Gen Psychiatry* 1996;53:595–606.
10. Bremner JD, Randall P, Scott TM, et al. MRI-based measurement of hippocampal volume in patients with combat-related posttraumatic stress disorder. *Am J Psychiatry* 1995;152:972–981.
11. Bremner JD, Randall P, Vermetten E, et al. Magnetic resonance imaging-based measurement of hippocampal volume in posttraumatic stress disorder related to childhood physical and sexual abuse—a preliminary report. *Biol Psychiatry* 1997;41:23–32.
12. Bremner JD, Innis RB, Salomon RM, et al. Positron emission tomography measurement of cerebral metabolic correlates of tryptophan depletion-induced depressive relapse. *Arch Gen Psychiatry* 1997;54:364–74.
13. Bremner JD, Narayan M, Anderson ER, et al. Hippocampal volume reduction in major depression. *Am J Psychiatry* 2000;157:115–118.
14. Büchel C, Dolan R, Armony JL, et al. Amygdala-hippocampal involvement in human aversive trace conditioning revealed through event-related functional magnetic resonance imaging. *J Neurosci* 1999;19:10869–10876.
15. Bush G, Whalen PJ, Rosen BR, et al. The counting Stroop: an interference task specialized for functional neuroimaging-validation study with functional MRI. *Hum Brain Mapp* 1998;6:270–282.
16. Bush G, Luu P, Posner MI. Cognitive and emotional influences in anterior cingulate cortex. *Trends Cogn Sci* 2000;4:215–222.
17. Carroll BJ, Curtis GC, Mendels J. Cerebrospinal fluid and plasma free cortisol concentrations in depression. *Psychol Med* 1976;6:235–244.

18. Carter CS, Botvinick MM, Cohen JD. The contribution of the anterior cingulate cortex to executive processes in cognition. *Rev Neurosci* 1999;10:49–57.
19. Carter CS, Macdonald AM, Botvinick M, et al. Parsing executive processes: strategic vs. evaluative functions of the anterior cingulate cortex. *Proc Natl Acad Sci U S A* 2000;97:1944–1948.
20. Chen G, Rajkowska G, Du F, et al. Enhancement of hippocampal neurogenesis by lithium. *J Neurochem* 2000;75:1729–1734.
21. Coffey CE, Wilkinson WE, Weiner RD, et al. Quantitative cerebral anatomy in depression. A controlled magnetic resonance imaging study. *Arch Gen Psychiatry* 1993;50:7–16.
22. Corcoran KA, Maren S. Hippocampal inactivation disrupts contextual retrieval of fear memory after extinction. *J Neurosci* 2001;21:1720–1726.
23. Cordes D, Haughton VM, Arfanakis K, et al. Mapping functionally related regions of brain with functional connectivity MR imaging. *Am J Neuroradiol* 2000;21:1636–1644.
24. Critchley HD, Mathias CJ, Dolan RJ. Neural activity in the human brain relating to uncertainty and arousal during anticipation. *Neuron* 2001;29:537–545.
25. Damasio AR. *Descartes-error: emotion, reason, and the human brain.* New York: Avon Books, 1994.
26. Davidson RJ, Abercrombie HC, Nitschke JB, et al. Regional brain function, emotion and disorders of emotion. *Curr Opin Neurobiol* 1999;9:228–234.
27. Davidson RJ, Irwin W. The functional neuroanatomy of emotion and affective style. *Trends Cogn Sci* 1999;3:11–21.
28. Davidson RJ. Affective style, psychopathology and resilience: brain mechanisms and plasticity. *Am Psychol* 2000;55:1193–1214.
29. Davidson RJ, Jackson DC, Kalin NH. Emotion, plasticity, context and regulation. *Psychol Bull* 2000;126:890–906.
30. Davidson RJ, Pizzagalli, D, Nitschke JB, et al. Parsing the subcomponents of emotion and disorders of emotion: perspectives from affective neuroscience. In: Davidson RJ, Scherer K, Goldsmith HH, eds. *Handbook of affective sciences.* New York: Oxford University Press 2002:8–24.
31. Davidson RJ, Putnam KM, Larson CL. Dysfunction in the neural circuitry of emotion regulation—a possible prelude to violence. *Science* 2000;289:591–594.
32. Davis M, Whalen PJ. The amygdala: vigilance and emotion. *Mol Psychiatry* 2001;6:13–34.
33. Devinsky O, Morrell MJ, Vogt BA. Contributions of anterior cingulate cortex to behaviour. *Brain* 1995;118:279–306.
34. Drevets WC, Videen TO, Price JL, et al. A functional anatomical study of unipolar depression. *J Neurosci* 1992;12:3628–3641.
35. Drevets WC, Price JL, Simpson JRJ, et al. Subgenual prefrontal cortex abnormalities in mood disorders. *Nature* 1997;386:824–827.
36. Drevets WC. Functional neuroimaging studies of depression: the anatomy of melancholia. *Annu Rev Med* 1998;49:341–361.
37. Drevets WC. Neuroimaging and neuropathological studies of depression: implications for the cognitive-emotional features of mood disorders. *Curr Opin Neurobiol* 2001;11:240–249.
38. Driessen M, Herrmann J, Stahl K, et al. Magnetic resonance imaging volumes of the hippocampus and the amygdala in women with borderline personality disorder and early traumatization. *Arch Gen Psychiatry* 2000;57:1115–1122.
39. Duman RS, Malberg J, Nakagawa S, et al. Neuronal plasticity and survival in mood disorders. *Biol Psychiatry* 2000;48:732–739.
40. Ebert D, Feistel H, Barocka A. Effects of sleep deprivation on the limbic system and the frontal lobes in affective disorders: a study with Tc-99m-HMPAO SPECT. *Psychiatry Res* 1991;40:247–251.
41. Ebert D, Ebmeier KP. The role of the cingulate gyrus in depression: from functional anatomy to neurochemistry. *Biol Psychiatry* 1996;39:1044–1050.
42. Elliott R, Dolan RJ. Differential neural responses during performance of matching and nonmatching to sample tasks at two delay intervals. *J Neurosci* 1999;19:5066–5073.
43. Eriksson PS, Perfilieva E, Bjork-Eriksson T, et al. Neurogenesis in the adult human hippocampus. *Nat Med* 1998;4:1313–1317.
44. Fanselow MS. Contextual fear, gestalt memories, and the hippocampus. *Behav Brain Res* 2000;110:73–81.
45. Ferry B, Roozendaal B, McGaugh JL. Role of norepinephrine in mediating stress hormone regulation of long-term memory storage: a critical involvement of the amygdala. *Biol Psychiatry* 1999;46:1140–1152.
46. Fink GR, Markowitsch HJ, Reinkemeier M, et al. Cerebral representation of one's own past: neural networks involved in autobiographical memory. *J Neurosci* 1996;16:4275–4282.
47. Garavan H, Ross RH, Stein EA. Right hemispheric dominance of inhibitory control: an event-related functional MRI study. *Proc Natl Acad Sci U S A* 1999;96:8301–8306.
48. Gold PW, Goodwin FK, Chrousos GP. Clinical and biochemical manifestations of depression: relation to the neurobiology of stress. *N Engl J Med* 1988;314:348–353.
49. Gould E, Tanapat P, Rydel T, et al. Regulation of hippocampal neurogenesis in adulthood. *Biol Psychiatry* 2000;48:715–720.
50. Hamann SB, Ely TD, Grafton ST, et al. Amygdala activity related to enhanced memory for pleasant and aversive stimuli. *Nat Neurosci* 1999;2:289–293.
51. Henriques JB, Glowacki JM, Davidson RJ. Reward fails to alter response bias in depression. *J Abnorm Psychol* 1994;103:460–466.
52. Henriques JB, Davidson RJ. Decreased responsiveness to reward in depression. *Cogn Emotion* 2000;15:711–724.
53. Ho AP, Gillin JC, Buchsbaum MS, et al. Brain glucose metabolism during non-rapid eye movement sleep in major depression. A positron emission tomography study. *Arch Gen Psychiatry* 1996;53:645–652.
54. Holland PC, Gallagher M. Amygdala circuitry in attentional and representational processes. *Trends Cogn Sci* 1999;3:65–73.
55. Hrdina PD, Demeter E, Vu TB, et al. 5-HT uptake sites and 5-HT2 receptors in brain of antidepressant-free suicide victims/depressives: increase in 5-HT2 sites in cortex and amygdala. *Brain Res* 1993;614:37–44.
56. Ikegaya Y, Saito H, Abe K. Attenuated hippocampal long-term potentiation in basolateral amygdala-lesioned rats. *Brain Res* 1994;656:157–164.
57. Ikegaya Y, Abe K, Saito H, et al. Medial amygdala enhances synaptic transmission and synaptic plasticity in the dentate gyrus of rats in vivo. *J Neurophysiol* 1995;74:2201–2203.
58. Ikegaya Y, Saito H, Abe K. Requirement of basolateral amygdala neuron activity for the induction of long-term potentiation in the dentate gyrus in vivo. *Brain Res* 1995;671:351–354.
59. Isenberg N, Silbersweig D, Engelien A, et al. Linguistic threat activates the human amygdala. *Proc Natl Acad Sci U S A* 1999;96:10456–10459.
60. Jacobs BL, Praag H, Gage FH. Adult brain neurogenesis and psychiatry: a novel theory of depression. *Mol Psychiatry* 2000;5:262–269.
61. Jacobson L, Sapolsky RM. The role of the hippocampus in feedback regulation of the hypothalamic-pituitary-adrenocortical axis. *Endocrinol Rev* 1991;12:118–134.
62. Javanmard M, Shlik J, Kennedy SH, et al. Neuroanatomic correlates of CCK-4-induced panic attacks in healthy humans: a comparison of two time points. *Biol Psychiatry* 1999;45:872–882.

63. Kahneman D. Objective happiness. In: Kahneman E, Diener E, Schwartz N, eds. *Well-being: the foundations of hedonic psychology.* New York: Russell Sage Foundation, 1999:3–25.

64. Kalin NH, Shelton SE. The regulation of defensive behaviors in rhesus monkeys: Implications for understanding anxiety disorders. In: Davidson RJ, ed. *Anxiety, depression and emotion.* New York: Oxford University Press, 2000:50–68.

65. Kawasaki H, Adolphs R, Kaufman O, et al. Single-neuron responses to emotional visual stimuli recorded in human ventral prefrontal cortex. *Nat Neurosci* 2001;4:15–16.

66. Kempermann G, Kuhn HG, Gage FH. More hippocampal neurons in adult mice living in an enriched environment. *Nature* 1997;386:493–495.

67. Ketter TA, Andreason PJ, George MS, et al. Anterior paralimbic mediation of procaine-induced emotional and psychosensory experiences. *Arch Gen Psychiatry* 1996;53:59–69.

68. Ketter TA, Kimbrell TA, George MS, et al. Effects of mood and subtype on cerebral glucose metabolism in treatment-resistant bipolar disorder. *Biol Psychiatry* 2001;49:97–109.

69. Konishi S, Nakajima K, Uchida I, et al. Common inhibitory mechanism in human inferior prefrontal cortex revealed by event-related functional MRI. *Brain* 1999;122:981–991.

70. Lane RD, Fink GR, Chau PM, et al. Neural activation during selective attention to subjective emotional responses. *Neuroreport* 1997;8:3969–3972.

71. Le Bihan D, Mangin JF, Poupon C, et al. Diffusion tensor imaging: concepts and applications. *J Magn Reson Imaging* 2001;13:534–546.

72. LeDoux JE. Emotion circuits in the brain. *Annu Rev Neurosci* 2000;23:155–184.

73. Leverenz JB, Wilkinson CW, Wamble M, et al. Effect of chronic high-dose exogenous cortisol on hippocampal neuronal number in aged nonhuman primates. *J Neurosci* 1999;19:2356–2361.

74. Liberzon I, Taylor SF, Amdur R, et al. Brain activation in PTSD in response to trauma-related stimuli. *Biol Psychiatry* 1999;45:817–826.

75. Lupien SJ, de Leon M, de Santi S, et al. Cortisol levels during human aging predict hippocampal atrophy and memory deficits. *Nat Neurosci* 1998;1:69–73.

76. MacDonald AW, Cohen JD, Stenger VA, et al. Dissociating the role of the dorsolateral prefrontal and anterior cingulate cortex in cognitive control. *Science* 2000;288:1835–1838.

77. Madhav TR, Pei Q, Grahame-Smith DG, et al. Repeated electroconvulsive shock promotes the sprouting of serotonergic axons in the lesioned rat hippocampus. *Neuroscience* 2000;97:677–683.

78. Magarinos AM, McEwen BS, Flugge G, et al. Chronic psychosocial stress causes apical dendritic atrophy of hippocampal CA3 pyramidal neurons in subordinate tree shrews. *J Neurosci* 1996;16:3534–3540.

79. Malberg JE, Eisch AJ, Nestler EJ, et al. Chronic antidepressant treatment increases neurogenesis in adult rat hippocampus. *J Neurosci* 2000;20:9104–9110.

80. Mayberg HS. Limbic-cortical dysregulation: a proposed model of depression. *J Neuropsychiatry Clin Neurosci* 1997;9:471–481.

81. Mayberg HS, Liotti M, Brannan SK, et al. Reciprocal limbic-cortical function and negative mood: converging PET findings in depression and normal sadness. *Am J Psychiatry* 1999;156:675–682.

82. McEwen BS. Protective and damaging effects of stress mediators. *N Engl J Med* 1998;338:171–179.

83. Mega MS, Cummings JL, Salloway S, et al. The limbic system: an anatomic, phylogenetic, and clinical perspective. *J Neuropsychiatry Clin Neurosci* 1997;9:315–330.

84. Mervaala E, Fohr J, Kononen M, et al. Quantitative MRI of the hippocampus and amygdala in severe depression. *Psychol Med* 2000;30:117–125.

85. Miller EK, Cohen JD. An integrative theory of prefrontal cortex function. *Annu Rev Neurosci* 2001;24:167–202.

86. Mirz F, Gjedde A, Sodkilde-Jorgensen H, et al. Functional brain imaging of tinnitus-like perception induced by aversive auditory stimuli. *Neuroreport* 2000;11:633–637.

87. Morris JS, Friston KJ, Dolan RJ. Neural responses to salient visual stimuli. *Proc R Soc Lond* 1997;264:769–775.

88. Morris JS, Ohman A, Dolan RJ. A subcortical pathway to the right amygdala mediating "unseen" fear. *Proc Natl Acad Sci U S A* 1999;96:1680–1685.

89. Nader K, Schafe GE, Le Doux JE. Fear memories require protein synthesis in the amygdala for reconsolidation after retrieval. *Nature* 2000;406:722–726.

90. Nauta WH. The problem of the frontal lobe: a reinterpretation. *J Psychiatr Res* 1971;8:167–187.

91. Nobre AC, Sebestyen GN, Gitelman DR, et al. Functional localization of the system for visuospatial attention using positron emission tomography. *Brain* 1997;120:515–533.

92. Nofzinger EA, Nichols TE, Meltzer CC, et al. Changes in forebrain function from waking to REM sleep in depression: preliminary analyses of [18F]FDG PET studies. *Psychiatry Res* 1999;91:59–78.

93. Noga JT, Vladar K, Torrey EF. A volumetric magnetic resonance imaging study of monozygotic twins discordant for bipolar disorder. *Psychiatry Res* 2001;106:25–34.

94. O'Doherty J, Kringelbach ML, Rolls ET, et al. Abstract reward and punishment representations in the human orbitofrontal cortex. *Nat Neurosci* 2001;4:95–102.

95. Ochsner KN, Barrett LF. A multiprocess perspective on the neuroscience of emotion. In: Mayne TJ, Bonanno GA, eds. *Emotions: current issues and future directions.* New York: Guilford Press, 2001:38–81.

96. Öngür D, An X, Price JL. Prefrontal cortical projections to the hypothalamus in macaque monkeys. *J Comp Neurol* 1998;401:480–505.

97. Öngür D, Drevets WC, et al. Glial reduction in the subgenual prefrontal cortex in mood disorders. *Proc Natl Acad Sci U S A* 1998;95:13290–13295.

98. Pantel J, Schroder J, Essig M, et al. Quantitative magnetic resonance imaging in geriatric depression and primary degenerative dementia. *J Affect Disord* 1997;42:69–83.

99. Pardo JV, Pardo PJ, Janer KW, et al. The anterior cingulate cortex mediates processing selection in the Stroop attentional conflict paradigm. *Proc Natl Acad Sci U S A* 1990;87:256–259.

100. Pariante CM, Miller AH. Glucocorticoid receptors in major depression: relevance to pathophysiology and treatment. *Biol Psychiatry* 2001;49:391–404.

101. Paus T, Zatorre RJ, Hofle N, et al. Time-related changes in neural systems underlying attention and arousal during the performance of an auditory vigilance task. *J Cogn Neurosci* 1997;9:392–408.

102. Phillips ML, Bullmore ET, Howard R, et al. Investigation of facial recognition memory and happy and sad facial expression perception: an fMRI study. *Psychiatry Res* 1998;83:127–138.

103. Pizzagalli D, Pascual-Marqui RD, Nitschke JB, et al. Anterior cingulate activity as a predictor of degree of treatment response in major depression: evidence from brain electrical tomography analysis. *Am J Psychiatry* 2001;158:405–415.

104. Pruessner JC, Collins DL, Pruessner M, et al. Age and gender predict volume decline in the anterior and posterior hippocampus in early adulthood. *J Neurosci* 2001;21:194–200.

105. Rauch SL, Savage CR, Alpert NM, et al. A positron emission tomographic study of simple phobic symptom provocation. *Arch Gen Psychiatry* 1995;52:20–28.

106. Rauch SL, van der Kolk BA, Fisler RE, et al. A symptom provocation study of posttraumatic stress disorder using positron emission tomography and script-driven imagery. *Arch Gen Psychiatry* 1996;53:380–387.

107. Rauch SL, Savage CR, Alpert NM, et al. A study of three disorders using positron emission tomography and symptom provocation. *Biol Psychiatry* 1997;42:446–452.

108. Rauch SL, Whalen PJ, Shin LM, et al. Exaggerated amygdala response to masked facial stimuli in posttraumatic stress disorder: a functional MRI study. *Biol Psychiatry* 2000;47:769–776.

109. Reiman EM. The application of positron emission tomography to the study of normal and pathologic emotions. *J Clin Psychiatry* 1997;58:4–12.

110. Reul JM, de Kloet ER. Anatomical resolution of two types of corticosterone receptor sites in rat brain with in vitro autoradiography and computerized image analysis. *J Steroid Biochem Mol Biol* 1986;24:269–272.

111. Rogers RD, Owen AM, Middleton HC, et al. Choosing between small, likely rewards and large, unlikely rewards activates inferior and orbital prefrontal cortex. *J Neurosci* 1999;20:9029–9038.

112. Rolls ET. The functions of the orbitofrontal cortex. *Neurocase* 1999;5:301–312.

113. Rusch BD, Abercrombie HC, Oakes TR, et al. Hippocampal morphometry in depressed patients and controls: relations to anxiety symptoms. *Biol Psychiatry* 2001;50:960–964.

114. Sapolsky RM, Krey LC, McEwan BS. The neuroendocrinology of stress and aging: the glucocorticoid cascade hypothesis. *Endocr Rev* 1986;7:284–301.

115. Sapolsky RM, Uno H, Rebert CS, et al. Hippocampal damage associated with prolonged glucocorticoid exposure in primates. *J Neurosci* 1990;10:2897–2902.

116. Sapolsky RM. Glucocorticoids and hippocampal atrophy in neuropsychiatric disorders. *Arch Gen Psychiatry* 2000;57:925–935.

117. Saxena S, Brody AL, Ho ML, et al. Cerebral metabolism in major depression and obsessive-compulsive disorder occurring separately and concurrently. *Biol Psychiatry* 2001;50:159–170.

118. Schaefer SM, Abercrombie HC, Lindgren KA, et al. Six-month test-retest reliability of MRI-defined PET measures of regional cerebral glucose metabolic rate in selected subcortical structures. *Hum Brain Mapp* 2000;10:1–9.

119. Schneider F, Weiss U, Kessler C, et al. Subcortical correlates of differential classical conditioning of aversive emotional reactions in social phobia. *Biol Psychiatry* 1999;45:863–871.

120. Schulkin J. Melancholic depression and the hormones of adversity—a role for the amygdala. *Curr Dir Psychol Sci* 1994;3:41–44.

121. Servan-Schreiber D, Perlstein WM. Pharmacologic activation of limbic structures and neuroimaging studies of emotions. *J Clin Psychiatry* 1997;58:13–15.

122. Semple WE, Goyer PF, McCormick R, et al. Higher brain blood flow at amygdala and lower frontal cortex blood flow in PTSD patients with comorbid cocaine and alcohol abuse compared with normals. *Psychiatry* 2000;63:65–74.

123. Shah PJ, Ebmeier KP, Glabus MF, et al. Cortical grey matter reductions associated with treatment-resistant chronic unipolar depression. Controlled magnetic resonance imaging study. *Br J Psychiatry* 1998;172:527–532.

124. Sheline YI, Wang PW, Gado MH, et al. Hippocampal atrophy in recurrent major depression. *Proc Natl Acad Sci U S A* 1996;93:3908–3913.

125. Sheline YI, Gado MH, Price JL. Amygdala core nuclei volumes are decreased in recurrent major depression. *Neuroreport* 1998;9:2023–2028.

126. Sheline YI, Sanghavi M, Mintun MA, et al. Depression dura-tion but not age predicts hippocampal volume loss in medically healthy women with recurrent major depression. *J Neurosci* 1999;19:5034–5043.

127. Sheline YI. 3D MRI studies of neuroanatomic changes in unipolar major depression: the role of stress and medical comorbidity. *Biol Psychiatry* 2000;48:791–800.

128. Shin LM, Kosslyn SM, McNally RJ, et al. Visual imagery and perception in posttraumatic stress disorder. A positron emission tomographic investigation. *Arch Gen Psychiatry* 1997;54:233–241.

129. Squire LR, Knowlton BJ. The medial temporal lobe, the hippocampus, and the memory systems of the brain. In: Gazzaniga MS, ed. *The new cognitive neurosciences.* Cambridge, MA: MIT Press, 2000:765–779.

130. Starkman MN, Giordani B, Gebarski SS, et al. Decrease in cortisol reverses human hippocampal atrophy following treatment of Cushing's disease. *Biol Psychiatry* 1999;46:1595–1602.

131. Steffens DC, Byrum CE, McQuoid DR, et al. Hippocampal volume in geriatric depression. *Biol Psychiatry* 2000;48:301–309.

132. Stein MB, Koverola C, Hanna C, et al. Hippocampal volume in women victimized by childhood sexual abuse. *Psychol Med* 1997;27:951–959.

133. Strakowski SM, DelBello MP, Sax KW, et al. Brain magnetic resonance imaging of structural abnormalities in bipolar disorder. *Arch Gen Psychiatry* 1999;56:254–260.

134. Taylor SE. Asymmetrical effects of positive and negative events: the mobilization-minimization hypothesis. *Psychol Bull* 1991;110:67–85.

135. Tebartz van Elst L, Woermann FG, Lemieux L, et al. Amygdala enlargement in dysthymia: a volumetric study of patients with temporal lobe epilepsy. *Biol Psychiatry* 1999;46:1614–1623.

136. Tebartz van Elst L, Woermann F, Lemieux L, et al. Increased amygdala volumes in female and depressed humans. A quantitative magnetic resonance imaging study. *Neurosci Lett* 2000;281:103–106.

137. Thayer JF, Lane RD. A model of neurovisceral integration in emotion regulation and dysregulation. *J Affect Disord* 2000;61:201–216.

138. Vakili K, Pillay SS, Lafer B, et al. Hippocampal volume in primary unipolar major depression: a magnetic resonance imaging study. *Biol Psychiatry* 2000;47:1087–1090.

139. Veith RC, Lewis N, Linares OA, et al. Sympathetic nervous system activity in major depression. Basal and desipramine-induced alterations in plasma norepinephrine kinetics. *Arch Gen Psychiatry* 1994;51:411–422.

140. Vogt BA, Finch DM, Olson CR. Functional heterogeneity in cingulate cortex: the anterior executive and posterior evaluative regions. *Cereb Cortex* 1992;2:435–443.

141. Vogt BA, Nimchinsky EA, Vogt LJ, et al. Human cingulate cortex: surface features, flat maps, and cytoarchitecture. *J Comp Neurol* 1995;359:490–506.

142. von Gunten A, Fox NC, Cipolotti L, et al. A volumetric study of hippocampus and amygdala in depressed patients with subjective memory problems. *J Neuropsychiatry Clin Neurosci* 2000;12:493–498.

143. Whalen PJ. Fear, vigilance, and ambiguity: initial neuroimaging studies of the human amygdala. *Curr Dir Psychol Sci* 1998;7:177–188.

144. Whalen PJ, Bush G, McNally RJ, et al. The emotional Stroop paradigm: a functional magnetic resonance imaging probe of the anterior cingulate affective division. *Biol Psychiatry* 1998;44:1219–1228.

145. Wu JC, Gillin JC, Buchsbaum MS, et al. Effect of sleep depriva-

tion on brain metabolism of depressed patients. *Am J Psychiatry* 1992;149:538–543.

146. Wu J, Buschbaum MS, Gillin JC, et al. Prediction of antidepressant effects of sleep deprivation by metabolic rates in the ventral anterior cingulate and medical prefrontal cortex. *Am J Psychiatry* 1999;156:1149–1158.

147. Yurgelun-Todd DA, Gruber SA, Kanayama G, et al. fMRI during affect discrimination in bipolar affective disorder. *Bipolar Disord* 2000;2:237–248.

148. Zald DH, Lee JT, Fluegel KW, et al. Aversive gustatory stimulation activates limbic circuits in humans. *Brain* 1998;121:1143–1154.

149. Zalla T, Koechlin E, Pietrini P, et al. Differential amygdala responses to winning and losing: a functional magnetic resonance imaging study in humans. *Eur J Neurosci* 2000;12:1764–1770.

150. Dolan RJ, Morris JS, de Gelder B. Crossmodal binding of fear in voice and face. *Proc Natl Acad Sci USA* 2001;98:10006–10010.

21

PERCEPTUAL DISTURBANCES IN NEUROLOGIC AND PSYCHIATRIC POPULATIONS

RUTH SALO
LYNN C. ROBERTSON

PERCEPTUAL DISTURBANCES IN PSYCHIATRIC AND NEUROLOGIC DISORDERS

With the emergence of the field of cognitive neuroscience, knowledge of how damage to different neural systems affects perception in humans has grown substantially. This can be attributed in part to the improvements in functional imaging, including positron emission tomography (PET) and functional magnetic resonance imaging (fMRI). Although these technologies have contributed a great deal to the field, converging behavioral evidence is essential to adequately capture and measure brain–behavior relationships.

Historically, the field of neuropsychology has focused on broad areas of behavior. Attempts to further dissect the behavioral deficits were more descriptive than analytical. For instance, visuospatial deficits could be divided into difficulties with parts versus wholes or gestalts or they could be described as unilateral neglect, meaning that patients would not respond to unilaterally presented stimuli in the contralesional field. A diagnosis of Balint syndrome meant that patients would only respond to one object (simultanagnosia), had difficulty with visually guided reaching (optic ataxia), and exhibited visual tracking deficits (optic apraxia). A diagnosis of Gerstmann syndrome meant that the patient had difficulty with left–right discrimination, calculation (acalculia), naming digits of the hand (finger agnosia), and writing (agraphia). These examples demonstrate that neuropsychological classifications do not elucidate the underlying cognitive processes but rather are descriptions of behavior that one is likely to see in these patients. Although these descriptions may be helpful clinically, they are not useful in elucidating the cognitive or neural mechanisms that underlie the behavioral deficits in question.

This chapter presents some selected advances in understanding the components of visuospatial and auditory perceptual functions and how they relate to neural systems. There are two focuses of this chapter: (a) anatomic corre-lates of visuoperceptual dysfunction in neurologic patients and (b) anatomic correlates of visual and auditory perceptual disturbances in psychiatric patients. This focus does not imply that psychopharmacologic or electrophysiologic correlates are not also important. Some evidence using these methods are noted within this chapter, but the emphasis is on perceptual deficits caused by structural damage to the human brain and perceptual disturbances associated with major psychiatric disorders.

PERCEPTUAL DISTURBANCES IN NEUROLOGIC PATIENTS

Lesion Method: Caveats and Credence

Much of neuropsychology's history is based on the lesion method. In clinical neuropsychology, there was a pragmatic reason for knowing how behavior related to an affected region. Before the advent of techniques to image the human brain, clinicians had to rely on behavioral data to determine the probable neural regions of damage. Despite well-known variability in functional damage between individuals, accuracy in determining the neural regions involved was rather impressive given the limited technologic tools of the time. Clearly, localization of deficit occurs in the sense that the performance of particular tasks is likely to be affected by a lesion in a particular location, whereas the performance of other tasks is not. Given the superiority of radiologic measures in localizing neurologic damage, the use of psychometric measures for this purpose has become increasingly less important. In clinical neuropsychology, concerns have shifted from an emphasis on locating the lesion to an emphasis on articulating the specific cognitive deficits and intact cognitive capacities that a patient possesses.

Despite its historical record, the lesion method has been challenged, in both animal and human research. One problem in relating damaged regions directly to a behavioral deficit is that the region of injury may be in the pathway of other regions that are integrally involved in the behavior. To take two simple examples, a visual hallucination might be the result of a lesion to the brainstem, bilateral occipital lesions, or a state of visual deterioration owing to macular degeneration. The inability to navigate one's way through large-scale environments may occur after lesions to the superior parietal lobe, damage to the retrosplenial cortex, or a lesion in the parahippocampal region. These types of relationship between cognition and brain regions have led some investigators to question the idea of specialization of function and to argue for a systems approach (1–4).

However, proposed systems or networks continue to include specialization within a network. For instance, Mesulam (3) proposed an attention network to account for unilateral neglect in which association cortices are involved in sensory representation of the stimulus, the frontal lobe in motor representation, the cingulate in motivation, and the reticular structures in arousal. All interact with parietal function. This approach is clearly superior to other approaches that make claims such as attention is a parietal lobe function, but it is important to note that localization of function remains within the network. One of the most important benefits of the network approach is that it demonstrates the need to consider the underlying operations associated with different regions that then interact with other areas (perhaps quite distant). When the network is intact, it results in behavior that we might call "attending," and damage to different parts of it will disrupt attending in different ways.

Another limitation of the lesion method is that neural reorganization and/or compensatory strategies occur after brain damage. Because of complications such as diaschisis and edema after acute insult, investigators who wish to specify the neural structures that correspond to a cognitive process must test subjects well after damage occurs, whether in humans or animals, leaving substantial opportunity for change. For instance, Merzenich and Kaas (5) showed that, in primates, neural reorganization occurs rapidly, even in the primary sensory cortex. When monkeys lose the use of one digit, the somatic cortex associated with that digit is reorganized to include adjacent digits. A similar reorganization in the visual cortex had been observed after the introduction of small retinal scotomas (6). The challenge posed by neural plasticity for scientific investigations is less of a concern in older than younger subjects, although it is still present. Conversely, there is no reason to believe that developing compensatory strategies is decreased in aged subjects.

One way to address problems associated with the lesion method is to collect converging evidence from different populations of subjects and to use various techniques to address the questions of interest. For example, if the prefrontal cortex is associated with a particular memory deficit in stroke patients and the frontal lobe shows increased activity during similar procedures in PET studies in normal subjects, then

confidence in the frontal lobe's role in memory is increased. If hemisphere laterality effects found in groups of patients with focal lesions are supported by converging evidence in normal subjects using lateralized visual field presentations or evoked potentials, then theories derived from patients with unilateral lesions are given substantial support. Additionally, if the distinct component operations that are affected with different lesions can be disassociated using measures of normal performance, then there is far more confidence that the deficits observed in patient groups are not owing to compensatory strategies.

So why perform studies using the lesion method at all if it is fraught with such difficulties? Perhaps the most important and obvious reason is to determine which deficits are caused by stroke and trauma, events that afflict thousands of people each day. A second reason is that for purposes of basic research, there is no foolproof method to replace it. All techniques have their strengths and weaknesses, but only the lesion method can reveal whether a specific region is necessary for normal function.

Another reason to use the lesion method in basic research is to study the division of labor between neural subregions. If a specific lesion affects one aspect of cognition while leaving others intact, we can be reasonably sure that there are areas of the brain that perform distinct operations and can study what those operations may be. The best evidence for separate systems remains the double dissociation. If damage to one region affects process X without affecting process Y, evidence for functional separation is strongly supported. For instance, damage to the left angular gyrus can affect language comprehension without affecting semantic comprehension of pictures, whereas damage to the right parietal regions can affect spatial orientation without disturbing language comprehension. This observation demonstrates that a component of one system associated with the left hemisphere is necessary for normal language function, whereas another involving the right hemisphere is necessary for spatial function. Note that this interpretation is not the same as saying that the left hemisphere is specialized for language and the right hemisphere is specialized for space. Rather, it poses the question of what different computations occur in the right and left hemispheres that will disrupt language, on the one hand, and spatial orientation on the other. Finally, the lesion method can be as valuable in studying deficits associated with damage in different areas as in studying the functions that remain intact. If a lesion does not disrupt a cognitive operation of interest, it suggests that the structures that support the process are intact.

Two Cortical Streams

Two major streams of processing traverse the cerebral cortex. One stream is associated with inferior occipitotemporal pathways and is involved in object identification or determining what is there. The other is associated with occipitoparietal pathways and is involved in analyzing spatial information or determining where objects are (7). The separation of these streams begins in the retina, continues through the geniculate and primary visual cortex, and is relatively separate (although by no means completely) as information flows anteriorly (8). The two major streams correspond roughly to the parvocellular and magnocellular systems, respectively. The receptive field size of cells in both the magnocellular and parvocellular systems increases as information is processed through progressively higher levels, but this increase in size is more pronounced in the magnocellular system that projects to the parietal lobe (8,9).

Cells in the parietal lobe often have receptive fields that include the ipsilateral and contralateral fields (8). Large receptive fields are well suited for spatial analysis to analyze motion, low spatial frequencies, and the relative location of objects. The parvocellular system has smaller receptive fields that respond to features such as color, brightness, shape, and higher spatial frequencies. These features provide some of the fundamental building blocks of object perception. A great deal of work with primates has revealed functionally separate areas within each cortical stream. The number of visual areas increases in vertebrates as the ratio of brain to body weight increases across species (10). Thus, humans are likely to have quite a large number of visual areas that respond to different features of a stimulus. A major question is how all these different areas rejoin to produce the unitary percept that we take for granted.

A clinical implication of this division of information is that there are several ways in which the system can go wrong. The good news is that a deficit in perceiving one feature (e.g., color) can be fairly circumscribed. The bad news is that if the wrong wiring takes place during development or reorganization, the number of abnormal connections that could end up providing false information is substantial. The study of neurologic patients with focal lesions has helped in understanding the ways in which the brain divides information and the mechanisms that are necessary to integrate or bind the information into an accurate (or at least relatively accurate) percept (Fig. 21.1).

Deficits in Perceiving Space

Balint Syndrome

On extensive neuropsychological testing of English war veterans with gunshot wounds to the head, Newcombe and Russell (11) showed that penetrating wounds in the parietal lobe produced difficulty in locating objects but not in identifying them, whereas occipitotemporal damage caused difficulty in identifying objects but not in locating them. Consistently, posterior ventral lesions produce object agnosias, whereas posterior dorsal lesions produce spatial deficits, as seen in unilateral visual neglect, visual extinction, or Balint syndrome (Fig. 21.2).

Despite the evidence that parietal lobes encode where things are, this can be only a first approximation. Studies of patients with Balint syndrome produced by bilateral parietal lobe damage have shown that the perception of spatial relationships between objects can be altered without affecting spatial relationships within objects. That is, these patients may know where a part of an object is in relation to other parts but not know where the object is in relation to other objects. In a simple task of reporting whether there were one or two colors in a display with many circles, Humphreys and Riddoch (12) found that a patient with Balint syndrome reported that there were two colors 75% of the time when the circles were unconnected. However, when lines were placed so that the circles on one end of the lines were one color and circles on the other end were another color, his accuracy increased to 90% (Fig. 21.3). Patients with Balint syndrome are rare and often mistaken as blind. They typically report their vision as hazy or blurred. However, in-depth testing reveals that these patients are able to recognize individual objects but have great difficulty seeing two objects placed next to each other, either side by side or one behind another along the same line of sight. In addition to deficits in perceiving spatial relations between objects, eye movements in patients with Balint syndrome appear fixed, and there is great difficulty tracking an object through space. These patients also have severe spatial reaching problems with no accompanying motor deficits.

We had the opportunity to test a patient diagnosed with Balint syndrome and found a similar profile (13,14). R.M. suffered embolic infarcts separated by approximately 7 months to each parietooccipital region. Testing began approximately 5 months after the second infarct. As with all Balint patients, R.M. could easily identify single objects in a visual scene but could not accurately reach for the object or track it. He could not accurately reach for an object when reaching was visually guided nor could he point to the location of an object.

Further experimental testing showed that when a single object was tilted away from or toward R.M., he was able to determine which side of the object was nearer to him, but when two objects were presented closer to or farther from him, he would guess which one was nearer to him. He could readily report the color of an object presented briefly in his peripheral vision but could not do so when a second object was presented foveally. He could name single pictures of everyday objects without difficulty and could do so even when critical features were covered over (e.g., the trunk of an elephant). When two pictures were presented, he named the two only infrequently and then only with a great deal of effort. This does not mean that he did not register the entire visual field. He also made many binding errors or what Treisman and Schmidt (15) call illusory conjunctions. If he saw the letter A, for instance, he often saw it as a color that was somewhere else in the visual field. He combined incorrectly different features such as shape and color, motion,

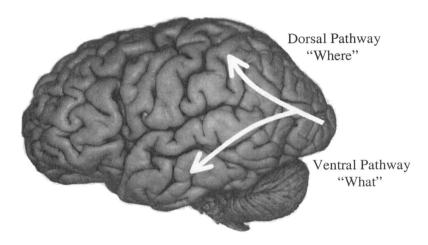

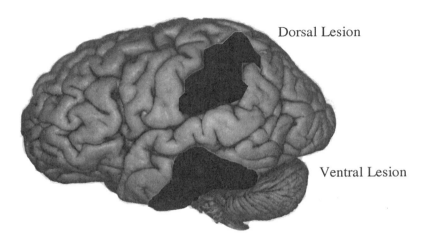

FIGURE 21.1. Two visual pathways in the human brain. The lower or what pathway extends from the occipital cortex to the temporal lobe and is believed to be involved in object recognition. The upper or where pathway extends from the occipital cortex to the parietal lobe and is more specialized for encoding object location. The lower part of the figure shows examples of brain lesions that might disrupt the what and where pathways.

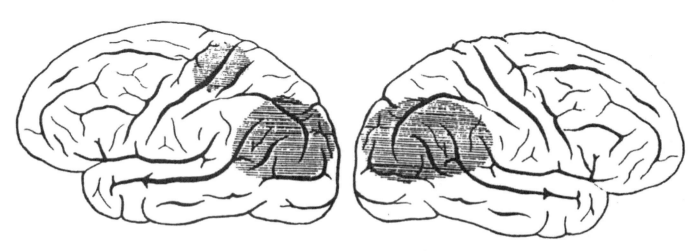

FIGURE 21.2. A reconstruction of bilateral parietal lesions from a Balint syndrome patient. (Reprinted from Balint R. Seelenlahmung des 'Schauens', optische Ataxie, raumliche Storung der Aufmerksamkeit. *Monatsschr Psychiatrie Neurologie* 1909;24:51–81, with permission.)

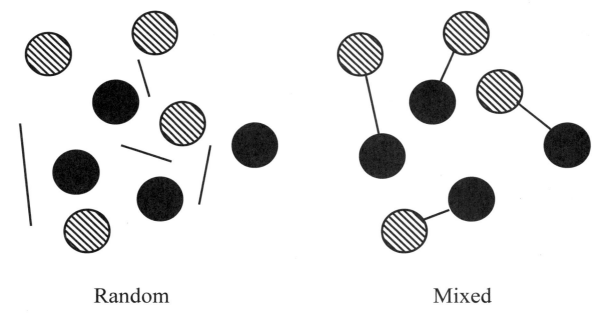

FIGURE 21.3. Example of stimuli. (Modified from Humphreys GW, Riddoch MJ. Interactions between object and space systems revealed through neuropsychology. In: Meyer DE, Kornblum S, eds. *Attention and performance XIV*. Hillsdale, NJ: Lawrence Erlbaum, 1993, with permission.)

and size, all of which are encoded by more ventral brain regions.

An MRI scan revealed bilateral parietooccipital involvement of nearly equal location and volume. There were also cerebellar infarcts that were unlikely to contribute to his perceptual deficits. Standard tests of his primary vision, including the perception of random dot stereograms, contrast sensitivity, color perception, acuity, shape from shading, and shape from motion, showed that his primary vision was intact. The spatial information of the parietal lobes was necessary to properly bind features encoded in more ventral object pathways. These data also demonstrate one instance of dorsal–ventral interactions that contribute to the perception of space and objects. They show an instance in which a lesion in one system (the "where" pathway) can create an illusory perception by disconnections to another pathway (the "what" pathway) (16–18).

Intact parietal lobes appear necessary for normal spatial vision between objects, whether presented side by side or one within another. Data from patients with Balint syndrome suggest that the what and where distinction for the respective dorsal and ventral systems is a good heuristic that captures some of the functional differences between the two systems but not all. It is not space *per se* that is disrupted but rather some interaction between overall spatial processing and objects that seems affected. Yet determining which features predict when spatial relationships between items will be perceived as part of an object and when they will be perceived as separate objects is not a trivial matter. In fact, one of the more perplexing problems in the study of pattern perception concerns what constitutes an object. Abrupt changes in color and contrast can distinguish one object from another, but

not always. A pattern carved in a table is not perceived as a separate object from the table. A contour change in a swatch of fabric does not divide the cloth into two perceptual objects. The screen of a computer monitor is not perceived as a completely independent object from the casing that houses it. It is part of the object that we perceive as a computer monitor.

To understand the dorsal and ventral streams fully, we need to understand the cognitive mechanisms involved in how the visual system determines when a luminance flux in the visual field is perceived as part of an object and when it is perceived as space between or within objects. These questions have begun to be addressed in neuropsychology, partly because the questions have only recently been posed in this manner. Whatever the answers ultimately may be, they must explain both normal vision and the way in which visual perception breaks down.

Topographic Disorientation

Several case reports have appeared over the years describing patients who have lost their ability to self-navigate their environment (19,20). The mechanisms underlying these navigational deficits may be heterogeneous in nature, with selective spatial deficits producing different forms of spatial disorientation. Aguirre and D'Esposito (19) describe these patients as being topographically disoriented and present a descriptive review of a number of case studies. Unlike neglect patients who can have a selective disruption of egocentric personal space, the exocentric space in some of these patients appears to be selectively damaged. That is, they are unable to use information from familiar landmarks in their environment to guide their direction or spatial orientation. These patients are unable to describe routes or draw maps between familiar landmarks and are unable to describe the spatial relation between one well-known or familiar place and another. Three groups of patients were described in the review by Aguirre and D'Esposito: (a) those with egocentric disorientation, (b) those with heading disorientation, and (c) those with landmark agnosia. Those patients with egocentric disorientation had lesions to the superior parietal lobe and appeared to have profound impairments in the perception of egocentric space, while having preserved visual recognition. Those with heading disorientation had lesions to the retrosplenial cortex. Their navigational impairment was especially intriguing because they could recognize landmarks but could not extract directional or heading information from those landmarks even though they were familiar with them. For instance, a native New Yorker with such damage could be in the middle of bustling New York City, clearly recognize Times Square, but from Times Square fail to determine which direction he should proceed to travel home. The third group of patients described as having landmark agnosia had damage to the medial occipital lobe, involving fusiform and lingual gyri and sometimes parahippocampal gyri. These patients exhibited a visual recognition deficit that appeared to be restricted to environmental features. Patients with such landmark recognition deficits are often unable to recognize photographs of famous buildings and landmarks but exhibit intact performance on tests of spatial representation (21).

The heterogeneity of the lesion sites that produce deficits in navigating the environment suggests a network of dorsal and ventral regions that work together to produce a spatial map that is critical for environmental navigation. Although it has been proposed that the dorsal stream encodes spatial information egocentrically and the ventral stream encodes spatial information from an allocentric perspective, both streams may be involved in object identification (22). Milner and Goodale (22) suggest that it is a disruption of the allocentric ventral stream that produces topographic disorientation. More research is needed to understand how the two visual streams act in concert to perceive objects, plan and execute actions, and navigate the environment. It is clearly not as simple as a where and what stream of perceiving.

In concurrence with the patient cases reported above, recent evidence from neuroimaging (23) suggests that activation within the parahippocampal region is selectively enhanced when viewing images of landscapes and scenes (discussed later in the chapter). It is possible that multiple regions contribute to the process of self-navigation, with the parahippocampal region playing a critical role in the consolidation of incoming information in relation to landmark objects.

Deficits in Perceiving Objects

The perception of objects can break down in highly specific ways. The investigation of deficits in recognizing objects (vi-

sual agnosias) has had a great influence on theories of normal and abnormal object perception (24). Studies of visual agnosia have also focused attention on specific brain regions as more important in object perception than others. Generally, visual agnosia occurs with ventral damage along the occipital and temporal lobes. Pure agnosias are seldom as specific as one would like for scientific investigations, and pure cases are difficult to find. Nevertheless, the purest cases demonstrate some interesting problems that limit the way in which investigators should conceptualize normal object perception.

Associative Agnosia

Normal individuals cannot look at a picture of a dog without perceiving a dog. Perceiving objects seems to happen effortlessly, and the concept of an object seems impossible to inhibit. When looking at or attending to a dog, one recognizes a dog. It appears that part of seeing a dog is knowing that what one sees is a dog (i.e., having a form concept). Neuropsychological evidence, however, does not support this simplistic view. There are patients who can see patterns that normal subjects perceive as a dog, draw what they see, distinguish one dog from another, match pictures of dogs but may not be able to report that the object is a dog or even that is it generally kept as a pet. In other words, these patients lose both the name and the meaning of the objects on visual presentation. This problem has been historically called associative agnosia. One striking aspect of this syndrome is that the meaning and the name of an object are readily available to the patient through other modalities. They will easily identify a dog by touch or when it barks.

Shallice (25) reviewed this literature and found only two patients with pure associative agnosia. That is, there were no other visual deficits that could account for the phenomenon. These patients had posterior left-hemisphere lesions with lesion extension into white matter areas that disrupted posterior callosal transfer. Commissurotomized subjects, in whom the corpus callosum is separated, do not exhibit associative agnosia. It appears that a left occipitotemporal lesion combined with disrupted interhemispheric communication between posterior regions is necessary to produce dissociation between an intact visual representation of the object and its meaning. Note that these patients are able to perceptually isolate objects. They see objects and perceive the spatial relations between objects. They have no reported spatial deficits. This is a case in which damage to the ventral processing stream that presumably identifies objects does not disrupt the perception of the objects. Rather, it disrupts the connection between the object's function and its meaning.

Integrative Agnosia and Simultanagnosia

Some patients have trouble perceiving an object as a whole but little difficulty seeing its parts. This syndrome is sometimes called integrative agnosia. These patients may perceive the parts such as the four feet, the body, and the trunk of an elephant, but they either cannot integrate the parts to form the internal representation of an elephant or are very slow to do so. If these patients do perceive a part that uniquely defines the object, such as the trunk, they are likely to guess correctly that the object is an elephant, but objects with few distinctive features are very difficult for such patients to recognize. Unlike patients with associative agnosia, these patients do have form concept access but do not have the ability to perceive the objects in the same way that normal subjects do (sometimes called integrative agnosia). Riddoch and Humphreys (26) were the first to use the term integrative agnosia as opposed to simultanagnosia to refer to this type of deficit. Wolpert (27) introduced the term simultaneous agnosia and referred to it as integration apperception. His patient could not comprehend the meaning of a scene but probably had a different underlying deficit. Although he could identify objects in a scene, he could not report the meaning implied in that scene. Individual objects, however, were readily apprehended. For instance, a picture of a boy crouching behind a bush with an angry man on the porch of his house with a broken front window at his side would be described accurately, but the idea that the boy had just thrown a ball and broken the window would not be comprehended.

Humphreys and Riddoch (28) reported an in-depth study of their patient H.J.A., who was previously categorized as having associative agnosia on neuropsychological examination. They found that the patient's deficit was an extreme slowness in integrating the pieces of objects. H.J.A. would laboriously trace a picture of an object and construct objects piece by piece. Once the object was perceptually integrated, the identity and function could be reported without difficulty. A computed tomography scan revealed bilateral deep lesions in the inferior, posterior occipitotemporal region (28).

A different way to think about H.J.A.'s deficits is that they reflect a problem in perceptual organization (in H.J.A.'s case, a specific organizational deficit that required the strategy of piecemeal processing). Once organization was achieved, he could readily access the meaning of each object. This patient showed abnormally slow perceptual organization for visually presented patterns but could recognize simple patterns if given enough time. He could report only one piece of the whole at a time, despite the fact that in-depth testing showed that primary vision was intact.

Lesions located in several different areas produce a deficit in seeing two objects simultaneously or, more accurately, in seeing two objects in the same amount of time that it takes normal subjects to see two objects. For instance, unilateral extinction is a deficit in recognizing or detecting contralesional stimuli in the presence of ipsilesional stimuli. Recognition is normal when the contralesional stimulus is presented in isolation. Global/local deficits also qualify because these deficits are defined as an abnormal ability to see either the local or global form within normal time limits. However, deficits of this sort have been associated with asymmetries in temporoparietal junction in the right or left hemisphere (29,30). Difficulties in perceiving two objects in patients with Balint syndrome could also be and have been described as simultanagnosia. Yet, this deficit is found in patients with bilateral parietooccipital damage.

It is not clear what constitutes perceiving a whole, and this question poses one of the challenges for investigators of object perception. One thread that does appear throughout the history of psychology is the observation that wholes are often perceived before the parts that constitute them (31–33). Until there is better understanding of how objects are perceived as objects and how an object can be perceived before seeing its parts, the underlying cause of visual integrative agnosias will remain unresolved.

Functional Imaging of Object Perception

Advances in neuroimaging have allowed the further exploration of cortical regions involved in object processing. In particular, fMRI studies have attempted to localize discrete areas that respond to unique properties and features of objects. Building on the lesion literature and the knowledge that deficits in particular aspects of object recognition appear after damage to specific areas, a wide range of functional imaging studies have emerged.

It has long been observed that face recognition deficits (prosopagnosia) appear after lesions to the ventral temporal lobe region. However, prosopagnosics often have intact object recognition. They may fail to recognize a picture of the President of the United States but can clearly identify the objects that adorn the picture such as the American flag and the presidential desk. This dissociation in visual deficits after lesions to the temporal lobe suggests that neural regions may be specialized in the processing of objects versus faces. In addition, those areas that are activated to a greater extent during object processing may be further divided into categories such as living versus nonliving things, functional versus nonfunctional objects, and landscapes and scenes. The identification of these specialized object processing areas within the temporal lobe has motivated a large number of functional imaging studies. The next section provides a brief overview of imaging studies in this area and discusses the relationship to the lesion literature.

One of the most fundamental properties of an object is its shape. Although other features such as color, shading, and texture are integral components of an object, shape appears to be the most important feature in object recognition. Recent fMRI studies have shown that the lateral occipital cortex (LOC) reliably shows increased activation when subjects are shown images of intact objects compared with scrambled images of these same objects (34). This suggests that the LOC is not just responding to randomly presented features but instead to a constellation of features bound together to form an object. The activation of the LOC does not appear to be modulated by familiarity of the object but rather responds to familiar and unique objects with the same level of activation. This activation is also robust for both line drawings and gray-shaded images of objects (34). What this pattern of activation suggests is that the response of the LOC is driven by the stimulus properties of an object *per se* and not by a stored memory of a previously viewed object.

Reports in the lesion literature provide compelling evidence that object processing is highly specialized and to some extent "modularized" within the human brain. Lesions to different areas of the cortex produce category-specific deficits that cause some patients to lose the ability to name living things while retaining the ability to identify man-made objects. These remarkable dissociations have given rise to a number of elegant imaging studies on object perception.

Category-specific object processing has been investigated in a number of imaging studies. In a study using PET, naming animal pictures activated a region within the left medial occipital cortex, whereas naming pictures of tools activated areas within the left premotor area and the left middle temporal gyrus (35). Later work also identified more robust activation for animals compared with that for tools in the lateral region of the fusiform gyrus. In contrast, tools elicited greater activation compared with animals in the medial portion of the fusiform gyrus (36). This different pattern of activity for the naming of tools versus animals suggests that the identification of tools may be modulated by knowledge about their functional properties and use, whereas the activation pattern for animals may be generated by more perceptual attributes.

In addition to neural regions that have been linked to the processing of natural versus man-made objects, recent work has also identified a region within the parahippocampal gyrus that responds more strongly to images of landscapes and scenes than to other visual stimuli. This region has been named the parahippocampal place area (PPA) and appears to play a role in the perception of the local environment (37). Interestingly, the activation of the PPA is just as strong when subjects are shown images of an empty room as when the subjects view a room filled with furniture and ornamental detail. This finding suggests that the PPA is responding to the spatial context of a scene and not to the local elements contained within it. For instance, when a photograph of a room is cut into pieces (similar to a puzzle), the PPA still responds to the image as long as the spatial configuration remains the same. If, however, the pieces of the photograph are rearranged so that the overall spatial configuration no longer resembles a visual scene, the activation in the PPA diminishes and is no longer significant. An important exception to note is that activation within the PPA increases when subjects view images of landmarks presented in isolation, in particular photographs of buildings. This exception to landmarks such as buildings may indicate that some landmarks elicit familiarity associated with the spatial surrounding of the building.

In other words, the images of buildings provide cues or landmarks that are essential in the development of spatial maps. When access to these cues are weakened or damaged, neurologic conditions often emerge that impair a person's ability to form spatial maps and navigate their environment (see previous discussion of topographic disorientation).

Prosopagnosia

Prosopagnosia refers to a recognition deficit limited to faces, although testing has revealed that other categories can also be affected, such as automobiles, breeds of dogs, and species of birds. Many investigators have argued that the deficit is one of accessing the meaning of the face as with pure associative agnosia for objects. However, unlike patients with associative agnosia for objects, patients with prosopagnosia can identify a face as a face, but they simply cannot classify it as a particular person or accurately report whether they have seen the face before. Lesions that produce this deficit are most commonly linked to inferior right occipitotemporal damage. However, there is much debate about whether a right-hemisphere lesion alone is sufficient to cause prosopagnosia or whether bilateral lesions are required (for a discussion, see reference 24). Damasio (38) rather convincingly argues that the fusiform gyrus of the mesial occipitotemporal region bilaterally must be affected.

Prosopagnosia may be a deficit in accessing the memory trace or structural description of a particular face (38,39). However, Farah (40) notes that visual discrimination problems with unfamiliar faces have accompanied prosopagnosia in every case in which it has been tested, consistent with a primary deficit in facial discrimination. She argues that there are higher-order perceptual deficits that may disproportionately affect face recognition. Identification of any one individual requires expertise in making subtle visual distinctions along a number of dimensions. Farah's argument can also account for cases like the prosopagnosic farmer who lost the ability to discriminate his cows (41) or another prosopagnosic patient who lost the ability to discriminate racehorses (42). Although Farah's conclusion makes a great deal of sense, it cannot account for all the data. Other evidence collected in prosopagnosic patients demonstrates that there is intact visual analysis of faces at a preconscious level. Tranel and Damasio (43) demonstrated an increased galvanic skin response to familiar faces in prosopagnosics. Although these patients could not accurately report which faces were familiar, physical responses demonstrated below-threshold knowledge of which faces had been seen and which had not. The galvanic skin response was greater to familiar faces than to unfamiliar faces. Visual discrimination must be intact at some level for such responses to occur.

Imaging Studies of Face Perception

The interest in how the human brain perceives faces has fueled an explosion of neuroimaging studies attempting to find an area within the human cortex specialized for face perception. This is a strong example of how neuropsychology and lesion studies have influenced the direction of neuroscience. The neurologic syndrome of prosopagnosia described earlier has provided one of the strongest bits of behavioral evidence for cortical specialization. Building on the evidence from the lesion literature that damage to the occipitotemporal cortex can result in prosopagnosia, imaging studies have isolated discrete areas within the temporal cortex that respond more selectively to faces than to other stimuli.

Kanwisher (44) studied a highly localized region in the ventral lateral portion of the fusiform gyrus that she named the fusiform face area (FFA). In a series of fMRI studies, responses to face stimuli were found to strongly activate this area compared with nonface stimuli. This area within the ventral region of the temporal cortex is more active during face viewing than object viewing and responds more actively to faces than scrambled pictures of faces (45). The FFA has also been found to respond to a variety of face stimuli including faces of cats and other animals (46). The FFA's response to animal faces is not surprising given the case histories of prosopagnosics whose deficits in recognizing faces can, but not always, extend to their household pets or farm animals. Again, this convergence of results between the lesion literature and the imaging findings illustrates the value of combining approaches. If there were no reports of prosopagnosia from the patient case literature, how would scientists and psychologists interpret the findings of the activation in the

FFA? It is one thing to observe a hot spot on a computerized picture of the brain when a specific category of stimuli is presented to the subject, but it is quite another to observe neural activation in a healthy subject and then discover that it maps onto a brain region that is linked to a neurologic deficit.

The spatial selectivity of the FFA region appears to be to some extent position invariant in that some but not all inverted images of faces can activate the FFA (47). As long as the inverted face is still recognizable as a face, the FFA exhibits increased activation. Furthermore, the FFA responds whether the face is consciously perceived or not. Using fMRI, Vuilleumier et al. (48) showed strong FFA activation in a patient with unilateral neglect even when the face was neglected (Fig. 21.4).

Agnosia for Features

Some individuals with neurologic damage in posterior regions involved in extrastriate occipital, temporal, and parietal lobes have difficulty with more fundamental visual analysis, including the perception of color (achromatopsia), orientation, and motion. These types of agnosias are considered primary deficits or "bottom-up" deficits and are generally classified as apperceptive agnosias. These agnosias represent basic visual deficits produced by cortical lesions. Each has been observed in isolation in patients and can be evident even when visual sensory function is normal (49). These dissociations are consistent with PET studies showing that

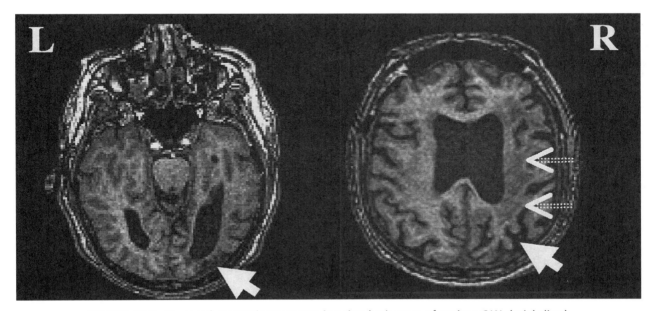

FIGURE 21.4. Structural magnetic resonance imaging brain scan of patient C.W. (axial slices) shows focal damage in the inferior posterior parietal cortex (angular gyrus, *thick arrows*), extending into the subcortical white matter of the right hemisphere (*dashed arrows*). (Reprinted from Vuilleumier P, Sagiv N, Hazeltine E, et al. Neural fate of seen and unseen faces in visuospatial neglect: a combined event-related functional MRI and event-related potential study. *Proc Natl Acad Sci U S A* 2001;98:3495–3500, with permission.)

identifying shape, color, and velocity has different profiles in the extrastriate cortex (for a discussion, see reference 50).

Color agnosia over the full visual field is associated with lesions similar to those that produce prosopagnosia. In fact, prosopagnosia typically occurs simultaneously with achromatopsia. However, achromatopsia can exist in isolation and can affect only one quadrant or one hemifield. Damasio (38) reports that patients with such deficits may describe what they see as "dirty" or "dulled," a washing out of colors. Some patients do not lose the ability to distinguish between monochromatic colors but cannot accurately report the color that is presented (51).

Achromatopsia has been associated with inferior occipitotemporal damage. Single-unit recordings from animal studies are consistent with the human literature. Although color *per se* (i.e., the basic colors from the light spectrum) is first extracted in the lateral geniculate nucleus of the thalamus (52), the color that is perceived is more complex and appears to be associated with higher visual analysis in occipitotemporal association areas. The absolute color (or more accurately the wavelength projected to the retina) may change as a function of lighting, but the relationship between wavelengths at contiguous locations remains perceptually constant as lighting changes. There is color constancy across various lighting conditions. Single-unit recordings in V4 in monkeys have revealed responses to relative color, which is consistent with color constancy (53). Thus, when damage occurs to areas that may include V4 in humans, discrimination of one monochromatic color from another need not suffer, but the experience of color may change in a way to disrupt color constancy. As noted before, identification of shape can also suffer with damage in similar regions. Again, the animal literature is consistent with these clinical observations. Single-unit responses in inferior occipitotemporal regions show sensitivity to color, brightness, orientation, and shape.

A rare case of motion agnosia has been reported by Zihl et al. (54) who studied a woman with difficulty detecting motion and reported that objects in motion looked as if they jumped from place to place. Continuous motion, as in running water, looked frozen. There is some controversy about whether motion perception deficits can occur in isolation (38). Animal work in areas MT and MTS in the posterior superior temporal sulcus of the monkey has shown convincingly that cells in this region are sensitive to the direction and velocity of movement (55). However, cells in parietal regions are also sensitive to movements, which may explain why movement deficits in isolation are extremely rare in humans.

In sum, the animal, human, and functional imaging literature demonstrates that inferior posterior association cortices are extensively involved in higher-order vision. Given the number of visual centers that lie in close proximity to one another and often abut one another, it is not surprising that only rare cases of isolated-feature agnosias occur. Clearly, patients with posterior temporoparietal and extrastriate occipital lesions should show some visual deficit, even if visual acuity is intact. The exact nature of the deficit will depend on which visual areas have been affected.

Deficits Associated with Visual Space

A great deal of evidence has shown that parietal lobes are extensively involved in spatial analysis. Animal data have demonstrated that cells in the parietal lobes are involved in the computation of abstract space. Parietal regions appear to be critical for maintaining a stable spatial structure to allow accurate movement throughout the environment (56). Cells in this area attend to locations in space independent of eye movements and are involved in the covert movement of attention across the visual field but only in particular ways, as described below. It is well established that parietal lobes are involved in location analysis and the analysis of spatial relationships. However, parietal lobes in humans are involved in covert attention processes that selectively attend to spatial locations but also to object features. The evidence for this claim is the subject of the next section.

Unilateral Visual Neglect

To neuropsychologists, unilateral neglect is considered the quintessential example of a spatial deficit. Patients with left neglect (owing to right-hemisphere damage) often act as if they are attempting to escape from their left side, sometimes to the point that they will deny left paresis or other problems on the left side of their body. On neuropsychological examination, patients with unilateral visual neglect will respond to the right side of drawings and bisect a horizontal line well to the right side, as if the length of the line toward the left side was misperceived (57). The magnitude of neglect varies tremendously between patients and more so in acute stages. Eye deviations to the right with left neglect are often observed but need not be present for left neglect to occur. Patients may interact normally with a person standing on their right side and be completely unaware of a person standing on their left side.

The bulk of the evidence from studies of unilateral visual neglect suggests that right inferior parietal lobe lesions produce more severe and long-lasting neglect than other areas of the brain (58). However, neglect has been found in patients with thalamic, basal ganglia, frontal, and cingulate damage (3). Left or right parietal lesions can also produce an extinctionlike deficit that is equal in frequency and severity, but the full-blown neglect syndrome is clearly more severe and more frequent with right-hemisphere damage (59). This asymmetry suggests that left neglect owing to right-hemisphere damage is a combination of an attention deficit (that produces an extinctionlike phenomenon) plus some other cognitive deficit that is more associated with right-hemisphere function (60).

Unilateral neglect typically resolves into extinction. Patients with extinction will see an object in their contralesional field if it is presented alone but will not see the same object if it is presented at the same time with another object in their ipsilesional field. Current neuropsychological theories of neglect and extinction disagree about whether these phenomena are a reflection of a direct deficit in spatial attention or a deficit in the visual representation of one side of space with secondary effects on attention (4,58,61–66). The resolution of this debate is important in understanding the neuropsychological syndrome itself, but it is also important in determining which cognitive functions involve the parietal lobe. For this reason, evidence related to this matter is discussed at length below.

Both human and animal research has shown that parietal lobes contribute to the covert movement of attention across the visual field (67,68). Early evidence by Morrow and Ratcliff (66) showed that right parietal damage produced more severe effects on covert attention (moving attention without accompanying eye movements) than left. In accordance with Posner et al. (61), they argue that the deficit was in one particular elementary operation that contributes to attention orienting, namely, the ability to disengage attention from an attended location to move to the contralesional side. In other words, the patient's attention gets stuck on the good side because they have difficulty disengaging attention from that side. Theoretically, the intact hemisphere (e.g., left) orients attention into the good field, but the signal to disengage to move attention in a leftward direction is disrupted by the damaged hemisphere, and attention remains fixed at a location on the right side.

Posner et al. (61) used a simple detection task in which subjects were asked to press a button when they detected a target. The targets could appear ipsilesionally or contralesionally and were preceded by a cue. The cue told the subjects whether the target would more likely to appear on the right or left of the screen. In 80% of the trials, the target appeared in the cued location (valid trials), and in 20%, it appeared in the opposite location (invalid trials). Groups of patients with right or left parietal damage responded to valid targets approximately equally well in the ipsilesional and contralesional fields. Both patient groups also had difficulty in responding to invalid targets when they occurred in the contralesional field. That is, for both groups, attention got stuck in the ipsilesional field. The right parietal group took longer to detect an invalidly cued target than the left parietal group in their contralesional field. Posner et al. (61) conclude that the parietal lobe is involved in the disengagement of attention from one location to move it to a new location. Covert attention orienting can be broken down into its components, and right parietal damage disrupts one component disproportionately. Posner et al. argue that successful covert orienting within the visual field requires (a) moving attention in the direction of an expected location, (b) engaging attention at that location, and (c) disengaging attention from that location to move it elsewhere. Patients with parietal lobe damage are especially deficient in the latter, or disengaging, process. Both hemi-

spheres appear to be involved in disengaging attention from its current location to move it in the contralateral direction (61). However, the strength of the disengagement process is asymmetrically represented (69). This conclusion is also supported using a similar paradigm in patients with clear signs of unilateral visual neglect (61,66). The cueing paradigm of Posner et al. (71) had been used in a variety of patient populations, including subjects with psychosis, to examine left/right differences in disengagement deficits (70,71). The evidence is consistent with a spatial attention deficit associated with left parietal function.

PERCEPTUAL DISTURBANCES IN PSYCHIATRIC PATIENTS

When people with normal brains perceive false information, we typically call it an illusion. When people with abnormal brains perceive false information, we often call it a hallucination. According to the *Diagnostic and Statistical Manual of Mental Disorders*, Fourth Edition, (*DSM-IV*) (72), hallucinations are sensory perceptions that have the compelling sense of reality of a true perception that occur without external stimulation of the relevant sensory organ. In contrast, an illusion is defined in *DSM-IV* as a misperception or misinterpretation of a real external stimulus. Abnormal perceptual experiences in psychiatric patients are most often called hallucinations. A large percentage of abnormal sensory experiences among psychiatric patients occurs in those individuals with a diagnosis of schizophrenia. Hallucinations in schizophrenia patients can occur in all modalities including auditory, visual, tactile, and olfactory. Auditory hallucinations, or hearing voices, are the most common perceptual disturbances reported in patients with schizophrenia, with hallucinations of vision, touch, and olfaction reported by only 20% of schizophrenia patients (73) (see further discussion of this issue later in this chapter). This next section focuses on the neural underpinnings associated with perceptual disturbances in schizophrenia. In the first two sections, we review and discuss some of the most recent neuroimaging findings on visual and auditory hallucinations and discuss how these findings may increase our understanding of the abnormal sensory experiences that occur in patients with schizophrenia. In the last section, we discuss the recent literature on global/local perceptual processing in schizophrenia and the implications of these studies for hemispheric contributions to visuospatial disturbances in schizophrenia.

Auditory Hallucinations

Hearing voices is one of the common features associated with schizophrenia, yet little is known about the underlying neuropathology. Although not all schizophrenia patients experience auditory hallucinations, a large percentage

(74%) of these patients does (74). Many have suggested that schizophrenia patients fail to monitor their own thoughts, which gives rise to a false perception that voices are talking to them (75). Although nonhallucinating individuals are able to monitor their own inner thoughts (i.e., thinking in words) and identify them as their own, schizophrenia patients who experience auditory hallucinations may not be able to identify when the source is internally generated or when it arises from an external source.

Frith (76) described three properties of these abnormal sensory experiences: (a) they occur in the absence of sensation, (b) self-generated activity is perceived to come from an external source, and (c) the voice (or auditory hallucination) is perceived to come from a source that is attempting to influence the patient. Patients with schizophrenia often report that the voice or voices comment on what they as individuals are doing, and, in some instances, the voices command the person to perform particular actions. An interesting example of an auditory hallucination was provided by a first-person narrative from the nineteenth century (77): "Whilst eating my breakfast, different spirits assailed me, trying me. One said, eat a piece of bread for my sake; another at the same time would say, refuse it for my sake or refuse that piece for my sake and take that; others, in like manner, would direct me to take or refuse my tea. I could seldom refuse one, without disobeying the other."

Although the example is quite benign in nature, auditory hallucinations can often be derogatory, frightening, and threatening in nature. Auditory hallucinations often contain very specific messages that are personally directed toward the person receiving them. These messages can at times direct or command the person to carry out specific behaviors. These actions can on occasion be dangerous to the person experiencing the hallucination or to others.

Auditory hallucinations have also been reported in congenitally deaf schizophrenia patients. Critchley et al. (78) reported that groups of prelingually deaf schizophrenia patients consistently reported seeing the arms and hands of persons signing to them. Several of these patients described the experiences as hearing, not seeing. Because prelingually deaf people have been noted to have some speech, these reports are not totally inconsistent with the idea that the hearing of signs could also be related to inner speech. Thus, the experience of a deaf person seeing a person signing may be similar to a hearing schizophrenia patient hearing voices.

Many attempts have been made to identify the underlying pathophysiology that contributes to auditory hallucinations. Early imaging research examining structural brain scans and volumetric measures found robust inverse correlations between regions in the superior temporal gyrus and the magnitude of auditory hallucinations (79,80). These studies reported that smaller brain volumes in the superior temporal gyrus correlated with increased frequency or magnitude of hallucinations. These were more lateralized to the left hemisphere. This lateralization or left-sided decrease sug-

gests that the brain areas associated with language (i.e., left temporal regions) may be abnormal in patients who experience a large number of hallucinations. This is not surprising given the finding that temporal lobe stimulation in neurosurgical patients produces a crude form of auditory hallucination (81). Although the content of auditory hallucinations in schizophrenia patients is usually more precise in content than those reported in patients during neurosurgery, there is nonetheless some overlap. This convergence of evidence suggests that disruption of processing within language-related areas in the temporal region may contribute to the occurrence of auditory hallucinations.

Recent advances in imaging techniques have allowed more precise localization of neural activity during hallucinatory behavior by allowing the measurement of brain activity while a person is actually experiencing an auditory hallucination or hearing voices. Early studies using single photon emission computed tomography reported increased activity in the Broca area during periods when schizophrenia subjects were actively hallucinating (82). The Broca area is located within the left inferior frontal cortex and is an area associated with speech. In the nonhallucinating state, the same subjects showed a significant decrease in activity within the Broca area, suggesting that the increased activity in the frontal lobe region correlated with auditory hallucinations.

Imaging studies using PET continued to explore the neural source of auditory hallucinations in schizophrenia patients. One PET study showed a relationship between activity in the Broca area and hallucinations, whereas another reported a positive relationship between metabolic activity in the anterior cingulate, left superior temporal lobe, and the presence of hallucinations (83,84).

More recent imaging studies using fMRI have pointed to the primary auditory cortex (PAC) as being a brain region involved in the production of voices in schizophrenia patients. One study reported increased activation of the PAC during periods that the schizophrenia patients were experiencing auditory hallucinations (85). It is important to note that the PAC normally shows a decrease in activation in normal subjects when perceiving inner speech (86) but an increase when perceiving sounds through earphones. This difference in activation of the PAC between healthy subjects and actively hallucinating schizophrenia patients is consistent with evidence that healthy subjects accurately perceive self-generated inner speech or subvocalizations as their own thoughts, whereas schizophrenia patients do not. Healthy individuals recognize that they are producing the subvocalizations themselves and are able to monitor their own thoughts. In contrast, the inner speech generated by schizophrenia patients is misperceived in such a way that the patients believe that the subvocalizations are produced by others, thus giving rise to hearing voices within their head, which sounds as real as if they were stimulated through sensory channels. Activity within the PAC of schizophrenia patients appears not to distinguish between externally and internally generated speech (87).

Are auditory hallucinations in schizophrenia patients a consequence of abnormal brain function in one particular region or are these perceptual disturbances a result of a disconnection between brain regions? Abnormal brain connectivity has been proposed to be a key problem in schizophrenia with reduced electrophysiologic activity having been recorded in the frontal and temporal lobe regions (88). If the frontal lobe is responsible for modulating and inhibiting activity in more posterior regions, could the abnormal perceptual experiences that plague schizophrenia patients be a result of a disconnection between the anterior and posterior cortices? One relatively rare neurologic condition associated with demyelination of frontal white matter (metachromatic leukodystrophy) has been reported to produce auditory hallucinations (89). Because demyelination of axons can disrupt neural connectivity, metachromatic leukodystrophy is a case in which abnormal connections between frontal and posterior cortices can produce perceptual disturbances in the auditory modality.

Is it possible, then, that the increased activity observed in the PAC (85) may actually be a result of the frontal cortex failing to modulate sensory input such that schizophrenia patients fail to recognize their own internally generated linguistic activity as their own and instead attribute them to outside voices? Because healthy, nonhallucinating individuals normally show a reduction in PAC activity when perceiving inner speech, this reduction may represent a modulation of activity that allows individuals to accurately monitor self-generated actions. The PAC and association auditory cortex have been repeatedly identified as neural regions involved in producing auditory hallucinations, but the findings from the different studies do not always agree on how these regions contribute to the production of abnormal auditory hallucinations. More imaging studies are needed to isolate the pathways that contribute to auditory hallucinations in patients with schizophrenia as well as in other disorders. The identification of these pathways will not only elucidate the brain mechanisms involved but may also lead to more targeted treatments that can alleviate these symptoms.

Visual Hallucinations

The pathology underlying visual hallucinations can be attributed to many causes. The presence of visual hallucinations spans the disciplines of neurology, psychiatry, ophthalmology, and addiction medicine. Neurologic conditions such as Alzheimer disease, epilepsy, brainstem lesions, and Parkinson disease can all produce visual hallucinations. Hallucinations in the visual modality have also been reported in various psychiatric conditions including mania, depression, and schizophrenia. In addition to these medical conditions, several drugs (e.g., LSD, mescaline, amphetamines) have also been strongly associated with visual perceptual hallucinations. The similarities in the visual disturbances produced by the conditions listed above are intriguing and might suggest that at some point in processing common mechanisms or pathways are affected in all disorders. The goal of the next section is to describe the phenomenology of visual hallucinations associated with one of the major psychiatric disorders, schizophrenia. Although numerous studies exist that examine the mechanisms underlying auditory hallucinations in schizophrenia, less is known about the frequency and mechanisms that produce hallucinations of the visual modality in schizophrenia.

Although reports of visual hallucinations in schizophrenia patients are far less common than auditory hallucinations, they are not rare. Estimates of visual hallucinations range from 15% (90) to as high as 62% (91). Recent research suggests that visual hallucinations may be an underreported phenomenon in schizophrenia (92,93). Patients who experience visual hallucinations almost always report having experienced auditory hallucinations at some point during their illness. This is in sharp contrast to visual hallucinations produced by illicit drug use and many neurologic illnesses. In those cases, visual hallucinations are usually reported to occur in isolation, and a co-occurrence of auditory hallucinations is not commonplace. Although it is difficult at times to clearly separate the qualities of visual hallucinations in schizophrenia from their qualities in other disorders, some differing patterns do emerge.

Visual hallucinations in schizophrenia patients have been reported to be both simple and complex, occur in black and white or color, and often converge with other modalities. Although hallucinations associated with Parkinson and other diseases often occur when the patient is drowsy, at the end of the day, visual hallucinations in schizophrenia patients frequently occur during the daytime hours when the patient is alert. Some schizophrenia patients report visual hallucinations of newly appearing objects, whereas others report distortions in shape or color of familiar objects. Figures 21.5 and 21.6 are drawings of such distortions done by a 23-year-old schizophrenia patient. This patient experienced vivid and uncontrollable visual hallucinations of faces, lips, limbs, and sexual organs that became distorted (94).

Visual hallucinations in schizophrenia patients often contain religious images, faces, demons, or lilliputian figures. A visual hallucination was vividly described by a 61-year-old French-Canadian laborer; he described the hallucination as a persistent vision of a little man who indulges in humorous and tormenting pranks (95). "The little man had been with him for seven years and varied in size from six inches to two feet. He could pass through windows, walls and closed doors. He was bald, gray-haired, seventy years of age and owns a brown poorly defined coat or gray vestments" (95).

Bracha et al. (91) examined a large group of severely ill schizophrenia patients both retrospectively and prospectively and found that more than half of the patients (56%) had experienced visual hallucinations at some point during their lifetime. The sample examined in this recent study was a group of severely ill, treatment-refractory inpatients. The

FIGURE 21.5. A drawing of a visual hallucination experienced by a 23-year-old schizophrenia patient. This drawing conveys the facial distortion (bulging eyes) that she experienced in her hallucinations. (Modified and redrawn from Guttman E, Maclay WS. Clinical observations on schizophrenic drawings. *Br J Med Psychol* 1937;16:184–205, with permission.)

high incidence of reported visual hallucinations in this group is in contrast to other studies that reported much lower rate of occurrence (18%) in less severely ill patients (96). Recent research suggests that visual hallucinations may be correlated with severity of illness because other studies have reported a higher rate of visual hallucinations among more severely ill schizophrenia patients (92).

This link between severity of illness and hallucinations in the visual modality brings into question the role of cultural influences in producing visual disturbances. Less industrialized societies are thought to have a higher incidence of visual hallucinations than industrialized societies owing to cultural practices and tolerance of deviant behavior. In Saudi Arabia and Africa, the reported rates of visual hallucinations in schizophrenia patients have always exceeded those from more industrial countries (more than 50%) (97–99). One description of a Saudi schizophrenia patient describes a woman who kept her face and mouth continuously covered with a handkerchief to prevent a visualized army of ants from entering her mouth (98). Another Saudi patient describes a vision of a woman who sat continuously by his side beating him with the palm of her hands (98).

FIGURE 21.6. A drawing of a visual hallucination by a schizophrenia patient depicting a visual distortion of the mouth and lips. (Modified and redrawn from Guttman E, Maclay WS. Clinical observations on schizophrenic drawings. *Br J Med Psychol* 1937;16:184–205, with permission.)

Because deviant behavior may be tolerated to a greater extent in nonindustrialized societies, it may be that only the most severely ill schizophrenia patients are seen in treatment settings where their visual hallucinations are then observed. If this is the case, then an underreporting of visual hallucinations in industrialized societies in contrast to the severely ill sample in nonindustrialized countries may contribute to the different prevalence of visual hallucinations reported in the literature. It is quite possible that if patient samples were matched for severity of illness, the rate of visual hallucinations among schizophrenia patients would not differ across cultures.

Neural Underpinnings of Visual Hallucinations

To the best of our knowledge, no imaging studies have examined the neurologic substrates of visual hallucinations in schizophrenia. This next section reviews a set of fMRI studies that examined visual hallucinations produced by damage to the cortex. These studies may generate hypotheses about the brain regions involved in visual hallucinations across disorders, including schizophrenia.

A vast body of research has shown that the visual cortex in humans is highly specialized (100), with discrete regions processing distinct features and properties of visual stimuli (101). If the cortex is specialized for normal visual processing, how might abnormalities in visual perception be localized? Will hallucinations that produce images of faces be linked to abnormal metabolism in brain regions that process faces? Will visual hallucinations of animals pathologically activate areas in the visual cortex that respond to images of animals? This is an intriguing question and one that has not yet been explored in schizophrenia patients.

A recent study, however, probed this question in persons with Charles Bonnet syndrome, a condition characterized by complex visual hallucinations in scotomas that appear in the visual field as a result of visual deterioration. This condition can arise with macular degeneration as well as bilateral lesions of the occipital lobes. Using fMRI, Santhouse et al. (102) measured cortical activity that occurred during active visual hallucinations. Their goal was to map out the distribution of these hallucinations and to determine whether the functional specificity associated with normal visual processing also extended to visual perceptual disturbances (i.e., visual hallucinations). Their findings revealed that the presence of category-specific visual hallucinations did map onto abnormally increased activity in specific regions. Patients who reported visual hallucinations of landscapes and objects showed increased activity in the cortex specialized for processing scenery (103), and visual hallucinations associated with faces corresponded with increased activity in fusiform gyrus in areas thought to be responsible for face processing (104). These findings suggest that cortical areas that show increased activation during normal object perception are also activated during abnormal perceptual experiences. It is the mechanism of this abnormal brain activity that is less understood, and future imaging studies may be able to reveal the source of this activity.

MISSING PARTS: GLOBAL/LOCAL PROCESSING IN SCHIZOPHRENIA

Although visual hallucinations may represent a vivid and sometimes bizarre form of perceptual disturbance in pa-

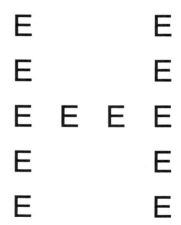

FIGURE 21.7. Example of a global H created from local E's.

tients with schizophrenia, more subtle visual spatial processing deficits have also been reported. Tasks that measure the ability to perceive the whole global configuration versus the local parts have been successfully employed to examine more fully the processing of complex shapes in patients with schizophrenia. In a part/whole or global/local task, a hierarchically organized figure is presented to the subject (Fig. 21.7) who is instructed to attend either to a specific level or to divide attention between the two levels and identify a predesignated target. These patterns represent a different type of spatial relationship between objects, with one object being a part of the other. For instance, the pattern in Figure 21.7 contains a global H created from a repetition of local E's. The local E's are parts of the global H.

Processing the global shape has been consistently associated with the right occipitotemporal lobes, whereas processing the local shape had been consistently associated with the left occipitotemporal lobes (105–109). Hemispheric abnormalities have long been reported in schizophrenia patients, with both hemispheres implicated in processing deficits. Although patients with schizophrenia have been reported to exhibit deficits in processing the gestalt or global properties of stimuli generally associated with right-hemisphere processing (110), the majority of studies provides evidence that schizophrenia affects the left hemisphere to a greater degree than the right hemisphere (71,111). Because local processing advantages have been linked to left-hemisphere processing and global processing advantages to right-hemisphere processing, one hypothesis that has been put forth is that schizophrenia patients exhibit deficits in local processing owing to greater left-hemisphere dysfunction.

This hypothesis of a local processing deficit has been tested using a divided attention task with hierarchical stimuli. Schizophrenia patients were asked to identify a predesignated target letter (H or S) that appeared as part of a classic Navon global/local figure (Fig. 21.7). The target letter could appear at either the local level (small H's in a big E) or at a global level (big H comprised of small E's). Schizophrenia patients showed an exaggerated global advantage in reaction times compared with controls (111–113). That is, the schizophrenia patients were faster to identify targets at the global level compared with the local level. Another way to think about the results is that the schizophrenia patients were slower to identify the target letters when they appeared as part of the local configuration of the stimuli compared with when they comprised the global form of the hierarchical stimuli. Local processing deficits in schizophrenia patients have also been correlated with both auditory hallucinations and abnormal left superior temporal gyral metabolism (111).

The schizophrenia studies in which attention was directed to either the global or local level in a blocked design reveal a different pattern of results. When attention was modulated by task directions and the schizophrenia patients were instructed to attend to one level over another, performance deficits either disappeared or shifted to a local advantage (112). These findings suggest that both attention resource demands and attention modulation play an important role in the perceptual processing abilities of patients with schizophrenia. Although global processing deficits reported in other perceptual tasks may represent a breakdown of "top-down" contextual operations in schizophrenia, the results from the global/local studies may represent more "bottom-up" or feature-processing operations. Deficits in local processing when attention resources are divided may be linked to left-hemisphere abnormalities. The disappearance of these local disadvantages through attention modulation suggests that perceptual encoding deficits in schizophrenia may be improved through attention control (112).

GENERAL CONCLUSION

The emergence of the field of cognitive neuroscience has brought together investigators from a variety of disciplines. This bridging across different levels of inquiry has proven essential for the growth of the field. This bridging of disciplines has inspired investigators to accept the concept of interdisciplinary thinking and collaboration. It has brought together scientists who test hypotheses from the cellular level all the way to those who test human behavior both at the hospital bedside and in the psychology laboratory. This cooperation between levels of analysis is essential to fully understand brain–behavior relationships.

Advances in neuroimaging techniques have added to our knowledge of brain regions involved in cognitive processing. These advances have allowed the further demarcation of regions involved in perception that is not always possible after brain injury. Lesions often encompass large areas of the brain, and regional analysis is often confounded by edema and rerouting of connections after traumatic brain injury. However, lesion studies are still essential for assessing whether a brain region is necessary to support a specific process. Lesion studies provide an important view of how

behavior is altered by damage to different areas. It is the continued union of neuroimaging and lesion studies that may produce the greatest advances in cognitive neuroscience. As imaging techniques become more refined and sensitive, imaging studies of both normal and patient populations can be used together to further our understanding of how the human brain processes the world around us.

The knowledge gained from this approach has proven useful in evaluating psychiatric problems and will continue to do so. Furthermore, knowledge gained from neurocognitive studies in psychiatric populations can advance the field of cognitive neuroscience in the same way that studies of neurologic patients have. We have presented only a small segment of how neuropsychiatry has benefited from studies of perceptual deficits in psychiatric and neurologic patients, augmented by behavioral and neuroimaging studies in normal healthy populations.

ACKNOWLEDGMENT

The preparation of this chapter was supported by a Senior Research Medical Scientist and Merit Award from the Office of Veterans Affairs to L.C.R. and University of California at Davis Department of Psychiatry support to R.S. We thank Ting Wong and Anne M. Salo for their help with the figures in this chapter and Thomas Nordahl and Martin Leamon for their helpful suggestions on an earlier draft.

REFERENCES

1. Efron R. *The decline and fall of hemispheric specialization.* Hillsdale NJ: Lawrence Erlbaum, 1990.
2. Goldman-Rakic PS. Circuitry of primate prefrontal cortex and regulation of behavior by representational memory. In: Mountcastle VB, Plum G, Geiger SR, eds. *Handbook of physiology I.* Bethesda, MD: American Physiological Society, 1987.
3. Mesulam M-M. Attention, confusional states and neglect. In: Mesulam M-M, ed. *Principles of behavioral neurology.* Philadelphia: FA Davis Co, 1985.
4. Rizzolatti G, Camarda R. Neural circuits for spatial attention and unilateral neglect. In: Jeannerod M, ed. *Neurophysiological and neuropsychological aspects of spatial neglect.* Amsterdam: Elsevier Science, 1987.
5. Merzenich MM, Kaas JH. Reorganization of mammalian somatosensory cortex following peripheral nerve injury. *Trends Neurosci* 1982;5:434–436.
6. Kaas JH, Krubitzer LA, Chino YM, et al. Reorganization of retinotopic cortical maps in adult mammal after lesions of the retina. *Science* 1990;248:229–231.
7. Ungerleider LG, Mishkin M. Two cortical visual streams. In: Ingle DJ, Goodale MA, Mansfield RJW, eds. *Analysis of visual behavior.* Cambridge, MA: MIT Press, 1982.
8. Livingstone M, Hubel D. Segregation of form, color, movement, and depth: anatomy, physiology, and perception. *Science* 1988;240:740–749.
9. Gattass R, Sousa APB, Covery E. Cortical visual areas of the macaque: possible substrates for pattern recognition

mechanisms. In: Chagas C, Gattass R, Gross C, eds. *Pattern recognition mechanisms.* Berlin: Springer-Verlag, 1985.

10. Kaas JH. Why does the brain have so many visual areas? *J Cogn Neurosci* 1989;1:121–135.

11. Newcombe F, Russell WR. Dissociated visual perceptual and spatial deficits in focal lesions of the right hemisphere. *J Neurol Neurosurg Psychiatry* 1969;32:73–81.

12. Humphreys GW, Riddoch MJ. Interactions between object and space systems revealed through neuropsychology. In: Meyer DE, Kornblum S, eds. *Attention and performance, XIV.* Hillsdale, NJ: Lawrence Erlbaum, 1992.

13. Grabowecky M, Egly R, Robertson LC, et al. Attentional control in a patient with bilateral parietal lesions. Paper presented at the annual meeting of the Society of Neuroscience, Washington, DC, November 1993.

14. Egly R, Robertson LC, Rafal R, Grabowecky M. Implicit processing of unreportable objects in Balint's syndrome. Poster presented at the 36th annual meeting of the Psychonomic Society, November, 1995, Los Angeles, California.

15. Treisman A, Schmidt H. Illusory conjunctions in the perception of objects. *Cognit Psychol* 1982;14:107–141.

16. Bernstein LJ, Robertson LC. Illusory conjunctions of color and motion with shape following bilateral parietal lesions. *Psychol Sci* 1998;9:167–175.

17. Friedman-Hill SR, Robertson LC, Treisman A. Parietal contributions to visual feature binding: Evidence from a patient with bilateral lesions. *Science* 1995;269:853–855.

18. Robertson LC, Treisman A, Friedman-Hill S, et al. The interaction of spatial and object pathways: evidence from Balint's syndrome. *J Cogn Neurosci* 1997;9:295–317.

19. Aguirre GK, D'Esposito M. Topographical disorientation: a synthesis and taxonomy. *Brain* 1999;122:1613–1628.

20. Farrell MJ. Topographical disorientation. *Neurocase* 1996;2:509–520.

21. McCarthy RA, Evans JJ, Hodges JR. Topographic amnesia: spatial memory disorder, perceptual dysfunction, or category specific semantic memory impairment? *J Neurol Neurosurg Psychiatry* 1996;60:318–325.

22. Milner AD, Goodale MA. *The visual brain in action.* Oxford: Oxford University Press, 1995.

23. Epstein R, Kanwisher N. A cortical representation of the local visual environment. *Nature* 1998;392:598–601.

24. Farah MJ, Brunn JL, Wong AB, et al. Frames of reference for allocation attention in space. Evidence from the neglect syndrome. *Neuropsychologia* 1990;28:335–347.

25. Shallice T. *From neuropsychology to mental structure.* Melbourne: Cambridge University Press, 1988.

26. Riddoch MJ, Humphreys GW. A case of integrative agnosia. *Brain* 1987;110:1431–1462.

27. Wolpert I. Die Simultanagnosie: Storung der Gesamtauffassung. Z.f.d. gesamte Neurol. U. Psychiatr., 93, 397–425 (as referenced by A. Benton [1985]). In: Heilman KM, Valenstein E, eds. *Clinical neuropsychology.* New York: Oxford University Press, 1924.

28. Humphreys GW, Riddoch MJ. *To see but not to see: a case study of visual agnosia.* Hillsdale, NJ: Lawrence Erlbaum, 1987.

29. Lamb MR, Robertson LC, Knight RT. Component mechanisms underlying the processing of hierarchically organized patterns: inferences from patients with unilateral cortical lesions. *J Exp Psychol Learn Mem Cogn* 1990;16:471–483.

30. Robertson LC, Lamb MR, Knight RT. Effects of lesions of temporal-parietal junction on perceptual and attentional processing in humans. *J Neurosci* 1988;8:3757–3769.

31. Krech D, Calvin A. Levels of perceptual organization and cognition. *J Abnorm Soc Psychol* 1953;48:394–400.

32. Krecehvsky I. An experimental investigation of the principle of proximity in the visual perception of the rat. *J Exp Psychol* 1938;22:497–523.

33. Navon D. Forest before trees: the precedence of global features in visual perception. *Cognit Psychol* 1977;9:353–383.

34. O'Craven KM, Downing PE, Kanwisher N. fMRI evidence for objects as the units of attentional selection. *Nature* 1999;401:584–587.

35. Martin A, Wiggs CL, Ungerleider LG, et al. Neural correlates of category-specific knowledge. *Nature* 1996;379:649–652.

36. Chao LL, Haxby JV, Martin A. Attribute-based neural substrates in temporal cortex for perceiving and knowing about objects. *Nat Neurosci* 1999;2:913–919.

37. Epstein R, Kanwisher N. A cortical representation of the local visual environment. *Nature* 1998;392:598–601.

38. Damasio AR. Disorders of complex visual processing: agnosias, achromatopsia, Balint's syndrome, and related difficulties of orientation and construction. In: Mesulam M-M, ed. *Principles of behavioral neurology.* Philadelphia: FA Davis Co, 1985:269–288.

39. Bruce V, Young A. Understanding face recognition. *Br J Psychol* 1986;77:305–327.

40. Farah MJ. *Visual agnosia: disorders of object recognition and what they tell us about normal vision.* New York: Academic Press, 1990.

41. Borenstein B, Stroka H, Munitz H. Prosopagnosia with animal face agnosia. *Cortex* 1969;5:164–169.

42. Newcombe F. The processing of visual information in prosopagnosia and acquired dyslexia: functional versus physiological interpretation. In: Osborne DJ, Gruneberg MM, Eiser JR, eds. *Research in psychology and medicine.* London: Academic Press, 1979.

43. Tranel D, Damasio AR. Knowledge without awareness: an autonomic index of facial recognition by prosopagnosics. *Science* 1985;228:1453–1454.

44. Kanwisher N. Domain specificity in face perception. *Nat Neurosci* 2000;3:759–763.

45. Puce A, Asgari M, Gore JC, et al. Differential sensitivity of human visual cortex to faces, letterstrings, and textures: a functional magnetic resonance imaging study. *J Neurosci* 1996;16:5205–5215.

46. Chao LL, Martin A, Haxby JV. Are face-responsive regions selective only for faces? *Neuroreport* 1999;10:2945–2950.

47. Kanwisher N, Tong F, Nakayama K. The effect of face inversion on the human fusiform face area. *Cognition* 1998;68:B1–B11.

48. Vuilleumier P, Sagiv N, Hazeltine E, et al. Neural fate of seen and unseen faces in visuospatial neglect: a combined event-related functional MRI and event-related potential study. *Proc Natl Acad Sci U S A* 2000;98:3495–3500.

49. Warrington EK. Visual deficits associated with occipital lobe lesions in man. *Exp Brain Res* 1986;11[Suppl]:247–261.

50. La Berge D. Thalamic and cortical mechanisms of attention suggested by recent positron emission tomographic experiments. *J Cogn Neurosci* 1990;4:358–372.

51. Davidoff J. *Cognition through color.* Cambridge, MA: MIT Press, 1991.

52. De Valois RL. Analysis and coding of color vision in the primate visual system. *Cold Spring Harbor Symp Quant Biol* 1965;30:567–579.

53. Desimone R, Schein SJ, Albright TD. Form, color and motion analysis in prestriate cortex of the macaque. In: Chagas C, Gattass R, Gross C, eds. *Pattern recognition mechanisms.* New York: Springer-Verlag, 1985.

54. Zihl J, Von Cramon D, Mai N. Selective disturbance of movement vision after bilateral brain damage. *Brain* 1983;106:313–340.

55. Maunsell JHR, Newsome WT. Visual processing in monkey extrastriate cortex. *Annu Rev Neurosci* 1987;10:363–401.

56. Anderson RA. Inferior parietal lobule function in spatial perception and visuomotor integration. In: Mountcastle VB, Plum F, Geiger SR, eds. *Handbook of physiology I.* Bethesda, MD: American Physiological Society, 1987:483–518.

57. Marshall JC, Halligan PW. When right goes left: an investigation of line bisection in a case of visual neglect. *Cortex* 1989;25:503–515.

58. Heilman KM, Watson RT, Valenstein E. Neglect and related disorders. In: Heilman KM, Valenstein E, eds. *Clinical neuropsychology,* 2nd ed. New York: Oxford University Press, 1985.

59. Ogden JA. The "neglected" left hemisphere and its contribution to visuospatial neglect. In: Jeannerod M, ed. *Neurophysiological and neuropsychological aspects of spatial neglect.* Amsterdam: Elsevier Science, 1987.

60. Robertson LC, Eglin M. Attentional search in unilateral neglect. In: Robertson I, Marshall J, eds. *Unilateral neglect: clinical and experimental studies.* London: Taylor & Francis, 1993:169–191.

61. Posner MI, Walker JA, Freidrich FJ, et al. Effects of parietal injury on covert orienting of attention. *J Neurosci* 1984;4:1863–1874.

62. Bisiach E, Luzzatti C. Unilateral neglect of representational space. *Cortex* 1978;14:129–133.

63. Bisiach E, Luzzatti C, Perani D. Unilateral neglect, representational schema and consciousness. *Brain* 1979;102:609–618.

64. Kinsbourne M. Mechanisms of unilateral neglect. In: Jeannerod M, ed. *Neurophysiological and neuropsychological aspects of spatial neglect.* Amsterdam: Elsevier Science, 1987.

65. Kinsbourne M. Orientational bias model of unilateral neglect: evidence from attentional gradients within hemispace. In: Robertson I, Marshall J, eds. *Unilateral neglect: clinical and experimental studies.* Hillsdale, NJ: Lawrence Erlbaum, 1993.

66. Morrow LA, Ratcliff G. Attentional mechanisms in clinical neglect. *J Clin Exp Neuropsychol* 1987;9:74–75.

67. Lynch JC. The functional organization of posterior parietal association cortex. *Behav Brain Sci* 1980;3:485–534.

68. Posner MI, Peterson SE. The attention system of the human brain. *Annu Rev Neurosci* 1990;13:25–42.

69. Corbetta M, Miezen FM, Shulman GL, et al. Shifts of visuospatial attention: a PET study. *J Neurosci* 1993;13:1202–1226.

70. Carter CS, Robertson LC, Chaderjian MR, et al. Attentional asymmetries in schizophrenia: the role of illness subtype and symptomatology. *Prog Neuropsychopharmacol Biol Psychiatry* 1994;18:661–683.

71. Posner MI, Early RS, Reiman E, et al. Asymmetries of attentional control in schizophrenia. *Arch Gen Psychiatry* 1988;45:814–821.

72. Spitzer RL, Williams JBW. *Structured clinical interview for DSM-IV—patient version.* New York: New York State Psychiatric Institute.

73. Frith CD. Positive symptoms, abnormal experiences. In: Frith CD, ed. *The cognitive neuropsychology of schizophrenia.* Hillsdale, NJ: Lawrence Erlbaum, 1992.

74. David AS. The neuropsychology of schizophrenia. In: David AS, Cutting JC, eds. *Neuropsychology of schizophrenia.* Hillsdale, NJ: Lawrence Erlbaum, 1994;269–313.

75. Frith CD. The positive and negative symptoms of schizophrenia reflect impairments in the perception and initiation of action. *Psychol Med* 1987;17:631–648.

76. Frith CD. The role of the prefrontal cortex in self-consciousness: the case of auditory hallucinations. In: Roberts AC, Robbins TW, Weiskrantz L, eds. *The prefrontal cortex: executive and cognitive functions.* Oxford, UK: Oxford University Press, 1998:181–194.

77. Peterson D, ed. *A mad people's history of madness.* Pittsburgh: University of Pittsburgh Press, 1984.

78. Critchley EM, Denmark JC, Warren F, et al. Hallucination experiences of prelingually profoundly deaf schizophrenics. *Br J Psychiatry* 1981;138:30–32.

79. McCarley RW, Shenton ME, O'Donnell BF, et al. Uniting Kraepelin and Bleuler: the psychology of schizophrenia and the biology of temporal lobe abnormalities. *Harv Rev Psychiatry* 1993;1:36–56.

80. Barta PE, Pearlson GD, Powers RE, et al. Auditory hallucinations and smaller superior temporal gyral volume in schizophrenia. *Am J Psychiatry* 1990;147:1457–1462.

81. Penfield W, Perot P. The brain's recording of auditory and visual experience. *Brain* 1963;86:655–663.

82. McGuire PK, Shah GMS, Murray RM. Increased blood flow in Broca's area during auditory hallucinations in schizophrenia. *Lancet* 1995;342:703–706.

83. McGuire PK, Silbersweig I, Murray RM, et al. The neural correlates of inner speech and auditory verbal imagery in schizophrenia. Relationship to auditory hallucinations. *Br J Psychiatry* 1996;169:148–159.

84. Silbersweig DA, Stern E, Frith CD, et al. A functional neuroanatomy of hallucinations in schizophrenia. *Nature* 1995;378:176–179.

85. Dierks T, Linden DE, Jangle M, et al. Activation of Heschl's gyrus during auditory hallucinations. *Neuron* 1999;22:615–621.

86. Frith CD Dolan RJ. Higher cognitive processes. In: Frackowiack RSJ, Friston K, Frith CD, et al., eds. *Human brain function.* Boston: Academic Press, 329–365.

87. Lennox BR, Park SB, Jones PB, et al. Spatial and temporal mapping of neural activity associated with auditory hallucinations. *Lancet* 1999;353:644–644.

88. Friston K. Schizophrenia and the disconnection hypothesis. *Acta Psychiatr Scand Suppl* 1999;395:68–79.

89. Hyde TM, Ziegler JC, Weinberger DR. Psychiatric disturbances in metachromatic leukodystrophy: insights into the neurobiology of psychosis. *Arch Neurol* 1992;49:401–406.

90. Goodwin DW, Alderson P, Rosenthal R. Clinical significance of hallucinations in psychiatric disorders. *Arch Gen Psychiatry* 1971;24:76–80.

91. Bracha HS, Wolkowitz OM, Lohr JB, et al. High prevalence of visual hallucinations in research subjects with chronic schizophrenia. *Am J Psychiatry* 1989;146:526–528.

92. Hendrickson J. Lilliputian hallucinations in schizophrenia: case report and review of literature. *Psychopathology* 1996;29:35–38.

93. Mueser KT, Bellack AS, Brady EU. Hallucinations in schizophrenia. *Acta Psychiatr Scand* 1990;82:26–29.

94. Guttman E, Maclay WS. Clinical observations on schizophrenic drawings. *Br J Med Psychol Soc* 1937;16:184–204.

95. Lewis DJ. Lilliputian hallucinations in the functional psychoses. *Can Psychiatr Assoc J* 1961;6:177–200.

96. Goldberg SC, Klerman GL, Cole JO. Changes in schizophrenic pathology and ward behavior as a function of phenothiazine treatment. *Br J Psychiatry* 1965;111:120–133.

97. Ndetei DM, Singh A. Hallucinations in Kenyan schizophrenic patients. *Acta Psychiatr Scand* 1983;67:144–147.

98. Zarroug EA. The frequency of visual hallucinations in schizophrenic patients in Saudi Arabia. *Br J Psychiatry* 1975;127:553–555.

99. Al-Issa I. Sociocultural factors in hallucinations. *Int J Soc Psychiatry* 1978;24:167–176.

100. Van Essen DC, Felleman DJ, Deyoe EA, et al. Probing the primate visual cortex: pathways and perspectives. In: Gulyas B,

Ottoson D, Roland PE, eds. *Functional organization of the human visual cortex.* Oxford: Pergamon Press, 1993:2941.

101. Ungerleider LG, Haxby JV. 'What' and 'where' in the human brain. *Curr Opin Neurobiol* 1994;4:157–165.

102. Santhouse AM, Howard RJ, Ffytche DH. Visual hallucinatory syndromes and the anatomy of the visual brain. *Brain* 2000;123:2055–2064.

103. Ffytche DH, Howard RJ. The perceptual consequences of visual loss: 'positive' pathologies of vision. *Brain* 1999;122:1247–1260.

104. Kanwisher N, McDermott J, Chun MM. The fusiform face area: a module in human extrastriate cortex specialized for face perception. *J Neurosci* 1997;17:4302–4311.

105. Martinez A, Moses P, Frank L, et al. Lateralized differences in spatial processing. Evidence from RT and fMRI. Paper presented at the Second International Conference in Functional Mapping of the Human Brain, Boston, 1996.

106. Heinze HJ, Johannes S, Munte TF, et al. The order of global and local-level information processing: electrophysiological evidence for parallel perceptual processes. In: Heinze HJ, Munte TF, Mangun GR, eds. *Cognitive electrophysiology.* Boston: Birkhauser, 1994:102–123.

107. Fink G, Halligan P, Marshall J, et al. Where in the brain does visual attention select the forest and the trees? *Nature* 1996;15:626–628.

108. Robertson LC, Lamb MR, Knight RT. Effects of lesions to temporo-parietal junction on perceptual and attentional processing in humans. *J Neurosci* 1988;10:3757–3769.

109. Robertson LC, Rafal RD. Disorders of visual attention. In: Gazzaniga MS, ed. *The new cognitive neurosciences.* Cambridge, MA: MIT Press, 2000.

110. Place EJS, Gilmore GC. Perceptual organization in schizophrenia. *J Abnorm Psychol* 1980;89:409–418.

111. Carter CS, Robertson LC, Nordahl TE, et al. Perceptual and attentional asymmetries in schizophrenia: further evidence for a left hemisphere deficit. *Psychiatry Res* 1996;62:111–119.

112. Granholm E, Perry W, Filoteo JV, et al. Hemispheric and attentional contributions to perceptual organization deficits on the global-local task in schizophrenia. *Neuropsychology* 1999;13:271–281.

113. Ferman TJ, Primeau M, Delis D, et al. Global-local processing in schizophrenia: hemispheric asymmetry and symptom-specific interference. *J Int Neuropsychol Soc* 1999;5:442–451.

SECTION III

SYNDROMES AND DISORDERS

22

AUTISM AND PERVASIVE DEVELOPMENTAL DISORDERS

LEEANNE GREEN
STEPHEN P. JOY
DIANA L. ROBINS
KARA M. BROOKLIER
LYNN H. WATERHOUSE
DEBORAH FEIN

Autism has drawn unprecedented attention in public and research arenas in recent years. Interest in the search for causes and treatments of its devastating impairments has gained momentum, in association with widespread reports of dramatic increases in its incidence. This review examines (a) the historical development of the concept of autism and pervasive developmental disorders (PDD); (b) diagnostic criteria; (c) findings and theories relative to social and cognitive characteristics; (d) the developmental course and epidemiology, including prevalence; (e) the boundaries of the autism/PDD concept and its comorbidity with other neuropsychiatric conditions; (f) genetics, associated medical conditions, and teratogens; (g) neuroanatomic findings; (h) neurochemical findings; (i) neurophysiologic findings; (j) clinical assessment; and (k) pharmacologic and psychological/educational treatments.

HISTORICAL DEVELOPMENT OF THE CONCEPT OF AUTISM

PDD is the term for a group of neurodevelopmental disorders characterized by a similar behavioral profile, more currently referred to as autistic spectrum disorders to reflect the heterogeneity known to this group (1). In clinical practice, PDD is often used to describe children who have mild autism, high-functioning autism (HFA), or an incomplete set of autistic features. DiLalla and Rogers (2) found that autistic disorder can best be differentiated from PDD by the degree of social impairment, although the spectrum concept reduces the need for categorization. The validity of nonautistic PDD syndromes was explored by Szatmari (1). Autism was first described by Kanner (3) and became known as infantile autism or autistic disorder. The concept has expanded since then, and the term Kanner autism is sometimes used to refer to cases presenting with symptoms similar to those of Kanner's original sample; such cases are a subset of PDD. Kanner's original description remains influential, and there has been a tendency in the literature to assume that persons with Kanner autism represent the nuclear or core form of PDD, an assumption that may or may not be warranted (4).

Kanner (3) presented case studies of 11 children (eight boys, three girls). He identified three fundamental characteristics common to the entire group: (a) "extreme autistic aloneness"—an inability to enter into affective contact with others, (2) failure to use language in a communicative fashion (muteness, literalness, echolalia, and/or pronoun reversal), and (3) "anxiously obsessive desire for the maintenance of sameness," with the children showing distress at any changes in their usual routines. Related to these factors were a severely limited repertoire of spontaneous activities and an associated tendency to repeat the same actions in a ritualized manner. Kanner argued that their extreme need for constancy led to a skillful enjoyment of inanimate physical objects.

Two more characteristics asserted by Kanner were normal intellectual potential (based mainly on their "intelligent physiognomies") and good physical health (although the case descriptions include neurologic abnormalities) (3). Kanner concluded that "these children have come into the world with innate inability to form the usual, biologically provided affective contact with people, just as other children come into the world with innate physical or intellectual handicaps."

With regard to intelligence, Kanner's group obviously sampled a higher functioning subset of persons with autism. Otherwise, his clinical description has held up remarkably well for more than half a century of research.

The major trend in thinking about autism during the two decades following Kanner's original paper must be regarded as an unfortunate digression. Influenced by popular psycho-dynamic ideas, several authors (5–7) claimed that autism was caused by pathologic parenting. Kanner (8) himself came to believe that lack of parental warmth contributed to the genesis of the disorder (the famous "refrigerator parent" hypothesis). Mahler's (9,10) writings on autistic and "symbiotic" psychoses of childhood figured prominently in this trend, although, as Hobson (11) points out, Mahler accepted the notion of a constitutional deficit in the child, which she identified as an inability to use the mother as a "beacon of emotional orientation." Although Mahler's clinical observations anticipated later research on social referencing, her views led researchers away from neurodevelopment. The erroneous attribution of responsibility to the parents, termed the "pernicious hypothesis" by Rimland (12), can only be understood in the context of the general psychiatry of that period, which usually minimized biologic formulations.

A shift toward neurodevelopmental theories began in the 1960's. Researchers outlined a variety of models of autistic neural impairment: (a) allocating attention to coordinate new stimuli with memories (12), (b) vestibular mediation of perceptual processes (13), and (c) left hemisphere–mediated linguistic abilities (14). Those advocating a primary cognitive deficit (15,16) and those favoring a primary socioaffective deficit (17,18) agree that the basic PDD deficit stems from a neurologic dysfunction.

Differential diagnosis was a conceptual problem for early autism research. Some clinicians believed that autism was a variant or precursor of schizophrenia. In the 1970's came an awareness that psychotic disorders beginning in infancy must be regarded as separate in kind from those with onset in later childhood, adolescence, or adulthood (19,20). Autisticlike disorders virtually always begin before 3 years of age, whereas schizophreniclike disorders virtually never begin before 7 years of age (19). This realization revitalized interest in infantile autism as a distinct nosologic entity, leading to the development of more operationally precise diagnostic criteria (20–22) and a reconceptualization of the syndrome as a PDD, under which label it was incorporated by the American Psychiatric Association in the third edition of the *Diagnostic and Statistical Manual of Mental Disorders* (*DSM-III*) (23).

DIAGNOSTIC CRITERIA

The first sets of systematic diagnostic criteria for autism appeared in 1978 (20,21).

Rutter (20) proposed four essential criteria:

1. Onset before the age of 30 months,
2. Impaired social development out of keeping with the child's intellectual level,
3. Delayed and deviant language development out of keeping with the child's intellectual level,

4. Insistence on sameness, as shown by stereotyped patterns, abnormal preoccupations, or resistance to change.

In addition, Rutter urged the adoption of a multiaxial classification system to include the behavioral diagnosis of autism, child's intellectual level, and neurologic/medical status.

Rutter's criteria of social and linguistic development "out of keeping with the child's intellectual level" represented an important advance in diagnostic identification and research (e.g., selection of appropriate control groups). Limiting the diagnosis to cases with the onset before 30 months of age was meant to differentiate maximally between autism and other symptomatically similar disorders such as Heller syndrome and Asperger syndrome (AS). Rutter recognized that some cases of autism arise at slightly older ages, but unfortunately, the 30-month guideline became a rigid rule in *DSM-III*. Rutter emphasized that social impairment often lessens after early childhood, although more subtle social ineptitude persists. In fact, Rutter's paper contains rich descriptions of developmental stages not captured in the summary diagnostic criteria that he proposed.

Ritvo and Freeman (21) outlined slightly different criteria for autism, which were adopted by the National Society for Autistic Children:

1. Disturbances of developmental rate and/or sequences (e.g., dissociations among motor, social, and cognitive lines of development),
2. Disturbances of response to sensory stimuli (perhaps including generalized hyperreactivity or hyporeactivity or an alternation between these extremes),
3. Disturbances of speech, language/cognition, and nonverbal communication (including delayed or absent speech, poor symbolic/abstract abilities, and poor use of gesture),
4. Disturbances of the capacity to relate appropriately to people, events, and objects (e.g., failure to manifest age-appropriate social behavior; stereotypic use of play objects).

Age of onset was stated to be typically before 30 months of age, but the system is descriptive and has no fixed, requisite criteria. Associated features are said to include mood lability without identifiable cause, failure to appreciate real danger, inappropriate fears, self-injurious behaviors, mental retardation, and abnormal electroencephalograms (EEG) with or without seizures.

Ritvo and Freeman placed strong emphasis on sensory abnormalities. They described abnormalities of every sensory modality, visual (prolonged staring, absence of eye contact), auditory (nonresponse or overresponse to sounds), tactile (over- or underresponse to pain or temperature change), vestibular (whirling and preoccupation with spinning objects), olfactory/gustatory (repetitive sniffing), and proprioceptive (posturing or hand flapping). All these behaviors are seen in many autistic children (a very low sensitivity to painful stimulation is, for example, a common finding). Whether they reflect an underlying deficit in sensory modulation is not known, and most sets of diagnostic criteria have not included sensory abnormalities as an essential feature. It should be noted that research using the criteria of Ritvo and Freeman would identify a somewhat different set of children as autistic than would Rutter's (20) criteria.

Wing and colleagues (24,25) took an empirical approach to the diagnosis, investigating a population of mentally handicapped children with deficits in social relatedness, communicative language, and repertoire of interests and behaviors. They identified three subgroups based on social behavior: aloof, passive, and active but odd.

Two aspects of the work of Wing et al. are relevant to diagnosis. First, they broadened the concept of PDD. Previously, researchers and clinicians had focused on Kanner's criteria. Wing et al. started with the fundamental, universally agreed-on symptom areas of autism and used them as the basis of classification. Hence, any child with serious social, communication, and imaginative/behavioral impairment greater than that expected for his or her mental age (MA) would be deemed part of the autistic spectrum, even if his or her presentation differed slightly from that of prototypical Kanner-type autism.

Second, Wing et al. described social behaviors associated with different types of functioning at different developmental stages. Behaviors of the aloof group, for example, most closely approximate Kanner autism, with aberrant absorption in their own pursuits and lack of interaction with others. Those in the passive group, conversely, tolerate interactions with others, although they do not initiate such interactions. Finally, those in the active-but-odd group initiate interactions with others, but these social bids tend to be awkward, one-sided affairs consisting of an attempt to draw another person into their own sphere of interest. Subgroup membership can change with development in the direction of greater sociability. Although many children remain in the same group over time, some become more sociable. An aloof young child may grow into a passive or active-but-odd older child. An active-but-odd young child, however, is unlikely to become either passive or aloof.

In *DSM-III* (23), infantile autism, now listed under the general heading of PPD, was diagnosed based on a monothetic criterion set (in which every major feature had to be present for the diagnosis to be made). The *DSM-III* criteria were as follows:

1. Onset before 30 months of age;
2. Pervasive lack of responsiveness to other people (autism);
3. Gross deficits in language development;
4. If speech is present, peculiar speech patterns such as immediate and delayed echolalia, metaphorical language, pronominal reversal;
5. Bizarre responses to various aspects of the environment, e.g., resistance to change, peculiar interests in or attachments to animate or inanimate objects;
6. Absence of delusions, hallucinations, loosening of associations, and incoherence as in schizophrenia.

Because failure to meet any one of these six criteria rules out a diagnosis of infantile autism, it is not surprising that use of *DSM-III* criteria yielded lower prevalence rates of autism. The sensitivity of the *DSM-III* criteria is relatively low, but their specificity is high, producing few false positives. In the *DSM-III,* older children and adults who no longer met all diagnostic criteria were assigned a new, separate diagnosis: infantile autism, residual state. The *DSM-III* also included a category termed childhood-onset PDD, which applied to few cases and was open to cases with diverse behavioral presentations and prognoses (e.g., Heller syndrome, AS). The *DSM-III* also included a category for atypical cases, which in some samples outnumbered those diagnosed as autistic.

Substantial changes were made in the revision of *DSM-III* (*DSM-III-R*) (26). The purely monothetic criteria were replaced by a partially polythetic and more fully operationalized set of criteria developed by Wing. The *DSM-III-R* criteria are as follows. At least eight of the following 16 items are present, including at least two items from A, one from B, and one from C. Note: Consider a criterion to be met only if the behavior is abnormal for the person's developmental level.

A. Qualitative impairment in reciprocal social interaction as manifested by the following:
   1. Marked lack of awareness of the existence or feelings of others
   2. No or abnormal seeking of comfort at times of distress
   3. No or impaired imitation
   4. No or abnormal social play
   5. Gross impairment in ability to make peer friendships
B. Qualitative impairment in verbal and nonverbal communication and imaginative activity, as manifested by the following:
   1. No mode of communication, such as communicative babbling, facial expression, gesture, mime, or spoken language
   2. Markedly abnormal nonverbal communication, as in the use of eye-to-eye gaze, facial expression, body posture, or gestures to initiate or modulate social interaction
   3. Absence of imaginative activity
   4. Marked abnormalities in the production of speech, including volume, pitch, stress, rate, rhythm, and intonation
   5. Marked abnormalities in the form or content of speech
   6. Marked impairment in the ability to initiate or sustain a conversation with others, despite adequate speech

C. Markedly restricted repertoire of activities and interests, as manifested by the following:
 1. Stereotyped body movements
 2. Persistent preoccupation with parts of objects
 3. Marked distress over changes in trivial aspects of environment
 4. Unreasonable insistence on following routines in precise detail
 5. Markedly restricted range of interests and a preoccupation with one narrow interest
D. Onset during infancy or childhood [specify if childhood onset (after 36 months of age)]

The *DSM-IV* (27) followed *DSM-III-R* categories for autistic disorder and PDD not otherwise specified but added three diagnostic categories to parallel the World Health Organization's *International Classification of Diseases,* 10th Revision (*ICD-10*): PDD subgroups Rett syndrome, other childhood disintegrative disorder (CDD), and AS (28). *DSM-IV* PDD include the following subgroups: autistic disorder, Rett syndrome, other CDD, AS, and PPD not otherwise specified (including atypical autism). It is important to note that Rett syndrome, AS, and PDD not otherwise specified all carry the same diagnostic code of 299.80. *DSM-IV* (27) and *ICD-10* (28) provide 12 parallel polythetic diagnostic criteria for autistic disorder and childhood autism. Six symptoms must be present, four of which are specified. At least two symptoms must come from a subset of four impaired social skills (nonverbal behavior, friendship, joint attention, and reciprocity). One must come from a subset of four impaired communication skills (delayed or absent language, abnormal conversation, perseverative speech, and abnormal play), and at least one of the six must come from a subset of four abnormal activities (obsessive interests, rigid rituals, stereotypies, and preoccupation with parts of objects). A study comparing *DSM-III, DSM-III-R,* and *ICD-10* diagnoses of autism with diagnoses determined by latent class analysis found that *ICD-10* best fit the latent diagnostic standard (29). However, this latent class model indicated that the latent standard was best approximated "by the sensitivity of *DSM-III-R* criteria and the specificity of the two other systems (*DSM-III* and *ICD-10*)" (29).

A study comparing autism and nonautistic PDD in these three systems plus the old *DSM-III* reported that all four diagnostic systems identified autistic groups that were significantly lower in IQ and adaptive functioning and expressed significantly more autistic symptomatology than the four systems' nonautistic PDD groups (30). However, all four nonautistic PDD groups expressed a pattern of autistic characteristics nearly identical to the pattern found for autism. It is true that "it has been difficult to develop a methodology for the evaluation and comparison of diagnostic criteria in psychiatry" (29).

SOCIOEMOTIONAL AND COGNITIVE CHARACTERISTICS

This section presents a selective review of the social and cognitive factors associated with autism/PDD and the current theories attempting to explain how these factors may play an etiologic role in the disorder. Most of the current theories implicate a specific social or cognitive deficit, and some offer an associated brain mechanism (neuroanatomic and neurochemical findings in autism/PDD are reviewed later in the chapter). Although the field has progressed from largely speculative notions regarding the social and cognitive deficits in autism to a clear understanding of the strengths and weaknesses present, none of the "core deficit" models proposed clearly explains the mechanism behind all the observed deficits. More integrative models postulate the interplay of several neural mechanisms to explain the deficits, although these models have yet to be systematically examined (31,32).

Comparison Groups

The study of differential deficit in autistic individuals poses the problem of identifying appropriate control groups. Normal children of similar chronologic age (CA) will score higher than autistic children on any standard measure. Normal children matched for MA are a better choice but will not control for general effects of psychopathology or for years of experience. Normal MA matches will be younger chronologically than the autistic children, so their sensorimotor systems may be too immature to be comparable. Psychiatric patients with non-PDD diagnoses also are likely to outperform autistic subjects on most measures. Mentally retarded controls may be matched with autistic individuals on both CA and MA and are the most frequently used comparison group. Mental retardation, however, is an even more heterogeneous classification than is PDD; retarded persons may have little in common beyond their general cognitive delay. Sometimes this heterogeneity is reduced by using a mentally retarded sample of known, homogeneous cause such as Down syndrome. Even this, however, limits inference. Most individuals with Down syndrome are highly social (above their MA, in many cases), and this will skew comparisons with autistic samples. Another comparison population is children with developmental language disorders because they have verbal and nonverbal skill patterns similar to those of autistic children. The choice of control group obviously will affect the areas in which autistic individuals appear relatively impaired. Some studies use more than one comparison group to offset the problems with each alternative. Other researchers study only autistic individuals with IQs above the retarded range. Although it is difficult to assemble samples meeting this criterion, it is a useful strategy in that it allows the study of autism uncomplicated by global intellectual impairment. The generalizability of results to the whole autistic spectrum, however, will be limited.

Another problem for research design is that because autistic individuals by definition have a specific deficit in language skill, their verbal IQ (VIQ) tend to be substantially lower than their non-VIQ. In most comparison samples (language disorders being an exception), verbal and nonverbal mental levels do not differ equally. Consequently, if a control group is matched to the autistic group on VIQ, then the autistic group will tend to be more capable overall because of their superior visuospatial skills. If a control group is matched on non-VIQ, the opposite situation will obtain. Some forms of PDD, such as AS, however, do not display this verbal/non-VIQ discrepancy.

In practice, all the comparison groups described above are used, as are a variety of matching strategies. Therefore, it is difficult to compare research findings across studies.

Social Impairment

Social impairment is a universally recognized core deficit in autism. It differs qualitatively from the impairments seen in other developmental or behavioral disorders. Autism is marked by a general inability to form relationships, failure to use nonverbal communicative behaviors such as eye contact, lack of reciprocity, lack of awareness of others, and failure to share experiences with others (27). The social impairments seen in autism may be categorized into four key areas: empathy, play, imitation, and joint attention. The deficit in joint attention appears to be the social deficit most specific to autism (33). Failure to use joint attention is seen in both dyadic (child–adult) and triadic (child–adult–object) levels of interaction (34). Additionally, autistic children are impaired in both gestural and procedural imitation (33). CA and MA are associated with both joint attention and imitation in autistic children but not in developmentally delayed children, in whom the skills emerge spontaneously regardless of developmental level (34). During play sessions with unfamiliar adults, autistic children are less likely to engage in joint attention and monitor the channel of communication less frequently than children with a developmental language disorder (35).

The most profound social deficit in autism is seen in face-to-face interaction skills, often examined through autistic children's play. In general, studies show that autistic children cannot orchestrate the give-and-take of social interactions, including conversation and play (36). Abnormalities are present in joint attention behaviors and in both initiations and responses to others, extending to the realm of peer play (37,38). These symptoms appear to vary depending on the interactive partner and setting involved (36,39) as well as on a cognitive level (36,37). In sum, findings suggest that autistic children are unlikely to initiate interactions but are more likely to respond in adult-initiated situations (36,40). In situations in which free play is observed, autistic children

are found to engage in play much less frequently and at a developmental level much lower than peers of a similar intellectual ability (41). When symbolic play is modeled, autistic children appear to be able to engage in a higher, albeit relatively deficient, level of play. When vigorously encouraged and elicited, normal levels of play have emerged in high-functioning autistic children (42). This is consistent with the finding of a large-scale study in which symbolic play deficits in high-functioning autistic children seemed more related to motivation than capacity (43). A review of studies affirms deficits in play relative to intellectual level and tendencies for sensorimotor play to the exclusion of symbolic play (44).

Anecdotally, it has been observed that social impairments in children with autism occur very early in life, and more recently early diagnostic instruments have allowed researchers to support this notion in observational studies (45). Charman et al. (45) found autistic children as young as 20 months old to have deficits in empathy, joint attention, and imitation compared with both normally developing children and developmentally delayed children. The developmentally delayed and autistic children both displayed nonverbal MA 3 months behind CA, although the developmentally delayed children had significantly higher verbal comprehension scores. Although spontaneous pretend play is limited in both of these groups and tends not to be a reliable discriminator, developmentally delayed children are more likely to engage when prompted. Although attachment is not absent, autistic children are more likely to have delayed development of attachment and are disproportionately prone toward insecure attachment styles (46). Extending these findings to adulthood, Howlin et al. (47) followed a group of autistic individuals from age 7 to 25 years and reported that deficits in stereotyped behavior, social relationships, jobs, and independence persisted to adulthood. Several mechanisms responsible for observed social deficits have been postulated, including both social cognitive and affective mechanisms.

In the past decade, the role of social cognition and socioemotional processing in autism has been extensively examined. Recent research into socioaffective communication in PDD has been influenced by Hobson (18,48,49). Hobson took as his starting point the notion that humans are neurobiologically programmed to display characteristic facial expressions, gestures, and vocalizations when experiencing specific affective states and are also predisposed to perceive and respond to those expressive actions on the part of others. Comprehension of emotional expressions requires one to (a) perceive qualitative differences among expressions, (b) become aware of which emotional expressions are meaningfully interrelated (express the same or similar affects), and (c) produce appropriate affective responses to expressions of the various emotions (50,51). In Hobson's view, this third aspect of emotional communication (Darwin's "instinct of sympathy") may be the basis of theory of mind rather than the other way around.

Hobson's initial studies (48,49) explored the hypothesis that autistic children should be more impaired in emotional recognition than their cognitive deficit would warrant. Hobson compared autistic children with several matched control groups for their ability to recognize facial, gestural, and vocal expressions of four basic emotions: happy, unhappy, angry, and scared. Subjects were also tested for recognition of movements, sounds, and contexts of inanimate objects (e.g., automobiles). All subjects obtained maximal scores on the object recognition task, but the autistic children alone were severely impaired on the emotion recognition measures. There were two methodologic problems with Hobson's early studies. The emotion and object recognition tasks were not of equal difficulty and controls were matched with autistic subjects on nonverbal MA, so the effects of differential verbal skills were not adequately controlled for (52). Hobson et al. (53) addressed these problems in a further study. The autistic sample continued to show a significantly greater discrepancy between their object and affect scores, but their performance on the affect recognition task was now not significantly lower than that of the controls. Thus, Hobson's original finding was only partly replicated. Several investigators have attempted to replicate and extend Hobson's findings. Tantam et al. (54) found that autistic children impaired relative to age and performance IQ matched controls on an affect recognition task, a finding that also holds for high-functioning autistic adults (55). Braverman et al. (56) found that PDD children were impaired relative to MA-matched controls on an affect-matching task and that the autistic children performed significantly less well on affect and face matching than on object matching. PDD children also are selectively impaired at matching emotion with appropriate contexts (57) and facial expressions (58) and at responding to affective stimuli (59). These findings do not seem to reflect a deficit in recognition of faces as such (60). They are, however, influenced by verbal MA. Like Hobson et al. (53), Prior et al. (61) reported a nonsignificant difference between the emotion recognition scores of autistic children and those of psychiatric controls matched for verbal MA, whereas Ozonoff et al. (62) reported a significant emotion perception deficit relative to nonverbal MA-matched controls, which shrank to nonsignificance relative to a verbal MA-matched control group. The extent to which emotion perception depends on verbal development remains uncertain. It may be that the experimental measures require more verbal processing than does everyday affective responsiveness. However, recent research implicates neural mechanisms responsible for this difference. During a facial processing task, high-functioning autistic individuals were found to have significant differences from matched controls in the function of brain areas that have been previously found to be impaired in autism (e.g., visual and auditory cortices, cerebellum, mesolimbic areas, lateral temporal lobe) (63).

Autistic subjects also seem to be impaired in the expression of emotion. Yirmiya et al. (64) coded the affect expressions of autistic children and several control groups, using Izard's (51) system for rating facial expressions. The autistic children displayed less facial affect, especially positive affect, than the other groups, but more incongruous, difficult to interpret expressions. MacDonald et al. (55) reported that high-functioning autistic adults displayed impaired emotional expressiveness and noted that, although no single measure differentiated individual autistic subjects from controls, a composite multitask score of socioemotional communication correctly classified 18 of 20 subjects as autistic or nonautistic.

Fine et al. (65) studied the extent to which high-functioning autistic, AS, and psychiatric control subjects employed intonation (prosody) in a meaningfully communicative fashion. High-functioning autistic subjects used communicative intonation less than the other groups but made greater use of noncommunicative patterns of intonation. AS subjects differed only slightly from controls in their use of different types of intonation, although they often seemed to use intonation inappropriately. The authors argued that because autistic individuals are able to perceive and produce patterns of vocal stress when asked, their failure to do so in conversation reflects an inability to map these vocal patterns onto appropriate verbal contexts.

In a recent examination of friendship and loneliness in autism, high-functioning autistic children reported more loneliness than normal controls and were more likely to have a cognitive as opposed to emotional definition of loneliness (66). Understanding loneliness through a cognitive mechanism suggests their loneliness is likely acquired through self-evaluation or social comparison of their experience with that of others or with their own past social experiences (66). The autistic children clearly failed to use emotion in their understanding of both friendship and loneliness, underscoring that the emotional deficit is primary.

The emerging consensus on a cognitive profile of PDD does not include an agreement as to which deficits are causal and which are epiphenomenal. Currently, the most influential theories focus on social cognition. A major debate is whether the affective or representational aspect of social cognition is primary. This reflects a fundamental question that goes far beyond autism: Is human emotion a result of cognitive evaluation or does it depend on the operation of phylogenetically older neural systems independent of cognition (67)? Affective primacy in autism has been proposed by Waterhouse and Fein (68), Dawson et al. (31), and Hobson (49). Representational primacy, reviewed in the theory of mind section, has been claimed by Baron-Cohen et al. (15).

Dawson et al. (31) examined the hypothesis that impaired medial temporal lobe functioning causes the core social and affective symptoms of autism as well as downstream impairment of prefrontal executive functions. In support of this hypothesis, social impairments "strongly and consistently correlated" with performance on the tasks sensitive to medial temporal lobe impairments and correlated less with the dorsolateral prefrontal cortex–related task. The authors concluded that the social and emotional deficits are primary to autism and are likely related to the functional integrity of the medial temporal lobe and limbic structures, which then may result in "downstream" consequences on the higher cortical functions (e.g., executive functions).

Waterhouse and Fein (68) argued that none of the reported cognitive deficits adequately explains the frequently observed autistic aloofness or disinterest in people during the first year of life. They pointed out that, although social and cognitive development is ordinarily highly correlated and presumably interdependent, they can be dissociated in disorders of development. Within the PDD population, social impairment occurs regardless of cognitive level and is the single feature both unique and universal to the disorders in the autistic spectrum. Autistic individuals with higher IQs may manifest their social deficit in less disruptive or less salient ways, but it is still apparent. Waterhouse et al. (32) presented a model of four dysfunctional neural systems with affiliated neurofunctional impairments resulting in the symptoms of autism. The authors proposed abnormalities in (a) the hippocampal system, resulting in canalization of sensory information such that integration of sensory information is disrupted; (b) the amygdaloid system, impairing the assignment of affective significance to stimuli; (c) oxytocin system, flattening social bonding and affiliative behavior; and (d) temporal and parietal polysensory regions, resulting in extended selective attention.

Kanner (3) proposed an innate, presumably biologically determined social deficit as the basis of autism. In the ensuing decades, the socioaffective deficit continued to occupy center stage in autism theory, but parents were postulated as causing the disorder. When the concept of a neurologic basis reemerged, the field turned away from both the discredited notion that cold parents created autism and the idea that the core dysfunction in autism was affective. Measurement tools have also constrained exploration of affective deficits. Almost the entire technology of neuropsychological assessment relates to cognitive functions, and only recently have ratings of social behavior in autism been standardized [e.g., the Autism Diagnostic Observation Schedule–Generic (69,70)], resulting in much of the research examining factors that are easily measured, whether or not they addressed the impairments most relevant to understanding autism.

Deficits in Theory of Mind

A specific social cognition deficit that has gained a lot of attention in relation to autism is that of theory of mind. Baron-Cohen et al. (15) proposed that individuals with autism have a specific impairment in the development of theory of mind; that is, they are impaired in the ability to attribute mental states (beliefs, desires, intentions) to themselves and others. Theory of mind has been extensively examined as a key underlying deficit in autism that can account for the triad of Wing and Gould (24) of impairments in communication, imagination, and socialization (71). In this view, the inability to represent the internal states of others leads to the autistic child's range of difficulty with socialization. Pretend play, pragmatic communication, and empathy, which are impaired in autism, require one to symbolize the surrounding world and be aware that other people have their own subjective frame of reference. Supporting the theory of mind hypothesis, autistic children have been shown to perform poorly across a variety of experimental tasks requiring theory of mind for successful solutions (15,61,72–75).

Metaanalyses on theory of mind studies (71) report significant differences between individuals with autism and individuals with mental retardation, individuals with autism and normally developing children, and normally developing children and children with mental retardation, suggesting that theory of mind deficits are not unique to autism but may be unique in severity. MA was a moderator for those with mental retardation and normally developing children (the older the MA, the better the performance) but not for those with autism. Consistent with this finding, recent studies have found high-functioning autistic individuals (IQ >70) and individuals with AS to be impaired on advanced theory of mind tasks, including the ability to use the language of the eyes to infer the mental state of another person (the eyes task) and attributing social meaning to ambiguous visual stimuli (the social attribution task) (76–78). Moreover, the individuals with HFA and AS performed similarly on the social attribution task, and scores on this task were not related to VIQ (77), disputing the notion that deficient theory of mind performance is owing to poor linguistic skills. High-functioning individuals with autism and AS have been found to be relatively unimpaired in the basic theory of mind tasks (16).

Despite the pervasiveness of the findings of theory of mind impairments in the autistic spectrum and the model's ability to account for some autistic behaviors and cognitive deficits, it is not a primary causal factor. For a deficit to be primary to autism, it must be both universal, manifested in all individuals with autism, and unique, not manifested by most individuals in other clinical populations. Although recent research appears to support the criteria of universality, it is clear that impairments in theory of mind are not unique to autism, as seen in the weak performance of individuals with mental retardation (71,75), language disorders (78), and deafness (79). Moreover, other researchers have argued that the deficient performance of theory of mind tasks may be the result of other proposed deficits in autism, such as weak central coherence (80) or executive function (81). Finally, the theory of mind hypothesis has also been criticized for its failure to account for the very early social impairments seen in autism before the development of mentalizing abilities (82), its failure to account for the presence of social behaviors hypothesized to be dependent on theory of mind (82), and its inability to account for the observed deficits in the nonsocial aspects of autism, such as restriction of repertoire (83). These findings suggest that, although autistic children are impaired in theory of mind tasks, their failure to develop a theory of mind is not the cause of the primary social symptoms of autism.

Language Deficits

Many aspects of verbal functioning are impaired in autistic children, many of whom are entirely mute. Verbal autistic children generally are able to acquire normal grammatical morphology and syntax, although onset is delayed (84,85). Some autistic children learn grapheme–phoneme correspondence, enabling them to write and decode words for reading (85). However, language comprehension is significantly impaired relative to expression, and deficits in the semantic and pragmatic aspects of language are common (85). Autistic children generally do not rely on meaning when comprehending or expressing acquired morphemes, grammatical structures, and vocabulary (85,86). They are also deficient in interactive communication, including conversational behavior, nonverbal communication, and speech prosody (87).

Even when language is relatively spared, there are usually deficits in comprehending complex language and formulating complex output (85). Minshew et al. (88) compared language skills of high-functioning autistic individuals (VIQ >70) with normal controls matched on IQ. No differences were found between those with autism and normal controls in tasks assessing basic procedural or mechanical skills; however, complex language ability was significantly impaired in the autistic group. Moreover, high-functioning autistic children have been shown to be more deficient in complex language comprehension and output than children with language disorders (85). Baron-Cohen et al. (76) argued that the pervasiveness of the language deficits in autism suggests that early social impairments may be affecting subsequent language development.

Rutter (14), however, argued that deficits in language-mediated functions are independent of social impairment and may play a causal role in the disorder. The findings of a 15-year follow-up study conducted by Mahwood et al. (89) demonstrated the importance of early language usage (not simply delay) in the outcome of those with autism. Specifically, early language usage was significantly related to social functioning. The study also suggested that individuals with autism can make substantial improvements in their language usage as they grow older; individuals with autism improved more than the language-delayed individuals on receptive language and VIQ, although they remained more impaired in

the language and social domains than the language delay–matched individuals. Rutter (90) suggested earlier that the language deficit might be only a special case of a more general impairment in the perception and expression of meaningfully patterned stimuli.

Deficits in Attention

Numerous studies have documented attention abnormalities in autistic individuals. Autistic individuals are found to be overly selective in their attention to stimuli, such that their focus of attention is overly narrow (91–93). Extended selective attention has been conceptually linked to the savant skills (exceptional ability in one domain) seen in a small subset of autistic individuals (32,94). Autistic individuals, even those with IQ greater than 70, display a general deficit in shifting attention (91,95–98). Sustained attention appears to be spared (91,99).

Plaisted et al. (100) found that high-functioning autistic individuals show a deficit in divided attention, but no deficit in selective attention. Specifically, the children with autism displayed a local advantage effect and a local interference effect (attending to a part of the stimuli rather than the whole), whereas the normally developing children displayed the opposite pattern (a global advantage effect and a global interference effect). The children with autism performed in the same manner as the normally developing children on the selective attention task; their reaction time data showed a global advantage and a global interference effect. Overall, these data support Frith's (101) notion of weak, but not absent, central coherence in children with autism. The weak central coherence in children with autism may be explained by (a) the inability to filter out information at the local level rather than a deficit in the ability to draw component information together, (b) children with autism voluntarily attending to the local level unless instructed to do otherwise, or (c) children with autism displaying a deficit in shifting attention. Performance consistent with that seen in the children with autism has also been seen in children with AS (102).

Several early cognitive theories of autism focused on abnormalities in mechanisms related to attention. Rimland (12) proposed that the autistic child failed to associate incoming stimuli with established memory. In his model, autistic social deficit arises from a failure to link the presence of mother with the memory of the biologic rewards that she has brought. Rimland hypothesized that this deficit might result from a hypoactive or underaroused reticular formation. Hutt et al. (103) developed a related but opposite theory: autistic children, they suggested, have a hyperactive or overaroused reticular formation and gate out all stimuli that threaten to further arouse them. Both of these theories accounted for many of the known perceptual and behavioral peculiarities of autism, although in different ways. For example, Rimland saw motor stereotypies as a method of increasing arousal; Hutt et al. saw them as a method of reducing arousal by screening out extrinsic stimuli. Other more recent theories also invoke neurologically based attention dysregulation as the cause of perseverative behavior and social deficits in autism (104–106) (see more extensive discussion of Courchesne's work below). Kinsbourne (106) and Dawson and Lewy (105) suggested that hypersensitivity to novelty and overactivation of brainstem mechanisms of arousal cause narrowed attention focus and stimulus overselectivity in autistic children. This could result in deployment of mechanisms for dearousal such as stereotypies and avoidance of social stimuli.

Courchesne et al. (98) demonstrated that the pattern of attention deficits in autism was consistent with the attention deficits observed in individuals with abnormalities in the cerebellum and parietal cortex. Further, structural abnormalities in the cerebellum and parietal cortex have been found in children with autism [for a review, see reference (98)]. The authors found a strong correlation between attention deficits and the degree of parietal abnormality in autistic adults (95). In an extension of this investigation (96), attention deficits were found to have a significant correlation with cerebellar hypoplasia but not with the size of other neuroanatomic regions in autistic children. These studies were among the first to link a specific behavioral deficit with the size of a specific structural abnormality in autism.

Memory Deficits

Several investigators have posited that autism is similar to medial temporal lobe amnesia (104), supported by both neuropsychological findings regarding memory skills (107–109) and neuroanatomic evidence of damage to the hippocampus and amygdala seen in individuals with autism and individuals with amnesia. However, recent studies have disputed this notion. Contrary to the pattern of intact implicit but impaired explicit memory seen in amnesic individuals, high-functioning autistic children (110) and adults with AS (111) have shown intact implicit and explicit (recognition and recall) memory. However, children with autism did not show the expected primacy effect for word list recall, suggesting that children with autism use different organizational strategies during encoding or retrieval of items from memory. Consistently, Fein et al. (112) reported verbal memory impairments increased as the semantic structure of the material increased. Moreover, Minshew et al. (113) reported that individuals with autism fail to group words into conceptual categories when given a random list of words. These results may be viewed as consistent with executive function deficits in planning and organization (114).

A consistent pattern of memory deficits has not been found across the studies examining memory in the autistic spectrum. The disparity in the literature may be partly explained by the cognitive functioning of the research partic-

ipants included in the study (110). Across studies, the most specific and consistent finding is that visual and rote auditory memory tends to be spared, whereas verbal memory for semantically organized material such as stories is quite impaired.

Bachevalier (115) proposed that impairments in memory (and mental retardation) seen in some individuals with autism are the result of lesions to the hippocampus, whereas the social impairments are the result of impairments in the amygdala. It is possible that some autistic individuals have only structural impairments in the amygdala, resulting in the social deficits but sparing memory and cognitive functioning. Bachevalier's hypothesis has been supported in animal models but has yet to be systematically tested in humans with autism.

Executive Function

Deficient executive functions (e.g., organization, planning, flexibility, impulse control, working memory) have been proposed to be a possible explanation for the core social and cognitive impairments seen in autism. Numerous studies have documented impairments in executive function tasks for individuals with autism (16,116–122) and AS (16,123). Specific deficits have been observed in set shifting and planning (122), and autistic children have been shown to make more perseverative errors than matched controls (117). Performance of executive function tasks is thought to be mediated by frontal cortical systems. Clearly, autism is not the only disorder reported to have executive function deficits. Several other developmental and neurologic disorders, such as schizophrenia, Tourette syndrome, and attention deficient/hyperactivity disorder (ADHD) have documented deficits in executive functions (116). However, a comparison of children with autism, ADHD, Tourette syndrome, and controls revealed a unique pattern of deficits for the autistic group. The children with autism displayed deficits in planning, organization, and cognitive flexibility, with intact inhibition; the opposite pattern was found for children with ADHD. No executive function deficits were seen in children with Tourette syndrome.

Consistent with the executive function hypothesis, Bennetto et al. (114) argued that the observed pattern of memory deficits seen in individuals with autism resembles the pattern seen in patients with frontal lobe lesions. Specifically, deficits are seen in temporal order memory for verbal information, supraspan verbal learning, and the ability to maintain the appropriate context of the information that has been learned. The authors suggest that these deficits may all arise from a general deficit in working memory, which could also explain the specific impairment in social cognition seen in autism. Social interaction depends on the integration of constantly changing, context-specific information, which is both subtle and complex in social interactions, placing too great a load on a defective working memory, resulting in observable social impairments.

To examine the executive function hypothesis, Griffith et al. (124) examined executive functioning in preschoolers with autism in a 1-year longitudinal study. The authors failed to find executive function deficits in the autistic children compared with developmentally delayed children. Further, when looking at the data longitudinally, the authors concluded that there was no evidence that children with autism were growing into an executive function deficit over time (instead the data appeared to indicate that the developmentally delayed children may be growing out of an executive function deficit). Nonetheless, the results of the study do not support an early autism-specific executive function deficit; therefore, it is unlikely that executive functions are responsible for the development of the social and behavioral symptoms characteristic of autism.

Overall, although executive function impairments have been consistently documented in high-functioning and older children with autism, they have been displayed less consistently for younger children. McEvoy et al. (125) found that executive function impairments in preschoolers related to measures of joint attention and social interaction, which may suggest that executive function deficits develop secondary to developmentally earlier social impairments.

Characteristics of Asperger Syndrome

As discussed above, individuals with AS have been shown to have visual memory deficits and intact basic theory of mind, but deficient advanced theory of mind, executive function impairments, difficulty shifting attention, and impairment in language pragmatics. Individuals with AS show fine and gross motor deficits (with a higher prevalence rate than autism) and are often described by parental report as clumsy (126).

There has been a debate in the literature regarding the uniqueness of AS from HFA. It is often cited that individuals with AS show a discrete pattern of intelligence scores, with significantly higher VIQ than Performance IQ (PIQ), higher verbal memory, and theory of mind performances and larger VIQ–PIQ splits than individuals with autism (127,117). Although some believe that these findings may simply be reflective of the definition of AS as a verbally able version of autism, these group differences have led some researchers to posit a specific pattern of deficits for AS similar to that seen in nonverbal learning disability (NLD). For example, Klin et al. (126) found that individuals with AS (but not HFA) displayed deficits in fine and gross motor skills, visuomotor integration, visuospatial perception, nonverbal concept formation, and visual memory. The HFA cases exhibited deficits in articulation, verbal output, auditory perception, vocabulary, and verbal memory, which appear spared in those with AS. Consistent with the social

deficits seen in the NLD profile, those with AS show deficits in social perception, social judgment, and interpretation of affect, which are believed to stem from difficulties understanding the nuances of nonverbal communication. This pattern of deficits is proposed to be associated with right-hemisphere dysfunction (41).

Other research supports the view that AS is not distinct from HFA. Szatmari et al. (123) compared cases of AS, HFA, and non-PDD psychiatric controls on a neuropsychological battery. AS and HFA were indistinguishable. High-functioning autistic subjects did less well on the Wisconsin Card Sorting Test and the similarities test of the Wechsler Intelligence Scale for Children-Revised (WISC-R), whereas AS patients did less well on a pegboard task using their nondominant hand, but these differences, although statistically significant, did not allow accurate group classification. Similarly, in a follow-up study of AS and HFA, Szatmari et al. (120) found few qualitative differences between the groups; high-functioning autistic subjects were more socially impaired and had more restricted interests, whereas AS patients had more associated psychiatric problems. Miller and Ozonoff (127) tested individuals with HFA and AS diagnosed by *DSM-IV* criteria on a number of neuropsychological measures and were unable to differentiate the two groups once they had controlled for full-scale IQ. Specifically, no group differences were found in measures of motor, visuospatial, or executive functions after controlling for full-scale IQ. The authors indicated that both groups contained individuals displaying the characteristics believed to be unique both to AS (normal language development, history of motor delays, and VIQ > PIQ discrepancies) and to autism (echolalia, pronoun reversal, and neologisms). Thus, the authors concluded that AS is not likely distinct from HFA and view AS as "high IQ autism" with deficits likely suggestive of widespread damage involving both hemispheres (32). Indeed, neurobiologic and neuropsychological research does not make a distinction between the two groups comprising subject samples.

Discussion

No cognitive impairment has yet to be shown to be specific to autism or present in all autistic individuals. The most specific and universal finding to the PDD spectrum is that of impaired basic relatedness. These findings have generated speculation about subtypes of autism/PDD with specific cognitive, social, and associated neural deficits (128,129). It is possible that there are multiple neural subsystems involved in social behavior, that these subsystems operate as parts of a complex, integrated whole, and that disruption of any one of these elements can lead to some form of PDD. If a dysfunctional brain region is involved in another function (e.g., perception, language, verbal memory), the subset of individuals with PDD with damage to that specific brain locus also would display impairment in the cognitive functions subserved by that anatomic site. Integrative models postulating dysfunction in multiple neural systems (31,32) seem to be the most promising avenue for understanding the complex presentation of autism.

DEVELOPMENTAL COURSE AND PROGNOSIS

Autism, as a developmental disorder, cannot accurately be described from a purely cross-sectional perspective. The aloofness and greater severity of symptoms sometimes characteristic of the preschool years can change with development. Even during the early years, there often are signs of increasing social relatedness (e.g., differential attachment to caregivers) and diminishing behavioral peculiarities. During middle childhood, autistic children often master daily living skills and make some accommodation to demands made by other people, and their ritualized behaviors and idiosyncratic preoccupations may diminish (130). School behavior may come to resemble that of hyperactive and/or retarded children rather than continuing to conform to classic autistic patterns.

Adolescence can be a difficult time. Besides the frequent onset of seizures during this period, a large minority of PDD children regresses behaviorally as adolescents. Paradoxically, one of the most difficult aspects of life for adolescents with autism is their increasing interest in close personal relationships during these years (130). Higher-functioning individuals (e.g., those with AS) are especially prone to psychological problems as they realize the extent of their ineptitude in the realm of social interaction. Their feelings of helplessness may engender a depressed mood, even depressive episodes. On a more positive note, both social and language skills usually do continue to develop during adolescence, and increased interest in relating to other people can make psychosocial intervention easier.

Older long-term follow-up studies from the 1960's and 1970's, summarized by Paul (130), estimated approximately 50% of autistic adults required residential care. These estimates obviously change with altered trends in care for psychiatric and developmentally disabled populations. It is generally accepted that many adults with autism will require caregiver support of varying degrees because only a minority is observed to achieve truly independent living including gainful employment.

All studies agree that the two most important predictors of good outcome in autism are IQ and language. The presence of associated neurologic disorders is a strong negative prognostic indicator. Those with IQ in the normal or near-normal range are observed to have a reasonably good chance of becoming more independent. Language development obviously is related to IQ but apparently makes an independent contribution to the prediction of long-term outcome. In particular, the emergence of communicative speech by 5 years of age is thought to be a critical indicator. According to Paul (130), there is a subset of PDD children whose speech is limited but who go on to improve socially; these children are characterized by relatively high IQ and good receptive language skills. Follow-up studies (131) generally support these findings. It is noteworthy that both IQ and adaptive behavior measures are as stable among autistic subjects as among other children. There seems to be a relatively low correlation between intelligence and the presence of maladaptive behaviors (132). Recent evidence suggests that there may be two distinct subgroups of autism varying in level of function, with the higher-functioning group improving over time and the lower group showing stable or declining functioning. Other studies support this bifurcation of outcome (133,134).

Given that only those autistic children with relatively high IQ and some use of language can be expected to do well over the long term and given that only a subset of these children will in fact enjoy good outcomes, there has been interest in identifying outcome predictors within the high-functioning autistic population. Follow-up studies (120,135) of such subjects showed even within this group that IQ, particularly VIQ, was the best predictor of adaptive behavior and achievement in adulthood, followed by other language measures.

EPIDEMIOLOGY
Prevalence

Long-standing convention previously placed the prevalence of Kanner autism at four per 10,000 children (136). For preschoolers, recent estimates from the United Kingdom have grown to roughly 16 per 10,000 for classic autism, and 62 per 10,000 for general PDD, which represents a four-fold increase (137). This may be attributable to any combination of the following factors: (a) improved procedures for identifying cases, (b) more lenient diagnostic practice, (c) actual increases in the proportion of individuals with the disorder, or (d) changes in the conceptualization of the disorder. All of these factors may have affected one or another of the recent epidemiologic surveys, but it seems likely that the actual prevalence of PDD has been greater than previously believed all along.

Prior studies embracing a broader spectrum definition foreshadowed current data. In an epidemiologic study of mentally retarded children with deficits in social relatedness, communication, and behavior, Wing and Gould (24) found that, although the prevalence of Kanner autism was in the four per 10,000 range, the prevalence of individuals with the Wing triad of deficits was in the range of 21 per 10,000. Similar diagnostic criteria applied in a Swedish population (138) also resulted in an estimated prevalence of 21 per 10,000 population. Neither of these studies included individuals displaying the triad of deficits who were not also mentally retarded.

Gillberg's research has provided a good perspective on the changing figures for PDD prevalence because his team has conducted three epidemiologic surveys over the course of a decade (139–141) and thus are in a position to look for the sources of the increase. In their earliest study, in the mid-1980's, the prevalence of autism was 2.0 per 10,000 children (another 1.9 per 10,000 were described as PDD but not classically autistic). Four years later, these figures became, respectively, 4.7 and 2.8 per 10,000. In 1991, in studying rates on the rise, they reported the prevalence of autism to be 7.8 per 10,000, with 3.4 per 10,000 fulfilling criteria for autistic spectrum disorders. The autisticlike conditions described in these studies were symptomatically very similar to autism, but either had a relatively late age of onset or some atypical symptom such as severe, frequent seizures of early onset. The total prevalence estimate from the last study (11.2 per 10,000) found agreement with a Canadian epidemiologic survey (142) that found a rate of 10.1 per 10,000. Even with inclusion of atypical autism, these figures are dramatically lower than current estimates.

Gillberg (4) claimed that part of the increased rate of autistic conditions in the Gothenburg, Sweden, area is attributable to a large number of cases among the children of new immigrants from distant, especially tropical, countries, possibly owing to maternal health problems during pregnancy. The remainder of the increase is accounted for by cases who are either mentally retarded or of normal range intelligence. Gillberg suggested that these populations previously were unlikely to be identified as autistic. Only with increasing sensitivity to the clinical phenomenology of the disorder did diagnostic practice begin consistently to look for evidence of PDD among institutionalized mentally impaired individuals or superficially intact children. (The prevalence of classic Kanner autism, which is associated with moderate mental retardation, has not increased over the years.) There is only speculation as to what may account for the remainder of the observed increase in today's prevalence rates, much of which centers on environmental factors (discussed below).

Sex Ratio

PDD are more common among males than among females. This is a universal finding dating back to Kanner's original clinical sample. Gillberg (4) reported that the male:female ratio in broadly defined PDD is between 2.0:1 and 2.9:1. The preponderance of males is greater in narrowly defined Kanner autism, with sex ratios ranging from 2.6:1 to 5.7:1, and greater still in AS, with sex ratios ranging from 7.1:1 to 10:1 across studies. The male:female ratio appears to be lower only when individuals with severe and profound retardation are included (143).

BOUNDARY CONDITIONS AND COMORBIDITY

Comorbidity is the co-occurrence of two discrete disorders in one individual. If two disorders occur together with a

frequency greater than that predicted by chance (i.e., by multiplying the base rates of the disorders), then it may be concluded that they are somehow related. The nature of this relationship is correlational; one cannot assume that one disorder causes the other or that a single underlying factor causes both, but these hypotheses do merit further investigation.

The assessment of comorbidity is complicated when one or both of the disorders lacks clearly defined boundaries. In such cases, it is possible that the two syndromes represent two aspects of a single disease process. As a practical matter, comorbidity cannot be ascertained when hierarchical diagnostic rules are applied. For example, if epilepsy were designated as a diagnostic entity superordinate to autism, such that the presence of seizures precluded a diagnosis of PDD, then clinical research would indicate zero overlap between the two conditions. The same would apply where autism was the superordinate category, e.g., if a PDD diagnosis ruled out consideration of Tourette syndrome. Thus, any discussion of comorbidity is inextricably intertwined with the issue of boundary conditions. Moreover, prevalence will vary depending on the narrowness of the defining criteria, including the number and nature of any diagnoses regarded as superordinate.

As noted earlier, it is generally acknowledged that Kanner autism is only one of a heterogeneous group of subtly differing, overlapping clinical pictures. Kanner autism could be the core disorder, and the other forms could be variants, but there is no *a priori* reason to accept this proposition. Alternatively, it is possible that (a) a common cause may lead to any of a number of more or less clearly demarcated outcomes, perhaps depending on the specific brain regions affected by a central nervous system (CNS) insult and/or (b) the autistic spectrum may represent a final common pathway for a number of different pathologic events, such as viral infections, toxic conditions, or genetically programmed neurodevelopmental deviations. Attempts to subtype the PDD spectrum are important and need not take as their starting point the assumption that Kanner autism is the prototypical disorder. Furthermore, there is no logical border between identification of subtypes, on the one hand, and identification of comorbid conditions on the other. We must recognize that at our current state of knowledge, the outer boundaries of the PDD spectrum (as well as boundaries within the spectrum) remain uncertain.

In our view, autism/PDD is a clinical diagnosis that should be made whenever a patient's symptoms meet diagnostic criteria, regardless of any associated conditions. Autism is a syndrome, like mental retardation; it is applied to persons with a wide range of etiologies or comorbid disorders. If, for example, a patient displays autistic symptoms, moderate mental retardation, and the fragile X abnormality, all three diagnoses should be made; failure to do so would impoverish the clinical portrayal of that patient. Diagnosis of associated conditions will also contribute to knowledge concerning the relationship of the disorders.

DISORDERS WIDELY REGARDED AS BELONGING TO THE PERVASIVE DEVELOPMENTAL DISORDER SPECTRUM

Childhood Disintegrative Disorder

One variant form of autism is the disorder formerly known as infantile dementia (144), disintegrative psychosis (19,145), or Heller syndrome. The disorder is characterized by 2 or more years of essentially normal development followed by marked regression into a state seemingly indistinguishable from autism, including the loss of already acquired social abilities and linguistic competence. The validity of this diagnostic category remains controversial, and CDD is poorly understood. Early studies cited few cases of CDD (146). Prognosis for CDD has been noted as particularly poor, and follow-up has shown static impairments to include the absence of language (145,147). Volkmar and Cohen (148) identified ten cases in a group of 165 autistic children. All had reached the point of speaking in sentences before regressing. At the time of the study, the subjects were in their teens and twenties. Most were severely retarded, four remained entirely mute, and all but one were in residential placements. They had IQ lower than those of comparison groups of PDD cases and exhibited stereotypies, self-injurious behavior, and aggression.

The validity of CDD remains in doubt. Considered by many to be very rare, questions remain as to whether CDD is distinct from the autistic regression at age 2 reported by many parents of children with autism. Kurita (149) reviewed 60 known cases and concluded that the case for its nosologic validity was insufficient. Volkmar (150), conversely, reviewed 77 cases (including those reported by Kurita) and argued for the distinctiveness of the syndrome, primarily by reason of its later age of onset (mean age was 3.36 years, with wide variability).

It has been argued that Heller syndrome should not be regarded as a developmental disorder because its defining feature is behavioral regression rather than failure to develop (hence disintegrative disorder). Yet the syndrome clearly involves developmental processes gone awry, and, although termed disintegrative, it is included as a PDD in *ICD-10* and *DSM-IV*.

In summary, CDD syndrome is indistinguishable from autism on cross-sectional inspection. Onset (by definition) is later, and the prognosis appears to be uniformly poor. It is not clear that it is a discrete clinical entity. The syndrome could represent the extreme tail of the distribution of age of onset of PDD. The label may be useful as a marker of poor prognosis and deserves further investigation.

Rett Syndrome

Another autistic spectrum condition is Rett syndrome, which is unusual among developmental disorders in that it affects females nearly exclusively. Recently, a few boys have been identified as meeting the criteria for Rett syndrome. First described by Andreas Rett in 1966, the syndrome remained largely unknown to the English-speaking world until Hagberg et al. (151) described 35 cases in the *Annals of Neurology*. Diagnostic criteria (152) and developmental stages (153) were soon proposed, and interest in the syndrome has since expanded.

Like CDD, Rett syndrome is characterized by a period of apparently normal development (6–18 months) followed by profound deterioration of social and psychomotor skills accompanying a progressive cerebral atrophy. The classic symptom involves loss of purposeful use of the hands accompanied by the development of complex stereotypic hand movements.

Rett syndrome follows a clear longitudinal course with four well-defined stages (153). Stage 1, referred to as early-onset stagnation, typically lasts several months. The child shows reduced interest in playing, may begin showing odd hand-waving behaviors at times, and head growth decelerates. There may be reductions in communicative abilities and eye contact.

Stage 2, the rapid destructive stage, can take place in a few weeks or several months. Cases of especially rapid onset may be mistaken for toxic reactions or encephalitic conditions. During this phase, the classic Rett syndrome stereotypies (hand wringing, hand clapping, and hand washing) appear, and purposeful hand use is lost. Other motor abilities are better preserved, but the child will be clumsy, with ataxia and apraxia. Breathing is often irregular, and hyperventilation may occur. Cognitively, the child displays severe dementia. Seizures sometimes develop. Behaviorally, classic autistic symptomatology is present.

Stage 3, the pseudostationary period, usually lasts several years. There is some cognitive recovery, and autistic symptoms diminish markedly. With the increased emotional contact of this period, the child's presentation is better described as severely mentally retarded than as autisticlike. However, gross motor skills often deteriorate further, with gait apraxia and jerky movements of the trunk common. Seizures are present in approximately 70% of cases, and abnormal EEG are apparently universal.

The final stage, stage 4, termed late motor deterioration, begins anywhere from 5 years of age through late adolescence. Epileptic symptoms diminish, and emotional contact with caregivers improves. Growth is retarded, but puberty occurs at the normal age. There are progressive wasting away and muscle weakness combined with spasticity; severe scoliosis and trophic foot disturbances are common, and most Rett syndrome sufferers become wheelchair bound.

The estimated prevalence of Rett syndrome is approximately one per 15,000 (152) female livebirths. Because Rett syndrome is (at different stages) easily mistaken for Kanner autism, cerebral palsy, mental retardation of unknown origin, or encephalopathy, diagnosis may be difficult.

Rett syndrome has been linked to the *MeCP2* gene at Xq28 in approximately 30% of cases studied (154).

Asperger Syndrome

Another condition assumed to lie on the autistic spectrum is AS (155). Where the syndromes described by Heller and Rett are severe variants of PDD, Asperger apparently intended to describe a milder variant, with normal or near-normal IQ and relatively good language skills.

Like Rett syndrome, AS was little known in the English-speaking world for many years. Early reports (156,157) attracted little attention, and it was not until the 1980's that the syndrome began to be referred to with any frequency. An influential article by Wing (158) undoubtedly contributed to this trend. Recently, AS has become very popular as a diagnosis among parents and some professionals. Confusion regarding the proper diagnosis of this disorder has led to overuse and has served to dilute its value as a diagnostic category. The criteria themselves in *DSM-IV* make "official" Asperger Disorder very rare by design; strictly normal language development is required for diagnosis. Conservative practice would leave AS a rarely used category. More common clinical practice appears to be to accept language delay and social difficulties in making the diagnosis, which, when applied too liberally and in disregard for degree of impairment, may result in overdiagnosis. For example, in the clinical setting, it is seen that many children with disrupted peer relationships, such as those with ADHD and even early antisocial personality, have been mistakenly diagnosed with AS. As discussed above, researchers also do not agree that AS exists as an entity unique and separate from autistic disorder. In fact, much evidence supports the contrary, and most researchers advocate the view of a true continuum of symptom expression.

There have been many descriptions of AS; Attwood (159) and Klin and Volkmar (160) have provided relatively recent and well-received volumes. Diagnoses of AS often are not made until after children enter school. In early childhood, their behavior is odd but often not enough to impel the parents to seek psychiatric advice. Anecdotal evidence suggests that many children with AS often are slow to walk and talk (despite diagnostic criteria requiring the contrary). Speech improves rapidly once begun. Parents report some children as having first spoken in complete sentences. There may be language problems such as pronoun reversal, but these are more transient than in cases of AS. Imaginative play may be limited, as in autism, but stereotypies are less common. Instead, the insistence on sameness takes the form of special interests. Children diagnosed with AS usually have one or two topics (such as trains or weather) to which they are passionately devoted, largely to the exclusion of all else. Often the same general topic continues to captivate them for many years. Occasionally, these develop into useful careers.

Once children with AS enter school, their deficits soon become apparent. Single-minded pursuit of their own interests, on their own schedule, conflicts with the demands of the structured classroom. Academic performance will be variable. Persons diagnosed with AS tend to excel in subjects requiring rote memorization but fail when problem solving or higher-order conceptualization is needed. In elementary grades, this may entail high grades in reading and spelling but poor performance in arithmetic. This pattern of performance may also relate to the NLD profile often observed in this group.

Descriptions of AS highlighted not social withdrawal but awkward and odd attempts at social interaction. Often these individuals try to tell others about their special subjects, without regard for the responses of the other person. Their speech tends to be stilted and pedantic, with flat or exaggerated prosody and gesture. They have great difficulty understanding or following the rules of games and tend to be motorically clumsy. These characteristics are conducive to social isolation and peer rejection in school. More recently, descriptions of AS have included shyness as a possible characteristic.

Persons diagnosed with AS may crave and pursue social contact and thus are keenly aware of the consequences of their social deficits. Sexual frustration is a major clinical issue in adolescence and adulthood, as is their frequent failure to achieve occupational goals in keeping with their academic work. They often seek clearly stated rules of conduct to guide them in their interactions, but their clumsy, rigid enactment of these rules seldom succeeds in winning them friends. Depression is a frequent complication, and there is evidence of a genetic link between AS and affective disorders (161–163). Unless school difficulties are extreme or other psychiatric conditions develop, persons with AS may not be brought to clinical attention. Tantam (164,165) suggested that many cases of lifelong social isolation combined with eccentric behavior and interests may be AS.

Kanner and Asperger worked independently and simultaneously, each describing what he believed to be a unique disorder, yet the two syndromes share an unmistakable family resemblance, and most authors assume that the syndromes are at the very least related (166). The differential diagnosis of AS can be difficult. Some (167,168) conceive of Asperger-type cases as schizoid personalities, and there is some evidence that people with AS may be more likely than those with other PDD to develop comorbid psychotic symptoms. In addition, Nagy and Szatmari (169) describe a group of children meeting criteria for schizotypal personality disorder who also could be diagnosed as having AS. Nagy and Szatmari regard these cases as exemplifying autistic spectrum disorders.

Verbal ability and motoric clumsiness are often cited as the main features differentiating AS from autism. As discussed above, the former may be in part an artifact of the definition of AS. In fact, AS does not involve normal use of language; pragmatic interactive skills are lacking.

Several sets of diagnostic criteria for AS have been proposed by various investigators (159). One study (170) suggests that the *ICD-10* criteria are more restrictive than are those unofficial criteria, chiefly because the diagnosis is disallowed in cases in which language development is delayed. This holds true for *DSM-IV* criteria as well.

DISORDERS USUALLY NOT REGARDED AS PART OF THE PERVASIVE DEVELOPMENTAL DISORDER SPECTRUM

Nonverbal Learning Disabilities

Shea and Mesibov (171) have noted the lack of a clear boundary between HFA and severe learning disabilities, and family studies suggested that some milder cognitive and social disabilities may be part of the PDD spectrum. The classification of learning disabilities continues to be fraught with controversy. One distinction of possible value is that between linguistic or academic skills disorders, on the one hand, and disturbances of social development on the other. Discussion of dyslexia is beyond the scope of this chapter, but the less familiar social or nonverbal developmental disorders deserve mention because of their sometimes striking resemblance to the core disturbances of autism.

Based on the right (nondominant) cerebral hemisphere's apparent role in the comprehension and expression of emotion, visuospatial skills, and directed attention, other researchers have sought to identify children with specific deficits in these functions. Thus, different authors speak of developmental learning disabilities of the right hemisphere (172), right-hemisphere deficit syndrome (173), and socioemotional learning disability (174,175), and some have described similarities with AS or HFA (126,172).

Rourke (176,177; see Chapter 25) used a psychometric approach to identify a subgroup of learning disabled children with a syndrome that he labeled NLD. These children displayed a pattern of cognitive strengths and weaknesses opposite to dyslexia. Children with NLD were reported to be strong in phoneme–grapheme matching, rote verbal learning, and related psycholinguistic skills but impaired on visuospatial and nonverbal problem-solving tasks, including arithmetic. Rourke found psychomotor and tactile/perceptual deficits as well. Based on the awkward social behavior of subjects with NLD, which includes disturbances in affect comprehension, gestural communication, and prosody, Rourke (176) suggested that the NLD syndrome represents the mild end of the autistic spectrum.

It is likely that the diagnostic categories of NLD/right-hemisphere syndrome and PDD/AS have been used to describe overlapping groups of children. The differential diagnosis of AS/PDD and NLD, when desired, is usually made depending on the severity of social impairment and restricted repertoire behaviors (i.e., unusual preoccupations and sensory issues) typified by the child with PDD.

Tourette Syndrome and Obsessive-Compulsive Disorder

Tourette syndrome is characterized by the presence of both motor and vocal tics, persisting for at least 1 year, and typically following a chronic course. Often as the disorder develops, the tics become increasingly complex and apparently meaningful. Some complex tics resemble the rituals and stereotypies found in autism and PDD. For example, common complex motor tics include seemingly ritualized walking behavior and self-injurious behaviors such as self-biting; common complex vocal tics include repetition of phrases out of context. Studies have suggested that Tourette syndrome and other chronic tic disorders are strongly influenced by genetic factors and that this familial basis is shared with that for obsessive-compulsive disorder (OCD).

Tourette syndrome is not regarded as a PDD spectrum disorder. Individuals with a tic disorder alone are not impaired in social relatedness, and onset of Tourette syndrome is later than autism (7 years of age is the mean). It appears likely that there is a relationship between at least some forms of PDD and Tourette syndrome (178,179). The incidence of PDD in patients with Tourette syndrome and, likewise, of tics in PDD patients exceeds that expected by chance (180–182). Reports tend to suggest that these patients are higher functioning. Clinicians will rarely have difficulty with the differential diagnosis of PDD and Tourette syndrome but may fail to identify cases in which tics develop in individuals previously diagnosed with PDD. Identification of comorbid Tourette syndrome in some cases of PDD may be difficult. The relation or overlap of complex tics and the repetitive motor mannerisms of the autistic spectrum is unclear. Because tics, unlike autistic symptoms, frequently respond to treatment, it is important to consider a comorbid diagnosis of Tourette syndrome. On a nosologic level, the association of the two syndromes deserves further investigation and may provide useful clues to the pathogenesis of both conditions.

Because Tourette syndrome is linked with OCD, the association between PDD and Tourette syndrome suggests the likelihood of a PDD–OCD connection. If complex motor tics can be difficult to distinguish from stereotypies and complex vocal tics from delayed echolalia or stereotyped language, then the distinction between autistic behaviors and OCD symptoms is even less clear. In OCD, unlike Tourette syndrome, the behaviors are performed voluntarily, albeit under internally mediated pressure (those with Tourette syndrome also feel mounting pressure when they resist their tics). Classically, OCD symptoms are differentiated from so-called psychotic ones largely by virtue of their ego-dystonic nature: the person resists engaging in them and has insight into their senselessness. These criteria, however, are not required for children with OCD. Obviously, there will be cases in which distinguishing between an overvalued idea and preoccupation with a special interest will be a difficult and probably unreliable procedure. It should be noted that the content of obsessions in OCD may involve intellectual material, making differentiation from high-functioning PDD potentially difficult. The differential diagnosis of OCD usually relies on the preservation of social relatedness (e.g., joint attention) and, to a lesser extent, the absence of bizarre content in preoccupations. The boundary between PDD and OCD is an area of interest for future research.

Childhood Schizophrenia

The issue of a possible connection between PDD and childhood schizophrenia has a long history. At one time, the two disorders were widely believed to be related, but the bimodal age of onset and different symptomatology of the disorders (19,183,184) have clearly indicated otherwise. Kolvin (19) found delusions, hallucinations, blunted or incongruent affect, and loose associations to be more common in cases with onset after 5 years of age (i.e., childhood schizophrenia). He found gaze avoidance, odd preoccupations, disinterest in people, impoverished play, stereotypies, echolalia, and overactivity to be more common in cases with onset before the age of 3 years (i.e., infantile autism). Green et al. (185) verified these findings in a study that used *DSM-III* diagnostic criteria for the two disorders.

There are case reports of comorbidity (186–189). However, other more recent data (190) suggest a very low rate of co-occurrence (one in 163 cases reviewed). Although prodromal phases of childhood schizophrenia may be difficult to distinguish from the autistic spectrum, proper diagnosis will attend to the presence of discernible delusions and hallucinations and later deteriorating course, none of which is typically observed in PDD. It seems reasonable to conclude, barring future evidence to the contrary, that a pathogenetic link between PDD and schizophrenia is unlikely.

Developmental Language Disorder

Another problematic differential diagnosis is that between PDD and the class of developmental dysphasias (or childhood aphasias). Language disturbance is, after all, one of the cardinal features of infantile autism and has even been posited to be the core deficit (14). The existence of PDD with apparently excellent command of language (as in AS) militates against this position, except that, even in these cases, the pragmatic use of language seems to be markedly impaired. The most useful distinguishing characteristic is that children with language disorders usually possess normal interpersonal relatedness and an interest in communicating with others via gestures, facial expressions, and intonation. At the same time, some children with deficient or absent comprehension of language do exhibit social deficits.

Early studies (191) found that aberrant behavior in the past or present distinguished effectively between the groups, as did the nature of both past and current language behavior. Autistic subjects did less well on the similarities, com-

prehension, and picture arrangement subtests of the WISC and were less able to comprehend the meaning of gestural communication. These findings make it appear that differentiating between PDD and developmental dysphasia is relatively straightforward; however, approximately 20% of the sample could not clearly be classified as belonging to either diagnostic group. These subjects appeared to be autistic on some variables and dysphasic on others. A more recent longitudinal study by Rapin (85) found somewhat different patterns of early language development between autistic and language-disordered children; delay of comprehension milestones was more differentiating than delay of expressive milestones. Within the group of expressive milestones, delay in appearance of "wh" questions and connected sentences was most discriminating. Regression of language skills was also much more typical of the autistic spectrum than language-disordered children. On formal language testing, somewhat different patterns of language skill appear; degree of deficit in comprehension is the single most differentiating feature of language, with the autistic children showing worse comprehension than that of the language-disordered children. The autistic children were also much less able to use meaning to assist in recall of verbal material. Although the language-disordered children were virtually normal in their pretend play, the high-functioning autistic children seemed able to engage in this kind of play but unwilling or unable to sustain it. Other clear differences in behavior, particularly social skills, also emerged.

Seizure Disorders

The link between autism and seizures dates back to Kanner's first sample of 18 children, two of whom went on to develop epileptic conditions. Estimates of the prevalence of seizures in autism traditionally have fallen in the range of 25% to 30% of autistic individuals (131,193–195). If autistic patients without seizures, but with abnormal EEG, are added, then approximately half of all persons with autism exhibit epileptoid conditions. A bimodal distribution of age of seizure onset is frequently observed; infancy and adolescence are both high-risk periods. Seizures occur more often in individuals with lower IQ.

Tuchman et al. (193) cited as a major risk factor for epilepsy the presence of severe mental deficiency and motor deficit. Perinatal maternal disorder, a difficult perinatal course, and family history of epilepsy were not risk factors. The most frequent types of seizures were generalized tonic-clonic and atypical absence, followed by myoclonic and partial seizures, followed by atonic seizures and infantile spasms.

Two other studies, one clinic based (194) and one population based (195), provided information about seizures in autism. Volkmar and Nelson found that 21% of the 194 autistic patients evaluated had seizure histories, with onset occurring in either early childhood or adolescence. Because many of the subjects had not yet passed the peak risk years (mean age, 14.1), the adjusted lifetime morbidity estimate was 29%. Those with full-scale IQ below 50 were particularly likely to develop seizures (84.1%), and females were slightly more likely than males to have seizures. Olsson et al. (195) found that 20% of prepubertal autistic children and 41% of prepubertal PDD children were experiencing seizures. Although autistic symptoms preceded seizures in all the infantile autism cases, onset of seizure activity actually preceded the development of autistic symptoms in several autisticlike PDD cases. This supports the hypothesis, consistent with clinical observations, that early-onset seizures can play a causal role in the development of autisticlike behavior. Landau-Kleffner syndrome, marked by language regression and specific abnormalities on EEG, can involve the development of autistic symptoms (196).

ETIOLOGIES OF AUTISM

There may be nearly as many different theories of the causes of autism as there are variations in expression of the disorder itself. Research into the etiology of autism has recently been reviewed by Tanguay (197), Buitelaar and Willemsen-Swinkels (198), and Trottier et al. (199), among others. The observed variability across the spectrum has contributed to the growing idea that there in fact may be multiple etiologies, the starting point for which is presumed by most to be genetic.

Genetics

Genetic Syndromes

Specific syndromes of known genetic origin have been noted to involve autistic symptoms. The most frequently cited genetic syndromes involving autistic symptoms are fragile X (in which 15% to 30% of patients have autistic traits), Rett syndrome, tuberous sclerosis (in which 40% of patients show autistic traits), neurofibromatosis, and phenylketonuria (see also discussion of medical factors below). The most common and well-known genetic syndrome linked to autism is the fragile X syndrome. The result of a mutation in the region near the *FMR-1* gene at Xq27, it is noted for the presence of mental retardation and autistic symptoms in addition to physical stigmata, all of which vary depending on the degree of expression of the mutation. The fra(x) PDD pattern is fairly distinctive (200). Compared with other autistic patients, fra(x) patients are less likely to show echolalia but more likely to display perseverative speech. Impairments of articulation, prosody, and general expressive language are usually present. Hand flapping, hand biting, and hyperactivity are more common in fra(x) than in other forms of PDD. The most interesting distinguishing characteristic of fra(x) autism is that fra(x) cases generally are not aloof; rather, they are interested in social interactions, but experience intense social anxiety. Cohen et al. (201) studied gaze patterns in fra(x) males and non-fra(x) autistic males and found that, although the non-fra(x) subjects tended to ignore parent-initiated eye contact, the fra(x) subjects were highly sensitive to eye contact but found it aversive. Nonautistic fra(x) males also display gaze avoidance while making a social approach (202). It has been estimated that 15% to 30% of fragile X patients show autistic symptoms (196,200). Conversely, from a diagnostic perspective, an estimated 2% to 10% of cases of autism may be associated with the fragile X mutation (200,203–205); usage of molecular techniques yields low prevalence rates. (See also the chapter on mental retardation by Joy et al. in this volume.)

Proximal to the fra(x) region, the *MeCP2* gene at Xq28 has recently been linked to Rett syndrome. As discussed above, the incidence of Rett syndrome is considered to be quite rare and perhaps unrelated to other types of autistic spectrum disorder. From the diagnostician's perspective, it is important to note that only approximately 30% of Rett syndrome cases show this particular mutation, leaving the majority of cases as yet unaccounted for. Another genetic condition, albeit a very rare one, merits a brief mention: Joubert syndrome, an autosomal recessive trait whose major feature is agenesis of the cerebellar vermis. Mental retardation, abnormal eye movements, poor muscle tone, ataxia, and tongue protrusion are typical. Holroyd et al. (206) reported two cases of Joubert syndrome, both of which demonstrated autistic traits. As discussed by Trottier et al. (199), a genetic syndrome's association with autism may reflect direct causation via CNS disruption or, alternately, proximity of the genes separately causing each disorder. This minority of so-called syndromic cases of autism has been distinguished from idiopathic cases, many of which are presumed also to have genetic bases yet undiscovered.

Heritability and Concordance

Twin and family studies have clearly demonstrated the heritability of autistic traits. Authors have cited concordance rates ranging from 36% for classic autism in monozygotic twins in older studies to 92% to 96% for what has been termed the broader phenotype of PDD (207) versus only 0% to 24% for dizygotic twins (208–212). Currently, the accepted estimate is commonly reported as approximately 90%. Siblings are commonly estimated as having a 2% to 3% concordance rate (208). Past studies reported sibling risk at 3% (213) to 4.5% (214). The recurrence risk of autism within families has been estimated between 3% and 8%, which is higher than the sibling concordance rate (208). According to Ritvo et al. (215), the recurrence risk was as high as 14.5% after the birth of a female proband. Interestingly, the recurrence rate of single autistic traits between siblings is as high as 50% (216).

Not only is the occurrence of autistic traits elevated in first-degree relatives (216–222), particularly siblings

(217,223), but families of autistic children are also reported to show higher rates of related impairments such as learning disability, language-based deficits, and mental retardation when compared with both base rates and study controls (212,213,216,224,225). One study (226) did not replicate this finding but reported deficits in abstract reasoning in relatives of probands. Bolton et al. (217) reported that in autistic probands with speech, the number of symptoms was significantly correlated with the number of relatives expressing PDD symptoms. Studies specifically indicate the presence of social and communicative difficulties, rigid behaviors, and executive functioning deficits in parents (227–229) and adult siblings of autistic subjects (213). Bailey et al. (222) reported that fathers show increased PDD traits compared with mothers.

Specific psychiatric disorders are associated with familial autism, which could reflect related etiologies. Associations between autism and affective illness have been reported, both in probands and relatives. Studies (213) reported approximately 16% of adult siblings of autistic individuals had been treated for a mood disorder, and autistic individuals themselves have received treatment for mood disorders, including bipolar disorder (230). A review of autism cases with comorbid affective illness found a 50% rate of family history for affective disorder, particularly for lower functioning subjects (231). AS has also been associated with affective illness (161), including bipolar disorder (162), and both AS and autism have been associated with schizoaffective disorder (232). An increase in social phobia in first-degree relatives is also discussed in the literature.

Heritability patterns for autistic traits in family studies support the concept of the autistic spectrum but have also led researchers to suspect the existence of multiple "susceptibility" genes, given, for example, the discordance rates of approximately 10% in monozygotic twins. Arguments for polygenic or multifactorial inheritance are based on (a) the very high (50%) family recurrence rate for single autistic traits, relative to the low (3% to 7%) recurrence of autistic disorder, suggesting that individual traits are heritable, (b) the "trough" between concordance rates for monozygotic twins versus dizygotic twins or sibs, (c) the failure to observe rates consistent with single dominant inheritance (50% chance) or single recessive inheritance (25% chance), (d) the finding that family members of autistic individuals can carry a common genetic feature without expression of the phenotype, and (e) the variety in genetic syndromes and defects linked to autism. Bolton et al. (217) have proposed that there exist at least two polygenic sources of autism. The search for candidate genes has exploded within the past 5 years.

Chromosomal Abnormalities

Maestrini et al. (208), Trottier et al. (199), Cook (233) and Gillberg (234) provide excellent reviews of genetic studies. Linkage studies seek to establish the occurrence and loca-

tion of genetic markers in families with autism. Findings from recent molecular genetics studies agree with previous genetic models in suggesting multiple loci (208); Pickles et al. (235) previously suggested that three interacting loci best fit the existing familial data. Genome screens have implicated different regions on several chromosomes as containing potential susceptibility genes. Maestrini et al. (208) reports that several groups have now all implicated a large region on chromosome 7q, potential linkage to which was first identified by the International Molecular Genetic Study of Autism Consortium (236,237–241).

The *HOXA1* gene located on chromosome 7 has been examined as a potential candidate gene for autism. Studies of the *HOXA1* gene in autism were inspired by Rodier's (242) observations of similarities between some autistic features (i.e., brainstem and cranial nerve abnormalities, ear anomalies) and HOXA1 knockout mice. *HOXA1* is thought to modulate other genes and may be active in brainstem development in the early prenatal period in mice. Allelic variants of HOXA1, indicating altered DNA sequences of the *HOXA1* gene, were found to be significantly increased in autistic subjects relative to controls and family members (243,244).

Although interesting in the context of previous findings, Rodier is careful to point out that the variant allele of interest shows variable expression and that it was both present in some nonautistic subjects and absent in many autistic subjects as well. She concludes that *HOXA1* may represent one susceptibility gene and that some allelic variants in fact may function to limit phenotypic expression, which would account for the variable expression of the autistic spectrum.

Region 15q11-q13 also has generated a great deal of interest. Genes in this region are associated with Prader-Willi and Angelman syndromes, both of which can exhibit autistic traits. Consistent with many previous case studies reporting abnormalities to chromosome 15 in cases of autism, including partial trisomy (245), partial tetrasomy (246,247), and duplication (248,249), this area also has shown evidence of linkage in genome screens. This region is of interest additionally because it is also the site of the gene for polypeptide 7B2, which modulates the enzyme (PC2) that may synthesize oxytocin, a behaviorally active peptide (see discussion below) (250). As the *PC2* gene itself is proximal to the *OT* gene (20p13), defects on chromosome 20 as well as chromosome 15 were hypothesized to contribute to autism (250).

Several other candidate genes have been studied, as reviewed by Trottier et al. (199). Frequencies of markers for the c-Harvey-*ras* gene, an important neurotrophic factor on chromosome 11, were elevated in autistic subjects (251). An association with autism was also found for one marker of the homeobox gene *EN2*, on chromosome 2, which is involved in cerebellar development (252).

Studies looking at serotonin (5-HT) transporter gene promoters are inconsistent in their findings regarding the presence and type of allelic variants associated with autism

(253–255). No association was found between autism and other genes related to the catecholamines, including the $5HT_{2A}$ receptor gene and the genes for tyrosine hydroxylase, dopamine (DA) β-hydroxylase, and tryptophan hydroxylase (256,257). Several studies of the gene responsible for fragile X have found no involvement of the *FMR* gene in samples of non–fragile X autistic subjects and their families (258–261).

One of the many fundamental questions remaining in genetics research is whether the autistic spectrum reflects heterogeneous genotypes that converge on a similar phenotype or a complex, multiloci genotype that is modulated by the environment and/or allelic variants. Trottier et al. (199) point out that family recurrence rates, which largely indicate polygenic inheritance, are not always consistent with findings from studies of mechanisms of inheritance, such as segregation analyses suggesting autosomal recessive transmission (262), reflecting that this question has not been resolved. As in other areas of PDD research, the issues of subtyping and establishing the boundaries of the PDD construct complicate efforts to model familial patterns. If some as yet undifferentiated subtypes of autism represent discrete disorders, with their own modes of transmission, then the mixture of different disorders within autism will confound investigations of causality. London and Etzel (244) refer to yet another source of complexity in genetics research in their discussion of the replication versus transcription functions of genes, the latter of which may be subject to modulation by other nongenetic factors.

Although no single known genetic defect is universal to PDD, several medical findings are associated with autism and may relate to the development of autistic impairments.

Medical Factors

Many medical conditions, including the genetic disorders discussed above, have been associated with the development of autistic symptoms. Gillberg and Coleman (263) report autism's association with marker chromosome syndrome; neurocutaneous disorders (tuberous sclerosis, neurofibromatosis, hypomelanosis of Ito); phenylketonuria; Goldenhar, Möbius, Lange, Landau-Kleffner, Lujan-Fryns, Sotos, Tourette, and Williams syndromes; Duchenne muscular dystrophy; hypothyroidism; the metabolic disorders of lactic acidosis, mucopolysaccharidoses, and lactic acidosis; free fatty acid abnormalities; and the viral illnesses rubella embryopathy, herpes encephalitis, cytomegalovirus infection, human immunodeficiency syndrome; and *Treponema pallidum*. Steffenburg and Gillberg's groups (263,264), who conducted a thorough neurobiologic workup on their sample, found that 38% had specific medical conditions. Gillberg (263) posits that if only the more standard diagnostic tests had been performed, the prevalence would have been only 18%.

Various attempts at finding other medical markers of autism have been made, ranging from plasma amino acids to γ-aminobutyric acid, with no universal or well-replicated

findings. Elevated neonatal plasma levels of neurotrophins may serve as a marker of later development of autism; however, elevations are also seen in babies who later develop mental retardation as well (265).

Of the viral processes linked to cases of autism, congenital rubella shows the strongest association, with a high prevalence of autism and autisticlike features. Older studies cited prevalence rates greater than 100 times that in the population at large (266,267). Follow-up at 8 to 9 years of age (267) showed six had made a full recovery, but four other children had begun to exhibit autistic features in the interim. Investigators suggested that the high recovery rate indicated that the autisticlike illness was part of a chronic infection running its course.

It has been hypothesized that pre- or postnatal viral infection may also result in diseases of autoimmunity by increasing brain antigens, ultimately causing disruption of CNS development (268). Some evidence of abnormal autoimmune reactivity is present in autism; a host of immune abnormalities in autistic individuals is reviewed by Comi et al. (269) and Trottier et al. (199). There are reports of partial C4bp deficiency, abnormal natural killer, B, and T cell activation, decreases in helper-induced lymphocytes, and increases in interferon-α as well as various antibodies (including antibodies to myelin basic protein, integral to CNS development) (199,269). Families of autistic individuals show a significantly elevated incidence of autoimmune disorders (such as rheumatoid arthritis, lupus, connective tissue disorders) compared with controls (269). Autistic individuals themselves are reported to be eight times more likely to have been born to a mother with an autoimmune disorder than nonautistic individuals. These findings may have been subject to bias in terms of retrospective report and possibly to selection bias but are nonetheless consistent with clinical anecdotes. Comi et al. (269) provide several hypotheses for the mechanisms by which autoimmune dysregulation may be involved in autism, including genetic defects or neurotransmitter effects on the immune system, which could result in the pathologic action of cytokines in the brain. They suggest that a genetic disposition for autoimmunity, expressed by the carrier mother as an autoimmune disorder, may be triggered by a critically timed medical or environmental factor to express the phenotype of autism in the affected child. Treatments of intravenous immunoglobulin, steroids, and adrenocorticotropic hormone have been discussed, with steroids showing short-lived success.

Relatedly, autistic subjects have been shown to display increased eosinophil and basophil responses to IgE-mediated reactions, which researchers claim can account for the increased incidence of allergies in autism (270,271). In contrast, Comi et al. (269) report a reduced incidence of allergy in autistic subjects relative to controls. Food allergies in autistic individuals have been observed by way of increased antibody reactivity to foods, and some suggest that perhaps the toxicity of food allergens promotes CNS disturbance.

Special diets, such as the casein- and gluten-free diet, have been advocated by some groups to eliminate the suspected allergic response.

It is currently estimated that 25% of cases of autism are caused by known diseases or genetic syndromes, leaving many cases as yet unexplained. The concept of susceptibility genes relies on the presence of environmental risk factors for expression of the broad autism phenotype and finds a growing opinion that a critically timed early CNS insult is involved in many cases.

Environmental Factors

Teratogens and Other Developmental Risk Factors

The increased presence of minor congenital physical anomalies in what are described as nonfamilial, nongenetic cases of autism was suggestive of an insult sustained early in gestation (212,272–276). London and Etzel (244) recently reviewed environmental factors in the development of autism.

The types of anomalies observed in autism were noted by some researchers to be reminiscent of those seen after exposure to specific teratogens. For example, thalidomide and valproic acid exposures both produce malformations to the ears and cranial nerves originating in the brainstem in animals; exposures in humans early in gestation have been associated with cases of autism (244). The known timing of the development of these structures, between 20 and 24 weeks, led Rodier (242) to propose that autism results from disruptions to neural tube closure and brainstem development. Similarly, retinoid exposure results in birth defects in both humans and animals; in animals, features include cerebellar and cranial nerve abnormalities, which have been paralleled with findings in autism (244). Prenatal alcohol exposure and the development of fetal alcohol syndrome has been recently cited as resulting in some cases in social deficits severe enough to earn autism diagnoses (244). Researchers have concluded variously that in cases of autism, the timing of an environmental insult is probably between 16 to 24 weeks of gestation.

Largely outside the mainstream of medical and research communities, parent and advocacy groups have raised consciousness regarding the issue of environmental toxins. Brick Township in New Jersey, for example, called for an investigation of the exceptionally high rates of autism found in their community. Although consistent covariation could not be demonstrated, increased levels of T-chloroethylenes and trihalomethanes were found in the water supply. This is potentially important, as London and Etzel (244) point out, because trihalomethanes are noted in the literature for causing neural tube defects. Environmental toxins have also been blamed for precipitating autoimmune disease by way of their entry to the brain. Preliminary evidence is provided for increased levels of toxins (such as trimethylbenzenes) and abnormal liver detoxification in autistic individuals; despite methodologic limitations, these studies are of interest in their attempt to cover a meagerly studied area (277). Further research in developmental neurotoxicity is critically needed.

Fears of possible neurotoxic effects and autoimmune reactions from vaccinations became a focus in the late 1990's, with the measles, mumps, and rubella (MMR) vaccine in particular the subject of numerous anecdotal reports of autistic regression at ages 1 to 2 years. There came observations of patients having supposed the onset of autistic regression with symptoms of colitis (termed autistic enterocolitis) after their MMR immunization, and reports of measles virus in the intestine of inflammatory bowel cases (278). Wakefield et al. (278) conjecture that MMR components, specifically measles, may produce inflammatory bowel disease, with the effect of increased permeability to gut peptides, resulting in CNS toxicity. Kawashima et al. (279) did find evidence of vaccine strains of measles virus (as opposed to wild-strain virus) in some patients with autistic enterocolitis but also in some patients with ulcerative colitis. Countering this link, a report in the *Bulletin of the World Health Organization* concluded in a review that a majority of studies could not confirm evidence of measles vaccine or virus in inflammatory bowel cases (280). Further, a recent review by the Institute of Medicine concluded that there is no link between the MMR vaccine and autism, pointing to the lack of correlation between rates of MMR usage and rates of autism. These trends were observed in both California and the United Kingdom. In addition, autistic individuals were reported in one study to show no differences from others in rates of traditional negative reactions to immunizations (269). The interaction with susceptibility genes in a subset of vulnerable susceptibility genes in a subset of vulnerable children have not been investigated.

Obstetric complications (281–285) are documented in association with autism, but researchers raise questions regarding the direction of causality. Regarding the baby and the delivery itself, increased incidences of breech delivery, low birth weight, low Apgar scores, delayed crying, elevated bilirubin, anemia, respiratory distress, emergency cesarean sections, clinical dysmaturity, and decreases in aggregated optimality measures are observed in autistic samples relative to controls (281–284). In the mother, higher incidences of midtrimester bleeding, maternal infection, and edema have been reported. A measure of aggregated optimality was a powerful discriminator between autistic and nonautistic groups both prenatally and neonatally but in at least one study was not universally depressed within the autistic group (282,284). One study showed that only those autistic children with reduced optimality in the prenatal period showed reduced optimality perinatally or as neonates, thus peri- and neonatal problems may be signals of nonoptimal pregnancy rather than independent risk factors (284). In addition, not all studies indicate differences between autistic samples and controls (286,287). A recent study found no differences in some prenatal complications (such as maternal hypertension), but trends in increases in other factors such as maternal infection, neonatal hypoxia, and prematurity (269). Although maternal infection is widely hypothesized elsewhere to present a risk of prenatal insult, the latter two factors are arguably secondary to what could be described as fetal vulnerability. In their review, Buitelaar and Willemsen-Swinkels (198) present findings to support their assertion that the observed obstetric complications are by-products of preexisting genetic susceptibility and not themselves causal factors.

The linkage of autism to multiple genetic loci as well as to heterogeneous medical conditions and environmental factors highlights the absence of a universal cause and lends support to the concept that autism may represent a "final common pathway" for heterogeneous etiologies. The ultimate effect of any or all of these factors is presumed to depend on their disruption of specific components of the vast anatomy and physiology of the brain.

STRUCTURAL AND FUNCTIONAL NEUROANATOMIC FINDINGS

Studies of structural brain differences in autism have relied on postmortem neuropathology examinations and imaging techniques such as computed tomography and magnetic resonance imaging (MRI). Differences in the function of specific brain regions have been studied with functional MRI (fMRI), magnetic resonance spectroscopy (MRS), single photon emission computed tomography (SPECT), and positron emission tomography (PET). Eliez and Reiss (288), Rumsey and Ernst (289), and Deb and Thompson (290) have reviewed the recent imaging research. It is important to note that imaging studies have varied greatly in terms of the methodologies and subjects employed, which renders comparison and generalization of results difficult. Despite inconsistencies across studies and apparent variability within the autistic population, some trends have emerged that highlight characteristics of the brain in autism.

A structural MRI study of monozygous twins discordant for classic autism revealed volumetric decreases in the amygdala, caudate, hippocampus, and cerebellar vermal lobules VI and VII in the autistic proband, in contrast to decreased superior temporal gyrus and frontal lobe volumes in the milder PDD twin. Although this study was limited by its lack of statistical power as well as the twins' prematurity, the differences observed are interesting as reflections of the possible neuroanatomic distinctions between the broader versus the classic phenotype of autism (291). The areas implicated in the case of classic autism parallel those broadly found in the literature (reviewed below).

Hindbrain

Studies of the cerebellum generated much interest and data in the 1980's and 1990's. Cerebellar Purkinje and granule cell loss, ectopias and immature cellular organization on postmortem examination (292–296), reduced cerebellar size on MRI (297,298), and hypoplasia of neocerebellar vermal lobules VI and VII were all reported in samples of autistic individuals (298,299). Hypoplasia of vermal lobules VI and VII was proposed to result from a circumscribed loss of Purkinje and granule cells during prenatal or early postnatal neural genesis and migration (294). In addition, case studies of Joubert syndrome, a genetic disorder marked by cerebellar hypoplasia, showed striking similarities to autism (300). Courchesne et al. (301,302) previously hypothesized that neocerebellar damage caused impairment in shifting of attention and to higher cognition and later reported findings of two subgroups with vermal hypo- and hyperplasia among autistic patients. Other studies failed to replicate the findings of Courchesne et al. when controlling for potential confounding factors such as age and IQ (303–306). In fact, Abell et al. (307) reported increased cerebellar volume on MRI.

Functional imaging studies have revived some support for involvement of the cerebellum in autism, although findings have not been unique or specific to that region alone. Chugani et al. (308) and Otsuka et al. (309) both found disruptions in neuronal metabolism in the cerebellum using MRS, which was hypothesized to relate to reductions in Purkinje and granule cells. Muller et al. (310) reported abnormally decreased cerebellar activation during nonverbal auditory perception tasks using ^{15}O PET. In addition, in an uncontrolled SPECT regional cerebral blood flow (rCBF) study, cerebellar hypoperfusion was found in nearly all of 23 young autistic subjects studied, despite an absence of gross structural findings on MRI (311). Although findings regarding the cerebellum have continued to surface, the lack of specificity in findings renders interpretation of their significance difficult.

The focus of some of the earliest theories of autism, the brainstem (12) has drawn renewed interest with the recent work of Rodier (242). Past MRI studies suggested diminished size of the brainstem and possible fourth ventricle abnormalities (297,312), and postmortem studies revealed ectopias, decreased neurons, and age-related changes in the inferior olive, the latter of which are indicative of the presence of an insult occurring at less than 30 weeks of gestation (294,296). A more recent single postmortem case revealed a "shortening" of the brainstem, involving significant diminishment of the facial nucleus and superior olive; these are structures that are formed during gestational days 20 through 24 during neural tube development (242). This finding is relevant to observations of aberrant facial expression, auditory responsiveness, and eye movement in autistic children and reminiscent of features of the HOXA1 knockout mouse, which held hopes of being an animal model of autism.

Cerebral Cortex

The most common structural finding is that of increased brain volume or megalencephaly, which is believed to be specific to autism (313). This observation has been attributed to the phenomenon of cortical thickening, particularly of the parietal and temporal lobes (313). The nature of abnormalities of the cerebral cortex itself is not consistent across autistic subjects (292,293,314–316). Citing diverse methodologies, various malformations have been reported (296), including smaller neurons and higher cell-packing density (294), cerebral lipidosis (317), microgyria and schizencephaly suggestive of migrational defects (318), enlargement of parietal and occipital regions (313,319), and slightly lower counts of primary auditory cortex glia in the left hemisphere and of auditory association cortex pyramidal neurons in the right hemisphere (315). Neuropathology studies concluded the absence of gliosis or signs of previous inflammation or anoxia (320).

Frontal Lobe

Imaging studies have pointed to differences in specific cerebral regions. Findings for frontal cortex typically emerge in the context of other regional abnormalities and may not be specific to autism. Interest in the frontal lobe arises in part from findings of executive function deficits and their association with theory of mind in autism. Recent neuropsychological studies by Dawson and colleagues (321) also have indirectly implicated ventromedial prefrontal cortex in the development of joint attention in autism.

Structural imaging and neuropathology studies have found no consistent frontal lobe abnormalities; however, functional imaging studies tend to suggest reduced frontal activation. Preschool-age autistic children show reduced frontal rCBF that is found to resolve by the elementary school years, possibly indicating delayed development of the frontal cortex (322). Although consistent with executive function deficits in autism, this finding is not specific to autism (323). MRS studies have found disruption of prefrontal metabolism (324) that correlates with the degree of autistic impairment (see below). Frontal DA processing, thought to be involved in working memory, may be decreased in autistic subjects as well (325). Functional imaging studies have implicated the prefrontal cortex specifically in theory of mind. Happe et al. (326) found that the prefrontal cortex is preferentially activated in normal subjects during theory of memory tasks, relative to logical inference or comprehension tasks. A PET study of rCBF activation suggested that both theory of mind and logical inference tasks ordinarily have in common the activation of the posterior cingulate and left superior temporal gyrus, whereas theory of mind uniquely activates the left medial frontal gyrus (326). Although high-functioning AS subjects showed the same shared activation patterns common to both theory of memory and inference tasks, they

failed to show the normal pattern of prefrontal cortical activation unique to theory of mind (326). Rumsey and Ernst (289) note that this study was limited by relatively restricted imaging resolution and the resulting uncertainty in regional specificity. Findings were consistent, however, with several previous studies showing left medial prefrontal activation during theory of mind tasks in normal subjects (for a review, see reference 327).

Temporal Lobe and Limbic Structures

Empirically supported as the seat of what has been called the "social brain" (328), limbic structures, in particular the amygdala, have been the focus of many recent structural and functional studies because damage to them can produce autistic symptomatology (329). Waterhouse et al. (32) and Fein and Waterhouse (330) have suggested that amygdala dysfunction produces what may be autism's primary impairment: disrupted assignment of affective significance to social and novel stimuli. Other ideas implicating the amygdala have been suggested by Bachevalier (329), Brothers (328), and Fotheringham (331), and, as of late, by Baron-Cohen et al. (332).

Case studies of both tumors and epileptogenic lesions of the temporal lobe, particularly involving the amygdala and hippocampus, support this region's involvement in the development of autistic symptoms (333,334). In fact, the occurrence of autism in cases of tuberous sclerosis appears to correspond with the presence of hamartomas specifically in the temporal lobes (335). Early imaging studies employing pneumoencephalography found enlargement of the temporal horn in autistic subjects, implicating disruption of the hippocampus (336). More recent MRI studies of the amygdala find increased volumes in high-functioning autistic subjects, with slight possible decreases in the hippocampus and parahippocampal gyri (337–339). It has been noted that enlargement of the amygdala may not be specific to autism because it is found in bipolar patients as well. Further, other recent MRI studies actually found decreases in the amygdala (337) and no differences in size of the hippocampus compared with controls (340,341), although the possibility remains that these discrepancies may be accounted for by methodologic differences. An MRI study of segments of the anterior cingulate in a small group of autistic adults found reduced volume in a segment corresponding to Brodmann area (BA) 24 and increases in BA 25 (342).

Postmortem investigation by Bauman and Kemper (292) found that selected nuclei of the amygdala showed abnormally increased cell-packing density and diminished neuron size and revealed these as well as other abnormalities in the hippocampus (341). Another neuropathology study did not replicate findings for the hippocampus (296). The studies of Bauman and Kemper's (292,294) and Raymond et al. (341) also have revealed limbic pathology, including abnormalities of the entorhinal cortex, anterior cingulate gyrus, septal nuclei, and mamillary body, all of which appeared to have occurred before 30 weeks of gestation. They summarize this set of findings as indicating stunted neural development in the forebrain limbic system.

Functional imaging studies have generally yielded findings of temporal abnormalities in autistic subjects and support the role of these abnormalities in the aberrant processing of sociocognitive and affective stimuli. These studies typically involve higher-functioning autistic spectrum individuals. Reduced limbic activation has been observed in several studies involving autistic subjects. An eye expression attribution task elicited activation of the frontotemporal cortex, left amygdala, hippocampus, bilateral insulae, and left striatum on fMRI in normal subjects (332). Proposed components of the social brain (the amygdala, orbitofrontal cortex, and superior temporal gyrus) failed to show normal activation in autistic subjects during this task (332). Abnormally increased activation of the superior temporal gyrus was observed in autistic subjects, consistent with previous studies associating this area with simple face perception. This study was confounded by the presence of other neuropsychological functions required by the task, particularly word recognition, which could activate areas unrelated to the task of mental state inference itself.

Facial discrimination, a known deficit for autistic individuals, can induce selective activation of the fusiform gyrus in normal subjects, as opposed to portions of the inferior temporal, parahippocampal, and occipital gyri, which are normally activated during common object perception. An fMRI facial discrimination task with high-functioning autistic and AS subjects yielded reduced activation of the right fusiform gyrus and abnormally increased activation of the inferior temporal gyrus (343). This pattern was consistent with the pattern of activation observed for normal subjects during an object perception trial. Investigators suggested that autistic individuals process faces abnormally using fragmented feature-based processing, perhaps as a result of disruptions to the amygdala, which would have downstream effects on its afferent the fusiform gyrus.

It should be noted that the inferior temporal lobe, especially the amygdala, has been cited as exceptionally difficult with regard to deriving a signal, which introduces the potential for distortion and error in measurement.

Other Functional Imaging Findings: Neuropsychology

Imaging studies have also yielded sets of findings specific to the various neuropsychological functions studied. Basic attention tasks show disrupted hemispheric lateralization in glucose metabolic rates in autistic subjects (344,345). Listening tasks, including receptive language tasks, elicit reversal of normal hemispheric and regional lateralization patterns in rCBF (346) and in ^{15}O PET (310,347) in autistic subjects. Reduced cerebellar activation during nonverbal

auditory tasks and reduced activation of thalamic, right dentate, and Broca areas during receptive and expressive tasks were also reported in the latter studies, possibly corresponding to the dentatothalamocortical circuit as proposed in Chugani's 5-HT studies (see below). Glucose hypometabolism was found in the anterior cingulate gyrus in a small PET activation study involving a verbal memory task; rates failed to correlate with observed MRI abnormalities (342).

Embedded figure tasks have elicited fMRI activation of the middle and inferior temporal gyrus, supramarginal gyrus, inferior frontal gyrus, and midoccipital gyrus in control subjects (348). Autistic individuals long have been noted for their exceptional ability in this type of task, which has been attributed to "weak central coherence" or difficulty with the perceptual ability of gestalt closure (101). During an experimental embedded figure task, a small group of autistic subjects showed reduced parietal and right dorsolateral prefrontal cortical activation and disproportionate activation of the right ventral occipitotemporal regions.

Evidence from Cerebral Metabolic and Blood Flow Studies

Findings of studies of cerebral activity or metabolism are more heterogeneous. Although some metabolic (PET glucose) and CBF (SPECT) studies with autistic subjects have specifically implicated temporal areas (327), the observed patterns of hypometabolism and hypoperfusion across studies are somewhat diffuse and variously observed for the whole brain, basal ganglia, cingulate, insular, frontal, parietal, occipital, cerebellar, and thalamic regions (289). This stands in contrast to older studies indicating increased rates of glucose metabolism in autistic subjects (349–352), similar to those found for patients with Down syndrome (353). Findings of interest from the SPECT literature include reversed perfusion lateralization patterns (346,354) and delayed maturation in frontal rCBF patterns in autistic children (322), although the latter finding is not specific to autism (323). Autistic children with comorbid mental retardation also have shown weaker correspondence in metabolic rates between regions compared with controls (355). Changes in metabolism and rCBF in typical children are thought to occur as a function of rapid synaptic proliferation and later pruning over the course of development, and abnormal findings in autistic samples are usually interpreted as evidence of developmental inefficiencies in energy utilization and neuronal activity (327). However, this body of findings is nonspecific and variable. Although a few studies have correlated metabolic and structural findings (e.g., PET glucose hypometabolism and structural lesions) (356), regional structural data do not always corroborate functional findings and vice versa (311). Further, metabolic abnormalities are observed more frequently in cases of autism associated with medical conditions (e.g., epilepsy) compared with idiopathic autism.

MRS metabolic studies report reductions in metabolic indices in several selected regions of study, namely, in phosphocreatine in the prefrontal cortex (324) and in *N*-acetyl-D-aspartate in the cerebellum (308,309). These indices are thought to reflect disruptions in metabolic processes such as energy utilization, membrane synthesis, and neuronal viability. More specifically, reduced high-energy phosphate in the prefrontal cortex is explained as a marker of a hypermetabolic state resulting from primary adenosine 5′-triphosphate depletion; its involvement in autism is tentatively supported by its correlation with autistic impairments (324). Abnormalities in cerebellar metabolism have been suggested to relate to reductions in Purkinje and granule cells, suggested by previous research (308). No abnormalities in the medial temporal lobe have been found. Likewise, signal intensities were normal in the corpus callosum (357), and parietal ratios of metabolic factors were no different from controls (358). The more general significance of MRS findings is unclear. MRS metabolic studies are particularly limited by the scope and variability in methodologies employed.

In light of inconsistencies in neuroanatomic data, investigators have argued for differentiation of gray versus white matter in volumetric studies, the results of which have suggested reductions in whole or regional corpus callosum volume (355,359,360). A recent gray matter volume study, employing the reputedly less biased whole-brain method of voxel-based morphometry, found decreased volumes in the right paracingulate sulcus, left occipitotemporal cortex, and left inferior frontal sulcus and increases in the left middle and right inferior temporal gyri, all of which are said to form a circuit with the amygdala (307).

Discussion

Aside from considerable support for involvement of the limbic system in autism, the significance of some sets of neuroanatomic findings is not yet clear given discrepancies across studies. Occasionally, correlations will be reported between structural or functional findings and behavioral data, for example, between thalamic perfusion rates and repetitive behaviors (355) and between autistic symptoms and amygdala abnormality (338). However, these demonstrations of external validity are the exceptions to an otherwise varied body of findings. Replications will benefit from attempts to standardize methodologies, including imaging technology, and design characteristics, such as the use of sedation, reduction of artifact, sample sizes, and matching of controls. [For an excellent treatment of these issues, see Eliez and Reiss (288).] The research in this area has pointed to the need for reducing the heterogeneity within samples through efforts to subtype. Subtyping may in turn be clearer with knowledge of relevant biologic markers, many of which are speculated to play a causal role in autism.

NEUROCHEMICAL CORRELATES

Studies of the neurochemical correlates of autism have focused on blood and cerebrospinal fluid (CSF) transmitter levels and functional neuroimaging of transmitter synthesis and metabolism and receptor function. Studies vary greatly in methods and samples, which necessitates caution in the comparison and interpretation of their findings.

Serotonin

The occurrence of abnormally high levels of whole blood, plasma, and platelet 5-HT, or hyperserotonemia, in approximately 30% autistic children is the most solidly documented neurochemical finding in autism research. 5-HT is known as a neurotrophic factor, regulating early neuronal differentiation and synaptic growth and connectivity in the sensory cortex as demonstrated by animal studies. It is implicated in learning, memory, sensory, and motor processes (361). 5-HT is shown in the animal literature to be associated with restricted activity and avoidance of novel situations (362–364), inhibited startle response (365,366), and disrupted attachment behavior (367,368). Increased blood 5-HT has been associated with the severity of autism (369) and verbal deficits in patients and their relatives (370). Hyperserotonemia appears to be a familial phenomenon because 5-HT levels are correlated between autistic children and their family members (371–373). In fact, approximately 40% of hyperserotonemic autistic patients have been found to have family members who also show hyperserotonemia (374). Hyperserotonemia has been suggested as a marker for familial genetic liability for autism/PDD because 5-HT levels have been found to be higher in autistic children who have siblings with PDD symptomology than in those with normal siblings (375).

However, there are some challenges to validating the role of 5-HT in autism. Hyperserotonemia itself is not unique to autism; it has been found in other behavior disorders (376) and in mental retardation (377–380) in five of six studies currently reviewed (381). 5-HT also has been found to be decreased in autistic subjects (369). Also, although increased 5-HT is shown to correlate with the severity of autism, reviews reveal that tryptophan depletion actually worsens symptoms and that increasing synaptic levels of 5-HT with selective 5-HT reuptake inhibitors is reported to improve symptoms (199). Further, the finding of autistic individuals' blunted responses to pharmacologic probes of the 5-HT system is difficult to assimilate with the body of research on hyperserotonemia (380,382). Initial findings of behavioral improvements with the 5-HT uptake inhibitor fenfluramine, long-term administration of which may reduce 5-HT, were thought to support the role of central 5-HT in autism (383–392). However, later findings not only raised concerns regarding side effects and limited efficacy but also highlighted that the direction of its effects may be in fact inconsistent

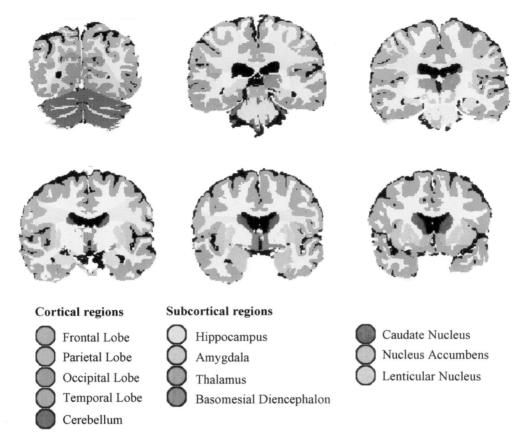

Cortical regions

Frontal Lobe
Parietal Lobe
Occipital Lobe
Temporal Lobe
Cerebellum

Subcortical regions

Hippocampus
Amygdala
Thalamus
Basomesial Diencephalon

Caudate Nucleus
Nucleus Accumbens
Lenticular Nucleus

FIGURE 3.4.

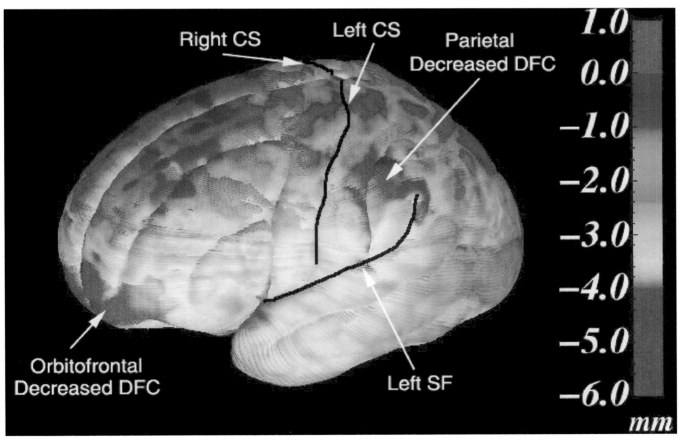

FIGURE 3.5.

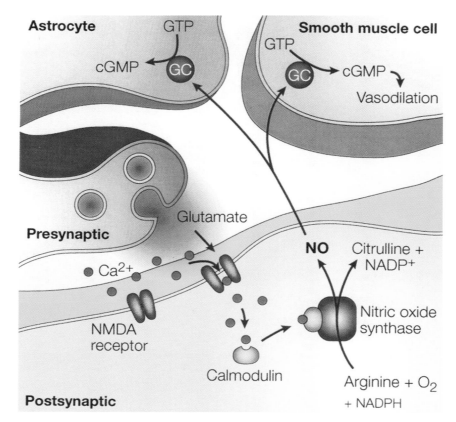

FIGURE 4.1.

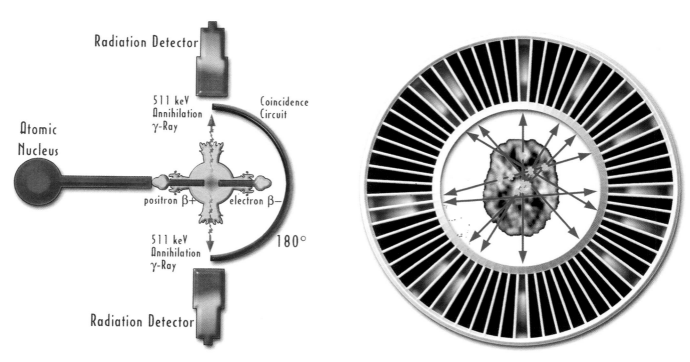

FIGURE 4.3.

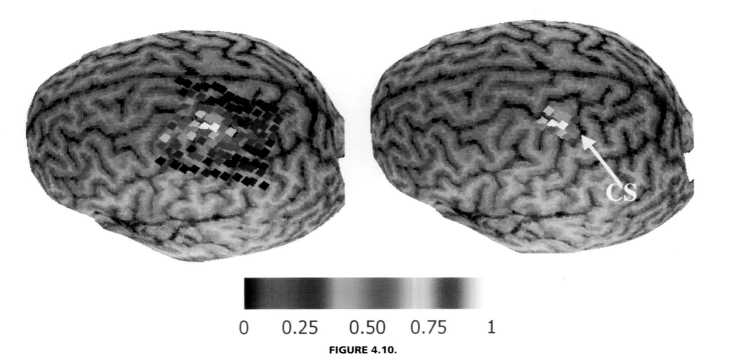

FIGURE 4.10.

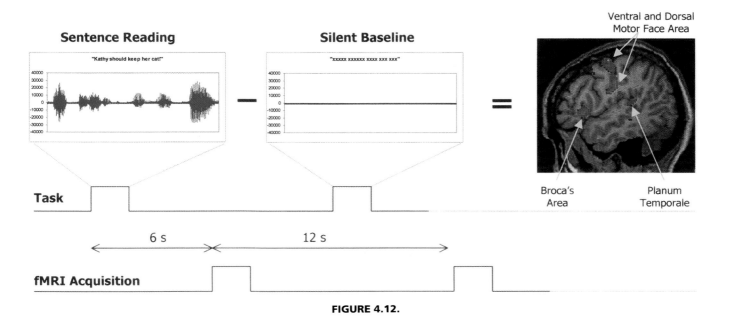

FIGURE 4.12.

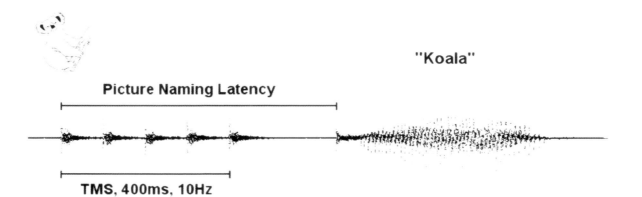

"Koala"

Picture Naming Latency

TMS, 400ms, 10Hz

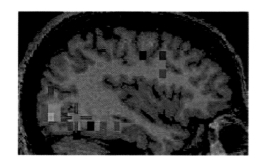

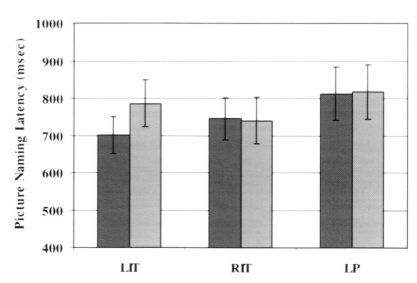

FIGURE 4.13.

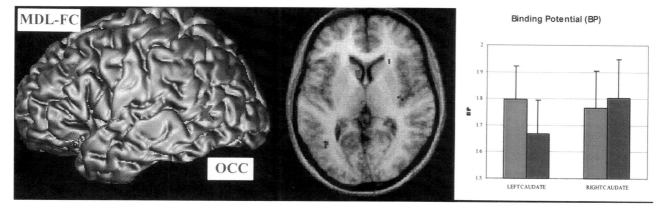

FIGURE 4.17.

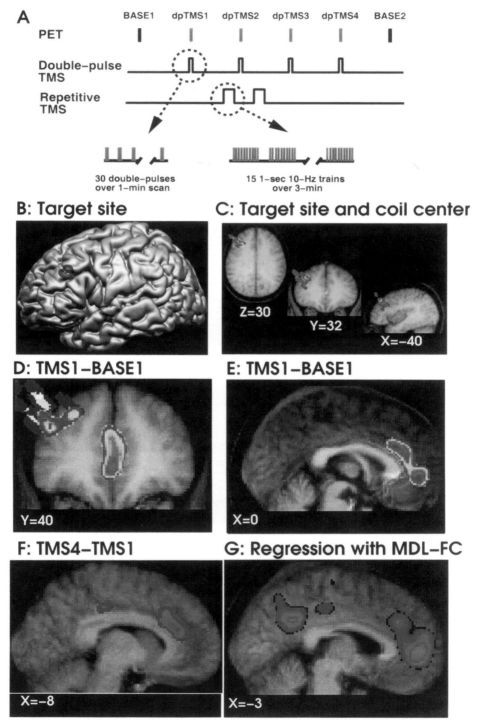

A

BASE1   dpTMS1   dpTMS2   dpTMS3   dpTMS4   BASE2

PET

Double-pulse
TMS

Repetitive
TMS

30 double-pulses
over 1-min scan

15 1-sec 10-Hz trains
over 3-min

B: Target site

C: Target site and coil center

Z=30   Y=32   X=−40

D: TMS1−BASE1

Y=40

E: TMS1−BASE1

X=0

F: TMS4−TMS1

X=−8

G: Regression with MDL−FC

X=−3

FIGURE 4.16.

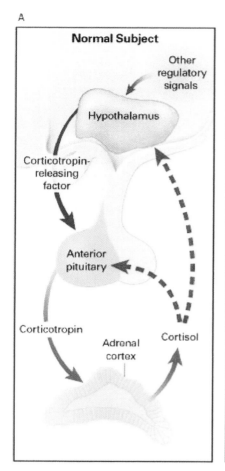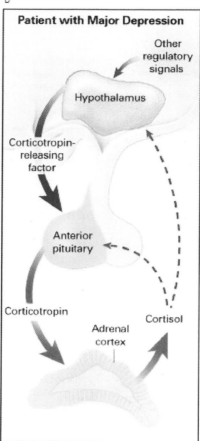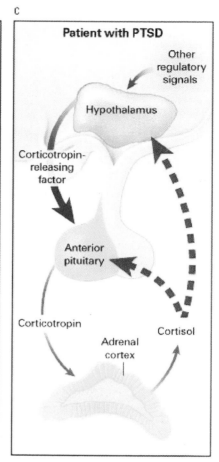

A — Normal Subject

B — Patient with Major Depression

C — Patient with PTSD

FIGURE 9.5.

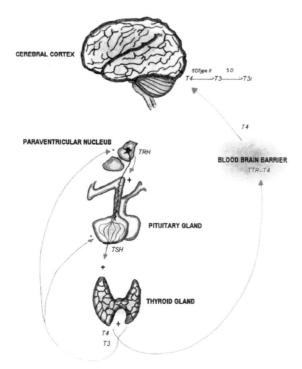

FIGURE 9.6.

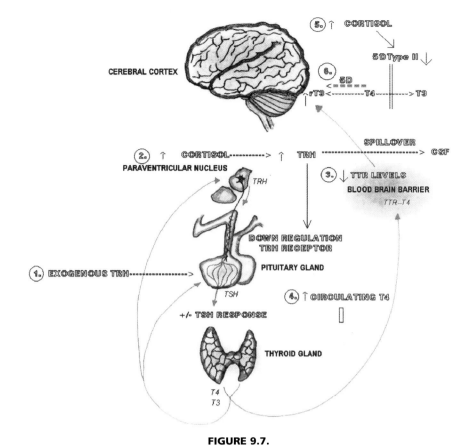

FIGURE 9.7.

FIGURE 9.7.

(5.) ↑ CORTISOL

5D Type II ↓

(6.) 5D
<=====
rT3 <------- T4 -------> T3

CEREBRAL CORTEX

SPILLOVER ------- > CSF

(2.) ↑ CORTISOL ------- > ↑ TRH
PARAVENTRICULAR NUCLEUS

(3.) ↓ TTR LEVELS
BLOOD BRAIN BARRIER
TTR–T4

TRH

DOWN REGULATION
TRH RECEPTOR

(1.) EXOGENOUS TRH ------- >

PITUITARY GLAND

TSH

(4.) ↑ CIRCULATING T4

+/- TSH RESPONSE

THYROID GLAND

T4
T3

FIGURE 9.7.

Preoptic
Area

Thalamus

Midbrain

Pons

Hypothalamus

BF → LH

DR
LDT/PPT
LC

VLPO

SCN

TMN

Optic Chiasm

Pituitary

FIGURE 15.3.

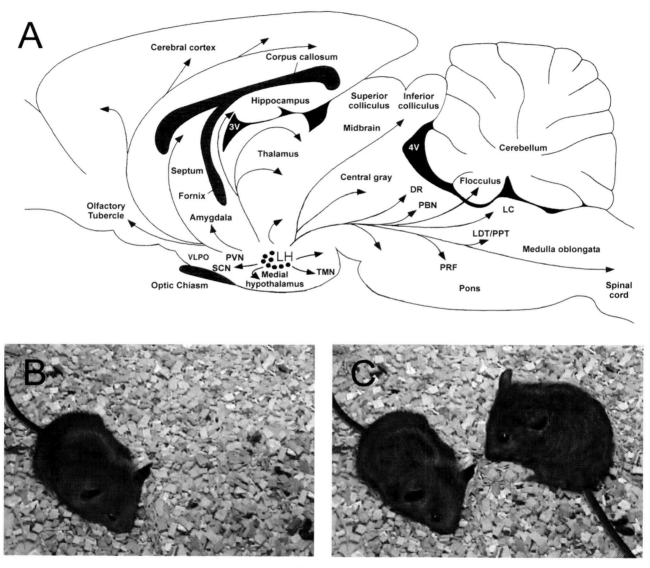

FIGURE 15.5.

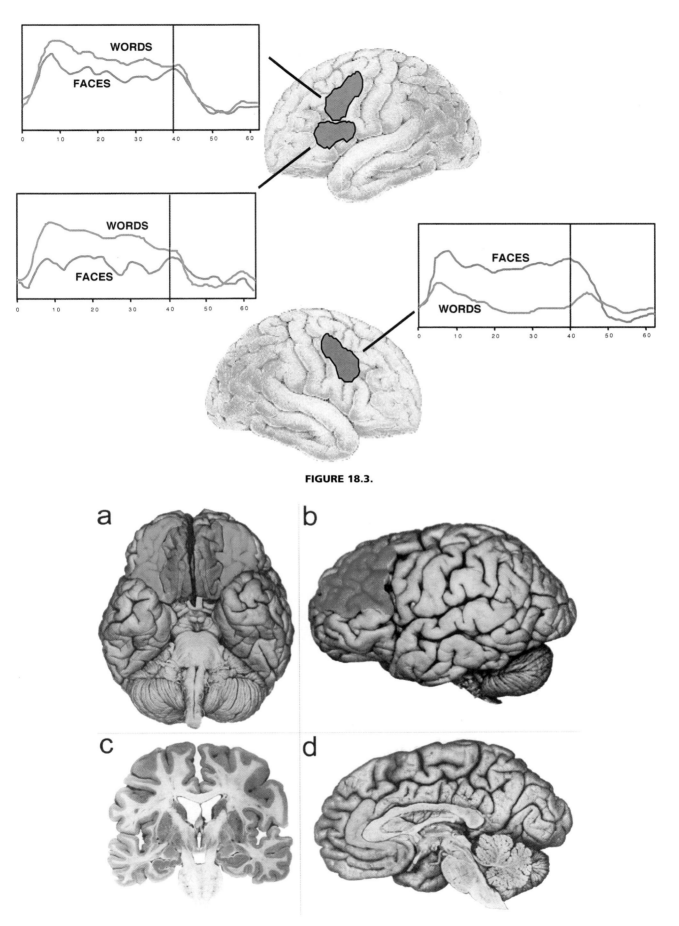

FIGURE 18.3.

FIGURE 20.1.

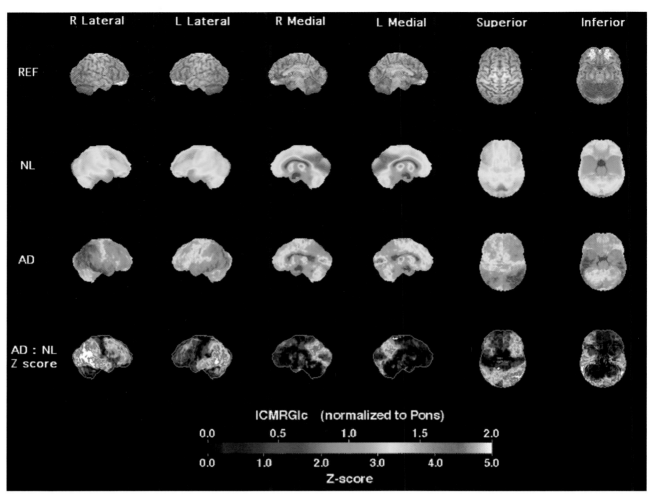

FIGURE 38.2.

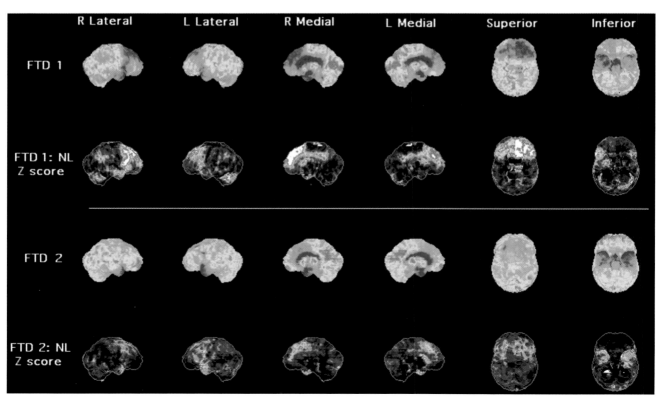

FIGURE 38.5.

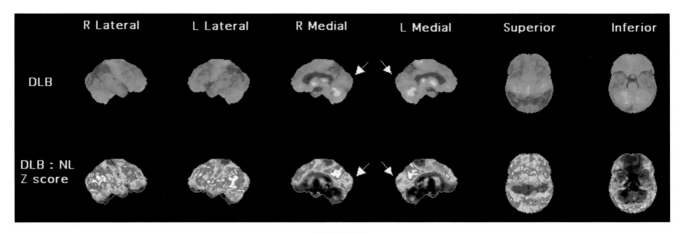

FIGURE 38.6.

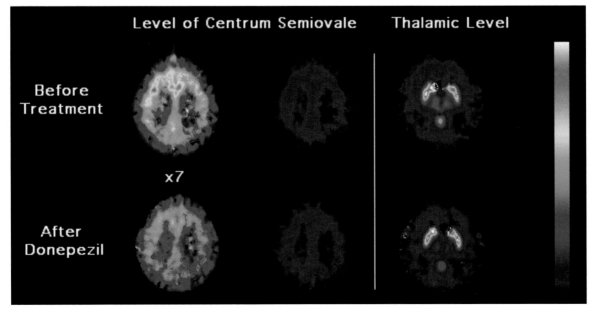

FIGURE 38.7.

with hyperserotonemia. The specificity of 5-HT's effects has also been questioned. Carlsson (393) points out the possible importance of glutamatergic systems with regard to 5-HT's role in autism. Research on the etiology of hyperserotonemia has likewise not offered consistent support.

The precise cause of hyperserotonemia in autism remains unknown. Blood 5-HT levels reflect peripheral mechanisms such as the platelet, which accounts for 99% of measured whole blood 5-HT (372). Past investigations of platelet uptake and release in autism have yielded very conflicting findings, a majority of which indicated normal uptake (394–402), perhaps attributable in part to the individual variability known to platelet 5-HT uptake (403). Studies of platelet 5-HT receptor binding sites have also yielded conflicting results of either normal or reduced numbers of sites (404,405). Investigations of platelet function have suggested various abnormalities in platelet aggregation in autistic subjects, including decreased plasma β-thromboglobulin and platelet factor 4, slight trends for decreased ability to aggregate in response to adenosine 5′-diphosphate and collagen, and higher adenosine 5′-diphosphate threshold concentrations (404,406,407). More recent platelet studies have examined the 5-HT transporter gene (408,409). The 5-HT transporter controlling platelet 5-HT uptake is identical in sequence to the 5-HT transporter expressed in 5-HT neurons. Cook et al. (408) found preferential transmission of the short variant of the transporter gene promoter 5HTTLPR in autism; in contrast, Klauck et al. (409) found higher frequency of transmission of the long variant in a broad autistic spectrum group. The long variant is believed to be responsible for increased 5-HT gene expression and 5-HT uptake. Most recently, a group from Italy (255) reported no linkage between 5-HT transporter gene promoter variants and autism and encouraged the search for associations between frequency of variants and autism subtypes based on 5-HT blood levels.

Hyperserotonemia could also be caused by increased synthesis or decreased metabolism of 5-HT; however, the evidence of this is inconsistent as well. Older findings of increased tryptophan levels and lack of correspondence between urine 5-hydroxyindoleacetic acid (5-HIAA) (a 5-HT metabolite) and platelet monoamine oxidase (which breaks 5-HT down into 5-HIAA) initially suggested disrupted metabolism and overactive synthesis (410). However, platelet levels of monoamine oxidase in autistic patients were consistently no different from those of normal individuals (397,411,412), and urine levels of 5-HT (413) and 5-HIAA (380,414) in autistic subjects were usually found to be normal. Although CSF 5-HIAA has been reported to be lower in autistic subjects (411), in general, CSF studies have revealed no consistent differences between normal and autistic subjects (411,415).

Recent studies by the Chugani et al. (416) examined regional brain 5-HT synthesis using PET with a tryptophan analogue tracer. They found significantly reduced 5-HT syn-

thesis in the left frontal cortex and left thalamus but increased values in the right dentate nucleus in a small number of autistic children relative to unaffected siblings. Two autistic subjects showed a reversed pattern with regard to lateralization of findings. Findings were thought to be consistent with dysfunction of the dentatothalamocortical pathway. A larger cross-sectional study by the same group reported that global 5-HT synthesis capacity increased with age in autistic children in contrast to the expected normal decreases observed in the nonautistic samples (417), which was taken as evidence of abnormal developmental regulation of 5-HT in autism.

Other studies of central 5-HT in autistic children have looked at receptors, finding decreased binding to the 5HT$_2$ receptor (418) and evidence of autoantibodies to 5-HT receptors that would block binding of 5-HT (419,420). However, binding inhibition has not been correlated with 5-HT levels (420,421), and immunoglobulins isolated from plasma showed no differences in inhibition of binding of a 5HT$_{1A}$ agonist in hippocampal membrane compared with controls (422).

Peripheral 5-HT may have an inverse relationship with central 5-HT function. Although some autistic children may possess overactive peripheral (e.g., platelet) 5-HT mechanisms, they may actually have concurrent decreases in central 5-HT. This would account for the clinical effectiveness of selective 5-HT reuptake inhibitors, the disparity between CSF 5-HIAA and blood 5-HT levels, and findings of diminished CSF 5-HIAA in an autistic sample (411). Further study of central function of 5-HT and its receptors is needed, particularly with attention to regional patterns as demonstrated by Chugani and colleagues.

Dopamine

The catecholamine DA has been implicated in such functions as stimulus-reward mechanisms, selective attention, motor activity, and cognition. Disruption to the nigrostriatal and mesolimbic DA systems was once hypothesized to result in motor stereotypies and other symptoms of autism (423). Hypoactivity of hypothalamic DA was also suggested (424,425). The effectiveness of neuroleptics (DA antagonists) in reducing autistic symptoms and the ability of amphetamine to exacerbate symptoms (i.e., stereotypies, self-injurious behavior, and even social deficits) in both animals and humans may be suggestive of the role of dopaminergic overactivity in autism. Clinical neurochemical studies themselves, however, have typically lent little support.

Plasma and platelet DA and urinary homovanillic acid [(HVA), the principal metabolite of DA] results are very conflicting (426–434). Plasma DA and HVA were found to show abnormal changes by age in autistic children (435). Aside from urinary HVA, the majority of which is believed to originate in the brain, most plasma HVA and plasma and urinary DA appear to originate peripherally and are not as reliable in indicating central DA activity as CSF levels of HVA. Studies appeared to implicate elevated CSF HVA in the severity of autistic symptoms (411,415,436,437), although this could not be replicated in one correlational study (438). In contrast, observations of delayed or blunted growth hormone response to levodopa have been taken as evidence of depressed central DA (425). In addition, a more recent PET study found decreased regional 6-[^{18}F]fluoro-L-DOPA in the anterior medial prefrontal cortex in subjects with autism, which may reflect reduced DA synthesis and storage (325).

In summary, although CSF studies showing overactive DA metabolism are consistent with animal and clinical drug studies, PET findings indicate regional depressed DA transmission.

Norepinephrine and Epinephrine

The catecholamines norepinephrine (NE) and epinephrine are rather ubiquitous and are involved in the regulation of arousal, attention, activity level, anxiety, response to stress, memory, and learning. Accordingly, findings are not consistent or specific to autism. One investigator found abnormal changes in plasma NE and epinephrine, as well as other transmitters, with age in autistic children (435). Increases have been found among autistic subjects in plasma NE (412,430,431) and platelet NE (430) but not in plasma 3-methoxy-4-hydroxy-phenylglycol (MHPG), NE's primary metabolite (439,440). This is in contrast to CSF MHPG's usual correspondence with plasma NE in the normal population (441,442). Findings of decreased urinary NE, epinephrine, and MHPG also are in contrast to plasma findings (441,443). Similarly, plasma measurement of DA β-hydroxylase, the enzyme that synthesizes NE from DA, has yielded mixed results (412,428). Later studies found disruptions in levels of catecholamine metabolites but no association with genetic markers coding for enzymes involved in catecholamine synthesis (444). One of the more recent investigations found no abnormalities in plasma and urine MHPG or urinary excretion rates of NE, epinephrine, and vanillylmandelic acid, leading investigators to conclude that there is no apparent adrenergic/noradrenergic dysfunction in autism (445).

Peptides

Opioids

Observations of the similarities between the symptoms of opiate addiction and autism (446) and the role of opioids in the attenuation of the perception of pain have led some researchers to implicate endorphins in autism. Elevated endorphin levels were hypothesized to contribute to the development of autistic symptoms, specifically deficits in attachment and social interaction and self-injurious behavior caused by elevated pain threshold (446,447). The endogenous opioids are distributed in areas of the brain known

to integrate sensation and emotion (448), and their role in these functions has been supported by animal models (449). Prenatal exposure to high levels of β-endorphin has been reported to result in disruptions of cognitive and developmental processes and adult social and sexual behavior in animals (448). Animals have been observed to exhibit autisticlike symptoms after administration of exogenous opiates and their agonists, including the absence of normal distress on separation from mothers, decreased affiliative behaviors, and high pain threshold (448). Blockade of opioid receptors through administration of opiate antagonists can result in the reversal of all these behaviors as well as facilitation of separation distress vocalizations, although this effect has been observed to vary slightly depending on the age of the subject, time of administration, and type of opioid receptor involved (448).

Although some findings of endorphin deficits in plasma (450,451) and excesses in CSF (452) seemed to tentatively suggest opioid system disruption and a reciprocal relationship between central and peripheral opioids in autism, the direction of findings has not been confirmed. More recently, inconsistencies in plasma and CSF β-endorphin findings were suggested (453,454). Also, nonautistic self-injuring individuals display low plasma and high CSF endorphin levels as well (452), which could suggest that opioid impairment is not central to autism or that opioids may contribute to autistic symptoms only through disruption of pain and reinforcement mechanisms specifically.

Although initial clinical pharmacologic trials were indirectly supportive of opioid involvement in the broad symptoms of autism (455–465), later studies with the opiate antagonists naloxone and naltrexone have been disappointing for any targets other than hyperactivity [for a review, see Buitelaar and Willemsen-Swinkels (298)]. In summary, the role of the opioid system in autism has compelling theoretical foundations with little clinical empirical support.

Oxytocin

Another peptide that has been implicated in the syndrome of autism is oxytocin (466–468), a centrally and peripherally active neurohormone that modulates affiliative behaviors among mammals (469). Oxytocin levels are positively associated with aspects of human attachment, including mother–infant bonding, sexual relations, and socially extroversive personalities (466). Modahl and Green's group (250) found that oxytocin is abnormally low and shows abnormal associations with sociality and age in autistic children, whereas its precursor prohormone OT-X is abnormally high. Insufficient metabolism of OT-X in autistic children was suggested.

Attempts to integrate peptide findings include Chamberlain and Herman's model (470) of melatonin's cascade of effects on 5-HT and β-endorphin. However, like a majority of the literature on the neurochemistry of autism, findings are neither universal nor consistent.

SENSORY FUNCTION: NEUROPHYSIOLOGIC STUDIES

Autism has been studied and treated by some as a disorder of sensory processing, from the 1960's to the present. Investigators have examined a range of sensory/perceptual functioning with measures of electrical activity in various anatomic regions. Included here are selected sleep studies.

Sleep and Ocular Studies

Based on a genetic model having concurrent cerebellar Purkinje cell depletions and retinal dysfunction, a 1980's study examined the retina using electroretinography and found b wave decreases in autistic subjects (471–473) that were proposed to result from dopaminergic dysfunction. Sleep studies have had mixed results, ranging from normal findings (474,475) to decreases or developmental delays in rapid eye movement (REM) in terms of oculomotor burst activity and REM frequency (475,476). A recent sleep study involving autistic children reported normal REM sleep but reduced sleep time, which correlated with behavioral variables (477). Autistic subjects were also found to have diminished phasic oculomotor activity during REM sleep in response to vestibular stimulation (478).

Vestibular Response

This finding and clinical observations of spinning and of nonresponsiveness to stimuli led to studies of the vestibular responses of autistic children, which are abnormal (479–483). Several researchers have demonstrated that autistic children show a suppression of postrotatory nystagmus, a vestibularly induced oculomotor response (479,481,483). Autistic children were found to have abnormally low oculomotor output in response to vestibular stimulation in the presence of visual stimulation (481). Studies also have found prolonged time constants of the nystagmus response to acceleration and deficiencies in recorded nystagmus beats (482). Ornitz et al. (481,482) suggest that abnormalities in nystagmus response may be owing to dysfunction of the vestibular nuclei in the brainstem.

Cerebral Electrical Recording

Brainstem dysfunction was a prevalent hypothesis in autism research in the 1960's and 1970's and led to studies examining waves derived from the EEG and event-related potentials (ERP) to reflect deficits in sensory processing. Clinical and empirical observations point to diffuse or variable focal abnormalities on EEG, with no patterns emerging as specific to autism. An exception to this observation is Landau-Kleffner syndrome, in which autistic symptoms may develop with specific EEG patterns involving temporal regions. Interestingly, specific behavioral features were recently found to cor-

relate with regional abnormalities on magnetoencephalography in subjects with autistic regression (484).

An early study examining ERP specifically found that autistic children younger than 5 years of age failed to display normal inhibition of wave N2 (reflecting basic sensory response) during REM (13). Later, abnormalities in brainstem auditory evoked responses were found in a subgroup of autistic subjects in the form of increased brainstem transmission time, high intertrial ERP variability, and delayed latencies for waves I, III, IV, and V, especially at lower stimulus intensities (485–493). Studies of that time finding no brainstem auditory evoked response abnormalities appear to have tested older children with higher IQ (494,495).

Cognitive or perceptual processes are thought to be reflected in middle to longer latency components, or waves, of the ERP. Earlier models of autism as a disorder of attention and information processing implicated the frontal cortex, basal ganglia (423), and limbic system (423,496) and found some tentative support in ERP studies (497–506). High-functioning autistic subjects of varying ages have usually shown abnormally small amplitudes for long latency, auditory evoked wave P3b, which is thought to reflect the cognitive processes of detection and classification of target stimuli (497–502,507). Dawson et al. (502) found P3 amplitudes diminished in response to phonetic but not musical stimuli at central cortical and left hemisphere recording sites. Interestingly, the P3 amplitude may be higher in response to affective/prosodic stimuli than to linguistic/prosodic stimuli in autistic subjects (508). However, decreases in P3 amplitudes are not unique to autism but are seen in ADHD as well (509).

Wave component Nc and the negative difference waves known as auditory Nd and N270 are thought to reflect frontally mediated, self-generated attention. Investigators have found abnormally small amplitudes of component Nc (494,501) and total absences of waves Nd and N270 (498) in autistic subjects.

Recent ERP studies showed that two components of the late positive ERP response, corresponding to functions of attention, were deficient during basic visuospatial processing tasks in autistic subjects, similar to patterns observed in cases of cerebellar lesions (510).

Other studies have found disrupted patterns of normal hemispheric lateralization (475,511,512) including inhibition of left-hemisphere EEG alpha attenuation (513) and right temporoparietal ERP N1 amplitude augmentation (514) in response to phonetic processing.

Summary of Sensory Findings

Data corroborate clinical observations of sensory processing deficits and vestibular dysfunction, which may suggest brainstem pathology. Auditory evoked ERP have revealed delays in basic brainstem auditory pathway processing for a subgroup of autistic patients, consistent with some previous neuroanatomic findings. Attenuated longer latency ERP in autistic subjects suggested disruptions in attention and information processing. Rapin and Katzman (196) caution, however, that ERP studies are limited by the electrode's lack of regional specificity and the confound of attention status during testing.

Discussions in the field of autism emphasize the need to synthesize the broad body of literature in the field of autism research (515). Integrative theoretical models are needed to attempt to bridge knowledge across the levels of neurobiologic research in autism. The model of Waterhouse et al. (32) of four proposed neurofunctional mechanisms of autism draws on evidence from neurobiologic, electrophysiologic, and neuropsychological studies. In it, they link the symptoms of asociality, impaired intrinsic reward (assignment of significance), fragmented perceptual integration, and perseveration (extended selective attention) to impairments of, respectively, oxytocin, the amygdala, hippocampus, and temporoparietal polysensory cortex. From their review of the data, they posit two polygenic sources of autism subtypes and comment that all etiologies ultimately share a common path of localized neurologic damage.

Discussion

Individual studies often lack complexity in scope in studying the nervous system. Research into the causes of autism is poised to begin to simultaneously examine genetics, neuroanatomy, neurochemistry, and neuropsychology. Despite technologic developments, neurobiologic studies in autism are still limited by a lack of standardized methodologies and other confounding factors that limit interpretation and generalization. [For discussion of these issues, see Trottier et al. (199).] Investigators variously argue for and against subtyping. Although acceptance of the concept of a spectrum of autism may be regarded as an advancement in autism research, the emergence of seminal markers or etiologies may continue to be obscured by a lack of homogeneous sampling. Variability in the biomedical and neuropsychological profiles of people with autism concur with genetic findings in suggesting that multiple etiologies may be responsible and that further efforts to subtype are warranted.

CLINICAL ASSESSMENT

For recent reviews of issues pertaining to the assessment and diagnosis of autistic spectrum disorders, see Filipek et al. (516), Newsom and Hovanitz (517), and Volkmar et al. (518).

Early Detection

Early detection of autistic spectrum disorders is possible at ages as young as 18 to 30 months and is critical for the implementation of early intervention. Several tools are under development. The most widely studied instrument is the Checklist for Autism in Toddlers (519–521), which consists of nine parent report items and five observations made by home health visitors in Great Britain. An American version, called the Modified Checklist for Autism in Toddlers (522), is currently being validated on a large population. Other instruments under development include the Pervasive Developmental Disorders Screening Test and the Screening Tool for Autism in Two-Year-Olds, which are reviewed by Filipek et al. (523). The Screening Tool for Autism in Two-Year-Olds is a second-level screen, designed for children already identified with developmental concerns, whereas the others can be used as a first-level screening (524).

It appears that the items pertaining to social relatedness (e.g., pointing, response to name, imitation) may be the best at discriminating very young children with autistic spectrum disorders from both typically developing children and children with other developmental disorders, such as language disorders and global cognitive impairments (522,525).

Medical Assessment

Issues in assessment, referrals, and management of children with autism and related syndromes are reviewed in a number of recent publications (526–534). In brief, the developmental history and behavioral and mental status examinations are the basis for the diagnosis of autism or PDD.

Once a diagnosis or tentative diagnosis of PDD or autism is made, assessments in specific areas should be done. The child's hearing must be assessed thoroughly and definitively (535). In any case in which behavioral assessment of hearing is not considered reliable, the child should be referred for brainstem auditory evoked response audiology.

Individuals with autism are subject to a variety of motor impairments, especially stereotypies, hypotonia, and apraxia, as well as frank movement disorders (536,537). A complete motor examination by a neurologist may reveal conditions that are remediable or that suggest a specific neurologic diagnosis.

Medical and family history and physical examination may suggest specific etiologies, such as hydrocephalus, or a genetic or neurocutaneous syndrome. [For reviews of the genetics involved in autism, see Cook (233) and Rutter (538).] The genetic syndrome most often reported in association with autism is fragile X syndrome (discussed previously) (539). Examination of the child's physiognomy can suggest the diagnoses of Williams syndrome, Prader-Willi syndrome, or other neurobehavioral entities and can dictate referral to a clinical geneticist. Macrocephaly is also common in autism (540). Allen et al. (528) conclude that, in general, screening for abnormalities in amino acids, organic acids, or other metabolites in blood and urine in the absence of a specific indication for such testing, has a very low yield.

However, as discussed above, Gillberg (263,530) claims that a high proportion of autistic individuals has associated conditions that can be uncovered by extensive medical investigation. He proposes the investigation of fragile X syndrome, tuberous sclerosis, neurofibromatosis, hypomelanosis of Ito, Rett syndrome, Möbius syndrome, Williams syndrome, de Lange syndrome, Laurence-Moon-Biedl syndrome, mucopolysaccharidosis, and Coffin-Lowry syndrome. Gillberg (530) suggests that other conditions worth investigating include phenylketonuria, rubella, lactic acidosis, congenital toxoplasmosis, cytomegalovirus infection, herpes encephalitis, hydrocephalus, hypothyroidism, infantile spasms, and purine and calcium disorders. Recent research suggests that tuberous sclerosis (541,542) and neurofibromatosis (543) may be more common in autistic spectrum disorders than in a nonselected population. Gillberg (530) also recommends a specific regimen of laboratory analyses. This includes blood work (chromosomes, phenylalanine, uric acid, lactic acid, pyruvic acid, herpes titer), urine studies (metabolic screen, uric acid, calcium), CSF examination for protein, EEG, and computed tomography or MRI to look for evidence of tuberous sclerosis, infection, neurofibromatosis, and hypomelanosis of Ito. Chromosome abnormalities may be present in some children with autism (544; for a review, see reference 234).

The high prevalence of epilepsy in association with autism supports the inclusion of EEG as part of the medical evaluation of children with autistic symptoms. Adolescent onset of seizures is more common in autism than in other developmental syndromes. Tuchman et al. (193) found agreement between the clinical diagnosis of epilepsy and epileptiform EEG and concluded that EEG may be useful in determining whether unusual stereotyped behaviors represent seizure activity. In young autistic children without stereotypies or seizures, EEG may be useful in cases with regressive loss of previously acquired language skills. When EEG are performed, they should include a sleep EEG. Although there is disagreement about the diagnosis of Landau-Kleffner syndrome (acquired epileptic aphasia) in children with PDD behaviors, an EEG may be useful in ruling out Landau-Kleffner syndrome when marked language regression is present. This is particularly important because children with Landau-Kleffner syndrome may respond to medications [the relationship of EEG abnormalities, seizures, Landau-Kleffner syndrome, and autism is reviewed by Tuchman et al. (545)].

The use of routine computed tomography or MRI remains controversial. Although the likelihood of finding an abnormality is low and a discovered abnormality is not likely to have treatment implications, the routine screening of autistic conditions with brain imaging is a common practice. Gillberg (530) recommends it as part of a standard medical workup of autistic children.

Psychiatric evaluation includes assessment for comorbid psychiatric disorders, including ADHD, depression, psychosis, anxiety, and OCD. Individuals with autistic spectrum disorders are at greater risk of anxiety and mood disorders (546), attention and sleep problems, hyperactivity, and tics (547). Family issues and interpersonal conflicts also should be assessed.

Neurologic assessment includes sufficient neurocognitive assessment to determine whether retardation is present. Rapin (526) found that neurologists missed the diagnosis of mental retardation in approximately 25% of children shown to have mental retardation by formal testing. IQ testing in autistic children has been shown to have the same predictive validity and stability as for normal children (548). Therefore, formal testing of intellectual function would be preferable to reliance on a clinical impression based on an incomplete neurocognitive assessment.

It is also important to look at the comorbidity rates of other psychiatric disorders in various subgroups of autism. Bonde (549) suggests that the children classified as aloof according to the Wing subtypes had fewer comorbidities than those children classified as active but odd.

Neuropsychological Assessment

Children and adolescents with autism or PDD should also have periodic neuropsychological assessments. These are necessary for describing the child's current level and profile of cognitive abilities, which will have implications for educational programs and long-range goals. Periodic reevaluations are also necessary to monitor the child's progress, detect any deterioration that might signal the onset of negative medical or psychological events, and evaluate the success of programs. Procedures and issues in neuropsychological evaluations of autistic children have been reviewed (532,550).

These evaluations should be done by a trained neuropsychologist experienced with autistic or PDD children. Multiple sessions at the office, home, or school may be necessary to complete the evaluation. With difficult children, the school can be an excellent place for testing because the children are accustomed to working there and the teachers can aid compliance. Whenever possible, a neuropsychological evaluation should thoroughly assess cognition in the areas of language (vocabulary, syntax, and pragmatics), visuospatial skills, abstract thinking and problem solving, memory, attention, and social cognition. Children with autistic spectrum disorders demonstrate a wide range of IQ and neuropsychological profiles (551,552); within IQ, speed of processing may be impaired in children with autism (553). Language, in particular, must be very thoroughly assessed. Usually a neuropsychologist can administer a complete language assessment. If not, a referral to a speech and language pathologist is advisable. Many clinicians would argue that such a referral is warranted in any case.

Watson and Marcus (554) reviewed a series of tests for use with young autistic children, including the Bayley Mental Scale of Infant Development, Merrill-Palmer Test of Mental

Abilities, McCarthy Scales of Children's Abilities, Wechsler tests, and the Kaufman Assessment Battery for Children. Additional tests are frequently used with procedures for the elicitation of play. The clinician should judge the degree of play deficit relative to the child's MA and not CA. A clear description of problem behaviors is important. Hadwin and Hutley (555) point out several behaviors (such as joint attention) that differentiate children with autism and comorbid learning difficulties from children who have severe learning difficulties without autism.

Behavioral Assessment

A thorough behavioral description is as important as assessment of cognition [for a review, see Harris et al. (556)]. This may be done by a neuropsychologist, psychiatrist, or other developmental, behavioral, or pediatric specialist. A profile of the individual's adaptive abilities should be included in the behavioral assessment (557). One clinical/psychoeducational approach to assessment of functional skills at home and in the classroom is described by Mesibov et al. (558). A standardized instrument for the assessment of adaptive skills is the revised Vineland Adaptive Behavior Scales (559); it has been shown to be a powerful, highly descriptive, "well-normed" instrument that works well with autistic children (82). It provides age equivalents and standard scores for communication, daily living, and socialization skills and motor scores for younger children. Because autistic children have more intradomain scatter in development than other children (560), it behooves the clinician to examine individual developmental items passed. Autistic children will often have their lowest age equivalent scores in the socialization domain. Daily living scores reflect self-help activities taught at home or school, so particularly low scores in this domain may suggest that the child has not been pushed to capacity or is particularly deficient in motivation to master these skills. Deficits in motor ability may also affect daily living skills. The Vineland Adaptive Behavior Scales may be better than tests of intellectual functioning (i.e., Stanford-Binet) at discriminating autism from mental retardation (561).

Play is an area of adaptive skill that warrants special attention. The development of play in normal children proceeds from simple sensorimotor play to functional play to the emergence of symbolic play, which then follow development along several lines of increasing abstraction and complexity (562). Particular impairment of symbolic play has been shown to characterize autistic children (563–566). The results of these investigations suggest that spontaneous free play is infrequent and immature in autism but that modeling can elicit free play in high-functioning autistic children, who are nevertheless generally uninterested in symbolic play activities. The clinician assessing play should be familiar with the development of play in normal children and with procedures for the elicitation of play. The clinician should judge the degree of play deficit relative to the child's overall MA and language items in particular and not CA.

A clear description of problem behaviors is important (567). Behaviors central to the syndrome (such as social incapacity and resistance to change), those associated with the syndrome (such as self-injury and abnormal motor behaviors), and those ancillary to it (such as hyperactivity, aggressiveness, and passivity) should be noted. Many instruments exist to document abnormal behaviors in autism; reviews of such instruments are found in Parks (568) and Powers (569). The Autism Diagnostic Interview (570) and the Autism Diagnostic Observation Schedule (571) are a pair of companion instruments for interviewing an informant and for direct interview of the autistic individual. These instruments are reliable and have good coverage of clinical content. However, administration of these instruments is quite lengthy, and training of examiners is required.

Often neglected in the description of problem behaviors is the analysis of antecedent conditions and consequences of the behaviors. These may clarify the role or function of the behavior for the particular autistic individual and may suggest changes in stimulus conditions and reinforcements to ameliorate problem behaviors as well as foster positive behaviors. Such behavioral assessment procedures are reviewed by Powers (569).

Family Assessment

Assessments of strengths, resources, and needs of families coping with autism are described in detail by Harris (572). Harris reviewed family assessment instruments and outlined the application of the McCubbin and Patterson (573) framework for assessment of the family, in which the clinician assesses personal financial resources, educational attainment, physical and emotional health, and psychological characteristics, family cohesion and adaptability, and social support of the family. Harris also describes how to arrive at the proper balance between the focus on the child's difficulties and the family's handling of the problems. A recent book offers parents a comprehensive guide for living with a child with autism (574). Parents and siblings of a child with autism may also have psychiatric difficulties as a result of the stresses of living with an autistic child; thus, adequate family assessment is critical for the well-being of the family.

TREATMENT

Pharmacologic Treatments

Pharmacologic treatment for autism is fairly prevalent; a survey (575) found that 50% of individuals with autism were being treated with medication or vitamins. Pharmacologic treatment is reviewed in several recent publications [197,576–578; see also *Journal of Autism and Developmental*

Disorders (special issue on treatments for people with PDD) 2000;30(5)]. Gringras (579) suggests choosing one of two approaches when considering medication for children with autism: (a) the patient-specific approach targets symptoms of comorbid conditions, such as hyperactivity and aggression, whereas (b) the target symptom approach focuses on symptoms specific to autism, including ritualistic behavior and resistance to change. Gualtieri (580) reviews many treatments for comorbid conditions, including epilepsy, self-injurious behavior, and compulsions, and Hellings (581) reviews pharmacologic treatment of comorbid mood disorders. Holm and Varley (582) caution that many studies of pharmacologic effectiveness are anecdotal, use inadequately defined populations, use inadequate instruments for measuring change, lack control groups, or report statistically but not clinically significant changes. Harty (583) argues that, although pharmacologic therapies do not fundamentally alter the natural course or core symptoms of autism, they can be useful in reducing aberrant behaviors.

Although attention difficulties and hyperactivity are often prominent in the clinical picture of autism, the autistic child's behavior usually does not improve with stimulant medications. Exacerbation or initiation of stereotypies and psychosis has in fact been reported in autistic children placed on stimulants. Therefore, Holm and Varley (582) conclude that stimulant medication is therapeutic for a few autistic children; these tend to be high-functioning autistic children with symptoms of attention deficit. Harty (583) recommended trials of stimulants for hyperactive autistic children without stereotypies. Aman and Langworthy (584) reviewed 41 studies targeting hyperactivity in autism, including studies of antipsychotics, selective 5-HT reuptake inhibitors, antianxiety drugs, stimulants, α-adrenergic agonists, and opiate blockers. They conclude that evidence is strongest in favor of antipsychotics, stimulants, and opiate blockers, but caution that these medications have been shown to affect hyperactivity more than attention functioning.

Research on the use of anxiolytics with anxious or sleep-disturbed autistic individuals is sparse. Harty (583) reports that agents such as buspirone will reduce agitated behavior in some autistic children, whereas King (585), in a review, concludes that barbiturates and benzodiazepines are rarely used and can worsen aggression and self-injury. β-Blockers have been reported to be useful in managing impulsive, aggressive, and self-abusive behavior in autistic adults (586). Ratey et al. (586) hypothesize that β-blockers reduce a chronic state of hyperarousal. Lipinski et al. (587) suggest that β-blockers may act by ameliorating akathisia induced by long-term neuroleptic administration.

The use of neuroleptics, especially haloperidol, is controversial. Although some researchers have reported positive effects (584,585), others (575) dispute its efficacy. Holm and Varley (582) conclude that, although haloperidol may have beneficial effects on agitation, hyperactivity, aggression,

stereotypies, and liability, it is less effective in improving social relationships and language. They advise that the risk of tardive dyskinesia (which occurs in 20% of treated autistic individuals) be weighed carefully. King (585), conversely, reviewed studies that suggest that DA antagonists can reduce maladaptive behavior and promote prosocial behavior beyond any sedative effects. Clozapine, an atypical neuroleptic, has received favorable review for its effectiveness (588,589).

Other drugs considered effective for self-injury include fluoxetine, clomipramine, buspirone, and opiate antagonists especially when combined with behavioral therapy (580,585,590), although King (585) notes that most studies of serotonergic drugs are not controlled trials. Although the opiate blockers naltrexone and naloxone were reported to have beneficial effects on self-injurious behavior, social withdrawal, and stereotypies, some authors (591) caution against overestimating the ameliorative effects of naltrexone. Harty (583) reviewed the differences between the two classes of agents and the symptoms targeted by each.

A recent review (592) of the use of tricyclic antidepressant medication in autism revealed that few studies have been conducted using double-blind, placebo-controlled methodology and that few studies show positive results. However, clinicians sometimes prescribe them, especially to autistic individuals who have a close relative who had depression that was responsive to tricyclics. Holm and Varley (582) note that, although tricyclics appear to improve language and social responsiveness in some autistic persons, many show adverse behavioral reactions, and aggravation of seizures is possible. Lithium carbonate has also been tried in autistic children (585,593); some children experienced amelioration of agitation and aggression. DeLong and Aldershof (593) report that high-functioning, perseverative children with hyperactivity unresponsive to stimulants and a family history of bipolar disorder are most likely to respond well.

Studies (585,594–597) have obtained encouraging results using clomipramine, a tricyclic compound that acts preferentially on the serotonergic system and that is widely used in the treatment of OCD. Two of these studies (594,597) found clomipramine superior to desipramine, using a crossover design and a double-blind comparison. Social relatedness improved, whereas aggression and ritualistic behaviors diminished. A recent open trial found reductions in adventitious movements and compulsions (596), and case reports indicate similar improvements (558). Recent reviews of pharmacologic treatment of repetitive thoughts and behaviors (598,599) suggest that selective 5-HT uptake inhibitors such as fluoxetine, clomipramine, and sertraline are promising and should be studied further to determine their efficacy. Similarly, case reports have begun to appear on the use of selective 5-HT reuptake inhibitors such as fluoxetine in autism (600,601), including autism comorbid with depression (602), trichotillomania (603), self-injurious behavior (580), and OCD (604). One study (605) found that fluoxetine led to clinical improvement in many subjects, although some subjects experienced undesirable side effects such as agitation. A recent 6-month, open-label trial with fluoxetine resulted in mild to moderate improvements in five children with autism, particularly in reducing ritualistic behavior and motor stereotypies and improving social relatedness (606). These results are encouraging but are preliminary and therefore mainly of heuristic significance.

The much more numerous studies on the effectiveness of fenfluramine were reviewed in detail by Holm and Varley (582). The cumulative results are inconsistent, but more studies show positive change than not. In general, where there is an effect, behaviors related to hyperactivity and stereotypies appear to show more consistent improvement than cognition, language, or social relatedness. Holm and Varley caution that many parents and clinicians have thought that the benefits were transitory, and Leventhal et al. (607) reported no substantial advantage to using fenfluramine for autism, particularly its social deficits.

Nutritional treatments, reviewed by Holm and Varley (582) and Raiten (608), have also been recommended by various autism researchers and clinicians (609–612). Holm and Varley (582) criticized these studies on methodologic grounds and cited evidence that large doses of vitamins such as B_6 may not be as safe as had been supposed. Kozlowski (613) asserts that megavitamins are by no means a proven success in the treatment of behavioral problems in mentally retarded autistic children. Page (614) reviewed the vitamin literature in autism from the perspective of effects on possible metabolic disorders and concludes that several well-designed studies show pyridoxine in combination with magnesium to have beneficial effects in treating autism, including positive effects on language, aggression, and interest in the environment. Tanguay (197), conversely, concludes that convincing evidence of the effectiveness of vitamin therapy in autism is lacking.

Folate has been tried with boys with fragile X syndrome (615), showing positive effects on behavior. However, Lowe et al. (616) found no positive effects of folate on an unselected group of autistic children. A preliminary crossover trial of supplemental ascorbic acid found improvements in the severity of symptoms over 10 weeks of treatment (617). Rimland's ongoing publication, *Autism Research Review International* (Autism Research Institute, 4182 Adams Avenue, San Diego, CA 92116), periodically reviews nutritional studies and advocates specific regimens of vitamins and minerals.

Antiepileptic drugs generally are given for control of documented seizure disorders in autistic children and not for the control of behavior. There is no firm evidence of their effectiveness in improving behavior, although some anecdotal evidence exists of improvement of depression and irritability (545,582,583).

Recent reports in the popular press have generated a great deal of interest in intravenous or transdermal administration of secretin to promote social and language improvement in children with autism or PDD. Well-controlled trials are just beginning to appear, but the initial publications are not promising [for controlled trials of secretin, see references 618–620, but see also Rimland (621)].

Recently, anecdotal testimony suggests that removing the proteins gluten and casein from the diet of individuals with autism improves social relatedness and reduces negative behaviors associated with the disorder. However, there have been few empirical studies examining the effects of eliminating these proteins from the diet, and almost all have been open trials (622). One controlled trial (623) with a placebo versus gluten challenge found negative results, although the subject sample was very small. Sponheim (623) concludes that not only were his results negative but that this highly restrictive diet contributes to further social isolation for these challenged families and children.

In sum, pharmacologic treatment is as individual as the presentation of autistic symptoms. Clinicians prescribing medication must monitor effects closely and be open to many options. Associated behaviors such as self-injury, aggression, and mood disturbances can be targeted specifically and may be more amenable to change than the core social and communication disturbances of these disorders.

Behavioral and Educational Treatments

The major treatments for autism are special education and behavioral programming. Early and extensive work on language, social interaction, preacademic and academic skills, and self-help skills are the autistic child's best hope for a positive outcome. There is a very extensive body of literature dating back 30 years on various aspects of behavior modification and other nonpharmacologic interventions for autism, but it is beyond the scope of this chapter to review this literature in detail (for selected reviews, see references 624–628). In brief, there are different approaches to intervention, but no large-scale, controlled study exists to compare their efficacy. The methodologic and practical problems inherent in attempting such a study seem almost insurmountable.

Representatives of several such approaches are the developmental program of Rogers and Lewis (629) in Denver, the incidental teaching approach (630), the behavioral program of Lovaas (631) at University of California at Los Angeles, and the Treatment and Education of Autistic and Related Communication Handicapped Children (TEACCH) approach developed by Schopler (632) and Mesibov (633). In the Rogers and Lewis approach, the child is stimulated to take the next step in relatedness, language, and conceptual knowledge, and social knowledge and social interaction are made as reinforcing as possible. Similarly, the Developmental Individual Difference Relationship Based Model approach of Greenspan and Wieder (634) focuses on promoting developmental progress. In the Lovaas approach, the child is the recipient of intensive behavior modification from a very early age. His group (635) reports that 47% of children in this program attained normal intellectual and academic functioning by first grade. Others, however, urge caution in accepting the full extent of their claims (636). Behavioral interventions are discussed by a number of authors (637–640). The TEACCH approach is a structured teaching program that places emphasis on the development of cognitive abilities and the use of visual skills (632,641).

There is good evidence that behavioral methods are the treatment of choice for problem behaviors, such as self-injury, and for establishing foundation skills such as attention, compliance, and basic language and can be successful in teaching components of even advanced social skills (637). A recent book (642) contains many extremely useful chapters with highly specific teaching programs and approaches. More naturalistic methods, such as play groups with typical peers, may effectively stimulate play and social development in children who have the foundational skills (643,644). A review of educational intervention in Great Britain (645) found that the common component to successful intervention is an early aggressive approach that includes teaching essential skills directly. Freeman (646) suggests general guidelines to follow when evaluating intervention plans.

Intellectual, language, and social status before treatment certainly contribute major variance to outcome; children with higher levels of communication, language, IQ, and social skills will have better outcomes in any type of program. It is probable that some types of children may do better with particular types of programs; for example, some clinicians refer children with severe social withdrawal, poor compliance, or poor attention to behavioral programs and refer children with odd but related social behavior and some degree of compliance and attention to more developmental or language-based programs. In the absence of any empirical data concerning which type of autistic child does well in which type of program, the clinician is forced to make a referral based on clinical judgment of the child and knowledge of available local resources. Information is lacking concerning which type of special education or behavioral treatment is most effective for which type of child, but there is general agreement that early aggressive intervention optimizes the chances for better outcome (525,647–649). There is a growing trend to refer autistic children for language and/or behavioral and other therapies as early as possible, family and community resources permitting (649).

At school age, there is a strong recent trend toward inclusion or integration in which the autistic child is placed, with or without an aide, into a normal public school class, sometimes with pull-out time for special services. Although there is ample evidence that autistic children's social interaction benefits from exposure to normal role models and peers who initiate and reinforce social interaction (650–652), it is too soon to tell whether the cohort of autistic children placed in mainstream settings will be able to cope with the increasingly complex social and academic activities of the later grade school years. An intermediate solution is the integrated classroom, especially in preschool, in which typical and atypical children are mixed in roughly equal proportions. Behavioral and educational treatments are covered in several very useful reviews in the *Journal of Autism and Developmental Disorders* [special issue on treatments for people with autism, 2000:3(5), see in particular references 653–655].

As part of the child's special education, and perhaps in addition to it, the child should receive aggressive language therapy. Except for children with severe articulatory dysfunction, this should focus more on the semantic and pragmatic use of language than on articulation. Prizant and Wetherby (649) discuss the analysis of the communicative functions of autistic utterances and outline a very pragmatically oriented approach to language therapy. Evidence shows that teaching functional communication skills not only increases interactive possibilities but may actually decrease problem behaviors (656). Teaching specific, pivotal social skills not only increases the occurrence of these skills but also generalizes to other social behaviors and to ratings of overall social appropriateness (657). Many autistic children have a variety of motor delays and impairments and may benefit from physical and/or occupational therapy. One approach, again lacking in well-controlled experimental validation but adhered to by many clinicians and parents, is the sensory integration school of occupational therapy, described by Cook (658). Dawson and Watling (659) review the scanty evidence concerning sensory integration and auditory integration training and conclude that evidence of their effectiveness is lacking.

So-called "facilitated communication," in which a facilitator assists the autistic individual to communicate by holding his or her hand or arm while he or she points to letters, is based on the notion that a severe motor apraxia interferes with the autistic individual's capacity to express him/herself. This treatment has led to claims of remarkable success with autistic children, adolescents, and adults thought to be nonverbal and severely retarded communicating via facilitation at a cognitively normal level. Controlled experiments have been done in which the autistic individual is exposed to information to which the facilitator has no access and then is questioned about this information. These studies show that the autistic individuals cannot communicate via facilitation at a level very discrepant from their usual verbal and nonverbal communicative capacity, unless the facilitator can influence the communication consciously or unconsciously (660,661). Tragically, several parents have been accused of abuse, supposedly by the child with autism, and at least one father has spent a significant amount of time in prison before the communication was disproved.

Finally, physicians, psychologists, and other clinicians must help families to obtain other necessary services, such as respite care, extended day programs, and summer programs to prevent the behavioral and cognitive regression that can occur within 2 months of unstructured time. They may also be able to suggest appropriate leisure activities, such as gymnastics, swimming, or play or social groups, that can provide constructive ways to spend after school hours and opportunities for social interaction with typical children and can promote self-esteem.

Prescription with therapies and services for the autistic individual must always include sensitivity to the often devastating effect of the disability on the family. Harris (572), Schopler and Mesibov (662), and Konstantareas (663) describe frameworks for assessing and helping families to cope with their disabled member. Harris (572) and Konstantareas (663) discuss how to recognize and try to address family stress, without undue attention to family dynamics that may imply adherence to the outdated notion that family pathology causes the autistic condition. Social support from other affected families and keeping abreast of the latest developments in treatment and other research can help families manage their affected children and their own emotional reactions. The Autism Society of America (8601 Georgia Ave., Suite 503, Silver Spring, MD 20910) publishes a regular newsletter with much information useful to parents.

SUMMARY AND CONCLUSIONS

PDD is a psychiatric diagnostic category that encompasses the autistic spectrum of disorders: autism, AS, CDD, Rett syndrome, and PDD not otherwise specified. All children diagnosed with PDD will have serious social impairments. Most will have some form of cognitive impairment, from profound retardation to rigidity of executive functions. Many will evince signs of neurologic impairment, such as motor dysfunction, seizures, and abnormal EEG. Very few will make near-normal adjustment in adult life, despite intense educational treatment programs and sensitive pharmacologic therapy, although the long-range outcome for children receiving intensive behavioral intervention at an early age, some of whom are doing extremely well, may be very positive.

Clinical practice rests on assessment; neurologic, medical, psychiatric, behavioral, neuropsychological, and language evaluations are crucial to treatment and management.

Future understanding of autism may depend on a change in research strategy. Despite evidence from all domains of research that autism is not a unitary disease entity, most investigations of the autistic spectrum are designed to search for a single cause for the entire spectrum, whether it be genetic, neuroanatomic, neurophysiologic, or psychological. Future research programs that are designed to explore various causal mechanisms within one framework will have a better chance to improve our understanding of the autistic spectrum of disorders.

The core deficit of the spectrum is social impairment. Neuroscience has begun exploration of the neurobiology of social behavior. Advances in our understanding of brain systems that serve sociability will also help us to understand the nature of the autistic spectrum.

REFERENCES

1. Szatmari P. The validity of autistic spectrum disorders: a literature review. *J Autism Dev Disord* 1992;22:583–600.
2. DiLalla DL, Rogers SJ. Domains of the Childhood Autism Rating Scale: relevance for diagnosis and treatment. *J Autism Dev Disord* 1994;24:115–128.
3. Kanner L. Autistic disturbances of affective contact. *Nerv Child* 1943;2:217–250.
4. Gillberg C. Autism and autistic-like conditions: subclasses among disorders of empathy. *J Child Psychol Psychiatry* 1992;33:813–842.
5. Bene E. A Rorschach investigation into the mothers of autistic children. *Br J Med Psychol* 1958;38:226–227.
6. Bettelheim B. *The empty fortress—infantile autism and the birth of the self.* New York: The Free Press, Collier-MacMillan, 1967.
7. Despert J. Some considerations relating to the genesis of autistic behavior in children. *Am J Orthopsychiatry* 1951;12:366–371.
8. Eisenberg L, Kanner L. Early infantile autism: 1943–55. *Am J Orthopsychiatry* 1956;26:55–65.
9. Mahler M. On child psychosis and schizophrenia: autistic and symbiotic infantile psychoses. *Psychoanal Study Child* 1952;7:286–305.
10. Mahler M, Gosliner B. On symbiotic child psychosis: genetic, dynamic and restitutive aspects. *Psychoanal Study Child* 1955;10:195–212.
11. Hobson RP. On psychoanalytic approaches to autism. *Am J Orthopsychiatry* 1990;60:324–336.
12. Rimland B. *Infantile autism: the syndrome and its implications.* New York: Appleton-Century-Crofts, 1964.
13. Ornitz E, Ritvo E. Perceptual inconstancy in early infantile autism. *Arch Gen Psychiatry* 1968;18:76–98.
14. Rutter M. Concepts of autism: a review of research. *J Child Psychol Psychiatry* 1968;9:1–25.
15. Baron-Cohen S, Leslie A, Frith U. Does the autistic child have a "theory of mind"? *Cognition* 1985;21:37–46.
16. Ozonoff S, Pennington B, Rogers S. Executive function deficits in high-functioning autistic individuals: relationship to theory of mind. *J Child Psychol Psychiatry* 1991;32:1081–1105.
17. Fein D, Pennington B, Markowitz P, et al. Toward a neuropsychological model of infantile autism: are the social deficits primary? *J Am Acad Child Psychiatry* 1986;25:198–212.
18. Hobson RP. Beyond cognition: a theory of autism. In: Dawson G, ed. *Autism: nature, diagnosis, and treatment.* New York: Guilford.
19. Kolvin I. Studies in the childhood psychoses: I. Diagnostic criteria and classification. *Br J Psychiatry* 1971;118:381–384.
20. Rutter M. Diagnosis and definition of childhood autism. *J Autism Child Schizophr* 1978;8:139–161.
21. Ritvo E, Freeman B. National Society for Autistic Children definition of the syndrome of autism. *J Autism Child Schizophr* 1978;8:162–169.
22. Schopler E. Diagnosis and definition of autism. *J Autism Child Schizophr* 1978;8:167–169.
23. American Psychiatric Association. *Diagnostic and statistical manual of mental disorders (DSM-III),* 3rd ed. Washington, DC: American Psychiatric Association, 1980.
24. Wing L, Gould J. Severe impairments of social interaction and associated abnormalities in children: epidemiology and classification. *J Autism Dev Disord* 1979;9:11–29.
25. Wing L, Attwood A. Syndromes of autism and atypical development. In: Cohen D, Donnellan A, Paul R, eds. *Handbook of autism and pervasive developmental disorder.* New York: John Wiley & Sons, 1987.

26. American Psychiatric Association. *Diagnostic and statistical manual of mental disorders, (DSM-III-R)*, 3rd ed., revised Washington, DC: American Psychiatric Association, 1987.

27. American Psychiatric Association. *Diagnostic and statistical manual of mental disorders*, 4th ed. Washington DC: American Psychiatric Association, 1994.

28. World Health Organization. *Mental disorders: the ICD-10 classification of mental and behavioral disorders. Diagnostic criteria for research.* Geneva, Switzerland: World Health Organization, 1993.

29. Szatmari P, Volkmar F, Walter S. Evaluation of diagnostic criteria for autism using latent class models. *J Am Acad Child Adolesc Psychiatry* 1995;34:216–222.

30. Waterhouse L, Morris R, Allen D, et al. Diagnosis and classification in autism. *J Autism Dev Disord* 1996;26:59–86.

31. Dawson G, Meltzoff A, Osterling J, et al. Neuropsychological correlates of early symptoms in autism. *Child Dev* 1998;69:1276–1285.

32. Waterhouse L, Fein D, Modahl C. Neurofunctional mechanisms in autism. *Psychol Rev* 1996;103:457–489.

33. Roeyers H, Van Oost P, Bothuyne S. Immediate imitation and joint attention in young children with autism. *Dev Psychopathol* 1998;10:441–450.

34. Leekam S, Lopez B, Moore C. Attention and joint attention in preschool children with autism. *Dev Psychol* 2000;36:261–273.

35. McArthur D, Adamson L. Joint attention in preverbal children: autism and developmental language disorder. *J Autism Dev Disord* 1996;26:481–496.

36. Hauck M, Fein D, Waterhouse L, et al. Social initiations by autistic children to adults and other children. *J Autism Dev Disord* 1995;25:579–595.

37. Stone W, Caro-Martinez L. Naturalistic observations of spontaneous communications in autistic children. *J Autism Dev Disord* 1990;20:437–453.

38. Stone W, Lemanek K. Parental report of social behaviors in autistic preschoolers. *J Autism Dev Disord* 1990;20:513–522.

39. Volkmar F. Social development. In: Cohen D, Donnellan A, eds. *Handbook of autism and pervasive developmental disorders.* New York: John Wiley & Sons, 1987.

40. Kasari C, Sigman M, Yirmiya N. Focused and social attention of autistic children in interactions with familiar and unfamiliar adults: a comparison of autistic, mentally retarded, and normal children. *Dev Psychopathol* 1993;5:403–414.

41. Riquet CB, Taylor ND, Benroya S, et al. Symbolic play in autistic, Downs and normal children of equivalent mental age. *J Autism Dev Disord* 1981;11:439–448.

42. Lewis V, Boucher J. Spontaneous, instructed and elicited play in relatively able autistic children. *Br J Dev Psychol* 1988;6:325–339.

43. Wainwright L, Fein D. Results of play analyses. In: Rapin I, ed. *Preschool children with inadequate communication: developmental language disorder, autism, low IQ. Clinics in Developmental Medicine*, No. 139. London: Mac Keith Press, 1998:173–189.

44. Roeyers H, Van Berckelaer O. Play in autistic children. *Commun Cogn* 1994;27:349–359.

45. Charman T, Swettenham J, Baron-Cohen S, et al. Infants with autism: an investigation of empathy, pretend play, joint attention, and imitation. *Dev Psychol* 1997;33:781–789.

46. Rogers SH, Ozonoff S, Maslin-Cole C. Developmental aspects of attachment behavior in young children and pervasive developmental disorder. *J Am Acad Child Adolesc Psychiatry* 1991;32:1274–1282.

47. Howlin P, Mawhood L, Rutter M. Autism and developmental receptive language disorder—a comparative follow-up in early

adult life. II: Social, behavioural, and psychiatric outcomes. *J Child Psychol Psychiatry* 2000;41:561–578.

48. Hobson RP. The autistic child's appraisal of expressions of emotion. *J Child Psychol Psychiatry* 1986;27:321–342.

49. Hobson RP. The autistic child's appraisal of expressions of emotion: a further study. *J Child Psychol Psychiatry* 1986;27:671–680.

50. Ekman P, Friesen W. *Unmasking the face: a guide to recognizing emotions from facial cues.* Englewood Cliffs, NJ: Prentice-Hall, 1975.

51. Izard C. *The face of emotion.* New York: Appleton-Century-Crofts, 1971.

52. Hobson R. Methodological issues for experiments on autistic individuals' perception and understanding of emotion. *J Child Psychol Psychiatry* 1991;32:1135–1158.

53. Hobson R, Ouston J, Lee A. Emotion recognition in autism: coordinating faces and voices. *Psychol Med* 1988;18:911–923.

54. Tantam D, Monaghan L, Nicholson H, et al. Autistic children's ability to interpret faces: a research note. *J Child Psychol Psychiatry* 1989;30:623–630.

55. MacDonald H, Rutter M, Howlin P, et al. Recognition and expression of emotional cues by autistic and normal adults. *J Child Psychol Psychiatry* 1989;30:865–877.

56. Braverman M, Fein D, Lucci D, et al. Affect comprehension in children with pervasive developmental disorders. *J Autism Dev Dis* 1989;19:301–315.

57. Fein D, Lucci D, Braverman M, et al. Comprehension of affect in context in children with pervasive developmental disorders. *J Child Psychol Psychiatry* 1992;33:1157–1157.

58. Boucher J, Lewis V, Collins G. Voice processing abilities in children with autism, children with specific language impairments, and young typically developing children. *J Child Psychol Psychiatry* 2000;41:847–857.

59. Hertzig M, Snow M, Sherman M. Affect and cognition in autism. *J Am Acad Child Adolesc Psychiatry* 1989;28:195–199.

60. Volkmar F, Sparrow S, Rende R, et al. Facial perception in autism. *J Child Psychol Psychiatry* 1989;30:591–598.

61. Prior M, Dahlstrom B, Squires T. Autistic children's knowledge of thinking and feeling states in other people. *J Child Psychol Psychiatry* 1990;31:587–601.

62. Ozonoff S, Pennington B, Rogers S. Are there emotion perception deficits in young autistic children? *J Child Psychol Psychiatry* 1990;31:343–361.

63. Critchley H, Daly E, Bullmore E, et al. The functional neuroanatomy of social behavior: changes in cerebral blood flow when people with autistic disorder process facial expressions. *Brain* 2000;123:2203–2213.

64. Yirmiya N, Kasari C, Sigman M, et al. Facial expressions of affect in autistic, mentally retarded and normal children. *J Child Psychol Psychiatry* 1989;30:725–735.

65. Fine J, Bartolucci G, Ginsberg G, et al. The use of intonation to communicate in pervasive developmental disorders. *J Child Psychol Psychiatry* 1991;32:771–782.

66. Bauminger N, Kasari C. Loneliness and friendship in high-functioning children with autism. *Child Dev* 2000;71:447–456.

67. Zajonc R. Feeling and thinking: preferences need no inferences. *Am Psychol* 1980;35:151–175.

68. Waterhouse L, Fein D. Social or cognitive or both? Crucial dysfunctions in autism. In: Gillberg C, ed. *Diagnosis and treatment of autism.* New York: Plenum, 1989.

69. Lord C, Risi S, Lambrecht L, et al. The Autism Diagnostic Observation Schedule—Generic: a standard measure of social and communication deficits associated with the spectrum of autism. *J Autism Dev Disord* 2000;30:205–223.

70. Lord C, Rutter M, DiLavore P, et al. *Autism Diagnostic Observation Schedule—WPS (ADOS-WPS)*. Los Angeles: Western Psychological Services, 1999.

71. Yirmiya N, Erel O, Shaked M, et al. Meta-analyses comparing theory of mind abilities of individuals with autism, individuals with mental retardation, and normally developing individuals. *Psychol Bull* 1998;124:283–307.

72. Happe FGE. An advanced test of theory of mind: understanding of story character's thoughts and feelings by able autistic, mentally handicapped and normal children and adults. *J Autism Dev Disord* 1994;24:129–154.

73. Tager-Flusberg H, Sullivan K. Predicting and explaining behavior: a comparison of autistic, mentally retarded and normal children. *J Child Psychol Psychiatry* 1994;35:1059–1075.

74. Baron-Cohen S, Goodhart F. The seeing leads to knowing deficit in autism: the Pratt and Bryant probe. *Br J Dev Psychol* 1994;12:397–401.

75. Yirmiya N, Solomonica-Levi D, Shulman C. The ability to manipulate behavior and to understand manipulation of beliefs: a comparison of individuals with autism, mental retardation, and normal development. *Dev Psychol* 1996;32:62–69.

76. Baron-Cohen S, Jolliffe T, Mortimore C, et al. Another advanced test of theory of mind: evidence from very high functioning adults with autism or Asperger syndrome. *J Child Psychol Psychiatry* 1997;38:813–822.

77. Klin A. Attributing social meaning to ambiguous visual stimuli in higher-functioning autism and Asperger syndrome: the social attribution task. *J Child Psychol Psychiatry* 2000;41:831–846.

78. Shields J, Varley R, Broks P, et al. Social cognition in developmental language disorders and high level autism. *Dev Med Child Neurol* 1996;38:487–495.

79. Peterson C, Siegal M. Deafness, conversation and theory of mind. *J Child Psychol Psychiatry* 1995;36:459–474.

80. Jarrold C, Butler D, Cottington E, et al. Linking theory of mind and central coherence bias in autism and in the general population. *Dev Psychol* 2000;36:126–138.

81. Russell J, Saltmarsh R, Hill E. What do executive factors contribute to the failure on false belief tasks by children with autism? *J Child Psychol Psychiatry* 1999;40:859–868.

82. Klin A, Volkmar F, Sparrow S. Autistic social dysfunction: some limitations of the theory of mind hypothesis. *J Child Psychol Psychiatry* 1992;33:861–876.

83. Happe F, Frith U. The neuropsychology of autism. *Brain* 1996;119:1377–1400.

84. Waterhouse L, Fein D. Language skills in developmentally disabled children. *Brain Lang* 1982;15:307–333.

85. Fein D, Dunn M, Allen D, et al. Neuropsychological and language data. In: Rapin I, ed. *Preschool children with inadequate communication: developmental language disorder, autism, low IQ*. Clinics in developmental medicine, No. 139. London: Mac Keith Press, 1996:123–154.

86. Tager-Flusberg H. A psycholinguistic perspective on language development in the autistic child. In: Dawson G, ed. *Autism: nature, diagnosis and treatment*. New York: Guilford Press, 1989.

87. Waterhouse L, Fein D. Language skills in developmentally disabled children. *Brain Lang* 1982;15:307–333.

88. Minshew N, Goldstein G, Siegel D. Speech and language in high-functioning autistic individuals. *Neuropsychology* 1995;9:255–261.

89. Mawhood L, Howlin P, Rutter M. Autism and developmental receptive language disorder—a comparative follow-up in early adult life. I: Cognitive and language outcomes. *J Child Psychol Psychiatry* 2000;41:547–559.

90. Rutter M. The description and classification of infantile autism. In: Churchill D, Alpern G, DeMyer M, eds. *Infantile autism: proceedings of the Indiana University Colloquium*. Springfield, IL: Charles C Thomas, 1971:8–28.

91. Wainwright JA, Bryson S. Visual-spatial orienting in autism. *J Autism Dev Disord* 1996;26:423–438.

92. Lovaas OI, Schreibman L, Koegel R, et al. Selective responding by autistic children to multiple sensory input. *J Abnorm Psychol* 1971;77:211–222.

93. Fein D, Tinder P, Waterhouse L. Stimulus generalization in autistic and normal children. *J Child Psychol Child Psychiatry* 1979;20:325–335.

94. Rimland B, Fein D. Special talents of autistic savants. In: Obler L, Fein D, eds. *The exceptional brain: neuropsychology of talent and special abilities*. New York: Guilford, 1988.

95. Townsend J, Courchesne E, Egaas B. Slowed orienting of covert visual-spatial attention in autism: specific deficits associated with cerebellar and parietal abnormality. *Dev Psychopathol* 1996;8:563–584.

96. Harris N, Courchesne E, Townsend J, et al. Neuroanatomic contributions to slowed orienting of attention in children with autism. *Cogn Brain Res* 1999;8:61–71.

97. Townsend J, Courchesne E, Eggas B. Visual attention deficits in autistic adults with cerebellar and parietal abnormalities. *Soc Neurosci Abstr* 1992;18:332.

98. Courchesne E, Townsend J, Akshoomoff N, et al. Impairment in shifting attention in autistic and cerebellar patients. *Behav Neurosci* 1994;105:848–865.

99. Pascualvaca D, Fantie B, Papageorgiou M, et al. Attentional capacities in children with autism: is there a general deficit in shifting focus? *J Autism Dev Disord* 1998;28:467–478.

100. Plaisted K, Swettenham J, Rees L. Children with autism show local precedence in a divided attention task and global precedence in a selective attention task. *J Child Psychol Psychiatry* 1999;40:733–742.

101. Frith U. *Autism: explaining the enigma*. London: Basil Blackwell, 1989.

102. Jolliffe T, Baron-Cohen S. Are people with autism and Asperger syndrome faster than normal on the embedded figures test? *J Child Psychol Psychiatry* 1997;38:527–534.

103. Hutt S, Hutt C, Lee D, et al. A behavioral and electroencephalographic study of autistic children. *J Psychiatr Res* 1965;3:181–197.

104. DeLong GR. Autism, amnesia, hippocampus and learning. *Neurosci Biobehav Rev* 1992;16:63–70.

105. Dawson G, Lewy A. Arousal, attention and the socioemotional impairments of individuals with autism. In: Dawson G, ed. *Autism: nature, diagnosis and treatment* New York: Guilford, 1989.

106. Kinsbourne M. Cerebral-brainstem relations in infantile autism. In: Dawson G, ed. *Autism: nature, diagnosis and treatment*. New York: Guilford Press, 1989.

107. Boucher J. Immediate free recall in early childhood autism: another point of behavioral similarity with the amnesic syndrome. *Br J Psychol* 1981;72:211–215.

108. Boucher J. Memory for recent events in amnesic children. *J Autism Dev Disord* 1981;11:293–302.

109. Minshew N, Goldstein G. Is autism an amnesic disorder? Evidence from the California Verbal Learning Test. *Neuropsychology* 1993;7:209–216.

110. Renner P, Klinger L, Klinger M. Implicit and explicit memory in autism: is autism an amnesic disorder? *J Autism Dev Disord* 2000;30:3–14.

111. Bowler D, Matthews N, Gardiner J. Asperger's syndrome and memory: similarity to autism but not amnesia. *Neuropsychologia* 1997;35:65–70.

112. Fein D, Dunn M, Allen D, et al. Language and neuropsychological findings. In: Wing IRL, ed. *Preschool children with inadequate communication: developmental language disorder, autism, low IQ.* London: Mac Keith, 1996:123–154.
113. Minshew NJ, Goldstein G, Muenz LR, et al. Neuropsychological functioning nonmentally retarded autistic individuals. *J Clin Exp Neuropsychol* 1992;14:749–761.
114. Bennetto L, Pennington B, Rogers S. Intact and impaired memory functions in autism. *Child Dev* 1996;67:1816–1835.
115. Bachevalier J. Medial temporal lobe structures and autism: a review of clinical and experimental findings. *Neuropsychologia* 1994;32:627–648.
116. Ozonoff SJ J. Brief report: specific executive function profiles in three neurodevelopmental disorders. *J Autism Dev Disord* 1999;29:171–177.
117. Ozonoff S, Rogers S, Pennington B. Asperger's syndrome: evidence of an empirical distinction from high-functioning autism. *J Child Psychol Psychiatry* 1991;32:1107–1122.
118. Rumsey J, Hamburger S. Neuropsychological findings in high-functioning men with infantile autism, residual state. *J Clin Exp Neuropsychol* 1988;10:210–221.
119. Rumsey J, Hamburger S. Neuropsychological divergence of high-level autism and severe dyslexia. *J Autism Dev Disord* 1990;20:155–168.
120. Szatmari P, Bartolucci G, Bremner R, et al. A follow-up study of high-functioning autistic children. *J Autism Dev Disord* 1989;19:213–225.
121. Ozonoff, S, McEvoy RE. A longitudinal study of executive function and theory of mind and development in autism. *Dev Psychopathol* 1994;6:415–431.
122. Hughes C, Russell J, Robbins TW. Evidence for executive dysfunction in autism. *Neuropsychologia* 1994;32:477–492.
123. Szatmari P, Tuff L, Finlayson A, et al. Asperger's syndrome and autism: neurocognitive aspects. *J Am Acad Child Adolesc Psychiatry* 1990;29:130–136.
124. Griffith E, Pennington B, Wehner E, et al. Executive functions in young children with autism. *Child Dev* 1999;70:817–832.
125. McEvoy R, Rogers S, Pennington B. Executive function and social communication deficits in young autistic children. *J Child Psychol Psychiatry* 1993;34:563–578.
126. Klin A, Volkmar F, Sparrow S, et al. Validity and neuropsychological characterization of Asperger syndrome: convergence with nonverbal learning disabilities syndrome. *J Child Psychol Psychiatry* 1995;36:1127–1140.
127. Miller J, Ozonoff S. The external validity of Asperger disorder: lack of evidence from the domain of neuropsychology. *J Abnorm Psychol* 2000;109:227–238.
128. O'Brien S. The validity and reliability of the Wing Subgroups Questionnaire. *J Autism Dev Disord* 1996;26:321–335.
129. Bonde E. Comorbidity and subgroups in childhood autism. *Eur Child Adolesc Psychiatry* 2000;9:7–10.
130. Paul R. Natural history. In: Cohen D, Donnellan A, Paul R, eds. *Handbook of autism and pervasive developmental disorder.* New York: John Wiley & Sons, 1987.
131. Gillberg C, Steffenburg S. Outcome and prognostic factors in infantile autism and similar conditions: a population-based study of 46 cases followed through puberty. *J Autism Dev Disord* 1987;17:273–287.
132. Freeman B, Rahbar B, Ritvo E, et al. The stability of cognitive and behavioral parameters in autism: a twelve-year prospective study. *J Am Acad Child Adolescent Psychiatry* 1991;30:479–482.
133. Kobayashi R, Murata T, Yoshinaga K. A follow-up study of 201 children with autism in Kyusha and Yamaguchi areas, Japan. *J Autism Dev Disord* 1992;22:395–411.
134. Waterhouse L, Fein D. Longitudinal trends in cognitive skills
135. Venter A, Lord C, Schopler E. A follow-up study of high-functioning autistic children. *J Child Psychol Psychiatry* 1992;33:489–507.
136. Lotter V. Epidemiology of autistic conditions in young children: I. Prevalence. *Soc Psychiatry* 1966;1:124–137.
137. Chakrabarti S, Fombonne E. Pervasive developmental disorders in preschool children. *JAMA* 2001;285:3141–3142.
138. Gillberg C, Persson U, Grufman M, et al. Psychiatric disorders in mildly and severely mentally retarded urban children and adolescents: epidemiological aspects. *Br J Psychiatry* 1986;149:68–74.
139. Gillberg C. Infantile autism and other childhood psychoses in a Swedish urban region. Epidemiological aspects. *J Child Psychol Psychiatry* 1984;25:377–403.
140. Steffenburg S, Gillberg C. Autism and autistic-like conditions in Swedish rural and urban areas: a population study. *Br J Psychiatry* 1986;149:81–87.
141. Gillberg C, Steffenburg S, Schaumann H. Is autism more common now than ten years ago? *Br J Psychiatry* 1991;158:403–409.
142. Bryson S, Clark B, Smith I. First report of a Canadian epidemiological study of autistic syndromes. *J Child Psychol Psychiatry* 1988;29:433–445.
143. Wing L. Sex ratios in early childhood autism and related conditions. *Psychiatry Res* 1981;5:129–137.
144. Hulse W. Dementia infantilis [translation of T. Heller's 1930 article]. *J Nerv Ment Dis* 1954;119:471–477.
145. Evans-Jones L, Rosenbloom L. Disintegrative psychosis in childhood. *Dev Med Child Neurol* 1978;20:462–470.
146. Burd L, Fisher W, Kerbeshian J. Childhood onset pervasive developmental disorder. *J Child Psychol Psychiatry* 1988;29:155–163.
147. Hill A, Rosenbloom L. Disintegrative psychosis of childhood: teenage follow-up. *Dev Med Child Neurol* 1986;28:34–40.
148. Volkmar F, Cohen D. Disintegrative disorder or "late onset" autism. *J Child Psychol Psychiatry* 1989;30:717–724.
149. Kurita H. The concept and nosology of Heller's syndrome: review of articles and report of two cases. *Jpn J Psychiatry Neurol* 1988;42:785–793.
150. Volkmar F. Childhood disintegrative disorder: issues for DSM-IV. *J Autism Dev Disord* 1992;22:625–642.
151. Hagberg B, Aicardi J, Dias K, et al. A progressive syndrome of autism, dementia, ataxia, and loss of purposeful hand use in girls: Rett's syndrome: report of 35 cases. *Ann Neurol* 1983;14:471–479.
152. Hagberg B, Goutieres F, Hanefeld F, et al. Rett syndrome: criteria for inclusion and exclusion. *Brain Dev* 1985;7:372–373.
153. Hagberg B, Witt-Engerstrom I. Rett syndrome: a suggested staging system for describing impairment profile with increasing age towards adolescence. *Am J Med Genet* 1986;24[Suppl 1]:47–59.
154. Amir RE, Van den Veyver IB, Wan M, et al. Rett syndrome is caused by mutations in X-linked MECP2 encoding methyl-CpG-binding protein 2. *Nat Genet* 1999;23:185–189.
155. Asperger H. 'Autistic psychopathy' in childhood. In: Frith U, ed. *Autism and Asperger syndrome.* Cambridge: Cambridge University Press, 1991 (original article published 1944). Frith U, translator.
156. Van Krevelen D. The psychopathology of autistic psychopathy. *Acta Paedopsychiatr* 1962;29:22–31.
157. Van Krevelen D. Early infantile autism and autistic psychopathy. *J Autism Child Schizophr* 1971;1:82–86.

158. Wing L. Asperger's syndrome: a clinical account. *Psychol Med* 1981;11:115–129.
159. Attwood T. *Asperger's syndrome.* London: Jessica Kingsley Publishers, 1998.
160. Klin A, Volkmar FR, eds. *Asperger syndrome.* New York: Guilford Press, 2000.
161. Tantam D. Asperger syndrome in adulthood. In: Frith U, ed. *Autism and Asperger syndrome.* Cambridge: Cambridge University Press, 1991.
162. Delong G, Dwyer J. Correlation of family history with specific autistic subgroups: Asperger's syndrome and bipolar affective disease. *J Autism Dev Disord* 1988;18:593–600.
163. Gillberg C. Clinical and neurobiological aspects of Asperger syndrome in six family studies. In: Frith U, ed. *Autism and Asperger syndrome.* Cambridge: Cambridge University Press, 1991.
164. Tantam D. Lifelong eccentricity and social isolation: I. Psychiatric, social, and forensic aspects. *Br J Psychiatry* 1988;153:777–782.
165. Tantam D. Lifelong eccentricity and social isolation: II. Asperger's syndrome or schizoid personality disorder? *Br J Psychiatry* 1988;153:783–791.
166. Wing L. The relationship between Asperger's syndrome and Kanner's autism. In: Frith U, ed. *Autism and Asperger syndrome.* Cambridge: Cambridge University Press, 1991.
167. Wolff S, Barlow A. Schizoid personality in childhood: a comparative study of schizoid, autistic and normal children. *J Child Psychol Psychiatry* 1979;20:29–46.
168. Wolff S, Chick J. Schizoid personality in childhood: a controlled follow-up study. *Psychol Med* 1980;10:85–100.
169. Nagy J, Szatmari P. A chart review of schizotypal personality disorders in children. *J Autism Dev Disord* 1986;16:351–367.
170. Ghaziuddin M, Tsai L, Ghaiuddin N. A comparison of the diagnostic criteria for Asperger syndrome. *J Autism Dev Disord* 1992;22:643–649.
171. Shea V, Mesibov G. Brief report: the relationship of learning disabilities and higher-level autism. *J Autism Dev Disord* 1985;15:425–435.
172. Weintraub S, Mesulam M. Developmental learning disabilities of the right hemisphere: emotional, interpersonal, and cognitive components. *Arch Neurol* 1983;40:463–468.
173. Voeller K. Right-hemisphere deficit syndrome in children. *Am J Psychiatry* 1986;143:1004–1009.
174. Denckla M. The neuropsychology of social-emotional learning disabilities. *Arch Neurol* 1983;40:461–462.
175. Morris M. Social-emotional learning disability. *J Exp Clin Neuropsychol* 1992;14:369 (abst).
176. Rourke B. The syndrome of nonverbal learning disabilities: developmental manifestation in neurological disease, disorder, and dysfunction. *Clin Neuropsychol* 1988;2:293–330.
177. Rourke B. *Nonverbal learning disabilities: the syndrome and the model.* New York: Guilford, 1989.
178. Marriage K, Miles T, Stokes D, et al. Clinical and research implications of the co-occurrence of Asperger's and Tourette's syndromes. *Aust N Z J Psychiatry* 1993;27:666–674.
179. Kerbeshian J, Burd L. Asperger's syndrome and Tourette syndrome: the case of the pinball wizard. *Br J Psychiatry* 1986;148:731–736.
180. Burd L, Fisher W, Kerbeshian J, et al. Is development of Tourette disorder a marker for improvement in patients with autism and other pervasive developmental disorders? *J Am Acad Child Adolesc Psychiatry* 1987;26:162–165.
181. Comings D, Comings B. Clinical and genetic relationships between autism-pervasive developmental disorder and Tourette

syndrome: a study of 19 cases. *Am J Med Genet* 1991;39:180–191.
182. Sverd J. Tourette syndrome and autistic disorder: a significant relationship. *Am J Med Genet* 1991;39:173–179.
183. Kolvin I, Ounsted C, Humphrey M, et al. Studies in the childhood psychoses: II. The phenomenology of childhood psychoses. *Br J Psychiatry* 1971;118:385–395.
184. Rutter M. Childhood schizophrenia reconsidered. *J Autism Child Schizophr* 1972;2:315–337.
185. Green W, Campbell M, Hardesty A, et al. A comparison of schizophrenic and autistic children. *J Am Acad Child Psychiatry* 1984;23:399–409.
186. Petty L, Ornitz E, Michelman J, et al. Autistic children who become schizophrenic. *Arch Gen Psychiatry* 1984;41:129–135.
187. Cantor S. *Childhood schizophrenia.* New York: Guilford, 1988.
188. Cantor S, Evans J, Pearce J, et al. Childhood schizophrenia: present but not accounted for. *Am J Psychiatry* 1982;139:758–762.
189. Tanguay P, Cantor S. Schizophrenia in children. *J Am Acad Child Psychiatry* 1986;25:591–594.
190. Volkmar F, Cohen D. Comorbid association of autism and schizophrenia. *Am J Psychiatry* 1991;148:1705–1707.
191. Bartak L, Rutter M, Cox A. A comparative study of infantile autism and specific developmental receptive language disorders: III. Discriminant function analysis. *J Autism Child Schizophr* 1977;7:383–396.
192. [Reserved.]
193. Tuchman RF, Rapin I, Shinnar S. Autistic and dysphasic children. II. Epilepsy. *Pediatrics* 1991;88:1219–1225.
194. Volkmar F, Nelson D. Seizure disorders in autism. *J Am Acad Child Adolesc Psychiatry* 1990;29:127–129.
195. Olsson I, Steffenburg S, Gillberg C. Epilepsy in autism and autisticlike conditions: a population-based study. *Arch Neurol* 1988;45:666–668.
196. Rapin I, Katzman R. Neurobiology of autism. *Ann Neurol* 1998;43:7–14.
197. Tanguay P. Pervasive developmental disorders: a 10-year review. *J Am Acad Child Adolesc Psychiatry* 2000;39:1079–1095.
198. Buitelaar JK, Willemsen-Swinkels SHN. Autism: current theories regarding its pathogenesis and implications for rational pharmacotherapy. *Paediatr Drugs* 2000;2:67–81.
199. Trottier G, Srivastava L, Walker C. Etiology of infantile autism: a review of recent advances in genetic and neurobiological research. *J Psychiatry Neurosci* 1999;24:103–115.
200. Hagerman R. Physical and behavioral phenotype. In: Hagerman R, Silverman A, eds. *Fragile X syndrome: diagnosis, treatment, and research.* Baltimore: Johns Hopkins University Press, 1991:3–68.
201. Cohen I, Vietze P, Sudhalter V, et al. Parent-child dyadic gaze patterns in fragile X males and fragile X females with autistic disorder. *J Child Psychol Psychiatry* 1989;30:845–856.
202. Cohen I, Fisch G, Sudhalter V, et al. Social gaze, social avoidance and repetitive behavior in fragile X males: a controlled study. *Am J Ment Retard* 1988;92:436–446.
203. Einfeld S, Molony H, Hall H. Autism is not associated with the fragile X syndrome. *Am J Med Genet* 1989;34:187–193.
204. Payton JB, Steele M, Wenger S, et al. The fragile X marker and autism in perspective. *J Am Acad Child Adolesc Psychiatry* 1989;28:417–421.
205. Piven J, Gayle J, Landa R, et al. The prevalence of fragile X in a sample of autistic individuals diagnosed using a standardized interview. *J Am Acad Child and Adolesc Psychiatry* 1991;30:825–830.
206. Holroyd S, Reiss A, Bryan N. Autistic features in Joubert

syndrome: a genetic disorder with agenesis of the cerebellar vermis. *Biol Psychiatry* 1991;29:287–294.

207. Bailey A, LeCouteur A, Gottesman I, et al. Autism as a strongly genetic disorder: evidence from a British twin study. *Psychol Med* 1995;25:63–77.

208. Maestrini E, Paul A, Monaco A, et al. Identifying autism susceptibility genes. *Neuron* 2000;28:19–24.

209. Folstein S, Rutter M. Infantile autism: a genetic study of 21 twin pairs. *J Child Psychol Psychiatry* 1977;18:297–321.

210. Ritvo E, Freeman B, Mason-Brothers A, et al. Concordance for the syndrome of autism in 40 pairs of afflicted twins. *Am J Psychiatry* 1985;142:74–77.

211. Steffenburg S, Gillberg C, Hellgren L, et al. A twin study of autism in Denmark, Finland, Iceland, Norway and Sweden. *J Child Psychol Psychiatry* 1989;30:405–416.

212. Smalley S, Asarnow R, Spence A. Autism and genetics: a decade of research. *Arch Gen Psychiatry* 1988;45:953–961.

213. Piven J, Gayle J, Chase G, et al. A family history study of neuropsychiatric disorders in the adult siblings of autistic individuals. *J Am Acad Child Adolesc Psychiatry* 1990;29:177–183.

214. Ritvo E, Freeman B, Pingree C, et al. The UCLA-University of Utah epidemiologic survey of autism: prevalence. *Am J Psychiatry* 1989;146:194–199.

215. Ritvo E, Jorde L, Mason-Brothers A, et al. The UCLA-University of Utah epidemiologic survey of autism: recurrence risk estimates and genetic counseling. *Am J Psychiatry* 1989;146:1032–1036.

216. Piven J, Palmer P, Jacobi D, et al. Broader autism phenotype. Evidence from a family history study of multiple-incidence autism families. *Am J Psychiatry* 1997;154:185–190.

217. Bolton P, MacDonald H, Pickles A, et al. A case-control family history study of autism. *J Child Psychol Psychiatry* 1994;35:877–900.

218. Delong G, Dwyer J. Correlation of family history with specific autistic subgroups: Asperger's syndrome and bipolar affective disease. *J Autism Dev Disord* 1988;18:593–600.

219. Evans-Jones L, Rosenbloom L. Disintegrative psychosis in childhood. *Dev Med Child Neurol* 1978;20:462–470.

220. Kurita H. Infantile autism with speech loss before the age of thirty months. *J Am Acad Child Psychiatry* 1985;24:191–196.

221. Piven J, Palmer P, Landa R, et al. Personality and language characteristics in parents from multiple-incidence autism families. *Am J Med Genet* 1997;74:398–411.

222. Bailey A, Palferman S, Heavey L, et al. Autism: the phenotype in relatives. *J Autism Dev Dis* 1998;28:439–445.

223. Szatmari P, Jones MB, Holden J, et al. High phenotypic correlations among siblings with autism and pervasive developmental disorders. *Am J Med Genet* 1996;67:354–360.

224. August G, Stewart M, Tsai L. The incidence of cognitive disabilities in the siblings of autistic children. *Br J Psychiatry* 1981;138:416–422.

225. Minton J, Campbell M, Green W. Cognitive assessment of siblings of autistic children. *J Am Acad Child Psychiatry* 1982;21:256–261.

226. Freeman B, Ritvo E, Mason-Brothers A, et al. Psychometric assessment of first-degree relatives of 62 autistic probands in Utah. *Am J Psychiatry* 1989;146:361–364.

227. Wolff S, Narayan S, Moyes B. Personality characteristics of parents of autistic children: a controlled study. *J Child Psychol Psychiatry* 1988;29:143–155.

228. Folstein S, Piven J. Etiology of autism: genetic influences. *Pediatrics* 1991;87:767–773.

229. Hughes C, Leboyer M, Bouvard M. Executive function in parents of children with autism. *Psychol Med* 1997;27:209–220.

230. Steingard R, Biederman J. Lithium responsive maniclike symptoms in two individuals with autism and mental retardation. *J Am Acad Child Adolesc Psychiatry* 1987;26:632–635.

231. Lainhart JE, Folstein SE. Affective disorders in people with autism: a review of published cases. *J Autism Dev Disord* 1994;24:587–601.

232. Gillberg C, Gillberg I, Steffenburg S. Siblings and parents of children with autism: a controlled population based study. *Dev Med Child Neurol* 1992;34:389–398.

233. Cook EHJ. Genetics of autism. *Ment Retard Dev Disabil* 1998;4:113–120.

234. Gillberg C. Chromosomal disorders and autism. *J Autism Dev Disord* 1998;28:415–425.

235. Pickles A, Bolton P, MacDonald H, et al. Latent-class analysis of recurrence risks for complex phenotypes with selection and measurement error: a twin and family history study of autism. *Am J Hum Genet* 1995;57:717–726.

236. Internation Molecular Genetic Study of Autism Consortium (IMFSAC). A full genome screen for autism with evidence for linkage to a region on chromosome 7q. *Hum Mol Genet* 1994;7:571–578.

237. Phillipe A, Martinez M, Bataille-Guillot M, et al. Genome-wide scan for autism susceptibility genes. *Hum Mol Genet* 1999;8:805–812.

238. Barrett S, Beck JC, Bernier R, et al. An autosomal gene screen for autism. Collaborative linkage study of autism. *Am J Med Genet* 1999;88:609–615.

239. Ashley-Koch A, Wolpert CM, Menold MM, et al. Genetic studies of autistic disorder and chromosome 7. *Genomics* 1999;61:227–236.

240. Bass MP, Menold MM, Wolpert CM, et al. Genetic studies in autistic disorder and chromosome 15. *Neurogenetics* 2000;2:219–226.

241. Risch N, Spiker D, Lotspeich L, et al. A genomic screen of autism: evidence for a multilocus etiology. *Am J Hum Genet* 1999;65:493–507.

242. Rodier PM. The early origins of autism. *Sci Am* 2000;282:56–63.

243. Ingram JL, Stodgell CJ, Hyman SL, Figlewicz DA, Weitkamp LR, Rodier PM. Discovery of allelic variants of HOXA1 and HOXB1: genetic susceptibility to autism spectrum disorders. *Teratology* 2000;62:393–405.

244. London E, Etzel R. The environment as an etiologic factor in autism: a new direction for research. *Environ Health Perspect* 2000;108:401–404.

245. Gillberg C, Steffenburg S, Wahlstrom J, et al. Autism associated with marker chromosome. *J Am Acad Child Adolesc Psychiatry* 1991;30:489–494.

246. Hotopf M, Bolton P. A case of autism associated with partial tetrasomy 15. *J Autism Dev Disord* 1995;25:41–49.

247. Ghaziuddin M, Sheldon S, Benkataraman S, et al. Autism associated with tetrasomy 15: a further report. *Eur Child Adolesc Psychiatry* 1993;2:226–230.

248. Bundy S, Hardy C, Vickers, S, et al. Duplication of the 15q 11-13 region in a patient with autism, epilepsy and ataxia. *Dev Med Child Neurol* 1994;36:736–742.

249. Baker P, Piven J, Schwartz S, et al. Duplication of chromosome 15q11-13 in two individuals with autistic disorder. *J Autism Dev Disord* 1994;24:529–535.

250. Green L, Fein D, Modahl C, et al. Oxytocin and autistic disorder: alterations in peptide forms. *Biol Psychiatry* 2001;50:609–613.

251. Herault J, Martineau J, Petit E, et al. Genetic markers in autism: association study on short arm of chromosome 11. *J Autism Dev Disord* 1994;24:233–235.

252. Petit E, Herault J, Martineau J, et al. Association study with two markers of a human homogene in autism. *J Med Genet* 1995;32:269–274.

253. Cook EHJ, Courchesne R, Lord C, et al. Evidence of linkage between the serotonin transporter and autistic disorder. *Mol Psychiatry* 1997;2:247–250.

254. Klauck SM, Poutska F, Benner A, et al. Serotonin transporter (5-HTT) gene variants associated with autism. *Hum Mol Genet* 1997;6:2233–2238.

255. Persico AM, Militerni R, Bravaccio C, et al. Lack of association between serotonin transporter gene promoter variants and autistic disorder in two ethnically distinct samples. *Am J Med Genet* 2000;96:123–127.

256. Herault J, Petit E, Martineau J, et al. Autism and genetics—clinical approach and association study with two markers of the HRAS gene. *Am J Med Genet* 1995;60:276–281.

257. Herault J, Petit E, Martineau J, et al. Serotonin and autism: biochemical and molecular biology features. *Psychiatry Res* 1996;65:33–43.

258. Hallmayer J, Pintado E, Lotspeich L, et al. Molecular analysis and test of linkage between the FMR-1 gene and infantile autism in multiplex families. *Am J Hum Genet* 1994;55:951–959.

259. Gurling HM, Bolton PF, Vincent J, et al. Molecular and cytogenetic investigations of the fragile X region including the Frax A and Frax ECGG trinucleotide repeat sequences in families multiplex for autism and related phenotypes. *Hum Hered* 1997;47:254–262.

260. Holden JJ, Wing M, Chalifoux M, et al. Lack of expansion of triplet repeats in the FMR1, FRAXE and FRAXF loci in male multiplex families with autism and pervasive developmental disorders. *Am J Med Genet* 1996;64:399–403.

261. Vincent JB, Konecki DS, Munstermann E, et al. Point mutation analysis of the FMR-1 gene in autism. *Mol Psychiatry* 1996;1:227–231.

262. Smalley SL. Genetic influences in autism. *Psychiatr Clin North Am* 1991;14:125–39.

263. Gillberg C, Coleman M. Autism and medical disorders: a review of the literature. *Dev Med Child Neurol* 1996;38:191–202.

264. Steffenburg S. Neuropsychiatric assessment of children with autism: a population-based study. *Dev Med Child Neurol* 1991;33:495–511.

265. Nelson KB, Grether JK, Croen LA, et al. Neuropeptides and neurotrophins in neonatal blood of children with autism or mental retardation. *Ann Neurol* 2001;49:597–606.

266. Chess S. Autism in children with congenital rubella. *J Autism Child Schizophr* 1971;1:33–47.

267. Chess S. Follow-up report on autism in congenital rubella. *J Autism Child Schizophr* 1977;7:69–81.

268. Singh VK, Warren RP, Odell JD, et al. Antibodies to myelin basic protein in children with autistic behavior. *Brain Behav Immun* 1993;7:97–103.

269. Comi AM, Zimmerman AW, Fyer VH, et al. Familial clustering of autoimmune disorders (AID's) and evaluation of medical risk factors in autism. *J Child Neurol* 1999;14:388–394.

270. Bidet B, Leboyer M, Descours B, et al. Allergic sensitization in infantile autism [Letter]. *J Autism Dev Disord* 1993;23:419–420.

271. Renzoni E, Beltrami V, Sestini P, et al. Brief report: allergological evaluation of children with autism. *J Autism Dev Disord* 1995;25:327–333.

272. Campbell M, Geller B, Small A, et al. Minor physical anomalies in young psychotic children. *Am J Psychiatry* 1978;135:573–575.

273. Gualtieri C, Adams A, Shen C, et al. Minor physical anomalies in alcoholic and schizophrenic adults and hyperactive and autistic children. *Am J Psychiatry* 1982;139:640–643.

274. Links P, Stockwell M, Abichandani F, et al. Minor physical anomalies in childhood autism: I. Their relationship to pre- and perinatal complications. *J Autism Dev Disord* 1980;10:273–292.

275. Steg J, Rapaport J. Minor physical anomalies in normal, neurotic, learning disabled, and severely disturbed children. *J Autism Child Schizophr* 1975;5:299–307.

276. Walker H. Incidence of minor physical anomaly in autism. *J Autism Dev Disord* 1977;7:165–176.

277. Edelson S, Cantor D. Autism: xenobiotic influences. *Toxicol Indust Health* 1998;14:553–563.

278. Wakefield AJ, Murch SH, Anthony A, et al. Ileal-lymphoid nodular hyperplasia, non-specific colitis, and pervasive developmental disorder in children. *Lancet* 1998;351:637–641.

279. Kawashima H, Mor T, Kashiwagi Y, et al. Detection and sequencing of measles virus from peripheral mononuclear cells from patients with inflammatory bowel disease and autism. *Dig Dis Sci* 2000;45:723–729.

280. Afzal MA, Minor PD, Schild GC. Clinical safety issues of measles, mumps and rubella vaccines. *Bull World Health Organ* 2000;78:199–204.

281. Deykin E, MacMahon B. Pregnancy, delivery, and neonatal complications among autistic children. *Am J Dis Child* 1980;134:860–864.

282. Bryson S, Smith I, Eastwood D. Obstetrical suboptimality in autistic children. *J Am Acad Child Adolesc Psychiatry* 1988;27:418–422.

283. Finegan J, Quarrington B. Pre-, peri-, and neonatal factors and infantile autism. *J Child Psychol Psychiatry* 1979;20:119–128.

284. Gillberg C, Gillberg IC. Infantile autism: a total population study of reduced optimality in the pre-, peri, and neonatal period. *J Autism Dev Disord* 1983;13:153–166.

285. Mason-Brothers A, Ritvo E, Guze B, et al. Pre-, peri-, and postnatal factors in 181 autistic patients from single and multiple incidence families. *J Am Acad Child Adolesc Psychiatry* 1987;26:39–42.

286. Levy S, Zoitak B, Saelens T. A comparison of obstetrical records of autistic and nonautistic referrals for psychoeducational evaluations. *J Autism Dev Disord* 1988;18:573–581.

287. Mason-Brothers A, Ritvo E, Pingree C, et al. The UCLA-University of Utah epidemiologic survey of autism: prenatal, perinatal, and postnatal factors. *Pediatrics* 1990;86:514–519.

288. Eliez S, Reiss AL. Annotation: MRI neuroimaging of childhood psychiatric disorders: a selective review. *J Child Psychol Psychiatry* 2000;41:679–694.

289. Rumsey J, Ernst M. Functional neuroimaging of autistic disorders. *Ment Retard Dev Disabil Res Rev* 2000;6:171–179.

290. Deb S, Thompson B. Neuroimaging in autism. *Br J Psychiatry* 1998;173:299–302.

291. Kates WR, Mostofsky SH, Zimmerman AW, et al. Neuroanatomical and neurocognitive differences in a pair of monozygous twins discordant for strictly defined autism. *Ann Neurol* 1998;43:782–791.

292. Bauman M, Kemper T. Histoanatomic observations of the brain in early infantile autism. *Neurology* 1985;35:866–874.

293. Williams RS, Hauser SL, Purpura DP, et al. Autism and mental retardation: neuropathologic studies performed on four retarded persons with autistic behavior. *Arch Neurol* 1980;37:749–753.

294. Bauman M, Kemper T. Limbic and cerebellar abnormalities: consistent findings in infantile autism. *J Neuropathol Exp Neurol* 1988;47:369.

295. Ritvo ER, Freeman BJ, Scheibel AB, et al. Lower Purkinje cell counts in the cerebella of four autistic subjects: initial findings

of the UCLA-NSAC autopsy research report. *Am J Psychiatry* 1986;143:862–866.

296. Bailey A, Luthert P, Dean A, et al. A clinicopathological study of autism. *Brain* 1998;121:889–905.

297. Gaffney GR, Tsai LY, Kuperman S, et al. Cerebellar structure in autism. *Am J Disabled Child* 1987;141:1330–1332.

298. Murakami JW, Courchesne E, Press GA, et al. Reduced cerebellar hemisphere size and its relationship to vermal hypoplasia in autism. *Arch Neurol* 1989;46:689–694.

299. Courchesne E, Yeung-Courchesne R, Press GA, et al. Hypoplasia of cerebellar vermal lobules VI and VII in autism. *N Engl J Med* 1988;318:1349–1354.

300. Holroyd S, Reiss A, Bryan N. Autistic features in Joubert syndrome: a genetic disorder with agenesis of the cerebellar vermis. *Biol Psychiatry* 1991;29:287–294.

301. Courchesne E, Townsend J, Aksoomoff N, et al. Impairment in shifting attention in autistic and cerebellar patients. *Behav Neurosci* 1994;108:848–865.

302. Courchesne E, Saitoh O, Yeung CR, et al. Abnormalities of cerebellar vermian lobules VI and VII in patients with infantile autism: identification of hypoplastic and hyperplastic subgroups with MR imagining. *AJR Am J Roentgenol* 1994;162:123–130.

303. Ritvo ER, Garber HJ. Cerebellar hypoplasia and autism. *N Engl J Med* 1988;319:1152.

304. Garber HJ, Ritvo ER. Magnetic resonance imaging of the posterior fossa in autistic adults. *Am J Psychiatry* 1992;149:245–247.

305. Kleiman MD, Neff S, Rosman NP. The brain in infantile autism: are posterior fossa structures abnormal? *Neurology* 1992;42:753–760.

306. Piven J, Saliba K, Bailey J, et al. An MRI study of autism: the cerebellum revisited. *Neurology* 1997;49:546–551.

307. Abell F, Krams H, Ashburner J, et al. The neuroanatomy of autism: a voxelbased whole brain analysis of structural scans. *Neuroreport* 1999;10:1647–1651.

308. Chugani DC, Sundram BS, Behen M, et al. Evidence of altered energy metabolism in autistic children. *Prog Neuropsychopharmacol Biol Psychiatry* 1999;23:635–641.

309. Otsuka H, Harada M, Mori K, et al. Brain metabolites in the hippocampus-amygdala region and cerebellum in autism: an 1H-MR spectroscopy study. *Neuroradiology* 1999;41:517–519.

310. Muller RA, Behen ME, Rothermel RD, et al. Brain mapping of language and auditory perception in high-functioning autistic adults: a PET study. *J Autism Dev Disord* 1999;29:19–31.

311. Ryu YH, Lee JD, Yoon PH, et al. Perfusion impairments in infantile autism on technetium-99m ethyl cysteinate dimer brain single-photon emission tomography: comparison with findings on magnetic resonance imaging. *Eur J Nucl Med* 1999;26:253–259.

312. Hashimoto T, Tayama M, Murakawa K, et al. Development of the brainstem and cerebellum in autistic patients. *J Autism Dev Disord* 1995;25:1–18.

313. Piven J, Arndt S, Bailey J, et al. Regional brain enlargement in autism: a magnetic resonance imaging study. *J Am Acad Child Adolesc Psychiatry* 1996;35:530–536.

314. Prior MR, Tress B, Hoffman WL, et al. Computed tomographic study of children with classic autism. *Arch Neurol* 1984;41:482–484.

315. Coleman P, Romano J, Lapham L, et al. Cell counts in cerebral cortex of an autistic patient. *J Autism Dev Disord* 1985;15:245–256.

316. Creasey H, Rumsey JM, Schwartz M, et al. Brain morphometry in autistic men as measured by volumetric computed tomography. *Arch Neurol* 1986;43:669–672.

317. Darby, JK. Neuropathologic aspects of psychosis in children. *J Autism Child Schizophr* 1976;6:339–352.

318. Piven J, Berthier ML, Starkstein SR, et al. Magnetic resonance imaging evidence for a defect of cerebral cortical development in autism. *Am J Psychiatry* 1990;147:734–739.

319. Hier DB, LeMay M, Rosenberger PB. Autism and unfavorable left-right asymmetries of the brain. *J Autism Dev Disord* 1979;9:153–159.

320. Bauman ML. Brief report: neuroanatomic observations of the brain in pervasive developmental disorders. *J Autism Dev Disord* 1996;26:199–203.

321. Dawson G, Munson J, Estes A, et al. Neurocognitive functioning and joint attention ability in young children with autistic spectrum disorder versus developmental delay. *Child Dev* 2002;73:345–358.

322. Zilbovicius M, Garreau B, Samson Y, et al. Delayed maturation of the frontal cortex in childhood autism. *Am J Psychiatry* 1995;152:248–252.

323. Deutsch G. The nonspecificity of frontal dysfunction in disease and altered states: cortical blood flow evidence. *Neuropsychiatry Neuropsychol Behav Neurol* 1992;5:301–307.

324. Minshew NJ, Goldstein G, Dombrowski SM, et al. A preliminary 31-P MRS study of autism: evidence for undersynthesis and increased degradation of brain membranes. *Biol Psychiatry* 1993;33:762–763.

325. Ernst M, Zametkin AJ, Matochik JA, et al. Low medial prefrontal dopaminergic activity in autistic children. *Lancet* 1997;350:638.

326. Happe P, Ehlers S, Fletcher P, et al. Theory of mind in the brain: evidence from a PET scan study of Asperger syndrome. *Neuroreport* 1996;8:197–201.

327. Ohnishi T, Matsuda H, Hashimoto T, et al. Abnormal rCBF in childhood autism. *Brain* 2000;123:1838–1844.

328. Brothers L. A biological perspective on empathy. *Am J Psychiatry* 1989;146:10–19.

329. Bachevalier J. Medial temporal lobe structures and autism: a review of clinical and experimental findings. *Neuropsychologia* 1994;32:627–648.

330. Fein D, Waterhouse L. Infantile autism: delineating the key deficits. Paper presented at a meeting of the International Neuropsychology Symposium, North Berwick, Scotland, 1985.

331. Fotheringham JB. Autism and its primary psychological and neurological deficit. *Can J Psychiatry* 1991;36:686–692.

332. Baron-Cohen S, Ring H, Wheelwright S, et al. Social intelligence in the normal and autistic brain: an fMRI study. *Eur J Neurosci* 1999;11:1891–1898.

333. Hoon AH, Reiss AL. The mesial-temporal lobe and autism: case report and review. *Dev Med Child Neurol* 1992;34:252–265.

334. Deonna T, Ziegler AL, Moura-Serra J, et al. Autistic regression in relation to limbic pathology and epilepsy: report of two cases. *Dev Med Child Neurol* 1993;35:166–176.

335. Bolton PF, Griffiths PD. Association of tuberous sclerosis of temporal lobes with autism and atypical autism. *Lancet* 1997;349:392–395.

336. Hauser SL, DeLong G, Rosman NP. Pneumographic findings in the infantile autism syndrome: a correlation with temporal lobe disease. *Brain* 1975;98:667–688.

337. Aylward EH, Minshew NJ, Goldstein G, et al. MRI volumes of amygdala and hippocampus in non-mentally retarded autistic adolescents and adults. *Neurology* 1999;53:2145–2150.

338. Howard MA, Cowell PE, Boucher J, et al. Convergent neuroanatomical and behavioral evidence of an amygdala hypothesis of autism. *Brain Imaging* 2000;11:2931–2935.

339. Saitoh O, Courchesne E, Egaas B, et al. Cross-sectional area of the posterior hippocampus in autistic patients with cerebellar and corpus callosum abnormalities. *Neurology* 1995;45:317–324.

340. Piven J, Baily J, Ranson RJ, et al. No difference in hippocampus volume detected on magnetic resonance imaging in autistic individuals. *J Autism Dev Disord* 1998;28:105–110.

341. Raymond GV, Bauman ML, Kemper TL. Hippocampus in autism—a Golgi analysis. *Acta Neuropathol (Berl)* 1996;91:117–119.

342. Haznedar MM, Buchsbaum MS, Metzger M, et al. Anterior cingulate gyrus volume and glucose metabolism in autistic disorder. *Am J Psychiatry* 1997;154:1047–1050.

343. Schultz RT, Gauthier I, Klin A, et al. Abnormal ventral temporal cortical activity during face discrimination among individuals with autism and Asperger syndrome. *Arch Gen Psychiatry* 2000;57:331–340.

344. Siegel BV, Asarnow RB, Tanguay P, et al. Regional cerebral glucose metabolism and attention in adults with a history of childhood autism. *J Neuropsychiatry Clin Neurosci* 1992;4:406–414.

345. Buschsbaum MS, Siegel BV, Wu JC, et al. Brief report: attention performance in autism and regional brain metabolic rate assessed by positron emission tomography. *J Autism Dev Dis* 1992;22:115–125.

346. Garreau B, Zilbovicius M, Guerin P, et al. Effects of auditory stimulation on regional cerebral blood flow in autistic children. *Dev Brain Dysfunct* 1994;7:119–128.

347. Muller RA, Chugani DC, Behen ME. Impairment of dentato-thalamo-cortical pathway in autistic men: language activation data from positron emission tomography. *Neurosci Lett* 1998;245:1–4.

348. Ring HA, Baron-Cohen S, Wheelwright S. Cerebral correlates of preserved cognitive skills in autism: a functional MRI study of embedded figures task performance. *Brain* 1999;122:1305–1315.

349. Rumsey JM, Duara R, Grady C, et al. Brain metabolism in autism: resting cerebral glucose utilization rates as measured with positron emission tomography. *Arch Gen Psychiatry* 1985;42:448–455.

350. Heh CWC, Smith R, Wu J, et al. Positron emission tomography of the cerebellum in autism. *Am J Psychiatry* 1989;146:242–245.

351. Small JG. EEG and neurophysiological studies of early infantile autism. *Biol Psychiatry* 1975;10:385–397.

352. Dawson G, Warrenburg S, Fuller P. Cerebral lateralization in individuals diagnosed as autistic in early childhood. *Brain Lang* 1982;15:353–368.

353. Schwartz M, Duara R, Haxby J, et al. Down's syndrome in adults: brain metabolism. *Science* 1983;221:781–785.

354. Chiron C, Leboyer M, Leon F. SPECT of the brain in childhood autism: evidence for a lack of normal hemispheric symmetry. *Dev Med Child Neurol* 1995;37:849–860.

355. Starkstein SE, Vasquez S, Vrancic D, et al. SPECT findings in mentally retarded autistic individuals. *J Neuropsychiatry Clin Neurosci* 2000;12:370–375.

356. Schifter T, Hoffman JM, Hatten JP. Neuroimaging in infantile autism. *J Child Neurol* 1994;9:155–161.

357. Belmonte M, Egaas B, Townsend J. NMR intensity of corpus callosum differs with age but not with diagnosis of autism. *Neuroreport* 1995;6:1253–1256.

358. Hashimoto TM, Tayama M, Miyazaki Y. Differences in brain metabolites between patients with autism and mental retardation as detected by in vivo localized proton magnetic resonance spectroscopy. *J Child Neurol* 1997;12:91–96.

359. Egaas B, Courchesne E, Saitoh O. Reduced size of corpus callosum in autism. *Arch Neurol* 1995;52:794–801.

360. Manes F, Piven J, Vrancic D, et al. An MRI study of the corpus callosum and cerebellum in mentally retarded autistic individuals. *J Neuropsychiatry Clin Neurosci* 1999;11:470–474.

361. Ciarenello RD, Vandenberg SR, Anders TF. Intrinsic and extrinsic determinants of neuronal development: relation to infantile autism. *J Autism Dev Disord* 1982;12:115–145.

362. Pucilowski O, Plaznik A, Kostowski W. Aggressive behavior inhibition by serotonin and quipazine injected into the amygdala in the rat. *Behav Neural Biol* 1985;43:58–68.

363. Shepard RA, Buxton DS, Broadhurst PL. Beta-adrenoceptor antagonists may attenuate hyponeophagia in the rat through serotonergic mechanism. *Pharmacol Biochem Behav* 1982;16:741–744.

364. Gately PF, Poon SL, Segal DS, et al. Depletion of brain serotonin by 5,7-dihydroxytryptamine alters the response to amphetamine and the habituation of locomotor activity in rats. *Psychopharmacology* 1985;87:400–405.

365. Davis M, Kehne JH, Commissaris RL. Antagonism of apomorphine-enhanced startle by alpha 1-adrenergic antagonists. *Eur J Pharmacol* 1985;108:233–241.

366. Kuperman S, Beeghly JHL, Burns TL, et al. Association of serotonin concentration to behavior and IQ in autistic children. *J Autism Dev Disord* 1987;17:133–140.

367. Panksepp J, Sivy SM, Normansell LA. Brain opioids and social emotions. In: Reite M, Field T, eds. *The psychobiology of attachment and separation.* New York: Academic Press, 1985.

368. Herman BH, Panksepp J. Ascending endorphin inhibition of distress vocalization. *Science* 1981;221:1060–1062.

369. Herault J, Petit E, Martineau J, et al. Serotonin and autism: biochemical and molecular biology features. *Psychiatry Res* 1996;65:33–43.

370. Cuccaro ML, Wright HH, Abramson RK, et al. Whole blood serotonin and cognitive functioning in autistic individuals and their first degree relatives. *J Neuropsychiatry Clin Neurosci* 1993;2:268–74.

371. Kuperman S, Beeghly J, Burns T, et al. Serotonin relationships of autistic probands and their first-degree relatives. *J Am Acad Child Adolesc Psychiatry* 1985;24:186–190.

372. Cook EH, Leventhal BL, Freedman DX. Free serotonin in plasma: autistic children and their first degree relatives. *Biol Psychiatry* 1988;24:488–491.

373. Cook EH, Leventhal BL, Heller W, et al. Autistic children and their first-degree relatives: relationship between serotonin and norepinephrine levels and intelligence. *J Neuropsychiatry Clin Neurosci* 1990;2:268–274.

374. Leventhal BL, Cook EH, Morford M, et al. Relationships of whole blood serotonin and plasma norepinephrine within families. *J Autism Dev Disord* 1990;20:499–511.

375. Piven J, Tsai G, Nehme E, et al. Platelet serotonin, a possible marker for familial autism. *J Autism Dev Disord* 1991;21:51–59.

376. Takahashi S, Kanai H, Miyamoto Y. Reassessment of elevated serotonin levels in blood platelet in early infantile autism. *J Autism Child Schizophr* 1976;6:317–326.

377. Hanley HG, Stahl SM, Freedman DX. Hyperserotonemia and amine metabolites in autistic and retarded children. *Arch Gen Psychiatry* 1977;34:521–531.

378. Pare CMB, Sandler M, Stacey RS. 5-Hydroxyindoles in mental deficiency. *J Neurol Neurosurg Psychiatry* 1960;23:341–346.

379. Schain RJ, Freedman DX. Studies on 5-hydroxyindoleamine metabolism in autistic and other mentally retarded children. *J Pediatrics* 1961;58:315–320.

380. Partington MW, Tu JB, Wong CY. Blood serotonin levels in severe mental retardation. *Dev Med Child Neurol* 1973;15:616–627.

381. McBride PA, Anderson GM, Hertzig ME, et al. Effects of diagnosis, race, and puberty on platelet serotonin levels in autism and mental retardation. *J Am Acad Child Adolesc Psychiatry* 1998;37:767–776.
382. Hoshino Y, Tachibana R, Watanabe M, et al. Serotonin metabolism and hypothalamic pituitary function in children with infantile autism and minimal brain dysfunction. *J Psychiatry Neurol* 1984;26:937–945.
383. Ritvo ER, Freeman BJ, Yuwiler A, et al. Fenfluramine treatment of autism: UCLA collaborative study of 81 patients at nine medical centers. *Psychopharmacol Bull* 1986;22:133–140.
384. Aman MG, Kern RA. Review of fenfluramine in the treatment of the developmental disabilities. *J Am Acad Child Adolesc Psychiatry* 1990;28:549–565.
385. August GJ, Raz N, Baird TD. Fenfluramine response in high and low functioning autistic children. *J Am Acad Child Adolescent Psychiatry* 1987;26:342–346.
386. Campbell M, Deutsch SI, Perry R, et al. Short-term efficacy and safety of fenfluramine in hospitalized preschool-age autistic children: an open study. *Psychopharmacol Bull* 1986;22:141–147.
387. Campbell M, Perry R, Polonsky B. Brief report: an open study of fenfluramine in hospitalized young autistic children. *J Autism Dev Disord* 1986;46:495–506.
388. Du Verglas G, Banks SR, Guyer KE. Clinical effects of fenfluramine on children with autism: a review of the research. *J Autism Dev Disord* 1988;18:297–308.
389. Geller E, Ritvo ER, Freeman BJ, et al. Preliminary observations on the effect of fenfluramine on blood serotonin and symptoms in three autistic boys. *N Engl J Med* 1982;307:165–169.
390. Groden G, Groden J, Dondey M, et al. Effects of fenfluramine on the behavior of autistic individuals. *Res Dev Disabil* 1987;8:203–211.
391. Ritvo ER, Freeman BJ, Yuwiler A, et al. Study of fenfluramine in outpatients with the syndrome of autism. *J Pediatrics* 1984;105:823–828.
392. Ritvo ER, Freeman BJ, Geller E, et al. Effects of fenfluramine on 14 outpatients with the syndrome of autism. *J Am Acad Child Psychiatry* 1983;22:549–558.
393. Carlsson ML. Hypothesis: is infantile autism a hypoglutamatergic disorder? Relevance of glutamate-serotonin interactions for pharmacotherapy. *J Neural Transm* 1998;105:525–535.
394. Sankar DVS. Biogenic amine uptake by blood platelets and RBC in childhood schizophrenia. *Acta Paedopsychiatr* 1970;37:174–182.
395. Sankar DVS. Uptake of 5-hydroxytryptamine by isolated platelets in childhood schizophrenia and autism. *Neuropsychobiology* 1977;3:234–239.
396. Boullin DJ, Coleman M, O'Brien RA. Abnormalities in platelet 5-hydroxy-tryptamine efflux in patients with infantile autism. *Nature* 1970;226:371–372.
397. Boullin DJ, Freeman BJ, Geller E, et al. Toward the resolution of conflicting findings. *J Autism Dev Disord* 1982;12:97–98.
398. Yuwiler A, Ritvo E, Geller E, et al. Uptake and efflux of serotonin from platelets of autistic and nonautistic children. *J Autism Dev Disord* 1975;5:83–98.
399. Katsui T, Okuda M, Usuda S, et al. Kinetics of sub-3-serotonin uptake by platelets in infantile autism and developmental language disorder (including five pairs of twins). *J Autism Dev Disord* 1986;16:69–76.
400. Rotman A, Caplan R, Szekely GA. Platelet uptake of serotonin in psychotic children. *Psychopharmacology* 1980;67:245–248.
401. Anderson GM, Minderaa RB, van Bentem PPG, et al. Platelet imipramine binding in autistic subjects. *Psychiatry Res* 1984;11:133–141.
402. Boullin DJ, Coleman M, O'Brien RA, et al. Laboratory predictions of infantile autism based on 5-hydroxy-tryptamine efflux from blood platelets and their correlation with the Rimland E-2 score. *J Autism Childhood Schizophr* 1971;1:63–71.
403. Halbreich U, Rojansky N, Zander KJ, et al. Influence of age, sex and diurnal variability on imipramine receptor binding and serotonin uptake in platelets of normal subjects. *J Psychiatr Res* 1991;25:7–18.
404. McBride PA, Anderson GM, Hertzig ME, et al. Serotonergic responsivity in male young adults with autistic disorder. *Arch Gen Psychiatry* 1989;46:205–212.
405. Perry BD, Cook EH, Leventhal BL, et al. Platelet 5-HT-sub-2 receptor binding sites in autistic children and their first-degree relatives. In: *Proceedings of the American Academy of Child and Adolescent Psychiatry.* New York, 1989:67–68.
406. Saffai-Kutti S, Denfors I, Kutti J, et al. In vitro platelet function in infantile autism. *Folia Haematol (Leipzig)* 1988[Suppl]:897–901(abst).
407. Saffai-Kutti S, Kutti J, Gillberg C. Impaired in vivo platelet reactivity in infantile autism. *Acta Paediatr Scand* 1985;74:799–800.
408. Cook EHJ, Courchesne R, Lord C. Evidence of linkage between the serotonin transporter and autistic disorder. *Mol Psychiatry* 1997;2:247–250.
409. Klauck SM, Poustka F, Benner A, et al. Serotonin transporter (5-HTT) gene variants associated with autism? *Hum Mol Genet* 1997;6:2233–2238.
410. Launay JM, Ferrari P, Haimart M, et al. Serotonin metabolism and other biochemical parameters in infantile autism. *Neuropsychobiology* 1988;20:1–11.
411. Cohen DJ, Caparulo BK, Shaywitz BA, et al. Dopamine and serotonin metabolism in neuropsychiatrically disturbed children. *Arch Gen Psychiatry* 1977;34:545–550.
412. Lake C, Ziegler M, Murphy D. Increased norepinephrine levels and decreased dopamine-beta-hydroxylase activity in primary autism. *Arch Gen Psychiatry* 1977;34:553–566.
413. Anderson GM, Minderaa RB, Cho SC, et al. The issue of hyperserotonemia and platelet serotonin exposure: a preliminary study. *J Autism Dev Disord* 1989;19:349–351.
414. Minderaa RB, Anderson GM, Volkmar FR, et al. Urinary 5-hydroxyindoleacetic acid and whole blood serotonin and tryptophan in autistic and normal subjects. *Biol Psychiatry* 1987;22:933–940.
415. Gillberg C, Svennerholm L, Hamilton-Hellberg C. Childhood psychosis and monoamine metabolites in spinal fluid. *J Autism Dev Disord* 1983;13:383–396.
416. Chugani DC, Muzik O, Rothermel R, et al. Altered serotonin synthesis in the dentatothalamocortical pathway in autistic boys. *Ann Neurol* 1997;42:666–669.
417. Chugani D, Muzik O, Behen M, et al. Developmental changes in brain serotonin synthesis capacity in autistic and nonautistic children. *Ann Neurol* 1999;45:287–295.
418. Cook ET, Leventhal BL. The serotonin system in autism. *Curr Opin Pediatr* 1996;8:348–354.
419. Todd RD, Ciaranello RD. Demonstration of inter- and intraspecies differences in serotonin binding sites by antibodies from an autistic child. *Proc Natl Acad Sci U S A* 1985;82:612–616.
420. Singh V, Singh E, Warren R. Hyperserotonemia and serotonin receptor antibodies in children with autism but not mental retardation. *Biol Psychiatry* 1997;41:753–755.
421. Yuwiler A, Shih JC, Chen CH, et al. Hyperserotonemia and antiserotonin antibodies in autism and other disorders. *J Autism Dev Disord* 1992;22:33–45.
422. Cook EH, Perry BD, Dawson G, et al. Receptor inhibition by immunoglobulins: specific inhibition by autistic children, their

relatives and control subjects. *J Autism Dev Disord* 1993;23: 67–78.

423. Damasio A, Maurer R. A neurological model for childhood autism. *Arch Neurol* 1978;35:777–786.

424. Meltzer HY, Busch D, Fang VS. Hormones, dopamine and schizophrenia. *Psychoneuroendocrinology* 1981;6:17–36.

425. Deutsch SI, Campbell M, Sachar EJ, et al. Plasma growth hormone response to oral l-dopa in infantile autism. *J Autism Dev Disord* 1985;15:205–212.

426. Barthelemy C, Bruneau N, Cottet-Eymard JM, et al. Urinary free and conjugated catecholamines and metabolites in autistic children. *J Autism Dev Disord* 1988;18:583–591.

427. Boullin DJ, O'Brien RA. Uptake and loss of sup-14-C-dopamine by platelets from children with infantile autism. *J Autism Dev Disord* 1972;12:97–98.

428. Garnier C, Comoy E, Barthelemy C, et al. Dopamine-beta-hydroxylase (DBH) and homovanillic acid (HVA) in autistic children. *J Autism Dev Disord* 1986;16:23–29.

429. Garreau B, Barthelemy C, Domenech J, et al. Disturbances in dopamine metabolism in autistic children: results of clinical tests and urinary dosages of homovanillic acid (HVA). *Acta Psychiatr Belg* 1980;80:249–265.

430. Iskrangkun PP, Newman HAI, Patel ST. Potential biochemical markers for infantile autism. *Neurochem Pathol* 1986;5:51–70.

431. Launay JM, Bursztje C, Ferrari P, et al. Catecholamines metabolism in infantile autism: a controlled study of 22 autistic children. *J Autism Dev Disord* 1987;17:333–347.

432. Minderaa RB, Anderson GM, Volkmar FR, et al. Neurochemical study of dopamine functioning in autistic and normal subjects. *J Am Acad Child Adolesc Psychiatry* 1989;28:190–194.

433. Piggot LR. Overview of selected basic research in autism. *J Autism Dev Disord* 1979;9:199–218.

434. Garreau B, Barthelemy C, Jouve J, et al. Urinary homovanillic acid levels of autistic children. *Dev Med Child Neurol* 1988;30:93–98.

435. Martineau J, Barthelemy C, Jouve J, et al. Monoamines (serotonin and catecholamines) and their derivatives in infantile autism: age-related changes and drug effects. *Dev Med Child Neurol* 1992;34:593–603.

436. Gillberg C, Svennerholm L. CSF monoamines in autistic syndromes and other pervasive developmental disorders of early childhood. *Br J Psychiatry* 1987;151:89–94.

437. Cohen DJ, Shaywitz BS, Johnson WT, et al. Biogenic amines in autistic and atypical children: cerebrospinal fluid measures of homovanillic acid and 5-hydroxyindoleacetic acid. *Arch Gen Psychiatry* 1974;31:845–853.

438. Ross DL, Klykylo WM, Anderson GM. Cerebrospinal fluid indoleamine and monoamine effects of fenfluramine treatment of infantile autism. *Ann Neurol* 1985;18:394.

439. Young JG, Cohen DJ, Kavanagh ME, et al. Cerebrospinal fluid, plasma, and urinary MHPG in children. *Life Sci* 1981;28:2837–2845.

440. Minderaa RB, Anderson GM, Volkmar FR, et al. Plasma levels of 3-methoxy-4-hydroxyphenylglycoll (MHPG) and urinary excretion of norepinephrine, epinephrine, and MHPG in autistic and normal subjects, 1988 (unpublished manuscript).

441. Roy A, Pickar D, De Jong J, et al. Norepinephrine and its metabolites in cerebrospinal fluid, plasma, and urine: relationship to hypothalamic-pituitary-adrenal axis function in depression. *Arch Gen Psychiatry* 1988;45:849–857.

442. Ziegler MG, Wood JH, Lake CR, et al. Norepinephrine and 3-methoxy-4-hydroxyphenyl glycol gradients in human cerebrospinal fluid. *Am J Psychiatry* 1977;134:565–568.

443. Young JG, et al. Decreased 24 hour urinary MHPG in childhood autism. *Am J Psychiatry* 1979;136:1055–1057.

444. Martineau J, Herault J, Petit E, et al. Catecholaminergic metabolism and autism. *Dev Med Child Neurol* 1994;36:688–697.

445. Minderaa RB, Anderson GM, Volkmar FR, et al. Noradrenergic and adrenergic functioning in autism. *Biol Psychiatry* 1994;36:237–241.

446. Kalat JW. Letter to the editor: speculations on similarities between autism and opiate addiction. *J Autism Child Schizophr* 1978;8:447–479.

447. Herman BH, Panksepp J. Effects of morphine and naloxone on separation distress and approach attachment: evidence for opiate mediation of social affect. *Pharmacol Biochem Behav* 1978;9:213–220.

448. Panksepp J, Sivy SM, Normansell LA. Brain opioids and social emotions. In: Reite M, Field T, eds. *The psychobiology of attachment and separation.* New York: Academic Press, 1985.

449. Panksepp J, Meeker R, Bean NJ. The neurochemical control of crying. *Pharmacol Biochem Behav* 1980;12:437–443.

450. Weizman R, Gil-Ad I, Dick J, et al. Low plasma immunoreactive b-endorphin levels in autism. *J Am Acad Child Adolesc Psychiatry* 1988;27:430–433.

451. Weizman R et al. Humoral endorphin blood levels in autistic, schizophrenic and healthy subjects. *Psychopharmacology* 1984;82:368–370.

452. Gillberg C, Terenius L, Lonnerholm G. Endorphin activity in childhood psychosis: spinal fluid levels in 24 cases. *Arch Gen Psychiatry* 1985;42:780–783.

453. Willemsen-Swinkels SHN, Buitelaar JK, Weijnen FG, et al. Plasma beta-endorphin concentrations in people with learning disability and self-injurious and/or autistic behavior. *Br J Psychiatry* 1996;168:105–109.

454. Tordjman S, Anderson GM, McBride PA, et al. Plasma beta-endorphin, adrenocorticotrophic hormone, and cortisol in autism. *J Child Psychol Psychiatry* 1997;38:705–715.

455. Barrett RP, Feinstein C, Hole WT. Effects of naloxone and naltrexone on self-injury: a double-blind, placebo-controlled analysis. *Am J Ment Retard* 1989;93:644–651.

456. Bernstein GA, Hughes JR, Mitchell JE, et al. Effects of narcotic antagonists on self-injurious behavior: a single case study. *J Am Acad Child Adolesc Psychiatry* 1987;26:886–889.

457. Herman BH, Hammock MK, Arthur-Smith A, et al. Naltrexone decreases self-injurious behavior in children. *Ann Neurol* 1987;22:550–552.

458. LeBoyer M, Bouvard MP, Dugas M. Effects of naltrexone on infantile autism. *Lancet* 1988;1:715.

459. Panksepp J, Lensing P. A synopsis of an open-trial of naltrexone treatment of autism with four children. *J Autism Dev Disord* 1991;21:243–249.

460. Sandman CA, Datta PC, Barron J, et al. Naloxone attenuates self-abusive behavior in developmentally disabled clients. *Appl Res Ment Retard* 1983;4:5–11.

461. Sandman CA. B-endorphin disregulation in autistic and self-injurious behavior: a neurodevelopmental hypothesis. *Synapse* 1988;2:193–199.

462. Sandyk R. Naloxone abolished self-injuring in a mentally retarded child. *Ann Neurol* 1985;17:520.

463. Walters A, Barrett RP, Feinstein C, et al. A case report of naltrexone treatment of self-injury and social withdrawal in autism. *J Autism Dev Disord* 1990;20:169–176.

464. Campbell M, Adams P, Small AM, et al. Naltrexone in infantile autism. *Psychopharmacol Bull* 1988;24:135–139.

465. Herman BH, Hammock Arthur-Smith, et al. Role of opioid peptides in autism: Effects of acute administration of naltrexone. *Soc Neurosci Abst* 1986;14:465.

466. Modahl C, Fein D, Waterhouse L, et al. Does oxytocin deficiency

mediate social deficits in autism? [Letter]. *J Autism Dev Disord* 1992;22:449–451.

467. Panksepp J. Oxytocin effects on emotional processes: separation distress, social bonding, and relationships to psychiatric disorders. *Ann N Y Acad Sci* 1992;652:243–252.

468. Insel TR, O'Brien DJ, Leckman JF. Oxytocin, vasopressin, and autism: is there a connection? *Biol Psychiatry* 1999;45: 145–157.

469. Insel TR. Oxytocin—a neuropeptide for affiliation: evidence from behavioral, receptor autoradiographic, and comparative studies. *Psychoneuroendocrinology* 1992;17:3–35.

470. Chamberlain RS, Herman BH. A novel biochemical model linking dysfunctions in brain melatonin, proopiomelanocortin peptides, and serotonin in autism. *Biol Psychiatry* 1990;28: 773–793.

471. Ritvo ER, Creel D, Realmuto G, et al. Electroretinograms in autism: a pilot study of b-wave amplitudes. *Am J Psychiatry* 1988;145:229–232.

472. Castrogiovanni P, Marazziti D. ERG b-wave amplitude and brain dopaminergic activity. *Am J Psychiatry* 1989;146:1085–1086.

473. Yuwiler A, Ritvo ER. ERG b-wave amplitude and brain dopaminergic activity: reply. *Am J Psychiatry* 1989;146:1086.

474. Ornitz E, Ritvo E, Walter R. Dreaming sleep in autistic twins. *Arch Gen Psychiatry* 1965;12:77–79.

475. Tanguay PE, Ornitz EM, Forsythe AB, et al. Rapid eye movement (REM) activity in normal and autistic children during REM sleep. *J Autism Child Schizophr* 1976;6:275–288.

476. Ornitz EM, Ritvo ER, Brown MB, et al. The EEG and rapid eye movements during REM sleep in normal and autistic children. *Electroencephalogr Clin Neurophysiol* 1969;26:167–175.

477. Elia M, Ferri R, Musumeci SA, et al. Sleep in subjects with autistic disorder: a neuropsychological and psychological study. *Brain Dev* 2000;22:88–92.

478. Ornitz EM, Forsythe AB, de la Pena A. Effect of vestibular and auditory stimulation on the REM's of REM sleep in autistic children. *Arch Gen Psychiatry* 1973;29:786–791.

479. Colbert G, Koegler RR, Markham CH. Vestibular dysfunction in childhood schizophrenia. *Arch Gen Psychiatry* 1959;1: 600–617.

480. Ornitz EM. Vestibular dysfunction in schizophrenia and childhood autism. *Comp Psychiatry* 1970;11:159–173.

481. Ornitz EM, Brown MB, Mason A, et al. Effect of visual input on vestibular nystagmus in autistic children. *Arch Gen Psychiatry* 1974;31:369–375.

482. Ornitz EM, Atwell CW, Kaplan AR, et al. Brain-stem dysfunction in autism. *Arch Gen Psychiatry* 1985;42:1018–1025.

483. Ritvo ER, Ornitz EM, Eviatar A, et al. Decreased postrotatory nystagmus in early infantile autism. *Neurology* 1969;19: 653–658.

484. Lewine JD, Andrews R, Chez M, et al. Magnetoencephalographic patterns of epileptiform activity in children with regressive autism spectrum disorders. *Pediatrics* 1999;104:405–418.

485. Fein D, Skoff B, Mirsky AF. Clinical correlates of brainstem dysfunction in autistic children. *J Autism Dev Disord* 1981;11: 303–315.

486. Narita T, Koga Y. Neuropsychological assessment of childhood autism. *Adv Biol Psychiatry* 1987;16:156–170.

487. Novick B, Vaughn HG, Kurtzberg D, et al. An electrophysiologic indication of auditory processing defects in autism. *Psychiatry Res* 1980;3:107–114.

488. Rosenblum SM, Arick JR, Krug DA, et al. Auditory brainstem evoked responses in autistic children. *J Autism Dev Disord* 1980;10:215–225.

489. Skoff B, Mirsky A, Turner D. Prolonged brainstem transmission time in autism. *Psychiatry Res* 1980;2:157–166.

490. Student M, Sohmer H. Evidence from auditory nerve and brainstem evoked responses for an organic brain lesion in children with autistic traits. *J Autism Child Schizophr* 1978;8:13–20.

491. Tanguay PE, Edwards RM, Buchwald J, et al. Auditory brainstem evoked responses in autistic children. *Arch Gen Psychiatry* 1982;39:174–180.

492. Thivierge J, Bedard C, Cote R, et al. Brainstem auditory evoked response and subcortical abnormalities in autism. *Am J Psychiatry* 1990;147:1609–1613.

493. Wong V, Wong SN. Brainstem auditory evoked potential study in children with autistic disorder. *J Autism Dev Disord* 1991;21:329–340.

494. Courchesne E, Courchesne RY, Hicks G, et al. Functioning of the brainstem auditory pathway in non-retarded autistic individuals. *Electroencephalogr Clin Neurophysiol* 1985;61:491–501.

495. Rumsey JM, Grimes AM, Pikus AM, et al. Auditory brainstem responses in pervasive developmental disorders. *Biol Psychiatry* 1984;19:1403–1418.

496. DeLong GR. A neuropsychological interpretation of infantile autism. In: Schopler E, Rutter M, eds. *Autism: a reappraisal of concepts and treatment.* New York: Plenum, 1978:207–218.

497. Novick B, Vaughn HG, Kurtzberg D, et al. An electrophysiologic indication of auditory processing defects in autism. *Psychiatry Res* 1980;3:107–114.

498. Cieselski KT, Courchesne E, Elmasian R. Effects of focused selective attention tasks on event-related potentials in autism and normal individuals. *Electroencephalogr Clin Neurophysiol* 1990;75:207–220.

499. Courchesne E, Kilman BA, Galambos R, et al. Autism: processing of novel auditory information assessed by event-related brain potentials. *Electroencephalogr Clin Neurophysiol* 1984;59: 238–248.

500. Courchesne E, Lincoln AJ, Kilman BA, et al. Event related brain potentials of the processing of novel visual and auditory information in autism. *J Autism Dev Disord* 1985;15:55–76.

501. Courchesne E, Lincoln AJ, Yeung-Courchesne R, et al. Pathophysiologic findings in nonretarded autism and receptive developmental language disorder. *J Autism Dev Disord* 1989;19: 1–18.

502. Dawson G, Finley C, Phillips S, et al. Reduced P3 amplitude of the event-related brain potential: its relationship to language ability in autism. *J Autism Dev Disord* 1988;18:493–504.

503. Martineau J, et al. Effects of vitamin B6 on averaged evoked potentials in infantile autism. *Biol Psychiatry* 1981;16:627–641.

504. Oades RD, Walker MK, Geffen LB, et al. Event-related potentials in autistic and healthy children on an auditory choice reaction time task. *Int J Psychophysiol* 1988;6:25–37.

505. Novick B, Kurtzberg D, Vaughn HG. An electrophysiologic indication of defective information storage in childhood autism. *Psychiatry Res* 1979;1:101–108.

506. Verbaten R, Van Engeland H. Autisme: psychofysiologisch benaderd [Autism: psychophysiologic assessment]. *Psycholoog* 1988;23:289–295(abst).

507. Niwa S, Ohta M, Yamazaki M. P300 and stimulus evaluation process in autistic subjects. *J Autism Dev Disord* 1983;13: 33–42.

508. Erwin RJ, Van Lancker D, Guthrie D, et al. P3 responses to prosodic stimuli in adult autistic subjects. *Electroencephalogr Clin Neurophysiol Evoked Potentials* 1991;80:561–571.

509. Kemner C, Verbaten MN, Koelega HS, et al. Are abnormal event-related potentials specific to children with ADHD? A comparison with two clinical groups. *Percept Mot Skills* 1998;87: 1083–1090.

510. Townsend J, Westerfield M, Leaver E, et al. Event-related brain response abnormalities in autism: evidence for impaired

cerebello-frontal spatial-attention. *Brain Res Cogn Brain Res* 2001;11:127–145.

511. Ogawa T, Sugiyama A, Ishiwa S, et al. Ontogenic development of EEG-asymmetry in early infantile autism. *Brain Dev* 1982;4:439–449.

512. Small JG. EEG and neurophysiological studies of early infantile autism. *Biol Psychiatry* 1975;10:385–397.

513. Dawson G, Warrenburg S, Fuller P. Cerebral lateralization in individuals diagnosed as autistic in early childhood. *Brain Lang* 1982;15:353–368.

514. Dawson G, Finley C, Phillips S, et al. Hemispheric specialization and the language abilities of autistic children. *Child Dev* 1986;57:1440–1453.

515. Bailey A, Philips W, Rutter M. Autism: towards an integration of clinical, genetic, neuropsychological and neurobiological perspectives. *J Child Psychol Psychiatry* 1996;37: 89–126.

516. Filipek PA, Accardo PJ, Ashwal S, et al. Practice parameter: screening and diagnosis of autism. *Neurology* 2000;55: 468–479.

517. Newsom C, Hovanitz CA. Autistic disorder. In: Mash EJ, ed. *Assessment of childhood disorders,* 3rd ed. New York: Guilford Press, 1997:408–452

518. Volkmar FR, Klin A, Marans W, et al. The pervasive developmental disorders: diagnosis and assessment. *Child Adolesc Psychiatr Clin North Am* 1996;5:963–977.

519. Baron-Cohen S, Allen J, Gillberg C. Can autism be detected at 18 months? The needle, the haystack, and the CHAT. *Br J Psychiatry* 1992;161:839–843.

520. Baron-Cohen S, Cox A, Baird G, et al. Psychological markers in the detection of autism in infancy in a large population. *Br J Psychiatry* 1996;168:158–163.

521. Cox A, Klein K, Charman T, et al. Autism spectrum disorders at 20 and 42 months of age: stability of clinical and ADI-R diagnosis. *J Child Psychol Psychiatry* 1999;40:719–732.

522. Robins D, Fein D, Barton M, et al. The Modified Checklist for Autism in Toddlers: an initial study investigating the early detection of autism and pervasive developmental disorders. *J Autism Dev Disord* 2001;31:131–144.

523. Filipek PA, Accardo PJ, Baranek GT, et al. The screening and diagnosis of autistic spectrum disorders. *J Autism Dev Disord* 1999;29:437–482.

524. Stone W, Coonrod E, Ousley O. Brief report: Screening Tool for Autism in Two-Year-Olds (STAT): development and preliminary data. *J Autism Dev Disord* 2000;30:607–612.

525. Lord C. Follow-up of two-year-olds referred for possible autism. *J Child Adolesc Psychiatry* 1995;36:1365–1382.

526. Rapin I. Children with inadequate language development: management guidelines for otolaryngologists. *Int J Pediatr Otorhinolaryngol* 1988;16:189–198.

527. Klein SK, Rapin I. Clinical assessment of pediatric disorders of higher cerebral function. *Curr Probl Pediatr* 1990;20:111–160.

528. Allen DA, Rapin I, Wiznitzer M. Communication disorders of preschool children: the physician's responsibility. *Dev Behav Pediatr* 1988;9:164–170.

529. Minshew N, Payton JB. New perspectives in autism, part 2: the differential diagnosis and neurobiology of autism. *Curr Probl Pediatr* 1988;988:615–694.

530. Gillberg C. Medical work-up in children with autism and Asperger syndrome. *Brain Dysfunction* 1990;3:249–260.

531. Marcus LM, Stone WL. Assessment of the young autistic child. In: Schopler E, Van Bourgondien ME, eds. *Preschool issues in autism. Current issues in autism.* New York: Plenum Press, 149–173.

532. Siegel DJ. Evaluation of high-functioning autism. In: Goldstein

G, ed. *Neuropsychology. Human brain function: assessment and rehabilitation.* New York: Plenum Press, 1998:107–134.

533. Volkmar F, Cook EH, Pomeroy J, et al. Practice parameters for the assessment and treatment of children, adolescents, and adults with autism and other pervasive developmental disorders. *J Am Acad Child Adolesc Psychiatry* 1999;38[Suppl]: 32S–54S.

534. Bucy JE, Smith T, Landau S. Assessment of preschool children with developmental disabilities and at-risk conditions. In: Nuttall EV, Romero I, eds. *Assessing and screening preschoolers: psychological and educational dimensions,* 2nd ed. Needham Heights, MA: Allyn & Bacon, 1999:318–339.

535. Rosenhall U, Nordin V, Sandstroem M, et al. Autism and hearing loss. *J Autism Dev Disord* 1999;29:349–357.

536. Bauman ML. Motor dysfunction in autism. In: Joseph AB, Young RR, eds. *Movement disorders in neurology and psychiatry.* Boston: Blackwell Scientific, 1992.

537. Rossi PG, Visconti P, Posar A. Stereotypies in autistic children. *J Mov Disord* 1992;7[Suppl 1]:166.

538. Rutter M. Genetic studies of autism: from the 1970s into the millennium. *J Abnorm Child Psychol* 2000;28:3–14.

539. Bailey DB, Mesibov GB, Hatton DD, et al. Autistic behavior in young boys with fragile X syndrome. *J Autism Dev Disord* 1998;28:499–508.

540. Lainhart JE, Piven J, Wzorek M, et al. Macrocephaly in children and adults with autism. *J Am Acad Child Adolesc Psychiatry* 1997;36:282–290.

541. Gillberg C, Uvebrant P, Carlsson G, et al. Case report: autism and epilepsy (and tuberous sclerosis?) in two pre-adolescent boys: neuropsychiatric aspects before and after epilepsy surgery. *J Intellect Disabil Res* 1996;40:75–81.

542. Smalley SL. Autism and tuberous sclerosis. *J Autism Dev Disord* 1998;28:407–414.

543. Williams PG, Hersh JH. The association of neurofibromatosis type 1 and autism. *J Autism Dev Disord* 1998;28:567–571.

544. Konstantareas MM, Homatidis S. Chromosomal abnormalities in a series of children with autistic disorder. *J Autism Dev Disord* 1999;29:275–285.

545. Tuchman R. Treatment of seizure disorders and EEG abnormalities in autism. *J Autism Dev Disord* 2000;30:485–489.

546. Kim JA, Szatmari P, Bryson SE, et al. The prevalence of anxiety and mood problems among children with autism and Asperger syndrome. *Autism* 2000;4:117–132.

547. Lainhart JE. Psychiatric problems in individuals with autism, their parents and siblings. *Int Rev Psychiatry* 1999;11: 278–298

548. Lord C, Schopler E. Intellectual and developmental assessment. In: Schopler E, Mesibov G, eds. *Diagnosis and assessment in autism.* New York: Plenum Press, 1980.

549. Bonde E. Comorbidity and subgroups in childhood autism. *Eur Child Adolesc Psychiatry* 2000;9:7–10.

550. Wainwright L, Fein D, Waterhouse L. Neuropsychological assessment of children with developmental disabilities. In: Amir N, Rapin I, Branski D, eds. *Pediatric neurology: behavior and cognition of the child with brain dysfunction.* Basel: Karger, 1991:146–163.

551. Siegel DJ, Minshew NJ, Goldstein G. Wechsler IQ profiles in diagnosis of high-functioning autism. *J Autism Dev Disord* 1996;26:389–406.

552. Minshew NJ, Goldstein G, Siegel DJ. Neuropsychologic functioning in autism: profile of a complex information processing disorder. *J Int Neuropsychol Soc* 1997;3:303–316.

553. Scheuffgen K, Happe F, Anderson M, et al. High "intelligence," low "IQ"? Speed of processing and measured IQ in children with autism. *Dev Psychopathol* 2000;12:83–90.

554. Watson LR, Marcus LM. Diagnosis and assessment of preschool children. In: Schopler E, Mesibov G, eds. *Diagnosis and assessment of autism.* New York: Plenum Press, 1988.

555. Hadwin J, Hutley G. Detecting features of autism in children with severe learning difficulties: a brief report. *Autism* 1998;2:269–280.

556. Harris SL, Belchic J, Blum L, et al. Behavioral assessment of autistic disorder. In: Matson JL ed. *Autism in children and adults: etiology, assessment, and intervention.* Pacific Grove, CA: Brooks/Cole Publishing, 127–146.

557. Carter AS, Gillham JE, Sparrow SS, et al. Adaptive behavior in autism. *Child Adolesc Psychiatr Clin North Am* 1996;5:945–961.

558. Mesibov GB, Troxler M, Boswell S. Assessment in the classroom. In: Schopler E, Mesibov G, eds. *Diagnosis and assessment of autism* New York: Plenum Press, 1988.

559. Sparrow S, Balla D, Cicchetti D. *Vineland Adaptive Behavior Scales.* Circle Pines, MN: American Guidance Service, 1984.

560. Van Meter L, Fein D, Morris R. Social behavior in autism: An analysis of intrasubtest scatter on the Vineland Adaptive Behavior Scales. *J Clin Exp Neuropsychol* 1992;14:390(abst).

561. Carpentieri S, Morgan SB. Adaptive and intellectual functioning in autistic and nonautistic retarded children. *J Autism Dev Disord* 1996;26:611–620.

562. McCune-Nicolich L. Toward symbolic functioning: structure of early pretend games and potential parallels with language. *Child Dev* 1981;52:785–797.

563. Lewis V, Boucher J. Spontaneous, instructed and elicited play in relatively able autistic children. *Br J Dev Psychol* 1988;6:325–339.

564. Mundy P, Sigman M, Ungerer J, et al. Nonverbal communication and play correlates of language development in autistic children. *J Autism Dev Disord* 1987;17:349–364.

565. Ungerer J, Sigman M. Symbolic play and language comprehension in autistic children. *J Am Acad Child Psychiatry* 1981;20:318–337.

566. Wainwright L, Fein D. Free play in autistic language disordered and mentally deficient children. Paper presented before the International Neuropsychological Society, Galveston, TX, 1993.

567. Groden G, Groden J, Stevenson S. Facilitating comprehensive behavioral assessments. *Focus Autism Other Dev Disabil* 1997;12:49–52.

568. Parks SL. Psychometric instruments available for the assessment of autistic children. In: Schopler E, Mesibov G, eds. *Diagnosis and assessment in autism.* New York: Plenum Press, 1988.

569. Powers M. Behavioral assessment of autism. In: Schopler E, Mesibov G, eds. *Diagnosis and assessment in autism.* New York: Plenum Press, 1988.

570. LeCouteur A, Rutter M, Lord C, et al. Autism Diagnostic Interview: a standardized investigator-based instrument. *J Autism Dev Disord* 1989;19:363–387.

571. Lord C, Rutter M, Good S, et al. Autism Diagnostic Observation Schedule: a standardized observation of communication and social behavior. *J Autism Dev Disord* 1989;19:185–212.

572. Harris S. Family assessment in autism. In: Schopler E, Mesibov G, eds. *Diagnosis and assessment in autism.* New York: Plenum Press, 1988.

573. McCubbin J, Patterson J. *Stress and the family. Vol. I: Coping with normative transitions.* New York: Brunner/Mazel, 1983.

574. Powers MD. Children with autism and their families. In: Powers M, ed. *Children with autism: a parent's guide.* Bethesda, MD: Woodbine House, 2000:119–153.

575. Aman MG, Van Bourgondien ME, Wolford PL, et al. Psychotropic and anticonvulsant drugs in subjects with autism: prevalence and patterns of use. *J Am Acad Child Adolesc Psychiatry* 1995;34:1672–1681.

576. Tsai LY. Psychopharmacology in autism. *Psychosom Med* 1999;61:651–665.

577. Martin A, Patzer DK, Volkmar FR. Psychopharmacological treatment of higher-functioning pervasive developmental disorders. In: Klin A, Volkmar FR, eds. *Asperger syndrome.* New York: Guilford Press, 2000:210–228.

578. McDougle CJ. Psychopharmacology. In: Volkmar FR, ed. *Autism and pervasive developmental disorders (Cambridge Monographs in Child and Adolescent Psychiatry).* New York: Cambridge University Press, 1998:169–194.

579. Gringras P. Practical paediatric psychopharmacological prescribing in autism: the potential and the pitfalls. *Autism* 2000;4:229–247.

580. Gualtieri CT. New developments in the psychopharmacology of autism. *Ital J Intellect Impair* 1992;5:127–136.

581. Hellings JA. Psychopharmacology of mood disorders in persons with mental retardation and autism. *Ment Retard Dev Disabil Res Rev* 1999;5:270–278.

582. Holm VA, Varley CK. Pharmacological treatment of autistic children. In: Dawson G, ed. *Autism: nature, diagnosis, and treatment.* New York: Guilford Press, 1989.

583. Harty JR. Pharmacotherapy in infantile autism. *Focus Autistic Behav* 1990;5:1–15.

584. Aman M, Langworthy K. Pharmacotherapy for hyperactivity in children with autism and other pervasive developmental disorders. *J Autism Pervasive Dev Disord* 2000;30:451–460.

585. King B. Pharmacological treatment of mood disturbances, aggression, and self-injury in persons with pervasive developmental disorders. *J Autism Dev Disord* 2000;30:439–446.

586. Ratey J, Mikkelson E, Sorgi P, et al. Autism: the treatment of aggressive behaviors. *J Clin Psychopharmacol* 1987;7:35–41.

587. Lipinski J, Keck P, McElroy SL. Beta-adrenergic antagonists in psychosis: is improvement due to treatment of neuroleptic-induced akathisia? *J Clin Psychopharmacol* 1988;8:409–416.

588. Rapoport JL. Clozapine and child psychiatry. *J Child Adolesc Psychopharmacol* 1994;4:1–3.

589. Posey DJ, McDougle CJ. The pharmacotherapy of target symptoms associated with autistic disorder and other pervasive developmental disorders. *Harv Rev Psychiatry* 2000;8:45–63.

590. Rothenberger A. Psychopharmacological treatment of self-injurious behavior in individuals with autism. *Acta Paedopsychiatr* 1993;56:999–104.

591. Buitelaar JK, Willemsen-Swinkels S, Van Engeland H, et al. Naltrexone in children with autism. *J Am Acad Child Adolesc Psychiatry* 1998;37:800–802.

592. Geller B, Reising D, Leonard HL, et al. Critical review of tricyclic antidepressant use in children and adolescents. *J Am Acad Child Adolesc Psychiatry* 1999;38:513–528.

593. DeLong G, Aldershof A. Long-term experience with lithium treatment in childhood: correlation with clinical diagnosis. *J Am Acad Child Adolesc Psychiatry* 1987;26:389–394.

594. Gordon C, Rapoport J, Hamburger S. Differential response of seven subjects with autistic disorder to clomipramine and desipramine. *Am J Psychiatry* 1992;149:363–366.

595. McDougle C, Price L, Volkmar F, et al. Clomipramine in autism: preliminary evidence of efficacy. *J Am Acad Child Adolesc Psychiatry* 1992;31:746–750.

596. Brasic JR, Barnett JY, Kaplan D, et al. Clomipramine ameliorates adventitious movements and compulsions in prepubertal boys with autistic disorder and severe mental retardation. *Neurology* 1994;44:1309–1312.

597. Gordon CT, State RC, Nelson JE, et al. A double blind comparison of clomipramine, desipramine, and placebo in the treatment of autistic disorder. *Arch Gen Psychiatry* 1993;50:441–447.

598. McDougle CJ. Repetitive thoughts and behavior in pervasive developmental disorders: phenomenology and pharmacotherapy. In: Scopler E, Mesibov GB, eds. *Asperger syndrome or high-functioning autism? Current issues in autism.* New York: Plenum Press, 1998:293–316.

599. McDougle C, Kresch L, Posey D. Repetitive thoughts and behavior in pervasive developmental disorders: treatment with serotonin reuptake inhibitors. *J Autism Dev Disord* 2000;30:427–436.

600. Mehlinger R, Scheftner W, Poznanski E. Fluoxetine and autism. *J Am Acad Child Adolesc Psychiatry* 1990;29:985.

601. Todd R. Fluoxetine in autism. *Am J Psychiatry* 1991;148:1089.

602. Ghaziuddin M, Tsai L, Ghaziuddin N. Fluoxetine in autism with depression. *J Am Acad Child Adolesc Psychiatry* 1991;30:508–509.

603. Hamdan-Allen G. Brief report: trichotillomania in an autistic male. *J Autism Dev Disord* 1991;21:79–82.

604. McDougle C, Price L, Goodman W. Fluvoxamine treatment of coincident autistic disorder and obsessive-compulsive disorder: a case report. *J Autism Dev Disord* 1990;20:537–543.

605. Cook E, Rowlett R, Jaselskis C. Fluoxetine treatment of children and adults with autistic disorder and mental retardation. *J Am Acad Child Adolesc Psychiatry* 1992;31:739–745.

606. Perel M, Alcami M, Gilaberti I. Fluoxetine in children with autism. *J Am Acad Child Adolesc Psychiatry* 1999;38:1472–1473

607. Leventhal BL, Cook EH, Morford M, et al. Clinical and neurochemical effects of fenfluramine in children with autism. *J Neuropsychiatry Clin Neurosci* 1993;5:307–315.

608. Raiten D. Nutrition and developmental disabilities: a clinical assessment. In: Schopler E, Mesibov G, eds. *Diagnosis and assessment of autism.* New York: Plenum Press, 1988.

609. Rimland B. Controversies in the treatment of autistic children: vitamin and drug therapy. *J Child Neurol* 1988;3[Suppl]:68–72.

610. Rimland B, Callaway E, Dreyfus P. The effects of high doses of vitamin B6 on autistic children: a double-blind crossover study. *Am J Psychiatry* 1978;135:472–475.

611. LeLord G, Muh J, Barthelemy C, et al. Effects of pyridoxine and magnesium on autistic symptoms: initial observations. *J Autism Dev Disord* 1981;11:219–230.

612. Martineau J, Barthelemy C, Garreau B, et al. Vitamin B6, magnesium, and combined B6-Mg: therapeutic effects in childhood autism. *Biol Psychiatry* 1985;20:476–478.

613. Kozlowski BW. Megavitamin treatment of mental retardation in children: a review of effects on behavior and cognition. *J Child Adolesc Psychopharmacol* 1992;2:307–320.

614. Page T. Metabolic approaches to the treatment of autism spectrum disorders. *J Autism Dev Disord* 2000;30:463–469.

615. Hagerman R, Jackson A, Levitas A, et al. Oral folic acid versus placebo in the treatment of males with the fragile X syndrome. *Am J Med Genet* 1986;23:241–262.

616. Lowe T, Cohen D, Miller S, et al. Folic acid and B12 in autism and neuropsychiatric disturbances of childhood. *J Am Acad Child Psychiatry* 1981;20:104–111.

617. Dolske MC, Spollen J, McKay S, et al. A preliminary trial of ascorbic acid as supplemental therapy for autism. *Prog Neuropsychopharmacol Biol Psychiatry* 1993;117:765–774.

618. Sandler A, Sutton K, DeWeese J, et al. Lack of benefit of a single dose of synthetic human secretin in the treatment of autism and pervasive developmental disorders. *N Engl J Med* 1999;341:1801–1806.

619. Dunn-Geier J, Ho H, Auersperg E, et al. Effect of secretin on children with autism: a randomized controlled trial. *Dev Med Child Neurol* 2000;42:796–802.

620. Chez M, Buchanan C, Bagan B, et al. Secretin and autism: a two-part clinical investigation. *J Autism Dev Disord* 2000;30:87–94.

621. Rimland B. Comments on "Secretin and autism: a two-part clinical investigation" by M. G. Chez et al. *J Autism Dev Disord* 2000;30:95.

622. Whitley P, Rodgers J, Savery D, et al. A gluten-free diet as an intervention for autism and associated spectrum disorders: preliminary findings. *Autism* 1999;3:45–65.

623. Sponheim E. Gluten-free diet in infantile autism. A therapeutic trial. *Tidsskr Nors Laegeforen* 1991;111:704–707.

624. Rogers SJ. Empirically supported comprehensive treatments for young children with autism. *J Clin Child Psychol* 1998;27:168–179.

625. Erba HW. Early intervention programs for children with autism: conceptual frameworks for implementation. *Am J Orthopsychiatry* 2000;70:82–94.

626. Gresham FM, Beebe-Frankenberger ME, MacMillan DL. A selective review of treatments for children with autism: description and methodological considerations. *School Psychol Rev* 1999;28:559–575.

627. Smith T. Outcome of early intervention for children with autism. *Clin Psychol Sci Pract* 1999;6:33–49.

628. Matson J, Benavidez D, Compton L, et al. Behavioral treatment of autistic persons: a review of research from 1980 to the present. *Res Dev Disabil* 1996;17:433–465.

629. Rogers S, Lewis H. An effective day treatment model for young children with pervasive developmental disorder. *J Am Acad Child Adolesc Psychiatry* 1989;28:207–214.

630. McGee GG, Morrier MJ, Daly T. An incidental teaching approach to early intervention for toddlers with autism. *J Assoc Persons Severe Handicap* 1999;24:133–146.

631. Lovaas OI. Behavioral treatment and normal educational and intellectual functioning in young autistic children. *J Consult Clin Psychol* 1987;55:3–9.

632. Schopler E. Prevention and management of behavior problems: the TEACCH approach. In: Sanavio E, ed. *Behavior and cognitive therapy today: essays in honor of Hans J. Eysenck.* West Point, New York: Elsevier Science, 1998:249–259.

633. Mesibov GB. Formal and informal measures on the effectiveness of the TEACCH programme. *Autism* 1997;1:25–35.

634. Greenspan SI, Wieder S. A functional developmental approach to autism spectrum disorders. *J Assoc Persons Severe Handicap* 1999;24:147–161.

635. McEachin JJ, Smith T, Lovaas OI. Long term outcome for children with autism who received early intensive behavioral treatment. *Am J Mental Retard* 1993;97:359–372.

636. Kazdin AE. Replication and extension of behavioral treatment of autistic disorder. *Am J Mental Retard* 1993;97:377–379.

637. Green G. Early behavioral intervention for autism: what does research tell us? In: Maurice C, Green G, Luce S, eds. *Behavioral intervention for young children with autism: a manual for parents and professionals.* Austin, TX: Pro-Ed, Inc., 1992:29–44.

638. Andersson SR, Romanczyk RG. Early intervention for young children with autism: continuum-based behavioral models. *J Assoc Persons Severe Handicap* 1999;24:162–173.

639. Harris SL, Weiss MJ. Right from the start: behavioral intervention for young children with autism. In: *Topics in autism.* Bethesda, MD: Woodbine House, 1998.

640. Harris SL. Behavioural and educational approaches to the pervasive developmental disorders. In: Volkmar FR, ed. *Autism and pervasive developmental disorders (Cambridge Monographs in*

Child and Adolescent Psychiatry). New York: Cambridge University Press, 1998:195–208.

641. Schopler E, Mesibov GB, Hearsey K. Structured teaching in the TEACCH system. In: Schopler E, Mesibov GB, eds. *Learning and cognition in autism. Current issues in autism.* New York: Plenum Press, 1995:243–268.

642. Maurice C, Green G, Foxx R. *Making a difference: behavioral intervention for autism.* Austin, TX: Pro-Ed, 2001.

643. Wolfberg PJ, Schuler AL. Integrated play groups: a model for promoting the social and cognitive dimensions of play in children with autism. *J Autism Dev Disord* 1993;23:467–489.

644. McGee GG, Paradis T, Feldman RS. Free effects of integration on levels of autistic behavior. *Top Early Child Special Educ* 1993;13:57–67.

645. Jordan R, Jones G. Review of research into educational interventions for children with autism in the UK. *Autism* 1999;3:101–110.

646. Freeman BJ. Guidelines for evaluating intervention programs for children with autism. *J Autism Dev Disord* 1997;27:641–651.

647. Mays RM, Gillon JE. Autism in young children: an update. *J Pediatr Health Care* 1993;7:17–23.

648. Siegel B, Pliner C, Eschler J, et al. How children with autism are diagnosed: difficulties in identification of children with multiple developmental delays. *J Dev Behav Pediatr* 1988;9:199–204.

649. Prizant B, Wetherby A. Providing services to children with autism (ages 0 to 2 years) and their families. *Focus Autistic Behav* 1988;4:1–16 (Reprinted from *Top in Lang Disord* 1988;9:1–23).

650. McGee G, Almeida M, Sulzer-Azaroff B, et al. Promoting reciprocal interactions via peer incidental teaching. *J Appl Behav Anal* 1992;25:117–126.

651. Lord D, Hopkins J. The social behavior of autistic children. *J Autism Dev Disord* 1986;16:449–462.

652. Odom S, Strain P. Comparison of peer initiation and teacher antecedent interventions. *J Appl Behav Anal* 1986;19:59–71.

653. Schreibman L. Intensive behavioral/psychoeducational treatment for autism: research needs and future directions. *J Autism Dev Disord* 2000;30:373–378.

654. Lord C. Commentary: achievements and future directions for intervention research in communication and autism spectrum disorders. *J Autism Dev Disord* 2000;30:393–398.

655. Rogers S. Interventions that facilitate socialization in children with autism. *J Autism Dev Disord* 2000;30:399–410.

656. Day HM, Horner RH, O Neill RE. Multiple functions of problem behaviors: assessment and intervention. *J Appl Behav Anal* 1994;27:279–289.

657. Koegel RL, Frea WS. Treatment of social behavior in autism through the modification of pivotal social skills. *J Appl Behav Anal* 1993;26:369–277.

658. Cook D. A sensory approach to the treatment and management of children with autism. *Focus Autistic Behav* 1990;5:1–19.

659. Dawson G, Watling R. Interventions to facilitate auditory, visual, and motor integration in autism: a review of the evidence. *J Autism Dev Disord* 2000;30:415–422.

660. Delmolino L, Romanczyk R. Facilitated communication: a critical review. *Behav Ther* 1995;18:27–30.

661. Jacobson J, Mulick J, Schwartz A. A history of facilitated communication: science, pseudoscience, and antiscience. *Am Psychol* 1995;50:750–765.

662. Schopler E, Mesibov G, eds. *The effects of autism on the family.* New York: Plenum Press, 1984.

663. Konstantareas M. A psychoeducational model for working with families of autistic children. *J Marital Fam Ther* 1990;16:59–70.

23

MENTAL RETARDATION AND DEVELOPMENTAL DISABILITIES

STEPHEN P. JOY
JESSICA STEWART LORD
LEEANNE GREEN
DEBORAH FEIN

Mental retardation refers to subnormal intellectual functioning that impacts negatively upon the individual's adjustment. The intellectual impairment is general in nature, affecting multiple cognitive functions, although in some cases there may be isolated islands of intact or even exceptional ability (so-called *savant syndrome*). When only one or two cognitive abilities are compromised in an otherwise intellectually intact individual, terms such as learning disability, learning disorder, and specific developmental disorder are used. *Developmental disability* refers to any handicapping condition that occurs as part of a child's natural developmental processes. This chapter provides (a) a descriptive overview of mental retardation (MR), together with brief comments on specific developmental disorders (communication disorders and learning disorders), including information relating to assessment instruments and treatment, and (b) detailed information concerning some of the more common, challenging, and/or interesting biomedical syndromes frequently associated with a mentally retarded or learning disabled presentation. Physical manifestations and biomedical complications will be discussed, but the emphasis will be on neuropsychological, behavioral, and psychiatric aspects of these disorders. Due to space limitations, developmental disabilities that are not strongly linked to neurocognitive impairment or abnormal behavior (e.g., cystic fibrosis and sickle cell anemia) are excluded. Pervasive developmental disorders (PDD; e.g., autistic disorder) and childhood epilepsy are considered in subsequent chapters.

DEFINING MENTAL RETARDATION AND LEARNING DISORDERS

The two most widely accepted sets of diagnostic criteria for MR are those of the American Association on Mental Retardation (AAMR) (1,2) and the American Psychiatric Association's *Diagnostic and Statistical Manual of Mental Disorders, Fourth Edition* (DSM-IV) (3). The two sets of criteria are generally similar, although there are differences in the particulars. MR is defined by the combination of (a) significantly subaverage intellectual functioning, (b) significant problems with adaptive functioning associated with the intellectual impairment, and (c) onset during the developmental years (i.e., before age 18).

With respect to intellectual functioning, DSM-IV mentions an approximate cutoff score of 70 on a standardized intelligence quotient (IQ) test but fails to specify the rating scale used. In practice, a score more than two standard deviations below the population mean is intended. The current AAMR criteria (unlike their 1983 criteria) offer a more lenient (and controversial) cutoff of 75.

With respect to adaptive functioning, the AAMR criteria require limitations in at least two of ten domains of adaptive functioning: communication, home living, community use, health and safety, leisure, self-care, social skills, self-direction, functional academic skills, and work. The DSM-IV list is similar, but health and safety are listed separately.

The age of onset criterion has one curious consequence. Adherence to the DSM-IV mandates applying a diagnosis of MR for children and even adolescents who suffer intellectual deterioration secondary to traumatic brain injuries, brain tumors, or degenerative conditions. This is not especially consistent with the idea of a *developmental* disorder; however, to date no better system has achieved widespread acceptance. In such cases, a diagnosis of dementia also is applied.

The current AAMR criteria have been criticized on two counts (4,5). First, because of the approximately normal distribution of IQ in the population, easing the IQ threshold threatens to bring far more children under the MR umbrella. Second, the domains of adaptive functioning listed in the criterion set have not been empirically validated, and in some cases, reliable instruments with which to evaluate them do not exist.

General intellectual functioning is operationally defined by scores on individually administered IQ tests. Alfred Binet and Theophile Simon developed the first such tests at the end of the nineteenth century, when universal education was introduced. Their purpose was to identify children who were unlikely to succeed in regular academic programs but who could benefit from modified forms of instruction (e.g., slower-paced classes or training in practical skills). Binet developed the concept of *mental age* (MA): a child's performance on mental tests as compared with the average for children of similar age. For example, if performance is equivalent to that of the "average" child of 6 years 6 months, then MA is 6.5, regardless of chronologic age (CA). When MA falls well short of CA, the child's mental development is said to be retarded. In Binet's view, the purpose of remedial instruction was to work upward from a child's current MA at a pace appropriate to the degree of slowing shown by a comparison of MA and CA—a relatively optimistic and still useful perspective.

Binet never accepted Stern's suggestion that MA be divided by CA, then multiplied by 100 to yield an "intelligence quotient"—presciently, he was concerned over the insidious reification of IQ, with its implied permanence, so different from his steadily improving MA—but to no avail; the IQ concept triumphed. Eventually, Stern's IQ formula was replaced by Wechsler's deviation IQ, which sets the mean at 100 and the standard deviation at 15 (a few tests use a standard deviation of 16 or an altogether different scale, but follow the same principle). The cutoff score defining MR is essentially arbitrary. The distribution of IQ scores is not dichotomous or even strongly bimodal, but it conforms fairly closely to the gaussian distribution ("bell-shaped curve") typical of multiply determined characteristics. At one time, people with IQ scores of more than one but less than two standard deviations below the mean were diagnosed with "borderline mental retardation," but this involved deeming almost 16% of the population to be retarded, which was far too many given that most of these people are perfectly capable of functioning adequately in ordinary human society. Today, individuals with IQ scores in the range from 70 to 84 are not diagnosed with a disorder, but the DSM-IV does include a "V" code (V62.89) for use when so-called *borderline intellectual functioning* is a focus of clinical attention.

Departments of MR (and of education) set cutoff scores for eligibility for services that may differ from agency to agency. For instance, one department may allow an MR diagnosis (and therefore fund special services) if *any* IQ score below 70 is obtained, whereas another may refuse to consider the diagnosis unless *all* IQ scores are below 70. Similar variability may be obtained with respect to level of impairment in adaptive functioning.

PREVALENCE AND LEVELS OF SEVERITY

Based on purely statistical considerations, the prevalence of IQ less than 70 should be about 2.3%, with increasing rarity as scores decline (e.g., 0.13% below 55), and these figures come close to the actual state of affairs. In fact, due to biomedical conditions that produce severe mental impairment, prevalence appears to be more than expected toward the bottom end of the distribution. On the other hand, many people with IQ close to 70 succeed in adapting to regular society and so need not be diagnosed with MR (or cease to be so diagnosed as adults). Actual prevalence, then, is probably less than 2%. A modest preponderance of males is consistently found, especially among those more seriously affected. In general, the lower the IQ, the greater the probabilities that (a) a biomedical condition caused the MR and (b) comorbid behavioral problems or psychiatric disorders exist.

Because of the wide range of ability involved, it is important to differentiate among levels of MR. As with the primary diagnosis, however, the cutoff scores between levels are essentially arbitrary, and clinical judgment should be exercised when diagnosing individuals whose IQ (and/or adaptive functioning) scores fall near the cusp between one level and another. In keeping with customary usage, DSM-IV distinguishes four levels of MR: mild, moderate, severe, and profound. The 1992 AAMR criteria also describe four levels of severity, but they do so in terms of the level of psychosocial support required in order to facilitate improved functioning: intermittent, limited, extensive, and pervasive.

Mild MR (coded 317 in DSM-IV) is diagnosed when IQ is in the range from 55 to 69. These individuals sometimes have been referred to as the *educable mentally retarded.* Nearly 90% of mentally retarded persons fall into this category. Most of these "cases" lack definable biomedical etiologies and may best be described as ordinary people who happen to fall in the lower range of intellectual endowment. They are not often recognized as having a problem until they reach school age, because they may reach many of the early childhood developmental milestones within normal limits, and other children who show the same mild developmental delays ultimately "catch up" with their age-mates. The mildly retarded, however, do not do well in school unless provided with special help. With appropriate educational services, they are capable of mastering basic academic skills. By the end of their high school years, they may be close to a sixth-grade level of scholarly attainment. Mildly retarded people are capable of enjoying hobbies and social activities and even of competitive employment in the regular workplace, although their jobs are likely to be relatively simple. Comorbid conditions (physical or psychiatric) may, however, preclude this level of adaptation.

Moderate MR (coded 318.0 in DSM-IV) applies when IQ is in the range from 41 to 54. These individuals sometimes have been referred to as the *trainable mentally retarded.* They are more likely to be identified during the preschool years, when they already lag behind their age-mates (e.g., not reaching developmental milestones such as walking or talking until about twice the average age). In time, however, moderately retarded people usually achieve reasonable linguistic competence (although as preschoolers they may use an amalgam of limited spoken language, signing, and other methods) and master basic self-care skills. Academically, they may achieve basic literacy and numeracy up to a second-grade level by the end of high school. Occupationally, they can acquire work-related skills and maintain employment, but generally they need considerable support in order to do so; many work in sheltered settings. Socially they need some degree of support and are unlikely to live independently. As with mild MR, comorbid conditions may adversely affect their adjustment.

Severe MR (coded 318.1 in DSM-IV) applies when IQ is in the range from 26 to 40. These individuals (and those with profound MR) are likely to "bottom out" on most standardized IQ tests, failing to complete even the simplest items correctly. Such cases are likely to be recognized very early in life and often bear physical stigmata (e.g., of a recognized medical syndrome associated with mental impairment). Communication skills are likely to remain rudimentary, but they are capable of mastering some spoken language, signs, and/or use of communication boards or similar devices. Intensive skill training using behavior modification methods can produce some competence in basic self-care, but neither academic nor occupational training is likely to be effective. A high level of personal supervision is required throughout the life span. This may be provided by devoted family members but is likely to involve residential treatment (e.g., at a training school or group home) for at least some portion of the life span.

Profound MR (coded 381.2 in DSM-IV) applies when IQ is 25 or lower. This level of impairment is rare (only about 1% of the MR population falls into this range) and almost always is produced by biomedical causes (e.g., severe neurologic conditions). Residential treatment and constant personal supervision are likely to be required throughout the life span. Some degree of motor development (perhaps including elementary self-care, e.g., toilet training) and primitive communicative competence may be established by training.

SPECIFIC DEVELOPMENTAL DISORDERS

The two groups of specific developmental disorders most relevant to understanding milder cases of neurocognitive impairment are *communication disorders* (also known as *language disorders* or *developmental dysphasias*) and *learning disorders* (LD) (also known as *learning disabilities*). The two principal language disorders are *expressive language disorder* (coded 315.31 in DSM-IV) and *mixed receptive-expressive language disorder* (also coded 315.31). Further information on language disorders may be found in the appropriate chap-

ter of this text. A substantial difference between receptive aphasia acquired in adulthood and its developmental equivalent is that the former is characterized by fluent language production (albeit of limited meaningfulness) whereas the latter involves reduced production of speech as well as reduced comprehension. Diagnosis is based upon discrepancy between relatively good performance on nonverbal tasks and poor performance on standardized language measures and other verbally mediated tasks.

Learning disorders are diagnosed based on discrepancies between IQ and academic achievement as measured by individually administered standardized tests. In essence, these are diagnoses of exclusion; if achievement is substantially lower than that expected based on the person's general ability level and the discrepancy cannot be explained based on other factors (e.g., inadequate schooling, sensory deficits, or chaotic home environment), then LD is deemed to be present. In general, one expects to find an uneven pattern of abilities, as well as the required discrepancy between ability and achievement. Presumably the relatively weak areas of cognitive functioning produce the learning impairment. Measures of attention, working memory, and/or executive functions are especially likely to show deficits. Such findings increase confidence in the diagnosis (e.g., differentiating between dyslexia and "garden-variety" poor reading, or among different forms of dyslexia) but are neither necessary nor sufficient in the eyes of many school systems.

The three major learning disorders are *reading disorder* (coded 315.00 in DSM-IV), *mathematics disorder* (coded 315.1), and *disorder of written expression* (coded 315.2). These obviously are intended to refer to educational demands rather than basic neuropsychological abilities. Of them, reading disorder *(dyslexia)* is by far the most extensively researched. In recent years, some attention also has been paid to nonverbal learning disabilities (6), which are discussed in Chapters 22 and 25. These are characterized chiefly by deficits in visual-spatial reasoning and social awareness in a context of strong rote verbal skills and may result from diffuse white matter pathology principally affecting the right cerebral hemisphere. This disorder is not recognized by DSM-IV, but a diagnosis of *learning disorder not otherwise specified* (coded 315.9) can be applied.

As with MR, specific cutoff scores for learning disorders may vary across school districts. In some, a 1-year lag may entitle a child to special services, in others, 2 or more years' delay may be required. A discrepancy of a standard deviation between IQ (or some component of IQ) and a specific achievement test score is another common criterion. Some may exclude the diagnosis of a learning disorder if IQ is in the retarded range or if academic achievement is at grade level, but there is no theoretical reason for doing so. In fact, mildly retarded students may also suffer from specific disorders of language or academic skills, and otherwise gifted students may be able partially to compensate for their learning disorders. In our view, members of both groups merit the diagnosis of a learning disorder.

Both expressive language disorder and mixed receptive-expressive language disorder affect approximately 3% to 5% of school-aged children. The prevalence of reading disorder is in the same range. All three of these disorders are two to three times more common in males than females. Prevalence data are lacking on the other learning disorders. No formal differentiations are made between different levels of severity, but the degree of disability produced by these disorders varies considerably. An expressive language disorder, for example, may present as word-finding difficulty and a tendency to speak in relatively short, telegraphic sentences; near total absence of speech; or anywhere in between.

COGNITIVE AND SOCIAL ASSESSMENT

A number of well-designed and standardized individually administered IQ tests are available for school-aged children. These should be administered, scored, and interpreted by clinical psychologists, school psychologists, or clinical neuropsychologists who are trained in standardized procedures and sensitivity to factors that may invalidate test performance (e.g., cultural differences or specific impairments in areas such as language or motor control). Comprehensive IQ tests consist of a number of different tasks, and the IQ score is only one small part of the information they yield. In fact, these tests constitute an essential component of a neuropsychological evaluation and can contribute much to the differential diagnosis of disorders associated with cognitive dysfunction as well as to treatment planning. The most widely used are the *Wechsler Intelligence Scale for Children, Third Edition* (WISC-III) (7), the *Stanford-Binet Intelligence Scale, Fourth Edition* (SB-IV) (8), the *Kaufman Assessment Battery for Children* (K-ABC) (9,10), and the *Differential Ability Scales* (DAS) (11). Each instrument involves a different set of tasks and so offers a somewhat different set of neuropsychological data, but IQ scores derived from different tests should not differ too markedly (correlating at $r = 0.80$–0.90 across tests). One exception to this rule derives from the fact that performance on IQ tests has been improving worldwide for over half a century (12,13). Because of this, scores on tests that were standardized more than 25–30 years ago will tend to be inflated. Reliability, criterion and construct validity, and representativeness of the standardization samples are very good to excellent for all of the instruments discussed in this section except where otherwise noted.

The WISC-III is the most popular test among neuropsychologists. It yields a verbal IQ (VIQ) based on language-mediated tasks (including arithmetic), a performance IQ (PIQ) based on visuospatial tasks, and a full-scale IQ that is a composite of the other two. Comparison of VIQ and PIQ helps to specify areas of relative strength and weakness. Anal-

ysis of scores on smaller groups of subtests provides information on a variety of cognitive domains: verbal comprehension, spatial organization, speed of processing, and working memory. Adults are evaluated using the highly comparable *Wechsler Adult Intelligence Scale, Third Edition* (WAIS-III) (14).

The Stanford-Binet, once the gold standard against which other IQ tests were judged, has slipped in popularity over recent decades but continues to be widely used, especially by school psychologists. SB-IV represents a fairly radical departure from previous editions. As with the Wechsler scales, a general intelligence score is derived from the aggregated subtest scores, with several more specific domains of cognitive functioning assessed by subsets of tests. Two of these, crystallized intelligence (linguistically mediated abilities built up through cultural transmission) and fluid intelligence (solution of abstract, novel problems) are derived from Cattell's theory of intelligence. The third, short-term memory, derives from cognitive science. The standardization sample was skewed toward higher socioeconomic strata, but this bias was corrected for statistically in the development of norms.

The K-ABC is based on a model of intelligence that posits two largely independent abilities: sequential processing and simultaneous processing. Sequential processing involves arranging stimuli (e.g., words) in serial order. Simultaneous processing involves the integration of multiple inputs (e.g., shapes) into larger wholes or gestalts. Although a mental processing composite equivalent to a full-scale IQ is derived, the separate scores for sequential and simultaneous processing are interpretively more important. The K-ABC also includes "achievement" tests that closely parallel some of the verbal subtests of the Wechsler scales.

The DAS is based on the British Ability Scales, long used in the United Kingdom (15). As with other tests, a global intelligence score (general cognitive ability) is supplemented by several lower-order composites. In the case of the DAS, these composites are verbal ability, nonverbal reasoning, and spatial ability. The latter two overlap substantially. The DAS may be better able to differentiate among more severe levels of MR than most IQ tests, but it is not yet widely used in the United States.

Intelligence theory is a complex and contentious field in psychology, and some of this complexity is reflected in the different terminology used by the various IQ tests described. There are, however, many points of similarity among these instruments. Each yields an estimate of overall ability (whether termed IQ or otherwise) in deference to Spearman's "g" factor and practical demands. Verbal IQ (WISC-III), crystallized intelligence (SB-IV), and verbal ability (DAS) all are conceptually related, whereas PIQ (WISC-III), fluid intelligence (SB-IV), and the overlapping spatial ability and nonverbal reasoning factors (DAS) are similarly related. When a third factor is present, it typically relates to some combination of processing speed, attention, working memory, and memory formation, although formal assessment of learning and memory requires utilization of other tests in addition to these IQ measures.

Assessment of infants, toddlers, and preschoolers poses special challenges. It can be difficult to elicit active collaboration, and, even assuming a willing examinee, limited language development may invalidate many procedures. Measurement of general intellectual functioning in infancy or early childhood is useful in that one may obtain a valid estimate of a child's level of intellectual development relative to his or her age-mates but problematic in that IQ scores are less stable over time in the very young. The primary purpose of evaluation, especially in this age range, is to delineate a pattern of cognitive strengths and weaknesses that can be used to guide treatment planning. In general, MA may be a more useful concept (and less distressing to parents) than IQ until the latter stabilizes at about age 6, and clinicians should apply MR diagnoses among preschoolers cautiously unless the impairment is obvious and severe.

The *Bayley Scales of Infant Development, Second Edition* (Bayley II) probably are the most widely used tests of general intellectual functioning for the very young (ages 1–42 months) (16). There are two main scores: the mental development index and the psychomotor development index. An infant behavior record supplies personality information. Mental and psychomotor development tend to be more highly correlated in infancy than in later childhood. Except in cases of serious mental deficiency, scores on the Bayley II have little predictive power with respect to IQ in later childhood or adulthood. Research involving children aged 0 to 3 years is increasingly using the *Mullen Scales of Early Learning,* because this instrument provides separate scores for visual concepts, receptive and expressive language, and fine and gross motor skills (17).

The *McCarthy Scales of Children's Abilities* consist of 18 subtests that contribute to five scales (verbal, perceptual, quantitative, memory, and motor) with means of 50 and standard deviations of 10, plus a general cognitive index (18). The McCarthy Scales are normed for children aged 2.5 to 8.5 years. Thus, they typically are used with children too old for the Bayley Scales but too young for most IQ tests. Alternatively, the *Wechsler Preschool and Primary Scale of Intelligence-Revised* (WPPSI-R) is useful because it includes simpler versions of the subtests used in other versions of the Wechsler scales and thus facilitates examination of intellectual development over the years (19). DAS norms also cover the preschool years.

Assessment of adaptive functioning typically is less systematic than intellectual assessment. There are, however, at least two reliable, standardized instruments for use in this area: the *Vineland Adaptive Behavior Scale-Revised* (20) and the *AAMR Adaptive Behavior Scales* (21). These rating scales must be completed by a trained professional using information supplied by parents, teachers, or other primary caregivers. Some state departments or other agencies will

also have their own locally developed rating scales for this purpose.

The Vineland provides an overall rating with a mean of 100 and a standard deviation of 15 (conveniently comparable to IQ scores), as well as separate scores for communication, daily living skills, socialization, and motor skills; maladaptive behaviors also may be assessed. The standardization sample consisted of 3,000 children ranging in age from infancy to 19 years and closely matched with US census data. The interview takes 20 to 60 minutes to complete, and interrater reliability is excellent. A Spanish edition is available.

The AAMR Adaptive Behavior Scale assesses nine domains of adaptive behaviors and 12 domains of maladaptive behaviors, takes about as long as the Vineland to administer, and should have adequate interrater reliability, although data on this issue are lacking. The standardization sample was large (6,523 children aged 3–17 years) and included healthy controls as well as cases of mild and moderate MR, but it was not stratified to match the census on key demographic variables.

A number of achievement tests are available for use in diagnosing learning disorders and for further evaluating the skills of MR children. All evaluate reading, writing, and mathematical skills, but with varying degrees of sophistication. The *Wide Range Achievement Test-Third Edition* (WRAT-3) is little more than a screening test, measuring only single-word reading, spelling, and arithmetic (22). As noted earlier, the DAS has its own achievement subtests, also best thought of as screening tests. The *Wechsler Individual Achievement Test-Third Edition* (WIAT-III) has the advantage of being standardized along with the WISC-III, facilitating score comparisons (23). The *Peabody Individual Achievement Test-Revised* (PIAT-R) (24) and *Kaufman Test of Educational Achievement* (K-TEA) (25) also are well regarded. Most school systems also use some sort of group-administered standardized achievement test as part of their evaluation of pupil progress and program effectiveness. Correlations between IQ and relevant achievement test scores typically are in the *r* range from 0.60 to 0.70 during the school years.

Special mention should be made of the *Woodcock-Johnson Psycho-educational Battery-Revised* (W-J-R), which evaluates a broad spectrum of cognitive skills and can be used to derive general intelligence scores as well as achievement levels from the preschool years through old age (26). This extensive and complex test battery is widely used for assessment of learning disorders, although the correlation between its basic cognitive ability score and IQ scores obtained using other instruments are relatively low (typically $r < 0.70$), which limits its popularity as a measure of general intelligence.

Proper evaluation of MR or otherwise learning disabled individuals should include more than assessment of general intelligence or even ability/achievement comparisons, especially when the degree of global impairment is mild to moderate. As noted earlier, IQ tests actually offer considerable neuropsychological information, although a floor effect limits their value in more severe cases. If administered in conjunction with tests of specific cognitive functions, the data provide a profile of neuropsychological abilities that can be useful in educational and habilitative treatment planning, as well as informative regarding the functional status of the central nervous system (CNS). Tests of verbal and nonverbal learning and memory, psychomotor abilities, and executive functions (e.g., planning and inhibition) are especially important. Lezak's classic text remains the definitive source of information on neuropsychological assessment procedures (27). Other sources provide further details on the neuropsychology of developmental disorders (28–30).

DIFFERENTIAL DIAGNOSIS

As noted earlier, MR is distinguished from communication disorders and learning disorders by virtue of the relatively global nature of intellectual deficits in the former as opposed to the more circumscribed impairment found in the latter. An individual with a reading disorder, for example, should perform much better on nonverbal tests and most poorly on those that require reading (or knowledge usually acquired through reading). This is not to say that the neuropsychological profile of any given mentally retarded person will be altogether "flat." To the contrary, intellectual functioning is apt to vary across domains for individuals who suffer from MR just as it does in the "normal" population, and some syndromes have identifiable cognitive profiles. Conversely, the selective cognitive deficits in a communication or learning disorder may, by restricting exposure to information, exert an increasingly negative influence upon measured IQ over the years. The differential, then, may be problematic when IQ is near 70 and an uneven profile of intellectual strengths and weaknesses is present.

Assuming IQ in the mid 60s to low 70s, the boundary between mild MR and normalcy (or *borderline intellectual functioning*) is defined by adaptive functioning. If standardized instruments are not used to evaluate adaptive functioning, clinical judgment must be exercised based on all available data. The boundary between a learning disorder and normalcy (e.g., dyslexia vs. "garden-variety" poor reading) is defined partly on the basis of test scores and partly by psychosocial factors. Scores on reading tests should be substantially below IQ before a diagnosis of dyslexia is considered. When an uncorrected sensory deficit, inadequate schooling, or a disruptive home environment is present, academic failure should not be attributed to a developmental disorder unless correction of such contributing factors fails to produce improved academic functioning.

Sensory deficits (especially reduced auditory acuity) reduce the amount of information a child can derive from the environment and can produce the appearance of MR,

a communication disorder, or a learning disorder—an appearance that ultimately may become a reality, owing to the resultant environmental deprivation, unless corrected. Audiologic and/or visual examinations should be ordered whenever there is a suspicion of sensory impairment. Indeed, deafness without exposure to sign language has been called the leading preventable cause of MR. Many syndromes linked to MR also are associated with frequent middle ear infections (otitis media) that may result in conductive hearing loss, further impeding language development and producing an impression of graver retardation than is in fact present.

Serious motor control deficits (e.g., cerebral palsy) and developmental dyspraxias are capable of artificially lowering scores on standardized tests, because the examinee simply cannot respond appropriately (e.g., copying designs). Dysarthric speech, spasticity, or other signs of problems with fine motor control or praxis signal the need for careful neurologic examination. In some cases, this may be a difficult differential, especially as MR can occur in company with disorders of fine motor control and/or praxis.

Grossly impoverished environments will, by failing to expose children to appropriate stimulation, invariably produce deficits on standardized testing. If the environmental deprivation is limited to academic material, then performance should be relatively better on nonverbal tests that do not call upon school-related knowledge. Pervasive deprivation during early childhood can, however, produce genuine MR. The arborization and pruning of synaptic connections in the developing brain are guided in part by appropriate experiences, and a dearth of such experience can preclude normal development. Cognitive impairment can be produced experimentally in animals by environmental (e.g., maternal grooming) deprivation and is clinically observed among humans. A cluster of related syndromes have been described in various literature (e.g., affect hunger, psychosocial dwarfism, hospitalism). All present with some combination of intellectual deficits, emotional disturbance, and retarded growth. Money (31) suggests the sobriquet Kaspar Hauser syndrome for all such cases.

COMORBID PSYCHIATRIC ISSUES

Evaluation of behavior problems in MR is complicated by three factors. First, the behavior problem (or symptom) may be (a) an inherent component of the syndrome that caused MR, (b) a secondary consequence of the stresses to which MR has exposed the individual interacting with inadequate coping skills, or (c) purely coincidental (i.e., resulting from a separate disorder unrelated to MR).

Second, evaluation of symptoms is complicated to the degree that verbal competence and insightfulness are compromised. Mildly retarded individuals can be interviewed in much the same fashion as other patients, but care must be taken to use appropriate vocabulary, simple phrasing, and concrete examples. Most formal self-report inventories [e.g., Beck Depression Inventory or Minnesota Multiphasic Personality Disorder-Second Edition (MMPI-2)] require too high a level of reading comprehension even for the best-functioning persons with MR. Those designed for children may be more appropriate but usually lack adult norms. Evaluation of moderately retarded persons requires supplementation of interview data with behavioral observations (perhaps conducted by regular caregivers). Where severely and profoundly retarded patients are concerned, only behavioral observation is useful. Observer-rated instruments such as the Child Behavior Checklist (CBCL) (32) and the Personality Inventory for Children (PIC) (33) are helpful in evaluating behavior problems in children and adolescents, whether mentally retarded or otherwise. A promising upward extension of the CBCL into young adulthood has been published (34). Some projective tests (e.g., human figure drawings) may circumvent the limited verbal skills of those with MR, but in addition to suffering from serious problems with scoring reliability these tests reflect intellectual ability and simply may provide another piece of evidence that the examinee is cognitively impaired.

Third, although standard diagnostic criteria are generally applicable, the symptoms of mental illness among the mentally retarded may differ from those expressed by the intellectually normal population. A person with limited cognitive abilities may respond to the discomfort produced by anxiety, for example, by engaging in hand flapping or rocking, which are not behaviors ordinarily associated with anxiety disorders. Careful analysis of the conditions under which symptomatic behaviors are elicited and the full spectrum of behavior changes wrought in an emotionally disturbed MR individual is a necessity if adequate treatment is to be prescribed.

TREATMENT

With rare and usually partial exceptions (e.g., low phenylalanine diets for children with phenylketonuria), MR is not a curable or reversible condition. Treatment involves maximizing the individual's intellectual potential and social adjustment. In principle, treatments may target (a) the core symptoms of MR (e.g., limited knowledge, information processing deficits); (b) abnormal behaviors produced by MR (e.g., delayed mastery of personal grooming and language); or (c) behavioral problems/psychiatric disorders comorbid with the MR.

Pharmacotherapy

Recent years have witnessed considerable interest in drugs that may function as cognitive enhancers. Acetylcholinesterase inhibitors (e.g., tacrine), for example, may improve learning and memory in cases of mild and moderate

dementia of the Alzheimer type, in which acetylcholine systems are among the first to deteriorate. With the exception of one animal study in which pyridostigmine was found to protect rat pups against the intellectual deficits usually produced by maternal deprivation (35), however, the effect of such drugs in MR has not been studied, although they may be appropriate when (e.g., in Down syndrome) a superimposed dementia develops. Nootropics (e.g., piracetam) also work (at least in part) on acetylcholine systems. These experimental agents are known to improve learning and memory in animal studies (36) and have occasionally been reported to produce positive results in cognitively impaired humans, including reading disordered students (37). Considerably more research is needed before they can be recommended for clinical use.

Psychostimulants (e.g., methylphenidate) offer some enhancement of intellectual function. They are the medications of choice for attention deficit/hyperactivity disorder (ADHD), symptoms of which are frequently observed in mentally retarded patients. Among nonretarded patients, psychostimulants are effective in nearly nine of ten cases; efficacy is somewhat reduced among the retarded, but about two thirds of patients respond favorably (38). In short, no psychotropic medication is known to offer specific amelioration of subnormal mental functioning of developmental origin.

For the most part, psychotropic medications are prescribed to mentally retarded patients for emotional or behavioral problems, whether they stem from the retardation or are coincidental. Self-injurious behavior (SIB) and aggressive behavior, for example, are common problems (especially among the more severely retarded) and often prompt psychiatric referrals (39). A variety of pharmacologic agents have been used for these indications, with mixed results. Practice in this area tends to be guided more by case reports and retrospective studies than by well-controlled clinical trials, a fact bemoaned by successive reviewers of the literature but as yet inadequately addressed (40–42). Treatment has long been empirical, but theoretical models of aggression and SIB (partly derived from animal studies) are placing it on a more rational footing. Dopamine, serotonin, and endorphin systems all have been implicated in the pathogenesis of aggression and/or SIB.

Neuroleptics (e.g., thioridazine and haloperidol) have long been used to treat aggression and, to a lesser extent, SIB (43), but their use for these purposes has fallen somewhat out of favor in recent years. They tend to suppress the target behaviors only at the expense of a more general behavioral suppression that, combined with possible drug-related cognitive deficits, does little to facilitate participation in training programs or other activities (39,44). The advent of second-generation antipsychotics associated with fewer extrapyramidal side effects (e.g., risperidone) has, however, rekindled interest in dopamine antagonists for the treatment of aggression.

The five classes of medications for SIB on which attention has been focused of late are (a) opioid antagonists (especially naltrexone); (b) serotonin reuptake inhibitors (clomipramine and fluoxetine); (c) buspirone; (d) β-adrenergic antagonists; and (e) anticonvulsants (especially valproic acid). Each usage is founded, at least in part, on theoretical models of aggression and self-injury; each shows some promise. Aggressive behavior also has been treated with lithium.

Naltrexone, a competitive opiate antagonist, presumably prevents reinforcement of SIB by endogenous opioids. Early reports on the efficacy of naloxone (a similar but shorter-acting agent), were mixed (45,46), but several small double-blinded, placebo-controlled studies of the longer-acting naltrexone reported substantial reduction in frequency and severity of SIB for eight of a cumulative 11 mentally retarded patients (47–49). Encouragingly, naltrexone appears to do this without changing the patient's general activity level (47,50). Several later studies also reported reduction of SIB by naltrexone treatment in at least 50% of patients (51–54), although a temporary increase in SIBs may occur (55), perhaps reflecting the extinction burst effect. Some reports suggest that the therapeutic benefit increases with long-term treatment (52) and even persists for years after such treatment is discontinued (56). However, the picture is not unmixed: naltrexone failed to produce positive effects in the largest double-blinded, placebo-controlled study to date (57).

The rationale for using serotonin reuptake inhibitors to treat SIB or aggression is twofold: reduced serotonin functioning has been linked to violence and suicidal behavior in many studies; and these drugs also are used to treat psychiatric disorders that could present with deliberate attempts at self-harm (depression) or potentially self-damaging stereotyped behavior [obsessive-compulsive disorder (OCD)]. Several open trials and case reports suggest that fluoxetine can reduce SIB and aggression by 20% to 88% in a majority of patients (58,59), and at least one study found a similar effect for sertraline (60), although paroxetine failed to reduce aggression in the one open trial reported to date (61) and fluoxetine was associated with a significant *increase* in aggression in another study (62). Clomipramine also has been reported to reduce SIB and stereotyped behaviors in at least one open trial (63) and in the only double-blinded, placebo-controlled study of serotonin reuptake inhibitors of which we are aware (64). In the latter study, clomipramine also was associated with improved teacher ratings on social withdrawal.

Buspirone is a partial serotonin agonist particularly active at 5-HT$_1$ receptors that also may affect several other receptor systems. Its use for SIB and aggression is justified by the serotonin model of aggressive behavior and, secondarily, by the possibility that underlying anxiety disorders may elicit SIB in the developmentally disabled. A substantial number of case reports and open trials report successful reduction of SIB and/or aggression in a majority of mentally retarded

patients treated (65–68). Encouragingly, this improvement took place without marked sedation. Unfortunately, despite more than a decade of active interest in buspirone's possibilities for this indication, no placebo-controlled, double-blinded studies have been conducted [although Ratey et al. (66) did use a placebo lead-in].

β-Adrenergic antagonists have been used for treatment of aggression (and, to a lesser extent, SIB) with generally favorable results. A series of case studies and open trials conducted in the 1980's produced varying degrees of improvement in 86% of patients with organic mental disorders, including MR (69). Further research, including some placebo-controlled, double-blinded studies, also were generally supportive (70). These drugs, of course, are not associated with either sedation or cognitive side effects. One limitation on this body of research is that most of the patients treated with β-blockers also were receiving other medications (e.g., neuroleptics and/or mood stabilizers), so it remains uncertain whether β-blockers alone can produce reductions in SIB or aggression.

Two studies have reported favorable results for valproic acid in the treatment of SIBs (71–72). Approximately 75% of the 46 patients in these samples were clinically improved, and virtually all of these individuals were able to discontinue taking other psychotropic medications (especially neuroleptics and phenobarbital). Neither study was placebo controlled, but the results are promising. These and a number of other trials involving patients with a variety of organic mental disorders, none well controlled but with a cumulative sample size of 164, have found that valproic acid reduces aggressive behaviors by 50% or more in approximately three fourths of all patients. Given the increasing popularity of valproic acid in the treatment of SIB and aggression, the lack of systematic evaluation of its therapeutic action is regrettable, but it remains one of the more promising pharmacologic agents for this indication.

Two retrospective studies using lithium report successful reduction of aggressive behaviors in approximately half of the mentally retarded patients treated (73,74). Again, though, well-controlled studies are lacking.

Finally, several studies have evaluated risperidone and other second-generation antipsychotics in the treatment of aggressive behaviors displayed by mentally retarded patients. Three open-label trials of risperidone, involving a total of 62 patients, reported significant reductions in target behaviors for a clear majority of patients (75–77). However, the samples for two of these studies consisted of patients also diagnosed with PDD, whereas in the third study, behavioral treatment was initiated simultaneously with risperidone. A small open trial of clozapine also resulted in reduced aggression and SIB (78). Whether these results will be replicated with larger samples or in double-blinded, placebo-controlled studies remains to be seen.

Cumulatively, recent research on pharmacologic management of SIB and aggression presents a fairly encouraging picture, but the prognosis for this approach remains guarded. Not only are most studies of the subject small in scope and relatively poorly controlled, but even the results reported are modest. To reduce SIB or aggressive outbursts by 40% to 50% in two thirds of patients is certainly a worthwhile accomplishment, but it leaves patients and caregivers alike suffering from all too many violent episodes. One promising avenue for further investigation might be the identification of subgroups of mentally retarded patients for whom one or another of the existing treatments might prove effective. For example, it is reasonable to suppose that naltrexone would be indicated in cases where SIB served a self-stimulatory function and/or elevated pain thresholds were observed, clomipramine if compulsive behaviors or depressive symptoms were present, or dopamine blockers if involuntary movements were in evidence. Surprisingly little work of this nature has been undertaken. One study reported that patients with elevated plasma β-endorphin activity (but not adrenocorticotropic hormone) after SIB were more responsive to naltrexone (79); this kind of research ultimately could lead to more adequate pharmacologic treatment of SIB and aggression in the mentally retarded.

Other psychiatric problems occur among the mentally retarded, as in the general population, and they typically are treated with the same medications used with nonretarded patients. For example, episodes of depression may be treated with tricyclic compounds or selective serotonin reuptake inhibitors. In some cases, solid research supports this clinical usage; more often, only case reports or retrospective studies have been published and the practice rests more upon anecdotal evidence.

Psychosocial Treatments

Given the relative paucity of effective pharmacologic treatments for MR, the mainstays of treatment are psychosocial. Special education, behavior modification, occupational therapy, speech and language therapy, and vocational habilitation all are widely used to treat the symptoms of MR, and each is associated with improved outcomes in some domains of functioning. Family and individual counseling are helpful in order to facilitate adjustment to stress, especially at transitional periods.

Children with MR and/or learning disorders are eligible for special services in school, and these services may constitute most of the psychosocial interventions used during the school years. The Education for all Handicapped Children Act of 1975 (P.L. 94-142), which mandates that public schools provide an appropriate education in the least restrictive environment possible for the needs of each student, greatly expanded the role played by the educational system in the treatment of mentally retarded and otherwise disabled children. Special education programs grew, and schools began providing speech therapy and other ancillary services. The Individuals with Disabilities in Education Act (IDEA)

of 1990 reauthorized and expanded upon this mandate, emphasizing the importance of maximizing the inclusion of handicapped children in as many aspects of regular society (and regular school) as possible.

Special education services sometimes are provided in regular classrooms (e.g., using a one-on-one aide) or in the form of extra academic assistance outside the regular classroom (e.g., a resource room staffed by special educators or developmental reading specialists). Sometimes services are provided in separate classrooms within the regular school (e.g., classes for the "educable mentally retarded"), typically small classes directed by special educators with the assistance of one or more aides. Depending upon the degree of intellectual impairment, a modified academic program or a skills-based training model may be used. If the behavioral (or medical) problems posed by a student are considered too grave for management within the regular school system, placement in a school for developmentally disabled students may be sought, and it will be the responsibility of the school system to provide funding for this service. For fiscal reasons, some school systems may be reluctant to acknowledge that a given child requires a particular service. Under these circumstances, it becomes one of the jobs of the child's clinical providers in general, and of pediatricians, psychiatrists, and psychologists in particular, to advocate for the child's genuine needs.

A major debate has raged within the educational system about the relative merits of segregated special education classrooms as opposed to maximal integration of mentally retarded students into regular classes. The latter approach (known as *mainstreaming*) has tended to predominate in recent years. Arguments in favor of mainstreaming include the least restrictive environment principle and the expanded opportunities for social development that are available in regular classes. Arguments against mainstreaming include concerns over the ability of regular classrooms to meet special educational needs and/or any social stigmatization faced by mentally retarded students, as well as the possibility that their disruptive behaviors may interfere with the education of other students. This issue is likely to be discussed and various solutions tried for some years to come. On the level of the individual case, factors such as degree of cognitive impairment and extent of disruptive behavior should be considered.

In a less publicized echo of the deinstitutionalization of psychiatric patients, many of the large state-operated training schools that once housed many of the mentally retarded for life have been closed; others have been drastically reduced in size. A variety of reasons may be suggested for this movement: disappointment with the results produced by those institutions, ethical concerns lest the human rights of mentally retarded persons be abrogated (connected to the least restrictive alternative doctrine), and/or financial considerations. As discussed earlier, the education system has taken on increased responsibility for school-aged children. "Birth to three" programs, typically managed by state departments of MR, coordinate services for younger children. These early intervention programs are expanding rapidly in scope and offer hope for improved functioning on the part of future generations of mentally retarded persons. Community-based care for mentally retarded adults relies on networks of public and private group homes, supervised apartments, and training programs similar to those providing care for deinstitutionalized psychiatric patients.

Behavior modification uses principles derived from the scientific study of learning to facilitate positive change in client behaviors. Operant conditioning (e.g., reinforcement of target behaviors) may be the most important aspect of this approach, but techniques derived from classical conditioning (e.g., systematic desensitization) and observational learning (modeling) also are applied. Applied behavior analysis involves the detailed description of target behaviors, the situations under which they occur or fail to occur (i.e., antecedent stimuli) and their reinforcing or punishing consequences. Understanding a behavior in context facilitates creation of plans whereby its frequency may be increased or decreased. Stimulus control, incompatible behaviors, reinforcement of desired behaviors, removal of existing reinforcers for undesirable behaviors, and (when necessary) punishment of undesirable behaviors are all used. Behavior modification can be applied to a wide range of problems (e.g., toilet training, concentrating on work) and applied behavior analysis should be a component of most treatment plans, but our discussion focuses upon those aggressive and SIBs that have been so recalcitrant to pharmacologic manipulation.

A behavioral analysis of SIB or aggression begins with observation of the behavior's frequency, the situations in which it appears, and the consequences of the behavior. Although some SIB may be produced directly by a disorder (e.g., Tourette syndrome), most instances perform operant functions, and even those that do reflect a disorder often are elicited by specific stressors. SIB or aggression may be associated with any of the following (not an exhaustive list):

1. Distress over frustration that cannot be expressed more adaptively due to cognitive limitations (e.g., an assigned task is difficult to complete satisfactorily)
2. Attempt at communication, signaling an unmet need (e.g., a client in an institution where staff attention is rarely offered needs assistance with toileting)
3. Instrumental attempt to escape from (or avoid) an annoying situation (e.g., a client who does not want to attend a scheduled activity)
4. Instrumental attempt to obtain some externally administered reinforcer (e.g., additional medication, one-on-one attention)
5. Attempt to self-generate reinforcement (e.g., by stimulating endorphin release)
6. Involuntary, or only partially voluntary, stereotyped behavior (e.g., tic or compulsion), possibly triggered by anxiety and/or psychosocial stress

7. Attempt to hurt, damage, or destroy oneself (e.g., suicidality due to depression) via primitive means

Clearly, neither aggression nor SIB can be understood in isolation. When such behaviors occur, they mandate careful examination of the patient and the situation in search of the mechanisms by which the behavior is produced and maintained. Specialists in behavior modification, who usually are psychologists by training, are particularly adept at this process of functional analysis. Implementation of behavior modification plans usually depends upon family members or the staff of a residential or educational institution, whose active collaboration must be solicited.

Several of the allied health professions are involved in treatment of the developmentally disabled. Physical and/or occupational therapy can be important in cases where hypotonia and/or poor development of gross motor skills are clinically significant. Occupational therapy is helpful in developing fine motor skills, visuomotor integration, and (to some degree) higher cognitive functions whose expression is dependent upon visuomotor abilities (e.g., constructional tasks). Speech and language therapy should be introduced whenever language development is significantly delayed, whether due to a specific communication disorder or to more general MR. Impairment of specific linguistic functions, or of articulation, often is present to a degree not fully accounted for by global retardation. Supplementation of standard language instruction by training in the use of sign language, communication boards, and/or picture cards can increase the communicative competence of mentally retarded individuals.

Family counseling may take various forms. Many families respond to a child's diagnosis as mentally retarded with a combination of shock, disbelief, mourning, guilt, and a search for answers. A sympathetic clinician can respond to each of these needs, facilitating acceptance of the situation and providing accurate information (e.g., prognostic). Psychoeducation may focus upon the special management needs of a mentally retarded child. While training parents to be more effective caregivers, one also should attend to the possibility of emotional exhaustion on the part of hard-working parents, especially if their best efforts produce only limited success. It is important to emphasize the fact that the child is, first and foremost, a human being, capable of experiencing many of the same joys and sorrows as are other people. To the extent that the child's clinical picture includes realistically favorable prognostic features, these too should be stressed.

Individuals with MR must negotiate many of the same social and emotional milestones as their peers, but they must do so with fewer intellectual resources. It is to be expected, therefore, that a significant proportion of them will need personal counseling at some point during their life span. Major mental illness aside, developmental issues (friendship, family conflicts, romance, occupational choice, aging) should be addressed as they arise. Specific issues wrought by developmental disabilities (e.g., physical limitations, stigmatization) are likely to arise. The areas where conflicts are most obviously accentuated by MR are independence and sexuality. Intellectual limitations constrain occupational choice and, in most cases, income, and even under optimal conditions persons with MR are likely to face delays in the establishment of independence. When the degree of retardation is more than mild, truly independent living is likely to be impossible. This exacerbates the difficulty of finding and competing for the love of potential partners.

Sexuality is a particularly thorny issue for those who work with the mentally retarded. For some, the old stereotype of "a child in a man's (or woman's) body" dies hard. Even those who accept the fact that the physical developments of puberty mandate some kind of social-emotional adaptation often are conflicted about sexual behavior on the part of people with MR. On the one hand, mating is a natural process for humans (as for other species) and certainly is not predicated upon academic intelligence. Likewise, persons with MR ought to enjoy the same right to pursuit of happiness enjoyed by their fellow humans. On the other hand, the potential for sexual victimization is clearly greater for those with limited verbal ability and/or problem-solving skills, and a person with a diminished capacity for informed decision making also must be regarded as having a reduced ability to offer informed consent. There is no easy universal solution to this dilemma. Some states have outlawed sexual contact with mentally retarded individuals, but this clearly does nothing to meet the legitimate sexual and romantic needs of those with MR. It is more in keeping with the old eugenic sterilization practices of the pre-World War II era than with any affirmation of the humanity of the individuals in question.

Whatever the specific issues for which a mentally retarded person seeks counseling, the nature of the counseling obviously must be tailored to the cognitive abilities of the client. For those (the majority) whose retardation is mild, many conventional techniques of counseling or psychotherapy may be applied, as long as the language is kept relatively simple and the concepts relatively concrete (80a,80b). The more serious the intellectual deficits, the more modification of standard practice will be required. Nonverbal channels of communication (e.g., facial expressions, vocal prosody) become increasingly important means of expressing empathy and encouraging growth of a therapeutic relationship. Engaging clients in nonverbal tasks (e.g., play or art activities) may be helpful. When talking treatments are not feasible, medication and behavior modification remain useful alternatives.

BIOMEDICAL DISORDERS ASSOCIATED WITH NEUROCOGNITIVE IMPAIRMENT

A variety of biomedical conditions can produce neurocognitive impairment. Some are strongly linked to MR; in other cases, milder impairment of general intellectual functioning and/or a learning disabled profile are more typical. For many of these diagnoses, clinical lore and research have identified a characteristic pattern of neuropsychological strengths and weaknesses (a cognitive phenotype) and/or a temperamental/personality profile (a behavioral phenotype). Frequently a syndrome is first identified among severely disabled, institutionalized patients and at first is believed to involve severe MR as a cardinal feature, but eventually milder forms of the disorder are recognized in outpatient samples. These milder types (sometimes more common than the "core" disorder) typically possess the physical characteristics of the syndrome in attenuated degree and are intellectually more able, although with a cognitive and behavioral phenotype akin to that presented by those more seriously affected.

Some of these syndromes involve (or appear to involve) specific genes that may be familial or represent *de novo* mutations. Some involve chromosomal anomalies such as added or deleted chromosomes or parts thereof. In either of the cases discussed earlier, there may be a known biochemical pathway by which the genotype is translated into a cognitively impaired phenotype. Still others are produced by teratogens, postnatally delivered neurotoxins, or infectious agents. Differential diagnosis is important; some of these disorders can be ameliorated via physical interventions (e.g., diet or surgery), many are associated with significant biomedical pathology that requires careful management, and some have specific psychiatric manifestations. Unfortunately, the rarity of many of these conditions implies that most psychiatrists and neuropsychologists will enjoy little opportunity to acquire familiarity with their various clinical presentations, rendering differential diagnosis difficult. As a general rule, the presence of seizures or marked physical (e.g., craniofacial) anomalies should raise a suspicion of biomedical etiology. If such peculiarities conform to those characteristic of a known syndrome, careful evaluation is mandated. The assistance of experienced dysmorphologists should be enlisted whenever possible. Likewise, if family history is positive for biomedical conditions associated with MR or prenatal exposure to teratogens or infectious agents is suspected, these possibilities should be explored even in the absence of obvious physical manifestations.

Because patients suffering from these disorders are liable to multiple handicaps, including (in some cases) potentially life-threatening physical disorders as well as intellectual limitations, the neuropsychiatrist, who combines standard medical training with expertise in mental and emotional functions, is ideally suited to assume a central position in their care. Likewise, clinical neuropsychologists can be crucial intercessors between the medical and educational components of patient care. Unfortunately, fiscal considerations often limit patient access to such highly trained care providers. Because mentally retarded patients are unlikely to be skilled negotiators for their own needs, a greater burden falls upon care providers to advocate for their patients.

In the following discussion, we try to strike a balance between breadth and depth of coverage. The disorders described do not comprise an exhaustive list of the biomedical conditions associated with cognitive impairment (of which there may be several hundred), but we attempt to include all of the more common conditions as well as several relatively rare disorders that present with interesting and/or psychiatrically challenging aspects. Similarly, exhaustive coverage of any one disorder is precluded, but we attempt to present an accurate, clinically useful synopsis of current knowledge regarding each. Aspects pertinent to general medical practice are mentioned briefly, with considerably more space allotted to their cognitive, behavioral, emotional, and social manifestations.

Down Syndrome

Down syndrome (trisomy 21) has been recognized since the nineteenth century. With an estimated incidence of approximately one in 1,000 live births, Down syndrome is one of the most common conditions associated with MR. Craniofacial stigmata include the well-known epicanthal folds that gave the disorder its original name of mongolism (although these folds are not, in fact, like those found in East Asian populations), small head, flat nose, and large, often protruding, tongue (81). The disorder usually is identifiable in infancy, although this may not be true when a mosaic pattern with only 50% or, less frequently, 25% of the individual's cells expressing the chromosomal anomaly is present.

People with Down syndrome are liable to a variety of physical problems. Structural abnormalities of the ear, combined with recurrent middle ear infections, produce hearing deficits in a majority of cases. Congenital heart abnormalities, present in up to half of all cases, may lead to hypertension, heart failure, or failure to thrive. Hypothyroidism, present in perhaps one sixth of cases, is easily missed but will exacerbate the learning problems that are already part of Down syndrome. Gastrointestinal disorders and seizure disorders are fairly common (82). All of these should be evaluated carefully and (if present) treated as for any other patient. The cumulative impact of these medical issues is that only about one fourth of those with Down syndrome survive beyond age 50 (83).

Down syndrome is unusual for its well-known association with maternal age. Risk of bearing a Down syndrome child accelerates as women approach middle age; the risk is about one in 384 at age 35, one in 112 at age 40, one in 30 at age 45, and one in 11 at age 48 (84a,84b,85). Generally, pregnant women over age 35 are advised to have the fetus tested for the presence of Down syndrome. Note that because most expecting mothers are younger than 35, this policy detects only about 20% of all cases (29). A weaker but still significant association with paternal age also exists (86). The risk of siring a Down syndrome child increases by approximately 25% if the father is aged 50 years or over (87).

The cognitive phenotype typical of Down syndrome is well described. Moderate MR is typical, but considerable variability exists. Some obtain IQ scores in the borderline range; this is especially likely if the trisomy 21 is mosaicized.

A specific disorder of expressive language is to be expected, over and above the effects of general intellectual delay (88–90). Average children undergo an acceleration of expressive language development beginning at roughly 18 months of age, typically increasing spoken vocabulary from 30 to 100 words over the next 6 months. This is followed by the even more pronounced acceleration associated with combining words into sentences. Children with Down syndrome keep pace linguistically with other children of the same MA until an MA of about 18 months, but they do not seem to undergo the first expected spurt of vocabulary growth (91). Thus, by the time they reach an MA of 24 months, they lag far behind. Children with Down syndrome may benefit to a degree from supplementary instruction in signing during this period, acquiring some words solely in vocal or sign form and thus increasing their total expressive vocabulary. Use of syntax is delayed relative to MA but continues to develop well into adolescence (88). Receptive language is *not* specifically impaired in Down syndrome; both vocabulary and syntax are well developed from the standpoint of comprehension, and people with the disorder typically understand considerably more than their speech makes it appear (88).

Some of the physical problems common in Down syndrome may exacerbate language problems. For example, recurrent middle ear infections may reduce auditory acuity and thus exposure to spoken language (91,92). Similarly, the oral-motor control deficits and enlarged tongue typical of the disorder produces articulation difficulties independent of the degree of language disability *per se* (91,93); these also may reduce the richness of early babbling, which is itself important in development of phonologic awareness (92).

Auditory-verbal short-term memory (the phonologic loop) appears to be specifically impaired in Down syndrome (89,94). Rote verbal learning is adversely affected by this deficit (91). Finally, most visuomotor skills are relatively strong in Down syndrome (94), although specific impairment of certain spatial representation abilities has been documented (95).

Children with Down syndrome are less likely than other MR children to suffer from behavioral disorders or other forms of psychopathology (89,90). They are at increased risk for depressive disorders during adulthood (90) but may be at reduced risk for bipolar disorder (96). The few reported cases of bipolar disorder in Down syndrome all have been males without a family history of the disorder (97). In general, the behavioral phenotype of Down syndrome has many assets. They often are perceived by others as warm, spontaneous, gentle, patient, and sensitive to interpersonal cues (82). Children and young adults with Down syndrome tend to function well, and this has become increasingly the case in recent years (98,99). Performance by these individuals in school and on the job tends to exceed that which would be expected based on IQ alone (82), especially if they received early intervention and intensive instruction in a supportive setting (100).

One of the most serious issues for adults with Down syndrome is a greatly increased risk of early-onset dementia of the Alzheimer type. The characteristic β-amyloid plaques and neurofibrillary tangles are present in virtually all Down syndrome brains after age 30, although most do not develop clinical signs of dementia until after age 50 (83,101), with prevalence approaching 50% (89). It is possible that subtle cognitive deficits appear before this, but many neuropsychological tests are not sufficiently sensitive to detect modest declines in cognitive functioning among the moderately retarded (102). Because other problems (e.g., depression) also are associated with functional impairment, a careful neuropsychiatric evaluation conducted around age 35 has been recommended to provide an adequate baseline for clinical evaluations that may be needed later on (103). Risk of dementia is *not* associated with degree of preexisting cognitive deficit (104). Gender *is* associated with dementia risk; males are up to three times more likely to develop dementia and have an earlier age of onset than do females (105). Dementia in Down syndrome rarely, if ever, includes psychotic features such as delusions or hallucinations (104), nor is it associated with increased risk of aggressive behavior (106).

Neuroimaging studies consistently document reduced volume of cortical structures, especially in the frontal and temporal regions, relative to subcortical nuclei of normal volume (94,107).

Fragile X Syndrome

Around 1970 a number of investigators realized that the preponderance of males in the mentally retarded population could be accounted for by X-linked recessive traits. Between seven and 19 different genetic syndromes of this type may exist; their cumulative prevalence in the general population probably is about 2.44 per 1,000 females (nearly all carriers) and 1.83 per 1,000 males (nearly all clinically affected) (108). One pedigree consistent with such a mode of transmission was identified in the 1940s, and the physical phenotype became known as Martin-Bell syndrome. In 1969 the fragile X anomaly, often abbreviated fra(X), was found in this family (109). The crucial site is in the Xq27.3 region. The *FMR1* gene whose mutation is responsible for fra(X) syndrome was sequenced in 1991.

The characteristic fra(X) face is long and narrow, with wide, prominent ears. Each of these features is present in about 80% of cases at some point, although the ears tend to grow less prominent and the face more elongated after puberty (110). Narrow palpebral fissures, puffiness around the eyes, a large jaw, and macrocephaly are fairly common features. Among postpubertal males, macroorchidism is present in approximately 80% of cases (110). Ovarian enlargement

has been reported in females with the disorder, but this is rarely assessed. A velvety quality of the skin often is seen.

The most frequent medical issue is recurrent otitis media, afflicting over 80% of cases (110). Strabismus is present in at least one third of all cases, and ophthalmologic examinations should be conducted during the preschool years (111). Cardiologic symptoms (e.g., heart murmurs) develop in up to one third of cases, usually during adolescence or early adulthood. Sinusitis afflicts up to one fourth of cases, and a majority of patients are flat footed. Finally, seizures develop in at least 20% of cases, usually with onset in childhood or adolescence (110). Pharmacotherapy (e.g., carbamazepine) is generally successful, and seizures often cease by adulthood (111). Neurologic "hard signs" are not expected, but hypotonia and (in males) a positive palmomental reflex (suggestive of frontal lobe involvement) are common (110).

The *FMR1* gene is a trinucleotide sequence of CGG repeats. Its normal function is to produce the FMR1 protein (110). Fewer than 30 of these CGG repeats are present in non-fra(X) individuals and are typically interspersed with about one AGG sequence for every ten CGG repeats (112).

The premutation form of fra(X) is produced when the number of CGG repeats increases and/or the AGG sequences are lost. People with this form are carriers and may have anywhere from 50 to 200 CGG repeats. This form is unstable, and the number of repeats is likely to be increased when transmitted to offspring (an anticipation effect) (112). The full mutation involves more than 200 CGG repeats, which usually are completely methylated and, as a result, incapable of producing FMR1 protein (110). There is, then, a wide spectrum of fra(X) involvement:

1. Premutation carriers are, as a rule, clinically unaffected, although prominent ears, attention deficits, and marked social anxiety are present in a subgroup.
2. Females with one fully mutated and one normal X chromosome may function as carriers. Only about one fourth are mentally retarded, but the remainder suffer from one or more of the following: borderline intellectual functioning, learning disabilities, or emotional problems.
3. Males with a mosaic pattern (e.g., only 50% of cells expressing the mutation) and males with incomplete methylation of the mutation produce some FMR1 protein, tend to be only mildly retarded, and bear fewer of the fra(X) physical stigmata.
4. Males with the full (and fully methylated) mutation, who produce no FMR1 protein, are most severely affected. This presumably would also be the case for a female with two mutated X chromosomes (e.g., the offspring of two carriers), but we are not aware of any such cases reported in the literature.

Knowledge of the *FMR1* gene has, among other things, led to an explanation of the hitherto baffling Sherman paradox (113). All the daughters of carrier males inherit the mutation, yet none are clinically affected. In contrast, although only half of the daughters of carrier females inherit the mutation, approximately 30% are clinically affected (this includes granddaughters of carrier males). The reason is that spermatozoa carry only the premutation, which is associated with carrier status.

It is remarkable that the fra(X) syndrome was not recognized until such a late date, for in addition to possessing a rather distinctive phenotype it is one of the most common causes (quite likely, the most common inherited cause) of MR. Exact prevalence figures have been difficult to establish for several reasons, including the fact that only retarded populations usually are tested, which omits an unknown number of less severely affected cases. Averaging the results of seven population studies of the retarded using cytogenetic tests yields an estimated prevalence of six per 10,000 males and four per 10,000 females in the general population. The more recently developed deoxyribonucleic acid (DNA) test for the *FMR1* gene tends to yield considerably lower estimates: 2.5 per 10,000 males and 1.2 per 10,000 females (112). Prevalence among the mentally retarded is substantially higher; approximately 1.9% of mentally retarded males and 0.3% of mentally retarded females have fra(X) syndrome (112).

Considerable research has been devoted to exploring the fra(X) cognitive phenotype or rather phenotypes, given that males differ substantially from females and carriers from those with the full mutation.

IQ among males is typically in the moderately retarded range, but mild and severe levels also are common. IQ appears to decline at some point in late childhood or early adolescence. Averaging across studies, mean IQ in younger samples is about 50, whereas that in older samples is about 34 (114). Only a fraction of adult fra(X) males have IQs in the borderline range or higher. Some studies suggest that a mosaic pattern or fewer CGG repeats are associated with higher IQ, but others have produced discrepant findings (115–117). Declining IQ may not transpire in incompletely methylated patients who produce some FMR1 protein (118), but the causes of the IQ decline are not yet clear. In general, fra(X) males do better on tasks requiring simultaneous (rather than sequential) processing, although visual-motor integration deficits impede performance on many spatial tasks. Both auditory and spatial memory are impaired, but object memory may be an area of relative strength. Vocabulary (expressive and receptive) also is relatively good, but because of the degree of overall retardation, speech usually is delayed and often is the initial presenting issue that brings the child to clinical attention. Articulation and syntactic usage are delayed, consistent with the degree of MR. The rate and prosody of speech, on the other hand, are deviant. In addition to a tendency toward cluttering, fra(X) speech has been described as possessing a "jocular" or "litany-like" quality. The pragmatic use of language tends to be an area of weakness. Stereotyped repetitive statements (including palilalia), one-sided conversation, and inappropriate inter-

jections are common issues. Pragmatic deviance is less pronounced than in autistic disorder in that fra(X) males seem to know that a conversation is supposed to be interactive, but it is more pronounced than in other forms of MR. A number of their deficits (e.g., impairment of working memory and pragmatic communication, impulsivity) are consistent with the hypothesis that executive functions are particularly impaired in fra(X), but most fra(X) males are too globally impaired for valid assessment using traditional executive function tests.

Averaged across studies, IQ among females with the full fra(X) mutation typically is in the low average to borderline range (mean = 81.6) (114). Although only about one fourth are in the mentally retarded range, at least half of the remainder suffer from learning disabilities (119). It is unclear whether females with the mutation suffer from a decline in IQ like that observed among males. Characteristic areas of weakness include visuospatial tasks, working memory (auditory and spatial), and arithmetic; vocabulary tends to be an area of relative strength. Linguistic competence is otherwise generally consistent with IQ, but subtle pragmatic deficits may exist. Because fra(X) females are more intelligent than fra(X) males, they can be tested with a wider array of neuropsychological instruments, including tests of executive functions. There is, at this point, reasonably compelling evidence for selective deficits in executive functions among fra(X) females (114).

Females carrying only the premutation appear to be unaffected cognitively. Mean IQ across studies is slightly over 100, with no clear pattern of strengths and weaknesses like those found among both males and females with the full mutation (114).

Behavior problems are common in fra(X) and second only to developmental delays as the cause of clinical consultation. Typical complaints include temper tantrums (about 50% of adults continue to exhibit episodic aggression), hyperactivity, autistic-like symptoms, behavioral stereotypies, and social anxiety.

Hyperactive and/or impulsive behaviors manifest in about 70% of males and 35% of females (110). Because attention problems are all but ubiquitous, ADHD will frequently be diagnosed. Psychostimulant response rate is about that expected for other mentally retarded patients.

Autistic disorder can be diagnosed in about 16% of fra(X) males (conversely, about 6.5% of those diagnosed with autistic disorder will test positive for the fragile X mutation); autistic features are far more common. The large majority of fra(X) males display poor eye contact, hand flapping, perseverative behavior, and tactile defensiveness; nearly two thirds also display hand biting (110). Interestingly, a specific fra(X) autism pattern has been described. It is characterized by active gaze avoidance (as opposed to gaze insensitivity), relatively good social relatedness, hyperactivity, perseverative speech, hand flapping, and hand biting. The presence of several of these features in an autistic patient should trigger cytogenetic or DNA testing. What is interesting about this pattern is the implication that fra(X) "autistics" are not aloof from human interaction. Rather, they appear to crave contact but to be overstimulated or even intimidated by such contact. A characteristic expression of this approach-avoidance conflict is their tendency to avert their gaze (and even body) while greeting a person and shaking hands. It also has been noted that hand flapping and hand biting are exacerbated in social settings.

The social anxiety and shyness commonly found in fra(X) syndrome may represent the mild end of a continuum with fra(X) "autistic" behaviors at the other end. Social anxiety, present in about one third of male patients (110), is the most frequent presenting problem for females with the disorder, up to two thirds of whom may be diagnosed with avoidant disorder in childhood, with dysthymia frequently comorbid (120). In general, shyness is more likely to be an issue in those with higher IQ (110). Interestingly, hyperactivity seems to offer some protection against the effects of this social anxiety, presumably because of impulsive social bids (120). Fluoxetine has demonstrated effectiveness in treating the combination of dysthymia and social anxiety in fra(X) syndrome (121). Social skills training (progressing from making eye contact to conversational skills) and provision of structured, low-pressure social opportunities (progressing from parallel play to more cooperative interactions) may be helpful (120).

A variety of perseverative stereotyped behaviors are common in fra(X). Up to 20% may be diagnosed with motor tic disorders (110); outbursts of pressured speech that resemble complex vocal tics and obsessive-compulsive behaviors are common. Perhaps 20% of first-degree relatives also manifest obsessive-compulsive behaviors.

Fetal Alcohol Syndrome

Alcohol has been suspected of teratogenic effects for thousands of years, but it was not until the latter part of the twentieth century that a specific syndrome secondary to prenatal alcohol exposure was described in the medical literature (122). This is surprising given its rather high incidence. Fetal alcohol syndrome (FAS) occurs in between one in 2,000 and one in 3,500 live births, and far more frequently among certain ethnic groups, such as Native Americans (123). It is estimated that approximately 12% of children born to heavy-drinking mothers will display the features of FAS (124). Less obvious but still frequently disabling fetal alcohol effects appear to be considerably more common (125), even among children of moderate drinkers (126). Thus, prenatal exposure to alcohol may be the single most frequent cause of MR, at least in the developed western world.

A diagnosis of FAS requires that three criteria be met: (a) prenatal and/or postnatal growth deficiency, (b) characteristic craniofacial malformations, and (c) evidence of CNS dysfunction. Small size often is obvious in infancy and persists throughout the developmental years. People with FAS typically fall below the tenth percentile on height and weight,

even as adults (127). The classic appearance involves microcephaly, short palpebral fissures, long, smooth philtrum, epicanthal folds, and flat midface. CNS dysfunction is demonstrated by one or more of the following: MR, learning disabilities, or symptoms akin to those of ADHD. Abnormal findings on electroencephalogram (EEG) are typical.

Many studies have compared children (or adults) with FAS to others who do not express the physical anomalies required for that diagnosis, but who do have a documented history of prenatal exposure to alcohol. Typically, non-FAS alcohol-exposed participants display cognitive deficits (global IQ and specific neuropsychological abilities) similar to those of participants with full-blown FAS (125,128,129). Not surprisingly, when differences are found FAS involves graver impairment. The behavioral profile also is similar, although less pathologic. Therefore, clinicians should not assume that a fetal alcohol effect is absent simply because a patient's appearance is normal. If prenatal exposure to alcohol is known or strongly suspected, psychiatric and neuropsychological evaluation is warranted.

Cognitive impairment has been documented in prospective studies of children exposed to alcohol (130), even in moderate amounts (126), and in retrospective studies of children of known alcoholics. The latter also tend to produce abnormal EEG findings (124). The amount of alcohol consumed is directly related to the degree of cognitive impairment (131). No perfectly safe level of alcohol ingestion during pregnancy has been established. Severe cognitive deficits are unlikely when daily consumption is no more than one drink (131), but subtle impairment may occur even at that modest level (132). The second trimester may be an especially risky period (132). Pattern of use may be important, with binge drinking more likely to produce fetal alcohol effects than more moderate regular use even though the absolute amount of alcohol consumed per week is similar (126,133).

Behavioral problems and psychiatric disorders are common. A majority of FAS children have problems in school and with authorities; up to half may be confined in hospitals or prisons by early adulthood (126). Parent and teacher ratings on the CBCL and Teacher Rating Form (TRF) often reveal attention and social problems; delinquency and/or aggression may be issues (128,134). The parent-completed PIC also reveals mean scores in the clinical range for intellectual problems, delinquency, thought disorder, lack of achievement, and developmental delays (135). Social skills, especially awareness of interpersonal cues, are a particular area of weakness, over and above any social deficits that might be expected based on intellectual limitations (136). Feeding and sleep problems are common in early childhood but tend to abate. Emotional disorders, however, tend to increase in frequency as children mature (134). Among the most common clinical diagnoses are ADHD (present in up to 20%) and speech and language disorders (present in up to 15%) (134).

Cognitive functioning in FAS has been studied extensively, although only recently has a detailed cognitive phenotype begun to emerge. Cognitive effects of prenatal exposure to alcohol in children who do not express physical features of FAS have been subjected to considerable research. Mattson and Riley (125) exhaustively reviewed the research published before 1997. The mean IQ across 79 case studies (mostly published in the 1970's) was 65.73, whereas that across 17 retrospective studies (cumulative n = 269) was 72.26. (The trend toward higher scores may reflect increased awareness of the disorder; milder cases would not have been identified in the 1970's.) In other words, either mild MR or borderline intellectual functioning is typical (134). More recent studies (128,135,136) continue to report mean IQ in the borderline range, except when mentally retarded cases are excluded so as to facilitate finer-grained neuropsychological analysis (137,138).

It often is stated that PIQ tends to be superior to VIQ in FAS, but the review by Mattson and Riley (125) disputes that claim. They identified 17 studies in which the two scores were similar: five reported the expected PIQ>VIQ discrepancy, and two reported the opposite trend. Of the more recent studies we reviewed, four reported PIQ>VIQ by at least four points (128,129,138,139), and one reported virtually identical scores (136). Thus, the issue remains unresolved, but it seems clear that simply comparing mean scores on batteries of verbal and visuospatial tasks is overly simplistic and fails to capture the quality of intellectual functioning in FAS.

Children with FAS often show deficits on visuospatial tasks, such as copying designs or drawing clocks (125,140), a finding inconsistent with the notion of preserved spatial abilities, although poor fine motor control (e.g., tremulousness) may contribute to these deficits (141). People with FAS often are able to capture the overall configuration or gestalt of a design but are inaccurate in their rendition of details (125).

Delayed language acquisition is typical, and clinically significant deficits in both expressive and receptive language may be present in a majority of FAS patients (142). A language disorder diagnosis often is justified. Comprehension of words (semantics), understanding and utilization of syntax, and fluency all may be affected. Poor articulation is common. Some, but not all, language problems in FAS may be secondary to impaired auditory acuity, which varies in nature (e.g., central vs. peripheral) but is found in a large majority of cases (142). Careful audiologic examination always is warranted but does not preclude a comorbid communication disorder.

Deficits in verbal learning have been reported (129,143). Children with FAS acquire less information on first presentation and make gains relatively slowly with practice. They retain the expected proportion of what is learned but make many false alarms on recognition tests. Results on visual memory tests are more mixed, but FAS may involve a partial dissociation of object memory (intact) and spatial memory (impaired) (144,145). In other words, people with FAS can learn *what* they have seen, but they have disproportionate difficulty remembering *where* something was. This is consistent with findings for animals exposed prenatally to alcohol (125,146).

Academic achievement in single-word reading and spelling tend to be consistent with IQ, whereas arithmetic achievement tends to be substantially lower (138,139). Mathematics may be the one subject in which FAS children most often receive special educational services (140). Thus, a diagnosis of mathematics disorder may be appropriate in many cases of FAS where IQ is in the normal range. Note that actual school performance often is considerably worse than would be expected based on IQ or achievement test scores.

It has been suggested that the combination of impulsivity, hyperactivity, cognitive deficits, and behavioral disorganization typical of FAS may be explicable in terms of a core impairment of frontally mediated executive functions: planning, self-monitoring, and inhibiting inappropriate impulses in the service of longer-term goals (147). Children with FAS usually (138,148), but not always (149), display impairment on the Wisconsin Card Sorting Test, a popular (albeit limited) measure of executive functioning, and on other tests requiring cognitive flexibility (138). An investigation of FAS children using the comprehensive Delis-Kaplan Executive Function Scale found selective deficits in concept formation, planning, cognitive flexibility, and inhibition that were not fully explicable in terms of general intellectual functioning or component skills (147). This proposed deficit in frontally mediated functions would be consistent with their reported difficulty with mathematics and spatial memory.

Research also has focused on neuroanatomic features associated with FAS or prenatal exposure to alcohol. Three lines of research are relevant here: (a) neuropathologic studies; (b) neuroimaging studies [especially magnetic resonance imaging (MRI)], both involving human FAS sufferers; and (c) animal research involving experimental manipulation of the dose and timing of alcohol. In general, all three forms of research produce congruent findings. Autopsy frequently reveals microcephaly, cerebral dysgenesis, and abnormal neuronal migration. Agenesis or marked thinning of the corpus callosum, cerebellar abnormalities (including agenesis of the cerebellar vermis), and partially fused frontal lobes have been reported in multiple cases (150). Of course, the subjects of these studies presumably suffered from relatively severe forms of FAS, but MRI studies have yielded similar results. Cerebrum, cerebellum, and corpus callosum typically are smaller in FAS patients than controls, and even agenesis of the corpus callosum is not uncommon. The basal ganglia tend to be reduced in size, even when controlling for total brain size (150). In most cases, the differences are subtle enough to be missed on clinical inspection and become apparent only when careful quantitative techniques are used (150).

Animal studies reveal abnormalities in neocortical structures (reduction in neurons and glial cells by 30% and disruption of neuronal migration patterns), hippocampus (reduction in N-methyl-D-aspartate receptor density and diminished long-term potentiation), and cerebellum (reduced numbers of Purkinje and granular cells) (146). The precise nature of the structural changes depends on the timing of alcohol delivery. Alcohol exposure during the first trimester produces a high incidence of structural brain anomalies, dysmorphic features reminiscent of FAS, and behavioral changes (e.g., hyperactivity and impaired spatial learning). Exposure during the second trimester produces similar behavioral changes and abnormal neuronal migration, but not structural changes in brain or facial features. Exposure during the third trimester produces microcephaly due to reduced proliferation of cells. The cerebellum is highly vulnerable to alcohol exposure at this time, and reduced cerebellar volume is associated with the motor control problems often seen in FAS (146).

Williams Syndrome

Williams syndrome, first fully identified in 1961, is one of the most neuropsychologically interesting syndromes associated with cognitive impairment. The disorder, which has an incidence of approximately one in 20,000 live births (equally common among males and females), nearly always is caused by deletion of a gene on the seventh chromosome in the 7q11.23 region (151). One exception was a mini-epidemic of mild cases produced in postwar Europe by excess vitamin D in infant food supplements. The diagnosis of idiopathic infantile hypercalcemia was applied at that time and occasionally still is used in cases of Williams syndrome. In addition to its cognitive manifestations, Williams syndrome presents with characteristic craniofacial features, cardiovascular abnormalities, small stature, and (in some cases) hypercalcemia. Up to half of patients may develop seizures.

Typical facial features include a small upturned nose; wide, full-lipped mouth; long philtrum; high wide cheeks; low-set ears; small pointed chin; and epicanthal folds (152). Stellate irides are common, although the starry pattern may be concealed or absent when the iris is heavily pigmented (153,154). The cumulative effect often is described as "elfin."

The most common cardiovascular defect is supravalvular aortic stenosis, with narrowing of the pulmonary arteries next in frequency. The former tends to be progressive, whereas the latter tends to improve over time (155). In severe cases, surgical correction may be necessary; even in mild cases, there is an increased risk of hypertension. Sudden death from cardiac problems occasionally occurs in Williams syndrome; anesthesia induction is often the immediate trigger (156).

Low birth weight is typical, and gains are made slowly for the first few years. Feeding problems (e.g., weak sucking, difficulty swallowing, vomiting, constipation) exacerbate this tendency; many cases of Williams syndrome may be diagnosed with failure to thrive. When present, hypercalcemia can be a contributing factor, and a low-calcium diet is prescribed (157). Growth accelerates after age 4, but even in adulthood relatively small stature is to be expected. Males average 65 inches in height and females 60 inches (158).

Common neurologic findings include gait abnormalities (but without cerebellar signs), poor fine motor control, abnormal deep tendon reflexes, and praxis deficits; both hypotonia and hypertonia are found among children, but only hypertonia is found among adults (159).

When severe cardiac abnormalities or hypercalcemia bring a case to medical attention, the diagnosis is often made in infancy. More commonly, the child is brought in for evaluation of developmental delays around age 4 or 5 years and the diagnosis is made based on dysmorphic features (157). Wide variability is (as always) the rule, but the average child with Williams syndrome walks independently and begins speaking single words at slightly older that 2 years. Fine motor skills are more delayed than gross motor abilities, and language comprehension develops slowly relative to expressive language (160). For example, children in the general population develop the ability to point to a picture in a book and to combine two words at about age 16 months; children with Williams syndrome begin combining words (expressive) at an average age of 26 months, but they are not able to point to a picture (receptive) until an average age of 37 months (160).

The cognitive profile of Williams syndrome has generated a substantial body of research. In terms of full-scale IQ, a typical individual with the syndrome functions in the mildly retarded range overall, with verbal abilities superior to visuospatial abilities. Smaller proportions function on a borderline or moderately retarded level (161,162). VIQ tends to be higher than PIQ, although the mean difference may be only about four points (161,163).

Although early language development tends to be delayed, by age 4 or 5 years language skills become a relative strength, albeit with unusual features. Semantics are not especially well developed, although vocabulary tends to be larger than among other children of like MA. Syntax, in contrast, is remarkably advanced. It is not altogether "normal" but is far superior to what is expected of mentally retarded individuals. Expressive language use is quite fluent and more remarkable for the sophistication of its syntactic organization than for meaning. Language comprehension, oddly, is less well developed, so it is all too easy for people (including educators and clinicians) to overestimate the true verbal ability of these patients. Many children with Williams syndrome tend to produce a large volume of speech organized into complex statements, using an adultlike vocabulary replete with social phrases or clichés, the classic pattern is known as "cocktail party speech." A pragmatic deficit sometimes is noted, with children tending to perseverate on topics in which they are interested regardless of the degree of interest shown by their listeners (164–167).

Visuospatial abilities are strikingly weak in Williams syndrome. Visuomotor integration and visuoconstructional abilities tend to be impaired. These deficits are exacerbated, in many cases, by poor fine motor skills. Typically, global analysis is more impaired than local analysis; a child may accurately represent certain details but miss the overall gestalt of a design. A striking exception to the poor performance typical of Williams syndrome on visual tasks is face recognition, which, like syntax, tends to be an area of surprising strength (168).

Learning and memory abilities also express the verbal versus visuospatial discrepancy. Children with Williams syndrome perform as well as, or better than, MA- or IQ-matched controls on short-term memory tasks involving digits (169,170) or words (171). Short-term spatial memory is relatively poor (170); findings on object memory are equivocal but consistent with a slight verbal superiority (172). Acquisition of long-term verbal memories proceeds slowly, although once established such memories are retained appropriately (169). Reduced speed of verbal learning appears to be due to a failure to use lexical-semantic information (as opposed to purely phonologic processing) during encoding (171). Acquisition of long-term visual-spatial memories is impaired, even after controlling for perceptual-motor difficulties (169).

We are unaware of any studies systematically evaluating executive functions in Williams syndrome.

The traditional clinical portrait of the Williams syndrome behavioral phenotype describes these people as affectionate, charming, loquacious, and gentle (173). Research has modified this picture somewhat. Temperament in infancy actually has many "difficult" features: colic, a low arousal threshold, intense responsiveness to stimuli, and slow adaptation to change (173,174), but unlike prototypical difficult children, those with Williams syndrome tend to approach, rather than withdraw from, others (173). Some of these features may reflect degree of MR rather than Williams syndrome *per se* (174). If hypercalcemia is present, temperament is likely to improve after initiation of a low-calcium diet (157).

Some temperamental characteristics persist in the form of personality traits. In fact, personality in Williams syndrome seems to be surprisingly unaffected by parental behaviors or social context (175). On the positive side, a friendly outgoing style is almost universally reported (176). This agreeableness is unusual in the retarded population (175) and is associated with increased willingness to help others (177), although it rarely translates into stable friendships (176,178). The social disinhibition and excessive (often one-sided) talk characteristic of Williams syndrome are not conducive to relationship development, although they can render these individuals vulnerable to exploitation, and they tend to be highly sensitive to rejection (177). Social problems are frequently reported (172). It is not uncommon for persons with Williams syndrome to develop (unreciprocated) infatuations with acquaintances or celebrities, and their behavior secondary to these "crushes" can be disruptive (178). Many children with Williams syndrome develop special interests (e.g., geology, electrical appliances, fabric textures) that may persist into adulthood (176,178). High levels of anxiety are common (179) and may manifest as phobias and/or hypochondriacal concerns, although anxiety may be subclinical in intensity despite a patient's overly dramatic way of reporting problems (178).

Unfortunately, the verbal abilities displayed by many persons with Williams syndrome do *not* translate into adaptive behavior. Standardized measures of adaptive behavior typically yield scores akin to, or somewhat lower than, IQ (172,180). Estimates of daily living skills vary. One study reports that very few are able to cook, clean house, or shop independently (181), whereas another study reports that most are able to handle those activities (154). Very few individuals are able to work competitively (176,181). Even in sheltered workshops, their attention problems, social peculiarities, and anxieties cause difficulty (181). Clinicians should be careful not to overestimate the capabilities of patients with Williams syndrome based on apparent sophistication of expressive language.

Phenylketonuria

Phenylketonuria (PKU) has the distinction of being one of the few largely preventable genetic causes of MR. The basic defect is an inability properly to oxidize dietary phenylalanine into tyrosine, which is converted instead into phenylpyruvic acid, the presence of which in the urine led to the discovery of the syndrome in 1934. This defect is transmitted as an autosomal recessive trait. Neonatal screening for the disorder was introduced in the 1960's (182). PKU has an overall incidence somewhat greater than one in 20,000 live births, suggesting that up to 1.5% of the population carries the gene responsible. The disorder is more common among persons of northern European descent (e.g., incidence in Germany is one in 10,000) (183) and rare in the African American and Jewish populations (182). The basic treatment is a special diet low in phenylalanine. Developmental course and outcome are altered dramatically by adherence to such a diet (184), necessitating separate discussion of treated and untreated forms of the disorder.

Under standard conditions, inability to transform phenylalanine into tyrosine has two consequences: tyrosine deficiency, and accumulation of phenylalanine and its metabolites. Both phenomena affect CNS neurochemistry. Tyrosine is the precursor to the catecholamines, and its scarcity in the PKU brain reduces synthesis of dopamine and norepinephrine. Phenylalanine also competes directly with tyrosine for transport into neurons, compounding this scarcity. Finally, high concentrations of phenylalanine metabolites disrupt the action of various enzymes required for monoamine synthesis, including tyrosine hydroxylase (dopamine) and 5-hydroxytryptophan decarboxylase (serotonin). All three neurotransmitter systems are affected in PKU; dopamine depletion is especially pronounced. Dietary treatment reduces the accumulation of phenylalanine and its metabolites, but it does not necessarily provide sufficient tyrosine to make up the deficiency. Attempts to remedy this via tyrosine supplementation have, to date, been disappointing (185).

Left untreated, PKU produces defective myelination, gliosis, and, ultimately, marked cortical atrophy. Based on MRI studies, some degree of abnormal myelination is typical of treated PKU. These MRI findings are unrelated to neurologic status or cognitive functioning. They appear to reflect current levels of phenylalanine and to remit spontaneously when phenylalanine concentration is reduced. These findings imply that only *prolonged* elevation of phenylalanine is associated with marked permanent structural damage to the CNS.

Untreated PKU is associated with progressive impairment over the first few years of life. Developmental milestones are delayed, abnormal hand movements are common, seizures occur in about one fourth of cases (with abnormal EEGs found in at least three fourths), and MR tends to be severe (186). One study reported a mean IQ of 25 (187). After age 3, no further intellectual decline is observed, but behavior continues to be problematic, characterized by hyperactivity, irritability, restlessness, and destructive tantrums (188).

The prognosis for treated PKU is far more benign, approaching normalcy of behavior and intelligence. Early initiation of dietary management (preferably before age 6 months) and strict adherence to the prescribed diet throughout childhood are associated with better outcomes. One early study found dietary initiation before 3 months of age associated with a mean IQ of 89; between 3 and 6 months, 74; between 6 and 13 months, 50; and after 13 months, 26 (189). Results from four studies agree that with appropriate treatment, mean IQ is about 98 (190–193); one other study reported a mean of 91 (194). These figures are safely in the average range but consistently lower than the mean IQ of unaffected siblings or other matched control groups. In two of these studies academic achievement also was in the average range but lower than that for siblings. Discontinuation of the special diet (i.e., increased phenylalanine levels) is consistently associated with lower IQ and achievement test scores. Discontinuation before age 6 is especially problematic, resulting in a mean ten-point drop in IQ relative to those who continue treatment (192,194). In general, mean plasma phenylalanine levels are correlated with IQ until about age 10 to 12 (191,195). If these levels are maintained below 400 μmol/L, outcome tends to be normal or near normal (196), but every 300 μmol/L increase is associated with an IQ decline of half a standard deviation (197).

The benefits of dietary management are not absolute. Early-treated PKU is associated with a characteristic pattern of neuropsychological strengths and deficits. Language skills are generally intact, except among those with poor dietary control (198). Visual-spatial and perceptual-motor abilities, on the other hand, tend to be relatively weak (194,199,200). Executive functions also tend to be deficient (201–205), and a specific deficit in this area has been proposed as the neuropsychological "core" of PKU (206,207). This prefrontal model of PKU is consistent with the neurochemical fact of reduced dopaminergic functioning in the mesocortical pathway of people with PKU. However, not all studies show executive function deficits in early treated PKU (208). This variability may be due to the specific "frontal tests" used or to sample characteristics such as age and degree of metabolic control (201).

Not all cases of PKU are identified through neonatal screening. When the disorder is allowed to develop, diagnosis may be made at any time after about age 6 months; the average is 2.5 years, at which time mean IQ has already declined to the low 50s (209). Dietary management is still helpful in these cases, partially reversing the cognitive decline and producing a mean IQ in adulthood of 79 (209). The earlier treatment is initiated, the better the results.

Behaviorally, PKU treated early often is associated with ADHD-like symptoms: impersistence, inattention, distractibility, and impulsivity frequently are reported (210–213). Up to half may be diagnosable with ADHD. These findings, suggestive of an attenuated but still significant impact of PKU upon behavior even after dietary management, are consistent with the prefrontal model of the neuropsychology of PKU proposed by Welsh and Pennington (206). In adolescence and adulthood, PKU treated early is associated with an increased incidence of internalizing disorders, especially depression, and a decreased incidence of externalizing behavior problems such as aggression (213–216). These problems are not associated with any biochemical factors but rather appear to reflect psychosocial influences (216) such as an overly restrictive, controlling parental style (214). Most patients, however, make a satisfactory adjustment to adult life (215).

Overall, the cognitive and behavioral problems in persons treated early for PKU are minor. By any standards, neonatal screening and dietary management have all but neutralized a disorder that once accounted for up to 1% of the institutionalized MR population. The near conquest of PKU stands as one of the triumphs of pediatric neuropsychiatry.

Lead Poisoning

Lead (Pb) is one of the best-known environmental causes of MR and milder learning problems. Both prenatal and postnatal lead exposure can exert negative effects upon cognitive development. Prenatal exposure to lead usually is estimated using either the mother's blood lead level or that of the umbilical cord; these tend to be strongly correlated (about $r = 0.80$) (217) but do not always correlate similarly with outcome variables. Blood lead level is the standard means of assessing postnatal exposure, although this approach measures only recent (past month) exposure. Blood lead level underestimates actual exposure if a child was heavily exposed in the more distant past but exposure has ceased. This is important, because the effect of low-level lead on cognitive abilities may emerge gradually, over the course of many months, and persist for years. Measuring dentine lead level is an alternative approach for children who are shedding their baby teeth; dentine lead and blood lead correlate at about $r = 0.50$ (218,219), a moderately strong relationship that is attenuated by the more durable nature of dentine lead levels. Other approaches (e.g., assessing tibia lead level via x-ray fluorescence spectroscopy) are used less frequently.

Blood lead levels above 70 μg/dL constitute a medical emergency, because levels much above this point are associated with increasing risk of encephalopathy, seizures, coma, and even death. Cases of frank lead poisoning (plumbism), usually defined by blood lead levels above 45 μg/dL, are rare nowadays but do occur. The level at which cognitive deficits and/or behavioral problems become likely is more controversial, but the trend for the past 20 years has been for the Centers for Disease Control periodically to lower the suggested cutoff. Previously, levels up to 25 to 30 μg/dL were considered acceptable, but today anything in excess of 10 μg/dL is deemed high enough to warrant intervention.

Concern over low-level lead exposure began in the mid-1970's, but the several studies conducted during that period produced mixed results and often failed to control for potentially confounding factors such as parental IQ, socioeconomic status (SES), home environment, and general health. Still, enough positive findings emerged to prepare the scientific community for the landmark 1979 study by Needleman et al. (220), who reported a 4.5-point difference in IQ score between groups of children with high versus low dentine lead levels, a difference that could not be accounted for by other variables. Lead-related deficits also emerged on tests of reaction time, auditory processing, and language skills. Furthermore, teacher behavior ratings showed the high-lead group to be more distractible, more easily frustrated, less persistent, and less organized, which are problems reminiscent of ADHD.

During the past 2 decades, several sizable cohorts of children have been followed longitudinally, in some cases since birth, with lead levels and cognitive functioning measured at intervals. These include children of mostly white, college-educated parents in Boston, of mostly single black mothers on public assistance in Cincinnati, and of black and white working-class urbanites in Cleveland. Children from lead-smelting communities in Australia and Yugoslavia, a complete birth cohort in a New Zealand region, a Danish sample, and the original Chelsea/Somerville sample of Needleman et al. also have been evaluated repeatedly. These may provide the most convincing evidence of the relationship between lead levels and cognitive functioning, but by no means do they exhaust the literature, which also includes cross-sectional studies from France, Italy, Germany, Scotland, the Netherlands, Finland, Croatia, Mexico, and India, as well as other American samples. We reviewed a total of 43 studies (involving 30 independent samples) of the association between lead level and cognitive functioning. The vast majority of these studies were carefully designed and executed. The large majority show an unequivocal adverse effect of lead at some point in the life span (218,220–249), although a few show no such effect (217,250–253), and several more show adverse effects under conditions that affect only a subset of the sample (219,254–258). Some of the studies reporting negative findings involved samples with very low mean lead levels (252,253).

Two important points are frequently made. First, the adverse effects of lead exposure often emerge gradually, reaching a maximum 1 or 2 years after peak blood lead level (222,225,249). Because blood lead level tends to peak at about age 2, this implies that preschool and early elementary school-aged children are apt to show the most noteworthy effects. Second, lead exposure may interact with other variables, especially lower SES (225,257,258), but also health-related factors such as neonatal jaundice (254). The developing brain may be able to compensate for low-level lead exposure if other aspects of life are optimized, but not when multiple environmental problems are present.

The main implication of this mass of evidence is quite clear: low-level lead exposure is associated with cognitive deficits. Beyond this point of agreement, however, lies continued controversy over the magnitude and precise nature of this effect. Based on a World Health Organization meta-analysis (259), some authorities maintain that the effect, although dose dependent and linear, is modest, amounting to only about two IQ points per 10 μg/dL increment in blood lead level above the official safety cutoff of 10 μg/dL. This may well, however, be an underestimate. To ensure that they are identifying a genuine effect of lead on IQ, most investigators since the 1979 study by Needleman et al. have controlled statistically for a number of other factors known to be associated with IQ (e.g., SES, home environment, maternal IQ, birth weight). Cumulatively, these other variables often account for a third or even half of the variance in IQ scores, rendering it difficult for any other factor to contribute substantially to the regression model. Furthermore, at least some of these control variables are confounded with lead exposure. Maternal IQ, which correlates quite strongly with child IQ, is a blatant example. The problem is that the mothers of lead-exposed children may themselves suffer from lead-produced cognitive deficits. Two studies reported a significant negative correlation between maternal IQ and prenatal lead levels (255,258). Similarly, because people of lower SES are more likely to live in high-lead environments, some of the IQ difference between populations of higher and lower SES may be attributable to lead. Higher prenatal lead levels also may cause lower birth weight (260). We suggest that the "true" effect of each 10 μg/dL increment in blood lead level may range from four to five points.

Even a five-point reduction in IQ may sound modest, and it would have little visible impact on a population whose mean IQ is in the high average range. In a hypothetical middle-class community with a "true" mean IQ of 115, an effect of this magnitude would increase the expected prevalence of nonspecific MR from 0.13% to 0.38%, which is still very low. (There might, of course, be more notable increases in learning problems and ADHD-type behavior). In contrast, in a hypothetical community of lower SES and a "true" mean IQ of 90, the same IQ effect would increase the expected prevalence of MR from 4.75% to 9.18%, with serious consequences for the school system and the future of the local economy. The evident fact that lead produces graver cognitive deficits in lower-SES children would further compound this problem.

The nature of the cognitive deficits associated with elevated lead varies across samples and is dependent upon the assessment instruments used. For example, some studies using the Wechsler scales report greater deficits on VIQ (218,244), whereas others report greater deficits on PIQ (230,257) or similar levels on both scales (222). On the K-ABC, simultaneous processing is more affected than sequential processing (229,258); similarly, the largest effects on the McCarthy scales are found for perceptual-performance tests (241,247). One commonly reported specific deficit involves visuomotor and visuoconstructional tasks, such as the Bender, the Rey Complex Figure, and block design tests (218,222,227,244,257,261). Another involves tasks tapping into sustained attention and reaction time (218,220,237–239,243,254,257). Thus, processing speed and perceptual organization and integration often emerge as the areas chiefly affected. However, measures of crystallized abilities, such as vocabulary and general knowledge, also represent major deficits in some samples (218,222). Age at exposure may determine what cognitive abilities are compromised (242). Alternatively, impairment of crystallized abilities may reflect the cumulative impact of reduced information processing efficiency over the years, especially the crucial late preschool and early elementary school years when the cognitive effects of lead exposure are at their peak.

Behavioral problems, especially externalizing behaviors (e.g., aggression, delinquency) but also some internalizing problems (e.g., somatic complaints), are commonly elevated among lead-exposed cohorts (218,220,232,238–240,262). These, too, may exert a cumulative impact over the years, resulting in poor reading test scores and a significantly increased dropout rate, even after controlling for IQ (234,240).

In addition to the many studies of lead levels in community-based samples, comparison of children with mild MR or borderline IQ and normal controls has revealed higher hair lead levels in the intellectually challenged (263,264). In addition, blood lead level correlates negatively with IQ within the MR population, except where other etiologies explain the MR (265). Similarly, children with ADHD have been found to have higher mean dentine lead

levels than other children (266). This implies that elevated lead levels should be ruled out as a factor in cases of mild MR, learning problems, and ADHD. The hypothesis that higher lead levels are a result, rather than a cause, of MR (due to increased rates of pica) is largely discredited. The fact remains, however, that members of high-lead groups are more likely than others to have a history of pica (220). Children with pica who reside in older housing are likely to develop cognitive deficits, and children with MR and pica are liable to increase the severity of their retardation as a result of disordered eating behavior. Effective management of pica in the MR population has led to lower levels of blood lead in that group (267).

It should be noted that low to moderate levels of lead exposure in adulthood exerts adverse effects upon cognitive functioning. Almost every adult study reports at least some performance deficits associated with lead levels. Most of these involve lead-exposed workers, usually spanning a wide age range (268–274). Findings for a cumulative 793 lead workers (and 186 controls) are rather convincing: lead-related occupations are associated with impairment of verbal learning and memory, visuomotor and constructional skills (e.g., design copying and block design), and fine motor speed and dexterity. Impairment persists for decades after discontinuation of lead exposure (269,270). Self-reported depression and somatic symptoms appear to increase with lead exposure (272,274). The effects of lead upon intellectual functioning may grow more severe with age, perhaps due to the increased vulnerability of the aging brain (269,270). Two unanswered questions are whether lead exposure produces cognitive impairment within the first year or so of lead-related employment (272,273) and whether lead exposure affects executive functions (269,274). Even in the general adult population (in which mean blood lead levels are well below 10 μg/dL), blood and tibia lead levels correlate with a pattern of neuropsychological deficits similar to that found among lead workers (275,276).

The blood lead levels of American children have declined markedly since the late 1970's, chiefly due to the elimination of leaded gasoline. In the late 1970's, the mean blood lead level was 15 μg/dL; by the early 1990s, this had dropped by three fourths to about 4 μg/dL (277,278). Lead paint (mainly inhalation of dust, but also eating of paint chips) became the principal vector of contamination. Progress continued into the late 1990's thanks to the gradual rehabilitation or replacement of older housing units that contained lead-based paint and/or lead plumbing. For example, 9.3% of children tested in Boston in 1994 had blood lead levels above 10 μg/dL (1.5% were >20 μg/dL); by 1999, these figures fell to 5.1% and 0.5%, respectively (279). At the present time, only children with identified risk factors need to be screened for lead. Children who live in older housing (where lead paint may remain) who also are African-American, Mexican-American, and/or living in poverty face greatly increased risk of elevated lead levels. Children enrolled in Medicaid are three times as likely to have elevated blood lead levels as are other children, and they account for 60% of all children with blood lead levels above 10 μg/dL [83% of those >20 μg/dL (279)]. Medicaid regulations technically require screening of blood lead levels, although in practice two thirds of these children have not been screened (277). It should be remembered that children of lower SES are at greater risk of incurring cognitive deficits as a result of relatively low levels of lead exposure. It is prudent to check the blood lead level of any mildly retarded child whose MR lacks other definable etiologies. Whether this should be extended to children with some combination of learning disorders, ADHD, and/or borderline IQ perhaps must be left to the discretion of the individual clinician.

The degree to which lead-induced cognitive deficits may be reversible is a subject of great interest. If low-level lead exposure inflicts structural damage, then little improvement is to be expected, but insofar as the toxic effects of lead reflect only current lead levels, considerable recovery is likely. Lead levels could drop either naturally or as a result of medical intervention (i.e., chelation). As noted earlier, mean lead levels tend to peak in early childhood and decline slowly thereafter. The extent of this decline varies from child to child. In at least one major prospective study, declining lead levels were only marginally associated with improved intellectual functioning (280). Other reports indicate a tendency toward resolution of cognitive problems within a few years (241).

Chelation is mandated if blood lead levels are above 45 μg/dL. Outcome studies of chelation treatment in more moderately exposed children are surprisingly scarce. One study combining parenteral edetate calcium disodium (EDTA), iron supplementation, and aggressive efforts to reduce exposure to lead-based paint provided mixed but generally positive results (281). The degree to which blood lead level dropped was related to improved IQ at 6-month follow-up. The effect size amounted to one point per 3.3 μg/dL reduction in blood lead. However, chelation treatment *per se* had no effect on cognitive functioning, and blood lead levels declined just as steeply among the children assigned to the no-chelation condition. A larger and better-controlled study (involving orally administered dimercaptosuccinic acid, or succimer) demonstrated significantly greater declines in blood lead level for the group receiving chelation but no difference in IQ between the succimer and placebo groups (282). Whether declining lead levels were associated with IQ changes in this group is unclear. In any case, despite its intuitive appeal, lead chelation for children with blood lead levels below 45 μg/dL remains an empirically unproved therapy (282). Lead abatement of homes or other sources of contamination is even more important. Note, too, that high lead levels often are associated with iron deficiency, which exerts an independent negative effect upon cognitive status. Iron supplementation tends to reverse anemia-related cognitive deficits relatively quickly,

producing clinical improvement in lead-exposed children (247,281).

Whatever the ultimate fate of chelation treatment for low-level lead exposure, the fact remains that this is one of the few common but preventable causes of MR. Public health efforts to reduce lead exposure at home and in school have reduced lead burden substantially and will continue to improve the environment for years to come. In addition, identifying and removing sources of lead exposure for individual children with elevated blood lead levels is an effective means of reducing lead level and improving cognitive status.

Neurocutaneous Disorders

Tuberous sclerosis, neurofibromatosis, and Sturge-Weber syndrome (SWS) together comprise the so-called *neurocutaneous disorders.* Each is sufficiently common and serious to warrant a separate section.

Neurofibromatosis

Neurofibromatosis comprises a subfamily of neurocutaneous disorders. The large majority of cases fall into two categories: type 1 (NF1, also known as peripheral neurofibromatosis or von Recklinghausen disease) and type 2 (NF2, also known as central neurofibromatosis) (283). These differ considerably in presentation and etiology; for space reasons, only NF1 is considered here.

NF1 is an autosomal dominant disorder with high penetrance, although about 50% of all cases reflect spontaneous mutations. The *NF1* gene, situated at 17q11.2, is coded to produce neurofibromin and evidently functions primarily as a tumor suppressor. With an incidence of up to one in 3,000 live births, NF1 is the most common single-gene condition known to afflict the human nervous system (284).

According to National Institutes of Health (NIH) criteria, diagnosis of NF1 requires two (or more) of the following symptoms: (a) family history of NF1; (b) six or more café au lait spots on the skin; (c) one or more neurofibromas; (d) freckling under the arms or in the groin; (e) Lisch nodules; (f) skeletal abnormalities; and (g) optic gliomas. These NIH criteria are associated with high specificity, but their sensitivity is limited, at least in early childhood. Nearly half of all patients fail to meet criteria during the first year of life, although more than 95% meet criteria by age 8 (285). Café au lait spots tend to appear first, followed (in order) by axillar freckling, Lisch nodules, and neurofibromas. A protein truncation test for NF1 has been developed, but its sensitivity and specificity remain unknown, so the diagnosis continues to be clinical (286).

NF1 was long believed to carry a very high risk of MR, but the early studies of this issue were compromised by ascertainment bias. NF1 also is associated with LD. Interest in the specific cognitive phenotype characteristic of NF1 was sparked in the mid-1980's by reports that children with the disorder tended to display a nonverbal learning disability profile, with specific weaknesses in visuospatial skills, organization, impulse control, and social awareness. This hypothesis has not been strongly supported.

In a consensus statement on cognitive functioning in NF1 as understood by the mid-1990's, a group of leading researchers reviewed ten studies (N = 505) of cognitive functioning in NF1 (287). Using the data presented in this review, we calculated the risk of MR as 6.9% (about three times that for the general population, but still relatively low) and the risk of LD as 33.5% (also considerably elevated). Mean full-scale IQ was 92.7 (significantly but not drastically below population levels), with no clear pattern of verbal or nonverbal dominance. Performance on a single neuropsychological test, judgment of line orientation, is clearly impaired in all NF1 samples studied; therefore, this test may be useful in screening for cognitive impairment in NF1.

More recent studies have modified the cognitive profile picture slightly (e.g., prevalence of LD may be somewhat higher) and suggested new ways of conceptualizing the NF1 cognitive profile (288,289). One study identified a bimodal distribution of IQ, with one peak at 100 and another at 85 (289). A similar finding was implicit in the figures presented by at least one earlier study but was passed unremarked at the time (290). This implies that there may be at least two subpopulations in NF1, one with and one without cognitive impairment. Some evidence also suggests that the LD group can be subdivided into a larger subgroup with global, but relatively mild, intellectual deficits and a smaller subgroup with outstanding visuospatial-visuoconstructional deficits (288). This, combined with the finding that NF1 patients being treated for reading disorders suffer from visuospatial deficits relative to other reading disordered students (291), suggests that the issue of nonverbal learning disabilities in NF1 is not altogether resolved in the negative. Finally, the specific intellectual deficits most often associated with NF1 (cognitive slowing, memory disturbances, strategic problem-solving deficits, and visuospatial deficits) have been noted to parallel those of so-called *subcortical dementias* (292). Because the latter typically are associated with lesions of the basal ganglia, thalamus, and/or brainstem—all of which have been implicated in NF1—the notion of a developmental equivalent is interesting.

Behaviorally, NF1 is frequently associated with a syndrome most accurately diagnosed as ADHD, primarily inattentive type: poor sustained attention, task organization, and impulse control combined with social problems, but not marked by aggressive, delinquent, or otherwise disruptive behaviors (289). Anecdotal evidence suggests that these behaviors respond well to standard treatment with psychostimulants (287,289).

NF1 is associated with an increased risk of psychiatric disorder in adulthood. In particular, the incidence of dysthymia is markedly elevated, affecting 20% or more of NF1 patients (293). No data are available on the efficacy of standard an-

tidepressant treatments in NF1, although there is no *prima facie* reason why they should not be as effective as in other cases.

The contribution of MRI to our understanding of NF1 is potentially important but plagued by controversy. Virtually all studies concur that MRI abnormalities are common among children with NF1, appearing in up to two thirds of cases (287). Interestingly, these abnormalities tend to disappear during late adolescence or early adulthood (287). The usual finding is one or more T2-weighted hyperintensities [also known as *unidentified bright objects* (UBO)], most frequently appearing in the basal ganglia, optic tract, brainstem, and/or cerebellum. There is some dispute even here: some UBOs in the optic tract may reflect optic gliomas, and some of those reportedly found in the basal ganglia may actually be situated in the thalamus (294), but the existence of these UBOs is universally agreed upon. More controversial is their significance. A connection between these evident brain lesions and cognitive dysfunction seems only logical, yet three early studies found no evidence of such a link (295–297). These studies were flawed by the inclusion of many adult subjects (among whom UBOs are rare) and/or by an inexplicable failure to conduct formal intellectual evaluations of all subjects. Several subsequent studies involving only children and in which detailed neuropsychological testing was performed obtained significant correlations between intellectual dysfunction and MRI abnormalities (298–301), but two other equally well-controlled studies reported null findings (302,303), so some doubt remains as to the strength of this association. Still, the consensus view is that lesions in the basal ganglia are associated with visuospatial deficits (especially judgment of line orientation) whereas lesions in the thalamus may be associated with attentional impairment and lowered IQ (287).

MRI clearly should not be used in lieu of detailed neuropsychological evaluation. Unless indicated for other reasons (e.g., focal neurologic deficits suggesting tumor growth), MRI has little clinical value in classic NF1 (287) but may help to make the diagnosis in cases that fail to meet full NIH criteria (304,305).

One study of the neuropathology of UBOs in NF1 found these sites were associated with glial proliferation and fluid buildup within the myelin sheath, suggestive of abnormal development of myelination (306). Conjecturally, the later disappearance of UBOs may result from resolution of these problems. The consensus statement on cognitive functioning in NF1 proposed an explanatory model of NF1-related cognitive deficits based on this finding (287): assuming a role of neurofibromin in early CNS development, NF1 may produce dysmyelination and gliosis that impede the formation of neural circuits designed to serve important cognitive functions.

Neither lowered IQ nor UBOs are associated with disease severity, SES, gender, or family history of NF1. Nor, interestingly, are they strongly associated with macrocephaly, although NF1 children with macrocephaly may be less linguistically competent (305), and enlargement of the corpus callosum may be associated with reduced visuospatial and fine motor skills (303).

Sturge-Weber Syndrome

First described by Sturge in 1879, SWS (occasionally referred to as *encephalotrigeminal angiomatosis*) is most prominently characterized by nevus flammeus, always including the area innervated by the first (ophthalmic) branch of the trigeminal nerve on at least one side of the face, and often including other areas as well. Typically roseate in hue at birth, the nevus deepens to a dark red, hence the sobriquet *port-wine stain* is frequently applied. Congenital buphthalmia or later-onset glaucoma with loss of vision (visual field defects) ipsilateral to the nevus, hemiparesis or hemiplegia contralateral to the nevus, seizures, and MR round out the prototypical clinical picture, although as with many other syndromes not all features are present in all cases. In 1922, Weber described cortical calcification and atrophy ipsilateral to the nevus as additional core features of the syndrome; these phenomena are secondary to leptomeningeal angiomatosis. Prevalence data are lacking, but the disorder may not be especially rare. Given that about one in 200 people are born with port-wine stains, then if as many as 10% of these people suffer from SWS, prevalence would be about one in 2,000.

We reviewed five clinical studies (307–311) that involved moderately large samples of patients with SWS to clarify the frequency with which the various clinical features are present, bringing in more specialized studies where appropriate.

Port-wine stains are almost universally present, although cases of SWS without facial nevus have been reported (312). The stains are unilateral in between one third (308,309) and one half (310) of cases and bilateral in the remainder. When unilateral, they are more frequently on the left side (309). When limited to the ophthalmic area of the trigeminal nerve, as in perhaps one in five cases, the stains are almost always unilateral (310). In somewhat more than one case in four, the stains extend over the maxillary trigeminal area as well; most of these cases are unilateral, but bilateral involvement is common. In about half of cases, the stains extend over both the maxillary and mandibular trigeminal areas; perhaps three fourths of such cases exhibit bilateral stains (310). Other patterns, such as a unilateral ophthalmic stain accompanied by a contralateral maxillary stain, are rare. Extracranial stains occur in a large minority, perhaps even half, of cases (310,311).

Glaucoma and visual field defects occur in between half and three fourths of patients with SWS (308,310,311,313–315). A majority of these cases develop during infancy, with the remainder developing after age 5 (314). However, fewer than one in four patients with unilateral port-wine stains limited to the ophthalmic region develop glaucoma; if they do, the glaucoma tends to be restricted to the eye ipsilateral to

the stain (310). Medical treatment (e.g., topical application of β-blockers) usually fails to normalize intraocular pressure (313,315), although it is successful often enough to be considered the initial treatment of choice (313). Glaucoma of later onset may be more likely to respond to medical treatment (316). A variety of surgical interventions have been reported to be effective in a majority of cases; these include goniotomy (313,317), trabeculotomy (313,317–319), cryocoagulation of the ciliary body (313,315), and the Ahmed glaucoma valve implant (313,320).

Seizures develop in approximately 80% of cases (307–311). Most studies agree that a majority of these seizure disorders develop during the first 1 to 2 years of life (308–310,321), and nearly all develop by age 5 (310), although some samples include larger numbers of cases with later-onset seizures (307). Partial motor seizures contralateral to the facial nevus are most typical (322), but generalized tonic-clonic or myoclonic seizures also are common; sometimes focal seizures later generalize. Infantile spasms are relatively rare but do occur (309). Approximately half of the seizure disorders in SWS can be controlled fully by standard anticonvulsant treatment (e.g., carbamazepine, valproic acid, or phenytoin) (308–310), but a sizable minority [between 11% (310) and 28% (309)] remain entirely uncontrolled.

A number of neuroimaging studies have been conducted. MRI enhanced with gadolinium (Gd-DPTA), which reveals not only cerebral lesions but also meningeal angiomatosis and intraocular lesions, is generally considered to be most revealing and clinically useful in confirming the diagnosis of SWS (309,323–326). The meningeal angiomas occur in the pia mater. Atrophy and calcification of underlying cortical and subcortical regions are nearly always evident before the end of the first year of life (326,327). The distribution of calcifications has been reported as occipital (92% of cases), parietal (61%), frontal (38%), and temporal (22%) (326,327). Both atrophy and calcification are progressive; the calcified regions expand in size and increase in density over the years (309). Interestingly, these calcified regions remain partially functional, being activated appropriately during performance of language and motor tasks, but as the calcification progresses other brain regions (both within and contralateral to the affected hemisphere) are activated by these tasks, reflecting functional reorganization (328).

Deficient cortical venous drainage secondary to the continued presence of sinusoidal vascular channels that ordinarily disappear early in the course of neurodevelopment has been hypothesized as the cause of cortical atrophy and calcification (327). A combination of decreased cerebral blood flow and glucose hypometabolism also has been proposed to underlie ongoing neurologic deterioration (329). Interestingly, whereas all SWS patients older than age 1 show evidence of hypoperfusion, before age 1 only those who have already developed seizures display this finding, with the remainder manifesting hyperperfusion during infancy (330).

Mean IQ may be in the low borderline range, about 75 (308), but this figure conceals wide variability. The two studies that reported the numbers of patients in different IQ ranges included very few cases of borderline intellectual functioning (307,309). To the contrary, the distribution of IQ seems to be bimodal, with many individuals testing in the average or above-average range and many others testing in the mentally retarded range. Averaging across studies, 50% to 60% of those with SWS function in the retarded range (307,309–311). About two thirds of these individuals may be diagnosed with mild to moderate MR, with the remaining third diagnosed with severe MR (307,309). There is a strong association between the presence of seizures and risk of MR; between 68% (307) and 84% (307) of SWS patients with seizures are diagnosed with MR. For those without seizures, intellectual functioning is generally in the normal range (307). Earlier onset of seizures is associated with more severe intellectual deficits (331). Similarly, duration of seizures correlates strongly and negatively with IQ (321). Presence of hemiparesis increases the risk of MR (307,331). The extent and nature of port-wine stains does not independently affect the risk of MR (307,309), but because those with bilateral stains are more likely to have seizures, there may be an indirect link with cognitive impairment (307).

Neuropsychological studies of SWS or even studies of intellectual functioning that go beyond the IQ level are lacking; hence, the cognitive phenotype (if, as seems probable, one exists) remains unknown. Given the large numbers of SWS patients with IQ in the normal range, it would seem relatively simple to conduct studies of this kind, which in turn would materially enhance our understanding of SWS pathology and the educational needs of SWS patients.

Concern over the medical and cognitive sequelae of uncontrolled seizures in SWS sometimes is sufficiently severe as to indicate neurosurgery. Hemispherectomy is the most common procedure, although occipital lobectomy and commissurotomy also have been used. When surgery is performed in the first few years of life (as soon after seizure onset as is compatible with a serious effort to establish pharmacologic control), long-term outcome tends to be remarkably good (322). Complete seizure control is the rule, often without further need for anticonvulsants. Hemiparesis is likely to resolve, although residual right-sided weakness and hypertonia may persist, and visual field defects (e.g., homonymous anopsia) are likely. Full-scale IQ tends to be in the low average range (322). When a (dominant) left cerebral hemisphere is removed during this age range, the right hemisphere functionally reorganizes itself to assume control of language. Typically, subsequent language development is grossly normal. Subtle morphosyntactic deficits may be elicited in some cases (332), but they may be an artifact of low IQ (333). Visuospatial and constructional abilities, however, are apt to be impaired, as evidenced by a VIQ>PIQ profile and poor performance on a variety of neuropsychological measures involving design copying or judgment of line orien-

tation (333), presumably because the right hemispheric circuits normally dedicated to such tasks have been reallocated to language. When surgery is delayed until later in childhood, IQ is likely to remain in the retarded range, although seizure control may still be achieved (322). In general, left hemispherectomy after about age 6 years is not followed by functional reorganization of the right hemisphere for language, although one 9-year-old boy displayed a dramatic first emergence of genuine language after hemispherectomy (334).

Few studies have examined psychosocial functioning or long-term outcome in SWS. Parents report elevated levels of social problems, somatic concerns, and (to a small degree) depressive affect, but not anxiety, withdrawal, delinquency, or hyperactivity (308). Social problems correlate with lower IQ, presence of seizures, and hemiparesis, and they may increase with age. Teachers also report social problems, emotional distress, and some oppositional-defiant and ADHD behaviors, especially in patients with seizure disorders (308). Note, however, that this study involved behavioral rating scales, not psychiatric diagnoses. At this time, it is not safe to assert that SWS patients are at elevated risk of ADHD or disruptive behavior disorders. Adults continue frequently to complain of depressive affect (quite possibly secondary to the social consequences of disfigurement). Among those functioning in the mentally retarded range, noncompliant and even violent behaviors are not uncommon, although SIBs and hyperactive behavior are rare (311).

Adjustment in adult life is variable. A majority of SWS patients are gainfully employed, and nearly half may achieve financial independence (311). Up to half may marry, and many will sire or bear children. Of 20 children of SWS parents in one study, only one suffered from a serious neurocutaneous syndrome (tuberous sclerosis), and two others had minor birthmarks (311). Patients suffering from seizures are less likely to maintain employment and/or suffer from emotional problems in adulthood (311).

Tuberous Sclerosis

Tuberous sclerosis was first described by Bourneville in 1880, although it was Vogt, a generation later, who completed the clinical picture of the disorder. The *Vogt triad* of facial angiofibroma, MR, and intractable epilepsy (plus the confirmatory evidence of numerous "potato-like" tumors in the brain, identified upon autopsy) remained the standard diagnostic criteria for many years (335). During that era, the disorder was thought to be quite rare (estimated at no more than one in 50,000). It gradually became evident, though, that none of the diagnostic signs were universally present; the complete Vogt triad may be present in fewer than one third of cases (336). As the diagnostic net broadened, prevalence estimates rose, and currently hover around one in 25,000 overall with an incidence of one in 10,000 live births (337). The higher prevalence among young children probably reflects both increased mortality rates and ascertainment bias. Even these estimates may omit many people who do not suffer from any clinical symptoms and fail to come to medical attention (337).

Transmission is autosomal dominant, although 60% to 70% of cases represent new mutations (337–339). Whether a given case, identified in early childhood, is familial makes a great difference from the standpoint of genetic counseling; if either parent carries the gene, recurrence risk is 50%, but if a new mutation is responsible, recurrence risk is only in the 1% to 2% range (still higher than the population risk, evidently because some genetic lines are more prone to this mutation than are others). Either of two separate genes will produce tuberous sclerosis. One, sometimes referred to as *TSC1*, is situated on chromosome 9 (in the 9q34 region); the other, *TSC2*, is on chromosome 16 (in the 16p13 region) (340). To date, no reliable differences in clinical presentation have been identified between the two genotypes.

Diagnostic criteria continued to be refined during the latter part of the twentieth century (335,341). The most recent revision abandons the notion that any feature is pathognomic, turning instead to a polythetic criterion set (342). Eleven major and nine minor features are listed. The major features are (a) facial angiofibromas or forehead plaque; (b) nontraumatic ungual or periungual fibroma; (c) three or more hypomelanotic macules; (d) shagreen patch; (e) multiple retinal nodular hemartomas; (f) cortical tuber; (g) subependymal nodule; (h) subependymal giant cell astrocytoma; (i) one or more cardiac rhabdomyomata; (j) lymphangiomyomatosis; and (k) renal angiomyolipoma. The minor features are (a) pitted dental enamel, (b) hamartomatous rectal polyps, (c) bone cysts, (d) cerebral white matter migration lines, (e) gingival fibromas, (f) nonrenal hamartoma, (g) retinal achromic patch, (h) "confetti" skin lesions, and (i) multiple renal cysts. With few exceptions, the presence of two major features or one major plus two minor features indicates definite tuberous sclerosis; the presence of one major and one minor feature indicates probable tuberous sclerosis; and the presence of one major or two minor features indicates possible tuberous sclerosis.

If a diagnosis of tuberous sclerosis has been made or is being seriously considered, at least five further evaluation procedures should be conducted (343). (a) Cranial computed tomographic (CT) scan or MRI provide diagnostic information. Results also help to predict neurologic and developmental course (although they should not be interpreted in isolation). In general, T2-weighted MRI is more revealing because it will identify both subependymal nodules and cortical tubers, whereas CT scan will identify only subependymal nodules (336). This should be repeated every 1 to 3 years until adulthood. (b) Neuropsychological assessment, including IQ and other tests, establishes the baseline level of intellectual functioning and facilitates educational and psychosocial treatment planning. Testing should be repeated at the time of school entry and whenever clinically indicated

(343). [Intellectual assessment may be omitted in the rare case of a newly diagnosed adult who enjoys a totally normal social and academic history (343).] (c) Ophthalmic examination helps diagnostically and identifies patients with poor vision due to macular lesions. About 75% of tuberous sclerosis patients have ophthalmic lesions, although they tend not to be progressive (343). (d) Renal ultrasonography will identify renal tumors and should be repeated every 1 to 3 years. By age 10, about 75% of tuberous sclerosis patients have at least one renal angiomyolipoma (343). (e) Electrocardiography will identify cardiac arrhythmias, often before clinical symptoms emerge. Cardiac rhabdomyomas and Wolff-Parkinson-White syndrome occur in tuberous sclerosis (343). If cardiac symptoms occur, then echocardiography should be considered. If seizures occur, then an EEG should be done. Dermatologic evaluation is indicated whenever facial angiofibromas or ungual fibromas are present. Finally, adult female patients with pulmonary complaints should undergo chest CT to rule out lymphangiomyomatosis.

Evaluation of family members (primarily for purposes of genetic counseling but also to identify hitherto unsuspected medical problems) need not be as comprehensive. Here, cranial CT scan may be preferable to MRI because of the higher rate of false-positive findings obtained using the latter approach (343). Renal ultrasonography is advised, but the other scanning procedures are not usually thought necessary.

The classic dermatologic manifestations of tuberous sclerosis remain important in making the diagnosis. Children under age 5 will nearly always display hypomelanic macules but may show no other features (337). These depigmented patches sometimes are called *ash leaf macules* because of their shape (306). Facial angiofibromas (pink or red nodules found on the butterfly area of the face) are present in the large majority of cases after age 5. Up to 50% may display ungual fibromas (growths in the nail beds), and almost as many have shagreen patches (which resemble untanned leather and usually are found in the lower back) (336,337). Like the growths sometimes found in the heart, kidneys, eyes, or elsewhere, all of these growths usually are benign.

Approximately 80% to 90% of patients suffer from seizure disorders (337,344,345), although the actual incidence of epilepsy in tuberous sclerosis may be as low as 62% due to unidentified mild cases (346). Onset of seizures is during the first year of life in between two thirds (344,345) and nine tenths (347) of cases, with most of the remaining cases developing by age 5. In general, earlier onset of seizures is associated with a more severe clinical course. Of those developing seizures in infancy, about one third present initially with infantile spasms, one third with partial motor or complex partial seizures, and one third with both infantile spasms and partial motor seizures (347). A majority of those presenting with infantile spasms go on to become persistently and severely epileptic; those presenting initially with partial seizures in infancy also are likely to develop epilepsy, but with a milder course (348). Later-onset cases nearly always present with partial seizures and tend to have the most favorable outcomes (347,348).

Epilepsy in tuberous sclerosis is notoriously difficult to manage. Fewer than one third of all patients achieve complete seizure control, and more than one third may never achieve even partial control (344). Infantile spasms usually are treated with adrenocorticotropic hormone or, more recently, with vigabatrin (349). Carbamazepine, valproic acid, and phenytoin all are used for partial seizures (336).

MR is common and often severe. Averaging across seven studies with a cumulative sample size of 459, risk of MR is 58.4% (337,338,345,346,350–352). [Note that this may greatly overestimate the actual risk, as large numbers of mild cases pass unnoticed. But even if only one third of those with tuberous sclerosis become mentally retarded (337), the higher figure continues to apply to those actually brought to clinical attention, including unaffected relatives of patients.] Most studies of cognitive functioning in tuberous sclerosis have reported only global imprecise data. The cognitive phenotype of the disorder, if one exists, remains to be elucidated by more systematic neuropsychological studies.

There is a strong association between seizures and MR in tuberous sclerosis. Patients who do not develop seizures almost never suffer from MR (337,344,345). For those with seizure disorders, risk of MR is approximately 70% (337,345). Onset of seizures before age 1 is most frequently associated with severe MR; onset during the next few years produces MR in about half of cases (345). Transient infantile spasms, however, do not necessarily indicate cognitive impairment (345,348). Late-onset partial seizures carry the most favorable intellectual prognosis; partial seizures with secondary generalization and partial seizures followed by spasms suggest the likelihood of MR (348).

Nearly all tuberous sclerosis patients manifest brain lesions, typically subependymal nodules and/or cortical tubers (335,336). Subependymal nodules consist of astrocytes, usually are located on ventricle walls, and tend to calcify in early childhood. They are generally benign, but they can develop into subependymal giant cell astrocytomas, which may block cerebrospinal fluid (CSF) circulation and necessitate surgical intervention (336). Cortical tubers (so termed because of their location and resemblance to tiny potatoes) also are composed of astrocytes. Their number varies widely across patients; remember that CT scan underestimates their number. One study of epileptic tuberous sclerosis patients found tubers in 94% of cases (347,348). Their frequency was bimodally distributed; a large minority of the patients had only one or two, whereas nearly all the rest had eight or more. When tubers are few, they are nearly always found in the parietal and occipital lobes, especially the former. When they are greater in number, they are found mainly in the parietal (and parietooccipital) and frontal regions. The number of tubers is inversely related to level of cognitive functioning;

few nonretarded patients have more than one tuber (348). Tubers also seem to function as epileptogenic foci (347).

Psychiatrically, tuberous sclerosis is remarkable for the frequency with which autistic features (often meriting a diagnosis of PDD) are present. The rate varies across samples, ranging from 50% (344,353) to 86% (351). The extent to which tuberous sclerosis presents with a unique form of PDD remains inadequately explored. Tuberous sclerosis patients who are severely retarded, have a history of infantile spasms that develop into intractable epilepsy, and manifest eight or more cortical tubers may be especially likely to meet criteria for PDD (348). Behaviors typical of ADHD are frequently observed, although these behaviors may be secondary to PDD (336). Other common behavioral problems include marked sleep disturbance, noncompliance, and aggression (344). Those functioning in the MR range are likely to have limited or no speech; to be unable to dress, wash, or toilet themselves; and (in some cases) to suffer from ataxic gait or even an inability to walk (344). It must be borne in mind that, contrary to persistent clinical lore based on the older literature (which suffered from a strong ascertainment bias), a great many people with tuberous sclerosis appear to function perfectly adequately.

Prader-Willi Syndrome and Angelman Syndrome

Prader-Willi syndrome (PWS) and Angelman syndrome (AS), although phenotypically distinct, share a common genetic basis (354) and so will be considered in one section, albeit sequentially. Each affects between one in 10,000 and one in 15,000 live births (354). Deletion of chromosome 15 was recognized as a cause of PWS by the early 1980's (355). The fact that AS also could result from deletion of chromosome 15 was not appreciated until several years later (356,357). At the same time, maternal disomy 15 (where the mother contributes both copies of the chromosome) was identified as an alternative source of PWS (358). It now is clear that both syndromes are produced by abnormalities in the q11-q13 region of chromosome 15. About 70% of cases involve deletion of the critical area. Nearly all remaining cases of PWS are produced by uniparental disomy (i.e., one parent contributing both copies of chromosome 15), with the few remaining cases resulting from imprinting center defects. In the case of AS, uniparental disomy and imprinting center defects are both relatively uncommon, with most nondeletion cases attributable to a single gene mutation (gene *UBE3A*) or unknown causes (354). In general, nondeletion cases present with milder clinical manifestations. The remarkable aspect of these syndromes is that PWS is always associated with an absence of *paternal* genetic material (i.e., paternal deletion or maternal disomy), whereas AS is almost always associated with an absence of *maternal* genetic material (i.e., maternal deletion or paternal disomy). This puzzling nonmendelian pattern was not recognized until the late 1980's and represents the first instance in which genetic imprinting has been associated with different neuropsychiatric manifestations (358–360).

PWS was first described in 1956 (361), although the first case reported in an English-language journal came 5 years later (362) and no review of the ensuing trickle of case reports appeared until 1968 (363). In those early years, the disorder sometimes was referred to in terms of its cardinal features: hypotonia-hypomentia-hypogonadism-obesity (HHHO) syndrome. Given the outstanding behavioral and physical manifestations of the disorder, it is remarkable that it passed unnoticed for so long.

Current diagnostic criteria for PWS syndrome are as follows (364): The major criteria are (a) neonatal and infantile hypotonia; (b) feeding problems in infancy (e.g., need for special feeding, failure to thrive); (c) excessive or rapid weight gain after age 12 months but before 6 years, with obesity resulting unless dietary interventions are imposed; (d) three or more of the characteristic facial features (almond-shaped eyes, small mouth with thin upper lip, downward-turning corners of mouth, narrow face or bifrontal diameter, dolichocephaly in infancy); (e) hypogonadism (e.g., cryptorchidism or testicular size below the fifth percentile for males; absence or severe hypoplasia of labia minora and/or clitoris for females; puberty delayed past age 16 years); (f) global developmental delay before age 6, mild to moderate MR or severe learning problems in older children; (g) hyperphagia, food foraging, and/or obsessions with food; and (h) 15q11-q13 deletion or other cytogenetic/molecular abnormality of this region. Minor criteria are (a) decreased fetal movements or infantile lethargy with weak cry, improving over time; (b) five or more characteristic behavior problems (temper tantrums, violent outbursts, obsessive-compulsive behavior, argumentative, oppositional, rigid, manipulative, stubborn, possessive, lying, stealing, perseverative); (c) sleep disturbance or sleep apnea; (d) short stature for genetic background by age 15; (e) hypopigmentation for genetic background; (f) small hands (below 25th percentile) or feet (below 10th percentile) for height and age; (g) narrow hands with straight ulnar border; (h) eye abnormalities (e.g., myopia, esotropia); (i) thick viscous saliva with crusting at the corners of the mouth; (j) speech or articulation deficits; and (k) skin picking. The diagnostic algorithm is as follows: one point is awarded for each major criterion met, and one-half point for each minor criterion. Children age 3 years and under must obtain five or more points, including at least four from the "major" group, to be diagnosed. Older children must obtain eight or more points, including at least five from the "major" group. Certain other findings, although not scored, increase diagnostic confidence. These are (a) high pain threshold, (b) decreased vomiting, (c) altered temperature sensitivity, (d) scoliosis and/or kyphosis, (e) early adrenarche, (f) unusual skill with jigsaw puzzles, and (g) normal neuromuscular studies (i.e., the hypotonia is deemed benign and treatable).

Many studies have reported on intellectual functioning in PWS, although few include finer-grained neuropsychological data (365–376). They display a striking uniformity: mean IQ is typically near 70 (on the cusp between mild MR and borderline intellectual functioning). There also is less variability in IQ than in the population generally; the large majority of PWS patients (probably >90%) function in the borderline or mild MR range, with diminishing numbers above or below this range. There may be a small superiority of linguistic over visuospatial processing (369,375). Certainly, PWS patients perform poorly on many neuropsychological tests sensitive to spatial organization (e.g., the Bender-Gestalt, Rey Complex Figure, and Hooper) (369), but relatively well on tests of receptive language (e.g., the Peabody Picture Vocabulary Test, PPVT, and the K-ABC "achievement" scales) (375). At the same time, however, they are known to display both interest in, and aptitude for, jigsaw puzzles. This suggests a dissociation between object recognition and abstract spatial processing, but this possibility has not been systematically explored. Sequential processing is another weak area (375). PWS is associated with learning disabilities, especially dyscalculia and dysgraphia. Some evidence suggests that cognitive impairment tends to be more severe in cases where excessive eating begins early (age 1), which also is associated with greater stature (377). One study found that deletion cases are more impaired than disomy cases (with mean IQs of 63 and 71, respectively) (370). Another study using a smaller sample reported no significant difference, but this may have been an artifact of low statistical power (378); the mean IQs for the two groups (63.3 and 69.8, respectively) were similar to those in the first study. The IQ difference may be produced by disparate VIQ, with PIQ similar across genetic subtypes (376). Unusual skill with jigsaw puzzles may be characteristic only of deletion cases (378).

Numerous studies have been devoted to elucidating the behavioral phenotype of PWS (365–367,369,371,373,374, 376,379–383). These also tend to yield remarkably consistent results. Behavior problems of one sort or another are common. For example, studies using the CBCL show 70% or more in the clinical range (compared with 40% or fewer of the general MR population) (367,369,373,376,378,379). Both internalizing and externalizing problems are common, sometimes equivalently so (369,373), but with externalizing symptoms often more serious (367,376,378). Evidence is mixed, but behavior problems (relative to other groups) may escalate during adolescence and early adulthood (373,379,384). Deletion cases exhibit more behavior problems than disomy cases (377). In general, IQ does not seem to be related to behavior problems in PWS.

Aside from overeating, the most common behavior problems in PWS are underactivity, excessive sleep, skin picking, argumentativeness, repetitive, inappropriate speech, temper tantrums, quick mood changes, being easily upset by minor hurts or changes in routine, being teased, stealing (especially at home, mainly involving food or money for food), obsessions, and compulsions. Lethargy obviously both contributes to, and is increased by, obesity.

SIBs characterize 80% to 90% or more of PWS patients, with skin picking (often leading to skin infections) by far the most common form. The shins and the nasal-oral area are most often injured, but practically any body site can be targeted. The head banging, hand biting, and hitting that typify SIBs in the retarded population generally are less common in PWS. The extent of SIBs tends to worsen with age (385). Deletion cases seem to injure more widely than disomy cases. It has been suggested that SIBs in PWS are related to compulsions. Consistent with this, at least one open-label trial reported that fluoxetine is efficacious in reducing their severity (386).

The social behavior characteristic of PWS is somewhat paradoxical. On the one hand, these individuals are perceived as naively friendly. They tend to have nurturing qualities, being especially fond of pets, babies, and young children. Unfortunately, their inappropriate speech may alienate others, and obesity invites teasing, especially during later childhood and adolescence. Perhaps these factors, plus the physical limitations imposed by obesity and their lethargic habits, produce the social withdrawal that tends to worsen over the years. At the same time, PWS patients are perceived as argumentative, intolerant of changes in routine, and subject to temper tantrums. A large proportion of these outbursts, as with the stealing that is their most prominent delinquent behavior, relate to the ongoing struggle between caregivers and PWS sufferers over access to food. In fact, PWS patients whose weight is kept within normal limits actually display *more* behavior problems than do their obese peers.

Without doubt, the most prominent behavior problem in PWS is the tendency toward gross overeating. Unless access to food is strictly controlled, virtually all PWS patients will become obese, often morbidly so. Group homes specializing in PWS tend to have the most success in keeping weight within reasonable limits; PWS patients dwelling at home tend to be the heaviest. Literally padlocking refrigerators and cupboards is a common expedient on the part of caregivers, although as noted earlier this may elicit escalated levels of foraging and stealing on the PWS sufferer's part. Obesity often produces medical complications. Nearly 20% of young adults with PWS suffer from diabetes that usually requires insulin treatment; almost as many suffer from hypertension; and nearly 50% develop scoliosis (383).

It was long believed that PWS patients ate indiscriminately, but they do in fact display food preferences (387). Unlike obese controls, those with PWS may prefer high-carbohydrate foods over those high in fat or protein (388) and show a liking for sour foods equivalent to that for sweet foods (389). Small amounts of high-preference foods can be used as reinforcers for target behaviors such as increased physical activity (387,389). Behavior modification techniques including contingency contracting, self-monitoring, differential reinforcement of other behaviors, and punishment

have been effective in reducing food intake and inappropriate foraging (e.g., stealing) in PWS (390–392) and are mainstays of treatment in group homes that specialize in PWS. An impaired satiety response similar to that produced experimentally by ventromedial hypothalamic lesions has been suggested as the source of hyperphagia in PWS (393,394).

Although hyperphagia produces rapid weight gain in PWS, another factor—growth hormone (GH) deficiency—may do much to perpetuate it. GH deficiency is associated with reduced levels of physical activity and resting energy expenditure, as well as body composition akin to that found in PWS. Once reduced GH secretion was documented in PWS (395), GH replacement therapy was tried (396). At least seven relatively well-controlled studies involving a cumulative 133 treated patients have established the efficacy of GH supplementation in PWS (397–403). Most of the gains appear to come during the first year of treatment. These improvements include increased lean mass, decreased fat mass, reduced body mass index, increased height growth velocity, improved ratio of low-density lipoprotein to high-density lipoprotein, increased physical strength and agility, improved respiratory function, and possibly increased resting energy expenditure. After the first year, however, little more can be expected on most of these parameters, although height and fat mass continue to benefit. Rarely, patients receiving GH show sharp weight increases; these may signal the onset of noninsulin-dependent diabetes mellitus, which may benefit from cessation of GH supplementation (398).

In patients with GH deficiency, GH replacement therapy also produces psychological improvement (e.g., reduced anxiety and irritability). Whether this also applies to PWS patients has not been adequately investigated. One study reported that GH produced reductions in depressive symptoms in PWS (especially among patients older than 11 years) but no change in attention, anxiety, obsessive-compulsive behaviors, or aggression (404).

From the psychiatric perspective, it is noteworthy that PWS patients often are reported to display elevated levels of obsessional ideation and compulsive behaviors (367,369,371,381). Possibly 25% to 45% of PWS patients meet criteria for OCD (372,382). This may take a rather specific form. Obsessions and compulsions related to hoarding are common (more so than among OCD sufferers generally), and concerns over needing to tell or ask, ordering or arranging, repeating rituals, and symmetry or exactness are common. On the other hand, PWS patients display relatively low levels of religious concerns and checking rituals (372). Hoarding behavior includes, but is not limited to, accumulation of food. Deletion cases seem to display higher levels of hoarding behavior than do disomy cases (377). OCD symptoms tend to worsen with age and to be more severe in patients with IQ in the MR range (372).

PWS may carry with it a modest increase in the risk of psychosis (405), more often in the affective spectrum but sometimes of a schizophrenic nature (406). Consistent with the nature of the psychopathology often found in PWS, at least one study has reported abnormal levels of the serotonin metabolite 5-hydroxyindoleacetic acid (5-HIAA); and, to a lesser extent, of dopamine metabolites in the CSF of patients with PWS (407).

Adaptive functioning in PWS tends to be poorer than would be expected based upon IQ. Up to half of individuals continue to reside with their parents in adulthood, with the remainder fairly evenly divided among specialty PWS group homes, generic MR group homes, and other treatment facilities (e.g., residential schools, psychiatric facilities). Only a tiny fraction live independently. In terms of work adjustment, up to 40% may perform no work at all, with 50% employed in sheltered workshops and smaller numbers volunteering or working out of the home. Fewer than 5% hold competitive employment. Social life tends to be restricted and sexual life minimal (383).

AS has received far less research attention than PWS. It was first described in the clinical literature in 1965, although it has been suggested that a famous Walt Disney character anticipated Angelman by several decades (408). Its classic form involves severe MR with seizures, absent speech, a characteristic jerky, ataxic gait with hands held high, and a cheerful disposition marked by frequent outbursts of laughter, hence the original name *happy puppet syndrome*. Current diagnostic criteria are as follows (354). *Consistent manifestations* are (a) functionally severe developmental delay; (b) severe speech impairment (none or minimal use of words), with relatively good receptive and nonverbal communication skills; (c) movement or balance disorder (e.g., gait ataxia and/or tremulous limb movements); and (d) a behavioral phenotype involving any combination of frequent smiling or laughter, happy demeanor, excitability, hand flapping, hyperkinetic behavior, and short attention span. *Frequent manifestations* (present in 80% or more of all cases) are (a) slow head growth, usually resulting in microcephaly by age 2; (b) seizures, usually with onset by age 3; and (c) abnormal EEG findings, especially large-amplitude slow-wave spikes. There also are many *associated features* that may help to confirm a tentative diagnosis or alert a clinician to the possibility of AS: uplifted flexed arms during walking; attraction to and fascination with water; tongue thrusting; a characteristic facies with wide mouth and widely spaced teeth; hypopigmentation of the skin and eyes; excessive chewing or mouthing; strabismus; brachycephaly; feeding problems in infancy; sleep disturbance; and hyperactive tendon reflexes.

AS obviously entails far graver impairment than PWS. The large majority of patients never speak more than two or three words (354,409,410). A specific oromotor apraxia contributes to this (354). Although most AS patients make some use of signs or gestures, only one in five masters American Sign Language. Beyond the expected ataxic gait, 20% to 40% never walk (310,411,412); females may be at

especially elevated risk of remaining wheelchair bound (411). Scoliosis frequently develops in adulthood (412).

Epilepsy, often severe, characterizes approximately 85% of AS patients (410–414), with onset usually by age 3. Initial presentation varies widely, although febrile convulsions are frequent (413). Absence, tonic-clonic, and myoclonic seizures and drop attacks are common seizure forms as epilepsy progresses (413,414). Parental reports suggest that valproate, clonazepam, and lamotrigine are generally effective and well tolerated, but that carbamazepine and vigabatrin produce less desirable results (414). At least one clinical report concurs that vigabatrin may induce or increase the frequency of seizures in AS (415).

Nearly all AS patients exhibit abnormal EEGs. One characteristic finding is triphasic δ waves with maximum amplitude over the frontal region (413). This may assist in diagnosis of cases lacking some of the characteristic clinical features (416). In addition, EEG findings can help make the differential between AS and Rett disorder (417). This distinction also can be made using molecular (methylation) analysis (418).

The diagnosis of AS is made most easily between early childhood and early adolescence. Infants often (but by no means always) exhibit head circumference below the 10th percentile, hypotonia, and feeding difficulties (410), but these are nonspecific findings. By late adolescence, some characteristics grow less obvious; for example, facial features typically coarsen (412). It is likely that many cases have gone undiagnosed (416,419). AS should be considered whenever a patient presents with MR and absent or near absent speech, especially if seizures occur. It would not be surprising to find incidentally that a milder variant exists, especially in families in which AS occurs as a result of an inherited imprinting center defect.

Klinefelter Syndrome

In 1942, Klinefelter, Reifenstein, and Albright identified a group of male patients who shared a number of characteristics, including tall stature with elongated limbs, infertility, and hypogonadism (420). Klinefelter syndrome (KS) is thought to affect between one in 700 and one in 900 men (421), although some reports have found an incidence as high as one in 426 (422). The syndrome has since been linked to abnormality of the sex chromosomes. Typical KS males exhibit an additional X chromosome (47,XXY karyotype). Other less frequent chromosomal patterns associated with KS include XXXY, XXXXY, or a combination of abnormal and normal cells (mosaicism). These abnormal patterns are thought to result from unsuccessful separation of the sex chromosomes during production of the egg or sperm (meiotic nondisjunction). Although less frequently observed, some genetic abnormalities also can occur as a result of an error in mitotic division in the zygote (423). Reports regarding the parental source of the additional chromosomal material and the role of parental age are inconsistent, but the additional X chromosome is maternal in origin in 50% or more of cases (424,425) and the likelihood of paternal or maternal nondisjunction may increase with age (424,426).

In addition to the physical features, individuals with KS face a variety of other physical and medical concerns. Obesity, diabetes mellitus, hyperlipidemia, hypercholesterolemia, gall bladder disease, chronic pulmonary disease, and peptic ulcer are of concern (427). Individuals with KS are at risk for renal, cardiac, and lymphatic conditions (428). At puberty, growth of facial, chest, and pubic hair may be diminished. Gynecomastia, eunuchoidism, and feminine fat distribution may be present in these men (429). By late puberty, 30% to 60% of boys with KS exhibit gynecomastia, and approximately nine in 1,000 individuals with KS will develop carcinoma of the breast (430). Depending upon the severity of the gynecomastia, mastectomy may be recommended.

KS is the most common cause of hypogonadism in males (430), and endocrinologic disturbances are a hallmark of the disorder. Males with KS exhibit reduced levels of testosterone and elevated levels of luteinizing hormone and follicle stimulating hormone at the beginning of adolescence (431). They may be at elevated risk of developing thyroid problems, especially hypothyroidism (430). The androgen deficiency associated with KS places these individuals at risk for osteoporosis (430).

The physical and cognitive characteristics associated with KS can be subtle and can appear at different stages in development. Therefore, many individuals are not diagnosed until puberty or adulthood. However, abnormal growth rate and elongated leg length are observable by age 5 (432). Therefore, diagnosis can be made earlier if physicians are aware of the specific symptoms associated with the disorder. Diagnosis is made through karyotyping, performed either prenatally using amniocentesis or chorionic villus sampling (429) or after birth if the physician has reason to suspect that the patient has KS.

Research on cognitive functioning in KS has consistently found a language-based disability associated with the disorder. Most individuals with KS exhibit IQ in the average range (433–436), although IQ may be depressed by 10 to 15 points relative to unaffected controls (437). Lower full-scale IQ scores appear to be due to discrepancies between verbal/linguistic (VIQ) and performance/visuospatial abilities (PIQ), with VIQ impaired and PIQ typically in the normal range (433,434). However, one study of KS adults challenges this generalization, indicating that as many adults with KS exhibit a VIQ>PIQ profile as the reverse (438). The authors proposed that the language deficits characteristic of children with KS sometimes remediate in adulthood as a result of postpubertal hypoandrogenic activity. These groups also displayed distinct patterns of neuropsychological performance. The VIQ>PIQ group exhibited reduced nonver-

bal information processing speed, nonverbal executive skills, visuospatial/constructional ability, and motor dexterity. In contrast, the PIQ>VIQ group was impaired on measures of verbal attention and verbal executive ability, whereas motor functions, visuospatial ability, and nonverbal executive functioning were intact. In addition to linguistic deficits, mirror movements, motor overflow, poor motor dexterity, and other abnormalities in motor functioning have been noted in KS (436,438).

Observations regarding attention are inconsistent. Some researchers report hyperactivity (433,435,439); others describe lower activity levels in males with KS (440,441). Impairment of sustained attention and working memory have been reported (442), although more empirical studies are needed before firm conclusions can be drawn.

Atypical hemispheric specialization may be common in the KS population. Because 97% to 99% of right-handed individuals demonstrate language localized in the left perisylvian region, handedness is frequently used as a measure of cerebral dominance (436). Reports of hand preference among individuals with KS are inconsistent, with some finding an increased incidence of sinistrality (435) and others finding no significant differences (443) or decreased right handedness only when handedness was assessed as a skill as opposed to a preference (436). KS also is associated with reduced right ear dominance on dichotic listening tasks (444), decreased left hemisphere specialization for language, and increased right hemisphere specialization for nonverbal processing (445).

The cognitive deficits typical of KS, although mild, do impact educational achievement. Due to their verbal-linguistic difficulties, boys with KS tend to fall four to five grade levels behind by adolescence (435) and are more likely to have failed a grade or to have received special education services than their typically developing peers (435). Many aspects of language have been implicated, including rate of auditory processing and verbal memory (446); comprehension and expression, word decoding and spelling, and written language skills (435); articulation, comprehension, verbal abstraction, sequencing, and expressing a story idea (447); and delayed language development, limited vocabulary and syntax, lack of fluency, and difficulty with concepts (448). Deficits in math problem solving, arithmetic, and acquisition of new knowledge have been observed (435).

Individuals with KS demonstrate a common behavioral and psychosocial profile. Children with KS tend to be less active, less assertive, and more susceptible to stress than typically developing boys (437). In addition to being pliant, withdrawn, and low in energy (433,445), younger individuals with KS also have been described as "shy," "reserved," and "having poor relationships" (449) or less teasing, less sarcastic, and more submissive than other boys (450). In adulthood, individuals with KS often are lonely, immature, and passive; lack friends (451); and suffer from low self-esteem, poor self-concept, and poor body image (421). They may date and become sexually involved at a later age than their peers (452). A supportive and stable environment is beneficial (449).

Neuroimaging studies of KS are scarce. One MRI investigation found reduced whole brain volumes and enlarged lateral ventricles in a group of young men with KS (453). A second MRI study reported a reduction in left temporal lobe gray matter in individuals with KS who had not undergone testosterone therapy (454). Given that the left temporal perisylvian areas are associated with many aspects of language functioning, this finding is consistent with the verbal and language processing deficits typical of KS. Importantly, the brain morphology of individuals with KS who had undergone testosterone supplementation did not differ significantly from that of typical controls (454). KS patients who had received testosterone supplementation also outperformed their nonsupplemented peers on a verbal fluency task. These findings support the hypothesis that gonadal hormones play an important role in brain development and suggest that testosterone supplementation can improve cognitive functioning in KS.

Currently, there are two hypotheses that explain the differences in brain function and structure found in individuals with KS. Cells with an additional X chromosome divide at a slower rate than cells with one X chromosome (437,455). Differential rates of cellular division in individuals with KS may have permanent effects on brain development. Alternatively, the low levels of testosterone characteristic of KS may affect brain development and, ultimately, verbal ability (437).

Turner Syndrome

Turner Syndrome (TS), first described in 1938, is another sex chromosome disorder associated with physical, cognitive, and psychosocial effects. TS affects between one in 2,000 and one in 5,000 live female births. As originally described, these patients (all phenotypically female) share a number of physical characteristics, including short stature, sexual infantilism, webbed neck, cubitus valgus (wide carrying angle of the elbow), and primary amenorrhea. The underlying chromosomal anomaly was discovered 20 years later (456). Approximately 50% of cases exhibit total X monosomy [45, X or XO karyotype (457)]resulting from meiotic nondisjunction or from an error in mitosis after conception (458). A variety of other karyotypes account for the remaining cases (459), including partial X monosomy (deletions of the short or long arm of the X chromosome), isochromosomes (duplication of the long arm of the second X chromosome), and mosaicism (only some cells express the XO anomaly).

Additional physical characteristics (not observed by Turner) include ovarian dysgenesis, low hairline, shield chest, lymphedema at birth, scoliosis and other skeletal abnormalities, cardiovascular abnormalities such as coarctation of the aorta and bicuspid aortic valve, recurrent otitis media with associated hearing loss, and renal abnormalities such as horseshoe kidney. Craniofacial abnormalities such as narrow palate and craniosynostosis may be present (437).

Because of the variety of karyotypes associated with TS, the relationship between the phenotypes and genotypes observed within this population has been the subject of recent investigations. Mutations or deletions around a homeobox-containing gene on the X chromosome are associated with the short stature phenotype (460). This gene, referred to as *SHOX,* is also thought to be related to Leri-Weill dyschondrosteosis and other skeletal abnormalities observed in this population (461–463). Many aspects of the TS phenotype, including premature ovarian failure, may be due to decreased gene dosage on the short arm of the X chromosome (464).

Because lack of sex chromatin material has both indirect and direct effects on the brain development, the structure and function of the brain of individuals with TS has been of interest (465). MRI studies consistently show decreased volume in the parietal lobes, especially posterior parietooccipital areas, typically more pronounced on the right (466–469). Sometimes MRI also reveals increased bilateral ventricular volume (466), increased CSF volume, decreased gray matter volume in the right prefrontal and left parietoperisylvian cortical areas (469), and/or decreased volume in the caudate, hippocampus, lenticular, and thalamic nuclei (467). In two case reports of agenesis of the corpus callosum in TS, both patients were mentally retarded (470,471).

Positron emission tomography has revealed bilaterally decreased glucose metabolism in the parietal and occipital cortex (472). In a functional MRI study, individuals with TS exhibited bilateral differences in activation of the supramarginal gyrus, dorsolateral prefrontal cortex, and caudate nucleus while performing visuospatial working memory tasks compared to controls (473).

The link between TS and neuropsychological impairment has been studied and debated for more than 40 years. TS was formerly thought to be strongly linked to MR (474–476). However, the early studies were confounded by ascertainment bias. As more patients were identified, it became clear that the effect of the disorder on cognitive functioning usually is milder and more specific. The most consistent finding is for individuals with the disorder to exhibit significant deficits in visuospatial and visuoperceptual processing. A large discrepancy between VIQ and PIQ is typical, with verbal abilities in the normal range and visuospatial abilities significantly below the average range (477,478). This has been interpreted as a visuospatial deficit or space-form blindness (479). Expanding upon this idea, subsequent research indicated that many specific functions related to spatial thinking, including left-right orientation (480), mental rotation (481–483), visuoperceptual and visual-conceptual skills (484), drawing (485), arithmetic (486–488), facial processing (480,481), and handwriting (489), are impaired in many individuals with TS. Two studies suggest that the working memory system associated with the spatial vision pathway is impaired relative to the working memory system associated with the object recognition vision pathway (480,481). Cumulatively, these findings suggest that cognitive deficits in TS may result from right hemisphere dysfunction (490), especially in the right parietal lobe (491), hypotheses consistent with neuroimaging findings.

The neuropsychological deficits of individuals with TS are not limited to the well-documented impairment in spatial thinking. Rovet (465) found that both VIQ and PIQ were significantly lowered in girls with TS, although the deficit in PIQ was more substantial. Specific verbal difficulties may account for the lowered VIQ. For example, a group of school-aged girls with TS exhibited significantly impaired auditory short-term memory and verbal fluency compared to typically developing control subjects (492). Others have reported similar difficulties on verbal tasks (480,483,493). Motor impairments and impairments in executive functioning have been observed (494,495).

The degree to which estrogen replacement therapy may enhance cognitive function in TS has received little attention. One study found superior motor functions among TS patients receiving estrogen supplementation (496). Unfortunately, cognitive deficits in the areas of spatial-perceptual skills, visuomotor integration, affect recognition, visual memory, attention, and executive functioning were unaffected.

Cognitive ability seems to be related to karyotype. Individuals with TS who have the cytogenetic variant known as the small X ring chromosome may be at increased risk for MR (497), and girls with complete X monosomy (45,X) have a tendency toward lower levels of overall cognitive functioning than do girls with a mosaic karyotype (498).

Girls with TS may be at risk for a variety of educational problems, but difficulties with arithmetic are the most consistent finding. For example, Rovet (499) found that school-aged girls with TS tended to perform one to two grade levels below their actual grade level on a test of basic arithmetic skills, even though 27% of the girls had repeated at least one grade. The educational challenges faced by some individuals with TS relate logically to their visuospatial impairment, difficulties with attention and working memory, and slower processing speed (498). Educators should be made aware of these issues, and special education and tutoring should be implemented for those girls who are having significant difficulties.

Some early studies suggested that TS might increase risk of disorders such as depression and anorexia nervosa (500–503), but it now appears that serious psychopathology is present in a relatively low proportion of women with the disorder. However, difficulties with psychological adaptation and specific personality styles that vary with developmental level have been observed. In girls 9 to 14 years old, immaturity and muted affect (504), as well as increased activity levels, difficulty with concentration, and distractibility, are common (505–507). As adolescence approaches, low self-esteem, anxiety, depression, and social withdrawal are more frequently observed (508,509). Adolescents with TS may have difficulty with social interaction and have fewer friends than their peers (510). Adult women with TS may be viewed as more mature, but they may be less active in intimate relationships than their peers and have difficulty understanding the subtler aspects of social interaction (496).

Some believe the psychosocial difficulties found in TS to result from neurodevelopmental factors (511,512), but the physical anomalies associated with TS suggest another interpretation. The short stature and other physical manifestations of the disorder may affect how others interact with those with TS, which in turn may negatively influence self-esteem and maturity levels (513). It is important that individuals with TS have a strong support system through family, friends, educators, and organizations such as the Turner Syndrome Society.

Cornelia de Lange Syndrome

Cornelia de Lange syndrome (CdLS, also known as Brachmann-de Lange syndrome), first described in 1933, is relatively uncommon. One population-based study estimates incidence as one in 50,000 live births (514), although it has been suggested that the actual incidence may be as high as one in 10,000 (515). The large majority of cases are sporadic, but a number of cases of familial CdLS have been reported. Controversy remains as to the mode of genetic transmission, but most studies suggest that it is an autosomal dominant trait (516). In the absence of an affected parent, risk of recurrence is low, probably less than 1% (515).

The diagnosis is made based upon physical stigmata. Thin lips, crescent-shaped mouth (with the corners down), elongated philtrum, anteverted nostrils, and well-defined, high-arched eyebrows with a "penciled on" look are the features most useful in distinguishing CdLS from other developmental disorders (517). Hirsutism (including confluent eyebrows, or synophrys, and long eyelashes) is typical but less specific to CdLS (517). Low-set ears are part of the traditional CdLS profile, but their frequency is debatable (515,517). Some of these facial features may disappear with maturation, at least among males (517). Limb abnormalities are common. Small hands and/or feet with short digits are typical; in about 25% of cases gross malformations of the upper limbs are present (515,517). Low birth weight is typical; the average is approximately 5 pounds. Physical growth continues to be slow postnatally.

Although there is some controversy on this point (518), most authorities agree that CdLS manifests in two forms: the "classic" CdLS phenotype and a "mild" type (519–523). Sometimes these are referred to as type I and type II, respectively; cases presenting with CdLS features due to other factors (e.g., fetal hydantoin syndrome) may be termed type III (523). Classic CdLS is associated with lower birth weight, slower postnatal growth, higher frequency of serious malformations, and severe psychomotor problems and MR.

The mild phenotype presents with similar facial features, but to a lesser degree, with only minor malformations (e.g., small hands), and somewhat better psychomotor control and intellectual functioning. On the other hand, the "mild" cases may, by virtue of their greater capacity for self-directed action, pose more serious behavioral problems.

Intellectual functioning varies but is generally in the retarded range. At one time it was believed that severe MR was the rule, but as clinicians began recognizing the milder phenotype it gradually became apparent that many persons with CdLS are much higher functioning. Cases of normal-range IQ began to be reported in the literature (524–526), and it now appears that up to 10% of patients may not be in the retarded range (527). Still, serious intellectual deficits are the rule; the average IQ is in the low to mid 50s, on the cusp between mild and moderate MR (528). Whether because of an increased probability of diagnosis of mild cases or improved psychosocial services to children with developmental disabilities, IQ tends to be higher in patients born after 1980 (528). Language development is more severely impaired than are perceptual organization or visuospatial memory (528). Within the linguistic domain, expressive language lags considerably behind receptive language (528,529), and syntax is less well developed than vocabulary (529). Only half of patients speak in sentences by age 4, and up to one third lack all spoken language except perhaps for 1 or 2 words (529). Even those with relatively large vocabularies use speech less than other children (529).

Better intellectual functioning, including language skills, is predicted by birth weight greater than 5.5 pounds and the absence of severe upper limb malformations [i.e., the milder phenotype (528)]. Better language development also is predicted by social relatedness and the absence of hearing impairment (529).

Classic descriptions of the CdLS behavioral phenotype stressed the high incidence of autistic features and SIB (530,531). Studies concur that SIBs occur in at least half of CdLS patients and are a severe problem for up to one fourth of patients. Daily acts of aggression also characterize roughly half of all patients, and symptoms of hyperactivity are common (527). As many as half display the combination of social deficits, communication deficits, and restricted range of behavior associated with PDDs (527). High frequencies of gastrointestinal problems (e.g., difficulty swallowing) and sleep disturbance are observed. Autistic behaviors, hyperactivity, and self-injury are distinctly more common among patients with the classic CdLS phenotype; to a lesser degree, this applies to disturbances of sleep and feeding as well. Aggression, however, is almost as common among those with the mild phenotype (527).

Cri du Chat Syndrome

Cri du chat syndrome (also known as Lejeune syndrome or 5p− syndrome), first described in 1963, is best known for the high-pitched (catlike) cry of affected infants. Poor muscle tone and weak feeding are common in infancy, as are certain craniofacial stigmata: microcephaly, micrognathia, a round face with wide-set eyes, and epicanthic folds. (As the child grows, the face becomes elongated and the epicanthic folds become less prominent.) Medical complications include frequent middle ear infections, upper respiratory tract infections, and digestive disturbances (532). Intellectual functioning typically is in the retarded range. Cri du chat syndrome is not common; it occurs in approximately one in 50,000 live births (533). The female-to-male ratio is approximately 2:1. The disorder is associated with a partial deletion (rarely, a translocation) of the fifth chromosome. Band 5p15.2 appears to be the critical locus, except that the distinctive cry appears to be related to band 5p5.3 (534). Early reports (based on institutionalized patients) suggested that cri du chat syndrome was almost always associated with a total lack of expressive speech, grossly retarded motor development (e.g., inability to walk), and extreme social withdrawal. As is commonly the case, subsequent research on more representative samples of patients revealed much more heterogeneity and evidence for the beneficial effects of early intervention upon both motor and language development (532). A majority of the patients can walk, although often less steadily than most children. Gross motor control usually is adequate, although fine motor control often is impaired. At least one fourth of patients use some speech. Excessive friendliness with strangers is at least as much of a problem as social withdrawal, although eye contact tends to be poor (535). Socially withdrawn behavior is more pronounced in cases caused by translocation, as opposed to deletion (536).

Individuals with cri du chat syndrome often are described as happy, even excessively so, but simultaneously as being irritable and subject to outbursts of anger (535). Level of irritability and aggression may not be more severe than among comparably retarded individuals with other disorders (537), but oppositionality and/or aggressive behavior are problematic in a majority of cases (535,536). SIB is also a problem in the clear majority of cases (535,536). Stereotypic behavior is another common problem (535,536), especially when IQ and/or adaptive behavior scores are low (536), and may be more characteristic of cri du chat syndrome than of similarly retarded individuals with different etiologies (537). Hyperactivity is an outstanding issue in most (536,537) but not all (535) samples studied. Psychostimulants have been used, but they may be less effective in cases where stereotyped behaviors are present (536).

Cri du chat syndrome almost always is associated with MR, usually in the moderate range; mean IQ is about 48 (538,539). There is no characteristic pattern of strengths and weaknesses, although this may be due to the relative insensitivity of most tests to fine gradations of ability at such low levels of intellectual functioning. Language development is particularly delayed, but receptive language tends to be superior to expressive language. By age 8 or 9 years, the average child will understand spoken language at about a

4-year-old level but will speak at less than a 2-year-old level (538–540). About half of all cases may present with echolalia. When present, speech is characterized by poor articulation (535). The weak vocalization characteristic of the syndrome in infancy may be a contributing factor to the extreme delay of speech; if so, instruction in sign language might be of some value. Larger chromosome 5 deletions are associated with lower IQ (532). Cases with only a 5p15.3 deletion display far less intellectual impairment (all in the average or borderline range of IQ), with VIQ consistently lower than PIQ, receptive language at or near age level, and distinctly delayed expressive language use (539).

Lesch-Nyhan Syndrome

Lesch-Nyhan syndrome was first described in 1964. With an estimated incidence of one in 380,000 live births, Lesch-Nyhan syndrome is exceedingly rare but has devastating psychiatric (as well as biomedical) consequences. It is an X-linked recessive trait; only two female patients have been described. The basic defect is deficiency of the enzyme hypoxanthine-guanine phosphoribosyltransferase (HPRT), the normal functions of which are to catalyze the conversion of hypoxanthine to inosinic acid and of guanine to guanylic acid. In its absence, uric acid is overproduced and accumulates to toxic levels (541). Degree of HPRT deficiency correlates with disease severity (542). The first sign of the disorder is often a crystallized deposit in the urine ("orange sand") (543). Physical manifestations of the disorder resemble those of gout and may include gouty arthritis and renal dysfunction. Renal failure is a common cause of death. Patients formerly died by age 6, but treatment with allopurinol has proved effective in controlling uric acid levels, and survival into early adulthood now is common (544). Neurologic manifestations include delayed development of motor skills plus some combination of choreiform movements (e.g., shrugging, grimacing), athetoid movements (e.g., arm and finger writhing), severe spasticity, and dysarthric speech. Behavioral manifestations include aggression, compulsive behaviors, and (most famously) SIB (541).

SIB typically begins at about age 2. Its most characteristic forms are self-biting (of the lips, tongue, and/or fingers) and head banging. Nose gouging and eye gouging also have been reported (543,545). As motor and cognitive abilities develop, children with the disorder discover new means by which to injure themselves. These injuries can be severe; for example, portions of the lips or fingers may be severed completely. SIB in Lesch-Nyhan syndrome is unique in quality. It rarely has an instrumental purpose beyond the actual infliction of painful injury. It does not appear to be associated with self-stimulation or endorphin release; rather, it is triggered by all sorts of stressful situations (including the discomfort attendant upon the disease itself). It is perceived by the patients themselves as unpleasant and undesirable. Aware of their behavior, they nonetheless perceive SIB as

involuntary (545). Treatment is problematic, although SIB sometimes diminishes spontaneously after age 10 (544). Traditional behavior modification produces modest, but not dramatic, improvement in many cases (546–548). Medications (e.g., risperidone) (549) have been tried, and animal research suggests that dopamine (especially D1) antagonists are promising (550). Punishing SIB via electrical shock has been reported to *increase* its frequency. An increased sense of personal control achieved by reducing perceived stress tends to decrease the frequency of SIB. Parents may allow their children to participate in decision making (e.g., choice of dinner menu) and teachers may attempt to present these students with tasks that will be viewed as moderately challenging but probably achievable. Another approach involves prevention of SIB by physical means. This is supported by the fact that Lesch-Nyhan sufferers do not desire to engage in SIB and many of these patients are able to anticipate onset of SIB and implement (or request) protective measures (545). For example, several case studies reported success using lower lip guards to prevent self-biting (551,552). In some intractable cases, tooth extraction has been implemented.

Aggressive behavior is a serious problem in Lesch-Nyhan syndrome. This seems to be closely related to SIB, elicited by perceived stress and experienced as involuntary and unpleasant by the patients themselves. Apparently sincere apologies often follow outbursts of aggression, but they do not reduce the probability of further attacks. As cognitive abilities develop, verbal aggression also manifests (e.g., criticism of people's appearance or behavior).

Perhaps because of the specific association of Lesch-Nyhan syndrome with aggression and self-injury, there has been considerable interest in identifying the pathways by which defective purine metabolism converts into behavioral dysfunction. Most research centers on abnormal development of dopamine systems, with attention also paid to abnormalities in norepinephrine and serotonin systems. For example, the motor disturbances include extrapyramidal symptoms consistent with involvement of the dopamine-rich basal ganglia. Rats whose nigrostriatal tracts are lesioned with 6-hydroxydopamine (6-OHDA) develop spasticity and engage in self-biting behavior reminiscent of Lesch-Nyhan syndrome (553,554). Dopamine (especially D1) receptor antagonists reduce this behavior (553), whereas dopamine agonists such as L-dopa or amphetamines exacerbate it (555). Most patients display reduced CSF levels of homovanillic acid (HVA) and 3-methoxy-4-hydroxyphenylglycol (MHPG), principal metabolites of dopamine and norepinephrine, respectively. Some also show elevated levels of 5-HIAA, the major metabolite of serotonin (556). MRI findings suggest substantial reductions in caudate and (to a lesser degree) putamen volume (557). Reduced dopamine receptor density in the basal ganglia also is likely. Two positron emission tomographic studies showed dramatic reductions of dopamine binding activity in the caudate and putamen (558,559); one of these studies also found substantial reductions in the frontal cortex (558). Neuropathology findings also suggest that it is in the basal ganglia, rather than the substantia nigra, that dopaminergic systems are disrupted (560). In sum, evidence for developmental disruption of the nigrostriatal (and perhaps the mesocortical) dopamine systems, and for a link between that disruption and the behavioral manifestations of Lesch-Nyhan syndrome, is reasonably consistent. The dopamine systems of the basal ganglia, in turn, appear to be closely associated with serotonin systems, which also may play a role in the SIB component of Lesch-Nyhan syndrome (561).

Few studies have examined cognitive functioning in Lesch-Nyhan syndrome, but the findings of those studies suggest that, contrary to early belief, the disorder is not necessarily associated with severe intellectual disability. This was first suggested by the results of a questionnaire distributed to parents of children with Lesch-Nyhan syndrome (562). Nearly all patients older than age 5 showed awareness of time and place (e.g., watching the clock for preferred television programs, offering directions when in the family car), memory skills (e.g., remembering facts about their friends, following sports statistics), comprehension (e.g., following the plots of television programs or the course of an athletic competition), and used language as a principal mode of communication. As these illustrative examples suggest, the children tended to be socially engaged, enjoying interactions with friends and family despite the intrusion of disease-related behaviors. They also tended to exhibit insight into their illness and awareness of impending SIB. Few, however, were functioning at grade level academically.

A study of seven hospitalized children largely confirmed these impressionistic findings (563). Scores on the various SB-IV domains ranged from moderately mentally retarded to low average. Six of these children were retested 4 years later, at which time they were an average age close to 18 years (564). They continued to make gains, but slowly, so standardized scores declined. Incidentally, it is possible that most test scores underestimate the cognitive abilities of these patients, whose spasticity and dysarthric speech render it difficult for them to respond properly.

Maple Syrup Urine Disease

Maple syrup urine disease (MSUD) is a disorder of amino acid metabolism. Deficiency of the branched chain α-keto acid dehydrogenase complex impedes metabolism of branched chain amino acids, leading to potentially toxic accumulations of these substances (e.g., leucine) and of branched chain α-keto acids (565). Recurrent severe episodes of ketoacidosis typically begin in early infancy and result in MR, neurologic symptoms, and often death (566). During acute crises, patients often become comatose and seizures are common (567). A milder variant, with only intermittent metabolic disruption, has been identified (568). The characteristic odor of the urine is that of sotolone, a flavor com-

pound found in maple syrup, fenugreek, and lovage (569). The disorder, which is transmitted as an autosomal recessive trait, is rare, occurring in approximately one in 200,000 live births (570). Neonatal screening has been implemented sporadically since the 1960's (570–573). Unfortunately, the intermittent variant often is missed by these screening tests (570).

The initial crisis (which typically occurs within weeks of birth) is treated aggressively, using peritoneal dialysis or continuous venovenous hemodialysis to reduce plasma leucine levels (574,575). Subsequent crises are treated in a similar fashion. Nasogastric administration of an amino acid solution devoid of leucine and other branched chain amino acids may accomplish the same effect, provided the needs for fluids and calories are met intravenously (576). Maintenance treatment for MSUD involves dietary management, with restricted intake of leucine, isoleucine, and valine (577). Thiamine supplementation may help to reduce plasma concentrations of these amino acids (578,579). Liver transplantation has successfully eliminated the MSUD defect in several cases, but it is considered too expensive and risky to replace dietary management (580).

The few brain scan and neuropathologic studies of MSUD have obtained generally consistent findings. The basic neurochemical defect may involve inadequate synthesis of glutamate and γ-aminobutyric acid (567). Computed axial tomographic scan, MRI, and cranial sonography reveal abnormalities of the white matter and pallidum, which may resolve after treatment (581–583). Postmortem studies also identify white matter pathology; abnormal neuronal orientation and dendritic branching have been observed, suggesting that disruption begins toward the end of the neuronal migration period (584). Abnormal EEG findings are typical in MSUD, even when the disorder is under control, and are associated with leucine levels (567).

Few neuropsychological investigations of MSUD have been undertaken. One study of 22 children found a mean (nonverbal) IQ of 74; individual scores ranged from 50 to 103 (585). Quality of metabolic control in infancy and after was correlated with intellectual outcome. Consistent with this, another study (n = 16) reported a mean IQ of 78 and noted a bimodal distribution of IQ scores (586). Those with normal-range IQ were diagnosed, on average, at age 3.5 days, whereas those with IQ in the retarded range were diagnosed, on average, at age 10 days. Again, quality of metabolic control over the long term was correlated with IQ. When the disorder is diagnosed early (preferably within 3 days of the emergence of symptoms) and managed successfully, IQ should be in the normal range and the child should be able to attend regular school (587,588). Little has been done to describe the cognitive phenotype, but verbal abilities may be relatively strong, with perceptual organization and visuomotor abilities more impaired (589). We are not aware of any systematic studies investigating behavioral or psychiatric problems in children or adults with MSUD.

Rubinstein-Taybi Syndrome

First described in 1962, this syndrome presents with short stature, MR, microcephaly, and broad thumbs and first toes (590). Prevalence may be approximately one in 125,000 live births (591). The disorder has been linked to microscopic deletions on the short arm of chromosome 16 (592), but such deletions account for only 12% of diagnosed cases (593). Patients often display characteristic facial features such as a prominent and beaked nose, downward slanting palpebral fissures, and a small mouth with a "pouting lower lip" (594). Medical concerns include congenital heart defects, ocular problems, dental problems, urinary tract infections, ear infections, palate abnormalities, and constipation.

Early reports of cognitive functioning indicated mean IQ scores in the range from 30 to 50, but this research suffered from ascertainment bias, having been conducted on institution-dwelling individuals (595). More recent research has found IQ scores ranging from 30 to 79, with a mean of 51 (596). VIQ may tend to be lower than PIQ (597). Delayed speech acquisition is almost universal (596). Although articulation and intelligibility usually are described as good, receptive skills are better than expressive skills, and pragmatic use of language is relatively strong (597). IQ, especially VIQ, may decline with increasing age (597). Note, however, that these findings are based on limited research and should be treated with caution.

The behavioral phenotype of Rubinstein-Taybi syndrome has been little studied. Anecdotally, these patients are described as loving, friendly, and happy, and they are likely to be interested in music (596,597). One study reported relatively good adaptive functioning, but short attention spans are typical. Most engage in self-stimulating behaviors, and up to half may exhibit heightened sensory sensitivity, especially auditory (596). In a study using the CBCL, parents of individuals with Rubinstein-Taybi syndrome expressed concern with concentration, attention seeking, impulsivity, and social withdrawal (597).

Rubinstein-Taybi syndrome may entail increased risk of mood disorders, tic/OCD spectrum disorders, and/or neuroleptic-induced movement disorders (598). Absolute level of risk for psychiatric disorders remains unknown, but when individuals with Rubinstein-Taybi syndrome develop such disorders, they are likely to fall into these categories.

Mucopolysaccharidoses: Hurler, Hunter, and Sanfilippo Syndromes

Hurler syndrome (occasionally known by the unfortunate name *gargoylism*) is one of several lysosomal storage disorders (mucopolysaccharidoses) associated with severe physical disability and deteriorating cognitive functioning and has been designated MPS I. Hunter syndrome (MPS II) and Sanfilippo syndrome (MPS III) also fall within this category. Hurler syndrome is an autosomal recessive disorder; the gene,

situated on chromosome 4 (599), results in a deficiency of an enzyme (iduronidase) and ultimately in toxic accumulation of GAGs and other substances throughout the body. Incidence may be close to one in 100,000 live births (600,601). The disorder can be recognized during the first year of life (602), although often it is not diagnosed until the second year, as the child's appearance grows more obviously abnormal. Multiple physical handicaps develop and frequently are lethal during childhood. Progressive skeletal abnormalities impair motor control, corneal clouding impairs vision, and fluid buildup in the middle ear produces a loss of auditory acuity that adversely affects language development (603). Intelligence as measured by the Bayley scales typically declines steeply during the second and third years of life, resulting in MR (604,605). Hydrocephalus, also common, can produce further cognitive impairment. On the positive side, children with Hurler syndrome (unlike those with the related disorders discussed later) are generally regarded as emotionally well adjusted, socially engaged, and pleasant, at least until the disorder progresses towards its end stage (603). A mild form, sometimes called Scheie syndrome, also exists: onset is later, physical impairment less pronounced, and cognitive deficits relatively mild (606,607). Patients with the more severe disorder (sometimes termed MPS IH) typically have died in childhood, whereas those with the mild variant (MPS IS), despite suffering from joint stiffness, corneal clouding, and aortic valve disorders, have lived out ordinary lives. Cases of intermediate severity sometimes are called Hurler-Scheie syndrome (MPS I H/S).

Bone marrow transplant, preferably before age 2, is associated with both physical and intellectual improvement, virtually a resumption of normal cognitive development except for mild attentional impairment (604,605,608–610). Skeletal deformities are not resolved by bone marrow transplant, and further surgical interventions may be needed (611,612). Enzyme replacement therapy (intravenous α-L-iduronidase) has shown promise in canines (613) and may soon be attempted with humans (614).

Hunter syndrome (MPS II) is an X-linked autosomal recessive disorder. The gene, situated at Xq28, causes deficient production of the enzyme iduronate sulfatase. As with Hurler syndrome, this leads to toxic accumulation of GAGs in multiple organ systems, including the CNS. Incidence is less than one in 100,000 (600,601) except in Israel, where the disorder is several times more common (615). There are two forms of the disorder: the more severe typically is associated with progressive MR, severe physical disability, and death during childhood or early adolescence, whereas the milder variant is associated with survival well into adulthood and far less intellectual impairment. Physically, Hunter syndrome closely resembles Hurler syndrome, except that it progresses more slowly and does not usually involve corneal clouding. Cognitively, too, the two disorders are similar, although there may be a specific language impairment in Hunter syndrome (603). Behaviorally, children with Hunter syndrome are more likely to display hyperactivity and/or aggression (616). Unfortunately, bone marrow transplant has only limited efficacy in Hunter syndrome (617).

Sanfilippo syndrome (MPS III) is an autosomal recessive disorder; the genes responsible for at least two of its four subtypes are located on chromosome 17. As with Hurler and Hunter syndromes, the proximal cause of intellectual deterioration (and the assorted physical manifestations) of the disorder is a toxic accumulation of GAGs secondary to enzyme deficiency. In Sanfilippo syndrome, this may involve any one of four enzymes. Although the four variants follow somewhat different trajectories, clinical presentation is similar. The first clear sign is macrocephaly with onset during the second year of life; enlargement of the liver develops concurrently (603). Diagnosis is possible at this time, but on average children are 6 years old before the disorder is recognized (618). Behavioral manifestations emerge at about age 5 (603). Memory problems and declining language skills come first, followed by symptoms resembling ADHD (but not responsive to psychostimulants) along with oppositionality and aggression (603,616). Bone marrow transplant may improve physical symptoms (which are less severe than those of Hurler or Hunter syndromes), but it does not produce intellectual or behavioral improvement (619).

CONCLUSIONS

It is clearly a misconception to treat MR, whatever its level of severity, as a homogeneous entity. Individual differences in specific neurocognitive abilities and personality traits are as important among the mentally retarded as among the remainder of the population; their evaluation is, if anything, more important. Children of average to above average ability can compensate for specific weaknesses, but optimization of the social and educational environment may be critically important for those with significantly subaverage intellectual ability. These considerations also apply to the learning disabled population, members of which are at elevated risk of school failure and behavioral and social problems that may culminate in delinquency, but who also possess significant intellectual strengths on which individualized educational programs can (and should) capitalize.

Granted that the majority of the MR population consists of idiopathic cases not inherently linked to medical issues, enough cases are caused by identifiable physical disorders to warrant careful investigation. A few of these syndromes (e.g., PKU) are treatable; a few (e.g., FAS) are, at least in principle, preventable; many are associated with specific biomedical and psychological issues. Knowing the diagnosis facilitates targeted assessment; knowing the syndrome (including any craniofacial anomalies, typical cognitive profiles) facilitates making the diagnosis. As stated earlier, neuropsychiatrists and neuropsychologists are unusually well situated to bring together the diverse strands of evidence and coordinate the

efforts of those providing medical care, psychosocial support, and educational stimulation to these individuals. Proper evaluation and clinical management requires wide-ranging knowledge of general medicine, neurocognitive functions, and adaptive behavior, as well as the interventions appropriate to each domain. One ought to be equally comfortable consulting with pediatricians, occupational therapists, special educators, parents, and a wide variety of other personnel.

Providing care to the mentally retarded is sometimes regarded as unrewarding, not likely to offer either the gratification of positive outcomes or high levels of intellectual stimulation. As a subject of research, it has even been referred to as psychiatry's "Cinderella" (620). It is true that many of the more severely retarded face limited prospects for adaptive functioning and that some of the metabolic disorders associated with MR reduce quality of life and shorten life expectancy. Still, functional outcome and quality of life are quite variable, especially for those with mild to moderate levels of MR, and appropriate medical, educational, and psychosocial interventions can make a tremendous difference. Again, this is also true for those suffering from milder, more specific cognitive limitations such as learning disorders, a category that includes many patients diagnosed with biomedical conditions (e.g., neurofibromatosis). The lives of family members and other care providers may be dramatically affected by apparently minor improvements in the behavior or health of their charges. Although the volume of research into MR remained fairly stable in the last quarter of the twentieth century, the advances made in this area are sufficiently important to lead one set of reviewers to suggest that this Cinderella is now becoming the "belle of the biopsychosocial ball" (621,622). It probably is safe to say that a majority of the psychiatric disorders identified in the last century have been in the area of developmental neuropsychiatry. Furthermore, several of these have turned out to be surprisingly common: consider the cases of FAS and fragile X syndrome. There almost certainly exist more such disorders awaiting the keen eye of an astute clinician. In short, work in this area offers the potential for alleviation of severe human suffering, intellectual challenge, and extension of our knowledge of the genetic, neurologic, endocrinologic, and psychological nature of the human being.

REFERENCES

1. Grossman HJ. *Manual of terminology and classification on mental retardation.* Washington, DC: American Association on Mental Deficiency, 1983.
2. Luckasson R, Coulter D, Poloway EA. *Mental retardation: definition, classification, and systems of support,* 9th ed. Washington, DC: American Association on Mental Retardation, 1992.
3. American Psychiatric Association. *Diagnostic and statistical manual of mental disorders, fourth edition (DSM-IV).* Washington, DC: Author, 1994.
4. MacMillan DL, Gresham FM, Siperstein GN. Conceptual and

psychometric concerns about the 1992 AAMR definition of mental retardation. *Am J Ment Retard* 1993;98:325–335.

5. MacMillan DL, Gresham FM, Siperstein GN. Heightened concerns over the 1992 AAMR definition: advocacy versus precision. *Am J Ment Retard* 1994;100:87–95.

6. Rourke BP. *Nonverbal learning disabilities: the syndrome and the model.* New York: Guilford, 1989.

7. Wechsler D. *Wechsler intelligence scale for children-third edition.* San Antonio, TX: The Psychological Corporation, 1991.

8. Thorndike RL, Hagen EP, Sattler JM. *Technical manual for the Stanford-Binet intelligence scale, fourth edition.* Chicago: Riverside, 1986.

9. Kaufman AS, Kaufman NL. *Administration and scoring manual for the Kaufman Assessment Battery for Children.* Circle Pines, MN: American Guidance Service, 1983.

10. Kaufman AS, Kaufman NL. *Interpretive manual for the Kaufman Assessment Battery for Children.* Circle Pines, MN: American Guidance Service, 1983.

11. Elliott CD. *Differential ability scales: introductory and technical handbook.* San Antonio, TX: The Psychological Corporation, 1990.

12. Flynn JR. Massive IQ tests in 14 nations: what IQ tests really measure. *Psychol Bull* 1987;101:171–191.

13. Neisser U, ed. *The rising curve.* Washington, DC: American Psychological Association, 1998.

14. Wechsler D. *Wechsler Adult Intelligence Scale-third edition (WAIS-III).* San Antonio, TX: The Psychological Corporation, 1997.

15. Elliott CD, Murray DJ, Pearson LS. *British ability scales.* Windsor, UK: National Foundation for Educational Research, 1979.

16. Bayley N. *Bayley scales of infant development, second edition.* San Antonio, TX: The Psychological Corporation, 1993.

17. Mullen E. *Mullen Scales of Early Learning: AGS edition.* Circle Pines, MN: American Guidance Service, 1995.

18. McCarthy D. *McCarthy scales of children's abilities.* San Antonio, TX: The Psychological Corporation, 1972.

19. Wechsler D. *Wechsler preschool and primary scales of intelligence-revised.* San Antonio, TX: The Psychological Corporation, 1989.

20. Sparrow SS, Balla DA, Cicchetti DV. *Vineland adaptive behavior scales.* Circle Pines, MN, American Guidance Service, 1984.

21. Lambert NM, Windmiller MB, Tharinger D, et al. *AAMR Adaptive Behavior Scale: school version.* Washington, DC: American Association on Mental Retardation, 1981.

22. Wilkinson GS. *Wide Range Achievement Test-third edition: administration manual.* Wilmington, DE: Wide Range, 1993.

23. The Psychological Corporation. *Wechsler Individual Achievement Test-third edition (WIAT-III).* San Antonio, TX: Author.

24. Markwardt FC. *Peabody Individual Achievement Test-revised.* Circle Pines, MN: American Guidance Service, 1989.

25. Kaufman AS, Kaufman NL. *Kaufman Test of Educational Achievement (K-TEA).* Circle Pines. MN: AGS Publishing, 1989.

26. Woodcock RW, Johnson MB. *Woodcock-Johnson Psychoeducational Battery-revised.* Allen, TX, DLM Teaching Resources, 1989.

27. Lezak M. *Neuropsychological assessment, third edition.* New York: Oxford University Press, 1995.

28. Hynd GW, Willis WG. *Pediatric neuropsychology.* Orlando, FL: Grune & Stratton, 1988.

29. Teeter P, Semrud-Clikeman M. *Child neuropsychology: assessment and interventions for neurodevelopmental disorders.* Needham Heights, MA: Allyn & Bacon, 1997.

30. Yeates KO, Ris MD, Yalor HG. *Pediatric neuropsychology: research, theory, and practice.* New York: Guilford, 2000.

31. Money J. *The Kaspar Hauser syndrome of psychosocial dwarfism.* Amherst, NY: Prometheus Books, 1992.

32. Achenbach TM. *Manual for the Child Behavior Checklist/4-18 and 1991 Profile.* Burlington, VT: University of Vermont, Department of Psychiatry, 1991.

33. Wirt RD, Lachar D, Klinedinst JE. *Personality Inventory for Children-revised.* Los Angeles, CA: Western Psychological Services, 1998.

34. Achenbach TM. *Manual for the Young Adult Self-Report and Young Adult Behavior Checklist.* Burlington, VT: University of Vermont, Department of Psychiatry, 1997.

35. Dorner G, Tonjes R, Hecht K, et al. Pyridostigmine administration in new-born rats prevents permanent mental ill-effects produced by maternal deprivation. *Endokrinologie* 1981;77:101–104.

36. Mondadori C. In search of the mechanism of action of the nootropics: new insights and potential clinical implications. *Life Sci* 1994;55:2171–2178.

37. Deberdt W. Interaction between psychological and pharmacological treatment in cognitive impairment. *Life Sci* 1994;55:2057–2066.

38. Mayes SD, Crites DL, Bixler EO, et al. Methylphenidate and ADHD: influence of age, IQ, and neurodevelopmental status. *Dev Med Child Neurol* 1994;36:1099–1107.

39. Schroeder SR, Schroeder CS, Smith B, et al. Prevalence of self-injurious behavior in a large state facility for the retarded: a three-year follow-up study. *J Autism Child Schizophr* 1978;8:261–269.

40. Fava M. Psychopharmacologic treatment of pathologic aggression. *Psychiatr Clin North Am* 1997;20:427–451.

41. Osman OT, Loschen EL. Self-injurious behavior in the developmentally disabled: pharmacologic treatment. *Psychopharmacol Bull* 1992;28:439–449.

42. Singh NN, Millichamp CJ. Pharmacological treatment of self-injurious behavior in mentally retarded persons. *J Autism Dev Disord* 1985;15:257–267.

43. Aman MG, Singh NN. The usefulness of thioridazine for treating childhood disorders—fact or folklore? *Am J Med Defic* 1980;84:331–338.

44. Baumeister AA, Todd ME, Sevin JA. Efficacy and specificity of pharmacological therapies for behavioral disorders in persons with mental retardation. *Clin Neuropharmacol* 1993;16:271–294.

45. Beckwith BE, Couk DI, Schumacher K. Failure of naloxone to reduce self-injurious behavior in two developmentally disabled females. *Appl Res Ment Retard* 1986;7:183–188.

46. Sandman CA, Datta PC, Barron J, et al. Naloxone attenuates self-injurious behavior in developmentally disabled clients. *Appl Res Ment Retard* 1983;4:5–11.

47. Kars H, Broekema W, Glaudemans-van Gelderen I, et al. Naltrexone attenuates self-injurious behavior in mentally retarded subjects. *Biol Psychiatry* 1990;27:741–746.

48. Sandman CA, Barron JL, Colman H. An orally administered opiate blocker, naltrexone, attenuates self-injurious behavior. *Am J Ment Retard* 1990;95:93–102.

49. Walters AS, Barrett RP, Feinstein C, et al. A care report of naltrexone treatment of self-injury and social withdrawal in autism. *J Autism Dev Disord* 1990;20:169–176.

50. Taylor DV, Hetrick WP, Neri CL, et al. Effect of naltrexone upon self-injurious behavior, learning, and activity: a case study. *Pharmacol Biochem Behav* 1991;40:79–82.

51. Buzan RD, Thomas M, Dubovsky SL, et al. The use of opiate antagonists for recurrent self-injurious behavior. *J Neuropsychiatry Clin Neurosci* 1995;7:437–444.

52. Buzan RD, Dubovsky SL, Treadway JT, et al. Opiate antagonists for recurrent self-injurious behavior in three mentally retarded adults. *Psychiatr Serv* 1995;46:511–512.

53. Casner JA, Weinheimer B, Gualtieri CT. Naltrexone and

self-injurious behavior: a retrospective population study. *J Clin Psychopharmacol* 1996;16:389–394.

54. Thompson T, Hackenburg T, Cerutti D, et al. Opioid antagonist effects on self-injury in adults with mental retardation: response form and location as determinants of medication effects. *Am J Ment Retard* 1994;99:85–102.

55. Benjamin S, Seek A, Tresise L, et al. Case study: paradoxical response to naltrexone treatment of self-injurious behaviors. *J Am Acad Child Adolesc Psychiatry* 1995;34:238–242.

56. Crews WD, Rhodes RD, Bonaventura SH, et al. Cessation of long-term naltrexone administration: longitudinal follow-up. *Rev Dev Disabil* 1999;20:23–30.

57. Willemsen-Swinkels SH, Buitelaar JK, Nijhof GJ, et al. Failure of naltrexone hydrochloride to reduce self-injurious and autistic behavior in mentally retarded adults: double-blind placebo-controlled studies. *Arch Gen Psychiatry* 1995;52:766–773.

58. Markowitz PI. Effect of fluoxetine on self-injurious behavior in the developmentally disabled: a preliminary study. *J Clin Psychopharmacol* 12:27–31.

59. Ricketts RW, Goza AB, Ellis CR, et al. Fluoxetine treatment of severe self-injury in young adults with mental retardation. *J Am Acad Child Adolesc Psychiatry* 1993;32:865–869.

60. Hellings JA, Kelley LA, Gabrielli WF, et al. Sertraline response in adults with mental retardation and autistic disorder. *J Clin Psychiatry* 1996;57:333–336.

61. Davanzo PA, Belin TR, Widawski MH, et al. Paroxetine treatment of aggression and self-injury in persons with mental retardation. *Am J Ment Retard* 1998;102:427–437.

62. Troisi A, Vicario E, Nuccetelli F, et al. Effects of fluoxetine on aggressive behavior of adult inpatients with mental retardation and epilepsy. *Pharmacopsychiatry* 1995;28:73–76.

63. Garber HJ, MocGonigle JJ, Slomka GT, et al. Clomipramine treatment of stereotypic behaviors and self-injury in patients with developmental disabilities. *J Am Acad Child Adolesc Psychiatry* 1992;31:1157–1160.

64. Lewis MH, Bodfish JW, Powell SB, et al. Clomipramine treatment for self-injurious behavior of individuals with mental retardation: a double-blind comparison with placebo. *Am J Ment Retard* 1996;100:654–665.

65. Ratey JJ, Sovner R, Mikkelsen E, et al. Buspirone therapy for maladaptive behavior and anxiety in developmentally disabled persons. *J Clin Psychiatry* 1989;50:382–384.

66. Ratey JJ, Sovner R, Parks A, et al. Buspirone treatment of aggression and anxiety in mentally retarded patients: a multiple-baseline, placebo lead-in study. *J Clin Psychiatry* 1991;52:159–162.

67. Ricketts RW, Goza AB, Ellis CR, et al. Clinical effects of buspirone on intractable self-injury in adults with mental retardation. *J Am Acad Child Adolesc Psychiatry* 1994;33:270–276.

68. Verhoeven WM, Tuinier S. The effect of buspirone on challenging behaviors in mentally retarded patients: an open prospective multiple-case study. *J Intellect Disabil Res* 1996;40:502–508.

69. Volavka J. Can aggressive behavior in humans be modified by beta blockers? *Postgrad Med* 1988;29:163–168.

70. Haspel T. Beta-blockers and the treatment of aggression. *Harv Rev Psychiatry* 1995;2:274–281.

71. Kastner T, Finesmith R, Walsh K. Long-term administration of valproic acid in the treatment of affective symptoms in people with mental retardation. *J Clin Psychopharmacol* 1993;13:448–451.

72. Ruedrich S, Swales TP, Fossaceca C, et al. Effect of divalproex sodium on aggression and self-injurious behaviour in adults with intellectual disability: a retrospective review. *J Intellect Disabil Res* 1999;43:105–111.

73. Langee HR. Retrospective study of lithium use for institutional-ized mentally retarded individuals with behavior disorders. *Am J Ment Retard* 1990;94:448–452.

74. Spreat S, Behar D, Reneski B, et al. Lithium carbonate for aggression in mentally retarded persons. *Comprehens Psychiatry* 1989;30:505–511.

75. Horrigan JP, Barnhill LJ. Risperidone and explosive aggressive autism. *J Autism Dev Disord* 1997;27:313–323.

76. Lott RS, Kerrick JM, Cohen SA. Clinical and economic aspects of risperidone treatment in adults with mental retardation and behavioral disturbance. *Psychopharmacol Bull* 1996;32:721–729.

77. McDougle CJ, Holmes JP, Bronson MR, et al. Risperidone treatment of children and adolescents with pervasive developmental disorders: a prospective open-label study. *J Am Acad Child Adolesc Psychiatry* 1997;36:685–693.

78. Cohen SA, Underwood MT. The use of clozapine in a mentally retarded and aggressive population. *J Clin Psychiatry* 1994;55:440–444.

79. Sandman CA, Hetrick W, Taylor DV, et al. Dissociation of POMC peptides after self-injury predicts responses to centrally acting opiate blockers. *Am J Ment Retard* 1997;102:182–199.

80a. Gaedt C. Psychotherapeutic approaches in the treatment of mental illness and behavioural disorders in mentally retarded people: the significance of a psychoanalytic perspective. *J Intellect Disabil Res* 1995;39:233–239.

80b. Johnston R, Kaslow NJ, Brooks A. Insight oriented psychotherapy with low IQ patients. *Psychother Bull* 1997;32:17–21.

81. Korenberg JR, Pulst S-M, Gerwehr S. Advances in the understanding of chromosome 21 and Down syndrome. In: Lott IT, McCoy EE, eds. *Down syndrome: advances in medical care.* New York: Wiley-Liss, 1992.

82. Pueschel SM. The person with Down syndrome: medical concerns and educational strategies. In: Lott IT, McCoy EB, eds. *Down syndrome: advances in medical care.* New York: Wiley-Liss, 1992.

83. Wisniewski KE, Hill AL, Wisniewski HM. Aging and Alzheimer's disease in people with Down syndrome. In: Lott IT, McCoy EB, eds. *Down syndrome: advances in medical care.* New York: Wiley-Liss, 1992.

84a. Hook EB, Cross PK, Jackson L, et al. Maternal age-specific rates of 47,+21 and other cytogenetic abnormalities diagnosed in the first trimester of pregnancy in chorionic villus biopsy specimens: comparison with rates expected from observations at amniocentesis. *Am J Hum Genet* 1988;42:797–807.

84b. Hook EB, Cross PK, Schreinenmachers DM. Chromosomal abnormality rates at amniocentesis and in live-born infants. *JAMA* 1983;249:2034–2038.

85. Dill FJ, Hayden MR, McGillivray B. *Genetics.* Baltimore, MD: Williams & Wilkins, 1992.

86. Zihni L. Raised paternal age and the occurrence of Down's syndrome. *Hist Psychiatry* 1994;5:71–88.

87. Erickson JD, Bjerkedal T. Down's syndrome associated with father's age in Norway. *J Med Genet* 1981;18:22–28.

88. Chapman RS. Language development in children and adolescents with Down syndrome. *Ment Retard Dev Disabil Res Rev* 1997;3:307–312.

89. Chapman RS, Heskith LJ. Behavioral phenotype of individuals with Down syndrome. *Ment Retard Dev Disabil Res Rev* 2000;6:84–95.

90. Hodapp RM. Down syndrome: developmental, psychiatric, and management issues. *Child Adolescent Psychiatr Clin N Am* 1996;5:881–894.

91. Miller JF. Development of speech and language in children with Down syndrome. In: Lott IT, McCoy EB, eds. *Down syndrome: advances in medical care.* New York: Wiley-Liss, 1992.

92. Stoel-Gammon C. Phonological development in Down syndrome. *Ment Retard Dev Disabil Res Rev* 1997;3:300–306.

93. Kumin L. Speech and language skills in children with Down syndrome. *Ment Retard Dev Disabil Res Rev* 1996;2:109–115.

94. Wang PP. A neuropsychological profile of Down syndrome: cognitive skills and brain morphology. *Ment Retard Dev Disabil Res Rev* 1996;2:102–108.

95. Uecker A, Mangan PA, Obrzut JE, et al. Down syndrome in neurobiological perspective: an emphasis on spatial cognition. *J Clin Child Psychol* 1993;22:266–276.

96. Craddock N, Owen M. Is there an inverse relationship between Down's syndrome and bipolar disorder? Literature review and genetic implications. *J Intellect Disabil Res* 1994;38:613–620.

97. Cooper S, Collacott R. Mania and Down's syndrome. *Br J Psychiatry* 1993;162:739–743.

98. Carr J. Annotation: long-term outcome for people with Down's syndrome. *J Child Psychol Psychiatry* 1994;35:425–439.

99. Nadel L. Learning and cognition in Down syndrome. In: Lott IT, McCoy EB, eds. *Down syndrome: advances in medical care.* New York: Wiley-Liss, 1992.

100. Cicchetti D, Beeghly M, eds. *Children with Down syndrome: a developmental perspective.* New York: Cambridge University Press, 1990.

101. Zigman W, Silverman W, Wisniewski HM. Aging and Alzheimer's disease in Down syndrome: clinical and pathological changes. *Ment Retard Dev Disabil Res Rev* 1996;2:73–79.

102. Dalton AK, Wisniewski HM. Down's syndrome and the dementia of Alzheimer's disease. *Int Rev Psychiatry* 1990;2:43–52.

103. McCreary BD. Functional regression in adults with Down's syndrome. *J Dev Disabil* 1992;1:34–39.

104. Prasher VP. Psychotic features and effect of severity of learning disability on dementia in adults with Down syndrome: review of the literature. *Br J Dev Disabil* 1997;43:85–92.

105. Schupf N, Kapell D, Nightingale B, et al. Earlier onset of Alzheimer's disease in men with Down syndrome. *Neurology* 1998;50:991–995.

106. Cosgrave MP, Tyrrell J, McCarron M, et al. Determinants of aggression, and adaptive and maladaptive behaviour in older people with Down's syndrome with and without dementia. *J Intellect Disabil Res* 1999;43:393–399.

107. Peterson B. Neuroimaging in child and adolescent neuropsychiatric disorders. *J Am Acad Child Adolesc Psychiatry* 1995;34:1560–1576.

108. Sherman S. Epidemiology. In: Hagerman RJ, Cronister A, eds. *Fragile X syndrome: diagnosis, treatment, and research,* 2nd ed. Baltimore, MD: Johns Hopkins University Press, 1996.

109. Lubs HA. A marker X chromosome. *Am J Hum Genet* 1969;21:231–244.

110. Hagerman RJ. Physical and behavioral phenotype. In: Hagerman RJ, Cronister A, eds. *Fragile X syndrome: diagnosis, treatment, and research,* 2nd ed. Baltimore, MD: Johns Hopkins University Press, 1996.

111. Hagerman RJ. Medical follow-up and pharmacotherapy. In: Hagerman RJ, Cronister A, eds. *Fragile X syndrome: diagnosis, treatment, and research,* 2nd ed. Baltimore, MD: Johns Hopkins University Press, 1996.

112. Sherman SL, Meadows KL, Ashley AE. Examination of factors that influence the expansion of the fragile X mutation in a sample of conceptuses from known carrier females. *Am J Med Genet* 1996;64:256–260.

113. Sherman S, Marton NE, Jacobs PA, et al. The marker (X) syndrome: a cytogenetic and genetic analysis. *Ann Hum Genet* 1984;48:21–37.

114. Pennington B. The neuropsychology of fragile X syndrome. In: Hagerman RJ, Cronister A, eds. *Fragile X syndrome: diagnosis,*

treatment, and research, 2nd ed. Baltimore, MD: Johns Hopkins University Press, 1996.

115. DeVries BB, Wiegers AM, DeGraaff E, et al. Mental status and fragile X expression in relation to FMR-1 gene mutation. *Eur J Hum Genet* 1993;1:72–79.

116. Merenstein SA, Sobesky WE, Taylor AK, et al. Molecular-clinical correlations in males with an expanded FMR1 mutation. *Am J Med Genet* 1996;64:388–394.

117. Rousseau F, Heitz D, Tarleton J, et al. A multicenter study on genotype-phenotype correlations in the fragile X syndrome, using direct diagnosis with probe StB12.3: the first 2,253 cases. *Am J Hum Genet* 1994;55:225–237.

118. Wright-Talamante C, Cheema A, Riddle JE, et al. A controlled study of longitudinal IQ changes in females and males with fragile X syndrome. *Am J Med Genet* 1996;64:350–355.

119. Cronister A, Hagerman RJ, Wittenberger M, et al. Mental impairment in cytogenetically positive fragile X females. *Am J Med Genet* 1991;38:503–504.

120. Sobesky WE. The treatment of emotional and behavioral problems. In: Hagerman RJ, Cronister A, eds. *Fragile X syndrome: diagnosis, treatment, and research, second edition.* Baltimore, MD: Johns Hopkins University Press, 1996.

121. Linden MG, Tassone F, Gane LW, et al. Compound heterozygous female with fragile X syndrome. *Am J Med Genet* 1999;83:318–321.

122. Jones KL, Smith DW. Recognition of the fetal alcohol syndrome in early infancy. *Lancet* 1973;2:999–1001.

123. Abel EL, Sokol RJ. A revised conservative estimate of the incidence of FAS and its economic impact. *Alcohol Clin Exp Res* 1991;15:514–524.

124. Johnson JL, Leff M. Children of substance abusers: overview of research findings. *Pediatrics* 1999;103:1085–1099.

125. Mattson SN, Riley EP. A review of the neurobehavioral deficits in children with fetal alcohol syndrome or prenatal exposure to alcohol. *Alcohol Clin Exp Res* 1998;22:279–294.

126. Streissguth AP, Barr HM, Sampson PD. Moderate prenatal alcohol exposure: effects on child IQ and learning problems at age 7½ years. *Alcohol Clin Exp Res* 1990;14:662–669.

127. Streissguth AP, Aase JM, Clarren SK, et al. Fetal alcohol syndrome in adolescents and adults. *JAMA* 1991;265:1961–1967.

128. Mattson SN, Riley EP. Parent ratings of behavior in children with heavy prenatal alcohol exposure and IQ-matched controls. *Alcohol Clin Exp Res* 2000;24:226–231.

129. Mattson SN, Riley EP, Gramling L, et al. Neuropsychological comparison of alcohol-exposed children with or without physical features of fetal alcohol syndrome. *Neuropsychology* 1998;12:146–153.

130. Mattson SN, Riley EP, Gramling L, et al. Heavy prenatal alcohol exposure with or without physical features of fetal alcohol syndrome leads to IQ deficits. *J Pediatr* 1997;131:718–721.

131. Larroque B, Kaminski M. Prenatal alcohol exposure and development at preschool age: main results of a French study. *Alcohol Clin Exp Res* 1998;22:295–303.

132. Goldschmidt L, Richardson G, Stoffer DS, et al. Prenatal alcohol exposure and academic achievement at age six: a nonlinear fit. *Alcohol Clin Exp Res* 1996;20:763–770.

133. Bonthius DJ, West JR. Alcohol-induced neuronal loss in developing rats: increased brain damage with binge exposure. *Alcohol Clin Exp Res* 1990;14:107–118.

134. Steinhausen HC, Spohr HL. Long-term outcome of children with fetal alcohol syndrome: psychopathology, behavior, and intelligence. *Alcohol Clin Exp Res* 1998;22:334–338.

135. Roebuck TM, Mattson SN, Riley EP. Behavioral and psychosocial profiles of alcohol-exposed children. *Alcohol Clin Exp Res* 1999;23:1070–1076.

136. Thomas SE, Kelly SJ, Mattson SN, et al. Comparison of social abilities of children with fetal alcohol syndrome to those of children with similar IQ scores and normal controls. *Alcohol Clin Exp Res* 1998;22:528–533.

137. Mattson SN, Goodman AM, Caine C, et al. Executive functioning in children with heavy prenatal alcohol exposure. *Alcohol Clin Exp Res* 1999;23:1808–1815.

138. Carmichael Olson H, Feldman JJ, Streissguth AP, et al. Neuropsychological deficits in adolescents with fetal alcohol syndrome: clinical findings. *Alcohol Clin Exp Res* 1998;22:1998–2012.

139. Kerns KA, Don A, Mateer CA, et al. Cognitive deficits in non-retarded adults with fetal alcohol syndrome. *J Learn Disabil* 1997;30:685–693.

140. Aronson M, Hagberg B. Neuropsychological disorders in children exposed to alcohol during pregnancy: a follow-up study of 24 children to alcoholic mothers in Goteborg, Sweden. *Alcohol Clin Exp Res* 1998;22:321–324.

141. Kyllerman M, Aronson M, Sabel K-G, et al. Children of alcoholic mothers: growth and motor performance compared to matched controls. *Acta Paediatr Scand* 1985;74:20–26.

142. Church MW, Kaltenbach JA. Hearing, speech, language, and vestibular disorders in the fetal alcohol syndrome: a literature review. *Alcohol Clin Exp Res* 1997;21:495–512.

143. Mattson SN, Riley EP, Delis DC, et al. Verbal learning and memory in children with fetal alcohol syndrome. *Alcohol Clin Exp Res* 1996;20:810–816.

144. Uecker A, Nadel L. Spatial locations gone awry: object and spatial memory deficits in children with fetal alcohol syndrome. *Neuropsychologia* 1996;34:209–223.

145. Uecker A, Nadel L. Spatial but not object memory impairments in children with fetal alcohol syndrome. *Am J Ment Retard* 1998;103:12–18.

146. Guerri C. Neuroanatomical and neurophysiological mechanisms involved in central nervous system dysfunctions induced by prenatal alcohol exposure. *Alcohol Clin Exp Res* 1998;22:304–312.

147. Mattson SN, Schoenfeld AM, Riley EP. Teratogenic effects of alcohol on brain and behavior. *Alcohol Res Health* 2001;25:185–191.

148. Kodituwakku PW, May PA, Clericozio CL, et al. Emotion-related learning in individuals prenatally exposed to alcohol: an investigation of the relation between set shifting, extinction of responses, and behavior. *Neuropsychologia* 2001;39:699–708.

149. Connor PD, Sampson PD, Bookstein FL, et al. Direct and indirect effects of prenatal alcohol damage on executive function. *Dev Neuropsychol* 2001;18:331–354.

150. Roebuck TM, Mattson SN, Riley EP. A review of the neuroanatomical findings in children with fetal alcohol syndrome or prenatal exposure to alcohol. *Alcohol Clin Exp Res* 1998;22:339–344.

151. Kotzot D, Bernasconi F, Brecevic L. Phenotype of the Williams Beuren syndrome associated with hemizygosity at the elastin locus. *Eur J Pediatr* 1995;154:477–482.

152. Mass E, Belostoky L. Craniofacial morphology of children with Williams syndrome. *Cleft Palate Craniofac J* 1993;30:343–349.

153. Holmstrom G, Almond G, Temple K, et al. The iris in Williams syndrome. *Arch Dis Child* 1990;65:987–989.

154. Lopez-Rangel E, Maurice M, McGillivray B, et al. Williams syndrome in adults. *Am J Med Genet* 1992;44:720–729.

155. Kim YM, Yoo SJ, Choi YJ, et al. Natural course of supravalvular aortic stenosis and peripheral pulmonary arterial stenosis in Williams syndrome. *Cardiol Young* 1999;9:37–41.

156. Bird LM, Billman GF, Lacro RV, et al. Sudden death in Williams syndrome. *J Pediatr* 1996;129:926–931.

157. Nicholson WR, Hockey KA. Williams syndrome: a clinical study of children and adults. *J Paediatr Child Health* 1993;29:468–472.

158. Partsch CJ, Dreyer G, Gosch A, et al. Longitudinal evaluation of growth, puberty, and bone maturation in children with Williams syndrome. *J Pediatr* 1999;134:82–89.

159. Chapman CA, du Plessis A, Pober BR. Neurologic findings in children and adults with Williams syndrome. *J Child Neurol* 1995;11:63–65.

160. Plissent L, Fryns JP. Early development (5 to 48 months) in Williams syndrome: a study of 14 children. *Genet Couns* 1999;10:151–156.

161. Howlin P, Davies M, Udwin O. Cognitive functioning in adults with Williams syndrome. *J Child Psychol Psychiatry* 1998;39:183–189.

162. Udwin O, Davies M, Howlin P. A longitudinal study of cognitive abilities and educational attainment in Williams syndrome. *Dev Med Child Neurol* 1996;38:1020–1029.

163. Jarrold C, Baddeley AD, Hewes AK. Verbal and nonverbal abilities in the Williams syndrome phenotype: evidence for diverging developmental trajectories. *J Child Psychol Psychiatry* 1998;39:511–523.

164. Arnold R, Yule W, Martin N. The psychological characteristics of infantile hypercalcemia: a preliminary investigation. *Dev Med Child Neurol* 1985;27:49–59.

165. Gosch A, Pankau G. Linguistic abilities in children with Williams-Beuren syndrome. *Am J Med Genet* 1994;52:291–296.

166. Karmiloff-Smith A, Grant J, Berthoud I, et al. Language and Williams syndrome: how intact is "intact?" *Child Dev* 1996;68:274–290.

167. Udwin O, Yule W. Expressive language of children with Williams syndrome. *Am J Med Genet* 1990;6:108–114.

168. Wang PP, Doherty S, Rourke SB, et al. Unique profile of visuospatial skills in a genetic syndrome. *Brain Cogn* 1995;29:54–65.

169. Vicari S, Brizzolara D, Carlesimo GA, et al. Memory abilities in children with Williams syndrome. *Cortex* 1996;32:503–514.

170. Wang PP, Bellugi U. Evidence from two genetic syndromes for a dissociation between verbal and visuo-spatial short-term memory. *J Clin Exp Neuropsychol* 1994;16:317–322.

171. Vicari S, Carlesimo GA, Brizzolara D, et al. Short-term memory in children with Williams syndrome: a reduced contribution of lexical-semantic knowledge to word span. *Neuropsychologia* 1996;34:919–925.

172. Greer MK, Brown FR, Pai GS, et al. Cognitive, adaptive, and behavioral characteristics of Williams syndrome. *Am J Med Genet* 1997;74:521–525.

173. Tomc SA, Williamson NK, Pauli RM. Temperament in Williams syndrome. *Am J Med Genet* 1990;36:345–352.

174. Plissart L, Borghgraef M, Fryns JP. Temperament in Williams syndrome. *Genet Couns* 1996;7:41–46.

175. van Lieshout CF, De Meyer RE, Curfs LM, et al. Family contexts, parental behaviour, and personality profiles of children and adolescents with Prader-Willi, fragile-X, or Williams syndrome. *J Child Psychol Psychiatry* 1998;39:699–710.

176. Gorsch A, Pankau R. Personality characteristics and behavior problems in individuals of different ages with Williams syndrome. *Dev Med Child Neurol* 1997;39:527–533.

177. Dykens EM, Rosner BA. Refining behavioral phenotypes: personality-motivation in Williams and Prader-Willi syndromes. *Am J Ment Retard* 1999;104:158–169.

178. Davies M, Udwin O, Howlin P. Adults with Williams syndrome: preliminary study of social, emotional, and behavioural difficulties. *Br J Psychiatry* 1998;172:273–276.

179. Udwin O, Yule W. A cognitive and behavioral phenotype in Williams syndrome. *J Clin Exp Neuropsychol* 1991;13:232–244.

180. Udwin O, Howlin P, Davies M, et al. Community care for adults with Williams syndrome: how families cope and the availability of support networks. *J Intellect Disabil Res* 1998;42:238–245.

181. Davies M, Howlin P, Udwin O. Independence and adaptive behavior in adults with Williams syndrome. *Am J Med Genet* 1997;70:188–195.

182. Knox WE. Phenylketonuria. In: Stanbury JB, Wyngaarden JB, Fredrickson DS, eds. *The metabolic basis of inherited disease.* New York: McGraw-Hill, 1972.

183. Matthias D, Bickel H. Follow-up study of 16 years neonatal screening for inborn errors of metabolism in West Germany. *Eur J Pediatr* 1986;145:310–312.

184. Levy HL. Nutritional therapy for selected inborn errors of metabolism. *J Am Coll Nutr* 1989;8[Suppl]:S54–S60.

185. Smith ML, Hamley WB, Clarke JT, et al. Randomised controlled trial of tryosine supplementation on neuropsychological performance in phenylketonuria. *Arch Dis Child* 1998;78:116–121.

186. Paine RS. The variability in manifestations of untreated patients with phenylketonuria (phenylpyruvic aciduria). *Pediatrics* 1957;20:290–331.

187. Dobson JC. Intellectual performance of 36 phenylketonuria patients and their non-affected siblings. *Pediatrics* 1976;58:53–58.

188. Wright SW, Tarjan G. Phenylketonuria. *Am J Dis Child* 1957;93:405.

189. Baumeister AA. The effects of dietary control on intelligence in phenylketonuria. *Am J Med Defic* 1967;71:840–847.

190. Berry HK, O'Grady DJ, Perlmutter LJ, et al. Intellectual development and academic achievement of children treated early for phenylketonuria. *Dev Med Child Neurol* 1979;21:311–320.

191. Pietz J, Dunckelmann R, Rupp A, et al. Neurological outcome in adult patients with early-treated phenylketonuria. *Eur J Pediatr* 1998;157:824–830.

192. Waisbren SE, Mahon BE, Schnell RR, et al. Predictors of intelligence quotient changes in persons treated for phenylketonuria early in life. *Pediatrics* 1987;79:351–355.

193. Williamson ML, Koch R, Azen C, et al. Correlations of intelligence test results in treated phenylketonuric children. *Pediatrics* 1981;68:161–167.

194. Griffiths PV, Demellweek C, Fay N, et al. Wechsler subscale IQ and subtest profile in early treated phenylketonuria. *Arch Dis Child* 2000;82:209–215.

195. Griffiths P. Neuropsychological approaches to treatment policy issues in phenylketonuria. *Eur J Pediatr* 2000;159[Suppl 2]:S82–S86.

196. Burgard P, Rey F, Rupp A, et al. Neuropsychologic functions of early treated patients with phenylketonuria, on and off diet: results of a cross-national and cross-sectional study. *Pediatr Res* 1997;41:368–374.

197. Burgard P. Development of intelligence in early treated phenylketonuria. *Eur J Pediatr* 2000;159[Suppl 2]:S74–S79.

198. Ozanne AE, Krimmer H, Murdoch BE. Speech and language skills in children with early treated phenylketonuria. *Am J Ment Retard* 1990;94:625–632.

199. Davis DD, McIntyre CW, Murray ME, et al. Cognitive styles in children with dietary treated phenylketonuria. *Educ Psychol Res* 1986;6:9–15.

200. Fischler K, Azen CG, Henderson R, et al. Psychoeducational findings among children treated for phenylketonuria. *Am J Ment Defic* 1987;92:65–73.

201. Arnold GL, Kramer BM, Kirby RS, et al. Factors affecting cognitive, motor, behavioral and executive functioning in children with phenylketonuria. *Acta Paediatr* 1998;87:565–570.

202. Diamond A, Revor MB, Callender G, et al. Prefrontal cortex cognitive deficits in children treated early and continuously for PKU. *Monogr Soc Res Child Dev* 1997;62:1–208.

203. Pennington BF, van Doorninck WJ, McCabe LL, et al. Neuropsychological deficits in early-treated phenylketonurics. *Am J Ment Defic* 1985;89:467–474.

204. Ris MD, Williams SE, Hunt MM, et al. Early-treated phenylketonuria: adult neuropsychologic outcome. *J Pediatr* 1994;124:388–392.

205. Welsh MC, Pennington BF, Ozonoff S, et al. Neuropsychology of early-treated phenylketonuria: specific executive function deficits. *Child Dev* 1990;61:1697–1713.

206. Welsh MC, Pennington BF. Assessing frontal lobe function in children: views from developmental psychology. *Dev Neuropsychol* 1988;4:199–230.

207. Welsh MC, Pennington BF. Phenylketonuria. In: Yeates KO, Ris MD, Taylor HG, eds. *Pediatric neuropsychology: research, theory, and practice.* New York: Guilford, 2000.

208. Griffiths PV, Campbell R, Robinson PR. Executive function in treated phenylketonuria as measured by the one-back and two-back versions of the continuous performance test. *J Inherit Metab Dis* 1998;21:125–135.

209. Trefz FK, Cipcic-Schmidt S, Koch R. Final intelligence in late treated persons with phenylketonuria. *Eur J Pediatr* 2000;159[Suppl 2]:S145–S148.

210. Fisch RO, Sines LK, Chang P. Personality characteristics of nonretarded phenylketonurics and their family members. *J Clin Psychiatry* 1981;42:106–113.

211. Realmuto GM, Garfinkel BD, Tuchman M, et al. Psychiatric diagnosis and behavioral characteristics of phenylketonuric children. *J Nerv Ment Dis* 1986;174:536–540.

212. Schor DP. PKU and temperament: rating children three through seven years old in PKU families. *Clin Pediatr* 1983;22:807–811.

213. Smith I, Knowles J. Behavior in early treated phenylketonuria: a systematic review. *Eur J Pediatr* 2000;159[Suppl 2]:S89–S93.

214. Pietz J, Fatkenheuer B, Burgard P, et al. Psychiatric disorders in adult patients with early-treated phenylketonuria. *Pediatrics* 1997;99:345–350.

215. Ris MD, Weber AM, Hunt MM, et al. Adult psychosocial outcomes in early-treated phenylketonuria. *J Inherit Metab Dis* 1997;20:499–508.

216. Weglage J, Grenzebach M, Pietsch M, et al. Behavioral and emotional problems in early-treated adolescents with phenylketonuria in comparison with diabetic patients and healthy controls. *J Inherit Metab Dis* 2000;23:487–496.

217. Ernhart CB, Morrow-Tlucak M, Marler MR, et al. Low level lead exposure in the prenatal and early postnatal periods: early preschool development. *Neurotoxicol Teratol* 1987;9:259–270.

218. Hansen ON, Trillingsgaard A, Beese I, et al. A neuropsychological study of children with elevated dentine lead level: assessment of the effect of lead in different socio-economic groups. *Neurotoxicol Teratol* 1989;11:205–213.

219. Winneke G, Kraemer U. Neuropsychological effects of lead in children: interactions with social background variables. *Neuropsychobiology* 1984;11:195–202.

220. Needleman HL, Gunnoe C, Leviton A, et al. Deficits in psychologic and classroom performance of children with elevated dentine lead levels. *N Engl J Med* 1979;300:689–694.

221. Altmuto L, Sveinsson K, Kraemer U, et al. Visual functions in 6-year-old children in relation to lead and mercury levels. *Neurotoxicol Teratol* 1998;20:9–17.

222. Baghurst PA, McMichael AJ, Wigg NR, et al. Environmental exposure to lead and children's intelligence at the age of seven years: the Port Pirie cohort study. *N Engl J Med* 1992;327:1279–1284.

223. Bellinger DC. Early sensory-motor development and prenatal

exposure to lead. *Neurobehav Toxicol Teratol* 1984;6:387–402.

224. Bellinger D, Leviton A, Waternaux C, et al. Longitudinal analyses of prenatal and postnatal lead exposure and early cognitive development. *N Engl J Med* 1987;316:1037–1043.

225. Bellinger D, Leviton A, Waterneaux C, et al. Low-level lead exposure, social class, and infant development. *Neurotoxicol Teratol* 1988;10:497–503.

226. Bergomi M, Borella P, Fantuzzi G, et al. Relationship between lead exposure indicators and neuropsychological performance in children. *Dev Med Child Neurol* 1989;31:181–190.

227. Bonithon-Kopp C, Huel G, Moreau T, et al. Prenatal exposure to lead and cadmium and psychomotor development of the child at 6 years. *Neurobehav Toxicol Teratol* 1986;8:307–310.

228. Campbell TF, Needleman HL, Riess JA, et al. Bone lead levels and language processing performance. *Dev Neuropsychol* 2001;18:171–186.

229. Dietrich KM, Succop PA, Berger OG, et al. Lead exposure and the central auditory processing abilities and cognitive development of urban children: the Cincinnati Lead Study cohort at age 5 years. *Neurotoxicol Teratol* 1992;14:51–56.

230. Dietrich KM, Berger OG, Succop PA, et al. The developmental consequences of low to moderate prenatal and postnatal lead exposure: intellectual attainment in the Cincinnati Lead Study Cohort following school entry. *Neurotoxicol Teratol* 1993;15:37–44.

231. Evans HL, Daniel SA, Marmor M. Reversal learning tasks may provide rapid determination of cognitive deficits in lead-exposed children. *Neurotoxicol Teratol* 1994;16:471–477.

232. Fergusson DM, Fergusson JE, Horwood LJ, et al. A longitudinal study of dentine lead levels, intelligence, school performance and behaviour: II. Dentine lead and cognitive ability. *J Child Psychol Psychiatry* 1988;29:793–809.

233. Fergusson DM, Horwood LJ, Lynskey MT. Early dentine lead levels and subsequent cognitive and behavioural development. *J Child Psychol Psychiatry* 1993;34:215–227.

234. Fergusson DM, Horwood LJ, Lynskey MT. Early dentine lead levels and educational outcomes at 18 years. *J Child Psychol Psychiatry* 1997;38:471–478.

235. Fulton M, Thomson G, Hunter R, et al. Influence of blood lead on the ability and attainment of children in Edinburgh. *Lancet* 1984;1:1221–1225.

236. Mendelsohn AL, Dreyer BP, Fieran AH, et al. Low-level lead exposure and cognitive development in early childhood. *J Dev Behav Pediatr* 1999;20:425–431.

237. Minder B, Das-Smaal EA, Brand E, et al. Exposure to lead and specific attentional problems in schoolchildren. *J Learn Disabil* 1994;27:393–399.

238. Needleman HL. The neurobehavioral consequences of low lead exposure in childhood. *Neurobehav Toxicol Teratol* 1982;4:729–732.

239. Needleman HL. Lead at low dose and the behavior of children. *Acta Psychiatr Scand* 1983;67[Suppl 303]:26–37.

240. Needleman HL, Schell A, Bellinger D, et al. The long-term effects of exposure to low doses of lead in childhood: an 11-year follow-up report. *N Engl J Med* 1990;322:83–88.

241. Schnaas L, Rothernberg SJ, Perroni E, et al. Temporal pattern in the effect of postnatal blood lead level on intellectual development of young children. *Neurotoxicol Teratol* 2000;22:805–810.

242. Shaheen SJ. Neuromaturation and behavior development: the case of childhood lead poisoning. *Dev Psychol* 1984;20:542–550.

243. Sinha SP, Anjama Sharma V. Intelligence and vigilance performance as related to lead exposure among children. *Ind J Clin Psychol* 1998;25:194–199.

244. Stiles KM, Bellinger DC. Neuropsychological correlates of low-level lead exposure in school-age children: a prospective study. *Neurotoxicol Teratol* 1993;15:27–35.

245. Thatcher RW. Intelligence and lead toxins in rural children. *J Learn Disabil* 1983;16:355–359.

246. Walkowiak J, Altmann L, Kraemer U, et al. Cognitive and sensorimotor functions in 6-year-old children in relation to lead and mercury levels: adjustment for intelligence and contrast sensitivity in computerized testing. *Neurotoxicol Teratol* 1998;20:511–521.

247. Wasserman GA, Graziano JH, Factor-Litvak P, et al. Consequences of lead exposure and iron supplementation on childhood development at age 4 years. *Neurotoxicol Teratol* 1994;16:233–240.

248. Wasserman GA, Liu X, Popovac D, et al. The Yugoslavia Prospective Lead Study: contributions of prenatal and postnatal lead exposure to early intelligence. *Neurotoxicol Teratol* 2000;22:811–818.

249. Wigg NR, Vimpani GV, McMichael AJ, et al. Port Pirie cohort study: childhood blood lead and neuropsychological development at age two years. *J Epidemiol Commun Health* 1988;42:213–219.

250. Ernhart CB, Landa B, Wolf AW. Subclinical lead level and developmental deficit: re-analyses of data. *J Learn Disabil* 1985;18:475–479.

251. Harvey PG, Hamlin MW, Kumar R, et al. Relationships between blood lead, behaviour, psychometric and neuropsychological test performance in young children. *Br J Dev Psychol* 1988;6:145–156.

252. Minder B, Das-Smaal EA, Orlebeke JF. Cognition in children does not suffer from very low lead exposure. *J Learn Disabil* 1998;31:494–502.

253. Prpic-Majic D, Bobic J, Simic D, et al. Lead absorption and psychological function in Zagreb (Croatia) school children. *Neurotoxicol Teratol* 2000;22:347–356.

254. Damm D, Grandjean P, Lyngbye T, et al. Early lead exposure and neonatal jaundice: relation to neurobehavioral performance at 15 years of age. *Neurotoxicol Teratol* 1993;15:173–181.

255. Ernhart CB, Morrow-Tlucak M, Wolf AW, et al. Low level lead exposure in the prenatal and early preschool periods: intelligence prior to school entry. *Neurotoxicol Teratol* 1989;11:162–170.

256. Moon C, Marlowe M, Stellern J, et al. Main and interaction effects of metallic pollutants on cognitive functioning. *J Learn Disabil* 1985;18:217–221.

257. Winneke G, Kraemer U. Neuropsychological effects of lead in children: interactions with social background variables. *Neuropsychobiology* 1984;11:195–202.

258. Dietrich KN, Succop PA, Berger OG, et al. Lead exposure and the cognitive development of urban preschool children: the Cincinnati Lead Study cohort at age 4 years. *Neurotoxicol Teratol* 1991;13:203–211.

259. World Health Organization. *Environmental health criteria 165—inorganic lead.* Geneva, Switzerland: Author, 1995.

260. McMichael AJ, Vimpani GV, Robertson EF, et al. The Port Pirie cohort study: blood lead and pregnancy outcomes. *J Epidemiol Commun Health* 1986;40:18–25.

261. Schwartz BS, Stewart WF, Bolla KI, et al. Past adult lead exposure is associated with longitudinal decline in cognitive function. *Neurology* 2000;55:1144–1150.

262. Yule W, Urbanowicz M, Lansdown R, et al. Teacher's ratings of children's behavior in relation to blood lead levels. *Br J Dev Psychol* 1984;2:295–305.

263. Marlowe M, Folio R, Hall D, et al. Increased lead burdens and trace-mineral status in mentally retarded children. *J Spec Educ* 1982;16:87–99.

264. Marlowe M, Errera J, Jacobs J. Increased lead and cadmium burdens among mentally retarded children and children with borderline intelligence. *Am J Med Defic* 1983;87:477–483.

265. David OJ, Grad G, McGann B, et al. Mental retardation and "nontoxic" lead levels. *Am J Psychiatry* 1982;139:806–809.

266. Gittelman R, Eskanazi B. Lead and hyperactivity revisited: an investigation of nondisadvantaged children. *Arch Gen Psychiatry* 1983;40:827–833.

267. Lohiya GS, Crinella FM, Figueroa LT, et al. Lead exposure of people with developmental disabilities: success of control measures. *Ment Retard* 1996;34:215–219.

268. Bleeker ML, Lindgren KN, Ford DP. Differential contributions of current and cumulative indices of lead dose to neuropsychological performance by age. *Neurology* 1997;48:639–645.

269. Schwartz BS, Stewart WF, Bolla KI, et al. Past adult lead exposure is associated with longitudinal decline in cognitive function. *Neurology* 2000;55:1144–1150.

270. Stewart WF, Schwartz BS, Simon D, et al. Neurobehavioral function and tibial and chelatable lead levels in 543 former organolead workers. *Neurology* 1999;52:1610–1617.

271. Valciukas JA. Central nervous system dysfunction due to lead exposure. *Science* 1978;201:465–467.

272. Bolla K, Rignani J. Clinical course of neuropsychological functioning after chronic exposure to organic and inorganic lead. *Arch Clin Neuropsychol* 1997;12:123–131.

273. Mantere P, Haenninen H, Hernberg S. Subclinical neurotoxic lead effects: two-year follow-up studies with psychological test methods. *Neurobehavioral Toxicology and Teratol* 1982;4:725–727.

274. Braun CM, Daigneault S. Sparing of cognitive executive functions and impairment of motor functions after industrial exposure to lead: a field study with a control group. *Neuropsychology* 1991;5:179–193.

275. Muldoon SB, Cauley JA, Kuller LH, et al. Effects of blood lead levels on cognitive function of older women. *Neuroepidemiology* 1996;15:62–72.

276. Payton M, Riggs KM, Spiro A, et al. Relations of bone and blood lead to cognitive function: the VA normative aging study. *Neurotoxicol Teratol* 1998;20:19–27.

277. Centers for Disease Control. Recommendations for blood lead screening of young children enrolled in Medicaid: targeting a group at high risk. Advisory committee on childhood lead poisoning prevention (ACCLPP). *Morb Mortal Wkly Rep* 2000;49: Whole #RR-14.

278. Centers for Disease Control. Trends in blood lead levels among children–Boston, Massachusetts, 1994–1999. *Morb Mortal Wkly Rep* 2001;50:337–339.

279. Centers for Disease Control. Blood lead levels in young children–United States and selected states, 1996–1999. *Morb Mortal Wkly Rep* 2000;49:1133–1137.

280. Tong S, Baghurst PA, Sawyer MG, et al. Declining blood lead levels and changes in cognitive function during childhood: the Port Pirie cohort study. *JAMA* 1998;280:1915–1919.

281. Ruff HA, Bijur PE, Markowitz M, et al. Declining blood lead levels and cognitive changes in moderately lead-poisoned children. *JAMA* 1993;269:1641–1646.

282. Rogan WJ, Dietrich KN, Ware JH, et al. The effect of chelation therapy with succimer on neuropsychological development in children exposed to lead. *N Engl J Med* 2001;344:1421–1426.

283. Ruggieri M. The different forms of neurofibromatosis. *Childs Nerv Syst* 1999;15:295–308.

284. Friedman JM. Epidemiology of neurofibromatosis type 1. *Am J Med Genet* 1999;89:1–6.

285. DeBella K, Poskitt K, Szudek J, et al. Use of "unidentified bright objects" on MRI for diagnosis of neurofibromatosis 1 in children. *Neurology* 2000;54:1646–1651.

286. Rasmussen SA, Freedman JM. BF1 gene and neurofibromatosis 1. *Am J Epidemiol* 2000;151:33–40.

287. North KN, Riccardi V, Samango-Sprouse C, et al. Cognitive function and academic performance in neurofibromatosis 1: consensus statement from the NF1 cognitive disorders task force. *Neurology* 1997;48:1121–1127.

288. Brewer VR, Moore BD, Hiscock M. Learning disability subtypes in children with neurofibromatosis. *J Learn Disabil* 1997;30:521–533.

289. North KN, Joy P, Yuille D, et al. Cognitive function and academic performance in children with neurofibromatosis type 1. *Dev Med Child Neurol* 2000;37:427–436.

290. Moore BD, Ater JL, Needle M, et al. Neuropsychological profile of children with neurofibromatosis, brain tumor, or both. *J Child Neurol* 1994;9:368–377.

291. Cutting LE, Koth CW, Denckla MB. How children with neurofibromatosis type 1 differ from "typical" learning disabled clinic attenders: nonverbal learning disabilities revisited. *Dev Neuropsychol* 2000;17:29–47.

292. Zoller MET, Rembeck B, Backman L. Neuropsychological deficits in adults with neurofibromatosis type 1. *Acta Neurol Scand* 1997;95:225–232.

293. Zoller MET, Rembeck B. A psychiatric 12-year follow-up of adult patients with neurofibromatosis type 1. *J Psychiatr Res* 1999;33:63–68.

294. Moore BD, Slopsis J, Schomer D, et al. Neuropsychological significance of areas of high signal intensity on brain MRIs of children with neurofibromatosis. *Neurology* 1996;46:1660–1668.

295. Duffner PK, Cohen ME, Seidel FG, et al. The significance of MRI abnormalities in children with neurofibromatosis. *Neurology* 1989;39:373–378.

296. Dunn DW, Roos KL. MRI evaluation of learning difficulties and incoordination in neurofibromatosis type 1. *Neurofibromatosis* 1989;2:1–5.

297. Ferner RE, Chaudhuri R, Bingham J, et al. MRI in neurofibromatosis 1. The nature and evolution of increased intensity Tx weighted lesions and their relationship to intellectual impairment. *J Neurol Neurosurg Psychiatry* 1993;56:492–495.

298. Denckla MB, Hofman K, Mazzocco M. Relationship between T2 weighted hyperintensities (unidentified bright objects) and lower IQs in children with neurofibromatosis type 1. *Am J Med Genet* 1996;67:98–102.

299. Hofman KJ, Harris EL, Bryan RN, et al. Neurofibromatosis type 1: the cognitive phenotype. *J Pediatr* 1994;124:51–58.

300. Mott S, Baumgardner T, Skryja PB. Neurofibromatosis type 1: association between volume of T2-weighted high intensity signals on MRI and impaired judgment of line orientation. *Pediatr Neurol* 1994;11:88(abst).

301. North KN, Joy P, Yuille D, et al. Specific learning disability in children with neurofibromatosis type 1: significance of MRI abnormalities. *Neurology* 1994;44:878–883.

302. Legius E, Desceemaeker MJ, Steyaert J. Neurofibromatosis type 1 in childhood: correlation of MRI findings with intelligence. *J Neurol Neurosurg Psychiatry* 1995;59:638–640.

303. Moore BD, Slopsis JM, Jackson EF, et al. Brain volume in children with neurofibromatosis type 1: relation to neuropsychological status. *Neurology* 2000;54:914–920.

304. Curless RG, Siatowski M, Glaser JS, et al. MRI diagnosis of NF-1 in children without cafe-au-lait skin lesions. *Pediatr Neurol* 1998;18:269–271.

305. DeBella K, Szudek J, Friedman JM. Use of the national institute of health criteria for diagnosis of neurofibromatosis 1 in children. *Pediatrics* 2000;105:608–614.

306. DiPaolo DP, Zimmerman RA, Rorke LB, et al. Neurofibromatosis type 1: pathologic substrate of high-signal of high-intensity foci in the brain. *Radiology* 1995;195:721–724.

307. Bebin EM, Gomez MR. Prognosis in Sturge-Weber disease: comparison of unihemispheric and bihemispheric involvement. *J Child Neurol* 1988;3:181–184.

308. Chapieski L, Friedman A, Lachar D. Psychological functioning in children and adolescents with Sturge-Weber syndrome. *J Child Neurol* 2000;15:660–665.

309. Pascual-Castroviejo I, Diaz-Gonzalez C, Garcia-Melian RM, et al. Sturge-Weber syndrome: study of 40 patients. *Pediatr Neurol* 1993;9:283–288.

310. Sujansky E, Conradi S. Sturge-Weber syndrome: age of onset of seizures and glaucoma and the prognosis for affected children. *J Child Neurol* 1995;10:49–58.

311. Sujansky E, Conradi S. Outcome of Sturge-Weber syndrome in 52 adults. *Am J Med Genet* 1995;57:35–45.

312. Aydin A, Cakmakci H, Kovanlikaya A, et al. Sturge-Weber syndrome without facial nevus. *Pediatr Neurol* 2000;22:400–402.

313. Awad AH, Mullaney PB, Al-Mesfer S, et al. Glaucoma in Sturge-Weber syndrome. *J AAPOS* 1999;3:40–45.

314. Sullivan TJ, Clarke MP, Morin JD. The ocular manifestations of the Sturge-Weber syndrome. *J Pediatr Ophthalmol Strabismus* 1992;29:349–356.

315. van Emelen C, Goethals M, Dralands L, et al. Treatment of glaucoma in children with Sturge-Weber syndrome. *J Pediatr Ophthalmol Strabismus* 2000;37:29–34.

316. Yang CB, Freedman SF, Myers JS, et al. Use of latanoprost in the treatment of glaucoma associated with Sturge-Weber syndrome. *Am J Ophthalmol* 1998;126:600–602.

317. Olsen KE, Huang AS, Wright MM. The efficacy of goniotomy/trabeculotomy in early-onset glaucoma associated with the Sturge-Weber syndrome. *J AAPOS* 1998;2:365–368.

318. Mandal AK. Primary combined trabeculotomy-trabeculectomy for early-onset glaucoma in Sturge-Weber syndrome. *Ophthalmology* 1999;106:1621–1627.

319. Irkec M, Kiratli H, Bilgic S. Results of trabeculotomy and guarded filtration procedure for glaucoma associated with Sturge-Weber syndrome. *Eur J Ophthalmol* 1999;9:99–102.

320. Hamush NG, Coleman AL, Wilson MR. Ahmed glaucoma valve implant for management of glaucoma in Sturge-Weber syndrome. *Am J Ophthalmol* 1999;128:758–760.

321. Lee JS, Asano E, Muzik O, et al. Sturge-Weber syndrome: correlation between clinical course and FDG PET findings. *Neurology* 2001;57:189–195.

322. Ogunmekan AO, Hwang PA, Hoffman HJ. Sturge-Weber-Dimitri disease: role of hemispherectomy in prognosis. *Can J Neurol Sci* 1989;16:78–80.

323. Benedikt RA, Brown DC, Walker R, et al. Sturge-Weber syndrome: cranial MR imaging with Gd-DTPA. *AJNR Am J Neuroradiol* 1993;14:409–415.

324. Marti-Bonmati L, Menor F, Poyatos C, et al. Diagnosis of Sturge-Weber syndrome: comparison of the efficacy of CT and MR imaging in 14 cases. *AJR Am J Roentgenol* 1992;158:867–871.

325. Sperner J, Schmauser I, Bittner R, et al. MR-imaging findings in children with Sturge-Weber syndrome. *Neuropediatrics* 1990;21:146–152.

326. Griffiths PD. Sturge-Weber syndrome revisited: the role of neuroradiology. *Neuropediatrics* 1996;27:284–294.

327. Herron J, Dorrah R, Quagheebeur G. Intra-cranial manifestations of the neurocutaneous syndromes. *Clin Radiol* 2000;55:82–98.

328. Muller R-A, Chugani HT, Muzik O, et al. Language and motor functions activate calcified hemisphere in patients with Sturge-Weber syndrome: a positron emission tomography study. *J Child Neurol* 1997;12:431–437.

329. Maria BL, Neufeld JA, Rosainz LC, et al. Central nervous system structure and function in Sturge-Weber syndrome: evidence of neurologic and radiologic progression. *J Child Neurol* 1998;13:606–618.

330. Pinton F, Chiron C, Enjolras O, et al. Early single photon emission computed tomography in Sturge-Weber syndrome. *J Neurol Neurosurg Psychiatry* 1997;63:616–621.

331. Arzimanoglou A, Aicardi J. The epilepsy of Sturge-Weber syndrome: clinical features and treatment in 23 patients. *Acta Neurol Scand* 1992;86[Suppl 140]:18–22.

332. Vargha-Khadem F, Isaacs EB, Papaleloudi H, et al. Development of language in six hemispherectomized patients. *Brain* 1991;473–495.

333. Marriotti P, Iuvone L, Torrioli MG, et al. Linguistic and non-linguistic abilities in a patient with early left hemispherectomy. *Neuropsychologia* 1998;36:1303–1312.

334. Vargha-Khadem F, Carr LJ, Isaacs E, et al. Onset of speech after left hemispherectomy in a nine-year-old boy. *Brain* 1997;120:159–182.

335. Gomez MR. *Tuberous sclerosis.* New York: Raven, 1988.

336. Harrison JE, Bolton PF. Annotation: tuberous sclerosis. *J Child Psychol Psychiatry* 1997;38:603–614.

337. Osborne JP, Freyer A, Webb D. Epidemiology of tuberous sclerosis. *Ann N Y Acad Sci* 1991;615:125–127.

338. Hunt A, Lindenbaum RH. Tuberous sclerosis—a new estimate of prevalence within the Oxford region. *J Med Genet* 1984;21:272–277.

339. Sampson JR, Scahill SJ, Stephenson JBP. Genetic aspects of tuberous sclerosis in the West of Scotland. *J Med Genet* 1989;26:28–31.

340. Povey S, Burley MW, Attwood J, et al. Two loci for tuberous sclerosis—one on 9q34 and one on 16p13. *Ann Hum Genet* 1994;58:107–127.

341. Osborne JP. Diagnosis of tuberous sclerosis. *Arch Dis Child* 1988;63:1423–1425.

342. Roach ES, Gomez MR, Northrup H. Tuberous sclerosis complex consensus conference: revised clinical diagnostic criteria. *J Child Neurol* 1999;13:624–628.

343. Roach ES, DiMario FJ, Kandt RS, et al. Tuberous sclerosis consensus conference: recommendations for diagnostic evaluation. *J Child Neurol* 1999;14:401–407.

344. Hunt A. Development, behavior, and seizures in 300 cases of tuberous sclerosis. *J Intellect Disabil Res* 1993;37:41–51.

345. Shepherd CW, Stephenson JBP. Seizures and intellectual disability associated with tuberous sclerosis complex in the west of Scotland. *Dev Med Child Neurol* 1992;34:766–774.

346. Webb DW, Fryer AE, Osborne JP. On the incidence of fits and mental retardation in tuberous sclerosis. *J Med Genet* 1991;28:395–397.

347. Curatolo P, Cusmai R, Cortesi F, et al. Neuropsychiatric aspects of tuberous sclerosis. *Ann N Y Acad Sci* 1991;615:8–16.

348. Jambaque I, Cusmai R, Curatolo P, et al. Neuropsychological aspects of tuberous sclerosis in relation to epilepsy and MRI findings. *Dev Med Child Neurol* 1991;33:698–705.

349. Curatolo P. Vigabatrin for refractory partial seizures in children with tuberous sclerosis. *Neuropediatrics* 1994;25:55.

350. Nevin MC, Pearce WG. Diagnostic and genetical aspects to tuberous sclerosis. *J Med Genet* 1968;5:273–279.

351. Gillberg IC, Gillberg C, Ahlsen G. Autistic behavior and attention deficits in tuberous sclerosis: a population-based study. *Dev Med Child Neurol* 1994;36:50–56.

352. Webb DW, Fryer AE, Osborne JP. Morbidity associated with tuberous sclerosis: a population study. *Dev Med Child Neurol* 1996;38:146–155.

353. Hunt A, Dennis J. Psychiatric disorder among children

with tuberous sclerosis. *Dev Med Child Neurol* 1987;29:190–198.

354. Cassidy SB, Dykens E, Williams CA. Prader-Willi and Angelman syndromes: sister imprinted disorders. *Am J Med Genet* 2000;97:136–146.

355. Ledbetter DH, Riccardi VM, Airhart SD, et al. Deletions of chromosome 15 as a cause of the Prader-Willi syndrome. *J Pediatr* 1981;81:286–293.

356. Magenis RE, Brown MG, Lacy DA, et al. Is Angelman syndrome an alternate result of del(15)(q11q13)? *Am J Med Genet* 1987;28:829–838.

357. Williams CA, Gray BA, Hendrickson JE, et al. Incidence of 15q deletions in the Angelman syndrome: a survey of twelve affected persons. *Am J Med Genet* 1989;32:339–345.

358. Nichols RD, Knoll JH, Butler MG, et al. Genetic imprinting suggested by maternal heterodisomy in nondeletion Prader-Willi syndrome. *Nature* 1989;342:281–285.

359. Knoll JH, Nichols RD, Magenis RE, et al. Angelman and Prader-Willi syndromes share a common chromosome 15 deletion but differ in parental origin of the deletion. *Am J Med Genet* 1989;32:285–290.

360. Williams CA, Zori RT, Stone JW, et al. Maternal origin of 15q11-13 deletions in Angelman syndrome suggests a role for genomic imprinting. *Am J Med Genet* 1990;35:495–497.

361. Prader A, Labhart A, Willi H. Ein syndrom von adipositas, kleinwuchs, kryptorchismus und oligophrenie nac myatonieartigem zustand in neugeborenenalter. *Schqeiz Med Wschr* 1956;86:120.

362. Lawrence BM. Hypotonia, obesity, hypogonadism, and mental retardation in childhood. *Arch Dis Child* 1961;36:690.

363. Zellweger H, Schneider HJ. Syndrome of hypotonia-hypomentia-hypogonadism-obesity (HHHO) or Prader-Willi syndrome. *Am J Dis Child* 1968;115:588–598.

364. Holm VA, Cassidy SB, Butler MG, et al. Prader-Willi syndrome: consensus diagnostic criteria. *Pediatrics* 1993;91:398–402.

365. Dykens EM, Rosner BA. Refining behavioral phenotypes: personality-motivation in Williams and Prader-Willi syndromes. *Am J Ment Retard* 1999;104:158–169.

366. Clarke DJ, Boer H. Problem behaviors associated with deletion Prader-Willi, Smith-Magenis, and cri du chat syndromes. *Am J Ment Retard* 1998;103:264–271.

367. Dykens EM, Kasari C. Maladaptive behavior in children with Prader-Willi syndrome, Down syndrome, and nonspecific mental retardation. *Am J Ment Retard* 1997;102:228–237.

368. Glover D, Maltzman I, Williams C. Food preferences among individuals with and without Prader-Willi syndrome. *Am J Ment Retard* 1996;101:195–205.

369. Gross-Tsur V, Landau YE, Benarroch F, et al. Cognition, attention, and behavior in Prader-Willi syndrome. *J Child Neurol* 2001;16:288–290.

370. Dykens EM, Cassidy SB, King BH. Maladaptive behavior differences in Prader-Willi syndrome due to paternal deletion versus maternal uniparental disomy. *Am J Ment Retard* 1999;104:67–77.

371. Einfeld SL, Smith A, Durvasula S, et al. Behavior and emotional disturbance in Prader-Willi syndrome. *Am J Med Genet* 1999;82:123–127.

372. Dykens EM, Leckman JF. Obsessions and compulsions in Prader-Willi syndrome. *J Child Psychol Psychiatry* 1996;37:995–1002.

373. Curfs LMG, Verhulst FC, Fryns JP. Behavioral and emotional problems in youngsters with Prader-Willi syndrome. *Genet Couns* 1991;2:33–41.

374. Borghgraef M, Fryns JP, Van Den Berghe H. Psychological profile and behavioral characteristics in 12 patients with Prader-Willi syndrome. *Genet Couns* 1990;38:141–150.

375. Dykens EM, Hodapp RM, Walsh K, et al. Profiles, correlates,

and trajectories of intelligence in Prader-Willi syndrome. *J Am Acad Child Adolesc Psychiatry* 1992;31:1125–1130.

376. Roof E, Stone W, MacLean W, et al. Intellectual characteristics of Prader-Willi syndrome: comparison of genetic subtypes. *J Intellect Disabil Res* 2000;44:25–30.

377. Laurance BM, Brito A, Wilkinson J. Prader-Willi syndrome after age 15 years. *Arch Dis Child* 1981;56:181–186.

378. Cassidy SB, Forsythe M, Heeger S, et al. Comparison of phenotype between patients with Prader-Willi syndrome due to deletion 15q and uniparental disomy 15. *Am J Med Genet* 1997;68:433–440.

379. Dykens EM, Hodapp RM, Walsh K, et al. Adaptive and maladaptive behaviors in Prader-Willi syndrome. *J Am Acad Child Adolesc Psychiatry* 1992;31:1131–1136.

380. Van Lieshaut CFM, deMeyer RE, Curfs LMG, et al. Problem behaviors and personality of children and adolescents with Prader-Willi syndrome. *J Pediatr Psychol* 1998;23:111–120.

381. Akefeldt A, Gillberg C. Behavior and personality characteristics of children and young adults with Prader-Willi syndrome: a controlled study. *J Am Acad Child Adolesc Psychiatry* 1999;38:761–769.

382. Whitman BY, Accardo P. Emotional symptoms in Prader-Willi syndrome adolescents. *Am J Med Genet* 1987;28:897–905.

383. Greenswag LR. Adults with Prader-Willi syndrome: a survey of 232 cases. *Dev Med Child Neurol* 1987;29:145–152.

384. Dykens EM, Cassidy SB. Correlates of maladaptive behavior in children and adults with Prader-Willi syndrome. *Am J Med Genet* 1995;60:546–549.

385. Symons FJ, Butler MG, Sanders MD, et al. Self-injurious behavior and Prader-Willi syndrome: behavioral forms and body locations. *Am J Ment Retard* 1999;104:260–269.

386. Hellings JA, Warnock JK. Self-injurious behavior and serotonin in Prader-Willi syndrome. *Psychopharmacol Bull* 1994;30:245–250.

387. Caldwell ML, Taylor RL, Bloom SR. An investigation of the use of high- and low-preference food as a reinforcer for increased activity of individuals with Prader-Willi syndrome. *J Ment Defic Res* 1986;30:347–354.

388. Fieldstone A, Zipf WB, Schwartz HC, et al. Food preferences in Prader-Willi syndrome, normal weight, and obese controls. *Int J Obes Rel Metab Dis* 1997;21:1046–1052.

389. Glover D, Maltzman I, Williams C. Food preferences among individuals with and without Prader-Willi syndrome. *Am J Ment Retard* 1996;101:195–205.

390. Altman K, Bondy A, Hirsch G. Behavioral treatment of obesity in patients with Prader-Willi syndrome. *J Behav Med* 1978;1:403–412.

391. Page TJ, Finney JW, Parrish JM, et al. Assessment and reduction of food stealing in Prader-Willi syndrome. *Appl Res Ment Retard* 1983;4:219–228.

392. Maglieri KA, DeLeon IG, Rodriguez-Catter V, et al. Treatment of covert food stealing in an individual with Prader-Willi syndrome. *J Appl Behav Anal* 2000;33:615–618.

393. Holland AJ, Treasure J, Coskeran P, et al. Characteristics of the eating disorder in Prader-Willi syndrome: implications for treatment. *J Intellect Disabil Res* 1995;39:373–381.

394. Lindgren AC, Barkeling B, Hagg A, et al. Eating behavior in Prader-Willi syndrome, normal weight, and obese control groups. *J Pediatr* 2000;137:50–55.

395. Costeff H, Holm VA, Ruvalcaba R, et al. Growth hormone secretion in Prader-Willi syndrome. *Acta Paediatr Scand* 1991;80:1254–1256.

396. Lee PD, Hwu K, Henson H, et al. Body composition studies in Prader-Willi syndrome: effects of growth hormone therapy. *Basic Life Sci* 1993;60:201–205.

397. Davies PS, Evans S, Broomhead S, et al. Effect of growth hormone on height, weight, and body composition in Prader-Willi syndrome. *Arch Dis Child* 1998;78:474–476.

398. Lindgren AC, Ritzen EM. Five years of growth hormone treatment in children with Prader-Willi syndrome. Swedish National Growth Hormone Advisory Group. *Acta Paediatr Suppl* 1999;88:109–111.

399. Myers SE, Carrel AL, Whitman BY, et al. Sustained benefit after 2 years of growth hormone on body composition, fat utilization, physical strength and agility, and growth in Prader-Willi syndrome. *J Pediatr* 2000;137:42–49.

400. Bosio L, Beccaria L, Benzi F, et al. Body composition during GH treatment in Prader-Labhardt-Willi syndrome. *J Pediatr Endocrinol Metab* 1999;12[Suppl 1]:351–353.

401. Eiholzer U, l'Allemand D, van der Sluis I, et al. Body composition abnormalities in children with Prader-Willi syndrome and long-term effects of growth hormone therapy. *Horm Res* 2000;53:200–206.

402. l'Allemand D, Eiholzer U, Schlumpf M, et al. Cardiovascular risk factors improve during 3 years of growth hormone therapy in Prader-Willi syndrome. *Eur J Pediatr* 2000;159:835–842.

403. Tauber M, Barbeau C, Jouret B, et al. Auxological and endocrine evolution of 28 children with Prader-Willi syndrome: effect of GH therapy in 14 children. *Horm Res* 2000;53:279–287.

404. Whitman BY, Myers S, Carrel A, et al. The behavioral impact of growth hormone treatment for children and adolescents with Prader-Willi syndrome: a 2-year, controlled study. *Pediatrics* 2002;109:E35.

405. Clarke D. Prader-Willi syndrome and psychotic symptoms. 2. A preliminary study of prevalence using the Psychopathology Assessment Schedule for Adults with Developmental Disability checklist. *J Intellect Disabil Res* 1998;42:451–454.

406. Beardsmore A, Dorman T, Cooper SA, et al. Affective psychosis and Prader-Willi syndrome. *J Intellect Disabil Res* 1998;42:463–471.

407. Akefeldt A, Ekman R, Gillberg C, et al. Cerebrospinal fluid monoamines in Prader-Willi syndrome. *Biol Psychiatry* 1998;44:1321–1328.

408. Dan B, Christiaens F. Dopey's seizure. *Seizure* 1999;8:238–240.

409. Clarke DJ, Marston G. Problem behaviors associated with 15q− Angelman syndrome. *Am J Ment Retard* 2000;105:25–31.

410. Smith A, Wiles C, Haan E, et al. Clinical features in 27 patients with Angelman syndrome resulting from DNA deletion. *J Med Genet* 1996;33:107–112.

411. Leitner RP, Smith A. An Angelman syndrome clinic: report on 24 patients. *J Paediatr Child Health* 1996;32:94–98.

412. Laan LA, den Boer AT, Hennekam RC, et al. Angelman syndrome in adulthood. *Am J Med Genet* 1996;66:356–360.

413. Laan LA, Renier WO, Arts WF, et al. Evolution of epilepsy and EEG findings in Angelman syndrome. *Epilepsia* 1997;38:195–199.

414. Ruggieri M, McShane MA. Parental view of epilepsy in Angelman syndrome: a questionnaire study. *Arch Dis Child* 1998;79:423–426.

415. Kuenzle C, Steinlin M, Wohlrab G, et al. Adverse effects of vigabatrin in Angelman syndrome. *Epilepsia* 1998;39:1213–1215.

416. Buoni S, Grosso S, Pucci L, et al. Diagnosis of Angelman syndrome: clinical and EEG criteria. *Brain Dev* 1999;21:296–302.

417. Laan LA, Brouwer OF, Begeer CH, et al. The diagnostic value of the EEG in Angelman and Rett syndrome at a young age. *Electroencephalogr Clin Neurophysiol* 1998;106:404–408.

418. Ellaway C, Buchholz T, Smith A, et al. Rett syndrome: significant clinical overlap with Angelman syndrome but not with methylation status. *J Child Neurol* 1998;13:448–451.

419. Buckley RH, Dinno N, Weber P. Angelman syndrome: are the estimates too low? *Am J Med Genet* 1998;80:385–390.

420. Klinefelter HF, Reifenstein EC, Albright F. Syndrome characterized by gynecomastia, aspermatogenesis without aleydigism, and increased excretion of follicle-stimulation hormone. *J Clin Endocrinol* 1942;2:615–627.

421. Cody H, Hynd G. Klinefelter syndrome. In: Goldstein S, Reynolds CR, eds. *Handbook of neurodevelopmental and genetic disorders in children.* New York: Guilford Press, 1999.

422. Nielson J, Wohlert M. Chromosome abnormalities among 34,910 newborn children: results from a 13-year incidence study in Aarhus, Denmark. *Hum Genet* 1991;87:81–83.

423. Jacobs PA, Hassold TJ, Whittington E, et al. Klinefelter's syndrome: an analysis of the origin of the additional sex chromosome using molecular probes. *Ann Hum Genet* 1988;52[Pt 2]:93–109.

424. Eskenazi B, Wyrobek AJ, Kidd SA, et al. Sperm aneuploidy in fathers of children with paternally and maternally inherited Klinefelter syndrome. *Hum Reprod* 2002;17:576–583.

425. Itsuka Y, Bock A, Nguyen DD, et al. Evidence of skewed X-chromosome inactivation in 47,XXY and 48,XXXY Klinefelter patients. *Am J Med Genet* 2001;98:25–31.

426. Drugan A, Isada NB, Johnson MP, et al. Parental chromosomal anomalies. In: Isada NB, Drugan A, Johnson MP, et al., eds. *Maternal genetic disease.* Stamford, CT: Appleton & Lange, 1996.

427. Zuppinger K, Engel E, Forbes AP, et al. Klinefelter's syndrome: a clinical and cytogenetic study in twenty-four cases. *Acta Endocrinol* 1967;54[Suppl 113]:5–48.

428. Evans MI, Drugan A, Pryde PG, et al. Genetics and the obstetrician. In: Isada NB, Drugan A, Johnson MP, et al., eds. *Maternal genetic disease.* Stamford, CT: Appleton & Lange, 1996.

429. Pierce BA. *The family genetic sourcebook.* New York: Wiley, 1990.

430. Schwartz ID, Root AW. The Klinefelter syndrome of testicular dysgenesis. *Endocrinol Metab Clin North Am* 1991;20:153–163.

431. Robinson A, Bender BG, Borelli JB, et al. Sex chromosomal aneuploidy: prospective and longitudinal studies. In: Ratcliffe SG, Paul N, eds. *Prospective studies on children with sex chromosome aneuploidy.* New York: Alan R. Liss, 1986.

432. Weatherall DJ. *The new genetics and clinical practice,* 3rd ed. New York: Oxford University Press, 1991.

433. Walzer S, Bashir AS, Silbert AR. Cognitive and behavioral factors in the learning disabilities of 47,XXY and 47,XYY boys. *Birth Defects* 1990;26:45–58.

434. Ratcliffe SG, Bancroft J, Axworthy D, et al. Klinefelter's syndrome in adolescence. *Arch Dis Child* 1982;57:6–12.

435. Rovet J, Netley C, Keenan M, et al. The psychoeducational profile of boys with Klinefelter syndrome. *J Learn Disabil* 1996;29:180–196.

436. Geschwind DH, Gregg J, Boone K, et al. Klinefelter's syndrome as a model of anomalous cerebral laterality: testing gene dosage in the X chromosome pseudoautosomal region using DNA microarray. *Dev Genet* 1998;23:215–229.

437. Bender BG, Berch DB. Overview: psychological phenotypes and sex chromosome abnormalities. In: Berch DB, Bender BG, eds. *Sex chromosome abnormalities and human behavior.* Boulder, CO: Westview Press, 1991.

438. Brauer-Boone KB, Swerdloff RS, Miller BL, et al. Neuropsychological profiles of adults with Klinefelter syndrome. *J Int Neuropsychol Soc* 2001;7:446–456.

439. Thielgaard A. A psychological study of the personalities of XYY and XXY men. *Acta Psychiatr Scand Suppl* 1984;315:1–13.

440. Nielsen J, Pelsen B, Sorenson K. Follow-up of 30 Klinefelter males treated with testosterone. *Clin Genet* 1988;33:262–269.

441. Stewart D, Bailey J, Netley C, et al. Growth and development

from early to misadolescence of children with X and Y chromosome aneuploidy: the Toronto study. *Birth Defects* 1986;22:119–182.

442. Robinson A, Bender BG, Borelli JB, et al. Sex chromosomal aneuploidy: prospective and longitudinal studies. In: Ratcliffe SG, Paul N, eds. *Prospective studies on children with sex chromosome aneuploidy.* New York: Alan R. Liss, 1986.

443. Bender BG, Puck MH, Salbenblatt JA, et al. Dyslexia in 47,XXY boys identified at birth. *Behav Genet* 1986;16:343–354.

444. Netley C, Rovet J. Hemispheric lateralization in 47,XXY Klinefelter's syndrome boys. *Brain Cogn* 1984;3:10–18.

445. Netley C. Behavior and extra X aneuploid states. In: Berch DB, Bender BG, eds. *Sex chromosome abnormalities and human behavior.* Boulder, CO: Westview Press, 1991.

446. Graham JM Jr, Bashir AS, Stark RE, et al. Oral and written language abilities of XXY boys: implications for anticipatory guidance. *Pediatrics* 1988;81:795–806.

447. Mandoki MW, Sumner GS, Hoffman RP, et al. A review of Klinefelter's syndrome in children and adolescents. *J Am Acad Child Adolesc Psychiatry* 1991;30:160–172.

448. Leonard M, Sparrow S. Prospective study of development of children with sex chromosome anomalies: New Haven Study IV. Adolescence. *Birth Defects* 1986;22:221–249.

449. Robinson A, Bender B, Linden M, et al. Sex chromosome aneuploidy: the Denver prospective study. *Birth Defects* 1990;26:59–115.

450. Schiavi RC, Thiegaard A, Owen DR, et al. Sex chromosome anomalies, hormones, and aggressivity. *Arch Gen Psychiatry* 1984;41:93–99.

451. Nielsen J, Johnsen SG, Sorensen K. Follow-up 10 years later of 34 Klinefelter males with karyotype 47,XXY and 16 hypogonadal males with karyotype 46,XY. *Psychol Med* 1980;10:345–352.

452. Raboch J, Mellan J, Starka L. Klinefelter's syndrome: sexual development and activity. *Arch Sex Behav* 1979;8:333–339.

453. Warwick MM, Doody GA, Lawrie SM, et al. Volumetric magnetic resonance imaging study of the brain in subjects with sex chromosome aneuploidy. *J Neurol Neurosurg Psychiatry* 1999;66:628–632.

454. Patwardhan AJ, Eliez S, Bender B, et al. Brain morphology in Klinefelter syndrome: extra X chromosome and testosterone supplementation. *Neurology* 2000;54:2218–2223.

455. Walzer S, Bashir AS, Graham JM, et al. Behavioral development of boys with X chromosome aneuploidy: impact of reactive style on the educational intervention for learning deficits. In: Ratcliffe SG, Paul N, eds. *Prospective studies on children with sex chromosome aneuploidy.* New York: Alan R. Liss, 1986.

456. Ford CE, Jones KW, Polani PE, et al. A sex-chromosome anomaly in a case of gonadal dysgenesis (Turner's syndrome). *Lancet* 1959;i:711.

457. Hall JG, Sybert VP, Williamson RA, et al. Turner's syndrome. *West J Med* 1982;137:32–44.

458. Ross JL. Disorders of the sex chromosomes: a medical overview. In: Holmes CS, ed. *Psychoneuroendocrinology: brain, behavior and hormonal interactions.* New York: Springer-Verlag, 1990:127–137.

459. Berch DB, Bender BG. Turner syndrome. In: Yeates KO, Ris MD, Taylor HG, eds. *Pediatric neuropsychology: research, theory, and practice.* New York: The Guilford Press, 2000.

460. Rao E, Weiss B, Fukami M, et al. Pseudoautosomal deletions encompassing a novel heterodox gene cause growth failure in idiopathic short stature and Turner syndrome. *Nat Genet* 1997;16:54–63.

461. Boucher CA, Sargent CA, Ogata T, et al. Breakpoint analysis of Turner patients with partial Xp deletions: implications for the lymphodema gene location. *J Med Genet* 2001;38:591–598.

462. Clement-Jones M, Schiller S, Rao E, et al. The short stature homeobox gene SHOX is involved in skeletal abnormalities in Turner syndrome. *Hum Mol Genet* 2000;9:695–702.

463. Palka G, Stuppia L, Guanciali-Franchi P, et al. Short arm rearrangements of sex chromosomes with haploinsufficiency of the SHOX gene are associated with Leri-Weill dyschondrosteosis. *Clin Genet* 2000;57:449–453.

464. Zinn AR, Ross JL. Molecular analysis of genes on Xp controlling Turner syndrome and premature ovarian failure (POF). *Semin Reprod Med* 2001;19:141–146.

465. Rovet JF. The cognitive and neuropsychological characteristics of females with Turner syndrome. In: Berch DB, Bender BG, eds. *Sex chromosome abnormalities and human behavior: psychological studies.* Boulder, CO: AAAS/Westview Press, 1990:38–77.

466. Ross JL, Reiss AL, Freund L, et al. Neurocognitive function and brain imaging in Turner syndrome—preliminary results. *Horm Res* 1993;39[Suppl 2]:65–69.

467. Murphy DGM, DeCarli C, Daly E, et al. X-chromosome effects on female brain: a magnetic resonance imaging study of Turner's syndrome. *Lancet* 1993;342:1197–1200.

468. Reiss AL, Mazzocco MMM, Greenlaw R, et al. Neurodevelopmental effects of X monosomy: a volumetric imaging study. *Ann Neurol* 1995;38:731–738.

469. Reiss AL, Freuns L, Plotnick L, et al. The effects of X chromosome monosomy on brain development: monozygotic twins discordant for Turner's syndrome. *Ann Neurol* 1993;34:95–107.

470. Araki K, Matsumoto K, Shiraishi T, et al. Turner's syndrome with agenesis of the corpus callosum, Hasimoto's thyroiditis and horseshoe kidney. *Acta Paediatr Jpn* 1987;29:622–626.

471. Kimura M, Nakajima M, Yoshino K. Ullrich-Turner syndrome with agenesis of the corpus callosum. *Am J Med Genet* 1990;37:227–228.

472. Clark C, Klonoff H, Hayden M. Regional cerebral glucose metabolism in Turner syndrome. *Can J Neurosci* 1990;17:140–144.

473. Haberecht MF, Menon V, Warsofsky IS, et al. Functional neuroanatomy of visuo-spatial working memory in Turner syndrome. *Hum Brain Mapp* 2001;14:96–107.

474. Grumbach MM, Van Wyck JJ, Wilkins L. Chromosomal sex in gonadal dysgenesis relationship to male pseudohermaphroditism and theories of human sex differentiation. *J Clin Endocrinol* 1955;15:1161–1193.

475. Haddad HM, Wilkins L. Congenital abnormalities associated with gonadal aplasia: review of 55 cases. *Pediatrics* 1959;23:885–902.

476. Polani PE. Turner syndrome and allied conditions. *Br Med Bull* 1961;17:200–205.

477. Cohen H. Psychological test findings in adolescents having ovarian dysgenesis. *Psychosom Med* 1962;24:249–256.

478. Shaffer JW. A specific cognitive deficit observed in gonadal aplasia (Turner's syndrome). *J Clin Psychol* 1962;18:403–406.

479. Money J. Cytogenetic and psychosexual incongruities with a note on space-form blindness. *Am J Psychiatry* 1963;119:820–827.

480. Waber DP. Neuropsychological aspects of Turner syndrome. *Dev Med Child Neurology* 1979;21:58–70.

481. Berch DB, Kirkendall KL. Spatial information processing in 45,X children. In: Robinson A, chair. *Cognitive and psychosocial dysfunctions associated with sex chromosome abnormalities.* Symposium presented at the meeting of the American Association for the Advancement of Science, Philadelphia, Pennsylvania, 1986.

482. Rovet J, Netley C. The mental rotation task performance of Turner syndrome subjects. *Behav Genet* 1980;10:437–443.

483. Rovet J, Netley C. Processing deficits in Turner's syndrome. *Dev Psychol* 1982;18:77–94.
484. Temple CM, Carney RA. Patterns of spatial functioning in Turner's syndrome. *Cortex* 1995;31:109–118.
485. Silbert A, Wolff PH, Lilienthal J. Spatial and temporal processing in patients with Turner's syndrome. *Behav Genet* 1977;7:11–21.
486. Buchanan L, Pavlovic J, Rovet J. A reexamination of the visuospatial deficit in Turner syndrome: contributions of working memory. *Dev Neuropsychol* 1998;14:341–367.
487. Buchanan L, Pavlovic J, Rovet J. The contribution of visuospatial working memory to impairments in facial processing and arithmetic in Turner syndrome. *Brain Cogn* 1998;37:72–75.
488. Garron DC. Intelligence among persons with Turner's syndrome. *Behav Genet* 1977;7:105–127.
489. Pennington BF, Bender B, Puck M, et al. Learning disabilities in children with sex chromosome anomalies. *Child Dev* 1982;53:1182–1192.
490. Alexander D, Money J. Turner's syndrome and Gerstmann's syndrome: neuropsychological comparisons. *Neuropsychologia* 1966;4:165–273.
491. Money J. Turner's syndrome and parietal lobe functions. *Cortex* 1973;385–393.
492. Bender BG, Puck M, Salbenblatt J, et al. Cognitive development of unselected girls with complete and partial X monosomy. *Pediatrics* 1984;73:175–182.
493. Pennington BF, Heaton RK, Karzmark P, et al. The neuropsychological phenotype in Turner syndrome. *Cortex* 1985;21:391–404.
494. Nijhuis-van der Sanden RWG, Smits-Engelsman BCM, Eling PATM. Motor performance in girls with Turner syndrome. *Dev Med Child Neurol* 2000;42:685–690.
495. Temple CM, Carney RA, Mullarkey S. Frontal lobe function and executive skills in children with Turner's syndrome. *Dev Neuropsychol* 1996;12:343–363.
496. Ross JL, Stafanatos GA, Kushner H, et al. Persistent cognitive deficits in adult women with Turner syndrome. *Neurology* 2002;58:218–225.
497. Van Dyke DL, Wiktor A, Roberson JR, et al. Mental retardation in Turner syndrome. *J Pediatr* 1991;118:415–417.
498. O'Connor J, Fitzgerald M, Hoey H. The relationship between karyotype and cognitive functioning in Turner syndrome. *Irish J Psychol Med* 2000;17:82–85.
499. Rovet JF. The psychoeducational characteristics of children with Turner syndrome. *J Learn Disabil* 1993;26:333–341.
500. Darby PL, Garfinkel PE, Vale JM, et al. Anorexia nervosa in Turner syndrome: cause or coincidence. *Psychol Med* 1981;11:141–145.
501. Halmi KA, DeBault LE. Gonosomal aneuploidy in anorexia nervosa. *Am J Hum Genet* 1974;26:195–198.
502. Raft D, Spencer RF, Toomey TC. Ambiguity of gender identity and fantasies and aspects of normality and pathology in hypopituitary dwarfism and Turner's syndrome: three cases. *J Sex Res* 1976;12:161–172.
503. Sabbath JC, Morris TA, Menzer-Benaron D, et al. Psychiatric observation in the adolescent girls lacking ovarian function. *Psychosom Med* 1961;23:224–231.
504. Rothchild E, Owens RP. Adolescent girls who lack functioning ovaries. *J Am Acad Child Psychiatry* 1972;11:88–113.
505. Rovet J. Processing deficits in 45,X females. In: Robinson A, chair. *Cognitive and psychosocial dysfunctions associated with sex chromosome abnormalities.* Symposium presented at the meeting of the American Association for the Advancement of Science, Philadelphia, Pennsylvania, 1986.
506. Rovet J, Ireland L. Behavioral phenotype in children with Turner syndrome. *J Pediatr Psychol* 1994;19:779–790.
507. Swillen A, Fryns JP, Kleczkowska A, et al. Intelligence, behavior, and psychosocial development in Turner syndrome. *Genet Couns* 1993;4:7–18.
508. Sonis WA, Levine-Ross J, Blue J, et al. *Hyperactivity of Turner's syndrome.* Paper presented at the meeting of the American Academy of Child Psychiatry, San Francisco, California, 1983.
509. McCauley E, Ross JL, Kushner H, et al. Self-esteem and behavior in girls with Turner syndrome. *J Dev Behav Pediatr* 1995;16:82–88.
510. McCauley E, Ito J, Kay T. Psychosocial functioning in girls with Turner syndrome and short stature. *J Am Acad Child Psychiatry* 1986;25:105–112.
511. Skuse DH, James RS, Bishop DVM, et al. Evidence from Turner's syndrome of an imprinted X-linked locus affecting cognitive function. *Nature* 1997;387:705–708.
512. Ross J, Zinn A, McCauley E. Neurodevelopmental and psychosocial aspects of Turner syndrome. *Ment Retard Dev Disabil Rev* 2000;6:135–141.
513. Bender BG, Linden MG, Robinson A. Neurocognitive and psychosocial phenotypes associated with Turner syndrome. In: Broman SH, Grafman J, eds. *Atypical cognitive deficits in developmental disorders: implications for brain function.* Hillsdale, NJ: Erlbaum, 1994:197–216.
514. Beck B, Fenger K. Mortality, pathological findings, and causes of death in the deLange syndrome. *Acta Paediatr Scand* 1985;74:765–769.
515. Jackson L, Kline AD, Barr MA, et al. deLange syndrome: a clinical review of 310 individuals. *Am J Med Genet* 1993;47:940–946.
516. Feingold M, Lin A. Familial Brachmann-de Lange syndrome: further evidence for autosomal dominant inheritance and review of the literature. *Am J Med Genet* 1993;47:1064–1067.
517. Ireland M, Donnai D, Burn J. Brachmann-de Lange syndrome. Delineation of the clinical phenotype. *Am J Med Genet* 1993;47:959–964.
518. Leroy JG, Persijn J, Van de Weghe V, et al. On the variability of the Brachmann-de Lange syndrome in seven patients. *Am J Med Genet* 1993;47:983–991.
519. Bay C, Mauk J, Radcliffe J, et al. Mild Brachmann-de Lange syndrome. Delineation of the clinical phenotype, and characteristic behaviors in a six-year-old boy. *Am J Med Genet* 1993;47:965–968.
520. Moeschler JB, Graham JM. Mild Brachmann-de Lange syndrome. Phenotypic and developmental characteristics of mildly affected individuals. *Am J Med Genet* 1993;47:969–976.
521. Saul RA, Curtis Rogers R, Phelan MC, et al. Brachmann-de Lange syndrome: diagnostic difficulties posed by the mild phenotype. *Am J Med Genet* 1993;47:999–1002.
522. Sellicorni A, Lalatta F, Livini E, et al. Variability of the Brachmann-de Lange syndrome. *Am J Med Genet* 1993;47:977–982.
523. Van Allen M, Fillippi G, Siegel-Bartelt J, et al. Clinical variability within Brachmann-de Lange syndrome: a proposed classification system. *Am J Med Genet* 1993;47:947–958.
524. Barr AN, Grabow JD, Matthews CG, et al. Neurologic and psychometric findings in the Brachmann-de Lange syndrome. *Neuropaediatrie* 1971;3:46–66.
525. Cameron TH, Kelly DP. Normal language skills and normal intelligence in a child with de Lange syndrome. *J Speech Hearing Disord* 1988;53:219–222.
526. Saal HM, Samango-Sprouse C, Rodnan L, et al. Brachmann-de Lange syndrome with normal IQ. *Am J Med Genet* 1993;47:995–998.
527. Berney TP, Ireland M, Burn J. Behavioral phenotype of Cornelia de Lange syndrome. *Arch Dis Child* 1999;81:333–336.

528. Kline AD, Stanley C, Belevich J, et al. Developmental data on individuals with the Brachmann-de Lange syndrome. *Am J Med Genet* 1993;47:1053–1058.

529. Goodban MT. Survey of speech and language skills with prognostic indicators in 116 patients with Cornelia de Lange syndrome. *Am J Med Genet* 1993;47:1059–1063.

530. Bryson Y, Sakati N, Nyhan WL, et al. Self-mutilative behavior in the Cornelia de Lange syndrome. *Am J Med Defic* 1971;76:319–324.

531. Johnson HG, Ekman P, Friesen W, et al. A behavioral phenotype in the de Lange syndrome. *J Pediatr* 1976;78:506–509.

532. Wilkins LE, Brown JA, Nance WE, et al. Clinical heterogeneity in 80 home-reared children with cri-du-chat syndrome. *J Pediatr* 1983;102:528–533.

533. Niebuhr E. The cri du chat syndrome: epidemiology, cytogenetics, and clinical features. *Hum Genet* 1978;16:227–275.

534. Gersh M, Goodart SA, Pasztor LM, et al. Evidence for a distinct region causing a cat-like cry in patients with 5p deletions. *Am J Hum Genet* 1995;56:1404–1410.

535. Cornish KM, Pigram J. Developmental and behavioural characteristics of cri du chat syndrome. *Arch Dis Child* 1996;75:448–450.

536. Dykens EM, Clarke DJ. Correlates of maladaptive behavior in individuals with 5p– (cri du chat) syndrome. *Dev Med Child Neurol* 1997;39:752–756.

537. Clarke DJ, Boer H. Problem behaviors associated with deletion Prader-Willi, Smith-Magenis, and cri du chat syndromes. *Am J Ment Retard* 1998;103:264–271.

538. Cornish KM, Bramble D, Munir F, et al. Cognitive functioning in children with typical cri du chat (5p–) syndrome. *Dev Med Child Neurol* 1999;41:263–266.

539. Cornish KM, Cross G, Green A, et al. A neuropsychological-genetic profile of atypical cri du chat syndrome: implications for prognosis. *J Med Genet* 1999;36:567–570.

540. Cornish KM, Munir F. Receptive and expressive language skills in children with cri-du-chat syndrome. *J Commun Disord* 1998;31:73–81.

541. Nyhan WL, Wong DF. New approaches to understanding Lesch-Nyhan disease. *N Engl J Med* 1996;334:1602–1604.

542. Nyhan WL. The recognition of Lesch-Nyhan syndrome as an inborn error of purine metabolism. *J Inherit Metab Dis* 1997;20:171–178.

543. Christie R, Bay C, Kaufman IA, et al. Lesch-Nyhan disease: clinical experience with nineteen patients. *Dev Med Child Neurol* 1982;24:293–306.

544. Mizuno T. Long-term follow-up of ten patients with Lesch-Nyhan syndrome. *Neuropediatrics* 1986;17:158–161.

545. Anderson LT, Ernst M. Self-injury in Lesch-Nyhan disease. *J Autism Dev Disord* 1994;24:67–81.

546. Gilbert S, Spellacy E, Watts RW. Problems in the behavioural treatment of self-injury in the Lesch-Nyhan syndrome. *Dev Med Child Neurol* 1979;21:795–800.

547. McGreevy P, Arthur M. Effective behavioral treatment of self-biting by a child with Lesch-Nyhan syndrome. *Dev Med Child Neurol* 1987;29:536–540.

548. Olson L, Houlihan D. A review of behavioral treatments used for Lesch-Nyhan syndrome. *Behav Modif* 2000;24:202–222.

549. Allen SM, Rice S. Risperidone antagonism of self-mutilation in a Lesch-Nyhan patient. *Prog Neuropsychopharmacol Biol Psychiatry* 1996;20:793–800.

550. Allen SM, Freeman JN, Davis WM. Evaluation of risperidone in the neonatal 6-OHDA model of Lesch-Nyhan syndrome. *Pharmacol Biochem Behav* 1998;59:327–330.

551. Evans J, Sirikumara M, Gregory M. Lesch-Nyhan syndrome and the lower lip guard. *Oral Surg Oral Med Oral Pathol* 1993;76:437–440.

552. Sugahara T, Mishima K, Mori Y. Lesch-Nyhan syndrome: successful prevention of lower lip ulceration caused by self-mutilation by use of mouth guard. *Int J Oral Maxillofac Surg* 1994;23:37–38.

553. Breese GR, Criswell HE, Duncan GE, et al. A dopamine deficiency model of Lesch-Nyhan disease-the neonatal 6-OHDA–lesioned rat. *Brain Res Bull* 1990;25:477–484.

554. Goldstein M. Dopaminergic mechanisms in self-inflicting biting behavior. *Psychopharmacol Bull* 1989;25:349–352.

555. Jinnah HA, Gage FH, Fruiedman T. Amphetamine-induced behavioral phenotype in a hypoxanthine-guanine phosphoribosyltransferase-deficient mouse model of Lesch-Nyhan syndrome. *Behav Neurosci* 1991;105:1004–1012.

556. Jankovic J, Caskey TC, Stout JT, et al. Lesch-Nyhan syndrome: a study of motor behavior and cerebrospinal fluid neurotransmitters. *Ann Neurol* 1988;23:466–469.

557. Harris JC, Les RR, Jinnah HA, et al. Craniocerebral magnetic resonance imaging measurement and findings in Lesch-Nyhan syndrome. *Arch Neurol* 1998;55:47–553.

558. Ernst M, Zametkin AJ, Matochik JA, et al. Presynaptic dopaminergic deficits in Lesch-Nyhan disease. *N Engl J Med* 1996;334:1568–1572.

559. Wong DF, Harris JC, Naidu S, et al. Dopamine transporters are markedly reduced in Lesch-Nyhan disease in vivo. *Proc Natl Acad Sci U S A* 1996;93:5534–5543.

560. Saito Y, Ito M, Hanaoka S, et al. Dopamine receptor upregulation in Lesch-Nyhan syndrome: a postmortem study. *Neuropediatrics* 1999;30:66–71.

561. Allen SM, Davis WM. Relationship of dopamine to serotonin in the neonatal 6-OHDA rat model of Lesch-Nyhan syndrome. *Behav Pharmacol* 1999;10:467–474.

562. Anderson LT, Ernst M, Davis SV. Cognitive abilities of patients with Lesch-Nyhan disease. *J Autism Dev Disord* 1992;22:189–203.

563. Matthews WS, Solan A, Barabas G. Cognitive functioning in Lesch-Nyhan syndrome. *Dev Med Child Neurol* 1995;37:715–722.

564. Matthews WS, Solan A, Barabas G, et al. Cognitive functioning in Lesch-Nyhan syndrome: a 4-year follow-up study. *Dev Med Child Neurol* 1999;41:260–262.

565. Chuang DT. Maple syrup urine disease: it has come a long way. *J Pediatr* 1998;132[Pt 2]:S17–S23.

566. Leonard JV, Daish P, Naughten ER, et al. The management and long term outcome of organic acidemias. *J Inherit Metab Dis* 1984;7[Suppl. 1]:13–17.

567. Korein J, Sansaricq C, Kalmijn M, et al. Maple syrup urine disease: clinical, EEG, and plasma amino acid correlations with a theoretical mechanism of acute neurotoxicity. *Int J Neurosci* 1994;21–45.

568. Kodama S, Seki A, Hanabusa M, et al. Mild variant of maple syrup urine disease. *Eur J Pediatr* 1976;124:31–36.

569. Podebrad F, Heil M, Reichert S, et al. 4,5-dimethyl-3-hydroxy-2[5H]-furanone (sotolone)—the odour of maple syrup urine disease. *J Inherit Metab Dis* 1999;22:107–114.

570. Naylor EW, Guthrie R. Newborn screening for maple syrup urine disease (branched-chain ketoaciduria). *Pediatrics* 1978;61:262–266.

571. Grover R, Wethers D, Shahidi S, et al. Evaluation of the expanded newborn screening program in New York City. *Pediatrics* 1978;61:740–749.

572. Irons M. Screening for metabolic disorders. How are we doing? *Pediatr Clin North Am* 1993;40:1073–1085.

573. Levy HL, Cornier AS. Current approaches to genetics—

metabolic screening in newborns. *Curr Opin Pediatr* 1994; 6:707–711.

574. Jouvet P, Poggi F, Rabier D, et al. Continuous venovenous haemodiafiltration in the acute phase of neonatal maple syrup urine disease. *Journal of Inherited Metabolic Diseases* 1997; 20:463–472.

575. Schaefer F, Straube E, Oh J, et al. Dialysis in neonates with inborn errors of metabolism. *Nephrol Dial Transplant* 1999;14:910–918.

576. Nyhan WL, Rice-Kelts M, Klein J, et al. Treatment of the acute crisis in maple syrup urine disease. *Arch Pediatr Adolesc Med* 1998;152:593–598.

577. Thomas E. Dietary management of inborn errors of amino acid metabolism with protein-modified diets. *J Child Neurol* 1992;7[Suppl]:S91–S111.

578. Duran M, Wadman SK. Thiamine-responsive inborn errors of metabolism. *J Inherit Metab Dis* 1985;8[Suppl 1]:70–75.

579. Fernhoff PM, Lubitz D, Danner DJ, et al. Thiamine response in maple syrup urine disease. *Pediatr Res* 1985;19:1011–1016.

580. Wendel U, Saudabray JM, Bodner A, et al. Liver transplantation in maple syrup urine disease. *Eur J Pediatr* 1999;158[Suppl 2]:S60–S64.

581. Fariello G, Dionisi-Vici C, Orazi C, et al. Cranial ultrasonography in maple syrup urine disease. *Am J Neuroradiol* 1996;17:311–315.

582. Taccone A, Schiaffino MC, Cerone R, et al. Computed tomography in maple syrup urine disease. *Eur J Radiology* 1992;14:207–212.

583. Uziel G, Savoiardo M, Nardocci N. CT and MRI in maple syrup urine disease. *Neurology* 1988;38:486–488.

584. Kamei A, Takashima S, Chan F, et al. Abnormal dendritic development in maple syrup urine disease. *Pediatr Neurol* 1992;8:145–147.

585. Hilliges C, Awiszus D, Wendel U. Intellectual performance of children with maple syrup urine disease. *Eur J Pediatr* 1993;152:144–147.

586. Kaplan P, Mazur A, Field M, et al. Intellectual outcome in children with maple syrup urine disease. *J Pediatr* 1991;119:46–50.

587. Clow CL, Reade TM, Scriver CR. Outcome of early and long-term management of classical maple syrup urine disease. *Pediatrics* 1981;68:856–862.

588. Naughten ER, Jenkins J, Francis DE, et al. Outcome of maple syrup urine disease. *Arch Dis Child* 1982;57:918–921.

589. Nord A, van Doorninck WJ, Greene C. Developmental profile of patients with maple syrup urine disease. *J Inherit Metab Dis* 1991;14:881–889.

590. Rubinstein JH, Taybi H. Broad thumbs and toes and facial abnormalities: a possible mental retardation syndrome. *Am J Dis Child* 1962;105:88–108.

591. Hennekam RCM, Boogaard VD, Doorne V. A cephalometric study in Rubinstein-Taybi syndrome. *J Craniofac Genet Dev Biol* 1991;20:33–40.

592. Lacombe D, Sabra R, Taint L, et al. Confirmation of an assignment of a locus for Rubinstein-Taybi syndrome gene to 16p13.3. *Am J Med Genet* 1992;44:126–128.

593. Wallerstein R, Anderson CE, Hay B, et al. Submicroscopic deletions at 16p13. in Rubinstein-Taybi syndrome: frequency and clinical manifestations in a North American population. *J Med Genet* 1997;34:203–206.

594. Rubinstein JH. Broad thumb-hallux (Rubinstein-Taybi) syndrome. *Am J Med Genet Suppl* 1990;6:3–16.

595. Padfield CJ, Partington MW, Simpson NE. The Rubinstein-Taybi syndrome. *Arch Dis Child* 1968;43:94–101.

596. Stevens CA, Carey JC, Blackburn BL. Rubinstein-Taybi syndrome: a natural history study. *Am J Med Genet Suppl* 1990;6:30–37.

597. Hennekam RCM, Baselier ACA, Beyaert E, et al. Psychological and speech studies in Rubinstein-Taybi syndrome. *Am J Ment Retard* 1992;96:645–660.

598. Levitas AS, Reid CS. Rubinstein-Taybi syndrome and psychiatric disorders. *J Intellect Disabil Res* 1998;42:284–292.

599. Scott HS, Litjens T, Hopwood JJ, et al. A common mutation for mucopolysaccharidosis type I associated with a severe Hurler syndrome phenotype. *Hum Mutat* 1992;1:103–108.

600. Lowry RB, Applegarth DA, Toone JR, et al. An update on the frequency of mucopolysaccharide syndromes in British Columbia. *Hum Genet* 1990;85:389–390.

601. Nelson J. Incidence of the mucopolysaccharidoses in Northern Ireland. *Hum Genet* 1997;101:355–358.

602. Shapiro E, Lockman L, Balthazor M, et al. Neuropsychological outcomes of several storage diseases with and without bone marrow transplantation. *J Inherit Metab Dis* 1995;18:413–429.

603. Shapiro E, Balthazor M. Metabolic and neurodegenerative disorders. In: Yeates KO, Ris MD, Taylor HG, eds. *Pediatric neuropsychology: research, theory, and practice.* New York: Guilford, 2000.

604. Peters C, Balthazor M, Shapiro E, et al. Outcome of unrelated bone marrow transplantation in forty children with Hurler syndrome. *Blood* 1996;87:4894–4902.

605. Peters C, Shapiro E, Anderson J, et al. Hurler's syndrome: II. Outcome of HLA-genotypically identical sibling and non-genotypically identical related donor bone marrow transplantation in fifty-four children. *Blood* 1998;91:2601–2608.

606. Schuchman EH, Desnick RJ. Mucopolysaccharidosis type I subtypes. Presence of immunologically cross-reactive material and in vitro enhancement of the residual alpha-L-iduronidase activities. *J Clin Invest* 1988;81:98–105.

607. Keith O, Scully C, Weidman GM. Orofacial features of Scheie (Hurler-Scheie) syndrome (alpha-L-iduronidase deficiency). *Oral Surg Oral Med Oral Pathol* 1990;70:70–74.

608. Vellodi A, Young E, Cooper A, et al. Bone marrow transplantation for mucopolysaccharidosis type I: experience of two British centers. *Arch Dis Child* 1997;76:92–99.

609. Guffon N, Souillet G, Maire I, et al. Follow-up of nine patients with Hurler syndrome after bone marrow transplantation. *J Pediatr* 1998;133:119–125.

610. Whitley CB, Belani KG, Chang PN, et al. Long-term outcome of Hurler syndrome following bone marrow transplantation. *Am J Med Genet* 1993;46:209–218.

611. Field RE, Buchanan JA, Copplemans MG, et al. Bone marrow transplantation in Hurler's syndrome: effect on skeletal development. *J Bone Joint Surg Br* 1994;76:975–981.

612. Tandon V, Williamson JB, Cowie RA, et al. Spinal problems in mucopolysaccharidosis I (Hurler syndrome). *J Bone Joint Surg Br* 1996;78:938–944.

613. Kakkis ED, McEntee MF, Schmidtchen A, et al. Long-term and high-dose trials of enzyme replacement therapy in the canine model of mucopolysaccharidosis I. *Biochem Mol Med* 1996;58:156–167.

614. Wraith JE. Enzyme replacement therapy in mucopolysaccharidosis type I: progress and emerging difficulties. *J Inherit Metab Dis* 2001;24:245–250.

615. Schaap T, Bach G. Incidence of mucopolysaccharidoses in Israel: is Hunter disease a "Jewish disease?" *Hum Genet* 1980;56:221–223.

616. Bax MC, Colville GA. Behavior in mucopolysaccharide disorders. *Arch Dis Child* 1995;73:77–81.

617. McKinnis EJ, Sulzbacher S, Rutledge JC, et al. Bone marrow

transplantation in Hunter syndrome. *J Pediatr* 1996;129:145–148.

618. Nidiffer FD, Kelly TE. Developmental and degenerative patterns associated with cognitive, behavioral, and motor difficulties in the Sanfilippo syndrome: an epidemiological study. *J Ment Defic Res* 1983;27:185–203.

619. Vellodi A, Young E, New M, et al. Bone marrow transplantation for Sanfilippo disease type B. *J Inherit Metab Dis* 1992;15:911–918.

620. Tarjan G. Cinderella and the prince: mental retardation and community psychiatry. *Am J Psychiatry* 1966;122:1057–1059.

621. King BH, State MW, Shah B, et al. Mental retardation: a review of the past 10 years. Part I. *J Am Acad Child Adolesc Psychiatry* 1997;36:1656–1663.

622. State MW, King BH, Dykens EM. Mental retardation: a review of the past 10 years. Part II. *J Am Acad Child Adolesc Psychiatry* 1997;36:1664–1671.

24

ATTENTION DEFICIT/HYPERACTIVITY DISORDER

JASMIN VASSILEVA
MARIELLEN FISCHER

Attention deficit/hyperactivity disorder (ADHD) is one of the most common childhood neurobehavioral disorders. It is characterized by developmentally inappropriate levels of inattention, hyperactivity, and impulsivity that occur across settings, cause functional impairment, and cannot be attributed to another disorder. It occurs in about 3% to 7% of children (1). It is much more common in boys than in girls, with estimated sex ratios varying between 3:1 and 9:1 (2). This ratio could be an inflated estimate, because ADHD of the purely inattentive type, which may be more prevalent in girls, may be overlooked.

The definition of, and diagnostic criteria for, ADHD have been modified on multiple occasions. The periodic changes in the definition of the disorder, reflected in the different names applied to the condition, follow the conceptualization of what has been considered central to the disorder at different times and reflect evolving theories of etiology of the major symptoms of the disorder. The first description of the disorder is attributed to George Still, who in 1902 described 43 children with aggression, disinhibition, limited sustained attention, and impaired rule-governed behavior (3). He attributed the central feature of the disorder to a "defect in moral control." From the 1930's to 1950's, the relationship of these symptoms to brain insults was highlighted, and the term *minimal brain damage* was coined, which was later changed to *minimal brain dysfunction* as there was no direct evidence for "damage" (3). Hyperactivity began to play a focal point in the diagnosis of the condition in the late 1960's, when the second edition of the *Diagnostic and Statistical Manual of Mental Disorders* (DSM-II) first gave official recognition to the disorder and called it *hyperkinetic reaction of childhood*. In the 1970's, the research pendulum shifted back to the central role of deficits in sustained attention in the disorder, which was accordingly renamed *attention deficit disorder* in the third revision of the DSM (DSM-III). In the late 1980's and 1990's, the primary role of inattention was questioned, emphasis reverted to the role of impulsivity/ hyperactivity as the core feature of the disorder (3), and consequently the disorder was retitled by the DSM-III-R as *attention deficit/hyperactivity disorder.*

Currently, ADHD is defined in behavioral terms and is diagnosed based on observable behavioral features, such as short attention span, distractibility, impulsivity, and overactivity. The most recent revision of the diagnosis of ADHD in the DSM-IV includes separate diagnostic criteria for symptoms of inattention and hyperactivity/impulsivity. Currently ADHD can be subtyped in one of three ways: ADHD predominantly inattentive type, ADHD predominantly hyperactive/impulsive type, and ADHD combined type. In addition to the heterogeneity in the presentation of the disorder, another source of variance is the frequent co-occurrence of other conditions (comorbidity). ADHD occurs commonly in association with conduct disorder (CD), oppositional defiant disorder (ODD), mood disorders, anxiety disorders, and many developmental disorders such as learning disabilities and speech and language delays (4–6). As many as one third of children with ADHD have one or more comorbid conditions. The prevalence of coexistent conditions in children with ADHD is estimated most recently to be 35% for ODD, 26% for CD, 26% for anxiety disorders, and 18% for depressive disorders (4). Similarly, up to 25% of children with ADHD also have specific learning disorders in reading, math, or written or spoken language (5). Comorbidity is associated with poorer outcome in adolescence and adulthood, as evidenced by significantly greater emotional, social, and psychological difficulties (6). Research suggests that the poorest outcomes for ADHD individuals in adolescence and young adulthood are associated with the presence of comorbid ODD and CD (7). Importantly, the presence of psychiatric comorbidity raises the question of whether the research findings reported in children with ADHD are related to the ADHD itself, the comorbid disorders, or the combination of both (6).

GENETICS

Family, twin, adoption, and molecular genetic studies suggest that ADHD has a significant genetic component (8,9). First- and second-degree relatives of probands with ADHD are at increased risk for the disorder (10,11). Parents of children with ADHD have a two- to eight-fold increase in the risk for ADHD, which provides support for the validity of diagnosing ADHD in adults. Heritability estimates range from 64% to 91% (12–14). More recently, technologic advances in the field of molecular genetics have made it possible to move beyond the quantification of heritability to attempt to isolate specific genes that may be associated with ADHD. Although a specific genetic pattern of inheritance has not yet been identified, one of the most promising research findings in this area is an association between ADHD and a specific form of the dopamine transporter gene (the DAT1 480 base pair allele) (15,16). The dopamine transporter is of particular interest because it is the site of pharmacologic action of methylphenidate and other dopamine reuptake inhibitors shown to be effective in treating ADHD. In addition, several studies have found an association between ADHD and a particular variant of dopamine type 4 receptor (DRD4-7) containing seven repeats of a specific noncoding region (17,18). Approximately 30% of ADHD individuals carry the seven repeat allele, whereas it occurs about half as frequently in the general population. More recently, research also has focused on other candidate susceptibility genes, such as the dopamine D5 receptor gene (DRD5) (19). Importantly, in 1999, an ADHD Molecular Genetics Network was formed, one of whose primary goals was to develop a standard assessment battery across various research groups. The operational criteria have been compiled in a new data-pooling measure, called *Hypescheme,* which already is being used by research groups conducting molecular genetics research in ADHD (20). Research in molecular genetics provides a promising avenue for clarifying the genetic contributions to ADHD, and future research will help determine whether one or a few genes are implicated in ADHD symptomatology or whether specific alleles of several genes occurring together in the same individual contribute to the development of the clinical syndrome.

ATTENTION DEFICIT/HYPERACTIVITY DISORDER IN ADULTS

In contrast to the widespread acceptance of ADHD as a childhood diagnosis, its prevalence in adults has been a source of much controversy. Until recently, it was believed that ADHD was limited to childhood and "outgrown" thereafter. In the early 1970's, Paul Wender noted that the parents of children with ADHD describe having similar behavioral problems in their own childhood, which continue throughout life for a large majority of them (21). This observation precipitated a systematic line of inquiry into the presentation and clinical symptoms of the disorder in adults. The consensus of more recent retrospective and prospective studies is that ADHD persists into adulthood in approximately 30% to 66% of children diagnosed with the disorder, who

continue to present either the full syndrome or significant residual symptoms (22–24). The prevalence of the disorder in the general adult population is estimated to be 1% to 6% (21). Other researchers have estimated the prevalence of ADHD in an adult sample at 4.7% (25) and the incidence at 2% (26). Nonetheless, the diagnosis of adult ADHD remains controversial, primarily because of the typically retrospective nature of the diagnostic process and its reliance on questionably accurate recollection of childhood events (27). For example, it has been noted that because the DSM-IV diagnosis of ADHD requires onset of symptoms before age 7, it assumes accurate recall of distant events. Because most patients evaluated for the adult form of the disorder were not psychiatrically evaluated as children, the retrospective nature of the adult diagnosis becomes a serious problem. Further, ADHD adults, like ADHD children, may deny or seriously underestimate their symptomatology (28). For example, a follow-up study of young adults revealed that self-report substantially underestimated the presence of the disorder in comparison to parent reports (24% vs. 58%, respectively) (23). This suggests that the current estimates may be underestimating the prevalence of the disorder in adults because most investigators depend almost exclusively on self-report. In addition, the field trial studies for the DSM-IV did not include adult subjects. Therefore, the current DSM diagnostic criteria were developed exclusively for children, do not include age-appropriate diagnostic items for adolescents and adults with ADHD, and thus become insensitive to the disorder with age (25). Ultimately, such diagnostic limitations contribute to underdiagnosis of the adult condition.

Fischer and colleagues developed an alternate method that they believe is more valid for diagnosing ADHD in adults (29,30). They collected ADHD symptom data based on parent report from a large group of normal control adults and a matched ADHD group, both of whom were followed longitudinally from childhood. They defined the disorder by a score that fell two standard deviations above the mean for the normal control group as adults. In applying this score to the formerly ADHD group, 66% of this group continued to exhibit the disorder in adulthood. In contrast, when DSM-III-R diagnostic criteria were applied, only 3% of this group could be diagnosed with ADHD in adulthood. This indicates that the use of developmentally sensitive diagnostic criteria may help to resolve the diagnostic dilemma about the validity of adult ADHD as a *bona fide* disorder in adulthood.

This diagnostic uncertainty impedes research aimed at elucidating the nature of the underlying deficits in adults with ADHD and whether they are consistent with the deficits observed in children with ADHD. The most notable characteristic of ADHD in children is motor hyperactivity, which tends to decrease over time independent of treatment (5,31). In contrast, symptoms of inattention show little decline over time, as evidenced by follow-up studies of children diagnosed with ADHD, which show that hyperactive/impulsive symptoms tend to decrease with age, whereas symptoms of inattention do not (31,32). Most often symptoms of inattention appear most prominently part way through elementary school, when academic challenges typically become more complex. Therefore, the ADHD subtype based on the current DSM-IV criteria is at least partially dependent on the age of the individual being diagnosed. Wender (2) also showed that hyperkinesis is observed to a lesser extent in adulthood than childhood and, if present, is manifested primarily as restlessness, personal disorganization, and inability to relax. Thus, although the core symptoms of ADHD may persist into adulthood, the expression of some symptoms may change over time. This suggests that some of the commonly used assessment methods for ADHD may need to incorporate unique procedures that will capture some of the distinct characteristics of adult ADHD.

DIAGNOSIS OF ATTENTION DEFICIT/HYPERACTIVITY DISORDER

Particularly with the advent of the DSM-IV, the diagnosis of ADHD for research and clinical purposes is increasingly driven by the DSM criteria. This provides a common standard or metric, as these criteria have come progressively closer to the state-of-the-art definition of the disorder agreed upon by research experts in the field. A current consensus definition holds that ADHD is a developmental disorder of inattention and/or hyperactivity/impulsivity, although many experts believe that the "predominantly inattentive" subtype of the disorder is not a true subtype but actually is a separate and distinct disorder independent of ADHD (33). The general consensus definition includes a developmentally inappropriate level of symptoms, a childhood onset, relative cross-setting occurrence of these symptoms, a resulting significant impairment in major life activities secondary to these symptoms, and exclusion of other disorders such as mental retardation, pervasive developmental disorder, and psychosis as a more primary cause of the symptoms.

Even though the DSM-IV diagnostic criteria are some of the most rigorous and most empirically derived criteria ever available in the history of clinical diagnosis for ADHD, several unresolved problems make them imperfect for research and clinical diagnosis. As discussed previously, the symptoms listed in the criteria are not developmentally scaled, and some are not suitable for adults. For example, "often leaves seat in classroom or in other situations in which remaining seated is expected" needs an appropriate adult analogue. As Barkley (33) points out, the DSM criteria need to reflect the *heterotypic continuity* present in ADHD across the life span, in which the construct remains the same but the symptoms change with age. Another problem with the DSM criteria is that cutoffs for number of symptoms required for diagnosis are not developmentally referenced and are not appropriate for persons younger than age 4 or older than age 16 (age groups outside those used in the field trial) (25,34).

The cutoffs also are not appropriate for females. Requiring that the symptoms have persisted 6 months probably is too short a duration for preschoolers (35). Similarly, defining the age of onset as before age 7 rather than as more generally prior to puberty has been shown to have little validity (36–38). Finally, the DSM-IV diagnostic criteria do not define what represents a developmentally inappropriate level of symptoms and require no corroboration by others when the diagnosis is made in an adult.

In addition to the clinical interview and the interview of parents or significant others, establishing a developmentally inappropriate level of ADHD symptoms typically is operationalized by the administration of behavior rating scales. Factor scores 1.5 to 2 standard deviations above the mean for same-age, same-sex normal children, adolescents, or adults on inattention, impulsivity, or hyperactivity factors on well-standardized behavior rating scales completed by parents and/or teachers and/or self-report remain the *sine qua non* of this aspect of diagnosis at present. Specific rating scales chosen may be partially a matter of clinician or researcher preference, ranging from narrow-band measures such as the *ADHD Rating Scale-IV* (39,40), which reflects actual DSM-IV criteria, through other often-used and well-respected measures such as the Conners Rating Scales (41,42), to the carefully researched and elaborate Achenbach System of Empirically Based Assessment (43), a group of broad-band behavior rating scales that encompass parent-, teacher-, self-report, and observational versions and span the preschool years through adulthood. Broad-band scales, although lengthier and sometimes more encumbering, have the virtue of often aiding in differential psychiatric diagnosis, a formidable task in ADHD assessment and diagnosis.

Neuropsychological studies in children and adults with ADHD are considered in some detail in the next section of this chapter. This research provides significant insight into the nature of the neurocognitive impairments inherent in ADHD. However, there has been some debate about whether neuropsychological testing plays a role in clinical diagnosis of the disorder. If the DSM-IV is considered the current "gold standard" for diagnosis, formal testing would not be considered to have a role. However, because comorbid learning disorders are relatively common in ADHD and because other developmental or acquired neurologic disorders can produce symptoms similar to those of ADHD, neuropsychological testing can play a valuable role in differential diagnosis, assessment of individual strengths and weaknesses, development of interventions, and treatment planning (44). If the testing is used more narrowly to rule in or rule out ADHD symptoms, it is important to note that findings are somewhat inconsistent with respect to sensitivity and not necessarily specific to ADHD, because impaired test scores can have many underlying causes and can be seen in a number of other disorders. Furthermore, the degree of test impairment does not correlate well with ratings of severity of symptoms, and neuropsychological test performance in a structured, nondistracting environment may not capture everyday difficulties, especially those in the cross-temporal organization of behavior (44,45).

NEUROPSYCHOLOGICAL STUDIES WITH CHILDREN

Faraono and Biederman (9) note that two of the main advantages to using neuropsychological tests are that many of the tests have been standardized on large normative populations, which makes it easier to define levels of impairment, and because the tests have been used extensively with brain-damaged patients, they provide a noninvasive and inexpensive means of generating hypotheses about brain pathophysiology, which can be tested with more direct neuroimaging measures.

Because neither specific deficits in attention nor motor control of behavior adequately portray the wide variety of ADHD symptoms, the construct of "executive functions" has gained increasing attention in the research literature and has been suggested to provide a more useful framework for describing ADHD symptoms (1,5,47). Deficits in executive functions purported to measure frontal lobe functioning have proved to be among the most consistent neuropsychological findings in children with ADHD. The executive functions tasks include, but are not limited to, the Continuous Performance Test (CPT) (7,48–50), Wisconsin Card Sorting Test (WCST) (50–61), Stroop test (49–51,53,57,62), Tower of Hanoi (63–65), hand movements test, and go/no go test (58,66,67). Generally, these tasks involve executive functions involving planning, organization, complex problem solving, and working memory (68), in addition to sustained attention/vigilance, response inhibition, motor inhibition, and verbal learning (69).

However, a comprehensive review of 22 neuropsychological studies of frontal lobe functions in children with ADHD (68) revealed highly inconsistent findings across studies, which appeared to be strongly related to the age of the subjects and the particular version of the test used. For example, whereas impairments on one of the most widely used tests of frontal function—the WCST—were common among children with ADHD, these deficits began to attenuate in adolescence. In addition, some studies found significant between-group differences on this test (52,58,59,70), whereas others failed to find any deficits in the ADHD group (49,51,54,61,71,72). A review of studies using the WCST with ADHD children indicated that although the task differentiated between ADHD and control children in 17 of 26 studies, it more consistently differentiated children with autism from normal children than it did children with ADHD from normal children (73). Overall, the review by Barkley et al. (68) revealed that the tests that were most reliable in detecting differences between children with ADHD and normal groups of children were the CPT, the Stroop test (interference score), and the go/no go test. As suggested by

the authors, a unifying feature of these tests is their assessment of the ability to inhibit motor responses on demand, which indicates a primary underlying deficit in behavioral inhibition (68).

Another neuropsychological review revealed that 15 of 18 studies found significant differences between ADHD and control children on one or more executive function measures (74). For 67% of the executive functions measures used, the performance of the ADHD group was significantly worse than that of the control group. None of the 60 executive measures was associated with significantly better performance in the ADHD group. The Tower of Hanoi, Trail-Making Test Part B, matching familiar figures test, and antisaccade task consistently yielded group differences. In addition, the ADHD group performed significantly worse on two working memory measures (sequential memory task and self-ordered pointing task). Importantly, consistent with the review by Barkley et al. (68), the Stroop test and tests of motor inhibition (go/no go and stopping task) emerged as measures that appeared especially sensitive to ADHD.

Other studies have found children with ADHD to be impaired on tests of interference control and resistance to distraction, such as the Stroop test (49–51,61,62,65), and on tests of inhibition of prepotent responses, such as the CPT and go/no go test (71). However, analysis of the types of errors performed by children with ADHD on CPTs revealed that whereas some of the studies report that children with ADHD commit significantly more errors of both omission and commission (7,49,71), others indicate that children with ADHD commit significantly more errors of omission only (50,51,56,57,68), which is evidence of inattention rather than of response disinhibition (75). In addition, a review of the ADHD literature using the CPT indicated that the task has sufficient sensitivity to ADHD in both children and adults with the disorder but inadequate specificity, which suggests that its ability to differentiate between ADHD and other neuropsychiatric conditions is suboptimal (76). In general, despite the inconsistent findings, neuropsychological studies of children with ADHD indicate that the most reliable findings are response disinhibition, inattention, and high variability in response speed on CPTs (77).

A number of methodologic differences among the studies may have contributed to the observed inconsistent findings. First, the use of small sample sizes may have resulted in insufficient power to detect differences among groups. Second, the selection criteria for defining ADHD were inconsistent. For example, some studies did not control for pervasiveness of ADHD and used only parent *or* teacher ratings. Third, some of the studies did not include a matched normal control group for comparison. Importantly, many studies failed to control for psychiatric comorbidity. It has been suggested that the comorbid conditions of CD or ODD may carry their own neuropsychological deficits that are similar to those observed in children with ADHD (78). For example, a number of studies found executive function deficits in CD children (79–82). As pointed out by Nigg et al. (78), of all the studies examining executive or effortful processes in ADHD, those controlling for comorbid ODD or CD (60,83) are the exception rather than the rule. This failure to control for comorbid CD or ODD becomes particularly problematic in light of the results from a recent metaanalysis of studies using one of the primary measures of response inhibition, namely the stop signal task (84), which revealed no significant difference between ADHD and CD children. To complicate matters even further, executive function deficits have been found in learning/reading disabled children (68,85,86), which further clouds the interpretation of the findings in ADHD, as many of the ADHD participants in the studies reviewed had comorbid learning disabilities.

In addition, the equivocal nature of the findings from neuropsychological studies in children with ADHD may be due to the limited validity of the tests of executive function for use with children, as the measures (e.g., the WCST) primarily have been validated in adults with focal brain lesions. Furthermore, the most common task used in the studies, the CPT, has been criticized for its limited neuropsychological validity as a measure of executive function in general (87) and for its insufficient specificity to ADHD (76). Finally, the relationship between performance on tests of executive function and the normal maturation of the central nervous system still is not well understood (88). Thus, examination on such tests of adults with ADHD whose central nervous system and frontal lobes in particular are fully developed may prove more informative about the underlying neuropsychological deficits of the disorder.

NEUROPSYCHOLOGICAL STUDIES WITH ADULTS

Because attentional and executive processes frequently are impaired in children with ADHD, demonstration of similar neuropsychological deficits in adulthood may serve as an additional external validation of the adult diagnosis (89). Neuropsychological measures have the added benefit of not being dependent on self-reports and the associated recall biases; thus, they provide a more direct assessment of performance. However, the different clinical manifestations of childhood and adult ADHD suggest that the adult form of the disorder may present with a distinct neuropsychological profile relative to the childhood form of ADHD. For example, it has been suggested that response inhibition is one of the most prominent features of childhood ADHD, whereas inattention and slowed information processing may be the key neuropsychological features in adults with ADHD (89–91).

Unfortunately, research into the cognitive and neuropsychological functioning of adults with ADHD has been relatively limited and characterized by inconsistent findings (11). To date, 21 studies have examined executive functioning in

adults with ADHD (57,89–108). It is of note that with three exceptions (92,99,106), all of the studies have been conducted within the last 7 years (i.e., since 1995). This suggests increased awareness among researchers of the frequent continuation of ADHD into adulthood and increased interest in the neuropsychological sequelae of the disorder in adults. Of all 21 studies, only one (99) diagnosed adult ADHD prospectively by following its subjects from childhood into adulthood, whereas the rest of the studies diagnosed ADHD retrospectively. Many of the studies indicate that some of the neuropsychological tests that have proved useful in discriminating between children with and without ADHD do not appear to do so in adults with ADHD. For example, the Stroop interference score, which is an index that has consistently differentiated children with ADHD from normal controls, does not reliably discriminate between adults with ADHD and those without. Of ten studies that used the Stroop paradigm (57,89–91,93,95,99,101,103,105), only four (93,99,103,105) report that adults with ADHD show deficient performance specifically on the interference condition. However, not one of the four studies provides any information about the performance of the ADHD group on the other two conditions of the Stroop task (i.e., word and color), which makes it difficult to discern whether the observed deficits are specific to the interference condition. The rest of the studies reveal either no group differences or overall slower information processing in the ADHD group on all Stroop conditions.

In accordance with the review of the neuropsychological literature in ADHD by Barkley et al. (48), which suggested that scores on the WCST are not reliably impaired in adolescents and adults with ADHD, only two of eight studies (57,89,97,98,100,105,106,108) found impaired performance in the ADHD group on the WCST. Specifically, Rugle and Melamed (106) reported that a group of adult gamblers who also had ADHD required more trials on the WCST to achieve the traditional criteria of six categories relative to a control group, and Seidman et al. (57) reported more perseverative errors and less categories achieved by young adults with ADHD. The other studies did not find evidence for impaired performance of adults with ADHD on the WCST.

Studies using the CPT tend to yield some of the most robust findings in discriminating between adults with ADHD and those without. Of 13 studies using the CPT or one of its variants (Gordon Diagnostic System, test of variables of attention) with adults having ADHD (57,89–94,97,98,101,102,105,108), only three (101,102,108) did not find group differences. Yet, an examination of the ten studies that reported group differences revealed that adults with ADHD tend to show more errors of omission (89–91,93,97,105), slower reaction times (57,89,98), and greater response variability (90–94) than normal controls, apparent evidence of inattention rather than of response disinhibition. Only four of the studies reported increased number of errors of commission, indicative of deficits in impulsivity/response inhibition (91,93,94,97). Similarly, adolescents with ADHD do not seem to make more errors of commission relative to controls (57,109). Importantly, Barkley (110) has suggested that the CPT's "ecologic validity" or relevance to the real world may be weaker for adults and adolescents than it is for younger children.

Of the neuropsychological studies with adults with ADHD reviewed, 13 used various measures of working (or short-term) memory for a total of 16 measures of working memory (89,91–93,95,97,100–103,105,107,108). Overall, the performance of adults with ADHD was significantly impaired on all but four of the 16 measures: digit span backward test (100), letter-number span test (105), Wechsler Adult Intelligence Scale-Revised (WAIS-R) Freedom from Distractibility Factor (89), and Tower of Hanoi (108). Adults with ADHD evidenced significant impairments on the remaining working memory measures, such as the pattern memory test and the serial digit learning test (92), test of auditory discrimination (95), paced auditory serial addition test (PASAT) (100,101,107), long delay of the auditory consonant trigrams test (97), digit backward subtest of the WAIS-III (91,93,102), freedom from distractibility index from the WAIS-R (103), and Simon test of nonverbal working memory (93). In addition, of three studies (98,101,103) that used the short delay free recall index of the California Verbal Learning Test (CVLT), which is considered to be an index of working or short-term memory, two studies (98,101) reported significant deficits in the ADHD group. These findings, together with the increased number of errors of omission that are more commonly observed in adults with ADHD, suggest that in contrast to the prominent response inhibition deficits in children with ADHD, adults with ADHD display more consistent impairments on measures of attention and the related construct of working memory.

WORKING MEMORY AND ATTENTION DEFICIT/HYPERACTIVITY DISORDER

Working memory has become the focus of increased research attention in the ADHD literature. Some have theorized that the consistent deficits in inhibition noted in children with ADHD could be attributed, in part, to deficits in working memory (111–113). Further, some models of working memory consider inhibition to be an intrinsic property of the working memory system (114,115). Increased activation of working memory processes is proposed to inhibit the activations required for competing response outcomes and to be generally involved in selecting appropriate action alternatives and inhibiting inappropriate ones (116).

A number of studies found deficits in working memory in children with ADHD using measures of mental arithmetic (117,118), backward digit span (112,117,119,120), memory for spatial location (117), and self-ordered pointing tasks (121,122). Similar deficits were observed in adolescents

with ADHD. For example, Oie et al. (123) found a group of adolescents with ADHD to be impaired on a digit span distractibility test with and without distractors relative to a normal control group. Importantly, on the condition without distractors, the performance of the ADHD group was worse than that of a group of adolescents with schizophrenia. Finally, as discussed earlier, deficits on tasks of working memory are some of the most consistent neuropsychological findings in adults with ADHD.

Barkley (1) proposed that evidence for deficits related to working memory in individuals with ADHD also comes from the following findings with persons with ADHD: (a) significantly greater variability in patterns of responding on tasks involving reaction time or CPTs; and (b) significantly greater problems with task performance when delays are imposed within the task and when these delays increase in duration. Barkley (47) also predicted that children with ADHD will have difficulties with commands or instructions that include a reference to a delayed or future performance of a behavior.

Barkley (124) also proposed that nonverbal working memory might be involved in the organization and reproduction of complex designs, such as in the Rey-Osterrieth Complex Figure Test. There is some evidence identifying organizational deficits in children with ADHD on this task (49,50,125). Nonverbal working memory in individuals with ADHD has been explored with more experimental research paradigms. For example, Frank et al. (126) used the Sternberg search paradigm with geometric shapes and found impaired performance in children with ADHD, as evidenced by significantly more errors and a larger variability of responding. Karatekin and Asarnow (112) used the dot test of visuospatial working memory and showed that, relative to a group of normal children, the performance of a group of children with ADHD was impaired to an extent that was equivalent to that of a group of children with schizophrenia. Another study investigated the performance of children with ADHD and normal control children on the Cambridge Neuropsychological Test Automated Battery (CANTAB), which includes the self-ordered pointing test of working memory, a spatial span test, delayed nonmatching to sample test, and an attentional set shifting task, all of which presumably tap working memory (87). The results revealed that the ADHD group had significantly shorter spatial span, committed significantly more errors on the delayed nonmatching to sample test, and were impaired on the attentional set shifting task and the self-ordered task. This indicated that in addition to having reduced working memory spans, the children with ADHD had problems when they had to manipulate information within working memory (as in the self-ordered task), which presumably involves the central executive component of working memory. In relation to the latter finding, Shue and Douglas (58) reported similarly impaired performance in a group of ADHD children on the self-ordered pointing task of working memory.

With regard to verbal working memory, the storage and recall of simple information within working memory do not appear to be impaired in individuals with ADHD (127–129). Deficits emerged, however, when larger and more complex amounts of verbal information must be held in mind over a delay period (50,128). In addition, both children and adults with ADHD are less proficient than controls when strategies that assist with organizing material so that it can be remembered more effectively are required (98,125). Norrelgen et al. (130) studied phonologic working memory in children with "pure" ADHD, children with ADHD and motor problems, and a control group. The findings revealed significant differences between controls and the group of children with ADHD and motor problems, but not between controls and the group with "pure" ADHD. The authors suggested that the motor problems in those children could be due to impaired phonologic verbal memory, which may affect the storage or retrieval of motor commands.

Overall, the extant literature indicates that both verbal and nonverbal working memory are affected in ADHD. Working memory is a cognitive domain that only recently has caught the attention of researchers in the field, yet the studies with children and particularly with adults with ADHD indicate that it may be the executive function that most reliably captures some of the neuropsychological impairments in ADHD.

ATTENTION DEFICIT/HYPERACTIVITY DISORDER AND THE BRAIN

The theory that the frontal lobes mediate some of the behavioral and cognitive manifestations of ADHD was first introduced in the 1930's, based on the observed behavioral and cognitive similarities between children with ADHD and patients with frontal lobe lesions. Satterfield and Dawson (131) were among the first to suggest that the symptoms of ADHD are related to frontolimbic dysfunction. Specifically, they posited that ADHD might be caused by weak frontal inhibitory control over limbic functions. The efficacy of stimulant medications, together with animal models implicating dopamine pathways in the pathophysiology of the disorder, provided support for this model. The systematic review of the neurologic literature by Mattes (132) further highlighted the similarities between children with ADHD and patients with frontal lobe lesions. Since then, neuropsychological impairments on different tasks purporting to measure frontal lobe function have been among the most consistent findings in the ADHD literature. Most of these cognitive impairments cluster in the domain of executive functions, associated with abnormalities of the prefrontal cortex and its connections with other brain regions. These neuropsychological impairments have been corroborated by evidence from neuroimaging studies, which implicate a frontostriatal dysfunction in the pathophysiology of ADHD.

Thus, the frontostriatal hypothesis of ADHD has become one of the most widely studied models of the neural correlates of ADHD. It has been tested by both neuropsychological and neuroimaging studies, which are reviewed in the next section.

NEUROIMAGING STUDIES WITH CHILDREN

Structural Imaging Studies

Imaging studies investigating structural brain differences in patients with ADHD typically use either computed axial tomography or magnetic resonance imaging (MRI). To date, 23 structural imaging studies have been conducted with ADHD (133–155). With one exception (151), all of the studies found evidence for structural brain abnormalities among ADHD patients. Twenty-two of the studies were conducted with children or adolescents with ADHD. The only study performed with young adults with ADHD showed sulcal widening and cerebellar atrophy in that group (154).

Total Brain Volume

Several studies found that children with ADHD have smaller total brain volume than control children (135–137,139). Using a large group of boys with ADHD and age-matched normal control subjects, Castellanos et al. (136) were the first to report that children with ADHD had on average 5% smaller total brain volumes compared to age-matched controls. The same investigators later replicated their findings with a larger group of children with ADHD (137). Similarly, Berquin et al. (135) report that total cerebral volume was smaller by 6.1% in a group of boys with ADHD relative to a control group. However, all three studies were conducted by the same research group and involved a significant overlap in subjects (135–137). Castellanos et al. (139) later replicated their earlier findings with an independent sample of 50 girls with ADHD. However, two studies by different teams of investigators failed to find significant differences in total brain volume between children with ADHD and healthy children (140,144). It is of note that the studies that did not find differences in total brain volume matched their ADHD and control subjects on intelligence quotient (IQ) (140,144), whereas the four studies that reported significant group differences in brain volume did not (135–137,139). Furthermore, when Castellanos et al. (137,139) and Berquin et al. (127) recomputed their analyses while controlling for IQ, the between-group difference in total brain volume was no longer significant. Importantly, it has been shown that IQ is positively correlated with total cerebral volume in both children and adults (151,156). Therefore, it is possible that decreased brain volume in children with ADHD might be a function of lower IQ rather than a result of the disorder itself. Thus, future studies investigating total brain volume in patients with ADHD will need to control for the effects of IQ to conclusively establish the presence of smaller total brain volumes in ADHD and should ideally use independent samples of research participants.

Frontal Lobes

The frontal lobes have been one of the most frequently investigated brain regions in ADHD. Among the structural imaging studies investigating the frontal lobes, there appears to be a consensus that the right frontal lobe is smaller in individuals with ADHD (137,140,148,157). For example, Hynd et al. (157) were among the first to reveal that the frontal lobes are symmetric in children with ADHD, in contrast to the normal L < R asymmetry reported in the literature. This was due to narrower right frontal widths in the ADHD group. Similarly, Filipek et al. (140) found that the volume of the anterior-superior frontal region (posterior prefrontal, motor association, and midanterior cingulate) was significantly (about 10%) smaller in adolescents with ADHD, particularly on the right. In addition, Filipek et al., who performed gray-white matter segmentation, also found smaller white matter volumes in bilateral prefrontal and retrocallosal (posterior parietal-occipital) hemispheric regions. Likewise, Castellanos et al. (137) found smaller right anterior frontal volumes in male children and adolescents with ADHD relative to controls, even after controlling for group differences in IQ. However, an attempt to replicate the findings with girls with ADHD revealed smaller volumes in the left rather than in the right frontal lobes, and the findings were no longer significant after controlling for verbal IQ (139). Fredricksen et al. (152) also found reduced left frontal lobe volumes in boys with ADHD, with a similar trend in the right hemisphere. Overmeyer et al. (148) found significant gray matter deficits in the right superior frontal gyrus in a group of hyperactive children.

In sum, of six studies investigating frontal lobe volumes in children with ADHD, four (137,140,148,157) found smaller right frontal volumes, whereas two (139,152) reported smaller left frontal volumes. However, Fredricksen et al. (152) found a trend for smaller frontal volumes on the right, which might have been significant if they had a larger sample size. The discrepant findings between the two studies of Castellanos et al. (137,139) may have been due to the different quantification techniques that they used.

Basal Ganglia

There are rich interconnections between the frontal regions and the basal ganglia (158). A number of studies attest to anomalies in some basal ganglia structures in ADHD. Among these structures, the caudate nucleus most reliably has been found to show volumetric structural abnormalities in individuals with ADHD (133,136,137,139, 140,143,145). However, findings from studies investigating asymmetry of the caudate have not been entirely consistent in terms of whether the right or left caudate is differentially affected. For example, Hynd et al. (143) performed a region

of interest analysis (ROI) of the caudate nucleus and found that in contrast to control children who evidenced an L > R asymmetry of the head of the caudate nucleus, children with ADHD were characterized by an R > L asymmetry. This was due to a relatively smaller left caudate nucleus in the children with ADHD, accounted for primarily by the male subjects. Notably, this difference was not due to significant differences between groups in overall brain size. Another study found that adolescents with ADHD had larger caudate nucleus areas on both the left and the right side (145). As in the study by Hynd et al., there was a tendency for the control group to show an L > R pattern of asymmetry of the head of the caudate, whereas the ADHD group showed the reverse pattern (R > L). Thus, the right caudate nucleus was significantly larger in patients with ADHD than in normal control individuals. However, there were no significant hemispheric differences in the right versus left caudate nuclei *within* groups. Filipek et al. (140) also found that a normal control group had an L > R caudate asymmetry; however, their ADHD subjects exhibited symmetric caudate volumes with smaller volumes of the total caudate (including right and left head and tail combined). This difference was accounted for by the left but not the right caudate, which was significantly smaller in the ADHD subjects. However, all of the ADHD subjects in this study were taking stimulant medication at the time of testing. When the group of ADHD subjects was divided into responders and nonresponders to the stimulants, it was revealed that the responders had more hyperactivity/ impulsivity symptoms than the nonresponders, whereas the nonresponders showed more inattention symptoms than the responders. Significant differences between groups also emerged when the caudate volumes were compared among the two ADHD groups and the controls. Specifically, in the controls, the left caudate was larger than the right (L > R); in the stimulant nonresponders, the left caudate was smaller than the right (R > L), whereas in the stimulant responders, the caudate volumes were symmetric (R = L). Another study also found R > L asymmetry in a group of children and adolescents with ADHD (150).

In contrast to the L > R asymmetry in the control subjects in the previously described studies (140,143,145), Castellanos et al. (136,137) found that control subjects exhibited significant R > L caudate asymmetry, whereas child and adolescent subjects with ADHD had symmetric right and left caudate volumes due to reduced right caudate volumes. Analysis by age revealed that caudate volumes decreased with increasing age in normal but not in ADHD subjects, which suggests that there could be a failure of the maturational processes that normally result in caudate volume reductions. Notably, the smaller total brain volumes found in the ADHD group did not account for the differences in caudate asymmetry in the two groups.

The studies reviewed suggest abnormalities in the caudate nucleus in children with ADHD; however, the findings are inconsistent with regard to the morphologic asymmetry of this brain structure in both ADHD and healthy control subjects. The inconsistencies might be partly related to methodologic differences among the studies. For example, two of the studies used only one-slice analysis of caudate volumes (143,145). In addition, whereas some studies (136,137) measured both the head and the body of the caudate, other studies (145) assessed only the head of the caudate. However, it should be noted that the subjects who participated in the two studies by Castellanos et al. overlapped significantly. In addition, whereas the ADHD and control groups in the study by Filipek et al. (140) were similar with regard to full-scale IQ, the controls in the studies by Castellanos et al. had significantly higher full-scale IQs than the ADHD subjects. Further, while the subjects in studies by Filipek et al. and Semrud-Clikeman et al. (150) had "pure" ADHD, many of the subjects in the studies by Castellanos et al., Hynd et al., and Mataro et al. (145) had comorbid conditions. All of the subjects in the study by Filipek et al. were receiving stimulant medication at the time of testing. Thus, to sort out the conflicting findings to date, future studies need to control for comorbidity, medication effects, IQ, and morphometric measures.

The putamen has not been implicated as consistently as the caudate in the pathophysiology of ADHD. For instance, of the few structural imaging studies of the putamen, two studies did not detect any structural abnormalities in this region in subjects with ADHD (133,137). However, a morphologic study of adolescents with ADHD revealed significant gray matter deficits in the right putamen and globus pallidus (148). In addition, a study of children with closed head injury revealed that children who develop ADHD secondary to a closed head injury had more lesions in the right putamen relative to those children who did not develop ADHD (159). Similarly, a study of children with a history of a single stroke and no prestroke manifestation of symptoms/traits of ADHD revealed that lesions involving the posterior putamen appear to lead to subsequent manifestation of ADHD symptoms in the absence of caudate damage (160). Because there are fewer morphometric studies of the putamen in individuals with ADHD compared with the number of morphometric studies of the caudate performed in this population, the relationship of the size of this basal ganglia structure to ADHD needs further exploration, particularly in light of the more recent reports suggesting an association between the putamen and primary or secondary symptoms of ADHD.

Similarly, few studies have investigated the size of the globus pallidus in patients with ADHD. To date, four studies have found significant reductions in the volume of this structure in ADHD, although similar inconsistencies exist with regard to the asymmetry of the reductions (133,137,138,148). Specifically, whereas both Castellanos et al. (137,138) and Overmeyer et al. (148) found that the globus pallidus was smaller on the right in children with ADHD, Aylward et al. (133) found that it was smaller on the left. Similarly, a study of children with Tourette syndrome with comorbid ADHD found a smaller globus pallidus on the left (155). It is of note that Castellanos et al.

(138) found that the globus pallidus was smaller on the right both in children with "pure" ADHD and in children who had ADHD with comorbid Tourette syndrome. The results of morphometric studies that investigated groups of children with ADHD with comorbid disorders should be interpreted with caution, because their findings may not be specific to the presence of ADHD *per se* but instead could be associated with the presence of comorbid conditions.

Corpus Callosum

A number of studies reported smaller anterior regions (rostrum, genu, rostral body) of the corpus callosum in children with ADHD (134,136,142,153). In contrast, other studies showed that posterior portions of the corpus callosum, such as the splenium, are smaller in children with ADHD (142,144,149). Castellanos et al. (137) did not find any significant differences in either the anterior or the posterior segment of the corpus callosum in children with ADHD. Similarly, Overmeyer et al. (147) compared children with ADHD to siblings of children with ADHD and did not find significant differences in the size of any part of the corpus callosum between the two groups. Therefore, to date, results on the nature of the morphologic anomalies in the corpus callosum in children with ADHD are not entirely concordant. The discrepancy may be related to the particular parcellation method used and the number of subregions defined by various research groups. For example, whereas some (134,142) used a five-segment division, others (136,144,147,149) implemented a seven-segment division. In sum, it remains unclear whether any regions of the corpus callosum are disproportionately affected in ADHD and, if so, whether it is the anterior or the posterior region.

Cerebellum

Until recently, the cerebellum was viewed primarily as a motor coordination center; however, research over the last decade suggests that it also may play a role in cognitive functioning (161). A number of studies reported higher executive dysfunction in patients with cerebellar lesions (146,162,163). In general, cerebellar lesions have been shown to impair perceptual timing and to result in deficits in learning, attention, planning of action, and working memory (135).

The first study reporting significantly reduced cerebellar volume in children with ADHD was published by Castellanos et al. (137), who indicated that cerebellar size was significantly reduced in 57 male children and adolescents with ADHD relative to 55 male controls. A subsequent study by the same research group (135) using a subsample of children from their initial study (137) indicated that the cerebellar vermis in particular was smaller in children and adolescents with ADHD. The reduction was localized specifically to the inferior portion of the posterior vermis (lobules VIII–X). This finding was replicated by Mostofsky et al. (146), who found identical reduction of the size of the posterior-inferior vermis (lobules VIII–X) in boys with ADHD. Castellanos et al. (139) found similarly reduced volumes in lobules VIII to X of the posterior-inferior vermis in girls with ADHD. The agreement between the four studies suggests that ADHD is associated with a specific abnormality of the inferior portion of the posterior vermis. As noted by Mostofsky et al., the presence of cerebellar anomalies in individuals with ADHD is not inconsistent with the frontostriatal hypothesis of the disorder because the inferior posterior vermis may be part of a larger frontal-subcortical network, as suggested by studies finding impaired executive and attentional functions in patients with cerebellar lesions. Cerebellar dysfunction also has been related to problems involving timing (164,165) that often are found in individuals with ADHD. In light of these findings, it has been suggested that a dysfunction of the cerebello-thalamic-basal ganglia-prefrontal circuits may be related to the executive, motor, and inhibitory deficits found in patients with ADHD (166).

Structural-Functional Relationships

A number of studies have examined the relationship between neuroanatomic structural brain differences and performance on neuropsychological measures in patients with ADHD (136,137,142,145,150,167). Castellanos et al. (136,137) found that scores on the vocabulary subtest of the Wechsler Intelligence Scale for Children-Revised (168) were significantly correlated with total brain volume in a group of children with ADHD. Semrud-Clikeman et al. (169) examined the relationship between neuropsychological measures and MRI measures in three groups of children: dyslexia, ADHD, and normal controls. They found a positive relationship between verbal comprehension and the size of the planum temporale in all three groups. More recently, the same research group studied ten children and adolescents with ADHD and 11 control subjects, and examined the relationship between structures compromised in children with ADHD and measures of inhibition and sustained attention. They found that children with reversed caudate asymmetry (R > L) performed more poorly on measures of inhibition (assessed by the Stroop test), regardless of group membership (150). The size of the left caudate (found to be smaller in the ADHD group) also was related to higher scores on the externalizing measure of the Child Behavioral Checklist. In contrast, there was a relationship between a measure of sustained attention and bilaterally smaller volumes of the white matter in anterior-superior frontal regions in both groups. It is of note that although this study found a relationship between inhibition and sustained attention and brain morphology, this relationship was not limited to the ADHD group but also was present in normal subjects exhibiting poorer performance on the neuropsychological measures.

Mataro et al. (145) assessed the correlations between performance on tests of attention/working memory and caudate volumes. Interestingly, whereas they found that a larger caudate was associated with worse performance on the CPT test and higher ratings on the Conners Teachers Rating Scale in healthy adolescents, there was a lack of significant associations in the ADHD group.

Casey et al. (167) performed correlational analyses examining the relation among attention, response inhibition, and MRI measures, while controlling for age, estimated full-scale IQ, and cerebral volume in groups of children with and without ADHD. They observed longer and more variable reaction times across both the control and the inhibitory trials in the ADHD group, and reduced accuracy of responding only on the inhibitory trials in the ADHD group. Volumetric measures of the prefrontal cortex, caudate nucleus, and globus pallidus correlated with task performance, whereas measures of the putamen did not. The significant correlations were predominantly in the right hemisphere. Only the prefrontal volumes were correlated specifically with the inhibitory conditions on the tasks, whereas the caudate and globus pallidus volumes were correlated with performance on both the control and the inhibitory trials. This suggests that prefrontal regions are associated predominantly with attentional/inhibitory control, whereas regions in the basal ganglia appear more related to general motor control. In addition, performance on the inhibitory tasks improved as a function of age for both groups.

Analyzing correlations between morphometric and neuropsychological measures is a relatively crude approach to assessing structure-function relationships. A more promising approach is to use functional imaging methods, which are reviewed in the next section.

Functional Imaging Studies

A number of techniques have been used to obtain functional images of the brain. Until recently, the most common technique was positron emission tomography (PET). This technique is less widely used in children because of the health risks associated with administering radioisotopes. For this reason, PET studies have been more common with adults than with children. Additional methods for functional imaging include single photon emission computed tomography (SPECT) and functional magnetic resonance imaging (fMRI). The latter holds the greatest promise because of the lack of ionizing radiation resulting in lesser risk for the participants and the superior spatial and temporal resolutions. In general, all of the functional imaging studies in ADHD completed to date provide some evidence for brain dysfunction in persons with ADHD.

Resting Functional Imaging Studies

Almost all of the resting functional imaging studies of ADHD were performed with children (170–177); only two studies (178,179) were conducted with adults.

Some of the earliest studies with children with ADHD were performed by Lou et al. (176,177) using xenon emission computed tomography. Lou et al. (172) demonstrated bilateral hypoperfusion centrally in the frontal lobes and in the caudate in a group of children with ADHD, some of whom had developmental dysphasia. In a later study, Lou et al. (176) added a relatively pure group of children with ADHD to the group who had additional comorbid disorders and found that both groups exhibited hypoperfusion in striatal regions. The hypoperfusion was statistically significant in the right striatum in the pure ADHD group, and bilateral in the striatum in the group of children with ADHD and comorbid conditions. In addition, the pure ADHD group showed evidence of hyperperfusion in left sensorimotor, primary auditory, and bilateral occipital cortices. In a third study by the same research group, similar hypoperfusion in striatal and posterior periventricular areas in a group of children with pure ADHD was seen (177).

A SPECT study found normal resting striatal perfusion and decreased perfusion in the left frontal and parietal regions in a group of children with ADHD who had comorbid psychiatric disorders, relative to a group of mixed psychiatric controls (173). However, because of ethical constraints in using ionizing radiation in normal control children, mixed psychiatric controls were used for comparison. In addition, the ADHD participants were significantly younger and had significantly lower IQ scores than the controls, which could have exerted a confounding effect.

In 1994, Ernst et al. (175) studied the combined effects of ADHD and gender on cerebral glucose metabolism. They found greater abnormalities in brain metabolism in girls with ADHD than in boys with the disorder. The reductions were in both global metabolism (15% reduction) and absolute regional metabolism of 27 brain regions (45% reduction of the examined regions), with the left prefrontal region being most affected. No differences were demonstrated in cerebral glucose metabolism between boys with ADHD and controls. This was interpreted as suggesting differential penetrance of the genes for ADHD by gender such that ADHD girls must have the vulnerability more severely before the symptom expression is seen. However, a study designed to replicate the findings of Ernst et al. (175) failed to find abnormally low global cerebral metabolic rates for glucose in girls with ADHD (170). Regional analyses revealed lower glucose metabolism in the left sylvian region in the parietal cortex, higher metabolism in the right anterior putamen, lower metabolism in the left anterior putamen, and higher regional cerebral metabolism in the limbic regions, particularly the hippocampus, of girls with ADHD.

In two other studies, the same group of researchers (171,179) used PET with fluorine-18 fluoro-DOPA ([18F]fluoro-DOPA) to compare presynaptic dopaminergic function in children (171) and adults (179) with ADHD. The studies revealed some intriguing differences between children and adults with ADHD. Specifically, in children

with ADHD, accumulation of [18F]fluoro-DOPA was 48% higher in the right midbrain relative to control children. In contrast, [18F]fluoro-DOPA was lower in medial and left prefrontal areas in adults with ADHD, where the [18F] ratios were, respectively, 52% and 51% significantly lower than those of controls. In addition, the [18F] ratios in the left prefrontal cortex, but not in the medial prefrontal cortex, were significantly correlated with the criteria for childhood ADHD. The results indicate that adults with ADHD may have abnormally low dopa decarboxylase activity in the prefrontal cortex, whereas children with ADHD tend to display abnormalities in the midbrain. This suggests that there might be a more extensive cortical involvement in adults with ADHD, whereas the subcortical dopaminergic nuclei could be more affected in children than in adults. However, a SPECT study of six adults (four women) showed that the dopamine transporter in the striatum of adults with ADHD was elevated by 70% relative to healthy controls (178). These findings merit further investigation.

Teicher et al. (174) used a new fMRI procedure, T2 relaxometry, which provides steady-state blood flow measures, to assess blood volume in the striatum (caudate and putamen) of boys with ADHD. They found that boys with ADHD had higher T2 relaxation time measures (indicating lower perfusion) than healthy controls bilaterally in the putamen. These measures were strongly correlated with the accuracy of their performance on a CPT task and with their capacity to sit still. The putamen has been implicated in the regulation of motor behavior, whereas the caudate also subserves higher cognitive functions (180), which may explain the greater relationship between motor activity and activation of the putamen rather than the caudate.

One caveat of performing resting functional imaging studies is that the resting state is inherently uncontrollable and variable at both an experiential and physiologic level (181); as a result, it may be less suitable to study brain function, especially in clinical groups. Specifically, there is evidence that subtle variations in what a subject attends to or thinks about during the resting state might substantially affect functional activation patterns (182). Therefore, given the individual differences in such uncontrollable factors during the resting state, it is difficult to determine the source and nature of the resulting signal.

Cognitive Activation Studies

Of the existing functional imaging studies that incorporate cognitive activation paradigms in their design, two (67,172) were conducted with children with ADHD, three (109,183,184) studied adolescents with the disorder, and four (107,185–187) were performed with adults.

Lou et al. (172) used a subsample of five children who had developmental dysphasia and attention deficit disorder to subtract their regional cerebral blood flow (rCBF) pattern at rest from the rCBF pattern while they were performing an object-naming task. They found that an object-naming task failed to produce an increased flow in expected brain regions (perisylvian, frontal, and occipital areas) in the ADHD group.

An fMRI study with two go/no go tasks with children with ADHD and healthy controls revealed that the ADHD group had greater frontal activation on a response-controlled go/no go task and reduced striatal activation on a stimulus-controlled go/no go task, relative to control children (67). Administration of methylphenidate affected activation only on the stimulus-controlled task in both groups, such that it increased striatal activation in the ADHD children but reduced it in the control children.

A SPECT study revealed that 22% of a group of 54 children and adolescents with ADHD showed prefrontal hypoperfusion at rest, and 65% showed prefrontal deactivation while performing a CPT task (183,188). However, a substantial number of the ADHD subjects (38%) also had increased focal left frontal dorsolateral activity at rest, and 35% had increased focal activity in the same area during the CPT task.

Another study with ADHD adolescents used fMRI to investigate the functional integrity of prefrontal brain regions in a group of seven males with ADHD and nine control subjects during two tasks of executive functions assessing motor response inhibition and motor timing (184). During a "stop" task requiring inhibition of a planned response, the ADHD group exhibited reduced brain activation in the right mesial frontal cortex (BA 8/32), right lateral prefrontal cortex (BA 45 and 9/46), and the left caudate nucleus, while showing greater activation in the right precentral and postcentral gyrus, right inferior parietal lobe, and right caudate nucleus relative to the control group. During the "delay" task of motor timing, which required synchronization of a motor response to an intermittently appearing visual stimulus, the ADHD participants showed greater activation bilaterally in the putamen, right supplementary motor area, and right extrastriate cortex, whereas the controls showed greater magnitude of response in the anterior and posterior cingulate gyrus relative to the ADHD group.

In a PET study, Zametkin et al. (187) found that global cerebral glucose metabolism was 8.1% lower in 25 adults with ADHD relative to 50 controls during the performance of a CPT task. The reduction in glucose metabolism was greater in women (12.7% reduction) than in men (6.0% reduction). Absolute glucose metabolism was reduced in 30 of 60 brain regions in the ADHD group. The greatest reductions were in bilateral premotor and superior prefrontal cortices; however, the reduced metabolism was not task specific. The study design was repeated in adolescents with ADHD, which revealed that neither global nor absolute measures of metabolism were significantly different in adolescents with ADHD versus normal adolescents (109). Only when the values for the individual regions of interest were normalized did ADHD adolescents exhibit significantly lower metabolism

in six regions, including three regions in the left frontal area (which correlated significantly with symptom severity), as well as the left thalamus, right temporal region, and right hippocampus in the ADHD subjects. Similarly to the earlier study (187), the reductions in metabolism were greater in girls than in boys and were most significant in girls with ADHD (17% reduction).

Ernst et al. (186) used PET to study 39 adults with ADHD and 56 healthy controls during the performance of a CPT task. The results revealed no group differences in cerebral glucose metabolism; however, there was a main effect of gender, such that metabolism was generally higher in women than in men.

Bush et al. (185) conducted an fMRI study using the Stroop paradigm to specifically examine the functional integrity of the anterior cingulate cortex in adults with ADHD. They found hypoactivation in the anterior cingulate in the ADHD group during the performance of the task, whereas significant activation in this area was exhibited by the matched normal control group. Importantly, the study revealed that the subjects with ADHD appear to recruit a different network during the task, which included BA 45 of the lateral prefrontal cortex bilaterally, the insular cortex bilaterally, and the left caudate, right putamen, right thalamus, and left pulvinar. In contrast, the healthy control subjects exhibited significant activations in the bilateral anterior cingulate, left middle frontal gyrus (BA 9), and left parietal and left medial/inferior occipital gyrus. The anterior cingulate has been implicated in attention and executive functions, such as working memory, complex motor control, motivation, novelty, error detection, and anticipation (185). Thus, the anterior cingulate, via its extensive reciprocal connections with the lateral prefrontal cortex, may be an additional structure involved in the frontostriatal network implicated in the pathophysiology of ADHD.

Schweitzer et al. (107) used PET while adults with ADHD and a normal control group performed a working memory paradigm (PASAT) and reported a lack of task-related frontal activations in the ADHD group. In contrast, the control group activated brain regions that were consistent with neural models of working memory.

One potential confound pertaining to functional imaging studies that use cognitive activation paradigms is the frequent presence of group differences in behavioral performance between the clinical group and the control group on the task that is being used. The presence of such differences creates significant circularity in the interpretation of the functional imaging results in that it is difficult to know whether an observed aberrant functional activation pattern is due to the inherent differences in behavioral performance, or conversely whether the differences in behavioral performance are due to the abnormal functional activation (189). It is of note that among the functional imaging studies of ADHD that have used cognitive activation paradigms, most of the reported reduced frontostriatal activation in ADHD individuals was observed within the context of poor behavioral performance relative to that of a control group (67,107,184). Therefore, it is possible that the reported hypoactivation may be epiphenomenal to the existing group differences in task performance.

The evidence from resting functional imaging studies and functional imaging studies incorporating cognitive activation paradigms suggests that cortical regions are mostly affected in adults (179,187) and adolescents with ADHD (107,109,175,184,185), whereas subcortical structures seem to be more affected in children (170–172,174). This may help to explain, in part, the clinical observations that adults who continue to meet the criteria for ADHD present with fewer symptoms of hyperactivity but with unchanged or more pronounced symptoms of inattention when compared to children with the disorder. It also may help explain the differential neuropsychological impairments characteristic of children and adults with the disorder, namely, that the primary deficit in children with ADHD is behavioral inhibition, whereas adults with ADHD show deficits predominantly in attention and working memory, processes that are heavily dependent on cortical involvement. This evolution of ADHD symptomatology may suggest that a functional normalization of the structures that control motor activity (such as the basal ganglia) may occur with age.

Review of Evidence for Frontostriatal Involvement in Attention Deficit/Hyperactivity Disorder

Frontal Effects

Among the structural imaging studies investigating the frontal lobes, there appears to be a consensus that the right frontal lobe is smaller in individuals with ADHD (137,140,148,157). However, the evidence does not appear to be as consistent with regard to the functional imaging studies of the frontal lobes in ADHD. For example, a number of studies report reduced activity in the frontal lobes of subjects with ADHD (107,109,173,175,179,183,184,187); yet many of the same studies also report *increased* activation in various frontal areas in ADHD subjects. For example, Amen and Carmichael (183) reported increased dorsolateral frontal activation, Ernst et al. (175) noted increased left posterior frontal activation, Rubia et al. (184) reported increased right supplementary motor area (SMA) and right precentral gyrus activation, Schweitzer et al. (107) observed increased activation in the right middle frontal gyrus, and Zametkin et al. (109) reported higher metabolism in the left posterior frontal cortex in ADHD subjects. In line with these findings, Vaidya et al. (67) reported that children with ADHD showed greater frontal activation and reduced striatal activation than controls during tasks of response inhibition, which was interpreted as being a function of a greater inhibitory effort exerted by the ADHD group in performing the task.

Based on the studies reviewed, there appears to be a much greater consensus in the structural imaging literature documenting abnormalities in the frontal lobes in individuals with ADHD than in the functional imaging literature, where the results appear to be less consistent in terms of the direction of the functional abnormalities in that population, that is, whether they consist of "hypoactivation" or "hyperactivation." It should be noted that increased metabolism or signal intensity in frontal regions in individuals with ADHD relative to controls does not in any way indicate that ADHD subjects do not have abnormalities in the functioning of their frontal lobes, because, under certain circumstances (high working memory load), having an increased rather than decreased response in the frontal lobes may be the aberrant response (181,190,191).

Striatal Effects

With regard to the basal ganglia and the striatum in particular, two xenon emission computed tomography studies (176,177) and an fMRI study (67) reported significantly reduced striatal activity in children with ADHD. Of nine functional neuroimaging studies that specifically examined the caudate nucleus in individuals with ADHD (109,170–172,174,175,184,186,187), only two (172,184) reported decreased activation of the caudate in subjects with ADHD. However, Lou et al. (172) did not use statistical analysis, which brings into questions the validity of the finding. Similarly, of eight functional imaging studies investigating the putamen in individuals with ADHD (109,170,171,174,175,184,186,187), only three (170,174,175) reported reduced activity in this structure in those with ADHD. In one of the studies (175), metabolism was decreased bilaterally in the putamen only in female and not in male children with ADHD. Ernst et al. (170) found that regional cerebral metabolic rates for glucose were higher in girls with ADHD in the right anterior putamen and lower in the left relative to control girls. In addition, even though the main focus of their study was not the basal ganglia, Bush et al. (185) reported increased activation in the right putamen in adults with ADHD while performing the Stroop task. It is of interest that structural imaging studies implicate most often the caudate rather than the putamen in the pathophysiology of ADHD. Yet, based on the functional imaging studies reviewed, it appears that abnormalities in the putamen are found at least as often as in the caudate in subjects with ADHD. This difference may reflect the relative paucity of structural imaging studies that examine the putamen compared with the number of studies that map the caudate.

In summary, reduced activation in the striatum in individuals with ADHD has not been demonstrated to a reliable degree of consistency, and it appears that the putamen is a structure that has not received sufficient attention in the existing ADHD imaging literature to date.

CURRENT THEORETICAL MODELS OF ATTENTION DEFICIT/HYPERACTIVITY DISORDER

Attention Deficit/Hyperactivity Disorder as a Deficit in Behavioral Inhibition

Based on the significant progress made during the last 2 decades in elucidating the neuropsychological functions of the prefrontal cortex and its implication in executive functions and on evidence suggesting that ADHD might arise from deficits in the development, structure, and function of the prefrontal cortex and its connections with other brain regions, Barkley (47) presented the most comprehensive neuropsychological model of ADHD to date, an attempt to account for these findings. The model postulates that the core mechanism of the disorder is a deficit in behavioral inhibition. This primary impairment leads to secondary deficits in four executive functions, which subserve self-controlled and goal-directed behavior and are related to the intact functioning of the prefrontal cortex: (a) *nonverbal working memory,* which represents the capacity to hold information in mind in order to control a response; (b) *verbal working memory,* which is composed of the internalization of speech, which progresses from public and other-directed speech to speech turned to the self, and ultimately to fully covert speech or verbal thought; (c) *self-regulation of affect, motivation, and arousal,* which refers to the capacity of the two forms of working memory to elicit and modulate affective and motivational states in unison with the internally represented information; and (d) *reconstitution,* or the capacity to take such internally represented information apart (analysis) and recombine it (synthesis) into novel, complex behavioral sequences. These secondary deficits, in turn, lead to reduced internal control of motor behavior, which ultimately is expressed in the symptoms of hyperactivity and impulsivity.

Barkley et al. (120) argue that "executive function is now seen as a special case of attention," which is reminiscent of what Alan Baddeley (192) has stated about working memory, namely, that it would be better conceptualized as "working attention" rather than "working memory." Barkley (193) posits that it is the behavior inhibition deficits that account for the appearance of inattention seen in individuals with ADHD. Thus, according to his theory, the "inattention" associated with ADHD is not a primary but a secondary symptom; that is, it is a consequence of the impairment that poor behavioral inhibition creates in the executive control of behavior. Barkley (194) also recommended eliminating the term *attention deficit* and replacing it with the term *behavioral inhibition deficit.*

Barkley's model is unarguably the most comprehensive model of ADHD developed so far and accounts to a large extent for the variety of deficits observed in individuals with this disorder. Importantly, many of the neuropsychological findings in the ADHD literature are in accord with the predictions of the model (47) reviewed in reference 1. For example, the increased number of errors of commission (i.e., inappropriately responding to incorrect stimuli) that are emblematic of the performance of children with ADHD are readily explained by a primary deficit in behavioral inhibition as suggested by the model. However, other common findings, such as increased number of errors of omission (i.e., failure to respond to correct stimuli) that also are characteristic of the performance of individuals with ADHD are not easily explained by a dominant deficit in behavioral inhibition. Similarly, the increased latencies to respond that are commonly reported in the neuropsychological literature on ADHD cannot be accounted for by a primary deficit in behavioral inhibition. If these deficits were due primarily to impaired behavioral inhibition, one would expect decreased latencies to respond (i.e., faster responding) rather than increased latencies. Of course, latency measures could be influenced by both children's typical rate of responding and the efficiency with which they process information (195). It is highly possible that individuals with ADHD could sacrifice speed in order to achieve a level of accuracy on par with that of normal individuals. The problem for the behavioral inhibition hypothesis arises in those instances when increased latencies to respond are the *only* apparent deficit exhibited by individuals with ADHD and are not accompanied by impaired quality of responding (evidenced by decreased accuracy).

In addition, some studies yielded results that contradict the behavioral inhibition hypothesis. For example, in a series of studies, Sonuga-Barke et al. (196,197) showed that inhibitory problems in ADHD occur only in certain situations and are more likely to reflect a sensitivity to delay rather than an underlying inhibitory deficit.

Furthermore, whereas Barkley (47) asserts that deficits in working memory in ADHD are secondary and due to a primary deficit in behavioral inhibition, other investigators have suggested the opposite, namely, that the consistent deficits in inhibition noted in children with ADHD could be secondary to deficits in working memory (74,111,112). Thus, whereas Barkley (47,198) posits that the working memory impairments in individuals with ADHD are due to their inability to inhibit prepotent responses to ongoing events, the reverse also could be true, that is, a deficit in behavioral inhibition could be due to impaired working memory that, as a result, cannot provide appropriate and timely feedback to the behavioral inhibition system about what particular response needs to be inhibited and what needs to be executed.

Barkley (47) also notes that behavioral inhibition provides the delay necessary for the four executive functions to occur and thus protects behavior from being determined and controlled by the immediate environment. Behavioral inhibitory processes presumably also help to bring behavior under the control of internally represented information, which directs and sustains behavior toward future hypothetical events or goals. Interestingly, this description of the functions of the behavioral inhibition system is highly reminiscent of the description of working memory described by other investigators (74,111,199). For example, one of the primary features of working memory is that it occurs during delays between attention/perception and the motor response (111). Some models of working memory consider inhibition to be an intrinsic property of a working memory system (114,115). Increased activation of working memory processes is suggested to inhibit the activations required for competing response outcomes and to be generally involved in selecting appropriate action alternatives and inhibiting inappropriate ones (116). Impaired or hypoactive working memory, in turn, may result in reduced inhibitory control and selection of inappropriate responses.

In summary, working memory and behavioral inhibition appear closely related both in Barkley's model of ADHD and in various models of working memory. The only difference between the models lies in the weight assigned to each function by different models. For example, in Barkley's model, behavioral inhibition is posited to be primary and working memory secondary, whereas in other models behavioral inhibition is but one subcomponent of working memory. The nature of the neuropsychological deficits in the majority of studies with children with ADHD suggests that both behavioral inhibition and working memory are impaired, which is in accordance with Barkley's model. However, the neuropsychological literature on adults with ADHD suggests that behavioral inhibition deficits tend to abate with age, whereas working memory/attentional deficits become more prominent in adulthood.

Attention Deficit/Hyperactivity Disorder as "Intention" Deficit Disorder

The model proposed by Denckla (111) posits that working memory and intention are the two executive functions that are impaired in children with ADHD. She distinguishes between attention, which generally precedes sensory detection/perception, and "intention," which occurs between sensation/perception and action. *Intention,* or *intentionality,* is defined as the "when" system in the brain, controlling when an action should be executed and thus is the state of preparedness to respond, which is closely affiliated to praxis and to the final common pathways of motor output. Thus, Denckla suggests that what is impaired in children with ADHD is not attention *per se,* but "intention," which reflects the failure of received directions to elicit desired actions. Another prominent construct in Denckla's model is the construct of working memory, which often is impaired in patients with ADHD. Working memory, as defined by Denckla (111), is strongly dependent on inhibition, because "if there was no inhibition, allowing delay between stimulus and response, there could be no working memory." Similarly, others also have proclaimed working memory as the key construct embodying executive functions. For example, Pennington (see

reference 200) has collapsed his two-factor model of executive functions (working memory and inhibition) into a single-factor model (working memory) "because as working memory is a limited capacity system, inhibition (or interference control) is intrinsic to its operation." Denckla suggests that impairments in rule-governed behavior, which often characterize children with ADHD, may be dependent on the phonologic component of working memory such that verbal rules do not "feed forward" to the central executive component of working memory. Thus, as children with ADHD grow older, they may be less overtly hyperactive or impulsive and may appear more similar to children with a verbal learning disability due to their deficits in the phonologic and central executive components of working memory.

Denckla's model focuses on the role of output dysfunction (due to impaired information processing, which includes impaired intention and working memory), as opposed to input dysfunction (inattention). She notes that the diagnosis of ADHD is vague with regard to which of the many aspects of the multifaceted construct of attention are impaired and which are spared in individuals with ADHD. She draws attention to the fact that one of the most widely used laboratory measures (CPT) that has been used to assess deficits in *sustained attention* in children with ADHD does not appear to adequately measure this construct. She indicates that CPT tasks reveal increased errors of commission, prolonged reaction times, and greater variability of time to respond correctly in children with ADHD, yet they fail to reveal any greater decline *over time* in accuracy of responses in individuals with ADHD relative to controls. That is, "what is abnormal in CPT scores in children with ADHD is abnormal from the very outset, but the rate of performance decrement (relative to controls) over the duration of the test is not abnormal" (111, p. 117). One of the few studies that has examined the performance of children with ADHD on the CPT task over an extended period of time did not find any decrement in the performance of the ADHD group (86). Sergeant (201) also notes that a decline in perceptual sensitivity with time-on-task is the classic index of a failure in sustained attention, yet most studies have used only the false-positive rates (i.e., errors of commission) as a measure of sustained attention. He suggests that false-negative rates (i.e., errors of omission) could have more validity as indices of failures of attention. These remarks question the validity of studies showing deficits in sustained attention in children with ADHD. Others have similarly questioned the widely accepted view of impulsive responsive style in children with ADHD, based on consistent findings of longer response latencies/slower reaction times in subjects with ADHD that are indicative of inefficient rather than of impulsive response style in this population (202,203).

In general, Denckla's account of the primary deficits in ADHD is similar to Barkley's model in that both postulate deficits in response inhibition and working memory. The difference stems from the priority that they assign to these constructs. Whereas Barkley posits that the primary deficit in ADHD is in behavioral inhibition and that working memory is a secondary deficit, Denckla claims that the primary deficit is in working memory (particularly the phonologic/verbal and central executive components) that, by being a superordinate construct, includes response inhibition.

Attention Deficit/Hyperactivity Disorder as a Disorder of Inattention (Sustained Attention)

Hyperactivity was considered the core feature of ADHD before 1980, when the disorder was described as "hyperkinetic reaction of childhood." Douglas (204–206) challenged the predominant clinical descriptions of the disorder as one of hyperactivity and hyperkinesis by recommending the inclusion of a separate domain of inattention in the symptom domains of ADHD. She emphasized the role of poor sustained attention, defined as the "maintenance of attention over time" (204, p. 285), over impulsivity and hyperactivity in her conceptualization of the disorder. Douglas (204,205) proposes that the primary deficit in hyperactive children is one of self-regulation/executive processes, which involves defective attentional, inhibitory, arousal, and reinforcement mechanisms. According to her theory, the self-regulatory deficit encompasses three principal aspects: (a) organization and monitoring of information processing, such as planning, executive functions, self-monitoring, and regulation of arousal and alertness to meet task demands; (b) mobilization of attention and effort throughout information processing, which refers to the deployment of sufficient "effortful" attention throughout all stages of information processing, including stimulus evaluation, response decision, and response execution, and maintaining it for as long as is required by the task; and (c) inhibition of inappropriate responding, which includes withholding inappropriate responses, responses to inappropriate stimuli, and responses to inappropriate reinforcers.

Consequently, she predicts that the deficits in children with ADHD would become most evident when demands for self-regulated, effortful, and sustained performance are high. Thus, the tasks that would be most sensitive to such deficits should be sufficiently demanding to engage the self-regulatory capabilities.

Notably, Douglas (204) argues that attentional problems in children with ADHD should not be overemphasized and inhibitory processes minimized, thereby underlining both the facilitatory and the inhibitory aspects of self-regulation. She acknowledges that the types of errors commonly associated with the performance of children with ADHD provide clear evidence for inhibitory problems, "provided it is remembered that they appear within the general context of tasks making heavy attentional demands" (204).

Some of the findings in the neuropsychological literature provide support for the predictions of Douglas's model. For example, she predicts that deficits will be observed on CPT tasks, which, although simple, do require sustained effortful attention for correct execution. In addition, as mentioned earlier, studies have found that storage and recall of simple information in memory is not impaired in individuals with ADHD (129,205,207). Deficits emerge only when larger and more complex amounts of verbal information must be processed or held in mind (205,208). In addition, individuals with ADHD are less proficient than controls when strategies that assist with organizing material so that it can be remembered more effectively are required (98,125).

Douglas's model of ADHD (204,205) also is similar to Barkley's model in that it includes impaired response inhibition and executive functions as some of the major deficits in the disorder. Again, the differences between the two models pertain to the primary emphasis that is being placed on various constructs. Douglas posits that inattention is the cardinal deficit and inhibitory deficits appear only within the general context of tasks that make heavy attentional demands, whereas Barkley claims the opposite—that deficits in attention are a function of the primary deficit in behavioral inhibition.

Attention Deficit/Hyperactivity Disorder as a Cognitive Energetic Deficit

The model of ADHD proposed by Sergeant (209) is based on their cognitive-energetic model of information processing, which posits that the overall efficiency of information processing is determined by both state/energetic factors, such as effort, arousal, and activation, and process/computational factors, such as stimulus encoding, memory search, and decision and motor processing. The first level of the model includes the process of attention, which has four stages: encoding, search, decision, and motor organization. The second level of the model contains three energetic pools: effort (the energy required to meet the demands of a task), arousal (phasic responding related to stimulus processing), and activation (tonic changes in physiologic activity). The third level of the model includes executive functions, which perform managerial or evaluative processes, such as strategic planning, monitoring, intention to inhibit a response, detection, and correction of errors.

Sergeant (209) suggests that two of the second level energetic pools, activation and effort, are particularly relevant to ADHD. Specifically, what appears to be impaired in ADHD is the activation component, which is necessary for the inhibition of a motor response to occur and is crucial in explaining disinhibition in ADHD. He suggests that ADHD may be related to a nonoptimal energetic state, which causes impaired motor processing that becomes evident in slow and inaccurate responding and high variability in the speed of responding. According to this model, individuals with ADHD suffer from a deficit in the energetic maintenance and allocation of resources, which leads to the secondary symptoms of disinhibited behavior. Thus, poor response inhibition in his model is but one aspect of the nonoptimal energetic state. In addition, it is proposed that ADHD children may have a deficit in the third level of the cognitive-energetic model, namely, in executive functions (84), but this part of the model does not seem to be very well developed.

Consequently, Sergeant (209) has argued that it is an oversimplification to conclude that ADHD children suffer uniquely from an inhibition deficit that accounts for all of the experimental findings on a variety of tasks. He has questioned Barkley's (47) proposal that the core deficit of ADHD is a behavioral inhibition deficit, based on findings that indicate that inhibitory dysfunction is not exclusively related to ADHD but also is present in other disorders such as CD and ODD (84,210), as well as in psychological states such as aggression (82,211). Such findings appear to question the discriminant validity of behavioral inhibition. Sergeant proposes that behavioral inhibition, although impaired in ADHD, is not uniquely associated with this disorder, but it is a characteristic of children with disruptive behavior in general. He acknowledges that there may be certain aspects of inhibition that are deficient in children with ADHD, but states that this is also dependent on the energetic state of the child and the allocation of energy to the task at hand. In particular, Sergeant proposes that the core deficit of ADHD is not a fixed "process" deficit such as behavioral disinhibition, but instead is a situational "state" deficit that depends on the task at hand and on the current state of the individual. The model suggests that the deficit in response inhibition observed in ADHD children is modulated by their inability to adjust their current state. In addition, the proposed state (activation) deficit in ADHD is suggested to selectively affect output stages rather than input stages of information processing.

In summary, Sergeant's model of ADHD has commonalities with Barkley's model in that it acknowledges that response inhibition could be impaired in ADHD. However, Sergeant proposes that the observed deficit in response inhibition in children with ADHD is modulated by their inability to adjust their current state, and, in this respect, response inhibition is but one aspect of a nonoptimal energetic state of children with ADHD. In addition, Sergeant posits that response inhibition is not unique to ADHD, as it cannot differentiate between children with ADHD and those with other disruptive disorders.

Neural Network Model of Attention Deficit/Hyperactivity Disorder

Swanson et al. (212) align the clinical symptom domains of ADHD (inattention, impulsivity, and hyperactivity) to cognitive concepts of attention and findings from the neurosciences. Their model is based on the neuroanatomic network theory of attention proposed by Posner and Raichle (213) positing that distinct neuroanatomic networks are in-

volved in three component processes of attention: alerting, orienting, and executive control.

The alerting network is involved in establishing a vigilant state and maintaining readiness to react. It is defined by a network of brain areas in the right frontal lobe, the right parietal lobe, and the locus caeruleus. The orienting network is involved in covert orienting to sensory stimuli and is defined by regions centered in the posterior parietal lobes, superior colliculus, and thalamus. The executive control network is related to the control of goal directed behavior, target and error detection, conflict resolution, and inhibition of automatic responses. It includes the midline frontal areas, such as the anterior cingulate, as well as the left lateral frontal lobe and basal ganglia. The alerting network is suggested to be involved in sustained attention, the orienting network in selective attention, and the executive control network in divided attention.

Swanson et al. (212) characterize ADHD as a combination of executive control and alerting deficits. Accordingly, they predict brain pathology in areas involved in these processes, such as the midline frontal cortex, basal ganglia, anterior prefrontal cortex, and anterior right parietal cortex. The neuropsychological and neuroimaging studies in ADHD reviewed appear to support these predictions. Because other disruptive behavior disorders, such as CD and ODD, also are related to hyperactivity and impulsivity, Swanson et al. suggest that a requirement to manifest symptoms of inattention (particularly deficits in sustained and divided attention) may reduce the source of heterogeneity and the source of misdiagnosis in ADHD.

An extension to the model of Swanson et al. is proposed by Berger and Posner (215), who argue that it is possible to conceptualize three of the current theoretical models of ADHD (1,47,212,216) under the umbrella of pathologies of attentional networks with primary impairments in the executive control and alerting networks.

TREATMENT OF ATTENTION DEFICIT/HYPERACTIVITY DISORDER

The National Institute of Mental Health (NIMH) Collaborative Multisite Multimodal Treatment Study of Children with Attention-Deficit/Hyperactivity Disorder (MTA) is the most carefully crafted and extensive study to date examining differential response to widely used, evidence-based interventions for childhood ADHD (217,218). This cooperative agreement treatment study conducted by the NIMH examined the long-term effectiveness of medication versus behavioral treatment versus both for treatment of ADHD and compared state-of-the-art treatment with routine community care. The subjects were 579 children (aged 7–9 years) with ADHD combined type across six different sites who were randomized to four different conditions: (a) medication titration followed by monthly visits; (b) intensive behavioral treatment (parent management training, strategies for home-based reinforcement for school behaviors and performance, individual psychotherapy, social skills training, academic study skills training, remedial tutoring as needed, with therapist involvement reduced over time); (c) an optimal combination of both approaches; or (d) standard community care (whatever treatments families chose to obtain through their local community providers, including medication if desired). The treatments took place for 14 months, and outcomes were assessed in multiple domains before, during, and at treatment endpoint, with periodic reassessment during the subsequent 24 months.

The results of the MTA study revealed that all four groups showed reductions in symptoms over time, but there were significant differences among them in the degree of change. For ADHD symptoms, carefully crafted medication management and combined treatment were superior to behavioral treatment and to routine community care (which included medication two thirds of the time). Combined treatment was not better than medication management for core ADHD symptoms. However, combined treatment did show a modest advantage over the other treatment conditions in improving non-ADHD symptoms (oppositional/aggressive symptoms, internalizing symptoms) and positive functioning outcomes (teacher-rated social skills, parent-child relations, reading achievement).

The MTA study confirms that stimulant medication remains the single most powerful intervention available for management of the core behavioral symptoms of the disorder. However, it also shows there is a significant increase in the efficacy of medication that is carefully titrated in the context of an ongoing therapeutic relationship, even one with brief contacts, over the medication management typically available from community practitioners. Nevertheless, the role of behavioral intervention in combination with carefully crafted medication management in the treatment of ADHD should not be downplayed. In addition to effecting some improvement in comorbid symptoms and positive functioning outcomes in the ADHD subjects, Vitiello et al. (219) note that children who received the combined medication and behavioral treatment conditions ended the study taking an average 20% lower medication dosages than children who were in the medication condition alone.

In addition to its sheer magnitude and careful implementation, the MTA study has a number of other strengths. It was able to examine treatment response not only in children with "pure" ADHD combined type, but also in those who could be grouped into three comorbidity profiles: ADHD + Anxiety, ADHD + ODD/CD, and ADHD + Anxiety + ODD/CD. The ADHD + Anxiety subjects responded equally well to all MTA treatments, belying the previous indication that they were more likely to have poor or adverse responses to stimulants (220–223). The "pure" ADHD and ADHD + ODD/CD subjects responded only to treatments that included medication, and the ADHD + Anxiety +

ODD/CD subjects responded best to combined treatment (medication and behavioral treatment) interventions.

Despite its strengths, there have been criticisms of the MTA study (224,225). One weakness of the study design, albeit an ethically necessary one, is that it did not include a no-treatment control group. Another possible weakness is that the behavioral treatment condition may not have been fully tested because the involvement of the therapist faded before the end of the study. Barkley (224) critiqued the behavioral intervention for failing to be guided by a theoretical hypothesis or new conceptual insight into the nature of the core deficit in the disorder. Yet Swanson et al. (226) argue that based on Barkley's theory of response disinhibition in ADHD (227), he would recommend intervention procedures consistent with the components of the MTA behavioral treatment (228).

With respect to medication intervention, the last several years have seen a burgeoning of new forms of stimulant medications and new systems for their delivery introduced to the market. Many of these provide longer-acting preparations that increase compliance, provide smoother control of symptoms across the day, and eliminate the need for children to take medication at school, which frequently is a socially stigmatizing event. Stimulant medications remain the first-line drugs for treatment of ADHD, including methylphenidate, which inhibits catecholamine reuptake, and dextroamphetamine, which both inhibits reuptake of dopamine and norepinephrine and facilitates their release. The older tricyclic antidepressants, imipramine and desipramine, which block norepinephrine and serotonin reuptake acutely but reduce output over the long run, continue to be used as well. Some persons respond positively to the newer noradrenergic/dopaminergic reuptake inhibitor, bupropion, or the serotonergic/noradrenergic reuptake inhibitor, venlafaxine. The noradrenergic reuptake inhibitors are the newest class of drugs that hold significant promise for treatment of ADHD. These drugs are expected to be available soon in the United States, with the first exemplar being atomoxetine (Eli Lilly). α- and α_2-Adrenoreceptor agonists (clonidine and guanfacine hydrochloride) are among the medications that can be used for treatment of ADHD that is comorbid with other conditions such as tics or Tourette syndrome, or aggression/CD.

Although the MTA study outlines intervention for ADHD children, there is no comparable research for ADHD adults beyond studies demonstrating the short-term efficacy of various medications, particularly those in the stimulant class. Other typical interventions for adults with ADHD include counseling, including marital or family counseling; psychotherapy; implementation of home, work, school, and lifestyle accommodations; better management of legal substances (avoiding caffeine/nicotine/alcohol abuse); substance abuse treatment as needed; management of other comorbid disorders; group therapy with other ADHD adults; vocational assessment and/or career counseling; and consultation to employer as needed. Assistance with time management, organization, and goal setting and attainment (use of a coaching model) has been especially popular and useful for many persons.

SUMMARY

ADHD is the most common neurobehavioral disorder of childhood, and it continues into adulthood in approximately one to two thirds of cases. The diagnosis of the adult form of the disorder is complicated by its often retrospective nature and remains somewhat controversial. ADHD has a substantial genetic component and is characterized by considerable comorbidity with other disorders/conditions. Clinical observations reveal that motor hyperactivity and impulsivity are the most prominent symptoms in children with ADHD, whereas symptoms of inattention appear more common in adults with ADHD. Therefore, although the core symptoms of ADHD persist into adulthood, symptoms may change over time and diagnostic criteria need to be developmentally referenced.

Neuropsychological findings suggest that the construct of "executive functions" best captures the wide variety of ADHD symptoms and provides the most useful framework for describing the neuropsychological deficits displayed by individuals with the disorder. Accordingly, most of the extant theoretical models of ADHD incorporate one or more executive functions in their major postulations, thus reflecting the major neuropsychological findings in the disorder. Specifically, the majority of neuropsychological studies reveal that deficits in response inhibition are particularly pronounced in children with ADHD, which is in accordance with the notable clinical symptoms of hyperactivity/impulsivity in children with the disorder. In contrast, adults with the disorder show deficits most significantly in the domains of attention and working memory, consistent with the clinical manifestation of the disorder in the adult population.

The neuropsychological findings with individuals with ADHD have been corroborated by results from neuroimaging studies attesting to structural and functional abnormalities in frontostriatal circuits, which are putative neuroanatomic substrates of various executive functions. The most salient and consistent structural imaging findings include smaller total brain volumes and reduced volumes in the right frontal lobes, caudate, and cerebellum. Functional imaging studies also reveal abnormalities in the frontal lobes and basal ganglia in individuals with ADHD. However, the direction of the abnormalities in the frontal lobes, that is, hypoactivation versus hyperactivation, appears to be inconsistent, with studies reporting somewhat conflicting results. In addition, whereas the caudate is the basal ganglia structure most consistently implicated in the pathophysiology of ADHD based on structural imaging studies (most of which were conducted with children with the disorder), in con-

trast, functional imaging studies suggest that abnormalities in the putamen are found at least as often as in the caudate in subjects with ADHD. It should be noted that the neuropsychological and neuroimaging literature continues to be plagued by a number of inconsistencies, primarily related to methodologic limitations, such as small sample sizes, presence of comorbid conditions, current and past medication use, and differences in cognitive abilities between the ADHD and the control groups.

Pharmacotherapy is the principal form of treatment for ADHD. Research supports the role of stimulants as a first-line treatment for both children and adults with ADHD. Yet the combination of pharmacologic and behavioral interventions appears to exert a synergistic effect with respect to some outcomes and may prove to be most beneficial overall.

ACKNOWLEDGMENTS

This work was supported in part by NIH Grant 5 R01 MH57836 and Fonds FCAR predoctoral fellowship.

REFERENCES

1. Barkley RA. *ADHD and the nature of self-control.* New York: Guilford Press, 1997.
2. Wender PH. Attention-deficit hyperactivity disorder in adults. *Psychiatr Clin North Am* 1998;21:761–774.
3. Mercugliano M. What is attention-deficit/hyperactivity disorder? *Pediatr Clin North Am* 1999;46:831–843.
4. American Academy of Pediatrics. Clinical practice guideline: diagnosis and evaluation of the child with attention-deficit/hyperactivity disorder. American Academy of Pediatrics. *Pediatrics* 2000;105:1158–1170.
5. Castellanos FX. Toward a pathophysiology of attention-deficit/hyperactivity disorder. *Clin Pediatr (Phila)* 1997;36:381–393.
6. Biederman J, Newcorn J, Sprich S. Comorbidity of attention deficit hyperactivity disorder with conduct, depressive, anxiety, and other disorders. *Am J Psychiatry* 1991;148:564–577.
7. Barkley RA, Fischer M, Edelbrock CS, et al. The adolescent outcome of hyperactive children diagnosed by research criteria: I. An 8-year prospective follow-up study. *J Am Acad Child Adolesc Psychiatry* 1990;29:546–557.
8. Faraone S, Biederman J, Monuteaux MC. Further evidence for the diagnostic continuity between child and adolescent ADHD. *J Atten Disord* 2002;6:5–13.
9. Faraone SV, Biederman J. Neurobiology of attention-deficit hyperactivity disorder. *Biol Psychiatry* 1998;44:951–958.
10. Milberger S, Faraone SV, Biederman J, et al. New phenotype definition of attention deficit hyperactivity disorder in relatives for genetic analyses. *Am J Med Genet* 1996;67:369–377.
11. Tannock R. Attention deficit hyperactivity disorder: advances in cognitive, neurobiological, and genetic research. *J Child Psychol Psychiatry* 1998;39:65–99.
12. Goodman R, Stevenson J. A twin study of hyperactivity—I. An examination of hyperactivity scores and categories derived from Rutter teacher and parent questionnaires. *J Child Psychol Psychiatry* 1989;30:671–689.

13. Goodman R, Stevenson J. A twin study of hyperactivity—II. The aetiological role of genes, family relationships and perinatal adversity. *J Child Psychol Psychiatry* 1989;30:691–709.

14. Levy F, Hay DA, McStephen M, et al. Attention-deficit hyperactivity disorder: a category or a continuum? Genetic analysis of a large-scale twin study. *J Am Acad Child Adolesc Psychiatry* 1997;36:737–744.

15. Cook EH Jr, Stein MA, Krasowski MD, et al. Association of attention-deficit disorder and the dopamine transporter gene. *Am J Hum Genet* 1995;56:993–998.

16. Gill M, Daly G, Heron S, et al. Confirmation of association between attention deficit hyperactivity disorder and a dopamine transporter polymorphism. *Mol Psychiatry* 1997;2:311–313.

17. LaHoste GJ, Swanson JM, Wigal SB, et al. Dopamine D4 receptor gene polymorphism is associated with attention deficit hyperactivity disorder. *Mol Psychiatry* 1996;1:121–124.

18. Swanson JM, Sunohara GA, Kennedy JL, et al. Association of the dopamine receptor D4 (DRD4) gene with a refined phenotype of attention deficit hyperactivity disorder (ADHD): a family-based approach. *Mol Psychiatry* 1998;3:38–41.

19. Faraone S. Report from the third international meeting of the attention-deficit hyperactivity disorder molecular genetics network. *Am J Med Genet* 2002;114:272–276.

20. Biederman J, Faraone SV, Monuteaux MC. Impact of exposure to parental attention-deficit hyperactivity disorder on clinical features and dysfunction in the offspring. *Psychol Med* 2002;32:817–827.

21. Wender PH, Wolf LE, Wasserstein J. Adults with ADHD. An overview. *Ann N Y Acad Sci* 2001;931:1–16.

22. Gittelman R, Mannuzza S. Diagnosing ADD-H in adolescents. *Psychopharmacol Bull* 1985;21:237–242.

23. Barkley RA. Advancing age, declining ADHD. *Am J Psychiatry* 1997;154:1323–1325.

24. Weiss G, Hechtman L, Milroy T, et al. Psychiatric status of hyperactives as adults: a controlled prospective 15-year follow-up of 63 hyperactive children. *J Am Acad Child Psychiatry* 1985;24:211–220.

25. Murphy K, Barkley RA. Attention deficit hyperactivity disorder adults: comorbidities and adaptive impairments. *Compr Psychiatry* 1996;37:393–401.

26. Wilens TE, Biederman J, Spencer TJ, et al. Pharmacotherapy of adult attention deficit/hyperactivity disorder: a review. *J Clin Psychopharmacol* 1995;15:270–279.

27. Mannuzza S, Klein RG, Bessler A, et al. Adult outcome of hyperactive boys. Educational achievement, occupational rank, and psychiatric status. *Arch Gen Psychiatry* 1993;50:565–576.

28. Wender PH, Reimherr FW, Wood DR. Attention deficit disorder ("minimal brain dysfunction") in adults. A replication study of diagnosis and drug treatment. *Arch Gen Psychiatry* 1981;38:449–456.

29. Barkley RA, Fischer M, Smallish L, et al. The persistence of attention-deficit/hyperactivity disorder into young adulthood as a function of reporting source and definition of disorder. *J Abnorm Psychol* 2002;111:279–289.

30. Fischer M. The persistence of ADHD into adulthood: it depends on whom you ask. *ADHD Rep* 1997;5:8–10.

31. Biederman J, Mick E, Faraone SV. Age-dependent decline of symptoms of attention deficit hyperactivity disorder: impact of remission definition and symptom type. *Am J Psychiatry* 2000;157:816–818.

32. Swanson JM, Sergeant JA, Taylor E, et al. Attention-deficit hyperactivity disorder and hyperkinetic disorder. *Lancet* 1998;351:429–433.

33. Barkley RA. Attention-deficit hyperactivity disorder. *Sci Am* 1998;279:66–71.

34. Hart EL, Lahey BB, Loeber R, et al. Developmental change in attention-deficit hyperactivity disorder in boys: a four-year longitudinal study. *J Abnorm Child Psychol* 1995;23:729–749.

35. Campbell LR, Cohen M. Management of attention deficit hyperactivity disorder (ADHD). A continuing dilemma for physicians and educators. *Clin Pediatr (Phila)* 1990;29:191–193.

36. Applegate B, Lahey BB, Hart EL, et al. Validity of the age-of-onset criterion for ADHD: a report from the DSM-IV field trials. *J Am Acad Child Adolesc Psychiatry* 1997;36:1211–1221.

37. Barkley RA, Biederman J. Toward a broader definition of the age-of-onset criterion for attention-deficit hyperactivity disorder. *J Am Acad Child Adolesc Psychiatry* 1997;36:1204–1210.

38. McGee R, Williams S, Feehan M. Attention deficit disorder and age of onset of problem behaviors. *J Abnorm Child Psychol* 1992;20:487–502.

39. DuPaul GJ, Power TJ, Anastopoulos AD, et al. *ADHD rating scale-IV.* New York: Guilford Press, 1998.

40. Reid R, DuPaul GJ, Power TJ, et al. Assessing culturally different students for attention deficit hyperactivity disorder using behavior rating scales. *J Abnorm Child Psychol* 1998;26:187–198.

41. Conners CK. *Conners' parent rating scale-revised.* North Tonawanda, NY: Multi-Health Systems, 1997.

42. Conners CK. *Conners' teacher rating scale-revised.* North Tonawanda, NY: Multi-Health Systems, 1997.

43. Achenbach TM, Rescorla LA. *Manual for the ASEBA school-age forms & profiles.* Burlington, VT: University of Vermont, Research Center for Children, Youth & Families, 2001.

44. Mapou RL. Assessment, diagnosis, and documentation of adult attention deficit hyperactivity disorder (ADHD). *J Int Neuropsychol Soc* 1999;5:112(abst).

45. Woods SP, Lovejoy DW, Ball JD. Neuropsychological characteristics of adults with ADHD: a comprehensive review of initial studies. *Clin Neuropsychol* 2002;16:12–34.

46. Wozniak J, Crawford MH, Biederman J, et al. Antecedents and complications of trauma in boys with ADHD: findings from a longitudinal study. *J Am Acad Child Adolesc Psychiatry* 1999;38:48–55.

47. Barkley RA. Behavioral inhibition, sustained attention, and executive functions: constructing a unifying theory of ADHD. *Psychol Bull* 1997;121:65–94.

48. Barkley RA, Grodzinsky G, DuPaul GJ. Frontal lobe functions in attention deficit disorder with and without hyperactivity: a review and research report. *J Abnorm Child Psychol* 1992;20:163–188.

49. Grodzinsky GM, Diamond R. Frontal lobe functioning in boys with attention-deficit hyperactivity disorder. *Dev Neuropsychol* 1992;8:427–445.

50. Seidman LJ, Benedict KB, Biederman J, et al. Performance of children with ADHD on the Rey-Osterrieth complex figure: a pilot neuropsychological study. *J Child Psychol Psychiatry* 1995;36:1459–1473.

51. Carter CS, Krener P, Chaderjian M, et al. Abnormal processing of irrelevant information in attention deficit hyperactivity disorder. *Psychiatry Res* 1995;56:59–70.

52. Chelune GJ, Ferguson W, Koon R, et al. Frontal lobe disinhibition in attention deficit disorder. *Child Psychiatry Hum Dev* 1986;16:221–234.

53. Gorenstein EE, Mammato CA, Sandy JM. Performance of inattentive-overactive children on selected measures of prefrontal-type function. *J Clin Psychol* 1989;45:619–632.

54. Loge DV, Staton RD, Beatty WW. Performance of children with ADHD on tests sensitive to frontal lobe dysfunction. *J Am Acad Child Adolesc Psychiatry* 1990;29:540–545.

55. McGee R, Williams S, Moffitt T, et al. A comparison of 13-year-old boys with attention deficit and/or reading disorder on

neuropsychological measures. *J Abnorm Child Psychol* 1989;17:37–53.

56. Reader MJ, Schuerholz LJ. Attention deficit hyperactivity disorder and executive dysfunction. *Dev Neuropsychol* 1994;10.

57. Seidman LJ, Biederman J, Faraone SV, et al. Toward defining a neuropsychology of attention deficit-hyperactivity disorder: performance of children and adolescents from a large clinically referred sample. *J Consult Clin Psychol* 1997;65:150–160.

58. Shue KL, Douglas VI. Attention deficit hyperactivity disorder and the frontal lobe syndrome. *Brain Cogn* 1992;20:104–124.

59. Boucugnani LL, Jones RW. Behaviors analogous to frontal lobe dysfunction in children with attention deficit disorder. *Arch Clin Neuropsychol* 1989;4:161–173.

60. McBurnett K, Harris SM, Swanson JM, et al. Neuropsychological and psychophysiological differentiation in attention/overactivity and aggression/defiance symptom groups. *J Clin Child Psychol* 1993;22:165–171.

61. Ozonoff S, Jensen J. Brief report: specific executive function profiles in three neurodevelopmental disorders. *J Autism Dev Disord* 1999;29:171–177.

62. Leung PW, Connolly KJ. Distractibility in hyperactive and conduct-disordered children. *J Child Psychol Psychiatry* 1996;37:305–312.

63. Aman CJ, Roberts RJ Jr, Pennington BF. A neuropsychological examination of the underlying deficit in attention deficit hyperactivity disorder: frontal lobe versus right parietal lobe theories. *Dev Psychol* 1998;34:956–969.

64. Klorman R, Hazel-Fernandez LA, Shaywitz SE, et al. Executive functioning deficits in attention-deficit/hyperactivity disorder are independent of oppositional defiant or reading disorder. *J Am Acad Child Adolesc Psychiatry* 1999;38:1148–1155.

65. Pennington BF, Grossier D, Welsh MC. Contrasting cognitive deficits in attention deficit disorder versus reading disability. *Developmental Psychology* 1993;29:511–523.

66. Iaboni F, Douglas VI, Baker AG. Effects of reward and response costs on inhibition in ADHD children. *J Abnorm Psychol* 1995;104:232–240.

67. Vaidya CJ, Austin G, Kirkorian G, et al. Selective effects of methylphenidate in attention deficit hyperactivity disorder: a functional magnetic resonance study. *Proc Natl Acad Sci U S A* 1998;95:14494–14499.

68. Barkley RA, Grodzinsky G, DuPaul GJ. Frontal lobe functions in attention deficit disorder with and without hyperactivity: a review and research report. *J Abnorm Child Psychol* 1992;20:163–188.

69. Breen MJ. Cognitive and behavioral differences in ADHD boys and girls. *J Child Psychol Psychiatry* 1989;30:711–716.

70. Shue KL, Douglas VI. Attention deficit hyperactivity disorder and the frontal lobe syndrome. *Can Psychol* 1989;30:498.

71. Fischer M, Barkley RA, Edelbrock CS, et al. The adolescent outcome of hyperactive children diagnosed by research criteria: II. Academic, attentional, and neuropsychological status. *J Consult Clin Psychol* 1990;58:580–588.

72. McGee R, Williams S, Moffitt T, et al. A comparison of 13-year-old boys with attention deficit and/or reading disorder on neuropsychological measures. *J Abnorm Child Psychol* 1983;17:37–53.

73. Sergeant JA, Geurts H, Oosterlaan J. How specific is a deficit of executive functioning for attention-deficit/hyperactivity disorder? *Behav Brain Res* 2002;130:3–28.

74. Pennington BF, Ozonoff S. Executive functions and developmental psychopathology. *J Child Psychol Psychiatry* 1996;37:51–87.

75. Halperin JM, Wolf LE, Pascualvaca DM, et al. Differential assessment of attention and impulsivity in children. *J Am Acad Child Adolesc Psychiatry* 1988;27:326–329.

76. Riccio CA, Reynolds CR. Continuous performance tests are sensitive to ADHD in adults but lack specificity. A review and critique for differential diagnosis. *Ann N Y Acad Sci* 2001;931:113–139.

77. Barkley RA, Grodzinsky G. Are neuropsychological tests of frontal lobe functions useful in the diagnosis of attention deficit disorders? *Clin Neuropsychol* 1994;8:121–139.

78. Nigg JT, Hinshaw SP, Carte ET, et al. Neuropsychological correlates of childhood attention-deficit/hyperactivity disorder: explainable by comorbid disruptive behavior or reading problems? *J Abnorm Psychol* 1998;107:468–480.

79. Giancola PR, Mezzich AC, Tarter RE. Executive cognitive functioning, temperament, and antisocial behavior in conduct-disordered adolescent females. *Journal of Abnormal Psychology* 1998;107:629–641.

80. Moffitt TE. Neuropsychological assessment of executive functions in self-reported delinquents. *Dev Psychopathol* 1989;1.

81. Seguin JR, Boulerice B, Harden PW, et al. Executive functions and physical aggression after controlling for attention deficit hyperactivity disorder, general memory, and IQ. *J Child Psychol Psychiatry* 1999;40:1197–1208.

82. Seguin JR, Pihl RO, Harden PW, et al. Cognitive and neuropsychological characteristics of physically aggressive boys. *J Abnorm Psychol* 1995;104:614–624.

83. Dykman RA, Ackerman PT. Attention deficit disorder and specific reading disability: separate but often overlapping disorders. *J Learn Disabil* 1991;24:96–103.

84. Oosterlaan J, Logan GD, Sergeant JA. Response inhibition in AD/HD, CD, comorbid AD/HD + CD, anxious, and control children: a meta-analysis of studies with the stop task. *J Child Psychol Psychiatry* 1998;39:411–425.

85. Kelly MS, Best CT, Kirk U. Cognitive processing deficits in reading disabilities: a prefrontal cortical hypothesis. *Brain Cogn* 1989;11:275–293.

86. Purvis KL, Tannock R. Phonological processing, not inhibitory control, differentiates ADHD and reading disability. *J Am Acad Child Adolesc Psychiatry* 2000;39:485–494.

87. Kempton S, Vance A, Maruff P, et al. Executive function and attention deficit hyperactivity disorder: stimulant medication and better executive function performance in children. *Psychol Med* 1999;29:527–538.

88. Diamond A. Abilities and neural mechanisms underlying AB performance. *Child Dev* 1988;59:523–527.

89. Seidman LJ, Biederman J, Weber W, et al. Neuropsychological function in adults with attention-deficit hyperactivity disorder. *Biol Psychiatry* 1998;44:260–268.

90. Downey KK, Stelson FW, Pomerleau OF, et al. Adult attention deficit hyperactivity disorder: psychological test profiles in a clinical population. *J Nerv Ment Dis* 1997;185:32–38.

91. Walker AJ, Shores EA, Trollor JN, et al. Neuropsychological functioning of adults with attention deficit hyperactivity disorder. *J Clin Exp Neuropsychol* 2000;22:115–124.

92. Arcia E, Gualtieri CT. Neurobehavioural performance of adults with closed-head injury, adults with attention deficit, and controls. *Brain Inj* 1994;8:395–404.

93. Murphy KR, Barkley RA, Bush T. Executive functioning and olfactory identification in young adults with attention deficit-hyperactivity disorder. *Neuropsychology* 2001;15:211–220.

94. Epstein JN, Johnson DE, Varia IM, et al. Neuropsychological assessment of response inhibition in adults with ADHD. *J Clin Exp Neuropsychol* 2001;23:362–371.

95. Corbett B, Stanczak DE. Neuropsychological performance of

adults evidencing attention-deficit hyperactivity disorder. *Arch Clin Neuropsychol* 1999;14:373–387.

96. Epstein JN, Conners CK, Erhardt D, et al. Asymmetrical hemispheric control of visual-spatial attention in adults with attention deficit hyperactivity disorder. *Neuropsychology* 1997;11:467–473.

97. Gansler DA, Fucetola R, Krengel M, et al. Are there cognitive subtypes in adult attention deficit/hyperactivity disorder? *J Nerv Ment Dis* 1998;186:776–781.

98. Holdnack JA, Moberg PJ, Arnold SE, et al. Speed of processing and verbal learning deficits in adults diagnosed with attention deficit disorder. *Neuropsychiatry Neuropsychol Behav Neurol* 1995;8:282–292.

99. Hopkins J, Perlman T, Hechtman L, et al. Cognitive style in adults originally diagnosed as hyperactives. *J Child Psychol Psychiatry* 1979;20:209–216.

100. Jenkins M, Cohen R, Malloy P, et al. Neuropsychological measures which discriminate among adults with residual symptoms of attention deficit disorder and other attentional complaints. *Clin Neuropsychol* 1998;74–83.

101. Katz LJ, Wood DS, Goldstein G, et al. The utility of neuropsychological tests in evaluation of attention-deficit/hyperactivity disorder (ADHD) versus depression in adults. *Assessment* 1998;5:45–52.

102. Kovner R, Budman C, Frank Y, et al. Neuropsychological testing in adult attention deficit hyperactivity disorder: a pilot study. *Int J Neurosci* 1998;96:225–235.

103. Lovejoy DW, Ball JD, Keats M, et al. Neuropsychological performance of adults with attention deficit hyperactivity disorder (ADHD): diagnostic classification estimates for measures of frontal lobe/executive functioning. *J Int Neuropsychol Soc* 1999;5:222–233.

104. Barkley RA, Murphy KR, Bush T. Time perception and reproduction in young adults with attention deficit hyperactivity disorder. *Neuropsychology* 2001;15:351–360.

105. Rapport LJ, Van Voorhis A, Tzelepis A, et al. Executive functioning in adult attention-deficit hyperactivity disorder. *Clin Neuropsychol* 2001;15:479–491.

106. Rugle L, Melamed L. Neuropsychological assessment of attention problems in pathological gamblers. *J Nerv Ment Dis* 1993;181:107–112.

107. Schweitzer JB, Faber TL, Grafton ST, et al. Alterations in the functional anatomy of working memory in adult attention deficit hyperactivity disorder. *Am J Psychiatry* 2000;157:278–280.

108. Weyandt LL, Rice JA, Linterman I, et al. Neuropsychological performance of a sample of adults with ADHD, developmental reading disorder, and controls. *Dev Neuropsychol* 1998;14:643–656.

109. Zametkin AJ, Liebenauer LL, Fitzgerald GA, et al. Brain metabolism in teenagers with attention-deficit hyperactivity disorder. *Arch Gen Psychiatry* 1993;50:333–340.

110. Barkley RA. The ecological validity of laboratory and analogue assessment methods of ADHD symptoms. *J Abnorm Child Psychol* 1991;19:149–178.

111. Denckla MB. Biological correlates of learning and attention: what is relevant to learning disability and attention-deficit hyperactivity disorder? *J Dev Behav Pediatr* 1996;17:114–119.

112. Karatekin C, Asarnow RF. Working memory in childhood-onset schizophrenia and attention-deficit/hyperactivity disorder. *Psychiatry Res* 1998;80:165–176.

113. Pennington B, Bennetto L, McAleer O, et al. Executive functions and working memory: theoretical and measurement issues. In: Lyon GR, Krasnegor NN, eds. *Attention, memory, and executive functions.* Baltimore, MD: Paul Brookes Publishing, 1996:327–348.

114. Cohen JD, Servan-Schreiber D. Context, cortex, and dopamine: a connectionist approach to behavior and biology in schizophrenia. *Psychol Rev* 1992;99:45–77.

115. Kimberg DY, Farah MJ. A unified account of cognitive impairments following frontal lobe damage: the role of working memory in complex, organized behavior. *J Exp Psychol Gen* 1993;122:411–428.

116. Roberts RJ Jr, Hager LD, Heron C. Prefrontal cognitive processes: working memory and inhibition in the antisaccade task. *J Exp Psychol Gen* 1994;123:374–393.

117. Mariani MA, Barkley R. Neuropsychological and academic functioning in preschool boys with attention deficit hyperactivity disorder. *Dev Neuropsychol* 1997;13:111–129.

118. Zentall SS, Smith YN. Mathematical performance and behavior of children with hyperactivity with and without coexisting aggression. *Behav Res Ther* 1993;31:701–710.

119. Lazar JW, Frank Y. Frontal systems dysfunction in children with attention-deficit/hyperactivity disorder and learning disabilities. *J Neuropsychiatry Clin Neurosci* 1998;10:160–167.

120. Barkley RA, Murphy KR, Kwasnik D. Psychological adjustment and adaptive impairments in young adults with ADHD. *J Atten Disord* 1996;1:41–54.

121. Wiers RW, Gunning WB, Sergeant JA. Is a mild deficit in executive functions in boys related to childhood ADHD or to parental multigenerational alcoholism? *J Abnorm Child Psychol* 1998;26:415–430.

122. Minshew NJ, Goldstein G, Muenz LR, et al. Neuropsychological functioning in nonmentally retarded autistic individuals. *J Clin Exp Neuropsychol* 1992;14:749–761.

123. Oie M, Sunde K, Rund BR. Contrasts in memory functions between adolescents with schizophrenia or ADHD. *Neuropsychologia* 1999;37:1351–1358.

124. Barkley RA. Behavioral inhibition, sustained attention, and executive functions: constructing a unifying theory of ADHD. *Psychological Bulletin* 1997;121:65–94.

125. Douglas VI, Benezra E. Supraspan verbal memory in attention deficit disorder with hyperactivity normal and reading-disabled boys. *J Abnorm Child Psychol* 1990;18:617–638.

126. Frank Y, Seiden J, Napolitano B. Visual event related potentials and reaction time in normal adults, normal children, and children with attention deficit hyperactivity disorder: differences in short-term memory processing. *Int J Neurosci* 1996;88:109–124.

127. Barkley RA, DuPaul GJ, McMurray MB. Comprehensive evaluation of attention deficit disorder with and without hyperactivity as defined by research criteria. *J Consult Clin Psychol* 1990;58:775–789.

128. Douglas VI, Barr RG, Amin K, et al. Dosage effects and individual responsivity to methylphenidate in attention deficit disorder. *J Child Psychol Psychiatry* 1988;29:453–475.

129. Cahn DA, Marcotte AC. Rates of forgetting in attention deficit hyperactivity disorder. *Child Neuropsychol* 1995;1:158–163.

130. Norrelgen F, Lacerda F, Forssberg H. Speech discrimination and phonological working memory in children with ADHD. *Dev Med Child Neurol* 1999;41:335–339.

131. Satterfield JH, Dawson ME. Electrodermal correlates of hyperactivity in children. *Psychophysiology* 1971;8:191–197.

132. Mattes JA. The role of frontal lobe dysfunction in childhood hyperkinesis. *Compr Psychiatry* 1980;21:358–369.

133. Aylward EH, Reiss AL, Reader MJ, et al. Basal ganglia volumes in children with attention-deficit hyperactivity disorder. *J Child Neurol* 1996;11:112–115.

134. Baumgardner TL, Singer HS, Denckla MB, et al. Corpus callosum morphology in children with Tourette syndrome and attention deficit hyperactivity disorder. *Neurology* 1996;47:477–482.

135. Berquin PC, Giedd JN, Jacobsen LK, et al. Cerebellum in attention-deficit hyperactivity disorder: a morphometric MRI study. *Neurology* 1998;50:1087–1093.

136. Castellanos FX, Giedd JN, Eckburg P, et al. Quantitative morphology of the caudate nucleus in attention deficit hyperactivity disorder. *Am J Psychiatry* 1994;151:1791–1796.

137. Castellanos FX, Giedd JN, Marsh WL, et al. Quantitative brain magnetic resonance imaging in attention-deficit hyperactivity disorder. *Arch Gen Psychiatry* 1996;53:607–616.

138. Castellanos FX, Giedd JN, Hamburger SD, et al. Brain morphometry in Tourette's syndrome: the influence of comorbid attention-deficit/hyperactivity disorder. *Neurology* 1996;47:1581–1583.

139. Castellanos FX, Giedd JN, Berquin PC, et al. Quantitative brain magnetic resonance imaging in girls with attention-deficit/hyperactivity disorder. *Arch Gen Psychiatry* 2001;58:289–295.

140. Filipek PA, Semrud-Clikeman M, Steingard RJ, et al. Volumetric MRI analysis comparing subjects having attention-deficit hyperactivity disorder with normal controls. *Neurology* 1997;48:589–601.

141. Foster LM, Hynd GW, Morgan AE, et al. Planum temporale asymmetry and ear advantage in dichotic listening in developmental dyslexia and attention-deficit/hyperactivity disorder (ADHD). *J Int Neuropsychol Soc* 2002;8:22–36.

142. Hynd GW, Semrud-Clikeman M, Lorys AR, et al. Corpus callosum morphology in attention deficit-hyperactivity disorder: morphometric analysis of MRI. *J Learn Disabil* 1991;24:141–146.

143. Hynd GW, Hern KL, Novey ES, et al. Attention deficit-hyperactivity disorder and asymmetry of the caudate nucleus. *J Child Neurol* 1993;8:339–347.

144. Lyoo IK, Noam GG, Lee CK, et al. The corpus callosum and lateral ventricles in children with attention-deficit hyperactivity disorder: a brain magnetic resonance imaging study. *Biol Psychiatry* 1996;40:1060–1063.

145. Mataro M, Garcia-Sanchez C, Junque C, et al. Magnetic resonance imaging measurement of the caudate nucleus in adolescents with attention-deficit hyperactivity disorder and its relationship with neuropsychological and behavioral measures. *Arch Neurol* 1997;54:963–968.

146. Mostofsky SH, Reiss AL, Lockhart P, et al. Evaluation of cerebellar size in attention-deficit hyperactivity disorder. *J Child Neurol* 1998;13:434–439.

147. Overmeyer S, Simmons A, Santosh J, et al. Corpus callosum may be similar in children with ADHD and siblings of children with ADHD. *Dev Med Child Neurol* 2000;42:8–13.

148. Overmeyer S, Bullmore ET, Suckling J, et al. Distributed grey and white matter deficits in hyperkinetic disorder: MRI evidence for anatomical abnormality in an attentional network. *Psychol Med* 2001;31:1425–1435.

149. Semrud-Clikeman M, Filipek PA, Biederman J, et al. Attention-deficit hyperactivity disorder: magnetic resonance imaging morphometric analysis of the corpus callosum. *J Am Acad Child Adolesc Psychiatry* 1994;33:875–881.

150. Semrud-Clikeman M, Steingard RJ, Filipek P, et al. Using MRI to examine brain-behavior relationships in males with attention deficit disorder with hyperactivity. *J Am Acad Child Adolesc Psychiatry* 2000;39:477–484.

151. Shaywitz BA, Shaywitz SE, Byrne T, et al. Attention deficit disorder: quantitative analysis of CT. *Neurology* 1983;33:1500–1503.

152. Fredricksen KA, Cutting LE, Katkin ES, et al. Disproportionate increases in white matter in right frontal lobe in Tourette syndrome. *Neurology* 2002;58:85–89.

153. Giedd JN, Castellanos FX, Casey BJ, et al. Quantitative morphology of the corpus callosum in attention deficit hyperactivity disorder. *Am J Psychiatry* 1994;151:665–669.

154. Nasrallah HA, Loney J, Olson SC, et al. Cortical atrophy in young adults with a history of hyperactivity in childhood. *Psychiatry Res* 1986;17:241–246.

155. Singer HS, Reiss AL, Brown JE, et al. Volumetric MRI changes in basal ganglia of children with Tourette's syndrome. *Neurology* 1993;43:950–956.

156. Reiss AL, Abrams MT, Singer HS, et al. Brain development, gender and IQ in children. A volumetric imaging study. *Brain* 1996;119[Pt 5]:1763–1774.

157. Hynd GW, Semrud-Clikeman M, Lorys AR, et al. Brain morphology in developmental dyslexia and attention deficit disorder/hyperactivity. *Arch Neurol* 1990;47:919–926.

158. Alexander GE, DeLong MR, Strick PL. Parallel organization of functionally segregated circuits linking basal ganglia and cortex. *Annu Rev Neurosci* 1986;9:357–381.

159. Herskovits EH, Megalooikonomou V, Davatzikos C, et al. Is the spatial distribution of brain lesions associated with closed-head injury predictive of subsequent development of attention-deficit/hyperactivity disorder? Analysis with brain-image database. *Radiology* 1999;213:389–394.

160. Max JE, Fox PT, Lancaster JL, et al. Putamen lesions and the development of attention-deficit/hyperactivity symptomatology. *J Am Acad Child Adolesc Psychiatry* 2002;41:563–571.

161. Fiez JA. Cerebellar contributions to cognition. *Neuron* 1996;16:13–15.

162. Burka K, Boscha S, Globasa C, et al. Executive dysfunction in spinocerebellar ataxia type 1. *Eur Neurol* 2001;46:43–48.

163. Riva D, Giorgi C. The cerebellum contributes to higher functions during development: evidence from a series of children surgically treated for posterior fossa tumours. *Brain* 2000;123[Pt 5]:1051–1061.

164. Ivry R. Cerebellar involvement in the explicit representation of temporal information. *Ann N Y Acad Sci* 1993;682:214–230.

165. Salman MS. The cerebellum: it's about time! But timing is not everything—new insights into the role of the cerebellum in timing motor and cognitive tasks. *J Child Neurol* 2002;17:1–9.

166. Giedd JN, Blumenthal J, Molloy E, et al. Brain imaging of attention deficit/hyperactivity disorder. *Ann N Y Acad Sci* 2001;931:33–49.

167. Casey BJ, Castellanos FX, Giedd JN, et al. Implication of right frontostriatal circuitry in response inhibition and attention-deficit/hyperactivity disorder. *J Am Acad Child Adolesc Psychiatry* 1997;36:374–383.

168. Wechsler D. *Manual for the Wechsler Intelligence Scale for Children-Revised.* New York: Psychological Corporation, 1974.

169. Semrud-Clikeman M, Hynd GW, Novey ES, et al. Dyslexia and brain morphology: relationships between neuroanatomical variation and neurolinguistic tasks. *Learn Individ Differ* 1991;3:205–223.

170. Ernst M, Cohen RM, Liebenauer LL, et al. Cerebral glucose metabolism in adolescent girls with attention-deficit/hyperactivity disorder. *J Am Acad Child Adolesc Psychiatry* 1997;36:1399–1406.

171. Ernst M, Zametkin AJ, Matochik JA, et al. High midbrain [18F]DOPA accumulation in children with attention deficit hyperactivity disorder. *Am J Psychiatry* 1999;156:1209–1215.

172. Lou HC, Henriksen L, Bruhn P. Focal cerebral hypoperfusion in children with dysphasia and/or attention deficit disorder. *Arch Neurol* 1984;41:825–829.

173. Sieg KG, Gaffney GR, Preston DF, et al. SPECT brain imaging abnormalities in attention deficit hyperactivity disorder. *Clin Nucl Med* 1995;20:55–60.

174. Teicher MH, Anderson CM, Polcari A, et al. Functional deficits in basal ganglia of children with attention-deficit/hyperactivity disorder shown with functional magnetic resonance imaging relaxometry. *Nat Med* 2000;6:470–473.

175. Ernst M, Liebenauer LL, King AC, et al. Reduced brain metabolism in hyperactive girls. *J Am Acad Child Adolesc Psychiatry* 1994;33:858–868.

176. Lou HC, Henriksen L, Bruhn P, et al. Striatal dysfunction in attention deficit and hyperkinetic disorder. *Arch Neurol* 1989;46:48–52.

177. Lou HC, Henriksen L, Bruhn P. Focal cerebral dysfunction in developmental learning disabilities. *Lancet* 1990;335:8–11.

178. Dougherty DD, Bonab AA, Spencer TJ, et al. Dopamine transporter density in patients with attention deficit hyperactivity disorder. *Lancet* 1999;354:2132–2133.

179. Ernst M, Zametkin AJ, Matochik JA, et al. DOPA decarboxylase activity in attention deficit hyperactivity disorder adults. A [fluorine-18]fluorodopa positron emission tomographic study. *J Neurosci* 1998;18:5901–5907.

180. Graybeal A, Apfel RJ, Flamini R, et al. The basal ganglia and adaptive motor control. *Science* 1994;265:1826–1831.

181. Goldberg TE, Berman KF, Fleming K, et al. Uncoupling cognitive workload and prefrontal cortical physiology: a PET rCBF study. *Neuroimage* 1998;7:296–303.

182. Binder JR, Frost JA, Hammeke TA, et al. Conceptual processing during the conscious resting state. A functional MRI study. *J Cogn Neurosci* 1999;11:80–95.

183. Amen DG, Carmichael BD. High-resolution brain SPECT imaging in ADHD. *Ann Clin Psychiatry* 1997;9:81–86.

184. Rubia K, Overmeyer S, Taylor E, et al. Hypofrontality in attention deficit hyperactivity disorder during higher-order motor control: a study with functional MRI. *Am J Psychiatry* 1999;156:891–896.

185. Bush G, Frazier JA, Rauch SL, et al. Anterior cingulate cortex dysfunction in attention-deficit/hyperactivity disorder revealed by fMRI and the counting Stroop. *Biol Psychiatry* 1999;45:1542–1552.

186. Ernst M, Zametkin AJ, Phillips RL, et al. Age-related changes in brain glucose metabolism in adults with attention-deficit/hyperactivity disorder and control subjects. *J Neuropsychiatry Clin Neurosci* 1998;10:168–177.

187. Zametkin AJ, Nordahl TE, Gross M, et al. Cerebral glucose metabolism in adults with hyperactivity of childhood onset. *N Engl J Med* 1990;323:1361–1366.

188. Amen DG, Paldi F, Thisted RA. Brain SPECT imaging. *J Am Acad Child Adolesc Psychiatry* 1993;32:1080–1081.

189. Price CJ, Friston KJ. Scanning patients with tasks they can perform. *Hum Brain Mapp* 1999;8:102–108.

190. Callicott JH, Bertolino A, Mattay VS, et al. Physiological dysfunction of the dorsolateral prefrontal cortex in schizophrenia revisited. *Cereb Cortex* 2000;10:1078–1092.

191. Callicott JH, Mattay VS, Bertolino A, et al. Physiological characteristics of capacity constraints in working memory as revealed by functional MRI. *Cereb Cortex* 1999;9:20–26.

192. Baddeley AD. Working memory or working attention? In: Baddeley AD, Weiskrantz L, eds. *Attention selection, awareness, and control. A tribute to Donald Broadbent.* New York: Oxford University Press, 1993:152–170.

193. Barkley RA. Attention-deficit/hyperactivity disorder, self-regulation, and time: toward a more comprehensive theory. *J Dev Behav Pediatr* 1997;18:271–279.

194. Barkley RA. Is there an attention deficit in ADHD? *ADHD Rep* 1995;3:1–3.

195. Douglas VI, Parry PA. Effects of reward on delayed reaction time task performance of hyperactive children. *J Abnorm Child Psychol* 1983;11:313–326.

196. Sonuga-Barke EJ, Houlberg K, Hall M. When is "impulsiveness" not impulsive? The case of hyperactive children's cognitive style. *J Child Psychol Psychiatry* 1994;35:1247–1253.

197. Sonuga-Barke EJ, Williams E, Hall M, et al. Hyperactivity and delay aversion. III: the effect on cognitive style of imposing delay after errors. *J Child Psychol Psychiatry* 1996;37:189–194.

198. Barkley RA. Attention-deficit/hyperactivity disorder, self-regulation, and time: toward a more comprehensive theory. *J Dev Behav Pediatr* 1997;18:271–279.

199. O'Reilly RC, Braver TS, Cohen JD. A biologically based computational model of working memory. In: Miyake A, Shah P, eds. *Models of working memory: mechanisms of active maintenance and executive control.* Cambridge: Cambridge University Press, 1999:375–411.

200. Denkla MB. A theory and model of executive function: a neuropsychological perspective. In: Lyon GR, Krasnegor NN, eds. *Attention memory, and executive function.* Baltimore, MD: Brookes Publishing, 1996:263–278.

201. Sergeant JA. A theory of attention: an information processing perspective. In: Lyon GR, Krasnegor NN, eds. *Attention, memory, and executive function.* Baltimore, MD: Brookes Publishing, 1996:57–69.

202. Carte ET, Nigg JT, Hinshaw SP. Neuropsychological functioning, motor speed, and language processing in boys with and without ADHD. *J Abnorm Child Psychol* 1996;24:481–498.

203. Swanson J, Castellanos FX, Murias M, et al. Cognitive neuroscience of attention deficit hyperactivity disorder and hyperkinetic disorder. *Curr Opin Neurobiol* 1998;8:263–271.

204. Douglas VI. Attentional and cognitive problems. In: Rutter M, ed. *Developmental neuropsychiatry.* New York: Guilford Press, 1983:280–328.

205. Douglas VI. Cognitive deficits in children with attention deficit disorder with hyperactivity. In: Bloomingdale LM, Sergeant JA, eds. *Attention deficit disorder: criteria, cognition, intervention.* London: Pergamon Press, 1988:65–82.

206. Douglas VI. Stop, look and listen: the problem of sustained attention and impulse control in hyperactive and normal children. *Can J Behav Sci* 1972;4:259–282.

207. Barkley RA, DuPaul GJ, McMurray MB. Comprehensive evaluation of attention deficit disorder with and without hyperactivity as defined by research criteria. *J Consult Clin Psychology* 1990;58:775–789.

208. Seidman LJ, Biederman J, Faraone SV, et al. Effects of family history and comorbidity on the neuropsychological performance of children with ADHD: preliminary findings. *J Am Acad Child Adolesc Psychiatry* 1995;34:1015–1024.

209. Sergeant J. The cognitive-energetic model: an empirical approach to attention-deficit hyperactivity disorder. *Neurosci Biobehav Rev* 2000;24:7–12.

210. Oosterlaan J, Sergeant JA. Effects of reward and response cost on response inhibition in AD/HD, disruptive, anxious, and normal children. *J Abnorm Child Psychol* 1998;26:161–174.

211. LeMarquand DG, Pihl RO, Young SN, et al. Tryptophan depletion, executive functions, and disinhibition in aggressive, adolescent males. *Neuropsychopharmacology* 1998;19:333–341.

212. Swanson J, Posner MI, Cantwell D, et al. Attention-deficit/hyperactivity disorder: symptom domains, cognitive processes, and neural networks. In: Parasuraman R, ed. *The attentive brain.* Cambridge: The MIT Press, 1998:445–460.

213. Posner MI, Raichle ME. *Images of mind.* New York: Scientific American Library, 1994.

214. [Reserved]

215. Berger A, Posner MI. Pathologies of brain attentional networks. *Neurosci Biobehav Rev* 2000;24:3–5.

216. Sergeant JA, Oosterlaan J, van der Meere J. Information

processing and energetic factors in attention-deficit/hyperactivity disorder. In: Quay HCE, Hogan AEE, et al., eds. *Handbook of disruptive behavior disorders.* New York: Kluwer Academic/Plenum Publishers, 1999:75–104.

217. A 14-month randomized clinical trial of treatment strategies for attention-deficit/hyperactivity disorder. The MTA Cooperative Group. Multimodal Treatment Study of Children with ADHD. *Arch Gen Psychiatry* 1999;56:1073–1086.

218. MTA Cooperative Group. Moderator and mediator challenges to the MTA study: effects of comorbid anxiety disorder, family poverty, session attendance, and community medication on treatment outcome. *Arch Gen Psychiatry* 1999;56:1088–1096.

219. Vitiello B, Severe JB, Greenhill LL, et al. Methylphenidate dosage for children with ADHD over time under controlled conditions: lessons from the MTA. *J Am Acad Child Adolesc Psychiatry* 2001;40:188–196.

220. DuPaul GJ, Barkley RA, McMurray MB. Response of children with ADHD to methylphenidate: interaction with internalizing symptoms. *J Am Acad Child Adolesc Psychiatry* 1994;33:894–903.

221. Pliszka SR. Effect of anxiety on cognition, behavior, and stimulant response in ADHD. *J Am Acad Child Adolesc Psychiatry* 1989;28:882–887.

222. Taylor EA. Drug response and diagnostic validation. In: Rutter M, ed. *Developmental neuropsychiatry.* New York: Guilford Press, 1983:348–368.

223. Voelker SL, Lachar D, Gdowski CL. The Personality Inventory for Children and response to methylphenidate: preliminary evidence for predictive validity. *J Pediatr Psychol* 1983;8:161–169.

224. Barkley RA. Commentary on the multimodal treatment study of children with ADHD. *J Abnorm Child Psychol* 2000;28:595–599.

225. Jensen PS. Fact versus fancy concerning the multimodal treatment study for attention-deficit hyperactivity disorder. *Can J Psychiatry* 1999;44:975–980.

226. Swanson JM, Arnold LE, Vitiello B, et al. Response to commentary on the multimodal treatment study of ADHD (MTA): mining the meaning of the MTA. *J Abnorm Child Psychol* 2002;30:327–332.

227. Barkley RA. Attention deficit hyperactivity disorder. In: Mash EJ, Wolfe DA, eds. *Abnormal child psychology.* Belmont, CA: Wadsworth, 1999.

228. Wells KC, Pelham WE, Kotkin RA, et al. Psychosocial treatment strategies in the MTA study: rationale, methods, and critical issues in design and implementation. *J Abnorm Child Psychol* 2000;28:483–505.

25

LEARNING DISABILITIES: A NEUROPSYCHOLOGICAL PERSPECTIVE

BYRON P. ROURKE
BRENT A. HAYMAN-ABELLO
DAVID W. COLLINS

This chapter is designed to introduce child clinical psychologists, child neurologists, and child psychiatrists to research and practice that relates to children with learning disabilities (LD). It is organized in terms of a frankly neuropsychological approach to the topic because, in our view, this constitutes the most comprehensive perspective from which to structure this field.

A summary and critique of the various viewpoints regarding research and professional practice within the field of LD is not presented because (a) these have been summarized many times elsewhere (13,24,25,75,139,146,152,163); and (b) it was deemed advantageous for the reader to be introduced to what we consider to be a coherent, comprehensive, and cohesive view of the field. The organizational format for this chapter is simple. It moves from a brief historical overview, continues with an explication of an approach to the specification of subtypes of LD, and concludes with clinical considerations. References for the interested reader who wishes to pursue these topics in greater depth are provided within each section. Before embarking on this agenda, however, it is necessary to offer some definitions to guide our review.

DEFINITIONS OF LEARNING DISABILITIES

The definitions that we offer are, essentially, of the "process" variety, are frankly neuropsychological in nature, and are designed to encompass reliable and valid characterizations of various LD subtypes (to be explained later). If we wish to retain the term *LD* as a generic descriptor of this heterogeneous group of disorders, then there are at least two issues with which we must deal, as follows:

1. There is a need for a general definition.
2. There is a need for specific definitions of specific subtypes. Some of these can be formulated now. Others will need to wait upon further research.

The following are our general and specific working definitions of LD. These definitions arise from that part of our research program that has focused on the delineation of reliable and valid subtypes of LD and their hypothesized etiologies (for reviews of this work, see references 146, 151, 152, 155–157, 161, 173).

General

LDs are specific patterns (subtypes) of neuropsychological assets and deficits that eventuate in specific patterns of formal (e.g., academic) and informal (e.g., social) learning assets and deficits. LD also may lead to specific patterns of psychosocial functioning. It is clear that these generalizations need to be construed and evaluated within the context of particular historical and sociocultural contexts.

Specific Subtypes: Formal and Informal Learning, and Psychosocial Functioning

Of particular interest to us are two subtypes of LD that we have identified in a reliable and valid manner: nonverbal learning disabilities (NLD) and basic phonologic processing disabilities (BPPD).

The NLD subtype (syndrome) is a specific pattern of neuropsychological assets and deficits that eventuates in the following: a specific pattern of relative assets and deficits in academic (well-developed single-word reading and spelling relative to mechanical arithmetic) and social (e.g., more efficient use of verbal than nonverbal information in social situations) learning; and specific, developmentally dependent patterns of psychosocial functioning. Typically, in children younger than age 4 years, psychosocial functioning is relatively normal or reflective of mild deficits. After this period, emerging manifestations of externalized psychopathology are frequent; the child may be characterized as "hyperactive" and "inattentive" during this period. The usual course with respect to activity level is one of perceived "hyperactivity" through evident normoactivity to hypoactivity with advancing years. By older childhood and early adolescence, the typical pattern of psychopathology in evidence is of the internalized variety, characterized by withdrawal, anxiety, depression, atypical behaviors, and social skill deficits.

The BPPD subtype is a specific pattern of neuropsychological assets and deficits that eventuates in the following: a specific pattern of relative assets and deficits in academic (poorly developed single-word reading and spelling relative to mechanical arithmetic) and social (e.g., more efficient use of nonverbal than verbal information in social situations) learning. The neuropsychological profile and outcomes of this subtype of LD stand in marked contrast to NLD. In particular, the BPPD subtype does not lead in a necessary manner to any particular configuration of difficulties in psychosocial/adaptive behavior.

We turn now to a consideration of the research that led to these formulations and to an explication of their scientific and clinical implications.

GENERAL HISTORICAL BACKGROUND

Learning Disabilities as a Univocal Homogeneous Disorder

Early work in the neuropsychology of LD focused almost exclusively on the study of reading disability (13,138). With a few notable exceptions, this research was characterized by the "contrasting groups" or "comparative populations" (5) investigative strategy, particularly among North American and European researchers. Studies involving comparisons of groups of children with LD and groups of age- and otherwise-matched "normal" controls were used to determine the deficit(s) that characterizes the neuropsychological ability structure of the LD child (33). Use of this method assumed, implicitly or explicitly, that children with LD formed a univocal, homogeneous diagnostic entity.

This approach also was characterized by narrow samplings of areas of human performance thought to be related to the LD. For example, some researchers who were convinced that a deficiency in memory formed the basis for LD in youngsters would measure and compare "memory" in groups of children with LD and matched control children. Almost without fail, the conclusion of such research was that children with LD were deficient in memory, leading to the inference that this deficit was the cause of LD. In fact, the only crucial variable that, typically, influenced the results of such studies was the severity of the learning deficit of the children with LD studied. The more severe the learning deficit, the more likely it was that the contrasting groups would differ [for an alternative strategy for studying memory in children with LD, see Brandys and Rourke (15)].

Such research was periodically seen as somewhat short-sighted and inconclusive when it could be demonstrated, for example, that disabled learners were more deficient in "auditory memory" than they were in "visual memory." This constituted something of an advance in the field, because a rather more comprehensive sampling of skills and abilities was undertaken. Nevertheless, the commitment to the contrasting-groups methodology remained, even when investigators considered another complicating factor: the developmental dimension. Specifically, it became clear to some that one "obtained" significant differences between contrasting groups of disabled and normal readers on some variables and not on others depending on the age at which the children were studied. Thus, an example of the next advance in this research tradition was to compare groups of younger and older children on auditory and visual memory tasks to determine if the main effects (for auditory and visual memory) and the interactions (with age) were significant.

Even with these advances, however, those who adopted this essentially narrow nomothetic approach remained reluctant to analyze the abilities and deficits of individual children with LD. For example, investigators rarely reported results of individual cases in their studies of auditory and visual memory, even though this is very simply accomplished (for examples of how individual results can be reported in such studies, see references 27 and 162). Thus, researchers were left with group results that do not necessarily reflect how all individuals within the group performed. Some critics of this nomothetic approach to the understanding of LD in children adopted a position that was antithetical to it, that is, one that is best described as an "idiosyncratic" or "idiographic" approach to the field.

Learning Disabilities as an Idiosyncratic Disorder

As was the case in the history of the study of frank brain damage in youngsters, there have long been investigators of LD who maintain that persons so designated exhibit very unique, idiosyncratic, nongeneralizable origins, characteristics, and reactions and, as such, must be treated in an idiographic manner. Investigators of this ilk suggest that the single-case study is the only viable means for discovering valid scientific information and argue that the important dimensions of this disorder are individualized to the extent that meaningful generalizations are virtually impossible.

It is clear that this point of view contrasts sharply and in almost all respects with the contrasting-groups position outlined earlier. This nomothetic/idiographic debate is not unique to the study of LD and frankly brain-damaged individuals. [A particularly poignant and insightful characterization of this nomothetic/idiographic controversy in the entire field of clinical neuropsychology has been presented by Reitan (132).]

Critique of the Contrasting-Groups and Idiosyncratic Approaches

Contrasting-Groups Approach

As suggested earlier, studies within this tradition served to do little more than produce a lengthy catalogue of variables that differentiate so-called "learning-disabled" from "normal" children (139). Considered as a whole, the results of this corpus of research made it abundantly evident that virtually any single measure of perceptual, psycholinguistic, or higher-order cognitive skill could be used to illustrate the superiority of normal children over comparable groups of children diagnosed as having LD.

Despite all of its limitations, many investigators still persist in using the comparative populations or contrasting-groups methodology with respect to research findings, model building, and theory development. Unfortunately, the results of numerous studies using this approach to study LD are trivial, contradict one another, and are not supported in replication attempts, and there is little to suggest that these "characteristic" factors of children with LD are related to one another in any meaningful fashion (146). In sum, the evidence arising from this literature is, at best, equivocal. A coherent and meaningful pattern of neuropsychological characteristics of children with LD does not emerge from the contrasting-groups approach. We maintain that this is the case because such a univocal pattern does not obtain (more on this point later).

In any event, there appear to be at least four major reasons why the results of these studies have not been very contributory to the investigation of the neuropsychological integrity of youngsters with LD. These are as follows:

1. *Lack of a conceptual model.* The vast majority of contrasting-groups studies do not use a conceptual model to elucidate the skills that are involved in perception, learning, memory, and cognition and that are deficient in children with LD. Most models developed in this genre of research represent either *post hoc* theorizing about a very limited spectrum of skills and abilities or "pet" theories derived from more "general" models of perception, learning, memory, and cognition. Research cast in terms of these very limited, narrow-band models and *a fortiori* research that is essentially atheoretical in nature is virtually certain to yield a less than comprehensive explanation of LD.

2. *Definition of LD.* There has been no consistent formulation of the criteria for LD in the contrasting-groups genre of study. Most studies of this type use vaguely defined or even undefined groups, and others use the ratings of teachers and other school personnel that remain otherwise unspecified. Some even use the guidelines of a particular political jurisdiction to define LD! This lack of clarity and consistency has a negative impact on the generalizability of the (limited) findings of such studies.

3. *Developmental considerations.* Part and parcel of the contrasting-groups approach is a gross insensitivity to age differences and developmental considerations. Several studies support the notion that the nature and patterning of the skill and ability deficits of (some subtypes of) children with LD vary with age (23,52,62,107,114,122,165). Because it seems reasonable to infer that the neuropsychological functioning of (some subtypes of) children with LD varies as a function of age (considered as one index of developmental change), the aforementioned inconsistencies in contrasting-groups research results could reflect differences in the ages of the subjects used. Although this possibility sometimes is acknowledged by investigators who use this type of methodology, such a realization has, until recently, been the exception rather than the rule.

4. *Treatment considerations.* The assumption of the contrasting-groups approach that children with LD form a homogenous entity implies that one particular method of treatment will be appropriate for all such children. However, practical experience with children who exhibit marked LD suggests clearly that only some of these individuals respond positively to particular forms of treatment. It also is the case that some forms of treatment for some children with LD actually are counterproductive (141,143).

Idiosyncratic Approach

A strictly individualistic standpoint in this field, although attractive from some clinical perspectives, is not viable on practical (applied/clinical) theoretical grounds. It is patently obvious that persons with LD are unique. It also is the case that most of them can be classified into homogeneous subtypes on the basis of sets of shared relative strengths and deficiencies in neuropsychological skills and abilities. That is, they have enough in common to be grouped (i.e., subtyped) for the purpose of some important aspects of treatment, as well as for specific model-testing purposes. The fact that models based upon subtype analysis can be shown to be internally consistent and externally (including ecologically) valid (146,151,153,156) suggests strongly that one need not adopt a position that maintains that each and every individual child with LD is so implacably unique that nothing can be said about his/her development, prognosis, and preferred mode of intervention/treatment. That treatment effects such as those demonstrated by Lyon (102) obtain is sufficient proof that strict individualism is not necessary.

To accomplish the desirable goals of scientific and clinical generalizability while ensuring the individual applicability of the "general" results of studies in this area, it is necessary to be able to demonstrate that each individual who falls within a particular LD subtype exhibits a close approximation to the pattern of neuropsychological assets and deficits that characterizes the subtype as a whole. This should "satisfy" the individualistic criterion while maintaining the generalizability of treatment and hypothesis-testing imperatives of scientific investigations within this area (155). The possibility of obtaining this degree of individuation of group results will be particularly apparent when we examine such issues in relationship to the NLD and BPPD subtypes.

The "subtype" approach can provide viable solutions to the four criticisms leveled at the contrasting-groups method. First, a componential analysis of academic and other types of learning can be of substantial aid in developing conceptual models of LD. For example, whereas some subtypes of children with LD may experience learning problems because of deficiencies in certain perceptual, cognitive, or behavioral skills, others may manifest such problems as a more direct result of attitudinal/motivational or vastly different perceptual, cognitive, or behavioral difficulties. In addition, these different patterns of neuropsychological assets and deficits may encourage different types or degrees of learning difficulties. Several theories emphasizing this type of comprehensive componential approach to the design of developmental neuropsychological explanatory models in this area have emerged (64). Our own efforts in this regard are described in subsequent sections of this chapter (142,147,149,151,155,156,169,172,173,175). Regarding the need for clear definitions of LD that are amenable to consensual validation, it seems patently obvious that the definitions themselves should be a *result* of sophisticated subtype analysis rather than the starting point of such inquiry (58–60,63,68,112,144–146,155,177).

Whereas the contrasting-groups approach is insensitive to possible developmental changes in the neuropsychological ability structures of children with LD, attention to these issues is an almost direct function of the sophistication in subtype analysis exhibited by neuropsychological investigators in this area [for an especially good example of perspicacious sensitivity to such issues, see Morris et al. (114)]. Finally, subtypal analysis can be of notable benefit in the area of treatment. For example, Lyon (102) demonstrated that specific subtypes of disabled readers respond differently to synthetic phonics and "sight-word" approaches to reading instruction. Similarly, in the study by Sweeney and Rourke (214), what appeared to be overreliance on the phonetic analysis of individual words may have hampered the reading speed of phonetically accurate disabled spellers.

We now turn to a brief historical overview of the "subtype" approach to the investigation of children with LD.

Learning Disabilities as a Heterogeneous Group of Disorders (Subtypes)

Scientific investigation of LD subtypes emerged primarily from careful clinical observations of children who were assessed because they were thought to be experiencing perceptual disabilities and/or LD. Astute observers of children

with LD, such as Johnson and Myklebust (89), noted that there are clear differences in patterns of assets and deficits exhibited by such children and that specific variations in developmental problems (e.g., learning to read) appeared to stem from these different patterns of assets and deficits. In other words, equally impaired *levels* of learning appeared to result from different "etiologies," as suggested in patterns of relative perceptual and cognitive strengths and weaknesses.

The search for subtypes of LD took on added importance when it was inferred that these potentially isolatable patterns of perceptual and cognitive strengths and weaknesses (which were thought to be responsible for learning difficulties) might be more or less amenable to different modes of intervention. Indeed, the "subtype by treatment interaction" hypothesis, that is, the notion that the tailoring of specific forms of treatment to the underlying abilities and deficits of children with LD would be advantageous, was alive and well early in the history of the LD field (92). Many researchers and professionals in the field virtually *assumed* that this was the most efficient and effective therapeutic course to follow for children experiencing outstanding difficulties in learning.

Positions of this sort were mirrored in the studies conducted by Johnson and Myklebust (89) and later by Mattis et al. (106), who used an essentially "clinical" approach to the identification of subtypes of LD. This research, which involved an emphasis on subtypal patterns, was accompanied in the late 1970's by research of investigators who sought to isolate reliable subtypes of children with LD through the use of statistical algorithms such as Q-type factor analysis and cluster analysis (34,35,52,126). By the 1980's, the use of statistical algorithms to assist in the generating of subtypes of LD was well under way (52,90,113). This type of organized search for reliable subtypes of children with LD clearly has begun to have a profoundly positive effect on investigative efforts in this area [for examples see Rourke (146,152,156)]. Systematic attempts at establishing the concurrent, predictive, and construct validity of such subtypes have been proceeding apace (21,41,53,81,102,152,153,161,170,182,207).

Systematic Developmental Neuropsychological Approach to the Study of Learning Disabilities

We now turn to a developmental neuropsychological approach to the study of LD that seeks to understand problems in learning by studying developmental change in behavior as seen through the perspective of brain–behavior models. This approach of combining the study of brain development with that of behavior development involves research strategies that pay as much attention to the development of an individual's approach to material to be learned as they do (say) to the electrophysiologic correlates at the level of the cerebral hemispheres that accompany that approach. In other words, it is a systematic attempt to fashion a complete understanding of brain–behavior relationships as they are mirrored in the development of central processing abilities and deficits throughout the life span.

Unfortunately, use of a neuropsychological framework such as this for investigation of LD in children often has been misinterpreted as reflecting an emphasis on static, intractable (and, therefore, limited) notions of the effects of brain impairment on behavior. Although specific statements to the contrary have been made on many occasions (137,139,140,141,155), many otherwise competent researchers and clinicians persist in the notion that such an approach *assumes* that brain damage, disorder, or dysfunction lies at the basis of LD. Nothing could be further from the truth, because the thrust of much work in this area has been to *demonstrate* whether and to what extent this might be the case (13,137,139,151,152,156, 217).

Be that as it may, the emphasis that we have brought to this enterprise is one that attempts to integrate dimensions of individual and social development on the one hand with relevant central processing features on the other, all in order to fashion useful models with which to study crucial aspects of perceptual, cognitive, and other dimensions of human development. That models and explanatory concepts developed with this aim in mind (137,143,145,147,149,151,155,156) contain explanations that are thought to apply both to some children with frank brain damage, to some types of children with LD, and to some aspects of normal human development should come as no surprise, because maximum generalizability is one goal of any scientific model or theory. Specifically with respect to the child with LD, the aspects of these concepts and models that are most relevant are those that have to do with proposed linkages between patterns of central processing abilities and deficits that may predispose a youngster to predictably different patterns of social as well as academic LD. In addition, these models are designed to encompass developmental change and outcome in patterns of learning and behavioral responsivity.

In the neuropsychology of LD, the important relationships to bear in mind are those that obtain between *patterns* of performance and *models* of developmental brain–behavior relationships. This is not meant to imply something as trite as that LD are the *result* of brain damage or dysfunction. The illogical leap of such an implication (assumption) is only one of its shortcomings. More important is the notion that, even were such a cause-and-effect (i.e., brain damage/dysfunction→LD) relationship shown to be the case, it would carry few, if any, corollaries for the understanding of developmental brain–behavior relationships or their treatment.

This is not to say that brain–behavior relationships are of little or no importance in the study of LD. Rather, it is simply the case that we currently understand more about developmental–neuro *psychological* performance interactions

than we do about developmental–*neuro* psychological performance interactions. The patterns of interaction among brain, development, and behavior that have been exquisitely hypothesized are, unfortunately, still largely unexplored and must be the subject of rigorous scientific test (215). Attempts to address this issue [as summarized by Dool et al. (41) and Stelmack et al. (207)] have suggested rather strongly that specific types of brain function/dysfunction/difference play important roles in the manifestations of some subtypes of LD.

Neurologic Correlates of Learning Disabilities

Over the last quarter century the study of the neurobiologic substrate of learning disorders has yielded a considerable volume of work. However, most of the work in this area has involved rather heterogeneous groupings of individuals with reading disabilities (dyslexia). As a consequence, much of what is known regarding the neurologic substrate of LD comes from studies of individuals said to have this disorder.

Postmortem studies have identified unusual cerebral symmetry and neuronal abnormalities, predominantly in the left hemisphere, as marked features of adult dyslexic brains (76,77,85). In particular, all autopsy studies of individuals with dyslexia thus far have revealed an unusual symmetry in the posterior aspect of the left and right superior temporal lobes, an area called the *planum temporale.* The plana temporale region in the majority of nonimpaired individuals shows an asymmetry with the left greater than right (11,50). Studies using computed tomographic scanning also have shown atypical cerebral symmetries in posterior regions in individuals with dyslexia (83,97,135).

Magnetic resonance imaging studies of brain morphology in individuals with dyslexia have focused on temporal lobe asymmetry (87,95,96,190), corpus callosum morphology (42,86,134), and brain language areas (42,46,87,88). Such studies have provided further evidence that the brains of individuals with dyslexia are more symmetric than control brains, particularly with respect to posterior regions. However, the evidence for other brain differences is equivocal.

Taken together, the only certain finding arising from studies of brain morphology is that no gross neuropathology is present in dyslexia. Otherwise, no robust brain differences between individuals with and without dyslexia are found in the literature. Rather, there may be subtle deviations in brain morphology in some individuals with dyslexia. One of the more frequent findings across studies is atypical cerebral asymmetry. Given the likely heterogeneity of reading disabilities, brain features such as atypical symmetries may be unique only to certain subtypes of dyslexia.

Studies of brain functioning in individuals with dyslexia have used either functional neuroimaging or event-related potential (ERP) methodologies. Rumsey and colleagues have conducted a series of studies of individuals with dyslexia using positron emission tomographic techniques

(84,189,191–193). Collectively, these studies provide evidence for dysfunctional left posterior cortical areas in dyslexia, typically reflected by a decrease or absence of brain activity in these areas relative to controls. Specifically, these researchers have identified the left angular gyrus as a probable site of dysfunction in dyslexia. Several positron emission tomographic studies have focused on the functional connectivity between the left angular gyrus region and other left hemisphere brain areas (84,191).

Functional magnetic resonance imaging studies of brain functioning in individuals with dyslexia have primarily targeted the visual system (29,30,44) and phonologic processing (78,200–222). Studies of visual brain areas provide consistent evidence of decreased or absent activity in magnocellular systems, with generally no lateralization pattern. Functional magnetic resonance imaging studies of phonologic processing in dyslexia have generally found reduced or absent activity in the left temporal and frontal regions during language-related tasks.

Electroencephalography (EEG) and magnetoencephalography are procedures that map ERPs, which are small signal fluctuations in brain activity that are time locked to specific events or stimuli. Such procedures are ideal for studying the temporal course of brain functioning. ERP studies in individuals with dyslexia using EEG have implicated generalized deficits that are not hemisphere specific (1,82,101). Further, the developmental EEG literature indicates that functional brain deficits in individuals with dyslexia may be present at birth and appear to be longstanding (100,110,205,219). Such findings underscore the importance of controlling for age.

Using magnetoencephalography, Simos et al. (201,202) showed that brain events during word recognition do not progress through the same brain regions in individuals with dyslexia as in nonimpaired readers. Generally, they found that children with dyslexia showed less activation confined to the left hemisphere compared to controls. It is noteworthy that Simos et al. (203) reported age-related changes in brain activation related to phonologic processing and word recognition in normal readers. Their findings suggested that children and adults differ in their respective spatiotemporal profiles of brain activation during reading. Some of these data suggest that as reading skills develop from childhood to adulthood, the underlying neurologic systems become increasingly localized within the left hemisphere.

One relatively consistent finding in the literature is that functional brain systems in individuals with dyslexia tend to be depressed, absent, or slow compared to controls. A general discrepancy in the literature is whether functional differences are found over large cortical areas or whether left hemisphere systems in particular are atypical. A unified view of these overall findings suggests that separate brain systems are functionally anomalous in dyslexia, only some of which involve left posterior regions. Specifically, individuals with dyslexia seem to show widespread functional deficits in magnocellular visual systems and in some aspects of central cognitive processing systems. In addition, many individuals with dyslexia appear to have functional deficits confined to linguistic processing systems in the left hemisphere.

Because individuals with LDs are a heterogeneous grouping, there is likely no universal deficit to explain poor reading (151). Rather, difficulties in reading may arise from distinct forms of brain dysfunction. According to this view, the behavioral identification of LD subtypes is key to understanding the underlying neurobiologic characteristics. Reviews of the literature have indicated that neurophysiologic differences in children with LD may be masked in studies neglecting to separate their samples into distinct subtypes (41,207). Inasmuch as LDs constitute a heterogeneous group, failure to control for LD subtypes may lead to inconsistent findings and erroneous conclusions.

Generally, few studies of brain functioning have taken into consideration the neuropsychological profile of individuals with LD. Nonetheless, several ERP (EEG) studies have differentiated two subtypes of reading disabilities based on the primary domain of dysfunction in each subtype, either phonologic or visual (2,9,65). Results from these studies have suggested that the dysfunctional brain regions are unique to specific subtypes.

Studies by Stelmack and colleagues (80,108,206,208) have yielded consistent neurophysiologic evidence for the differentiation of LD into subtypes that are consistent with the theoretical model developed by Rourke (142). In such a model (described in this chapter), distinct subtypes of LD are defined by patterns of neuropsychological assets and deficits. Collectively, the studies by Stelmack and colleagues provided support for the view that particular areas of brain dysfunction are associated with distinct patterns of neuropsychological functioning. Specifically, left hemisphere dysfunction was associated with linguistic deficits in one subtype (i.e., group R-S), and right hemisphere dysfunction was associated with visual-spatial analysis deficits in another subtype (i.e., group A). The findings of between-group differences in electrophysiologic brain activity support the utility of establishing LD subtypes based on neuropsychological and academic performance.

Overall, it seems clear that brain differences are found between individuals with LD and individuals without such disabilities. The nature of these brain differences, although still in question, may include microstructural abnormalities, unusual cerebral symmetries, functional anomalies in left hemisphere language systems, dysfunctional right hemisphere processes, and impaired magnosystems. A key obstacle to identifying reliable brain differences in LD is not so much a limitation of technology as a problem of nosology. Many of the inconclusive and discrepant findings in the field of LD may be due to the lack of precision in definition. For example, much of the earlier research focused on dyslexia as a single, presumably unitary, disorder. As the cited series of studies demonstrates, a simplified conceptualization of LD is not adequate to the task of understanding the unique strengths and weaknesses of these children. Unique patterns of brain activation are important correlates of LD subtypes. Results from electrophysiologic studies suggest differences in brain activity across LD subtypes. In particular, Rourke's typology has been well validated with electrophysiologic studies (142). The question arises whether other techniques to measure brain activity will show distinct patterns of functioning across these LD subtypes in a similar manner.

We now turn to a discussion of selected investigations carried out in our laboratory that should serve to illustrate the heuristic properties of the systematic developmental neuropsychological approach to the study of LD. For a compendium of our efforts in this area, see Rourke (146,151,152,156).

THE WINDSOR STUDIES

The focus of the studies discussed in this section is on the academic and psychosocial dimensions of LD. The perspective used is one that has emerged from our study of these and related problems over some 30 years in the University of Windsor Laboratory.

To appreciate the main issues involved in this presentation, it is necessary to reflect on the crucial historical issues that bear upon them, as well as the relationships that obtain between LD on the one hand and deficits in academic and psychosocial functioning on the other. In this connection, it should be made clear from the outset that the principal hypothesis that we have attempted to test in our work is the following: LD are manifestations of basic neuropsychological assets and deficits; the subtypes of LD in question may lead to problems in academic functioning and/or psychosocial functioning; and the relationships among neuropsychological deficits, LD, and academic and social learning deficits can be understood fully only within a neurodevelopmental framework that takes into consideration the changing nature of the academic, socioemotional, and vocational demands confronting humans in a particular sociocultural milieu.

To illustrate the research and model development strategy used in our laboratory that speaks to the issues of concern in this chapter, we describe in some detail the nature of two of the subtypes of children with LD who have been the subject of intensive investigation in our laboratory.

Two Subtypes of Learning Disabilities

Although there are several subtypes of LD that have been isolated in our investigations, we have been interested in the study of two in particular. Children in one group, those with BPPD (151), are those who exhibit many relatively poor psycholinguistic skills in conjunction with very well-developed visual-spatial-organizational, tactile-perceptual, psychomotor, and nonverbal problem-solving skills. They exhibit very

poor reading and spelling skills and significantly better, although still impaired, mechanical arithmetic competence. The other group, those with NLD (151), exhibits outstanding problems in visual-spatial-organizational, tactile-perceptual, psychomotor, and nonverbal problem-solving skills, within a context of clear strengths in psycholinguistic skills such as rote verbal learning, regular phoneme-grapheme matching, amount of verbal output, and verbal classification. Children who exhibit the syndrome of NLD experience their major academic learning difficulties in mechanical arithmetic while exhibiting advanced levels of word recognition and spelling. Both of these subtypes of children with LD, especially the second subtype of child characterized as having NLD (116,151), have been the subject of much scrutiny in our laboratory (for reviews, see references 137,140,142,147,149,151,155,156,180,209–211).

Examination of Figures 25.1 and 25.2 will elucidate the models that we have used in our attempts to explain the developmental dynamics of the neuropsychological and adaptive dimensions of these two subtypes of LD. For example, the summaries of the assets and deficits of the NLD subtype should be viewed within a specific context of cause-and-effect relationships, that is, the primary neuropsychological assets and deficits are thought to lead to the secondary neuropsychological assets and deficits, and so on within the four categories of neuropsychological dimensions, and the foregoing are seen as causative, that is, the academic and socioemotional/adaptive aspects of this subtype. In this sense, the latter dimensions are, essentially, *dependent* variables (i.e., effects rather than causes) in the NLD subtype. The same applies, *mutatis mutandis,* to the summary of the assets and deficits of the BPPD subtype.

Basic Phonologic Processing Disabilities

Figure 25.1 contains a description of this subtype. The following is offered as a summary of the assets and deficits that characterize this subtype.

Neuropsychological Assets

Tactile-perceptual, visual-spatial-organizational, psychomotor, and nonverbal problem-solving and concept-formation skills and abilities are developed to an average to above average degree. The capacity to deal with novelty and the amount and quality of exploratory behavior are average. Attention to tactile and visual input is normal.

Neuropsychological Deficits

Disordered phonemic hearing, segmenting, and blending are paramount. Attention to and memory for auditory-verbal material are clearly impaired. Poor verbal reception, repetition, and storage are evident. Amount and quality of verbal associations are clearly underdeveloped. There is a less than average amount of verbal output.

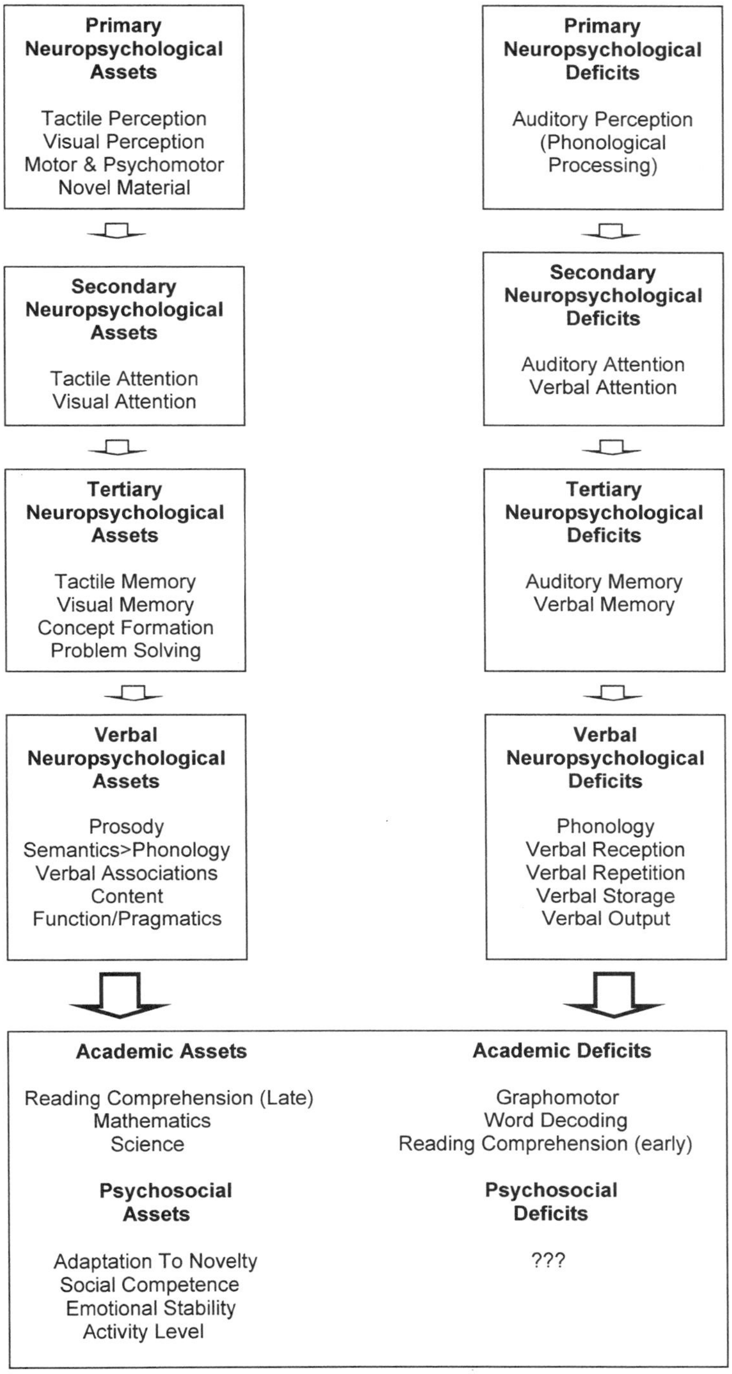

FIGURE 25.1. Content and dynamics of Basic Phonologic Processing Disabilities.

Academic Assets and Deficits

Reading and spelling are affected, as are those aspects of arithmetic performance that require reading and writing. The symbolic aspects of writing are affected. The nonverbal aspects of arithmetic and mathematics are left unaffected. The prognosis for advances in reading and spelling and the verbal-symbolic aspects of writing and arithmetic must be very guarded.

Psychosocial Assets and Deficits

Psychosocial disturbance may occur if parents, teachers, and other caretakers establish unattainable goals for the child and/or if antisocial models and lifestyles hold reinforcing properties for the child that the school is unable to provide. Otherwise, there would appear to be no very good reason to infer that such children are at any particular risk for development of serious psychopathology. Our investigations

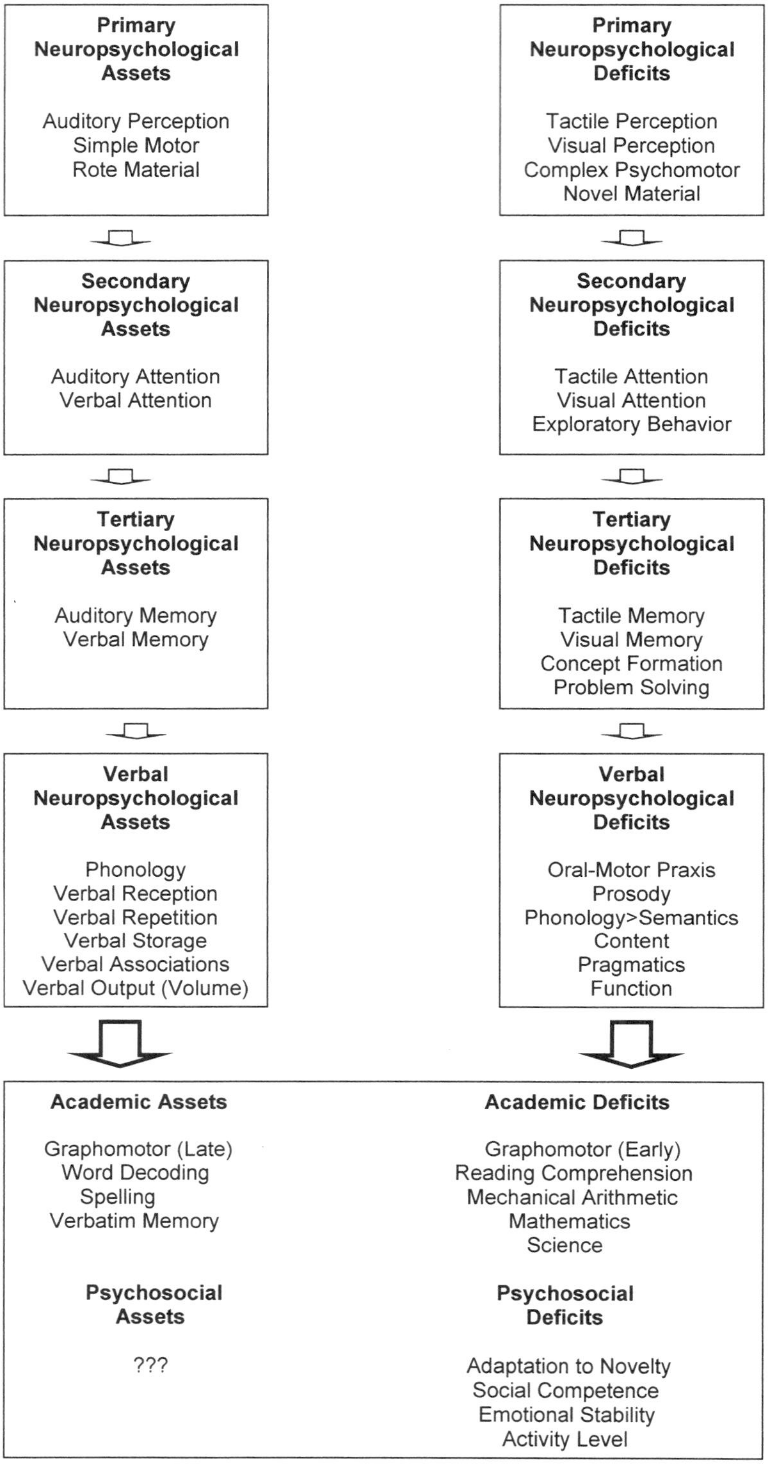

**Primary
Neuropsychological
Assets**

Auditory Perception
Simple Motor
Rote Material

**Primary
Neuropsychological
Deficits**

Tactile Perception
Visual Perception
Complex Psychomotor
Novel Material

**Secondary
Neuropsychological
Assets**

Auditory Attention
Verbal Attention

**Secondary
Neuropsychological
Deficits**

Tactile Attention
Visual Attention
Exploratory Behavior

**Tertiary
Neuropsychological
Assets**

Auditory Memory
Verbal Memory

**Tertiary
Neuropsychological
Deficits**

Tactile Memory
Visual Memory
Concept Formation
Problem Solving

**Verbal
Neuropsychological
Assets**

Phonology
Verbal Reception
Verbal Repetition
Verbal Storage
Verbal Associations
Verbal Output (Volume)

**Verbal
Neuropsychological
Deficits**

Oral-Motor Praxis
Prosody
Phonology>Semantics
Content
Pragmatics
Function

Academic Assets

Graphomotor (Late)
Word Decoding
Spelling
Verbatim Memory

Academic Deficits

Graphomotor (Early)
Reading Comprehension
Mechanical Arithmetic
Mathematics
Science

**Psychosocial
Assets**

???

**Psychosocial
Deficits**

Adaptation to Novelty
Social Competence
Emotional Stability
Activity Level

FIGURE 25.2. Content and dynamics of Nonverbal Learning Disabilities Syndrome.

in this area appear to support this view (69,70,73,125, 230).

Nonverbal Learning Disabilities

Speculation regarding the content of the NLD syndrome began with the seminal work of Myklebust (116). What we have been able to do is to flesh out this content, mostly as a result of our study of subtypes of LD. We have approached this investigative effort from the perspective of developmental neuropsychology, that is, the study of brain–behavior relationships within a developmental context. A review of these studies is available elsewhere (151,155,156). For our current purposes, the relevant dimensions of the conclusions of these investigations are what we take to be the content and dynamics of NLD.

Two features of this description of the NLD syndrome characteristics should be borne in mind:

1. Because of our focus on children and adolescents, manifestations of NLD in persons within these developmental stages are emphasized.

2. This description is couched in terms of the "developmental" manifestation of the NLD syndrome, that is, in terms of its characteristics in a child who has been so afflicted since his/her earliest developmental stages. A schematic outline of these neurodevelopmental dynamics is shown in Figure 25.2. Some modifications in the manifestations of the syndrome are necessary when we turn to considerations of the onset of the syndrome in an older child, adolescent, or adult who has enjoyed a normal early developmental course.

NEUROPSYCHOLOGICAL ASSETS

Primary Assets

Simple Motor Skills

Simple, repetitive motoric skills are generally intact, especially at older age levels (middle childhood and beyond).

Auditory Perception

After a very early developmental period when such skills appear to be lagging, auditory-perceptual capacities become very well developed.

Rote Material

Repetition and/or constancy of stimulus input, especially through the auditory modality but not confined to it, is well appreciated. Repetitious motoric acts, including some aspects of speech and well-practiced skills such as handwriting, eventually develop to average or above average levels.

Secondary Assets

Attention Skills

Deployment of selective and sustained attention for simple, repetitive verbal material (especially that delivered through the auditory modality) becomes very well developed.

Tertiary Assets

Memory

Rote verbal memory and memory for material that is readily coded in a rote verbal fashion becomes extremely well developed.

Verbal Assets

Speech and Language

After an early developmental period when linguistic skills appear to be lagging, a number of such skills emerge and develop in a rapid fashion. Excellent phonemic hearing, segmentation, blending, and repetition and very well-developed receptive language skills and rote verbal capacities are evident. A large store of rote verbal material and verbal associations, and a very high volume of speech output, is present. All of these characteristics tend to become more prominent with advancing years.

Academic Assets

After initial problems with the visual-motor aspects of writing and much practice with a writing instrument, graphomotor skills (for words) reach good to excellent levels. After initial problems with the development of the visual-spatial feature analysis skills necessary for reading, good to excellent single-word reading skills develop to above average levels. Misspellings are almost exclusively of the phonetically accurate variety. Verbatim memory for oral and written verbal material can be outstanding in the middle to late elementary school years and thereafter.

NEUROPSYCHOLOGICAL DEFICITS

Primary Deficits

Tactile Perception

Bilateral tactile-perceptual deficits are evident, usually more marked on the left side of the body. Deficits in simple tactile imperception and suppression tend to become less prominent with advancing years, but problems in dealing with complex tactile input tend to persist.

Visual Perception

There is impaired discrimination and recognition of visual detail and visual relationships, and there are outstanding deficiencies in visual-spatial-organizational abilities. Simple visual discrimination, especially for material that is verbalizable, usually approaches normal levels with advancing years. Complex visual-spatial-organizational skills, especially when required within a novel framework, tend to worsen relative to age-based norms.

Complex Psychomotor Skills

Bilateral psychomotor coordination deficiencies are prominent and often are more marked on the left side of the body. These deficits, except for well-practiced skills such as handwriting, tend to increase in severity with age, especially when they are required within a novel framework.

Novel Material

As long as stimulus configurations remain novel, they are dealt with very poorly and inappropriately. Difficulties in age-appropriate accommodation to, and a marked tendency toward overassimilation of, novel events increase with advancing years. There is an overreliance on prosaic rote (and, in consequence, inappropriate) behaviors in such situations. The capacity to deal with novel experiences often remains poor or even worsens with age.

Secondary Deficits
Attention

Attention to tactile and visual input is poor. Deficiencies in visual attention tend to increase over the course of development, except for material that is programmatic and overlearned (e.g., printed text). Deployment of selective and sustained attention is much better for simple, repetitive verbal material (especially that delivered through the auditory modality) than for complex, novel nonverbal material (especially that delivered through the visual or tactile modalities). The disparity between attentional deployment capacities for these two sets of materials tends to increase with age.

Exploratory Behavior

There is little physical exploration of any kind. This is the case even for objects that are immediately within reach and could be explored through visual or tactile means. A tendency toward sedentary and physically limited modes of functioning increases with age.

Tertiary Deficits
Memory

Memory for tactile and visual input is poor. Deficiencies in these areas tend to increase over the course of development, except for material that is programmatic and overlearned (e.g., spoken natural language). Memory for nonverbal ma-

terial, whether presented through the auditory, visual, or tactile modalities, is poor if such material is not readily coded in a verbal fashion. Relatively poor memory for complex, meaningful, and/or novel verbal and nonverbal material is typical. Differences between good to excellent memory for rote material and impaired memory for complex material and/or that which is not readily coded in a verbal fashion tend to increase with age.

Concept Formation, Problem Solving, Strategy Generation, and Hypothesis Testing/Appreciation of Informational Feedback

Marked deficits in all of these areas are apparent, especially when the concept to be formed, the problem to be solved, and/or the problem-solving milieu(s) is/are novel or complex. Also evident are significant difficulties in dealing with cause-and-effect relationships and marked deficiencies in the appreciation of incongruities (e.g., age-appropriate sensitivity to humor). Most noticeable as formal operational thought becomes a developmental demand (i.e., in late childhood and early adolescence), deficits in these areas tend to increase markedly with advancing years, as does the gap between performance on rote (overlearned) and novel tasks.

Verbal Deficits
Speech and Language

Mildly deficient oral-motor praxis, little or no speech prosody, and much verbosity of a repetitive, straightforward, rote nature are characteristic. When paraphasic errors are in evidence, they are much more likely to be of the phonologic than of the semantic variety. Also typical are content disorders of language, characterized by very poor psycholinguistic content and pragmatics (e.g., "cocktail party" speech) and reliance upon language as a principal means for social relating, information gathering, and relief from anxiety. "Memory" for complex verbal material usually is very poor, probably as a result of poor initial comprehension of such material. All of these characteristics, except oral-motor praxis difficulties, tend to become more prominent with advancing years.

ACADEMIC DEFICITS
Graphomotor

In the early school years, there is much difficulty with printing and cursive script; with considerable practice, handwriting often becomes quite good. However, some avoid practice and remain deficient in this skill.

Reading Comprehension

Reading comprehension is much poorer than single-word reading (decoding). Deficits in reading comprehension, especially for novel material, tend to increase with advancing years.

Mechanical Arithmetic and Mathematics

There are outstanding relative deficiencies in mechanical arithmetic compared to proficiencies in reading (word recognition) and spelling. With advancing years, the gap between good to excellent single-word reading and spelling and deficient mechanical arithmetic performance widens. Absolute level of mechanical arithmetic performance only rarely exceeds the grade 5 level; mathematical reasoning, as opposed to programmatic arithmetic calculation, remains poorly developed.

Science

Persistent difficulties in academic subjects involving problem solving and complex concept formation (e.g., physics) are prominent. Problems in dealing with scientific concepts and theories become apparent by early adolescence. The gap between deficiencies in this type of complex academic endeavor and other, more rote, programmatic academic pursuits widens with age.

PSYCHOSOCIAL/ADAPTATIONAL DEFICITS

Adaptation in Novel Situations

There is extreme difficulty in adapting to (i.e., countenancing, organizing, analyzing, and synthesizing) novel and otherwise complex situations. An over-reliance on prosaic rote (and, in consequence, inappropriate) behaviors in such situations is common. These characteristics tend to become more prominent with advancing years.

Social Competence

Significant deficits are apparent in social perception, social judgment, and social interaction skills and become more prominent as age increases. There is a marked tendency toward social withdrawal and even social isolation with advancing years.

Psychosocial Disturbance

Often characterized during early childhood as afflicted with some type of acting out or conduct disorder, such children are at high risk for development of "internalized" forms of psychopathology. Indications of excessive anxiety, depression, and associated internalized forms of psychosocial disturbance tend to increase with advancing years.

Activity Level

Children who exhibit the syndrome are frequently perceived as hyperactive during early childhood. With advanc-

ing years, they tend to become normoactive and eventually hypoactive.

Time Sense

Extremely Distorted Sense of Time

This may be reflected in poor estimation of elapsed time during common activities and poor estimation of time of day. (This deficit may not appear spontaneously; it usually requires a direct attempt to elicit it. We include this dimension provisionally because we have not yet examined it in depth.)

Specific Comparisons of Basic Phonologic Processing Disabilities and Nonverbal Learning Disabilities

In a series of studies, we investigated the neuropsychological assets and deficits of these two subtypes of children with LD (166,179,209). Based upon these and related analyses (122,123), there are several specific conclusions that apply to these two subtypes, as follows. (In the following, "older" refers to 9- to 14-year-old children and "younger" to 7- and 8-year-old children.)

1. Older children with BPPD exhibit some (often no more than mild) deficiencies in the more rote aspects of psycholinguistic skills, such as recall of information and word definitions. Children with NLD exhibit average to superior skills in these areas.

2. Older children with BPPD exhibit outstanding deficiencies in the more complex semantic-acoustic aspects of psycholinguistic skills, such as sentence memory and auditory analysis of common words. Children with NLD exhibit less than normal performances in these areas, but their levels of performance are superior to those with BPPD. Children with NLD tend to perform least well on those tests of semantic-acoustic processing that place an emphasis upon the processing of novel, complex, and/or meaningful material.

3. Older children with BPPD exhibit normal levels of performance on visual-spatial-organizational, psychomotor, and tactile-perceptual tasks. Children with NLD have outstanding difficulties on such tasks. The deficiencies exhibited by children with NLD on tactile-perceptual and psychomotor tasks are in evidence bilaterally. When there is evidence of lateralized impairment on such tasks for the child with NLD, it is almost always a relative deficiency in performance on the left side of the body. In general, the more novel the visual-spatial, psychomotor, and tactile-perceptual task, the more impairment, relative to age-based norms, exhibited by the child with NLD.

4. Older children with BPPD perform normally on nonverbal problem-solving tasks. They have no difficulty in benefiting from experience with such tasks. They are particularly

adept at using nonverbal informational feedback to modify their performance to meet the demands of such tasks. Children with NLD exhibit moderate to severe problems on nonverbal problem-solving tasks. They benefit very little from informational feedback and continued experience with such tasks, even when the information provided would be expected to be well within estimates based upon their largely rote verbal learning capacity (51).

5. Younger children with BPPD and those with NLD exhibit intergroup and intragroup patterns of relative abilities and deficits on automatic (rote) verbal, semantic-acoustic, and visual-spatial-organizational tasks that are similar to those exhibited by older children in these two groups. The exception is that the more rote aspects of verbal skills tend to be more deficient relative to age-based norms in younger than in older children with BPPD.

6. Differentiations in terms of psychomotor and tactile-perceptual skills and abilities are less marked in younger than in older children of these two groups.

7. Although there is some evidence suggesting that younger children with NLD have more difficulty in adapting to novel problem-solving situations than do their BPPD counterparts, the precise measurement of such suspected difficulties has yet to be completed.

8. The adaptive implications of the relative assets and deficits of older children with BPPD and NLD extend well beyond the confines of the academic setting. For example, children with BPPD tend to fare far better in social situations than do children with NLD (168). More precise delineations of these extraacademic learning considerations are discussed in Rourke (151) and in other investigations (28,70,99,121,123,164,172,174,230).

Note on Research Methodology

These conclusions and generalizations should be sufficient to make the desired point that children with LD chosen solely on the basis of specific variations in patterns of academic performance (with adequate controls for age, full-scale IQ, and other important attributes) can be shown to exhibit different patterns of neuropsychological abilities and deficits. In this connection, it should be noted that both of the subtypes under consideration exhibited deficient mechanical arithmetic performance. Thus, members of each of these groups could have been included in the "LD" sample of a contrasting-groups type of study that is designed to compare "arithmetic-disabled" and "normal" children. It should be clear from the conclusions and generalizations that such an "arithmetic-disabled" group could be made up of at least two very distinct arithmetic-disabled subtypes. It is also the case that (a) these subtypes of children share virtually nothing in common, from a neuropsychological standpoint, with each other; (b) their similarly deficient levels of performance in mechanical arithmetic reflect these quite different "etiologies"; and (c) their needs for academic habilitation/rehabilitation with

respect to mechanical arithmetic differ markedly from one another (151,153,161,180,210,211).

NEUROPSYCHOLOGICAL ASSESSMENT PROCEDURES FOR CHILDREN WITH LEARNING DISABILITIES

Content and Rationale

Extensive treatments of the neuropsychological assessment procedures that we have found useful for children and adolescents can be found in Rourke (138,141,154), Rourke et al. (160, especially Chapter 5), and Rourke et al. (171, especially Chapters 1 and 2). Only a few points are emphasized here, as follows.

1. *Comprehensiveness.* A neuropsychological assessment of a child who is thought to have LD should be comprehensive in nature. Our experience with procedures that aim simply to "screen" for LD is that they typically do more harm than good. Even disregarding for the moment the very high levels of false-positive and false-negative results that such procedures typically generate, it should be pointed out that it is rarely, if ever, useful to know simply that a child does or does not have LD. Such knowledge, especially when a positive "diagnosis" of LD is made on the basis of such an assessment procedure, may very well be counterproductive for the child. This is so because there are patterns of assets and deficits evident in various subtypes of children with LD that differ markedly from one another (see, for example, Fig. 25.1 and 25.2 for comparison of NLD and BPPD profiles). Thus, lumping them under the common rubric of LD may lead to gross misunderstanding on the part of professionals and other caretakers involved with the child, not to mention the potentially dramatic negative ramifications that such a homogeneous designation might have for treatment.

It usually is essential to know at least as much about what a child with LD can do as it is to know what he/she *cannot* do, if one's goal is to design an appropriate treatment/intervention plan for the youngster. A designation of LD very often accentuates the disabilities and deficits that the child exhibits while virtually ignoring the assets and strengths that probably will be useful in compensatory intervention strategies. Most treatment plans for children with LD usually involve a combination of direct "attack" on the deficits and exploitation of the child's assets for compensatory modes of adaptation. In general, the older and the more impaired the child, the more likely it is that compensatory techniques accentuating the use of the child's assets will be found to be effective and will tend to dominate the treatment picture. Comprehensive neuropsychological assessment procedures are designed with this "accentuate the positive" rule in mind.

What constitutes a comprehensive assessment? As explained in detail in the references cited earlier, a comprehensive neuropsychological assessment is one that involves the measurement of the principal skills and abilities that are thought to be subserved by the brain. Thus, a fairly broad sampling of tasks involving sensory, perceptual, motor/psychomotor, attentional, mnestic, linguistic, and concept formation/problem solving/hypothesis testing would need to be used. In addition, it would be well to vary the levels of complexity of such tasks (from quite simple to quite complex) and to present tasks that vary along the dimensions of rote and novel requirements. Inclusion of tasks that vary from those that involve information processing within a single modality to those that involve the coordination and execution of response requirements within several modalities would be desirable. It is clear that the aforementioned continua of tests and measures are not mutually exclusive. Finally, it is important in the analysis of children with LD to have available fairly comprehensive personality and "behavioral" data on the child. Standardized tests of important dimensions of psychopathology, activity level, and common problem behaviors are useful—sometimes essential—for this purpose.

A collection of tests and measures that includes these dimensions would meet the minimal requirements for comprehensiveness. For a list of the tests that fulfill such criteria and that we use routinely in our evaluation of children referred for neuropsychological assessment, the interested reader should examine the Appendix section of Rourke et al. (160) or Rourke et al. (171). Some specification of these procedures is contained within the next section of this chapter.

2. *Standardized administration procedures and the availability of norms.* We have found that testing procedures that are standardized and that have norms available for them are essential in the study of children who exhibit LD and any number of other neuropsychological difficulties. This is so for a variety of reasons, not the least of which is the fact that the manifestations of LD may change in many ways over time. The predictability that can be seen to obtain in such circumstances is best couched in terms of deviations from age-expected performances that decline, remain stable, or increase over time. For example, without the availability of developmental norms and standardized tests, it would not have been possible to determine the predictable developmental manifestations of the NLD syndrome. Furthermore, the use of unstandardized procedures would have made it all but impossible to assess its fulminations, responses to therapy, and other aspects of its course in the individual patient. In the last analysis, standardized assessment procedures and their attendant benefits are what make it possible to translate clinical lore into rigorous, testable clinical generalizations (e.g., the NLD syndrome and model).

3. *Procedures amenable to a variety of methods of inference.* The methods of inference for comprehensive neuropsychological assessment procedures first proposed by Reitan (131) are as important now as they were when he articulated them. It is quite crucial that one (a) be able to apply *level of performance* interpretations (especially within the context of developmental norms as outlined in point 2); (b) have a suffi-

ciently broad sample of performances so that *pathognomonic signs* of brain impairment may emerge; (c) have data in a variety of realms that are amenable to the *differential score* approach; and (d) be able to carry out systematic *comparisons of performance on the two sides of the body* (137). In the analysis of NLD and other LD subtypes, all of these methods of inference are used. As an example of a battery of tests that can be used and for which these methods of inference are useful, we present the following recommended protocols for assessment of children with known or suspected NLD.

These are the protocols we find most helpful in the determination of the presence of NLD syndrome. There is one protocol for older children and adolescents (9–15 years) and one for younger children (5–8 years). The older children/adolescent protocol is one about which we know a great deal and in which we feel quite confident. The protocol for younger children currently is under review and so it is not couched in the same terms as is that for older children.

4. There are a number of other features of comprehensive, systematic, neuropsychological assessment that are important considerations in the analysis of the behavior of children with LD (158). Some of these arise in the next section. More comprehensive treatments of such issues are contained in the references cited earlier. We now turn to a description of a comprehensive model of neuropsychological assessment and intervention.

DEVELOPMENTAL NEUROPSYCHOLOGICAL REMEDIATION/HABILITATION MODEL

Figure 25.3 contains a modification of a remediation/ habilitation model that was developed by Rourke et al. (160, 171) for general neuropsychological assessment. It is modified slightly in Figure 25.3 to take into consideration the special dimensions of LD. Note that the model is couched in terms of "steps." In explaining the content and dynamics of the model, we consider each step in order.

Step 1: Levels and Patterns of Assets and Deficits

This step appears, at first blush, to be simple. However, it is meant to imply a number of dimensions that are fairly complex and sometimes difficult to determine in the assessment process. In general, the pictorial representation of this step is meant to imply that the levels and patterns of neuropsychological assets and deficits that constitute the child's LD may have a differential impact on the child's capacities for academic and social learning. The impact may be small or large; it may be in academic and not in social areas; and it may have more impact on some areas of academic functioning than in others. Furthermore, it is taken for granted that factors other than LD may have a bearing on academic and social learning. Thus, the initial goal of the assessment

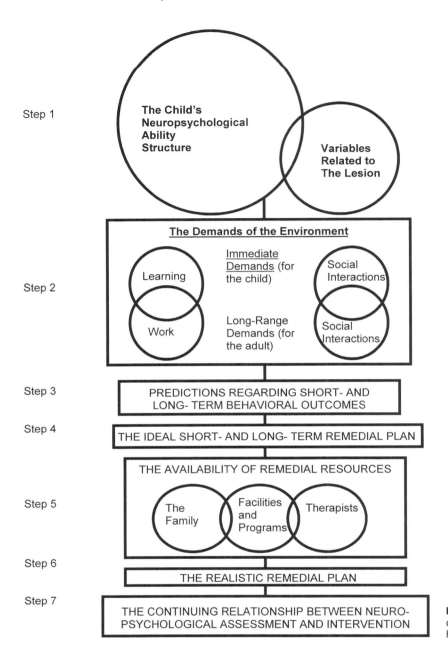

Step 1

Step 2

Step 3

Step 4

Step 5

Step 6

Step 7

FIGURE 25.3. Formulating the remedial plan: developmental neuropsychological remediation/habilitation model.

process is to make a determination of whether and to what extent LD impacts on various academic and social learning/adaptation capacities.

To accomplish this purpose, the neuropsychological assessment procedures should be reliable (18) and valid (152). The principal dimensions of validity in the consideration of the clinical dimensions of assessment are those relating to content, concurrent, predictive, and ecologic/clinical validity. The issues of content (or coverage) and concurrent validity are discussed later. Issues surrounding predictive validity are particularly relevant within the context of steps 2 and 3. Ecologic/clinical validity is basic to all of the steps, but most specifically to considerations relevant to steps 4 through 7.

Content validity, or what we prefer to view as "coverage," is a simple concept but one with far-reaching ramifications. Simply put, a neuropsychological assessment has adequate content validity or coverage to the extent that it meets the following criteria: (a) the skill and ability domains tapped are sufficient to mirror the principal areas of functioning that are thought to be mediated by the brain; (b) the data gathered are sufficient to deal with the clinical (sometimes the referral) problems presented by the child. Failure to meet these criteria, that is, having poor content/coverage validity, can, and often does, severely limit our knowledge regarding the child's present neuropsychological status (concurrent validity), his/her probable prognosis with and without intervention (predictive validity), and the impact of the child's neuropsychological status on a reasonably broad spectrum of his/her developmental/adaptive task demands (ecologic validity).

If the clinician is interested merely in determining if a child has LD, the particular approach to assessment advocated by us is, for the most part, irrelevant. Adopting a more reasonable perspective, for example, one that invokes criteria of probable therapeutic relevance for determining the validity of a neuropsychological assessment, places the evaluation of coverage within a vastly different and potentially more exciting framework. Thus, the evaluation of content/coverage validity within the context of ecologic validity is tantamount to requiring that the content of the neuropsychological assessment (assuming for the moment that such content is adequately interpreted) must be fairly broad, pervasive and comprehensive. It should be clear that such a goal will be achieved only through a reasonably painstaking analysis of a wide variety of functions, with attention given to their interrelatedness and their dissociations.

To this end, we propose that the ability and skill areas outlined in Table 25.1 constitute a bare minimum with respect to the dimensions that must be addressed in order to provide reasonable, relevant, and comprehensive answers to developmentally important clinical questions for children with LD. The reader should bear in mind that the specific tests and measures mentioned in Table 25.2 are of secondary importance. The crucial issues in the present context are related to the ability and skill domains that are sampled and developmental considerations that bear upon levels of performance and complexity. The actual tests and measures used must be evaluated in relation to these issues and considerations. We should be quick to add that this categorization of skills and abilities lacks a firm scientific basis. However, the continuation of our work regarding the construct validity (dimensionality) of this test battery (66,67) is expected to yield such a basis.

Step 2: Demands of the Environment

It is clear and evident that the results and dimensions of the neuropsychological assessment described, no matter how thorough and insightful, remain virtually useless until such time as they are related in a meaningful way to the developmental demands (behavioral tasks, skill developments) that the child faces. This necessitates an appreciation on the part of the assessor of the socio-historical-cultural milieu within which the child is expected to function. In North American society and, for the most part, that of Europe, the demands for learning in both academic and extraacademic milieux are well known. Some of these are immediate (short term, in the model); others are more far flung (long term, in the model). Some have to do with formal learning environments such as the schoolroom, others with informal learning situations such as play. All are important; hence, all four sectors within the matrix associated with step 2 (Fig. 25.3) need to be addressed. Failure to do so can, and often does, lead to terribly

limited and even counterproductive forms of intervention for the child [for examples of such outcomes see Rourke et al. (171), especially Chapters 1, 5, and 6].

Step 3: Predictions Regarding Short-Term and Long-Term Behavioral Outcomes

This step involves, essentially, the formulation of prognostications for the individual child that flow from a melding of the conclusions arrived at in steps 1 and 2. Thus, the child's neuropsychological assessment results are viewed within the context of the short-term and long-term developmental demands that he/she is expected to face so that the relative ease or difficulty of these demands for the child can be seen within the context of the particular neuropsychological assets and deficits exhibited by the youngster. The formulation of confident prognostic statements is a complicated step that requires considerable experience with (a) those deficits in children at different ages that are expected to diminish or increase with the passage of time, with and without treatment; and (b) the relative impact of various forms of treatment on those deficits. It should be clear that the processes involved at this level of the model are intimately connected with those required to deal effectively with the next step in the process.

Step 4: Ideal Short-Term and Long-Term Remedial Plans

Although seldom possible to put into practice, the formulation of ideal remedial plans offers an opportunity to "flesh out" the implications of steps 1, 2, and 3. More important from a clinical perspective is the necessity to formulate a prognosis that fits the neuropsychological status of the child. For example, it might be anticipated that a particular form of intervention (e.g., the institution of a synthetic phonics approach to single-word reading) would be beneficial in the short term but counterproductive over the long haul. Such prognoses regarding the advisability of starting and stopping appropriate treatments/interventions at particular times often are essential to the judgment of treatment efficacy. This approach to this phase of the assessment process also implies that the assessor will proffer some advice regarding the necessity for the type of assessment process that should be applied in order to determine whether a change of therapeutic/intervention tack is advisable. Such an assessment could be as simple as the application of a standardized achievement test to something as complex as a comprehensive examination (see step 7).

Step 5: Availability of Remedial Resources

The best therapeutic plans in the world cannot be applied without appropriate therapeutic resources. Hence, it is necessary for the psychologist to be able to evaluate the therapeutic/intervention resources of the family, specialized "therapists," and the community. By evaluation we mean the fairly precise specification of the types of therapies and forms of intervention that the child with a specific subtype of LD needs and the likelihood that particular families, therapists, and other caregivers can and will provide.

Perhaps the importance of this principle is best seen in an example of its breach. In our experience, the conclusions of standard psychoeducational assessments often read like excerpts from the services offered by a particular school district. The following suggestions are common: placement in

TABLE 25.1. NONVERBAL LEARNING DISABILITIES ASSESSMENT PROTOCOL

NLD Assessment Protocol (9–15 Years)		
	Tests and Measures	
Primary Assets and Deficits	**Simple (Rote)**	**Complex**
Motor and psychomotor	Grip strength	**Maze**
	Static steadiness	**Grooved pegboard**
Tactile-perceptual	Simple tactile perception	**Fingertip number writing**
		Tactile form recognition
Visual-spatial-organizational	Simple visual perception	**WISC object assembly**
	Trail Making Test (part A)	**WISC block design**
Auditory-perceptual	Speech-sounds perception	**Sentence memory**
	Auditory closure	**WISC digit span (backward)**
	Auditory	**Visual**
Attention/memory	WISC digit span (forward)	**Target**
	Seashore rhythm	**Rey-Osterreith Complex**
	Serial digits	**Figure**
	Complex	
Problem solving, hypothesis testing, etc.	**Category test**	
	Wisconsin Card Sorting Test	
	Tactual performance test	
	Trail Making Test (part B)	
	Simple (Rote)	**Complex**
Language	WISC similarities	**WISC comprehension**
	WISC vocabulary	**Verbal fluency (phonemically cued)**
Academic achievement	WRAT reading	**WRAT arithmetic**
	WRAT spelling	**Reading comprehension**
Psychosocial functioning	**Vineland Adaptive Behavior Scale**	
	Personality Inventory for Children	

NLD Assessment Protocol (5–8 Years)[a]	
Primary Assets and Deficits	**Tests and Measures**
Motor and psychomotor	Grip strength
	Grooved pegboard
Tactile-perceptual	Simple tactile perception
	Finger agnosia
	Fingertip number writing
Visual-spatial-organizational	WISC object assembly
	WISC block design
	Target test
Auditory-perceptual	Speech-sounds perception
	Auditory closure
	WISC digit span

(continued)

TABLE 25.1. (*continued*)

	Simple (Rote)	Complex
Language	WISC similarities WISC vocabulary	Sentence memory Verbal fluency (phonemically cued)
Problem solving, hypothesis testing, etc.	Children's word finding test Tactual performance test Matching pictures test	
Academic achievement	WRAT reading WRAT spelling WRAT arithmetic	
Psychosocial functioning	Vineland Adaptive Behavior Scale Personality Inventory for Children	

Note: Tests and measures listed in bold are those in which relative deficits are evident.
The tests are arranged to coincide with the developmental dimensions of the content and dynamics of the nonverbal learning disabilities (NLD) syndrome (238).
Descriptions of the tests, except for the Rey-Osterreith Complex Figure, the serial digits test, and the Wisconsin Card Sorting Test, can be found in Rourke (151), Rourke et al. (160), and Rourke et al. (171).
"Reading comprehension" is not a test, but an area of functioning that is measured in any number of ways.
[a]The relative assets and deficits for the 5 to 8 year age range will be specified at a later date.
WISC, Wechsler Intelligence Scale for Children, Revised; WRAT, Wide Range Achievement Test, Revised.

TABLE 25.2. TESTS INCLUDED IN THE NEUROPSYCHOLOGICAL TEST BATTERY

Tactile-perceptual
1. Reitan-Kløve Tactile-Perceptual and Tactile-Form Recognition Tests
   A. Tactile imperception and suppression
   B. Finger agnosia
   C. Fingertip number writing perception (9–15 yr)
      Fingertip symbol writing perception (5–8 yr)
   D. Coin recognition (9–15 yr)
      Tactile-forms recognition (5–8 yr)

Visual-perceptual
1. Reitan-Kløve Visual-Perceptual Tests
2. Target test
3. Constructional dyspraxia items, Halstead-Wepman Aphasia Screening Test
4. Wechsler picture completion, picture arrangement, block design, object assembly subtests
5. Trail Making Test for Children, part A (9–15 yr)
6. Color form test (5–8 yr)
7. Progressive figures test (5–8 yr)
8. Individual performance test (5–8 yr)
   A. Matching figures
   B. Star drawing
   C. Matching V's
   D. Concentric squares drawing

Auditory-perceptual and language related
1. Reitan-Kløve Auditory-Perceptual Tests
2. Seashore rhythm test (9–15 yr)
3. Auditory closure test (91)
4. Auditory analysis test (136)

5. Peabody Picture Vocabulary Test (43)
6. Speech-sounds perception test
7. Sentence memory test (12)
8. Verbal fluency test (212)
9. Wechsler information, comprehension, similarities, vocabulary, digit span subtests
10. Aphasoid items, aphasia screening test

Problem solving, concept formation, reasoning
1. Halstead Category Test
2. Children's word finding test (167)
3. Wechsler arithmetic subtest
4. Matching pictures test (5–8 yr)

Motor and psychomotor
1. Reitan-Kløve Lateral Dominance Examination
2. Dynamometer
3. Finger tapping test
4. Foot tapping test
5. Kløve-Matthews Motor Steadiness Battery
   A. Maze coordination test
   B. Static steadiness test
   C. Grooved pegboard test

Other
1. Underlining test (32, 176, 178)
2. Wechsler coding subtest
3. Tactual performance test
4. Trail Making Test for Children, part B (9–15 yr)

Note: Unless otherwise indicated, these measures are described in Reitan and Davison (133) and in manuals for the Wechsler scales. [In these descriptions, we refer to the Wechsler scales rather than to the Wechsler Intelligence Scale for Children (WISC) or Wechsler Intelligence Scale for Children, Revised (WISC-R) because some of the child cases presented were administered on one or both of these scales. In the text, we simply refer to the "Wechsler" scale(s) rather than making a distinction between the two for interpretation purposes.] All are described extensively by Rourke (151).

a specific learning disabilities class; phonics training; referral to a speech pathologist. The results of any and, indeed, all such recommendations may turn out to be quite productive. However, it is far more common that such recommendations—lacking, as they do, much in specifications regarding precise modes of intervention, intensity, goals, and prognoses—form a poor guide for the caregiver. What usually happens is that the therapists (e.g., teacher, speech pathologist, physiotherapist) must carry out their own assessment and relate those findings to what they view as their own therapeutic facilities. This is an unproductive state of affairs, because it wastes time, effort, and resources, and—of far more importance—causes unnecessary delays in the application of efficacious modes of intervention. Even worse, such shilly-shallying can lead to grossly inappropriate modes of intervention being applied.

Step 6: Realistic Remedial Plan

The differences between steps 4 and 5 constitute the broad outline for the realistic remedial plan. When such differences are consistently large and/or consistently within one particular area (e.g., speech/language therapy), they constitute parts of a "laundry list" of services required to address the needs of children with LD.

Step 7: Persisting Relationship Between Neuropsychological Assessment and Intervention

In view of the changes that maturation and formal and informal learning experiences can effect, it is necessary to point out that neuropsychological and other forms of comprehensive assessment are not "one-shot" affairs. It should come as no surprise that repeated neuropsychological assessments often are required in order to fine tune, or even grossly alter, remedial plans for children with LD.

Follow-up neuropsychological assessments afford an opportunity to evaluate the effectiveness of particular modes of remedial intervention. Given our current (fairly abysmal) state of knowledge regarding the therapeutic efficacy of various forms of intervention for children with LD, a rigorous, objective, comprehensive, standardized assessment appears to be a minimal requirement for any protocol designed to assess the effectiveness of these complex undertakings.

With these issues regarding assessment as background, we now turn to a consideration of issues relating to prognosis and treatment for the child who exhibits the NLD syndrome (subtype). We present these general principles as an example of the sort of treatment program that can emerge from the systematic comprehensive analysis of children who exhibit a specific subtype of LD.

TREATMENT PROGRAMS FOR CHILDREN WITH DIFFERENT SUBTYPES OF LEARNING DISABILITY

Identification of subtypes of LD in combination with intensive and comprehensive neuropsychological examination of the children with LD should provide insight into their disposition and treatment. One of the main objectives of subtyping efforts is to delineate with greater precision the sorts of deficiencies that may account for a child's inability to acquire normal reading, spelling, arithmetic, or psychosocial skills so that they might be addressed in a treatment program. The comprehensive neuropsychological assessment further individualizes this process. The aim of this double-barreled process is to fashion remedial programs tailored to the individual's specific assets and deficits. It has been our clinical experience that a remedial management intervention that fails to "fit" the neurocognitive ability makeup of the child can, in effect, be counterproductive with respect to the acquisition of basic academic-related and psychosocial skills (171).

Subtyping methods have an additional long-term economy in permitting the use of multivariate classification schemes that have some hope of being replicated in other clinical and research settings. In our experience, such a systematic collection of data has proved helpful in the design, implementation, and validation of various treatment approaches. Some of the reasons for failure to replicate findings by independent investigators are that there has not been sufficient denotation of the measurement operations or that the investigators have used different sets of performance measures. One can hardly expect to develop efficacious and generalizable treatment procedures when the assessment data have been collected by one-of-a-kind techniques.

The problem of translating neuropsychological assessment data into viable and effective educational and therapeutic procedures is not a simple one and has not yet been satisfactorily solved. Although a lack of experimental verification of treatment approaches is partly responsible for this state of affairs, there also has been the tendency for the educational community to favor the more traditional psychoeducational assessment for purposes of remedial planning. Unfortunately, children who have been subjected to remedial regimens prescribed by this approach often make gains in skills that they practice and yet exhibit little or no transfer of training to the classroom situation (216).

As we suggested in the past (141), perhaps one of the main problems in bridging the gap between assessment information and intervention strategies lies in the lack of a sophisticated method for determining distinct groups of individual differences among youngsters with LD that would allow for an adequate test of the presumed relationship between patterns of abilities and deficits and patterns of failure and success in the classroom. In the absence of any reliable and valid model of the child's cognitive ability structure, it is no wonder that most, if not all, academic remedial prescriptions can be viewed as speculative and having little hope of enjoying the luxury of a definitive empirical test. To complicate matters further, many educators seem to be more attached to the defense of professional "turf" rather than to the progress of scientific knowledge.

Still another factor of considerable importance in the treatment of children with LD involves the relationship between the child's ability-related strengths and weaknesses and his/her adaptive capacities. There has been a tendency in the design of remedial programs to focus almost exclusively on the immediate demands of the academic milieu. Most often, scant attention is paid to the child's capacities to adapt to informal learning situations (e.g., play), social relationship requirements (within the family and with age-mates), and long-term learning and social developmental demands (171). It is not enough simply to focus on promoting the child's academic growth while giving short shrift to issues of personal integration and psychosocial adjustment and adaptation.

Although a number of treatment programs for LD have been reported in the literature, few have been demonstrated to be effective in well-controlled and replicated research (216). An excellent summary of several of the factors critical in assessing the interaction between instructional methods and outcome for children with LD has been provided by Lyon and Flynn (103). According to these investigators, the degree of relationship among LD subtype, instructional treatment method, and outcome is difficult to interpret because of variations and/or limitations in variables such as the following: (a) measuring adherence to, or compliance with, a particular treatment strategy; (b) teacher-related variables, such as preparation or style; (c) specific classroom climate; and (d) the LD child's previous and concurrent instructional experiences. To complicate matters further, limited follow-up compliance, initial refusal to "experiment" with any given form of instruction, and delayed recognition of the effects of remediation pose additional problems when measuring the efficacy of remedial efforts. At the same time, the work of Fiedorowicz and Trites (49), Bakker (6,7), and Bakker et al. (8) seems to hold some promise in the effective treatment of subtypes of reading disorders in particular.

Our own work in this area has used more in the way of clinical inferential rather than empirically based investigative strategies. For example, several single-case follow-up examinations of distinct subgroups of reading- and spelling-disordered children have shown them to be less amenable to instructional strategies that tend to be more "deficit driven" than "compensatory" in nature. This appears to be the case in children who exhibit marked deficits in phonologic, symbolic, sequential, and/or mnemonic processing. In addition, neither age nor sex of the child appears to be a contributing factor in differentiating outcome. For example, a standard

remediation program that emphasizes complex semantic-acoustic aspects of language (e.g., phonics) likely would not be very helpful for a child with BPPD. However, it may be the case that very intensive "phonics" training would be of benefit at some level (124). Nevertheless, standard remediation approaches using only phonics training probably would result primarily in frustration or even resentment on the part of the child, and as such he/she could even become resistant to remediation in general. Alternatively, use of compensatory methods that take advantage of the visual-spatial perception developing sight-word recognition strategies for reading instruction should have much more favorable results in cases of BPPD. For some children, this might be integrated with phonics as well, but in most cases the emphasis on visual cues and learning to recognize words as whole units is key. This would foster development of a "sight word" vocabulary and could be accomplished in several ways, such as using flash cards, learning to associate words with pictures, and presenting novel words in the context of familiar sentences. Similarly, for spelling instruction children could be taught to try to memorize graphical visual features of words (particularly irregularly spelled words) so that they can examine whether a particular word "looks" correct after spelling it.

Although the results of relevant research to date suggests that children whose LD seems due primarily to auditory-perceptual and/or psycholinguistic-related processing deficiencies (i.e., BPPD) tend to be generally well adjusted (71,72,99,121,123,129,172,230), those children who experience some degree of emotional maladjustment seem to profit little from verbally biased psychotherapeutic techniques. This condition most likely reflects compromised verbal comprehension competencies and may lead such children to avoid situations that are verbally interactive. Nevertheless, the fact remains that the deficits exhibited by such children do not have a necessary negative impact on psychosocial functioning (172, 230).

As is the case of the psycholinguistically impaired youngster, most of the habilitational considerations for children with NLD are garnered from single-case follow-up investigations and reflect general principles of treatment intervention. We have attempted to relate the unique pattern of neuropsychological assets and deficits of such children to modes of intervention that are primarily, although not exclusively, compensatory in nature.

Our formulation of treatment strategies with children with NLD is based, in part, on a model of brain functioning that implicates the involvement of white matter disease or dysfunction in its manifestation (151,156). This is of both theoretical and clinical importance in selecting the appropriate "mode" of intervention. For example, the earlier in development that white matter disease occurs, the more propitious it would be to use a treatment approach that attacks the deficits. This notion is predicated on the assumption that such an approach will serve to stimulate the functioning of the remaining white matter in order to foster the development of centers of gray matter. On the other hand, the longer the NLD syndrome persists and the later its psychosocial manifestations, the more likely that compensatory techniques will prove to be efficacious in pursuit of a therapeutic nature. For a complete explication of the specifics of this particular program, the interested reader is referred to Rourke (151,156), but some techniques are summarized here.

In general, compensatory techniques for children with NLD should emphasize rote, easily memorized strategies, preferably those that are easily verbalized. Whenever possible, these strategies should be broken down into simple, step-by-step sequences that can be individually learned in small portions. Given the problem-solving and organizational difficulties in children with NLD, this method will be most successful if each part of the sequence is presented in the proper order and the child is given adequate time to fully master each step through a slow, highly redundant, and repetitive approach. The emphasis on verbalizable steps is important because it will allow the child to "talk through" them when solving problems or encountering novel situations. This could be vocal at first but eventually become an internalized list that would allow the child with NLD to make use of his/her strengths in verbal abilities and rote memorization. This piece-by-piece training approach should be used for specific areas and skills such as mechanical arithmetic and more general problem-solving techniques. Thus, the child should be taught to examine a situation systematically, generate possible approaches (or more likely attempt to recall several possible means of solving similar problems previously encountered), consider potential consequences of each action, weigh the alternatives, and then choose the best approach. Most persons have learned through experience to conduct these steps automatically and often will not realize they are specifically following them. For the individual with NLD, though, the steps must be made explicit and consciously followed in novel and complex situations.

With regard to psychosocial problems in children with NLD, the tendency for anxiety, depression, and social isolation generally result from their consistent difficulties in social interactions. Their impairments in producing and reading both subtle and more obvious nonverbal information in social situations and their difficulties dealing with complex or novel situations and learning from experience tend to result in serious problems with interpersonal relations. As these children grow older, their psychosocial skills fail to improve and they often become withdrawn, isolated, and consequently demonstrate symptoms of notable internalized psychopathology (172).

The best means of intervention for psychosocial difficulties in children with NLD is to address the area at the root of these problems—impaired social interaction skills. As with academic interventions, social skills training for a child with NLD should be done using compensatory methods and in a systematic, sequential manner using verbalizable

steps and procedures as much as possible. What may seem to be very basic social skills (e.g., eye contact during conversation, taking turns during interchanges) should be taught in the same methodical repetitive manner as more complex skills until they are mastered through memorization. Once the child with NLD can interact more appropriately and comfortably with others, he/she should be able to participate more frequently and successfully in social situations. However, this should be done in a very gradual manner with multiple opportunities for feedback and few consequences initially. Therapists could present videotapes of social interactions that can be paused, allowing interpretations of nonverbal cues and discussion of possible actions/consequences with the child. Eventually, structured social activities, such as in intramural sports, clubs, or community groups, could be used for further opportunities to practice appropriate verbal and nonverbal social interaction skills.

Generalization of training in a remedial setting to outside contexts can be difficult with any intervention, but it is particularly problematic for the child with NLD. Therefore, the training of various methods and sequential strategies should include a number of rules and specific descriptions of when to follow various scripts and strategies. These rules should be made as explicit as possible, with time spent teaching the child with NLD to examine all aspects of a situation in a systematic fashion, consider alternative actions, and choose an appropriate response. This often will be necessary even in what may seem to be very straightforward situations to an outside observer, and in areas of both deficit (e.g., social skills, problem-solving) and relative strength, such as verbal expression. For example, the tendency of a child with NLD to engage in excessive verbal output, often tangential in nature, can be curtailed through instruction of rules for language pragmatics (what to say, when to say it, and how it should be said).

For some children, it may be beneficial to help them gain insight into their own patterns of abilities and deficits, which could aid them in learning what compensatory methods to use and the situations in which they would be most appropriate. However, this approach would be very difficult for children with NLD. The difficulties with abstraction, dealing with novel or unusual information, and overreliance on highly routinized responses that characterize this subtype of LD can be substantial barriers to such techniques. This is not to say that such insight is impossible. In fact, frequent appropriate feedback may help children with NLD comprehend their areas of cognitive strengths and learn independently when they should apply certain strategies. Working toward this insight, though, likely will not be as successful in other areas such as social skills training. The emphasis there should remain on memorizing particular responses and rules for applying them in various situations.

Finally, providing age-appropriate aids for children with subtypes of LD can be very productive. For example, a hand calculator could be used to check the accuracy of written arithmetic for younger children with NLD so they may immediately have feedback on their performance and look for where they made errors when they occur. This practice may be required even more frequently at the secondary school level so as to allow greater time to focus on learning common mathematics concepts. Both children with NLD and those with BPPD could make extensive use of computers as therapeutic devices. The ability to present visual cues in multiple forms that can be linked to various words would be of notable benefit to a child with BPPD and reduce the burden of constantly creating multiple visual aids. In the case of NLD, programs that present material in a systematic way could be useful in teaching academic material and in problem-solving and social situations. The provision of immediate corrective feedback for each response that this format allows is crucial if learning is to proceed apace.

DIRECTIONS FOR FUTURE STUDY

Although some of the shortcomings of existing theoretical formulations and experimental procedures in the study of LD are beginning to be understood, it is clear that the investigations conducted to date represent at best only a small fraction of the work yet to be done in this area. The results of multivariate classification studies have made us cognizant of the fact that LD in children can be analyzed at a number of levels and with rather sophisticated statistical precision. Ultimately, we hope for an integration of these findings with other levels of analyses such as those provided by more in-depth educational, psychosocial, and neuropsychological models. To accomplish this, the following areas of investigation appear to be of considerable importance.

First, there remains a state of confusion regarding the definition of LD. There are those who continue to "diagnose" LD, and yet professionals vary greatly when asked to state what it is they are diagnosing. In a climate of definitional—and, as one consequence—etiologic confusion, it is easy to see why the determination of the types of intervention that would be most appropriate for different subtypes of children with LD has been elusive. We have proposed some specific definitions and are in the process of determining their validity. The goal of this exercise is to arrive at consensually validated characterizations of LD.

With the exception of NLD, subtypal definitions have received little investigative attention. There has been an increasing amount of empirical study aimed at determining the reliability and validity of LD subtypes [for a review of these investigations, see Rourke (152,156)], but still much needs to be done to relate these subtypes to considerations of ecologic validity (e.g., areas of psychosocial and adaptive functioning). We believe that there has been sufficient work in this area to this point (71,72,148,150,172,183,194,230) to justify some optimism with respect to the fruitfulness of research in this area.

There is the issue of comparisons of subtypes of LD with comorbid conditions, such as attention deficit disorder (22), and the generalizability of psychosocial subtypes of LD to other areas of neuropsychological interest, such as traumatic brain injury (20). Even more generally, the incidence of subtypes of LD, such as NLD, in conditions such as Asperger syndrome (93), as well as a host of other neurologic syndromes (156,159), is an area that is ripe for neuropsychological investigation (228).

Finally, it is clear that we need to establish whether and to what extent the specificity of subtype identification leads to specificity of efficacious treatments. Vapid debates regarding the relative utility of deficit- or compensatory-driven strategies at different stages of development will continue to be the rule until we make considerably more headway in this area. To this end, we need to focus on investigations in which preferred modes of intervention are deduced from models of neuropsychological relationships that have been shown to have concurrent and predictive validity. The work of Bakker and associates (6,7,8) is one example of this approach, and that of Lyon and Flynn (103) is another. Both of these approaches have shown considerable promise, and we can expect to see more developments along these lines in the near future.

ADDENDUM

NLD has been referred to as the *final common pathway* of white matter disease/dysfunction (147). What is meant by this? What is the most recent classification and hierarchy of diseases and disorders wherein the NLD syndrome is more or less manifest?

Referring to NLD as the "final common pathway" is meant to suggest that NLD is not thought to be a cause or a mediating variable in some types of neurologic disease and disorder. Rather, it is meant to denote that some types of neurologic disease and disorder (Table 25.3) appear to have one common set of characteristics of their neuropsychological phenotype (i.e., the content of the NLD syndrome). Implied is the notion that these disorders would have other phenotypical characteristics in addition to the set of neuropsychological assets and deficits that constitute the NLD syndrome.

These forms of neurologic disease, disorder, and dysfunction in which NLD is the "final common pathway" are arranged in terms of a hierarchy. The levels within this hierarchy denote decreasing phenotypic similarity to the set of neuropsychological assets and deficits that constitute the manifestations of the NLD syndrome.

As investigators pursue the clarification of the many dimensions of this question, the hierarchy originally proposed (156,228) has changed in some significant ways. Our current characterization of this system is shown in Table 25.3.

TABLE 25.3. NONVERBAL LEARNING DISABILITIES: OVERVIEW OF MANIFESTATIONS IN NEUROLOGIC DISEASE, DISORDER, AND DYSFUNCTION

Disease/Disorder/Dysfunction	References/Notes
Level 1 (virtually all of the NLD assets and deficits are manifest)	
Callosal agenesis (uncomplicated)	Rourke (147); Smith and Rourke (204)
Asperger syndrome	Ellis and Gunter (47); Klin et al. (93); Rourke and Tsatsanis (181)
Velocardiofacial syndrome	Fuerst et al. (74); Golding-Kushner et al. (79); Moss et al. (115)
Williams syndrome	Anderson and Rourke (3); Don et al. (37); MacDonald and Roy (104); Morris and Mervis (111); Udwin and Yule (231)
de Lange syndrome	Tsatsanis and Rourke (229)
Hydrocephalus (early; shunted)	Donders et al. (38); Fletcher et al. (59); Fletcher et al. (56); Fletcher et al. (54,55); Rourke et al. (160, pp. 290–297)
Turner syndrome (45,X)	Powell and Schulte (130); Rovet (186)
Significant damage or dysfunction of the right cerebral hemisphere	Rourke et al. (160, pp. 230–253)
Level 2 (a considerable majority of the NLD assets and deficits are evident)	
Sotos syndrome	Dool et al. (40); Buono et al. (19); Fletcher and Copeland (57); Lesnik et al. (98); Picard and Rourke (128)
Prophylactic treatment for acute lymphocytic leukemia (ALL); (long-term survivors) and treatment of children with some forms of brain cancer (long-term survivors)	*Note:* Prophylactic treatment, of course, is not a form of neurologic disease. It is included at level 2 because children who are long-term survivors of ALL (and some other forms of cancer affecting the brain), and who have received very high doses of whole brain cranial irradiation and some other types of therapies, frequently exhibit a considerable majority of the NLD assets and deficits.
Metachromatic leukodystrophy (early in disease progression)	Dool et al. (39); Shapiro and Balthazor (197); Shapiro et al. (198); Weber et al. (233)
Congenital hypothyroidism	Rovet (185)
Fetal alcohol syndrome (high functioning)	Don and Rourke (36); Streissguth et al. (213)
Level 3 (fairly clear evidence of NLD; many of the NLD assets and deficits are manifested by a significant subset of children with these disorders)	
Multiple sclerosis (early to middle stages)	White and Krengel (235)
Traumatic brain injury (diffuse white matter perturbations)	Ewing-Cobbs et al. (48); Fletcher and Levin (61)
Toxicant-induced encephalopathy (affecting white matter) and teratology	Magee et al. (105); White and Krengel (236)
Children with human immunodeficiency virus and white matter disease	Brouwers et al. (17)
Fragile X (high functioning)	Crowe and Hay (26)
Triple X syndrome	Ryan et al. (195)
Leukodystrophies other than metachromatic (early in disease)	Shapiro and Balthazor (197); Shapiro et al. (199)
Haemophilus influenzae meningitis	Anderson and Taylor (4)
Early-treated phenylketonuria	Welsh and Pennington (234); Waisbren (232)
Intraventricular hemorrhage (early)	Landry et al. (94); Selzer et al. (196)
Children with cardiac disease treated with extracorporeal membrane oxygenation	Teeter (220); Tindall et al. (225)
Children with very low birth weight	Picard et al. (127); Taylor et al. (218)
Congenital adrenal hyperplasia	Nass et al. (117)
Insulin-dependent diabetes mellitus (very early onset)	Rovet (184, 187); Rovet et al. (188)
Fahr syndrome	Blackburn (14)
Level 4 (research evidence is ambiguous with respect to the phenotype of NLD in these disorders)	
Neurofibromatosis (early to middle stages of disease progression)	Bawden et al. (10); Dilts et al. (31); Eliason (45); Nilsson and Bradford (118); North et al. (119, 120)
Noonan syndrome	Teeter (220); Troyer and Joschko (227)
Difficult to classify	
Cerebral palsies of perinatal origin	*Note:* Many children with cerebral palsy of perinatal origin exhibit a considerable majority of the NLD assets and deficits. However, because of the wide variety of etiologies and manifestations considered under this rubric, the classification by level of NLD manifestations is rendered problematic.

(continued)

TABLE 25.3. (*continued*)

Disease/Disorder/Dysfunction	References/Notes
Similar, but basically different	
Tourette syndrome	Brookshire et al. (16); Yeates and Bornstein (237)
	Note: Tourette syndrome is one example of a neurologic disorder in which several of the NLD manifestations are evident. However, there are many basic differences that suggest strongly that Tourette syndrome is not a syndrome that should be considered within the group of neurologic disorders that can be characterized in terms of the NLD spectrum [Brookshire et al. (16)].
Autism (high functioning)	Klin et al. (93); Minshew et al. (109)
	Note: As with Tourette syndrome, a similar state of affairs obtains with respect to "high-functioning" autism [Klin et al. (93)].

NLD, nonverbal learning disability.

Some notes regarding the hierarchy:

1. It is probable that this hierarchy will change somewhat as more becomes known about the neuropsychological manifestations of the diseases in question.
2. It is likely that other forms of neurologic disease, disorder, and dysfunction will be added to this hierarchy.
3. It appears highly probable that further advances in neuroimaging of white matter functioning, neuropathologic findings regarding white matter perturbations, and other advances in the specification of the developmental and functional neuroanatomy of myelination will throw considerable light upon the underpinnings of this NLD hierarchy.

It should be clear that this characterization of the *content* of NLD as a phenotype does not speak to the issues of *developmental dynamics* (causes and effects) that may transpire in these forms of neuropathology. Specific investigations of the developmental dynamics that transpire in each of them would be expected to yield interesting information regarding behavioral cause-and-effect relationships therein. In tandem with investigations of the behavioral dimensions of the dynamics in these diseases, it would be desirable to test the tenets of the white matter model (151) in each of them.

In this exercise, the following would be expected to arise from the investigation of these disorders:

1. The NLD *content* of the phenotypes would be virtually identical, even though other characteristics of the phenotype of each disorder may vary widely.
2. The *developmental dynamics* would be similar but by no means uniform.
3. The role of the type and degree of *white matter* maldevelopment, dysfunction, and/or destruction would yield different neurodevelopmental dynamics and trajectories.

Viewing all of this within the context of the testing of the developmental tenets of the white matter model also would be expected to yield important information with respect to the neuropsychological consequences of perturbations in white matter. For example, the model would suggest that one would expect to see significant differences in neuropsychological outcome in children whose disease affected white matter development and function during early uterine versus neonatal/very early infancy versus late infancy/early childhood developmental epochs.

Final Note

These diseases are being examined from the perspective of the NLD/white matter model (151) in our own and a number of other laboratories. The interested reader may wish to consult Thatcher (223,224), Fuerst and Rourke, Tsatsanis and Rourke (228), and Ellis and Gunter (47) for some implications and ramifications of white matter disease, disorder, and dysfunction, especially with regard to the hypothesized etiology of the NLD syndrome.

REFERENCES

1. Ackerman PT, Dykman RA, Oglesby DM. Visual event-related potentials of dyslexic children to rhyming and nonrhyming stimuli. *Clin Exp Neuropsychol* 1994;16:136–154.
2. Ackerman PT, McPherson WB, Oglesby DM, et al. EEG power spectra of adolescent poor readers. *J Learn Disabil* 1998;31:83–90.
3. Anderson P, Rourke BP. Williams syndrome. In: Rourke BP, ed. *Syndrome of nonverbal learning disabilities: neurodevelopmental manifestations.* New York: Guilford Press, 1995:138–170.
4. Anderson V, Taylor HG. Meningitis. In: Yeates KO, Ris RD, Taylor HG, eds. *Pediatric neuropsychology: research, theory, and practice.* New York: Guilford Press, 1999:117–148.
5. Applebee AN. Research in reading retardation: two critical problems. *J Child Psychol Psychiatry* 1971;12:91–113.
6. Bakker DJ. Electrophysiological validation of L- and P-type dyslexia. *J Clin Exp Neuropsychol* 1986;8:133.
7. Bakker DJ. *Neuropsychological treatment of dyslexia.* New York: Oxford University Press, 1990.
8. Bakker DJ, Licht R, Van Strien J. Biopsychological validation of L- and P-type dyslexia. In: Rourke BP, ed. *Neuropsychological*

validation of learning disability subtypes. New York: Guilford Press, 1991:124–139.

9. Bakker DJ, Vinke J. Effects of hemisphere-specific stimulation on brain activity and reading in dyslexics. *J Clin Exp Neuropsychol* 1985;7:505–525.

10. Bawden H, Dooley J, Buckley D, et al. MRI and nonverbal cognitive deficits in children with neurofibromatosis 1. *J Clin Exp Neuropsychol* 1996;18:784–792.

11. Beaton AA. The relation of planum temporale asymmetry and morphology of the corpus callosum to handedness, gender, and dyslexia: a review of the evidence. *Brain Lang* 1997;60:255–322.

12. Benton AL. *Sentence memory test.* Iowa City, IA: Author, 1965.

13. Benton AL. Developmental dyslexia: neurological aspects. In: Friedlander WJ, ed. *Advances in neurology, vol. 7.* New York: Raven Press, 1975:1–41.

14. Blackburn LB. Neurodevelopmental course in pediatric onset progressive calcification of the basal ganglion: a case report. *Arch Clin Neuropsychol* 1996;11:369–370.

15. Brandys CF, Rourke BP. Differential memory capacities in reading- and arithmetic-disabled children. In: Rourke BP, ed. *Neuropsychological validation of learning disability subtypes.* New York: Guilford Press, 1991:73–96.

16. Brookshire B, Butler IJ, Ewing-Cobbs L, et al. Neuropsychological characteristics of children with Tourette syndrome: evidence for a nonverbal learning disability? *J Clin Exp Neuropsychol* 1994;16:289–302.

17. Brouwers P, van der Vlugt H, Moss H, et al. White matter changes on CT brain scan are associated with neurobehavioral dysfunction in children with symptomatic HIV disease. *Child Neuropsychol* 1995;1:93–105.

18. Brown SJ, Rourke BP, Cicchetti DV. Reliability of tests and measures used in the neuropsychological assessment of children. *Clin Neuropsychol* 1989;3:353–368.

19. Buono LA, Morris MK, Morris RD, et al. Evidence for the syndrome of nonverbal learning disabilities in children with brain tumors. *Child Neuropsychol* 1998;4:144–157.

20. Butler K, Rourke BP, Fuerst DR, et al. A typology of psychosocial functioning in pediatric closed-head injury. *Child Neuropsychol* 1997;3:98–133.

21. Casey JE, Rourke BP. Construct validation of the nonverbal learning disabilities syndrome and model. In: Rourke BP, ed. *Neuropsychological validation of learning disability subtypes.* New York: Guilford Press, 1991:271–292.

22. Casey JE, Rourke BP, Del Dotto JE. Learning disabilities in children with attention deficit disorder with and without hyperactivity. *Child Neuropsychol* 1996;2:83–98.

23. Casey JE, Rourke BP, Picard EM. Syndrome of nonverbal learning disabilities: age differences in neuropsychological, academic, and socioemotional functioning. *Dev Psychopathol* 1991;3:331–347.

24. Ceci SJ. *Handbook of cognitive, social, and neuropsychological aspects of learning disabilities, vol. 1.* Hillsdale, NJ: Lawrence Erlbaum, 1986.

25. Ceci SJ. *Handbook of cognitive, social, and neuropsychological aspects of learning disabilities, vol. 2.* Hillsdale, NJ: Lawrence Erlbaum, 1987.

26. Crowe SF, Hay DA. Neuropsychological dimensions of the fragile X syndrome: support for a non-dominant hemisphere dysfunction hypothesis. *Neuropsychologia* 1990;28:9–16.

27. Czudner G, Rourke BP. Age differences in visual reaction time in "brain-damaged" and normal children under regular and irregular preparatory interval conditions. *J Exp Child Psychol* 1972;13:516–526.

28. Del Dotto JE, Fisk JL, McFadden GT, et al. Developmental analysis of children/adolescents with nonverbal learning disabili-

ties: long-term impact on personality adjustment and patterns of adaptive functioning. In: Rourke BP, ed. *Neuropsychological validation of learning disability subtypes.* New York: Guilford Press, 1991:293–308.

29. Demb JB, Boynton GM, Heeger DJ. Brain activity in visual cortex predicts individual differences in reading performance. *Proc Natl Acad Sci USA* 1997;94:13363–3366.

30. Demb JB, Boynton GM, Heeger DJ. Functional magnetic resonance imaging of early visual pathways in dyslexia. *J Neurosci* 1998;18:6939–6951.

31. Dilts C, Carey J, Kircher J, Hoffman R, et al. Children and adolescents with neurofibromatosis 1: a behavioral phenotype. *J Dev Behav Pediatr* 1996;17:229–241.

32. Doehring DG. *Patterns of impairment in specific learning disability.* Bloomington, IN: Indiana University Press. 1968.

33. Doehring DG. The tangled web of behavioral research on developmental dyslexia. In: Benton AL, Pearl D, eds. *Dyslexia: an appraisal of current knowledge.* New York: Oxford University Press, 1978:125–135.

34. Doehring DG, Hoshko IM. Classification of reading problems by the Q-technique of factor analysis. *Cortex* 1977;13:281–294.

35. Doehring DG, Hoshko IM, Bryans BN. Statistical classification of children with reading problems. *J Clin Neuropsychol* 1979;1:5–16.

36. Don A, Rourke BP. Fetal alcohol syndrome. In: Rourke BP, ed. *Syndrome of nonverbal learning disabilities: neurodevelopmental manifestations.* New York: Guilford Press, 1995:372–406.

37. Don A, Schellenberg G, Rourke BP. Auditory pattern perception in children with Williams syndrome. *Child Neuropsychol* 1999;5:154–170.

38. Donders J, Rourke BP, Canady AI. Neuropsychological functioning of hydrocephalic children. *J Clin Exp Neuropsychol* 1991;13:607–613.

39. Dool CB, Fuerst KB, Rourke BP. Metachromatic leukodystrophy. In: Rourke BP, ed. *Syndrome of nonverbal learning disabilities: neurodevelopmental manifestations.* New York: Guilford Press, 1995:331–350.

40. Dool CB, Fuerst KB, Rourke BP. Sotos syndrome. In: Rourke BP, ed. *Syndrome of nonverbal learning disabilities: neurodevelopmental manifestations.* New York: Guilford Press, 1995.

41. Dool CB, Stelmack RM, Rourke BP. Event-related potentials in children with learning disabilities. *J Clin Child Psychol* 1993;22:387–398.

42. Duara R, Kusch A, Gross-Glenn K, et al. Neuroanatomic differences between dyslexic and normal readers on magnetic resonance imaging scans. *Arch Neurol* 1991;48:410–416.

43. Dunn LM. *Expanded manual for the Peabody Picture Vocabulary Test.* Minneapolis, MN: American Guidance Services, 1965.

44. Eden GF, VanMeter JW, Rumsey JM, et al. Abnormal processing of visual motion in dyslexia revealed by functional brain imaging. *Nature* 1996;382:66–69.

45. Eliason MJ. Neurofibromatosis: implications for learning and behavior. *J Dev Behav Pediatr* 1986;7:175–179.

46. Eliez S, Rumsey JM, Giedd JN, et al. Morphological alteration of temporal lobe gray matter in dyslexia: an MRI study. *J Child Psychol Psychiatry* 2000;41:637–644.

47. Ellis HD, Gunter HL. Asperger syndrome: a simple matter of white matter? *Trends Cogn Sci* 1999;3:192–200.

48. Ewing-Cobbs L, Fletcher JM, Levin HS. Traumatic brain injury. In: Rourke BP, ed. *Syndrome of nonverbal learning disabilities: neurodevelopmental manifestations.* New York: Guilford Press, 1995:433–459.

49. Fiedorowicz CAM, Trites RL. From theory to practice with subtypes of reading disabilities. In: Rourke BP, ed. *Neuropsychological*

validation of learning disability subtypes. New York: Guilford Press, 1991:243–266.

50. Filipek PA. Neurobiologic correlates of developmental dyslexia: how do dyslexics' brains differ from those of normal readers? *J Child Neurol* 1995;10[Suppl 1]:S62–S69.

51. Fisher NJ, DeLuca JW, Rourke BP. Wisconsin Card Sorting Test and Halstead Category Test performances of children and adolescents who exhibit the syndrome of nonverbal learning disabilities. *Child Neuropsychol* 1997;3:61–70.

52. Fisk JL, Rourke BP. Identification of subtypes of learning-disabled children at three age levels: a neuropsychological, multivariate approach. *J Clin Neuropsychol* 1979;1:289–310.

53. Fletcher JM. External validation of learning disability typologies. In: Rourke BP, ed. *Neuropsychology of learning disabilities: essentials of subtype analysis.* New York: Guilford Press, 1985:187–211.

54. Fletcher JM, Bohan TP, Brandt ME, et al. Cerebral white matter and cognition in hydrocephalic children. *Arch Neurol* 1992;49:818–825.

55. Fletcher JM, Bohan TP, Brandt M, et al. Morphometric evaluation of the hydrocephalic brain: relationships with cognitive development. *Childs Nerv Syst* 1996;12:192–199.

56. Fletcher JM, Brookshire BL, Bohan TP, et al. Early hydrocephalus. In: Rourke BP, ed. *Syndrome of nonverbal learning disabilities: neurodevelopmental manifestations.* New York: Guilford Press, 1995:206–238.

57. Fletcher JM, Copeland DR. Neurobehavioral effects of central nervous system prophylactic treatment of cancer in children. *J Clin Exp Neuropsychol* 1988;10:495–537.

58. Fletcher JM, Espy KA, Francis DJ, et al. Comparisons of cut-off score and regression-based definitions of reading disabilities. *J Learn Disabil* 1989;22:334–338.

59. Fletcher JM, Francis DJ, Rourke BP, et al. The validity of discrepancy-based definitions of reading disabilities. *J Learn Disabil* 1992;25:555–561.

60. Fletcher JM, Francis DJ, Rourke BP, et al. Classification of learning disabilities: Relationships with other childhood disorders. In: Lyon GR, Gray DB, Kavanagh JF, et al., eds. *Better understanding learning disabilities: new views from research and their implications for education and public policies.* Baltimore, MD: Paul H. Brookes, 1993:27–55.

61. Fletcher JM, Levin H. Neurobehavioral effects of brain injury in children. In: Routh DK, ed. *Handbook of pediatric psychology.* New York: Guilford Press, 1988:258–295.

62. Fletcher JM, Satz P. Developmental changes in the neuropsychological correlates of reading achievement: a six-year longitudinal follow-up. *J Clin Neuropsychol* 1980;2:23–37.

63. Fletcher JM, Stuebing KK, Shaywitz BA, et al. Validity of the concept of dyslexia: Alternative approaches to definition and classification. In: van den Bos KP, Siegel LS, Bakker DJ, et al., eds. *Current directions in dyslexia research.* Amsterdam: Swets & Zeitlinger, 1994:31–43.

64. Fletcher JM, Taylor HG. Neuropsychological approaches to children: towards a developmental neuropsychology. *J Clin Neuropsychol* 1984;6:39–56.

65. Flynn JM, Deering W, Goldstein M, et al. Electrophysiological correlates of dyslexic subtypes. *J Learn Disabil* 1992;25:133–141.

66. Francis DJ, Fletcher JM, Rourke BP. Discriminant validity of lateral sensorimotor measures in children. *J Clin Exp Neuropsychol* 1988;10:779–799.

67. Francis DJ, Fletcher JM, Rourke BP, et al. A five-factor model for motor, psychomotor, and visual-spatial tests used in the neuropsychological assessment of children. *J Clin Exp Neuropsychol* 1992;14:625–637.

68. Francis DJ, Fletcher JM, Shaywitz BA, et al. Defining learning and language disabilities: conceptual and psychometric issues with the use of IQ tests. *Lang Speech Hearing Sci Schools* 1996;27:132–143.

69. Fuerst DR, Rourke BP. Psychosocial functioning of children: relations between personality subtypes and academic achievement. *J Abnorm Child Psychol* 1993;21:597–607.

70. Fuerst DR, Rourke BP. Psychosocial functioning of children with learning disabilities at three age levels. *Child Neuropsychol* 1995;1:38–55.

71. Fuerst DR, Fisk JL, Rourke BP. Psychosocial functioning of learning-disabled children: replicability of statistically derived subtypes. *J Consult Clin Psychol* 1989;57:275–280.

72. Fuerst DR, Fisk JL, Rourke BP. Psychosocial functioning of learning-disabled children: relations between WISC verbal IQ-performance IQ discrepancies and personality subtypes. *J Consult Clin Psychol* 1993;58:657–660.

73. Fuerst KB, Rourke BP. White matter physiology and pathology. In: Rourke BP, ed. *Syndrome of nonverbal learning disabilities: neurodevelopmental manifestations.* New York: Guilford Press, 1995:27–44.

74. Fuerst KB, Dool CB, Rourke BP. Velocardiofacial syndrome. In: Rourke BP, ed. *Syndrome of nonverbal learning disabilities: neurodevelopmental manifestations.* New York: Guilford Press, 1995:119–137.

75. Gaddes WH. *Learning disabilities and brain function,* 2nd ed. New York: Springer-Verlag, 1985.

76. Galaburda AM. Ordinary and extraordinary brain development: anatomical variation in developmental dyslexia. *Ann Dyslexia* 1989;39:67–80.

77. Galaburda AM, Sherman GF, Rosen GD, et al. Developmental dyslexia: four consecutive patients with cortical anomalies. *Ann Neurol* 1985;18:222–233.

78. Georgiewa P, Rzanny R, Hopf JM, et al. FMRI during word processing in dyslexic and normal reading children. *Neuroreport* 1999;10:3459–3465.

79. Golding-Kushner KJ, Weller G, Shprintzen RJ. Velo-cardio-facial syndrome: language and psychological profiles. *J Craniofac Genet Dev Biol* 1985;5:259–266.

80. Greenham SL, Stelmack RM, van der Vlugt H. Learning disability subtypes and the role of attention during the naming of pictures and words: an event-related potential analysis. *Dev Neuropsychol (in press).*

81. Harnadek MCS, Rourke BP. Principal identifying features of the syndrome of nonverbal learning disabilities in children. *J Learn Disabil* 1994;27:144–154.

82. Harter MR, Anllo-Vento L, Wood FB. Event-related potentials, spatial orienting, and reading disabilities. *Psychophysiology* 1989;26:404–421.

83. Hier DB, LeMay M, Rosenberger PB, et al. Developmental dyslexia. Evidence for a subgroup with a reversal of cerebral asymmetry. *Arch Neurol* 1978;35:90–92.

84. Horwitz B, Rumsey JM, Donohue BC. Functional connectivity of the angular gyrus in normal reading and dyslexia. *Proc Natl Acad Sci USA* 1998;95:8939–8944.

85. Humphreys P, Kaufman WE, Galaburda AM. Developmental dyslexia in women: neuropathological findings in three patients. *Ann Neurol* 1990;28:727–738.

86. Hynd GW, Hall J, Novey ES, et al. Dyslexia and corpus callosum morphology. *Arch Neurol* 1995;52:32–38.

87. Hynd GW, Semrud-Clikeman M, Lorys AR, et al. Brain morphology in developmental dyslexia and attention deficit disorder/hyperactivity. *Arch Neurol* 1990;47:919–926.

88. Jernigan TL, Hesselink JR, Sowell E, et al. Cerebral structure on magnetic resonance imaging in language- and learning-impaired children. *Arch Neurol* 1991;48:539–545.

89. Johnson DJ, Myklebust HR. *Learning disabilities.* New York: Grune & Stratton, 1967.

90. Joschko M, Rourke BP. Neuropsychological subtypes of learning-disabled children who exhibit the ACID pattern on the WISC. In: Rourke BP, ed. *Neuropsychology of learning disabilities: essentials of subtype analysis.* New York: Guilford Press, 1985:65–88.

91. Kass CE. Auditory closure test. In: Olson JJ, Olson JJ, eds. *Validity studies on the Illinois Test of Psycholinguistic Abilities.* Madison, WI: Authors, 1964.

92. Kirk SA, McCarthy JJ. The Illinois Test of Psycholinguistic Abilities: an approach to differential diagnosis. *Am J Ment Defic* 1961;66:399–412.

93. Klin A, Volkmar FR, Sparrow SS, et al. Validity and neuropsychological characterization of Asperger syndrome: convergence with nonverbal learning disabilities syndrome. *J Child Psychol Psychiatry* 1995;36:1127–1140.

94. Landry SH, Fletcher JM, Denson SE. Longitudinal outcome for low birth weight infants: effects of intraventricular hemorrhage and bronchopulmonary dysplasia. *J Clin Exp Neuropsychol* 1993;15:205–218.

95. Larsen JP, Höien T, Ödegaard H. Magnetic resonance imaging of the corpus callosum in developmental dyslexia. *Cogn Neuropsychol* 1992;9:123–134.

96. Larsen JP, Höien T, Lundberg I, et al. MRI evaluation of the size and symmetry of the planum temporale in adolescents with developmental dyslexia. *Brain Lang* 1990;39:289–301.

97. Leisman G, Ashkenazi M. Aetiological factors in dyslexia: IV. Cerebral hemispheres are functionally equivalent. *Int J Neurosci* 1980;11:157–164.

98. Lesnik PG, Ciesielski KT, Hart BL, et al. Evidence for cerebellar-frontal subsystem changes in children treated with intrathecal chemotherapy for leukemia: enhanced data analysis using an effect size model. *Arch Neurol* 1998;52:1561–1568.

99. Loveland KA, Fletcher JM, Bailey V. Nonverbal communication of events in learning-disability subtypes. *J Clin Exp Neuropsychol* 1990;12:433–447.

100. Lovrich D, Cheng JC, Velting DM, et al. Auditory ERPs during rhyme and semantic processing: effects of reading ability in college students. *J Clin Exp Neuropsychol* 1997;19:313–330.

101. Lovrich D, Stamm JS. Event-related potential and behavioral correlates of attention in reading retardation. *J Clin Neuropsychol* 1983;5:13–37.

102. Lyon GR. Educational validation studies of learning disability subtypes. In: Rourke BP, ed. *Neuropsychology of learning disabilities: essentials of subtype analysis.* New York: Guilford Press, 1985:257–280.

103. Lyon GR, Flynn JM. Educational validation studies with subtypes of learning-disabled readers. In: Rourke BP, ed. *Neuropsychological validation of learning disability subtypes.* New York: Guilford Press, 1991:223–242.

104. MacDonald GW, Roy DL. Williams syndrome: a neuropsychological profile. *J Clin Exp Neuropsychol* 1988;10:125–131.

105. Magee LA, Nulman I, Rovet JF, et al. Neurodevelopment after in utero amiodarone exposure. *Neurotoxicol Teratol* 1999;21:261–265.

106. Mattis S, French JH, Rapin I. Dyslexia in children and young adults: three independent neuropsychological syndromes. *Dev Med Child Neurol* 1975;17:150–163.

107. McKinney JD, Short EJ, Feagans L. Academic consequences of perceptual-linguistic subtypes of learning-disabled children. *Learn Disabil Res* 1985;1:6–17.

108. Miles J, Stelmack RM. Learning disability subtypes and the effects of auditory and visual priming on visual event-related potentials to words. *J Clin Exp Neuropsychol* 1994;16:43–64.

109. Minshew NJ, Goldstein G, Muenz LR, et al. Neuropsychological functioning in non-mentally retarded autistic individuals. *J Clin Exp Neuropsychol* 1992;14:749–761.

110. Molfese DL. Predicting dyslexia at 8 years of age using neonatal brain responses. *Brain Lang* 2000;72:238–245.

111. Morris CA, Mervis CB. Williams syndrome. In: Goldstein S, Reynolds CR, eds. *Handbook of neurodevelopmental and genetic disorders in children.* New York: Guilford Press, 1999:555–590.

112. Morris RD, Fletcher JM. Classification in neuropsychology: a theoretical framework and research paradigm. *J Clin Exp Neuropsychol* 1988;10:640–658.

113. Morris R, Blashfield R, Satz P. Neuropsychology and cluster analysis: potentials and problems. *J Clin Neuropsychol* 1981;3:79–99.

114. Morris R, Blashfield R, Satz P. Developmental classification of reading-disabled children. *J Clin Exp Neuropsychol* 1986;8:371–392.

115. Moss EM, Batshaw ML, Solot CB, et al. Psychoeducational profile of the 22q11. 2 microdeletion: a complex pattern. *J Pediatr* 1999;134:193–198.

116. Myklebust HR. Nonverbal learning disabilities: assessment and intervention. In: Myklebust HR, ed. *Progress in learning disabilities, vol. 3.* New York: Grune & Stratton, 1975:85–121.

117. Nass R, Speiser P, Heier L, et al. White matter abnormalities in congenital adrenal hyperplasia. *Ann Neurol* 1990;28:470.

118. Nilsson DE, Bradford LW. Neurofibromatosis. In: Goldstein S, Reynolds CR, eds. *Handbook of neurodevelopmental and genetic disorders in children.* New York: Guilford Press, 1999:350–367.

119. North KN, Joy P, Yuille D, et al. Specific learning disability in children with neurofibromatosis type 1: significance of MRI abnormalities, *Neurology* 1994;44:878–883.

120. North KN, Riccardi V, Samango-Sprouse C, et al. Cognitive function and academic performance in neurofibromatosis 1: consensus statement from the NF-1 Cognitive Disorders Task Force. *Neurology* 1997;48:1121–1127.

121. Ozols EJ, Rourke BP. Dimensions of social sensitivity in two types of learning-disabled children. In: Rourke BP, ed. *Neuropsychology of learning disabilities: essentials of subtype analysis.* New York: Guilford Press, 1985:281–301.

122. Ozols EJ, Rourke BP. Characteristics of young learning-disabled children classified according to patterns of academic achievement: auditory-perceptual and visual-perceptual disabilities. *J Clin Child Psychol* 1988;17:44–52.

123. Ozols EJ, Rourke BP. Classification of young learning-disabled children according to patterns of academic achievement: validity studies. In: Rourke BP, ed. *Neuropsychological validation of learning disability subtypes.* New York: Guilford Press, 1991:97–123.

124. Papanicolaou AC, Simos PG, Breier JI, et al. Brain mechanisms for reading in children with and without dyslexia: a review of studies of normal development and plasticity. *Dev Neuropsychol (in press).*

125. Pelletier PM, Ahmad SA, Rourke BP. Classification rules for basic phonological processing disabilities and nonverbal learning disabilities: formulation and external validity. *Child Neuropsychology* 2001;7:84–98.

126. Petrauskas RJ, Rourke BP. Identification of subtypes of retarded readers: a neuropsychological, multivariate approach. *J Clin Neuropsychol* 1979;1:17–37.

127. Picard EM, Del Dotto JE, Breslau N. Very low birth weight. In: Yeates KO, Ris RD, Taylor HG, eds. *Pediatric neuropsychology: research, theory, and practice.* New York: Guilford Press, 1999:275–299.

128. Picard EM, Rourke BP. Neuropsychological consequences of prophylactic treatment for acute lymphocytic leukaemia. In: Rourke BP, ed. *Syndrome of nonverbal learning disabilities: neurodevelopmental manifestations.* New York: Guilford Press, 1995:282–330.

129. Porter JE, Rourke BP. Personality and socioemotional dimensions of learning disabilities in children. In: Rourke BP, ed. *Neuropsychology of learning disabilities: essentials of subtype analysis*. New York: Guilford Press, 1985:257–280.

130. Powell MP, Schulte T. Turner syndrome. In: Goldstein S, Reynolds CR, eds. *Handbook of neurodevelopmental and genetic disorders in children*. New York: Guilford Press, 1999:277–297.

131. Reitan RM. A research program on the psychological effects of brain lesions in human beings. In: Ellis NR, ed. *International review of research in mental retardation, vol. 1*. New York: Academic Press, 1966:1534–218.

132. Reitan RM. Methodological problems in clinical neuropsychology. In: Reitan RM, Davison LA, eds. *Clinical neuropsychology: current status and applications*. New York: Wiley, 1974:19–46.

133. Reitan RM, Davison LA. *Clinical neuropsychology: current status and applications*. Washington, DC: VH Winston, 1974.

134. Robichon F, Bouchard P, Démonet JF, et al. Developmental dyslexia: re-evaluation of the corpus callosum in male adults. *Eur Neurol* 2000;43:233–237.

135. Rosenberger PB, Hier DB. Cerebral asymmetry and verbal intellectual deficits. *Ann Neurol* 1980;8:300–304.

136. Rosner J, Simon DP. The Auditory Analysis Test: an initial report. *J Learn Disabil* 1971;4:40–48.

137. Rourke BP. Brain-behavior relationships in children with learning disabilities. *Am Psychol* 1975;30:911–920.

138. Rourke BP. Issues in the neuropsychological assessment of children with learning disabilities. *Can Psychol Rev* 1976;17:89–102.

139. Rourke BP. Neuropsychological research in reading retardation: a review. In: Benton AL, Pearl D, eds. *Dyslexia: an appraisal of current knowledge*. New York: Oxford University Press, 1978:141–171.

140. Rourke BP. Reading, spelling, arithmetic disabilities: a neuropsychologic perspective. In: Myklebust HR, ed. *Progress in learning disabilities, vol. 4*. New York: Grune & Stratton, 1978:97–120.

141. Rourke BP. Neuropsychological assessment of children with learning disabilities. In: Filskov SB, Boll TJ, eds. *Handbook of clinical neuropsychology*. New York: Wiley-Interscience, 1981:453–478.

142. Rourke BP. Central processing deficiencies in children: toward a developmental neuropsychological model. *J Clin Neuropsychol* 1982;4:1–18.

143. Rourke BP. Child-clinical neuropsychology: assessment and intervention with the disabled child. In: de Wit J, Benton AL, eds. *Perspectives in child study: integration of theory and practice*. Lisse, The Netherlands: Swets & Zeitlinger, 1982:62–72.

144. Rourke BP. Outstanding issues in research on learning disabilities. In: Rutter M, ed. *Developmental neuropsychiatry*. New York: Guilford Press, 1983:564–574.

145. Rourke BP. Reading and spelling disabilities: a developmental neuropsychological perspective. In: Kirk U, ed. *Neuropsychology of language, reading, and spelling*. New York: Academic Press, 1983:209–234.

146. Rourke BP, Ed. *Neuropsychology of learning disabilities: essentials of subtype analysis*. New York: Guilford Press, 1985.

147. Rourke BP. Syndrome of nonverbal learning disabilities: the final common pathway of white-matter disease/dysfunction? *Clin Neuropsychol* 1987;1:209–234.

148. Rourke BP. Socioemotional disturbances of learning-disabled children. *J Consult Clin Psychol* 1988;56:801–810.

149. Rourke BP. The syndrome of nonverbal learning disabilities: developmental manifestations in neurological disease, disorder, and dysfunction. *Clin Neuropsychol* 1988;2:293–330.

150. Rourke BP. Nonverbal learning disabilities, socioemotional disturbance, and suicide: a reply to Fletcher, Kowalchuk, King, and Bigler. *J Learn Disabil* 1989;21:186–187.

151. Rourke BP. *Nonverbal learning disabilities: the syndrome and the model*. New York: Guilford Press, 1989.

152. Rourke BP, ed. *Neuropsychological validation of learning disability subtypes*. New York: Guilford Press, 1991.

153. Rourke BP. Arithmetic disabilities, specific and otherwise: a neuropsychological perspective. *J Learn Disabil* 1993;26:214–226.

154. Rourke BP. Neuropsychological assessment of children with learning disabilities: measurement issues. In: Lyon GR, ed. *Frames of reference for the assessment of learning disabilities: new views on measurement issues*. Baltimore, MD: Paul H. Brookes, 1994:475–514.

155. Rourke BP. The science of practice and the practice of science: the scientist-practitioner model in clinical neuropsychology. *Can Psychol* 1995;36:259–287.

156. Rourke BP, ed. *Syndrome of nonverbal learning disabilities: neurodevelopmental manifestations*. New York: The Guilford Press, 1995.

157. Rourke BP. Neuropsychological and psychosocial subtyping: a review of investigations within the University of Windsor laboratory. *Can Psychol* 2000;41:34–51.

158. Rourke BP, Adams KM. Quantitative approaches to the neuropsychological assessment of children. In: Tarter RE, Goldstein G, eds. *Advances in clinical neuropsychology, vol. 2*. New York: Plenum, 1984:79–108.

159. Rourke BP, Ahmad SA, Collins DW, et al. Child-clinical/pediatric neuropsychology: some recent advances. *Annu Rev Psychol* 2002;53:309–339.

160. Rourke BP, Bakker DJ, Fisk JL, et al. *Child neuropsychology: an introduction to theory, research, and clinical practice*. New York: Guilford Press, 1983.

161. Rourke BP, Conway JA. Disabilities of arithmetic and mathematical reasoning: perspectives from neurology and neuropsychology. *J Learn Disabil* 1997;30:34–46.

162. Rourke BP, Czudner G. Age differences in auditory reaction time in "brain-damaged" and normal children under regular and irregular preparatory interval conditions. *J Exp Child Psychol* 1972;14:372–378.

163. Rourke BP, Del Dotto JE. *Learning disabilities: a neuropsychological perspective*. Thousand Oaks, CA: Sage, 1994.

164. Rourke BP, Del Dotto JE, Rourke SB, et al. Nonverbal learning disabilities: the syndrome and a case study. *J School Psychol* 1990;28:361–385.

165. Rourke BP, Dietrich DM, Young GC. Significance of WISC verbal-performance discrepancies for younger children with learning disabilities. *Percept Motor Skills* 1973;36:275–282.

166. Rourke BP, Finlayson MAJ. Neuropsychological significance of variations in patterns of academic performance: verbal and visual abilities. *J Abnorm Child Psychol* 1978;6:121–133.

167. Rourke BP, Fisk JL. *Children's Word-Finding Test, revised*. University of Windsor, Department of Psychology, Windsor, Ontario, 1976.

168. Rourke BP, Fisk JL. Socio-emotional disturbances of learning disabled children: the role of central processing deficits. *Bull Orton Soc* 1981;31:77–88.

169. Rourke BP, Fisk JL. Subtypes of learning-disabled children: implications for a neurodevelopmental model of differential hemispheric processing. In: Molfese DL, Segalowitz SJ, eds. *Developmental implications of brain lateralization*. New York: Guilford Press, 1988:547–565.

170. Rourke BP, Fisk JL. Adult presentations of learning disabilities. In: White RF, ed. *Clinical syndromes in adult neuropsychology: the practitioner's handbook*. Amsterdam: Elsevier, 1992:451–473.

171. Rourke BP, Fisk JL, Strang JD. *Neuropsychological assessment of children: a treatment-oriented approach*. New York: Guilford Press, 1986.

172. Rourke BP, Fuerst DR. *Learning disabilities and psychosocial functioning: a neuropsychological perspective.* New York: Guilford Press, 1991.

173. Rourke BP, Fuerst DR. Psychosocial dimensions of learning disability subtypes: neuropsychological studies in the Windsor Laboratory. *School Psychol Rev* 1992;21:360–373.

174. Rourke BP, Fuerst DR.Cognitive processing, academic achievement, and psychosocial functioning: a neuropsychological perspective. In: Cicchetti D, Cohen D, eds. *Developmental psychopathology, vol. 1.* New York: Wiley, 1995:391–423.

175. Rourke BP, Fuerst DR. Psychosocial dimensions of learning disability subtypes. *Assessment* 1996;3:277–290.

176. Rourke BP, Gates RD.*Underlining test: preliminary norms.* University of Windsor, Department of Psychology, Windsor, Ontario, 1980.

177. Rourke BP, Gates RD.Neuropsychological research and school psychology. In: Hynd GW, Obrzut JE, eds. *Neuropsychological assessment and the school-age child: issues and procedures.* New York: Grune & Stratton, 1981:3–25.

178. Rourke BP, Petrauskas RJ.*Underlining test, revised.* University of Windsor, Department of Psychology, Windsor, Ontario, 1977.

179. Rourke BP, Strang JD. Neuropsychological significance of variations in patterns of academic performance: motor, psychomotor, and tactile-perceptual abilities. *J Pediatr Psychol* 1978;3:62–66.

180. Rourke BP, Strang JD.Subtypes of reading and arithmetical disabilities: a neuropsychological analysis. In: Rutter M, ed. *Developmental neuropsychiatry.* New York: Guilford Press, 1983:473–488.

181. Rourke BP, Tsatsanis KD.Syndrome of nonverbal learning disabilities and Asperger syndrome. In: Klin A, Volkmar F, Sparrow SS, eds. *Asperger syndrome.* New York: Guilford Press, 2000:231–253.

182. Rourke BP, Tsatsanis KD. Syndrome of nonverbal learning disabilities: psycholinguistic assets and deficits. *Topics Lang Disord* 1996;16:30–44.

183. Rourke BP, Young GC, Leenaars A. A childhood learning disability that predisposes those afflicted to adolescent and adult depression and suicide risk. *J Learn Disabil* 1989;21:169–175.

184. Rovet JF. Learning disabilities profiles in four endocrine disorders. *J Clin Exp Neuropsychol* 1991;13:58–59.

185. Rovet J.Congential hypothyroidism. In: Rourke BP, ed. *Syndrome of nonverbal learning disabilities: neurodevelopmental manifestations.* New York: Guilford Press, 1995:255–281.

186. Rovet J.Turner syndrome. In: Rourke BP, ed. *Syndrome of nonverbal learning disabilities: neurodevelopmental manifestations.* New York: Guilford Press, 1995:351–371.

187. Rovet JF. Diabetes. In: Yeates KO, Ris RD, Taylor HG, eds. *Pediatric neuropsychology: research, theory, and practice.* New York: Guilford Press, 1999:336–365.

188. Rovet JF, Ehrlich RM, Czuchta D, et al. Psychoeducational characteristics of children and adolescents with insulin dependent diabetes mellitus. *J Learn Disabil* 1993;26:7–22.

189. Rumsey JM, Andreason P, Zametkin, et al. Failure to activate the left temporoparietal cortex in dyslexia: an oxygen 15 positron emission tomography study. *Arch Neurol* 1992;49:527–534.

190. Rumsey JM, Dorwart R, Vermess M, et al. Magnetic resonance imaging of brain anatomy in severe developmental dyslexia. *Arch Neurol* 1986;43:1045–1046.

191. Rumsey JM, Horwitz B, Donohue BC, et al. A functional lesion in developmental dyslexia: left angular gyrus blood flow predicts severity. *Brain Lang* 1999;70:187–204.

192. Rumsey JM, Nace K, Donohue B, et al. A positron emission tomographic study of impaired word recognition and phonological processing in dyslexic men. *Arch Neurol* 1997;54:562–573.

193. Rumsey JM, Zametkin AJ, Andreason P, et al. Normal activation of frontotemporal language cortex in dyslexia, as measured with oxygen 15 positron emission tomography. *Arch Neurol* 1994;51:27–38.

194. Russell DL, Rourke BP. Concurrent and predictive validity of phonetic accuracy of misspellings in normal and disabled readers and spellers. In: Rourke BP, ed. *Neuropsychological validation of learning disability subtypes.* New York: Guilford Press, 1991:57–72.

195. Ryan TV, Crews WD Jr, Cowan L, et al. A case of triple X syndrome manifesting with the syndrome of nonverbal learning disabilities. *Child Neuropsychol* 1998;4:225–232.

196. Selzer SC, Lindgren SD, Blackman JA. Long-term neuropsychological outcome of high risk infants with intracranial hemorrhage. *J Pediatr Psychol* 1992;17:407–422.

197. Shapiro EG, Balthazor M.Metabolic and neurodegenerative disorders. In: Yeates KO, Ris RD, Taylor HG, eds. *Pediatric neuropsychology: research, theory, and practice.* New York: Guilford Press, 1999:171–205.

198. Shapiro EG, Lipton ME, Krivit W. White matter dysfunction and its neuropsychological correlates: a longitudinal study of a case of metachromatic leukodystrophy. *J Clin Exp Neuropsychol* 1992;14:610–624.

199. Shapiro E, Lockman L, Balthazor M, et al. Neuropsychological outcomes of several storage diseases with and without bone marrow transplantation. *J Inherit Metab Dis* 1995;18:413–429.

200. Shaywitz SE, Shaywitz BA, Pugh KR, et al. Functional disruption in the organization of the brain for reading in dyslexia. *Proc Natl Acad Sci USA* 1998;95:2636–2641.

201. Simos PG, Breier JL, Fletcher JM, et al. Cerebral mechanisms involved in word reading in dyslexic children: a magnetic source imaging approach. *Cereb Cortex* 2000;10:809–816.

202. Simos PG, Breier JL, Fletcher JM, et al. Brain activation profiles in dyslexic children during non-reading: a magnetic source imaging study. *Neurosci Lett* 2000;290:61–65.

203. Simos PG, Breier JL, Fletcher JM, et al. Age-related changes in regional brain activation during phonological decoding and printed word recognition. *Dev Neuropsychol* 2001;19:191–210.

204. Smith LA, Rourke BP. Callosal agenesis. In: Rourke BP, ed. *Syndrome of nonverbal learning disabilities: neurodevelopmental manifestations.* New York: Guilford Press, 1995:45–92.

205. Sobotka KR, May JG. Visual evoked potentials and reaction time in normal and dyslexic children. *Psychophysiology* 1977;14:18–24.

206. Stelmack RM, Miles J. The effect of picture priming on event-related potentials of normal and disabled readers during a word recognition memory task. *J Clin Exp Neuropsychol* 1990;12:887–903.

207. Stelmack RM, Rourke BP, van der Vlugt H. Intelligence, learning disabilities, and event-related potentials. *Dev Neuropsychol* 1995;11:445–465.

208. Stelmack RM, Saxe BJ, Noldy-Cullum N, et al. Recognition memory for words and event-related potentials: a comparison of normal and disabled readers. *J Clin Exp Neuropsychol* 1988;10:185–200.

209. Strang JD, Rourke BP. Concept-formation/non-verbal reasoning abilities of children who exhibit specific academic problems with arithmetic. *J Clin Child Psychol* 1983;12:33–39.

210. Strang JD, Rourke BP. Adaptive behavior of children with specific arithmetic disabilities and associated neuropsychological abilities and deficits. In: Rourke BP, ed. *Neuropsychology of learning disabilities: essentials of subtype analysis.* New York: Guilford Press, 1985:302–328.

211. Strang JD, Rourke BP. Arithmetic disability subtypes: the neuropsychological significance of specific arithmetical impairment

in childhood. In: Rourke BP, ed. *Neuropsychology of learning disabilities: essentials of subtype analysis.* New York: Guilford Press, 1985:167–183.

212. Strong RT Jr. *Intellectual deficits associated with minimal brain disorders in primary school children.* Columbus, OH: Columbus State School, 1963.

213. Streissguth AP, Aase JM, Clarren SK, et al. Fetal alcohol syndrome in adolescents and adults. *JAMA* 1991;265:1961–1967.

214. Sweeney JE, Rourke BP. Spelling disability subtypes. In: Rourke BP, ed. *Neuropsychology of learning disabilities: essentials of subtype analysis.* New York: Guilford Press, 1985:147–166.

215. Taylor HG. MBD: meanings and misconceptions. *J Clin Neuropsychol* 1983;5:271–287.

216. Taylor HG. Learning disabilities. In: Mash EJ, Barkley RA, eds. *Behavioral treatment of childhood disorders.* New York: Guilford, 1989:347–380.

217. Taylor HG, Fletcher JM. Biological foundations of specific developmental disorders: methods, findings, and future directions. *J Clin Child Psychol* 1983;12:46–65.

218. Taylor HG, Hack M, Klein N, et al. Achievement of <750 gm birthweight children with normal cognitive abilities: evidence for specific learning problems. *J Pediatr Psychol* 1995;20:703–719.

219. Taylor MJ, Keenan NK. Event-related potentials to visual and language stimuli in normal and dyslexic children. *Psychophysiology* 1990;27:318–327.

220. Teeter PA. Noonan syndrome. In: Goldstein S, Reynolds CR, eds. *Handbook of neurodevelopmental and genetic disorders in children.* New York: Guilford Press, 1999:337–349.

221. Temple E, Poldrack RA, Protopapas A, et al. Disruption of the neural response to rapid acoustic stimuli in dyslexia: evidence from functional fMRI. *Proc Natl Acad Sci USA* 2000;97:13907–13912.

222. Temple E, Poldrack RA, Salidis J, et al. Disrupted neural responses to phonological and orthographic processing in dyslexic children: an fMRI study. *Neuroreport* 2001;12:299–307.

223. Thatcher RW. Cyclic cortical reorganization: origins of human cognitive development. In: Dawson G, Fischer KW, eds. *Human behavior and the developing brain.* New York: Guilford Press, 1994:232–266.

224. Thatcher RW. Neuroimaging of cyclic cortical reorganization during human development. In: Thatcher RW, ed. *Developmental neuroimaging: mapping the development of brain and behavior.* San Diego: Academic Press, 1997:91–106.

225. Tindall S, Rothermel RR, Delamater A, et al. Neuropsychological abilities of children with cardiac disease treated with extracorporeal membrane oxygenation. *Dev Neuropsychol* 1999;15:101–115.

226. [Reserved]

227. Troyer AK, Joschko M. Cognitive characteristics associated with Noonan syndrome: two case reports. *Child Neuropsychol* 1997;3:199–205.

228. Tsatsanis KD, Rourke BP. Conclusions and future directions. In: Rourke BP, ed. *Syndrome of nonverbal learning disabilities: neurodevelopmental manifestations.* New York: Guilford Press, 1995:476–496.

229. Tsatsanis KD, Rourke BP. de Lange syndrome. In: Rourke BP, ed. *Syndrome of nonverbal learning disabilities: neurodevelopmental manifestations* New York: Guilford Press, 1995:171–205.

230. Tsatsanis KD, Fuerst DR, Rourke BP. Psychosocial dimensions of learning disabilities: external validation and relationship with age and academic functioning. *J Learn Disabil* 1997;30:490–502.

231. Udwin O, Yule W. A cognitive and behavioural phenotype in Williams syndrome. *J Clin Exp Neuropsychol* 1991;13:232–244.

232. Waisbren SE. Phenylketonuria. In: Goldstein S, Reynolds CR, eds. *Handbook of neurodevelopmental and genetic disorders in children.* New York: Guilford Press, 1999:433–458.

233. Weber AM, McKellop JM, Gyato K, et al. Metachromatic leukodystrophy and nonverbal learning disability: neuropsychological and neuroradiological findings in heterozygosity and pseudodeficiency. *J Int Neuropsychol Soc* 1998;4:42.

234. Welsh MC, Pennington BF. Phenylketonuria. In: Yeates KO, Ris RD, Taylor HG, eds. *Pediatric neuropsychology: research, theory, and practice.* New York: Guilford Press, 1999:275–299.

235. White RF, Krengel M. Multiple sclerosis. In: Rourke BP, ed. *Syndrome of nonverbal learning disabilities: neurodevelopmental manifestations.* New York: Guilford Press, 1995.

236. White RF, Krengel M. Toxicant-induced encephalopathy. In: Rourke BP, ed. *Syndrome of nonverbal learning disabilities: neurodevelopmental manifestations.* New York: Guilford Press, 1995:460–475.

237. Yeates KO, Bornstein RA. Psychosocial correlates of learning disability subtypes in children with Tourette's syndrome. *Child Neuropsychol* 1996;2:193–203.

26

THE NEUROPSYCHIATRY OF AGGRESSION

SHARON S. ISHIKAWA
ADRIAN RAINE

Set against a backdrop of several highly publicized school shootings, violence in the United States has emerged as a major public health concern. A review of headlines from the past decade highlights the loss inflicted by aggressive individuals: the Jonesboro massacre, Arkansas shootings, Columbine High School massacre, and Oklahoma City and World Trade Center bombings represent only a small sampling. These events not only aggrieved the victims and their families, but they also mobilized a nation into regarding violence as a significant public threat.

Ironically, this intensified focus has occurred when violent crime and homicide have decreased steadily among adults 25 years or older (1). Nevertheless, consistent with the media hype on school shootings, murders perpetrated by individuals 14 to 17 years old increased 172% between 1985 and 1994 (2). Although homicides committed by those between ages 14 and 24 began to drop in the early 1990's, the rates remained much higher than before the 1980's upswing. In 1998, homicide remained the second leading cause of death for 15- to 24-year-olds and the third leading cause of death for 1- to 14-year-olds (3). In addition, juvenile violence is anticipated to increase into the twenty-first century coincident with an increase in the number of children maturing into adolescence (4). Thus, understanding the risk and protective factors in the development of violence remains a pressing health issue.

CONCEPTUAL BACKGROUND

Prevention and management of aggressive behavior falls within the domain of several disciplines, including law, neurology, psychiatry, psychology, education, social work, and public health. Within and between these fields, much controversy on the causes of aggression and its appropriate treatment has been generated. This controversy stems partially from rivalry and/or unfamiliarity across disciplines and partially from early attempts to develop a unitary model of aggressive behavior. In reality, however, aggression is a multimodal construct. Not only does the expression of aggression differ from person to person, but the same person may aggress in different ways across situations. Scientists have made initial attempts to acknowledge this heterogeneity by identifying subtypes of aggressive behavior.

Aggression Subtypes

Based on extensive neuroanatomic and neurochemical studies, seven subtypes of animal aggression have been validated: predatory, intermale, territorial, maternal, irritable, defensive, and instrumental. The studies reviewed in this chapter focus primarily on predatory, intermale, and defensive aggression, with predatory aggression referring to stereotyped interspecific aggression intended to result in efficient and rapid killing of the intended victim; intermale aggression referring to intraspecific aggression intended to ensure male dominance within the group; and defensive aggression referring to aggressive behavior elicited in response to a perceived threat or attack.

Following the lead of animal research, researchers have begun to examine subtypes of human aggression. The most common distinctions include impulsive versus predatory aggression, proactive versus reactive aggression, and life-course persistent versus adolescent-limited antisocial behavior (5–7). Impulsive aggression may be directed toward self or others and often is marked by behavioral disinhibition and affective instability. It is observed frequently in psychiatric or neurologic patients and is not necessarily indicative of an antisocial personality. In contrast, predatory aggression is instrumental planned aggression used to obtain some goal other than harming the victim and is more likely to be observed in psychopathic antisocial individuals (8). A related concept used frequently in studies with children is proactive aggression, which refers to behavior such as coercion, direct threats, or attacks designed to achieve a goal. Reactive aggression, on the other hand, refers to an aggressive response to provocation or perceived threat. Finally, life-course persistent antisocial behavior is stable across the life span and typically begins in childhood or early adolescence, whereas adolescent-limited antisocial behavior refers to "typical" delinquent behavior limited to the adolescent developmental period. Although the latter subgroupings do not ostensibly refer to aggression, accumulating research indicates that life-course persistent antisocial individuals are more prone to aggression and serious offending than are adolescent-limited delinquents.

Nature Versus Nurture Debate

Historically, aggressive human behavior has been considered the domain of criminologists and sociologists rather than psychiatrists, neurologists, and other medical and mental health professionals. As such, many theories of aggressive behavior focus exclusively on social factors. Poverty, bad parenting, and child abuse all have been fingered as cultivators of aggression (9). Although research has generally supported the links between a disruptive family environment and aggressive behavior, these social models fail to account fully for the occurrence of aggression. Similarly, scientific inquiry geared toward elucidating biology as the sole determinant of aggression has not fared well empirically, theoretically, or politically. Theoretical and empirical attempts have been made to integrate biologic and social factors in the study of aggression, thus serving as a catalyst for furthering understanding of this complex human behavior.

This chapter explores biologic risk factors of aggressive behavior in humans, paying particular attention to biosocial interactions. In addition, although aggression has adaptive functions and in certain situations is a healthy response, the primary focus rests on "pathologic" aggression, or aggression that is extreme, repetitive, and/or not necessarily elicited by an immediate threat to survival. As appropriate, research on the biologic correlates of antisocial personality are reviewed in light of findings that chronically antisocial individuals are particularly prone to engage in aggression.

NEUROCOGNITIVE AND NEUROANATOMIC CORRELATES

Neuropsychological Research

Some of the earliest work on brain–behavior relationships in aggressive populations used neurologic and neuropsychological testing. Depending on the test used, poor performance is assumed to infer general brain damage or impairment of specific brain regions. In the study of aggressive behavior, general cognitive functioning [e.g., learning disabilities, intelligence, verbal-performance intelligence quotient (IQ) splits] and executive functions have been the most intensively investigated.

General Neurocognitive Functioning

Generalized brain damage as assessed by standardized neuropsychological test batteries relates to the commission of violent crimes (10,11). Violent offenders, compared to nonviolent offenders, show significantly greater impairment on the Halstead-Reitan and/or Luria neuropsychological test batteries (12,13), with a trend for greater neuropsychological impairment to be related to greater aggression (i.e., homicidal > nonhomicidal > nonviolent offenders) (13).

IQ deficits are reliably associated with delinquency. Significantly lower full-scale IQs are observed in violent offenders versus first-year college students (14) and in violent versus nonviolent offenders (15,16). It is estimated that adult criminals show an approximate ten-point deficit compared to controls (17), and juvenile delinquents show an approximate eight-point deficit (18,19). Although full-scale IQs of violent criminals are significantly lower compared to various control groups, it should be noted that they still generally fall within the low average to average range.

Reduced Hemispheric Lateralization

An alternative conceptualization of these neurocognitive findings is that they reflect reduced hemispheric lateralization of speech processes (11). A review of the literature indicates that delinquents and criminals—particularly those with aggressive psychopathic traits—not only show lower full-scale IQs, but also consistently show lower verbal IQs relative to performance IQs (17,19). Consistent with a reduced lateralization hypothesis, results from extensive neuropsychological test batteries indicate that dysfunction is significantly more likely to be localized to the left frontotemporal brain regions among violent adult criminals (20) and to the left temporal lobe among violent adolescent offenders (21). Similarly, adolescent and adult psychopathic criminals show less lateralization on a verbal dichotic listening task than do nonpsychopathic, age-matched criminals (22,23).

Although an indirect marker of lateralization, learning disabilities, which reflect a significantly lower level of academic achievement despite adequate ability (i.e., IQs), are frequently observed in delinquent populations. For example, formal diagnosis of both a developmental reading disability and childhood conduct disorder is associated with an earlier age of conviction for violent crime [i.e., late adolescence or early adulthood (24)]. Increased rates of developmental dyslexia also appear to be specific to impulsive and not premeditated aggression (25). Similarly, findings that recidivistic offenders are more likely to be left handed than the general population provide indirect support for a reduced lateralization hypothesis.

Interestingly, reduced hemispheric lateralization in life-course persistent antisocial individuals may actually result from early *right* hemispheric dysfunction, whereas left hemispheric deficits observed in adolescence and adulthood may be acquired. One study found that although life-course persistent antisocials have significantly lower verbal and spatial IQs at age 11 compared to non-antisocial controls, they demonstrate spatial but not verbal IQ deficits at age 3 (26). In addition, poor scholastic ability, in part, mediates the age-11 IQ deficits, whereas age-3 spatial IQ deficits are directly associated with antisocial behavior. Early spatial deficits may relate to persistent antisocial behavior in that they interfere with proper emotion regulation and socialization (26). Consistent with this speculation, healthy infants are right hemisphere dominant, presumably due to the importance of visual spatial skills, facial recognition, and emotion processing prior to the development of language (26). As such, verbal deficits in adolescence or adulthood may reflect (a) early interhemispheric reorganization prompted by spatial deficits that impeded the development of verbal abilities and/or (b) the cumulative effects of poor socialization that interfered with language-based academic and social functioning (26).

Executive Functioning

Another line of neuropsychological research focuses on impairment in executive functions among violent individuals. Executive functions refer to a cluster of higher-order cognitive processes involving initiation, planning, cognitive flexibility, abstraction, and decision making that together allow the execution of contextually appropriate behavior (27). When executive dysfunction is observed, the prefrontal cortex (PFC) traditionally is thought to be involved, although dysfunction may reflect disruption in any of the neural pathways connected to the PFC (28). Executive dysfunction is thought to relate to antisocial and/or aggressive behavior due to decreased behavioral inhibition and inability to generate socially acceptable responses in challenging situations (29).

Studies controlling for methodologic weaknesses present in the initial work generally support the executive dysfunction hypothesis in aggressive antisocial individuals. Poorer executive functions are correlated with increased antisocial aggressive behavior in preschool boys and girls (30) and differentiate conduct disordered or aggressive children from nonaggressive controls (31–33). Adult psychiatric inpatients with a history of community violence also exhibit significantly greater executive and psychomotor dysfunction than inpatients without such a history (34). Importantly, executive dysfunction is observed after controlling for potential confounds such as attention deficit/hyperactivity disorder (ADHD) (31,33), family socioeconomic status (SES) (31), hard-to-manage temperament (30), IQ (32–34), early family psychosocial adversity (32,33), age, race, gender, and other measures of neurocognitive function (32–34).

Although the majority of studies examine only boys, similar findings of executive dysfunction are observed in aggressive girls. After controlling for age, SES, ADHD diagnosis, and verbal knowledge, executive dysfunction is more strongly related to number of aggressive behaviors in adolescent girls, whereas difficult temperament is more strongly related to number of *non*-aggressive antisocial behaviors (35). Similarly, after controlling for age and SES, impaired executive function is more strongly related to current and lifetime history of aggression, whereas drug use is more strongly associated with nonaggressive delinquency (36).

Theoretical Implications

With regard to reduced lateralization, the accompanying poor verbal comprehension and communication skills may predispose to aggression by (a) simultaneously increasing the likelihood of interpersonal misunderstandings and decreasing one's ability to talk oneself out of conflict, and/or (b) reflecting neurologic markers of poor emotion regulation (26). Executive dysfunction, on the other hand, is thought to predispose one toward antisocial behavior through a reduction in impulse control and ability to direct current behavior based on the anticipation of future consequences. Although rarely considered, it also is possible that antisocial personality predisposes one to do poorly in the classroom setting, thus impairing one's general test-taking ability (9,37).

Brain Imaging Studies

Although neuropsychological tests can only infer the location of a hypothesized brain-behavior association, functional and structural brain imaging have revolutionized the examination of such relationships. Imaging studies of various aggressive populations using positron emission tomography, single photon emission computerized tomography, and magnetic resonance imaging have focused primarily on the frontal lobes, temporal lobes, and related subcortical structures, which are reviewed in the following.

Frontal Dysfunction

The frontal lobe is regarded as central to the processing of executive functions (38) and, as such, has been an area of intensive focus in the study of human aggression. When specific subareas of the frontal lobes are studied with regard to aggressive behavior, the orbitofrontal-ventromedial and dorsolateral divisions of the PFC are typically implicated. The orbitofrontal division is extensively connected to the amygdala and hippocampus (39). It plays a role in behavioral inhibition; autonomic reactivity; social and self-awareness (39,40); and regulation of negative affect (41,42). The dorsolateral division represents the primary connection with the cortical association areas and is critical to information processing and working memory (39). It also is hypothesized to be involved in emotional processing (41).

Studies using different functional brain imaging techniques exhibit frontal deficits in various classes of aggressive individuals. For example, compared to normal controls, repetitively violent psychiatric inpatients have reduced neuronal density and abnormal phosphate metabolism in the PFC (43) and reduced glucose metabolism in bilateral prefrontal and medial temporal regions (44). Murderers, relative to controls, showed reduced glucose metabolism specifically in the PFC, with no other frontal lobe regions implicated (45). These frontal and frontotemporal deficits are not simply an artifact of comparing patients to normal controls, as reduced prefrontal metabolic activity is observed when violent psychiatric patients are compared to nonviolent psychiatric controls (46).

Aggression in various neurologic conditions also appears to involve the frontal lobes. For example, repetitive violence in patients with temporal lobe epilepsy is related to frontal but not temporal abnormalities (47). Similarly, patients diagnosed with frontotemporal dementia are significantly more likely to engage in inappropriate aggressive, sexual, and antisocial behavior than are patients diagnosed with Alzheimer disease (48). In addition, aggressive dementia patients (i.e., Alzheimer disease, vascular dementia; no frontotemporal dementia) show significant hypoperfusion in the left and right dorsolateral frontal areas, left anterior temporal cortex, and right superior parietal areas compared to nonaggressive dementia patients (49). Importantly, these differences are not due to factors such as age, gender, SES, premorbid education, Mini-Mental State Examination score, severity of cognitive impairment, or current medication use (48,49).

Although the majority of imaging studies are conducted on hospitalized incarcerated subjects, frontal deficits also are observed in antisocial individuals from the community. Males who meet the *Diagnostic and Statistical Manual of Mental Disorders, Fourth Edition* (DSM-IV) criteria for antisocial personality disorder (APD)—the majority of whom have committed severe violence such as physical assault resulting in injury to the victim, armed robbery, rape, firing a gun at an intended victim—demonstrate significantly reduced prefrontal gray matter volume compared to healthy controls, substance-dependent non-antisocial personality disorder controls, and psychiatric controls matched on affective and schizophrenia spectrum disorders (50). Furthermore, these prefrontal structural deficits are independent of psychosocial deficits.

Additional research suggests that the link between frontal deficit and aggression depends on different personality and social factors. First, frontal abnormalities may be more pronounced in individuals engaging in impulsive rather than premeditated aggression. Support for this argument comes from studies showing that frequency of impulsive aggression is associated with reduced glucose metabolism in the frontal cortex of patients with personality disorders (51). Similarly, pretrial forensic patients arrested for impulsive violent offenses exhibit reduced cerebral blood flow in the frontal and temporal lobes compared to healthy controls (52) and reduced glucose metabolism in the prefrontal area compared to predatory instrumental murderers (53).

Second, frontal deficits may be particularly pronounced in violent individuals who have not been exposed to significant social stressors. When Raine et al. (54) separated the psychiatric forensic sample into subjects who had and those who had not experienced early psychosocial deprivation, nondeprived murderers showed a 14.2% reduction in functioning of the right orbitofrontal cortex relative to deprived murderers (54). Their findings run counter to popular opinion that criminals from abusive backgrounds are more likely to show biologic deficits, and they speculated that, in this sample, individuals without a psychosocial "push" toward violence required a greater neurobiologic "push." This is not to say, however, that among violent individuals, those who suffer early trauma do not show brain deficits. Violent offenders severely abused in childhood showed more extensive abnormal brain activation patterns than did nonabused violent offenders (55). Thus, frontal deficits and physical abuse may independently and additively relate to later aggressive behavior.

Finally, it should be noted that imaging studies observe frontal abnormalities in a number of antisocial populations (e.g., alcoholics, 56; psychopaths, 57), as well as in patients with affective disorders (58). As such, frontal dysfunction cannot be regarded as specific to aggression. However, it can be safely stated that frontal dysfunction represents one predisposing factor to antisocial, aggressive, and/or impulsive behavior. Furthermore, it is clear that the PFC is but one of several brain structures implicated in violence (42); other involved structures are reviewed in the following.

Temporal Lobes, Amygdala, and Subcortex

The temporal lobes are generally involved in learning, memory, and emotion. They are interconnected with the frontal lobes and subcortical limbic structures (38). Within this complex, the amygdala has been particularly implicated in aggressive behavior, which is likely due to its role in the processing of emotional and social cues. The amygdala receives information from the frontal lobe, hippocampus, thalamus, and other limbic structures (38). It is critical to the recognition of facial expressions (59), particularly those connoting threat (60), and is involved in social judgment (61). The amygdala also is crucial to the initial learning of responses to negative or aversive stimuli (61,62) but does not appear necessary for the expression of previously acquired emotional behavior (41). However, until recently, empirical research on aggressive behavior has not specifically examined the amygdala (instead focusing on the temporal area or simultaneously examining several subcortical limbic structures); thus, the explicit role of the amygdala in human aggression can only be cursorily addressed at present.

The classic work of Klüver and Bucy helped first establish the connection between the temporal lobes and aggression.

In experiments with monkeys, they demonstrated that bilateral removal of the temporal lobes, hippocampal formation, amygdala, and non-limbic temporal cortex resulted in a dramatic behavioral shift to passivity, flattened affect, hyperorality, and inappropriate sexual behavior (63). Subsequent research using various animal species presumably localized the taming effect to the amygdala. Lesions to the amygdaloid nucleus in monkeys (64), cats (65), and rats (66) generally resulted in taming, whereas stimulation of the amygdala elicited aggression in hamsters (67). However, much of this early lesion work also resulted in damage to surrounding areas. Lesions localized specifically in the amygdala have not resulted in the Klüver-Bucy syndrome, indicating that the taming effect occurs elsewhere in the medial temporal lobe (41).

Support for a link between temporal/limbic system damage and aggressive behavior has been found in various human populations. The number of localized temporal lobe changes identified through electroencephalography (EEG) and computed tomography (CT) is positively associated with severity ratings of preadmission violence among mentally disordered offenders (68). Temporal lobe abnormalities are more likely to be found in aggressive versus nonaggressive psychiatric patients (46,69). Compared to healthy controls, impulsive and planned murderers both show increased subcortical activity [i.e., midbrain, amygdala, hippocampus, thalamus (45); and abnormal subcortical asymmetries, right > left hemisphere (53)]. Among epileptics, prevalence rates of aggression range from under 5% to 50% depending on sampling issues, although well-controlled epidemiologic studies are needed to further assess whether epileptics engage in higher rates of aggression compared to other medical populations and/or the general population (70).

Paralleling the ambiguity of animal research, the role of the amygdala in aggressive human behavior remains to be clarified. For example, violent inpatients exhibit abnormal phosphate metabolism in the amygdala-hippocampal complex (43), and epileptics with intractable seizures generally show a reduction in the severity and/or frequency of aggressive behaviors after bilateral amygdalotomy (71). However, aggressive and nonaggressive temporal lobe epilepsy patients do not differ on amygdala volume or amygdala pathology as measured by magnetic resonance imaging (70), suggesting that amygdala function is more important than structural integrity in understanding aggressive behavior.

Two imaging studies bear out the functional importance of the amygdala and other related structures. For example, individuals with APD and healthy controls exhibit different cerebral activation patterns during exposure to pairings of conditioned stimuli (i.e., neutral faces) and unconditioned stimuli (noxious odor, unscented air). During initial exposures to the noxious unconditioned stimuli of the conditioned stimuli-unconditioned stimuli pairs, controls exhibit activation of the amygdala, whereas persons with antisocial personality disorder demonstrate deactiva-

tion (72). However, after several exposures to the conditioned stimuli-unconditioned stimuli pairs, controls then demonstrate decreased amygdala and dorsolateral prefrontal activation, whereas persons with antisocial personality disorder show *increased* activation in these areas and a tendency toward decreased orbitofrontal activation. Similarly, substance-abusing psychopaths, compared to nonpsychopathic substance abusers, demonstrate increased bilateral activation in frontotemporal regions during the processing of emotional words (57). Although one intuitively might expect antisocial individuals to show reduced subcortical activation, this increased frontotemporal/amygdala activation may result because antisocial individuals must expend significantly greater effort to execute emotionally valenced laboratory tasks (57,72). In contrast, antisocials would exhibit the expected relative inactivation of the amygdala and temporolimbic structures in naturally occurring situations where they can direct their attention at will. Partial support for this interpretation is provided by the decreased amygdala activation exhibited by persons with antisocial personality disorder during initial exposures to a noxious stimulus (72).

As might be inferred from the importance of the amygdala in emotional learning, animal and human research suggests that the temporal-amygdala area interacts with social context in modulating aggression. After the three most dominant monkeys in a group of eight underwent bilateral amygdalectomies, they engaged in increased aggressive displays, with two monkeys falling to the bottom of the social hierarchy (73). This suggests that the amygdala may not control the absolute threshold for exhibiting aggressive behavior but instead may regulate previously learned associations between stimuli and aggressive responses (74). Thus, whether damage to the amygdala results in decreased or increased aggression may depend on prior punishment of, or reinforcement for, aggressive displays.

Although not a direct assessment of prior reinforcement and punishment of aggression, an interaction of childhood abuse and biologic vulnerability is related to aggression in adult humans. Although community-based subjects who suffered childhood abuse generally exhibit reduced left hemisphere (frontal, temporal, occipital) activation during a working memory task, only abused subjects who perpetrate severe violent crime in adulthood exhibit significantly reduced right temporal and frontal activation (55). In contrast, physically abused, nonviolent subjects show activation of the superior temporal gyrus comparable to controls. Intact right hemisphere activation may afford a biologic protective factor against the development of violent behavior by virtue of sparing a person's ability to process negative emotions, whereas someone with right temporal abnormalities may fail to perceive emotional cues and escalate an encounter into an aggressive response (55).

Finally, it should be noted that many temporal lobe findings in aggressive populations reflect frontotemporal dysfunction, as evidenced by the fact that all of the studies reviewed above except two found coexisting frontal deficits (43,45,46,53,55,57,72). Although Wong et al. (68,69) failed to observe frontal deficits, this may be related to the fact that they based their findings on a single clinical evaluation of the EEG/CT scans, whereas all the other studies used either objective measurement techniques or multiple clinical ratings for each scan.

Hypothalamus

The hypothalamus is extensively interconnected with the limbic system and the neocortex and is responsible for coordinating inputs and regulating the autonomic system and emotional behavior. As such, it plays an important yet complex role in the modulation of aggressive behavior. To date, this body of research has primarily been conducted on animals. For example, stimulation of the ventral hypothalamus facilitates predatory attack behavior in cats and stimulation of the dorsal hypothalamus delays such behavior (75). Stimulation of, and lesions to, the lateral hypothalamus in cats elicits behavioral displays of anger [i.e., piloerection, pupillary constriction, arching of the back, increased blood pressure, raising of the tail (76)] and taming (77), respectively. Moreover, "sham rage" can only be elicited in a decorticated cat as long as the posterior lateral hypothalamus has not been ablated (38). In addition, lesions to the medial hypothalamus in the cat and other species result in increased irritability and aggressive responding to neutral stimuli (78). Although case studies of humans with tumors in the medial hypothalamus have documented dramatic increases in aggression (79), whether there is a general relationship between the hypothalamus and aggressive behavior in humans has not been routinely investigated. However, with the advances in, and increased availability of, high-resolution functional brain imaging, this gap may be addressed in future research.

Corpus Callosum

Although damage to the corpus callosum has long been hypothesized to represent a neurologic predisposition to violence (80), only one imaging study to date has tested for, and found, that murderers exhibit decreased metabolic activity in the corpus callosum compared to healthy controls (45). The authors speculate that this poor interhemispheric connection might account for the neuropsychological findings of abnormal cerebral asymmetries and reduced hemispheric lateralization of language processes in violent offenders. Alternatively, they hypothesize that the right hemisphere, which is involved in the generation of negative emotion (81), may experience less regulation and control by the inhibitory processes of the more dominant left hemisphere, thus contributing to the expression of anger and aggression. In support of this, patients who had their corpus callosum surgically severed and patients with callosal lesions often exhibit poor

emotion regulation, "imperviousness," and apathy (82). Thus, it is possible that the inappropriate emotion regulation and poor long-term planning exhibited by life-course persistent antisocials may be mediated, in part, through the corpus callosum. However, because these callosal patients and forensic murderers also have accompanying frontal dysfunction, the relative contribution of each structure to aggression has yet to be explored.

Theoretical Implications: Drawing Upon Affective Neuroscience

As alluded to earlier, the field of affective neuroscience has been working to identify the neural circuitry involved in emotion regulation, or the "processes that amplify, attenuate, or maintain an emotion" (42). Davidson et al. (41,42) elucidated a general model of emotion regulation focusing on the PFC, amygdala, hippocampus, hypothalamus, anterior cingulate cortex, insular cortex, and ventral striatum, with particular attention paid to the PFC, amygdala, and anterior cingulate in impulsive aggression. They make a compelling argument that the orbitofrontal PFC regulates emotion by inhibiting negative emotion generated by the amygdala. When the orbitofrontal PFC fails to exert its inhibitory effects, negative affect is sustained, thus increasing one's propensity for engaging in impulsive, affective aggression (42). Such a theory is consistent with the findings demonstrating frontotemporal deficits and increased subcortical activation in impulsively aggressive individuals.

With regard to premeditated aggression, no formal integrated theories have been extended to date, and it is more than likely that different neural circuitry underlies this aggressive behavior typology (42). Based on the currently available research, extensive prefrontal structural damage does not seem to be involved. Rather, it appears that abnormal activation patterns are localized in the subcortical limbic structures (53). Abnormal subcortical functioning may predispose an individual to chronic antisocial behavior and/or premeditated aggression due to poor recognition of threatening cues (60), faulty social judgment (61), and reduced conditioning to aversive emotional stimuli (72). This pattern of findings would be consistent with previous speculations that excessive (and/or abnormal) subcortical activity predisposes one toward aggressive behavior, but how these impulses are regulated depends, in part, on the integrity of prefrontal functioning (53). In contrast, intact subcortical function may be important for the inhibition of aggressive responses among healthy and at-risk (i.e., physically abused) individuals (55).

Given that aggressive behavior likely has its basis in emotional arousal and regulation, it is no surprise that studies of human aggression parallel the theory and research of affective neuroscience. Current understanding of aggression can only benefit from more systematic integration of the two fields. In addition to placing greater emphasis on the study of interconnected structures (as opposed to a single structure), future research will benefit from examining aggressive subgroups (e.g., impulsive vs. premeditated, antisocial vs. psychiatric/medical, males vs. females) and the incorporation of social factors (e.g., social deprivation, child abuse, academic failure, presence of positive adult role models) as potential moderators and mediators of brain–behavior relationships.

PRENATAL AND PERINATAL RISK FACTORS

In addition to neuroanatomic and cognitive abnormalities, prenatal or perinatal difficulties such as obstetric complications, prenatal substance abuse, and minor physical anomalies are elevated in aggressive populations. Although it is frequently assumed that such complications result in subtle brain damage that predisposes an individual toward aggression, direct tests of this hypothesis have not yet been examined.

Obstetric Complications

Several studies show that children who suffer obstetric complications such as anoxia, forceps delivery, and preeclampsia are more likely to develop antisocial behavior in childhood and adulthood than are children who do not suffer from birth complications. Specifically, serious perinatal complications are related to violent offenses, particularly recidivistic violence, in young adulthood but are not related to antisocial personality (83). In addition, the interaction of prenatal or perinatal complications and family stress (e.g., disadvantaged family background, maternal rejection) significantly increases the risk for later violent offending; this interaction is not related to nonviolent offending (84,85). Moreover, the combination of maternal rejection and obstetric complications is uniquely related to (a) serious violence (severe physical aggression/assault) rather than minor violence (possession of weapons, threats of violence); and (b) early-onset violence (convicted before age 18) (86).

Drug and Alcohol Exposure

It has been well documented that prenatal exposure to alcohol and other drugs can have long-term effects on offspring (87). Human fetal alcohol exposure is linked to cognitive deficits across the life span (87). In addition, it is associated with social deficits such as disrupted attachment behavior and emotional dysregulation in infancy (87); increased anger, aggression, and distractibility in early childhood (88); and inappropriate sexual behavior, increased legal problems, depression, suicide, and poor caretaking of one's own children in adulthood (87). Similar cognitive and social deficits are found in children exposed to methadone (89) and cocaine (90); however, the extent to which these effects occur in-

dependently of concomitant prenatal alcohol exposure and other confounding variables (91) is unclear.

Large-scale studies provide strong evidence that smoking during pregnancy is associated with later conduct disorder and violent offending (92–97). Importantly, a number of these studies control for many third factors (e.g., low SES, family size, poor parenting, obstetric complications, birth weight, family problems, parental psychiatric diagnoses, offspring smoking, other drug use during pregnancy) that could account for the relationship. In addition, there is some suggestion that later antisocial behavior may result from an interaction between prenatal exposure to nicotine and other complicating factors. For example, maternal smoking is associated with later antisocial aggressive behavior only if the offspring also experienced delivery complications (94); was born to a teenaged, single mother; was an unwanted pregnancy; and/or experienced developmental motor lags (95).

Animal experiments have demonstrated findings consistent with the human longitudinal, correlational studies, which lend support to the causal influence of prenatal substance exposure on aggression by controlling for potential confounds (e.g., rearing, prenatal and postnatal nutrition, documentation of birthing difficulties). Prenatal alcohol and cocaine exposure alters aggressive behavior during play, grooming, encountering a strange animal, and/or electric shock in juvenile rats (98–101). Prenatal alcohol or drug exposure is associated with deficits in communication, parenting, and sexual behavior in adult rats (87). Prenatal nicotine exposure disrupts the noradrenergic (102) and cardiac M2-muscarinic cholinergic receptors (103), which regulate autonomic functions. Thus, prenatal exposure to these substances elevates the later risk for several cognitive, mood, and social problems. However, whether increased aggression is observed in prenatally substance-exposed children likely depends on the presence of other social and biologic risk factors.

Minor Physical Anomalies

Minor physical anomalies (MPAs) are thought to reflect maldevelopment of the fetus toward the end of the first 3 months of pregnancy. Although MPAs may have a genetic basis, they also may be caused by environmental stress (e.g., anoxia, bleeding, infection) to which the fetus is exposed (104).

MPAs relate to aggressive antisocial behavior in children. They are linked to peer aggression at age 3 (105), aggressive and impulsive behavior in preschool boys (106), and problem school behavior in male elementary schoolchildren (107). MPAs also predict violent but not nonviolent delinquency during adolescence (108). However, it is of note that MPAs are generally elevated in aggressive children when compared to children drawn from the normal population (109), thus limiting any assertions of specificity to aggression.

Although not uniformly observed (108), there is some suggestion that MPAs interact with social factors to predispose an individual toward aggression. MPAs at 12 years relate to violent offending at 21 years only in individuals raised in unstable (vs. stable) home environments (110). MPAs are not associated with nonviolent property offending, regardless of home environment. This interaction is similar to those on birth complications and prenatal nicotine exposure that suggest a negative psychosocial factor is required to "trigger" the biologic risk factor. Moreover, the biosocial interaction is specific to violent offending. Further studies are needed to explore which environmental or social stressors interact with MPAs to predispose to violence.

NEUROCHEMICAL AND HORMONAL CORRELATES

The neuroanatomic and neurocognitive factors reviewed likely involve disruptions in the neurotransmitters innervating and enervating these sites. Although the influence of neurotransmitters in psychiatric disorders such as schizophrenia and depression has been intensively studied, extensive examination of aggressive antisocial behavior did not really take place until the early 1990's. This review focuses on serotonin, dopamine (DA), norepinephrine (NE), acetylcholine (ACh), and γ-aminobutyric acid (GABA). Although not neurotransmitters, testosterone and cortisol will also be discussed as they have been consistently associated with aggression.

Serotonin

Serotonin (5-HT) is the most widely studied neurotransmitter with regard to violence and aggression. In addition to its role in sleep regulation, appetite, and sexual behavior, serotonin is regarded as a behavioral inhibitor (111), which partly accounts for its intensive scrutiny in the study of aggression. Its pathways originate in the raphe nuclei of the upper brainstem. Rostral projections diffusely innervate the prefrontal, frontal, hypothalamic, hippocampal amygdaloid, striatal, and cerebellar areas, and caudal projections extend to the dorsal horns in the spinal cord (38,112).

Decades of research have established that an inverse relationship between serotonin and aggressive/violent behavior generally exists in humans (112–114), with a metaanalysis indicating the overall cerebrospinal fluid (CSF) serotonin–antisocial behavior relationship is moderate in magnitude (115). Markers of low serotonin are associated with aggression in naval recruits (116,117) and incarcerated juvenile offenders (118). Lower serotonin level also is observed in incarcerated murderers diagnosed with antisocial personality disorder versus age-matched, nonincarcerated healthy controls (119), aggressive versus nonaggressive psychiatric patients (113,120), and depressed persons with

a history of suicide attempts or completions versus those without such a history (113).

Scientists further postulate that 5-HT is primarily related to poor impulse control, and it is this impulsivity that predisposes an individual to engage in aggressive behavior (113,114). Support for this position comes from research finding lower 5-HT in repeat rather than one-time violent offenders (121), impulsive rather than premeditated violent offenders (121), impulsively aggressive psychiatric patients rather than nonaggressive controls (122), and depressed patients with, rather than without, anger attacks (123). In addition, not only do alcoholic, impulsive, violent offenders have significantly lower CSF 5-hydroxyindoleacetic acid (5-HIAA) than do nonimpulsive violent offenders, but nonimpulsive offenders have significantly higher CSF 5-HIAA levels than do normal controls (124).

Researchers are investigating in what populations, or under what environmental conditions, the serotonin–aggression relationship is most likely to be observed. For example, a consistent effect appears for men but not women (115,125). Within antisocial and criminal populations, serotonergic dysfunction appears to be specific to violent offending and not general crime (125–127). Although replication of findings has been considerably more mixed in child than in adult populations (115), particularly around puberty (128), abnormal serotonin function in preadolescent boys may be more prevalent in those with aggressive parents (129).

The complexity of the serotonin–aggression relationship is further evidenced by the fact that manipulating serotonin levels does not always result in expected alterations in aggressive behavior. D,L-Fenfluramine, a serotonin-releasing drug, decreases experimentally elicited aggressive responses in males with a history of conduct disorder (130). However, acute dietary depletion of tryptophan, a 5-HT precursor, does not increase impulsive behavior in aggressive adolescent boys (131). In contrast, tryptophan depletion does increase aggressive behavior and hostility in normal males (132,133), particularly if they have an aggressive trait disposition (134).

In addition, serotonin activity in certain regions of the brain may be more important to the manifestation of aggressive behavior than are whole brain or blood circulating measures of serotonin. Fenfluramine administration significantly increases cortical metabolism and/or cerebral blood flow in the PFC of healthy humans (135,136). However, impulsively aggressive, personality-disordered patients who are administered D,L-fenfluramine show a blunted response in the prefrontal (particularly orbital) and cingulate cortex compared to controls (137). Similar findings of diminished serotonergic stimulation in the PFC, as well as in the left superior temporal gyrus and right insular cortex, are observed in patients with borderline personality disorder (138).

Animal research is isolating specific serotonin receptor subtypes (5-HT$_{1A}$ and 5-HT$_{1B}$) central to the serotonin-aggression link. 5-HT$_{1A}$ receptors inhibit aggressive behavior in mice, particularly those that become significantly more aggressive after drinking alcohol (139). In addition, it appears that the somatodendritic 5-HT$_{1A}$ autoreceptors are involved in the aggression-heightening effects of alcohol, indicating that reduced aggression in this sample may be associated with decreased—not increased—serotonin transmission (139).

After being housed in isolation, mice genetically altered to specifically lack the 5-HT$_{1B}$ receptor attack an intruder more quickly and intensely than unaltered mice experiencing the same stress manipulation (140). However, there is some discrepancy as to whether the 5-HT$_{1B}$ receptor relates to aggression in general or to defensive impulsive aggression in particular. For example, the mutant mice do not differ in aggression while housed in a group, suggesting that 5-HT$_{1B}$ receptors are specific to defensive aggression. In contrast, mice treated with a 5-HT$_{1B}$ receptor selective agonist demonstrate significant decreases in unprovoked, species-typical aggression, as well as alcohol-heightened and territorial aggression (141). Importantly, decreased aggression is not the result of sedation, generally decreased locomotor activity, or increased anxiety, indicating that, despite variations in behavioral effects, the 5-HT$_{1B}$ receptors play an important role in modulating aggression. Additional research should attempt to clarify which aggressive behavior subtypes are modulated by the 5-HT$_{1B}$ receptor.

The modulation effects of the 5-HT$_{1A}$ and 5-HT$_{1B}$ receptors may depend on hormones (specifically androgens) and/or the particular brain region involved. Reduced aggression in animals treated with 5-HT$_{1A}$ and 5-HT$_{1B}$ receptor agonists is observed in male but not female hamsters (142). In addition, administration of 5-HT$_{1A}$ and 5-HT$_{1B}$ receptor agonists administered directly to the lateral septum in male gonadectomized mice reduces attack behavior only in those treated with synthetic androgen; those treated with estrogen are not affected (143). In contrast, all agonist treatments administered to the medial preoptic area decreases aggression, regardless of androgen or estrogen treatment. Nevertheless, the specificity of the 5-HT$_{1B}$ receptor to human impulsive aggression remains to be determined. Postmortem brain tissue analysis of various groups of patients (pathologic aggression, substance abuse, depression) did not find significant differences in measures of prefrontal 5-HT$_{1B}$ receptor binding or 5-HT$_{1B}$ genotype (144).

Research in animals and humans has demonstrated that the reciprocal regulatory effects of the vasopressin-serotonin system are important to the control of aggression (145,146). Serotonin inversely relates to, and vasopressin directly relates to, aggressive behavior in hamsters (145) and in personality-disordered humans (146). Moreover, it appears that environmental threat plays an important role in disrupting the vasopressin-serotonin system. Young adult hamsters that were attacked daily by adult hamsters during adolescence show an increased density of 5-HT terminals and decreased density of vasopressin fibers and neurons in the hypothalamus compared to isolation-stressed and nonstressed controls (147). In addition, compared to isolation-stressed controls, the abused hamsters exhibit exaggerated inappropriate aggression toward smaller males. However, compared to nonstressed controls, the abused hamsters exhibit increased submission to same-size males (147).

Finally, increasing evidence from human (127,148) and nonhuman primate (149,150) research suggests that the serotonin–aggression relationship holds for pathologic populations but not normal subjects. Male inpatients who die before age 40 have significantly lower CSF 5-HIAA, the main metabolite of 5-HT, than those still living at the time of follow-up, with the early deaths resulting from homicide, suicide, or suspicious accidents suggesting suicide (151). Free-ranging male rhesus monkeys with the lowest CSF 5-HIAA levels are significantly more likely to initiate excessive, unrestrained, injurious aggression; engage in risk-taking behavior; and suffer from premature death during a 4-year period than are those with average or above average levels (150). Similarly, low CSF 5-HIAA is found in female rhesus macaques that engage in inappropriate spontaneous aggression, but CSF-HIAA is unrelated to noninjurious aggression involved in maintaining social dominance status (149). Thus, whereas average or above average serotonin corresponds with competent social behavior, abnormally low serotonin may increase one's mortality through exposure to dangerous aggressive behavior (150).

Dopamine

DA is considered a behavioral facilitator, and its midbrain neurotransmitter system acts in opposition of the serotonergic system (152). Although human research on the DA-aggression link has been fairly limited, animal studies provide compelling evidence that dopaminergic function is related both to predatory aggression and exaggerated defenses (153). Pharmacologic challenge studies in rats demonstrate that selective D$_1$ and D$_2$ agonists enhance shock-induced aggression and indiscriminate biting (113,153) but diminish predatory aggression (153). In addition, D$_2$ antagonists decrease aggressive behavior in animals (113) and humans (153), although this appears to be due to general sedating effects rather than an effect specific to aggression (153).

Brain levels of DA in mice that have just engaged in aggressive behavior are consistently found to be elevated (154,155), with elevations most noticeable in the frontal and limbic portions of the brain associated with the initiation and execution of behavior and with behavioral reinforcement (153). Thus, the mesolimbic and mesocortical dopaminergic systems may influence aggressive behavior, in part, through positive reinforcement of successfully executed aggression (156). Consistent with this hypothesis, haloperidol, a nonselective DA antagonist, only decreases aggressive or defensive behavior in rats without a prior history of aggression; it shows no affect in mice with a history of repetitive aggression (157). Similarly, when confronted with a nonaggressive conspecific, mice that have experienced repeated defeat in prior aggressive encounters exhibit significantly greater DA metabolism in the PFC and more submissive and defensive behaviors than do mice with prior nonaggressive social encounters (158). Alterations in social exploration and activation of the nucleus accumbens septi further depend on the gender of the prior aggressor and current conspecific.

It appears that different DA receptor subtypes mediate locomotor (e.g., aggression, escape, jump) versus nonlocomotor defensive behaviors (e.g., freezing, upright defensive posturing) (159). For example, although both D_2 and D_3 agonists increase nonlocomotor responses, only agonists selective to the D_3 receptor (a subtype of the D_2-like receptor densely concentrated in limbic areas) enhance locomotor responses (159). Thus, the D_3 receptor may be particularly important in modulating aggression.

Although a general DA-aggression link is not supported in humans, DA does appear to be associated with aggression in certain patient populations. CSF and plasma measures of DA are positively associated with aggression in patients with Alzheimer disease (160,161), and postmortem analysis indicates that dopaminergic tracts in the substantia nigra are spared in aggressive but not nonaggressive patients with Alzheimer disease (162). Moreover, research suggests that genetic variation of the *DRD1* gene, in particular, may predispose those with Alzheimer disease toward aggression (163).

Alcohol-, cocaine-, and amphetamine-related aggression in animals and humans also is mediated, in part, by dopaminergic activation (164). Brain imaging data of striatal dopaminergic sites indicate that alcoholics prone to violence differ structurally from nonviolent alcoholics. Compared to controls, violent alcoholics have significantly increased densities, whereas nonviolent alcoholics have significantly reduced densities (164,165).

Norepinephrine

NE also acts as a facilitator of aggressive behavior in animals and humans, with the α- and β-adrenergic receptor subtypes in the hypothalamus being particularly involved (113,153). NE correlates positively with aggression in mice (113), rhesus monkeys (166), and healthy male humans (167,168) using brain, CSF, and blood measures of NE. In addition, the most aggressive of three strains of mice demonstrate significantly higher steady-state levels and turnover rate of NE in the frontal cortex, caudate nucleus, and hypothalamus than the least aggressive strain (169).

Consistent with a facilitation hypothesis, administration of psychopharmacologic agents that increase NE levels (i.e., tricyclic and monoamine oxidase inhibitor antidepressants) results in increased aggression in mice (170). Moreover, this effect can be abolished through noradrenergic denervation (171). In addition, β-adrenergic blockers such as propanolol reduce aggressive behavior in animals, as well as in

aggressive psychiatric patients (172–174). However, because β-blockers subsequently have been discovered to act on 5-HT_{1A}, it is unclear whether these drugs reduce aggression through their effect on the androgenergic system, serotonergic system, or both (153).

The relationship between NE and aggression may depend on personality and behavioral characteristics of the individual. Quantitative analysis of 29 CSF NE studies in humans found that, although NE is not related to antisocial behavior in general, it is significantly lower in specific subtypes of aggressive individuals (175): alcoholic versus nonalcoholic; those with borderline personality disorder versus those without BPD; and those with histories of depression versus those without such histories. In addition, NE is reduced in children with conduct disorders (176). Because of the conflicting findings in this overall body of research, it has not yet been possible to delineate a coherent noradrenergic system of aggression (153).

Acetylcholine

Brain ACh acts on two basic receptors, muscarinic and nicotinic (38), with the largest concentrations of cell bodies found in the caudate nucleus, hippocampus, amygdala, select brain stem nuclei, and frontal lobe (111). ACh is important to behavioral and regulatory functions such as arousal, attention, learning, memory, and mood (38,111). Animal research reveals that cholinergic agonists, particularly muscarinic (177), facilitate aggressive behavior (178), whereas cholinergic antagonists decrease aggression (178). In addition, injections of nicotine decrease aggressive behavior in a variety of nonhuman species (153). However, due to the toxicity of most agents, a cholinergic hypothesis of aggression in humans has only been examined in nicotinic receptors (153). This limited body of research indicates that aggressive responses toward a confederate in a laboratory task decrease after smoking cigarettes (179), and aggression-prone smokers naturally titrate their intake of nicotine in order to reduce anger and hostility (180).

γ-Aminobutyric Acid

GABA serves as a major inhibitor of neurotransmission (111), with activation of the $GABA_A$ receptor subtype resulting in hyperpolarization of the affected membrane (111). Evidence that anxiolytics act on the benzodiazepine receptor–$GABA_A$ complex spurred research on its potential modulatory role for aggressive behavior (153), with the initial speculation that GABA exerted an inhibitory influence (74). For example, decreased GABA is associated with increased aggressive behavior in mice (181) or killing in rats (182). In rats and cats, benzodiazepines reduce defensive aggression elicited by an approaching experimenter, electric shock, or stimulation of the hypothalamus (153). In humans, acute administration of benzodiazepines and related

GABA agonists effectively reduces aggressive and agitated behavior (111). However, the effects appear to be the result of general sedation and muscle relaxation and therefore are not aggression specific (153).

Considerable research has contradicted an inhibitory influence of GABA. Long-term use of benzodiazepines can paradoxically result in increased aggressive behavior in some patients (183). Knockout of the mouse *GAD65* gene, which is essential for GABA synthesis, results in reduced intermale aggression and, to a lesser extent, reduced anxiety (184). In addition, naloxone, an opioid antagonist, significantly decreases aggression in rats rendered disinhibited through 5-HT lesions (185) via antagonism of the GABA$_A$ receptor. In addition, certain aggressive strains of mice show increased levels of GABA in the hypothalamus and limbic structures (153).

Temporal aspects of prior aggressive experience are related to GABA facilitation of aggressive behavior. Perfusion of a glutamate agonist/GABA$_A$ antagonist combination into the hypothalamic attack area of rats only elicits aggression in those with recent aggressive experiences. Rats without any fighting experience and rats with temporally distant fighting experiences only respond to the treatment with increased grooming (186). At present, it appears that GABA$_A$ receptors modulate the expression of aggressive behavior, but whether they facilitate or decrease aggression depends on the region of the brain stimulated, the type of aggression studied (e.g., maternal, defensive, killing), and prior aggression experience. Clearly, more research in this area is needed.

Testosterone and Androgens

Contrary to popular opinion, a linear relationship between testosterone and aggression in humans is not well supported (187). Although animal research has established a causal link between increased testosterone and certain forms of aggression (187,188), direct application of these nonprimate models to humans has not met with success (187). Elevated serum or salivary testosterone is observed in select populations such as violent offenders (124,189), abusers of anabolic steroids (190), clinic-referred children with aggressive behavior problems (191), and female prisoners convicted of unprovoked assault (192). Elevated dehydroepiandrosterone sulfate (DHEAS), an adrenal androgen that is synthesized by the brain and has antagonistic effects on central GABA functions, is observed in children with conduct or oppositional defiant disorder (ODD) (193). Despite these findings, however, a comparable number of studies fail to observe such relationships despite using similar sample populations and research methodologies (187). Pharmacologic manipulations of testosterone in medical, criminal, and normal populations also consistently fail to alter aggression in expected directions (187).

To address these conflicts, research is attempting to elucidate in what social contexts testosterone is likely to relate to aggression. Studies with monkeys find that, in newly formed groups, males that became dominant exhibit the most aggressive behavior and show a significant increase in testosterone from baseline, whereas the most subordinate monkeys show a significant decrease in testosterone (194,195). Once the dominance hierarchy is stabilized, however, dominant males no longer exhibit elevated testosterone or aggression (194). Similarly, in human pubescent males, social dominance and aggression manifested as peer-rated leadership and "toughness" are associated with high testosterone levels (196,197). In contrast, persistent physical aggression exhibited throughout childhood is associated with lower testosterone levels and peer rejection (196). Thus, testosterone appears to be positively correlated with aggressive displays that secure social dominance and negatively correlated with extreme aggression that encourages subordination or social rejection.

Cortisol

Scientists have begun to turn their attention to the role of cortisol in aggressive behavior. Cortisol is generally accepted as a peripheral measure of hypothalamic–pituitary–adrenal (HPA) axis activity, which is thought to mobilize an individual's response to stressful events. Assuming that a propensity to engage in aggression may result from autonomic underarousal, it has been suggested that antisocial aggressive individuals should show decreased cortisol levels or reactivity compared to nonaggressive individuals.

As with the testosterone literature, empirical tests of this theory have met with conflicting results. Lower cortisol levels are observed in aggressive schoolchildren (198), adolescents with conduct problems (199), boys with disruptive behavior (200) or ODD (201), and disinhibited children (202). However, this relationship is not always observed in aggressive adolescent boys (203) or noninstitutionalized aggressive adult males (167). Similarly, the findings of cortisol stress reactivity in antisocial individuals are conflicting; some studies found decreased reactivity (201) and other studies found increased reactivity (204).

Paralleling the work with serotonin, studies of violent and psychiatrically disordered subgroups suggest that reduced cortisol may occur only in the most extreme populations. For example, cortisol is lower in habitually violent adult offenders compared to nonhabitually violent offenders, nonviolent offenders, and normal controls (205), and in nonanxious, conduct-disordered boys compared to anxious conduct-disordered boys (206). Similarly decreased cortisol is present in aggressive children with persistent ADHD diagnoses but not in children who fail to meet ADHD diagnostic criteria between kindergarten and second grade (207). Future studies should integrate social, environmental, and individual difference variables that likely influence HPA response to further clarify the cortisol-aggression link.

Interaction of Systems

Although neurochemistry has provided several important clues regarding aggression, a continued limitation is that most research examines a single neurotransmitter rather than considering the complex interplay of multiple neurotransmitter systems (74). For example, a number of studies demonstrated that serotonin interacts with vasopressin and DA or that androgens modulate GABA activity. Thus, future research that measures multiple neurotransmitter systems, considers regional brain differences, and examines the influence of social experience on neurochemical function will help develop a more complete picture of how neurotransmitters are related to aggression.

PSYCHOPHYSIOLOGIC CORRELATES

Since the 1940's an extensive body of research has been built up on the psychophysiologic basis of antisocial and aggressive behavior. This chapter focuses on low physiologic arousal as measured by EEG and heart rate because they encompass the strongest psychophysiologic findings in this area. In addition, although cardiovascular underarousal research has long been conducted independently from neurobiologic research, autonomic nervous system functions clearly interact with and are regulated by the central nervous system, including some of the neuroanatomic structures reviewed [i.e., prefrontal, temporal areas (208,209)].

Electroencephalogram

An extensive literature base demonstrates that central nervous system underarousal as indicated by EEG abnormalities—particularly excessive slow wave activity—is overrepresented among violent recidivistic offenders (210–213). Similarly, increased θ activity and decreased α activity are associated with aggression in a sample of adult male drug abusers (214), and increased δ and decreased α activity are associated with violent acting out by psychiatric inpatients (215). Compared to depressed and headache control patients, patients with either intermittent explosive disorder or episodic dyscontrol are disproportionately more likely to have abnormal diffuse or focal slowing on EEG (216). In addition, the bulk of the research implicates EEG arousal deficits in the frontal regions of the brain for both general and violent crime (217,218).

There is some suggestion that age and gender moderate the relationship between EEG activity and antisocial behavior. In a longitudinal study spanning preschool to elementary school, children with DSM-IV ODD were compared to

healthy children on frontal brain activation and aggression (219). ODD children, regardless of age and gender, were significantly more aggressive than healthy children. With regard to EEG activity, ODD girls showed a pattern of more right than left frontal activation at ages $4\frac{1}{2}$ and 8, whereas healthy girls showed no asymmetry at age $4\frac{1}{2}$ and more left than right frontal activation at age 8. In contrast, healthy boys exhibited greater right than left frontal activation at both ages, whereas ODD boys failed to show frontal asymmetry at either age. The authors speculate that these atypical patterns of activation in ODD children may reflect an affective style that predisposes them to behavioral problems. Atypical frontal activation on EEG also was associated with behavioral problems in another preschool sample (220). Interestingly, whether preschoolers with atypical frontal activation exhibited internalizing (anxiety and/or depression) or externalizing behavior problems (inattention, impulsivity, and/or aggression) depended on whether the child had a shy or sociable temperament, respectively.

Heart Rate

Low resting heart rate is the best replicated biologic marker of antisocial and aggressive behavior in community samples. Low resting heart rate at age 3 is related to increased aggressive behavior but not general delinquency at age 11 (221). Low heart rate at age 12 is associated with more frequent fighting between childhood and preadolescence in boys with general disruptive behavior problems (222). Cardiac underarousal in preadolescent or late-teenaged boys predicts criminal offending across the life span, with the lowest heart rates observed in those who commit violent offenses (223–225) and in life-course persistent antisocial individuals (224). Although most studies have examined only males, low heart rate also characterizes female antisocial individuals (176,221,226,227).

Unlike most other biologic correlates, an unusual and important feature of the heart rate–aggression relationship is its diagnostic specificity. No other psychiatric condition has been linked to low resting heart rate. Other conditions such as alcoholism, depression, schizophrenia, and anxiety disorder have, if anything, been linked to *higher* resting heart rate. In addition, whereas serotonin and cortisol most strongly relate to aggressive behavior in psychopathologic populations, cardiovascular underarousal is routinely observed in uninstitutionalized community samples. The heart rate–aggression relationship is also not the result of potential confounds such as height, weight, body bulk, physical development, and muscle tone (221,223,228), scholastic ability and IQ (228,229), excess motor activity and inattention (221,228), drug and alcohol use (221), physical exercise and sports activity (223,228), or low social class, divorce, family size, teenage pregnancy, and other aspects of psychosocial adversity (223,228,229). Consistent heart rate findings are found across different countries (e.g., England, Germany, New Zealand, United States, Mauritius, Canada) using different measurement techniques.

Interpretations of Low Arousal: Fearlessness and Stimulation-Seeking Theories

Two main theoretical interpretations attempt to explain the link between psychophysiologic underarousal and aggressive antisocial behavior. First, fearlessness theory presumes that subjects are not actually at "rest" in research settings but instead are reacting to a mildly stressful, novel situation. Thus, autonomic underarousal reflects lack of anxiety and fear, and this lack of fear is assumed to predispose one to antisocial behavior because such behavior requires a degree of fearlessness to execute (9,230). Fearlessness, especially in childhood, might account for poor socialization because low fear of punishment would likely reduce the effectiveness of behavioral conditioning. A second theory explaining reduced arousal is stimulation-seeking theory (9,231–233). This theory argues that low autonomic arousal represents an unpleasant physiologic state, and antisocials seek stimulation in order to increase their arousal to an optimal or normal level.

Although conceptualized independently, it is possible that these two theories are complementary rather than competing. That is, underarousal may predispose to crime because it both results in fearlessness and encourages antisocial stimulation-seeking. Consistent with this, behavioral indicators of stimulation-seeking behavior and fearlessness measured at age 3 each predicts aggressive behavior at age 11 (233), thus suggesting that their combined effect may be more important in explaining antisocial behavior than either influence taken alone.

INTEGRATED NEURAL THEORY OF EMOTION AND AGGRESSION

Although neuroanatomic, neurotransmitter, and psychophysiologic research traditionally has been conducted independently, there have been attempts to link them into a coherent model reflecting a biologic basis to human aggression (41,42,234). Damasio et al. (235) posit that the ventromedial PFC regulates behavior, in part, by the generation of somatic markers (i.e., skin conductance responses to aversive stimuli). Somatic markers alert individuals to risky or threatening situations, thus allowing them to maintain homeostasis by intuitively guiding their behavior toward advantageous decision making (234,235). The original model, based on patients with selective lesions (236–238), placed primary emphasis on the ventromedial PFC. However, subsequent research has implicated the PFC and amygdala (238), as well as their connections with additional subcortical structures

involved in emotion and homeostasis regulation [e.g., cingulate, hypothalamus, motor cortex, dorsal pons, midbrain, insula, cerebellum (42,235)], thus suggesting a broadening of the original model. Recent research on community-based psychopaths suggests that excessive cardiovascular stress reactivity may also reflect a somatic marker (239).

Drawing upon this theory and related research, the risk for impulsive aggression may be elevated when an individual is unable to generate or appreciate the significance of somatic markers. Specifically, PFC abnormalities may directly interfere with the generation of somatic markers [i.e., skin conductance is related to the orbitofrontal PFC (209,234)]. PFC damage also may render one unable to utilize affective information by interfering with "affective working memory" or the ability to use emotional memory to guide current behavior based on anticipation of future consequences (41). In addition, abnormal function of the amygdala, hypothalamus, and other subcortical structures involved in emotion and psychophysiologic responses may result in a failure to generate somatic markers, thus increasing one's propensity for aggression. As reviewed earlier, abnormalities in the PFC, temporal area, amygdala, and autonomic nervous system arousal have been routinely observed in aggressive populations from the community, prison, and psychiatric hospitals, thus lending support to such an interpretation. Research on neurotransmitter abnormalities in aggressive hospitalized individuals fits well within this framework. When neurotransmitter anomalies have been localized to specific brain regions (PFC, hypothalamus, striatum), these regions generally correspond to the neuroanatomic findings.

Despite the general convergence of these different biologic correlates into a viable neural model of impulsive aggression, the current state of research leaves many questions unanswered. Undoubtedly, different models will need to be generated for different subtypes of aggression, with the most obvious gap being premeditated aggression (42). In addition, the neural circuitry underlying aggressive behavior in men and women is likely to be different. Finally, research will need to address the issue of developmental neural plasticity, temperamental differences, the environmental context, and their combined association with aggression across the life span (41). Nevertheless, the current research not only demonstrates how the complex interplay of neurobiologic systems can create a "behavioral capacity" (188) for aggression, but it also highlights how this dynamic system can influence, and be influenced by, the environment.

REFERENCES

1. Fox JA, Zawitz MW. Homicide trends in the United States: 1998 update. *Bureau of Justice Statistics Crime Data Brief* 2000.
2. Fox JA. Trends in juvenile violence: a report to the United States Attorney General on current and future rates of juvenile offending. *Bureau of Justice Statistics* 1996.

3. National Vital Statistics. *Death and death rates for the 10 leading causes of death in specified age groups, by race and sex: United States.* 1998.

4. Fox JA. Trends in juvenile violence: 1997 update. *Bureau of Justice Statistics* 1997.

5. Vitiello B, Stoff DM. Subtypes of aggression and their relevance to child psychiatry. *J Am Acad Child Adolesc Psychiatry* 1997;36:307–315.

6. Loeber R, Burke JD, Lahey BB, et al. Oppositional defiant and conduct disorder: a review of the past 10 years, part I. *J Am Acad Child Adolesc Psychiatry* 2000;39:1468–1484.

7. Moffitt, TE. The neuropsychology of conduct disorder. Special issue: toward a developmental perspective on conduct disorder. *Dev Psychopathol* 1993;5:135–151.

8. Williamson S, Hare RD, Wong S. Violence: criminal psychopaths and their victims. *Can J Behav Sci* 1987;19:454–462.

9. Raine A. *The psychopathology of crime: criminal behavior as a clinical disorder.* San Diego, CA: Academic Press, 1993.

10. Golden CJ, Jackson ML, Peterson-Rohne, et al. Neuropsychological correlates of violence and aggression: a review of the clinical literature. *Aggression Violent Behav* 1996;1:3–25.

11. Raine A, Buchsbaum MS. Violence, brain imaging, and neuropsychology. In: Stoff DM, Cairns RB, eds. *Aggression and violence: genetic, neurobiological, and biosocial perspectives.* Mahwah, NJ: Lawrence Erlbaum Associates, 1996;195–218.

12. Bryant ET, Scott ML, Tori C, et al. Neuropsychological deficits, learning disability, and violent behavior. *J Consult Clin Psychol* 1984;52:323–324.

13. Langevin R, Ben-Aron M, Wortzman G, et al. Brain damage, diagnosis, and substance abuse among violent offenders. *Behav Sci Law* 1987;5:77–94.

14. Vaillant PM, Asu ME, Cooper D, et al. Profile of dangerous and nondangerous offenders referred for pretrial psychiatric assessment. *Psychol Rep* 1984;54:411–418.

15. Syverson KL, Romney DM. A further attempt to differentiate violent from nonviolent offenders by means of a battery of psychological tests. *Can J Behav Sci* 1985;17:87–92.

16. Holland TR, Beckett GE, Levi M. Intelligence, personality, and criminal violence: a multivariate analysis. *J Consult Clin Psychol* 1981;49:106–111.

17. Wilson JQ, Hernnstein R. *Crime and human nature.* New York: Simon & Schuster, 1985.

18. Hirschi T, Hindelang MJ. Intelligence and delinquency: a revisionist review. *Am Sociol Rev* 1977;42:571–587.

19. Quay HC. Intelligence. In: Quay HC, ed. *Handbook of juvenile delinquency.* New York: Wiley, 1987:106–117.

20. Yeudall LT, Flor-Henry P. Lateralized neuropsychological impairments in depression and criminal psychopathy. Paper presented at the Conference of the Psychiatric Association of Alberta, Calgary, Alberta, 1975.

21. Yeudall LT. *The neuropsychology of aggression.* Clarence Hinks Memorial Lecture, University of Western Ontario, Canada, 1978.

22. Hare RD, McPherson LM. Psychopathy and perceptual asymmetry during verbal dichotic listening. *J Abnorm Psychol* 1984;93:141–149.

23. Raine A, O'Brien M, Smiley N, et al. Reduced lateralization in verbal dichotic listening in adolescent psychopaths. *J Abnorm Psychol* 1990;99:272–277.

24. Nestor PG. Neuropsychological and clinical correlates of murder and other forms of extreme violence in a forensic psychiatric population. *J Nerv Ment Dis* 1992;180:418–423.

25. Barratt ES, Stanford MS, Kent TA, et al. Neuropsychological and cognitive psychophysiological substrates of impulsive aggression. *Biol Psychiatry* 1997;41:1045–1061.

26. Raine A, Yaralian PS, Reynolds C, et al. Spatial but not verbal cognitive deficits at age 3 years in persistently antisocial individuals: a prospective, longitudinal study. *Development and Psychopathology* 2002;14(1):25–44.

27. Spreen O, Strauss E. *A compendium of neuropsychological tests: administration, norms, and commentary,* 2nd ed. New York: Oxford University Press, 1998.

28. Lezak, MD. *Neuropsychological assessment,* 3rd ed. New York: Oxford University Press, 1995.

29. Giancola P. Evidence of dorsolateral and orbital prefrontal cortical involvement in the expression of aggressive behavior. *Aggress Behav* 1995;21:431–450.

30. Hughes C, White A, Sharpen J, et al. Antisocial, angry, and unsympathetic: "hard-to-manage" preschoolers' peer problems and possible cognitive influences. *J Child Psychol Psychiatry* 2000;41:169–179.

31. Toupin J, Dery M, Pauze R, et al. Cognitive and familial contributions to conduct disorder in children. *J Child Psychol Psychiatry* 2000;41:333–344.

32. Seguin JR, Pihl RO, Harden PW, et al. Cognitive and neuropsychological characteristics of physically aggressive boys. *J Abnorm Psychol* 1995;104:614–624.

33. Seguin JR, Boulerice B, Harden PW, et al. Executive functions and physical aggression after controlling for attention deficit hyperactivity disorder, general memory, and IQ. *J Child Psychol Psychiatry* 1999;40:1197–1208.

34. Krakowski M, Czobor P, Carpenter MD, et al. Community violence and inpatient assaults: neurobiological deficits. *J Neuropsychiatry Clin Neurosci* 1997;9:549–555.

35. Giancola PR, Mezzich AC, Tarter RE. Executive cognitive functioning, temperament, and antisocial behavior in conduct-disordered adolescent females. *J Abnorm Psychol* 1998;107:629–641.

36. Giancola PR, Mezzich AC, Tarter RE. Disruptive, delinquent and aggressive behavior in female adolescents with a psychoactive substance use disorder: relation to executive cognitive functioning. *J Stud Alcohol* 1998;59:560–567.

37. Rutter M, Giller H, Hagell A. *Antisocial behavior by young people.* New York: Cambridge University Press, 1997.

38. Kandel ER, Schwartz JH, Jessell TM. *Principles of neural science,* 3rd ed. Norwalk, CT: Appleton & Lange, 1991.

39. Damasio AR, Anderson SW. The frontal lobes. In: Heilman KM, Valenstein E, eds. *Clinical neuropsychology,* 3rd ed. New York: Oxford University Press, 1993:410–460.

40. LaPierre D, Braun, CMJ, Hodgins S. Ventral frontal deficits in psychopathy: neuropsychological test findings. *Neuropsychologia* 1995;33:139–151.

41. Davidson RJ, Jackson DC, Kalin NH. Emotion, plasticity, context and regulation: perspectives from affective neuroscience. *Psychol Bull* 2000;126:890–909.

42. Davidson RJ, Putnam KM, Larson CL. Dysfunction in the neural circuitry of emotion regulation: a possible prelude to violence. *Science* 2000;289:591–594.

43. Critchley HD, Simmons A, Daly E, et al. Prefrontal and medial temporal correlates of repetitive violence to self and others. *Biol Psychiatry* 2000;47:928–934.

44. Volkow ND, Tancredi LR, Grant C, et al. Brain glucose metabolism in violent psychiatric patients: a preliminary study. *Psychiatry Res Neuroimag* 1995;61:243–253.

45. Raine A, Buchsbaum M, LaCasse L. Brain abnormalities in murderers indicated by positron emission tomography. *Biol Psychiatry* 1997;42:495–508.

46. Amen DG, Stubblefield M, Carmichael B, et al. Brain SPECT findings and aggressiveness. *Ann Clin Psychiatry* 1996;8:129–137.

47. Woermann FG, van Elst LT, Koepp MJ, et al. Reduction of frontal neocortical grey matter associated with affective

aggression in patients with temporal lobe epilepsy: an objective voxel by voxel analysis of automatically segmented MRI. *J Neurol Neurosurg Psychiatry* 2000;68:162–169.

48. Miller BL, Darby A, Benson DF, et al. Aggressive, socially disruptive and antisocial behaviour associated with fronto-temporal dementia. *Br J Psychiatry* 1997;170:150–154.

49. Hirono N, Mega MS, Dinov ID, et al. Left frontotemporal hypoperfusion is associated with aggression in patients with dementia. *Arch Neurol* 2000;57:861–866.

50. Raine A, Lencz T, Bihrle S, et al. Reduced prefrontal gray matter volume and reduced autonomic activity in antisocial personality disorder. *Arch Gen Psychiatry* 2000;57:119–127.

51. Goyer PF, Andreason PJ, Semple WE, et al. Positron-emission tomography and personality disorders. *Neuropsychopharmacology* 1994;10:21–28.

52. Soderstrom H, Tullberg M, Wikkelso C, et al. Reduced regional blood flow in non-psychotic violent offenders. *Psychiatry Res* 2000;98:29–41.

53. Raine A, Meloy JR, Bihrle S, et al. Reduced prefrontal and increased subcortical brain functioning assessed using positron emission tomography in predatory and affective murderers. Special issue: impulsive aggression. *Behav Sci Law* 1998;16:319–332.

54. Raine A, Stoddard J, Bihrle S, et al. Prefrontal glucose deficits in murderers lacking psychosocial deprivation. *Neuropsychiatry Neuropsychol Behav Neurol* 1998;11:1–7.

55. Raine A, Park S, Lencz T, et al. Reduced right hemisphere activation in severely abused violent offenders during a working memory task: an fMRI study. *Aggress Behav* 2001;27(2):111–129.

56. Kuruoglu AC, Arikan Z, Vural G, et al. Single photon emission computerised tomography in chronic alcoholism: antisocial personality disorder may be associated with decreased frontal perfusion. *Br J Psychiatry* 1996;169:348–354.

57. Intrator J, Hare R, Stritzke P, et al. A brain mapping (single positron emission computerized tomography) study of semantic and affective processing in psychopaths. *Biol Psychiatry* 1997;42:96–103.

58. Powell KB, Miklowitz DJ. Frontal lobe dysfunction in the affective disorders. *Clin Psychol Rev* 1994;14:525–546.

59. Adolphs R, Tranel D, Damasio H, et al. Impaired recognition of emotion in facial expression following bilateral damage to human amygdala. *Nature* 1994;372:669–672.

60. LeDoux JE. Emotion circuits in the brain. *Annu Rev Neurosci* 2000;23:155–184.

61. Adolphs R, Tranel D, Damasio AR. The human amygdala in social judgment. *Nature* 1998;393:470–474.

62. Schneider F, Habel U, Kessler C, et al. Functional imaging of conditioned aversive emotional responses in antisocial personality disorder. *Neuropsychobiology* 2000;42:192–201.

63. Kluver H, Bucy PC. Preliminary analysis of the functions of the temporal lobe in monkeys. *Arch Neurol Psychiatry* 1939;42:979–1000.

64. Kling A, Lancaster J, Benitone J. Amygdalectomy in the free-ranging vervet (Cercopithecus Aethiops). *J Psychiatr Res* 1970;7:191–199.

65. Schreiner L, Kling A. Behavioral changes following rhinencephalic injury in cat. *J Neurophysiol* 1953;16:643–659.

66. Albert DJ, Walsh ML. Neural systems and the inhibitory modulation of agonistic behavior: a comparison of mammalian species. *Neurosci Biobehav Rev* 1984;8:5–24.

67. Potegal M, Hebert M, DeCoster M, et al. Brief, high-frequency stimulation of the corticomedial amygdala induces a delayed and prolonged increase of aggressiveness in male Syrian golden hamsters. *Behav Neurosci* 1996;110:401–412.

68. Wong MTH, Lumsden J, Fenton GQ, et al. Electroencephalog-

raphy, computed tomography and violence ratings of male patients in a maximum-security mental hospital. *Acta Psychiatr Scand* 1994;90:97–101.

69. Wong M, Fenwick P, Fenton G, et al. Repetitive and non-repetitive violent offending behaviour in male patients in a maximum-security mental hospital–clinical and neuroimaging findings. *Med Sci Law* 1997;37:150–160.

70. Van Elst LT, Woermann FG, Lemieux L, et al. Affective aggression in patients with temporal lobe epilepsy: a quantitative MRI study of the amygdala. *Brain* 2000;123:234–243.

71. Lee GP, Bechara A, Adolphs R, et al. Clinical and physiological effects of stereotaxic bilateral amygdalotomy for intractable aggression. *J Neuropsychiatry Clin Neurosci* 1998;10:413–420.

72. Schneider F, Habel U, Kessler C, et al. Functional imaging of conditioned aversive emotional responses in antisocial personality disorder. *Neuropsychobiology* 2000;42:192–201.

73. Rosvold H, Enger H, Mirsky AF, et al. Influence of amygdalectomy on social behavior in monkeys. *J Comp Physiol Psychol* 1954;47:173–178.

74. Saver JL, Salloway SP, Devinsky O, et al. Neuropsychiatry of aggression. In: Fogel BS, Schiffer RB, Rao SM, eds. *Neuropsychiatry.* Baltimore, MD: Lippincott Williams & Wilkins, 1997:523–548.

75. Watson RE, Edinger HM, Siegel A. An analysis of the mechanisms underlying hippocampal control of hypothalamically-elicited aggression in the cat. *Brain Res* 1983;269:327–345.

76. Mirsky AF, Siegel A. The neurobiology of violence and aggression. In: Reiss AJ, Miczek KA, Roth JA, eds. *Understanding and preventing violence: volume 2, biobehavioral influences.* Washington, DC: National Academy Press, 1994:59–172.

77. Hess WR. *Diencephalon: autonomic and extrapyramidal functions.* New York: Grune & Stratton, 1954.

78. Albert DJ, Walsh ML, Jonik RH. Aggression in humans: what is its biological foundation? *Neurosci Biobehav Rev* 1993;17:405–445.

79. Reeves AG, Plum F. Hyperphagia, rage, and dementia accompanying a ventromedial hypothalamic neoplasm. *Arch Neurol* 1969;20:616–624.

80. Yeudall LT. Neuropsychological assessment of forensic disorders. *Can Ment Health* 1977;25:7–16.

81. Davidson RJ, Fox NA. Frontal brain asymmetry predicts infants' response to maternal separation. *J Abnorm Psychol* 1989;98:127–131.

82. Bogen JE. The callosal syndromes. In: Heilman KM, Valenstein E, eds. *Clinical neuropsychology,* 3rd ed. New York: Oxford University Press, 1993:337–408.

83. Kandel E. Biology, violence, and antisocial personality. *J Forens Sci* 1992;37:912–918.

84. Piquero A, Tibbetts S. The impact of pre/perinatal disturbances and disadvantaged familial environment in predicting criminal offending. *Stud Crime Crime Prev* 1999;8:52–70.

85. Raine A, Brennan P, Mednick SA. Birth complications combined with early maternal rejection at age 1 year predispose to violent crime at age 18 years. *Arch Gen Psychiatry* 1994;51:984–988.

86. Raine A, Brennan PA, Mednick SA. Interaction between birth complications and early maternal rejection in predisposing to adult violence: specificity to serious, early onset violence. *Am J Psychiatry* 1997;154:1265–1271.

87. Kelly SJ, Day N, Streissguth AP. Effects of prenatal alcohol exposure on social behavior in humans and other species. *Neurotoxicol Teratol* 2000;22:143–149.

88. Cohen S, Erwin EJ. Characteristics of children with prenatal drug exposure being served in preschool special education programs in New York City. *Topics Early Child Special Educ* 1994;14:232–253.

89. De Cubas, Mercedes M, Field T. Children of methadone-dependent women: developmental outcomes. *Am J Orthopsychiatry* 1993;63:266–276.

90. Lewis M, Bendersky M, eds. *Mothers, babies, and cocaine: the role of toxins in development.* Hillsdale, NJ: Lawrence Erlbaum Associates, 1997.

91. Neuspiel DR. The problem of confounding in research on prenatal cocaine effects on behavior and development. In: Lewis M, Bendersky M, eds. *Mothers, babies, and cocaine: the role of toxins in development.* Hillsdale, NJ: Lawrence Erlbaum Associates, 1995:95–110.

92. Weissman MM, Warner V, Wickramaratne PJ, et al. Maternal smoking during pregnancy and psychopathology in offspring followed to adulthood. *J Am Acad Child Adolesc Psychiatry* 1999;38:892–899.

93. Fergusson DM, Woodward LJ, Horwood J. Maternal smoking during pregnancy and psychiatric adjustment in late adolescence. *Arch Gen Psychiatry* 1998;55:721–727.

94. Brennan PA, Grekin ER, Mednick SA. Maternal smoking during pregnancy and adult male criminal outcomes. *Arch Gen Psychiatry* 1999;56:215–219.

95. Rasanen P, Hakko H, Isohanni M, et al. Maternal smoking during pregnancy and risk of criminal behavior among adult male offspring in the northern Finland 1996 birth cohort. *Am J Psychiatry* 1999;156:857–862.

96. Wakschlag LS, Lahey BB, Loeber R, et al. Maternal smoking during pregnancy and the risk of conduct disorder in boys. *Arch Gen Psychiatry* 1997;54:670–676.

97. Day NL, Richardson GA, Goldschmidt L, et al. Effects of prenatal tobacco exposure on preschoolers' behavior. *J Dev Behav Pediatr* 2000;21:180–188.

98. Royalty J. Effects of prenatal ethanol exposure on juvenile play-fighting and postpubertal aggression in rats. *Psychol Rep* 1990;66:551–560.

99. Davis SF, Nielson LD, Weaver MS, et al. Shock-elicited aggression as a function of early ethanol exposure. *J Gen Psychol* 1984;110:93–98.

100. Johns JM, Noonan LR. Prenatal cocaine exposure affects social behavior in Sprague-Dawley rats. *Neurotoxicol Teratol* 1995;17:569–576.

101. Wood RD, Spear LP. Prenatal cocaine alters social competition of infant, adolescent, and adult rats. *Behav Neurosci* 1998;112:419–431.

102. Levin ED, Wilkerson A, Jones JP, et al. Prenatal nicotine effects on memory in rats: pharmacological and behavioral challenges. *Dev Brain Res* 1996;97:207–215.

103. Slotkin TA, Epps TA, Stenger ML, et al. Cholinergic receptors in heart and brainstem of rats exposed to nicotine during development: implications for hypoxia tolerance and perinatal mortality. *Brain Res* 1999;113:1–12.

104. Guy JD, Majorski LV, Wallace CJ, et al. The incidence of minor physical anomalies in adult male schizophrenics. *Schizophr Bull* 1983;9:571–582.

105. Waldrop MF, Bell RQ, McLaughlin B, et al. Newborn minor physical anomalies predict short attention span, peer aggression, and impulsivity at age 3. *Science* 1978;199:563–564.

106. Paulus DL, Martin CL. Predicting adult temperament from minor physical anomalies. *J Pers Soc Psychol* 1986;50:1235–1239.

107. Halverson CF, Victor JB. Minor physical anomalies and problem behavior in elementary schoolchildren. *Child Dev* 1976;47:281–285.

108. Arseneault L, Tremblay RE, Boulerice B, et al. Minor physical anomalies and family adversity as risk factors for violent delinquency in adolescence. *Am J Psychiatry* 2000;157:917–923.

109. Pomeroy JC, Sprafkin J, Gadow KD. Minor physical anomalies as a biological marker for behavior disorders. *J Am Acad Child Adolesc Psychiatry* 1988;27:466–473.

110. Mednick SA, Kandel E. Genetic and perinatal factors in violence. In: Mednick SA, Moffitt T, eds. *Biological contributions to crime causation.* Dordrecht, Holland: Martinus Nijhoff, 1988:121–134.

111. Julien RM. *A primer of drug action,* 7th ed. New York: WH Freeman and Co., 1996.

112. Miczek KA, Weerts E, Haney M, et al. Neurobiological mechanisms controlling aggression: preclinical developments for pharmacotherapeutic interventions. *Neurosci Rev* 1994;18:97–110.

113. Coccaro EF, Kavoussi RJ. Neurotransmitter correlates of impulsive aggression. In: Stoff DM, Cairns RB, eds. *Aggression and violence: genetic, neurobiological, and biosocial perspectives.* Mahwah, NJ: Lawrence Erlbaum Associates, 1996:67–99.

114. Linnoila VM, Virkkunen M. Aggression, suicidality, and serotonin. *J Clin Psychol* 1992;53:46–51.

115. Moore TM, Scarpa A, Raine A. A meta-analysis of serotonin metabolite 5-HIAA and antisocial behavior. *Aggress Behav* 2002;28:299–316.

116. Brown GL, Goodwin FK, Ballenger JC, et al. Aggression in humans correlates with cerebrospinal fluid amine metabolites. *Psychiatry Res* 1979;1:131–139.

117. Brown GL, Ebert MH, Goyer PF, et al. Aggression, suicide, and serotonin: relationships to CSF amine metabolites. *Am J Psychiatry* 1982;139:741–746.

118. Unis AS, Cook EH, Vincent JG, et al. Platelet serotonin measures in adolescents with conduct disorder. *Biol Psychiatry* 1997;42:553–559.

119. O'Keane V, Moloney E, O'Neill H, et al. Blunted prolactin responses to d-Fenfluramine in sociopathy: evidence for subsensitivity of central serotonergic function. *Br J Psychiatry* 1992;160:643–646.

120. Stanley B, Molcho A, Stanley M, et al. Association of aggressive behavior with altered serotonergic function in patients who are not suicidal. *Am J Psychiatry* 2000;157:609–614.

121. Linnoila M, Virkkunen M, Scheinin M, et al. Low cerebrospinal fluid 5 hydroxyindoleacetic acid concentration differentiates impulsive from nonimpulsive violent behavior. *Life Sci* 1983;33:2609–2614.

122. Hibbeln JR, Umhau JC, Linnoila M, et al. A replication study of violent and nonviolent subjects: cerebrospinal fluid metabolites of serotonin and dopamine are predicted by plasma essential fatty acids. *Biol Psychiatry* 1998;44:243–249.

123. Fava M, Vuolo RD, Wright EC, et al. Fenfluramine challenge in unipolar depression with and without anger attacks. *Psychiatry Res* 2000;94:9–18.

124. Virkkunen M, Rawlings R, Tokola R, et al. CSF biochemistries, glucose metabolism, and diurnal activity rhythms in alcoholic, violent offenders, fire setters, and healthy volunteers. *Arch Gen Psychiatry* 1994;51:20–27.

125. Moffitt TE, Brammer GL, Caspi A, et al. Whole blood serotonin relates to violence in an epidemiological study. *Biol Psychiatry* 1998;43:446–457.

126. Cherek DR, Moeller FG, Khan-Dawood F, et al. Prolactin response to buspirone was reduced in violent compared to nonviolent parolees. *Psychopharmacology* 1999;142:144–148.

127. Sarne Y, Mandel J, Goncalves MH, et al. Imipramine binding to blood platelets and aggressive behavior in offenders, schizophrenics and normal volunteers. *Biol Psychiatry* 1995;31:120–124.

128. Halperin JM, Newcorn JH, Schwartz ST, et al. Age-related changes in the association between serotonergic function and

aggression in boys with ADHD. *Biol Psychiatry* 1997;41:682–689.

129. Halperin JM, Newcorn JH, Kopstein I, et al. Serotonin, aggression, and parental psychopathology in children with attention-deficit hyperactivity disorder. *J Am Acad Child Dev* 1997;36:1391–1398.

130. Cherek DR, Lane SD. Effects of d,l-fenfluramine on aggressive and impulsive responding in adult males with a history of conduct disorder. *Psychopharmacology* 1999;146:473–481.

131. LeMarquand DG, Pihl RO, Young SN, et al. Tryptophan depletion, executive functions, and disinhibition in aggressive, adolescent males. *Neuropsychopharmacology* 1998;19:333–341.

132. Moeller FG, Dougherty DM, Swann AC, et al. Tryptophan depletion and aggressive responding in healthy males. *Psychopharmacology* 1996;126:97–103.

133. Ellenbogen MA, Young SN, Dean P, et al. Mood response to acute tryptophan depletion in healthy volunteers: sex differences and temporal stability. *Neuropsychopharmacology* 1996;15:465–474.

134. Cleare AJ, Bond AJ. The effect of tryptophan depletion and enhancement on subjective and behavioural aggression in normal male subjects. *Psychopharmacology* 1995;118:72–81.

135. Mann JJ, Malone KM, Deihl DJ, et al. Positron emission tomography imaging of serotonin activation effects on prefrontal cortex in healthy volunteers. *J Cereb Blood Flow Metab* 1996;16:418–426.

136. Kapur S, Meyer J, Wilson AA, et al. Modulation of cortical neuronal activity by a serotonergic agent: a PET study in humans. *Brain Res* 1994;646:292–294.

137. Siever LJ, Buchsbaum MS, New AS, et al. d,l-fenfluramine response in impulsive personality disorder assessed with [18F]fluorodeoxyglucose positron emission tomography. *Neuropsychopharmacology* 1999;20:413–423.

138. Soloff PH, Meltzer CC, Greer PJ, et al. A fenfluramine-activated FDG-PET study of borderline personality disorder. *Biol Psychiatry* 2000;47:540–547.

139. Miczek KA, Hussain S, Faccidomo S. Alcohol-heightened aggression in mice: attenuation by 5-HT1A receptor agonists. *Psychopharmacology* 1998;139:160–168.

140. Ramboz S, Saudou F, Amara DA, et al. 5-TH1B receptor knock out—behavioral consequences. *Behav Brain Res* 1996;73:305–312.

141. Fish EW, Faccidomo S, Miczek KA. Aggression heightened by alcohol or social instigation in mice: reduction by the 5-HT(1B) receptor agonist CP-94,253. *Psychopharmacology* 1999;146:391–399.

142. Joppa MA, Rowe RK, Meisel RL. Effects of serotonin 1A or 1B receptor agonists on social aggression in male and female Syrian hamsters. *Pharmacol Biochem Behav* 1997;58:349–353.

143. Cologer-Clifford A, Simon NG, Lu SF, et al. Serotonin agonist-induced decreases in intermale aggression are dependent on brain region and receptor subtype. *Pharmacol Biochem Behav* 1997;58:425–430.

144. Huang YY, Grailhe R, Arango V, et al. Relationship of psychopathology to the human serotonin 1B genotype and receptor binding kinetics in postmortem brain tissue. *Neuropsychopharmacology* 1999;21:238–246.

145. Ferris CF, Melloni RH, Koppel G, et al. Vasopressin/serotonin interactions in the anterior hypothalamus control aggressive behavior in golden hamsters. *J Neurosci* 1997;17:4331–4340.

146. Coccaro EF, Kavoussi RJ, Hauger RL, et al. Cerebrospinal fluid vasopressin levels: correlates with aggression and serotonin function in personality-disordered subjects. *Arch Gen Psychiatry* 1998;55:708–714.

147. Ferris CF. Adolescent stress and neural plasticity in hamsters: a vasopressin-serotonin model of inappropriate aggressive behaviour. *Exp Physiol* 2000;85:85–90.

148. Evans J, Platts H, Lightman S, et al. Impulsiveness and the prolactin response to d-fenfluramine. *Psychopharmacology* 2000;149:147–152.

149. Higley JD, King ST, Hasert MF, et al. Stability of interindividual differences in serotonin function and its relationship to severe aggression and competent social behavior in rhesus macaque females. *Neuropsychopharmacology* 1996;14:67–76.

150. Higley JD, Mehlman PT, Higley SB, et al. Excessive mortality in young free-ranging male nonhuman primates with low cerebrospinal fluid 5-hydroxyindoleacetic acid concentrations. *Arch Gen Psychiatry* 1996;53:537–543.

151. Faustman WO, Ringo DL, Faull KF. An association between low levels of 5-HIAA and HVA in cerebrospinal fluid and early mortality in a diagnostically mixed psychiatric sample. *Br J Psychiatry* 1993;163:519–521.

152. Kelland MD, Chiodo LA. Serotonergic modulation of midbrain dopamine systems. In: Ashby CR Jr, ed. *The modulation of dopaminergic neurotransmission by other neurotransmitters*. Boca Raton, FL: CRC Press, 1996:87–122.

153. Miczek KA, Haney M, Tidey J, et al. Neurochemistry and pharmacotherapeutic management of aggression and violence. In: Reiss AJ, Miczek KA, Roth JA, eds: *Understanding and preventing violence: volume 2, biobehavioral influences*. Washington, DC: National Academy Press, 1994:245–514.

154. Bernard BK, Finkelstein ER, Everett GM. Alterations in mouse aggressive behavior and brain monamine dynamics as a function of age. *Physiol Behav* 1975;15:731–736.

155. Modigh K. Effects of isolation and fighting in mice on the rate of synthesis of noradrenaline, dopamine, and 5-hydroxytryptamine in the brain. *Psychopharmacologia* 1973;33:1–17.

156. Miczek KA, Mirsky AF, Carey G, et al. An overview of biological influences on violent behavior. In: Reiss AJ, Miczek KA, Roth JA, eds. *Understanding and preventing violence: volume 2, biobehavioral influences*. Washington, DC: National Academy Press, 1994:1–20.

157. Kudryavtseva NN, Lipina TV, Koryakina LA. Effects of haloperidol on communicative and aggressive behavior in male mice with different experiences of aggression. *Pharmacol Biochem Behav* 1999;63:229–236.

158. Cabib S, D'Amato FR, Puglisi-Allegra S, et al. Behavioral and mesocorticolimbic dopamine responses to non aggressive social interactions depend on previous social experiences and on the opponent's sex. *Behav Brain Res* 2000;112:13–22.

159. Gendreau PL, Petitto JM, Petrova A, et al. D$_3$ and D$_2$ dopamine receptor agonists differentially modulate isolation-induced social-emotional reactivity in mice. *Behav Brain Res* 2000;114:107–117.

160. Lopez OL, Kaufer D, Reiter CT, et al. Relationship between CSF neurotransmitter metabolites and aggressive behavior in Alzheimer's disease. *Eur J Neurol* 1996;3:153–155.

161. Sweet RZ, Pollack BG, Mulsant BH. Association of plasma homovanillate with behavioral symptoms in patients diagnosed with dementia: a preliminary report. *Biol Psychiatry* 1997;42:1016–1023.

162. Victoroff J, Zarow C, Mack WJ, et al. Physical aggression is associated with preservation of substantia nigra pars compacta in Alzheimer disease. *Arch Neurol* 1996;53:428–434.

163. Sweet RA, Nimgaonkar VL, Vishwajit L, et al. Dopamine receptor genetic variation, psychosis, and aggression in Alzheimer disease. *Arch Neurol* 1998;55:1335–1340.

164. Kuikka JT, Tiihonen J, Bergstrom KA, et al. Abnormal structure of human striatal dopamine re-uptake sites in habitually

violent alcoholic offenders: a fractal analysis. *Neurosci Lett* 1998;253:195–1973

165. Tiihonen J, Kuikka J, Bergstrom K, et al. Altered striatal dopamine reuptake site densities in habitually violent and non-violent alcoholics. *Nat Med* 1995;1:645–657.

166. Higley JD, Mehlman PT, Taub DM, et al. Cerebrospinal fluid monamine and adrenal correlates of aggression in free-ranging rhesus monkeys. *Arch Gen Psychiatry* 1992;49:436–441.

167. Gerra G, Avanzini P, Zaimovic A, et al. Neurotransmitter and endocrine modulation of aggressive behavior and its components in normal humans. *Behav Brain Res* 1996;81:19–24.

168. Gerra G, Zaimovic A, Avanzini P, et al. Neurotransmitter-neuroendocrine responses to experimentally induced aggression in humans: influence of personality variable. *Psychiatry Res* 1997;66:33–43.

169. Tizabi Y, Thoa NB, Maengywn-Davies GD, et al. Behavioral correlation of catecholamine concentration and turnover in discrete brain areas of the three strains of mice. *Brain Res* 1979;166:199–205.

170. Eichelman BS, Barchas J. Facilitated shock-induced aggression following anti-depressive medication in the rat. *Pharmacol Biochem Behav* 1975;3:601–604.

171. Matsumoto K, Ojima K, Watanabe H. Noradrenergic denervation attenuates desipramine enhancement of aggressive behavior on isolated mice. *Pharmacol Biochem Behav* 1995;50:481–484.

172. Yudofsky S, Williams D, Gorma J. Propanolol in the treatment of rage and violent behavior in patients with chronic brain syndromes. *Am J Psychiatry* 1981;138:218–220.

173. Yudofsky SC, Silver JM, Scheider SE. Propanolol in the treatment of rage and violent behavior associated with Korsakoff's psychosis. *Am J Psychiatry* 1987;141:114–115.

174. Greendyke RM, Schuster DB, Wooton JA. Propanolol in the treatment of assaultive patients with organic brain disease. *J Clin Psychopharmacol* 1984;4:282–285.

175. Scerbo A. Violence and antisocial behavior: a meta-analysis of neurotransmitters in humans (unpublished manuscript).

176. Rogeness GA, Javors MA, Mass JW, et al. Catecholamines and diagnoses in children. *J Am Acad Child Adolesc Psychiatry* 1990;29:234–241.

177. Bell R, Warburton DM, Brown K. Drugs as research tools in psychology: cholinergic drugs and aggression. *Neuropsychobiology* 1985;14:181–192.

178. Herbut M, Rolinsky Z. The cholinergic influences on aggression in mice. *Pol J Pharmacol Pharm* 1985;37:1–10.

179. Cherek DR. Effects of smoking different doses of nicotine on human aggressive behavior. *Psychopharmacology* 1981;75:339–345.

180. Warburton DM. Nicotine and the smoker. *Rev Environ Health* 1985;5:343–390.

181. Simler S, Puglisi-Allegra S, Mandel P. Gamma-aminobutyric acid in brain areas of isolated aggressive or non-aggressive in-bred strains of mice. *Pharmacol Biochem Behav* 1982;16:57–61.

182. Mandel P, Mack G, Kempf E. Molecular basis of some models of aggressive behavior. In: Sandler M, ed. *Psychopharmacology of aggression.* New York: Raven Press, 1979:95–110.

183. Lipman RS, Covi L, Rickels K, et al. Imipramine and chlordiazepoxide in depressive and anxiety disorders. *Arch Gen Psychiatry* 1986;43:68–77.

184. Stork O, Ji FY, Kaneko K, et al. Postnatal development of a GABA deficit and disturbance of neural functions in mice lacking GAD65. *Brain Res* 2000;865:45–58.

185. Soderpalm B. Svensson AI. Naloxone reverses disinhibitory/aggressive behavior in 5,7-DHT-lesioned rats; involvement

186. Haller J, Abraham I, Zelena D, et al. Aggressive experience affects the sensitivity of neurons towards pharmacological treatment in the hypothalamic attack area. *Behav Pharmacol* 1998;9:469–475.

187. Albert DJ, Walsh ML, Jonik RH. Aggression in humans: what is its biological foundation? *Neurosci Biobehav Rev* 1993;17:405–425.

188. Rubinow DR, Schmidt PJ. Androgens, brain, and behavior. *Am J Psychiatry* 1996;153:974–984.

189. Kreuz LE, Rose RM. Assessment of aggressive behavior and plasma testosterone in a young criminal population. *Psychosom Med* 1972;36:469–475.

190. Bahrke MS, Yesalis CE, Wright JE. Psychological and behavioral effects of endogenous testosterone and anabolic-androgenic steroids among males: a review. *Sports Med* 1990;10:303–337.

191. Scerbo AS, Kolko DJ. Salivary testosterone and cortisol in disruptive children: relationship to aggressive, hyperactive, and internalizing behaviors. *J Am Acad Child Adolesc Psychiatry* 1994;33:1174–1184.

192. Dabbs JM, Ruback RB, Frady RL, et al. Saliva testosterone and criminal violence among women. *Pers Individ Differ* 1988;9:269–275.

193. Van Goozen SHM, Van Den Ban E, Matthys W, et al. Increased adrenal androgen functioning in children with oppositional defiant disorder: a comparison with psychiatric and normal controls. *J Am Acad Child Adolesc Psychiatry* 2000;39:1446–1451.

194. Eberhart JA, Keverne EB, Meller RE. Social influences on plasma testosterone levels in male talopoin monkeys. *Horm Behav* 1980;14:247–266.

195. Rose RM, Bernstein IS, Gordon TP. Consequences of social conflict on plasma testosterone levels in rhesus monkeys. *Psychosom Med* 1975;37:50–61.

196. Schaal B, Tremblay RE, Soussignan R, et al. Male testosterone linked to high social dominance but low physical aggression in early adolescence. *J Am Acad Child Adolesc Psychiatry* 1996;35:1322–1330.

197. Tremblay RE, Schaal B, Boulerice B, et al. Testosterone, physical aggression, dominance, and physical development in early adolescence. *Int J Behav Dev* 1998;22:753–777.

198. Tennes K, Kreye M. Childrens' adrenocortical response to classroom activities in elementary school. *Psychosom Med* 1985;47:451–460.

199. Susman EJ, Petersen AC. Hormones and behavior in adolescence. In: McAnarney RE, Kreipe DP, Comerci GD, eds. *Textbook of adolescent medicine.* New York: Saunders Publication, 1992;125–130.

200. McBurnett K, Lahey BB, Rathouz PJ, et al. Low salivary cortisol and persistent aggression in boys referred for disruptive behavior. *Arch Gen Psychiatry* 2000;57:38–43.

201. Van Goozen SHM, Matthys W, Cohenkettenis PT, et al. Salivary cortisol and cardiovascular activity during stress in oppositional-defiant disorder boys and normal controls. *Biol Psychiatry* 1998;43:531–539.

202. Kagan J, Reznick SJ, Snidman N, et al. Temperamental variation in response to the unfamiliar. In: Krasnegor NA, Blass EM, eds. *Perinatal development: a psychobiological perspective.* New York: Brunner/Mazel, 1987:421–440.

203. Gerra G, Giucastro G, Folli F, et al. Neurotransmitter-hormonal responses to psychological stress in peripubertal subjects: relationship to aggressive behavior. *Life Sci* 1998;62:617–625.

204. Susman EJ, Dorn LD, Inoff-Germain G, et al. Cortisol reactivity, distress behavior, behavior problems, and emotionality in young

adolescents: a longitudinal perspective. *J Res Adolesc* 1997;7:81–105.

205. Virkunnen M. Urinary free cortisol secretion in habitually violent offenders. *Acta Psychiatr Scand* 1985;72:40–44.

206. McBurnett K, Lahey BB, Frick PJ, et al. Anxiety, inhibition, and conduct disorder in children: II. Relation to salivary cortisol. *J Am Acad Child Adolesc Psychiatry* 1991;30:192–196.

207. King JA, Barkley RA, Barrett S. Attention-deficit hyperactivity disorder and the stress response. *Biol Psychiatry* 1998;44:72–74.

208. Critchley HD, Corfield DR, Chandler MP, et al. Cerebral correlates of autonomic cardiovascular arousal: a functional neuroimaging investigation of humans. *J Physiol* 2000;523:259–270.

209. Critchley HD, Elliott R, Mathias CJ, et al. Neural activity relating to generation and representation of galvanic skin conductance responses: a functional magnetic resonance imaging study. *J Neurosci* 2000;20:3033–3040.

210. Bach-y-Rita G, Lion JR, Climent CE, et al. Episodic dyscontrol: a study of 139 violent patients. *Am J Psychiatry* 1971;127:1473–1478.

211. Hill D, Pond DA. Reflections on 100 capital cases submitted for electroencephalography. *J Ment Sci* 1952;98:23–43.

212. Mark VH, Ervin FR. *Violence and the brain.* New York: Harper Row, 1970.

213. Williams D. Neural factors related to habitual aggression: consideration of those differences between those habitually aggressive and others who have committed crimes of violence. *Brain* 1969;92:503–520.

214. Fishbein DH, Herning RI, Pickworth WB, et al. EEG and brainstem auditory evoked response potentials in adult male drug abusers with self-reported histories of aggressive behavior. *Biol Psychiatry* 1989;26:595–611.

215. Convit A, Czobor P, Volavka J. Lateralized abnormality in the EEG of persistently violent psychiatric inpatients. *Biol Psychiatry* 1991;30:363–370.

216. Drake ME, Hietter SA, Pakalnis A. EEG and evoked potentials in episodic-dyscontrol syndrome. *Neuropsychobiology* 1992;26:125–128.

217. Volavka J. Electroencephalogram among criminals. In: Mednick SA, Moffitt TE, Stack S, eds. *The causes of crime: new biological approaches.* Cambridge: Cambridge University Press, 1987:137–145.

218. Milstein V. EEG topography in patients with aggressive violent behavior. In: Moffitt TE, Mednick SA, eds. *Biological contributions to crime causation.* Dordrecht, the Netherlands: Kluwer, 1988.

219. Baving L, Laucht M, Schmidt MH. Oppositional children differ from healthy children in frontal brain activation. *J Abnorm Psychol* 2000;28:267–275.

220. Fox NA, Schmidt LA, Calkins SD, et al. The role of frontal activation in the regulation and dysregulation of social behavior during the preschool years. *Dev Psychopathol* 1996;8:89–102.

221. Raine A, Venables PH, Mednick SA. Low resting heart rate at age 3 years predisposes to aggression at age 11 years: findings from the Mauritius Joint Child Health Project. *J Am Acad Child Adolesc Psychiatry* 1997;36:1457–1464.

222. Kindlon DJ, Tremblay RE, Mezzacappa E, et al. Longitudinal patterns of heart rate and fighting behavior in 9- through 12-year-old boys. *J Am Acad Child Adolesc Psychiatry* 1995;34:371–377.

223. Moffitt TE, Caspi A. Childhood predictors differentiate life-course persistent and adolescent limited pathways among males and females. *Dev Psychopathol* 2001;13:355–375.

224. Wadsworth MEJ. Delinquency, pulse rate and early emotional deprivation. *Br J Criminol* 1976;16:245–256.

225. West DJ, Farrington D. *The delinquent way of life.* London: Heinemann, 1977.

226. Farrington DP. Implications of biological findings for criminological research. In: Mednick SA, Moffitt TE, Stack SA, eds. *The causes of crime: new biological approaches.* New York: Cambridge University Press, 1987:42–64.

227. Maliphant R, Hume F, Furnham A. Autonomic nervous system (ANS) activity, personality characteristics and disruptive behavior in girls. *J Child Psychol Psychiatry* 1990;31:619–628.

228. Farrington DP. The relationship between low resting heart rate and violence. In: Raine A, Brennan PA, Farrington DP, et al., eds. *Biosocial bases of violence.* New York: Plenum, 1997:89–106.

229. Raine A, Venables PH, Williams M. Relationships between CNS and ANS measures of arousal at age 15 and criminality at age 24. *Arch Gen Psychiatry* 1990;47:1003–1007.

230. Raine A. Psychophysiology and antisocial behavior. In: Stoff D, Breiling J, Maser J, eds. *Handbook of antisocial behavior.* New York: Wiley, 1997:289–304

231. Eysenck HJ. *Crime and personality.* London: Methuen, 1964.

232. Quay HC. Psychopathic personality as pathological stimulation-seeking. Special issue: toward a developmental perspective on conduct disorder. *Am J Psychiatry* 1965;122:180–183.

233. Raine A, Reynolds C, Venables PH, et al. Fearlessness, stimulation-seeking, and large body size at age 3 years as early predispositions to childhood aggression at age 11 years. *Arch Gen Psychiatry* 1998;55:745–751.

234. Damasio AR. *Descartes' error: emotion, reason, and the human brain.* New York: Grosset/Putnam, 1994.

235. Damasio AR, Grabowski TJ, Bechara A, et al. Subcortical and cortical brain activity during the feeling of self-generated emotions. *Nat Neurosci* 2000;3:1049–1056.

236. Bechara A, Tranel D, Damasio H, et al. Failure to respond autonomically to anticipated future outcomes following damage to prefrontal cortex. *Cereb Cortex* 1996;6:215–225.

237. Bechara A, Damasio H, Damasio AR. Emotion, decision making, and the orbitofrontal cortex. *Cereb Cortex* 2000;10:295–307.

238. Bechara A, Damasio H, Damasio AR, et al. Different contributions of the human amygdala and ventromedial prefrontal cortex to decision-making. *J Neurosci* 1999;19:5473–5481.

239. Ishikawa SS, Raine A, Lencz T, et al. Autonomic stress reactivity and executive functions in successful and unsuccessful criminal psychopaths from the community. *J Abnorm Psychol* 2001;110:423–432.

27

THE NEUROPSYCHIATRY OF SUICIDE

YEATES CONWELL
AMER BURHAN

OVERVIEW

Although the word *suicide* did not come into regular use until the late seventeenth century, references to intentional self-destructive behavior are as old as recorded time (1). Sacred oriental writings suggest an ambivalent acceptance of suicide by Brahminism and Buddhism consistent with their philosophies of the acquisition of knowledge through the denial of life's passions. Muhammadanism vigorously condemned suicide, as did Judaism. During Greek and Roman periods, suicide gained gradual acceptance, paralleled by the growing autonomy of the individual expressed in the philosophies of Seneca, the Stoics, Cynics, and Epicureans. Suicide was initially embraced by early Christian martyrs. The church's subsequent antisuicide stance was first articulated by St. Augustine and later by St. Thomas Aquinas in the thirteenth century. Subsequent modification of those attitudes has continued through recent centuries, shaped by a broad range of social, religious, and cultural influences. Indeed, the phenomenon of self-destruction can be understood for any individual, as for any society, only in terms of the underlying influence of these and other factors.

Great advances in the neurosciences in the past two decades have further shaped our understanding of suicide and self-injury. This chapter reviews the evidence that links suicidal and self-injurious behaviors to central nervous system (CNS) function. It first provides an overview of the epidemiology and demography of suicidal behaviors as well as social and psychological risk factors. After a review of animal models of self-injury, we consider the neurobiologic studies of human populations that implicate specific neurochemical or neuroanatomic systems in the etiology of suicide. We review the known associations between suicide and specific psychiatric, medical, and neuropsychiatric syndromes and their implications for the diagnosis and treatment of patients at risk of self-injury.

DEFINITIONS

The terms used in the literature to describe suicidal behavior and the populations studied are often ill-defined and loosely applied, making the interpretation and generalization of findings difficult. Although the need for a more precise typology of aggressive and self-destructive behaviors in humans has been well recognized, consensus has been difficult to achieve (2). A common system of classification is to distinguish people who commit suicide, those who make suicide attempts, and those with suicidal ideation. The relationships between these three groups are, however, uncertain (3). Striking demographic differences between attempters and completers suggest that they are distinct but overlapping populations. Whereas the risk for completed suicide is highest for elderly males, the risk for attempted suicide is highest in young females. Suicidal ideation is common to both groups as well as the general population.

Most authorities advocate distinguishing direct from indirect self-destructive behavior based on conscious intent to do self-harm (4,5). Kreitman (6) coined the term *parasuicide,* referring to all forms of intentional self-destructive behavior, which is in turn distinguished from self-mutilation (self-harm without stated or apparent intent to die). Some authors have even suggested that deliberate self-harm, in the absence of suicidal intent, is sufficiently distinctive to warrant classification as a separate diagnostic syndrome (7). The most recent effort to standardize the nomenclature of suicide is gaining more widespread acceptance (5). It categorizes suicide-related behaviors into groups based on the intent to die and the extent of injury, as depicted in Table 27.1.

Recognizing the limitations of categorical nosologies, other authors advocate classifying self-destructive behavior on continua of intent (8) and lethality, either of medical risk (lethality of implementation) or extent of the will to die (lethality of intent) (3,9). Many variables of interest in neurobiologic studies are continuous, such as the level of

TABLE 27.1. PROPOSED NOMENCLATURE FOR SUICIDE-RELATED BEHAVIORS

Suicide-related Behavior	Definition
1. Instrumental behavior a. With injuries b. Without injuries c. With fatal outcome	Potentially self-injurious behavior for which there is evidence (implicit or explicit) that (a) the person had no intent to die and (b) and wished to use the appearance of intent to die to achieve some other end

2. Suicidal acts
a. Suicide attempt
   With injuries
   Without injuries
	Self-injurious behavior with nonfatal outcome for which there is evidence that the person intended at some (nonzero) level to die

b. Completed suicide	Fatal self-injurious behavior with evidence of some (nonzero) intent to die

From O'Carroll PW, Berman AL. Beyond the tower of Babel: a nomenclature for suicidology, *Suicide Life Threat Behav* 1996;26:237–252, with permission.

monoamine metabolites in the cerebrospinal fluid (CSF), whereas the populations studied are classified in categorical terms.

This chapter reviews only those behaviors that have conscious self-destructive intent, for which we use the terms *attempted* suicide and *completed* suicide. There is stronger emphasis on completed suicides for two reasons. First, it is a more clearly defined behavior for which constructs such as lethality and impulsivity can be taken as associated variables rather than components of the definition. Second, much of the compelling evidence of a neurobiologic basis for suicide is derived from postmortem studies of suicide victims.

EPIDEMIOLOGY

Attempted Suicide

The largest study to date of the prevalence of psychiatric illness in the general community is the Epidemiological Catchment Area Study, in which more than 18,000 individuals 18 years of age and older from five representative communities across the United States were administered structured diagnostic interviews. Of the entire sample, 2.9% reported that they had attempted suicide at some time in their lives. By far, those at highest risk of having made an attempt were individuals with a lifetime diagnosis of psychiatric disorder. Additional risk factors included female gender, being separated or divorced, and a lower socioeconomic status. Whites had significantly higher rates of suicide attempts than blacks, and with increasing age, a smaller proportion had a history of suicidal behavior (10). Other studies have estimated a prevalence of from 120 to 730 attempts per 100,000 population per year compared with rates of approximately 11

per 100,000 per year for completed suicide in the United States.

Estimates of the ratio of attempts to completed suicides range from 8:1 to as high as 20:1 for the general population and vary markedly in subgroups defined by age and gender. Parkin and Stengel (11) found, for example, only four attempts for each completed suicide victim age 60 years and older, whereas young white women have been estimated to make 200 attempts per completed suicide (12). In contrast to completed suicides, suicide attempts are more common in women than men at all ages (13). Of those who attempt suicide, approximately two-thirds make no further attempt, 10% to 20% make an additional attempt within 1 year, 1% to 2% will complete suicide within the year of an attempt, and approximately 10% to 15% of suicide attempters will eventually take their own lives (12).

Completed Suicide

Statistics regarding completed suicides in the United States are compiled from death certificates submitted by medical examiners and coroners from each state and the District of Columbia. There is general agreement that suicide rates are underreported owing to variations in the training and reporting practices of the responsible officials. However, studies differ on the extent of this phenomenon and its implications (14,15). By comparing suicide rates with rates of undetermined and suicide deaths combined, Speechley and Stavraky (16) estimated that the average potential underreporting for suicide in Canada was 17.5% for females and 12% for males. Adopting a similar strategy, Barraclough (17) compared rank order of suicide rates in 22 countries in 1968 with the rank order of combined rates for suicide plus undetermined deaths. Finding few differences, he concluded that, although underreporting likely occurred, the differences between nations in suicide rates could not simply reflect differences in methods of ascertainment. These and other studies (18,19) suggest that underreporting of completed suicide does not invalidate the observations made by epidemiologists in comparing rates between and within countries over time.

Reflecting period effects, suicide rates in the United States fluctuated widely in the first half of the twentieth century. Period effects are phenomena occurring at a particular time that influence suicide rates in a population. Reproducible period effects include elevations in suicide rates in the years during and immediately after a significant economic downturn, exemplified by the Great Depression, and decreasing rates during times of war (World Wars I and II). In the years after War II, fluctuation in suicide rates diminished considerably, a function of more uniform reporting practices and a larger database, stabilizing at a rate of approximately 11 per 100,000 per year.

In 1998, 30,575 people died of suicide in the United States, a rate of 11.3 per 100,000. For the general population, suicide was the eighth leading cause of death. It was the third

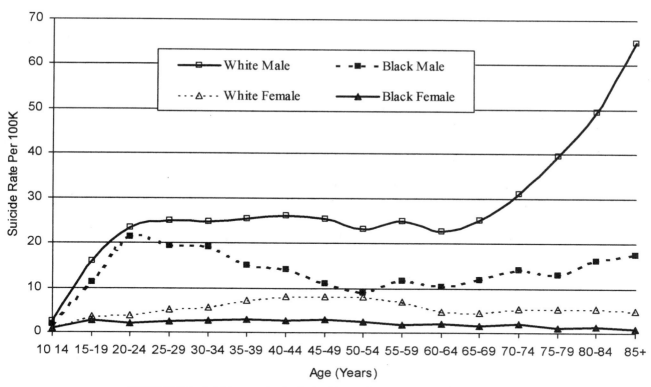

FIGURE 27.1. Suicide rates in the United States by age, race, and gender, 1997.

leading cause of death for young adults, and the fourteenth leading cause of death in the elderly (20). As depicted in Figure 27.1, suicide rates differ dramatically by age, gender, and race. In the general population, rates rise through young adulthood to an initial peak of approximately 15 per 100,000 in the 35- to 44-year age group, plateau through midlife, and rise again to a high of more 20 per 100,000 in 75 to 84 year olds. Thus, in contrast to the popular perception that young adults are the group at highest risk, it is the elderly who kill themselves at higher rates than any other age group.

In 1998, the suicide rate for white Americans was more two times that for blacks (12.4 vs. 5.7 per 100,000). This differential is present at all ages across the life course, although the relative proportion changes. In young adulthood, rates for blacks approximate those for whites, whereas in late life, whites are at approximately six times greater risk (20).

At all ages, men are at greater risk than women within each racial subgroup. Although women make approximately three times as many suicide attempts as men, the overall ratio of female-to-male suicides is 1:4. This difference has been ascribed in large part to men's characteristic choice of more violent, and hence potentially lethal, means. For example, approximately one-half of the suicides in the United States are committed by men using firearms. Although women also use firearms in substantial numbers, poisonings constitute a far greater proportion of suicides than for men (Table 27.2). Firearms are the method of choice for an even larger proportion of suicides in older adults, reflecting the greater lethality of suicidal behavior in later life (21).

In addition to period, age, race, and gender effects on suicide rates, epidemiologic studies have demonstrated potent cohort effects as well. Solomon and Hellon (22) were the

TABLE 27.2. PERCENTAGE OF SUICIDES, BY GENDER AND AGE CATEGORY, USING SPECIFIC METHODS (UNITED STATES, 1998)

Method	Men	Women	15–24 yr	≥65 yr
Firearms	61.9	38.4	60.7	70.9
Hanging, suffocation	19.2	16.8	25.4	11.2
Gas poisons	5.4	6.7	3.1	4.0
Solid and liquid poisons	7.0	27.1	4.1	6.8
All other	6.9	11.0	6.7	7.1

Reprinted from Murphy SL. Deaths: final data for 1998. Hyattsville, MD: National Center for Health Statistics; 2000; DHHS 2000-1120, with permission.

first to apply cohort analysis to death records, demonstrating that suicide rates in Canada were birth cohort specific. That is, within any given birth cohort, suicide rates rise with age. Across the course of the twentieth century, however, patterns of suicide rates for young adults and the elderly differed greatly. Whereas rates for the population aged 65 years and older have dropped substantially since a peak in 1930, rates rose more than 300% between 1950 and 1980 for white males of ages 15 to 19 years and 200% for males of ages 20 to 24 years. These changes have been ascribed to a variety of factors including, in the case of the elderly population, increased economic security resulting from the implementation of Social Security and Medicare legislation (23), and in the young adult population, the more widespread use of alcohol and drugs (24). Beyond these factors, however, cohort effects are clearly operative, as demonstrated by Solomon and Hellon (22) in Alberta, Canada, and Murphy and Wetzel (25) and Manton et al. (26) in the United States. However, studies in Great Britain (27) and Australia (28) have not supported the ability of a cohort's suicide rates in young adulthood to predict their risk in later years.

Although the overall suicide rate has remained stable, this complex interaction of period and cohort effects with other demographic factors continued in the past decade to cause significant fluctuations in suicide rates in specific demographic subgroups. Meehan et al. (29) noted significant increases in suicide rates between 1980 and 1986 among elderly white (23%), black (42%), and divorced (38%) men. Other more recent reports have noted substantial increases among elderly women (30) and blacks 18 to 24 years of age (31).

Regional and national differences in suicide rates are worth noting as well because, like age and gender effects, they may shed light on the relative contribution of social, cultural, and perhaps biologic underpinnings of suicide. As depicted in Figure 27.2, rates within the United States are lowest in the northeastern region. They increase as one moves south and west, peaking in the mountain region and decreasing again in the Pacific states. Satisfactory explanations for these sometimes substantial differences have not been established.

Circumannual variation has been identified in which rates characteristically peak in spring and early summer months and decline to a nadir in early winter. This pattern is present

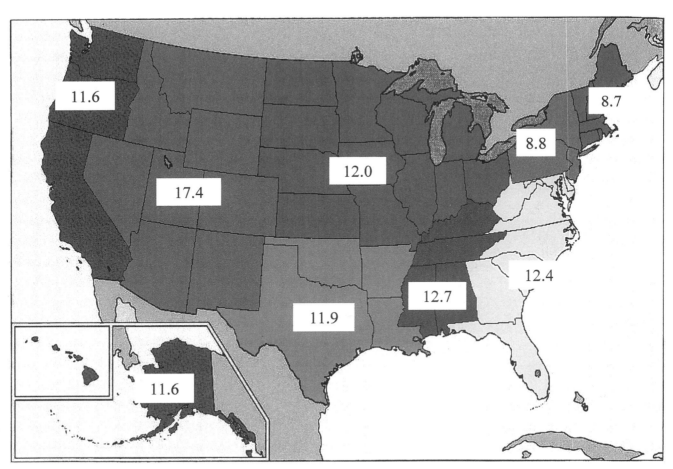

FIGURE 27.2. Suicide rates per 100,000 by region of the United States, 1997. (Source: National Center for Health Statistics.)

TABLE 27.3. SUICIDE RATES (PER 100,000/YEAR)

Country	Year	Males	Females
Australia	1995	19.0	5.1
Austria	1998	30.0	9.2
Belarus	1998	63.4	10.1
Brazil	1992	5.6	1.6
Canada	1997	19.6	5.1
China (mainland)	1994	14.3	17.9
Czech Republic	1998	25.3	6.5
Denmark	1996	24.3	9.8
Egypt	1987	0.1	0.0
Estonia	1998	59.4	10.5
Finland	1996	38.7	10.7
France	1997	28.4	10.1
Germany	1998	21.5	7.3
Hungary	1998	51.1	14.7
India	1995	11.4	8.0
Israel	1996	8.2	2.6
Latvia	1998	59.8	12.2
Lithuania	1998	73.7	13.7
Mexico	1995	5.4	1.0
Netherlands	1997	13.5	6.7
Peru	1989	0.6	0.4
Philippines	1993	2.5	1.7
Puerto Rico	1992	16.0	1.9
Russian Federation	1997	66.4	12.3
Spain	1996	12.8	4.3
Sweden	1996	20.0	8.5
United Kingdom	1997	11.0	3.2
United States of America	1997	18.7	4.4
Zimbabwe	1990	10.6	5.2

Reprinted from World Health Organization. *Suicide rates per 100,000.* Geneva: World Health Organization, 2000, with permission.

in a wide range of industrialized and agrarian nations (32). In addition, McCleary et al. (33) demonstrated peaks in the number of suicides occurring in the United States on the first day of each week and in the first week of each month. More detailed analysis by age and gender, however, demonstrated these cycles occurring only in select subpopulations, supporting a combination of psychosocial and biologic etiologic influences on suicide.

Comparison of rates across nations further underscores the complex determinants of suicide. Table 27.3 demonstrates the wide variation in suicide rates per 100,000 population in nations across the world. Stillion et al. (34) reviewed suicide rates by age and gender for 12 countries that reported suicide statistics between 1983 and 1985, demonstrating remarkable consistency in age- and gender-specific patterns. Eleven of the 12 reporting countries showed suicide rates rising progressively with age in males. Only Poland showed a peak rate for males in midlife. Rates for women, however, were more variable. In the United States, Australia, Denmark, Poland, and Sweden, rates for women peaked for the period from 45 to 54 years of age. In Canada and the United Kingdom, rates for women peaked somewhat later, from 55 to 64 years of age, whereas in the remaining five countries (Austria, France, Italy, Japan, and the Netherlands),

the suicide rate for women, like men, peaked at age 65 years or older.

The large differences in suicide rates between countries and across ethnic subgroups within a country reflect the complex interplay of social, cultural, and religious influences. Particularly relevant to this chapter, however, is that fairly consistent associations between suicide rates, age, and gender expressed across cultures may suggest further support for the neurobiologic hypotheses developed further below. Indeed, any adequate neurobiologic model of suicide must account for gender and age effects.

RISK FACTORS

The Multideterminant Nature of Suicide

Havens (35) characterized suicide as "the final common pathway of diverse circumstances, of an interdependent network rather than an isolated cause, a know of circumstances tightening around a single time and place." Often, the popular understanding of suicide is oversimplified, ascribed to a single factor such as depression or physical illness. The reality is far more complex; there is no single cause for any suicide, and no two suicides have the same cause. As no single factor is universally causal, no single intervention will prevent all suicide deaths. The multidetermination of suicide presents great challenges but also diverse opportunities for prevention (36).

Demographic Risk Factors

In addition to the epidemiologically defined suicide risk factors of age, gender, and race, a number of other demographic influences should be noted. A married conjugal status has been consistently associated with lower suicide risk, with those who are single, separated, divorced, or widowed at higher risk across the life course. Individuals widowed in young adulthood may be especially vulnerable (37), at greatest risk in the first year after the loss of a spouse, with gradually decreasing risk thereafter (38,39).

As it has throughout history, religion continues to play a powerful role in determining suicide risk. Observing that rates are generally higher among Protestants than Catholics and Jews, Durkheim (40) suggested that this pattern reflects the possible influence of the doctrinal prohibitions of each religion against suicide. Indeed, those countries with a greater proportion of Catholics in the population tend to have lower suicide rates. In the United States, Templer and Veleber (41) found an inverse relationship between the proportion of the population who were Catholic and the suicide rate in each state. Others suggest, however, that it may be the degree of religious involvement rather than the religion and its doctrines that is the vital factor (42).

The association between suicide and socioeconomic status is complex. Some studies have found a link between unemployment and suicide in the general population, whereas others have found that this association held only for men and only on a national rather than regional level (43). Several studies have found that people with blue collar occupations were more likely than white collar employees to commit suicide (44,45), whereas others have failed to find any such association (46).

Stressful Life Events

In general, studies of suicide attempters have found that an increased number of stressful life events occur in the months preceding the self-destructive act (47). High rates of stressful events have been noted in the lives of completed suicides as well (48–50). More specifically, the nature of the stressful life events immediately preceding suicide differs as a function of both age and diagnosis. Job, financial, and interpersonal difficulties, often the result of psychiatric illness and substance abuse, are the events most commonly associated with suicide in young adulthood and middle age. The elderly most frequently commit suicide in the context of physical ill health and loss. Bereavement (38), retirement (51), and threatened loss of autonomy, such as functional disability or impending nursing home placement (52), are losses that commonly precede late-life suicide.

Physical Illness

Physical illness is a stressor far more often associated with suicide in the elderly than in younger populations. Sainsbury (18), for example, estimated that physical illness contributed to suicide in 10% of younger cases, 25% of middle age cases, and 35% of elderly cases. Others have made similar observations (53–56). However, few investigations have included an adequate control group with which to compare the prevalence of specific physical conditions in completed suicides. The concerns about physical ill health expressed by people in the days and weeks before they take their own lives are for some a reflection of distorted or delusional ideation rather than objectively evident organic disease (57). Murphy, for example, observed in 1,300 consecutive forensic autopsies a pattern in which suicide victims often believed that they had cancer but in fact did not (58).

Whitlock (59) found in a review of the literature that reports of the prevalence of physical illness in completed suicide varied from 25% to 70% of cases and that it appeared to be an important contributory cause of the patient's death in 10.9% to 51%. Malignant neoplasms and diseases of the CNS were overrepresented in suicide deaths. In his study of 35 completed suicides, ages 65 years and older, Barraclough (60) found physical illness in 56% of cases, a considerably higher rate than in a matched control group of accident victims. In a review of 235 prospective studies with at least two years of follow-up and no more than 10% attrition, Harris and Barraclough (61) calculated standardized mortality ratios for suicide in more than 60 medical disorders and treatments. They concluded that human immunodeficiency

virus/acquired immunodeficiency syndrome, head and neck cancers, Huntington disease (HD), multiple sclerosis (MS), peptic ulcer, renal disease, spinal cord injury, and systemic lupus erythematosus conferred increased suicide risk. The data on Parkinson disease (PD) and several other disorders were suggestive but ultimately inconclusive.

Given the limitations of the data reviewed by Harris and Barraclough, the mortality ratios could not be adjusted for the influence of other factors, such as comorbid affective disorders or personality traits that may have interacted with the physical illness in determining the suicide outcome. Yet it is clear that the associations of suicide, physical illness, functional impairment, pain, and affective disorder are complex (62,63). As demonstrated by Chochinov et al. (64) and by Brown et al. (65), suicidal ideation among seriously ill people is rare in the absence of clinically significant mood disturbance. It remains unclear, therefore, whether physical illnesses represent risk factors for suicide independent of comorbid affective disorders or other potential mediators.

The literature regarding specific neurologic and psychiatric disorders most highly associated with suicide is reviewed in detail in the following sections.

Neurologic Illness

Most of the studies that have focused on completed suicide in neurologic illness have been epidemiologic, seeking to examine the prevalence and risk of suicide in known neurodegenerative disorders such as HD and MS. Increased suicide risk in these disorders relative to the general population is suggestive, in part, of a biologic basis for the behavior (either primary or secondary). Supplementation of these data by clinicopathologic/correlative studies, most often in the form of single case design, allows more specific examination of structural or neurochemical brain abnormalities.

The neurologic conditions in which suicide has been studied in greatest detail include HD, MS, PD, epilepsy, and traumatic spinal cord injury. For the most part, these investigations have been retrospective, using archival databases such as death certificates, hospital records, and federal or state disease rosters. Less frequently, the information gathered has been supplemented by seeking interviews with identified patients or with family members when the patient is deceased. On rare occasions, longitudinal research has been conducted, the goal of which has been to study the natural history of a specific disease. In such cases, evaluation of suicide and related behaviors has had a secondary or incidental focus. The study of attempted suicide in neurologic illness has received less attention than completed suicide and can best be described, with the exception of the epilepsy literature, as anecdotal.

Huntington Disease

HD is a neurodegenerative, autosomally dominant genetic disorder of adult onset. The major neuropathologic changes include neuronal degeneration in the caudate nucleus, putamen, and associated pathways (66). In addition to progressive choreiform movement abnormalities, psychiatric sequelae, including mood disorder, personality change, and dementia, are considered hallmarks of this disorder (67–69). Perhaps because of its clear neuropsychiatric nature, the cognitive and emotional symptoms, including suicide, have been of interest to study since the first description of this disease by George Huntington in 1872 (70). Early work focused on case histories, which included anecdotal reference to completed or attempted suicide. Two investigations were carried out to investigate the course, progression, and inherited transmission patterns of the disease. The first of these studies (71), a comprehensive review of identified HD cases in England, revealed that four subjects had attempted suicide. Psychiatric syndromes found in the group included schizophrenia, paranoid reaction, affective disorder, and alcoholism. The method of psychiatric diagnosis and the description of suicidal behavior were not clearly defined, nor was the relationship between suicidal behavior and psychiatric diagnosis explored.

The survey of Chandler et al. (72) of 761 individuals with HD in Michigan found that 7.8% of male and 6.4% of female deaths were suicides. Because subjects were not routinely followed until death, these figures may be underestimates. The authors concluded that suicide was an "important cause of death in non-institutionalized HD patients," especially in early-onset cases (ages 15–40 years).

To examine the sociopsychiatric consequences associated with HD, Dewhurst et al. (69) studied 102 HD patients and their families. They found ten completed suicides, 11 individuals who had attempted suicide, and 13 self-mutilators among subjects with a diagnosis of HD.

In a study designed to evaluate gender-related differences in psychopathology, Tamir et al. (73) retrospectively reviewed the clinical records of 32 psychiatrically hospitalized HD patients. The authors identified one of 13 women and eight of 19 men who were depressed, with no men and three women who were suicidal during the time of hospitalization. They concluded that there was a greater prevalence of depression and suicidality in women, with greater aggression in men with HD. Methods and definition of psychiatric terms were not specified.

Of 199 patients with HD admitted to psychiatric departments in Norway during a 59-year period, one individual subsequently attempted and one completed suicide (74). When compared with non-HD psychiatric admissions, there was no evidence to suggest increased risk of completed suicide in the HD group.

The primary purpose of two studies was to examine the prevalence of completed suicide in HD. Schoenfeld et al. (75) conducted record reviews and family interviews for 506 cases of HD deaths in the New England area. Known cause of death could be established in 157 cases. Twenty suicides were identified, making it the third leading cause of death in this group. Adjusting for age and gender to com-

pare their findings with the Massachusetts census, the authors concluded that HD patients age 50 years and older were at eight times higher risk of suicide than the general population. There was no increased risk for HD patients in the younger age group.

Using the National HD Research Roster, Farrer (76) examined the proportion of deaths in 831 patients that were owing to suicide using a detailed questionnaire sent to families. Of 452 individuals identified as deceased, 25 had died of suicide, four times greater than would be expected in a general population sample. Of the suicide victims, 27.6% had attempted suicide at least once (18.5% of males, 22.2% of females). Interestingly, age at onset of HD was not correlated with completed suicide. However, suicide did tend to occur in the early to middle stages of the illness. Cautions regarding this work include the lack of age and gender adjustment and a relatively high dropout rate.

The advent of preclinical genetic testing has fostered examination of psychiatric sequelae, including suicidal behavior, in asymptomatic persons at risk of expression of the *HD* gene. Kessler et al. (77) conducted semistructured interviews and administered multiple psychological symptom inventories in their investigation of attitudes toward predicative testing. Of 69 subjects, 37% reported a history of suicide attempts in family members, with 34% identified as having a family member who had been psychiatrically hospitalized. Eleven percent reported they would consider suicide if they tested positive. Similarly, 29% of 131 at-risk individuals surveyed by Mastromauro et al. (78) stated that they would consider suicide if testing revealed that they carried the *HD* gene. These latter two studies highlight the psychological impact of the illness on an individual and his/her family (79), complicating efforts to correlate suicide with the brain lesions of HD.

Only two case studies of HD patients have attempted to correlate neurobiologic measures with suicidal behavior. Albin et al. (80) reported the results of an immunohistochemical analysis, in conjunction with a structured family interview, of a 32-year-old presymptomatic (i.e., no movement abnormalities) woman whose death was attributable to suicide. Her history was also notable for two prior suicide attempts, beginning at 21 years of age. Psychiatrically, she was diagnosed as having cyclothymia with superimposed major depression. Neurochemical analysis revealed selective impairment of striatal enkephalinergic neurons projecting to the globus pallidus. In addition, substance P neurons projecting to the substantia nigra were of reduced density.

The second case study described a woman in her 30's who attempted suicide after positive genetic testing results (81). She also had made multiple suicide attempts in the remote past, although no significant psychiatric symptoms had been reported for the past 13 years. Positron emission tomography scan results obtained before the index suicide attempt revealed diminished glucose metabolism in both caudate nuclei. As in the previous study, the relationship between suicidal behavior, other disease variables, and the observed neurochemical abnormalities is unclear. It is an avenue of study worthy of further pursuit.

In summary, multiple investigations of completed and attempted suicide in HD are suggestive of increased prevalence and risk in these patients, with most studies focusing on completed suicide. Calculated risk has varied significantly across studies, ranging from one to eight times that of normals, with the most consistent estimate approximately four times that of the general population. An increased incidence of other psychopathologic symptoms is also consistently reported, although the comorbidity of other psychiatric phenomena within the context of suicidal behavior has not been addressed. These findings must be considered with caution, given the diversity of populations and control groups used across studies and the clear methodologic difficulties inherent to most investigations. Specific study of neuropathologic abnormalities in relation to suicide in HD is in its infancy, with no research as yet conducted of which suicidal behavior serves as a primary focus.

Multiple Sclerosis

MS is a chronic demyelinating disease of the CNS, more common in females than males (1.7:1) (82). The course is progressive but unpredictable, and patients may have multiple periods of illness exacerbation and remission. Although there is a site preference for the periventricular white matter, particularly surrounding the anterior horns, lesions may occur virtually anywhere in the brain and spinal cord, causing considerable variability across patients in clinical presentation. Onset usually occurs in the 20's and 30's. Mental disorder, including depression, has long been associated with this disease (83). These symptoms may precede the onset of focal neurologic abnormality, leading to misdiagnosis of primary psychiatric illness in some patients. To date, the most comprehensive evaluation of coexisting psychopathology in MS patients was conducted by Joffe et al. (84). Through structured clinical interviews and affective rating scales administered to 100 MS clinic patients, they reported a 42% lifetime prevalence of depressive illness, with 13% fulfilling the criteria for a major depressive episode. Thirteen percent of patients received a diagnosis of bipolar affective disorder. The authors suggested that MS and affective disorder may share a common pathobiology, citing as additional evidence the lack of any correlation between level of functional disability and the presence of mood disorder. Unfortunately, suicidal behavior was not examined in this well-designed study. Given the high prevalence of depressive illness within the context of a chronic debilitating illness, the potential for suicide and its rate in this population are clearly important.

Of four studies that attempted to establish the prevalence of suicide in patients with MS, all but one found patients to be at increased risk. Schwartz and Pierron (85), who examined death certificates only, found four suicides among 408 MS-related deaths in Michigan, a proportion of suicides (1%) no greater than is seen in the general population. Sadovnick et al. (86) found 18 of 119 deaths in MS patients for which a cause could be established were owing to suicide, estimated to be a rate 7.5 times higher than expected.

The most recent and carefully conducted study was an epidemiologic investigation (87) of 6,088 Danish patients with MS. Adjusting for age, gender, and effects of early mortality in MS from causes other than suicide, the expected suicide rate in this population was significantly lower than observed (2.0% vs. 3.7%). The risk of suicide in MS was greatest for males, individuals with histories of suicidal behavior and mental illness, and those with recent deterioration of their condition (88). Furthermore, in two studies of the physical illnesses found in victims of completed suicide, MS was present in numbers significantly greater than would be expected in the general population (59,89a).

Given the close association of affective illness with both MS and suicide, one would expect to find depressive symptomatology in a large proportion of MS patients who took their lives. Only one preliminary report has directly addressed this issue. Of 18 MS patient suicides investigated by psychological autopsy, 83% had a major depressive episode (89b).

There is no research with an MS population, to our knowledge, in which the focus is attempted suicide. One study did examine a series of 50 outpatients with concurrent psychiatric and physical illness who were referred after a suicide attempt (90). These subjects were compared with 85 suicide attempters without concurrent physical illness. Forty percent of the concurrent illness group had a neurologic disease, of which MS represented 2%. Major affective illness was identified as the most common comorbid psychiatric syndrome. The physical illness attempter group had a greater incidence of organic and major depression, was prone to more violent attempts, and had profiles more similar to completers than did the attempters without physical illness.

In summary, there is clear evidence to support the findings of increased suicide risk in this population, with estimates that 3.7% to 5.7% of MS patients take their own lives. Affective illness, primarily major depressive syndromes and bipolar illness, is also present in increased numbers, although the relationship between mood disorder and suicidal behavior in this population has yet to be systematically studied. In addition, direct studies of brain abnormalities in MS patients with psychiatric symptoms are almost nonexistent, with no study focusing primarily on either attempted or completed suicide.

Parkinson Disease

PD is a progressive neurodegenerative disorder that affects the nigrostriatal system, resulting in movement abnormalities including tremor, gait disturbance, rigidity, and bradykinesia. Sensory symptoms and cognitive phenomena, including dementia, may also be part of the disease. Primary PD, considered idiopathic in its etiology, commonly has onset in the sixth or seventh decade of life. It is surprising that suicidal behaviors have rarely been studied in PD because depressive disorders are estimated to occur simultaneously in 20% to 90% of cases, the majority of studies suggesting a 40% to 60% prevalence rate (91). Furthermore, PD is associated with degradation of both central dopaminergic and serotonergic function (92,93), systems that the pre- and postmortem neurobiologic evidence suggests may be dysregulated in suicidal states (reviewed below). One study, the focus of which was to examine the prevalence of physical disease in victims of completed suicide, found PD present at a rate four times greater than expected based on the prevalence of PD in the general population (89a).

In contrast, Hoehn and Yahr (94) reviewed clinical records and death certificates of 672 idiopathic PD cases. Of 340 deaths, only three were attributed to suicide. In 60 cases, the cause of death could not be determined. Although age, gender, and the percentage of deaths were compared with statistics for the general population of New York City for several causes of death, suicide was not included in the comparative analysis. No study, to our knowledge, has examined the prevalence of attempted suicide in PD.

Epilepsy

Epilepsy is the term used to describe a heterogeneous group of disorders whose common feature is multiple seizure episodes. Single or multiple brain regions may be involved. Problems with the use of epilepsy as a neurobiologic model include the heterogeneity of symptoms, etiologies, and brain regions affected. In addition, in the cases of childhood-onset seizures, the degree to which early CNS insult may affect the developing brain is not well understood.

Comorbid psychological variables, including suicide risk, have been a topic of considerable interest in the literature. Depressive symptoms are common in epilepsy (95), and severe depressive illness has been estimated to occur in 19% of temporal lobe epileptics (96). In addition, an increased prevalence of psychosis has been identified. It is not surprising, therefore, that estimates of suicide risk in individuals with epilepsy are three to four times higher than in the general population (97).

The most comprehensive and methodologically sound study of completed suicide in epilepsy was conducted by White et al. (98). Evaluating the records of 2,099 epilepsy patients admitted to a British epilepsy center, they concluded that suicide was among the chief causes of excess death in this population. Of 636 deaths that had occurred, 21 were by suicide, a rate 5.4 times greater than expected.

Mendez and Doss (99) studied the factors associated with four completed suicides that had occurred among 1,611 patients with epilepsy seen in the neurology clinic. Three of the suicide victims were male, three were actively psychotic, and all four had a temporal lobe seizure focus.

In a review of 16 studies, Barraclough (100a) concluded that severe epileptics were at five times greater risk of suicide than the general population, with as much as a 25 times greater risk in individuals with temporal lobe epilepsy. Matthews and Barabas (100b), in their review of five studies, arrived at a similar estimate that approximately 5% of patients with epilepsy die by suicide. Because the articles reviewed in both papers were diverse in their methodology and populations, the overall findings should be considered tentative.

Like completed suicide, attempted suicide rates among those with epilepsy is very high, four to five times greater than in the general population (101,102a). Factors distinguishing attempters with epilepsy from attempters with other diagnoses include recent alcohol use (102a), psychotic symptomatology, borderline personality disorder, and a history of impulsive attempts (97). Comparing 325 epileptic patients who had attempted suicide with 166 nonattempter epileptic controls, Batzel and Dodrill (102b) found an increased incidence of anxiety, general psychopathology, and diminished ego strength on the Minnesota Multiphasic Personality Inventory profile of the suicide attempter group. No difference was found on the depression or schizophrenia subscales. Unfortunately, neither formal diagnoses nor specific symptoms were considered.

Spinal Cord Trauma

Investigation of suicidal behavior in traumatic spinal cord injury is of potential research interest for a variety of reasons. As with increased prevalence of completed suicide in other populations, the identification of risk factors is critical to effective diagnosis and treatment of vulnerable patients. Moreover, given that these static lesions are confined to the spinal cord, this group may represent a suitable control group when trying to establish neurobiologic models of suicide in other neurologic disease entities.

A series of rather large investigations studied mortality and associated cause of death in spinal injury cases. Increased mortality in this group has been well established (103,104), as has increased frequency of death by suicide (105–109). Risk estimates range from two to six times that of the general population (109), with the most methodologically sound investigations reporting a four- to five-fold increase (105,106). Younger age (105,106,108) and quadriplegia (104–106) place patients with spinal cord injury at increased risk, particularly in the first few years after injury.

Charlifue and Gerhart (109) compared 42 completed suicides with spinal cord lesions with a spine-injured group matched for gender, age at injury, and lesion location. Preinjury variables associated with suicide included a history of alcohol, drug abuse, or depression. Postinjury variables of import included despondency, shame, apathy and helplessness, anger, alcohol abuse, destructive behaviors, and attempted suicide. Interestingly, 7% of completed suicides suffered their spinal cord injury during an attempt on their lives, whereas no control group patients acquired their injury in that manner. Judd and Brown (110) identified six suicides among 342 spinal cord patients occurring over a 5-year period. Clinical aspects common to the completed suicide group included male gender; schizoid, depressive, or narcissistic personality traits; alcohol or drug abuse; and the presence of significant depression. These findings should be considered tentative, given their descriptive nature.

In summary, completed suicide in traumatic spinal cord injury patients appears to occur at a rate that is four to five times that in the general population when mortality and demographic factors are considered. Those at greatest risk include male quadriplegics in the first few years after injury. Although the prevalence of affective illness in this population has not been established, depressive symptomatology is common. Studies of attempted suicide in traumatic spinal injury have yet to be conducted.

Other Neurologic Disorders

Cerebral neoplasms and cerebral vascular accidents are other neurologic disorders of potential interest in the study of attempted suicide, particularly given the reportedly high incidence of depressive illness in the latter group (111). No focused research on attempted or completed suicide has been conducted in either of these groups of disorders, although anecdotal case illustrations are reported (83). Studies of the prevalence of specific physical illnesses in suicide have suggested a possible increased incidence of completed suicide in individuals with CNS neoplasms. In these studies, illness prevalence rates were compared with prevalence rates for the disease in the general population (59,89a). The results are less consistent for cerebral vascular accidents, with one study reporting a higher rate of stroke in completed suicide (59) and another finding no suggestion of increased illness representation in a completed suicide population (89a).

Summary

Increased risk of suicide has been fairly well established in MS, epilepsy, and traumatic spinal cord injury, with probable increased risk associated with HD as well. The study of suicidal behavior in PD has been too limited to allow definite conclusions. Phenomenologic description of specific suicidal behaviors in these disorders is scant.

The relative contributions of a high incidence of comorbid psychiatric sequelae, particularly depressive illness, have also been well documented. The degree to which depressive illness or other psychiatric phenomena such as psychosis serve to mediate suicidal behavior in neurologic illness versus suicidal behavior existing as an independent disease-related behavior is unclear. In addition, a comprehensive assessment

of pertinent demographic and socioenvironmental variables such as age, gender, functional disability, and disease prognosis issues has not been adequately performed.

Clearly, the development of systematic research that addresses and rigorously defines this comprehensive array of factors within specific neurologic illnesses will represent an important step toward the development of neurobiologic models of completed and attempted suicide. At the very least, the application of such carefully designed investigations will serve to provide greater clarity in the identification of pertinent risk factors for suicidal behavior in these disorders.

Psychiatric Illness

Of the postulated risk factors for suicide, none is more potent than mental illness. In describing the association between emotional disorders and suicide, one must take two complementary approaches. Initially, studies of general population samples are made in which retrospective psychiatric diagnoses are determined for all suicides occurring in a particular area and time period. Although such studies are important in establishing the prevalence of particular diagnoses in a sample of completed suicides, they do not specify the risk inherent in a given psychiatric or neuropsychiatric disorder. These data must be derived from retrospective or prospective studies of particular diagnostic groups. Each of these approaches is reviewed in the sections that follow.

Studies of Suicide in the General Population

Since the mid-1950's, more than a dozen retrospective studies of suicide in the general population have been conducted worldwide. Using a method known as the psychological autopsy, investigators construct a detailed picture of the victim's symptomatology, behaviors, and life circumstances in the weeks and months before death by reviewing records and conducting detailed interviews with knowledgeable informants (112). Psychiatric diagnoses established for victims in 11 of the most comprehensive of these studies are listed in Table 27.4 (24,53,55,113–120). They differ in sample size, the years during which they were conducted, and the criteria by which diagnoses were made. Nonetheless, all these investigations show a remarkable and important consistency in finding diagnosable psychopathology in 90% or more of cases. In general, affective disorders were most common, present in 30% to 87% of cases, closely followed by substance use disorders in 19% to 63% of cases. Primary psychotic illnesses were diagnosed in 2% to 17% of completed suicides in these studies, whereas personality disorders were diagnosed in 3% to 44%.

To explore the marked variation in suicide rates as a function of age and gender (Fig. 27.1), several authors examined differences between males and females who completed suicide. Rich et al. (121) compared 143 men and 61 women

TABLE 27.4. PSYCHOLOGICAL AUTOPSY STUDIES OF GENERAL POPULATION SAMPLES: PERCENTAGE OF COMPLETED SUICIDES WITH SELECTED DIAGNOSES

Study	N	Site	Age	Affective Disorder	Substance Use	Schizophrenia	Axis II	None
				| % With Diagnosis |				
Rich et al., 1986 (24)	283	United States	13–88	44	60	14	5	6
Robins et al., 1959 (53)	134	United States		45	25	2	—	6
Chynoweth et al., 1980 (55)	135	Australia	≥15	55	34	4	3	11
Dorpat and Ripley, 1960 (113)	108	United States	—	30	27	12	9	0
Barraclough et al., 1974 (114)	100	Great Britain	—	80	19	3	—	7
Conwell et al., 1996 (115)	141	United States	21–92	47	63	17	—	10
Vijayakumar and Rajkumar, 1997 (118)	100	India	≥14	35	34	4	20	12
Foster et al., 1997 (117)	118	Northern Ireland	≥14	36	37	11	44	10
Henriksson et al., 1993 (116)	229	Finland	>10	59	43	10	31	2
Cheng, 1995 (119)	116	Taiwan	≥15	87	44	7	—	2
Gustafsson and Jacobsson, 2000 (120)	100	Sweden	18–87	64	37	15	22	3

in their sample from the San Diego suicide study and found that women had a significantly older mean age at death. There was no difference in the proportion in each group with a substance use disorder diagnosis. However, women were significantly more likely to have had a major depressive disorder. Men used firearms significantly more often and drugs or poison significantly less often as a means of death than did women.

A study by Asgard (122) of 104 women who committed suicide in Sweden helps to further characterize gender differences in the psychopathology of victims. They found, for example, that 59% met the criteria for a mood disorder in the last month of life, a percentage similar to that reported for men. Substance use disorders were present in only 12% of cases. They further found that 63% of their victims had attempted suicide at some time in their lives, 36% having done so in the year preceding death. Seventy-one percent had a history of psychiatric care, with almost 60% having received treatment in the last year of life. There are no data on Swedish men with which to compare these figures.

Several investigators have noted differences in the distribution of psychiatric diagnoses by age. Dorpat and Ripley (113) found schizophrenia to be the most common diagnosis in completed suicides in those younger than the age of 40 years, alcoholism the most common in those between the ages of 40 and 60, and psychotic depression the most common in suicide victims older than the age of 60 years. Carlson and Cantwell (123) reanalyzed Robins's detailed case histories of suicides occurring in St. Louis in the 1950's (53) to show that affective disorder diagnoses were more common and alcoholism less frequent in those older than 65 years of age than in younger victims. In a more contemporaneous sample, Rich et al. (24) found significantly more substance use disorders and antisocial personality in suicide victims

age 30 years and younger than in the older cohort, who were more likely to have an affective illness.

The examination of differences in psychiatric diagnosis by age in completed suicides was the principal objective of a psychological autopsy study by Conwell et al. (115). In 141 victims of completed suicide of ages 21 to 92 years, multiple regression analyses were used to establish whether age, gender, or their interaction could predict specific diagnoses in suicide victims. They found that increasing age was a significant predictor of major depressive disorder. Younger age predicted schizophrenic illness and substance use disorders. In those subjects with major depressive disorder, increasing age significantly predicted a single episode rather than recurrent illness. Finally, younger victims were significantly more likely to have multiple comorbid diagnoses than their older counterparts, whose affective illness was much less likely to be complicated by substance abuse or dependence.

We now turn to more detailed consideration of suicide in specific diagnostic subgroups.

Affective Disorder

Suicide rates in patients with affective illness are far higher than in the general population. Miles (124) reviewed 30 studies in which patients with primary depressive disorders were followed longitudinally. In agreement with Guze and Robins (125), he concluded that approximately 15% of affected individuals will ultimately die by suicide. Several studies have attempted to calculate the rate of suicide in patients with depressive disorder. Pokorny (126) followed 4,800 male veterans for a mean of 5 years, during which time 67 took their own lives. Whereas the overall suicide rate was 279 per 100,000 per year, patients with affective illness diagnoses died by suicide at the rate of 695 per 100,000 per year. Again, age may be a factor. By linking psychiatric case register data with death certificates, Gardner et al. (127) calculated a suicide rate for psychiatric patients with a diagnosis of affective psychoses of 351 per 100,000. However, when they divided the population at age 55 years, younger depressives had a rate of 207 per 100,000 compared with a rate of 475 per 100,000 for older patients. Furthermore, they observed that although 2% of the patient population who attempted suicide went on to take their own lives within one year, 6% of attempters older than age 55 completed suicide within the same period. These data suggest that elderly people with affective illness may be at higher risk of suicide than both normal elderly and younger depressive individuals as well (115).

In an attempt to define the clinical characteristics unique to those depressed patients at highest risk of suicide, several studies compared depressed completers with living patients. Barraclough and Pallis (128) found that depressed suicides studied by the psychological autopsy method were more likely to be male, single, and living alone than the comparison group of ambulatory patients with affective disorder diagnoses. Symptoms of insomnia, self-neglect, impaired memory, and a history of suicide attempts also were distinguishing features. Roy (129) matched psychiatric patients with recurrent depression who were known to have committed suicide with depressed inpatients who had not taken their own lives and found that significantly more of those who committed suicide had lived alone, were young when they lost their parent(s), and had histories of suicide attempts. Berglund and Nilsson (130) followed more than 1,200 psychiatric inpatients with severe affective illness for as long as 27 years. Comparing male subjects who committed suicide during the follow-up period with those males who had not, they found an acute onset of illness, marital problems, and lower ratings for psychomotor retardation distinguished the suicide group. Women completers with depressive disorders had a significantly higher frequency of attempted suicide than other women.

Fawcett et al. (131) examined predictors of suicide in 954 patients with major affective disorder followed longitudinally for an average of 10 years. They divided the 32 subjects who committed suicide in the follow-up period into those who had died within 1 year of initial evaluation and those who had died 2 to 10 years later in an effort to distinguish short- from long-term predictors. They found that short- but not long-term suicide was associated with severe anhedonia and psychic anxiety, obsessive-compulsive features, global insomnia, acute alcohol intoxication, impairments in concentration, and panic attacks. Symptoms associated with long-term suicide (2 to 10 years) were hopelessness, suicidal ideation, and a history of suicide attempts. The authors suggest that recognition of and intervention in short-term risk factors, such as severe anxiety and panic attacks, through appropriate pharmacotherapy, psychotherapy, and environmental manipulation may help to prevent imminent suicide.

Although consensus is lacking, there is no clear association between any subtype of affective illness and suicide. Whereas some studies found patients with bipolar affective illness to be at lower risk than those with unipolar disorders (132,133), others found no difference in suicide rates between unipolar and bipolar patients (134,135) or even higher risk in the bipolar type (136).

Roose et al. (137) found that delusionally depressed patients were overrepresented in suicides committed in an inpatient facility. Others, however, found no increased risk of suicide in psychotic depressed patients in either short- (138) or long-term (139) follow-up.

The posthospitalization period is the time of greatest risk of suicide in affective illness (124), decreasing progressively after the first several years postdischarge. Thirty-two percent of the suicides among patients followed longitudinally by Fawcett et al. (131) occurred within 6 months of entry into their study, and 52% within 1 year. Hospital discharge, therefore, represents a time for special vigilance on the part of patients and their caretakers.

Schizophrenia

Epidemiologic studies in the United States and other countries estimate that the prevalence of schizophrenia in the general population is approximately 1%. Yet approximately 5% to 10% of completed suicides carry that diagnosis. The implication that schizophrenics are at a much increased risk of completed suicide has been confirmed in a number of longterm follow-up studies (140). Based on suicides in follow-up intervals from 4 to 40 years, suicide rates for schizophrenics have been estimated to be as high as 615 per 100,000 (141), more than 50 times greater than that of the general population. An estimated 10% of schizophrenics ultimately end their lives by suicide (134,142).

Although some studies indicated that the standardized mortality ratios for suicide are greater for schizophrenic women than men (143–145), other investigators noted a heavy preponderance of men among schizophrenic suicide victims (146–148).

Younger age appears to be a risk factor for suicide in schizophrenics. Breier and Astrachan (146) found that the mean age of schizophrenic suicide victims of 30.3 years to be significantly younger than that of schizophrenic patient controls, consistent with an earlier report by Roy (147). Similarly, schizophrenic suicide victims have been shown to be significantly younger than suicide victims with other diagnoses (149,150). With increasing age, however, risk seems to diminish. In a record linkage study, Newman and Bland (151) found the highest risk of suicide among schizophrenic patients age 29 years or less, with a standardized mortality ratio of approximately 33. Between the ages of 30 and 49 years, however, relative risk decreased to 18 times that of the general population, whereas in schizophrenic patients aged 50 to 69 years, risk was only seven times that of the general population. Others reported similar trends, with the exception that the peak risk for women occurs approximately 10 years later (145), consistent with their characteristically later onset of illness. Perhaps more relevant than age at death is the total duration of illness, which is similar for men and women—typically 5 to 10 years after initial diagnosis (147,148).

Like patients with other psychiatric disorders, the schizophrenic patient is at highest risk of suicide in the first weeks and months after admission and discharge from the hospital. In studies of suicide committed while in the hospital, schizophrenia is the most common diagnosis (152). Roy (153) reviewed seven studies revealing that as many as 50% of schizophrenic patients who commit suicide do so in the first few weeks and months after discharge. Contrary to popular belief, suicide is less common in the acutely psychotic state than at a time when, as psychosis resolves, the patient is more vulnerable to a depressive syndrome. In such situations, the proper diagnosis and treatment of depression, which may be difficult to differentiate from medication effects or the negative symptomatology of the psychotic process, are critical.

Other factors associated with risk of suicide in people with schizophrenia include positive symptoms, the paranoid subtype (154), unemployment, and a relapsing, deteriorating course (147), particularly in individuals with higher levels of educational achievement (148), which result in feelings of inadequacy and hopelessness regarding their future (155,156). As with depressed patients, a history of serious attempts to take their lives also places schizophrenics at greater risk of completed suicide (156).

Substance Use Disorders

Through long-term follow-up of hospitalized patients, Black et al. (157) and others clearly demonstrated that patients with alcohol and drug abuse have excess mortality not only because of suicide, but also physical causes, accidental, and other violent deaths as well. Lindberg and Agren (158), for example, found in a 2- to 20-year follow-up that the relative risk of death from all causes was three times greater for male alcoholics and more than five times greater for female alcoholics than the general Swedish population. Furthermore, male alcoholics were 7.9 times and female alcoholics 15.2 times more likely than expected to die by suicide. Murphy and Wetzel (159) reviewed the literature in critical examination of the frequently quoted estimate that the lifetime risk of suicide and alcoholism was as high as 15%. They found this figure untenable based on suicide rates reported in both short- and long-term follow-up studies and on the estimation that the alcoholic population at risk of suicide is approximately 5.6 million people in the United States. They calculated a lifetime risk of suicide of 2% to 3.4% in alcoholics in the United States and other Western countries with similar suicide rates or 60 to 120 times higher than that of the nonpsychiatrically ill.

Whereas depression is more common among completed suicides with increasing age, substance use disorders are less prevalent. The most frequently cited reason for this observation is early mortality among alcoholics from other causes (160), leaving a smaller absolute number subject to suicide risk in later life. However, suicide rates among alcoholics may increase with age (127), with the duration of abusive drinking being an important variable. Murphy (161) calculated that the mean duration of alcoholism before death was 19.8 years for men and 12.3 years for women. However, alcoholics with an older age at death were more likely to have had a late onset of the disorder and thus a duration of illness similar to that of younger alcoholic victims.

Far fewer data are available on the suicidal behavior and risk of other substance users. Allgulander et al. (162) noted four suicides among 40 patients with sedative hypnotic dependence in a 4- to 6-year follow-up. Social deterioration and a prevalence of depressive symptoms were characteristic. Suicide rates for heroin addicts have been estimated to range from 82 to 350 per 100,000, with 2.5% to 7% dying by their own hand. These age-adjusted rates are five to 20 times greater than that for the general population. Murphy et al. (163) found that 17.3% of 533 opiate addicts treated in a drug dependency unit had a history of at least one suicide attempt. Marzuk et al. (164) reported that cocaine metabolites were detectable in more than 20% of New York City suicide victims age 60 years and younger, with the highest prevalence (45.7%) in young Hispanic males. Seven percent of all cocaine-positive deaths in New York City in 1986 were suicides. In the San Diego Suicide Study, Fowler et al. (165) found that 30% of suicide victims younger than 30 years of age had a history of cocaine use. The authors speculate that the striking rise in suicide rates in young adults over the past three decades is in part a consequence of increasing substance use over the same period.

Because drug and alcohol abuse have such widespread effects both in the short and long term on the individual's physical health, social network, and psychological equilibrium, the mechanism of an association between suicide and substance abuse is obscure. Comorbidity is common. Data from the Epidemiologic Catchment Area Study show that individuals with either an alcohol or drug use disorder were more than seven times more likely than the rest of the population to have a second addictive disorder and that 37% of those with alcoholism and 53% of those with other drug addictions had a comorbid mental illness (166). Among alcoholics who completed suicide, Murphy et al. (167) found that 72% met the criteria for a definite or probable major affective disorder. Citing five other studies that yielded qualitatively similar results, they speculate that the presence of depression in alcoholics is a major risk factor for suicide in that group.

Additional data from studies of alcoholics (49,168) and other substance abusers (50,169) demonstrated that a disproportionate number of suicide victims with substance use disorders had experienced the loss of a close interpersonal relationship in the last 6 weeks of life. This concentration was markedly statistically different from suicides with major depressive disorders in the absence of substance use comorbidity.

In a subsequent analysis, Murphy et al. (167) compared alcoholic suicides studied by the psychological autopsy method with data from living alcoholic controls. They identified seven features that appeared to be related to suicide: continued drinking, development of a major depressive episode, communication of suicidal intent, poor social supports, unemployment, living alone, and a serious intervening medical illness. Their findings underscored the need in the treatment of alcohol and drug abuse to achieve and sustain abstinence, diagnose and aggressively treat depressive symptoms, and to be especially alert for suicide at times of disruption and further impoverishment of the patient's social support network.

Biologic factors require further clarification. Prolonged exposure to alcohol has been shown to produce progressive depressive symptoms, in both normals and abusive drinkers (170). The toxic effects of alcohol on the CNS are well known. The length of abstinence from alcohol is correlated with CSF 5-hydroxyindoleacetic acid (5-HIAA) levels: alcoholics have deficits in central serotonin (5-HT) function, and chronic alcohol consumption serves only to further deplete those levels (171). Decreased tryptophan to total amino acid ratios have been demonstrated in a subgroup of alcoholics at risk of depression, suicide, and aggressive behavior as well (172). Like major depressives, alcoholics have been shown to have a blunted adrenocorticotropic hormone response to corticotropin releasing factor (CRF) infusion and to escape from dexamethasone suppression in the acute withdrawal (173). These findings suggest a neurobiologic vulnerability to suicide created, or at least exacerbated, by heavy alcohol use.

Other Psychiatric Disorders

Based on their review of eight cohort and 11 case-based studies of axis II disorders and completed suicide, Duberstein (174) concluded that approximately 30% to 40% of suicides had a personality disorder. The diagnoses most closely associated were borderline and antisocial personality disorder, with suggestive evidence for risk associated with avoidant and schizoid personality disorders as well.

Borderline personality disorder seems to confer the highest risk of the personality disorders, with estimates of from 5% to 9% taking their own lives (174). Perry (175) concluded that the period immediately after intake to psychiatric care was most hazardous. Other characteristics associated with suicide in borderline personality disorder include a history of attempts, greater educational attainment, and fewer psychotic symptoms (176).

Many personality traits have been associated with completed suicide, including hostility, hopelessness, helplessness,

dependency, neuroticism, and low openness to experience (177). This literature is important because it suggests that personality traits, in addition to personality disorders, can increase suicide risk. For example, Marttunen et al. (178) noted that antisocial symptoms, even those of relatively brief duration and not severe enough to meet diagnostic criteria for antisocial personality disorder, may still indicate suicide risk.

Individuals with panic disorder may be at significantly greater risk of suicidal ideation and attempts than the general population (179). Some studies, however, indicate that the elevated risk results from high rates of the comorbidity of panic with other psychiatric illnesses such as major depression (180). Anxiety disorder diagnoses are rarely made in psychological autopsy studies (181). It remains unclear whether this discrepancy represents an artifact of the psychological autopsy method (difficulty obtaining information from informants about the subjective experience of anxiety) or whether suicide attempts in panic disorder signify less risk of subsequent completed suicide than does suicidal behavior in major depression, schizophrenia, or substance use disorders.

NEUROBIOLOGY OF SUICIDE

Having reviewed the factors associated with suicide risk, we now turn to a review of studies concerning the neurobiology of suicidal behavior. These include the consideration of relevant animal models of self-destruction, evidence from genetic studies of a heritable component to suicide that may be independent of social and environmental factors, and data derived from both pre- and postmortem studies in humans regarding specific neurochemical systems.

Animal Models

The development of animal models of psychiatric illness is an important objective for three reasons. Animal models allow investigators to control experimental factors and thus manipulate the system in ways that enable far more precise testing of hypotheses than is feasible in humans. Invasive studies that could not be conducted in people for obvious ethical reasons can be conducted in animals to both generate and test hypotheses concerning the mechanisms underlying disorders or behaviors, such as suicide. These same models can then be used to develop and test treatments and screening tools based on biologic measures to identify individuals at high risk.

Although animal models may be useful for the development and testing of neurobiologic hypotheses, their applicability to suicide research has clear limitations as well. For example, if one requires conscious intent to end one's life as a defining feature of suicide, then animal models of self-destructive behaviors may be of limited or no relevance

(182). The migratory patterns of the Norwegian lemming are a case in point. Their plunge into the sea is commonly interpreted as self-destruction, whereas for many ethologists, it is an incidental outcome of emigration away from an area of overcrowding. Hamilton (183) offers numerous similar examples from which he draws the conclusion that suicide-like behavior is in fact rare in animals and that, where found, it probably has its basis in parental behavior or kin selection.

Self-sacrifice as an altruistic behavior is common among social insects and other animals. Crawley et al. (182) offer as an example the self-sacrifice of Hymenoptera, including wasps, ants, and bees, in which members of the sterile soldier caste die after using their stinger once in the defense of the community. Similar soldiering behaviors are seen in other mammals, including primates and in schools of fish and flocks of birds. When considered as behaviors of each individual organism, they may have some relevance to our understanding of suicide. When seen, however, as constituents of a larger social organism, an individual's sacrifice has less immediate relevance to human suicide.

Stress-related behaviors in animals have more inherent similarity to self-destructive behaviors in humans. Frequently cited examples include self-mutilation in animals confined to small cages and removed from social stimuli. Bach-Y-Rita and Vero (184) compared confinement-induced self-injury in animals with self-destructive behavior in human prison populations. This similarity in behavior and social context can be taken as evidence to support noncognitive determinants of human self-injury. Among laboratory models, learned helplessness in rats and social separation in primates elicit behaviors reminiscent of depression in humans, which respond in turn to known human antidepressant therapies. Each has an incidence of self-destructive behavior associated as well.

Although it is beyond the scope of this chapter, extensive studies concerning animal models of externally aggressive behavior may well be relevant to suicide in humans. Neurotransmitters that have been associated with aggression in animals include 5-HT, γ-aminobutyric acid (GABA), dopamine (DA), norepinephrine (NE), acetylcholine, cyclic nucleotides, and neurohormones (185). Indeed, the clinical correlate of altered neurotransmitter function in studies of the psychobiology of suicide in humans may be impulsiveness or aggression rather than self-destructive intent. Whether that aggression or impulse to injure is directed inwardly or externally may be determined by the dynamic balance of neurobiologic systems (186), culture, and cognition (187–189). Animal models offer the opportunity to test hypotheses as they become better formulated by studies of self-destructive behavior in humans.

Human Studies

The study of biologic correlates for suicidal behavior has been complicated by several limitations; some are general and others are specific to the particular technique used. Addressing the limitations of the specific techniques is beyond the scope of this text. Among the general limitations are small sample sizes, the retrospective nature of data from postmortem studies, lack of control for important confounders [e.g., diagnosis, medication exposure, toxic exposure including substance use, method of suicide (violent vs. nonviolent), demographic and seasonal variation in some of measures used, postmortem delay].

Genetics of Suicide

It has long been recognized that suicidal behaviors tend to cluster in vulnerable families, suggesting a role for genetic factors in determining the behavior. Sainsbury (51), for example, found that 2% of the deceased relatives of completed suicides had also died by that same means. Farberow and Simon (190) reported that 6% of suicide victims had a parent who had died by suicide. Psychiatric patients with a family history of suicide are at greater risk of suicidal behavior as well. Roy (191) found that a family history of suicide significantly increased the risk of a suicide attempt in patients with diagnoses of schizophrenia, major affective disorders, depressive neurosis, and personality disorders.

Tsuang (134,192) reported on the long-term follow-up of three subject groups: 525 patients with diagnoses of schizophrenia or manic-depressive illness, 160 patient controls admitted to the hospital for routine surgical procedures, and 5,721 of their first-degree relatives. He found significantly more suicides among patients than controls during the 30- to 40-year follow-up period and a significantly greater risk of suicide in the relatives of patients than in the relatives of surgical controls. Furthermore, the relatives of patients who had committed suicide showed a significantly higher risk of suicide than did the relatives of patients who did not commit suicide in the follow-up period. Notably, the increased risk was most pronounced for relatives of patients with affective illness, leaving open the question of whether the vulnerability was for the depressive disorder rather than suicide more directly.

Scheftner et al. (193) followed 955 affectively disordered probands for 5 years, comparing the frequency of suicide in first-degree relatives of patients who committed suicide during the follow-up period with relatives of nonsuicidal patients. They found no significant difference either for completed or attempted suicides. Although the shorter follow-up period could explain these negative findings, they would also be expected if the transmissible vulnerability related to the affective disorder were common to both groups in this comparison. Conversely, Brent et al. (194) found a four-fold elevated risk of suicidal behavior in the first-degree relatives of adolescent suicide victims compared with control patients, even after controlling for the increased rates of psychiatric disorder and assaultive behavior in the relatives. They believe that exposure to parental substance abuse and depression may have added an environmental risk factor to the genetic risk of suicide.

Egeland and Sussex (195) saw an opportunity to study familial patterns of suicide in the absence of several important confounding variables in the Old Order Amish community. The Amish are by tradition nonviolent, abstain from drug and alcohol use, provide extensive community support throughout life, and have negligible unemployment. Meticulously kept records of multigenerational families enabled the researchers to ascertain all suicides in the community over a 100-year period. Ninety-two percent of the 26 documented suicides occurred in individuals with retrospectively diagnosed major affective illness. Furthermore, the suicides clustered in a distribution closely matching that of affective illness in these Amish pedigrees. Although this study suggested the role of inheritance for both suicide and major affective disorders, it could neither distinguish genetic from environmental influences nor determine whether the heritability of suicide was distinct from that for affective illness.

Additional evidence for a genetic component to suicide has been provided by twin studies that demonstrate a higher concordance for suicide among monozygotic than dizygotic twins (196). Again, however, separating the genetic predisposition to psychiatric illness from a genetic risk of suicide is problematic. Of ten twin pairs concordant for suicide reported in the literature reviewed by Roy (197), all were monozygotic. In five of these twin pairs, however, the twins were also concordant for either depression or schizophrenia. In a later study, Roy et al. (198) examined the prevalence of suicide attempts among living co-twins whose twin had committed suicide. Suicidal behavior was significantly more common among monozygotic (ten of 26) than dizygotic (zero of nine) co-twins. No such difference was evident in another comparison of monozygotic and dizygotic twins whose co-twins had died by other means, effectively ruling out the role of grief rather than the suicidal behavior *per se* (199).

Twin studies are unable to distinguish with confidence the effects of nature, or genetic predisposition, from nurture, or environmental influences. Although presumably raised under comparable conditions, identical twins may be subject to different social, environmental, and cultural influences than fraternal twins. Danish records of individuals adopted between the years 1925 and 1948 offered an opportunity to control for these confounds. Wender et al. (200) matched 71 adoptees with major or minor depressive disorders with 71 adoptive controls and searched death records for indications of suicide in biologic and adoptive relatives. They found that a significantly greater proportion of biologic relatives of depressed adoptees had committed suicide than had biologic relatives of controls. Schulsinger et al. (201) conducted a similar search for suicides in the biologic and adoptive relatives of 57 adoptees who had themselves committed suicide and 57 nonsuicidal control adoptees matched

for gender, age, social class of adoptive parents, and time spent with biologic parents. Twelve of 269 biologic relatives of suicide adoptees (4.5%) had themselves committed suicide, a significantly greater proportion than of the biologic relatives of control adoptees (0.7%). None of the adoptive relatives in either group had committed suicide. Given that proband adoptees had been reared separately from, and thus unexposed to, the environmental influences of their biologic families, the data strongly favor a genetic determinant of suicide. However, given the well-established heritability of psychiatric disorders that are themselves highly correlated with suicidal behavior, the available data cannot at this point distinguish a genetic factor for suicide that is distinct from a genetic predisposition to major mental disorders.

Recent advances in molecular genetics techniques have enabled new lines of research. Of most interest to investigators have been genes that are involved in the transcription of proteins related to the serotonergic system: tryptophan hydroxylase (TPH), the 5-HT transporter (5-HTT), and 5-HT receptors.

TPH is the rate-limiting enzyme in the biosynthesis of 5-HT. Abbar et al. (202) were the first to test the association between suicide attempts and the *TPH* gene but found no difference in allelic frequencies at the TPH locus between suicidal patients and controls. Nielsen et al. (203) also observed no difference in the allelic distribution at the TPH locus between Finnish violent alcoholic offenders with a history of suicidal behaviors and controls. However, they did report a significant association between TRH polymorphism and a history of suicide attempts in their study group, independent of impulsivity and CSF 5-HIAA concentration. Other studies report both negative (204–208) and positive results (209–212), possibly reflecting differences in the racial and diagnostic composition of the samples, variations in the polymorphism studied, and other methodologic issues.

Rotondo et al. (213) screened the human TPH promoter for genetic variants that could modulate TPH gene transcription in 260 individuals from Finnish, Italian, white American, and Native American populations. Four common polymorphisms were identified, one of which was significantly associated with suicidal behavior in 167 unrelated Finnish offenders compared with controls. They concluded that these four polymorphisms can be used for haplotype-based analysis to localize functional TPH alleles influencing behavior.

Genetic studies of the 5-HTT have generally failed to show any difference between suicides and controls (214–216). Neither have studies of the *5HT$_{2A}$* (212,217–220) or the *5-HT$_{1B}$* receptor gene (221) shown a consistent pattern.

In summary, the initial evidence from family history, twin, and adoption studies of possible heritability of suicidal behavior has stimulated research using molecular genetics techniques in an attempt to identify genetic markers for suicide risk. Limited success thus far must be evaluated in light of the general limitations of suicide research as a whole and the more specific limitations to genetic studies: genetic variations between different populations, the problem of distinguishing the genetics of different diagnoses and suicide *per se,* the influence of environmental factors on gene expression, and other issues. This invites a systematic, comprehensive approach that allows better identification of the population studied, diagnostic issues, medication and toxic exposures, and psychosocial factors.

Neurochemical Systems and Suicide in Humans

The emphasis in genetic studies on 5-HT stems from an extensive body of literature on its role in the pathogenesis of suicidal behavior. Indeed, abnormalities in central 5-HT function have been implicated in the pathobiology of depression for more than 30 years. Based on the observation that reserpine, an alkaloid that depletes brain stores of monoamines, causes severe depression and on the discovery of antidepressant drugs whose effects are mediated through augmentation of monoaminergic systems, noradrenergic (222) and serotonergic (223) hypotheses of affective illness have been proposed. Among the numerous studies that ensued from the formulation of these hypotheses, many researchers attempted to correlate abnormalities in monoamine function with specific depressive symptoms, including suicidality. These studies in turn lead to the rapidly expanding literature on the psychobiology of suicidal behavior, in which abnormalities of numerous neurochemical systems have been implicated. It is the 5-HT system that has received the most attention.

Serotonin

In addition to genetic marker studies, the serotonergic system in suicide can be studied in the brain tissue of victims obtained postmortem; the CSF of individuals who have attempted suicide or those at high risk followed longitudinally until they attempt or complete suicide; in other peripheral tissues of living subjects, principally platelets; and through neuroendocrine challenge paradigms. Here we examine the evidence of a role for 5-HT in modulating suicidal behavior from each of these sources.

Postmortem Studies

Tissue Levels of Serotonin and 5-Hydroxyindoleacetic Acid. The first evidence linking neurochemical changes in the brain and suicidal behavior was provided by studies of 5-HT and 5-HIAA in tissue obtained postmortem from suicide victims. Table 27.5 lists studies in which indolamine and metabolite levels from a variety of brain regions were compared in suicides and controls. In many of these studies, investigators could find no differences in either 5-HIAA (224–229) or 5-HT levels between groups (224,225,228–232) in a variety of brain regions. 5-HT levels were significantly decreased in specific brain regions of suicides in

five studies, including the brainstem (227,233), raphe nuclei (234), putamen (224), and hypothalamus of victims with diagnoses other than schizophrenia (235). Only two studies showed significant reduction in 5-HIAA levels in brains of suicides relative to controls. In both cases, long postmortem intervals complicate interpretation (230,232), as does the finding of *elevated* 5-HT levels in the basal ganglia (235) and *elevated* 5-HIAA levels in the amygdala (224) and hippocampus (236) of suicide victims.

Underwood et al. (237) examined brainstem dorsal raphe nucleus serotonergic neurons for number and morphometry in the postmortem brains of seven suicide victims and six controls. They found that serotonergic neuron number and density were higher in suicide victims than controls, concluding that impaired serotonergic transmission in suicide is not owing to fewer neurons.

At most, these studies suggest a trend for decreased indoleamine and metabolite levels in the brainstem nuclei of suicide victims. The inconsistency of findings may be attributed to the influence on monoamine levels of age, gender, postmortem interval, antemortem exposure to centrally active drugs, diurnal and seasonal variation, and the duration and nature of the agonal state (238). Lack of detail regarding diagnosis and treatment histories of subjects further limits the interpretation of these data.

Serotonin Transporter Studies

As a methodology for studying neurochemical systems in suicide, the assay of receptor binding in postmortem tissue has great advantages over measurement of monoamine levels. In general, receptor binding is far less subject to the influences of postmortem interval and acute (but not long-term) antidepressant exposure. Presynaptically, 5-HTT binding has been studied extensively using imipramine and paroxetine ligands and presumably reflects the number and functional status of 5-HT nerve terminals.

Table 27.6 lists results of studies comparing measures of binding at the 5-HT reuptake site in the brains of suicide victims and controls. Although the greatest number of studies found no difference in the frontal cortex (236,239–242) or other brain regions (236,241–246), others found significant decreases in imipramine binding (225,239,241,247–253) in the brains of suicide victims. One study that showed *greater* imipramine binding in the frontal cortex of suicides than in controls ascribed this inconsistent finding to premorbid drug exposure in the suicide group (254). Other explanations offered for the discrepancies include the observation that imipramine is a relatively nonspecific ligand that binds to muscarinic and α-adrenergic receptors in addition to the 5-HT reuptake site (244,255).

In general, paroxetine binding studies have yielded negative findings in various brain regions as well. Rosel et al. (250) reported that imipramine and paroxetine bind to distinct sites in the 5-HTT macromolecule complex and that

TABLE 27.5. STUDIES COMPARING LEVELS OF SEROTONIN AND 5-HYDROXYINDOLEACETIC ACID IN BRAINS OF SUICIDE VICTIMS AND CONTROLS

Study	Subjects	Tissue	Findings
Crow et al., 1984 (226)	10 suicides (7 depressed), 9 depressives with natural deaths, 20 Alzheimer disease, 19 controls	Frontal cortex	Nonsignificant trend for ↓ 5-HIAA in suicides
Bourne et al., 1968 (230)	23 suicides, 28 controls	Brainstem	↓ 5-HIAA in suicides; no difference in 5-HT
Cochran et al., 1976 (231)	10 depressed suicides, 9 alcoholic suicides, 12 controls	33 brain regions	No difference in 5-HT
Beskow et al., 1976 (232)	23 suicides, 62 controls	7 brain regions	No difference in 5-HT; ↓ 5-HIAA in suicides no difference after controlling for postmortem interval
Shaw et al., 1967 (233)	28 suicides, 17 controls	Hindbrain	↓ 5-HT in depressed suicides
Cheetham et al., 1989 (224)	19 suicides, 19 controls	Cortex, hippocampus, brainstem	↑ 5-HIAA in amygdala of suicides; ↓ 5-HT in putamen
Pare et al., 1969 (227)	24 suicides, 15 controls	Brainstem, hypothalamus, caudate	No difference in 5-HIAA; ↓ 5-HT in brainstem of suicides
Owen et al., 1986 (236)	19 suicides, 19 controls	Occipital and frontal cortex, hippocampus	↑ 5-HIAA in hippocampus of suicides
Owen et al., 1983 (261)	17 suicides, 20 controls	Frontal cortex	No difference in 5-HIAA
Lloyd et al., 1974 (234)	5 suicides, 5 controls	Raphe nuclei	↓ 5-HT in nucleus dorsalis and centralis inferior
Korpi et al., 1986 (235)	30 schizophrenics, 14 nonschizophrenic suicides, 29 normal controls	14 brain regions	↓ 5-HT in hypothalamus of nonschizophrenic suicides
Arato et al., 1987 (225)	13 suicides, 14 controls	Frontal cortex, hippocampus	No difference in 5-HT or 5-HIAA
Mann et al., 1996 (228)	24 suicides, 24 matched controls	Prefrontal cortex, temporal cortex	No difference in 5-HT, 5-HIAA, 5-HTP, or TRY
Arranz et al., 1997 (229)	18 suicides, 23 controls	Frontal cortex, gyrus cinguli, hypothalamus	No difference in 5-HT, NE, DA, 5-HIAA, DOPAC, HAV, or 5-hydroxy-L-tryptophan

5-HIAA, 5-hydroxyindoleacetic acid; 5-HT, Serotonin; 5-HTP, 5-hydroxytryptophan; TRY, tryptophan; NE, norepinephrine; DA, dopamine; DOPAC, dihydroxyphenly acetic acid; HVA, homovanillic acid.

imipramine is a better biologic marker for 5-HTT function in suicides with affective disorder.

Regional differences in neurochemical measures may also help to explain the variability between studies. Gross-Isseroff et al. (239) found increased imipramine binding in pyramidal and molecular cell layers of some hippocampal fields and decreased imipramine binding in tissues of the claustrum, postcentral gyrus, and insular cortex. Similarly, Arato et al. (225,256) found that in suicides, imipramine binding was significantly greater in the left compared with the right frontal cortex, in marked contrast to controls in whom binding was greater on the right. Although others have failed to replicate this pattern of interhemispheric differences

(240,244), it serves to stress the need for specificity in future studies regarding not only descriptive characteristics of the population but of the neuroanatomic region sampled as well.

Postsynaptic Serotonin Receptor Binding Studies

In addition to the 5-HTT site, five distinct 5-HT receptors have been identified. Here we review receptor types that have been studied in suicide, including 5HT$_1$ (and its 5HT$_{1A,1B,1D,1E/1F}$ subtypes), 5HT$_2$ (and its subtype 5HT$_{2A}$), and 5HT$_3$.

The 5HT$_2$ receptor is located on the postsynaptic neuron, with its greatest density in the frontal cortex projections

TABLE 27.6. STUDIES COMPARING BINDING AT THE SEROTONIN REUPTAKE SITE IN BRAINS OF SUICIDE VICTIMS AND CONTROLS

Study	Subjects	Tissue	Findings
Crow et al., 1984 (226)	10 suicides, (7 depressed), 9 depressives natural deaths, 20 Alzheimer disease, 19 controls	Frontal cortex	↓ [3H]imipramine binding only in depressed suicides
Owen et al., 1986 (236)	19 suicides, 19 controls	Frontal cortex, occipital cortex, hippocampus	[3H]imipramine binding—no difference between groups
Gross-Isseroff et al., 1989 (239)	12 suicides, 12 controls	Multiple cortical and subcortical regions	↑ [3H]imipramine binding in specific cell layers of hippocampus fields; ↓ imipramine binding in postcentral and insular cortex and claustrum; no difference in prefrontal cortex or brainstem
Arato et al., 1987 (225)	13 suicides, 14 controls	Frontal cortex, hippocampus	[3H]imipramine binding in frontal cortex of suicides left > right hemisphere; in controls right > left hemisphere; no difference in hippocampus
Lawrence et al., 1990 (243)	22 suicides, 20 controls	10 brain regions	No difference in [3H]paroxetine binding in any region
Lawrence et al., 1990 (244)	8 suicides, 8 controls	Frontal cortex, putamen, substantia nigra	[3H]paroxetine binding—no interhemispheric differences; no differences in either hemisphere between suicides and controls
Meyerson et al., 1982 (254)	8 suicides, 10 controls	Frontal cortex	↑ [3H]imipramine binding in suicides
Stanley et al., 1982 (248)	9 suicides, 9 controls	Frontal cortex	↓ [3H]imipramine binding in suicides
Arora and Meltzer, 1991 (240)	6 suicides, 10 controls	Frontal cortex	No interhemispheric differences; no differences between suicides and controls
Arora and Meltzer, 1989 (477)	28 suicides, 28 controls	Frontal cortex	[3H]imipramine binding—no differences between groups
Lawrence et al., 1998 (253)	22 depressed suicides, antidepressant free; 17 depressed suicides, ADT	Frontal cortex, putamen, substantia nigra	↓ DMI-defined [3H]imipramine binding in the putamen of ADT suicides
Andersson and Marcusson, 1992 (242)	19 suicides, 22 controls	Frontal cortex, cingulate cortex, hypothalamus	[3H]paroxetine binding—no difference between groups
Little et al., 1997 (478)	8 suicides with MDD, 8 matched controls	Frontal cortex, midbrain, hippocampus	[125]RTI-55 binding assay to the 5-HTT and 5-HTT mRNA level—no difference between groups
Bligh-Glover et al., 2000 (246)	10 suicides with MDD, 10 matched controls	Subnuclei of the dorsal raphe	[3H]paroxetine binding—no significant differences
Arango et al., 1995 (252)	22 suicides, 22 controls	Prefrontal cortex	↓ 5-HTT and ↑ 5-HT$_{1A}$, mainly in ventrolateral prefrontal cortex of suicides
Rosel et al., 1997 (250)	11 depressed suicides, 11 controls	Frontal, temporal, cingulate cortex; hypothalamus; hippocampus; amygdala	↓ B$_{max}$ of [3H]imipramine without change in K$_d$ in the hippocampus of suicides; no difference in [3H]paroxetine binding

(continued)

TABLE 27.6. (*continued*)

Study	Subjects	Tissue	Findings
Rosel et al., 1998 (251)	17 depressed suicides, 17 controls	Frontal and cingulate cortex, hippocampus, amygdala	↓ B_{max} of [^{3}H]imipramine with no change in K_d in the hippocampus of suicides; no differences in [^{3}H]paroxetine binding
Mann et al., 1996 (228)	24 suicides, 24 matched controls	Prefrontal and temporal cortex	No difference in 5-HTT sites in Brodmann area 9 between groups
Hrdina et al., 1993 (245)	10 depressed suicides, 10 controls	Prefrontal cortex, amygdala	[^{3}H]paroxetine, [^{3}H]ketanserin binding—↑ in B_{max} of 5-HT$_2$; no difference in paroxetine binding
Arato et al., 1991 (256)	23 suicides, 23 controls	Frontal cortex	↑ [^{3}H]imipramine binding in left hemisphere of suicides vs. controls; ↓ binding in right hemisphere of suicides vs. controls; differences more pronounced in violent suicides

DMI, desmethylimipramine; ADT, antidepressant treated; MDD, major depressive disorder; 5-HTT, serotonin transporter.

of the ascending dorsal raphe nucleus. Based on the observation that the lesioning of serotonergic neurons in the raphe nuclei leads to up-regulation of postsynaptic 5HT$_2$ receptors and decreased imipramine binding presynaptically (257,258), Stanley and Mann (247) predicted that if the 5-HT deficiency hypothesis is operative, then a similar pattern would be found in victims of completed suicide. They found that the density (B_{max}) of 5HT$_2$ receptors was 44% greater in the frontal cortex of 11 suicide victims than in that of controls matched for age, gender, and postmortem interval. As shown in Table 27.7, these results were subsequently replicated by some, but not all, groups. Mann et al. found a 28% increase in 5HT$_2$ receptor binding in the frontal cortex of 21 suicides compared with matched controls (259), and Arora and Meltzer (260) found 35% greater B_{max} in suicides versus controls. This latter study, which also used [^{3}H]spiroperidol, found a 35% increase in an even larger group of suicide victims compared with controls, noting higher values in men than women, but no difference in 5HT$_2$ binding between suicides who died by violent and nonviolent means. Four other studies investigated this question using [^{3}H]ketanserin as the ligand, all of which failed to show a significant difference between suicide victims and controls in the frontal cortex (226,236,261,262).

Cheetham et al. (262) studied 19 suicides, ages 16 to 57 years, all of whom had documented depressive disorders. Although they found no difference in ketanserin binding between suicides and controls in the frontal cortex, they noted an inverse correlation between receptor density and age and that 5HT$_2$ receptor binding was significantly decreased in the hippocampus of depressed suicide victims who had not been treated with antidepressant medications compared with controls. They concluded that reports of increased 5HT$_2$ binding in the frontal cortex of suicide victims may be owing to drug treatments or the quality of

aggression rather than to depression or suicide *per se*. Other studies, however, have found increased 5HT$_2$ receptor binding in subjects with major depression who died from natural causes compared with controls (263), and Mann et al. (255) noted that chronic antidepressant therapies down-regulate 5HT$_2$ receptors. The meaning of these discrepant findings, therefore, remains unclear.

Some studies have used the technique of *in vitro* quantitative receptor autoradiography to compare receptor binding in suicides and controls. Gross-Isseroff et al. (264), using [^{3}H]ketanserin as the ligand, found decreased binding in the frontal cortex of victims of suicide, ages 50 years and younger, compared with controls, whereas there was no such difference in subjects older than age 50 years. Furthermore, they found differing patterns of age dependence of ketanserin binding in different brain regions. In the frontal cortex of controls and, to a lesser extent, suicides, age was inversely correlated with ketanserin binding. In the hippocampal formation, however, young suicide victims had higher binding and older victims had lower binding than controls. Thus, the impact of aging on 5HT$_2$ receptor function may differ between groups as a function of anatomic site. Lower binding in the hippocampus of suicide victims was reported by Rosel et al. (251), although age was not considered in the analysis.

In autoradiographic studies, Arango et al. (265,266) used the more specific 5HT$_2$ receptor ligand lysergic acid diethylamide-125 to demonstrate significantly greater binding in the prefrontal, but not the temporal, cortex of suicides compared with controls. They further demonstrated a laminar pattern in the cortex, similar in both suicides and controls, in which binding was greatest in intermediate cell layers. Negative results were reported by Stockmeier et al. (267) when they examined suicides with depression specifically in the right frontal cortex and hippocampus using [^{3}H]ketanserin and also by Arranz et al. (268), even

TABLE 27.7. STUDIES COMPARING 5HT$_2$ RECEPTOR BINDING IN BRAINS OF SUICIDE VICTIMS AND CONTROLS

Study	Subjects	Tissue	Findings
Owen et al., 1983 (261)	17 suicides, 20 controls	Frontal cortex	[^{3}H]ketanserin binding—trend for ↓ in suicides
Owen et al., 1986 (236)	19 suicides, 19 controls	Frontal cortex, occipital cortex, hippocampus	[^{3}H]ketanserin binding—no difference at any site
Mann et al., 1986 (259)	21 suicides, 21 controls	Frontal cortex	↑ Spiroperidol binding in suicides
Arora and Meltzer, 1989 (260)	32 suicides, 37 controls	Frontal cortex	↑ [^{3}H]spiperone binding in suicides (males > females; violent > nonviolent)
Gross-Isseroff et al., 1990 (264)	12 suicides, 12 controls	Frontal cortex, hippocampus, other	↓ [^{3}H]ketanserin binding in frontal cortex of young suicides vs. controls; no difference for suicides age >50 years vs. controls
Hrdina et al., 1993 (245)	10 depressed suicides, 10 controls	Prefrontal cortex, amygdala	↑ [^{3}H]ketanserin binding in B$_{max}$ of 5-HT$_2$ in prefrontal cortex and amygdala
Rosel et al., 1998 (251)	17 depressed suicides, 17 controls	Frontal and cingulate cortex, hippocampus, amygdala	[^{3}H]ketanserin, ↓ 5-HT$_{2A}$ binding sites in hippocampus of suicides vs. controls
Arranz et al., 1994 (268)	18 suicides, 23 controls	Frontal cortex	[^{3}H]ketanserin—no difference in 5HT$_2$ binding between groups
Stanley and Mann, 1983 (247)	11 suicides, 11 controls	Frontal cortex	↑ [^{3}H]spiroperidol binding in suicides
Arango et al., 1992 (266)	13 suicides, 13 controls	Prefrontal cortex, temporal	↑ ^{125}I-LSD binding in suicides in frontal cortex only
Cheetham et al., 1988 (262)	19 suicides, 19 controls	Frontal, temporal, occipital cortex; amygdala; hippocampus	↓ [^{3}H]ketanserin binding in hippocampus, no difference in other regions
Arango et al., 1990 (265)	11 suicides, 11 controls	Prefrontal cortex, temporal cortex	↑ ^{125}I-LSD binding in suicides in frontal cortex only
Crow et al., 1984 (226)	10 suicides, 9 depressive controls, 20 Alzheimers disease, 19 normal controls	Frontal cortex	[^{3}H]ketanserin binding—no difference between groups

when they took into account the presence or absence of depressive symptoms and the violence of the suicide method. The variability in findings, therefore, can be accounted for by many factors, including variation in the specificity of ligands, age of subjects, and brain region studied, in addition to the effects previously noted of antemortem drug exposures, manner of death (violent vs. nonviolent, prolonged vs. brief agonal state), and other factors. Decrease in 5HT$_2$ receptors has been observed in schizophrenics who die of natural causes (269) but not suicide (241), underscoring the importance of diagnosis in neurobiologic research on suicide as well.

5HT$_1$ receptors have both pre- and postsynaptic subtypes, with highest density in areas of the hippocampus and basal ganglia receiving projections from the median raphe nucleus. Table 27.8 lists studies examining 5HT$_1$ receptor binding in brains of suicide victims and controls assayed by a variety of methods. Using [^{3}H]5-HT, several studies have found

no difference between suicides and controls in the frontal (226,236,259,261,270), occipital (236), or temporal cortex (270). However, Cheetham et al. (270) found a 20% decrease in [^{3}H]5-HT binding in the hippocampus of suicides.

Using a selective 5HT$_{1A}$ receptor agonist, hydroxy-2-(di-*n*-propylamino) tetralin, as a ligand, investigators found no (267,268,270,271) or equivocal differences between groups. Matsubara et al. (272), for example, found that suicides who died by nonviolent means had 25% higher 5-HT$_{1A}$ receptor binding than did violent suicides or controls, possibly owing to the effect on receptors of carbon monoxide poisoning and drugs taken in overdose, the principal means of death in the nonviolent group. They also reported that 5HT$_{1A}$ receptor binding was negatively correlated with age in male controls but not females, whereas no such relationship could be demonstrated for suicide victims of either gender.

Quantitative autoradiography for 5HT$_{1A}$ was performed on coronal sections of the prefrontal cortex of suicide

TABLE 27.8. STUDIES COMPARING 5HT₁ AND 5HT₁ₐ RECEPTOR BINDING IN BRAINS OF SUICIDE VICTIMS AND CONTROLS

Study	Subjects	Tissue	Findings
Owen et al., 1983 (261)	17 suicides, 20 controls	Frontal cortex	[^{3}H]5-HT receptor binding—no difference between groups
Owen et al., 1986 (236)	19 suicides, 19 controls	Frontal and occipital cortex, hippocampus	[^{3}H]5-HT receptor binding—no difference between groups
Dillon et al., 1991 (271)	14 suicides, 14 controls	Numerous cortical and subcortical brain regions	OH-DPAT binding—no difference between groups; negative correlation with age in males
Matsubara et al., 1991 (272)	23 suicides, 40 controls	Frontal cortex	OH-DPAT receptor binding—negative correlation with age in male controls but not females or suicides of either sex; ↑ binding in nonviolent suicides
Mann et al., 1986 (259)	21 suicides, 21 controls	Frontal cortex	[^{3}H]5-HT receptor binding—no difference between groups
Cheetham et al., 1990 (270)	19 suicides, 19 controls	Frontal and temporal cortex, hippocampus	↓ [^{3}H]5-HT receptor binding in hippocampus of suicides; no difference in cortical regions
Arango et al., 1995 (252)	22 suicides, 22 controls	Prefrontal cortex	↑ OH-DPAT receptor binding; 5HT₁ₐ receptor binding in suicides, mainly in the ventrolateral prefrontal cortex
Arranz et al., 1994 (268)	18 suicides, 23 controls	Frontal cortex	OH-DPAT receptor binding—no difference in 5HT₁ₐ binding
Stockmeier et al., 1997 (267)	Depressed suicides vs. controls	Right prefrontal cortex, hippocampus	OH-DPAT—no difference in 5HT₁ₐ binding
Crow et al., 1984 (226)	10 suicides, 9 depressives, 20 Alzheimer disease, 19 controls	Frontal cortex	[^{3}H]5HT receptor binding—no differences between groups

OH-DPAT, hydroxypropylaminotetralin.

victims and controls by Arango et al. (252). They found higher 5HT₁ₐ receptor binding in the suicide group, an effect most pronounced in the ventrolateral prefrontal cortex. Stockmeier et al. (273) reported enhanced 5HT₁ₐ binding in the dorsal raphe nucleus in suicides with depression compared with controls but did not examine the prefrontal cortex in these subjects.

5HT₁D and 5HT₁E/1F receptors were examined in the basal ganglia and cortex of depressed suicides and matched controls by Lowther et al. (274). They found a significant increase in 5HT₁D binding in the globus pallidus of suicide victims who died by violent methods and were reported to be antidepressant free but no differences between groups in other brain regions in 5HT₁D receptor binding and no significant differences in any brain regions for the 5HT₁E/1F receptor. A significant increase in 5HT₁D receptor binding was reported by Arranz et al. (268) in the frontal cortex of suicides regardless of their depression history.

In summary, evidence from studies of both suicide attempters and victims of completed suicide supports a potential role for abnormalities in central 5-HT function in the expression of suicidal behavior. This evidence is derived from CSF and brain tissue levels of tryptophan, 5-HT, and 5-HIAA, 5-HT uptake sites using imipramine and paroxetine binding, and 5-HT receptor binding studies. In general, these findings are consistent with a decrease in presynaptic 5-HT function, reflected in decreased 5-HT levels, 5-HT turnover, and presynaptic receptor binding, coupled with postsynaptic receptor changes as a compensatory response. Frontal cortex is the brain region most intensively studied, with few significant abnormalities reported in temporal cortex tissue. Cheetham et al. (262,270) noted decrements in hippocampal 5HT₁ and 5HT₂ receptor binding in the absence of differences in these measures in the frontal cortex of depressed suicides and controls, leading them to suggest the need for more detailed studies of this brain region as a mediator of suicidal behavior in affective illness.

Despite the weight of these findings, they are far from unanimous. The inconsistency between studies has been ascribed to many factors, including the postmortem interval, season of the year, premorbid exposure to prescribed and illicit drugs, manner of death (including its violent nature, which may be a central feature of the purported biologic abnormality), age and gender of subjects, specificity of ligands used, and regional differences within the brain (interhemispheric and intrahemispheric).

Cerebrospinal Fluid Studies

Because of the availability of brain tissue for postmortem analysis, CSF 5-HT measures have not been studied in completed suicide. However, Stanley et al. (275) simultaneously sampled CSF 5-HIAA drawn from the spinal cistern postmortem with levels of 5-HIAA in the cerebral cortex. Demonstrating a strong correlation, they concluded that CSF metabolite concentrations are a valid reflection of brain 5-HT activity. It is with greater confidence, then, that the findings of many studies of serotonergic indices from the lumbar CSF of suicide attempters can be assumed to reflect brain function.

CSF sampling has the great advantage of easy accessibility. Furthermore, because the fluid is obtained from a living subject, simultaneous measurements of symptoms and behaviors can be made, and the subjects can be followed longitudinally to establish the predictive value of CSF measures. There are, nonetheless, numerous limitations to such studies that should be considered in their interpretation. As previously noted, suicide attempters are not uniformly representative of completed suicides. The subjects in CSF studies likely represent a heterogeneous group, only some of whom may share any postulated neurobiologic profile with suicide completers. Studies frequently fail to define adequately the subjects' diagnoses or history of suicidal behavior and to distinguish adequately, at times, among suicidal ideation, low lethality attempts, and high lethality attempts.

Measuring CSF monoamines and metabolites is technically difficult because of numerous factors that influence their concentration. These include age (increasing concentrations of 5-HIAA with aging), gender (lower concentrations in men), body height, diet, level of physical activity before the lumbar puncture, the intervertebral space used and position of the subject, the amount of CSF drawn, the time of day and season of the year, and recent history of exposure to psychotropic medications. Given these cautions, a large and relatively consistent body of work offers support to the postmortem studies suggesting altered central 5-HT function in attempted suicide.

The first observation of an association between CSF 5-HT measures and suicidal behavior was made during the course of investigations of CSF monoamines in major depression. Asberg et al. (276) noted that depressives with low CSF levels of 5-HIAA had a higher incidence of suicide attempts. Of 68 patients with depressive illness in their study, 15 had attempted suicide during the index episode. Concentrations of CSF 5-HIAA were bimodally distributed in the sample, with 40% of the low 5-HIAA group having attempted suicide compared with 15% of patients with normal CSF

5-HIAA levels. Furthermore, all suicide attempts by violent means clustered in the low 5-HIAA group, two of whom subsequently killed themselves during the study period.

Several subsequent studies confirmed the original finding of Asberg et al., whereas others failed to do so. Negative results may be explained by methodologic differences or differences between patient groups. For example, two of the studies that found no association between CSF 5-HIAA and suicidal behavior included substantial numbers of patients with bipolar disorders in their sample (277,278). Although Roy-Byrne et al. (278) found no difference in mean CSF 5-HIAA between patients with affective disorders with and without a history of suicide attempts, separate analyses for unipolar and bipolar patients were more revealing. They found no difference between suicidal and nonsuicidal bipolars, but suicidal unipolar depressives tended to have lower CSF 5-HIAA than the other groups. Agren (279) also found a significant negative correlation between 5-HIAA levels and a history of suicide attempts among unipolar depressives only, with no such difference evident for bipolar patients.

The frequent need for maintenance antidepressant medications in treated depressives makes follow-up studies of CSF monoamines in suicide attempters problematic. Nonetheless, the few available data suggest the possibility that low CSF 5-HIAA is a trait rather than solely a state characteristic. In approximately 50% of patients, a low CSF 5-HIAA concentration persists after recovery from the depressive episode. Abnormal values have been associated with both past (276,279,280) and future (276,281) attempts. Asberg et al. (238) suggest the existence of a subgroup of depressives whose CSF concentrations of 5-HIAA are both lower and more unstable over time than normal. Further reductions at times of illness exacerbation may augment suicide risk. Traskman et al. (281) reported on the follow-up of 119 patients for whom CSF 5-HIAA determinations had been made. Seven of these patients went on to commit suicide, six of whom had a history of attempt. A 5-HIAA level below the median for the entire patient group was a significant predictor of subsequent suicide. Among attempters, all six who went on to complete suicide had 5-HIAA levels below the median. Subsequent reports support the notion that low CSF 5-HIAA may be a useful indicator of subsequent attempted (282) or completed suicide.

Diagnoses Other than Depression

To further explore whether abnormalities in CSF 5-HIAA in suicide attempters were a result solely of their mood disturbance rather than, for example, a dimension of psychopathology that crosses diagnostic boundaries, researchers expanded studies to include patients with a variety of psychiatric illnesses. Schizophrenics with a history of suicide attempts have lower levels of CSF 5-HIAA than either controls or schizophrenics without a history of suicidal behavior (283–286), a finding confirmed in a long-term follow-up study comparing schizophrenics who eventually attempted suicide with those who did not (287). Banki et al. (286) found a significant association between recent suicide attempts and CSF 5-HIAA levels in both patients with adjustment disorders and subjects with alcoholism.

In further support of the notion that CSF 5-HIAA measures a dimension of psychopathology independent of affective disorder, Brown et al. (280) studied 22 young active-duty military men with diagnoses of personality disorder in the absence of affective symptomatology. Subjects with a history of suicide attempt had lower CSF 5-HIAA than those without such a history. In a subsequent study of 12 military men with borderline personality disorder and no affective illness, their findings were the same (288). In both studies, however, the authors went on to make associations between 5-HIAA levels and measures of aggression and impulsivity, demonstrating a strong and significant inverse relationship. Subjects with more impulsive personality disorders, higher scores on a measure of lifetime history of aggression/impulsive behavior, and higher T-scores on the psychopathic deviate scale of the Minnesota Multiphasic Personality Inventory had lower levels of CSF 5-HIAA.

These findings spurred other investigators to question whether similar correlations could be found in other subject groups with histories characterized by aggressiveness or impulsivity rather than suicidal behavior *per se*. Linnoila et al. (289) studied individuals with personality disorders incarcerated after violent acts. Dividing the group into impulsive and nonimpulsive offenders, they found significantly lower CSF 5-HIAA concentrations in the former group. Some subjects, however, had a history of a suicide attempt; these individuals had the lowest metabolite levels. Subsequently Virkkunen et al. (290) compared CSF monoamine metabolite levels in normal controls with 20 habitually violent offenders and 20 arsonists, a group thought to represent an extreme of impulsivity relatively free from traits of aggressiveness and violence. 5-HIAA levels were significantly lower in arsonists than comparison groups. Controls had the highest levels, with violent offenders showing intermediate levels of 5-HIAA. The results were interpreted to support the hypothesis that poor impulse control is associated with low levels of monoamine metabolites. A 3-year follow-up study of 36 violent offenders and 22 arsonists subsequently demonstrated lower CSF 5-HIAA levels at baseline in subjects who went on during the follow-up period to commit new violent offenses or arson (291). Again, interpretation of results is complicated by the fact that 27 of 58 subjects had a history of a suicide attempt, 25 of whom fulfilled criteria for an alcohol abuse diagnosis (292). Based on the additional observation that CSF 5-HIAA levels were significantly lower among the 35 subjects with alcoholic fathers than among those without, Linnoila et al. (293) suggested that low CSF 5-HIAA, in addition to impulsivity, may be characteristic of type 2 alcoholism, explaining some of the inconsistency in the literature regarding associations between alcoholism, suicidality, and CSF monoamine measures.

Other studies have demonstrated altered 5-HT metabolism in aggressive patients with an XYY chromosomal pattern (294), murder-suicides (295), children and adolescents with disruptive behavior disorders (296), and individuals who had murdered a sexual partner (297). In this latter study, the finding that CSF measures were significantly lower in crimes of passion than for other homicide offenders was taken as an indication of impulsivity in the former group. To rule out suicidal behavior as a potential confounder of the relationship between low CSF 5-HIAA and aggression, Stanley et al. (298) studied 64 mixed diagnosis patients who had no past suicidal behavior. They found that subjects with a history of adult aggression had significantly lower CSF 5-HIAA concentrations than the nonaggressive group. Aggressive individuals also scored significantly higher on self-report measures of hostility, impulsiveness, and sensation seeking.

Other Biologic Measures of Serotonin Function

In addition to CNS tissue assays and CSF measures, a variety of other strategies have been used to explore 5-HT function in suicidal people.

Glucose Tolerance Test

In addition to measuring CSF 5-HIAA in his populations of violent and impulsive offenders, Virkkunen et al. (290, 299,300) conducted glucose tolerance tests to demonstrate that the pattern of response differentiated impulsive from nonimpulsive groups. Arsonists and subjects with antisocial personality disorder and intermittent explosive disorder demonstrated blood glucose nadirs significantly below those of controls. They speculate that altered 5-HT input to the suprachiasmatic nucleus alters the regulatory mechanisms demonstrated in rat models to regulate insulin and hyperglycemic responses to oral or intravenous glucose loads (301).

Blood and Platelet Studies

Peripheral blood and platelet 5-HT indices are easily accessible, allow the study of living subjects, and thus may have clinical utility in suicide risk assessment. Studies have examined whole blood tryptophan and 5-HT measures, platelet 5-HT content, monoamine oxidase (MAO) enzyme activity, 5-HT uptake sites, and $5HT_2$ receptors, postulating that these peripheral measures reflect central 5-HT activity. For example, human platelets have long been recognized to have qualities of central presynaptic neurons (302).

Studies of whole blood 5-HT show reduced levels in acutely suicidal inpatients relative to controls (303). However, that there is no difference when subjects with and without a history of suicide attempts (304) are compared suggests that whole blood 5-HT is a state rather than trait measure of suicide risk. Whole blood tryptophan and platelet 5-HT levels have been found to be lower in prepubertal children with a recent suicide attempt (305). In contrast, platelet 5-HT levels were elevated in inpatients with borderline personality disorder after a suicide attempt (306,307).

Further, an increase in the baseline level of platelet 5-HT significantly predicted the recurrence of suicide attempts in a 1-year follow-up (308), suggesting the possibility of a trait relationship between increased platelet 5-HT level and recurrent suicidal behavior in patients with this high-risk illness.

Although results have been mixed in studies of imipramine binding and 5-HT uptake in platelets of major depressives (309), several investigators have found significant results in studies of depressed suicide attempters. Wagner et al. (310) found a trend for depressives who had made a violent suicide attempt to have higher platelet imipramine binding values than depressed nonattempters. Roy et al. (309) compared patients with affective disorders with controls, finding differences only between the women in each group. However, four of the 51 depressed patients committed suicide in a 1-year follow-up. In retrospect, a history of suicide attempt in combination with either decreased platelet imipramine binding or 5-HT uptake appeared to place patients at especially high risk of future suicide. In subsequent studies, both positive (311) and negative (312) findings of association between platelet imipramine binding and suicidal behavior have been reported. Marazziti et al. (313) found decreased density of both platelet [^{3}H]imipramine binding and of 5-HT uptake sites, with no change in sulfotransferase activity (a marker of catecholamine activity) in platelets of suicide attempters relative to healthy controls. A higher affinity constant (K_d) of platelet [^{3}H]paroxetine binding was related to a higher risk of short-term recurrence of a suicide attempt, suggesting a state relationship, whereas the maximal number of binding sites (B_{max}) did not correlate with suicidality (308).

Roy (314) examined platelet 5-HT uptake in relation to suicidal behavior in depression, finding a significantly higher dissociation constant (K_m) among depressed patients who had a lifetime history of a suicide attempt than either depressed patients who had never attempted suicide or controls. Patients with suicidal ideation at the index admission had significantly higher K_m values than patients without suicidal ideation. In addition, patients who reattempted or committed suicide during a 5-year follow-up period had significantly higher K_m values than controls.

Platelet MAO enzyme activity showed no correlation with history of suicide attempts in a study of inpatients with no axis I diagnosis. However, in a separate study by the same investigators in a similar patient population, a low MAO activity level was associated with impulsive suicide attempts (307). Although one study reported lower platelet MAO activity in patients with a history of suicidal behavior (304), another reported higher than average mean MAO platelet activity in patients with history of chronic suicidality, a finding that persisted even after resolution of the depression. The investigators concluded that platelet MAO activity may be a trait marker of suicide risk (315).

In addition to imipramine and paroxetine binding sites and 5-HT uptake sites, human platelets also have $5HT_2$ receptors involved in the mediation of 5-HT–amplified platelet aggregation (316). Biegon et al. (317) compared the platelets of suicidal men with controls and found increased $5HT_2$ receptors, whereas Pandey et al. (316) found significantly higher $5HT_2$ binding in a subgroup of depressives with recent attempts or serious ideation compared with nonsuicidal depressives and controls. McBride et al. (318) found a direct association among depressed attempters between $5HT_2$ receptor binding on platelets and lethality of suicidal behavior as well. Neuger et al. (319) showed a strong positive correlation between $5-HT_2$ receptor binding (but not density) and suicidal behavior in depressives. Similar findings have been reported regardless of the diagnosis (304).

In summary, studies of peripheral 5-HT measures in suicide complement other lines of CNS 5-HT studies in pointing out the association between 5-HT dysfunction and suicidal behavior. There is a trend toward lower 5-HT as a state measure in acutely suicidal patients and higher baseline 5-HT measures in patients with recurrent suicidality, a trait measure. Studies of $5HT_2$ receptors are relatively consistent in showing decreased levels and, hypothetically, decreased function, whereas studies of imipramine binding are truly mixed. Further work is needed to better define the predictive value of these measures for suicidal risk. Continued attention to seasonal variation, medication exposure history, and diagnoses is warranted in future studies.

Neuroendocrine Studies

Altered levels of 5-HT and metabolites in the CSF and peripheral tissues leave unresolved the question of where in this dynamic system the deficit lies. Presynaptic lesions may result in decreased 5-HIAA owing to diminished synthesis and release of 5-HT, whereas *increased* presynaptic function could result in decreased turnover as a compensatory postsynaptic response. Various challenge paradigms have been designed that take advantage of the regulatory interaction between 5-HT and neuroendocrine systems to test their integrity. Among the agents most frequently used are 5-HT precursors and agonists.

Meltzer et al. (320) administered 5-hydroxytryptophan to 40 patients with major affective disorder, observing a significant correlation between the degree of cortisol response and a history of suicidal behavior. They interpreted the finding to reflect postsynaptic $5HT_2$ receptor up-regulation in the hypothalamus and thus presynaptic *hypo*function.

Concerned with the possibility that cortisol response to 5-hydroxytryptophan does not specifically reflect $5HT_2$ receptor function, however, Coccaro et al. (321) chose fenfluramine as their challenge agent. Fenfluramine both augments the release and blocks the reuptake of endogenous stores of 5-HT, resulting in increased serum prolactin levels in the peripheral circulation. They found that compared with normal controls and male patients with no history of suicidal behavior, subjects with either major depression or a personality disorder diagnosis who had made a suicide attempt in the past demonstrated a significantly blunted prolactin response to fenfluramine infusion. They interpreted these findings as suggesting an overall reduction in 5-HT neurotransmission in the limbic–hypothalamic–pituitary axis in affective and/or personality disordered patients with a history of parasuicide.

Extending this line of research, Malone et al. (322) demonstrated that depressed patients with a history of a higher lethality suicide attempt had a significantly lower prolactin response to fenfluramine compared with those with low lethality suicidal behavior, even when controlling for a wide range of potential confounds.

How useful can the fenfluramine challenge test be? Studying a range of 5-HT indices, Mann et al. (323) found a positive correlation of CSF 5-HIAA with a maximal prolactin response to fenfluramine but not with platelet 5-HT$_2$ receptor indices. The fenfluramine-stimulated maximal prolactin response correlated with platelet 5-HT$_2$ receptor number, particularly in older patients. They concluded that less invasive procedures, such as a fenfluramine challenge test or platelet 5-HT$_2$ measures, cannot replace but only complement CSF-5-HIAA in suicide research with live patients.

Summary

In summary, data from studies using a broad range of technologies and tissues support the existence of an association between suicidal behavior and functioning of the central 5-HT system. Mann et al. (324) proposed a stress diathesis model in which the risk of suicidal acts is determined by the interplay between a psychiatric illness (the stressor) and predisposing traits, such as aggression and impulsivity, that, as reviewed above, are closely linked to serotonergic function. Much more work is necessary, however, owing to the complexity of this issue. The methodologies for measuring both clinical and neurobiologic parameters are intricate, easily confounded by difficulties controlling the numerous variables that independently affect them. Despite these limitations, it seems most likely that 5-HT plays its role in suicidal behavior through the mediation of impulse control and aggression. Deficits in this system may predispose individuals to violent behavior, as has been demonstrated in animal models, under specific circumstances, such as the development of a comorbid depressive disorder or individually significant stressful life circumstances. Whether the emergent behavior is destructive to self or others may depend in turn on complex interactions with other neurochemical systems, reviewed below.

Dopamine

The catecholamines DA and NE have received far less attention than 5-HT in studies of suicidal behavior. There is an association in the CSF between levels of 5-HIAA and the DA metabolite homovanillic acid (HVA), but it is unclear whether the close correlation is owing to a shared transport mechanism or a functional connection between the parent amines. Several authors have reported low levels of CSF HVA paralleling decreases in 5-HIAA in suicidal depressives (281,325,326). For example, Roy et al. (326) studied 27 patients with major depressive disorders, finding lower CSF HVA in those with a history of suicide attempts. They then followed these patients for 5 years (282) and continued to find an association between decreased CSF HVA levels and further suicide attempts during the follow-up period. In fact, in some studies of depressed patients, the association has been stronger between suicide attempts and low CSF HVA than with low CSF 5-HIAA (279,327, 328).

Some authors suggest that the association between central dopaminergic function and suicidal behavior is specific to affective illness (281). The failure by Ninan et al. (329) to find any difference in CSF HVA levels between suicidal and nonsuicidal schizophrenics supports this possibility. More recently, however, Engstrom et al. (328) showed that suicide attempters in a group of hospitalized patients of mixed diagnoses had significantly lower HVA levels, HVA/5-HIAA ratios, and HVA/3-methoxy-4-hydroxyphenylglycol (MPHG) ratios than controls.

Roy et al. (330) also measured the urinary output of DA metabolites in patients with major depression and normal controls to explore whether DA dysregulation differentiated those with suicidal behavior. Not only did patients with a history of suicide attempts have lower urinary outputs of DA metabolites than nonattempters or controls, but patients who made repeat attempts during the 5-year follow-up period had significantly lower values than other groups. Conversely, Tripodianakis et al. (331) failed to show any difference in urinary 5-HIAA and HVA between patients with adjustment disorder admitted after a suicidal attempt and healthy controls. In combination, these data indicate that decreased CSF HVA levels may signal vulnerability to suicide attempts in patients with major affective illness; however, its association with suicidality in patients with other diagnoses seems less likely.

Several studies showed no difference between suicides and controls in DA content of brain tissue assayed postmortem from any brain region (227,229,232). Others, however, did find differences. Crow et al. (226) found an insignificant trend for suicides to have lower levels of HVA in the frontal cortex than normal controls, whereas Bowden et al. (332) found evidence of reduced DA turnover in depressed suicides compared with controls. They noted, however, that the finding could be explained by the subjects' ingestion of toxic agents (332). No studies that have examined D$_1$ or D$_2$ receptors in the caudate, putamen, or nucleus accumbens of depressed suicide victims and controls could yet demonstrate a difference between groups (333,334).

Despite the lack of findings, several other lines of evidence do support the need for future research in this area. In reviewing the epidemiology of suicide, we note the striking rise in rate with age in men. As in other monoaminergic systems, there are well-documented reductions in dopaminergic neurons, the function of receptors and uptake sites, and synthetic enzyme activity with increasing age (334–337). Furthermore, as discussed elsewhere in this chapter, suicide risk is elevated in both HD and PD, disorders in which degenerative changes of the central dopaminergic systems have been implicated (75,76,338,339). Its role, therefore, may be more important in segments of the population at greater risk owing to aging and/or the neurobiologic effects of specific pathologic processes.

Norepinephrine/Epinephrine

As for DA, there have been relatively few studies examining the noradrenergic system in suicide, and those that have been conducted provide equivocal results. Positive findings include the observation of an inverse correlation between CSF levels of the NE metabolite MPHG and ratings of suicidal intent in unipolar and bipolar depressives (325) and arsonists and violent offenders (292). MHPG in the CSF is negatively correlated with a lifetime history of aggression in children and adolescents with disruptive behavior disorders as well (296). In contrast, Brown et al. (280) found a significant positive correlation between CSF MHPG levels and aggression scores among 26 military recruits with personality disorders and significantly higher levels in those with a history of suicidal behavior than those who had not attempted suicide. These investigators were unable to replicate this finding, however, in a subsequent study of 12 men with borderline personality disorder but no depressive illness (288). Roy et al. (340) found no difference between depressives with and without suicide attempts in CSF, plasma, or urinary indices of noradrenergic function, but Ostroff et al. (186,341) showed in two studies that a low ratio of urinary NE to epinephrine was predictive of suicidal behavior in patients with mixed psychiatric diagnoses. Traskman-Bendz et al. (342) found that violent suicide attempters had elevated CSF concentrations of MHPG but no significant differences between violent and nonviolent attempters in 24-hour urinary NE-epinephrine. In prospective follow-up, four patients completed suicides, all of whom had CSF MHPG concentrations above the median.

Studies that have examined postmortem tissue levels of NE and MHPG have failed to show differences between suicides and controls in the prefrontal cortex (343), frontal cortex, gyrus cinguli, hypothalamus (229), and locus caeruleus (344). However, each study was influenced by limitations of the high-performance liquid chromatography methodology and diagnostic and medication exposure issues.

Studies that used membrane preparations and/or autoradiography to measure β-adrenergic receptors in postmortem tissue are listed in Table 27.9. Three reported significantly increased binding in the frontal cortex of suicides compared with controls (3,265,345). Arango et al. (265) found no apparent correlation with age, gender, or 5HT₂ receptor binding in the same subject group. Conversely, three studies found no difference (226,254,346), and one found *decreased* β-receptor density in the frontal cortex of suicides compared with controls (347). Using the highly specific ligand H3CGP12177 in nine brain regions in drug-free depressed suicides and controls, De Paermentier et al. (348) found significantly decreased β-receptor binding in the temporal

TABLE 27.9. STUDIES COMPARING MEASURES OF β-ADRENERGIC RECEPTOR BINDING IN BRAINS OF SUICIDE VICTIMS AND CONTROLS

Study	Subjects	Tissue	Findings
Arango et al., 1990 (265)	11 suicides, 11 controls	Frontal and temporal cortex	↑ [¹²⁵I]pindolol binding in prefrontal and temporal cortex in suicides
Stockmeier and Meltzer, 1991 (346)	22 suicides, 22 controls	Frontal cortex	DHA binding—no difference between groups
Meyerson et al., 1982 (254)	8 suicides, 10 controls	Frontal cortex	DHA binding—no difference between groups
De Paermentier et al., 1990 (348)	21 suicides, 20 controls	9 brain regions	↓ CGP 12177 binding in temporal and frontal cortex of suicides (violent > nonviolent)
Biegon and Israeli, 1988 (345)	14 suicides, 14 controls	Prefrontal cortex	↑ DHA binding in suicides
Mann et al., 1986 (259)	21 suicides, 21 controls	Frontal cortex	↑ DHA binding in suicides
Crow et al., 1984 (226)	10 suicides, 19 controls	Frontal cortex	DHA binding—no difference between groups
Little et al., 1997 (349)	7 depressed suicides, 8 controls	Pineal gland	[¹²⁵I]pindalol binding—no difference between groups
Little et al., 1990 (347)	15 suicides, 15 controls	Frontal cortex	↓ [¹²⁵I]pindolol binding in suicides

DHA, dihydroalprenolol.

cortex of suicide victims. Dividing depressives into those with violent and nonviolent suicides, they found that the victims by violent means had significantly lower β-receptor binding in the frontal cortex than either controls or nonviolent suicides, whereas the nonviolent suicides had significantly lower β-receptor binding compared with controls in the occipital and temporal cortex. There was no difference in subcortical structures. Little et al. (349) found no difference in β and β_1 autoradiography in the pineal gland of depressed suicides compared with controls. As with postmortem examination of other receptor populations, variation in results may be explained by differences in the region studied, ligand specificity, and the extent to which the effects of age, postmortem interval, drug effects, and circadian rhythms were accounted for in study design.

Fewer studies have examined α_1-adrenergic receptor function (Table 27.10). Whereas one study using *in vitro* quantitative autoradiography with tritiated prazosin found significantly decreased binding in the prefrontal cortex of suicides compared with controls (350), Arango et al. (343), in a study with similar methodology, found increased binding specific to layers IV and V of the prefrontal cortex in suicides.

α_2-Adrenergic receptors have been examined in suicide victims and controls as well. Finding no difference in iodoclonidine binding on autoradiography of the locus ceruleus of suicides and controls, Ordway et al. (351)

concluded that if a difference exists in α_2-adrenergic receptor binding, it must be in the terminal fields. Arango et al. (343), however, found no difference in any cortical layer when comparing α_2-adrenergic receptor binding in suicides and controls. Meana et al. (352) found increased α_2-adrenergic receptor density in the frontal cortex, amygdala, and head of the caudate for depressed suicides compared with controls. Unable to demonstrate any differences between schizophrenic suicides and controls, their findings suggest a role for noradrenergic receptor dysfunction in affective illness rather than suicidal behavior *per se*.

Ordway et al. (344) showed that binding of the α_2-adrenoreceptor agonist ligand p-[^{125}I]iodoclonidine was significantly higher in the locus ceruleus of suicide victims than controls, whereas binding of the antagonist ligand [^{3}H]yohimbine was no different. Because p-[^{125}I]iodoclonidine binds preferably to the high-affinity state of the receptor, they concluded that the difference in suicide is observable only in that state.

Other Neurochemical Systems

Although not systematically investigated, several neurochemical systems other than monoamines have been implicated in suicidal behavior as well. In light of the proposed relationship of changes in cholinergic function to affective

TABLE 27.10. STUDIES COMPARING α_1- AND α_2-ADRENERGIC BINDING IN BRAINS OF SUICIDE VICTIMS AND CONTROLS

Study	Subjects	Tissue	Findings
α_1-Receptor studies			
Gross-Isseroff et al., 1990 (350)	12 suicides, 12 controls	Prefrontal cortex, temporal cortex, amygdala, hippocampus, caudate	↓ [^{3}H]prazosin binding in cortical regions and caudate of suicides
Arango et al., 1993 (343)	17 suicides, 17 controls	Areas 9 + 38, prefrontal cortex	↑ [^{3}H]prazosin binding in prefrontal cortex (layers IV–V) of suicides
Arango et al., 1989 (479)	10 suicides, 10 controls	Prefrontal and temporal cortex	↑ [^{3}H]prazosin binding in prefrontal cortex (layers IV–V) of suicides
α_2-Receptor studies			
Ordway et al., 1994 (344)	10 suicides, 10 controls	Locus caeruleus	[^{125}I]iodoclonidine (α_2 agonist) and [^{3}H]yohimbine (α_2 antagonist)— ↑ only in iodoclonidine binding (high-affinity state receptors)
Ordway et al., 1990 (351)	13 suicides, 13 controls	Locus caeruleus	Iodoclonidine binding—no difference between groups
Arango et al., 1993 (343)	17 suicides, 17 controls	Areas 9 + 38, prefrontal cortex	[^{3}H]p-aminoclonidine binding—no difference in α_2 binding
Arango et al., 1990 (265)	12 suicides, 12 controls	Prefrontal cortex	[^{3}H]clonidine binding—no difference between groups
Meana et al., 1992 (352)	45 suicides, 42 controls	7 brain regions	[^{3}H]clonidine binding—↑ in frontal cortex, amygdala, and caudate in depressed suicides; no difference in schizophrenic suicides

disorders (353), three groups have measured receptor ligand binding for [^{3}H]quinuclidinyl benzilate, a muscarinic antagonist, in the brains of suicide victims and controls. Using the frontal cortex, Meyerson et al. (254) found a 47% increase in receptor binding in suicides compared with controls. Neither Kaufman et al. (354) nor Stanley (355) could replicate this finding in the frontal cortex or other tissues. This discrepancy may be explainable by differences between studies in the proportion of suicide victims with affective disorder.

Findings from a variety of studies suggest that the inhibitory neurotransmitter GABA plays a role in anxiety and affective disorders (356). Moreover, anxiety has been reported by some authors to be a powerful short-term predictor of completed suicide (131). Therefore, several groups of investigators have explored various aspects of the GABA system in the brains of suicide victims. None, however, has demonstrated a difference between suicides and controls in the function of GABA's principal synthetic (356) or degradative (357) enzymes in cortical sites, GABA receptor binding in cortex or hippocampus (358), GABA uptake sites (359), or GABA$_B$ binding in the frontal cortex (360). Although no difference was noted between suicide and controls in benzodiazepine receptor binding (to which the GABA receptor complex is linked) in subcortical tissue (361,362), one study found an increase in the affinity to benzodiazepine binding sites in the hippocampus of a small number of drug-free suicides by violent methods (363). Korpi et al. (364) found no difference between suicides and controls in GABA levels by reversed-phase high-performance liquid chromatography in the frontal cortex or four other subcortical sites. Only Cheetham et al. (356) found a significant (18%) increase in benzodiazepine binding in the frontal cortex of depressed suicide victims compared with controls. One study examined CSF concentrations of GABA in patients, finding no significant difference between depressed patients who attempted suicide and those who did not (365).

It is possible that abnormalities in GABA function may be more relevant to suicide in later life than in young adulthood and middle age. In comparing elderly depressed patients who died of natural causes with controls, Perry et al. (366) found a marked reduction in the GABA synthetic enzyme glutamic acid decarboxylase in elderly depressives, individuals who may be at higher suicide risk than any other demographic or diagnostic subgroup. Furthermore, glutamic acid decarboxylase was markedly reduced in terminal illness, particularly when associated with hypoxia, a circumstance associated with a much increased suicide risk.

Neuropeptides have been the focus in a number of recent studies. For example, Charlton et al. (367) found no difference in concentrations of cortical somatostatin-I immunoreactivity between depressed suicides and controls. Among living suicidal depressives, CSF levels of somatostatin were significantly reduced in one study (368) but not in another (365).

Neuropeptide Y, which coexists in the brain with monoamines and GABA, was significantly reduced in the frontal cortex, temporal cortex, and caudate of suicide victims compared with controls in one study (369), but another showed no difference between groups in the frontal cortex, cingulate cortex, and hypothalamus (370). Neuropeptide Y measured in plasma was shown to be lower in patients with recent suicide attempts and even lower in those who had repeatedly attempted suicide (371).

Cholecystokinin (CCK) is another peptide that has received some attention in the suicide literature. Harro et al. (372) analyzed CCK and benzodiazepine receptor binding characteristics in postmortem tissue samples from 19 suicide victims and 23 age-matched controls. In the frontal cortex, significantly more CCK receptors and higher-affinity constants were found in the suicide victims. Younger suicide victims had a higher density of CCK receptors in the cingulate cortex, whereas in older suicides, the value was lower compared with age-matched controls.

Bachus et al. (373) measured CCK mRNA in postmortem samples of the dorsolateral prefrontal cortex, anterior cingulate cortex, and entorhinal cortex of seven schizophrenics, nine nonpsychotic suicides, and seven normal controls. The suicide victims had elevated CCK mRNA in the dorsolateral prefrontal cortex, where schizophrenics do not differ from normals, and increased cellular density of CCK mRNA relative to both normals and schizophrenics. Loefberg et al. (374) measured levels of CCK peptides in the CSF of 105 patients with major depressive disorders. Patients who had made one or more suicide attempts tended to have higher CSF CCK levels than those who had made none. No correlations were found between CSF CCK and 5-HIAA, HVA, or with plasma cortisol.

A large body of literature has accumulated to suggest that disturbances of the endogenous opioid system may be involved in the pathophysiology of self-mutilatory behaviors in individuals with mental retardation (375–378), Tourette syndrome (379), Lesch-Nyhan syndrome (380), and personality disorders (381,382). Very few studies, however, have considered this issue in patients whose intent was clearly suicide. Frecska et al. (383) administered the opioid agonist fentanyl to ten women with major depression and measured their prolactin secretory response. Although they found no difference between healthy volunteers and depressed patients, four of the depressed women showed the most blunted response, three of whom committed suicide within a 1 year follow-up. Gross-Isseroff et al. (384) used quantitative autoradiography to examine regional differences in μ-opioid receptors between the brains of 14 pairs of suicide victims and age- and gender-matched controls. Finding a significant increase in μ-opioid receptor density in frontal and temporal cortical gyri in suicides, the authors speculate that greater postsynaptic receptor density is a compensatory response to diminished presynaptic release of endogenous peptides. They also noted significant age effects, most pronounced in temporal gyri, in which receptor binding was correlated with age in both suicides and controls. This finding was replicated by Gabilondo et al. (385) who found significantly greater μ-opioid receptor density in the frontal cortex and caudate but not the thalamus of suicide victims and also confirmed the direct correlation with age.

Another postmortem study of the opiate system examined β-endorphin levels in the frontal and temporal cortex, caudate nucleus, hypothalamus, and thalamus in seven suicide victims and controls, showing both intergroup and interhemispheric differences (386). β-Endorphin concentrations were decreased in the temporal and frontal cortex and caudate nucleus of suicides compared with controls. However, the reductions were observed only in samples taken from the left side of the brain. Suicide victims demonstrated a marked asymmetry in the distribution of β-endorphin, lower on the left side of the brain than on the right, a pattern that was not observed in controls. Although consistent with the findings of Gross-Isseroff et al. (384), lower β-endorphin levels may be owing to either decreased production of the endogenous peptide or a greatly increased turnover. These data suggest the need for studies that measure the integrity and function of neuropeptide-containing neurons and additional data concerning regional specificity and lateralization of neurochemical function as it relates to the regulation of mood and impulse.

Scattered reports are also available concerning other neurochemicals such as δ sleep–inducing peptidelike immunoreactivity and vasopressin, but with the investigators reporting no significant findings (368).

Neuroendocrine Systems and Suicide

Of the neuroendocrine systems implicated in the pathophysiology of psychiatric illness, two have been investigated in some detail with regard to their role in suicidal behavior: the hypothalamic–pituitary–thyroid axis and the hypothalamic–pituitary–adrenal (HPA) axis.

Thyroid Axis

Individuals with depressive disorders have a blunted thyroid stimulating hormone (TSH) response to the infusion of thyrotropin releasing hormone (387). Retrospective analysis in two studies demonstrated an association between blunted TSH response and attempted (388) and completed (389) suicide. In contrast, Banki et al. (286) found an augmented TSH response to thyrotropin releasing hormone in suicidal versus nonsuicidal psychiatric patients of mixed diagnoses. As with other neurobiologic measures, the distinction between violent and nonviolent suicidal behavior may be important because a number of studies have found associations between thyroid axis abnormalities and suicide in those who used violent methods only (388,390,391). Corrigan et al. (392) reported that the symptom constellation of panic, agitation, and suicidality in depression may correlate with the

greatest reduction in TSH response in a study of 27 euthyroid primary unipolar depressed inpatient women.

Hypothalamic–Pituitary–Adrenal Axis and Suicide

Many lines of evidence suggest that disturbances of the HPA axis may be associated with suicide as well. The studies of this system have looked at serum and/or urine cortisol and its metabolites, adrenocorticotropic hormone, CRF, and their receptors. The earliest observations were of elevated urinary excretion of cortisol (341) and 17-hydroxy corticosteroids in patients who attempted or completed suicide (393,394). Elevated plasma cortisol has also been associated with completed suicide at follow-up (395). Traskman-Bendz et al. (342) reported no significant difference in plasma or urine cortisol between suicide attempters and nonattempters at baseline. However, those who completed suicide in a later stage had significantly higher urine cortisol than those who did not. In postmortem studies, the size and weight of the adrenal glands of suicides were shown to be significantly greater in suicides than controls by several groups (396–398), a finding hypothesized to reflect stress-related hypertrophy. CSF CRF levels were significantly higher (399) and CRF binding sites in the frontal cortex were significantly reduced (400) in suicides compared with sudden death controls. These findings are important because of the known role of CRF in the regulation of a range of affective and neurovegetative functions. Others, however, failed to replicate theses findings (401–405).

Escape from dexamethasone suppression predicted both attempted (406–412) and completed (413,414) suicide, in some, but not all, studies (342,395,415–417). Numerous explanations for these mixed findings have been offered, including differences between studies in diagnostic composition, definition of suicidal behavior, and dexamethasone suppression test methodology. Age and gender may be factors as well. Although there are few overt changes in HPA axis functioning with normal aging (418), subtle abnormalities of the axis's regulatory response emerge in stressful conditions or pathologic states (419,420). For example, Stangl et al. (421) found that in two-thirds of 50 studies of the dexamethasone suppression test in major affective disorder, nonsuppressors were older than suppressors. Furthermore, in men, postdexamethasone cortisol was positively correlated with age in both depressives and normals, albeit at higher levels across the age spectrum in depressives. In women, conversely, only depressives showed the correlation between age and postdexamethasone cortisol. Unfortunately, no analyses were reported regarding the subjects' histories of suicidal behavior, and we are aware of no reported studies of the dexamethasone suppression test response in suicidal elders.

Taken together, studies support the notion of chronic stimulation of the HPA axis as a risk factor for suicide. However, the relative contribution of mood disorders, independent of suicidality *per se,* is unclear.

Other Neuroendocrine Systems

Finally, preliminary evidence linking a number of other biologic variables to suicidal behavior is noteworthy. Reports have pointed out the gender difference in teenage suicide and the rise in suicidality during the years when hypothalamic–pituitary–gonadal axis changes are dramatic (422). In addition, there is an apparent decrease in suicidality during days 5 through 21 of the menstrual cycle (423). Despite such observations, the association of the hypothalamic–pituitary–gonadotropin axis with suicidal behavior has received less attention from investigators. In one of the few such studies, Martin et al. (424) found a significant correlation between progesterone level and suicidal ideation only in male subjects. This correlation was even stronger in the presence of alcohol or marijuana use within the past month.

Testosterone has long been associated with aggression in both animal models and humans (425), yet few studies examine its role in suicidal behavior. Ehrenreich et al. (426) reported a rare case of bilateral cryptorchism that did not come to medical attention until much later in adult life. The patient developed severe suicidal depression that responded solely to testosterone. Bergman and Brismar (427) observed an association between elevated levels of serum testosterone and violence in suicidal alcoholics.

Banki et al. (428) observed that psychiatric patients who had attempted suicide by either violent or nonviolent means had significantly lower mean CSF magnesium levels, regardless of diagnosis, than nonsuicidal patients or controls. CSF magnesium and 5-HIAA levels were highly correlated.

Exploring the possibility of a "low melatonin syndrome" in patients with major depressive disorders, Beck-Friis et al. (429) found that patients with a history of suicidal behavior had significantly higher mean nocturnal serum melatonin levels than did those without a history of suicidal behavior. In contrast, however, another report indicates that the mean nocturnal melatonin level was significantly lower in a suicidal subgroup of MS patients compared with the nonsuicidal patients from the same sample (430).

Cholesterol and Suicide

An association between low serum cholesterol concentrations and suicide provides another example of the complex interplay of neurobiologic factors in determining risk of self-injury. Evidence for this association comes from a variety of sources: cross-sectional studies of the relationships between suicidal behavior and cholesterol level, longitudinal cohort studies, and primary and secondary prevention studies designed to reduce the prevalence of cardiovascular disease and its associated morbidity and mortality by lowering serum cholesterol (431).

In a review of six randomized, controlled, primary prevention trials in which the serum cholesterol of middle-age subjects was lowered by diet, drugs, or both strategies, Muldoon et al. (432) found that, although mortality from coronary heart disease was reduced, deaths from other causes were not. A subsequent Swedish study (433) linked that excess mortality to violent deaths, including suicide. In more than 50,000 people followed for 20 years, mortality from injuries was strongly negatively correlated with baseline cholesterol level in men. Those individuals in the lowest quartile of the cholesterol distribution had a 4.2 times greater risk of suicide than other male subjects. No such relationship was observed for women. Complicating interpretation of the results, investigators had no information regarding the psychological state of their subjects. They also could not account for other potential confounds, such as alcohol use and nutritional state, in the follow-up period, which may have been modified as a function of the development of affective or other psychopathology.

Partonen and Loennqvist (434) investigated the association of low serum total cholesterol with major depression and suicide in a total of 29,133 Finnish men of ages 50 to 69 years followed prospectively. They found that low serum total cholesterol was associated with low mood and a heightened risk of suicide as well as hospital treatment owing to major depressive disorder. In a separate Finnish study, Tanskanen et al. (435) examined total serum cholesterol levels of 37,635 adults from five independent population surveys conducted between 1972 and 1992. Serum cholesterol concentration was positively associated with the risk of violent suicide, and the violent/nonviolent suicide ratio increased linearly with increasing cholesterol category (435). Numerous studies have shown an association between low cholesterol and suicidal ideation or attempt in psychiatric patient samples (436–440).

In an effort to explain these observations, Engelberg (441) postulates that a reduction in brain cell cholesterol may cause dysfunction at the 5-HT reuptake site through alterations in membrane lipid viscosity. Lower presynaptic levels of 5-HT would result, causing disruptions in mood and impulse control and suicidal behavior.

Interactions such as these at the neuronal level are likely to characterize biologic abnormalities in suicide more generally. Monoaminergic, GABA-ergic, peptidergic, and neuroendocrine systems have complex interrelationships that are beyond the scope of this text. For example, as reviewed by Sulser (442), an intact serotonergic neuronal input is required for the proper functioning of β-adrenergic receptors. The secretion of cortisol and prolactin is regulated in part through stimulation of 5HT$_1$ receptors in addition to renin, vasopressin, and adrenocorticotropic hormone. In addition, the HPA axis participates in the modulation of both serotonergic and catecholamine function, the relationships of which may be altered during episodes of illness. It is most likely, therefore, that the expression of suicidal behavior in an individual represents, in addition to psychological and social factors, subtle alterations in the balance of numerous neurochemical systems. These alterations, in turn, result from a wide range of intrinsic or extrinsic factors, such as genetic programming, the aging of neurotransmitter and hormonal systems, structural lesions, exposure to toxins, nutritional influences, and others. Personality, life circumstances, and the impact of these factors on individual neurochemical or neuroanatomic systems determine whether the resulting violent behavior is directed outward toward others or inward toward the self.

TREATMENT AND PREVENTION

Treatment

Assessment

Despite decades of research and progress in the recognition of risk factors for suicide, efforts to construct a useful and sufficiently sensitive and specific predictive scale have thus far failed (126,443). Although many hope that the development of biologic markers will greatly enhance our ability to identify individuals at risk, the body of knowledge that we now possess can serve only in a nonspecific manner to enhance our clinical judgment and guide best practice (Table 27.11). Assessment relies heavily on the recognition of risk factors previously noted. As with any medical tool, however, risk factors are only as useful as the skill with which they are applied in any particular case. For more detailed discussions of assessment and interview techniques, the reader is referred to several excellent texts in this area (444–446).

Beyond acknowledgment of an individual's sociodemographic characteristics, risk assessment begins with an understanding of the patient's perceived stressors and his/her psychological reactions to them. Screening for both depressive symptoms and substance abuse should be a part of routine primary care and mental health practice. Wherever indicated by positive or ambiguous responses, the alert clinician should initiate a direct, yet sensitive, discussion with the patient about the nature and extent of his/her suicidal ideation, including the potential lethality of any plans he/she has considered. One should carefully define the nature and extent of the patient's social support network and the elements in his/her environment that may be mobilized to ameliorate a suicidal crisis. Most important, one must determine the patient's willingness and ability to work with his/her caretakers in finding alternative solutions to suicide.

In the presence of more highly developed suicidal ideation and planning, the absence of helpful supports in the patient's social network, and with resistance by the patient to available interventions, hospitalization may be lifesaving. In the great majority of cases, suicidal thoughts are transient, a symptom of illness that emerges at times of episodic crisis and resolves with treatment. Therefore, the maintenance of the individual's safety—even by involuntary hospitalization or restraint—will usually lead, after appropriate treatment, to the resolution of suicidal thoughts. In less emergent sit-

TABLE 27.11. SUMMARY OF PUTATIVE RISK FACTORS FOR SUICIDE

Older
Male
White
Living alone
History of suicide attempt(s)
Psychological characteristics
  Neuroticism
  Low openness to experience (older adults)
  Impulsivity/aggression
  Hopelessness

Psychiatric illness
  Affective disorder
  Substance abuse/dependence
  Schizophrenia
  Personality disorder (borderline, antisocial)

Medical Illness
  Human immunodeficiency virus/acquired immunodeficiency syndrome
  Cancer
  Peptic ulcer
  Renal disease
  Systemic lupus erythematosus
  Other

Neurologic illness
  Multiple sclerosis
  Epilepsy
  Traumatic spinal cord injury
  Huntington disease

Life stressors
  Recent or threatened loss

Lack of available support
Inability to accept help

uations, outpatient treatment may be a viable alternative, particularly if family and friends can be responsibly engaged in a supportive supervisory role.

Psychiatric consultation should be considered for all suicidal patients to help assess the extent of risk and establish a psychiatric diagnosis and formulation. This latter element of the database includes a thorough assessment of the individual's intrapersonal (psychological/cognitive) resources, most prominent defense mechanisms, and the life stressors that have led to the current situation. With an understanding of these factors, one can better define with the patient the source of his/her intolerable pain and thereby a means to ameliorate it. The wish for death is *always* ambivalently held by the suicidal person (444). As articulated by Shneidman (444), "The clinical rule is: Reduce the level of suffering, often just a little bit, and the individual will choose to live." This strategy, in addition to the maintenance of the patient's safety, is the short-term goal of treatment.

Having established the patient's safety and defused the acute suicidal crisis, one can afford to shift the focus of treatment to longer-term goals: the evaluation and treatment of psychiatric illness, resolution of conflicts, and augmentation of internal defenses and external supports so that the individual can cope more effectively with new and ongoing stressors in his/her life. The choice of treatment modalities, which depends of course on the specific needs of the patient, may include a wide range of effective somatic and/or psychological therapies.

Pharmacotherapy

Conventional practice holds that pharmacotherapeutic interventions are primarily directed at the psychiatric condition of which suicidal ideation and behavior may be symptomatic manifestations. However, the emerging evidence that neurobiologic abnormalities underlie suicide suggests that psychotropic medications (in particular, serotonergic agents) may have a role in the treatment of the suicidal person independent of their antidepressant effects (447).

In one of the few randomized, controlled trials of the effectiveness of a medication in reducing recurrent self-destructive behavior, Montgomery et al. (448) compared intramuscular flupentixol with placebo over 6 months in patients admitted after suicide attempts. Patients treated with antipsychotics were significantly less likely to repeat attempts (21% vs. 75%). The rationale for treatment of suicidal patients with neuroleptic drugs in the absence of psychosis remains poorly defined, however, and in the absence of replication, the finding has not translated to standard care.

Because of ethical concerns over withholding antidepressant therapies from affectively ill, suicidal individuals, evidence is lacking from carefully controlled outcome studies to support the belief that antidepressants decrease suicide risk (449). However, cohort studies of bipolar patients provide suggestive evidence that maintenance on lithium may be protective. Tondo and Baldessarini (450) reviewed 22 studies and additional unpublished data concerning the suicidal behavior of manic depressive patients with and without lithium treatment. Relative to patients who were untreated or discontinued lithium treatment, they found an overall reduction of risk by nearly seven-fold among those taking the drug, from 1.78 to 0.26 suicide attempts and completed suicides per 100 patient-years at risk. Discontinuation resulted in sharp increases in risk (451). The smaller body of available evidence does not support a similar protective effect with other thymoleptic agents (452). Further, preliminary indications are that electroconvulsive therapy may decrease suicidal outcomes in depressed patients followed prospectively (453–455). Much more research is needed on the neurobiologic correlates of suicidal behavior before specific recommendations can be formulated.

Psychotherapy

In addition to somatic therapies, suicidal people may respond to a range of psychological, cognitive, and behav-

ioral interventions. As reviewed by Hawton et al. (449), the effectiveness of problem-solving therapy (a task-centered, cognitive/behavioral approach) in reducing recurrent suicidal behavior has been compared with care as usual in five randomized trials. The experimental intervention was more effective in all five, but when the studies were combined in a metaanalysis, the summary odds ratio did not reach significance. Neither was there a significantly reduced odds ratio of self-harming behavior compared with treatment as usual among patients in six studies who were provided with intensive interventions (urgent and after-hours appointments, 24-hour telephone access) and outreach services.

Linehan et al. (456) designed dialectical behavior therapy, also derived from the cognitive/behavioral model, specifically to address the self-destructive acts of high-risk patients with borderline personality disorder. In an important 1991 study, they demonstrated a significantly lower rate of repeated self-harming behavior among patients treated with dialectical behavior therapy than among those randomized to care as usual (26% vs. 60%).

Psychoanalytically oriented therapy offered in the context of a partial hospital program may also decrease risk of parasuicide (457). A review of all behavioral and psychosocial therapies for suicidal states suggests that the best predictor of response is the inclusion of patients with extensive histories of deliberate self-harm (458). None has been tested in a prospective design in which completed suicide is a measured outcome. It remains unclear, therefore, how applicable the findings of these trials are to individuals without a history of suicide attempts or to the prevention of completed suicide.

Prevention

Strategies for prevention of suicide follow two theoretical approaches: population-based strategies and high-risk models (459).

Population-based or universal strategies include interventions or initiatives that have a potential impact on large segments of a society independent of risk status. Gun control, both by education and legislative action, has been proposed as one such approach. Sixty percent of people who die by suicide each year in the United States use a firearm to kill themselves. Almost 50% of all firearm-related deaths are suicides. Although the subject remains politically controversial, several studies have provided evidence to suggest that more restrictive gun control legislation would lower suicide rates. In a carefully conducted psychological autopsy study comparing adolescent suicide victims with psychiatric inpatients and normal controls, Brent et al. (460) found that the presence of a firearm in the home was significantly associated with a completed suicide. The manner in which the gun was stored (e.g., in a locked place without ammunition) had no impact on the likelihood that it would be used for suicide. Sloan et al. (461) compared suicide rates of two locales with markedly different gun control regulations. Although risk of death from suicide did not significantly differ between the two regions, the rate of suicide by handguns was far higher in the region with less restrictive laws. Loftin et al. (462) compared rates of gun-related death in the District of Columbia before and after the adoption of restrictive hand gun policies and found a 25% decrease in firearm homicides and suicides after the law went into effect. There was no such drop in the surrounding metropolitan area, where the laws remain unchanged, and no compensatory increase in suicides by other means. Although it appears that access to means of suicide with greater potential lethality is associated with higher rates of completed suicide (463), there is ongoing debate about whether removal of more lethal means will lead only to substitution of other methods (461,464).

A second population-based approach is the education of the public and their health care providers about the warning signs of suicide and its close association with treatable psychiatric illness. A great many myths about suicide are commonly held by the public. One such misconception is that if a person truly wants to commit suicide, there is little that can be done to prevent him/her. Other myths are that people who threaten suicide will not follow through or that discussing suicide with a depressed person may lead him/her to the act (12). Ageist attitudes may also contribute to the belief that depression suicidal thoughts are "normal" among older adults facing losses or nearing the end of life. Given these misconceptions, people may fail to recognize the need for intervention or understand its potential benefits.

High-risk approaches target segments of the population recognized to be at increased risk of suicide owing to the presence of one or more vulnerabilities, such as clinical depressive illness. In addition to the general public, education of health care professionals about the diagnosis and treatment of affective disorder must be a high priority, particularly for late-life suicide prevention. Numerous studies have shown that as many as 75% of elderly suicide victims had seen a physician in the last month of life and as many as half in their final week (60,465–467). However, physicians in general have limited knowledge of suicide risk (468,469) and often fail to recognize and adequately treat depressive disorders in primary care settings (470). Given the association between affective illness and suicide, therefore, it is reasonable to conclude that a program to educate primary care providers about the recognition and aggressive treatment of depression would decrease suicide rates in that subgroup of the population. This hypothesis was validated in one Swedish study (471,472).

Postvention

Despite the best efforts of all involved, suicides will continue to occur. Although empirical studies are few (473), there is a growing consensus that the grief after a death by suicide involves a complex combination of depression, guilt, and anger that is distinct from normal bereavement. It is a grief that may place the survivor at increased risk of psychiatric morbidity and suicide mortality (474). Therefore, treatment should not end with the death of a patient by suicide. The important tasks that remain include attention to the impact of the suicide on the patient's survivors—family, friends, and health care providers, including the physician him/herself.

The recognition that survivors are a large group in need of understanding and services has led to the development of a broad network of self-help and support groups (475). As with almost every other aspect of suicide, however, there remains a great need for carefully conducted, ethically sound, and clinically sensitive research. As we learn more about the differential roles played by social, psychological, and biologic determinants, suicide, like other illnesses, may give up the mystery and stigma in which it is shrouded.

REFERENCES

1. Farberow NL. Cultural history of suicide. In: Farberow NL, ed. *Suicide in different cultures.* Baltimore: University Park Press, 1975:1–15.
2. Eichelman B. Aggressive behavior: from laboratory to clinic. Quo vadit? *Arch Gen Psychiatry* 1992;49:488–492.
3. Linehan MM. Suicidal people: one population or two? In: Mann JJ, Stanley M, eds. *The psychobiology of suicidal behavior.* New York: New York Academy of Sciences, 1986:16–33.
4. Farberow NL. Indirect self-destructive behavior: classification and characteristics. In: Farberow NL, ed. *The many faces of suicide.* New York: McGraw-Hill, 1980:15–27.
5. O'Carroll PW, Berman AL. Beyond the tower of Babel: a nomenclature for suicidology. *Suicide Life Threat Behav* 1996;26:237–252.
6. Kreitman N. *Parasuicide.* London: John Wiley & Sons, 1977.
7. Pattison EM, Kahan J. The deliberate self-harm syndrome. *Am J Psychiatry* 1983;140:867–872.
8. Neeleman J, Wessely S, Wadsworth M. Predictors of suicide, accidental death, and premature natural death in a general-population birth cohort. *Lancet* 1998;351:93–97.
9. Litman RE. Psychodynamics of indirect self-destructive behavior. In: Farberow NL, ed. *The many faces of suicide.* New York: McGraw-Hill, 1980:28–40.
10. Moscicki EK, O'Carroll P, Rae DS, et al. Suicide attempts in the Epidemiologic Catchment Area Study. *Yale J Biol Med* 1988;61:259–268.
11. Parkin D, Stengel E. Incidence of suicidal attempts in an urban community. *BMJ* 1965;2:133–138.
12. Fremouw WJ, dePerczel M, Ellis TE. *Suicide risk: assessment and response guidelines.* New York: Pergamon Press, 1990.
13. Moscicki EK. Identification of suicide risk factors using epidemiologic studies. *Psychiatric Clin North Am* 1997;3:499–517.
14. O'Donnell I, Farmer R. The limitations of official suicide statistics. *Br J Psychiatry* 1995;166:458–461.
15. McCarthy PD, Walsh D. Suicide in Dublin: I. The under-reporting of suicide and the consequences for national statistics. *Br J Psychiatry* 1975;126:301–308.
16. Speechley M, Stavraky KM. The adequacy of suicide statistics

for use in epidemiology and public health. *Can J Public Health* 1991;82:38–42.

17. Barraclough BM. Differences between national suicide rates. *Br J Psychiatry* 1973;122:95–96.
18. Sainsbury P. *Suicide in London.* Maudsley Monographs, 1955.
19. Atkinson MW, Kessel N, Dalgaard JB. The comparability of suicide rates. *Br J Psychiatry* 1975;127:247–256.
20. Murphy SL. *Deaths: final data for 1998.* Hyattsville, MD: National Center for Health Statistics, 2000.
21. Conwell Y, Duberstein PR, Cox C, et al. Age differences in behaviors leading to completed suicide. *Am J Geriatr Psychiatry* 1998;6:122–126.
22. Solomon MI, Hellon CP. Suicide and age in Alberta, Canada, 1951 to 1977. A cohort analysis. *Arch Gen Psychiatry* 1980;37:511–513.
23. Busse EW. Geropsychiatry: social dimensions. In: Maletta GJ, ed. *Survey reports on the aging nervous system* (DHEW publ. no. 74-296). Washington, DC: U.S. Government Printing Office, 1974:195–225.
24. Rich CL, Young D, Fowler RC. San Diego suicide study I. Young vs old subjects. *Arch Gen Psychiatry* 1986;43:577–582.
25. Murphy GE, Wetzel RD. Suicide risk by birth cohort in the United States, 1949 to 1974. *Arch Gen Psychiatry* 1980;37:519–523.
26. Manton KG, Blazer DG, Woodbury MA. Suicide in middle age and later life: sex- and race-specific life table and cohort analyses. *J Gerontol* 1987;42:219–227.
27. Murphy E, Lindesay J, Grundy E. 60 years of suicide in England and Wales. A cohort study. *Arch Gen Psychiatry* 1986;43:969–976.
28. Goldney RD, Katsikitis M. Cohort analysis of suicide rates in Australia. *Arch Gen Psychiatry* 1983;40:71–74.
29. Meehan PJ, Saltzman LE, Sattin RW. Suicides among older United States residents: epidemiologic characteristics and trends. *Am J Public Health* 1991;81:1198–1200.
30. Suicide among older persons—United States, 1980–1992. *MMWR Morb Mortal Wkly Rep* 1996;45:3–6.
31. Centers for Disease Control and Prevention. Suicide among black youths—United States, 1980–1995. *JAMA* 1998;279:1431.
32. Chew KS, McCleary R. The spring peak in suicides: a cross-national analysis. *Soc Sci Med* 1995;40:223–230.
33. McCleary R, Chew KS, Hellsten JJ, et al. Age- and sex-specific cycles in United States suicides, 1973 to 1985. *Am J Public Health* 1991;81:1494–1497.
34. Stillion JM, McDowell EE, May JH. Suicide across the lifespan—premature exits. New York: Hemisphere Publishing, 1989.
35. Havens L. The anatomy of a suicide. *N Engl J Med* 1965;272:401–406.
36. O'Carroll PW. Suicide causation: pies, paths, and pointless polemics. *Suicide Life Threat Behav* 1993;23:27–36.
37. Kreitman N. Suicide, age and marital status. *Psychol Med* 1988;18:121–128.
38. Duberstein PR, Conwell Y, Cox C. Suicide in widowed persons. A psychological autopsy comparison of recently and remotely bereaved older subjects. *Am J Geriatr Psychiatry* 1998;6:328–334.
39. MacMahon B, Pugh TF. Suicide in the widowed. *Am J Epidemiol* 1965;81:23–31.
40. Durkheim E. *Suicide.* New York: Free Press, 1966.
41. Templer DI, Veleber DM. Suicide rate and religion within the United States. *Psychol Rep* 1980;47:898.
42. Maris RW. *Pathways to suicide: a survey of self-destructive behavior.* Baltimore: Johns Hopkins University Press, 1981.
43. Platt S. Parasuicide and unemployment. *Br J Psychiatry* 1986;149:401–405.
44. Kung HC, Liu X, Juon HS. Risk factors for suicide in Caucasians and in African-Americans: a matched case-control study. *Soc Psychiatry* 1998;33:155–161.
45. Drever F, Whitehead M, Roden M. Current patterns and trends in male mortality by social class (based on occupation). *Pop Trends* 1996;86:15–20.
46. Kagamimori S, Matsubara I, Sokejima S, et al. The comparative study on occupational mortality, 1980 between Japan and Great Britain. *Indust Health* 1998;36:252–257.
47. Paykel ES, Prusoff BA, Myers JK. Suicide attempts and recent life events. A controlled comparison. *Arch Gen Psychiatry* 1975;32:327–333.
48. Conwell Y, Rotenberg M, Caine ED. Completed suicide at age 50 and over. *J Am Geriatr Soc* 1990;38:640–644.
49. Murphy GE, Armstrong JWJ, Hermele SL, et al. Suicide and alcoholism. Interpersonal loss confirmed as a predictor. *Arch Gen Psychiatry* 1979;36:65–69.
50. Rich CL, Fowler RC, Fogarty LA, et al. San Diego Suicide Study. III. Relationships between diagnoses and stressors. *Arch Gen Psychiatry* 1988;45:589–592.
51. Sainsbury P. Suicide and depression. In: Coppen A, Walk A, eds. *Recent developments in affective disorders: a symposium.* Ashford, England: Headley Bros., 1968:1–13.
52. Loebel JP, Loebel JS, Dager SR, et al. Anticipation of nursing home placement may be a precipitant of suicide among the elderly. *J Am Geriatr Soc* 1991;39:407–408.
53. Robins E, Murphy GE, Wilkinson RH Jr, et al. Some clinical considerations in the prevention of suicide based on a study of 134 successful suicides. *Am J Public Health* 1959;49:888–899.
54. Mackenzie TB, Popkin MK. Suicide in the medical patient. *Int J Psychiatry Med* 1987;17:3–22.
55. Chynoweth R, Tonge JI, Armstrong J. Suicide in Brisbane—a retrospective psychosocial study. *Aust N Z J Psychiatry* 1980;14:37–45.
56. Dorpat TL, Anderson WF, Ripley HS. The relationship of physical illness to suicide. In: Resnik HPL, ed. *Suicidal behaviors: diagnosis and management.* Boston: Little, Brown, 1968.
57. Conwell Y, Caine ED, Olsen K. Suicide and cancer in late life. *Hosp Commun Psychiatry* 1990;41:1334–1339.
58. Murphy GK. Cancer and the coroner. *JAMA* 1977;237:786–788.
59. Whitlock FA. Suicide and physical illness. In: Roy A, ed. *Suicide.* Baltimore: Williams & Wilkins, 1986:151–170.
60. Barraclough BM. Suicide in the elderly: recent developments in psychogeriatrics. *Br J Psychiatry* 1971[Suppl 6]:87–97.
61. Harris EC, Barraclough BM. Suicide as an outcome for medical disorders. *Medicine* 1994;73:281–296.
62. Magni E, DeLeo D, Schifano F. Depression in geriatric and adult medical inpatients. *J Clin Psychology* 1985;41:337–344.
63. Magni E. Depression and somatic symptoms in the elderly: the role of cognitive function. *Int J Geriatr Psychiatry* 1996;11:517–522.
64. Chochinov HM, Wilson KG, Enns M, et al. Desire for death in the terminally ill. *Am J Psychiatry* 1995;152:1185–1191.
65. Brown JH, Henteleff P, Barakat S, et al. Is it normal for terminally ill patients to desire death? *Am J Psychiatry* 1986;143:208–211.
66. Vonsattel JP, Myers RH, Stevens TJ, et al. Neuropathological classification of Huntington's disease. *J Neuropathol Exp Neurol* 1985;44:559–577.
67. Boll TJ, Heaton R, Reitan RM. Neuropsychological and emotional correlates of Huntington's chorea. *J Nerv Ment Dis* 1974;158:61–69.

68. Heathfield KW. Huntington's chorea: a centenary review. *Postgrad Med J* 1973;49:32–45.

69. Dewhurst K, Oliver JE, McKnight AL. Socio-psychiatric consequences of Huntington's disease. *Br J Psychiatry* 1970;116:255–258.

70. Huntington G. On chorea. *Med Surg Rep* 1872;26:317–321.

71. Minski L, Guttmann E. Huntington's chorea: a study of thirty-four families. *J Ment Sci* 1938;84:21–96.

72. Chandler JH, Reed E, DeJong RN. Huntington's chorea in Michigan. III. Clinical observations. *Neurology* 1960;10:148–153.

73. Tamir A, Whittier J, Korenyi C. Huntington's chorea: a sex difference in psychopathological symptoms. *Dis Nerv Syst* 1969;30:103.

74. Saugstad L, Odegard O. Huntington's chorea in Norway. *Psychol Med* 1986;16:39–48.

75. Schoenfeld M, Myers RH, Cupples LA, et al. Increased rate of suicide among patients with Huntington's disease. *J Neurol Neurosurg Psychiatry* 1984;47:1283–1287.

76. Farrer LA. Suicide and attempted suicide in Huntington disease: implications for preclinical testing of persons at risk. *Am J Med Genet* 1986;24:305–311.

77. Kessler S, Field T, Worth L, et al. Attitudes of persons at risk for Huntington disease toward predictive testing. *Am J Med Genet* 1987;26:259–270.

78. Mastromauro C, Myers RH, Berkman B. Attitudes toward presymptomatic testing in Huntington disease. *Am J Med Genet* 1987;26:271–282.

79. Kessler S, Bloch M. Social system responses to Huntington disease. *Fam Process* 1989;28:59–68.

80. Albin RL, Young AB, Penney JB, et al. Abnormalities of striatal projection neurons and N-methyl-D-aspartate receptors in presymptomatic Huntington's disease. *N Engl J Med* 1990;322:1293–1298.

81. Lam RW, Bloch M, Jones BD, et al. Psychiatric morbidity associated with early clinical diagnosis of Huntington disease in a predictive testing program. *J Clin Psychiatry* 1988;49:444–447.

82. Rao SM. Neuropsychology of multiple sclerosis: a critical review. *J Clin Exp Neuropsychol* 1986;8:503–542.

83. Stenager EN, Stenager E. Suicide and patients with neurologic diseases. Methodologic problems. *Arch Neurol* 1992;49:1296–1303.

84. Joffe RT, Lippert GP, Gray TA, et al. Mood disorder and multiple sclerosis. *Arch Neurol* 1987;44:376–378.

85. Schwartz ML, Pierron M. Suicide and fatal accidents in multiple sclerosis. *Omega* 1972;3:291–293.

86. Sadovnick AD, Eisen K, Ebers GC, et al. Cause of death in patients attending multiple sclerosis clinics. *Neurology* 1991;41:1193–1196.

87. Stenager EN, Stenager E, Koch-Henriksen N, et al. Suicide and multiple sclerosis: an epidemiological investigation. *J Neurol Neurosurg Psychiatry* 1992;55:542–545.

88. Stenager EN, Koch-Henriksen N, Stenager E. Risk factors for suicide in multiple sclerosis. *Psychother Psychosom* 1996;65:86–90.

89a. Stensman R, Sundqvist-Stensman UB. Physical disease and disability among 416 suicide cases in Sweden. *Scand J Soc Med* 1988;16:149–153.

89b. Berman A, Samuel L. Suicide among people with multiple sclerosis. *J Neurolog Rehab* 1993;7:53–62.

90. Kontaxakis VP, Christodoulou GN, Mavreas VG, et al. Attempted suicide in psychiatric outpatients with concurrent physical illness. *Psychother Psychosom* 1988;50:201–206.

91. Santamaria J, Tolosa E. Clinical subtypes of Parkinson's disease and depression. In: Huber SJ, Cummings JL, eds. *Parkinson's disease: neurobehavioral aspects.* New York: Oxford University Press, 1992:217–228.

92. Enna SJ. Brain serotonin receptors and neuropsychiatric disorders. *Adv Exp Med Biol* 1981;133:347–357.

93. Cash R, Raisman R, Ploska A, et al. High and low affinity [3H]imipramine binding sites in control and parkinsonian brains. *Eur J Pharmacol* 1985;117:71–80.

94. Hoehn MM, Yahr MD. Parkinsonism: onset, progression, and mortality. 1967 [classic article]. *Neurology* 1998;50:318.

95. Lambert MV, Robertson MM. Depression in epilepsy: etiology, phenomenology, and treatment. *Epilepsia* 1999;40[Suppl 10]:S21–S47.

96. Currie S, Heathfield KW, Henson RA, et al. Clinical course and prognosis of temporal lobe epilepsy. A survey of 666 patients. *Brain* 1971;94:173–190.

97. Mendez MF, Lanska DJ, Manon-Espaillat R, et al. Causative factors for suicide attempts by overdose in epileptics. *Arch Neurol* 1989;46:1065–1068.

98. White SJ, McLean AE, Howland C. Anticonvulsant drugs and cancer. A cohort study in patients with severe epilepsy. *Lancet* 1979;2:458–461.

99. Mendez MF, Doss RC. Ictal and psychiatric aspects of suicide in epileptic patients. *Int J Psychiatry Med* 1992;22:231–237.

100a. Barraclough BM. The suicide rate of epilepsy. *Acta Psychiatr Scand* 1987;76:339–345.

100b. Matthews WS, Barabas G. Suicide and epilepsy: a review of the literature. *Psychosomatics* 1981;22:515–524.

101. Mendez MF, Cummings JL, Benson DF. Depression in epilepsy. Significance and phenomenology. *Arch Neurol* 1986;43:766–770.

102a. Hawton K, Fagg J, Marsack P. Association between epilepsy and attempted suicide. *J Neurol Neurosurg Psychiatry* 1980;43:168–170.

102b. Batzel LW, Dodrill CB. Emotional and intellectual correlates of unsuccessful suicide attempts in people with epilepsy. *J Clin Psychology* 1986;4:699–702.

103. Frankel HL, Coll JR, Charlifue SW, et al. Long-term survival in spinal cord injury: a fifty year investigation. *Spinal Cord* 1998;36:266–274.

104. Le CT, Price M. Survival from spinal cord injury. *J Chronic Dis* 1982;35:487–492.

105. DeVivo MJ, Black KJ, Richards JS, et al. Suicide following spinal cord injury. *Paraplegia* 1991;29:620–627.

106. Frisbie JH, Kache A. Increasing survival and changing causes of death in myelopathy patients. *J Am Paraplegia Soc* 1983;6:51–56.

107. Geisler WO, Jousse AT, Wynne-Jones M. Survival in traumatic transverse myelitis. *Paraplegia* 1977;14:262–275.

108. Nyquist RH, Bors E. Mortality and survival in traumatic myelopathy during nineteen years, from 1946 to 1965. *Paraplegia* 1967;5:22–48.

109. Charlifue SW, Gerhart KA. Behavioral and demographic predictors of suicide after traumatic spinal cord injury. *Arch Phys Med Rehabil* 1991;72:488–492.

110. Judd FK, Brown DJ. Suicide following acute traumatic spinal cord injury. *Paraplegia* 1992;30:173–77.

111. Chemerinski E, Robinson RG. The neuropsychiatry of stroke. *Psychosomatics* 2000;41:5–14.

112. Beskow J, Runeson B, Asgard U. Psychological autopsies: methods and ethics. *Suicide Life Threat Behav* 1990;20:307–323.

113. Dorpat TL, Ripley HS. A study of suicide in the Seattle area. *Comp Psychiatry* 1960;1:349–359.

114. Barraclough B, Bunch J, Nelson B, et al. A hundred cases of suicide: clinical aspects. *Br J Psychiatry* 1974;125:355–373.

115. Conwell Y, Duberstein PR, Cox C, et al. Relationships of age and axis I diagnoses in victims of completed suicide: a psychological autopsy study. *Am J Psychiatry* 1996;153:1001–1008.

116. Henriksson MM, Aro HM, Marttunen MJ, et al. Mental disorders and comorbidity in suicide. *Am J Psychiatry* 1993;150:935–940.

117. Foster T, Gillespie K, McClelland R. Mental disorders and suicide in Northern Ireland. *Br J Psychiatry* 1997;170:447–452.

118. Vijayakumar L, Rajkumar S. Are risk factors for suicide universal? A case-control study in India. *Acta Psychiatr Scand* 1999;99:407–411.

119. Cheng AT. Mental illness and suicide. A case-control study in east Taiwan. *Arch Gen Psychiatry* 1995;52:594–603.

120. Gustafsson L, Jacobsson L. On mental disorder and somatic disease in suicide: a psychological autopsy study of 100 suicides in northern Sweden. *Nord J Psychiatry* 2000;54:383–395.

121. Rich CL, Ricketts JE, Fowler RC, et al. Some differences between men and women who commit suicide. *Am J Psychiatry* 1988;145:718–722.

122. Asgard U. A psychiatric study of suicide among urban Swedish women. *Acta Psychiatr Scand* 1990;82:115–124.

123. Carlson G, Cantwell D. Suicidal behavior and depression in children and adolescents. *J Am Acad Child Adol Psychiatry* 1982;21:361–368.

124. Miles CP. Conditions predisposing to suicide: a review. *J Nerv Ment Dis* 1977;164:231–246.

125. Guze SB, Robins E. Suicide and primary affective disorders. *Br J Psychiatry* 1970;117:437–438.

126. Pokorny AD. Prediction of suicide in psychiatric patients. Report of a prospective study. *Arch Gen Psychiatry* 1983;40:249–257.

127. Gardner EA, Bahn AK, Mack M. Suicide and psychiatric care in the aging. *Arch Gen Psychiatry* 1964;10:547–553.

128. Barraclough BM, Pallis DJ. Depression followed by suicide: a comparison of depressed suicides with living depressives. *Psychol Med* 1975;5:55–61.

129. Roy A. Suicide in recurrent affective disorder patients. *Can J Psychiatry* 1984;29:319–322.

130. Berglund M, Nilsson K. Mortality in severe depression. A prospective study including 103 suicides. *Acta Psychiatr Scand* 1987;76:372–380.

131. Fawcett J, Scheftner WA, Fogg L, et al. Time-related predictors of suicide in major affective disorder. *Am J Psychiatry* 1990;147:1189–1194.

132. McGlashan TH. The Chestnut Lodge follow-up study. II. Long-term outcome of schizophrenia and the affective disorders. *Arch Gen Psychiatry* 1984;41:586–601.

133. Black DW, Winokur G, Nasrallah A. Suicide in subtypes of major affective disorder. A comparison with general population suicide mortality. *Arch Gen Psychiatry* 1987;44:878–880.

134. Tsuang MT. Suicide in schizophrenics, manics, depressives, and surgical controls. A comparison with general population suicide mortality. *Arch Gen Psychiatry* 1978;35:153–155.

135. Weeke A, Vaeth M. Excess mortality of bipolar and unipolar manic-depressive patients. *J Affect Disord* 1986;11:227–234.

136. Morrison JR. Suicide in a psychiatric practice population. *J Clin Psychiatry* 1982;43:348–352.

137. Roose SP, Glassman AH, Walsh BT, et al. Depression, delusions, and suicide. *Am J Psychiatry* 1983;140:1159–1162.

138. Wolfersdorf M, Keller F, Steiner B, et al. Delusional depression and suicide. *Acta Psychiatr Scand* 1987;76:359–363.

139. Coryell W, Tsuang MT. Primary unipolar depression and the prognostic importance of delusions. *Arch Gen Psychiatry* 1982;39:1181–1184.

140. Brown S. Excess mortality of schizophrenia. A meta-analysis. *Br J Psychiatry* 1997;171:502–508.

141. Wilkinson DG. The suicide rate in schizophrenia. *Br J Psychiatry* 1982;140:138–141.

142. Harris EC, Barraclough B. Excess mortality of mental disorder. *Br J Psychiatry* 1998;173:11–53.

143. Allebeck P, Bolund C. Suicides and suicide attempts in cancer patients. *Psychol Med* 1991;21:979–984.

144. Allebeck P, Wistedt B. Mortality in schizophrenia. A ten-year follow-up based on the Stockholm County inpatient register. *Arch Gen Psychiatry* 1986;43:650–653.

145. Black DW, Winokur G, Warrack G. Suicide in schizophrenia: the Iowa Record Linkage Study. *J Clin Psychiatry* 1985;46:14–17.

146. Breier A, Astrachan BM. Characterization of schizophrenic patients who commit suicide. *Am J Psychiatry* 1984;141:206–209.

147. Roy A. Suicide in chronic schizophrenia. *Br J Psychiatry* 1982;141:171–177.

148. Drake RE, Gates C, Cotton PG, et al. Suicide among schizophrenics. Who is at risk? *J Nerv Ment Dis* 1984;172:613–617.

149. Langley GE, Bayatti NN. Suicides in Exe Vale Hospital, 1972–1981. *Br J Psychiatry* 1984;145:463–467.

150. Virkkunen M. Suicides in schizophrenia and paranoid psychoses. *Acta Psychiatr Scand Suppl* 1974;250:1–305.

151. Newman SC, Bland RC. Mortality in a cohort of patients with schizophrenia: a record linkage study. *Can J Psychiatry* 1991;36:239–245.

152. Caldwell CB, Gottesman II. Schizophrenia—a high-risk factor for suicide: clues to risk reduction. *Suicide Life Threat Behav* 1992;22:479–493.

153. Roy A. Suicide in schizophrenia. In: Roy A, ed. *Suicide.* Baltimore: Williams & Wilkins; 1986:97–112.

154. Fenton WS, McGlashan TH, Victor BJ, et al. Symptoms, subtype, and suicidality in patients with schizophrenia spectrum disorders. *Am J Psychiatry* 1997;154:199–204.

155. Drake RE, Cotton PG. Depression, hopelessness and suicide in chronic schizophrenia. *Br J Psychiatry* 1986;148:554–559.

156. Warnes H. Suicide in schizophrenics. *Dis Nerv Syst* 1968;29[Suppl]:40.

157. Black DW, Warrack G, Winokur G. The Iowa record-linkage study. I. Suicides and accidental deaths among psychiatric patients. *Arch Gen Psychiatry* 1985;42:71–75.

158. Lindberg S, Agren G. Mortality among male and female hospitalized alcoholics in Stockholm 1962–1983 [published erratum appears in *Br J Addict* 1989;84:706]. *Br J Addict* 1988;83:1193–1200.

159. Murphy GE, Wetzel RD. The lifetime risk of suicide in alcoholism. *Arch Gen Psychiatry* 1990;47:383–392.

160. Goodwin DW. Alcohol in suicide and homicide. *Q J Stud Alcohol* 1973;34:144–156.

161. Murphy GE. *Suicide in alcoholism.* New York: Oxford University Press, 1992.

162. Allgulander C, Borg S, Vikander B. A 4–6-year follow-up of 50 patients with primary dependence on sedative and hypnotic drugs. *Am J Psychiatry* 1984;141:1580–1582.

163. Murphy SL, Rounsaville BJ, Eyre S, et al. Suicide attempts in treated opiate addicts. *Compr Psychiatry* 1983;24:79–89.

164. Marzuk PM, Tardiff K, Leon AC, et al. Prevalence of cocaine use among residents of New York City who committed suicide during a one-year period. *Am J Psychiatry* 1992;149:371–375.

165. Fowler RC, Rich CL, Young D. San Diego Suicide Study. II. Substance abuse in young cases. *Arch Gen Psychiatry* 1986;43:962–965.

166. Regier DA, Farmer ME, Rae DS, et al. Comorbidity of

mental disorders with alcohol and other drug abuse. Results from the Epidemiologic Catchment Area (ECA) Study. *JAMA* 1990;264:2511–2518.

167. Murphy GE, Wetzel RD, Robins E, et al. Multiple risk factors predict suicide in alcoholism. *Arch Gen Psychiatry* 1992;49:459–463.

168. Murphy GE, Robins E. Social factors in suicide. *JAMA* 1967;199:303–308.

169. Duberstein PR, Conwell Y, Caine ED. Interpersonal stressors, substance abuse, and suicide. *J Nerv Ment Dis* 1993;181:80–85.

170. Tamerin JS, Mendelson JH. The psychodynamics of chronic inebriation: observations of alcoholics during the process of drinking in an experimental group setting. *Am J Psychiatry* 1969;125:886–899.

171. Ballenger JC, Goodwin FK, Major LF, et al. Alcohol and central serotonin metabolism in man. *Arch Gen Psychiatry* 1979;36:224–227.

172. Branchey L, Branchey M, Shaw S, et al. Depression, suicide, and aggression in alcoholics and their relationship to plasma amino acids. *Psychiatry Res* 1984;12:219–226.

173. Buydens-Branchey L, Branchey MH, Noumair D, et al. Age of alcoholism onset. II. Relationship to susceptibility to serotonin precursor availability. *Arch Gen Psychiatry* 1989;46:231–236.

174. Duberstein PR. Personality disorders and completed suicide: a methodological and conceptual review. *Clin Psychol Sci Pract* 1997;359:4.

175. Perry JC. Longitudinal studies of personality disorders. *J Pers Disord* 1993;[Suppl 1]:63–85.

176. Stone MH, Hurt SW, Stone DK. The PI 500: long-term follow-up of borderline inpatients meeting DSM-III criteria: I. Global outcome. *J Pers Disord* 1987;1:291–298.

177. Duberstein PR, Seidlitz L, Conwell Y. Reconsidering the role of hostility in completed suicide: a life-course perspective. In: Masling J, Bornstein RF, eds. *Psychoanalytic perspectives on developmental psychology.* Washington, DC: American Psychological Association, 1996:57–323.

178. Marttunen MJ, Aro HM, Henriksson MM, et al. Antisocial behaviour in adolescent suicide. *Acta Psychiatr Scand* 1994;89:167–173.

179. Weissman MM, Klerman GL, Markowitz JS, et al. Suicidal ideation and suicide attempts in panic disorder and attacks. *N Engl J Med* 1989;321:1209–1214.

180. Hornig CD, McNally RJ. Panic disorder and suicide attempt. A reanalysis of data from the Epidemiologic Catchment Area study. *Br J Psychiatry* 1995;167:76–79.

181. Johnson J, Weissman MM, Klerman GL. Panic disorder, comorbidity, and suicide attempts. *Arch Gen Psychiatry* 1990;47:805–808.

182. Crawley JN, Sutton ME, Pickar D. Animal models of self-destructive behavior and suicide. *Psychiatric Clin North Am* 1985;8:299–310.

183. Hamilton WJ. Do nonhuman animals commit suicide? *Behav Brain Sci* 1980;3:278–279.

184. Bach-Y-Rita G, Veno A. Habitual violence: a profile of 62 men. *Am J Psychiatry* 1974;131:1015–1017.

185. Brown GL, Goodwin FK. Cerebrospinal fluid correlates of suicide attempts and aggression. *Ann N Y Acad Sci* 1986;487:175–188.

186. Ostroff RB, Giller E, Harkness L, et al. The norepinephrine-to-epinephrine ratio in patients with a history of suicide attempts. *Am J Psychiatry* 1985;142:224–227.

187. Jones IH. Self-injury: toward a biological basis. *Perspect Biol Med* 1982;26:137–150.

188. Weiger WA, Bear DM. An approach to the neurology of aggression. *J Psychiatr Res* 1988;22:85–98.

189. Sheard MH. Clinical pharmacology of aggressive behavior. *Clin Neuropharmacol* 1988;11:483–492.

190. Farberow NL, Simon MD. Suicide in Los Angeles and Vienna: an intercultural study of two cities. *Public Health Rep* 1969;84:389–403.

191. Roy A. Family history of suicide. *Arch Gen Psychiatry* 1983;40:971–974.

192. Tsuang MT. Risk of suicide in the relatives of schizophrenics, manics, depressives, and controls. *J Clin Psychiatry* 398;44:396–397.

193. Scheftner WA, Young MA, Endicott J, et al. Family history and five-year suicide risk. *Br J Psychiatry* 1988;153:805–809.

194. Brent DA, Perper JA, Moritz G, et al. Familial risk factors for adolescent suicide: a case-control study. *Acta Psychiatr Scand* 1994;89:52–58.

195. Egeland JA, Sussex JN. Suicide and family loading for affective disorders. *JAMA* 1985;254:915–918.

196. Roy A, Segal NL, Centerwall BS, et al. Suicide in twins. *Arch Gen Psychiatry* 1991;48:29–32.

197. Roy A. Genetics of suicide. *Ann N Y Acad Sci* 1986;487:97–105.

198. Roy A, Segal NL, Sarchiapone M. Attempted suicide among living co-twins of twin suicide victims. *Am J Psychiatry* 1995;152:1075–1076.

199. Segal NL. Suicide attempts in twins whose co-twins' deaths were non-suicides. *Pers Individ Diff* 1995;19:937–940.

200. Wender PH, Kety SS, Rosenthal D, et al. Psychiatric disorders in the biological and adoptive families of adopted individuals with affective disorders. *Arch Gen Psychiatry* 1986;43:923–929.

201. Schulsinger F, Kety SS, Rosenthal D, et al. A family study of suicide. In: Schou M, Stromgren E, eds. *Origin, prevention, and treatment of affective disorders.* New York: Academic Press, 1979:277–287.

202. Abbar M, Amadeo S, Malafosse A, et al. An association study between suicidal behavior and tryptophan hydroxylase markers. *Clin Neuropharmacol* 1992;15:299.

203. Nielsen DA, Goldman D, Virkkunen M, et al. Suicidality and 5-hydroxyindoleacetic acid concentration associated with a tryptophan hydroxylase polymorphism. *Arch Gen Psychiatry* 1994;51:34–38.

204. Abbar M, Courtet P, Amadeo S, et al. Suicidal behaviors and the tryptophan hydroxylase gene. *Arch Gen Psychiatry* 1995;52:846–849.

205. Kunugi H, Ishida S, Kato T, et al. No evidence for an association of polymorphisms of the tryptophan hydroxylase gene with affective disorders or attempted suicide among Japanese patients. *Am J Psychiatry* 1999;156:774–776.

206. Bennett PJ, McMahon WM, Watabe J, et al. Tryptophan hydroxylase polymorphisms in suicide victims. *Psychiatr Genet* 2000;10:13–17.

207. Bellivier F, Leboyer M, Courtet P, et al. Association between the tryptophan hydroxylase gene and manic-depressive illness. *Arch Gen Psychiatry* 1998;55:33–37.

208. Furlong RA, Ho L, Rubinsztein JS, et al. No association of the tryptophan hydroxylase gene with bipolar affective disorder, unipolar affective disorder, or suicidal behaviour in major affective disorder. *Am J Med Genet* 1998;81:245–247.

209. Geijer T, Frisch A, Persson ML, et al. Search for association between suicide attempt and serotonergic polymorphisms. *Psychiatr Genet* 2000;10:19–26.

210. Mann JJ, Malone KM, Nielsen DA, et al. Possible association of a polymorphism of the tryptophan hydroxylase gene with suicidal behavior in depressed patients. *Am J Psychiatry* 1997;154:1451–1453.

211. Tsai SJ, Hong CJ, Wang YC. Tryptophan hydroxylase gene polymorphism (A218C) and suicidal behaviors. *Neuroreport* 1999;10:3773–3775.
212. Du L, Faludi G, Palkovits M, et al. Frequency of long allele in serotonin transporter gene is increased in depressed suicide victims. *Biol Psychiatry* 1999;46:196–201.
213. Rotondo A, Schuebel K, Bergen A, et al. Identification of four variants in the tryptophan hydroxylase promoter and association to behavior. *Mol Psychiatry* 1999;4:360–368.
214. Bellivier F, Laplanche JL, Leboyer M, et al. Serotonin transporter gene and manic depressive illness: an association study. *Biol Psychiatry* 1997;41:750–752.
215. Russ MJ, Lachman HM, Kashdan T, et al. Analysis of catechol-O-methyltransferase and 5-hydroxytryptamine transporter polymorphisms in patients at risk for suicide. *Psychiatry Res* 2000;93:73–78.
216. Mann JJ, Huang YY, Underwood MD, et al. A serotonin transporter gene promoter polymorphism (5-HTTLPR) and prefrontal cortical binding in major depression and suicide. *Arch Gen Psychiatry* 2000;57:729–738.
217. Turecki G, Briere R, Dewar K, et al. Prediction of level of serotonin 2A receptor binding by serotonin receptor 2A genetic variation in postmortem brain samples from subjects who did or did not commit suicide. *Am J Psychiatry* 1999;156:1456–458.
218. Ohara K, Nagai M, Tsukamoto T, et al. 5-HT2A receptor gene promoter polymorphism—1438G/A and mood disorders. *Neuroreport* 1998;9:1139–1141.
219. Du L, Bakish D, Lapierre YD, et al. Association of polymorphism of serotonin 2A receptor gene with suicidal ideation in major depressive disorder. *Am J Med Genet* 2000;96:56–60.
220. Alda M, Hrdina PD. Distribution of platelet 5-HT (2A) receptor densities in suicidal and non-suicidal depressives and control subjects. *Psychiatry Res* 2000;94:273–277.
221. Huang YY, Grailhe R, Arango V, et al. Relationship of psychopathology to the human serotonin1B genotype and receptor binding kinetics in postmortem brain tissue. *Neuropsychopharmacology* 1999;21:238–246.
222. Schildkraut JJ. The catecholamine hypothesis of affective disorders: a review of supporting evidence [1965 classic article]. *J Neuropsychiatry Clin Neurosci* 1995;7:524–533.
223. Lapin IP, Oxenkrug GF. Intensification of the central serotoninergic processes as a possible determinant of the thymoleptic effect. *Lancet* 1969;1:132–136.
224. Cheetham SC, Crompton MR, Czudek C, et al. Serotonin concentrations and turnover in brains of depressed suicides. *Brain Res* 1989;502:332–340.
225. Arato M, Tekes K, Tothfalusi L, et al. Serotonergic split brain and suicide. *Psychiatry Res* 1987;21:355–356.
226. Crow TJ, Cross AJ, Cooper SJ, et al. Neurotransmitter receptors and monoamine metabolites in the brains of patients with Alzheimer-type dementia and depression, and suicides. *Neuropharmacology* 1984;23:1561–1569.
227. Pare CM, Yeung DP, Price K, et al. 5-hydroxytryptamine, noradrenaline, and dopamine in brainstem, hypothalamus, and caudate nucleus of controls and of patients committing suicide by coal-gas poisoning. *Lancet* 1969;2:133–135.
228. Mann JJ, Henteleff RA, Lagattuta TF, et al. Lower 3H-paroxetine binding in cerebral cortex of suicide victims is partly due to fewer high affinity, non-transporter sites. *J Neural Transm* 1996;103:1337–1350.
229. Arranz B, Blennow K, Eriksson A, et al. Serotonergic, noradrenergic, and dopaminergic measures in suicide brains. *Biol Psychiatry* 1997;41:1000–1009.
230. Bourne HR, Bunney WE Jr, Colburn RW, et al. Noradrenaline, 5-hydroxytryptamine, and 5-hydroxyindoleacetic acid in hindbrains of suicidal patients. *Lancet* 1968;2:805–808.
231. Cochran E, Robins E, Grote S. Regional serotonin levels in brain: a comparison of depressive suicides and alcoholic suicides with controls. *Biol Psychiatry* 1976;11:283–294.
232. Beskow J, Gottfries CG, Roos BE, et al. Determination of monoamine and monoamine metabolites in the human brain: post mortem studies in a group of suicides and in a control group. *Acta Psychiatr Scand* 1976;53:7–20.
233. Shaw DM, Camps FE, Eccleston EG. 5-Hydroxytryptamine in the hind-brain of depressive suicides. *Br J Psychiatry* 1967;113:1407–1411.
234. Lloyd KG, Farley IJ, Deck JH, et al. Serotonin and 5-hydroxyindoleacetic acid in discrete areas of the brainstem of suicide victims and control patients. *Adv Biochem Psychopharmacol* 1974;11:387–397.
235. Korpi ER, Kleinman JE, Goodman SI, et al. Serotonin and 5-hydroxyindoleacetic acid in brains of suicide victims. Comparison in chronic schizophrenic patients with suicide as cause of death. *Arch Gen Psychiatry* 1986;43:594–600.
236. Owen F, Chambers DR, Cooper SJ, et al. Serotonergic mechanisms in brains of suicide victims. *Brain Res* 1986;362:185–188.
237. Underwood MD, Khaibulina AA, Ellis SP, et al. Morphometry of the dorsal raphe nucleus serotonergic neurons in suicide victims. *Biol Psychiatry* 1999;46:473–483.
238. Asberg M, Nordstrom P, Traskman-Bendz L. Cerebrospinal fluid studies in suicide. An overview. *Ann N Y Acad Sci* 1986;487:243–255.
239. Gross-Isseroff R, Israeli M, Biegon A. Autoradiographic analysis of tritiated imipramine binding in the human brain post mortem: effects of suicide. *Arch Gen Psychiatry* 1989;46:237–241.
240. Arora RC, Meltzer HY. Laterality and 3H-imipramine binding: studies in the frontal cortex of normal controls and suicide victims. *Biol Psychiatry* 1991;29:1016–1022.
241. Joyce JN, Shane A, Lexow N, et al. Serotonin uptake sites and serotonin receptors are altered in the limbic system of schizophrenics. *Neuropsychopharmacology* 1993;8:315–336.
242. Andersson A, Eriksson A, Marcusson J. Unaltered number of brain serotonin uptake sites in suicide victims. *J Psychopharmacol* 1992;6:509–513.
243. Lawrence KM, De Paermentier F, Cheetham SC, et al. Brain 5-HT uptake sites, labelled with [3H]paroxetine, in antidepressant-free depressed suicides. *Brain Res* 1990;526:17–22.
244. Lawrence KM, De Paermentier F, Cheetham SC, et al. Symmetrical hemispheric distribution of 3H-paroxetine binding sites in postmortem human brain from controls and suicides. *Biol Psychiatry* 1990;28:544–546.
245. Hrdina PD, Demeter E, Vu TB, et al. 5-HT uptake sites and 5-HT2 receptors in brain of antidepressant-free suicide victims/depressives: increase in 5-HT2 sites in cortex and amygdala. *Brain Res* 1993;614:37–44.
246. Bligh-Glover W, Kolli TN, Shapiro K, et al. The serotonin transporter in the midbrain of suicide victims with major depression. *Biol Psychiatry* 2000;47:1015–1024.
247. Stanley M, Mann JJ. Increased serotonin-2 binding sites in frontal cortex of suicide victims. *Lancet* 1983;1:214–216.
248. Stanley M, Virgilio J, Gershon S. Tritiated imipramine binding sites are decreased in the frontal cortex of suicides. *Science* 1982;216:1337–1339.
249. Rosel P, Menchon JM, Vallejo J, et al. Platelet [3H]imipramine and [3H]paroxetine binding in depressed patients. *J Affect Disord* 1997;44:79–85.
250. Rosel P, Arranz B, Vallejo J, et al. High affinity [3H]imipramine

and [3H]paroxetine binding sites in suicide brains. *J Neural Transm* 1997;104:921–929.

251. Rosel P, Arranz B, Vallejo J, et al. Variations in [3H]imipramine and 5-HT2A but not [3H]paroxetine binding sites in suicide brains. *Psychiatry Res* 1998;82:161–170.

252. Arango V, Underwood MD, Gubbi AV, et al. Localized alterations in pre- and postsynaptic serotonin binding sites in the ventrolateral prefrontal cortex of suicide victims. *Brain Res* 1995;688:121–133.

253. Lawrence KM, Kanagasundaram M, Lowther S, et al. [3H] imipramine binding in brain samples from depressed suicides and controls: 5-HT uptake sites compared with sites defined by desmethylimipramine. *J Affect Disord* 1998;47:105–112.

254. Meyerson LR, Wennogle LP, Abel MS, et al. Human brain receptor alterations in suicide victims. *Pharmacol Biochem Behav* 1982;17:159–163.

255. Mann JJ, Arango V, Marzuk PM, et al. Evidence for the 5-HT hypothesis of suicide. A review of post-mortem studies. *Br J Psychiatry Suppl* 1989:7–14.

256. Arato M, Tekes K, Tothfalusi L, et al. Reversed hemispheric asymmetry of imipramine binding in suicide victims. *Biol Psychiatry* 1991;29:699–702.

257. Brunello N, Chuang DM, Costa E. Different synaptic location of mianserin and imipramine binding sites. *Science* 1982;215: 1112–1115.

258. Roth BL, McLean S, Zhu XZ, et al. Characterization of two [3H]ketanserin recognition sites in rat striatum. *J Neurochem* 1987;49:1833–1838.

259. Mann JJ, Stanley M, McBride PA, et al. Increased serotonin 2 and beta-adrenergic receptor binding in the frontal cortices of suicide victims. *Arch Gen Psychiatry* 1986;43:954–959.

260. Arora RC, Meltzer HY. Serotonergic measures in the brains of suicide victims: 5-HT2 binding sites in the frontal cortex of suicide victims and control subjects. *Am J Psychiatry* 1989;146:730–736.

261. Owen F, Cross AJ, Crow TJ, et al. Brain 5-HT-2 receptors and suicide [Letter]. *Lancet* 1983;2:1256.

262. Cheetham SC, Crompton MR, Katona CL, et al. Brain 5-HT2 receptor binding sites in depressed suicide victims. *Brain Res* 1988;443:272–280.

263. Yates M, Leake A, Candy JM, et al. 5HT2 receptor changes in major depression. *Biol Psychiatry* 1990;27:489–496.

264. Gross-Isseroff R, Salama D, Israeli M, et al. Autoradiographic analysis of [3H]ketanserin binding in the human brain postmortem: effect of suicide. *Brain Res* 1990;507:208–215.

265. Arango V, Ernsberger P, Marzuk PM, et al. Autoradiographic demonstration of increased serotonin 5-HT2 and beta-adrenergic receptor binding sites in the brain of suicide victims. *Arch Gen Psychiatry* 1990;47:1038–1047.

266. Arango V, Underwood MD, Mann JJ. Alterations in monoamine receptors in the brain of suicide victims. *J Clin Psychopharmacol* 1992;12:8S–12S.

267. Stockmeier CA, Dilley GE, Shapiro LA, et al. Serotonin receptors in suicide victims with major depression. *Neuropsychopharmacology* 1997;16:162–173.

268. Arranz B, Eriksson A, Mellerup E, et al. Brain 5-HT1A, 5-HT1D, and 5-HT2 receptors in suicide victims. *Biol Psychiatry* 1994;35:457–463.

269. Arora RC, Meltzer HY. Serotonin2 (5-HT2) receptor binding in the frontal cortex of schizophrenic patients. *J Neural Transm* 1991;85:19–29.

270. Cheetham SC, Crompton MR, Katona CL, et al. Brain 5-HT1 binding sites in depressed suicides. *Psychopharmacology* 1990;102:544–548.

271. Dillon KA, Gross-Isseroff R, Israeli M, et al. Autoradiographic

analysis of serotonin 5-HT1A receptor binding in the human brain postmortem: effects of age and alcohol. *Brain Res* 1991; 554:56–64.

272. Matsubara S, Arora RC, Meltzer HY. Serotonergic measures in suicide brain: 5-HT1A binding sites in frontal cortex of suicide victims. *J Neural Transm* 1991;85:181–194.

273. Stockmeier CA, Shapiro LA, Dilley GE, et al. Increase in serotonin-1A autoreceptors in the midbrain of suicide victims with major depression-postmortem evidence for decreased serotonin activity. *J Neurosci* 1998;18:7394–7401.

274. Lowther S, Katona CL, Crompton MR, et al. 5-HT1D and 5-HT1E/1F binding sites in depressed suicides: increased 5-HT1D binding in globus pallidus but not cortex. *Mol Psychiatry* 1997;2:314–321.

275. Stanley M, Traskman-Bendz L, Dorovini-Zis K. Correlations between aminergic metabolites simultaneously obtained from human CSF and brain. *Life Sci* 1985;37:1279–1286.

276. Asberg M, Thoren P, Traskman L, et al. "Serotonin depression"—a biochemical subgroup within the affective disorders? *Science* 1976;191:478–480.

277. Vestergaard P, Sorensen T, Hoppe E, et al. Biogenic amine metabolites in cerebrospinal fluid of patients with affective disorders. *Acta Psychiatr Scand* 1978;58:88–96.

278. Roy-Byrne P, Post RM, Rubinow DR, et al. CSF 5HIAA and personal and family history of suicide in affectively ill patients: a negative study. *Psychiatry Res* 1983;10:263–274.

279. Agren H. Life at risk: markers of suicidality in depression. *Psychiatr Dev* 1983;1:87–103.

280. Brown GL, Goodwin FK, Ballenger JC, et al. Aggression in humans correlates with cerebrospinal fluid amine metabolites. *Psychiatry Res* 1979;1:131–139.

281. Traskman L, Asberg M, Bertilsson L, et al. Monoamine metabolites in CSF and suicidal behavior. *Arch Gen Psychiatry* 1981; 38:631–636.

282. Roy A, De Jong J, Linnoila M. Cerebrospinal fluid monoamine metabolites and suicidal behavior in depressed patients. A 5-year follow-up study. *Arch Gen Psychiatry* 1989;46:609–612.

283. van Praag HM. CSF 5-HIAA and suicide in non-depressed schizophrenics [Letter]. *Lancet* 1983;2:977–978.

284. van Praag HM. (Auto)aggression and CSF 5-HIAA in depression and schizophrenia. *Psychopharmacol Bull* 1986;22:669–673.

285. Ninan PT, van Kammen DP, Scheinin M, et al. CSF 5-hydroxyindoleacetic acid levels in suicidal schizophrenic patients. *Am J Psychiatry* 1984;141:566–569.

286. Banki CM, Arato M, Papp Z, et al. Biochemical markers in suicidal patients. Investigations with cerebrospinal fluid amine metabolites and neuroendocrine tests. *J Affect Disord* 1984;6: 341–350.

287. Cooper SJ, Kelly CB, King DJ. 5-Hydroxyindoleacetic acid in cerebrospinal fluid and prediction of suicidal behaviour in schizophrenia. *Lancet* 1992;340:940–941.

288. Brown GL, Ebert MH, Goyer PF, et al. Aggression, suicide, and serotonin: relationships to CSF amine metabolites. *Am J Psychiatry* 1982;139:741–746.

289. Linnoila M, Virkkunen M, Scheinin M, et al. Low cerebrospinal fluid 5-hydroxyindoleacetic acid concentration differentiates impulsive from nonimpulsive violent behavior. *Life Sci* 1983;33:2609–2614.

290. Virkkunen M, Nuutila A, Goodwin FK, et al. Cerebrospinal fluid monoamine metabolite levels in male arsonists [published erratum appears in *Arch Gen Psychiatry* 1989;46:960]. *Arch Gen Psychiatry* 1987;44:241–247.

291. Virkkunen M, De Jong J, Bartko J, et al. Relationship of psychobiological variables to recidivism in violent offenders and

impulsive fire setters. A follow-up study [published erratum appears in *Arch Gen Psychiatry* 1989;46:913]. *Arch Gen Psychiatry* 1989;46:600–603.

292. Virkkunen M, De Jong J, Bartko J, et al. Psychobiological concomitants of history of suicide attempts among violent offenders and impulsive fire setters [published erratum appears in *Arch Gen Psychiatry* 1989;46:913]. *Arch Gen Psychiatry* 1989;46:604–606.

293. Linnoila M, De Jong J, Virkkunen M. Family history of alcoholism in violent offenders and impulsive fire setters. *Arch Gen Psychiatry* 1989;46:613–616.

294. Bioulac B, Benezech M, Renaud B, et al. Serotoninergic dysfunction in the 47, XYY syndrome. *Biol Psychiatry* 1980;15:917–923.

295. Lidberg L, Asberg M, Sundqvist-Stensman UB. 5-Hydroxyindoleacetic acid levels in attempted suicides who have killed their children [Letter]. *Lancet* 1984;2:928.

296. Kruesi MJ, Rapoport JL, Hamburger S, et al. Cerebrospinal fluid monoamine metabolites, aggression, and impulsivity in disruptive behavior disorders of children and adolescents. *Arch Gen Psychiatry* 1990;47:419–426.

297. Lidberg L, Tuck JR, Asberg M, et al. Homicide, suicide and CSF 5-HIAA. *Acta Psychiatr Scand* 1985;71:230–236.

298. Stanley B, Molcho A, Stanley M, et al. Association of aggressive behavior with altered serotonergic function in patients who are not suicidal. *Am J Psychiatry* 2000;157:609–614.

299. Virkkunen M. Reactive hypoglycemic tendency among habitually violent offenders. A further study by means of the glucose tolerance test. *Neuropsychobiology* 1982;8:35–40.

300. Virkkunen M. Reactive hypoglycemic tendency among arsonists. *Acta Psychiatr Scand* 1984;69:445–452.

301. Roy A, Virkkunen M, Linnoila M. Monoamines, glucose metabolism, aggression towards self and others. *Int J Neurosci* 1988;41:261–264.

302. Stahl SM. The human platelet. A diagnostic and research tool for the study of biogenic amines in psychiatric and neurologic disorders. *Arch Gen Psychiatry* 1977;34:509–516.

303. Rao ML, Hawellek B, Papassotiropoulos A, et al. Upregulation of the platelet serotonin2A receptor and low blood serotonin in suicidal psychiatric patients. *Neuropsychobiology* 1998;38:84–89.

304. Simonsson P, Traskman-Bendz L, Alling C, et al. Peripheral serotonergic markers in patients with suicidal behavior. *Eur Neuropsychopharmacol* 1991;1:503–510.

305. Pfeffer CR, McBride PA, Anderson GM, et al. Peripheral serotonin measures in prepubertal psychiatric inpatients and normal children: associations with suicidal behavior and its risk factors. *Biol Psychiatry* 1998;44:568–577.

306. Mann JJ, McBride PA, Anderson GM, et al. Platelet and whole blood serotonin content in depressed inpatients: correlations with acute and life-time psychopathology. *Biol Psychiatry* 1992;32:243–257.

307. Verkes RJ, Van der Mast RC, Kerkhof AJ, et al. Platelet serotonin, monoamine oxidase activity, and [3H]paroxetine binding related to impulsive suicide attempts and borderline personality disorder. *Biol Psychiatry* 1998;43:740–746.

308. Verkes RJ, Fekkes D, Zwinderman AH, et al. Platelet serotonin and [3H]paroxetine binding correlate with recurrence of suicidal behavior. *Psychopharmacology* 1997;132:89–94.

309. Roy A, Everett D, Pickar D, et al. Platelet tritiated imipramine binding and serotonin uptake in depressed patients and controls. Relationship to plasma cortisol levels before and after dexamethasone administration. *Arch Gen Psychiatry* 1987;44:320–327.

310. Wagner A, Aberg-Wistedt A, Asberg M, et al. Lower

311. Marazziti D, De Leo D, Conti L. Further evidence supporting the role of the serotonin system in suicidal behavior: a preliminary study of suicide attempters. *Acta Psychiatr Scand* 1989;80:322–324.

312. Theodorou AE, Katona CL, Davies SL, et al. 3H-imipramine binding to freshly prepared platelet membranes in depression. *Psychiatry Res* 1989;29:87–103.

313. Marazziti D, Presta S, Silvestri S, et al. Platelet markers in suicide attempters. *Prog Neuropsychopharmacol Biol Psychiatry* 1995;19:375–383.

314. Roy A. Suicidal behavior in depression: relationship to platelet serotonin transporter. *Neuropsychobiology* 1999;39:71–75.

315. Van Kempen GM, Notten P, Hengeveld MW. Repeated measures of platelet MAO activity and 5-HT in a group of suicidal women. *Biol Psychiatry* 1992;31:529–530.

316. Pandey GN, Pandey SC, Janicak PG, et al. Platelet serotonin-2 receptor binding sites in depression and suicide. *Biol Psychiatry* 1990;28:215–222.

317. Biegon A, Grinspoon A, Blumenfeld B, et al. Increased serotonin 5-HT2 receptor binding on blood platelets of suicidal men. *Psychopharmacology* 1990;100:165–167.

318. McBride PA, Brown RP, DeMeo M, et al. The relationship of platelet 5-HT2 receptor indices to major depressive disorder, personality traits, and suicidal behavior. *Biol Psychiatry* 1994;35:295–308.

319. Neuger J, El Khoury A, Kjellman BF, et al. Platelet serotonin functions in untreated major depression. *Psychiatry Res* 1999;85:189–198.

320. Meltzer HY, Perline R, Tricou BJ, et al. Effect of 5-hydroxytryptophan on serum cortisol levels in major affective disorders. II. Relation to suicide, psychosis, and depressive symptoms. *Arch Gen Psychiatry* 1984;41:379–387.

321. Coccaro EF, Siever LJ, Klar HM, et al. Serotonergic studies in patients with affective and personality disorders. Correlates with suicidal and impulsive aggressive behavior [published erratum appears in *Arch Gen Psychiatry* 1990;47:124]. *Arch Gen Psychiatry* 1989;46:587–599.

322. Malone KM, Corbitt EM, Li S, et al. Prolactin response to fenfluramine and suicide attempt lethality in major depression. *Br J Psychiatry* 1996;168:324–329.

323. Mann JJ, McBride PA, Brown RP, et al. Relationship between central and peripheral serotonin indexes in depressed and suicidal psychiatric inpatients. *Arch Gen Psychiatry* 1992;49:442–446.

324. Mann JJ, Waternaux C, Haas GL, et al. Toward a clinical model of suicidal behavior in psychiatric patients. *Am J Psychiatry* 1999;156:181–189.

325. Agren H. Symptom patterns in unipolar and bipolar depression correlating with monoamine metabolites in the cerebrospinal fluid: II. Suicide. *Psychiatry Res* 1980;3:225–236.

326. Roy A, Agren H, Pickar D, et al. Reduced CSF concentrations of homovanillic acid and homovanillic acid to 5-hydroxyindoleacetic acid ratios in depressed patients: relationship to suicidal behavior and dexamethasone nonsuppression. *Am J Psychiatry* 1986;143:1539–1545.

327. Montgomery SA, Montgomery D. Pharmacological prevention of suicidal behaviour. *J Affect Disord* 1982;4:291–298.

328. Engstrom G, Alling C, Blennow K, et al. Reduced cerebrospinal HVA concentrations and HVA/5-HIAA ratios in suicide attempters. Monoamine metabolites in 120 suicide attempters and 47 controls. *Eur Neuropsychopharmacol* 1999;9:399–405.

3H-imipramine binding in platelets from untreated depressed patients compared to healthy controls. *Psychiatry Res* 1985;16:131–139.

329. Ninan PT, van Kammen DP, Linnoila M. Letter to the editor. *Am J Psychiatry* 1985;142:148.

330. Roy A, Karoum F, Pollack S. Marked reduction in indexes of dopamine metabolism among patients with depression who attempt suicide. *Arch Gen Psychiatry* 1992;49:447–450.

331. Tripodianakis J, Markianos M, Sarantidis D, et al. Neurochemical variables in subjects with adjustment disorder after suicide attempts. *Eur Psychiatry* 2000;15:190–195.

332. Bowden C, Cheetham SC, Lowther S, et al. Reduced dopamine turnover in the basal ganglia of depressed suicides. *Brain Res* 1997;769:135–140.

333. Allard P, Norlen M. Unchanged density of caudate nucleus dopamine uptake sites in depressed suicide victims. *J Neural Transm* 1997;104:1353–1360.

334. Bowden C, Lowther SW. Dopamine uptake sites, labelled with [3H]GBR12935, in brain samples from depressed suicides and controls. *Eur Neuropsychopharmacol* 1997;7:247–252.

335. Govoni S, Rius RA, Battaini F, et al. The central dopaminergic system: susceptibility to risk factors for accelerated aging. *Gerontology* 1988;34:29–34.

336. Morgan DG. The dopamine and serotonin systems during aging in human and rodent brain. A brief review. *Prog Neuropsychopharmacol Biol Psychiatry* 1987;11:153–157.

337. De Keyser J, Ebinger G, Vauquelin G. Age-related changes in the human nigrostriatal dopaminergic system. *Ann Neurol* 1990;27:157–161.

338. Richfield EK, O'Brien CF, Eskin T, et al. Heterogeneous dopamine receptor changes in early and late Huntington's disease. *Neurosci Lett* 1991;132:121–126.

339. Mayeux R. Depression in the patient with Parkinson's disease. *J Clin Psychiatry* 1990;51[Suppl]:20–23.

340. Roy A, Pickar D, De Jong J, et al. Suicidal behavior in depression: relationship to noradrenergic function. *Biol Psychiatry* 1989;25:341–350.

341. Ostroff R, Giller E, Bonese K, et al. Neuroendocrine risk factors of suicidal behavior. *Am J Psychiatry* 1982;139:1323–1325.

342. Traskman-Bendz L, Alling C, Oreland L, et al. Prediction of suicidal behavior from biological tests. *J Clin Psychopharmacol* 1992;12:21S–6S.

343. Arango V, Ernsberger P, Sved AF, et al. Quantitative autoradiography of alpha 1- and alpha 2-adrenergic receptors in the cerebral cortex of controls and suicide victims. *Brain Res* 1993;630:271–282.

344. Ordway GA, Widdowson PS, Smith KS, et al. Agonist binding to alpha 2-adrenoceptors is elevated in the locus coeruleus from victims of suicide. *J Neurochem* 1994;63:617–624.

345. Biegon A, Israeli M. Regionally selective increases in beta-adrenergic receptor density in the brains of suicide victims. *Brain Res* 1988;442:199–203.

346. Stockmeier CA, Meltzer HY. Beta-adrenergic receptor binding in frontal cortex of suicide victims. *Biol Psychiatry* 1991;29:183–191.

347. Little KY, Duncan GE, Breese GR. Beta-adrenergic binding in suicide victims. *Soc Neurosci Abst* 1990;16:140(abst).

348. De Paermentier F, Cheetham SC, Crompton MR, et al. Brain beta-adrenoceptor binding sites in antidepressant-free depressed suicide victims. *Brain Res* 1990;525:71–77.

349. Little KY, Ranc J, Gilmore J, et al. Lack of pineal beta-adrenergic receptor alterations in suicide victims with major depression. *Psychoneuroendocrinology* 1997;22:53–62.

350. Gross-Isseroff R, Dillon KA, Fieldust SJ, et al. Autoradiographic analysis of alpha 1-noradrenergic receptors in the human brain postmortem. Effect of suicide [published erratum appears in *Arch Gen Psychiatry* 1991;48):862]. *Arch Gen Psychiatry* 1990;47:1049–1053.

351. Ordway GA, Widdowson PS, Streator-Smith K, et al. Neurochemistry of the human locus coeruleus in suicide and depression. *Soc Neurosci Abst* 1990;16:140(abst).

352. Meana JJ, Barturen F, Garcia-Sevilla JA. Alpha 2-adrenoceptors in the brain of suicide victims: increased receptor density associated with major depression. *Biol Psychiatry* 1992;31:471–490.

353. Janowsky DS, el Yousef MK, Davis JM, et al. A cholinergic-adrenergic hypothesis of mania and depression. *Lancet* 1972;2:632–635.

354. Kaufmann CA, Gillin JC, Hill B, et al. Muscarinic binding in suicides. *Psychiatry Res* 1984;12:47–55.

355. Stanley M. Cholinergic receptor binding in the frontal cortex of suicide victims. *Am J Psychiatry* 1984;141:1432–1436.

356. Cheetham SC, Crompton MR, Katona CL, et al. Brain GABAA/benzodiazepine binding sites and glutamic acid decarboxylase activity in depressed suicide victims. *Brain Res* 1988;460:114–123.

357. Sherif F, Marcusson J, Oreland L. Brain gamma-aminobutyrate transaminase and monoamine oxidase activities in suicide victims. *Eur Arch Psychiatry Clin Neurosci* 1991;241:139–144.

358. Cross JA, Cheetham SC, Crompton MR, et al. Brain GABAB binding sites in depressed suicide victims. *Psychiatry Res* 1988;26:119–129.

359. Sundman I, Allard P, Eriksson A, et al. GABA uptake sites in frontal cortex from suicide victims and in aging. *Neuropsychobiology* 1997;35:11–15.

360. Arranz B, Cowburn R, Eriksson A, et al. Gamma-aminobutyric acid-B (GABAB) binding sites in postmortem suicide brains. *Neuropsychobiology* 1992;26:33–36.

361. Manchon M, Kopp N, Rouzioux JJ, et al. Benzodiazepine receptor and neurotransmitter studies in the brain of suicides. *Life Sci* 1987;41:2623–2630.

362. Stocks GM, Cheetham SC, Crompton MR, et al. Benzodiazepine binding sites in amygdala and hippocampus of depressed suicide victims. *J Affect Disord* 1990;18:11–15.

363. Rochet T, Kopp N, Vedrinne J, et al. Benzodiazepine binding sites and their modulators in hippocampus of violent suicide victims. *Biol Psychiatry* 1992;32:922–931.

364. Korpi ER, Kleinman JE, Wyatt RJ. GABA concentrations in forebrain areas of suicide victims. *Biol Psychiatry* 1988;23:109–114.

365. Roy A. Neuropeptides in relation to suicidal behavior in depression. *Neuropsychobiology* 1993;28:184–186.

366. Perry EK, Gibson PH, Blessed G, et al. Neurotransmitter enzyme abnormalities in senile dementia. Choline acetyltransferase and glutamic acid decarboxylase activities in necropsy brain tissue. *J Neurol Sci* 1977;34:247–265.

367. Charlton BG, Wright C, Leake A, et al. Somatostatin immunoreactivity in postmortem brain from depressed suicides [Letter]. *Arch Gen Psychiatry* 1988;45:597–598.

368. Traskman-Bendz L, Ekman R, Regnell G, et al. HPA-related CSF neuropeptides in suicide attempters. *Eur Neuropsychopharmacol* 1992;2:99–106.

369. Widdowson PS, Ordway GA, Halaris A. Neuropeptide Y concentrations in postmortem brain from victims of suicide and controls. *Soc Neurosci Abst* 1990:16:140(abst).

370. Arranz B, Blennow K, Ekman R, et al. Brain neuropeptidergic function in suicide victims. *Hum Psychopharmacol* 1996;11:451–461.

371. Westrin A, Ekman R, Traskman-Bendz L. Alterations of corticotropin releasing hormone (CRH) and neuropeptide Y (NPY) plasma levels in mood disorder patients with a recent suicide attempt. *Eur Neuropsychopharmacol* 1999;9:205–211.

372. Harro J, Marcusson J, Oreland L. Alterations in brain

cholecystokinin receptors in suicide victims. *Eur Neuropsychopharmacol* 1992;2:57–63.

373. Bachus SE, Hyde TM, Herman MM, et al. Abnormal cholecystokinin mRNA levels in entorhinal cortex of schizophrenics. *J Psychiatr Res* 1997;31:233–256.

374. Loefberg C, Agren H, Harro, et al. Cholecystokinin in CSF from depressed patients: possible relations to severity of depression and suicidal behaviour. *Eur Neuropsychopharmacol* 1998;8:153–157.

375. Kars H, Broekema W, Glaudemans-van Gelderen I, et al. Naltrexone attenuates self-injurious behavior in mentally retarded subjects. *Biol Psychiatry* 1990;27:741–746.

376. Sandman CA, Barron JL, Chicz-DeMet A, et al. Plasma B-endorphin levels in patients with self-injurious behavior and stereotypy. *Am J Ment Retard* 1990;95:84–92.

377. Barrett RP, Feinstein C, Hole WT. Effects of naloxone and naltrexone on self-injury: a double-blind, placebo-controlled analysis. *Am J Ment Retard* 1989;93:644–651.

378. Taylor DV, Hetrick WP, Neri CL, et al. Effect of naltrexone upon self-injurious behavior, learning and activity: a case study. *Pharmacol Biochem Behav* 1991;40:79–82.

379. Bruun R, Kurlan R. Opiate therapy and self-harming behavior in Tourette's syndrome [Letter]. *Mov Disord* 1991;6:184–185.

380. Richardson JS, Zaleski WA. Naloxone and self-mutilation. *Biol Psychiatry* 1983;18:99–101.

381. McGee MD. Cessation of self-mutilation in a patient with borderline personality disorder treated with naltrexone [Letter]. *J Clin Psychiatry* 1997;58:32–33.

382. Konicki PE, Schulz SC. Rationale for clinical trials of opiate antagonists in treating patients with personality disorders and self-injurious behavior. *Psychopharmacol Bull* 1989;25:556–563.

383. Frecska E, Arato M, Banki CM, et al. Prolactin response to fentanyl in depression. *Biol Psychiatry* 1989;25:692–696.

384. Gross-Isseroff R, Dillon KA, Israeli M, et al. Regionally selective increases in mu opioid receptor density in the brains of suicide victims. *Brain Res* 1990;530:312–316.

385. Gabilondo AM, Meana JJ, Garcia-Sevilla JA. Increased density of mu-opioid receptors in the postmortem brain of suicide victims. *Brain Res* 1995;682:245–250.

386. Scarone S, Gambini O, Calabrese G, et al. Asymmetrical distribution of beta-endorphin in cerebral hemispheres of suicides: preliminary data. *Psychiatry Res* 1990;32:159–166.

387. Loosen PT, Prange AJ Jr. Serum thyrotropin response to thyrotropin-releasing hormone in psychiatric patients: a review. *Am J Psychiatry* 1982;139:405–416.

388. Linkowski P, Van Wettere JP, Kerkhofs M, et al. Thyrotrophin response to thyrostimulin in affectively ill women relationship to suicidal behaviour. *Br J Psychiatry* 1983;143:401–405.

389. Linkowski P, Van Wettere JP, Kerkhofs M, et al. Violent suicidal behavior and the thyrotropin-releasing hormone-thyroid-stimulating hormone test: a clinical outcome study. *Neuropsychobiology* 1984;12:19–22.

390. Banki CM, Bissette G, Arato M, et al. Elevation of immunoreactive CSF TRH in depressed patients. *Am J Psychiatry* 1988;145:1526–1531.

391. Prasad AJ. Neuroendocrine differences between violent and non-violent parasuicides. *Neuropsychobiology* 1985;13:157–159.

392. Corrigan MH, Gillette GM, Quade D, et al. Panic, suicide, and agitation: independent correlates of the TSH response to TRH in depression. *Biol Psychiatry* 1992;31:984–992.

393. Bunney WE, Fawcett J. Possibility of a biochemical test for suicidal potential. *Arch Gen Psychiatry* 1965;13:232–239.

394. Bunney WE, Fawcett J, Davis JM, et al. Further evaluation of urinary 17-hydroxy-corticosteroid in suicidal patients. *Arch Gen Psychiatry* 1969;21:138–150.

395. Krieger G. The plasma level of cortisol as a predictor of suicide. *Dis Nerv Syst* 1979;35:237–240.

396. Dorovini-Zis K, Zis AP. Increased adrenal weight in victims of violent suicide. *Am J Psychiatry* 1987;144:1214–1215.

397. Szigethy E, Conwell Y, Forbes NT, et al. Adrenal weight and morphology in victims of completed suicide. *Biol Psychiatry* 1994;36:374–380.

398. Dumser T, Barocka A, Schubert E. Weight of adrenal glands may be increased in persons who commit suicide. *Am J Forensic Med Pathol* 1998;19:72–76.

399. Arato M, Banki CM, Bissette G, et al. Elevated CSF CRF in suicide victims. *Biol Psychiatry* 1989;25:355–359.

400. Nemeroff CB, Owens MJ, Bissette G, et al. Reduced corticotropin releasing factor binding sites in the frontal cortex of suicide victims. *Arch Gen Psychiatry* 1988;45:577–579.

401. Krieger G. Biochemical predictors of suicide. *Dis Nerv Syst* 1970;31:478–482.

402. Traskman L, Tybring G, Asberg M, et al. Cortisol in the CSF of depressed and suicidal patients. *Arch Gen Psychiatry* 1980;37:761–767.

403. Banki CM, Bissette G, Arato M, et al. CSF corticotropin-releasing factor-like immunoreactivity in depression and schizophrenia. *Am J Psychiatry* 1987;144:873–877.

404. Kocsis JH, Kennedy S, Brown RP, et al. Suicide and adrenocortical function. *Psychopharmacol Bull* 1986;22:650–655.

405. Stein E, McCrank E, Schaefer B, et al. Adrenal gland weight and suicide. *Can J Psychiatry* 1993;38:563–566.

406. Lester D. The dexamethasone suppression test as an indicator of suicide: a meta-analysis. *Pharmacopsychiatry* 1992;25:265–270.

407. Norman WH, Brown WA, Miller IW, et al. The dexamethasone suppression test and completed suicide. *Acta Psychiatr Scand* 1990;81:120–125.

408. Coryell W, Schlesser MA. Suicide and the dexamethasone suppression test in unipolar depression. *Am J Psychiatry* 1981;138:1120–1121.

409. Targum SD, Rosen L, Capodanno AE. The dexamethasone suppression test in suicidal patients with unipolar depression. *Am J Psychiatry* 1983;140:877–879.

410. Robbins DR, Alessi NE. Suicide and the dexamethasone suppression test in adolescence. *Biol Psychiatry* 1985;20:107–110.

411. Pfeffer CR, Stokes P, Shindledecker R. Suicidal behavior and hypothalamic-pituitary-adrenocortical axis indices in child psychiatric inpatients. *Biol Psychiatry* 1991;29:909–917.

412. Jones JS, Stein DJ, Stanley B, et al. Negative and depressive symptoms in suicidal schizophrenics. *Acta Psychiatr Scand* 1994;89:81–87.

413. Yerevanian BI, Olafsdottir H, Milanese E, et al. Normalization of the dexamethasone suppression test at discharge from hospital. Its prognostic value. *J Affect Disord* 1983;5:191–197.

414. Greden JF, Albala AA, Haskett RF, et al. Normalization of dexamethasone suppression test: a laboratory index of recovery from endogenous depression. *Biol Psychiatry* 1980;15:449–458.

415. Schmidtke A, Fleckenstein P, Beckmann H. The dexamethasone suppression test and suicide attempts. *Acta Psychiatr Scand* 1989;79:276–282.

416. Zimmerman M, Coryell W, Pfohl B. The validity of the dexamethasone suppression test as a marker for endogenous depression. *Arch Gen Psychiatry* 1986;43:347–355.

417. Ayuso-Gutierrez JL, Cabranes JA, Garcia-Camba E, et al. Pituitary-adrenal disinhibition and suicide attempts in depressed patients. *Biol Psychiatry* 1987;22:1409–1412.

418. Blackman MR. Pituitary hormones and aging. *Endocrinol Metab Clin North Am* 1987;16:981–994.

419. Greenwald BS, Mathe AA, Mohs RC, et al. Cortisol and Alzheimer's disease, II: dexamethasone suppression, dementia severity, and affective symptoms. *Am J Psychiatry* 1986;143:442–446.

420. Jacobs S, Mason J, Kosten T, et al. Urinary-free cortisol excretion in relation to age in acutely stressed persons with depressive symptoms. *Psychosom Med* 1984;46:213–221.

421. Stangl D, Pfohl B, Zimmerman M, et al. The relationship between age and post-dexamethasone cortisol: a test of three hypotheses. *J Affect Disord* 1986;11:185–197.

422. Shaffer D. The epidemiology of teen suicide: an examination of risk factors. *J Clin Psychiatry* 1988;49[Suppl]:36–41.

423. Lester D. Suicide and the menstrual cycle. *Med Hypotheses* 1990;31:197–199.

424. Martin CA, Mainous AG III, Mainous RO, et al. Progesterone and adolescent suicidality. *Biol Psychiatry* 1997;42:956–958.

425. Olweus D, Mattsson A, Schalling D, et al. Circulating testosterone levels and aggression in adolescent males: a causal analysis. *Psychosom Med* 1988;50:261–272.

426. Ehrenreich H, Halaris A, Ruether E, et al. Psychoendocrine sequelae of chronic testosterone deficiency. *J Psychiatric Res* 1999;33:379–387.

427. Bergman B, Brismar B. Hormone levels and personality traits in abusive and suicidal male alcoholics. *Alcohol Clin Exp Res* 1994;18:311–316.

428. Banki CM, Vojnik M, Papp Z, et al. Cerebrospinal fluid magnesium and calcium related to amine metabolites, diagnosis, and suicide attempts. *Biol Psychiatry* 1985;20:163–171.

429. Beck-Friis J, Kjellman BF, Aperia B, et al. Serum melatonin in relation to clinical variables in patients with major depressive disorder and a hypothesis of a low melatonin syndrome. *Acta Psychiatr Scand* 1985;71:319–330.

430. Sandyk R, Awerbuch GI. Nocturnal melatonin secretion in suicidal patients with multiple sclerosis. *Int J Neurosci* 1993;71:173–182.

431. Boston PF, Dursun SM, Reveley MA. Cholesterol and mental disorder. *Br J Psychiatry* 1996;169:682–689.

432. Muldoon MF, Manuck SB, Matthews KA. Lowering cholesterol concentrations and mortality: a quantitative review of primary prevention trials. *BMJ* 1990;301:309–314.

433. Lindberg G, Rastam L, Gullberg B, et al. Low serum cholesterol concentration and short term mortality from injuries in men and women. *BMJ* 1992;305:277–279.

434. Partonen T, Loennqvist J. Association of low serum total cholesterol with major depression and suicide. *Br J Psychiatry* 1999;175:259–262.

435. Tanskanen A, Vartiainen E, Tuomilehto J, et al. High serum cholesterol and risk of suicide. *Am J Psychiatry* 2000;157:648–650.

436. Gallerani M, Manfredini R, Caracciolo S, et al. Serum cholesterol concentrations in parasuicide. *BMJ* 1995;310:1632–1636.

437. Golier JA, Marzuk PM, Leon AC, et al. Low serum cholesterol level and attempted suicide. *Am J Psychiatry.* 1995;152:419–423.

438. Modai I, Valevski A, Dror S, et al. Serum cholesterol levels and suicidal tendencies in psychiatric inpatients. *J Clin Psychiatry* 1994;55:252–254.

439. Sullivan PF, Joyce PR, Bulik CM, et al. Total cholesterol and suicidality in depression. *Biol Psychiatry* 1994;36:472–477.

440. Takei N, Kunugi H, Nanko S, et al. Low serum cholesterol and suicide attempts [Letter]. *Br J Psychiatry* 1994;164:702–703.

441. Engelberg H. Low serum cholesterol and suicide. *Lancet* 1992;339:727–729.

442. Sulser F. Serotonin-norepinephrine receptor interactions in the brain: implications for the pharmacology and pathophysiology of affective disorders. *J Clin Psychiatry* 1987;48[Suppl]:12–18.

443. Goldstein RB, Black DW, Nasrallah A, et al. The prediction of suicide. Sensitivity, specificity, and predictive value of a multivariate model applied to suicide among 1906 patients with affective disorders. *Arch Gen Psychiatry* 1991;48:418–422.

444. Shneidman E. Some essentials of suicide and some implications for response. In: Roy A, ed. *Suicide.* Baltimore: Williams & Wilkins, 1986:1–16.

445. Maltsberger JT. *Suicide risk—the formulation of clinical judgement.* New York: New York University Press, 1986.

446. Shea SC. *The practical art of suicide assessment: a guide for mental health professionals and substance abuse counselors.* New York: John Wiley & Sons, 1999.

447. Montgomery SA, Montgomery DB, Green M, et al. Pharmacotherapy in the prevention of suicidal behavior. *J Clin Psychopharmacol* 1992;12:27S–31S.

448. Montgomery SA, Montgomery DB, Jayanthi-Rani S, et al. Maintenance therapy in repeat suicidal behavior: a placebo controlled trial. In: Proceedings of the 10th International Congress for Suicide Prevention and Crisis Intervention, Ottawa. 1979:227–229.

449. Hawton K, Townsend E, Arensman E, et al. Psychosocial versus pharmacological treatments for deliberate self harm. *Cochrane Database Syst Rev* 2000;CD001764.

450. Tondo L, Baldessarini RJ. Reduced suicide risk during lithium maintenance treatment. *J Clin Psychiatry* 2000;61[Suppl 9]:97–104.

451. Tondo L, Baldessarini RJ, Hennen J, et al. Lithium treatment and risk of suicidal behavior in bipolar disorder patients. *J Clin Psychiatry* 1998;59:405–414.

452. Schou M. The effect of prophylactic lithium treatment on mortality and suicidal behavior: a review for clinicians. *J Affect Disord* 1998;50:253–259.

453. Avery D, Winokur G. Suicide, attempted suicide, and relapse rates in depression. *Arch Gen Psychiatry* 1978;35:749–753.

454. Avery D, Winokur G. Mortality in depressed patients treated with electroconvulsive therapy and antidepressants. *Arch Gen Psychiatry* 1976;33:1029–1037.

455. Huston PE, Locher LM. Manic depressive psychosis. Course when treated and untreated with electric shock. *Arch Neurol Psychiatry* 1948;60:37–48.

456. Linehan MM, Armstrong HE, Suarez A, et al. Cognitive-behavioral treatment of chronically parasuicidal borderline patients. *Arch Gen Psychiatry* 1991;48:1060–1064.

457. Bateman A, Fonagy P. Effectiveness of partial hospitalization in the treatment of borderline personality disorder: a randomized controlled trial. *Am J Psychiatry* 1999;156:1563–1569.

458. Linehan MM. Behavioral treatments of suicidal behaviors. Definitional obfuscation and treatment outcomes. *Ann N Y Acad Sci* 1997;836:302–328.

459. Lewis G, Hawton K, Jones P. Strategies for preventing suicide. *Br J Psychiatry* 1997;171:351–354.

460. Brent DA, Perper JA, Allman CJ, et al. The presence and accessibility of firearms in the homes of adolescent suicides. A case-control study. *JAMA* 1991;266:2989–2995.

461. Sloan JH, Rivara FP, Reay DT, et al. Firearm regulations and rates of suicide. A comparison of two metropolitan areas. *N Engl J Med* 1990;322:369–373.

462. Loftin C, McDowall D, Wiersema B, et al. Effects of restrictive

licensing of handguns on homicide and suicide in the District of Columbia. *N Engl J Med* 1991;325:1615–1620.

463. Marzuk PM, Leon AC, Tardiff K, et al. The effect of access to lethal methods of injury on suicide rates. *Arch Gen Psychiatry* 1992;49:451–458.

464. Rich CL, Young JG, Fowler RC, et al. Guns and suicide: possible effects of some specific legislation. *Am J Psychiatry* 1990;147:342–346.

465. Conwell Y. Suicide in elderly patients. In: Schneider LS, Lebowitz BD, Friedhoff AJ, eds. *Diagnosis and treatment of depression in late life: rresults of the NIH Consensus Development Conference.* Washington, DC: American Psychiatric Press, 1994:397–418.

466. Carney SS, Rich CL, Burke PA, et al. Suicide over 60: the San Diego study. *J Am Geriatr Soc.* 1994;42:174–180.

467. Cattell G, Jolley DJ. One hundred cases of suicide in elderly people. *Br J Psychiatry* 1995;166:451–457.

468. Ben-Arie O, Welman M, Teggin AF. The depressed elderly living in the community. A follow-up study. *Br J Psychiatry* 1990;157:425–427.

469. Diekstra RF, van Egmond M. Suicide and attempted suicide in general practice, 1979–1986. *Acta Psychiatr Scand* 1989;79:268–275.

470. Nielsen AC, Williams TA. Depression in ambulatory medical patients. *Arch Gen Psychiatry* 1980;37:999–1004.

471. Rihmer Z, Rutz W, Pihlgren H. Depression and suicide on Gotland. An intensive study of all suicides before and after a depression-training programme for general practitioners. *J Affect Disord* 1995;35:147–152.

472. Rutz W, von Knorring L, Walinder J, et al. Effect of an educational program for general practitioners on Gotland on the pattern of prescription of psychotropic drugs. *Acta Psychiatr Scand* 1990;82:399–403.

473. Farberow NL, Gallagher DE, Gilewski MJ, et al. An examination of the early impact of bereavement on psychological distress in survivors of suicide. *Gerontologist* 1987;27:592–598.

474. Ness DE, Pfeffer CR. Sequelae of bereavement resulting from suicide. *Am J Psychiatry* 1990;147:279–285.

475. Appel YH, Wrobleski A. Self-help and support groups: mutual aid for survivors. In: Dunne EJ, McIntosh JL, Dunne-Maxim K, eds. *Suicide and its aftermath: understanding and counseling the survivors.* New York: WW Norton, 1987:215–233.

476. World Health Organization. *Suicide rates per 100,000.* Geneva: World Health Organization, 2001.

477. Arora RC, Meltzer HY. 3H-imipramine binding in the frontal cortex of suicides. *Psychiatry Res* 1989;30:125–135.

478. Little KY, McLauglin DP, Ranc J, et al. Serotonin transporter binding sites and mRNA levels in depressed persons committing suicide. *Biol Psychiatry* 1997;41:1156–1164.

479. Arango V, Hoffman L, Ernsberger P, et al. Quantitative autoradiography of cortical alpha-adrenergic receptors in suicide victims. *Soc Neurosci Abst* 1989;15:585 (abst).

28

THE NEUROPSYCHIATRY
OF MOOD DISORDER

ROBERT G. ROBINSON

Although the neuropsychiatry of mood disorders would logically include the biochemical neurophysiologic and neuroanatomic correlates of mood disorders not associated with brain injury, that topic with an enormous literature would be substantially more than can be covered in this chapter. Rather, disorders characterized by demonstrable brain lesions that are associated with mood disorders are the focus. This is itself a huge undertaking because of the large number of neuropsychiatric disorders.

The neuropsychiatric disorders associated with mood disorders are neurologic disorders characterized by structural brain lesions such as stroke, traumatic brain injury (TBI), Parkinson disease (PD), spinocerebellar degenerations, Alzheimer' disease and other degenerative dementias, Huntington disease, multiple sclerosis, and brain tumors, to name only a few. Similarly, the clinical manifestations of affective disorder include a wide range of emotional and cognitive disturbances and can be divided based on the type of mood disorder and the underlying neuropsychiatric disorder. Although studies providing empirical data about individual neuropsychiatric disorders have only begun to emerge within the past few years, these kinds of investigations are essential to build a firm empirical database for our understanding of the clinical manifestations, treatments, and mechanisms of these disorders.

In this chapter, the historical development of neuropsychiatric aspects of affective disorder is briefly reviewed and then the disturbances (depressive, mania, and bipolar disorders) associated with selected neuropsychiatric disorders, specifically stroke, TBI, and PD, are presented.

A confounding factor in our understanding of affective disorders associated with brain disease is the tendency of investigators to intermix different types of brain disorders when studying emotional problems in patients with brain injury. For example, the early work of Babinski (1) or Denny-Brown et al. (2), as well as the systematic study of emotional disorders in patients with brain injury by Gainotti (3), included patients with various types of brain injuries such as traumatic closed head injury, penetrating head injury, thromboembolic stroke, surgical incision, and intracerebral hemorrhage. Although it is generally assumed that neuronal death produced by a variety of mechanisms will result in similar clinical symptoms, depending on the size and location of the lesion, that is not necessarily the case. The fact that cognitive impairment is associated with major depression after stroke, but not TBI, is an example. Moreover, rarely will two conditions producing brain injury result in identical types of lesions. For example, closed head injury generally produces widespread brain injury with multiple small areas of shear or torsion injury, whereas cerebral embolism produces a focal lesion with an area of transient peripheral ischemia. Thus, much of the early information about emotional disorders associated with brain disease has drawn on data obtained from a heterogeneous group of patients, some with clearly defined and localized brain disease and others without it.

HISTORICAL BACKGROUND

Meyer (4) was one of the first neuropsychiatrists to examine the relationship of psychopathology to brain damage. Although he warned that new discoveries of cerebral localization in the 1880's and early 1900's, such as language function, led to an excess of identification of brain centers, he singled out several disorders, such as delirium, dementia, and aphasia, that he thought were the direct result of brain injury. In keeping with his biopsychosocial view of most mental "reactions," however, he saw disorders such as depression arising from a combination of head injury (specifically citing the left frontal lobe and cortical convexities), as well as family history and prior life events as leading to the psychiatric disorder. He did, nevertheless, suggest that some traumatic insanities may be produced by specific causes and locations of brain injury.

Babinski (1) noted that patients with right-hemisphere disease frequently displayed symptoms of anosognosia, euphoria, and indifference. Bleuler (5) wrote that after stroke, "melancholic moods lasting for months and sometimes longer appear frequently." Kraepelin (6) recognized an association between manic depressive insanity and cerebrovascular disease.

Goldstein (7) was the first to describe an emotional mood disorder thought to be uniquely associated with brain disease: the catastrophic reaction. The catastrophic reaction is an emotional outburst involving various degrees of anger, frustration, depression, tearfulness, refusal, shouting, swearing, and sometimes aggressive behavior. In his extensive studies of brain injury in war, Goldstein (8) described two symptom clusters: those related directly to the physical damage of a circumscribed area of the brain and those related secondarily to the organism's psychological response to injury. Emotional symptoms in his view represented the latter category (i.e., the psychological response of an organism struggling with physical or cognitive impairments).

A second abnormality unique to brain injury, also involving a disturbance of (this time the absence of) mood, was the indifference reaction described by Hecaen et al. (9) and Denny-Brown et al. (2). The indifference reaction associated with right-hemisphere lesions consists of symptoms of indifference toward failures, lack of interest in family and friends, enjoyment of foolish jokes, and minimizing of physical difficulties. It is associated with neglect of the opposite half of the body and space.

A third "mood" disorder associated with brain injury, such as cerebral infarction, is pathologic laughter or crying. Ironside (10) described the clinical manifestations of this disorder. Such patients' emotional displays are characteristically unrelated to their inner emotional state. Crying, for example, may occur spontaneously or after some seemingly minor provocation. This phenomenon has been given various names such as emotional incontinence, emotional lability, pseudobulbar affect, and pathologic emotionalism.

Despite the assertions by Kraepelin (6) and others (11) that mood disorder may be produced directly by focal brain injury, many investigators have adopted "psychological" explanations for the affective disorders associated with brain injury. Studies associating emotional symptoms with psychological reactions to the losses accompanying cerebrovascular disease began to appear in the early 1960's. Ullman and Gruen (12) reported that stroke was a particularly severe stress to the organism, as Goldstein (8) had suggested, because the organ governing the emotional response to injury had itself been damaged. Fisher (13) described depression associated with cerebrovascular disease as reactive and understandable because "the brain is the most cherished organ of humanity." Thus, depression was viewed as a natural emotional response to a decrease in self-esteem from a life-threatening injury and the resulting disability and dependence.

Systematic studies led other investigators who were impressed by the frequency of association between brain injury and mood disorders to hypothesize more direct causal links. In a study of 100 elderly patients with affective disorders,

Post (11) stated that the high frequency of brain ischemia associated with first episodes of depressive disorder suggested that the causes for atherosclerotic disease and depression may be linked. Folstein et al. (14) compared 20 stroke patients with ten orthopedic patients. Although the functional disability in both groups was comparable, more of the stroke patients were depressed. The authors concluded that "mood disorder is a more specific complication of stroke than simply a response to motor disability."

In summary, there have been two primary lines of thought in the study of affective disorders that are associated with structural brain disease. One attributes mood disorders to an understandable psychological reaction to the associated impairment; the other, based on a lack of association between severity of impairment and severity of emotional disorder, suggests a direct causal connection between mood disorders and structural brain damage.

DEPRESSION ASSOCIATED WITH STROKE

Prevalence

Depression is probably the most common emotional disorder associated with stroke. The prevalence of depression has been studied by investigators all over the world. Investigators who have used structured psychiatric interviews and established diagnostic criteria have usually identified two forms of depressive disorder associated with brain disease. One type is depression owing to stroke with a major depressivelike episode as defined by the *Diagnostic and Statistical Manual of Mental Disorders, Fourth Edition* (*DSM-IV*). The second type of depression is minor depression as defined by *DSM-IV* research criteria (this is a subsyndromal form of major depression).

The prevalence of these depressions varies, depending on whether patients were examined in a hospital or community setting (Table 28.1). The overall mean frequency of major depression is 20% among hospitalized patients and 13% among community samples. The mean prevalence of minor depression is 19% among hospital patients and 11% among community samples.

Diagnosis

Investigators of depression associated with physical illness have debated the most appropriate method for diagnosis of depression when some symptoms (e.g., sleep or appetite disturbance) could result from the physical illness.

Cohen-Cole and Stoudemire (15) reported that four approaches have been used to assess depression in the physically ill. These approaches are the inclusive approach in which depressive diagnostic symptoms are counted regardless of whether they may be related to physical illness (16), the etiologic approach in which a symptom is counted only if the diagnostician thinks that it is not caused by the physical

TABLE 28.1. PREVALENCE STUDIES OF POSTSTROKE DEPRESSION

Investigators	Patient Population	N	Criteria	% Major	% Minor	Total %
Finklestein, 1982 (140)	Rehab hosp	25	Cutoff score	—	—	48
Sinyor, 1986 (141)	Rehab hosp	35	Cutoff score	—	—	36
Finset, 1989 (142)	Rehab hosp	42	Cutoff score	—	—	36
Eastwood, 1989	Rehab hosp	87	SADS-RDC	10	40	50
Morris, 1990	Rehab hosp	99	CIDI-*DSM-III*	14	21	35
Shubert, 1992 (143)	Rehab hosp	18	*DSM-III-R*	28	44	72
Schwartz, 1993 (144)	Rehab hosp	91	*DSM-III*	40	—	40[a]
Feibel, 1982 (145)	Outpatient 6 mo	91	Nursing evaluation	—	—	26
Robinson, 1982	Outpatient 6 mo–10 yr	103	Cutoff score	—	—	29
Collin, 1987 (146)	Outpatient	111	Cutoff score	—	—	42
Astrom, 1993	Outpatient					
	3 mo	73	*DSM-III*	31	NR	31[a]
	1 yr	73	*DSM-III*	16	NR	16[a]
	2 yr	57	*DSM-III*	19	NR	19[a]
	3 yr	49	*DSM-III*	29	NR	29[a]
Castillo, 1995 (147)	Outpatient					
	3 mo	77	PSE-*DSM-III*	20	13	33
	6 mo	80	PSE-*DSM-III*	21	21	42
	1 yr	70	PSE-*DSM-III*	11	16	27
	2 yr	67	PSE-*DSM-III*	18	17	35
Wade, 1987 (148)	Community	379	Cutoff score	—	—	30
House, 1991	Community	89	PSE-*DSM-III*	11	12	23
Burvill, 1995	Community	294	PSE-*DSM-III*	15	8	23
Kotila, 1998 (149)	Community					
	3 mo	321	Beck cutoff score	—	—	48
	1 yr	311	—	—	—	49
Pohjasvaara, 1998 (150)	Community	277	*DSM-III-R*	26	14	40
Dennis, 2000 (151)	Outpatient 6 mo	309	Cutoff score	—	—	38
Robinson, 1983	Acute hosp	103	PSE-*DSM-III*	27	20	47
Ebrahim, 1987 (152)	Acute hosp	149	Cutoff score	—	—	23
Fedoroff, 1991 (153)	Acute hosp	205	PSE-*DSM-III*	22	19	41
Castillo, 1995 (147)	Acute hosp	291	PSE-*DSM-III*	20	18	38
Starkstein, 1992 (154)	Acute hosp	80	PSE-*DSM-III*	16	13	29
Astrom, 1993	Acute hosp	80	*DSM-III*	25	NR	25[a]
Herrmann, 1993	Acute hosp	21	RDC	24	14	38
Andersen, 1994	Acute hosp or outpatient	285	HDRS cutoff	10	11	21
Herrmann, 1998 (155)	Acute hosp					
	3 mo	150	MADRS cutoff	—	—	27
	1 yr	136	MADRS cutoff	—	—	22
Gainotti, 1999	Acute hosp	58	*DSM-III-R*	27	NR	27+[a]
	2–4 mo	52	*DSM-III-R*	27	NR	27+[a]
	>4 mo	43	*DSM-III-R*	40	NR	40+[a]
			Mean	22	21	36

[a]Because minor depression was not included, these values may be low.
Rehab hosp, rehabilitation hospital; PSE, present state examination; SADS, schedule for affective disorders and schizophrenia; RDC, research diagnostic criteria; *DSM-III, Diagnostic and Statistical Manual of Mental Disorders, Third Edition; DSM-III-R, Diagnostic and Statistical Manual of Mental Disorders, Revised Third Edition;* CIDI, composite international diagnostic interview; HDRS, Hamilton Depression Rating Scale.

illness (17), the substitutive approach of Endicott (18) in which other psychological symptoms of depression replace the vegetative symptoms, and the exclusive approach in which symptoms are removed from the diagnostic criteria if they are not found to be more frequent in depressed than nondepressed patients (19).

Paradiso et al. (20) examined the utility of these methods in the diagnosis of poststroke depression (PSD) during the first 2 years after stroke. Among 205 patients with acute stroke; 142 patients were followed for examination at 3, 6, 12, or 24 months after stroke. There were no significant differences in the background characteristics between patients with a depressed mood and nondepressed patients except that the depressed group was significantly younger ($p = 0.006$) and had a significantly higher frequency of personal history of psychiatric disorder ($p = 0.04$).

Throughout the 2-year follow-up, depressed patients showed a higher frequency both of vegetative and psychological symptoms compared with the nondepressed patients. The only symptoms that were not more frequent in the depressed compared with nondepressed patients were weight loss and early morning awakening at the initial evaluation; weight loss and early morning awakening at 6 months; weight loss, early morning awakening, anxious foreboding, and loss of libido at 1 year; and weight loss and loss of libido at 2 years. Among the psychological symptoms, the depressed patients had a higher frequency of most psychological symptoms throughout the 2-year follow-up. The only psychological symptoms that were not significantly more frequent in the depressed than in the nondepressed group were suicide plans, simple ideas of reference, and pathologic guilt at 3 months; pathologic guilt at 6 months; pathologic guilt, suicide plans, guilty ideas of reference, and irritability at 1 year; and pathologic guilt and self-depreciation at 2 years.

The effect of using each of the proposed alternative diagnostic methods for PSD using *DSM-IV* criteria was examined. The symptoms were obtained using the inclusive approach (i.e., symptoms that the patients acknowledged were included as positive even if there was some suspicion that the symptom may have been related to the physical illness). During the in-hospital evaluation, 26 patients (18%) met *DSM-IV* diagnostic criteria for major depression. Compared with diagnoses based solely on the existence of five or more specific symptoms of the diagnosis of *DSM-IV* major depression, diagnoses based on unmodified symptoms (i.e., early awakening and weight loss included) had a specificity of 98% and a sensitivity of 100%.

Modified *DSM-IV* criteria are then used to examine the substitutive approach (i.e., all vegetative symptoms were eliminated and the presence of four psychological symptoms plus depressed mood was required for the diagnosis of major depression). Using this approach, none of the original 27 patients with major depression was excluded. Similar results were found at 3-, 6-, 12-, and 24-month follow-up. Thus, one could reasonably conclude that modifying *DSM-IV*

criteria because of the existence of an acute medical illness is unnecessary and the inclusive approach can be used with confidence that patients are not being over- or underdiagnosed.

These findings also suggest that the nature of PSD may be changing over time. Because the symptoms that were specific to depression changed over time, this may reflect an alteration in the underling etiology of PSD associated with early-onset depression compared with the late or chronic poststroke period.

Phenomenology

Lipsey et al. (21) examined the frequency of depressive symptoms in a group of 43 patients with major PSD compared with that in a group of 43 age-matched patients with "functional" (i.e., no known brain pathology) depression. The main finding was that both groups showed almost identical profiles of symptoms, including those that were not part of the diagnostic criteria (Fig. 28.1). More than 50% of the patients who met the diagnostic criteria for major PSD reported sadness, anxiety, tension, loss of interest and concentration, sleep disturbances with early morning awakening, loss of appetite with weight loss, difficulty concentrating and thinking, and thoughts of death.

Gainotti et al. (22) examined the phenomenology of PSD using the Poststroke Depression Rating Scale that they devised. The scale includes 10 items: depressed mood, guilt feelings, thoughts of death or suicide, vegetative symptoms, apathy and loss of interest, anxiety, catastrophic reactions, hyperemotionalism, anhedonia, and diurnal mood variations. The last section on diurnal mood variations is scored between a +2 and a −2 score, +2 indicating a motivated depression associated with situational stresses, handicaps, or disabilities and −2 indicating a lack of associated motivation with depression being more prominent in the early morning.

Gainotti et al. compared patients with poststroke major depression who were less than 2 months (n = 58), 2 to 4 months (n = 52), and more than 4 months (n = 43) poststroke on the scale compared with 30 patients admitted to the psychiatric hospital with a diagnosis of endogenous major depression. Although statistical adjustments controlling for the large number of comparisons were not provided, the data were interpreted to indicate that patients with endogenous depression had higher scores on suicide and anhedonia, whereas patients with PSD had higher scores on catastrophic reactions, hyperemotionalism, and diurnal mood variation, which indicates an association with disability.

Gainotti et al. (22) asserted that failure to assess these aspects of depression (included in the Poststroke Depression Rating Scale) indicates methodologic errors in the assessment of depression by Robinson et al. (23). There are clearly established criteria for the diagnosis of major depression as validated through numerous studies supporting *DSM-IV* (23). The inclusion of catastrophic reactions, hyperemotionalism,

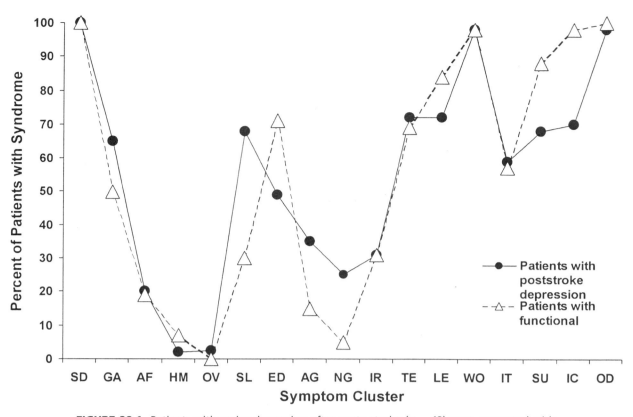

FIGURE 28.1. Patients with major depression after acute stroke (n = 43) were compared with age-comparable patients hospitalized for functional primary depression (n = 43). The symptom clusters are "syndromes" derived from the semistructured interview of the present state examination. SD, simple depression; GA, general anxiety; AF, affective flattening; HM, hypomania; OV, overactivity; SL, slowness; ED, special features of depression; AG, agitation; NG, self-neglect; IR, ideas of reference; TE, tension; LE, lack of energy; WO, worrying; IT, irritability; SU, social unease; IC, loss of interest and concentration; OD, other symptoms of depression. Patients with primary and poststroke depressions showed the same frequency of all syndromes except slowness (stroke patients showed a higher frequency) and loss of interest and concentration (primary depression patients showed a higher frequency). (Reprinted from Robinson RG. The clinical neuropsychiatry of stroke. Cambridge, UK: Cambridge University Press, 1998;61, with permission.)

and the patients' attribution of their depression to life stressors and disability is clearly idiosyncratic criteria for the diagnosis of depression, arbitrarily added to the diagnostic criteria to show differences with primary depression. Both catastrophic reactions and hyperemotionalism occur in patients without depression, indicating the comorbid nature of these conditions and not symptoms that are integral to the diagnosis of depression. The Gainotti et al. data do not validate the alteration of diagnostic criteria for PSD; they only support a preconceived hypothesis about the etiology of PSD.

Longitudinal Course of Depression

Among a group of 103 acute stroke patients in a 2-year longitudinal study of PSD (24), both major and minor depressive disorders were still present in 86% of patients at a 6-month follow-up evaluation. However, only one of five patients with major depression continued to have major depression at

1-year follow-up (two patients had minor depression). Patients with minor depression had a less favorable prognosis; only 40% had no depression at 1-year follow-up, and 30% had no depression at 2-year follow-up.

Morris et al. (25) found that among a group of 99 patients in a stroke rehabilitation hospital in Australia, those with major depression had a duration of major depression of 40 weeks, whereas those with adjustment disorders (minor depression) had a mean duration of depression of only 12 weeks. Astrom et al. (26) found that among 80 patients with acute stroke, 27 developed major depression in-hospital or at 3-month follow-up. Of these patients with major depression, 15 (60%) had recovered by 1-year follow-up, but by 3-year follow-up, only one more patient had recovered. Burvill et al. (27) similarly found that 30% of patients with major depression identified in a community survey were still depressed at 1-year follow-up. The percentages of patients with major depression who had recovered by 1-year follow-up without treatment are summarized in Figure 28.2

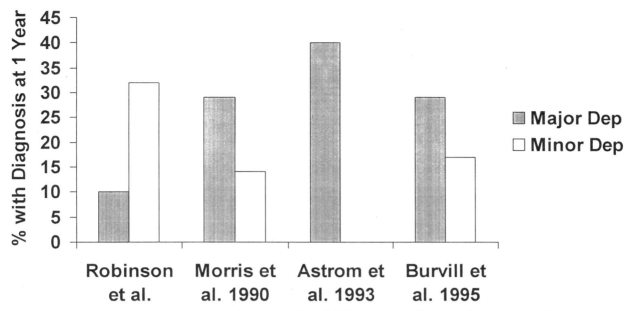

FIGURE 28.2. The percentage of patients with an initial assessment diagnosis of major poststroke depression who continued to have a diagnosis of major depression or had improved to a diagnosis of minor depression at 1-year follow-up. Note that the number of chronic cases varies between studies, probably reflecting a mixture of etiologies among the group with an in-hospital diagnosis of major poststroke depression. The mean frequency of persistent major depression at 1-year follow-up across all studies was 26%. (Reprinted from Robinson RG. The clinical neuropsychiatry of stroke. Cambridge, UK: Cambridge University Press, 1998;61, with permission.)

for all published longitudinal studies. The mean frequency of major depression that persisted beyond 1 year was 26%.

Two factors have been identified that can influence the natural course of PSD. One is treatment of depression with antidepressant medications (discussed below). The second factor is lesion location. Starkstein et al. (28) compared two groups of depressed patients: one group (n = 6) had spontaneously recovered from depression by 6 months after stroke, whereas the other group (n = 10) remained depressed at this point. There were no significant between-group differences in important demographic variables, such as age, gender, and education, and both groups had similar levels of social functioning and degrees of cognitive dysfunction. There were two significant between-group differences. One was lesion location: The recovered group had a higher frequency of subcortical and cerebellar/brainstem lesions; the nonrecovered group had a higher frequency of cortical lesions (*p* < 0.01). Impairments in activities of daily living (ADL) were also significantly different between the two groups: The nonrecovered group had significantly more severe impairments in ADL in-hospital than did the recovered group (*p* < 0.01).

The available data suggest that PSD is not transient but is usually a long-standing disorder with a natural course of approximately 9 to 10 months for most major depressions. Depressions lasting more than 2 years do occur in some patients with major or minor depression. Lesion location

and severity of associated impairments may influence the longitudinal evolution of PSD.

Relationship to Lesion Variables

The relationship between depressive disorder and lesion location has been, perhaps, the most controversial area of research in the field of poststroke mood disorder. Although the association between specific clinical symptomatology and lesion location is one of the fundamental goals of clinical practice in neurology, this has rarely been the case with psychiatric disorders. Cognitive functions, speech impairment, and the extent and severity of motor or sensory deficits are all symptoms of stroke that are commonly used by clinicians to localize lesions to particular brain regions. There is no known neuropathology consistently associated with primary mood disorders (i.e., mood disorders without known brain injury) or secondary mood disorders associated with a physical illness). The idea that there may be a neuropathology associated with the development of major depression has led to both surprise and skepticism.

The first study to report a significant clinicopathologic correlation in PSD was an investigation by Robinson and Szetela (29) of 29 patients with left-hemisphere brain injury secondary to stroke (n = 18) or to TBI (n = 11). Based on localization of the lesion by computed tomography (CT) scan, there was a significant inverse correlation between the severity of depression and the distance of the anterior border of the lesion from the frontal pole ($r = -0.76$). This surprising finding led to a number of subsequent examinations of this phenomenon in other populations. This finding of a significant correlation between severity of depression and proximity of the lesion to the left (and sometimes the right) frontal pole has been replicated by Eastwood et al. (30), House et al. (31), Morris et al. (32), Hermann and Walesch (33), and several times by Robinson et al. (34–38) using different populations of patients, and Starkstein et al. (39).

In addition lesion location has also been found to influence the frequency of depression. In a study of 45 patients with a single lesion restricted to either cortical or subcortical structures in the left or right hemisphere, Starkstein et al. (40) found that among patients with anterior lesions, five of five patients with left cortical lesions involving the frontal lobe had depression compared with two of 11 patients with left cortical posterior lesions. Moreover, four of six patients with left subcortical anterior lesions had depression compared with one of seven patients with left subcortical posterior lesions. Finally, correlations between depression scores and the distance of the lesion from the frontal pole were significant for patients with left cortical lesions and patients with left subcortical lesions. These relationships were not significant for patients with right-hemisphere lesions.

In a subsequent study, Starkstein et al. (41) examined the relationship between lesions of specific subcortical nuclei and depression. Basal ganglia (caudate and/or putamen) lesions produced major PSD in seven of eight patients with left-sided lesions compared with only one of seven patients with right-sided lesions and zero of ten with thalamic lesions ($p < 0.001$).

Astrom et al. (26) similarly found that among patients with acute stroke, 12 of 14 with left anterior lesions had major depression compared with only two of seven patients with left posterior lesions ($p = 0.017$) and two of 23 with right-hemisphere lesions ($p < 0.001$). Numerous studies, however, have failed to replicate these findings. This failure to replicate findings that were reported in at least four other studies was referred to by Gainotti et al. (22) as a "factual error of Robinson et al." Negative findings, of course, happen all the time, and failure to replicate is far from proving the null hypothesis.

Shimoda and Robinson (42) examined the relationship between lesion location and time since stroke using a longitudinally studied patient population. This study examined 60 patients with single lesions involving either the right or left middle cerebral artery distribution that were visible on CT scan and who had follow-up at 3 or 6 months (short-term follow-up) and at 12 or 24 months (long-term follow-up). There were no statistically significant differences between the patients with right- and left-hemisphere lesions in age, gender, race, marital status, or other background characteristics. The frequency of depression during the initial evaluation was significantly higher for both major and minor depres-

sion among patients with left-hemisphere stroke compared with patients with right-hemisphere stroke ($p = 0.0006$) (Fig. 28.3). At short- and long-term follow-up there were no significant differences between right- and left-hemisphere lesion groups in terms of the frequency of major or minor depression.

This study suggests that the failure of other investigators to replicate the association of left anterior lesion location with increased frequency of depression in most cases may be related to time since stroke. The lateralized effect of left anterior lesions on both major and minor depression is a phenomenon of the acute poststroke period when the patients are less than 1 or 2 months poststroke. This lateralized effect is lost after these first few weeks, and subsequently there is an equal likelihood of patients with right- or left-hemisphere lesions having PSD.

Although it is uncertain why this temporal dynamic occurs in the relationship between the severity of depression and lesion location, it suggests that the mechanism of depression (e.g., depletion of biogenic amines) in patients with left anterior lesions are hemisphere specific for only a few weeks after cerebral infarction. By 2 to 3 months poststroke, similar or perhaps alternative mechanisms occur in patients with right frontal lesions that lead to depression and eliminate any demonstrable hemispheric association.

Relationship to Impairment in Activities of Daily Living

Although many clinicians have assumed that the most powerful determinant of depression after stroke was the severity of associated physical impairment, empirical studies have consistently failed to find a strong relationship between severity of depression and severity of physical impairment (25,30,43). This is not to say that there is no relationship. Numerous studies have demonstrated that severity of physical impairment is one of several factors that contributes to depression and, in some subpopulations, may be strongly associated with depression (26,30,43,44).

Although the effect of impairment on depression appears to be fairly weak, there is a significant influence of PSD on recovery in ADL. Parikh et al. (45) compared 25 patients with PSD (either major or minor depression) and 38 stroke patients with no mood disorders who were matched for severity of ADL impairments in-hospital. After controlling for all the variables that have been shown to influence stroke outcome, such as acute treatment on a stroke unit, the size, nature, and location of the brain injury, age, education, and duration of rehabilitation services, patients with in-hospital PSD were found to have a significantly poorer recovery than nondepressed stroke patients, even after their depression had subsided.

A study by Chemerinski et al. (46) included a consecutive series of patients with PSD (n = 55) divided into those whose mood improved at 3- to 6-month follow-up

Depression and Hemispheric Lesion Location

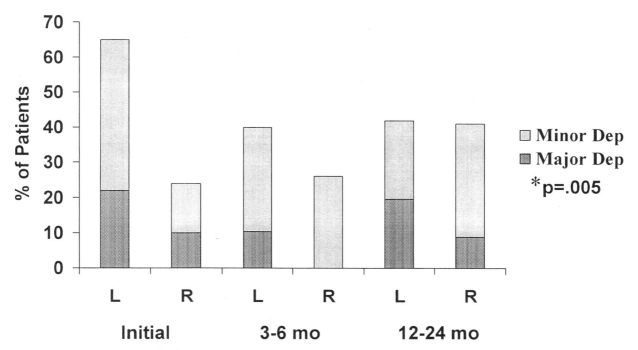

FIGURE 28.3. The frequency of major and minor depression defined by the *Diagnostic and Statistical Manual of Mental Disorders, Fourth Edition* criteria associated with single lesions of the right or left hemisphere during the acute stroke period and at follow-up. The lateralized effect of left-hemisphere lesions on both major and minor depression was found only during the acute stroke period. At short- and long-term follow-up, there were no hemispheric lesion effects on the frequency of depression. (Reprinted from Robinson RG. The clinical neuropsychiatry of stroke. Cambridge, UK: Cambridge University Press, 1998;61, with permission.)

and those whose mood had not improved (n = 34). This study found significantly greater improvement in ADL score among patients whose major or minor depression improved compared with those whose depression did not (Fig. 28.4) (46). There were no significant differences in the background characteristics of patients whose mood did and did not improve including lesion characteristics, neurologic symptoms, age, and education. In addition, patients with minor depression showed the same degree of improvement with remission of their depression as patients with major depression. The fact that both major and minor depression showed an equal degree of recovery in ADL with remission of their depression suggests that the effect of depression on physical impairment may be mediated by psychological or physiologic mechanisms that are independent of diagnosis. For example, depressed patients may be hopeless about the future and thus less motivated to put an effort into rehabilitation exercises. Alternatively, physiologic changes in the brain may lead to low energy or apathy in both major and minor depression.

Relationship to Cognitive Impairment

Several investigators have found that poststroke patients with major depression have intellectual deficits that are more severe than nondepressed patients (47–49). This issue was first examined in patients with PSD by Robinson et al. (47). Patients with major depression after a left-hemisphere infarct were found to have significantly lower (i.e., more impaired) scores on the Mini-Mental State Examination (MMSE) (50) than did a comparable group of nondepressed patients. Both the size of the patients' lesions and their depression scores correlated independently with severity of cognitive impairment.

In a second study (51), stroke patients with and without major depression were matched for lesion location and

In-hospital and at 3- or 6-months follow-up of ADL in post-stroke patients with and without remission of depression

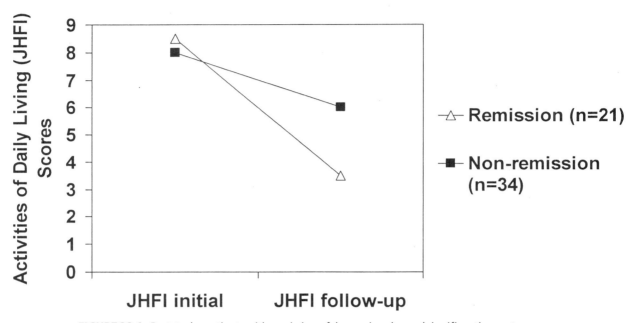

FIGURE 28.4. Poststroke patients with remission of depression showed significantly greater recovery in activities of daily living than nonremitted patients at the 3- or 6-month follow-up. (Reprinted from Chemerinski et al., 2001 (46), with permission.)

volume. Of 13 patients with major PSD, ten had an MMSE score lower than that of their matched control subjects, two had the same score, and only one patient had a higher score ($p < 0.001$). Thus, even when patients were matched for lesion size and location, depressed patients were more cognitively impaired.

In a follow-up study, Bolla-Wilson et al. (52) administered a comprehensive neuropsychologic battery and found that patients with major depression and left-hemisphere lesions had significantly greater cognitive impairments than did nondepressed patients with comparable left-hemisphere lesions ($p < 0.05$). These cognitive deficits primarily involved tasks of temporal orientation, language, and executive motor and frontal lobe functions. Conversely, among patients with right-hemisphere lesions, patients with major depression did not differ from nondepressed patients on any of the measures of cognitive impairment.

In a 2-year follow-up study of 140 patients after the occurrence of stroke, Downhill and Robinson (53) found that major depression was associated with a greater degree of cognitive impairment, as measured by the MMSE, than minor depression or no depression for 1 year after stroke. The intellectual deficit, however, was most prominent among patients

with major depression after a left-hemisphere lesion. Patients who had major depression after a right-hemisphere stroke or patients with minor depression or in whom stroke had occurred 2 years previously did not show an effect of major depression on cognitive impairment.

Treatment studies of PSD have consistently failed to show an improvement in cognitive function even when poststroke mood disorders responded to antidepressant therapy (49). Kimura et al. (54) examined this issue in a study comparing nortriptyline and placebo using a double-blind treatment methodology among patients with major ($n = 33$) or minor ($n = 14$) PSD. Although there were no significant differences between the nortriptyline- and placebo-treated groups in the change in MMSE scores from beginning to end of the treatment study, when patients were divided into those who responded to treatment (i.e., >50% decline in Hamilton Rating Scale for Depression score and no longer meeting depression diagnosis criteria) and those who failed to respond, there was a significantly greater improvement in MMSE among patients who responded to treatment ($n = 24$) compared with patients who failed to respond to treatment ($n = 23$) (Fig. 28.5). The responder group included 16 patients treated with nortriptyline and nine with placebo.

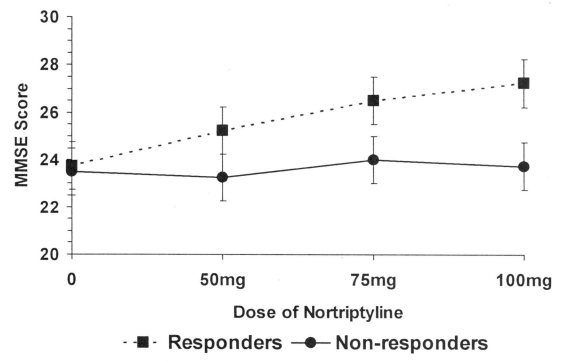

FIGURE 28.5. Change of Mini-Mental State Examination scores in patients with poststroke major depression during treatment study. The mood improvement group (n = 15) showed significantly greater improvement in cognitive function than the no mood improvement group (n = 18) ($p = 0.0087$). Error bars represent standard error. (Reprinted from Kimura M, Robinson RG, Kosier T. Treatment of cognitive impairment after poststroke depression. *Stroke* 2000;31:1482–1486, with permission.)

Although the nonresponder group included five patients treated with nortriptyline and 18 with placebo, there were no significant differences between the two groups in baseline Hamilton Rating Scale for Depression scores, demographic characteristics, stroke characteristics, and neurologic findings. A repeated-measures analysis of variance demonstrated a significant group by time interaction ($p = 0.005$), and planned *post hoc* comparisons demonstrated that the responders had significantly less impaired MMSE scores than the nonresponders at nortriptyline doses of 75 ($p = 0.036$) and 100 mg ($p = 0.024$). Thus, the failure to demonstrate cognitive improvement in prior studies was not the result of nortriptyline drug effects such as sedation or impaired attention owing to anticholinergic effects. When the effect of major versus minor depression was examined, patients with major depression who responded to treatment (n = 15) showed significantly greater improvement in MMSE scores than patients with major depression who did not respond (n = 18) ($p = 0.0087$). Among patients with minor depression (nine responders and five nonresponders), repeated-measures analysis of variance of MMSE scores showed no significant group by time interaction.

The reason that prior treatment studies did not show a significant effect of treatment of depression on cognitive function was the result of effect size. When patients treated

with nortriptyline, some of whom responded to treatment and some of whom did not, were compared with placebo-treated patients, some of whom responded to treatment and some of whom did not, the effect size was only 0.16. When patients were divided into those who responded and those who did not respond, the effect size increased to 0.96, thus allowing a significant difference to be demonstrated with a much smaller group size.

Relationship to Aphasia

The diagnosis of depression among patients with severe comprehension deficits is very difficult, and most investigators have excluded these patients from studies of PSD. Some investigators (55) suggested that a diagnosis of depression should be based on behavioral observations (i.e., diminished sleep, food intake, restlessness and agitation, retarded or tearful behavior). However, the sensitivity and specificity of those methods for detecting depression have not yet been demonstrated.

Robinson and Benson (56), in a study of depression in patients with fluent or nonfluent aphasias, found that 53% were depressed and that those findings were similar to the frequencies of major and minor depression seen in nonaphasic patients with comparable-sized left-hemisphere lesions (34).

This study also found that patients with nonfluent aphasia had a significantly higher frequency of depression than patients with fluent aphasia (56).

Perhaps the higher frequency of depression among nonfluent aphasic patients is attributable to the greater awareness related to the language impairment. Starkstein and Robinson (57) concluded that the most important variable in the association between PSD and nonfluent aphasia was lesion location, suggesting that nonfluent language impairment and depression may not be casually related but may be independent outcomes of the same left frontal lesion.

Relationship to Vascular Territory of Stroke

Starkstein et al. (58) reported that patients with middle cerebral artery lesions showed a higher incidence and longer course of PSD compared with those with posterior circulation lesions (i.e., temporooccipital and cerebellar/brainstem region). Major or minor depression occurred in 48% of the patients in the middle cerebral artery lesion and in 35% of patients in the posterior circulation lesion groups. At 6-month follow-up, frequencies of depression among the patients with in-hospital depression were 82% and 20%, respectively in patients with middle cerebral artery and patients with posterior circulation infarcts. At follow-up 1 to 2 years poststroke, frequencies of depression were 68% and 0%, respectively.

This finding suggests that the mechanism of depression after middle cerebral artery infarcts may differ from that after posterior lesions. It has been suggested that the shorter duration may be related to the smaller size and to the possibility that cerebellar/brainstem lesions produce less injury to the biogenic amine pathways than middle cerebral artery distribution lesions.

Mechanism of Depression after Stroke

Although the cause of PSD is not known, it has been hypothesized that disruption of the amine pathways by the stroke lesion may play an etiologic role (34). The noradrenergic and serotonergic cell bodies are located in the brainstem and send ascending projections through the median forebrain bundle to the frontal cortex. The ascending axons then arc posteriorly and run longitudinally through the deep layers of the cortex, arborizing and sending terminal projections into the superficial cortical layers (59). Lesions that disrupt these pathways in the frontal cortex or basal ganglia may affect many downstream fibers. Based on these neuroanatomic facts and the clinical findings that the severity of depression correlates with the proximity of the lesion to the frontal pole, one could hypothesize that post-stroke depression is caused in part by depletion of norepinephrine and/or serotonin produced by lesions in the frontal lobe or basal ganglia.

In support of this hypothesis, a lateralized biochemical response to ischemia in human subjects was reported by Mayberg et al. (60). Patients with stroke lesions in the right

hemisphere had significantly higher ratios of ipsilateral to contralateral spiperone binding (presumably 5HT$_2$ type receptor binding) in noninjured temporal and parietal cortex than patients with comparable left-hemisphere strokes. Patients with left-hemisphere lesions showed a significant inverse correlation between the amount of spiperone binding in the left temporal cortex and depression scores (i.e., higher depression scores were associated with lower serotonin receptor binding).

Thus, a greater depletion of biogenic amines in patients with right-hemisphere lesions than in those with left-hemisphere lesions could lead to a compensatory up-regulation of receptors that might protect against depression. Conversely, patients with left-hemisphere lesions may have moderate depletion of biogenic amines but without a compensatory up-regulation of serotonin receptors and, therefore, a dysfunction of biogenic amine systems in the left hemisphere. This dysfunction may ultimately lead to the clinical manifestations of depression.

TREATMENT OF POSTSTROKE DEPRESSION

At present, there are four placebo-controlled, randomized, double-blind treatment studies on the efficacy of single antidepressant treatment of PSD. In the first study, Lipsey et al. (61) examined 14 patients treated with nortriptyline and 20 patients given placebo. The 11 patients treated with nortriptyline who completed the 6-week study showed significantly greater improvement in their scores on the Hamilton Rating Scale for Depression (62) than did 15 placebo-treated patients ($p < 0.01$). Successfully treated patients had serum nortriptyline levels of 50 to 150 ng/mL. Three patients experienced side effects (including delirium, confusion, drowsiness, and agitation) that were severe enough to require the discontinuation of nortriptyline. Similarly, Reding et al. (63) reported that patients with PSD (defined as having an abnormal dexamethasone suppression test) taking trazodone had greater improvement in Barthel ADL scores (64) than did placebo-treated control subjects ($p < 0.05$). Another double-blind, controlled trial in which the selective serotonin reuptake inhibitor citalopram was used found that Hamilton Rating Scale for Depression scores were significantly more improved over 6 weeks in patients receiving active treatment (n = 27) than placebo (n = 32) (65). At both 3 and 6 weeks, the group receiving active treatment had significantly lower Hamilton depression scores than did the group receiving placebo. This study established for the first time the efficacy of a selective serotonin reuptake inhibitor in the treatment of PSD.

The most recent treatment study was conducted by Robinson et al. (66). This study compared depressed patients treated with fluoxetine (n = 23), nortriptyline (n = 16), or placebo (n = 17) in a double-blind, randomized treatment design. Patients were enrolled if they had a diagnosis of either major or minor PSD and had no contraindication to the use of fluoxetine or nortriptyline such as intracerebral hemorrhage (fluoxetine) or cardiac conduction abnormalities (nortriptyline). Patients were treated with 10-mg doses of fluoxetine for the first 3 weeks, 20 mg for weeks 4 through 6, 30 mg for weeks 6 through 9, and 40 mg for the last 4 weeks (9 through 12). The nortriptyline-treated patients were given 25 mg for the first week, 50 mg for weeks 2 and 3, 75 mg for weeks 3 through 6, and 100 mg for weeks 6 through 12. Patients treated with placebo were given identical capsules of the same number used for the active-treatment patients. Intention-to-treat analysis demonstrated significant time by treatment interaction with patients treated with nortriptyline, showing a significantly greater decline in Hamilton depression scores than either the placebo- or fluoxetine-treated patients at 12 weeks of treatment. There were no significant differences between the fluoxetine and the placebo groups (Fig. 28.6). Nortriptyline led to a significantly higher response rate (response is defined as >50% drop in Hamilton depression score and the patient no longer meeting diagnos-

tic criteria for major or minor depression) [ten of 16 (62%)] than either fluoxetine [two of 23 (9%)] or placebo [four of 17 (24%)] (Fisher exact, $p = 0.001$). In addition, fluoxetine treatment was associated with a mean weight loss of 15.1 lb or 8.5% of initial body weight from the beginning to the end of the 12-week treatment trial, which was not seen among patients treated with placebo or nortriptyline. Of the 12 patients treated with fluoxetine, ten lost 10 lb or more, whereas only two of 13 nortriptyline- and one of 11 placebo-treated patients lost this amount of weight (Fisher exact, $p = 0.004$).

Based on the available data, if there are no contraindications to nortriptyline such as heart block, cardiac arrhythmia, narrow angle glaucoma, sedation, or orthostatic hypotension, nortriptyline remains the first-line treatment for PSD. Doses of nortriptyline should be increased slowly, and blood levels should be monitored with a goal of achieving serum concentrations between 50 and 150 ng/mL. If there are contraindications to the use of nortriptyline, citalopram (20 mg

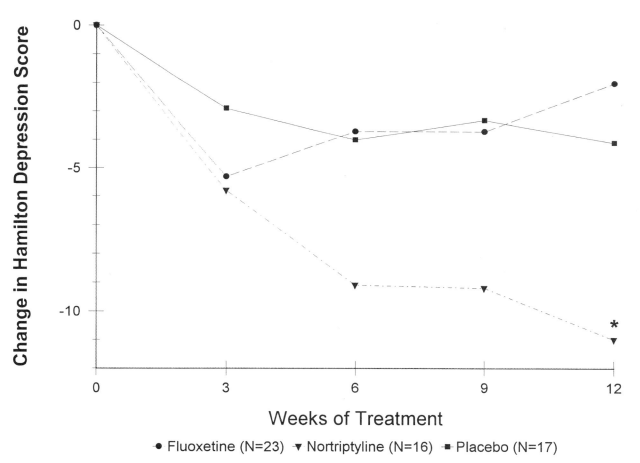

FIGURE 28.6. Changes in Hamilton Rating Scale for Depression score (28 items) over 12 weeks of treatment for all patients who were entered in the study (i.e., intention to treat analysis) [significant group by time interaction ($F = 3.45$; df = 8, 212; $p = 0.0035$) and *post hoc* significantly greater change in patients treated with nortriptyline compared with those treated with fluoxetine or placebo at 12 weeks]. (From Robinson RG, Schultz SK, Castillo C, et al. Nortriptyline versus fluoxetine in the treatment of depression and in short term recovery after stroke: a placebo controlled, double-blind study. *Am J Psychiatry* 2000;157:351–359, with permission.)

for those younger than age 66 and 10 mg age for those 66 and older) would be the next choice.

Electroconvulsive therapy has also been reported to be effective for treating PSD (67). It causes few side effects and no neurologic deterioration. Psychostimulants have also been reported in one double-blind trial of stroke patients (not all were depressed) to significantly lower Hamilton depression scores and significantly improve ADL scores compared with placebo (68). Finally, psychological treatment, including cognitive/behavioral therapy (69) and group and family therapy, has also been reported to be useful (70,71). However, controlled studies for these psychosocial treatments have not been conducted.

Depression Associated with Parkinson Disease

As with stroke, depression is a frequent finding in patients with PD. Although the frequent association of emotional disorders with PD was recognized more than 50 years ago (72), it has only been within the past several years that investigators have begun to empirically examine the nature of the relationship. Some investigators have suggested that the high frequency of depression in PD is the understandable consequence of progressive physical impairment (73). Other investigators have not found a significant correlation between the severity of depression and the severity of physical impairment and have suggested that depression may be a consequence of neurochemical changes in specific brain areas (74,75).

Prevalence

In several studies, the frequency of depression was reported to be approximately 40% (74,76,77). In a prospective study of a consecutive series of 105 patients, Starkstein et al. (78) found that 21% of outpatients in a PD clinic met *DSM-III* (*DSM, Third Edition*) for criteria for major depression, whereas 20% met *DSM-III* criteria for dysthymic (minor) depression. The highest frequency of depression was found in the early and late stages of PD.

Relationship to Cognitive Impairments

In PD, cognitive impairments may range from subtle deficits in frontal lobe–related tasks to an overt dementia (79). Mayeux et al. (80), using a modified MMSE, reported a significant correlation between cognitive deficits and severity of depression (i.e., severe depression was associated with severe cognitive impairments). This relationship was also reported by Starkstein et al. in three studies (81–83). In the first study, the association between depression, cognitive impairments, and stage of PD was examined (81). Patients in the late stages showed significantly greater overall cognitive impairments than patients in the early stages, and those impairments were restricted to tasks involving motor-related

functions. Depressed patients in the late stages of the disease showed the most significant impairments (82). Taken together, these findings suggest that cognitive deficits may primarily be a result of motor impairments, but when depression also occurs, the cognitive deficits are greater and increase in severity as the PD progresses (81).

In the second study, the association between cognitive impairments and type or severity of depression (major or minor) among patients with PD was examined (82). No differences were found on cognitive tasks between minor depression and nondepressed patients, but patients with major depression showed the worst cognitive performance. This impairment was greatest on frontal lobe–related tasks, such as the Wisconsin Card Sorting Test (82).

In the third study, the influence of depression on the longitudinal evolution of cognitive deficits was examined in a 3- to 4-year follow-up (83). Both groups, depressed and nondepressed patients, showed significant decreases in MMSE scores over time, but depressed patients had significantly greater cognitive decline than nondepressed subjects (Fig. 28.7) (78). These findings demonstrate that depression may not only be associated with cognitive impairments at the time depression is present but may also be associated with more rapid cognitive deterioration. These findings support the speculation of Sano et al. (83) that depression may be an early finding in patients with PD who later show dementia.

Mechanism

Several studies have demonstrated that patients with PD and dementia may show senile plaques and neurofibrillary tangles compatible with the diagnosis of Alzheimer disease, as well as severe depletion of cholinergic neurons in the nucleus basalis of Meynert or Lewy bodies in cortical regions (84). Few neuropathologic studies have been carried out in patients with PD and depression. Torack and Morris (85) reported a marked loss of pigmented neurons in the ventral tegmental area in a small group of patients with PD, dementia, and depression. The ventral tegmental area contains cell bodies of dopaminergic neurons that provide most of the dopaminergic innervation to the prefrontal cortex, nucleus accumbens, and amygdala. Thus, dysfunction of the mesocorticolimbic dopaminergic system may play an important role in the production of deficits in frontal lobe tasks among patients with PD and major depression.

Depression in PD may also be related to changes in other biogenic amines. Mayeux et al. (74) showed that patients with PD and depression had significantly lower 5-hydroxyindoleacetic acid (a metabolite of serotonin) levels in the cerebrospinal fluid than patients with PD without depression. However, patients with PD and both dementia and depression had the lowest 5-hydroxyindoleacetic acid cerebrospinal fluid values (83).

The metabolic abnormalities associated with depression in PD have also been examined using neuroimaging

Neuropsychological Findings
Mini Mental Status

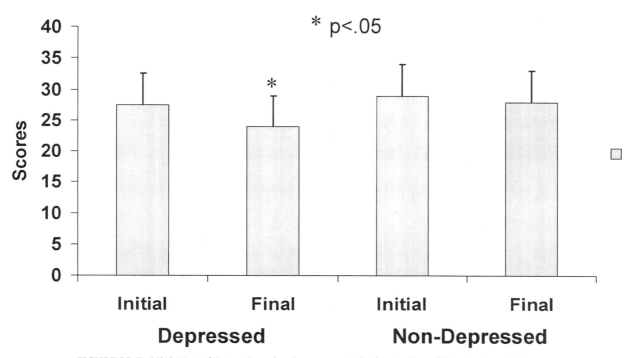

FIGURE 28.7. Mini-Mental Status Examination scores at the beginning of the study and between 3 and 4 years of follow-up. Parkinson disease patients who were depressed at the initial evaluation demonstrated a significant decline in their cognitive function during the follow-up period, whereas the nondepressed patients showed no significant change.

techniques (86). Regional cerebral glucose metabolism was determined in five depressed and four nondepressed patients with PD using [^{18}F]fluoro-1-deoxy-D-glucose positron emission tomography. Patients with PD and major depression had significantly lower metabolic activity in the head of the caudate and inferior frontal cortex than nondepressed PD patients of comparable age, duration, and stage of illness. Moreover, there was a significant correlation between Hamilton depression scores and the relative regional metabolism in the inferior frontal cortex ($r = 0.73$, $p < 0.05$) (i.e., the lower the relative regional metabolic activity in the inferior frontal cortex, the more severe the depression).

Mayberg et al. (86) proposed that dopaminergic projections from the ventral tegmental area show regional specificity for the orbitofrontal cortex (87), and the serotonergic projections are more diffusely distributed (88). Conversely, the major cortical outflow back to the mesencephalon originates in the orbitofrontal cortex (89). Thus, degeneration of the mesocorticolimbic dopaminergic systems in patients with PD may lead to dysfunction of the orbitofrontal cortex, which may secondarily affect cell bodies in the dorsal raphe.

This combination of dopamine and serotonergic deficits may account for the presence of cognitive impairments and depression in patients with PD.

Treatment

The first double-blind treatment study of depression in PD was conducted by Andersen et al. (90). This crossover design study compared 15 patients treated with nortriptyline then placebo with nine patients treated with placebo then nortriptyline. During the first leg of the study, patients given nortriptyline had significantly lower (more improved scores on a measurement scale designed for PD) than placebo patients over 4 weeks of treatment (repeated-measures analysis of variance, $p < 0.05$). This study did not use structured interviews and standard rating scales. The most recent studies were conducted by Hauser and Zesiewicz (91) and Tesei et al. (92). Hauser and Zesiewicz conducted an open-label trial of sertraline in 15 patients. Patients treated with 25 mg sertraline for 1 week and 50 mg for 6 weeks showed significantly lower scores on the Beck Depression Inventory compared with their scores at the beginning of the treatment trial. The authors concluded that a placebo-controlled trial of sertraline in PD with depression should be conducted. Tesi et al. treated 65 patients with PD and depression in an open-label study with paroxetine for 3 months. Patients were given 10 mg daily for 4 weeks and then 20 mg daily. There were 13 dropouts, and the study completers showed a significant decrease in Hamilton depression scores from 21.7 ± 6.4 to 13.8 ± 5.8 ($p < 0.001$).

A metaanalysis conducted in 1995 by Klaassen et al. (93) examined 12 studies that carried out treatment trials of depression in PD. The authors concluded that all the treatment trials were flawed either by not being conducted double blind or not using standardized diagnostic criteria or standardized depression rating scales. They concluded by stating that there are virtually no empirical data on the treatment of depression in PD.

A question that remains unanswered is whether the treatment of depression may influence the progression of intellectual impairments in patients with PD. In the longitudinal 3- to 4-year follow-up study described previously (78), six patients with depression who had received treatment for depression had only an 11% decrease in cognitive scores compared with a 23% decrease in cognitive scores among nontreated depressed patients. Moreover, the two patients who were receiving the highest doses of tricyclics did not show decline in their MMSE scores. These preliminary findings suggest the need for prospective double-blind treatment studies to determine whether the use of antidepressants may delay the progression of cognitive impairment in patients with PD.

DEPRESSION ASSOCIATED WITH TRAUMATIC BRAIN INJURY

TBI has an annual incidence of 2 million cases and represents the most common cause of brain injury and the most common cause of death in people younger than age 45 in the United States (94). Although there is an extensive literature on the neurobehavioral consequences of head injury, very few studies have examined the frequency or course of mood disorders that occur after TBI.

Prevalence

Most of the studies of emotional or depressive symptoms among patients with TBI have relied on rating scales or relatives' reports rather than on structured interviews and established diagnostic criteria [e.g., *DSM-III-R* (*DSM-III Revised*). Perhaps as a result, the frequency of depression has varied from one study to the next. For example, Rutherford et al. (95) found that 6% of 145 patients with minor head injury had a significant number of depressive symptoms within 6 weeks after injury. Conversely, McKinlay et al. (96) reported indirect evidence of a depressed mood in

approximately 50% of their patients at 3, 6, or 12 months after severe head injury. Kinsella et al. (97) reported that in a series of 39 patients, 33% were classified as depressed and 26% as suffering from anxiety within 2 years of severe head injury.

Studies by Fedeoroff et al. (98) and Jorge et al. (99) found that 28 (42%) of 66 patients admitted to a head trauma unit developed major depression at some point during a 1-year follow-up. Among the 66 patients admitted to a specialized trauma unit with acute closed head injury without significant spinal cord or other organ system injury, 17 (26%) met diagnostic criteria for major depression at the time of the initial in-hospital evaluation. In addition, 3% met criteria for minor (dysthymic) depressive disorder. This frequency is consistent with the findings of several other investigators (100,101). These data suggest that major depression constitutes a significant psychiatric complication in this population.

Longitudinal Course of Depression

Mood disorders after TBI may be transient syndromes lasting for a few weeks or persistent disorders lasting for many months (102). Other authors have suggested that transient disorders may be the result of neurochemical changes provoked by brain injury, whereas prolonged depressive disorders may be of a more complex nature and may be reactive to physical or cognitive impairment (103–105).

Empirical data supporting these suggestions have been reported by Jorge et al. (99). Diagnoses of depression were based on a semistructured psychiatric interview (106) and *DSM-III-R* criteria for major or minor (dysthymic) depression (107). Of the original 66 patients evaluated with acute TBI, 54 were reevaluated at 3 months, 43 at 6 months, and 43 at 1 year. The prevalence of depression was 30% at 3 months, 26% at 6 months, and 26% at 1 year (108). The mean duration of major depression was 4.7 months. There was, however, a group of seven patients (41% of the depressed group) who had transient depressions lasting 1.5 months, whereas the depressions of the nine remaining depressed patients had a mean duration of 7 months. The patients with transient depressions showed a strong association with left anterior lesion location (Fisher exact, $p = 0.006$), whereas prolonged depressions were associated with impaired social functioning. These findings support the hypothesis that biologic factors may lead to transient depression, whereas prolonged depressions may result from psychological factors.

Risk Factors for Depression after Traumatic Brain Injury

Several premorbid factors may influence patients' emotional responses to acute TBI and may therefore be relevant to the etiology of depressive disorders after TBI (109). In the study of 66 patients with acute TBI, there was significantly greater frequency of a personal history of psychiatric disorder in

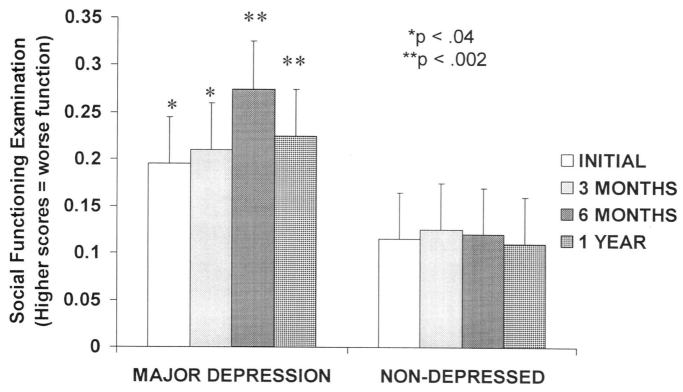

FIGURE 28.8. Social functioning examination scores for patients with traumatic brain injury during 1 year of follow-up. Patients with major depression at the initial evaluation and at each of the follow-up examinations had significantly more impaired social function than the nondepressed patients. (Data taken from Jorge RE, Robinson RG, Arndt SV, et al. Depression following traumatic brain injury: a 1 year longitudinal study. *J Affect Disord* 1993;27:233–243, with permission.)

the major depressed group compared with the nondepressed patients (98). In addition, the depressed group had significantly more impaired social functioning as measured by the Social Functioning Examination (Fig. 28.8) (110). The Social Functioning Examination, during the initial evaluation, measured the patient's quality of and personal satisfaction with social functioning during the period before the brain injury. This suggests, as other investigators have reported, that patients with poor social adjustment and social dissatisfaction before the brain injury were more prone to develop depression than patients with adequate social support.

Relationship to Lesion Location

There is some empirical evidence supporting an association between post-TBI depression and specific lesion locations. Lishman (111) reported that several years after penetrating brain injury, depressive symptoms were more common among patients with right-hemisphere lesions. Depressive symptoms were also more frequent among patients with frontal and parietal lesions compared with other lesion locations. Grafman et al. (112) also reported that several years after head injury, depressive symptoms were more frequently associated with penetrating injuries involving the right-hemisphere (right orbitofrontal) lesions than any other lesion location.

TBI is characterized by the presence of diffuse and focal lesions that may be the direct result of traumatic shear injury or contusions or may be secondary to ischemic complications (113). Of the 66 patients previously described, 42 (64%) had a diffuse pattern of brain injury on their CT scans, and 24 (36%) presented with focal lesions. There were no significant differences between major depressed and nondepressed groups in the frequency of diffuse or focal patterns of injury. In addition, no significant differences were found in the frequency of extraparenchymal hemorrhages, contusions, intracerebral or intraventricular hemorrhages, hydrocephalus, or CT findings suggestive of brain atrophy. There was a significant association between lesion location and the development of major depression. The presence of left anterior hemisphere lesions (i.e., left dorsolateral frontal cortex or left basal ganglia) was the strongest correlate of major depression (Fig. 28.9). Other frontal lesions (i.e., left, right, or bilateral frontal lesions, including the orbitofrontal cortex) were associated with a lower probability of developing major depression (98).

These results are consistent with previous findings in acute stroke patients of an increased frequency of depression

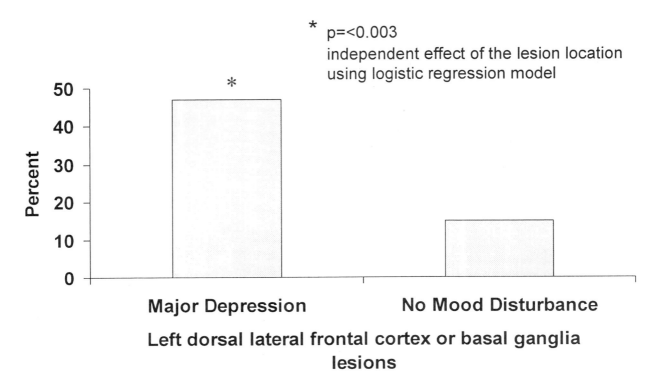

*p=<0.003
independent effect of the lesion location
using logistic regression model

Left dorsal lateral frontal cortex or basal ganglia lesions

FIGURE 28.9. The percentage of patients having no mood disturbance or major depression after traumatic brain injury. A significantly higher percentage of patients with major depression had a left frontal or left basal ganglia lesion than nondepressed patients. A logistic regression examining impairment and demographic variables, as well as lesion location, demonstrated a significant independent effect of lesion location on the existence of major depression. (Data taken from Fedoroff JP, Starkstein SE, Forrester AW, et al. Depression in patients with acute traumatic brain injury. *Am J Psychiatry* 1992;149:918–923, with permission.)

among patients with left dorsolateral cortical or left basal ganglia lesions (39). These findings are also consistent with a previous study in which patients with anterior left-hemisphere lesions were found to have more severe depressive symptoms than patients with left posterior–hemisphere lesions in both stroke and TBI (56). This clinicopathologic relationship applies only to the initial in-hospital evaluation. This finding is consistent with stroke patients, in whom the association between lesion location and major depression is also a phenomenon of the acute stroke period (42). The difference between stroke and TBI depression in the duration of depression (i.e., 9 months in stroke, 5 months after TBI) may be the result of differences in the nature of the brain injury with TBI, leading to reorganization (e.g., pruning, reactive synaptogenesis, regenerative sprouting) (114,115) and perhaps clinical recovery from depression.

In conclusion, some acute-onset depressions appear to be related to lesion characteristics and may have their etiology in biologic responses such as neurochemical changes. Left dorsolateral frontal and left basal ganglia lesions were significantly associated with major depression during the initial in-hospital evaluation and may represent a strategic lesion location that elicits biochemical responses that ultimately lead to depression. In addition, other major correlates of depression were a history of a psychiatric disorder and impaired social functioning, suggesting psychosocial factors in the mechanism of some post-TBI depressions.

Relationship to Impairment Variables

Empirical studies have reported conflicting findings concerning the relationship between impairment and depressive symptoms after TBI (116,117). In the previously described study of 66 patients with TBI, there was no significant association between the depression and the severity of intellectual impairment (i.e., MMSE score) or ADL (98). Social functioning was the clinical variable that had the most consistent relationship with depression throughout the whole follow-up period (118). Although it is likely that lack of social support contributed to the development of depression, the existence of depression may have negatively influenced social functioning. This suggestion is supported by the finding that depression scores correlated with the Social Functioning Examination only at the onset of depression and not with the initial Social Functioning Examination scores (108). One might infer from these findings that social intervention, as well as the treatment of depression, may be necessary to alleviate these severe and long-lasting mood disorders.

Treatment of Traumatic Brain Injury Depression

There have been only two placebo-controlled studies of the efficacy of pharmacologic treatments of depression in TBI patients. The first controlled treatment study was conducted by Wroblewski et al. (119) in which ten patients with severe TBI and depressions lasting 2 years or longer were randomly assigned to desipramine (n = 6) or placebo (n = 4) for 1 month. Patients were evaluated blind to medication, but after the first month, all placebo patients were crossed over to desipramine and the desipramine patients continued on active treatment. Results found that seven of ten patients completed 2 to 3 months of treatment and six of seven had resolution of depression. There was a statistically significantly greater improvement in the affect/mood scale (nurse rated) among the active-treatment compared with the placebo-treated patients over the first month. Although this study was not double blind throughout, it is the first study to demonstrate improvement of depression in patients with TBI using a prospective, placebo-controlled design and blinded ratings.

The most recent studies of depression in TBI were reported by Fann et al. (120) and Newburn (121). The Fann et al. study administered sertraline (25–200 mg daily) to 15 patients with major depression after TBI after 1 week of single-blind treatment with placebo. Using the Hamilton depression score as the outcome, 13 patients had a greater than 50% reduction and ten had a score of 7 or less (remission) by week 8 of active treatment. Newburn et al. treated 26 patients with major depression 4.7 years post-TBI with moclobemide, a monoamine oxidase A inhibitor (450–600 mg daily). There were seven dropouts, and the patients were treated for 6 weeks. The mean initial Hamilton depression score was 23.4 and Hamilton Anxiety score was 21.2. The mean reduction was 81% in both the Hamilton depression score and the Hamilton Anxiety score. Onset of response was rapid, with 17 patients responding by day 3. This was an open-label study without a control group.

These are also case reports of successful treatments of post-TBI depression with psychostimulants (122). These include dextroamphetamine (8–60 mg daily), methylphenidate (10–60 mg daily), and pemoline (56–75 mg daily). They are given twice daily, with the last dose at least 6 hours before sleep to prevent initial insomnia. Treatment is begun at lower doses, which are then gradually increased. Patients taking stimulants need close medical monitoring to prevent abuse or toxic effects. The most common side effects are anxiety, dysphoria, headaches, irritability, anorexia, insomnia, cardiovascular symptoms, dyskinesias, or even psychotic symptoms (123).

Buspirone, a drug that has an agonist effect on $5HT_{1A}$ receptors and an antagonist effect on D_2 dopaminergic receptors, has proved to be a safe and efficacious anxiolytic. Initial dosing is 15 mg daily given in three divided doses, and it may gradually be increased (5 mg every 4 days) to as much as 60 mg daily. The most common side effects are dizziness and headaches (122,123).

Electroconvulsive therapy is not contraindicated in TBI patients and may be considered if other methods of treatment prove to be unsuccessful. Finally, the role of social interventions and adequate psychotherapeutic support may be included in the treatment of depression. Psychological treatments need to be examined in controlled treatment trials.

POSTSTROKE MANIA

Although poststroke mania occurs much less frequently than depression (we have observed only three cases among a consecutive series of more than 300 stroke patients), manic syndromes are sometimes associated with stroke.

Starkstein et al. (41) examined a series of 12 consecutive patients who met *DSM-III* criteria for an organic affective syndrome, manic type. These patients, who developed mania after a stroke, TBI, or tumors, were compared with patients with functional (i.e., no known neuropathology) mania (40). Both groups of patients showed similar frequencies of elation, pressured speech, flight of ideas, grandiose thoughts, insomnia, hallucinations, and paranoid delusions. Thus, the symptoms of mania that occurred after brain damage (secondary mania) appeared to be the same as those found in patients with mania without brain damage (primary mania).

Lesion Location

Several studies of patients with brain damage have found that patients who develop secondary mania have a significantly greater frequency of lesions in the right hemisphere than patients with depression or no mood disturbance (124,125). The right-hemisphere lesions that lead to mania tend to be in specific right-hemisphere structures that have connections to the limbic system. The right basotemporal cortex appears to be particularly important because direct lesions, as well as distant hypometabolic effects (diaschisis) of this cortical region, are frequently associated with secondary mania.

Cummings and Mendez (124) reported on two patients who developed mania after right thalamic stroke lesions. After review of the literature, these authors suggested a specific association between secondary mania and lesions in the limbic system or in limbic-related areas of the right hemisphere.

Robinson et al. (125) reported on 17 patients with secondary mania. Most of the patients had right-hemisphere lesions involving either cortical limbic areas, such as the orbitofrontal cortex and basotemporal cortex, or subcortical nuclei, such as the head of the caudate or thalamus. The frequency of right-hemisphere lesions was significantly greater than in patients with major depression, who tended to have left frontal or basal ganglia lesions (Fig. 28.10).

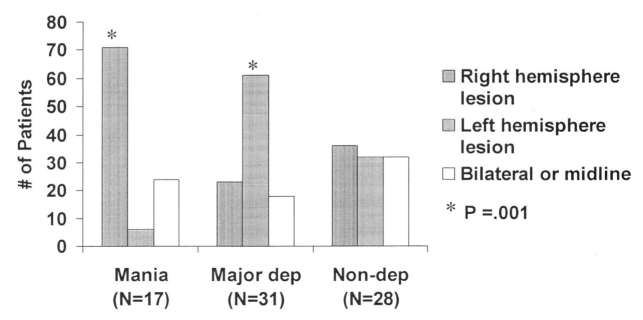

FIGURE 28.10. Frequency of right- or left-hemisphere lesion location in patients with mania after stroke (n = 9), tumors (n = 6), or traumatic brain injury (n = 2) compared with patients with acute poststroke major depression (Major dep) or no mood disturbance (Nondep) after stroke. Mania was strongly associated with a right-hemisphere lesion location, whereas major depression after acute stroke was associated with left hemisphere lesions. The association of diagnosis with lesion location was highly significant ($p = 0.001$). (Reprinted from Robinson RG. The clinical neuropsychiatry of stroke. Cambridge, UK: Cambridge University Press, 1998;61, with permission.)

These findings have been replicated in another study of eight patients with secondary mania (126). All eight patients had right-hemisphere lesions (seven unilateral and one bilateral injury). Lesions were either cortical (basotemporal cortex in four patients and orbitofrontal cortex in one patient) or subcortical (frontal white matter, head of the caudate, and anterior limb of the internal capsule in three patients, respectively). Positron emission tomography scans with [18F]fluorodeoxyglucose were carried out in the three patients with purely subcortical lesions. They all showed a focal hypometabolic deficit in the right basotemporal cortex (Fig. 28.11).

Risk Factors

Not every patient with a lesion in limbic areas of the right hemisphere will develop secondary mania. Therefore, there must be risk factors for this disorder. In one study, patients with secondary mania were compared with patients with secondary major depression (125). Results indicated that patients with secondary mania had a significantly higher frequency of positive family history of affective disorders than did depressed patients or patients with no mood disturbance ($p < 0.05$). Therefore, it appears that genetic predisposition to affective disorders may constitute a risk factor for mania.

In another study, patients with secondary mania were compared with patients with no mood disturbance who were matched for the size, location, and etiology of the brain lesion (61). The groups were also compared with patients with primary mania and control subjects. No significant between-group differences were found in either demographic variables or neurologic evaluation. Patients with secondary mania had a significantly greater degree of subcortical atrophy, as measured by bifrontal- and third ventricular-to-brain ratio ($p < 0.001$). Moreover, of the patients who developed secondary mania, those with a positive family history of psychiatric disorders had significantly less atrophy than those without such a family history ($p < 0.05$), suggesting that genetic predisposition to affective disorders and brain atrophy may be independent risk factors.

The relatively rare occurrence of mania after stroke suggests that there are premorbid risk factors that have an impact on the expression of this disorder. Studies thus far have identified two such factors. One is a genetic vulnerability for affective disorder, and the other is a mild degree of subcortical atrophy. The subcortical atrophy probably preceded the stroke, but its cause remains unknown (61).

Mechanism of Secondary Mania

Several studies have demonstrated that the amygdala (located in the medial portion of the temporal lobe) has an important role in the production of instinctive reactions and the association between stimulus and emotional response (127). The amygdala receives its main

Regional Brain Glucose Metabolic Rates

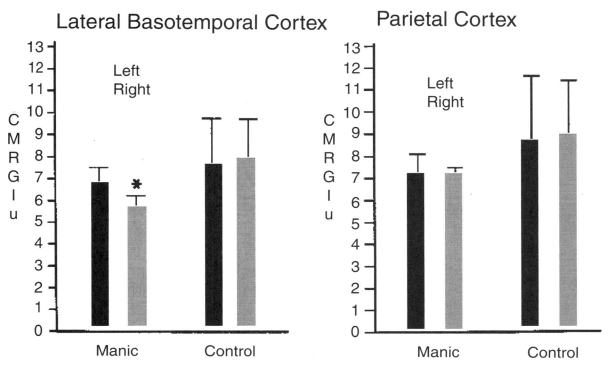

FIGURE 28.11. Regional brain glucose metabolic rates among patients with mania after stroke and age comparable controls (n = 18). Although there were no significant differences between metabolic rates in symmetric areas of the right and left parietal cortex, among patients with poststroke mania, there was a significantly lower metabolic rate in the uninjured areas of the basotemporal cortex in the right compared with the left hemisphere. Control patients showed no lateralized effect on basotemporal glucose metabolic rate. These findings are consistent with the association of lesions of the right basotemporal cortex with mania.

afferents from the basal diencephalon (which in turn receives psychosensory and psychomotor information from the reticular formation) and the temporopolar and basolateral cortex (which receive main afferents from heteromodal association areas) (128,129). The basotemporal cortex receives afferents from association cortical areas and the orbitofrontal cortex and sends efferent projections to the entorhinal cortex, hippocampus, and amygdala. By virtue of these connections, the basotemporal cortex may represent a cortical link between sensory afferents and instinctive reactions (130).

The orbitofrontal cortex may be subdivided into two regions: a posterior one, which is restricted to limbic functions and should be considered part of the limbic system, and an anterior one, which exerts tonic inhibitory control over the amygdala by means of its connection through the uncinate fasciculus with the basotemporal cortex (89). Thus, the uncinate fasciculus and the basotemporal cortex may mediate connections between psychomotor and volitional processes generated in the frontal lobe and vital pro-

cesses and instinctive behaviors generated in the amygdala (41).

A case report by Starkstein et al. (131) suggested that the mechanism of secondary mania is not related to the release of transcallosal inhibitory fibers (i.e., the release of left limbic areas from tonic inhibition owing to a right-hemisphere lesion). A patient who developed secondary mania after bleeding from a right basotemporal arteriovenous malformation underwent a Wada test before the therapeutic embolization of the malformation. Injection of amobarbital in the left carotid artery did not abolish the manic symptoms (which would be expected if the "release" theory were correct).

Although the mechanism of secondary mania remains unknown, both lesion studies and metabolic studies suggest that the right basotemporal cortex may play an important role. A combination of biogenic amine system dysfunction and release of tonic inhibitory input into the basotemporal cortex and lateral limbic system may lead to the production of mania.

Treatment of Secondary Mania

Although no systematic treatment studies of secondary mania have been conducted, one report suggested several potentially useful treatment modalities. Bakchine et al. (132) carried out a double-blind, placebo-controlled treatment study in a patient with secondary mania. Clonidine (0.6 mg daily) rapidly reversed the manic symptoms, whereas carbamazepine (1,200 mg daily) was associated with no mood changes and levodopa (375 mg daily) was associated with an increase in manic symptoms. In other treatment studies, the anticonvulsants valproic acid and carbamazepine, as well as neuroleptics and lithium therapy, have been reported to be useful in treating secondary mania (133). None of these treatments has been evaluated in double-blind, placebo-controlled studies.

Mania Associated with Traumatic Brain Injury

Prevalence

Mania is more frequent among patients with TBI than among patients with stroke lesions (125,134,135). In the previously described study of 66 patients with acute TBI, six cases (9%) with secondary mania were found (135). One of these patients presented a bipolar course. The manic episodes were short-lived, with a mean duration of 2 months. The mean duration of the elevated mood (without meeting other diagnostic criteria for mania) was 5.7 months. In addition, three of the six secondary manic patients developed brief episodes of violent behavior at some point during the 1-year follow-up. Aggressive behavior was significantly more frequent in the secondary mania group than among those who did not experience an affective disorder. Although at the time of the diagnosis, three patients were receiving drug treatment (two patients received lorazepam and one patient haloperidol), the duration of mania did not appear to be significantly different from those who were not treated.

Relationship to Impairment Variables

In the previously described study of 66 patients with mania after TBI, the severity of mania was not associated with severity of brain injury, degree of physical or cognitive impairment, personal or family history of psychiatric disorder, or the availability of social support or quality of social functioning (135). Thus, although further studies of the relationships between impairment or other risk factors and the development of mania need to be conducted, the present data suggest that mania is not a response to the associated impairments.

Relationship to Lesion Location

The cortical areas most frequently affected by closed head injury are the ventral aspects of the frontal and temporal lobes. There is also evidence that damage to subcortical, diencephalic, and brainstem structures is involved (136).

The Jorge et al. (135) study of secondary manic syndromes found that the major correlate of mania was the presence of anterior temporal lesions. This finding is consistent with the finding in patients with stroke that mania is associated with right basotemporal lesions. The trauma study did not have sufficient numbers of patients with unilateral lesions to examine the right- versus left-hemisphere effect. Factors such as personal history of mood disorders or posttraumatic epilepsy did not appear to significantly influence the frequency of secondary mania in this group of patients.

Mechanism

It has been suggested that the development of abnormal activation patterns in limbic networks, functional changes in aminergic inhibitory systems, and the presence of aberrant regeneration pathways may play an important role in the genesis of these syndromes (137,138).

BIPOLAR DISORDER

Bipolar Disorder Associated with Stroke or Trauma

Starkstein et al. (133) studied 19 patients with the diagnosis of secondary mania in an effort to determine which factors played a role in the development of bipolar or unipolar affective disorder. The bipolar (manic-depressive) group consisted of seven patients who, after the brain lesion, met the *DSM-III-R* criteria for organic mood syndrome, mania, followed or preceded by organic mood syndrome, depression. The unipolar mania (mania only) consisted of 12 patients who met the criteria for mania just described, not followed or preceded by depression. All the patients had CT scan evidence of vascular, neoplastic, or traumatic brain lesion and a history of other neurologic, toxic, or metabolic conditions.

There were no significant between-group differences in age, gender, race, education, handedness, or personal history of psychiatric disease. In addition, no significant differences were found on neurologic examination. On psychiatric examination, which was carried out during the index manic episode, no significant differences were observed in the type or frequency of manic symptoms. The bipolar group, however, showed significantly greater intellectual impairment as measured by MMSE scores ($p < 0.05$).

Patients with bipolar disorder had lesions restricted to the right hemisphere, which involved the head of the caudate or thalamus (Fig. 28.12). One patient developed a bipolar illness after surgical removal of a pituitary adenoma. In contrast to these subcortical lesions, the unipolar mania group had lesions involving the right basotemporal and orbitofrontal cortex.

This suggests that patients with bipolar disorder tend to have subcortical lesions (mainly involving the right head of the caudate or right thalamus), although patients with pure

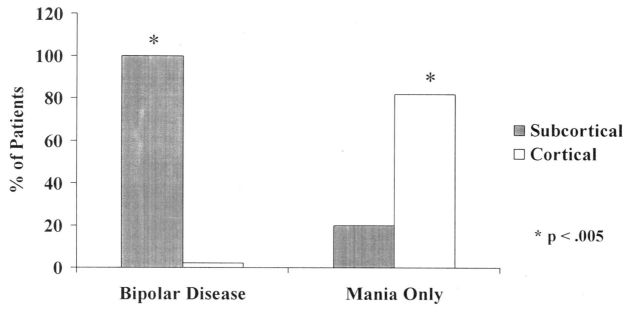

FIGURE 28.12. The frequency of subcortical and cortical lesions in patients with bipolar disorder or mania only after brain lesions. Bipolar disorder was strongly associated with lesions of the basal ganglia or thalamus in the right hemisphere, whereas mania only was associated with right cortical lesions. (Reprinted from Robinson RG. The clinical neuropsychiatry of stroke. Cambridge, UK: Cambridge University Press, 1998;61, with permission.)

mania tend to show a higher frequency of cortical lesions, particularly in the right orbitofrontal and right basotemporal cortex.

The question that now arises is why subcortical lesions might produce a bipolar disease. The cause of both bipolar and unipolar mood disorders remains unknown and is a major goal of future investigation. Numerous hypotheses have been proposed to explain these disorders. Pappata et al. (139) reported that subcortical lesions induce hypometabolic effects in many regions, including contralateral brain areas (i.e., crossed-hemisphere and crossed-cerebellar diaschisis). It is therefore possible that subcortical lesions will induce metabolic changes in left frontocortical regions that are associated with depression. Mania may develop at a later stage, when these changes become restructured to the orbitofrontal and basotemporal cortex of the right hemisphere.

SUMMARY AND CONCLUSIONS

The neuropsychiatry of mood disorders involves the investigation of similarities and differences in mood disorders associated with several neurologic disorders. This chapter focused on mood disorders associated with stroke, PD, and TBI.

Depression occurs in approximately 40% of patients with acute stroke lesions, and its natural evolution is from 1 to 2 years, although patients with subcortical or brainstem lesions may show depression of shorter duration. Both intrahemispheric and interhemispheric lesion locations appear to contribute to the development of depression. Major depression is significantly more frequent during the acute stroke period (i.e., 1 to 2 months poststroke) among patients with left-hemisphere lesions involving anterior cortical (frontal) or subcortical (basal ganglia) regions than any other lesion location. Conversely, depressions after right-hemisphere lesions are sometimes associated with a genetic vulnerability and frontal or parietal lobe damage. Risk factors for the development of PSD includes the severity of cognitive and physical impairment, as well as site of lesion location.

Mania that develops after stroke has a phenomenology similar to that of mania without known neuropathology. Secondary mania, a rare complication in stroke patients, is usually the consequence of lesions involving right cortical (orbitofrontal or basotemporal) or right subcortical (head of the caudate or thalamus) limbic-related regions. Among patients with the right cortical or right subcortical lesions, dysfunction of the basotemporal cortex seems to be particularly important to the development of secondary mania, and basotemporal dysfunction may be produced by direct or indirect (diaschisis) damage. Risk factors for secondary mania include a familial history of psychiatric disorders or prior subcortical atrophy.

Among patients with PD, the frequency of depression is approximately 40% with equal numbers of cases of major and minor depression. Depression in patients with PD is associated with early and late-stage PD, greater cognitive impairment, more rapid progression of the disease, and abnormality of both serotonin production and orbitofrontal cortical function.

Among patients with TBI, the frequency of major depression is approximately 26%. Although some patients remain depressed for more than a year, the average duration of depression is 4 to 5 months. Risk factors for depression include a history of psychiatric disorder and poor social functioning. Intellectual or physical impairment is not associated with the degree of depression. In addition, transiently depressed patients are more likely to have sustained injury to left dorsolateral frontal or left basal ganglia structures than patients with more prolonged depressions. Similarly, manic patients are significantly more likely to have temporal basopolar lesions than depressed or nondepressed patients.

Many areas are ripe for future research. The most important elements of social functioning that contribute to depression need to be explored, as well as the effect of social intervention. The role of antidepressants in treating these depressive disorders has not been systematically explored and deserves study.

Finally, the mechanism of these depressions, both those associated with psychosocial factors and those associated with neurobiologic variables (e.g., strategic lesion locations) need to be investigated. It is only through the discovery of their mechanisms that specific and rational treatment strategies for these disorders will be developed.

REFERENCES

1. Babinski J. Contribution a l'etude des troubles mentaux dans l'hemiplegie organique cerebrale (anosognosie). *Rev Neurol (Paris)* 1914;27:845–848.
2. Denny-Brown D, Meyer JS, Horenstein S. The significance of perceptual rivalry resulting from parietal lesions. *Brain* 1952;75:434–471.
3. Gainotti G. Emotional behavior and hemispheric side of the brain. *Cortex* 1972;8:41–55.
4. Meyer A. The neuroanatomical facts and clinical varieties of traumatic insanity. *Am J Insanity* 1904;60:373.
5. Bleuler EP. *Textbook of psychiatry.* New York: Macmillan, 1951:131–197.
6. Kraepelin E. *Manic depressive insanity and paranoia.* Edinburgh: E&S Livingstone, 1921.
7. Goldstein K. *The organism: a holistic approach to biology derived from pathological data in man.* New York: American Books, 1939:35.
8. Goldstein K. *After effects of brain injuries in war.* New York: Grune & Stratton, 1942.
9. Heacen H, de Ajuriaguerra J, Massonet J. Les troubles visoconstructifs para lesion parieto occipitale droit. *Encephale* 1951;40:122–179.
10. Ironside R. Disorders of laughter due to brain lesions. *Brain* 1956;79:589–609.
11. Post F. *The significance of affective symptoms in old age* (Maudsley Monograph 10). London: Oxford University Press, 1962.
12. Ullman M, Gruen A. Behavioral changes in patients with stroke. *Am J Psychiatry* 1960;117:1004–1009.
13. Fisher SH. Psychiatric considerations of cerebral vascular disease. *Am J Cardiol* 1961;7:379–385.
14. Folstein MF, Maiberger R, McHugh PR. Mood disorder as a specific complication of stroke. *J Neurol Neurosurg Psychiatry* 1977;40:1018–1020.
15. Cohen-Cole SA, Stoudemire A. Major depression and physical illness: special considerations in diagnosis and biologic treatment. *Psychiatr Clin North Am* 1987;10:1–17.
16. Rifkin A, Reardon G, Siris S, et al. Trimipramine in physical illness with depression. *J Clin Psychiatry* 1985;46:4–8.
17. Rapp SR, Vrana S. Substituting nonsomatic for somatic symptoms in the diagnosis of depression in elderly male medical patients. *Am J Psychiatry* 1989;146:1197–1200.
18. Endicott J. Measurement of depression in patients with cancer. *Cancer* 1984;53[Suppl]:2243–2248.
19. Bukberg J, Penman D, Holland JC. Depression in hospitalized cancer patients. *Psychosom Med* 1984;46:199–212.
20. Paradiso S, Ohkubo T, Robinson RG. Vegetative and psychological symptoms associated with depressed mood over the first two years after stroke. *Int J Psychiatry Med* 1997;27:137–157.
21. Lipsey JR, Spencer WC, Rabins PV, et al. Phenomenological comparison of functional and post-stroke depression. *Am J Psychiatry* 1986;143:527–529.
22. Gainotti G, Azzoni A, Marra C. Frequency, phenomenology and anatomical-clinical correlates of major post-stroke depression. *Br J Psychiatry* 1999;175:163–167.
23. American Psychiatric Association. *Diagnostic and statistical manual of mental disorders, fourth edition.* Washington, DC: American Psychiatric Press, 1994.
24. Robinson RG, Bolduc P, Price TR. A two year longitudinal study of post-stroke depression: diagnosis and outcome at one and two year follow-up. *Stroke* 1987;18:837–843.
25. Morris PLP, Robinson RG, Raphael B. Prevalence and course of depressive disorders in hospitalized stroke patients. *Int J Psychiatry Med* 1990;20:349–364.
26. Astrom M, Adolfsson R, Asplund K. Major depression in stroke patients: a 3-year longitudinal study. *Stroke* 1993;24:976–982.
27. Burvill PW, Johnson GA, Jamrozik KD, et al. Prevalence of depression after stroke: the Perth Community Stroke Study. *Br J Psychiatry* 1995;166:320–327.
28. Starkstein SE, Moran TH, Bowersox JA, et al. Behavioral abnormalities induced by frontal cortical and nucleus accumbens lesions. *Brain Res* 1988;473:74–80.
29. Robinson RG, Szetela B. Mood change following left hemispheric brain injury. *Ann Neurol* 1981;9:447–453.
30. Eastwood MR, Rifat SL, Nobbs H, et al. Mood disorder following cerebrovascular accident. *Br J Psychiatry* 1989;154:195–200.
31. House A, Dennis M, Mogridge L, et al. Mood disorders in the year after first stroke. *Br J Psychiatry* 1991;158:83–92.
32. Morris PLP, Robinson RG, Raphael B. Lesion location and depression in hospitalized stroke patients: evidence supporting a specific relationship in the left hemisphere. *Neuropsychiatry Neuropsychol Behav Neurol* 1992;3:75–82.
33. Herrmann M, Walesch C-W. Depressive changes in stroke patients. *Disabil Rehabil* 1993;15:55–66.
34. Robinson RG, Kubos KL, Starr LB, et al. Mood disorders in stroke patients: importance of location of lesion. *Brain* 1984;107:81–93.
35. Robinson RG, Starr LB, Price TR. A two-year longitudinal study of post-stroke mood disorders: dynamic changes in associated variables over the first six months of follow-up. *Stroke* 1984;15:510–517.
36. Robinson RG, Starr LB, Price TR. A two year longitudinal study of mood disorders following stroke. Prevalence and duration at six months follow-up. *Br J Psychiatry* 1984;144:256–262.
37. Robinson RG, Lipsey JR, Bolla-Wilson K, et al. Mood disorders

in left handed stroke patients. *Am J Psychiatry* 1985;142:1424–1429.

38. Robinson RG, Bolduc PL, Kubos KL, et al. Social functioning assessment in stroke patients. *Arch Phys Med Rehabil* 1985;66:496–500.
39. Starkstein SE, Robinson RG, Price TR. Comparison of cortical and subcortical lesions in the production of post-stroke mood disorders. *Brain* 1987;110:1045–1059.
40. Starkstein SE, Pearlson GD, Boston J, et al. Mania after brain injury: a controlled study of causative factors. *Arch Neurol* 1987;44:1069–1073.
41. Starkstein SE, Boston JD, Robinson RG. Mechanisms of mania after brain injury: 12 case reports and review of the literature. *J Nerv Ment Dis* 1988;176:87–100.
42. Shimoda K, Robinson RG. The relationship between post-stroke depression and lesion location in long-term follow-up. *Biol Psychiatry* 1999;45:187–192.
43. Robinson RG, Starr LB, Kubos KL, et al. A two year longitudinal study of post-stroke mood disorders: findings during the initial evaluation. *Stroke* 1983;14:736–744.
44. Morris PLP, Robinson RG, Raphael B, et al. The relationship between risk factors for affective disorder and post-stroke depression in hospitalized stroke patients. *Aust N Z J Psychiatry* 1992;26:208–217.
45. Parikh RM, Robinson RG, Lipsey JR, et al. The impact of post-stroke depression on recovery in activities of daily living over two year follow-up. *Arch Neurol* 1990;47:785–789.
46. Chemerinski E, Robinson RG, Kosier JT. Improved recovery in activities of daily living associated with remission of post-stroke depression. *Stroke* 2001;32:113–117.
47. Robinson RG, Bolla-Wilson K, Kaplan E, et al. Depression influences intellectual impairment in stroke patients. *Br J Psychiatry* 1986;148:541–547.
48. House A, Dennis M, Warlow C, et al. The relationship between intellectual impairment and mood disorder in the first year after stroke. *Psychol Med* 1990;20:805–814.
49. Andersen G, Vestergaard K, Riis JO, et al. Dementia of depression or depression of dementia in stroke? *Acta Psychiatr Scand* 1996;94:272–278.
50. Folstein MF, Folstein SE, McHugh PR. Mini-Mental State: a practical method for grading the cognitive state of patients for the clinician. *J Psychiatr Res* 1975;12:189–198.
51. Starkstein SE, Robinson RG, Price TR. Comparison of patients with and without post-stroke major depression matched for size and location of lesion. *Arch Gen Psychiatry* 1988;45:247–252.
52. Bolla-Wilson K, Robinson RG, Starkstein SE, et al. Lateralization of dementia of depression in stroke patients. *Am J Psychiatry* 1989;146:627–634.
53. Downhill JE Jr, Robinson RG. Longitudinal assessment of depression and cognitive impairment following stroke. *J Nerv Ment Dis* 1994;182:425–431.
54. Kimura M, Robinson RG, Kosier T. Treatment of cognitive impairment after poststroke depression. *Stroke* 2000;31:1482–1486.
55. Ross ED, Rush AJ. Diagnosis and neuroanatomical correlates of depression brain damaged patients. *Arch Gen Psychiatry* 1981;38:1344–1354.
56. Robinson RG, Benson DF. Depression in aphasic patients: frequency, severity and clinical-pathological correlations. *Brain Lang* 1981;14:282–291.
57. Starkstein SE, Robinson RG. Aphasia and depression. *Aphasiology* 1988;2:1–20.
58. Starkstein SE, Robinson RG, Berthier ML, et al. Depressive disorder following posterior circulation compared with middle cerebral artery infarcts. *Brain* 1988;111:375–387.
59. Morrison JH, Molliver ME, Grzanna R. Noradrenergic inner-

vation of the cerebral cortex: widespread effects of local cortical lesions. *Science* 1979;205:313–316.
60. Mayberg HS, Robinson RG, Wong DF, et al. PET imaging of cortical S$_2$-serotonin receptors after stroke: lateralized changes and relationship to depression. *Am J Psychiatry* 1988;145:937–943.
61. Lipsey JR, Robinson RG, Pearlson GD, et al. Nortriptyline treatment of post-stroke depression: a double-blind study. *Lancet* 1984;1:297–300.
62. Hamilton MA. A rating scale for depression. *J Neurol Neurosurg Psychiatry* 1960;23:56–62.
63. Reding MJ, Orto LA, Winter SW, et al. Antidepressant therapy after stroke: a double-blind trial. *Arch Neurol* 1986;43:763–765.
64. Granger CV, Denis LS, Peters NC, et al. Stroke rehabilitation: analysis of repeated Barthel Index measures. *Arch Phys Med Rehabil* 1979;60:14–17.
65. Andersen G, Vestergaard K, Lauritzen L. Effective treatment of poststroke depression with the selective serotonin reuptake inhibitor citalopram. *Stroke* 1994;25:1099–1104.
66. Robinson RG, Schultz SK, Castillo C, et al. Nortriptyline versus fluoxetine in the treatment of depression and in short term recovery after stroke: a placebo controlled, double-blind study. *Am J Psychiatry* 2000;157:351–359.
67. Murray GB, Shea V, Conn DR. Electroconvulsive therapy for post-stroke depression. *J Clin Psychiatry* 1987;47:458–360.
68. Grade C, Redford B, Chrostowski J, et al. Methylphenidate in early poststroke recovery: a double-blind, placebo-controlled study. *Arch Phys Med Rehabil* 1998;79:1047–1050.
69. Hibbard MR, Grober SE, Gordon WA, et al. Modification of cognitive psychotherapy for the treatment of post-stroke depression. *Behav Ther* 1990;13:15–17.
70. Oradei DM, Waite NS. Group psychotherapy with stroke patients during the immediate recovery phase. *Am J Orthopsychiatry* 1974;44:386–395.
71. Watzlawick P, Coyne JC. Depression following stroke: brief, problem-focused family treatment. *Fam Process* 1980;19:13–18.
72. Lewy FH. *Die Lehre von Tonus und der Bewegung.* Berlin: Springer, 1923.
73. Mindham RHS. Psychiatric symptoms in Parkinsonism. *J Neurol Neurosurg Psychiatry* 1970;33:188–191.
74. Mayeux R, Williams JBW, Stern Y. Depression in Parkinson disease. *Adv Neurol* 1984;40:242–250.
75. Mayeux R, Stern Y, William JBW. Clinical and biochemical features of depression in Parkinson's disease. *Am J Psychiatry* 1986;143:756–759.
76. Celesia GG, Wanamaker WM. Psychiatric disturbances in Parkinson's disease. *Dis Nerv Syst* 1972;33:577–583.
77. Gothan AM, Brown RG, Marsden CD. Depression in Parkinson's disease: a quantitative and qualitative analysis. *J Neurol Neurosurg Psychiatry* 1986;49:381–389.
78. Starkstein SE, Bolduc PL, Mayberg HS, et al. Cognitive impairment and depression in Parkinson's disease: a follow-up study. *J Neurol Neurosurg Psychiatry* 1990;53:597–602.
79. El-Awar M, Bekcer JT, Hammond KM. Learning deficits in Parkinson's disease: comparison with Alzheimer's disease and normal aging. *Arch Neurol* 1987;44:180–184.
80. Mayeux R, Stern Y, Rose J. Depression, intellectual impairment and Parkinson disease. *Neurology* 1981;32:645–650.
81. Starkstein SE, Colduc PL, Preziosi TJ. Cognitive impairments in different states of Parkinson disease. *J Neuropsychiatry Clin Neurosci* 1989;1:243–248.
82. Starkstein SE, Preziosi TJ, Berthier ML, et al. Depression and cognitive impairments in Parkinson's disease. *Brain* 1989;112:1141–1153.

83. Sano M, Stern Y, William J. Coexisting dementia and depression in Parkinson's disease. *Arch Neurol* 1989;46:1284–1286.

84. Perry EK, Curtis M, Dick DJ. Cholinergic correlates of cognitive impairment in Parkinson disease: comparisons with Alzheimer's disease. *J Neurol Neurosurg Psychiatry* 1985;48:413–421.

85. Torack RM, Morris JC. The association of ventral tegmental area histopathology with adult dementia. *Arch Neurol* 1988;45:211–218.

86. Mayberg HS, Starkstein SE, Sadzot B, et al. Selective hypometabolism in the inferior frontal lobe in depressed patients with Parkinson's disease. *Ann Neurol* 1990;28:57–64.

87. Simon H, LeMoal M, Calas A. Efferents and afferents of the ventral tegmental A-10 region studied after local injection of [3H]-leucine and horseradish peroxidase. *Brain Res* 1979;178:17–40.

88. Azmitia EC, Gannon PJ. Primate serotonergic system: a review of human and animal studies and a report on *Macacca fascicularis*. *Adv Neurol* 1986;43:407–468.

89. Nauta WJH. The problem of the frontal lobe: a reinterpretation. *J Psychol Res* 1971;8:167–187.

90. Andersen J, Aabro E, Gulmann N, et al. Antidepressive treatment in Parkinson's disease. *Acta Neurol Scand* 1980;62:210–219.

91. Hauser RA, Zesiewicz TA. Sertraline for the treatment of depression in Parkinson's disease. *Mov Disord* 1997;12:756–759.

92. Tesei S, Antonini A, Canesi M, et al. Tolerability of paroxetine in Parkinson's disease: a prospective study. *Mov Disord* 2000;15:986–989.

93. Klaassen T, Verhey FR, Sneijders GH, et al. Treatment of depression in Parkinson's disease: a meta-analysis. *J Neuropsychiatry Clin Neurosci* 1995;7:281–286.

94. Silver JM, Yudofsky SC, Hales RE. Depression in traumatic brain injury. *Neuropsychiatry Neuropsychol Behav Neurol* 1991;4:12–23.

95. Rutherford WH, Merrett JD, McDonald JR. Sequelae of concussion caused by minor head injuries. *Lancet* 1977;1:1–4.

96. McKinlay WW, Brooks DN, Bond MR, et al. The short-term outcome of severe blunt head injury as reported by the relatives of the head injury person. *J Neurol Neurosurg Psychiatry* 1981;44:527–533.

97. Kinsella G, Moran C, Ford B. Emotional disorders and its assessment within the severe head injured population. *Psychol Med* 1988;18:57–63.

98. Fedoroff JP, Starkstein SE, Forrester AW, et al. Depression in patients with acute traumatic brain injury. *Am J Psychiatry* 1992;149:918–923.

99. Jorge RE, Robinson RG, Arndt SV, et al. Depression following traumatic brain injury: a 1 year longitudinal study. *J Affect Disord* 1993;27:233–243.

100. Gualtieri CT, Cox DR. The delayed neurobehavioral sequelae of traumatic brain injury. *Brain Inj* 1991;5:219–232.

101. Brooks N, Campsie L, Symington C. The five year outcome of severe blunt head injury: a relative's view. *J Neurol Neurosurg Psychiatry* 1986;49:764–770.

102. Grant I, Alwes W. Psychiatric and psychological disturbances in head injury. In: Levin HS, Grafman J, Eisenberg HM, eds. *Neurobehavioral recovery from head injury.* London: Oxford University Press, 1987:215–232.

103. Prigatano GP. Disturbances of self-awareness of deficit after traumatic brain injury. In: Prigatano GP, Schacter DL, eds. *Awareness of deficit after brain injury: clinical and theoretical issues.* New York: Oxford University Press, 1991.

104. Lishman WA. Physiogenesis and psychogenesis in the postconcussional syndrome. *Br J Psychiatry* 1988;153:460–469.

105. VanZomeren AH, Saan RJ. Psychological and social sequelae of severe head injury. In: Braakman R, ed. *Handbook of clinical neurology. Volume 13.* Amsterdam: Elsevier Science, 1990:397–420.

106. Wing JK, Cooper JE, Sartorius N. *The measurement and classification of psychiatric symptoms: an instructional manual for the PSE and CATEGO programs.* New York: Cambridge University Press, 1974.

107. American Psychiatric Association. *Diagnostic and statistical manual of mental disorders, revised third edition.* Washington, DC: American Psychiatric Press, 1987.

108. Jorge RE, Robinson RG, Arndt SV, et al. Comparison between acute and delayed onset depression following traumatic brain injury. *J Neuropsychiatry* 1993;5:43–49.

109. Lishman WA. The psychiatric sequelae of head injury: a review. *Psychol Med* 1973;3:304–318.

110. Starr LB, Robinson RG, Price TR. The social functioning exam: an assessment of stroke patients. *Soc Work Res Abst* 1983;18:28–33.

111. Lishman WA. Brain damage in relation to psychiatric disability after head injury. *Br J Psychiatry* 1968;114:373–410.

112. Grafman J, Vance SC, Swingartner H. The effects of lateralized frontal lesions on mood regulation. *Brain* 1986;109:1127–1148.

113. Katz DI. Neuropathology and neurobehavioral recovery from closed head injury. *J Head Trauma Rehabil* 1992;7:1–15.

114. Chollet F, DiPiero V, Wise RJS, et al. The functional anatomy of motor recovery after stroke in humans: a study with positron emission tomography. *Ann Neurol* 1991;29:63–71.

115. Steward O. Reorganization of neuronal connections following CNS trauma: principles and experimental paradigms. *J Neurotrauma* 1989;6:99–152.

116. Prigatano GP. *Neuropsychological rehabilitation after brain injury.* Baltimore: Johns Hopkins University Press, 1986.

117. Bornstein RA, Miller HB, Van Schoor JT. Neuropsychological deficit and emotional disturbance in head injured patients. *J Neurosurg* 1989;70:509–513.

118. Jorge RE, Robinson RG, Starkstein SE, et al. Depression and anxiety following traumatic brain injury. *J Neuropsychiatry Clin Neurosci* 1993;5:43–49.

119. Wroblewski BA, Joseph AB, Cornblatt RR. Antidepressant pharmacotherapy and the treatment of depression in patients with severe traumatic brain injury: a controlled, prospective study. *J Clin Psychiatry* 1996;57:582–587.

120. Fann JR, Uomoto JM, Katon WJ. Sertraline in the treatment of major depression following mild traumatic brain injury. *J Neuropsychiatry Clin Neurosci* 2000;12:226–232.

121. Newburn G, Edwards R, Thomas H, et al. Moclobemide in the treatment of major depressive disorder (DSM-3) following traumatic brain injury. *Brain Inj* 1999;13:637–642.

122. Gualtieri CT. Pharmacotherapy and the neurobehavioral sequelae of traumatic brain injury. *Brain Inj* 1988;2:101–129.

123. Zasler ND. Advances in neuropharmacological rehabilitation for brain dysfunction. *Brain Inj* 1992;6:1–14.

124. Cummings JL, Mendez MF. Secondary mania with focal cerebrovascular lesions. *Am J Psychiatry* 1984;141:1084–1087.

125. Robinson RG, Boston JD, Starkstein SE, et al. Comparison of mania with depression following brain injury: causal factors. *Am J Psychiatry* 1988;145:172–178.

126. Starkstein SE, Mayberg HS, Berthier ML, et al. Secondary mania: neuroradiological and metabolic findings. *Ann Neurol* 1990;27:652–659.

127. Gloor P. Role of the human limbic system in perception, memory and affect: lessons for temporal lobe epilepsy. In: Doane BK, Livingston KE, eds. *The limbic system: functional organization and clinical disorders.* New York: Raven Press, 1986.

128. Beck E. A cytoarchitectural investigation into the boundaries of cortical areas 13 and 14 in the human brain. *J Anat* 1949;83:145–147.

129. Crosby E, Humphrey T, Laner E. *Correlative anatomy of the nervous system.* New York: MacMillan, 1962.

130. Goldar JC, Outes DL. Fisiopatologia de la desinhibicion instintiva. *Acta Psiquiatr Psicol Am Lat* 1972;18:177–185.

131. Starkstein SE, Berthier PL, Lylyk A, et al. Emotional behavior after a WADA test in a patient with secondary mania. *J Neuropsychiatry Clin Neurosci* 1989;1:408–412.

132. Bakchine S, Lacomblez L, Benoit N, et al. Manic-like state after orbitofrontal and right temporoparietal injury: efficacy of clonidine. *Neurology* 1989;39:778–781.

133. Starkstein SE, Fedoroff JP, Berthier MD, et al. Manic depressive and pure manic states after brain lesions. *Biol Psychiatry* 1991;29:149–158.

134. Shukla S, Cook BL, Mukherjee S, et al. Mania following head trauma. *Am J Psychiatry* 1987;144:93–96.

135. Jorge RE, Robinson RG, Starkstein SE, et al. Secondary mania following traumatic brain injury. *Am J Psychiatry* 1993;150:916–921.

136. Teasdale G, Mendelow D. Pathophysiology of head injuries. In: Brooks N, ed. *Closed head injury: psychological, social, and family consequences.* London: Oxford University Press, 1984:4–36.

137. Stevens JR. Psychiatric consequences of temporal lobectomy for intractable seizures. *Psychol Med* 1990;20:529–545.

138. Csernansky JG, Mellentin J, Beauclair L, et al. Mesolimbic dopaminergic supersensitivity following electrical kindling of the amygdala. *Biol Psychiatry* 1988;23:285–294.

139. Pappata S, Tran Dinh S, Baron JC, et al. Remote metabolic effects of cerebrovascular lesions: magnetic resonance and positron tomography imaging. *Neuroradiology* 1987;29:1–6.

140. Finklestein S, Benowitz LI, Baldessarini RJ, et al. Mood, vegetative disturbance, and dexamethasone suppression test after stroke. *Ann Neurol* 1982;12:463–468.

141. Sinyor D, Amato P, Kaloupek P. Post-stroke depression: relationship to functional impairment, coping strategies, and rehabilitation outcome. *Stroke* 1986;17:112–117.

142. Finset A, Goffeng L, Landro NI, Haakonsen M. Depressed mood and intra-hemispheric location of lesion in right hemisphere stroke patients. *Scand J Rehabil Med* 1989;21:1–6.

143. Schubert DSP, Taylor C, Lee S, et al. Physical consequences of depression in the stroke patient. *Gen Hosp Psychiatry* 1992;14:69–76.

144. Schwartz JA, Speed NM, Brunberg JA, et al. Depression in stroke rehabilitation. *Biol Psychiatry* 1993;33.

145. Feibel JH, Springer CJ. Depression and failure to resume social activities after stroke. *Arch Phys Med Rehabil* 1982;63:276–278.

146. Collin SJ, Tinson D, Lincoln NB. Depression after stroke. *Clin Rehabil* 1987;1:27–32.

147. Castillo CS, Schultz SK, Robinson RG. Clinical correlates of early-onset and late-onset poststroke generalized anxiety. *Am J Psychiatry* 1995;152:1174–1179.

148. Wade DT, Legh-Smith J, Hewer RA. Depressed mood after stroke, a community study of its frequency. *Br J Psychiatry* 1987;151:200–205.

149. Kotila M, Numminen H, Waltimo O, Kaste M. Depression after stroke. Results of the FINNSTROKE study. *Stroke* 1998;29:368–372.

150. Pohjasvaara T, Leppavuori A, Siira I, et al. Frequency and clinical determinants of poststroke depression. *Stroke* 1998;29:2311–2317.

151. Dennis M, O'Rourke S, Lewis S, et al. Emotional outcomes after stroke: factors associated with poor outcome. *J Neurol Neurosurg Psychiatry* 2000;68:47–52.

152. Ebrahim S, Barer D, Nouri F. Affective illness after stroke. *Br J Psychiatry* 1987;151:52–56.

153. Fedoroff JP, Starkstein SE, Parikh RM, et al. Are depressive symptoms non-specific in patients with acute stroke? *Am J Psychiatry* 1991;148:1172–1176.

154. Starkstein SE, Fedoroff JP, Price TR, et al. Anosognosia in patients with cerebrovascular lesions. A study of causative factors. *Stroke* 1992;23:1446–1453.

155. Herrmann N, Black SE, Lawrence J, et al. The Sunnybrook stroke study. A prospective study of depressive symptoms and functional outcome. *Stroke* 1998;29:618–624.

29

ANXIETY

ALEXANDER NEUMEISTER
MEENA VYTHILINGAM
OMER BONNE
JOHANNA SARA KAPLAN
DENNIS S. CHARNEY

The past decade provided us with tremendous progress in the acquisition of knowledge about the molecular, cellular, and anatomic correlates of fear and anxiety. Recent research has advanced our knowledge about the neurotransmitter systems that play roles in fear and anxiety. The anatomic circuitries where these transmitters participate in mediating and modulating fear and anxiety have also been illuminated through improvements in neurotoxic techniques, which have enhanced the selectivity of lesion analyses in experimental animal studies, and advances in neuroimaging technology, which have permitted mapping of the neurophysiologic correlates of emotions in humans. This chapter reviews preclinical and clinical data regarding the neural mechanisms underlying normal and pathologic anxiety and may help to explain the pathophysiologic mechanisms that appear to be involved in the development of anxiety disorders, such as panic disorder (PD) with and without agoraphobia, social phobia, specific phobia, posttraumatic stress disorder (PTSD), acute stress disorder, and generalized anxiety disorder (GAD). Initially, the neurobiologic foundation needed to understand the etiology of the major anxiety disorders is described. Subsequently, epidemiologic data, a basic description of the syndrome, comparisons with other psychiatric and nonpsychiatric disorders, and current etiologic theories for each disorder are reviewed.

FUNCTIONAL NEUROANATOMIC CORRELATES OF ANXIETY AND FEAR

Evidence accumulated from a large body of preclinical studies provides a basis for proposing a functional neuroanatomy of anxiety and fear. The brain structures constituting a neural circuit of anxiety or fear should have several features:

1. There must be sufficient afferent sensory input to permit assessment of the fear- or anxiety-producing nature of external or internal stimuli.

2. The neuronal interactions among the brain structures must be capable of incorporating an individual's prior experience (memory) into the cognitive appraisal of stimuli. These interactions are important in the attachment of affective significance to specific stimuli and the mobilization of adaptive behavioral responses.

3. The efferent projections from the brain structures should be able to mediate an individual's neuroendocrine, autonomic, and skeletal motor responses to threat to facilitate survival and account for the pathologic reactions that result in anxiety-related signs and symptoms.

The major afferent arms of the neural circuitry of anxiety include the exteroceptive sensory systems of the brain (auditory, visual, somatosensory), consisting of serially organized relay channels that convey directly or through multisynaptic pathways information relevant to the experience of fear or anxiety. The sensory information contained in a fear- or anxiety-inducing stimulus is transmitted from peripheral receptor cells to the dorsal thalamus (1). An exception is the olfactory system, which does not relay information through the thalamus and whose principal targets in the brain are the amygdala and entorhinal cortex (2). Visceral afferent pathways alter the function of the locus caeruleus (LC) and amygdala, either through direct connections or via the nucleus paragigantocellularis and nucleus tractus solitarius (3–5).

The thalamus relays sensory information to the primary sensory receptive areas of the cortex. In turn, these primary sensory regions project to adjacent unimodal and polymodal cortical association areas (6–8). The cortical association areas of visual, auditory, and somatosensory systems send projections to other brain structures, including the amygdala, entorhinal cortex, orbitofrontal cortex, and cingulate gyrus (9–11). The hippocampus receives convergent, integrated inputs from all sensory systems by way of projections from entorhinal cortex (12).

Thus, much of the sensory information of fear- and anxiety-inducing stimuli are first processed in the sensory cortex before transfer to subcortical structures, which are more involved in affective, behavioral, and somatic responses. It is noteworthy that the amygdala also receives sensory information directly from the thalamus. The medial geniculate nuclei of the thalamus (acoustic thalamus) sends projections to the amygdala and hypothalamus. The thalamic areas associated with the visual system also innervate the amygdala. These data support a pivotal role for the amygdala in the transmission and interpretation of fear- and anxiety-inducing sensory information because it receives afferents from thalamic and cortical exteroceptive systems and subcortical visceral afferent pathways (13). The neuronal interactions between the amygdala and cortical regions, such as the orbitofrontal cortex, enable the individual to initiate adaptive behaviors to threat based on the nature of the threat and prior experience.

The efferent pathways of the anxiety–fear circuit mediate autonomic, neuroendocrine, and skeletal-motor responses. The structures involved in these responses include the amygdala, LC, hypothalamus, periaqueductal gray, and striatum.

Many of the autonomic changes produced by anxiety- and fear-inducing stimuli are produced by the sympathetic and parasympathetic neural systems. Stimulation of the lateral hypothalamus results in sympathetic system activation—increases in blood pressure and heart rate, sweating, piloerection, and pupil dilation. Activation of the paraventricular nucleus of the hypothalamus promotes the release of various hormones and peptides. The hypothalamus integrates information that it receives from various brain structures into a coordinated pattern of sympathetic responses. The sympathetic activation and hormonal release associated with anxiety and fear are probably mediated in part by stimulation of the hypothalamus via projections from the amygdala and LC (14–16). In addition, the paragigantocellularis also plays an important role in regulating sympathetic function and may account for the parallel activation of the peripheral sympathetic system and LC.

The vagus and splanchnic nerves are major projections of the parasympathetic nervous system. Afferents to the vagus nerves include the lateral hypothalamus, paraventricular nucleus, LC, and amygdala. There are afferent connections to the splanchnic nerves from the LC (17,18). This innervation of the parasympathetic nervous system may relate to visceral symptoms associated with anxiety, such as gastrointestinal and genitourinary disturbances.

The regulatory control of skeletal muscle by the brain in response to emotions is complex. Both subtle movements involving a few muscle groups (facial muscles) and fully integrated responses requiring the entire musculoskeletal system for fight or flight may be required. Adaptive mobilization of the skeletal motor system to respond to threat probably involves pathways between the cortical association areas and the motor cortex, the cortical association areas and the striatum, and the amygdala and the striatum.

The amygdala has strong projections to most areas of the striatum, including the nucleus accumbens, olfactory tubercle, and parts of the caudate and putamen. The portion of the striatum that is innervated by the amygdala also receives efferents from the orbitofrontal cortex and the ventral tegmental area. The amygdalocortical and amygdalostriatal projections are topographically organized and occur in register. Individual areas of the amygdala, and in some cases individual amygdaloid neurons, can integrate information from the corticostriatopallidal systems. The dense innervation of the striatum and prefrontal cortex by the amygdala indicates that the amygdala can powerfully regulate both of these systems (19,20). These interactions between the amygdala and the extrapyramidal motor system may be very important for generating motor responses to threatening stimuli, especially those related to prior adverse experiences.

Effects of Prior Experience

Memories and previously learned behaviors influence the responses to anxiety- and fear-inducing stimuli via such neural mechanisms as fear conditioning, extinction, and sensitization (a discussion follows). Although within the medial temporal lobe memory system, emotional responsiveness (amygdala) and memory (hippocampus) may be separately organized, there is considerable interaction between storage and recall of memory and affect. This is exemplified by the critical role of the amygdala in conditioned fear acquisition, sensitization, extinction, and the attachment of affective significance to neutral stimuli (21,22).

The hippocampus and amygdala are sites of convergent reciprocal projections from widespread unimodal and polymodal cortical association areas. It is probably through these interactions, as well as cortical–cortical connections, that memories stored in the cortex, which are continually being reinforced by ongoing experience, are intensified and develop greater coherence (23).

The hippocampal memory system is essential to short-term memory. However, long-term memory storage may be organized such that, as time passes, with subsequent additional retrieval opportunities and the acquisition of related material, the role of the hippocampus diminishes until it may no longer be necessary for memory. The repository of long-term memory may be in the same areas of the cortex where the initial sensory impressions take place (24). The shift in memory storage to the cortex may represent a shift from conscious representational memory to unconscious memory processes that indirectly affect behavior.

Therefore, once a fear- or anxiety-inducing sensory stimulus is relayed through the thalamus into neural circuits involving the cortex, hippocampus, and amygdala, relevant memory traces of posttraumatic experiences are stimulated. Most likely, the potency of the cognitive and somatic responses to the stimulus will be strongly correlated with prior experiences, owing to the strengthening of neural connections within the circuit. These functional neuroanatomic relationships can explain how a single sensory stimulus such as a sight or sound can elicit a specific memory. Moreover, if the sight or sound is associated with a particular traumatic event, a cascade of anxiety- and fear-related symptoms will ensue, probably mediated by the efferent arm of the proposed circuit.

NEUROTRANSMITTERS IMPLICATED IN ANXIETY AND FEAR

Numerous preclinical and clinical studies suggest that central monoaminergic transmitter systems and neuropeptides mediate the behavioral responses associated with anxiety and fear. The next section briefly describes the evidence implicating specific neurotransmitter and neuropeptide systems in anxiety and fear.

Noradrenergic System

Effects of Stressful and Fear-Inducing Stimuli on Biochemical Indices of Noradrenergic Function

Stressful stimuli of many types produce marked increases in brain noradrenergic function. Stress produces regional selective increases in norepinephrine (NE) turnover in brain regions identified as part of the neural circuitry of anxiety, including the LC, hypothalamus, hippocampus, amygdala, and cerebral cortex. It has recently been demonstrated that immobilization stress, footshock stress, tailpinch stress, and conditioned fear increase noradrenergic metabolism in the hypothalamus and amygdala (62–64). Stress also increases tyrosine hydroxylase levels in the LC (65). Anxiolytic agents reverse the effects of stress on noradrenergic metabolism (62,66,67). Consistent with these findings, acute cold-restraint stress results in decreased density of α_2-adrenergic receptors in the hippocampus and amygdala (68). The stress-induced decrease in NE turnover is also associated with a decrease in postsynaptic β-receptor density (69). The heightened responsiveness of the noradrenergic system to stress is consistent with the notion that the elevated sense of fear or anxiety connected to stress may be a critical factor in the neurochemical effects observed.

Locus Caeruleus Activity and Behavioral States Associated with Stress and Fear

In laboratory rats, chronic stress results in increased firing of the LC (70,71). Animals exposed to chronic inescapable shock, which is associated with learned helplessness, have an increase in responsiveness of the LC to an excitatory stimulus compared with animals exposed to escapable shock (71).

The effect of stressful and fear-inducing stimuli on LC activity has been assessed in freely moving cats (72,73).

Conditions that are behaviorally activating but not stressful, such as visual exposure to other rats or to food that is physically inaccessible, do not increase LC firing. In contrast, stressful and fear-inducing stimuli, such as loud white noise, air puff, restraint, and confrontation with a dog, produce a rapid, robust, and sustained increase in LC activity (72). Interestingly, these increases in LC function are accompanied by sympathetic activation. Generally, the greater the sympathetic activation in response to the stressor, as indicated by heart rate, the greater the correlation observed. Thus, a stimulus intensity threshold for coactivation of central and peripheral NE systems may exist.

A parallel activation of LC neurons and splanchnic sympathetic nerves is produced by noxious stimuli. The LC, like sympathetic splanchnic activity, is highly responsive to various peripheral cardiovascular events, such as alterations in blood volume or blood pressure. Internal events that must be responded to for survival, such as thermoregulatory disturbance, hypoglycemia, blood loss, an increase in P_{CO_2}, or a marked reduction in blood pressure, cause robust and long-lasting increases in LC activity (67).

There are also peripheral visceral influences on LC activity. In rats, distention of the urinary bladder, distal colon, or rectum activates LC neurons. These findings suggest that changes in autonomic or visceral function may result in specific behavioral responses via the brain LC–NE system. The LC–NE network may help to determine whether, under threat, an individual turns attention toward external sensory stimuli or internal vegetative events. The system, when functioning normally, may be important in facilitating the planning and execution of behaviors important for survival (67).

Behavioral Effects of Locus Caeruleus Stimulation

Electrical stimulation of the LC produces a series of behavioral responses similar to those observed in naturally occurring or experimentally induced fear (74). These behaviors are also elicited by administration of drugs, such as yohimbine and piperoxan, that activate the LC by blocking α_2-adrenergic autoreceptors. Drugs that decrease the function of the LC by interacting with inhibitory opiate (morphine), benzodiazepine (BZD) (diazepam), and α_2-(clonidine) receptors on the LC decrease fearful behavior and partially antagonize the effects of electrical stimulation of the LC in the monkey (75).

Altogether, the studies suggest that single or recurrent exposures to aversive stimuli or some pharmacologic agents can increase the behavioral sensitivity to subsequent stressors (76). It has been suggested that such phenomena account for clinical observations that patients with anxiety disorders report experiencing exaggerated sensitivity to psychosocial stress. Neural models for the pathophysiology of anxiety disorders built on sensitization phenomena thus hold that repeated exposure to traumatic events resulting in stress comprises a risk factor for the subsequent development of anxiety disorders, particularly PTSD.

Dopaminergic System

Acute stress increases dopamine release and metabolism in a number of specific brain areas. However, the dopamine innervation of the medial prefrontal cortex (mPFC) appears to be particularly vulnerable to stress; sufficiently low-intensity stress (such as that associated with conditioned fear) or brief exposure to stress increases dopamine release and metabolism in the prefrontal cortex in the absence of overt changes in other mesotelencephalic dopamine regions (77,78). Low-intensity electric footshock increases *in vivo* tyrosine hydroxylase and dopamine turnover in the mPFC but not the nucleus accumbens or striatum (79–81). BZD anxiolytics prevent selective increases in dopamine utilization in mPFC after mild stress (81). Anxiogenic BZD inverse agonists exert an opposite effect (81,82). Selective activation of mPFC dopamine neurons can also be induced by intracerebroventricular injection of corticotropin releasing factor, which has anxiogenic properties (80).

Stress can enhance dopamine release and metabolism in other areas receiving dopamine innervation, provided greater intensity or longer duration stress is used (77,83). Thus, the mPFC dopamine innervation is preferentially activated by stress compared with mesolimbic and nigrostriatal systems, and the mesolimbic dopamine innervation appears to be more sensitive to stress than the striatal dopamine innervation (77,84,85).

Serotonergic Function

The effects of stress on serotonin (5-HT) systems have been studied less thoroughly than noradrenergic and dopamine systems, and the data are somewhat contradictory. Animals exposed to a variety of stressors including footshock, tailshock, tailpinch, and restraint stress have all been shown to produce an increase in 5-HT turnover in the mPFC (80, 86–91).

Conversely, inescapable stress paradigms producing "learned helplessness" behavioral deficits have been associated with reduced *in vivo* release of 5-HT in the cerebral cortex (92,93). 5-HT antagonists produce behavioral deficits resembling those seen after inescapable shock (94). Drugs that enhance 5-HT neurotransmission [selective 5-HT reuptake inhibitors (SSRI)] are effective in reversing learned helplessness (95).

The effect of stress to active 5-HT turnover may stimulate a system that has both anxiogenic and anxiolytic pathways in the forebrain (96,97). A primary distinction in the qualitative effects of 5-HT may be between the dorsal and median raphe nuclei—the two midbrain nuclei that produce most of the forebrain 5-HT. The serotonergic innervation of the amygdala and hippocampus by the dorsal raphe is believed to mediate anxiogenic effects via $5HT_{2A}$ receptors. In contrast, the median raphe innervation of hippocampal $5HT_{1A}$ receptors has been hypothesized to facilitate the disconnection of previously learned associations with aversive events or to suppress the formation of new associations, thus providing a resilience to aversive events (96,98). Chronic stress increases cortical $5HT_2$ receptors (99–101) and reduces hippocampal $5HT_{1A}$ receptors (101–106). Potentially compatible with this hypothesis, $5HT_{1A}$ receptor knockout mice exhibit behaviors consistent with increased fear and anxiety, and administration of $5HT_{1A}$ receptor partial agonists exert anxiolytic effects in GAD (107).

Benzodiazepine Receptor Systems

Several lines of evidence support the hypothesis that BZD receptor function is altered in anxiety disorders. Central BZD receptors are present throughout the brain but are most densely concentrated in the gray matter. The BZD and γ-aminobutyric acid A ($GABA_A$) receptors form parts of the same macromolecular complex, and although they constitute distinct binding sites, they are functionally coupled and regulate each other in an allosteric manner (108). Central BZD receptor agonists have been shown to prolong and potentiate the synaptic actions of GABA by increasing the frequency of GABA-mediated chloride channel openings (108).

Microinjection of BZD receptor agonists in limbic and brainstem regions, e.g., the amygdala, exert antianxiety effects in animal models of anxiety and fear (109). In contrast, administration of BZD receptor inverse agonists, e.g., β-carboline-3-carboxylic acid ester, induces behaviors and increases in heart rate, blood pressure, plasma cortisol, and catecholamines similar to those seen in anxiety and stress (110). These effects have been shown to be effectively blocked by administration of BZD receptor agonists (111). Animal data suggest that the anxiolytic effects of BZD receptor agonists are mediated at least in part by the $GABA_A$ receptor α_2 subunit, which is largely expressed in the limbic system, but not in the α_1 subunit, which is implicated in mediating the sedative, amnestic, and anticonvulsive effects of BZD (112,113), or the α_3 subunit, which is predominantly expressed in the reticular activating system (112,114).

Other agents with anxiolytic effects appear to modulate the function of the $GABA_A$/BZD receptor–chloride ionophore complex via mechanisms distinct from those of BZD agonists. The neurosteroid allopregnanolone has been shown to exert antianxiety effects in conflict paradigms that serve as a model of anxiety. However, allopregnanolone seems not to bind at the BZD site but appears to stimulate chloride channels in $GABA_A$ receptors by binding at the picrotoxinin site on the $GABA_A$ receptor or at a site specific for RO15-4513 (a BZD receptor inverse agonist that inhibits $GABA_A$-activated chloride flux in neuronal membranes).

The anxiolytic effects of antidepressants with their primary mode of action on monoamine uptake may also include, at least in part, GABA-ergic mechanisms. These agents have been shown to be effective for different diagnostic subtypes of anxiety disorders, such as GAD, social anxiety disorder, PD, and PTSD. Their mechanisms of action include a potentiation of GABA-ergic function (115) mediated by an increase of brain GABA concentrations, in contrast to classic BZD, which instead increases the affinity of $GABA_A$ receptors for GABA.

The animal model of inescapable stress has been used as a model for human anxiety disorders and for the development of medications for these disorders. Animals exposed to inescapable stress during a period between 7 days and several months developed a number of behaviors that were noted to parallel symptoms of anxiety disorders in humans. It was shown that these animals develop a 20% to 30% decrease in BZD receptor binding in the frontal (116,117) and cerebral cortex (118–121) and with some studies showing reductions in the hippocampus (118,119,121). However, it remains unclear how fast these changes begin after exposure to stress. One study that used a different model of stress (defeat stress) showed an increase in BZD receptor binding in the cortex (122). A decrease of BZD receptor density was found in the Maudsley genetically fearful rat strain in several brain structures including the hippocampus (123), consistent with the effects of chronic stress on BZD receptors. Also, administration of corticosterone for 10 days has been shown to induce complex changes in the mRNA levels of multiple $GABA_A$ receptor subunits (dentate gyrus: α_1, α_2, and β_1 mRNA decrease, whereas β_2 and β_3 subunits increased; cingulate gyrus: α_1, β_1, and β_2 levels increased) (124).

Neuropeptides Implicated in Anxiety and Fear

Hypothalamic–Pituitary–Adrenal Axis and Corticotropin Releasing Hormone

Corticosteroid hormones (mainly cortisol in humans and corticosterone in rodents) are lipophilic molecules secreted by the adrenal gland after a circadian pattern and in response to stress. Hypothalamic–pituitary–adrenal (HPA) axis function is regulated via feedback actions of glucocorticoids at the level of the pituitary, hypothalamus, and suprahypothalamic limbic compartments such as the hippocampus, mPFC, and amygdala, structures also involved in the regulation of mood and cognition (125). Hormonal signals are transduced via heterogeneous corticosteroid nuclear receptors (known as type 1) or mineralocorticoid receptors and type 2 or glucocorticoid receptors as homo- or heterodimers. Negative or stressful events and social isolation increase HPA axis activity, but individuals vary considerably in their hormonal response to events (126). The hippocampus, amygdala, and mPFC are plastic limbic structures linked anatomically and are targets and modulators of adrenal steroids. Neurotoxic damage to the hippocampus and suppression of ongoing neurogenesis has been found to be owing to a direct action of glucocorticoids (126–128). Other mechanisms include calcium influx, N-methyl-D-aspartate subunit expression, and increased synaptic glutamate levels (130,131).

Exposure to various types of stress results in a release of corticotropin releasing hormone (CRH), adrenocorticotropic hormone (ACTH), and cortisol. This HPA axis activation during episodes of acute stress can result in transient elevation of the plasma concentration of cortisol and partial resistance to feedback inhibition of cortisol release that persists during and shortly after the duration of the stressful stimulus. This phenomenon may result in a rapid down-regulation of glucocorticoid receptors (132), resulting in increased corticosterone secretion and feedback resistance. After stress termination, glucocorticoid receptor density increases and feedback sensitivity normalizes.

Glucocorticoid receptors are present on both neurons and glia and in almost every region of the brain but are in particularly high density in areas related to stress responses such as the hypothalamus, serotonergic and noradrenergic cell bodies, and hippocampus. Mineralocorticoid expression is mainly in the limbic brain regions, such as the hippocampus, septum, and amygdala in rodents (133,134). Although the anatomic distribution of glucocorticoid receptors is well described in rodents, less is known about either mineralocorticoid or glucocorticoid distribution in the human/primate brain. Patel et al. (135) reported that mineralocorticoid and glucocorticoid receptors in the nonhuman primate hippocampus and amygdala were expressed in proportions expected from rodent studies. Glucocorticoid mRNA levels have been reported in the hippocampal formation of humans (136–139). Brooke et al. (140) confirmed the presence of glucocorticoid binding in the monkey hippocampus; however, it was less abundant compared with the cortex, cerebellum, or hypothalamus. Patel et al. (135) also found glucocorticoid receptor mRNA in all prefrontal cortical regions. A recent study questioned the presence of glucocorticoid receptors in the hippocampal subfields excluding the dentate gyrus and subiculum (125). A similar reduction in the glucocorticoid receptors was also reported in the amygdala, whereas the distribution in the prefrontal, pituitary, and entorhinal cortices were abundant and similar to the rodent data (125).

Some stressors experienced within critical periods of neurodevelopment exert long-term effects on HPA axis function. In rats exposed to either severe prenatal (*in utero*) stress or early maternal deprivation stress (141,142), the plasma concentrations of corticosterone achieved in response to subsequent stressors are increased, and this tendency to show exaggerated glucocorticoid responses to stress persists into adulthood. Early postnatal adverse experiences, such as maternal separation, are associated with long-lasting alterations in the basal concentrations of hypothalamic CRH mRNA, hippocampal glucocorticoid receptor mRNA, and median eminence CRH and in the magnitude of stress-induced CRH, corticosterone, and ACTH release (143–145). In nonhuman primates, adverse early experiences induced by variable maternal foraging requirements reportedly result in alterations in juvenile and adult social behavior, such that animals are more timid, less socially interactive, and more subordinate (146). Adult monkeys who were raised in such a maternal environment are also hyperresponsive to yohimbine and have elevated CRH concentrations and decreased cortisol levels in the cerebrospinal fluid, findings that parallel those in humans with PTSD (146).

Conversely, positive early life experiences during critical developmental periods may have beneficial long-term consequences on the ability to mount adaptive responses to stress or threat. For example, daily postnatal handling of rat pups by human experimenters within the first few weeks of life has been shown to produce persistent (throughout life) increases in the density of type 2 glucocorticoid receptors. This increase was associated with enhanced feedback sensitivity to glucocorticoid exposure and reduced glucocorticoid-mediated hippocampal damage in later life (147,148). These effects are hypothesized to comprise a type of a stress inoculation induced by the mothers repeatedly licking the pups after they were handled by humans. Taken together with the data reviewed in the preceding paragraph, these data indicate that a high degree of plasticity exists in stress-responsive neural systems during the prenatal and early postnatal periods that programs future biologic responses to stressful stimuli (143).

Another level through which the CRH–glucocorticoid system maintains homeostasis and provides mechanisms for modulating stress or anxiety responses involves functional differences between CRH receptor subtypes. The CRH_1 and CRH_2 receptors appear to play reciprocal roles in mediating stress responsiveness and anxietylike behaviors (149). Mice genetically deficient in CRH_1 receptor expression exhibit diminished anxiety and stress responses to threat or stress (150,151). In contrast, mice deficient in CRH_2 receptors display heightened anxiety in response to stress (152,153). The affinity of CRH is higher for CRH_1 than CRH_2 receptors, consistent with evidence that CRH elicits anxiogenic effects either when exogenously administered to native animals or when endogenously released in mice genetically altered to overexpress CRH (149). Also consistent with the hypothesis that CRH_1 receptor stimulation facilitates anxiety responses, oral administration of the CRH_1 receptor antagonist antalarmin inhibits the behavioral, sympathetic autonomic, and neuroendocrine responses (i.e., attenuating increases in the cerebrospinal fluid CRH concentration and the pituitary—adrenal and adrenal–medullary activity) to acute social stress in monkeys (154).

Regional differences in the anatomic distribution of CRH_1 and CRH_2 receptors likely play a role in balancing facilitatory versus modulatory effects of CRH receptor stimulation on stress responses. In monkeys, the CRH_1 receptor density is high in most amygdaloid nuclei, cingulate cortex, the remainder of the PFC (medial > dorsolateral > orbital), insular cortex, parietal cortex, dentate gyrus, and entorhinal cortex and moderate in the CeA and LC. The CRH_2 receptor

density is high in the cingulate cortex, mPFC, CeA, CA-1 region of the hippocampus, and paraventricular nuclei and supraoptic nucleus of the hypothalamus. An important avenue of future research will involve assessments of the homeostatic balance between CRH_1 and CRH_2 receptor systems in anxiety disorders.

Cholecystokinin

Recently, preclinical investigations support the involvement of the peptide cholecystokinin (CCK) in anxiety and fear. CCK, an octapeptide originally discovered in the gastrointestinal tract, has been found to be present in high concentrations in the cerebral cortex, amygdala, and hippocampus in mammalian brain (155). CCK has anxiogenic effects in laboratory animals and appears to act as a neurotransmitter or neuromodulator in the brain. Interactions with BZD receptors are of particular interest. BZD anxiolytics antagonize the excitatory neuronal and anxiogenic effects of CCK, and CCK_β antagonists have anxiolytic actions in animal models (156). Studies in healthy human subjects have demonstrated that CCK4 induces severe anxiety or short-lived panic attacks. This effect is reduced by lorazepam (157).

Neuropeptide Y

Low doses of neuropeptide Y administered intraventricularly have anxiolytic effects in several animal models of anxiety (158,159). These actions may be mediated by neuropeptide Y in the amygdala. Administration of an antisense oligonucleotide targeted at the neuropeptide Y_1 receptor message, which led to a 60% decrease in the B_{max} of Y_1 receptors, was accompanied by marked anxiogeniclike effects in the elevated plus-maze, a test in which neuropeptide Y itself is markedly anxiolytic (160). This suggests that endogenous neuropeptide Y could be anxiolytic by activating Y_1 receptors and that disturbed neuropeptide Y transmission might have a role in the symptoms of anxiety (161).

ANXIETY DISORDERS: DESCRIPTION AND PATHOPHYSIOLOGY

Panic Disorder and Agoraphobia

PD is an anxiety disorder characterized by recurrent unexpected panic attacks, with at least one of the attacks being followed by one or more of the following symptoms: (a) persistent concern about having additional attacks, (b) worry about the implication of the attack or its consequences, and (c) a significant change in behavior related to the attacks. Usually the condition is accompanied by agoraphobia, which consists of excessive fear (and often avoidance) of situations, such as driving, crowded places, stores, or being alone, in which escape or obtaining help would be difficult. Current *Diagnostic and Statistical Manual of Mental Disorders, Fourth Edition* (*DSM-IV*) classifications include "panic disorder without agoraphobia," "panic disorder with agoraphobia," and "agoraphobia without history of panic disorder." The life-time and 12-month prevalence rates of panic and depression have been reported in the National Comorbidity Survey and include, respectively, 7.2% and 4.2% for panic attack, 3.4% and 2.2% for PD, and 16.9 and 10% for depression (162). Age of onset is typically in the late teens to early 30's and is unusual after the age of 40 years. The majority (78%) of patients describe the initial attack as spontaneous (occurring without an environmental trigger). In the remainder, the first attack is precipitated by confrontation with a phobic stimulus or use of a psychoactive drug. Onset of the disorder often follows within 6 months of a major stressful life event, such as marital separation, occupational change, or pregnancy. Patients with PD have a high rate of comorbid major depression, with a 12-month comorbidity of 80.2% and a lifetime comorbidity of 55.6% (163). PD has been shown to be heterogeneous in its pattern of onset and recovery (164) with a subset of patients with comorbid depression having a chronic course with high rates of relapse (165). Furthermore, comorbidity significantly contributes to psychological distress, interpersonal impairment, and overall impairment (166). The substantial cost on society as a result of untreated PD may be alleviated by more widespread awareness, recognition, and appropriate early treatment (167).

Description and Differential Diagnosis

PD usually begins with a spontaneous panic attack that often leads the individual to seek medical treatment, such as presenting to an emergency department believing that he/she is having a heart attack, stroke, losing his/her mind, or experiencing some other serious medical event. Some time may pass before another attack, or the patient may continue to have frequent attacks. Patients may feel constantly fearful and anxious after the first attack, wondering what is wrong and fearing that it will happen again. Some patients experience nocturnal attacks that awaken them from sleep. Usually, patients gradually become fearful of situations (a) that they associate with the attacks; (b) in which they would be unable to flee if the attack occurred; (c) in which help would not be readily available; or (d) in which they would be embarrassed if others noticed that they are experiencing an attack (although attacks are not usually evident to others). Less frequently, a history of phobia may precede the first panic attack. Before patients are educated about the symptoms of the disorder, they believe that they have a serious medical condition. They are often embarrassed about their symptoms and try to hide them from others, often making excuses not to attend functions or enter phobic situations.

The differential diagnosis of PD and agoraphobia includes secondary anxiety disorder owing to medical conditions; substances such as caffeine, cocaine, or amphetamines; withdrawal from alcohol, sedative-hypnotics, or BZD; and other phobic conditions, GAD, and psychosis. Medical illness that can produce symptoms similar to those of panic attacks must be excluded. Endocrine disturbances, such as pheochromocytoma, hyperthyroidism, or hypoglycemia, can produce similar symptoms and can be excluded with appropriate clinical history and laboratory evaluations.

When gastrointestinal symptoms of attacks are prominent, the diagnosis of colitis may need to be excluded. Symptoms of tachycardia, palpitations, chest pain or pressure, and dyspnea may be confused with cardiac or respiratory conditions. Lightheadedness, faintness, dizziness, derealization, shaking numbness, and tingling may suggest a neurologic condition. The association of mitral valve prolapse with PD is controversial. The presence of mitral valve prolapse in PD patients does not appear to alter treatment response or course, so the diagnosis of PD should be made independently of mitral valve prolapse.

PD differs from GAD in that the former is distinguished by recurrent discrete, intense episodes of panic symptoms, although in both disorders anticipatory anxiety and generalized feelings of anxiety may be present. Although some of the same situations may be feared, agoraphobia differs from social and simple phobias in that the fear is related to feeling trapped or being unable to escape and that the fears often become generalized. Agoraphobics may additionally have a history of other phobias.

PD is frequently associated with major depression, other anxiety disorders, and alcohol and substance abuse. In clinical samples, as many as two-thirds of panic patients report experiencing a major depressive episode at some time in their lives. Similarly, studies of patients seeking treatment for major depression report high rates of panic in these patients and their relatives. Once symptoms begin, patients often describe becoming demoralized as a result of fear related to the symptoms and their imagined causes and because of impairment when their activities are restricted by their agoraphobia. Unlike depressed patients, PD patients usually lack vegetative symptoms and have a normal desire to engage in activities but avoid them because of their phobias.

Attempts to self-medicate the intolerable anxiety may increase the risk of alcoholism and substance abuse. Patients may require a drink before entering phobic situations. Approximately 20% of patients report a history of alcohol abuse, but the onset of alcoholism precedes the first attack in almost all patients (168). Alcoholism may also alter the course of the disorder. Preliminary data suggest that PD precipitated by cocaine use may be less likely to respond well to the usual pharmacologic treatments and may have less association with a family history of PD.

Genetic Studies in Panic Disorder

Genetic epidemiologic studies have consistently demonstrated increased rates of PD among first- and second-degree relatives of PD probands. Family studies suggest that first-degree relatives of patients with PD have a 2.6- to 20-fold relative risk of developing the disorder (169). The data published by Kendler et al. (170) imply that heritable factors contribute approximately 30% to 40% of the diathesis toward PD. These genetic factors have been the focus of recent work employing genomic (171) and candidate gene (172–177) strategies. Comparison of concordance rates in monozygotic versus dizygotic twins is used to differentiate between genetic and environmental factors in families because monozygotic twins share 100% of their genetic material and dizygotic twins on average share half. Torgersen (178) found concordance for anxiety disorder with panic attacks in four of 13 monozygotic and zero of 16 dizygotic twin pairs. A recent genetic linkage study provides strong evidence for a susceptibility locus for PD either within the catechol-*O*-methyltransferase gene or in a nearby region of chromosome 22 (179).

Pathophysiology

Studies of Noradrenergic Function in Panic Disorder

Among all the anxiety disorders, the evidence of an abnormality in noradrenergic function is most compelling for PD and PTSD. PD and PTSD patients frequently report cardiovascular, gastrointestinal, and respiratory symptoms. Because the LC is responsive to peripheral alterations in the function of these systems, minor physiologic changes in these patients may result in abnormal activation of LC neurons and, consequently, panic attacks and flashbacks. These functional interactions may explain the association of anxiety symptoms with tachycardia, tachypnea, hypoglycemia, and visceral and organ distention, as well as the marked sensitivity of PD and PTSD patients to interoceptive stimuli. The important role of the noradrenergic system in fear conditioning may account for the development of phobic symptoms in these patients. Finally, the involvement of noradrenergic neurons in learning and memory may relate to the persistence of traumatic memories in both disorders.

Peripheral Catecholamine Levels

Generally, measurement of peripheral NE and its metabolites have revealed concentrations in PD patients similar to those in controls (180–185).

Regulation of Noradrenergic Function in Panic Disorder

The regulation of noradrenergic neuronal function has been examined by determining the behavioral, biochemical, and cardiovascular effects of oral and intravenous yohimbine, an α_2-adrenergic receptor antagonist, in a spectrum of psychiatric disorders, including schizophrenia, major depres-

sion, obsessive-compulsive disorder, GAD, PD, and PTSD (186). Specific abnormalities have been identified in PD and PTSD. Approximately 60% to 70% of PD patients experience yohimbine-induced panic attacks. These patients have larger yohimbine-induced increases in plasma 3-methoxy-4-hydroxy-phenylglycol (MHPG), blood pressure, and heart rate compared with healthy subjects and other psychiatric disorders (187–189). The effects of yohimbine on regional cerebral blood flow and metabolism have been evaluated in PD and PTSD patients. In PD patients, yohimbine significantly reduced frontal regional cerebral blood flow rates in patients compared with healthy subjects (190).

A consistent finding in the literature is that the growth hormone rise induced by clonidine is blunted in PD patients (191). In a recent investigation, a blunted growth hormone response was found primarily in the patients who experienced yohimbine-induced panic attacks (189). This suggests that the diminished postsynaptic α_2-adrenergic receptor function reflected by the blunted clonidine–growth hormone response may relate to presynaptic noradrenergic neuronal hyperactivity.

Several previous investigations observed that clonidine produced greater decreases in plasma MHPG and blood pressure in PD patients compared with healthy subjects (189,192). The clonidine-induced decreases in plasma MHPG may be greatest in the PD patients who experienced yohimbine-induced panic attacks (189), suggesting that there is a distinct subgroup of PD patients who manifest noradrenergic neuronal dysfunction.

β-Adrenergic Receptor Function

Infusion of isoproterenol, a peripherally acting compound that is selective for the β-adrenoceptor, has been reported to trigger anxiety responses in patients with PD compared with controls (193). Successful treatment of PD patients with tricyclic antidepressants blunted isoproterenol-induced anxiety and systolic blood pressure responses (194). These studies are consistent with the hypothesis of increased β_1-adrenoceptor sensitivity in PD, which is normalized by effective pharmacotherapy (194).

If the noradrenergic system is dysregulated in PD, the mechanism of action of antipanic therapy may be its ability to decrease the wide and unpredictable fluctuations in noradrenergic activity. Likewise, it can improve efficiency by reducing basal activity (decreasing noise) while effecting a more specific responsiveness to specific stimuli (increasing signal-to-noise ratio).

Dopamine Function in Panic Disorder

The potential role of dopamine function in PD is unclear, but, thus far, there is little evidence that dopaminergic dysfunction plays a primary role in the pathophysiology of human anxiety disorders. It has been shown that homovanillic acid levels are higher in symptomatic patients with PD relative to healthy controls (195). However, Ericksson et al. (196) found no evidence for alterations in the cerebrospinal fluid homovanillic acid levels in PD patients. Genetic studies exploring associations between PD and gene polymorphisms revealed a susceptibility locus for PD either within the catechol-*O*-methyltransferase gene or in a nearby region of chromosome 22 (179). In contrast, no association between PD and gene polymorphisms was found for the dopamine D_4 receptor and the dopamine transporter (175).

Serotonin Function in Panic Disorder

Peripheral Serotonin Function
Platelet imipramine binding (a marker of the 5-HT reuptake site), which is generally reduced in depression, has been found to be normal in PD (197,198), whereas platelet 5-HT uptake in PD has been reported to be elevated (199), normal (200), or reduced (201). One study found that PD patients had lower levels of circulating 5-HT compared with controls (202). Thus, no clear pattern of abnormality in 5-HT function in PD has emerged from analysis of peripheral blood elements.

$5HT_{1A}$ Receptor Function in Panic Disorder
As shown earlier in preclinical studies, there is substantial evidence available that $5HT_{1A}$ receptors are involved in the development of symptoms of anxiety and that exposure to stress plays a critical role. Potentially compatible with this hypothesis, $5HT_{1A}$ receptor knockout mice exhibit behaviors consistent with increased anxiety and fear (107). In humans with PD, pharmacologic challenge studies exploring the responsivity of the $5HT_{1A}$ receptor assessed hypothermic, neuroendocrine, and behavioral responses to the $5HT_{1A}$ partial agonist ipsapirone. Such studies showed significantly attenuated ACTH/cortisol and thermoregulatory responses to ipsapirone in PD subjects relative to healthy controls (203–205). Finally, $5HT_{1A}$ partial agonists have been shown to be effective in the treatment of PD patients (206).

Possible Role of Cortisol Hypersecretion and $5HT_{1A}$ Receptor Abnormalities

One factor that may contribute to the reduction in $5HT_{1A}$ receptor binding in anxiety disorders is increased cortisol secretion because postsynaptic $5HT_{1A}$ receptor mRNA expression is under tonic inhibition by corticosteroid receptor stimulation in some brain regions (207,208). The magnitude of the reduction in $5HT_{1A}$ receptor density and mRNA levels induced by stress-induced glucocorticoid secretion in rodents is similar to that of the differences between depressed and healthy humans (208). For example, in rats, chronic, unpredictable stress reduced $5HT_{1A}$ receptor density an average of 22% across hippocampal subfields (208), similar to the 25% reduction in hippocampal $5HT_{1A}$ receptor binding potential (BP) documented in depression.

Similarly, in tree shrews, chronic social subordination stress (for 28 days) decreased the density of $5HT_{1A}$ receptors in the posterior cingulate, parietal cortex, prefrontal cortex, and hippocampus (by 11%–34%) (209), similar to the magnitude of reduced $5HT_{1A}$ receptor BP found by Sargent et al. (210) and Drevets et al. (211) in these regions. This response is mediated primarily by stimulation of mineralocorticoid receptors (208), which are concentrated in the hippocampus in rats but extend throughout the cortex in primates (Juan Lopez, 2001, *personal communication*). Thus, elevated cortisol concentrations may conceivably induce a relatively widespread reduction of $5HT_{1A}$ expression in humans (208).

Hypothalamic–Pituitary–Adrenal Axis Function and Corticotropin Releasing Hormone Release in Panic Disorder

In PD, the results of studies examining CRH and HPA axis function have been inconsistent. One study (212), but not another (213), reported elevated cortisol levels. In addition, results of studies assessing urinary free cortisol have been inconsistent (214,215). One study reported elevations of nocturnal ACTH and cortisol secretion and greater amplitude of ultradian secretory episodes in PD patients relative to healthy controls (216). Both normal and elevated rates of cortisol nonsuppression after dexamethasone (DEX) administration have been reported in PD (217). After DEX/CRH challenge, the HPA axis response was higher in PD subjects than in controls (218,219). The ACTH response to CRH was blunted in some studies (213,218), but not others (219), although cerebrospinal fluid levels of CRH did not differ between PD patients and controls (220).

Benzodiazepine Receptor Function in Panic Disorder

BZD are present throughout the brain (221), with the highest concentration in cortical gray matter (222). BZD agonists have anxiolytic effects, whereas inverse agonists have anxiogenic properties. The major behavioral effects of BZD agonists and antagonists are mediated through saturable high-affinity BZD receptor sites located on a subunit of the $GABA_A$ receptor (112,113). Some anxiety disorder subjects show reduced sensitivity to BZD agonists (223,224). Hypotheses regarding the role of $GABA_A$/BZD receptor function in anxiety disorders have proposed either changes in the $GABA_A$/BZD macromolecular complex or that alterations in the concentration account for the pathologic anxiety seen in anxiety disorder patients. It must be acknowledged that these hypotheses have so far not been tested at all by *in vivo* or postmortem studies of anxiety disorder patients.

Administration of the BZD receptor antagonist flumazenil to patients with PD results in an increase in panic attacks and subjective anxiety compared with controls (224,225). Both oral (225) and intravenous (224) flumazenil have been shown to produce panic in a subgroup of PD patients but not in healthy subjects. The BZD receptor inverse agonist FG7142 induces severe anxiety resembling panic attacks and biologic characteristics of anxiety in healthy subjects (226). This observation raises a question regarding the existence of endogenous equivalents of FG7142 that might be released to provoke panic attacks. One candidate for such an endogenous ligand is tribulin, a substance with BZD receptor binding and monoamine oxidase–inhibiting properties, found in human urine (227). Increased levels of tribulin have been found in patients with anxiety and after lactate-induced panic attacks (228).

The BZD receptors are highly concentrated in gray matter structures, including cortical areas, such as the temporal, parietal, and frontal cortex, and subcortical areas, such as the hippocampus. Patients with PD may have abnormalities of the right temporal lobe, with some studies suggesting atrophy and areas of increased signal activity detectable by magnetic resonance imaging (MRI) (229,230).

Receptor imaging studies using single photo emission computed tomography (SPECT) have assessed BZD receptor binding in patients with PD. SPECT studies have reported reduced uptake of the selective BZD receptor ligand [^{123}I]iomazenil in the frontal (231–233), temporal (231,232), and occipital (231) cortex of PD patients relative to controls. However, interpretation of data is limited owing to methodologic problems, such as small sample sizes, absence of medication-free patients, lack of suitable control groups, and their dependence on nonquantitative methods for estimating BZD binding. A SPECT iomazenil study that quantitated BZD receptor binding by derivation of distribution volumes found a decrease in left hippocampal and precuneus in PD patients (234). Conclusively, these studies support hypotheses that stress down-regulates BZD receptor binding and that this may be an important pathway in the pathogenesis of PD.

Two studies assessed central BZD receptor binding in PD using positron emission tomography (PET) and [^{11}C]flumazenil (235). In a small study including seven patients and healthy controls, Malizia et al. (235) reported a global reduction in BZD site binding throughout the brain, with the most prominent reductions in the right orbitofrontal cortex and right insula (regions that are consistently activated during normal anxiety processing) in PD patients relative to healthy controls. However, in a second study using a similar study design, those findings were not replicated. Abadie et al. (236) found no differences in the B_{max}, K_d, or bound/free values for [^{11}C]flumazenil in any brain region in ten unmedicated PD patients relative to healthy controls.

Cholecystokinin Hypothesis of Panic Disorder

Preclinical studies have shown that CCK agonists are anxiogenic in laboratory animals. DeMontigny (157) demonstrated that in healthy volunteers intravenous administration of CCK4 (a tetrapeptide that crosses the blood–brain barrier more readily than CCK8) can induce severe anxiety or short-lived panic attacks. The anxiogenic effect of CCK was blocked by the BZD lorazepam, although this may merely be pharmacologic opposition and not true antagonism.

Recently, several investigations showed that PD patients are more sensitive to the anxiogenic effects of CCK4 and a closely related peptide, pentagastrin, and that these effects are blocked by CCK antagonists. The mechanism responsible for the enhanced sensitivity to CCK4 has not been elucidated. Patients may have an elevated production or turnover of CCK or increased sensitivity of CCK receptors. Because CCK has important functional interactions with other systems implicated in anxiety and fear (noradrenergic, dopaminergic, BZD), these interactions need to be evaluated in PD patients. CCK_{β} antagonists are now being tested as antipanic drugs (237–239).

Pharmacologic Treatment of Patients with Panic Disorder

Antidepressants and BZD are mainstays in the pharmacologic treatment of patients with PD. Among the antidepressants, both tricyclic antidepressants and SSRI are effective in the treatment of PD (240). Although tricyclic antidepressants are effective in the treatment of PD, their unfavorable side effect profile limits their extensive clinical use (241). Antidepressants such as SSRI have not only been shown to be effective in the treatment of patients with PD but have a better side effect profile compared with the tricyclic antidepressants (242,243). Fluvoxamine (244,245), sertraline (246,247), citalopram (248), fluoxetine (249), and paroxetine (250,251) have been shown to be effective in placebo-controlled trials in patients with PD. A metaanalysis suggests that SSRI may be even more effective than tricyclic antidepressants or BZD for the disorder (252). For paroxetine, a minimal dose demonstrated to be significantly superior to placebo was 40 mg per day, although some patients did respond at lower doses (250).

Although SSRI have proven efficacy in the treatment of PD, there is delay of as long as 4 weeks for the onset of antipanic effects. Furthermore, using stringent response criteria, defined as 50% reduction in the number of panic attacks, there is a substantial number of patients with PD who do not show full response to the SSRI alone (250). Moreover, tricyclic antidepressants and SSRI have been shown to induce anxietylike symptoms in the early treatment phase of the disorder.

There is also evidence available from the literature and clinical practice that BZD, administered either alone (253) or in combination with antidepressants (246), are effective in the acute treatment of PD. Clonazepam, a BZD with a long half-life, has been shown to be an effective and safe, short-term treatment for PD (253). However, the potential of BZD to cause abuse and dependence and rebound panic attacks limits their use as a long-term treatment option.

Combining SSRI or other classes of antidepressants with BZD during the course of antipanic treatment appears to be a natural psychopharmacologic approach. This strategy was designed to take advantage of the rapid anxiolytic effects of the BZD before the delayed onset of SSRI treatment, thereby accelerating and possibly enhancing overall treatment effects. However, there is only a very small number of placebo-controlled treatment studies supporting such an approach in PD. One group showed a rapid onset of antipanic efficacy of a combination of tricyclic antidepressant, imipramine, and a short-acting BZD, alprazolam, in the treatment of PD (254). However, there was a relapse of panic symptoms after tapering the alprazolam. A recent randomized, double-blind study showed the superior efficacy of an SSRI, sertraline, and a long-acting BZD, clonazepam, over the SSRI alone (255). Given the high comorbidity of PD with major depression, there is a need to extend the findings from treatment studies evaluating the combination of an SSRI and a long-acting BZD in patients with PD to patients with PD and comorbid major depression.

POSTTRAUMATIC STRESS DISORDER

PTSD has been classified as an anxiety disorder in the *DSM-IV*. It is characterized by three subsets of symptoms:

1. Intrusive symptoms: flashbacks, nightmares, intrusive recollections of the associated traumatic event;
2. Avoidance symptoms: social withdrawal, emotional numbing, avoiding thoughts, feelings, activities, places, or people associated with the trauma;
3. Hyperarousal symptoms: sleep disturbance, outbursts of anger, and exaggerated startle response.

PTSD can either be acute (duration of symptoms is less than 3 months) or chronic (duration of symptoms of 3 months or longer) and can also have a delayed onset (at least 6 months between the traumatic event and the onset of symptoms).

Epidemiologic studies in the United States have estimated that the lifetime prevalence of PTSD is between 8% and 12% (256–259). Several factors contribute to the high variation in the prevalence rates, including severity, duration, frequency of trauma, and the age when traumatized. If untreated, approximately 10% to 50% of survivors of severe trauma will develop chronic PTSD that may persist for years. The majority (as many as 91%) of patients with PTSD have at least one additional diagnosis (e.g., affective disorders, anxiety disorders, major depressive disorder, alcoholism, drug abuse) (260).

PTSD is one of the most pervasively disabling of the anxiety disorders, causing significant impairment of social and occupational functioning, increased risk of suicidality, and increased burden to spouse, family, and community (261).

Pathophysiology

Neuroimaging Abnormalities in Posttraumatic Stress Disorder

Hippocampal Volumetric Studies in Posttraumatic Stress Disorder

Clinical investigation of hippocampal volume in PTSD was largely stimulated by numerous preclinical studies reporting hippocampal neuronal loss and dendritic atrophy after exposure to hydrocortisone or psychosocial stress in primates (262,263). The first report of Bremner et al. (264) of a small but significant decrease (8%) in the body of the right hippocampus in patients with combat-related PTSD was confirmed by three studies from two independent groups (265–267). The decrease in hippocampal volume was seen in patients with PTSD related to either combat or childhood sexual and/or physical abuse. We recently found a significant 18% reduction in the left hippocampal volume in patients with early childhood abuse and current major depressive disorder compared with patients with current major depressive disorder without abuse after controlling for age, race, education, whole brain volume, alcohol use, and PTSD (268).

More recent studies failed to find a reduction in hippocampal volume in holocaust victims (269), combat veterans (270), or women with PTSD secondary to domestic violence. Longitudinal MRI studies in adult subjects with PTSD immediately and 6 months after a motor vehicle accident (271) and a 2-year study of children with sexual and/or physical abuse–related PTSD (272) did not find reductions in hippocampal volume. The largest volumetric study to date of 44 abused children with PTSD and 61 healthy children found a significant decrease in intracranial volume and corpus callosal volume but not hippocampal volume (273). A *post hoc* analysis in patients with alcohol dependence showed that the volume of the hippocampus in women with alcohol dependence and PTSD was similar to women with alcohol dependence without PTSD (274).

Several factors could explain the contradictory findings in hippocampal volume studies in PTSD. Trauma variables include differences in the kinds of trauma (e.g., sexual abuse/rape, physical abuse, witnessing violence, motor vehicle accident, combat, victim of mugging), duration of trauma (repeated episodes over a period of years versus a single episode), severity of trauma, and the timing of trauma in development (prepubertal versus postpubertal). Differences in the prevalence of comorbid disorders such as major depression and alcohol and substance use could also explain the variance in hippocampal volume. Possibly, exposure to antidepressants could enhance dendritic branching (275,276) and contribute to differences in hippocampal volume in humans. Ongoing twin studies in PTSD have just begun to elucidate the contribution of genetic differences to hippocampal volumetric changes in patients with PTSD. Future studies in patients with PTSD should take into account the factors that could explain heterogeneity in biologic markers and study more homogeneous groups of traumatized patients and compare them with control subjects who are exposed to similar kinds of trauma without current PTSD and healthy subjects.

Putative pathophysiologic mechanisms for smaller hippocampal volume include increased levels of cortisol at the time of the trauma and increased levels of excitatory amino acids such as glutamate. Evaluating whether patients with PTSD have increased central glucocorticoid sensitivity may help to address whether this phenomenon is related to smaller hippocampal volume. If patients with PTSD have increased central glucocorticoid sensitivity and/or frequent episodes of stress-induced hypercortisolemia, cumulative hippocampal neuronal loss may ensue over time.

Abnormalities in the Amygdala and Prefrontal Cortex in Posttraumatic Stress Disorder

Several functional imaging studies have confirmed abnormalities in the amygdala and prefrontal cortex in patients with PTSD. Because glucocorticoid receptors are abundantly distributed in these regions, it is important to determine whether glucose utilization in the amygdala and prefrontal cortical regions is inhibited in addition to the hippocampus in patients with PTSD after hydrocortisone administration.

Although the role of the amygdala in fear conditioning has been repeatedly demonstrated in preclinical studies, few human studies have demonstrated activation of the amygdala during acquisition and extinction phases of conditioning (277). Studies of amygdala activation in PTSD are mixed and are confounded by different behavioral paradigms of symptom provocation used in the various studies. Right amygdala activation was reported in combat-related PTSD when patients and controls were exposed to traumatic imagery and combat pictures (278,279), whereas the left amygdala was activated in response to combat sounds (280). However, other studies that used similar paradigms were unable to replicate the finding of greater amygdala activation in patients with PTSD (281–283).

Because early and transient activation of the amygdala may not be detected by techniques such as PET or SPECT, which have poor temporal resolution in contrast to functional MRI (fMRI), it is possible that transient activation was missed by these techniques. In contrast, fMRI has a much greater temporal resolution, and similar studies using this technique in larger samples of patients and controls may provide definitive answers. A recent fMRI study by Rauch et al. (284) extended the masked faces paradigm to patients with PTSD. Whalen et al. (285) demonstrated activation of the amygdala on viewing pictures of fearful versus neutral or happy faces presented below the level of awareness. In contrast, overt presentation of emotional faces result in significant medial frontal activation (285,286). Rauch et al. found exaggerated amygdala responses to mask-fearful versus mask-happy faces in patients with PTSD compared with combat-exposed veterans without PTSD, suggesting that these patients exhibit exaggerated amygdala activation to

general threat-related stimuli presented at the subliminal level. Studies using similar symptom provocation designs, but using fMRI techniques in larger samples of patients and controls, may provide definitive answers with respect to the role of amygdala in PTSD symptoms.

Abnormalities in the functioning of subregions of the mPFC have been shown in PET and SPECT studies, using personalized scripts, combat slides, or sounds, of patients with PTSD. The prefrontal cortex is reciprocally connected to the amygdala and inhibits acquisition of the fear response and promotes extinction of behavioral response to fear-conditioned stimuli that are no longer reinforced (287,288). Various subregions of the mPFC mediate different responses. Lesions of the ventral mPFC or the orbital cortex prolong the extinction phase (289), whereas lesions of the dorsal mPFC (anterior cingulate) facilitated the fear response during the acquisition and extinction phases of fear conditioning, resulting in a generalized increase in fear response (287). Suppression of the neuronal firing in the subgenual prefrontal cortical equivalent in the rat (prelimbic cortex) is inversely correlated with an increase in amygdala neuronal activity (290). Based on the lesion studies, it can be hypothesized that dysfunction in the mPFC can result in a facilitation of acquisition of a fear response to traumatic stimuli with a failure of extinction of the fear response despite the lack of continuing trauma.

Functional imaging studies in PTSD are confounded by the number of various paradigms used for recreating symptoms, small sample sizes, and the presence of comorbid alcohol and/or substance dependence and major depression. However, a consistent finding among the various PET studies is the failure of activation of the anterior cingulate cortex in response to personalized scripts of the abuse (282,291) or exposure to combat slides and sounds (281). In contrast, cerebral blood flow increased in the right pregenual anterior cingulate cortex in response to combat sound– and script–driven imagery (278,280), suggesting laterality of mPFC regulation of emotional behavior in patients with PTSD. Further confirmation of this function of the anterior cingulate cortex comes from evidence supporting impaired performance on a Stroop test containing emotional combat words (285). In keeping with the hypothesis that a dysfunctional mPFC, particularly the anterior cingulate cortex, inadequately inhibits the amygdala, increased activation of the amygdala has been demonstrated in response to trauma-related stimuli (278–280). Although cerebral blood flow in other limbic and paralimbic cortical structures, such as the posterior orbital cortex, anterior temporal lobe, and the anterior insula, increased both in patients with PTSD and those with trauma without symptoms of PTSD, these findings did not distinguish these two groups.

These findings support the possibility that neural processes mediating extinction to trauma-related stimuli may be impaired in patients with PTSD. Recent psychophysiologic data confirmed that, compared with subjects without PTSD, patients with PTSD acquire conditioned responses more readily and take a longer time to extinguish the response, despite the absence of the primary stimulus (292,293).

Hypothalamic–Pituitary–Adrenal Axis Function in Posttraumatic Stress Disorder

In PTSD, abnormalities of CRH or HPA axis function have been inconsistent. Basal plasma or 24-hour urine cortisol levels have been reported to be decreased (294), no different (295,296), or abnormally increased (297–300) in PTSD patients compared with healthy or trauma-matched controls. These inconsistencies may reflect differences in gender, age of onset of the illness, and trauma type or duration. Behavioral challenges did not result in differences in cortisol secretion in PTSD subjects relative to controls (301). One group found that PTSD patients have an increased density of glucocorticoid receptors on peripheral lymphocytes (302) and are hypersensitive to low-dose DEX (303). Other alterations in HPA axis function include abnormally increased CRH levels (304) and blunted responses to CRH in PTSD patients compared with controls (305). There is evidence available that cortisol elevations after occurrence of a trauma are transient (306) and that loss of brain tissue is reversible if the exposure to elevated glucocorticoid levels does not exceed a critical time period (307).

Benzodiazepine Receptor Function in Posttraumatic Stress Disorder

Platelet peripheral-type BZD receptors have been shown to be decreased in patients with increased anxiety (308) and PTSD (309,310) and increased during diazepam treatment. Perhaps the most convincing piece of evidence linking BZD receptor function to PD is the efficacy of BZD in its treatment.

Although BZD administered as single agents have anxiolytic and antipanic effects, their role in the treatment of PTSD is limited. Braun et al. (311) demonstrated a modest, albeit significant improvement in anxiety and depressive symptoms in ten Israeli patients with PTSD using alprazolam, a short-acting BZD, in a double-blind crossover study. However, there was no significant change in core PTSD symptoms. Furthermore, administration of a BZD (clonazepam or alprazolam) immediately after a traumatic event did not alter the onset of either anxiety or PTSD symptoms at 1 and 6 months compared with placebo (312). However, PTSD patients randomized to BZD had a decrease in heart rate over time.

Serotonin Function and Posttraumatic Stress Disorder

Although only two studies directly examined the 5-HT system in PTSD, there is a large body of indirect evidence

suggesting that this neurotransmitter may be important in the pathophysiology of trauma-related symptomatology. In humans, low 5-HT functioning has been associated with aggression (313), impulsivity, and suicidal behavior (314). Patients with PTSD are frequently described as aggressive or impulsive and often have depression, suicidal tendencies, and intrusive thoughts that have been likened to obsessions.

More direct evidence of serotonergic dysregulation in PTSD comes from the results of electrophysiologic studies by Paige et al. (315). They demonstrated a predominance of "reducing" patterns to auditory evoked potential paradigms in combat veterans with PTSD compared with normal controls. Meanwhile, Hegerl et al. (316) suggested that the slope of the stimulus–response curve to auditory tones is positively associated with serotonergic activity. Based on their findings, it is possible that the discoveries of Paige et al. are, at least in part, related to low serotonergic activity in PTSD. This hypothesis gains further support from the observation that SSRI have been found to be partially effective in treating PTSD symptoms such as intrusive memories and avoidance symptoms (317,318).

The first report of serotonergic function in PTSD is a study examining paroxetine binding in blood platelets of 20 combat veterans with PTSD under baseline conditions (319). In this study, platelet 5-HT uptake was significantly decreased in PTSD patients compared with normals, with PTSD patients meeting criteria for comorbid major depressive disorder. Because decreased platelet 5-HT uptake has also been reported in patients with depression and alcoholism, the specificity of these findings has yet to be determined.

More recently, the behavioral effects of *m*-chlorophenylpiperazine were examined in a preliminary study of 14 combat veterans with PTSD (320). Five of the 14 patients with PTSD had a panic attack, and four had a flashback after *m*-chlorophenylpiperazine administration. In contrast, no patient had a panic attack and one patient experienced a flashback after the infusion of placebo saline. Thus, a subgroup of patients with PTSD exhibited a marked behavioral sensitivity to serotonergic provocation, raising the possibility of pathophysiologic subtypes among traumatized combat veterans. Clearly, further studies are needed to delineate possible serotonergic alterations in PTSD.

Noradrenergic Function in Posttraumatic Stress Disorder

Considering the broad functions of the noradrenergic system, abnormal regulation of this system could account for many of the clinical features of PTSD. Two studies found significantly elevated 24-hour urine NE excretion in combat veterans with PTSD compared with healthy subjects or patients with schizophrenia or major depression (321,322). The density of platelet α_2-adrenergic receptors is reduced in PTSD, perhaps reflecting adaptive down-regulation in response to chronically elevated levels of circulating endogenous catecholamines (323). The relevance of these studies to brain α-receptor function is questionable because of the lack of evidence demonstrating that platelet α-adrenergic receptors reflect brain α-receptor function (324).

Plasma lymphocyte β-adrenergic receptors are downregulated in PTSD (325,326); however, the functional significance of these findings is unclear.

Most of the early clinical investigations of the pathophysiology of PTSD identified a relationship between severe stress exposure, increased peripheral sympathetic nervous system activity, and conditioned physiologic, and emotional responses (327–329). Psychophysiologic studies have found hyperreactive responses to combat-associated stimuli but not to other stressful noncombat-related stimuli. These data are consistent with the hypothesis that noradrenergic hyperreactivity in patients with PTSD may relate to the conditioned or sensitized responses to specific traumatic stimuli. Studies evaluating the efficacy of psychotherapeutic techniques emphasizing desensitization to reduce hyperarousal responses to stimuli associated with the psychological trauma represent a current focus of investigation (330,331).

Similar to the PD patients, approximately two-thirds of PTSD patients experience yohimbine-induced panic attacks. In addition, 40% of PTSD patients report flashbacks after yohimbine. As a group, PTSD patients also have greater yohimbine-induced increases in plasma MHPG, sitting systolic blood pressure, and heart rate than healthy subjects. In the PTSD patients, yohimbine induced significant increases in core PTSD symptoms, such as intrusive traumatic thoughts, emotional numbing, and grief (332).

Indeed, the strength of traumatic memories may relate to the degree to which particular neuromodulatory systems are activated by the traumatic experience. Experimental and clinical investigations have demonstrated that memory processes are susceptible to modulating influences after the information has been acquired. Stimulation of noradrenergic receptors on the amygdala after a learning experience has memory-enhancing effects (333,334). Activating the LC–NE system that projects to the amygdala by frightening and traumatic experiences may facilitate the encoding of memories associated with the experiences. Moreover, it is possible that reproducing a neurobiologic state (noradrenergic hyperactivity in specific brain regions), similar to the one that existed at the time of the memory encoding, can elicit the traumatic memory.

PHOBIAS
Social Phobia

Social fears are commonly experienced by healthy subjects, especially in initial public-speaking experiences. For some people, this fear becomes persistent and overwhelming, limiting their social or occupational functioning because of intense anxiety and often avoidance. Social phobia is an area of active research; therefore, our knowledge base will be expanding rapidly in the next several years.

Epidemiology

Social phobia is a highly prevalent disorder. There is a wide range of lifetime prevalence rates, with the lowest levels reported in Asian study populations [0.5% (335) and 0.6% (336)], and the highest rates in the United States [13.3% (162)] and Switzerland [16% (337)]. In a recent prospective longitudinal community study, Merikangas et al. (338) reported that 6% of participants met lifetime criteria for social phobia, 12% at the subthreshold level and 24% with social phobia symptoms alone. It was confirmed that women had a greater risk of developing the disorder than men, whereas there was an equal gender ratio of social phobia symptoms. Onset is usually between 15 and 20 years of age, and the course tends to be chronic and unremitting. Complications include interference with work or school, social isolation, and abuse of alcohol or drugs. In inpatient alcoholism treatment programs, 20% to 25% report social phobia beginning before the onset of alcoholism or persisting after 1 year of abstinence. A substantial portion of patients also has at least one comorbid mental disorder, particularly major depression, any other anxiety disorder, or substance abuse/dependence. Results from recent twin studies suggest that at least 30% of the variance underlying familial concordance is attributable to genetic influence (339,340).

Description and Differential Diagnosis

Social phobia is characterized by a persistent and exaggerated fear of humiliation or embarrassment in social situations, leading to high levels of distress and possibly avoidance of those situations. Patients may become fearful that their anxiety will be evident to others, which may intensify their symptoms or even produce a situational panic attack. The fear may be of speaking, meeting people, or eating or writing in public, and relates to the fear of appearing nervous or foolish, making mistakes, being criticized, or being laughed at. Often, physical symptoms of anxiety such as blushing, trembling, sweating, and tachycardia are triggered when the patient feels that he/she is being evaluated or scrutinized.

Probably the most difficult diagnostic distinctions are between social phobia and normal performance anxiety or social phobia and PD. Normal fear of public speaking usually diminishes as the individual is speaking or with additional experience, whereas in social phobia, the anxiety may worsen or fail to attenuate with rehearsal. Social phobics may experience situational panic attacks resulting from anticipation or exposure to the feared social situation. Some PD/agoraphobia patients avoid social situations because of fear of embarrassment if a panic attack should occur, but usually their initial panic attack is unexpected (occurs in a situation that they previously did not fear), and the subsequent development of phobias is generalized beyond social phobia situations. Sometimes social phobia and PD coexist. Social phobia can be differentiated from simple phobias in that the latter do not involve social situations involving scrutiny, humiliation, or embarrassment. In major depression, social avoidance may develop from apathy rather than fear and resolves with remission of the depressive episode. In schizoid personality disorder, social isolation results from a lack of interest rather than from fear. In avoidant personality disorder, the avoidance is of personal relationships; however, if the patient develops a marked anxiety about and avoidance of most social situations, the additional diagnosis of social phobia should be made.

Specific Phobia

Specific phobia shares many of the basic features of the general phobias, but the fear is limited to a specific object or situation, such as dogs or heights, so the extent of interference in a patient's life tends to be mild.

Epidemiology

Six-month prevalence rates of specific phobia reported in the Epidemiological Catchment Area Study are between 4.5% and 11.8%; rates are higher for females than for males. The onset of animal phobia is usually in childhood. Blood-injury phobia usually begins in adolescence or early adulthood and may be associated with vasovagal fainting on exposure to the phobic stimulus. Age of onset may be more variable for other simple phobias. Many childhood-onset phobias may remit spontaneously. Impairment depends on the extent to which the phobic object or situation is routinely encountered in the individual's life. Simple phobias may coexist with social phobia and PD, but they are believed to be unrelated.

Description and Differential Diagnosis

Specific phobia is usually a circumscribed fear of a specific object or situation. As with the other phobias, the fear is excessive and unrealistic, exposure to the phobic stimulus produces an anxiety response, expectation of exposure may produce anticipatory anxiety, and the object or situation is either avoided or endured with considerable discomfort. However, unlike social phobia, the fear does not involve scrutiny or embarrassment, and, unlike agoraphobia, the fear is not of being trapped or having a panic attack. The nature of the fear is specific to the phobia, such as a fear of falling or loss of visual support in height phobia or fear of crashing in a flying phobia. Isolated fears are common in the general population; a diagnosis of simple phobia is reserved for situations in which the phobia results in marked distress or some degree of impairment in activities or relationships.

Pathophysiologic Studies of Social and Specific Phobias

Noradrenergic Function

Few studies have examined noradrenergic function in patients with phobic disorders. In patients with specific phobias, increases in subjective anxiety and increased heart rate, blood pressure, and plasma NE and epinephrine have been associated with exposure to the phobic stimulus (341). This finding may be of interest from the standpoint of the model of conditioned fear, reviewed earlier, in which a potentiated release of NE occurs in response to a reexposure to the original stressful stimulus.

Patients with social phobia have been found to have greater increases in plasma NE on orthostatic challenge compared with healthy controls and patients with PD (342). In contrast to PD patients, the density of lymphocyte β-adrenoceptors is normal in social phobia patients (343).

Dopaminergic Function

Brain imaging studies using SPECT reported abnormal reductions in dopamine receptor and transporter binding. Tiihonen et al. (344) found a significant reduction in striatal β-CIT binding in patients with social phobia compared with controls, suggesting a reduction in dopamine transporter binding. A more recent study by Schneier et al. (345) reported reduced uptake of the D_2/D_3 receptor ligand [^{123}I]IBZM in patients compared with controls. Both findings await replication.

GENERALIZED ANXIETY DISORDER

GAD is characterized by excessive and uncontrollable worry about multiple life circumstances. It is accompanied by symptoms of muscle tension, restlessness, fatigue, concentration problems, difficulty falling or staying asleep, and irritability. The anxiety is unrelated to panic attacks, phobic stimuli, obsessions, illness, or traumatic events (in PTSD). The validity of GAD as a diagnosis distinct from other anxiety disorders or depression and whether it is a homogeneous category are still being examined.

Epidemiology

GAD is a very common disorder with similar high prevalence rates. The mean onset of GAD has generally been reported as relatively late as 35 years compared with most forms of anxiety disorders that usually start in adolescence. This may reflect to some extent the delay in recognition of GAD. GAD in children and young adolescents is recognized in the *DSM-IV* criteria, which include the diagnosis of "overanxious anxiety disorder of childhood," but the prevalence of *DSM-IV* GAD among subjects younger than

20 years of age is comparatively very low, with estimates well below 1%. Early-onset GAD has an almost equal gender distribution, but this separates sharply after the age of 20 years when the female predominance becomes obvious (346).

The estimates of lifetime prevalence of *DSM-III-R* (*DSM, Revised Third Edition*) and *DSM-IV* GAD reported in various population surveys around the world have been generally in line with the 5.1% rate reported in the National Comorbidity Survey (347). The reported 12-month prevalence rates vary between 2% and 4%. When the important levels of subthreshold GAD are also taken into account, it becomes evident that GAD is among the more frequent mental disorders.

There is a high ratio of current-to-lifetime rates, indicating that GAD mostly runs a very chronic course. This is reflected in the mean duration of 20 years reported in epidemiologic studies (348). Clinical experience also suggests that those who develop GAD are likely to still have the disorder 20 or 30 years later.

Description and Differential Diagnosis

The symptom of anxiety is prominent in a number of conditions, including depressive, psychotic, substance use, and somatoform disorders as well as some medical conditions and medication side effects, so careful questioning is necessary to differentiate among the multiple causes of anxiety.

GAD is characterized by chronic excessive worry that is out of proportion to the likelihood or impact of the feared events, is pervasive (focused on many life circumstances), is difficult to control, and is not related to hypochondriacal concerns or part of PTSD; anxiety secondary to substance-induced or nonpsychiatric medical etiologies will be excluded. This is accompanied by tension or nervousness as manifested by at least three of the following symptoms: restlessness, quick fatigue, feeling keyed up or on edge, difficulty concentrating/mind going blank, and irritability. In addition, significant functional impairment or marked distress is required for the diagnosis.

Generalized persistent anxiety may develop between attacks in PD. GAD symptoms are often present in episodes of depression. In patients with somatization disorder, the thrust of concern regards health and physical symptoms rather than apprehensive worry about life circumstances. As with PD, medical conditions that may produce anxiety symptoms, such as hyperthyroidism or caffeinism, must be excluded. If an anxiety disorder is present that does not fit the criteria of any of the anxiety disorders, somatization, psychoactive substance related, or medical conditions, then the diagnosis of anxiety disorder not otherwise specified may be considered. However, careful assessment will usually reveal the presence of a more specific diagnosis. If there is doubt, an anxiety disorders specialist should be consulted.

Pathophysiology

Noradrenergic Function

Plasma MHPG levels have been shown to be both increased (349,350) and not different (351) compared with normal controls. Similarly, increases in resting plasma NE in GAD patients have been reported in some studies (350,352,353) but not others (349,352).

α_2-Adrenergic receptor number, as measured by specific binding of tritiated yohimbine on platelet membranes, has been found to be reduced (350) or unchanged (354) in GAD patients. Growth hormone response to clonidine has been found to be blunted in GAD patients (355). Patients with GAD have been found to have normal responses to the α_2 antagonist yohimbine (351).

Benzodiazepine Function

Patients with GAD have been found to have decreases in peripheral-type BZD receptor binding as assessed by [3H]PK11195 binding to lymphocyte membranes, although the relationship to central BZD receptor function is unclear (356). In addition, reduced binding to [3H]PK11195 has been reversed with BZD therapy in patients with anxiety disorders (357).

Brain Imaging Studies in Generalized Anxiety Disorder

Patients with GAD have been found to have a decrease in metabolism at baseline measured with PET in basal ganglia and white matter and an increase in normalized left occipital cortex, right posterior temporal lobe, and right precentral frontal gyrus metabolism compared with healthy controls (358). Administration of BZD therapy in patients with GAD results in a decrease in glucose metabolism in the occipital cortex, a brain region high in BZD receptors (359).

CONCLUDING COMMENTS

The neural circuits and neural mechanisms of anxiety and fear described in this chapter may have implications for the development of novel treatment strategies for patients with anxiety disorders. It must be taken into account that there exist inconsistencies in the results of biologic investigations of anxiety disorders that highlight the importance of addressing the neurobiologic heterogeneity inherent in criteria-based psychiatric diagnoses. Understanding this heterogeneity will be facilitated by the continued development and application of genetic, neuroimaging, and neurochemical approaches that can refine anxiety disorder phenotypes and elucidate the genotypes associated with these disorders. Application of these experimental approaches may also help to develop research protocols aimed at elucidating the mechanisms of antianxiety therapies, including both pharmacotherapy and psychotherapy and their underlying mechanisms of action.

Anxiolytic treatments appear to inhibit neuronal activity in the structures mediating fear expression and behavioral sensitization and/or facilitate endogenous mechanisms for modulating the neural transmission of information about aversive stimuli and responses to such stimuli. Novel treatments being developed to exploit the former type of mechanisms include pharmacologic agents that selectively target subcortical and brainstem pathways supporting specific components of emotional expression (e.g., CRH receptor antagonists). In contrast, nonpharmacologic treatments for anxiety may augment the brain systems for modulating anxiety responses by facilitating the extinction of putative fear-conditioned responses or directing the reinterpretation of anxiety-related thoughts and somatic sensations (so that they produce less subjective distress). Informed by increasingly detailed knowledge about the pathophysiology of specific anxiety disorders and the neural pathways involved in anxiety and fear processing, the development of therapeutic strategies that combine both types of approaches may ultimately provide the optimal means for reducing the morbidity of anxiety disorders.

REFERENCES

1. LeDeux JE. Nervous System V. Emotion. In: Blum F, ed. *Handbook of physiology.* Washington, DC: American Physiological Society, 1987:419–459.
2. Turner B, Gupta KC, Mishkin M. The locus and cytoarchitecture of the projection areas of the olfactory bulb in Macaca mulatta. *J Comp Neurol* 1978;177:381–396.
3. Whitlock DG, Nauta WJH. Subcortical projections from the temporal neocortex in Macaca mulatta. *J Comp Neurol* 1956;106:183–212.
4. Saper CB. Convergence of autonomic and limbic connections in the insular cortex of the rat. *J Comp Neurol* 1982;210:163–173.
5. Elam M, Svensson THE, Thoren P. Locus coeruleus neurons in sympathetic nerves: activation by visceral afferents. *Brain Res* 1986;375:117–125.
6. Jones EG. The thalamus. In: Emson P, ed. *Chemical neuroanatomy.* New York: Raven Press, 1983:257–293.
7. Jones EG, Powell TPSS. An experimental study of converging sensory pathways within the cerebral cortex of the monkey. *Brain* 1970;93:793–820.
8. Mesulam MM, Van Hoesen G, Pandya DN, et al. Limbic and sensory connections of the IPL in the rhesus monkey. *Brain Res* 1977;136:393–414.
9. Turner BH, Mishkin M, Knapp M. Organization of the amygdalopetal projections from modality-specific cortical association areas in the monkey. *J Comp Neurol* 1980;191:515–543.
10. VanHoesen GW, Pandya DN, Butters N. Cortical afferents to the entorhinal cortex of the rhesus monkey. *Science* 1972;175:1471–1473.
11. Vogt BA, Miller MW. Cortical connections between rat cingulate cortex and visual, motor, and postsubicular cortices. *J Comp Neurol* 1983;216:192–210.
12. Swanson LW. The hippocampus and the concept of the limbic system. In: Seifert W, ed. *Neurobiology of the hippocampus.* London: Academic Press, 1983:3–19.
13. Amaral DG, Price JL, Pitanken A, et al. Anatomical organization of the primate amygdala complex. In: Aggleton JP, ed. *The amygdala: neurobiological aspects of emotion, memory and mental dysfunction.* New York: Wiley-Liss, 1992:1–66.
14. LeDoux JE, Iwata J, Cicchetti P, et al. Different projections of the central amygdaloid nucleus mediate autonomic and behavioral correlates of conditioned fear. *J Neurosci* 1988;8:2517–2529.
15. Sawchenko PE, Swanson LW. The organization of forebrain afferents to the paraventricular and supraoptic nucleus of the rat. *J Comp Neurol* 1983;218:121–144.
16. Sawchenko PE, Swanson LWW. Central noradrenergic pathways for the integration of hypothalamic neuroendocrine and autonomic responses. *Science* 1982;214:685–687.
17. Westlund KN, Coulter JDD. Descending projections of the locus coeruleus and subcoeruleus/medial brachial nuclei in monkey: axonal transport studies and dopamine beat hydroxylase immunocytochemistry. *Brain Res Rev* 1980;2:235–264.
18. Clark FM, Proudfit HK. The projection of locus coeruleus neurons to the spinal cord in the rat determined by anterograde tracing combined with immunocytochemistry. *Brain Res* 1991;538:231–245.
19. McDonald AJ. Organization of amygdaloid projections to prefrontal cortex and associated striatum in the rat. *Neuroscience* 1991;44:1–14.
20. McDonald AJ. Topographical organization of amygdaloid projections to the caudatoputamen, nucleus accumbens, and related striatal-like areas of the rat brain. *Neuroscience* 1991;44:15–33.
21. Veening JG, Swanson LW, Sawchenko PE. The organization of projections from the central nucleus of the amygdala to brain stem sites involved in central autonomic regulation: a retrograde transport-immunohistochemical study. *Brain Res* 1984;303:337–357.
22. Charney DS, Deutch AY, Krystal JH, et al. Psychobiologic mechanisms of posttraumatic stress disorder. *Arch Gen Psychiatry* 1993;50:294–305.
23. Squire LR. Memory and the hippocampus: a synthesis from findings with rats, monkeys, and humans. *Psychol Rev* 1992;99:195–231.
24. Squire LR, Zola-Morgan S. The medial temporal lobe memory system. *Science* 1991;253:2380–2386.
25–61. [Reserved.]
62. Ida Y, Tanaka M, Tsuda A, et al. Attenuating effect of diazepam on stress-induced increases in noradrenaline turnover in specific brain regions of rats: antagonism by Ro 15-1788. *Life Sci* 1985;37:2491–2498.
63. Tanaka T, Yokoo H, Mizoguchi K, et al. Noradrenaline release in rat amygdala is increased by stress: studies with intracerebral microdialysis. *Brain Res* 1991;544:174–181.
64. Tanaka T, Yokoo H, Tsuda A, et al. Stress increased hypothalamic noradrenaline release studied by intracerebral dialysis method. *Neuroscience* 1990;16:293–300.
65. Melia KR, Nestler EJ, Duman RS. Chronic imipramine treatment normalized levels of tyrosine hydroxylase in the locus coeruleus of chronically stressed rats. *Psychopharmacology* 1992;108:23–26.
66. Shirao I, Tsuda A, Yoshisshige I, et al. Effect of acute ethanol administration on noradrenaline metabolism in brain regions of stressed and nonstressed rats. *Pharmacol Biochem Behav* 1988;30:769–773.

67. Svensson TH. Peripheral, autonomic regulation of locus coeruleus noradrenergic neurons in brain. Putative implications for psychiatry and psychopharmacology. *Psychopharmacology* 1987;92:1–10.

68. Torda T, Kvetnansky R, Petrikova M. Effect of repeated immobilization stress on rat central and peripheral adrenoceptors. In: Usdin E, Kvetnansky R, Axelrod J, eds. *Stress: the role of catecholamines and other neurotransmitters.* New York: Gordon & Breach, 1984:691–701.

69. U'Prichard DC, Kvethansky R. Central and peripheral adrenergic receptors in acute and repeated immobilization stress. In: Usdin E, Kvetnansky R, Kopin IJ, eds. *Catecholamines and stress: recent advances.* Amsterdam: Elsevier, 1980:299–308.

70. Pavcovich LA, Cancela LM, Volosin M, et al. Chronic stress-induced changes in locus coeruleus neuronal activity. *Brain Res Bull* 1990;24:293–296.

71. Simson PE, Weiss JM. Altered activity of the locus coeruleus in an animal model of depression. *Neuropsychopharmacology* 1988;1:287–295.

72. Levine ES, Litto WJ, Jacobs BL. Activity of cat locus coeruleus noradrenergic neurons during the defense reaction. *Brain Res* 1990;531:189–195.

73. Rasmussen K, Morilak DA, Jacobs BL. Single unit activity of locus coeruleus neurons in the freely moving cat. I. During naturalistic behaviors and in response to simple and complex stimuli. *Brain Res* 1986;371:324–334.

74. Redmond DE Jr, Huang YH, Snyder DR, et al. Behavioral effects of stimulation of the locus coeruleus in the stumptail monkey (Macaca arctoides). *Brain Res* 1976;116:502–507.

75. Redmond DE Jr. Studies of the nucleus locus coeruleus in monkeys and hypotheses for neuropsychopharmacology. In: Meltzer HY, ed. *Psychopharmacology: the third generation of progress.* New York: Raven Press, 1987:467–974.

76. Charney DS, Woods SW, Price LH, et al. Noradrenergic dysregulation in panic disorder. In: Ballenger JC, ed. *Neurobiology of panic disorders.* New York: Alan R. Liss, 1990:91–105.

77. Deutch AY, Roth RH. The determinants of stress induced activation of the prefrontal cortical dopamine system. *Prog Brain Res* 1990;85:367–403.

78. Herman J-P, Guilloneau PP, Dantzer R, et al. Differential effects of inescapable footshocks and of stimuli previously paired with inescapable footshocks on DA turnover in cortical and limbic areas of the rat. *Science* 1982;23:1549–1556.

79. Fadda F, Argiolas A, Melis MR, et al. Stress-induced increase in 3,4-dihydroxyphenylacetic acid (DOPAC) levels in the cerebral cortex and in n. accumbens: reversal by diazepam. *Life Sci* 1978;23:2219–2224.

80. Dunn AJ. Stress-related activation of cerebral dopaminergic systems. *Ann N Y Acad Sci* 1988;537:188–205.

81. Roth RH, Tam S-Y, Ida Y, et al. Stress and the mesocorticolimbic dopamine systems. *Ann N Y Acad Sci* 1988;537:138–147.

82. Bradberry CW, Lory JS, Roth RH. The anxiogenic beta-carboline FG 7142 selectively increases dopamine release in rat prefrontal cortex as measured by microdialysis. *J Neurochem* 1991;56:748–752.

83. Roth RH, Tam S-Y, Ida Y, et al. Stress and the mesocorticolimbic dopamine systems. *Ann N Y Acad Sci* 1988;537:138–147.

84. Abercrombie ED, Keefe KA, DiFrischia DS, et al. Differential effect of stress on in vivo dopamine release in striatum, nucleus accumbens, and medial frontal cortex. *J Neurochem* 1989;52:1655–1658.

85. Mantz J, Thierry AM, Glowinski J. Effect of noxious tail pinch on the discharge rate of mesocortical and mesolimbic dopamine neurons: selective activation of the mesocortical system. *Brain Res* 1989;476:377–381.

86. Adell A, Trullas R, Gelpi E. Time course of changes in serotonin and noradrenaline in rat brain after predictable or unpredictable shock. *Brain Res* 1988;459:54–59.

87. Pei Q, Zetterstrom T, Fillenz M. Tail pinch-induced changes in the turnover and release of dopamine and 5-hydroxytryptamine in different brain regions of the rat. *Neuroscience* 1990;35:133–138.

88. Dunn AJ. Stress-related activation of cerebral dopaminergic systems. *Ann N Y Acad Sci* 1988;537:188–205.

89. Heinsbroek RPW, Van Haaven F, Fecustra MGP, et al. Controllable and uncontrollable footshock and monoaminergic activity in the frontal cortex of male and female rats. *Brain Res* 1991;551:247–255.

90. Kramer GL, Petty F. Inhibition of stress-induced increase in dopamine and serotonin metabolism in frontal neocortex by diazepam in vivo using brain microdialysis perfusion. *J Neurochem* 1989;52:S155.

91. Rasmussen AM, Goldstein LE, Bunney BS, et al. Nonsedative dose lorazepam dissociates the neurochemical and behavioral correlates of fear conditioning in the rat. *Soc Neurosci Abst* 1992;18:813.

92. Petty F, Sherman AD. Learned helplessness induction decreases in vivo cortical serotonin release. *Pharmacol Biochem Behav* 1983;18:649–650.

93. Hellhammer DH, Rea MA, Bell L, et al. Learned helplessness effects on brain monoamines and the pituitary gonadal axis. *Pharmacol Biochem Behav* 1984;21:481–485.

94. Petty F, Sherman AD. Reversal of learned helplessness by imipramine. *Commun Psychopharmacol* 1980;3:371–375.

95. Martin P, Soubrie P, Pueuch AJ. Reversal of helpless behavior by serotonin uptake blocker. *Psychopharmacology* 1990;101:403–407.

96. Graeff F. Role of 5-HT in defensive behavior and anxiety. *Rev Neurosci* 1993;4:181–211.

97. Watanabe Y, Gould E, Daniels D, et al. Tianeptine attenuates stress-induced morphological changes in the hippocampus. *Eur J Pharmacol* 1992;222:157–162.

98. Deakin W, Graeff F. 5-HT and mechanisms of defense. *J Psychopharmacol* 1991;5:305–315.

99. Kuroda Y, Mikuni M, Ogawa T, et al. Effect of ACTH, adrenalectomy and the combination treatment on the density of $5-HT_2$ receptor binding sites in neocortex of rat forebrain and $5-HT_2$ receptor-mediated wet-dog shake behaviors. *Psychopharmacology* 1992;108:27–32.

100. Kuroda Y, Mikuni M, Nomura N, et al. Differential effect of subchronic dexamethasone treatment on serotonin-$_2$ and β-adrenergic receptors in the rat cerebral cortex and hippocampus. *Neurosci Lett* 1993;155:195–198.

101. Fernandes C, McKittrick CR, File SE, McEwen BS. Decreased 5-HT1A and increased 5-HT2A receptor binding after chronic corticosterone associated with a behavioral indication of depression but not anxiety. *Psychoneuroendocrinology* 1997;22(7):447–491.

102. Martire M, Navarra P, Pistritto G, et al. Adrenal steroid-induced chances in serotonin receptors in rat hippocampus and hypothalamus. *Pharmacol Res* 1988;20:415–416.

103. Mendelson S, McEwen BS. Autoradiographic analyses of the effects of restraint-induced stress on $5-HT_{1A}$, $5-HT_{1C}$ and

5-HT$_2$ receptors in the dorsal hippocampus of male and female rats. *Neuroendocrinology* 1991;54:454–461.

104. Mendelson S, McEwen BS. Autoradiographic analyses of the effects of adrenalectomy and corticosterone on 5-HT$_{1A}$ and 5-HT$_{1B}$ receptors in the dorsal hippocampus and cortex of the rat. *Neuroendocrinology* 1992;55:444–450.

105. Chalmers D, Kwak S, Mansour A, et al. Corticosteroids regulate brain hippocampal 5-HT$_{1A}$ receptor mRNA expression. *J Neurosci* 1993;13:914–923.

106. Burnet P, Mefford I, Smith C, et al. Hippocampal 8-[3H]hydroxy-2-(di-n-propylamino) tetralin binding site densities, serotonin receptor (5-HT$_{1A}$) messenger ribonucleic acid abundance and serotonin levels parallel the activity of the hypothalamopituitary-adrenal axis in rat. *J Neurochem* 1992;59:1062–1070.

107. Ramboz S, Oosting R, Amara DA, et al. Serotonin receptor 1A knockout: an animal model of anxiety-related disorder. *Proc Natl Acad Sci U S A* 1998;95:14476–14481.

108. Choi DW, Farb DH, Fischbach GD. Chlordiazepoxide selectively potentiates GABA conductance of spinal cord and sensory neurons in cell culture. *J Neurophysiol* 1981;45:621–631.

109. Graeff FG, Silveira MC, Nogueira RL, et al. Role of the amygdala and periaqueductal gray in anxiety and panic. *Behav Brain Res* 1993;58:123–131.

110. Braestrup C, Schmiechen R, Neef G, et al. Interaction of convulsive ligands with benzodiazepine receptors. *Science* 1982;216:1241–1243.

111. Ninan PT, Insel TM, Cohen RM, et al. Benzodiazepine receptor mediated experimental "anxiety" in primates. *Science* 1982;218:1332–1334.

112. McKernan RM, Rosahl TW, Reynolds DS, et al. Sedative but not anxiolytic properties of benzodiazepines are mediated by the GABA$_A$ receptor α_1 subtype. *Nat Neurosci* 2000;3:587–592.

113. Rudolph U, et al. Benzodiazepine actions mediated by specific α-aminobutyric acid A receptor subtypes. *Nature* 1999;410:796–800.

114. Low K, Crestani F, Keist R, et al. Molecular and neuronal substrate for the selective attenuation of anxiety. *Science* 2000;290:131–134.

115. Paslawski T, Treit D, Baker GB, et al. The antidepressant drug phenelzine produces antianxiety effects in the plus-maze and increases in rat brain GABA. *Psychopharmacology* 1996;127:19–24.

116. Lippa AS, Klepner CA, Yunger L, et al. Relationship between benzodiazepine receptors and experimental anxiety in rats. *Pharmacol Biochem Behav* 1978;9:853–856.

117. Weizman R, Weizman A, Kook KA, et al. Repeated swim stress alters brain benzodiazepine receptors measured in vivo. *J Pharmacol Exp Ther* 1989;249:701–707.

118. Medina JH, Novas ML, de Robertis E. Changes in benzodiazepine receptors by acute stress: different effects of chronic diazepam or RO 15-1788 treatment. *Eur J Pharmacol* 1983;96:181–185.

119. Medina JH, Novas ML, Wolfman CNV, et al. Benzodiazepine receptors in rat cerebral cortex and hippocampus undergo rapid and reversible changes after acute stress. *Neuroscience* 1983;9:331–335.

120. Drugan RC, Basile AC, Deutsch SI, et al. Inescapable shock reduces [3H]Ro 5-4864 binding to peripheral type benzodiazepine receptors in the rat. *Pharmacol Biochem Behav* 1986;24:1673–1677.

121. Drugan RC, Morrow AL, Weizman R, et al. Stress-induced behavioral depression in the rat is associated with a decrease

in GABA receptor-mediated chloride ion flux and benzodiazepine receptor occupancy. *Brain Res* 1989;487:45–51.

122. Miller LG, Thompson ML, Greenblatt DJ, et al. Rapid increase in brain benzodiazepine receptor binding following defeat stress in mice. *Brain Res* 1987;414:395–400.

123. Robertson HA, Martin IL, Candy JM. Differences in benzodiazepine receptor binding in Maudsley-reactive and nonreactive rats. *Eur J Pharmacol* 1978;50:455–457.

124. Orchinik M, Weiland NG, McEwen NG. Chronic exposure to stress levels of corticosterone alters GABA$_A$ receptor subunit mRNA levels in rat hippocampus. *Brain Res Mol Brain Res* 1995;34:29–34.

125. Sanchez MM, Young LJ, Plotsky PM, et al. Distribution of corticosteroid receptors in the rhesus brain: relative absence of glucocorticoid receptors in the hippocampal formation. *J Neurosci* 2000;20:4657–4668.

126. McEwen BS. From molecules to mind. Stress, individual differences, and the social environment. *Ann N Y Acad Sci* 2001;935:42–49.

127. Cameron HA, Gould E. Adult neurogenesis is regulated by adrenal steroids in the dentate gyrus. *Neuroscience* 1994;61:203–209.

128. Sapolsky RM. Glucocorticoids and hippocampal atrophy in neuropsychiatric disorders. *Arch Gen Psychiatry* 2000;57:925–935.

129. Moghaddam B. Stress preferentially increases extraneuronal levels of excitatory amino acids in the prefrontal cortex: comparison to hippocampus and basal ganglia. *J Neurochem* 1993;60:1650–1670.

130. Lowy MT, Gault L, Yamamoto BK. Adrenalectomy attenuates stress-induced elevations in extracellular glutamate concentrations in the hippocampus. *J Neurochem* 1993;61:1957–1960.

131. Moghaddam B. Recent basic findings in support of excitatory amino acid hypotheses of schizophrenia. *Prog Neuropsychopharmacol Biol Psychiatry* 1994;18:859–870.

132. Sapolsky RM, Plotsky PM. Hypercortisolism and its possible neural bases. *Biol Psychiatry* 1990;27:937–952.

133. Veldhuis HD, De Kloet ER. Significance of ACTH4-10 in the control of hippocampal corticosterone receptor capacity of hypophysectomized rats. *Neuroendocrinology* 1982;34:374–380.

134. Reul JM, de Kloet ER. Two receptor systems for corticosterone in rat brain: microdistribution and differential occupation. *Endocrinology* 1985;117:2505–2511.

135. Patel PD, Lopez JF, Lyons DM, et al. Glucocorticoid and mineralocorticoid receptor mRNA expression in squirrel monkey brain. *J Psychiatr Res* 2000;34:383–392.

136. Seckl JR, Dickson KL, Yates C, et al. Distribution of glucocorticoid and mineralocorticoid receptor messenger RNA expression in human postmortem hippocampus. *Brain Res* 1991;561:332–337.

137. Wetzel DM, Bohn MC, Kazee AM, et al. Glucocorticoid receptor mRNA in Alzheimer's diseased hippocampus. *Brain Res* 1995;679:72–81.

138. Watzka M, Bidlingmaier F, Beyenburg S, et al. Corticosteroid receptor mRNA expression in the brains of patients with epilepsy. *Steroids* 2000;65:895–901.

139. Watzka M, Beyenburg S, Blumcke I, et al. Expression of mineralocorticoid and glucocorticoid receptor mRNA in the human hippocampus. *Neurosci Lett* 2000;290:121–124.

140. Brooke SM, de Haas-Johnson AM, Kaplan JR, et al. Characterization of mineralocorticoid and glucocorticoid receptors in primate brain. *Brain Res* 1994;637:303–307.

141. Stanton, ME, Gutierrez YR, Levine S. Maternal deprivation

potentiates pituitary-adrenal stress responses in infant rats. *Behav Neurosci* 1988;102:69–70.

142. Levine S, Wiener SG, Coe CL. Temporal and social factors influencing behavioral and hormonal responses to separation in mother and infant squirrel monkey. *Psychoneuroendocrinology* 1993;18:297–306.

143. Liu D, Diorio J, Tannenbaum B, et al. Maternal care, hippocampal glucocorticoid receptors and hypothalamic-pituitary adrenal responses to stress. *Science* 1997;277:1654–1662.

144. Plotsky PM, Meaney MJ. Early postnatal experience alters hypothalamic corticotropin releasing factor MRNA medial CRF context, and stress induced release in adult rats. *Mol Brain Res* 1993;18:195–200.

145. Heit C, Woens MJ, Plotsky PM, et al. Persistent changes in corticotropin releasing factor systems due to early life stress: relationship to pathophysiology of major depression and post-traumatic stress disorder. *Psychopharmacol Bull* 1997;33:185–192.

146. Copplan JD, Andrews MW, Rosemblum LA, et al. Persistent elevations of cerebrospinal fluid concentrations of corticotropin releasing factor in adult nonhuman primates exposed to early life stressors: implications for the pathophysiology of mood and anxiety disorders. *Proc Natl Acad Sci U S A* 1996;93:1619–1623.

147. Meaney MJ, Airken DH, van Berkel C, et al. Effect of neonatal handling on age-related impairments associated with the hippocampus. *Science* 1988;239:766–768.

148. Meaney MJ, Aitken DH, Sharma S, et al. Neonatal handling alters adrenocortical negative feedback sensitivity and hippocampal type II glucocorticoid receptor binding in the rat. *Neuroendocrinology* 1989;50:597–604.

149. Koob GF, Heinrichs SC. A role for corticotropin releasing factor and urocortin in behavioral responses to stressors. *Brain Res* 1999;848:141–152.

150. Smith GW, et al. Corticotropin releasing factor 1-deficient mice display decreased anxiety, impaired stress response, and aberrant neuroendocrine development. *Neuron* 1998;20:1093–1102.

151. Timpl P, Spanagel R, Sillaber I, et al. Impaired stress response and reduced anxiety in mice lacking a functional corticotropin-releasing hormone receptor. *Nat Genet* 1998;19:162–166.

152. Bale TL, Contarino A, Smith GW, et al. Mice deficient for corticotropin-releasing hormone receptor-2 display anxiety-like behaviour and are hypersensitive to stress. *Nat Genet* 2000;24:410–414.

153. Kiskimoto T, Radulovic J, Radulovic M, et al. Deletion of Chr2 reveals an anxiolytic role for corticotropin-releasing hormone receptor-2. *Nat Genet* 2000;24:415–419.

154. Habib KE, Weld KP, Rice KC, et al. Oral administration of a cortico-releasing hormone receptor antagonist significantly attentuates behavioral, neuroendocrine, and autonomic responses to stress in primates. *Proc Natl Acad Sci U S A* 2000;97:6079–6084.

155. Woodruff GN, Hill DR, Boden P, et al. Functional role of brain CCK receptors. *Neuropeptides* 1991;19[Suppl]:45–56.

156. Bradwejn J, Koszyck D, Couetoux du Terte A, et al. The cholecystokinin hypothesis of panic and anxiety disorders. *J Psychopharmacol* 1992;6:345–351.

157. DeMontigny C. Cholecystokinin tetrapeptide induces panic like attacks in healthy volunteers. *Arch Gen Psychiatry* 1989;46:511–517.

158. Heilig M, Soderpalm B, Engel JA, et al. Centrally adminis-tered neuropeptide Y (NPY) produces anxiolytic-like effects in animal anxiety models. *Psychopharmacology* 1989;98:524–529.

159. Heilig M, McLeod S, Koob GF, et al. Anxiolytic-like effect of neuropeptide Y (NPY), but not other peptides in an operant conflict test. *Regul Pept* 1992;41:61–69.

160. Wahlestedt C, Pich EM, Koob GF, et al. Modulation of anxiety and neuropeptide Y-Y₁ receptors by antisense oligodeoxynucleotides. *Science* 1993;259:528–531.

161. Heilig M, Koob GF, Ekman R, et al. Corticotropin releasing factor and neuropeptide Y: role in emotional integration. *Trends Neurosci* 1994;17:80–85.

162. Kessler R, McGonagle K, Zhao S, et al. Lifetime and 12-month prevalence of DSM-III-R psychiatric disorders in the United States: results from the National Comorbidity Study. *Arch Gen Psychiatry* 1994;51:8–19.

163. Roy-Byrne, PP, Stang P, Wittchen HU, et al. Lifetime panic-depression comorbidity in the National Comorbidity Survey. *Br J Psychiatry* 2000;176:229–235.

164. Eaton WW, Anthony JC, Romanoski A, et al. Onset and recovery from panic disorder in the Baltimore Epidemiologic Catchment Area follow-up. *Br J Psychiatry* 1998;173:501–507.

165. Keller MB, Yonkers KA, Warshaw MG, et al. Remission and relapse in subjects with panic disorder and panic with agoraphobia: a prospective short-interval naturalistic follow-up. *J Nerv Ment Dis* 1994;182:290–296.

166. Apfeldorf WJ, Spielman LA, Cloitre M, et al. Morbidity of comorbid psychiatric diagnoses in the clinical presentation of panic disorder. *Depress Anxiety* 2000;12:78–84.

167. Greenberg PE, Sisitsky T, Kessler RC, et al. The economic burden of anxiety disorders in the 1990s. *J Clin Psychiatry* 1999;60:427–435.

168. Breier A, Charney DS, Heninger GR. Agoraphobia with panic attacks: development, diagnostic stability, and course of illness. *Arch Gen Psychiatry* 1986;43:1029–1036.

169. Knowles JA, Weissman MM. Panic disorder and agoraphobia. In: Oldham JM, Riba MB, eds. *Reviews of psychiatry, volume 14.* Washington, DC: American Psychiatric Press, 1996:383–404.

170. Kendler KS, Neale MC, Kessler RC, et al. Panic disorder in women: a population-based twin study. *Psychol Med* 1993;23:397–406.

171. Knowles JA, Fyer AJ, Vieland VJ, et al. Results of a genome-wide genetic screen for panic disorder. *Am J Med Genet* 1998;81:139–147.

172. Crowe RR, Wang Z, Noyes RJ, et al. Candidate gene study of eight GABAA receptor subunits in panic disorder. *Am J Psychiatry* 1997;154:1096–1100.

173. Deckert J, Catalano M, Syagailo YV, et al. Excess of high activity monoamine oxidase A gene promoter alleles in female patients with panic disorder. *Hum Mol Genet* 1999;8:621–624.

174. Hamilton SP, Heiman GA, Haghighi F, et al. Lack of genetic linkage or association between a functional serotonin transporter polymorphism and panic disorder. *Psychiatr Genet* 1999;9:1–6.

175. Hamilton SP, Haghighi F, Heiman GA, et al. Investigation of dopamine receptor (DRD4) and dopamine transporter (DAT) polymorphisms for genetic linkage or association to panic disorder. *Am J Med Genet* 2000;96:324–330.

176. Hamilton SP, Slager SL, Hellby L, et al. No association or linkage between polymorphisms in the genes encoding cholecystokinin and cholecystokinin B receptor and panic disorder. *Mol Psychiatry* 2001;6:59–65.

177. Kennedy JL, Bradwejn J, Koszycki D, et al. Investigation of cholecystokinin system genes in panic disorder. *Mol Psychiatry* 1999;4:284–285.

178. Torgersen S. Genetic factors in anxiety disorders. *Arch Gen Psychiatry* 1983;40:1085–1089.

179. Hamilton SP, Slager SL, Heiman GA, et al. Evidence for a susceptibility locus for panic disorder near the catechol-o-methyltransferase gene on chromosome 22. *Biol Psychiatry* 2002;51:591–601.

180. Cameron OG, Lee MA, Curtis GC, et al. Endocrine and physiological changes during spontaneous panic attacks. *Psychoneuroendocrinology* 1987;12:321–331.

181. Woods SW, Charney DS, McPherson CA, et al. Situational panic attacks: behavioral, physiological, and biochemical characterization. *Arch Gen Psychiatry* 1987;44:365–375.

182. Nesse RM, Cameron OG, Buda AJ, et al. Urinary catecholamines and mitral valve prolapse in panic anxiety patients. *Psychiatry Res* 1985;14:67–75.

183. Nesse RM, Cameron OG, Curtis GC, et al. Adrenergic function in patients with panic anxiety. *Arch Gen Psychiatry* 1984;41:771.

184. Edlund MJ, Swann AC, Davis CM. Plasma MHPG in untreated panic disorder. *Biol Psychiatry* 1987;22:1491–1495.

185. Finaly JM, Abercrombie ED. Stress induced sensitization of norepinephrine release in the medial prefrontal cortex. *Soc Neurosci Abst* 1991;17:151.

186. Charney DS, Woods SW, Price LH, et al. Noradrenergic dysregulation in panic disorder. In: Ballenger JC, ed. *Neurobiology of panic disorders.* New York: Alan R. Liss, 1990:91–105.

187. Charney DS, Heninger GR, Breier A. Noradrenergic function in panic anxiety: effects of yohimbine in healthy subjects and patients with agoraphobia and panic disorder. *Arch Gen Psychiatry* 1984;41:751–763.

188. Charney DS, Woods SW, Goodman WK, et al. Neurobiological mechanisms of panic anxiety: biochemical and behavioral correlates of yohimbine-induced panic attacks. *Am J Psychiatry* 1987;144:1030–1036.

189. Charney DS, Woods SW, Krystal JH, et al. Noradrenergic neuronal dysregulation in panic disorder: the effects of intravenous yohimbine and clonidine in panic disorder patients. *Acta Psychiatr Scand* 1992;86:273–282.

190. Woods SW, Hoffer PB, McDougle CJ, et al. Cerebral noradrenergic function in panic disorder. *Biol Psychiatry (in press).*

191. Uhde TW, Murray MB, Vittone BJ, et al. Behavioral and physiological effects of short-term and long-term administration of clonidine in panic disorder. *Arch Gen Psychiatry* 1989;46:170–177.

192. Nutt DJ. Altered alpha$_2$-adrenoceptor sensitivity in panic disorder. *Arch Gen Psychiatry* 1989;46:165–169.

193. Pohl R, Yeragani VK, Balon R, et al. Isoproterenol induced panic attacks. *Biol Psychiatry* 1988;24:891–902.

194. Pohl R, Yeragani VK, Balon R. Effects of isoproterenol in panic disorder patients after antidepressant treatment. *Biol Psychiatry* 1990;28:203–214.

195. Roy-Byrne PP, Udhe TW, Sack DA, et al. Plasma HVA and anxiety in patients with panic disorder. *Biol Psychiatry* 1986;21:847–849.

196. Eriksson E, Westberg L, Alling C, et al. Cerebrospinal fluid levels of monoamine metabolites in panic disorder. *Psychiatry Res* 1991;36:243–251.

197. Uhde TW, Berrettini WH, Roy-Byrne PP, et al. Platelet 3$_H$-imipramine binding in patients with panic disorder. *Biol Psychiatry* 1987;22:52–58.

198. Innis RB, Charney DS, Heninger GR. Differential 3$_H$-imipramine platelet binding in patients with panic disorder and depression. *Psychiatry Res* 1987;21:33–41.

199. Norman TR, Judd FK, Gregory M, et al. Platelet serotonin uptake in panic disorder. *J Affect Disord* 1986;11:69–72.

200. Balon R, Poh R, Yeragani V, et al. Platelet serotonin levels in panic disorder. *Acta Psychiatr Scand* 1987;75:315.

201. Pecknold JC, Suranyi-Cadotte B, Chang H, et al. Serotonin uptake in panic disorder and agoraphobia. *Neuropsychopharmacology* 1988;1:173–176.

202. Schneider LS, Munjack D, Severson JA, et al. Platelet 3$_H$imipramine binding in generalized anxiety disorder, panic disorder, and agoraphobia with panic attacks. *Biol Psychiatry* 1987;21:33–41.

203. Lesch K. The ipsapirone/5-HT1A receptor challenge in anxiety disorders and depression. In: Stahl S, Hesselink JK, Gastpar M, et al., eds. *Serotonin 1A receptors in depression and anxiety.* New York: Raven Press, 1992:135–162.

204. Lesch KP, Wiesmann M, Hoh A, et al. 5-HT1A receptor-effector system responsivity in panic disorder. *Psychopharmacology* 1992;106:111–117.

205. Broocks A, Bandelow B, George A, et al. Increased psychological responses and divergent neuroendocrine responses to m-CPP and ipsapirone in patients with panic disorder. *Int Clin Psychopharmacol* 2000;15:153–161.

206. Van Vliet IM, Westenberg HG, den Boer JA. Effects of the 5-HT1A receptor agonist flesinoxan in panic disorder. *Psychopharmacology* 1996;127:74–180.

207. Meijer OC, Van Oosten RV, De Kloet ER. Elevated basal trough levels of corticosterone suppress hippocampal 5-hydroxytryptamine(1A) receptor expression in adrenally intact rats: implication for the pathogenesis of depression. *Neuroscience* 1997;80:419–426.

208. Lopez JF, Chalmers DT, Little KY, et al. Regulation of serotonin 1A, glucocorticoid, and mineralocorticoid receptor in rat and human hippocampus: implications for neurobiology of depression. *Biol Psychiatry* 1998;43:547–573.

209. Flugge G. Dynamics of central nervous 5-HT1A-receptors under psychosocial stress. *J Neurosci* 1995;15:7132–7140.

210. Sargent PA, Kjaer KH, Bench CJ, et al. Brain serotonin1A receptor binding measured by positron emission tomography with [11C]WAY-100635: effects of depression and antidepressant treatment. *Arch Gen Psychiatry* 2000;57:174–180.

211. Drevets WC, Frank E, Price JC, et al. PET image of serotonin 1A receptor binding in depression. *Biol Psychiatry* 1999;46:1375–1387.

212. Goldstein S, Halbreich U, Asnis G, et al. The hypothalamic pituitary-adrenal system in panic disorder. *Am J Psychiatry* 1987;144:1320–1323.

213. Holsboer F, vanBardeleben U, Buller R, et al. Stimulation response to corticotropin-releasing hormone (CRH) in patients with depression alcoholism in panic disorder. *Horm Metab Res* 1987;16[Suppl]:80–88.

214. Kathol RG, Anton R, Noyes R, et al. Relationship of urinary free cortisol levels in patients with panic disorder to symptoms of depression and agoraphobia. *Psychiatry Res* 1988;24:211–221.

215. Uhde T, Joffe RT, Jimerson DC, et al. Normal urinary free cortisol and plasma MHPG in panic disorder: clinical and theoretical implications. *Biol Psychiatry* 1988;23:575–585.

216. Abelson JL, Curtis GC. Hypothalamic-pituitary-adrenal axis

activity in panic disorder. *Arch Gen Psychiatry* 1996;53:323–332.

217. Corynell W, Noyes R. HPA axis disturbance and treatment outcome in panic disorder. *Biol Psychiatry* 1988;24:762–755.

218. Roy-Byrne PP, Udhe TW, Post RW, et al. The corticotropin-releasing hormone stimulation test in patients with panic disorder. *Am J Psychiatry* 1986;143:896–899.

219. Rapaport MH, Risch SC, Golshan S, et al. Neuroendocrine effects of ovine corticotropin-releasing hormone in panic disorder patients. *Biol Psychiatry* 1989;26:344–348.

220. Jolkkonen J, Lepola V, Bissette G, et al. CSF corticotropin-releasing factor is not affected in panic disorder. *Biol Psychiatry* 1993;33:136–138.

221. Mohler H, Okada T. Benzodiazepine receptor demonstration in the central nervous system. *Science* 1977;198:849–851.

222. Hirsch JD, Garrett KM, Beer B. Heterogeneity of benzodiazepine binding sites. A review of recent research. *Pharmacol Biochem Behav* 1985;23:681–685.

223. Roy-Byrne PP, Cowley DS, Greenblatt DJ, et al. Reduced benzodiazepine sensitivity in panic disorder. *Arch Gen Psychiatry* 1990;47:917–925.

224. Nutt DJ, Glue P, Lawson C, et al. Flumazenil provocation of panic attacks: evidence for altered benzodiazepine receptor sensitivity in panic disorder. *Arch Gen Psychiatry* 1990;47:917–925.

225. Woods SW, Charney DS, Silver JM, et al. Behavioral, biochemical and cardiovascular responses to benzodiazepine receptor antagonist flumazenil in panic disorder. *Psychol Res* 1991;36:115–124.

226. Dorow R, Horowski R, Paschelke G, et al. Severe anxiety induced by FG 7142, a β-carboline ligand for benzodiazepine receptors. *Lancet* 1983;7:98–99.

227. Clow A, Glover V, Armando I, et al. New endogenous benzodiazepine receptor ligand in human urine. Identity with endogenous monoamine oxidase inhibitor? *Life Sci* 1983;33:735–741.

228. Clow A, Glover V, Sandler M, et al. Increased urinary tribulin output in generalized anxiety disorder. *Psychopharmacology* 1988;95:378–380.

229. Ontiveros A, Fontaine R, Breton G, et al. Correlation of severity of panic disorder and neuroanatomical changes on magnetic resonance imaging. *J Neuropsychiatry Clin Neurosci* 1989;1:404–408.

230. Fontaine R, Breton G, Dery R, et al. Temporal lobe abnormalities in panic disorder. An MRI study. *Biol Psychiatry* 1990;27:304–310.

231. Schlegel S, Teinert H, Bockisch A, et al. Decreased benzodiazepine receptor binding in panic disorder measured by iomazenil SPECT. A preliminary report. *Eur Arch Psychiatry Clin Neurosci* 1994;244:49–51.

232. Kaschka W, Feistel H, Ebert D. Reduced benzodiazepine receptor binding in panic disorders measured by iomazenil SPECT. *J Psychiatry Res* 1995;29:427–434.

233. Kuikka JT, Pitkanen A, Lepola U, et al. Abnormal regional benzodiazepine receptor uptake in the prefrontal cortex in patients with panic disorder. *Nucl Med Commun* 1995;16:273–280.

234. Bremner JD, Innis RB, White T, et al. SPECT [123I]iomazenil measurements of the benzodiazepine receptor in panic disorder. *Biol Psychiatry* 2000;47:96–106.

235. Malizia AL, Cunningham VJ, Bell JB, et al. Decreased brain GABA$_A$-benzodiazepine receptor binding in panic disorder. *Arch Gen Psychiatry* 1998;55:715–720.

236. Abadie P, Boulenger JP, Benali K, et al. Relationship between trait and state anxiety and the central benzodiazepine receptor: a PET study. *Eur J Neurosci* 1999;11:1470–1478.

237. Bradwejn J, Koszyck D, Couetoux du Terte A, et al. The cholecystokinin hypothesis of panic and anxiety disorders. *J Psychopharmacol* 1992;6:345–351.

238. Bradwejn J, Koszycki D, Annable L, et al. A dose-ranging study of the behavioral and cardiovascular effects of CCK-tetrapeptide in panic disorder. *Biol Psychiatry* 1992;32:903–912.

239. Bradwejn J, Koszycki MA. Imipramine antagonism of the panicogenic effects of the cholecystokinin tetrapeptide in panic disorder. *Arch Gen Psychiatry* 1994;151:261–263.

240. Zohar J, Westenberg HG. Anxiety disorders: a review of tricyclic antidepressants and selective serotonin reuptake inhibitors. *Acta Psychiatr Scand Suppl* 2000;403:39–40.

241. Yeragani VK, Pohl R, Jampala VC, et al. Effects of nortriptyline and paroxetine on QT variability in patients with panic disorder. *Depress Anxiety* 2000;11:16–130.

242. Sheehan DV, Harnett-Sheehan K. The role of SSRIs in panic disorder. *J Clin Psychiatry* 1996;57:51–58.

243. Davidson J, Lydiard RB, McCann U, et al. Algorithm for the treatment of panic disorder with agoraphobia. *Psychopharmacol Bull* 1995;31:457–507.

244. Black DW, Wesner R, Bowers W, et al. A comparison of fluvoxamine, cognitive therapy, and placebo in the treatment of panic disorder. *Arch Gen Psychiatry* 1993;50:44–50.

245. Hoehn-Saric R, McLeod DR, Hipsley PA. Effect of fluvoxamine on panic disorder. *J Clin Psychopharmacol* 1987;7:329–332.

246. Rapaport MH, Pollack MH, Clary CM, et al. Panic disorder and response to sertraline: the effect of previous treatment with benzodiazepines. *J Clin Psychopharmacol* 2001;21:104–107.

247. Pollack MH, Rapaport MH, Clary CM, et al. Sertraline treatment of panic disorder: response in patients at risk for poor outcome. *J Clin Psychiatry* 2000;61:922–927.

248. Lepola UM, Wade AG, Leininen EV, et al. A controlled, prospective, 1-year trial of citalopram in the treatment of panic disorder. *J Clin Psychiatry* 1998;59:528–534.

249. Michelson D, Lydiard RB, Pollack MH, et al. Outcome assessment and clinical improvement in panic disorder: evidence from a randomized controlled trial of fluoxetine and placebo. The Fluoxetine Panic Disorder Study Group. *Am J Psychiatry* 1998;155:1570–1577.

250. Ballenger JC, Wheadon DE, Steiner M, et al. Double-blind, fixed-dose, placebo-controlled study of paroxetine in the treatment of panic disorder. *Am J Psychiatry* 1998;155:36–42.

251. Oehrberg S, Christiansen PE, Behnke K, et al. Paroxetine in the treatment of panic disorder: a randomised, double-blind, placebo-controlled study. *Br J Psychiatry* 1995;167:374–379.

252. Boyer W. Serotonin reuptake inhibitors are superior to imipramine and alprazolam in alleviating panic attacks: a meta-analysis. *Int Clin Psychopharmacol* 1995;10:45–49.

253. Moroz G, Rosenbaum JF. Efficacy, safety, and gradual discontinuation of clonazepam in panic disorder: a placebo-controlled, multicenter study using optimized dosages. *J Clin Psychiatry* 1999;60:604–612.

254. Woods SW, Kosten K, Krystal JH, et al. Yohimbine alters regional cerebral blood flow in panic disorder. *Lancet* 1988;2:678.

255. Goddard AW, Brouette T, Almai A, et al. Early coadministration of clonazepam with sertraline for panic disorder. *Arch Gen Psychiatry* 2001;58:681–686.

256. Breslau N, Davis GC, Andreski P, et al. Traumatic events and posttraumatic stress disorder in an urban population of young adults. *Arch Gen Psychiatry* 1991;48:216–222.

257. Davidson JR, Hughes D, Blazer DG, et al. Post-traumatic stress disorder in the community: an epidemiological study. *Psychol Med* 1991;21:713–721.

258. Kessler RC, Sonnega A, Bromet E, et al. Posttraumatic stress disorder in the National Comorbidity Survey. *Arch Gen Psychiatry* 1995;52:1048–1060.

259. Margolin G, Gordis EB. The effects of family and community violence on children. *Annu Rev Psychol* 2000;51:445–479.

260. Blanchard EB, Buckley TC, Hickling EJ, et al. Posttraumatic stress disorder and comorbid major depression: is the correlation an illusion? *J Anxiety Disord* 1998;12:21–37.

261. Beckham JC, Lytle BL, Feldman ME. Caregiver burden in partners of Vietnam War veterans with posttraumatic stress disorder. *J Consult Clin Psychol* 1996;64:1068–1072.

262. Uno H, Tarara R, Else JG, et al. Hippocampal damage associated with prolonged and fatal stress in primates. *J Neurosci* 1989;9:1705–1711.

263. Watanabe Y, Gould E, McEwen BS. Stress induces atrophy of apical dendrites of hippocampal CA3 pyramidal neurons. *Brain Res* 1992;588:341–345.

264. Bremner JD, Randall P, Scott TM, et al. MRI-based measurement of hippocampal volume in patients with combat-related posttraumatic stress disorder. *Am J Psychiatry* 1995;152:973–981.

265. Bremner JD, Randall P, Scott TM, et al. MRI-based measurement of hippocampal volume in patients with combat-related posttraumatic stress disorder. *Am J Psychiatry* 1995;152:973–981.

266. Stein MB, Koverola C, Hanna C, et al. Hippocampal volume in women victimized by childhood sexual abuse. *Psychol Med* 1997;27:951–959.

267. Gurvits TV, Shenton ME, Hokama H, et al. Magnetic resonance imaging study of hippocampal volume in chronic, combat-related posttraumatic stress disorder. *Biol Psychiatry* 1996;40:1091–1099.

268. Vythilingam M, Heim C, Newport D, et al. Childhood trauma associated with smaller hippocampal volume in women with major depression. *Am J Psychiatry* 2002;159:2072–2080.

269. Golier J, Yehuda R, Grossman R, et al. Hippocampal volume and memory performance in Holocaust survivors with and without PTSD. Presented at the annual meeting of the American College of Neuropsychopharmacology, San Juan, PR, 2000.

270. Schuff N, Marmar CR, Weiss DS, et al. Reduced hippocampal volume and n-acetyl aspartate in posttraumatic stress disorder. *Ann N Y Acad Sci* 1997;821:516–520.

271. Bonne O, Brandes D, Gilboa A, et al. Longitudinal MRI study of hippocampal volume in trauma survivors with PTSD. *Am J Psychiatry* 2001;158:1248–1251.

272. De Bellis MD, Hall J, Boring AM, et al. A pilot longitudinal study of hippocampal volumes in pediatric maltreatment-related posttraumatic stress disorder. *Biol Psychiatry* 2001;50:305–309.

273. De Bellis MD, Keshavan MS, Clark DB, et al. A.E. Bennett Research Award. Developmental traumatology. Part II: brain development. *Biol Psychiatry* 1999;45:1271–1284.

274. Agartz I, Momenan R, Rawlings RR, et al. Hippocampal volume in patients with alcohol dependence. *Arch Gen Psychiatry* 1999;56:356–363.

275. Duman RS, Malberg J, Thome J. Neural plasticity to stress and antidepressant treatment. *Biol Psychiatry* 1999;46:1181–1191.

276. Duman RS, Malberg J, Nakagawa S, et al. Neuronal plasticity and survival in mood disorders. *Biol Psychiatry* 2000;48:732–739.

277. LaBar KS, Gatenby JC, Gore JC, et al. Human amygdala activation during conditioned fear acquisition and extinction: a mixed-trial fMRI study. *Neuron* 1998;20:937–945.

278. Rauch SL, van der Kolk BA, Fisler RE, et al. A symptom provocation study of posttraumatic stress disorder using positron emission tomography and script-driven imagery. *Arch Gen Psychiatry* 1996;53:380–387.

279. Shin LM, Kosslyn SM, McNally RJ, et al. Visual imagery and perception in posttraumatic stress disorder. A positron emission tomographic investigation. *Arch Gen Psychiatry* 1997;54:233–241.

280. Liberzon I, Taylor SF, Amdur R, et al. Brain activation in PTSD in response to trauma-related stimuli. *Biol Psychiatry* 1999;45:817–826.

281. Bremner JD, Staib LH, Kaloupek D, et al. Neural correlates of exposure to traumatic pictures and sound in Vietnam combat veterans with and without posttraumatic stress disorder: a positron emission tomography study. *Biol Psychiatry* 1999;45:806–816.

282. Bremner JD, Narayan M, Staib LH, et al. Neural correlates of memories of childhood sexual abuse in women with and without posttraumatic stress disorder. *Am J Psychiatry* 1999;156:1787–1795.

283. Zubieta JK, Chinitz JA, Lombardi U, et al. Medial frontal cortex involvement in PTSD symptoms: a SPECT study. *J Psychiatr Res* 1999;33:259–264.

284. Rauch SL, Whalen PJ, Shin LM, et al. Exaggerated amygdala response to masked facial stimuli in posttraumatic stress disorder: a functional MRI study. *Biol Psychiatry* 2000;47:769–776.

285. Whalen PJ, Rauch SL, Etcoff NL, et al. Masked presentations of emotional facial expressions modulate amygdala activity without explicit knowledge. *J Neurosci* 1998;18:411–418.

286. Morris JS, Friston KJ, Buchel C, et al. A neuromodulatory role for the human amygdala in processing emotional facial expressions. *Brain* 1998;121:47–57.

287. Morgan MA, LeDoux JE. Differential contribution of dorsal and ventral medial prefrontal cortex to the acquisition and extinction of conditioned fear in rats. *Behav Neurosci* 1995;109:681–688.

288. Quirk GJ, Russo GK, Barron JL, et al. The role of ventromedial prefrontal cortex in the recovery of extinguished fear. *J Neurosci* 2000;20:6225–6231.

289. Morgan MA, Romanski LM, LeDoux JE. Extinction of emotional learning: contribution of medial prefrontal cortex. *Neurosci Lett* 1993;163:109–113.

290. Garcia R, Vouimba RM, Baudry M, et al. The amygdala modulates prefrontal cortex activity relative to conditioned fear. *Nature* 1999;402:294–296.

291. Shin LM, McNally RJ, Kosslyn SM, et al. Regional cerebral blood flow during script-driven imagery in childhood sexual abuse-related PTSD: a PET investigation. *Am J Psychiatry* 1999;156:575–584.

292. Orr SP, Metzger LJ, Lasko NB, et al. De novo conditioning in trauma-exposed individuals with and without posttraumatic stress disorder. *J Abnorm Psychol* 2000;109:290–298.

293. Peri T, Ben-Shakhar G, Orr SP, et al. Psychophysiologic assessment of aversive conditioning in posttraumatic stress disorder. *Biol Psychiatry* 2000;47:512–519.

294. Yehuda R, Boisoneau D, Lowy MT, et al. Dose-response

changes in plasma cortisol and lymphocyte glucocorticoid receptors following dexamethasone administration in combat veterans with and without posttraumatic stress disorder. *Arch Gen Psychiatry* 1995;52:583–593.

295. Pitman RK, van der Kolk BA, Orr SP, et al. Naloxone-reversible analgesic response to combat-related stimuli in posttraumatic stress disorder. *Arch Gen Psychiatry* 1990;47:541–544.

296. Yehuda R, Boiseneau S, Mason JW, et al. Glucocorticoid receptor number and cortisol excretion in mood, anxiety, and psychotic disorder. *Biol Psychiatry* 1993;34:18–25.

297. DeBellis MD, Lefter L, Trickett PK, et al. Urinary catecholamine excretion in sexually abused girls. *J Am Acad Child Adoles Psychiatry* 1994;33:320–327.

298. Lemieux AM, Coe CL. Abuse-related PTSD: evidence for chronic neuroendocrine activation in women. *Psychosom Med* 1995;57:105–115.

299. Liberzon I, Abelson JL, Flagel SB, et al. Neuroendocrine and psychophysiological responses in PTSD: a symptom provocation study. *Neuropsychopharmacology* 1999;21:40–50.

300. Maes M, Lin A, Bonaccorso S, et al. Increased 24-hour urinary cortisol excretion in patients with posttraumatic stress disorder and patients with major depression, but not in patients with fibromyalgia. *Acta Psychiatr Scand* 1998;90:325–328.

301. Liberzon I, Taylor SF, Amdur R, et al. Brain activation in PTSD in response to trauma-related stimuli. *Biol Psychiatry* 1999;45:817–826.

302. Yehuda R, Lowy MT, Southwick SM, et al. Lymphocyte glucocorticoid receptor number in posttraumatic stress disorder. *Am J Psychiatry* 1991;148:499–504.

303. Yehuda R, Southwick SM, Krystal JH, et al. Enhanced suppression of cortisol following dexamethasone administration in posttraumatic stress disorder. *Am J Psychiatry* 1993;150:83–86.

304. Bremner JD, Licinio J, Darnell A, et al. Elevated CSF corticotropin-releasing factor concentrations in posttraumatic stress disorder. *Am J Psychiatry* 1997;154:624–629.

305. Smith MA, Davidson J, Ritchie JC, et al. The corticotropin-releasing homone-test in patients with PTSD. *Biol Psychiatry* 1989;26:349–355.

306. Hawk LW, Dougall AL, Orsano RJ, et al. Urinary catecholamines and cortisol in recent-onset posttraumatic stress disorder after motor vehicle accidents. *Psychosom Med* 2000;62:423–434.

307. McEwen BS. Stress and hippocampal plasticity. *Annu Rev Neurosci* 1999;22:105–122.

308. Weizman R, Tanne Z, Granek M, et al. Peripheral benzodiazepine binding sites on platelet membranes are increased during diazepam treatment of anxious patients. *Eur J Pharmacol* 1987;138:289–292.

309. Weizman R, Laor N, Karp L, et al. Alteration of platelet benzodiazepine receptors by stress of war. *Am J Psychiatry* 1994;151:766–767.

310. Gavish M, Laor N, Bidder M, et al. Altered platelet peripheral-type benzodiazepine receptors in posttraumatic stress disorder. *Neuropsychopharmacology* 1996;14:181–186.

311. Braun P, Greenberg D, Dasberg H, et al. Core symptoms of posttraumatic stress disorder unimproved by alprazolam treatment. *J Clin Psychiatry* 1990;50:236–238.

312. Gelpin E, Bonne O, Peri T, et al. Treatment of recent trauma survivors with benzodiazepines: a prospective study. *J Clin Psychiatry* 1996;57:390–394.

313. Brown GL, Linnoila MI. CSF serotonin metabolite (5-HIAA) studies in depression, impulsivity, and violence. *J Clin Psychiatry* 1990;51:31–41.

314. Stanley M, Stanley B. Postmortem evidence for serotonin's role in suicide. *J Clin Psychiatry* 1990;51:22–28.

315. Paige S, Reid G, Allen M. Psychophysiological correlates of post traumatic stress disorder. *Biol Psychiatry* 1990;27:419–430.

316. Hegerl U, Ulrich G, Muller-Oerlinghausen B. Auditory evoked potentials and response to lithium prophylaxis. *Pharmacopsychiatry* 1987;20:213–216.

317. McDougle C, Southwick SM, Charney DS, et al. An open trial of fluoxetine in the treatment of post traumatic stress disorder. *J Clin Psychopharmacol* 1992;1:325–327.

318. Nagy LM, Morgan CA, Southwick SM, et al. Open prospective trial of fluoxetine for posttraumatic stress disorder. *J Clin Psychopharmacol* 1993;13:107–113.

319. Arora RC, Fitchuer CT, O'Connor F. Paroxetine binding in the blood platelets of posttraumatic stress disordered patients. *Life Sci* 1993;53:919–928.

320. Southwick SM, Krystal JH, Morgan CA, et al. Yohimbine and m-chloro-phenylpiperazine in PTSD. New Research Abstracts of the American Psychiatric Association 144th Annual Meeting, 1991:NR3481.

321. Kosten TR, Mason JW, Giller EL, et al. Sustained urinary norepinephrine and epinephrine elevation in posttraumatic stress disorder. *Psychoneuroendocrinology* 1987;12:13–20.

322. Yehuda R, Southwick SM, Giller EL. Urinary catecholamine excretion and severity of post-traumatic stress disorder in Vietnam combat veterans. *J Nerv Ment Dis* 1992;180:321–325.

323. Perry BD, Giller EL, Southwick SM. Altered platelet alpha2-adrenergic binding sites in posttraumatic stress disorder. *Am J Psychiatry* 1987;144:1511–1512.

324. Nutt DJ. Increased central alpha-2 adrenoceptor sensitivity in panic disorder. *Psychopharmacology* 1986;90:268–269.

325. Brown SL, Charney DS, Woods SW, et al. Lymphocyte beta-adrenergic receptor binding in panic disorder. *Psychopharmacology* 1988;94:2428–2434.

326. Krystal JH, Kosten TR, Perry BD, et al. Neurobiological aspects of PTSD: review of clinical and preclinical studies. *Behav Ther* 1989;20:177–198.

327. Grinker RR, Spiegle JJ. *Men under stress.* New York: McGraw-Hill, 1945:219.

328. Meakes JC, Wilson RM. The effect of certain sensory stimulations on respiratory and heart rate in case of "so-called" irritable heart. *Heart* 1918;7:17–22.

329. Peabody FW, Clough HD, Sturgis LC, et al. Effects of injection of norepinephrine in soldiers with "irritable heart." *JAMA* 1918;21:1912–1917.

330. Boudewyns PA, Hyer L. Psychophysiological response to combat memories and preliminary treatment outcome in Vietnam veteran PTSD patients treated with direct therapeutic exposure. *Behav Ther* 1990;21:63–87.

331. Keane T, Fairbank J, Caddell JM, et al. Implosive therapy reduces symptoms of PTSD in Vietnam combat veterans. *Behav Ther* 1989;20:245–260.

332. Southwick SM, Krystal JH, Morgan CA, et al. Abnormal noradrenergic function in post-traumatic stress disorder. *Arch Gen Psychiatry* 1993;181:31–37.

333. McGaugh JL. Involvement of hormonal and neuromodulatory systems in the regulation of memory storage: endogenous modulation of memory storage. *Annu Rev Neurosci* 1989;12:255–287.

334. McGaugh JL. Significance and remembrance: the role of neuromodulatory systems. *Psychol Sci* 1990;1:15–25.

335. Hwu H-G, Yeh E-K, Chang L-Y. Prevalence of psychiatric disorders in Taiwan defined by Chinese Diagnostic Interview Schedule. *Acta Psychiatr Scand* 1989;79:136–147.

336. Lee CK, Kwak YS, Yamamoto J, et al. Psychiatric epidemiology in Korea: part 1. Gender and age differences in Seoul. *J Nerv Ment Dis* 1990;178:242–246.

337. Wacker H, Mullejans R, Kelin K, et al. Identification of cases of anxiety disorders and affective disorders in the community according to the ICD-10 and the DSM-III-R by using the Composite International Diagnostic Interview (CIDI). *Int J Methods Psychiatr Res* 1992;2:91–100.

338. Merikangas K, Avenevoli S, Acharyya S, et al. The spectrum of social phobia in the zurich cohort study of young adults. *Biol Psychiatry* 2002;51:81–91.

339. Kendler KS, Neale MC, Kessler RC, et al. The genetic epidemiology of phobias in women. *Arch Gen Psychiatry* 1992;49:273–281.

340. Nelson EC, Grant JD, Bucholz KK, et al. Social phobia in a population-based female adolescent twin sample: Comorbidity and associated suicide-related symptoms. *Psychol Med* 2000;30:797–804.

341. Nesse RM, Curtis GC, Thyer BA, et al. Endocrine and cardiovascular responses during phobic anxiety. *Psychosom Med* 1985;47:320–332.

342. Stein MB, Tancer ME, Uhde TW. Heart rate and plasma norepinephrine responsivity to orthostatic challenge in anxiety disorders. Comparison of patients with panic disorder and social phobia and normal control subjects. *Arch Gen Psychiatry* 1992;49:311–317.

343. Stein MB, Huzel LL, Delaney SM. Lymphocyte β-adrenoceptors in social phobia. *Biol Psychiatry* 1993;34:45–50.

344. Tiihonen J, Kuikka J, Bergstrom K, et al. Dopamine reuptake site densities in patients with social phobia. *Am J Psychiatry* 1997;154:239–242.

345. Schneier FR, Liebowitz MR, Abi-Dargham A, et al. Low dopamine D(2) receptor binding potential in social phobia. *Am J Psychiatry* 2000;158:327–328.

346. Pigott TA. Gender differences in the epidemiology and treatment of anxiety disorders. *J Clin Psychiatry* 1999;18:4–15.

347. Wittchen HU, Zhao S, Kessler RC, et al. DSM-III-R generalized anxiety disorder in the National Comorbidity Survey. *Arch Gen Psychiatry* 1994;51:355–364.

348. Yonkers KA, Warshaw MG, Massion AO, et al. Phenomenology and course of generalised anxiety disorder. *Br J Psychiatry* 1996;168:308–313.

349. Cedarbaum JM, Aghajanian GK. Afferent projections to the rat locus coeruleus as determined by a retrograde tracing technique. *J Comp Neurol* 1978;178:1–16.

350. Sevy S, Papadimitriou GN, Surmont DW, et al. Noradrenergic function in generalized anxiety disorder, major depressive disorder, and healthy subjects. *Biol Psychiatry* 1989;25:141–152.

351. Charney DS, Woods SW, Heninger GR. Noradrenergic function in generalized anxiety disorder. Effects of yohimbine in healthy subjects and patients with generalized anxiety disorder. *Psychiatry Res* 1987;27:173–182.

352. Matthew RJ, Ho BT, Francis DJ, et al. Catecholamines and anxiety. *Acta Psychiatr Scand* 1982;65:142–147.

353. Matthew RJ, Ho BT, Kralik P, et al. Catecholamines and monoamine oxidase activity in anxiety. *Acta Psychiatr Scand* 1981;63:245–252.

354. Cameron OG, Smith CB, Lee MA, et al. Adrenergic status in anxiety disorders. Platelet alpha-2-adrenergic receptor binding, blood pressure, pulse, and plasma catecholamines in panic and generalized anxiety disorder patients and in normal subjects. *Biol Psychiatry* 1990;28:3–20.

355. Curtis G, Lee MA, Glitz DA, et al. Growth hormone response to clonidine in anxiety disorders. *Biol Psychiatry* 1989;25: 6A.

356. Rocca P, Ferrero P, Gualerzi A, et al. Peripheral type benzodiazepine receptors in anxiety disorders. *Acta Psychiatr Scand* 1992;84:537–544.

357. Weizman R, Tanne Z, Granek M, et al. Peripheral benzodiazepine binding sites on platelet membranes are increased during diazepam treatment of anxious patients. *Eur J Pharmacol* 1987;138:289–292.

358. Wu JC, Buchsbaum MS, Hershey TG, et al. PET in generalized anxiety disorder. *Biol Psychiatry* 1991;29:1181–1199.

359. Buchsbaum MS, Wu J, Haier R, et al. Positron emission tomography assessment of the effect of benzodiazepines on regional glucose metabolic rate in patients with anxiety disorder. *Life Sci* 1987;40:2392–2400.

SUGGESTED READINGS

Litz BT, Keane TM. Information processing in anxiety disorders: application to the understanding of post-traumatic stress disorder. *Clin Psychol Rev* 1989;9:243–257.

McNally RJ, Luedke DL, Besyner JK, et al. Sensitivity to stress relevant stimuli in post traumatic stress disorder. *J Anxiety Disord* 1987;1:105–116.

Phillips RG, LeDoux JE. Differential contribution of amygdala and hippocampus to cued and contextual fear conditioning. *Behav Neurosci* 1992;106:274–285.

Hoffman HS, Selekman W, Fleishler M. Stimulus aspects of aversive controls: long term effects of suppression procedures. *J Exp Anal Behav* 1966;9:659–662.

Davis M. Animal models of anxiety based upon classical conditioning: the conditioned emotional response and fear potentiated startle effect. *Pharmacol Ther* 1990;47:147–165.

Kim JJ, Fanselow MS. Modality-specific retrograde amnesia of fear. *Science* 1992;256:675–677.

Cassens G, Kuruc A, Roffman M, et al. Alterations in brain norepinephrine metabolism and behavior induced by environmental stimuli previously paired with inescapable shock. *Behav Brain Res* 1981;2:387–407.

Tanaka M, Ida Y, Tsuda A, et al. Involvement of brain noradrenaline and opioid peptides in emotional changes induced by stress in rats. In: Oomura Y, ed. *Emotions: neural and chemical control.* Tokyo: Scientific Societies Press, 1986:417–427.

Rasmussen K, Marilak DA, Jacobs BL. Single unit activity of the locus coeruleus in the freely moving cat. I: during naturalistic behaviors and in response to simple and complex stimuli. *Brain Res* 1986;371:324–334.

Tsaltas E, Gray JA, Fillenz M. Alleviation of response suppression to conditioned aversive stimuli by lesions of the dorsal noradrenergic bundle. *Behav Brain Res* 1987;13:115–127.

Cole BJ, Robbins TW. Dissociable effects of lesions to dorsal and ventral noradrenergic bundle on the acquisition performance, and extinction of aversive conditioning. *Behav Neurosci* 1987;101:476–488.

Estes WK. Statistical theory of spontaneous recovery and regression. *Psychol Rev* 1955;62:145–154.

Konorski J. *Conditioned reflexes and neuronal organization.* London: Cambridge University Press, 1948.

Bouton ME, Bulles RC. Contexts, event memories, and extinction.

In: Balsam PDD, Tomic KA, eds. *Context and learning* Hillsdale, NJ: Lawrence Erlbaum, 1985:133–166.

Bouton ME, King DA. Contextual control of conditioned fear: tests for the associative value of the context. *J Exp Psychol Anim Behav Process* 1983;9:248–256.

Bouton ME, King DA. Effect of context with mixed histories of reinforcement and nonreinforcement. *J Exp Psychol Anim Behav Process* 1986;12:4–15.

Bouton ME, Bolles RC. Role of contextual stimuli in reinstatement of extinguished fear. *J Exp Psychol Anim Behav Process* 1979;5:368–378.

Pavlov JP. *Conditioned reflexes.* Oxford, England: Oxford University Press, 1927.

Rescola RA, Heth CD. Reinstatement of fear to extinguished conditioned stimulus. *J Psychol Anim Behav Process* 1975;104:88–96.

McAllister WR, McAllister DE. Reconditioning of extinguished fear after a one-year delay. *Bull Psychonom Soc* 1988;26:463–466.

VanDyke C, Zilberg NJ, MacKinnon JA. Post traumatic stress disorder: a thirty year delay in a World War II veteran. *Am J Psychiatry* 1985;142:1070–1073.

Solomen Z, Garb K, Bleich A, et al. Reactivation of combat related post traumatic stress disorder. *Am J Psychiatry* 1987;144:51–55.

Falls WA, Miserendino MJD, Davis M. Excitatory amino acid antagonists infused into the amygdala block extinction of fear-potentiated startle. *Soc Neurosci Abst* 1990;16:767.

Teich AH, McCabe PM, Gentile CC, et al. Auditory cortex lesions prevent extinction of pavlovian differential heart rate conditioning to tonal stimuli in rabbits. *Brain Res* 1989;480:210–218.

Post RM. Transduction of psychosocial stress into the neurobiology of recurrent affective disorders. *Am J Psychiatry* 1992;149:999–1010.

Weiss SRB, Post RM, Pert A, et al. Context-dependent cocaine sensitization: differential effect of haloperidol on development versus expression. *Pharmacol Biochem Behav* 1989;34:655–661.

King GR, Joyner C, Lee T, et al. Intermittent and continuous cocaine administration: residual behavioral states during withdrawal. *Pharmacol Biochem Behav* 1992;43:243–248.

Antelman SM. Time dependent sensitization as the cornerstone for a new approach to pharmacotherapy: drugs as foreign or stressful stimuli. *Drug Dev Res* 1988;14:1–30.

Sorg B, Kalivas PW. Behavioral and neurochemical cross-sensitization between footshock stress and cocaine. *Brain Res* 1991;559:29–36.

Caggiula AR, Antelman SM, Aul E, et al. Prior stress attenuates the analgesic response but sensitizes the corticosterone and cortical dopamine responses to stress 10 days later. *Psychopharmacology* 1989;99:233–237.

Kalivas PW, Duffy P. Similar effects of daily cocaine and stress on mesocorticolimbic dopamine neurotransmission in the rat. *Biol Psychiatry* 1989;25:913–928.

Kalivas PW, Duffy P, Abhold R, et al. Sensitization of mesolimbic dopamine neurons by neuropeptides and stress. In: Kalivas PW, Barnes CD, eds. *Sensitization in the nervous system.* Caldwell, NJ: Telford Press, 1990:119–124.

Kalivas PW, Striplin CD, Steketee JD, et al. Cellular mechanisms of behavioral sensitization to drugs of abuse. *Ann N Y Acad Sci* 1992;654:128–135.

Criswell HE, Mueller RA, Breese GR. Long-term D1-dopamine receptor sensitization in neonatal 6-OHD-lesioned rats is blocked by an NMDA antagonist. *Brain Res* 1990;512:284–290.

Fontana DJ, Post RM, Pert A. Conditioned increase in mesolimbic dopamine overflow by stimuli associated with cocaine. *Brain Res* 1993;629(1):31–39.

Nisenbaum LK, Zigmand MJ, Sved AF, et al. Prior exposure to chronic stress results in enhanced synthesis and release of hippocampal norepinephrine in response to a novel stressor. *J Neurosci* 1991;11:1478–1484.

Finlay JM, Abercrombie ED. Stress induced sensitization of norepinephrine release in the medial prefrontal cortex. *Soc Neurosci Abst* 1991;17:151.

30

NEUROPSYCHIATRIC ASPECTS
OF SCHIZOPHRENIA

ELIZABETH W. TWAMLEY
CHRISTIAN R. DOLDER
JODY COREY-BLOOM
DILIP V. JESTE

Schizophrenia is a serious psychiatric illness that can involve massive disruptions of thinking, perception, emotion, and behavior. Many experts consider schizophrenia to be the most severe expression of mental dysfunction in the spectrum of psychopathology. With a worldwide lifetime prevalence rate of 1%, schizophrenia is a lifelong psychiatric disability, often tragically affecting patients' family relationships and social and occupational functioning and frequently requiring periodic or even chronic hospitalization. The symptoms of schizophrenia are variable across patients and fall into three groups: positive symptoms (e.g., hallucinations, delusions), negative symptoms (e.g., affective blunting, anhedonia, poverty of speech, impaired initiation of behavior), and disorganized symptoms (e.g., formal thought disorder, bizarre behavior) (1).

Schizophrenia is, in all probability, as ancient as human civilization (2). Kraepelin provided the first thorough description of the constellation of symptoms that comprise schizophrenia, which he called dementia praecox (literally, early dementia). At the turn of the century, Kraepelin distinguished the more pervasive perceptual and cognitive dysfunctions of dementia praecox from that of manic-depressive psychoses and noted that the two conditions differed in both course and outcome (3). Dementia praecox had an earlier onset and a progressively deteriorating course with no return to the premorbid level of functioning, whereas the mood disorders had episodic periods of severe psychopathology followed by periods of normal functioning. Kraepelin also noted that no single pathognomonic symptom or cluster of symptoms served to characterize dementia praecox, an illness that he considered so mysterious that he referred to it as a disorder whose causes were shrouded in "impenetrable darkness."

Although many questions remain about the causes and consequences of schizophrenia, it is a less enigmatic condition than it once was. In the past decades, scientific understanding of the risk factors for schizophrenia, as well as its neuropathology and neurochemistry, has grown exponentially. Contemporary conceptualizations of schizophrenia concur that there are progressive features of the disorder (e.g., decline in functioning from premorbid levels) but emphasize the neurodevelopmental nature of the illness. McGlashan and Hoffman (4), for example, describe schizophrenia as a "progressive neurodevelopmental disorder." Today, schizophrenia is viewed as a brain disease of disrupted neuronal connections. Neuronal dysfunction leads to multiple neurocognitive deficits, and these deficits in turn lead to striking functional impairment and disability. According to this view, the dramatic positive symptoms of schizophrenia (hallucinations and delusions) are present during active phases of the disorder but are less central to the conceptualization of schizophrenia. Positive symptoms, so long a focus of treatment, are merely "along for the ride," explaining little about the persistent cognitive deficits, negative symptoms, and functional impairment that characterize schizophrenia. During the past several years, the research literature has identified a number of cognitive and neurologic deficits (e.g., in attention, perception, eye tracking, sensory gating, and evoked potentials) that appear to be markers for predisposition to schizophrenia (5).

Where does schizophrenia come from? Many decades of genetic and epidemiologic research have led to the conclusion that schizophrenia is a complex, multiply determined disorder. Risk for schizophrenia is, to some extent, genetically transmitted and is probably determined by multiple genes; a recent review of the molecular genetics of schizophrenia (6) cited linkage studies implicating chromosomes 6, 8, 10, 13, and 22. Association studies have investigated not only the neurotransmission of dopamine but also the roles of serotonin, acetylcholine, γ-aminobutyric acid (GABA), and glutamate in susceptibility to schizophrenia. Environmental risk factors, such as malnutrition in the first trimester of pregnancy or influenza in the second trimester, also have been identified. Genetic and environmental factors appear to interact, leading to dysfunction in cell migration and pruning in neurodevelopment. This suboptimal cytoarchitecture is thought to be the cause of suboptimal neuronal connectivity, which then results in the cognitive, affective, and social impairments of schizophrenia. As stated by Green (5), "The illness starts with disconnection at the neural level and leads to disconnection at the interpersonal level."

Because it is recognized as a neurobiologic disorder, current approaches to schizophrenia have focused on neurophysiologic explanations of its etiology (7,8). Recent technologic innovations in brain imaging have improved the ability to map and examine the structure and function of the brain and have resulted in improved understanding of the neuropathologic underpinnings of schizophrenia. However, schizophrenia presents no clear-cut neuropathology, unlike Alzheimer disease or brain tumors, and the exact localization and nature of the neuronal dysfunctions that produce schizophrenia remain uncertain. Partly because there is no consistent neuropathology of schizophrenia, current theories of the disorder have moved away from the view that it is a single entity and toward a conceptualization of schizophrenia as a collection of etiologically disparate disorders with similar clinical presentation—as a syndrome with heterogeneous symptoms, implicating a diffuse and diverse neuropathology (9,10). Thus, in the past decade, research on schizophrenia has focused increasingly on the different symptom complexes of the disorder (positive, negative, and disorganized) and their distinct neuroanatomic substrates. For example, positive symptoms may involve the anterior cingulate–basal ganglia–thalamocortical and language pathways, whereas negative symptoms may involve the dorsolateral prefrontal–basal ganglia–thalamocortical circuit (11–13).

In this chapter, we consider the following aspects of schizophrenia: neuropsychology, brain imaging, neuropathology, neurochemistry, late-onset schizophrenia (LOS), secondary psychoses, and management, including neuroleptic-induced tardive dyskinesia (TD).

NEUROPSYCHOLOGY

The neuropsychological approach assumes that the cognitive functions measured by specific tests are associated with specific brain regions or systems and, thus, impairment on a specific test is taken as evidence of dysfunction in the specific brain region or system associated with the test (see Chapter 1). Several extensive reviews (14–20) have described numerous neuropsychological deficits in patients with schizophrenia. The most consistently replicated deficits in patients with schizophrenia have been observed on tasks measuring attention, information-processing speed, working memory, learning, and executive (i.e., frontal systems) functions (further discussed below) (21). Although patients with schizophrenia frequently show movement disorders and psychomotor slowing, impairment on motor tasks is often attributable to

attention, information-processing speed, or executive function deficits or medication effects (18,22). Similarly, findings of impairment in basic language and visuospatial functions have been inconsistent, and deficits in these domains may be owing to increased demands on attention or executive functions (18). With regard to language functions, it is clear that patients with schizophrenia produce abnormal utterances [e.g., neologisms (words that are made up by the individual)]; however, an area of debate is whether these utterances are the result of disordered language or disordered thought (23).

The brief selective review below focuses on the more robust findings of deficits in attention and information-processing speed, working memory, learning/memory, and executive functions. These four domains are not only the most commonly impaired in schizophrenia but are also the most strongly associated with functional outcomes (i.e., social and community functioning) (21). This pattern of neuropsychological impairments in patients with schizophrenia has led to the hypothesis that specific frontal–subcortical brain systems are involved in the pathophysiology of schizophrenia. However, given the widespread cognitive impairments characteristic of schizophrenia, it is not likely that schizophrenia can be attributed to one locus in the brain or one central cognitive deficit. It should be noted that the cognitive deficits that characterize schizophrenia are stable, rather than progressive, and tend to remain stable regardless of aging or fluctuations in clinical state (24,25). Additionally, it has been shown that a sizable minority (approximately one-fourth) of schizophrenia patients is "neuropsychologically normal," i.e., their cognitive performances are not significantly different from those of demographically matched normal comparison subjects (26). Thus, a focus of ongoing research has been the derivation of meaningful subtypes of schizophrenia based on cognitive profiles (10).

Attention and Information-Processing Speed

Multiple types of attention exist, including selective attention (ability to focus attention on a single source of incoming information), divided attention (ability to divide attentional resources among multiple sources of incoming information), and vigilance (ability to sustain attention over time). Preattentive abilities, which involve perception of information to be attended to, are frequently studied along with attention. Information-processing speed also is considered along with attention because most tasks measuring processing speed necessarily involve attention to various stimuli. Individuals with schizophrenia often exhibit impaired performances in these domains, particularly when the processing demands of the tasks are increased (e.g., by increasing the number or complexity of cognitive operations needed to complete the task) (19,27,28).

Preattentive Abilities

A common task used to measure this perceptual domain is the span of apprehension task, assessing the number of items that can be perceived at one time. For example, subjects must report whether a T or F target letter is among a group of distractor letters flashed briefly in a visual display. In the span of apprehension task studies reviewed (29), schizophrenia patients accurately reported significantly fewer target letters than nonpsychiatric individuals with 8-, 10-, or 12-letter arrays (higher processing demands), but the patients' detection accuracy did not differ significantly from nonpsychiatric subjects with smaller array sizes (lower processing demands).

Selective Attention

Tasks in this domain are designed to measure the ability to select one stimulus or channel (e.g., ear) over another for processing. Studies using dichotic listening tasks, for example, have shown that schizophrenia patients are abnormally susceptible to the effects of distraction. On these tasks, subjects are asked to listen to a message and repeat (shadow) it out loud, while ignoring an irrelevant message played simultaneously either in a different ear or in a different voice. When shadowing random word lists, schizophrenia patients commonly make more shadowing omission errors than normal subjects when irrelevant word lists are introduced (30,31). These impairments are, however, reduced or eliminated when shadowing semantically and syntactically structured textual information (32,33), which has inherent syntactic organization that can be more automatically deciphered (reduced processing demand) (32). Impaired shadowing during distraction in schizophrenia patients has also been shown with faster (50 items per minute), but not slower (25 items per minute), presentation rates (31), which again suggests that higher processing demands lead to information-processing overload and cognitive impairment in schizophrenia.

Vigilance

On the Continuous Performance Test (CPT), which measures sustained attention (for 8–30 min), subjects are typically required to press a response key every time a critical target stimulus (e.g., an X or a 7) appears in a random sequence of individually presented letter or digit distractors, which are shown for 40 to 100 ms each. Schizophrenia patients obtain significantly lower target hit rates than normal control subjects on this version of the CPT (34). Deficits are not always observed on the CPT, however, unless distraction or other factors that increase burden on processing are present. For example, the CPT deficits in schizophrenia patients are more clearly evidenced when subjects are required

to actively ignore distracting stimuli (digits presented aurally), when a sequential target version of the CPT is used (requiring a response every time two targets appear sequentially, e.g., a 5 followed by a 9), or when stimuli are degraded (e.g., blurred or partially obscured) (19).

Processing Speed

One of the most consistently replicated impairments in schizophrenia patients has been found on reaction time tasks (35). Individuals with schizophrenia show slower reaction time relative to normal individuals on tasks that require a rapid response to a single stimulus (e.g., a tone). In addition, unlike normal subjects, individuals with schizophrenia do not benefit (i.e., react faster) when a warning stimulus is presented at regular, *longer* preparatory intervals before the stimulus. This finding may suggest a failure to establish a preparatory set or a failure to sustain attention over longer, more attention-demanding intervals (35).

Working Memory

Working memory is defined as a system of "temporary storage and manipulation of the information necessary for such complex cognitive tasks as language comprehension, learning, and reasoning" (36). Goldman-Rakic and Selemon (37) proposed that working memory is the core deficit of schizophrenia, arising from dysregulation of the dopamine system in the prefrontal cortex and explaining multiple cognitive impairments, negative symptoms, and disorganized behavior. Working memory tasks may be classified in different ways, e.g., as auditory or visuospatial. For example, Park and Holzman (38) reported that their schizophrenia sample had differential impairment in visuospatial, but not auditory, working memory. Perry et al. (39) proposed that there are two kinds of working memory: (a) transient online "storage and retrieval working memory" (e.g., forward digit span, which requires repetition of a string of numbers) and (b) "executive-functioning working memory," which additionally involves manipulation of that information (e.g., backward digit span, which requires reordering the numbers and repeating them in reverse order). In a series of four experiments, these authors found that schizophrenia patients tended to exhibit a similar level of impairment on both types of tasks and that they were impaired on both auditory and visuospatial tasks. Their working memory deficits, however, were neither dramatic nor significantly worse than their deficits in other domains. Rather, the patients appeared to have more generalized neuropsychological impairment. Some researchers have suggested that the correlations among working memory measures should be stronger in schizophrenia patients than in normal comparison subjects because of a selective working memory deficit in schizophrenia (40,41). The results of Perry et al. did not support this assumption and argue against working memory being a "core" deficit of schizophrenia.

Learning and Memory

A common finding is that recall, but not recognition, performance is deficient in patients with schizophrenia, suggesting a pattern of frontal–subcortical memory impairment (27,42,43). Using a discriminant function analysis of performances on the California Verbal Learning Test, Paulsen et al. (42) classified 50% of schizophrenia patients in their sample as having a subcortical memory profile, 15% as having a cortical memory profile with rapid forgetting, and 35% as having a normal memory profile.

Several studies (44–46) provide evidence of a similar level of impairment on recall and recognition tasks in patients with schizophrenia when tasks require subjects to use semantic or other strategic encoding operations to improve performance. Unlike normal subjects, schizophrenia patients do not make normal use of semantic categorical clustering of word lists; benefit from affective clustering of word lists based on "pleasantness"; show normal levels of idiosyncratic, subjective organization of random word lists; or show normal release from proactive interference (43,45). These findings have led to the conclusion that schizophrenia patients fail to spontaneously carry out semantic or other organizational mnemonic processes necessary for effective memory encoding and retrieval (43). Memory for visual stimuli has received less attention in neuropsychological studies of schizophrenia, but these patients typically show comparable levels of impairment on verbal and nonverbal memory tasks (18,47).

Several studies have examined the learning (or automation) of perceptual-motor skills with practice (i.e., procedural learning) in schizophrenia patients. Intact procedural learning has been reported on the pursuit rotor task, which requires subjects to learn perceptual-motor skills involved in keeping a handheld stylus on a target stimulus rotating on a turntable (48,49), by Schwartz et al. (50); on the multiple-frame visual search task, which requires subjects to automate target detection operations on a task similar to the span of attention task described above (51); and on a motor-sequencing reaction time task, which requires subjects to learn a repeating pattern of key presses (52). This consistent pattern of results suggests that, despite impairments in the ability to learn verbal and nonverbal factual (declarative) information, schizophrenia patients do learn procedural skills involving motor and attentional functions.

In summary, the learning and memory deficits in schizophrenia appear to be related to a failure to spontaneously use contextual cues (e.g., semantic category) and strategic processes to organize their encoding and retrieval of information (43). Because patients with frontal lobe dysfunction show similar impairments in the spontaneous use of strategic mnemonic processes and patients with temporal hippocampal dysfunction show similar impairment in the encoding of semantic information, the memory impairments in schizophrenia have been interpreted as being consistent with frontal and/or temporolimbic abnormalities (45,47,48). In addition, because better recognition than recall memory is commonly observed in patients with primarily subcortical (e.g., Huntington disease) rather than cortical (e.g., Alzheimer disease) neuropathology (53), the general finding of superior recognition over recall performance in schizophrenia may reflect subcortical pathology (54).

Executive Functions

Numerous studies have reported impairments on tasks measuring executive functions that are mediated by frontal lobe systems (i.e., neural circuitry involving the frontal cortex and its basal ganglia, limbic, and reticular connections). Executive functions are those neuropsychological processes important for adapting to the environment, such as preparation, initiation, and modulation of action; maintenance of arousal; cognitive set maintenance and set shifting; abstract reasoning; hypothesis testing; and monitoring of ongoing purposeful behavior. The executive function deficits shown by schizophrenia patients include sorting fewer categories and making more perseverative errors than normal subjects on the Wisconsin Card Sorting Test (55–57); increased errors on the Halstead Category Test (56), reduced verbal fluency on the Benton Word Generation Test (58), and slower performance on the Trail-Making Test, Trail B (56), which requires set shifting. These executive impairments are consistent with frontal system dysfunction in schizophrenia patients. Additional evidence for such dysfunction in schizophrenia is found in functional brain imaging studies, as discussed in the next section.

Conclusions and Caveats

The neuropsychological impairments discussed above suggest that inefficient or inadequate brain resources play a central role in schizophrenia. Since the time of Kraepelin and Bleuler, many researchers have viewed schizophrenia as resulting from frontal lobe pathophysiology. The modern version of this hypothesis considers the frontal cortex in the context of its subcortical connections with basal ganglia and reticular structures. Robbins (59a) suggested that dysfunction in the flow of information through frontostriatal cognitive and motor loops is involved in schizophrenia. Weinberger et al. (60) proposed a diffuse lesion in schizophrenia, involving the periventricular limbic and diencephalic nuclei and connections between these structures and the dorsolateral prefrontal cortex, which occurs early in development but results in symptoms later when the brain areas involved reach maturity. Andreasen (61) posited that pathways between the prefrontal cortex, thalamic nuclei, and cerebellum result in the "cognitive dysmetria" of schizophrenia (i.e., impairments in information processing, retrieval, and prioritization of cognitive tasks). Carlsson and Carlsson (59b) described a negative feedback loop involving frontal, striatal, thalamic, and reticular structures, which together modulate cortical sensory input (through thalamic nucleus reticularis sensory gating) and cortical tone (through arousal centers in the midbrain reticular formation). They further postulated that disruption in the balance of glutamate and dopamine in this system leads to the inability to modulate the influx and processing of information, which may lead to psychosis. As discussed in the previous sections, the neuropsychological deficits in schizophrenia patients are consistent with the hypothesis that the pathophysiology of schizophrenia involves dysfunction in the complex interactions between the frontal cortex and limbic, basal ganglia, and reticular structures connected with the frontal lobes.

This finding that schizophrenia patients are relatively easily overloaded by the information-processing demands of cognitive tasks has led some researchers to conceptualize the cognitive impairments of schizophrenia patients within the framework of processing "resource" (or "capacity") theories from contemporary cognitive psychology (19,27,28). Processing resources are broadly defined as the limited pool of fuels, processes, and structures that are available at a given moment for performance of cognitive tasks. The use of this concept simply provides a means of describing the human brain as a resource-dependent system, i.e., there are clear limits on its capacity to perform. Resource models of schizophrenia emphasize that patients with schizophrenia perform poorly only when processing loads (i.e., resource demands) are higher. That schizophrenia patients are more adversely affected by higher processing resource demands than normal subjects is consistent with the hypothesis that these patients deplete their available resources at lower processing loads (i.e., have fewer resources mobilized and available) than do normal individuals (19,27,28).

The brain systems that govern the availability of processing resources involve frontal–basal ganglia–reticular circuits (62), which are similar to those circuits frequently postulated to be involved in the pathophysiology of schizophrenia (62,63). Kinsbourne and Hicks (63) suggested that different functional regions of the cerebral cortex constitute different processing resource pools that can be differentially activated for performance of different cognitive tasks. The frontal lobes recruit (directly or through basal ganglia connections) reticular nuclei to modulate cortical tone (activation) in specific regions in accordance with task-processing demands (64). The dynamic interactions between the frontal cortex, basal ganglia, and reticular nuclei have been implicated as the mechanism through which cortical resource pools are modulated to provide the resources appropriate to the processing demands of a task (65). A failure to mobilize appropriate cortical resources when processing demands increase in attention and information-processing tasks may occur in

patients with schizophrenia because of dysfunction in this circuit.

Impairments on some of these attention, information-processing, working memory, learning, and executive functioning tasks have been observed in individuals with remitted schizophrenia, nonpsychotic first-degree relatives of schizophrenia patients, and individuals with schizotypal personality characteristics (such as unusual beliefs, constricted affect, and the lack of close interpersonal relationships) (19,29). Particular psychophysiologic (e.g., sensory gating and prepulse inhibition) and electrophysiologic tests (e.g., the P50 component of event-related potential waveforms) show similar patterns (66). These consistent findings of impairment in individuals across the schizophrenia spectrum suggest that some of these tasks are sensitive, not only to the psychotic state but also to subtler core deficits of schizophrenia that may mark a genetic vulnerability, and possibly a genetically transmitted pathophysiology, of schizophrenia (67).

Several important caveats should be noted. The localization of brain pathophysiology based on neuropsychological test results is complicated by the fact that impairment on a specific neuropsychological test does not necessarily implicate pathophysiology in a specific region of the brain. The validity of this association between impaired test performance and localized brain pathophysiology is based on the assumption that diffuse brain damage, motivational and arousal factors, variations in task difficulty, and other nonspecific factors have not produced the impairment. In addition, psychiatric medications taken by schizophrenia patients may also affect neuropsychological test performance. Long-term treatment with antipsychotic medications tends to normalize schizophrenia patients' performance on several neuropsychological measures (especially attention/information-processing tasks), whereas anticholinergic medications tend to impair performance on many memory tasks (16,68,69). Finally, although the neuropsychological deficits in schizophrenia in this brief review were interpreted as being consistent with frontal system dysfunction, alternative interpretations of the neuropsychological findings exist. Some researchers (15) have described a more generalized deficit syndrome in schizophrenia patients involving significantly lower performance and verbal IQ, more diffuse impairments, and possibly a deteriorating course. Other researchers (14,58) have suggested that left-hemisphere deficits are more common than right-hemisphere deficits in schizophrenia patients.

The possibility of different profiles of neuropsychological deficits (e.g., more frontal, lateralized, or diffuse deficits) in schizophrenia patients, as well as findings that a considerable number may demonstrate intact performance on neuropsychological tests (26,29,56,58), has led to the notion of various subgroups of schizophrenia patients. Three commonly discussed subtypes of schizophrenia are the negative subtype, characterized by a psychomotor poverty syndrome; the positive subtype, characterized by a reality distortion syndrome; and the disorganized subtype, characterized by thought disorder. One subgrouping hypothesis with a fair amount of support is that neuropsychological impairment is more common in patients with predominant negative rather than positive symptoms (10,22,70). Patients with predominantly negative symptoms exhibit general difficulty with self-generated mental activity, whereas those in the disorganized group seem to have more trouble regulating their cognitive activity; the positive subtype is less associated with neuropsychological functioning (10). The possible identification of different subgroups of schizophrenia patients based on different neuropsychological test profiles has important implications for treatment/outcome prediction. Neuropsychological test performance has been reported to predict medication response (29,71), success in social and community functioning (72,73), and effectiveness of social skills training (74). Future neuropsychological research will have etiologic and treatment implications if these subgrouping issues are examined further.

BRAIN IMAGING

There is a wealth of neuroimaging evidence to suggest that structural and functional cerebral pathology exists in schizophrenia. The localization of the site(s) responsible for the pathophysiology of schizophrenia has remained elusive, however. The following sections summarize what is known from structural and functional neuroimaging studies of schizophrenia.

Structural Neuroimaging

The most reproducible finding in the myriad computed tomography (CT) and magnetic resonance imaging (MRI) studies of schizophrenia has been that of enlarged ventricular spaces, particularly in the lateral ventricle and third ventricle (75). It remains unclear, however, what the significance of the enlarged ventricles is to the pathophysiology of the illness and whether the ventricular enlargement reflects tissue loss, as is generally suggested. If so, the nature of this tissue loss, i.e., generalized or focal, remains uncertain. Most CT and many early MRI studies of individuals with schizophrenia have relied on visual inspection or linear and area measurements of carefully selected sections. Now, current MRI techniques make it possible to quantify the volume of both whole brain and specific small brain structures. The advantages of quantitative volumetric measurements are many. Even so, methodology for the semiautomated computerized approaches varies greatly across investigative groups, and it is often difficult to relate results from different laboratories. Sagittal, axial, and coronal sections of the brain have been utilized by various investigators to compare volumes of different cerebral structures. In addition, differences in MRI scanners, selected scanning sequences, section thickness, and image analysis algorithms have probably contributed to variations in results.

Enlargement of the lateral ventricles of the brain in a group of chronic schizophrenia patients was first demonstrated on CT in 1976 by Johnstone et al. (76). This was confirmed by Weinberger et al. (77) and again by Owens et al. (78), who were careful to use demographically matched control groups. Additionally, there appears to be a relationship between brain asymmetry and the disease process (79,80), especially when handedness is taken into account. When unilateral abnormalities have been reported, they have typically involved the left hemisphere (79).

MRI, with its ability to detect relatively small structural brain abnormalities, has not only confirmed the ventricular findings of CT (81–85) but, additionally, has revealed evidence of significantly reduced volumes of important cerebral structures. In MRI studies of schizophrenia, the most frequently described morphologic abnormalities have involved the temporal, frontal, and parietal lobes and subcortical structures. However, evidence to support a specific neuroanatomic pathophysiology has been less than compelling.

The structural neuroimaging literature has addressed whether the brain abnormalities in schizophrenia are related to neuroleptic treatment and whether they are neurodevelopmental or progressive or both. Typical neuroleptics tend to increase, but atypical neuroleptics decrease, basal ganglia volumes (75). Most studies, however, show no dramatic effects of pharmacologic treatment on cerebral volumes. Most of the abnormalities discussed above appear to be developmental in nature; however, subtle progressive volume reductions in the frontal lobes (and possibly in the parietal lobes, superior temporal gyrus, and lateral ventricles) seem to occur (75).

Relatively little attention has been paid to the question of associations between regional abnormalities and specific symptoms or subtypes of schizophrenia. The most robust findings so far in this area point toward relationships between negative symptoms and the prefrontal cortex, positive symptoms and the temporal lobes, and thought disorder and the planum temporale (86). Schroder et al. (87) found that both patients with predominantly negative symptoms and those with mostly positive symptoms had increased frontal interhemispheric fissure widths; the positive subgroup also had generalized sulcal widening. Patients in the disorganized group had enlarged ventricles in general and particularly enlarged third ventricles. Allen et al. (88), comparing subgroups hypothesized to be similar, found third ventricle enlargement in a negative symptom subgroup and sulcal widening in a severely cognitively impaired group. Thus, there was no common neuroradiologic correlate for negative symptoms and cognitive impairment.

In summary, structural neuroimaging findings, particularly MRI findings, have dramatically increased understanding of the neuropathologic substrates of schizophrenia. These studies, like the neuropsychological literature, point to diffuse abnormalities and suggest dysfunctional connections between brain regions. They also have spawned new theories of schizophrenia, such as the "cognitive dysmetria" theory (89) (implicating the thalamus and its cortical and cerebellar connections), the "disconnection model" (90) (implicating temporolimbic and prefrontal disconnections), and the "asynchronous neural firing model" (91). The emphasis has shifted from studying structures and regions of interest to investigating networks that are affected in schizophrenia. This emphasis has also become prominent in functional brain imaging studies, as discussed below.

Functional Neuroimaging

Positron emission tomography (PET), single photon emission computed tomography, and functional MRI are the three main functional neuroimaging techniques used to map neuronal activity in the brain. These techniques measure either cerebral perfusion [referred to as regional cerebral blood flow (rCBF)] or glucose metabolism, both of which are believed to reflect neuronal activity and thus serve as markers of brain function. Abnormalities in these markers have been reported by a number of investigators, with most studies finding less activation on cognitive challenge in schizophrenia subjects compared with normal controls (92). The literature that has emerged from functional brain imaging studies has provided a number of important themes about the neural basis of schizophrenia, with special emphasis on frontal, temporal, and basal ganglia dysfunction.

Since the seminal studies of Ingvar and Franzen (93) using intracarotid xenon-133 to measure rCBF, several functional imaging studies have reported diminished cerebral activity in the frontal lobes of schizophrenia patients. Several rCBF studies have reported hypofrontality in both medicated (55,94–96) and drug-free schizophrenia patients (60,97). Most were evaluated in the resting condition only; however, some (55,60) also included executive functioning and control tasks. For example, Stevens et al. (98) found frontal hypoactivity in the context of an auditory working memory task. Numerous PET studies of (mostly chronic) schizophrenia patients have confirmed the finding of hypofrontality (96,99–107). However, findings of *increased* frontal metabolism have been reported in acute, unmedicated patients (108–110). Thus, although the literature is not entirely consistent, hypofrontality remains a frequently observed functional abnormality in schizophrenia.

In parallel with the structural imaging findings, functional imaging studies have also found abnormalities in the temporal lobes of schizophrenia patients. For example, O'Leary and Webb (111) found decreased rCBF in the left superior temporal gyrus during a dichotic listening task. Eyler Zorrilla et al. (112,113) reported both frontal and medial temporal abnormalities in response to a visual learning task. Glucose metabolism in the left temporal lobe has been found to be greater than that in the right temporal lobe in many studies (114). Hallucinations appear to activate the hip-

pocampal region as well as the thalamus and basal ganglia (115).

The basal ganglia have also been implicated in the pathophysiology of schizophrenia. Sheppard et al. (106) noted decreased basal ganglia metabolism in 12 schizophrenia patients, half of whom were drug naïve. Decreased absolute (102,107,116) and relative (100,102) basal ganglia activity has been reported by a number of investigators, primarily in patients who were never medicated or who had not taken medications for some time. Conversely, increased absolute basal ganglia activity has been described in 18 chronic schizophrenia patients on medication (117). Finally, several investigators reported an increase in relative basal ganglia metabolism in schizophrenia (107,116–118). Except for one study (118), most patients either were on medication at the time of scanning or had been on neuroleptics for various intervals before examination. It may, therefore, be that striatal metabolism in schizophrenia is more related to neuroleptic treatment than to disease-related pathophysiology.

The nature of the clinical presentation of schizophrenia patients may help to explain some of the variations in results across studies. Whether there is a relationship between frontal lobe hypometabolism and negative symptoms, as reported by some investigators, remains unresolved. Schroder et al. (119) reported medial frontal lobe changes (hypoactivity in the anterior cingulate and medial frontal gyrus) in patients with predominantly negative symptoms as well as in those with predominantly positive symptoms. However, the negative symptom patients had prominent, generalized frontal hypoactivity and increased left temporal activity; positive symptom or delusional patients had decreased hippocampal function; and disorganized patients had parietal and motor cortex hyperactivity and decreased activity in the corpus callosum. A review by Liddle (120) also identified different patterns of metabolism and blood flow depending on clinical subtype. Across imaging studies, negative symptom subjects had decreased activity in the left prefrontal cortex, whereas disorganized patients exhibited decreased activity in the right ventral prefrontal region. Positive symptom patients showed increased medial temporal activity but had increased perfusion of the left medial temporal lobe and decreased perfusion in the left lateral temporal lobe. Reviews of total hemispheric activity as indexed by psychophysiologic measures also revealed a pattern of greater left- than right-sided activity in positive symptom patients and greater right- than left-sided activity in negative symptom patients (121).

Functional neuroimaging also can be used to examine neuroreceptors *in vivo*, for example, using radioligands for the dopamine receptor with PET and single photon emission computed tomography technology. This is an exciting and important application of functional brain imaging because alterations in dopamine receptors have been suggested in this illness.

It has long been known that antipsychotic medications work by blocking dopamine receptors, particularly the D_2 receptor (122). The newer atypical antipsychotics appear to have an even greater affinity for the $5HT_{2A}$ receptor (123). Using single photon emission computed tomography, Crawley et al. (124) found elevated values of dopamine receptors in a small group of schizophrenia subjects off medication for 6 months. Whether a 6-month washout period is enough to nullify the effects of long-term use of neuroleptics on striatal receptors is uncertain. Using PET, Wong et al. (125) found highly elevated dopamine receptor densities in the caudates of both treated and nontreated schizophrenia patients before and after administration of 7.5 mg haloperidol. These findings, however, were not confirmed by other PET receptor ligand studies. Farde et al. (126), for example, found no increase in dopamine receptor binding in 18 neuroleptic-naïve schizophrenia patients compared with normal subjects.

Conclusions

Neuroimaging techniques have provided unique opportunities for exploring brain anatomy and function in schizophrenia. The precise morphologic and physiologic substrates of this disabling disorder remain ambiguous, but evidence suggests that the frontal lobes, temporal lobe structures, and basal ganglia, either alone or in concert, are involved in the pathophysiology of schizophrenia. Current theories suggest that disruptions in cortical–subcortical circuitry (e.g., connections between the hippocampus and frontal cortex) play a central role in the neuropathology of schizophrenia. As the technology improves and measurements are standardized and refined, neuroimaging techniques will probably become an even more powerful aid in understanding the primary brain lesions responsible for producing the symptoms and signs of schizophrenia. Future studies will continue to involve combined structural and functional modalities in larger cohorts of well-characterized and carefully selected subjects and appropriate controls.

NEUROPATHOLOGY AND NEUROCHEMISTRY

In parallel with neuroimaging studies, neuropathologic examination of the brains of schizophrenia patients has provided clear evidence of structural abnormalities. These findings do not yet form a clearly understood chain of events, and in many instances have been hotly debated, but they provide clues that should help to focus further research. Studies have addressed two important questions: Where are the abnormalities in the brains of schizophrenia patients? What type of pathologic process is likely to result in the changes that are found?

Early studies of brain weight, gross brain structure, and histologic appearance in schizophrenia were inconclusive (127), attributable to flaws in methodology and sampling.

Efforts were hampered by a lack of quantitative methods, confounding factors such as postmortem shrinkage of tissue, brain changes owing to neurologic diseases such as stroke, and failure to match with controls for potential confounding factors such as age and gender. Nonetheless, most such studies have shown that brain abnormalities in schizophrenia do exist, providing an organic basis for the psychosis. Three sets of studies that controlled for variables such as age, gender, weight, and height, found that the overall brain weight was decreased in schizophrenia patients (128–130). The extent of the decrease was modest but consistent (approximately 5%–8%). In one of these studies, the decrease in volume of the hemispheres in schizophrenia patients compared with normal subjects was accounted for more strongly by a loss of cortical gray matter (12%) than of central gray matter (6%), whereas white matter volume did not differ (130).

Congruent with neuroimaging studies of living patients, autopsy studies of the brains of schizophrenia patients have generally shown ventricular enlargement beyond that attributable to normal aging alone (79,128). The increase mainly affects the lateral ventricles, especially the temporal horn, and the left temporal horn may be more severely affected than the right (131). In contrast to Alzheimer disease, in which enlargement of the lateral ventricles is generalized, in schizophrenia, the temporal horns are enlarged to a much greater extent than the rest of the ventricles (131). This selective enlargement suggests that structures adjacent to the temporal horn bear the brunt of the pathology, including the temporal cortex, hippocampus, and amygdala.

Planimetric studies (128,132–135) have shown decreased volumes or areas on cross section of temporal lobe structures, notably the hippocampus, parahippocampal gyrus, amygdala, and inner pallidal segment. The thinness of the parahippocampal cortex in schizophrenia patients is correlated with the extent of ventricular enlargement (36). Studies have found no differences in the volumes of basal ganglia structures or the cingulate or insular cortex (128,134,136). However, in patients on long-term neuroleptic treatment, striatal volumes may be slightly increased (137). A quantitative study of the major subnuclei of the thalamus showed no significant difference between schizophrenia patients and normal subjects, apart from volume loss in gray matter adjacent to the third ventricle (138), although other studies have noted neuronal loss in the mediodorsal nucleus of the thalamus (139). A report of an altered gyral pattern of the temporal lobe in some schizophrenia patients (140) was not confirmed in a larger study (129).

Histologic studies have used quantitative microscopic methods and a variety of staining and immunohistochemical techniques to study the number and arrangement of neurons in various brain areas in schizophrenia. Apart from being technically demanding, these studies have a number of potential clinical confounds. Cases and controls need to be carefully matched for age and gender. Many schizophrenia patients die at a relatively advanced age, when the effects of aging, long-term treatment with medications, comorbidity, and factors such as alcohol abuse may need to be considered. Indeed, studies have found an increased rate of neuropathologic lesions owing to Alzheimer disease, Parkinson disease, or infarcts in schizophrenia patients compared with elderly control subjects (129,141), further confounding the findings.

More recent studies have paid special attention to matching cases with controls, studying brains while "blind" to clinical diagnoses, and using computer-assisted cell-counting methods of serial thin sections. Several investigators have reported decreased numbers and disordered arrangement of neurons in temporal lobe structures (135,140,142–144), especially the parahippocampal gyrus, where neuronal loss as high as 20% has been reported (135). Pyramidal cell density and number appear to be especially affected (144). Other studies have found a more modest degree of neuronal loss in several layers of the prefrontal and cingulate cortex (145,146) and to a lesser extent in other brain areas. Only rarely do studies not concur on these findings (147). Although the picture and extent of neuronal loss described in studies are not entirely consistent across studies, this is likely to be due to different methods of histologic analysis. The balance of evidence from these controlled studies supports the presence of nonspecific pathology in temporolimbic areas in the brains of schizophrenia patients.

The reported changes are relatively subtle, and, unlike neurodegenerative disorders such as Alzheimer or Parkinson disease, there are no characteristic neuronal ultrastructural changes or inclusion bodies. Instead, a variety of alterations of neuronal size and/or arrangement has been noted. This includes loss of small interneurons (146) or large pyramidal cells (143), especially in layer III, neuronal disarray (148–152), and decreased neuronal size in some areas (142). For example, increased neuronal density in the context of reduced cortical thickness has been reported, which may implicate reduced cortical neuropil (153). The disorganized orientation of neurons described is subtle and not unique to the brains of schizophrenia patients but is more common in patients than in normal subjects. The severity of neuronal loss or other changes does not seem to increase in proportion to the duration of illness. This implies that the pathology is established at the time of initial clinical presentation and does not necessarily progress. The concept of schizophrenia as a neurodevelopmental disorder that manifests clinically in adolescence or adulthood is thus supported.

A study that shed further light on the timing and nature of neuronal changes examined the brains of schizophrenia patients and normal subjects for nicotinamide adenine dinucleotide phosphate diaphorase, which labels a class of neurons resistant to the neurodegeneration of Alzheimer disease and other conditions. In the schizophrenia patients, diaphorase-positive neurons were decreased in the cortical gray matter and increased in the white matter deep to the cor-

tex in the frontal lobes (154), temporal lobes, and hippocampus (155). This suggests failure of migration toward the cortical plate, which normally occurs between 6 and 14 weeks of gestation. How these out-of-position cells affect cerebral dysfunction, later brain development, and the etiology of schizophrenia remains to be determined, but again the findings indicate frontal and temporolimbic sites of pathology.

Many studies have consistently reported that inflammation or gliosis does not accompany neuronal abnormalities or atrophy in the brains of schizophrenia patients (79,135,139,156,157). Studies that excluded cases with acquired pathology such as infarcts supported this conclusion most clearly. This further supports the idea that the pathology reflects a developmental event occurring *in utero* because acquired lesions generally elicit a glial response. Gliosis occurs from the third trimester onward, again narrowing the window during which abnormalities may arise. Obstetric injuries and perinatal infections should leave a trail identified by gliosis and are therefore less strongly favored than genetic or developmental factors.

Although the events that incite the pathology of schizophrenia remain elusive, clues and theories abound. Infectious agents have been sought with great effort, including molecular biologic techniques of testing for viral DNA, without avail. Some current theories attribute neuronal misarrangements to abnormal migration of developing neurons in cortical columns. This migration is genetically controlled, and it is conceivable that a disturbance in the genetic regulation of brain development may result in the types of abnormalities cited above. From twin studies, it is apparent that although heredity does increase the risk of schizophrenia, the concordance rate in monozygotic twins is substantially less than 100%. This suggests that environmental factors are probably important in many cases.

The neurochemical basis of schizophrenia symptoms has been a separate focus of inquiry. Until recently, the focus was on the dopamine hypothesis, which attributed the illness to an excess of dopaminergic neurotransmission. The dopamine hypothesis rests largely on two pillars. First, neuroleptics, which block D_2 receptors, decrease the positive symptoms of schizophrenia. Their clinical efficacy is roughly proportional to their affinity for D_2 receptors (158,159). Second, amphetamine and other aminergic agents, which increase the activity of transmission of dopamine and other catecholamines, can provoke a psychotic state or worsen the symptoms of some schizophrenia patients, especially in those who are more treatment resistant (160–162). Although these observations indicate that manipulating the dopaminergic system can regulate psychotic symptoms, they do not directly implicate a central excess of dopamine as the cardinal cause of schizophrenia symptoms. In fact, Davis et al. (163) proposed that a chronically low level of striatal dopamine release causes negative symptoms (owing to low levels of dopamine in the prefrontal cortex); the diminished dopamine release also leads to up-regulation of dopamine receptors, resulting in supersensitivity to phasic dopamine release in the context of environmental stressors (exhibited behaviorally as positive symptoms). This theory would explain why dopamine-blocking drugs would improve positive but not negative symptoms.

More direct evidence to support the dopamine hypothesis has been sought from studies of brain tissue, serum, and cerebrospinal fluid (CSF) aminergic neurotransmitters and neuroimaging techniques such as PET. The least ambiguous evidence should come from brains of patients not exposed to neuroleptics, with tissue frozen within a few hours after death. In the absence of large numbers of such patients, it is not surprising that neurochemical studies of catecholamine concentrations in their brains have been difficult to interpret. Modest increases in dopamine and its metabolites were reported in several early studies of the caudate, putamen, nucleus accumbens (164–166), and amygdala (167). There were also trends toward increased levels in the temporal and frontal cortex. Later studies did not, however, confirm these findings (168,169). In general, increased levels of dopamine and its metabolites were more apparent in younger patients and in patients who had received neuroleptics (163). In neuroleptic-free patients, there was no consistent finding of an overall hyperdopaminergic state.

The technique of radioligand receptor binding studies was used to examine the distribution and levels of dopamine receptors in schizophrenia. Findings of increased D_2 receptors in the caudate and nucleus accumbens, targets of mesolimbic catecholaminergic projections, have been attributed to the effect of treatment with neuroleptics rather than to the illness itself (165,170,171). Nonetheless, patients with Alzheimer or Huntington disease treated with neuroleptics show lesser degrees of increased D_2 receptor binding, implying indirectly that medications do not fully account for the increased receptor binding found in schizophrenia.

Studies of dopamine and its metabolites in the CSF, especially homovanillic acid (HVA), have yielded inconsistent findings. Reported results have included an increase in schizophrenia patients compared with normal subjects, no difference, and a decrease (172,173). Inconsistencies may be related to variables such as neuroleptic treatment, diet, and activity that affect CSF neurotransmitters as well as to assay variability, but even those studies with positive results have shown considerable overlap between concentrations in schizophrenia patients and normal subjects. Plasma amines, especially HVA (pHVA), have been extensively studied in patients before and after neuroleptic treatment (163,172). Although no clear and consistent difference at baseline distinguishes schizophrenia patients from normal subjects, pretreatment pHVA may predict patients who are more likely to respond to neuroleptics.

In light of these inconclusive findings and the realization that the important negative or deficit symptoms of apathy and withdrawal are not readily explained by augmented dopamine transmission, the original theory has undergone several revisions (163,174). These have tried to explain both the positive and negative symptoms associated with schizophrenia by alterations in limbic dopaminergic circuits, particularly the projections of the ventral tegmental area or mesolimbic projections. These neurons have extensive connections to the frontal lobes, nucleus accumbens, and amygdala. Damage to the ventral tegmental area system may potentially explain hypofrontality, thought to be an important basis of the deficit syndrome of schizophrenia. Prefrontal neurons inhibit subcortical dopaminergic pathways, which may possibly trigger positive phenomena such as delusions and hallucinations. Thus, as suggested by Davis et al. (163), frontal hypodopaminergic function, accompanied by phasic hyperdopaminergic function in temporolimbic areas, may account more closely for the clinical features of schizophrenia. The introduction of clozapine, the first atypical antipsychotic, shed light on other important neurochemical pathways in schizophrenia because it relieves psychotic symptoms with minimal effect on D_2 dopaminergic receptors and has been effective in treating patients refractory to neuroleptics. The cloning of several subtypes of dopamine receptors indicates that the regulation and effects of aminergic neurotransmission are more complicated than previously thought. Clozapine binds strongly to D_4 receptors (a variant of the D_2 class) but poorly to other subtypes (175), whereas typical neuroleptics show relatively greater binding to D_2 than to D_4 receptors. Additionally, clozapine is a partial agonist at the $5HT_{1A}$ receptor (176) and may also affect the glutamate system (177).

The monoamines serotonin and norepinephrine, long the focus of attention in depressive disorders, also seem to play a role in schizophrenia. Most research has focused on the $5HT_{2A}$ receptor as related to hallucinatory phenomena but, as noted above, the $5HT_{1A}$ receptor may also be an appropriate target for treatment of schizophrenia (176). Alteration in monoamine activity in the limbic circuit has been proposed as a possible link to the dopamine hypothesis of schizophrenia (178). For example, van Kammen et al. (179) reported that CSF norepinephrine levels are associated with both the severity of negative symptoms and psychosis ratings.

Several neuropeptide and amino acid transmitters have also been studied in schizophrenia. Normal levels of GABA, the major inhibitory neurotransmitter in the brain, have been found in cortical and subcortical areas, making it unlikely that positive symptoms result from inadequate inhibition of neuronal circuits. Decreased density of GABA uptake sites in the hippocampus has been reported, possibly reflecting neuronal loss (149). Benes et al. (142,180), in fact, proposed that a loss of GABA neurons and excessive dopamine transmission to other inhibitory neurons are the key neuronal components of schizophrenia. Several neuropeptides have been measured in the brains of schizophrenia patients, the rationale being that these either colocalized with dopamine or interacted with aminergic pathways. Although neurotensin levels are normal (181), slight reductions of the neuropeptides somatostatin and cholecystokinin have been found in the hippocampus, amygdala, and frontal cortex (182–184). The implications of these findings are unclear, but they may reflect loss of neurons.

Several studies of the role of glutamate (the primary excitatory neurotransmitter) in schizophrenia have been carried out. Glutamatergic neurons link the cortex, limbic system, and thalamus. Impaired functioning of the *N*-methyl-D-aspartate subtype of the glutamate receptor may be especially important in schizophrenia (185). Blocking this receptor produces negative symptoms and cognitive deficits in normal subjects while simultaneously increasing mesolimbic dopamine; in individuals with schizophrenia, enhancement of *N*-methyl-D-aspartate receptor function reduces negative symptoms and cognitive deficits (186). Binding to the kainate subtype of glutamate receptor in schizophrenia patients has been reported as decreased in the left hemisphere (187) and increased in the frontal lobe (188). Thus, drugs that affect glutamate function may be promising for alleviating the symptoms that are not improved by antipsychotic medications.

Substantial progress has been made in neuropathologic and neuroanatomic studies of schizophrenia. Insights from physiologic imaging studies of living patients, studies of brain development and its genetic control, and pharmacologic studies using newer atypical antipsychotics are promising stepping stones that may lead to an explanation of the pathogenesis of schizophrenia.

LATE-ONSET SCHIZOPHRENIA

The possible development of schizophrenia in mid- or late life has been a topic of continuing controversy with important implications for the understanding of schizophrenia. Historically, schizophrenia has been defined as a condition for which early onset is an intrinsic criterion (189–191). In the past, the etiology of late-onset psychoses has been attributed almost exclusively to organic illnesses or brain lesions, substance abuse, mood disorders, and sensory deficits.

The current (fourth) edition of the *Diagnostic and Statistical Manual of Mental Disorders* (*DSM-IV*) (192) does not recognize LOS as a separate entity but no longer prohibits the diagnosis of schizophrenia when the age of onset is over age 45 (as *DSM-III* did). Although LOS has been recognized since the 1940's (Bleuler, cited in reference 193), the concept generally has not been included in formal diagnostic criteria. Research in the past decade, however, has clarified the nature of LOS and its distinction from earlier, typical onset schizophrenia. The consensus statement of the International Late-Onset Schizophrenia Group recently suggested that schizophrenia with onset between the ages of 40 and 60 be labeled LOS, whereas after age 60, it should be called very late onset schizophrenialike psychosis (193).

Despite problems in methodology, inconsistent or incompatible terminology and nosologic criteria, and difficulties in accurately diagnosing schizophrenia and determining a date of onset in any clinical instance, it is clear that a particular proportion of patients demonstrates their first recognizable symptoms of schizophrenia only after age 40 or 45 (194). A literature review by Harris and Jeste (194) found that approximately 13% of hospitalized schizophrenia patients had onset in the fifth decade, 7% in the sixth decade, and 3% thereafter. These findings provide evidence that schizophrenia with onset after age 40 is not as rare as previously believed.

LOS is characterized by many of the typical symptoms of schizophrenia, especially of the paranoid type: persecutory delusions, auditory hallucinations, schizoid or paranoid traits in premorbid personality, a tendency toward chronicity, and improvement in positive symptoms with neuroleptics. The disorganized subtype is much less common than the paranoid subtype in LOS patients (195). Additionally, LOS patients seem to exhibit similar levels of positive symptoms but less severe negative symptoms than their earlier-onset schizophrenia (EOS) counterparts (195,196). In contrast to EOS, there is a preponderance of women among LOS patients (197–200), which has led to hypotheses regarding the protective role of endogenous estrogen premenopausally (201).

Investigations of the course and prognosis of LOS have concluded that the course tends to be chronic, but the prognosis may not be as unfavorable as in EOS (194,198,202–205). There have been only a few studies of the use of neuroleptics in treating LOS patients. Jeste et al. (206) suggest that treatment with neuroleptics is as effective in counteracting the symptoms of schizophrenia in LOS as in EOS and that the risk of relapse after neuroleptic withdrawal is similar in both groups. LOS patients, however, tend to require lower daily doses of antipsychotic medications (195). There is a relatively high risk of side effects associated with neuroleptic treatment in older patients, including TD (207,208). It must be remembered that pharmacokinetic and pharmacodynamic changes associated with aging require that any clinical treatment of LOS patients always weigh the significant risks involved. It is also important to gauge the extent to which the sedative or anticholinergic properties of different neuroleptics may adversely affect patients with other medical problems.

An analysis of family histories shows that both EOS and LOS patients have greater numbers of first-degree relatives with schizophrenia than do normal subjects. Some studies suggest that a positive family history may be more common in EOS than in LOS, but others have found similar rates (10%–15% with positive family histories) in both groups (195). The premorbid personality of LOS patients sometimes includes schizoid or paranoid traits, and LOS patients have elevated rates of minor physical anomalies, as do EOS patients (209). A diagnosis of schizophrenia in the LOS patients is also supported by the numerous similarities between the EOS and LOS groups on neurocognitive and psychopathologic measures (194,210,211). However, despite similar levels of childhood maladjustment (195), the proportion of LOS patients who have married is higher than that in the EOS group, suggesting somewhat better premorbid social functioning in LOS than in EOS.

There has been a modest number of CT and MRI studies of later life–onset schizophrenia (212–215). Although the relationship of this entity to EOS remains uncertain, in general, these investigations have demonstrated similar increases in ventricular size in LOS and EOS patients compared with elderly normal subjects. A higher prevalence of white matter abnormalities in this patient group was described by Breitner et al. (216) and Miller et al. (217). In one study, LOS patients had larger thalamic volumes than demographically similar EOS patients (218). However, in another study, most LOS patients had clinically normal MRI findings and no increased prevalence of volume loss, strokes, tumors, or white matter hyperintensity when compared with EOS subjects (219).

Neurocognitive similarities and differences between LOS and EOS have been studied extensively in the past decade. LOS and EOS patients generally exhibit similar levels of cognitive impairment, with one exception: Those with LOS tend to have less impairment in learning, abstraction, and cognitive flexibility (195). Further studies have shown that community-dwelling patients with LOS show no evidence of cognitive decline over time beyond that associated with normal aging (25,220). Together, the neuropsychological and neuroimaging data support the notion of LOS as a late-onset neurodevelopmental condition rather than a neurodegenerative disorder.

SECONDARY PSYCHOSES

Secondary psychosis is diagnosed in patients with brain lesions who have prominent delusions or hallucinations. Typically in secondary psychoses, fragments of the schizophrenia syndrome, especially delusions, occur rather than the full repertoire of positive and negative symptoms. Many reports of secondary psychosis have been incompletely documented and lack details of the precise symptoms and their time course or the presence of previous psychopathology, affective symptoms, or a family history of schizophrenia. Most reports take the form of case reports or small series, and it is difficult to assess how frequently psychotic symptoms occur in various neurologic conditions. With these reservations in mind, the association between psychotic symptoms and brain lesions is discussed below, with conditions being grouped as neurodegenerative diseases, infections, focal lesions, and toxic or metabolic encephalopathies. For more complete lists of disorders reported in combination with delusions or hallucinations, the reader is referred to references 221,222.

Neurodegenerative diseases have been relatively well studied in terms of psychotic symptoms. In dementing conditions, delusions and hallucinations may be prominent. Psychosis of Alzheimer disease is now recognized as a distinct syndrome with proposed diagnostic criteria (223). In Alzheimer disease, the prototypical cortical dementia, approximately 30% to 40% of patients have delusions at some point during the illness (224). These are most common during early and intermediate stages and less so in advanced dementia, presumably because functioning cortical areas are necessary to generate delusions. Usually the delusions in Alzheimer disease are simpler than those in schizophrenia and are not systematized. They often have a paranoid quality and may be related to memory impairment, e.g., blaming people for stealing items that the patient misplaces. Spousal infidelity, Capgras syndrome (the belief that a significant person, commonly the spouse, has been replaced by an identical impostor), beliefs that strangers are entering the house, and paranoid delusions that people are plotting to hurt the patient or are spying are also common. Vascular dementia is characterized by a similar spectrum of delusions. Psychotic symptoms occur less often in Pick disease and Creutzfeldt-Jakob disease.

Idiopathic Parkinson disease is associated with psychotic symptoms mainly in two situations: in patients who develop dementia, usually late in the course of the illness, and as a result of overstimulation with L-dopa (225). The latter occurs more commonly in patients with severe parkinsonism that is less responsive to L-dopa and tends to produce hallucinations rather than delusions. Postencephalitic parkinsonism, now rare, was linked to a variety of psychotic symptoms, especially delusions. Atypical antipsychotics have been reported as beneficial for psychosis in Parkinson disease, a situation in which typical neuroleptic agents (especially the high-potency ones such as haloperidol) are not favored because they worsen extrapyramidal function (226).

Approximately 50% of Huntington disease patients develop psychosis, sometimes as the presenting feature. Rarer conditions involving the basal ganglia, such as Wilson disease and idiopathic calcification of the basal ganglia, and cerebellar or multisystem degenerations may also produce psychosis (227).

Delusions are somewhat less common in demyelinating diseases. They have been reported in multiple sclerosis and inherited demyelinating conditions such as metachromatic leukodystrophy and adrenoleukodystrophy. Although the result of an infectious agent, human immunodeficiency virus encephalopathy is convenient to group along with neurodegenerative conditions because of its association with dementia. Psychotic symptoms occur in a small proportion of subjects with human immunodeficiency virus encephalopathy (228,229) and respond symptomatically to relatively low doses of neuroleptics.

Many focal brain lesions have been reported as potential causes of psychotic symptoms, which often are transient. The highest risk occurs with temporolimbic lesions. Left-sided lesions are relatively more likely to produce persecutory delusions, whereas right-sided lesions tend to cause visual hallucinations and sometimes delusions. Infarcts, arteriovenous malformations, tumors, trauma, herpes simplex encephalitis (and other infections), hydrocephalus, and anoxic encephalopathy are among the list of culprits, although more common than any of these is the psychosis associated with epilepsy. The relationship between epilepsy and psychosis has been extensively studied and comprehensively reviewed (228,230) and is dealt with in another chapter in this volume. In brief, complex partial seizures are more likely than other types of seizures to manifest with psychosis (which may be ictal, interictal, or postictal) and bear a poor relationship to the adequacy of seizure control (231,232). Interictal psychosis can be a major management problem in patients with complex partial seizures. The content of delusions and hallucinations may closely resemble that of schizophrenia (233). Rarely, marked improvement in seizure control by medications may precipitate psychotic symptoms (234). Left-sided seizure foci and structural lesions of the temporal lobe are risk factors for interictal psychosis (235–237).

Most systemic illnesses and metabolic disorders have the potential to produce psychotic symptoms, although delirium is far more common. Psychosis may appear in the setting of delirium or fluctuation of consciousness. Renal, hepatic, or pulmonary failure, hypoxia, disorders of the thyroid, parathyroid or adrenal glands, collagen-vascular diseases such as lupus and temporal arteritis, and vitamin deficiencies (B_{12}, folate, thiamine) are examples of such conditions. The psychoses seen in intensive care units or postoperatively are usually multifactorial.

Toxic or medication-induced encephalopathies may present with prominent delusions or hallucinations. An extensive list of medications reported as being potentially associated with delusions or hallucinations has been compiled by Cummings (221). Therapeutic drugs affecting neurotransmitter function, such as anticholinergics, dopaminergic agents, antidepressants, anticonvulsants, antihistamines, and antihypertensive agents, may provoke psychotic symptoms, as may many other drugs, such as cimetidine, benzodiazepines, corticosteroids, and digitalis. Many drugs of abuse may produce psychiatric symptoms that include hallucinations or delusions. Lysergic acid diethylamide, phencyclidine, psilocybin, and cocaine may cause acute excited states with hallucinations. More sustained psychotic symptoms that sometimes persist after withdrawal of the drug occur with amphetamines, lysergic acid diethylamide, phencyclidine, and mescaline and with inhalation of glue or other organic solvents. Withdrawal from alcohol, barbiturates, opiates, and, more rarely, other drugs may precipitate psychosis.

In the face of such a large array of lesions and disorders that can result in psychotic symptoms, it is difficult to extract general principles regarding secondary psychoses. Clinically, there are several important distinctions between these secondary psychoses and schizophrenia. The secondary psy-

choses usually have a later age of onset, and a family history of schizophrenia is usually absent. The delusions are often less chronic and intractable than those in schizophrenia and often respond well to low doses of neuroleptics. First-rank symptoms can occur but are rare in many of the conditions listed above. In patients whose overall level of cognition is impaired, the content of delusions is less complex than that in schizophrenia.

From the plethora of conditions associated with secondary psychotic symptoms, some patterns emerge that link brain sites of lesions and specific symptoms. Delusions associated with disease processes affecting the cortex tend to be less elaborate than those related to subcortical or limbic lesions. Left-hemisphere lesions are overrepresented in the series and case reports as are temporal lobe lesions. This is especially evident in conditions that can arise in various areas of the brain, such as seizures or tumors, where delusions are far more common if the temporal lobe is affected. Most of the neurologic conditions listed above display prominent neurologic and cognitive deficits and usually do not pose diagnostic problems in the setting of psychosis. Many toxic or metabolic disturbances produce psychotic symptoms in association with delirium. These patients may need a period of observation and relevant blood and urine tests to establish the diagnosis. Neuroimaging studies are needed in patients in whom the underlying disease process is not obvious on clinical examination.

MANAGEMENT OF SCHIZOPHRENIA

There is no cure for schizophrenia, and the goal of any treatment program must be to alleviate the suffering of the patient and those around him/her and to help the patient function better in the world in terms of cognitive, affective, and social performance. Medication alone is insufficient; management of schizophrenia requires psychosocial care conjoined with pharmacotherapy.

Any management of schizophrenia must be based on as accurate a diagnosis as possible. Before diagnosing schizophrenia, a clinician must perform an appropriate diagnostic workup, both to rule out conditions in the differential diagnosis and to identify any coexisting medical conditions that could affect treatment. Given the lack of insight that characterizes schizophrenia patients, it is important that the most complete psychiatric and other medical history possible be gathered from the patient's relatives, friends, significant others, colleagues, and medical records. Thorough neurologic and medical examinations are also critical. It is especially important that the diagnostician rule out primary affective disorders, delusional disorder, and secondary psychoses. This may involve the use of brain imaging, blood chemistry, and neuropsychological testing.

Pharmacologic treatments are an essential component of the management for patients with schizophrenia, of which antipsychotics are the primary medication (238). In most schizophrenia patients, many of the positive symptoms and some negative symptoms of the disorder can be brought under control by the use of antipsychotics, although functioning may not improve to the premorbid level of social adjustment. In approximately two-thirds of acute schizophrenia patients, antipsychotic medication can reduce the psychotic symptoms, sometimes within a matter of weeks. A maintenance program of continuous neuroleptic medication is necessary for preventing psychotic relapses in remitting patients.

The development of chlorpromazine and related conventional antipsychotics ushered in the modern era of pharmacologic treatment for schizophrenia. Numerous clinical trials showed that conventional agents are effective for both acute exacerbations and long-term maintenance, but decades of experience with these agents highlighted important deficiencies (239). Such drawbacks include the high incidence of acute and chronic neurologic effects [e.g., extrapyramidal symptoms (EPS), TD], frequency of only partial or poor response of positive and negative symptoms, high rates of antipsychotic nonadherence, and unanswered questions regarding the neuroleptics' effects on the long-term course of schizophrenia (240). Furthermore, negative symptoms that often characterize a patient's chronic course tend to be relatively refractory to typical antipsychotic treatment (239). Thus, treating only positive symptoms leaves the majority of patients with schizophrenia significantly functionally disabled (241).

The development of atypical antipsychotics and emerging findings from studies of these agents have demonstrated important improvements with these medications compared with typical antipsychotics. For example, large, randomized, controlled trials have demonstrated reductions in parkinsonian symptoms (242–246) and a reduced risk of TD (244,247–250) in patients treated with atypical antipsychotics. Further study is needed regarding other promising therapeutic aspects of atypical antipsychotics, including their effects on negative symptoms and cognitive functioning (240).

In the past decade, studies have shown that the atypical antipsychotic drugs have some effects on cognitive performance. For example, improvements in verbal fluency (251–253), working memory and learning (254,255), and motor skill (256) have been reported. However, the improvements, which are often attributed to the lower incidence of sedation in atypical antipsychotics, are not dramatic and residual cognitive deficits remain (257).

The benefits of atypical antipsychotics over typical agents have important clinical implications. Despite the increased costs of newer atypical agents, it is hard to justify initiating therapy with conventional agents or continuing typical antipsychotics in patients who are experiencing bothersome side effects such as neuroleptic-induced movement disorders. Consideration should also be given to switching those patients who are at increased risk of long-term side effects, such

as the elderly. Stable patients who are doing well on conventional agents, without TD or EPS, probably can remain on their regimen (239).

All antipsychotic medications carry a risk of side effects, including but not limited to sedation, hypotension, dryness of the mouth, blurred vision, tachycardia, cardiac effects, amenorrhea, galactorrhea, hyperpyrexia, pigmentary retinopathy, weight gain and associated metabolic changes, allergic reactions, and seizures (239). The propensity to cause particular side effects varies among individual agents and the type of antipsychotic (conventional vs. atypical). A small number of patients may develop neuroleptic malignant syndrome, a potentially fatal condition (258,259). Pharmacokinetic and pharmacodynamic changes attributable to aging may significantly intensify the response to antipsychotic medications and the likelihood of side effects, necessitating considerable caution in prolonged use of antipsychotics in elderly patients (239).

Given the potential side effects of antipsychotics, the current clinical strategy is to minimize the total lifetime amount of antipsychotic medication administered to a patient by prescribing the lowest effective dose and avoiding the unnecessary use of antipsychotics. Because commonly prescribed antipsychotic medications seem to be equally efficacious in controlling the active symptoms of schizophrenia, the selection of a specific agent in a given instance is often dictated by careful consideration of the medication's side effects and the patient's particular medical condition. Atypical antipsychotics, such as clozapine, risperidone, quetiapine, olanzapine, ziprasidone, and aripiprazole, are useful in controlling the symptoms of schizophrenia and may carry a reduced risk of TD, although they have their own potential long-term side effects (see next section). The ultimate decision on suitability of a given antipsychotic must be determined in each case after evaluating the risk:benefit ratio of each drug.

In a recent study of schizophrenia and aging (24,260), older age was associated with decreased psychopathology. There was no evidence of accelerated aging-related decline in various outcomes, but elderly patients were more impaired than age-matched controls on various measures. Thus, the course of schizophrenia in late life appears stable, but most elderly patients remain symptomatic and impaired, disputing notions of either progressive deterioration or marked improvement in aging schizophrenia patients. These findings underscore the need for lifelong care for most individuals with schizophrenia.

Schizophrenia has an organic, neurophysiologic basis, and pharmacotherapy is an essential part of its treatment, but pharmacotherapy must be supplemented by psychosocial support. Successful treatment requires that each patient be engaged in a therapeutic process fashioned to his/her personal situation and directed toward providing the patient with an expanded repertoire of functional skills. The therapy should involve behavior modification directed toward helping the patient to develop social or adaptive skills. A trusting therapist–patient relationship can facilitate the success of other modes of treatment, providing a secure environment in which the patient can learn and can practice strategies of interpersonal relationship and communication. Evidence-based practice guidelines for psychosocial treatment of schizophrenia now include psychosocial treatments such as social skills training (261), cognitive/cognitive-behavioral therapy (262,263), family psychoeducation (264), and vocational rehabilitation (265,266). Newer treatments, such as cognitive training and adaptive skills training (e.g., medication management, transportation use), may also help to improve functional outcomes in patients with schizophrenia.

The future treatment of schizophrenia will focus increasingly on interventions tailored to each individual's symptoms. For example, in addition to the medications currently used, there may become available specific medications for negative symptoms and cognitive symptoms. Psychotherapeutic techniques, too, can become more individualized and tailored to each patient's needs and personal goals.

LONG-TERM SIDE EFFECTS OF ANTIPSYCHOTICS

In the treatment of schizophrenia, antipsychotic medications usually must be administered for long periods. Such long-term treatment exposes the patient to increased risks of deleterious side effects, the most troublesome of which is TD. Neuroleptic-induced TD is a serious problem in the psychopharmacology of schizophrenia. The TD syndrome consists of abnormal, involuntary movements, usually of the choreoathetoid type, sometimes stereotyped (128). Many different hyperkinetic movements, excluding tremor, have been reported as manifestations of TD. Typically, however, orofacial and upper extremity musculature is involved, with orofacial dyskinesias occurring in approximately 80% of affected patients. In some patients, the trunk, lower extremities, pharynx, and diaphragm are also affected (157). Other tardive syndromes such as tardive dystonia and tardive akathisia may be dominated by single particular forms of hyperkinesis (267–270), but TD usually presents multiple and disparate hyperkinesias simultaneously. Whether these manifestations imply disparate syndromes or are variants within a unitary TD syndrome remains unclear.

Antipsychotic treatment is a primary and necessary factor in the etiology and development of TD. (Non-TD dyskinesia may be secondary to different causes or may be spontaneous dyskinesia.) Usually, the syndrome appears after at least 3 months of neuroleptic treatment (except in older patients, in whom 1 month of treatment may be sufficient), and, other than withdrawal from antipsychotics, there are no consistently reliable treatments for TD. Several trials examined the effects of vitamin E with mixed results, and other small experimental trials were conducted, but more investigation is warranted (271).

Differential diagnosis of TD from other disorders with hyperkinetic and hypokinetic manifestations is critical (272). TD has also been found to coexist with many other disorders that present with movement-related symptoms (272–277). It can be difficult to separate the stereotypies seen in schizophrenia, autism, catatonia, and mental retardation from those of TD or from illnesses such as Huntington disease and Wilson disease, in which psychiatric symptoms manifested in the early stage of the disease may be treated with neuroleptics. Separating TD from akathisia is important, and Munetz and Cornes (278) proposed a set of clinical guidelines. Several studies have reported that TD and neuroleptic-induced parkinsonism can coexist (279–281).

Epidemiology and Risk Factors

The reported prevalence of TD has varied as a result of differences in populations of patients and the methods used (282). Yassa and Jeste (283) reviewed 76 studies of the prevalence of TD published from 1960 to 1990. In a total population of approximately 40,000 patients, the overall prevalence of TD was 24.2%.

A few studies measured the incidence of TD. Kane et al. (284) prospectively studied more than 850 patients (mean age, 29 years) and determined the incidence of TD after cumulative exposure to conventional antipsychotics to be 5% after 1 year, 18.5% after 4 years, and 40% after 8 years. The incidence in older populations has been found to be much higher. Saltz et al. (208) reported an incidence of 31% after 43 weeks of conventional antipsychotic treatment in a population of elderly patients. Jeste et al. (285) evaluated 266 psychiatric patients with a mean age of 66 years and found that 26.1% of the sample met criteria for TD during the first 12 months of study treatment, 51.7% had TD by the end of 24 months, and 59.8% had TD by the end of 36 months. Total exposure to typical antipsychotic agents has been shown to correlate with TD risk (286), and in the elderly, the cumulative amount of typical antipsychotics has also been associated with TD risk, especially with high-potency conventional agents (285).

Older age is the most important patient-related risk factor for TD. Patients older than age 45 are several times more likely to develop TD than are younger patients, and the prevalence, severity of dyskinetic symptoms, and intractability of the course of the disease increase with advancing age (206,287). Gender may be another risk factor for TD. Researchers have reported significantly greater incidence of TD in women. In their review of the published literature, Yassa and Jeste (283) calculated a global mean value of 27% for women and 22% in men. Moreover, women tended to have more severe TD and a higher prevalence of spontaneous dyskinesia than did men. Mood disorders (especially unipolar depression) have been reported to be risk factors for TD in a number of publications, although findings have been mixed (208,284,285,288,289). The evidence for organic mental syndromes as risk factors for TD so far has been inconclusive (290–294). Patients who react to antipsychotic drugs with acute or subacute EPS may be at greater risk of developing TD with continued treatment (287).

Risk Reduction with Atypical Antipsychotic Medications

Evidence supporting a reduced risk of TD with atypical antipsychotics is beginning to emerge. The lower risk of EPS with atypical agents has led to the conclusion that these agents will also have reduced TD risk. The low risk of TD in clozapine-treated individuals has been clearly demonstrated (247). In addition, a lower incidence of TD has been reported with risperidone (244,248,249). Furthermore, studies have shown a reduction in TD symptoms in patients with existing TD after switching from a conventional to atypical antipsychotic (247,295–298).

Neurochemistry

TD has frequently been thought to result from supersensitivity of striatal dopamine receptors. However, all antipsychotic-treated patients develop some degree of supersensitivity to dopamine, but only some develop TD. Jeste and Wyatt (299) summarized the evidence against the supersensitivity hypothesis of TD and concluded that it may be more likely that a number of separate neurotransmitter systems are involved in the pathogenesis of TD, leading to different subtypes of TD, each involving a unique profile of neurochemical imbalance.

Other Long-Term Side Effects

Although the risk of TD appears to be significantly reduced with the use of atypical antipsychotics, concern regarding other long-term side effects (e.g., weight gain, diabetes, and hyperlipidemia) with atypical agents is growing. Significant weight gain may negatively affect long-term health by contributing to comorbid conditions such as diabetes, coronary artery disease, and hypertension (300). Weight gain has been reported with conventional antipsychotics, but even greater weight gain has been seen with such atypical agents as clozapine, olanzapine, and, to a lesser extent, quetiapine (301). Changes in weight also have an important effect on an individual's lipid profile. Specifically, weight gain leads to an increase in triglycerides, which leads to an increased risk of serious events such as coronary artery disease and stroke (302). The accumulation of even relatively small amounts of visceral adipose tissue can cause insulin resistance (303), thus worsening existing diabetes or potentially leading to the development of diabetes in susceptible individuals. Although the exact mechanism is unclear, new-onset diabetes, hyperglycemia, and exacerbation of existing diabetes have been associated with atypical antipsychotic treatment, most frequently reported with clozapine and olanzapine (302,304–

306). The long-term consequences of weight gain, diabetes, and hyperlipidemia should prompt providers to monitor the weight of patients initiated on atypical agents and perform glucose monitoring as appropriate.

REFERENCES

1. Buchanan RW, Carpenter WT. Domains of psychopathology: an approach to the reduction of heterogeneity in schizophrenia. *J Nerv Ment Dis* 1994;182:193–204.
2. Jeste DV, del Carmen R, Lohr JB, et al. Did schizophrenia exist before the eighteenth century? *Compr Psychiatry* 1985;26:493–503.
3. Jellinger K. Neuropathological findings after neuroleptic long-term therapy. In: Roizen L, Sharki H, Grevil N, eds. *Neurotoxicology.* New York: Raven Press, 1977.
4. McGlashan TH, Hoffman RE. Schizophrenia as a disorder of developmentally reduced synaptic connectivity. *Arch Gen Psychiatry* 2000;57:637–648.
5. Green MF. *Schizophrenia revealed: from neurons to social interactions.* New York: Norton, 2001.
6. Pulver AE. Search for schizophrenia susceptibility genes. *Biol Psychiatry* 2000;47:221–230.
7. Nasrallah HA, ed. *Handbook of schizophrenia, volume 1: the neurology of schizophrenia.* Amsterdam: Elsevier Science, 1986.
8. Gottesman II, Shields J. *Schizophrenia: the epigenetic puzzle.* Cambridge, UK: Cambridge University, 1982.
9. Andreasen NC, Carpenter WT. Diagnosis and classification of schizophrenia. *Schizophr Bull* 1993;19:199–214.
10. Seaton BE, Goldstein G, Allen DN. Sources of heterogeneity in schizophrenia: the role of neuropsychological functioning. *Neuropsychol Rev* 2001;11:45–67.
11. Tamminga CA, Thaker GK, Buchanan R, et al. Limbic system abnormalities identified in schizophrenia using positron emission tomography with fluorodeoxyglucose and neocortical alterations with deficit syndrome. *Arch Gen Psychiatry* 1992;49:522–530.
12. Cleghorn JM, Franco S, Szechtman B, et al. Toward a brain map of auditory hallucinations. *Am J Psychiatry* 1992;149:1062–1069.
13. Liddle PF, Friston KJ, Frith CD, et al. Patterns of cerebral blood flow in schizophrenia. *Br J Psychiatry* 1992;160:179–186.
14. Flor-Henry P. Influence of gender in schizophrenia as related to other psychopathological syndromes. *Schizophr Bull* 1990;16:211–227.
15. Goldstein G. The neuropsychology of schizophrenia. In: Grant I, Adams KM, eds. *Neuropsychological assessment of neuropsychiatric disorders.* Oxford, UK: Oxford University Press, 1986.
16. Heaton RK, Crowley TJ. Effects of psychiatric disorders and their somatic treatments on neuropsychological test results. In: Filskov SB, Bell TJ, eds. *Handbook of clinical neurology.* New York: Wiley-Interscience, 1981.
17. Isaacman DJ, Verdile VP, Kohen FP, et al. Pediatric telephone advice in the emergency department: results of a mock scenario. *Pediatrics* 1992;89:35–39.
18. Levin S, Yurgelun-Todd D, Craft S. Contributions of clinical neuropsychology to the study of schizophrenia. *J Abnorm Psychol* 1989;98:341–356.
19. Nuechterlein KH, Dawson ME. Information processing and attentional functioning in the course of schizophrenic disorder. *Schizophr Bull* 1984;10:160–203.
20. Seidman L. Schizophrenia and brain dysfunction: an

integration of recent neurodiagnostic findings. *Psychol Bull* 1983;94:195–238.

21. Green MF, Kern RS, Braff DL, et al. Neurocognitive deficits and functional outcome in schizophrenia: are we measuring the "right stuff"? *Schizophr Bull* 2000;26:119–136.

22. Nuechterlein KH, Edell WS, Norris M, et al. Attentional vulnerability indicators, thought disorder, and negative symptoms. *Schizophr Bull* 1986;12:408–426.

23. Maher BA. Language and schizophrenia. In: Steinhauer S, Gruzelier JH, Zubin J, eds. *Handbook of schizophrenia, volume 4: neuropsychology, psychophysiology, and information processing.* Amsterdam: Elsevier Science, 1991.

24. Eyler Zorrilla LT, Heaton RK, McAdams LA, et al. Cross-sectional study of older outpatients with schizophrenia and healthy comparison subjects: no differences in age-related cognitive decline. *Am J Psychiatry* 2000;157:1324–1326.

25. Heaton RK, Gladsjo JA, Palmer BW, et al. Stability and course of neuropsychological deficits in schizophrenia. *Arch Gen Psychiatry* 2001;58:24–32.

26. Palmer BW, Heaton RK, Paulsen JS, et al. Is it possible to be schizophrenic yet neuropsychologically normal? *Neuropsychology* 1997;11:437–446.

27. Gjerde PF. Attentional capacity dysfunction and arousal in schizophrenia. *Psychol Bull* 1983;93:57–72.

28. Granholm E. Processing resource limitations in schizophrenia: implications for predicting medication response and planning attentional training. In: Margolin DI, ed. *Cognitive neuropsychology in clinical practice.* New York: Oxford University Press, 1992.

29. Asarnow RF, Granholm E, Sherman T. Span of apprehension in schizophrenia. In: Steinhauer S, Gruzelier JH, Zubin J, eds. *Handbook of schizophrenia, volume 4: neuropsychology, psychophysiology, and information processing.* Amsterdam: Elsevier Science, 1991.

30. Payne RW, Hochberg AC, Hawks DV. Dichotic stimulation as a method of assessing the disorder of attention in overinclusive schizophrenic patients. *J Abnorm Psychol* 1970;76:185–193.

31. Wishner J, Wahl O. Dichotic listening in schizophrenia. *J Consult Clin Psychol* 1974;42:538–546.

32. Pogue-Geile MF, Oltmanns TF. Sentence perception and distractibility in schizophrenic, manic and depressed patients. *J Abnorm Psychol* 1980;89:115–124.

33. Wielgus MS, Harvey PD. Dichotic listening and recall in schizophrenia and mania. *Schizophr Bull* 1988;14:689–700.

34. Orzak MH, Kornetsky C. Attention dysfunction in chronic schizophrenia. *Arch Gen Psychiatry* 1966;14:323–326.

35. Nuechterlein KH. Reaction time and attention in schizophrenia: a critical evaluation of the data and the theories. *Schizophr Bull* 1977;3:373–428.

36. Baddeley A. Working memory. *Science* 1992;255:556–559.

37. Goldman-Rakic PS, Selemon LD. Functional and anatomical aspects of prefrontal pathology in schizophrenia. *Schizophr Bull* 1997;23:437–458.

38. Park S, Holzman PS. Schizophrenics show spatial working memory deficits. *Arch Gen Psychiatry* 1992;49:975–982.

39. Perry W, Heaton RK, Potterat E, et al. Working memory in schizophrenia: transient "online" storage versus executive functioning. *Schizophr Bull* 2001;27:157–176.

40. Gold JM, Carpenter C, Randolph C, et al. Auditory working memory and Wisconsin Card Sorting Test performance in schizophrenia. *Arch Gen Psychiatry* 1997;54:159–165.

41. Goldman-Rakic PS. Prefrontal cortical dysfunction in schizophrenia: the relevance of working memory. In: Carroll BJ, Barnett JE, eds. *Psychopathology and the brain.* New York: Raven Press, 1991.

42. Paulsen JS, Salmon DP, Monsch AU, et al. Discrimination of cortical from subcortical dementias on the basis of memory and problem-solving tests. *J Clin Psychol* 1995;51:48–58.

43. Koh SD. Remembering of verbal materials by schizophrenic young adults. In: Schwartz S, ed. *Language and cognition in schizophrenia.* Hillsdale, NJ: Lawrence Erlbaum, 1978.

44. Bauman E. Schizophrenic short-term memory: the role of organization at input. *J Consult Clin Psychol* 1971;36:4–19.

45. Gold JM, Randolph C, Carpenter CJ, et al. Forms of memory failure in schizophrenia. *J Abnorm Psychol* 1992;101:487–494.

46. Russell PN, Bannatyne PA, Smith JF. Associative strength as a mode of organization in recall and recognition: a comparison of schizophrenics and normals. *J Abnorm Psychol* 1975;84:122–128.

47. Saykin AJ, Gur RC, Gur RE, et al. Neuropsychological function in schizophrenia: selective impairment in memory and learning. *Arch Gen Psychiatry* 1991;48:618–624.

48. Goldberg TE, Saint-Cyr JA, Weinberger DR. Assessment of procedural learning and problem solving in schizophrenia patients by Tower of Hanoi type tasks. *J Neuropsychiatry* 1990;2:165–173.

49. Granholm E, Bartzokis G, Asarnow RF, et al. Preliminary associations between motor procedural learning, basal ganglia T2 relaxation times, and tardive dyskinesia in schizophrenia. *Psychiatry Res* 1993;50:33–44.

50. Schwartz BL, Rosse RB, Deutsch SI. Toward a neuropsychology of memory in schizophrenia. *Psychopharmacol Bull* 1992;28:341–351.

51. Granholm E, Asarnow RF, Marder SR. Controlled information processing resources and the development of automatic detection responses in schizophrenia. *J Abnorm Psychol* 1991;100:22–30.

52. Schmand B, Brand N, Kuipers T. Procedural learning of cognitive and motor skills in psychotic patients. *Schizophr Res* 1992;8:157–170.

53. Butters N, Salmon DP, Granholm E, et al. Neuropsychological differentiation of amnesiac and dementing states. In: Stahl S, Iversen S, Goodman E, eds. *Cognitive neurochemistry.* Oxford: Oxford University Press, 1987.

54. Pantelis C, Barnes RRE, Nelson HE. Is the concept of frontal-subcortical dementia relevant to schizophrenia? *Br J Psychiatry* 1970;76:185–193.

55. Berman KF, Zec RF, Weinberger DR. Physiologic dysfunction of dorsolateral prefrontal cortex in schizophrenia: II. Role of neuroleptic treatment. *Arch Gen Psychiatry* 1986;43:126–135.

56. Braff DL, Heaton R, Kuck J, et al. The generalized pattern of neuropsychological deficits in outpatients with chronic schizophrenia with heterogeneous Wisconsin Card Sorting Test results. *Arch Gen Psychiatry* 1991;48:891–898.

57. Weinberger D. The pathogenesis of schizophrenia: a neurodevelopment theory. In: Nasrallah HA, Weinberger DR, eds. *Handbook of schizophrenia, volume I: the neurology of schizophrenia.* New York: Elsevier Science, 1986.

58. Gruzelier J, Seymour K, Wilson L, et al. Impairments on neuropsychologic tests of temporohippocampal and frontohippocampal functions and word fluency in remitting schizophrenia and affective disorders. *Arch Gen Psychiatry* 1988;45:623–629.

59a. Robbins TW. The case of frontostriatal dysfunction in schizophrenia. *Schizophrenia Bulletin* 1990;16:391–402.

59b. Carlsson M, Carlsson A. Schizophrenia: a subcortical neurotransmitter imbalance syndrome? *Schizophr Bull* 1990;16:425–432.

60. Weinberger DR, Berman KF, Zec RF. Physiological dysfunction of dorsolateral prefrontal cortex in schizophrenia, I: Regional cerebral blood flow evidence. *Arch Gen Psychiatry* 1986;43:114–124.
61. Andreasen NC. Understanding schizophrenia: a silent spring? *Am J Psychiatry* 1998;155:1657–1659.
62. Goldberg TE, Ragland JD, Torrey EF, et al. Neuropsychological assessment of monozygotic twins discordant for schizophrenia. *Arch Gen Psychiatry* 1990;47:1066–1072.
63. Kinsbourne M, Hicks RE. Functional cerebral space: a model for overflow, transfer and interference effects in human performance. In: Requin J, ed. *Attention and performance, VII.* Hillsdale, NJ: Lawrence Erlbaum, 1978.
64. Luria RR. *The working brain.* New York: Basic Books, 1973.
65. Beatty J. Task-evoked pupillary responses, processing load and the structure of processing resources. *Psychol Bull* 1982;91:276–292.
66. Swerdlow NR, Geyer MA. Using an animal model of deficient sensorimotor gating to study the pathophysiology and new treatments of schizophrenia. *Schizophr Bull* 1998;24:285–301.
67. Asarnow RF, Granholm E. The contributions of cognitive psychology to vulnerability models. In: Hafner H, Gattaz WF, eds. *Search for the causes of schizophrenia, volume II.* Heidelberg: Springer-Verlag, 1991.
68. Bilder RM, Turkel E, Lipschutz-Broch L, et al. Antipsychotic medication effects on neuropsychological functions. *Psychopharmacol Bull* 1992;28:353–366.
69. Spohn HE, Strauss ME. Relation of neuroleptic and anticholinergic medication to cognitive functions in schizophrenia. *J Abnorm Psychol* 1989;98:367–380.
70. Cornblatt BA, Lenzenweger MF, Dworkin RH, et al. Positive and negative schizophrenia symptoms, attention, and information processing. *Schizophr Bull* 1985;11:397–408.
71. Smith RC, Largen J, Vroulis G, et al. Neuropsychological test scores and clinical response to neuroleptic drugs in schizophrenic patients. *Compr Psychiatry* 1992;33:139–145.
72. Breier A, Schreiber JL, Dyer J, et al. National Institute of Mental Health longitudinal study of chronic schizophrenia—prognosis and predictors of outcome. *Arch Gen Psychiatry* 1991;48:239–246.
73. Perlick D, Mattis S, Stastny P, et al. Neuropsychological discriminators of long-term inpatient or outpatient status in chronic schizophrenia. *J Neuropsychiatry Clin Neurosci* 1992;4:428–434.
74. Kern RS, Green MF, Satz P. Neuropsychological predictors of skills training for chronic psychiatric patients. *Psychiatry Res* 1992;43:223–230.
75. Shenton ME, Dickey CC, Frumin M, et al. A review of MRI findings in schizophrenia. *Schizophr Res* 2001;49:1–52.
76. Johnstone EV, Crow TJ, Frith CD, et al. Cerebral ventricular size and cognitive impairment in chronic schizophrenia. *Lancet* 1976;2:924–926.
77. Weinberger D, Torrey E, Neophytides A, et al. Lateral cerebral ventricular enlargement in chronic schizophrenia. *Arch Gen Psychiatry* 1979;36:735–739.
78. Owen D, Johnstone E, Crow T, et al. Cerebral ventricular enlargement in schizophrenia: Relationship to the disease process and its clinical correlates. *Psychol Med* 1985;15:27–41.
79. Crow T, Ball J, Bloom S, et al. Schizophrenia as an anomaly of development of cerebral asymmetry: a post mortem study and a proposal concerning the genetic basis of the disease. *Arch Gen Psychiatry* 1989;46:1145–1150.
80. Daniel D, Myslobodsky M, Ingraham L, et al. The relationship of occipital skull asymmetry to brain parenchymal measures in schizophrenia. *Schizophr Res* 1989;2:465–472.
81. Andreasen NC, Ehrhardt JC, Swayze VW, et al. Magnetic resonance imaging of the brain in schizophrenia. *Arch Gen Psychiatry* 1990;47:35–44.
82. Gur R, Mozley P, Resnick S. Magnetic resonance imaging in schizophrenia: I. Volumetric analysis of brain and cerebrospinal fluid. *Arch Gen Psychiatry* 1991;48:407–412.
83. Kelsoe JR, Cadet JL, Pickar D, et al. Quantitative neuroanatomy in schizophrenia: a controlled magnetic resonance imaging study. *Arch Gen Psychiatry* 1988;45:533–541.
84. Suddath RL, Casanova MF, Goldberg TE, et al. Temporal lobe pathology in schizophrenia: a quantitative magnetic resonance imaging study. *Am J Psychiatry* 1989;146:464–472.
85. Suddath RL, Christison GW, Torrey EF, et al. Anatomical abnormalities in the brains of monozygotic twins discordant for schizophrenia. *N Engl J Med* 1990;322:789–794.
86. Andreasen NC, Paradiso S, O'Leary DS. "Cognitive dysmetria" as an integrative theory of schizophrenia: a dysfunction in cortical-subcortical-cerebellar circuitry? *Schizophr Bull* 1998;24:203–218.
87. Schroder J, Buchsbaum MS, Siegel BV, et al. Structural and functional correlates of subsyndromes in chronic schizophrenia. *Psychopathology* 1995;28:38–45.
88. Allen DN, Seaton BE, Goldstein G, et al. Neuroanatomic differences among cognitive and symptom subtypes of schizophrenia. *J Nerv Ment Dis* 2000;188:381–384.
89. Andreasen NC, Nopoulos P, O'Leary DS, et al. Defining the phenotype of schizophrenia: cognitive dysmetria and its neural mechanisms. *Biol Psychiatry* 1999;46:908–920.
90. Weinberger DR. Neurodevelopmental perspectives on schizophrenia. In: Bloom FE, Kupfer DJ, eds. *Psychopharmacology: the fourth generation of progress.* New York: Raven Press, 1995.
91. Green MF, Neuchterlein KH. Should schizophrenia be treated as a neurocognitive disorder? *Schizophr Bull* 1999;25:309–318.
92. Kindermann SS, Karimi A, Symonds L, et al. Review of functional magnetic resonance imaging in schizophrenia. *Schizophr Res* 1997;27:143–156.
93. Ingvar D, Franze G. Abnormalities of cerebral blood flow distribution in patients with chronic schizophrenia. *Acta Psychiatr Scand* 1974;50:425–462.
94. Ariel RN, Golden CJ, Quaife MA, et al. Regional cerebral blood flow in schizophrenics. *Arch Gen Psychiatry* 1983;40:258–263.
95. Kurachi M, Kobayashi K, Malsubara R. Regional cerebral blood flow in schizophrenic disorders. *Eur Neurol* 1985;24:176–181.
96. Mubrin Z, Knezevic S, Koretic D, et al. Regional cerebral blood flow patterns in schizophrenic patients. *Cereb Blood Flow Bull* 1982;3:43–46.
97. Chabrol H, Guell A, Bes A, et al. Cerebral blood flow in schizophrenic adolescents. *Am J Psychiatry* 1986;143:130.
98. Stevens AA, Goldman-Rakic PS, Gore JC, et al. Cortical dysfunction in schizophrenia during auditory word and tone working memory demonstrated by functional magnetic resonance imaging. *Arch Gen Psychiatry* 1998;55:1097–1103.
99. Buchsbaum M, DeLisi L, Holcomb H. Anterior gradients in cerebral glucose use in schizophrenia and affective disorders. *Arch Gen Psychiatry* 1984;41:1159–1166.
100. Buchsbaum M, Ingvar D, Kessler R. Cerebral glucography with positron emission tomography. *Arch Gen Psychiatry* 1982;39:251–259.

101. Buchsbaum MS. The frontal lobes, basal ganglia, and temporal lobes as sites for schizophrenia. *Schizophr Bull* 1990;16:379–389.

102. Buchsbaum MS, Haier RJ, Potkin SG, et al. Frontostriatal disorder if cerebral metabolism in never-medicated schizophrenics. *Arch Gen Psychiatry* 1992;49:935–942.

103. Cohen R, Semple W, Gross M. Dysfunction in a prefrontal substrate of sustained attention in schizophrenia. *Life Sci* 1987;40:2031–2039.

104. Early TS, Reiman EM, Raichle ME, et al. Left globus pallidus abnormality in never-medicated patients with schizophrenia. *Proc Natl Acad Sci U S A* 1987;84:561–563.

105. Farkas T, Wolf A, Jaeger J, et al. Regional brain glucose metabolism in chronic schizophrenia: a positron emission transaxial tomographic study. *Arch Gen Psychiatry* 1984;41:293–300.

106. Sheppard G, Gruzelier J, Manchanda R, et al. O-15 Positron emission tomographic scanning in predominantly never-treated acute schizophrenic patients. *Lancet* 1983;2:1448–1452.

107. Wolkin A, Jaeger J, Brodie J. Persistence of cerebral metabolic abnormalities in chronic schizophrenia as determined by positron emission tomography. *Am J Psychiatry* 1985;142:564–571.

108. Cleghorn J, Garnett E, Nahmias C. Increased frontal and reduced parietal glucose metabolism in acute untreated schizophrenia. *Psychiatry Res* 1989;28:119–133.

109. Szechtman H, Nahmias C, Garnett E. Effects of neuroleptics on altered cerebral glucose metabolism in schizophrenia. *Arch Gen Psychiatry* 1988;45:523–532.

110. Warkentin S, Nilsson A, Risberg J. Regional cerebral blood flow in schizophrenia: repeated studies during a psychotic episode. *Psychiatry Res* 1990;35:27–38.

111. O'Leary D, Webb M. The need for care assessment—a longitudinal approach. *Psychiatr Bull* 1996;20:134–136.

112. Eyler Zorrilla LT, Jeste DV, Paulus M, Brown GG. Functional abnormalities of medial temporal cortex during novel picture learning among patients with chronic schizophrenia. *Schizophr Res* 2003;59:187–198.

113. Eyler Zorrilla LT, Jeste DV, Brown GG. Functional MRI and novel picture learning among older patients with chronic schizophrenia: abnormal correlations between recognition memory and medial temporal brain response. *Am J Geriatr Psychiatry* 2002;10:52–61.

114. Buchsbaum MS, Hazlett EA. Positron emission tomography studies of abnormal glucose metabolism in schizophrenia. *Schizophr Bull* 1998;24:343–364.

115. Silbersweig DA, Stern E, Frith C, et al. A functional neuroanatomy of hallucinations in schizophrenia. *Nature* 1995;378:176–179.

116. Gur RE, Resnick SM, Alavi A, et al. Regional brain function in schizophrenia, I: a positron emission tomography study. *Arch Gen Psychiatry* 1987;44:119–125.

117. Volkow N, Wolf A, Van Gelder P. Phenomenological correlates of metabolic activity in 18 patients with chronic schizophrenia. *Am J Psychiatry* 1987;144:151–158.

118. Cleghorn J, Szechtman H, Garnett E. Apomorphine effects on brain metabolism in neuroleptic naive schizophrenic patients. *Psychiatry Res* 1991;40:135–153.

119. Schroder J, Wenz F, Schad LR, et al. Sensorimotor cortex and supplementary motor area changes in schizophrenia. *Br J Psychiatry* 1995;167:197–201.

120. Liddle PF. Functional imaging—schizophrenia. In: Johnstone EC, ed. *Biological psychiatry.* Oxford, UK: Royal Society of Medicine Press, 1996.

121. Gruzelier JH. Functional neuropsychological asymmetry in schizophrenia: a review and reorientation. *Schizophr Bull* 1999;25:91–120.

122. Wilson JM, Sanyal S, Van Tol HH. Dopamine D2 and D4 receptor ligands: relation to antipsychotic action. *Eur J Pharmacol* 1998;351:273–286.

123. Meltzer HY, McGurk SR. The effects of clozapine, risperidone, and olanzapine on cognitive function in schizophrenia. *Schizophr Bull* 1999;25:233–255.

124. Crawley J, Crow T, Johnstone E. Dopamine D2 receptors in schizophrenia studied in vivo. *Lancet* 1986;2:224–225.

125. Wong D, Wagner H, Tune L. Positron emission tomography reveals elevated D2 dopamine receptors in drug-naive schizophrenic patients. *Science* 1986;234:1558–1563.

126. Farde L, Wiesel F, Stone-Elander S. D2 dopamine receptors in neuroleptic-naive schizophrenic patients: a positron emission tomography study with [11C] raclopride. *Arch Gen Psychiatry* 1990;47:213–219.

127. Corsellis J. Psychoses of obscure pathology. In: Blackwood W, Corsellis J, ed. *Greenfield's neuropathology.* London: E. Arnold, 1976.

128. Brown R, Colter N, Corsellis J, et al. Postmortem evidence of structural brain changes in schizophrenia: differences in brain weight, temporal horn area, and parahippocampal gyrus compared with affective disorder. *Arch Gen Psychiatry* 1986;43:36–42.

129. Bruton CJ, Crow TJ, Frith CD, et al. Schizophrenia and the brain: a prospective clinico-neuropathological study. *Psychol Med* 1990;20:285–304.

130. Pakkenberg B. Post-mortem study of chronic schizophrenic brains. *Br J Psychiatry* 1987;151:744–752.

131. Crow T, Colter N, Brown R, et al. Lateralized asymmetry of temporal horn enlargement in schizophrenia. *Schizophr Res* 1988;1:155–156.

132. Atlshuler L, Casanova M, Goldberg T, et al. The hippocampus and parahippocampus in schizophrenic, suicide and control brains. *Arch Gen Psychiatry* 1990;47:1029–1034.

133. Bogerts B, Falkai P, Haupts M, et al. Post-mortem volume measurements of limbic systems and basal ganglia structures in chronic schizophrenics. *Schizophr Res* 1990;3:295–301.

134. Bogerts B, Meertz E, Schonfeldt-Bausch R. Basal ganglia and limbic system pathology in schizophrenia: a morphometric study of brain volume and shrinkage. *Arch Gen Psychiatry* 1985;42:784–791.

135. Falkai P, Bogerts B, Rozumek M. Limbic pathology in schizophrenia: the entorhinal region: a morphometric study. *Biol Psychiatry* 1988;24:515–521.

136. Stevens J. Clinicopathological correlations in schizophrenia. *Arch Gen Psychiatry* 1986;43:715–716.

137. Heckers S, Heinsen H, Heinsen Y, et al. Cortex white matter, and basal ganglia in schizophrenia: a volumetric postmortem study. *Biol Psychiatry* 1991;29:556–566.

138. Lesch A, Bogerts B. The diencephalon in schizophrenia: evidence for reduced thickness of the periventricular gray matter. *Eur Arch Psychiatry Neurol Sci* 1984;234:212–219.

139. Pakkenberg B. Pronounced reduction of total neuron number in mediodorsal thalamic nucleus and nucleus accumbens in schizophrenia. *Arch Gen Psychiatry* 1990;47:1023–1028.

140. Jakob H, Beckmann H. Gross and histological criteria for developmental disorders in brains of schizophrenics. *J R Soc Med* 1989;39:1131–1139.

141. Stevens JR. Neuropathology of schizophrenia. *Arch Gen Psychiatry* 1982;39:1131–1139.

142. Benes F, Sorensen I, Bird E. Reduced neuronal size in posterior hippocampus of schizophrenic patients. *Schizophr Bull* 1991;17:597–608.

143. Falkai P, Bogerts B. Cell loss in the hippocampus of schizophrenics. *Eur Arch Psychiatry Neurol Sci* 1986;236:154–161.
144. Jeste DV, Lohr JB. Hippocampal pathologic findings in schizophrenia: a morphometric study. *Arch Gen Psychiatry* 1989;46:1019–1024.
145. Benes F, Majocha R, Bird E, et al. Increased vertical axon numbers in the cingulate cortex of schizophrenics. *Arch Gen Psychiatry* 1987;4417:1017–1021.
146. Benes F, McSparren J, Bird E, et al. Deficits in small interneurons in prefrontal and cingulate cortices of schizophrenic and schizoaffective patients. *Arch Gen Psychiatry* 1993;48:996–1001.
147. Heckers S, Heinsen H, Geiger B, et al. Hippocampal neuron number in schizophrenia—a stereological study. *Arch Gen Psychiatry* 1993;48:1002–1008.
148. Benes F, Bird E. An analysis of the arrangement of neurons in the cingulate cortex of schizophrenic patients. *Arch Gen Psychiatry* 1987;44:608–616.
149. Benes F, Davidson J, Bird E. Quantitative cytoarchitectural studies of the cerebral cortex of schizophrenics. *Arch Gen Psychiatry* 1986;43:31–35.
150. Christian G, Casanova M, Weinberger D, et al. A quantitative investigation of parahippocampal pyramidal cell size, shape and variability of orientation in schizophrenia. *Arch Gen Psychiatry* 1989;46:1027–1032.
151. Conrad A, Abebe T, Austin R, et al. Hippocampal pyramidal cell disarray in schizophrenia. *Arch Gen Psychiatry* 1993;48:413–417.
152. Kovelman J, Scheibel A. A neurohistological correlate of schizophrenia. *Biol Psychiatry* 1984;19:1601–1621.
153. Selemon LD, Rajkowska G, Goldman-Rakic PS. Abnormally high neuronal density in the schizophrenic cortex. *Arch Gen Psychiatry* 1995;52:805–818.
154. Akbarian S, Bunney WE Jr, Potkin SG, et al. Altered distribution of nicotinamide-adenine dinucleotide phosphate-diaphorase cells in frontal lobe of schizophrenics implies disturbances of cortical development. *Arch Gen Psychiatry* 1993;50:169–177.
155. Akbarian S, Vinuela A, Kim J, et al. Distorted distribution of nicotinamide-adenine dinucleotide phosphate-diaphorase neurons in temporal lobe of schizophrenics implies anomalous cortical development. *Arch Gen Psychiatry* 1993;40:178–187.
156. Casanova MF, Stevens JR, Kleinman JE. Astrocytosis in the molecular layer of the dentate gyrus: a study in Alzheimer's disease and schizophrenia. *Psychiatry Res* 1990;35:149–166.
157. Roberts G, Colter N, Lofthouse R, et al. Is there gliosis in schizophrenia? Investigation of the temporal lobe. *Biol Psychiatry* 1987;22:1459–1486.
158. Creese I, Burt D, Snyder S. Dopamine receptor binding predicts clinical and pharmacological potencies of antischizophrenic drugs. *Science* 1976;192:481–483.
159. Seeman P, Lee T, Chau Wong K. Antipsychotic drug doses and neuroleptic/dopamine receptors. *Nature* 1976;261:717–719.
160. Davidson M, Keefe RS, Mohs RC, et al. L-Dopa challenge and relapse in schizophrenia. *Am J Psychiatry* 1987;144:934–938.
161. Lieberman J, Kane J, Gadaleta D, et al. Methylphenidate challenge as a predictor of relapse in schizophrenia. *Am J Psychiatry* 1984;141:633–638.
162. Snyder S. Catecholamines in the brain as mediators of amphetamine psychosis. *Arch Gen Psychiatry* 1972;27:169–179.
163. Davis KL, Kahn RS, Ko G, et al. Dopamine in schizophrenia: a review and reconceptualization. *Am J Psychiatry* 1991;148:1474–1486.
164. Bird E, Barnes J, Iversen L, et al. Increased brain dopamine and reduced glutamic acid decarboxylase and choline acetyltransferase activity in schizophrenia and related psychoses. *Lancet* 1977;2:1157–1159.
165. Mackay AVP, Iversen LL, Rosser M, et al. Increased brain dopamine and dopamine receptors in schizophrenia. *Arch Gen Psychiatry* 1982;39:991–997.
166. Owen F, Crow T, Poulter M, et al. Increased dopamine-receptor sensitivity in schizophrenia. *Lancet* 1978;2:223–225.
167. Reynolds G. Increased concentrations and lateral asymmetry of amygdala dopamine in schizophrenia. *Nature* 1983;305:527–529.
168. Crow TJ, Baker HF, Cross AJ, et al. Monamine mechanisms in chronic schizophrenia: post-mortem neurochemical findings. *Br J Psychiatry* 1979;134:249–256.
169. Reynolds G, Czudek C, Bzowej N, et al. Dopamine receptor asymmetry in schizophrenia. *Lancet* 1987;1:979.
170. Lee T, Seeman P, Tourtellotte W, et al. Binding of 3H-neuroleptics and 3H-apomorphine in schizophrenic brains. *Nature* 1978;274:897–900.
171. Mackay A, Bird E, Spokes E, et al. Dopamine receptors and schizophrenia: drug effect or illness? *Lancet* 1980;2:925–926.
172. Lieberman JA, Koreen AR. Neurochemistry and neuroendocrinology of schizophrenia: a selective review. *Schizophr Bull* 1993;19:371–429.
173. Widerlov E. A critical appraisal of CSF monoamine metabolite studies in schizophrenia. *Ann N Y Acad Sci* 1988;537:309–323.
174. Weinberger DR. Implications of normal brain development for the pathogenesis of schizophrenia. *Arch Gen Psychiatry* 1987;44:660–669.
175. Van Tol H, Bunjow J, Guan H, et al. Cloning of the gene for a human dopamine D4 receptor with high affinity for the antipsychotic clozapine. *Nature* 1991;350:610–614.
176. Millan MJ. Improving the treatment of schizophrenia: focus on serotonin (5-HT)(1A) receptors. *J Pharmacol Exp Ther* 2000;295:853–861.
177. Goff DC, Wine L. Glutamate in schizophrenia: clinical and research implications. *Schizophr Res* 1997;2:157–158.
178. Joyce JN. The dopamine hypothesis of schizophrenia: limbic interactions with serotonin and norepinephrine. *Psychopharmacology* 1993;112:S16–S34.
179. van Kammen DP, Peters J, Yao J, et al. Norepinephrine in acute exacerbations of chronic schizophrenia. Negative symptoms revisited. *Arch Gen Psychiatry* 1990;47:161–168.
180. Benes FM. Altered glutamatergic and GABAergic mechanisms in the cingulate cortex of the schizophrenic brain. *Arch Gen Psychiatry* 1995;52:1015–1018.
181. Bissette G, Nemeroff C, Mackay A. Peptides in schizophrenia. In: Enson P, Rosser M, Tohyama M, eds. *Progress in brain research*. Amsterdam: Elsevier Science, 1986.
182. Farmery SM, Owen F, Poulter M, et al. Reduced high affinity cholecystokinin binding in hippocampus and frontal cortex of schizophrenic patients. *Life Sci* 1985;36:473–477.
183. Ferrier IN, Roberts GW, Crow TJ, et al. Reduced cholecystokinin-like and somatostatin-like immunoreactivity in limbic lobe is associated with negative symptoms in schizophrenia. *Life Sci* 1983;33:475–482.
184. Roberts G, Ferrier I, Lee Y, et al. Peptides, the limbic lobe and schizophrenia. *Brain Res* 1983;288:199–211.
185. Olney JW, Farber NB. Glutamate receptor dysfunction and schizophrenia. *Arch Gen Psychiatry* 1995;52:998–1007.
186. Goff DC, Coyle JT. The emerging role of glutamate in the pathophysiology and treatment of schizophrenia. *Am J Psychiatry* 2001;158:1367–1377.

187. Kerwin R, Patel S, Maldrum B, et al. Asymmetrical loss of glutamate receptor subtype in left hippocampus in schizophrenia. *Lancet* 1988;1:583–584.

188. Nishikawa T, Takashima M, Toru M. Increased (3H) kainic acid binding in the prefrontal cortex in schizophrenia. *Neurosci Lett* 1983;40:245–250.

189. American Psychiatric Association. *Diagnostic and Statistical Manual of Mental Disorders, Third Edition.* Washington, DC: American Psychiatric Press, 1980.

190. Feighner JP, Robins E, Guze SB, et al. Diagnostic criteria for use in psychiatric research. *Arch Gen Psychiatry* 1972;26:57–63.

191. Kraepelin E. *Dementia praecox and paraphrenia, 1919.* Huntington, NY: Krieger, 1971.Barclay RM, Robertson GM, translators.

192. American Psychiatric Association. *Diagnostic Criteria from DSM-IV.* Washington, DC: American Psychiatric Association, 1994.

193. Howard R, Rabins PV, Seeman MV, et al. Late-onset schizophrenia and very-late-onset schizophrenia-like psychosis: an international consensus. *Am J Psychiatry* 2000;157:172–178.

194. Harris MJ, Jeste DV. Late-onset schizophrenia: an overview. *Schizophr Bull* 1988;14:39–55.

195. Jeste DV, Symonds LL, Harris MJ, et al. Non-dementia nonpraecox dementia praecox?: late-onset schizophrenia. *Am J Geriatr Psychiatry* 1997;5:302–317.

196. Palmer BW, McClure F, Jeste DV. Schizophrenia in late-life: findings challenge traditional concepts. *Harv Rev Psychiatry* 2001;9:51–58.

197. Bleuler M. *The schizophrenic disorders: long-term patient and family studies.* New Haven, CT and London: Yale University Press, 1978.Clemens SM, translator.

198. Castle DJ, Murray RM. The neurodevelopmental basis of sex differences in schizophrenia. *Psychol Med* 1991;21:565–575.

199. Kay DWK, Beamish P, Roth M. Old age mental disorders in Newcastle-Upon-Tyne. *Br J Psychiatry* 1964;110:146–158.

200. Pearlson GD, Kreger L, Rabins RV, et al. A chart review study of late-onset and early-onset schizophrenia. *Am J Psychiatry* 1989;146:1568–1574.

201. Lindamer LA, Lohr JB, Harris MJ, et al. Gender, estrogen, and schizophrenia. *Psychopharmacol Bull* 1997;33:221–228.

202. Herbert ME, Jacobson S. Late paraphrenia. *Br J Psychiatry* 1967;113:461–469.

203. Kay DWK, Roth M. Environmental and hereditary factors in the schizophrenias of old age ("late paraphrenia") and their bearing on the general problem of causation in schizophrenia. *J Ment Sci* 1961;107:649–686.

204. Post F. *Persistent persecutory states of the elderly.* London: Pergamon Press, 1966.

205. Rabins P, Pauker S, Thomas J. Can schizophrenia begin after age 44? *Compr Psychiatry* 1984;25:290–293.

206. Jeste DV, Lacro JP, Gilbert PL, et al. Treatment of late-life schizophrenia with neuroleptics. *Schizophr Bull* 1993;19:817–830.

207. Jeste DV, Caligiuri MP. Tardive dyskinesia. *Schizophr Bull* 1993;19:303–315.

208. Saltz BL, Woerner MG, Kane JM, et al. Prospective study of tardive dyskinesia incidence in the elderly. *JAMA* 1991;266:2402–2406.

209. Lohr JB, Alder M, Flynn K, et al. Minor physical anomalies in older patients with late-onset schizophrenia, early-onset schizophrenia, depression, and Alzheimer's disease. *Am J Geriatr Psychiatry* 1997;5:318–323.

210. Harris MJ, Cullum CM, Jeste DV. Clinical presentation of late-onset schizophrenia. *J Clin Psychiatry* 1988;49:356–360.

211. Jeste DV, Harris MJ, Zweifach M.Late-onset schizophrenia. In: Michels R, Cavenar JO Jr, Brodie NKH, et al., eds. *Psychiatry,* rev. ed. Philadelphia: JB Lippincott, 1988.

212. Burns A, Carrick J, Ames D, et al. The cerebral cortical appearance in late paraphrenia. *Int J Geriatr Psychiatry* 1989;4:31–34.

213. Krull AJ, Press G, Dupont R, et al. Brain imaging in late-onset schizophrenia and related psychoses. *Int J Geriatr Psychiatry* 1991;6:651–658.

214. Naguib M, Levy R. Neuropsychological impairment and structural brain abnormalities on computed tomography. *Int J Geriatr Psychiatry* 1987;2:83–90.

215. Rabins P, Pearlson G, Jayaram G, et al. Elevated VBR in late-onset schizophrenia. *Am J Psychiatry* 1987;144:1216–1218.

216. Breitner J, Husain M, Figiel G, et al. Cerebral white matter disease in late-onset psychosis. *Biol Psychiatry* 1990;28:266–274.

217. Miller BL, Lesser IM, Boone KB, et al. Brain lesions and cognitive function in late-life psychosis. *Br J Psychiatry* 1991;158:76–82.

218. Corey-Bloom J, Jernigan T, Archibald S, et al. Quantitative magnetic resonance imaging of the brain in late-life schizophrenia. *Am J Psychiatry* 1995;152:447–449.

219. Symonds LL, Olichney JM, Jernigan TL, et al. Lack of clinically significant structural abnormalities in MRIs of older patients with schizophrenia and related psychoses. *J Neuropsychiatry Clin Neurosci* 1997;9:251–258.

220. Palmer BW, Bondi MW, Twamley EW, et al. Are late-onset schizophrenia-spectrum disorders neurodegenerative conditions?: annual rates of change on two dementia measures. *J Neuropsychiatry Clin Neurosci* 2003 *(in press).*

221. Cummings JL. Organic delusions: phenomenology, anatomical correlations, and review. *Br J Psychiatry* 1985;146:184–197.

222. Davison K, Bagley C. Schizophrenia-like psychoses associated with organic disorders of the central nervous system—a review of the literature. In: Hetherington R, ed. *Current problems in neuropsychiatry,* 1969.

223. Jeste DV, Finkel SI. Psychosis of Alzheimer's disease and related dementias: diagnostic criteria for a distinct syndrome. *Am J Geriatr Psychiatry* 2000;8:29–34.

224. Jeste DV, Wragg RE, Salmon DP, et al. Cognitive deficits of patients with Alzheimer's disease with and without delusions. *Am J Psychiatry* 1992;149:184–189.

225. Peyser CE, Naimark D, Langston JW, et al. Psychotic syndromes in Parkinson's disease. *Semin Clin Neuropsychiatry* 1998;3:41–50.

226. Friedman J, Lannon M. Clozapine in the treatment of psychosis in Parkinson's disease. *Neurology* 1989;39:1219–1221.

227. Jeste DV, Wyatt RJ. *Neuropsychiatric movement disorders.* Washington, DC: American Psychiatric Press, 1984.

228. Harris MJ, Jeste DV, Gleghorn A, et al. New-onset psychosis in HIV-infected patients. *J Clin Psychiatry* 1991;52:369–376.

229. Sewell DD, Jeste DV, Atkinson JH, et al. HIV-associated psychosis: a longitudinal study 20 cases. *Am J Psychiatry* 1994;151:237–242.

230. Trimble M. Interictal psychosis. In: Trimble M, ed. *The psychoses of epilepsy.* New York: Raven Press, 1991.

231. Bear D, Levin K, Blumer D, et al. Interictal behavior in hospitalized temporal lobe epileptics: relationship to idiopathic psychiatric syndromes. *J Neurol Neurosurg Psychiatry* 1993;45:481–488.

232. Gibbs A. Ictal and non-ictal psychiatric disorders in temporal lobe epilepsy. *J Nerv Ment Dis* 1951;113:522–528.

233. Perez M, Trimble M, Michael R. Epileptic psychosis:

diagnostic comparison with process schizophrenia. *Br J Psychiatry* 1980;137:245–249.

234. Pakalnis A, Drake M, John K, et al. Forced normalization: acute psychosis after seizure control in seven patients. *Arch Neurol* 1987;44:289–292.

235. Mendez M, Grau R, Doss R, et al. Schizophrenia in epilepsy: seizure and psychosis variables. *Neurology* 1993;43:1073–1077.

236. Roberts GW, Done DJ, Bruton C, et al.: A "mock up" of schizophrenia: temporal lobe epilepsy and schizophrenia-like psychosis. *Biol Psychiatry* 1990;28:127–143.

237. Sherwin I, Peron-Magnan P, Bancaud J, et al. Prevalence of psychosis in epilepsy as a function of the laterality of the epileptogenic lesion. *Arch Neurol* 1982;39:621–625.

238. American Psychiatric Association. *Practice guidelines for the treatment of patients with schizophrenia.* Washington, DC: American Psychiatric Association, 1997:1–63.

239. Arana GW, Rosenbaum JF. *Handbook of psychiatric drug therapy* 4th ed. Philadelphia:, Lippincott Williams & Wilkins, 2000.

240. Kane JM. Pharmacologic treatment of schizophrenia. *Biol Psychiatry* 1999;46:1396–1408.

241. Meltzer HY. Treatment-resistant schizophrenia-the role of clozapine. *Curr Med Res Opin* 1997;14:1–20.

242. Arvanitis LA, Miller BG, the Seroquel Trial 13 Study Group. Multiple fixed doses of "seroquel" (quetiapine) in patients with acute exacerbation of schizophrenia: a comparison with haloperidol and placebo. *Biol Psychiatry* 1997;42:233–246.

243. Beasley CM, Hamilton SH, Crawford AM, et al. Olanzapine versus haloperidol: acute phase results of the international double-blind olanzapine trial. *Eur J Neuropsychopharmacol* 1997;7:125–137.

244. Chouinard G, Jones B, Remington G, et al. A Canadian multicenter placebo-controlled study of fixed doses of risperidone and haloperidol in the treatment of chronic schizophrenic patients. *J Clin Psychopharmacol* 1993;13:25–40.

245. Kane JM, Honigfeld G, Singer J, et al. Clozapine for the treatment resistant schizophrenic: a double-blind comparison with chlorpromazine. *Arch Gen Psychiatry* 1988;45:789–796.

246. Simpson GM, Lindenmayer JP. Extrapyramidal symptoms in patients treated with risperidone. *J Clin Psychopharmacol* 1997;17:194–201.

247. Kane JM, Woerner MG, Pollack S, et al. Does clozapine cause tardive dyskinesia? *J Clin Psychiatry* 1993;54:327–330.

248. Jeste DV, Lacro JP, Palmer BW, et al. Incidence of tardive dyskinesia in early stages of low-dose treatment with typical neuroleptics in older patients. *Am J Psychiatry* 1999;156:309–311.

249. Jeste DV, Lacro JP, Bailey A, et al. Lower incidence of tardive dyskinesia with risperidone compared with haloperidol in older patients. *J Am Geriatr Soc* 1999;47:716–719.

250. Tollefson GD, Beasley CM, Tran PV, et al. Olanzapine versus haloperidol in the treatment of schizophrenia and schizoaffective and schizophreniform disorders: results of an international collaborative trial. *Am J Psychiatry* 1997;154:457–465.

251. Buchanan RW, Holstein C, Breier A. The comparative efficacy and long-term effect of clozapine treatment on neuropsychological test performance. *Biol Psychiatry* 1994;36:717–725.

252. Hagger C, Buckley P, Kenny JT, et al. Improvement in cognitive functions and psychiatric symptoms in treatment-refractory schizophrenic patients receiving clozapine. *Biol Psychiatry* 1993;34:702–712.

253. Hoff AL, Faustman WO, Wieneke M, et al. The effects of clozapine on symptom reduction, neurocognitive function, and clinical management in treatment-refractory state hospital schizophrenic inpatients. *Neuropsychopharmacology* 1996;15:361–369.

254. Green MF, Marshall BD, Wirshing WC, et al. Does risperidone improve verbal working memory in treatment-resistant schizophrenia? *Am J Psychiatry* 1997;154:799–804.

255. Kern RS, Green MF, Barringer DM Jr, et al. Risperidone versus haloperidol on secondary memory: can newer medications aid learning? *Schizophr Bull* 1999;25:223–232.

256. Purdon SE, Jones BDW, Stip E, et al. Neuropsychological changes in early phase schizophrenia during 12 months of treatment with olanzapine, risperidone, or haloperidol. *Arch Gen Psychiatry* 2000;57:249–258.

257. Purdon SE. Cognitive improvement in schizophrenia with novel antipsychotic medications. *Schizophr Res* 1999;35:S51–S60.

258. Levensen JL. Neuroleptic malignant syndrome. *Am J Psychiatry* 1985;142:1137–1145.

259. Sewell DD, Jeste DV. Neuroleptic malignant syndrome: clinical presentation, pathophysiology, and treatment. In: Stoudemire A, Fogel BS, eds. *Medical psychiatric practice, volume 1.* Washington, DC: American Psychiatric Press, 1991.

260. Jeste DV, Twamley EW, Eyler Zorrilla LT, et al. Aging and outcome in schizophrenia. *Acta Psychiatr Scand* 2003;107:1–8.

261. Heinssen RK, Liberman RP, Kopelowicz A. Psychosocial skills training for schizophrenia: lessons from the laboratory. *Schizophr Bull* 2000;26:21–46.

262. Beck AT, Rector NA. Cognitive therapy for schizophrenia patients. *Harv Ment Health Lett* 1998;15:4–6.

263. Garety PA, Fowler D, Kuipers E. Cognitive-behavioral therapy for medication-resistant symptoms. *Schizophr Bull* 2000;26:73–86.

264. Dixon L, Adams C, Lucksted A. Update on family psychoeducation for schizophrenia. *Schizophr Bull* 2000;26:5–20.

265. Bond GR, Resnick SG, Drake RE, et al. Does competitive employment improve nonvocational outcomes for people with severe mental illness? *J Consult Clin Psychol* 2001;69:489–501.

266. Cook JA, Razzano L. Vocational rehabilitation for persons with schizophrenia: recent research and implications for practice. *Schizophr Bull* 2000;26:87–103.

267. Burke RF, Fahn S, Jankovic J, et al. Tardive dystonia: late-onset and persistent dystonia caused by antipsychotic drugs. *Neurology* 1982;32:1335.

268. Gimenez-Roldan S, Mateo D, Bartolome P. Tardive dystonia and severe tardive dyskinesia: a comparison of risk factors and prognosis. *Acta Psychiatr Scand* 1985;44:417.

269. Stahl SM. Tardive Tourette syndrome in an autistic patient after long-term neuroleptic administration. *Am J Psychiatry* 1980;137:1267.

270. Weiner WJ, Luby ED. Tardive akathisia. *J Clin Psychiatry* 1983;44:417.

271. Gupta S, Mosnik D, Black DW, et al. Tardive dyskinesia: review of treatments past, present, and future. *Ann Clin Psychiatry* 1999;11:257–266.

272. Lohr J, Wisniewski A, Jeste DV. Neurological aspects of tardive dyskinesia. In: Nasrallah HA, Weinberger DR, eds. *The neurology of schizophrenia.* Amsterdam: Elsevier Science, 1986.

273. Klawans HL, Barr A. Prevalence of spontaneous lingual-facial-buccal dyskinesias in the elderly. *Neurology* 1982;32:558–559.

274. Koller WC. Eduntulous orodyskinesia. *Ann Neurol* 1983;13:97.

275. Nasrallah HA, Pappas NJ, Crowe RR. Oculogyric dystonia in tardive dyskinesia. *Am J Psychiatry* 1980;137:850–851.

276. Weiner WJ, Nausieda PA, Glantz RH. Meige syndrome (blepharospasmoromandibular dystonia) after long-term neuroleptic therapy. *Neurology* 1981;31:1555.

277. Weiss KJ, Ciraulo DA, Shader RI. Physostigmine test in the rabbit syndrome and tardive dyskinesia. *Am J Psychiatry* 1980;137:627–628.

278. Munetz MR, Cornes CL. Distinguishing akathisia and tardive dyskinesia: a review of the literature. *J Clin Psychopharmacol* 1983;3:343.

279. Britton V, Melamed E. Coexistence of severe parkinsonism and tardive dyskinesia as side effects of neuroleptic therapy. *J Clin Psychiatry* 1984;45:28.

280. Crane GE. Pseudoparkinsonism and tardive dyskinesia. *Arch Neurol* 1972;27:426–430.

281. Richardson MA, Craig TJ. The coexistence of parkinsonism-like symptoms and tardive dyskinesia. *Am J Psychiatry* 1982;139:341.

282. Jeste DV, Wyatt RJ. Changing epidemiology of tardive dyskinesia—an overview. *Am J Psychiatry* 1981;138:297–309.

283. Yassa R, Jeste DV. Gender differences in tardive dyskinesia: a critical review of the literature. *Schizophr Bull* 1992;18:701–715.

284. Kane JM, Woerner M, Lieberman J. Tardive dyskinesia: prevalence, incidence, and risk factors. *J Clin Psychopharmacol* 1988;8:52S–56S.

285. Jeste DV, Caligiuri MP, Paulsen JS, et al. Risk of tardive dyskinesia in older patients: a prospective longitudinal study of 266 patients. *Arch Gen Psychiatry* 1995;52:756–765.

286. Casey DE. Will the new antipsychotics bring hope of reducing the risk of developing extrapyramidal syndromes and tardive dyskinesia? *Int Clin Psychopharmacol* 1997;12:S19–S27.

287. Kane JM, Jeste DV, Barnes TRE, et al. *Tardive dyskinesia: a task force report of the American Psychiatric Association.* Washington, DC: American Psychiatric Association, 1992.

288. Casey DE. Affective disorders and tardive dyskinesia. *Encephale* 1988;14:221–226.

289. Casey DE, Keepers GA. Neuroleptic side effects: acute extrapyramidal syndromes and tardive dyskinesia. In: Casey DE, Christensen AV, eds. *Psychopharmacology: current trends.* Berlin: Springer-Verlag, 1988.

290. Brown KW, White T. The influence of topography on the cognitive and psychopathological effects of tardive dyskinesia. *Am J Psychiatry* 1992;149:1385–1389.

291. Gold JM, Egan MF, Kirch DG, et al. Tardive dyskinesia: neuropsychological, computerized tomographic, and psychiatric symptom findings. *Biol Psychiatry* 1991;30:587–599.

292. Manschreck TC, Keuthen NJ, Schneyer ML, et al. Abnormal involuntary movements and chronic schizophrenic disorders. *Biol Psychiatry* 1990;27:150–158.

293. Mion CC, Andreasen NC, Arndt S, et al. MRI abnormalities in tardive dyskinesia. *Psychiatry Res* 1991;40:157–166.

294. Yassa R, Nair V, Schwartz G. Tardive dyskinesia: a two-year follow-up study. *Psychosomatics* 1984;25:852–855.

295. Simpson GM, Lee JM, Shrivastava RK. Clozapine in tardive dyskinesia. *Psychopharmacology* 1978;56:75–80.

296. Littrell KH, Johnson CG, Littrell S, et al. Marked reduction of tardive dyskinesia with olanzapine. *Arch Gen Psychiatry* 1998;55:279–280.

297. Street JS, Tollefson GD, Tohen M, et al. Olanzapine for psychotic conditions in the elderly. *Psychiatr Ann* 2000;30:191–196.

298. Jeste DV, Klausner M, Brecher M, et al. A clinical evaluation of risperidone in the treatment of schizophrenia: a 10-week, open-label, multicenter trial involving 945 patients. *Psychopharmacology* 1997;131:239–247.

299. Jeste DV, Wyatt RJ. *Understanding and treating tardive dyskinesia.* New York: Guilford Press, 1982.

300. Must A, Spadano J, Coakley EH, et al. The disease burden associated with overweight and obesity. *JAMA* 1999;282:1523–1529.

301. Meyer JM. A retrospective comparison of lipid, glucose and weight changes at one year between olanzapine and risperidone treated inpatients. *J Clin Psychiatry* 2002;63:425–433.

302. Wirshing DA, Pierre JM, Eyeler J, et al. Risperidone-associated new-onset diabetes. *Biol Psychiatry* 2001;50:148–149.

303. Lebovitz HE. Diagnosis, classification, and pathogenesis of diabetes mellitus. *J Clin Psychiatry* 2001;62:5–9.

304. Haupt DW, Newcomer JW. Hyperglycemia and antipsychotic medications. *J Clin Psychiatry* 2001;62:15–26.

305. Jin H, Meyer JM, Jeste DV. Phenomenology of and risk factors for new-onset diabetes mellitus and diabetic ketoacidosis: an analysis of 45 published cases. *Ann Clin Psychiatry* 2002;14:59–64.

306. Wirshing DA, Spellberg BJ, Erhart SM, et al. Novel antipsychotics and new onset diabetes. *Biol Psychiatry* 1998;44:778–783.

31

THE PSYCHOPHARMACOLOGIC TREATMENT OF CHRONIC NONNEUROPATHIC PAIN AND ITS ASSOCIATED PSYCHOPATHOLOGY

DAVID A. FISHBAIN

MEASUREMENT OF PAIN

Pain is measured in two situations: experimental and clinical (1). Experimental pain is easier to study because one can measure the intensity of the pain-inducing stimulus. In clinical pain, the nature of the stimulus is often unknown, and its intensity can only be measured indirectly (2). In addition, because pain is a complex multidimensional subjective experience (3), a single dimension such as intensity fails to capture the variations in pain associated with a pinprick, toothache, or burn. Additionally, in any given dimension, such as intensity, the reported severity of the pain is related to numerous variables: cultural background, past experience, the meaning of the situation, personality variables, attention, arousal level, the prevailing contingencies of reinforcement, individual pain threshold and pain tolerance differences, pain responsivity differences, inability to habituate differences, suggestion, and environmental cues (1,3–5). Because of these factors, the clinical measurement of pain has been a difficult problem, infrequently evaluated as an end point of treatment (2). Currently, the following indirect methods are available for the measurement of clinical pain (2,6,7):

1. Rating scale methods in which patients rate pain experiences on structured scales with clearly defined limits;
2. Psychophysical methods that attempt to define pain threshold and pain tolerance in terms of experimentally induced pain and then ask the patient to match the perceived experimentally induced pain to his/her current clinical pain;
3. Measurement of drug dose required to relieve the pain;
4. Measurement of observed pain behaviors, such as moving in bed, bracing, rubbing an affected part;
5. Magnitude estimation procedures in which judgments of perceived pain are translated into cross-modality matching techniques such as handgrip force;

6. Measurement of performance ability on laboratory tasks;
7. Human physiologic correlates, such as direct recording from peripheral nerves and evoked potentials.

Each of these general methods has inherent limitations (2). Thus, it is almost mandatory to use a combination of methods in most pain measurement studies. Currently, methods 5, 6, and 7 are investigational only, whereas 1, 2, and 3 are the most widely researched methods. Some specific approaches to the utilization of methods 1, 2, and 4 are described in the sections that follow.

Rating Scale Methods

Visual Analogue Scale

Of the rating scale methods, the visual analogue scale (VAS) is the most researched and accepted method and the most widely used in measuring clinical pain (1,2,8–10). The VAS consists of a 10-cm line anchored by two extremes of pain, "no pain" and "pain as bad as it could be." Patients are asked to make a mark on the line that represents their level of perceived pain intensity, and the scale is scored by measuring the distance from the "no pain" to the patient's mark. There are various types of VAS, e.g., scaled from 0 to 10, but the most practical index uses a 101-point numerical scale (9). An example of this scale is shown in Figure 31.1.

A common problem with the VAS is that it assumes pain to be a one-dimensional experience that varies only in intensity (2). Although the VAS is subject to response biases (2), it has been shown to be internally consistent both in experimentally induced pain and chronic clinical pain, thereby demonstrating validity (2). The VAS has been shown to be more sensitive than verbal rating scales (9) and may be more reliable (2). In addition, VAS scores appear to predict chronic pain patient (CPP) status at 6 months (11). Of the rating scale methods, the VAS is preferable for clinical application.

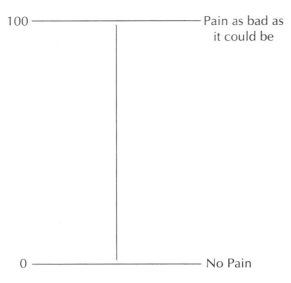

Please indicate on the line above the number between zero (0) and one hundred (100) that best describes your pain. A zero (0) would mean "no pain" and a one hundred (100) would mean "pain as bad as it could be."

FIGURE 31.1. The Visual Analogue Scale for pain (10 cm).

However, even the VAS may have poor sensitivity to treatment effects (2). Therefore, it has been recommended that, when measuring treatment effects, two types of VAS be employed: (a) the absolute VAS just described and (b) a comparative VAS (8). The comparative VAS is 20 cm in length and has three points: less severe, unchanged (at the midline), and more severe (Fig. 31.2).

The VAS can also be used in other ways. Depending on the directions given to the patient completing the VAS, a different aspects of pain can be measured. Thus, the patient can be asked to rate his/her pain at "this point in time" or as "average pain level over the past 24 hours" or "average pain over the last week." By combining visual analog scales based on different directions, the intensity, frequency, and duration of clinical pain can be indirectly assessed (2). For example, if one wanted to measure the relationship between occurrence of pain symptoms and drug use, one could construct four VAS with the different directions below and give them to the same patient at the same time:

1. "Rate your average pain over the past 24 hours."
2. "What level of pain do you consider intolerable?"
3. "For what level of pain would you consider taking medication?"
4. "What level of pain do you consider disabling?"

Description Differential Scale

The Description Differential Scale consists of a list of adjectives describing different levels of pain intensity (Fig. 31.3) (12). CPP are asked to rate the intensity of their pain

Directions

Please tell us how your pain has changed in the last month by making a line on the scale. Zero (0) represents no change, +100 represents most severe increase while −100 represents the greatest possible decrease.

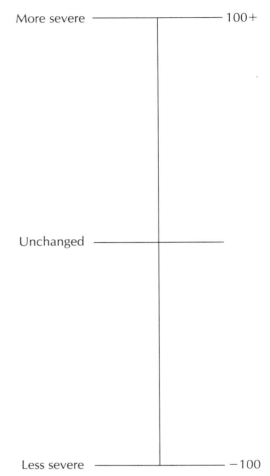

FIGURE 31.2. The comparative Visual Analogue Scale (20 cm, total).

as more or less than the word on the list by placing a check mark either to the left (less) or right (more) of the word being rated. If the word describes their pain level, they place a check mark under the word. There are 10 points to the right and left of each word, giving a 21-point rating scale. Pain intensity is defined as the mean of each rating and ranges from 0 to 20. The Description Differential Scale has high internal consistency (12). The major advantage of this scale is the fact that CPP can be checked to see how consistently they are using the scale by examining the relationship among different items. As displayed in Figure 31.3, terms describing the intensity of pain are presented in random order. If pain is "fair," it cannot also be "very intense." Responses are checked to ensure that they are consistent in this sense (Fig. 31.3).

TABLE 31.1. BEHAVIORS USED IN PAIN BEHAVIOR RATING SCALES

1. Stiffness on sitting
2. Slow movement
3. Limping
4. Frequent shifts in position
5. Stands bent forward
6. Walks bent forward
7. Distorted gait
8. Facial grimacing
9. Holds painful part
10. Rubs painful part
11. Moans
12. Groans
13. Writhes
14. Uses cane
15. Takes medications for pain
16. Moves in a guarded fashion
17. Uses heat or ice
18. Uses prosthetic devices
19. Frequent rests (lying down)
20. Avoids physical activity

Pain affect can be defined as the emotional arousal and disruption engendered by the pain experience. The pain affect component may be conceptually and empirically distinct from pain intensity (12). Multiple-item measures may be more reliable in measuring pain affect than single-item measures. Therefore, the Description Differential Scale was modified to measure pain affect and can be used in this fashion. This scale is demonstrated in Figure 31.4.

Pain Measurement Inventories

McGill Pain Questionnaire

The McGill Pain Questionnaire (MPQ) (13) is an inventory rather than a rating scale. It is designed to quantify three dimensions of pain: sensory, affective, and valuative. The MPQ is made up of 20 sets of pain word descriptions, each set containing up to six words (2). Pain patients are asked to circle words in each set that are relevant. The investigator scores the number of words chosen on the total number of word sets that apply to the pain. Because the words within each set have been ranked, one can compute a total score.

Instructions: Each word represents an amount of sensation. Rate your sensation in relation to each word with a checkmark.

FIGURE 31.3. Description Differential Scale of pain intensity.

Instructions: Each word represents an amount of sensation. Rate your sensation in relation to each word with a checkmark.

FIGURE **31.4.** Description Differential Scale of pain affect.

Although the MPQ is the most widely used instrument for pain measurement (14), there is little consensus over the scaling of the words (2). In addition, patients often have difficulty with the complexity of the vocabulary used (2), and the MPQ weights sensory aspects of pain more heavily than the affective and valuative aspects. This last issue may bias MPQ scores (2), but, most important because of this bias, it does not focus on the most important aspect of pain treatment: the response to treatment in the valuative and affective dimension (2). In the past, caution has been urged in using the MPQ for the measurement of pain intensity (15), and doubt has been cast on the discriminate validity of the MPQ pain rating index (16). However, a short form of the MPQ has been developed and found to be reliable and valid (14) but has not yet been used widely.

Million Scale

There is now convincing evidence that chronic pain intensity levels and measures of functional impairment and disability are correlated (17–26). From these data, one may conclude that pain itself results in functional impairment and/or dis-

ability or that the pain is perceived by the pain patient as a disability or both. This evidence has been used in developing rating scales and inventories that tap both the perceived pain and perceived functional impairment aspects of the pain experience. The Million Scale is an example of this concept. This is a 15-item inventory scored on a VAS concerning the association of pain with several activities, and self-perceived functional impairment (27). It appears to be valid and reliable with chronic low back pain (CLBP) patients. Low scores on this scale appear to predict return to work (28). VonKorff et al. (26) also developed a scale using the pain/disability concept. However, the reliability and validity of this scale have not yet been determined.

Pain Behavior Methods

Pain behavior has been defined as "any and all outputs of the patient that a reasonable observer would characterize as suggesting pain, such as (but not limited to) posture, facial expression, verbalizing, lying down, taking medications, seeking medical assistance and receiving compensation" (29). A list of identified pain behaviors is presented in Table 31.1.

Pain behaviors can often be elicited during physical examination (30) and correlate with physical examination findings, number of operations, and longer pain histories (30). In addition, pain behaviors correlate with perceived severity of pain (31,32), extent of functional impairments (31), and scores on the Illness Behavior Questionnaire (33). As pain improves, pain behavior diminishes (2,32,34). Evidence from pain stimulation experiments indicates that pain behavior may be either a consequence of CLBP or an important risk factor for the development of CLBP (35). These studies highlight the importance of pain behavior measurements and the possible interrelationship of pain behavior to functional status.

To measure pain behavior in a systematic fashion, the University of Alabama (UAB) Pain Behavior Scale was developed (34). The UAB Pain Behavior Scale quantifies the observed pain behavior and has been shown to be reliable (34). In addition, ratings of pain behavior using the UAB Pain Behavior Scale appear to be significantly related to both VAS sensory and VAS affective ratings (36). The concept of pain behavior has been criticized from a number of standpoints (37). However, the aforementioned studies indicate that the pain behavior concept can be operationalized as a powerful pain observational clinical tool that is particularly useful in more severely affected patients.

Pain-Matching Methods

Although the VAS is efficient and simple to use in pain measurement (2), it has one major problem: pain threshold (level of stimulation at which the subject begins to perceive pain in a pain stimulation experiment) and pain tolerance (level of stimulation at which the subject can no longer tolerate the induced pain in a pain stimulation experiment) vary among individuals (1). Although a pain stimulus may approach pain tolerance or "pain as bad as it could be" for one individual, another may perceive it as mild. Pain threshold and pain tolerance mediate a patient's rating of pain intensity. As an example, CLBP patients seem to have a higher pain threshold and lower pain tolerance than pain-free controls, and they may not be able to habituate pain threshold like pain-free controls (38). Therefore, the VAS is most effective in measuring pain intensity and pain change in the same individual but is less reliable when it is used for between-subject pain comparisons. In an attempt to solve this dilemma, the pain-matching method was developed. This method allows a more direct measurement of the actual clinical pain a patient experiences and comes closest to the true experimental situation. Various methods can be used in humans to induce clinical experimental pain, including heat, electric shock, noise, and pressure (2). Of these, pressure has been noted to most closely approximate chronic, gnawing pain (10). A pressure algometer was developed for this type of pain induction (39) and improved (40) to give greater reliability while producing sensations more closely related to clinical pain. This stimulator has been modified by the addition of a motor to increase the pressure in a linear manner

over a short period of time, thereby improving reliability (41).

Before using this pain stimulation method, the CLBP patient is given these instructions: "Tell us when you first feel the pain (pain threshold), then tell us when the pain that you feel in your finger is equal to the pain in your lower back (matching pain), and finally tell us when you can no longer tolerate the pain (pain tolerance)." In this way, the pain patient's current pain can be matched to his/her pain threshold and tolerance and thereby matched across other pain patients. Although the use of a pain-matching technique raises some ethical questions, this procedure is necessary to adequately assess pain threshold/tolerance. Again, this is the only method currently available by which pain comparisons can be made across patients.

Dilemma of the Severity of Pain and its Relationship to Function and Disability

The major problem in the evaluation and treatment of chronic pain is the frequent discrepancy between the patient's reported level of pain and the resultant functional status versus the physician's perception of what the patient should be able to do functionally. This discrepancy in perceptions is attributable to the following complex problems:

1. CPP perceive their pain as a disability that limits their functional status (25). The perception of pain as a disability was investigated by Riley et al. (25). To address this problem, they developed the Pain and Impairment Relationship Scale. This scale was shown to covary in CLBP patients with four standardized measures of physical and overall impairment that were not intercorrelated (25). The authors interpreted this finding as indicating that patients who are more impaired are, indeed, more impaired as a function of their pain (25). The great number of patients who claim disability based on pain alone has forced the United States Social Security Administration Commission on the Evaluation of Pain to recommend the development of a listing based on "impairment due primarily to pain" (42).

2. The pathology model does not predict back pain (43), making the reliability and validity of this model questionable. As a result, objective medical impairment may vary markedly from subjective disability (44). Also, CPP often lack documentation of clear structural organic pathology (45), thereby having minimal "medical impairment" as currently rated by the American Medical Association's (AMA) *Guides to the Evaluation of Permanent Impairment* (46). At the same time, many CPP will demonstrate higher disability than medical impairment ratings (47).

3. Medical impairment ratings vary widely from one physician to another (48,49) as do measures of disability (50). These differences in impairment ratings have been blamed on evaluation schedules that are not scientifically

based and that do not take functional status into consideration (49).

4. In cases in which patients report severe pain, physicians have great difficulty in estimating patients' functional status (51). In addition, physicians have great difficulty in estimating anxiety, pain, and activity limitation from patients' self-report and physical examination. These three dimensions are underestimated 35% of the time, and the activity limitation dimension is the one most often underestimated (52). This is because the interobserver reliability decreases as functional status decreases (53).

5. In CLBP, physical findings have low reliability and reproducibility (54), creating a problem in the determination of medical impairment. In addition, in CLBP, physical findings are not predictive of disability status, treatment outcome, and return to work (55,56).

6. Patient-generated statements about severity of pain and functional impairment from pain have been shown to correlate (57,58). However, the measures show few relationships to other measures of severity of pain: medication consumption, health care utilization, activity level, and frequency counts of engaging in a set of commonplace activities (57).

7. CPP display a reduced activity level attributable to pain, resulting in a consistent negative relationship between exercise and pain behavior, i.e., the more exercise performance, the fewer the pain behaviors (59).

8. It appears that pain behavior and movement ratings are strongly influenced by self-efficacy for movement, which in turn is determined by response expectancies for pain. Thus, the patients' movement is determined by how much pain they believe that movement will produce, how much pain they wish to avoid, and whether patients can accurately predict the pain involved in each movement (60). It has been postulated that the prediction of pain in movement promotes pain avoidance behavior and that inaccurate prediction could promote excessive avoidance (61).

The previously reviewed studies indicate that there is little relationship between the physician's prediction of functional status based on the physical examination and the CPP's demonstrated functional status. This discrepancy relates to the inability of the physician to gauge the patient's pain and the CPP's perception of pain as a disability and differences in the CPP's willingness to risk increases in pain by increasing activities.

Dilemma of the Discrepancy between Chronic Pain Patients' Self-Report of Pain and Observed Pain Behavior

Pain researchers and clinicians have in the past been troubled by a concern that pain self-report and observed pain behavior

might be inconsistent (62,63). However, there is little evidence that this is a problem. Indeed, pain behavior ratings have been noted to correlate significantly with reported perceived pain (31,32,36,62). As perceived pain improves, pain behavior diminishes (2,32,34). The question remains then: Why do some clinicians believe that in CPP there is little correlation between reported perceived pain and observed pain behavior? The probable reasons for this perception are as follows:

1. As pointed out previously, there are discrepancies between what the CPP should be able to do according to the clinician and what he/she actually is willing to do. Such a situation leads to this perception.

2. For a proportion of patients presenting with chronic pain, nonverbal expression may be discordant with self-reports (63). Reports of acute and severe distress may be contradicted by a calm and dispassionate manner, or displays of agony may be accompanied by denials of discomfort (63). Verbal behavior is more likely to be discounted than nonverbal behavior when the two are discordant (63). Thus, a calm CPP who demonstrates little pain behavior but reported having severe pain would lead to the "discrepancy" observation.

Conclusion

At present, there is no reliable direct method for measuring clinical pain. The VAS appears to be the simplest and most reliable tool for that purpose. In addition, pain behavior appears to be an extremely important concept in chronic pain. The UAB Pain Behavior Scale is recommended for the measurement of pain behavior. For measurement of change in pain over time, the comparative VAS is recommended along with the UAB Pain Behavior Scale.

New Ways to Study Cerebral Pain Response

In the past decade, there has been an explosion in the use and application of new cerebral imaging technologies. These have included positron emission tomography, single photon emission computed tomography, and functional magnetic resonance imaging. Because of the cost of these machines, applications to the study of pain have been slow. Nevertheless, because these technologies allow comparison of brain function during rest with that during exposure to noxious stimuli, the early results are exciting. Tables 31.2 through 31.5 outline some of the main current findings and the significance of these findings in this line of research. The details of the findings of these studies are presented in these tables. However, a number of general conclusions can be drawn from this work:

1. Noxious stimuli are registered in specific brain areas, and therefore this technology may eventually solve the pain

TABLE 31.2. FINDINGS WITH POSITRON EMISSION TOMOGRAPHY, SINGLE PHOTON EMISSION COMPUTED TOMOGRAPHY, AND FUNCTIONAL MAGNETIC RESONANCE IMAGING ON NOXIOUS STIMULI PROCESSING IN THE BRAIN

Type of Pain Imaging Study	Research Question	Findings	Significance
PET, SPECT, fMRI	How are noxious stimuli processed in the brain (416–426)?	1. Noxious (painful stimuli) primarily activate the following brain areas: thalamus, primary somatosensory cortex, secondary somatosensory, cortex (52), and ACC 2. Noxious stimuli can also activate the following brain areas: insula, lentiform nucleus, prefrontal cortex, inferior parietal cortex, primary and secondary motor cortex, cerebellum, and brainstem 3. Many regions of brain engaged but thalamus activated at pain threshold whereas cortical structures involved higher intensities 4. S1 and S2 are only areas activated on contralateral side 5. There is subject-to-subject variability (ACC more consistently activated than S1) 6. Females will respond with more intense activation than males to same noxious stimuli	1. Human cerebral cortex involved perception, arousal, cognitive evaluation process, and, hence, affective reactions to pain 2. Activation of these areas represents the sensory, emotional, and cognitive components of the pain experience as follows: A. Sensory = S1 and S2 B. Affective = ACC and insula C. Cognitive = prefrontal and parietal cortex D. Inhibited motor response = lentiform nucleus and motor cortices 3. May be basis for individual differences in pain perception 4. May be basis for sex differences in pain perception
PET	How is visceral pain perceived? (429)	Acute noxious stimulation of rectum in normal subjects evokes brain activity in ACC	Visceral pain perceived in a region associated with affective qualities of pain

PET, positron emission tomography; SPECT, single photon emission tomography; fMRI, functional magnetic resonance imaging; ACC, anterior cingulate cortex.

measurement problem outlined above. However, there is no highly localized "pain center" in the brain. Pain is processed in multiple brain regions.

2. Noxious stimuli are transmitted to areas of the brain responsible for emotional response, and this may explain the concept of "suffering" from pain.

3. There are individual and gender differences in pain processing, and this may explain individual differences in pain response.

4. CPP process pain differently from normal controls, and this pattern of response could eventually be used as a diagnostic test to identify the CPP.

5. CPP diagnosed with fibromyalgia do process pain differently from controls. Such activation pattern could serve as a diagnostic test for fibromyalgia and should finally serve to dispel the historical suspicion around this diagnosis.

6. Mental activity training can apparently have an impact on the activation pattern to noxious stimuli.

Although in its infancy, the cerebral imaging of pain is, therefore, revolutionizing our thinking about pain processing. It is expected that this technology will eventually solve some of the controversies surrounding pain and its percep-

tion. In particular, brain imaging in the course of pharmacologic manipulation will refine our approach to the treatment of acute and chronic pain.

Pain Memory

Previously experienced visceral pain associated with a strong affective dimension (angina, labor pain) has been reproduced by stimulation of those posteroinferior regions of the human thalamus that project to parietal operculum and insular cortex (427). This type of pain is only elicited in patients with a previous experience of such pain, whereas similar pain but without a strong affective component can be elicited in patients without the previous experience of such pain (427). These results suggest that the strong affective component of pain elicited by thalamic stimulation is the result of previous experience and learning and that thalamic stimulation apparently activates a memory trace. This has been called pain memory. The important structure related to memory here is the insula (428). The insula may contribute to memory and learning of events related to painful stimuli (428) and demonstrates the existence of long-term neural mechanisms capable of storing the results of previous pain experience.

TABLE 31.3. FINDINGS WITH POSITRON EMISSION TOMOGRAPHY, SINGLE PHOTON EMISSION COMPUTED TO-MOGRAPHY, FUNCTIONAL MAGNETIC RESONANCE IMAGING, AND PROTON MAGNETIC RESONANCE SPEC-TROSCOPY ON DIFFERENCES IN CEREBRAL PROCESSING OF PAIN BETWEEN NORMAL SUBJECTS AND CHRONIC PAIN PATIENTS

Type of Pain Imaging Study	Research Question	Findings	Significance
PET	What is the difference in response patterns to noxious stimuli between those with neuropathic pain and cluster headache and without chronic pain (419,423)?	Experimentally induced unilateral pain in normals results in contralateral activation (S1, insula, ACC, thalamus); in individuals with neuropathic pain and cluster headache, response in right hemisphere (area 24) and dorsal medial prefrontal cortex (area 9 x 32)	Contralateral response represents the perception and localization of stimulus and intention to initiate escape motor response; cortical response in nondominant hemisphere represents emotional suffering associated with painful response
PET, fMRI, SPECT	Are there different activation patterns in neuropathic pain (418,422)?	Reduced thalamic activity without stimulation in chronic pain patients with chronic peripheral neuropathic pain and central neuropathic pain; thalamic hyperactivity in central neuropathic pain when patient's allodynic side stimulated but not normal since	Pathologic thalamus hyperactivity in resting hemithalamus plus underlying hyperresponsive-ness to noxious stimulation that in turn could be owing to loss of resting inhibitory activity in thalamus
PET	What is the effect on activation patterns in reflex sympathetic dystrophy, neuropathic pain patients during pathologic pain (allodynia) in their baseline pain state, and in a pain-free state (after blocks) (423)?	Activation pattern in baseline pain with pathologic pain state very different from that in controls; activation pattern with pathologic pain (allodynia) in a pain-free state similar to that in controls	Abnormal activation pattern can be reversed by pain-free state
Proton magnetic resonance spectroscopy	Does chronic pain cause brain chemistry changes?	Chronic low back pain patients compared with controls demonstrated reduced *N*-acetyl-D-aspartate and glucose reduction in dorsolateral prefrontal cortex	Abnormal brain chemistry in chronic pain potentially can be used as a diagnostic tool

PET, positron emission tomography; fMRI, functional magnetic resonance imaging; SPECT, single photon emission computed tomography; ACC, anterior cingulate cortex.

TABLE 31.4. FINDINGS WITH POSITRON EMISSION TOMOGRAPHY, SINGLE PHOTON EMISSION COMPUTED TO-MOGRAPHY, AND FUNCTIONAL MAGNETIC RESONANCE IMAGING ON DIFFERENCES IN CEREBRAL PROCESS-ING OF PAIN BETWEEN NORMAL SUBJECTS AND PATIENTS WITH FIBROMYALGIA

Type of Pain Imaging Study	Research Question	Findings	Significance
SPECT PET	Do FMS patients demonstrate resting activation abnormalities compared with controls (417,423)?	Low resting state rCBF in left and right thalamus and caudate nucleus	Potential diagnostic test for FMS
SPECT	Do FMS patients demonstrate abnormal activation with noxious stimuli (424)?	FMS patients demonstrate abnormal ipsilateral rCBF responses in the anterior cingulate cortex in response to noxious stimuli	Potential diagnostic test for FMS

SPECT, single photon emission computed tomography; PET, positron emission tomography; FMS, fibromyalgia; rCBF, regional cerebral blood flow.

TABLE 31.5. FINDINGS WITH POSITRON EMISSION TOMOGRAPHY AND FUNCTIONAL MAGNETIC RESONANCE IMAGING THAT INDICATE THAT THE EMOTIONAL RESPONSE TO PAIN CAN BE MODIFIED BY MENTAL ACTIVITY

Type of Pain Imaging Study	Research Question	Findings	Significance
fMRI	Can instructions for mental imagery during noxious stimuli modify the pain activation response (425)?	Instructions for distraction (pleasant imagery, away state) can modify the activation response, i.e., increase the number and volume of activated regions during noxious stimuli	May be the neuropsychological basis of the relaxation response in reference to perceived pain
fMRI	Can pain perception be separated from pain anticipation (422)?	Activation of some regions better correlated with anticipation of pain than with pain perception	Anticipation of pain may be a factor in some chronic pain patients
PET	Can hypnotic suggestion diminish the "unpleasantness" of heat stimulation (426)?	Activation of the ACC correlates with unpleasantness of heat and is diminished by hypnotic suggestion	Hypnosis may reduce the unpleasantness associated with pain perceived in the ACC (part of older limbic cortex believed to be associated with emotional content)

fMRI, functional magnetic resonance imaging; PET, positron emission tomography; ACC, anterior cingulate cortex.

CHRONIC PAIN AND PSYCHIATRIC COMORBIDITY

What is Chronic Pain?

There is no consensus definition of chronic pain (64). The current definitions can be summarized as follows:

1. The Institute of Medicine Committee Report on Pain and Disability (65) recognized chronic pain as a syndrome and defined it as: "pain lasting for long periods of time... more than 6 months may be associated with a residual structural defect... may be pain persisting past healing time without objective physical findings of residual structural defect... may be pain that recurs regularly and frequently over long periods." In addition, the committee pointed out that there was no agreed-on operational definition of chronic pain in the studies reviewed.

2. The most recent edition of the AMA's *Guides to the Evaluation of Permanent Impairment* (46) recognized chronic pain as a syndrome and pointed out that the medical profession had been slow to identify chronic pain as a specific medical disorder. The AMA defined acute pain, acute recurrent pain, and chronic pain. Acute recurrent pain was defined as "episodic noxious sensations resulting from tissue damage in chronic disorders, e.g. arthritis, tic douloureux." Chronic pain was defined as "not a symptom of an underlying acute somatic injury... a pathological disorder in its own right... chronic... long-lived... progressive... tissue damage has healed and does not serve as a generator of pain... although applied to pain of greater than 6 months duration a chronic pain syndrome can be diagnosed two to four weeks after onset." In addition, the AMA awarded a 5% impairment (lumbar) for discal herniation or other soft-tissue lesions with a minimum of 6 months medically documented pain associated with muscle spasm and rigidity.

3. In the classification of chronic pain (67) by the International Association for the Study of Pain Subcommittee on Taxonomy, chronic pain was supposed to be coded on axis IV as 1 month, 1 to 6 months, or more than 6 months. In the its definitions, the subcommittee advised that "many people report pain in the absence of tissue damage... if they regard their experience as pain, it should be accepted as pain... this definition would avoid tying pain to the stimulus."

Based on the previous statements, a number of conclusions can be drawn: (a) chronic pain is now recognized as a distinct syndrome; (b) the quoted authorities appear to consider pain chronic if it has lasted for more than 6 months, but they point out that chronic pain may begin earlier; (c) there is disagreement about the importance of the presence or absence of continued tissue damage and presence or absence of objective physical findings in the definition; (d) overall, there appears to be no universally accepted operational definition of chronic pain.

In an article (64), this author defined chronic pain as "continuous noxious output, like that of acute pain, but modulated and compounded by the prolonged or recurrent nature of the chronic state and complicated by a multitude of economic and psychosocial factors." However, the problem with the chronic pain concept is not so much the lack of a definition of chronic pain but the lack of an operational diagnosis for chronic pain. An operational diagnosis for chronic pain would be useful for both clinical practice and research. This author made the following recommendations for the operational criteria for the diagnosis of chronic pain:

1. That the suggestion of the International Association for the Study of Pain Subcommittee on Taxonomy not to tie the definition of chronic pain to the presence or absence of noxious stimulus (67) be accepted for the operational diagnosis of chronic pain.

2. That the operational diagnosis of chronic pain should not be based on any psychiatric or psychological criteria (psychopathology, if present, should be investigated and diagnosed separately).

3. That the temporal criteria of the diagnosis be flexible and defined in each patient either by the physician making the diagnosis of chronic pain or the patient identifying him/herself as having chronic pain. The last point is based on the approach to case identification as put forward in the Diagnostic Interview Schedule for psychopathology (68). Individuals were identified as cases if they viewed their symptoms as interfering with their lives or if the symptoms led to a consultation. In other words, presumably patients disturbed enough by their symptom will seek treatment for that symptom.

4. That the category of acute recurrent pain as put forward by the AMA be eliminated and placed under acute pain (66). Those patients whose acute recurrent pain was frequent enough and severe enough as to be identified by either the physician or the patient as developing into chronic pain would be diagnosed as a CPP.

5. That the operational diagnosis of chronic pain contain some patient-perceived functional impairment (69) and patient-perceived disability criteria (70). These two criteria are suggested because many CPP link perceived pain, perceived functional impairment, and perceived disability, believing that they cannot live normal lives as long as they have pain because their level of functioning is inversely proportional to their perceived pain (69). These two criteria (perceived functional impairment, perceived disability) become necessary to adequately distinguish between patient groups (using the aforementioned criteria). As an example, the patient who has chronic pain but is able to manage the pain and remain functional—perhaps because he/she has control over the pace of work at his/her job—may not seek treatment and, therefore, may not be identified as a CPP. The perceived functional impairment and perceived disability criteria are, therefore, necessary.

In summary, patients demonstrate having chronic pain through pain behavior, impaired function, and disability perception. Patients define themselves as suffering from chronic pain if the periods of what they perceive as intolerable pain are frequent enough to interfere with normal function. However, as noted earlier, there is neither a totally satisfactory definition for chronic pain nor an operational diagnosis for chronic pain. This major problem has interfered with de-lineating the frequency of chronic pain within the general population and within the psychiatric population.

Epidemiology of Pain and Chronic Pain in the General Population

Epidemiologic data on pain are incomplete (71) but indicate that the problem is enormous. The prevalence of persistent pain in Canada is 11%, and the age-specific morbidity rate for persistent pain increases with age (72). In Sweden, 40% of the general population reported "obvious pain" lasting more than 6 months (73). A statistical model applied to this population found that 25% had moderate pain and that 12% required pain treatment (73). The U.S. Nuprin Pain Report found that 5% to 10% of those studied reported various types of pain lasting more than 3 months (74), with a reduction with age and the prevalence of chronic pain in all sites other than joints (74). The percentage of prevalent cases of low back pain with persistent pain is thought to be 16% to 29% (74,75).

A New Zealand study indicated that 81.7% of the general population has had a pain experience severe enough to have led to a consultation with a doctor or other health professional, or led to the use of medications for the pain, or that interfered with life or activities "a lot" (76). In general, the prevalence of pain increased with age, except for headache and abdominal pain (76). In a suburban Australian community, 4.3% of the population were in constant pain (77). Twenty percent of hospitalized patients reported pain lasting for more than 6 months (78). Finally, it has been reported that 14.4% of the U.S. population between the ages of 65 and 74 may suffer from chronic pain related to the joints and musculoskeletal system (79). Based on these reports, the following observations can be drawn:

1. Pain is extremely common in the general population and is reported in 4.3% to 40% of the patients.

2. The lifetime prevalence of persistent pain may be as high as 80%.

3. The percentage of the reported patients who had chronic pain remains unclear. This is because of the definition variations described before. However, probably 4.3% to 12% of the general population has chronic pain.

4. It is claimed that 12% of the general population require pain treatment (73). However, even if one requires pain treatment, one may not be a CPP of the type that chronic pain treatment centers treat. These patients have been shown to be distinctly different from other pain patients on many pain behavior and emotional variables (80–82). This makes them more difficult to treat, necessitating specialized treatments such as psychiatric care.

These observations lead to two major conclusions:

1. The prevalence of CPP with severe disability in the general population is unknown.

2. The prevalence of pain complaints within the general population is uncertain. This makes comparisons with specialized populations, such as the population with overt psychiatric disorders, extremely difficult.

Pain in Psychiatric Disorders

Pain is a common documented symptom in psychiatric patients (83), but there are few studies of pain complaints in these patients (84). France et al. (84) described a number of characteristics of the pain in psychiatric disorders. To delineate the differences between psychiatric patients with pain and CPP seen in pain centers, the pain characteristics of France et al. have been put into table form (Table 31.6).

It is clear from this table that the pain seen in psychiatric patients differs from the chronic pain seen in pain centers. In addition, although information is lacking on this point, France et al. (84) claim that chronic pain is rarely seen in the psychiatric patient group. This has also been my own clinical experience.

France et al. (84), from their extrapolation of the literature, also claim that pain complaints are common in the following psychiatric diagnostic groups: major depression, dysthymia, generalized anxiety disorder, somatoform disorder, atypical somatoform disorder, alcohol dependence, opiate dependence, conversion disorder, somatoform pain disorder, factitious disorder with physical symptoms, hypochondriasis, and personality disorders.

However, few incidence and prevalence data for pain are available for these diagnostic groups. Available incidence data

TABLE 31.6. SIMILARITIES AND DIFFERENCES IN PAIN CHARACTERISTICS BETWEEN PSYCHIATRIC PATIENTS AND CHRONIC PAIN PATIENTS

Characteristic	Psychiatric Patients	Chronic Pain Patients
Pain as a presenting complaint	Rarely	Always
Consistency of pain	Transient	Chronic
Location	Cephalic, truncal	Low back, neck
Onset of pain with injury	Insidious	Usually associated
Pain sites	Multiple	Generally low back and/or neck
Neuroanatomic correlation	Poor	Poor to good
Exacerbating factors	Nonspecific	For low back pain, usually lifting, bending, sitting, or standing
Alleviating factors	Nonspecific	For low back pain, usually walking, lying down, changing positions, ice, or heat

indicate that 75% of nonpsychotic psychiatric inpatients had been bothered with pain in the past 3 months (85). In this study (85), pain was most frequent in patients with neurosis and personality disorders and was associated with unskilled work. The prevalence rate for chronic pain in a psychiatric outpatient clinic was reported to be 14.37% (86,87). These patients were usually urban, middle-aged housewives and were more likely to have a *Diagnostic and Statistical Manual of Mental Disorders, Third Edition* (*DSM-III*) diagnosis of anxiety disorder, dysthymic disorder, or no diagnosis than a pain-free psychiatric control group (86,87). Interestingly, psychiatric patients without pain were as likely to have a *DSM-III* diagnosis of major depression or conversion disorder (86,87).

Mood disorders as a group appear to have been most intensively studied for the association of a specific psychiatric disorder with pain. Here the prevalence of pain complaints is reported to vary from 30% to 84% (84). Data on the occurrence of pain complaints in different well-defined subtypes of depression are not available (84). It is interesting to note that pain thresholds are increased in depressed patients both with and without pain complaints (88). Finally, most psychiatric patients with pain complaints report that pain developed either before or at the same time as the depression (89,90).

Relationship between Perceived Disability and Depression

People with physical disabilities have elevated rates of both depressive symptomatology and major depression (91–93). In older adults (60 years and older), there is a significant relationship between disability and most mental health measures (94). Older adults with severe disabilities experience higher levels of anxiety, suicidal ideation, and overall distress than those with moderate disabilities (94). Severity of perceived disability correlates with severity of psychiatric symptoms (95).

In addition, disability is a frequent accompaniment of late-onset geriatric depression (96). Conversely, psychiatric disorders (affective and anxiety) are independently associated with both acute and chronic limitations in physical functioning (97). A longitudinal study demonstrated a synchronous relationship between depression and disability (93). In rheumatoid arthritis patients, depression predicts functional status ratings (98). Finally, in CPP with headache, depression and perceived disability are significantly associated (99). Thus, there is an association between disability perception and depression.

The Comorbidity of Pain and Depression

In pain treatment facilities, most CPP are depressed (79,100,101). The reported point prevalence of major depression in the chronic pain population varies between 1.5%

and 54.5%; the reported lifetime prevalence of major depression among CPP varies between 20% and 71% (79,102). The reported prevalence of dysthymia ranges from 0.0% to 43.3% (79). Using the Million Clinical Multiaxial Inventory, more than 50% of CPP had clinically elevated scores for depression (103). Adjustment disorder with depressed mood has been reported in 28.3% of CPP (100). Wells et al. (97) reported that 56.2% of CPP had some form of mood disorder (depression) as delineated by *DSM-III* criteria. Conversely, psychiatric outpatients who report chronic pain (14.8%) more frequently have dysthymia and anxiety disorders (104), and individuals with two or more pain conditions are at increased risk of a diagnosis of major depression (105). The differences in the reported prevalence of major depression in CPP may be related to differences in pain center CPP selection criteria (100) and the lack of operationally specific procedures for determining whether depressive symptoms are owing to an organic factor, an exclusion criterion for *DSM-III-R* depressive disorders (106).

In addition to the reported association between pain and depression, there is strong evidence that pain affects mood (107) and the severity of depression. Pain severity has been found to be associated with negative mood (108). It also appears that negative mood increases and becomes fixed as pain continues and becomes persistent (109,110). Several studies (111–114) reported a significant relationship between the level or degree of perceived pain and the degree of depression and the relationship between depression, pain, and pain behavior (115). Commonly used measurements of depression, such as the Beck Depression Inventory, are compounded by pain symptomatology (114). Finally, the level of pain appears to be more important in predicting the level of depression than physical dependency (116).

Major differences between authors in the reported prevalence of depression in CPP are likely to arise from problems in attributing symptoms to depression versus physical illness *per se*. For example, a high percentage of CPP have a sleep disorder (117–119), and many gain weight because of inactivity (100). Such a situation has confounded the utilization of the *DSM* criteria for major depression in this group (100). Furthermore, CPP may demonstrate a distinctive depressive syndrome (119). This is compatible with the observation that dysthymia and atypical depression are associated with greater severity of pain than major depression or adjustment disorder with depressed mood (120).

Does the Depression Seen in Chronic Pain Patients Precede or Follow the Development of Chronic Pain?

Although depression can precede pain as an independent phenomenon, there is empirical review evidence that persistent pain causes depression (168). A longitudinal study (121) of rheumatoid arthritis patients, for example, found that pain severity predicted subsequent depression. The causal rela-

tionship between pain severity and depression occurred only after the first 12 months of the study. Zarkowska and Philips (109) also determined that as pain persists, pain intensity becomes more closely related to a number of subjective and behavioral dimensions. In two published studies, Gamsa (122,123) concluded that emotional disturbance in CPP is more likely to be a consequence than a cause of chronic pain (122), although psychological events are risk factors in the development of chronic pain (123). Finally, a study (124) demonstrated that CLBP patients had significantly higher lifetime rates of major depression, alcohol use disorders, and major anxiety disorder. However, the first episode of major depression generally followed pain onset. This current evidence indicates that psychological events may be risk factors for the development of chronic pain, but that emotional disturbance is likely the result of chronic pain.

Pain Comorbidity with Other Psychiatric Diagnoses

Only a limited number of studies have addressed the issue of the prevalence of *DSM-III* diagnoses other than affective disorders in CPP. These studies and their results are presented in Table 31.7. The studies support the following conclusions:

1. A wide range of psychiatric disorders besides affective disorders is present in CPP.
2. There are wide discrepancies between research reports on the prevalence of some *DSM-III* disorders.
3. The discrepancy in the prevalence of some *DSM-III* diagnoses in CPP is greatest for disorders that have questionable reliability.
4. The majority of CPP will have an axis I diagnosis.
5. Much work is needed in this area to resolve the conflicting data.

Is There Evidence for Preexisting Psychiatric Pathology in Chronic Pain Patients?

This issue has not been extensively explored. To my knowledge, only three studies addressed this concern. An excellent study using the Diagnostic Interview Schedule found that in 81% of CPP, alcohol use disorders preceded pain onset. CPP, when compared with age-matched controls, had significantly higher prepain rates of alcohol use disorders but not of depression (124). The second study used a self-designed questionnaire. In this study, 46% of the CPP had prepain stress-related illness, 34% had a history of psychiatric illness, and 17% had been previously disabled (125). In the third study, the structured clinical interview for the diagnostic statistical manual (SCID-I) was administered to 98 CLBP patients with a diagnosis of somatoform pain disorder. Thirty-nine percent of the CPP admitted having a preexisting

TABLE 31.7. PREVALENCE OF *DIAGNOSTIC AND STATISTICAL MANUAL OF MENTAL DISORDERS, THIRD EDITION,* DIAGNOSES OTHER THAN AFFECTIVE DISORDER AND DRUG ABUSE/DEPENDENCE IN CHRONIC PAIN PATIENTS

Category and Diagnosis	Fishbain et al. (100)			Reich et al. (127)	Katon et al. (251)	Large (252)	Muse (253)	Fishbain et al. (254)
	Males (n = 156) (%)	Females (n = 127) (%)	Total (N = 283) (%)					
Somatoform disorders								
Somatization disorder	0.6[a]	7.9[a]	3.9	12% (F) 0% (M) 5% (T)	16.2% (T)	8% (T)	—	—
Conversion disorder	42.3	32.3	37.8	2% (T)	—	8% (T)	—	—
Psychogenic pain	0.6	0.0	0.3	32% (T)	—	—	—	—
Hypochondriasis	0.6	0.8	0.7	—	—	—	—	—
No diagnosis	5.7	4.7	—	—	—	—	—	—
Anxiety disorders								
Panic disorder	—	—	—	—	11.0% (T)	—	—	—
Agoraphobia with panic attacks and simple phobia	1.2	3.2	2.1	—	—	—	—	—
Generalized anxiety disorder	15.4	15.0	15.2	—	—	—	—	—
Obsessive-compulsive disorder	0.6	1.6	1.1	—	—	—	—	—
Posttraumatic stress disorder, acute and chronic	1.2	0.8	1.1	—	—	—	10% (T)	—
Adjustment disorder with anxious mood	40.4	45.7	42.8	—	—	—	—	—
Total number of patients with anxiety (anxiety disorders and adjustment disorder with anxious mood)	58.8[a]	66.3[a]	62.5	7% (T)	—	8% (T)	—	—
Organic mental disorders								
Delirium	0.6	0.0	0.4	—	—	—	—	—
Dementia	5.1	11.0	7.8	—	—	—	—	—
Substance abuse disorders								
Current alcohol abuse/ dependence	5.7	2.4	4.3	2% (T)	5.4% (T)	—	—	—
Alcohol abuse/ dependence in remission	0.3	3.9	7.4	—	35.1% (T)	—	—	—
Current drug dependence (opioids, barbiturates, sedatives, and cannabinoid)	14.7[b]	5.5[b]	10.6	25.5% (T)	24.3% (T)	—	—	—
Opioid dependence in remission	0.6	0.0	0.4	—	—	—	—	—
Total current alcohol and other drug dependence	20.4[a]	7.9[a]	14.9	—	—	—	—	—
Illicit drug abuse current	—	—	—	—	—	—	—	—
Schizophrenia	—	—	0%	0% (T)	—	—	—	—
Intermittent explosive disorder	16.7	1.6	9.9	—	—	—	—	—
Factitious disorder	—	—	—	2% (T)	—	—	—	0.14 (T)
Adjustment disorder with work inhibition	17.9[a]	7.1[a]	13.0	5% (T)	—	—	—	—

(continued)

TABLE 31.7. (continued)

Category and Diagnosis	Fishbain et al. (100)			Reich et al. (127)	Katon et al. (251)	Large (252)	Muse (253)	Fishbain et al. (254)
	Males (n = 156) (%)	Females (n = 127) (%)	Total (N = 283) (%)					
Psychological factors affecting physical condition	—	—	0%	19% (T)	0% (T)	34%	—	—
Uncomplicated bereavement	2.6	4.7	3.5	—	—	—	—	—
Marital problem	7.7	8.7	8.2	7% (T)	—	—	—	—
Personality disorders	62.3	55.1	59.0	37% (T)	—	—	—	—
Paranoid	5.1[b]	0.0[b]	2.8	—	—	—	—	—
Schizoid	3.2	0.0	1.7	—	—	—	—	—
Compulsive	7.7	5.5	6.7	—	—	—	—	—
Histrionic	4.5	20.5	11.7	—	—	—	—	—
Dependent	21.2	12.6	17.4	—	—	—	—	—
Narcissistic	4.5[b]	0.0[b]	2.4	—	—	—	—	—
Borderline	0.0	2.4	1.0	—	—	—	—	—
Passive-aggressive	15.4	14.2	14.9	—	—	—	—	—

[a] *p* <0.01.
[b] *p* <0.05.
F, female; M, male; T, total.

substance abuse disorder (41% of the males, 33% of the females). Twenty-nine percent had at least one episode of major depression before the onset of chronic pain (25% of the males, 36% of the females). Twenty-one percent had had preexisting symptoms consistent with an anxiety disorder (126). Finally, two studies (100,127) addressed this issue indirectly. Both studies found a high prevalence of personality disorders in CPP: 59% (100) and 37% (127). These findings on personality disorders could be significant because apparently axis II disorders may be less influenced by state phenomena than axis I disorders (128).

Conclusion

The general conclusions that can be drawn from this section are:

1. Chronic pain is a common complaint in the general population.
2. Psychiatric patients are unlikely to have a greater prevalence of chronic pain than the general population.
3. A wide range of psychiatric disorders are present in the chronic pain population, the most frequent being mood disorders.
4. Psychiatric disorders, especially depression, usually follow the development of chronic pain (168).
5. There is some evidence that CPP may have preexisting psychiatric pathology, particularly alcohol-related problems, episodes of major depression, and personality disorders.

MULTIDISCIPLINARY PAIN CENTER TREATMENT OF CHRONIC PAIN

A multidisciplinary pain center (MPC) was defined by Aronoff (129) as "a facility that offers multidisciplinary evaluation and treatment, a cohesive team approach directed towards the modification of pain and drug seeking behavior and towards the interruption of the disability process." MPC goals in the treatment of the CPP have been stated as follows (130–132): to reduce pain; reduce medication intake; reduce psychiatric/psychological impairment; correct posture, gait, and range of motion abnormalities; educate patients in the roles that emotions, behaviors, and attitudes play in chronic pain; remove rewards for pain behavior while encouraging healthy behavior; improve "up time" and thereby improve activities of daily living; improve level of function in social, familial, and household roles; improve strength and functional status; and restore occupational role function. Initially, there were many questions about the efficacy of MPC in treating chronic pain (133). However, evidence from well-designed outcome studies indicates that MPC do indeed fulfill some of their goals. In their review, Aronoff et al. (133) concluded that MPC do indeed return CPP to the workplace. Our group completed two review studies of this outcome literature, including an MA for return to work (134,135). Both studies concluded that MPC do return CPP to work, that the increased rates of return to work are the result of treatment, and that the benefits of treatment are not temporary. At issue, then, is not so much whether MPC treatment of chronic pain is efficacious but by what mechanism it works.

Operant Conditioning

The operant conditioning model states that behavior is controlled by its consequences. Thus, when pain behavior is reinforced in a positive fashion, it is likely that pain behavior will increase (136). Fordyce et al. (137) were the first to propose that the behavior of CPP fits an operant conditioning model. In a landmark study, they demonstrated that operant conditioning treatment could modify pain behavior. In this study, nurses withheld social reinforcement when CPP displayed pain behaviors and provided attention when CPP displayed "well behaviors," such as exercise. In addition, the reinforcing effects of pain medications were removed, using instead "pain cocktail" detoxification. The results demonstrated a dramatic increase in exercise tolerance and activity and a decrease in pain ratings and medication intake (137).

Since that study, the assertion of Fordyce et al. that operant conditioning is the cause of pain behavior has been severely criticized on three fronts: (a) there are questions as to what degree pain behavior is affected by social contingencies; (b) there are assertions that operant methods do not treat pain but instead teach CPP to be more stoic about pain they experience; and (c) positive treatment outcome for operant treatment programs is not evidence of the importance of operant factors to the etiology and maintenance of chronic pain, i.e., pain behavior is not necessarily learned or acquired just because it can be subsequently modified (138). In addition, it was pointed out that empirical data for this theory are lacking (138,139) and cannot be supported experimentally (140). There are now difficulties in determining the unique contribution of contingency management techniques to modify chronic pain behavior because these techniques are usually integrated into MPC (138). Fordyce et al. were accused of jumping to conclusions when they stated that pain behavior is influenced by factors other than pain (138).

Fordyce (141) has deemphasized the role of social contingencies and now believes that CPP are characterized by avoidance behavior. CPP display pain behaviors because they anticipate that rapid movement will increase pain, and thus pain behaviors are avoidance behaviors (141). The avoidance behaviors, he believes, are self-reinforcing, i.e., the avoidance of anticipated pain is self-reinforced (141). Fordyce (141) then adds that the pain behaviors may lead directly to reinforcing consequences, e.g., special attention. The avoidance learning concept for pain behaviors has some experimental support (142). These data, however, indicate that behavior treatments such as exercise quota systems are effective not because of reinforcement but because of a deconditioning process (142). At present, one can conclude that pain behaviors are probably not a consequence of operant conditioning but can be modified by operant conditioning. It is unclear, however, what the operant conditioning is acting on; that is, we may have targeted pain behavior for the operant conditioning but what is changed are the patient's beliefs about the pain, which may in turn lead to a change in pain behavior. Thus, it has been proposed that all forms of behavioral interventions, including operant conditioning, may exert an influence on chronic pain by changing the way in which the CPP thinks about his/her pain (143).

Cognitive/Behavioral Chronic Pain Treatment

Methods

Table 31.8 summarizes the pain management techniques that are currently being used in MPC. Cognitive/behavioral

TABLE 31.8. PAIN MANAGEMENT TECHNIQUES

A. Suggestion
   Information
   Direct verbal suggestion
   Programmed suggestion
   Hypnosis
   Self-hypnosis
B. Distraction
   Internal (imagination)
   External (ceiling TV)
C. Cognitive awareness
   Detailed information
   Mental preparation
   Rehearsal of strain to reduce pain
   Stimulus control (identification of stimulus and
      rearrangement to minimize exposure)
D. Anxiety-reduction techniques
   Relaxation
   Desensitization
E. Behavior skills training
   Social skills
   Stress management
   Self-management
   Self-improvement training
   Self-control
   Self-efficacy
   Anxiety management
   Decision making
   Social intervention
   Psychoeducation
F. Self-control
   Biofeedback
   Autogenic training
   Progressive muscle relaxation
   Imagery relaxation
   Breathing exercises
   Tension control
G. Cognitive therapy
   Transactional analysis
   Specific therapy to reevaluate meaning of pain, i.e., subjective
      ideas about possible aversive consequences of pain
H. Operant techniques
   Decrease attention in environment to pain behavior
I. Conditioning
J. Total push programs
   All the above techniques

References 260, 261.

methods are made up of a variety of these techniques and generally encompass items A through E on Table 31.8 (144). Reviewers concluded that cognitive/behavioral methods do indeed show that various cognitive strategies can increase pain tolerance levels (136,144). In addition, cognitive coping strategies, a subcategory of cognitive/behavioral methods, have been shown to be more effective in alleviating pain compared with either no treatment or expectancy controls (145).

Perhaps there is no difference in the efficacy of treatment results between the various pain management techniques. For example, one study showed operant conditioning pain treatment to be equal in efficacy to cognitive/behavioral group treatment (146). In addition, individual cognitive/behavioral therapy was equal in efficacy to cognitive/behavioral group therapy (147). Cognitive/behavioral treatment effects may be additive to the effects of physical therapy. The combined condition of cognitive/behavioral group treatment with physical therapy has been shown to be superior to physical therapy alone (148). This study offers empirical support for combining various treatment disciplines in an MPC.

Indications for Multidisciplinary Pain Center Referral

There are several approaches for the treatment of chronic pain. These approaches can be categorized as A, surgical; B, nerve blocks; C, psychopharmacologic; D, drug detoxification; E, physical therapy; F, occupational therapy; and G, behavioral pain management techniques, as outlined in Table 31.8. MPC use any combination of these approaches, depending on philosophy, the medical specialty providing leadership, and the size of the center. In addition, most MPC use the "total push program" for the behavioral pain management techniques used. This author believes that the effectiveness of the treatment provided by MPC rests in their ability to integrate these various treatment approaches in one setting. The larger MPC will generally combine most of these approaches, usually in the following combinations: B-G, A and C-G, or C-G.

Indications for referring a CPP to an MPC are provided in Table 31.9. Keep in mind that this table represents this author's opinions. The author believes that a CPP should be referred to an MPC if he/she fulfills criterion A and any one of the criteria in list B (Table 31.9). However, if the CPP fulfills any of the criteria in list C, then he/she should be excluded from MPC treatment because it is likely that the CPP will not be treated successfully. The referring doctor may be able to overcome these exclusion criteria by proper education and/or counseling of the CPP. Often the CPP is fearful or anxious about MPC treatment because he/she believes that the increased activity may increase the pain. This should be addressed directly as a risk, with the possibility that the chronic pain will eventu-

TABLE 31.9. INDICATIONS FOR MULTIDISCIPLINARY PAIN CENTER TREATMENT

Inclusion criterion A
  Chronic pain or chronic benign pain >6 mo in duration

Inclusion criteria B
  Surgical failure (failed back surgery syndrome)
  Few physical findings
  High-level pain behavior
  High-level functional disability
  Extended vocational disability (>6 mo)
  Drug dependence
  Severe suffering
  Significant psychopathology as a result of the chronic pain
  Failure of other modes of treatment (e.g., physical therapy alone)

Exclusion criteria C
  Chronic pain patient's unwillingness to participate in a pain management program
  Unrealistic expectation of what can be accomplished (e.g., that pain will be completely alleviated)

See reference 169.

ally decrease with MPC treatment. The referring physician should also encourage realistic expectations because many CPP wish to undergo treatments that will guarantee the removal of all their pain. Care should be taken to refer the CPP to an MPC that provides the full range of disciplines and treatments.

Conclusions

A number of conclusions can be drawn from the previous discussion:

1. It is likely that operant factors are not the reason for pain behaviors.
2. Operant conditioning can, however, aid in the treatment of the CPP.
3. The most effective CPP treatment includes behavioral or other treatments that provide a reduction in pain intensity. This gives the CPP an increased perception of being in personal control of the chronic pain (149).
4. The most effective CPP treatments are combinations of techniques that give the CPP this perception.

PSYCHOTROPIC DRUGS FOR THE TREATMENT OF NONNEUROPATHIC CHRONIC PAIN

This section deals with the use of psychotropic drugs in the treatment of chronic pain. Only the treatment of non-neuropathic chronic pain is discussed, i.e., nociceptive pain (150). As such, this review does not present data on disease entities that appear to have a clear neuropathic component: central pain, postherpetic neuralgia, avulsion of plexus, neuroma formation, phantom limb, lancinating neuralgias, nerve compression, painful polyneuropathy, and reflex sympathetic dystrophy/causalgia (151). The treatment of neuropathic pain is addressed in Chapter 16. The utility of the following psychotropic drug groups (the World Health Organization classification) (152) in the treatment of nonneuropathic chronic pain is discussed: antidepressants (AD) [tricyclics, heterocyclics, serotonin reuptake inhibitors, monoamine oxidase inhibitors (MAOI)], neuroleptics, antihistaminics, psychostimulants, and antiepileptic drugs.

Antidepressants (Tricyclic, Heterocyclics, and Selective Serotonin Reuptake Inhibitors): Evidence for Analgesic Efficacy in Chronic Pain

Besides the neuropathic pain conditions, for which there is excellent evidence of the efficacy (153,154,430) of AD, these drugs have been used for numerous other syndromes for which chronic pain is believed to be either nociceptive or owing to a psychological condition or both (155,493). These include headache, facial pain, arthritis/rheumatic pain, ulcer healing, fibromyalgia, CLBP, pelvic pain, cancer-associated pain, and psychogenic pain. AD have also been used in stud-

ies in which the pain conditions were of mixed etiology (156). The results of the treatment studies for these various etiologies are presented in Tables 31.10 through 31.21. The tables list all placebo-controlled treatment studies available in the literature, organized according to the number of CPP in the study, AD used, dose in milligrams per day, and/or whether the drug was statistically better than placebo. The tables also identify the AD as either serotonergic, noradrenergic, or serotonergic-noradrenergic. This division is used to ascertain whether some class of AD is superior to another for pain effects.

Chronic Low Back Pain

Table 31.10 summarizes the trials for CLBP. Ten trials (264–271,432) used serotonergic-noradrenergic AD. Five trials used noradrenergic AD (270,271,431–433), and two trials used a serotonergic AD (263,432). Of the ten serotonergic-noradrenergic trials, seven (70%) reported the AD to have an antinociceptive effect. Of the five noradrenergic AD trials, four (80%) reported the AD to have an antinociceptive effect. The two serotonergic AD trials did not find the AD effective. Overall, these data indicate that the serotonergic-noradrenergic AD and the noradrenergic AD are relatively consistently effective for CLBP.

TABLE 31.10. PLACEBO-CONTROLLED ANTIDEPRESSANT STUDIES FOR CHRONIC LOW BACK PAIN

Study	No. of Patients	Type of Drug	Dose (mg/d)	Active Drug Significantly More Effective Than Placebo?
Atkinson et al., 1998 (431)	57	Nortriptyline (NA)	50–150	Yes
Atkinson et al., 1999 (434)	74	Maprotiline (NA)	150	Yes
		Paroxetine (S)	30	No
Goodkin et al., 1990 (263)	42	Trazodone (S)	201 average	No
Hameroff et al., 1982 (264)	30	Doxepin (S-NA)	50–300	Yes
Hameroff et al., 1982 (266)	27	Doxepin (S-NA)	2.5 mg/kg	Yes
Hameroff et al., 1984 (265)	51	Doxepin (S-NA)	300	Yes
Jenkins et al., 1976 (267)	44	Imipramine (S-NA)	75	No
Pheasant et al., 1983 (268)	9	Amitriptyline (S-NA)	150	Yes
Sternbach et al., 1976 (269)	9	Clomipramine (S-NA)	150	Yes
		Amitriptyline (S-NA)	150	No
Storch and Steck, 1982 (433)	35	Maprotiline (NA)	Variable	No
Treves et al., 1991 (432)	68	Clomipramine (S-NA)	75	Yes
Ward et al., 1984 (271)	26	Doxepin (S-NA)	188	Yes, 50% overall (doxepin reduced pain severity significantly more than desipramine)
		Desipramine (NA)	173	
		Compared placebo responders dropped from study		
Ward, 1986 (270)	35	Doxepin (S-NA)	?	Yes (no significant difference in pain response noted between drugs, so patient groups were collapsed. Retests indicated a statistically significant drop in pain by wk 1)
		Desipramine (NA)	?	
		Compared placebo responders dropped from study		

NA, noradrenergic; S, serotonergic; S-NA, serotonergic/noradrenergic.
Abstracted from Fishbain DA. Evidence based data on pain relief with antidepressants. *Ann Med* 2000;32:305–316.

TABLE 31.11. PLACEBO-CONTROLLED ANTIDEPRESSANT STUDIES FOR CHRONIC PELVIC PAIN

Study	No. of Patients	Type of Drug	Dose (mg/d)	Active Drug Significantly More Effective Than Placebo?
Engel et al., 1998 (435)	23	Sertraline (S)	20	No

S, serotonergic.
Abstracted from Fishbain DA. Evidence based data on pain relief with antidepressants. *Ann Med* 2000;32:305–316.

Chronic Pelvic Pain and Pain Associated with Cancer

Table 31.11 presents the data available on chronic pelvic pain. One study found that the serotonergic AD sertraline was not effective (435).

Table 31.12 summarizes the placebo-controlled trials for cancer-associated pain. The two trials of a serotonergic-noradrenergic AD provided inconsistent results (436,437).

Osteoarthritis and Rheumatoid Arthritis

Table 31.13 summarizes the placebo-controlled treatment trials for osteoarthritis or rheumatoid arthritis pain.

In 11 of these trials (300,301,303–305,308,309,437–440), a serotonergic-noradrenergic AD was used, in four trials, a serotonergic AD was used (300,315,438,440), and in one trial, a noradrenergic AD was used (300). Of the 11 trials using serotonergic-noradrenergic AD, seven (63.6%) reported the AD to have an antinociceptive effect. Of the four serotonergic trials, two (50.0%) reported the AD to have an antinociceptive effect. The trial with a noradrenergic agent showed no antinociceptive effect (300). Data from the table indicate that AD are not consistently effective for osteoarthritic or rheumatoid pain but that the serotonergic-noradrenergic agents are more consistently effective than the serotonergic agents.

Fibromyalgia

Table 31.14 summarizes the placebo-controlled treatment trials for fibromyalgia pain. In ten of these trials, a serotonergic-noradrenergic AD was used (317,320, 321,441,442,445–448), and in three trials, the drug was serotonergic (442–444). All ten of these serotonergic-noradrenergic AD trials reported the AD to have an antinociceptive effect. Of the three serotonergic trials (442–444), one (33.3%) reported the AD to have an antinociceptive effect. Data from this table indicate that serotonergic-noradrenergic AD are consistently effective for fibrositis or fibromyalgia pain, whereas serotonergic AD are not.

Headache

Table 31.15 summarizes placebo-controlled treatment trials for headache including 18 trials using serotonergic-noradrenergic AD (170,193,336,339,340,343,344,346–349,352,449–464, 15 trials using serotonergic AD (343,346,347,352,450,452–458,461,463), and one trial with a noradrenergic AD (340). Of the 18 serotonergic-noradrenergic AD trials, 15 (77.7%) reported the AD to have an antinociceptive effect. Of the 15 serotonergic AD trials, eight (53.3%) reported the AD to have an antinociceptive effect. The one trial using a noradrenergic AD reported a positive antinociceptive effect. Data in Table 31.15 indicate that AD are not consistently effective for headache pain; however, the serotonergic-noradrenergic AD are more consistently effective than the serotonergic AD.

Facial Pain

Table 31.16 summarizes placebo-controlled treatment trials for facial pain. There were two trials, one with a serotonergic-noradrenergic AD (163) and the other with a serotonergic AD (465). Both trials reported a positive antinociceptive effect.

TABLE 31.12. PLACEBO-CONTROLLED ANTIDEPRESSANT STUDIES FOR PAIN ASSOCIATED WITH CANCER

Study	No. of Patients	Type of Drug	Dose (mg/d)	Active Drug Significantly More Effective Than Placebo?
Fiorentino, 1967 (285)	40	Imipramine (S-NA)	150	No (nearly significant $p < 0.06$)
Walsh, 1986 (436)	47	Imipramine (S-NA)	25–75	Yes (in reducing morphine requirements)

S-NA, serotonergic/noradrenergic.

TABLE 31.13. PLACEBO-CONTROLLED ANTIDEPRESSANT STUDIES FOR PAIN ASSOCIATED WITH OSTEOARTHRI-TIS/RHEUMATOID ARTHRITIS

Study	No. of Patients	Type of Drug	Dose (mg/d)	Active Drug Significantly More Effective Than Placebo?
Caruso et al., 1987 (298)	734 OA	S-Adenosyl-L-methionine	200 i.m.	Yes
Frank et al., 1988 (300)	47 RA	Amitriptyline (S-NA)	1.5 mg/kg	Yes
		Desipramine (NA)	1.5 mg/kg	No
		Trazodone (S)	3 mg/kg	No
Ganvir et al., 1980 (301)	49 OA	Clomipramine (S-NA)	25	No
Glick and Fowler, 1979 (303)	11 OA	Imipramine (S-NA)	75	Yes
Grace et al., 1985 (304)	36 RA	Amitriptyline (S-NA)	75	No
Gringas, 1976 (305)	55 RA, OA	Imipramine (S-NA)	75	Yes
MacFarlane et al., 1986 (308)	27 RA	Trimipramine (S-NA)	75	Yes
McDonald-Scott, 1969 (309)	22 OA	Imipramine (S-NA)	75	Yes
MacNeil and Dick, 1976 (437)	45 OA	Imipramine (S-NA)	75	No
Rani et al., 1996 (438)	59 RA	Fluoxetine (S)	20 mg	Yes
		Amitriptyline (S-NA)	25 mg	Yes
Singh et al., 1976 (439)	20	Imipramine (S-NA)	?	No
Wheatley, 1986 (315)	65	Mianserin (S)	Various	No

OA, osteoarthritis; RA, rheumatoid arthritis; S, serotonergic; NA, noradrenergic; S-NA, serotonergic/noradrenergic; i.m., intramuscularly.
Abstracted from Fishbain DA. Evidence based data on pain relief with antidepressants. *Ann Med* 2000;32:305–316.

Irritable Bowel Syndrome

Table 31.17 summarizes placebo-controlled trials for the treatment of pain associated with irritable bowel syndrome with AD. There were only two trials, one with a serotonergic-noradrenergic AD (466) and one with a noradrenergic AD (467). Both studies demonstrated a positive antinociceptive effect. The data in this table indicate a consistent trend for efficacy, but too few studies have been performed to draw any general conclusions.

Ulcer Healing

Table 31.18 summarizes the eight placebo-controlled trials for ulcer healing using serotonergic-noradrenergic AD (330–334,468–471). Of these, six trials (75%) reported the AD to have an ulcer-healing effect (331–334,449,468,470,471). Data from this table indicate that serotonergic-noradrenergic AD appear to demonstrate a relatively consistent ulcer-healing effect.

TABLE 31.14. PLACEBO-CONTROLLED ANTIDEPRESSANT STUDIES FOR PAIN ASSOCIATED WITH FIBROSITIS/FIBROMYALGIA

Study	No. of Patients	Type of Drug	Dose (mg/d)	Active Drug Significantly More Effective Than Placebo?
Bennett et al., 1988 (446)	120	Cyclobenzaprine (S-NA)	40	Yes
Carette et al., 1986 (317)	59	Amitriptyline (S-NA)	50	Yes
Carette et al., 1994 (445)	208	Amitriptyline (S-NA)	50	Yes
		Cyclobenzaprine (S-NA)	30	Yes
Ginsberg et al., 1996 (448)	46	Amitriptyline (S-NA)	25 SR	Yes
Goldenberg et al., 1986 (320)	58	Amitriptyline (S-NA)	25	Yes
Goldenberg et al., 1996 (442)	19	Amitriptyline (S-NA)	25	Yes
		Fluoxetine (S)	20	Yes
Hannonen et al., 1998 (441)	130	Amitriptyline (S-NA)	25–37.5	Yes
Kempenaers et al., 1994 (447)	36	Amitriptyline (S-NA)	50	Yes
Narregaard et al., 1995 (443)	22	Citalopram (S)	20	No
Scudds et al., 1989 (321)	36	Amitriptyline (S-NA)	50	Yes
Tavoni et al., 1987 (322)	25	S-Adenosyl-L-methionine	200 i.m.	Yes (significantly decreased number of tender points)
Wolfe et al., 1994 (444)	42	Fluoxetine (S)	20	No

SR, sustained release; S, serotonergic; S-A, serotonergic/noradrenergic; i.m., intramuscular.
Abstracted from Fishbain DA. Evidence based data on pain relief with antidepressants. *Ann Med* 2000;32:305–316.

TABLE 31.15. PLACEBO-CONTROLLED STUDIES ON THE USE OF ANTIDEPRESSANT FOR HEADACHES

Study	No. of Patients	Type of Drug	Dose (mg/d)	Active Drug Significantly More Effective Than Placebo?
Couch, 1979 (337)	100 (M)	Amitriptyline (S-NA)	100	Yes
Diamond and Baltes, 1971 (339)	56 (TH)	Amitriptyline (S-NA)	10–25	Yes
Fogelholm and Murros, 1985 (340)	30 (TH)	Maprotiline (NA)	75	Yes
Indaco and Carrieri, 1988 (193)	15 (TH)	Amitriptyline (S-NA)	50	Yes
Lance and Curran, 1964 (170)	280 (TH)	Amitriptyline (S-NA)	30–75	Yes
		Imipramine (S-NA)	30–75	Yes
Langemark et al., 1990 (343)	82 (TH)	Clomipramine (S-NA)	Variable	Yes
		Mianserin (S)	Variable	No
Langohr et al., 1985 (344)	63	Clomipramine (S-NA)	100	No
Martucci et al., 1985 (346)	20	Mianserin (S) (migraine headaches)	30	No
Monro et al., 1985 (347)	60	Mianserin (S)	60	No
Morland et al., 1979 (348)	14	Doxepin (S-NA)	100	Yes
Noone, 1980 (349)	10	Clomipramine (S-NA)	30	No
Saper et al., 1994 (450)	64 (TH), 58 (M)	Fluoxetine (S)	40	Yes
Sjaastad, 1983 (352)	16 (TH)	Femoxetine (S)	400	Yes
Ziegler, 1987 (449)	30 (M)	Amitriptyline (S-NA)	50–150	Yes
Adly et al., 1992 (458)	16 (M)	Fluoxetine (S)	40	Yes
Battistella et al., 1993 (453)	40 (M)	Trazodone (S)	1 mg/kg	Yes
Bendtsen et al., 1996 (454)	40 (TH)	Amitriptyline (S-NA)	75	Yes
		Citalopram (S)	20	No
Gobel et al., 1994 (451)	78 (TH)	Amitriptyline (S-NA)	75	Yes
Kangashemi et al., 1983 (457)	24 (M)	Femoxetine (S)	400	No
Loldrup et al., 1987 (463)	?(M,C)	Clomipramine (S-NA)	75–100	Yes
		Miaserin (S)	30–60	Yes
Manna et al., 1994 (461)	41 (TH)	Mianserin (S)	30–60	Yes
		Fluvoxamine (S)	50–100	Yes
Orholm et al., 1986 (456)	65 (M)	Femoxetine (S)	600	No
Pfaffenrath et al., 1993 (459)	211 (TH)	Amitriptyline (S-NA)	50–90	No
Pfaffenrath et al., 1994 (462)	197 (TH)	Amitriptyline oxide (S-NA)	60–90	Yes
		Amitriptyline (S-NA)	50–75	No
Steiner et al., 1998 (452)	53 (M)	S-Fluoxetine (S)	20	Yes
Ward et al., 1979 (464)	16 (?)	Doxepin (S-NA)	75	Yes
Zeeberg et al., 1981 (455)	59 (M)	Femoxetine (S)	200–300	No
Ziegler et al., 1993 (460)	30 (M)	Amitriptyline (S-NA)	40–150	Yes

M, migraine; S-NA, serotonergic/noradrenergic; TH, tension headache; NA, noradrenergic; S, serotonergic; C, cluster.
Abstracted from Fishbain DA. Evidence based data on pain relief with antidepressants. *Ann Med* 2000;32:305–316.

Mixed Etiologies

Table 31.19 summarizes placebo-controlled trials for AD treatment of pain of mixed etiology. These included five trials using serotonergic-noradrenergic AD (175,189,357,363,366) and two using serotonergic AD (358,360). Of the five trials using a serotonergic-noradrenergic AD, three (60.0%) reported the AD to have antinociceptive effect. Of the two trials using serotonergic AD (103,358), one (50%) reported the AD to have an

TABLE 31.16. PLACEBO-CONTROLLED ANTIDEPRESSANT STUDIES FOR FACIAL PAIN

Study	No. of Patients	Type of Drug	Dose (mg/d)	Active Drug Significantly More Effective Than Placebo?
Harrison et al., 1997 (465)	178	Fluoxetine (S)	20	Yes
Sharav et al., 1987 (163)	28	Amitriptyline (S-NA)	129 (mean)	Yes

S, serotonergic; S-NA, serotonergic/noradrenergic.
Abstracted from Fishbain DA. Evidence based data on pain relief with antidepressants. *Ann Med* 2000;32:305–316.

TABLE 31.17. PLACEBO-CONTROLLED ANTIDEPRESSANT STUDIES ASSOCIATED WITH IRRITABLE BOWEL SYNDROME

Study	No. of Patients	Type of Drug	Dose (mg/d)	Active Drug Significantly More Effective Than Placebo?
Greenbaum et al., 1987 (466)	28	Desipramine (NA)	Variable	Yes
Rajagopalan et al., 1998 (467)	40	Amitriptyline (S-NA)	25–75	Yes

NA, noradrenergic; S-NA, serotonergic/noradrenergic.
Abstracted from Fishbain DA. Evidence based data on pain relief with antidepressants. *Ann Med* 2000;32:305–316.

antinociceptive effect. Data from this table indicate that neither the serotonergic-noradrenergic nor the serotonergic AD is consistently effective.

Acute Pain

Table 31.20 summarizes the data on placebo-controlled AD studies for human acute pain. Serotonergic-noradrenergic AD were used in five studies (472–475,477), whereas a noradrenergic AD was used in one study (473) and a serotonergic AD in one study (476). Overall, this group of studies indicates that the serotonergic-noradrenergic AD have a consistent antinociceptive effect for human acute pain. Similarly, the noradrenergic AD may have such an effect, whereas the serotonergic ones do not.

Metaanalytic Studies

Table 31.21 summarizes the metaanalytic studies evaluating AD in pain treatment. Four metaanalyses (MA) were performed: two on various types of chronic pain (181,480), one on neuropathic pain (430), and one on alleged psychogenic or somatoform disorder pain (479). These MA are consistent in reporting that AD do have an antinociceptive effect. However, a closer inspection of these MA (Table 31.21) indicates that they included pooled studies in which different types of antidepressants were used. Thus,

these metaanalytic studies do not help in determining which types of AD are most effective in terms of an antinociceptive effect.

Antinociception and Antidepressant Blood Levels

Table 31.22 summarizes studies that addressed the relationship between the blood level of AD and pain relief in humans and rats. There were six studies in this group (480–485), and only one of them (481) established a relationship between the blood level of the AD and pain relief. This group of studies indicates that there is no relationship between the blood level and the antinociceptive effect of an AD.

Tables 31.10 through 31.22 present data for AD pain efficacy by presenting three levels of evidence. This evidence includes human studies on acute pain, individual placebo-controlled studies for the treatment of specific chronic pain syndromes, and metaanalytic studies. The results can be summarized as follows. Overall, evidence from human studies on acute pain, individual placebo-controlled studies on the treatment of specific chronic pain syndromes, and four MA (181,430,479,480) is consistent in indicating that AD do have an antinociceptive effect. In addition, evidence from one MA indicates that AD have an antinociceptive effect for neuropathic pain (430). This MA also demonstrated that AD could have an antinociceptive effect on psychogenic pain or somatoform pain disorder (430). Finally, the evidence from

TABLE 31.18. PLACEBO-CONTROLLED ANTIDEPRESSANT STUDIES FOR ULCER HEALING

Study	No. of Patients	Type of Drug	Dose (mg/d)	Active Drug Significantly More Effective Than Placebo?
Anderson, 1984 (470)	23	Doxepin (S-NA)	50	Yes
Berstand, 1980 (330)	95	Trimipramine (S-NA)	25	No
Don, 1980 (471)	65	Trimipramine (S-NA)	25	No
Guldhal, 1977 (331)	8	Trimipramine (S-NA)	25–75	Yes
Moshal, 1981 (468)	20	Trimipramine (S-NA)	50	Yes
Niller et al., 1977 (333)	19	Trimipramine (S-NA)	50	Yes
Shrivastava, 1985 (449)	10	Doxepin (S-NA)	50–100	Yes
Wetterhus et al., 1977 (334)	14	Trimipramine (S-NA)	50	Yes

S-NA, serotonergic/noradrenergic.
Abstracted from Fishbain DA. Evidence based data on pain relief with antidepressants. *Ann Med* 2000;32:305–316.

TABLE 31.19. PLACEBO-CONTROLLED ANTIDEPRESSANT STUDIES FOR PAIN OF MIXED ETIOLOGIES

Study	No. of Patients	Type of Drug	Dose (mg/d)	Active Drug Significantly More Effective Than Placebo?
Evans, 1973 (357)	22	Doxepin (S-NA)	150	No
Gourlay, 1986 (358)	19	Zimeldine (S)	300	No
Johansson and Von Knorring, 1979 (360)	32	Zimeldine (S)	200	Yes
McQuay et al., 1992 (189)	25	Amitriptyline (S-NA)	25	Yes
Pilowsky et al., 1982 (363)	32	Amitriptyline (S-NA)	150	No
Raft et al., 1981 (175)	23	Amitriptyline (S-NA)	3.5 mg/kg	Yes
Zitman et al., 1990 (366)	49	Amitriptyline (S-NA)	75	Yes

S-NA, serotonergic/noradrenergic; S, serotonergic.
Abstracted from Fishbain DA. Evidence based data on pain relief with antidepressants. *Ann Med* 2000;32:305–316.

TABLE 31.20. PLACEBO-CONTROLLED ANTIDEPRESSANT STUDIES FOR ACUTE PAIN

Study	No. of Patients	Type of Drug	Dose (mg/d)	Active Drug Significantly More Effective Than Placebo?
Cannon et al., 1994 (477)	60 (chest pain with normal coronaries)	Imipramine (S-NA)	50	Yes (reduced frequency of chest pain)
Gordon et al., 1994 (476)	70 (postoperative pain)	Fluoxetine (S) preoperatively	10	No (antagonized morphine analgesia)
Kerrick et al., 1993 (474)	28 (postoperative pain)	Amitriptyline (S-NA)	50	Yes
Levine et al., 1986 (473)	? (Preoperative)	Amitriptyline (S-NA)	25	No
		Desipramine (NA)	25	Yes (increased and prolonged morphine analgesia)
Stein et al., 1996 (472)	39 (acute low back pain)	Amitriptyline (S-NA)	150	Yes
Tiengo et al., 1987 (475)	40 (postoperative pain)	Clomipramine (S-NA) vs. pentazocine	Variable	No difference to pentazocine for pain control

S-NA, serotonergic/noradrenergic; S, serotonergic; NA, noradrenergic.
Abstracted from Fishbain DA. Evidence based data on pain relief with antidepressants. *Ann Med* 2000;32:305–316.

TABLE 31.21. METAANALYSES OF ANTIDEPRESSANT TREATMENT FOR CHRONIC PAIN

Ref.	Type of Pain	No. of Studies Pooled and No. of Trials of Each Type of Antidepressant	Author Clinical Interpretation of Metaanalysis Results
Fishbain et al., 1998 (478)	Psychogenic pain and/or somatoform pain disorder	11 trials: 4 S, 1 NA, 5 S-NA, 2 MAO	Antidepressants effective for psychogenic pain/somatoform pain disorder
McQuay et al., 1996 (430)	Neuropathic pain (diabetic neuropathy, postherpetic neuralgia, atypical facial pain)	20 trials: 4 S, 6 NA, 13 S-NA, 2 MAO	Antidepressants much more effective than placebo
Onghena and Van Houdenhove 1992 (181)	Chronic pain	48 trials: 10 S, 2 NA, 32 S-NA, 4 MAO	Average chronic pain patient receiving antidepressants had less pain than 74% of chronic pain patients receiving placebo
Phillipp and Fickinger 1993 (479)	Various types chronic pain	47 trials: 1 S, 3 NA, 42 S-NA, 1 MAO	For 24 studies pooling the data gave a global improvement rate of 57% for the active drug versus 31% for placebo; difference was highly significant at ($p = 0.007$)

S, serotonergic; NA, noradrenergic; S-NA, serotonergic/noradrenergic; MAO, monoamine oxidase inhibitor.
Abstracted from Fishbain DA. Evidence based data on pain relief with antidepressants. *Ann Med* 2000;32:305–316.

TABLE 31.22. RELATIONSHIP BETWEEN ANTIDEPRESSANT BLOOD LEVELS AND PAIN RELIEF OR OTHER OUTCOME VARIABLE IN HUMANS AND RATS

Drug and Ref.	Outcome Variable	Relationship Established?
AMI (480) (H)	Pain improvement	No
AMI (481) (H)	Pain improvement	Yes
AMI (482) (H)	Pain improvement	No
CL (483) (H)	Pain improvement	No
CL (484) (R)	Pain improvement	No
AMI (485) (H)	Pain improvement	No

AMI, amitriptyline; H, humans; CL, clomipramine; R, rats.
Abstracted from Fishbain DA. Evidence based data on pain relief with antidepressants. *Ann Med* 2000;32:305–316.

human acute pain studies and individual placebo-controlled studies for CLBP, headache pain, osteoarthritic or rheumatoid arthritis pain, and fibrositis or fibromyalgia pain indicates that the serotonergic-noradrenergic AD have a more consistent antinociceptive effect than the serotonergic AD.

AD treatment of chronic pain was assessed in four MA. Two of these combined various types of pain syndromes (479,480), one was performed on studies addressing neuropathic pain (430), and one (181) was performed on studies addressing psychogenic pain or somatoform pain disorder (430). Because quantitative techniques are used in MA, any two sets of investigators should come up with roughly the same results with the same universe of available data (486), i.e., there should be replicate nonvariability. This is demonstrated in the case of the above four MA: They all consistently

TABLE 31.23. HYPOTHESES FOR THE MODE OF ACTION OF THE ANTIDEPRESSANT ANTINOCICEPTIVE EFFECT

Antidepressant effect (manifest or masked depression)
Stabilizing aberrantly conducting neurons or inhibiting their afferent transmission at the spinal cord (migraine headaches, neuropathic pain)
Antiepileptic (suppressing epileptiform activity in deafferented neurons)
Selective serotonin inhibition
Facilitating analgesia by facilitating central monoamine transmission (inhibiting serotonin and norepinephrine reuptake at the synapse)
Facilitating the descending pain modulation system (resultant decrease in afferent input through the spinothalamic tract)
Interaction with opioid receptors (altering binding characteristics of the receptor to morphine and enkephalins)
Affecting a core disorder (common psychobiologic abnormalities but multiple clinical and diagnostic presentations, e.g., chronic pain and depression)
Peripheral antiinflammatory action (inflammatory disorders)
Central skeletal muscle relaxant action (depressing polysynaptic reflexes causing muscle spasm or tension)
Sedative effect
Placebo effect

From references 153, 181, 367, 373, and 374.

indicate that AD have an antinociceptive effect. These four studies provide the strongest evidence of the antinociceptive efficacy of AD. However, the reliability of this evidence depends on the quality of the MA. There can be reliability problems with MA. Results of MA depend on the quality of the data within the selected studies. Pain treatment studies have often been criticized for the following methodologic problems: poor definition of outcome variables; time sampling differences between studies; different treatment populations within studies; different outcome variables, making comparisons difficult between studies; different selection criteria, making comparisons between studies problematic; different patient adherence patterns, again making comparison difficult; different and poorly described methods of treatment; and poor control for nontreatment factors that affect outcome variables such as return to work (486–488). These methodologic problems could have affected the results of the MA (Table 31.22). In addition, there has been some recent concern about the quality of pain studies and the impact of this on the results of MA (489,490). There are also technical factors not reviewed here in performing MA procedures that can be sources of error (486). These technical factors can prevent the recognition of poor-quality studies, which may thus be included in the MA. All the above-listed factors can lead to poor-quality MA (486). In an attempt to address this problem, Fishbain et al. (486) recently reviewed pain treatment MA for quality and demonstrated that MA evaluating the treatment of chronic pain do indeed demonstrate procedure problems that could lead to developing a poor-quality MA.

This chapter includes three of the MA in Table 31.21 (181,430,479). Their quality was found to vary but was, however, above average (486). In addition to the above evidence of the antinociceptive effect of AD in human acute and chronic pain, there is a significant body of evidence from animal studies for this antinociceptive effect. This literature was recently reviewed by Fishbain et al. (491). They isolated 22 animal studies in which different AD had been used. They divided the studies into serotonergic-noradrenergic AD, serotonergic AD, and noradrenergic AD trials. Overall, 92% of the serotonergic-noradrenergic trials, 25% of the serotonergic trials, and 88.9% of the noradrenergic trials reported an antinociceptive effect (491). Thus, the evidence from animal studies supports the evidence from human acute and chronic pain treatment trials indicating that (a) AD do have an antinociceptive effect, (b) the antinociceptive effect may differ according to the AD class, and (c) in general, the serotonergic-noradrenergic and noradrenergic AD have a stronger antinociceptive effect than the serotonergic AD.

Tables 31.21 and 31.22 and previous reviews also provide the following in reference to the antinociceptive effect of AD:

1. The beneficial effect from AD for most pain conditions is not related to mood (181). AD can relieve pain even if they have no significant effect on depressed mood (164).

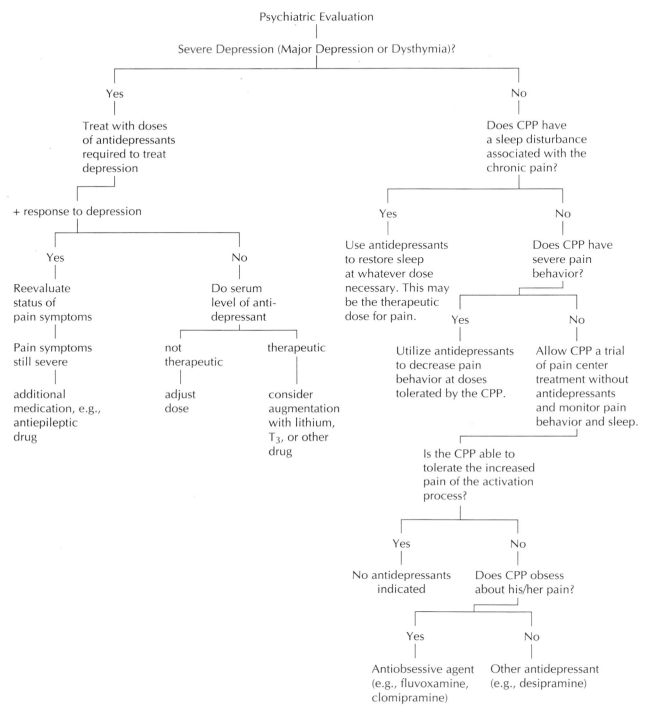

FIGURE 31.5. Algorithm for the use or nonuse of antidepressants in chronic pain patients.

The presence of depression is not required for an analgesic response (151,170,193).

2. The beneficial effect appears to be mild to moderate (172,184).

3. CPP may have an analgesic response to lower doses of AD than those used to treat depression (181,184,191–193).

4. The delay in the onset of action of the pain effect varies from a few days to several weeks (184) but seems to be more rapid than the antidepressant effect (189).

5. The MA of Onghena and Van Houdenhove (181) and Fishbain et al. (478) suggest that the antinociceptive effect of AD is not much different for pain having an organic

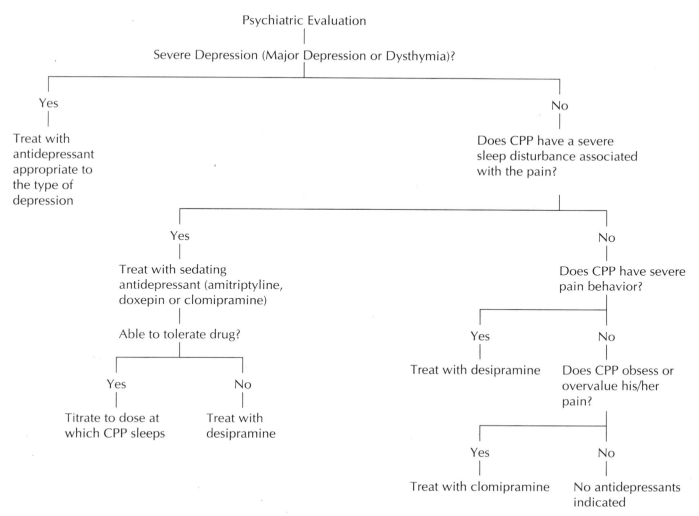

FIGURE 31.6. Algorithm for type of antidepressant to be used for chronic pain patients.

versus psychogenic etiology. If idiopathic pain were truly psychogenic, one would expect the following outcomes to occur with AD treatment: (a) no response at all, (b) a response equal to a placebo response, or (c) a response dependent on the antidepressant properties of the drug, i.e., the masked depression hypothesis (182). Hypothesis a, b, or c has not been shown to be true in studies of AD treatment of idiopathic pain. Thus, the analgesic response to AD in idiopathic pain disorder questions the psychogenic concept. As pointed out by Davis (183), these studies suggest that these CPP have an organically based pain disorder.

6. In general, there is no correlation between AD serum levels and the antinociceptive effect (Table 31.22). There is also no correlation between AD serum levels and changes in depression scores in those patients reporting an antinociceptive effect (196,197). Serum levels of those patients reporting an antinociceptive effect are lower than AD serum levels required for an AD effect (174,194,195).

AD serum levels, therefore, cannot be used as a guide for the antinociceptive effect.

Mechanism of Action of the Antinociceptive Effect of Antidepressants

Based on the data given in Tables 31.10 through 31.22, it appears that there is a large body of evidence indicating that AD do have an antinociceptive effect. Several hypotheses have been advanced as a potential explanation for the mode of action of the antinociceptive effect of AD. These hypotheses are summarized in Table 31.23. The accuracy of these hypotheses in light of the current evidence is discussed below.

From MA results and data from human and animal studies, Onghena and Van Houdenhove (181) conclude that the manifest depression, masked depression, and sedation hypotheses for the analgesic mode of action of AD were not correct (Table 31.23). Thus, there is no support for the commonly held opinion that the response of a patient's pain to an AD implies that the pain is a manifestation of some form of depression (210–212). Also, there is now enough evidence to conclude that the manifest depression, masked depression, and sedation hypotheses should be eliminated as explanations of the analgesic action of AD.

Onghena and Van Houdenhove (181) now believe that a biochemical hypothesis is correct, that antidepressants have hitherto unsuspected intrinsic analgesic properties. Max (492) postulates that serotonergic AD are only weakly antinociceptive but augment noradrenergic antinociception. As a result, the serotonergic-noradrenergic AD should be superior to the other two groups (serotonergic and noradrenergic). This has proven to be the case, as indicated above. Thus, the selective serotonin inhibition theory of antinociception (Table 31.23) is no longer a viable hypothesis. Finally, most studies have found that there is a significant difference between the analgesic effect of the active drug and placebo, which would rule out the placebo hypothesis.

In summary, manifest depression, masked depression, placebo, sedation, and selective serotonin inhibition are no longer viable hypotheses for the antinociceptive effect of AD. The other hypotheses presented in Table 31.23 are still viable alternatives.

Practical Guidelines for the Use of Antidepressants in Chronic Pain

As discussed earlier, it is now clear that AD have a significant antinociceptive effect independent of their antidepressant effect. In addition, although it was initially suspected that there may be a therapeutic blood level for the antinociceptive effects, i.e., a therapeutic window (214), methodologically sound studies (197) indicate that there is no relationship between serum AD levels and their antinociceptive effect.

Thus, this author believes that any practical guidelines for the treatment of the CPP with AD should depend on the following variables determined during the psychiatric examination: the level and severity of depression, the level and severity of the sleep disturbance and whether it is related to the chronic pain, the degree of pain perceived by the CPP, the level of pain behavior; and the tolerance of the CPP of his/her painful condition. The treatment approach taken by the clinician for the AD treatment of the CPP should use the clinician's knowledge of these variables in terms of how they affect the individual CPP and the clinician's knowledge that the AD dose used to treat pain can vary but in general will be less than the dose usually required to treat depression. Based on these concepts, the author has developed two algorithms for the AD treatment of chronic pain (Figs. 31.5 and 31.6).

When using the algorithm in Fig. 31.5, the following factors should be taken into account:

1. The presence of depression, which, if severe, should be treated with AD at the usual therapeutic doses for depression.
2. If sleep is disturbed, insomnia should be treated by AD at whatever dose facilitates sleep.
3. For pain, AD should be used at a dose tolerated by the patient but adequate to reduce pain behavior.

A final issue is which AD should be used in the treatment of the CPP (Fig. 31.6). Here the evidence appears to favor the tricyclic AD: amitriptyline, doxepin, clomipramine, and desipramine. Based on the previous discussion and the evidence of the efficacy of the tricyclic AD, this author advocates the algorithm presented in Figure 31.6. In this algorithm, the factors are the same as in Figure 31.5, but the drugs are chosen for their specific properties or side effects that address the problems identified in the psychiatric evaluation.

TABLE 31.24. MONOAMINE OXIDASE INHIBITORS IN NONNEUROPATHIC PAIN

Study	N	Type of Pain	Type of Drug	Dose (mg/d)	Controlled Study	Percentage of Patients Reporting Pain Relief
Anthony and Lance, 1969 (335)	25	Headache	Phenelzine	45	No	80
Lascelles, 1966 (162)	40	Facial pain	Phenelzine	45	Yes	[a]
Raft et al., 1981 (175)	23	Various	Phenelzine vs. amitriptyline vs. placebo	1.5 mg/kg 3.5 mg/kg	Yes	[a] Phenelzine superior to amitriptyline and placebo for pain relief
Raskin, 1982 (375)	1	Chest pain	Tranylcypromine	30	No	Marked decrease in chest pain
Rees & Harris, 1978–79 (329)	4	Facial pain	Phenelzine	?	No	100% complete pain relief or mild brief attacks

[a] Drug more effective than placebo or other comparison drug.

TABLE 31.25. NEUROLEPTICS IN NONNEUROPATHIC PAIN

Study	N	Type of Pain	Type of Drug	Dose (mg/d)	Controlled Study	Percentage of Patients Reporting Pain Relief
Beaver et al., 1966 (218)	40	Cancer pain	Morphine vs. methotrimeprazine	8 7.5, 15	No	Methotrimeprazine gave 15% more pain relief than morphine
Bell et al., 1990 (222)	24 26 26	Migraine	Chlorpromazine vs. dihydroergotamine vs. lidocaine	12.5–37.5 i.v.	No	Chlorpromazine more effective than the other two drugs
Bloomfield et al., 1964 (217)	19	Various types, chronic pain	Placebo vs. morphine vs. methotrimeprazine	15	Yes	[a]
						Morphine and methotrimeprazine superior to placebo and indistinguishable from each other
Caviness and O'Brien, 1980 (224)	13	Cluster headache	Chlorpromazine	75–700	No	93% headache free
Clarke, 1981 (376)	120	Various, many with neuropathic pain	Amitriptyline Perphenazine combined	25 2	No	33.68% pain free, best results with neuropathic pain
Duthie, 1977 (377)	12	Various	Amitriptyline Trifluoperazine combined	75 3	No	67% complete relief
Hakkarainen, 1977 (223)	50	Tension headache	Fluphenazine	1	Yes	[a]
						Duration and severity of headache and analgesic use decreases by drug
Kast, 1966 (216)	51	Various types of chronic pain	Morphine vs. methotrimeprazine	15	Yes	No difference between drugs for pain relief
Lane et al., 1989 (220)	24 22	Acute Migraine	Comparison i.v. chlorpromazine and i.v. meperidine with dimenhydrinate	Variable by body weight (0.1 mg/kg chlorpromazine and 0.4 mg/kg meperidine)	Yes	[a] Only 2 in chlorpromazine group vs. 11 in the meperedine group received inadequate pain relief
Moertel et al., 1974 (378)	100	Cancer pain	Placebo vs. aspirin vs. aspirin plus promazine	25	Yes	No significant difference between groups for pain relief
Montilla et al., 1963 (219)	95	Various types	Morphine vs. methotrimeprazine	—	No	Similar pain relief
Peters and Friedman, 1983 (379)	4	Severe burn pain	Methotrimeprazine	37.5	No	Control of pain
Raft et al., 1979 (380)	16	Atypical facial pain	Haloperidol	2–6	No	15 patients had 85% improvement over baseline pain
Sherwin, 1979 (381)	14	Headache	Amitriptyline Perphenazine combined	100–200 8–64	No	71% stopped complaining of headache
Stiell et al., 1991 (221)	37	Migraine	Methotrimeprazine vs. meperidine	37.5 i.m.	Yes	Methotrimeprazine as effective as meperidine in controlling pain (no statistical difference)
Zitman et al., 1991 (372)	37 34	Psychogenic pain	Flupentixol	75 i.m. 3	Yes	[a] Flupentixol did not add to the analgesic activity of amitriptyline but had a statistically significant effect on pain

[a] Drug more effective than placebo or other comparison drug.
i.v., intravenous; i.m., intramuscular.

TABLE 31.26. ANTIHISTAMINICS IN NONNEUROPATHIC PAIN

Study	N	Type of Pain	Type of Drug	Dose (mg/d)	Controlled Study	Percentage of Patients Reporting Pain Relief
Bakris et al., 1982 (382)	38	Muscle contraction headache	Orphenadrine vs. diazepam	200 15	No	After 6 mo, 79% of the diazepam and 74% of the orphenadrine group reported decreased symptoms
Batterman, 1965 (383)	78	Arthritic patients, musculoskeletal disorders	Orphenadrine vs. APC vs. orphenadrine + APC	25 225 + 160 + 30	Yes	In a 2–4 wk trial, orphenadrine not more effective than placebo
Birkeland and Clawson, 1968 (384)	325	Muscle spasm, gunshot wounds, fractures, low back syndromes	Orphenadrine + APC vs. orphenadrine vs. APC	25 25	Yes	Orphenadrine plus APC superior to single drugs and all significantly greater than placebo in pain effect
Cass and Brederick, 1964 (385)	40	Chronic arthritis, phlebitis	Orphenadrine vs. APC vs. orphenadrine + APC	25 25	No	Combination superior to two others, whereas orphenadrine was superior to APC
Gilbert, 1976 (386)	160	Musculoskeletal, nonrheumatic	Phenyltoloxamine vs. acetaminophen vs. both drugs	60 325 60 + 325	Yes	Phenyltoloxamine had no significant effect on pain but added to acetaminophen
Gilbert, 1976 (387)	208	Headache	Acetaminophen vs. phenyltoloxamine vs. both drugs	650 60 650 + 60	Yes	[a]
Gold, 1978 (388)	60	Low back pain	Orphenadrine vs. aspirin	200 2,600	Yes	Phenyltoloxamine effective analgesic and additive to acetaminophen [a]
Hingorani, 1971 (389)	99	Low back pain	Orphenadrine + acetaminophen vs. aspirin	210 270 1,800	No	Orphenadrine superior to aspirin and placebo Equal pain relief for both treatments
Krabbe and Olesen, 1980 (390)	35	Muscle contraction headache, migraine	Mepyramine	0.5 mg/kg i.v.	No	Headache completely abolished
McGuinness, 1983 (391)	28	Sprains, low back pain, sports injuries	Orphenadrine + acetaminophen vs. acetaminophen	35 450 450	No	Combination superior to acetaminophen alone
Nanda, 1980 (392)	34	Migraine	Cimetidine vs. chlorpheniramine + cimetidine	200 2 200	No	No response in either medication group
Stambaugh and Lane, 1983 (393)	30	Cancer	Meperidine vs. hydroxyzine vs. meperidine + hydroxyzine	50 i.m. 100 i.m. 50 i.m. 100 i.m.	No	Hydroxyzine alone provided pain relief; 70% had excellent or very good response
Tek and Mellon, 1987 (228)	94	Headache	Nalbuphine i.m. vs. nalbuphine + hydroxyzine i.m.	10 10 50	Yes	Combination not superior to nalbuphine alone
Tervo et al., 1976 (394)	50	Acute lumbago	Orphenadrine + acetaminophen vs. acetaminophen	210 270 270	No	Combination significantly superior

[a] Drug more effective than placebo.
APC, aspirin, phenacetin, caffeine; i.m., intramuscular; i.v., intravenous.

Monoamine Oxidase Inhibitors

Table 31.24 reveals that only a few studies have evaluated MAOI for the treatment of pain. Of these, two were controlled and demonstrated a significant analgesic effect (162,175). One study demonstrated an analgesic effect superior to that of amitriptyline (175). However, one study used facial pain patients (162), whereas the other used patients with various pains (175). Thus, it is not yet established that MAOI have an analgesic effect, although it is likely. In addition, although empirical data do not imply that MAOI should be prescribed first, CPP with facial pain unresponsive to a tricyclic AD should have a trial of an MAOI.

Neuroleptics

Although neuroleptics such as fluphenazine and perphenazine appear to be effective in neuropathic pain (215), the situation is less clear for nonneuropathic pain. Table 31.25 delineates the available studies for the treatment of nonneuropathic chronic pain and some forms of intermittent pain, e.g., headache. This table indicates that there is strong evidence that some neuroleptics may have a significant analgesic effect. In a number of well-designed studies (216–219), methotrimeprazine has been shown to have as strong or stronger an analgesic effect as morphine. In addition, neuroleptics have been demonstrated to have a strong analgesic effect on various types of headaches: acute migraine (220–222), tension (223), and cluster (224).

However, there continues to be controversy (225) over whether neuroleptics do indeed have a true analgesic effect because data from experimentally induced pain do not indicate an analgesic or analgesia-potentiating effect for neuroleptics. This controversy is not yet settled. At this point, methotrimeprazine is the only neuroleptic with well-established analgesic properties. Haloperidol, however, does have a molecular structure resembling that of morphine and meperidine (226). This molecular similarity could be the basis of the analgesic effect of some neuroleptics. The risk of tardive dyskinesia associated with neuroleptics is a major deterrent to the use of neuroleptics for the treatment of chronic pain. This author recommends that neuroleptics should not be used as first-line drugs in the treatment of nonneuropathic chronic pain. These drugs should be used only when other drugs have failed and only with the informed consent of the patient, the latter including specific mention of the risk of tardive dyskinesia.

Antihistamines

Several studies have attempted to use antihistamines for the treatment of nonneuropathic pain. These studies and their results are presented in Table 31.26. A review of this table shows that, in general, antihistamines could have an analgesic effect on nonneuropathic pain and that this effect can be additive to that of other analgesics. This was also the conclusion of Rumore and Schlichting (227), who, in a review of 27 controlled clinical trials of antihistamines, concluded that there is evidence of a direct analgesic effect of these drugs. Diphenhydramine, hydroxyzine, orphenadrine, and pyrilamine were shown to produce analgesia (227). This, however, was not the case for chlorpheniramine and phenyltoloxamine (227). It has been claimed that 100 mg hydroxyzine administered intramuscularly demonstrates the analgesic efficacy of 5 to 8 mg morphine (227). Based on these reports, it has become a common practice to combine an opioid with an antihistamine such as hydroxyzine. However, like the neuroleptics, the use of antihistamines for their analgesic properties is mired in controversy. Evidence indicates that antihistamines may not potentiate opiate analgesia. For example, the combination of hydroxyzine and nalbuphine was not more effective than nalbuphine alone for the treatment of acute headache (228). A reviewer therefore concluded that the data do not confirm the purported clinical benefits of hydroxyzine–opioid combinations in comparison with appropriate regimens of opioids alone. Further research is needed to clarify this matter.

Psychostimulants

A limited number of studies has been conducted on the treatment of nonneuropathic pain with psychostimulants (Table 31.27). The studies to date indicate that, at least in

TABLE 31.27. PSYCHOSTIMULANTS IN NONNEUROPATHIC PAIN

Study	N	Type of Pain	Type of Drug	Dose (mg/d)	Controlled Study	Percentage of Patients Reporting Pain Relief
Bruera et al., 1987 (395)	32	Cancer	Methylphenidate	15	Yes	[a]
Kaiko et al., 1987 (396)	19	Chronic nonmalignant pain	Cocaine	10	Yes	Not significantly better than placebo
Portenoy, 1989 (397)	48	Cancer	Methylphenidate	15	No	In most cases, a significant improvement in pain was observed

[a]Drug more effective than placebo.

TABLE 31.28. ANTIEPILEPTIC DRUGS IN THE TREATMENT OF NONNEUROPATHIC PAIN

Study	N	Type of Pain	Type of Drug	Dose (mg/d)	Controlled Study	Percentage of Patients Reporting Pain Relief
Ashley, 1984 (398)	43	Discogenic chronic pain	Clonazepam + trifluoperazine	1 / 2	No	42 pts had a positive response
Bowsher, 1987 (399)	1	Tabes dorsalis with lightning pains	Valproate	?	No	Total relief
Caccia, 1975 (400)	2	Cluster headache	Clonazepam	4–8	Yes	Negligible effect
Dunsker, 1976 (243)	5	Postsurgical (e.g., laminectomy) "flashing" pain	Carbamazepine	Variable	No	100% responded
Emhjellen and Skjelbred, 1990 (401)	20	Chronic orofacial pain	Clonazepam vs. amitriptyline	1 / 25	No	Clonazepam produced significantly more pain relief
Fishbain et al., 1991 (402)	40	Chronic myofascial pain	Clonazepam	0.5–4	No	85% had partial pain relief
Hering and Kuritzky, 1989 (403)	15	Chronic cluster headache (2 pts) and episodic cluster headache (13 pts)	Valproate	600–2,000	No	73.3% had positive response, 9 no pain, 2 marked improvement
Martin, 1980 (244)	14	Postlaminectomy persistent sciatic pain characterized by lightning or lancinating pains	Carbamazepine (8 pts) Clonazepam (2 pts)	Variable	No	50% responded in each drug group
Mathew and Ali, 1991 (404)	30	Chronic headache	Divalproex (valproate)	100–200	No	2 of 3 improved significantly
Stensrud and Sjaastad, 1979 (240)	38	Migraine	Clonazepam	2	Yes	Headache days and headache index differed significantly from pretest status; however, no difference from placebo; 8 pts continued with beneficial effects
Swerdlow, 1980 (246)	70	Various types of conditions, neuropathic, characterized by lancinating, stabbing, or shooting pains with a burning background	Phenytoin or valproate	150 + / 600 +	No	67% improved on one or the other drug
Swerdlow and Cundill, 1981 (245)	170	Various types of conditions including neuropathic patients characterized by lancinating, stabbing, or shooting pains with a burning background	Carbamazepine (37 pts) Clonazepam (35 pts) Phenytoin (47 pts) Valproate (51 pts)	Variable	No	Proportion of pts relieved by each drug as follows: carbamazepine, 30%; clonazepam, 66%; phenytoin, 47%; valproate, 39%
Yajnik et al., 1992 (405)	75	Cancer Various etiologies	Phenytoin vs. buprenorphine vs. both combined	200 / 0.4 SL	No	Proportion of pts relieved at least 50%: 72% for phenytoin alone; 84% for buprenorphine alone; 88% for combined treatment

Editor's note: Many of the pains treated had qualitative features of neuropathic pain, even though the investigators did not label the pains as neuropathic. See Chapter 16.

pts, patients.

cancer pain, methylphenidate may have an analgesic effect. However, in a postoperative pain study, methylphenidate did not demonstrate an analgesic effect (230). Methylphenidate does, however, appear to improve cognitive functions in cancer patients on opioids (231), thus permitting an increase in the daily opioid dose (232). Other psychostimulants may also have an analgesic effect. Dextroamphetamine has been found to potentiate the analgesic effect of morphine in acute clinical pain (233). Fenfluramine was also demonstrated to significantly increase the analgesic potency of morphine and had a mild analgesic effect alone (234).

As a psychostimulant, cocaine presents a special case. Cocaine has been included in the Brompton cocktail for the treatment of cancer pain (235). However, there has been much controversy over the inclusion of cocaine in the mixture. Evidence indicates that cocaine has an analgesic effect on experimental pain (236) and potentiates opiate analgesia (237).

The reviewed studies indicate that there may be a place for the use of psychostimulants as adjuncts in the treatment of cancer-associated chronic pain, especially when sedation limits opiate dose.

Antiepileptic Drugs

It is now generally accepted that antiepileptic drugs do have some efficacy in neuropathic pain syndromes (238,239). However, the issue is less clear for nonneuropathic pain because there has only been a limited number of studies of the treatment of nonneuropathic pain with antiepileptic drugs. These studies are presented in Table 31.28. Note that only two of these studies were controlled (240,241). In addition, few of the studies used comparable diagnoses; thus, it is difficult to compare them. However, overall, these studies indicate that antiepileptic drugs could have an analgesic effect in some pain conditions. Experimentally, phenytoin has been demonstrated to have an analgesic effect against suxamethonium-induced myalgia (242). However, its mechanism of action is still unclear. In addition, phenytoin, carbamazepine, valproate, and clonazepam have different modes of action (242). Therefore, it is difficult to postulate one mechanism of action of the analgesic effect of these drugs in nonneuropathic pain.

Finally, as pointed out previously, it is difficult to make a distinction between neuropathic pain and nonneuropathic

TABLE 31.29. LITHIUM IN NONNEUROPATHIC PAIN

Study	N	Type of Pain	Type of Drug	Dose (mg/d)	Controlled Study	Percentage of Patients Reporting Pain Relief
Bussone et al., 1990 (406)	50	Cluster headache	Lithium vs. verapamil	300–1,800	Yes	Both drugs were effective in preventing chronic cluster headache attacks
Chazot et al., 1979 (407)	25	Migraine	Lithium	300–1,800	No	50% had decreased number of attacks
Ekbom, 1977 (408)	5	Cluster headache	Lithium	300–1,800	No	3 pts with chronic cluster had 70% improvement
Ekbom, 1981 (249)	19	Cluster headache	Lithium	300–1,800	No	8 pts with chronic cluster improved
Kudrow, 1978 (409)	15	Cluster headache	Lithium vs. prednisone vs. methylsergide	300–1,800	No	87% had ≥ 75% improvement with lithium; 41% improved with methylsergide; 50% improved with prednisone
Mathew, 1978 (410)	31	Cluster headache	Lithium	300–1,800	No	Chronic cluster: 80% improved; episodic cluster; 84% improved
Medina et al., 1978 (411)	12	Cluster headache	Lithium	300–1,800	No	All improved
Medina, 1982 (412)	22	Migraine	Lithium	300–1,800	No	19 pts with cyclic form responded
Nieper, 1978 (413)	44	Migraine	Lithium	300–1,800	No	39 pts had decreased severity and frequency of attacks
Peatfield and Rose, 1981 (414)	5	Migraine	Lithium	300–1,800	No	All worse
Peatfield and Rose, 1981 (415)	31	Cluster headache	Lithium	300–1,800	No	14 pts markedly improved, 10 showed some improvement
Tyber, 1990 (250)	3	Fibromyalgia	Lithium	300–1,800	No	All had sustained pain relief

pts, patients.

pain based on diagnosis. It is unclear whether all lancinating, shooting, stabbing, and burning pains have a neuropathic component. Some authors consider these pains neuropathic in nature (238,239,243–246), but these pains may not fit the diagnostic nomenclature for neuropathic pain, e.g., as diabetic neuropathy and herpes zoster do. In the author's experience, the preponderance of CPP has these kinds of pain and yet does not have a neuropathic diagnosis. These patients are usually diagnosed as suffering from failed back surgery syndrome or myofascial pain syndrome, among others. It is therefore possible that some or many CPP have more than one type of pain, i.e., both neuropathic and nonneuropathic. The pain clinician needs to keep this in mind when treating CPP. Drug choice should not only be made based on pain diagnostic category, e.g., peripheral neuropathy, but also on the pain description. This approach has been advocated by Jensen (247).

Lithium

Lithium has been used extensively for the treatment of migraine and cluster headache, as demonstrated in Table 31.29. One reviewer concluded that lithium aggravates or exacerbates the symptoms of migraine except in cyclic migraine, in which it possibly has a positive prophylactic effect (248). The reviewer concluded that lithium is effective for the prophylaxis of chronic cluster headache (248). Patients with chronic cluster headache may respond to lithium in the first 2 weeks of therapy. They usually respond to typical mood-stabilizing serum levels (e.g., 0.8–1.5 E/L) (249).

This author could find only one report of lithium use in nonneuropathic pain other than headache. In this case report (250), lithium appeared to be effective for the chronic pain of fibromyalgia.

Conclusion

There is strong evidence of the analgesic effect of tricyclic AD on nonneuropathic pain. The evidence for the analgesic effect of neuroleptics, MAOI, antihistamines, psychostimulants, and antiepileptic drugs for nonneuropathic pain is not as strong. More extensive, well-designed research is needed to determine the appropriate role of these psychotropic drugs in pain treatment.

REFERENCES

1. Price DDR, McGrath PA, Rafii A, et al. The validity of visual analogue scales as ration scale measures for chronic and experimental pain. *Pain* 1983;17:45–56.
2. Chapman CR, Casey KL, Dubner R, et al. Pain measurement: an overview. *Pain* 1985;22:131.
3. Frederiksen LW, Lynce RS, Ros J. Methodology in the measurement of pain. *Behav Ther* 1978;9:486–488.
4. Bayer TL, Baer PE, Early C. Situational and psychophysiological factors in psychologically induced pain. *Pain* 1991;44:45–50.
5. Chen ACN, Dworkin SF, Haug J, et al. Human pain responsivity in a tonic pain model: psychological determinants. *Pain* 1989;37:143–160.
6. Bromm B. Modern techniques to measure pain in healthy man. *Methods Find Exp Clin Pharmacol* 1985;7:161–169.
7. Chen ACN, Dworkin SF, Haug J, et al. Topographic brain measures of human pain and pain responsivity. *Pain* 1989;37:129–141.
8. Huskisson EC. Measurement of pain. *Lancet* 1974;1127–1132.
9. Jensen MP, Karoly P, Braver S. Measurement of clinical pain intensity: a comparison of six methods. *Pain* 1986;27:117–126.
10. Malow RM, Olson RE. Changes in pain perception after treatment for chronic pain. *Pain* 1981;11:65–72.
11. Yang JC, Clark WC, Janal MN. Sensory decision theory and visual analogue scale indices predict status of chronic pain patients six months later. *J Pain Symptom Manage* 1991;6:58–64.
12. Gracely RH, Kwilosz DM. The descriptor differential scale: applying psychophysical principles to clinical pain assessment. *Pain* 1988;35:279–288.
13. Melzack R. The McGill Pain Questionnaire: major properties and scoring methods. *Pain* 1975;7:275–299.
14. Melzack R. The short form McGill Pain Questionnaire. *Pain* 1987;30:191–197.
15. Skevington SM. Activities as indices of illness behaviour in chronic pain. *Pain* 1983;15:295–307.
16. Holroyd KA, Holm JE, Keefe FJ, et al. A multi-center evaluation of the McGill Pain Questionnaire: results from more than 1700 chronic pain patients. *Pain* 1992;48:301–311.
17. Gronbold M, Lukinmaa A, Konttinen YT. Chronic low back pain: intercorrelation of repeated measures for pain and disability. *Scand J Rehabil Med* 1990;22:73–77.
18. Tooney TC, Mann JD, Abashian S, et al. Description of a brief scale to measure functional impairment in chronic pain patients. *Pain* 1990;S5:A578
19. Millard RW, Palt RB. A comparison of measures for low back pain disability. *Pain* 1990;S5:A573
20. Waddell G, Newton M, Henderson WI. Pain and disability. *Pain* 1990;S5:S966.
21. Sonty N, Tart RC, Chibnall JT. Use of the psychosomatic symptom checklist (PSC) as a screening instrument in patients with chronic pain. Presented at the 10th Annual Meeting of the American Pain Society, New Orleans, 1991:A91404, 124.
22. Put CL, Witkower A. Pain and impairment, beliefs in patients treated in an interdisciplinary inpatient pain program. Presented at the 10th Annual Meeting of the American Pain Society, New Orleans, 1991:A91408, 126.
23. Turk DC. Associations among impairment, pain perception, and disability: results of a national survey. Presented at the 10th Annual Meeting of the American Pain Society, New Orleans, 1991:A91230, 51.
24. VonKorff M, LeResche L, Whitney CW, et al. Prediction of temporomandibular disorder (TMD) pain disability. *Pain* 1990;S5:S329(A631).
25. Riley JF, Ahern DK, Follick MJ. Chronic low back pain and functional improvement: assessing beliefs about their relationship. *Arch Phys Med Rehabil* 1988;69:579–582.
26. VonKorff M, Ormel J, Keefe FJ, et al. Grading the severity of chronic pain. *Pain* 1992;50:133–149.
27. Deyo RA. Measuring the functional status of patients with low back pain. *Arch Phys Med Rehabil* 1988;69:1044–1053.
28. Barnes D, Smith D, Gatchel RJ, et al. Psychosocioeconomic predictors of treatment success/failure in chronic low back pain patients. *Spine* 1989;14:427–430.
29. Turk DC, Matyas TA. Pain-related behaviors—communication of pain. *Am Pain Soc J* 1992;1:109–111.

30. Keefe FJ, Wilkins RH, Cook WA. Direct observation of pain behavior in low back pain patients during physical examination. *Pain* 1984;20:59–68.
31. Fordyce WE, Lansky D, Calsyn DA, et al. Pain measurement and pain behavior. *Pain* 1984;18:53–69.
32. Romano JM, Syrjala KL, Levy RL, et al. Overt pain behaviors: relationship to patient functioning and treatment outcome. *Behav Ther* 1988;19:191–201.
33. Keefe FJ, Crisson JE, Maltbie A, et al. Illness behavior as a predictor of pain and overt behavior patterns in chronic low back pain patients. *J Psychosom Res* 1986;30:543–551.
34. Richards JS, Nepomuceno C, Riles R, et al. Assessing pain behavior: the UAB Pain Behavior Scale. *Pain* 1982;14:393–398.
35. Brands Anne-Mieke EF, Schmidt JM. Learning processes in the persistence behavior of CLBP patients with repeated acute pain stimulation. *Pain* 1987;30:329–337.
36. Gramling SE, Elliott TR. Efficient pain assessment in clinical settings. *Behav Res Ther* 1992;30:71–73.
37. Keefe FJ, Dunsmore J. Pain behavior—concepts and controversies. *Am Pain Soc J* 1992;1:92–100.
38. Peters ML, Schmidt AJM, VandenHout MA. Chronic low back pain and the reaction to repeated acute pain stimulation. *Pain* 1989;39:69–76.
39. Merskey H, Spear FG. The reliability of the pressure algometer. *Br J Soc Clin Psychol* 1964;3:130–136.
40. Forgione AG, Barber TX. A strain gauge pain stimulator. *Psychophysiology* 1971;8:102–106.
41. Asfour SS, Khalil TM, Sipes AJ, et al. Quantitative measurement of pain: a new design of a pressure algometer. Presented at the 6th Annual Meeting of the American Pain Society, Washington, DC, 1986.
42. Atkinson JH, Slater MA, Grant I, et al. Depressed mood in chronic low back pain: relationship with stressful life events. *Pain* 1988;35:47–55.
42a. Turk DC, Rudy TE, Stieg RL. The disability determination dilemma: towards a multi-axial solution. *Pain* 1988;34:217–229.
43. Haldeman S. Failure of the pathology model to predict back pain. *Spine* 1990;15:718–724.
44. Waddell G, Main CJ. Assessment of severity in low back pain. *Spine* 1984;9:204–208.
45. Spektor S. Chronic pain and pain related disabilities. *J Disabil* 1990;1:98–102.
46. American Medical Association. *Guides to the evaluation of permanent impairment*, 3rd ed. rev. Chicago: American Medical Association, 1990.
47. Brena SF, Chapman SL, Stegal PG, et al. Chronic pain states: their relationship to impairment and disability. *Arch Phys Med Rehabil* 1979;60:387–389.
48. Brand RA, Lehmann TR. Low back impairment rating practices of orthopaedic surgeons. *Spine* 1983;8:75–78.
49. Clark WC, Haldeman S, Johnson P, et al. Back impairment and disability determination another attempt at objective, reliable rating. *Spine* 1988;13:332–341.
50. Carey TS, Haltler NM, Gillings D, et al. Medical disability assessment of the back pain patient for the Social Security Administration: the weighing of presenting clinical features. *J Clin Epidemiol* 1988;41:691–697.
51. Pettingill BF. Physicians' estimates of disability vs. patients' reports of pain. *Psychosomatics* 1979;20:827–830.
52. Wartman SA, Morlock LL, Malitz FE, et al. Impact of divergent evaluations by physicians and patients of patients' complaints. *Public Health Rep* 1983;98:141–145.
53. Jette AM, Deniston OL. Inter-observer reliability of a functional status assessment instrument. *J Chronic Dis* 1978;31:573–580.
54. Waddell G, Main CJ, Morris EW, et al. Normality and reliability in the clinical assessment of backache. *BMJ* 1982;284:1519–1523.
55. Milhous RL, Hough LD, Frymoyer JW, et al. Determinants of vocational disability in patients with low back pain. *Arch Phys Med Rehabil* 1989;70:589–593.
56. Deyo RA, Diehe AK. Psychosocial predictors of disability in patients with low back pain. *J Rheumatol* 1988;15:1557–1564.
57. Fordyce WE, Lansky D, Calsyn DA, et al. Pain measurement and pain behavior. *Pain* 1984;18:53–69.
58. Linton SJ. The relationship between activity and chronic back pain. *Pain* 1985;21:289–294.
59. Fordyce W, McMahon R, Rainwater G, et al. Pain complaint-exercise performance relationship in chronic pain. *Pain* 1981;10:311–321.
60. Council JR, Ahern DK, Follick CL, et al. Expectancies and function improvement in chronic low back pain. *Pain* 1988;33:323–331.
61. Rachman S, Lopatka C. Accurate and inaccurate predictions of pain. *Behav Res Ther* 1988;26:291–296.
62. Teske K, Daut RL, Cleeland CS. Relationships between nurses' observations and patients' self-reports on pain. *Pain* 1983;16:289–296.
63. Craig KD. The facial expression of pain. *Am Pain Soc J* 1992;1:153–162.
64. Fishbain DA, Rosomoff HL. What is chronic pain? *Clin J Pain* 1990;6:164–166.
65. Osterweiss M, Kleinman A, Mechanic D, eds. *Pain and disability. Clinical, behavioral, and public policy perspectives.* Washington, DC: National Academy Press, Institute of Medicine Committee on Pain, Disability and Chronic Illness Behavior, 1987.
66. [Reserved.]
67. International Association for the Study of Pain Subcommittee on Taxonomy. Classification of chronic pain. *Pain* 1986[Suppl]:S3.
68. Robins L, Helze J, Croughan J, et al. The National Institute of Mental Health Division Diagnostic Interview Schedule: its history and characteristics and validity. *Arch Gen Psychiatry* 1981;38:381–389.
69. Riley JF, Adhera DIT, Follick MJ. Chronic low back pain and functional improvement: assessing beliefs about their relationship. *Arch Phys Med Rehabil* 1988;69:579–584.
70. Tait RC, Pallard CA, Margolis RB. The pain disability index: psychometric and validity data. *Arch Psychometric Med Rehabil* 1987;68:430–441.
71. Melding PS. Is there such a thing as geriatric pain? *Pain* 1991;46:119–121.
72. Crook J, Rideout EL, Brown G. The prevalence of pain complaints in a general population. *Pain* 1984;18:299–314.
73. Brattberg G, Thorslunce M, Wilman A. The prevalence of pain in a general population. The results of a postal survey in a country of Sweden. *Pain* 1989;37:215–222.
74. Sternbach RA. Pain and "hassles" in the United States. Findings of the Nuprin pain report. *Pain* 1986;27:69–80.
75. VonKorff M, Dworkin SF, LeResche L, et al. An epidemiologic comparison of pain complaints. *Pain* 1988;32:173–183.
76. James FR, Large RG, Bushnell JA, et al. Epidemiology of pain in New Zealand. *Pain* 1991;44:279–283.
77. Baum FE, Cooke RD, Kaluncy E. The prevalence of pain in a suburban Australian community. *Pain* 1990;S5:S335(A642).
78. Abbott FV, Gray-Donald K, Sewitch MJ, et al. The prevalence of pain in hospitalized patients and resolution over six months. *Pain* 1992;50:15–28.
79. Magni G, Caldieron C, Regatti-Luchini S, et al. Chronic musculoskeletal pain and depressive symptoms in the general

population. An analysis of the First National Health and Nutrition Examination Survey Data. *Pain* 1990;43:299–307.

80. Crook J, Weir R, Tunks E. An epidemiological follow-up survey of persistent pain sufferers in a group family practice and specialty pain clinic. *Pain* 1989;36:49–61.

81. Merskey H, Lau CL, Russell ES, et al. Screening for psychiatric morbidity. The pattern of psychological illness and premorbid characteristics in four chronic pain populations. *Pain* 1987;30:141–157.

82. Crook J, Tinks E, Kalaher S, et al. Coping with persistent pain: a comparison of persistent pain sufferers in a speciality clinic and in a family practice clinic. *Pain* 1988;34:175–184.

83. Devine R, Merskey H. The description of pain in psychiatric and general medical patients. *J Psychosom Res* 1965;9:311–316.

84. France RD, Rama Krishnan KR, eds. *Chronic pain.* Washington, DC: American Psychiatric Press, 1988.

85. Jensen J. Pain in non-psychotic psychiatric patients: life events, symptomatology and personality traits. *Acta Psychiatr Scand* 1988;78:201–207.

86. Chaturvedi SK, Michael A. Chronic pain in a psychiatric clinic. *J Psychosom Res* 1986;30:347–354.

87. Chaturvedi SK. Prevalence of chronic pain in psychiatric patients. *Pain* 1987;29:231–237.

88. Davis GC, Buchsbaum MS, Bunney WE. Analgesia to painful stimuli in affective illness. *Am J Psychiatry* 1979;136:1148–1151.

89. Linsey PG, Wyckoff M. The depression-pain syndrome and its response to antidepressants. *Psychosomatics* 1981;22:571–577.

90. Bradley JJ. Severe localized pain associated with the depressive syndrome. *Br J Psychiatry* 1963;109:741–745.

91. Turner R, Beiser M. Major depression and depressive symptomatology among the physically disabled. Assessing the role of chronic stress. *J Nerv Ment Dis* 1990;178:343–350.

92. Turner RJ, Noh S. Physical disability and depression: a longitudinal analysis. *J Health Soc Behav* 1988;29:23–27.

93. VonKorff M, Ormel J, Katon W, et al. Disability and depression among high utilizers of health care. A longitudinal analysis. *Arch Gen Psychiatry* 1992;49:91–99.

94. Zautra AJ, Maxwell BM, Reich JW. Relationship among physical impairment, distress, and well being in older adults. *J Behav Med* 1989;12:543–557.

95. Viney LL, Westbrook MT. Psychological reactions to chronic illness-related disability as a function of its severity and type. *J Psychosom Res* 1981;25:513–523.

96. Alexopoulos GS, Meyers BS, Young RC, et al. Disability and environment in late-onset depression. Presented at the 144th Annual Meeting of the American Psychiatric Association, New Orleans, 1991:NR643.

97. Wells KB, Golding JM, Burnam MA. Psychiatric disorder and limitations in physical functioning of a Los Angeles general population. *Am J Psychiatry* 1988;145:712–716.

98. Anderson KO, Keefe FJ, Bradley LA, et al. Prediction of pain behavior and functional status of rheumatoid arthritis patients using medical status and psychological variables. *Pain* 1988;33:25–32.

99. Tschannen TA, Duckro PN, Margolis RB, et al. The relationship of anger, depression, and perceived disability among headache patients. *Headache* 1992;32:501–503.

100. Fishbain DA, Goldberg M, Meagher BR, et al. Male and female chronic pain patients categorized by DSM-III Psychiatric Diagnostic Criteria. *Pain* 1986;26:181–197.

101. Romano JM, Turner JA. Chronic pain and depression: does the evidence support a relationship? *Psychol Bull* 1985;97:18–34.

102. Dworkin RH, Gitlin MJ. Clinical aspects of depression in chronic pain patients. *Clin J Pain* 1991;7:79–94.

103. Marshall M, Helmes E, Deathe AB. A comparison of psychosocial functioning and personality in amputee and chronic pain populations. *Clin J Pain* 1992;8:351–357.

104. Chaturvedi SK, Michael A. Chronic pain in a psychiatric clinic. *J Psychosom Res* 1986;30:347–354.

105. Dworkin-Samuel F, Von-Korff M, LeResche L. Multiple pains and psychiatric disturbance: an epidemiologic investigation. *Arch Gen Psychiatry* 1990;47:239–244.

106. Fogel BS. Major depression versus organic mood disorder: a questionable distinction. *J Clin Psychiatry* 1990;51:2:53–56.

107. Devins GM, Armstrong SJ, Mandin H, et al. Recurrent pain, illness intrusiveness, and quality of life in end-stage renal disease. *Pain* 1990;42:279–285.

108. Shacham S, Dar R, Cleeland CS. The relationship of mood state to the severity of clinical pain. *Pain* 1984;18:187–197.

109. Zarkowska E, Philips HC. Recent onset vs. persistent pain: evidence for a distinction. *Pain* 1986;25:365–372.

110. Dworkin RH, Hartstein G, Rosner HL, et al. Distinguishing psychological antecedents from psychological consequences of chronic pain: the prospective investigation of herpes zoster. Presented at the 9th Annual Meeting of the American Pain Society, St. Louis, 1990:A278, 57.

111. Haythornthwaite JA, Sieber WJ, Kerns RD. Depression and the chronic pain experience. *Pain* 1991;46:177–184.

112. Parmelee PA, Katz IR, Lawton MP. The relation of pain to depression among institutionalized aged. *J Gerontol* 1991;46:15–21.

113. Doan BD, Wadden NP. Relationships between depressive symptoms and descriptions of chronic pain. *Pain* 1989;36:75–84.

114. Wesley AL, Gatchel RJ, Polatin PB, et al. Differentiation between somatic and cognitive/affective components in commonly used measurements of depression in patients with chronic low-back pain. Let's not mix apples and oranges. *Spine* 1991;16:S213–S215.

115. Keefe FJ, Wilkins RH, Cook WA, et al. Depression, pain, and pain behavior. *J Consult Clin Psychol* 1986;54:665–669.

116. Williams AK, Schulz R. Association of pain and physical dependency with depression in physically ill middle-aged and elderly persons. *Phys Ther* 1988;68:1226–1230.

117. Pilwosky I, Creltenden I, Townley M. Sleep disturbance in pain clinic patients. *Pain* 1985;23:27–33.

118. Wittig RM, Zorick FJ, Blumer D, et al. Disturbed sleep in patients complaining of chronic pain. *J Nerv Ment Dis* 1982;170:429–435.

119. Bacon SF, Klapow JC, Slater MA. Characterizing subsyndromal depression in chronic low back pain. Presented at the 10th Annual Meeting of the American Pain Society, New Orleans, 1991:A91251, 61.

120. Magni-Guido SR, Schifano-Fabrizio LD. Pain as a symptom in elderly depressed patients: relationship to diagnostic subgroups. *Eur Arch Psychiatry Clin Neurol Sci* 1985;235:143–145.

121. Brown GK. A causal analysis of chronic pain and depression. *J Abnorm Psychol* 1990;99:127–137.

122. Gamsa A. Is emotional disturbance a precipitator or a consequence of chronic pain? *Pain* 1990;42:183–195.

123. Gamsa A, Vikis-Freibergs V. Psychological events are both risk factors in, and consequences of, chronic pain. *Pain* 1991;44:271–277.

124. Atkinson JH, Slater MA, Patterson TL, et al. Prevalence, onset, and risk of psychiatric disorders in men with chronic low back pain: a controlled study. *Pain* 1991;45:111–121.

125. Ciccone DS, Grzesiak RC. Psychological vulnerability to chronic back and neck pain. Presented at the 9th Annual Meeting American Pain Society, St. Louis, 1990:A248, 60.

126. Polatin PB, Kinney RK, Gatchel RJ. Premorbid psychopathology

in somatoform pain syndrome. Presented at the 144th Annual Meeting of the American Psychiatric Association, New Orleans, 1991:NR553, 181.

127. Reich J, Rosenblatt RM, Tupen J. DSM-III: a new nomenclature for classifying patients with chronic pain. *Pain* 1983;16:201–206.

128. Loranger AW, Lenzenweger MF, Gartner AF, et al. Trait-state artifacts and the diagnosis of personality disorders. *Arch Gen Psychiatry* 1991;48:720–728.

129. Aronoff GM. Role of the pain center in the treatment of intractable suffering and disability resulting from chronic pain. *Semin Neurol* 1982;3:377.

130. Aronoff GM, Evans WO, Enders PL. A review of follow-up studies of multidisciplinary pain units. *Pain* 1982;16:1–11.

131. Chapman SL, Brena SF, Bradford LA. Treatment outcome in a chronic pain rehabilitation program. *Pain* 1981;11:255–268.

132. Roy R. Pain clinics: reassessment of objectives and outcomes. *Arch Phys Med Rehabil* 1984;65:448–451.

133. Aronoff GM, McAlary PW, Witkower A, et al. Pain treatment programs: do they return workers to the workplace? *Occup Med* 1988;3:123–136.

134. Fishbain DA, Rosomoff HL, Goldberg M, et al. The prediction of return to work after pain center treatment. A review. *Clin J Pain* 1993;9:3–15.

135. Cutler RB, Fishbain DA, Rosomoff HL, et al. Does non-surgical pain center treatment of chronic pain return patients to work? A review and meta-analysis of the literature. *Spine* 1994;19:643–652.

136. Keefe FJ, Bradley LA. Behavioral and psychological approaches to the assessment and treatment of chronic pain. *Gen Hosp Psychiatry* 1984;6:49–54.

137. Fordyce WE, Fowler RS, Lehmann JF, et al. Operant conditioning in the treatment of chronic pain. *Arch Phys Med Rehabil* 1973;54:399–408.

138. Schmidt AJM. The behavioral management of pain: a criticism of a response. *Pain* 1987;30:285–291.

139. Latimer PR. External contingency management for chronic pain: critical review of the evidence. *Am J Psychiatry* 1982;139:1308–1312.

140. [Reserved.]

141. Fordyce WE. The cognitive/behavioral perspective on clinical pain. In: Loeser JD, Egan KJ, eds. *Managing the chronic pain patient.* New York: Raven Press, 1989:51–64.

142. Dolce JJ, Crocker MF, Moletteire C, et al. Exercise quotas, anticipatory concern and self-efficacy expectancies in chronic pain: a preliminary report. *Pain* 1986;24:365–372.

143. Ciccone DS, Grzesjak RC. Cognitive dimensions of chronic pain. *Soc Sci Med* 1984;19:1339–1345.

144. Pearce S. A review of cognitive-behavioral methods for the treatment of chronic pain. *J Psychosom Res* 1983;27:431–440.

145. Fernandez E, Turk DC. The utility of cognitive coping strategies for altering pain perception: a meta-analysis. *Pain* 1989;38:123–135.

146. Turner JA, Clancy S. Comparison of operant behavioral and cognitive-behavioral group treatment for chronic low back pain. *Clin Psychol* 1988;56:261–266.

147. Spence SH. Cognitive-behavior therapy in the management of chronic, occupational pain of the upper limbs. *Behav Res Ther* 1989;27:435–446.

148. Nicholas MK, Wilson PH, Goyen J. Comparison of cognitive-behavioral group treatment as an alternative non-psychological treatment for chronic low back pain. *Pain* 1992;48:339–347.

149. Toomey TC, Mann JD, Abashian S, et al. Relationship between perceived self-control of pain, pain description and functioning. *Pain* 1991;45:129–133.

150. Ward SJ, Portenoy RK, Yaksh TL. Nociceptive models: relevance to clinical pain states. In: Basbaum AI, Besson JM, eds. *Towards a new pharmacotherapy of pain.* New York: John Wiley & Sons, 1991:381–392.

151. Feinmann C. Pain relief by antidepressants: possible modes of action. *Pain* 1985;23:1–8.

152. Krishnan KRR, France RD. Antidepressants in chronic pain syndromes. *Am Fam Pract* 1989;39:233–237.

153. Satterthwaite JR, Tollison CD, Kriegel ML. The use of tricyclic antidepressants for the treatment of intractable pain. *Compr Ther* 1990;16:10–15.

154. Clifford DB. Treatment of pain with antidepressants. *Am Fam Pract* 1985;31:181–185.

155. France RD, Houpt JL, Ellinwood EH. Therapeutic effects of antidepressants in chronic pain. *Gen Hosp Psychiatry* 1984;6:55–63.

156. France RD. The future for antidepressants: treatment of pain. *Psychopathology* 1987;20[Suppl 1]:99–113.

157. Egbunike IG, Chaffee BJ. Antidepressants in the management of chronic pain syndromes. *Pharmacotherapy* 1990;10:262–270.

158. Walsh TD. Antidepressants in chronic pain. *Clin Neuropharmacol* 1983;6:271–295.

159. Magni G, Arsie D, DeLeo D. Antidepressants in the treatment of cancer pain. A survey in Italy. *Pain* 1987;29:347–353.

160. Ashby MA, Fleming BG, Brooksbank M, et al. Description of a mechanistic approach to pain management in advanced cancer. Preliminary report. *Pain* 1992;51:153–161.

161. Tasker RR. The problem of deafferentation pain in the management of the patient with cancer. *J Palliat Care* 1987;2:8–12.

162. Lascelles RG. Atypical facial pain and depression. *Br J Psychiatry* 1966;112:651–659.

163. Sharav Y, Singer E, Schmidt E, et al. The analgesic effect of amitriptyline on chronic facial pain. *Pain* 1987;31:199–209.

164. Feinmann C, Harris M. Psychogenic facial pain. Part 1: management and prognosis. *Br Dent J* 1984;156:205–208.

165. Brown RS, Bottomley WK. The utilization and mechanism of action of tricyclic antidepressants in the treatment of chronic facial pain: a review of the literature. *Anesth Prog* 1990;37:223–229.

166. Hoff GS, Ruud RE, Tonder M, et al. Doxepin in the treatment of duodenal ulcer. An open clinical and endoscopic study comparing doxepin and cimetidine. *J Clin Psychiatry* 1982;43:56–60.

167. France RD, Krishnan KRR. Assessment of chronic pain. In: France RD, Krishnan KRR, eds. *Chronic pain.* Washington, DC: Psychiatric Press, 1988:265–297.

168. Fishbain DA, Cutler R, Rosornoff HL, et al. Chronic pain associated depression: antecedent or consequence of chronic pain? A review. *Clin J Pain* 1997;13:197–205.

169. Fishbain DA, Rosornoff HL, Steele-Rosornoff R, et al. What is a multidisciplinary pain center and what are the indicators for treatment at such centers? *Archives Family Med* 1995;4:58–66.

170. Lance JW, Curran DA. Treatment of chronic tension headache. *Lancet* 1964;1:1236–1239.

171. [Reserved.]

172. Zitman FG, Linssen ACG, Edelbroek PM, et al. Low dose amitriptyline in chronic pain: the gait is modest. *Pain* 1990;42:35–42.

173. [Reserved.]

174. [Reserved.]

175. Raft D, Davidson J, Wasik J, et al. Relationship between response to phenelzine and MAO inhibition in a clinical trial of phenelzine, amitriptyline and placebo. *Neuropsychobiology* 1981;7:122–126.

176. Ward GN, Bloom VL, Friedel RO. The effectiveness of

tricyclic antidepressants in the treatment of coexisting pain and depression. *Pain* 1979;7:331–341.

177. Pilowsky I, Barrow CG. A controlled study of psychotherapy and amitriptyline used individually and in combination in the treatment of chronic intractable, "psychogenic" pain. *Pain* 1990;40:3–19.

178. Eberhard G, VonKnorring L, Nilsson HL, et al. A double-blind randomized study of clomipramine versus maprotiline in patients with idiopathic pain syndromes. *Neuropsychobiology* 1988;19:25–34.

179. Valdes M, Garcia L, Treserra J, et al. Psychogenic pain and depressive disorders: an empirical study. *J Affect Disord* 1989;16:21–25.

180. Pilowsky I, Barrow G. Predictors of outcome in the treatment of chronic "psychogenic" pain with amitriptyline and brief psychotherapy. *Clin J Pain* 1992;8:358–362.

181. Onghena P, Van Houdenhove B. Antidepressant-induced analgesia in chronic non-malignant pain: a meta-analysis of 39 placebo-controlled studies. *Pain* 1992;49:205–219.

182. [Reserved.]

183. Davis RW. Comments on "A controlled study of psychotherapy and amitriptyline used individually and in combination in the treatment of chronic intractable, psychogenic pain." *Pain* 1990;40:3–19.

184. Magni G. The use of antidepressants in the treatment of chronic pain. A review of the current evidence. *Drugs* 1991;42:730–748.

185. Goodkin K, Gullion CM. Antidepressants for the relief of chronic pain: do they work? *Ann Behav Med* 1989;11:83–101.

186. Rosenblatt RM, Reich J, Dehring D. Tricyclic antidepressants in treatment of depression and chronic pain. *Anesth Analg* 1984;63:1025–1032.

187. Lance JW, Curran DA, Anthony M. Investigations into the mechanism and treatment of chronic headache. *Med J Aust* 1965;2:909–914.

188. Loldrup D, Langemark M, Hansen HJ, et al. Clomipramine and mianserin in chronic idiopathic pain syndrome: a placebo controlled study. *Psychopharmacology* 1989;99:1–7.

189. McQuay HJ, Carroll D, Glynn CJ. Low dose amitriptyline in the treatment of chronic pain. *Anaesthesia* 1992;47:646–652.

190. Sindrup SH, Brosen K, Gram LF. Antidepressants in pain treatment: antidepressant or analgesic effect? *Clin Neuropharmacol* 1992;15[Suppl 1]:636A–637A.

191. [Reserved.]

192. [Reserved.]

193. Indaco A, Carrieri PB. Amitriptyline in the treatment of headache in patients with Parkinson's disease: a double-blind placebo-controlled study. *Neurology* 1988;28:1720–1722.

194. Watson CP, Evans RJ, Reed K, et al. Amitriptyline versus placebo in postherpetic neuralgia. *Neurology* 1982;32:671–673.

195. Kvinesdal B, Molin J, Froland A, et al. Imipramine treatment of painful diabetic neuropathy. *JAMA* 1984;251:1727–1730.

196. Rascol O, Tran MA, Bonnevaille P, et al. Lack of correlation between plasma levels of amitriptyline (and nortriptyline) and clinical improvement of chronic pain of peripheral neurologic origin. *Clin Neuropharmacol* 1987;10:560–564.

197. Edelbroek PM, Linssen CG, Frans MA, et al. Analgesic and antidepressive effects of low-dose amitriptyline in relation to its metabolism in patients with chronic pain. *Clin Pharmacol Ther* 1986;39:156–162.

198. [Reserved.]

199. [Reserved.]

200. [Reserved.]

201. [Reserved.]

202. [Reserved.]

203. [Reserved.]

204. [Reserved.]

205. [Reserved.]

206. [Reserved.]

207. [Reserved.]

208. [Reserved.]

209. [Reserved.]

210. Gupta MA. Is chronic pain a variant of depressive illness? A critical review. *Can J Psychiatry* 1986;31:241–248.

211. [Reserved.]

212. [Reserved.]

213. Taiwo YO, Fabian A, Pazoles CJ, et al. Potentiation of morphine antinociception by monoamine reuptake inhibitors in the rat spinal cord. *Pain* 1985;21:329–337.

214. Watson CPN. Therapeutic window for amitriptyline analgesia. *Can Med Assoc J* 1984;130:105–106.

215. Getto CJ, Sorkness CA, Howell T. Antidepressants and chronic nonmalignant pain. A review. *J Pain Symptom Manage* 1987;2:9–18.

216. Kast EC. An understanding of pain and its measurement. *Med Times* 1966;94:1501–1513.

217. Bloomfield S, Simard-Savoie S, Bernier J, et al. Comparative analgesic activity of levomepromazine and morphine in patients with chronic pain. *Can Med Assoc J* 1964;90:1156–1159.

218. Beaver WT, Wallenstein SL, Houde RS, et al. A comparison of the analgesic effect of methotrimeprazine and morphine in patients with cancer. *Clin Pharmacol Ther* 1966;7:436–446.

219. Montilla E, Frederik WS, Cass IJ. Analgesic effects of methotrimeprazine and morphine. *Arch Intern Med* 1963;111:725–728.

220. Lane PL, McLellan BA, Baggoley CJ. Comparative efficacy of chlorpromazine and meperidine with dimenhydrate in migraine headache. *Ann Emerg Med* 1989;18:53–58.

221. Stiell IG, Dufour DG, Moher D, et al. Methotrimeprazine versus meperidine and dimenhydrate in the treatment of severe migraine: a randomized, controlled trial. *Ann Emerg Med* 1991;20:1201–1205.

222. Bell R, Montoya D, Shuaib A, et al. A comparative trial of three agents in the treatment of acute migraine headache. *Ann Emerg Med* 1990;19:1079–1082.

223. Hakkarainen H. Fluphenazine for tension headache; double-blind study. *Headache* 1977;5:216–218.

224. Caviness VS Jr, O'Brien P. Cluster headache: response to chlorpromazine. *Headache* 1980;20:128–131.

225. McGee JL, Alexander MR. Phenothiazine analgesia—fact or fantasy? *Am J Hosp Pharm* 1979;36:633–638.

226. Maltbie AA, Cavenar JO, Sullivan JL, et al. Analgesia and haloperidol: a hypothesis. *Clin J Psychiatry* 1979;57–63.

227. Rumore MM, Schlichting DA. Clinical efficacy of antihistaminics as analgesics. *Pain* 1986;25:7–22.

228. Tek D, Mellon M. The effectiveness of nalbuphine and hydroxyzine for the emergency treatment of severe headache. *Ann Emerg Med* 1987;16:308–313.

229. Glazier HS. Potentiation of pain relief with hydroxyzine: a therapeutic myth? *DICP* 1990;24:484–488.

230. Dodson ME, Fryer JM. Postoperative effects of methylphenidate. *Br J Anaesth* 1980;52:1265–1270.

231. Bruera E, Miller MJ, Macmillan K, et al. Neuropsychological effects of methylphenidate in patients receiving a continuous infusion of narcotics for cancer pain. *Pain* 1992;48:163–166.

232. Bruera E, Fainsinger R, MacEachern T, et al. The use of methylphenidate in patients with incident cancer pain receiving regular opiates. A preliminary report. *Pain* 1992;50:75–77.

233. Forrest WH, Brown BW, Brown CR, et al. Dextroamphetamine

with morphine for the treatment of postoperative pain. *N Engl J Med* 1977;296:712–715.

234. Coda BA, Hill HF, Schaffer RL, et al. Enhancement of morphine analgesia by fenfluramine in subjects receiving tailored opioid infusions. *Pain* 1993;52:85–91.

235. Twycross R. Value of cocaine in opiate-containing elixirs. *BMJ* 1977;2:1348.

236. Yang JC, Clark WC, Dooley JC, et al. Effect of intranasal cocaine on experimental pain in man. *Anesth Analg* 1982;61:358–361.

237. Misra AL, Pontani RB, Vadlamani NL. Stereospecific potentiation of opiate analgesia by cocaine: predominant role of noradrenaline. *Pain* 1987;28:129–138.

238. Swerdlow M. Anticonvulsant drugs and chronic pain. *Clin Neuropharmacol* 1984;7:51–82.

239. McQuay HJ. Pharmacological treatment of neuralgic and neuropathic pain. *Cancer Surv* 1988;7:141–159.

240. Stensrud P, Sjaastad O. Clonazepam (Rivotril) in migraine prophylaxis. *Headache* 1979;19:333–334.

241. [Reserved.]

242. Hatta V, Saxena A, Kaul HL. Phenytoin reduces suxamethonium-induced myalgia. *Anaesthesia* 1992;47:664–667.

243. Dunsker SB, Mayfield FH. Carbamazepine in the treatment of the flashing pain syndrome. *J Neurosurg* 1976;45:49–51.

244. Martin B. Recurrent pain of a pseudotabetic variety after laminectomy for lumbar disc lesion. *J Neurol Neurosurg Psychiatry* 1980;43:283–284.

245. Swerdlow M, Cundill JG. Anticonvulsant drugs used in the treatment of lancinating pain. A comparison. *Anaesthesia* 1981;36:1129–1132.

246. Swerdlow M. The treatment of "shooting" pain. *Postgrad Med J* 1980;56:159–161.

247. Jensen NH. Accurate diagnosis and drug selection in chronic pain patients. *Postgrad Med J* 1991;67[Suppl]:S2–S8.

248. Yung CY. A review of clinical trials of lithium in neurology. *Pharmacol Biochem Behav* 1984;21[Suppl 1]:57–64.

249. Ekbom K. Lithium for cluster headache: review of the literature and preliminary results of long-term treatment. *Headache* 1981;21:132–139.

250. Tyber MA. Lithium for persistent fibromyalgia. *Can Med Assoc J* 1990;143:902–904.

251. Katon W, Egan K, Millder D. Chronic pain: lifetime psychiatric diagnoses and family history. *Am J Psychiatry* 1985;142:1156–1160.

252. Large RG. DSM-III diagnoses in chronic pain—confusion or clarity? *J Nerv Ment Dis* 1986;174:295–302.

253. Muse M. Stress-related, posttraumatic chronic pain syndrome: criteria for diagnosis, and preliminary report on prevalence. *Pain* 1985;24:295–300.

254. Fishbain DA, Goldberg M, Steele-Rosomoff R, et al. More Munchausen with chronic pain. *Clin J Pain* 1991;7:237–244.

255. [Reserved.]

256. Steele-Rosomoff R, Fishbain DA, Goldberg M, et al. Chronic pain patients who lie in this psychiatric examination about current drug/alcohol use. *Pain* 1990;5[Suppl]:S299.

257. Rafii A, Haller DL, Poklis A. Incidence of recreational drug use among chronic pain clinic patients. Presented at the Ninth Annual Meeting of the American Pain Society, 1990:A33.

258. Evans PJD. Narcotic addiction in patients with chronic pain. *Anaesthesia* 1981;36:597–602.

259. [Reserved.]

260. Haward LRC. The stress and strain of pain. *Stress Med* 1985;1:41–46.

261. McKegney FP, Schwartz CE. Behavioral medicine: treatment and organization issues. *Gen Hosp Psychiatry* 1986;8:330–339.

262. [Reserved.]

263. Goodkin K, Gullion CM, Agras WS. A randomized, double-blind, placebo-controlled trial of trazodone hydrochloride in chronic low back pain syndrome. *J Clin Psychopharmacol* 1990;10:269–278.

264. Hameroff SR, Cork RC, Scherer K, et al. Doxepin effects on chronic pain, depression and plasma opioids. *J Clin Psychiatry* 1982;43:22–26.

265. Hameroff SR, Weiss JL, Lerman JC, et al. Doxepin's effects on chronic pain and depression: a controlled study. *J Clin Psychiatry* 1984;45:47–62.

266. Hameroff SR, Cork RC, Scherer K, et al. Doxepin effects on chronic pain, depression and plasma opioids. *J Clin Psychiatry* 1982;43:22–26.

267. Jenkins DG, Ebbutt AF, Evans CD. Tofranil in the treatment of low back pain. *J Int Med Res* 1976;4:28–40.

268. Pheasant H, Bursk A, Goldfarb J, et al. Amitriptyline and chronic low back pain: a randomized double-blind crossover study. *Spine* 1983;8:552–557.

269. Sternbach RA, Janowsky DS, Huey LY, et al. Effects of altering brain serotonin activity on human chronic pain. In: *Advances in pain research and therapy, volume 1.* New York: Raven Press, 1976.

270. Ward NG. Tricyclic antidepressants for chronic low-back pain. Mechanisms of action and predictors of response. *Spine* 1986;11:661–665.

271. Ward N, Bokan JA, Phillips M, et al. Antidepressants in concomitant chronic back pain and depression: doxepin and desipramine compared. *J Clin Psychiatry* 1984;45:54–57.

272. [Reserved.]

273. [Reserved.]

274. [Reserved.]

275. [Reserved.]

276. [Reserved.]

277. [Reserved.]

278. [Reserved.]

279. [Reserved.]

280. [Reserved.]

281. [Reserved.]

282. [Reserved.]

283. [Reserved.]

284. [Reserved.]

285. Fiorentino M. Sperimentazione controllata dell'imipramina come analgesico maggiore in oncologia. *Riv Med Trentina* 1967;4:387–397.

286. [Reserved.]

287. [Reserved.]

288. [Reserved.]

289. [Reserved.]

290. [Reserved.]

291. [Reserved.]

292. [Reserved.]

293. [Reserved.]

294. [Reserved.]

295. [Reserved.]

296. [Reserved.]

297. [Reserved.]

298. Caruso I, Pietro Grande V. Italian double blind multicenter study comparing S-adenosyl-methionine, naproxen and placebo in the treatment of degenerative joint disease. *Am J Med* 1987;83[Suppl 5A]:66–71.

299. [Reserved.]

300. Frank RG, Kashani JH, Parker JC, et al. Antidepressant analgesia in rheumatoid arthritis. *J Rheumatol* 1988;15:1632–1638.

301. Ganvir P, Beaumont G, Seldrug J. A comparative trial of

clomipramine and placebo as adjunctive therapy in arthralgia. *J Int Med Res* 1980;8[Suppl 3]:60–66.

302. [Reserved.]
303. Glick EN, Fowler PD. Imipramine in chronic arthritis. *Pharmacol Med* 1979;1:94–96.
304. Grace EM, Bellamy N, Kassam Y, et al. Controlled double blind, randomized trial of amitriptyline in relieving articular pain and tenderness in patients with rheumatoid arthritis. *Curr Med Res Opin* 1985;9:426–429.
305. Gringas M. A clinical trial of Tofranil in rheumatic pain in general practice. *J Intern Med Res* 1976;4:41–49.
306. [Reserved.]
307. [Reserved.]
308. McFarland JG, Jalali S, Grace EM. Trimipramine in rheumatoid arthritis: a double-blind trial in relieving pain and joint tenderness. *Curr Med Res Opin* 1986;10:89–93.
309. McDonald-Scott WA. The relief of pain with an antidepressant in arthritis. *Practitioner* 1969;202:802–807.
310. [Reserved.]
311. [Reserved.]
312. [Reserved.]
313. [Reserved.]
314. [Reserved.]
315. Wheatley D. Antidepressants in elderly arthritis. *Practitioner* 1986;230:477–481.
316. [Reserved.]
317. Carette S, McCain GA, Bell DA, et al. Evaluation of amitriptyline in primary fibrositis: a double blind, placebo controlled study. *Arthritis Rheum* 1986;29:655–659.
318. [Reserved.]
319. [Reserved.]
320. Goldenberg DL, Felson DT, Dinerman H. A randomized controlled trial of amitriptyline and naproxen in the treatment of patients with fibromyalgia. *Arthritis Rheum* 1986;29:1371–1377.
321. Scudds RA, McCain GA, Rollman GB, et al. Improvements in pain responsiveness in patients with fibrositis after successful treatment with amitriptyline. *J Rheumatol* 1989;16:98–105.
322. Tavoni A, Vitali C, Bombardieri S, et al. Evaluation of S-adenasemethionine in primary fibromyalgia: a double blind crossover study. *Am J Med* 1987;83[Suppl 5A]:107–110.
323. [Reserved.]
324. [Reserved.]
325. [Reserved.]
326. [Reserved.]
327. [Reserved.]
328. [Reserved.]
329. Nees RT, Harris M. A typical odontalgia. *Br J Oral Surg* 1978;16:212–218.
330. Berstad A, et al. Treatment of duodenal ulcer with antacids in combination with trimipramine and cimetidine. *Scand J Gastroenterol Suppl* 1980;58:46–52.
331. Guldahl M. The effect of trimipramine on marked depression in patients with duodenal ulcer. *Scand J Gastroenterol Suppl* 1977;43:27–31.
332. [Reserved.]
333. Niller L, Haraoldsson A, Holck P, et al. The effect of trimipramine on the healing of peptic ulcer. *Scand J Gastroenterol* 1977;12[Suppl]:39–45.
334. Wetterhus S, Aubert E, Berg CE, et al. The effect of trimipramine on symptoms and healing of peptic ulcer. *Scand J Gastroenterol* 1977;12[Suppl]:33–38.
335. Anthony M, Lance JW. Monoamine oxidase inhibition in the treatment of migraine. *Arch Neurol* 1969;21(3):263–268.

336. [Reserved.]
337. Couch JR, Hassanein RS. Amitriptyline in migraine prophylaxis. *Arch Neurol* 1979;36:695–699.
338. [Reserved.]
339. Diamond S, Baltes BJ. Chronic tension headaches: treated with amitriptyline—a double-blind study. *Headache* 1971;11:110–116.
340. Fogelholm R, Murros K. Maprotiline in chronic tension headache: a double-blind cross-over study. *Headache* 1985;25:273–275.
341. [Reserved.]
342. [Reserved.]
343. Langemark M, Loldrup D, Bech P, et al. Clomipramine and mianserin in the treatment of chronic tension headache. A double-blind, controlled study. *Headache* 1990;30:118–121.
344. Langohr HD, Gerber WD, Koletzki E, et al. Clomipramine and metoprolol in migraine prophylaxis: a double blind crossover study. *Headache* 1985;25:107–113.
345. [Reserved.]
346. Martucci N, Manna V, Porto C, et al. Migraine and the noradrenergic control of vasomotricity: a study with alpha-2 stimulant and alpha-2 blocker drugs. *Headache* 1985;25:95–100.
347. Monro P, Swade C, Coppen A. Mianserin in the prophylaxis of migraine: a double-blind study. *Acta Psychiatr Scand* 1985;72[Suppl]:98–103.
348. Morland TJ, Storli OV, Mogstad TE. Doxepin in the treatment of mixed vascular and tension headaches. *Headache* 1979;19:382–383.
349. Noone JF. Clomipramine in the prevention of migraine. *J Int Med Res* 1980;8[Suppl]:49–52.
350. [Reserved.]
351. [Reserved.]
352. [Reserved.]
353. [Reserved.]
354. [Reserved.]
355. [Reserved.]
356. [Reserved.]
357. Evans W, Gensler F, Blackwell B, Galbrecht C. The effects of antidepressant drugs on pain relief and mood in the chronically ill. A double-blind study. *Psychosomatics* 1973;14(4):214–219.
358. Gourlay GK, et al. A controlled study of the serotonin reuptake blocker, 3 imelidine, in the treatment of chronic pain. *Pain* 1987;25:35–52.
359. [Reserved.]
360. Johansson F, Von Knorring L. A double-blind controlled study of a serotonin uptake inhibitor (zimeldine) versus placebo in chronic pain patients. *Pain* 1979;7:69–78.
361. [Reserved.]
362. [Reserved.]
363. Pilowsky I, Hallett EC, Bassett DL, et al. A controlled study of amitriptyline in the treatment of chronic pain. *Pain* 1982;14:169–179.
364. [Reserved.]
365. [Reserved.]
366. Zitman FG, Linssen ACG, Edelbroek PM, et al. Low dose amitriptyline in chronic pain: the gain is modest. *Pain* 1990;42:35–42.
367. Dahl LE, Decker SJ, Lundin N. A double blind study of dothiepin (Prothiaden) and amitriptyline in outpatients with masked depression. *J Int Med Res* 1981;9:103–107.
368. Lindsay PG, Wyckoff M. The depression-pain syndrome and its response to antidepressants. *Psychosomatics* 1981;22:571–577.
369. Eisendrath SJ, Kodama KT. Fluoxetine management of chronic abdominal pain. *Psychosomatics* 1992;33:227–229.

370. Feinmann C, Harris M. Psychogenic facial pain: the clinical presentation. *Br Dent J* 1984;156:165–168.
371. Loldrup D, Langemark M, Hansen HJ, et al. Clomipramine and mianserin in chronic idiopathic pain syndrome. A placebo controlled study. *Psychopharmacology (Berl)* 1989;99:1–7.
372. Zitman FG, Linssen ACG, Edelbroek PM, et al. Does addition of low-dose flupentixol enhance the analgetic effects of low-dose amitriptyline in somatoform pain disorder? *Pain* 1991;47:25–40.
373. Murphy DL, Siever LJ, Insel TR. Therapeutic responses to tricyclic antidepressants and related drugs in non-affective disorder patient populations. *Prog Neuropsychopharmacol Biol Psychiatry* 1985;9:3–13.
374. Bigeon A, Samuel D. Interaction of tricyclic antidepressants with opiate receptors. *Biochem Pharmacol* 1980;29:460–462.
375. Raskin DE. MAO inhibitors in chronic pain and depression. *J Clin Psychiatry* 1982;43:3.
376. Clarke IMC. Amitriptyline and perphenazine (Triptafen DA) in chronic pain. *Anaesthesia* 1981;36:210–212.
377. Duthie AM. The use of phenothiazines and tricyclic antidepressants in the treatment of intractable pain. *S Afr Med J* 1977;51:246–247.
378. Moertel CG, Ahamann DL, Taylor WF, et al. Relief of pain by oral medications. *JAMA* 1974;229:55–59.
379. Peters WJ, Friedman J. Methotrimeprazine for burn patients. *Can Med Assoc J* 1983;128:894–897.
380. Raft D, Toomey T, Gregg JM. Behavior modification and haloperidol in chronic facial pain. *South Med J* 1979;72:155–159.
381. Sherwin D. A new method for treating headaches. *Am J Psychiatry* 1979;136:1181–1183.
382. Bakris GL, Mulopulos BB, Tiwari S, et al. Orphenadrine citrate: an effective alternative for muscle contraction headaches. *Ill Med J* 1982;161:106–108.
383. Batterman RC. Methodology of analgesic evaluation: experience with orphenadrine citrate compound. *Curr Ther Res* 1965;7:639–647.
384. Birkeland IW, Clawson DK. Drug combinations with orphenadrine for pain relief associated with muscle spasm. *Clin Pharmacol Ther* 1968;9:639–646.
385. Cass LJ, Brederick WS. An evaluation of orphenadrine citrate in combination with APC as an analgesic. *Curr Ther Res* 1964;6:400–408.
386. Gilbert MM. Efficacy of Percogesic in relief of musculoskeletal pain associated with anxiety. *Psychosomatics* 1976;17:190–193.
387. Gilbert MM. Analgesic/calmative effects of acetaminophen and phenyltoloxamine in treatment of simple nervous tension accompanied by headache. *Curr Ther Res* 1976;20:53–58.
388. Gold RH. Treatment of low back syndrome with oral orphenadrine citrate. *Curr Ther Res* 1978;23:271–276.
389. Hingorani K. Orphenadrine/paracetamol in backache—a double blind controlled trial. *Br J Clin Pract* 1971;25:227–231.
390. Krabbe AA, Olesen J. Headache provocation by continuous intravenous infusion of histamine. Clinical results and receptor mechanism. *Pain* 1980;8:253–259.
391. McGuinness BW. A double-blind comparison in general practice of a combination tablet containing orphenadrine citrate and paracetamol with paracetamol alone. *J Intern Med Res* 1983;11:42–45.
392. Nanda RN. Cimetidine in the prophylaxis of migraine. *Acta Neurol Scand* 1980;62:90–95.
393. Stambaugh JE, Lane C. Analgesic efficacy and pharmacokinetic evaluation of meperidine and hydroxyzine, alone and in combination. *Cancer Invest* 1983;1:111–117.
394. Tervo T, Petaja L, Lepisto P. A controlled clinical trial of a muscle relaxant analgesic combination in the treatment of acute lumbago. *Br J Clin Pract* 1976;30:62–64.
395. Bruera E, Chadwick S, Brenneis C, et al. Methylphenidate associated with narcotics for the treatment of cancer pain. *Cancer Treat Rep* 1987;71:67–70.
396. Kaiko RF, Kanner R, Foley KM, et al. Cocaine and morphine interaction in acute and chronic cancer pain. *Pain* 1987;31:35–45.
397. Portenoy RK. Use of methylphenidate as an adjuvant to narcotic. *J Pain Symptom Manage* 1989;4[Suppl]:2, 4.
398. Ashley JJ. Chronic intractable pain. *N Z Med J* 1984;97:23.
399. Bowsher D. A case of tabes dorsalis with tonec pulis and lightning pairs relieved by sodium valproate. *J Neurol Neurosurg Psychiatry* 1987;50:239–241.
400. Caccia MR. Clonazepam in facial neuralgia and cluster headache. Clinical and electrophysiological study. *Eur Neurol* 1975;13:560–563.
401. Emhjellen S, Skjelbred P. Clonazepam versus amitriptyline in patients with chronic orofacial pain. *Pain* 1990;S5:S42(A78).
402. Fishbain DA, Goldberg M, Rosomoff H, et al.Clonazepam open clinical trial for chronic pain of myofascial pain syndrome origin refractory to pain unit treatment. Presented at the 144th Annual Meeting of the American Psychiatric Association, New Orleans, 1991:NR538, 117.
403. Hering R, Kuritzky A. Sodium valproate in the treatment of cluster headache: an open clinical trial. *Cephalalgia* 1989;9:195–198.
404. Mathew NT, Ali S. Valproate in the treatment of persistent chronic daily headache. An open label study. *Headache* 1991;31:71–74.
405. Yajnik S, Singh GP, Singh G, et al. Phenytoin as a coanalgesic in cancer pain. *J Pain Symptom Manage* 1992;7:209–213.
406. Bussone G, Leone M, Peccarisi C, et al. Double blind comparison of lithium and verapamil in cluster headache prophylaxis. *Headache* 1990;30:411–417.
407. Chazot G, Chauplannaz G, Biron A, et al. Migraines: traitement par lithium. *Nouvelle Presse Med* 1979;8;2836–2837.
408. Ekbom K. Lithium in the treatment of chronic cluster headache. *Headache* 1977;17:39–40.
409. Kudrow L.Comparative results of prednisone, methysergide and lithium therapy in cluster headache. In: Green R, ed. *Current concepts in migraine research.* New York: Raven Press, 1978:159–163.
410. Mathew NT. Clinical subtypes of cluster headaches and response to lithium therapy. *Headache* 1978;18:26–30.
411. Medina JL, Fareed J, Diamond S. Blood amines and platelet changes during treatment of cluster headache with lithium and other drugs. *Headache* 1978;19:112.
412. Medina JL. Cyclic migraine: a disorder responsive to lithium carbonate. *Psychosomatics* 1982;23:625–637.
413. Nieper HA. The clinical application of lithium orotate. A two-year study. *Agressologie* 1978;14:407–411.
414. Peatfield RC, Rose FC. Exacerbation of migraine by treatment with lithium. *Headache* 1981;21:140–142.
415. Peatfield RC. Lithium in migraine and cluster headache: a review. *J R Soc Med* 1981;74:432–436.
416. Grachev ID, Fredrickson BE, Apkarian AV. Abnormal brain chemistry in chronic back pain: an in vivo proton magnetic resonance spectroscopy study. *Pain* 2000;80:7–18.
417. Mountz JM, Bradley LA, Modell JG, et al. Fibromyalgia in women. Abnormalities of regional cerebral blood flow in the thalamus and the caudate nucleus are associated with low pain threshold levels. *Arthritis Rheum* 1995;38:926–938.
418. Hsieh JC, Belfrage M, Stone-Elander S, et al. Central

representation of chronic ongoing neuropathic pain studies by positron emission tomography. *Pain* 1995;63:225–236.

419. Iadarola MJ, Max MB, Berman KF, et al. Unilateral decrease in thalamic activity observed with positron emission tomography in patients with chronic neuropathic pain. *Pain* 1995;63:55–64.

420. Hsieh JC, Stahle-Backdahl M, Hagermark O, et al. Traumatic nociceptive pain activates the hypothalamus and the periaqueductal gray: a positron emission tomography study. *Pain* 1995;64:303–314.

421. Derbyshire SWG. Imaging the brain in pain. *APS Bull* 1999;May/June:7–9.

422. Casey KL, Bushnell MC. Why image the brain during pain? *Int Assoc Study Pain Clin Updates* 2000;VIII:1–4.

423. Gelnar PA, Apkarian V. Imaging pain, science and technology. *APS Bull* 1997;March/April:3–11.

424. Bradley LA, Alarcon GS, Mountz JM, et al. Acute pain produces abnormal regional cerebral blood flow (RCBF) in the anterior cingulate (AC) cortex in patients with fibromyalgia (FM). American Pain Society Meeting, Atlanta, GA, 2000:170(abst 812).

425. Jackson EF, Cleeland CS, Anderson KO, et al. Functional MRI study of pressure pain and its modulation using mental imagery. American Pain Society Meeting, Atlanta, GA, 2000:169(abst 810).

426. Rainville P. Pain affect encoded in human anterior cingulate but not somatosensory cortex. *Science* 1997;277:968–971.

427. Treede RD, Apkarian AV, Bromm B, et al. Cortical representation of pain: functional characterization of nociceptive areas near the lateral sulcus. *Pain* 2000;87:113–119.

428. Augustine JR. Circuitry and functional aspects of the insular lobe in primates including humans. *Brain Res Rev* 1996;22:229–244.

429. Silverman DHS, Munakata JA, Ennes H, et al. Regional cerebral activity in normal and pathological perception of visceral pain. *Gastroenterology* 1997;112:64–72.

430. McQuay HJ, Tramer M, Nye BA, et al. A systematic review of antidepressants in neuropathic pain. *Pain* 1996;68:217–227.

431. Atkinson JH, Slater MA, Williams RA, et al. A placebo-controlled randomized clinical trial of nortriptyline for chronic low back pain. *Pain* 1998;76:278–296.

432. Treves R, Montaine de la Roque P, Dumond JJ, et al. Prospective study of the analgesic action of clomipramine versus placebo in refractory lumbosciatica (68 cases) [in French]. *Rev Rhum Mal Osteoartic* 1991;58:549–552.

433. Storch H, Steck P. Concomitant thymoleptic therapy in the frame of a controlled study with maprotiline (Ludiomil) in the treatment of back pain [in German]. *Nervenarzt* 1982;53:445–450.

434. Atkinson JH, Slater MA, Wahlgren DR, et al. Effects of noradrenergic and serotonergic antidepressants on chronic low back pain. *Pain* 1999;83:137–145.

435. Engel CC Jr, Walker EA, Engel AL, et al. Randomized, double-blind crossover trial of sertraline in women with chronic pelvic pain. *J Psychosom Res* 1998;44:203–207.

436. Walsh TD. Controlled study of imipramine (IM) and morphine (M) in chronic pain due to advanced cancer. In: *Proceedings of the American Society of Clinical Oncology Annual Meeting 1986.* Alexandria, VA: ASCO, 1986:237.

437. MacNeill AL, Dick WC. Imipramine and rheumatoid factor. *J Int Med Res* 1976;83:371–375.

438. Rani PU, Naidu MUR, Prasad VBN, et al. An evaluation of antidepressants in rheumatic pain conditions. *Anesth Analg* 1996;83:371–375.

439. Singh AN, Saxena B, Gent M, et al. Maprotiline (Ludiomil, Ciba 34, 276-BA) and imipramine in depressed outpatients: a double-

blind clinical study. *Curr Ther Res Clin Exp* 1976;19:451–462.

440. [Reserved.]

441. Hannonen P, Malminiemi K, Yli-Kerttula U, et al. A randomized, double-blind, placebo-controlled study of moclobemide and amitriptyline in the treatment of fibromyalgia in females without psychiatric disorder. *Br J Rheumatol* 1998;37:1279–1286.

442. Goldenberg D, Mayskiy M, Mossey C, et al. A randomized, double blind crossover trial of fluoxetine and amitriptyline in the treatment of fibromyalgia. *Arthritis Rheum* 1996;39:1852–1859.

443. Norregaard J, Volkmann H, Danneskiold-Samsoe B. A randomized controlled trial of citalopram in the treatment of fibromyalgia. *Pain* 1995;61:445–459.

444. Wolfe F, Cathey MA, Hawley DJ. A double-blind placebo controlled trial of fluoxetine in fibromyalgia. *Scand J Rheum* 1994;23:255–259.

445. Carette S, Bell MJ, Reynolds WJ, et al. Comparison of amitriptyline, cyclobenzaprine, and placebo in the treatment of fibromyalgia. A randomized, double blind clinical trial. *Arthritis Rheum* 1994;37:32–40.

446. Bennett RM, Gatter RA, Campbel SM, et al. A comparison of cyclobenzaprine and placebo in the management of fibrositis. A double-blind controlled study. *Arthritis Rheum* 1988;31:1535–1542.

447. Kempanaers CH, Simenon G, Vander Elst M, et al. Effect of an antidiencephalon immune serum on pain and sleep in primary fibromyalgia. *Neuropsychobiology* 1994;30:66–72.

448. Ginsberg F, Mancaux A, Joos E, et al. A randomized placebo-controlled trial of sustained-release amitriptyline in primary fibromyalgia. *J Musculo Pain* 1996;14:37–47.

449. Ziegler DK, et al. Migraine prophylaxis: a comparison of propranolol and amitriptyline. *Arch Neurol* 1987;44:486–489.

450. Saper JR, Silberstein SD, Lake AE 3rd, et al. Double-blind trial of fluoxetine: chronic daily headache and migraine. *Headache* 1994;34:497–502.

451. Gobel H, et al. Chronic tension-type headache: amitriptyline reduces clinical headache-duration and experimental pain sensitivity but does not alter pericranial muscle activity readings. *Pain* 1994;59:2419.

452. Steiner TJ, Ahbme F, Findley LJ, et al. S-fluoxetine in the prophylaxis of migraine: a phase 11 double-blind randomized placebo-controlled study. *Cephalalgia* 1998;18:283–286.

453. Battisstella PA, Ruffilli R, Cernetti R, et al. A placebo controlled crossover trial using trazodone in pediatric migraine. *Headache* 1993;33:36–39.

454. Bendtsen L, Jensen R, Olesen J. A non-selective (amitriptyline), but not a selective (citalopram), serotonin reuptake inhibitor is effective in the prophylactic treatment of chronic tension-type headache. *J Neurol Neurosurg Psychiatry* 1996;61:285–290.

455. Zeeberg IB, Orholm M, Nielsen JD, et al. Femoxetine in the prophylaxis of migraine—randomized comparison with placebo. *Acta Neurol Scand* 1981;64:452–459.

456. Orholm M, Honore PF, Zeeberg I. A randomized general practice group—comparative study of femoxetine and placebo in the prophylaxis of migraine. *Acta Neurol Scand* 1986;74:235–239.

457. Kangasniemi PJ, Nyrke T, Lang AH, et al. Femoxetine—a new 5 HT uptake inhibitor—and prophylactic treatment of migraine. *Acta Neurol Scand* 1983;68:262–267.

458. Adly C, Straumanis J, Chesson A. Fluoxetine prophylaxis of migraine. *Headache* 1992;32:101–104.

459. Pfaffenrath V, et al. Effectiveness and tolerance of amitriptyline oxide in chronic tension headache: a multicenter double-blind

study versus amitriptyline versus placebo [in German]. *Nerve-narzt* 1993;64:114–120.

460. Ziegler DK, Hurwitz A, Preskorn S, et al. Propranolol and amitriptyline in prophylaxis of migraine. Pharmacokinetics and therapeutic effects. *Arch Neurol* 1993;50:825–830.

461. Mama V, Bolino F, DiCicco L. Chronic tension-type headache, mood depression and serotonin: therapeutic effects of fluvoxamine and mianserin. *Headache* 1994;34:44–49.

462. Pfaffenrath V, Diener HC, Isler H, et al. Efficacy and tolerability of amitriptylinoxide in the treatment of chronic tension-type headache multicenter controlled study. *Cephalalgia* 1994;14:149–155.

463. Poulsen DL, Hansen HJ, Langemark M, et al. Discomfort or disability in patients with chronic pain syndrome. *Psychother Psychosom* 1987;48:60–62.

464. Ward NG, Bloom VL, Friedel RO. The effectiveness of tricyclic antidepressants in the treatment of coexisting pain and depression. *Pain* 1979;7:331–341.

465. Harrison SD, Glover L, Feinmann C, et al. A comparison of antidepressant medication alone and in conjunction with cognitive behavioral therapy for chronic idiopathic facial pain. *Prog Pain Res Manage* 1997;8:663–671.

466. Greenbaum DS, Mayle JE, Vanegeren IE, et al. Effects of desipramine on irritable bowel syndrome compared with atropine and placebo. *Dig Dis Sci* 1987;32:257–266.

467. [Reserved.]

468. Moshal MG, Khan F. Trimipramine in the treatment of active duodenal ulceration. *Scand J Gastroenterol* 1981;16(2):295–298.

469. [Reserved.]

470. Anderson OK, et al. Doxepan in the treatment of duodenal ulcer. A double-blind clinical study comparing doxepan and placebo. *Scand J Gastroenterol* 1984;19:923–925.

471. Don GJ, Shearman DJ, Hecker R. Double-blind trial of trimipramine in the treatment of duodenal ulcer. *Med J Aust* 1980;2(10):567–968.

472. Stein D, Peri T, Edelstein E, et al. The efficacy of amitriptyline and acetaminophen in the management of acute low back pain. *Psychosomatics* 1996;37:63–69.

473. Levine JD, Gordon NC, Smith R, et al. Desipramine enhances opiate postoperative analgesia. *Pain* 1986;27:45.

474. Kerrick JM, Fine PG, Lipman AG, et al. Low-dose amitriptyline as an adjunct to opioids for postoperative orthopedic pain: a placebo-controlled trial. *Pain* 1993;52:325–330.

475. Tiengo M, Pagnoni B, Calmi A, et al. Clomipramine compared with pentazocine as a unique treatment in postoperative pain. *Int J Clin Pharmacol* 1987;7:1141–1143.

476. Gordon NC, Heller PH, Gear RW, et al. Interactions between fluoxetine and opiate analgesia for postoperative dental pain. *Pain* 1994;58:85–88.

477. Cannon RO, Quyyumi AA, Mincemoyer R, et al. Imipramine in patients with chest pain despite normal coronary angiograms. *N Engl J Med* 1994;330:1411–1417.

478. Fishbain DA, Cutler RB, Rosomoff HL, et al. Do antidepressants have an analgesic effect in psychogenic pain and somatoform pain disorder: a meta-analysis. *J Psychosom Med* 1998;60:5039.

479. Phillipp M, Fickinger M. Psychotropic drugs in the management of chronic pain syndromes. *Pharmacopsychiatry* 1993;26:221–234.

480. Guiraud-Chaumeil B, et al. Lack of correlation between plasma levels of amitriptyline (and nortriptyline) and clinical improvement of chronic pain of peripheral neurologic origin. *Clin Neuropharmacol* 1987;10:560–564.

481. Watson CPN, Smythe IE. Therapeutic window for amitriptyline analgesia. *Can Med Assoc J* 1984;130:105.

482. Edelbroek PM, Linssen CG, Zitman FG, et al. Analgesic and antidepressive effects of low-dose amitriptyline in relation to its metabolism in patients with chronic pain. *Clin Pharmacol Ther* 1986;39:156–161.

483. Montastruc JL, Tran MA, Charlet JP, et al. Analgesic properties and plasma concentrations of clomipramine in chronic pain. *Rev Neurol* 1983;139:583–587.

484. Guiol C, Tran MA, Blac M, et al. Is there any relationship between antinociception, doses and plasma levels of clomipramine during subchronic treatment in rats? *Biomed Pharmacother* 1984;38:59–62.

485. Brenne E, van der Hagen K, Maehlum E, et al. Treatment of chronic pain with amitriptyline. A double-blind dosage study with determination of serum levels [in Norwegian]. *Tidsskr Nor Laegeforen* 1997;117:3491–3494.

486. Fishbain DA, Cutler RB, Rosomoff HL, et al. What is the quality of implemented meta-analytic procedures in chronic pain treatment meta-analyses? *Clin J Pain* 2000;16:73–85.

487. Fishbain DA, Rosomoff HL, Goldberg M, et al. The prediction of return to the workplace after multidisciplinary pain center treatment. *Clin J Pain* 1993;9:3–15.

488. Wells KB. Treatment research at the crossroads: the scientific interface of clinical trials and effectiveness research. *Am J Psychiatry* 1999;156:5–10.

489. Goodkin K, Feaster DJ. Meta-analysis and the need for study quality control. *Pain Forum* 1998;7:90–94.

490. Olkin I. Quality and diagnostics in meta-analysis. *Pain Forum* 1998;7:95–97.

491. Fishbain DA, Cutler R, Rosomoff HL, et al. Evidence-based data from animal and human experimental studies on pain relief with antidepressants: a structured review. *Pain Med* 2000;1:310–316.

492. Max MB. Antidepressants as analgesics. In: Fields HL, Liebeskind JC, eds. *Progress in pain research and management.* Seattle: IASP Press, 1994:229–246.

493. Fishbain DA. Evidence based data on pain relief with antidepressants. *Ann Med* 2000;32:305–316.

32

NEUROBIOLOGY OF DRUG ABUSE

R. DAYNE MAYFIELD
R. LISA POPP
TINA K. MACHU
STEVEN J. GRANT
MICHAEL JOHN MORGAN
STEPHEN R. ZUKIN
EDYTHE D. LONDON

Drug abuse has important social and legal dimensions; however, from a biologic standpoint, abuse potential is an attribute of a drug, and it derives from the reinforcing properties of the drug. Drugs of abuse comprise a chemically heterogeneous group that represents a very small percentage of all drugs. The neurobiology of substance abuse encompasses both drug-specific mechanisms in the brain and commonalities of action that explain how such diverse compounds share the property of abuse liability. This chapter reviews the biologic aspects of the acute and chronic effects of drug abuse. Information about neurochemical and anatomic substrates of these effects is presented, along with a discussion of the techniques used to elucidate them. Clinical aspects of drug abuse are the subject of Chapter 33.

RECEPTORS: INITIAL TARGETS FOR ACTIONS OF ABUSED DRUGS

Given the complexity of brain circuitry, it is unlikely that the action of a drug at a single target *per se* could produce the constellation of behaviors, such as reward, conditioning, and compulsive self-administration, that characterize addiction. Nonetheless, despite the propagation of pharmacologic effects to locations that are remote from the initial loci of action, drugs of abuse interact with specific neuronal receptors (Table 32.1). There is a considerable and rapidly expanding body of data on the central target sites at which drugs of abuse initiate their effects.

This chapter is updated by Drs. Mayfield, Popp, and Machu from the original chapter by Drs. Grant, Morgan, Zukin, and London in *Neuropsychiatry,* 1st ed., 1996.

Dopamine Receptors and the Dopamine Transporter

Although drugs of abuse generally do not interact directly with dopamine (DA) receptors, as shown later, activation of the mesolimbic DA system is critically important to the rewarding effects and perhaps to other behavioral actions of these drugs. Therefore, it is useful to review the following information on DA receptors and the DA transporter (DAT).

Dopamine Receptors

By the early 1980's, biochemical studies had revealed heterogeneity among DA receptors. All subtypes of DA receptors belong to a class of receptors that manifest relatively slow (seconds) responses, consistent with a modulatory function (1). Two DA receptor subtypes were originally proposed based on their ability to either stimulate (D_1 receptors) or inhibit (D_2 receptors) adenylyl cyclase activity (2–4). Both D_1- and D_2-like receptors are coupled to adenylate cyclase by guanine nucleotide regulatory proteins (G proteins). The D_1 receptor is coupled via a G_s subtype (5), and the D_2 receptor is coupled via a G_i or G_o subtype (5). Inhibitory D_2-like autoreceptors localized on DA neuronal terminals in the brain modulate DA synthesis (6,7) and release (8). In addition, both synergistic and opposing interactions between D_1 and D_2 receptors occur (9–11). Numerous studies have attempted to associate the D_2 receptor gene as a marker for vulnerability to substance abuse. The results of these studies have been equivocal and have created considerable controversy (12–19).

To date, five distinct DA receptors have been cloned: D_1 and D_5 receptors ("D_1-like"); and D_2, D_3, and D_4 receptors ("D_2-like") (20–24). The two products of D_1 receptor genes are known as D_{1A} and D_{1B}. The D_{1B} gene in rats may be homologous to the D_5 receptor gene identified in humans (25). The D_{1B}/D_5 receptor appears to have a higher

TABLE 32.1. RECEPTORS FOR DRUGS OF ABUSE

Prototype Compounds	Chemical/ Pharmacologic Classification	Receptor	Receptor Family	Drug Effect	Endogenous Ligands of Drug Binding Sites
Morphine	Opiate alkaloid	μ-Opiate receptor	G-protein–coupled receptors	Inhibits adenylate cyclase	Enkephalins, Endorphins, Morphine (?)
Δ^9-Tetrahydro-cannabinol	Cannabinoid	Cannabinoid receptor	G-protein–coupled receptors	Inhibits adenylate cyclase	Anandamide (?)
Diazepam	Benzodiazepine	Benzodiazepine site of $GABA_A$ receptor complex	Ligand gated anion channels	Enhances GABA effect; increases Cl^- influx	Diazepam binding inhibitor?
Phencyclidine (PCP)	Arylcyclohexylamine (dissociative anesthetic)	PCP site within channel of NMDA receptor complex	Ligand gated cation channels	Blocks channel; noncompetitive NMDA antagonist	?
Nicotine	Alkaloid	Nicotinic cholinergic receptor	Ligand gated cation channels	Activates nicotinic receptor	Acetylcholine
Cocaine	Alkaloid (psycho-motor stimulant)	DA transporter	Na^+/Cl^--dependent neurotransmitter transporters	Blocks DA uptake	Dopamine ?
Amphetamine	Psychomotor stimulant	DA terminal (DA transporter monoamine oxidase)	Na^+/Cl^--dependent neurotransmitter transporters	Promotes DA release	?
Lysergic acid diethylamide	Hallucinogen	$5HT_2$-serotonin receptor	Second messenger-coupled receptor	Partial agonist; increases phospho-inositol hydrolysis	Serotonin

DA, dopamine; GABA, γ-aminobutyric acid; NMDA, N-methyl-D-aspartate.

affinity for DA than the D_{1A} receptor (26). The D_2 receptor gene family includes D_2, D_3, and D_4 receptors. All of these receptor genes have considerable amino acid sequence homology, and they belong to a superfamily of receptors with seven membrane-spanning regions. The major differences in the amino acid sequences of the receptors occur in the intracellular portions believed to be responsible for the G-protein coupling to adenylate cyclase. In addition, two isoforms of the D_2 receptor that are encoded by a single gene and differentially spliced to include (long form) or exclude (short form) 29 amino acids within the third intracellular loop of the receptor (27,28) have been identified.

Neurochemical markers for DA receptor subtypes exhibit differential distributions in the brain (21,23,29). DA receptors were initially mapped in the brain by assay of binding in dissected regions (30); the findings were extended using light microscopic autoradiography (31–34). The development of hybridization probes for receptor ribonucleic acid (RNA) and antibodies against nonhomologous receptor protein segments has allowed more precise mapping of new receptor subtypes (35–40). The highest levels of D_{1A} receptors are found in dopaminergic terminal fields, including the striatum, nucleus accumbens, and olfactory tubercle in the forebrain. Lower densities occur in the cortex, amygdala, hypothalamus, and thalamus. In these regions, the receptors are located primarily on postsynaptic neurons (41,42). Although D_1 receptors are about ten-fold more abundant

than D_2 receptors in the cortex of most species, including primates, the electrophysiologic effects of DA in rats have a profile more consistent with D_2 receptor interactions (43–45). There is evidence that D_1 and D_2 receptors may be located on different functional subsets of pyramidal cells in the primate prefrontal cortex (46). In primate and human cortex, staining for antibodies against the D_{1A} receptor has been found at extrasynaptic locations on spines of pyramidal cell dendrites (47). Extrasynaptic DA receptors in both the cortex and striatum may have an important role in regulating input from afferents that use excitatory amino acids as transmitters (47,48). Such extrasynaptic receptors may be important to the actions of drugs of abuse because they enhance DA output and thereby increase the overflow of DA out of the synaptic cleft (48). In contrast, the D_{1B} receptor is confined primarily to limbic regions, mainly the mammillary nuclei and hippocampus. It has little or no presence in the striatum (23,29).

DA D_2 receptors are not only expressed in high concentrations in postsynaptic neurons localized to DA terminal regions. Unlike other DA receptors, they also are located presynaptically on both axonal terminals and somatodendritic portions of dopaminergic neurons (21,23,29,41,42). Presynaptic D_2 receptors are referred to as autoreceptors (49–51). Autoreceptors are not unique to dopaminergic neurons, but they are found in all monoamine-containing systems, on cholinergic neurons, and perhaps on other chemically defined neurons. D_2-like autoreceptors localized on the soma and dendrites of DA neurons inhibit impulse flow by activating G-protein–coupled inwardly rectifying potassium channels and thereby hyperpolarizing the neurons (52,53). The D_2 DA autoreceptor exerts inhibitory feedback onto dopaminergic neurons. As concentrations of DA in the synapse increase, autoreceptors inhibit firing rates, decrease transmitter release, and inhibit transmitter synthesis. Compelling evidence that D_2-like autoreceptors are D_2 receptors comes from reports that autoreceptor functions are lost in D_2-receptor–deficient mice (54,55) but not in D_3-receptor–deficient mice (56). Not all dopaminergic neurons contain autoreceptors. In the rat, a subpopulation of dopaminergic neurons that project to the medial portions of the prefrontal cortex lacks release or impulse-modulating autoreceptors (51,57). In human and nonhuman primate brains, a larger proportion of DA neurons lacks autoreceptors. A previous report indicated that virtually all dopaminergic neurons in the ventral tegmental area (VTA) in human and monkey brains fail to express D_2 autoreceptors (58).

Both D_3 and D_4 receptors are much less abundant than D_1 and D_2 receptors; they are present in concentrations about one tenth those of the other receptors (21,23,29,41,42). In contrast to D_1 and D_2 receptors, D_3 receptors have a restricted distribution in the brain and are found primarily in mesolimbic DA terminal regions, such as amygdala, olfactory tubercle, and the shell region of the nucleus accumbens (59–61). Neurochemical studies have identified the shell of the nucleus accumbens as an important anatomic site of action for drugs of abuse (62). There also is evidence that D_3 receptors are autoreceptors on a subpopulation of dopaminergic neurons (63). The D_4 receptor is highly expressed in the frontal cortex, amygdala, hippocampus, hypothalamus, and mesencephalon (64,65). Low levels of messenger RNA (mRNA) have been found in the basal ganglia. Specific antibodies directed against the D_4 receptor have been developed, and immunohistochemical studies have revealed that, in the cerebral cortex and hippocampus, D_4 receptors are present in γ-aminobutyric acid (GABA)-ergic pyramidal and nonpyramidal neurons (66). This suggests that D_4 receptors modulate GABA-ergic transmission in these brain regions. D_4 receptors also have been found in GABA-ergic neurons of the globus pallidus, substantia nigra pars reticulata, and reticular nucleus of the thalamus (66). Substantial interest in the D_4 receptor has been prompted by the finding that the atypical antipsychotic drug clozapine is a relatively selective ligand for this receptor, and that D_4 receptors are present in a higher concentration in postmortem striatal tissue from schizophrenic patients than from controls (67,68). Several polymorphic variants of this receptor that could result in physiologic or biologic response heterogeneity have been identified (69–72). Such differences in receptor function could affect susceptibility to schizophrenia and variability in drug response.

Dopamine Transporter

It is well established that cocaine and other drugs of abuse block the reuptake of DA and other amine and indoleamine transmitters into their synaptic terminals (49). The DAT belongs to a large family of neurotransmitter and amino acid transporters that are functionally related by their requirement for extracellular Na^+ and Cl^- (73,74). DAT couples the translocation of DA (or other substrates) to the driving force of these ions down their electrochemical gradients. This transporter, which is expressed exclusively in DA neurons (75,76), plays a critical role in regulating synaptic concentrations of DA (77). DAT-mediated reaccumulation of DA into the presynaptic terminals terminates the synaptic actions of DA at its receptors, thus regulating the intensity, duration, and extent of dopaminergic neurotransmission. Hence, abnormal DAT function has a profound impact on DA-mediated presynaptic and postsynaptic signaling. The development of new ligands that are highly selective for DAT has permitted detailed study of the binding properties of the transporter (78–82) and has identified more than one binding site on the transporter. These findings suggest that different drugs that inhibit DA uptake may have differential influences on transporter functions (83).

DAT has been sequenced and cloned (84,85). It is a single subunit membrane-bound protein consisting of 12 putative membrane-spanning regions with the amino and carboxyl termini located cytosolically (86,87). Between transmembrane regions three and four, DAT has a large extracellular loop containing a number of glycosylation sites. In addition, the protein contains cytosolic consensus sites for phosphorylation, which suggests that intracellular signaling regulates transporter function. A number of studies have demonstrated that DAT function can be dynamically regulated by various drugs, substrates, receptors, and signaling systems (86). The structure of DAT is similar to that of other Na^+/Cl^--dependent neurotransmitter transporters, including the norepinephrine (NE) and serotonin [5-hydroxytryptamine (5-HT)] transporters, which suggests that these molecules form a superfamily of proteins analogous to the G-protein–linked receptor superfamily (88).

The distribution of the DAT has been studied using receptor autoradiography, *in situ* hybridization, and specific antibodies (89–92). Ligands for the DAT have been used for positron emission tomography (PET) and single photon emission computed tomography (93,94). In binding studies, the DAT is found in nearly all of the DA terminal fields. The highest concentration is found in the striatal DA terminal fields, including the ventral striatal components (nucleus accumbens and olfactory tubercle) (90, 95–97). Moderate concentrations are found in the regions containing DA neurons, including the substantia nigra, VTA, hypothalamus, and anterior cingulated cortex. Low levels of DAT are found in other DA terminal areas, such as the cerebral cortex, globus pallidus, thalamus, and amygdala. *In situ* hybridization reveals detectable amounts of DAT mRNA only in dopaminergic cell bodies (98). Consistent with the data from ligand binding studies, dopaminergic neurons in the VTA contain lower concentrations of DAT mRNA than neurons in the substantia nigra. Some regions that are known to contain DA neurons, such as the retina, olfactory bulb, and arcuate nucleus, have virtually no detectable levels of DAT mRNA. This pattern of DAT localization is consistent with that determined with immunochemical methods (75,91,99).

Advances in molecular biologic techniques have added significant insight to our understanding of the neuronal substrates involved in mediating the effects of various drugs and their involvement in reward and reinforcement. For example, the psychomotor stimulant effects of cocaine and amphetamine are completely absent in genetically altered mice in which DAT has been deleted (77). Such "knockout" mice display a number of profound neurochemical adaptations, such as elevated extracellular DA levels and spontaneous hyperlocomotion, which are consistent with the hypothesis that DAT plays a key role in regulating synaptic DA and ultimately DA-mediated behavioral responses. It was commonly believed that in addition to hyperlocomotor effects, the reinforcing and addictive properties of cocaine depended solely on its ability to block DAT and increase synaptic DA in specific brain regions (100). However, DAT knockout mice were shown to paradoxically self-administer cocaine (101). Mapping of neuronal activity in these mice suggests that, in addition to DA, serotonergic systems may play a role in the reinforcing effects of psychomotor stimulants. Support for this theory has been obtained from mice in which both DAT and the serotonin transporter have been deleted (102).

Opiate Receptors

About 30 years ago, biochemical evidence of the existence of stereospecific binding sites for opiates in the brain was reported by several laboratories (103–105). These sites bound opiate agonists and antagonists in a rank order of affinities consistent with the analgesic potencies of the drugs. In a chronic spinal dog preparation, Martin and co-workers demonstrated distinct physiologic profiles for the prototypical opiate compounds morphine, ketocyclazocine, and SKF 10,047 (*N*-allylnormetazocine). They proposed that these disparate profiles reflected selective interactions of the drugs with corresponding distinct μ-, κ-, and σ-opiate receptors (106,107). It subsequently was demonstrated that σ receptors were unaffected by naltrexone (108). Given this lack of sensitivity to naltrexone, the σ receptor was no longer classified as an opiate receptor (109). The initial confusion may have arisen from the observation that σ receptors possessed

high-affinity binding sites for drugs with psychotropic activity (reviewed in reference 110). It then became apparent that at least some of the behavioral properties of SKF 10,047 and related compounds were mediated through phencyclidine (PCP)-like action at the *N*-methyl-D-aspartate (NMDA) receptor (reviewed in reference 111), as well as μ- and κ-opioid receptors (112). Although sigma receptors are not opioid receptors, they have been implicated as having important functions outside the central nervous system (CNS). The two σ receptor subtypes, σ_1 and σ_2, have been implicated in cell differentiation and development and may be useful reagents to treat the side effects produced by antipsychotic drugs (113).

Additional pharmacologic studies led to the identification of a third opioid receptor. The endogenous enkephalin neuropeptides were more efficacious and potent at a receptor that differed from the μ and κ receptors; thus, a third opioid receptor was proposed and named the δ receptor (114). Of the endogenous opioid neuropeptides, the dynorphins interact preferentially with κ receptors (115,116), the enkephalins with δ receptors (117,118) and the endorphins with μ receptors (119,120). Two recently identified endogenous neuropeptides, endomorphin-1 and endomorphin-2, also are selective for μ receptors (121). With the exception of the endomorphins, all of the other neuropeptides are cleavage products from three larger precursor peptides: dynorphins from prodynorphin (122), enkephalins from proenkephalin A (123), and endorphins from proopiomelanocortin (124). Differences in their pharmacologic profiles led to the speculation that the three main opioid receptors could be divided further into subtypes: κ_1, κ_2, and κ_3; δ_1 and δ_2; and μ_1, μ_2, and μ_3 (for review see references 125 and 126).

The validity of classifying opioid receptors into μ, δ, and κ classes with distinct patterns of ligand selectivity, stereospecificity (127,128), and neuroanatomic distribution has been confirmed by molecular cloning of brain μ (129,130), δ (131–133), and κ (133–135) receptors. All of the cloned receptors manifest structures consistent with members of the superfamily of G-protein–coupled receptors and the subfamily of rhodopsin receptors. Thus, opioid receptors have the putative seven transmembrane regions in which the amino terminus is located extracellularly and contains multiple glycosylation sites. The third intracellular loop contains multiple amphiphatic α helixes, and the putative palmitoylation sites at the carboxy tails form the fourth intracellular loop. Among the three receptors there is 60% sequence homology: 73% to 76% in the transmembrane domains and 86% to 100% in the intracellular loops. The greatest divergent areas exist in the amino terminus (9%–10%), extracellular loops (14%–72%), and carboxy terminus (14%–20%) (136). Prototypically, opiate receptor subtypes couple to G_i/G_o. Similar to other receptors that use the G_i subfamily opioid receptors, they inhibit adenylate cyclase (137); inhibit various different ion channels, including N-type (138) and L-type (139,140) Ca^{2+} channels; stimulate inwardly rectifying potassium channels (141); increase intracellular calcium levels (142,143); and regulate the mitogen-activated protein kinases ERK1 and ERK2 (144–146).

There are a large number of opioid ligands that differ widely in their efficacy, potency, and selectivity for the opioid receptors (for review see reference 147). Naturally occurring opioid peptides lack absolute specificity for any single class of opiate receptor; therefore, their behavioral effects represent a summation of their actions at μ, δ, and κ receptors. Pharmacologic determination of the predominant receptor type governing the effects of a particular opiate can be guided by criteria, such as the sensitivity of its effects to the prototypical opiate antagonist naloxone, its ability to act as a discriminative stimulus in experimental animals trained to distinguish injections of highly subtype-specific opiates from inactive treatments, and its binding affinity in membrane preparations labeled with a highly subtype-specific radioligand (148). Opiates acting preferentially at μ or δ receptors display reinforcing properties in animal models and human subjects (149). In addition to producing analgesia, agonists at κ receptors can produce undesirable effects such as dysphoria, psychotomimetic effects, and sedation (150). κ-Receptor agonists also appear to have a lower abuse potential than μ agonists (151).

The anatomic distribution of opioid receptors has been the subject of numerous autoradiographic studies of rodent brain (152,153). These studies have shown that μ, δ, and κ receptors display sharply different distribution patterns in the CNS. High levels of μ receptors labeled with the μ-selective radioligand [D-Ala^2, methyl-Phe^4, Gly^5-ol]-enkephalin (DAMGO) occur (in approximate rank order) in striatal "patches" and "streaks," accessory olfactory bulb, nucleus accumbens, ventral subiculum and dentate gyrus of the hippocampal formation, amygdala, central gray, superior and inferior colliculi, geniculate bodies, thalamic nuclei, and substantia nigra. These observations have been confirmed by *in vitro* autoradiography of receptor-activated G proteins by DAMGO (agonist)-stimulated guanylyl 5'-[γ-[^{35}S]thiol]-triphosphate binding (154). This assay allows for assessment of functional activity combined with autoradiographic identification of receptor G-protein coupling. In contrast, high levels of δ receptors labeled with the selective δ radioligand [^{3}H][D-Pen^2,D-Pen^5]-enkephalin ([^{3}H]DPDPE) are found in nucleus accumbens, external plexiform layer of olfactory bulb, olfactory tubercle, striatum, amygdala, and layers of I–II and V–VI of cortex. The distribution pattern of κ receptors, as identified with [^{125}I]dynorphin (1–8) in the presence of specific blockers of μ and δ receptors was much different than that observed for μ and δ receptors (155). Such comparative studies revealed a narrower range of densities than did those of μ and δ receptors, with high densities observed in the tail of the striatum, hypothalamus (medial preoptic area), suprachiasmatic nucleus, globus pallidus, and

nucleus accumbens (155). More recent studies with the κ-preferring drug [^{3}H]U-69,593 confirmed the initial results. In the rat brain there is a lower concentration of κ_1 receptors compared to μ and δ receptors, and they are located in discrete brain regions, namely, the claustrum endopiriform nucleus, caudate putamen, nucleus accumbens, midline nuclear group of the thalamus, and superficial gray layer of the superior colliculus and central. In contrast, κ_2 sites are more widely distributed throughout the brain gray layer (156). However, κ_1-receptor densities are higher in the guinea pig brain than in the rat brain gray layer (156). These higher densities correlate well with the enhanced analgesic effects observed by κ agonists in guinea pigs compared to rats (156). Opiate receptors in sensory and accessory sensory brain areas presumably mediate the analgesic effects of opiate drugs, whereas limbic, extrapyramidal, and hypothalamic/neuropituitary opiate receptors may be associated with reinforcing, motor, and endocrine effects of opiates, respectively.

Nicotinic Acetylcholine Receptor

The actions of nicotine are mediated by receptors for the natural transmitter acetylcholine in the central and peripheral nervous systems. Cholinergic receptors are classified as either muscarinic or nicotinic, depending on their sensitivities to the natural alkaloids muscarine and nicotine. Nicotinic acetylcholine receptors (nAChRs) represent a class of heterogeneous receptors, including those found in the electric organ of *Torpedo* skeletal muscle, autonomic ganglia, and the brain. Unlike muscarinic receptors, which are coupled to G proteins (1), nAChRs (which give a faster response to agonists) are rapidly responding ligand gated sodium ion (Na$^+$) channels. The five subunits of the nAChR belong to a ligand gated channel superfamily that also includes the GABA$_A$ and GABA$_C$ subtypes of the GABA receptor, the glycine receptor, and the 5-HT$_3$ receptor (157). Muscle-type nAChRs are composed of α, β, γ, and δ or ϵ subunits, whereas brain nAChRs are composed of α or α and β subunits (158).

The availability of radioligands that can bind with high affinity to central nicotinic receptors has allowed mapping of the distribution of nAChR in the brain (159). Quantitative *in vitro* autoradiographic studies of [^{3}H]nicotine to slices of rat brain have shown densest labeling in the interpeduncular nucleus and medial habenula; dense labeling in thalamic nuclei, brain areas related to sensory function, and cerebral cortex; moderate labeling in the molecular layer of the dentate gyrus and the subiculum; sparse specific binding in the hypothalamus and caudate putamen; and no detectable specific binding in Ammon's horn of the hippocampus or in the periaqueductal gray matter (160).

A series of critical studies of nAChR from *Electrophorus* and *Torpedo* electric organs rendered nAChR the best characterized neurotransmitter receptor almost two decades ago

(161,162). This work was advanced by characterization of nAChR in detergent extracts (163–165) and its subsequent purification (166–169). Additional critical information was derived from knowledge of the subunit composition (170–173) and the primary structure of the subunits (171,174). Purified nAChR from *Torpedo* has been characterized as a membrane glycoprotein formed by five subunits that surround the cationic channel (161,162). Each subunit contains a long extracellular N-terminal domain, four transmembrane domains, a long intracellular loop between transmembranes 3 and 4, and a short C terminus. The second transmembrane domain of each subunit lines the channel pore. There are two binding sites per receptor for acetylcholine and other agonists (e.g., nicotine, carbamylcholine, methyl-carbamylcholine). The binding sites for these agonists (as well as competitive antagonists) are at subunit interfaces: α-γ and α-δ/ϵ. A group of structurally unrelated drugs, including mecamylamine, PCP, chlorpromazine, and local anesthetics, act as noncompetitive inhibitors by interacting at sites within the channel. Prolonged exposure to agonists or allosteric effectors causes decrements in the response amplitude of the receptor. This process is termed *desensitization.*

Neuronal nAChRs from a number of species have been revealed through molecular biologic investigations that have allowed cloning and sequence analysis of the various subunits. The genes that encode neuronal nAChR subunits have been designated α_2–α_{10} and β_2–β_4 based on the primary structure of the encoded proteins (175–181). The subunits can combine to form functional receptors, as demonstrated by injection of complementary RNAs encoding the α_2 and α_4 subunits in *Xenopus* oocytes (175,181). Furthermore, *in situ* hybridization studies have revealed that although the β_2 gene and the α_2 subunit genes, as a group, are expressed in most regions of the brain, each gene is localized to a unique but partially overlapping set of neuronal structures (182).

N-Methyl-D-Aspartate Receptor

A stereospecific brain receptor site for PCP and related drugs was demonstrated biochemically in 1979 (Fig. 32.1) (183,184). Considerable biochemical, neuroanatomic, functional, and clinical evidence indicates that this receptor represents the central target site at which PCP-type drugs initiate their unique effects (185,186). The rank order of potencies of drugs in eliciting PCP-like behavioral effects correlates highly with the potencies of the drugs in inhibiting specific binding of radioligands for the PCP receptor and in inhibiting NMDA receptor–mediated increases in calcium (Ca^{2+}) conductances in electrophysiologic assays (187). The PCP receptor is located within the ion channel gated by the NMDA class of glutamate receptors. When PCP-type ligands bind, they block the channel and act as uncompetitive NMDA receptor antagonists.

Glutamate is the primary excitatory amino acid neurotransmitter in the CNS. Glutamatergic neurotransmission is mediated by both metabotropic and ionotropic receptors (for review see reference 188). The ionotropic receptors can be subdivided by agonist preference into two groups: NMDA receptors and non-NMDA receptors [α-amino-3-hydroxy-5-methyl-4-isoxazole-propionate (AMPA) and kainate receptors]. The NMDA receptor requires the co-agonist glycine (189) and an endogenous excitatory amino acid for activation (190,191). The NMDA receptor possesses unusual characteristics of strong voltage dependence (open channel block by Mg^{2+}). Although it is

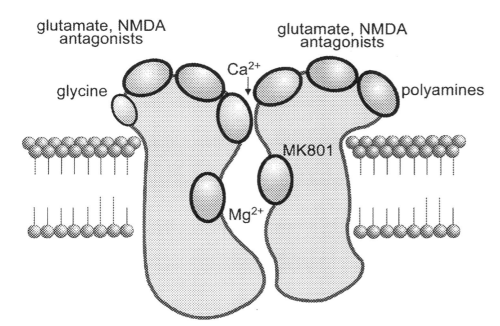

FIGURE 32.1. Schematic diagram of the *N*-methyl-D-aspartate (NMDA) receptor. The receptor is a heteromeric complex that is highly permeant to cations. Native functional receptors require the presence of the NR1 subunit, which contains the binding site for the co-agonist glycine, in combination with the NR2/NR3 subunits. The NR2 subunit contains the binding site for excitatory amino acids. Competitive antagonists also can bind to these two sites. Sites for various modulators, including protons, polyamines, and Zn^{2+}, exist on the channel and, when bound, lead to further modulation of receptor function. Phenylcyclidine and Mg^{2+} bind within the channel to block the influx of Ca^{2+}.

permeant to both Na^{2+} and Ca^{2+} ions, permeability for Ca^{2+} is ten-fold greater than that for Na^{2+} (192).

Identification and molecular cloning of the NMDA receptor has identified the receptor as a multimeric assembly of at least two to three subunits named NMDA NR1 (ζ subunits, mouse homologue); NMDA NR2A-D (ϵ subunits, mouse homologue) (193–197); and NMDA NR3, which is composed of NR3A (198–200) and NR3B subunits (201). Eight functional splice variants of NR1 arise from the splicing in or out of a 21-amino-acid insert at the N terminus or two adjacent deletions (37 and 38 amino acids, one of each or both combined) at the C terminus (194,202). The NR3 subunits are believed to coassemble with the NR1 and NR2 subunits *in vivo* (200,201,203). Native receptors are believed to be heterotetramers (204) and appear to be composed of subunits from the three gene families. Electrophysiologic studies report that the presence of NR3 subunits combined with the other two subunits produces a dominant, negative effect in that whole-cell currents are attenuated and single-channel conductance is decreased (200,205). There is some controversy that NR1 subunits can self-assemble to form functional channels in *Xenopus* oocytes (193). However, these functional receptors have a different physiology than native receptors, and they may be formed by the combination of the transiently expressed NR1 with an endogenous oocyte NR2-like protein UR-1 (206).

In adult rats, mRNA that encodes the NR1 subunit is ubiquitously expressed in neuronal cells throughout the CNS (195,207). Studies measuring mRNA levels have indicated that the NR1 splice variants are differentially expressed in different regions of the CNS (207). The dominant variant in cortex and hippocampus is NR1a (splice variants without the N-terminal cassettes), which represents 85% to 90% of total NR1 mRNA (207,208). In the cerebellum, NR1a and NR1b variants are expressed equally (207). The studies also showed that the NR1-1 variants that contain both the C1- and C2-terminal cassettes were dominantly expressed in cortex and hippocampus but were lower in cerebellum. NR1–4 variants that lack both the C1- and C2-terminal cassettes were highly expressed in the cerebellum (208). However, Zong et al. (207) reported that the hippocampus and cortex also contain significant amounts of the NR1–4 variants. These changes in splice variant expression in different brain regions may underlie some of the changes in the functional and pharmacologic properties of the NMDA receptor. It has been suggested that the C1-terminal cassettes are involved in cytoskeletal association (209) and endoplasmic reticulum retention of the NMDA receptors (210). Furthermore, NR1 splice variant studies suggest that the selective functions of the NMDA receptor and its postsynaptic location may be based on the particular NR1 splice variants present in the complex.

The four NR2 subunits (NR2A–D) play a major role in determining the function and pharmacology of the NMDA receptor complex. The NR2 receptor subunits are 55% to 70% identical in sequence. NMDA NR2 subunits are differentially expressed in various brain regions as determined by *in situ* hybridization and histo-blot analysis (211,212). The NR2A subunit is distributed throughout the adult brain, including the forebrain, cerebellum, and thalamus. The NR2B subunit is expressed mainly in the forebrain. NR2C and NR2D are expressed primarily in the adult cerebellum and brainstem, respectively (211,212).

The NR3 subunits have a much different distribution pattern than the NR1 and NR2 subunits. Although NR3A is ubiquitously expressed throughout the CNS at birth, it is found only in discrete nuclei in the thalamus, amygdala, and nucleus of the lateral olfactory tract in the adult rodent brain (198,199). The NR3B subunit is even more limited in expression; it has been identified only in somatic motor neurons of the brainstem and spinal cord (201). Both NR3 subunits are present in primary cultured cerebrocortical neurons (200,203).

Observations from behavioral studies provide strong support that the site of action of PCP in the CNS is through the NMDA receptor (213–215). These behavioral experimental paradigms include disruption of prepulse inhibition of the startle reflex in rats by PCP and other NMDA receptor antagonists (213); impairment of latent learning in mice by PCP administration (214); and the observation that NMDA receptor antagonists can substitute for PCP in drug discrimination tests (215).

γ-Aminobutyric Acid Receptor

A molecular target of benzodiazepines, ethanol, barbiturates, and other general anesthetics, the GABA ($GABA_A$) receptor is an inhibitory, ligand gated, receptor-channel complex that has significant sequence and structural homology to the nAChR. The native $GABA_A$ receptor likely is composed of α, β, and γ subunits; minor subunits are the δ, ϵ, and π subunits. Because each subunit exists in multiple isoforms, there are several thousand possible heteropentameric complexes of the $GABA_A$ receptor; however, the theoretical number is limited by the brain regional tissue distribution of the subunits and their subtypes (216,217).

The $GABA_A$ complex is composed of multiple functional domains (Fig. 32.2). The GABA recognition site binds up to two molecules of agonist, resulting in conformational changes that permit Cl^- ion flux through the channel (218). Ligand binding to the benzodiazepine recognition site enhances the affinity of GABA for its own recognition site and thus alters the sensitivity of the entire complex to agonist. Ligands for the benzodiazepine site can be divided into three groups based on their effects. Clinically used anxiolytic, sedative, and anticonvulsant benzodiazepines (such as diazepam), which are classified as *agonists* of the benzodiazepine binding site, increase the sensitivity of the complex to GABA. *Inverse agonists,* such as the β-carbolines, decrease the sensitivity of the complex to GABA and thus exert anxiogenic effects

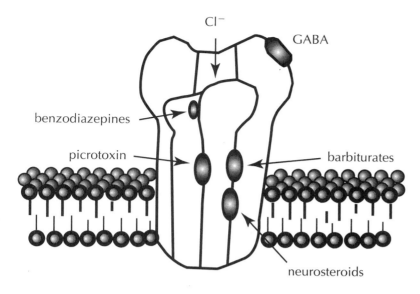

FIGURE 32.2. Schematic diagram of the γ-amino-butyric acid (GABA)-benzodiazepine receptor. The receptor is a polymeric complex that gates a Cl^- channel. It bears binding sites for GABA and agonistic drugs, such as muscimol, and for a variety of drugs (benzodiazepines, barbiturates, and convulsants such as picrotoxin). In addition, the activity of the receptor is modulated through the actions of endogenous neurosteroids, such as pregnenolone sulfate, dehydroepiandrosterone sulfate, and desoxycorticosterone sulfate.

(219). *Antagonists,* such as flumazenil, lack a significant intrinsic effect, but they block the effects of benzodiazepine agonists or of inverse agonists (220).

Barbiturates also enhance GABA$_A$ receptor function. They act at a domain of the complex independent of the benzodiazepine binding site. At lower concentrations barbiturates act primarily by enhancing the effects of GABA; at higher concentrations they may activate the channel even in the absence of GABA (221). Other general anesthetics, including volatile anesthetics such as sevoflurane, and intravenously administered anesthetics such as propofol also enhance GABA-mediated currents. In contrast with barbiturates, these drugs do not activate GABA$_A$ receptors in the absence of GABA (222). Ethanol, which is categorized as a general anesthetic, also interacts with the GABA$_A$ receptor. In CNS-derived tissues and in cells that express recombinant receptors, ethanol potentiates GABA-activated chloride (Cl^-) currents in a concentration-dependent manner (223).

The neurosteroids dehydroepiandrosterone sulfate (224,225) and pregnenolone sulfate (226) act as negative allosteric modulators of GABA$_A$ receptor function, whereas tetrahydroprogesterone and tetrahydrodeoxycorticosterone manifest allosteric GABA-agonistic features (227,228). These observations suggest that these steroids may serve as endogenous regulators of GABA$_A$ receptor function. The rate and type of neurosteroidogenesis may be regulated in part by binding of an endogenous peptide, termed the *diazepam binding inhibitor,* to mitochondrial benzodiazepine receptors on glial cells (229).

Cannabinoid Receptor

The active principle of marijuana, Δ^9-tetrahydrocannabinol (Δ^9-THC) (230), proved to be a challenging probe for a CNS recognition site because of its extreme hydrophobicity (231). Identification of the cannabinoid receptor was fa-

cilitated by demonstration of cannabinoid enantioselectivity (232) and by synthesis of a series of novel cannabinoids (233). These developments made it possible to demonstrate that cannabinoids inhibit adenylate cyclase activity in rank order that correlates with their physiologic effects (234,235). One of the new cannabinoids was found to bind to a unique site that displayed biochemical and neuroanatomic characteristics consistent with those of a receptor coupled to a G protein (236–238). A complementary deoxyribonucleic acid (DNA) isolated from a rat cerebral cortex library and later named cannabinoid receptor 1 (CB1) was found to encode a 473-amino-acid protein structurally belonging to the family of G-protein–coupled receptors (239). The CB1 receptors are located in both the central and peripheral nervous systems. A second cannabinoid receptor, CB2, was cloned and discovered to be localized in non-nervous tissues, primarily immune cells (240).

Although the euphoric and sedative/hypnotic effects of marijuana are the most well known, the identification of endogenous cannabinoid ligands raised the issue of the function of cannabinoid receptors in the nervous system. To date, two endogenous ligands for CB1 receptors in the brain have been identified. A novel derivative of arachidonic acid, arachidonylethanolamide (anandamide), was isolated from porcine brain (241) and was proven to exert cannabinoid-like behavioral effects (242). Another endogenous ligand, 2-arachidonoylglycerol, is present in larger concentrations in the brain than is anandamide (243). In addition to its actions at CB1 receptors, anandamide is an agonist at capsaicin and resiniferatoxin (vanilloid) receptors, which play a role in pain transmission (244). Anandamide is transported into neurons and astrocytes via a membrane-bound cannabinoid transporter, and it activates vanilloid receptors intracellularly (245). These actions demonstrate the importance of endogenous cannabinoid receptor ligands in mediating pain transmission. The endogenous cannabinoid system also affects memory, cognition, appetite, vomiting, and movement (245).

THEORIES OF ADDICTION AND THE SELF-ADMINISTRATION PARADIGM

Until the mid-1950's, it was widely assumed that addictive behavior was uniquely human and was defined primarily by drug withdrawal syndromes. According to this physical dependence model, drug abusers were driven by predisposing internal states and were soon trapped in a cycle of self-administration, tolerance, physical dependence, and readministration. The drug abuser was believed to be motivated primarily by the desire to avoid the negative consequences of withdrawal. By the 1930's, it had been found that monkeys could be made physically dependent if they were forced to consume a drug, but the monkeys had never been observed to make themselves physically dependent simply as a result of free access to the drug (246). It was believed that animals were not capable of learning the association between an injection and subsequent relief (15 to 20 minutes later) from withdrawal symptoms (247). These findings and interpretation resulted in considerable skepticism about the possibility of developing an animal model of drug abuse.

During the 1950's, researchers began to find evidence that laboratory animals would learn to perform behavior that resulted in drug injections. This work was boosted by the development of technology that allowed intravenous delivery of drug infusions to the freely moving animal. Reflecting the influence of the physical dependence doctrine, early studies of drug self-administration in animals were performed in subjects that had been made physically dependent by repeated injections of morphine prior to experimentation. The animals were placed in an operant chamber and given the opportunity to self-administer morphine via an intravenous catheter. They quickly learned to self-administer morphine under these conditions, and it soon became apparent that they were responding to the drug infusion as if it were a positive reinforcer (248). In subsequent studies, animals self-administered doses of morphine that were too low to produce physical dependence (249). They also self-administered stimulants, such as cocaine, but no obvious withdrawal syndrome upon cessation of stimulant use had been described (250,251). Further studies demonstrated that animals even self-administered certain abused substances without any previous experience of having received the drug (252). Studies such as these had an immense impact on the concept of motivation underlying drug abuse in humans. They resulted in a shift in emphasis away from the physical dependence model toward the view that drug abuse is largely driven by the motivation to seek and experience the incentive (rewarding or reinforcing) properties of abused drugs.

In addition to the roles played by negative and positive primary reinforcement in drug addiction, secondary reinforce-

ment has been found to play a significant role. Secondary reinforcers or incentives are otherwise neutral stimuli present in the environment that become associated with drug taking after repeated administration through a process of incentive learning. Once acquired, secondary reinforcement can become an extremely powerful source of motivation for drug-seeking behavior. The recent incentive-sensitization theory posits that after incentive learning has occurred, "wanting" a drug (drug craving) can become more important than "liking" a drug (drug reward) in determining drug-seeking behavior (253). The authors of this theory cite the example of cigarette smoking in experienced smokers, in whom the intensity of cravings for cigarettes (produced by abstinence and/or the presence of secondary reinforcers) appears to be out of proportion to the reward produced by smoking. They also cite a report that low doses of opiates, which produce no subjective effects, can maintain a response in human volunteers with histories of intravenous opiate abuse (254) as evidence that drug taking can become independent of positive subjective effects. Currently, addiction is defined as a behavior that ultimately leads to harmful consequences over which an individual has impaired control (255,256). Individuals are aware that a particular behavior is harmful to them or those they care about, but they find themselves unable to halt the behavior even when they try to do so (257). Based on this definition, it is clear that the medical, social, and psychological harm caused by addiction warrants it to be considered a psychological disorder (258).

Brain Reward Systems

One of the most basic principles of behavior is the *law of effect,* which states that behaviors that lead to a desired outcome are repeated (259,260). These outcomes are referred to as *reinforcers.* A precise definition of the term *reinforcer* has been difficult to formulate and is beyond the scope of this review. In general, however, a reinforcer is an event that facilitates learning. In the literature on animal research, reinforcers often are events or objects that satisfy biologic needs. For humans, reinforcers also can be objects (rewards) that produce pleasurable (hedonic) subjective experiences.

The terms *reward* and *hedonic,* which is used by physiologic psychologists, are not synonymous with the term *reinforcement.* For example, when animals are trained on a conditioned emotional response in which they acquire an association between a tone and a withdrawal response produced by mild footshock, the posttraining noncontingent consumption of sucrose (261) or a brief application of intense footshock (262) has identical memory-improving effects. Thus, both rewarding and aversive events can produce reinforcement. Rewarding events can produce positive reinforcement, and aversive events can produce negative reinforcement. Furthermore, in the paradigm just outlined, memory is not improved by posttraining consumption of saccharine solutions in concentrations that are equally preferred to those of retention-improving sucrose solutions (261,263). Thus, equally rewarding events do not necessarily produce the same degree of reinforcement, and the property of being rewarding is neither necessary nor sufficient for reinforcement to occur. The difference between reinforcement and reward, and the more subtle distinction between positive reinforcement and reward, has resulted in some confusion in the literature on the neuroanatomic and neurochemical substrates of these phenomena.

The notion that the brain contains circuits that mediate reward or pleasure is central to our current understanding of the neural basis of drug abuse. Identification of the specific neural systems involved in reward originated with the work of Olds and Milner (264), who discovered that laboratory animals avidly self-administered electrical stimulation delivered via electrodes to specific deep regions of the brain. This finding was of great significance for several reasons. First, this apparently rewarding brain stimulation acted as a positive reinforcer that could be used to selectively strengthen any behavior contingently linked to it. Second, only a limited number of sites were found to support brain stimulation reward (265). This observation suggested that there are anatomically specific circuits in the brain that mediate reward or pleasure (266). Third, it subsequently was found that rewarding brain stimulation could elicit natural consummatory responses (267–271). Thus, rewarding electrical stimulation appeared to activate the same neural systems that are involved in natural reward and motivation. Only a few years after the original discovery of brain stimulation reward, it was found that drugs that are abused by humans facilitated brain stimulation reward when the drugs were administered to animals (272). This finding suggested that drugs of abuse influence the same brain reward system that is activated by electrical stimulation or by natural rewards. The acute enhancement of brain reward mechanisms by abusable drugs now is considered to constitute their single most important property and provides the basis for the hypothesis that abused substances produce reward by acting at specific neural systems. The remainder of this section reviews evidence on the neuroanatomy of brain reward systems derived from two main paradigms: self-administration and electrical brain stimulation reward.

Self-Administration Paradigm

The self-administration procedure generally uses the intravenous route of administration, although many other routes, including oral, intramuscular, and intracerebral, have been used. Animals that are surgically implanted with chronic indwelling catheters are placed in operant chambers and given the opportunity to press a lever that causes delivery of the drug through the catheter. Typically rhesus monkeys and rats are used, although many other species, such as squirrel monkeys, baboons, dogs, cats, pigs, and mice, have been used successfully (273). Such operant conditioning proce-

dures have demonstrated that abusable drugs have positive reinforcing properties that are very similar to those of natural rewards, such as food and water. Unlike natural rewards, however, the drugs can have nonspecific effects on motor activity and, under certain reinforcement schedules, can accumulate in the body and cause changes in responding over time.

The most striking feature of drug self-administration is how closely the classes of drugs that are self-administered by animals correspond with those that are abused by humans. With ratio schedules of operant reinforcement, laboratory animals have been observed to respond for amphetamine (274–278), cocaine (274,279–281), methamphetamine (275,278,282,283), methylphenidate (284,285), nicotine (273,286–289), caffeine (290), opiates (291–293), ethanol (294–297), barbiturates (298–300), benzodiazepines (301,302), PCP (279,303–306), and cannabinoid receptor agonists (307). With interval schedules of reinforcement, responding for cocaine (247), opiates (308), ethanol (309), barbiturates (298,310,311), and other abusable substances has been demonstrated. The list of drugs that are self-administered is remarkable because it is composed of compounds that differ markedly in their chemical structure and spectra of pharmacologic activity. Perhaps even more significant is the observation that laboratory animals will not self-administer drugs from the myriad of other classes available. Many drugs, including opioid antagonists, neuroleptics, and tricyclic antidepressants, either do not maintain self-administration in animals (312,313) or stimulate voluntary termination of drug infusion (314,315). These results suggest that drugs that are self-administered must share some underlying commonalities that distinguish them from the much larger number of drugs that are not self-administered.

Furthermore, drugs that have not been found to support self-administration in animals are, almost without exception, not abused by humans. Lysergic acid diethylamide (LSD) is one of the few exceptions. It is not self-administered by laboratory animals; in fact, it is actively avoided by rhesus monkeys (273,314). This disparity between animal and human drug self-administration may be partly explained by a peculiarly human urge to seek altered states of consciousness. The fact that most drugs, other than hallucinogens, that are abused by humans are also self-administered by animals provides general support to the hypothesis that drug self-administration in animals is a reliable predictor of potential abuse liability in humans.

Intravenous drug self-administration behavior in animals is maintained across a wide range of doses (316). Although specific patterns of intake vary with drug class, generally the rate of drug self-administration is inversely related to the dose or duration of the drug effect. Low-to-intermediate drug doses increase the rate of responding, but higher doses decrease response rate. It has been a long-standing assertion that this pattern of responding results from attempts by the subject to titrate blood levels of the drug (317). According to this hypothesis, when the injection dose is increased, the concentration of drug in the blood declines at a slower rate and results in longer interinjection intervals and decreased rates of responding. Conversely, when the injection dose is decreased, the level of drug in the blood falls more rapidly and results in shorter interinjection intervals and increased rates of responding.

Neural Substrates of Drug Self-Administration Reward

Systemic Pharmacologic Challenge

Pharmacologic challenge is one approach that has been used in an attempt to elucidate the neurochemical substrates of drug-induced reward. This technique involves the noncontingent coadministration of an agonist or antagonist of a particular neurotransmitter and examination of the effect of this manipulation on drug self-administration responding. The rationale for challenge with an agonist of a particular neurotransmitter is that if the neurotransmitter facilitates reward, the agonist should substitute for the self-administered drug and temporarily decrease self-administration of the drug. Conversely, challenge with an antagonist should produce a selective increase in responding to compensate for the reduced effectiveness of the self-administered drug. However, higher doses would further reduce rewarding efficacy of the self-administered drug and therefore would reduce responding.

Examples of such pharmacologic challenge paradigms can be drawn from studies of cocaine self-administration. For example, treatment with bromocriptine, a DA receptor agonist, decreased cocaine self-administration in the rat (318). In contrast, pretreatment with low doses of a variety of DA receptor antagonists, including pimozide (319,320), α-flupenthixol (320,321), haloperidol (320), perphenazine (322), and chlorpromazine (320), increased the rate of responding for intravenous cocaine by rats. When a high dose of a DA receptor antagonist was given, the rate of responding was reduced, which suggests that the reinforcing effects of cocaine were attenuated (319). In subsequent experiments, SCH 23390 and spiperone, which are antagonists that are relatively selective for D_1-like and D_2-like DA receptors, respectively, increased the rate of responding for cocaine on a fixed-ratio schedule of reinforcement at doses up to 10 mg/kg and reduced the rate at higher doses (323). In the same study, both drugs decreased the highest ratio completed in rats self-administering cocaine on a progressive ratio. These findings suggested that both D_1-like and D_2-like DA receptors contribute to the reinforcing effects of cocaine.

More recently, similar pharmacologic approaches have been used to implicate the DA D_3 receptor in cocaine self-administration (324). As outlined in an earlier section, D_3 receptors have restricted localization in the brain and are found primarily in the shell of the nucleus accumbens (325–327), which is a region known to be an important site of action

for drugs of abuse (328). In rats trained to self-administer intravenous cocaine, pretreatment with D_3 receptor agonists decreased cocaine self administration (329,330), which suggests that D_3 receptor agonists enhance the reinforcing properties of cocaine. In addition, the potency of these agonists to decrease cocaine self-administration is correlated with their potency at D_3, but not D_2, receptors (329). In contrast, D_3 receptor antagonists increase cocaine self-administration with a potency consistent with D_3 receptor blockade (329). These findings suggest that D_3 receptor activation enhances the reinforcing effects of cocaine.

One major problem encountered with pharmacologic challenges of self-administration is related to the nonspecific pharmacologic effects of the coadministered drug. A compound might affect the ability of an animal to press a lever or bar for drug injection by stimulating or inhibiting motor activity generally, rather than by selectively increasing or reducing the rewarding properties of the self-administered drug. Confounding by such nonspecific effects must be guarded against by ensuring that the dose of the drug being used to challenge self-administration responding has minimal effects on performance.

Lesion Studies

Another approach has been used to map the neuroanatomic and neurochemical substrates of reward associated with systemic self-administration of various abusable drugs. This method examines the effect of selective lesions of specific neurotransmitter systems on drug responding. Particular neurotransmitter systems can be rendered dysfunctional by surgery (e.g., knife cuts), electrolytic lesions, thermocoagulative lesions, or treatment with neurotransmitter-specific toxins, such as 6-hydroxydopamine (6-OHDA), kainate, quinolinate, and ibotenate. Effects of the loss of contribution of a particular neurotransmitter system on drug self-administration then can be investigated.

Lesion studies have provided considerable information about the localization of central sites that are important to drug-induced reward. Such studies have demonstrated that the mesolimbic DA system plays a critical role (Fig. 32.3). The cell bodies of this system originate in the VTA. They project rostrally to the nucleus accumbens and to more anterior regions, including the amygdala, olfactory tubercle, and prefrontal cortex. Selective lesions of the nucleus accumbens, produced by the catecholamine-specific neurotoxin 6-OHDA, disrupt self-administration of cocaine (331–334) and amphetamine (335). Both drugs act presynaptically to augment concentrations of intrasynaptic DA. Similarly, selective 6-OHDA lesions of the VTA disrupt self-administration of cocaine (336,337). In contrast, self-administration of the direct DA receptor agonist apomorphine, which acts at postsynaptic sites in the mesolimbic DA system, is unaffected by such lesions (331,336). In general, 6-OHDA lesions of other brain regions have not been found to affect cocaine self-administration (331,332,338).

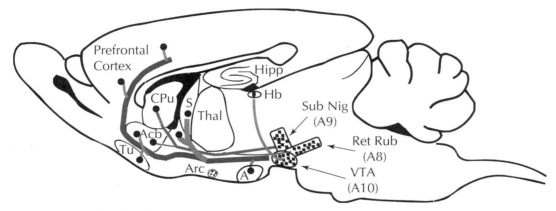

FIGURE 32.3. Chemical neuroanatomy of rat dopamine (DA) systems relevant to drug abuse. Cell bodies of DA-containing neurons are indicated by *dotted regions*. The mesocorticolimbic dopaminergic pathway *(heavy stippling)* has been specifically implicated in the rewarding properties of drugs. The mesocorticolimbic DA system consists of projections from dopaminergic neurons found mainly in the ventral tegmental area (VTA; A10) to the nucleus accumbens (Acb), olfactory tubercle (Tu), bed nucleus of the stria terminalis and septum (S), amygdala (A), lateral habenula (Hb), and portions of the prefrontal cortex. The nigrostriatal system *(light stippling)* consists of dopaminergic neurons in the substantia nigra (A9) that project primarily to the dorsal portions of the caudate and putamen (CPu). DA neurons in the retrorubral (Ret Rub) region (A8) contribute to both systems. The tuberoinfundibular DA neurons (A12) in the arcuate nucleus (Arc) are of interest because of their regulation of neuroendocrine function, such as inhibition of prolactin release. (Modified from Cooper JR, Bloom FE, Roth RH, eds. *The biochemical basis of neuropharmacology*, 6th ed. New York: Oxford University Press, 1991.)

It has been reported that 6-OHDA lesions of the prefrontal cortex had no effect on self-administration at higher doses of cocaine. At lower doses, however, significant increases in the number of responses made to obtain the last reinforcement of a session were observed, which suggests that dopaminergic innervation of the prefrontal cortex may have a role in cocaine self-administration (339). Whether these findings are contingent on accumbens DA function cannot be ruled out.

Interpretation of some of the earlier of these findings (331) was complicated because 6-OHDA produces lesions of both DA and NE systems. Later studies that used pretreatment with desmethylimipramine to prevent depletion of NE found similar reductions in cocaine self-administration (332).

Another concern is related to the anatomic specificity of 6-OHDA lesions of the nucleus accumbens, because such lesions deplete DA and NE in more anterior structures. This issue was addressed by examining the effect of kainic acid lesion of the nucleus accumbens on cocaine self-administration. Kainic acid selectively destroys cell bodies but preserves the fibers of passage to, in this case, more anterior regions, such as the olfactory tubercle, amygdala, and prefrontal cortex. Placement of kainic acid lesions in the nucleus accumbens significantly reduced responding for cocaine but did not affect catecholamine levels in more anterior structures (336). Thus, studies using kainic acid and those using 6-OHDA lesions provided compelling evidence that the nucleus accumbens plays a critical role in self-administration of psychomotor stimulants by laboratory mammals.

Lesion studies have demonstrated that the mesolimbic DA system is critical for self-administration of other classes of abusable drugs. Selective 6-OHDA lesions of this system have been found to disrupt self-administration of heroin and morphine (340–343). Furthermore, selective lesions of the core and shell subregions of the nucleus accumbens using the excitotoxins quinolinate and ibotenate, respectively, demonstrated anatomic specificity within this brain region (344). Lesions of the core but not of the shell reduced levels of responding during heroin self-administration (345,346). These results suggest that the core region of the nucleus accumbens plays an important role in the acquisition of heroin-seeking behavior. However, other studies have shown that 6-OHDA lesions of accumbens had no effect on initiation of heroin self-administration or on lever pressing preceding the behavior test session (334). Clearly, the mesolimbic DA system appears to comprise a common neuronal substrate that mediates both opioid and psychomotor stimulant reward (347–349); however, this system may not be solely responsible for the general motivational aspects of drug self-administration (334).

The ventral pallidum appears to play a role in mediating reward produced by self-administration of cocaine as well as opioids. This region receives afferent projections from the nucleus accumbens (350). Lesions produced by injections of ibotenic acid, which, like kainic acid, is an axon-sparing excitotoxin, into the ventral pallidum of rats reduce cocaine and heroin self-administration behavior maintained on a fixed-ratio schedule of reinforcement (351,352). At the same time, they decrease the highest ratio sustained in progressive-ratio procedures. These findings suggest that a circuit involving the nucleus accumbens and ventral pallidum may be a common pathway in stimulant and opioid reward.

Intracranial Self-Administration of Drugs

Challenge of systemic self-administration of drugs by coadministration of selective agonists and antagonists for particular receptors, or by production of selective lesions, has considerably advanced our knowledge of the neural bases of reward. Nonetheless, both of these techniques are inherently indirect. A more direct method of exploring the neural systems underlying self-administration reward involves microinjection of abusable substances into the brain itself. The rationale behind this technique is straightforward. If there are discrete brain circuits that mediate drug reward, then animals should self-administer abusable drugs directly into these regions of the brain, and not into other regions that are not involved in reward. In practice, however, many methodologic problems are encountered. Some of these problems are associated with slow infusion rates of drugs and minute injection volumes; both of these conditions are necessary to minimize nonspecific neuronal responses. Furthermore, problems similar to those encountered with systemic self-administration of drugs arise; therefore, it is not always possible to conclude that a brain region is not involved in reward simply because it does not support self-administration. It is possible that the dose of drug in question is activating competing behaviors that can inhibit responding and mask the reward behavior.

The results of intracranial self-administration studies suggesting that the nucleus accumbens is an important anatomic substrate for drugs of abuse are, in general, consistent with those from lesion studies. For example, animals will voluntarily self-administer microinjections of amphetamine into the nucleus accumbens and prefrontal cortex (orbitofrontal cortex of the rhesus monkey) (353–355) but not into other brain regions. In addition, intracranial self-administration studies have provided evidence for the involvement of opioid (356), NMDA (357), muscarinic (358), and neuropeptide Y (356) receptors within the nucleus accumbens in reinforcement behaviors. Cocaine is voluntarily self-administered into the prefrontal cortex (359–361). Morphine is self-administered into the VTA (362–365), lateral hypothalamus (366,367), and nucleus accumbens (356,367,368), all of which are either nuclei or terminal projection regions of the mesolimbic DA system. Other synthetic and endogenous opioids also are self-administered intracranially into the mesolimbic system. Fentanyl is self-administered into the VTA (369), metenkephalin into the nucleus accumbens (370), and the metenkephalin analogue D-ala^2-metenkephalinamide into the lateral hypothalamus (371).

A variant of the intracranial microinjection procedure has provided further evidence that the mesolimbic DA system plays a critical role in opioid self-administration, although the periaqueductal gray also appears to be involved. In these studies, intravenous self-administration of opioids was challenged with an intracranially administered quaternary opioid antagonist. Microinjection of diallyl-normorphinium bromide into the VTA, but not into other brain regions, produced dose-dependent increases in rates of responding for heroin self-administration (372). In contrast, administration of a low dose of methylnaloxonium chloride into the nucleus accumbens increased heroin self-administration, whereas injections into the VTA were not effective (373). These results support a role of the mesolimbic DA system in opioid reward, but they imply that opioid receptors in the VTA are not essential in this regard. The first evidence that opiate receptors in the vicinity of the periaqueductal gray may play a role in intravenous self-administration of opioid drugs derives from the observation that intracranial microinjections of methylnaltrexone into this brain region produced dose-related increases in responding for heroin, but not for cocaine, by rats on a continuous reinforcement schedule (374). In the same study, microinjections of methylnaltrexone in the nucleus accumbens also increased responding for heroin, thus confirming the involvement of this DA terminal region in opiate self-administration.

Self-Administration of Direct and Indirect Dopaminergic Agonists

There is now a considerable body of evidence that implicates DA as a critical neurochemical substrate of drug self-administration reward. The most obvious is that the psychomotor stimulants amphetamine and cocaine, which are indirect DA agonists that prolong the actions of DA in the synaptic cleft, are avidly self-administered by laboratory animals. Further support for the latter feature of this theory is provided by reports that laboratory animals also avidly self-administer direct DA receptor agonists, such as apomorphine and piribedil (375–379). In addition, the intracranial self-administration of selective D_1 and D_2 agonists has been used to demonstrate that coactivation of D_1 and D_2 receptors in the nucleus accumbens is reinforcing (380). Neither the D_1 agonist SKF 38393 nor the D_2 agonist (−)-quinpirole was self-administered into the shell region of the accumbens. However, a combination of each agonist was self-administered into this brain region. These DA agonists alone or in combination were not self-administered in the core region of the accumbens. Coadministration of either SCH 23390, a D_1 antagonist, or (−)-sulpiride, a D_2 antagonist, blocked the self-administration of the DA agonists into the shell, which suggests that coactivation of D_1 and D_2 receptor subtypes had cooperative effects on DA-mediated reward. In addition, DA reuptake blockers, such as bupropion, mazindol, nomifensine, GBR 12909, and RTI 111, which prolong the action of DA in the synaptic cleft, are readily self-administered (322,381–390).

It has been known for over 2 decades that cocaine blocks the reuptake of DA in the brain (391). Although cocaine also blocks the reuptake of NE and serotonin, the potencies of a series of cocaine analogues in self-administration studies in monkeys were highly correlated with potencies of the same compounds for inhibiting binding to uptake sites for DA but not to NE or 5-HT uptake sites (78). This observation indicated that the rewarding properties of cocaine were due to interaction of the drug with the DAT. However, recent studies with DAT knockout mice demonstrate that multiple neurotransmitter systems may be involved in some of the behavioral actions of psychomotor stimulants.

Self-Administration of Nicotine

In addition to the evidence implicating dopaminergic neurotransmission in the self-administration of direct and indirect dopaminergic agonists, studies of nicotine self-administration provide support for the view that nicotine-induced reward requires intact functioning of the mesolimbic DA system (392–396). Rats given D_1- and D_2-selective DA receptor antagonists have reduced self-administration of nicotine (397,398). In selective lesioning studies, rats subjected to injections of 6-OHDA into the nucleus accumbens show reductions of DA in the nucleus accumbens, caudate putamen, and olfactory tubercle, compared with levels in sham-injected control rats (399,400). When lesioned in this manner, rats showed reduced self-administration of nicotine over a 10-day acquisition period in one study (399) and over a 3-week period of testing in a continuous reinforcement schedule in other studies (397,400).

Neurochemical evidence supports the feasibility of a role of mesolimbic DA in reinforcement due to nicotine. Nicotinic receptors are present in the midbrain area containing the substantia nigra and the VTA (160) and in the terminal fields (nucleus accumbens, olfactory tubercle) of the mesolimbic DA system (182,401–403). Systemic nicotine increases the firing rate of neurons in the VTA, and intracellular recording from presumed DA-containing cells of the VTA *in vitro* reveals that nicotine directly depolarizes the neurons (404–406). Likewise, studies on intact rats demonstrate that systemic nicotine stimulates the regional cerebral metabolic rate for glucose (rCMRglc), which is an index of local brain function, in the VTA (407). Nicotine stimulates DA release from slices of the nucleus accumbens (408) and increases the concentration of extracellular DA in the nucleus accumbens as measured by *in vivo* microdialysis (409–411).

Electrical Brain Stimulation Reward Paradigm

Since the seminal discovery by Olds and Milner (264) that laboratory animals will self-administer electrical stimulation to specific brain regions, the electrical brain stimulation reward paradigm and the relationships between the type of reward it produces and drug and natural rewards have been well researched. The technique of intracranial self-stimulation (ICSS) involves surgically implanting chronic indwelling stimulating electrodes into specific brain nuclei of animals and training the animals to self-administer the rewarding electrical stimulation by pressing a bar or lever. Generally, animals rapidly learn to press the lever for brain stimulation, with very little prompting, and then maintain extremely high levels of responding. The reward derived from the electrical stimulation of certain brain regions can be sufficiently powerful that animals ignore natural rewards or endure pain to receive it.

Human studies of the effects of electrical stimulation of the brain have confirmed that stimulation of certain brain regions can produce intense feelings of euphoria (412). Thus, the reward produced by electrical self-stimulation of the brain is an extremely potent phenomenon that is matched only by that produced by the most avidly self-administered drugs.

A notable common feature of drugs such as opiates, stimulants, barbiturates, benzodiazepines, PCP, alcohol, and marijuana, which are self-administered systemically by laboratory animals, is that they enhance brain stimulation reward (347,348,413–415). When coadministered, abusable substances from different pharmacologic classes have synergistic effects on brain stimulation reward thresholds (416,417). These findings provide support for the hypothesis that abused drugs activate the same reward system that is activated by electrical stimulation of specific anatomic sites in the brain.

Characteristics of Intracranial Self-Stimulation

A detailed discussion of all of the characteristics of ICSS is beyond the scope of this chapter (see reference 415). Nonetheless, it is important to discuss some characteristics that, at first glance, appear to distinguish responding for ICSS from responding for either abusable drugs or natural rewards. In the case of simple operant schedules of reinforcement for abusable drugs or natural rewards, termination of reward elicits a temporary increase in responding. This increase characterizes the frustrative, nonreward response phase that precedes extinction. In the case of conventional ICSS, however, termination of the stimulation results in an almost instantaneous extinction of responding. Furthermore, drugs that are believed to block the rewarding properties of brain stimulation depress the rate of ICSS or lead to extinction of the response. In contrast, as noted earlier, administration of a pharmacologic antagonist of a drug generally produces a compensatory increase in self-administration responding.

The apparent differences between the characteristics of ICSS and other types of reward have puzzled investigators and have been taken by some as evidence that the neural substrates underlying these types of reward are dissociable. An alternative hypothesis is that these discrepancies are artifacts that arise from differences between the kinetics of brain stimulation and other types of reward. Thus, even the most rapidly acting drugs, such as heroin or cocaine, have a time course of action that is at least an order of magnitude longer than that of a maximally effective train of brain stimulation. Furthermore, animals respond reliably for drug reward even when the time course of the effect of a single dose lasts for many minutes or even hours (418,419).

The latter hypothesis was investigated in an important series of studies in which drug kinetics were modeled with frequency-modulated trains of brain stimulation (420). Under these conditions, animals self-administer brain stimulation in a manner that closely resembles drug self-administration. Termination of stimulation results in a frustrative nonreward increase in lever pressing, followed by a slow decline and ultimate cessation of responding. Similarly, treatment with drugs that block reward produces an increase in responding. This work suggests that brain stimulation reward is not as easily distinguishable from pharmacologic and natural rewards as it was once believed to be. Further direct support for the hypothesis that ICSS activates, albeit indirectly, the same neural substrates acted upon by abusable drugs and natural rewards is provided by neuroanatomic investigations of brain stimulation reward.

Neuroanatomic and Neurochemical Substrates

Shortly after his discovery of ICSS, Olds found that brain stimulation reward is strongly attenuated by drugs that block catecholaminergic transmission (266,421). He subsequently identified the medial forebrain bundle (MFB) in the region of the lateral hypothalamus as a highly effective locus for ICSS (265). In the mid 1960's, the use of histofluorescence microscopy, which then was a newly developed technique for mapping neurotransmitter systems, demonstrated that most ICSS-positive sites in the MFB were proximal to the mesotelencephalic DA system. This anatomic correspondence led to investigations of the effects of pharmacologic manipulations of dopaminergic neurotransmission on brain stimulation reward. By the end of the 1980's, extensive research suggested that DA is the most important catecholamine involved in reward produced by stimulation of the MFB (422). The main evidence for this conclusion was that low doses of specific DA antagonists mimicked the effect of decreasing the intensity of rewarding brain stimulation, whereas higher doses eliminated ICSS. Microinjections of DA antagonists into various sites have shown that the relevant site of dopaminergic action for ICSS is the nucleus accumbens (423–427). A similar conclusion can be drawn from studies examining the effects of DA agonists on ICSS (428,429). There are two major dopaminergic systems in the MFB. They are the nigrostriatal system, which terminates in the caudate putamen; and the mesolimbic system, which innervates several forebrain regions, including the nucleus accumbens, olfactory tubercle, ventral pallidum, and parts of the frontal cortex (Fig. 32.3) (430). Discovery of dopaminergic pathways in the MFB led many investigators in the 1960's to assume that these must be the paths activated by rewarding MFB

stimulation. However, the axons of dopaminergic neurons in the MFB are unmyelinated. Although the stimulation most commonly used at the time was a 60-Hz sine wave current, which is capable of stimulating unmyelinated fibers, it was known that animals would work to receive pulses of stimulation of 0.1 millisecond or shorter duration, and measurements of refractory period have suggested that such short pulses stimulate mostly myelinated fibers (431–434). More recent studies confirm that the neurophysiologic properties of the primary MFB substrate that is directly activated by electrical brain stimulation reward do not match those of the mesotelencephalic DA system (435–437). Instead, rewarding electrical stimulation appears primarily to activate a nondopaminergic, myelinated, caudally projecting fiber system whose neurons have absolute refractory periods of 0.4 to 1.2 milliseconds.

To explain the inconsistency between the physiologic properties of dopaminergic neurons and those activated by rewarding stimulation, it was proposed that directly stimulated, non-DA "first-stage" neurons synapse with "second-stage" dopaminergic neurons in the midbrain tegmentum and stimulate ascending DA pathways indirectly (349,438). The neurochemistry of the "first-stage," myelinated, caudally projecting fiber system remains an open question. There is some evidence that at least some of these neurons may be cholinergic. Cholinergic neurons in the pedunculopontine and laterodorsal tegmental nucleus make synaptic contact with dopaminergic neurons (439–442). The axons with the shortest refractory periods (0.4–0.7 millisecond) are blocked by peripherally injected atropine (443). However, axons with longer refractory periods (0.7–1.2 millisecond) are insensitive to this manipulation (444).

Effects of intracranial microinjections of abusable drugs on electrical brain stimulation reward suggest that these substances produce reward by direct actions on a small subset of "second-stage" dopaminergic, unmyelinated axons (422,445). Studies using this paradigm have demonstrated that neural substrates responsible for enhancement of brain stimulation reward by amphetamine lie within the nucleus accumbens and neostriatal forebrain terminal projections of mesotelencephalic dopaminergic neurons (446). Similarly, the substrate for the enhancing effect of morphine on brain stimulation reward was localized to cell body regions of the mesotelencephalic DA system, including the posterior hypothalamus, and, less consistently, the VTA (447). Finally, microinjection of haloperidol, a DA receptor antagonist, into the nucleus accumbens and other dopaminergic neostriatal projection sites inhibits brain stimulation reward. This finding provides more general support for the view that the mesotelencephalic DA system plays an essential role in electrical brain stimulation reward (448).

Microdialysis experiments have provided direct evidence that extracellular DA concentrations in the nucleus accumbens are increased during ICSS (449–451). Given the low time resolution of the microdialysis technique, it is not possible to determine the correlation between DA levels and patterns of ICSS in real time. In contrast, *in vivo* voltametry can measure evoked DA release at millisecond time intervals. Studies using this technique have demonstrated that stimulation mimicking ICSS induced different patterns of DA release, depending on the schedule of reinforcement (452–454). Although activation of DA-releasing neurons appears to be required for ICSS, evoked release actually is decreased during ICSS (455). Hence, DA may be more important for novelty (456) or reward expectation (457), rather than the reward itself.

Virtually all substances that are self-administered by animals increase basal neuronal firing and/or basal neurotransmitter release in brain circuits that are relevant to reward. As outlined earlier, studies using *in vivo* brain microdialysis and *in vivo* brain voltametry have demonstrated that monoamines, in general, and the neurotransmitter DA, in particular, participate in the neurochemical actions of abused drugs. Enhancement of extracellular mesotelencephalic DA has been reported after administration of amphetamine (458,459), cocaine (459,460), opiates (461), ethanol (462), nicotine (409,411), barbiturates (463), PCP (459,464), and Δ^9-THC (465,466). Although benzodiazepines are self-administered as well, neurochemical evidence does not support the view that this class of abused drugs increases the activity of dopaminergic neurons (467–471). This common feature among all drugs that are self-administered by animals provides the basis for the theory that drug self-administration activates the same reward circuits as those activated by electrical brain stimulation reward, and that the mechanisms involve an important dopaminergic component.

In addition to the wealth of information pointing to a major role of DA in reward secondary to the administration of drugs of abuse, pharmacologic interventions have demonstrated an involvement of serotonergic systems as well. Treatments that would be expected to enhance serotonergic tone *reduce* self-administration of drugs of abuse. For example, pretreatment with L-tryptophan, a precursor of 5-HT (472), or feeding with a diet enriched in tryptophan (473) attenuated amphetamine self-administration. Administration of inhibitors of 5-HT uptake, such as fluoxetine (474,475), or 5-HT receptor agonists (472,476,477) also reduced amphetamine self-administration in rats. The 5-HT uptake inhibitors likewise reduced the self-administration of ethanol (478,479) and cocaine (480) in rats. Another 5-HT uptake inhibitor, zimelidine, reduced morphine intake in dependent rats (481) and reduced ethanol consumption in humans (482). Furthermore, the potencies of amphetamine and related phenethylamines in operant drug self-administration studies were inversely correlated with their potencies in inhibiting [³H]paroxetine binding to the 5-HT transporter (483). This finding supported the view that inhibition of 5-HT uptake opposes reinforcing effects of amphetamine and related drugs. In addition to these studies of self-administration, a serotonergic role in drug-induced reward was shown using the conditioned place preference paradigm. In this regard, ICS 205-930 and MDL 72222, which are specific antagonists of 5-HT₃ receptors, reduced place preference produced by morphine and nicotine, but not amphetamine, in rats (484). These results indicated an involvement of 5-HT₃ receptors in the reinforcing properties of morphine and nicotine.

Another system that has been implicated in drug abuse-related reward is glutamatergic transmission involving the NMDA subtype of glutamate receptor. When 2-amino-5-phosphonovaleric acid, an antagonist of the NMDA receptor, was microinfused into the nucleus accumbens of rats trained to press a lever for intravenous administration of cocaine or heroin, the intake of cocaine, but not of heroin, was increased (485). Thus, it appeared that blocking glutamatergic transmission at NMDA receptors of the nucleus accumbens reduced the rewarding value of cocaine but not of heroin.

Finally, a role of Ca^{2+} in the rewarding properties of both morphine and cocaine has been implied by studies in which dihydropyridine Ca^{2+} channel antagonists reduced self-administration of both cocaine and morphine in mice (486). The effects of Ca^{2+} antagonists may be related to the DA system, as impulse-driven release of DA is dependent upon Ca^{2+}, and removal of Ca^{2+} from the perfusion medium abolishes cocaine-stimulated increases in extracellular DA in the nucleus accumbens (459). Human studies support the view that Ca^{2+} antagonists interfere with positive affective responses that may be related to reward. In this regard, verapamil reduced morphine-elevated scores on the morphine-Benzedrine group subscale of the Addiction Research Center Inventory while potentiating the effect of morphine to elevate pain threshold (487). These results support the view that Ca^{2+} has a differential role in mediating or influencing actions of opioids.

Dopamine in Natural Reward

The view that DA is a mediator of drug-induced reward is consonant with the literature implicating brain DA in the maintenance of positively rewarded behaviors in general. The techniques used to generate these findings include systemic treatment of animals with DA agonist and antagonist drugs, microinjections into specific brain loci, microdialysis, selective lesioning of the DA system, and genetically altered animals.

Free feeding, operant feeding, and stimulation of the perifornical lateral hypothalamus all increase DA turnover in the nucleus accumbens, as measured by microdialysis in rats (488). DA agonist drugs increase rates of operant responding by rats for food, water, or rewarding brain stimulation, whereas DA receptor antagonists reduce responding at doses below those that would reduce appetite or general motor activity (489,490). For example, when the effects of DA receptor antagonists on water-rewarded operant lever pressing and on nonconditioned water intake were studied in parallel, the drugs were more potent in attenuating lever pressing than in affecting consummatory water intake (491,492). Destruction of DA terminals in the nucleus accumbens and olfactory tubercle disrupted schedule-induced drinking (493). In developing rat pups, treatment with either SCH 23390, a D_1-like DA receptor antagonist, or with raclopride, a D_2-like DA receptor antagonist, was more potent in reducing intake of sucrose in an independent ingestion test than in a test in which sucrose was continually infused into the mouth in developing rats (494). Therefore, both D_1-like and D_2-like DA receptors appear to be important to sucrose reward.

Studies of conditioned reinforcement also suggest a role of DA, particularly in the nucleus accumbens, in supporting reward. Microinjections of D-amphetamine into the nucleus accumbens elicit dose-dependent, selective enhancement of responding for conditioned reinforcement, whereas injections into other striatal regions produce effects that are ineffectual, nonreproducible, or not selective for conditioned reward (495). Additional experiments designed to elucidate the neurochemical specificity of the effect of intra-accumbens D-amphetamine on conditioned reinforcement show that DA, but not NE, microinjected into the nucleus accumbens selectively increases responding for conditioned reinforcement, and that injections into the caudate putamen are ineffective (496).

Neurochemical evidence for the effects of natural rewards on dopaminergic activity has been obtained from postmortem studies of the striata of rats that responded over 30 minutes for water (497) and of the nucleus accumbens and striatum after 1 hour of feeding by food-deprived rats. However, observations that the type of food ingested can influence the response, measured in terms of DA metabolism, argue against an exclusive role of reward or motor processes in determining dopaminergic responses. Using *in vivo* microdialysis, it was shown that either free feeding or lever pressing for food reward produced an increase in extracellular DA, followed by an increase in homovanillic acid (HVA, a metabolite of DA) in the nucleus accumbens but not in the ventral striatum (488). A subsequent study that used linear sweep voltammetry at electrodes implanted intracranially demonstrated that 30-minute sessions of responding for food reward increased uric acid and then HVA in the caudate putamen, with a similar effect on HVA in the nucleus accumbens (498). Thus, it appears that an increase of DA metabolism accompanies food-induced reward.

LONG-TERM EFFECTS OF DRUGS OF ABUSE ON THE CENTRAL NERVOUS SYSTEM

The long-term consequences of chronic drug administration are defining features of drug abuse and addiction. The most prominent and persistent of these effects include alterations in the magnitude of the response to the drug, effects that emerge upon cessation of drug taking, and responses to environmental cues that become associated with taking the drug. Over the past decade, there has been substantial progress made in our understanding of the neuronal mechanisms by which drugs act to effect these consequences.

The term *neuroadaptation* is now commonly used to describe the neural mechanisms underlying alteration in magnitude of drug effect, as both decreases (tolerance) and increases (sensitization) in the response can accompany chronic exposure to a drug. The emergence of withdrawal symptoms when the drug is no longer present is thought to reflect the consequences of such enduring neuroadaptations. In addition, a distinction has been made between neuroadaptive mechanisms that occur at receptors, which are the primary sites of drug action (within-system changes), and the secondary consequences that are propagated throughout the target neuronal networks (between-system changes) (499).

Tolerance, or a diminution in the response to a given dose of the drug, is the classic consequence of chronic drug administration. Although the development of tolerance to a particular drug, by definition, requires that higher doses of drug are needed to obtain the same level of effect, the presence of tolerance to a particular drug does not necessarily imply a potential for abuse of the drug. In fact, the ability to produce tolerance is a nearly universal property of all drugs (500). For example, tolerance develops rapidly to antihistamines, yet these agents are not commonly considered to be drugs of abuse. Furthermore, tolerance is not a unitary phenomenon. The mechanisms underlying tolerance are varied, and tolerance can develop differentially to various actions of a drug (501). For example, individuals who use opioid drugs and are highly tolerant to respiratory depressant actions still manifest opioid-induced miosis and experience constipation (502).

Classic pharmacology attributed tolerance to altered availability of a drug at the relevant receptor site (bioavailability) or to changes in the receptor (503). Altered bioavailability constitutes the simplest mechanism of neuroadaptation. In this case, tolerance can result from reduced absorption, increased metabolism or excretion, or impaired distribution of the drug (e.g., through reduced blood flow). With chronic use of some drugs of abuse, these mechanisms contribute appreciably to the development of tolerance. Induction of hepatic microsomal enzymes by alcohol and barbiturates provides a prototypical example of metabolic tolerance (502). Nonetheless, altered bioavailability is not a primary reason for tolerance to most drugs of abuse.

Repeated exposure to a drug can lead to an increase in the response to the drug. This phenomenon is termed *sensitization* or *reverse tolerance*. Sensitization has been increasingly emphasized in neurobiologic hypotheses of addiction, and it is thought to contribute to the escalation of drug taking or relapse after a period of abstinence (504,505). For example, locomotor activity tends to increase with repeated administration of psychomotor stimulants or opioids. The neurobiologic substrates of sensitization are not characterized as well as those of tolerance, and they may involve complex interactions between multiple neuronal systems (505,506). There is evidence in rodents that increased bioavailability contributes to enhancement of the locomotor effects of cocaine (507,508).

Long-term changes in the specific receptors that are targets of the drugs of abuse have proved to be complex. Both up- and down-regulation of receptor function accompanying chronic drug exposure have been documented. In addition to changes at the ligand binding site, alterations can occur along signal transduction pathways. (In fact, at the cellular level, adaptation to a particular drug may depend on the signal transduction mechanism used by the receptor to which the drug binds.) Tolerance to drugs that act on receptors linked to second messengers (e.g., opioids and stimulants) appears to involve modulation of signal transduction mechanisms, with little or no change in ligand binding affinity or receptor density. On the other hand, tolerance to drugs, such as nicotine and sedative/hypnotics, which act on ionotropic receptors, appears to involve a paradoxical increase in ligand binding associated with functional desensitization of the relevant receptor. Some common mechanisms may contribute to tolerance produced by drugs of abuse belonging to different chemical classes. Evidence for this view derives from observations of cross-tolerance (i.e., the phenomenon observed when one drug induces tolerance to another one). Such cross-tolerance has been observed between nicotine and ethanol (509,510).

The presence of a withdrawal syndrome was long been considered the defining feature of addiction (502). It is currently thought that the cellular mechanisms underlying tolerance and withdrawal often are related. The specific features of a given withdrawal syndrome often are expressions of a rebound from the homeostatic adaptations associated with the development of tolerance. The cellular mechanisms underlying withdrawal symptoms induced by opioids are best understood; less is known about the basis of withdrawal from psychostimulants and sedative/hypnotics.

Opioids

Over the past 3 decades, substantial progress has been made in resolving the cellular mechanisms underlying opioid tolerance and dependence. It is now believed that adaptations involving changes in signal transduction pathways, rather than in the binding sites of receptors, mediate tolerance and dependence (511). As proof for this hypothesis, no consistent relation has been found between physical dependence and the number or affinity of opioid binding sites or the level of endogenous opioids in the brain (512). It originally was proposed that tolerance and dependence rely on specific opioid receptors or subtypes based on the observation that subtypes of μ receptors make independent contributions to analgesic tolerance and physical dependence (513).

Concurrent administration of naloxonazine, an antagonist at the μ_1-receptor subtype, with morphine prevented the development of analgesic tolerance but not the expression of withdrawal signs (514). This finding implied that analgesic tolerance is caused by changes in μ_1 receptors, whereas physical withdrawal involves another type of μ receptor. However, CXBK mice, which are deficient in μ_1 receptors, exhibited a less severe withdrawal syndrome than C57BL/6 mice, which suggests that μ_1 receptors make a partial contribution to the expression of opioid withdrawal (515). Although there is evidence supporting the existence of opioid receptor subtypes (516,517), these subtypes sometimes do not differ from the original receptor in their pharmacology or function. Furthermore, despite efforts to clone these different subtypes, no complementary DNAs corresponding to all proposed receptor subtypes have been found (518). This has led to the speculation that the observed differences in opioid ligand binding properties may be due to opioid receptor dimerization (518). Finally, cellular changes that occur in a specific neuronal system may be responsible for adaptations to only one aspect of opioid action. For example, local injection of morphine into the VTA supports self-administration of morphine, but it does not lead to the development of physical dependence (519).

Tolerance

The molecular and cellular adaptation best identified that occurs due to repeated opioid administration is the upregulation of the cyclic adenosine 3',5'-monophosphate (cAMP) signal transduction cascade (520–523). These cellular adaptive changes appear to be mediated through μ receptors on noradrenergic neurons in the locus caeruleus (LC). The effects of opioids on these neurons have been studied intensively because of their critical contribution to the expression of opioid withdrawal symptoms (511,524–526). Most of the noradrenergic cell bodies in the brain are found in the LC, a nucleus adjacent to the floor of the fourth ventricle at the pontine-medullary junction (527,528). Both μ and κ receptors are found in this nucleus. The somata of LC neurons contain μ-opioid receptors (153,503), which are thought to mediate the direct effects of opiates on LC neurons (153,503). In contrast, κ receptors probably are located on the presynaptic terminals, given that κ agonists attenuate excitatory input without altering spontaneous firing rates (529,530).

Acute administration of morphine or selective μ-opioid agonists produces dose-dependent decreases in the spontaneous activity of LC neurons (524,526,531–535). These decreases are blocked by specific μ antagonists, such as β-chlornaltrexamine and β-funaltrexamine, as well as by prototypical opiate antagonists, such as naloxone and naltrexone (526,535,536). Intracellular studies indicate that the decrease in impulse activity is accompanied by a hyperpolarization generated by increased outward K^+ conductance and

ACUTE OPIATE ACTION IN THE LC

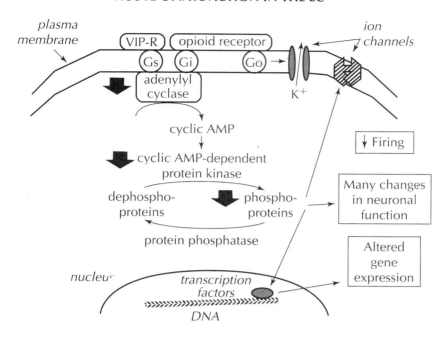

CHRONIC OPIATE ACTION IN THE LC

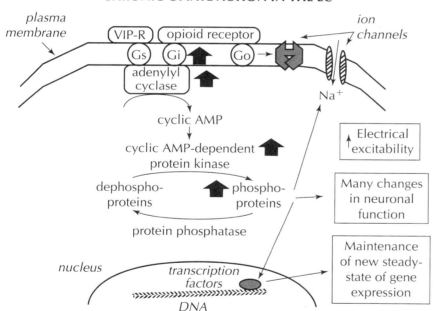

FIGURE 32.4. Sequence of cellular events underlying development of tolerance to opiates in locus caeruleus (LC) neurons. **A:** Upon acute administration, opiates decrease the spontaneous firing rate of LC neurons through activation of G proteins associated with the μ-opioid receptor. One G protein (G_i) decreases cyclic adenosine monophosphate (cAMP) synthesis, and the fall in cAMP levels results in closing of nonspecific cation channels, which contributes to the pacemaker potentials that regulate spontaneous firing. Another G protein (G_o) directly opens K^+ ion channels, and this action hyperpolarizes the cell. Decreased cAMP synthesis impacts upon other biochemical processes, including deoxyribonucleic acid (DNA) transcription, which is reduced. **B:** After chronic administration of opiates, the G proteins are desensitized, which leads to a decrease in K^+ conductance but a net up-regulation of cAMP synthesis and inward cation current. Other proteins dependent on cAMP phosphorylation also are up-regulated, as is expression of immediate early genes (e.g., c-*fos*). When the tonic opioid levels drop, as occurs during abstinence or antagonist-precipitated withdrawal, these adaptive changes leave the neuron in a state of increased excitability. This state is responsible for the hyperactivity seen in these neurons during withdrawal. Clonidine can normalize firing, as the α_2 receptor is coupled to the same set of G proteins as the μ-opioid receptor. See text for details. (Modified from Nestler EJ. Molecular mechanisms of drug addiction. *J Neurosci* 1992;12:2439–2450.)

decreased slow inward cation current (524,526,535,537). Electrophysiologic studies on LC neurons indicate that both μ- and δ-opioid receptors can activate K_G channels via a pertussis-sensitive G protein. Opiate receptors expressed in LC neurons couple to PTX-sensitive G proteins, which then results in inhibition of cAMP synthesis (Fig. 32.4) (524,537,538–540).

Neurons in the LC develop tolerance to opioids in rats chronically treated with morphine. After 3 to 5 days of chronic treatment, spontaneous firing rates of LC neurons are no longer suppressed and are equivalent to those in drug-naive animals (535,536,538). Larger doses of opiates then are required to suppress firing (535,536). This tolerance appears to result from an up-regulation involving increases in the components of the cAMP-dependent signal transduction cascade (Fig. 32.4B) (522,540,541). Up-regulation of this pathway negates the inhibitory effects produced by acute opiate exposure, and the resultant physical manifestation is tolerance. In support of this hypothesis, chronic opiate treatment inactivates G_i proteins through ribosylation of adenosine diphosphate (ADP), which produces a net up-regulation of cAMP synthesis and restores protein kinase A (PKA) activity to predrug levels (540,542,543). This up-regulation of the cAMP pathway results in an increased intrinsic firing rate of the LC neurons by a nonselective cation channel (544,545). As predicted from earlier studies of whole brain (546,547), cAMP levels in the LC increase above pre-drug levels during precipitated withdrawal (548).

Chronic opiate treatment increases type 1 and type IV adenylate cyclase, PKA regulatory and catalytic type II subunits, and several phosphoproteins, including cAMP response element-binding protein (CREB) (549,550). All of these changes can lead to alterations in gene regulation (Fig. 32.4 B). Acute administration of opiates decreases the expression of the immediate early genes c-*fos* and c-*jun*. Conversely, expression of these genes increases several-fold during naltrexone-induced withdrawal (522). Because the time course for gene regulation is much longer than the decrease in spontaneous activity of the cell in response to acute opiates, decreased expression of these immediate early genes may play a role in adaptation of the cells to chronic treatment.

In addition to opioids themselves, other endogenous substances modulate tolerance to the chronic effects of morphine. The NMDA receptor, and the novel neuronal messenger nitric oxide (NO), both have been implicated in the development of tolerance. Coadministration of either an NMDA receptor antagonist (551) or an inhibitor of NO synthetase (552) blocks the development of analgesic tolerance to morphine and the development of locomotor sensitization (553). NO synthesis attenuates the development of tolerance to morphine at the cellular level (554). When rats were given an inhibitor of NO synthase during chronic morphine treatment, LC neurons in brain slices of morphine-dependent animals were as sensitive to challenge doses of either morphine or the μ-specific agonist DAMGO as were neurons from animals that had never received morphine. The direct effects of opioids on LC neurons are mediated exclusively through μ receptors; thus, cellular effects are consistent with the ability of NO synthesis inhibitors to block tolerance to μ- but not κ-receptor agonists (555). The molecular bases of these modulatory effects are not known. It has been proposed that the modulatory actions of NMDA receptors and NO may be causally linked because NMDA cytotoxicity apparently depends on synthesis of NO (556). NO may be critical to the ribosylation of G proteins, a process that has been proposed as the molecular mechanism of tolerance for μ receptors (see previous discussion). Results of a study

showing that NO promotes ADP ribosylation (557) supports this hypothesis.

Physical Withdrawal

The classic hallmark of addiction to opioid drugs is the emergence of a well-defined cluster of physiologic signs and symptoms, the "physical" withdrawal syndrome, upon discontinuation of opioid drug use. The appearance of this syndrome indicates that the organism depends on the presence of the opioid for normal physiologic functioning. The severity of the syndrome is dependent on the dosage and duration of use. When the withdrawal syndrome is precipitated by administration of an opioid antagonist, it has a more rapid onset and is more severe than the syndrome produced by "spontaneous" withdrawal. Otherwise, spontaneous and precipitated withdrawal are qualitatively similar, and it is commonly acknowledged that they involve an identical neural mechanism (558).

The opioid withdrawal syndrome is similar across species, although the exact time course and specific constellation of signs and symptoms are different. The general characteristics of the physical withdrawal syndrome resemble those of influenza, and they are similar in humans and animals (246,502,559,560). Behavioral and physiologic signs include pronounced secretory activity (lacrimation, rhinorrhea, sweating); piloerection (goose pimples); mastication (teeth chattering and chewing); elevated heart rate, blood pressure, respiratory rate, and body temperature; irritability; tremors and "wet dog shakes"; and gastrointestinal disturbances (intestinal spasms, nausea, diarrhea). Subjective symptoms in humans include agitation and anxiety, dysphoria, drug craving, aches in joints and muscles, loss of appetite, and insomnia. In humans, the symptoms begin to appear 8 to 12 hours after the last dose of an opioid agonist, depending on the pharmacokinetics of the individual drug (502). The syndrome reaches a peak over the next 48 to 72 hours and is virtually absent by 7 to 10 days. In rodents, the time course is more rapid (559). Opioid withdrawal in rodents includes increased locomotor activity, escape behavior (digging, jumping), and ptosis.

With the development of selective antagonists, it has become possible to assess the differential contribution made by specific opioid receptor subtypes to physical dependence. Because morphine and other commonly abused opioids are primarily μ agonists, it is not surprising that physical dependence appears to be mediated primarily by μ-opioid receptors (561), as are the reinforcing effects of heroin (562). Substantially higher doses of κ and δ antagonists are required to elicit even a mild withdrawal syndrome compared with doses of selective μ antagonists (561).

Many brain areas, including thalamic, hypothalamic, limbic, and hindbrain areas, are activated during physical withdrawal from opiates (563). This activation can be attributed not only to adaptive changes within the opioid system itself, but also to nonopioid neurotransmitter systems (564). Much biochemical and pharmacologic evidence indicates that hyperactivity of noradrenergic innervation in the CNS contributes substantially to the expression of many of the somatic symptoms seen during opiate withdrawal (525,564,565). This hypothesis is based on both preclinical animal research and clinical treatment research.

In vivo microdialysis studies have corroborated findings from initial studies in which changes in brain catecholamine levels during withdrawal were observed (566). However, some of the most compelling evidence for a noradrenergic role in opiate withdrawal is that pharmacologic suppression of noradrenergic activity attenuates morphine withdrawal in both animals and humans (525,567–569). Systemic and local (intracerebral) administration of clonidine, a specific α_2-adrenergic receptor agonist that decreases the spontaneous activity of LC neurons, attenuates both the increase in LC firing due to withdrawal (524,532) and most of the accompanying physiologic signs (568–571). Unfortunately, use of clonidine, which is the nonopioid treatment of choice for treatment of opiate addiction, is somewhat controversial (564). Clonidine cannot account for all of the withdrawal symptoms observed in humans and animals, and a lack of correlation between some biochemical and behavioral responses suggests that other neurotransmitter systems may be involved in opiate withdrawal effects (564).

LC neurons become hyperactive during antagonist-precipitated withdrawal, as evidenced by increased impulse activity, transmitter release, and metabolite levels (571–576). The hyperactivity is correlated with the intensity of the physical signs (524,532,573,577). Behaviorally, the effects of electrical or pharmacologic stimulation of the LC, which is the major source of noradrenergic innervation in the CNS, resemble opiate withdrawal (578). Opiate withdrawal can be induced by local injections of naloxone (571) or of methylnaloxonium, a quaternary derivative of naloxone that has more limited diffusion, directly into the LC (579). A pattern of widely distributed increases in rCMRglc, similar to that observed after systemic injection of naloxone in morphine-dependent rats, is obtained by intracerebral injections of methylnaloxonium directly into the LC (563). Finally, lesions of the LC greatly attenuate the withdrawal behaviors precipitated by intraventricular injection of opiate antagonists (580).

The hyperactivity associated with withdrawal is generated by a rebound of the mechanisms that underlie the development of tolerance (Fig. 32.4B). After cessation of chronic opiate administration (abstinence) or blockade of opiate receptors by naloxone (precipitated withdrawal), cAMP levels rise as the down-regulated G_i proteins cease to inhibit up-regulated adenylate cyclase as well as increases in PKA expression (360,523,546–548,581,582). Because cAMP normally inactivates an outward K^+ channel and activates an inward nonspecific cation conductance, the rise in cAMP levels dur-

ing withdrawal produces a net depolarization of the neuron and a profound increase in firing rates (536,538,541). These cellular mechanisms account for the ability of cAMP and pharmacologic agents that elevate cAMP to increase the spontaneous firing rate of LC neurons and to mimic the behavioral signs of opiate withdrawal (quasi-morphine withdrawal syndrome) (541,544,572,576,577,583).

Withdrawal hyperactivity may involve altered glutamatergic afferent input and intrinsic changes in LC neurons. Intraventricular or direct administration of nonselective glutamate antagonist into the LC blocks the withdrawal-induced activation of LC neurons (584–587). However, this does not occur when the drugs are administered systemically (586). Systemic administration of antagonists for either NMDA or non-NMDA glutamate receptors does attenuate withdrawal behaviors (551,588), which suggests that mechanisms other than LC activation are involved. Increased release of glutamate and aspartate adds excitatory drive to LC neurons during withdrawal (589), but opioids do not change the sensitivity of excitatory amino acid receptors on LC neurons (576,590). The nucleus paragigantocellularis (Pgi) in the ventrolateral medulla is the most prominent source of excitatory input to the LC, and hyperactivity of these neurons during opioid withdrawal could be the source of increased excitatory drive (591). Lesions of the Pgi attenuate, but do not completely abolish, the increase in basal firing rates after induced withdrawal (584). It is not known whether neurons in the Pgi become hyperactive during opioid withdrawal, yet chronic morphine treatment produces changes in the levels of cAMP in the Pgi (522). Unlike the LC and other regions, the Pgi does not show an increase in c-*fos* expression, which is indicative of cellular activation during opioid withdrawal (592). This discrepancy suggests that increased neuronal activity in other regions, such as the prefrontal cortex, which provides excitatory amino acid input to the LC, may contribute to withdrawal-induced hyperactivity in the LC (592).

Coadministration of MK-801, a noncompetitive NMDA receptor antagonist, or of 5,7-dichlorokynurenic acid, an antagonist at the glycine co-agonist site of the NMDA receptor, during chronic morphine treatment has been observed to attenuate a behavioral sign (jumping) of precipitated withdrawal. These findings suggest that NMDA receptor activation is involved in the development of opiate dependence (551,558). Acute administration of either a competitive or noncompetitive NMDA receptor antagonist immediately before precipitating withdrawal also attenuated withdrawal behaviors, but it was less effective than subchronic treatment (551,593). However, the attenuation of withdrawal behaviors after acute administration of noncompetitive NMDA antagonists may be the result of nonspecific motoric effects of these drugs (551,594). This interpretation is consistent with the inability of acute administration of either competitive or noncompetitive NMDA antagonists to block the withdrawal-associated hyperactivity of LC neurons (593), even though LC neurons contain NMDA receptors (595,596). Alternatively, it is possible that the relevant NMDA receptors are located on neurons in regions that receive primary or secondary projections from the LC. Despite the numerous reports implicating a role of glutamatergic transmission in drug addiction, the interaction between NMDA receptor antagonists and opiates warrants further inspection. It has been observed that PCP and MK801, which are NMDA receptor antagonists, have powerful and reinforcing actions of their own, and they may potentiate the activating and reinforcing effects of opiates (597). Therefore, additional study is required to elucidate the interactions between these two neuronal systems and their involvement in adaptive changes due to chronic opiate use.

NO has been implicated in the development of opioid tolerance and physical dependence. Coadministration of NOS inhibitors with morphine or other opioid agonists has been shown to prevent the development of tolerance to morphine-induced antinociception (598). The inhibition of NO synthesis was shown to attenuate the behavioral signs of opioid withdrawal (26,599–604). This observation reflects the evidence reviewed previously proposing a critical role of NO in the development of morphine tolerance at both the behavioral and cellular levels. Furthermore, immunohistochemical studies have identified the presence of NOS-like immunoreactivity in the LC (605,606). There also is voltametric (607) and electrophysiologic data (608) supporting functional evidence for the role of NO in opiate withdrawal in the LC. Finally, given that NO synthase can generate NO upon NMDA receptor activation, this provides another possible mechanism by which changes in glutamatergic neurotransmission lead to the development of opiate withdrawal symptoms (609).

Affective Aspects of Withdrawal

Opioid withdrawal has an affective component that represents a rebound from reinforcing effects of opioids, much as classic physical withdrawal signs represent the converse of acute physiologic effects of opioids (e.g., diarrhea is the opposite of constipation) (506). Dysphoria is a prominent part of opioid withdrawal in human subjects (568,610). Opioid withdrawal interferes with motivational states, as seen by reductions in the performance of operant schedules for food reward (611). Opioid withdrawal also acts as a negative unconditioned stimulus in conditioning paradigms. Presentation of a normally preferred tastant solution (e.g., saccharine) during induction of opioid withdrawal results in taste aversion conditioning (i.e., persistent avoidance of the tastant upon later presentations) (612). The place-conditioning paradigm has been used to demonstrate that opioid withdrawal induces a negative affective state (613). Place conditioning consists of repeatedly administering a drug in a distinctive physical environment. If the subjects choose to spend more time in that environment, it is taken as an index of the reinforcing

properties of a drug. The finding that subjects will avoid an environment where they have undergone opioid withdrawal (place aversion) indicates that the environment was associated with an unpleasant experience. The data supporting the affective component of opiate abuse are extensive. Of importance is the role that this component plays in the development of addiction. If the affective component is not responsible for the initial development of compulsive drug use (614,615), then it likely accounts for the contribution of these negative reinforcements to the continued drug use required to alleviate withdrawal symptoms, which results in the maintenance of drug use (616).

Although it is possible that these changes in motivated behaviors are generated by peripheral mechanisms (e.g., gastrointestinal disturbance), they appear to be separate from the physiologic malaise associated with classic physical withdrawal. Much lower doses of naloxone are required to affect motivated behaviors than to cause expression of classic withdrawal signs (617). Place aversion associated with opioid withdrawal can be produced by intraventricular but not subcutaneous administration of methylnaloxonium (613). Because methylnaloxonium does not readily cross the blood–brain barrier, the results indicate that affective components of withdrawal are generated by central mechanisms and are not due to peripherally mediated physiologic malaise.

The nucleus accumbens appears to play a critical role in the affective components of withdrawal, complementary to its role in opioid reinforcement. Local injection of methylnaloxonium disrupts performance on operant schedules for food reward and produces place-aversion conditioning (263,618). In contrast to local injections of methylnaloxonium into the LC, the affective withdrawal signs induced by nucleus accumbens injections are not accompanied by expression of physical signs of withdrawal (263). Furthermore, clonidine is less effective in alleviating dysphoria than other signs in humans during opioid withdrawal (568).

Chronic morphine produces an increase in adenylate cyclase activity and in the levels of PKA in the nucleus accumbens, similar to that seen in the LC, and may be attributed to a reduction in the levels of $G_{i/alpha}$ (619–621). The mechanism of opioid action in the nucleus accumbens is more complex than that in the LC. Many afferents from different brain regions synapse onto a plethora of cell types in the nucleus accumbens to modify neuronal output by mechanisms still not clearly understood (622). This is in contrast to the situation in the LC, where opioids act primarily through postsynaptic μ receptors. For example, opioid-induced changes in other transmitter systems, especially in dopaminergic afferents, influence the activity of nucleus accumbens neurons (623,624). Reports have indicated that DA release in the nucleus accumbens decreases (575,625–627) but acetylcholine release increases during opioid withdrawal.

In conclusion, expression of opioid withdrawal is not a unitary phenomenon. There are at least two distinct components, each mediated through distinct neural systems.

Classic physical withdrawal signs involve a μ-receptor–mediated hyperactivity of brain noradrenergic neurons. In contrast, the same mesolimbic system regions involved in opioid reinforcement (e.g., nucleus accumbens) probably mediate affective and motivational aspects of opioid withdrawal.

Sedative/Hypnotics

Tolerance

Continued use of sedative/hypnotic drugs (e.g., benzodiazepines, barbiturates, ethanol, aldehydes) produces tolerance to their sedative, anticonvulsant, and anxiolytic effects (628–630). However, tolerance develops differentially to each of these effects. Tolerance to the sedative and anticonvulsive effects develops within days. Whether tolerance to the anxiolytic effects occurs is still unclear. Tolerance to the anxiolytic effects has been difficult to demonstrate, especially in the clinical setting except under specific experimental conditions (628,631). This finding suggests that different neuronal mechanisms could be responsible for tolerance to these individual effects. The discovery of various specific binding sites associated with GABA receptors suggests that the primary consequence of chronic administration of sedative/hypnotic drugs, such as benzodiazepines and barbiturates (which bind to these sites), involves alterations in the GABA-benzodiazepine-ionophore receptor complex. These changes are best understood with respect to the benzodiazepine family.

After repeated administration, benzodiazepines produce less potentiation of the electrophysiologic inhibition by GABA of cerebellar Purkinje cells (632) and of serotonergic neurons in the dorsal raphe (633,634). The diminution in effect does not appear to reflect changes in benzodiazepine receptors. Decreases in the number of benzodiazepine binding sites, but not their affinity, after chronic benzodiazepine treatment were found by some investigators (635–638), but others observed little or no changes in benzodiazepine binding (633,639,640). These discrepancies may result from differences in the specific drug administered (e.g., diazepam, lorazepam, alprazolam, flurazepam), in routes of administration (oral vs. injected), or in dose levels.

Chronic administration of benzodiazepines has been reported to have no effect on, or to decrease, the actions of benzodiazepines; however, there is general agreement that chronic benzodiazepine treatment induces a reduction in the actions of GABA. This view is based on observations that both GABA-potentiated benzodiazepine binding (633) and electrophysiologic responses to direct application to GABA alone are diminished after chronic benzodiazepine treatment (634,641). These changes occur whether drugs are injected intermittently (633) or delivered continuously (634). The decreased electrophysiologic response to GABA is associated with decreases in GABA-stimulated Cl^- flux (642), even though chronic benzodiazepine treatment does not decrease the number of GABA receptor-linked Cl^- channels (643). Consistent with the development of GABA subsensitivity, the threshold for seizures induced by bicuculline, a convulsant that binds to the $GABA_A$ receptor complex, was reduced during chronic benzodiazepine treatment (644).

These alterations in GABA sensitivity are not seen in all neurons responsive to benzodiazepines. Neurons in the substantia nigra pars reticulata, which also use GABA as a transmitter, exhibit tolerance to benzodiazepine administration without changes in GABA sensitivity (645–647), and cerebellar Purkinje cells do not exhibit decreases in GABA-stimulated Cl^- influx (642).

The development of GABA-ergic subsensitivity may be due to changes in the molecular composition of the GABA-benzodiazepine-ionophore complex. Although chronic benzodiazepine treatment fails to alter the total number of GABA binding sites or the high-affinity component of GABA binding, the affinity for GABA increases at low-affinity GABA binding sites (648). This change may involve alterations in the protein structure of the receptor complex. Chronic, but not acute, administration of benzodiazepines decreases the expression of mRNA that codes for the α_1 (649) and $\alpha 5$ (650) subunits of the GABA-receptor complex. Changes in the protein structure of the receptor complex are consistent with the time-dependent nature of the changes in GABA sensitivity (644). GABA subsensitivity does not start to develop until after a week of chronic treatment, but it persists for at least 5 days after discontinuation of drug administration.

Based on this combination of electrophysiologic and biochemical data, some researchers have argued that subsensitivity to GABA underlies tolerance to benzodiazepines. In turn, this subsensitivity is due to conversion of low-affinity GABA binding sites to a desensitized state that exhibits higher affinity (651). Because nicotinic receptors also exhibit a paradoxical increase in binding coupled to a functional desensitization, this pattern may represent a general neuroadaptive mechanism underlying tolerance in the superfamily of ionotropic receptors.

Tolerance to other sedative/hypnotic drugs may involve changes to the GABA-benzodiazepine receptor complex. This mechanism most likely is the basis of cross-tolerance between these drug classes (652). It has been suggested that chronic ethanol and barbiturate administration reduces the coupling between the Cl^- channel and the GABA and benzodiazepine binding sites (653–655). Similar to treatment with benzodiazepines, chronic administration of barbiturates or ethanol induces a functional subsensitivity of the receptor complex by reducing GABA- or benzodiazepine-stimulated Cl^- influx (639,655). Unlike the benzodiazepines, however, chronic ethanol has generally been reported as having no effect on $GABA_A$ receptor number (656). However, alterations of $GABA_A$ receptor subunit mRNAs and their encoded proteins have been reported as a result of chronic ethanol treatment (657–660). Thus, alterations of the composition of expressed $GABA_A$ receptors and their affinities for GABA may be related to the development of tolerance and dependence to ethanol.

Tolerance to ethanol involves direct changes in the function of the NMDA receptor (661). Acutely, ethanol decreases NMDA receptor function (653,662), most likely by interacting directly within the ion channel (662,663). Chronic exposure to ethanol produces up-regulation of NMDA receptors, in contrast to the subsensitivity of the GABA receptors induced by ethanol (653,664). Enhanced NMDA receptor function may mediate cell death that can occur during chronic ethanol use (661,665). Changes in NMDA receptors during chronic ethanol treatment may be involved in conditioned drug effects (666).

Dependence and Withdrawal

It is now well accepted that withdrawal syndromes can be produced by abrupt cessation of treatment with any anxiolytic/sedative drug (628,629,630,667–669). Drugs in this class produce withdrawal syndromes that are qualitatively similar but not identical (670). Because characterization of the benzodiazepine withdrawal syndrome has led to recognition of common cellular substrates underlying withdrawal from this diverse class of drugs (667–669), the following section focuses on benzodiazepine withdrawal as prototypical of the response to abstinence from chronic treatment with sedative/hypnotics.

Behaviorally, the predominant symptoms are the converse of the primary effects of the sedative/hypnotics. They include anxiety, increased locomotor activity, insomnia, tremors, and convulsions (671–678). As with opioid withdrawal, the severity of the syndrome is dependent on the dose, duration of use, and especially the rate of elimination of the drug (668,670,672,679–681).

Recognition of a benzodiazepine withdrawal syndrome was hampered by its relatively slow temporal development. For nearly 20 years after the clinical introduction of the benzodiazepines, the existence of a benzodiazepine withdrawal syndrome was questioned, even though withdrawal from other anxiolytic/sedative drugs (e.g., barbiturates and ethanol) was well known (628,629,668,682,683). Barbiturate and ethanol withdrawal symptoms appear soon after discontinuation of drug use, even after short-term use of moderate doses (670,671,677). Benzodiazepine withdrawal symptoms can take several days or weeks to emerge, although faster onsets have been noted after very long durations of use (675,684). This delayed development of symptoms has been attributed to the slow decline in plasma levels of most benzodiazepines in humans, even when drug administration is abruptly discontinued (667,685). Thus, the full severity of benzodiazepine withdrawal was not seen until specific antagonists that could displace benzodiazepines from their receptors were developed (672,673,679,681,686,687). Similarly,

the introduction and widespread use of shorter-acting benzodiazepines (e.g., lorazepam and triazolam) with more rapid elimination rates accelerated the acceptance of the concept of a benzodiazepine withdrawal syndrome (667,668).

The neurobiologic substrates of anxiolytic/sedative withdrawal are not as well established as those that mediate opioid withdrawal, but development of a persistent subsensitivity to GABA appears to be the most proximal factor. Again, chronic exposure to anxiolytic/sedative drugs has the common effect of decreasing the sensitivity of receptors to GABA. Thus, during chronic administration, the normally potentiating effects of benzodiazepines or barbiturates on the action of GABA are offset by the attenuated response to the transmitter itself; however, the subsensitivity persists for days after benzodiazepines are no longer present in the brain (644,688). Because GABA is the primary inhibitory transmitter in the CNS, the net GABA-ergic subsensitivity increases the general excitability of the nervous system. This postulated release from inhibition is consistent with the appearance of anxiety, seizures, and other behavioral signs of CNS hyperexcitability associated with withdrawal from anxiolytic/sedative drugs.

Withdrawal from ethanol involves increased NMDA receptor function, in addition to changes in GABA-ergic function (653). Upon discontinuation of ethanol treatment, up-regulated NMDA receptors exacerbate the overall neural hyperexcitability, including the seizure activity associated with ethanol withdrawal (664). Stimulation of NMDA receptors also mediates the increase in c-*fos* expression that accompanies ethanol withdrawal (689).

Although anxiolytic/sedative withdrawal could reflect a global increase in excitability throughout the neuraxis, individual behavioral signs probably reflect hyperactivity in specific brain systems. A variety of cortical and subcortical sites exhibit significant alterations in glucose metabolism during mild diazepam withdrawal (690,691). Interestingly, the pattern of glucose utilization was not entirely consistent with the known distribution of benzodiazepine receptors, which suggests that some regions were activated transsynaptically (691,692). In some cases, reasonable connections between specific withdrawal signs and individual brain structures can be made. It is widely accepted that decreased GABA-ergic function in the hippocampus and cerebral cortical regions leads to seizures (49,693). Likewise, the amygdala, which is thought to be critically involved in fear and anxiety, has high concentrations of benzodiazepine receptors and putative benzodiazepine-like peptides (229,692). On the other hand, dysfunction in any number of the myriad regions implicated in sleep and arousal may contribute to the insomnia associated with anxiolytic/sedative withdrawal (694,695).

Hyperactivity of the central noradrenergic system may contribute to the anxiolytic/sedative abstinence syndrome, although evidence for noradrenergic involvement is less well established in this situation than it is for opioid withdrawal. The potential involvement of noradrenergic systems in withdrawal from opiates and anxiolytic/sedative drugs may explain the commonalities between the two withdrawal syndromes. These include gastrointestinal distress (decreased food intake, retching, and vomiting), weight loss, wet dog shakes, increased startle responses, vocalizations, and piloerection (672,674,676,680,681,687). Both antagonist-precipitated withdrawal from diazepam in primates (687) and discontinuation of chronic use of diazepam by humans (696) increase the concentrations of NE metabolites in the brain. Furthermore, activation of the noradrenergic system by yohimbine exacerbates benzodiazepine withdrawal (697). There also is evidence for noradrenergic hyperactivity during withdrawal from barbiturates and alcohol (698). Firing rates of LC neurons increase during ethanol withdrawal, and this hyperactivity is mediated through up-regulated NMDA receptors (699). These changes may not be simply correlative, given that lesions of the noradrenergic system diminish the effects of benzodiazepine withdrawal on the firing rate of neurons in the cerebellum (651).

A noradrenergic role in sedative/hypnotic withdrawal was predicted from the hypothesis that hyperactivity of brain noradrenergic systems contributes to anxiety states, especially panic in humans (700). Acute administration of diazepam decreases the firing rate of noradrenergic neurons in the LC of rats (701,702), attenuates the behavioral effects of LC stimulation in primates (700,703), and decreases the concentrations of noradrenergic metabolites in human subjects (704). Alcohol and barbiturates decrease central noradrenergic activity (629,705,706), and lesions of noradrenergic neurons decrease tolerance to barbiturates (707,708). Conversely, inverse agonists at the benzodiazepine receptor increase firing rates of LC neurons (709) and mimic the behavioral effects of LC stimulation (574,700,710).

If hyperactivity of the noradrenergic system is involved in withdrawal from anxiolytic/sedative drugs, treatment that reduces noradrenergic activity should attenuate this withdrawal syndrome, as it does the opioid abstinence syndrome. A preliminary clinical study demonstrated that clonidine decreased the severity of withdrawal from alcohol (711) and the time required for tapered withdrawal from alprazolam (712). Consistent with the known anticonvulsant actions of clonidine (713), preliminary studies in monkeys indicated that clonidine attenuated seizures associated with benzodiazepine withdrawal. In contrast, preclinical studies in rats have not shown a consistent action of clonidine against sedative/hypnotic withdrawal (714–716).

It is likely that non-noradrenergic transmitter systems contribute to the withdrawal syndrome. However, the functional contributions of serotonergic or dopaminergic systems to anxiolytic/sedative withdrawal have not been well characterized (626,717–720).

Stimulants

The existence and nature of long-term consequences of chronic stimulant use have been subjects of considerable research over the past decade. The prototypical stimulants are amphetamine and cocaine. Because recent research on stimulant abuse has been driven largely by the widespread abuse of cocaine, especially in the form of "crack," this section focuses on the sequelae of chronic cocaine use on brain function. Many of the findings, however, can be generalized to other stimulants, such as amphetamine, which act as indirect aminergic agonists (721,722). In addition, changes in the mesolimbic DA system will be emphasized because this system is the primary target of stimulant action.

The long-term consequences of psychostimulant abuse differ from those of other abused drugs, such as opioids and sedative/hypnotics. In addition to the psychological problems engendered by addiction, the detrimental consequences of prolonged use of these drugs include cardiovascular pathology (hypertension, cardiac infarctions, arrhythmias, and neuropathy) (723–726), hepatotoxicity (727), neurotoxicity (728–730), seizures (731), and psychopathology (paranoid psychosis) (723,729,732). On the other hand, long-term use of stimulants does not invariably lead to tolerance, nor does cessation of chronic use produce clear withdrawal signs.

Tolerance and Sensitization

After repeated administration of stimulants, the response to a challenge dose of the drug can either be attenuated (tolerance) or enhanced (sensitization). Classically, tolerance was thought to underlie the progressive increase in drug intake associated with addiction. As the response to a drug diminishes, larger amounts of the drug must be taken to obtain the same level of effect. Some cellular and behavioral effects of stimulants, such as induction of immediate early genes or discriminative stimulus properties, exhibit tolerance (733–736). Nonetheless, long-term use of stimulants is now recognized to lead to an enhancement of other effects of these drugs. These effects include increased locomotor activity and seizure susceptibility (731,737–739). Because sensitization can persist long after discontinuation of drug administration, it has been proposed that sensitization is the basis not only for the pathologic consequences of long-term stimulant use but also for other essential features of addiction (e.g., craving and relapse) (58,253,740).

It is unclear whether tolerance or sensitization develops to the rewarding effects of stimulants. Tolerance to the subjective and cardiovascular effects of cocaine in human subjects was reported to occur within hours after a single dose, but it had dissipated by the next day (741). On the other hand, studies in animals suggest that preexposure to stimulants produces sensitization to the reinforcing effects of these drugs (742–744).

One difficulty in assessing the chronic effects of stimulants is that the same drug effect can show tolerance under one set of experimental conditions but exhibit sensitization under a different set of conditions (742). Sensitization of locomotor activity depends critically on the treatment regimen, including dosage and intervals between drug administrations (733). For example, if the drug is administered intermittently, locomotor activity exhibits sensitization, but if the drug is continuously infused, tolerance occurs (745). The interval between the last administration of the drug and the challenge dose is also an important factor. Tolerance is more evident during the first 24 to 72 hours after the last drug administration, especially when high doses are used, whereas sensitization can take several days to emerge (746). The differential development between sensitization and tolerance can be influenced by psychological variables, such as method of drug administration (noncontingent vs. self-administered) or environmental stimuli associated with drug administration (conditioned cues) (31,742,744,747,748).

Initial hypotheses on the long-term effects of stimulants assumed that tolerance would develop as the release and metabolism of DA outpaced its synthesis and the presynaptic stores of DA became depleted (49,749). Blockade of DA reuptake into the presynaptic terminal would eliminate a major source of releasable transmitter. Under normal conditions, D_2-like dopaminergic autoreceptors located on dopaminergic cell bodies, dendrites, and, axons limit increases in intrasynaptic DA levels by inhibiting impulse activity and transmitter release (750). Increases in intrasynaptic DA levels would stimulate these presynaptic autoreceptors, resulting in inhibition of transmitter synthesis and impulse-dependent transmitter release. The combination of these regulatory mechanisms was expected to counteract the rise in intrasynaptic DA levels by diminishing overall neurotransmission and, thus, was expected to produce tolerance. These adaptive mechanisms were long cited as the basis for tolerance to the euphorigenic effects of stimulants and the rapid escalation in doses of stimulants self-administered by addicts (751). What was not taken into account was the fact that the homeostatic mechanisms themselves could be altered by continued use of a stimulant.

Recognizing that increased dopaminergic transmission in the mesolimbic system was responsible for the behaviors that exhibit sensitization (e.g., locomotor activity), researchers hypothesized that sensitization is due to a progressive enhancement of the ability of stimulant drugs to increase the levels of synaptic DA in the nucleus accumbens (459,461,463). Despite the apparent simplicity of the hypothesis that synaptic DA is augmented by chronic stimulant use, the empirical results have not yielded a clear picture. Evidently, whether sensitization to cocaine-induced increases in extracellular DA is measurable depends on whether the absolute concentration of DA—or rather the change in DA concentration relative to basal levels—is assayed (752). Whereas rats treated with saline for 10 days exhibited larger increases in extracellular DA in the nucleus accumbens in response to a challenge dose of cocaine, absolute DA concentrations in response to cocaine challenge were larger in rats that had received the subchronic cocaine treatment. These results reflected an elevation in basal levels of extracellular DA in the cocaine-treated rats. Furthermore, the increase in basal level of extracellular DA and in the levels of DA assayed after cocaine challenge were no longer present 1 week after discontinuation of daily cocaine treatments, even though sensitization of locomotor responses persisted for several weeks after the final dose of cocaine. A similar pattern of effect on extracellular DA was observed in the VTA (753). Both basal levels of DA in this cell body region and the relative increase in extracellular DA evoked by a challenge dose of cocaine were elevated the day after discontinuation of daily cocaine administration; however, as in the terminal regions, the enhancing effects on DA levels were no longer present 2 weeks after discontinuation of daily cocaine, despite persistence of sensitization of locomotor responses to cocaine.

Paradoxically, sensitization of the increase in extracellular DA levels produced by cocaine may be inversely related to the dose of daily cocaine during the pretreatment period (754). When rats were pretreated with half the dose of cocaine used in the studies just cited, the relative increase in extracellular DA in the nucleus accumbens in response to a challenge dose of cocaine was enhanced, and the enhancement was correlated with sensitization of locomotor activity. However, when the daily dose of cocaine was increased to the amount used in these same studies, enhancement of the increase in DA levels by cocaine challenge the day after discontinuation of daily cocaine was less than in the control group, even though an alteration in basal DA levels, as seen in the previous studies, was not observed. In contrast, higher daily doses of cocaine produced more sensitization of the locomotor responses. Furthermore, the DA response to cocaine challenge continued to increase rather than diminish over a 21-day period in both of the cocaine-treated groups.

Although the process of sensitization does not depend on a single mechanism (755), it is believed that sensitization is initiated by an action on dopaminergic neurons. Repeated injection of amphetamine directly into the VTA of rats is sufficient to produce sensitization to a subsequent systemic administration of amphetamine or cocaine (756). This finding suggests that changes in the dopaminergic cell bodies mediate sensitization (746). After repeated administration of cocaine or amphetamine, these drugs have a decreased ability to inhibit the firing rate of dopaminergic neurons due to a decrease in the sensitivity of the autoreceptors (757–759). Therefore, the net increase in DA output at the terminal regions would increase because the accumulation of synaptic DA would be less effective in exerting inhibitory feedback through the autoreceptors. A correlation between behavioral sensitization and molecular changes in dopaminergic neurons also has been reported. The functional significance of these changes is not clear (619,760,761) because the molecular effects (e.g., changes in DA autoreceptor sensitivity and G-protein ribosylation in the VTA) are fairly transient and cannot account for the persistence of sensitization (757,760). Also, increased inhibitory feedback from terminal regions appears to contribute more than activation of somatodendritic autoreceptors to the inhibition of DA impulse activity by acute amphetamine and cocaine (762).

It is now believed that persistent changes in DA terminal regions are responsible for the maintenance of sensitized responses, but also that dopaminergic transmission may contribute only minimally to the sensitized response once sensitization is established. Chronic injections of amphetamine directly into the VTA can induce sensitization to a subsequent systemic administration of amphetamine (756), but a challenge dose of amphetamine injected directly into the VTA does not elicit a sensitized response (763). In contrast, a sensitized response is seen if the challenge injection is made directly into the nucleus accumbens (763). At present, the two most likely alternative interpretations are that adaptations occurring either in the presynaptic dopaminergic terminals or at nondopaminergic neurons or terminals could be critical to sensitization. Changes in the number of presynaptic DATs do not seem to be well correlated with the degree of sensitization (764,765). On the other hand, electrophysiologic studies have shown that neurons in the nucleus accumbens become supersensitive to DA, and that this supersensitivity is mediated by D_1 receptors (758,766). Unlike the changes in the autoreceptors on DA cell bodies, D_1 supersensitivity can persist for several weeks.

The role of D_1 receptors has been examined with D_1 receptor drugs and with D_1 receptor knockout mice. Reversal of cocaine sensitization has been achieved with injection of a D_1 receptor antagonist (767). Similarly, pretreatment with a D_1 receptor antagonist prevents the development of sensitization (768). Mice that lack the D1 receptor show reduction, but not elimination, of sensitization to cocaine (769), which suggests that the D_1 receptor plays a role in this phenomenon, but it is not the sole determinant. Recent studies suggest that the D_2 receptor also may influence sensitization (770). Finally, it has been suggested that sensitization may be due in part to up-regulation of adenylyl cyclase activity, which is regulated by D_1 receptors (771).

If sensitization is caused by changes in nondopaminergic neurons, other transmitters must be involved in maintaining sensitization. Endogenous opioids appear to contribute to sensitization, and there is evidence for cross-sensitization between stimulants and opioids (772). Coadministration of naltrindole, a δ-opioid antagonist, with cocaine blocks sensitization (773). In addition, repeated doses of cocaine increase the expression of precursors for opioid peptides in the striatum (736). Excitatory amino acid transmitters, most likely originating from cortical inputs to the striatum and nucleus accumbens, also are involved in sensitization. Coadministration of the noncompetitive NMDA receptor antagonist dizocilpine prevents the development of behavioral sensitization (774,775). Dizocilpine also prevents development of autoreceptor desensitization in dopaminergic neurons and D_1 receptor supersensitivity in nucleus accumbens neurons associated with sensitization (776). It is likely that the di-

rect effects of stimulants on noradrenergic and serotonergic systems are important contributors to the chronic effects of these drugs (777–779). Finally, activation of two transcription factors, CREB (780) and *Delta* FosB (781), have been linked to long-term cocaine use.

Stimulant Withdrawal

The epidemic of cocaine use during the last 2 decades has provided overwhelming evidence for the addictive properties of cocaine and other psychostimulants. Unlike withdrawal syndromes from opioids and sedative/hypnotic drugs, abstinence from cocaine and other psychostimulants is believed to be characterized primarily by emotional and motivational disturbances (782).

The description of a cocaine abstinence syndrome by Gawin and Kleber (782,783) has provided a provocative heuristic framework for both clinical and preclinical research. Based on clinical interviews with an outpatient population, these authors proposed that cocaine withdrawal consists of three distinct temporal phases, each marked by progressive changes in mood and drug craving. The first stage ("crash") is analogous to an alcohol "hangover." The "crash" phase lasts for 1 to 4 days after cessation of cocaine use and is characterized by exhaustion and intense craving for cocaine. The true withdrawal syndrome was considered to start about 1 week after the last dose of cocaine and to last for 1 to 10 weeks, during which there was a high probability of relapse. The second phase is marked by a subtle dysphoric syndrome, which consists of decreased activity, loss of motivation, increased boredom, and anhedonia, punctuated by episodes of drug craving. If abstinence is maintained for 1 to 2 months, mood, although somewhat labile, begins to normalize. This normalization of mood represents the beginning of a protracted extinction phase, during which the craving response to cues gradually decreases if abstinence is maintained, although intermittent episodes of drug craving and relapse can occur.

Although this description of cocaine withdrawal has substantial intuitive appeal, it has been difficult to validate these phases under controlled conditions. Data from a sample of six inpatients suggested no consistent disturbances in mood (anxiety or depression) during a 22-day detoxification period, but there was a sharp increase in craving for cocaine 1 to 2 weeks after admission (784). However, two subsequent studies using a larger sample of subjects failed to demonstrate a phasic pattern of "crash" and "withdrawal" (782,783). In a 28-day residential study, mood disturbances and drug craving were highest upon admission and declined steadily over the following 4-week period (785). Furthermore, mood disturbances were relatively mild. Cocaine abusers and normal control subjects exhibited similar sleep patterns throughout the study period (i.e., cocaine addicts did not exhibit definite sleep disturbances), and there were no consistent cardiovascular or other physiologic alterations during the study period. Three independent residential studies confirmed that mood disturbances and drug craving decrease sharply over the first week of abstinence and normalize over the next 3 weeks, although there was electroencephalographic evidence of protracted sleep disturbances (786–788).

The failure to observe discrete phases of cocaine withdrawal, especially with respect to craving, in these residential studies may be related to differences between inpatient and outpatient settings. In particular, these findings have led to an emphasis on the importance of environmental stimuli (e.g., people, places, paraphernalia) related to drug use in triggering craving (782,783,786,789–791). Even though craving is central to descriptions of stimulant withdrawal and is a major factor in relapse, the neural basis of craving is currently unknown (782,783,789,790). One problem with this line of investigation is that craving does not have a widely accepted definition or physiologic measure (789,792–794). The subjective nature of craving has impeded the development of animal models that could be used to assess the physiologic basis for craving (795). Robinson and Berridge (253) have argued that sensitization of an incentive-motivational system centered in the mesolimbic DA system may serve as the neural basis for craving. Although only a heuristic model, it would be consistent with the evidence for increased dopaminergic activity during presentation of cues that predict delivery of cocaine or food rewards (796–798).

Despite difficulties in demonstrating a distinct clinical cocaine abstinence syndrome, there has been substantial progress in characterizing neuronal phenomena following chronic use of stimulants. Cocaine and other psychostimulants exert their acute behavioral effects through enhancement of dopaminergic transmission, particularly in the mesolimbic system. Accordingly, behavioral changes seen upon abstinence from cocaine and other psychostimulants were presumably the result of depletion of DA and/or down-regulation of dopaminergic neurotransmission (782,783,799). However, neurochemical evidence for dopaminergic depletion has been difficult to demonstrate in clinical studies (784,786,799). A postmortem binding study reported reductions in the number of DAT sites in the prefrontal cortex of cocaine users, but these changes may have been due to residual cocaine present at the time of death (800).

Evidence for altered dopaminergic function has been found in PET studies of abstinent cocaine abusers. A transient decrease in availability of D_2-like DA receptors labeled with [^{18}F]N-methylspiroperidol in the striatum was observed during the first week of abstinence, but labeling of these receptors was at control levels when they were assayed 1 month later (801). In one report, similar transient alterations were seen when glucose metabolism was assayed in PET scans of abstinent cocaine users (802). Compared to controls who were not taking cocaine, rCMRglc of cocaine users was higher during the first week of cocaine abstinence in the basal ganglia and a ventral medial portion of the pre-

frontal cortex—areas that receive dopaminergic projections. However, there was no difference in rCMRglc in any region between subjects who had been abstinent for 2 weeks and controls who were not taking cocaine. In contrast, a subsequent study from the same group reported that cocaine users who had been abstinent for 1 to 6 weeks had lower rCMRglc in dorsal prefrontal regions than controls who were not taking cocaine (803). This difference in rCMRglc was still present in a subset of subjects who were retested after 3 to 4 months of abstinence.

More substantial evidence has been obtained from animal studies. Behavioral changes consistent with down-regulation of the DA system have been noted in animal studies during the first few hours to days of abstinence. There is behavioral evidence for increased anxiety and anhedonia during cocaine withdrawal, but these effects normalize within several days (804,805). A more general disruption of operant responding, which persists for up to 2 weeks and perhaps is indicative of motivational dysregulation, also occurs (806).

Direct evidence for progressive changes in dopaminergic transmission during each of the proposed phases of cocaine withdrawal has been obtained from animal studies. Although acute administration of cocaine produced selective stimulation of rCMRglc in extrapyramidal motor and limbic regions (807,808) immediately after 12 hours of self-administration by rats, there was a decrease in rCMRglc in the mesolimbic and anterior striatal regions, including the nucleus accumbens, olfactory tubercle, amygdala, septum, limbic cortex (cingulate and piriform cortices), and rostral caudate putamen (809). Decreases also occurred in motor, somatosensory, and auditory regions of the neocortex. The decrements in rCMRglc were inversely correlated to the amount of self-administered cocaine. Spontaneous firing rates of DA neurons in the VTA were increased during the first day after 1 to 2 weeks of cocaine or amphetamine administration (759,810,811). The increased impulse activity was attributed to a subsensitivity of inhibitory autoreceptors. In a different study, autoreceptor subsensitivity was seen in DA neurons of the substantia nigra, but the subsensitivity was not accompanied by changes in the spontaneous firing rates (812). The increase in extracellular DA levels, observed 24 hours after the cessation of chronic cocaine administration in one study, is consistent with an increase in firing rate (752). However, three other studies reported that basal extracellular levels of DA measured by *in vivo* dialysis in the nucleus accumbens were normal during the early stages of withdrawal (754,813,814). The discrepancy between increased impulse activity and normal extrasynaptic DA levels could be explained by enhanced clearance of synaptic DA due to an increase in the number of DATs or enhancement of their efficiency. An elevation in DAT binding sites was reported in one study (815), but three other studies found that the number of DAT binding sites in the nucleus accumbens was unchanged either immediately after or up to 3 days after cessation of chronic cocaine administration in rats or mice

(764,765,816). However, a study reporting an increase in the velocity of DA transport and turnover rate in rats revealed a functional activation of the DAT 24 hours after cessation of chronic cocaine administration that persisted for at least 2 weeks (817). An increase in the efficiency of the transporter presumably would lead to a normalization of synaptic DA levels.

Studies of postsynaptic receptor densities during the early phases of abstinence have yielded equivocal results. A decrease in the number of D_1-DA receptors in rats was observed 2 hours after the final cocaine dose in one study (818) but not at 4 hours in a second study (819). Paradoxically, electrophysiologic studies found that neurons of the nucleus accumbens exhibit an enhanced response to DA and D_1 agonists the day after cessation of 2 weeks of cocaine treatment (758,766). An increase in the number of D_2 receptors in the striatum, but a decrease in the nucleus accumbens, immediately after 2 weeks of cocaine administration have been observed. As with the induction of sensitization (see previous discussion), differences in the method of cocaine administration may underlie the discrepant findings on cocaine withdrawal. In most of the studies cited, cocaine was noncontingently administered by the investigators one or more times during the day; however, rats in other studies self-administered cocaine (809,815,818) or were given a cocaine infusion continuously (812).

About 1 week after the last cocaine administration, more reliable alterations in dopaminergic function can be detected, but these changes do not clearly support a pattern of either dopaminergic hypoactivity or hyperactivity (799,820). There is a marked decline in the number of DAT sites in the nucleus accumbens but not in the more dorsal portions of the striatal complex (764,765,815,816,821). This decline is accompanied by a decrease in expression of mRNA for the transporter in ventral tegmental DA neurons, but not in DA neurons of the substantia nigra (821). Although a decline in transporter binding should result in less DA reuptake and thus an elevation of extracellular DA concentrations, most studies have found either normal (821–823) or decreased DA levels in the nucleus accumbens 1 week after abstinence (814,824). The lack of elevated DA levels may be due to a persistent increase in the velocity of turnover or transport (800). The reduction in the spontaneous activity of DA neurons in the VTA along with normalized autoreceptor sensitivity would further compensate for increased DA release (811,812). Despite a reduction in the number of D_1 receptors (818,819), neurons in the nucleus accumbens continue to exhibit a paradoxical increase in electrophysiologic responses to D_1 agonists. In contrast, D_2 receptor density in the nucleus accumbens is normal, as is the response to D_2 receptor agonists.

The picture that emerges from the contradictory changes seen during this period is that of a progressive restabilization of the interactive mechanisms that regulate dopaminergic neurotransmission. Consistent with the clinical model, many of these changes persist for several weeks. For example, the reduction in DAT binding in the rat nucleus accumbens persists for up to 2 months after the final dose of cocaine (765). The increased sensitivity of nucleus accumbens neurons to D_1 agonists also is present after 2 months of abstinence (766).

Nicotine

Tolerance

The effects of repeated administration of nicotine involve a variety of physiologic and behavioral functions. Chronic administration of nicotine produces tolerance, although evidence for sensitization to the effects of nicotine has also been obtained in animal studies. Furthermore, it is now clear that deprivation of nicotine, in subjects that have been exposed chronically to the drug, also leads to the emergence of withdrawal signs, indicating dependence upon the drug. The mechanisms for producing these effects appear to involve cellular adaptation rather than dispositional changes.

Tolerance to the chronic effects of nicotine has been demonstrated in human investigations (825). In addition, studies in animals have demonstrated a rapid development of tolerance, albeit incomplete, to many of the effects of nicotine (826). Reversible tolerance to some of the effects of nicotine can develop in minutes (827). Tolerance is not explained by pharmacokinetics. In this regard, human studies demonstrated higher blood levels of nicotine despite a diminution in heart rate acceleration and subjective responses to a second intravenous infusion of nicotine given 1 or 2 hours after the first infusion (828).

In rodents, nicotine produces biphasic effects, including a depression in locomotor activity, followed by stimulation. Both of these phenomena have been used to characterize the effects of chronic or subchronic treatment with nicotine on the response to an acute nicotine challenge. The initial depression of spontaneous motor activity by nicotine was used to demonstrate tolerance to this central effect of nicotine (829,830). In rats given nicotine for about 1 month, motor activity, measured using a T maze, and several parameters of exploratory activity showed tolerance to a nicotine challenge that was presented 24 hours after withdrawal (831). Tolerance to the depressant effects of nicotine on motor behavior in rats persisted 80 days after withdrawal (830). Frequent dosing with nicotine was not required for the production of tolerance to the depressant effect on locomotor behavior. Tolerance developed when nicotine was given once every 3 days, and a single administration of nicotine produced acute tolerance 2 hours later (832).

Another central effect of nicotine that shows the development of tolerance is hypothermia. Tolerance to the hypothermic effect of a challenge dose of nicotine was seen in mice that were treated subchronically (4 days) with nicotine three times per day (833). At times after nicotine challenge when cerebral nicotine levels were the same in mice given the subchronic treatment as in naive mice, naive mice showed maximum hypothermia but pretreated mice showed none. These data provided no support for a pharmacokinetic explanation of nicotine tolerance but suggested a cellular mechanism. Other experiments in which tolerance to hypothermia in rats was attenuated by the concomitant administration of mecamylamine pointed to a mechanism of tolerance that involves central nicotinic receptors (834).

Sensitization

Whereas tolerance has been the predominant effect of chronic nicotine on nicotine-induced hypothermia and suppression of locomotor activity, chronic nicotine enhances behavioral activation, which follows the early suppression of locomotor behavior by nicotine. Such stimulation has been observed 20 to 80 minutes after a nicotine challenge in rats given repeated administration of low doses of nicotine (835).

Dependence and Withdrawal

Observations of long-term neurochemical changes in the brain following chronic administration of nicotine are consistent with the view that nicotine can produce physical dependence. Acceptance of the view that nicotine produces such dependence derives from observations of withdrawal signs and symptoms in human subjects who were previously exposed chronically and then deprived of cigarettes or smokeless tobacco products (836). The syndrome is characterized by anxiety, irritability, difficulty in concentration, sleep disturbances, increases in appetite, and craving for tobacco that may last from 1 to 10 weeks (837,838). The withdrawal induced by cessation of tobacco use has been hypothesized to create a diminished interest in pleasurable activities, which also is a hallmark of depression (839). The overwhelming desire to continue smoking may be an attempt to alleviate these affective symptoms. Withdrawal symptoms may be alleviated by use of nicotine-containing products, such as chewable gums and transdermal patches; these products have enabled some smokers to quit (840). However, abrupt termination of nicotine gum usage has been shown to produce withdrawal symptoms that vary with the dosage used. Abstinence from chronic treatment with 4-mg nicotine gum produces more signs and symptoms than withdrawal from 2-mg nicotine gum, and more severe symptoms are seen after withdrawal from cigarettes than from 4-mg gum (841).

The greater severity of the withdrawal syndrome following abstinence from cigarette smoking compared with abstinence from chewing nicotine gum likely reflects differences in the bioavailability of nicotine administered by the two routes. Unlike nicotine gum, which provides a more constant concentration of the drug in the bloodstream, cigarette smoking produces rapid peaks in the arterial concentration of nicotine (842,843).

Because the signs and symptoms of withdrawal are similar whether subjects are withdrawn from tobacco or from nicotine gum (844,845), it appears that nicotine is the active ingredient in tobacco that produces dependence. Other evidence that attributes dependence on the consumption of tobacco products to nicotine could include production of withdrawal signs by antagonists of nAChR activation and amelioration of withdrawal signs by administration of nicotine. In this regard, results of studies in which mecamylamine was used to test whether antagonism of nAChRs produces signs and symptoms of withdrawal have been equivocal (846–848). On the other hand, nicotine gum alleviates some of the symptoms of withdrawal (849–853). These include anxiety, irritability, and decreased ability to concentrate, but not craving for tobacco. Therefore, although interaction with nAChRs appears to be a key feature in the production and amelioration of dependence upon nicotine, long-term changes in nonnicotinic neurotransmitter systems also may play a role in the behavioral response to nicotine abstinence.

Neurochemical Mechanisms in Long-Term Effects of Nicotine

Studies on the effects of chronic or subchronic nicotine on the binding of radioligands to nAChRs in the brain have supported the view that nAChR is at least one of the cellular targets that contribute to the long-term changes in behavioral responses to nicotine. Despite some discrepancies in reports on the effects of chronic nicotine on the binding of radioligands to nAChR in different regions of the rodent brain, the findings generally indicate that chronic nicotine produces up-regulation of nAChRs, especially in the cerebral cortex (854–858). In particular, α_4-subunit–containing nicotinic receptors are up-regulated after chronic nicotine treatment (859). In addition, administration of the selective $\alpha_4\beta_2$ nicotinic receptor antagonist DHβE to chronic nicotine-treated rats precipitated affective aspects of withdrawal (860). The α_7 nicotinic receptor subtype appears to play a role in the rewarding aspects of nicotine actions; however, blockade of the receptor subtype does not precipitate withdrawal (861).

Alterations in nAChRs produced by chronic treatment with nicotine in parallel with changes in the behavioral response to the drug have implicated nAChR as critical in neuroadaptive mechanisms. For example, a 14-day nicotine treatment regimen produced tolerance to the acute effects of nicotine on body temperature and locomotor depressant activity in rats and increased the binding of [^{3}H]acetylcholine (in the presence of an antagonist to block muscarinic sites) in the midbrain and hippocampus (856). This observation suggested that the tolerance was related to up-regulation of nAChRs in the brain. However, in the same study, chronic nicotine treatment had no effect on the binding of [³H]nicotine in the brain, which suggests that the two radioligands used had labeled different sites.

In another study, repeated administration of nicotine for 5 days enhanced both the behavioral stimulant effects of nicotine in rats and the binding of [³H]acetylcholine to nAChRs measured in postmortem brain tissue (862). By 21 days after the last exposure to nicotine in the same study, both the behavioral effect as well as the effect on [³H]acetylcholine binding were no longer seen, which suggests that the two effects were functionally linked. Ksir et al. (863) suggested several interpretations of the augmentation of the stimulant properties of nicotine by chronic administration of the drug. One view purports that the depressant and stimulant effects of nicotine are caused by actions of the drug at different receptor types, and that stimulation is "unmasked" when tolerance develops to the depressant actions. Another interpretation is that biochemical events in the brain lead to sensitization. Subsequently, Ksir et al. (863) showed that repeated administration of 0.03 to 0.3 mg/kg/day (but not 0.01 mg/kg/day) nicotine enhanced the behavioral stimulant effect of nicotine and increased the binding of [³H]acetylcholine in the brain. The parallelism in the behavioral and biochemical responses again indicated that these two phenomena were linked. However, in the same study, rats that received 1.6 mg/kg nicotine twice daily showed no increase in behavioral stimulation due to a test dose of nicotine, although receptor binding was increased even more than at the lower doses of nicotine. The authors suggested that the dissociation between receptor up-regulation and the behavioral response to repeated high-dose treatment with nicotine reflects actions on multiple receptor types, additional tolerance mechanisms, and prolonged receptor desensitization.

The finding of an increased density of nAChRs in animals chronically or subchronically treated with nicotine does little to explain the mechanism for an alteration in the behavioral response. One theory of the neuroadaptational change in response to chronic nicotine involves desensitization of cholinergic mechanisms (864). In this model of neuroadaptation, a desensitized receptor can bind a drug, but the binding is not coupled with a biologic response (864). Current information about nAChR is consistent with this model because exposure to an agonist (such as nicotine) induces a conformational change in the receptor, which leads to increased affinity but reduced sensitivity.

Aside from the focus on nAChR as the prime mediator of long-term adaptational changes to chronic nicotine, noncholinergic systems also have been implicated. For example, evidence for opioid involvement in the response to nicotine withdrawal derives from the use of a rodent model of the nicotine abstinence syndrome (865). The model uses continuous subcutaneous infusion of nicotine tartrate via osmotic minipump. After termination of nicotine infusion, a syndrome characterized by teeth chatter, chewing, gasps, writhes, body shakes, ptosis, and seminal ejaculation is seen. These signs, which are similar to those seen in opiate with-

drawal, can be reversed by administration of nicotine. In rats treated with continuous nicotine infusion, the withdrawal syndrome can be elicited by the injection of naloxone, and morphine can reduce spontaneous abstinence signs observed after termination of nicotine infusion (866). These findings support the view that changes in endogenous opioid systems are involved in nicotine dependence.

Acute nicotine use enhances DA overflow in the mesolimbic system, similar to other drugs of abuse. However, due to nicotine-induced nAChR desensitization, this enhanced activity is reduced or eliminated with subsequent nicotine use (405,409,411,867). This phenomenon may contribute in part to the acute tolerance observed with cigarette smoking. Acutely, nicotine acts at presynaptic α_7 nicotinic receptors to enhance glutamate release in the VTA (868). The role of glutamate transmission in withdrawal, however, is equivocal and needs to be pursued further (861).

Some studies of the chronic effects of nicotine have involved metabolic mapping with 2-deoxy-D-[1-¹⁴C]glucose method. One of these experiments provided an example of a dissociation between the behavioral response to chronic nicotine administration and a biochemical marker in the brain (869). When rats were subjected to a twice daily regimen of subcutaneous nicotine injection for 10 days, the percentage of animals manifesting certain behaviors after the nicotine injection either increased (Straub tail, hyperkinesia, tremor) or did not change (ataxia) over the 10-day period. In contrast, none of the 45 brain regions showed evidence of sensitization to the stimulant effects of nicotine on rCMRglc, but tolerance to nicotine challenge was seen in the metabolic responses of the VTA, some components of visual pathways, the cerebellum, and vestibular nuclei. The discrepancies between the behavioral and neurochemical findings in this study may be due in part to involvement of brain regions, other than those that showed tolerance, in the behavioral responses to nicotine that were enhanced by repeated administration. In addition, because rCMRglc is an indication of activity in a heterogeneous population of neurons within a region, it may not discriminate an enhanced response to nicotine in a discrete neurochemical system within the region. Another study on the effects of chronic nicotine on rCMRglc was performed in rats (870). Using a subcutaneous infusion method (osmotic minipumps), this study showed stimulation of rCMRglc in regions that previously showed stimulation in response to acute nicotine. There was no evidence of tolerance, although a challenge dose was not given to test for tolerance.

CONCLUSION

The last 3 decades have seen remarkable advances in our understanding of the mechanisms of action and drug abuse. Receptor pharmacology has provided major insights into the initial sites of drug action in the brain. It is now recognized that drugs of abuse act on classic receptors or transporter molecules associated with specific neurotransmitter systems. Starting with the first assays of opiate receptor binding using radiolabeled naloxone in the 1970's, relevant receptors for every major drug of abuse, except ethanol, have been identified in the brain. The impact of chronic administration of drugs of abuse on receptor function has been characterized in increasing detail and has helped to explain tolerance and other long-term drug effects at a cellular level. For example, observation of changes in expression of mRNAs and their encoded proteins in response to drug treatment has shed light on receptors and proteins that may play a role in tolerance and dependence. Analysis by gene-chip microarray systems of genes in animals and in humans that are up- or down-regulated by drugs of abuse has provided us with information regarding thousands of genes that are affected by long-term drug use.

Equally important are advances in our understanding of how receptor-mediated events influence neural systems and how the effects on these systems produce the behavioral manifestations of addiction. One critical feature of drugs of abuse is their rewarding properties, which are required for the initiation and maintenance of drug use. It is now well established that the mesolimbic DA system serves as a common substrate for this phase of addiction. On the other hand, a plethora of evidence has implicated noradrenergic systems in the physical aspects of withdrawal. The neurobiologic substrates of other aspects of the addictive process, such as sensitization, conditioning, and predisposition, are not as well characterized. Current investigations are uncovering mechanisms that include involvement of serotonergic, glutamatergic, and other neurotransmitter systems in these processes. Gene knockout or knockin technology is aiding in this discovery process and has allowed for the identification of transporters and receptors that are critical for drug action.

The propensity to abuse drugs, such as alcohol, has long been recognized as a heritable trait. Gene linkage studies are underway to identify allelic variants that predispose individuals to abuse. Once uncovered, these genes may not only be a marker for identifying risk for disease, but they ultimately may lead to better treatments for addictive diseases.

REFERENCES

1. Strange PG. The structure and mechanism of neurotransmitter receptors. Implications for the structure and function of the central nervous system. *Biochem J* 1988;249:309.
2. Creese I, Sibley DR, Hamblin MW et al. The classification of dopamine receptors: relationship to radioligand binding. *Annu Rev Neurosci* 1983;6:43.
3. Kebabian JW, Calne DB. Multiple receptors for dopamine. *Nature* 1979;277:93.
4. Stoof JC, Kebabian JW. Two dopamine receptors: biochemistry, physiology and pharmacology. *Life Sci* 1984;35:2281.
5. Freissmuth M, Casey PJ, Gilman AG. G proteins control diverse pathways of transmembrane signaling. *FASEB J* 1989;3:2125–2131.
6. Tissari AH, Atzori L, Galdieri MT. Inhibition of dopamine synthesis in striatal synaptosomes by lisuride: stereospecific reversal by (-)-sulpiride. *Naunyn Schmiedebergs Arch Pharmacol* 1983;322:89–91.
7. Wachtel SR, Hu X-T, Galloway MP, et al. D1 dopamine receptor stimulation enables the postsynaptic, but not autoreceptor, effects of D2 dopamine agonists in nigrostriatal and mesoaccumbens dopamine systems. *Synapse* 1989;4:327–346.
8. Starke K, Göthert M, Kilbinger H. Modulation of neurotransmitter release by presynaptic autoreceptors. *Physiol Rev* 1989;69:864–989.
9. Carlson JJ, Bergstom DA, Walters JR. Stimulation of both D$_1$ and D$_2$ dopamine receptors appears necessary for full expression of postsynaptic effects of dopamine agonists: a neurophysiological study. *Brain Res* 1987;400:205.
10. Walters JR, Bergstrom DA, Carlson JH, et al. D$_1$ dopamine receptors activation required for postsynaptic expression of D$_2$ agonist effects. *Science* 1987;236:719.
11. Clark D, White FJ. Review: D1 dopamine receptor—the search for a function: a critical evaluation of the D1/D2 dopamine receptor classification and its functional implications. *Synapse* 1987;1:347.
12. Blum K, Noble EP, Sheridan PJ, et al. Allelic association of human dopamine D$_2$ receptor gene in alcoholism. *JAMA* 1990;263:2055.
13. Smith SS, O'Hara BF, Persico A, et al. Genetic vulnerability to drug abuse. The D$_2$ dopamine receptor Taq B1 restriction fragment length polymorphism appears to move frequently in polysubstance abusers. *Arch Gen Psychiatry* 1992;49:723.
14. Gejman PV, Ram A, Gelernter J, et al. No structural mutation in the dopamine D$_2$ receptor gene in alcoholism or schizophrenia. *JAMA* 1994;271:204.
15. Matsushita S, Muramatsu T, Murayama M, et al. Alcoholism, ALDH2*2 allele and the A1 allele of the dopamine D2 receptor gene: an association study. *Psychiatry Res* 2001;104:19–26.
16. Lawford BR, Young RM, Rowell JA, et al. Association of the D2 dopamine receptor A1 allele with alcoholism: medical severity of alcoholism and type of controls. *Biol Psychiatry* 1997;41:386–393.
17. Neiswanger K, Hill SY, Kaplan BB. Association and linkage studies of the TAQI A1 allele at the dopamine D2 receptor gene in samples of female and male alcoholics. *Am J Med Genet* 1995;60:267–271.
18. Chen WJ, Lu ML, Hsu YP, et al. Dopamine D2 receptor gene and alcoholism among four aboriginal groups and Han in Taiwan. *Am J Med Genet* 1997;74:129–136.
19. Edenberg HJ, Foroud T, Koller DL, et al. A family-based analysis of the association of the dopamine D2 receptor (DRD2) with alcoholism. *Alcohol Clin Exp Res* 1998;22:505–512.
20. Sokoloff P, Giros B, Martres M-P, et al. A new dopamine receptor: the gain falls mainly in the brain. Discovery of a third dopamine receptor stimulates new studies of neurotransmitter systems and signal transduction pathways in the brain. *J NIH Res* 1990;2:59.
21. Schwartz J-C, Giros B, Martres M-P, et al. The dopamine receptor family: molecular biology and pharmacology. *Semin Neurosci* 1992;4:99.
22. Van Tol HHM, Bunzow JR, Guan H-C, et al. Cloning of the gene for a human dopamine D$_4$ receptor with high affinity for the antipsychotic clozapine. *Nature* 1991;350:610.
23. Gingrich JA, Caron MG. Recent advances in the molecular biology of dopamine receptors. *Annu Rev Neurosci* 1993;16:229.
24. Van Tol HHM. The dopamine-D$_4$ receptor. In: Niznik HB, ed. *Dopamine receptors and transporters. Pharmacology, structure, and function*. New York: Marcel Dekker, 1994:189.

25. Grandy DK, Zhang Y, Bouvier C, et al. Multiple human D5 dopamine receptor genes: a functional receptor and two pseudogenes. *Proc Natl Acad Sci U S A* 1991;88:9175.

26. Sunahara RK, Guan H-C, O'Dowd BF, et al. Cloning of the gene for a human dopamine D₅ receptor with higher affinity for dopamine that D₁. *Nature* 1991;350:614.

27. Giros B, Sokoloff P, Martres MP, et al. Alternative splicing directs the expression of two D2 dopamine receptor isoforms. *Nature* 1989;342:923–926.

28. Monsma FJ, McVittie LD, Gerfen CR, et al. Multiple D2 dopamine receptors produced by alternative RNA splicing. *Nature* 1989;342:926–929.

29. Meador-Woodruff JH, Mansour A, Saul J, et al. Neuroanatomical distribution of dopamine receptor messenger RNAs. In: Niznik HB, ed. *Dopamine receptors and transporters. Pharmacology, structure, and function.* New York: Marcel Dekker, 1994:401.

30. Snyder SH, Creese I, Burt DR. The brain's dopamine receptor: labeling with [³H]dopamine and [³H]haloperidol. *Psychopharmacol Commun* 1975;1:663.

31. Hollt V, Schubert P. Demonstration of neuroleptic sites in mouse brain by autoradiography. *Brain Res* 1978;151:149.

32. Kuhar MJ, Murrin LC, Malouf AT, et al. Dopamine receptor binding *in vivo:* the feasibility of autoradiographic studies. *Life Sci* 1978;22:203.

33. Murrin LC, Kuhar MJ. Dopamine receptors in the rat frontal cortex: an autoradiography study. *Brain Res* 1979;177:279.

34. Palacios JM, Niehoff DL, Kuhar MJ. ³H-Spiperone binding sites in brain: autoradiographic localization of multiple receptors. *Brain Res* 1981;213:277.

35. Mansour A, Meador-Woodruff JH, Bunzow JR, et al. Localization of dopamine D2 receptor mRNA and D1 and D2 receptor binding in the rat brain and pituitary: an in situ hybridization-receptor autoradiographic analysis. *J Neurosci* 1990;10:2587–2600.

36. Meador-Woodruff JH, Mansour A, Bunzow JR, et al. Distribution of D2 dopamine receptor mRNA in rat brain. *Proc Natl Acad Sci U S A* 1989;86:7625–7628.

37. Meador-Woodruff JH, Mansour A, Healy DJ, et al. Comparison of the distributions of D1 and D2 dopamine receptor mRNAs in rat brain. *Neuropsychopharmacology* 1991;5:231–242.

38. Mengod G, Martinez-Mir MI, Vilaro MT, et al. Localization of the mRNA for the dopamine D2 receptor in the rat brain by in situ hybridization histochemistry. *Proc Natl Acad Sci U S A* 1989;86:8560–8564.

39. Hurd YL, Suzuki M, Sedvall GC. D1 and D2 dopamine receptor mRNA expression in whole hemisphere sections of the human brain. *J Chem Neuroanat* 2001;22:127–137.

40. Maltais S, C te S, Drolet G, Falardeau P. Cellular colocalization of dopamine D1 mRNA and D2 receptor in rat brain using a D2 dopamine receptor specific polyclonal antibody. *Prog Neuropsychopharmacol Biol Psychiatry* 2000;24:1127–1149.

41. Madras BK, Fahey MA, Canfield DR, et al. D₁ and D₂ dopamine receptors in caudate-putamen of nonhuman primates (*Macaca fascicularis*). *J Neurochem* 1988;51:934.

42. Beckstead RM. Association of dopamine D₁ and D₂ receptors with specific cellular elements in the basal ganglia of the cat: the uneven topography of dopamine receptors in the striatum is determined by intrinsic striatal cells, not nigrostriatal axons. *Neuroscience* 1988;27:851.

43. Richfield EK, Young AB, Penney JB. Comparative distributions of dopamine D-1 and D-2 receptors in the cerebral cortex of rats, cats, and monkeys. *J Comp Neurol* 1989;286:409.

44. Lidow MS, Goldman-Rakic PS, Gallager DW, et al. Distribution of dopaminergic receptors in the primate cerebral cortex: quantitative autoradiographic analysis using

[³H]raclopride, [³H]spiperone and [³H]SCH23390. *Neuroscience* 1991;40:657.

45. Sesack SR, Bunney BS. Pharmacological characterization of the receptor mediating electrophysiological responses to dopamine in the rat medial prefrontal cortex: a microiontophoretic study. *J Pharmacol Exp Ther* 1989;248:1323.

46. Goldman-Rakic PS. Dopamine-mediated mechanisms of the prefrontal cortex. *Semin Neurosci* 1992;4:149.

47. Smiley JF, Levey AI, Ciliax BJ, et al. D₁ dopamine receptor immunoreactivity in human and monkey cerebral cortex: predominant and extrasynaptic localization in dendritic spines. *Proc Natl Acad Sci U S A* 1994;91:5270.

48. Grace AA. Phasic versus tonic dopamine release and the modulation of dopamine system responsibility: a hypothesis for the etiology of schizophrenia. *Neuroscience* 1991;41:1.

49. Cooper JR, Bloom FE, Roth RH, eds. *The biochemical basis of neuropharmacology,* 6th ed. New York: Oxford University Press, 1991.

50. Bunney BS, Sesack SR, Silva NL. Midbrain dopaminergic systems: neurophysiology and electrophysiological pharmacology. In: Meltzer HY, ed. *Psychopharmacology: the third generation of progress.* New York: Raven Press, 1994:113.

51. Weinrieb RM, O'Brien CP. Persistent cognitive deficits attributed to substance abuse. *Neurol Clin* 1993;11:663.

52. White FJ. Synaptic regulation of mesocorticolimbic dopamine neurons. *Annu Rev Neurosci* 1996;19:405–436.

53. Lacey MG. Neurotransmitter receptors and ionic conductances regulating the activity of neurones in substantia nigra pars compacta and ventral tegmental area. *Prog Brain Res* 1993;99:251–276.

54. L'Hirondel M, Cheramy A, Godeheu G, et al. Lack of autoreceptor-mediated inhibitory control of dopamine release in striatal synaptosomes of D2 receptor-deficient mice. *Brain Res* 1998;792:253–262.

55. Mercuri NB, Saiardi A, Bonci A, et al. Loss of autoreceptor function in dopaminergic neurons from dopamine D2 receptor deficient mice. *Neuroscience* 1997;79:323–327.

56. Koeltzow TE, Xu M, Cooper DC, et al. Alterations in dopamine release but not dopamine autoreceptor function in dopamine D3 receptor mutant mice. *J Neurosci* 1998;18:2231–2238.

57. Chiodo LA, Bannon MJ, Grace AA, et al. Evidence for the absence of impulse-regulating somatodendritic and synthesis-modulating nerve terminal autoreceptors on subpopulations of mesocortical dopamine neurons. *Neuroscience* 1984;12:1.

58. Meador-Woodruff JH, Damask SP, Watson SJ Jr. Differential expression of autoreceptors in the ascending dopamine systems of the human brain. *Proc Natl Acad Sci U S A* 1994;91:8297.

59. Diaz J, Levesque D, Lammers CH, et al. Phenotypical characterization of neurons expressing the dopamine D3 receptor in the rat brain. *Neuroscience* 1995;65:731–745.

60. Levesque D, Diaz J, Pilon C, et al. Identification, characterization, and localization of the dopamine D3 receptor in rat brain using 7-[³H]hydroxy-N,N-di-n-propyl-2-aminotetralin. *Proc Natl Acad Sci U S A* 1992;89:8155–8159.

61. Bouthenet ML, Souil E, Martres MP, et al. Localization of dopamine D3 receptor mRNA in the rat brain using in situ hybridization histochemistry: comparison with dopamine D2 receptor mRNA. *Brain Res* 1991;564:203–219.

62. Altman J, Everitt BJ, Glautier S, et al. The biological, social and clinical bases of drug addiction: commentary and debate. *Psychopharmacology (Berl)* 1996;125:285–345.

63. Sokoloff P, Giros B, Martres M-P, et al. Molecular cloning and characterization of a novel dopamine receptor (D₃) as a target for neuroleptics: *Nature* 1990;347:146.

64. O'Malley KL, Harmon S, Tang L, et al. The rat dopamine

D4 receptor: sequence, gene structure, and demonstration of expression in the cardiovascular system. *New Biol* 1992;4: 137–146.

65. Van Tol HH, Bunzow JR, Guan HC, et al. Cloning of the gene for a human dopamine D4 receptor with high affinity for the antipsychotic clozapine. *Nature* 1991;350:610–614.

66. Mrzljak L, Bergson C, Pappy M, et al. Localization of dopamine D4 receptors in GABAergic neurons of the primate brain. *Nature* 1996;381:245–248.

67. Seeman P. Dopamine receptor sequences. Therapeutic levels of neuroleptics occupy D$_2$ receptors, clozapine occupies D$_4$. *Neuropsychopharmacology* 1992;7:261.

68. Seeman P, Guan H-C, Van Tol HHM. Dopamine D4 receptors elevated in schizophrenia. *Nature* 1993;365:441.

69. Lichter JB, Barr CL, Kennedy JL, et al. A hypervariable segment in the human dopamine receptor D4 (DRD4) gene. *Hum Mol Genet* 1993;2:767–773.

70. Asghari V, Schoots O, van Kats S, et al. Dopamine D4 receptor repeat: analysis of different native and mutant forms of the human and rat genes. *Mol Pharmacol* 1994;46:364–373.

71. Catalano M, Nobile M, Novelli E, et al. Distribution of a novel mutation in the first exon of the human dopamine D4 receptor gene in psychotic patients. *Biol Psychiatry* 1993;34:459–464.

72. Petronis A, O'Hara K, Barr CL, et al. (G)n-mononucleotide polymorphism in the human D4 dopamine receptor (DRD4) gene. *Hum Genet* 1994;93:719.

73. Amara SG, Kuhar MJ. Neurotransmitter transporters: recent progress. *Annu Rev Neurosci* 1993;16:73–93.

74. Uhl GR, Hartig PR. Transporter explosion: update on uptake. *Trends Pharmacol Sci* 1992;13:421–425.

75. Ciliax BJ, Heilman C, Demchyshyn LL, et al. The dopamine transporter: immunochemical characterization and localization in brain. *J Neurosci* 1995;15:1714–1723.

76. Kuhar MJ, Vaughan R, Uhl G, et al. Localization of dopamine transporter protein by microscopic histochemistry. *Adv Pharmacol* 1998;42:168–170.

77. Giros B, Jaber M, Jones SR, et al. Hyperlocomotion and indifference to cocaine and amphetamine in mice lacking the dopamine transporter. *Nature* 1996;379:606–612.

78. Ritz MC, Lamb RJ, Goldberg SR, et al. Cocaine receptors on dopamine transporters are related to self-administration of cocaine. *Science* 1987;237:1219.

79. Boja JW, Vaughan R, Patel A, et al. The dopamine transporter. In: Niznik HB, ed. *Dopamine receptors and transporters. Pharmacology, structure, and function.* New York: Marcel Dekker Inc., 1994:611.

80. Dutta AK, Reith ME, Madras BK. Synthesis and preliminary characterization of a high-affinity novel radioligand for the dopamine transporter. *Synapse* 2001;39:175–181.

81. Reith ME, Xu C, Coffey LL. Binding domains for blockers and substrates on the cloned human dopamine transporter studied by protection against N-ethylmaleimide-induced reduction of 2 beta-carbomethoxy-3 beta-(4-fluorophenyl)[³H]tropane ([³H]WIN 35,428) binding. *Biochem Pharmacol* 1996;52:1435–1446.

82. Xu C, Coffey LL, Reith ME. Binding domains for blockers and substrates on the dopamine transporter in rat striatal membranes studied by protection against N-ethylmaleimide-induced reduction of [³H]WIN 35,428 binding. *Naunyn Schmiedebergs Arch Pharmacol* 1997;355:64–73.

83. Dersch CM, Akunne HC, Partilla JS, et al. Studies of the biogenic amine (amine) transporters. I. Dopamine reuptake blockers inhibit [³H]mazindol binding to the dopamine transporter by a competitive mechanism: preliminary evidence for different binding domains. *Neurochem Res* 1994;19:201.

84. Shimada S, Kitayama, S, Lin C-L, et al. Cloning and expression of a cocaine-sensitive dopamine transporter complementary DNA. *Science* 1991;254:576.

85. Pacholczyk T, Blakely RD, Amara SG. Expression cloning of a cocaine and antidepressant-sensitive human noradrenaline transporter. *Nature* 1991;350:350.

86. Zahniser NR, Doolen S. Chronic and acute regulation of Na$^+$/Cl$^-$-dependent neurotransmitter transporters: drugs, substrates, presynaptic receptors, and signaling systems. *Pharmacol Ther* 2001;92:21–55.

87. Chen N, Reith ME. Structure and function of the dopamine transporter. *Eur J Pharmacol* 2000;405:329–339.

88. Uhl GR. Neurotransmitter transporters (plus): a promising new gene family. *Trends Neurosci* 1992;15:265.

89. Kaufman MJ, Spealman RD, Madras BK. Distribution of cocaine recognition sites in monkey brain: I. In vitro autoradiography with [³H]CFT. *Synapse* 1991;9:177.

90. Letchworth SR, Smith HR, Porrino LJ, et al. Characterization of a tropane radioligand, [(3)H]2beta-propanoyl-3beta-(4-tolyl) tropane ([(3)H]PTT), for dopamine transport sites in rat brain. *J Pharmacol Exp Ther* 2000;293:686–696.

91. Freed C, Revay R, Vaughan RA, et al. Dopamine transporter immunoreactivity in rat brain. *J Comp Neurol* 1995;359:340–349.

92. Lorang D, Amara SG, Simerly RB. Cell-type-specific expression of catecholamine transporters in the rat brain. *J Neurosci* 1994;14:4903–4914.

93. Scheffel U, Dannals RF, Wong DF, et al. Dopamine transporter imaging with novel, selective cocaine analogs. *Neuroreport* 1992;3:969.

94. Wong DF, Yung B, Dannals RF, et al. *In vivo* imaging of baboon and human dopamine transporters by positron emission tomography using [¹¹C]WIN 35,428. *Synapse* 1993;15:130.

95. Madras BK, Spealman RD, Fahey MA, et al. Cocaine receptors labeled by [³H]2 beta-carbomethoxy-3 beta-(4-fluorophenyl)tropane. *Mol Pharmacol* 1989;36:518–524.

96. Boja JW, Patel A, Carroll FI, et al. [¹²⁵I]RTI-55: a potent ligand for dopamine transporters. *Eur J Pharmacol* 1991;194:133–134.

97. Richfield EK. Quantitative autoradiography of the dopamine uptake complex in rat brain using [³H]GBR 12935: binding characteristics. *Brain Res* 1991;540:1–13.

98. Cerruti C, Walther DM, Kuhar MJ, et al. Dopamine transporter mRNA expression is intense in rat midbrain neurons and modest outside midbrain. *Mol Brain Res* 1993;18:181.

99. Nirenberg MJ, Chan J, Vaughan RA, et al. Immunogold localization of the dopamine transporter: an ultrastructural study of the rat ventral tegmental area. *J Neurosci* 1997;17:4037–4044.

100. Kuhar MJ, Ritz MC, Boja JW. The dopamine hypothesis of the reinforcing properties of cocaine. *Trends Neurosci* 1991;14:299–302.

101. Rocha BA, Fumagalli F, Gainetdinov RR, et al. Cocaine self-administration in dopamine-transporter knockout mice. *Nat Neurosci* 1998;1:132–137.

102. Sora I, Hall FS, Andrews AM, et al. Molecular mechanisms of cocaine reward: combined dopamine and serotonin transporter knockouts eliminate cocaine place preference. *Proc Natl Acad Sci U S A* 2001;98:5300–5305.

103. Simon EJ, Hiller JM, Edelman I. Stereospecific binding of the potent narcotic analgesic [³H]etorphine to rat brain homogenate. *Proc Natl Acad Sci U S A* 1973;70:1947.

104. Terenius L. Stereospecific interaction between narcotic analgesics and a synaptic plasma membrane fraction of rat brain cortex. *Acta Pharmacol Toxicol (Copenh)* 1973;32:317.

105. Pert CB, Snyder SH. Opiate receptor: demonstration in nervous tissue. *Science* 1973;179:1011.

106. Martin WR, Eades CG, Thompson JA, et al. The effects of morphine- and nalorphine-like drugs in the nondependent and morphine-dependent chronic spinal dog. *J Pharmacol Exp Ther* 1976;197:517.

107. Gilbert PE, Martin WR. The effects of morphine- and nalorphine-like drugs in the nondependent, morphine-dependent and cyclazocine-dependent chronic spinal dog. *J Pharmacol Exp Ther* 1976;198:66.

108. Vaupel DB. Naltrexone fails to antagonize the σ effects of PCP and SKF 10,047 in the dog. *Eur J Pharmacol* 1983;92:269.

109. Quirion R, Chicheportiche R, Contreras PC, et al. Classification and nomenclature of phencyclidine and *sigma* receptor sites. *Trends Neurosci* 1987;10:444.

110. Walker JM, Bowen WD, Walker FO, et al. Sigma receptors: biology and function. *Pharmacol Rev* 1990;42:355.

111. Zukin RS, Zukin SR. The σ receptor. In: Pasternak GW, ed. The opiate receptors. Clifton, NJ: Humana, 1988:143.

112. Magnan J, Paterson SJ, Tavani A, et al. The binding spectrum of narcotic analgesic drugs with different agonist and antagonist properties. *Naunyn Schmiedebergs Arch Pharmacol* 1982;319:197.

113. Bowen WD. Sigma receptors: recent advances and new clinical potentials. *Pharm Acta Helv* 2000;74:211.

114. Lord JA, Waterfield AA, Hughes J, et al. Endogenous opioid peptides: multiple agonists and receptors. *Nature* 1977;267:495.

115. Goldstein A, Fischli W, Lowney LI, et al. Porcine pituitary dynorphin: complete amino acid sequence of the biologically active heptadecapeptide. *Proc Natl Acad Sci U S A* 1981;78:7219.

116. Chavkin C, James IF, Goldstein A. Dynorphin is a specific endogenous ligand of the κ opioid receptor. *Science* 1982;215:413.

117. Hughes J, Smith TW, Kosterlitz HW, et al. Identification of two related pentapeptides from the brain with potent opiate agonist activity. *Nature* 1975;258:577.

118. Chang KJ, Cuatrecasas P. Multiple opiate receptors. Enkephalins and morphine bind to receptors of different specificity. *J Biol Chem* 1979;254:2610.

119. Bradbury AF, Smyth DG, Snell CR. Biosynthetic origin and receptor conformation of methionine enkephalin. *Nature* 1976;260:165.

120. Cox BM, Goldstein A, Hi CH. Opioid activity of a peptide, beta-lipotropin-(61-91), derived from beta-lipotropin. *Proc Natl Acad Sci U S A* 1976;73:1821.

121. Zadina JE, Hackler L, Ge LJ, et al. A potent and selective endogenous agonist for the mu-opiate receptor. *Nature* 1997;386:499.

122. Kakidani H, Furutani Y, Takahashi H, et al. Cloning and sequence analysis of cDNA for porcine beta-neo-endorphin/dynorphin precursor. *Nature* 1982;298:245.

123. Noda M, Furutani Y, Takahashi H, et al. Cloning and sequence analysis of cDNA for bovine adrenal preproenkephalin. *Nature* 1982;295:202.

124. Nakanishi S, Inoue A, Kita T, et al. Nucleotide sequence of cloned cDNA for bovine corticotropin-beta-lipotropin precursor. *Nature* 1979;278:423.

125. Traynor J. Subtypes of the kappa-opioid receptor: fact or fiction? *Trends Pharmacol Sci* 1989;10:52.

126. Traynor JR, Elliott J. delta-Opioid receptor subtypes and cross-talk with mu-receptors. *Trends Pharmacol Sci* 1993;14:84.

127. Zukin SR, Brady KT, Slifer BL, et al. Behavioral and biochemical stereoselectivity of *sigma* opiate/PCP receptors. *Brain Res* 1984;294:174.

128. Zukin SR. Differing stereospecificities distinguish opiate receptor subtypes. *Life Sci* 1982;31:1307.

129. Chen Y, Mestek A, Liu J, et al. Molecular cloning and functional

130. Wang JB, Imai Y, Eppler CM, et al. μ opiate receptor: cDNA cloning and expression. *Proc Natl Acad Sci U S A* 1993;90:10230.

131. Evans CJ, Keith DE Jr, Morrison H, et al. Cloning of a *delta* opioid receptor by functional expression. *Science* 1992;258:1952.

132. Kieffer BL, Befort K, Gaveriaux-Ruff C, et al. The delta-opioid receptor: isolation of a cDNA by expression cloning and pharmacological characterization. *Proc Natl Acad Sci U S A* 1992;89:12048–12052.

133. Yasuda K, Raynor K, Kong H, et al. Cloning and functional comparison of κ and σ opioids receptors from mouse brain. *Proc Natl Acad Sci U S A* 1993;90:6736.

134. Meng F, Xie GX, Thompson RC, et al. Cloning and pharmacological characterization of a rat kappa opioid receptor. *Proc Natl Acad Sci U S A.* 1993;90:9954.

135. Li S, Zhu J, Chen C, et al. Molecular cloning and expression of a rat kappa opioid receptor. *Biochem J* 1993;295:629.

136. Chen Y, Mestek A, Liu J, et al. Molecular cloning of a rat kappa opioid receptor reveals sequence similarities to the mu and delta opioid receptors. *Biochem J* 1993;295:625.

137. Sharma SK, Klee WA, Nirenberg M. Opiate-dependent modulation of adenylate cyclase. *Proc Natl Acad Sci U S A* 1977;74:3365.

138. Tallent M, Dichter MA, Bell GI, et al. The cloned kappa opioid receptor couples to an N-type calcium current in undifferentiated PC-12 cells. *Neuroscience* 1994;63:1033.

139. Piros ET, Prather PL, Loh HH, et al. Ca^{2+} channel and adenylyl cyclase modulation by cloned mu-opioid receptors in GH3 cells. *Mol Pharmacol* 1995;47:1041.

140. Piros ET, Prather PL, Law PY, et al. Voltage-dependent inhibition of Ca^{2+} channels in GH3 cells by cloned mu- and delta-opioid receptors. *Mol Pharmacol* 1996;50:947.

141. Henry DJ, Grandy DK, Lester HA, et al. Kappa-opioid receptors couple to inwardly rectifying potassium channels when coexpressed by Xenopus oocytes. *Mol Pharmacol* 1995;47:551.

142. Johnson PS, Wang JB, Wang WF, et al. Expressed mu opiate receptor couples to adenylate cyclase and phosphatidyl inositol turnover. *Neuroreport* 1994;5:507.

143. Spencer RJ, Jin W, Thayer SA, et al. Mobilization of Ca^{2+} from intracellular stores in transfected neuro2a cells by activation of multiple opioid receptor subtypes. *Biochem Pharmacol* 1997;54:809.

144. Li LY, Chang KJ. The stimulatory effect of opioids on mitogen-activated protein kinase in Chinese hamster ovary cells transfected to express mu-opioid receptors. *Mol Pharmacol* 1996;50:599.

145. Fukuda K, Kato S, Morikawa H, et al. Functional coupling of the delta-, mu-, and kappa-opioid receptors to mitogen-activated protein kinase and arachidonate release in Chinese hamster ovary cells. *J Neurochem* 1996;67:1309.

146. Burt AR, Carr IC, Mullaney I, et al. Agonist activation of p42 and p44 mitogen-activated protein kinases following expression of the mouse delta opioid receptor in Rat-1 fibroblasts: effects of receptor expression levels and comparisons with G-protein activation. *Biochem J* 1996;320:227.

147. Jordan BA, Cvejic S, Devi LA. Opioids and their complicated receptor complexes. *Neuropsychopharmacol* 2000;23[Suppl]:S5.

148. Zukin RS, Zukin SR. The case for multiple opiate receptors. *Trends Pharmacol Sci* 1984;7:160.

149. Simon EJ. Opiates: neurobiology. In: Lowinson JH, Ruiz P, Millman RB, et al., eds. *Substance abuse. A comprehensive textbook.* Baltimore: Williams & Wilkins, 1992:195.

150. Pfeiffer A, Brantl V, Herz A, et al. Psychotomimesis mediated by kappa opiate receptors. *Science* 1986;233:774.

151. Unterwald E, Sasson S, Kornetsky C. Evaluation of the supraspinal analgesic activity and abuse liability of ethylketocyclazocine. *Eur J Pharmacol* 1987;133:275.

152. Mansour A, Khachaturian H, Lewis ME, et al. Anatomy of CNS opioid receptors. *Trends Neurosci* 1988;11:308.

153. Knapp RJ, Hunt M, Wamsley JK, et al. CNS receptors for opioids. In: London ED, ed. *Imaging drug action in the brain.* Boca Raton, FL: CRC Press, 1993:119.

154. Sim LJ, Selley DE, Childers SR. In vitro autoradiography of receptor-activated G proteins in rat brain by agonist-stimulated guanylyl 5′-[gamma-[^{35}S]thio]-triphosphate binding. *Proc Natl Acad Sci U S A* 1995;92:7242.

155. Sharif NA, Hughes J. Discrete mapping of brain *mu* and *delta* opioid receptors using selective peptides: quantitative autoradiography, species differences and comparison with *kappa* receptors. *Peptides* 1989;10:499.

156. Unterwald EM, Knapp C, Zukin RS. Neuroanatomical localization of kappa 1 and kappa 2 opioid receptors in rat and guinea pig brain. *Brain Res* 1991;562:57.

157. Tobin AJ, Khrestchatisky M.Gene expression in the mammalian nervous system. In: Siegel G, Agranoff B, Albers RW, et al., eds. *Basic neurochemistry.* Raven Press: New York, 1989:417.

158. Arias HR. Localization of agonist and competitive antagonist binding sites on nicotinic acetylcholine receptors. *Neurochem Int* 2000;36:595.

159. Clarke PBS, Pert CB, Pert A. Autoradiographic distribution of nicotine receptors in rat brain. *Brain Res* 1984;323:390.

160. London ED, Waller SB, Wamsley JK. Autoradiographic localization of [^{3}H]nicotine binding sites in the rat brain. *Neurosci Lett* 1985;53:179.

161. Changeux J-P, Devillers-Thiéry A, Chemouilli P. Acetylcholine receptor: an allosteric protein. *Science* 1984;225:1335.

162. Changeux J-P, Giraudat J, Dennis M. The nicotinic acetylcholine receptor: molecular architecture of a ligand-regulated ion channel. *Trends Pharmacol Sci* 1987;8:459.

163. Changeux JP, Kasai M, Lee CY. Use of a snake venom toxin to characterize the cholinergic receptor protein. *Proc Natl Acad Sci U S A* 1970;67:1241.

164. Miledi R, Molinoff P, Potter LT. Isolation of the cholinergic receptor protein of *Torpedo* electric tissue. *Nature* 1971;229:554.

165. Changeux JP. The acetylcholine receptor: an allosteric membrane protein. *Harvey Lect* 1979–80;75:85.

166. Olsen RW, Meunier JC, Changeux JP. Progress in the purification of the cholinergic receptor protein from *Electrophorus electricus* by affinity chromatography. *FEBS Lett* 1972;28:96.

167. Karlin A, Cowburn D. The affinity-labeling of partially purified acetylcholine receptor from electric tissue of *Electrophorus*. *Proc Natl Acad Sci U S A* 1973;70:3636.

168. Chang HW. Purification and characterization of acetylcholine receptor-I from *Electrophorus electricus*. *Proc Natl Acad Sci U S A* 1974;71:2113.

169. Meunier JC, Sealock R, Olsen R, et al. Purification and properties of the cholinergic receptor protein from *Electrophorus electricus* electric tissue. *Proc Natl Acad Sci U S A* 1974;45:371.

170. Weill CL, McNamee MG, Karlin A. Affinity-labeling of purified acetylcholine receptor from *Torpedo californica*. *Biochem Biophys Res Commun* 1974;61:997.

171. Raftery MA, Hunkapiller MW, Strader CD, et al. Acetylcholine receptor: complex of homologous subunits. *Science* 1980;208:1454.

172. Raftery MA, Vandlen R, Michaelson D, et al. The biochemistry of an acetylcholine receptor. *J Supramol Struct* 1974;2:582.

173. Reynolds JA, Karlin A. Molecular weight in detergent solution of acetylcholine receptor from *Torpedo californica*. *Biochem* 1978;17:2035.

174. Devillers-Thiéry A, Changeux JP, Paroutaud P, et al. The amino-terminal sequence of the 40,000 molecular weight subunit of the acetylcholine receptor protein from *Torpedo marmorata*. *FEBS Lett* 1979;104:99.

175. Wada K, Ballivet M, Boulter J, et al. Functional expression of a new pharmacological subtype of brain nicotinic acetylcholine receptor. *Science* 1988;240:330.

176. Boulter J, Evans K, Goldman D, et al. Isolation of a cDNA clone coding for a possible neural nicotinic acetylcholine receptor α-subunit. *Nature* 1986;319:368.

177. Goldman D, Deneris E, Luyten W, et al. Members of a nicotinic acetylcholine receptor gene family are expressed in different regions of the mammalian central nervous system. *Cell* 1987;48:965.

178. Deneris ES, Connolly J, Boulter J, et al. Primary structure and expression of β_2: a novel subunit of neuronal nicotinic acetylcholine receptors. *Neuron* 1988;1:45.

179. Deneris ES, Boulter J, Swanson LW, et al. β_3: a new member of nicotinic acetylcholine receptor gene family is expressed in brain. *J Biol Chem* 1989;264:6268.

180. Boulter J, O'Shea-Greenfield A, Duvoisin RM, et al. α_3, α_5, and β_4: three members of the rat neuronal nicotinic acetylcholine receptor-related gene family form a gene cluster. *J Biol Chem* 1990;265:4472.

181. Boulter J, Connolly J, Deneris E, et al. Functional expression of two neuronal nicotinic acetylcholine receptors from cDNA clones identifies a gene family. *Proc Natl Acad Sci U S A* 1987;84:7763.

182. Wada E, Wada K, Boulter J, et al. Distribution of *alpha*$_2$, *alpha*$_3$, *alpha*$_4$, and *beta*$_2$ neuronal nicotinic receptor subunit mRNAs in the central nervous system: a hybridization histochemical study in the rat. *J Comp Neurol* 1989;284:314.

183. Zukin SR, Zukin RS. Specific [^{3}H]phencyclidine binding in rat central nervous system. *Proc Natl Acad Sci U S A* 1979;76:5372.

184. Vincent JP, Kartalovski B, Geneste P, et al. Interaction of phencyclidine ("angel dust") with a specific receptor in rat brain membranes. *Proc Natl Acad Sci U S A* 1979;76:4678.

185. Javitt DC, Zukin SR. The role of excitatory amino acids in neuropsychiatric illness. *J Neuropsychiatry Clin Neurosci* 1990;2:44.

186. Javitt DC, Zukin SR. Recent advances in the phencyclidine model of schizophrenia. *Am J Psychiatry* 1991;148:1301.

187. Zukin SR, Zukin RS.Phencyclidine. In: Lowinson JH, Ruiz P, Millman RB, et al., eds. *Substance abuse. A comprehensive textbook.* Baltimore: Williams & Wilkins, 1992:290.

188. Monaghan DT, Bridges RJ, Coteman CW. The excitatory amino acid receptors: their classes pharmacology, and distinct properties in the function of the central nervous system. *Annu Rev Pharm Toxicol* 1989;29:365.

189. Kleckner NW, Dingledine R. Requirement for glycine in activation of NMDA-receptors expressed in *Xenopus* oocytes. *Science* 1988;241:835.

190. Javitt DC, Frusciante MJ, Zukin SR. Rat brain *N*-methyl-D-aspartate receptors require multiple molecules of agonist for activation. *Mol Pharmacol* 1990;37:603.

191. Benveniste M, Mayer ML. Kinetic analysis of antagonist action at *N*-methyl-D-aspartic acid receptors. Two binding sites each for glutamate and glycine. *Biophys J* 1991;59:560.

192. Mayer ML, Westbrook GL. Permeation and block of N-methyl-D-aspartic acid receptor channels by divalent cations in mouse cultured central neurones. *J Physiol* 1987;394:501.

193. Moriyoshi K, Masu M, Ishii T, et al. Molecular cloning and characterization of the rat NMDA receptor. *Nature* 1991;354:31.

194. Durand GM, Gregor P, Zheng X, et al. Cloning of an apparent

splice variant of the *N*-methyl-D-aspartate receptor NMDAR1 with altered sensitivity to polyamines and activators of protein kinase C. *Proc Natl Acad Sci U S A* 1992;89:9359.

195. Monyer H, Sprengel R, Schoepfer R, et al. Heteromeric NMDA receptors: molecular and functional distinction of subtypes. *Science* 1992;256:1217.

196. Meguro H, Mori H, Araki K, et al. Functional characterization of a heteromeric NMDA receptor channel expressed from cloned cDNAs. *Nature* 1992;357:70

197. Kutsuwada T, Kashiwabuchi N, Mori H, et al. Molecular diversity of the NMDA receptor channel. *Nature* 1992;358:36.

198. Ciabarra AM, Sullivan JM, Gahn LG, et al. Cloning and characterization of chi-1: a developmentally regulated member of a novel class of the inotropic glutamate receptor family. *J Neurosci* 1995;15:6498.

199. Sucher NJ, Akbarian S, Chi CL, et al. Developmental and regional expression pattern of a novel NMDA receptor-like subunit (NMDAR-L) in the rodent brain. *J Neurosci* 1995;15:6509.

200. Das S, Sasaki YF, Rothe T, et al. Increased NMDA current and spine density in mice lacking the NMDA receptor subunit NR3A. *Nature* 1998;393:377.

201. Nishi M, Hinds H, Lu HP, et al. Motoneuron-specific expression of NR3B, a novel NMDA-type glutamate receptor subunit that works in a dominant-negative manner. *J Neurosci* 2001;21:RC185.

202. Ishii T, Moriyoshi K, Sugihara H, et al. Molecular characterization of the family of the *N*-methyl-D-aspartate receptor subunits. *J Biol Chem* 1993;268:2836.

203. Chatterton JE, Awobuluyi M, Premkumar LS, et al. Excitatory glycine receptors containing the NR3 family of NMDA receptor subunits. *Nature* 2002;415:793.

204. Laube B, Kuhse J, Betz H. Evidence for a tetrameric structure of recombinant NMDA receptors. *J Neurosci* 1998;18:2954.

205. Perez-Otano I, Schulteis CT, Contractor A, et al. Assembly with the NR1 subunit is required for surface expression of NR3A-containing NMDA receptors. *J Neurosci* 2001;21:1228.

206. Soloviev MM, Brierley MJ, Shao ZY, et al. Functional expression of a recombinant unitary glutamate receptor from Xenopus, which contains N-methyl-D-aspartate (NMDA) and non-NMDA receptor subunits. *J Biol Chem* 1996;271:32572.

207. Zhong J, Carrozza DP, Williams K, et al. Expression of mRNAs encoding subunits of the NMDA receptor in developing rat brain. *J Neurochem* 1995;64:531.

208. Laurie DJ, Seeburg PH. Ligand affinities at recombinant N-methyl-D-aspartate receptors depend on subunit composition. *Eur J Pharmacol* 1994;268:335.

209. Ehlers MD, Fung ET, O'Brien RJ, et al. Splice variant-specific interaction of the NMDA receptor subunit NR1 with neuronal intermediate filaments. *J Neurosci* 1998;18:720.

210. Scott DB, Blanpied TA, Swanson GT, et al. An NMDA receptor ER retention signal regulated by phosphorylation and alternative splicing. *J Neurosci* 2001;21:3063.

211. Monyer H, Burnashev N, Laurie DJ, et al. Developmental and regional expression in the rat brain and functional properties of four NMDA receptors. *Neuron* 1994;12:529.

212. Wenzel A, Scheurer L, Künzi R, et al. Distribution of NMDA receptor subunit proteins NR2A, 2B, 2C and 2D in rat brain. *Neuroreport* 1995;7:45.

213. Depoortere R, Perrault G, Sanger DJ. Prepulse inhibition of the startle reflex in rats: effects of compounds acting at various sites on the NMDA receptor complex. *Behav Pharmacol* 1999;10:51.

214. Noda A, Noda Y, Kamei H, et al. Phencyclidine impairs latent learning in mice: interaction between glutamatergic systems and sigma(1) receptors. *Neuropsychopharmacology* 2001;24:451.

215. Mori A, Noda Y, Mamiya T, et al. Phencyclidine-induced discriminative stimulus is mediated via phencyclidine binding sites on the N-methyl-D-aspartate receptor-ion channel complex, not via sigma(1) receptors. *Behav Brain Res* 2001;119:33.

216. Burt DR, Kamatchi GL. GABA$_A$ receptor subtypes: from pharmacology to molecular biology. *FASEB J* 1991;5:2916.

217. Barnard EA, Skolnick P, Olsen RW, et al. Subtypes of gamma-aminobutyric acid A receptors: classification on the basis of subunit structure and receptor function. International Union of Pharmacology. XV. *Pharmacol Rev* 1998;50:291.

218. Macdonald RL, Twyman RE. Kinetic properties and regulation of GABA$_A$ receptor channels. In: Narahashi T, ed. *Ion channels, volume 3*. New York: Plenum Press, 1992:315.

219. Doble A, Martin IL. Multiple benzodiazepine receptors: no reason for anxiety. *Trends Pharmacol Sci* 1992;13:76.

220. Luddens H, Wisden W. Function and pharmacology of multiple GABA$_A$ receptor subtypes. *Trends Pharmacol Sci* 1991;12:49.

221. Bormann J. Electrophysiology of GABA$_A$ and GABA$_B$ receptor subtypes. *Trends Neurosci* 1988;11:112.

222. Yamakura T, Bertaccini E, Trudell JR, et al. Anesthetics and ion channels: molecular models and sites of action. *Annu Rev Pharmacol Toxicol* 2001;41:23.

223. Weight FF, Aguayo LG, White G, et al.GABA- and glutamate-gated ion channels as molecular sites of alcohol and anesthetic action. In: Biggio G, Concas A, Costa E, eds. *GABAergic synaptic transmission*. New York: Raven Press, 1992:335.

224. Demirgören S, Majewska MD, Spivak CE, et al. Receptor binding and electrophysiological effects of dehydroepiandrosterone sulfate, an antagonist of the GABA$_A$ receptor. *Neuroscience* 1991;45:127.

225. Majewska MD, Demirgören S, London ED. Binding of pregnenolone sulfate to rat brain membranes suggest multiple sites of steroid action at the GABA$_A$ receptor. *Eur J Pharmacol* 1990;189:307.

226. Majewska MD, Demirgören S, Spivak CE, et al. The neurosteroid dehydroepiandrosterone sulfate is an allosteric antagonist of the GABA$_A$ receptor. *Brain Res* 1990;526:143.

227. Harrison NL, Majewska MD, Harrington JW, et al. Structure-activity relationships for steroid interaction with the γ-aminobutyric acid$_A$ receptor complex. *J Pharmacol Exp Ther* 1987;241:346.

228. Majewska MD, Harrison NL, Schwartz RD, et al. Steroid hormone metabolites are barbiturate-like modulators of the GABA receptor. *Science* 1986;232:1004.

229. Costa E, Guidotti A. Diazepam binding inhibitor (DBI): a peptide with multiple biological actions. *Life Sci* 1991;49:325.

230. Gaoni Y, Mechoulam R. Isolation, structure, and partial synthesis of an active constituent of hashish. *J Am Chem Soc* 1964;86:1646.

231. Garrett ER, Hunt CA. Physiochemical properties, solubility, and protein binding of Δ^9-tetrahydrocannabinol. *J Pharm Sci* 1974;63:1056.

232. Martin BR, Balster RL, Razdan RK, et al. Behavioral comparisons of the stereoisomers of tetrahydrocannabinols. *Life Sci* 1981;29:565.

233. Johnson MR, Melvin LS. The discovery of nonclassical cannabinoid analgetics. In: Mechoulam R, ed. *Cannabinoids as therapeutic agents*. Boca Raton, FL: CRC Press, 1986:121.

234. Howlett AC, Johnson MR, Melvin LS, et al. Nonclassical cannabinoid analgesics inhibit adenylate cyclase: development of a cannabinoid receptor model. *Mol Pharmacol* 1988;33:297.

235. Little PJ, Compton DR, Johnson MR, et al. Pharmacology and stereoselectivity of structurally novel cannabinoids in mice. *J Pharmacol Exp Ther* 1988;247:1046.

236. Devane WA, Dysarz FA III, Johnson MR, et al. Determination

and characterization of a cannabinoid receptor in rat brain. *Mol Pharmacol* 1988;34:605.

237. Herkenham M, Lynn AB, Little MB, et al. Cannabinoid receptor localization in brain. *Proc Natl Acad Sci U S A* 1990;87:1932.

238. Herkenham M, Lynn AB, Johnson MR, et al. Characterization and localization of cannabinoid receptor in rat brain: a quantitative *in vitro* autoradiographic study. *J Neurosci* 1991;11:563.

239. Matsuda LA, Lolait JS, Brownstein M, et al. Structure of a cannabinoid receptor and functional expression of the cloned cDNA. *Nature* 1990;346:561.

240. Munro S, Thomas KL, Abu-Shaar M. Molecular characterization of a peripheral receptor for cannabinoids. *Nature* 1993;365:61.

241. Devane WA, Hanus L, Breuer A, et al. Isolation and structure of a brain constituent that binds to the cannabinoid receptor. *Science* 1992;258:1946.

242. Fride E, Mechoulam R. Pharmacological activity of the cannabinoid receptor agonist, anandamide, a brain constituent. *Eur J Pharmacol* 1993;231:313.

243. Mechoulam R, Fride E, Di Marzo V. Endocannabinoids. *Eur J Pharmacol* 1998;359:1.

244. Kumar RN, Chambers WA, Pertwee RG. Pharmacological actions and therapeutic uses of cannabis and cannabinoids. *Anaesthesia* 2001;56:1059.

245. De Petrocellis L, Bisogno T, Maccarrone M, et al. The activity of anandamide at vanilloid VR1 receptors requires facilitated transport across the cell membrane and is limited by intracellular metabolism. *J Biol Chem* 2001;276:12856.

246. Seevers MH. Opiate addiction in the monkey I. Methods of study. *J Pharmacol Exp Ther* 1935;56:147.

247. Goldberg SR, Kelleher RT. Behavior controlled by scheduled injections of cocaine in squirrel and rhesus monkeys. *J Exp Anal Behav* 1976;25:93.

248. Thompson T, Schuster CR. Morphine self-administration, food-reinforced, and avoidance behaviors in rhesus monkeys. *Psychopharmacologia* 1964;5:87.

249. Schuster CR. Psychological approaches to opiate dependence and self-administration by laboratory animals. *Fed Proc* 1970;29:1.

250. Pickens R, Thompson T. Cocaine-reinforced behavior in rats: effects of reinforcement magnitude and fixed-ratio size. *J Pharmacol Exp Ther* 1968;161:122.

251. Thompson T, Pickens R. Stimulant self-administration by animals: some comparisons with opiate self-administration. *Fed Proc* 1970;29:6.

252. Gardner EL. Brain reward mechanisms. In: Lowinson JH, Ruiz P, Millman RB, et al., eds. *Substance abuse. A comprehensive textbook.* Baltimore: Williams & Wilkins, 1992:70.

253. Robinson TE, Berridge KC. The neural basis of drug craving: an incentive-sensitization theory of addiction. *Brain Res Rev* 1993;18:247.

254. Lamb RJ, Preston KL, Schindler CW, et al. The reinforcing and subjective effects of morphine in post-addicts: a dose-response study. *J Pharmacol Exp Ther* 1991;259:1165.

255. Cottler LB. Comparing DSM-III-R and ICD-10 substance use disorders. *Addiction* 1993;88:689–696.

256. Rounsaville BJ, Bryant K, Babor T, et al. Cross system agreement for substance use disorders: DSM-III-R, DSM-IV and ICD-10. *Addiction* 1993;88:337–348.

257. Heather N. A conceptual framework for explaining drug addiction. *J Psychopharmacol* 1998;12:3–7.

258. West R. Theories of addiction. *Addiction* 2001;96:3–13.

259. Thorndike EL, ed. *Animal intelligence.* New York: MacMillan, 1911.

260. Skinner BF, ed. *The behavior of organisms: an experimental analysis.* New York: Appleton-Century-Crofts, 1938.

261. Messier C, White NM. Contingent and non-contingent actions

262. of sucrose and saccharin reinforcers: effects on taste preference and memory. *Physiol Behav* 1984;32:195.

262. White N, Legree P. Effect of post-training exposure to an aversive stimulus on retention. *Physiol Psychol* 1984;12:233.

263. Stinus L, Le Moal M, Koob GF. Nucleus accumbens and amygdala are possible substrates for the aversive stimulus effects of opiate withdrawal. *Neuroscience* 1990;37:767.

264. Olds J, Milner P. Positive reinforcement produced by electrical stimulation of septal area and other regions of rat brain. *J Comp Physiol Psychol* 1954;47:419.

265. Olds ME, Olds J. Approach-avoidance analysis of rat diencephalon. *J Comp Neurol* 1963;120:259.

266. Olds J. Pleasure centers in the brain. *Sci Am* 1956;195:105.

267. Miller NE. Motivational effects of brain stimulation and drugs. *Fed Proc* 1960;19:846.

268. Margules DL, Olds J. Identical "feeding" and "rewarding" systems in the lateral hypothalamus of rats. *Science* 1962;135:374.

269. Hoebel BG, Teitelbaum P. Hypothalamic control of feeding and self-stimulation. *Science* 1962;132:375.

270. Coons EE, Levak M, Miller NE. Lateral hypothalamus: learning of food-seeking response motivated by electrical stimulation. *Science* 1965;150:1320.

271. Hoebel BG. Feeding and self-stimulation. *Ann N Y Acad Sci* 1969;157:758.

272. Killam KF, Olds J, Sinclair J. Further studies on the effects of centrally acting drugs on self-stimulation. *J Pharmacol Exp Ther* 1957;119:157

273. Yokel RA. Intravenous self-administration: response rates, the effects of pharmacological challenges, and drug preference. In: Bozarth MA, ed. *Methods of assessing the reinforcing properties of abused drugs.* New York: Springer-Verlag, 1987:1.

274. Goldberg SR. Comparable behavior maintained under fixed-ratio and second-order schedules of food presentation, cocaine injection or *d*-amphetamine injection in the squirrel monkey. *J Pharmacol Exp Ther* 1973;186:18.

275. Yokel RA, Pickens R. Self-administration of optical isomers of amphetamine and methylamphetamine by rats. *J Pharmacol Exp Ther* 1973;187:27.

276. Fletcher PJ, Azampanah A, Korth KM. Activation of 5-HT(1B) receptors in the nucleus accumbens reduces self-administration of amphetamine on a progressive ratio schedule. *Pharmacol Biochem Behav* 2002;71:717–725.

277. Klebaur JE, Phillips SB, Kelly TH, et al. Exposure to novel environmental stimuli decreases amphetamine self-administration in rats. *Exp Clin Psychopharmacol* 2001;9:372–379.

278. Harrod SB, Dwoskin LP, Crooks PA, et al. Lobeline attenuates d-methamphetamine self-administration in rats. *J Pharmacol Exp Ther* 2001;298:172–179.

279. Martin-Fardon R, Weiss F. BTCP is a potent reinforcer in rats: comparison of behavior maintained on fixed- and progressive-ratio schedules. *Pharmacol Biochem Behav* 2002;72:343–353.

280. Ranaldi R, Wise RA. Blockade of D1 dopamine receptors in the ventral tegmental area decreases cocaine reward: possible role for dendritically released dopamine. *J Neurosci* 2001;21:5841–5846.

281. Rowlett JK. A labor-supply analysis of cocaine self-administration under progressive-ratio schedules: antecedents, methodologies, and perspectives. *Psychopharmacology (Berl)* 2000;153:1–16.

282. Munzar P, Goldberg SR. Dopaminergic involvement in the discriminative-stimulus effects of methamphetamine in rats. *Psychopharmacology (Berl)* 2000;148:209–216.

283. McMillan DE, Li M. Drug discrimination under a concurrent fixed-ratio schedule. *J Exp Anal Behav* 1999;72:187–204.

284. Johanson CE, Schuster CR. A choice procedure for drug reinforcers: cocaine and methylphenidate in the rhesus monkey. *J Pharmacol Exp Ther* 1975;193:676.

285. Rush CR, Essman WD, Simpson CA, et al. Reinforcing and subject-rated effects of methylphenidate and d-amphetamine in non-drug-abusing humans. *J Clin Psychopharmacol* 2001;21:273–286.

286. Goldberg SR, Spealman RD, Goldberg DM. Persistent behavior at high rates maintained by intravenous self-administration of nicotine. *Science* 1981;214:573.

287. Shoaib M, Schindler CW, Goldberg SR. Nicotine self-administration in rats: strain and nicotine pre-exposure effects on acquisition. *Psychopharmacology (Berl)* 1997;129:35–43.

288. Suto N, Austin JD, Vezina P. Locomotor response to novelty predicts a rat's propensity to self-administer nicotine. *Psychopharmacology (Berl)* 2001;158:175–180.

289. Donny EC, Caggiula AR, Mielke MM, et al. Nicotine self-administration in rats on a progressive ratio schedule of reinforcement. *Psychopharmacology (Berl)* 1999;147:135–142.

290. Deneau G, Yanagita T, Seevers MH. Self-administration of psychoactive substances by the monkey. *Psychopharmacologia* 1969;16:30.

291. Weeks JR. Experimental morphine addiction: method for automatic intravenous injections in unrestrained rats. *Science* 1962;138:143.

292. Schulteis G, Stinus L, Risbrough VB, et al. Clonidine blocks acquisition but not expression of conditioned opiate withdrawal in rats. *Neuropsychopharmacology* 1998;19:406–416.

293. Alderson HL, Robbins TW, Everitt BJ. Heroin self-administration under a second-order schedule of reinforcement: acquisition and maintenance of heroin-seeking behaviour in rats. *Psychopharmacology (Berl)* 2000;153:120–133.

294. Woods JH, Ikomi F, Winger G. The reinforcing property of ethanol. In: Roach MK, McIsaac WM, Creaven PJ, eds. *Biological aspects of alcohol.* Austin, TX: University of Texas Press, 1971:371.

295. Rodd-Henricks ZA, Melendez RI, Zaffaroni A, et al. The reinforcing effects of acetaldehyde in the posterior ventral tegmental area of alcohol-preferring rats. *Pharmacol Biochem Behav* 2002;72:55–64.

296. Elmor GI, George FR. Genetic differences in the operant rate-depressant effects of ethanol between four inbred mouse strains. *Behav Pharmacol* 1995;6:794–800.

297. Czachowski CL, Chappell AM, Samson HH. Effects of raclopride in the nucleus accumbens on ethanol seeking and consumption. *Alcohol Clin Exp Res* 2001;25:1431–1440.

298. Davis JD, Lulenski GC, Miller NE. Comparative studies of barbiturate self-administration. *Int J Addict* 1968;3:207.

299. Munzar P, Yasar S, Redhi GH, et al. High rates of midazolam self-administration in squirrel monkeys. *Behav Pharmacol* 2001;12:257–265.

300. Weerts EM, Ator NA, Griffiths RR. Comparison of the intravenous reinforcing effects of propofol and methohexital in baboons. *Drug Alcohol Depend* 1999;57:51–60.

301. Findley JD, Robinson WW, Peregrino L. Addiction to secobarbital and chlordiazepoxide in the rhesus monkey by means of a self-infusion preference procedure. *Psychopharmacologia* 1972;26:93.

302. Weerts EM, Griffiths RR. Zolpidem self-injection with concurrent physical dependence under conditions of long-term continuous availability in baboons. *Behav Pharmacol* 1998;9:285–297.

303. Balster RL, Johanson CE, Harris RT. Phencyclidine self-administration in the rhesus monkey. *Pharmacol Biochem Behav* 1973;1:167.

304. Pilotto R, Singer G, Overstreet D. Self-injection of diazepam in naive rats: effects of dose, schedule and blockade of different receptors. *Psychopharmacology (Berl)* 1984;84:174.

305. Rodefer JS, Carroll ME. A comparison of progressive ratio schedules versus behavioral economic measures: effect of an alternative reinforcer on the reinforcing efficacy of phencyclidine. *Psychopharmacology (Berl)* 1997;132:95–103.

306. Rodefer JS, Carroll ME. Concurrent progressive-ratio schedules to compare reinforcing effectiveness of different phencyclidine (PCP) concentrations in rhesus monkeys. *Psychopharmacology (Berl)* 1999;144:163–174.

307. Fattore L, Cossu G, Martellotta CM, et al. Intravenous self-administration of the cannabinoid CB1 receptor agonist WIN 55,212-2 in rats. *Psychopharmacology (Berl)* 2001;156:410–416.

308. Woods JH, Schuster CR. Reinforcement properties of morphine, cocaine, and SPA as a function of unit dose. *Int J Addict* 1968;3:231.

309. Carney JM, Llewellyn ME, Woods JH. Variable interval responding maintained by intravenous codeine and ethanol injections in the rhesus monkey. *Pharmacol Biochem Behav* 1977;5:577.

310. Winger G, Stitzer ML, Woods JH. Barbiturate-reinforced responding in rhesus monkeys: comparisons of drugs with different durations of action. *J Pharmacol Exp Ther* 1975;195:505.

311. Kelleher RT. Characteristics of behavior controlled by scheduled injections of drugs. *Pharmacol Rev* 1975;27:307.

312. Hoffmeister F, Goldberg SR. A comparison of chlorpromazine, imipramine, morphine and d-amphetamine self-administration in cocaine-dependent rhesus monkeys. *J Pharmacol Exp Ther* 1973;187:8.

313. Collins RJ, Weeks JR, Cooper MM, et al. Prediction of abuse liability of drugs using IV self-administration by rats. *Psychopharmacology (Berl)* 1984;82:6.

314. Hoffmeister F, Wuttke W. Psychotropic drugs as negative reinforcers. *Pharmacol Rev* 1976;27:419.

315. Kandel DA, Schuster CR. An investigation of nalorphine and perphenazine as negative reinforcers in an escape paradigm. *Pharmacol Biochem Behav* 1977;6:61.

316. Herming RI, Jones RT, Benowitz NL, et al. How a cigarette is smoked determines blood nicotine levels. *Clin Pharmacol Ther* 1983;33:84.

317. Yokel RA, Pickens R. Drug levels of *d* and *l*-amphetamine during intravenous self-administration. *Psychopharmacologia* 1974;34:255.

318. Hubner CB, Koob GF. Bromocriptine produces decreases in cocaine self-administration in the rat. *Neuropsychopharmacology* 1990;3:101.

319. de Wit H, Wise RA. Blockade of cocaine reinforcement in rats with dopamine receptor blocker pimozide, but not with the noradrenergic blockers phentolamine or phenoxybenzamine. *Can J Psychol* 1977;31:195.

320. Roberts DCS, Vickers G. Atypical neuroleptics increase self-administration of cocaine: an evaluation of a behavioural screen for antipsychotic activity. *Psychopharmacology (Berl)* 1984;82:135.

321. Ettenberg A, Pettit HO, Bloom FE, et al. Heroin and cocaine intravenous self-administration in rats: mediation by separate neural systems. *Psychopharmacology (Berl)* 1982;78:204.

322. Johanson CE, Kandel DA, Bonese K. The effects of perphenazine on self-administration behavior. *Pharmacol Biochem Behav* 1976;4:427.

323. Hubner CB, Moreton JE. Effects of selective D_1 and D_2 dopamine antagonists on cocaine self-administration in the rat. *Psychopharmacology (Berl)* 1991;105:151.

324. Le Foll B, Schwartz JC, Sokoloff P. Dopamine D3 receptor agents

as potential new medications for drug addiction. *Eur Psychiatry* 2000;15:140–146.

325. Bouthenet ML, Souil E, Martres MP, et al. Localization of dopamine D3 receptor mRNA in the rat brain using in situ hybridization histochemistry: comparison with dopamine D2 receptor mRNA. *Brain Res* 1991;564:203–219.

326. Diaz J, Levesque D, Lammers CH, et al. Phenotypical characterization of neurons expressing the dopamine D3 receptor in the rat brain. *Neuroscience* 1995;65:731–745.

327. Levesque D, Diaz J, Pilon C, et al. Identification, characterization, and localization of the dopamine D3 receptor in rat brain using 7-[^{3}H]hydroxy-N,N-di-n-propyl-2-aminotetralin. *Proc Natl Acad Sci U S A* 1992;89:8155–8159.

328. Altman J, Everitt BJ, Glautier S, et al. The biological, social and clinical bases of drug addiction: commentary and debate. *Psychopharmacology (Berl)* 1996;125:285–345.

329. Caine SB, Koob GF, Parsons LH, et al. D3 receptor test in vitro predicts decreased cocaine self-administration in rats. *Neuroreport* 1997;8:2373–2377.

330. Caine SB, Koob GF. Modulation of cocaine self-administration in the rat through D-3 dopamine receptors. *Science* 1993;260:1814–1816.

331. Roberts DCS, Koob GF, Klonoff P, et al. Extinction and recovery of cocaine self-administration following 6-hydroxydopamine lesions of the nucleus accumbens. *Pharmacol Biochem Behav* 1980;12:781.

332. Roberts DCS, Corcoran ME, Fibiger HC. On the role of ascending catecholaminergic systems in intravenous self-administration of cocaine. *Pharmacol Biochem Behav* 1977;6:615.

333. Pettit HO, Ettenberg A, Bloom FE, et al. Destruction of dopamine in the nucleus accumbens selectively attenuates cocaine but not heroin self-administration in rats. *Psychopharmacology (Berl)* 1984;84:167.

334. Gerrits MA, Van Ree JM. Effect of nucleus accumbens dopamine depletion on motivational aspects involved in initiation of cocaine and heroin self-administration in rats. *Brain Res* 1996;713:114–124.

335. Lyness WH, Friedle NM, Moore KE. Destruction of dopaminergic nerve terminals in nucleus accumbens: effect on d-amphetamine self-administration. *Pharmacol Biochem Behav* 1979;11:553.

336. Zito KA, Vickers G, Roberts DCS. Disruption of cocaine and heroin self-administration following kainic acid lesions of the nucleus accumbens. *Pharmacol Biochem Behav* 1985;23:1029.

337. Roberts DCS, Koob GF. Disruption of cocaine self-administration following 6-hydroxydopamine lesions of the ventral tegmental area in rats. *Pharmacol Biochem Behav* 1982;17:901.

338. Roberts DCS, Zito KA. Interpretation of lesion effects on stimulant self-administration. In: Bozarth MA, ed. *Methods of assessing the reinforcing properties of abused drugs.* New York: Springer-Verlag, 1987:87.

339. McGregor A, Baker G, Roberts DC. Effect of 6-hydroxydopamine lesions of the medial prefrontal cortex on intravenous cocaine self-administration under a progressive ratio schedule of reinforcement. *Pharmacol Biochem Behav* 1996;53:5–9.

340. Glick SD, Cox RD. Dopaminergic and cholinergic influences on morphine self-administration in rats. *Res Commun Chem Pathol Pharmacol* 1975;12:17.

341. Bozarth MA, Wise RA. Heroin reward is dependent on a dopaminergic substrate. *Life Sci* 1981;29:1881.

342. Bozarth MA. Neural basis of psychomotor stimulant and opiate reward: evidence suggesting the involvement of a common dopaminergic system. *Behav Brain Res* 1986;22:107.

343. Spyraki C, Fibiger HC, Phillips AG. Attenuation of heroin reward in rats by disruption of the mesolimbic dopamine system. *Psychopharmacology (Berl)* 1983;79:278.

344. Parkinson JA, Olmstead MC, Burns LH, et al. Dissociation in effects of lesions of the nucleus accumbens core and shell on appetitive pavlovian approach behavior and the potentiation of conditioned reinforcement and locomotor activity by D-amphetamine. *J Neurosci* 1999;19:2401–2411.

345. Alderson HL, Parkinson JA, Robbins TW, et al. The effects of excitotoxic lesions of the nucleus accumbens core or shell regions on intravenous heroin self-administration in rats. *Psychopharmacology (Berl)* 2001;153:455–463.

346. Hutcheson DM, Parkinson JA, Robbins TW, et al. The effects of nucleus accumbens core and shell lesions on intravenous heroin self-administration and the acquisition of drug-seeking behaviour under a second-order schedule of heroin reinforcement. *Psychopharmacology (Berl)* 2001;153:464–472.

347. Wise RA. Action of drugs of abuse on brain reward systems. *Pharmacol Biochem Behav* 1980;13:213.

348. Wise RA. The dopamine synapse and the notion of "pleasure centers" in the brain. *Trends Neurosci* 1980;3:91.

349. Wise RA, Bozarth MA. Brain reward circuitry: four circuit elements "wired" in apparent series. *Brain Res Bull* 1984;12:203.

350. Zaborszky L, Alheid GF, Alones VE, et al. Afferents of the ventral pallidum studies with a combined immunohistochemical-anterograde degeneration method. *Soc Neurosci Abstr* 1982;8:218.

351. Hubner CB, Koob GF. The ventral pallidum plays a role in mediating cocaine and heroin self-administration in the rat. *Brain Res* 1989;508:20.

352. Robledo P, Koob GF. Two discrete nucleus accumbens projection areas differentially mediate cocaine self-administration in the rat. *Behav Brain Res* 1993;55:159–166.

353. Phillips AG, Mora F, Rolls ET. Intracerebral self-administration of amphetamine by Rhesus monkeys. *Neurosci Lett* 1981; 24:81.

354. Monaco AP, Hernandez L, Hoebel BG. Nucleus accumbens: site of amphetamine self-injection: comparison with lateral ventricle. In: Chronister RB, DeFrance JF, eds. *The neurobiology of the nucleus accumbens.* Brunswick, ME: Haer Institute for Electrophysiological Research, 1981:33.

355. Hoebel BG, Monaco AP, Hernandez L, et al. Self-injections of amphetamine directly into the brain. *Psychopharmacology (Berl)* 1983;81:158.

356. McBride WJ, Murphy JM, Ikemoto S. Localization of brain reinforcement mechanisms: intracranial self-administration and intracranial place-conditioning studies. *Behav Brain Res* 1999;101:129–152.

357. Carlezon WA Jr, Wise RA. Rewarding actions of phencyclidine and related drugs in nucleus accumbens shell and frontal cortex. *J Neurosci* 1996;16:3112–3122.

358. Ikemoto S, Glazier BS, Murphy JM, et al. Rats self-administer carbachol directly into the nucleus accumbens. *Physiol Behav* 1998;63:811–814.

359. Goeders NE, Smith JE. Cortical dopaminergic involvement in cocaine reinforcement. *Science* 1983;221:773.

360. Goeders NE, Smith JE. Intracranial cocaine self-administration into the medial prefrontal cortex increases dopamine turnover in the nucleus accumbens. *J Pharmacol Exp Ther* 1993;265:592–600.

361. Tzschentke TM. The medial prefrontal cortex as a part of the brain reward system. *Amino Acids* 2000;19:211–219.

362. Bozarth MA, Wise RA. Intracranial self-administration of morphine in to the ventral tegmental area in rats. *Life Sci* 1981;28:551.

363. Bedford JA, Bailey LP, Wilson MC. Cocaine reinforced progressive ratio performance in the rhesus monkey. *Pharmacol Biochem Behav* 1978;9:631.

364. David V, Durkin TP, Cazala P. Differential effects of the dopamine D(2)/D(3) receptor antagonist sulpiride on self-administration of morphine into the ventral tegmental area or the nucleus accumbens. *Psychopharmacology (Berl)* 2002;160:307–317.

365. Devine DP, Wise RA. Self-administration of morphine, DAMGO, and DPDPE into the ventral tegmental area of rats. *J Neurosci* 1994;14:1978–1984.

366. Stein EA, Olds J. Direct intracerebral self-administration of opiates in the rat. *Soc Neurosci Abstr* 1977;3:302

367. David V, Cazala P. Differentiation of intracranial morphine self-administration behavior among five brain regions in mice. *Pharmacol Biochem Behav* 1994;48:625–633.

368. Olds ME. Reinforcing effects of morphine in the nucleus accumbens. *Brain Res* 1982;237:429.

369. van Ree JM, de Wied D. Involvement of neurohypophyseal peptides in drug-mediated adaptive responses. *Pharmacol Biochem Behav* 1980;13[Suppl 1]:257.

370. Goeders NE, Lane JD, Smith JE. Self-administration of methionine enkephalin into the nucleus accumbens. *Pharmacol Biochem Behav* 1984;20:451.

371. Olds ME, Williams KN. Self-administration of D-Ala^2-Met-enkephalinamide at hypothalamic self-stimulation sites. *Brain Res* 1980;194:155.

372. Britt MD, Wise RA. Ventral tegmental site of opiate reward: antagonism by a hydrophilic opiate receptor blocker. *Brain Res* 1983;258:105.

373. Vaccarino FJ, Bloom FE, Koob GF. Blockade of nucleus accumbens opiate receptors attenuates intravenous heroin reward in the rat. *Psychopharmacology (Berl)* 1985;86:37.

374. Corrigall WA, Vaccarino FJ. Antagonist treatment in nucleus accumbens or periaqueductal grey affects heroin self-administration. *Pharmacol Biochem Behav* 1988;30:443.

375. Gill CA, Holz WC, Zirkle CL, et al. Pharmacological modification of cocaine and apomorphine self-administration in the squirrel monkey. In: Denker P, Radouco-Thomas C, Villeneuve A, eds. *Proceedings of the Tenth Congress of the Collegium Internationale Neuro-Psychopharmacologicum.* New York: Pergamon Press, 1978:1477.

376. Yokel RA, Wise RA. Amphetamine-type reinforcement by dopaminergic agonists in the rat. *Psychopharmacology (Berl)* 1978;58:289.

377. Woolverton WL, Goldberg LI, Ginos JZ. Intravenous self-administration of dopamine receptor agonists by rhesus monkeys. *J Pharmacol Exp Ther* 1984;230:678.

378. Baxter BL, Gluckman MI, Stein L, et al. Self-injection of apomorphine in the rat: positive reinforcement by a dopamine receptor stimulant. *Pharmacol Biochem Behav* 1974;2:387.

379. Davis WM, Smith SG. Catecholaminergic mechanisms of reinforcement: direct assessment by drug self-administration. *Life Sci* 1977;20:483.

380. Ikemoto S, Glazier BS, Murphy JM, et al. Role of dopamine D1 and D2 receptors in the nucleus accumbens in mediating reward. *J Neurosci* 1997;17:8580–8587.

381. Winger G, Woods JH. Comparison of fixed-ratio and progressive-ratio schedules of maintenance of stimulant drug-reinforced responding. *Drug Alcohol Depend* 1985;15:123.

382. Risner ME, Silcox DL. Psychostimulant self-administration by beagle dogs in a progressive-ratio paradigm. *Psychopharmacology (Berl)* 1981;75:25.

383. Corwin RL, Woolverton WL, Schuster CR, et al. Anorectics: effects on food intake and self-administration in rhesus monkeys. *Alcohol Drug Res* 1987;7:351.

384. van der Zee P, Koger HS, Gootjes J, et al. 1,4-dialk(en)ylpeperazines as selective and very potent inhibitors of dopamine uptake. *Eur J Med Chem* 1980;15:363.

385. Heikkila RE, Manzino L. Behavioral properties of GBR 12909, GBR 13609 and GBR 13098: specific inhibitors of dopamine uptake. *Eur J Pharmacol* 1984;103:241.

386. Woods JH, Katz JL, Medzihradsky F, et al. Evaluation of new compounds for opioid activity: 1982 annual report. *NIDA Res Monogr* 1983;43:457.

387. Wilson MC, Schuster CR. Mazindol self-administration in the rhesus monkey. *Pharmacol Biochem Behav* 1976;4:207.

388. Bergman J, Madras BK, Johnson SE, et al. Effects of cocaine and related drugs in nonhuman primates. III. Self-administration by squirrel monkeys. *J Pharmacol Exp Ther* 1989;251:150.

389. Ranaldi R, Anderson KG, Carroll FI, et al. Reinforcing and discriminative stimulus effects of RTI 111, a 3-phenyltropane analog, in rhesus monkeys: interaction with methamphetamine. *Psychopharmacology (Berl)* 2000;153:103–110.

390. Carlezon WA Jr, Devine DP, Wise RA. Habit-forming actions of nomifensine in nucleus accumbens. *Psychopharmacology (Berl)* 1995;122:194–197.

391. Heikkila RE, Cabbat FS, Manzino L, et al. Rotational behavior induced by cocaine analogs in rats with unilateral 6-hydroxydopamine lesions of the substantia nigra: dependence upon dopamine uptake inhibition. *J Pharmacol Exp Ther* 1979;211:189.

392. Dwoskin LP, Crooks PA. A novel mechanism of action and potential use for lobeline as a treatment for psychostimulant abuse. *Biochem Pharmacol* 2002;63:89–98.

393. Pontieri FE, Passarelli F, Calo L, et al. Functional correlates of nicotine administration: similarity with drugs of abuse. *J Mol Med* 1998;76:193–201.

394. Balfour DJ, Benwell ME, Birrell CE, et al. Sensitization of the mesoaccumbens dopamine response to nicotine. *Pharmacol Biochem Behav* 1998;59:1021–1030.

395. Sziraki I, Sershen H, Benuck M, et al. Receptor systems participating in nicotine-specific effects. *Neurochem Int* 1998;33:445–457.

396. Di Chiara G. Role of dopamine in the behavioural actions of nicotine related to addiction. *Eur J Pharmacol* 2000;393:295–314.

397. Corrigall WA. Regulation of intravenous nicotine self-administration—dopamine mechanisms. In: Adlkofer F, Thurau K, eds. *Effects of nicotine on biological systems.* Basel: Birkhauser Verlag, 1991:423.

398. Corrigall WA, Coen KM. Selective dopamine antagonists reduce nicotine self-administration. *Psychopharmacology (Berl)* 1991;104:171.

399. Singer G, Wallace M, Hall R. Effects of dopaminergic nucleus accumbens lesions on the acquisition of schedule induced self injection of nicotine in the rat. *Pharmacol Biochem Behav* 1982;17:579.

400. Corrigall WA, Franklin KBJ, Coen KM, et al. The mesolimbic dopaminergic system is implicated in the reinforcing effects of nicotine. *Psychopharmacology (Berl)* 1992;107:285.

401. Clarke PBS, Pert A. Autoradiographic evidence for nicotine receptors on nigrostriatal and mesolimbic dopaminergic neurons. *Brain Res* 1985;348:355.

402. Deutch AY, Holliday J, Roth RH, et al. Immunohistochemical localization of a neuronal nicotinic acetylcholine receptor in mammalian brain. *Proc Natl Acad Sci U S A* 1987;84:8697.

403. Swanson LW, Simmons DM, Whiting PJ, et al. Immunohis-

tochemical localization of neuronal nicotinic receptors in the rodent central nervous system. *J Neurosci* 1987;7:3334.

404. Calabresi P, Lacey MG, North RA. Nicotinic excitation of rat ventral tegmental neurones *in vitro* studied by intracellular recording. *Br J Pharmacol* 1989;98:135.

405. Grenhoff J, Aston-Jones G, Svensson TH. Nicotinic effects on the firing pattern of midbrain dopamine neurons. *Acta Physiol Scand* 1986;128:351.

406. Clarke PBS, Hommer DW, Pert A, et al. Electrophysiological actions of nicotine on substantia nigra single units. *Br J Pharmacol* 1985;85:827.

407. London ED, Connolly RJ, Szikszay M, et al. Effects of nicotine on local cerebral glucose utilization in the rat. *J Neurosci* 1988;8:3920.

408. Rowell PP, Carr LA, Garner AC. Stimulation of [^{3}H]dopamine release by nicotine in rat nucleus accumbens. *J Neurochem* 1987;49:1449.

409. Imperato A, Mulas A, Di Chiara G. Nicotine preferentially stimulates dopamine release in the limbic system of freely moving rats. *Eur J Pharmacol* 1986;132:337.

410. Mifsud J-C, Hernandez L, Hoebel BG. Nicotine infused into the nucleus accumbens increases synaptic dopamine as measured by *in vivo* microdialysis. *Brain Res* 1989;478:365.

411. Nisell M, Nomikos GG, Svensson TH. Systemic nicotine-induced dopamine release in the rat nucleus accumbens is regulated by nicotinic receptors in the ventral tegmental area. *Synapse* 1994;16:36.

412. Heath RG, Gallant DM. Activity of the human brain during emotional thought. In: Heath RG, ed. *The role of pleasure in behavior.* New York: Harper & Row, 1964:83.

413. Izenwasser S, Kornetsky C. Brain-stimulation reward: a method for assessing the neurochemical bases of drug-induced euphoria. In: Watson RR, ed. *Drugs of abuse and neurobiology.* Boca Raton, FL: CRC Press, 1992:1.

414. Esposito RU, Kornetsky C. Opioids and rewarding brain stimulation. *Neurosci Biobehav Rev* 1978;2:115.

415. Esposito RU, Porrino LJ, Seeger TF. Brain stimulation reward: measurement and mapping by psychophysical techniques and quantitative 2-[^{14}C]deoxyglucose autoradiography. In: Bozarth MA, ed. *Methods of assessing the reinforcing properties of abused drugs.* New York: Springer-Verlag, 1987: 421.

416. Seeger TF, Carlson KR. Amphetamine and morphine: additive effects on ICSS threshold. *Soc Neurosci Abstr* 1981;7:974

417. Hubner CB, Bain GT, Kornetsky C. Morphine and amphetamine: effect on brain stimulation reward. *Soc Neurosci Abstr* 1983;9:893

418. Weeks JR, Collins RJ. Factors affecting voluntary morphine intake in self-maintained addicted rats. *Psychopharmacologia* 1964;6:267.

419. Pickens R. Self-administration of stimulants by rats. *Int J Addict* 1968;3:215.

420. Lepore M, Franklin KBJ. Modelling drug kinetics with brain stimulation: dopamine antagonists increase self-stimulation. *Pharmacol Biochem Behav* 1992;41:489.

421. Olds J, Travis RP. Effects of chlorpromazine, meprobamate, pentobarbital and morphine on self-stimulation. *J Pharmacol Exp Ther* 1960;128:397.

422. Wise RA, Rompre PP. Brain dopamine and reward. *Annu Rev Psychol* 1989;40:191.

423. Robertson A, Mogenson GJ. Evidence for a role for dopamine in self-stimulation of the nucleus accumbens of the rat. *Can J Psychol* 1978;32:67.

424. Stellar JR, Kelley AE, Corbett D. Effects of peripheral and central dopamine blockade on lateral hypothalamic self-stimulation:

425. Stellar JR, Corbett D. Regional neuroleptic microinjections indicate a role for nucleus accumbens in lateral hypothalamic self-stimulation reward. *Brain Res* 1989;477:126.

426. Mora F, Sanguinetti AM, Rolls ET, et al. Differential effects of self-stimulation and motor behaviour produced by microintracranial injections of a dopamine-receptor blocking agent. *Neurosci Lett* 1975;1:179.

427. Kurumiya S, Nakajima S. Dopamine D$_1$ receptors in the nucleus accumbens: involvement in the reinforcing effect of tegmental stimulation. *Brain Res* 1988;448:1.

428. Singh J, Desiraju T, Raju TR. Dose-response functions of apomorphine, SKF 38393, LY 171555, haloperidol and clonidine on the self-stimulation evoked from lateral hypothalamus and ventral tegmentum. *Indian J Physiol Pharmacol* 1996;40:15–22.

429. Singh J, Desiraju T, Raju TR. Dopamine receptor sub-types involvement in nucleus accumbens and ventral tegmentum but not in medial prefrontal cortex: on self-stimulation of lateral hypothalamus and ventral mesencephalon. *Behav Brain Res* 1997;86:171–179.

430. Ungerstedt U. Stereotaxic mapping of the monoamine pathways in the rat brain. *Acta Physiol Scand* 1971;367[Suppl]:1.

431. Deutsch JA. Behavioral measurement of the neural refractory period and its application to intracranial self-stimulation. *J Comp Physiol Psychol* 1964;58:1.

432. Gallistel CR. Self-stimulation: failure of pretrial stimulation to affect rats' electrode preference. *J Comp Physiol Psychol* 1969;69:722.

433. Gallistel CR. The incentive of brain-stimulation reward. *J Comp Physiol Psychol* 1969;69:713.

434. Gallistel CR, Rolls E, Greene D. Neuron function inferred from behavioral and electrophysiological estimates of refractory period. *Science* 1969;166:1028.

435. Gallistel CR. Self-stimulation in the rat: quantitative characteristics of the reward pathway. *J Comp Physiol Psychol* 1978;92:977.

436. Gallistel CR, Shizgal P, Yeomans JS. A portrait of the substrate for self-stimulation. *Psychol Rev* 1981;88:228.

437. Shizgal P, Bielajew C, Corbett D, et al. Behavioral methods for inferring anatomical linkage between rewarding brain stimulation sites. *J Comp Physiol Psychol* 1980;94:227.

438. Gallistel CR. Self stimulation. In: Deutsch JA, ed. *The physiological basis of memory.* New York: Academic Press, 1983:269.

439. Beninato M, Spencer RF. The cholinergic innervation of the rat substantia nigra: a light and electron microscopic immunohistochemical study. *Exp Brain Res* 1988;72:178.

440. Clarke PBS, Hommer DW, Pert A, et al. Innervation of substantia nigra neurons by cholinergic afferents from pedunculopontine nucleus in the rat: neurochemical and electrophysiological evidence. *Neuroscience* 1987;23:1011.

441. Gould E, Woolf NJ, Butcher LL. Cholinergic projections to the substantia nigra from the pedunculopontine and lateral dorsal tegmental nuclei. *Neuroscience* 1989;28:611.

442. Scarnati E, Prioria A, Campana E, et al. A microiontophoretic study of the nature of the putative synaptic neurotransmitter involved in the pedunculopontine substantia nigra pars compacta excitatory pathway in the rat. *Exp Brain Res* 1986;62: 470.

443. Gratton A, Wise RA. Hypothalamic reward mechanism: two first-stage fiber populations with cholinergic component. *Science* 1985;227:545.

444. Yeomans JS. Two substrates for medial forebrain bundle self-stimulation: myelinated axons and dopamine axons. *Neurosci Biobehav Rev* 1989;13:91.

445. Wise RA, Bozarth MA. Brain substrates for reinforcement and

evidence for both reward and motor deficits. *Pharmacol Biochem Behav* 1983;18:433.

drug self-administration. *Prog Neuropsychopharmacol Biol Psychiatry* 1981;5:467.

446. Broekkamp CLE, Pijnenburg AJJ, Cools AR, et al. The effect of microinjections of amphetamine into the neostriatum and the nucleus accumbens on self-stimulation behaviour. *Psychopharmacologia* 1975;42:179.

447. Broekkamp CL, van den Bogaard JH, Heijnen HJ, et al. Separation of inhibiting and stimulating effects of morphine on self-stimulation behaviour by intracerebral microinjections. *Eur J Pharmacol* 1976;36:443.

448. Broekkamp CLE, Van Rossum JM. The effect of microinjections of morphine and haloperidol into the neostriatum and the nucleus accumbens on self-stimulation behaviour. *Arch Int Pharmacodyn Ther* 1975;217:110.

449. Fiorino DF, Coury A, Fibiger HC, et al. Electrical stimulation of reward sites in the ventral tegmental area increases dopamine transmission in the nucleus accumbens of the rat. *Behav Brain Res* 1993;55:131–141.

450. Miliaressis E, Emond C, Merali Z. Re-evaluation of the role of dopamine in intracranial self-stimulation using in vivo microdialysis. *Behav Brain Res* 1991;46:43–48.

451. Nakahara D, Fuchikami K, Ozaki N, et al. Differential effect of self-stimulation on dopamine release and metabolism in the rat medial frontal cortex, nucleus accumbens and striatum studied by in vivo microdialysis. *Brain Res* 1992;574:164–170.

452. Gilbert DB, Millar J, Cooper SJ. The putative dopamine D3 agonist, 7-OH-DPAT, reduces dopamine release in the nucleus accumbens and electrical self-stimulation to the ventral tegmentum. *Brain Res* 1995;681:1–7.

453. Yavich L, Tiihonen J. In vivo voltammetry with removable carbon fibre electrodes in freely-moving mice: dopamine release during intracranial self-stimulation. *J Neurosci Methods* 2000;104: 55–63.

454. Young SD, Michael AC. Voltammetry of extracellular dopamine in rat striatum during ICSS-like electrical stimulation of the medial forebrain bundle. *Brain Res* 1993;600:305–307.

455. Garris PA, Kilpatrick M, Bunin MA, et al. Dissociation of dopamine release in the nucleus accumbens from intracranial self-stimulation. *Nature* 1999;398:67–69.

456. Rebec GV, Grabner CP, Johnson M, et al. Transient increases in catecholaminergic activity in medial prefrontal cortex and nucleus accumbens shell during novelty. *Neuroscience* 1997;76:707–714.

457. Schultz W, Dayan P, Montague PR. A neural substrate of prediction and reward. *Science* 1997;275:1593–1599.

458. Zetterström T, Sharp T, Marsden CA, et al. *In vivo* measurement of dopamine and its metabolites by intracerebral dialysis: changes after *d*-amphetamine. *J Neurochem* 1983;41:1769.

459. Carboni E, Imperato A, Perezzani L, et al. Amphetamine, cocaine, phencyclidine and nomifensine increase extracellular dopamine concentrations preferentially in the nucleus accumbens of freely moving rats. *Neuroscience* 1989;28: 653.

460. Bradberry CW, Roth RH. Cocaine increases extracellular dopamine in rat nucleus accumbens and ventral tegmental area as shown by *in vivo* microdialysis. *Neurosci Lett* 1989;103:97.

461. Di Chiara G, Imperato A. Drugs abused by humans preferentially increase synaptic dopamine concentrations in the mesolimbic system of freely moving rats. *Proc Natl Acad Sci U S A* 1988;85:5274.

462. Imperato A, Di Chiara G. Preferential stimulation of dopamine release in the nucleus accumbens of freely moving rats by ethanol. *J Pharmacol Exp Ther* 1986;239:219.

463. Di Chiara G, Imperato A. Preferential stimulation of dopamine release in the nucleus accumbens by opiates, alcohol, and barbi-

turates: studies with transcerebral dialysis in freely moving rats. *Ann N Y Acad Sci* 1986;473:367.

464. Hernandez L, Auerbach S, Hoebel BG. Phencyclidine (PCP) injected in the nucleus accumbens increases extracellular dopamine and serotonin as measured by microdialysis. *Life Sci* 1988;42:1713.

465. Chen J, Paredes W, Li J, et al. Δ^9-Tetrahydrocannabinol produces naloxone-blockable enhancement of presynaptic basal dopamine efflux in nucleus accumbens of conscious, freely-moving rats as measured by intracerebral microdialysis. *Psychopharmacology (Berl)* 1990;102:156.

466. Chen J, Paredes W, Lowinson J, et al. Δ^9-Tetrahydrocannabinol enhances presynaptic dopamine efflux in medial prefrontal cortex. *Eur J Pharmacol* 1990;190:259.

467. Imperato A, Puglisi-Allegra S, Zocchi A, et al. Stress activation of limbic and cortical dopamine release is prevented by ICS 205-930 but not by diazepam. *Eur J Pharmacol* 1990;175:211.

468. Zetterström T, Fillenz M. Local administration of flurazepam has different effects on dopamine release in striatum and nucleus accumbens: a microdialysis study. *Neuropharmacology* 1990;29:129.

469. Finlay JM, Damsma G, Fibiger HC. Benzodiazepine-induced decreases in extracellular concentrations of dopamine in the nucleus accumbens after acute and repeated administration. *Psychopharmacology (Berl)* 1992;106:202.

470. Fuxe K, Agnati LF, Bolme P, et al. The possible involvement of GABA mechanisms in the action of benzodiazepines on central catecholamine neurons. In: Costa E, Greengard P, eds. *Mechanism of action of benzodiazepines*. New York: Raven Press, 1975:45.

471. Ishiko J, Inagaski C, Takaori S. Effects of diazepam, nitrazepam and brotizolam on dopamine turnover in the olfactory tubercle, nucleus accumbens and caudate nucleus of rats. *Jpn J Pharmacol* 1983;33:706.

472. Lyness WH. Effect of L-tryptophan pretreatment on *d*-amphetamine self administration. *Subst Alcohol Actions Misuse* 1983;4:305.

473. Cone EJ, Risner ME, Neidert GL. Concentrations of phenethylamine in dog following single doses and during intravenous self administration. *Res Commun Chem Pathol Pharmacol* 1978;22:211.

474. Yu DSL, Smith FL, Smith DG, et al. Fluoxetine-induced attenuation of amphetamine self-administration in rats. *Life Sci* 1986;39:1383.

475. Leccese AP, Lyness WH. The effects of putative 5-hydroxytryptamine receptor active agents on *d*-amphetamine self-administration in controls and rats with 5,7-dihydroxytryptamine median forebrain bundle lesions. *Brain Res* 1984;303:153.

476. Fletcher PJ, Azampanah A, Korth KM. Activation of 5-HT(1B) receptors in the nucleus accumbens reduces self-administration of amphetamine on a progressive ratio schedule. *Pharmacol Biochem Behav* 2002;71:717–725.

477. Fletcher PJ, Korth KM. RU-24969 disrupts d-amphetamine self-administration and responding for conditioned reward via stimulation of 5-HT1B receptors. *Behav Pharmacol* 1999;10:183–193.

478. Zabik JE, Roache JD, Sidor R, et al. The effects of fluoxetine on ethanol preference in the rat. *Pharmacologist* 1982;24:204.

479. Lamb RJ, Jarbe TU. Effects of fluvoxamine on ethanol-reinforced behavior in the rat. *J Pharmacol Exp Ther* 2001;297:1001–1009.

480. Carroll ME, Lac ST, Asencio M, et al. Fluoxetine reduces intravenous cocaine self-administration in rats. *Pharmacol Biochem Behav* 1990;35:237.

481. Rockman GE, Amit Z, Bourque C, et al. Reduction of voluntary morphine consumption following treatment with zimelidine. *Arch Int Pharmacodyn* 1980;244:124.
482. Naranjo CA, Sellers EM, Roach CA, et al. Zimelidine-induced variations in ethanol intake in non-depressed drinkers. *Clin Pharmacol Ther* 1984;35:374.
483. Ritz MC, Kuhar MJ. Relationship between self-administration of amphetamine and monoamine receptors in brain: comparison with cocaine. *J Pharmacol Exp Ther* 1989;248:1010.
484. Carboni E, Acquas E, Leone P, et al. 5HT₃ receptor antagonists block morphine- and nicotine- but not amphetamine-induced reward. *Psychopharmacology (Berl)* 1989;97:175.
485. Pulvirenti L, Maldonado-Lopez R, Koob GF. NMDA receptors in the nucleus accumbens modulate intravenous cocaine but not heroin self-administration in the rat. *Brain Res* 1992;594:327.
486. Panksepp J, Herman BH, Vilberg T, et al. Endogenous opioids and social behavior. *Neurosci Biobehav Rev* 1978;4:473.
487. Vaupel DB, Lange WR, London ED. Effects of verapamil on morphine-induced euphoria, analgesia and respiratory depression in humans. *J Pharmacol Exp Ther* 1993;267:1386.
488. Hernandez L, Hoebel BG. Feeding and hypothalamic stimulation increase dopamine turnover in the accumbens. *Physiol Behav* 1988;44:599.
489. Robbins TW. Behavioural determinants of drug action: rate-dependence revisited. In: Cooper SJ, ed. *Theory in psychopharmacology.* London: Academic Press, 1981:1.
490. Fibiger HC, Phillips AG. Mesocorticolimbic dopamine systems and reward. In: Kalivas PW, Nemeroff CB, eds. *The mesocorticolimbic dopamine system.* Bethesda, MD: New York Academy of Science, 1988:206.
491. Ljungberg T. Blockade by neuroleptics of water intake and operant responding for water in the rat: anhedonia, motor deficit, or both? *Pharmacol Biochem Behav* 1987;27:341.
492. Ljungberg T. Effects of the dopamine D-1 antagonist SCH 23390 on water intake, water-rewarded operant responding and apomorphine-induced decrease of water intake in rats. *Pharmacol Biochem Behav* 1989;33:709.
493. Negus SS, Dykstra LA. κ Antagonist properties of buprenorphine in the shock titration procedure. *Eur J Pharmacol* 1988;156:77.
494. Tyrka A, Gayle C, Smith GP. Raclopride decreases sucrose intake of rat pups in independent ingestion test. *Pharmacol Biochem Behav* 1992;43:863.
495. Kelley AE, Delfs JM. Dopamine and conditioned reinforcement: I. Differential effects of amphetamine microinjections into striatal subregions. *Psychopharmacology (Berl)* 1991;103:187.
496. Cador M, Taylor JR, Robbins TW. Potentiation of the effects of reward-related stimuli by dopaminergic-dependent mechanisms in the nucleus accumbens. *Psychopharmacology (Berl)* 1991;104:377.
497. Heffner TG, Luttinger D, Hartman JA, et al. Regional changes in brain catecholamine turnover in the rat during performance on fixed ratio and variable interval schedules of reinforcement. *Brain Res* 1981;214:215.
498. Joseph MH, Hodges H. Lever pressing for food reward and changes in dopamine turnover and uric acid in rat caudate and nucleus accumbens studied chronically by *in vivo* voltametry. *J Neurosci Methods* 1990;34:143.
499. Pert CB, Hill JM, Ruff MR, et al. Octapeptides deduced from the neuropeptide receptor-like pattern of antigen T4 in brain potently inhibit human immunodeficiency virus receptor binding and T-cell infectivity. *Proc Natl Acad Sci U S A* 1986;83:9254.
500. Browne RG, Welch WM. Stereoselective antagonism of phencyclidine's discriminative properties by adenosine receptor agonists. *Science* 1982;217:1157.
501. Nies AS. Principles of therapeutics. In: Gilman AG, Rall TW, Nies AS, et al., eds. *The pharmacological basis of therapeutics.* New York: Pergamon Press, 1990:62.
502. Jaffe JH. Drug addiction and drug abuse. In: Gilman AG, Rall TW, Nies AS, et al., eds. *The pharmacological basis of therapeutics.* New York: Pergamon Press, 1990:522.
503. Goodman RR, Adler BA, Pasternak GW. Regional distribution of opioid receptors. In: Pasternak GW, ed. *The opiate receptors.* Clifton, NJ: Humana, 1988:197.
504. White FJ, Wolf ME. Psychomotor stimulants. In: Pratt JA, ed. *The biological bases of drug tolerance.* London: Academic Press, 1991:153.
505. Wolf ME, White FJ, Hu X-T. Behavioral sensitization to MK-801 (dizocilpine): neurochemical and electrophysiological correlates in the mesoaccumbens dopamine system. *Behav Pharmacol* 1993;4:429.
506. Koob GF, Bloom FE. Cellular and molecular mechanisms of drug dependence. *Science* 1988;242:715.
507. Falk JL, Fang MA, Lau CE. Chronic oral cocaine self-administration: pharmacokinetics and effects or spontaneous and discriminative motor functions. *J Pharmacol Exp Ther* 1991;257:457.
508. Reith ME, Benuck M, Lajtha A. Cocaine disposition in the brain after continuous or intermittent treatment and locomotor stimulation in mice. *J Pharmacol Exp Ther* 1991;243:281.
509. Collins AC, Burch JB, de Fiebre CM, et al. Tolerance to and cross tolerance between ethanol and nicotine. *Pharmacol Biochem Behav* 1988;29:365.
510. Burch JB, de Fiebre CM, Marks MJ, et al. Chronic ethanol or nicotine treatment results in partial cross-tolerance between these agents. *Psychopharmacology (Berl)* 1988;95:452.
511. Nestler EJ. Molecular mechanisms of drug addiction. *J Neurosci* 1992;12:2439.
512. Loh HH, Smith AP. Molecular characterization of opioid receptors. *Annu Rev Pharmacol Toxicol* 1990;30:123.
513. Wuster M, Schulz R, Herz A. Opioid tolerance and dependence: re-evaluating the unitary hypothesis. *Trends Pharmacol Sci* 1985;64.
514. Ling GSF, MacLeod JM, Lee S, et al. Separation of morphine analgesia from physical dependence. *Science* 1984;226:462.
515. Suzuki T, Hayashi Y, Misawa M. The role of *mu*₁ receptor in physical dependence on morphine using the *mu* receptor deficient CXBK mouse. *Life Sci* 1992;50:849.
516. Zimprich A, Simon T, Hollt V. Cloning and expression of an isoform of the rat mu opioid receptor (rMOR1B) which differs in agonist induced desensitization from rMOR1. *FEBS Lett* 1995;359:142.
517. Jordan BA, Devi LA. G-protein-coupled receptor heterodimerization modulates receptor function. *Nature* 1999;399:697.
518. Jordan BA, Cvejic S, Devi LA. Opioids and their complicated receptor complexes. *Neuropsychopharmacology* 2000;23:S5.
519. Bozarth MA, Wise RA. Anatomically distinct opiate receptor fields mediate reward and physical dependence. *Science* 1984;224:516.
520. Sharma SK, Klee WA, Nirenberg M. Dual regulation of adenylate cyclase accounts for narcotic dependence and tolerance. *Proc Natl Acad Sci U S A* 1975;72:3092.
521. Collier HO. Cellular site of opiate dependence. *Nature* 1980;283:625.
522. Nestler EJ. Molecular mechanisms of drug addiction. *J Neurosci* 1992;12:2439.
523. Nestler EJ, Hope BT, Widnell KL. Drug addiction: a model for the molecular basis of neural plasticity. *Neuron* 1993;11:995.
524. Aghajanian GK, Wang Y-Y. Common α₂- and opiate effector mechanisms in the locus coeruleus: intracellular studies in brain slices. *Neuropharmacology* 1987;26:793.

525. Redmond DE Jr, Krystal JH. Multiple mechanism of withdrawal from opioid drugs. *Annu Rev Neurosci* 1984;7:443.

526. Williams JT, North RA. Opiate-receptor interactions on single locus coeruleus neurones. *Mol Pharmacol* 1984;26:489.

527. Grant SJ, Redmond DE Jr. The neuroanatomy and pharmacology of the nucleus locus coeruleus. In: Lal H, Fielding S, eds. *Psychopharmacology of clonidine.* New York: Alan R. Liss, 1981:5.

528. Foote SL, Bloom FE, Aston-Jones G. Nucleus locus coeruleus: new evidence of anatomical and physiological specificity. *Physiology* 1983;63:844.

529. Pinnock RD. A highly selective kappa-opioid receptor agonist, CI-977, reduces excitatory synaptic potentials in the rat locus coeruleus in vitro. *Neuroscience* 1992;47:87.

530. McFadzean I, Lacey MG, Hill RG, et al. Kappa opioid receptor activation depresses excitatory synaptic input to rat locus coeruleus neurons *in vitro. Neuroscience* 1987;20:231.

531. Korf J, Bunney BS, Aghajanian GK. Noradrenergic neurons: morphine inhibition of spontaneous activity. *Eur J Pharmacol* 1974;25:165.

532. Aghajanian GK. Tolerance of locus coeruleus neurones to morphine and suppression of withdrawal response by clonidine. *Nature* 1978;276:186.

533. Bird SJ, Kuhar MJ. Iontophoretic application of opiates to the locus coeruleus. *Brain Res* 1977;122:523.

534. Strahlendorf HK, Strahlendorf JC, Barnes CD. Endorphin-mediated inhibition of locus coeruleus neurons. *Brain Res* 1980;191:284.

535. Andrade R, Vandermaelen CP, Aghajanian GK. Morphine tolerance and dependence in the locus coeruleus: single cell studies in the brain slices. *Eur J Pharmacol* 1983;91:161.

536. Christie MJ, Williams JT, North RA. Cellular mechanisms of opioid tolerance: studies in single brain neurons. *Mol Pharmacol* 1987;32:633.

537. North RA, Williams JT. On the potassium conductance increased by opioids in rat locus coeruleus neurones. *J Physiol (Lond)* 1985;364:265.

538. Aghajanian GK, Wang Y-Y. Pertussis toxin blocks the outward currents evoked by opiate and α_2 agonists in locus coeruleus neurons. *Brain Res* 1986;371:390.

539. Pepper CM, Henderson G. Opiates and opioid peptides hyperpolarize locus coeruleus neurons *in vitro. Science* 1980;209:394.

540. Childers SR. Opioid receptor-coupled second messenger systems. *Life Sci* 1991;48:1991.

541. Wang Y-Y, Aghajanian GK. Intracellular GTPγS restores the ability of morphine to hyperpolarize rat locus coeruleus neurons after the blockade by pertussis toxin. *Brain Res* 1987;436:396.

542. Nestler EJ, Erdos JJ, Terwilligner R, et al. Regulation of G-proteins by chronic morphine in the rat locus coeruleus. *Brain Res* 1989;476:230.

543. Nestler EJ, Tallman JF. Chronic morphine treatment increases cyclic AMP-dependent protein kinase activity in the rat locus coeruleus. *Mol Pharmacol* 1988;33:127.

544. Alreja M, Aghajanian GK. Opiates suppress a resting sodium-dependent inward current and activate an outward potassium current in locus coeruleus neurons. *J Neurosci* 1993;13:3525.

545. Alreja M, Aghajanian GK. QX-314 blocks the potassium but not the sodium-dependent component of the opiate response in locus coeruleus neurons. *Brain Res* 1994;639:320.

546. Collier HOJ. Cellular site of opiate dependence. *Nature* 1980;283:625.

547. Collier HOJ, Francis DL. Morphine abstinence is associated with increased brain cyclic AMP. *Nature* 1975;255:159.

548. Duman RS, Tallman JF, Nestler EJ. Acute and chronic opiate-regulation of adenylate cyclase in brain: specific effects in locus coeruleus. *J Pharmacol Exp Ther* 1988;246:1033.

549. Matsuoka I, Maldonado R, Defer N, et al. Chronic morphine administration causes region-specific increase of brain type VIII adenylyl cyclase mRNA. *Eur J Pharmacol* 1994;268:215.

550. Lane-Ladd SB, Pineda J, Boundy VA, et al. CREB (cAMP response element-binding protein) in the locus coeruleus: biochemical, physiological, and behavioral evidence for a role in opiate dependence. *J Neurosci* 1997;17:7890.

551. Trujillo KA, Akil H. Inhibition of morphine tolerance and dependence by the NMDA receptor antagonist MK-801. *Science* 1991;251:85.

552. Kolesnikov YA, Pick CG, Pasternak GW. N^G-nitro-l-arginine prevents morphine tolerance. *Eur J Pharmacol* 1992;221:399.

553. Vaupel DB, Kimes AS, London ED. Nitric oxide synthase inhibitors. Preclinical studies of potential use for treatment of opioid withdrawal. *Neuropsychopharmacology* 1995;13:315.

554. Highfield DA, Grant S. Ng-nitro-L-arginine, an NOS inhibitor, reduces tolerance to morphine in the rat locus coeruleus. *Synapse* 1998;29:233.

555. Kolesnikov YA, Pick CG, Ciszewska G, et al. Blockade of tolerance to morphine but not to κ opioids by a nitric oxide synthase inhibitor. *Proc Natl Acad Sci U S A* 1993;90:5162.

556. Dawson VL, Dawson TM, London ED, et al. Nitric oxide mediates glutamate neurotoxicity in primary cortical cultures. *Proc Natl Acad Sci U S A* 1991;88:6368.

557. Zhang J, Dawson VL, Dawson TM, et al. Nitric oxide activation of poly(ADP-ribose) synthetase in neurotoxicity. *Science* 1994;263:687.

558. Geary WA, II, Wooten GF. Similar functional anatomy of spontaneous and precipitated morphine withdrawal. *Brain Res* 1985;334:183.

559. Blasig J, Hetz A, Reinhold K, et al. Development of physical dependence on morphine in respect to time and dosage and quantification of the precipitated withdrawal syndrome in rats. *Psychopharmacologia* 1973;33:19.

560. Tatum AL, Seevers MH, Collins KH. Morphine addiction and its physiological interpretation based on experimental evidences. *J Pharmacol Exp Ther* 1992;36:447.

561. Jiang ZG, North RA. Pre- and postsynaptic inhibition by opioids in rat striatum. *J Neurosci* 1992;12:356.

562. Negus SS, Henriksen SJ, Mattox A, et al. Effect of antagonists selective for *mu, delta,* and *kappa* opioid receptors on the reinforcing effects of heroin in rats. *J Pharmacol Exp Ther* 1993;265:1245.

563. Kimes AS, London ED. Glucose utilization in the rat brain during chronic morphine treatment and naloxone-precipitated morphine withdrawal. *J Pharmacol Exp Ther* 1989;248:538.

564. Maldonado R. Participation of noradrenergic pathways in the expression of opiate withdrawal: biochemical and pharmacological evidence. *Neurosci Biobehav Rev* 1997;21:91.

565. Koob GF, Maldonado R, Stinus L. Neural substrates of opiate withdrawal. *Trends Neurosci* 1992;15:186.

566. Rossetti ZL, Longu G, Mercuro G, et al. Extraneuronal noradrenaline in the prefrontal cortex of morphine-dependent rats: tolerance and withdrawal mechanisms. *Brain Res* 1993;609:316.

567. Bowen WD, Kirschner BN, Newman AH, et al. σ receptors negatively modulate agonist-stimulated phosphoinositide metabolism in rat brain. *Eur J Pharmacol* 1988;149:399.

568. Charney DS, Sternberg DE, Kleber HD, et al. The clinical use of clonidine in abrupt withdrawal from methadone. Effects on blood pressure and specific signs and symptoms. *Arch Gen Psychiatry* 1981;38:1273.

569. Taylor JR, Elsworth JD, Garcia EJ, et al. Clonidine infusions

into the locus coeruleus attenuate behavioral and neurochemical changes associated with naloxone-precipitated withdrawal. *Psychopharmacology (Berl)* 1988;96:121.

570. Klonoff DC, Andrews BT, Obana WG. Stroke associated with cocaine use. *Arch Neurol* 1989;46:989.

571. Esposito E, Kruszewska A, Ossowska G, et al. Noradrenergic and behavioural effects of naloxone injected in the locus coeruleus of morphine-dependent rats and their control by clonidine. *Psychopharmacology (Berl)* 1987;93:393.

572. Valentino RJ, Wehby RG. Locus ceruleus discharge characteristics of morphine-dependent rats: effects of naltrexone. *Brain Res* 1989;488:126.

573. Swann AC, Elsworth JD, Charney DS, et al. Brain catecholamine metabolites and behavior in morphine withdrawal. *Eur J Pharmacol* 1983;86:167.

574. Crawley JN, Ninan PT, Pickar D, et al. Neuropharmacological antagonism of the β carboline-induced "anxiety" response in rhesus monkeys. *J Neurosci* 1985;5:477.

575. Silverstone PH, Done C, Sharp T. In vivo monoamine release during naloxone-precipitated morphine withdrawal. *Neuroreport* 1993;4:1043.

576. Kogan JH, Nestler EJ, Aghajanian GK. Elevated basal firing rates and enhanced responses to 8-Br-cAMP in locus coeruleus neurons in brain slices from opiate-dependent rats. *Eur J Pharmacol* 1992;211:47.

577. Rasmussen K, Beitner-Johnson D, Krystal JH, et al. Opiate withdrawal and the rat locus coeruleus: behavioral, electrophysiological, and biochemical correlates. *J Neurosci* 1990;10:2308.

578. Grant SJ, Aston-Jones G, Redmond DE Jr. Responses of primate locus coeruleus neurons to simple and complex sensory stimuli. *Brain Res Bull* 1988;21:401.

579. Maldonado R, Stinus L, Gold LH, et al. Role of different brain structures in the expression of the physical morphine withdrawal syndrome. *J Pharmacol Exp Ther* 1992;261:669.

580. Maldonado R, Koob GF. Destruction of the locus coeruleus decreases physical signs of opiate withdrawal. *Brain Res* 1993;605:128.

581. Nestler EJ, Alreja M, Aghajanian GK. Molecular and cellular mechanisms of opiate action: studies in the rat locus coeruleus. *Brain Res Bull* 1994;35:521.

582. Matsuoka I, Maldonado R, Defer N, et al. Chronic morphine administration causes region-specific increase of brain type VIII adenylyl cyclase mRNA. *Eur J Pharmacol* 1994;268:215.

583. Valentino RJ, Aston-Jones G. Activation of locus coeruleus neurons in the rat by a benzazocine derivative (UM 1046) that mimics opiate withdrawal. *Neuropharmacology* 1983;22:1363.

584. Rasmussen K, Aghajanian GK. Withdrawal-induced activation of locus coeruleus neurons in opiate-dependent rats: attenuation by lesions of the nucleus paragigantocellularis. *Brain Res* 1989;505:346.

585. Akaoka H, Aston-Jones G. Opiate withdrawal-induced hyperactivity of locus coeruleus neurons is substantially mediated by augmented excitatory amino acid input. *J Neurosci* 1991;11:3830.

586. Rasmussen K, Krystal JH, Aghajanian GK. Excitatory amino acids and morphine withdrawal: differential effects of central and peripheral kynurenic acid administration. *Psychopharmacology* 1991;105:508.

587. Tung CS, Grenhoff J, Svensson TH. Morphine withdrawal responses of rat locus coeruleus neurons are blocked by an excitatory amino-acid antagonist. *Acta Physiol Scand* 1990;138:581.

588. Cappendijk SL, de Vries R, Dzoljic MR. Excitatory amino acid receptor antagonists and naloxone-precipitated withdrawal syndrome in morphine-dependent mice. *Eur Neuropsychopharmacol* 1993;3:111.

589. Aghajanian GK, Kogan JH, Moghaddam B. Opiate withdrawal increases glutamate and aspartate efflux in the locus coeruleus: an in vivo microdialysis study. *Brain Res* 1994;636:126.

590. Oleskevich S, Clements JD, Williams JT. Opioid-glutamate interactions in rat locus coeruleus neurons. *J Neurophysiol* 1993;70:931.

591. Saper CB. Function of the locus coeruleus. *Trends Neurosci* 1987;10:343.

592. Hayward MD, Duman RS, Nestler EJ. Induction of the c-*fos* proto-oncogene during opiate withdrawal in the locus coeruleus and other regions of rat brain. *Brain Res* 1990;525:256.

593. Murase S, Nisell M, Grenhoff J, et al. Decreased sensory responsiveness of noradrenergic neurons in the rat locus coeruleus following phencyclidine or dizocilpine (MK-801): role of NMDA antagonism. *Psychopharmacology (Berl)* 1992;109:271.

594. Rasmussen K, Fuller RW, Stockton ME, et al. NMDA receptor antagonists suppress behaviors but not norepinephrine turnover or locus coeruleus unit activity induced by opiate withdrawal. *Eur J Pharmacol* 1991;197:9.

595. Trujillo KA, Akil H. The NMDA receptor antagonist MK-801 increases morphine catalepsy and lethality. *Pharmacol Biochem Behav* 1991;38:673.

596. Shiekhattar R, Aston-Jones G. NMDA-receptor mediated sensory responses of brain noradrenergic neurons are suppressed by in vivo concentrations of extracellular magnesium. *Synapse* 1992;10:103.

597. Carlezon WA Jr, Wise RA. Morphine-induced potentiation of brain stimulation reward is enhanced by MK-801. *Brain Res* 1993;620:339.

598. Majeed NH, Przewlocka B, Machelska H, et al. Inhibition of nitric oxide synthase attenuates the development of morphine tolerance and dependence in mice. *Neuropharmacology* 1994;33:189–92.

599. Kimes AS, Vaupel DB, London ED. Attenuation of some signs of opioid withdrawal by inhibitors of nitric oxide synthase. *Psychopharmacology (Berl)* 1993;112:521.

600. Vaupel DB, Kimes AS, London ED. Comparison of 7-nitroindazole with other nitric oxide synthase inhibitors as attenuators of opioid withdrawal. *Psychopharmacology (Berl)* 1995;118:361.

601. Adams ML, Kalicki JM, Meyer ER, et al. Inhibition of the morphine withdrawal syndrome by a nitric oxide synthase inhibitor, N^G-nitro-L-arginine methyl ester. *Life Sci* 1993;52:PL245.

602. Bhargava HN, Thorat SN. Evidence for a role of nitric oxide of the central nervous system in morphine abstinence syndrome. *Pharmacology* 1996;52:86.

603. Cappendijk SL, Duval SY, de Vries R, et al. Comparative study of normotensive and hypertensive nitric oxide synthase inhibitors on morphine withdrawal syndrome in rats. *Neurosci Lett* 1995;183:67.

604. Hall S, Milne B, Jhamandas K. Nitric oxide synthase inhibitors attenuate acute and chronic morphine withdrawal response in the rat locus coeruleus: an in vivo voltametric study. *Brain Res* 1996;739:182.

605. Rodrigo J, Springall DR, Uttenthal O, et al. Localization of nitric oxide synthase in the adult rat brain. *Philos Trans R Soc Lond B Biol Sci* 1994;345:175.

606. Xu ZQ, Pieribone VA, Zhang X, et al. A functional role for nitric oxide in locus coeruleus: immunohistochemical and electrophysiological studies. *Exp Brain Res* 1994;98:75.

607. Hall S, Milne B, Jhamandas K. Nitric oxide synthase inhibitors attenuate acute and chronic morphine withdrawal response in the rat locus coeruleus: an in vivo voltametric study. *Brain Res* 1996;739:182.

608. Pineda J, Torrecilla M, Martin-Ruiz R, et al. Attenuation of

withdrawal-induced hyperactivity of locus coeruleus neurones by inhibitors of nitric oxide synthase in morphine-dependent rats. *Neuropharmacology* 1998;37:759.

609. Sullivan ME, Hall SR, Milne B, et al. Suppression of acute and chronic opioid withdrawal by a selective soluble guanylyl cyclase inhibitor. *Brain Res* 2000;859:45.

610. Kanof PD, Handelsman L, Aronson MJ, et al. Clinical characteristics of naloxone-precipitated withdrawal in human opioid-dependent subjects. *J Pharmacol Exp Ther* 1992;260:355.

611. Gellert VF, Sparber SB. A comparison of the effects of naloxone upon body weight loss and suppression of fixed-ratio operant behavior in morphine-dependent rats. *J Pharmacol Exp Ther* 1977;201:44.

612. Manning FJ, Jackson MC Jr. Enduring effects of morphine pellets revealed by conditioned taste aversion. *Psychopharmacology (Berl)* 1977;51:279.

613. Hand TH, Koob GF, Stinus L, et al. Aversive properties of opiate receptor blockade: evidence for exclusively central mediation in naive and morphine-dependent rats. *Brain Res* 1988;474:364.

614. Solomon RL. The opponent-process theory of acquired motivation: the affective dynamics of addiction. In: Maser JD, Seligman MEP, eds. *Psychopathology: experimental models.* San Francisco, CA: WH Freeman, 1997:124–145.

615. Koob GF, Le Moal M. Drug abuse: hedonic homeostatic dysregulation. *Science* 1997;278:52.

616. Koob GF. Neurobiology of addiction. Toward the development of new therapies. *Ann N Y Acad Sci* 2000;909:170.

617. Higgins GA, Sellers EM. Antagonist precipitated opioid withdrawal in rats: evidence for dissociation between physical and motivational signs. *Pharmacol Biochem Behav* 1994;48:1.

618. Koob GF, Wall TL, Bloom FE. Nucleus accumbens as a substrate for the aversive stimulus effects of opiate withdrawal. *Psychopharmacology (Berl)* 1989;98:530.

619. Terwilliger RZ, Beitner-Johnson D, Sevarino KA, et al. A general role for adaptations in G-proteins and the cyclic AMP system in mediating the chronic actions of morphine and cocaine on neuronal function. *Brain Res* 1991;548:100.

620. Nestler EJ, Hope BT, Widnell KL. Drug addiction: a model for the molecular basis of neural plasticity. *Neuron* 1993;11:995.

621. Self DW, Nestler EJ. Molecular mechanisms of drug reinforcement and addiction. *Annu Rev Neurosci* 1995;18:463.

622. Nestler EJ. Under siege: the brain on opiates. *Neuron* 1996;16:897.

623. Hakan RL, Henriksen SJ. Systemic opiate administration has heterogenous effects on activity recorded from nucleus accumbens neurons *in vivo. Neurosci Lett* 1987;83:307.

624. Hakan RL, Henriksen SJ. Opiate influences on nucleus accumbens neuronal electrophysiology: dopamine and non-dopamine mechanisms. *J Neurosci* 1989;8:3538.

625. Pothos E, Rada P, Mark GP, et al. Dopamine microdialysis in the nucleus accumbens during acute and chronic morphine, naloxone-precipitated withdrawal and clonidine treatment. *Brain Res* 1991;566:348.

626. Rossetti ZL, Hmaidan Y, Gessa GL. Marked inhibition of mesolimbic dopamine release: a common feature of ethanol, morphine, cocaine and amphetamine abstinence in rats. *Eur J Pharmacol* 1992;221:227.

627. Rada P, Pothos E, Mark GP, et al. Microdialysis evidence that acetylcholine in the nucleus accumbens is involved in morphine withdrawal and its treatment with clonidine. *Brain Res* 1991;561:354.

628. File SE, Pellow S. Behavioral pharmacology of minor tranquilizers. In: Balfour DJK, ed. *Psychotropic drugs of abuse.* New York: Pergamon Press, 1990:147.

629. Pohorecky LA, Brick J. Pharmacology of ethanol. In: Balfour DJK, ed. *Psychotropic drugs of abuse.* New York: Pergamon Press, 1990:189.

630. Okamoto M, Boisse NR. Sedative-hypnotic tolerance and physical dependence. *Trends Pharmacol Sci* 1981;2:9.

631. Tallman JF, Gallager DW. The GABA-ergic system: a locus of benzodiazepine action. *Annu Rev Neurosci* 1985;8:21.

632. Waterhouse BD, Moises HC, Yeh HH, et al. Comparison of norepinephrine- and benzodiazepine-induced augmentation of Purkinje cell response to γ-aminobutyric acid (GABA). *J Pharmacol Exp Ther* 1984;228:257.

633. Gallager DW, Lakoski JM, Gonsalves SF, et al. Chronic benzodiazepine treatment decreases postsynaptic GABA sensitivity. *Nature* 1984;308:74.

634. Gallager DW, Malcolm AB, Anderson SA, et al. Continuous release of diazepam: electrophysiological, biochemical and behavioral consequences. *Brain Res* 1985;342:26.

635. Miller LG, Woolverton S, Greenblatt DJ, et al. Chronic benzodiazepine administration. IV. Rapid development of tolerance and receptor downregulation associated with alprazolam administration. *Biochem Pharmacol* 1989;38:3773.

636. Rosenberg HC, Chiu TH. Regional specificity of benzodiazepine receptor down-regulation during chronic treatment of rats with flurazepam. *Neurosci Lett* 1981;24:4.

637. Chiu TH, Rosenberg HC. Reduced diazepam binding following chronic benzodiazepine treatment. *Life Sci* 1978;23:1153.

638. Rosenberg HC, Chiu TH. Decreased ^{3}H-diazepam binding is a specific response to chronic benzodiazepine treatment. *Life Sci* 1979;24:803.

639. Allan AM, Baier LD, Zhang X. Effects of lorazepam tolerance and withdrawal on GABA-$_A$ receptor-operated chloride channels. *J Pharmacol Exp Ther* 1992;261:395.

640. Lamb RJ, Griffiths RR. Effects of Ro 15-1788 and CGS 8216 in diazepam-dependent baboons. *Eur J Pharmacol* 1987;143:205.

641. Wilson MA, Gallager DW. GABAergic subsensitivity of dorsal raphe neurons *in vitro* after chronic benzodiazepine treatment *in vivo. Brain Res* 1988;473:198.

642. Marley R, Gallager DW. Chronic diazepam treatment produces regionally specific changes in GABA-stimulated chloride influx. *Eur J Pharmacol* 1989;159:217.

643. Heninger C, Gallager DW. Altered gamma-aminobutyric acid/benzodiazepine interaction after chronic diazepam exposure. *Neuropharmacology* 1988;27:1073.

644. Gonsalves SF, Gallager DW. Time course for development of anticonvulsant tolerance and GABAergic subsensitivity after chronic diazepam. *Brain Res* 1987;405:94.

645. Wilson MA, Gallager DW. Effects of chronic diazepam exposure on GABA sensitivity and on benzodiazepine potentiation of GABA-mediated responses of substantia nigra pars reticulata neurons of rats. *Eur J Pharmacol* 1987;136:333.

646. Wilson MA, Gallager DW. Responses of substantia nigra pars reticulata neurons to benzodiazepine ligands after acute and prolonged diazepam exposure. II. Modulation of firing rate. *J Pharmacol Exp Ther* 1989;248:886.

647. Wilson MA, Gallager DW. Responses of substantia nigra pars reticulata neurons to benzodiazepine ligands after acute and prolonged diazepam exposure. I. Modulation of γ-aminobutyric acid sensitivity. *J Pharmacol Exp Ther* 1989;248:879.

648. Gallager DW, Rauch SL, Malcolm AB. Alterations in a low affinity GABA recognition site following chronic benzodiazepine treatment. *Eur J Pharmacol* 1984;98:159.

649. Heninger C, Saito N, Tallman JF, et al. Effects of continuous diazepam administration on GABA-A subunit mRNA in rat brain. *J Mol Neurosci* 1990;2:101.

650. Li M, Szabo A, Rosenberg HC. Down-regulation of benzodiazepine binding to alpha 5 subunit-containing

gamma-aminobutyric acid(A) receptors in tolerant rat brain indicates particular involvement of the hippocampal CA1 region. *J Pharmacol Exp Ther* 2000;295:689.

651. Bell J, Bickford-Wimer PC, de la Garza R, et al. Increased central noradrenergic activity during benzodiazepine withdrawal: an electrophysiological study. *Neuropharmacology* 1988;27:1187.

652. Morrow AL, Suzdak PD, Karanian JW, et al. Chronic ethanol administration alters γ-aminobutyric acid, pentobarbital, and ethanol-mediated ³⁶Cl-uptake in cerebral cortical synaptoneurosomes. *J Pharmacol Exp Ther* 1988;246:158.

653. Sanna E, Harris RA. Recent developments in alcoholism: neuronal ion channels. *Recent Dev Alcohol* 1993;11:169.

654. Im WB, Blakeman DP. Correlation between γ-aminobutyric acid$_A$ receptor ligand-induced changes in *t*-butylbicyclophosphoro [³⁵S]thionate binding and ³⁶Cl-uptake in rat cerebrocortical membranes. *Mol Pharmacol* 1991;39:394.

655. Saunders PA, Ho IK. Barbiturates and the GABA-A receptor complex. *Prog Drug Res* 1990;34:261.

656. Rastogi SK, Thyagarajan R, Clothier J, et al. Effect of chronic treatment of ethanol on benzodiazepine and picrotoxin sites on the GABA receptor complex in regions of the brain of the rat. *Neuropharmacology* 198;25:1179.

657. Montpied P, Ginns EI, Martin BM, et al. Gamma-aminobutyric acid (GABA) induces a receptor-mediated reduction in GABAA receptor alpha subunit messenger RNAs in embryonic chick neurons in culture. *J Biol Chem* 1991;5:6011.

658. Mhatre MC, Ticku MK. Chronic ethanol administration alters gamma-aminobutyric acidA receptor gene expression. *Mol Pharmacol* 1992;42:415.

659. Devaud LL, Fritschy JM, Sieghart W, et al. Bidirectional alterations of GABA(A) receptor subunit peptide levels in rat cortex during chronic ethanol consumption and withdrawal. *J Neurochem* 1997;69:126.

660. Papadeas S, Grobin AC, Morrow AL. Chronic ethanol consumption differentially alters GABA(A) receptor alpha1 and alpha4 subunit peptide expression and GABA(A) receptor-mediated ³⁶Cl(−) uptake in mesocorticolimbic regions of rat brain. *Alcohol Clin Exp Res* 2001;25:1270.

661. Wu PH, Mihic SJ, Liu JF, et al. Blockade of chronic tolerance to ethanol by the NMDA antagonist, (+)-MK-801. *Eur J Pharmacol* 1993;231:157.

662. Weight FF, Lovinger DM, White G. Alcohol inhibition of NMDA channel function. *Alcohol Alcohol Suppl* 1991;1:163.

663. Snell LD, Tabakoff B, Hoffman PL. Radioligand binding to the N-methyl-D-aspartate receptor/ionophore complex: alterations by ethanol *in vitro* and by chronic *in vivo* ethanol injections. *Brain Res* 1993;602:91.

664. Tabakoff B, Hoffman PL. Ethanol, sedative hypnotics, and glutamate receptor function in brain and cultured cells. *Behav Genet* 1993;23:231.

665. Davidson MD, Wilce P, Shanley BC. Increased sensitivity of the hippocampus in ethanol-dependent rats to toxic effect of N-methyl-D-aspartic acid *in vivo*. *Brain Res* 1993;606:5.

666. Szabo G, Tabakoff B, Hoffman PL. The NMDA receptor antagonist dizocilpine differentially affects environment-dependent and environment-independent ethanol tolerance. *Psychopharmacology (Berl)* 1994;113:511.

667. Tyrer P. Dependence as a limiting factor in the clinical use of minor tranquilizers. In: Balfour DJK, ed. *Psychotropic drugs of abuse*. New York: Pergamon Press, 1990:173.

668. Griffiths RR, Sannerud CA. Abuse of and dependence on benzodiazepines and other anxiolytic/sedative drugs. In: Meltzer HY, ed. *Psychopharmacology: the third generation of progress*. New York: Raven Press, 1987:1535.

669. Tall JF, Gallager DW. The GAB-ergic system: a locus of benzodiazepine action. *Annu Rev Neurosci* 1985;8:21.

670. Martin WR, McNicholas LF, Cherian S. Diazepam and pentobarbital dependence in the rat. *Life Sci* 1982;31:721.

671. Wulff MH. The barbiturate withdrawal syndrome. A clinical and electroencephalographic study. *Electroencephalogr Clin Neurophysiol* 1960;14(Suppl):1.

672. Lukas SE, Griffiths RR. Precipitated withdrawal by a benzodiazepine receptor antagonist (Ro 15-1788) after 7 days of diazepam. *Science* 1982;217:1161.

673. Emmett-Oglesby M, Spencer DG Jr, Lewis MW, et al. Anxiogenic aspects of diazepam withdrawal can be detected in animals. *Eur J Pharmacol* 1983;92:127.

674. Martin WR, Sloan JW, Wala EP. Precipitated abstinence in the diazepam-dependent rat. *Pharmacol Biochem Behav* 1993;46:683.

675. Winokur A, Rickels K, Greenblatt DJ, et al. Withdrawal reaction from long-term, low-dosage administration of diazepam. A double-blind, placebo-controlled case study. *Arch Gen Psychiatry* 1990;37:101.

676. Miczek KA, Vivian JA. Automatic quantification of withdrawal from 5-day diazepam in rats: ultrasonic distress vocalizations and hyperreflexia to acoustic startle stimuli. *Psychopharmacology (Berl)* 1993;110:379.

677. Essig CF. Barbiturate withdrawal in white rats. *Int J Neuropharmacol* 1966;5:103.

678. Essig CF. Clinical and experimental aspects of barbiturate withdrawal convulsions. *Epilepsia* 1967;8:21.

679. Lukas SE, Griffiths RR. Precipitated diazepam withdrawal in baboons: effects of dose and duration of diazepam exposure. *Eur J Pharmacol* 1984;100:163.

680. Ryan GP, Boisse NR. Experimental induction of benzodiazepine tolerance and physical dependence. *J Pharmacol Exp Ther* 1983;226:100.

681. McNicholas LF, Martin WR. The effect of a benzodiazepine antagonist, Ro 15-1788, in diazepam dependent rats. *Life Sci* 1982;31:731.

682. Balfour DJK. Nicotine as the basis of the smoking habit. In: Balfour DJK, ed. *Psychotropic drugs of abuse*. New York: Pergamon Press, 1990:453.

683. Dituri B, Gilman LH. Withdrawal symptoms of barbiturate addiction: a little-known syndrome. *J Phila Gen Hosp* 1951;2:2.

684. Pevnick JS, Jasinski DR, Haertzen CA. Abrupt withdrawal from therapeutically administered diazepam. Report of a case. *Arch Gen Psychiatry* 1978;35:995.

685. Greenblatt DJ, Shader RI. Pharmacokinetics of antianxiety agents. In: Meltzer HY, ed. *Psychopharmacology: the third generation of progress*. New York: Raven Press, 1987:1377.

686. Rosenberg HW, Chiu TH. An antagonist-induced benzodiazepine abstinence syndrome. *Eur J Pharmacol* 1982;81:153.

687. Grant SJ, Golloway MP, Maynor R, et al. Precipitated diazepam withdrawal elevates noradrenergic metabolism in primate brain. *Eur J Pharmacol* 1985;107:127.

688. Gonsalves SF, Gallager DW. Spontaneous and Ro 15-1788-induced reversal of subsensitivity to GABA following chronic benzodiazepines. *Eur J Pharmacol* 1985;110:163.

689. Morgan PF, Nadi NS, Karanian J, et al. Mapping rat brain structures activated during ethanol withdrawal: role of glutamate and NMDA receptors. *Eur J Pharmacol* 1992;225:217.

690. Ableitner A, Wüster M, Herz A. Specific changes in local cerebral glucose utilization in the rat brain induced by acute and chronic diazepam. *Brain Res* 1985;359:49.

691. Marietta CA, Eckardt MJ, Zbicz KL, et al. Cerebral glucose utilization during diazepam withdrawal in rats. *Brain Res* 1990;511:192.

692. Young WS, III, Kuhar MJ. Radiohistochemical localization of benzodiazepine receptors in rat brain. *J Pharmacol Exp Ther* 1980;212:337.

693. Rall TW, Schleifer LS. Drugs effective in the therapy of the epilepsies. In: Gilman AG, Rall TW, Nies AS, et al., eds. *The pharmacological basis of therapeutics.* New York: Pergamon Press, 1990:436.

694. Hobson JA, Lydic R, Baghdoyan HA. Evolving concepts of sleep cycle generation: from brain centers to neuronal populations. *Behav Brain Sci* 1986;9:371.

695. Steriade M, McCarley RW, ed. *Brainstem control of wakefulness and sleep.* New York: Plenum Press, 1990.

696. Nutt D, Molyneux S. Benzodiazepines, plasma MHPG and *alpha-2* adrenoceptor function in man. *Int Clin Psychopharmacol* 1987;2:151.

697. Idemudia SO, Mathis DA, Lal H. Enhancement of a diazepam withdrawal symptom by bicuculline and yohimbine. *Neuropharmacology* 1987;26:1739.

698. Wilkins JN, Gorelick DA. Clinical neuroendocrinology and neuropharmacology of alcohol withdrawal. *Recent Dev Alcohol* 1986;4:241.

699. Engberg G, Hajos M. Alcohol withdrawal reaction as a result of adaptive changes of excitatory amino acid receptors. *Naunyn Schmiedebergs Arch Pharmacol* 1992;346:437.

700. Redmond DE Jr. Studies of the nucleus locus coeruleus in monkeys and hypotheses for neuropsychopharmacology. In: Meltzer HY, ed. *Psychopharmacology: the third generation of progress.* New York: Raven Press, 1987:967.

701. Grant SJ, Huang YH, Redmond DE. Benzodiazepines attenuate single unit activity in the locus coeruleus. *Life Sci* 1980;27:2331.

702. Simon PE, Weiss JM. Peripheral, but not local or intracerebroventricular, administration of benzodiazepines attenuates evoked activity of locus coeruleus neurons. *Brain Res* 1989;490:236.

703. Redmond DE Jr, Huang YH. New evidence for a locus coeruleus-norepinephrine connection with anxiety. *Life Sci* 1979;25:2149.

704. Duka T, Ackenheil M, Noderer J, et al. Changes in noradrenaline plasma levels and behavioural responses induced by benzodiazepine agonists with the benzodiazepine antagonist Ro 15-1788. *Psychopharmacology (Berl)* 1986;90:351.

705. Aston-Jones G, Foote SL, Bloom FE. Low doses of ethanol disrupt sensory responses of brain noradrenergic neurones. *Nature* 1982;296:857.

706. Pohorecky LA, Brick J. Activity of neurons in the locus coeruleus of the rat: inhibition by ethanol. *Brain Res* 1977;131:174.

707. Osmanovic SS, Shefner SA. Enhancement of current induced by superfusion of GABA in locus coeruleus neurons by pentobarbital, but not ethanol. *Brain Res* 1990;517:324.

708. Verbanck PMP, Seutin V, Massotte L, et al. Differential effects of picrotoxin and RO 15-1788 on high and low ethanol concentrations on rat locus coeruleus *in vitro. Eur J Pharmacol* 1992;211:15.

709. Grant SJ, Mayor R, Redmond DE Jr. Effects of alprazolam, a novel triazolobenzodiazepine on locus coeruleus unity activity. *Soc Neurosci Abstr* 1984;10:952.

710. Insel TR, Ninan PT, Aloi J, et al. A benzodiazepine receptor-mediated model of anxiety. Studies in nonhuman primates and clinical implications. *Arch Gen Psychiatry* 1984;41:741.

711. Wilkins AJ, Jenkins WJ, Steiner JA. Efficacy of clonidine in treatment of alcohol withdrawal state. *Psychopharmacology (Berl)* 1983;83:78.

712. Fyer AJ, Leibowitz MR, Gorman JM, et al. Effects of clonidine on alprazolam discontinuation in panic patients: a pilot study. *J Clin Psychopharmacol* 1994;8:270.

713. Papanicolaou J, Summers RJ, Vajda FJE, et al. The relationship between α_2-adrenoceptor selectivity and anticonvulsant effect in a series of clonidine-like drugs. *Brain Res* 1982;241:393.

714. Baldwin HA, Hitchcott PK, File SE. Evidence that the increased anxiety detected in the elevated plus-maze during chlordiazepoxide withdrawal is not due to enhanced noradrenergic activity. *Pharmacol Biochem Behav* 1989;34:931.

715. Siegel EG, Bonfiglio JF. Investigation of clonidine and lofexidine for the treatment of barbiturate withdrawal in mice. *Vet Human Toxicol* 1985;27:503.

716. Kunchandy J, Kulkarni SK. Reversal by *alpha-2* agonists of diazepam withdrawal hyperactivity in rats. *Psychopharmacology (Berl)* 1986;90:198.

717. Gil E, Colado I, Lopez F, et al. Effects of chronic treatment with ethanol and withdrawal of ethanol on levels of dopamine, 2,4-dihydroxyphenylacetic acid and homovanillic acid in the striatum of the rat. Influence of benzodiazepines, barbiturate and somatostatin. *Neuropharmacology* 1992;31:1151.

718. Nutt DJ, Cowen PJ. Diazepam alters brain 5HT function in man: implications for the acute and chronic effects of benzodiazepines. *Psychol Med* 1987;17:601.

719. Ida Y, Roth RH. The activation of mesoprefrontal dopamine neurons by FG 7142 is absent in rats treated chronically with diazepam. *Eur J Pharmacol* 1987;137:185.

720. Rastogi RB, Lapierre YD, Singhal RL. Evidence for the role of brain norepinephrine and dopamine in "rebound" phenomenon seen during withdrawal after repeated exposure to benzodiazepines. *J Psychiatr Res* 1976;13:65.

721. Creese I, ed. *Stimulants: neurochemical, behavioral, and clinical perspectives.* New York: Raven Press, 1983.

722. Cho AK, Segal DS, eds. *Amphetamine and its analogs. Psychopharmacology, toxicology, and abuse.* San Diego, CA: Academic Press, 1994.

723. Doweiko HE. ed. *Concepts of chemical dependency,* 2nd ed. Belmont, CA: Brooks/Cole Publishing Company, 1993.

724. Gillis RA, Quest JA, Wilkerson RD. Mechanisms responsible for cardiovascular disorders associated with cocaine abuse. In: Lakoski JM, Galloway MP, White FJ, eds. *Cocaine: pharmacology, physiology, and clinical strategies.* Boca Raton, FL: CRC Press, 1992:371.

725. Nademanee K. Cardiovascular effects and toxicities of cocaine. *J Addict Dis* 1992;11:71.

726. Gould L, Gopalaswamy C, Patel C, et al. Cocaine-induced myocardial infarction. *N Y State J Med* 1985;11:660.

727. Roberts SM, James RC, Harbison RD. Cocaine-induced hepatotoxicity. In: Lakoski JM, Galloway MP, White FJ, eds. *Cocaine: pharmacology, physiology, and clinical strategies.* Boca Raton, FL: CRC Press, 1992:15.

728. Gibb JW, Hanson GR, Johnson M. Neurochemical mechanisms of toxicity. In: Cho AK, Segal DS, eds. *Amphetamine and its analogs. Psychopharmacology, toxicology, and abuse.* San Diego, CA: Academic Press, 1994:269.

729. Ellison GD. Paranoid psychosis following continuous amphetamine or cocaine: relationship to selective neurotoxicity. In: Korenman SG, Barchas JD, eds. *Biological basis of substance abuse.* New York: Oxford University Press, 1993:355.

730. Marshall JF, O'Dell SJ, Weihmuller FB. Dopamine-glutamate interactions in methamphetamine-induced neurotoxicity. *J Neural Transm Gen Sec* 1993;91:241.

731. Post RM, Weiss SRB, Pert A. Sensitization and kindling effects of chronic cocaine administration. In: Lakoski JM, Galloway MP, White FJ, eds. *Cocaine: pharmacology, physiology, and clinical strategies.* Boca Raton, FL: CRC Press, 1992:115.

732. Angrist B. Psychoses induced by central nervous system stimulants and related drugs. In: Creese I, ed. *Stimulants: neurochemical, behavioral, and clinical perspectives.* New York: Raven Press, 1983:1.

733. Wood DM, Emmett-Oglesby MW. Characteristics of tolerance, recovery from tolerance and cross-tolerance for cocaine used as a discriminative stimulus. *J Pharmacol Exp Ther* 1986;237: 120.

734. Bhat RV, Cole AJ, Baraban JM. Chronic cocaine treatment suppresses basal expression of *zif268* in rat forebrain: *in situ* hybridization studies. *J Pharmacol Exp Ther* 1992;263:343.

735. Hope B, Kosofsky B, Hyman SE, et al. Regulation of immediate early gene expression and AP-1 binding in the rat nucleus accumbens by chronic cocaine. *Proc Natl Acad Sci U S A* 1992;89:5674.

736. Daunais JB, McGinty JF. Acute and chronic cocaine administration differentially alters striatal opioid and nuclear transcription factor mRNAs. *Synapse* 1994;18:35.

737. Segal DS, Mandell AJ. Long-term administration of d-amphetamine: progressive augmentation of motor activity and stereotypy. *Pharmacol Biochem Behav* 1974;2:249.

738. Post RM, Rose H. Increasing effects of repetitive cocaine administration in the rat. *Nature* 1976;260:731.

739. Downs AW, Eddy NB. The effect of repeated doses of cocaine on the rat. *J Pharmacol Exp Ther* 1932;46:199.

740. Segal DS, Schuckit MA. Animal models of stimulant-induced psychosis. In: Creese I, ed. *Stimulants: neurochemical, behavioral, and clinical perspectives.* New York: Raven Press, 1983:131.

741. Fischman MW, Schuster CR, Hatano Y. A comparison of the subjective and cardiovascular effects of cocaine and lidocaine in humans. *Pharmacol Biochem Behav* 1983;18:123.

742. Stewart J, Badiani A. Tolerance and sensitization to the behavioral effects of drugs. *Behav Pharmacol* 1993;4:289.

743. Horger BA, Shelton K, Schenk S. Preexposure sensitizes rats to the rewarding effects of cocaine. *Pharmacol Biochem Behav* 1990;37:707.

744. Kalivas PW, Sorg BA, Hooks MS. The pharmacology and neural circuitry of sensitization to psychostimulants. *Behav Pharmacol* 1993;4:315.

745. Reith MEA, Benuck M, Lajtha A. Cocaine disposition in the brain after continuous or intermittent treatment and locomotor stimulation in mice. *J Pharmacol Exp Ther* 1987;243: 281.

746. Kalivas PW, Stewart J. Dopamine transmission in the initiation and expression of drug- and stress-induced sensitization of motor activity. *Brain Res Rev* 1991;16:223.

747. Kalivas PW, Barnes CD, eds. *Sensitization in the nervous system.* Caldwell, NJ: Telford Press, 1988.

748. Segal DS, Kuczenski R. Behavioral pharmacology of amphetamine. In: Cho AK, Segal DS, eds. *Amphetamine and its analogs. Psychopharmacology, toxicology, and abuse.* San Diego, CA: Academic Press, 1994:115.

749. Mandell AJ, Knapp S. Acute versus chronic effects of psychotropic drugs: adaptive responses in brain amine systems and their clinical implications. Evidence for multiphasic presynaptic adaptation: studies with cocaine and lithium. *Psychopharmacol Bull* 1977;13:40.

750. Roth RH, Wolf ME, Deutch AY. Neurochemistry of midbrain dopamine systems. In: Meltzer HY, ed. *Psychopharmacology: the third generation of progress.* New York: Raven Press, 1987: 81.

751. Mandell AJ. Neurobiological barriers to euphoria. *Am Sci* 1993;61:565.

752. Weiss F, Paulus MP, Lorang MT, et al. Increases in extracellular dopamine in the nucleus accumbens by cocaine are inversely re-

lated to basal levels: effects of acute and repeated administration. *J Neurosci* 1992;12:4372.

753. Weber E, Sonders M, Quarum M, et al. 1,3-Di(2-[5-^{3}H]tolyl)guanidine: a selective ligand that labels σ-type receptors for psychotomimetic opiates and antipsychotic drugs. *Proc Natl Acad Sci U S A* 1986;83:8784.

754. Kalivas PW, Duffy P. Time course of extracellular dopamine and behavioral sensitization to cocaine. I. Dopamine axon terminals. *J Neurosci* 1993;13:266.

755. Wise RA, Leeb K. Psychomotor-stimulant sensitization: a unitary phenomenon? *Behav Pharmacol* 1993;4:339.

756. Kalivas PW, Weber B. Amphetamine injection into the ventral mesencephalon sensitizes rats to peripheral amphetamine and cocaine. *J Pharmacol Exp Ther* 1988;245:1095.

757. Ackerman JM, White FJ. A10 somatodendritic dopamine autoreceptor sensitivity following withdrawal from repeated cocaine treatment. *Neurosci Lett* 1990;117:181.

758. White FJ, Henry DJ, Hu X-T, et al. Electrophysiological effects of cocaine in the mesoaccumbens dopamine system. In: Lakoski JM, Galloway MP, White FJ, eds. *Cocaine: pharmacology, physiology, and clinical strategies.* Boca Raton, FL: CRC Press, 1992:261.

759. White FJ, Wang RY. Electrophysiological evidence for A10 dopamine autoreceptor subsensitivity following chronic d-amphetamine treatment. *Brain Res* 1984;309:283.

760. Striplin CD, Kalivas PW. Correlation between behavioral sensitization to cocaine and G protein ADP-ribosylation in the ventral tegmental area. *Brain Res* 1992;579:181.

761. Nestler EJ, Guitart X, Beiter-Johnson D. Second-messenger and protein phosphorylation mechanisms underlying opiate and cocaine addiction. In: Korenman SG, Barchas JD, eds. *Biological basis of substance abuse.* New York: Oxford University Press, 1993:49.

762. Einhorn LC, Johansen PA, White FJ. Electrophysiological effects of cocaine in the mesoaccumbens dopamine system: studies in the ventral tegmental area. *J Neurosci* 1988;8:100.

763. Perugini M, Vezina P. Amphetamine administered to the ventral tegmental area sensitizes rats to the locomotor effects of nucleus accumbens amphetamine. *J Pharmacol Exp Ther* 1994;270:690.

764. Koff JM, Shuster L, Miller LG. Chronic cocaine administration is associated with behavioral sensitization and time-dependent changes in striatal dopamine transporter binding. *J Pharmacol Exp Ther* 1994;268:277.

765. Pilotte NS, Sharpe LG, Kuhar MJ. Withdrawal of repeated intravenous infusions of cocaine persistently reduces binding to dopamine transporters in the nucleus accumbens of Lewis rats. *J Pharmacol Exp Ther* 1994;269:963.

766. Henry DJ, White FJ. Repeated cocaine administration causes persistent enhancement of D_1 dopamine receptor sensitivity within the rat nucleus accumbens. *J Pharmacol Exp Ther* 1991;258:882.

767. Henry DJ, White FJ. The persistence of behavioral sensitization to cocaine parallels enhanced inhibition of nucleus accumbens neurons. *J Neurosci* 1995;15:6287.

768. Li Y, White FJ, Wolf ME. Pharmacological reversal of behavioral and cellular indices of cocaine sensitization in the rat. *Psychopharmacology (Berl)* 2000;151:175.

769. Xu M, Guo Y, Vorhees CV, et al. Behavioral responses to cocaine and amphetamine administration in mice lacking the dopamine D1 receptor. *Brain Res* 2000;852:198.

770. Beyer CE, Steketee JD. Cocaine sensitization: modulation by dopamine d(2) receptors. *Cereb Cortex* 2002;12:526.

771. Self DW, Nestler EJ. Molecular mechanisms of drug reinforcement and addiction. *Annu Rev Neurosci* 1995;18:463.

772. Cunningham ST, Kelley AE. Evidence for opiate-dopamine cross-sensitization in nucleus accumbens: studies of conditioned reward. *Brain Res Bull* 1992;29:675.

773. Heidbreder C, Goldberg SR, Shippenberg TS. Inhibition of cocaine-induced sensitization by the γ-opioid receptor antagonist naltrindole. *Eur J Pharmacol* 1993;243:123.

774. Wolf ME, Jeziorski M. Coadministration of MK-801 with amphetamine, cocaine or morphine prevents rather than transiently masks the development of behavioral sensitization. *Brain Res* 1993;613:291.

775. Gur RC, Gur RE, Resnick SM, et al. The effect of anxiety on cortical cerebral blood flow and metabolism. *J Cereb Blood Flow Metab* 1987;7:173.

776. Wolf ME, White FJ, Hu X-T. MK-801 prevents alterations in the mesoaccumbens dopamine system associated with behavioral sensitization to amphetamine. *J Neurosci* 1994;14:1735.

777. White FJ, Hu X-T, Henry DJ. Electrophysiological effects of cocaine in the rat nucleus accumbens: microiontophoretic studies. *J Pharmacol Exp Ther* 1993;266:1075.

778. Lakoski JM, Black ER, Moday HJ. Electrophysiological effects of cocaine on serotonin neuronal systems. In: Lakoski JM, Galloway MP, White FJ, eds. *Cocaine: pharmacology, physiology, and clinical strategies.* Boca Raton, FL: CRC Press, 1991:295.

779. Harris GC, Williams JT. Sensitization of locus ceruleus neurons during withdrawal from chronic stimulants and antidepressants. *J Pharmacol Exp Ther* 1992;261:476.

780. Konradi C, Cole RL, Heckers S, et al. Amphetamine regulates gene expression in rat striatum via transcription factor CREB. *J Neurosci* 1994;14:5623.

781. Nestler EJ, Barrot M, Self DW. DeltaFosB: a sustained molecular switch for addiction. *Proc Natl Acad Sci U S A* 2001;98:11042.

782. Gawin FH, Kleber HD. Abstinence symptomatology and psychiatric diagnosis in cocaine abusers: clinical observations. *Arch Gen Psychiatry* 1986;43:107.

783. Gawin FH. Cocaine addiction: psychology and neurophysiology. *Science* 1991;251:1580.

784. Martin SD, Yeragani VK, Lodhi R, et al. Clinical ratings and plasma HVA during cocaine abstinence. *Biol Psychiatry* 1989;26:356.

785. Weddington WW, Brown BS, Haertzen CA, et al. Changes in mood, craving, and sleep during short-term abstinence reported by male cocaine addicts: a controlled, residential study. *Arch Gen Psychiatry* 1990;47:861.

786. Satel SL, Price LH, Palumbo JM, et al. Clinical phenomenology and neurobiology of cocaine abstinence: a prospective inpatient study. *Am J Psychiatry* 1991;148:1712.

787. Flowers Q, Elder IR, Voris J, et al. Daily cocaine craving in a 3-week inpatient treatment program. *J Clin Psychol* 1993;49:292.

788. Kowatch RA, Schnoll SS, Knisely JS, et al. Electroencephalographic sleep and mood during cocaine withdrawal. *J Addict Dis* 1992;11:21.

789. Ehrman RN, Robbins SJ, Childress AR, et al. Conditioned responses to cocaine-related stimuli in cocaine abuse patients. *Psychopharmacology (Berl)* 1992;107:523.

790. Kosten TR. Can cocaine craving be a medication development outcome? *Am J Addict* 1992;1:230.

791. Kranzler HR, Bauer LO. Bromocriptine and cocaine cue reactivity in cocaine-dependent patients. *Br J Addict* 1992;87:1537.

792. Koob GF, Stinus L, Le Moal M, et al. Opponent process theory of motivation: neurobiological evidence from studies of opiate dependence. *Neurosci Biobehav Res* 1989;13:135.

793. Newlin DB. A comparison of drug conditioning and craving for alcohol and cocaine. *Recent Dev Alcohol* 1992;10:147.

794. Kozlowski LT, Mann RE, Wilkinson DA, et al. "Cravings" are ambiguous: ask about urges or desires. *Addict Behav* 1989;14:443.

795. Markou A, Weiss F, Gold LH, et al. Animal models of drug craving. *Psychopharmacology (Berl)* 1993;112:163.

796. Gratton A, Wise RA. Drug- and behavior-associated changes in dopamine-related electrochemical signals during intravenous cocaine self-administration in rats. *J Neurosci* 1994;14:4130.

797. Schultz W. Activity of dopamine neurons in the behaving primate. *Semin Neurosci* 1992;4:129.

798. Schultz W, Romo R. Dopamine neurons of the monkey midbrain: contingencies of responses to stimuli eliciting immediate behavioral reactions. *J Neurophysiol* 1990;63:607.

799. Dackis CA, Gold MS. New concepts in cocaine addiction: the dopamine depletion hypothesis. *Neurosci Biobehav Rev* 1985;9:469.

800. Hitri A, Casanova MF, Kleinman JE, et al. Fewer dopamine transporter receptors in the prefrontal cortex of cocaine users. *Am J Psychiatry* 1994;15:1074.

801. Volkow ND, Fowler JS, Wolf AP, et al. Effects of chronic cocaine abuse on postsynaptic dopamine receptors. *Am J Psychiatry* 1990;147:719.

802. Volkow ND, Fowler JS, Wolf AP, et al. Changes in brain glucose metabolism in cocaine dependence and withdrawal. *Am J Psychiatry* 1991;148:621.

803. Volkow ND, Hitzemann R, Wang G-J, et al. Long-term frontal brain metabolic changes in cocaine abusers. *Synapse* 1992;11:184.

804. Wood DM, Lal H. Anxiogenic properties of cocaine withdrawal. *Life Sci* 1987;41:1431.

805. Markou A, Koob GF. Postcocaine anhedonia: an animal model of cocaine withdrawal. *Neuropsychopharmacology* 1991;4:17.

806. Carroll ME, Lac ST. Cocaine withdrawal produces behavioral disruptions in rats. *Life Sci* 1987;40:2183.

807. London ED, Wilkerson G, Goldberg SR, et al. Effects of *l*-cocaine on local cerebral glucose utilization in the rat. *Neurosci Lett* 1986;68:73.

808. Porrino LJ, Domer FR, Crane AM, et al. Selective alterations in cerebral metabolism within the mesocorticolimbic dopaminergic system produced by acute cocaine administration in rats. *Neuropsychopharmacology* 1988;1:109.

809. Hammer RP Jr, Pires WS, Markou A, et al. Withdrawal following cocaine self-administration decreases regional cerebral metabolic rate in critical brain reward regions. *Synapse* 1993;14:73.

810. Henry DJ, Greene MA, White FJ. Electrophysiological effects of cocaine in the mesoaccumbens dopamine system: repeated administration. *J Pharmacol Exp Ther* 1989;251:833–839.

811. Ackerman JM, White FJ. Decreased activity of rat A10 dopamine neurons following withdrawal from repeated cocaine. *Eur J Pharmacol* 1992;218:171.

812. Zhang H, Lee TH, Ellinwood EH Jr. The progressive changes of neuronal activities of the nigral dopaminergic neurons upon withdrawal from continuous infusion of cocaine. *Brain Res* 1992;594:315.

813. Pettit HO, Pan H, Parsons LH, et al. Extracellular concentrations of cocaine and dopamine are enhanced during chronic cocaine administration. *J Neurochem* 1990;55:798.

814. Parsons LH, Smith AD, Justice JB Jr. Basal extracellular dopamine is decreased in the rat nucleus accumbens during abstinence from chronic cocaine. *Synapse* 1991;9:60.

815. Wilson JM, Nobrega JN, Carroll ME, et al. Heterogeneous subregional binding patterns of ^{3}H-WIN 35,428 and ^{3}H-GBR 12,935 are differentially regulated by chronic cocaine self-administration. *J Neurosci* 1994;14:2966.

816. Sharpe LG, Pilotte NS, Mitchell WM, et al. Withdrawal of repeated cocaine decreases autoradiographic [^{3}H]mazindol-

labelling of dopamine transporter in rat nucleus accumbens. *Eur J Pharmacol* 1991;203:141.

817. Meiergerd SM, McElvain JS, Schenk JO. Effects of cocaine and repeated cocaine followed by withdrawal. Alterations of dopaminergic transporter turnover with no changes in kinetics of substrate recognition. *Biochem Pharmacol* 1994;47:1627.

818. Laurier LG, Corrigall WA, George SR. Dopamine receptor density, sensitivity, and mRNA levels are altered following self-administration of cocaine in the rat. *Brain Res* 1994;634: 31.

819. Neisewander JL, Lucki I, McGonigle P. Time-dependent changes in sensitivity to apomorphine and monoamine receptors following withdrawal from continuous cocaine administration in rats. *Synapse* 1994;16:1.

820. Gawin FH, Kleber HD. Evolving conceptualizations of cocaine dependence. *Yale J Biol Med* 1988;61:123.

821. Cerruti C, Pilotte NS, Uhl G, et al. Reduction in dopamine transporter mRNA after cessation of repeated cocaine administration. *Brain Res Mol Brain Res* 1994;22:132.

822. King GR, Kuhn C, Ellinwood EH Jr. Dopamine efflux during withdrawal from continuous or intermittent cocaine. *Psychopharmacology (Berl)* 1993;111:179.

823. Kalivas PW, Duffy P. Time course of extracellular dopamine and behavioral sensitization to cocaine. II. Dopamine perikarya. *J Neurosci* 1993;13:276.

824. Robertson MW, Leslie CA, Bennett JP Jr. Apparent synaptic dopamine deficiency induced by withdrawal from chronic cocaine treatment. *Brain Res* 1991;538:337.

825. Larson PS, Silvette H. eds. *Tobacco: experimental and clinical studies, supplement I.* Baltimore: Williams & Wilkins, 1968.

826. Marks MJ, Stitzel JA, Collins AC. Time course study of the effects of chronic nicotine infusion on drug response and brain receptors. *J Pharmacol Exp Ther* 1985;235:619.

827. Benowitz NL, Porchet H, Jacob P, III. Nicotine dependence and tolerance in man: pharmacokinetic and pharmacodynamic investigations. In: Nordberg A, Fuxe K, Holmsted B, et al., eds. *Progress in brain research, volume 79.* Amsterdam: Elsevier, 1989: 279.

828. Porchet HC, Benowitz NL, Sheiner LB. Pharmacodynamic model of tolerance: application to nicotine. *J Pharmacol Exp Ther* 1988;244:231.

829. Morrison CF, Stephenson JA. The occurrence of tolerance to a central depressant effect of nicotine. *Br J Pharmacol* 1972;46:151.

830. Stolerman IP, Fink R, Jarvik ME. Acute and chronic tolerance to nicotine measured by activity in rats. *Psychopharmacologia* 1973;30:329.

831. Falkeborn Y, Larsson C, Nordberg A. Chronic nicotine exposure in rat: a behavioural and biochemical study of tolerance. *Drug Alcohol Depend* 1981;8:51.

832. Stolerman IP, Bunker P, Jarvik ME. Nicotine tolerance in rats: role of dose and dose interval. *Psychopharmacologia* 1974;34:317.

833. Mansner R, Alhava E, Klinge E. Nicotine hypothermia and brain nicotine and catecholamine levels in the mouse. *Med Biol* 1974;52:390.

834. Horstmann M. Influence of mecamylamine and atropine on tolerance development to nicotine hypothermia in rats. *J Pharm Pharmacol* 1984;36:770.

835. Clarke PBS, Kumar R. The effects of nicotine on locomotor activity in non-tolerant and tolerant rats. *Br J Pharmacol* 1983;78:329.

836. US Department of Health and Human Services. In: Benowitz NL, Grunberg NE, Henningfield JE, et al., eds. *The health consequences of smoking: nicotine addiction.* A report of the Surgeon General. Washington, DC: US Department of Health and Human Services, Public Health Service, Centers for Disease Control, Center for Health Promotion and Education, Office on Smoking and Health, DHHS Publication No. (CDC) 88–8406, 1988.

837. Hughes JR, Higgins ST, Hatsukami D. Effects of abstinence from tobacco: a critical review. In: Kozlowski LT, Annis HM, Cappell HD, et al., eds. *Research advances in alcohol and drug problems, volume 10.* New York: Plenum Press, 1990:317.

838. Hughes JR, Gust SW, Skoog K, et al. Symptoms of tobacco withdrawal: a replication and extension. *Arch Gen Psychiatry* 1991;48:52.

839. Markou A, Kosten TR, Koob GF. Neurobiological similarities in depression and drug dependence: a self-medication hypothesis. *Neuropsychopharmacology* 1998;18:135.

840. Fagerstrom KO, Ramstrom LM, Svensson TH. Health in tobacco control. *Lancet* 1992;339:934.

841. Hatsukami DK, Skoog K, Huber M, et al. Signs and symptoms from nicotine gum abstinence. *Psychopharmacology (Berl)* 1991;104:496.

842. Henningfield JE, London ED, Benowitz NL. Arterio-venous differences in plasma concentrations of nicotine after cigarette smoking. *JAMA* 1990;263:2049.

843. Henningfield JE, Stapleton JM, Benowitz NL, et al. Higher levels of nicotine in arterial than in venous blood after cigarette smoking. *Drug Alcohol Depend* 1993;33:23.

844. West RJ, Russell MA. Effects of withdrawal from long-term nicotine gum use. *Psychol Med* 1985;15:891.

845. Hughes JR, Hatsukami DK, Skoog KP. Physical dependence on nicotine in gum. A placebo substitution trial. *JAMA* 1986;255:3277.

846. Stolerman IP, Goldfarb T, Fink R, et al. Influencing cigarette smoking with nicotine antagonists. *Psychopharmacologia* 1973;28:247.

847. Tennant FS Jr, Tarver AL, Rawson RA. Clinical evaluation of mecamylamine for withdrawal from nicotine dependence. *NIDA Res Monogr* 1984;49:239.

848. Nemeth-Coslett R, Henningfield JE, O'Keeffe MK, et al. Effects of mecamylamine on human cigarette smoking and subjective ratings. *Psychopharmacology (Berl)* 1986;88:420.

849. Jarvis MJ, Raw M, Russell MAH, et al. Randomized control trial of nicotine chewing gum. *Br Med J (Clin Res Ed)* 1982;285:537.

850. Hughes JR, Hatsukami DK, Pickens RW, et al. Effect of nicotine on the tobacco withdrawal syndrome. *Psychopharmacology (Berl)* 1984;83:82.

851. Schneider NG, Jarvik ME. Nicotine vs. placebo gum in the alleviation of withdrawal during smoking cessation. *Addict Behav* 1984;9:149.

852. West RJ, Jarvis MJ, Phil M, et al. Effect of nicotine replacement on the cigarette withdrawal syndrome. *Br J Addict* 1984;79:215.

853. Gross J, Stitzer ML. Nicotine replacement: ten-week effects on tobacco withdrawal symptoms. *Psychopharmacology (Berl)* 1989;98:334.

854. Schwartz RD, Kellar KJ. Nicotinic cholinergic receptor binding sites in the brain: regulation in vivo. *Science* 1983;220:214.

855. Ksir C, Hakan R, Hall DP Jr, et al. Exposure to nicotine enhances the behavioral stimulant effect of nicotine and increases binding of [³H]acetylcholine to nicotinic receptors. *Neuropharmacology* 1985;24:527.

856. Larsson C, Nilsson L, Halén A, et al. Subchronic treatment of rats with nicotine: effects on tolerance and on [³H]acetylcholine and [³H]nicotine binding in the brain. *Drug Alcohol Depend* 1986;17:37.

857. Marks MJ, Burch JB, Collins AC. Genetics of nicotine response in four inbred strains of mice. *J Pharmacol Exp Ther* 1983;226:291.

858. Nordberg A, Wahlström G, Arnelo U, et al. Effect of long-term nicotine treatment on [^{3}H]nicotine binding sites in the rats brain. *Drug Alcohol Depend* 1985;16:9.

859. Flores CM, Davila-Garcia MI, Ulrich YM, et al. Differential regulation of neuronal nicotinic receptor binding sites following chronic nicotine administration. *J Neurochem* 1997;69:2216.

860. Epping-Jordan MP, Watkins SS, Koob GF, et al. Dramatic decreases in brain reward function during nicotine withdrawal. *Nature* 1998;393:76.

861. Kenny PJ, Markou A. Neurobiology of the nicotine withdrawal syndrome. *Pharmacol Biochem Behav* 2001;70:531.

862. Price RW, Brew B, Sidtis J, et al. The brain in AIDS: central nervous system HIV-1 infection and AIDS dementia complex. *Science* 1988;239:586.

863. Ksir C, Hakan RL, Kellar KJ. Chronic nicotine and locomotor activity: influences of exposure dose and test dose. *Psychopharmacology (Berl)* 1987;92:25.

864. Ochoa ELM, Li L, McNamee MG. Desensitization of central cholinergic mechanisms and neuroadaptation to nicotine. *Mol Neurobiol* 1990;4:251.

865. Malin DH, Lake JR, Newlin-Maultsby P, et al. A rodent model of nicotine abstinence syndrome. *Pharmacol Biochem Behav* 1992;43:119.

866. Malin DH, Lake JR, Carter VA, et al. Naloxone precipitates nicotine abstinence syndrome in the rat. *Psychopharmacology (Berl)* 1993;112:339.

867. Tassin JP, Vezina P, Trovero F, et al. Cortico-subcortical interactions in behavioral sensitization: differential effects of daily nicotine and morphine. *Ann N Y Acad Sci* 1992;654:101.

868. Mansvelder HD, McGehee DS. Long-term potentiation of excitatory inputs to brain reward areas by nicotine. *Neuron* 2000;27:349.

869. London ED, Fanelli RJ, Kimes AS, et al. Effects of chronic nicotine on cerebral glucose utilization in the rat. *Brain Res* 1990;520:208.

870. Grnüwald F, Schröck H, Theilen H, et al. Local cerebral glucose utilization of the awake rat during chronic administration of nicotine. *Brain Res* 1988;456:350.

33

NEUROPSYCHIATRY OF ALCOHOL AND DRUG ABUSE

WALTER LING
PEGGY COMPTON
RICHARD RAWSON
DONALD R. WESSON

Although addiction medicine draws its membership from a diverse group of medical specialties, a chapter on substance abuse is especially fitting for a textbook on neuropsychiatry. Substance dependence and abuse are clinically defined in behavioral terms, and their description and classification constitute a formal part of the current psychiatric disease nomenclature. The generic disease of substance dependence and abuse, the clinical syndromes of substance intoxication and withdrawal, and the substance-induced psychiatric syndromes, many of which resemble primary psychiatric disorders, clinically fall within the realm of psychiatry. Conversely, neurologists see some of the most dramatic and devastating complications of addictive diseases, ranging from trauma to strokes, seizures, and the neurologic manifestations of acquired immunodeficiency syndrome (AIDS). Indeed, few medical events rival the drama of a patient's ostensible instant recovery from heroin-induced coma after administration of naloxone or the tragedy of severe head injury resulting from a drunk driving accident. Moreover, some medications used by neurologists to treat chronic painful neurologic disorders, such as chronic headache and painful neuropathies, may contribute to the development of an independent substance abuse disorder.

Thus, substance dependence and abuse apparently should be an area of common interest for neurologists and psychiatrists and, in particular, for neuropsychiatrists. Unfortunately, the area of common interest often becomes an area of mutual neglect. Few psychiatrists involved in the treatment of substance abuse disorders do in fact see the catastrophic neurologic complications of drug abuse, and few neurologists who treat neurologic complications are involved in the management of the underlying addictive disorder beyond recovery of the acute neurologic illness. The goal of this chapter is to provide psychiatrically trained neuropsychiatrists with some understanding of the neurologic syndromes related to substance abuse and neurologically trained neuropsychiatrists with some understanding of the pharmaco-

logic and nonpharmacologic approaches to treating the underlying substance-related disorders.

We begin with some definitions of terms, followed by a discussion of the general approach to patient assessment, and then formal clinical diagnosis with laboratory confirmations. We next discuss specific neuropsychiatric syndromes induced by alcohol, cocaine, other stimulants, opiates, sedative-hypnotics, and hallucinogens, and the pharmacologic management of the three major classes of drugs of abuse: opiates, stimulants, and alcohol. We then address some specific problems of neuropsychological function relating to drug abuse, issues of comorbidity, human immunodeficiency virus (HIV) and pain, and their implications for pharmacologic and nonpharmacologic treatment strategies.

It is not our intent to write a neurologic primer for psychiatrists or a psychiatric primer for neurologists; instead, the focus is on the more commonly seen neuropsychiatric syndromes of interest to both groups. Some of these are common knowledge, such as those related to acute and chronic alcohol abuse. Others have been more recently described, such as seizures and strokes in cocaine and other stimulant abusers. We do not include any discussion relating to caffeine and nicotine dependence, even though both are listed in the American Psychiatric Association's *Diagnostic and Statistical Manual of Mental Disorders*, Fourth Edition (*DSM-IV*), and we touch only lightly on the neurologic manifestations of AIDS, mainly to serve as a reminder that AIDS, deserving full treatment in its own right, is an increasingly common medical complication among substance abusers.

DEFINITION OF TERMS

Drug abuse. Broadly defined, this term refers to the use of a drug for nonmedicinal purposes, with intent to affect consciousness and is distinct from *misuse,* which implies inappropriate medicinal use (e.g., for the wrong indication, at the wrong dose, for the wrong amount of time). The term drug abuse is deeply embedded in cultural and social values and often implies social disapproval or illegality. The use of alcohol in any amount is prohibited in Muslim cultures, whereas the use of marijuana and other hallucinogens for religious purposes is quite acceptable in many Native American tribal cultures. Contemporary American psychiatry views abuse as a maladaptive pattern of use, leading to recurrent, significant adverse physical, legal, occupational, and/or social consequences. It is thus embedded within the larger context of dependence, although not necessarily less serious. The diagnosis of substance dependence preempts the diagnosis of abuse in the same drug class.

Dependence. This term includes both psychic dependence, characterized by compulsive drug-seeking behavior, and physical dependence, characterized by the presence of withdrawal symptoms on abrupt discontinuation of the drug. Physical dependence is best understood in terms of the body's adjustment to a new level of homeostasis during the period of drug use. This new state of equilibrium manifests itself by symptoms of withdrawal when the drug is discontinued or its action is reversed by administration of specific antagonists, as exemplified by naloxone-precipitated opiate withdrawal.

Addiction. This term refers to a state of physical and psychic dependence, although it can refer to psychic dependence alone when physical dependence is not apparent.

Tolerance. Tolerance is the body's compensatory response to the effect of the drug and refers to the decreased response to drug effects such that an increasingly larger dose is required to achieve the same effect. It is closely related to physical dependence and can be of several types. *Metabolic* tolerance refers to increased disposition of the drug from increased metabolism with decreased availability of the drug at its site of action. *Cellular* tolerance refers to diminished response without concomitant reduction in drug concentration or availability. *Functional* tolerance refers to the compensatory changes owing to receptor mechanisms, whereas *behavioral* tolerance refers to the clinical ability to compensate for drug effects. Tolerance invariably leads to the need for larger doses of a drug to achieve the same effect, but it is unclear at present how tolerance relates to drug craving and drug seeking.

CLINICAL ASSESSMENT OF PATIENTS

Patient assessment can serve clinical, administrative, research, or medicolegal purposes. In general, the goals of assessment include detecting substances of abuse, formulating diagnoses, establishing the severity of the disorder and the baseline status of the patient for purposes of treatment planning, monitoring treatment progress, and formulating a prognosis.

Clinical evaluation should lead to a formal diagnosis with laboratory confirmation. It should also consider the context within which the assessment is made. An assessment performed in a general medical setting may differ from one performed for admission to a substance abuse treatment program.

Substance abuse disorders are underrecognized in general medical and neuropsychiatric practices. This is suggested by the high incidence of substance abuse disorders noted in surveys of general medical and neuropsychiatric patients, in contrast to hospital chart reviews, which show a rather low documentation of these disorders (1–4). Failure of recognition, however, may only partially explain the underreporting of substance abuse disorders. Indeed, most physicians are aware that substance abuse is one of the most frequently encountered problems in their practice but may be reluctant to report it because of the current structure of medical care and reimbursement.

Although the general approach to assessing drug abuse disorders follows traditional lines of history taking and physical and mental status examinations supported by laboratory tests, the circumstance under which evaluation takes place merits special consideration. This is because it can greatly influence the patient's attitude toward the process of evaluation and the information to be gathered.

An empathetic, nonjudgmental attitude on the part of the evaluator is necessary but does not come easily for many physicians whose training has largely exposed them to the most medically frustrating aspects of drug abuse. The term addict itself conjures up images of the most undesirable of patients. Physicians therefore need to be aware of their own attitude toward the addict. A nonjudgmental attitude encourages self-disclosure and makes it easier to establish a working relationship. Patients referred from a medical service are likely to be apprehensive and are often embarrassed by the referral. The addict's seeming lack of candor in disclosing information should not automatically be interpreted as resistance, denial, or lack of motivation. Sometimes patients are simply too embarrassed by the implications of the referral, or there may be a genuine and not totally groundless fear of losing a job or professional licensure. Addicts are largely ambivalent but not necessarily in denial about their disease. The nature of addictive disease is such that it often cuts the patient off from all social contacts. The addiction itself becomes the only meaningful relationship left in life, and this is not easy for the addict to give up. In addition, some patients simply are unable to cooperate or remember details of events because they are still under the influence of drugs or have suffered from their chronic effects.

Patient interviews should be conducted in private and without interruption. Sensitivity to the patient's need to avoid feelings of shame and guilt promotes candid disclosure. It is not that addicts need special consideration. It is simply that many physicians do not give addicts the same consideration given to other patients with illnesses that cause potential embarrassment and shame. By ensuring privacy, the physician shows respect and sensitivity, essential first steps toward establishing a therapeutic rapport.

Patients should be assured at the outset of the confidential nature of the information they are about to give. Their understanding of the confidentiality will enhance the validity of the information obtained. Physicians should be aware that substance abuse treatment information is, under the law, treated with a degree of confidentiality more stringent than general medical information (5).

Although the general approach to patient inquiry needs to be empathetic, the questioning itself should be direct and straightforward. It is better to use simple language and avoid street terms, unless one is certain of the meaning. Misuse of street terms tends to give the addict a feeling of superiority over the physician, which may in turn encourage manipulative behavior. Most addicts will give a reliable enough history if some allowances are made and if inquiry is limited to information without dwelling excessively on reasoning and motivation. Keep in mind that addicts often rationalize as a defense mechanism against their own sense of guilt and shame; their explanations are mostly for themselves and they do not expect the inquirer to believe them. It is sufficient, for instance, in exploring legal complications, to ascertain how many times the patient has gone to jail and for what charges, without being too concerned with matters of justice. An addict will often volunteer a reason, such as bad luck or prejudice from the legal system, while admitting the fact of incarceration. There is no need to echo the patient's sentiment or contradict the patient's claims.

Remember that addicts do not always want to deny or minimize the extent of their drug problems. A man referred by his employee assistance program may well want to minimize his drinking, but a heroin addict seeking detoxification commonly exaggerates the size of his habit.

The history is the cornerstone of clinical evaluation. Begin with the suspected drug of abuse that brought the patient to medical attention and explore each commonly abused drug systematically, one at a time. Keep in mind local fads that may be unique to the population under evaluation. For some drugs of abuse, even the history of use on a single occasion may be considered abuse. For others, like alcohol, the pattern and extent of use determine whether abuse or dependence is present. Because substance abuse and dependence disorders are clinically defined as maladaptive patterns of use, it is important to explore such issues as the amount of use, presence of withdrawal on cessation of use, amount of time spent in procurement and use, past attempts to control or cut down, neglect of social responsibility, and continuous use in the face of adverse physical and social consequences.

For each drug class, the age at first use and age at first regular use or intoxication must be determined. Most drug abusers begin their careers in their early teens. Late adult onset of drug or alcohol abuse suggests the presence of a coexisting or preexisting psychiatric disease, in particular an affective disorder.

Past periods of heavy use, periods of abstinence, and the surrounding life circumstances that may give clues to events precipitating relapse or motivation for abstinence should also be ascertained. This information can also aid in planning treatment. The pattern of use during the 3 to 4 weeks before evaluation, including the amount and time of last use, should be explored in detail. This information is crucial in deciding whether a period of hospitalization will be necessary to establish abstinence because withdrawal from some substances, such as alcohol, sedative-hypnotics, and other central nervous system (CNS) depressants, may have serious medical complications. The current illness should be supplemented with a detailed review of systems, history, family history, and social and occupational history. Whenever possible, an independent source should be sought to confirm the information obtained. Even when an addict wants to be candid, he/she is likely to distort out of habit.

Screening for substance abuse disorders in a general medical or neuropsychiatric context differs from evaluating patients seeking treatment. In the former, the subject matter can sometimes be approached by means of a questionnaire, such as the Michigan Alcohol Screening Test (6), the CAGE (a four-item screener) (7) for alcoholism, or the Drug Abuse Screening Test for other drugs of abuse. These instruments are easy to administer and ensure that busy clinicians do not omit important questions. However, these instruments lack the flexibility of a live interview; they tend to be constructed around a specific drug and miss the local fads. The mechanical application of a questionnaire can detract from the process of building a therapeutic rapport. In practice, these instruments are of greatest value to supplement general medical and psychiatric evaluations for the clinician whose practice does not focus on addictive disease. For clinicians who devote a substantial portion of time to the practice of addiction medicine, the points covered by these instruments become almost second nature to general history taking, and their use is unnecessary.

The history should be supplemented by careful observation and a thorough physical examination. Common physical findings of injection drug abuse include fresh needle marks and old scars, thrombosed veins, abscesses and congested nasal mucosa, enlarged liver, and local lymph nodes. Spider nevi and gynecomastia are common in alcoholism, as are cardiac murmur and arrhythmia in cocaine and other stimulant addiction. Signs of intoxication or withdrawal may be present, depending on the time of the last "fix." The mental status examination should include observation for any unusual behaviors or mannerisms, speech, orientation, attention, concentration, thought content, simple calculation, and recent and remote memory. The patient's response and behavior during the history taking and physical examination usually give sufficient clues to the mental status, but sometimes it is useful to employ a standardized cognitive screening test, such as the Mini-Mental State Examination (8). Formal neurocognitive evaluation is usually reserved for patients suspected of having drug-induced dementia. Taken together, these components of patient evaluation should lead naturally to a formal diagnosis.

CLINICAL DIAGNOSIS

The formal diagnosis of drug-related neuropsychiatric syndromes, covered under Substance-Related Disorders in *DSM-IV* (9), contains three distinct but related diagnostic categories: (a) the generic diagnosis of substance dependence and substance abuse (substance use disorders), (b) the clinical syndromes directly related to intoxication and withdrawal (substance-induced disorders), and (c) the drug-induced psychiatric disorders phenomenologically related to other specific psychiatric disorders (substance-induced mental disorders).

Substance dependence and substance abuse are characterized by a cluster of cognitive, behavioral, and physiologic symptoms occurring within a specific time frame. Central to substance dependence are compulsive drug-seeking behavior and continuing drug use despite adverse consequences, with or without tolerance and withdrawal, whereas substance abuse emphasizes repeated use under hazardous conditions and/or despite harmful consequences in an individual who does not meet criteria for dependence on the substance.

Table 33.1 lists the salient features of substance dependence and substance abuse. For dependence, three of seven criteria must be met within a 12-month period, and for abuse, one of four criteria must be met and the individual must not be dependent on the substance. Most neurologic complications related to substance use disorders fall outside the primary concerns of psychiatry and are only briefly mentioned in *DSM-IV* as associated general medical conditions.

Laboratory Testing

Several laboratory tests serve important functions in the evaluation and treatment of substance-related disorders.

Screening for drugs of abuse is invaluable in investigating patients suspected of drug overdose, confirming the diagnosis of drug dependence for planning and initiating treatment, differentiating drug-induced psychopathology from syndromes of other causes, detecting multiple drugs of abuse, monitoring compliance and treatment progress, and detecting relapse. Results of drug testing can also be a powerful tool in counseling and, in some circumstances, such as in the military and the workplace, an effective deterrent for drug use (10). Drug testing has been used for other purposes outside the therapeutic setting. It is commonly used for screening prospective employees, detecting drug use by workers in particular professions, monitoring compliance of those on parole and probation, prosecuting suspected drunk drivers, excluding athletes from competitions, confirming or disputing information given on employment or insurance applications, and investigating some unexplained or suspicious deaths.

Drug testing can be performed on several body fluids and tissues, including blood, urine, saliva, sweat, and hair. By far the most commonly used is urine because of the large amounts of drugs and metabolites excreted in the urine and because urine is easily obtained. Unfortunately, addicts, for a variety of reasons, do not want their drug use made known and resort to many ways of cheating on urine tests. At first, this created the need for direct observation for urine collection, which became one of the most demeaning aspects of substance abuse treatment. The problem increased over the years as more women entered treatment.

Today, drug-testing kits have become widely available for laboratory and in-clinic urine testing. These kits include a collection device incorporating a temperature-sensitive strip to register the temperature of the urine at the time of

TABLE 33.1. DIAGNOSTIC AND STATISTICAL MANUAL OF MENTAL DISORDERS, FOURTH EDITION CRITERIA FOR SUBSTANCE DEPENDENCE AND SUBSTANCE ABUSE

Dependence (Three or More in a 12-mo Period)	Abuse (One or More in a 12-mo Period)[a]
Tolerance (marked increase in amount, marked decrease in effect)	Recurrent use resulting in failure to fulfill major role obligations at work, home, or school
Characteristic withdrawal symptoms/substance taken to relieve withdrawal	Recurrent use in physically hazardous situations
Substance taken in larger amount and for longer period than intended	Recurrent substance-related legal problems
Persistent desire or repeated unsuccessful attempts to quit	Continued use despite persistent or recurrent social or interpersonal problems caused or exacerbated by substance
Much time/activity to obtain, use, recover	
Important social, occupational, or recreational activities given up or reduced	
Use continues despite knowledge of adverse consequences (e.g., failure to fulfill role obligation, use when physically hazardous)	

[a] Symptoms must never have met criteria for substance dependence for this class of substance.

collection. Because the temperature of freshly voided urine falls within a narrow range, this device is virtually tamper-proof and has obviated direct observation of urine collection in most cases (11,12).

Routine use of blood for drug testing is not practical because of the need for venipuncture and because drug concentrations are often too low for routine laboratory detection. Drug-testing technologies have been developed that use saliva, sweat, and hair, although these methods have not yet reached general clinical utility.

Several test methods are available for drug screening. They vary in sensitivity, specificity, and cost, and each has its own advantages and disadvantages. Thin-layer chromatography (TLC), the least expensive, is also the least sensitive. Its primary advantage besides cost is that a single test can detect multiple drugs. It is highly suitable for detecting high-dose drug use of recent origin, as in cases of suspected overdose in emergency departments where the types of drugs may be unknown but the need to detect low levels of drugs is not an issue. TLC is also widely used for routine therapeutic monitoring of patients on methadone maintenance. Because of the low sensitivity, TLC tends to give false-negative results. It is not suitable, therefore, for detecting illicit drugs of high potency, such as fentanyl, lysergic acid diethylamide (LSD), or some of the designer drugs. Many laboratories confirm results of TLC by an alternative method, and TLC alone is insufficient for forensic purposes.

Several immunoassay methods are available, based on antigen–antibody interactions. These include the enzyme multiplied immunoassay test, radioimmunoassay, and fluorescent polarization immunoassay. These methods are highly sensitive, and the tests are quick and quite easy to perform. Unfortunately, however, they can test for only one drug at a time; thus, a drug that is not thought of is not detected. However, because no extraction is required for these tests, they lend themselves to automation for large-volume testing.

More highly sophisticated assay methods using gas-liquid chromatography and gas chromatography with mass spectrometry are available in some laboratories. These tests give the best sensitivity and specificity but are also costly and labor intensive. They are used mainly to confirm results of less expensive tests.

Interpreting results of drug screening is no simple matter for the clinician. The detectability of a drug depends not only on its characteristics but also on a host of other factors, including the size of the dose, frequency of its use, time of the last use, time of the sample collection, metabolites of the drug, and, of course, type of body fluid used for testing and the sensitivity and specificity of the analytic methods. For the busy clinician, the best strategy is to become acquainted with the personnel of the local laboratory, become familiar with their general methodology, and find out what a comprehensive panel means, so that if a particular drug of abuse is suspected that does not appear on the panel, a specific request can be made. Most laboratory directors are quite re-

sponsive to questions and special requests. It may be helpful to have available a copy of the laboratory's report form listing the drugs contained in its routine comprehensive panel, information about other drug tests that the laboratory can perform, knowledge about the sensitivity of the tests, and the cutoff level at which results are reported as positive or negative (13).

In addition to drug screening, several other laboratory tests are useful in the evaluation and treatment of substance-related disorders. A simple breath test can estimate a blood-alcohol level (BAL). Increased serum γ-glutamyltransferase, absent other evidence of liver disease or the use of hepatotoxic prescription drugs, is a fairly sensitive indicator of heavy drinking in alcoholics, although it does not distinguish alcoholic liver disease from other forms of liver disease, and its specificity for heavy alcohol consumption diminishes in the presence of chronic hepatitis. A high aspartate aminotransferase (AST) level can also indicate alcohol abuse, and the ratio of mitochondrial AST to total AST can differentiate chronic alcoholics from other patients (14). Changes in γ-glutamyltransferase and AST levels are more useful when taken in conjunction with changes in mean corpuscular volume, which increases after a period of heavy drinking (15). These laboratory findings tend to reverse with abstinence and can be used serially to monitor treatment progress. Other laboratory changes, such as increased uric acid, increased triglycerides, and decreased blood urea nitrogen, also suggest alcohol abuse and also tend to return toward normal with abstinence.

Injection drug abusers commonly show elevated liver enzymes that, under treatment, improve or become stable. In the absence of acute clinical liver disease, these elevated enzymes should not be cause for withholding substance abuse intervention. Any indication of active hepatitis, which, unfortunately, is also very common among addicts, requires appropriate evaluation and treatment. Recent advances in the neurobiology of drugs of abuse and in neuroimaging techniques have allowed the visualization of many acute and chronic drug effects in the living human brain. Such advances offer exciting possibilities in the development of new medications to treat substance-related disorders and may help to define subgroups of patients who may respond preferentially to different therapeutic interventions. Meanwhile, an unexpected clinical bonus has been realized from sharing with the patient the colorful and intriguing pictures from these neuroimaging studies. Patients are often fascinated by them and are motivated to enter or remain in treatment because they show drug-related brain dysfunction.

CLINICAL SYNDROMES

Some clinical syndromes, such as acute alcohol intoxication and withdrawal, result directly from the effects of the drug on the CNS; others, such as seizures and strokes in stimulant

TABLE 33.2. NEUROPSYCHIATRIC SYNDROMES ASSOCIATED WITH DRUG ABUSE

Syndrome	Drug-Related Etiology
Seizure	Alcohol intoxication
	Cocaine, CNS stimulant toxicity
	Alcohol, sedative-hypnotic withdrawal
Stroke	Cocaine, CNS stimulant intoxication, and toxicity
	Opiate toxicity
	Alcohol abuse
	Intravenous drug use
Coma	Opiate toxicity
	Alcohol, sedative-hypnotic toxicity
	Drug-induced stroke or seizure
	Lifestyle related (head trauma, hypoglycemia)
Delirium	Cocaine, CNS stimulant intoxication
	Alcohol, sedative-hypnotic withdrawal
CNS infection	Intravenous drug use
Peripheral neuropathy	Alcohol abuse
	Injection drug use

CNS, central nervous system.

abusers, may result from the drug's direct effect on the CNS or be secondary to drug effects on the cardiovascular system. Still others, such as meningitis, brain abscess, peripheral neuropathy, and AIDS, arise from more distant factors relating to the use of contaminated needles and from the addict's high-risk lifestyle.

Each class of drugs of abuse can produce a number of neuropsychiatric syndromes. Alcohol, for example, leads to intoxication, withdrawal, seizures, and dementia. Conversely, each syndrome can be produced by more than one class of drugs of abuse. For example, seizures can result from acute stimulant abuse or from withdrawal from alcohol or sedative-hypnotics. Consequently, without adequate information, making a differential diagnosis can be difficult.

Clinical syndromes can be organized from the perspective of the class of drugs of abuse or the clinical syndromes that they produce. The following sections present the more commonly seen clinical syndromes associated with each major class of drugs of abuse, followed by tables summarizing each one by drug class and clinical syndrome. Table 33.2 shows the common syndromes seen with each class of drugs of abuse.

Neuropsychiatric Syndromes by Drug Class
Alcohol

Alcohol is the most commonly used psychoactive drug in the United States. The 2001 National Household Survey on Drug Abuse found that almost half of all Americans age 12 or older (48.3% or 109 million) used alcohol in the month before the survey. This was approximately five million persons more than the year 2000 estimates when 46.6% of those 12 or older reported recent use (16). Its popularity

is owing in no small part to its legal and sanctioned status, making it readily available at a relatively low cost to virtually all Americans of drinking age and younger. Of the 109 million people reporting current alcohol use in 2001, approximately 10.1 million persons were aged 12 to 20, and for them, alcohol is an illicit substance. It is safe to say that one of the most frequently occurring neurologic syndromes in our society is alcohol intoxication, followed closely (in terms of both prevalence and time) by uncomplicated alcohol withdrawal or hangover.

In 2001, there were 13.4 million Americans classified with dependence on or abuse of alcohol (5.9% of the total population). Among past year users of alcohol, 9.3% were classified with alcohol dependence or abuse. Treatment admissions for alcohol have decreased over the past several years. However, this probably does not reflect a decline in the numbers of alcohol abusers (17). Alcohol-related neurologic syndromes occur in relationship to the last drink consumed, as well as to drinking over a lifetime, and differ markedly among individuals and between persons who are physically dependent and those who are not.

Alcohol Intoxication

At low doses, alcohol acts primarily as a CNS intoxicant, whereas at high doses, its sedative and anesthetic actions predominate (19–21). Thus, neurologic syndromes associated with acute alcohol ingestion vary with alcohol dose. Individual response to acute alcohol ingestion is highly variable and dependent on the amount of alcohol ingested, the rate at which blood levels of alcohol rise, individual tolerance to the psychological and physical effects of alcohol, genetic factors that determine the responsiveness of the nervous system to the effects of alcohol, mood state of the individual while drinking, and the setting or environment in which the drinking occurs. Metabolism of alcohol is decreased in women, which effectively increases their BAL (22).

Alcohol acutely depresses the integrating and inhibitory functions of the cerebral cortex and reticular activating system (23,24). Miller and Gold (25) refer to this effect as a "descending paralysis," affecting the cortex first, then the limbic system, and finally the brainstem. The action of alcohol in the CNS is nonspecific, which accounts in part for the high doses (24) needed to achieve, relative to other psychoactive drugs, comparable levels of intoxication (26). The sensitivity with which neurotransmitter systems respond to alcohol appears to have a genetic basis (27–32).

The average adult metabolizes 7 to 10 g (1 oz) of alcohol per hour (33–35). After this clearance rate is exceeded, the BAL begins to rise and CNS manifestations of intoxication occur. The BAL, a commonly used measure of intoxication, accurately reflects the CNS effects of alcohol in nontolerant persons. The severity of intoxication at a given blood level is typically greater when the BAL is rising as opposed to falling.

At BAL between 0.01% and 10%, mild declines in coordination and cognition are experienced, and individuals present with disinhibited behavior, euphoric mood, and decreased anxiety. At concentrations slightly above this level, persons are notably intoxicated, demonstrating dysarthria, ataxia, poor coordination, decreased sensory function, impaired judgment and psychomotor skills, decreased attention span, and mood swings. At 0.20%, individuals are severely intoxicated, manifesting nausea and vomiting, diplopia, gross motor incoordination, incoherence, and confusion (36–39).

Other neurologic findings in alcohol intoxication reflect CNS stimulation and include facial flushing, increased heart rate, increased respiratory rate, and hyperreflexia (24,35,38,40). Behavior may be boisterous, belligerent, or dysphoric. Examination of the eyes and third cranial nerve reveal mydriatic or normal-sized pupils, sluggish pupillary light reflex, nystagmus and diplopia, redness of sclera, corneal glazing, and nonconvergence (41). With increasing alcohol dependence, individuals become tolerant to its intoxicating effects and no longer complain of getting drunk when they drink. Alcoholics have been shown to perform quite well on psychomotor or cognitive tasks at BAL between 0.20% and 0.30% (42). Marked intoxication is rarely noted in alcoholics at this BAL.

Memory disturbances are common with alcohol intoxication (43). Anterograde amnesia has been demonstrated at BAL as low as 0.04%. During an alcohol blackout, a classic example of anterograde amnesia, the intoxicated individual appears alert and is able to ambulate, carry on a conversation, and perform previously learned tasks. No impairment of long-term memory or immediate recall is noted. On becoming sober, however, the person has total and permanent amnesia for the period. These amnestic episodes are self-limiting, rarely lasting longer than 48 hours, and can occur both in persons dependent and nondependent on alcohol. They tend to occur when the slope of increasing BAL over time is steep and at BAL of greater than 0.25% (44–46). Anterograde memory function is usually recovered fully when the individual is sober, but the chance of incurring residual memory deficits increases with the chronicity of alcohol abuse (45,47).

At higher BAL, the depressant and anesthetic actions of alcohol predominate. At 0.30%, individuals become stuporous, and by 0.40%, they may lose consciousness, thus protecting themselves from further increases in BAL. A BAL of 0.50% to 0.70% results in coma, respiratory depression, and eventual death (21,37). Typically, the chronic alcoholic is protected from alcohol overdose by his/her extreme tolerance to the depressant effects of the drug, although Mendelson and Mello (35) report no elevation in the lethal dose of alcohol for alcoholics, and severe respiratory depression can occur with chronic alcohol intoxication.

Coma induced by high-dose alcohol is a metabolic coma characterized by an absence of focal neurologic signs and a fluctuating level of consciousness. The patient is typically un-

responsive but may show a purposeful motor response with vigorous or painful stimulation. Respiration is depressed and noisy, pupils are normal to dilated and react sluggishly to light, and hypothermia is common, further depressing the neurologic response (48).

Alcohol frequently plays a role in overdose of other drugs, especially CNS depressants. In one study, three-fourths of comatose patients admitted to the emergency department tested positive for alcohol, 65% of them having a BAL higher than 0.08% (49). In another study of suspected drug overdose in 492 patients, serum toxicology was positive for alcohol in 50% of the sample, and 38% had a BAL of 0.10% or greater (50). Alcohol increases the overdose potential of other psychoactive drugs because of its sedative effects as well as its disinhibitory effects on behavior.

Alcohol Withdrawal

As with intoxication, the nature and course of alcohol withdrawal depend on whether the individual is alcohol dependent (Table 33.3). Alcohol withdrawal is characterized by rebound effects in those physiologic systems that were initially modified by its ingestion. Alcohol withdrawal results in adrenergic hyperactivity at the level of the cortex, limbic system, and brainstem (51,52).

Uncomplicated alcohol withdrawal or hangover is commonly experienced by the nonalcohol-dependent individual after an episode of heavy drinking (53). It is characterized by tremulousness, mild autonomic hyperactivity, nausea and vomiting, photophobia, malaise, irritability, and vascular headache. The syndrome typically occurs 4 to 6 hours after the last drink and subsides within 48 hours (54). It appears to be partly genetically determined. Sons of alcoholics have been shown to suffer a more severe hangover than control males drinking the same amount of alcohol (55).

For the individual physically dependent on alcohol, sudden abstinence or a significant decrease in alcohol intake precipitates a withdrawal syndrome of varying severity (38,56). CNS hyperstimulation results from both decreased inhibitory and increased excitatory neurotransmission (51). In their classic work, Victor and Adams (39) describe early and late stages of acute withdrawal in alcoholics that, although temporally related, do not necessarily co-occur. The severity of withdrawal correlates partly with the chronicity of alcoholism and the degree of tolerance (57), the genetic predisposition of the individual (29), the number and severity of prior withdrawal experiences, and age (57,58). A history of severe withdrawal is one of the best predictors of the severity of the subsequent withdrawal syndrome (59). The entire acute withdrawal syndrome typically subsides within 5 to 7 days.

Alcoholic tremulousness, alcoholic hallucinosis, and withdrawal seizures appear within 6 to 24 hours after the last drink. Although described as if progressive in nature, their sequence and interdependence are inconsistent

TABLE 33.3. NEUROLOGIC SYNDROMES ASSOCIATED WITH ALCOHOL WITHDRAWAL

Syndrome	Alcohol Dependent?	Time Since Last Drink	Duration	Neurologic Findings
Uncomplicated withdrawal	No	4–6 hr	24–48 hr	Tremors, vascular headache, photophobia, mild autonomic excitation, irritability
Acute withdrawal Early stage Alcohol tremulousness	Yes	6–24 hr	24–48 hr	Tremors, headache, irritability, anxiety, mild to moderate autonomic excitation, hyperreflexia, potential transient hallucinations, clear sensorium
Alcoholic hallucinosis				Vivid, threatening, auditory hallucinations; clear sensorium; tremors; moderate autonomic excitation
Withdrawal seizures				Generalized tonic-clonic seizures, occurring in groups of two to six, without EEG focus, postictal confusion, and disorientation
Acute withdrawal Late stage Alcohol withdrawal delirium	Yes	36–72 hr	24–72 hr	Profound disorientation; fluctuating level of consciousness; poor remote and immediate memory; visual, auditory or tactile hallucinations; gross tremor; dilated pupils; extreme autonomic hyperactivity
Protracted withdrawal	Yes	2–3 wk	6–24 mo	Autonomic dysfunction, sleep disturbance, EEG changes, impaired short-term memory, fatigue

EEG, electroencephalogram.

(21,38,39,47,60). Most alcoholics experience alcoholic tremulousness, the mildest of these syndromes, as the BAL drops to between 0.10% and 0.20% (21,38). This is similar to uncomplicated withdrawal in the nondependent individual but with a somewhat greater amount of autonomic excitation and the potential for mild and transient hallucinations. Postural tremors of the distal upper extremities and tongue, mild tachycardia, headache, irritability, anxiety, anorexia, insomnia, moderate diaphoresis, hyperreflexia, and systolic hypertension are common (35,39,60). Typically, this is the extent of alcohol withdrawal for the alcohol-dependent individual; only 10% of alcoholics progress to more severe withdrawal symptomatology (35,39).

Alcoholic hallucinosis, in which the individual experiences vivid, typically auditory, and often threatening hallucinations, also occurs in early withdrawal (61,62). The individual's level of consciousness and orientation does not change dramatically, and hallucinations are recognized as such (63,64). There is mixed evidence that hallucinosis can become chronic, persisting for weeks to years after late-stage alcohol withdrawal, although an underlying schizophrenia needs to be ruled out (61,64).

Alcohol withdrawal seizures also take place during the early stage of withdrawal, often within several hours after the last drink. They are characteristically generalized tonic-clonic seizures, occurring in groups of two to six (35,39,60), with 90% occurring within the first 48 hours of alcohol abstinence (52,65–67). They are more likely to occur in individuals with previous alcohol withdrawal seizures, an effect some authors have likened to the "kindling" effects of epileptic seizures (68). Of the persons who suffer withdrawal seizures, 88% have less than five such episodes. Rarely does status epilepticus develop (35,59,69). These individuals are not

epileptic and have a normal electroencephalogram (EEG) (67,70–72).

Management of these seizures must include ruling out structural diseases and other causes of seizures. Treatment should consider the entirety of the alcohol withdrawal syndrome because other signs of full-blown withdrawal usually develop quickly. Thiamine, 100 mg intravenously, should be part of the initial fluid and electrolyte management. There is no convincing evidence, however, to warrant prophylactic anticonvulsants to prevent alcohol withdrawal seizures (73,74).

The late stage of alcohol withdrawal syndrome, known as alcohol withdrawal delirium or, more commonly, delirium tremens (DT), occurs 36 to 72 hours after the last drink. It is relatively rare, occurring in approximately 5% of alcoholics (35). Seizure activity increases the likelihood of late-stage withdrawal; between 12% and 33% of alcoholics who experience seizures on withdrawal develop DT (35,65). Its presence suggests concomitant medical problems, such as pneumonia, subdural hematoma, pancreatitis, fractures, liver disease, and malnutrition.

Alcohol withdrawal delirium is characterized by extreme autonomic hyperactivity, profound disorientation, gross disturbances of recent and remote memory, fluctuating levels of consciousness, incoherence, and confusion (64). The individual is agitated and restless, with terrifying visual, auditory, and tactile hallucinations that cannot be differentiated from reality. Autonomic hyperactivity is evidenced by hyperventilation, tachycardia (commonly 130 to 150 beats per minute), hypertension (greater than 150/100), hyperthermia (38°–39.5°C), pronounced diaphoresis, facial flushing, gross tremor, and dilated pupils. The development of profuse diarrhea and hyperpyrexia (temperature higher than 40°C) portend vascular collapse and death (59). Mortality rates of alcohol withdrawal delirium range from 5% to 15% (75).

Although the most severe withdrawal symptoms for alcoholics occur within the first week of abstinence, there is increasing evidence that persons continue to suffer residual effects of chronic CNS intoxication for weeks to years after their last drink. This is referred to as protracted abstinence and appears to follow acute withdrawal from many drugs of abuse (e.g., opiates, cocaine) (76). The syndrome is characterized by the persistence of mild to moderate subjective feelings of discomfort or distress, which may play an important role in relapse. The syndrome, which varies considerably across individuals, appears to slowly clear with continued abstinence. The most commonly reported neurologic finding associated with protracted alcohol abstinence is some degree of autonomic dysfunction (irregular respiration, labile blood pressure, and pulse), which may persist for as long as 2 years (77). Sleep disturbances, EEG changes, impaired short-term memory, fatigue, tremor, and increased muscle tension may persist for 6 to 24 months after abstinence (47,78,79). More important, depression and anxiety also tend to persist.

Syndromes Associated with Chronic Alcohol Abuse

These syndromes typically present after at least a decade of heavy drinking and, in many cases, are permanent conditions, improving only slightly (if at all) with continued abstinence (64,80). Acute intoxication and withdrawal in persons who continue to drink often complicate the clinical picture.

Wernicke Encephalopathy

The onset of Wernicke encephalopathy is typically abrupt but may develop over several days to weeks and often is accompanied by drowsiness, headache, nausea, and vomiting. The so-called classic triad of ataxia, encephalopathy, and ophthalmoplegia was present in only one-third of persons in whom Wernicke lesions were found on autopsy; conversely, lesions have been found in a number of patients who were asymptomatic during life (80,81). Profound confusion and disorientation may be confounded by withdrawal delirium (64). Oculomotor palsies of the third and sixth cranial nerves are common, as is horizontal or vertical nystagmus. Pupils react sluggishly and may be unequal. Ataxia is typically cerebellar, evident in the gait and lower extremities, with the upper extremities and speech being relatively spared (82). Ataxia may be aggravated by peripheral polyneuropathy and cerebellar degeneration, which are common in Wernicke encephalopathy (83). Untreated, Wernicke encephalopathy will progress to stupor and coma. Fortunately, it is responsive to parenteral thiamine therapy, with improvement noted (especially ocular symptoms) within days to weeks (64). If memory deficits persist, a diagnosis of alcohol amnestic disorder must be considered.

Alcohol appears to interact with thiamine deficiency to accelerate the progression of neurologic dysfunction (84,85), although genetic factors also play a role, which explains why only a subset of malnourished thiamine-deficient alcoholics develop Wernicke encephalopathy and why there are large individual differences. Measurement of serum thiamine monophosphate may be a more accurate indicator of chronic alcoholism compared with measurement of free thiamine and thiamine diphosphate and may help to identify patients predisposed to Wernicke encephalopathy (86).

Alcohol Amnestic Disorder

More commonly known as Korsakoff psychosis, alcohol amnestic disorder appears to result from cumulative effects of Wernicke encephalopathy, hence the term Wernicke-Korsakoff syndrome (80,82). Alcohol amnestic disorder is primarily a gross memory disorder of both anterograde and retrograde function, although anterograde deficits predominate. Frequently, the sufferer cannot retain new information for more than a few seconds. Intelligence, language, and speech are unaffected, and sensorium is clear. Related to profound memory deficits, the patient is usually disoriented to time and place, confabulates to reconstruct forgotten events, and may perseverate. These patients are typically unaware of and unconcerned about their amnesia. The sufferer becomes emotionally flat, inattentive, placid, and apathetic (64,80,87). Typically, drinking is stopped spontaneously, which can result in mild memory improvement to complete recovery in 20% of cases. Based on the observation of a relative deficit of norepinephrine metabolites in the cerebrospinal fluid of a small number of alcoholic Korsakoff syndrome patients, McEntee and Mair treated two small groups of such patients with clonidine and reported improvement of memory performance after 2 weeks of treatment. O'Carroll et al. (88) and McEntee et al. (89–91), in subsequent studies, however, failed to replicate the results.

Alcoholic Dementia

The diagnosis of alcoholic dementia is made when slowly progressive but stable cognitive dysfunction, attributed to metabolic or organic disease, is noted in alcoholics with extensive drinking histories (80,92). Current thought is that the dementia relates to the neuropathology of Wernicke encephalopathy in that these lesions are commonly found on autopsy of persons diagnosed with alcoholic dementia (80,93,94). Factors to which alcohol dementia has been linked include nutritional deficits, history of head trauma, and direct alcohol neurotoxicity (95–98). Conflicting evidence exists of the correlation between the dementia and frontal lobe atrophy and ventricular enlargement noted on magnetic resonance imaging and computed tomography of alcoholics' brains.

Presenting symptoms are heterogeneous, but typically a progressive lack of interest in the environment and the appearance of oneself, impaired judgment, attention deficits, and slowed thought processes are noted. Symptomatology resembles frontal lobe dysfunction, but emotional lability and grandiosity are absent. Unlike most dementias, there is no disruption of language or loss of primary motor and sensory function. Cognitive deficits typically center on the areas of nonverbal intelligence and visuospatial abilities. Memory deficits are present but never to the extent noted in alcohol amnestic disorder (99–101). Sensorium is typically clear, and aphasia is not present. If the patient continues to drink, alcoholic dementia will progress. DT exacerbates its course. Slight improvement, absence of progression, and even complete recovery have been reported in abstinent persons (64).

Cerebellar Degeneration

Degeneration of the Purkinje cells in the cerebellar cortex has been noted in some alcoholics (80). Symptoms related to these lesions tend to develop subacutely over a period of months and worsen with an uneven trajectory as drinking continues. Symptoms include gait ataxia and may be accompanied by upper extremity ataxia and dysarthria. Cerebellar incoordination is typically bilateral (36,102). Unlike Wernicke encephalopathy, oculomotor and mental deficits are absent, although Victor et al. (80) presented evidence that alcoholic cerebellar degeneration represents the same disease process. The cerebellum is noted to have high thiamine turnover rates, and stance and gait may respond to thiamine therapy, implicating thiamine deficiency as a possible etiology (103,104). As with alcoholic dementia, mild improvement of symptoms is reported with abstinence.

Neuropathy

Inadequate nutrition has been shown to play an important role in the development of polyneuropathy in alcoholics, perhaps related to general vitamin B deficiencies as opposed to thiamine deficiency specifically (80,95,103). Alcohol-induced neurotoxic effects on axonal transport may be a contributing factor as well. Alcoholic neuropathy tends to progress gradually and affects sensory and motor nerves more than autonomic nerves. Deficits are distal, bilateral, and symmetric, in a glove and stocking distribution, and affect all sensory modalities. Motor weakness, atrophy, and absent or decreased distal deep tendon reflexes signify lower

motor neuron deficit (36). Improvement occurs with more adequate nutrition in conjunction with abstinence.

Head Trauma

Alcohol abuse, common among patients with traumatic brain injuries, affects the morbidity and recovery of these patients. Behavior disinhibition, prone to occur in such patients, is exaggerated by alcohol (105,106), and even moderate amounts of alcohol can profoundly affect their balance, reaction time, and cognitive functions (107,108). Moreover, alcohol lowers the seizure threshold in these patients, who are already at risk from their brain injury (109). Recognition and management of alcohol abuse should be an integral part of managing patients with traumatic brain injury (110,111).

Cocaine

Although the neuropsychiatric sequelae of cocaine abuse were well described nearly 80 years ago (112), the cocaine epidemic of the 1980s, combined with the switch to the potent smoked (alkaloidal) form, or crack, have resulted in a better clinical description of the intoxication, withdrawal, and cerebrovascular syndromes associated with cocaine and other stimulant use (Table 33.4) (113,114). In 2001, approximately 25 million Americans aged 12 and older reported having at some time used cocaine, with an estimated 1.7 million (0.7%) current cocaine users and 406,000 current crack users (115). Among past year users of cocaine, 24.9% (1.0 million) were classified with dependence or abuse of cocaine. Cocaine treatment admission trends show stable or declining rates for primary cocaine abuse in most states (116). The more dramatic neurologic complications of cocaine abuse tend to be related to acute cocaine toxicity, whereas withdrawal-related neurologic symptoms appear to be relatively mild. Neurologic impairment related to long-term use is less well defined and may be evident as frontotemporal dysfunction (117), cerebral atrophy (118), or subtle neuropsychological deficits.

Cocaine Intoxication

The acute effects of cocaine use are dependent on dose and route of administration and often confounded by concurrent

TABLE 33.4. SYNDROMES ASSOCIATED WITH STIMULANT INTOXICATION AND WITHDRAWAL

Clinical Syndromes	Clinical Findings	Other Associated Features
Intoxication	Euphoria or affective, blunting, sociability, hypervigilance, anxiety, tremor, stereotyped movements, dyskinesia and dystonia, impaired judgment	Tachycardia, pupillary dilation, elevated or lowered blood pressure, sweating or chills, nausea and vomiting, weight loss, psychomotor agitation or retardation, muscle weakness, respiratory depression, cardiac arrhythmia, confusion, seizure
Withdrawal	Phase 1: crash depression, fatigue, craving, anxiety, suspiciousness, paranoia, exhaustion, sleep Phase 2: dysphoria, anhedonia, craving (cue-induced) Phase 3: intermittent craving	

use of marijuana, heroin, alcohol, or other CNS depressants. Oral ingestion of cocaine hydrochloride results in a delayed onset (10–30 minutes) and prolonged duration (45–90 minutes) of action, with relatively low peak plasma levels; intranasal use results in similar peak plasma levels but with a more rapid onset (2–3 minutes) and shorter duration (30–45 minutes) of drug effects. Smoking and intravenous use result in rapid onset of cocaine effects, within 5 to 45 seconds, lasting between 5 and 20 minutes, with peak plasma levels five to ten times those achieved by oral and intranasal routes (119).

The plasma half-life of cocaine, whichever the route of administration, is between 40 and 60 minutes. Benzoylecgonine, the primary metabolite of liver enzymatic degradation, is itself a potent CNS stimulant. Cocaine is detectable in the urine for as long as 36 hours after use, whereas benzoylecgonine may be present in heavy users for as long as 3 weeks.

At low doses, cocaine produces euphoria and a sense of enhanced self-image. Users report increased energy and mental acuity, heightened sensory (including sexual) awareness and self-confidence, and diminished appetite. The euphoric "rush" after intravenous or crack cocaine use is described as intense and orgasmic, lasting from seconds to several minutes. The sympathomimetic effects of cocaine result in increased motor activity and tremor, heart rate, blood pressure, and temperature. At higher doses, taken by chronic users to counteract tolerance to cocaine's euphoric effects, psychological symptoms tend to be more dysphoric and physiologic symptoms more indicative of sympathetic overactivity. Anxiety, affective instability, and suspiciousness accompany tachycardia, hypertension, and tachypnea.

Psychotic disorders associated with cocaine abuse range from delusional disorders to schizophrenia-like symptoms and are characterized by anxiety, paranoia, agitation, and impaired judgment, with relative sparing of general cognitive function (119,120). Cocaine-induced delusions are commonly persecutory, jealous, or somatic, with the latter including parasitosis or cocaine "bugs." These psychoses correlate with high cocaine plasma levels and are believed to be the result of sustained dopamine release. Symptoms gradually resolve as plasma levels fall, and clear within 24 to 48 hours in relatively inexperienced cocaine users and in several days to weeks in chronic abusers. Increasingly, a hyperarousal delirium syndrome has been noted secondary to cocaine use, characterized by sudden onset of disturbances in perception and attention, disorientation, and cognitive impairment (121). The delirium, theorized to be a variant of neuroleptic malignant syndrome (122) after a prolonged cocaine binge, is frequently accompanied by extremely violent behavior (123). Untreated, the delirium can progress to include hyperthermia, extreme psychosis, autonomic instability, and fatal respiratory collapse (124).

Cocaine-induced seizures have been well described (125–128) and are believed to arise from lowered seizure threshold by the local anesthetic effects of the drug, although

cerebral anoxia may also play a role (123,129,130). Seizures typically occur immediately after or within hours of cocaine use and are more likely to occur when cocaine is injected intravenously or smoked as crack. Most often seizures develop in persons with no history of convulsive disorders. Although chronic cocaine use has been implicated as having a kindling effect (131) on seizure activity, seizures are not uncommon in first-time cocaine users. Typically, these seizures are single and generalized and not associated with lasting neurologic deficits. Multiple or focal seizures suggest underlying brain pathology and other substance abuse. Rarely, cocaine-induced partial complex seizures with clouded sensorium and stereotyped movements are misdiagnosed as cocaine psychosis (132). Individuals with preexisting convulsive disorders and women are most vulnerable to cocaine-induced seizures (128). Status epilepticus and death have occurred when large amounts of cocaine are ingested, as happens when cocaine-filled latex condoms rupture in the body of "body packers" (133). Death from cocaine-related seizures is believed to result from cardiac arrhythmias (123) and hyperthermia (134), mechanisms quite apart from underlying cocaine-induced seizures.

Cerebrovascular accidents in cocaine abusers result from specific effects of cocaine on the cerebral vasculature in combination with the sympathomimetic effects of cocaine on vascular tone. It has been estimated that the relative risk of stroke among drug abusers to the general population after controlling for other stroke risk factors is 6.5 to 1 (135), making drug abuse the most common cause of stroke in persons younger than the age of 35; cocaine abuse accounts for the overwhelming majority of these events (136). In addition, cocaine abusers frequently abuse alcohol, which itself is a risk factor for stroke.

Most cocaine-related strokes have a sudden onset and occur within 72 hours of cocaine use. Hemorrhagic and ischemic strokes occur with approximately equal frequency in cocaine abusers compared with the general population, in whom infarcts predominate (123), and there is some evidence that the use of alkaloidal cocaine is more likely to result in ischemic strokes than the use of cocaine hydrochloride (137). In almost one-half of hemorrhagic strokes, underlying vascular abnormalities (saccular aneurysms and arteriovenous malformations) are present; in the remainder of cases, thalamic and basal ganglia bleeds are common (138). Cocaine-induced hypertension is believed to be the primary cause of intracerebral and subarachnoid hemorrhage (139).

The mechanisms responsible for ischemic cerebrovascular accidents in cocaine abusers remain unclear, but several possible mechanisms have been described. Vasospasm of the cerebral vessels may be induced by the catecholaminergic effect of cocaine, deranged cerebral autoregulation related to sudden changes in perfusion pressure, increased release of serotonin, or the vasoconstrictive effects of cocaine's primary metabolite, benzoylecgonine (114,123,130,137, 140–142). Increased platelet responsivity and cardiogenic emboli are other possible causes of occlusion (123,143). Cocaine-induced cerebral hypoperfusion, specifically in the frontal lobe and other superficial cortical areas, contribute to these ischemic events. Transient ischemic attacks have also been reported in cocaine abusers (144,145), many of which escape medical attention because of their spontaneous resolution.

Cocaine Withdrawal

Physiologic dependence on cocaine is evidenced by the development of tolerance to its euphorigenic effects and by the emergence of distinct withdrawal phenomena within a few hours of the last cocaine use. Withdrawal symptoms are commonly recognized after heavy and prolonged cocaine use, although a substantial number of individuals dependent on cocaine has few clinically evident withdrawal symptoms on cessation of use (146). Symptoms characteristic of cocaine withdrawal attributed to acute dopamine depletion and receptor up-regulation include dysphoric mood; fatigue; vivid, unpleasant dreams; insomnia or hypersomnia; increased appetite; and either psychomotor agitation or retardation. Muscle pains, chills, twitching, and tremors have also been reported (147). However, the most serious problems associated with cocaine withdrawal are suicidal ideation and suicide attempts, associated with profound withdrawal depression.

Three phases of cocaine abstinence have been described (148). The first phase, commonly referred to as the "crash," occurs within hours of the last cocaine use and lasts 3 to 4 days. Agitation, anorexia, and intense cocaine craving appear early in the crash but are quickly replaced with fatigue, hypersomnolence, and hyperphagia as craving subsides. Depression persists throughout this first phase. During the second phase, which occurs over the next 8 to 10 weeks, patients experience an early return to normalized mood and sleep pattern with little cocaine craving, although as the anhedonia, anergia, anxiety, and craving for cocaine gradually intensify, patients are at high risk of relapse. The third, or final, extinction phase occurs with continued cocaine abstinence. During this period, the risk of relapse decreases as mood and hedonic response normalize, although cocaine craving continues to emerge intermittently, especially in the presence of conditioned cues. Such craving, resembling protracted alcohol and opiate withdrawal syndromes, can persist indefinitely.

Cocaine- and Stimulant-Induced Movement Disorders

Cocaine has been reported to induce chorea, dystonia, choreoathetoid movements, akathisia, buccolingual dyskinesia, eye blinking, and lip smacking (149–152). These movements, referred to as "crack dancing" within some drug-using communities, have typical onset within 2 hours of use and appear to be self-limited, although follow-up data are sparse and a few appear to persist. They are similar to those described with amphetamine abuse (153,154) and have recently been associated with norpseudoephedrine use (155).

Most likely, these symptoms result from accumulated high levels of dopamine in the synaptic cleft within the basal ganglia.

Amphetamines

Amphetamines made up less than 5% of the 1.6 million treatment admissions at federally funded substance abuse programs in 1999. However, this represented a significant increase from the 2% in 1993. Methamphetamine is the primary form of amphetamine seen in the United States and can be smoked, injected, or inhaled (156). Compared with amphetamine, methamphetamine has increased central nervous system penetration and a more prolonged duration of action (6–24 hours) (157). Methamphetamine use has periodically produced problems in the United States, and its abuse has become widespread in various countries abroad (158,159).

Methamphetamine Intoxication

Acutely, methamphetamine produces physiologic effects similar to those of cocaine: tachycardia, at times with arrhythmia, although apparently less often than with cocaine, hypertension, hyperthermia, increased respiratory rate, and dilation of the pupils. The onset of acute effects depends on the route of administration, from seconds after smoking and injection to a few minutes after oral ingestion. Maximal effect is reached in 15 minutes and lasts for 10 to 12 hours, followed by aftereffects lasting approximately 48 hours (160). This long duration of action distinguishes amphetamine effects from those of cocaine.

Metamphetamine's psychological effects include a heightened sense of well-being or euphoria, increased alertness and vigor, and decreased appetite and sleep. Acute administration has been shown to increase socialization in humans. However, mood changes are common, with the user rapidly changing from friendly to hostile. High doses may produce repetitive and stereotypic acts accompanied by irritability, excitement, auditory hallucinations, paranoia, and violent behavior. Dangerously elevated body temperature and convulsions occur with methamphetamine overdoses and if untreated can result in death. Tolerance and sensitization to the behavioral effects develop with repeated exposure. The underlying mechanism of methamphetamine action, similar to those of cocaine, appear to be related to increased levels of free dopamine in the brain's limbic reward system (161).

Methamphetamine Withdrawal

Methamphetamine withdrawal effects nearly exactly inversely recapitulate intoxication, a finding often ignored or unrecognized by the abuser. These effects, like those of cocaine, include fatigue, unpleasant dreams, insomnia or restless hypersomnia, hyperphagia, psychomotor agitation/retardation, dysphoria, and fragmented attention span. The symptoms are intense and protracted because of methamphetamine's long duration of action. A period of sleep that may be promoted by the use of secondary substances such as alcohol, barbiturates, and benzodiazepines, finally begins several hours after the last use and may last for several days. On awakening, the user may experience severe depression, lasting from a few days to several weeks. While in this depressed state, the user is at increased risk of suicide. After users feel that they have "recovered" from a binge episode, cravings set in, and the cycle often begins again (161).

Craving frequently occurs in response to exposure to conditioned cues (stimuli present during past episodes of methamphetamine use and euphoria). Such cues evoke a powerful craving for methamphetamines via classic conditioning principles. The likelihood of continued methamphetamine smoking or injecting appears to be related in part to the strength of the craving experienced from these craving-generating cues. The withdrawal dysphoria present in the context of ubiquitous methamphetamine availability and conditioned cues can produce a very pernicious dependence; indeed, inpatient hospitalization may be indicated to treat long-term methamphetamine dependence, at least in initial stages of detoxification. Medically managed inpatient care is expensive, however, and widespread methamphetamine abuse has appeared in impoverished populations with very limited access to such inpatient resources. Other than the actual presence of the drug and environmental cues, stress also appears to be a major cause for relapse (162).

Neuropsychiatric Syndromes Associated with Methamphetamine Abuse

The most common CNS syndromes associated with methamphetamine toxicity are acute and chronic psychoses, strokes, and seizures. Compared with cocaine abusers, who often present to emergency departments with acute medical problems relating to cardiovascular, pulmonary, and cerebrovascular complications, methamphetamine abusers are more likely to come to medical attention because of their behavioral manifestations, such as trauma relating to physical violence or vehicular accidents and acute psychosis, and are thus more likely to be seen by a psychiatrist than a neurologist. Recent animal and human neuroimaging studies suggest that methamphetamine neurotoxicity is both profound and prolonged (163). However, considering the large number of people using amphetamine worldwide, it is remarkable that there are not more reports of adverse events. This is largely owing to tolerance to the effects of methamphetamine.

Methamphetamine-Related Psychosis

Amphetamine-related psychosis was recognized soon after the drug was first introduced, with the first paper on the subject published 65 years ago (164). Amphetamine-related psychosis bears a striking resemblance to schizophrenia, and recent papers have evaluated the notion that amphetamine-induced psychosis may be a synthetic drug

model for schizophrenia (165). It has been hypothesized that amphetamine-related psychosis is more common than psychosis in cocaine users because of the high affinity of methamphetamine for σ receptors. σ Receptor binding is connected with the occurrence of psychosis. Amphetamine-related psychosis has been reported in large numbers in Japan (166) and Thailand (167) but has not been reported on such a large scale in the United States. Although the methamphetamine-related psychosis mostly resolves in a matter of days to a few weeks, it may persist in some patients for months or even years. This too appears to occur more frequently in some parts of the world, such as Japan and some countries in Southeast Asia. The reasons are not clear, but it is unlikely that these are simply cases with concomitant schizophrenic illness because the incidence of the latter is much the same in different parts of the world.

Even when not frankly psychotic, methamphetamine users are often restless, tense, and fearful. Some develop delusions of persecution; ideas of reference; and auditory, visual, and tactile hallucinations, usually with clear sensorium and recognition of the drug-induced nature of their symptoms. The predominance of visual hallucinations and the relative preservation of orientation are believed by some researchers to differentiate amphetamine psychosis from true schizophrenia (166).

Amphetamine-Related Stroke

There is good documentation to support the association of methamphetamine use and hemorrhagic and ischemic strokes. However, the mechanism of stroke in these patients is not always clear. The strokes have been reported in patients who used methamphetamine via smoking (168–170), oral ingestion (171,172), and intravenous injection (169,173–181). There have been no experimental histologic studies, and only a few autopsies of stroke victims have been reported (182).

Opiates

The extent to which Americans abuse opiates is difficult to quantify because most opiates ingested are legally dispensed for analgesic use; illicit use of street opiates accounts for a relatively small proportion of the total amount of opiates consumed. Precise estimates of heroin use remain difficult to determine. Standard methods of measuring prevalence such as household surveys are not adequate because only a small number of users would be included in a household survey. Survey-based estimates substantially underestimate the prevalence because of difficulties in locating heroin abusers and obtaining accurate reports because heroin use is illegal. To partially account for underestimation owing to underreporting and undercoverage, a 1996 study made adjustments based on treatment clients and counts of arrests. These adjustments resulted in estimates of approximately

TABLE 33.5. SYNDROMES OF OPIATE INTOXICATION AND WITHDRAWAL[a]

Syndrome (Onset and Duration)	Characteristics
Opiate intoxication	Conscious, sedated, "nodding," mood normal to euphoric, pinhole pupils, history of recent opiate use
Acute overdose	Unconscious; pinhole pupils; slow, shallow respirations
Opiate withdrawal,	
Anticipatory[b] (3–4 hr after last "fix")	Fear of withdrawal, anxiety, drug craving, drug-seeking behavior
Early (8–10 hr after last "fix")	Anxiety, restlessness, yawning, nausea, sweating, nasal stuffiness, rhinorrhea, lacrimation, dilated pupils, stomach cramps, drug-seeking behavior
Fully developed (1–3 d after last "fix")	Severe anxiety, tremor, restlessness, piloerection,[c] vomiting, diarrhea, muscle spasms,[d] muscle pain, increased blood pressure, tachycardia, fever, chills, impulse-driven drug-seeking behavior
Protracted abstinence (may last as long as 6 mo)	Hypotension, bradycardia, insomnia, loss of energy or appetite, stimulus-driven opiate cravings

[a] The times given refer to heroin. Withdrawal will develop more slowly with long-acting opiates.
[b] Anticipatory symptoms begin as the acute effects of heroin begin to subside.
[c] The piloerection gave rise to the term "cold turkey."
[d] The sudden muscle spasms in the legs gave rise to the term "kicking the habit."

3.0 million lifetime users in the United States and approximately 663,000 past year users (183). In the 1990's, there were approximately 1.6 million annual treatment admissions in publicly funded substance abuse treatment facilities across the nation. Of these, the proportion who sought treatment for heroin addiction increased from 12% in 1993 to 14% in 1999. Historically, people treated for heroin addiction have injected the drug. In 1998, 66% of the people treated for heroin addiction were injectors. This estimate reflects a decline from 74% in 1993. Meanwhile, inhalation of heroin increased in 1998 to 29% of the people treated for heroin addiction from 23% in 1993 (184).

The number of person reporting use of oxycodone (OxyContin) for nonmedical purposes at least once in their lifetime increased from 221,000 in 1999 to 399,000 in 2000 to 957,000 in 2001. The annual number of new users of pain relievers nonmedically has also been increasing since the mid-1980's when there were approximately 400,000 initiates. In 2000, there were an estimated 2.0 million (185).

As with alcohol, syndromes commonly associated with opiate use are temporally related to the last opiate use and chronicity of use (Table 33.5). The duration of opiate intoxication and onset of withdrawal vary with the pharmacologic half-life of the particular opiate. For example, heroin and meperidine have relatively short half-lives, and withdrawal symptoms from these opiates can emerge within 8 hours of drug abstinence. Tolerance and physical dependence develop quite rapidly after repeated opiate exposure; mild physical dependence has been demonstrated after short-term postoperative opiate analgesic administration. Tolerance, however, does not develop uniformly to all opiate effects; notable tolerance develops to the analgesia, euphoria, sedation, and respiratory depression and little to the gastrointestinal, pupillary, and endocrine effects (186). Cross-tolerance to the CNS effects of opiates exists across opiate medications.

Opiate Intoxication

The CNS effects of opiates include analgesia, drowsiness, changes in mood, and mental clouding. Respiratory depression, miosis, nausea and vomiting, and cough reflex suppression are other opiate effects mediated centrally (187). Acute subjective effects vary depending on whether the individual is in pain, is dependent on opiates, or is without either pain or addiction. In normal, pain-free individuals, the acute effects of opiates are frequently described as unpleasant and include nausea, vomiting, inability to concentrate, lethargy, and reduced visual acuity. For those experiencing pain, analgesia is the predominant response and is accompanied by drowsiness, pruritus, and occasionally euphoria. For drug-free exaddicts, mental clouding is less severe and euphoria is pronounced, typically peaking with the rush after intravenous injection.

Opiate overdose occurs when the respiratory depressant effects result in a decreased level of consciousness. Despite the notable tolerance that opiate addicts develop to opiate respiratory depression, overdose is not uncommon, partly because tolerance is not complete and partly because addicts tend to combine opiates with alcohol or other CNS depressants. Pinhole pupils, slow and shallow respiration, and coma are the classic triad of acute opiate overdose. If hypoxia persists, pulmonary edema, hypotension, and cardiovascular collapse occur, and eventually death supervenes. Establishment of an airway and intravenous administration of naloxone, an opiate antagonist, are the key elements of treatment of opiate overdose (188).

Opiate Withdrawal

CNS hyperexcitability characterizes the opiate withdrawal syndrome. The severity of withdrawal depends on the degree of physical dependence, the particular opiate used, the health and personality of the addict, and the setting in which the withdrawal occurs (186,188). Early symptoms include lacrimation, rhinorrhea, yawning, and sweating, with progression of symptoms to include dilated pupils, anorexia, gooseflesh, restlessness, irritability, and tremor. In fully developed opiate withdrawal, increased irritability, insomnia, marked anorexia, violent yawning, severe sneezing, lacrimation, nausea, vomiting, diarrhea, weakness, pronounced depressed mood, tachycardia, and hypertension are noted. Fluid and electrolyte imbalances occur with prolonged anorexia, vomiting, diaphoresis, and diarrhea, although rarely does opiate withdrawal require medical intervention. The most common outcome of opiate withdrawal is relapse to opiate use.

As with alcohol, a protracted abstinence syndrome has been described in drug-free opiate addicts, emerging after the patient has been completely detoxified and lasting from weeks to months. Although subtle, the symptoms of protracted abstinence, including hypotension, hyperalgesia, and poor stress tolerance, have been identified as important reasons for relapse after opiate withdrawal (187).

Neuropsychiatric Syndromes Associated with Opiate Abuse

Opiates themselves do not directly cause either neuropathology or psychopathology; rather, it is the common routes of administration (subcutaneous and intravenous), the adulterants used to cut or prepare abused opiates (e.g., quinine, lead, chloroquine, *N*-methyl-4-phenyl-1,2,3,6-tetrahydropyridine) (189,190), and the lifestyle associated with illicit opiate use that account for the majority of neurologic complications of opiate abuse. Other than those associated with HIV infection and AIDS (see later discussion of HIV), most infective neurologic complications of intravenous drug use are those that accompany endocarditis, which is typically right-sided with tricuspid valve involvement (191). Infectious material filtered in the pulmonary tree are circulated systemically when arteriovenous shunts develop secondary to pulmonary hypertension and can result in intraparenchymal and extraparenchymal abscess of the brain and spinal cord, meningitis, embolic cerebral infarction, diffuse vasculitis, and septic aneurysm (144,192). Other neurologic complications of infectious origin in street opiate addicts include hepatitis, encephalopathy, CNS syphilis, tetanus, and botulism (193).

A neurologic complication of noninfective origin associated with parenteral opiate use is cerebral or spinal embolic stroke as the result of relatively insoluble foreign materials or adulterants being injected directly into the vasculature along with opiate drugs (144); other possible causes of cerebral infarction in opiate addicts include hypoperfusion with opiate overdose and allergic or toxic vasculitis (194). Hemorrhagic strokes in heroin addicts follow hypertension secondary to nephropathy or clotting derangements secondary to liver disease. Peripheral neuropathy may occur owing to nerve damage at common injection sites or pressure palsies during overdose (195,196), and fibrotic myositis

has resulted from repeated needle trauma and local toxic responses (197).

Sedative-Hypnotics

The sedative-hypnotics include a chemically diverse group of medications that relieve anxiety, produce sedation, and induce sleep. Many Americans report having used sedatives for nonmedical purposes; however, most sedative-hypnotics are not primary drugs of abuse. Of the 7.0 million current users of illicit drugs other than marijuana, 4.8 million were current users of psychotherapeutic drugs. Of those who reported current use of any psychotherapeutics, 3.5 million used pain relievers, 1.4 million used tranquilizers, 1.0 million used stimulants (including methamphetamine or amphetamines), and 0.3 million used sedatives (115). Some of the short-acting barbiturates are injected for the rush or taken orally to produce a state of disinhibition similar to alcohol, whereas benzodiazepines are often used by substance abusers to relieve opiate withdrawal, to augment the effects of methadone, or to ameliorate the adverse effects of cocaine or methamphetamine. Unlike other drugs of abuse, most sedative-hypnotics are manufactured by legitimate pharmaceutical companies and diverted to the illicit street-drug market through medical channels.

Sedative-Hypnotic Intoxication

Most people do not find the subjective effects of sedative-hypnotics pleasant or appealing beyond their therapeutic effects (i.e., relief of anxiety or facilitation of sleep), whereas persons predisposed to addiction find the effects of sedative-hypnotics reinforcing (198). The acute effects of sedative-hypnotics consist of slurred speech, incoordination, ataxia, sustained nystagmus, impaired judgment, and mood lability. In large amounts, sedative-hypnotics produce progressive respiratory depression and coma. The amount of respiratory depression produced by the benzodiazepines is much less than that produced by the barbiturates and other sedative-hypnotics.

Sedative-Hypnotic Withdrawal

Barbiturates produce tolerance and physical dependence, which can be induced within several days of continuous infusion of anesthetic doses. The withdrawal syndrome from sedative-hypnotics is similar to that from alcohol. Signs and symptoms include anxiety, tremors, nightmares, insomnia, anorexia, nausea, vomiting, postural hypotension, seizures, delirium, and hyperpyrexia. The syndrome is qualitatively similar for all sedative-hypnotics; however, the time course depends on the particular drug. With short-acting sedative-hypnotics (e.g., pentobarbital, secobarbital, meprobamate, oxazepam, alprazolam, triazolam), withdrawal symptoms typically begin 12 to 24 hours after the last dose and peak in intensity between 24 and 72 hours. Symptoms may develop more slowly in patients with liver disease or the elderly because of decreased drug metabolism. With long-acting drugs (e.g., phenobarbital, diazepam, chlordiazepoxide), withdrawal symptoms peak on the fifth to eighth day.

During untreated sedative-hypnotic withdrawal, the EEG may show paroxysmal bursts of high-voltage, slow-frequency activity that precede the development of seizures. The withdrawal delirium may include confusion and visual and auditory hallucinations. Some patients may have only delirium or only seizures, whereas some may have both. Benzodiazepines can also produce a protracted withdrawal syndrome in some patients after cessation of long-term therapeutic use.

Hallucinogens

Hallucinogens are a chemically diverse group of compounds that produce perceptual distortion and prominent visual hallucinations. Perceptual changes include depersonalization, derealization, illusions, and synesthesias in which sensory stimuli in one sense produce changes in another (e.g., a beating drum is seen as a flashing light). Many drugs, such as nicotine and atropine, can produce visual hallucinogens as a facet of high-dose toxicity and in some cultures are ritualistically used for their hallucinatory effects. The hallucinogenic drugs considered here are those commonly used by drug abusers in the United States. Hallucinogens do not produce physical dependence, and because tolerance to the psychological effects develops rapidly, they are typically used episodically.

Hallucinogens are used for a variety of reasons, ranging from a means of spiritual quest to enhancing the enjoyment of dancing or simply experiencing the alterations of perception. Recently, 3,4-methylenedioxymethamphetamine (MDMA, known as Ecstasy) has increased in popularity at dance clubs in the United States (199) and England (200,201). In the 1960's, LSD was the most commonly used hallucinogen, whereas MDMA was more popular in the 1990's (see Club Drugs section). The incidence of hallucinogen use exhibited two prominent periods of increase. Between 1965 and 1971, the number of new hallucinogen initiates rose ten-fold, from 90,000 to 900,000. The second period of increase began in 1990 when there were approximately 600,000 new hallucinogen users. By 2000, the number rose nearly three-fold to 1.5 million initiates (202). Approximately 1.3 million (0.6%) of the population aged 12 and older reported that they were current users of hallucinogens in 2001 (203).

Club Drugs

The term "club drugs" came from the association of these substances with dance clubs called "raves." Generally, the following five drugs are included in this category: GHB (γ-hydroxybutyrate), ketamine, LSD, MDMA (Ecstasy), and methamphetamine. A 2001 update reports data on emergency department visits related to the abuse of club drugs, the Drug Abuse Warning Network report. According to this report, methamphetamine continues to account for the largest share of emergency department mentions (15,000 mentions in 2001) in large metropolitan areas across the United States. In 2001, MDMA (Ecstasy) was the next most frequent club drug in mentions (5,500 mentions) (203). The number of persons who had ever tried MDMA increased from 6.5 million in 2000 to 8.1 million in 2001. There were 786,000 current users in 2001. In 2000, an estimated 1.9 million persons used MDMA for the first time compared with 700,000 in 1998. This change represents a tripling in incidence in just 2 years (115).

Emergency department visits involving club drugs usually involve multiple substances, especially alcohol, marijuana, and other club drugs. Co-occurring cocaine and heroin also appear in substantial numbers. The long-term trends for GHB, ketamine, and MDMA have been upward, whereas LSD mentions have declined significantly since 1994. Emergency department visits involving GHB and MDMA demonstrated remarkably similar upward trends from 1994 to 1999, but from 1999 to 2001, MDMA mentions continued upward, and GHB mentions leveled off (203).

Hallucinogen Intoxication

The effects of hallucinogens depend on the drug, the dose, the setting, and the circumstances of use. Typically, CNS effects include dilated pupils, tachycardia, sweating, palpitations, and tremors. Experienced hallucinogenic users accept these signs and symptoms as an expected part of the drug experience, although novice users sometimes find the symptoms frightening. Perceptual alterations associated with intoxication include changes in mood, intensification of perceptions, depersonalization, and derealization. Hallucinations are usually visual, but auditory and tactile hallucinations sometimes occur. When users become frightened by the content of the hallucinations or by paranoid ideation, they may become acutely anxious and are said to have a "bad trip." These bad trips rarely come to medical attention because experienced users are skilled at "talk-down" techniques, which usually take the form of reassurance or redirecting attention to another topic or image. The combined effects of perceptual distortion, impaired judgment, and excitation sometimes result in serious adverse consequences. For instance, a user might jump off a building not because of suicidal ideation, but because they believe that they can fly.

Neuropsychiatric Syndromes Associated with Hallucinogen Abuse

Sporadic case reports suggest that LSD can cause significant vasospasm and stroke (204). Subarachnoid hemorrhage has been reported in association with MDMA abuse (205). 3,4-Methylenedioxyamphetamine (MDA) and MDMA have been implicated in degeneration of serotonergic neurons in animals, but the clinical significance is uncertain (206).

Neuropsychological impairment has been reported in heavy users of MDMA (207).

Flashbacks are transient recurrences of perceptual alterations like those that occur under the influence of the hallucinogen. Perceptual alterations may include flashes of color, peripheral field images, after images, halos, or macropsia and micropsia. Flashbacks may last seconds to several minutes and may be triggered by other drugs, anxiety, or fatigue. Flashbacks can occur as long as 5 years or more after abstinence. The person is aware that the perceptual alteration is a drug effect. LSD and MDMA can cause flashbacks and have also been reported to induce panic disorder (208–211) and chronic paranoid psychosis (212,213).

Marijuana

Marijuana is the most commonly used illicit drug. In 2001, it was used by 76% of current illicit drug users. An estimated 2.4 million Americans used marijuana for the first time in 2000. The annual number of new marijuana users has varied considerably since 1965 when there were an estimated 0.6 million new users. The number of new marijuana users reached a peak in 1976 and 1977 at approximately 3.2 million. Between 1990 and 1996, the estimated number of new users increased from 1.4 million to 2.5 million and has remained at this level (115).

Marijuana intoxication has been demonstrated to impair learning and short-term memory (214,215) and the performance of complex motor tasks (216–218). However, it is fairly well established that chronic marijuana use produces no notable neurologic impairment. In an extensive review of the literature, Wert and Raulin (219) found no significant differences between volunteer and heavy user samples and normal controls on computed tomography or EEG measures nor on detailed neurologic examination. Furthermore, most studies reviewed provided no evidence that long-term marijuana use resulted in diminished psychological or neuropsychological performance (220). Yet the authors caution that undetected subtle neurologic impairments may exist, and these may become more pronounced with age, noting that few of the subjects studied were older than the age of 35.

Marijuana is illegal under federal law but legal under California law and San Francisco ordinance if it is part of a medical treatment. In 1996, California became the first state to approve the use of marijuana for medical ailments. Since then, eight other states have passed medicinal marijuana laws (221). Proponents of medical marijuana describe it as useful in long-term pain management, mood elevation, and appetite stimulation for patients who may have AIDS, cancer, or other serious medical conditions.

It is pertinent to add that the cannabinoids found in marijuana have been reported to have both antiepileptic and antispasmodic activity. Before the 1937 passage of the Marijuana Tax Act, cannabis was listed in the *National Formulary and Pharmacopoeia* for these clinical indications and was available

in more than 28 different pharmaceuticals (222). Although virtually never used today as an antiepileptic agent, animal data have provided evidence that the cannabinoid cannabidiol depresses epileptiform electrophysiologic responses with little to no development of tolerance to these effects (223–226). In a small, placebo-controlled clinical trial with 15 patients with epilepsy refractory to traditional anticonvulsant pharmacotherapy, improvement was noted in 85% of patients receiving 200 to 300 mg of cannabinoid daily compared with only one patient on placebo (227).

There is much anecdotal but little published evidence of the use of marijuana as an antispasmodic agent for the treatment of muscle spasms that occur with spinal cord injury and multiple sclerosis (228), which is attributed to the inhibitory effects of the tetrahydrocannabinol cannabinoid (THC) on polysynaptic reflexes (229). In one small, double-blind clinical trial (230), nine patients with multiple sclerosis–related spasticity were treated with two doses of THC or placebo on three separate occasions. Both THC doses (5 and 10 mg) significantly improved spasticity as measured by deep tendon reflex tone, resistance to stretch, and the presence of abnormal reflexes. Clinical populations frequently report self-medicating with marijuana to lessen muscle spasticity and support recent efforts to make THC more readily available to clinicians and patients (231). There is increasing evidence that cannabinoid may be analgesic.

Specific Neuropsychiatric Syndromes Associated with Drug Abuse

Just as a particular drug of abuse can result in a number of neuropsychiatric syndromes, a particular neuropsychiatric syndrome can be associated with a number of drugs of abuse. The most common neuropsychiatric syndromes that occur across drug classes include seizures, stroke, coma, delirium, CNS infection, and peripheral neuropathies. When substance dependent patients present these complaints for medical care, it is essential to rule out primary neurologic diseases and to include long-term drug abuse treatment in the total care plan.

Seizures

Seizures are sudden alterations in cerebral function produced by excessive discharge of groups of neurons in the brain. They can be generalized or focal, depending on whether the entire brain or localized brain areas are involved; not uncommonly, seizures begin focally and quickly progress to generalized seizures as neuronal discharge spreads throughout the brain. Generalized seizures are the most frequent type encountered in drug abusers and are characterized by loss of consciousness, loss of motor control, incontinence, and tonic-clonic jerking of the extremities. Generalized seizures typically last 1 to 5 minutes. Those that recur without the patient regaining consciousness between seizures or

TABLE 33.6. SEIZURE ACTIVITY ASSOCIATED WITH DRUGS OF ABUSE

Drug Class	Intoxication	Withdrawal	Early Abstinence
Alcohol	Infrequent; may potentiate seizures in patients with underlying seizure disorder	Common	Infrequent; seizure threshold may be lowered
Opiates	May occur with meperidine and propoxyphene use	Absent except in the case of a newborn	Rare; seizure threshold may be lowered in persons with intoxication seizures
Stimulants	Common	Absent	Rare; seizure threshold may be lowered in persons with intoxication seizures
Sedative-hypnotics	Rare; these agents act as anticonvulsants	Common in abrupt withdrawal	Infrequent; seizure threshold may be lowered
Hallucinogens	Rare	Absent	Absent

continue unabated for more than 5 minutes are termed status epilepticus, which requires immediate anticonvulsant pharmacotherapy, although cocaine-induced status epilepticus may be refractory to standard anticonvulsant agents. Focal seizures in drug abusers suggest the presence of a focal lesion, such as brain abscess or stroke.

Clinical experience suggests that seizures are much more common among drug abusers than in the general population. Drug-related seizures occur most frequently between the ages of 25 and 40 years. Diagnosis of drug-related seizures, usually made based on patient history or witness report, can be difficult unless the incidents are observed because addicts may feign seizures to obtain disability status or medications, and EEG, although helpful when abnormal, may be normal between seizures.

Drug intoxication or withdrawal should not automatically be assumed to be the cause of seizures in drug abusers. Brain abscess, embolic stroke, and meningitis, all common complications of drug abuse, can first manifest as generalized or focal seizure activity, as can neurologic diseases of nondrug abuse origin. Other causes of seizures must be considered, especially in patients whose seizures cannot temporally be related to drug use. When drug-related seizures seem likely, it should be determined whether they arise from intoxication or occur during withdrawal (Table 33.6).

Stroke

Strokes are a sudden, focal disruption of brain function produced by interruption of blood flow to brain tissue. They can be ischemic, occurring from obstruction of blood flow, or hemorrhagic, resulting from rupture and bleeding within the cerebrovasculature. In either case, neurons distal to the blood flow interruption become anoxic and die. Hemorrhagic strokes also induce vasospasm of nearby vessels, further aggravating cerebral anoxia. The nature, severity, and duration of neurologic deficit after stroke vary with the location, size, and number of affected vessels; strokes originating in the circle of Willis can result in massive motor and sensory deficits, coma, or death, whereas multiple infarcts in small vessels may result in less obvious deficits that may become

evident only on careful neurologic or neuropsychological testing. Deficits arising from embolic stroke may be transient (lasting less than 24 hours) if the embolus dissolves or if collateral blood flow is sufficient to prevent infarct.

The relationship between drug abuse and stroke has been recognized for more than two decades (192), and abuse of drugs has been identified as a predisposing factor in 47% of stroke patients younger than the age of 35 (135). Stroke in stimulant abusers is likely to be hemorrhagic and related to drug-induced hypertensive bleeds in cerebral vessels with underlying weakness, such as aneurysms or arteriovenous malformations (130,144). Strokes in alcoholics also tend to be hemorrhagic secondary to hypertension from chronic alcohol abuse (232). Stroke in opiate abusers, conversely, is likely to be ischemic and related to inadequate cerebral perfusion during opiate overdose (233). Intravenous use of any drug puts individuals at risk for ischemic stroke owing to emboli arising from an infected heart valve or foreign particulate matter injected into the bloodstream with the drug (144,233) (Table 33.7).

Coma

Coma can complicate opiate, alcohol, and sedative-hypnotic overdose owing to profound respiratory depression and direct CNS effects associated with toxic levels of these drugs, singly or in combination. In opiate overdose, the respiratory rate is extremely low, and cyanosis may be present. Pupils are typically pinholes bilaterally, unless hypoxia is severe, in which case they are fixed and dilated. Blood pressure falls as hypoxia persists, the body temperature is low, the skin is cool and clammy, and the muscle tone is hypotonic. Pulmonary edema and shock commonly complicate opiate-induced coma, and respiratory failure is the typical cause of death (234). Maintenance of an adequate airway and administration of naloxone result in dramatic improvement, which can occur even in the presence of alcohol or sedative-hypnotics because of antagonism of opiate-induced respiratory depression.

The coma from toxic levels of barbiturates or alcohol is similar, with very slow or Cheyne-Stokes respiration and low

TABLE 33.7. CAUSES OF CEREBRAL EMBOLISM IN INTRAVENOUS DRUG ABUSERS

Cause	Mechanism
Injection of particulate matter into the carotid artery	While attempting to inject into the jugular vein, abuser accidentally punctures the carotid artery, sending particulate matter directly into cerebral circulation
Reverse blood flow across the patent foramen ovale	Recurrent embolization of particulate matter in the lungs causes pulmonary hypertension, shunting venous blood to the left ventricle and directly into the cerebral circulation
Formation of arteriovenous shunts in the lung	Recurrent embolization of particulate matter in lungs produces granulomas and arteriovenous shunts in lung tissue, introducing venous blood into the arterial circulation to the brain
Endocarditis	Repeated injection of infectious matter into the venous circulation results in friable right-sided heart valve tissue; valvular emboli reach the arterial circulation and brain via the patent foramen ovale or pulmonary arteriovenous shunts described above

body temperature. Deep tendon reflexes are generally preserved, and the Babinski sign is often positive. Pupils are constricted but remain reactive unless hypoxia is severe. Death results from severe hypotension and shock, renal failure, or pulmonary complications. Treatment is primarily supportive, consisting of airway maintenance, adequate supply of oxygen, and maintenance of blood pressure with intravenous fluid and dopamine administration. If drug ingestion occurred within 24 hours, gastric lavage may be initiated; rarely, hemodialysis may be indicated (235).

Overdose of benzodiazepines alone rarely causes coma; however, alcohol increases the absorption of benzodiazepines and potentiates its CNS depression (235). Other causes of coma in drug abusers include direct neurologic sequelae from drug use (e.g., stroke, seizure) and from risky lifestyle and health behaviors (e.g., hypoglycemia, head trauma).

Delirium

The delirium that accompanies cocaine intoxication and alcohol withdrawal is typical of delirium of neurologic or metabolic origin. Clinical features include inattention, rambling irrelevant speech, clouded sensorium, disorientation, hallucinations, memory impairment, disturbances of the sleep/wake cycle, and psychomotor activity. Typically, substance-induced delirium is accompanied by hyperthermia, psychomotor hyperactivity or agitation, and auditory or tactile hallucinations that are recognized by the patient as such until delirium becomes severe. Transient paranoid psychosis, brief panic attacks, and marked physical agitation (including violent behavior) have been reported with delirium associated with cocaine intoxication (121,236–238).

Management of substance-induced delirium includes provision of a calm, quiet, dimly lit, reassuring environment. Restraints, which should be used sparingly and only when the patient's safety cannot be ensured, can lead to symptom escalation in some cases. Benzodiazepines are the pharmacotherapy of choice for substance-induced delirium;

short-acting agents are preferred for stimulant-induced delirium and longer acting ones for the delirium associated with alcohol withdrawal (239).

Central Nervous System Infection

The immunosuppression that accompanies cocaine and opiate abuse, apart from that associated with HIV disease, puts intravenous users of these drugs at high risk for neurologic infections. Direct injection of pathogens, which are harbored on the heart valves in endocarditis and subsequently carried into the CNS as infective emboli or as disseminated sepsis, can result in brain and epidural abscess (240) and meningitis. Subarachnoid hemorrhage occurs from the rupture of mycotic aneurysms (241,242). Users who inject into the jugular vein are at particular risk of vertebral osteomyelitis (240,243,244), with secondary epidural and extradural spinal infections. Common bacterial and fungal pathogens include hemolytic streptococci, *Staphylococcus aureus* (245–248), *Pseudomonas* (249–251), and *Candida* (252), although more exotic organisms have also been identified (253, 254).

Peripheral Neuropathy

Peripheral neuropathy in substance abusers can result from direct trauma to the nerve from injection of toxic adulterants, from local infection caused by contaminated paraphernalia, or from pressure palsies during periods of unconsciousness or heavy sleep from intoxication (196). The disorder affects both motor and sensory nerves and is characterized by pain, weakness, and numbness. Alcohol neuropathies tend to be bilateral and symmetric, affecting the distal limbs most profoundly. Neuropathies from trauma, traction, and pressure tend to be focal and affect nerves prone to pressure, such as radial nerves in "Saturday night palsy" and brachial plexopathy from traction.

PHARMACOLOGIC TREATMENT OF DRUG ABUSE

Management of the neuropsychiatric complications of drug abuse includes treatment of the underlying addiction, which often begins with management of withdrawal symptoms accompanying abstinence. Although psychosocial and psychoeducational therapies have remained critical components of the overall recovery program, pharmacologic strategies have assumed an increasingly important role in the management of opiate, cocaine, methamphetamine, and alcohol abuse and dependence. Pharmacotherapy can be short term and narrowly focused, using specific medications for targeted clinical manifestations, such as benzodiazepines for alcohol withdrawal and naloxone for acute opiate overdose, or it can be more sustained and long term with more distal and global outcome implications. Treatment goals differ from the perspective of the physician, the patient, and the society. The physician wants the addict to stop using drugs and get healthier; the patient wants to avoid physical sickness from withdrawal and hassles from law enforcement but not necessarily to stop using drugs; society wants crime reduced, neighborhoods made safe, and addicts converted from tax eaters to taxpayers. These divergent goals can lead to frustration and disappointment when the patient and clinician fail to appreciate each other's expectations.

It is also important to distinguish pharmacologic efficacy from clinical efficacy. Because pharmacotherapy primarily affects physiology, only some changes can be attributed to the direct effects of the medication. For example, in substituting methadone for heroin, the prevention of withdrawal and blockade of the effects of subsequently administered street heroin can be attributed to the direct pharmacologic effects of methadone. Retention in treatment and reduction in craving and drug use can also be reasonably linked to methadone's pharmacologic effects, although less directly. Reduction in criminal activities and increase in employment, although no less important clinical outcomes, are nonetheless more distal effects of methadone treatment and subject to many intervening events. Pharmacologic efficacy cannot always be directly translated into clinical efficacy. Naltrexone, for instance, completely blocks the effects of heroin and is therefore pharmacologically highly efficacious but has only limited clinical efficacy because few addicts want to take it. Conversely, clinical efficacy can only be attributed to medication effects where pharmacologic efficacy has been shown.

Opiate Pharmacotherapies

Pharmacotherapy for opiate dependence includes treatment of acute overdose, detoxification, and long-term maintenance (Table 33.8). Naloxone, a highly potent, virtually pure opiate antagonist, is the specific treatment for acute opiate overdose, characterized by slow, shallow respiration, pinhole pupils, and coma. The response is both dramatic and immediate. The patient, who moments earlier appeared virtually dead, awakens after an intravenous injection of 0.4 to 0.8 mg naloxone, even before the needle is completely withdrawn from the injection site. This response is so specific that anything short of a clear reversal of coma should arouse suspicion of irreversible coma from severe, prolonged anoxia, mixed overdose, overdose from nonopiates, or presence of serious underlying neurologic complications. A transient response should suggest overdose from long-acting opiates, such as L-alpha acetyl methadol (LAAM) or methadone. Coma may return in these patients after the effects of naloxone wear off. Such patients require a longer period of observation or hospitalization overnight. Because of its short duration of action, naloxone may need to be given repeatedly or administered by an intravenous drip. It is especially important to keep in mind that polydrug abuse is now the rule rather than the exception.

Opiate Detoxification (Gradual Withdrawal)

Until recently, methadone and clonidine were the two most commonly used medications for opiate detoxification in the United States. A synthetic full opiate agonist, methadone effectively suppresses the symptoms of opiate withdrawal

TABLE 33.8. OPIATE ADDICTION PHARMACOTHERAPIES

Medication	Drug Class	Indication	Clinical Action
Methadone	Opiate agonist	Maintenance therapy, detoxification	Suppresses withdrawal, blocks effects of subsequently administered opiates
L-alpha acetylmethadol	Opiate agonist	Maintenance therapy	Suppresses withdrawal; blocks effects of subsequently administered opiates
Buprenorphine	Partial opiate agonist	Maintenance therapy, detoxification	Suppresses withdrawal; blocks effects of subsequently administered opiates; produces less physical dependence than full agonists
Naltrexone	Opiate antagonist	Postdetoxification	Blocks effects of subsequently administered opiates
Clonidine, lofexidine[a]	α_2-Adrenergic agonist	Withdrawal	Suppresses withdrawal-induced noradrenergic hyperactivity

[a] Not approved for use in the United States but is approved in the United Kingdom for opiate withdrawal.

and can be conveniently administered once daily in an oral preparation, usually diluted with juice. The starting dose for detoxification of street heroin addicts is usually 30 to 40 mg. After several days of stabilization, the dose is tapered to zero over 2 to 3 weeks. In practice, because methadone is a Schedule II narcotic under strict legal control and the law prohibits physicians from prescribing a Schedule II narcotic for treatment of opiate addiction, methadone detoxification is available only in specially licensed outpatient methadone clinics. Also, because federal regulations had limited detoxification to 21 days, there was a very high rate of treatment failure. Most patients do not complete the full 21-day program and drop out of treatment as the dose is reduced. For those few who manage to complete detoxification, the recidivism rate has been high. Recent changes in the federal regulations have extended methadone detoxification to 180 days, but because of the chronic, relapsing nature of opiate addiction, this has not appeared to substantially improve treatment results (255).

Clonidine, an α-adrenergic agonist marketed for treatment of hypertension, suppresses the physiologic manifestation of opiate withdrawal related to rebound sympathetic hyperactivity. It is not a scheduled medication and is therefore available to physicians in general. Although it has not received U.S. Food and Drug Administration (FDA) approval for the treatment of opiate withdrawal, it has been used for this purpose extensively, and several dosing strategies have been described (186). Its major drawbacks are orthostatic hypotension and sedation, which limit the amount of medication that can be safely administered. In addition, clonidine suppression of opiate withdrawal symptoms is incomplete. It has little effect on craving, and ancillary medication is often needed for aches and pains and for insomnia.

Lofexidine, an analogue of clonidine, has been available in the United Kingdom for several years as an opiate detoxification agent and is currently undergoing evaluation in the United States. It produces clinical results similar to clonidine, but with fewer hypotensive effects (256–258). Its clinical utility as a detoxification medication, at least in the United States, remains uncertain.

Buprenorphine alone and in combination with naloxone (in tablet form) has just been approved by the FDA for treatment of opiate dependence and will likely be superior to clonidine and lofexidine. Its availability in the general medical setting would also make it advantageous compared with methadone. It has already been shown to be efficacious in outpatient detoxification and appears to have wide patient acceptance (259,260).

The severity of opiate withdrawal depends on the degree of dependence and environmental influences. Experience in therapeutic communities and other closed environments suggest that in such settings, the physical symptoms of withdrawal are physiologically benign and clear rapidly without drastic medical intervention. However, there is a high rate of relapse after patients leave the controlled environment. Hence, a major challenge in opiate detoxification has been to find a medication that can be used with good patient acceptance, without unnecessarily burdensome legal constraints, by general physicians in their office practice while their patients live and work in the real world. The recent FDA approval of buprenorphine is a major step toward these goals.

There has also been interest in rapid and ultrarapid detoxification strategies, performed while the patient is under anesthesia or heavy sedation. These procedures, however, are not standardized and are often considered by practitioners to be trade secrets. Many patients continue to experience withdrawal symptoms after being discharged, and it is unclear whether these patients have actually achieved physiologic detoxification or have, as is often the case, simply been inducted on naltrexone. Long-term follow-up has not thus far demonstrated the advantages of this procedure, and the risks involved, albeit small, remain an issue of contention (259–261).

The major drawback of detoxification is what happens afterward. It is not particularly difficult to get off opiates, but staying off is hard. All the follow-up data show that detoxification by itself has a very low rate of completion and a high rate of relapse even for the completers. In view of the considerable risks of HIV and hepatitis from continued opiate use, detoxification must be considered a transitional strategy toward long-term treatment and rehabilitation.

Opiate Maintenance

Two relatively long-acting opiate agonists, methadone and LAAM, and a partial agonist, buprenorphine, are now available for maintenance treatment for opiate dependence. Buprenorphine, a partial agonist, was approved and made available to clinicians in 2002. Naltrexone, an orally effective, long-acting antagonist, has been available for nearly two decades for the prevention of relapse but has had only limited clinical success.

Methadone

Methadone maintenance is the prototype for opiate maintenance pharmacotherapy, having been administered in specially licensed clinics for more than 30 years. It is a full opiate agonist that, when administered orally, suppresses the emergence of withdrawal symptoms in opiate-dependent individuals and blocks the effects of subsequently administered opiates for as long as 72 hours (262,263). Methadone has a slower onset and longer duration of action than heroin and other short-acting, abused opiates and thus does not provide a reinforcing rush on administration. Most patients can be effectively maintained with once-daily dosing, and because of its blocking action at the μ-opiate receptor, addicts no longer need to use drugs to counteract or avoid withdrawal. Because it is orally administered, secondary reinforcements related to needle use are obviated. Methadone maintenance frees patients from having to engage in illegal and risky activities (including intravenous drug use) to obtain and use opiates and provides them the opportunity to participate in rehabilitative activities. High-dose methadone maintenance (70–90 mg daily) has proven more effective than low-dose maintenance in reducing injection drug use, which in turn is important in reducing HIV transmission among the drug-abusing population (264–266).

The upsurge in heroin use in the late 1990s, and the increased risks of HIV, hepatitis, and other infections have prompted consideration for changes in the methadone delivery system to allow more flexibility in meeting patient needs (267). These changes transfer oversight of methadone treatment from the FDA to the Substance Abuse and Mental Health Services Administration and a shift from regulation to accreditation for treatment programs. These measures should improve the outcome of methadone treatment (268) and make methadone available to a larger segment of the patients needing treatment (269). Recent work continues to confirm the efficacy of methadone treatment and the added benefit of psychosocial intervention, including contingency management (270,271).

L-Alpha Acetyl Methadol

A derivative of methadone, LAAM was first synthesized by German scientists in 1948. Its slow onset and long duration of action, unique among the group of related compounds, suggested the presence of active metabolites. This finding was subsequently confirmed by laboratory and human studies. Shortly after its discovery, Fraser and Isbell (272), in a series of studies with past addicts, and addicts maintained on morphine, showed that orally administered LAAM produced euphoria, pupillary constriction, and constipation and cross-substituted for morphine, preventing the emergence of abstinence in subjects maintained on 240 mg morphine daily in dosing intervals as long as 96 hours. Clinical interest in LAAM as a maintenance treatment for opiate addiction emerged in the late 1960's with the growing success of methadone maintenance. Its efficacy in reducing illicit opiate use and keeping patients in treatment as well as its safety with chronic administration were demonstrated convincingly in a series of studies, including a labeling assessment study, leading to its approval by the FDA in 1993 for treatment of opiate dependence (273,274), thus making LAAM the first available alternative to methadone in more than three decades.

After oral administration, LAAM is well absorbed in the gastrointestinal tract. It is metabolized by *N*-demethylation to nor-LAAM and dinor-LAAM, both potent opiate agonists, with the former being three to six times more active than methadone and the latter approximately equivalent to methadone. Opiate activities, representing the combined effects of the parent drug and its active metabolites, begin in

1 to 2 hours, peak at 4 to 8 hours, and last as long as 72 hours. It is widely distributed throughout the body and, unlike methadone, is excreted largely in the feces as nor-LAAM and dinor-LAAM. Less than 20% is excreted in the urine, mostly as conjugates.

The slow onset and long duration of action confer on LAAM a unique pharmacologic profile that offers both pharmacologic and logistic advantages in opiate maintenance pharmacotherapy. Its slow onset of action makes it less subject to abuse, greatly decreases its street value, and thus minimizes the risk for street diversion. Its long duration of action provides a more stable blood level and allows three times weekly dosing. Thus, clinics can serve larger numbers of patients by reducing the amount of paperwork and manpower needed for daily dosing preparation and record keeping (272,275–279).

LAAM may be particularly appealing to those who have difficulty attending clinics on a daily basis, find methadone's duration of action too short owing to rapid metabolism, or find methadone's effects too sedating because LAAM appears to have less sedative effect than methadone. Patients fearing methadone's stigmatization may also find LAAM an attractive alternative. Despite these potential advantages, the introduction of LAAM into clinical practice has not met with its anticipated success (280).

Because LAAM is also a Schedule II narcotic, its use is governed by regulations similar to those for methadone. Recent concerns about its cardiac effects led to its withdrawal from clinical use in Europe and an FDA "black box" labeling caution in the United States. Clinicians, however, continue to find it advantageous for some patients, and the Center for Substance Abuse Treatment is developing clinical guidelines for its use (281).

Buprenorphine

Buprenorphine is a potent partial opioid agonist at the μ receptor and an antagonist at the κ receptor. Its agonist activities, such as analgesia and respiratory depression, have a ceiling effect even at full receptor occupancy. Its κ antagonist effect renders it devoid of psychomimetic effects. It is a more potent analgesic than morphine, its equivalent parenteral analgesic dose being 0.3 mg. It blocks the effects of subsequently administered opiates and precipitates withdrawal in patients dependent on full agonists. Buprenorphine binds tightly to its receptors and dissociates slowly, conferring a long duration of action and poor irreversibility by naloxone. It is poorly absorbed after oral administration and has a large first-pass effect but is well absorbed sublingually, having approximately two-thirds bioavailability after parenteral administration. Buprenorphine is widely distributed throughout the body, with the highest concentration seen in the liver, kidneys, lungs, and heart. It possesses high affinity to brain tissue relative to plasma concentration, which peaks approximately 90 minutes after sublingual administration.

Its terminal half-life is approximately 45 hours and requires 10 days to reach steady state, accounting for the slow onset of relatively mild withdrawal symptoms after cessation of dosing. Buprenorphine is highly bound to α- and κ-globulin. It is metabolized by conjugation and *N*-dealkylation and is predominantly excreted by the fecal route as the buprenorphine conjugate *N*-dealkylbuprenorphine and its conjugate, with renal excretion accounting for less than 5% of the administered drug in 48 hours.

Buprenorphine's potential for the treatment of heroin addiction was first recognized in the late 1970's by Jasinski et al. (282), who showed that it substitutes for heroin and suppresses opiate withdrawal with few withdrawal symptoms of its own. Its ceiling effect, beyond which dose increases prolong its duration of action without increasing its agonist effect, confers on it a high clinical safety profile with limited effects on euphoria and respiratory depression. Human laboratory studies and early clinical observations showed that buprenorphine suppresses heroin self-administration (283–285) and provided evidence for its utility and general safety in clinical settings (286,287).

A series of controlled clinical trials affirmed buprenorphine's safety and efficacy, and it has been available for clinical use for several years in some European countries, notably France and the United Kingdom, and in Australia (214F). Several recently completed and ongoing studies in the United States have continued to support buprenorphine's safety and clinical utility. In October 2002, the FDA approved the use of buprenorphine in the treatment of opiate dependence in the United States. This comes against the background of the Drug Addiction Treatment Act of 2000 enacted by the U.S. Congress, which, for the first time in nearly eight decades, authorizes physicians to prescribe an opiate-based medication for the treatment of opiate dependence. This will enable physicians to treat opiate abuse patients in the same manner that they treat their other patients. Addiction medicine specialists have taken note that it literally took an act of Congress to make buprenorphine available, but it is undeniably a defining moment in the history of substance abuse treatment and, in particular, the treatment of opiate dependence.

In response to reports of buprenorphine abuse in some parts of the world, a buprenorphine-naloxone combination tablet was developed in concert with the buprenorphine-only formulation (288,289). When the combination tablet is taken sublingually, buprenorphine is readily absorbed but naloxone is not, resulting in the buprenorphine effect only. If injected, however, the naloxone (as an opiate antagonist) induces immediate withdrawal, thus deterring intravenous abuse. Studies indicated that a 4:1 combination of buprenorphine/naloxone was adequate to discourage abuse while retaining most of the opiate agonist effects of buprenorphine (290) and maintain acceptance by patients. This formulation was used in a large National Institute on Drug Abuse (NIDA)–sponsored multicenter trial that demonstrated that its clinical efficacy is comparable with the buprenorphine-only tablet (291). Together, these studies support the effectiveness and safety of buprenorphine for both short- and long-term use.

In clinical practice, patients can be inducted onto buprenorphine with relative ease with the exception of patients who are maintained on methadone because precipitated withdrawal can occur. Several induction strategies are available and are discussed in detail in most training programs provided by such addiction medicine specialist groups as California Society of Addiction Medicine and American Society of Addiction Medicine. Once stabilized, both patients and clinicians find maintenance buprenorphine treatment satisfactory. A daily dose of 4 to 32 mg is sufficient to suppress opiate withdrawal and prevent abuse. In contrast to methadone, the required maintenance dose tends to decrease with time. Buprenorphine's long duration of action permits dosing frequency to be reduced to two or three times weekly, although most patients prefer daily dosing, often with a small nighttime dose. Discontinuation of buprenorphine after maintenance can be achieved by gradual dose reduction, but follow-up treatment strategies must be made available because of the chronic and frequent tendency to relapse in opiate dependence.

The unique pharmacologic properties of buprenorphine make it an ideal first-line treatment agent for opiate dependence. For the few patients who can benefit from short-term detoxification, buprenorphine permits this to be done in a general medical setting with relative ease. For the majority of patients, a moderate to long-term treatment strategy is more appropriate. Patients with a high level of physical dependence may require a full opiate agonist such as methadone or LAAM. Patients can usually be moved from buprenorphine to these agents, whereas the reverse is more difficult because of the potential for precipitated withdrawal.

The integration of buprenorphine into mainstream medical practice is not simply an addition to available opiate pharmacotherapies, but rather it signals a major shift in opiate-dependence treatment strategy within a social and political context. Over time, it should encourage clinicians to view individuals with addictive diseases in the same light as all other patients.

Naltrexone

Produced by *N*-allyl substitution of naloxone with the cyclopropyl-methyl group of cyclazocine, naltrexone combines the relatively pure antagonist property of the former with the long duration of action of the latter and comes close to being the ideal opiate antagonist. It is absorbed quickly, after a single oral dose of 50 mg, reaching peak plasma concentration within an hour and lasting for as long as 24 hours. At doses of 150 mg, it provides blockade for subsequently administered opiates for as long as 72 hours and needs, therefore, to be given only once every 3 days. It produces no euphoria or dysphoria, is not addicting, has no street value, and is not abused by addicts. Clinical studies in the 1970's and early 1980's demonstrated its value as an adjunct to treatment of opiate dependence (292,293). It was approved by the FDA in 1984 and had been available under the trade name Trexan until 1995 when, after being approved for use in alcohol relapse prevention, its trade name was changed to ReVia.

Within the context of opiate maintenance pharmacotherapy, naltrexone has only found limited success. One major drawback relates to its potent and virtually pure antagonist property that precipitates acute abstinence in patients with any degree of physical dependence. It cannot, therefore, be administered to addicts until detoxification is complete, which, unfortunately, has generally been very difficult or impossible for the majority of cases. Detoxification in an inpatient setting, assisted by ancillary medication and other measures, has been somewhat more successful. Buprenorphine in this context may also prove to be a useful adjunct (294).

Another drawback of naltrexone is its low patient acceptance, perhaps also because of its lack of appreciable agonist activity. Patients often discontinue the medication after only a short time, and because naltrexone has no reinforcing property, patients rarely, if ever, return to treatment on their own. Some subgroups of patients, such as physicians and other professionals, do considerably better with naltrexone treatment compared with street addicts (295). The reasons for the discrepancy are not always clear, but the first group generally has a great deal more to lose and are under considerable outside pressure. Quite possibly, naltrexone works best when there is significant external motivation, such as the threat of loss of a job or professional license.

Because a major shortcoming of the currently available form of naltrexone is lack of patient compliance, efforts to develop a depo formulation that can be administered to patients once every several weeks have been ongoing for many years. Still, there is no such preparation approved by the FDA, although several are in different stages of clinical trial. Interest in this area of medication development has been bolstered by naltrexone's approval for use in treatment of alcoholism, which greatly expands its potential market, and by the recently increased interest in its use as an adjunctive medication in conjunction with rapid and ultrarapid opiate detoxification.

Stimulant Pharmacotherapies

Efforts to develop effective pharmacotherapies for cocaine, methamphetamine, and other stimulants have been ongoing for more than two decades; however, no effective medications have yet received FDA approval. At the height of the cocaine epidemic in the 1980's, physicians were apt to prescribe medications to manage the acute and chronic symptoms of withdrawal and to prevent relapse. Many of these attempts were based on clinical hunches, intuition, and common sense, but often they were also based on fads and "hypes" and often out of desperation. The list of medications that have been purported to be effective for cocaine addiction is lengthy, but none has stood up to the test of controlled clinical trials (296–298).

The reinforcing psychostimulant effects of cocaine are attributed to its actions on several neurotransmitter systems, notably dopamine, serotonin, and norepinephrine. Cocaine withdrawal, a significant precipitator of relapse, is believed to arise from neurobiologic adaptations in these systems as a result of chronic cocaine use. The primary outcome of the withdrawal was initially attributed to dopamine depletion, and the first pharmacotherapeutic agents evaluated for the treatment of cocaine dependence were those that either block cocaine's reinforcing effects at the dopamine receptor or increase the availability of dopamine within the mesolimbic pathway to minimize withdrawal discomfort, cocaine craving, and dysphoria. These agents, which increase the availability of dopamine and decrease withdrawal severity, include dopamine receptor antagonists (which make cocaine use less rewarding), dopamine receptor agonists, dopamine reuptake and degradation inhibitors, and dopamine releasers and precursors.

As with dopamine, the reuptake of serotonin is also blocked with acute cocaine use, and relative serotonin depletion appears to play a part in the withdrawal syndrome. Thus, medications that block the serotonin receptor have been used in attempts to attenuate cocaine reinforcement, and functional serotonin agonists have been used to minimize the severity of withdrawal. Because of the limited demonstrated efficacy of medications acting singly within the dopamine or serotonin systems, multiple alternate mechanisms, including concurrent serotonergic and dopaminergic therapy, stimulant substitution, and antiepileptic drug therapy, are also being explored.

Research efforts to develop medications to aid in the treatment of methamphetamine-related disorders are at an even earlier stage of development. There are at present no medications that can quickly and safely reverse life-threatening methamphetamine overdoses. Similarly, there are no medications that can reliably reduce the paranoia and psychotic symptoms that frequently contribute to episodes of dangerous and violent behavior associated with methamphetamine use, and as clinicians will attest, it would be tremendously helpful to have medications that could help methamphetamine users recover more quickly from the effects of chronic use. Medications that could reduce symptoms in the early days and weeks of recovery could be extremely valuable in promoting engagement and retention in behavioral and psychosocial treatments (299).

Relapse to methamphetamine use is a complex process. However, one important set of contributing factors is the unpleasant emotional and cognitive impairments that accompany the protracted abstinence syndrome for months after methamphetamine use is discontinued. To date, there have been fewer than ten placebo-controlled, double-blind efficacy trials of potential methamphetamine pharmacotherapies. In response to the need for more research to find an effective medication, the NIDA has recently established the Methamphetamine Clinical Trials Group, a network designed to provide new clinical research teams and sites in geographic areas where methamphetamine use is a major public health problem (300). Within NIDA, the establishment of the Medication Development Division, now the Division for Treatment Research and Development, has made possible a two-pronged approach toward pharmacotherapy development for stimulant abuse. One approach is to examine selected promising marketed medications taking advantage of available human data; the other builds on discoveries of basic science research bearing on the mechanisms of drug use, reinforcement, abstinence, craving, and relapse.

Alcohol Pharmacotherapies

Pharmacotherapy, apart from the treatment of medical complications and psychiatric comorbidity, may be directed toward treatment of acute intoxication, acute withdrawal, or relapse prevention.

Intoxication

In theory, the intoxicating effects of alcohol can be reduced or reversed by inhibiting its absorption, antagonizing its effect at the receptor sites, altering or counteracting its physiologic effects, or enhancing its elimination from the body. In practice, the search for a sobering agent has been singularly unsuccessful (301). Contrary to popular belief, caffeine and other stimulants do not reverse the intoxicating effects of alcohol, nor do they increase elimination of alcohol from the body. Ingestion of large amounts of fructose does modestly increase alcohol metabolism but has not proved clinically useful (302). Thus far, neither γ-aminobutyric acid antagonists nor reverse agonists have been shown to block the effects of alcohol (303).

Acute Alcohol Withdrawal

Treatment of acute withdrawal has enjoyed considerably more success. Most currently used medications for suppression of withdrawal are benzodiazepines. Some examples are diazepam (5 to 20 mg orally every 4–6 hours), chlordiazepoxide (25–100 mg orally every 4–6 hours), and oxazepam (15–60 mg orally every 4–6 hours). Benzodiazepines with a long duration of action and slow elimination are, in general, preferable. An added advantage of benzodiazepines is that they are also anticonvulsants and may prevent the emergence of alcohol withdrawal seizures (304,305). Where psychotic symptoms or extreme agitation occurs, however, an antipsychotic medication such as haloperidol may need to be used. Seizures complicating withdrawal require separate consideration and are dealt with elsewhere. Clonidine, an α-adrenergic agonist, has been useful in reducing the neurophysiologic effects of withdrawal, as has propranolol, which counteracts the tachycardia, anxiety, and tremor of acute withdrawal (306–308).

Relapse Prevention Pharmacotherapies

Pharmacologic strategies for relapse prevention have involved medications that produce an adverse reaction when alcohol is consumed and medications that modify craving and consumptive behavior. Disulfiram (Antabuse) irreversibly inactivates acetaldehyde dehydrogenase, a step in alcohol metabolism. This leads to the accumulation of acetaldehyde when alcohol is consumed and the appearance of the disulfiram-alcohol reaction, characterized by flushing, throbbing headache, nausea, vomiting, thirst, sweating, palpitation, chest pain, tachycardia, confusion, agitation, and, when severe, respiratory depression and cardiovascular collapse (309). The reaction can occur with BAL as low as 50 to 100 mg per 100 mL and last from 30 to 60 minutes. The dose used ranges from 125 to 1,000 mg per day, although most patients are treated with 250 to 500 mg daily. Alcoholics have characterized disulfiram as a "3-day insurance policy" because it takes about that long for the medication to be eliminated from the body after last dose. Some patients, however, are able to drink even while taking disulfiram without experiencing any disulfiram-alcohol reaction, whereas others are so sensitive that inadvertent use of over-the-counter medications, toiletries such as aftershave lotions, and foods that contain small amounts of alcohol may bring on a reaction. Patients must be cautioned, therefore, about the use of these products while on disulfiram.

Medications that decrease alcohol craving and consumption aim at manipulating the brain neurotransmitter systems mediating reward, mood, and appetite behavior. Several medications acting on several neurotransmitter systems have shown various degrees of success in clinical use. Among these are the selective serotonin reuptake inhibitors fluoxetine and citalopram, narcotic antagonists naltrexone and nalmefene, and acamprosate, a γ-aminobutyric acid agonist that also has an effect on glutamate neurotransmission (310). Clinical studies with selective serotonin reuptake inhibitors on alcohol consumption and relapse prevention have yielded mixed results, and none has as yet received official approval for clinical use (311). Naltrexone, originally developed for relapse prevention of opiate dependence, received FDA approval in 1994 as a recommended adjunct to alcohol relapse prevention in the context of comprehensive treatment programs. Although the exact mechanisms of naltrexone's action in preventing alcohol relapse remain unclear, animal studies have demonstrated empirically that it reduces alcohol consumption (312). It is believed that the reinforcing effects of alcohol may be mediated at least in part via the opioid receptor mechanisms. Two initial double-blind, placebo-controlled clinical studies showed that patients receiving naltrexone in the context of comprehensive treatment programs had lower rates of relapse, fewer drinking episodes, and longer time to relapse and were less likely to suffer full blown relapse after a "slip" (313,314). Subsequent studies and clinical experience suggest that naltrexone's therapeutic effect is modest in general clinical settings.

Nalmefene is a universal opioid antagonist that works on all opiate receptors. In a 1999 Miami study, patients who received nalmefene were 2.4 times less likely to relapse to heavy drinking than those who received a placebo. Although one-third of the nalmefene patients relapsed to heavy drinking at least once during the 12-week trial, significantly more nalmefene than placebo patients avoided relapse. Of those who did relapse, patients on nalmefene had fewer subsequent heavy drinking episodes (315).

Several controlled and uncontrolled studies conducted in several European countries, including France, Italy, and Belgium, have shown varying degrees of clinical efficacy with acamprosate in the treatment of alcohol dependence. The medication has been available in Europe for several years. A multicenter study was completed in the United States; however, the New Drug Application, in May 2002, failed FDA approval, and acamprosate remains unavailable for clinical use in the United States (311).

Psychosocial/Behavioral Treatment for Substance Abuse Disorders

The NIDA has produced several empirically tested treatment manuals for substance abuse, including manuals for cognitive behavioral therapy and contingency management. This material was developed and tested primarily with cocaine and crack users, although there is evidence to suggest that methamphetamine users respond similarly to these strategies (316–319). A recently completed seven-site, Center for Substance Abuse Treatment (CSAT)-funded evaluation of the Matrix Neurobehavioral Model suggested that such an approach can be effective across a varied group of treatment settings in a wide range of methamphetamine users (320–324). This manualized, 16-week, nonresidential psychosocial approach had been successfully used for more than a decade to treat other drug dependence. To meet future demands and the treatment needs of some special patient groups, continued efforts are essential to improve existing treatment protocols and establish new ones.

Special Populations of Methamphetamine Abusers

At present, intensive outpatient treatment protocols are the primary treatment paradigm for most methamphetamine users; however, several groups require other treatment considerations. Patients with severe psychiatric impairment require medically supervised inpatient treatment of varying duration and may require ongoing treatment with antipsychotic medications. Pregnant women and women with small children require increased levels of care, in particular, proper prenatal care. Women with small children frequently require a residential or intensive day treatment setting.

The needs of gay male methamphetamine users, especially those in some of the large gay communities on the West Coast, may require special treatment programs because their engaging in methamphetamine use generally becomes intertwined with high-risk sexual and social behaviors. These behaviors are often not effectively discussed in mixed patient groups that include heterosexuals, and this often contributes to poor treatment engagement and early treatment dropout. Successful and improved treatment is critical in this group because it is a critical vector in the spread of HIV (325).

SPECIAL ISSUES
Comorbidity of Psychiatric Illnesses in Substance Abuse

A substantial body of literature has accumulated documenting the frequency of coexisting substance abuse disorders among different diagnostic categories of psychiatric patients. The identification and management of substance-abusing patients with coexistent psychopathology require an awareness of the characteristics of specific substance abuse disorders and how these characteristics are manifested in the context of psychiatric illness.

Neuropsychiatric effects of psychoactive substances are most clearly manifested in intoxication and withdrawal syndromes. Yet a patient's cognitive development, interpersonal relationships, perception of self and others, and personality organization can be profoundly influenced by the use of psychoactive drugs. Consequences of substance abuse include the development of socially deviant, irresponsible, self-damaging, and antisocial patterns of behavior. Clearly, one of the challenges in diagnosing substance use disorders among psychiatric patients is to distinguish symptoms that are the result of psychiatric illness from those resulting from substance abuse or dependence. For example, the paranoia that occurs during and after extended stimulant abuse is difficult to distinguish from some forms of schizophrenia. Similarly, the socially and morally deviant behaviors associated with long-term drug dependence disorders overlap those of antisocial personality disorder.

The distinction between behaviors associated with psychiatric illness and psychoactive substance abuse is further obscured by the complex temporal relationship between these two classes of disorders. Psychopathology can be a risk factor or modifier of an addictive disorder (326). It can occur in the course of chronic intoxication or emerge as a consequence and persist into the period of remission. Over time, psychopathology and addictive disorders can become meaningfully linked. Finally, each can occur independent of the other. A large amount of literature has established some important patterns of coexistence between psychopathology and addictive disorders.

Comorbidity of Mental Health and Substance Abuse Disorders in the General Population

The 2001 National Household Survey on Drug Abuse presents national estimates of the prevalence and characteristics of persons with serious mental illness (SMI) and comorbid substance use and substance abuse/dependence. Those included in the description are adults aged 18 or older who had SMI in 2001. SMI is defined for that report as having at some time during the past year a diagnosable mental, behavioral, or emotional disorder that met the criteria specified in the *DSM-IV* and resulted in functional impairment that substantially interfered with or limited one or more major life activities. A scale consisting of six National Household Survey on Drug Abuse questions is used to measure SMI.

Adults who used illicit drugs were twice as likely to have SMI than adults who did not use an illicit drug. In 2001, among adults who used an illicit drug in the past year, 16.6% also had SMI in that year, whereas among adults who did not use an illicit drug, the rate of SMI was 6.1%. Conversely, among adults with SMI in 2001, 20.3% (3.0 million adults) were dependent on or abused alcohol or illicit drugs, whereas the rate among adults without SMI was only 6.3% (327). The Epidemiologic Catchment Areas study of a decade ago (18) found a lifetime prevalence of 32.7% for coexisting mental health and substance dependence disorders with alcohol abuse/dependence, drug abuse/dependence, and nonalcohol-related mental disorders diagnosed in 13.5%, 6.1%, and 22.5% of the sample, respectively. Higher rates were found in some special populations. For example, 82.1% of a sample of prisoners had coexistent mental health and substance abuse/dependence disorders, 56.2% had alcoholism, and 53.7% had substance abuse/dependence.

Alcoholism was the single most prevalent lifetime diagnosis across all categories of psychiatric diagnoses. The frequency of secondary mental health diagnoses among alcoholics (36.6%) almost doubles the rate for mental disorders in the general population (19.9%), and the prevalence of other drug disorders (21.5%) is nearly six times that (3.7%) of nonalcoholics. Men are more prone to alcoholism than women, but women alcoholics are more likely to have concurrent psychiatric disorders, especially depression. In men, alcoholism most often precedes the onset of depression, whereas depression precedes alcoholism in most women (328). Alcoholism frequently co-occurs with personality disorder. In these patients, onset of alcoholism is earlier, the history is longer, the symptoms more severe, and the treatment more difficult (328).

The prevalence of mental disorders among patients with drug abuse/dependence disorders is 53.1% or 4.5 times the rate in the nondrug-abusing/dependent population. Similarly, the prevalence of alcohol abuse or dependence among drug abusers is 47.3% or 7.1 times the rate in the nondrug-abusing/dependent population. In combination, the prevalence of any mental or alcohol disorder among drug abusers is 71.6% or 6.5 times the nondrug-abusing/dependent population (18). As with alcohol disorders, the rate of antisocial personality in the drug-dependent population is much higher (17.8%) than that found in nondrug-dependent groups, with an odds ratio of 13.4. This is differentially associated with different categories of drug use. For cocaine users, the rate of antisocial personality is 42.7% or 29.2 times that of noncocaine users; for opiate users, it is 36.7% or 24.3 times that of nonopiate users; and for marijuana users, it is 14.7% or 8.3 times that of the nonmarijuana-using group.

The co-occurrence of mental disorders and substance abuse disorders is substantially higher among patients who have sought treatment for either category of disorder than among untreated populations. The presence of a co-occurring disorder may exacerbate the symptoms of the other disorder and thus increase the probability of seeking treatment. Overall, the prevalence rates for psychiatric disorders among substance abusers in treatment were three times that of the general population (18). In one group of patients presenting for substance-abuse treatment, 68% had a concurrent nonsubstance-abuse psychiatric diagnosis and 43% had two or more (329).

Treatment of these patients is complicated because clinicians are typically trained to treat substance abuse disorders *or* psychiatric disorders. Consequently, treatment of comorbid disorders has generally been sequential rather than simultaneous. Clinicians who work with substance abusers often rely on a chemical dependency model of treatment that sees substance abuse as a life-long affliction, whereas other psychopathologies, even those that preceded drug use, are viewed as part of the addiction disease process (330). In this perspective, the cure for addiction and its concomitant problems, including psychiatric disorders, is believed to be abstinence. Clinicians who rely solely on this perspective are often unprepared when abstinence unmasks psychiatric symptoms that need attention (330). Conversely, clinicians who treat individuals with psychiatric disorders often see alcoholism or drug dependence as simply an attempt to treat an underlying disorder; this self-medication hypothesis implies that successful treatment of the psychiatric disorder will eliminate the addiction. Raskin and Miller (331) note that the opposite is usually true: Treatment of psychiatric disorders may only be possible after the addiction is under control. In general, treating individuals with comorbid diagnoses requires a broad conceptualization of addiction and psychopathology, encompassing the interactive and integral nature of the disorders. Treatment results are also less favorable among opiate addicts with depression (332).

Although uncommon, cocaine and other stimulants can induce panic attacks that persist after individuals become abstinent (236,330,333). These attacks have been reported to improve after treatment with clonazepam or carbamazepine (334).

Neuropsychological Correlates of Substance Abuse

In the final analysis, an altered neuropsychological state is the desired effect of drug abuse. Thus, it hardly seems surprising that long-term use of these drugs shows altered performance on neuropsychological testing. Specific abnormal findings in motor dexterity, sensory processing, attention and concentration, language, visuospatial analysis, verbal and nonverbal memory, abstraction, and problem-solving skills have been demonstrated with chronic alcohol, cocaine, opiate, and polysubstance abuse. The temporal relationships between neuropsychological findings, drug use, and abstinence have significant implications for treatment and recovery.

Although neuropsychological impairments have frequently been demonstrated in chronic substance abusers, relating specific findings to the use of a particular drug has been difficult. Polydrug abuse is common, and typically there are other confounding factors, such as head injury, nutritional deficits, liver disease, concurrent psychiatric and medical illness (notably HIV disease), and impoverished family and social environments, all of which can affect neuropsychological functioning. Demographic factors are also at play in neuropsychological performance. For example, older alcoholics are generally more impaired than younger alcoholics (335), and educational differences among subjects may result in erroneous conclusions about neuropsychological abilities (336). Although brain function is known to be less lateralized in females than in males, few studies have explored the influence of gender on substance abuse and neuropsychological function (337–340).

Most information regarding neuropsychological functioning in substance abusers, gathered over the past 20 years, has been derived from studies of alcoholics. These studies typically use recently detoxified (between 1 week and 2 months sober) alcoholic males compared with matched nonalcoholic samples. Deficits in abstraction and problem-solving abilities, learning and memory, and perceptual motor tasks are the three most frequently described areas of impairment (341–343) and appear to correlate with the lifetime amount of alcohol consumed (344). Impaired visual learning and memory in recently detoxified alcoholics have been noted. Conversely, Grant and Reed (342) found that alcoholics retain their capacity for learning but that more time and effort are required to master new material.

Despite the cocaine epidemic of the 1980's and 1990's, there have been few controlled studies on the neuropsychological sequelae of cocaine abuse (345). These preliminary studies describe mild to moderate levels of neuropsychological impairment in short-term memory and attention in chronic, heavy cocaine abusers. Deficits persisting for 6 months or more were specifically linked to cocaine-induced cerebral hypoperfusion in frontal, periventricular, and temporoparietal areas in a sample of cocaine abusers with an average use history of 2.5 years (346). Neuropsychological assessment in recently abstinent cocaine abusers may be complicated by depression, which accompanies acute cocaine withdrawal (345).

Uniquely, long-term opiate abusers seem not to suffer neuropsychological deficits, although empirical literature is sparse. Guerra et al. (347) compared attention, memory, and verbal fluency in 93 opiate addicts before and 1 week after detoxification with a group of matched controls. Although the addict group was significantly more impaired than the control group before detoxification, no differences between groups were detected at the second time point. Others have found that polysubstance abusers with a history of opiate abuse are significantly more likely to be identified as impaired on a blind clinical neuropsychological rating, although the potential effects of polysubstance abuse on performance could not be ruled out (348). In fact, neuropsychological deficits in most drug abusers result from a combination of substances. The Collaborative Neuropsychological Study of Polydrug Users (348,349) continues to provide the bulk of our current knowledge on the neuropsychological effects of polysubstance abuse. In general, perceptual/motor deficits were present in one-third of the chronic polysubstance abusers, but no definitive data were obtained regarding abstracting and language skills. Neuropsychological impairment in polysubstance abusers cannot be predicted by what is known about impairments unique to specific drugs of abuse but may reflect the individual's pattern of drug use and other factors, such as HIV/AIDS.

In NIDA technical review on the residual behavioral effects of abused drugs, Reed and Grant (351) cited available literature as indicative of possible long-term, slow recovery from neuropsychological impairment, especially among alcoholics, but also among polysubstance abusers. The researchers proposed an intermediate-duration neurobehavioral disorder associated with substance abuse, with a definitive diagnosis based on improvement in neurobehavioral testing with drug abstinence or reduced consumption.

A methodologic problem inherent in assessing persistence of substance abuse–related deficits is the stability and consistency of neuropsychological test scores over time in a population subject to the variable effects of drug intoxication, withdrawal, motivation, and compliance. Richards et al. (352) found good test–retest reliability on a large number of neuropsychological assessments over a mean test interval of 10.4 days with a small sample of parenteral substance abusers. Tests of verbal memory (the Selective Reminding Test) and motor functioning (the Perdue Pegboard) were less reliable in this same group over the testing interval.

The role of premorbid neuropsychological deficits in the development of substance use disorders remains an area of much empirical interest. Hesselbrock et al. (354) caution that the current evidence supporting cognitive deficits as risk factors for the development of alcoholism is not strong and that future studies should focus on specific versus general measures of neuropsychological functioning. These studies should consider the different subtypes of alcoholics for whom the predictive strength of neuropsychological functioning as a risk factor in the development of substance abuse may vary.

Implications for Treatment and Recovery

Neuropsychological impairment in treatment-seeking substance abusers necessitates consideration of how these impairments affect treatment and recovery. Becker and Kaplan (350) recommend that interventions with newly abstinent drug abusers should minimize cognitively demanding approaches. This is especially true in the case of methamphetamine abusers in whom cognitive impairment is often profound and persistent. Complex skill training and vocational and educational interventions should not be introduced until several weeks of abstinence have been achieved.

The degree of neuropsychological impairment may be a significant predictor of treatment outcome. Substance abusers with cognitive deficits are at greater risk of relapse than those without (353,355). Gregson and Taylor (356–358) found a positive correlation between performance on neuropsychological testing and counselor-assessed clinical progress in therapy in two groups of alcoholic men. In a large sample (n = 495) of substance abusers entering a drug-free therapeutic community, measures of diffuse cognitive impairment (Wechsler Adult Intelligence Scale Block Design and Digit-Symbol Subtests) predicted poor treatment retention, a finding attributed to the high level of complex information presented at treatment admission, on which subsequent treatment performance was based (359). In general, substance abusers with less neuropsychological compromise, whether predating or resulting from substance abuse, are better able to benefit from treatment interventions and achieve successful outcomes than those with greater neuropsychological compromise (360).

In summary, predictable patterns of neuropsychological impairment frequently accompany chronic substance abuse, with specific deficits varying according to the drug of abuse. The interaction of HIV disease with substance abuse complicates neuropsychological presentation but does not appear to predict increased severity or scope of impairment. Although there is evidence that neuropsychological deficits may predispose individuals to substance use disorders or make them more susceptible to the neurotoxic effects of drugs of abuse, there is also evidence that some deficits may clear slowly with continued abstinence. Neuropsychological assessment on treatment entry can help to structure treatment interventions and can predict the patient's ability to remain abstinent. Despite a wealth of research literature on neuropsychological correlates of substance abuse, questions remain about individual differences in the development of impairment, the link between deficits and drug effects, and the time necessary for neuropsychological recovery. Advances in and standardization of measurement techniques and increased emphasis on demographics, individual drug use history, and premorbid neuropsychological capacity will contribute to a better understanding of the relationship between substance abuse and neuropsychological function and have implications for treatment and prognosis.

PAIN AND SUBSTANCE ABUSE

Substance abusers frequently develop painful conditions from the adverse effects of abused drugs. In addition to the direct toxic effects of the drugs themselves, trauma and infection are common causes of acute and chronic pain in substance abusers. Pain in the patient with a history of substance abuse must be managed appropriately to minimize suffering and promote health and to establish a trusting relationship between the patient and the clinician, a crucial step in initiating or sustaining recovery from substance abuse.

Opiates provide one of the best pain relief options available to clinicians. Unfortunately, these medications have a clear abuse liability, and clinicians often are reluctant to provide them in sufficient amounts to their patients for fear of iatrogenic addiction (361–363). These fears are magnified in the case of patients with a history of substance abuse (364). The fear of "feeding the addiction" or triggering a relapse tends to outweigh concerns about managing discomfort. Moreover, current legislation governing prescription of opiates to known substance abusers puts prescribing physicians at considerable risk of prosecution.

Strategies for management of pain in the substance abuser vary according to whether the pain is acute or chronic and whether the patient is currently abusing opiates or other drugs, on opiate maintenance, or in drug-free recovery (365). Assessment of patterns of pain and opiate use provides the clinician with information on whether use patterns resemble those of abuse, pseudoaddiction, or appropriate pain self-medication.

Acute Versus Chronic Pain

Acute pain is typically sudden in onset, well localized, and easily recalled and described by the patient. Its duration is time limited and generally brief (typically less than 1 week), and it may occur in a single episode (e.g., appendicitis) or be recurrent in nature (e.g., migraine headaches). The origin of the pain is usually obvious and can be attributed to dysfunction in a specific organ or to systemic disease.

It is associated with signs of sympathetic hyperactivity and immobilization or guarding of the affected body region. The primary psychological response to acute pain is anxiety, with noticeable concern for determining its origin and seeking relief. Acute pain is extremely responsive to opiate analgesia, which may be the only intervention required to provide the patient adequate relief (366).

Chronic pain may or may not be preceded by an episode of acute pain. It can be of malignant (e.g., cancer) or nonmalignant (e.g., back pain) origin and may last for months to years. It is typically described as constant but may fluctuate in intensity. Psychological responses to chronic pain tend to be depression and insomnia, both of which can intensify the perception of pain. Opiate analgesics, although generally quite effective in managing chronic pain, are usually prescribed in combination with nonsteroidal analgesics and nonpharmacologic interventions, especially if the pain is nonmalignant in origin.

Concern for the development of opiate abuse in nonsubstance-abusing persons receiving opiate analgesia is exaggerated. In both single-episode and recurrent acute pain, the risk of iatrogenic addiction is low because the duration of opiate treatment is short and typically in response to intense pain. Patients with recurrent acute pain may be at slightly higher risk of developing abuse behaviors because of their increased access to opiate analgesics over prolonged periods, but empirical evidence of this occurrence is lacking. Evidence of opiate abuse is generally not a clinical issue in patients with chronic malignant or cancer-related pain, as maximal relief of discomfort and avoiding opiate toxicity are overriding concerns. Conversely, for persons with chronic pain of nonmalignant origin, opiates have not been generally considered a first-line strategy for pain management, and nonpharmacologic strategies are frequently emphasized. As clinicians have learned more about chronic nonmalignant pain syndromes, it is increasingly recognized that opiate maintenance may be required to provide adequate relief (367,368). Under these circumstances, development of tolerance and subsequent withdrawal on opiate discontinuation, characteristics of physical dependence, should not be confused with substance abuse.

Pain in Addicted Patients

Pain management strategies for persons with a history of substance abuse depend on whether the history is remote or current and whether the primary drug of abuse is an opiate or a nonopiate substance. For persons in drug-free recovery from opiate addiction, clinicians must recognize that unrelieved pain is itself a risk factor for relapse, and therefore aggressive attempts must be made to provide relief. The fear of relapse is likely to be strong in such patients. The clinician must be especially supportive and reinforce in the patient the distinction between use of opiate analgesics in the context of pain and his/her use in the absence of pain. The patient should be encouraged to intensify or reinitiate participation in recovery activities (e.g., 12-step meetings, relapse prevention sessions) (369). Certainly, the use of nonopiate analgesics and nonpharmacologic strategies is indicated, but their effectiveness may be less than adequate, and provision of an opiate analgesic may be necessary. Ineffective pain management and the accompanying high levels of anxiety can induce craving for the drug(s) that had provided relief for the patient in the past and increase the risk of relapse.

Tapering of opiate analgesics as pain diminishes should proceed at a more gradual rate with the recovering addict than with the opiate-naïve patient to minimize the emergence of withdrawal symptoms. A fixed dosing interval should be used. A partial opiate agonist (e.g., buprenorphine) may be preferable to full agonist analgesics because the level of physical dependence may be lower and subjective feelings of withdrawal (which may induce opiate craving) will be minimized. After completely withdrawn from opiates, the patient may be offered opiate antagonist maintenance therapy (e.g., naltrexone) for a short period to prevent relapse.

Patients, and even street addicts, on an opiate maintenance treatment (e.g., LAAM, methadone, buprenorphine) require higher opiate analgesic doses than those required by opiate-naïve patients because of tolerance. Maintenance opiate medication should be continued at the usual dose and not considered as contributing to analgesic requirement (370–374). Maintenance medications chronically administered to prevent withdrawal and to block the psychoactive effects of subsequently administered opiates have no demonstrated efficacy to provide analgesia. Instead, patients maintained on opiates for the treatment of addiction have been shown to have decreased tolerance to standard painful stimuli (375) and are highly tolerant to added opiate agonists, such as morphine (376–378). An effective maintenance dose of a long-acting opiate, such as LAAM, methadone, and buprenorphine, should be prescribed for opiate abusers not in treatment and, again, should be considered distinct from opiates prescribed to provide analgesia.

Mixed opiate agonist-antagonists or full antagonists cannot be used in opiate-dependent persons because they precipitate withdrawal, further diminish pain tolerance (379), and increase analgesic need. For patients with a history of opiate addiction, oral opiate analgesics with a slow onset to minimize psychoactive effects are preferred. It is best to avoid prescribing the patient's identified opiate of abuse. Any evidence that visitors may be supplying the patient with illicit opiates should trigger a review of the adequacy of pain relief or the presence of withdrawal instead of punishing or policing the patient as a matter of automatic response.

Guidelines for managing acute pain in known or suspected substance abusers have been developed by the Agency for Health Care Policy and Research Guideline Panel (380).

TABLE 33.9. CLINICAL FEATURES THAT DISTINGUISH PAIN PATIENTS AND ADDICTS

Clinical Features	Pain Patients	Addicts
Compulsive use	Rare	Common
Craving	Rare	Common
Obtain or purchase drugs from nonmedical sources	Rare	Common
Illegal activities to procure drugs	Absent	Common
Escalation of opiate dose without medical instruction	Rare	Common
Supplement with other opiate drugs	Unusual	Frequent
Demands specific opiate agent	Rare	Common
Ability to stop opiates when effective alternative treatments available	Successful	Usually unsuccessful
Requires unusually high dose	Rare	Common
Preference for specific routes of administration	No	Yes
Able to regulate use according to supply	Yes	No

During an episode of acute pain, detoxification should never be attempted. Remember that pain relief is the priority; discussion of drug treatment options can follow after pain is adequately managed. With chronic nonmalignant pain, detoxification may be warranted to get baseline assessments of pain, addiction, and analgesic need and, where possible, should be initiated in an inpatient environment where both pain and withdrawal can be actively assessed and managed. Detoxified chronic pain patients anecdotally report less pain relative to that experienced before detoxification. After the patient is opiate free, the clinician and patient can explore controlled opiate maintenance as an option to effectively manage pain. Recent studies in methadone-maintained patients found that they are very intolerant to cold pressor pain but are much more tolerant to electrically induced pain, suggesting that their pain experience is genuine (378).

A similar strategy can be implemented for patients with a history of abusing substances other than opiates. A history of drug abuse, regardless of specific drug, predisposes the patient to becoming addicted to another drug (opiates). If opiate dependence is not an issue, mixed opiate agonist-antagonists and partial agonists may minimize the risk of physical dependence (381). Benzodiazepines should be avoided to augment analgesia in persons with a history of alcohol, sedative-hypnotic, or CNS depressant abuse, as should stimulants in persons with a history of cocaine or CNS stimulant abuse.

Patterns of Analgesic Use

It is often necessary to determine whether a patient with chronic nonmalignant pain is using opiate analgesics in an abusive or addictive manner. True substance abuse behaviors in the chronic pain patient often preclude adequate pain management and place the physician at significant legal risk of prescribing opiates for an opiate addict. Distinguishing

between substance abuse and pain behaviors is complicated by the fact that pain patients who appropriately use opiate analgesics can meet *DSM-IV* diagnostic criteria for substance dependence.

Table 33.9 provides clinical features that further distinguish pain patients from addicts. It is necessary to rule out pseudoaddiction in pain patients manifesting addictive behaviors. Patients whose pain is not being effectively managed may demonstrate drug-seeking behaviors in legitimate attempts to achieve relief (e.g., using multiple physicians or obtaining opiates from street sources). For this reason, it is important to ensure that all patients have achieved adequate analgesia before assuming that drug-seeking behaviors are indicative of addiction.

The clinician treating the substance abuser on opiate maintenance for analgesia must anticipate that relapse to abuse can and does occur, especially early in treatment. Relapse is an expected outcome of attempts to stop drug use in substance abuse treatment and, in this context, should be viewed as an integral part of the substance abuser's attempt to recover. When treating for either the pain or the substance abuse, it is counterproductive to threaten the patient with withholding opiate analgesics when a relapse is detected. Rather, relapse must be treated aggressively, as such, in the context of chronic pain.

The following guidelines, abstracted and modified from Portenoy and Payne (366), are applicable to all patients, including those with a history of substance abuse:

1. Opiate maintenance should be considered only after careful evaluation of all available treatment options.

2. A single practitioner should take primary responsibility for treatment.

3. After drug selection, doses should be given on a fixed interval schedule.

4. The initial goal is to obtain at least partial but clinically meaningful analgesia. Failure to achieve this within

TABLE 33.10. COMMON NEUROLOGIC COMPLICATIONS OF HUMAN IMMUNODEFICIENCY VIRUS

	Central Nervous System	Peripheral Nervous System[a]
Human immunodeficiency virus neurotoxicity	AIDS dementia (subacute encephalitis, AIDS encephalopathy), vacuolar myelopathy	Distal symmetric polyneuropathy, progressive lumbosacral polyradiculopathy, mononeuropathy multiplex, chronic inflammatory polyneuropathy, autonomic neuropathy, herpes zoster radiculitis, toxoplasmosis myopathy, varicella zoster neuropathy, cytomegalovirus neuropathy
Secondary opportunistic infections	Cerebral toxoplasmosis, cryptococcal meningitis, progressive multifocal leukoencephalopathy, herpes simplex virus encephalitis, neurosyphillis, cytomegalovirus encephalitis, *Mycobacterium tuberculosis,* varicella zoster virus encephalitis or vasculitis	

[a] Additionally, toxic neuropathy or myopathy may arise as a side effect of therapeutic drugs (e.g., vincristine, isoniacid, dapsone, azidothymidine). AIDS, acquired immunodeficiency syndrome.

the first few weeks of maintenance at the initial doses (acknowledging tolerance) may indicate that the pain is not responsive to opiate maintenance therapy.

5. Emphasis should be given to measuring improved analgesia by gains in physical and social functioning.
6. In addition to the daily dose, patients should be permitted to escalate the dose in small amounts transiently on days of increased pain, under tightly controlled conditions.
7. Exacerbations in pain not effectively treated by transient, small increases of dose should be treated in the hospital.
8. Patients must be seen and drugs prescribed at frequent intervals.
9. Evidence of drug hoarding, drug diversion, or relapse should be immediately addressed with substance abuse treatment interventions.

A history of substance abuse does not preclude the use of opiate analgesics to provide pain relief. A clear understanding of the differences between physical dependence and addictive behaviors and of the abuse potential of opiates in the context of pain enables clinicians to effectively address the challenging problem of pain in substance abusers. Clinicians should keep in mind that a chronic pain patient with an addictive disorder does not have one insurmountable condition but two difficult yet treatable medical illnesses.

HUMAN IMMUNODEFICIENCY VIRUS AND THE NERVOUS SYSTEM

Use of contaminated paraphernalia and sharing of dirty needles put the intravenous drug user at an inordinate risk of contracting HIV. This risk is increased, especially for women,

when unsafe sexual behaviors are engaged in to support the drug habit. The recent increase in methamphetamine and club drug abuse, with the associated high-risk sexual behavior, increases the HIV risk in these groups, especially among men who have sex with men, even for those not engaged in intravenous drug use.

HIV infection has a wide range of effects on the central and peripheral nervous systems. The virus appears to invade the nervous system early (perhaps even before seroconversion occurs) (382) and is currently one of the most common causes of neurologic disorders (predominantly dementia, myelopathy, and neuropathy) in young persons (383). Furthermore, neurologic complications may result not only from secondary opportunistic infections and malignancies but from the neurodegenerative effects of the virus itself (Table 33.10). Cells in the nervous system, including macrophages, microglia, astrocytes, oligodendrocytes, and neurons, can be directly infected with HIV, although by different mechanisms from those identified in lymphocyte infection (384). Centrally, basal ganglia and temporolimbic brain structures are specifically susceptible, although secondary viral coinfections can produce cerebral vessel vasculitides, which can in turn lead to strokes and multiinfarcts in other brain areas.

Several psychiatric complications may also arise as a result of HIV infection. Most commonly, these include anxiety and depression, which are primarily reactive responses to the diagnosis itself, and dementia and delirium, which relate directly to the systemic or CNS disease (385). This symbiotic relationship between the effects of intravenous drug use and HIV infection presents an important clinical challenge for the health care provider who is treating HIV-infected substance abusers. Distinction must be made between the neurologic and psychological findings associated with HIV and those related to the chronic effects of substance abuse so that appropriate therapies can be instituted.

ACKNOWLEDGMENT

Preparation of this revised manuscript was supported in part by National Institute on Drug Abuse grants DA13706, DA13045, and DA12755. The authors gratefully acknowledge the assistance of Mrs. Sandy Dow and Ms. Deborah Ling, M.B.A.

REFERENCES

1. Beresford TD, Lowe D, Hall RC, et al. Alcoholism in the general hospital. *Psychiatr Med* 1984;2:139–148.
2. Moore RD, Bone LR, Geller G, et al. Prevalence, detection and treatment of alcoholism in hospitalized patients. *JAMA* 1989;261:403–407.
3. Cleary PD, Miller M, Bush PT, et al. Prevalence and recognition of alcohol abuse in the primary care population. *Am J Med* 1988;85:466–471.
4. Buchsbaum DG, Buchanan RG, Schnoll SH, et al. Screening for alcohol abuse using C.A.G.E. scores and likelihood ratios. *Ann Intern Med* 1991;115:744–777.
5. Center for Substance Abuse Treatment, U.S. Department of Health and Human Services. *State methadone maintenance guidelines,* 1992:337–369.
6. Selzer ML. The Michigan Alcoholism Screening Test: the quest for a new diagnostic instrument. *Am J Psychol* 1971;127:1653–1658.
7. Ewing J. Detecting alcoholism: the CAGE questionnaire. *JAMA* 1984;252:1905–1907.
8. Folstein M, Folstein S, McHugh P. The Mini-Mental State Examination. *J Psychiatr Res* 1975;12:189–198.
9. American Psychiatric Association. *Diagnostic and statistical manual of mental disorders,* 4th ed., revised. Washington, DC: American Psychiatric Association, 1994.
10. Willette E. Drug testing programs. *NIDA Research Monogr* 1986;73:5–12.
11. Person N, Ehrenkranz J. Evaluation of urine temperature methods to screen urine specimens for drug testing. *Clin Chem* 1989;35:1181–1189.
12. Herridge P, Ehrenkranz J, Pottash A, et al. The clinical laboratory. In: Sederer L, ed. *Inpatient psychiatry.* Baltimore: Williams & Wilkins, 1991:338–359.
13. Council on Scientific Affairs. Scientific issues in drug testing. *JAMA* 1987;257:3110–3114.
14. Nalpas B, Vassault A, Le Guillou A, et al. Serum activity of mitochondrial aspartate aminotransferase: a sensitive marker of alcoholism with or without alcohol hepatitis. *Hepatology* 1984;4:893.
15. Bernadt MW, Taylor C, Mumford J, et al. Comparison of questionnaire and laboratory tests in the detection of excessive drinking and alcoholism. *Lancet* 1982;1:525.
16. SAMHSA. 2001 National Household Survey on Drug Abuse (NHSDA): highlights, volume I, 2001 summary of findings, 2001.
17. SAMHSA. 2001 NHSDA: chapter 7: substance dependence, abuse, and treatment, 2001.
18. Regier DA, Farmer ME, Rae DS, et al. Comorbidity of mental disorders with alcohol and other drug abuse. Results from the epidemiologic catchment area (ECA) study. *JAMA* 1990;264:2511–2518.
19. Kalant H, Woo N. Electrophysiological effects of ethanol on the nervous system. *Pharmacol Ther* 1981;14:431.
20. Majchrowicz E. Biologic properties of ethanol and the biphasic nature of the ethanol withdrawal syndrome. In: Tarter RE, Van Thiel DH, eds. *Alcohol and the brain: chronic effects.* New York: Plenum, 1985:315–338.
21. Rubino FA. Neurologic complications of alcoholism. *Psychiatr Clin North Am* 1992;15:359–372.
22. Frezza M, DiPadova C, Pazzato G, et al. High blood alcohol levels in women: the role of decreased gastric alcohol dehydrogenase activity and first pass metabolism. *N Engl J Med* 1990;322:95–99.
23. Delin CR, Lee TH. Drinking and the brain: current evidence. *Alcohol* 1992;27:117–126.
24. Victor M. The effects of alcohol on the nervous system: clinical features, pathogenesis and treatment. In: Liever CS, ed. *Medical and nutritional complications of alcoholism.* New York: Plenum, 1992:413–457.
25. Miller NS, Gold MS. *Alcohol.* New York: Plenum, 1991.
26. Light WJH. *The neurobiology of alcohol abuse.* Springfield, IL: Charles C Thomas, 1986.
27. Crabbe JC, Feller DJ, Terdal ES, et al. Genetic components of ethanol responses. *Alcohol* 1990;7:245–248.
28. Crabbe JC, Harris RA. *The genetic basis of alcohol and drug actions.* New York: Plenum Press, 1991.
29. Crabbe JC, Phillips TJ, Cunningham CL, et al. Genetic determinants of ethanol reinforcement. *Ann N Y Acad Sci* 1992;654:302–310.
30. Harris RA, Allan AM. Alcohol intoxication: ion channels and genetics. *FASEB J* 1989;3:1689–1695.
31. Kiianmaa K, Helevuo K. The alcohol tolerant and alcohol nontolerant rat line selected for differential sensitivity to ethanol: a tool to study mechanisms of the actions of ethanol. *Ann Intern Med* 1990;22:283–287.
32. Palmer MR. Neurophysiological mechanisms in the genetics of ethanol sensitivity. *Soc Biol* 1985;32:241–254.
33. Fernandez-del Moral R, Dawid-Milner S, Diaz-Calvia JE. Pharmacology of acute alcoholic intoxication. *Rev Esp Fisiol* 1989;45[Suppl]:337–346.
34. Jaffe J. Drug addiction and drug abuse. In: Gilman AG, Rolf TW, Nies AS, et al., eds. *Goodman and Gilman's the pharmacological basis of therapeutics,* 8th ed. New York: Pergamon Press, 1990:522–523.
35. Mendelson JH, Mello NK. *The diagnosis and treatment of alcoholism.* New York: McGraw-Hill, 1979.
36. Charness ME, Simon RP, Greenberg DA. Ethanol and the nervous system. *N Engl J Med* 1989;321:442–454.
37. Little HJ. Mechanisms that may underlie the behavioral effects of ethanol. *Prog Neurobiol* 1991;36:171–194.
38. Majchrowicz E. Biologic properties of ethanol and the biphasic nature of the ethanol withdrawal syndrome. In: Tarter RE, Van Thiel DH, eds. *Alcohol and the brain: chronic effects.* New York: Plenum, 1985:315–338.
39. Victor M, Adams RD. The effects of alcohol on the nervous system. *Res Publ Assoc Nerv Ment Dis* 1953;32:526–573.
40. Naranjo CA, Bremner KE. Behavioral correlates of alcohol intoxication. *Addiction* 1993;88:25–35.
41. Tennant F. The rapid eye test to detect drug abuse. *Postgrad Med* 1988;84:108–114.
42. Mello NK, Mendelson JH. Alcohol and human behavior. In: Iversen LL, Eversen SD, Snyder SH, eds. *Handbook of psychopharmacology.* New York: Plenum, 1978:235–317.
43. Tamerin JS, Weiner S, Poppen R, et al. Alcohol and memory:

amnesia and short-term memory function during experimentally induced intoxication. *Am J Psychiatry* 1971;127:1659–1664.

44. Goodwin DW, Othmer E, Halikas JA, et al. Loss of short-term memory as a predictor of the alcoholic "blackout." *Nature* 1971;227:201–202.
45. Goodwin DW. Blackouts and alcohol induced memory dysfunction. In: Mello NK, Mendelson JH, eds. *Recent advances in studies of alcoholism.* Bethesda: NIMH, 1971:508–536.
46. Sweeney DF. Alcohol versus mnemosyne—blackouts. *J Subst Abuse Treat* 1989;6:159–162.
47. Gallant DM. *Alcoholism: a guide to diagnosis, intervention and treatment.* New York: WW Norton, 1987.
48. Taylor WA, Slaby AE. Acute treatment of alcohol and cocaine emergencies. *Recent Dev Alcohol* 1992;10:179–191.
49. Holt S, Steward IC, Dixon JW, et al. Alcohol and the emergency service patient. *BMJ* 1980;281:638–640.
50. Kellermann AL, Fihn SD, LoGerfo JP, et al. Impact of drug screening in suspected overdose. *Ann Emerg Med* 1987;16:1206–1216.
51. Nutt DJ, Glue P. Neuropharmacological and clinical aspects of alcohol withdrawal. *Ann Intern Med* 1990;22:275–281.
52. Turner RC, Lichstein PR, Peden JG, et al. Alcohol withdrawal syndromes: a review of pathophysiology, clinical presentation, and treatment. *J Gen Intern Med* 1989;4:432–444.
53. Gauvin DV, Youngblood BD, Holloway FA. The discriminative stimulus properties of acute ethanol withdrawal (hangover) in rats. *Alcohol Clin Exp Res* 1992;16:336–341.
54. Missouri Medicine. Alcohol consumption and hangover. *Mo Med* 1990;87:875–876.
55. Newlin DB, Pretorius MB. Sons of alcoholics report greater hangover symptoms than sons of nonalcoholics: a pilot study. *Alcohol Clin Exp Res* 1990;14:713–716.
56. Rubino C, Butters N. Cognitive effects in alcohol abuse. In: Kissin B, Begleiter H, eds. *The biology of alcoholism.* New York: Plenum, 1982:485–538.
57. Edwards G. Withdrawal symptoms and alcohol dependence: fruitful mysteries. *Br J Addict* 1990;85:447–461.
58. Maier DM, Pohorecky LA. The effects of repeated withdrawal episodes on subsequent withdrawal severity in ethanol-treated rats. *Drug Alcohol Depend* 1989;23:103–110.
59. Cushman P. Delirium tremens: update on an old disorder. *Postgrad Med* 1987;5:117–122.
60. Alpert MA. Modern management of delirium tremens. *Hosp Med* 1990;26:111–136.
61. Glass IB. Alcoholic hallucinosis: a psychiatric enigma. *Br J Addict* 1989;84:29–41.
62. Surawicz FG. Alcoholic hallucinosis: a missed diagnosis. *Can J Psychiatry* 1980;25:57–63.
63. McMicken DB. Alcohol withdrawal syndromes. *Emerg Med Clin North Am* 1990;8:805–819.
64. Strub RL, Black FW, eds. *Neurobehavioral disorders: a clinical approach.* Philadelphia: FA Davis, 1988.
65. Gillman MA, Lichtigfeld FJ. The drug management of severe alcohol withdrawal syndrome. *Postgrad Med J* 1990;66:1005–1009.
66. Romach MK, Seller EM. Management of the alcohol withdrawal syndrome. *Annu Rev Med* 1991;42:323–340.
67. Young GP. Seizures in the alcoholic patient. *Emerg Med Clin North Am* 1990;8:821–833.
68. Lechtenberg R, Worner TM. Total ethanol consumption as a seizure risk factor in alcoholics. *Acta Neurol Scand* 1992;85:90–94.
69. Lowenstein DH, Alldredge BK. Status of epilepticus at an urban public hospital in the 1980s. *Neurology* 1993;43:483–488.
70. Krauss GL, Niedermeyer E. Neuropharmacology of cocaine and ethanol dependence. *Recent Dev Alcohol* 1992;10:201–233.
71. Loiseau P, Duche B, Loiseau J. Classification of epilepsies and epileptic syndromes in two different samples of patients. *Epilepsia* 1991;32:303–309.
72. Verma NP, Policheria H, Buber BA. Prior head injury accounts for the heterogeneity of the alcohol-epilepsy relationship. *Clin Electroencephalogr* 1992;23:147–151.
73. Alldredge BK, Lowenstein DH, Simon RP. Placebo-controlled trial of intravenous diphenylhydrate for short-term treatment of alcohol withdrawal seizures. *Am J Med* 1989;87:645–648.
74. Hillborn M, Tokola R, Kuusela V, et al. Prevention of alcohol withdrawal seizures with carbamazepine and valproic acid. *Alcohol* 1989;6:223–226.
75. Thompson WL, Johnson AD, Maddrey WL. Diazepam and paraldehyde for treatment of severe delirium tremens. *Ann Intern Med* 1975;82:175–180.
76. Satel SL, Kosten TR, Schuckit MA, et al. Should protracted withdrawal from drugs be included in DSM-IV? *Am J Psychiatry* 1993;150:695–704.
77. Kissin B. Biologic investigations in alcohol research. *J Stud Alcohol* 1979;8:146–181.
78. Alling C, Balldin M, Bokstrom K, et al. Studies on duration of a late recovery period after chronic abuse of ethanol. *Acta Psychiatr Scand* 1982;66:384–397.
79. De Soto CB, O'Donnell WE, Allred LJ, et al. Symptomatology in alcoholics at various stages of abstinence. *Alcohol Clin Exp Res* 1985;9:505–512.
80. Victor M, Adams RD, Collins GH, eds. *The Wernicke-Korsakoff syndrome and related neurologic disorders due to alcoholism and malnutrition.* Philadelphia: FA Davis, 1989.
81. Carlin AS, Strauss FF, Adams KM, et al. The prediction of neuropsychological impairment in polydrug abusers. *Addict Behav* 1978;3:5–12.
82. Greenberg DA, Diamond I. Wernicke-Korsakoff syndrome. In: Tarter RE, Van Thiel DH, eds. *Alcohol and the brain: chronic effects.* New York: Plenum, 1986:295–314.
83. Reuler JB, Girard DE, Cooney TG. Wernicke's encephalopathy. *N Engl J Med* 1985;312:1035–1040.
84. Zimitat C, Kril J, Harper CG, et al. Progression of neurological disease in thiamin-deficient rats is enhanced by ethanol. *Alcohol* 1990;7:493–501.
85. Manzo L, Locatelli C, Candura SM, et al. Nutrition and alcohol neurotoxicity. *Neurotoxicology* 1994;15:555–566.
86. Tallaksen CM, Bohmer T, Bell H. Blood serum thiamin and thiamin phosphate esters concentrations in patients with alcohol dependence syndrome before and after thiamin treatment. *Alcohol Clin Exp Res* 1992;16:320–325.
87. Irle E, Kaiser P, Naumann-Stoll G. Differential patterns of memory loss in patients with Alzheimer's disease and Korsakoff's disease. *Int J Neurosci* 1990;52:67–77.
88. O'Carroll RE, Moffoot A, Ebmeier KP, et al. Korsakoff's syndrome, cognition and clonidine. *Psychol Med* 1993;23:341–347.
89. McEntee WJ, Mair RG. Memory impairment in Korsakoff's psychosis: a correlation with brain noradrenergic activity. *Science* 1978;202:905–907.
90. McEntree WJ, Mair RG. Memory enhancement in Korsakoff's psychosis by clonidine: further evidence for a noradrenergic deficit. *Ann Neurol* 1980;7:466–470.
91. McEntree WJ, Miar RG, Langlais PJ. Neurochemical pathology in Korsakoff's psychosis: implications for other cognitive disorders. *Neurology* 1984;34:648–652.
92. Benson DF, Cummings JL. A scheme to differentiate the

dementias. In: Jeste DV, ed. *Neuropsychiatric dementias.* Washington, DC: American Psychological Association, 1986:1–25.

93. Akai J. Anatomo-pathological studies on alcohol dementia: a review and up to date research. *Drug Alcohol Depend* 1991;26:134.

94. Cutting J. The relationship between Korsakoff syndrome and "alcoholic dementia." *Br J Psychiatry* 1978;132:240–245.

95. Blusewicz MJ. Neuropsychological correlates of chronic alcoholism and aging. *J Nerv Ment Dis* 1977;165:348–356.

96. Goldstein G. Dementia associated with alcoholism. In: Tarter RE, Van Thiel DH, eds. *Alcohol and the brain: chronic effects.* New York: Plenum, 1985:283–294.

97. Page RD, Shaub LH. Intellectual functioning in alcoholics during six months abstinence. *J Stud Alcohol* 1977;38:1240–1248.

98. Victor M, Banker BQ. Alcohol and dementia. In: Katzman R, Terry RD, Bick KL, eds. *Alzheimer's disease: senile dementia and related disorders.* New York: Raven Press, 1978:149–170.

99. Jones B, Parsons OA. Impaired abstraction ability in chronic alcoholics. *Arch Gen Psychiatry* 1971;24:431–457.

100. Tarter R. An analysis of cognitive deficits in chronic alcoholics. *J Nerv Ment Dis* 1973;157:138–147.

101. Tarter R, Alterman A. Neuropsychological deficits in chronic alcoholics: etiological considerations. *J Stud Alcohol* 1984;45:1–9.

102. Neiman J, Lang AE, Fornazzari L, et al. Movement disorders in alcoholism: a review. *Neurology* 1990;40:741–746.

103. Butterworth RF, D'Amour M, Bruneau J, et al. Role of thiamine deficiency in the pathogenesis of alcoholic peripheral neuropathy and the Wernicke-Korsakoff syndrome: an update. In: Palmer TN, ed. *Alcoholism: a molecular perspective.* New York: Plenum, 1991:269–277.

104. Graham JR, Woodhouse D, Read FH. Massive thiamine dosage in an alcoholic with cerebellar degeneration. *Lancet* 1971;11:107.

105. Karol RL, Halla PD. *Brain injury and alcohol: a workbook for making decisions after injury.* Minneapolis: Thompson, 1987.

106. Elliot FA. Neuroanatomy and neurology of aggression. *Psychol Ann* 1987;17:385–387.

107. Corthell DW, Tooman M. *Twelfth Institute on Rehabilitation issues: rehabilitation of traumatic brain injury.* Stout, WI: Research and Training Center, University of Wisconsin, 1985.

108. Bombardier CH. Alcohol use and traumatic brain injury. *West J Med* 1995;162:150–151.

109. Heikkinen ER, Ronty HS, Tolonen U, et al. Development of posttraumatic epilepsy. *Stereotact Funct Neurosurg* 1990;54/55:25–33.

110. Solomon DA, Sparadeo F. The effects of substance abuse on persons with traumatic brain injury. *Neurorehabilitation* 1992;2:16–26.

111. Langley MJ, Lindsay WP, Lam CS, et al. Programme development. A comprehensive alcohol abuse treatment programme for persons with traumatic brain injury. *Brain Inj* 1990;4:77–86.

112. Mayer E. The toxic effects following the use of local anesthetics. *JAMA* 1924;82:876–878.

113. Mendoza R, Miller BL, Mena I. Emergency room evaluation of cocaine-associated neuropsychiatric disorders. In: Galantar M, ed. *Recent developments in alcoholism, vol. 10* New York: Plenum, 1992:73–87.

114. Levine SR, Brust JCM, Futrell N, et al. Cerebrovascular complications of the use of the crack form of alkaloidal cocaine. *N Engl J Med* 1990;323:699–704.

115. SAMHSA. 2001 NHSDA: chapter 2: illicit drug use, 2001.

116. Office of Applied Studies, SAMHSA. The Drug and Alcohol Services Information System (DASIS) report, cocaine treatment admissions decrease: 1993–1999, Jan. 25, 2002.

117. Pascual-Leone A, Dhuna A, Anderson DC. Longterm neurological complications of chronic, habitual cocaine abuse. *Neurotoxicology* 1991;12:393–400.

118. Pascual-Leone A, Dhuna A, Anderson DC. Cerebral atrophy in habitual cocaine abusers: a planimetric CT study. *Neurology* 1991;41:34–38.

119. Gold MS. Cocaine (and crack): clinical aspects. In: Lowinson JH, Ruiz P, Millman RB, eds. *Substance abuse: a comprehensive textbook,* 2nd ed. Baltimore: Williams & Wilkins, 1992:205–221.

120. Mendoza R, Miller BL. Neuropsychiatric disorders associated with cocaine use. *Hosp Community Psychiatry* 1992;43:677–678.

121. Wetli CV, Fishbain DA. Cocaine-induced psychosis and sudden death in recreational cocaine users. *J Forensic Sci* 1985;30:873–879.

122. Kosten TR, Kleber HD. Rapid death during cocaine abuse: a variant of the neuroleptic malignant syndrome? *Am J Drug Alcohol Abuse* 1988;14:335–346.

123. Karch SB. *The pathology of drug abuse.* Boca Raton, FL: CRC Press, 1993.

124. Karch SB, Stephens BG. Drug abusers who die during arrest or in custody. *J R Soc Med* 1999;92:110–113.

125. Allredge BK, Lowenstein DH, Simon RP. Seizures associated with recreational drug abuse. *Neurology* 1989;39:1037–1039.

126. Chong-Kwong M, Lipton RB. Seizures in hospitalized cocaine users. *Neurology* 1989;39:425–427.

127. Kramer LD, Locke GE, Ogunyemi A, et al. Cocaine-related seizures in adults. *Am J Drug Alcohol Abuse* 1990;16:309–317.

128. Pascual-Leone A, Dhuna A, Altafullah I, et al. Cocaine-induced seizures. *Neurology* 1990;40:404–407.

129. Root RK, Rowbotham M. Neurologic aspects of cocaine abuse. *West J Med* 1988;149:442–448.

130. Rowbotham MC, Lowenstein DH. Neurologic consequences of cocaine use. *Am Rev Med* 1990;41:417–422.

131. Post RM, Kopanda RT. Cocaine, kindling, and psychosis. *Am J Psychiatry* 1976;133:327–334.

132. Merriam A, Medalia A, Levine B. Partial complex status epilepticus associated with cocaine abuse. *Biol Psychiatry* 1988;23:515–518.

133. Wetli C, Mittleman RE. The "body packer syndrome"—toxicity following ingestion of illicit drugs packaged for transportation. *J Forensic Sci* 1981;26:492–500.

134. Catravas JD, Waters IW. Acute cocaine intoxication in the conscious dog: studies on the mechanism of lethality. *J Pharmacol Exp Ther* 1981;217:350–356.

135. Kaku DA, Lowenstein DH. Emergence of recreational drug abuse as a major risk factor for stroke in young adults. *Ann Intern Med* 1990;113:821–827.

136. Sloan MA, Kittner SU, Rigamonti D, et al. Occurrence of stroke associated with use/abuse of drugs. *Neurology* 1991;41:1358–1364.

137. Levine SR, Brust JCM, Futrell N, et al. A comparative study of the cerebrovascular complications of cocaine: alkaloidal versus hydrochloride—a review. *Neurology* 1991;41:1173–1177.

138. Brust JCM. *Neurological aspects of substance abuse.* Boston: Butterworth-Heineman, 1993.

139. Kibayashi K, Mastri AR, Hirsch CS. Cocaine induced intracerebral hemorrhage: analysis of predisposing factors and mechanisms causing hemorrhagic strokes. *Hum Pathol* 1995;26:659–663.

140. Madden J, Powers R. Effects of cocaine and cocaine metabolites on cerebral arteries in vitro. *Life Sci* 1990;47:1109–1114.

141. Levine S, Washington J, Jefferson M, et al. "Crack" cocaine associated stroke. *Neurology* 1987;37:1849–1853.

142. Bartzokis G, Beckson M, Lu PH, et al. Age-related brain volume reductions in amphetamine and cocaine addicts and normal controls: implications for addiction research. *Psychiatry Res* 2000;98:93–102.

143. Tongna G, Tempesta E, Tongna AR, et al. Platelet responsiveness and biosynthesis of thromboxane and prostacyclin in response to in vitro cocaine treatment. *Haemostasis* 1985;15:100–107.

144. Kokkinos J, Levine SR. Stroke. *Neurol Clin* 1993;II:577–590.

145. Mendoza RP, Miller BL, Mena I. Emergency room evaluation of cocaine-associated neuropsychiatric disorders. In: Galanter M, ed. *Recent developments in alcoholism, vol 10: alcohol and cocaine: similarities and differences.* New York: Plenum, 1992.

146. American Psychiatric Association. Substance related disorders. In: *Diagnostic and statistical manual of mental disorders,* 4th ed. Washington, DC: American Psychiatric Association, 1994.

147. Brower KJ, Paredes A. Cocaine withdrawal [Editorial]. *Arch Gen Psychiatry* 1987;44:297.

148. Gawin FH, Kleber HD. Abstinence symptomatology and psychiatric diagnosis in cocaine abusers. *Arch Gen Psychiatry* 1986;43:107–113.

149. Daras M, Koppel BS, Atos-Radzion E. Cocaine-induced choreoathetoid movements ("crack dancing"). *Neurology* 1994;44:751–752.

150. Habal R, Sauter D, Olowe O, et al. Cocaine and chorea. *Am J Emerg Med* 1991;9:618–620.

151. Catalano G, Catalano MC, Rodriguez R. Dystonia associated with crack cocaine use. *South Med J* 1997;90:1050–1052.

152. van Harten PN, van Trier JC, Horwitz EH, et al. Cocaine as a risk factor for neuroleptic-induced acute dystonia. *J Clin Psychiatry* 1998;59:128–130.

153. Lundh H, Tunving K. An extrapyramidal choreiform syndrome caused by amphetamine addiction. *J Neurol Neurosurg Psychiatry* 1981;44:728–730.

154. Rhee KJ, Albertson TE, Douglas JC. Choreoathetoid disorder associated with amphetamine-like drugs. *Am J Emerg Med* 1988;6:131–133.

155. Thief A, Dressler D. Dyskinesias possibly induced by norpseudoephedrine. *J Neurol* 1994;241:167–169.

156. Office of Applied Studies, SAMHSA. The DASIS Report, amphetamine treatment admissions increase: 1993–1999, Nov. 16, 2001.

157. Division of Mental Health and Prevention of Substance Abuse, World Health Organization. Programme on substance abuse, amphetamine-type stimulants, Geneva, 1997.

158. Stimmel B. Editorial. *J Addict Dis* 2002;21:1.

159. Rawson RA, Anglin MD, Ling W. Will the methamphetamine problem go away. *J Addict Dis* 2002;21:5–19.

160. Koda LY, Gibb JW. Adrenal and striatal tyrosine hydroxylase activity after methamphetamine. *J Pharmacol Exp Ther* 1973;185:42–48.

161. SAMHSA, Center for Substance Abuse Treatment. Treatment for stimulant use disorders, treatment improvement protocol (TIP) series no. 33, 1999;29–32.

162. Wang B, Luo F, Zhang WT, et al. Stress or drug priming induces reinstatement of extinguished conditioned place preference. *Neuroreport* 2000;11:2781–2784.

163. Cho AK, Melega WP. Patterns of methamphetamine abuse and their consequences. *J Addict Dis* 2002;21:21–34.

164. Young D, Scoville W. Paranoid psychosis in narcolepsy and the possible danger of Benzedrine treatment. *Med Clin North Am* 1938;22:637–639.

165. Sato M, Numachi Y, Haamura T. Relapse of paranoid psychotic state in methamphetamine model of schizophrenia. *Schizophrenia Bull* 1992;18:115–122.

166. Iwanami A, Sugiyama A, et al. Patient with methamphetamine psychosis admitted to a psychiatric hospital in Japan. A preliminary report. *Acta Psychiatr Scand* 1994;89:428–432.

167. Verachai V, Dechongkit S, Patarakorn A, et al. Drug addicts treatment for ten years in Thanyarak Hospital. *J Med Assoc Thai* 2001;84:24–29.

168. Rothrock JF, Rubenstein R, et al. Ischemic stroke associated with methamphetamine inhalation. *Neurology* 1988;38:589–592.

169. Yen DJ, Wang SJ, et al. Stroke associated with amphetamine inhalation. *Eur Neurol* 1994;34:16–22.

170. Perez Jr JA, Asura EL, et al. Methamphetamine-related stroke: four cases. *J Emerg Med* 1999;17:469–471.

171. Delaney P, Estes M. Intracranial hemorrhage with amphetamine abuse. *Neurology* 1980;30:1125–1128.

172. Yu YJ, Cooper DR, et al. Cerebral angiitis and intracerebral hemorrhage associated with methamphetamine abuse: case report. *J Neurosurg* 1983;58:109–111.

173. D'Souza T, Shraberg D. Intracranial hemorrhage associated with amphetamine use. *Neurology* 1981;31:922–923.

174. Caplan LR, Hier DB, et al. Current concepts of cerebrovascular disease—stroke: stroke and drug abuse. *Stroke* 1982;13:869–872.

175. Lukes SA. Intracerebral hemorrhage from an arteriovenous malformation after amphetamine injection. *Arch Neurol* 1983;49:60–61.

176. Lessing MP, Hyman NM. Intracranial hemorrhage caused by amphetamine abuse. *J R Soc Med* 1980;82:766–777.

177. Imanse J, Vanneste J. Intraventricular hemorrhage following amphetamine abuse. *Neurology* 1990;40:1318–1319.

178. Goplen AK, Berg-Johnson J, et al. Fatal cerebral hemorrhage in young amphetamine addicts. *Tidsskr Nor Laegoforen* 1995;115:832–834.

179. O'Brien CP. Stroke in young women who use cocaine or amphetamines. *Epidemiology* 1998;9:587–588.

180. Petitti DB, Sidney S, et al. Stroke and cocaine or amphetamine use. *Epidemiology* 1998;9:596–600.

181. Bakheit AM. Intracerebral haemorrhage in previously healthy young adults. *Postgrad Med J* 1999;75:499–500.

182. Bostwick DG. Amphetamine induced cerebral vasculitis. *Hum Pathol* 1981;12:1031–1033.

183. Epstein JF, Gfroerer JC. SAMHSA: heroin abuse in the United States: need for treatment, 1994–1996.

184. Office of Applied Studies, SAMHSA. The Drug and Alcohol Services Information System (DASIS) report, heroin—changes in how it is used. July 20, 2001.

185. U.S. Department of Health and Human Services. NHSDA 2001, Press release. Annual household survey finds millions of Americans in denial about drug abuse. Sept 5, 2002.

186. Ling W, Wesson DR. Drugs of abuse-opiates. *West J Med* 1990;152:565–572.

187. Jaffe JH, Martin WR. Opioid analgesics and antagonists. In: Gilman AG, Rall TW, Nies AS, et al., eds. *Goodman and Gilman's the pharmacological basis of therapeutics,* 8th ed. New York: McGraw-Hill, 1993:21.

188. Jaffe JH. Opiates: clinical aspects. In: Lowinson JH, Ruiz JH, Millman RB, et al., eds. *Substance abuse: a comprehensive textbook,* 2nd ed. Baltimore: Williams & Wilkins, 1992:186–194.

189. Brust JCM, Richter RW. Quinine amblyopia related to heroin addiction. *Ann Intern Med* 1971;74:84.

190. Dalessandro-Gandolfo L, Macci A, Biolcati G, et al. Inconsueta modaity d'intossicazione da piombo. Presentazione di un caso. *Recenti Prog Med* 1989;80:140.

191. Novick DM. The medically ill substance abuser. In:

Lowinson JH, Ruiz P, Millman RB, et al., eds. *Substance abuse: a comprehensive textbook,* 2nd ed. Baltimore: Williams & Wilkins, 1992:657–674.

192. Caplan L, Hier D, Banks G. Current concepts of cerebrovascular disease stroke: stroke and drug abuse. *Stroke* 1982;13:869–872.

193. Brust JCM. *Neurological aspects of substance abuse.* Boca Raton, FL: CRC Press, 1993.

194. Caplan LR, Hier DB, Banks G. Stroke and drug abuse. *Stroke* 1982;13:869.

195. Ammueilaph R, Boongird P, Leechawengwongs M, et al. Heroin neuropathy. *Lancet* 1973;1:1517.

196. Ritland D, Butterfield W. Extremity complications of drug abuse. *Am J Surg* 1973;126:639.

197. Chen SS, Chien CH, Yu HS. Syndrome of deltoid and gluteal fibrotic contracture: an injection myopathy. *Acta Neurol Scand* 1988;78:167.

198. Griffiths RR, Roache JD. Abuse liability of benzodiazepines: a review of human studies evaluation subjective and/or reinforcing effects. In: Smith DE, Wesson DR, eds. *The benzodiazepines: current standards for medical practice.* Hingham, MA: MTP Press, 1985:209–225.

199. Randall T. "Rave" scene, ecstasy use, leap Atlantic. *JAMA* 1992;268:1506.

200. Randall T. Ecstasy-fueled "rave" parties become dances of death for English youths. *JAMA* 1992;268:1505–1506.

201. Henry JA. Ecstasy and the dance of death. *BMJ* 1992;305:5–6.

202. SAMHSA. 2001 NHSDA: chapter 5: trends of initiation of substance abuse, 2001.

203. Office of Applied Studies, SAMHSA. The drug abuse warning network, club drugs, 2001 update, Oct. 2002.

204. Sobel J, Espinas O, Friedman, S. Carotid artery obstruction following LSD capsule ingestion. *Arch Intern Med* 1971;127:290–291.

205. Gledhill JA, Moore DF, Bell D, et al. Subarachnoid haemorrhage associated with MDMA abuse [Letter]. *J Neurol Neurosurg Psychiatry* 1993;56:1036–1037.

206. Allen RP, McCann UD, Ricaurte GA. Persistent effects of (+) 3,4-methylenedioxymethamphetamine (MDMA) (ecstasy) on human sleep. *Sleep* 1993;16:560–564.

207. Krystal JH, Price LH, Opsahl C, et al. Chronic 3,4-methylenedioxymethamphetamine (MDMA) use: effects on mood and neuropsychological function. *Am J Drug Alcohol Abuse* 1992;18:331–341.

208. Creighton FJ, Black DL, Hyde CE. "Ecstasy" psychosis and flashbacks. *Br J Psychiatry* 1991;159:713–715.

209. Pallanti S, Mazzi D. MDMA (ecstasy) precipitation of panic disorder. *Biol Psychiatry* 1992;32:91–95.

210. McCann UD, Ricaurte GA. MDMA (ecstasy) and panic disorder: induction by single dose. *Biol Psychiatry* 1992;32:950–953.

211. Whitaker-Azmitia PM, Aronson TA. "Ecstasy" (3,4-methylenedioxymethamphetamine)-induced panic. *Am J Psychiatry* 1989;146:119.

212. Winstock AR. Chronic paranoid psychosis after misuse of 4,3-methylenedioxymethamphetamine. *BMJ* 1991;302:1150–1151.

213. Schifano F. Chronic atypical psychosis associated with 3,4-methylenedioxymethamphetamine ("ecstasy") abuse. *Lancet* 1991;338:1335.

214. Varma VK, Malhotra AK, Dang R. Cannabis and cognitive functions: a prospective study. *Drug Alcohol Depend* 1988;21:147–152.

215. Schwartz RH, Gruenewald PJ, Klitzner M, et al. Short-term memory impairment in cannabis-dependent adolescents. *Am J Dis Child* 1989;143:1214–1219.

216. Leirer VO, Yesavage JA, Morrow DG. Marijuana carryover effects on aircraft pilot performance. *Aviat Space Environ Med* 1991;62:221–227.

217. Janowsky DS, Meacham MP, Blaine JD, et al. Simulated flying performance after marijuana intoxication. *Aviat Space Environ Med* 1976;47:124–128.

218. Leirer VO, Yesavage JA, Morrow DG. Marijuana, aging and task difficulty effects on pilot performance. *Aviat Space Environ Med* 1989;60:1145–1152.

219. Wert RC, Raulin ML. The chronic cerebral effects of cannabis use. I. Methodological issues and neurological findings. *Int J Addict* 1986;21:605–628.

220. Wert RC, Raulin ML. The chronic cerebral effects of cannabis use. II. Psychological findings and conclusions. *Int J Addict* 1986;21:629–642.

221. Mendoza M. San Francisco considers growing and distributing marijuana. Associated Press, Nov. 2, 2002.

222. Mikuriya TH, ed. *Marijuana: medical papers 1939–1972.* Oakland, CA: Medi-Comp Press, 1973.

223. Turkanis SA, Karler R. Electrophysiologic properties of the cannabinoids. *J Clin Pharmacol* 1981;21:449S–463S.

224. Karler R, Turkanis SA. The cannabinoids as potential antiepileptics. *J Clin Pharmacol* 1981;21:437S–448S.

225. Consroe P, Martin A, Singh V. Antiepileptic potential of cannabidiol analogs. *J Clin Pharmacol* 1981;21:428S–436S.

226. Lemberger L. Potential therapeutic usefulness of marijuana. *Annu Rev Pharmacol Toxicol* 1980;20:151–172.

227. Carlini EA, Cunha JM. Hypnotic and antiepileptic effects of cannabidiol. *J Clin Pharmacol* 1981;21:417S–427S.

228. Doyle E, Spence AA. Cannabis as a medicine? *Br J Anaesth* 1995;74:359–360.

229. Dagirmangian R, Boyd ES. Some pharmacological effects of two tetrahydrocannabinols. *J Pharmacol Exp Ther* 1962;135:25–33.

230. Petro DJ, Ellenberger C. Treatment of human spasticity with delta⁹-tetrahydrocannabinol. *J Clin Pharmacol* 1981;21:413S–416S.

231. Grinspoon L, Bakalar JB. Marihuana as medicine. A plea for reconsideration. *JAMA* 1995;273:1875–1876.

232. Gorelick DA. Serotonin uptake blockers and the treatment of alcoholism. In: Galanter ed. *Alcoholism treatment research, vol. 7.* New York: Plenum, 1989:267.

233. Lerner PI. Neurological complications of infective endocarditis. *Med Clin North Am* 1985;69:385–398.

234. Jaffe JH, Martin WR. Opioid analgesics and antagonists. In: Gilman AG, Rall TW, Nies AS, et al., eds. *Goodman and Gilman's the pharmacological basis of therapeutics,* 8th ed. New York: McGraw-Hill, 1993:485–521.

235. Rall TW. Hypnotics and sedatives. In: Gilman AG, Rall TW, Nies AS, et al., eds. *Goodman and Gilman's the pharmacological basis of therapeutics,* 8th ed. New York: McGraw-Hill, 1993:345–382.

236. Aronson TA, Craig TJ. Cocaine precipitation of panic disorder. *Am J Psychiatry* 1986;143:643–645.

237. Gawin FH, Ellinwood EH Jr. Cocaine and other stimulants. Actions, abuse, and treatment. *N Engl J Med* 1988;318:1173–1182.

238. Gold MS, Washton AM, Dackis CA. Cocaine abuse: neurochemistry, pharmacology, and treatment. *NIDA Res Monogr* 1985;61:130–150.

239. Litten RZ, Allen JP. Pharmacological therapies of alcohol addiction. In: Miller NS, Gold MS, eds. *Pharmacological therapies for drug and alcohol addictions.* New York: Marcel Dekker, 1995:127–141.

240. Koppel BS, Tuchman AJ, Mangiardi JR, et al. Epidural spinal infection in intravenous drug abusers. *Arch Neurol* 1988;45:1331–1337.

241. Amine AB. Neurosurgical complications of heroin addiction: brain abscess and mycotic aneurysm. *Surg Neurol* 1977;7:385.

242. Gilroy J, Andaya L, Thomas VJ. Intracranial mycotic aneurysms and subacute endocarditis in heroin addiction. *Neurology* 1973;23:1193.

243. Holzman RS, Bishko F. Osteomyelitis in heroin addicts. *Ann Intern Med* 1971;75:693–696.

244. Siao PTC, McCabe P, Yagnik P. Nocardial spinal epidural abscess. *Neurology* 1989;39:996.

245. Cherubin CE. The medical sequelae of narcotic addition. *Ann Intern Med* 1967;67:23.

246. Gattell JM, Miro JM, Para C, et al. Infective endocarditis in drug addicts. *Lancet* 1984;1:228.

247. Hubbell G, Cheitlin MD, Rapaport E. Presentation, management, and follow-up of infective endocarditis in drug addicts. *Am Heart J* 1981;138:85.

248. Louria DB, Hensle T, Rose J. The major medical complications of heroin addiction. *Ann Intern Med* 1967;67:1.

249. Jabbari B, Pierce JF. Spinal cord compression due to *Pseudomonas* in a heroin addict. *Neurology* 1977;27:1034.

250. Kasplan SS. *Pseudomonas* disc space infection in an occasional heroin user. *Ariz Med* 1974;31:916.

251. Reyes MP, Palutke WA, Wylin RF, et al. *Pseudomonas* endocarditis in the Detroit Medical Center, 1969–1972. *Medicine* 1973;52:173.

252. Harris PD, Yeoh CB, Breault J, et al. Fungal endocarditis secondary to drug addiction. Recent concepts in diagnosis and therapy. *J Thorac Cardiovasc Surg* 1972;6:980.

253. Pollack S, Magtader A, Lange M. *Neisseria subflava* endocarditis. Case report and review of the literature. *Am J Med* 1984;76:752.

254. Vartian CV, Shlaes DM, Padhye AA, et al. *Wangiella dermatitides* endocarditis in an intravenous drug user. *Am J Med* 1985;78:703.

255. Sees KL, Delucchi KL, Carmen M, et al. Methadone maintenance vs. 180-day psychosocially enriched detoxification in treatment of opioid dependence. *JAMA* 2000;283:1303–1310.

256. Akhurst JS. The use of lofexidine by drug dependency units in the United Kingdom. *Eur Addict Res* 1999;5:43–49.

257. Strang J, Bearn J, Gossop M. Lofexidine for opiate detoxification: review of recent randomized and open controlled trials. *Am J Addict* 1999;8:337–348.

258. Gerra G, Zaimovic A, Giusti F, et al. Lofexidine versus clonidine in rapid opiate detoxification. *J Subst Abuse Treat* 2000;21:11–17.

259. Gowing L, Ali R, White J. Buprenorphine for the management of opioid withdrawal. *Cochrane Database Syst Rev* 2000;3:CD002025.

260. Ling W, Huber A, Rawson RA. New trends in opiate pharmacotherapy. *Drug Alcohol Rev* 2001;20:79–94.

261. McGregor C, Ali R, White JM, et al. A comparison of antagonist-precipitated withdrawal under anesthesia to standard inpatient withdrawal as a precursor to maintenance naltrexone treatment in heroin users: outcomes at 6 to 12 months. *Drug Alcohol Depend* 2002;68:5–14.

262. Dole VP, Nyswander ME. A medical treatment for diacetylmorphine (heroin) addiction. *JAMA* 1965;193:646.

263. Dole VP, Nyswander ME, Kreek MJ. Narcotic blockade. *Arch Intern Med* 1966;118:304–309.

264. Ball JC, Lange WR, Myers CP, et al. Reducing the risk of AIDS through methadone maintenance treatment. *J Health Soc Behav* 1988;29:214–226.

265. Cooper JR. Methadone treatment and acquired immunodeficiency syndrome. *JAMA* 1989;262:1664–1668.

266. Des Jarlais DC, Friedman SR, Hopkins W. Risk reduction for acquired immunodeficiency syndrome among intravenous drug users. *Ann Intern Med* 1985;103:755–759.

267. U.S. Department of Health and Human Services, Food and Drug Administration. Narcotic drugs in maintenance and detoxification treatment of narcotic dependence; repeal of current regulations and proposal to adopt new regulations. *Federal Register* 1999 Jul 22;64:39810–39857 (codified at 21 CFR 291, 42 CFR 8).

268. Lewis DC. Access to narcotic addiction treatment and medical care prospects for the expansion of methadone maintenance treatment. *J Addict Dis* 1999;18:5–21.

269. Rounsaville BJ, Kosten TR. Treatment for opioid dependence: quality and access. *JAMA* 2000;283:1337–1338.

270. Simpson DD, Joe GW, Rowan-Szal G, et al. Client engagement and change during drug abuse treatment. *J Subst Abuse* 1995;7:117–134.

271. Hartz DT, Meek P, Piotrowski NA, et al. A cost-effective and cost-benefit analysis of contingency contracting-enhanced methadone detoxification treatment. *Am J Drug Alcohol Abuse* 1999;25:207–218.

272. Fraser HF, Isbell H. Actions and addiction liabilities of alpha-acetylmethadols in man. *J Pharmacol Exp Ther* 1952;105:458–465.

273. Blaine JD, Renault PF. Clinical use of LAAM. *Ann N Y Acad Sci* 1978;311:214–231.

274. Fudala PJ, Vocci F, Montgomery A, et al. Unpublished, LAAM New Drug Application submission, 1993.

275. Levine R, Zaks A, Fink M, et al. Levomethadyl acetate: prolonged duration of opioid effects, including cross-tolerance to heroin. *NIDA Res Monogr* 1976;8.

276. Ling W, Blaine JD. The use of LAAM in treatment. In: Dupont RL, Goldstein A, O'Donnell J, eds. *Handbook on drug abuse.* Washington, DC: U.S. Government Printing Office, 1976.

277. Ling W, Charuvastra VC, Kaim SC, et al. Methadyl acetate and methadone maintenance treatments for heroin addicts: a Veterans Administration cooperative study. *Arch Gen Psychiatry* 1976;33:709–720.

278. Ling W, Klett CJ, Gillis R. A cooperative clinical study of methadyl acetate. *Arch Gen Psychiatry* 1978;35:345–353.

279. Tennant FS, Rawson RA, Pumphrey E, et al. Clinical experience with 959 opioid-dependent patients treated with levo-alpha-acetylmethadol (LAAM). *J Subst Abuse Treat* 1986;3:195–202.

280. Ling W, Rawson RA, Anglin MD. Pharmacology, practice, and politics: a tale of two opiate pharmacotherapies. In: Sorenson J, Rawson RA, Guydish J, Zweben J, eds. *Research to practice, practice to research: promoting scientific-clinical interchange in drug abuse treatment.* Washington, DC: American Psychological Association, 2002;207–220.

281. Expert Panel Clinical guidelines on LAAM In Opiate Agonist Therapy (OAT). U.S. Department of Health and Human Services, Public Health Service, Substance Abuse and Mental Health Services Administration, Center for Substance Abuse Treatment, Rockville, MD *(in press).*

282. Jasinski DR, Pevnick JS, Griffith JD. Human pharmacology and abuse potential of the analgesic buprenorphine. *Arch Gen Psychiatry* 1978;35:501–516.

283. Mello NK, Bree MP, Mendelson JH. Comparison of buprenorphine and methadone effects on opiate self-administration in primates. *J Pharmacol Exp Ther* 1983;225:378–386.

284. Mello MK, Mendelson JH. Buprenorphine suppresses heroin use by heroin addicts. *Science* 1980;27:657–659.

285. Mello MK, Mendelson JH, Kuehnle JC. Buprenorphine effects on human heroin self-administration: an operant analysis. *J Pharmacol Exp Ther* 1982;223:30–39.

286. Bickel WK, Stitzer ML, Bigelow GE, et al. A clinical trial of buprenorphine: comparison with methadone in the detoxification of heroin addicts. *Clin Pharmacol Ther* 1988;43:72–78.

287. Kosten TR, Morgan C, Kleber HD. Treatment of heroin addict using buprenorphine. *Am J Drug Alcohol Abuse* 1991;17:119–128.

288. Singh RA, Mattoo SK, Malhotra A, et al. Cases of buprenorphine abuse in India. *Acta Psychiatr Scand* 1992;86:46–48.

289. Quigley AJ, Bredemeyer DE, Seow SS. A case of buprenorphine abuse in Australia. *Med J Aust* 1984;140:425–426.

290. Mendelson J, Jones RT, Wlem S, et al. Buprenorphine and naloxone combinations: the effects of three dose rations in morphine-stabilized, opiate-dependent volunteers. *Psychopharmacology* 1999;141:37–46.

291. Fudala PJ, Bridge TP, Herbert S, et al. Office-based treatment of opiate dependence: efficacy and safety of buprenorphine/naloxone sublingual tablets *(submitted)*.

292. Martin WR, Jasinski DR, Mansley PA. Naltrexone, an antagonist for the treatment of heroin dependence. *Arch Gen Psychiatry* 1973;28:784–791.

293. Report of the National Research Council Committee on Clinical Evaluation of Narcotics Antagonists. Clinical evaluation of naltrexone treatment of opiate dependent individuals. *Arch Gen Psychiatry* 1978;35:335–340.

294. Kosten TR, Kleber HD. Buprenorphine detoxification from opioid dependence: a pilot study. *Life Sci* 1988;42:635–641.

295. Ling W, Wesson DR. Naltrexone treatment for addicted healthcare professionals: a collaborative private practice experience. *J Clin Psychiatry* 1984;45:46–48.

296. Halikas JA, Nugent SM, Crosby RD, et al. 1990–1991 survey of pharmacotherapies used in the treatment of cocaine abuse. *J Addict Dis* 1993;12:129–139.

297. Meyers RE. New pharmacotherapies for cocaine dependence revisited. *Arch Gen Psychiatry* 1992;49:900–904.

298. Ling W, Shoptaw S. Research in pharmacotherapy for substance abuse: Where are we? Where are we going? *J Addict Dis* 1997;16:83–102.

299. Vocci F. Medication development for methamphetamine-related disorders. May 1998. Presented to the Methamphetamine Advisory Group to Attorney General Reno. Washington, DC, 1998.

300. Rawson RA. The NIDA Methamphetamine Clinical Trials Group (MCTG): taking research into the field. Presented at the Annual ASAM Meeting, Los Angeles, April 2002.

301. Gorelick DA. Overview of pharmacological treatment approaches for alcohol and other drug addiction: intoxication, withdrawal, relapse prevention. *Psychiatr Clin North Am* 1993.

302. Seller EM, Kalant H. Alcohol intoxication and withdrawal. *N Engl J Med* 1976;294:757–762.

303. Tabakoff B, Hoffman PL. Adaptive responses to ethanol in the central nervous system. In: Goedde HE, Agarwal DP, eds. *Alcoholism: biomedical and genetic aspects.* New York: Pergamon Press, 1989:99–112.

304. Mayo-Smith MF. Pharmacological management of alcohol withdrawal. A meta-analysis of evidence-based practice guidelines. American Society of Addiction Medicine Working Group on Pharmacological Management of Alcohol Withdrawal. *JAMA* 1997;278:144–151.

305. Temkin NR. Antiepileptogenesis and seizure prevention trials with antiepileptic drugs: meta-analysis of controlled trials. *Epilepsia* 2001;42:515–524.

306. Litten RZ, Allen JP. Pharmacotherapies for alcoholism. Promising agents and clinical issues. *Alcoholism* 1991;15:620–633.

307. Wilkins AJ, Jerkins WJ, Steiner JA. Efficacy of clonidine in treatment of alcohol withdrawal state. *Psychopharmacology* 1983;81:78–80.

308. Sellers EM, Zilm DH, Degani NC. Comparative efficacy of propranolol and chlordiazepoxide in alcohol withdrawal. *J Stud Alcohol* 1977;38:2096–2108.

309. Peachy JE, Annis HM. Pharmacologic treatment of chronic alcoholism. *Psychiatry Clin North Am* 1984;7:745–756.

310. Zeise ML, Kasparov S, Capogna M, et al. Acamprosate (calciumacetylhomotaurinate) decreases postsynaptic potentials in the rat neocortex: possible involvement of excitatory amino acid receptors. *Eur J Pharmacol* 1993;231:47–52.

311. Kranzler HR. Pharmacotherapy of alcoholism: gaps in knowledge and opportunities for research. *Alcohol Alcohol* 2000;35:537–547.

312. Tabakoff B, Hoffman PL. Alcohol: neurobiology. In: Lowinson JH, Ruiz P, Millman RB, et al., eds. *Substance abuse: a comprehensive textbook,* 2nd ed. Baltimore: Williams & Wilkins, 1992:152–185.

313. O'Malley SS, Jaffe AJ, Chang G, et al. Naltrexone and coping skills therapy for alcohol dependence. A controlled study. *Arch Gen Psychiatry* 1992;49:881–887.

314. Volpicelli JR, Alterman AI, Hayashida M, et al. Naltrexone in the treatment of alcohol dependence. *Arch Gen Psychiatry* 1992;49:876–880.

315. Mason BJ, et al. A double-blind, placebo-controlled study of oral nalmefene for alcohol dependence. *Arch Gen Psychiatry* 1999;56:719–724.

316. Huber A, Ling W, Shoptaw S, et al. Integrating treatments for methamphetamine abuse: a psychosocial perspective. *J Addict Dis* 1997;16:41–50.

317. Rawson RA, Huber A, Brethen PB, et al. Methamphetamine and cocaine users: differences in characteristics and treatment retention. *J Psychoactive Drugs* 2000;32:233–240.

318. Rawson RA, McCann MJ, Shoptaw S, et al. A comparison of contingency management and cognitive-behavioral approaches for cocaine- and methamphetamine-dependent individuals *(submitted)*.

319. Rawson RA, McCann MJ, Huber A, et al. A comparison of contingency management and cognitive-behavioral approaches for cocaine dependent methadone-maintained individuals. *Arch Gen Psychiatry.* 2002;59(9):817–824.

320. Anglin MD, Rawson RA, eds. The CSAT methamphetamine treatment project: moving research into the "real world." *J Psychoactive Drugs* 2000;32.

321. Rawson RA, Obert JL, McCann MJ, et al. *The neurobehavioral treatment manual* Beverly Hills, CA: Matrix, 1989.

322. Rawson RA, Shoptaw S, Obert JL, et al. An intensive outpatient approach for cocaine abuse treatment: the matrix model. *J Subst Abuse Treat* 1995;12:117–127.

323. Shoptaw S, Rawson RA, McCann MJ, et al. The matrix model of outpatient stimulant abuse treatment: evidence of efficacy. *J Addict Dis* 1994;13:129–141.

324. Rawson RA, Huber A, Brethen P, et al. Status of methamphetamine users 2–5 years after outpatient treatment. *J Addict Dis* 2002;21:107–119.

325. Reback CJ. The social construction of a gay drug: methamphetamine use among gay and bisexual males in Los Angeles (report 93427). City of Los Angeles AIDS Coordinator, 1997.

326. Meyer RE. How to understand the relationship between psychopathology and addictive disorders: another example of the chicken and the egg. In: Meyer RE, ed. *Psychopathology and addictive disorders.* New York: Guilford, 1986:3–16.

327. SAMHSA. 2001 NHSDA: chapter 8: prevalence and treatment of mental health problems, 2001.

328. Helzer JE, Pryzbeck TR. The co-occurrence of alcoholism with other psychiatric disorders in the general population and its impact on treatment. *J Stud Alcohol* 1988;49:219–224.

329. Ross HE, Glaser FB, Germanson T. The prevalence of psychiatric

disorders in patients with alcohol and other drug problems. *Arch Gen Psychiatry* 1988;45:1023–1031.

330. Beeder AB, Millman RB. In: Lowinson JH, Ruiz P, Millman RB, et al., eds. *Substance abuse: a comprehensive textbook,* 2nd ed. Baltimore: Williams & Wilkins, 1992:51.

331. Raskin VD, Miller NS. The epidemiology of the comorbidity of psychiatric and addictive disorders: a critical review. *J Addict Dis* 1993;12:45–57.

332. Khantzian EJ, Treece C. DSM-III psychiatric diagnosis of narcotic addicts: recent findings. *Arch Gen Psychiatry* 1985;42:1067–1071.

333. Geracioti TD, Post RM. Onset of panic disorder associated with rare use of cocaine. *Biol Psychiatry* 1991;29:403–406.

334. Louie AK, Lannon RA, Ketter TA. Treatment of cocaine-induced panic disorder. *Am J Psychiatry* 1989;146:40–44.

335. Klisz DK, Parsons OA. Hypothesis testing in younger and older alcoholics. *J Stud Alcohol* 1977;38:1718–1729.

336. Grant I, Reed R. Neuropsychology of alcohol and drug abuse. In: Alterman AI, ed. *Substance abuse and psychopathology.* New York: Plenum, 1985:289–341.

337. Fabian MS, Parsons OA, Sheldon MD. Effects of gender and alcoholism on verbal and visuo-spatial learning. *J Nerv Ment Dis* 1985;172:16–20.

338. Fabian MS, Parsons OA, Silberstein JA. Impaired perceptual-cognitive functioning in women alcoholics: cross-validated findings. *J Stud Alcohol* 1981;42:217–229.

339. Glenn SW, Parsons OA. Neuropsychological efficiency measures in male and female alcoholics. *J Stud Alcohol* 1992;53:546–552.

340. Hewett LJ, Nixon SJ, Glenn SW, et al. Verbal fluency deficits in female alcoholics. *J Clin Psychol* 1991;47:716–720.

341. Grant I. Alcohol and the brain. Neuropsychological correlates. *J Consult Clin Psychol* 1987;545:310–324.

342. Grant I, Reed R. Neuropsychology of alcohol and drug abuse. In: Alterman AI, ed. *Substance abuse and psychopathology.* New York: Plenum Press, 1985:289–341.

343. Patterson BW, Sinha R, Williams HL, et al. The relationship between neuropsychological and late component evoked potential measures in chronic alcoholics. *Int J Neurosci* 1989;49:319–327.

344. Williams CM, Skinner AEG. The cognitive effects of alcohol abuse: a controlled study. *Br J Addict* 1990;85:911–917.

345. Berry J, Van Gorp WG, Herzberg DS, et al. Neuropsychological deficits in abstinent cocaine abusers: preliminary findings after two weeks of abstinence. *Drug Alcohol Depend* 1993;32:231–237.

346. Strickland TL, Mena I, Villaneueva-Meyer J. Cerebral perfusion and neuropsychological consequences of chronic cocaine use. *J Neuropsychiatry Clin Neurosci* 1993;5:419–427.

347. Guerra D, Sole A, Cami J, et al. Neuropsychological performance in opiate addicts after rapid detoxification. *Drug Alcohol Depend* 1987;20:261–270.

348. Grant I, Adams KM, Carlin AS, et al. The collaborative neuropsychological study of polydrug users. *Arch Gen Psychiatry* 1978;35:1063–1074.

349. Grant I, Judd L. Neuropsychological and EEG disturbances in polydrug users. *Am J Psychiatry* 1976;133:1039–1042.

350. Becker JT, Kaplan RF. Neurophysiological and neuropsychological concomitants of brain dysfunction in alcoholics. In: Meyers RE, ed. *Psychopathology and addictive disorders.* New York: Guilford, 1986:1–25.

351. Reed RJ, Grant I. The long-term neurobehavioral consequences of substance abuse: conceptual and methodological challenges for future research. Residual effects of abused drugs on behavior. *NIDA Res Monogr* 1990;101:10–56.

352. Richards M, Sano M, Goldstein S, et al. The stability of neuropsychological test performance in a group of parenteral drug users. *J Subst Abuse Treat* 1992;9:371–377.

353. Parsons OA. Do neuropsychological deficits predict alcoholics' treatment course and posttreatment recovery? In: Parsons OA, Butters N, Nathan PE, eds. *Neuropsychology of alcoholism: implications for diagnosis and treatment.* New York: Guilford, 1987:273–290.

354. Hesselbrock V, Bauer LO, Hesselbrock MN, et al. Neuropsychological factors in individuals as high risk for alcoholism. *Recent Dev Alcohol* 1991;9:21–39.

355. Yohman JR, Parsons OA, Leber WR. Lack of recovery in male alcoholics' neuropsychological performance one year after treatment. *Alcohol Clin Exp Res* 1985;9:114–117.

356. Gregson RAM, Taylor GM. Prediction of relapse in men alcoholics. *J Stud Alcohol* 1977;38:1749–1760.

357. Leber WR, Parsons OA, Nichols N. Neuropsychological test results are related to ratings of men alcoholics' therapeutic progress: a replicated study. *J Stud Alcohol* 1985;46:116–121.

358. Kupke T, O'Brien W. Neuropsychological impairment and behavioral limitations exhibited within an alcohol treatment program. *J Clin Exp Neuropsychol* 1985;7:292–304.

359. Fals-Stewart W, Schafer J, Fals-Stewart W, et al. The relationship between length of stay in drug-free therapeutic communities and neurocognitive functioning. *J Clin Psychol* 1992;48:539–543.

360. Smith DE, McCrady BS. Cognitive impairment among alcoholics: impact on drink refusal skill acquisition and treatment outcome. *Addict Behav* 1991;16:265–274.

361. Morgan JP. American opiophobia: customary underutilization of opioid analgesics. *Adv Alcohol Subst Abuse* 1986;5:163–173.

362. Melzack R. The tragedy of needless pain. *Sci Am* 1990;262:27–33.

363. Friedman DP. Perspectives on the medical use of drugs of abuse. *J Pain Symptom Manage* 1990;5[Suppl 1]:S2–S5.

364. Perry SW. Irrational attitudes toward addicts and narcotics. *Bull N Y Acad Med* 1985;61:706–727.

365. Wesson DR, Ling W, Smith DE. Prescription of opioids for treatment of pain in patients with addictive disease. *J Pain Symptom Manage* 1993;8:428–435.

366. Portenoy RK, Payne R. Acute and chronic pain. In: Lowinson J, Ruiz P, Millman RB, eds. *Substance abuse: a comprehensive textbook,* 2nd ed. Baltimore: Williams & Wilkins, 1992.

367. Portenoy RK. Chronic opioid therapy in nonmalignant pain. *J Pain Symptom Manage* 1990;5[Suppl 1]:S46–S62.

368. Portenoy RK, Foley KM. Chronic use of opioid analgesics in non-malignant pain: report of 38 cases. *Pain* 1986;25:171–186.

369. Dunbar SA, Katz NP. Chronic opioid therapy for nonmalignant pain in patients with a history of substance abuse: report of 20 cases. *J Pain Symptom Manage* 1996;11:163–171.

370. Hicks RD. Pain management in the chemically dependent patient. *Hawaii Med J* 1989;48:491–495.

371. Scimeca MM, Savage SR, Portenoy R, et al. Treatment of pain in methadone-maintained patients. *Mt Sinai J Med* 2000;67:412–422.

372. Senay EC, Becker CE, Schnoll SH. Management of addicts in the general hospital. Emergency treatment of the drug abusing patient for treatment staff physicians. Rockville, MD: 1980:35–39.

373. Rubenstein RB, Spira I, Wolff WI. Management of surgical problems in patients on methadone maintenance. *Am J Surg* 1976;131:566–569.

374. Kantor TG, Cantor R, Tom E. A study of hospitalized surgical patients on methadone maintenance. *NIDA Res Monogr* 1981;34:243–249.

375. Compton M. Cold-pressor pain tolerance in opiate and cocaine abusers: correlates of drug type and use status. *J Pain Symptom Manage* 1994;9:462–473.

376. Compton P, Charuvastra VC, Ling W. Pain intolerance in opioid-maintained former opiate addicts: effect of long-acting maintenance agent. *Drug Alcohol Depend* 2001;63:139–146.

377. Doverty M, Somogyi AA, White JM, et al. Methadone maintenance patients are cross-tolerant to the antinociceptive effects of morphine. *Pain* 2001;93:155–163.

378. Doverty M, White JM, Somogyi AA, et al. Hyperalgesic responses in methadone maintenance patients. *Pain* 2001;90:91–96.

379. Tilson HA, Rech RH, Stolman S. Hyperalgesia during withdrawal as a means of measuring the degree of dependence in morphine dependent rats. *Psychopharmacology* 1973;28:287–300.

380. U. S. Department of Health and Human Services. *Acute pain management: operative or medical procedures and trauma.* Rockville, MD: Agency for Health Care Policy and Research, 1992.

381. Hughes JR, Bickel WK, Higgins ST. Buprenorphine for pain relief in a patient with drug abuse. *Am J Drug Alcohol Abuse* 1991;17:451–455.

382. Johnson RT. Pathogenesis of HIV infections. American Academy of Neurology Annual Courses, vol. 5: infections, emergency, critical care, 1993;140:169–177.

383. McArthur JC. HIV-associated CNS syndromes. American Academy of Neurology Annual Courses, vol. 5: infections, emergency, critical care, 1993;140:15–29.

384. Gonzales-Scarano F. Basic biology of HIV-1. American Academy of Neurology Annual Courses, vol. 5: infections, emergency, critical care, 1993;140:3–14.

385. Perry S. Treatments of psychiatric complications. American Academy of Neurology Annual Courses, vol. 5: infections, emergency, critical care, 1993;140:133–143.

34

BASAL GANGLIA AND BEHAVIORAL DISORDERS

JOSEPH JANKOVIC
MARIA L. DE LEÓN

During the past century, neurologists, psychiatrists, and neuroscientists have sought to understand the role and function of the basal ganglia, not only in control of motor function but also in cognitive function and behavior (1). Despite the enormous increase in the knowledge of anatomy, chemistry, physiology, and clinical correlates, the function of the basal ganglia remains a mystery. Long considered to be primarily involved with motor control, accumulating clinical and scientific evidence supports the role of basal ganglia in nonmotor (e.g., behavioral, cognitive, sensory) functions. The clinical hallmark of basal ganglia disease or dysfunction is paucity of movement (hypokinesia) or excessive abnormal involuntary movement (hyperkinesias). In addition, dysfunction of the basal ganglia and their connections frequently is associated with apraxia (2). Aside from motor dysfunction, the basal ganglia have been increasingly recognized to play a major role in the production and modification of behavioral and cognitive disorders such as obsessive-compulsive disorder (OCD), attention deficit/hyperactivity disorder (ADHD), depression, set shifting, and executive dysfunction (Table 34.1). In this chapter, we briefly review the current concepts of basal ganglia and their role in behavior and cognition. The first part of the chapter focuses on the functional anatomy of the basal ganglia, particularly their connections with cortical areas involved with behavior and cognition; the second part reviews various syndromes related to basal ganglia dysfunction; and the third part examines specific anatomic-behavioral correlations.

FUNCTIONAL ANATOMY OF THE BASAL GANGLIA

Basal ganglia comprise a complex of subcortical nuclei that include the caudate and putamen (striatum), globus pallidus interna (GPi), globus pallidus externa (GPe), subthalamic nucleus, claustrum, substantia nigra compacta (SNc), substantia nigra reticulata (SNr), and ventral tegmental area.

Although these portions of the basal ganglia have been studied extensively because of their role in motor control, various anatomic, physiologic, and behavioral studies have demonstrated that certain cognitive and behavioral disorders result from a dysfunction in the cortico-striato-pallido-thalamo-cortical circuit (3–7). In the early 1990's, Alexander et al. (8) proposed the existence of five circuits that were thought to be responsible for modulating the expression of behavior. The motor circuit is composed of direct connections among the striatum, GPi, and SNr. There is also an indirect pathway that connects the striatum to the subthalamic nucleus via the GPe and GPi/SNr. The output nuclei, GPi and SNr, project to the thalamus and brainstem. The thalamus in turn projects to the cortex. The basal ganglia are closely integrated with the cerebral cortex via multiple parallel loops. One of the loops connects the basal ganglia with the somatosensory cortex, one with oculomotor areas, and the other three with nonmotor areas of the frontal cortex (dorsolateral prefrontal area 46, lateral orbitofrontal area 12, and anterior cingulate/medial orbitofrontal areas 24 and 13). These frontal areas are involved in planning, organizing, working memory, learning, attention, and other executive functions, as well as emotional expression (contentment and euphoria) (9). Renewed interest in the nonmotor function of the basal ganglia has been fueled by advances in anatomic studies using neurotropic viruses as transneuronal tracers and later supplemented by physiologic studies, including single-unit recordings and functional imaging studies (7).

The role of basal ganglia in mood and behavior has been suspected since the early twentieth century, when it was recognized that diseases of the basal ganglia, such as Parkinson disease (PD), progressive supranuclear palsy (PSP), multiple system atrophy (MSA), dementia with Lewy bodies, and Huntington disease (HD), often were associated with apathy, depression, and other behavioral problems (10–12). Subsequent studies also recognized that properly functioning basal ganglia suppress not only unwanted movements (e.g., chorea, tics), but also inappropriate or impulsive behavior,

TABLE 34.1. BEHAVIORAL COGNITIVE DISORDERS RELATED TO BASAL GANGLIA DYSFUNCTION

Brain Lesion	Abnormal Behavior	Movement Disorder
Caudate	Obsessive-compulsive disorder (compulsive repetitive behaviors), Tourette syndrome, depression, dementia, inability shifting sets	Tourette syndrome, Sydenham chorea
Putamen	Tourette syndrome, inertia, depression, PANDAS	Extrapyramidal symptoms, Huntington disease, tics, Sydenham chorea
Globus pallidus	Inertia, depression, apathy, loss of motivation, stuttering, aggressive behavior	Tics (tongue thrusting, shoulder shrugging, arm/leg jerking), parkinsonism
Thalamus	Mood dysfunction, memory loss, language perception	Hemichorea
Cingulate gyrus	Obsessive-compulsive disorder	Phonic tics
Frontal lobe/prefrontal	Impulsive behavior, poor planning, lack of foresight, apathy, executive dysfunction, inability shifting sets	
Substantia nigra (compacta and reticulata)	Worthlessness, depression, morbid preoccupation with death, peduncular hallucinosis	Parkinson disease
Subthalamic nucleus		Hemichorea, hemiballismus

PANDAS, pediatric autoimmune neuropsychiatric disorders associated with streptococcal infections.

obsessive thoughts, and other intrusions (9,13). The study of "negative symptoms" associated with various basal ganglia disorders, manifested not only by poverty of movement but also by slowness and paucity of cognition (bradyphrenia) and emotion (apathy), may provide insights into the mechanisms of "goal-directed" and "reward" behaviors (14,15). Neurobehavioral and neuropsychiatric assessment of patients with movement disorders is beyond the scope of this chapter and is reviewed elsewhere (16,17).

PARKINSON DISEASE AND OTHER PARKINSONIAN DISORDERS

PD is the most common cause of hypokinetic disorders, but there are many other disorders in which parkinsonism, manifested chiefly by hypokinesia (paucity of movement) and bradykinesia (slowness of movement), is the predominant feature (18). SNc is the site of the brunt of the pathology in PD, characterized by loss of dopaminergic neurons. The neuronal loss is not confined to this area; it extends diffusely into areas of the ventral tegmentum and involves noradrenergic neurons in the locus caeruleus, cholinergic neurons in the nucleus basalis of Meynert, and serotonergic neurons in the dorsal raphe. Depending on the nature and distribution of biochemical and pathologic changes, these parkinsonian disorders manifest a variety of cognitive and behavioral disorders, including deficits in attention set shifting, working memory, planning, and problem solving; mood changes; and abnormalities in visual and auditory perception (Table 34.2) (16,19). Some of the cognitive difficulties seen in patients with PD and related basal ganglia dysfunction are due to impaired attention (20–22). According to one

study, PD patients are more easily distracted in visual and auditory attention tasks compared to normal controls (20). Other studies have shown that PD patients have subtle visual exploratory directional biases (23). The findings add to the complexity of visuospatial abnormalities encountered in PD patients and further support the role of basal ganglia in the control of attention (22). PD patients may have limited perception of large spatial configurations (they may see the trees but not the forest) (24).

In parkinsonian syndromes, the function of the frontosubcortical circuits that run parallel to the motor circuits and enter the striatum at the caudate is thought to be impaired. Thus, disruption of these projections results in executive dysfunction manifested by perseveration, difficulty in set shifting, impaired higher-order attention, and poor abstraction and concept formation (25). On the other hand, a disruption of the motor circuit that involves the striatum via the putamen is thought to result primarily in bradykinesia or akinesia, such as impaired movement initiation, loss of associated movements, slowed execution of motor movements, and prolonged time in the planning of intended movements. Clinically, patients demonstrate paucity of gestures, decreased spontaneity, and decline in verbal output. In addition to these motor and cognitive deficits, other behavioral deficits have been observed in the early stages of parkinsonism (26). Moreover, similar nonmotor deficits have been seen in other parkinsonian disorders, such as PSP and MSA (27,28). For example, one may find prominent symptoms of apathy and social disinhibition, along with impaired planning skills, in patients with PSP.

Some studies designed to evaluate executive dysfunction in patients with parkinsonism have revealed consistent activation (based on measures of regional cerebral blood flow)

TABLE 34.2. COMMON PARKINSONISM-DEMENTIA SYNDROMES

Syndrome	Clinical Features
Parkinson disease with dementia	Progressive memory deficits, with/without visuospatial deficits; difficulty set shifting
Dementia with Lewy bodies	Early hallucinations unrelated to therapy; cortical dementia (executive dysfunction); marked decline in cognitive abilities; parkinsonism
Progressive supranuclear palsy	Decreased fluency; difficulty shifting concepts greater than in Parkinson disease; impaired planning, behavioral disturbances (e.g., apathy, disinhibition), with/without mild memory disturbances; visuospatial deficits; parkinsonism; postural instability; pseudobulbar palsy and supranuclear gaze palsy
Corticobasal degeneration	Ideomotor apraxia; alien hand; aphasia; sensory/visual neglect; progressive focal motor dysfunction (e.g., parkinsonism, dystonia, myoclonus)
Vascular dementia	Stepwise appearance of motor and cognitive deficits; history of acute neurologic deficits
Alzheimer disease	Progressive anterograde memory loss; apraxia; aphasia; agnosia; executive dysfunction; may have parkinsonism in late stages
Pick disease	Progressive "frontal lobe personality"; disinhibition; impulsivity; social misconduct; difficulty planning; sequencing deficits; aphasia; echolalia; parkinsonism develops in late stages

of the putamen, along with other regions in the cortex and cerebellum, while patients are performing learned sequential finger movements. This activation pattern was seen repeatedly, irrespective of sequence or length of movements. The putamen apparently serves an executive role in the process of carrying out specific motor tasks (in this case, finger movements), whereas the right dorsal premotor region of the brain may be more involved with memory, given that it was found to activate only in direct relation to the sequence length of finger movements (29). Therefore, some of the behavioral impairments associated with basal ganglia abnormalities or disorders are directly related to executive dysfunction.

According to several studies, about 40% (range 10%–90%) of parkinsonian patients have clinical depression that appears to be independent of age and length of illness (30). Depression precedes the onset of motor symptoms in an estimated 40% of patients with PD. A community-based study showed that 7.7% of PD patients met the criteria for major depression, 5.1% for moderate-to-severe depression, and 45% for mild depression (31). Aarsland et al. (32) found at least one psychiatric symptom in 61% of 139 patients with PD. These symptoms included depression (38%), hallucinations (27%), and a variety of other behavioral and cognitive changes. In another study, the authors found that the presence of hallucinations was the strongest predictor of placement in a nursing home (33). Minor hallucinations may occur in as many as 40% of patients with PD, illusions in 25%, formed visual hallucinations in 22%, and auditory hallucinations in 10% (34). Risk factors for hallu-

cinations include older age, duration of illness, depression, cognitive disorder, daytime somnolence, poor visual acuity, and dopaminergic drugs (35–37). Hallucinations seem to correlate with daytime episodes of rapid eye movement (REM) sleep, suggesting that hallucinations and psychosis are associated with narcolepsy-like REM sleep disorder (38). In a community-based, prospective study, Aarsland et al. (39) found that patients with PD have an almost six-fold increased risk of dementia. In one study, 38% of patients who presented with idiopathic REM sleep behavior disorder (RBD) eventually developed parkinsonism (40). In another study, 86% of patients with RBD had associated parkinsonism (PD 47%, MSA 26%, PSP 2%) (41). In this study, RBD preceded the onset of parkinsonism in 52% of patients with PD. Several other studies reported this association of RBD and parkinsonism (42–44).

Single photon emission computed tomography (SPECT) studies have revealed correlations between depression and impaired dorsolateral frontal lobe perfusion, as well as in idiopathic depression (30,45). There is evidence of decreased perfusion in caudate nuclei, orbital frontal cortex, and anterior temporal and inferomedial frontal areas during bouts of depression as well (46). Ring et al. (47) found a correlation between depression and reduced activity in the anteromedial aspect of the medial frontal cortex and cingulate cortex. Some of these regions may be affected in PD at various stages, which may account for the variability in frequency of depression in PD and other parkinsonian syndromes. Temporoparietal cortical hypometabolism is present in patients with PD and may be a useful predictor of future cognitive impairment (48). Another predictor of cognitive dysfunction appears to be reduced [^{18}F]-fluorodopa uptake in the caudate nucleus and frontal cortex (49). In a group of 94 patients with primary depression, Starkstein et al. (50) found that 20% had parkinsonism that could be reversed by treatment of the depression in some of the patients.

Aside from depression, dementia often may be seen in later stages of PD. In one study, the risk of dementia in PD was estimated to be two times greater compared to the risk in age-matched, nondemented elderly controls (51). Siblings of demented PD patients may have an increased risk of developing Alzheimer disease (52). PD patients often exhibit many cognitive deficits, such as decreased ability to process information, impaired verbal fluency, lack of insight, faulty reasoning, difficulty executing sequential actions, and problems shifting between tasks. According to Jacobs et al. (53), poor performance on verbal fluency tests appears to correlate with incipient dementia in early stages of PD. Cognitive abnormalities are not excluded from other atypical parkinsonian syndromes. According to one author, as many as three of the five frontosubcortical circuits that unite regions of the frontal lobe with GP, striatum, and thalamus may be interrupted by lesions in the basal ganglia. These areas are responsible for mediating cognition, motivation, and behavior. Axial impairment, probably mediated predominantly by nondopaminergic systems, is associated with incident dementia (54,55). One study found that executive function is impaired in both familial and sporadic PD, but explicit memory recall is more impaired in the sporadic form of PD (56). The reader is referred to previous reviews of cognitive and behavioral abnormalities associated with non-PD parkinsonian disorders, including PSP, MSA, dementia with Lewy bodies, and various tauopathies (17,57–60).

HUNTINGTON DISEASE

HD is an autosomal dominant neurodegenerative disease. It is a prototype of a neurobehavioral disorder predominantly involving the basal ganglia that manifests itself as dementia, chorea, and a broad spectrum of behavioral abnormalities. The disease results from an abnormal expansion of a trinucleotide (CAG) repeat in the huntingtin gene on the short arm of chromosome 4p16.3 (61).

Psychiatric abnormalities are well known in HD patients and occur to a variable degree in up to 95% of all patients. The neurobehavioral symptoms typically consist of personality changes, agitation, irritability, anxiety, apathy, social withdrawal, impulsiveness, depression, mania, paranoia, delusions, hostility, hallucinations, psychosis, and various sexual disorders (16,62–64). In a study of 52 patients with HD, Paulsen et al. (65) found the following neuropsychiatric symptoms (in descending order of frequency): dysphoria,

agitation, irritability, apathy, anxiety, disinhibition, euphoria, delusions, and hallucinations. Irritability and mood swings are seen in 58% of HD patients, apathy in 48%, and aggressive behavior in 59% (66). Mood disorder, mainly depression (32%–44%) and mania (9%–10%), also may dominate the clinical expression of HD patients (64,67,68). In some patients, affective disorders precede HD by 2 to 20 years (69). Criminal behavior, which is closely linked to the personality changes, depression, and alcohol abuse, has been reported to be present more frequently in patients with HD compared to nonaffected, first-degree relatives (70). Such behavior may be a manifestation of an impulsive disorder as part of disinhibition, which is seen not only in HD but also in other frontal lobe–basal ganglia disorders, particularly Tourette syndrome (TS; see later) (71).

According to some authors, early onset of depression in HD suggests that the medial aspect of the caudate nucleus is the area responsible for HD-related mood disorders, because this region is affected early in the disease (72). The pathologic hallmark of HD is atrophy of the caudate nucleus, putamen, and globus pallidus (73). Other disorders (discussed elsewhere in this chapter) associated with dysfunction of the caudate nucleus also are associated with depression, which suggests a key role of the caudate nucleus in mediating mood disturbances (30,74). Behaviors such as agitation, anxiety, and irritability have been related to hyperactivity of the medial and orbitofrontal cortical circuitry (16). This is in contrast to the apathy and hypoactive behaviors seen in PSP, which are attributed to a dysfunction in the frontal cortex and associated circuitry. Involvement of the caudate nucleus in HD may lead not only to chorea and depression, but also to OCD, as highlighted in a report of HD patients who manifested classic OCD findings (75). We also reported a patient in whom familial tics and OCD were the presenting symptoms of HD (76).

Cognitive changes, which are manifested chiefly by loss of recent memory, poor judgment, and impaired concentration and acquisition, occur in nearly all patients with HD (16,77). In one study, dementia was found in 66% of 35 HD patients (78). Tasks that require psychomotor or visuospatial processing (e.g., Trail-Making B Test or Stroop Interference Test) are impaired early in the course of the disease and deteriorate at a more rapid rate than memory impairment (79). Although neurobehavioral symptoms may precede motor disturbances in some cases, de Boo et al. (80) showed that motor symptoms are more evident than cognitive symptoms in early stages of HD. Most studies have found that neuropsychological tests do not differentiate between presymptomatic persons who are positive for the HD gene from those who are negative (80), but some studies have found that cognitive changes may be the first symptoms of HD (81). In a longitudinal study performed by the Huntington Study Group of 260 persons considered "at risk" for HD, Paulsen et al. (65) found that this group had worse scores on the cognitive section of the UHDRS at baseline, an average two years before the development of motor manifestations of the disease.

TOURETTE SYNDROME AND COMORBID BEHAVIORAL DISORDERS

Tics

Tics, which are the hallmark of TS, are brief, abrupt involuntary movements (motor tics) or sounds (phonic tics) that may resemble fragments of normal motor behavior or gestures (1,82). In addition to TS, tics may be seen in the setting of a variety of neurologic disorders, including Sydenham chorea, HD, neuroacanthocytosis, tardive dyskinesia, brain tumors, and brain injury (83,84). Although considered a genetic disorder, no gene-specific locus has yet been identified. The disorder is thought to be inherited through a bilineal transmission in which both parents exhibit to a variable degree certain components of TS, such as tics, traits of OCD, or ADHD (85). That OCD is an integral part of the neurobehavioral spectrum of TS has been known since the disorder was first described by Georges Gilles de la Tourette in 1885 (86). The "just right" perception, which is manifested in patients with TS, is thought to be a manifestation of comorbid OCD (87,88). An important connection between motor and behavioral manifestations of TS is the loss of impulse control, one of the most disturbing symptoms of TS. The overlap between tics and other behavioral symptoms, including loss of impulse control, suggests that TS is a "disinhibition" disorder. The frequently observed lack of control of impulses in TS has been attributed to a faulty editor of intentions (89). Neurobehavioral disorders, such as TS, have been attributed to a dysfunction in the basal ganglia and the associated cortico-striato-thalamo-cortical circuits (1,90).

Obsessive-Compulsive Disorder

The *Diagnostic and Statistical Manual of Mental Disorders, Fourth Edition* (91) defines OCD as the presence of obsessions or compulsions causing marked distress interfering with normal functioning. *Obsessions* are persistent recurrent images, impulses, or thoughts that are inappropriate and intrusive. *Compulsions* are the behavior equivalents of thoughts that an individual feels compelled to perform, often in response to an obsession. Common examples of obsessions include need for symmetry, fear of communication, somatic preoccupations, and aggressive sexual thoughts. Common compulsions include checking, washing, counting, evening up, and hoarding. TS has been reported to have the highest comorbidity rate for OCD, with rates ranging from 50% to 80% (92).

Several studies have supported the notion of basal ganglia involvement in the pathogenesis of OCD, particularly

the caudate nucleus and inferior prefrontal cortex (93–95). Further evidence comes from the functional neuroimaging studies of Rauch et al. (96) showing bilateral medial temporal activation in patients with OCD as compared to normal controls. Functional magnetic resonance imaging (MRI) revealed decreased neuronal activity during periods of tic suppression in the ventral globus pallidus, putamen, and thalamus, along with increased activity in the right caudate nucleus, right frontal cortex, and other cortical areas normally involved in inhibition of unwanted impulses (prefrontal, parietal, temporal, and cingulate gyri) (97). Transcranial magnetic stimulation studies have shown shortened cortical silent periods and defective intracortical inhibition in patients with OCD (98) and TS (99). The latter finding may account for the clinically intrusive phenomenon of the compulsive thoughts and motor findings in TS patients. In addition to the cortico-striato-thalamo-cortical circuit, the cingulate gyrus is thought to play an important role in the production of tics, particularly phonic tics, and coexistent behavioral disorders. This notion is based in part on the observation that cingulotomy leads to resolution of phonic tics, whereas stimulation of the cingulate cortex produces vocalizations (100). Similarly, stimulation of supplementary motor areas of the brain in some instances has led to phonic tics (90).

One hypothesis based on the work of Ikonomidou et al. (101) suggests that there is increased dopaminergic innervation in the ventral striatum and associated limbic system in TS. These two regions are not only anatomically linked to one another but also functionally linked. The connection between the two may be responsible for the frequent association between tics and behavioral disturbances. A dysfunction in this system can easily account for the apparent loss of impulse control characteristic of TS, ADHD, and OCD. In one study, 59% of 54 patients with OCD had a life time history of tics (102).

There is now a substantial body of evidence that implicates basal ganglia dysfunction in the pathogenesis of OCD (69,103,104). In studies of patients with bilateral basal ganglia lesions secondary to trauma or to anoxic or toxic encephalopathy, stereotyped and obsessive-compulsive behaviors were seen in many of the patients after the reported injuries (Table 34.3). We and others have observed patients with PD and atypical parkinsonism who developed features of OCD, such as obsessions with bowels, compulsive gambling, and other obsessive-compulsive behaviors, as their disease progressed (105,106). LaPlane et al. (107) reported obsessive-compulsive behavior in seven patients with bilateral basal ganglia lesions, a majority of whom had an involvement of the globus pallidus (107). Bilateral globus pallidus infarction presenting as OCD also highlights the importance of this nucleus in OCD (108). There have been numerous other reports of subjects with lesions in the caudate nucleus who manifested symptoms of compulsive and repetitive behaviors (104). Baxter et al. (109) found bilaterally decreased

TABLE 34.3. BASAL GANGLIA LESIONS ASSOCIATED WITH BEHAVIORAL DISORDERS[a]

Author Reference	Lesion Site	Etiology	Behavior
LaPlane et al. (107)	Bilateral lentiform (putamen and globus pallidus) lesions	Unknown	Inertia, apathy
Strub (148)	Bilateral globus pallidus lesion	Hemorrhage	Apathy, depression, loss of motivation
Trautner et al. (149)	Bilateral basal ganglia lesions, caudate lesions	Calcifications	Mood disorder, obsessive-compulsive disorder
Bejjani et al. (152)	Left substantia nigra lesion	High frequency stimulation	Acute depression
Steffens et al. (154)	Substantia nigra lesion	Cerebrovascular accident	Depression
McKee et al. (155)	Bilateral medial substantia nigra, pars reticulata lesions	Occlusion of paramedian branches of rostral basilar artery	Peduncular hallucinosis
Demirkol et al. (151)	Bilateral globus pallidus lesions	Unknown	Aggressive and compulsive behavior
Joseph (163)	Frontal lobe lesions	Trauma	Mania, depression, obsessive-compulsive disorder, confabulation, perseveration

[a]For additional review, see Kwak and Jankovic (104).

caudate volume by volumetric computed tomography in patients with known OCD.

Attention Deficit Disorder

Attention deficit disorder (ADD) is one of the most frequent comorbidities in patients with TS, but it is by no means limited to this disorder. PD patients and patients with atypical parkinsonism also may manifest features of ADD, which may be linked to executive dysfunction. Executive function includes the ability to monitor, anticipate, inhibit, and thus control mental processing, whereas impulsivity, distractibility, difficulty maintaining or selecting activities (sets), and lack of self-concern are viewed as evidence of executive dysfunction. D'Esposito and Grossman (110) describe executive function as a medium by which the "highest order of cognitive abilities" is captured and subserves cognitive abilities such as "sustained attention, fluency, flexibility of thought in the generation of solutions to novel problems, planning and regulation of adaptive and goal directed behavior." In patients with parkinsonism, inflexibility of thought and action to novel events and ideas is one of the most characteristic behavioral disorders. These patients often can engage in activities initiated and sustained by others, but they are incapable of doing either one on their own because of apathy and lack of initiative. Auto-activation deficit, also known as "psychic akinesia," is characterized by a deficit in spontaneous activation of mental processing that can be dramatically reversed by external simulation, similar to "kinesia paradoxica" (111). This is in contrast to the abulia observed in patients with frontal lobe lesions. Such executive impairment may be construed by some as ADD. Because basal ganglia damage can result in "psychic akinesia," a disconnection of input from output was proposed by Brown and Marsden (5) in which "neither thought nor sensory information are linked to mental or physical action." They further speculated that a type of

"focused attention" was necessary for the automatic binding of input to output. It is believed that attention exists at a subconscious level and is only brought into consciousness during taxing episodes in which an individual may feel "superattentive" or have a sense of "focused concentration." Further, it is believed that such attention is required in order to set into motion the operation of a particular motor plan that otherwise would be under automatic control, such as walking. This disruption manifests itself clearly in patients with PD in whom paradoxical kinesis is observed in the face of danger, for example, "fleeing from an oncoming car" at normal or increased speed yet being unable to walk under normal circumstances. The same is seen in patients who talk or walk during sleep. It appears that the basal ganglia are responsible for facilitating the selection and grouping of neurons required for the execution of an intended movement by focusing attention to those things relevant for carrying out that one function. If there is a disruption in the system, motor processing takes longer and results in bradykinesia, which is a salient feature of parkinsonism. Similarly, physiologic studies and recordings in PD patients during stereotactic surgery have provided evidence that the basal ganglia play a crucial role in the initiation, preparation, and suppression of inappropriate actions (112). These studies found that the basal ganglia become active only when a stimulus is attended or a movement is volitionally implemented. There appears to be a segregated group of neurons responsible for assessing actions and carrying out the motor aspect of the action. There is also an inhibitory, opposing neuronal mechanism that is activated when initiating and suppressing inappropriate actions. Finally, when attention to a task is required, a separate group in the striatum mediates this preparatory stage. In a study of parkinsonian and normal primates, it was observed that the "breakdown of the independent processing is the hallmark of PD" because the normal dopaminergic system supports segregation of the various functional subcircuits (113).

ADHD is one of the most common childhood behavioral disorders; it is estimated to affect 5% to 8% of boys and 2% to 4% of girls (114,115). About half of TS patients are speculated to have evidence of ADHD (82,116). Furthermore, attention disturbances have been observed commonly in children with various autistic disorders, including Asperger syndrome (117,118). Aylward et al. (119) found that children with ADHD had significantly smaller globus pallidus volumes, particularly in the left side, compared to age-matched controls. In another volumetric MRI study, Frederickson et al. (120) found evidence of smaller gray matter volumes in the left frontal lobes of patients with TS, further supporting the findings of loss of normal left greater than right asymmetry. In one study of ADHD, researchers found increased frontal activation and reduced striatal activation via functional MRI while the patients were performing various tasks. However, after treatment with methylphenidate, they noticed increased striatal function while the patients were performing the same tasks, presumably because of greater ability to concentrate and focus on the task at hand (121).

Aggressive and Self-Injurious Behavior

Aggressive behavior is seen commonly in patients with TS. It may be manifested as oppositional, argumentative, defiant disorder, and even violent. Comings and Comings (122) noted that 61% of children with TS in a cohort of 250 school-aged children had significant problems with violence and anger. Aggression may be directed by the patients not only to others but also to themselves. Self-injurious behavior (SIB) consists of repetitive behaviors that can be viewed as extreme cases of stereotypies with harmful consequences. SIB is a well-recognized symptom of TS (123). TS-related SIB includes biting the lips, picking at sores, poking at the eye, banging the head, and even carving words into the skin (123). Such SIB seems to be highly correlated with the severity of obsessive personality, hostility, and aggression (124), and even with violent and criminal behavior (71).

SIB is most frequently present in patients with various mental and psychiatric disorders. One study of patients with psychiatric disorders reports that 8% of patients had SIB and 6.5% had stereotyped behavior (125). The incidence appears to be higher in males than in females, particularly among patients with handicaps (126). Studies of institutionalized individuals have shown a prevalence between 10% to 15%; the rate is as high as 38% in patients in public facilities but is between 1% and 2.5% in mentally retarded individuals living outside institutions (127). In addition to TS, other movement disorders associated with SIB include Lesch-Nyhan syndrome and neuroacanthocytosis (84,128). Lesch-Nyhan syndrome is an X-linked recessive disease caused by a defect in hypoxanthine-guanine phosphoribosyltransferase (HPRT) gene. The enzymatic defect in this condition results in uric acid accumulation in all body fluid compartments, as well as gout secondary to urate crystal deposition in the joints. Self-mutilating behavior, which is the hallmark of this condition, begins in the second year of life. Lip biting is the most distressful of these behaviors, but they also may include biting of other body parts. Neuroacanthocytosis is a rare syndrome that presents with a variety of behaviors, including parkinsonism, chorea, tics, dystonia, and biting of lips and cheeks. In addition to these movement and behavioral symptoms, patients with this familial disorder have amyotrophy and an elevated creatine kinase level.

SYDENHAM CHOREA

The term chorea is derived from the Greek *chorea* meaning *dance*. This condition, formerly known as St. Vitus dance or chorea minor, is a brief, jerky, unpredictable movement that moves from one body part to another in random fashion. In SC, the movements occur as a consequence of an immune response to group A β-hemolyic streptococcal infection with neuronal antibodies directed against the basal ganglia. In 50 consecutive patients with rheumatic fever, 26% developed chorea, but arthritis was more frequent in patients without chorea (84%) than in those with chorea (31%) (129). Although the majority of patients have bilateral involvement, the distribution of chorea usually is asymmetric, and pure hemichorea can be seen in 20% of all Sydenham patients. Neurobehavioral symptoms, such as irritability, emotional lability, obsessive-compulsive behavior, attention deficit, and anxiety, usually begin within 2 to 4 weeks after the onset of the choreic movements. The disorder tends to spontaneously resolve in 3 to 4 months, but it may persist in half of the patients during 3-year follow-up (130). Serial brain single photon emission computed tomographic images show hyperperfusion in the regions of the striatum and thalamus bilaterally during the stage of active chorea. Perfusion returns to baseline levels after the chorea resolves (131).

Several reports have drawn attention to the association between SC and OCD (132–136). Swedo et al. (133) found that patients with SC had significantly more OCD symptoms than patients who were afflicted with only rheumatic fever. In a prospective 6-month study of obsessive-compulsive symptoms in a total of 50 children and adolescents with rheumatic fever with (n = 30) and without (n = 20) chorea, Asbahr et al. (135) found that the onset of obsessive-compulsive behavior occurred either at the same time as or after the onset of SC in over 70% of the 30 patients who progressed to develop SC. Obsessive-compulsive symptoms previously have been associated with SC but not with rheumatic fever without chorea, which provides further evidence of basal ganglia involvement in both chorea and OCD.

A number of studies have drawn attention to a possible link between exacerbation of tics and TS symptoms preceded by streptococcal infection. This was confirmed by antistreptolysin O (ASO) titers (which peak 3–6 weeks after acute infection) and anti-DNAse B antibodies (which peak at 6–8 weeks) (137). Higher titers of antineuronal antibodies directed against putamen have been demonstrated in TS children and adults compared to controls, but these higher values did not correlate with the severity of symptoms (138). One study found higher antibody titers in ADHD patients with larger volumes of the putamen and globus pallidus nuclei (139). The occurrence or exacerbation of tics and OCD after streptococcal infection has been known as pediatric autoimmune neuropsychiatric disorders associated with streptococcal infections (PANDAS) (140). However, this concept is still highly controversial and is not accepted by many experts in the field.

BEHAVIORAL AND CLINICAL SYNDROMES ASSOCIATED WITH FOCAL LESIONS

Ventral Striatum

The concept of ventral striatum originally was used to describe the extension of the basal ganglia into the olfactory tubercle (141). The ventral striatum includes the nucleus accumbens, which has been closely connected to reward and motivation behaviors (142). Studies of rats with lesions of the nucleus accumbens core, the brain region noted for reward and reinforcement, showed that the rats preferred small immediate rewards over larger delayed rewards (143). In addition to the ventromedial prefrontal cortices, lesions in the amygdala have been known to result in altered decision-making processes and a disregard for future consequences (144). Neuropathologic changes in the ventral striatum, caudate nucleus, and mediodorsal nucleus of the thalamus have been associated with catatonia (145), immobility, mutism, and withdrawal (refusal to eat or drink). A wide array of additional behavioral and motor disorders have been described, including staring, rigidity, posturing, negativism, waxy flexibility (or catalepsy), echophenomena, verbigeration, mannerisms, grimacing, increased muscular tension, and other findings on neurologic examination. We described a patient with catatonic symptoms, including mutism, but instead of immobility she had bilateral ballism associated with basal ganglia calcification (146).

Lentiform Nucleus (Putamen and Globus Pallidus) and Caudate Nucleus

Many reports have described focal lesions in the globus pallidus leading to specific neuropsychiatric disturbances such as apathy and depression. One report documented bilateral lentiform lesions resulting in behavioral inertia with only mild extrapyramidal findings (147). Another case illustrating involvement of the lentiform nucleus in the development of depressed mood and affect is that of a previously asymptomatic 60-year-old man who, after developing bi-

lateral globus pallidus hemorrhages, suddenly experienced severe apathy and loss of motivation (148). In a separate series, five patients were reported to have mood disorders secondary to bilateral calcifications of the basal ganglia predominantly in the region of the pallidum (149). It is important to note that cognitive impairments (slow incremental learning of nonmotor tasks, loss of drive, loss of flexibility of thought, poor perception) appear earlier than motor deficits (22,150). Demirkol et al. (151) presented the case of a 17-year-old boy with history of stuttering, aggressive behavior, tics (tongue thrusting, shoulder shrugging, arm and leg jerking), and compulsive behaviors (e.g., turning on and off lights) in whom radiologic evidence revealed bilateral lesions restricted to the globus pallidus.

Substantia Nigra

One of the most compelling pieces of evidence that the substantia nigra (SN) is involved in regulation of mood comes from a report of an acute bout of depression that occurred after stimulation of the midbrain nucleus (152). In this report, a 65-year-old woman with advanced PD was treated with continuous high-frequency (130 Hz) stimulation of the left SN. After a 7-minute stimulation, she developed feelings of sadness, emptiness, fatigue, guilt, worthlessness, and morbid preoccupation with death. The episodes of depression occurred almost immediately (within 5 seconds) after onset of stimulation, and the symptoms were duplicated each time stimulation was carried out up to 8 months later. Because similar stimulation in the subthalamic nucleus did not produce symptoms of depression, the authors concluded that depression was due to a site-specific stimulus. When the appropriate target is stimulated, high-frequency deep brain stimulation usually does not produce meaningful cognitive or behavioral changes. This procedure currently is being investigated as a possible treatment for a variety of neuropsychiatric disorders, including OCD and schizophrenia (153). Results of other studies have supported the notion that SN is involved in control of mood. In a sample of 3,660 patients older than 65 years, lesions in the SN (smaller than 3 mm) noted on MRI corresponded statistically with a higher frequency of depressive symptoms (154). Other reports have directly linked behavioral dysfunction to lesions of the SN. One such report involves an 83-year-old man with peduncular hallucinosis found to have bilateral symmetric lesions in the medial SN pars reticulata (155). The lesions were the result of occlusion of paramedian branches of the rostral basilar artery.

Subthalamic Nucleus

The behavioral correlates of subthalamic lesions are not well understood, but insight might be gained from reports of uncontrollable laughter and hilarity in two patients with PD treated with acute subthalamic deep brain stimulation (156).

Thalamus

The thalamus is not considered part of the basal ganglia, but because of its connection to the basal ganglia, a lesion or dysfunction of the thalamus may have an influence on the basal ganglia and on behavior. Several reports in the literature support the notion that disruptions in the anterior and dorsomedial thalamus result in dysfunction in mood, language, perception, memory, and drive (157–161). One example is the case of a 56-year-old man who exhibited profound apathy and amnesia after suffering bilateral thalamic infarcts in the paramedian arterial distribution (158). He also showed significant disturbance in learning both verbal and nonverbal information, and he developed a destructive/aggressive personality. Other reports of confusion, apathy, and memory loss have been attributed to disruption of fibers from the anterior thalamic peduncle connecting the thalamus to the prefrontal cortex. Interruption of communication between the inferior thalamic peduncle and anterior thalamic nucleus with insula, amygdala, and medial temporal lobes may lead to similar deficits (162).

In summary, basal ganglia and their connections are being increasingly recognized as playing a critical role not only in control of motor function but also in human behavior. Better understanding of the basal ganglia circuitry in various neurobehavioral and psychiatric disorders undoubtedly will lead to improved behavioral, pharmacologic, and surgical management of patients affected by these disorders.

REFERENCES

1. Leckman J, Cohen D, Goetz C, et al. Tourette syndrome: pieces of the puzzle. In: Cohen DJ, Jankovic J, Goetz CG, eds. *Tourette syndrome. Advances in neurology, volume 85.* Philadelphia: Lippincott Williams & Wilkins, 2001:369–390.
2. Roy EA. Apraxia in disease of the basal ganglia. *Mov Disord* 2000;15:598–600.
3. Alexander GE, Crutcher MD. Functional architecture of basal ganglia circuits: neural substrates of parallel processing. *Trends Neurosci* 1990;13:266–271.
4. Hoover JE, Strick PL. Multiple output channels in the basal ganglia. *Science* 1993;259:819–21.
5. Brown P, Marsden CD. What do the basal ganglia do? *Lancet* 1998;351:1801–1804.
6. Parent A, Cicchetti F. The current model of basal ganglia organization under scrutiny. *Mov Disord* 1998;13:199–202.
7. Middleton FA, Strick PL. Basal ganglia output and cognition: evidence from anatomical, behavioral, and clinical studies. *Brain Cogn* 2000;42:183–200.
8. Alexander GE, Crutcher MD, De Long MR. Basal ganglia-thalamocortical circuits: parallel substrates for motor, oculomotor, "prefrontal" and "limbic" functions. *Prog Brain Res* 1990;85:119–146.
9. Palumbo D, Maughan A, Kurlan R. Hypothesis III: Tourette syndrome is only one of several causes of a developmental basal ganglia syndrome. *Arch Neurol* 1997;54:475–483.
10. Folstein SE, Brandt J, Folstein MF. The subcortical dementia of Huntington's disease. In: Cummings J, ed. *Subcortical dementia.* Oxford, UK: Oxford University Press, 1990.

11. Brandt J. Cognitive impairment in Huntington's disease: insights into the neuropsychology of the striatum. In: Buller F, Grafman J, eds. *Handbook of neuropsychology, volume 5.* Amsterdam: Elsevier Science 1991:241–264.
12. Lange KW, Sahakian BJ, Quinn NP et al. Comparison of executive and visuospatial memory function in Huntington's disease and dementia of Alzheimer's type matched for degree of dementia. *J Neurol Neurosurg Psychiatry* 1995;58:598–606.
13. Mink JW. Neurobiology of basal ganglia circuits in Tourette syndrome: faulty inhibition of unwanted motor patterns? In: Cohen DJ, Jankovic J, Goetz CG, eds. *Tourette syndrome. Advances in neurology, volume 85.* Philadelphia: Lippincott Williams & Wilkins, 2001:113–122.
14. Brown RG, Pluck G. Negative symptoms: the "pathology" of motivation and goal-directed behaviour. *Trends Neurosci* 2000;23:412–417.
15. Graybiel AM, Canales JJ. The neurobiology of repetitive behaviors: clues to the neurobiology of Tourette syndrome. In: Cohen DJ, Jankovic J, Goetz CG, eds. *Tourette syndrome. Advances in neurology, volume 85.* Philadelphia: Lippincott Williams & Wilkins, 2001:123–132.
16. Litvan I, Paulsen JS, Mega MS, et al. Neuropsychiatric assessment of patients with hyperkinetic and hypokinetic movement disorders. *Arch Neurol* 1998;55:1313–1319.
17. Ring HA, Serra-Mestres J. Neuropsychiatry of the basal ganglia. *J Neurol Neurosurg Psychiatry* 2002 72:12–21.
18. Jankovic J. Parkinsonian syndromes. In: Kurlan R, ed. *Treatment of movement disorders.* Philadelphia: JB Lippincott Company, 1995:95–114.
19. Cooper JA, Sagar HJ, Doherty SM, et al. Different effects of dopaminergic and anticholinergic therapies on cognitive and motor function in Parkinson's disease. A follow-up study of untreated patients. *Brain* 1992;115:1701–1725.
20. Sharpe MH. Auditory attention in early Parkinson's disease: an impairment in focused attention. *Neuropsychologia* 1992;30:101–106.
21. Lange KW, Robbins TW, Marsden CD, et al. L-dopa withdrawal in Parkinson's disease selectively impairs cognitive performance in tests sensitive to frontal lobe dysfunction. *Psychopharmacology* 1992;107:394–404.
22. Brown LL, Schneider JS, Lidsky TI. Sensory and cognitive functions of the basal ganglia. *Curr Opin Neurobiol* 1997;7:157–163.
23. Ebersbach G, Trottenburg T, Hattig H, et al. Directional bias of initial visual exploration. A symptom of neglect in Parkinson's disease. *Brain* 1996;119:79–87.
24. Barrett AM, Crucian GP, Schwartz R, et al. Seeing trees but not the forest. Limited perception of large configurations in PD. *Neurology* 2001;56:724–729.
25. Levy ML, Cummings JL. Parkinson's disease and parkinsonism. In: Joseph AB, Young RR, eds. *Movement disorders in neurology and neuropsychiatry,* 2nd ed. Cambridge, Massachusetts: Blackwell Science, 1999:171–179.
26. Dubois B, Pillon B. Do cognitive changes of Parkinson's disease result from dopamine depletion? *J Neural Transm* 1995;45[Suppl]:27–34.
27. Robbins TW, James M, Owen AM, et al. Cognitive deficits in progressive supranuclear palsy, Parkinson's disease, and multisystem atrophy in tests sensitive to frontal lobe dysfunction. *J Neurol Neurosurg Psychiatry* 1994;57:79–88.
28. Partiot A, Verin M, Pillon B, et al. Delayed response task in basal ganglia lesions in man: further evidence for a striato-frontal cooperation in behavioral adaptation. *Neuropsychologia* 1996;34:709–721.
29. Sadato N, Campbell G, Ibanez V, et al. Complexity affects

regional cerebral blood flow change during sequential finger movements. *J Neurosci* 1996;16:2691–2700.

30. Cummings JL. Depression and Parkinson's disease. A review. *Am J Psychiatry* 1992;149:443–454.

31. Tandberg E, Larsen JP, Aarsland D, et al. The occurrence of depression in Parkinson's disease. *Arch Neurol* 1996;53:175–179.

32. Aarsland D, Larsen JP, Lim NG, et al. Range of neuropsychiatric disturbances in patients with Parkinson's disease. *J Neurol Neurosurg Psychiatry* 1999;67:492–496.

33. Aarsland D, Larsen JP, Tandberg E, et al. Predictors of nursing home placement in Parkinson's disease: a population-based, prospective study. *J Am Geriatr Soc* 2000;48:1–5.

34. Fénelon G, Mahieux F, Huon R, et al. Hallucinations in Parkinson's disease. Prevalence, phenomenology and risk factors. *Brain* 2000;123:733–745.

35. Barnes J, David AS. Visual hallucinations in Parkinson's disease: a review and phenomenological survey. *J Neurol Neurosurg Psychiatry* 2001;70:727–733.

36. Holroyd S, Currie L, Wooten GF. Prospective study of hallucinations and delusions in Parkinson's disease. *J Neurol Neurosurg Psychiatry* 2001;70:734–738.

37. Goetz CG, Leurgans S, Pappert EJ, et al. Prospective longitudinal assessment of hallucinations in Parkinson's disease. *Neurology* 2001;57:2078–2082.

38. Arnulf I, Bonnet A-M, Damier P, et al. Hallucinations, REM sleep, and Parkinson's disease. A medical hypothesis. *Neurology* 2000;55:281–288.

39. Aarsland D, Andersen K, Larsen JP, et al. Risk of dementia in Parkinson's disease: a community-based, prospective study. *Neurology* 2001;56:730–736.

40. Ferini-Strambi L, Zucconi M. REM sleep behavior disorder. *Clin Neurophysiol* 2000;111[Suppl 2]:S136–S140.

41. Olson EJ, Boeve BF, Silber MH. Rapid eye movement sleep behaviour disorder: demographic, clinical and laboratory findings in 93 cases. *Brain* 2000;123:331–339.

42. Plazzi G, Corsini R, Provini F, et al. REM sleep behavior disorders in multiple system atrophy. *Neurology* 1997;48:1094–1097.

43. Comella CL, Nardine TM, Diederich NJ, et al. Sleep-related violence, injury, and REM sleep behavior disorder in Parkinson's disease. *Neurology* 1998;51:526–529.

44. Wetter TC, Collado-Seidel V, Pollmacher T, et al. Sleep and periodic leg movement patterns in drug-free patients with Parkinson's disease and multiple system atrophy. *Sleep* 2000;23:361–367.

45. Jagust WJ, Reed BR, Martin EM, et al. Cognitive function and regional cerebral blood flow in Parkinson's disease. *Brain* 1992;115:521–537.

46. Mayberg HS, Starkstein SE, Sadzot B, et al. Selective hypometabolism in the inferior frontal lobe in depressed patients with Parkinson's disease. *Ann Neurol* 1990;28:57–64.

47. Ring HA, Bench CJ, Trimble MR, et al. Depression in Parkinson's disease: a positron emission study. *Br J Psychiatry* 1994;165:333–339.

48. Hu MTM, Taylor-Robinson SD, Chaudhuri KR, et al. Cortical dysfunction in non-demented Parkinson's disease patients. A combined 31P-MRS and 18FDG-PET study. *Brain* 2000;123:340–352.

49. Rinne JO, Portin R, Ruottinen H, et al. Cognitive impairment and the brain dopaminergic system in Parkinson disease. [18F]fluorodopa positron emission tomographic study. *Arch Neurol* 2000;57:470–475.

50. Starkstein SE, Petracca G, Chemerinski E, et al. Prevalence and correlates of parkinsonism in patients with primary depression. *Neurology* 2001;57:553–555.

51. Marder K, Cote L, Tang, et al. The risk and predictive factors associated with dementia in Parkinson's disease. In: Korczyn A, ed. *Dementia in Parkinson's disease.* Bologna, Italy: Monduzzi Editore, 1994:51–54.

52. Marder K, Tang MX, Alfaro B, et al. Risk of Alzheimer's disease in relatives of Parkinson's disease patients with and without dementia. *Neurology* 1999;52:719–724.

53. Jacobs DM, Marder K, Cote LJ, et al. Neuropsychological characteristics of preclinical dementia in Parkinson's disease. *Neurology* 1995;45:1691–1696.

54. Jankovic J, McDermott M, Carter J, et al. Variable expression of Parkinson's disease: a base-line analysis of the DATATOP cohort. *Neurology* 1990;41:1529–1534.

55. Levy G, Tang M-X, Cote LJ. Motor impairment in PD. Relationship to incident dementia and age. *Neurology* 2000;55:539–544.

56. Dujardin K, Defbvre L, Grunberg C, et al. Memory and executive function in sporadic and familial Parkinson's disease. *Brain* 2001;124:389–398.

57. Litvan I. Parkinsonism-dementia syndromes. In: Jankovic J, Tolosa E, eds. *Parkinson's disease and movement disorders,* 3rd ed. Baltimore: Williams & Wilkins, 1998:819–836.

58. Hohl U, Tiraboschi P, Hansen LA, et al. Diagnostic accuracy of dementia with Lewy bodies. *Arch Neurol* 2000;57:347–351.

59. Kertesz A, Kawarai T, Rogaeva E, et al. Familial frontotemporal dementia with ubiquitin-positive, tau-negative inclusions. *Neurology* 2000;54:818–827.

60. Kurlan R, Richard IH, Papka M, et al. Movement disorders in Alzheimer's disease: more rigidity of definitions needed. *Mov Disord* 2000;15:24–29.

61. Cattaneo E, Rigamonti D, Coffredo D, et al. Loss of normal huntingtin function: new developments in Huntington's disease research. *Trends Neurosci* 2001;24:182–188.

62. Cummings JL. Behavioral and psychiatric symptoms associated with Huntington's disease. In: Weiner WJ, Lang AE, eds. *Behavioral neurology of movement disorders. Advances in neurology, volume 65.* New York: Raven Press, 1995:179.

63. Folstein SE, Peyser CE, Starkstein SE, et al. Subcortical triad of Huntington's disease: a model for neuropathology of depression, dementia, and dyskinesia. In: Carroll BJ, Barrett JE, eds. *Psychopathology and the brain.* New York: Raven Press, 1991.

64. Woodcock JH. Behavioral aspects of Huntington's disease. In: Joseph AB, Young RR, eds. *Movement disorders in neurology and neuropsychiatry,* 2nd ed. Cambridge, Massachusetts: Blackwell Science, 1999:155–160.

65. Paulsen JS, Ready RE, Hamilton JM, et al. Neuropsychiatric aspects of Huntington's disease. *J Neurol Neurosurg Psychiatry* 2001;71:310–314.

66. Burns A, Folstein S, Brandt J, et al. Clinical assessment of irritability, aggression, and apathy in Huntington and Alzheimer disease. *J Nerv Ment Dis* 1990;178:20.

67. Cummings JL. Anatomic and behavioral aspects of frontal-subcortical circuits. *Ann N Y Acad Sci* 1995;769:1–13.

68. Folstein SE. The psychopathology of Huntington's disease. In: McHugh PR, McKusick VA, eds. *Genes, brain, and behavior.* New York: Raven Press, 1991:181.

69. Salloway S, Cummings J. Subcortical structures and neuropsychiatric illness. *Neuroscientist* 1996;2:66–75.

70. Jensen P, Fenger K, Bowling TG, et al. Crime in Huntington's disease: a study of registered offences among patients, relatives, and controls. *J Neurol Neurosurg Psychiatry* 1998;65:467–471.

71. Brower M, Price B. Neuropsychiatry of frontal lobe

dysfunction in violent and criminal behavior: a critical review. *J Neurol Neurosurg Psychiatry* 2001;71:720–726.

72. Haddad MS, Cummings JL. Huntington's disease. In: Miguel EC, Rauch SL, Leckman JF, eds. *The psychiatric clinics of North America: neuropsychiatry of the basal ganglia.* Philadelphia: WB Saunders, 1997;20:791–808.

73. Vonsattel JPG. Huntington's disease neuropathology. In: Joseph AB, Young RR, eds. *Movement disorders in neurology and neuropsychiatry,* 2nd ed. Cambridge, Massachusetts: Blackwell Science, 1999:161–170.

74. Marques-Dias MJ, Mercadente MT, Tucker D, et al. Sydenham's chorea. In: Miguel EC, Rauch SL, Leckman JF, eds. *The psychiatric clinics of North America: neuropsychiatry of the basal ganglia.* Philadelphia: WB Saunders, 1997;20:809–820.

75. Cummings JL, Cunningham K. Obsessive-compulsive disorder in Huntington's disease. *Biol Psychiatry* 1992;31:263.

76. Jankovic J, Ashizawa T. Tourettism associated with Huntington's disease. *Mov Disord* 1995;10:103–105.

77. Paulsen JS, Zhao H, Staout JC, et al. Clinical markers of early disease in persons near onset of Huntington's disease. *Neurology* 2001;57:658–662.

78. Pillon B, Dubois B, Ploska A, et al. Severity and specificity of cognitive impairment in Alzheimer's, Huntington's, and Parkinson's diseases and progressive supranuclear palsy. *Neurology* 1991;41:634–643.

79. Bamford KA, Caine ED, Kido DK, et al. A prospective evaluation of cognitive decline in early Huntington's disease. *Neurology* 1995;45:1867–1873.

80. de Boo GM, Tibben A, Lanser JB, et al. Early cognitive and motor symptoms in identified carriers of the gene for Huntington's disease. *Arch Neurol* 1997;54:1353–1357.

81. Hahn-Barma V, Deweer B, Dürr A, et al. Are cognitive changes the first symptoms of Huntington's disease? A study of gene carriers. *J Neurol Neurosurg Psychiatry* 1998;64:172–177.

82. Jankovic J. Tourette's syndrome. *N Engl J Med* 2001;345:1184–1192.

83. Krauss JK, Jankovic J. Tics secondary to craniocerebral trauma. *Mov Disord* 1997;12:776–782.

84. Jankovic J. Differential diagnosis and etiology of tics. In: Cohen DJ, Jankovic J, Goetz CG, eds. *Tourette syndrome. Advances in neurology, volume 85.* Philadelphia: Lippincott Williams & Wilkins, 2001:15–29.

85. Hanna PA, Janjua FN, Contant CF, et al. Bilineal transmission in Tourette syndrome. *Neurology* 1999;53:813–818.

86. Stein D. The neurobiology of obsessive-compulsive disorder. *Neuroscientist* 1996;2:300–305.

87. Leckman JF, Walker DE, Goodman WK, et al. "Just right" perceptions associated with compulsive behavior in Tourette's syndrome. *Am J Psychiatry* 1994;151:675–680.

88. Eapen V, Robertson MM, Alsobrook JP, et al. Obsessive-compulsive symptoms in Gilles de la Tourette syndrome and obsessive-compulsive: differences by diagnosis and family history. *Am J Med Genet* 1997;74:432–438.

89. Baron-Cohen S, Cross P, Crowson M, et al. Can children with Gilles de la Tourette syndrome edit their intentions? *Psychol Med* 1994;24:29–40.

90. Morris HR, Thacker AJ, Newman PK, et al. Sign language tics in a prelingually deaf man. *Mov Disord* 2000;15:318–320.

91. *Diagnostic and statistical manual of mental disorders, fourth edition.* Washington, DC: American Psychiatric Association, 1994:100–105.

92. Saba PR, Dastur K, Keshavan MS, et al. Obsessive-compulsive disorder, Tourette's syndrome, and basal ganglia pathology on MRI. *J Neuropsychiatry Clin Neurosci* 1998;10:116–117.

93. Mc Guire PK. The brain in obsessive-compulsive disorder. *J Neurol Neurosurg Psychiatry* 1995;59:457–459.

94. Swoboda KJ, Jenike MA. Frontal abnormalities in a patient with obsessive-compulsive disorder. The role of structural lesions in obsessive-compulsive behavior. *Neurology* 1995;45:2130–2134.

95. Micallef J, Blin O. Neurobiology and clinical pharmacology of obsessive-compulsive disorder. *Clin Neuropharmacol* 2001;24:191–207.

96. Rauch SL, Whalen PJ, Curran T, et al. Probing striato-thalamic function in obsessive-compulsive disorder and Tourette syndrome using neuroimaging methods. In: Cohen DJ, Jankovic J, Goetz CG, eds. *Tourette syndrome. Advances in neurology, volume 85.* Philadelphia: Lippincott Williams & Wilkins, 2001:207–224.

97. Peterson BS, Skudlarski P, Anderson AW, et al. A functional magnetic resonance imaging study of tic suppression in Tourette syndrome. *Arch Gen Psychiatry* 1998;54:326–333.

98. Greenberg BD, Ziemann U, Harmon A, et al. Decreased neuronal inhibition in cerebral cortex in obsessive-compulsive disorder on trans cranial magnetic stimulation. *Lancet* 1998;352:881–882.

99. Ziemann U, Paulus W, Rothenberger A. Decreased motor inhibition in Tourette's disorder: evidence from transcranial magnetic stimulation. *Am J Psychiatry* 1997;154:1277–1284.

100. Weeks RA, Turjanski N, Brooks DJ. Tourette's syndrome: a disorder of cingulate and orbitofrontal function? *Q J Med* 1996;89:401–408.

101. Ikonomidou C, Bosch F, Miksa M, et al. Blockade of NMDA receptors and apoptotic neurodegeneration in the developing brain. *Science* 1999;283:70–74.

102. Leonard H. Tourette syndrome and obsessive compulsive disorder. In: Chase T, Friedhoff A, Cohen DJ, eds. *Tourette syndrome. Advances in neurology, volume 85.* New York, Raven Press, 1992:83–94.

103. Berthier ML, Kulisevsky J, Gironell A, et al. Obsessive-compulsive disorder associated with brain lesions: clinical phenomenology, cognitive function, and anatomic correlates. *Neurology* 1996;47:353–361.

104. Kwak C, Jankovic J. Tourettism and dystonia after subcortical stroke. *Mov Disord* 2002;17:821–825.

105. Alegret M, Junque C, Valldeoriola F, et al. Obsessive-compulsive symptoms in Parkinson's disease. *J Neurol Neurosurg Psychiatry* 2001;70:394–396.

106. Gschwandtner U, Aston J, Renaud S, et al. Pathologic gambling in patients with Parkinson's disease. *Clin Neuropharmacol* 2001;24:170–172.

107. LaPlane D, Lavasseur M, Pillon B, et al. Obsessive-compulsive and other behavioral changes with bilateral basal ganglia lesions. *Brain* 1989;112:699–725.

108. Escalona PR, Adair JC, Roberts BB, et al. Obsessive-compulsive disorder following bilateral globus pallidus infarction. *Biol Psychiatry* 1997;42:410–412.

109. Baxter LR Jr, Schwartz JM, Bergman KS, et al. Caudate glucose metabolic rate changes with both drug and behavior therapy for obsessive-compulsive disorder. *Arch Gen Psychiatry* 1992;49:681–689.

110. D'Esposito M, Grossman M. The psychological basis of executive function and working memory. *Neuroscientist* 1996;2:345–352.

111. Laplane D, Dubois B. Auto-activation deficit: a basal ganglia related syndrome. *Mov Disord* 2001;16:810–814.

112. Kropotov JD, Etlinger SC. Selection of actions in the basal ganglia-thalamocortical circuits: review and model. *Int J Psychophysiol* 1999;31:197–217.

113. Bergman H, Feingold A, Nini A, et al. The physiological aspects of information processing in the basal ganglia of normal and parkinsonian primates. *Trends Neurosci* 1998;21:32–38.

114. Barkley RA. A critique of current diagnostic criteria for attention-deficit hyperactivity disorder: clinical and research implications. *J Dev Behav Pediatr* 1990;11:343–352.

115. Goldman LS, Genel M, Bezman RJ, et al. Diagnosis and treatment of attention-deficit/hyperactivity disorder in children and adolescents. *JAMA* 1998;279:1100–1107.

116. Comings DE, Himes JA, Comings BG. An epidemiologic study of Tourette's syndrome in a single school district. *J Clin Psychiatry* 1990;51:463–469.

117. Saint-Cyr JA, Taylor AE, Nicholson K. Behavior and the basal ganglia. In: Weiner WJ, Lang AE, eds. *Behavioral neurology of movement disorders. Advances in neurology, volume 65.* New York: Raven Press, 1995:1–28.

118. Ringman JM, Jankovic J. The occurrence of tics in Asperger syndrome and autistic disorder. *J Child Neurol* 2000;15:394–400.

119. Aylward EH, Reiss AL, Reader MJ, et al. Basal ganglia volumes in children with attention-deficit hyperactivity disorder. *J Child Neurol* 1996;11:112–115.

120. Frederickson KA, Cutting LE, Kates WR, et al. Disproportionate increases of white matter in right frontal lobe in Tourette syndrome. *Neurology* 2002;58:85–89.

121. Vaidya C, Austin G, Kirkorian G, et al. Selective effects of methylphenidate in attention deficit hyperactivity disorder: a functional magnetic resonance study. *Proc Natl Acad Sci U S A* 1998;95:14494–14499.

122. Comings DE, Comings BG. Tourette syndrome: clinical and psychological aspects of 250 cases. *Am J Hum Genet* 1985;37:435–450.

123. Jankovic J, Sekula SL, Milas D. Dermatological manifestations of TS and OCD. *Arch Dermatol* 1998;134:113–114.

124. Maughan A, Palumbo D, Kurlan R. Comorbidity in Tourette syndrome: evidence of a developmental basal ganglia syndrome. Presented at the Eighth Annual New York State Office of Mental Health Research Conference, December 2, 1995, Albany, New York.

125. Bojahn J, Borthwick-Duffy SA, Jacobson JW. The association between psychiatric diagnosis and severe behavior problems in mental retardation. *Ann Clin Psychiatry* 1993;5:163–170.

126. Johnson S. Epidemiology. In: Luiselli JK, Matson JL, Singh NN, eds. *Self-injurious behavior: analysis, assessment, and treatment.* New York: Springer-Verlag, 1992:21–57.

127. Aman MG. Efficacy for psychotropic drugs for reducing self-injurious behavior in the developmental disabilities. *Ann Clin Psychiatry* 1993;5:171–188.

128. Robertson MM, Yakeley JW. Obsessive-compulsive disorder and self-injurious behavior. In: Kurlan R, ed. *Handbook of Tourette's syndrome and related tic and behavioral disorders.* New York: Marcel Dekker, 1993:45–88.

129. Cardoso F, Eduardo C, Silva AP, et al. Chorea in 50 consecutive patients with rheumatic fever. *Mov Disord* 1997;12:701–703.

130. Cardoso F, Vargas AP, Oliveira LD, et al. Persistent Sydenham's chorea. *Mov Disord* 1999;14:805–807.

131. Lee PH, Nam HS, Lee KY, et al. Serial brain SPECT images in a case of Sydenham chorea. *Arch Neurol* 1999;56:237–240.

132. Pauls D, Leckman J. The inheritance of Gilles de la Tourette's syndrome and associated behaviors. *N Engl J Med* 1986;315:933–997.

133. Swedo SE, Leonard HL, Schapiro MB, et al. Sydenham's chorea: physical and psychological symptoms of St. Vitus dance. *Pediatrics* 1993;91:706–713.

134. Moore DP. Neuropsychiatric aspects of Sydenham's chorea: a comprehensive review. *J Clin Psychiatry* 1996;57:407–414.

135. Asbahr FR, Negrao AB, Gentil V, et al. Obsessive-compulsive and related symptoms in children and adolescents with rheumatic fever with and without chorea: a prospective 6-month study. *Am J Psychiatry* 1998;155:1122–1124.

136. Murphy TK, Petitto JM, Voeller KK, et al. Obsessive compulsive disorder: is there an association with childhood streptococcal infections and altered immune function? *Semin Clin Neuropsychiatry* 2001;6:266–276.

137. Kurlan R. Tourette's syndrome and "PANDAS": will the relation bear out? *Neurology* 1998;50:1530–1534.

138. Singer HS, Giuliano JD, Hansen BH, et al. Antibodies against human putamen in children with Tourette syndrome. *Neurology* 1998;50:1618–1624.

139. Peterson BS, Leckman JF, Tucker D, et al. Preliminary findings of antistreptococcal titers and basal ganglia volumes in tic, obsessive-compulsive, and attention deficit/hyperactivity disorders. *Arch Gen Psychiatry* 2000;57:364–372.

140. Perlmutter SJ, Leitman SF, Garvey MA, et al. Therapeutic plasma exchange and intravenous immunoglobulin for obsessive-compulsive disorder and tic disorders in childhood. *Lancet* 1999;354:1153–58.

141. Haber SN, McFarland NR. The concept of the ventral striatum in nonhuman primates. *Ann N Y Acad Sci* 1999;877:33–48.

142. Schultz W, Tremblay L, Hollerman Jr. Reward prediction in primate basal ganglia and frontal cortex. *Rev Neuropharmacol* 1998;37:421–429.

143. Cardinal R, Pennicott D, Sugathapala C, et al. Impulsive choice induced in rats by lesions of the nucleus accumbens core. *Science* 2001;292:2499–2501.

144. Bechara A, Tranel D, Damasio H. Characterization of the decision-making deficit of patients with ventromedial prefrontal cortex lesions. *Brain* 2000;123:2189–2202.

145. Northoff G. Brain imaging and catatonia: current findings and a pathophysiological model. *CNS Spectrums* 2000;5:34–46.

146. Inbody S, Jankovic J. Hyperkinetic mutism: bilateral ballism and basal ganglia calcification. *Neurology* 1986;36:825–827.

147. La Plane D. Is loss of psychic self activation a heuristic concept? *Behav Neurol* 1990;3:27–38.

148. Strub RL. Frontal lobe syndrome in a patient with bilateral globus pallidus lesions. *Arch Neurol* 1989;46:1024–1027.

149. Trautner RJ, Cummings JL, Read SL, et al. Idiopathic basal ganglia calcifications and organic mood disorder. *Am J Psychiatry* 1988;145:350–353.

150. Schneider JS, Pope-Coleman A. Cognitive deficits precede motor deficits in a slowly progressing model of parkinsonism in the monkey. *Neurodegeneration* 1995;4:245–255.

151. Demirkol A, Erdem H, Inan L, et al. Bilateral globus pallidus lesions in a patient with Tourette syndrome and related disorders. *Biol Psychiatry* 1999;46:863–867.

152. Bejjani BP, Damier P, Arnulf I, et al. Transient acute depression induced by high frequency deep brain stimulation. *N Engl J Med* 1999:340:1475–1480.

153. Roth RM, Flashman LA, Saykin AJ, et al. Deep brain stimulation in neuropsychiatric disorders. *Curr Psychiatry Reports* 2001;3:366–372.

154. Steffens DC, Helms MJ, Krishnan KRR, et al. Cerebrovascular disease and depression symptoms in the cardiovascular health study. *Stroke* 1999;30:2159–2166.

155. McKee AC, Levine DN, Kowall NW, et al. Peduncular hallucinosis associated with isolated infarction of the substantia nigra pars reticulata. *Ann Neurol* 1990;27:500–504.

156. Krack P, Kumar R, Ardouin C, et al. Mirthful laughter induced by subthalamic nucleus stimulation. *Mov Disord* 2001;16:867–875.

157. Eslinger PJ, Warner GC, Grattan LM, et al. "Frontal lobe" utilization behavior associated with paramedian thalamic infarction. *Neurology* 1991;41:450–452.

158. Salloway S. Diagnosis and treatment of patients with frontal lobe syndromes. *J Neuropsychiatry Clin Neurosci* 1994;6:388–398.

159. Sandson TA, Daffner KR, Carvalho PA, et al. Frontal lobe dysfunction following infarction of the left-sided medial thalamus. *Arch Neurol* 1991;48:1300–1303.

160. Malamut BL, Graff-Radford N, Chawluk J, et al. Memory in a case of bilateral thalamic infarction. *Neurology* 1992;42:163–169.

161. Campbell J, Duffy J, Salloway S. Apathy and memory loss following bilateral anteromedial thalamic infarctions. *J Neuropsychiatry Clin Neurosci* 1993;5:446.

162. Tatemichi TK, Desmond DW, Prohovnik I, et al. Confusion and memory loss from capsular, genu infarction. *Neurology* 1992;42:1966–1979.

163. Joseph J. Frontal lobe psychopathology: mania, depression, confabulation, catatonia, perservation, obsessive compulsions, and schizophrenia. *Psychiatry* 1999;62[2]:138–172.

35

GILLES DE LA TOURETTE SYNDROME AND OBSESSIVE-COMPULSIVE DISORDER

VALSAMMA EAPEN
JESSICA W. YAKELEY
MARY MAY ROBERTSON

DEFINITION AND CLINICAL CHARACTERISTICS

The generally accepted diagnostic criteria for Gilles de la Tourette syndrome (GTS) are included in the *Diagnostic and Statistical Manual of Mental Disorders, Fourth Edition* (DSM-IV) of the American Psychiatric Association (1) and the 10th Edition of the *International Statistical Classification of Diseases and Related Health Problems* of the World Health Organization (2). The criteria include both multiple motor tics and one or more vocal tics that do not necessarily present concurrently. They occur many times a day (usually in bouts) and last for more than 1 year. The anatomic location, number, frequency, complexity, and severity of the tics classically change over time (1,2).

The first clear medical description of GTS was made in 1825, when Itard (3) reported the case of a French noblewoman, the Marquise de Dampierre, who developed the first symptoms of GTS at age 7 years. Georges Gilles de la Tourette (4), the prominent French neuropsychiatrist and pupil of Charcot, later described nine cases of GTS and emphasized the triad of multiple tics, coprolalia, and echolalia; this earned him eponymous fame. The first reported case of GTS in the United Kingdom may have been Mary Hall of Gadsden, who was reported in 1663 by William Drage (5). Dr. Samuel Johnson, the prominent eighteenth-century literary figure and genius, most likely was afflicted with GTS (6,7). Some investigators have suggested that Mozart had GTS, although the evidence is insufficient and disputed (8).

Obsessive-compulsive disorder (OCD) is characterized by recurrent obsessions or compulsions that are sufficiently severe to (a) cause marked distress that is time consuming or (b) significantly interfere with the person's normal routine, functioning, social activities, or relationships (1). Obsessions are recurrent ideas, thoughts, images, or impulses that enter the mind and are persistent, intrusive, and unwelcome.

Attempts are made to ignore or suppress these thoughts or to neutralize them with other thoughts or actions. The individual recognizes them as a product of his/her own mind. Compulsions are repetitive purposeful behaviors performed in response to an obsession and are designed to neutralize or prevent discomfort or a dreaded event or situation. However, the activity is excessive or is not connected realistically with what the activity is designed to prevent. The affected person recognizes that his/her behavior is unreasonable.

OCD has been described by authors for centuries. Since the Medieval Period, Latin terms such as obsession, compulsion, and impulsion have appeared in the European medical literature to describe OCD-like behavior and phenomena. In his comprehensive review of the conceptual history of OCD during the nineteenth century, Berrios (9) focuses on French psychiatrists who were instrumental in developing the classification of OCD through insanity, neurosis, and psychosis to the final position of OCD as part of the new class of neuroses. OCD was formally recognized in the 1830's as a cluster of related symptoms. Over the next 40 years, the components of OCD were differentiated and defined, with separation of the terms obsession from delusion, and compulsion from impulsion. OCD now is classified as an anxiety disorder, and it has a significant overlap with other anxiety disorders. OCD originally was viewed as resulting from psychodynamic factors. Our understanding of the pathogenesis of OCD has been revolutionized over the years, and it currently is considered a neurobiologic disorder.

The fact that GTS and OCD are linked is undisputed, but the precise relationship is complex. The relationship between GTS and OCD was documented for the first time by Itard (3) in 1825, when he described the Marquise de Dampierre. Many years later, Charcot saw the Marquise, and his observations were recorded by his student Georges Gilles de la Tourette in 1885. In this account, Gilles de la Tourette described the obsessive thoughts that plagued the

Marquise, as well as her motor tics and vocalizations (10). Charcot, however, was the first neurologist to identify the involuntary "impulsive" ideas, such as doubting mania, double checking, touching, and arithmomania (an obsession with counting and numbers), as part of GTS and to link them to the impulsive movements (11). Some of these "impulsions" would now be classified as compulsions (current definitions specify that executing an impulsive deed gives the patient some form of pleasure, satisfaction, or excitement, whereas carrying out the compulsive activity provides the patient only relief from tension; thus, impulsive acts are ego-syntonic, whereas compulsive acts are ego-dystonic) (12).

Following the studies by Charcot and Gilles de la Tourette, several other nineteenth-century neurologists and physicians became interested in the relationship between these psychological aspects of GTS and its motor and vocal manifestations. In 1899, Gilles de la Tourette (13) documented the anxieties and phobias of his patients and acknowledged the ideas of Guinon (14), who suggested that "tiqueurs" (individuals with tics or GTS) nearly always had associated psychiatric problems characterized by multiple phobias and agoraphobia. Soon after in 1890, Grasset (15) referred to his patients' obsessions and phobias, which to him were an accompaniment of the tic disorder and represented psychical tics. Robertson and Reinstein (16) translated the writings of Gilles de la Tourette, Guinon, and Grasset and illustrated how they described the psychopathology of people with "compulsive tic disorder" with particular reference to OCD, including checking rituals, arithmomania, folie du doute, delire de toucher (forced touching), folie du pourqoui (the irresistible habit of seeking explanations for the most commonplace, insignificant facts by asking perpetual questions), and "mania" for order.

In their classic text, "Confessions of a Victim to Tic," Meige and Feindel (17) described a patient whom they considered to be the prototype of a tic patient. It now is clear that this patient had GTS. He had motor tics that began at age 11 years, echophenomena (copying behaviors), a vocal tic ("tic of phonation dating back to his 15th year"), and coprolalia ("an impulse to use slang"). He was impulsive, and he had suicidal tendencies and OCD. Meige and Feindel (17) stated that "the frequency with which obsessions, or at least a proclivity for them, and tics are associated, cannot be a simple coincidence." They described case histories of their tic patients who had typical features of OCD, including the relief of anxiety that came from carrying out a particular motor act. They also noted that there often was no direct connection between a patient's obsessions and the tics, the former occurring in the form of extraordinary scrupulousness, phobias, and excessive punctiliousness in their actions. They mentioned in particular arithmomania, onomatomania (the dread of uttering a forbidden word or the impulse to intercollate another), and folie du pourquoi.

Kinnear-Wilson (18) acknowledged a relationship between tics and OCD: "no feature is more prominent in tics than its irresistibility.... The element of compulsion links the condition intimately to the vast group of obsessions and fixed ideas." Ascher (19) later reported that all of the five GTS patients he reported had obsessive personalities, and Bockner (20) noted that the majority of GTS cases described in the literature had obsessive-compulsive (OC) neurosis.

These historical findings and speculations were based primarily on anecdotal descriptions of single case studies. The last 2 decades have seen a wealth of evidence revealing a more definite link between OC symptomatology and GTS. Evidence for this connection continues to grow. The link can be viewed from several perspectives, including historical, epidemiologic, phenomenologic, genetic, neurobiologic, and neurochemical, which are discussed in detail later in the chapter.

Clinical Characteristics of Gilles de la Tourette Syndrome

The clinical characteristics of patients with GTS seem to be independent of culture; they occur with some degree of uniformity irrespective of the country of origin. The age at symptom onset ranges from 2 to 15 years; a mean of 7 years is the most common. The most frequent initial symptoms involve the eyes (e.g., eye blinking, eye rolling). Although often referred to as a tic disorder, patients with GTS usually demonstrate a variety of complicated movements, including licking, hitting, jumping, smelling, spitting, squatting, abnormalities of gait, and forced touching (21).

The onset of vocalizations usually is later than that of the motor tics. The mean age at onset is 11 years. Throat clearing, sniffing, grunting, coughing, barking, snorting, humming, clicking, colloquial emotional exclamations, low- and high-pitched noises, and inarticulate sounds are the usual utterances (21). Vocalizations have been reported as initial symptoms in 12% to 37% of subjects; the most frequent vocalization is repeated throat clearing (22,23). Many GTS individuals describe premonitory "sensory" experiences that are distinct from the actual motor or vocal tic (24).

Coprolalia (the inappropriate and involuntary uttering of obscenities) occurs in less than one third of clinic patient populations (25) and in very few children (26) or mildly affected individuals, such as affected relatives of a proband, in whom it is encountered in only 2% to 4% (27,28). Coprolalia has been reported in 22% of GTS individuals in an epidemiologic study (29). The mean age at onset is 13 to 14 years, but it later disappears in up to one third of cases (21). There is some suggestion that coprolalia is culturally determined; only 4% of patients in Japan have true coprolalia (30) versus 26% in Denmark (31). Some researchers have noted that coprophenomena occur infrequently in certain GTS patients from middle-class and strict religious backgrounds (31,32), whereas coprolalia has been shown to occur more frequently in those with moderate or severe GTS symptoms (26). *Copropraxia* (involuntary and inappropriate obscene gestures) has been reported to occur in 1% to 21% of clinic sample populations (21).

Echolalia (the imitation of sounds or words of others) and *echopraxia* (the imitation of movements or actions of others) occur in 11% to 44% of patients. *Palilalia* (the repetition of the last word or phrase in a sentence or the last syllable of a word uttered by the patient) occurs in 6% to 15% of patients (21). It has been reported that these symptoms are common as GTS develops into its fullest form; other subjective symptoms, such as sensory tics, mental coprolalia, coprographia, and mental palilalia, are only discovered when they are inquired about directly (25). Although these clinical features are not essential to make the diagnosis of GTS, the presence of any of these behaviors increases the clinician's diagnostic confidence.

Tics and vocalizations are invariably aggravated by anxiety, stress, boredom, fatigue, and excitement, whereas sleep, alcohol, orgasm, fever, relaxation, and concentrating on an enjoyable task usually lead to temporary disappearance of symptoms (21). Premenstrual stress and exogenous stimulants, such as caffeine, methylphenidate, and amphetamines, have been implicated in tic exacerbation (25,33). Characteristically, the course of GTS over the person's lifetime is punctuated by the appearance of new tics and the disappearance of older ones (21). During adolescence, the symptoms tend to be more unpredictable from day to day. It is estimated that the tic symptoms will remit completely by late adolescence in one third of patients. An additional third will show significant improvement in symptoms, and the remaining third will continue to be symptomatic during adulthood (25). Spontaneous remissions have been reported in 3% to 5% of patients followed-up for 6 months to 3 years (11,34).

Abuzzahab and Anderson (34) reported significant cross-cultural symptom differences among France, Italy, the United Kingdom, and the United States. They noted fewer eye and neck tics and more echokinesis in France; more eye tics in Italy; less echophenomena in the United Kingdom; and more neck tics, coprolalia, and younger age at onset in the United States. A German sample did not differ significantly from any of the other countries. Many of the reported differences have not been replicated in later studies (5,11).

It once was thought that movements disappeared during sleep, but this has now been shown not to be the case (35,36). Tics of the eyes, face, and head often are persistent and remain the most refractory of pharmacologic interventions (37).

A problematic area for parents, teachers, and some physicians is the idea that GTS is a so-called "involuntary" disorder of movement, because patients can suppress symptoms during an interview or while they are involved in certain tasks, for example, those performed at school or while playing a sport (32).

Clinical Characteristics of Obsessive-Compulsive Disorder

OCD is classified as an anxiety disorder in DSM-IV (1). It is characterized by obsessive thinking and compulsive behavior. Obsessions and compulsions can take various forms. *Obses-sional thoughts* are intrusive words, ideas, and beliefs that often are upsetting and unwelcome, even though the person recognizes them to be a product of his/her own mind. *Obsessional doubts* involve excessive and inappropriate concern about previous actions and their consequences. For example, patients might fixate on the possibility that they left their front door open, which would allow a burglar to enter their home, even though they can clearly remember locking up the house. *Obsessional fears* are feelings of uneasiness and dread about imagined events that might happen but which may be highly improbable or unreasonable, such as the death of a friend or being unfairly blamed for something. They also may be concerned with the fear of harming themselves or others. *Obsessional ruminations* are recurring thoughts of a complex nature, such as the ending of the world. *Obsessional images* are unwelcome mental pictures that often suddenly intrude upon the mind and are of an unacceptable nature, possibly involving violent or sexual scenes. *Obsessional impulses* are urges to carry out an act that usually is socially unacceptable, that the patient does not want to do, and that he/she will resist doing, such as shouting obscenities, carrying out a crime (e.g., shoplifting or stabbing), or causing self-harm.

Compulsions (also called *obsessional rituals*) are repeated and meaningless rituals carried out in a purposeful and stereotyped way that may be performed to neutralize an obsessional thought that preceded the compulsion. For example, the repeated washing of hands may be an action performed to allay obsessional fears of contamination with germs, or the obsessional doubt of failing to turn off the gas may lead to compulsive checking. Although the compulsion may seem to reduce the anxiety resulting from the obsession and produce a tension release in its performance, the person strongly resists carrying out the ritual and does not derive pleasure from it. The compulsion occasionally may have no apparent connection with an obsession, except for an obsessional urge to carry out the act, such as ordering things in a particular way. Patients often will go to great lengths to hide their compulsive acts and rituals because of embarrassment and fear of being considered mad. This often leads to a delay in diagnosis.

Most obsessions and compulsions involve issues of cleanliness and contamination, safety and aggression, order, sex, and religion. In a survey that reported the obsessions and compulsions of 70 child and adolescent patients with OCD, 43% had obsessions involving disgust with bodily wastes or secretions, dirt, germs, and chemical and environmental contamination; 24% had obsessional fears that something terrible might happen; and 17% had obsessions of symmetry and exactness, such as the way things should be arranged or organized (38). Thirteen percent had obsessions of a religious nature; and 4% had forbidden or perverse sexual thoughts, images, or impulses, involving bestiality, pedophilia, incest, or homosexuality. Other obsessions may involve seemingly neutral images, colors with special significance, and intrusive sounds, words, or music (39). Somatic obsessions involve preoccupation with parts of the body, such as the shape of one's nose.

Compulsions mostly take the form of cleaning, checking, repeating, counting, and ordering. In the study cited earlier, 85% of subjects had rituals of excessive hand washing, bathing, showering, tooth brushing, or grooming; 51% had repeated rituals, such as going in and out of doors or getting up and down from a chair; 46% had compulsions of checking doors, stoves, and car brakes; 23% had cleaning rituals involving contact with contaminants; 20% had rituals involving the need to touch; 18% had counting rituals; and 17% had ordering and arranging compulsions, such as rearranging drawers or packing and unpacking suitcases (38). Other compulsions involved hoarding and collecting; the need to tell, ask, or confess; and the need to dress a special way.

Habit Disorders, Dysmorphophobia, and Other Disorders

The so-called *habit disorders,* such as trichotillomania (characterized by chronic severe hair pulling) and onychophagia (severe nail biting), are abnormal behaviors that bear some resemblance to the compulsions seen in OCD in that they are repetitive, purposeful, and stereotyped, and they relieve anxiety. These similarities and the evidence that trichotillomania and onychophagia, like OCD, respond to selective serotonin reuptake inhibitors (SSRIs), such as clomipramine (40,41) and fluoxetine (42,43), led to the theory that these disorders are variants of OCD. However, other researchers have indicated that this class of drugs is not as effective for hair pulling as previously thought. An 18-week, placebo-controlled, double-blinded, crossover study that investigated the efficacy of fluoxetine in 21 adult patients with trichotillomania showed no significant short-term benefit (44). Reports have indicated that relapse may occur during chronic treatment of trichotillomania with clomipramine (45). Neuroimaging studies and studies of the clinical features of the two conditions provide other evidence that trichotillomania is a disorder distinct from OCD. Positron emission tomographic (PET) studies have shown a different pattern of regional cerebral glucose metabolism in women with trichotillomania from that seen in patients with OCD (46). Comparison of the clinical features between a group of eight trichotillomania patients and 13 OCD patients showed that the former group experienced a greater degree of pleasure in hair pulling than did the OCD patients in enacting their compulsions (47). The study also showed that the two groups differed in terms of associated depression, anxiety, and personality characteristics (47).

Body dysmorphic disorder is a somatoform disorder in which a normal-appearing person has a nondelusional preoccupation with an imagined defect in appearance. The patient's concerns are excessive, distressing, persistent, and disruptive to his/her lifestyle; thus, they are similar to the obsessions of OCD. Reports in the literature have illustrated

the diagnostic overlap between OCD and body dysmorphic syndrome (48,49) and response of the latter disorder to treatment with SSRIs (50).

Trichotillomania, onychophagia, and dysmorphophobia are included in a group of associated disorders that resemble OCD, called the *OC spectrum disorders* (51). Other conditions included in this group are hypochondriasis, GTS, eating disorders, and impulse control disorders such as kleptomania, compulsive gambling, pyromania, and the paraphilias (52). Many clinicians have empirically treated these OC spectrum disorders with SSRIs without having firm evidence that these agents are effective because of the lack of controlled trials. Further phenomenologic, family, biologic, and treatment studies are needed to determine whether these disorders are actual variants of OCD, comorbid conditions, or separate syndromes with overlapping characteristics that complicate the differential diagnosis.

Clinical Characteristics of Gilles de la Tourette Syndrome Plus Obsessive-Compulsive Disorder

Of importance is the finding that the obsessive-compulsive behaviors (OCB) seen in GTS are phenomenologically and clinically different from those seen in primary OCD. Obsessions related to violent/sexual/religious themes and compulsions such as those involving touching, symmetry, and repeating behaviors until the patients gets it "just right" are commonly reported in GTS, whereas contamination fears, cleaning, and washing are commonly encountered in primary OCD (53–56). It also was noted that primary OCD patients who shared a symptom profile similar to that of GTS had a positive family history and that female relatives were more likely to be affected (53). The gender difference was noted by other investigators for GTS as well; female relatives of GTS probands are more likely to have OCB, whereas male relatives more often exhibit tic symptoms (57,58). Santangelo et al. (58a) observed that onset more often is rage in males and compulsive tics in females, and that perinatal brain injury may cause or exacerbate the development of OCD in TS subjects. Iida et al. (59) suggested that GTS+OCD is more strongly associated with organic cerebral disorders and greater severity of symptoms. Zohar et al. (60) noted in an epidemiologic study that adolescents with tics were more prone to aggressive and sexual images and obsessions than adolescents without tics and that these differences could not be wholly attributed to sex differences.

Researchers have attempted to define the precise phenomenology of the OC thoughts and behaviors that occur in GTS patients, to relate them to other variables, and to compare them to the symptomatology that occurs in OCD patients without a tic disorder. Holzer et al. (61) found that OCD patients with tics had more touching, tapping, rubbing, blinking, and staring rituals but fewer cleaning rituals compared to OCD patients without tics. Baer (62) found in a factor analysis study that the only factor related to co-morbid tics and OCD was "symmetry/hoarding." Pitman et al. (63) reported that certain kinds of compulsive behaviors, such as touching and symmetry behaviors ("evening-up" rituals to ensure that the body is symmetric or balanced), occurred more often in GTS patients than in OCD patients, although some GTS patients exhibited compulsions such as checking, washing, and counting. Similarly, the types of tics that occurred in OCD patients with a tic disorder, such as eye blinking and noise making (e.g., humming, sniffing, throat clearing), were not unlike those commonly experienced by GTS patients. They also noted other phenomena that occurred at much higher rates in both the GTS and OCD groups than in the control sample, phenomena such as pathologic doubt, slowness, and depersonalization. Both groups shared high rates of unipolar depressive illness and generalized anxiety disorder compared with controls. These results led the authors to propose "the notion of a symptomatic continuum from simple tic to complex tic to compulsion" (63). However, despite the large symptomatic overlap, the two conditions were not phenomenologically identical. Echophenomena, a history of attention deficit disorder (ADD), and self-injurious behaviors (SIB) occurred frequently in GTS but not OCD patients, whereas phobic and panic disorders were much more common in the OCD group (63).

Frankel et al. (64) reported an overlap in the type of OC symptoms (including checking and fear of contamination) experienced by GTS and non-GTS OCD patients. They also documented that symptomatology changed with age in GTS patients. Younger individuals exhibited compulsive behaviors related to impulse control, and older patients manifested behaviors more classically associated with OCD, such as checking, arranging, and fear of contamination (64). George et al. (65) prospectively studied ten subjects with OCD and 15 with OCD and comorbid GTS using the Yale-Brown Obsessive Compulsive Scale (Y-BOCS), the Leyton Obsessional Inventory (LOI), and a new questionnaire designed to emphasize the differences in symptoms between the two groups. Results showed that subjects with comorbid OCD and GTS had significantly more violent, sexual, and symmetric obsessions, and more touching, blinking, counting, and self-damaging compulsions. The group with OCD alone had more obsessions involving dirt or germs and more cleaning compulsions. The subjects who had both disorders reported that their compulsions arose spontaneously, whereas the subjects with OCD alone reported that their compulsions frequently were preceded by cognitions (65).

Leckman et al. (66) examined OC symptoms in 177 patients with OCD; 56 had tic-related OCD and 121 had non–tic-related OCD. Patients with tic-related OCD reported more OC symptoms (including more aggressive, religious, and sexual obsessions, and checking, counting, ordering, touching, and hoarding compulsions) than did patients with non–tic-related OCD.

Eapen et al. (53) examined patients with primary GTS and compared them with patients with GTS and OCD. Patients with OCD had significantly more obsessions involving dirt, germs, contamination, and fear of something bad happening, and compulsions concerned with washing and cleaning. Cath et al. (67) found that patients with GTS+OCD reported more Tourette-related impulsions, such as mental play, echophenomena, impulsive behavior, and SIB, and less overall symptomatic obsessions and symptomatic washing than patients with OCD without tics. Miguel et al. (68) highlighted the importance of sensory phenomena in differentiating GTS+OCD from primary OCD. They found that sensory phenomena, including bodily and mental sensations, are encountered more frequently in patients with GTS+OCD than in patients with OCD alone. On the other hand, intentional repetitive behaviors are associated with cognitive phenomena and autonomic anxiety in primary OCD but not in GTS+OCD (69, 70).

Other investigators have suggested that the frequency of OC symptoms increases with the duration of GTS (71). However, Robertson et al. (72) failed to demonstrate a significant relationship of OC symptoms with either age or duration of GTS. Robertson et al. (72) compared adults having GTS with depressed adults and normal controls using questionnaires that measured obsessionality, depression, and anxiety. The GTS and depressed groups both scored significantly higher than normal controls on all measures. The GTS subjects, however, had scores on measures of obsessionality similar to those of the depressed subjects but significantly lower scores on measures of depression and anxiety. This suggests that obsessionality is a prominent feature of GTS, and that the psychopathologic profile is different from that of patients with major depressive disorder. Coffey et al. (73) found that the rates of bipolar disorder, attention deficit/hyperactivity disorder (ADHD), social phobia, body dysmorphic disorder, and substance use disorders were elevated in the GTS+OCD group compared with the GTS only and OCD only groups. These authors concluded that GTS+OCD is a more severe disorder.

Coprolalia has rarely been described in OCD patients. Pitman and Jenike (74) presented a single case history of a man who fulfilled DSM-III-R criteria for OCD. The patient had manifested coprolalia since age 6 years but otherwise did not satisfy criteria for diagnosis of GTS. The authors emphasized that this case further blurred the phenomenologic distinction between the two disorders, and they remind us that both Janet (75) and Meige and Feindel (17) observed an association between coprolalia and obsessions almost a century ago.

Other investigators observed that GTS and OCD share clinical features, such as waxing and waning of symptoms, early age at onset, lifelong course, ego-dystonic behavior, worsening with depression, anxiety, and their occurrence in the same families (54,76).

ASSOCIATED PSYCHOPATHOLOGY

It is perhaps no coincidence that some of the first comments on the psychopathology of GTS patients were made by Georges Gilles de la Tourette himself in 1899 (13), when he commented on the anxieties and phobias of his patients. In that paper, he noted the ideas of Guinon (14), who suggested that "tiqueurs" nearly always had associated psychiatric disorders characterized by multiple phobias, arithmomania, and agoraphobia. Robertson and Reinstein (16) translated and documented the original accounts by the early French physicians Georges Gilles de la Tourette, Grasset, and Guinon of the psychopathology associated with GTS. Meige and Feindel (17), in the "Confessions of a Victim to Tic," described the psychopathology of their tic patient who almost certainly had GTS and drew attention to "the fundamental importance of the psychical element that precedes the motor reaction."

The GTS literature from 1900 to 1965 was predominantly psychoanalytic, with many classic descriptions by early writers (77–83). Although some investigators, such as Shapiro and colleagues (11,83,84), claimed that no specific psychopathology was found in association with GTS, Shapiro et al. (11) reported that only one of their cohort of 34 patients was free from psychiatric illness; the majority were diagnosed with various types of "personality disorders." They pointed out that there may be a subgroup of GTS patients who have a great deal of difficulty with compulsive ritualistic behaviors. In a controlled study using the Minnesota Multiphasic Personality Inventory, a group of GTS patients failed to differ significantly from general psychiatric outpatients on factors such as overt and underlying psychosis, OC traits, inhibition of hostility, hysteria, and general maladjustment (11). They found OC traits in 12.8% of GTS patients and 14.6% of controls (11). They also pointed out that there may be a subgroup of GTS patients who have substantial difficulty with OC rituals (11).

Other investigators, such as the Comingses (85–88), suggest that GTS is associated with a wide variety of psychopathologies, including learning disorders, dyslexia, conduct disorder, phobias, panic attacks, mania and manic depressive disorder, schizoid behaviors, alcoholism, drug abuse, and pathologic gambling. Most investigators report and agree that the psychopathology is more specific, and it is generally accepted that OCB, anxiety, depression, ADHD, and SIB are found in a large number of GTS patients.

Several studies found that GTS patients are more depressed than control groups (89–91), but other studies suggest that the depression is related to the duration of the GTS (72). This might be attributable to the stress of having a chronic, socially disabling, and stigmatizing disease (21). GTS patients have demonstrated more anxiety than controls (89), although Coffey et al. (92) suggest that there is a subgroup of GTS patients with non-OCD pure anxiety disorders. Depression and anxiety certainly are associated

with GTS, but we maintain that they likely are secondary phenomena.

The relationship between GTS and ADHD was studied extensively by the Comingses (88,93) and reviewed by Robertson and Eapen (33) and Towbin and Riddle (94). Evidently, ADHD occurs in a substantial proportion of GTS patients (21%–90% of GTS clinic populations), but the precise relationship between the two disorders is complex and remains unclear (33,94). There have been suggestions that GTS and ADHD are genetically related (88,91,95–97), but they have been followed by refutations (98) and debate (99). Pauls et al. (100) found that the rates of ADD in relatives of GTS patients were not significantly increased compared to controls. However, the results also suggest that there are two types of individuals with GTS and ADD: one in whom the ADD is independent of GTS and the second in whom the ADD is secondary to the GTS. The relationship between the disorders is complex and merits further study.

An association between some types of SIB and GTS has been suggested (39), but the exact nature of this association is unclear (for reviews, see references 101 and 102). In epidemiologic (29) and pedigree (27) studies, individuals with mild GTS were noted to exhibit SIB, which suggests that SIB is integral in some individuals with GTS and may not necessarily be associated with the severity of GTS or referral bias. SIB does, however, seem to be related to psychopathology, especially hostility and obsessionality (103). Robertson et al. (104) found that 71% of GTS patients studied had one or more personality disorders compared to 15% of the control group. However, this comorbidity may be the result of ADHD or other behaviors that often coexist with GTS. In a study of adult psychopathology among GTS patients, Eapen et al. (in preparation) performed principal components factor analysis using varimax rotation on psychopathology scores. This analysis yielded two components that accounted for 72% of the variance. The first component was interpreted as the "obsessionality component." The second was interpreted as the "anxiety/depression component," which had Eysenck extroversion negatively loaded on it. Although there is compelling evidence that obsessionality is etiologically related to GTS (53,105), the etiology of anxiety and depression in the context of GTS is less certain. The latter may be of multifactorial origin, as it is in the non-GTS populations. Earlier studies suggested that this may be due to the severe and socially handicapping nature of GTS, the effects of medication, or the effects of comorbidity and ascertainment bias within the clinic population (106). A possible genetic mechanism (107) and coexistence by chance due to the high lifetime prevalence of anxiety and depression (108) are other possible explanations.

There is no association between psychosis and GTS, apart from a few case studies reporting major depression (107), mania (109), bipolar disorder (110–113), schizophreniform psychosis (27,114,115), and psychosis in the setting of learning disability (116). There have been two reports of children with visual impairment, Ganser syndrome, and GTS, and the authors note that a number of their patients (all children) with GTS experienced "brief auditory hallucinations" that varied between true hallucinations and intense eidetic auditory imagery (117,118). The authors later described "schizophreniform" symptoms in 11 cases of GTS (although only two were typical and uncomplicated) (119). This demonstrates how childhood schizophrenia can be misdiagnosed, because all of the children actually had GTS. Six cases of concurrent GTS and Asperger syndrome were reported by the same group (120). Baron-Cohen et al. (121) reported a higher prevalence of GTS (4%) and tics (34%) in a large-scale study of 447 children with autism. Robertson et al. (72) reported no psychotic patients in a cohort of 90, although three patients exhibited ideas of reference.

It appears that patients with GTS are especially prone to depression, which may be related to the duration of the GTS and is partly connected to the reality that GTS sufferers have a chronic, socially disabling, and stigmatizing disease. Anxiety is high among patients with GTS, but depression may account for some of these cases. ADHD and SIB occur in a substantial number of GTS individuals, but the precise relationship between them is unclear. There is no generally accepted association between GTS and psychosis.

Family Psychopathology

If psychopathology is an important and distinctive feature of GTS, then a positive family history of psychiatric illness might be expected, but this aspect has not received much attention. Gilles de la Tourette (13) had already noted in 1899 that the family history of patients with GTS was almost invariably "loaded for nervous disorder." Samuel Johnson's father, Michael, clearly suffered from depression, "a general sensation of gloomy wretchedness," and it is from him that his famous son "inherited...a vile melancholy" (122). Comings and Comings (93) assert that a wide variety of psychopathologies are found in relatives of GTS probands. Other than Comings, only case reports and small studies have noted a positive family history of psychopathology (123–127). Corbett and colleagues (125) studied the parents of tiqueurs and found that 57 of 184 children (31%) of tiqueurs had one or both parents who were psychosomatically ill; over half of these were mothers with affective illness. This was a significantly higher number compared to an out-of-clinic sample and a psychiatric hospital population (125). Montgomery et al. (71) specifically addressed the question and found that 70% of 30 first-degree relatives (FDRs) of 15 GTS patients satisfied Feighner criteria for psychiatric illness. The most common diagnoses were unipolar depression, obsessive-compulsive illness, and panic disorder. Robertson et al. (72) reported that 48% of 90 probands had a positive family history of psychiatric illness; the most common disorders were depression, schizophrenia, and obsessional disorder. Robertson and Gourdie (27) reported a wide range of

psychopathology in 85 family members of a GTS proband. Individuals completed standardized psychiatric rating scales and, with the exception of the Leyton Obsessional Inventory, there were no statistical differences between cases and non-cases. All individuals also were diagnosed using the Schedule for Affective Disorders and Schizophrenia-Lifetime Version. Twenty-one family members had exhibited psychiatric illness at some time, but their occurrences were not in excess of the lifetime risks for major psychiatric illness, such as major depressive disorder, alcohol abuse, schizophrenia, and other "neurotic" disorders (27).

Pauls et al. (128) examined the relationship between GTS and psychopathology in 338 biologic FDRs of 86 GTS probands, 92 biologic FDRs of 27 unaffected control probands, and 21 nonbiologic FDRs of six adopted GTS probands, mostly by direct interview. Results showed that anxiety, panic disorder, phobias, affective disorders, substance abuse, and psychotic disorders are not genetically related to GTS.

PREVALENCE

The exact prevalence of GTS is unknown. The currently accepted figure is 0.5 in 1,000 (129), but this probably will prove to be an underestimate (130). A population-based study performed in Israel reported an estimated prevalence of 4.28 in 10,000, with rates of 4.90 in 10,000 observed in boys and 3.10 in 10,000 in girls (131). Reports have suggested that the prevalence rate is as high as 1% (132) to 2.9% (133) in some groups.

Lifetime prevalence of OCD ranges from 1.9% to 3.2%, despite problems with the definition of OCD and its differentiation from other psychiatric diagnoses (especially anxiety disorders and depression). The prevalence of OCD in patients with GTS is much greater than would be expected by chance alone. In this context, since 1987 the diagnoses of OCD and GTS are no longer mutually exclusive according to accepted diagnostic conventions. In its definition of OCD, DSM-III published in 1980 stated that the obsessions or compulsions must not be due to another mental disorder, such as GTS, schizophrenia, major depression, or organic mental disorder (134). The current DSM-IV published in 1994 states that in some people with GTS, OCD is an associated diagnosis (1).

With regard to GTS+OCD, acknowledgment of the prevalence of OCD in the context of GTS dates back to the first description in 1825. Shapiro et al. (11) divided the history of GTS into seven periods. The fourth period, called "Epidemiology and Reviews," commenced in 1954 with the retrospective studies of Zausmer (135), but it was not until 1962 that Torup (136) documented the frequency of obsessional symptoms, traits, or illness in a population of GTS patients. In a study of 237 children treated for tics in Denmark between 1946 and 1947, 12% were judged to have a compulsive behavior pattern (136). Since that investigation, many epidemiologic studies have revealed significant percentages of GTS and tic patient populations experiencing some form of OC phenomena. The reported rates have tended to increase over the years. Studies performed in the 1960's reported rates of 10% to 30% (124,137); by the 1970's and 1980's, investigators recorded far higher rates of 30% to 80% (103). The earlier lower figures may have resulted from a variety of causes, including the use of less specific measures to assess OCD, differences in the definition of OCD, and an unusually high frequency of OCD in control studies (11).

More recent investigations have used specially designed inventories to assess the incidence of OCD among GTS patients and normal controls. Frankel and colleagues (64) used a questionnaire derived from the LOI in their controlled study of British and American patients with GTS, American patients with OCD, and normal control American subjects. They reported that 47% of GTS males and 86% of GTS females met diagnostic criteria for OCD. Green and Pitman (138) found more OC problems in GTS patients compared with controls. Caine et al. (29) reported that 49% of the 41 patients in their epidemiologic study of school children in Monroe County (New York) had obsessional ideas or associated ritualistic behavior, although only three were significantly impaired to warrant a diagnosis of OCD. This study is limited because it included only children.

In a controlled study, Van de Wetering et al. (90) used a Dutch version of the LOI on 66 Dutch GTS patients and reported that 47% admitted to obsessive rituals, 18% obsessive thoughts, and 4% obsessive imaging. Robertson et al. (72) used both the LOI and the obsessional scale of the Crown Crisp Experiential Index (previously known as the Middlesex Hospital Questionnaire) and found that 37% of 90 GTS patients in the United Kingdom reported OCB. In this study, coprolalia and echophenomena were significantly associated with OC phenomena. Use of different questionnaires or scales in these studies, the varying sample sizes, and cultural differences all may contribute to the diversity of figures obtained.

Far fewer investigations have demonstrated the occurrence of tics and other associated features of GTS, such as coprolalia, echophenomena, and ADD in OCD patients. Rasmussen and Tsuang (139) reported a GTS prevalence of 5% in OCD patients, which still is greater than the generally accepted figure of 0.5 in 1,000 for GTS in the general population (129). In their study of 16 GTS outpatients, 16 OCD outpatients, and 16 normal controls, Pitman et al. (63) reported that 37% (six patients) of the OCD group satisfied criteria for any tic disorder (only one patient met criteria for GTS), compared with only one subject in the control group. Ten patients (63%) in the GTS group met the criteria for OCD. These figures suggest that although OCD might now be considered an integral part of GTS (6), the majority of OCD subjects do not suffer from a tic disorder and instead

may represent a separate etiologic group. Rappoport (140a) suggests that about 20% of OCD patients have tics.

The reported prevalence of GTS depends on the definition and ascertainment method used. Other reasons why the exact numbers of people with GTS is unknown include misdiagnosis and the reluctance of relatives to divulge family histories, as suggested by the case reports of Hajal and Leach (141). The apparent increase in the number of patients seen in recent years may be partly explained by the fact that many cases now being diagnosed tend to be those with milder symptoms. Most of the reported figures are thought to be underestimates because many patients with mild GTS who never consulted a doctor for their symptomatology have been identified in family studies (27). Some researchers suggest that GTS is very common in special education school populations (142), with a prevalence as high as 1,200 in 10,000 (143). Undoubtedly, more cases are coming to medical attention either through clinical or research sources. With more epidemiologic studies, the true prevalence rates will become clearer.

GTS is found in all cultures and racial groups, but it is rare among the U.S. black population, in which it has been represented in varying but small proportions ranging from 0.5% to 8.7% (11,144). Most large reports have come from the United States, but significant numbers of patients also have been reported from Europe, including the United Kingdom, Germany, the Netherlands, Denmark, and France, as well as New Zealand, Canada, the former Soviet Union, Japan, Korea, Hong Kong, China, and India. Case reports from Sri Lanka, Puerto Rico, Australia, Finland, South America, Malta, and the Middle East highlight the worldwide distribution of the disorder (for review see references 21, 106, 145, and 146).

Some early studies found an unexpectedly large proportion of their GTS subjects have an Eastern European/Ashkenazi-Jewish background (11,147,148), but this finding has not been reported by other investigators (5,23,26,54,72,149–152). It should be noted that two of the three studies reporting such a preponderance were from the metropolitan New York City area (11,147), which has a large Jewish population.

The majority of studies indicate that GTS occurs three to four times more often in males than in females (106). The syndrome is found in all social classes (4,11,54,149), although two studies found a clustering in the lower social class (11,54). Some studies have reported that around 60% of GTS patients (classified according to the Registrar General's classification) failed to attain their parental social class, thus suggesting that patients with GTS underachieve socially.

NEUROPHYSIOLOGY

Shapiro et al. (11) reviewed the literature on electroencephalographic (EEG) findings from 11 studies on GTS.

Of 127 patients reported, 66% were said to have abnormal EEGs. In their own cohort of 79 patients, 47% had EEG abnormalities, which were more common in children (71%) than in adults (25%). Other investigations reported lower rates, with some degree of consistency. Abnormalities have been found in 12.5% (153), 13% (5), 17% (154,155), 20% (156), and 37% (72) of GTS clinic patient populations. Abnormalities detected are nonspecific, and there is no evidence of any paroxysmal activity time locked to the tics (155). In summary, EEG findings in GTS are mostly normal; abnormalities are minor and nonspecific. It has been suggested that among GTS patients who are more psychologically impaired and who have more learning disabilities, there are more dysrhythmias on the EEG (157), but this suggestion was based on a small number of patients. Semerci (158) conducted EEG studies on 40 children and adolescents with GTS and found nonspecific EEG abnormalities and neurologic soft signs in 12% and 57.5% of cases, respectively. A statistically significant association between EEG abnormalities, neurologic soft signs, and low-performance intelligence quotient (IQ) was observed. Hyde et al. (159) studied EEG in 11 twin pairs, of which at least one twin had GTS. In nine of the 11 twin pairs that differed with regard to clinical severity, the twin with the more severe course of illness had a significantly more abnormal EEG by qualitative visual analysis. Most of the differences were due to excessive frontocentral θ activity, which suggests dysfunction outside the basal ganglia. It also was noted that the twin with the lower score on neuropsychological testing and lower birth weight had a more abnormal EEG. The same effect was found in twin pairs that had been free of medication for at least 6 months before entry into the study. The authors suggested that the slowness was related to an interaction between environmental insults to the central nervous system (CNS) and the genetic component of GTS that led to damage to the cortex, thalamus, or both. Stevens et al. (160) found abnormal topography of EEG microstates as evidenced by an abnormal increase in fields with a right frontal/left posterior configuration in GTS patients compared to matched controls. The temporal descriptors of the EEG were found to be unaffected in patients with GTS, and the abnormal EEG patterns were not similar to those elicited by simple or complex movements. The authors believe that the presence of abnormally facilitated, near threshold motor activity in GTS patients is not a likely explanation. Gunther et al. (161) compared EEGs from GTS patients and controls during resting and manumotor/music perception activation conditions. Resting EEGs did not show any differences, but subtle differences in α frequency were noted in patients with GTS during simple and complex hand movements, as well as during music perception tasks. The authors suggested there is reduced brain activation in frontal and central regions during motor tasks and in temporal and parietal regions during music perception.

In view of the wide spectrum of clinical manifestations of the tics associated with GTS, it is not surprising to find an equally wide variation in electromyographic (EMG) patterns recorded from muscles. One study has addressed the subject comprehensively (155). In EMG recordings of simple tics, cocontraction of prime mover and antagonist muscles was seen more frequently, and more complex tics were associated with a variety of EMG patterns. The same investigators reported that the buildup of a negative potential over the half second before EMG activation of an involved muscle (the premovement, readiness, or Bereitschafts potential, which probably represents the summed activity of changing firing patterns in cortical neurons in preparation for a voluntary movement) was absent in six cases of GTS. This suggests that the simple tics of GTS are not generated through the normal cortical motor pathways used for willed human movements.

Using transcranial magnetic stimulation, Ziemann et al. (162) found decreased motor inhibition in GTS patients compared to healthy controls. Motor threshold and peripheral motor excitability were normal in the GTS group, but the cortical silent period was shortened and the intracortical inhibition was reduced. Subgroup analysis revealed that these abnormalities were mainly seen when tics were present in the EMG target muscle or in patients without neuroleptic treatment. Age, sex, ADHD, OCD, and sensory urges had no significant effect on motor excitability. The authors concluded that tics originate from a primarily subcortical disorder affecting the motor cortex through disinhibited afferent signals, impaired inhibition directly at the level of the motor cortex, or both.

Gironell et al. (163) found abnormalities in the acoustic startle reflex and reaction time in patients with GTS. Compared to controls, GTS patients showed a significantly higher amplitude, major degree of spread, and fewer habituation phenomena of the startle reflex. GTS patients also showed poorer nonstatistically significant reaction time (RT) (four blocks of 50 trails) performance, with a significant correlation between RT and severity of disease. In addition, the start-react effect was significantly less pronounced in GTS patients.

Obeso et al. (155) and Krumholz et al. (153) investigated visual and sensory evoked potentials in GTS patients, but no consistent abnormalities were observed. Late components of event-related auditory evoked potentials were investigated in a controlled study (164). It was concluded that GTS subjects have no abnormalities in early or late components, but the components in the range from 90 to 280 msec are affected and may reflect specific attention deficits that can occur in some GTS patients.

Dursun et al. (165) reported antisaccade eye movement abnormalities in patients with GTS. Compared to controls, GTS patients showed significantly higher error rates and increased latencies in the antisaccade task, which suggests a dysfunction of the fronto-cortico-striatal network.

Georgiou et al. (166) found that GTS patients require external sensory cues to sequence a motor program effectively. When there was a high level of reduction in advance information, that is, the visual pathway to be followed was extinguished well in advance of each successive movement, executions progressively slowed as the sequence was traversed. Similarly, if no advance information was provided before each move, movement execution was slower than that of controls. The movement initiation times were not different from those of the controls, as were their movement execution times when advance visual information was available. Thus, with limited visual guidance, patients with GTS, regardless of medication or depression state, may require more time to plan and program each next submovement. It seems that, like other basal ganglia disorders such as Parkinson disease, GTS patients may be dysfunctional in internal switching mechanisms.

Tolosa et al. (167) evaluated neuronal excitability in 23 GTS patients by studying the effect of a conditioning stimulus on blink reflexes. Studies were repeated during maximal voluntary tic inhibition. Results indicated that there is increased brainstem interneuron excitability in patients with GTS, with reduction of this excitability during voluntary tic inhibition.

Electrophysiologic studies have yielded somewhat conflicting results with regard to OCD. The overall evidence points to the orbitofrontal and anterior cingulate-basal ganglia-thalamocortical circuits as the neuroanatomic substrate for OCD. Drake et al. (168) compared EEG spectral measures in OCD patients who were not taking medication and 12 controls. The modal α frequency and maximal α frequency were reduced in the frontal regions in patients compared to controls, which suggests frontal lobe disturbance in the pathophysiology of obsessions and compulsions.

NEUROIMAGING

Structural abnormalities of the brain in patients with GTS have been investigated by a variety of neuroimaging techniques, including computed tomography (CT) and magnetic resonance imaging (MRI). Although abnormalities have been few and only small numbers of patients have been examined, there are implications that abnormalities of the basal ganglia are found in GTS.

In investigations using CT scans, only 22 of 176 (13%) documented CT scans were abnormal, and only some of the abnormalities appeared to be of possible etiologic significance. Robertson et al. (72) documented that 71 of 73 GTS patients had normal CT scans (the two abnormalities were cavum septum pellucidum cavities in patients who banged their heads). In a controlled study, CT scans were performed on 19 patients with GTS and compared with those of patients with infantile autism, ADD, and a language disorder, and a control group of 20 medical patients (169). No significant differences were found between the groups or controls with respect to total ventricular volume, ratio of right ventricular to left ventricular volume, ventricular asymmetries,

ratio of ventricle to brain, or brain density. Regeur et al. (31a) reported that 47 of 53 CT scans in GTS patients were normal. Abnormalities included a small arachnoid cyst in the occipital region, a suprasellar epidermoid, a large defect in the right temporoparietal region, slight cortical atrophy, and asymmetry of the ventricles (in two patients). In contrast to the findings of these studies, Caparulo et al. (170) reported abnormalities such as mild ventricular dilation, prominent sylvian fissures, and cortical sulci in 38% (6/16) of patients. Chase et al. (171) reported 7 of 9 normal CT scans; abnormalities were mild ventricular dilation and mild, diffuse cortical atrophy. Other researchers have reported markedly enlarged occipital horns of the lateral ventricles bilaterally (172), a large porencephalic cyst in the right hemisphere involving the right basal ganglia, and contrast enhancement in the region of the left basal ganglia (173). In other CT reports, Yeragani et al. (174) documented a mild increase in the density of the caudate nuclei thought to be due to calcification, and Lakke and Wilmink (175) reported a pineal tumor. Vieregge (176) reported twins concordant for GTS who also had ventricular asymmetry on CT scan.

MRI scans in GTS patients suggest that there are minor abnormalities in the basal ganglia. However, some scans are normal even when GTS subjects are compared with controls (171,177), but these studies are limited because of the small numbers of patients involved (n = 13) and the lack of quantitative analysis. A few abnormalities on MRI have been described. Sandyk (178) reported a 7-year-old boy whose MRI revealed asymmetric cerebral peduncles. Robertson et al. (179) reported a patient with severe GTS who required psychosurgery (limbic leukotomy) and who had a high-signal lesion in the right globus pallidus on MRI. Demeter (177) documented two patients who showed focal abnormalities involving the basal ganglia. Two controlled MRI studies of 51 GTS patients and 32 controls indicate that the basal ganglia are abnormal in patients with GTS (180,181). In particular, the lenticular region in the left hemisphere is reduced in volume compared to the volume in controls. Likewise, in GTS patients there is an attenuation of the left side prominence relative to the right side observed in control groups (182).

Peterson et al. (183) suggested that structural interhemispheric connectivity is aberrant in the CNS of GTS patients, and that this may indirectly support altered lateral cerebralization in GTS. In a study of magnetic resonance images of the corpus callosum (CC) from 14 unmedicated GTS patients and 14 matched controls, the authors found statistically significant reductions in the overall CC area and all subregional areas. Measures of mean callosal curvature suggested that the CC in GTS patients is less rounded than in controls. Worst-ever motor tic symptoms showed the strongest significant correlation with the length of the CC center line in GTS patients. In an MRI study of 17 GTS patients and eight controls, Moriarty et al. (184) found no absolute differences in caudate nucleus volumes, but they observed an increase in CC cross-sectional area and a loss of normal asymmetry of the caudate nucleus in GTS. Loss of the normal correlation between cross-sectional area of CC and whole brain index was noted in the patient group.

Hyde et al. (185) performed morphometric analysis of magnetic resonance images from ten monozygotic (MZ) twin pairs discordant for severity of GTS but concordant for the presence of tics. Significantly reduced anterior right caudate volume was observed in the more severely affected twin in nine of the ten twin sets. The mean volume of the left lateral ventricle was significantly smaller in the more severely affected twins. The normal asymmetry of the lateral ventricles (left greater than right) was not present in the more severely affected twins, who had a trend toward a larger right lateral ventricle. Thus, subtle structural abnormalities in the CNS, particularly the caudate, may play a role in the pathophysiology of GTS.

Zimmerman et al. (186) performed MRI-based subcortical assessment on 19 GTS girls aged 7 to 15 years (11 with GTS only and 8 with GTS+ADHD), and 21 age- and sex-matched controls. Two group comparisons showed no robust differences between GTS and controls. Three group comparisons showed significantly smaller lateral ventricles in GTS only subjects compared to GTS+ADHD and control subjects. In view of the previously reported finding of putamen asymmetry index as a marker for GTS, retrospective comparisons were made with data from GTS boys. These comparisons demonstrated that girls with GTS had putamen asymmetry indices similar to those of boys with GTS; however, control girls also showed similar patterns. The authors concluded that gender differences confound the association between putamen asymmetry and GTS.

To investigate the pathophysiology, Biswal et al. (187) used functional MRI to study activation in the sensorimotor cortex of five GTS patients and five controls. In the GTS patients, the average number of pixels activated during the motor task of finger tapping in the sensorimotor cortices and supplementary motor areas significantly exceeded that in the controls. The area over which the pixels was distributed was significantly larger. The investigators suggested that motor function is organized differently in GTS patients. Peterson et al. (188) used MRI to evaluate the effect of tic suppression in 22 GTS patients. Significant changes in signal intensity were seen in the basal ganglia and thalamus and in anatomically connected cortical regions involved in attention-demanding tasks. The magnitudes of regional signal change in the basal ganglia and thalamus correlated inversely with the severity of tic symptoms. These findings suggest the involvement of impaired modulation of neuronal activity in subcortical neuronal circuits.

More abnormalities in brain structure of GTS patients are being reported using MRI than CT technique. This probably reflects the diagnostic accuracy of MRI and the ability to detect subtle abnormalities.

Functional imaging using PET and single photon emission computed tomography (SPECT) has yielded interesting abnormalities, especially considering the abnormalities found in OCD.

One study using PET showed abnormalities in five GTS patients compared with controls (189). There was a fairly close positive association between metabolism in the basal ganglia (particularly the corpus striatum) and metabolism throughout the cerebral cortex in the GTS patients. Regional glucose metabolism seemed to have a close inverse association with the severity of vocal tics in the middle and inferior parts of the frontal lobes bilaterally, extending posteriorly from the frontal poles to the central gyrus. Coprolalia, in contrast, was inversely correlated with hypometabolism in the left parasylvian region (189). Continuing their work in the area, Chase et al. (171) assessed 12 untreated GTS patients with matched normal controls using an improved PET scanner that had higher resolution and sensitivity than the earlier scanner. At horizontal levels from 8.4 to 8.8 cm caudal to the vertex, nonnormalized glucose utilization rates were approximately 15% below control values in the region of the frontal cingulate and possibly insular cortex and in the inferior corpus striatum ($p < 0.01$). In a controlled study, Stoetter et al. (190) examined 18 GTS patients and found that they had lower relative metabolic rates in inferior limbic regions of the cortex, striatum, and subcortical limbic structures and higher relative metabolic rates in superior sensorimotor cortices.

In a study of presynaptic functional integrity of dopaminergic terminals using [^{18}F]-DOPA PET in ten GTS patients, the striatal metabolism of exogenous levodopa and the density of striatal D_2 receptors were found to be normal (191). These investigators concluded that GTS does not arise from a primary dysfunction of dopaminergic terminals.

Wong et al. (192) studied 21 adult GTS patients and age- and sex-matched controls by PET with [^{11}C]3-*N*-methylspiperone ([^{11}C]NMSP). They used two methods, namely, caudate-to-cerebellar ratio and two-PET scan procedure, for absolute measure of receptor density (B_{max}). Neither group showed significant differences from their control group for caudate-to-cerebellar ratio. However, the two-PET scan B_{max} measurement showed that four of the 20 patients had significantly elevated D_2-like receptors. The authors also found a trend between the severity of vocal tics and B_{max} values and a significant association between B_{max} measure and the Wisconsin Card Sorting Test. These authors believe that not all GTS patients have an abnormality of D_2-like receptors but that a subgroup has a significant D_2-like dopamine receptor elevation.

Braun et al. (193) analyzed [^{18}F]fluoro-deoxyglucose PET scans from 18 medication-free GTS patients. They found that cognitive and behavioral features such as obsessions and compulsions, impulsivity, coprolalia, SIB, echophenomena, depression, and measures of attentional and visuospatial dysfunction were associated with significant increases in metabolic activity in the orbitofrontal cortices. Similar increases, although less robust, were observed in the putamen and, in the case of attentional and visuospatial dysfunction, in the inferior portions of insula. Behavioral and cognitive features were not associated with metabolic rates in other subcortical, paralimbic, and sensorimotor areas in which metabolism had, in some cases, distinguished between GTS and controls (194).

SPECT imaging has shown hypoperfusion in the basal ganglia, thalamus, and frontal and temporal cortical areas (195), as well as elevated frontal cortex blood flow (relative to basal ganglia, inclusive of the caudate nucleus and putamen) in GTS patients (196). Moriarty et al. (197) examined 50 GTS patients and 20 controls using hexamethyl-propyleneamine oxime (HMPAO) SPECT. Patients differed from controls on measures of relative blood flow to the left caudate, anterior cingulate cortex, and left dorsolateral prefrontal cortex. Severity of tics was related to hypoperfusion of the left caudate and cingulate and the left medial temporal region. Hypoperfusion in the left dorsolateral prefrontal region was related to mood. The authors related their findings to known functions of the frontal lobe and striatum. A wide range of perfusion patterns was seen, and no characteristic patterns for behavioral subgroups have been documented using HMPAO SPECT (197). Riddle et al. (198) found hypoperfusion in basal ganglia and frontal areas that reached statistical significance in the left putamen–globus pallidus (relative technetium-99m-HMPAO uptake was 4% lower in GTS compared to controls). Klieger et al. (199) performed SPECT scans with HMPAO in six GTS patients and nine controls. Five of the six GTS patients showed a significant decrease in right basal ganglia activity compared with none of the controls.

George et al. (200) performed single-slice dynamic SPECT with [123I]-iodo-6-benzamide ([123I]-IBZM) in 15 GTS patients (eight unmedicated) and six healthy volunteers. Unmedicated GTS patients showed no differences from control subjects, whereas GTS patients taking D_2-blocking medications had significantly decreased [123I]-IBZM binding in both right and left basal ganglia compared with normal control subjects. These results suggest that D_2 receptor availability as measured by [123I]-IBZM SPECT is not abnormal in GTS. Muller-Vahl et al. (201) demonstrated an increase in dopamine transporter (DAT) activity in a study of 12 GTS patients and nine control subjects using SPECT and [123I]-labeled 2β-carbomethoxy-3β-(4-iodophenyl)tropane ([123I]-β-CIT). They found significantly higher striatal activity ratios in GTS patients than in controls and an association between binding ratios and "self-injurious behavior" and "lack of impulse control." Similar higher striatal [123I]-β-CIT in GTS patients compared to age- and gender-matched comparison subjects was found by other investigators as well (202). These studies support the hypothesis of a dysregulation in presynaptic dopamine function in GTS. In another study using SPECT and [123I]-IBZM, Muller-Vahl et al. (203) examined post-

synaptic dopamine D_2 receptors in 17 GTS patients and a control group. In neuroleptic-treated patients, [123I]-IBZM binding was significantly reduced compared to both normal controls and unmedicated patients. In unmedicated patients, the mean binding ratio did not differ from controls. However, patients in an advanced stage of GTS showed significantly reduced relative striatal binding compared to patients in earlier stages and controls. The authors suggested that the spontaneous recovery of tics noted in early childhood in some cases is associated with reduced receptor binding capacity.

In a family study using SPECT imaging, Moriarty et al. (204) noted that the "affected" family members of GTS probands showed hypoperfusion in striatal, frontal, and temporal areas, irrespective of whether they had tics, GTS, or OCD. Whereas most of the earlier studies of primary OCD patients noted hyperperfusion (205), none of the affected members in this study, including those who had only OC symptoms, showed hyperperfusion. Furthermore, the "affected" family members were noted to have lower anxiety scores irrespective of their clinical presentation. This result, coupled with the finding of a correlation between improvement in anxiety and a reduction in the orbitofrontal metabolism in OCD subjects (206), suggests that the difference in perfusion findings between OCD subjects and TS+OCD patients is related to the difference in anxiety levels.

Structural abnormalities of the brain in OCD have been sought by various brain imaging techniques, but the results have been inconsistent. Using CT scans, Insel et al. (207) found no significant anatomic differences compared with normal controls, whereas other researchers found that adolescent patients with OCD had a significantly higher mean ventricular brain ratio than controls (208). Luxenberg et al. (209) suggested involvement of the basal ganglia in the pathogenesis of OCD because CT scanning of young male patients showed reduced caudate volumes. There have been several reports of patients showing compulsive stereotypical behavior: two patients after carbon monoxide poisoning, one after a wasp sting (210), and one of unknown etiology (211). In this last patient, CT scans showed bilateral cavitation of the basal ganglia. It also has been hypothesized that basal ganglia developmental processes (developmental basal ganglia syndrome) can produce clinical symptoms of GTS and OCD (212). The association of OCD symptoms with a wide range of genetic and environmental conditions that cause dysfunction of the basal ganglia, such as postencephalitic parkinsonism and Sydenham chorea, provides further evidence supporting the involvement of the basal ganglia in OCD (213).

MRI studies have provided similar results. Kellner et al. (214) detected no consistent gross brain anatomic abnormalities on MRI in 12 patients with OCD. Garber et al. (215) found no structural abnormality unique to OCD, but they observed regional tissue abnormalities, particularly in the orbitofrontal cortex, which correlated strongly with symptom severity in patients with OCD.

Neuroimaging techniques such as PET, which allow functional rather than anatomic assessment of the brain, have extended the findings of neuroanatomic studies by indicating that orbitofrontal-basal circuitry may mediate OC behavior. PET studies of cerebral glucose metabolism in patients with OCD found increased rates in the caudate nuclei and left orbital gyri (216,217); in the left orbitofrontal, right sensorimotor, and bilateral prefrontal and anterior cingulate regions (218); and in both orbital gyri (219). The patients in the latter study who had not been taking medication were rescanned after treatment with clomipramine (220). Symptomatic improvement coincided with a return of glucose metabolism to more normal levels in orbitofrontal regions and the basal ganglia compared with normal controls. Patients who responded well to treatment showed more marked changes in cerebral glucose metabolism than those who responded poorly.

Functional imaging using PET and SPECT has demonstrated metabolic and perfusion abnormalities in the basal ganglia and frontotemporal areas (221). Whereas studies involving GTS patients have predominantly found hypoperfusion in striatal and frontal areas, studies of primary OCD patients have found hyperperfusion of similar areas. It seems that, although the basal ganglia abnormalities are specific, the hypofrontality found in GTS may be a nonspecific finding that also is found in depression and schizophrenia (222–224).

NEUROANATOMY

Frontal-subcortical circuits, dopaminergic neurotransmission, and second messenger systems have attracted much attention as the neurobiologic substrates of GTS.

No abnormalities were found in the first case of "idiopathic" GTS carefully studied postmortem (225). Balthasar (226) subsequently suggested that there was pathologic development of the striatum after a careful cell count of the brain of a patient with GTS showed that the number of small striatal cells matched the cell count in a 1-year-old child. Richardson (227) reviewed the two findings and concluded that there were no distinctive histopathologic abnormalities.

More recently, postmortem investigations have suggested decreased levels of 5-hydroxytryptamine (5-HT; serotonin) and glutamate (especially in the subthalamus) in many areas of the basal ganglia (228), a reduction in the second messenger cyclic adenosine monophosphate, and an increased number of dopamine uptake carrier sites in the striatum (229). Haber and colleagues (230,231) reported detailed neuropathologic and immunohistochemical postmortem examinations of five GTS patients and found a reduction of dynorphin-like immunoreactivity in the globus pallidum. These findings have excited much interest but require replication.

Dysfunction centered in the nucleus accumbens (232) and the cingulate and orbitofrontal cortex (233) has been hypothesized to be the neurobiologic basis of GTS. Aberrant activity in the interrelated sensorimotor, language, executive, and paralimbic circuits has been suggested to be involved in the initiation and execution of diverse motor and vocal behaviors in GTS. Using event-related PET techniques, Stern et al. (234) demonstrated that brain regions in which activity was significantly correlated with tic occurrence were the medial and lateral premotor cortices, anterior cingulate cortex, dorsolateral-rostral premotor cortex, inferior parietal cortex, putamen, caudate, primary motor cortex, the Broca area, superior temporal gyrus, insula, and claustrum. In one patient with significant coprolalia, vocal tics were found to be associated with activity in the prerolandic and postrolandic language regions, insula, caudate, thalamus, and cerebellum, and motor tics were found to be associated with activity in the sensorimotor cortex.

The neuroanatomic basis of OCD has been inferred from the results of new radiologic techniques, neuropsychological and electrophysiologic studies in patients with OCD, and the occurrence of OC behavior in patients with certain brain lesions or neurologic disorders. In this regard, the basal ganglia-thalamocortical circuits appear to be important in both GTS and OCD. Saint-Cyr et al. (235) suggested that diverse lesions, depending on the site, can result in problems with the development and maintenance of behavioral sets ("hypophrenic") versus problems in relinquishing preferential sets ("hyperphrenic"). In OCD, the "hyperphrenic" pattern seems to apply only to obsessional rituals. Using problem-solving and cognitive tasks, OCD patients were found to require more practice and/or the provision of external guidelines to facilitate habit formation. Thus, as in other disorders of the basal ganglia, the establishment of useful heuristics by which to direct adaptive behavior suffers. These investigators concluded that OCD patients have at least two types of basal ganglia dysfunction: ritualistic obsessions and compulsions, and heuristic inefficiency.

At the molecular level, the biochemical mechanisms underlying the disorder are being elucidated by investigating abnormalities of neurotransmission in OCD patients, which has led to the so-called serotonin hypothesis of OCD. The evidence for this hypothesis and the involvement of other neurotransmitter systems in the modulation of OC behavior are reviewed in greater detail in the following sections.

Although GTS and OCD both involve neuropathology of the basal ganglia-thalamocortical pathways, differential involvement of sensorimotor and limbic circuits in GTS and prefrontal and limbic pathways in OCD has been suggested (236). Wolf et al. (237) studied MZ twins discordant for GTS severity and found that differences in D_2 receptor binding in the head of the caudate nucleus predicted differences in phenotypic severity, but this relation was not observed in putamen. The authors suggest that there is a link between GTS and a spectrum of neuropsychiatric disorders that involve associative striatal circuitry.

NEUROCHEMISTRY

Dopaminergic System

The neurochemical basis for GTS is not yet known. A thorough review has examined the evidence for biochemical abnormalities, and the main hypothesis is an imbalance of CNS neurotransmitter agents (238). Dopamine has received the most support, because haloperidol and other dopamine antagonists reduce the symptoms in a large number of patients, whereas stimulants such as pemoline and methylphenidate exacerbate the symptoms. Homovanillic acid (HVA) has been found to be decreased in the cerebrospinal fluid (CSF) of some GTS patients, although the proposed methods involved have been criticized (238). Friedhoff (239) addressed the possible role of D_1 and D_2 receptors in GTS, highlighting a distinction between the two. Thus, the D_2 system, unlike the D_1, is sensitive to "adaptive unregulation." When blocked by D_2 antagonists, such as haloperidol and pimozide, it increases the number of receptors, which serves to overcome the blockade. A similar phenomenon does not occur in the D_1 system unless there are very long periods of blockade. The implications of this difference and its relevance to GTS are discussed in detail by Friedhoff (239), including issues such as epigenetic transmission and the male-to-female ratio (240). The eyeblink rate, a supposed correlate of central dopamine activity, has been found by some investigators (77,241) but not others (242) to be increased in patients with GTS.

The integrity of presynaptic dopaminergic function in GTS children was studied by Ernst et al. (243) using PET and [^{18}F]-fluoro-DOPA (FDOPA). Subjects with GTS showed higher FDOPA accumulation in the left caudate nucleus and right midbrain than seen in controls, which indicates an up-regulation of DOPA decarboxylase activity. This suggests that dopaminergic dysfunction in GTS affects both cell nuclei and nerve terminals. Rabey et al. (244) measured the uptake of [^{3}H]-DA into platelet storage capsules and found that DA uptake by platelet storage capsules was significantly lower in GTS patients compared to controls. The authors suggested that this reflects compensatory presynaptic changes that reduce DA activity.

Other Transmitter Mechanisms

Investigations implicating noradrenergic systems and acetylcholine have been equivocal, and there is little support for the involvement of serotonin or γ-aminobutyric acid in the pathophysiology of GTS (238). Other investigators have suggested an underactivity of the endogenous opioid system (245).

With no one transmitter convincingly implicated to date, the importance of animal models and further postmortem studies is stressed. Cohen et al. (246) stated that GTS reflects the interaction among genetic, neurophysiologic, behavioral, and environmental factors. Eldridge and Denckla (247) suggest that a complex interaction of androgenic and immunologic factors is involved in susceptibility to neurodevelopmental disability and, therefore, GTS. They further note that, when viewed in this context, the observation of Balthasar (226) of an increased number of neurons in the caudate and putamen, which suggests the persistence of an immature neuronal pattern, may be more understandable (247). Moreover, it has been hypothesized that the brain regions involved in GTS are those that participate in primitive reproductive behavior and whose development are under sex hormone control (248).

Bonnet (241) addressed the anatomic localization of GTS and concluded that the biochemical structure of the limbic forebrain structures, particularly the anterior cingulate cortex, and their interrelationships with other specific nuclei suggest they are the anatomic site for GTS. Trimble and Robertson (249) agree with Bonnet. Others have suggested that pathology in the periaqueductal gray matter and midbrain tegmentum (175,250) and in the amygdaloid complex (251) is implicated in GTS.

Other Biochemical Evidence

Robertson et al. (252) reported that ten of 80 GTS patients had an abnormally low serum copper level. Two of the ten patients were investigated in detail with copper radioisotope studies. These two patients exhibited abnormalities of copper handling, as noted by fast disappearance of copper from the plasma, slow uptake by the liver, and low incorporation of copper into ceruloplasmin. Wilson disease was excluded in all 80 subjects based on detailed clinical examination. The abnormalities found carried no etiologic or treatment implications.

Biochemical Abnormalities in Obsessive-Compulsive Disorder

Serotonin Hypothesis

The neurotransmitter serotonin (5-HT) has been implicated as playing an important role in the pathogenesis of OCD. Convincing evidence from different lines of research has led to a serotonergic neurochemical hypothesis for OCD. The studies to date that have investigated such a role for serotonin can be grouped under three main separate headings (253): (a) drug treatment studies, (b) peripheral markers of 5-HT function, and (c) pharmacologic challenge studies.

Drug Treatment Studies. The most impressive evidence for serotonergic involvement in this disorder comes from numerous drug response studies demonstrating the efficacy of drugs that inhibit neuronal uptake of 5-HT in ameliorating the obsessions and compulsions of OCD. It was first noted in the 1960's that the tricyclic antidepressant clomipramine had antiobsessional properties, even when evidence of an underlying depressive illness was lacking (254–257). Since 1980, clomipramine has been found to be superior to placebo in terms of antiobsessive effects. Unlike antidepressive drugs, such as amitriptyline, desipramine, and clorgyline, which preferentially block uptake of norepinephrine over serotonin, clomipramine has a potent, though nonselective, action on serotonin and has been confirmed to have antiobsessional effects in double-blinded, placebo-controlled trials (251,258–265). Its efficacy for treatment of patients with OCD was proved in a large multicenter trial (266). Although conventional norepinephrine reuptake inhibitor antidepressants do not have significant antiobsessional effects (266), the pharmacologically active metabolite of clomipramine, desmethylclomipramine, has noradrenergic reuptake properties, which casts doubt on the belief that serotonin is essential for clomipramine's mechanism of action in the therapy of OCD. One study demonstrated a positive correlation between levels of this metabolite and improvement of symptoms (267), but other researchers have refuted this finding (258,268).

Additional evidence for the role of serotonin in OCD came with the development of more selective and potent 5-HT uptake inhibitors. Various studies confirmed their effectiveness in OCD. Thus, double-blinded, placebo-controlled clinical trials of fluvoxamine (269–271), fluoxetine (272), and sertraline (273) have confirmed their antiobsessional action and their superiority to desipramine (274). Zimeldine and trazodone, a serotonin 5-HT$_2$ antagonist (275,276), also may have antiobsessional properties.

The mechanism by which these drugs produce their therapeutic effects in OCD and their interaction with the brain's 5-HT system have not been fully elucidated. Significant reuptake inhibition of 5-HT can be detected after a single administration of a 5-HT reuptake inhibitor, yet it takes several weeks for the antiobsessional therapeutic action to become clinically apparent. There is ample evidence that the long-term administration of 5-HT uptake inhibitors produces changes in 5-HT function, but there is dispute over the direction of such changes. Thus, electrophysiologic studies in animals show enhanced postsynaptic 5-HT neurotransmission with the chronic administration of such drugs (277,278). The drugs appear to inhibit presynaptic autoreceptor 5-HT release. Conversely, down-regulation of the 5-HT$_2$ receptor subtype has been demonstrated after long-term administration of 5-HT uptake inhibitors (279,280), which suggests that they cause an overall decrease in serotonergic transmission. These differences may be explained by differential actions on separate subtypes of 5-HT receptor, of which there are at least eight (281), some of which possess reciprocal functional interaction. Furthermore, these drugs may, by their serotonergic action, induce compensatory alterations in other neurotransmitter systems that are more directly involved in the pathogenesis of OCD (253).

Peripheral Markers. Peripheral measurement of serotonin and its metabolites may reflect CNS 5-HT neurotransmission. Hence, abnormalities of putative peripheral markers of 5-HT, including whole blood (282,283) and platelet serotonin concentration (284), imipramine binding in platelets (285–290), and CSF levels of the 5-HT metabolite 5-hydroxyindoleacetic acid (5-HIAA) (261,285,291), have been sought in OCD. Although no coherent differences between patient groups and controls have emerged from these peripheral marker studies, some subgroups of OCD patients showed evidence of abnormal 5-HT turnover. In one study, OCD patients with a family history of OCD had significantly higher blood 5-HT levels than either those with no family history or normal controls (459). This finding suggests that increased serotonergic activity occurs in this subset of patients. On the other hand, very low levels of 5-HIAA have been detected in CSF from patients with recurrent violent obsessions, which implies a decrease in 5-HT turnover (292). de Groot et al. (293) attempted to identify clusters of OC characteristics in GTS and to explore their neurochemical correlates. Overall, 5-HIAA, the primary metabolite of serotonin, was found to be most highly correlated with the individual OC symptoms. Leckman et al. (294) studied CSF biogenic amines in OCD, GTS, and controls. Whereas CSF tyrosine concentration was found to be reduced in OCD than controls, CSF norepinephrine was found to be 55% higher in GTS and 35% higher in OCD than in controls. CSF HVA levels were reduced in OCD compared to GTS but not compared to controls. No mean differences were noted in CSF MHPG, tryptophan, and 5-HIAA across the three groups. The CSF norepinephrine data seem to suggest that noradrenergic mechanisms are involved in GTS, whereas alterations in the balance of dopaminergic, noradrenergic, and serotonergic systems are likely to be involved in the pathobiology of OCD. In a study of serotonergic mechanisms in GTS, Cath et al. (295) reported that their data did not support a predominant role of 5-HT in the tics associated with GTS. They noted a trend for amelioration of impulsions in response to *m*-chlorphenylpiperazine (m-CPP), a relatively selective 5-HT$_{2c}$ agonist, and concluded that this can be interpreted as circumstantial evidence for impulsivity-related 5-HT hypofunctionality in GTS. However, the authors cautioned that in view of the large variability of m-CPP plasma concentrations, the reliability of m-CPP as a probe for challenge studies is doubtful.

Pharmacologic Challenge Studies. A more direct approach to the investigation of 5-HT receptor function disturbance in OCD was anticipated with the introduction of neuroendocrine probe research. In pharmacologic challenge studies, the neuroendocrine, physiologic, or behavioral responses to selective serotonergic agents can be interpreted as a measure of the integrity of the hypothalamic serotonergic system. For example, the relative rise or fall in plasma levels of pituitary hormones or cortisol after administration of such a probe can be monitored, and the observed blunting or enhancement of the normal neuroendocrine response can be interpreted as underactivity or overactivity of 5-HT transmission, respectively. Similarly, the exacerbation or amelioration of OC symptoms in patients with OCD, as well as in healthy controls, after serotonergic pharmacologic challenge can be viewed as behavioral markers of changes in central 5-HT activity.

The results of such studies have been confusing and inconclusive. The most frequently used serotonergic probe is the metabolite of trazodone, m-CPP, which acts as an agonist for certain 5-HT receptor subtypes and may act as an antagonist for others. Administration of this agent reliably causes increases in body temperature and levels of serum prolactin and cortisol in laboratory animals (296) and produces behavioral responses akin to anxiety seen in humans. m-CPP has been given to normal volunteers and patients with OCD, with variable results. Some studies in untreated OCD patients reported a blunting of the normal prolactin response but no change in the cortisol response (297,298), which suggests hyposensitivity of some 5-HT receptors. Another study reported a blunting of the cortisol rather than the prolactin response (299) and no change in these responses in patients whose symptoms had remitted with SSRI treatment (300). However, another study found enhancement of the cortisol response after treatment.

The discovery that m-CPP was not as selective a 5-HT probe as originally thought, having action at noradrenergic, dopaminergic, muscarinic receptors, and serotonergic receptors, led researchers to examine the clinical effects of putative, more selective 5-HT agonists and antagonists. Studies using the 5-HT$_{1A}$ agonist ipsapirone (301) and the serotonin agonist and reuptake inhibitor fenfluramine (297,302) showed no difference in prolactin and cortisol between OCD patients and controls. An increase in the prolactin response, however, was observed with intravenous L-tryptophan (298), which is a 5-HT precursor, but it also may alter dopamine transmission (303).

The results of pharmacologic challenge studies examining the behavioral responses of OCD patients and normal controls have been equally inconsistent. m-CPP given to OCD patients induced exacerbations of OC symptoms in some studies (297,299) but not in others (253,298). On the other hand, differential anxiety-related responses between subject populations after m-CPP challenge that distinguishes OCD from panic disorder have been more consistently observed. In these cases, m-CPP failed to produce heightened anxiety in OCD patients (297,298) but caused panic attacks in patients with panic disorders (304,305).

L-Tryptophan, fenfluramine, and ipsapirone have been shown to produce negligible behavioral effects in OCD patients (297,298,301,302). Metergoline, a 5-HT$_1$ and 5-HT$_2$ receptor antagonist, apparently improved obsessional symptoms in untreated patients when it was given as a single dose (299). When administered in repeated doses, metergoline reversed the therapeutic effect of clomipramine in OCD patients (306), although these latter results may not have been significant.

The lack of consensus from these serotonergic challenge studies may be due in part to inherent methodologic flaws in their design. The symptomatology measured in the studies after a single dose of the pharmacologic probe may bear little resemblance to the obsessions and compulsions seen in the clinical setting of OCD, where response to intervention is observed over a much longer period of days or weeks. Furthermore, the behavioral responses observed may have been influenced by environmental triggers, which can be difficult to control. Other variables, such as menstrual cycle or seasonal changes, may alter 5-HT function. Results of earlier neuroendocrine and behavioral challenge studies should be reinterpreted in light of more advanced understanding of the complexity of brain 5-HT pathways and their multiple subsystems and receptors. Barr et al. (253) attempted to reconcile the disparate findings of these studies by proposing that OCD is characterized by neuroendocrine hyposensitivity linked with behavioral hypersensitivity to serotonergic stimulation. Baldwin et al. (307) observed enhanced prolactin and growth hormone response to L-tryptophan. They hypothesized that dysfunction of 5-HT receptors is important in the pathogenesis of OCD. Future double-blinded, placebo-controlled studies using more selective test compounds are required to delineate the precise serotonin deficit underlying this disorder.

Difficulties in the interpretation of studies implicating 5-HT in the pathophysiology of OCD must be due, at least in part, to the fact that there are many 5-HT receptor subtypes (5-HT$_{1A-E}$, 5-HT$_2$, 5-HT$_3$, 5-HT$_4$) (253). It has been stated that the ongoing identification of new subtypes of 5-HT receptors makes conclusions based on the specificity of a particular challenge agent problematic (253). We extend that cautioning to any treatment studies and investigations of peripheral 5-HT receptors.

Dopamine and Other Neurotransmitters

Other neurotransmitter systems may be more directly involved in OCD. Challenge studies of the noradrenergic agents clonidine (308–310) and yohimbine (311) provide no firm evidence that significant dysfunction of the noradrenergic system occurs in OCD. However, there is some evidence that the clinically relevant action of SSRIs in OCD is related to the down-regulation of α-adrenergic receptors (312–314). Exacerbated obsessions and compulsions in two patients given the opiate antagonist naloxone (315) has promoted speculation that opiate receptors are operational in OC symptomatology.

A more convincing case has been made for the role of dopamine in some forms of OCD following the results of preclinical studies (316). Animal studies have shown that some dopaminergic agents, such as L-dopa, bromocriptine, and the stimulant amphetamine, produce stereotypical behavior analogous to the compulsive rituals seen in patients with OCD (317–320). Moreover, the use of such stimulants in humans has been documented to induce purposeful and repetitive actions that closely resemble the compulsions that occurred in OCD in previously healthy subjects (321–325). As a result of the brain imaging and neuropsychological studies just reviewed, the basal ganglia, which is an area richly innervated by dopamine-containing and 5-HT–containing neurons, has been implicated as forming part of the neuroanatomic substrate for OCD. Similarly, dopamine neurotransmission probably is affected in both OCD and GTS. GTS is a disorder considered to be due in part to disturbed dopamine metabolism, and the presence of obsessions and compulsions in some neurologic diseases involving the basal ganglia supports the notion that dopamine neurotransmission is affected in OCD.

Drug response data have strengthened a dopamine-serotonin hypothesis of OCD. SSRIs have been shown to have dopamine-blocking activity (326), which may contribute to their antiobsessional effect. In addition, there are reports that neuroleptic dopamine antagonists have been used successfully in conjunction with a 5-HT uptake inhibitor for treatment of some cases of GTS concomitant with OCD (327,328) and for OCD unresponsive to the 5-HT uptake inhibitor alone (329). Fluoxetine has been associated with extrapyramidal symptoms similar to those seen with neuroleptic treatment (330–333), and decreased CSF levels of the principal dopamine metabolite HVA have been found after fluoxetine treatment (334). These findings have been interpreted as fluoxetine-facilitating serotonergic inhibition of striatal dopamine neurons. However, clomipramine administration has been associated with increased levels of HVA in the CSF (261). A study of CSF neurochemistry in children and adolescents with OCD showed that CSF HVA levels did not correlate with OCD symptoms (335). These conflicting results may be explained by the existence of several forms of OCD, some of which may involve dysregulation of dopamine neurotransmission.

Neurobiologic Link Between Gilles de la Tourette Syndrome and Obsessive-Compulsive Disorder

There is growing evidence suggesting that GTS and OCD have some common neurochemical and neuroanatomic bases. As stated earlier, in GTS patients, postmortem and neuroimaging studies suggest structural and functional abnormalities in the basal ganglia, cingulum, and frontal areas.

Also, the absence of premovement potentials prior to the tics indicates that the movements are subcortical in origin (155). Similar studies in OCD suggest orbitofrontal, anterior cingulate-basal ganglia-thalamocortical circuits as the neuroanatomic substrate.

Direct evidence to support this hypothesis is scarce. To date, a unique neurochemical abnormality has not been established in either condition. Dopamine is thought to be the main neurotransmitter involved in GTS, whereas serotonin, not dopamine, is considered the major neurotransmitter involved in OCD. Results from treatment and electrophysiologic studies do not exclude the possibility that dopamine abnormalities also are involved in OCD (336). Although OCD responds poorly to dopamine antagonist drugs alone, McDougle et al. (337) found in an open trial that nine of 17 fluvoxamine-resistant OCD patients responded to combined fluvoxamine-dopamine antagonist treatment. Several studies examining the efficacy of combined SSRIs and dopamine antagonists for treatment of OCD are underway. It is possible that the OCD patients who respond to this combined treatment represent the subgroup associated with a family history of GTS proposed earlier.

Although researchers on the whole have not concluded that serotonin is involved in the pathophysiology of GTS (238), other investigators (338) suggest that it is a hyposerotonergic condition. Successful treatment of OC symptoms in GTS patients using both clomipramine and fluoxetine has been reported, with the latter drug also decreasing the severity of tics in one patient (339). Use of other forms of treatment of GTS and OCD highlights the similarities and differences between the two conditions.

Despite differences in the effectiveness of pharmacologic treatment in the two disorders, the tics, vocalizations, and OCD in GTS patients and their relatives may be products of common neurochemical disturbances. The striatum and limbic system receive extensive projections from both dopamine and serotonin transmitter systems. Disturbances in these parts of the brain could be responsible for dopamine-mediated tics and vocalizations and for serotonin-medicated obsessions and compulsions in these patients.

NEUROPSYCHOLOGY

Early neuropsychological studies were reviewed by Golden (144) and Robertson (21). They noted an average IQ in GTS patient cohorts, with verbal/performance discrepancies of 15 points (performance being the lower); specific deficits in reading, writing, and arithmetic; and dysfunction on the Halstead-Reitan Neuropsychological Assessment Battery. Specific learning problems were found in 36% of 200 children with GTS (26). Learning disability and GTS have been reported to coexist in some individuals (116,340); the combination has been reported in 10% of one GTS cohort (151).

Language skills appear to be largely unimpaired, whereas consistent deficits in visual-motor performance have been documented by a number of authors. In a study of 31 GTS children, 20 normal siblings, and ten children with arithmetic disabilities (AD), Brookshire et al. (341) found evidence of nonverbal learning disabilities. The three groups generally performed within the average range on academic and cognitive measures, but the GTS children and the AD children demonstrated poorer performance on written arithmetic tasks but comparatively better performance on word reading and written word spelling. Unlike the AD group, which had generalized nonverbal deficiencies, GTS children had reduced performance on visual-motor and expressive language measures and measures of complex cognition. The investigators suggested that GTS is an "output" subtype of learning disability, perhaps involving the mesocortical dopaminergic system.

Controlled studies have suggested that the intention editor is impaired in GTS (342). This editor, which is a key mechanism that underlies the will and which begins to function in early childhood, is triggered whenever there are several intentions competing in parallel with each other. It is hypothesized to be a subcomponent of the supervisory attentional system, which serves inhibition and is subserved by frontal circuits (342). There also appear to be deficits in attention, especially on more complex tasks, including serial addition, block sequence span (forward), the Trail-Making Test, and letter cancellation vigilance tasks (343). Because GTS subjects and controls did not differ significantly in IQ (which was in the normal range), it was suggested that these findings represent selective deficits rather than global impairments in functioning for the GTS group (243). In this regard, Schultz et al. (344) observed that the integration of visual inputs and organized motor output is a specific area affected in GTS. In a study of 50 GTS children and 23 age-matched controls, the authors found that GTS children performed significantly worse than controls on the Berry Visual-Motor Integration Test. It also was noted that visuoperceptual and fine motor coordination subprocesses were significant predictors of visual motor integration scores.

Oades (345) recorded signal detection measures on four tests of "sustained attention," with increasing working memory requirements in 14 healthy children, 14 with ADHD, and 11 with a tic syndrome. Clinical associations also were made from 24-hour urinary measures of monoamine activity. The cancellation paper-and-pencil test revealed no group differences in errors or signal detection measures. On the computerized continuous performance test (CPT), ADHD children made more commission and omission errors than controls, but GTS children made mostly omission errors. This suggested poor perceptual sensitivity (d') for ADHD and conservative response criteria (beta) for GTS. This group difference extended to the CPTax (working memory), which was evident on regression analysis to test for putative working memory-related abilities and concentration. In all the

children, immediate response feedback reduced omissions and mostly improved d'. CPTax performance was negatively related to dopamine metabolism in controls and to serotonin metabolism in ADHD children. Independent of difficulty, CPT tasks demonstrated a perceptual-based impairment in ADHD and response conservation in GTS. Catecholamine activity was implicated in the promotion of perceptual processing in normal and ADHD children, but serotonin activity may contribute to poor CPTax (working memory) performance in ADHD patients.

de Groot et al. (346) studied 92 GTS children grouped by the presence or absence of obsessive-compulsive and/or attention deficit symptoms. After statistical control for complex motor symptoms, impaired performance on measures of achievement and executive functioning was found to be correlated with obsessive and obsessive/attentional symptoms but not with attentional symptoms alone. Similarly, in a study that compared executive function abilities in nonretarded autistic children, GTS children, and normal controls, Ozonoff et al. (347) found that GTS children did not differ from controls on any of the tasks, but the autistic sample was significantly impaired on a measure of cognitive flexibility. In another study, Ozonoff et al. (348) examined central inhibitory function using a negative priming task in 46 GTS children and 22 control children matched for age, gender, and IQ. As a group, the GTS children did not differ from controls in inhibitory function. Subgroup analysis of GTS children (one that had no evidence of comorbidity and the other that met criteria for ADHD, OCD or both) revealed that those with numerous and severe symptoms of GTS, ADHD, and OCD performed significantly less well than controls and GTS subjects having fewer and less severe symptoms. It appears that neuropsychological impairment that occurs in the context of GTS is a function of comorbidity and symptom severity.

Many researchers have investigated the effects of medication on cognitive function in patients with GTS. Apparently, haloperidol has no consistent global effect on performance (144). One study indicated that GTS patients who received medication, especially butyrophenones, had lower IQ scores (72). It remains to be shown, however, that individual patients are not impaired and that there is not a deleterious effect on specific areas of cognitive function (144).

Neuropsychological Evidence in Obsessive-Compulsive Disorder

Encouraged by the neuroradiologic findings, researchers have looked for correlating neuropsychological deficits in frontal lobe, basal ganglia, or limbic system functioning in patients with OCD. Earlier studies gave inconsistent results but were methodologically flawed by small sample size, questionable selection criteria for both patients and controls, and choice of test instruments. Neuropsychological testing of OCD patients has detected frontal lobe dysfunction

(208,349–351), memory impairment (350,352), and reduction of visual-spatial skills, performance IQ, and motor functioning, which are predominantly right-sided activities (207,349,350,353). Studies designed to avoid the methodologic problems encountered previously found cognitive deficits consistent with basal ganglia or right hemisphere disturbance but no frontal lobe dysfunction. Boone et al. (354) revealed subtle visual-spatial and visual-memory deficits in 20 nondepressed OCD patients. The deficits were most pronounced in patients with a family history of OCD, but the researchers documented no difference in frontal lobe functioning, verbal memory, attention, or intelligence compared with matched normal controls. Zielinski et al. (355) also noted visual-spatial and nonverbal processing deficits consistent with subcortical or right hemisphere pathology in 21 patients with OCD who actually scored as well or better than normal controls on frontal lobe tasks.

Christensen et al. (356) did not find a significant decrease in visual-spatial performance on neuropsychological testing of 18 nondepressed OCD patients compared with 18 nondepressed controls, but they did identify a recent nonverbal memory deficit that correlated to abnormalities of the limbic and paralimbic areas of the right hemisphere.

The variations in frontal lobe performance observed between the later studies and previously reported evidence for frontal lobe impairment in OCD may be due to the inclusion in earlier studies of depressed OCD subjects who may perform poorly on frontal lobe tests. However, it is possible that the frontal lobe tests used were not sufficiently sensitive to detect abnormalities within specific regions of the frontal lobes, such as the orbitofrontal area, in which abnormalities have been detected on radiologic studies (215,218,219).

NEUROIMMUNOLOGY

Swedo et al. (357) suggested that tics and associated behaviors, including OCD, develop from streptococcal infection by molecular mimicry, whereby antibodies directed against bacterial antigens cross-react with brain targets. This spectrum of neurobehavioral disorders has been termed pediatric autoimmune neuropsychiatric disorders associated with streptococcal infection (PANDAS). There is debate as to whether the autoimmune process following streptococcal infection plays a direct etiologic role or acts indirectly to modulate the phenotypic expression of GTS. In this regard, antineuronal antibodies in association with involuntary repetitive movement disorders have been demonstrated (358). These investigators performed immunoassays of antineuronal antibodies in children with GTS or Sydenham chorea. Sensitivity, specificity, positive predictive value, and negative predictive value of this assay for diagnosis of GTS were 79.1%, 61.2%, 61.6%, and 78.8% and for diagnosis of chorea were 71.1%, 60.9%, 68.6%, and 63.6%, respectively.

Singer et al. (359) studied serum antineuronal antibodies against neuron-like HTB-10 neuroblastoma cells in children with GTS and found that GTS children have higher median, but not mean, levels of antineuronal antibodies. They observed antibodies in both control and clinical groups, but they failed to identify a relationship between antibodies and clinical phenotypes or one-time markers for streptococcal infection.

Murphy et al. (360) investigated whether monoclonal antibody D18/17 levels are higher than normal in some forms of OCD and GTS. They found that the average percentage of B cells expressing the D8/17 antigen was significantly higher in the patients than in controls. All patients but only one comparison subject was positive for D8/17. The authors suggested that D8/17 serves as a marker for susceptibility in some forms of OCD and GTS.

Kiessling et al. (361) examined whether children with OCD show evidence of caudate nucleus involvement. They studied antineuronal antibodies from sera of clinical cases with OCD and ADHD with or without concomitant tics and found that antibodies were directed against caudate, putamen, or both, at rates significantly higher than in controls. They concluded that there is no differential effect against caudate but there is more generalized basal ganglia involvement in OCD. Singer et al. (362) found that, compared with controls, GTS subjects had a significant increase in the mean and median enzyme-linked immunosorbent assay optical density levels of serum antibodies against putamen but not against caudate or globus pallidus.

Muller et al. (363) investigated whether an increase of titers of streptococcal antibodies could be reproduced in GTS patients. They found that the mean values of antistreptolysin O and anti-deoxyribonuclease B titers were significantly higher in children and adults with GTS compared to control children and adults and to schizophrenic patients. These findings support the theory that infections or postinfectious phenomena play a role in the pathogenesis of GTS.

Preliminary serologic evidence suggests that tic disorders and OCD are sequelae of prior streptococcal infection. Peterson et al. (364) examined this issue further to determine whether this association was related to comorbidity rather than the condition itself. In a study of 105 people diagnosed as having chronic tic disorder (CTD), OCD, or ADHD and 37 controls, they found that a diagnosis of ADHD was significantly associated with two distinct antistreptococcal antibodies: antistreptolysin O and anti-deoxyribonuclease B. These associations remained significant after controlling for the effects of CTD and OCD comorbidity. Furthermore, they noted that the relationships between antibody titers and basal ganglia volumes were significantly different in OCD and ADHD subjects compared with other diagnostic groups. Higher antibody titers were associated with larger volumes of putamen and globus pallidus nuclei. The investigators concluded that the prior reports of an association between antistreptococcal antibodies and CTD or OCD may have been confounded by the presence of ADHD. They also suggested that in susceptible persons with ADHD or OCD, chronic or recurrent streptococcal infections are associated with structural alterations in basal ganglia nuclei. Hallett et al. (365) also suggested that the binding of antineuronal antibody from some children with GTS induced striatal dysfunction. Using MRI, Giedd et al. (366) studied 34 children with streptococcus-associated OCD and/or tics and 82 healthy controls and found that the patients with OCD and/or tics had enlarged basal ganglia.

Lougee et al. (367) studied psychiatric disorders in the FDRs of children with PANDAS and found that the rates of tic disorders and OCD in the relatives of these children were higher than those reported in the general population but similar to those reported in tic disorders and OCD. They suggested further studies are needed to determine the nature of the relationship between genetic and environmental factors in PANDAS.

In a review of immune abnormalities in GTS, OCD, and PANDAS, Trifiletti and Packard (368) reported that that there is a subset of patients with GTS and OCD, perhaps 10%, for whom there is a clear streptococcal trigger. This may have implications for diagnosis and treatment, particularly in children who do not respond to conventional therapies. In this regard, Perlmutter et al. (369) found that plasma exchange and intravenous immunoglobulin were both effective in lessening symptom severity in children with infection-triggered OCD and tic disorders. Plasma exchange is of no benefit in OCD patients who do not have streptococcus-related exacerbations (370). Well-designed and adequately controlled studies are needed to determine whether there is a true etiologic role for these streptococcal infections and whether the use of immune-modifying therapies are justified in these situations (371).

There is a widely held belief among both the lay public and GTS patients that GTS symptoms are associated with allergy. Bruun (129) reviewed 300 patients with GTS and stated that although there is no evidence that allergies cause GTS, her clinical experience has shown that symptom exacerbation often is associated with seasonal allergy responses, ingestion of allergens in food, and drugs used to treat allergies. Although other investigators also report exacerbation of symptoms upon exposure to allergens (372,373), there has been little scientific evidence for the involvement of allergy in GTS (374). A controlled study has indicated that GTS individuals do not have more allergies than controls, but they do have more hyperactivity (375). Ho et al. (376) studied 72 GTS patients, aged 4 to 17 years, using multiple allergens simultaneous tests (MAST) and compared the results with levels from a local population. When the MAST-positive group of GTS patients was compared with controls, the difference was significant (56.9% vs. 44.3%, respectively). The authors concluded that the prevalence of allergy in GTS patients is significantly higher than in the general population.

LABORATORY AND GENETIC STUDIES

Genetic Studies

In the quest for an etiologic factor for GTS, interest in the genetics of the disorder has mushroomed, and many groups and laboratories have become involved. A number of investigations have addressed the question of a genetic predisposition in GTS, but the precise genetics are unclear (377), because it can be sporadic (378), and 50% of the human genome has now been excluded (379). At present, the evidence is mostly in support of a major autosomal dominant gene.

Convincing evidence for a genetic factor comes from twin data in which MZ twins have been concordant for GTS (11,148,176,380–382). However, a male GTS whose MZ twin was discordant has been reported (383). In the large twin study involving 43 pairs of same-sex twins performed by Price et al. (384), concordance rates were only 77% and 23% for MZ and dizygotic (DZ) pairs, respectively. These findings suggest that nongenetic factors play a role in the expression of GTS. The original data from the large study (384) were reexamined, and it was found that in each case of discordant twins, the unaffected co-twin had a higher birth weight than the affected twin (385). It was speculated that some prenatal events or exposures, such as maternal stress, antiemetic medication, or other unknown agents, lead to changes in the sensitivity of some dopaminergic receptors, and this could partially determine the eventual severity of expression of the GTS diathesis (385).

Numerous family studies provide stronger evidence for the genetic factors involved in GTS, given that many relatives of probands with GTS may present with either GTS or motor or vocal tics only (386,387). Several large multiply affected families have been reported (388,389), and in all there is an increased incidence of tics (simple or chronic) and GTS in the relatives of probands with GTS.

Studies using complex segregation analysis on independent samples (390–394) and on large families (395,396) suggest the presence of a major autosomal dominant gene with varying but usually high penetrance. Some studies have shown that OCB is a phenotypic variation of the putative GTS gene (390). However, other studies have differed and suggested intermediate inheritance in which some heterozygotes manifest the disorder (397); mixed model of inheritance (398); polygenic inheritance (399); and no evidence for mendelian inheritance (400).

A range of chromosomal anomalies in GTS has been reported (for review see reference 401), which has led to investigations for linkage. Reports have tentatively assigned the gene to chromosomes 3 (402), 11 [D_2 receptors (239)], and 18 (403,404), and the dopamine D_3 receptor gene (405), but there has been much debate (406–408). Chromosomes 7 and 18 (409), chromosome 11 (410), and the D_1 dopamine receptor (411) have been excluded. In a linkage analysis of a large French-Canadian family consisting of 127 members with 20 cases of GTS and 20 cases of related tic disorder, Merette et al. (412) observed an LOD score of 3.24 on chromosome 11 (11q23). This result was obtained using marker D11S1377, the marker for which significant linkage disequilibrium with GTS was detected in an Afrikaner population by Simonic et al. (413). The investigators found the strongest evidence for association for markers within the chromosomal regions encompassing D2S1790 near the chromosome 2 centromere, D6S477 on distal 6p, D8S257 on 8q, D11S933 on 11q, D14S1003 on proximal 14q, D20S1085 on distal 20q, and D21S1252 on 21q.

Notably, Robertson and Trimble (401) reported that 65 of 68 (96%) consecutive GTS patients had entirely normal chromosomes. Their findings, plus a review of the literature, suggest that currently no single chromosomal abnormality can be said to be absolutely characteristic of GTS. Bilineality (i.e., affected individuals on both maternal and paternal sides of index cases) has been reported (414), which suggests that homozygosity is not uncommon in GTS and might further explain difficulties in localizing the gene defect by linkage analysis. Kurlan et al. (415) and Hanna et al. (416) also found evidence for bilineal transmission in at least one fourth of GTS families.

Eapen et al. (417), Lichter et al. (418), and others have suggested that the parent of origin effect caused by genomic imprinting may influence the phenotypic expression of GTS.

Two studies analyzed human leukocyte antigen (HLA) typing in GTS, and both found no association between GTS and HLA-A or HLA-B antigens or haplotypes (419,420). The latter study also found no association between GTS and HLA-C or HLA-DR antigens.

Evidence from Family and Genetic Studies

As noted previously, researchers have suggested that genetic factors are involved not only in GTS but also in OCD. Having established that they are inherited disorders, these investigations went on to look for a genetic relationship between GTS and OCD.

Among the studies that reported concordant MZ twin sets was the work of Jenkins and Ashby (381). In that study, both twins were described as obsessional. A triplet study in which the triplets were reared apart showed 100% concordance for GTS but not OCD (421).

In several family studies of GTS, relatives with OCB but without tics or GTS have been reported (27,422–424). As part of a large family study of GTS, Pauls et al. (425) interviewed 90% of all FDRs of 32 GTS probands, and they obtained family history reports about each family member from all available data obtained. Although the authors acknowledge the difficulties in making an accurate diagnosis of OCD, they found the frequency of OCD diagnoses among the FDRs were significantly higher (nine to 13 times) than the frequency of OCD estimated in the general population

from the epidemiologic catchment area study in which similar methodology was used (426). The rates of GTS and OCD were practically the same in families of GTS probands with OCD as the rates in families of GTS probands without OCD. In addition, the frequency of OCD without GTS among FDRs was significantly elevated, particularly among the female relatives, in both groups of families. Subsequent segregation analysis performed on these results gave further support to the autosomal dominant model for GTS, but analysis also indicated that in at least a proportion of cases, GTS and OCD constitute alternate expressions of the same autosomal dominant gene and that this expression may be sex specific (424).

Eapen et al. (390) performed complex segregation analyses on families ascertained through 40 unselected consecutive GTS patients (168 FDRs, all of whom were directly interviewed). Results were consistent with an autosomal dominant gene with high penetrance. Their results also showed that OCB was an integral part of the expression of GTS, whereas within the families, motor tics (chronic or transient) were not always etiologically or genetically related to GTS. Thus, twin, family, and, more importantly, complex segregation analysis studies strongly suggest that there is a genetic relationship between GTS and OCB.

The few studies that have looked for evidence of a genetic basis to OCD have involved only small numbers of patients. They also used different methods and diagnostic criteria, so the results have been difficult to interpret. Nevertheless, the data obtained are in favor of an underlying genetic defect in OCD. Twin studies have revealed higher concordance rates for OCD among MZ twins than for DZ twins (427), as well as OC traits among twins (428–430). In no cases, however, did concordance approach unity, which indicates that expression of OCD probably is influenced by nongenetic factors. Most family studies have shown that there is a significantly higher incidence of OCD among FDRs of patients with OCD than in the general population (431,432). Two recent studies, although uncontrolled, used stricter diagnostic criteria and more stringent methodology than previous studies by directly interviewing all FDRs of OCD patients (433,434). The rates of OCD in the parents of such patients were significantly greater than those obtained in the epidemiologic catchment area studies that determined the prevalence of OCD in the general population (426). Presenting OC symptoms of probands and their parents usually were dissimilar, which argues against any simple social or cultural transmission. Nestadt et al. (435) reported a significantly higher lifetime prevalence in cases compared to control relatives (11.7% vs. 2.7%). Case relatives had higher rates of both obsessions and compulsions, but the finding was more robust for obsessions, which suggests that obsessions are more specific to the phenotype than compulsions. Also, age at onset of less than 18 years in the case proband was significantly related to familiality.

There is convincing evidence that GTS and OCD share a genetic etiology, but not all cases of OCD are associated with GTS. The findings from the genetic studies and data described earlier in this chapter suggest that OCD can be divided into at least three categories: (a) sporadic (no family history); (b) familial (positive family history); and (c) OCD associated with a family history of tics or GTS (423). This division into categories challenges traditional assumptions of etiologic homogeneity in OCD.

In a segregation analysis study of OCD families, Cavallini et al. (436) suggested a dominant model of transmission. When the phenotype was widened to include GTS and chronic motor tics, an unrestricted model of transmission became the best fit, and they proposed that the OCD phenotype probably presents a higher level of heterogeneity than the GTS phenotype. Furthermore, the clinical and phenotypic heterogeneity may be linked to genetic heterogeneity, with some individuals inheriting the "GTS+OCD genotype" and the others the "classic OCD genotype." In a family study of OCD probands and GTS+OCD probands, Eapen et al. (53) found that all the OCD probands who shared a similar symptom profile to that of TS probands had at least one FDR with OCD, whereas none of the OCD probands with classic OCD symptoms had a positive family history. The authors concluded that the latter could be regarded as sporadic or nonfamilial cases. These observations may be consistent with genetic heterogeneity within both OCD and GTS. Alsobrook et al. (437) used factor-analytic symptom dimensions to subset the family sample based upon the symptom factor scores of the probands. They found that families high in the symptom score of symmetry and ordering had a higher risk of OCD in the relatives and suggested that this constitutes a genetically significant subtype of OCD. These investigators also found that although bilineal transmission of tics is relatively infrequent in consecutive TS pedigrees, cotransmission of OCB from an otherwise unaffected parent is common and significantly influences the development of OCB and SIB, but not tics, in the offspring.

According to the collective knowledge, it seems that although there is little doubt about a familial relationship between OCD and GTS, not all cases of OCD are either familial or etiologically related to GTS. It is hoped that further refinements at various levels of diagnostic hierarchies will lead to an improved definition of what constitutes the genetic spectrum of GTS. Molecular genetics offers a promising method for exploring the underlying etiologic heterogeneity of GTS and related conditions such as OCD.

Several studies have reported exclusion of various candidate genes, including D_5 receptor locus *(DRD5)* (438); D_1, D_2, D_3, D_4, and D_5 neuroreceptor gene loci; genes encoding dopamine β-hydroxylase, tyrosine, and tyrosine hydroxylase (439); and the norepinephrine transporter gene (440). On the other hand, Cruz et al. (441) reported increased prevalence of the seven-repeat variant of D_4 receptor gene in patients with OCD with tics and suggested that the *DRD4* gene is a factor in the phenotypic variance of tics in OCD patients.

In a review, Blum et al. (442) reported a number of independent metaanalyses that confirm an association of dopamine D_2 receptor gene *(DRD2)* polymorphisms and impulsive-additive-compulsive behaviors. Vandenbergh et al. (443) failed to identify any common protein coding DAT sequence variant in more than 150 individuals free of any neuropsychiatric disease, 109 individuals with GTS, 64 individuals with alcohol dependence, and 15 individuals with ADHD. The authors suggested that gene variants that alter levels of DAT expression provide the best candidate mechanism for reported DAT gene markers, ADHD, and other associated neuropsychiatric disorders.

The results of complete genome scan in 76 affected sib pair families with a total of 110 sib pairs were summarized in a report by the Tourette Syndrome Association International Consortium for Genetics (444). Although there was no statistically significant results, the multipoint maximum-likelihood scores for two regions (4q and 8p) were found to be suggestive.

Although genetic linkage studies are being conducted in GTS to try to establish a relationship between a marker and the hypothetical gene for the disorder, this methodology has not yet been used in OCD. It is hoped that these techniques will be applied to future research, but until biologic markers are identified, the family history of OCD is the strongest evidence suggesting a genetic foundation underlying at least some forms of OCD.

Despite conflicting evidence, the dysregulation of cerebral neurotransmitter systems involving serotonin and dopamine is strongly implicated in the pathophysiology of this disorder. The anatomic, functional, and biochemical findings may be consolidated and become clearer as neuroradiologic techniques, such as PET and SPECT, are refined.

It has been suggested, however, that some prenatal events or exposures, such as maternal stress, antiemetic medication, or other unknown agents, lead to changes in the sensitivity of some dopaminergic receptors. This could partially determine the eventual severity of expression of the GTS diathesis. Further studies on the mechanism of inheritance may be assisted by more precise definition of the phenotypes involved. Future studies likely will show that the mechanism is more complicated than a single gene, because previous studies have not entirely ruled out the possibility of other alleles, multifactorial inheritance, or genetic heterogeneity.

Clinical Evaluation

As with all neuropsychiatric patients, the clinician must take a full and detailed medical and psychiatric history, examine the mental state, and perform thorough physical and neurologic examinations. However, there are methods or aspects of assessment that require special mention.

Assessment and Rating of Tic Severity

Both clinicians and researchers need to make an accurate description and measure the severity of GTS. To this end, there are several schedules currently in use. The problems with rating tic severity, the statistical considerations for quantitative assessment of tics, and an in-depth discussion of available instruments have been elegantly reviewed by Kurlan and McDermott (445).

Briefly, the precise measurement of tic frequency nearly always poses a problem, because the severity often is variable and dependent on several factors. An attempt to overcome some of the difficulties has been made by using video recordings; in addition, recordings of patients' movements while performing prescribed tasks have been made using the large-scale integrated motor-activity monitor, which looks like and is worn as a wristwatch (446).

There are some self-report schedules that are used for evaluation of GTS. The Tourette Syndrome Questionnaire (447a) is a structured parental and/or self-report schedule that originally was developed for an epidemiologic survey by Cohen and co-workers. The questionnaire offers a systematic way of obtaining relevant information, such as personal and demographic data, developmental history, family history, general medical and treatment history, course of tic behaviors, and impact of GTS on the person's life. A second parent and/or self-report instrument was developed by Stefl and colleagues (cited in reference 445) for a needs assessment survey conducted for the Tourette Syndrome Association of Ohio. Another schedule, the Tourette Syndrome Symptom Checklist, was devised by Cohen et al. (447b) to assist parents in making daily or weekly ratings of tic severity. The schedule, which takes into account the frequency and disruption of both tics and behavioral symptoms, has been used successfully to monitor the longitudinal course of GTS and to document changes during medication trials (448).

There are several clinician or observer schedules available for assessment of tics and GTS. Tanner et al. (449) developed a scale that enables an observer to count the patient's tics under several different conditions that are videotaped (patient alone, with the examiner, sitting quietly, and performing a task). The Tourette Syndrome Severity Scale was developed by Shapiro and colleagues (450–452) for use in a clinical trial evaluating pimozide for treatment of GTS. It includes a composite rating of severity composed of five items or factors, the scores of which are totaled and converted to a global severity rating. The items are as follows: (a) degree to which tics are noticeable to observers; (b) whether the tics elicit comments or curiosity; (c) whether others consider the patient odd or bizarre; (d) whether the tics interfere with functioning; and (e) whether the patient is incapacitated or hospitalized

because of the tics. The Tourette Syndrome Global Scale (TSGS), which combines a variety of ratings for tic symptoms and social functioning into an overall global score for severity, was developed by Cohen and colleagues (453). The section on tics rates frequency and disruption for both simple and complex motor and vocal tics, which are combined in a complicated mathematical formula. Social functioning includes ratings of behavior, restlessness, learning, school, and occupational problems. A problems with the TSGS is that social functioning tends to be underweighted because of the multiplication of frequency by disruption scores (454).

The Yale Global Tic Severity Rating Scale was developed to refine measurement of GTS symptoms, building on the developers' experience with the TSGS (454). When using the Yale Global Tic Severity Rating Scale, examiners rate the number, frequency, intensity, complexity, and interference of both motor and vocal tics, and then generate a total tic score, an overall impairment rating, and a global severity score. The Yale group (454) also has produced a Global Clinical Impression Scale for tics. It is a seven-point ordinal scale ranging from "normal" to "extremely severe" for a rating of the impact of GTS symptoms on daily functioning.

Goetz et al. (455) devised a scale by which examiners can rate patient videotaped sessions recorded under standardized conditions in three settings (seated quietly with the examiner in the room; reading aloud with the examiner in the room; seated quietly without the examiner in the room). The videotapes are reviewed, and examiners assess various body regions, rate both motor and vocal tic severity (based on a scale from 0–5), and count both motor and vocal tics.

Shapiro et al. (450) devised the Shapiro TS Severity Scale, and they use videotaped recordings under three conditions.

1. Computation condition: subject is asked to add up numbers for 2.5 minutes
2. Reading condition: subject reads three paragraphs for 2.5 minutes
3. No-stimulus condition: subject is seated alone after the examiner has left the room.

After the videotape session, examiners again rate tic number, type, and severity.

A schedule that is widely used internationally for research is the Schedule for Tourette's Syndrome and Other Behavioral Syndromes. Developed at Yale by Pauls and Hurst (456), it is comprehensive and allows collection of information (tics and behaviors) about both the patient and family members. It takes approximately 4 hours to complete. Robertson and Eapen (457) developed a similar schedule (The National Hospital Interview Schedule for the Assessment of GTS and Related Behaviors), which takes about 1.5 hours to administer and has been widely used in the United Kingdom. It has been shown to be both reliable and valid.

It has been argued that the structured diagnostic interview methodology can reduce errors of omission. Coffey et al. (458) compared youths with GTS identified through a specialized program who had both a structured diagnostic interview-derived diagnosis of GTS and an expert evaluation of GTS (N = 103) with youths ascertained through a non-GTS specialized program who only had a structured diagnostic interview-derived diagnosis of GTS (N = 92). It was found that children who met diagnostic criteria for GTS on the structured diagnostic interviews shared similarities and patterns of clinical correlates, irrespective of whether they were ascertained through a GTS specialized or a non-GTS specialized clinic. These findings support the usefulness of the structured diagnostic interview methodology as a diagnostic aid for identification of GTS cases irrespective of the setting.

Clinical Evaluation of Obsessive-Compulsive Disorder

Because of the secretive nature of the disease and the reluctance of sufferers to reveal their symptoms, OCD can be difficult to recognize and diagnose by even the most experienced physicians. Two simple screening questions have been designed to isolate symptoms of the disorder. These questions can be asked routinely in suspected cases while taking a history, as follows: "Are you bothered by thoughts coming into your mind that make you anxious and that you are unable to get rid of?" "Are there certain behaviors that you do over and over that may seem silly to you or to others but that you feel you just have to do?" (459).

If the answers to these questions are affirmative, the clinician can proceed to inquire further about the nature, frequency, onset, duration, and severity of the obsessions and compulsions expressed by the patient. Thorough assessment of the psychological and physical status of the patient is desirable in view of the manifold presentation of OCD. Physical symptoms may include skin problems in obsessive hand washers; respiratory and gastrointestinal allergies in obsessive hoarders; hair loss in patients with trichotillomania; and poor personal hygiene or nutritional status in patients with a fear of contamination of poisoning (460). Accurate evaluation of these patients' psychological status will reveal the presence of comorbid disorders such as depression and anxiety, the symptoms of which may be the presenting features.

A number of standardized inventories or rating scales have been constructed to assess OCD. These scales have been used both in clinical settings and for research purposes, particularly to document treatment response. The LOI was the first rating scale used to quantify subjective reports of obsessive feelings and behavior (461). The original inventory consisted of 69 questions and was designed as a supervised card sorting procedure that gives information about feelings of resistance and interference with other activities, in addition to the straightforward answers to the questions. It subsequently has been administered as a questionnaire (462). Scores differentiate well between obsessional patients and normal volunteer subjects, but it has been criticized for its lack of reliability and validity data and because it originally was designed to study houseproud housewives (who are excessively concerned with cleaning and tidying their homes) and not for OC patients (338). The LOI in its card sorting form has been adapted for use in the assessment of childhood OCD and is termed the LOI-Child Version. It has been found to reliably distinguish OCD adolescents from normal controls and from other psychiatric patients with severe OC symptoms. It also can be used as a valid instrument in treatment studies (463).

Other rating scales used to tap OC symptomatology are the Sandler-Hazari Scale (464), the Obsessive-Compulsive Check List (465), and the Maudsley-Obsessive Comprehensive Inventory (466). The Y-BOCS has been designed to provide a specific measure of the severity of OCD symptoms independent of the particular type of obsession or compulsion present (467,468). The Y-BOCS consists of a scale of ten items, each rated from 0 (no symptoms) to 4 (extreme symptoms), with separate subtotals for severity of obsessions and compulsions. Because it is a clinician-rated instrument, it avoids the problems of self-assessment, such as resistance to changing self-ratings over a period of time (260). The Y-BOCS has been assessed as a valid instrument suitable for use as an outcome measure in drug trials of OCD (271).

Differential Diagnosis of Gilles de la Tourette Syndrome

Most of the conditions that should be considered in the differential diagnosis of GTS are described in detail by Bruun and Shapiro (469), Sacks (470), and Robertson (21). They include the athetoid type of cerebral palsy, which is seen commonly among learning disabled populations with an age at onset between birth and 3 years. These patients have evident neurologic deficits and a static course after age 3 years. Dystonia musculorum deformans usually presents with a torsion dystonia, often in the legs, and usually is progressive. A crippling state results 10 to 15 years after onset, and remissions are rare. Encephalitis lethargica, in its chronic form, and the more rare klazomania, an acute postencephalitic state, may mimic the symptoms of GTS, but usually there is other evidence or history of encephalitis and parkinsonian symptoms are associated with these states. Huntington chorea usually begins in the third to fifth decade, but 1% of cases have an onset in early childhood. A positive family history of the disorder, with attendant dementia and progression to death within 10 to 20 years, is the norm.

OC symptoms can occur in a number of neurologic conditions that should be excluded before a diagnosis of GTS is made. Such cases represent only a small proportion of the total number of patients with GTS+OCD (471).

It was observed as early as 1924 that OC symptoms occurred in patients with postencephalitic parkinsonism (472), a finding that was confirmed by subsequent investigators in the years following the pandemic of encephalitis lethargica after World War I (471–473). OC traits but not symptoms are common in Parkinson disease (474), although, paradoxically, oral levodopa for treatment of parkinsonism can cause obsessions and compulsions (358). OC symptoms can occur following head trauma, either at birth (475) or as a result of accidents (476). Up to one third of patients with Sydenham chorea have OC symptoms (477,478), which also have been reported in patients with Huntington chorea (479). OC symptoms are rarely reported in patients with diabetes insipidus (480). OC symptoms have been noted to develop in cases of epilepsy. One notable individual is Napoleon, who was epileptic and had a compulsive ritual of counting the number of windows in the buildings that he passed (471). There are a number of case reports of OC symptoms occurring in patients with petit mal (481), grand mal (482), and temporal lobe epilepsy (365), and as part of the epileptic aura (483,484). However, because both OCD and epilepsy are common conditions, the observed association between the two disorders may not be significant (485).

A condition similar to GTS has been reported following both short-term and long-term use of neuroleptics (486). Sacks (470) distinguishes GTS from cases of "acquired Tourettism," in which symptoms identical to those of GTS are seen in the setting of an acute or chronic cerebral insult, occurring most commonly in postencephalitic patients but also found after head trauma and carbon monoxide poisoning (for review see reference 487). A condition similar to GTS has been reported following both short-term and long-term use of neuroleptics. Stahl (488) coined the term "tardive Tourette syndrome" to describe such cases, which occurred in patients treated with long-term neuroleptic drugs and arose either as an elaboration of lower-level dyskinesia or *de novo*. Tics also have been reported in relation to the use of the antiepileptic drug lamotrigine (489). Eapen et al. (490) described adult-onset GTS cases that, compared with DSM-classified younger-onset GTS, more often were associated with severe symptoms, a potential trigger event, increased sensitivity, and poor response to neuroleptic medication.

One of the most difficult differential diagnoses is that of tics of childhood, which commence between ages 5 to 10 years. Many remit spontaneously and usually improve with age. For a review of the diagnostic criteria for transient tic disorder, chronic motor tic disorder, and GTS, see Woody and Laney (491) or DSM-IV (1).

Shapiro et al. (11) reported subtle neurologic deficits in 57% of patients studied; 20% were left handed or ambidextrous, which led the authors to argue an organic etiology. Most of the patients (78%) had minor motor asymmetry, and 20% had chorea or choreoathetoid movements. In contrast, Lees et al. (5) used a standardized handedness questionnaire and found that 87% of their sample of 53 patients were right handed. Other investigators reported only minor nonspecific

neurologic abnormalities in a few patients (5,22,26,29,72). Reported abnormalities include chorea, dystonia, torticollis, dysphonia, dysdiadochokinesis, postural abnormalities, reflex asymmetry, motor incoordination, nystagmus, and unilateral Babinski reflexes.

Spasmodic torticollis usually presents between ages 30 and 50 years, although it can commence at any age. It is a form of focal torsion dystonia that may be associated with more widespread disease and spastic speech. Approximately half of the cases are progressive; some remit spontaneously, whereas others remain static. Sydenham chorea occurs more frequently in females, and 75% of cases are associated with rheumatic fever, eosinophilia, and electrocardiographic (ECG) abnormalities. The choreiform movements usually are self-limiting. Wilson disease usually presents between ages 10 and 25 years. The classic signs of the disease are Kayser-Fleischer rings, cirrhosis or hepatitis, dementia, and associated copper abnormalities in the serum and urine. Wilson disease usually is progressive and fatal if left untreated for several years. Kayser-Fleischer rings almost always are detectable once neurologic signs, such as abnormal movements, are present (21).

Uncommon disorders, such as Hallervorden-Spatz disease, Pelizaeus-Merzbacher disease, status dysmyelinatus, and Jakob-Creutzfeldt disease, should be considered in the differential diagnosis of GTS, but most of these conditions have distinctive clinical features, classic courses, and characteristic types of movements that usually make it possible to differentiate them from GTS (469). Other conditions to be excluded are hypoparathyroidism and epilepsy, especially myoclonus epilepsy.

Comings and Comings (23) stress that incorrect diagnoses of GTS are common. They suggest the main reasons for misdiagnosis include unfamiliarity with the syndrome, suppressibility of the symptoms, and the erroneous belief that coprolalia must be present. It is noteworthy that despite the usually typical presentation of GTS, diagnostic difficulties do occur (492,493).

Differential Diagnosis of Obsessive-Compulsive Disorder

The differential diagnosis of OCD is large and includes all of the associated psychiatric and neurologic conditions, such as the other anxiety disorders, depression, GTS, eating disorders, trichotillomania, body dysmorphic syndrome, and personality disorders discussed previously. Misdiagnosis is common because of the overlapping characteristics of OCD and these comorbid conditions. For example, the morbid preoccupations and depressive ruminations that can occur in major depression may be confused with the obsessional thoughts of OCD. The former, however, are seen to be related to depressive symptomatology, such as thoughts of guilt, and there are few attempts to suppress them, in contrast with the obsessions seen in OCD. Likewise, anxiety disorders can be distinguished from OCD by the content of the anxious thoughts, which are realistic to patients with anxiety and irrational to patients with OCD. The extreme avoidance behavior of some patients with OCD differs from that of the phobic patient in that the behavior is less effective in providing relief. Other disorders, however, may be more difficult to differentiate from OCD. For example, the complex tics of GTS may be identical to the compulsions seen in OCD, such as the need to touch or the need for symmetry. Diagnostic accuracy may be paramount in distinguishing whether a GTS patient with complex tics also has comorbid OCD. Recent evidence suggests that patients who are resistant to treatment with SSRIs alone may respond to a combination of fluvoxamine and a neuroleptic (51,329).

The diagnosis of OCD is an area of confusion and controversy, and the present diagnostic criteria are under review (494). Recent research has led investigators to question the traditional views that obsessions are cognitive and compulsions are behavioral; that obsessions and compulsions can occur independently of each other; that they are always egodystonic; and that the sufferer always recognizes that they are senseless. A functional model has been proposed in which a dynamic relationship exists between obsessions and compulsions in which compulsions always occur with obsessions they are designed to neutralize. These compulsions may be thoughts that are formulated to neutralize other thoughts or obsessions and are thus termed "cognitive compulsions." The other main area of concern is how to establish precise definitions that accurately distinguish OCD from the many conditions that are associated with it. This task cannot be satisfactorily completed until the relationships between OCD and these disorders are fully understood. In the future, OCD most likely will be regarded as a heterogeneous disorder with separate phenomenologic subgroups that require different treatment.

Drug Treatment of Gilles de la Tourette Syndrome

At present, drug therapy is the mainstay of treatment for the motor and vocal symptoms and some of the associated behaviors of GTS. The medications most commonly used are dopamine antagonists, but other medications are useful for treatment of the motor and vocal tics and the associated behaviors.

The butyrophenone haloperidol was first used successfully for GTS by Seignot (495) in 1961. Since then, dopamine antagonists such as haloperidol, pimozide (496,497), and sulpiride (498) have been widely and successfully used. Doses are relatively small. For example, haloperidol therapy begins at 0.25 to 0.5 mg daily and is increased 0.5 mg per week; often 2 to 3 mg daily is sufficient (499). Extrapyramidal side effects, sedation, and dysphoric states are common with haloperidol, but less so with pimozide (499). In some patients, the sleepiness associated with haloperidol

was thought to adversely affect behavioral manifestations of GTS (500). School phobia and avoidance have been reported as unwanted side effects of haloperidol (501) and pimozide; these symptoms were successfully reversed by the addition of a tricyclic antidepressant (502). Butyrophenones may impair concentration and scholastic achievement (503), may be associated with lower IQs on formal testing (72), and may cause tardive dyskinesia (504). ECG changes (significantly prolonged QTc interval) during treatment with pimozide have been reported (505). Accordingly, some clinicians consider ECG monitoring prudent for patients receiving pimozide.

Sulpiride causes fewer extrapyramidal problems, including tardive dyskinesia and dystonia, and fewer cognitive and sedative side effects compared with haloperidol and pimozide. However, gynecomastia, galactorrhea, menstrual irregularities, and possible depression have been reported (498). Tiapride, which has a profile similar to sulpiride, has been used (506).

Atypical neuroleptics have been administered in GTS, with some success (507). Risperidone, an atypical neuroleptic, is primarily a 5-HT$_2$ receptor antagonist, but it also is a weaker D$_2$ antagonist. Available evidence suggests that although the therapeutic effectiveness of risperidone in controlling tic symptoms is less than that of conventional neuroleptic drugs, it does have a role in the treatment of GTS (508,509). Most studies have used a mean daily dose of 1.5 mg/day (range 0.5–4 mg/day). Drowsiness, dizziness, weight gain, anxiety, and headache are the common side effects. Extrapyramidal side effects, such as acute dystonia and akathisia, have been reported but at much lower frequencies (range 0%–20%) than with conventional neuroleptic drugs. Risperidone has been reported to be effective as adjunctive therapy in children with ADHD (510). Clozapine, another atypical neuroleptic, has been found to be ineffective in treating GTS symptoms (511).

Olanzapine, another atypical neuroleptic, has been reported to be of some benefit in controlling GTS (512). Stamenkovic et al. (513) in an open-label treatment and Onofrj et al. (514) in a double-blinded, crossover study compared 5 to 10 mg olanzapine with 2 to 4 mg pimozide and found that the reduction in GTS symptoms was highly significant with olanzapine compared to pimozide. Sallee et al. (515) found that ziprasidone was significantly more effective in GTS compared to placebo. The dose was started at 5 mg/day and gradually titrated up to a maximum of 40 mg/day.

Bruun (129) notes that GTS patients can be highly sensitive to side effects from neuroleptic agents, mainly because GTS patients (including children) developed tardive dyskinesia when they were treated with neuroleptics (516,517), especially when they had a family history of movement disorders, including GTS (518). Jeste et al. (519) suggest low-dose bromocriptine for treatment of dyskinesias secondary to neuroleptics in GTS. Tardive dyskinesia has been reported in a GTS patient who was receiving sulpiride (520). A cautionary note to prescribing neuroleptics in children was made by Silverstein and Johnston (521), who reported that although these drugs were indicated at times, they were not without problems. Syndromes similar to tardive dyskinesia were found in 140 of 410 (35%) children with a variety of psychiatric disorders whom these researchers treated.

Uhr et al. (522) successfully treated four patients with GTS using piquindone, which preferentially blocks dopamine receptors in the mesolimbic system rather than the nigrostriatal system and selectively blocks D_2 receptors. Motor tics responded to lower doses than vocal tics.

Clonidine has been used successfully in GTS patients (84,129,523,524). Bruun (129) noted that about 15% of patients had a better response to a combination of clonidine and haloperidol than they had to haloperidol alone. Clonidine may be the agent of choice if a child has GTS and associated ADHD (33).

Specific SSRIs, such as fluoxetine (327), can be used to treat the OC aspects of GTS. Augmentation of the antiobsessional effects of SSRIs (e.g., fluvoxamine) by neuroleptics (e.g., pimozide) has been reported in patients with OCD and GTS (328,329). Treatment of comorbid GTS and OCD was reviewed by Eapen and Robertson (507), who concluded that the best treatment strategy is the combination of a dopamine antagonist with a serotonergic drug.

Mesulam and Petersen (525) reviewed the pharmacologic treatment of 58 patients treated over an 8-year period. They noted that differences in response patterns were common and thus required individualized tailoring of management. Generally, dopamine-blocking neuroleptics were the mainstay of therapy, but frequently midcourse alterations were required because previously successful drugs stopped working or their side effects became intolerable. Clonidine proved to be inferior to neuroleptics for treatment of motor and vocal tics, but they suggested it may have a role in some patients with prominent OC symptomatology. Interestingly, tardive dyskinesia never occurred in their cohort (525).

Other drugs have been tried with varying success, including neuroleptics such as fluphenazine; penfluridol; the antidepressant clomipramine; the anticonvulsant clonazepam; the α_2 noradrenergic agonist guanfacine; the GABA-ergic drugs carbamazepine and progabide; the calcium antagonists nifedipine and verapamil; naloxone; lithium carbonate; tetrabenazine; muscarinic compound RS86; the cholinergic agent physostigmine; lecithin; nicotine transdermal patches; metoclopramide; the dopamine agonist pergolide; the antiandrogen flutamide; and botulinum toxin by local injection (507).

From the reviewers' clinical practice, allergy may have an effect on many GTS patients, manifested as worsening of their symptoms. Dietary manipulation may be an area worth pursuing in the future, especially in the case of children in whom one is hesitant to use medication unless absolutely necessary. There are no controlled trials of sulpiride versus haloperidol or pimozide, and such studies would be useful.

Treatment of Obsessive-Compulsive Disorder

There is little evidence that psychoanalysis and dynamic psychotherapy are effective in managing OCD (526,527), and these approaches to treatment of this disorder have largely been abandoned. Advances in behavioral theories and techniques, and more recently in pharmacotherapy, have led to rational and integrated psychopharmacologic treatment regimes for a condition that until a decade ago was considered mainly resistant to treatment. Psychosurgery now is reserved as a final option for only the most refractory cases.

Behavior Therapy

Meyer (528) was the first to demonstrate that exposure and response prevention was a successful treatment for OCD inpatients. In the past 2 decades, many studies have confirmed the efficacy of behavioral treatments for OCD but have also highlighted their limitations. The most widely evaluated and useful techniques are exposure and response prevention (529). Exposure *in vivo* therapy requires the patient maintain contact with the stimuli that provoke the obsessions or compulsions until habituation occurs. Response prevention instructs the patients to refrain from ritualistic behavior despite often overwhelming urges to do so. For example, patients with contamination fears or compulsive washing may be exposed in real life to dirt, blood, or excrement and then prevented from washing. When the anticipated dreaded consequences fail to materialize, the patient's anxiety and symptomatology improve. More recently, exposure in fantasy, or imaginal exposure, in which the patient is asked to think about the feared stimulus without being exposed to it in reality, has been added to reinforce the gains achieved with exposure *in vivo*, particularly for patients with catastrophic thinking.

Several follow-up studies have demonstrated significant improvements in OCD symptomatology after exposure therapy for various periods of time ranging from 1 to 6 years, depending on the study (530–532). Foa et al. (532) found that 90% of a group of patients treated intensively with exposure and response prevention therapy for 3 weeks experienced some immediate symptomatic improvement; 80% continued receiving these benefits after 1 year. Marks (533) reported improvement that persisted in up to 75% of patients 2 to 4.5 years after completion of therapy. O'Sullivan et al. (531) showed that gains lasted 6 years after treatment and that the best predictor of long-term outcome was improvement at the end of treatment. They also demonstrated that exposure periods of long duration were more effective than short sessions; patients who had undergone 6 weeks of therapy experienced more long-lasting improvement than those who had undergone therapy for only 3 weeks. The treatment period can be effective with limited therapist contact time.

These encouraging results disguise the fact that a substantial proportion of patients with OCD (up to 30%) refuse or fail to complete behavior therapy because of its arduous time-consuming and often frightening tasks. The high response rates reported in the aforementioned studies reflect the outcome of the most motivated patients. Poor response to behavior therapy is seen most often in patients with concomitant depression, those who use CNS-depressing drugs, and patients who are obsessional ruminators with no obvious compulsive behaviors (534). Based on theories that obsessions and compulsions originate in thoughts, some authors advocate cognitive therapy as an adjunct to existing behavioral techniques for treatment of OCD, but controlled trials are lacking.

Studies have shown that the combination of behavior and drug therapies may be the optimum management strategy for OCD. Although O'Sullivan et al. (531) found no benefit of clomipramine over placebo in the long-term outcome of OCD patients who also had been treated with exposure therapy, others found that clomipramine and behavior techniques have additive effects, at least early during treatment (262,535). Trials that combined the newer SSRI agents fluvoxamine and fluoxetine with behavioral therapy suggest that one drug enhances the effectiveness of the other (535, 536).

Numerous psychotropic drugs have been tried in the past 25 years to control the symptoms of OCD, but only the potent SSRIs have consistently proven beneficial. The tricyclic antidepressant clomipramine was first reported in the 1960's to be an effective drug for some patients with obsessional symptoms (254). During the next decade, a number of uncontrolled studies confirmed that it was useful for treatment of patients with primary OCD (541,542). Since then, many controlled, double-blinded studies have demonstrated the superiority of clomipramine over placebo (507). In two large, multicenter, double-blinded, placebo-controlled studies of more than 500 outpatients with OCD at 21 different centers in North America (537), clomipramine was found to be significantly superior to placebo on all measures of OC symptom severity. Fewer studies have compared the efficacy of clomipramine to that of other drugs classified as antidepressants for treatment of OCD, but the trend emerges that agents with less potent serotonin reuptake blocking activity are ineffective in reducing obsessions and compulsions.

Thus, double-blinded drug treatment studies have demonstrated that the antiobsessional effect of clomipramine is superior to that of nortriptyline (261), amitriptyline (263), clorgyline (260), imipramine (538), and desipramine (393). These studies also explored the issue that success with clomipramine in patients with OCD could simply reflect an antidepressant effect rather than specific antiobsessional and anticompulsive effects. Ultimately, the findings demonstrated that clomipramine ameliorated OC symptomatology in nondepressed patients (258,260,261,263,265,538).

The antiobsessional effect of clomipramine could be the result of its alteration of serotonin transmission. With this theory in mind, in the past 10 years researchers have experimented with other nontricyclic serotonin reuptake inhibitors that would not (unlike clomipramine) lose their selectivity for blocking serotonin reuptake *in vivo* and lack significant activity for histaminic, cholinergic, and α-adrenergic receptors. The most rigorously studied agent has been fluvoxamine, which has been shown in single- and double-blinded studies to be more effective than placebo in OCD patients (539–542).

Few studies directly compare the efficacies of the different SSRIs in OCD. In one controlled comparative study, the therapeutic effect of fluoxetine was found to be equivalent to that of clomipramine in OCD patients (543). In another multicenter, double-blinded study, sertraline was more effective than placebo in reducing OC symptoms, but its drug effect was not as large as that observed in studies of clomipramine and fluvoxamine (544). Comparing the response rates in independent trials of clomipramine, fluvoxamine, and fluoxetine in OCD patients, no drug emerges as significantly superior in efficacy. Clinician choice therefore depends more on matching the patient with the anticipated side effect profile than on drug efficacy. Clomipramine, like other tricyclic antidepressants, has anticholinergic side effects, such as weight gain, dry mouth, blurred vision, nausea, and constipation, which may be troublesome. These side effects are encountered less often with the newer SSRIs, although nausea is common with fluvoxamine, and insomnia and weight loss can occur with fluoxetine (507).

The empirical use of higher doses of antidepressant drugs in OCD than in depression is common, but there is little evidence that this practice is necessary. A 10-week study showed no advantage for administration of a daily dose of 40 or 60 mg over 20 mg fluoxetine in OCD patients (545). The antiobsessional effect of these drugs tends to take longer to emerge than the antidepressant effect observed during treatment of depression. The initial response may not be seen until 4 to 6 weeks after starting medication (261,271,546); therefore, a trial of therapy should continue for at least 10 weeks. It is unclear how long maintenance therapy should be continued. One double-blinded, placebo-controlled study of clomipramine found 90% rate of relapse within 2 months of discontinuing treatment, regardless of whether the duration had been for 4 to 8 months, 8 to 12 months, or more than 12 months (547). Most clinicians treat for a period of 6 to 12 months and then gradually taper the dose, but in view of the high relapse rates reported, patients may need medication indefinitely. There is emerging evidence that treatment with doses lower than those used initially to produce a response may be effective in maintaining OCD patients free of symptoms (548).

Although SSRIs can be very successful in improving the symptoms of some OCD patients, 40% to 60% of patients exhibit minimal to no change after treatment with an adequate trial of an SSRI alone. These treatment-refractory patients represent a large proportion of the population of OCD sufferers, and a number of different biologic therapeutic strategies have been advocated for this group of patients. Although debatable, there may be some value in substituting one SSRI for another if the patient fails to respond to an adequate trial of the first. Combination therapies have been recommended (546). If a purely antiobsessional regimen is ineffective, several different augmenting agents have been used, mostly based on anecdotal reports of their use in patients who did not respond to more conventional treatment.

Thus, other agents that modify serotonergic function, such as the serotonin releaser and reuptake blocker fenfluramine (549) and the serotonin type 1A agonist buspirone (214,219,350), have been added to ongoing SSRI therapy with variable success. Studies have documented improvement in OC symptoms with augmentation therapy using lithium and tryptophan with SSRIs (507). Nevertheless, double-blinded trials of addition of lithium to ongoing clomipramine and fluvoxamine treatment, as well as buspirone addition to clomipramine, fluoxetine, and fluvoxamine treatment, have failed to show significant benefit (507). In light of new evidence that brain dopamine transmission also may be involved in mediating OC manifestations of some patients with OCD, particularly those with a personal or family history of tics, the role of neuroleptics in the treatment of OCD has been reevaluated. Neuroleptics alone do not seem to be effective, but combination therapy involving treatment with SSRIs and low-dose dopamine-blocking agents may be of use (329), although the hazards of long-term treatment with the latter must be appreciated.

Although OCD is classified as an anxiety disorder, there have been only a few reports of idiosyncratic response to anxiolytics, such as clonazepam (550) and alprazolam (551), in OCD patients. There are no systematic studies of their role as augmenting agents, but their use in conjunction with SSRIs is common, despite the risks of producing dependency with long-term administration. A few uncontrolled studies have reported improvement in OCD symptoms after treatment with monoamine oxidase inhibitors (552). However, a more recent controlled study comparing clomipramine and the selective monoamine oxidase inhibitor clorgyline in OCD patients failed to find any difference between the two groups (260). A few case reports have indicated that trazodone (276), clonidine (553), and antiandrogens (549) produce a positive response in patients with OCD, but these results must be considered with caution. Until well-designed controlled studies involving large numbers of patients are undertaken, the use of drugs other than SSRIs for treatment of OCD remains speculative.

Neurosurgical Treatment

Nonpharmacologic biologic treatments of OCD have included electroconvulsive therapy (ECT) phototherapy, and psychosurgery. ECT is not generally effective in ameliorating obsessions and compulsions, but it may be indicated for treatment of severe depressive or suicidal symptoms in medication-refractory OCD patients (546). Bright light therapy in a study investigating the role of seasonal symptomatic variation in a small group of OCD patients had no significant effect on symptoms (554).

Neurosurgery is reserved for the minority of patients who fail to respond to behavior therapy and SSRIs and who have chronic unremitting illness for at least 2 years with severe life disruption. Few "psychosurgical" procedures are now performed, following their unjustifiable overuse with often disastrous consequences in the 1940's and 1950's. It is not possible to perform double-blinded controlled trials. Still, follow-up studies show that of all the disorders treated by neurosurgery, OCD consistently responds best (555). With a greater understanding of functional neuroanatomy and the development of more precise and accurate techniques (e.g., stereotactic cryosurgery, thermocoagulation, and multifocal leukocoagulation), adverse effects (e.g., epilepsy, cognitive deficits, personality change) are rare, and mortality approaches zero. Procedures that interrupt the pathways from the frontal cortex to the basal ganglia, such as cingulotomy, anterior capsulotomy, and orbitofrontal leukotomy, are most effective in ameliorating disabling OC symptoms in such patients (556). Follow-up studies show that at least two thirds of patients improve following psychosurgery (556,557). The most extensive study interviewed 90 patients with OCD who had undergone operation between 1960 and 1974 at a mean 12-year follow-up (555). Thirty percent had completely recovered with no symptoms and required no treatment; 38% were well with mild residual symptoms; 26% had improved but still had significant interfering symptoms; only 6% remained unchanged; and one was worse. Postoperative epilepsy occurred in only one case; no personality change was detectable to the patients themselves or to others in 85%; and operative morbidity was minimal and lasted only a few days after surgery. Relapse was rare unless recurrent depression also was present. The results confirm the efficacy of neurosurgery in chronic intractable cases of OCD and provide further research data for a neuroanatomic model of this condition.

The mainstay of treatment of OCD is an 8- to 12-week trial of pharmacotherapy with one of the SSRIs, which have been shown in double-blinded controlled trials to be superior to placebo in improving the symptoms of OCD. There is no convincing evidence suggesting that one SSRI is more efficacious than another, so clinician choice depends on side effect profile, availability, and patient preference. The duration of treatment should be at least 6 months and often longer, considering the high rates of relapse following cessation of treatment. Although published controlled research support for combination treatments is scarce, augmentation of SSRI treatment with agents such as lithium, buspirone, or a neuroleptic may be useful for patients who have not responded to an SSRI alone, particularly if there is concurrent depression, anxiety disorder, or tics. Neurosurgery is reserved for the most severe refractory cases. Evidence suggests that behavior therapy and SSRIs are complementary, and this combination may be the optimum treatment available for the OCD patient. Finally, the importance of education, reassurance, and self-help must not be underestimated. Family therapy and helping the OCD patient to cope at work or school may be as crucial to the outcome of the patient as pharmacobehavioral intervention.

MANAGEMENT AND REHABILITATION

Successful management of the person with GTS requires both psychosocial measures and pharmacologic intervention for the individual, as well as various strategies for dealing with the patient's family. For many adults with mild GTS, explanation and reassurance often are sufficient for their peace of mind. Several such patients have an initial appointment and require only one or two immediate follow-up sessions to be given feedback about investigations that are routinely performed, information about self-help groups, and booklets for relatives and general practitioners. In a similar way, parents of mildly affected children often can be reassured by the diagnosis, explanation about the nature of the disorder, information about self-help groups, and booklets for teachers. For the moderate to severely afflicted patient who may have the associated features of OCB, ADHD, SIB, and possibly antisocial behavior, the management is more complex.

Behavioral therapy has been the first line of treatment in OCD for more than 20 years, despite the 25% failure rate, high rates of refusal, dropout among otherwise suitable patients, and recent evidence that combined multimodal behavioral therapy and pharmacotherapy may be more effective than behavioral therapy alone (535,536). Behavioral therapy is no longer thought to be effective in ameliorating the tics and vocalizations of GTS, and this practice has largely been abandoned (21). Combined drug therapy and behavior therapy appears to be the treatment of choice for OCD with GTS. For the tic symptoms, massed practice—over rehearsal of the target tic—and other forms of behavior therapy (including EEG sensorimotor rhythm biofeedback training) can be helpful, but behavioral approaches in general are of limited value (558).

Formal psychoanalytic psychotherapy is now widely considered to have little place in the treatment of either disorder (507). What is vitally important is supportive psychotherapy, counseling, and advice to both the patient and family, as the illness can be disabling to the affected individual but distressing to relatives. Support in helping the patient and family to cope is important, as is guidance to teachers (559). In many instances, as the doctors in charge of the patient, we write letters to teachers and principals informing them about GTS and its associated behaviors (especially ADHD, which may simultaneously affect concentration and disrupt

a class). We may suggest individual tuition, extra explanation, extra time in examinations, allowing the use of word processors or computers in the classroom, or stating that the child is unlikely to be of danger to others, for example, in a chemistry class or swimming pool.

For the severely affected patient who may have the associated features of ADHD, SIB, and aggressive behavior, the management is complex. It is best handled by doctors and health professionals who are well acquainted with the disorder. In these cases, management includes counseling, regular assessment of mood and danger to the individual (depression and SIB), and often the prescribing of more than one medication, such as an SSRI and sulpiride. When the GTS symptomatology and associated behaviors are very severe and life threatening, neurosurgical treatment has been used, but it is not a common practice.

Adolescence is a difficult period in the lives of most people; it is a time of both emotional and physical unpredictability. It is associated with growth spurts, hormonal changes, and the development of secondary sexual characteristics. Not surprisingly, tics and associated TS symptoms seem to worsen and become more problematic at this time. There is some evidence that temper tantrums, aggression, and explosiveness appear in the preadolescent period; become severe in teenage years; and gradually recede thereafter. The tic symptomatology of GTS also ameliorates with age (560).

CONCLUSIONS

GTS is no longer the rarity it was once thought to be; both case reports and large cohorts have been reported worldwide. The clinical characteristics are well established, and the core symptomatology is uniform and genetically determined (motor and vocal tics). The associated symptoms are variable (SIB, ADHD) and likely result from of a variety of genetic and environmental factors in predisposed individuals. An important aspect of GTS is OCB, which may be a phenotype of the putative GTS gene(s). While the search continues for the exact genetic mechanism in GTS, there are several factors that complicate the process (561,562). These include uncertainties with diagnosis and phenotypic definitions, assortative mating, genomic imprinting, bilineality, genetic heterogeneity, and unclear mode of inheritance. Future research will emphasize the more precise definition of the phenotype. Once genetic linkage is established, it will be possible to identify the actual gene and its causative mutations. This will be followed by prenatal diagnosis and genetic counseling, as well as the development of new treatments based on knowledge of the disease pathways.

Postmortem and neuroimaging studies suggest both structural changes and functional abnormalities in the basal ganglia, cingulum, and frontal areas, although the specificity of the latter finding is questioned. MRI studies might shed light on subtle structural changes that may, in turn, correlate with changes in function assessed by various tests, including SPECT and MRI. Identification of an endophenotype would spark much interest and may well be found using a combination of genetic and neuroimaging techniques.

The evidence we presented suggests that both GTS and OCD have a genetic basis, which in some cases is linked and results in common neurochemical abnormalities. The phenotype of the anticipated GTS gene(s) may be expressed as a spectrum of symptoms, with chronic multiple tics (CMTs) or GTS alone at one end and some types of OCD at the other. This GTS-associated OCD appears to represent a separate subcategory of the OCD defined in DSM-III-R, which indicates that OCD in its entirety may consist of a heterogeneous group of conditions of different etiologies. If so, it would be necessary to reevaluate both the classification and treatment of OCD. Future studies are needed to confirm these conclusions, and the results from the ongoing genetic linkage studies and the combined treatment studies are eagerly awaited.

In conclusion, at least some types of OCD (e.g., OCB) probably are an integral part of GTS. In this context, it is interesting to note that in 1903, in his treatise *Les Obsessions et la Psychiasthenie*, Pierre Janet (75) described three stages of psychasthenic illness: the first was the "psychasthenic state"; the second was "forced agitations," which included motor tics; and the third was obsessions and compulsions (563).

REFERENCES

1. American Psychiatric Association. *Diagnostic and statistical manual of mental disorders, fourth edition.* Washington, DC: American Psychiatric Association, 1994.
2. World Health Organization. *International classification of diseases and health related problems, 10th revision.* Geneva, Switzerland: World Health Organization, 1992.
3. Itard JMG. Memoire sur quelques fonctions involuntaires des appareils de la locomotion de la prehension et de la voix. *Arch Gen Med* 1825;8:385–407.
4. Gilles de la Tourette G. Etude sur une affection nerveuse caracterisee par de l'incoordination motrice accompagnee d-echolalie et de copralalie. *Arch Neurol* 1885;9:19–24;158–200.
5. Lees AJ, Robertson MM, Trimble MR, et al. A clinical study of Gilles de la Tourette syndrome in the United Kingdom. *J Neurol Neurosurg Psychiatry* 1984;47:1–8.
6. McHenry LC Jr. Samuel Johnson's tics and gesticulations. *J Hist Med Allied Sci* 1967;22:152–168.
7. Murray TJ. Dr. Samuel Johnson's movement disorders. *Br Med J* 1979;i:1610–1614.
8. Robertson MM. Tourette syndrome, associated conditions and the complexities of treatment. *Brain* 2000;123:425–462.
9. Berrios GE. Obsessive-compulsive disorder: its conceptual history in France during the 19th century. *Compr Psychiatry* 1989;30:283–295.
10. Stevens H. The syndrome of Gilles de la Tourette and its treatment. *Med Ann DC* 1964;33:277–279.
11. Shapiro AK, Shapiro E, Bruun RD, et al. *Gilles de la Tourette syndrome.* New York: Raven Press, 1978.
12. Hoogduin K. On the diagnosis of obsessive-compulsive disorder. *Am J Psychother* 1986;40:36–51.
13. Gilles de la Tourette G. La maladie des tics convulsifs. *La Sem Med* 1899;19:153–156.
14. Guinon G. Sur la maladie des tics convulsifs. *Rev Med* 1886;6:50–80.
15. Grasset J. Lecons sur un cas de maladie des tics et un cas de tremblement singulier de la tete et des membres gauches. *Arch Neurol* 1890;20:27–45, 187–211.
16. Robertson MM, Reinstein DZ. Convulsive tic disorder Georges Gilles de la Tourette, Guinon and Grasset on the phenomenology and psychopathology of Gilles de la Tourette syndrome. *Behav Neurol* 1991;4:29–56.
17. Meige H, Feindel E. *Tics and their treatment.* New York: William Wood and Co., 1907. Wilson SAK, editor and translator.
18. Kinnear-Wilson SAK. Tics and allied conditions. *J Neurol Psychopathol* 1927;8:93–109.
19. Ascher E. Psychodynamic consideration in Gilles de la Tourette's disease (maladie des tics): with a report of five cases and discussion of the literature. *Am J Psychiatry* 1948;105:267–276.
20. Bockner S. Gilles de la Tourette's disease. *J Ment Sci* 1959;105:1078–1081.
21. Robertson MM. The Gilles de la Tourette syndrome: the current status. *Br J Psychiatry* 1989;154:147–169.
22. Regeur L, Pakkenberg B, Fog R, et al. Clinical features and long-term treatment with pimozide in 65 patients with Gilles de la Tourette's syndrome. *J Neurol Neurosurg Psychiatry* 1986;49:791–795.
23. Comings DE, Comings BG. Tourette syndrome: clinical and psychological aspects of 250 cases. *Am J Hum Genet* 1985;37:435–450.
24. Lang AE. Premonitory ("sensory") experiences. In: Kurlan R, ed. *Handbook of Tourette's and related disorders.* New York: Marcel Dekker, 1993:17–26.
25. Bruun RD, Budman CL. The natural history of Tourette's syndrome. In: Chase TN, Friedhoff AJ, Cohen DJ, eds. *Tourette's syndrome: genetics, neurobiology and treatment. Advances in neurology, volume 58.* New York: Raven Press, 1992:1–6.
26. Erenberger G, Cruse RP, Rothner AD. Tourette syndrome: an analysis of 200 pediatric and adolescent cases. *Cleve Clin J Med* 1986;53:127–131.
27. Robertson MM, Gourdie A. Familial Tourette's syndrome in a large British pedigree: associated psychopathology, severity and potential for linkage analysis. *Br J Psychiatry* 1990;156:515–521.
28. McMahon WM, Leppert M, Filloux F, et al. Tourette's syndrome in 161 related family members. *Adv Neurol* 1992;58:159–165.
29. Caine ED, McBride MC, Chiverton P, et al. Tourette syndrome in Monroe County school children. *Neurology* 1988;38:472–475.
30. Nomura Y, Segawa M. Tourette syndrome in Oriental children: clinical and pathophysiological considerations. *Adv Neurol* 1982;36:277–280.
31. Robertson MM, Stern JS. The Gilles de la Tourette syndrome. *Crit Rev Neurobiol* 1997;11:1–19.
31a. Regeur L, Korbo L, Bang N, et al. Decreased volume of the cerebral ventricles on CT images in Gilles de la Tourette's syndrome. *Behav Neurol* 1998;11(3):139–147.
32. Butler IJ. Tourette's syndrome. Some new concepts. *Neurol Clin* 1984;2:571–580.
33. Robertson MM, Eapen V. Pharmacologic controversy of CNS stimulants in Gilles de la Tourette syndrome. *Clin Neuropharmacol* 1992;15:408–425.
34. Abuzzahab FE, Anderson FO. Gilles de la Tourette's syndrome. *Minn Med* 1973;56:492–496.

35. Glaze DG, Jankovic J, Frost JD. Sleep in Gilles de la Tourette syndrome: disorder of arousal. *Neurology* 1982;32:586–592.

36. Incagnoli T, Kane R. Developmental perspective of the Gilles de la Tourette syndrome. *Percept Mot Skills* 1983;57:1271–1281.

37. Leckman JF, Cohen DJ. Recent advances in Gilles de la Tourette syndrome: implications for clinical practice and future research. *Psychiatr Dev* 1983;3:301–316.

38. Rapoport JL. The neurology of obsessive-compulsive disorder. *JAMA* 1988;260:2888–2890.

39. Rapoport JL. *The boy who couldn't stop washing.* Melbourne: Collins, 1990.

40. Swedo SE, Leonard HL, Rapoport JL, et al. A double-blind comparison of clomipramine and desipramine in the treatment of trichotillomania (hair-pulling). *N Engl J Med* 1989;321:497–501.

41. Leonard HL, Lenane M, Swedo S, et al. A double-blind comparison of clomipramine and desipramine treatment of severe onychophagia (nail biting). *Arch Gen Psychiatry* 1991;48:821–827.

42. Winchel RM, Stanley B, Guido J, et al. An open trial of fluoxetine for trichotillomania (hair pulling). Presented at the 28th annual meeting of the American College of Neuropsychopharmacology, Maui, Hawaii, December 13, 1989.

43. Stanley MA, Bowers TC, Taylor DJ. Treatment of trichotillomania with fluoxetine [Letter]. *J Clin Psychiatry* 1991;52:282.

44. Christenson GA, Popkin MK, Mackenzie TB, et al. Lithium treatment of chronic hair pulling. *J Clin Psychiatry* 1991;52:116–120.

45. Pollard CA, Ibe O, Krojanker DN, et al. Clomipramine treatment of trichotillomania: a follow-up report on four cases. *J Clin Psychiatry* 1991;52:128–130.

46. Swedo SE, Rapoport JL, Leonard HL, et al. Regional cerebral glucose metabolism of women with trichotillomania. *Arch Gen Psychiatry* 1991;48:828–833.

47. Stanley MA, Swann AC, Bowers TC, et al. A comparison of clinical feature in trichotillomania and obsessive-compulsive disorder. *Behav Res Ther* 1992;30:39–44.

48. Brady KT, Austin L, Lydiard RB. Body dysmorphic disorder: the relationship to obsessive-compulsive disorder. *J Nerv Ment Dis* 1990;178:538–540.

49. Hollander E, Neville D, Frankel M, et al. Body dysmorphic disorder. Diagnostic issues and related disorders. *Psychosomatics* 1992;33:156–165.

50. Hollander E, Liebowitz MR, Winchel R, et al. Treatment of body dysmorphic disorder with serotonin reuptake blockers. *Am J Psychiatry* 1989;145:786–770.

51. Rasmusen SA, Eisen JL. The epidemiology and differential diagnosis of obsessive compulsive disorder. *J Clin Psychiatry* 1992;53[Suppl 4]:4–10.

52. McElroy SL, Hudson JI, Pope H Jr, et al. The DSM-III-R impulse control disorders not elsewhere classified: clinical characteristics and relationship to other psychiatric disorders. *Am J Psychiatry* 1992;149:318–327.

53. Eapen V, Robertson MM, Alsobrook JP, et al. Obsessive compulsive symptoms in Gilles de la Tourette syndrome and obsessive compulsive disorder: differences by diagnosis and family history. *Am J Med Genet* 1997;74:432–438.

54. Leckman JF, Walker WK, Goodman WK, et al. "Just right" perceptions associated with compulsive behaviors in Tourette's syndrome. *Am J Psychiatry* 1994;151:675–680.

55. Muller N, Putz A, Kathmann N, et al. Characteristics of obsessive compulsive symptoms in Tourette's syndrome, obsessive compulsive disorder and Parkinson's disease. *Psychiatry Res* 1997;70:105–114.

56. Petter T, Richter MA, Sandor P. Clinical features distinguishing patients with Tourette's syndrome and obsessive compulsive disorder from patients with obsessive compulsive disorder without tics. *J Clin Psychiatry* 1998;59:456–459.

57. Pauls DL, Leckman JF. The inheritance of Gilles de la Tourette's syndrome and associated behaviours: evidence for autosomal dominant transmission. *N Engl J Med* 1986;315:993–997.

58. Pauls DL, Towbin KE, Leckman JF, et al. Gilles de la Tourette syndrome and obsessive compulsive disorder. Evidence supporting an etiological relationship. *Arch Gen Psychiatry* 1986;43:1180–1182.

58a. Santangelo SL, Pauls DL, Goldstein JM, et al. Tourette's syndrome: what are the influences of gender and comorbid obsessive-compulsive disorder? *J Am Acad Child Adolesc Psychiatry* 1994;33(6):795–804.

59. Iida J, Sakiyama S, Iwasaka H, et al. The clinical features of Tourette's disorder with obsessive-compulsive symptoms. *Psychiatry Clin Neurosci* 1996;50(4):185–189.

60. Zohar AH, Pauls DL, Ratzoni G, et al. Obsessive-compulsive disorder with and without tics in an epidemiological sample of adolescents. *Am J Psychiatry* 1997;154(2):274–276.

61. Holzer JC, Goodman WK, McDougle CJ, et al. Obsessive-compulsive disorder with and without a chronic tic disorder. A comparison of symptoms in 70 patients. *Br J Psychiatry* 1994;164(4):469–473.

62. Baer L. Factor analysis of symptom subtypes of obsessive-compulsive disorder and their relation to personality and tic disorders. *J Clin Psychiatry* 1994;(55):18–23.

63. Pitman RK, Green RC, Jenike MA, et al. Clinical comparison of Tourette's disorder and obsessive-compulsive disorder. *Am J Psychiatry* 1987;144:1166–1171.

64. Frankel M, Cummings JL, Robertson MM, et al. Obsessions and compulsions in Gilles de la Tourette's syndrome. *Neurology* 1986;36:378–382.

65. George MS, Trimble MR, Ring HA, et al. Obsessions in obsessive-compulsive disorder with and without Gilles de la Tourette's syndrome. *Am J Psychiatry* 1993;150:93–97.

66. Leckman JF, Grice DE, Barr LC, et al. Tic-related vs. non-tic-related obsessive compulsive disorder. *Anxiety* 1995;1:20800–15.

67. Cath DC, Spinhoven P, van Woerkom TC, et al. Gilles de la Tourette's syndrome with and without obsessive-compulsive disorder compared with obsessive-compulsive disorder without tics: which symptoms discriminate? *J Nerv Ment Dis* 2001;189(4):219–228.

68. Miguel EC, do Rosario-Campos MC, Prado HS, et al. Sensory phenomena in obsessive-compulsive disorder and Tourette's disorder. *J Clin Psychiatry* 2000;61(2):150–156.

69. Miguel EC, Baer L, Coffey BJ, et al. Phenomenological differences appearing with repetitive behaviours in obsessive-compulsive disorder and Gilles de la Tourette's syndrome. *Br J Psychiatry* 1997;170:140–145.

70. Miguel EC, Coffey BJ, Baer L, et al. Phenomenology of intentional repetitive behaviors in obsessive-compulsive disorder and Tourette's disorder. *J Clin Psychiatry* 1995;56(6):246–255.

71. Montgomery MA, Clayton PJ, Friedhoff AJ. Psychiatric illness in Tourette syndrome patients and first-degree relatives. In: Friedhoff AJ, Chase TN, eds. *Gilles de la Tourette syndrome. Advances in neurology, volume 35.* New York: Raven Press, 1982:335–339.

72. Robertson MM, Trimble MR, Lees AJ. The psychopathology of Gilles de la Tourette syndrome: a phenomenological analysis. *Br J Psychiatry* 1988;152:383–390.

73. Coffey BJ, Miguel EC, Biederman J, et al. Tourette's disorder

with and without obsessive-compulsive disorder in adults: are they different? *J Nerv Ment Dis* 1998;186(4):201–206.

74. Pitman RK, Jenike MA. Coprolalia in obsessive-compulsive disorder: a missing link. *J Nerv Ment Dis* 1988;176:311–313.

75. Janet P. *Les obsessions et la psychiasthenie, volume 1.* Paris: Felix Alcan, 1903.

76. Cummings JL, Frankel M. Gilles de la Tourette syndrome and the neurological basis of obsessions and compulsions. *Biol Psychiatry* 1985;20:1117–1126.

77. Ferenczi S. Psycho-analytic observations on tic. *Int J Psychoanal* 1921;2:1–30.

78. Mahler MS, Rangell L. A psychosomatic study of maladie des tics (Gilles de la Tourette's disease). *Psychiatry Q* 1943;17:579–603.

79. Mahler MS. Tics and impulsions in children: a study of motility. *Psychoanal* 1944;13:430–444.

80. Mahler MS, Luke JA, Daltroff W. Clinical and follow up study of the tic syndrome in children. *Am J Orthopsychiatry* 1945;15:631–647.

81. Mahler MS, Luke JA. Outcome of the tic syndrome. *J Nerv Ment Dis* 1946;103:433–445.

82. Fenichel O. *The psychoanalytic theory of neurosis.* New York: Norton and Company, 1945.

83. Heuscher JE. Intermediate states of consciousness in patients with generalized tics. *J Nerv Ment Dis* 1953;117:29–38.

84. Shapiro AK, Shapiro ES. Tourette syndrome: history and present status. In: Friedhoff AJ, Chase TN, eds. *The Gilles de la Tourette syndrome. Advances in neurology, volume 35.* New York: Raven Press, 1982:17–23.

85. Comings DE. A controlled study of Tourette syndrome. VII. Summary. A common genetic disorder due to disinhibition of the limbic system. *Am J Hum Genet* 1987;41:839–836.

86. Comings DE, Comings BG. A controlled study of Tourette syndrome, I-VII. *Am J Hum Gen* 1987;41:701–866.

87. Comings DE. *Tourette syndrome and human behavior.* Duarte, CA: Hope Press, 1990.

88. Comings DE, Comings BG. Comorbid behavioral disorders. In: Kurlan R, ed. *Handbook of Tourette's syndrome and related tic and behavioral disorders.* New York: Marcel Dekker, 1993:111–150.

89. Robertson MM, Channon S, Baker J, et al. The psychopathology of Gilles de la Tourette's syndrome: a controlled study. *Br J Psychiatry* 1993;162:114–117.

90. Van de Wetering BJM, Cohen AP, Minderaa RB, et al. Het syndroom van Gilles de la Tourette: Klinische Bevindigen. *Ned Tijdschr Geneeskd* 1988;132:21–25.

91. Robertson MM, Stern JS. Tic disorders: new developments in Tourette syndrome and related disorders. *Curr Opin Neurol* 1998;11:373–380.

92. Coffey B, Frazier J, Chen S. Comorbidity, Tourette's syndrome and anxiety disorders. In: Chase TN, Friedhoff AJ, Cohen DJ, eds. *Tourette's syndrome: genetics, neurobiology and treatment. Advances in neurology, volume 58.* New York: Raven Press, 1992:95–104.

93. Comings DE, Comings BG. A controlled family history study of Tourette syndrome 1: ADHD and learning disorders. *J Clin Psychiatr* 1990;51:275–280.

94. Towbin KE, Riddle MA. Attention deficit hyperactivity disorder. In: Kurlan R, ed. *Handbook of Tourette syndrome and related tic disorders.* New York: Marcel Dekker, 1993:89–110.

95. Comings DE, Comings BG. Tourette's syndrome and attention deficit disorder with hyperactivity: are they genetically related? *J Am Acad Child Psychiatry* 1984;23:138–146.

96. Comings DE, Comings BG. A controlled family history study of Tourette's syndrome, I: attention deficit hyperactivity disorder and learning disorders. *J Clin Psychiatry* 1990;51:275–280.

97. Knell E, Comings D. The sex ratio for tics, and attention deficit disorder (ADD) in persons with tics, is approximately 1:1 in relatives of Tourette's syndrome (TS) probands. 41st Annual Meeting of American Society of Human Genetics. *Clin Genet* 1990;1(244):126(abst).

98. Pauls DL, Cohen DJ, Kidd KK, et al. The Gilles de la Tourette syndrome [Letter]. *Am J Hum Genet* 1988;43:206–209.

99. Comings DE, Comings BG. Tourette's syndrome and attention deficit disorder. In: Cohen DJ, Bruun RD, Leckman JF, eds. *Tourette's syndrome and tic disorders: clinical understanding and treatment.* New York: John Wiley and Sons, 1988:119–135.

100. Pauls DL, Leckman JF, Cohen DJ. The familial relationship between Gilles de la Tourette syndrome, attention deficit disorder, learning disabilities, speech disorders and stuttering. *J Am Acad Child Adolesc Psychiatry* 1993;32:1044–1050.

101. Robertson MM, Trimble MR, Lees AJ. Self-injurious behavior and the Gilles de la Tourette syndrome. A clinical study and review of the literature. *Psychol Med* 1989;19:611–625.

102. Robertson MM. Self injurious behavior and Tourette syndrome. In: Chase TN, Friedhoff AJ, Cohen DJ, eds. *Tourette's syndrome: genetics, neurobiology and treatment. Advances in neurology, volume 58.* New York: Raven Press, 1992:105–114.

103. Robertson MM, Yakeley JW. Obsessive-compulsive disorder and self-injurious behavior. In: Kurlan R, ed. *Handbook of Tourette's syndrome and related tic and behavioral disorders.* New York: Marcel Dekker, 1993:45–87.

104. Robertson MM, Banerjee S, Fox-Hiley PJ, et al. Personality disorder and psychopathology in Tourette's syndrome: a controlled study. *Br J Psychiatry* 1997;171:283–286.

105. Pauls DL, Alsobrook JP II, Goodman W, et al. A family study of obsessive compulsive disorder. *Am J Psychiatry* 1995;152:76–84.

106. Robertson MM. Tourette syndrome, associated conditions and the complexities of treatment. *Brain* 2000;123:425–462.

107. Cumings DE. Role of genetic factors in depression based on studies of Tourette syndrome and ADHD probands and their relatives. *Am J Genet* 1995;60:111–121.

108. Keller MB, Hirshfield RMA, Hanks D. Double depression: a distance subtype of unipolar depression. *J Affect Disord* 1997;45:65–73.

109. Bleich A, Bernout E, Apter A, et al. Gilles de la Tourette syndrome and mania in an adolescent. *Br J Psychiatry* 1985;146:664–665.

110. Burd L, Kerbeshian J. Gilles de la Tourette's syndrome and bipolar disorder. *Arch Neurol* 1984;41:1236.

111. Berthier ML, Kulisevsky J, Campos VM. Bipolar disorder in adult patients with Tourette syndrome: a clinical study. *Biol Psychiatry* 1998;43:364–370.

112. Kerbeshian J, Burd L. Case study: co-morbidity among Tourette's syndrome, autistic disorder, and bipolar disorder. *J Am Acad Child Adolesc Psychiatry* 1996;35:681–685.

113. Kerbeshian J, Burd L, Klug MG. Comorbid Tourette's disorder and bipolar disorder: an aetiologic perspective. *Am J Psychiatry* 1995;152:1646–1651.

114. Caine ED, Margolin DI, Brown GL, et al. Gilles de la Tourette's syndrome, tardive dyskinesia and psychosis in an adolescent. *Am J Psychiatry* 1978;135:241–243.

115. Takevchi K, Yamashita M, Morikiyo M, et al. Gilles de la Tourette's syndrome and schizophrenia. *J Nerv Ment Dis* 1986:174–248.

116. Reid AH. Gilles de la Tourette syndrome in mental handicap. *J Med Defic Res* 1984;28:81–83.

117. Burd L, Kerbeshian J. Tourette syndrome, atypical pervasive developmental disorder and Ganser syndrome in a 15 year old, visually impaired, mentally retarded boy. *Can J Psychiatry* 1985;30:74–76.

118. Kerbeshian J, Burd L. A second visually impaired, mentally retarded male with pervasive developmental disorder, Tourette disorder, Ganser's syndrome: diagnostic classification and treatment. *Int J Psychiatry Med* 1986;6:67–75.

119. Kerbeshian J, Burd L. Are schizophreniform symptoms present in attenuated form in children with Tourette disorder and other developmental disorders? *Can J Psychiatry* 1987;32:123–135.

120. Kerbeshian J, Burd L. Asperger's syndrome and Tourette syndrome: the case of the pinball wizard. *Br J Psychiatry* 1986;148:731–736.

121. Baron-Cohen S, Martimore C, Moriarty J, et al. Prevalence of Gilles de la Tourette syndrome in children and adolescents with autism: a large scale study. *Psychol Med* 1999;29:1151–1159.

122. Boswell J. *The life of Samuel Johnson LLD.* London: George Routledge & Sons, 1867.

123. Lieh Mak F, Chung SY, Lee P, et al. Tourette syndrome in the Chinese: a follow up of fifteen cases. *Adv Neurol* 1982;35:281–284.

124. Fernando SJM. Gilles de la Tourette's syndrome: a report on four cases and a review of published case reports. *Br J Psychiatry* 1967;113:607–617.

125. Corbett JA, Mathews AM, Connell PH, et al. Tics and Gilles de la Tourette's syndrome: a follow-up study and critical review. *Br J Psychiatry* 1969;115:1229–1241.

126. Field JR, Corbin KB, Goldstein NP, et al. Gilles de la Tourette syndrome. *Neurology* 1966;16:453–462.

127. Morphew JA, Sim M. Gilles de la Tourette's syndrome: a clinical and psychopathological study. *Br J Med Psychol* 1969;42:293–301.

128. Pauls DL, Leckman JF, Cohen DJ. Evidence against a genetic relationship between Gilles de la Tourette syndrome and anxiety, depression, panic and phobic disorders. *Br J Psychiatry* 1994;164:215–221.

129. Bruun RD. Gilles de la Tourette's syndrome: an overview of clinical experience. *J Am Acad Child Adolesc Psychiatry* 1984;23:126–133.

130. Tanner CM, Goldman SM. Epidemiology of Tourette syndrome. *Neurol Clin* 1997;15:295–402.

131. Apter A, Pauls DL, Bleich A, et al. An epidemiological study of Gilles de la Tourette syndrome in Israel. *Arch Gen Psychiatry* 1993;50:734–8.

132. Kadesjo B, Gillberg C. Tourette's disorder: epidemiology and comorbidity in primary school children. *J Am Acad Child Adolesc Psychiatry* 2000;39:548–555.

133. Mason A, Banerjee S, Eapen V, et al. The prevalence of Gilles de la Tourette syndrome in a mainstream school population: a pilot study. *Dev Med Child Neurol* 1998;40:292–296.

134. American Psychiatric Association. *Diagnostic and statistical manual of mental disorders, third ed.* Washington, DC: American Psychiatric Association, 1980.

135. Zausmer DM. The treatment of tics in childhood: a review and follow-up study. *Arch Dis Child* 1954;29:537–542.

136. Torup E. A follow-up study of children with tics. *Acta Pediatr Scand* 1962;51:261–268.

137. Corbett JA, Matthews AM, Connell PH. Tics and Gilles de la Tourette's syndrome: a follow-up study and critical review. *Br J Psychiatry* 1969;115:1229–1241.

138. Green RC, Pitman RK. Tourette syndrome and obsessive-compulsive disorder. In: Jenike MA, Baer L, Minichiello WO, eds. *Obsessive-compulsive disorder: theory and management.* Littleton, MA: PSGP Publishing, 1986.

139. Rasmussen SA, Tsuang MT. Clinical characteristics and family history in DSM III obsessive-compulsive disorder. *Am J Psychiatry* 1986;143:317–322.

140. Swedo SE, Henriette L, Leonard MD. Childhood movement disorders and obsessive compulsive disorders. *J Clin Psychiatry* 1994;55[3 Suppl]:32–37.

140a. Rapoport JL, Swedo SE, Leonard HL. Childhood obsessive compulsive disorder. *J Clin Psychiatry* 1992;53(Suppl):11–16.

141. Hajal F, Leach AM. Familial aspects of Gilles de la Tourette syndrome. *Am J Psychiatr* 1981;138:90–92.

142. Kurlan R. Tourette's syndrome in a special education population. Hypotheses. *Adv Neurol* 1992;58:75–81.

143. Comings DE, Himes JA, Comings BG. An epidemiologic study of Tourette's syndrome in a single school district. *J Clin Psychiatr* 1990;51:463–469.

144. Golden GS. Psychologic and neuropsychologic aspects of Tourette's syndrome. *Neurol Clin* 1984;21:91–102.

145. Robertson MM. Annotation: Gilles de la Tourette syndrome—an update. *J Child Psychol Psychiatry* 1994;35:597–611.

146. Staley D, Wand R, Shady G. Tourette Disorder: a cross-cultural review. *Compr Psychiatry* 1997;38:6–16.

147. Eldridge R, Sweet R, Lake CR, et al. Gilles de la Tourette's syndrome: clinical, genetic, psychologic, and biochemical aspects in 21 selected families. *Neurology* 1977;27:115–124.

148. Wassman ER, Eldridge R, Abuzzahab IS Sr, et al. Gilles de la Tourette syndrome: clinical and genetic studies in a mid western city. *Neurology* 1978;28:304–307.

149. Moldofsky H, Tullis C, Lamon R. Multiple tics syndrome (Gilles de la Tourette's syndrome). *J Nerv Ment Dis* 1974;15:282–292.

150. Erenberger G, Rothner AD. Tourette syndrome: a childhood disorder. *Cleve Clin J Med* 1978;45:207–212.

151. Golden GS, Hood OJ. Tics and tremors. *Pediatr Clin North Am* 1982;29:95–103.

152. Lucas AR, Beard CM, Rajput AH, et al. Tourette syndrome in Rochester, Minnesota, 1968-1979. In: Friedhoff AJ, Chase TM, eds. *Gilles de la Tourette syndrome. Advances in neurology, volume 35.* New York: Raven Press, 1982:267–269.

153. Krumholz A, Singer HS, Niedermeyer E, et al. Electrophysiological studies in Tourette's syndrome. *Ann Neurol* 1983;14:638–641.

154. Bergen D, Tanner CM, Wilson R. The electroencephalogram in Tourette syndrome. *Ann Neurol* 1981;11:382–383.

155. Obeso JA, Rothwell JC, Marsden CD. The neurophysiology of Tourette's syndrome. In: Friedhoff AJ, Chase TN, eds. *Gilles de la Tourette syndrome. Advances in neurology, volume 35.* New York: Raven Press, 1982:105–114.

156. Verma NP, Syrigou-Papavasiliou A, Lewitt PA. Electroencephalographic findings in unmedicated, neurologically and intellectually intact Tourette syndrome patients. *Electroencephalogr Clin Neurophysiol* 1986;64:12–20.

157. Lucas AR, Rodin EA. Electroencephalogram in Gilles de la Tourette's disease. Fifth World Congress of Psychiatry, Mexico City, 1973:85–89.

158. Semerci ZB. Neurological soft signs and EEC findings in children and adolescents with Gilles de la Tourette syndrome. *Turk J Pediatr* 2000;42:53–55.

159. Hyde TM, Emsellem HA, Randolph C, et al. Electroencephalographic abnormalities in monozygotic twins with Tourette's syndrome. *Br J Psychiatry* 1994;164:811–817.

160. Stevens A, Gunther W, Lutzenberger W, et al. Abnormal topography of EEG microstates in Gilles de la Tourette syndrome. *Eur Arch Psychiatry Clin Neurosci* 1996;246:310–316.

161. Gunther W, Muller N, Trapp W, et al. Quantitative EEG

analysis during motor function and music perception in Tourette's syndrome. *Eur Arch Psychiatry Clin Neurosci* 1996;246:197–202.

162. Ziemann U, Paulus W, Rothenberger A. Decreased motor inhibition in Tourette's disorder: evidence from transcranial magnetic stimulation. *Am J Psychiatry* 1997;154:1277–1284.

163. Gironell A, Rodriguez-Fornells A, Kulisevsky J, et al. Abnormalities of the acoustic startle reflex and reaction time in Gilles de la Tourette syndrome. *Clin Neurophysiol* 2000;111:1366–1371.

164. Van de Wetering BJM, Martens CMC, Fortgens C, et al. Late components of the auditory evoked potentials in Gilles de la Tourette syndrome. *Clin Neurol Neurosurg* 1985;87:181–186.

165. Dursun SM, Burke JG, Reveley MA. Antisaccade eye movement abnormalities in Tourette syndrome: evidence for corticostriatal network dysfunction? *J Psychopharmacol* 2000;14:37–39.

166. Georgiou N, Bradshaw JL, Phillips JG, et al. Advance information and movement sequencing in Gilles de la Tourette's syndrome. *J Neurol Neurosurg Psychiatry* 1995;58:184–191.

167. Tolosa ES, Montserrat L, Bayes A. Reduction of brainstem interneuron excitability during voluntary tic inhibition in Tourette's syndrome. *Neurology* 1986;36[Suppl]:118–119.

168. Drake ME Jr, Pakalnis A, Newell SA. EEG frequency analysis in obsessive-compulsive disorder. *Neuropsychobiology* 1996;33:97–99.

169. Harcherik DF, Cohen DJ, Ort S, et al. Computed tomographic brain scanning in four neuropsychiatric disorders of childhood. *Am J Psychiatry* 1985;142:731–734.

170. Caparulo BK, Cohen DJ, Rothman SL, et al. Computed tomographic brain scanning in children with developmental neuropsychiatric disorders. *J Am Acad Child Psychiatry* 1981;20:338–357.

171. Chase TN, Geoffrey V, Gillespie M, et al. Structural and functional studies of Gilles de la Tourette syndrome. *Rev Neurol (Paris)* 1986;142:851–855.

172. Shaenboen MJ, Nigro MA, Martocci RJ. Colpocephali and Gilles de la Tourette's syndrome. *Arch Neurol* 1984;41:1023.

173. Kjaer M, Boris P, Hansen LG. Abnormal CT scan in a patient with Gilles de la Tourette syndrome. *Neuroradiology* 1986;28:362–363.

174. Yeragani VK, Blackman M, Barker GB. Biological and psychological aspects of a case of Gilles de la Tourette syndrome. *J Clin Psych* 1983;44:27–29.

175. Lakke JPWF, Wilmink JT. A case of Gilles de la Tourette's syndrome with midbrain involvement. *J Neurol Neurosurg Psychiatry* 1985;48:1293–1296.

176. Vieregge P. Gilles de la Tourette syndrome in monozygotic twins. *J Neurol Neurosurg Psychiatry* 1987;50:1554–1556.

177. Demeter S. Structural imaging in Tourette's syndrome. In: Chase TN, Friedhoff AJ, Cohen DJ, eds. *Tourette's syndrome: genetics, neurobiology and treatment. Advances in neurology, volume 58.* New York: Raven Press, 1992:201–206.

178. Sandyk R. A case of Tourette syndrome with midbrain involvement. *Int J Neurosci* 1988;43:171–175.

179. Robertson M, Doran M, Trimble M, et al. The treatment of Gilles de la Tourette syndrome by limbic leucotomy. *J Neurol Neurosurg Psychiatry* 1990;53:691–694.

180. Peterson B, Riddle MA, Cohen DJ, et al. Reduced basal ganglia volumes in Tourette's syndrome using three-dimensional reconstruction techniques from magnetic resonance images. *Neurology* 1993;43:941–949.

181. Singer HS, Reiss AL, Brown JE, et al. Volumetric MRI changes in basal ganglia of children with Tourette's syndrome. *Neurology* 1993;43:950–956.

182. Witelson SF. Clinical neurology as data for basic neuroscience: Tourette's syndrome and the human motor system. *Neurology* 1993;43:859–861.

183. Peterson BS, Leckman JF, Duncan JS, et al. Corpus callosum morphology from magnetic resonance images in Tourette's syndrome. *Psychiatry Res* 1994;55:85–99.

184. Moriarty J, Varma AR, Stevens J, et al. A volumetric MRI study of Gilles de la Tourette's syndrome. *Neurology* 1997;49:410–415.

185. Hyde TM, Stacey ME, Coppola R, et al. Cerebral morphometric abnormalities in Tourette's syndrome: a quantitative MRI study of monozygotic twins. *Neurology* 1995;45:1176–1182.

186. Zimmerman AM, Abrams MT, Giuliano JD, et al. Subcortical volumes in girls with Tourette syndrome: support for a gender effect. *Neurology* 2000;54:2224–2229.

187. Biswal B, Ulmer JL, Krippendorf RL, et al. Abnormal cerebral activation associated with a motor task in Tourette syndrome. *AJNR Am J Neuroradiol* 1998;19:1509–1512.

188. Peterson BS, Skudlarski P, Anderson AW. A functional magnetic resonance imaging study of tic suppression in Tourette syndrome. *Arch Gen Psychiatry* 1998;55:326–333.

189. Chase TN, Foster NL, Fedio P, et al. Gilles de la Tourette syndrome: studies with the fluorine-18-labelled fluorodeoxyglucose positron emission tomographic method. *Ann Neurol* 1984;15[Suppl]:S175.

190. Stoetter B, Braun AR, Randolph C, et al. Functional neuroanatomy of Tourette syndrome: limbic motor interactions studied with FDG PET. *Adv Neurol* 1992;58:213–226.

191. Turjanski N, Sawle GV, Playford ED, et al. PET studies of the presynaptic and postsynaptic dopaminergic system in Tourette's syndrome. *J Neurol Neurosurg Psychiatry* 1994;57:688–692.

192. Wong DF, Singer HS, Brandt J. D2-like dopamine receptor density in Tourette syndrome measured by PET. *J Nucl Med* 1997;38:1243–1247.

193. Braun AR, Randolph C, Stoetter B, et al. The functional neuroanatomy of Tourette's syndrome: an FDG-PET Study. II: relationships between regional cerebral metabolism and associated behavioral and cognitive features of the illness. *Neuropsychopharmacology* 1995;13:151–1568.

194. Braun AR, Stoetter B, Randolph C, et al. The functional neuroanatomy of Tourette's syndrome: an FDG-PET study. I. Regional changes in cerebral glucose metabolism differentiating patients and controls. *Neuropsychopharmacology* 1993;9:277–291.

195. Hall M, Costa DC, Shields J, et al. Brain perfusion patterns with ^{99m}Tc-HMPAO/SPET in patients with Gilles de la Tourette syndrome—short report. *Nucl Med* 1990;27[Suppl]:243–245.

196. George MS, Trimble MR, Costa DC, et al. Elevated frontal cerebral blood flow in Gilles de la Tourette syndrome A ^{99m}Tc-HMPAO SPECT study. *Psychiatry Res* 1992;45:143–151.

197. Moriarty J, Costa DC, Schmitz B, et al. Brain perfusion abnormalities in Gilles de la Tourette syndrome. *Br J Psychiatry* 1995;167:249–254.

198. Riddle M, Rasmussen AM, Woods SW, et al. SPECT imaging of cerebral blood flow in Tourette's syndrome. *Adv Neurol* 1992;58:207–212.

199. Klieger PS, Fett KA, Dimitsopulos T, et al. Asymmetry of basal ganglia perfusion in Tourette's syndrome shown by technetium-99m-HMPAO SPECT. *J Nucl Med* 1997;38:188–191.

200. George MS, Robertson MM, Costa DC, et al. Dopamine receptor availability in Tourette's syndrome. *Psychiatry Res Neuroimaging* 1994;55:193–203.

201. Muller-Vahl KR, Berding G, Brucke T, et al. Dopamine transporter binding in Gilles de la Tourette syndrome. *J Neurol* 2000;247:514–20.

202. Malison RT, McDougle CJ, van Dyck CH, et al. [^{123}I] beta-CIT SPECT imaging of striatal dopamine transporter binding in Tourette's disorder. *Am J Psychiatry* 1995;152:1359–1361.

203. Muller-Vahl KR, Berding G, Kolbe H, et al. Dopamine D2 receptor imaging in Gilles de la Tourette syndrome. *Acta Neurol Scand* 2000;101:165–171.

204. Moriarty J, Eapen V, Costa DC, et al. HMPAO SPECT does not distinguish obsessive-compulsive and tic syndromes in families multiply affected with Gilles de la Tourette's syndrome. *Psychol Med* 1997;27:737–740.

205. Insel TR. Towards a neuroanatomy of obsessive compulsive. *Arch Gen Psychiatry* 1992;49:739–744.

206. Swedo SE, Pictrini P, Leonard HL, et al. Cerebral glucose metabolism in childhood onset obsessive compulsive disorder: revisualization during pharmacotherapy. *Arch Gen Psychiatry* 1992;49:690–694.

207. Insel TR, Donnelly EF, Lalakea ML, et al. Neurological and neuropsychological studies of patients with obsessive-compulsive disorder. *Biol Psychiatry* 1983;18:741–751.

208. Behar D, Rapoport JL, Berg MA, et al. Computerized tomography and neuropsychological test measures in adolescents with obsessive-compulsive disorder. *Am J Psychiatry* 1984;141:363–369.

209. Luxenberg JS, Swedo SE, Flament MF, et al. Neuroanatomical abnormalities in obsessive-compulsive disorder detected with quantitative x-ray computed tomography. *Am J Psychiatry* 1988;145:1089–1093.

210. Laplane D, Baulac M, Widlocher D, et al. Pure psychic akinesia with bilateral lesions of basal ganglia. *J Neurol Neurosurg Psychiatry* 1984;47:377–385.

211. Williams AC, Owen C, Heath DA. A compulsive movement disorder with cavitation of caudate nucleus. *J Neurol Neurosurg Psychiatry* 1988;51:447–448.

212. Palumbo D, Maughan A, Kurlan R. Hypothesis III. Tourette syndrome is only one of several causes of a developmental basal ganglia syndrome. *Arch Neurol* 1997;54:475–483.

213. Swedo SE, Leonard HL. Childhood movement disorders and obsessive compulsive disorder. *J Clin Psychiatry* 1994;55[Suppl]:32–37.

214. Kellner CH, Jolley RR, Holgate RC, et al. Brain MRI in obsessive-compulsive disorder. *Psychiatry Res* 1991;36:45–49.

215. Garber HJ, Ananth JV, Chiu LC, et al. Nuclear magnetic resonance study of obsessive-compulsive disorder. *Am J Psychiatry* 1989;146:1001–1005.

216. Baxter LR, Phelps ME, Mazziotta JC, et al. Local cerebral glucose metabolic rates in obsessive-compulsive disorder. A comparison with rates in unipolar depression and in normal controls. *Arch Gen Psychiatry* 1987;44:211–218.

217. Baxter LR Jr, Schwartz JM, Phelps ME, et al. Cerebral glucose metabolic rates in nondepressed patients with obsessive-compulsive disorder. *Am J Psychiatry* 1988;145:1560–1563.

218. Swedo SE, Schapiro MB, Grady CL, et al. Cerebral glucose metabolism in childhood-onset obsessive-compulsive disorder. *Arch Gen Psychiatry* 1989;46:518–523.

219. Nordahal T, Benkelfat C, Semple W. Cerebral glucose metabolic rates in obsessive-compulsive disorder. *Neuropsychopharmacology* 1989;2:23–28.

220. Benkelfat C, Nordahal TE, Semple WE, et al. Local cerebral glucose metabolic rates in obsessive-compulsive disorder. Patients treated with clomipramine. *Arch Gen Psychiatry* 1990;47:840–848.

221. Baxter LR, Guze BH. Neuro-imaging. In: Kurlan R, ed. *Handbook of Tourette's syndrome and related tic and behavioral disorders.* New York: Marcel Dekker, 1993:289–304.

222. Baxter LR Jr, Schwartz JM, Phelps ME, et al. Reduction of prefrontal cortex glucose metabolism is common to three types of depression. *Arch Gen Psychiatry* 1989;46:243–250.

223. Deutsch G. The non-specificity of frontal dysfunction in disease and altered states: cortical blood flow evidence. *Behav Neurol* 1992;5:301–307.

224. Berman KF, Doran AR, Pickar D, et al. Is the mechanism of prefrontal hypofunction in depression the same as in schizophrenia? Regional cerebral blood flow during cognitive activation. *Br J Psychiatry* 1993;162:183–192.

225. Dewulf A, Van Bogaert L. Etudes anatomo-cliniques de syndromes hypercinetiques complexes-partie 3. Une observation anatomo-clinique de maladie des tics (Gilles de la Tourette). *Monatsschr Psychiatr Neurol* 1941;104:53–61.

226. Balthasar K. Uber das anatomische substrat der generalisierten Tic-Krankheit (maladie des tics, Gilles de la Tourette): Entwicklungshemmung des corpus striatum. *Arch Psychiatr Nervenkrankheiten (Berl)* 1957;195:531–549.

227. Richardson EP. Neuropathological studies of Tourette syndrome. In: Friedhoff AJ, Chase TN, eds. *Gilles de la Tourette syndrome. Advances in neurology, volume 35.* New York: Raven Press, 1982:83–87.

228. Anderson GM, Pollak ES, Chatterjee D. Brain monamines and amino acids in Gilles de la Tourette syndrome: a preliminary study of subcortical regions [Letter]. *Arch Gen Psychiatry* 1992;49:584–586.

229. Singer HS. Neurochemical analysis of post-mortem cortical and striatal brain tissue in patients with Tourette syndrome. *Adv Neurol* 1992;58:135–144.

230. Haber SN, Kowall NW, Vonsattel JP, et al. Gilles de la Tourette's syndrome: a postmortem neuropathological and immunohistochemical study. *J Neurol Sci* 1986;75:225–241.

231. Haber SN, Wolfer D. Basal ganglia peptidergic staining in Tourette's syndrome: a follow up study. In: Chase TN, Friedhoff AJ, Cohen DJ, eds. *Tourette's syndrome: genetics, neurobiology and treatment. Advances in neurology, volume 58.* New York: Raven Press, 1992:145–150.

232. Brito GN. A neurobiological model for Tourette syndrome centered on the nucleus accumbens. *Med Hypothesis* 1997;49:133–42.

233. Weeks RA, Turjanski N, Brooks DJ. Tourette's syndrome: a disorder of cingulate and orbitofrontal function? *QJM* 1996;89:401–408.

234. Stern E, Silbersweig DA, Chee KY, et al. A functional neuroanatomy of tics in Tourette syndrome. *Arch Gen Psychiatry* 2000;57:741–748.

235. Saint-Cyr JA, Taylor AE, Nicholson K. Behavior and the basal ganglia. *Adv Neurol* 1995;65:1–28.

236. Sheppard DM, Bradshaw JL, Purcell R, et al. Tourette's and comorbid syndromes: obsessive compulsive and attention deficit hyperactivity disorder. A common etiology? *Clin Psychol Rev* 1999;19:531–52.

237. Wolf SS, Jones DW, Knable MB, et al. Tourette syndrome: prediction of phenotypic variation in monozygotic twins by caudate nucleus D2 receptor binding. *Science* 1996;273:1225–1227.

238. Caine ED. Gilles de la Tourette's syndrome: a review of clinical and research studies and consideration of future directions for investigation. *Arch Neurol* 1985;42:393–397.

239. Friedhoff AJ. Insights into the pathophysiology and pathogenesis of Gilles de la Tourette syndrome. *Rev Neurol (Paris)* 1986;142:860–864.

240. Pauls DL, van de Wetering BJM. The genetics of tics and related disorders. In: Robertson MM, Eapen V, eds. *Movement and allied disorders in childhood.* West Sussex, UK: John Wiley and Sons, 1995:13–29.

241. Bonnet KA. Neurobiological dissection of Tourette syndrome: a neurochemical focus on a human neuroanatomical model. In: Friedhoff AJ, Chase TN, eds. *Gilles de la Tourette syndrome. Advances in neurology, volume 35.* New York: Raven Press, 1982:77–82.

242. Karson CN, Kaufmann CA, Shapiro AK, et al. Eye-blink rate in Tourette's syndrome. *J Nerv Ment Dis* 1985;173:566–568.

243. Ernst M, Zametkin AJ, Jons PH, et al. High presynaptic dopaminergic activity in children with Tourette's disorder. *J Am Acad Child Adolesc Psychiatry* 1999;38:86–94.

244. Rabey JM, Oberman Z, Graff E, et al. Decreased dopamine uptake into platelet storage granules in Gilles de la Tourette disease. *Biol Psychiatry* 1995; 38:112–115.

245. Merikangas JR, Merikangas KR, Kopp U, et al. Blood choline and response to clonazepam and haloperidol in Tourette's syndrome. *Acta Psychiatr Scand* 1985;72:395–399.

246. Cohen DJ, Detlor J, Shaywitz B, et al. Interaction of biological and psychological factors in the natural history of Tourette syndrome: a paradigm for childhood neuropsychiatric disorders. In: Friedhoff AJ, Chase TN, eds. *Gilles de la Tourette syndrome. Advances in neurology, volume 35.* New York: Raven Press, 1982:31–40.

247. Eldridge R, Denckla MB. The inheritance of Gilles de la Tourette's syndrome. *N Engl J Med* 1987;317:1346–1347.

248. Kurlan R. The pathogenesis of Tourette's syndrome. A possible role of hormonal and excitatory neurotransmitter influences in brain development. *Arch Neurol* 1992b;49:874–876.

249. Trimble MR, Robertson MM. The psychopathology of tics. In: Marsden CD, Fahn S. *Modern trends in neurology. Movement disorders, volume 2.* London: Butterworths, 1987:406–422.

250. Devinsky O. Neuroanatomy of Gilles de la Tourette's syndrome: possible midbrain involvement. *Arch Neurol* 1983;40:508–514.

251. Jadresic D. The role of the amygdaloid complex in Gilles de la Tourette syndrome. *Br J Psychiatry* 1992;161:532–534.

252. Robertson M, Evans K, Robins A, et al. Abnormalities of copper in Gilles de la Tourette syndrome. *Biol Psychiatry* 1987;22968–978.

253. Barr LC, Goodman WK, Price LH, et al. The serotonin hypothesis of obsessive compulsive disorder: implications of pharmacologic challenge studies. *J Clin Psychiatry* 1992;53[Suppl 4]:17–28.

254. Fernandez CE, Lopez Ibor AJ. La monoclorimipramina en enfernos psiquiatricos resistentes a otros tratamientos. *Actas Luso Esp Neurol Psiquiat* 1969;26:119–147.

255. Van Renynghe-de-Voxvrie G. Use of anafranil (G34586) in obsessive neuroses. *Acta Neurol Psychiatr Belg* 1968;68:787–792.

256. Capstick N. Chlorimipramine in obsessional states. *Psychosomatics* 1971;12:332–335.

257. Leonard H, Swedo SE, Rapoport J, et al. Treatment of obsessive-compulsive disorder with clomipramine and desipramine in children and adolescents. A double blind crossover comparison. *Arch Gen Psychiatry* 1989;46:1088–1092.

258. Flament MF, Rapoport JL, Berg CJ, et al. Clomipramine treatment of childhood obsessive-compulsive disorder. A double-blind controlled study. *Arch Gen Psychiatry* 1985;42:977–983.

259. de Veaugh Geiss J, Landau P, Katz R. Treatment of obsessive compulsive disorder with clomipramine. *Psychiatr Ann* 1989;19:97–101.

260. Insel TR, Murphy DL, Cohen RM, et al. A double-blind trial of clomipramine and clorgyline. *Arch Gen Psychiatry* 1983;40:605–612.

261. Thoren R, Asberg M, Cronholm B, et al. Clomipramine treatment of obsessive-compulsive disorder: I. A controlled study. *Arch Gen Psychiatry* 1980;37:1281–1285.

262. Marks IM, Stern RS, Mawson D, et al. Clomipramine and exposure for obsessive compulsive rituals: I. *Br J Psychiatry* 1980;136:1–25.

263. Ananth J, Pecknold JC, Van Den Steen N, et al. Double-blind comparative study of clomipramine and amitriptyline in obsessive neurosis. *Biol Psychiatry* 1981;5:257–262.

264. Zohar J, Insel T. Obsessive-compulsive disorder: psychobiological approaches to diagnosis, treatment and pathophysiology. *Biol Psychiatry* 1987;22:667–687.

265. Mavissakalian M, Turner SM, Michelson L, et al. Tricyclic antidepressants in obsessive-compulsive disorder: antiobsessional or antidepressant agents? II. *Am J Psychiatry* 1985;142:572–576.

266. DeVeaugh-Geiss J, Katz R, Landau P, et al. Clinical predictors of treatment response in obsessive-compulsive disorder: exploratory analyses from multicenter trials of clomipramine. *Psychopharmacol Bull* 1990;26:45–49.

267. DeVeaugh-Geiss J, Katz R, Landau P, et al. Clomipramine in the treatment of patients with obsessive-compulsive disorder: the clomipramine collaborative study group. *Arch Gen Psychiatry* 1991;48:730–738.

268. DeVeaugh-Geiss J, et al. Clomipramine hydrochloride in childhood and adolescent obsessive-compulsive disorder—a multicenter trial. *J Am Acad Child Adolesc Psychiatry* 1992;31:45–49.

269. Perse TL, Greist JH, Jefferson JW. Fluvoxamine treatment of obsessive compulsive disorder. *Am J Psychiatry* 1987;144:1543–1548.

270. Cottraux J, Mollard E, Bouvard M. A controlled study of fluvoxamine and exposure in obsessive compulsive disorder. *Arch Gen Psychiatry* 1990;46:36–44.

271. Goodman WK, Price LH, Rasmussen SA. Efficacy of fluvoxamine in obsessive compulsive disorder. *Arch Gen Psychiatry* 1989;46:36–44.

272. Pigott TA, Pato MT, Bernstein SE, et al. Controlled comparisons of clomipramine and fluoxetine in the treatment of obsessive-compulsive disorder. Behavioral and biological results. *Arch Gen Psychiatry* 1990;47:926–932.

273. Chouinard G, Goodman WK, Greist J, et al. Results of a double-blind placebo controlled trial of a new serotonin uptake inhibitor, sertraline, in the treatment of obsessive-compulsive disorder. *Psychopharmacol Bull* 1990;26:279–284.

274. Goodman WK, Price LH, Delgado PL, et al. Specificity of serotonin reuptake inhibitors in the treatment of obsessive-compulsive disorder. Comparison of fluvoxamine and desipramine (see comments). *Arch Gen Psychiatry* 1990;47:577–585.

275. Kahn RS, Westenberg HG, Jolles J. Zimeldine treatment of obsessive-compulsive disorder. Biological and neuropsychological aspects. *Acta Psychiatr Scand* 1984;69:259–261.

276. Prasad, AJ. Obsessive-compulsive disorder and trazodone [Letter]. *Am J Psychiatry* 1984;141:612–613.

277. Blier P, Chaput Y, de Montigny C. Long-term 5-HT reuptake blockade, but not monoamine oxidase inhibition, decreases the function of terminal 5-HT autoreceptors: an electrophysiological study in the rat brain. *Naunyn Schmiedebergs Arch Pharmacol* 1988;337:246–254.

278. Blier P, de Montigny C, Chaput Y. A role for the serotonin system in the mechanism of action of antidepressant treatments:

preclinical evidence. *J Clin Psychiatry* 1990;51[Suppl 4]:14–20.

279. Peroutka SJ, Snyder SH. Chronic antidepressant treatment decreases spiroperidol-labeled serotonin receptor binding. *Science* 1980;210:88–90.

280. Zohar J, Murphy DL, Zohar-Kadouch RC, et al. *Serotonin in major psychiatric disorders.* Washington, DC: American Psychiatric Association Press, 1990.

281. Anonymous. Drugs affecting 5-hydroxytryptamine function. *Drug Ther Bull* 1993;31:25–27.

282. Yaryura-Tobias JA, Bebirian RJ, Neziroglu FA, et al. Obsessive-compulsive disorders as a serotonergic defect. *Res Commun Psychol Psychiatry Behav* 1977;2:279–286.

283. Hanna GL, Yuwiler A, Cantwell DP. Whole blood serotonin in juvenile obsessive-compulsive disorder. *Biol Psychiatry* 1991;29:738–744.

284. Flament MF, Rapoport JL, Murphy DL, et al. Biochemical changes during clomipramine treatment of childhood obsessive-compulsive disorder (published erratum appears in *Arch Gen Psychiatry* 1987;44:548). *Arch Gen Psychiatry* 1987;44:219– 225.

285. Insel TR, Mueller EA, Alterman I, et al. Obsessive-compulsive disorder and serotonin: is there a connection? *Biol Psychiatry* 1985;20:1174–1188.

286. Weizman A, Carmi M, Hermesh H, et al. High affinity imipramine binding and serotonin uptake in platelets of eight adolescent and ten adult obsessive-compulsive patients. *Am J Psychiatry* 1986;143:335–359.

287. Black DW, Kelly M, Myers C, et al. Tritiated imipramine binding in obsessive-compulsive volunteers and psychiatrically normal controls. *Biol Psychiatry* 1990;27:319–327.

288. Bastani B, Arora RC, Meltzer HY. Serotonin uptake and imipramine binding in the blood platelets of obsessive-compulsive disorder patients. *Biol Psychiatry* 1991;30:131–139.

289. Vitello BH, Shimon H, Behar D, et al. Platelet imipramine binding and serotonin uptake in obsessive-compulsive patients. *Acta Psychiatr Scand* 1991;84:29–32.

290. Kim· SW, Dysken MW, Pandey GN, et al. Platelet ^{3}H-imipramine binding sites in obsessive-compulsive behavior. *Biol Psychiatry* 1991;30:467–474.

291. Lydiard RB, Ballenger JC, Ellinwood E, et al. CSF monoamine metabolites in obsessive-compulsive disorder. Presented at the 143rd annual meeting of the American Psychiatric Association, New York, 1990. Cited in Barr LC, Goodman WK, Price LH, et al. The serotonin hypothesis of obsessive compulsive disorder: implications of pharmacologic challenge studies. *J Clin Psychiatry* 1992;53[4 Suppl]:17–28.

292. Leckman JF, Goodman WK, Riddle MA, et al. Low CSF 5HIAA and obsessions of violence: report of two cases. *Psychiatry Res* 1990;35:95–99.

293. de Groot CM, Bornstein RA, Baker GB. Obsessive-compulsive symptom clusters and urinary amine correlates in Tourette syndrome. *J Nerv Ment Dis* 1995;183:224–230.

294. Leckman JF, Goodman WK, Anderson GM, et al. Cerebrospinal fluid biogenic amines in obsessive compulsive disorder, Tourette's syndrome, and healthy controls. *Neuropsychopharmacology* 1995;12:73–86.

295. Cath DC, Gijsman HJ, Schoemaker RC, et al. The effect of m-CPP on tics and obsessive-compulsive phenomena in Gilles de la Tourette syndrome. 1999;144:137–43.

296. Aloi JA, Insel TR, Mueller EA, et al. Neuroendocrine and behavioral effects of m-chlorophenylpiperazine administration in rhesus monkeys. *Life Sci* 1984;43:1325–1331.

297. Hollander E, DeCaria CM, Nitescu A, et al. Serotonergic function in obsessive-compulsive disorder: behavioral and neuroendocrine responses to oral m-chlorophenylpiperazine and fenfluramine in patients and healthy volunteers. *Arch Gen Psychiatry* 1992;49:21–28.

298. Charney DS, Goodman WK, Price LH, et al. Serotonin function in obsessive-compulsive disorder. A comparison of the effects of tryptophan and m-chlorophenylpiperazine in patients and healthy subjects. *Arch Gen Psychiatry* 1988;45:177–185.

299. Zohar J, Mueller A, Insel TR, et al. Serotonergic responsivity in obsessive-compulsive disorder. Comparison of patients and healthy controls. *Arch Gen Psychiatry* 1987;44:946–951.

300. Zohar J, Insel TR, Zohar-Kadouch RC, et al. Serotonergic responsivity in obsessive-compulsive disorder. Effects of chronic clomipramine treatment. *Arch Gen Psychiatry* 1988;46:167–172.

301. Lesch KP, Hoh A, Disselkamp-Tietze J, et al. 5-Hydroxytryptamine receptor responsivity in obsessive-compulsive disorder: comparison of patients and controls. *Arch Gen Psychiatry* 1991;48:540–547.

302. McBridge PA, DeMeo MD, Sweeney JA, et al. Neuroendocrine and behavioral responses to challenge with the indirect serotonin agonist DL-fenfluramine in adults with obsessive-compulsive disorder. *Biol Psychiatry* 1992;31:19–34.

303. Van Praag HM, Lemus C, Kahn R. Hormonal probes of central serotonergic activity: do they really exist? *Biol Psychiatry* 1987;22:86–98.

304. Kahn RS, Wetzler S, Van Praag HM, et al. Behavioral indications for serotonin receptor hypersensitivity in panic disorder. *Psychiatry Res* 1988;25:101–104.

305. Charney DS, Woods SW, Goodman WK, et al. Serotonin function in anxiety. II: effects of the serotonin agonist, MCPP, in panic disorder patients and healthy subjects. *Psychopharmacology* 1987;92:14–24.

306. Benkelfat C, Murphy DL, Zohar J, et al. Clomipramine in obsessive-compulsive disorder. Further evidence for a serotonergic mechanism of action. *Arch Gen Psychiatry* 1989;46:23–28.

307. Baldwin D, Fineberg N, Bullock T, et al. Serotonin 1A receptors and obsessive-compulsive disorder. In: Gastpar M, et al., eds. *Serotonin 1A receptors in depression and anxiety.* New York: Raven Press, 1992:193–200.

308. Siever LJ, Insel TR, Jumerson DC, et al. Growth hormone response to clonidine in obsessive-compulsive patients. *Br J Psychiatry* 1983;142:184–187.

309. Lee MA, Cameron OG, Gurguis GN, et al. Alpha 2-adrenoreceptor status in obsessive-compulsive disorder. *Biol Psychiatry* 1990;27:1983–1093.

310. Hollander E, DeCaria C, Nitescu A, et al. Noradrenergic function in obsessive-compulsive disorder: behavioral and neuroendocrine responses to clonidine and comparison to healthy controls. *Psychiatry Res* 1991;37:161–177.

311. Rasmussen SA, Goodman WK, Woods SW, et al. Effects of yohimbine in obsessive compulsive disorder. *Psychopharmacology* 1987;93:308–313.

312. Janowsky A, Okada F, Manier DH, et al. Role of serotonergic input in the regulation of the β-adrenergic receptor-coupled adenylate cyclase system. *Science* 1982;219:900–901.

313. Blier P, de Montigny C, Chaput Y. Modifications of the serotonin system by anti-depressant treatments: implications for the therapeutic response in major depression. *J Clin Psychopharmacol* 1987;7:24S–35S.

314. Racagni G, Bradford D. Proceedings of the 14th Collegium Internationale Neuro-Psychopharmacologicum Congress, Florence, Italy, 1984: 733.

315. Insel TR, Pickar D. Naloxone administration in obsessive-compulsive disorder: report of two cases. *Am J Psychiatry* 1983;140:1219–1220.

316. Goodman WK, McDougle CJ, Price LH, et al. Beyond the serotonin hypothesis: a role for dopamine in some forms of obsessive-compulsive disorder. *J Clin Psychiatry* 1990;51[Suppl]:36–43, 55–58.

317. Creese I, Iversen SD. The role of forebrain dopamine systems in amphetamine induced stereotyped behavior in the rat. *Psychopharmacologia* 1974;39:345–357.

318. Loew DM, Vigouret J-M, Jaton A. Neuropharmacology of bromocriptine and dihydroergotoxine (Hydergine). In: Goldstein M, ed. *Ergot compounds and brain function: neuroendocrine and neuropsychiatric aspects.* New York: Raven Press, 1980:63–74.

319. Wallach MB, Gershon S. A neuropsychopharmacological comparison of D-amphetamine, L-dopa and cocaine. *Neuropharmacology* 1971;10:743–752.

320. Willner JH, Samach M, Angrist BM, et al. Drug-induced stereotyped behavior and its antagonism in dogs. *Commun Behav Biol* 1970;5:135–142.

321. Randrup A, Munkvad HM. Stereotyped activities produced by amphetamine in several animal species and man. *Psychopharmacologia* 1967;11:300–310.

322. Koizumi HM. Obsessive-compulsive symptoms following stimulants. *Biol Psychiatry* 1985;20:1332–1337.

323. Frye PE, Arnold LE. Persistent amphetamine-induced compulsive rituals: response to pyridoxine (B6). *Biol Psychiatry* 1981;16:583–587.

324. Leonard HL, Rapoport JL. Relief of obsessive-compulsive symptoms following stimulants. *Biol Psychiatry* 1987;20:1332–1337.

325. McDougle CJ, Goodman WK, Delgado PL, et al. Pathophysiology of obsessive compulsive disorder [Letter]. *Am J Psychiatry* 1989;146:1350–1351.

326. Austin LS, Lydiard RB, Ballenger JC, et al. Dopamine blocking activity of clomipramine in patients with obsessive-compulsive disorder. *Biol Psychiatry* 1991;30:225–232.

327. Riddle MA, Hardin MT, King R, et al. Fluoxetine treatment of children and adolescents with Tourette's and obsessive-compulsive disorders: preliminary clinical experience. *J Am Acad Child Adolesc Psychiatry* 1990;29:45–48.

328. Delgado PL, Goodman WK, Price LH, et al. Fluvoxamine/pimozide treatment of concurrent Tourette's and obsessive compulsive disorder. *Br J Psychiatry* 1990;157:762–765.

329. McDougle CJ, Goodman WK, Price LH, et al. Neuroleptic addition in fluvoxamine-refractory obsessive-compulsive disorder. *Am J Psychiatry* 1990;147:652–654.

330. Lipinski JF Jr, Mallya G, Zimmerman P, et al. Fluoxetine-induced akathisia: clinical and theoretical implications. *J Clin Psychiatry* 1989;50:339–342.

331. Tate JL. Extrapyramidal symptoms in a patient taking haloperidol and fluoxetine. *Am J Psychiatry* 1989;146:399–400.

332. Bouchard RH, Pourcher E, Vincent P. Fluoxetine and extrapyramidal side effects. *Am J Psychiatry* 1989;146:1352–1353.

333. Brod TM. Fluoxetine and extrapyramidal side effects. *Am J Psychiatry* 1989;146:1353.

334. Meltzer HY, Young M, Metz J, et al. Extrapyramidal side effects and increased serum prolactin following fluoxetine, a new antidepressant. *J Neural Transm* 1979;45:165–175.

335. Swedo SE, Leonard HL, Kruesi MJP, et al. Cerebrospinal fluid neurochemistry in children and adolescents with obsessive-compulsive disorder. *Arch Gen Psychiatry* 1992;49:29–36.

336. Crespi F, Martin KF, Marsden C. Simultaneous in vivo voltametric measurement of striatal extracellular DOPAC and 5-HIAA levels: effect of electrical stimulation of DA and 5HT neuronal pathways. *Neurosci Lett* 1988;90:285–291.

337. McDougle CJ, Goodman WK, Price LH, et al. Neuroleptic addition in fluvoxamine refractory OCD. *Am Psychiatr Assoc New Res Abstracts* 1989:NR 350(abst).

338. Yaryura-Tobias JA, Neziroglu F. *Obsessive compulsive disorders: pathogenesis, diagnosis, and treatment.* New York: Marcel Dekker, 1983.

339. Riddle MA, Leckman JF, Hardin MT, et al. Fluoxetine treatment of obsessions and compulsions in patients with Tourette's syndrome [Letter]. *Am J Psychiatry* 1988;145:1173–1174.

340. Golden GS, Greenhill L. Tourette syndrome in mentally retarded children. *Ment Retard* 1981;19:17.

341. Brookshire BL, Butler IJ, Ewing-Cobbs L, et al. Neuropsychological characteristics of children with Tourette syndrome: evidence for a nonverbal learning disability? *J Clin Exp Neuropsychol* 1994;16:289–302.

342. Baron-Cohen S, Cross P, Crowson M, et al. Can children with Gilles de la Tourette syndrome edit their intentions? *Psychol Med* 1994;24:29–40.

343. Channon S, Flynn D, Robertson MM. Attentional deficits in Gilles de la Tourette syndrome. *Neuropsychiatry Neuropsychol Behav Neurol* 1992;5:170–177.

344. Schultz RT, Carter AS, Gladstone M, et al. Visual-motor integration functioning in children with Tourette syndrome. *Neuropsychology* 1998;12:134–45.

345. Oades RD. Differential measures of "sustained attention" in children with attention-deficit/hyperactivity or tic disorders: relations to monoamine metabolism. *Psychiatry Res* 2000;93:165–78.

346. de Groot CM, Yeates KO, Baker GB, et al. Impaired neuropsychological functioning in Tourette's syndrome subjects with co-occurring obsessive-compulsive and attention deficit symptoms. *J Neuropsychiatry Clin Neurosci* 1997;9:267–272.

347. Ozonoff S, Strayer DL, McMahon WM, et al. Executive function abilities in autism and Tourette syndrome: an information processing approach. *J Child Psychol Psychiatry* 1994;35:1015–32.

348. Ozonoff S, Strayer DL, McMahon WM, et al. Inhibitory deficits in Tourette syndrome: a function of comorbidity and symptom severity. *J Child Psychol Psychiatry* 1998;39:1109–1118.

349. Head D, Bolton D, Hymas N. Deficit in cognitive shifting ability in patients with obsessive-compulsive disorder. *Biol Psychiatry* 1989;25:929–937.

350. Cox CS, Fedio P, Rapoport JL. Neuropsychological testing of obsessive-compulsive adolescents. In: Rapoport JL, ed. *Obsessive-compulsive disorder in children and adolescents.* Washington, DC: American Psychiatric Press, 1989:73–85.

351. Malloy P. Frontal lobe dysfunction in obsessive-compulsive disorder. In: Perecman E, ed. *The frontal lobes revisited.* New York: The IRBN Press, 1987:207–223.

352. Sher KJ, Frost RO, Kushner M, et al. Memory deficits in compulsive checkers: replication and extension in a clinical sample. *Behav Res Ther* 1989;27:65–69.

353. Hollander E, Schiffman E, Cohen B, et al. Signs of central nervous system dysfunction in obsessive-compulsive disorder. *Arch Gen Psychiatry* 1990;47:27–32.

354. Boone KB, Ananth J, Philpott L, et al. Neuropsychological characteristics of nondepressed adults with obsessive-compulsive disorder. *Neuropsychiatry Neuropsychol Behav Neurol* 1991;4:96–109.

355. Zielinski CM, Taylor MA, Juzwin KR. Neuropsychological deficits in obsessive-compulsive disorder. *Neuropsychiatry Neuropsychol Behav Neurol* 1991;4:110–126.

356. Christensen GA, Mackenzie TB, Mitchell JE, et al. A placebo-controlled, double-blind crossover study of fluoxetine in trichotillomania. *Am J Psychiatry* 1991;148:1566–1571.

357. Swedo SE, Leonard HL, Garvey M, et al. Pediatric autoimmune neuropsychiatric disorders associated with streptococcal infections: clinical description of the first 50 cases. *Am J Psychiatry* 1998;155:264–71.

358. Laurino JP, Hallett J, Kiessling LS, et al. An immunoassay for anti-neuronal antibodies associated with involuntary repetitive movement disorders. *Ann Clin Lab Sci* 1997;27:230–5.

359. Singer HS, Giuliano JD, Hansen BH, et al. Antibodies against a neuron-like (HTB-10 neuroblastoma) cell in children with Tourette syndrome. *Biol Psychiatry* 1999 46:775–780.

360. Murphy TK, Goodman WK, Fudge MW, et al. B lymphocyte antigen D8/17: a peripheral marker for childhood-onset obsessive-compulsive disorder and Tourette's syndrome? *Am J Psychiatry* 1997;154:402–407.

361. Kiessling LS, Marcotte AC, Culpepper L. Antineuronal antibodies: tics and obsessive-compulsive symptoms. *J Dev Behav Pediatr* 1994;15:421–425.

362. Singer HS, Giuliano JD, Hansen BH, et al. Antibodies against human putamen in children with Tourette syndrome. *Neurology* 1998;50:1618–1624.

363. Muller N, Riedel M, Straube A, et al. Increased anti-streptococcal antibodies in patients with Tourette's syndrome. *Psychiatry Res* 2000;94:43–49.

364. Peterson BS, Leckman JF, Tucker D, et al. Preliminary findings of antistreptococcal antibody titers and basal ganglia volumes in tic, obsessive-compulsive, and attention deficit/hyperactivity disorders. *Arch Gen Psychiatry* 2000;57:364–372.

365. Hallett JJ, Harling-Berg CJ, Knopf PM, et al. Anti-striatal antibodies in Tourette syndrome cause neuronal dysfunction. *J Neuroimmunol* 2000;111:195–202.

366. Giedd JN, Rapoport JL, Garvey MA, et al. MRI assessment of children with obsessive-compulsive disorder or tics associated with streptococcal infection. *Am J Psychiatry* 2000;157:281–283.

367. Lougee L, Perlmutter SJ, Nicolson R, et al. Psychiatric disorders in first-degree relatives of children with pediatric autoimmune neuropsychiatric disorders associated with streptococcal infections (PANDAS). *J Am Acad Child Psychiatry* 2000;39:1120–1126.

368. Trifiletti RR, Packard AM. Immune mechanisms in pediatric neuropsychiatric disorders. Tourette's syndrome, OCD, and PANDAS. *Child Adolesc Psychiatr Clin N Am* 1999;8:767–775.

369. Perlmutter SJ, Leitman SF, Garvey MA, et al. Therapeutic plasma exchange and intravenous immunoglobulin for obsessive-compulsive disorder and tic disorders in childhood. *Lancet* 1999;354:1153–1158.

370. Nicolson R, Swedo SE, Lenane M, et al. An open trial of plasma exchange in childhood-onset obsessive-compulsive disorder without poststreptococcal exacerbations. *J Am Acad Child Adolesc Psychiatry* 2000;39:1313–1315.

371. Kurlan R. Investigating Tourette syndrome as a neurological sequelae of rheumatic fever. *CNS Spectrums* 1999;4:62–67.

372. Rapp DJ. Allergy and Tourette's syndrome. *Ann Allergy* 1986;56:507.

373. Mandell M. Allergy and Tourette's syndrome. *Ann Allergy* 1986;56:507–508.

374. Finegold I. Allergy and Tourette's syndrome. *Ann Allergy* 1985;55:119–121.

375. Robertson MM, Kalali A, Brostoff J. Cited in Carroll A, Robertson M. *Tourette syndrome: a guide for teachers, parents and carers.* London: David Fulton Publishers, 2000.

376. Ho CS, Shen Ey, Shyur SD, et al. Association of allergy with Tourette's syndrome. *J Formos Med Assoc* 1999;98:492–5.

377. Zausmer DM, Dewey ME. Tics and heredity. *Br J Psychiatry* 1987;150:628–634.

378. Baraitser M. *The genetics of neurological disorders.* London: Oxford University Press, 1982.

379. Pakstis AJ, Heutink P, Pauls DL, et al. Progress in the search for genetic linkage with Tourette syndrome: an exclusion map covering more than 50% of the autosomal genome. *Am J Hum Genet* 1991;48:281–294.

380. Escalar G, Majeron MA, Finavera L, et al. Contributo alla conoscenza della sindrome Gilles de la Tourette. *Minerva Med* 1972;63:3517–3522.

381. Jenkins RL, Ashby HB. Gilles de la Tourette syndrome in identical twins. *Arch Neurol* 1983;40:249–251.

382. Waserman J, Lal S, Gauthier S. Gilles de la Tourette syndrome in monozygotic twins. *J Neurol Neurosurg Psychiatry* 1983;46:75–77.

383. Ellison RM. Gilles de la Tourette syndrome. *Med J Aust* 1964;1:153–155.

384. Price RA, Kidd KK, Cohen DJ, et al. A twin study of Tourette syndrome. *Arch Gen Psychiatry* 1985;42:815–820.

385. Leckman JF, Price RV, Walkup JT, et al. Nongenetic factors in Gilles de la Tourette's syndrome. *Arch Gen Psychiatry* 1987;44:100.

386. Dunlap JR. A case of Gilles de la Tourette's disease (maladie des tics): a study of the intrafamily dynamics. *J Nerv Ment Dis* 1960;130:340–344.

387. Friel PB. Familial incidence of Gilles de la Tourette's disease, with observations on aetiology and treatment. *Br J Psychiatry* 1963;122:655–658.

388. Guggenheim MA. Familial Tourette syndrome. *Ann Neurol* 1979;5:104.

389. American Psychiatric Association. *Diagnostic and statistical manual of mental disorders, revised third edition.* Washington, DC: American Psychiatric Association, 1987.

390. Eapen V, Pauls DL, Robertson MM. Evidence of autosomal dominant transmission in Tourette's syndrome. United Kingdom cohort study. *Br J Psychiatry* 1993;162:593–596.

391. Comings DE, Devor EJ, Cloninger CR. Detection of a major gene for Gilles de la Tourette syndrome. *Am J Hum Genet* 1984;36:586–600.

392. Devor EJ. Complex segregation analysis of Gilles de la Tourette syndrome: further evidence for a major locus mode of transmission. *Am J Human Genet* 1984;36:704–709.

393. Price RA, Pauls DL, Caine ED. Pedigree and segregation analysis of clinically defined subgroups of Tourette syndrome. *Am J Hum Genet* 1984;36:178.

394. Pauls DL, Leckman JF. The inheritance of Gilles de la Tourette syndrome and associated behaviors. Evidence for autosomal dominant transmission. *N Engl J Med* 1986;315:993–997.

395. Pauls DL, Pakstis AJ, Kurlan R, et al. Segregation and linkage analyses of Tourette's syndrome and related disorders. *J Am Acad Child Adolesc Psychiatry* 1990;29:195–203.

396. Curtis D, Robertson MM, Gurling HMD. Autosomal dominant gene transmission in a large kindred with Gilles de la Tourette syndrome. *Br J Psychiatry* 1992;160:845–849.

397. Hasstedt SJ, Leppert M, Filloux F, et al. Intermediate inheritance of Tourette syndrome, assuming assortative mating. *Am J Hum Genet* 1995;57:682–9.

398. Singer HS, Brown J, Riddle MA, et al. Family study and segregation analysis of Tourette syndrome: evidence for a mixed model of inheritance. *Am J Hum Genet* 1996;59:684–693.

399. Comings DE, Wu S, Chiu C, et al. Polygenic inheritance of

Tourette syndrome, stuttering, attention deficit hyperactivity, conduct, and oppositional defiant disorder: the additive and subtractive effect of the three dopaminergic genes—DRD2, D beta H, and DAT1. *Am J Med Genet* 1996 67:264–88.

400. Seuchter SA, Hebebrand J, Klug B, et al. Complex segregation analysis of families ascertained through Gilles de la Tourette syndrome. *Genet Epidemiol* 2000;18:33–47.

401. Robertson MM, Trimble MR. Normal chromosomal findings in Gilles de la Tourette syndrome. *Psychiatr Genet* 1993;3:95–99.

402. Brett P, Curtis D, Gourdie A, et al. Possible linkage of Tourette syndrome to markers on short arm of chromosome 3 (C3p21-14). *Lancet* 1990;336:1076.

403. Comings DE, Comings BG, Dietz G, et al. Evidence of the Tourette syndrome gene is at 18g22.1. 7th International Congress of Human Genetics, Berlin, Germany, 1986: Abstract M,11,23:620.

404. Donnai D. Gene localization in Tourette's syndrome. *Lancet* 1987;i:627.

405. Comings DE, Muhleman D, Dibtz G, et al. Association between Tourette's syndrome and homozygosity at the dopamine D3 receptor gene. *Lancet* 1993;341:906.

406. Heutink P, Sandkuyl LA, van de Wetering BJM, et al. Linkage and Tourette syndrome. *Lancet* 1991;337:122–123.

407. Brett P, Schneiden V, Jackson G, et al. Chromosome markers in Tourette's syndrome. *Lancet* 1991;337:184–185.

408. Brett P, Robertson MM, Gurling HMD, et al. Failure to find linkage and increased homozygosity for the dopamine D3 receptor gene in Tourette's syndrome. *Lancet* 1993;341:1225.

409. Heutink P, van de Wetering BJM, Breedveld GJ, et al. No evidence for genetic linkage of Gilles de la Tourette syndrome on chromosomes 7 and 18. *J Med Genet* 1990;27:433–436.

410. Devor EJ, Grandy DK, Civelli O, et al. Genetic linkage is excluded for the D2-dopamine receptor lambda HD2G1 and flanking loci on chromosome 11q22-q23 in Tourette syndrome. *Hum Hered* 1990;40:105–108.

411. Gerlenter J, Kennedy JL, Grandy GK, et al. Exclusion of close linkage of Tourette's syndrome to D1 dopamine receptor. *Am J Psychiatry* 1993;150:449–453.

412. Merette C, Brassard A, Potvin A, et al. Significant linkage for Tourette syndrome in a large French Canadian family. *Am J Hum Genet* 2000;67:1008–1013.

413. Simonic I, Gericke GS, Ott J, et al. Identification of genetic markers associated with Gilles de la Tourette syndrome in an Afrikaner population. *Am J Hum Genet* 1998;63:839–846.

414. Kurlan R, Eapen V, Stern J, et al. Bilineal transmission in Tourette's syndrome families. *Neurology* 1994;44:2336–2342.

415. Kurlan R, Eapen V, Stern J, et al. Bilineal transmission in Tourette syndrome families. *Neurology* 1994;44:2236–2242.

416. Hanna PA, Janjua FN, Contant CF, et al. Bilineal transmission in Tourette syndrome. *Neurology* 1999;53:813–818.

417. Eapen V, O'Neill J, Gurling HM, et al. Sex of parent transmission effect in Tourette's syndrome: evidence for earlier age at onset in maternally transmitted cases suggests a genomic imprinting effect. *Neurology* 1997;48:934–937.

418. Lichter DG, Jackson LA, Schachter M. Clinical evidence of genomic imprinting in Tourette's syndrome. *Neurology* 1995;45:924–928.

419. Comings DE, Gursey B, Hecht T, et al. HLA typing in Tourette syndrome. In: Friedhoff AJ, Chase TN, eds. *Gilles de la Tourette syndrome. Advances in neurology, volume 35.* New York: Raven Press, 1982:251–253.

420. Caine ED, Lowell RW, Chiverton P, et al. Tourette syndrome and HLA. *J Neurol Sci* 1985;69:201–206.

421. Segal NL, Dysken MW, Bouchard TJ Jr, et al. Tourette's disorder in a set of reared apart triplets: genetic and environmental influences. *Am J Psychiatry* 1990;147:196–199.

422. Kurlan R, Behr J, Medved L, et al. Familial Tourette's syndrome: report of a large pedigree and potential for linkage analysis. *Neurology* 1986;36:772–776.

423. Comings DE, Comings BG. Hereditary agoraphobia and obsessive-compulsive behavior in relatives of patients with Gilles de la Tourette's syndrome. *Br J Psychiatry* 1987;151:195–199.

424. Pauls DL, Leckman J, Towbin KE, et al. A possible genetic relationship exists between Tourette's syndrome and obsessive-compulsive disorder. *Psychopharmacol Bull* 1986a;22:730–733.

425. Pauls DL, Towbin KE, Leckman JF, et al. Gilles de la Tourette's syndrome and obsessive compulsive disorder. Evidence supporting a genetic relationship. *Arch Gen Psychiatry* 1986;43:1180–1182.

426. Robins LN, Helzer JE, Weissman MM, et al. Lifetime prevalence of specific disorders in three sites. *Arch Gen Psychiatry* 1984;41:949–958.

427. Inouye E. Similar and dissimilar manifestations of obsessive-compulsive neurosis in monozygotic twins. *Am J Psychiatry* 1965;121:1171–1175.

428. Torgerson S. Genetic factors in anxiety disorders. *Arch Gen Psychiatry* 1983;40:1085–1089.

429. Carey G, Gottesman I. Twin and family studies of anxiety, phobic and obsessive disorders. In: Klein DF, Rabin JG. *Anxiety: new research and current concepts.* New York: Raven Press, 1981.

430. Clifford CA, Murray RM, Fulker DW. Genetic and environmental influences on obsessional traits and symptoms. *Psychol Med* 1984;14:791–800.

431. Kringlen E. Obsessional neurotics: a long-term follow-up. *Br J Psychiatry* 1965;111:709–722.

432. Lewis A. Problems of obsessional illness. *Proc R Soc Lond* 1935;29:325–336.

433. Pauls DL, Raymond CL, Hurst CR, et al. Transmission of obsessive compulsive disorder and associated behaviors. Proceedings of the 43rd Meeting of the Society of Biological Psychiatry, Montreal, Canada, 1988.

434. Lenane MC, Swedo SE, Leonard H, et al. Psychiatric disorders in first degree relatives of children and adolescents with obsessive compulsive disorders. *J Am Acad Child Adolesc Psychiatry* 1990;29:407–412.

435. Nestadt G, Samuels J, Riddle M, et al. A family study of obsessive-compulsive disorder. *Arch Gen Psychiatry* 2000;57:358–363.

436. Cavallini MC, Pasquale L, Bellodi L. Complex segregation analysis for obsessive compulsive disorder and related disorders. *Am J Med Genet* 1999;88:38–43.

437. Alsobrook JP II, Leckman JF, Goodman WK, et al. Segregation analysis of obsessive-compulsive disorder using symptom-based factor scores. *Am J Med Genet* 1999;15;88:669–75.

438. Barr CL, Wigg KG, Zovko E, et al. Linkage study of the dopamine D5 receptor gene and Gilles de la Tourette syndrome. *Am J Med Genet* 1997;74:58–61.

439. Brett PM, Curtis D, Robertson MM, et al. The genetic susceptibility to Gilles de la Tourette syndrome in a large multiple affected British kindred: linkage analysis excludes a role for the genes coding for dopamine D1, D2, D3, D4, D5 receptors, dopamine beta hydroxylase, tyrosinase, and tyrosine hydroxylase. *Biol Psychiatry* 1995;37:533–540.

440. Stober G, Hebebrand J, Cichon S, et al. Tourette syndrome and the norepinephrine transporter gene: results of a systematic mutation screening. *Am J Med Genet* 1999;88:158–63.

441. Cruz C, Camarena B, King N, et al. Increased prevalence of the seven-repeat variant of the dopamine D4 receptor gene in patients with obsessive-compulsive disorder with tics. *Neurosci Lett* 1997;231:1–4.

442. Blum K, Sheridan PJ, Wood RC, et al. Dopamine D2 receptor gene variants: association and linkage studies in impulsive-addictive-compulsive behaviour. *Pharmacogenetics* 1995;5:121–141.

443. Vandenbergh DJ, Thompson MD, Cook EH, et al. Human dopamine transporter gene: coding region conservation among normal, Tourette's disorder, alcohol dependence and attention-deficit hyperactivity disorder populations. *Mol Psychiatry* 2000;5:283–292.

444. Anonymous. A complete genome screen in sib pairs affected by Gilles de la Tourette syndrome. The Tourette Syndrome Association International Consortium for Genetics. *Am J Hum Genet* 1999;65:1428–1436.

445. Kurlan R, McDermott M.Rating tic severity. In: Kurlan R, ed. *Handbook of Tourette's and related disorders.* New York: Marcel Dekker, 1993:199–220.

446. Gillies D, Forsythe W. Treatment of multiple tics and the Tourette syndrome. *Dev Med Child Neurol* 1984;26:822–833.

447a. Jagger J, Prusoff BA, Cohen DJ, et al. The epidemiology of Tourette's syndrome: a pilot study. *Schizophr Bull* 1982;8:267–278.

447b. Cohen DJ, Leckman JF. Tourette syndrome: advances in treatment and research. *J Am Acad Child Adolesc Psychiatry* 1984;23:123–125.

448. Cohen DJ, Detlor J, Young JG, et al. Clonidine ameliorates Gilles de la Tourette syndrome. *Arch Gen Psychiatry* 1980;37:1350–1357.

449. Tanner CM, Goetz CG, Klawans HL. Cholinergic mechanisms in Tourette syndrome. *Neurology* 1982;32:1315–1317.

450. Shapiro AK, Shapiro ES, Young JG, et al.*Gilles de la Tourette syndrome.* New York: Raven Press, 1988.

451. Shapiro AK, Wayne HL, Eisenkraft G. Semiology, nosology and criteria for tic disorders. *Rev Neurol (Paris)* 1983;142:824–832.

452. Shapiro AK, Shapiro ES. Controlled study of pimozide versus placebo in Tourette's syndrome. *J Am Acad Child Adolesc Psychiatry* 1984;23:161–173.

453. Harcherik D, Leckman J, Detlor J, et al. A new instrument for clinical studies of Tourette's syndrome. *J Am Acad Child Adolesc Psychiatry* 1984;23:153–160.

454. Leckman JF, Riddle MA, Pradin MT, et al. The Yale Global Tic Severity Scale: initial testing of clinician rated scale of tic severity. *J Am Acad Child Adolesc Psychiatry* 1989;28:566–573.

455. Goetz CG, Tanner CM, Wilson RS, et al. Clonidine and Gilles de la Tourette syndrome: double blind study using objective rating methods. *Ann Neurol* 1987;21:307–310.

456. Pauls DL, Hurst CR. *Schedule for Tourette and other behavioral syndromes.* New Haven, CT: Child Study Center, Yale University of Medicine, 1981.

457. Robertson MM, Eapen V. The National Hospital Interview Schedule for the assessment of Gilles de la Tourette syndrome and related behaviours. *Int J Methods Psychiatr Res* 1996;6:203–226.

458. Coffey BJ, Miguel EC, Biederman J, et al. Tourette's disorder with and without obsessive-compulsive disorder in adults: are they different? *J Nerv Ment Dis* 1998;186:201–6.

459. Rasmussen SA, Eisen JL. Clinical features and phenomenology of obsessive-compulsive disorder. *Psychiatr Ann* 1989;19:67–73.

460. Alarcon RD. How to recognize obsessive-compulsive disorder. *Postgrad Med* 1991;90:131–143.

461. Cooper J. The Leyton Obsessional Inventory. *Psychol Med* 1970;1:46–64.

462. Snowdon J. A comparison of written and postbox forms of the Leyton Obsessional Inventory. *Psychol Med* 1980;10:165–170.

463. Berg CJ, Rapoport JL, Flament M. The Leyton Obsessional Inventory-child version. *J Am Acad Child Adolesc Psychiatry* 1986;25:84–91.

464. Mears R. Obsessionality: the Sandler-Hazari scale and spasmodic torticollis. *Br J Med Psychol* 1971;44:181–182.

465. Philpot R. Recent advances in the behavioral measurement of obsessional illness. *Scott Med J* 1975;20[Suppl 1]:33–40.

466. Rackman C, Hodgson R. *Obsessions and compulsions.* Englewood Cliffs, NJ: Prentice-Hall, 1980.

467. Goodman WK, Price LH, Rasmussen SA, et al. The Yale-Brown Obsessive-Compulsive Scale. II. Validity. *Arch Gen Psychiatry* 1989;46:1012–1016.

468. Goodman WK, Price LH, Rasmussen SA, et al. The Yale-Brown Obsessive-Compulsive Scale. I. Development, use, and reliability. *Arch Gen Psychiatry* 1989;46:1006–1011.

469. Bruun RD, Shapiro AK. Differential diagnosis of Gilles de la Tourette's syndrome. *J Nerv Ment Dis* 1972;155:328–332.

470. Sacks OW. Acquired Tourettism in adult life. In: Friedhoff AJ, Chase TN, eds. *Gilles de la Tourette syndrome. Advances in neurology, volume 35.* New York: Raven Press, 1982:89–92.

471. Kettl PA, Marks IM. Neurological factors in obsessive-compulsive disorder. Two case reports and a review of the literature. *Br J Psychiatry* 1986;149:315–319.

472. Mayer-Gross W, Steiner G. Z. ges. Encephalitis lethargica. *Neurol Psychiatr* 1921;73:283.

473. Von Economo C. *Encephalitis lethargica, its sequelae and treatment.* New York: Oxford University Press, 1931. Newman KO, translator.

474. Trimble MR. *Neuropsychiatry.* London: John Wiley and Sons, 1981.

475. Capstick N, Seldrup U. Obsessional states: a study in the relationship between abnormalities occurring at birth and subsequent development of obsessional symptoms. *Acta Psychiatr Scand* 1977;56:427–439.

476. McKeon J, McGuffin P, Robinson P. Obsessive-compulsive neurosis following head injury: a report of 4 cases. *Br J Psychiatry* 1984;144:190–192.

477. Freeman J, Aron A, Collard D. The emotional correlates of Sydenham's chorea. *Pediatrics* 1965;35:42–49.

478. Swedo SE, Rapoport JL, Cheslow DL. High prevalence of obsessive-compulsive symptoms in patients with Sydenham's chorea. *Am J Psychiatry* 1989;146:246–249.

479. Jenike M. Update on obsessive-compulsive disorder. *Curr Affect III* 1990;9:5–12.

480. Barton R. Diabetes insipidus and obsessional neurosis: a syndrome. *Lancet* 1965;1:133–135.

481. Pacella BL, Polatin P, Nagle SH. Clinical and EEG studies in obsessive compulsive states. *Am J Psychiatry* 1944;100:830–838.

482. Garmany G. Obsessional states in epileptics. *J Ment Sci* 1947;93:639–643.

483. Brickner RM. A human cortical area producing repetitive phenomena when stimulated. *J Neurophysiol* 1940;3:128–130.

484. Penfield W, Jasper H. *Epilepsy and the functional anatomy of the human brain.* London: Churchill Livingstone, 1954.

485. Guerrant J, Anderson W, Fischer A, et al. *Personality of epilepsy.* Springfield, IL: Charles C Thomas, 1962:43–54.

486. Cunningham-Owens DG. Drug related movement disorders.

In: Robertson MM, Eapen V, eds. *Movement and allied disorders in childhood.* West Sussex, UK: John Wiley Publications, 1995.

487. Jankovic J. Diagnosis and classification of tics and Tourette syndrome. *Adv Neurol* 1992;58:7–14.

488. Stahl SM. Tardive Tourette syndrome in an autistic patient after long-term neuroleptic administration. *Am J Psychiatry* 1980;137:1267–1269.

489. Sotero de Menezes MA, Rho JM, Murphy P, et al. Lamotrigine-induced tic disorder: report of five pediatric cases. *Epilepsia* 2000;41:862–867.

490. Eapen V, Lees A, Lakke JPWF, et al. Adult onset Gilles de la Tourette syndrome. *Mov Disord* 2002;17:735–740.

491. Woody RC, Laney M. Tics and Tourette's syndrome: a review. *J Arkansas Med Soc* 1986;83:53–55.

492. Feinberg TE, Shapiro AK, Shapiro E. Paroxysmal myoclonic dystonia with vocalizations: new entity of variant of preexisting syndromes? *J Neurol Neurosurg Psychiatry* 1986;49:52–57.

493. Fahn S. Paroxysmal myoclonic dystonia with vocalizations. *J Neurol Neurosurg Psychiatry* 1987;50:117–118.

494. Foa EB, Kozak MJ. Diagnostic criteria for obsessive-compulsive disorder. *Hosp Commun Psychiatry* 1991;42:679–680, 684.

495. Seignot MJN. Un cas de maladie des tics de Gilles de la Tourette gueri par le R 1625. *Ann Med Psychol (Paris)* 1961;119:578–579.

496. Shapiro AK, Shapiro E, Fulop G. Pimozide treatment of tic and Tourette disorders. *Pediatrics* 1987;79:1032–1039.

497. Shapiro E, Shapiro AK, Fulop G, et al. Controlled study of haloperidol, pimozide and placebo for the treatment of Gilles de la Tourette syndrome. *Arch Gen Psychiatry* 1989;46:722–730.

498. Robertson MM, Schnieden V, Lees AJ. Management of Gilles de la Tourette syndrome using sulpiride. *Clin Neuropharmacol* 1990;13:229–235.

499. Ross MS, Moldofsky H. A comparison of pimozide and haloperidol in the treatment of Gilles de la Tourette's syndrome. *Am J Psychiatry* 1978;135:585–587.

500. O'Quinn AN, Thompson RJ. Tourette's syndrome: an expanded view. *Pediatrics* 1980;66:420–424.

501. Mikkelsen EJ, Detlor J, Cohen DJ. School avoidance and social phobia triggered by haloperidol in patients with Tourette's disorder. *Am J Psychiatry* 1981;138:1572–1575.

502. Linet LS. Tourette syndrome, pimozide and school phobia: the neuroleptic separation anxiety syndrome. *Am J Psychiatry* 1985;142:613–615.

503. Bruun RD. The natural history of Tourette's syndrome. In: Cohen DJ, Bruun RD, Leckman JF, eds. *Tourette's syndrome and tic disorders: clinical understanding and treatment.* New York: John Wiley and Sons, 1988:21–39.

504. Korczyn AD. Pathophysiology of drug induced dyskinesias. *Neuropharmacology* 1972;11:601–607.

505. Fulop G, Phillips RA, Shapiro AK, et al. ECG changes during haloperidol and pimozide treatment of Tourette's disorder. *Am J Psychiatry* 1987;144:673–675.

506. Eggers C, Rothenberger A, Berghaus V. Clinical and neurobiological findings in children suffering from tic disease following treatment with tiapride. *Eur Arch Psychiatr Neurol Sci* 1988;237:223–229.

507. Eapen V, Robertson MM. Tourette syndrome and co-morbid obsessive compulsive disorder: therapeutic interventions. *CNS Drugs* 2000;13:173–183.

508. Robertson MM, Scull DA, Eapen V, et al. Risperidone in the treatment of Tourette syndrome: a retrospective case note study. *J Psychopharmacol* 1996;10:317–320.

509. Peterson BS. Considerations of natural history and pathophys-

510. Cosgrove F. Recent advances in paediatric psychopharmacology. *Hum Psychopharmacol* 1994;9:381–382.

511. Caine ED, Polinsky RJ, Ebert MH, et al. Trial of clomipramine and desipramine for Gilles de la Tourette syndrome. *Ann Neurol* 1979;6:305–306.

512. Bhandrinath BR. Olanzapine in Tourette syndrome [Letter]. *Br J Psychiatry* 1998;172:366.

513. Stamenkovic M, Schindler SD, Aschauer HN, et al. Effective open-label of Tourette's disorder. Olanzapine. *Int Clin Psychopharmacol* 2000;Jan 15:23–28.

514. Onofrj M, Paci C, D'Andreamatteo G, et al. Olanzapine in severe Gilles de la Tourette syndrome: a 52-week double-blind cross-over vs. low-dose pimozide. *J Neurol* 2000;247:443–446.

515. Sallee FR, Kurlan R, Goetz CG, et al. Ziprasidone treatment of children and adolescents with Tourette's syndrome: a pilot study. *J Am Acad Child Adolesc Psychiatry* 2000;39:292–299.

516. Mizrahi EM, Holtzman D, Tharp B. Haloperidol-induced tardive dyskinesia in a child with Gilles de la Tourette's disease. *Arch Neurol* 1980;37:380.

517. Caine ED, Polinsky RJ. Tardive dyskinesia in persons with Gilles de la Tourette's disease. *Arch Neurol* 1981;38:471–472.

518. Golden GS. Tardive dyskinesia in Tourette syndrome. *Pediatr Neurol* 1985;1:192–194.

519. Jeste DV, Cutler NR, Kaufmann CA, et al. Low-dose apomorphine and bromocriptine in neuroleptic-induced movement disorders. *Biol Psychiatry* 1983;18:1085–1091.

520. Eapen V, Katona CLE, Barnes TRE, et al. Sulpiride-induced tardive dyskinesia in a person with Gilles de la Tourette syndrome. *J Psychopharmacol* 1993;7:290–292.

521. Silverstein F, Johnston MV. Risks of neuroleptic drugs in children. *J Child Neurol* 1987;2:41–43.

522. Uhr SB, Pruitt B, Berger PA, et al. Improvement of symptoms in Tourette syndrome by piquindone, a novel dopamine-2 receptor antagonist. *Int Clin Psychopharmacol* 1986;1:216–220.

523. Leckman JF, Detlor J, Harcherik DF, et al. Short and long term treatment of Tourette's syndrome with clonidine: a clinical perspective. *Neurology* 1985;35:343–351.

524. Singer HS, Gammon K, Quaskey S. Haloperidol, fluphenazine and clonidine in Tourette syndrome: controversies in treatment. *Pediatr Neurosci* 1986;12:71–74.

525. Mesulam MM, Petersen RC. Treatment of Gilles de la Tourette's syndrome: eight-year, practice-based experience in a predominantly adult population. *Neurology* 1987;37:1828–1833.

526. Greist JH. Treatment of obsessive compulsive disorder psychotherapies, drugs, and other somatic treatment. *J Clin Psychiatry* 1990;51[Suppl]:44–50.

527. Malan DH. *Individual psychotherapy and the science of psychodynamics.* London: Butterworths, 1979:218–219.

528. Meyer V. Modification of expectations in cases with obsessive rituals. *Behav Res Ther* 1966;4:270–280.

529. Piacentini J, Jaffer M, Gitow A, et al. Psychopharmacologic treatment of child and adolescent obsessive compulsive disorder. *Pediatr Psychopharmacol* 1992;15:87–107.

530. Kasvikis Y, Marks IM. Clomipramine in obsessive-compulsive ritualizers treated with exposure therapy: relations between dose, plasma levels outcome and side effects. *Psychopharmacology* 1988;95:113–118.

531. O'Sullivan G, Noshirvani H, Marks I, et al. Six-year follow-up after exposure and clomipramine therapy for obsessive-compulsive disorder. *J Clin Psychiatry* 1991;52(4):150–155.

532. Foa E, Steketee G, Grayson JB, et al. Deliberate exposure and

blocking of obsessive compulsive rituals: immediate and long-term effects. *Behav Ther* 1984;15:450–472.

533. Marks IM. *Fears, phobias, and rituals: panic, anxiety, and their disorders.* New York: Oxford University Press, 1987:497–498.

534. Greist JH. An integrated approach to treatment of obsessive compulsive disorder. *J Clin Psychiatry* 1992;53[Suppl 4]:38–41.

535. Marks I, Lelliott P, Basoglu M, et al. Clomipramine, self-exposure and therapist aided exposure in obsessive-compulsive ritualizers. *Br J Psychiatry* 1988;152:522–534.

536. Orloff LM, Battle MA, Baer L, et al. Long-term follow-up of 85 patients with obsessive-compulsive disorder. *Am J Psychiatry* 1994;151:441–442.

537. Clomipramine Collaborative Study Group. Clomipramine in the treatment of patients with obsessive-compulsive disorder. *Arch Gen Psychiatry* 1991;48:730–738.

538. Volavka JF, Neziroglu FA, Yaryura-Tobias JA. Clomipramine and imipramine in obsessive-compulsive disorder. *Psychiatry Res* 1985;14:85–93.

539. Fuller RW, Wong DT. Serotonin reuptake blockers in vitro and in vivo. *J Clin Psychopharmacol* 1987;7:36S–43S.

540. Hollander E, Stein DJ, Decaria CM, et al. Disorders related to OCD-neurobiology. *Clin Neuropharmacol* 1992;15[Suppl 1]:259A–260A.

541. Montgomery SA, McIntyre A, Osterheider M, et al. A double-blind, placebo-controlled study of fluoxetine in patients with DSM-III-R obsessive-compulsive disorder. The Lilly European OCD Study Group. *Eur Neuropsychopharmacol* 1993;3:143–152.

542. Leonard HL, Topol D, Bukstein O, et al. Clonazepam as an augmenting agent in the treatment of childhood-onset obsessive-compulsive disorder. *J Am Acad Child Adolesc Psychiatry* 1994;33:792–794.

543. Stern RS, Marks IM, Wright J, et al. Clomipramine: plasma levels, side effects and outcome in obsessive compulsive neurosis. *Postgrad Med J* 1980;56[1 Suppl]:134–139.

544. Chouinard G, Goodman W, Greist J, et al. Results of a double-blind placebo controlled trial of a new serotonin uptake inhibitor, sertraline, in the treatment of obsessive-compulsive disorder. *Psychopharmacol Bull* 1990;26:279–284.

545. Wheadon DE. Placebo controlled multi-center trial of fluoxetine in OCD. Presented at the 5th World Congress of Biological Psychiatry, Florence, Italy, 1991. Cited in Barr CL, Goodman WK, Price LH, et al. The serotonin hypothesis of obsessive compulsive disorder: implications of pharmacologic challenge studies. *J Clin Psychiatry* 1992;56[4 Suppl]:17–28.

546. Zohar J, Zohar-Kadouch RC, Kindler S. Current concepts in the pharmacological treatment of obsessive compulsive disorder. *Pract Ther Drugs* 1992;43:210–218.

547. Pato MT, Zohar KR, Zohar J, et al. Return of symptoms after discontinuation of clomipramine in patients with obsessive-compulsive disorder. *Am J Psychiatry* 1988;145:1521–1525.

548. Pato MT, Hill JL, Murphy DL. A clomipramine dosage reduction study in the course of long-term treatment of obsessive-compulsive disorder patients. *Psychopharmacol Bull* 1990;26:211–214.

549. Hollander E, De Cariar CM, Schneier FR, et al. Fenfluramine augmentation of serotonin reuptake blockade antiobsessional treatment. *J Clin Psychiatry* 1990;51:119–123.

550. Hewlett WA, Vinogradov S, Agras WS. Clonazepam treatment of obsessions and compulsions. *J Clin Psychiatry* 1990;51:158–161.

551. Tollefson G. Alprazolam in the treatment of obsessive symptoms. *J Clin Psychopharmacol* 1985;5:39–42.

552. Jenike MA. Rapid response of severe obsessive compulsive disorder to tranylcypromine. *Am J Psychiatry* 1981;138:1249–1250.

553. Knesevich JW. Successful treatment of obsessive compulsive disorder with clonidine hydrochloride. *Am J Psychiatry* 1982;139:364–365.

554. Yoney TH, Pigott TA, L'Heureux F, et al. Seasonal variation in obsessive compulsive disorder: preliminary experience with light treatment. *Am J Psychiatry* 1991;148:1727–1729.

555. Bird JM, Crow CD. Psychosurgery in obsessional-compulsive disorder: old techniques and new data. Current approaches. In: Montgomery SA, Goodman WK, Goeting N, eds. *Obsessive compulsive disorder.* Southamptom, England: Ashford Colonial Press, Gosport Harris Duphar Medical Relations, 1990:82–92.

556. Mindus P. Capsulotomy in anxiety disorders: a multidisciplinary study. Kongl Carolinska Medico Chirurgiska Institute, Stockholm, Sweden, 1991. Cited in Goodman WK, McDougle CJ, Price LH. Pharmacotherapy of obsessive compulsive disorder. *J Clin Psychiatry* 1992;53[4 Suppl]:29–37.

557. Rack PH. Clinical experience in the treatment of obsessional states. *J Int Med Res* 1977;5:81–96.

558. Savicki V, Carlin AS. Behavioral treatment of Gilles de la Tourette syndrome. *Int J Child Psychother* 1972;1:97–109.

559. Stefl ME, Rubin M. Tourette syndrome in the classroom: special problems, special needs. *J Sch Health* 1985;55:72–75.

560. Freeman RD, Fast DK, Burd L, et al. An international perspective on Tourette syndrome: selected findings from 3,500 individuals in 22 countries. *Dev Med Child Neurol* 2000;42:436–447.

561. Alsobrook JP 2nd, Pauls DL. The genetics of Tourette syndrome. *Neurol Clin* 1997;15:381–393.

562. Barr CL, Sandor P. Current status of genetic studies of Gilles de la Tourette syndrome. *Can J Psychiatry* 1998;43:351–357.

563. Pitman RK. Pierre Janet on obsessive compulsive disorder (1903). *Arch Gen Psychiatry* 1987;44:226–232.

36

WHITE MATTER DISORDERS

CHRISTOPHER M. FILLEY

White matter has an important role in the study of brain–behavior relationships, although gray matter, particularly of the cerebral cortex, most often is regarded as the repository of higher functions. The assumption that the singular phenomena of human behavior require the activities of the cortical gray matter is one of the most pervasive in all of neuroscience. Whereas an impressive body of evidence exists to support this belief, the cerebral cortex consists of only the outermost 3 mm of the brain, and a wealth of clinical experience suggests that disorders affecting structures below the cortex, much of which is white matter, reliably and significantly alter mental functions.

A wide range of syndromes involving both cognitive decline and emotional dysfunction have been linked with structural involvement of the brain white matter. Clinical observations of patients with white matter disorders generate the essential data to support this claim, and magnetic resonance imaging (MRI) has provided detailed views of the white matter and permitted correlations with neurobehavioral syndromes. These syndromes may equal or surpass in clinical importance the deficits in motor and sensory function of white matter lesions well known from classic neurology.

Cognitive and emotional disturbances were recognized in patients with multiple sclerosis (MS) by Charcot in the nineteenth century (1). Other neurologists of the era developed brain–behavior relationships based on the study of focal white matter lesions, including Dejerine, who demonstrated involvement of the splenium of the corpus callosum in pure alexia, and Liepmann, who included the anterior corpus callosum in his explanation of unilateral apraxia (2). These disorders came to be known as disconnection syndromes because disruption of a tract connecting cortical areas was essential for their appearance. Disconnection of the hemispheres by corpus callosotomy later attracted considerable attention (3).

In the twentieth century, a different line of inquiry developed that would prove relevant to the study of cerebral white matter. Although it was generally believed that cortical damage was the major neuropathologic substrate of mental dysfunction (4), some investigators examined the possibility that disorders involving selective lesions of subcortical structures, such as the basal ganglia and thalamus, could disrupt cognition and behavior. This idea flourished in the 1970's after observations of a specific pattern of cognitive impairment in patients with progressive supranuclear palsy (5) and Huntington disease (6). In general, subcortical dementia was said to be characterized by cognitive slowing, forgetfulness, and personality and emotional changes, in contrast to the amnesia, aphasia, apraxia, and agnosia traditionally associated with cortical dementias. Subcortical dementia was theorized to disrupt the "fundamental" functions of arousal, attention, motivation, and mood that provide for the timing and activation of cortical processes, whereas cortical dementia was seen as interfering with the "instrumental" functions of memory, language, praxis, and perception associated with the neocortex (7). An analogous distinction was drawn between the concept of "channel" functions, which are the specific contents of cognition, and "state" functions, which maintain the state of information processing in the brain (8). Based on these formulations, descriptive work in the dementias flourished, and subcortical dementia became a popular but controversial term for the dementia seen in a variety of subcortical disorders (9,10).

Although not initially included in the list of subcortical dementias, disorders of white matter soon were recognized as being capable of producing similar disturbances. MS was the most prominent of these, and considerable effort was devoted to investigating its neurobehavioral sequelae (11). Other diseases with major white matter neuropathology were found to manifest similar deficits, including Binswanger disease (BD) (12) and the acquired immunodeficiency syndrome (AIDS) dementia complex (ADC) (13), and white matter disorders began to be regarded as etiologies of subcortical dementia (14).

Convergence of clinical and MRI evidence in the 1980's led to a reconsideration of the role of cerebral white matter. The most important syndrome to emerge from early studies was cognitive impairment, and many patients with cognitive loss or dementia were increasingly recognized. In addition, focal neurobehavioral syndromes and neuropsychiatric disorders became better appreciated. To highlight the dementia that could be ascribed to dysfunction of cerebral white matter, the term "white matter dementia" was proposed (15). The primary goal of this designation was to call attention to potential neurobehavioral issues in these patients, and evidence suggesting a specific neuropsychological profile lent credence to the syndrome (16). In addition to serving as an exhortation for clinical vigilance in evaluating individuals with white matter disorders, the concept of white matter dementia had a theoretical dimension (17).

Although white matter neuroanatomy is poorly understood in humans compared to monkeys, it is clear that white matter connects cortical and subcortical regions within and between the hemispheres in a variety of distributed neural networks that subserve higher cerebral functions (8). Gray matter areas operate in concert to mediate these activities, and white matter forms the connections that link these areas into coherent neural assemblies. Lesions in these tracts produce dysfunction by disconnecting neural networks.

Many patients with white matter disorders are evaluated without due regard for the potential existence of neurobehavioral syndromes. The prevalence of neurobehavioral syndromes caused by white matter disorders is difficult to determine, but the large number of individuals affected by these diseases and injuries strongly suggests that the burden of neurobehavioral disability is enormous and underappreciated. Moreover, because white matter lesions often involve damage to myelin alone without axonal damage, the potential for spontaneous improvement or effective treatment is higher than in disorders of gray matter that typically destroy neurons. Substantial clinical benefit may result from a better understanding of these issues.

Ambiguity may exist about whether a given disorder represents sufficiently selective white matter involvement to enable neurobehavioral correlation. Lesions in gray matter also occur in some primary white matter disorders. The problem of neuroanatomic specificity should not deter an attempt to identify the impact of white matter lesions on behavior. The cerebral localization of many well-known neurobehavioral disorders is still debated, and it is a process that energizes research and refines understanding. The intent of this chapter is to provide a broad survey of white matter disorders and the wide variety of neurobehavioral disturbances with which they can be associated. As the organ of the mind, the brain in all of its neuroanatomic and functional complexity deserves careful attention.

WHITE MATTER STRUCTURE AND FUNCTION

The white matter occupies about 50% of the adult cerebral volume (18). Axons invested with myelin constitute the white matter, and millions of these fibers combine to form the many tracts that travel within and between the hemispheres, as well as to more caudal brainstem and cerebellar regions. Myelinated fibers perform a critical role in normal brain function by virtue of this vast and intricate connectivity.

Although the great majority of the white matter is located within these tracts, myelinated fibers also can be found in gray matter areas. Thus, the distinction between gray and white matter is only relative, and disorders of myelin can affect gray matter regions as well.

Neuroanatomy

The human brain is a collection of neurons, glial cells, and vasculature weighing about 1,400 g in the adult (range 1,100–1,700), or roughly 2% of total body weight (19). An estimated 100 billion neurons are found in the brain, each of which may make contact with up to 10,000 others and at least ten times this many glial cells (20). Many neurons feature myelinated axons that course throughout the brain and provide essential connections between cortical and subcortical gray matter areas.

Myelin

Myelin is a complex lipid-rich substance that surrounds axons in a concentric sheath. In the freshly cut brain, myelin is easily visible because of its glistening white appearance that derives from its preponderance of lipid. Lipid accounts for approximately 70% of the dry weight of myelin, and protein accounts for about 30% (21). Cholesterol is the most abundant lipid, and most of the protein in myelin is proteolipid protein and myelin basic protein (21).

Oligodendrocytes

Of the four types of glial cells in the central nervous system (CNS)—oligodendrocytes, astrocytes, ependymal cells, and microglia—oligodendrocytes are the most important in determining the structure and function of white matter. Oligodendrocytes are responsible for the formation of myelin, which is wrapped around the axon to form the myelin sheath. An important feature of the myelin sheath is that it is discontinuous, thus creating short segments of the axon called the nodes of Ranvier.

White Matter Tracts

White matter consists of large collections of myelinated fibers that are grossly discernible but substantially interdigitated with each other. The word *tract* is the most commonly used descriptor for white matter pathways, but other similar words are fasciculus, funiculus, lemniscus, peduncle, and bundle (19). Cerebral white matter tracts are classified as projection, commissural, and association fibers.

Projection fibers consist of long ascending (corticopetal) and descending (corticofugal) tracts. Corticopetal (afferent) tracts connect structures lower in the brain and spinal cord with the cortex; corticofugal (efferent) tracts proceed in the opposite direction. Projection fiber systems include the thalamocortical radiations that link thalamic nuclei with sensory and visual cortices, and the corticospinal and corticobulbar tracts that connect motor cortices with lower motor areas via the internal capsule and cerebral peduncle. Commissural fibers connect the two halves of the cerebrum. The most important of these is the corpus callosum, a prominent structure that contains some 300 million myelinated axons (19). This fiber tract connects nearly all cortical regions in one hemisphere with homologous areas in the other. It consists of the posterior splenium, central body, anterior genu, and ventrally directed rostrum. Other commissures are the anterior commissure, which connects olfactory and temporal regions; and the hippocampal or fornical commissure, which links the two crura of the fornices. Association fibers join cerebral areas within each hemisphere. First, there are short association fibers, the U or arcuate fibers, which link adjacent cortical gyri. Second, a number of long association fiber tracts connect more distant cerebral areas: the arcuate (superior longitudinal) fasciculus, superior occipitofrontal fasciculus (subcallosal bundle), inferior occipitofrontal fasciculus, cingulum, and uncinate fasciculus. Association systems are generally arranged so that they are bidirectional, which allows for extensive reciprocal communication between cerebral regions. Long association fibers share the feature of having one terminus in the frontal lobe. The white matter is structurally organized to facilitate frontal lobe interaction with all other regions of the cerebrum. This pattern of connectivity provides a neuroanatomic basis for the essential role of white matter in human behavior (8,22).

White matter tracts coalesce with each other in the cerebrum and form a dense mass within each hemisphere. Above the internal capsule, through which nearly all the neural traffic to and from the cerebral cortex passes, lies a collection of fibers that fans out laterally; this is the corona radiata (19). Still higher is found the centrum semiovale, which is the white matter located subjacent to the cortical mantle (19). The "periventricular white matter" conveniently refers to the white matter that lies immediately adjacent to the lateral ventricles. Less specific is the "subcortical white matter," a term used by some authors to refer to the white matter just below the cortex and by others to designate the aggregate of tracts located within the hemispheres. A number of smaller tracts can be found deep within the brain, the clinical significance of which is slowly evolving. The fimbria emanates from the hippocampus and merges into the fornix as it curves posteriorly and dorsally to terminate in a nucleus of the hypothalamus called the mamillary body (19). This tract serves as an important link in the limbic system, and it has a role in both memory and emotion. The medial forebrain bundle connects the hypothalamus with the brainstem below and the cerebral cortex above (20). It conveys various biogenic amines (dopamine, norepinephrine, and serotonin) to their cortical destinations. The external capsule courses lateral to the lenticular nucleus (caudate and putamen), and it contains cholinergic fibers also destined for the cerebral cortex (23).

The neurotransmitters carried within these tracts enable the major modulatory influences on the cortex that originate from the ascending reticular activating system in the brainstem (8). The extreme capsule lies lateral to the claustrum and medial to the insula. Its fibers, like those of the arcuate fasciculus, connect the temporal and frontal lobes (24).

Laterality differences have been noted in the distribution of white matter. The right hemisphere has been reported to contain a larger relative proportion of white matter than the left hemisphere, a difference that is particularly apparent in the frontal lobes (25). This distribution of cerebral white matter may be important in terms of the functional specializations of the two hemispheres and even of individual regions within the hemispheres.

Blood Supply

The arterial supply of the cerebral white matter comes from many perforating arteries that arise from larger arteries at the base of the brain. Most prominent are the lenticulostriate arteries, which originate from the middle cerebral artery and follow a long course as they penetrate the deep structures of the cerebrum (19). The U or arcuate fibers, the extreme capsule, and the external capsule enjoy a rich blood supply provided by many short and interdigitated cortical arterioles. The corpus callosum is supplied by short, small-caliber arterioles that arise from the pial plexus. The white matter of the brainstem comes from penetrating arteries that arise from the basilar artery, and the cerebellar white matter is supplied by penetrating branches of the superior cerebellar, posterior inferior cerebellar, and anterior inferior cerebellar arteries (19).

Neurophysiology

The brain is an electrical organ, and its function depends on the capacity to transmit electrical signals. Neurons operate by conducting an electrical impulse known as the action potential. This impulse is propagated along the axon in an "all or none" fashion, and the neuron can influence others to which it is connected. The speed and efficiency of this process are significantly influenced by the degree of axonal myelination. In the absence of normal myelination and white matter integrity, brain function is dramatically compromised.

The Action Potential

This event represents a departure from the normal resting membrane potential of the neuron (approximately −65 mV), which is maintained by a charge difference across the cell membrane (20). The resting potential is determined by the separation of extracellular positive charges from intracellular negative charges resulting from the activity of the adenosine triphosphate (ATP)-dependent sodium/potassium pump. The action potential, or "spike," is generated by the rapid influx of positively charged sodium ions through voltage-gated sodium channels that temporarily depolarize the membrane so that the membrane potential briefly becomes positive. The depolarization so produced is quickly reversed by a rapid efflux of potassium ions that restores the resting potential. After a refractory period during which no action potential can be propagated, the axon is again prepared to conduct another impulse. Neurons in the brain conduct electrical impulses at a velocity ranging from 1 to 120 m/s. The axon diameter in part determines conduction velocity; therefore, larger fibers conduct electricity faster than smaller fibers (20).

Saltatory Conduction

An increase of neuronal conduction velocity also is conferred by the phenomenon of saltatory conduction. The myelin sheath is interrupted every 1 to 2 mm by the unmyelinated nodes of Ranvier (20). This arrangement permits the action potential to travel from one node to the next without the need for the entire axonal membrane to be depolarized. This kind of conduction is known as saltatory, derived from the Latin verb *saltare,* meaning "to leap." It is estimated that large myelinated fibers conduct impulses as much as 100 times faster than small unmyelinated fibers (20), an increase largely due to myelination and the advantage of saltatory conduction.

Clinical Neurophysiology

In clinical practice, the function of white matter tracts can be assessed by *in vivo* neurophysiologic techniques. In contrast to electroencephalography (EEG), which primarily evaluates cortical function, evoked potentials (EPs) can offer limited insight into the integrity of white matter tracts. The most familiar EPs are the conventional visual (VEP), auditory (AEP) or brainstem auditory (BAEP), and somatosensory (SEP) evoked potentials (26). These are scalp potentials that reflect the function of primary sensory pathways as they traverse the central nervous system. EPs initially found the most utility in the diagnosis of MS, but their usefulness has been diminished by the much improved depiction of neuroanatomy provided by MRI. The application of conventional EPs to behavioral neurology has not proved feasible because of the complex representation of cognition and emotion in comparison to elemental sensory function.

A special variety of EP is the event-related potential (ERP), in which cognition is engaged in the generation of the electrophysiologic response (27). ERPs also are known as "endogenous" potentials, to distinguish them from "exogenous" conventional EPs, and they hold considerable promise for the study of behavior. In general, ERPs are long-latency waves related to the cognitive processing of stimuli. The most familiar ERP is the P300 or P3. Because of the multiple neuroanatomic levels of processing involved, both gray and white matter structures are presumed to contribute to the generation of ERPs. Although ERP abnormalities may reflect disturbances in neural systems relevant to behavior, these potentials are limited in the degree to which they can localize areas of dysfunction. It is not surprising that prolonged P300 latencies can be found in a variety of dementing diseases, including Alzheimer disease (AD), vascular dementia, and Parkinson disease (28).

Preliminary studies of ERPs have been performed in patients with white matter disorders. In MS, P300 abnormalities have been correlated with cognitive dysfunction and MRI lesion burden by some investigators (29) but not others (30). Another cognitive evoked potential, the P50 (31), has been found to be abnormal in traumatic brain injury (TBI) patients (32). Although it remains difficult to establish the neuroanatomy responsible for the generation of ERP data, the technique may contribute to the study of cognitive function in patients with white matter disorders.

ONTOGENY OF WHITE MATTER

The human brain is a dynamic organ that undergoes a constant process of structural change during the life span. As development, maturity, and then aging occur, the brain continually remodels both its fine and its gross structure. Microscopically, this remodeling occurs in the gray matter at the level of the synapse, where constant coupling and uncoupling of dendrites takes place in parallel with the processes of learning and cognition. At the macroscopic level, changes in the brain appear to be most evident in the white matter.

Development

Most of the axons in the adult brain are myelinated. The process by which this is accomplished requires a prolonged period that begins *in utero* and continues for many years. The first information on this topic was derived from classic neuroanatomic studies on the sequence of myelination (33), followed by more recent investigations (34). The MRI era has added substantially to this field by enabling the *in vivo* brain imaging of normal neonates, infants, and children. MRI has become a powerful tool for assessment of the myelination—and hence maturation—of the young brain (35). The findings of neuroanatomy and neuroradiology have been consistent in describing an orderly progression of brain myelination.

Gray matter and white matter differ significantly in their patterns of development. The entire complement of central nervous system neurons is formed before birth. Development of gray matter involves continual pruning of inessential neurons by programmed cell death and the simultaneous establishment of synaptic contacts between the neurons that remain (20). In contrast, the white matter does not begin to form until the middle trimester of gestation. The process is only partially underway at birth and is still just

90% complete by age 2 years (35). The remainder of myelination requires many more years (33). The exact duration of this process is unclear, but evidence from a large series of normal brains studied post mortem suggests that myelination proceeds throughout the end of the sixth decade (34).

As a general rule, myelination follows a pattern that begins with more caudal regions and advances to more rostral structures in the brain (33). Thus, the brainstem and cerebellum are myelinated first, followed by the diencephalon and the cerebral hemispheres. This ontogenetic sequence mirrors the phylogenetic background of the brain, as more recently acquired brain structures require a longer time to myelinate than more ancient ones. The occipital and parietal lobes mature sooner than the temporal and frontal lobes (35), again in keeping with this principle. The major association and commissural fibers are the last to myelinate (33).

The clinical significance of the sequence of brain myelination has long been debated (33). The relative importance of white matter versus gray matter development has not been entirely clear. Recently, neuroradiologists have increasingly interpreted delayed myelination on MRI as indicative of neurologic dysfunction (35). Delayed myelination in children with congenital hydrocephalus, for example, has been correlated with cognitive impairment (36). Moreover, hydrocephalic children have reductions in the size of the corpus callosum and other cerebral white matter tracts that correlate with cognitive dysfunction (37).

One intriguing notion to arise from study of this area is the possibility that acquisition of the mature personality in young adulthood depends to a substantial extent on frontal lobe myelination (17). Normal personality development requires the acquisition of traits such as comportment, impulse control, and judgment, which traditionally are associated with frontal lobe function. Because myelination of the frontal lobes occurs late in development—at a time when gray matter is relatively stable—the arrival of the adult personality may require completion of this myelogenetic phase. Moreover, subtle personality change in late adulthood may be related to continuing myelination in the fifth and sixth decades (33). Understanding these potential correlations could help establish a foundation for considering the neural organization of personality throughout the life span.

Aging

Until recently, it was widely held that aging in the brain was characterized by the widespread death of neurons in the neocortex and hippocampus. With recent improvements in estimating neuron number, however, cortical gray matter loss in aging is less pronounced than previously thought (38), an insight that has helped stimulate a search for other structural changes in aging. Several lines of evidence now support the idea that cerebral white matter undergoes significant alterations in aging and that these changes have important functional consequences.

Postmortem studies generally leave little doubt that the brain decreases in weight and volume with advancing years (39). However, the assumption that this phenomenon, often called "cortical" atrophy, results from gray matter cell loss may be premature. Image analyzer techniques have been used to demonstrate that the relative proportion of cerebral white matter to gray matter varies considerably at different ages (18). Thus, in comparison to gray matter, there appears to be an early deficiency of white matter, followed by relative parity in midlife, and then again a deficiency in old age. MRI studies in normal individuals indicate a loss of white matter in the aging brain that exceeds that of cortical gray matter (40). Diminished white rather than gray matter predicts an increase in sulcal fluid volume in aging (41), which supports the notion that age-related "cortical atrophy" on MRI is related to white matter loss alone. Moreover, autopsy studies of normal older individuals indicate a decline of white matter in the brain exceeding that of cortical gray matter (42).

The origin of this white matter attrition is uncertain. The white matter hyperintensities commonly seen on MRI scans of older persons may account in part for the loss of white matter volume, but preliminary studies have found that these changes are not correlated with white matter volume loss (40). Alternatively, a decrease in subcortical myelin associated with an increase in unsaturated acyl chains has been demonstrated with aging. This relative desaturation of myelin lipid implies an instability of the white matter in the aging brain (43). Free radicals are believed by many to have a role in normal aging by virtue of their preferential attack on phospholipids that are vulnerable to peroxidation; these photolipids are abundant in myelin (44).

The selective loss of white matter in the aging brain has important neurobehavioral implications. Metaanalysis of structural neuroimaging studies of normal older people has concluded that cerebral white matter abnormalities are associated with attenuated performance on tasks of processing speed, immediate and delayed memory, executive functions, and measures of global cognitive function (45). More specifically, the relative abundance of white matter in the frontal lobes and the right hemisphere offers a basis for the hypothesis that normal aging reflects particular dysfunction in these areas. Many of the cognitive features of normal aging closely resemble changes associated with frontal lobe involvement seen clinically (22). Aging confers many cognitive advantages, such as the broad experience and expansive knowledge commonly known as "wisdom," but the mental slowing and rambling garrulousness of some older people is reminiscent of younger patients with frontal lobe dysfunction who are inattentive, distractible, and poorly organized. Large studies of elderly populations have found robust correlations between cognitive changes and periventricular white matter changes on MRI, with cognitive speed being the most affected domain (46). The "frontal aging hypothesis" postulates that cognitive functions dependent on the frontal lobes are selectively vulnerable in aging (47). However, the evidence for selective frontal lobe atrophy in aging is not conclusive, and other nonfrontal cognitive skills also are affected (48). Resolution of this issue may be provided by considering the neuroanatomy of white matter. Because white matter tracts course throughout the brain, alterations in myelinated systems in aging would be expected to impact all cognitive functions to some extent, but because the frontal lobes have the largest concentration of white matter, this effect would be most apparent in the performance of frontally mediated tasks. This explanation is consistent with a "myelin-based theory of aging" (48), which necessarily includes the idea of distributed neural networks (8). There also is evidence for a selective decline in right hemisphere function in aging that may in part reflect loss of cerebral white matter. The "classic aging pattern" has long been recognized by neuropsychologists as the relative stability of the verbal intelligence quotient (IQ) on the Wechsler Adult Intelligence Scale with a decline

in the performance IQ (PIQ) (49). Although the PIQ is not necessarily a measure of right hemisphere function, it is reflective of so-called "fluid" abilities that are largely nonverbal in nature (49). In this light, it is of interest that a study of healthy octogenarians found a selective decline in PIQ that correlated with increasing severity of white matter hyperintensities on MRI (50).

DISORDERS OF WHITE MATTER

A wide range of diseases and injuries can affect the white matter of the brain (Table 36.1). These disorders all feature prominent or exclusive white matter involvement as demonstrated by neuropathology or neuroimaging techniques, and for each disorder there is evidence for a role of white matter in neurobehavioral dysfunction. Complete discussions of

TABLE 36.1. DISORDERS OF WHITE MATTER

Genetic	Metabolic
Leukodystrophies	Cobalamin deficiency
Aminoacidurias	Folate deficiency
Phakomatoses	Central pontine myelinolysis
Mucopolysaccharidoses	Hypoxia
Muscular dystrophy	Hypertensive encephalopathy
Callosal agenesis	Eclampsia
Demyelinative	High-altitude cerebral edema
Multiple sclerosis	Vascular
Acute disseminated encephalomyelitis	Binswanger disease
Acute hemorrhagic encephalomyelitis	Cerebral autosomal dominant
Schilder disease	arteriopathy with subcortical infarcts
Marburg disease	and leukoencephalopathy (CADASIL)
Baló concentric sclerosis	Leukoaraiosis
Infectious diseases	Cerebral amyloid angiopathy
Acquired immunodeficiency syndrome	White matter disease of prematurity
dementia complex	Migraine
Progressive multifocal leukoencephalopathy	Traumatic
Subacute sclerosing panencephalitis	Traumatic brain injury
Progressive rubella panencephalitis	Shaken baby syndrome
Varicella zoster encephalitis	Corpus callosotomy
Cytomegalovirus encephalitis	Neoplasms
Lyme encephalopathy	Gliomatosis cerebri
Inflammatory	Diffusely infiltrative gliomas
Systemic lupus erythematosus	Primary cerebral lymphoma
Behçet-disease	Focal white matter tumors
Sjögren syndrome	Hydrocephalus
Wegener granulomatosis	Early hydrocephalus
Temporal arteritis	Normal pressure hydrocephalus
Polyarteritis nodosa	
Scleroderma	
Isolated angiitis of the central nervous system	
Sarcoidosis	
Toxic	
Radiation	
Therapeutic drugs	
Drugs of abuse	
Environmental toxins	

these disorders and their treatment can be found in standard neurology textbooks. In this section, each category shown in Table 36.1 will be considered using one representative disorder that best illustrates the importance of white matter dysfunction in brain–behavior relationships.

Genetic Diseases

These diseases mostly appear in infancy or childhood, but in some cases they come to attention in adulthood. At any age, neurobehavioral manifestations are prominent. Genetic white matter diseases highlight the importance of myelinated tracts in the development of normal behavior. Many are characterized by dysmyelination, which is the abnormal formation of myelin because of an inborn error of metabolism that interferes with the sequence of events leading to the establishment and maintenance of the myelin sheath. The most familiar examples of dysmyelination are the leukodystrophies.

Metachromatic Leukodystrophy

Metachromatic leukodystrophy (MLD) is the most common of the leukodystrophies. The usual age at onset of this autosomal recessive disease is 2 or 3 years, but MLD also may present in older children and adults. Common presenting manifestations include developmental delay, intellectual deterioration, gait disorder, strabismus, and spasticity. Peripheral dysmyelination produces neuropathy. In young children, steady deterioration progresses toward a vegetative state and death within a few years, but later ages of onset portend a somewhat less severe course. MLD results from a deficiency of the enzyme arylsulfatase A, which converts sulfatide to cerebroside, a major component of myelin (51). Sulfatide accumulation in oligodendrocyte lysosomes is visible microscopically as metachromatically staining granules, and the eventual death of oligodendrocytes precludes the possibility of any remyelination. Diagnosis is confirmed by the demonstration of reduced arylsulfatase A activity in leukocytes. MRI provides a detailed view of the characteristic cerebral dysmyelination (Fig. 36.1) (52).

In older children and adults, the longer course of the disease has permitted more detailed study of its neurobehavioral features. Dementia is the predominant manifestation. Neuropsychological testing has disclosed a pattern of deficits consistent with primary involvement of the white matter, including inattention, poor vigilance, impaired memory, relatively intact language, impaired visuospatial function, and executive dysfunction (53). A frequent tendency for psychosis to herald the onset of the disease has been noted (52,54), possibly because of disrupted corticocortical connections between frontal and temporal lobes (54). Dementia then follows as more extensive dysmyelination proceeds to disrupt other connections and produce more widespread neuronal dysfunction (52).

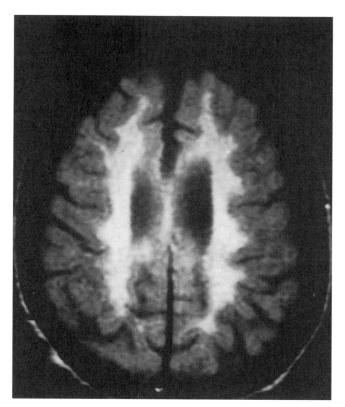

FIGURE 36.1. Magnetic resonance imaging scan of a patient with metachromatic leukodystrophy showing diffuse and symmetric white matter hyperintensity indicating cerebral dysmyelination.

Treatment of MLD has been limited, but the introduction of bone marrow transplantation has raised the hope that normal arylsulfatase A activity and clinical benefit can be achieved (55). Preliminary reports indicate that this procedure can restore the activity of the enzyme, stabilize cognitive deterioration, and improve the MRI appearance of white matter (55). Success in treating MLD with bone marrow transplantation provides evidence for the neurobehavioral importance of white matter in that improvements in cognition and cerebral neuroimaging occur in parallel.

Demyelinative Diseases

In contrast to dysmyelination, demyelination refers to a stripping away of myelin from the axon by an inflammatory attack on the myelin sheath. The most familiar example of the demyelinative diseases is MS, the most common nontraumatic disabling neurologic disease of young adults. Other demyelinative diseases are usually regarded as variants of MS.

Multiple Sclerosis

MS remains a perplexing disease of unknown etiology. Among many questions requiring further explication are the characterization, significance, and treatment of neurobehavioral dysfunction. As first noted by Charcot (1), both cognitive and emotional disturbances are recognized, and the wide range of neurobehavioral disturbances presents a challenge to clinicians and an opportunity for researchers.

Cognitive impairment is an important problem that affects many patients with MS (11,56). This syndrome can range from subtle cognitive loss that may easily escape clinical detection to frank dementia that mandates total care. An overall prevalence figure of 40% to 70% has been suggested, and more severe disease generally predicts more significant cognitive loss (56). Cognitive dysfunction in MS may not be associated with more obvious features of neurologic disease and by itself may constitute the major source of disability. Physical disability as measured by the Extended Disability Status Scale (57), for example, may be minimal in the presence of major neurobehavioral impairment (58). Clinicians also may be misled by a pattern of cognitive impairment that differs from that of more familiar dementia syndromes such as AD. The typical profile of AD, for example, is one of prominent amnesia and aphasia, whereas MS patients are likely to manifest greater impairment in attentional function (59). These distinctions mean that the use of routine cognitive screening instruments may be inadequate. Because the dementia of MS, which is similar to many other white matter disorders, does not significantly disrupt language function, heavily language-weighted tests such as the Mini-Mental Status Examination (60) are relatively insensitive to the cognitive loss that may be present (61). Neuropsychological testing may be required to confirm a clinical suspicion of cognitive dysfunction.

Understanding the cognitive impairment associated with MS begins with the neuropathology of the disease. Although cerebral plaques in patients with MS are most common in the periventricular regions, left and right hemispheres are equally affected, and plaques are distributed proportionately throughout the white matter (62). These findings, which are well known to clinicians, are demonstrated by MRI studies of MS patients showing a pattern of hyperintensities consistent with these observations (Fig. 36.2).

Advances in MRI have clarified the neuropathologic significance of white matter lesions in MS (63). White matter hyperintensities generally reflect an increase in the water content of the affected region; therefore, they are nonspecific findings. Identical MRI lesions may range in severity from simple interstitial edema, to demyelination with astrogliosis, and finally axonal destruction. The acute lesion of MS is an inflammatory process revealed by contrast enhancement that transiently increases the MRI signal until the inflammation subsides, after which variable degrees of tissue loss remain. With time, these lesions conspire to produce brain atrophy, which may be most prominent around the third ventricle (63).

Compelling evidence from MRI studies has demonstrated that the burden of white matter disease correlates with

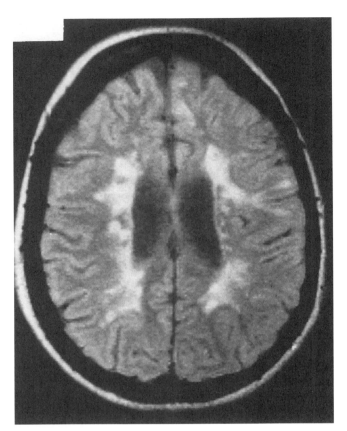

FIGURE 36.2. Magnetic resonance imaging scan of a patient with multiple sclerosis demonstrating large patchy areas of periventricular demyelination.

cognitive dysfunction (61,64–67). An area of about 30 cm^2 has been suggested as the threshold above which cognitive impairment is likely (64). MS plaques that occur in the cortex, judged to be 5% of the total by neuropathologic study (62), do not appear to cause this impairment, as cortical plaques revealed by MRI accounted for only 6% of the total lesion volume and did not correlate with neurocognitive test results (68). More advanced MRI techniques have added additional information. Magnetization transfer imaging studies have found that cognitive impairment in MS patients is correlated with neuropathologic changes in normal appearing white matter (69). Both magnetic resonance spectroscopy (70) and postmortem studies (71) of MS brains also have detected axonal loss in areas of white matter that appear to be normal. These findings suggest that more subtle white matter neuropathology can be detected with advanced neuroimaging techniques and correlated with cognitive dysfunction.

Characterization of neuropsychological deficits in MS patients has led to the conclusion that the disease qualifies as a subcortical dementia (11). This view is based on the prominence of deficits in attention, concentration, memory, executive function, and psychiatric status combined with the absence of significant language disturbance and other cortical deficits (11). Additional data suggest the possibility that MS has unique neuropsychological features that separate it not only from cortical diseases but also from subcortical gray matter diseases. First, the memory disturbance in MS has been described as a retrieval rather than an encoding deficit (72). Another feature that seemed to characterize MS was the sparing of procedural memory, which is known to be affected in subcortical gray matter diseases (73). These characteristics were postulated to distinguish MS from the cortical dementias, which involve an encoding deficit, and the traditional subcortical dementias, in which procedural memory is affected (17). Together with other findings indicating prominent frontal lobe and right hemisphere white matter dysfunction, these observations support the notion that MS is a prototype white matter dementia (17).

Treatment of cognitive dysfunction in MS is only beginning to be explored. Although the mainstay of conventional treatment remains corticosteroids for acute exacerbations, cognitive decline may be effectively treated by immunomodulatory drugs, three of which (interferon-β 1b, interferon-β 1a, and glatiramer) have been approved within the last decade. The rationale for the use of these drugs for relapsing MS is persuasive. This form of the disease, if untreated, is associated with progressive brain atrophy (63), and all three drugs appear to reduce relapse rate and improve white matter disease burden as noted on MRI (74). It is conceivable that therapy of this kind could improve or stabilize cognitive function. Improvement in neuropsychological function has been reported after 4 years of interferon-β 1b therapy (75).

MS is a diffuse disease that typically disturbs the function of multiple cerebral regions simultaneously. In relatively few situations can a single plaque or region of white matter involvement be securely correlated with a focal neurobehavioral syndrome, but some focal syndromes have been identified. These include amnesia (76), Broca aphasia (77), transcortical motor aphasia (78), conduction aphasia (79), global aphasia (80), mixed transcortical aphasia (78), pure alexia (81), alexia with agraphia (82), left hemineglect (83), visual agnosia (84), left tactile anomia, agraphia, apraxia (85), and executive dysfunction (66). The lesions responsible for the appearance of these familiar neurobehavioral syndromes are consistent with those expected on the basis of classic teachings on localization in behavioral neurology.

A number of neuropsychiatric syndromes have been described in patients with MS. These include depression, bipolar disorder, psychosis, emotional incontinence, euphoria, and fatigue. As might be expected, the site(s) of white matter involved in producing these syndromes is largely unknown, and other factors in addition to structural brain disease are likely to play a role in pathogenesis (56). MS, like other chronic neurologic diseases, frequently produces major disturbances that often can be ascribed to psychological distress, but white matter lesions probably also contribute to the development of neuropsychiatric distress. Depression is

the most significant neuropsychiatric syndrome, seen in 50% of MS patients at some point (56). MS patients have been found to be seven times as likely to commit suicide than age-matched control subjects (56). Bipolar disorder is less common than depression in patients with MS, but more so than expected in the normal population (56). Psychosis has been a rare complication, and frontal-limbic disconnection or temporal lobe demyelination has been suggested to be pathogenetic (52). Emotional incontinence, also known as pathologic laughter and crying or pseudobulbar affect, may occur in some patients, and demyelination of the corticobulbar tracts originating in the frontal lobes has been speculated to be responsible (56). A state of sustained and undue cheerfulness called euphoria has been described in patients with advanced MS, and support for frontal lobe involvement has appeared (56). Finally, fatigue is a problem for many MS patients and may be related to difficulty maintaining sustained cognitive activity (86).

Infectious Diseases

The white matter of the brain can be affected by a number of infectious diseases. The majority of these infections are caused by viruses, and many clinical and immunopathologic similarities to demyelinative diseases have invited comparison with MS. As is often the case with white matter disorders, infectious diseases may also involve gray matter regions, and the selectivity for white matter may only be relative.

Acquired Immunodeficiency Syndrome Dementia Complex

Infection with the human immunodeficiency virus type 1 (HIV-1), commonly known as HIV, is a major public health problem worldwide. Despite rapid progress in determining its etiology, prevention, and treatment, AIDS continues to be an incurable disease. Soon after the epidemic began, involvement of the brain was recognized, and the ADC was identified as a major manifestation of this infection (13,87). The ADC affects approximately 30% of AIDS patients at some point in the disease (88), and it takes the form of severe dementia in about half of these patients (87). The severity of ADC generally parallels that of AIDS itself. Left untreated, the onset of dementia predicts a mean survival of 6 months (88).

Since its identification, the most common neuropathologic feature of ADC has been diffuse myelin pallor in the cerebral white matter (13). HIV does not infect neurons. The virus is most concentrated in subcortical regions (88), where white matter pallor is accompanied by gliosis, multinucleated giant cells, microglial nodules, and increased numbers of perivascular macrophages (89). Studies of the cortex in HIV infection have found that either there is no cortical neuronal loss in patients with ADC (90) or that the loss of cortical neurons is not correlated with dementia severity (91). White matter changes can be seen on computed tomography and especially MRI, where patchy or diffuse hyperintensity and cerebral atrophy may be observed (87). The subcortical and cortical gray matter are not as significantly affected as the white matter.

In clinical terms, ADC has been characterized as a subcortical dementia (87,88). Patients typically present with impairments in attention, concentration, memory, and personality. Deficits in cognitive speed and mental flexibility may be striking. Language typically is normal, but dysarthria may be seen (13). Memory is uniformly impaired, with a pattern of dysfunction closely reminiscent of the hypothesized profile of white matter dementia (17). First, there appears to be a retrieval deficit in declarative memory (92); second, procedural memory appears to be spared (93). A general correlation between white matter pallor and degree of cognitive impairment has been observed (13,94).

Because the neuropathology of ADC also involves the subcortical gray matter to some extent (13,88), it is uncertain how much of the neurobehavioral picture can be attributed to white matter involvement alone. Nevertheless, further indirect evidence of white matter dysfunction comes from studies of AIDS treatment. A marked reduction in the incidence of ADC has accompanied the widespread use of antiretroviral therapy (95), possibly through its effects on white matter involvement in AIDS. The antiretroviral drug zidovudine (AZT), a mainstay of AIDS pharmacotherapy, has been shown to improve cognitive function in ADC (96) in parallel with a reduction in white matter lesion burden on MRI (97). Similarly, use of protease inhibitors in ADC may be associated with improvement in both cognitive function and extent of MRI white matter involvement (98). These observations support the hypothesis that ADC is associated with primary involvement of cerebral white matter, but the data do not exclude the possibility that gray matter changes also play a role.

Inflammatory Diseases

A diverse group of noninfectious inflammatory diseases can affect the white matter of the brain. These disorders, often referred to as connective tissue or collagen vascular diseases, are autoimmune diseases in which the immune system mounts an assault on various tissues of the body. Widespread lesions in the brain are typical, but specific damage to white matter often occurs and may have significant consequences.

Systemic Lupus Erythematosus

Systemic lupus erythematosus (SLE) is the best known systemic connective tissue disease that affects the nervous system. The central nervous system is implicated in up to two thirds of patients (99). When the brain is involved, the terms lupus cerebritis and, more commonly, neuropsychiatric lupus often are used (99). These designations are intended to include the whole range of neurologic presentations, in-

cluding the acute confusional state, cognitive dysfunction, dementia, and a variety of neuropsychiatric syndromes.

The neuropathology of SLE is dominated by a vasculopathy with hyalinization of vessel walls and perivascular inflammation; a true vasculitis is uncommon (100). Multiple ischemic and hemorrhagic lesions in both gray and white matter are thought to arise from this process (99). Neuroradiologic findings in SLE include cerebral atrophy, the presence of which is confounded by the fact the steroid therapy may produce this change (101). White matter lesions noted on CT and especially MRI may be the most prominent neuroradiologic finding (101).

The origin of neuropsychiatric disturbances in SLE remains unresolved. A number of neuropathologic processes may be involved, and it is plausible that a combination of structural injury from vascular occlusion and neuronal damage from antineuronal antibodies and cytokines contributes (99). Neuropsychological reports of the profile of cognitive dysfunction in SLE note that, despite some clinical heterogeneity, deficits in attention, concentration, visuospatial skills, and cognitive speed without major language involvement are typical (102). This pattern is reminiscent of that produced by other white matter disorders (17). Moreover, language-based screening instruments, such as the Mini-Mental Status Examination (60), often are insensitive to the cognitive impairments of SLE patients.

There is considerable indirect evidence for white matter involvement in producing neurobehavioral dysfunction in SLE. However, a mild degree of white matter hyperintensity as noted on MRI does not appear to produce neuropsychological deficits (103). One explanation for this finding may be that cognitive impairment may be detectable only when a certain threshold burden of disease is present on MRI (103). Additionally, there may be damage in the white matter below the level of detection of conventional MRI. Magnetic resonance spectroscopic measures of neuronal dysfunction in normal appearing white matter were found to have the strongest correlation with cognitive impairment in one series (104).

Toxic Leukoencephalopathies

The nervous system can be damaged by a wide range of toxins, and the selective vulnerability of the white matter delineates a subset of neurotoxicology called toxic leukoencephalopathy (105). Adverse effects on white matter resulting from radiation and chemotherapy are the most familiar (105). With the advent of new therapeutic agents for cancer and other diseases, as well as the appearance of new drugs of abuse, toxic leukoencephalopathies are being increasingly recognized (105).

Toluene Dementia

Injury to the nervous system caused by drugs of abuse has been difficult to characterize because drug abusers often are exposed to more than one agent, and neuropathologic studies of individuals with single exposures are rare. However, with the use of MRI and neuropathologic study, toluene has been identified as a selective white matter toxin. Toluene (methylbenzene) is the major solvent in spray paints, and it is also found in many other readily obtainable household products. Exposure to toluene, in addition to many other organic solvents, occurs in workers in many occupational settings and in the general public. High-level exposure to toluene has provided firm data on the neurotoxic effects of this solvent (106). Abuse of toluene is practiced by the intentional inhalation of solvent vapors derived mainly from spray paint; this inhalation induces euphoria without a dramatic withdrawal state. With prolonged exposure, dementia dominates a striking neurologic syndrome that includes ataxia, corticospinal dysfunction, and cranial nerve abnormalities (106). These effects persist in many abusers even after abstinence is achieved. The pattern of dementia is consistent with that of subcortical dementia (106) and, more specifically, of white matter dementia (17).

The first MRI studies of toluene dementia proved invaluable in documenting diffuse leukoencephalopathy in the cerebrum and cerebellum (Fig. 36.3). Findings included increased periventricular white matter signal intensity, loss of differentiation between gray and white matter, and diffuse cerebral atrophy (107). The white matter changes accounted for cognitive loss, as the severity of cerebral white matter involvement on MRI was strongly correlated with the degree of neuropsychological impairment (108).

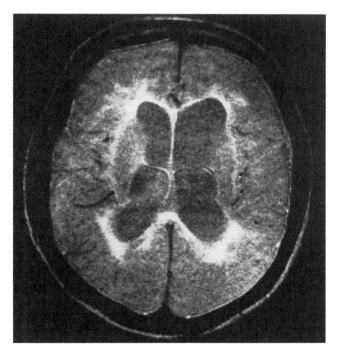

FIGURE 36.3. Magnetic resonance imaging scan of a patient with toluene dementia showing diffuse white matter hyperintensity and ventriculomegaly.

Neuropathologic studies of toluene dementia were able to confirm the selectivity of white matter involvement. Postmortem studies consistently disclosed widespread white matter changes in the brain with sparing of cortical and subcortical gray matter and axons (109). True demyelination was not observed, and an increase in very-long-chain fatty acids in the white matter suggested a neuropathologic commonality with adrenoleukodystrophy (110).

The clinical sequelae of extended toluene abuse are clear, but the impact of low-level occupational exposure to toluene and other solvents remains uncertain. Workers exposed in industrial settings often have many neurobehavioral complaints that could be due to solvent exposure. However, these symptoms, which typically include fatigue, poor concentration, memory loss, depression, and sleep disturbance, are nonspecific and may not correlate with neurologic findings or neuropsychological dysfunction. Determining a cause-and-effect relationship is difficult because many individuals are exposed to multiple solvents, experience depression or anxiety, have concurrent alcohol or other drug issues, or are involved in litigation. The issue has been controversial since the first description of so-called "chronic painters' syndrome" (111). Since then, much has been written about this condition, which also is called chronic toxic encephalopathy and the psychoorganic syndrome, and opinions on its existence range from supportive (112) to skeptical (113). Consistent patterns of neuropsychological impairment in attention, memory, and visuospatial dysfunction have been noted in many studies (112), but methodologic problems with this research have been stressed. CT scans may be normal in solvent-exposed workers, although MRI may show diffuse white matter hyperintensity in workers exposed to industrial solvents compared to age-matched controls (114). Accurate diagnosis of individuals in this setting is not straightforward, and many cases of alleged cognitive impairment after occupational solvent exposure are unconvincing after careful neurobehavioral evaluation. Thus, although the dementia of toluene abuse remains the best example of solvent-induced neurobehavioral dysfunction and is one of the most instructive varieties of white matter dementia (17), similar effects from low-level exposure to toluene or other solvents remain to be substantiated.

Metabolic Disorders

A number of disorders in the general category of metabolic dysfunction can result in white matter disease of the brain. Although there is considerable overlap between metabolic and toxic disorders, many conditions can be seen as resulting from a metabolic derangement in which neuroimaging or neuropathologic evidence of leukoencephalopathy has been obtained. Many are reversible if the metabolic derangement is corrected before irreversible damage develops. Neurobehavioral aspects of these disorders have been gradually characterized and general findings have emerged, although precise correlations between white matter lesions and behavioral features are not yet available.

Cobalamin Deficiency

One of the most commonly sought causes of reversible dementia is deficiency of cobalamin (vitamin B_{12}). Cobalamin deficiency has been noted in up to 40% of older individuals, and as many as 50% of those who are deficient have cerebral symptoms (115). Although classically associated with pernicious anemia, it is clear that neurologically significant cobalamin deficiency can occur in patients who have no anemia or macrocytosis (116). Cobalamin levels below 100 pg/mL regularly produce neurologic manifestations; levels above 300 pg/mL are considered normal. Vitamin B_{12} levels that fall within this range are difficult to interpret, and many recommend the measurement of serum homocysteine and methylmalonic acid in these patients. If one or both of these metabolites are elevated above the normal range, clinically significant cobalamin deficiency can be assumed (117). Peripheral neuropathy and subacute combined degeneration are well known sequelae in patients who are cobalamin deficient, but relatively little is known about the cerebral manifestations of this disorder.

The neurobehavioral manifestations of cobalamin deficiency are protean. Neuropsychiatric dysfunction has been frequently described, and psychosis appears to be particularly common (118). A study of community-dwelling older women found that metabolically significant cobalamin deficiency conferred a two-fold increase in the risk of severe depression (119). Cognitive loss and dementia have been documented (120), and the pattern of deficits in these syndromes, which often include cognitive slowing and confusion along with depression, has been regarded as consistent with subcortical dementia (120).

Neuropathologic observations in cobalamin deficiency have demonstrated the presence of white matter lesions in the brain that are identical to those in the spinal cord (121). Cerebral lesions usually develop later in the disease course and are characterized by perivascular degeneration of myelinated fibers with sparing of cortical and subcortical gray matter (121). The pathophysiology of cobalamin deficiency is unknown, but a disturbance of fatty acid synthesis leading to abnormal myelin has been postulated (122).

The cerebral white matter lesions seen neuropathologically have been interpreted as responsible for the mental changes in individuals with cobalamin deficiency (121). Case studies with MRI scans have supported this claim, as clinical and neuroradiologic improvement of leukoencephalopathy may occur in parallel with cobalamin replacement (123,124). However, it must be recognized that a variety of biochemical disturbances also occur in the brain and may contribute to the neurobehavioral picture (119). The biochemistry of cobalamin metabolism is complex, and many factors may contribute to dementia. One area that deserves further attention is the effect of hyperhomocysteine-

mia, which is associated with clinically significant cobalamin deficiency (116). Homocysteine is a risk factor for cerebrovascular disease, and vascular white matter lesions related to elevated homocysteine levels may occur in cobalamin deficiency.

Treatment with cobalamin has been observed to benefit neurobehavioral syndromes in some cases (116). Well-documented examples of significant cognitive and neuroimaging recovery with cobalamin treatment (123,124) support the notion that dementia results from white matter involvement in this nutritional disorder.

Vascular Diseases

Vascular disease is one of the major concerns of clinical neuroscience and medicine. Stroke continues to be the third leading cause of death in the United States. It represents an enormous source of disability resulting from neurologic and neurobehavioral dysfunction. Many controversies persist in the area of vascular disease, one of which centers on the origin and significance of white matter changes frequently seen on the neuroimaging scans of older individuals. These findings have prompted a renewed examination of the contribution of vascular white matter disease to neurobehavioral dysfunction.

Binswanger Disease

BD is a form of vascular dementia characterized by prominent involvement of the cerebral white matter. Autopsy studies have suggested that about 4% of the general population and 35% of dementia patients have the lesions of BD at postmortem examination (125). Clinically, the disease is associated with hypertension or other vascular risk factors. It presents in mid to late life with progressive neurologic and neurobehavioral features, often, but not always, with a stepwise course (12,126). A similar clinical picture in the absence of hypertension may indicate the presence of cerebral autosomal dominant arteriopathy with subcortical infarcts and leukoencephalopathy (CADASIL), another etiology of vascular white matter dementia (127). In BD, elemental neurologic features include subacute onset of focal pyramidal or extrapyramidal signs, acute lacunar syndromes, gait disorder, pseudobulbar signs, depression, and sometimes seizures. Neurobehavioral manifestations include apathy, inertia, abulia, memory impairment, visuospatial dysfunction, and poor judgment and insight. The diagnosis may be difficult in early stages of BD, when all of these features may be subtle. Apathy and inertia may be mistaken for the cognitive slowing that is a common feature of normal aging. In addition, depression is common in BD (128), but it may not be appreciated as a feature of the disease in the absence of prominent neurologic dysfunction.

Neuropathologic observations form the foundation for understanding the origin of dementia in BD. The long penetrating arterioles of the deep cerebral white matter are invariably affected by thickening and hyalinization of the vessel walls that narrow but usually do not occlude the lumen (12). Modern formulations have acknowledged that lacunar infarction is present in cerebral areas as well, and the finding of lacunar infarction closely links BD with the lacunar state (129). In any case, the cortex is spared from this process, and the subcortical gray matter is less affected than the white matter (126). Microscopically, findings in early cases may be limited to myelin pallor. In advanced cases, loss of myelin, axons, and oligodendrocytes is common, as are astrocytic gliosis and frank cavitation of the white matter. The subcortical U fibers typically are spared (12). These findings are similar to those of leukoaraiosis (130), a term given to white matter hyperintensities seen on MRI that are considered ischemic in origin (131). Because these lesions have cognitive sequelae (46), possibly when a certain threshold of white matter involvement is reached (132), it has been suggested that leukoaraiosis is a precursor to BD (15,129).

The neuroradiology of BD is controversial because of lingering doubt about the nosologic status and diagnosis of the disease. MRI scans typically show widespread white matter hyperintensity (Fig. 36.4), often with scattered lacunar infarctions (133). In view of the nonspecific nature of these findings, the diagnosis of BD is made based on clinical and neuropathologic grounds, and MRI evidence is considered supportive.

Neurobehavioral dysfunction in BD has been steadily linked with white matter vascular disease. Common clinical features include memory loss, confusion, apathy, and changes in mood and personality that usually are unaccompanied by aphasia, apraxia, and movement disorder (12,126,128,133). The disease is in contrast to the cortical syndrome often seen in patients with multiinfarct dementia, and its features are consistent with white matter dementia (17). Studies in stroke patients have suggested that even single lacunar infarctions produce functionally limiting neurobehavioral deficits, thus supporting the role of vascular white matter disease as a cause of dementia (134). When combined with other ischemic changes (46,132), the impact of slowly accumulating white matter lacunes appears likely to result in increasing cognitive impairment. In a quantitative study of vascular dementia patients using MRI, Liu et al. (135) found a strong correlation between white matter lesion area and dementia. Babikian and Ropper (12) pointed out a resemblance between the clinical features of BD and normal-pressure hydrocephalus (NPH), which is another disease that primarily affects cerebral white matter. Although further study of neurobehavioral features is needed, the pattern of deficits and preserved areas is consistent with primary white matter involvement (17). Quantitative comparisons of white matter involvement with cortical and particularly subcortical gray matter disease are required to determine the relative contributions of these tissues to dementia.

The many controversies surrounding BD highlight the uncertainties about the neurobehavioral implications of white matter disease in general. This issue is well illustrated by cerebrovascular disease. Although neurologists have long understood the potential for cortical infarcts to impact cognitive function, this principle has not been as readily applicable to lesions of the white matter. However, the much debated BD is gaining increasing credibility as a disease of the cerebral white matter that can have dramatic effects on behavior; therefore, it merits attention as a prototype white matter dementia (17).

Traumatic Disorders

Trauma to the brain can occur as a result of accidents, assaults, sporting events, or intentional therapeutic maneuvers. The neuropathologic changes in the brain caused by trauma are complex and multifaceted, and determining the clinical consequences of TBI is a formidable task. Nevertheless, the white matter is significantly damaged in TBI, and considering this category of injury from a neurobehavioral perspective is instructive.

Traumatic Brain Injury

TBI ranks as a major cause of neurologic disability in the United States. The problems faced by many TBI patients and their families and caregivers are particularly burdensome given the high incidence of TBI in young adults, who then usually survive and may be required to cope with long-term neurobehavioral sequelae for decades (22). These deficits in

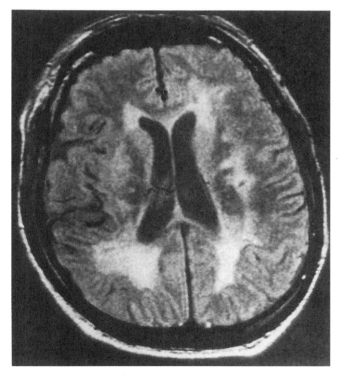

FIGURE 36.4. Magnetic resonance imaging scan of a patient with Binswanger disease demonstrating widespread ischemic white matter disease in the hemispheres.

cognition, and especially emotional status, are typically the most problematic, far outpacing physical disability (22). Evidence developed over the last half century has established that the most important factor underlying the neurobehavioral sequelae of TBI is injury to the cerebral white matter.

A consideration of TBI as a white matter disorder requires a review of fundamental neuropathology and pathophysiology. Although the range of neuropathology found in TBI is broad and includes focal cortical contusions, hypoxic-ischemic lesions, and extraaxial hemorrhages, clinical and experimental studies have implicated injury to white matter as the most prominent (136,137). The currently preferred label for this observation is diffuse axonal injury (DAI) (136). Although this term identifies the axon as the major site of injury, DAI primarily implies widespread structural lesions in the white matter of the brain.

Variable degrees of DAI have been shown in both severe and mild TBI. Neuropathologically, there are axonal retraction balls, microglial clusters, and wallerian degeneration in white matter fiber tracts (138). The pathophysiology of DAI involves shearing forces produced in the brain by sudden acceleration and deceleration (137). Rotational forces appear to be most deleterious. These shearing forces act mainly upon long fiber systems in the brain that are most vulnerable to mechanical disruption. Injury to blood vessels with multiple hemorrhagic foci also is common. The extent of DAI correlates with clinical measures of severity, including the Glasgow Coma Scale, the length of unconsciousness, and the duration of posttraumatic amnesia (137). Thus, the essential difference between mild and severe forms of TBI is the degree of DAI (137). Areas most prominently affected by DAI are the dorsal midbrain, corpus callosum, and hemispheric white matter (22).

Neuroimaging studies have generally supported neuropathologic findings highlighting the importance of white matter damage in TBI. In a prospective study, DAI was found to be the most common primary lesion in TBI (139). DAI lesions on MRI are seen in the hemispheric white matter, corpus callosum, and dorsal midbrain (139), which are locations consistent with neuropathologic studies of DAI (22). Correlations between white matter changes on MRI and neuropsychological function have been reported (140), and newer, more sensitive MRI techniques are likely to strengthen these relationships (141).

The clinical impact of DAI depends to a large extent on the severity of DAI, although other neurologic and systemic pathologies also contribute, and a profile of neurobehavioral deficits can be tentatively ascribed to DAI. As a general rule, attention, memory, and executive function are prominently affected in TBI, reflecting damage to frontal and temporal lobes (22,137). TBI patients display relative preservation of procedural memory in comparison to declarative memory (142), as well as specific difficulties with memory retrieval (143), a pattern consistent with white matter dementia (17). Executive dysfunction has been documented in TBI and may

reflect damage to white matter connections of the dorsolateral prefrontal cortices to posterior structures (22). Language is relatively preserved after TBI, but personality and emotional changes are common; depression affects nearly half of all TBI patients (144). A complex amalgam of neurobehavioral deficits is typical of TBI, and because of the common denominator of DAI and a clinical profile that tentatively matches that of other white matter disorders (17), white matter neuropathology likely is responsible for a significant component of the neurobehavioral disability.

Neoplasms

A discussion of brain tumors that selectively damage white matter may appear to be limited. As a rule, brain tumors do not affect one discrete region; instead they show a predilection for involving widespread areas of both gray and white matter. This characteristic, combined with the associated edema and mass effect that frequently occur, often renders correlations of tumor location with neurobehavioral status tentative at best. However, many neoplasms actually arise from the white matter and exert major effects on the function of these areas.

Gliomatosis Cerebri

Gliomatosis cerebri is a diffusely infiltrative glial cell neoplasm of the brain. Although rare, this disease serves as the best example of a neoplastic disorder that is confined to the cerebral white matter throughout most of its course. Gliomatosis cerebri usually appears in adulthood, but it can arise at any age. The insidious onset of mental status changes is the most frequent presentation; headache, motor dysfunction, and seizures also may occur (145,146). Diagnosis in life is difficult because the clinical presentation is consistent with a wide range of diffuse white matter disorders, and even with cerebral biopsy there may be confusion about the classification of this lesion. The clinical course is variable, with survival durations reported from weeks to many years (145,146), but a fatal outcome has been typical. Surgery, chemotherapy, and radiotherapy may be partially effective, but they are not curative.

In contrast to gliomas that tend to be single or multicentric, gliomatosis cerebri involves infiltration of contiguous areas of the cerebrum. Although gray matter structures may be affected, the major neuropathologic burden falls on the white matter, where there is destruction of the myelin sheath but little damage to neurons and axons (146). Thus, there is widespread white matter infiltration with relatively preserved cerebral architecture. Periventricular white matter often undergoes additional damage because of the hydrocephalus and increased intracranial pressure that can develop from aqueductal stenosis or tumor overgrowth (145). MRI has generally proven more sensitive than CT in detecting white matter changes in this disease, and MRI studies have documented high signal prominently involving the frontal lobes and corpus callosum (147).

The neurobehavioral changes associated with gliomatosis cerebri have not been well characterized, and there is no reported study of neuropsychological function in affected patients. This paucity of information is due largely to the rarity of the disease. A 1985 review located only 48 cases and an additional 10 from which to draw conclusions about the mental status changes (146). In addition, few reports contain substantial detail on neurobehavioral aspects of this disease. Despite these shortcomings, personality and mental status changes are repeatedly stated to be the most striking findings in patients with gliomatosis cerebri, whether in the initial or later stages of the disease (145,146). The mental changes typically are described as confusion, disorientation, and memory loss; focal cortical signs including aphasia are rarely noted. Neuropsychiatric dysfunction, typically described as personality change, is frequently cited as an initial or early manifestation. Although gray matter involvement may occur in some cases, these clinical features are consistent with those that would be expected with diffuse white matter involvement, and the neuropsychiatric dysfunction can plausibly be attributed to involvement of the frontal lobes.

Hydrocephalus

The term hydrocephalus refers to the accumulation of excessive water in the brain. In many cases, the higher ventricular volume of cerebrospinal fluid (CSF) in hydrocephalus exerts a prominent effect on the cerebral white matter, producing significant neurobehavioral manifestations. Hydrocephalus may develop at any age, but in older adults the etiology and management may be particularly problematic.

Normal-Pressure Hydrocephalus

Since its description in the 1960's (148), this disease has alternately been heralded as a reversible form of dementia and an entity whose very existence is open to debate. The opportunity to detect and effectively treat a cause of dementia is appealing, but the risks of treatment are not insubstantial. In this section, NPH is regarded as a disorder of cerebral white matter in an attempt to clarify this continuing clinical conundrum.

NPH presents with the classic clinical triad of dementia, gait disorder, and incontinence (148). The dementia has been noted to have frontal features, with cognitive slowing and apathy. The gait disorder and incontinence also seem to result from frontal lobe damage (148). Some cases are thought to follow meningitis, TBI, or subarachnoid hemorrhage, but many remain idiopathic. Neuroimaging scans show ventricular enlargement out of proportion to gyral atrophy. MRI may show high signal lesions in the deep cerebral white matter that have been suggested to represent transependymal flow that predicts a good outcome with

shunting (149). Diagnosis is made by identifying the clinical triad, finding a consistent neuroimaging pattern, and ruling out other possible causes. Many clinicians also perform a high-volume CSF "tap test" because a temporary improvement in gait and sometimes cognition may help predict a good surgical outcome (150).

The pathophysiology of NPH continues to be an enigma. Adams et al. (148) offered the first and still most familiar theory, which holds that obstruction of CSF outflow causes ventricular enlargement without cortical atrophy. The outflow obstruction is assumed to be at the level of the arachnoid villi. Other researchers have suggested another major hypothesis, which posits that ischemia and infarction in the cerebral white matter lead to ventriculomegaly, and the reduced tensile strength of the white matter causes further ventricular enlargement under the stress of intraventricular pulse pressure (151). This theory suggests that there may be an overlap of NPH with BD, or even that the two may be the same disease. Most authorities, however, favor the existence of NPH as a discrete entity, recognizing that it may coexist with other neuropathologic processes. From a clinical perspective, the beneficial outcome of surgery in some individuals justifies the conclusion that NPH may exist in isolation, but white matter ischemia and infarction often are present as well and limit its reversibility.

Regardless of the mechanism of NPH, the neuropathologic findings are generally consistent with primary white matter involvement. Details of the neuropathology are debated, but the white matter of the brain seems to bear the brunt of the damage, due to either mechanical compression or ischemia (152). Gray matter, both of subcortical nuclear structures and the cortex, is less affected, although with severe and prolonged hydrocephalus there may be some neuronal damage in these areas (152). Unrelated cortical pathology can be found in patients suspected of having NPH. One study found that 10 of 38 patients clinically thought to have NPH actually had biopsy-proven AD (153). These data further highlight the difficulty in diagnosing NPH, suggest that cortical pathology may be present in many cases, and offer another explanation as to why some patients have a poor outcome.

The neurobehavioral profile of NPH has been categorized as typical of subcortical dementia (154). Frontal signs, including cognitive slowing, inattention, and perseveration, are prominent, and memory loss occurs as well (154). Importantly, there typically is little or no aphasia, apraxia, or agnosia, except for gait disorder, which sometimes is called gait apraxia (154). In contrast, selective executive function deficits have been documented in patients with NPH that were not present in AD patients with comparable dementia severity (155). In light of these findings and the neuropathology of the disease, NPH has been classified as a white matter dementia (17).

Treatment of NPH involves the placement of a diversionary shunt to reduce ventricular volume. These shunts are rubber tubes with one-way valves that can be positioned to provide ventriculoperitoneal, ventriculoatrial, or lumboperitoneal CSF drainage. The critical therapeutic decision involves determining which patients are most likely to benefit from a shunt procedure, which is difficult because of the high incidence of shunt complications (156). Although clinicians have differing preferences, many refer patients for surgery who have the full clinical triad, consistent neuroimaging studies, no evidence of other cerebral diseases, and no significant surgical risk. Additional favorable prognostic signs are the presence of dementia for less than 2 years, the gait abnormality preceding the dementia, and a known cause of NPH (157).

WHITE MATTER DISORDERS AND HIGHER FUNCTION

The disorders discussed in this chapter demonstrate that damage or injury to the white matter of the brain can disrupt higher functions. From these observations it is possible to make some generalizations about the white matter disorders as a whole. Cognitive impairment emerges as the most frequent neurobehavioral syndrome in patients with white matter disorders. Focal neurobehavioral syndromes and neuropsychiatric syndromes also are encountered. In this section, selected disorders will serve to illustrate these various disturbances.

White Matter Dementia

Cognitive impairment may take the form of relatively mild cognitive dysfunction or the unmistakable syndrome of dementia. Both involve disturbances in a wide range of higher functions that each is mediated by a distributed neural network. The term white matter dementia was conceived to refer to the dementia syndrome of patients with white matter disorders, but it also implies that more subtle degrees of cognitive loss occur, which in many cases are precursors of frank dementia.

The degree of cognitive impairment in white matter disorders likely mirrors the extent of neuropathology, as suggested by the parallel literature. In MS, a lesion area greater than 30 cm^2 may predict cognitive impairment, which implies that the burden of disease determines the severity of cognitive change (64). Similarly, in patients with leukoaraiosis, a lesion area greater than 10 cm^2 has been cited as predictive of cognitive impairment (132). In toluene dementia, the severity of white matter change on MRI correlates with the degree of cognitive loss (108), which reflects a general tendency in toxic leukoencephalopathy (105).

Dementia is most usefully defined as an acquired and persistent disorder of intellectual function with deficits in at least three of the following: memory, language, visuospatial skills, complex cognition, and emotion or personality (158). This definition differs from that of the widely cited American Psychiatric Association's *Diagnostic and Statistical Manual of Mental Disorders, Fourth Edition* (DSM-IV), which emphasizes aphasia, apraxia, and agnosia as core disturbances (159). Analysis of patients with white matter disorder and dementia discloses that whereas disabling deficits in various domains arise, the cortical syndromes of aphasia, apraxia, and agnosia are encountered only exceptionally. Thus, the DSM-IV definition, based as it is on the cognitive profile of cortical dementias such as AD, is less useful than the more inclusive definition offered here (158).

The idea of white matter dementia was developed based on clinical experience with selected white matter disorders, particularly MS (61) and toluene leukoencephalopathy (106). Literature review of many other disorders confirmed that damage to white matter produces a consistent neurobehavioral pattern (15). The hypothesis was put forth in hopes of stimulating further work that would assist in our understanding and treatment of many neurologic disorders and in advancing knowledge of the operations of the human brain. In general, white matter dementia was thought to be characterized by inattention, forgetfulness, emotional changes, and the absence of aphasia, apraxia, and a movement disorder (15). Although this profile is similar to that of subcortical dementia (9), two clinical features seemed sufficiently prominent to support the creation of a separate dementia category. The first was the saliency of attentional dysfunction, which is closely allied with cognitive slowing (15). The second was the typical absence of a movement disorder, which distinguished white matter dementia from subcortical dementias that regularly feature an extrapyramidal component (15).

In the years since this proposal, further investigation pertaining to this hypothesis has been conducted (52,53,59,73,105,108,160), and critical reviews have dealt with conceptual issues (16,56). Although systematic study of the distinctions between these dementia categories has only rarely been undertaken (73), many relevant observations illuminate this topic. As a result, further refinement of the idea has been facilitated, and a more complete profile of deficits and preserved strengths in white matter dementia can be proposed. This formulation supports the concept that white matter dementia is a consistent clinical syndrome.

Sustained Attention Deficit

Most generally, attention refers to the ability to focus on stimuli while excluding competing distractors. This capacity, often called selective attention, operates over a period of seconds and is commonly investigated with the digit span test (8). When selective attention operates over a period of minutes, sustained attention, also known as concentration or vigilance, is engaged, and a variety of continuous performance tasks are suitable for assessment (8). A comparison of neuropsychological features in MS and AD revealed that sustained attention was significantly affected in the former but was relatively normal in the latter (59). Because of contrasting profiles of attention and concentration dysfunction in

MS patients versus memory and language impairment in patients with AD, it was argued that these two diseases represent prototype white matter and cortical dementias (59). Similar distinctions may separate white matter dementia from subcortical dementia. In one study, MS patients had better memory function than those with HD, but they exhibited slowed and less efficient cognition (161). Disruption of frontal white matter connections to more posterior structures, particularly on the right (22), has been implicated in sustained attention impairment (8).

Memory Retrieval Deficit with Normal Procedural Memory

Two distinctions in the realm of memory function prove most useful for this discussion. The first lies within the category of declarative memory and involves the separable processes of encoding versus retrieval. The other is between declarative and procedural memory, or the difference between fact and skill memory. In this light, it is helpful to review some neuropsychological studies of memory that suggest a unique pattern in patients with white matter disorders. Individuals with MS have been evaluated with this objective in mind. They were found to have a retrieval deficit in declarative memory (72) but a preservation of procedural memory (73). Retrieval deficits also have been documented in several other diseases with prominent white matter involvement, including ADC (92,93) and TBI (143), which suggests that this type of declarative memory loss can be generalized to all the white matter disorders. Preserved procedural memory also has been observed in ADC (92,93) and TBI (143). This pattern stands in contrast to patterns of memory loss seen in both cortical disease, in which there is an encoding deficit in declarative memory and normal procedural memory, and subcortical gray matter diseases, in which there is a retrieval deficit but impaired procedural memory (17). Clinical reports and functional neuroimaging data suggest that the uncinate fasciculus may be involved in memory retrieval (162). Because retrieval requires the engagement of working memory systems in the frontal lobes to recall information stored in the temporal lobes, it is plausible that the essential structure in this process is the uncinate fasciculus, which connects these two lobes (162). Furthermore, whereas episodic memories may be retrieved by this system in the right hemisphere, semantic memories may be retrieved in the left hemisphere (162).

Normal Language

One of the most robust observations in the white matter disorders is that language usually is well preserved. In this respect, the clinical lore of classic neurology is entirely accurate. In MS, for example, aphasia is rare and frequently is an isolated syndrome related to a focal plaque in an eloquent white matter tract such as the arcuate fasciculus (79). In comparison to patients with AD, MS patients have little

linguistic difficulty (59). Aphasia is rarely described in other white matter disorders, although reduced verbal fluency has been noted (154). In contrast, speech disorders are frequent in white matter disorders, probably related to involvement of corticobulbar fibers subserving articulation. Dysarthria is well known in MS (11) and sometimes can assume a scanning quality. Articulation deficits also are described in ADC (87), toluene leukoencephalopathy (106), and BD (12).

Visuospatial Dysfunction

Visuospatial dysfunction has been documented in patients with white matter disorders. In MS patients, scores on the performance subtests of the Wechsler Adult Intelligence Scale are about 10 points lower than those of the verbal subtests, and specific visuospatial deficits have been shown on a variety of standard tests of right hemisphere function (163). In MLD (53) and toluene leukoencephalopathy (160), nonverbal abilities are more impaired than verbal skills (53).

Frontal Lobe Impairment

Many mental operations can be reasonably subsumed under the heading of complex cognition, but among the most important are those referred to as frontal lobe functions. Deficits in tests measuring executive function have been noted in MS patients, and MRI studies have correlated these deficits with plaques in the frontal lobe white matter (66). In patients with MLD, a frontal lobe syndrome with poor attentional function has been documented neuropsychologically (53). Apathy is a prominent feature of toluene leukoencephalopathy and probably stems from frontal lobe myelin loss (106). In contrast to patients with AD, those with NPH have a pattern of executive dysfunction related to frontal lobe involvement (155).

Psychiatric Disturbance

Emotional and personality aspects of white matter dementia have often been recognized. Psychosis as an early feature of adult-onset MLD was noted as a frequent trend in this disease (52,54), and the development of this syndrome was interpreted as an early component of a sequential progression to dementia seen in these patients (52,53). In MS, depression has received considerable and well-deserved attention (56). Depression also has been reported to be more common in BD patients than in AD patients who are comparably demented (128).

Normal Extrapyramidal Function

An initial criticism of the white matter dementia hypothesis questioned the usefulness of the absence of movement disorders based on the observation that these phenomena may be encountered in white matter diseases (17). However, despite exceptions to the rule, white matter disorders generally do not involve significant disorders of movement, at least early in the clinical course (17). Movement disorders reflecting basal ganglia involvement can occur in white matter disorders that have reached a late stage when the basal ganglia may be affected, as in the case of myoclonus in ADC (87).

Summary

Based on the available information, the most likely neurobehavioral profile of white matter dementia is now a combination of deficits in sustained attention, memory retrieval, visuospatial skills, frontal lobe function, and psychiatric status, with preserved language, procedural memory, and extrapyramidal function (17). This pattern differs from those associated with both the cortical and subcortical gray matter diseases and is consistent with what would be expected from the prominence of white matter in the frontal lobes and right hemisphere.

Focal Neurobehavioral Syndromes

Focal neurobehavioral syndromes are rarely described in the white matter disorders because isolated white matter lesions are relatively uncommon, the syndrome may occur only transiently, and these problems, even when present, may be underappreciated because attention is focused on other clinical and basic aspects of the disorder. Focal syndromes have been recognized for many years, however, and their identification has been dramatically accelerated with the advent of MRI.

Amnesia

Whereas a retrieval deficit is most characteristic of white matter dementia, amnesia is a different type of memory disturbance that also can follow white matter lesions. Amnesia is a disorder of new learning in which the encoding of information is primarily deficient (22). One recurrent question has been whether damage to the fornix, a prominent limbic tract, can cause amnesia. Support for this idea comes from reports of recent memory disturbance occurring after damage to the fornices resulting from neoplasia (164), infarction (165), trauma (166), and surgical section (167). A left fornical neoplasm led to the prominence verbal memory loss, which is consistent with current theories on the lateralization of verbal and nonverbal memory in the brain (22). Amnesia also has been noted after infarction of the internal capsule, particularly the genu; in these cases, the lesion was interpreted as interfering with the connections of the declarative memory system while not directly damaging its gray matter components (168).

Aphasia

Aphasia, commonly defined as a disturbance of the symbolic communication system of language resulting from brain damage, is the quintessential cortical syndrome (22). How-

ever, as recognized by classic neurologists for more than a century (2), white matter lesions also are capable of producing this syndrome.

Broca Aphasia

Vascular white matter lesions have increasingly been seen as a cause of Broca aphasia. Comprehensive CT studies of left hemisphere stroke patients have concluded that white matter areas—specifically the subcallosal fasciculus (a branch of the superior occipitofrontal fasciculus) and the periventricular white matter—are crucial for language fluency (169). CT study of the brain of Leborgne, Broca's patient who had severe nonfluency, showed a similar pattern of white matter involvement (169). Acute exacerbations of demyelinative disease have provided useful information on aphasia in white matter disorders. Broca aphasia, also called motor aphasia, is the most common language disturbance in MS (77,78) and has been associated with left frontal white matter lesions (77).

Transcortical Motor Aphasia

Infarction of the left anterior periventricular white matter has been observed with transcortical motor aphasia (170). The initiation of speech is primarily compromised, and the responsible lesion can be in the white matter connections between the left supplementary motor area and the perisylvian language zone. Transcortical motor aphasia has been noted in patients with MS (78).

Conduction Aphasia

Classic neurology holds that a lesion in the arcuate fasciculus underlying the left inferior parietal cortex is responsible for conduction aphasia (2). However, lesions that cause conduction aphasia, most of which are strokes, typically damage overlying perisylvian cortex as well as the arcuate fasciculus. Several observations have confirmed that white matter disconnection can cause conduction aphasia. A patient with MS developed the syndrome and had an MRI-proven large plaque in the white matter subjacent to the left supramarginal gyrus (79). Conduction aphasia was noted in a patient with a left parietal infarct that spared the cortex on MRI (171). This syndrome may also occur with damage to the left extreme capsule, which contains other fibers connecting the temporal and frontal lobes (24).

Wernicke Aphasia

This syndrome appears to be extremely rare in the setting of white matter disorders. One case of fluent aphasia with impaired comprehension and repetition was reported by Day et al. (82) in a patient with MS. The MRI scan of the patient showed a large area of demyelination in the left temporoparietal region. The aphasia improved significantly with treatment of MS.

Global Aphasia

The case of a woman with MS who had global aphasia with right hemiparesis and homonymous hemianopia has been reported; CT demonstrated a large white matter lesion in the left periventricular region (80). The report concluded it was likely that the arcuate fasciculus and all connections from Broca and Wernicke areas were affected and thus produced this syndrome, which was substantially improved 1 year later (80).

Mixed Transcortical Aphasia

This aphasia has rarely been described with neuropathology of any kind, but one case was reported in a patient with a left subangular white matter lesion that followed a left parietal hemorrhage (172). This lesion was interpreted as disconnecting the left auditory cortex from other cortical regions where semantic information is represented (172). Mixed transcortical aphasia has been observed in MS (78).

Alexia

Acquired disorders of reading have been recognized for more than a century. The enduring works of Dejerine established the accepted distinction between alexia with agraphia and pure alexia (2). Pure alexia is a classic disconnection syndrome stemming from strategically placed white matter damage, and alexia with agraphia may occur on occasion from appropriately placed white matter lesions.

Alexia with Agraphia

The patient of Day et al. (82) cited earlier also had an acquired disorder of reading and writing. The large plaque in the left temporoparietal white matter was responsible for both the fluent aphasia and the alexia with agraphia. Thus, the single white matter lesion produced a clinical picture similar to that often seen after a lesion in the temporoparietal cortex immediately overlying this area.

Pure Alexia

The usual cause of pure alexia is ischemic cerebrovascular disease, and the presence of a left occipital lesion combined with a lesion in the splenium of the corpus callosum is widely thought to be responsible (2). One of the most elegant neurobehavioral syndromes, pure alexia represents a disconnection between the visual and language systems that disturbs reading but not the capacity to write (22). Pure alexia has been described in MS and related to plaques in the left occipital lobe and the splenium (81).

Developmental Dyslexia

This disorder, which affects about 10% of children, is an impairment of the ability to read despite adequate intelligence and access to instruction. A study using diffusion tensor MRI found a significant correlation between reading scores and lower anisotropy in the left temporoparietal white matter of dyslexic adults (173).

Gerstmann Syndrome

The tetrad of agraphia, acalculia, finger agnosia, and right–left confusion is classically associated with left inferior parietal lesions (22). These lesions typically are cortical infarctions. A case of Gerstmann syndrome was described in a patient with a vascular white matter lesion subjacent to the left angular gyrus (174).

Agnosia

Agnosia is a disorder of recognition most often linked with cortical damage in the relevant areas of sensory association cortex (22). With the advent of MRI, it has been possible to detect agnosic syndromes in patients with white matter lesions as well.

Visual Agnosia

Three cases of visual object agnosia analyzed with CT revealed that the left inferior longitudinal (occipitofrontal) fasciculus was the critical structure involved (175). Thus, associative visual agnosia can represent a unilateral left temporal disconnection syndrome involving selective damage to white matter. A case of visual form agnosia in a patient with MS has been described (84). This patient had bilateral occipitotemporal and callosal lesions that were interpreted as interrupting the ventral stream of the visual association system (84).

Auditory Agnosia

This syndrome has been divided into two categories: auditory verbal agnosia (pure word deafness) and auditory sound agnosia (22). A case of pure word deafness from a left thalamic hemorrhage primarily damaging white matter fibers of the auditory system has been reported (176). In this instance, the lesion was interpreted as interrupting the auditory input from both primary auditory regions to the Wernicke area (176). Auditory sound agnosia from a white matter lesion has not been described.

Neglect

The literature on neglect generally points to the cortex of the right hemisphere as responsible, and the right parietal lobe is most often implicated. More recent formulations of neglect have postulated a distributed right hemisphere network for directed attention that includes the parietal lobe, frontal cortex, cingulate gyrus, and subcortical regions (22). White matter tracts are implicated as well because of the necessary connections that link these areas. Subcortical neglect has been described in patients with right internal capsule infarcts that were thought to deactivate the ipsilateral parietal and frontal cortices (177).

Visuospatial Dysfunction

Visuospatial dysfunction has been observed in patients with right cerebral white matter lesions. A case of topographic disorientation was associated with a lesion in the posterior limb of the right internal capsule, and the lesion was thought to disrupt cortical metabolism in the overlying parietal lobe (178). In another report, spatial delirium with reduplicative paramnesia was ascribed to a white matter infarct in the right corona radiata that extended into the retrolenticular portion of the internal capsule (179).

Akinetic Mutism

This syndrome occasionally has been ascribed to white matter lesions. In one patient, akinetic mutism developed after removal of a tumor in the anterior hypothalamus; this lesion was thought to destroy the medial forebrain bundle (180). Treatment with dopamine agonists was successful, thus suggesting a role of dopamine depletion in the medial frontal lobes. In another case, a patient with MS developed transient akinetic mutism after the appearance of a midbrain plaque on MRI (181). Although the presence of cerebral demyelination complicated interpretation, involvement of dopaminergic transmission in the medial forebrain bundle is possible in this case. These observations introduce the possibility that isolated lesions of subcortical white matter tracts conveying neurotransmitters to the neocortex can result in specific neurobehavioral syndromes.

Executive Dysfunction

Executive dysfunction is most securely associated with frontal lobe lesions, including those confined to white matter. In MS, for example, executive dysfunction can dominate the clinical picture and by itself cause major disability (182). Arnett et al. (66) examined the relationship between focal white matter involvement in MS and performance on the Wisconsin Card Sorting Test, which is a test of conceptual reasoning generally regarded as a measure of executive function. The authors found a significant correlation between frontal lesion area and impaired performance on the Wisconsin Card Sorting Test (66). Other white matter lesions can interfere with frontal lobe function. Apathy, abulia, and other frontal lobe features occurred in patients with acute capsular genu infarction that was thought to cause ipsilateral frontal lobe deactivation (168).

Callosal Disconnection

The corpus callosum is the largest white matter tract in the brain. Although its clinical importance is debated because of the relative paucity of significant neurobehavioral deficits in many individuals with corpus callosum lesions, many examples of hemispheric disconnection have been described (2,3). Cerebrovascular disease is a commonly reported cause of focal callosal damage. It may cause ideomotor apraxia of the left hand (22), the alien hand syndrome (183), and left hemialexia (184). Demyelinative diseases may cause callosal disconnection, as reported in an MS patient with an atrophic callosum who had tactile anomia, agraphia, and apraxia that affected only the left hand (85).

Neuropsychiatric Syndromes

Neurobehavioral manifestations of white matter disorders are not limited to cognitive syndromes. Emotional dysfunction can be seen, often arising earlier in the course of the disorder and sometimes proving to be more problematic than cognitive impairment. In this respect, white matter disorders resemble the subcortical gray matter diseases, which also feature prominent neuropsychiatric manifestations (14).

Depression

In the combined literature describing all the white matter disorders discussed in this chapter, depression appears to be the most common neuropsychiatric syndrome. Preliminary observations suggest that although reactive depression probably is one contributory factor, white matter neuropathology is also likely to participate in the pathogenesis.

Depression is uncommon in dysmyelinative diseases, largely because most of these diseases begin early in life and are dominated throughout their course by devastating cognitive effects. Even with later-onset cases, however, personality change and psychosis tend to be more prominent. In contrast, demyelination, particularly in MS, is strongly associated with depression (56). The relationship of depression and demyelination in MS has been debated, and evidence for a relationship between MRI-proven white matter disease and depression idea is accumulating (56). For example, MRI studies have correlated severity of depression with frontal atrophy and T1-weighted hypointensities ("black holes") in the frontal, temporal, and parietal lobe white matter (185).

Depression occurs in the infectious, inflammatory, toxic, and metabolic white matter disorders. About 10% of patients with HIV infection experience depression (186). In patients with ADC, depression commingled with cognitive impairment can produce a complex clinical picture. In many cases, apathy dominates the mental status of AIDS patients, and it is difficult to determine whether depression is present. Both stimulant medications and antidepressants may be helpful in HIV-infected patients (186), which implies that apathy and depression can coexist. Depression is common in SLE, and a combination of structural and immunologic abnormalities likely is involved (99). In toluene abusers, depression has been interpreted as a consequence, rather than a cause, of drug abuse because some abusers have no premorbid psychiatric history (187). Depression has been observed as an important feature of both cobalamin (119) and folate (118) deficiency.

The relationship between depression and vascular white matter changes on neuroimaging scans in older people has been intensively studied. In patients with BD, depression is more frequent than in comparably demented patients with AD (128). Many reports have also focused on depression in individuals with leukoaraiosis. A comprehensive review of neuroimaging studies of mood disorders concluded that the best replicated structural brain abnormality in depression was an increased rate of white matter and periventricular hyperintensities (188).

A depressed mood may afflict individuals with TBI, tumors of the white matter, and hydrocephalus. Major depression is the most common psychiatric disorder following TBI, occurring in 44% of all patients across many studies (144). Tumors of the cerebral white matter may be associated with depression, and the occurrence of mood change before the diagnosis is made indicates that cerebral neuropathology may be causative (189). Depression and other psychiatric syndromes are more likely when the tumor involves the frontal or temporal lobes (189). Finally, depression can be seen as a presenting or prominent feature of NPH (190). This syndrome may be difficult to distinguish from the apathy or abulia that more often characterizes affected patients, but the beneficial response of depression to shunting in NPH (190) indicates that depression may occur in NPH as a result of damage to white matter tracts.

Mania

Mania, or the full syndrome of bipolar disorder, often is reported in the context of white matter disorders. The etiology of these syndromes is unknown; the possibility exists that they coexist with white matter disorders by chance alone. However, the association of mania and bipolar disorder with white matter dysfunction implies an etiologic connection and offers a new perspective on the neurobiology of mood disorders.

As is true of depression, mania has not been noted in many cases of dysmyelinative diseases. Both mania and bipolar disorder, however, have often been observed in patients with demyelinative disease. There is a higher than expected occurrence of bipolar disorder in patients with MS, and some evidence supports an association with temporal lobe plaques (56). Manic behavior has been observed with greater than chance frequency in patients with AIDS (191). Mania or hypomania develops in some patients with SLE, although exact prevalence figures are unavailable (99). In toxic disorders, mania is not commonly observed, but the syndrome appears fairly often in metabolic disorders, as cobalamin and folate deficiency have both been associated with mania (118). As in the case of depression, bipolar disease has frequently been associated with vascular white matter lesions (188). In patients with TBI, mania can occur either as an isolated feature or as part of bipolar disorder; the latter may occur four times more often in TBI patients than in the general pop-

ulation (144). Mania has been observed with involvement of the temporal lobe white matter with glioblastoma multiforme (189). NPH is uncommonly associated with mania, but improvement after neurosurgical intervention suggests that structural injury to white matter may produce this syndrome (192).

Psychosis

Psychosis has been observed in many different cerebral white matter disorders. In MLD, psychosis is one of the most frequent early clinical features, often preceding cognitive impairment and dementia in the clinical course (52,54). Temporal and/or frontal lobe involvement has been implicated (52). Psychosis is relatively rare in MS, but as in the case of mania, it can be an initial manifestation (56). The presence of psychotic symptoms has been linked with bilateral temporal lobe plaques on MRI scans (56). Psychosis is rare in HIV infection (87), but it can be seen in 5% to 15% of patients with SLE (99). Paranoid psychosis has been noted as a persistent problem in some individuals with toluene abuse (193). Psychosis may develop in cobalamin deficiency and pernicious anemia, and folate deficiency may actually be more closely associated with psychosis than cobalamin deficiency (118). Late-onset psychosis has been noted in association with vascular white matter disease. In BD, psychosis may appear as an early clinical feature, preceding the onset of dementia (194). In TBI, there appears to be only a slightly increased risk of psychosis as a late sequela (144). Cerebral tumors have often been associated with psychosis, mostly related to frontal and temporolimbic sites of origin (189). Finally, psychosis, often with a paranoid flavor, can be seen in patients with NPH (190). Shunt placement reportedly effects complete or substantial recovery in these individuals (190).

Personality Change

A change in personality can occur with white matter disorders, but the symptoms and signs often are vague and nonspecific. Informants may only yield the description that a person is not his/her "normal self." Apathy, irritability, lassitude, inattention, drowsiness, memory lapses, and emotional lability all may contribute to this picture.

Personality changes with more recognizable implications, however, may be seen in white matter disorders. Emotional incontinence, also called pathologic laughter and weeping or pseudobulbar affect, is recognized as a component of the syndrome of pseudobulbar palsy. In this disorder of emotional control that is seen often in MS and advanced BD, patients are plagued by uncontrollable laughter or weeping out of proportion to the actual degree of happiness or sadness. Classic teaching holds that emotional incontinence is due to bilateral damage to the corticobulbar tracts, lesions of which presumably release brainstem motor centers from frontal cortical control. In MS, emotional incontinence is generally associated with more severe cognitive impairment, which implies that affected patients have greater white matter involvement (56), but the exact locations of responsible lesions are unknown.

Euphoria, which is encountered mainly in patients with advanced MS, is a state of remarkable unconcern and even elation in the setting of severe physical disability. As such, this syndrome has much in common with the lack of insight that typifies patients with frontal lobe damage from any etiology (22). Structural neuroimaging studies have consistently suggested an association with cerebral white matter disease. Correlations between euphoria and frontotemporal white matter lesion burden on MRI have been reported (195).

Fatigue

In MS, fatigue is a common and frequently disabling symptom. Little is known about the origin of this complaint, which may occur independently of disease relapses. Three causes of fatigue in MS have been proposed: depression, motor dysfunction, and cognitive impairment. Krupp et al. (196) found no relationship between fatigue and depression, but Feinstein (56) expressed reservations about whether fatigue could exist apart from psychiatric symptoms. Motor dysfunction has been suggested by MRI studies documenting a correlation between fatigue and plaque burden in descending motor tracts (197). Some support for "cognitive fatigue" has been provided by MS patients who demonstrate a decline in performance after a continuous effortful cognitive task (86).

CONCLUSION

The study of white matter expands the clinical spectrum of brain–behavior relationships. The neurologic diseases and injuries discussed in this chapter convincingly demonstrate that white matter disorders are regularly associated with specific neurobehavioral syndromes. Cognitive impairment is well documented. It may be diffuse, resulting in syndromes ranging from mild cognitive dysfunction to severe dementia, or focal, causing a variety of neurobehavioral syndromes that may closely resemble classic syndromes traditionally ascribed to cortical lesions. Emotional dysfunction also occurs and assumes the form of many diverse neuropsychiatric syndromes. Just as cortical diseases produce recurrent clinical patterns, so do the many white matter disorders. A sufficient clinical database exists to justify a behavioral neurology of white matter (17).

In clinical practice, the diagnosis of white matter disorders is a major concern, as clinical features may be most elusive early in the course of the disorder when treatment may be most efficacious. Awareness of typical clinical manifestations and use of MRI can improve the detection of both disorders of the white matter and their neurobehavioral syndromes.

Despite the frequent difficulty with diagnosis, these disorders offer the refreshing prospect of a relatively good prognosis in many cases. Although the outcome of many patients remains poor, white matter disorders frequently have a less pervasive impact than gray matter disorders, including AD. Because neuronal elements other than myelin may be completely spared in white matter disorders, spontaneous recovery may be complete or nearly so if the neuropathologic process is identified and corrected early enough in the course. Even if myelin is damaged, axons are frequently preserved, and remyelination or other functional compensation can occur in many disorders. As for treatment, again the prospects appear to be favorable. Prevention holds considerable promise because many white matter disorders are acquired through lifestyle, injury, or iatrogenesis. This approach may avert any significant white matter involvement, and recognition of the potential injury will assist in primary care and counseling. Even after early involvement has taken place, intervention to prevent further damage often may be successful, particularly if axonal loss has not occurred. When the syndrome is fully established, a wide range of treatment options have been attempted, and the therapeutic armamentarium for these patients undoubtedly will expand.

In theoretical terms, the study of brain white matter implicates the notion of distributed neural networks. This concept postulates that a multitude of neural assemblies exist in the brain that are widely distributed anatomically, yet are structurally interconnected and functionally integrated to subserve specific neurobehavioral domains (8). By invoking the idea of multifocal but dedicated networks, this theory stands as a modern resolution of the old dispute in behavioral neurology between localizationists and equipotential theorists. In this time-worn controversy, those who believe higher functions are strictly localized in specific brain regions differ from those who contend that no such specificity exists. Despite much debate, evidence from the last several decades has suggested that a middle ground can be reasonably held; that is, whereas many higher functions are represented to a variable extent in specific brain regions, these functions also depend on other areas that support the cognitive or emotional operation. The white matter clearly participates in all of these networks.

Disconnection of cerebral regions offers the most plausible organizing principle for the behavioral neurology of white matter. The idea of cerebral disconnection has an honored position in behavioral neurology (2), but details of the pathogenesis of this phenomenon are complex and poorly understood. An improved understanding of disconnection can be achieved by a more directed emphasis on white matter dysfunction and on the interference of the normal operations of distributed neural networks by white matter lesions. In most white matter disorders, it is likely that a combination of lesions and lesion effects occurs that is unique to each patient. Unlike the discrete and localized lesions that can be produced under experimental conditions, nature's experiments are not so easily interpretable. This reality introduces an imposing degree of complexity in the interpretation of clinical data, but sufficient evidence exists to justify a major focus on the white matter disorders and the many syndromes they may produce. As the study of brain and mind continues to flourish in the new century (198), the white matter of the brain cannot be neglected.

REFERENCES

1. Charcot JM. *Lectures on the diseases of the nervous system delivered at La Salpetriere.* London: New Sydenham Society, 1877.
2. Geschwind N. Disconnexion syndromes in animals and man. *Brain* 1965;88:237–294, 585–644.
3. Gazzaniga MS. Cerebral specialization and interhemispheric communication: does the corpus callosum enable the human condition? *Brain* 2000;123:1293–1326.
4. Mandell AM, Albert ML. History of subcortical dementia. In: Cummings JL, ed. *Subcortical dementia.* New York: Oxford University Press, 1990:17–30.
5. Albert ML, Feldman RG, Willis AL. The "subcortical dementia" of progressive supranuclear palsy. *J Neurol Neurosurg Psychiatry* 1974;37:121–130.
6. McHugh PR, Folstein MF. Psychiatric syndromes of Huntington's chorea: a clinical and phenomenologic study. In: Benson DF, Blumer D, eds. *Psychiatric aspects of neurologic disease, volume 1.* New York: Grune and Stratton, 1975:267–285.
7. Albert ML. Subcortical dementia. In: Katzman R, Terry RD, Bick KL, eds. *Alzheimer's disease: senile dementia and related disorders.* New York: Raven Press, 1978:173–180.
8. Mesulam M-M. Behavioral neuroanatomy. Large-scale neural networks, association cortex, frontal systems, the limbic system, and hemispheric specializations. In: Mesulam M-M, ed. *Principles of behavioral and cognitive neurology,* 2nd ed. New York: Oxford University Press, 2000:1–120.
9. Cummings JL, Benson DF. Subcortical dementia. Review of an emerging concept. *Arch Neurol* 1984;41:874–879.
10. Whitehouse PJ. The concept of subcortical and cortical dementia: another look. *Ann Neurol* 1986;19:1–6.
11. Rao SM. Neuropsychology of multiple sclerosis. A critical review. *J Clin Exp Neuropsychol* 1986;8:503–542.
12. Babikian V, Ropper AH. Binswanger's disease: a review. *Stroke* 1987;18:2–12.
13. Navia BA, Cho E-S, Petito CK, et al. The AIDS dementia complex: II. Neuropathology. *Ann Neurol* 1986;19:525–535.
14. Cummings JL, ed. *Subcortical dementia.* New York: Oxford University Press, 1990.
15. Filley CM, Franklin GM, Heaton RK, et al. White matter dementia: clinical disorders and implications. *Neuropsychiatry Neuropsychol Behav Neurol* 1988;1:239–254.
16. Rao SM. White matter disease and dementia. *Brain Cogn* 1996;31:250–268.
17. Filley CM. The behavioral neurology of cerebral white matter. *Neurology* 1998;50:1535–1540.
18. Miller AKH, Alston RL, Corsellis JAN. Variation with age in the volumes of grey and white matter in the cerebral hemispheres of man: measurements with an image analyser. *Neuropathol Appl Neurobiol* 1980;6:119–132.
19. Nolte J. *The human brain,* 4th ed. An introduction to its functional anatomy. St. Louis, MO: Mosby, 1999.
20. Kandel ER, Schwartz JH, Jessell TM, eds. *Principles of neural science,* 4th ed. New York: McGraw-Hill, 2000.
21. McLaurin J, Yong VW. Oligodendrocytes and myelin. *Neurol Clin* 1995;13:23–49.
22. Filley CM. Neurobehavioral anatomy, 2nd ed. Boulder, CO: University Press of Colorado, 2001.
23. Selden NR, Gitelman DR, Salamon-Murayama N, et al. Trajectories of cholinergic pathways within the cerebral hemispheres of the human brain. *Brain* 1998;121:2249–2257.
24. Damasio H, Damasio AR. The anatomical basis of conduction aphasia. *Brain* 1980;103:337–350.
25. Gur RC, Packer IK, Hungerbuhler JP, et al. Differences in the distribution of gray and white matter in the human cerebral hemispheres. *Science* 1980;207:1226–1228.
26. Chiappa KH. Pattern shift visual, brainstem auditory, and short-latency somatosensory evoked potentials in multiple sclerosis. *Neurology* 1980;30:110–123.
27. Knight RT. Electrophysiological methods in behavioral neurology and neuropsychology. In: Feinberg TE, Farah MJ, eds. *Behavioral neurology and neuropsychology.* New York: McGraw-Hill, 1997:101–119.
28. Ito J. Somatosensory event-related potentials (ERPs) in patients with different types of dementia. *J Neurol Sci* 1994;121:139–146.
29. Newton MR, Barrett G, Callanan MM, et al. Cognitive event-related potentials in multiple sclerosis. *Brain* 1989;112:1637–1660.
30. Van Dijk JG, Jennekens-Schinkel A, Caekebeke JFV, et al. Are event-related potentials in multiple sclerosis indicative of cognitive impairment? Evoked and event-related potentials, psychometric testing and response speed: a controlled study. *J Neurol Sci* 1992;109:18–24.
31. Freedman R, Adler LE, Myles-Worsley M, et al. Inhibitory gating of an evoked response to repeated auditory stimuli in schizophrenic and normal subjects. Human recordings, computer simulation, and an animal model. *Arch Gen Psychiatry* 1996;53:1114–1121.
32. Arciniegas D, Adler L, Topkoff J, et al. Attention and memory dysfunction after traumatic brain injury: cholinergic mechanisms, sensory gating, and a hypothesis for further investigation. *Brain Injury* 1999;13:1–13.
33. Yakovlev PI, Lecours AR. The myelogenetic cycles of regional maturation of the brain. In: Minkowski A, ed. *Regional development of the brain in early life.* Oxford: Blackwell Scientific Publications, 1967:3–79.
34. Benes FM, Turtle M, Khan Y, et al. Myelination of a key relay zone in the hippocampal formation occurs in the human brain during childhood, adolescence, and adulthood. *Arch Gen Psychiatry* 1994;51:477–484.
35. Byrd SE, Darling CF, Wilczynski MA. White matter of the brain: maturation and myelination on magnetic resonance in infants and children. *Neuroimaging Clin N Am* 1993;3:247–266.
36. Van der Knaap MS, Valk J, Bakker CJ, et al. Myelination as an expression of the functional maturity of the brain. *Dev Med Child Neurol* 1991;33:849–857.
37. Fletcher JM, Bohan TP, Brandt ME, et al. Cerebral white matter and cognition in hydrocephalic children. *Arch Neurol* 1992;49:818–824.
38. Morrison JH, Hof PR. Life and death of neurons in the aging brain. *Science* 1997;278:412–419.
39. Creasey H, Rapoport SI. The aging human brain. *Ann Neurol* 1985;17:2–10.
40. Salat DH, Kaye JA, Janowsky JS. Prefrontal gray and white

matter volumes in healthy aging and Alzheimer disease. *Arch Neurol* 1999;56:338–344.

41. Symonds LL, Archibald SL, Grant I, et al. Does an increase in sulcal or ventricular fluid predict where brain tissue is lost? *J Neuroimaging* 1999;9:201–209.

42. Double KL, Halliday GM, Kril JJ, et al. Topography of brain atrophy during normal aging and Alzheimer's disease. *Neurobiol Aging* 1996;17:513–521.

43. Malone MJ, Szoke MC. Neurochemical changes in white matter. Aged human brain and Alzheimer's disease. *Arch Neurol* 1985;42:1063–1066.

44. Weber GF. The pathophysiology of reactive oxygen intermediates in the central nervous system. *Med Hypotheses* 1994;43:223–230.

45. Gunning-Dixon FM, Raz N. The cognitive correlates of white matter abnormalities in normal aging: a quantitative review. *Neuropsychology* 2000;14:224–232.

46. De Groot JC, de Leeuw F-E, Oudkerk M, et al. Cerebral white matter lesions and cognitive function: the Rotterdam scan study. *Ann Neurol* 2000;47:145–151.

47. West RL. An application of prefrontal cortex function theory to cognitive aging. *Psychol Bull* 1996;120:272–292.

48. Greenwood PM. The frontal aging hypothesis revisited. *J Int Neuropsychol Soc* 2000;6:705–726.

49. Weintraub S. Neuropsychological assessment of mental state. In: Mesulam M-M, ed. *Principles of behavioral and cognitive neurology,* 2nd ed. New York: Oxford University Press, 2000:121–173.

50. Garde E, Mortensen EL, Krabbe K, et al. Relation between age-related decline in intelligence and cerebral white-matter hyperintensities in healthy octogenarians: a longitudinal study. *Lancet* 2000;356:628–634.

51. Austin J, Armstrong D, Fouch S, et al. Metachromatic leukodystrophy (MLD). VIII. MLD in adults: diagnosis and pathogenesis. *Arch Neurol* 1968;18:225–240.

52. Filley CM, Gross KF. Psychosis with cerebral white matter disease. *Neuropsychiatry Neuropsychol Behav Neurol* 1992;5:119–125.

53. Shapiro EG, Lockman LA, Knopman D, et al. Characteristics of the dementia in late-onset metachromatic leukodystrophy. *Neurology* 1994;44:662–665.

54. Hyde TM, Ziegler JC, Weinberger DR. Psychiatric disturbances in metachromatic leukodystrophy. Insights into the neurobiology of psychosis. *Arch Neurol* 1992;49:401–406.

55. Krivit W, Peters C, Shapiro EG. Bone marrow transplantation as effective treatment of central nervous system disease in globoid cell leukodystrophy, metachromatic leukodystrophy, adrenoleukodystrophy, mannosidosis, fucosidosis, aspartylglucosaminuria, Hurler, Maroteaux-Lamy, and Sly syndromes, and Gaucher disease type III. *Curr Opin Neurol* 1999;12:167–176.

56. Feinstein A. *The clinical neuropsychiatry of multiple sclerosis.* Cambridge: Cambridge University Press, 1999.

57. Kurtzke JF. Rating neurologic impairment in multiple sclerosis: an expanded disability scale. *Neurology* 1983;33:1444–1452.

58. Franklin GM, Nelson LM, Filley CM, et al. Cognitive loss in multiple sclerosis. Case reports and review of the literature. *Arch Neurol* 1989;46:162–167.

59. Filley CM, Heaton RK, Nelson LM, et al. A comparison of dementia in Alzheimer's Disease and multiple sclerosis. *Arch Neurol* 1989;46:157–161.

60. Folstein MF, Folstein SE, McHugh PR. "Mini-Mental State": a practical method of grading the cognitive state of patients for the clinician. *J Psychiatr Res* 1976;12:189–198.

61. Franklin GM, Heaton RK, Nelson LM, et al. Correlation of neuropsychological and MRI findings in chronic/progressive multiple sclerosis. *Neurology* 1988;38:1826–1829.

62. Brownell B, Hughes JT. The distribution of plaques in the cerebrum in multiple sclerosis. *J Neurol Neurosurg Psychiatry* 1962;25:315–320.

63. Simon JH. From enhancing lesions to brain atrophy in relapsing MS. *J Neuroimmunol* 1999;98:7–15.

64. Swirsky-Sacchetti T, Mitchell DR, Seward J, et al. Neuropsychological and structural brain lesions in multiple sclerosis: a regional analysis. *Neurology* 1992;42:1291–1295.

65. Feinstein A, Ron MA, Thompson A. A serial study of psychometric and magnetic resonance imaging changes in multiple sclerosis. *Brain* 1993;116:569–602.

66. Arnett PA, Rao SM, Bernardin L, et al. Relationship between frontal lobe lesions and Wisconsin Card Sorting Test performance in patients with multiple sclerosis. *Neurology* 1994;44:420–425.

67. Hohol MJ, Guttmann CRG, Orav J, et al. Serial neuropsychological assessment and magnetic resonance imaging analysis in multiple sclerosis. *Arch Neurol* 1997;54:1018–1025.

68. Catalaa I, Fulton JC, Zhang X, et al. MR imaging quantitation of gray matter involvement in multiple sclerosis and its correlation with disability measures and neurocognitive testing. *AJNR Am J Neuroradiol* 1999;20:1613–1618.

69. Fillippi M, Tortorella C, Rovaris M, et al. Changes in the normal appearing brain tissue and cognitive impairment in multiple sclerosis. *J Neurol Neurosurg Psychiatry* 2000;68:157–161.

70. Sarchielli P, Presciutti O, Pelliccioli G, et al. Absolute quantification of brain metabolites by proton magnetic spectroscopy in normal-appearing white matter of multiple sclerosis patients. *Brain* 1999;122:513–521.

71. Evangelou N, Esiri MM, Smith S, et al. Quantitative pathological evidence for axonal loss in normal appearing white matter in multiple sclerosis. *Ann Neurol* 2000;47:391–395.

72. Rao SM, Leo GJ, Aubin-Faubert P. On the nature of memory disturbance in multiple sclerosis. *J Clin Exp Neuropsychol* 1989;11:699–712.

73. Rao SM, Grafman J, DiGiulio D, et al. Memory dysfunction in multiple sclerosis: its relation to working memory, semantic encoding, and implicit learning. *Neuropsychology* 1993;7:364–374.

74. Rudick RA, Cohen JA, Weinstock-Guttman B, et al. Management of multiple sclerosis. *N Engl J Med* 1997;337:1604–1611.

75. Pliskin NH, Hamer DP, Goldstein DS, et al. Improved delayed visual reproduction test performance in multiple sclerosis patients receiving interferon β-1-b. *Neurology* 1996;47:1463–1468.

76. Pozzilli C, Passfiume D, Bernardi S, et al. SPECT, MRI and cognitive functions in multiple sclerosis. *J Neurol Neurosurg Psychiatry* 1991;54:110–115.

77. Achiron A, Ziv I, Djaldetti R, et al. Aphasia in multiple sclerosis: clinical and radiologic correlations. *Neurology* 1992;42:2195–2197.

78. Devere TR, Trotter JL, Cross AH. Acute aphasia in multiple sclerosis. *Arch Neurol* 2000;57:1207–1209.

79. Arnett PA, Rao SM, Hussain M, et al. Conduction aphasia in multiple sclerosis: a case report with MRI findings. *Neurology* 1996;47:576–579.

80. Friedman JH, Brem H, Mayeux R. Global aphasia in multiple sclerosis. *Ann Neurol* 1983;222–223.

81. Dogulu CF, Kansu T, Karabudak R. Alexia without agraphia in multiple sclerosis. *J Neurol Neurosurg Psychiatry* 1996;61:528.

82. Day TJ, Fisher AG, Mastaglia FL. Alexia with agraphia in multiple sclerosis. *J Neurol Sci* 1987;78:343–348.

83. Graff-Radford NR, Rizzo M. Neglect in a patient with multiple sclerosis. *Eur Neurol* 1987;26:100–103.

84. Okuda B, Tanaka H, Tachibana H, et al. Visual form agnosia in multiple sclerosis. *Acta Neurol Scand* 1996;94:38–44.

85. Schnider A, Benson DF, Rosner LJ. Callosal disconnection in multiple sclerosis. *Neurology* 1993;43:1243–1245.

86. Krupp L, Elkins LE. Fatigue and declines in cognitive functioning in multiple sclerosis. *Neurology* 2000;55:934–939.

87. Navia BA, Jordan BD, Price RW. The AIDS dementia complex: I. Clinical features. *Ann Neurol* 1986;19:517–524.

88. McArthur JC, Sacktor N, Selnes O. Human immunodeficiency virus-associated dementia. *Semin Neurol* 1999;19:129–150.

89. Sharer L. Pathology of HIV-1 infection of the nervous system. A review. *J Neuropathol Exp Neurol* 1992;51:3–11.

90. Seilhean D, Duyckaerts C, Vazeux R, et al. HIV-1-associated cognitive/motor complex: absence of neuronal loss in the cerebral neocortex. *Neurology* 1993;43:1492–1499.

91. Everall IP, Glass JD, McArthur J, et al. Neuronal density in the superior frontal and temporal gyri does not correlate with the degree of human immunodeficiency virus-associated dementia. *Acta Neuropathol* 1994;88:538–544.

92. White DA, Taylor MJ, Butters N, et al. Memory for verbal information in individuals with HIV-associated dementia complex. HNRC Group. *J Clin Exp Neuropsychol* 1997;19:357–366.

93. Jones RD, Tranel D. Preservation of procedural memory in HIV-positive patients with subcortical dementia. *J Clin Exp Neuropsychol* 1991;13:74.

94. Bencherif B, Rottenberg DA. Neuroimaging of the AIDS dementia complex. *AIDS* 1998;12:233–244.

95. Clifford DB. Human immunodeficiency virus-associated dementia. *Arch Neurol* 2000;57:321–324.

96. Sidtis JJ, Gatsonis C, Price RW, et al. Zidovudine treatment of the AIDS dementia complex: results of a placebo-controlled trial. *Ann Neurol* 1993;33:343–349.

97. Tozzi V, Narciso P, Galgani S, et al. Effects of zidovudine in 30 patients with mild to endstage AIDS dementia complex. *AIDS* 1993;7:683–692.

98. Thurnher MM, Schindler EG, Thurnher SA, et al. Highly active antiretroviral therapy for patients with AIDS dementia complex: effect on MR findings and clinical course. *AJNR Am J Neuroradiol* 2000;21:670–678.

99. West SG. Neuropsychiatric lupus. *Rheum Dis Clin North Am* 1994;20:129–158.

100. Johnson RT, Richardson EP. The neurological manifestations of systemic lupus erythematosus. *Medicine* 1968;47:337–369.

101. Jacobs L, Kinkel PR, Costello PB, et al. Central nervous system lupus erythematosus: the value of magnetic resonance imaging. *J Rheumatol* 1988;15:601–606.

102. Denburg SD, Carbotte RM, Denburg JA. Cognition and mood in systemic lupus erythematosus. Evaluation and pathogenesis. *Ann N Y Acad Sci* 1997;823:44–59.

103. Kozora E, West SG, Kotzin B, et al. Magnetic resonance imaging abnormalities and cognitive deficits in systemic lupus erythematosus patients without overt central nervous system disease. *Arthritis Rheum* 1998;41:41–47.

104. Brooks WM, Jung RE, Ford CC, et al. Relationship between neurometabolite derangement and neurocognitive dysfunction in systemic lupus erythematosus. *J Rheumatol* 1999;26:81–85.

105. Filley CM, Kleinschmidt-DeMasters BK. Toxic leukoencephalopathy. *N Engl J Med* 2001;345:425–432.

106. Hormes JT, Filley CM, Rosenberg NL. Neurologic sequelae of chronic solvent vapor abuse. *Neurology* 1986;36:698–702.

107. Rosenberg NL, Spitz MC, Filley CM, et al. Central nervous system effects of chronic toluene abuse—clinical, brainstem evoked response and magnetic resonance imaging studies. *Neurotoxicol Teratol* 1988;10:489–495.

108. Filley CM, Rosenberg NL, Heaton RK. White matter dementia in chronic toluene abuse. *Neurology* 1990;40:532–534.

109. Rosenberg NL, Kleinschmidt-DeMasters BK, Davis KA, et al. Toluene abuse causes diffuse central nervous system white matter changes. *Ann Neurol* 1988;23:611–614.

110. Kornfeld M, Moser AB, Moser HW, et al. Solvent vapor abuse leukoencephalopathy. Comparison to adrenoleukodystrophy. *J Neuropathol Exp Neurol* 1994;53:389–398.

111. Arlien-Soborg P, Bruhn P, Glydensted C, et al. Chronic painters' syndrome. Chronic toxic encephalopathy in house painters. *Acta Neurol Scand* 1979;60:149–156.

112. Baker EL. A review of recent research on health effects of human occupational exposure to organic solvents. *J Occup Med* 1994;36:1079–1092.

113. Rosenberg NL. Neurotoxicity of organic solvents. In: Rosenberg NL, ed. *Occupational and environmental neurology.* Stoneham, MA: Butterworth-Heinemann, 1995:71–113.

114. Thuomas K-A, Moller C, Odkvist LM, et al. MR imaging in solvent-induced chronic toxic encephalopathy. *Acta Radiol* 1966;37:177–179.

115. Goebels N, Soyka M. Dementia associated with vitamin B12 deficiency: presentation of two cases and review of the literature. *J Neuropsychiatry Clin Neurosci* 2000;12:389–394.

116. Lindenbaum J, Healton EB, Savage DG, et al. Neuropsychiatric disorders caused by cobalamin deficiency in the absence of anemia or macrocytosis. *N Engl J Med* 1988;318:1720–1728.

117. Pennypacker LC, Allen RH, Kelly JP, et al. High prevalence of cobalamin deficiency in elderly outpatients. *J Am Geriatr Soc* 1992;40:1197–1204.

118. Hutto BR. Folate and cobalamin in psychiatric illness. *Comp Psychiatry* 1997;38:305–314.

119. Penninx BWJH, Guralnik JM, Ferrucci L, et al. Vitamin B12 deficiency and depression in physically disabled older women: epidemiologic evidence from the Women's Health and Aging Study. *Am J Psychiatry* 2000;157:715–721.

120. Larner AJ, Janssen JC, Cipolotti L, et al. Cognitive profile in dementia associated with B12 deficiency due to pernicious anaemia. *J Neurol* 1999;246:317–319.

121. Adams RD, Kubik CS. Subacute degeneration of the brain in pernicious anemia. *N Engl J Med* 1944;231:1–9.

122. Shevell MI, Rosenblatt DS. The neurology of cobalamin. *Can J Neurol Sci* 1992;19:472–486.

123. Chatterjee A, Yapundich R, Palmer CA, et al. Leukoencephalopathy associated with cobalamin deficiency. *Neurology* 1996;46:832–834.

124. Stojsavljevic N, Levic Z, Drulovic J, et al. A 44-month clinical-brain MRI follow-up in a patient with B12 deficiency. *Neurology* 1997;49:878–881.

125. Santamaria Ortiz J, Knight PV. Review. Binswanger's disease, leukoaraiosis and dementia. *Age Ageing* 1994;23:75–81.

126. Caplan LR. Binswanger's disease—revisited. *Neurology* 1995;45:626–633.

127. Filley CM, Thompson LL, Sze C-I, et al. White matter dementia in CADASIL. *J Neurol Sci* 1999;163:163–167.

128. Bennett DA, Gilley DW, Lee S, et al. White matter changes: neurobehavioral manifestations of Binswanger's disease and clinical correlates in Alzheimer's disease. *Dementia* 1994;5:148–152.

129. Roman GC. From UBOs to Binswanger's disease. Impact of magnetic resonance imaging on vascular dementia research. *Stroke* 1996;27:1269–1273.

130. Hachinski VC, Potter P, Merskey H. Leuko-araiosis. *Arch Neurol* 1987;44:21–23.

131. Pantoni L, Garcia JH. Pathogenesis of leukoaraiosis. A review. *Stroke* 1997;28:652–659.
132. Boone KB, Miller BL, Lesser IM, et al. Neuropsychological correlates of white-matter lesions in healthy elderly subjects. A threshold effect. *Arch Neurol* 1992;49:549–554.
133. Kinkel WR, Jacobs L, Polachini I, et al. Subcortical arteriosclerotic encephalopathy (Binswanger's disease). *Arch Neurol* 1985;42:951–959.
134. Van Zandvoort MJE, Kapelle LJ, Algra A, et al. Decreased capacity for mental effort after single supratentorial lacunar infarct may affect performance in everyday life. *J Neurol Neurosurg Psychiatry* 1998;65:697–702.
135. Liu CK, Miller BL, Cummings JL, et al. A quantitative MRI study of vascular dementia. *Neurology* 1992;42:138–143.
136. Adams JH, Graham DI, Murray, LS, et al. Diffuse axonal injury due to nonmissile head injury: an analysis of 45 cases. *Ann Neurol* 1982;12:557–563.
137. Alexander MP. Mild traumatic brain injury: pathophysiology, natural history, and clinical management. *Neurology* 1995;45:1252–1260.
138. Gennarelli TA, Thibault LE, Adams JH, et al. Diffuse axonal injury and traumatic coma in the primate. *Ann Neurol* 1982;12:564–574.
139. Gentry LR, Godersky JC, Thompson B. MR imaging of head trauma: review of the distribution and radiopathologic features of traumatic lesions. *AJR Am J Roentgenol* 1988;150:663–672.
140. Levin HS, Williams DH, Eisenberg HM, et al. Serial MRI and neurobehavioral findings after mild to moderate closed head injury. *J Neurol Neurosurg Psychiatry* 1992;55:255–262.
141. Smith DH, Meaney DF, Lenkinski RE, et al. New magnetic resonance imaging techniques for the evaluation of traumatic brain injury. *J Neurotrauma* 1995;12:573–577.
142. Ewert L, Levin HS, Watson MG, et al. Procedural memory during posttraumatic amnesia in survivors of severe closed head injury. Implications for rehabilitation. *Arch Neurol* 1989;46:911–916.
143. Timmerman ME, Brouwer WH. Slow information processing after very severe closed head injury: impaired access to declarative knowledge and intact application and acquisition of procedural knowledge. *Neuropsychologia* 1999;37:467–478.
144. Van Reekum R, Cohen T, Wong J. Can traumatic brain injury cause psychiatric disorders? *J Neuropsychiatry Clin Neurosci* 2000;12:316–327.
145. Artigas J, Cervos-Navarro J, Iglesia JR, et al. Gliomatosis cerebri: clinical and histological findings. *Clin Neuropathol* 1985;4:135–148.
146. Couch JR, Weiss SA. Gliomatosis cerebri. Report of four cases and review of the literature. *Neurology* 1974;24:504–511.
147. Keene DL, Jimenez C, Hsu E. MRI diagnosis of gliomatosis cerebri. *Pediatr Neurol* 1999;20:148–151.
148. Adams RD, Fisher CM, Hakim S, et al. Symptomatic occult hydrocephalus with "normal" cerebrospinal-fluid pressure. *N Engl J Med* 1965;273:117–126.
149. Jack CR, Mokri B, Laws ER, et al. MR findings in normal-pressure hydrocephalus: significance and comparison with other forms of dementia. *J Comp Assist Tomogr* 1987;11:923–931.
150. Fisher CM. Communicating hydrocephalus. *Lancet* 1978;1:37.
151. Earnest MP, Fahn S, Karp JH, et al. Normal pressure hydrocephalus and hypertensive cerebrovascular disease. *Arch Neurol* 1974;31:262–266.
152. Del Bigio MR. Neuropathological changes caused by hydrocephalus. *Acta Neuropathol* 1993;85:573–585.
153. Bech RA, Juhler M, Waldemar G, et al. Frontal brain and leptomeningeal biopsy specimens correlated with cerebrospinal fluid outflow resistance and B-wave activity in patients suspected of normal-pressure hydrocephalus. *Neurosurgery* 1997;40:497–502.
154. Derix MMA. *Neuropsychological differentiation of dementia syndromes.* Lisse, The Netherlands: Swets and Zeitlinger, 1994.
155. Iddon JL, Pickard JD, Cross JJL, et al. Specific patterns of cognitive impairment in patients with idiopathic normal pressure hydrocephalus and Alzheimer's disease: a pilot study. *J Neurol Neurosurg Psychiatry* 1999;67:723–732.
156. Vanneste J, Augustijn P, Dirven C, et al. Shunting normal-pressure hydrocephalus: do the benefits outweigh the risks? A multicenter study and literature review. *Neurology* 1992;42:54–59.
157. Graff-Radford NR. Normal pressure hydrocephalus. *Neurologist* 1999;5:194–204.
158. Cummings JL, Benson DF. *Dementia: a clinical approach.* Boston: Butterworths, 1994.
159. American Psychiatric Association. *Diagnostic and statistical manual of mental disorders, fourth edition.* Washington, DC: American Psychiatric Association Press, 1994:134–135.
160. Yamanouchi N, Okada S, Kodama K, et al. Effects of MRI abnormalities in WAIS-R performance in solvent abusers. *Acta Neurol Scand* 1997;96:34–39.
161. Caine ED, Bamford KA, Schiffer RB, et al. A controlled neuropsychological comparison of Huntington's disease and multiple sclerosis. *Arch Neurol* 1986;43:249–254.
162. Markowitsch HJ. Which brain regions are critically involved in the retrieval of old episodic memory? *Brain Res Rev* 1995;21:117–127.
163. Rao SM, Leo GJ, Bernardin L, et al. Cognitive dysfunction in multiple sclerosis. I. Frequency, patterns, and prediction. *Neurology* 1991;41:685–691.
164. Tucker DM, Roeltgen DP, Tully R, et al. Memory dysfunction following unilateral transection of the fornix: a hippocampal disconnection syndrome. *Cortex* 1988:;24:465–472.
165. Park SA, Hahn JH, Kim JI, et al. Memory deficits after bilateral fornix infarction. *Neurology* 2000;54:1379–1382.
166. D'Esposito M, Verfaellie M, Alexander MP, et al. Amnesia following traumatic bilateral fornix transection. *Neurology* 1995;45:1546–1550.
167. Gaffan D, Gaffan EA. Amnesia in man following transection of the fornix. A review. *Brain* 1991;114:2611–2618.
168. Tatemichi TK, Desmond DW, Prohovnik I, et al. Confusion and memory loss from capsular genu infarction: a thalamocortical disconnection syndrome? *Neurology* 1992;42:1966–1979.
169. Naeser MA, Palumbo CL, Helm-Estabrooks N, et al. Severe nonfluency in aphasia. Role of the medial subcallosal fasciculus and other white matter pathways in recovery of spontaneous speech. *Brain* 1989;112:1–38.
170. Alexander MP, Naeser MA, Palumbo CL. Correlations of subcortical CT lesion sites and aphasia profiles. *Brain* 1987;110:961–991.
171. Poncet M, Habib M, Robillard A. Deep parietal lobe syndrome: conduction aphasia and other neurobehavioral disorders due to a small subcortical lesion. *J Neurol Neurosurg Psychiatry* 1987;50:709–713.
172. Pirozzolo FJ, Kerr KL, Obrzut JE, et al. Neurolinguistic analysis of the language abilities of a patient with a "double disconnection syndrome": a case of subangular alexia in the presence of mixed transcortical aphasia. *J Neurol Neurosurg Psychiatry* 1981;44:152–155.
173. Klingberg T, Hedehus M, Temple E, et al. Microstructure of temporo-parietal white matter as a basis for reading ability:

evidence from diffusion tensor magnetic resonance imaging. *Neuron* 2000;25:493–500.

174. Mayer E, Martory M-D, Pegna AJ, et al. A pure case of Gerstmann syndrome with a subangular lesion. *Brain* 1999;122:1107–1120.

175. Feinberg TE, Schindler RJ, Ochoa E, et al. Associative visual agnosia and alexia without prosopagnosia. *Cortex* 1994;30:395–411.

176. Takahashi N, Kawamura M, Shinotou H, et al. Pure word deafness due to left hemisphere damage. *Cortex* 1992;28:295–303.

177. Bogousslavsky J, Miklossy J, Regli F, et al. Subcortical neglect: neuropsychological, SPECT, and neuropathological correlates with anterior choroidal territory artery infarctions. *Ann Neurol* 1988;23:448–452.

178. Hublet C, Demeurisse G. Pure topographical disorientation due to a deep-seated lesion with cortical remote effects. *Cortex* 1992;28:123–128.

179. Nighoghossian N, Trouillas P, Vighetto A, et al. Spatial delirium following a right subcortical infarct with frontal deactivation. *J Neurol Neurosurg Psychiatry* 1992;55:334–335.

180. Ross ED, Stewart RM. Akinetic mutism from hypothalamic damage: successful treatment with dopamine agonists. *Neurology* 1981;31:1435–1439.

181. Scott TF, Lang D, Girgis RM, et al. Prolonged akinetic mutism due to multiple sclerosis. *J Neuropsychiatry Clin Neurosci* 1995;7:90–92.

182. Filley CM. Clinical neurology and executive dysfunction. *Semin Speech Lang* 2000;21:95–108.

183. Geschwind DH, Iacoboni M, Mega M, et al. Alien hand syndrome: interhemispheric motor disconnection due to a lesion of the midbody of the corpus callosum. *Neurology* 1995;45:802–808.

184. Suzuki K, Yamadori A, Endo K, et al. Dissociation of letter and picture naming resulting from callosal disconnection. *Neurology* 1998;51:1390–1394.

185. Bakshi R, Czarnecki D, Shaikh ZA, et al. Brain MRI lesions and atrophy are related to depression in multiple sclerosis. *Neuroreport* 2000;11:1153–1158.

186. Perry SW. HIV-related depression. *Res Publ Assoc Res Nerv Ment Dis* 1994;72:223–238.

187. Zur J, Yule W. Chronic solvent abuse. 2. Relationship with depression. *Child Care Heath Dev* 1990;16:21–34.

188. Soares JC, Mann JJ. The anatomy of mood disorders—review of structural neuroimaging studies. *Biol Psychiatry* 1997;41:86–106.

189. Filley CM, Kleinschmidt-DeMasters BK. Neurobehavioral presentations of brain neoplasms. *West J Med* 1995;163:19–25.

190. Rice E, Gendelmann S. Psychiatric aspects of normal pressure hydrocephalus. *JAMA* 1973;223:409–412.

191. Kieburtz K, Zettelmaier AE, Ketonen L, et al. Manic syndrome in AIDS. *Am J Psychiatry* 1991;148:1068–1070.

192. Kwentus JA, Hart RP. Normal pressure hydrocephalus presenting as mania. *J Nerv Ment Dis* 1987;175:500–502.

193. Byrne A, Kirby B, Zibin T, et al. Psychiatric and neurological effects of chronic solvent abuse. *Can J Psychiatry* 1991;36:735–738.

194. Lawrence RM, Hillam JC. Psychiatric symptomatology in early-onset Binswanger's disease: two case reports. *Behav Neurol* 1995;8:43–46.

195. Diaz-Olavarrieta C, Cummings JL, Velasquez J, et al. Neuropsychiatric manifestations of multiple sclerosis. *J Neuropsychiatry Clin Neurosci* 1999;11:51–57.

196. Krupp LB, Alvarez LA, LaRocca NG, et al. Fatigue in multiple sclerosis. *Arch Neurol* 1988;45:435–437.

197. Colombo B, Boneschi FM, Rossi P, et al. MRI and motor evoked potential findings in nondisabled multiple sclerosis patients with and without symptoms of fatigue. *J Neurol* 2000;247:506–509.

198. Kandel ER, Squire LR. Neuroscience: breaking down barriers to the study of brain and mind. *Science* 2000;290:1113–1120.

37

NEUROBEHAVIORAL COMPLICATIONS OF HUMAN IMMUNODEFICIENCY VIRUS INFECTION AND ACQUIRED IMMUNODEFICIENCY SYNDROME

JORGE LOUIS MALDONADO
FRANCISCO FERNANDEZ

The epidemic of the human immunodeficiency virus (HIV) infection and acquired immunodeficiency syndrome (AIDS) continues to grow. It has been more than 20 years since the first cases were reported, and through the years, progress has been made. Zidovudine was the first antiretroviral medication used, but resistance was found to be a significant problem. After combination therapy was introduced, the prognosis for patients improved, and, since 1996, a continuous decline in the mortality and incidence of AIDS has been achieved (1,2). From being the leading cause of death in those aged 25 to 44 in 1995, HIV-related deaths have been the fifth leading cause since 1997 (3,4). Multiple prevention programs have been successful in decreasing the rate of infection in targeted populations. The epidemic is still present, however, as new persons are infected all the time with HIV and the prevalence of persons living with AIDS continues to rise. According to the Centers for Disease Control and Prevention, the estimate of persons living with AIDS in the United States is 331,518 as of the end of June 2001, and 457,667 have died from AIDS through the years (5).

Worldwide, the epidemic is still well out of control. The underdeveloped countries suffer the most because the implementation of prevention programs is difficult and access to medications is limited (6). The infection continues to spread at a rate of 14,000 new persons infected per day worldwide, and the number of deaths continues to rise as well, leaving more than 14 million orphans because of the AIDS epidemic. The Joint United Nations Program on HIV/AIDS and the World Health Organization estimate that at the end of 2001, there were 40 million persons living with HIV, with the epidemic having already taken 20 million lives up to that moment, three million of those deaths occurring during 2001 (7).

The HIV has been identified as affecting the immunologic system primarily. The central nervous system (CNS), however, is also one of its targets, and HIV and AIDS have been associated with an array of neurologic and neuropsychiatric disorders. Some of the populations at risk have been found to have a higher prevalence of psychopathology than the average population (8). Unfortunately, chronically mentally ill patients also are at increased risk of becoming infected because they get involved in high-risk behaviors more frequently than average persons (9). Therefore, psychiatrists are involved in the care of HIV-infected patients in many ways.

The HIV induces cognitive and motor disorders by infecting the brain and inducing a cascade of biochemical reactions with a result of neuronal death. It is estimated that 90% of terminal AIDS patients will present a cognitive and/or motor disorder (10). The HIV-infected CNS is also the host of several opportunistic infections and malignancies and has an increased rate of cerebrovascular accidents. All these may present as, or be associated with, neuropsychiatric disorders.

Vaccines have been under investigation without success. Until one is found and massive immunization campaigns are undertaken, we will continue to see more and more persons suffering from HIV infection and AIDS. We need a thorough understanding of the illness and the neuropsychiatric presentations, complications, and treatment modalities for these patients.

CENTRAL NERVOUS SYSTEM PATHOLOGY

The HIV enters the brain early in the course of the infection and replicates within the CNS, perpetuating the infection and serving as an anatomic reservoir for the disease (11). The direct effect is mainly through glycoprotein gp120 of the HIV, but more significantly, indirect effects are involved in the development of neuropathology. Multiple steps

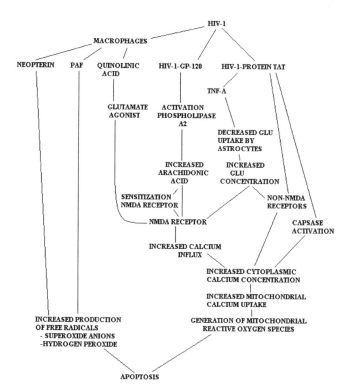

FIGURE 37.1. Neuropathogenesis of human immunodeficiency virus–associated dementia.

in a complex cascade of biochemical reactions have been identified (Fig. 37.1), and it is now understood that they may work independently as well as potentiate the neurotoxic effects of each other (12,13).

Only macrophages and microglial cells are significantly infected in the CNS (14). They are responsible for the production of neurotoxins and the HIV nonstructural regulatory proteins implicated in the neuropathogenesis of the HIV-associated dementia complex (HAD) (15). Noninfected cells are also stimulated by glycoprotein gp120 of HIV and are involved in the secretion of neurotoxic substances (16).

The final process of apoptosis is related to increased intracellular calcium (17,18), which is mediated at large by the *N*-methyl-D-aspartate (NMDA) receptor. It is sensitized by arachidonic acid to the excitatory amino acids aspartate and glutamate (19–21). Increased release of arachidonic acid is induced by phospholipase A_2, which is activated by glycoprotein gp120 of the HIV. Glutamate stimulates the NMDA receptor, and glutamate concentrations are increased as its uptake by astrocytes is diminished by tumor necrosis factor-α (TNF-α). TNF-α is activated by the HIV nonstructural regulatory protein tat (22,23), which also activates capsase, leading to an increase in intracellular calcium level as well (24). Glutamate also induces an increase in intracellular calcium by stimulating non-NMDA receptors. Another factor for increasing intracellular calcium is the presence of

quinolinic acid owing to hyperactivation of the kynurenine pathway of tryptophan metabolism in the brain, which results from enzymatic induction by cytokines released from the HIV-infected macrophage (25,26).

The last steps seem to involve dysregulation of metabolic pathways by free radicals and reactive oxygen species in the mitochondria that comes from increased mitochondrial calcium uptake (24,27) and from macrophage production of platelet-activating factor (16,28). Interferon-γ stimulates biosynthesis of neopterin, which has been useful as an immune activation marker in cerebrospinal fluid (CSF) and may also be involved in the production of oxygen free radicals (29).

Pathologic changes that are usually found in patients with HIV infection include microglial and glial changes and the presence of multinucleated giant cells, which are considered a pathognomonic finding in patients with HAD (30). White matter pallor with diffuse astrogliosis of the cerebral hemispheres, basal ganglia, and brainstem; vacuolation associated with myelin or axonal loss (31); cortical atrophy; and diminished brain weight are also possible findings, although they are less commonly observed (32–35).

The most affected areas in the brains of infected individuals are the basal ganglia and temporolimbic structures, which show a higher HIV load in the brain (36). Support cells such as astrocytes, which share similar CD4 receptors with the lymphocytes, are also significantly infected in the brain (37).

ASSESSMENT OF HUMAN IMMUNODEFICIENCY VIRUS–RELATED CENTRAL NERVOUS SYSTEM DYSFUNCTION

The assessment of the patient with HIV infection must include a detailed cognitive history (38). Some patients will have specific complaints that resemble the subcortical nature of HAD. Patients with HIV-associated minor cognitive motor disorder (HMCMD) are able to perform all aspects of work and activities of their daily life, except for the most demanding (39), and a detailed history is essential for a comprehensive evaluation. Asymptomatic patients have been found to have an elevated rate of cognitive dysfunction compared with HIV-negative individuals (40–46), and the degree of dysfunction is unrelated to the level of immunosuppression or other psychiatric disorders (44,45,47,48). Other studies have failed to show a correlation between cognitive symptoms and abnormalities in neuropsychological testing in otherwise asymptomatic HIV-infected patients (49–51). In the evaluation of the cognitive decline associated with HIV infection, neuropsychological tests show deficits in attention, concentration, learning and memory, abstract thinking and reasoning, visuospatial tasks, speech and language, and slowing of information processing (39,44,52). These become more severe as the disease progresses and may also be associated with neurologic symptoms such as incoordination of limbs, abnormal gait, hyperreflexia, hypertonia or weakness, and double incontinence.

Screening instruments such as the Mini-Mental State Examination (53) are not sensitive for HAD unless the disease is advanced and is also affecting cortical functions. Screening tests have been developed for the detection of HAD (54–56) and show reliable sensitivity and specificity for patients with dementia. They have not, however, proved to be of any value in detecting early stages of cognitive decline. Table 37.1 shows the full neuropsychological battery recommended by the National Institute of Mental Health for the evaluation of HIV-infected individuals (57). Modifications according to specific patients' limitations and complaints are more useful than the entire battery.

Electrophysiologic studies have produced contradictory results about the early electroencephalographic changes in the asymptomatic patient (58,59). In advanced dementia,

TABLE 37.1. NATIONAL INSTITUTE OF MENTAL HEALTH NEUROPSYCHOLOGICAL BATTERY

A. Indication of premorbid intelligence
  1. Vocabulary (Wechsler Adult Intelligence Scale-Revised)
  2. National Adult Reading Test
B. Attention
  1. Digit span (Wechsler Memory Scale-Revised)
  2. Visual span (Wechsler Memory Scale-Revised)
C. Speed of processing
  1. Sternberg search task
  2. Simple and choice reaction time
  3. Paced Auditory Serial Addition Test
D. Memory
  1. California Verbal Learning Test
  2. Memory working test
  3. Modified Visual Reproduction Test
E. Abstraction
  1. Category fluency test
  2. Trail-Making Test, Trails A and B
F. Language
  1. Boston Naming Test
  2. Letter and category fluency test
G. Visuospatial
  1. Embedded figures test
  2. Money's Standardized Road-map Test of Direction Sense
  3. Digit symbol substitution
H. Construction abilities
  1. Block design test
  2. Tactual performance test
I. Motor abilities
  1. Grooved pegboard
  2. Finger tapping test
  3. Grip strength
J. Psychiatric assessment
  1. Diagnostic Interview Schedule
  2. Hamilton Depression Scale
  3. State-Trait Anxiety Scale
  4. Mini-Mental State Examination

Adapted from Sacktor NC, Bacellar H, Hoover DR, et al. Psychomotor slowing in HIV infection: a predictor of dementia, AIDS, and death. *J Neurovirol* 1996; 2:404–410, with permission.

a low-amplitude pattern is commonly found (60). Evoked potentials are abnormal in physically asymptomatic patients (61), but their predictive value as indicators of cognitive disease has not been proven.

Viral Load

Plasma viral load measures have become a standard method in the diagnosis of severity of systemic HIV infection and monitoring response to treatment (62). Its usefulness in the evaluation of the presence or severity of CNS infection and the associated cognitive dysfunction has been very limited. Several studies have shown that there is no correlation between plasma and CSF viral load (63–68) because it is now understood that they have different dynamics (68). HIV replicates within the brain independently where it causes CNS pathology and from where it spreads to the CSF compartment (36,64). CSF viral load correlates with the severity of CNS infection and neurocognitive dysfunction and has a role in monitoring the effect of antiretroviral treatment in the CNS (65,66,69,70).

Other abnormalities found in the CSF include elevation of immunoglobulin levels, HIV antibodies, mononuclear cells, oligoclonal bands, β_2-microglobulin, and myelin basic protein (71–73). Their clinical value has decreased since the measurement of viral load was improved.

Neuroimaging

Structural imaging is most useful in detecting CNS opportunistic infections and malignancies when assessing the HIV-infected patient (9). Changes observed in patients with HAD are limited and usually more significant as the disease progresses (74) and include bilateral cortical atrophy with ventricular enlargement and diffuse white matter changes (75,76), and in children, calcification of the basal ganglia with delayed myelination can be found (77,78). In the context of a patient with neuropsychological symptoms, these findings add to the diagnosis but are not very sensitive and have limited clinical value in the diagnosis of cognitive dysfunction.

Increased metabolism in the basal ganglia has been observed early in HIV infection with positron emission tomography studies, even when structural imaging fails to detect any abnormalities (79,80), whereas hypometabolism of the basal ganglia and cortex is a late finding in HAD (80). Wiseman et al. (81) described abnormalities of brain activation associated with effortful retrieval and organizational process in HIV seropositive asymptomatic patients and patients with minor cognitive motor complex, but no difference from the control group was found when patients were at rest.

Many abnormalities have been found in single photon emission computed tomography studies (82); however, they have failed to consistently correlate with abnormalities in neuropsychological testing or cognitive symptoms in

HIV-infected individuals. Patients who have impairment in motor speed performance had decreased quantitative global cerebral uptake in one study (83). Extensive cortical cerebral blood flow changes with predominance for frontal and parietal lobes have been found to correlate with the severity of HAD (82). Mild changes in cortical and periventricular regions in patients with HAD correlated with improvement and/or further deterioration of the clinical status, suggesting that it may have a role in monitoring the response to treatment of the HIV-associated CNS infection (84).

Studies have shown the usefulness of magnetic resonance spectroscopy in detecting early changes in patients with HMCMD, differentiating them from HIV seropositive asymptomatic patients and those with more advanced HAD (85–91). Abnormalities in frontal white matter are an early sign in HMCMD, whereas abnormalities are also found in the basal ganglia and frontal cortex in patients with HAD (85). Magnetic resonance spectroscopy provides better information than single photon emission computed tomography in evaluating brain injury in early HMCMD (86), and it has also been suggested that MRS may have a role in monitoring response to treatment in HAD (89,90).

TREATMENT OF HUMAN IMMUNODEFICIENCY VIRUS INFECTION

It is now accepted that treatment of HIV infection should be started soon after the diagnosis because it is associated with virologic, immunologic, and clinical benefits. However, the decision to start treatment has to be individualized because it involves a large number of medications (92), and side effects usually arise. The result may imply changes in the quality of life and requires a long-term commitment, which is essential for the treatment to be effective and to minimize the emergence of resistant virus (93). Because cognitive disorders are related to the presence of HIV in the CNS, treatment of HIV infection is necessary to stop deterioration and to recover cognitive functions.

There are 16 antiretroviral drugs with proven efficacy against HIV that are approved by the U.S. Food and Drug Administration (FDA) for clinical use (Table 37.2). They work by inhibition of two different enzymes involved in separate stages of the HIV life cycle. The first two groups inhibit the reverse transcriptase (nucleoside reverse transcriptase inhibitors and nonnucleoside reverse transcriptase inhibitors) and the third group inhibits the protease [protease inhibitors (PI)].

Combination therapy with antiretroviral medications, also referred to as highly active antiretroviral therapy, was introduced as a common practice after the Delta trial showed a decrease in mortality of patients receiving two antiretrovirals compared with treatment with one only (94,95). The study showed that treatment-naïve patients (Delta 1) and pa-

TABLE 37.2. ANTIVIRAL MEDICATIONS APPROVED BY THE U.S. FOOD AND DRUG ADMINISTRATION AND AVAILABLE INFORMATION ON THEIR CSF CONCENTRATIONS

Antiviral Medication	CSF Concentration
NRTI	
Zidovudine	Plasma/CSF ratio = 0.6 (98)
Didanosine	Plasma/CSF ratio = 0.2 (98)
Lamivudine	Plasma/CSF ratio = 0.06 (98) to 0.41 (109)
Stavudine	Plasma/CSF ratio = 0.4 (98)
Zalcitabine	Plasma/CSF ratio = 0.20 (108)
Abacavir	Plasma/CSF ratio = 0.18 (112)
Tenofovir	N/A
nNRTI	
Delavirdine	Plasma/CSF ratio = 0.072 Desalkyl delvirdine plasma/CSF = 1.68 (112)
Nevirapine	Plasma/CSF (unbound to protein) ratio = 1 (110)
Efavirenz	Plasma/CSF (unbound to protein) ratio = 1 (111)
PI	
Indinavir	Plasma/CSF (unbound to protein) ratio >1 (113)
Nelfinavir	Not detected in CSF (114, 115)
Ritonavir	Undetectable or >60 times lower than in plasma (114)
Saquinavir	Not detected in CSF (116)
Amprenavir	Undetectable (117)
Lopinavir	Undetectable (118)

CSF, cerebrospinal fluid; NRTI, nucleoside reverse transcriptase inhibitors; N/A, information not available; nNRTI, nonnucleoside reverse transcriptase inhibitors; PI, protease inhibitors.

tients previously treated with zidovudine (ZDV) (Delta 2) had increased survival, regardless of the stage of the disease. Two other trials comparing monotherapy versus combination therapy showed similar results (96,97). It is currently recommended to initiate treatment with two to four antiviral medications (93).

TREATMENT OF THE CENTRAL NERVOUS SYSTEM HUMAN IMMUNODEFICIENCY VIRUS INFECTION

The CNS is both the target of HIV infection and its anatomic reservoir (11). Macrophages and macrophage-related microglial cells are the primary sites of CNS infection (98), and the virus replicates independently in the CNS, showing different viral load dynamics between plasma and CSF (99) as well as the independent emergence of virus resistant to specific drugs (100). It is important, therefore, that the antivirals achieve a good CSF level to suppress the replication of the virus and to minimize resistance. Table 37.2 summarizes the available information on the penetrance of the different drugs of the CNS. ZDV was initially used as the drug of choice for patients who had cognitive dysfunction because it showed good penetrance to the CNS with a plasma/CSF ratio of 0.6 (101). Its use was associated with improved cognitive function in patients who had minor cognitive changes and were otherwise asymptomatic. It also reduced cognitive deficits in patients who took it long term and reduced the overall risk of HAD (102–104). Didanosine was also tried in monotherapy. Progression to HAD was common in adults receiving treatment with didanosine at both low and high doses (105); however, it improved cognitive functioning in children with AIDS (106) and, when used in combination with ZDV, showed better results in preventing progression to HIV encephalopathy than ZDV alone (107). Similarly, zalcitabine in combination with ZDV was effective in reversing neurocognitive dysfunction in patients who had developed abnormalities despite treatment with ZDV alone (108). The combinations of lamivudine and ZDV and lamivudine and stavudine were equally effective in decreasing CSF HIV-RNA concentrations, and the CSF drug levels were consistent over time in patients without neurologic compromise (109). CSF levels of other medications have been determined only when used in combination therapy, but they do not necessarily correlate with clinical efficacy. Some of the medications achieve very low levels in the CSF, but they are comparable with the portion not bound to proteins in plasma (110,111) and therefore can be efficacious drugs for the treatment of HIV infection in the CNS. Delavirdine has low penetrance of CSF, but its metabolite desalkyl-delavirdine achieves a good concentration in the CSF (112). From the PI, only indinavir has been consistently found in CSF at levels similar to plasma concentration (113). Ritonavir is found in CSF only in one-third of patients taking it (114), and nelfinavir, saquinavir, amprenavir, and lopinavir are not detectable in CSF (114–118). However, nelfinavir was associated with good CSF viral load suppression (115), and saquinavir in combination with ritonavir suppressed the CSF viral load at 18 weeks, when they were used alone or in combination with stavudine (119). PI have also been found to improve neurocognitive function in patients with dementia when used in combination therapy (120).

NEUROBEHAVIORAL DISORDERS AND THEIR TREATMENTS

Cognitive Disorders

Cognitive/motor disorders are the most frequent neuropsychiatric complications in patients infected with HIV. As many as 90% of individuals in the late symptomatic phase of HIV infection will present with at least mild cognitive/motor impairment (9), and it is estimated that 25% will develop HAD (121), which is an AIDS-defining illness (122) and may be its initial presentation. The current nomenclature (39) divides the cognitive disorders associated with HIV into two categories (Table 37.3): HMCMD and HAD. Whether they are a continuum of the same pathology or separate disorders is still not known.

Patients with HMCMD may have symptoms of memory loss, difficulty with attention and concentration, and slowness of thought and movement. These symptoms may be subtle, and abnormalities may be apparent only on neuropsychological testing. Changes in mood and personality may be present. In patients with HAD, more pronounced cognitive symptoms are common, and neurologic impairment may be significant in patients with seizures, difficulty with gait and speech, and incontinence of the bladder and bowel, and on examination, ataxia, spasticity, hyperreflexia, or hypertonia may be found. As the disease progresses, the initial findings that characterize HAD as subcortical are exacerbated, and impairment of cortical functions such as aphasia and apraxia may also be present.

HIV cognitive disorders comprise a diagnosis of exclusion because the CNS may be host to opportunistic infections and malignancies that could present with similar symptomatology. Therefore, the workup should include studies to rule out other CNS pathology.

Besides treatment with antiviral medications, stimulants have been used to improve cognitive and affective symptoms in HIV-infected patients (123,124). Methylphenidate, dextroamphetamine, and pemoline are useful as adjuvant medications for patients with cognitive disorders. It is recommended to start with an initial dose of 5 to 10 mg methylphenidate or its equivalent with other drugs and adjust according to the patient's response. The extended-release formulations of the stimulants have not been formally studied in patients with HIV infection; however, it is expected that the benefit observed in terms of a lower side effect profile and prolonged efficacy seen in other disorders will also apply to this population. Medications that block the toxicity caused by HIV through the NMDA receptor have been studied. These include the calcium channel blockers nifedipine and nimodipine and the NMDA receptor antagonists memantine, nitroglycerin, 2-amino-5-phosphonopentatoic acid (AP5), and dizocilpine (125–127). Nimodipine has been studied in a clinical trial, with some improvement in neuropsychological testing being suggested in the group that received high doses (60 mg five times daily); however, it was not statistically significant (128). Memantine blocked completely the neurotoxicity caused by protein Tat and gp120 in patients with HIV encephalitis (129). AP5, memantine, and dizocilpine lowered gp120-induced increase in calcium in human fetal neurons (127). More clinical trials will be necessary before establishing its usefulness.

Delirium

Delirium is the most common neuropsychiatric disorder in hospitalized HIV-infected patients (130), with 30% to 40% prevalence in this population. In nursing homes, 14% to 46% of the AIDS patients had a diagnosis of delirium in consultation (131) and retrospectively (132). Delirium is also a predictor of outcome in hospitalized AIDS patients.

TABLE 37.3. AMERICAN ACADEMY OF NEUROLOGY AIDS TASK FORCE: NOMENCLATURE AND RESEARCH CASE DEFINITIONS FOR NEUROLOGIC MANIFESTATIONS OF HUMAN IMMUNODEFICIENCY VIRUS TYPE 1 INFECTION

A. Criteria for clinical diagnosis of HIV-1–associated minor cognitive motor disorder
  1. Cognitive/motor/behavioral abnormalities (each of the following)
    a. At least two of the following present for at least 1 month
      1. Impaired attention or concentration
      2. Mental slowing
      3. Impaired memory
      4. Slowed movements
      5. Incoordination
      6. Personality change, irritability, or emotional lability
    b. Acquired cognitive/motor abnormality verified by clinical neurologic examination or neuropsychological testing (e.g., fine motor speed, manual dexterity, perceptual motor skills, attention/concentration, speed of processing information, abstraction/reasoning, visuospatial skills, memory/learning, or speech/language)
  2. Disturbance from criterion 1 causes mild impairment of work or activities of daily living
  3. Does not meet criteria for HIV-1–associated dementia complex or HIV-1–associated myelopathy
  4. No evidence of another etiology, including active CNS opportunistic infection or malignancy, severe systemic illness, active alcohol or substance use, acute or chronic substance withdrawal, adjustment disorder, or other psychiatric disorders
  5. HIV seropositivity (ELISA confirmed by Western blot, polymerase chain reaction, or culture)
B. Criteria for clinical diagnosis of HIV-1–associated dementia complex (each of the following)
    a. Acquired abnormality in at least two of the following cognitive abilities for at least 1 month: attention/concentration, speed of processing information, abstraction/reasoning, visuospatial skills, memory and speech/language (cognitive dysfunction causing impairment of work or activities of daily living should not be attributable solely to severe systemic illness)
    b. At least one of the following
      i. Acquired abnormality in motor function or performance verified by clinical examination, neuropsychological testing, or both
      ii. Decline in motivation or emotional control or change in social behavior
    c. Absence of clouding of consciousness during a period long enough to establish the presence of criterion 1
    d. No evidence of another etiology, including active CNS opportunistic infection or malignancy, other psychiatric disorders (e.g., depression), active alcohol or substance use, or acute or chronic substance withdrawal
    e. HIV seropositivity (ELISA test confirmed by Western blot, polymerase chain reaction, or culture)

HIV-1, human immunodeficiency virus-1; CNS, central nervous system; ELISA, enzyme-linked immunosorbent assay.
Adapted from Levy JK, Fernandez F. HIV infection of the CNS: implications for neuropsychiatry. In: Yudofsky SC, Hales RE, eds. *Textbook of neuropsychiatry*. Washington, DC: American Psychiatric Press, 1999; 663–692, with permission.

Compared with AIDS patients with similar demographics and markers of medical morbidity, those who had delirium had higher mortality rates, longer hospitalizations, or need for long-term care if discharged (133). As treatments for HIV infection improve and patients are able to live longer, even when diagnosed with AIDS, it is possible that delirium in the outpatient setting will become more prevalent.

Delirium reflects a diffuse cerebral cellular metabolic dysfunction (134), and, clinically, the patients present a prodromal phase with difficulty in thinking, restlessness, irritability, insomnia, or disruption of sleep with vivid nightmares. As the delirium process continues, impairment in arousal, attention, concentration, short-term memory, and orientation becomes more evident. Delirium has a fluctuating course, with diurnal variations and usually worsening symptoms during the night. Motor abnormalities may be present and include tremor, myoclonus, asterixis, and picking at clothing or the air. Its etiology cannot always be established. In terminally ill hospitalized AIDS patients, Fernandez et al. (135) found CNS opportunistic infections and lymphomas as the cause of the delirium in 68% of the cases. Other causes included sepsis, metabolic abnormalities owing to hypoxemia, electrolyte imbalance, toxicity from analgesics, and multisystem failure. Delirium was associated with medications in 44%

of the cases in a nursing home (132), and this may become more common as more complex medical regimens are being used.

The resolution of the delirious process depends on the identification and treatment of the underlying causes. However, because this is not always possible, pharmacologic treatment is necessary. Traditional high-potency neuroleptics have been extensively used. Breitbart et al. (136) used low doses of oral and intramuscular haloperidol (2.8 ± 2.4 mg daily) and chlorpromazine (50 ± 23 mg daily) to control delirium in hospitalized AIDS patients. They also studied lorazepam as a single agent, which was not effective and poorly tolerated. Molindone has been reported as a safer alternative when oral high-potency antipsychotics are poorly tolerated by HIV-infected patients (137).

Although not approved by the FDA for intravenous use, haloperidol has been highly effective in controlling the agitated delirious patient when administered intravenously (138). Fewer treatment-emergent side effects are observed than with the oral agents, even when used at high doses (139). In their population, Fernandez et al. (139) used doses of 12 to 132 mg daily of haloperidol in combination with 4.5 to 37.5 mg lorazepam daily. Because the HIV-infected patients are more sensitive to the extrapyramidal reactions

(135,140) or neuroleptic malignant syndrome (141), careful adjustment to the minimal effective dose of the antipsychotic should be made. Combination intravenous chemotherapy with haloperidol, lorazepam, and hydromorphone (138) or buprenorphine (142) is also effective in the management of the agitated delirious patient, with lower doses of the neuroleptic required to control symptoms.

The newer atypical antipsychotics offer another alternative in the treatment of delirium in HIV-infected patients. Risperidone has been used in HIV patients with psychosis and was effective and well tolerated with no serious adverse effects associated (143). It has also been reported effective in delirious medically compromised and elderly patients (144) and in psychosis induced by medical illnesses (145). Olanzapine has also been used in the treatment of delirium in cancer patients. It showed good efficacy and safety profile in this population (146) and, when compared with haloperidol, was equally effective and better tolerated in delirious patients (147). According to one report, a patient with AIDS with poor tolerance to previous treatments developed akathisia when treated with olanzapine, confirming the sensitivity of such patients to side effects (148). Quetiapine has been studied in patients with Parkinson disease, and its use at low doses was shown to be efficacious and safe in this population (149). No studies of other medically ill patients or HIV-related delirium have been reported with the use of either oral or the now commercially available preparation of parenteral ziprasidone.

The treatment of the hypoactive delirium is still controversial; however, Platt et al. (150) found improvement in hypoactive delirious AIDS patients who received low doses of haloperidol or chlorpromazine, equivalent to those who had hyperactive delirium. Hypoactive delirium was one of the factors associated with poor response to olanzapine in the delirious cancer patients (146).

Mood Disorders: Depression

Mood disorders in HIV-infected patients are common, particularly depression. However, manic and hypomanic disorders have also been described (151,152). Substance-induced mania in patients treated with ZDV (153,154), didanosine (155), ganciclovir, or antidepressants (152) has been reported.

The presentation of manic disorders is different according to the stage of HIV infection. When mania presents early in the course of HIV infection, with CD4 counts greater than 200, patients are more likely to have a personal or family history of mood disorder. In late-onset mania [AIDS mania (156)] with CD4 less than 200, patients usually do not have history of mood disorder, manic symptoms are linked to cognitive impairment and executive dysfunction with psychomotor slowing, and they have more manic symptoms, with more pronounced irritability and less talkativeness than those who present with early-onset mania (151,156).

It was found in early studies that patients infected with HIV had a high incidence and prevalence of depressive disorders, comparable with other medically ill populations (157). It has been confirmed over the years as the HIV/AIDS epidemic continues to grow but is now understood that some HIV-negative populations at risk (i.e., homosexual men and intravenous drug abusers) also have a higher incidence and prevalence of depressive disorders, comparable with infected HIV individuals with similar risk factors (7,158). Different studies have found 3% to 8% of individuals reporting an episode of a major depressive disorder the month before the study evaluation, and the lifetime history of depression was 28% in HIV-positive patients and 45% in HIV-negative homosexual men (7,159), with most of the HIV-infected individuals having the onset of the depressive illness before the seroconversion (158,160).

Suicide risk has been found to be 16 to 24 times higher in HIV-infected individuals compared with the general population (161), with the highest risk shortly after the notification of seropositivity (161,162). As many as 14% of HIV-infected individuals have attempted suicide within the first 3 months (162). However, the number of completed suicides was not different in military service applicants who were disqualified owing to HIV seropositivity or other medical problems (163), and as the illness progresses and patients develop AIDS, their suicidality becomes comparable with other medically ill hospitalized patients (164).

Another question raised and still debated is whether progression of HIV infection is related to increased rates of depression. It has been accepted that early in the HIV infection, depressive symptoms do not increase during the first several months of the asymptomatic period (165,166). Prospective studies with long-term, semiannual follow-up of HIV-infected patients have given contradictory information. One study found stability of mood despite progression of HIV illness (167), whereas another found increased depression starting 12 to 18 months before the AIDS diagnosis, which reached a plateau 6 months before the AIDS diagnosis (166). In both studies, somatic symptoms, which may be associated with HIV, were analyzed independently. It is possible that results differ because they studied patients during different stages of HIV infection. In one study (167), not all patients progressed to an AIDS diagnosis, and the highest percentage of patients with a diagnosis of AIDS was 19% in follow-up visit number 6, whereas the other study (166) analyzed patients whose HIV infection progressed to a diagnosis of AIDS in all cases. Different methodology and interpretation of data may have also played a role because the first study (167) used the *Diagnostic and Statistical Manual for Mental Disorders, Revised Third Edition* structured clinical interview, whereas the second one (166) used the Center for Epidemiologic Studies of Depression Scale for the assessment of depression in HIV-infected patients.

The course of HIV infection in depressed patients has also been studied. Depressed patients report AIDS-related somatic symptoms more frequently than nondepressed patients with a similar immunologic and serologic status (168–170), but no difference in the progression of the immunologic compromise to an AIDS diagnosis and no difference in mortality rate has been found in most studies (170–172). Only one prospective study found a higher risk of progression to an AIDS diagnosis in a population of depressed homosexual and bisexual men (168), and another found a higher mortality rate in the same population (173). A more rapid decline of the immune system, as measured by CD4 cell count, was found in depressed versus nondepressed HIV-infected men (169); however, this difference was not sustained when this population was included in a metaanalysis with 18 other studies (171).

The diagnosis of depressive disorders in HIV-infected individuals is usually complicated because many of the symptoms of depression are also manifestations of HIV infection (174). Persistent alterations in the sleep architecture have been found in asymptomatic HIV-infected patients that cannot be attributed to primary mood or sleep disorders or to medication side effects (175,176). Fatigue, which is a common symptom in HIV-infected individuals, may or may not be related to depressed mood (177–179). Mental and motor slowness in HIV-infected patients may reflect neurologic and cognitive dysfunction (180). Apathy may be an indicator of CNS involvement in HIV patients (181). Weight loss is a manifestation of the wasting syndrome, which reflects metabolic disturbances (182). Cognitive symptoms may be related more to abnormalities in neuropsychological testing than to depression (48). Conversely, studies have shown a relationship between the somatic symptoms of fatigue and insomnia and depression but not neuropsychological abnormalities or CD4 counts (179), whereas depressed patients had significantly higher scores on the self-rating slowness scale (180). Other studies found a positive correlation between cognitive symptoms and levels of depression but not abnormalities in neuropsychological testing (48–51). We must also remember that depression in all medically ill patients is both underdiagnosed and undertreated (183), and, therefore, we recommend an inclusive approach in the diagnostic process in which symptoms of questionable physical or psychological origin are considered valid indicators of depressive disorders.

The treatment of depression in HIV-infected patients is similar to the treatment of depression in the noninfected patients, with few exceptions.

In double-blind, placebo-controlled studies, the response rates of patients treated with placebo varied from 26% to 46% (184–186), which is significantly lower than those for any antidepressant studied.

Tricyclic antidepressants (TCA) are effective and safe in HIV-infected patients. Similar efficacy has been found when compared with paroxetine (187), fluoxetine (188), and dextroamphetamine (189). Hintz et al. (188) compared different TCA with fluoxetine and trazodone and found no

significant difference in efficacy rates except for trazodone used at doses lower than the usual therapeutic dose for antidepressant effect (50–100 mg daily). Nortriptyline and imipramine (IMI) were the most efficacious, but IMI was better tolerated than amitriptyline, desipramine, and nortriptyline. In a comparative analysis (185), the IMI response rate was similar to that of fluoxetine and sertraline (70%–74%), but less than that of dextroamphetamine (93%) and testosterone (81%). Tolerability was similar in all trials, except in one in which the dropout rate was significantly greater with IMI compared with paroxetine (187). Although the risk of precipitating delirium or unmasking the cognitive disorder is potentially associated with the anticholinergic effect of TCA, it was not found in the studies mentioned. It is advisable to use TCA with caution for this reason as well as for the high risk of suicide among the HIV population.

Selective serotonin reuptake inhibitors also have proven efficacy and tolerability in HIV-infected individuals. Fluoxetine is superior to placebo (186) and supportive group psychotherapy (190) and is effective when IMI is either not tolerated or not efficacious (191). When compared with HIV-negative depressed patients, it shows similar efficacy, but the response to the medication becomes apparent later for the HIV-positive group (192). Sertraline was comparable with fluoxetine and IMI (185,193), and paroxetine has been found efficacious (187,194) and better tolerated than IMI (187). Fluvoxamine showed efficacy but poor tolerability in a small sample (195). No reports of citalopram are found in the literature, but we have used it extensively in the clinical setting with good results and infrequent adverse effects.

Nefazodone may be useful in the HIV population because it improves sleep architecture and has a low incidence of side effects. Elliot et al. (196) reported a small trial, finding good tolerance with only one dropout owing to side effects (7%) and no sexual side effects in the remaining 14 patients. However, PI produce strong inhibition of cytochrome P-450, with indinavir, ritonavir, and amprenavir inhibiting cytochrome P-450 2D6, which may lead to the accumulation and toxicity of *m*-chlorophenylpiperazine, the metabolite of nefazodone. The recommendation of Elliot et al. is to use doses less than or equal to 300 mg daily when coadministering with indinavir and less than or equal to 100 mg daily when coadministering with ritonavir. Usual doses (300–600 mg daily) are recommended when combined with saquinavir or nelfinavir. There is no clinical experience with the coadministration of nefazodone with amprenavir.

The limited experience with venlafaxine suggests that it has qualitative energizing effects akin to those of the psychomotor stimulants. An increase in *β* wave activity on electroencephalography has been observed, suggesting that it may also have cognitive-enhancing effects (J. K. Levy, *personal communication*, 2002).

Bupropion may be beneficial in the withdrawn and anhedonic patient owing to its activating effect (197) and can be an alternative to psychostimulants, but it must be used with caution in patients with significant CNS pathology because seizures may be precipitated or aggravated. Use of the newer antidepressants mirtazapine and escitalopram in the HIV-infected population has not been reported.

The psychostimulants methylphenidate and dextroamphetamine may be helpful for depressed patients who are symptomatic of HIV infection, when other antidepressants have not been tolerated or have failed, and especially for depressed patients who have also cognitive impairment (123,189,198–200). They have also been effective agents when used for augmentation therapy in depressed HIV-infected patients (184,191).

Patients who have been taking monoamine oxidase inhibitors or lithium carbonate before HIV infection can usually continue to take them, with the appropriate dose adjustment because patients with diarrhea may be more sensitive to lithium toxicity. Monoamine oxidase inhibitors are theoretically incompatible with ZDV treatment because they have a catechol-*O*-methyltransferase-inhibiting effect (197), and this combination should be avoided.

Rabkin et al. (201,202) reported the use of testosterone therapy for HIV-infected patients with deficient or low normal levels of testosterone. In their samples, 64% to 79% of those who had depressive disorders showed improvement at week 8. Because low testosterone levels are common in HIV patients (203), it should be tested as part of the workup for depressed patients.

In patients with treatment-resistant depression, electroconvulsive therapy may be an option (204). Although not approved by the FDA, intravenous amitriptyline may be another alternative for the medically ill patient if approved by one's pharmacy or institutional review board for compassionate use in patients who cannot take oral medications or when electroconvulsive therapy might be the treatment of choice but is refused or contraindicated (205).

Individual and group psychotherapies have been used in the treatment of depressive disorders in HIV patients. Individual interpersonal therapy has proven to be more effective than supportive or cognitive/behavioral therapies (206–208). Of the group therapies, cognitive/behavioral and support group therapies were effective in reducing depression, hostility, and somatization relative to a comparison group (209). However, supportive therapy alone was less effective than when used in combination with pharmacologic treatment with fluoxetine (190).

Mood Disorders: Mania

In the treatment of manic states in HIV patients, lithium is both effective and safe but needs to be monitored closely, especially in patients who have diarrhea or wasting syndrome, to prevent toxicity. The use of anticonvulsants for manic states is now widely recognized in the psychiatric literature. Carbamazepine is effective in the treatment of acute mania as well as maintenance therapy (210). In the HIV population,

it presents specific problems because the interaction with PI may result in carbamazepine toxicity owing to impaired metabolism through the cytochrome P-450 3A enzymatic pathway (211). More important, however, enzymatic induction by carbamazepine may result in a decrease in the levels of PI, which may lead to viral resistance (212). Valproic acid has proven effective for acute mania and is now FDA approved for this indication (210). In HIV-infected patients, ZDV clearance is reduced by 50% when coadministered with valproic acid owing to inhibition of ZDV glucuronidation (213). Increased peak and trough plasma and CSF levels of ZDV were reported when valproic acid was added to the treatment of a patient (214). On a negative note, valproic acid has enhanced viral replication *in vitro* (215), but there are no *in vivo* studies demonstrating a similar clinical effect. When used, close monitoring of free valproic acid is recommended because the use of multiple medications bound to serum albumin may increase the free fraction of valproic acid levels (216). The newer anticonvulsants gabapentin and lamotrigine have proven efficacy and tolerability for patients with mania/hypomania (217) and bipolar depressed disorders (218). They have not been used in the HIV population for mania, but gabapentin has been used in the treatment of neuropathic pain (219,220), and in one report, lamotrigine was used for cocaine abuse in HIV patients (221). Oxcarbazepine, topiramate, and zonisamide are proving to be useful in the treatment of bipolar disorder in the general population (222–224); however, their use in the HIV population has not been reported, and, therefore, they should not be first-line agents in the treatment of HIV bipolar patients.

Anxiety Disorders

Anxiety disorders are present in 17% to 36% of HIV-infected patients (158). Adjustment disorder is probably the most common and is present shortly after a patient is notified of seropositivity, with progression of the HIV infection, or with other psychosocial crisis. Panic disorder, generalized anxiety disorder with worries of related and unrelated HIV causes, and symptoms similar to those of posttraumatic stress syndrome, with nightmares and recurrent intrusive thoughts of the notification of seropositivity are also frequent in HIV-infected patients. With more advanced disease, secondary anxiety is more common and may be owing to toxicity or withdrawal from drugs (prescribed, illicit, or over-the-counter), hypoxia, pain, and CNS opportunistic infections or malignancies.

Treatment of acute and severe anxiety involves the use of short- to intermediate-acting benzodiazepines (197,225). Long-term benzodiazepines should be avoided to prevent development of amnesia, confusion, delirium, frontal lobe dysfunction, and disinhibition. Buspirone in doses of 15 to 60 mg daily is effective, either alone or in combination with benzodiazepines, and antidepressants are usually effective as well. Trazodone has been used for the treatment of anxiety and insomnia and is better tolerated than the benzo-diazepines in patients with cognitive impairment. Combination therapy should be started when the anxiety symptoms are prominent, with a benzodiazepine and buspirone or an antidepressant, and after the first 2 weeks, the benzodiazepine should be tapered. The use of atypical antipsychotics for the management of anxiety has been suggested (226–228). Their use in HIV patients with anxiety has not been studied, and caution needs to be exercised owing to potential side effects.

Psychotic Disorders

HIV-induced psychotic disorders are usually seen late in the course of the HIV infection when the patient already has cognitive impairment (229). HIV-infected individuals with new-onset psychosis frequently have a psychiatric history, no antiretroviral therapy, and lower cognitive performance (230). Conversely, chronically mentally ill patients may get involved in high-risk behaviors and thus may have a higher prevalence of HIV than the general population (8,231).

The treatment with neuroleptic medications has been discussed. The same precautions as when treating delirium are recommended. Because of the increased sensitivity to extrapyramidal reactions by HIV-infected patients, atypical neuroleptics should be the first choice, and, in rare cases, clozapine has been used with significant improvement at low doses (mean dose, 27 mg daily) with no associated extrapyramidal reactions (232).

CONCLUSIONS

The HIV/AIDS epidemic remains out of control around the world. Better understanding of the illness, improved technology in the assessment, and new drugs to treat the virus have allowed patients to live longer and to progress more slowly to an AIDS diagnosis. However, new individuals are infected continuously, and patients continue to die of the infection and its complications. Efforts will continue to find better treatments and vaccines that may prevent further spread of the disease.

The brain is affected in a large percentage of infected individuals, and the associated cognitive dysfunction becomes more prevalent as the illness progresses. Better diagnosis and assessment of the response to treatment are now possible. Patients are susceptible to psychiatric disorders and behavioral problems that may or may not be associated with the infection but that require the attention of mental health professionals. Specific problems are presented by this population, and specific treatment modalities are required because these patients may be medically compromised and receive a large number of medications with multiple drug interactions.

As the treatments for HIV infection improve, the prevalence of the neuropsychiatry-associated disorders may increase and their course may change. Research continues to improve treatments and outcomes of HIV-associated neuropsychiatric disorders.

REFERENCES

1. Centers for Disease Control and Prevention. *HIV/AIDS surveillance report.* Washington, DC: U.S. Department of Health and Human Services, 1997;9:1–37.
2. Centers for Disease Control and Prevention. *HIV/AIDS surveillance report.* Washington, DC: U.S. Department of Health and Human Services, 1998;10:1–40.
3. Hoyert DL, Kochanek KD, Murphy SL. Deaths: final data for 1997. *Nat Vital Stat Rep* 1998;47:27–28.
4. Minino AM, Smith BL. Deaths: preliminary data for 2000. *Nat Vital Stat Rep* 2001;49:25–26.
5. Centers for Disease Control and Prevention. *HIV/AIDS surveillance report.* Washington, DC: U.S. Department of Health and Human Services, 2001;13:1–41.
6. Pecoul B, Chirac P, Trouiller P, et al. Access to essential drugs in poor countries. A lost battle? *JAMA* 1999;281:361–367.
7. UNAIDS. Report on the global HIV/AIDS epidemic. "The Barcelona report." XIV International Conference on AIDS. Barcelona, July 2002.
8. Perkins DO, Stern RA, Golden RN, et al. Mood disorders in HIV infection: prevalence and risk factors in a nonepicenter of the AIDS epidemic. *Am J Psychiatry* 1994;151:233–236.
9. McKinnon K, Cournos F, Herman R. HIV among people with chronic mental illness. *Psychiatr Q* 2002;73:17–31.
10. Goodkin K, Wilkie FL, Concha M, et al. Subtle neuropsychological impairment and minor cognitive-motor disorder in HIV-1 infection. Neuroradiological, neurophysiological, neuroimmunological, and virological correlates. *Neuroimaging Clin N Am* 1997;7:561–579.
11. Schrager LK, D'Souza MP. Cellular and anatomical reservoirs of HIV-1 in patients receiving potent antiretroviral combination therapy. *JAMA* 1998;280:67–71.
12. Shi B, Raina J, Lorenzo A, et al. Neuronal apoptosis induced by HIV-1 Tat protein and TNF-alpha: potentiation of neurotoxicity mediated by oxidative stress and implications for HIV-1 dementia. *J Neurovirol* 1998;4:281–290.
13. Heyes MP, Brew BJ, Saito K, et al. Inter-relationships between quinolinic acid, neuroactive kynurenines, neopterin and beta 2-microglobulin in cerebrospinal fluid and serum of HIV-1-infected patients. *J Neuroimmunol* 1992;40:71–80.
14. Tyor WR, Wesselingh SL, Griffin JW, et al. Unifying hypothesis for the pathogenesis of HIV-associated dementia complex, vacuolar myelopathy, and sensory neuropathy. *J Acquir Immune Defic Syndr Hum Retrovirol* 1995;9:379–388.
15. Heyes MP. Potential mechanisms of neurologic disease in HIV infection. In: Bloom FE, Kupfer DJ, eds. *Psychopharmacology: the fourth generation of progress,* 4th ed. New York: Raven Press, 1995:1559–1566.
16. Lipton SA, Yeh M, Dreyer EB. Update on current models of HIV-related neuronal injury: platelet-activating factor, arachidonic acid and nitric oxide. *Adv Neruoimmunol* 1994;4:181–188.
17. Nath A, Padua RA, Geiger JD. HIV-1 coat protein gp120-induced increases in levels of intrasynaptosomal calcium. *Brain Res* 1995;678:200–206.
18. Codazzi F, Menegon A, Zacchetti D, et al. HIV-1 gp120 glycoprotein induces (Ca2+)i responses not only in type-2 but also type-1 astrocytes and oligodendrocytes of the rat cerebellum. *Eur J Neurosci* 1995;7:1333–1341.

19. Ushijima H, Nishio O, Klocking R, et al. Exposure to gp 120 of HIV-1 induces an increased release of arachidonic acid in rat primary neuronal cell culture followed by NMDA receptor-mediated neurotoxicity. *Eur J Neurosci* 1995;7:1353–1359.

20. Barks JDE, Sun R, Malinak C, et al. gp 120, an HIV-1 protein, increases susceptibility to hypoglycemic and ischemic brain injury in perinatal rats. *Exp Neurol* 1995;132:123–133.

21. Muller WEG, Pergande G, Ushijima H, et al. Neurotoxicity in rat cortical cells caused by N-methyl-D-aspartate (NMDA) and gp120 of HIV-1: induction and pharmacological intervention. *Prog Mol Subcell Biol* 1996;16:44–57.

22. Fine SM, Angel RA, Perry SW, et al. Tumor necrosis factor a inhibits glutamate uptake by primary human astrocytes. Implications for pathogenesis of HIV5-1 dementia. *J Biol Chem* 1996;271:15303–15306.

23. New DR, Maggirwar SB, Epstein LG, et al. HIV-1 Tat induces neuronal death via tumor necrosis factor-alpha and activation of non-N-methyl-D-aspartate receptors by a NFkappaB-independent mechanism. *J Biol Chem* 1998;273:17852–17858.

24. Kruman II, Nath A, Mattson MP. HIV-1 protein Tat induces apoptosis of hippocampal neurons by a mechanism involving caspase activation, calcium overload, and oxidative stress. *Exp Neurol* 1998;154:276–288.

25. Dursun SM, Reveley MA. Serotonin hypothesis of psychiatric disorders during HIV infection. *Med Hypotheses* 1995;44:263–267.

26. Shizuko S, Kuniaki S, Stewart SK, et al. Increased human immunodeficiency virus (HIV) type 1 DNA content and quinolinic acid concentration in brain tissues from patients with HIV encephalopathy. *J Infect Dis* 1995;172:638–647.

27. Nakamura S, Nagano I, Yoshioka M, et al. Detection of tumor necrosis factor-α-positive cells in cerebrospinal fluid of patients with HTLV-I-associated myelopathy. *J Neuroimmunol* 1993;42:127–130.

28. Lafon-Cazal M, Pietri S, Culcasi M, et al. NMDA-dependent superoxide production and neurotoxicity. *Nature* 1993;364:535–537.

29. Baier-Bitterlich G, Fuchs D, Wachter H. Chronic immune stimulation, oxidative stress, and apoptosis in HIV infection. *Biochem Pharmacol* 1997;53:755–763.

30. Sharer LR. Pathology of HIV-1 infection of the central nervous system. A review. *J Neuropathol Exp Neurol* 1992;51:3–11.

31. Smith TW, DeGirolami U, Hénin D, et al. Human immunodeficiency virus (HIV) leukoencephalopathy and the microcirculation. *J Neuropathol Exp Neurol* 1990;49:357–370.

32. Wiley CA, Masliah E, Morey M, et al. Neocortical damage during HIV infection. *Ann Neurol* 1991;29:651–657.

33. Aylward EH, Brettschneider PD, McArthur JC, et al. Magnetic resonance imaging measurement of gray matter volume reductions in HIV dementia. *Am J Psychiatry* 1995;152:987–994.

34. de la Monte SM, Ho DD, Schooley RT, et al. Subacute encephalomyelitis of AIDS and its relation to HTLV-III infection. *Neurology* 1987;37:562–569.

35. Mintz M. Clinical comparison of adult and pediatric neuroAIDS. *Adv Neuroimmunol* 1994;4:207–221.

36. Wiley CA, Soontornniyomkij V, Radhakrishnan L, et al. Distribution of brain HIV load in AIDS. *Brain Pathol* 1998;8:277–284.

37. Hill JM, Farrar WL, Pert CB. Autoradiographic localization of T4 antigen, the HIV receptor in human brain. *Int J Neurosci* 1987;32:687–693.

38. Levy JK, Fernandez F. HIV infection of the CNS: implications for neuropsychiatry. In: Yudofsky SC, Hales RE, eds. *Textbook of neuropsychiatry*, 3rd ed. Washington, DC: American Psychiatric Press, 1997:663–692.

39. Janssen RS, Saykin AJ, Cannon L, et al. American Academy of Neurology AIDS Task Force. Nomenclature and research case definitions for neurologic manifestations of human immunodeficiency virus-type 1 (HIV-1) infection. *Neurology* 1991;41:778–785.

40. Grant I, Atkinson JH, Hesselink JR, et al. Evidence for early central nevous system involvement in the immunodeficiency syndrome (AIDS) and other human immunodeficiency virus (HIV) infections: studies with neuropsychological testing and magnetic resonance imaging. *Ann Intern Med* 1987;107:828–836.

41. Stern Y, Marder K, Bell K, et al. Multidisciplinary baseline assessment of homosexual men with and without human immunodeficiency virus infection. *Arch Gen Psychiatry* 1991;48:131–138.

42. Lunn S, Skydsbjerg M, Schulsinger H, et al. A preliminary report on the neuropsychologic sequelae of human immunodeficiency virus. *Arch Gen Psychiatry* 1991;48:139–142.

43. Heaton R, Kirson D, Velin RA, et al. The utility of clinical ratings for detecting cognitive change in HIV infection. In: Grant I, Martin A, eds. *Neuropsychology of HIV infection*. New York: Oxford University Press, 1994:188–206.

44. Bornstein RA, Nasrallah HA, Para MF, et al. Neuropsychological performance in symptomatic and asymptomatic HIV infection. *J Acquir Immune Defic Syndr Hum Retrovirol* 1993;7:519–524.

45. Wilkie FL, Morgan R, Fletcher MA, et al. Cognition and immune function in HIV-1 infection. *J Acquir Immune Defic Syndr Hum Retrovirol* 1992;6:977–981.

46. Martin EM, Robertson LC, Edelstein HE, et al. Performance of patients with early HIV-1 infection on the Stroop task. *J Clin Exp Neuropsychol* 1992;14:857–868.

47. Podraza AM, Bornstein RA, Whitacre CC, et al. Neuropsychological performance and CD4 levels in HIV-1 asymptomatic infection. *J Clin Exp Neuropsychol* 1994;16:777–783.

48. Beason-Hazen S, Nasrallah HA, Bornstein RA. Self-report of symptoms and neuropsychological performance in asymptomatic HIV-positive individuals. *J Neuropsychiatry Clin Neurosci* 1994;6:43–49.

49. van Gorp WG, Satz P, Hinkin C, et al. Metacognition in HIV-1 seropositive asymptomatic individuals: self-ratings versus objective neuropsychological performance. Multicenter AIDS cohort study (MACS). *J Clin Exp Neuropsychol* 1991;13:812–819.

50. Moore LH, van Gorp WG, Hinkin CH, et al. Subjective complaints versus actual cognitive deficits in predominantly symptomatic HIV-1 seropositive individuals. *J Neuropsychiatry Clin Neurosci* 1997;9:37–44.

51. Wilkins JW, Robertson KR, Snyder CR, et al. Implications of self-reported cognitive and motor dysfunction in HIV-positive patients. *Am J Psychiatry* 1991;148:641–643.

52. Van Gorp WG, Miller E, Satz P, et al. Neuropsychological performance in HIV-1 immunocompromised patients. *J Clin Exp Neuropsychol* 1989;11:35.

53. Folstein MF, Folstein SE, McHugh PR. Mini-Mental State: a practical method for grading the cognitive state of patients for the clinician. *J Psychiatr Res* 1975;12:189–198.

54. Power C, Selnes OA, Grim JA, et al. HIV dementia scale: a rapid screening test. *J Acquir Immune Defic Syndr Hum Retrovirol* 1995;8:273–378.

55. Desi M, Seivel N, Korwezgliu J, et al. Cognition disorders in HIV infection. Validation of a brief neuropsychological evaluation battery. *Encephale* 1995;21:289–294.

56. Sacktor NC, Bacellar H, Hoover DR, et al. Psychomotor slowing in HIV infection: a predictor of dementia, AIDS and death. *J Neurovirol* 1996;2:404–410.

57. Butters N, Grant I, Haxby J, et al. Assessment of AIDS-related cognitive changes: recommendations of the NIMH workgroup

on neuropsychological assessment approaches. *J Clin Exp Neuropsychol* 1990;12:963–978.

58. Tinuper P, de Carolis P, Galeotti M, et al. Electroencephalography and HIV infection [Letter]. *Lancet* 1989;1:554.
59. Parisi A, Strosselli M, DiPerri G, et al. Electroencephalography in the early diagnosis of HIV-related subacute encephalitis: analysis of 185 patients. *Clin Electroencephalogr* 1989;20:1–5.
60. Harden CL, Daras M, Tuchman AJ, et al. Low amplitude EEGs in demented AIDS patients. *Electroencephalogr Clin Neurophysiol* 1993;87:54–56.
61. Smith T, Jakobsen J, Gaub J, et al. Clinical and electrophysiological studies of human immunodeficiency virus-seropositive men with AIDS. *Ann Neurol* 1988;23:295–297.
62. Mellors J, Rinaldo C, Gupta P, et al. Prognosis in HIV-1 infection predicted by the quantity of virus in plasma. *Science* 1996;271:1167–1170.
63. Robertson K, Fiscus S, Kapoor C, et al. CSF, plasma viral load and HIV associated dementia. *J Neurovirol* 1998;4:90–94.
64. McArthur JC, McClernon DR, Cronin MF, et al. Relationship between human immunodeficiency virus-associated dementia and viral load in cerebrospinal fluid and brain. *Ann Neurol* 1997;42:689–698.
65. Brew B, Pemberton L, Cunningham P, et al. Levels of human immunodeficiency virus type 1 RNA in cerebrospinal fluid correlate with AIDS dementia stage. *J Infect Dis* 1997;175:963–966.
66. Cinque P, Vago L, Ceresa D, et al. Cerebrospinal fluid HIV-1 RNA levels: correlation with HIV encephalitis. *AIDS* 1998;12:389–394.
67. Ellis RJ, Hsia K, Spector SA, et al. Cerebrospinal fluid human immunodeficiency virus type 1 RNA levels are elevated in neurocognitively impaired individuals with acquired immunodeficiency syndrome. *Ann Neurol* 1997;42:679–688.
68. Gisslen M, Hagberg L, Fuchs D, et al. Cerebrospinal fluid viral load in HIV-1-infected patients without antiretroviral treatment. A longitudinal study. *J Acquir Immune Defic Syndr Hum Retrovirol* 1998;17:291–295.
69. Conrad AJ, Schmid P, Syndulko K, et al. Quantifying HIV-1 RNA using the polymerase chain reaction on cerebrospinal fluid and serum of seropositive individuals with and without neurological abnormalities. *J Acquir Immune Defic Syndr Hum Retrovirol* 1995;10:425–435.
70. De Luca A, Ciancio BC, Larussa D, et al. Correlates of independent HIV-1 replication in the CNS and of its control by antiretrovirals. *Neurology* 2002;59:342–347.
71. Marshall DW, Brey RL, Cahill WT, et al. Spectrum of cerebrospinal fluid findings in various stages of human immunodeficiency virus infection. *Arch Neurol* 1988;45:954–958.
72. Brew BH, Bhalla RB, Fleisher M, et al. Cerebrospinal fluid 2 microglobulin in patients infected with human immunodeficiency virus. *Neurology* 1989;39:830–834.
73. Luizzi GM, Mastroianni CM, Fanelli M, et al. Myelin degrading activity in the CSF of HIV-1-infected patients with neurological diseases. *Neuroreport* 1994;6:157–160.
74. Raininko R, Elovaara I, Virta A, et al. Radiological study of the brain at various stages of human immunodeficiency virus infection: early development of brain atrophy. *Neuroradiology* 1992;34:190–196.
75. Syndulko K, Singer EJ, Nogales-Gaete J, et al. Laboratory evaluations in HIV-1-associated cognitive/motor complex. *Psychiatric Clin North Am* 1994;17:91–123.
76. Chrysikopoulos HS, Press GA, Grafe MR, et al. Encephalitis caused by human immunodeficiency virus: CT and MR imaging manifestations with clinical and pathologic correlation. *Radiology* 1990;175:185–191.
77. Chamberlain MC. Pediatric AIDS: a longitudinal comparative

MRI and CT brain imaging study. *J Child Neurol* 1993;8:175–181.
78. DeCarli C, Civitello LA, Brouwers P, et al. The prevalence of computed tomographic abnormalities of the cerebrum in 100 consecutive children symptomatic with the human immunodeficiency virus. *Ann Neurol* 1993;34:198–205.
79. Rottenberg DA, Sidtis JJ, Strother SC, et al. Abnormal cerebral glucose metabolism in HIV-1 seropositive subjects with and without dementia. *J Nucl Med* 1996;37:1133–1141.
80. Rottenberg D, Moeller J, Strother S, et al. The metabolic pathway of the AIDS dementia complex. *Ann Neurol* 1987;22:700–706.
81. Wiseman MB, Sanchez JA, Buechel C, et al. Patterns of relative blood flow in minor cognitive motor disorder in human immunodeficiency virus infection. *J Neuropsychiatry Clin Neurosci* 1999;11:222–233.
82. Maini CL, Pigorini F, Pau FM, et al. Cortical cerebral blood flow in HIV-1-related dementia complex. *Nucl Med Commun* 1990;11:639–648.
83. Sacktor N, Van Heertum RL, Dooneief G, et al. A comparison of cerebral SPECT abnormalities in HIV-positive homosexual men with and without cognitive impairment. *Arch Neurol* 1995;52:1170–1173.
84. Szeto ER, Freund J, Brew BJ, et al. Cerebral perfusion scanning in treating AIDS dementia: a pilot study. *J Nucl Med* 1998;39:298–302.
85. Chang L, Ernst T, Leonido-Yee M, et al. Cerebral metabolite abnormalities correlate with clinical severity of HIV-1 cognitive motor complex. *Neurology* 1999;52:100–108.
86. Ernst T, Itti E, Itti L, et al. Changes in cerebral metabolism are detected prior to perfusion changes in early HIV-CMC: a coregistered (1)H MRS and SPECT study. *J Magn Reson Imaging* 2000;12:859–865.
87. Salvan AM, Lamoreaux S, Michel G, et al. Localized proton magnetic resonance spectroscopy of the brain in children infected with human immunodeficiency virus with and without encephalopathy. *Pediatr Res* 1998;44:755–762.
88. Harrison MJ, Newman SP, Hall-Craggs MA, et al. Evidence of CNS impairment in HIV infection: clinical, neuropsychological, EEG, and MRI/MRS study. *J Neurol Neurosurg Psychiatry* 1998;65:301–307.
89. Wilkinson ID, Cunn S, Miszkiel KA, et al. Proton MRS and quantitative MRI assessment of the short term neurological response to antiretroviral therapy in AIDS. *J Neurol Neurosurg Psychiatry* 1997;63:477–482.
90. Salvan AM, Vion-Dury J, Confort-Gouny S, et al. Brain proton magnetic resonance spectroscopy in HIV-related encephalopathy: identification of evolving metabolic patterns in relation to dementia and therapy. *AIDS Res Hum Retroviruses* 1997;13:1055–1066.
91. Stankoff B, Tourdan A, Suarez S, et al. Clinical and spectroscopic improvement in HIV-associated cognitive impairment. *Neurology* 2001;56:112–115.
92. Von Bargen J, Moorman A, Holmberg S. How many pills do patients with HIV infection take? *JAMA* 1998;280:29.
93. Carperter CCJ, Fischl MA, Hammer SM, et al. Antiretroviral therapy for HIV infection in 1998. *JAMA* 1998;280:78–86.
94. Anonymous. Delta: a randomised double-blind controlled trial comparing combinations of zidovudine plus didanosine or zalcitabine with zidovudine alone in HIV-infected individuals. Delta Coordinating Committee. *Lancet* 1996;348:283–291.
95. Choo V. Combination superior to zidovudine in delta trial. *Lancet* 1995;346:895.
96. Randall P. CPCRA 007: preliminary results of combination antiretroviral study. *NIAID AIDS Agenda,* 1996:2.

97. Cooper EC. Antiretroviral combination treatment prolongs life in people with HIV/AIDS. *AMFAR Rep* 1996;1–5.

98. Kure K, Llena JF, Lyman WD, et al. Human immunodeficiency virus-1 infection of the nervous system: an autopsy study of 268 adult, pediatric, and fetal brains. *Hum Pathol* 1991;22:700–710.

99. Pratt RD, Nichols S, McKinney N, et al. Virologic markers of human immunodeficiency virus type 1 in cerebrospinal fluid of infected children. *J Infect Dis* 1996;174:288–293.

100. Cunningham P, Smith D, Satchell C, et al. Evidence for independent development of RT inhibitor (RTI) resistance patterns in cerebrospinal fluid compartment. International Conference on AIDS, 1998;12:577(abst 563/32284).

101. Portegies P. HIV-1, the brain, and combination therapy. *Lancet* 1995;346:1244–1245.

102. Elovaara I, Poutiainen E, Lahdevirta J, et al. Zidovudine reduces intrathecal immunoactivation in patients with early human immunodeficiency virus type I infection. *Arch Neurol* 1994;51:943–950.

103. Schmitt FA, Bigley JW, McKinnis R, et al. Neuropsychological outcome of zidovudine (AZT) treatment of patients with AIDS and AIDS-related complex. *N Engl J Med* 1988;319:1573–1578.

104. Baldeweg T, Catalan J, Lovett E, et al. Long-term zidovudine reduces neurocognitive deficits in HIV-1 infection. *J Acquir Immune Defic Syndr Hum Retrovirol* 1995;9:589–596.

105. Portegies P, Enting RH, de Jong MD, et al. AIDS dementia complex and didanosine. *Lancet* 1994;344:759.

106. Butler KM, Husson RN, Balis FM, et al. Dideoxyinosine in children with symptomatic human immunodeficiency virus infection. *N Engl J Med* 1991;324:137–144.

107. Charreaus I, Chemlal K, Yeni P, et al. Reduced risk of progression to HIV encephalopathy with ddI or ddC in combination to AZT compared to AZT alone. 5th Conference on the Retrovirology of Opportunistic Infections, 1998;165(abst 458).

108. McIntyre K, Torres R, Luck D, et al. Pilot study of zidovudine (AZT) and zalcitabine (ddC) combination in HIV-associated dementia. International Conference on AIDS, 1994;10:201(abst PB0233).

109. Foudraine NA, Hoetelmans RMW, Lange JMA, et al. Cerebrospinal-fluid HIV-1 RNA and drug concentrations after treatment with lamivudine plus zidovudine or stavudine. *Lancet* 1998;351:1547–1551.

110. Yazdanian M, Ratigan S, Joseph D, et al. Nevirapine, a nonnucleoside RT inhibitor, readily permeates the blood brain barrier. 4th Conference on the Retrovirology of Opportunistic Infections, 1997;169(abst 567).

111. Fiske WD, Brennan JM, Haines PJ, et al. Efavirenz (DMP 266) cerebrospinal fluid (CSF) concentrations after chronic oral administration to cynomolgus monkeys. 5th Conference on the Retrovirology of Opportunistic Infections, 1998;200(abst 640).

112. McDowell JA, Chittick GE, Ravitch JR, et al. Pharmacokinetics of [14C]abacavir, a human immunodeficiency virus type 1 (HIV-1) reverse transcriptase inhibitor, administered in a single oral dose to HIV-1 infected adults: a mass balance. *Antimicrob Agents Chemother* 1999;43:2855–2861.

113. Stahle L, Martin C, Svensson JO, et al. Indinavir in cerebrospinal fluid of HIV-1-infected patients. *Lancet* 1997;350:1823.

114. Gisolf EH, Portegies P, Hoetelmans R, et al. Effect of ritonavir (RTV)/saquinavir (SQV) versus RTV/SQV/stavudine (d4T) on cerebrospinal fluid (CSF) HIV-RNA levels: preliminary results. International Conference on AIDS, 1998;12:560(abst 32197).

115. Aweeka F, Jayewardene A, Staprans S, et al. Failure to detect nelfinavir in the cerebrospinal fluid of HIV-1-infected patients with and without AIDS dementia complex. *J Acquir Immune Defic Syndr Hum Retrovirol* 1999;1:39–43.

116. Tashima KT. Cerebrospinal fluid levels of antiretroviral medications. *JAMA* 1998;280:879–880.

117. Chang M, Sood VK, Wilson GJ, et al. Metabolism of the HIV-1 reverse transcriptase inhibitor delavirdine in mice. *Drug Metab Dispos* 1997;25:828–839.

118. Sadler BM, Chittick GE, Polk RE, et al. Metabolic disposition and pharmacokinetics of [14C]-amprenavir, a human immunodeficiency virus type 1 (HIV-1) protease inhibitor, administered as a single oral dose to healthy male subjects. *J Clin Pharmacol* 2001;41:386–396.

119. Gisolf E, Colebunders R, Can Wanzeele F, et al. Treatment with ritonavir/saquinavir versus ritonavir/saquinavir/stavudine. 5th Conference on the Retrovirology of Opportunistic Infections, 1998;152(abst 389).

120. Skolnick AA. Protease inhibitors may reverse AIDS dementia. *JAMA* 1998;279:419.

121. Forstein M, Cournos F, Etemad JG, et al. Policy guideline on the recognition and management of HIV-related neuropsychiatric findings and associated impairments. *Am J Psychiatry* 1998;155:1647.

122. Gold JWM. The diagnosis and management of HIV infection. In: Sande MA, Volberding PA, eds. *The medical management of AIDS,* 4th ed. Philadelphia: WB Saunders, 1995:1283–1307.

123. Fernandez F, Adams F, Levy JK, et al. Cognitive impairment due to AIDS-related complex and its response to psychostimulants. *Psychosomatics* 1988;9:38–46.

124. van Dyck CH, McMahon TJ, Rosen MI, et al. Sustained-release methylphenidate for cognitive impairment in HIV-1-infected drug abusers: a pilot study. *J Neuropsychiatry Clin Neurosci* 1997;9:29–36.

125. Lipton SA. Calcium channel antagonists and human immunodeficiency virus coat protein-mediated neuronal injury. *Ann Neurol* 1991;30:110–114.

126. Lipton SA, Gendelman HE. Dementia associated with the acquired immunodeficiency syndrome. *N Engl J Med* 1995;332:934–940.

127. Holden CP, Haughey NJ, Nath A, et al. Role of Na+/H+ exchangers, excitatory amino acid receptors and voltage-operated Ca2+ channels in human immunodeficiency virus type 1 gp 120-mediated increases in intracellular Ca2+ in human neurons and astrocytes. *Neuroscience* 1999;91:1369–1378.

128. Navia BA, Dafni U, Simpson D, et al. A phase I/II trial of nimodipine for HIV-related neurologic complications. *Neurology* 1998;51:221–228.

129. Nath A, Haughey NJ, Jones M, et al. Synergistic neurotoxicity by human immunodeficiency virus proteins Tat and gp 120: protection by memantine. *Ann Neurol* 2000;47:186–194.

130. Fernandez D, Holmes VF, Levy JK, et al. Consultation-liaison psychiatry and HIV-related disorders. *Hosp Community Psychiatry* 1989;40:146–153.

131. Cohen MA. Psychiatric care in an AIDS nursing home. *Psychosomatics* 1998;39:154–161.

132. Uldall KK, Berguis JP. Delirium in AIDS patients: recognition and medication factors. *AIDS Patient Care STD* 1997;11:435–441.

133. Uldall KK, Harris VL, Lalonde B. Outcomes associated with delirium in acutely hospitalized acquired immune deficiency syndrome patients. *Compr Psychiatry* 2000;41:88–91.

134. Lipowski ZJ. Delirium (acute confusional states). *JAMA* 1987;258:1789–1792.

135. Fernandez F, Levy JK, Mansell PWA. Management of delirium

in terminally ill AIDS patients. *Int J Psychiatry Med* 1989;19: 165–172.

136. Breitbart W, Marotta R, Platt MM, et al. A double-blind trial of haloperidol, chlorpromazine and lorazepam in the treatment of delirium in hospitalized AIDS patients. *Am J Psychiatry* 1996;153:231–237.

137. Fernandez F, Levy JK. The use of molindone in the treatment of psychotic and delirious patients infected with the human immunodeficiency virus. *Gen Hosp Psychiatry* 1993;15: 31–35.

138. Adams F. Emergency intravenous sedation of the delirious medically ill patient. *J Clin Psychiatry* 1988;49[Suppl]:22–26.

139. Fernandez F, Holmes VF, Adams F, et al. Treatment of severe, refractory agitation with a haloperidol drip. *J Clin Psychiatry* 1988;49:239–241.

140. Sewell DD, Jeste DV, McAdams LA, et al. Neuroleptic treatment of HIV-associated psychosis. HNRC group. *Neuropsychopharmacology* 1994;10:223–229.

141. Breitbart W, Marotta RF, Call P. AIDS and neuroleptic malignant syndrome. *Lancet* 1988;2:1488–1489.

142. Maldonado JL, Fernandez F. Management of neuropsychiatric complications in HIV infection. *Med Psychiatr* 1998;1:22–29.

143. Singh AN, Golledge H, Catalan J. Treatment of HIV-related psychotic disorders with risperidone: a series of 21 cases. *J Psychosom Res* 1997;42:489–493.

144. Sipahimalani A, Masand PS. Use of risperidone in delirium: case reports. *Ann Clin Psychiatry* 1997;9:105–107.

145. Furmaga KM, DeLeon OA, Sinha SB, et al. Psychosis in medical conditions: response to risperidone. *Gen Hosp Psychiatry* 1997;19:223–228.

146. Breitbart W, Tremblay A, Gibson C. An open trial of olanzapine for the treatment of delirium in hospitalized cancer patients. *Psychosomatics* 2002;43:175–182.

147. Sipahimalani A, Mahand PS. Olanzapine in the treatment of delirium. *Psychosomatics* 1998;39:422–430.

148. Meyer JM, Marsh J, Simpsom G. Differential sensitivities to risperidone and olanzapine in a human immunodeficiency virus patient. *Biol Psychiatry* 1998;44:791–794.

149. Reddy S, Factor SA, Molho ES, et al. The effect of quetiapine on psychosis and motor function in parkinsonian patients with and without dementia. *Mov Disord* 2002;17:676–681.

150. Platt MM, Breitbart W, Smith M, et al. Efficacy of neuroleptics for hypoactive delirium. *J Neuropsychiatry Clin Neurosci* 1994;6:66–67.

151. Lyketsos CG, Hanson AL, Fishman M, et al. Manic syndrome early and late in the course of HIV. *Am J Psychiatry* 1993;150:326–327.

152. Holmes VF, Fricchione GL. Hypomania in an AIDS patient receiving amitriptyline for neuropathic pain. *Neurology* 1989;39:305.

153. Maxwell S, Scheftner WA, Kessler HA, et al. Manic syndrome associated with zidovudine treatment. *JAMA* 1988;259:3406–3407.

154. O'Dowd MA, McKegney FP. Manic syndrome associated with zidovudine. *JAMA* 1988;260:3587–3588.

155. Brouillette MJ, Chouinard G, Lalonde R. Didanosine-induced mania in HIV infection. *Am J Psychiatry* 1994;151:1839–1840.

156. Lyketsos CG, Schwartz J, Fishman M, et al. AIDS mania. *J Neuropsychiatry Clin Neurosci* 1997;9:277–279.

157. Perry SW, Tross S. Psychiatric problems of AIDS inpatients at the New York Hospital: preliminary report. *Public Health Rep* 1984;99:200–205.

158. Atkinson JH, Grant I, Kennedy CJ, et al. Prevalence of psychiatric disorders among men infected with human immun-

odeficiency virus. A controlled study. *Arch Gen Psychiatry* 1988;45:859–864.

159. Leserman J, DiSantostefano R, Perkins DO. Gay identification and psychological health in HIV-positive and HIV-negative gay men. *J Appl Soc Psychol* 1994;24:2193–2208.

160. Kelly B, Raphael B, Judd F, et al. Psychiatric disorder in HIV infection. *Aust N Z J Psychiatry* 1998;32:441–453.

161. Beckett A, Shenson D. Suicide risk in patients with human immunodeficiency virus infection and acquired immunodeficiency syndrome. *Harv Rev Psychiatry* 1993;1:27–35.

162. Chandra PS, Ravi V, Desai A, et al. Anxiety and depression among HIV-infected heterosexuals. A report from India. *J Psychosom Res* 1998;45:401–409.

163. Dannenberg AL, McNeil JG, Brundage JF, et al. Suicide and HIV infection. Mortality follow-up of 4147 HIV seropositive military service applicants. *JAMA* 1996;276:1743–1746.

164. McKegney FP, O'Dowd MA. Suicidality and HIV status. *Am J Psychiatry* 1992;149:396–398.

165. Perry S, Jacobsberg L, Card CAL, et al. Severity of psychiatric symptoms after HIV testing. *Am J Psychiatry* 1993;150:775–779.

166. Lyketsos CG, Hoover DR, Guccione M, et al. Changes in depressive symptoms as AIDS develops. *Am J Psychiatry* 1996;153:1430–1437.

167. Rabkin JG, Goetz RR, Remien RH, et al. Stability of mood despite HIV illness progression in a group of homosexual men. *Am J Psychiatry* 1997;154:231–238.

168. Page-Shafer K, Delorenze GN, Satariano, et al. Comorbidity and survival in HIV-infected men in the San Francisco men's health survey. *Ann Epidemiol* 1996;6:420–430.

169. Burack JH, Barrett DC, Stall RD, et al. Depressive symptoms and CD4 lymphocyte decline among HIV-infected men. *JAMA* 1993;270:2568–2573.

170. Lyketsos CG, Hoover DR, Guccione M, et al. Depressive symptoms as predictors of medical outcomes in HIV infection. *JAMA* 1993;270:2563–2567.

171. Zorrilla EP, McKay JR, Luborsky L, et al. Relation of stressors and depressive symptoms to clinical progression of viral illness. *Am J Psychiatry* 1996;153:626–635.

172. Vedhara K, Schifitto G, McDermott M. Disease progression in HIV-positive women with moderate to severe immunosuppression: the role of depression. Dana consortium on therapy for HIV dementia and related cognitive disorders. *Behav Med* 1999;25:43–47.

173. Mayne TJ, Bittinghoff E, Chesney MA, et al. Depressive affect and survival among gay and bisexual men infected with HIV. *Arch Intern Med* 1996;156:2233–2238.

174. Ostrow D, Grant I, Atkinson H. Assessment and management of the AIDS patient with neuropsychiatric disturbances. *J Clin Psychiatry* 1988;49[Suppl]:14–22.

175. Norman SE, Chediak AD, Kiel M, et al. Sleep disturbances in HIV-infected homosexual men. *AIDS* 1990;4:775–781.

176. Norman SE, Chediak AD, Freeman C, et al. Sleep disturbances in men with asymptomatic human immunodeficiency (HIV) infection. *Sleep* 1992;15:150–155.

177. Darko DF, McCutchan JA, Kripke DF, et al. Fatigue, sleep disturbance, disability, and indices of progression of HIV infection. *Am J Psychiatry* 1992;149:514–520.

178. Ferrando S, Evans S, Goggin K, et al. Fatigue in HIV illness: relation to depression, physical limitations and disability. *Psychosom Med* 1998;60:759–764.

179. Perkins DO, Leserman J, Stern RA, et al. Somatic symptoms and HIV infection: relationship to depressive symptoms and indicators of HIV disease. *Am J Psychiatry* 1995;152:1776–1781.

180. Lopez OL, Wess J, Sanchez J, et al. Neurobehavioral correlates

of perceived mental and motor slowness in HIV infection and AIDS. *J Neuropsychiatry Clin Neurosci* 1998;10:343–350.

181. Castellon SA, Hinkin CH, Wood S, et al. Apathy, depression, and cognitive performance in HIV-1 infection. *J Neuropsychiatry Clin Neurosci* 1998;10:320–329.

182. Shambelan M, Grunfeld C. Endocrinologic manifestations of HIV infection. In: Sande MA, Volberding PA, eds. *The medical management of AIDS*, 4th ed. Philadelphia: WB Saunders, 1995:345–357.

183. Cohen-Cole SA, Stoudemire A. Major depression and physical illness. Special considerations in diagnosis and biologic treatment. *Psychiatric Clin North Am* 1987;10:1–17.

184. Rabkin JG, Rabkin R, Harrison W, et al. Effect of imipramine on mood and enumerative measures of immune status in depressed patients with HIV illness. *Am J Psychiatry* 1994;151:516–523.

185. Wagner GJ, Rabkin JG, Rabkin R. A comparative analysis of standard and alternative antidepressants in the treatment of human immunodeficiency virus patients. *Compr Psychiatry* 1996;37:402–408.

186. Rabkin JG, Wagner GJ, Rabkin R. Fluoxetine treatment for depression in patients with HIV and AIDS: a randomized, placebo-controlled trial. *Am J Psychiatry* 1999;156:101–107.

187. Elliot AJ, Uldall KK, Bergam K, et al. Randomized, placebo-controlled trial of paroxetine versus imipramine in depressed HIV-positive outpatients. *Am J Psychiatry* 1998;155:367–372.

188. Hintz S, Kuck J, Peterkin JJ, et al. Depression in the context of human immunodeficiency virus infection: implications for treatment. *J Clin Psychiatry* 1990;51:497–501.

189. Fernandez F, Levy JK, Samley HR, et al. Effects of methylphenidate in HIV-related depression: a comparative trial with desipramine. *Int J Psychiatry Med* 1995;25:53–67.

190. Zisook S, Peterkin J, Goggin KJ, et al. Treatment of major depression in HIV-seropositive men. *J Clin Psychiatry* 1998;59:217–224.

191. Rabkin JG, Rabkin R, Wagner G. Effects of fluoxetine on mood and immune status in depressed patients with HIV illness. *J Clin Psychiatry* 1994;55:92–97.

192. Cazzullo CL, Bessone E, Bertrando P, et al. Treatment of depression in HIV-infected patients. *J Psychiatry Neurosci* 1998;23:293–297.

193. Rabkin JG, Wagner G, Rabkin R. Effects of sertraline on mood and immune status in patients with major depression and HIV illness: an open trial. *J Clin Psychiatry* 1994;55:433–439.

194. Grassi B, Gambini O, Garghentini G, et al. Efficacy of paroxetine for the treatment of depression in the context of HIV infection. *Pharmacopsychiatry* 1997;30:70–71.

195. Grassi B, Gambini O, Scarone S. Notes of the use of fluvoxamine as treatment of depression in HIV-1-infected subjects. *Pharmacopsychiatry* 1995;28:93–94.

196. Elliot AJ, Russo J, Bergam K, et al. Antidepressant efficacy in HIV-seropositive outpatients with major depressive disorder: an open trial of nefazodone. *J Clin Psychiatry* 1999;60:226–231.

197. Fernandez F, Levy JK. Psychopharmacotherapy of psychiatric syndromes in asymptomatic and symptomatic HIV infection. *Psychiatry Med* 1991;9:377–394.

198. Wagner GJ, Rabkin JG, Rabkin R. Dextroamphetamine as a treatment for depression and low energy in AIDS patients: a pilot study. *J Psychosom Res* 1997;42:407–411.

199. Fernandez F, Levy JK, Galizzi H. Response of HIV-related depression to psychostimulants: case reports. *Hosp Community Psychiatry* 1988;39:628–631.

200. Holmes VF, Fernandez F, Levy JK. Psychostimulant response in AIDS-related complex patients. *J Clin Psychiatry* 1989;50:5–8.

201. Rabkin JG, Wagner GJ, Rabkin R. Testosterone therapy for human immunodeficiency virus-positive men with and without hypogonadism. *J Clin Psychopharmacol* 1999;19:19–27.

202. Rabkin JG, Rabkin R, Wagner G. Testosterone replacement therapy in HIV illness. *Gen Hosp Psychiatry* 1995;17:37–42.

203. Wagner G, Rabkin JG, Rabkin R. Illness stage, concurrent medications, and other correlates of low testosterone in men with HIV illness. *J Acquir Immune Defic Syndr Hum Retrovirol* 1995;8:204–207.

204. Schaerf FW, Miller RR, Lipsey JR, et al. ECT for major depression in four patients infected with human immunodeficiency virus. *Am J Psychiatry* 1989;146:782–784.

205. Olson D, Maldonado JL, Pipkin ML, et al. Use of intravenous amitriptyline in depressed medically ill patients: two case reports and a review of the literature. *Med Psychiatry* 1998;1:80–85.

206. Markowitz JC, Kocsis JH, Fishman B, et al. Treatment of depressive symptoms in human immunodeficiency virus-positive patients. *Arch Gen Psychiatry* 1998;55:452–457.

207. Markowitz JC, Klerman GL, Clougherty KF, et al. Individual psychotherapies for depressed HIV-positive patients. *Am J Psychiatry* 1995;152:1504–1509.

208. Markowitz JC, Klerman GL, Perry SW. Interpersonal psychotherapy of depressed HIV-positive outpatients. *Hosp Community Psychiatry* 1992;43:885–890.

209. Kell JA, Murphy DA, Bahr R, et al. Outcome of cognitive-behavioral and support group brief therapies for depressed, HIV-infected persons. *Am J Psychiatry* 1993;150:1679–1686.

210. Keck PE, McElroy SL, Strakowski SM: Anticonvulsants and antipsychotics in the treatment of bipolar disorder. *J Clin Psychiatry* 1998;59[Suppl 6]:74–81.

211. Brooks J, Daily J, Schwamm L. Protease inhibitors and anticonvulsants. *AIDS Clin Care* 1997;9:87–90.

212. Barry M, Gibbons S, Back D, et al. Protease inhibitors in patients with HIV disease. Clinically important pharmacokinetic considerations. *Clin Pharmacokinet* 1997;32:194–209.

213. Lertora JJ, Rege AB, Greenspan DL, et al. Pharmacokinetic interaction between zidovudine and valproic acid in patients infected with human immunodeficiency virus. *Clin Pharmacol Ther* 1994;56:272–278.

214. Akula SK, Rege AB, Dreisbach AW, et al. Valproic acid increases cerebrospinal fluid zidovudine levels in a patient with AIDS. *Am J Med Sci* 1997;313:244–246.

215. Moog C, Kuntz-Simon G, Caussin-Schwemling C, et al. Sodium valproate, an anticonvulsant drug, stimulates human immunodeficiency virus type 1 replication independently of glutathione levels. *J Gen Virol* 1996;77:1993–1999.

216. Dasgupta A, McLemore JL. Elevated free phenytoin and free valproic acid concentrations in sera of patients infected with human immunodeficiency virus. *Ther Drug Monit* 1998;20:63–67.

217. Cabras PL, Hardoy MJ, Hardoy MC, et al. Clinical experience with gabapentin in patients with bipolar or schizoaffective disorder: results of an open-label study. *J Clin Psychiatry* 1999;60:245–248.

218. Calabrese JL, Bowden CL, Sachs GS, et al. A double-blind placebo-controlled study of lamotrigine monotherapy in outpatients with bipolar I depression. *J Clin Psychiatry* 1999;60:79–88.

219. Vadivelu N, Berger J. Neuropathic pain after anti-HIV gene therapy successfully treated with gabapentin. *J Pain Symptom Manage* 1999;17:155–156.

220. Newshan G. HIV neuropathy treated with gabapentin. *AIDS* 1998;22:219–221.

221. Margoli A, Avants SK, DePhilippis D, et al. A preliminary

investigation of lamotrigine for cocaine abuse in HIV-sero-positive patients. *Am J Drug Alcohol Abuse* 1998;24:85–101.

222. Emrich HM, Altmann H, Dose M, et al. Therapeutic effects of GABA-ergic drugs in affective disorders. A preliminary report. *Pharmacol Biochem Behav* 1983;19:369–372.

223. Maidment ID. The use of topiramate in mood stabilization. *Ann Pharmacother* 2002;36:1277–1281.

224. Kanba S, Yagi G, Kamijima K, et al. The first open study of zonisamide, a novel anticonvulsant, shows efficacy in mania. *Prog Neuropsychopharmacol Biol Psychiatry* 1994;18:707–715.

225. Fernandez F, Levy JK. Psychopharmacology in HIV spectrum disorders. *Psychiatric Clin North Am* 1994;17:135–148.

226. Wilner KD, Anziano RJ, Johnson AC, et al. The anxiolytic effect of the novel antipsychotic ziprasidone compared with diazepam in subjects anxious before dental surgery. *J Clin Psychopharmacol* 2002;22:206–210.

227. Bruhwyler J, Chleide E, Liegeois JF, et al. Anxiolytic potential of sulpiride, clozapine and derivatives in the open-field test. *Pharmacol Biochem Behav* 1990;36:57–61.

228. Sewell DD, Jeste DV, Atkinson JH, et al. HIV-associated psychosis: a study of 20 cases. *Am J Psychiatry* 1994;151:237–242.

229. de Ronchi D, Faranca I, Forti P, et al. Development of acute psychotic disorders and HIV-1 infection. *Int J Psychiatry Med* 2000;30:173–183.

230. Meyer I, Empfield M, Engel D, et al. Characteristics of HIV-positive chronically mentally ill inpatients. *Psychiatr Q* 1995;66:201–207.

231. Lera G, Zirulnik J. Pilot study with clozapine in patients with HIV-associated psychosis and drug-induced parkinsonism. *Mov Disord* 1999;14:128–131.

232. Blaney SM, Daniel MJ, Harker AJ, et al. Pharmacokinetics of lamivudine and BCH-189 in plasma and cerebrospinal fluid of nonhuman primates. *Antimicrob Agents Chemother* 1995;39:2779–2782.

233. Tomaszewski JE, Gieshaber CK, Balzarini J, et al. Toxicologic and pharmacokinetic evaluation of 2'3'-dideoxycytidine, a potential drug to treat AIDS. *Proc Am Assoc Cancer Res* 1987;28:440.

38

NEUROPSYCHIATRY OF DEMENTING DISORDERS

NORMAN L. FOSTER

Dementia is an increasingly important focus of neuropsychiatric practice. The number of individuals with dementia is expected to increase dramatically, and neuropsychiatrists are particularly well suited to evaluate and treat dementia because of its complex combination of cognitive deficits and disordered behavior. The experience and skills of neuropsychiatrists are needed for the difficult tasks of recognizing specific dementing diseases and disentangling the effects of coexisting medical, neurologic, and psychiatric illnesses. Although dementia can occur at all ages, dementing diseases are most frequent in the elderly. Consequently, it is important for neuropsychiatrists to understand the changes in cognition and behavior that occur with normal aging and to be familiar with common illnesses of the elderly.

There is a great opportunity to improve the care of patients with dementia and to enhance the quality of life for families who provide the majority of care. Nearly 60 years passed between the initial pathologic descriptions of Alzheimer disease (AD) and Pick disease in the early 1900's and when dementia again gained the interest of a substantial number of research investigators (1). During the period of great advances in other areas of medicine, research on dementing illnesses languished. They were either ignored completely or vaguely attributed to aging. We are still burdened by the misconceptions and habits of this heritage, but fortunately the situation has changed immensely. There is a virtual explosion of information about AD and related disorders, and the study of dementing diseases is now in the vanguard of neuroscience research. The new insights and conceptual advances generated by this research are beginning to significantly change the evaluation and management of dementia, and there is hope for future treatment breakthroughs. Awareness and application of these recent advances provide great opportunities to make a difference for patients and society.

INCIDENCE AND PREVALENCE

The risk of dementia increases rapidly with age, and, thus, the number of individuals with dementia will grow dramatically as our population gets older. This number will expand most rapidly over the next decades in the developing world, but dementia will have the greatest social impact in Japan, Europe, and North America where the elderly constitute the highest proportion of the population.

It is difficult to measure accurately the incidence and prevalence of dementia and even more difficult to determine the incidence and prevalence of specific dementing diseases. The usual methods that are used to establish disease prevalence are unreliable for dementia. Dementia is infrequently recorded in hospital charts. Outpatient billing records markedly underestimate the number of visits attributable to dementia care, in part because reimbursement rules discourage coding for dementia (2). Even though dementia frequently necessitates institutional long-term care, medical records in nursing home and even nursing home admission forms often fail to indicate when dementia is present and even less frequently state its cause (3,4). Documentation of dementia is even worse for minority nursing home residents (5). Death certificates also significantly underestimate the incidence of dementia (6,7), particularly for minorities and women (8). As a result, we are dependent on epidemiologic studies to determine the prevalence and incidence of dementia. Unfortunately, there are no generally accepted methods for identifying dementia. Thus, there can be widely divergent rates depending on the criteria used, and studies have variable sensitivity to mild symptoms (9). Epidemiologic studies rarely include a sufficiently detailed evaluation to determine the specific cause of dementia. Furthermore, these studies generally use geographically circumscribed populations that may not be representative of the population as a whole. For all these reasons, the measured values for prevalence and incidence vary widely and must be considered rough estimates.

The incidence rates for dementia (new cases in a specified period) are approximately 0.3% per year at age 65 and increase to 5% per year at age 85 (10). AD accounts for one-half to two-thirds of the incident cases of dementia and most of the increasing incidence of dementia with age. However, many other dementing diseases, such as multiinfarct dementia and dementia with Lewy bodies (DLB), also increase with age and contribute to this trend. Incidence rates for AD are approximately 0.2% per year at age 65 and 3% per year at age 85.

Because of the prolonged course of most dementing illnesses, prevalence rates for dementia (cases at a specific time) are much greater than incidence rates. Approximately 5% of individuals older than age 65 have severe dementia and 10% have mild dementia. With increasing age, dementia increases dramatically, with prevalence rates doubling approximately every 5 years (11). By age 85, nearly 30% are at least mildly demented. Physicians encounter dementia in even a higher proportion of their elderly patients than would be suggested by prevalence rates. Dementia itself causes many medical visits, and more medical attention is necessary when dementia complicates medical illnesses and drug treatment. Dementia also accounts for a high proportion of admissions to long-term care facilities with their associated physician oversight. The prevalence of dementia in medical practice is so frequent that there is a considerable risk that physicians will inaccurately stereotype all elderly as having cognitive impairment. Such an underestimate of normal cognitive abilities in the elderly probably contributes to the underrecognition of mild dementia in clinical practice (12).

The epidemiology of dementia has significant public health implications. Unless we are able to find more effective treatments, the prevalence of dementia in the United States will nearly quadruple in the next 50 years (13). Conversely, if we can delay the development of AD by 5 years, we can decrease its prevalence by 50% (14). Estimates of dementia prevalence are important for the planning and distribution of health services, although changing preferences and health care financing may have unexpected effects. For example, the number of long-term nursing home residents has not increased as much as expected, probably because of the expansion of assisted living facilities (15).

DEFINITION OF TERMS

Dementia

Dementia is a decline of intellectual function from a previous level of performance that causes an altered pattern of everyday activity in someone who is alert and cooperative. For the purposes of this definition, intellectual function means two or more areas of cognition, such as language, visuospatial processing, abstraction, and judgment. Memory also is usually affected. Impairment of memory (both verbal and visuospatial) is an important sign in dementia because memory function is not localizable to a single cerebral hemisphere. By contrast, more limited cognitive deficits in isolation, such as aphasia, apraxia, or alexia, have a discrete localization to a single cerebral hemisphere and should be distinguished from dementia, which is more pervasive. Single cognitive deficits can be confused with dementia unless a careful mental status

examination is done that examines multiple areas of cognition and uses several methods to assess deficits. For example, comprehension and memory should be assessed using both verbal and nonverbal stimuli. Intellectual function is multifaceted, and occasionally the attention of the physician and patient is so drawn to a single component of cognition, such as language disturbance or memory loss, that the full extent of the patient's impairment is missed.

To establish that there has been a decline in intellectual function, the examiner must have a sense of the patient's prior intellectual attainment. Premorbid intellectual abilities obviously vary widely among individuals. However, a careful social history provides the examiner with important clues. Academic achievements and occupational activities can be used to gauge expectations, whereas psychiatric and social problems may suggest premorbid limitations. Detailed questioning of those who have known the individual for several years is particularly helpful. The difficulty of the testing should be gauged to the individual's expected abilities and background. More challenging tasks are needed to detect mild impairments in the highly educated or skilled. Questions should be chosen that are relevant to a person's individual situation. For example, a nurse should be familiar with proper administration of medications and a carpenter should be able to describe the purpose and selection of tools that are used in woodworking. The timing and degree of changes in intellectual abilities should be documented as much as possible to better understand the course of the illness. When insufficient historical information is available, direct longitudinal observation of the patient may be necessary to establish that there is truly a decline in cognitive ability.

The definition of dementia further requires an investigation of the functional implications of any decline in intellectual abilities. Current and premorbid activities must be compared with the help of an informant. Changes in day-to-day activities caused by cognitive impairment must be distinguished from changes that can be explained by social preference, physical disability, or psychiatric illness. Social roles and personal choice are important in determining what someone does, but activities should be judged as objectively as possible. Sometimes a change in social role, such as a spouse's death or change in residence, will reveal previously unsuspected functional deficits. In other cases, significant change will be manifested by a qualitative decline in performance, such as the inability to use a recipe while still being able to cook, or deterioration in previously immaculate housekeeping or personal hygiene. It is particularly important not to accept without question the reasons for functional loss offered by the patient or family. They naturally develop their own explanations and may misattribute changes to physical problems or lack of motivation. The physician must judge whether physical deficits are a sufficient explanation for functional decline based on findings on the physical examination. Likewise, the role of psychiatric illness in functional deficits must be based on the assessment of the presence and severity of depression or other definable mental illness.

A final requirement for the recognition of dementia is that the subject be fully alert and cooperative. Otherwise, the assessment of cognition is unreliable. As a consequence, acute illness is often an inappropriate time for assessing dementia, although observations in a hospital justifiably can trigger a subsequent outpatient dementia evaluation. Inconsistent responses, refusal or equal inability to answer both simple and complicated questions, and somnolence during the examination should raise the suspicion that the subject is not fully alert and cooperative. Fluctuating cooperation and alertness suggest delirium rather than dementia. In these situations, it is easy to underestimate an individual's true abilities with the risk of misjudging the presence of dementia.

It is important to note that dementia can be static or progressive, acute or chronic, reversible or irreversible. Furthermore, dementia can occur at any age. Although the prevalence of dementia is much greater in the elderly, dementia at younger ages should not be overlooked. Age alters the likely causes of dementia. In children, dementia is often caused by inherited metabolic diseases or focal mass lesions or is the result of central nervous system infections. In young adulthood, head trauma, acquired immunodeficiency syndrome, toxins, and tumors are major causes of dementia.

It must be emphasized that dementia is only a symptom, not a disease. It is never sufficient to merely identify a symptom; its cause must be found. Yet unlike most symptoms, dementia is not always easy for the physician to recognize. Patients often do not complain of memory loss or cognitive decline themselves and may even actively try to hide or deny that there is a problem. As a result, the physician must actively seek evidence of dementia; casual observation is not sufficient.

Delirium

Dementia should be distinguished from delirium because both cause similar changes in thinking and behavior and they frequently occur together. Delirium is primarily a disorder of attention. It therefore also reflects dysfunction of both cerebral hemispheres. However, unlike dementia, delirium characteristically has a fluctuating level of consciousness. As a result, dementia cannot be accurately accessed when delirium is also present. Delirium generally develops acutely over a period of minutes or hours in association with medical illness, infection, or the use of centrally active medication. Delirium varies in intensity throughout the day and is associated with altered perception, inappropriate and fluctuating psychomotor activity, disorientation, and memory impairment. It causes an inability to shift focus and sustain attention. Both external and internal stimuli fail to elicit a coherent plan of action, and the individual is easily distracted by irrelevant stimuli. It may be impossible for the patient to maintain conversation, visual contact, or performance of a

single sustained task on the physical examination because of wandering attention. There may be dramatic alterations in behavior depending on the environment, such as whether it is day or night or there are other activities in the room. The patient with delirium often becomes more combative and upset in the dark and at night. There also may be lucid intervals. Fluctuations may occur rather suddenly, and the patient may go rapidly from hypersomnolence to alertness and back. At times, the patient may be stuporous or even comatose and then become hypervigilant. Almost invariably, sleep/wake cycles are disturbed, and daytime drowsiness and nighttime insomnia are frequent. Perceptual disturbances include misinterpretations of the environment or the meaning of actions. For example, patients may mistake a knock on the door for a gunshot. There may be visual and auditory hallucinations or allusions, such as thinking a person is behind a curtain when it is simply blowing in the breeze. Delirious patients can be either sluggish or have extreme agitation. Patients may cry for help, scream, demonstrate euphoria, or have deep depression. Fear is a very common experience. Delirium has also been called acute brain syndrome, acute brain failure, acute confusion, and metabolic encephalopathy. The lack of agreement on terminology has hampered education, research, and even ordinary professional communication about this condition. The evaluation and treatment of delirium are beyond the scope of this chapter, but neuropsychiatrists should be familiar with published recommendations (16).

Dementia is a significant risk factor for the development of delirium. Furthermore, an episode of acute mental confusion is a common presenting symptom of dementia. One study found that 40% of adults with dementia had a superimposed delirium on admission to the hospital. Conversely, only 25% of those with delirium were found to have dementia once their acute mental status changes had resolved (17). Delirium is more likely when patients have more severe dementia (18). An important benefit of recognizing dementia is the ability to institute measures that reduce the risk of delirium during hospitalization (19).

Mild Cognitive Impairment or Isolated Memory Impairment

There is increasing interest in the earliest clinical manifestations of dementia. As with other medical problems, early interventions are likely to be most successful in limiting the extent and duration of disability. There is also a need to understand and manage elderly individuals who seek medical attention because they are concerned about having AD but who are found not to have dementia. When such nondemented individuals have clear and otherwise unexplained cognitive deficits on examination, they have been referred to in the literature as having mild cognitive impairment (MCI), isolated memory impairment (IMI), or a variety of similar terms. Understandably, because memory loss is the typical initial symptom in AD, a predominant deficit in memory has been emphasized in most formulations of this condition. The initial symptoms of other dementias are less well understood. MCI/IMI can be defined clinically as persistently impaired memory in an individual older than age 50 without a decline in intellectual abilities or cognitively limited everyday activities and that is unexplained by head injury, focal neurologic disease, medical or psychiatric illness, medication, or depression. Because only memory is significantly impaired and everyday activities are not affected, MCI/IMI can be distinguished from dementia.

The recognition of MCI/IMI is difficult because the boundaries of normal function are difficult to define. Neuropsychological testing is appropriate to quantify cognitive performance and to permit the use of age-based norms to assess deficits. Uniform criteria and terminology for this condition have not been agreed on yet. Some have suggested that deficient memory performance should be defined as 1.5 standard deviations (SD) or more below average values for elderly normals (20). Others have suggested that deficiency be defined as 1.5 SD or more below average values for young normals (21). Adjustments for educational attainment and premorbid intellectual ability determined from reading vocabulary also have been proposed (22). Appropriate criteria for research depend on their purpose. Criteria that are liberal are bound to include neurologically normal individuals, whereas criteria that are restrictive will have less sensitivity. In any case, there is general agreement that both clinical judgment and neuropsychological test performance must be taken into account with any criteria (20,23). Fortunately, the precise neuropsychological criteria have relatively little effect on observed outcomes once clinical judgment is factored in. Consequently, in clinical practice, physician judgment is critical rather than the technicalities of neuropsychological testing. Clinical assessment is most important in identifying and excluding other conditions that cause memory complaints and impair performance sufficiently to meet psychometric criteria for MCI/IMI. Particularly of concern are disorders of attention and concentration caused by medical illnesses, sleep disorders, medication side effects, and mood disturbances, all common in this age group. Limited cooperation or effort can also cause misleading test performance. Most criteria for MCI/IMI require a memory complaint, and it is interesting that this complaint itself is somewhat predictive of the later development of dementia (24). Clinicians should therefore be attentive to patient concerns about memory.

Epidemiologic studies have used somewhat different definitions and terminology, reflecting the uncertainty when medical evaluations are less extensive (25). Much remains to be learned about the natural history of MCI/IMI and its prevalence. It is unclear how frequently memory complaints are important in those who do not seek medical advice and how often memory deficits meeting proposed criteria are transient.

The clinical relevance of MCI/IMI is based on the possibility that symptoms will progress. Individuals with MCI/IMI are heterogeneous but are at a significantly increased risk of developing AD of approximately 15% per year (26,27). Dementia is quickly apparent in some individuals, whereas others may continue to have memory problems yet otherwise change little over 5 or more years of observation. Many studies have sought to identify factors that would predict which patients progress to dementia. Older age, abnormal cerebral glucose metabolism measured by positron emission tomography (PET), hippocampal atrophy on magnetic resonance imaging (MRI) scans, and evidence of even slight nonmemory impairments increase the risk of dementia (21,28,29). Conversely, apolipoprotein E genotype is not a predictor of dementia in those with MCI/IMI, although it is otherwise a risk factor for AD (30). MCI/IMI is also now the subject of several clinical drug trials. Treatment would be warranted if it could decrease the chance that symptoms would progress, but nothing yet has been shown to be effective. It is, therefore, currently appropriate to evaluate and monitor patients with MCI/IMI through regular clinic visits without drug treatment and advise patients about their risk of dementia so that they can make appropriate legal and social plans (23). While we await the results of further research and the proof of effective interventions, it is wise for patients to keep mentally and physically active and to optimize medical health.

EVALUATION AND DIAGNOSIS

The evaluation of dementia is composed of two distinct components: (a) recognition and initial assessment, and (b) a diagnostic determination of its cause. Because understanding and management of dementia have improved, considerable effort is now being directed at recognizing dementia early, while symptoms are mild and interventions clearly are most effective. It is relatively easy to recognize severe dementia, but considerable effort may be required to determine whether individuals with subtle deficits have dementia. Neuropsychiatrists should have a reasoned and prepared strategy for evaluation and diagnosis of dementia. It is most beneficial to consider the possibility of dementia in a broad group of patients and then target only select individuals with clear evidence of impairment for further testing to avoid unnecessary effort and expense.

Recognition and Initial Assessment of Dementia

The clinical practice guideline developed by the Agency for Health Care Policy and Research (now known as the Agency for Healthcare Research and Quality) provides a systematic approach to the early recognition and initial assessment of dementia (31). This guideline nicely summarizes the steps needed to decide whether a patient has dementia (Fig. 38.1). It is impractical to evaluate all elderly individuals

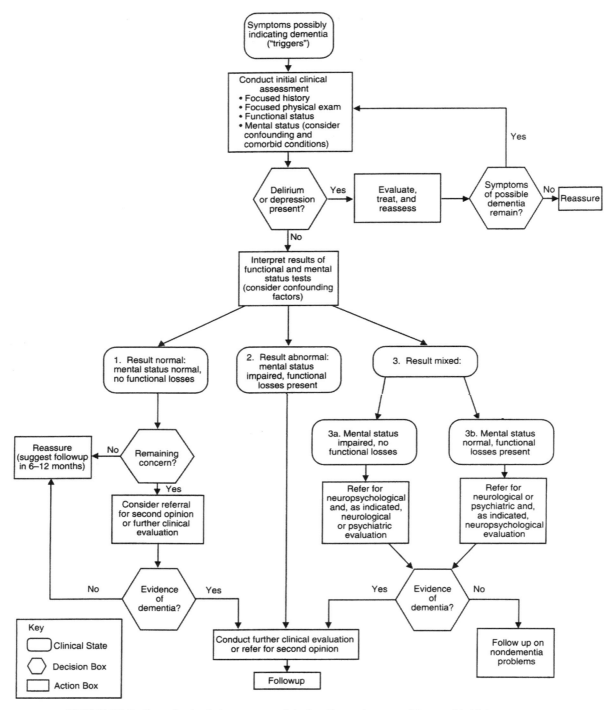

FIGURE 38.1. Flow chart of steps appropriate for the early recognition and initial assessment of dementia. (Reprinted from Costa PT Jr, Williams TF, Somerfield M, et al. *Recognition and initial assessment of Alzheimer's disease and related dementias. Clinical practice guideline no. 19.* Rockville, MD: U.S. Department of Health and Human Services, Public Health Service, Agency for Health Care Policy and Research; 1996; AHCPR publication 97-0702, in public domain.)

TABLE 38.1. SITUATIONS THAT SHOULD TRIGGER A DEMENTIA EVALUATION

Patient complaints or concerns of others
  Persistent and progressive memory loss
  Inability to learn and retain new information
  Difficulty handling complex tasks
  Poor reasoning ability
  Trouble with spatial ability and orientation
  Difficulty with language
  Change in behavior
Observations during medical care
  Difficulty following directions
  Inability to provide an accurate medical history
  Vague or evasive responses to direct questions
  Appears dependent on others for information or financial assistance
  Noncompliance or confusion about appointments and treatment
  Irritable or indifferent when asked direct questions

for dementia because its incidence (as opposed to prevalence) is relatively low and available screening tests are nonspecific and insensitive. Consequently, evaluations should be targeted at those who are at highest risk. This can be done by identifying triggers that suggest dementia based on patient or family complaints and from observations made by physicians, other health professionals, and staff during the course of medical care (Table 38.1). Educating staff and posting these dementia warning signs in a prominent location can increase awareness and encourage nurses and office personnel to share their relevant observations with the physician. It also is appropriate to have staff or the physician simply and directly ask elderly patients and their family members whether they have noticed any trouble with their memory and whether memory loss has affected their activities. Such questions offer an opportunity to those who might otherwise be embarrassed to discuss their concerns.

Once a trigger is identified, the physician obtains a medical history from the patient and at least one collateral source, such as a family member, friend, or employer. A collateral source is important because patients may not be able to adequately recognize or report their problems. When the reliability of other responses is uncertain, it may be efficient to defer the complete history until another source is available to confirm facts and avoid misinformation. The medical history should focus on cognitive and behavioral symptoms, changes in functional abilities, family history of dementia, medical illnesses, and use of medications that can affect cognition. Depression and other psychiatric illnesses cause poor concentration and inattention that can be mistaken for dementia and must be specifically considered. Depressed patients often perceive more memory loss than others recognize that they have, while the reverse is more typical of dementia (32). A history of psychiatric illness and inconsistent responses suggest a psychiatric illness.

Next, a careful medical examination is needed that searches for coexisting medical illnesses and focuses partic-

ularly on potential physical explanations of functional impairment, especially vision and hearing loss. Because neurologic diseases are the most common cause of dementia, a complete neurologic examination including a mental status examination is needed. Physical deficits, hearing loss, and vision problems should be corrected or taken into account when conducting mental status testing.

From the results of the medical history and examination, the physician must make three decisions. The first decision is whether the patient has depression or delirium. If so, these should be treated and the patient then reevaluated after the treatment (Fig. 38.1). If the patient has neither depression nor delirium, the physician must make two further judgments: (a) whether there is cognitive decline, and (b) whether there is functional impairment that is not explained by physical disability. The patient's premorbid state should be considered for both judgments (e.g., literacy, education level, usual activities).

Further action depends on the result of these two judgments. If both new cognitive impairment and functional impairment are present, the patient has dementia and a full diagnostic evaluation is warranted to determine its cause. If only one of the two is present, further observation or additional information is needed, such as neuropsychological testing or social work assessment. Patients with only memory impairment and no functional deficit may have MCI/IMI (see above). A specific cognitive deficit without a functional deficit suggests a focal brain lesion. Conversely, only functional disability without clear-cut cognitive deficits suggests depression, drug abuse, or an unfavorable social situation. These warrant further interventions before concluding that dementia is present.

If neither cognitive decline nor functional impairment is confirmed, despite the presence of the trigger, the patient can be reassured, but a reassessment is warranted in 6 to 12 months before the patient can be considered normal. Reevaluation is needed because patients and family members may be more sensitive to subtle changes than a physician during a single examination. When the patient is reexamined, the results of the earlier testing can be used to detect subtle changes in individual performance. When appropriate, relying on serial observations rather than making a definitive decision about dementia at a single visit will make certain that a mild dementia is not missed. Patients and families also appreciate this approach and find it both reassuring and convincing.

Diagnostic Assessment of Dementia

If the initial dementia assessment described above concludes that the patient has dementia, the cause of dementia must be determined. Further medical history and testing are needed to make a specific diagnosis, which can then be used to develop an appropriate treatment plan based on the patient's individual needs. A detailed history is used to describe the course of the illness. Depending on its cause,

TABLE 38.2. ROUTINE LABORATORY STUDIES FOR MEDICAL ILLNESSES IN DEMENTIA

Test	Rationale
Complete blood count	Severe anemia, systemic infection, leukemia
Electrolyte panel	Electrolyte disturbances
Glucose and calcium	Hypoglycemia, hypercalcemia
Chemistry panel	Hepatic or renal disease
Thyroid function tests	Hypothyroidism or hyperthyroidism
Vitamin B_{12} level	B_{12} deficiency, pernicious anemia

the onset of symptoms may be insidious or sudden, and the dementia may be stable or progress in either a smooth and continuous or stepwise manner. The order of appearance, rate of progression, and relative severity of different cognitive and behavioral changes should be charted. The possible role of medications should also be reconsidered. Centrally active drugs are the most common reversible cause of dementia. Psychotropic medications, particularly sedatives, hypnotics, tranquilizers, antidepressants, and anxiolytics such as benzodiazepines, are the most common offenders. However, drugs taken for other purposes, such as antihistamines, prescription, and over-the-counter medications that are anticholinergic, should not be overlooked.

A detailed neurologic examination is particularly relevant to identify subtle motor signs and changes in gait. The mental status examination is used to evaluate specific cognitive domains, including memory, language, visuospatial abilities, and executive function. These observations are then compared with the pattern of deficits described in the medical history. This comparison helps to illuminate the validity of the historical information.

Based on these clinical examinations, further studies are ordered to confirm or refute diagnostic impressions. Neuropsychological testing and an evaluation by a speech pathologist can further clarify deficits and are often helpful. These assessments provide independent, quantitative measures to

further characterize a patient's precise cognitive and behavioral deficits. Standardization permits these tests to compare the relative severity of deficits in one domain to another and in a patient to a population of normals.

In addition, laboratory testing should be obtained to identify medical illnesses that can cause or amplify the dementia (33). Blood tests for common medical illnesses should be routine and obtained in all cases (Table 38.2). Testing for other medical illnesses is needed only when they are clinically suspected (Table 38.3). It is important to identify medical illnesses causing dementia because many of these illnesses are potentially reversible. Routine testing and a high index of suspicion are warranted because in the elderly, medical disorders often cause cognitive symptoms before more typical symptoms are apparent. However, it is more common for medical illnesses to exacerbate an underlying dementing illness than to be the complete cause for dementia. The physician, therefore, should not be discouraged when successful treatment of medical problems fails to resolve the patient's dementia. Patients still benefit significantly when the cognitive changes caused by medical diseases improve. The resolution of this excess disability fully justifies the effort involved.

Neurologic diseases are the most common cause of dementia (Table 38.4). Consequently, additional testing for neurologic diseases is needed when evaluating the cause of dementia. Structural brain imaging, particularly MRI, is very sensitive to focal lesions, and either computed tomography (CT) or MRI should be obtained in all patients with dementia (33). Although clinicians naturally focus on mass lesions when interpreting these studies, widening of sulci and ventricular enlargement are very relevant in dementia and deserve comment. Because it is often overlooked, the possibility of focal atrophy deserves particular attention. Single photon emission computed tomography and PET are functional imaging methods that provide biochemical information not revealed by structural imaging studies. Although single photon emission computed tomography and PET are not always needed, they can help to distinguish specific neurodegenerative diseases and provide valuable clinical information that help to explain a patient's symptoms. Additional

TABLE 38.3. ANCILLARY LABORATORY STUDIES FOR DEMENTIA WHEN MEDICAL ILLNESSES ARE SUSPECTED

Test	Rationale and Indications
Human immunodeficiency virus	Acquired immunodeficiency syndrome (risk factors identified)
Tests for syphilis	Neurosyphilis (exposure, unreactive pupils)
Urinalysis	Urinary tract infection (incontinence or urinary complaints)
Sedimentation rate	Vasculitis, temporal arteritis (headache, seizures)
Blood gases	Hypoxia, hypercarbia (lung disease, exposure)
Drug screen	Alcohol, prescription and recreational drugs (exposure)
Heavy metals	Lead and arsenic poisoning (exposure, peripheral neuropathy)
Hu antibody	Paraneoplastic limbic encephalitis (history of relevant cancer)
Ceruloplasmin	Wilson disease (liver disease, young age, dystonia)

TABLE 38.4. SOME NEUROLOGIC DISEASES CAUSING DEMENTIA

Neurodegenerative Diseases
    Alzheimer disease and its mimics
    Parkinson disease
    Huntington disease
Neoplastic diseases
    Primary and metastatic brain tumors
    Paraneoplastic limbic encephalitis
Vascular diseases
    Multiple strokes
    Sequelae of subarachnoid hemorrhage
    Subdural hematoma
    CNS vasculitis
Metabolic and Demyelinating Disorders
    Multiple sclerosis
    Metachromatic leukodystrophy
    Cerebrotendinous xanthomatosis
Infectious disorders
    Neurosyphilis
    Central nervous system acquired immunodeficiency syndrome
    Brain abscess
    Progressive multifocal leukoencephalopathy
    Sequelae of encephalitis and meningitis
    Creutzfeldt-Jakob disease
Trauma
    Closed head injury

testing may be appropriate for rapidly progressive dementias and in familial dementias as described below.

Determining the cause of dementia is one of the most complex diagnostic challenges facing a physician. It is important not to underestimate the effort required for an accurate diagnostic assessment and to plan adequate time to obtain a detailed history and examination. Diagnosis relies heavily on reconstructing the clinical course of the illness and the appearance and timing of symptoms. This information, along with results of laboratory testing and brain imaging, is then compared with characteristic clinical features of individual dementing illnesses to arrive at a diagnosis. The best match provides the specific diagnosis, with the degree of congruence generally indicating the level of diagnostic certainty.

Although it is challenging, determining the cause of dementia is also intellectually rewarding and beneficial for the patient and family. Successful diagnosis requires both a good knowledge of the many possible dementing disorders and accurate, detailed information about the patient. Consensus clinical criteria are very helpful, but they cannot address all clinical situations. Clinical diagnosis is always a hypothesis until confirmed by pathologic examination or a definitive biomarker. This fact is particularly evident with dementia because pathologic information is usually unavailable during life and definitive biomarkers have been identified only for the very few patients with known genetic mutations. Despite unavoidable uncertainty, a specific diagnosis is important because it provides the basis for deciding on treatment, foretells

prognosis, and determines the kind of care that is needed. In deciding on the cause of dementia, the physician must have conviction but humility. Diagnosis must be open for review and revision, if necessary. Physicians should recognize the limits of current knowledge. It is appropriate to recommend that families seek confirmation of clinical diagnoses by postmortem examination, particularly because the results may have substantial implications for other family members.

Direct observations over time can be very helpful in reaching a better understanding of the illness and a more accurate diagnosis. Medical histories are often incomplete and inaccurate because of the insidious and complex nature of dementia. Serial examinations can compensate for these deficiencies. An illness that initially seems to be rapidly progressive may not be on repeated medical visits. The appearance of new motor symptoms may reveal a previously unsuspected cause of dementia. Unfortunately, most current clinical criteria have been developed for use in a single evaluation and do not take into account the great value of information obtained in repeated direct observations.

Serial examinations are particularly relevant when there is more than one disease contributing to the patient's symptoms. Such mixed dementias are not uncommon. The most frequent combinations are AD with strokes, AD with Parkinson disease, and AD accompanied by DLB. Repeated observations may disentangle symptoms and reveal which condition is the predominant cause of the dementia. For instance, in a patient with mixed dementia owing to AD and stroke, gradual progression of dementia without further strokes on MRI suggests that AD has caused the worsening of symptoms. By contrast, if there is sudden worsening of dementia associated with new focal deficits, stroke is implicated as the cause of worsening and is likely to be more important. Serial neuropsychological testing also can provide quantitative measures of change in specific cognitive domains. Using this method, even the most complex situations can be understood and attributed to specific diseases.

Diagnostic Assessment of Rapidly Progressive Dementia

Dementia that develops over less than 1 year requires urgent evaluation and warrants additional testing. Because outcome may depend on how soon treatment is initiated, a specific diagnosis should be reached in these cases as soon as possible and testing completed within a week. The diseases causing rapidly progressive dementia differ from those with a more indolent course (Table 38.5). The possibility of delirium should be carefully considered and laboratory testing obtained to exclude an unsuspected medical problem. Electroencephalography (EEG) should be performed quickly to identify nonconvulsive status epilepticus or periodic discharges typical of Creutzfeldt-Jakob disease (CJD). If CJD continues to be a consideration later, EEG can be helpful for comparison with subsequent studies to identify the

TABLE 38.5. CAUSES OF RAPIDLY PROGRESSIVE DEMENTIA

Creutzfeldt-Jakob disease
Meningitis/encephalitis (including tuberculous, fungal, herpes, carcinomatous, paraneoplastic)
Vasculitis
Central nervous system lymphoma, primary and metastatic brain tumors
Subdural hematoma
Nonconvulsive status epilepticus
Multiinfarct dementia (occasionally)
Alzheimer disease (rarely)

typical progression of abnormalities in CJD. MRI is preferable to CT for evaluating rapidly progressive dementia. It can detect subtle meningeal changes and small, disseminated lesions that can be missed on CT. Furthermore, focal abnormalities can be found on MRI in CJD, often involving the striatum (34). If a diagnosis still is not apparent after blood testing and brain imaging, a lumbar puncture should be performed. Cerebrospinal fluid (CSF) should be tested for cell count; glucose, protein, viral, and fungal antigens; and cytology. Cultures for opportunistic bacterial and fungal organisms and tuberculosis should also be obtained. In the appropriate setting, 14-3-3 protein should also be tested.

Creutzfeldt-Jakob Disease

It is important to alert other health professionals when evaluating patients with rapidly progressive dementia. All neuropsychiatrists should be familiar with and use proper precautions to prevent the potential iatrogenic transmission of CJD (35). CJD is caused by the replication of a normal protein that has assumed an abnormal conformation (prion). This prion is unaffected by routine sterilization but can be inactivated by denaturing procedures. Transmission occurs only under limited conditions so that family members and others providing ordinary care are not at risk and usual universal precautions are generally sufficient. However, neuropsychiatrists should help to ensure that written infectious disease precautions for CJD are in place and adhered to, particularly for neurosurgical procedures and autopsy. This simple step can avoid irrational responses by staff and unnecessary risks to others.

CJD can have a surprising variety of clinical presentations, including initial memory loss, focal cognitive deficits, and gait disorder. It has been possible to relate specific clinical courses to prion glycotype and genotype (36). Clinical criteria can be useful in making diagnosis (37). 14-3-3 Protein usually can be found in the CSF of patients with CJD and can be helpful, but it also occurs in several other disorders associated with rapid brain destruction and should not be viewed as definitive (38,39). There has been great public health concern raised by the appearance of new variant CJD, likely caused by transmission of bovine spongiform

encephalopathy. This form of CJD affects younger individuals than sporadic CJD and has distinctive clinical features (40). Although new variant CJD is thus far limited to the United Kingdom and Western Europe, potential reservoirs of ruminant spongiform encephalopathy exist in North America and elsewhere, and vigilance is needed. Autopsy examination and established surveillance programs should be utilized for all patients with suspected CJD (www.cjdsurveillance.com; www.cjd.ed.ac.uk).

Diagnostic Evaluation of Familial Dementia

A review of the patient's family history should be a part of each dementia evaluation, but additional steps are needed when there is evidence that dementia has affected other family members. The patient should have the usual evaluation procedures and diagnosis. The next step is to further investigate whether other family members have had a similar disorder and construct a pedigree chart. Affected individuals in two or more generations are generally needed to determine the pattern of inheritance. Most familial dementias are autosomal dominant. If there is sufficient evidence to establish that the same disease has caused a recognizable mendelian pattern of inheritance in the family, a familial dementia is confirmed. However, this process is not as straightforward in dementia as in other genetic disorders because diagnostic information is often not as reliable. Usually several family members must agree to provide information and assist in obtaining medical records for review. Books that provide families with guidance on how to construct a family medical history are available and can be quite helpful (41). The physician then must assess carefully the quality of the diagnostic information available and decide whether the same disorder is truly present in family members. The congruence of the diagnosis in the patient and other family members is often uncertain. Autopsy data are the most convincing. Unfortunately, often postmortem examinations have not been performed or are of poor quality. Next most convincing are similar clinical symptoms and signs, preferably from medical records. Unfortunately, dementia evaluations often have not been performed, and there remains some uncertainty.

When a familial dementia has been confirmed, it may be possible to obtain genetic testing that definitively establishes the cause of the dementia and that also can be used for diagnostic and predictive testing of other family members. Such testing requires informed consent and should be conducted only after careful consideration of its clinical, ethical, and legal implications (42).

Huntington Disease

Huntington disease is an exceptional form of familial dementia because it causes choreoathetosis and thus is usually easily recognized. However, patients with an earlier onset, particularly those with affected fathers, may have a large number of

CAG repeats in the Huntington gene and clinically present with a rigid akinetic form of the disease that can be confused with Parkinson disease (43). Moreover, psychiatric symptoms that may suggest a frontotemporal dementia (FTD) may appear before chorea. The neuropsychological profile of Huntington disease differs from that of AD and predominantly involves frontal lobe functions (44). The diagnosis of Huntington disease usually is not difficult because it is easy to recognize in a family history. The genetic mutation for Huntington disease rarely arises *de novo* and is easy to test for, even in isolated cases without a history of affected relatives (45).

Outcomes of a Dementia Evaluation

A complete dementia evaluation has benefits for patients and their families and physicians. It provides a specific diagnosis that explains the cause of symptoms and permits answers to questions about prognosis and family risk. It provides the physician with a rational approach to treatment and a plan for how the patient should be assessed over time. However, to take full advantage of the evaluation, more than a diagnosis is needed. The evaluation should meet the expectations of patients and their families and address the reasons that they sought a dementia evaluation (Table 38.6). Not only will this increase satisfaction, but it is also an important first step in eliciting the involvement of the patient and family in providing care. Many patient and family concerns are about the future. Physicians should provide hope and the assurance that someone knowledgeable about dementia will be available. Referring physicians also will be interested in more than a specific diagnosis. The complete evaluation should include a detailed management plan based on the diagnosis and dementia severity (Table 38.7). It should include patient education and address social and legal issues that are raised by the diagnosis. This plan should always include referral to community resources, such as the area agency on aging and the Alzheimer's Association. This will establish an approach that minimizes future demands on the primary care physician and the neuropsychiatrist. The evaluation must be considered incomplete if outcomes that address the needs of patients and their families and long-term management are not realized. Usually this will lead to further referrals and repeated evaluations that would be unnecessary if the initial evaluation had been adequate.

The value of the dementia evaluation is not fully realized unless it is shared with the patient and others whom the patient chooses to involve. There are many advantages of telling patients their diagnosis (Table 38.8). In the unusual situation in which there is concern about a serious psychological risk, counseling and psychiatric treatment should be provided before sharing the diagnosis.

THE NEUROBIOLOGY OF DEMENTIA

The explosion of knowledge about dementing disorders is overwhelming, but it has led to a unifying understanding of the neurobiology of dementia. Dementia is always the result of impaired function of the cerebral cortex. In most cases, there is direct damage to cortical neurons. However, in some diseases, there is sufficient damage to afferent inputs to impair the function of the cerebral cortex and cause dementia. Damage to afferents can occur from loss of subcortical projection neurons or from injury to axons and dendrites that connect to cortical neurons, such as in demyelinating diseases. We know that cerebral cortical function is critical for dementia even with damage to afferents because it always correlates most closely with dementia in these situations. For example, cerebral cortical hypometabolism is the best predictor of dementia in subcortical cerebral infarction and multiple sclerosis (46,47). Frequently, there is a combination

TABLE 38.7. EXPECTED OUTCOMES OF A COMPLETE DEMENTIA EVALUATION

A specific diagnosis
A management plan for current needs that includes
    Drug treatments specific to the disease
    Symptomatic drug treatments
    Appropriate environmental support
    Referral to community-based dementia services
A plan for future needs that
    Addresses social and legal issues
    Plans for long-term care
    Recommends continuing medical care and review of the
        management plan
Education about dementia and additional resources

TABLE 38.6. WHAT PATIENTS AND THEIR FAMILIES WANT FROM A DEMENTIA EVALUATION

A specific explanation for the patient's symptoms
What should be done now for the patient
What can be expected in the future, and how to prepare for the
    future now
Whether other family members are at risk of developing similar
    problems
When and how to obtain help in the future
Hope for future treatment and support

TABLE 38.8. ADVANTAGES OF TELLING PATIENTS WITH DEMENTIA THEIR DIAGNOSIS

Respects the autonomy and rights of patients
Helps patients understand the reason for treatment
    recommendations
Involves patients in their own treatment
Permits open discussion about treatment and research options
Permits patients to plan for their own future
Enhances communication within the family
Knowing what is wrong provides reassurance and hope for
    future support and treatment

of these mechanisms. For instance, although most damage in AD occurs in the cerebral cortex, afferents originating in the nucleus basalis of Meynert also are lost as these neurons fail (48,49). Likewise, even though subcortical neurons are most affected in progressive supranuclear palsy (PSP), there also is loss of neurons in the cerebral cortex (50).

Dementing diseases do not affect the cerebral cortex uniformly but instead cause selective neuronal damage. Specific dementing diseases distinguish themselves by the regional distribution of this damage. Initially, dementing diseases are focal or multifocal and then spread to other cortical regions in a predictable hierarchy that varies in different disorders and can be helpful diagnostically. Eventually dementias become pervasive and involve large areas of the cerebral cortex. As a result, it is easier to distinguish the cause of dementia early in the course of the illness than when it is severe and extensive neuronal loss obscures selective injury.

Neurodegenerative diseases causing dementia share several characteristics. They all appear to be explained by minimal imbalances in cellular metabolism whose effects thus become apparent only after many years. This accounts for their onset in late life and increasing incidence with age. This characteristic also explains their predilection for the central nervous system, even though metabolic derangement may be more widespread. Neurons persist throughout life, whereas most other tissues experience rapid cellular turnover. As a consequence, minor imbalances that have only cumulative effects over a long time can be devastating to the nervous system even though they are unimportant in other tissues. Moreover, the high metabolic activity of neurons may put them at particular risk by exaggerating the impact of minor abnormalities. Frequently, these metabolic derangements result in the accumulation of cellular by-products that develop "fatal attractions" and appear as inclusions (51).

Individual cell failure in most neurodegenerative disease appears to be gradual rather than catastrophic. Often there is functional impairment before cell death. For example, in AD, acetylcholinesterase activity decreases more than vesamicol, a marker of cholinergic neuron integrity (52). This offers the possibility that recovery of some function might be possible. Furthermore, neurodegenerative diseases evolve over several years. This provides hope that even modest long-term treatments might slow or even prevent progression.

Neurodegenerative diseases are the most common cause of dementia. Each neurodegenerative disease has a hierarchy of cellular vulnerability that presumably reflects cellular specialization and characteristics such as size, neurotransmitter type, and active metabolic pathways reflected in genetic expression. It is also likely that the location of neurons in the brain is significant because many neurodegenerative diseases have specific regional brain pathology. Because local environment influences neuronal differentiation and functional development, it is logical that local factors may be critical in the pathophysiology of neurodegenerative disease. We need to know a great deal more about neurodegenerative diseases and how they affect the underlying mechanisms of cognition. Nevertheless, these general principles provide the basis for understanding specific dementing disorders in the elderly.

COMMON CAUSES OF DEMENTIA IN THE ELDERLY

A review of all diseases causing dementia is beyond the scope of this chapter. Dementia developing in childhood and adolescence is usually due to trauma, sequelae of central nervous system infections, or inherited metabolic disorders and requires different considerations than dementia in adults (53). Helpful reviews of unusual metabolic disorders causing dementia in adults and less common medical and neurologic dementing disorders are available (54,55). The remainder of this chapter focuses on common causes of dementia in the elderly, particularly neurodegenerative disorders and stroke. These disorders can conveniently be grouped as they appear to the clinician (Table 38.9). Some patients have no evident focal features or a gait disorder (AD and AD mimics). Others have dementia and symptoms of parkinsonism. Still other patients have focal deficits on the neurologic examination or a nonparkinsonian gait disturbance. A final group of patients has a distinctive medical history or neurologic findings that distinguish them. It is evident from the table that the clinical presentation is not sufficient for diagnosis because several diseases can have different presentations.

Alzheimer Disease

AD is the most common cause of dementia in the elderly, accounting for 60% or more of all dementia. AD increases nearly exponentially with age after age 60, which is its most important risk factor. Age is a risk factor in most neurodegenerative dementing diseases, but other risk factors are not shared (Table 38.10).

TABLE 38.9. COMMON PRESENTATIONS OF DEMENTIA IN ADULTS

Alzheimer disease and Alzheimer mimics
  Alzheimer disease, frontotemporal dementia, including Pick disease, familial CJD, dementia with Lewy bodies, multiinfarct dementia, CBD, progressive supranuclear palsy
Dementia with parkinsonism
  Alzheimer disease, dementia with Lewy bodies, Parkinson disease, multiinfarct dementia, progressive supranuclear palsy, CBD, frontotemporal dementia with parkinsonism linked to chromosome 17
Dementia with focal findings or gait disturbance
  Multiinfarct dementia, subdural hematoma, CBD, traumatic brain injury, normal pressure hydrocephalus, CJD, sequelae of intracerebral hemorrhage

CJD, Creutzfeldt-Jakob disease; CBD, corticobasal degeneration.

TABLE 38.10. RISK FACTORS FOR NEURODEGENERA-TIVE DEMENTING DISEASES

Alzheimer disease
    Increasing age
    Family history of Alzheimer disease
    Inheriting the apolipoprotein E4 allele
    Low educational and occupational attainment (160)
    Severe head injury (161,162)
Parkinson disease with dementia
    Increasing age
Progressive supranuclear palsy
    Increasing age
    Inheriting A0/A0 and H1/H1 alleles of the *tau* gene (163,164)
Creutzfeldt-Jakob disease
    Homozygosity at codon 129 of the *PrP* gene
    Family history of Creutzfeldt-Jakob disease

Clinical Features and Diagnosis

AD initially causes the insidious onset and progressive worsening of memory loss. Patients have difficulty learning new information. Memory loss usually remains the most prominent symptom throughout the illness, but as the disease progresses, additional changes in cognition become evident and may obscure the memory loss. Patients develop progressive difficulty naming objects, expressing thoughts, and comprehending speech (aphasia). They also develop difficulty with visuospatial skills and become apraxic. They have difficulty recognizing faces and understanding the meaning of circumstances and concepts. This pattern of symptoms reflects the distribution of pathologic damage, which begins in the hippocampus and entorhinal cortex, and the regional distribution of glucose hypometabolism and hypoperfusion, which is most apparent in posterior temporoparietal association areas and the posterior cingulate gyrus (Fig. 38.2). Patients with AD have variable degrees of impaired judgment and insight, which probably reflects involvement of frontal association cortex. Behavioral changes reflecting frontal lobe damage are common but are typically less severe than other deficits and tend to occur in more severely impaired patients. Indeed, patients with AD usually maintain their social skills and social etiquette until they are severely impaired. Although many patients eventually lose insight into their deficits, the majority are aware of failing abilities early in their illness and may be very defensive when challenged. Even in these situations, however, patients are less aware of their deficits than others are.

The clinical diagnosis of AD is based on the National Institute of Neurological and Communicable Disease–Alzheimer's Disease and Related Disorders Association criteria (Table 38.11), which have been pathologically validated in several studies (56,57). These criteria describe three degrees of diagnostic certainty. Because pathologic information is not generally available, clinical diagnoses are limited to either probable or possible AD. Probable AD represents the highest degree of diagnostic certainty (85% to 90% validation by subsequent postmortem examinations) and is reserved for patients with typical onset and no complicating illnesses. Possible AD is used when there is an atypical onset or the patient has complicating illnesses that could account for some or all of the patient's cognitive symptoms. Understandably, this situation has less diagnostic certainty than in a patient who meets probable AD criteria. An important part of these criteria is that definite AD requires both clinical and pathologic data. Both the clinician and pathologist are needed to establish a diagnosis of definite AD.

Neurobiology

AD causes characteristic changes that are used to establish the pathologic diagnosis, including neuritic plaques, neurofibrillary tangles, and selective synaptic and neuronal loss. Neuritic plaques are extracellular inclusions in the cerebral cortex composed of a central amyloid core surrounded by abnormal neuronal processes that appear to be attracted to the plaque. The amyloid core is an aggregate of Aβ, a fragment of amyloid precursor protein (APP), a transmembrane protein present in cells throughout the body. Aβ in neuritic plaques is primarily found as a 42-amino-acid protein formed from APP by β- and γ-secretase (Fig. 38.3). A more common 40-amino-acid form of this protein causes vascular amyloid deposits. Initially, amyloid aggregates without any response, but as the deposit matures, it attracts neuritic processes, microglia, and other substances, including inflammatory markers, metal ions, and several plasma proteins.

Neurofibrillary tangles are intracellular inclusions composed of aggregates of tau, a cytoskeletal protein that binds and regulates microtubule assembly. Tau has six isoforms depending on the presence of an insert and the number of repeat regions that are transcribed (Fig. 38.4). For unknown reasons, in AD, all isoforms of tau becomes hyperphosphorylated and aggregate into paired helical filaments that are most evident in the axon hillock of large neurons. This undoubtedly disrupts microtubular transport and eventually leads to neuronal death.

Synapse loss and neuronal death partially, but not completely, reflect the distribution of these inclusions. Neuronal loss and neurofibrillary tangles are most prominent in the entorhinal cortex and hippocampus. The amygdala and temporoparietal association cortex are particularly affected with prominent neuronal loss and neuritic plaques. The location of inclusions corresponds to the distribution of synaptic loss, which correlates most closely with severity of dementia and is the major determinant of glucose hypometabolism (58). Not all neurons are equally susceptible within a brain region. There is a distinctive laminar distribution of neuronal loss particularly affecting large neurons (59,60).

Cholinergic neurons originating in the nucleus basalis of Meynert and projecting to the cerebral cortex appear to

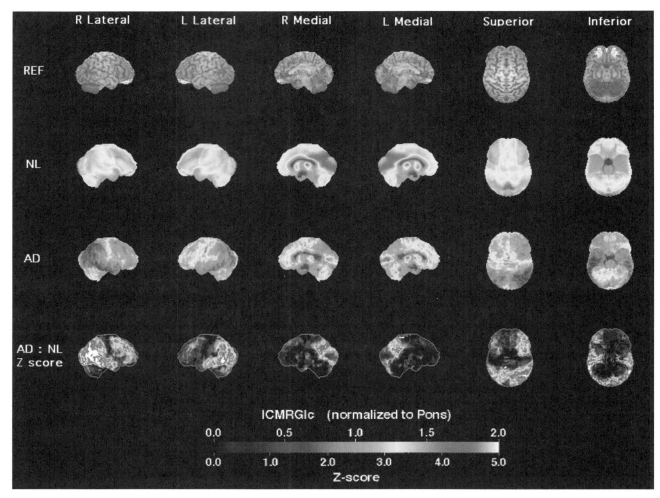

FIGURE 38.2. Cerebral glucose metabolism in a patient with autopsy-confirmed Alzheimer disease (AD) using stereotactic surface projection (SSP) analysis (183). The first row of images shows the right lateral, left lateral, right medial, left medial, superior, and inferior views of the standard reference atlas brain corresponding to the subsequent fluorodeoxyglucose positron emission tomography (FDG-PET) images. Cerebral glucose metabolism relative to that in the pons is shown in 33 normal elderly control subjects (row 2) and in a patient with autopsy-confirmed AD (row 3). The images for controls are more uniform than that for the patient because they represent values averaged pixel-by-pixel over 33 subjects. In the last row are statistical SSP maps showing pixel-by-pixel z-scores where metabolism is lower in the AD patient compared with the controls. Colors in the FDG-PET images correspond to the values as shown in the color bar at the bottom of the figure (white > red > yellow > green > blue). In the statistical maps, yellow corresponds to a z-score of 3. Note that glucose metabolism in the patient with AD is lower than in controls, particularly in association areas of the cerebral cortex. These decreases are greatest in the temporoparietal association cortex and to a lesser degree in the frontal cortex. This accentuates the relatively preserved metabolism of the primary motor and sensory strips of the cerebral cortex in both lateral views and the superior view. The medial views demonstrate the marked decline of metabolism in the posterior cingulate gyrus and relative preservation of the anterior cingulate gyrus, visual cortex, thalamus, and striatum in the AD patient. The inferior view demonstrates that the anterior temporal cortex is not statistically different in the patient with AD. In this particular patient, glucose metabolism is more affected in the right hemisphere. (See Color Figure 38.2 following page 526.)

be particularly susceptible to injury in AD. This is apparent on microscopic examination and by measurement of enzymes found in cholinergic neurons. Choline acetyltransferase, the enzyme that synthesizes acetylcholine, is 60% to 90% lost by the time of autopsy (61,62). Another important marker of cholinergic function is the acetylcholinesterase, the primary degradative enzyme for acetylcholine in the brain. This enzyme occurs in cholinergic and cholinoceptive neurons and thus is a marker for both pre- and postsynaptic components of cholinergic pathways. Acetylcholinesterase is present in both membrane-bound and soluble forms and is decreased 60% to 85% by the time of death (63). Another

TABLE 38.11. NATIONAL INSTITUTE OF NEUROLOGICAL AND COMMUNICABLE DISEASE AND STROKE–ALZHEIMER'S DISEASE AND RELATED DISORDERS ASSOCIATION CRITERIA FOR ALZHEIMER DISEASE

I. The criteria for the clinical diagnosis of PROBABLE Alzheimer disease include
   Dementia established by clinical examination and documented by the Mini-Mental State Examination, Blessed Dementia Scale, or some similar examination and confirmed by neuropsychological tests
   Deficits in two or more areas of cognition
   Progressive worsening of memory and other cognitive functions
   No disturbance of consciousness
   Onset between ages 40 and 90, most often after age 65
   Absence of systemic disorders or other brain diseases that in and of themselves could account for the progressive deficits in memory and cognition

II. The diagnosis of PROBABLE Alzheimer disease is supported by
   Progressive deterioration of specific cognitive functions, such as language (aphasia), motor skills (apraxia), and perception (agnosia)
   Impaired activities of daily living and altered patterns of behavior
   Family history of similar disorders, particularly if confirmed neuropathologically
   Laboratory results of normal lumbar puncture as evaluated by standard techniques, normal pattern or nonspecific changes in electroencephalogram such as increased slow-wave activity, and evidence of cerebral atrophy on computed tomography with progression documented by serial observation

III. Other clinical features consistent with the diagnosis of PROBABLE Alzheimer disease, after exclusion of causes of dementia other than Alzheimer disease, include
   Plateaus in the course of progression of the illness
   Associated symptoms of depression, insomnia, incontinence, delusions, illusions, hallucinations, catastrophic verbal, emotional, or physical outbursts, sexual disorders, and weight loss
   Other neurologic abnormalities in some patients, especially with more advanced disease and including motor signs such as increased muscle tone, myoclonus, and gait disorder
   Seizures in advanced disease
   Computed tomography normal for age

IV. Features that make the diagnosis of PROBABLE Alzheimer disease uncertain or unlikely include
   Sudden, apoplectic onset
   Focal neurologic findings, such as hemiparesis, sensory loss, visual field deficits, and incoordination early in the course of the illness
   Seizures or gait disturbances at the onset or very early in the course of the illness

V. Clinical diagnosis of POSSIBLE Alzheimer disease
   May be made on the basis of the dementia syndrome in the absence of other neurologic, psychiatric, or systemic disorders sufficient to cause dementia and in the presence of variations in the onset, presentation, or clinical course
   May be made in the presence of a second systemic or brain disorder sufficient to produce dementia, which is not considered to be the cause of the dementia
   Should be used in research studies when a single, gradually progressive, severe cognitive deficit is identified in the absence of other identifiable causes

VI. Criteria for diagnosis of DEFINITE Alzheimer disease are
   The clinical criteria for probable Alzheimer disease
   Histopathologic evidence obtained from a biopsy or autopsy

VII. Classification of Alzheimer disease for research purposes should specify features that may differentiate subtypes of the disorder such as
   Familial occurrence
   Onset before age of 65
   Presence of trisomy 21
   Coexistence of other relevant conditions such as Parkinson disease

From McKhann G, et al. Clinical diagnosis of Alzheimer's disease: report of the NINCDS-ADRDA Work Group under the auspices of Department of Health and Human Services Task Force on Alzheimer's Disease. *Neurology* 1984; 34:939–944, in the public domain.

enzyme, butyrylcholinesterase, normally plays a minor role in degradation of acetylcholine in the brain, but it is increased in AD and associated with neuritic plaques (64). The time course of cholinergic changes is not well understood. Cortical biopsies in one study showed 64% loss of choline acetyltransferase, significantly less than in autopsy specimens (62). A postmortem study of patients who died while only mildly demented found little change in either choline acetyltransferase or acetylcholinesterase (65). Other

studies have suggested that enzymatic deficits occur as a relatively early sign of injury before neurons die (66). This is consistent with functional imaging studies using cholinergic markers (49,52).

Familial Alzheimer Disease

The cause of most AD is not yet known, but it is now generally agreed that alterations of both APP and tau processing

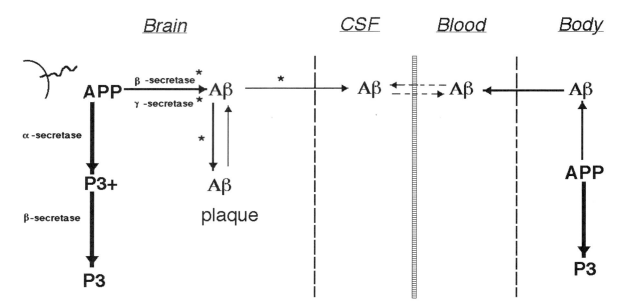

FIGURE 38.3. Schematic diagram of amyloid precursor protein (APP) processing. APP is a linear transmembrane protein found in neural and nonneural tissues. It has a short cytoplasmic C-terminal segment and a much longer extracellular N-terminal tail (schematically represented at the upper left corner of the figure). APP primarily is metabolized at the cell membrane through the α-secretase pathway. APP is cleaved between amino acids 16 and 17 by α-secretase to form a secreted product (sAPPa) and p3+ that is then further broken down by γ-secretase to p3. Aβ protein results from a much less active alternative pathway in which APP is catabolized first by β-secretase and then by γ-secretase. Some APP is normally catabolized in the lysosome, but its role in the formation of extracellular Aβ is poorly understood. When present in sufficiently high concentrations, Aβ self-assembles into β-pleated sheets of insoluble plaque. Although this plaque was long thought to be permanent, there is now evidence that dissolution can occur. Aβ is kept soluble and cleared into the CSF with the help of chaperone proteins such as apolipoprotein E. Aβ is present normally in both the CSF and blood; however, the origin of Aβ in the blood and the degree to which it is affected by brain Aβ is controversial. Slightly different forms of Aβ predominate in plaques (Aβ 1-42) and in vascular amyloid deposits (Aβ 1-40). Proposed targets for treatment *(asterisk)* include inhibition of β- and γ-secretase to decrease Aβ production through the alternative pathway, prevention of Aβ aggregation, and mechanisms that would increase clearance of free Aβ in the hope of decreasing the initial aggregation of Aβ and promote dissolution of already established plaques.

are involved. Evidence from genetic studies of familial AD has shown that abnormalities in APP processing are sufficient to cause all the features of AD. Approximately 10% of AD is familial (defined as one or more first-degree relatives with AD). Genetic mutations affecting APP metabolism have been identified in some cases of early-onset familial AD (defined as dementia onset before age 65). No genetic mutations have yet been identified in additional cases of early-onset familial AD or in any cases of late-onset familial AD. Although early-onset familial AD is uncommon, the identification of these mutations indicates that APP processing abnormalities are sufficient to cause all the features of AD. Early-onset familial AD is most commonly caused by a mutation in presenilin 1 found on chromosome 14. Less common mutations are on presenilin 2 (chromosome 1) and APP (chromosome 21). All these mutations have the effect of increasing the ratio of Aβ 1-42 over Aβ 1-40. Pathologic APP mutations are close to the Aβ fragment of APP and apparently alter splicing by secretases. The presenilins belong to an uncommon class of intramembranous enzymes and are themselves γ-secretases that cleave APP and help form Aβ protein.

Biomarkers for Alzheimer Disease

The identification of these genetic causes of AD has provided new approaches to AD treatment, as described later in this chapter, and also affected diagnostic evaluations. It is now appropriate to look at whether symptomatic patients with early-onset familial AD have one of the known genetic mutations. Although not reflected by current clinical criteria, identifying a known mutation of presenilin 1, presenilin 2, or APP provides a definitive diagnosis and obviates the need for more extensive investigations. At present, only genetic testing for presenilin 1 is commercially available, although it is usually not covered by medical insurance and is very expensive because it is necessary to sequence the entire length of the gene.

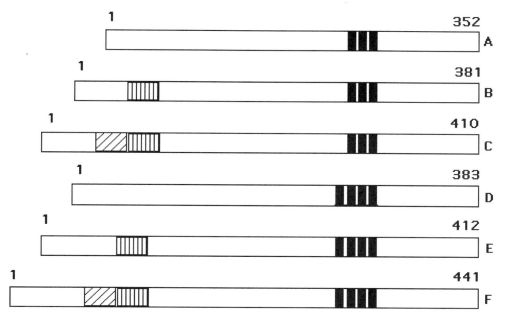

FIGURE 38.4. Isoforms of tau protein. There are six isoforms of tau **(A–F)**. Each isoform has multiple sites that can be phosphorylated. The degree of phosphorylation is tightly regulated and controls tau function. Isoforms contain either three- **(A–C)** or four- **(D–F)** tandem repeat sequences that serve as microtubule binding sites. Four-repeat isoforms are expressed when exon 10 is transcribed. Isoforms are further distinguished by the absence **(A, D)** or presence of an insert of either 29 **(B, E)** or 59 amino acids **(C, F)**. Only isoform A is expressed in fetal human brain. Normal adult brain has all six isoforms with an equal ratio of three- and four-tandem repeat isoforms. In Alzheimer disease, neurofibrillary tangles are composed of all six isoforms of tau, which are hyperphosphorylated. Pick disease involves only three-repeat isoforms. Neurofibrillary tangles in progressive supranuclear palsy and corticobasal degeneration consist of only four-repeat isoforms (184). Inclusions in frontotemporal dementia with parkinsonism linked to chromosome 17 can include all isoforms or only four-repeat forms, when the causative mutations affect the exon 10 splice site (184,185).

Our improved understanding of AD has led to the possibility of using biochemical disease markers for diagnostic purposes. $A\beta$ protein is found in both CSF and peripheral blood. Studies have shown that $A\beta$ 1-42 is decreased in the CSF of patients with AD and probably first increases and then decreases in plasma. Tau also can be measured in CSF. Tau increases in the CSF of patients with AD. Commercial tests are available to measure CSF tau and $A\beta$ 1-42, but these tests are not specific and have not been validated sufficiently to use in diagnosis (67). Apolipoprotein E also has been shown to affect the risk of AD (68). Inheritance of the apoE4 allele increases the risk of AD and is more frequently found in AD patients than in the general population. However, this risk factor appears to be primarily important for patients who are 60 to 70 years of age, and apolipoprotein E is inappropriate for presymptomatic diagnosis and provides little diagnostic help in a specific patient (68,69).

Alzheimer Mimics

AD was not recognized as the most common cause of dementia until systematic autopsy studies were performed in the 1960's and 1970's (70,71). These early autopsy studies found that the clinical diagnosis of dementia was frequently wrong. With the development of explicit clinical criteria and systematic evaluation methods, the clinical diagnosis of dementia now is significantly more accurate (56,57). Laboratory tests effectively identify all medical illnesses that cause dementia, and brain imaging recognizes all structural lesions. Yet diagnostic errors remain. Even in the most experienced and after excluding patients with complicating conditions, 10% to 15% of patients with a clinical diagnosis of probable AD do not have the typical neuritic plaques and neurofibrillary tangles of AD at autopsy. These diagnostic errors are owing to Alzheimer mimics, disorders that can share so many of the characteristics of AD that they may not be distinguishable from AD (Table 38.9). Better understanding of the typical clinical features and course of AD and other dementing disorders and the use of neuropsychological testing have further improved the accuracy of dementia diagnosis and permitted earlier diagnosis. Despite the improvements in diagnosis, 10% to 15% of patients with a clinical diagnosis of probable AD at academic centers do not have the typical neuritic plaques and neurofibrillary tangles of AD. Even after extensive evaluation, several disorders have many characteristics in common with AD and are difficult to differentiate. Recent intensive study of these individuals has led to the identification of several newly characterized disorders that are truly Alzheimer mimics: disorders that meet criteria for AD and fail to have evidence of another disorder on typical laboratory tests. There is a great deal of interest in improving our recognition of these mimics because they are likely to respond to different treatments and have different genetic implications for family members.

We now have a very good understanding of the natural history and clinical features of AD. This has permitted a retrospective comparison of patients with AD and Alzheimer mimics. Although there is a wide variation in the clinical features of patients with AD, unusual presenting features, predominant cognitive impairments other than memory loss, and recognizable variations in clinical course do seem to differ. These findings are now being used to improve our recognition of Alzheimer mimics. In some cases, this has led to proposed clinical criteria or diagnostic guidelines. Their validity in clinical practice is not yet established, and none is widely enough accepted to cause the modification of current AD criteria.

Several other dementing diseases can have clinical features that are indistinguishable from AD. Familial CJD may not have periodic discharges on EEG and an atypical long course of many years (72). Occasionally, eye movement abnormalities develop only late in PSP, making dementia a more prominent feature than usual and leading to a misdiagnosis of AD (73,74). Generally, these mimics can be recognized as patients are reexamined throughout the course of their illness, but it is always appropriate to recommend a postmortem examination to families to confirm the clinical diagnosis. An autopsy can provide invaluable information to assess the risk of other family members and contribute to further research.

Frontotemporal Dementia

Structural, and particularly, functional brain imaging has aided in the clinical recognition of dementias that predominantly affect the frontal and anterior temporal lobes with relative sparing of the temporoparietal cortex (Fig. 38.5). Such patients typically are characterized by the early onset of behavioral symptoms or language impairment. These characteristics remain prominent throughout the illness. Patients can exhibit a wide range of behavioral symptoms, including disinhibition, apathy, and perseverative behavior (Table 38.12). The speech disorder of FTD usually causes a loss of verbal fluency and spontaneity with stereotypy of speech, perseveration, and echolalia. Patients may become mute while appearing to retain significant comprehension.

The diagnosis of FTD remains challenging. Consensus diagnostic criteria have been developed, but symptoms of FTD are not unique and also occur in other dementias. The major distinguishing features are brain imaging and clinical course. Although frontal lobe function is also affected in AD,

this usually occurs only when there is more severe dementia and mutism, if it occurs at all, and appears only in those who are already completely dependent on others. Several clinical syndromes can be distinguished: a dysexecutive syndrome with or without motor neuron disease, progressive nonfluent aphasia, and semantic dementia (75). Semantic dementia has a severe naming and word comprehension impairment in the context of fluent, effortless, and grammatical speech. Parkinsonian signs often emerge late in the illness.

Several different disorders can be found on histologic examination of the brains of patients with a clinical diagnosis of FTD. These include Pick disease, progressive subcortical gliosis, PSP, and, probably most frequently, nonspecific neuronal loss and gliosis with or without vacuolization (76). There are insufficient clinicopathologic studies to permit the different pathologic conditions underlying FTD to be recognized clinically. Family history, age at onset, and other findings differ but remain unreliable for clinical diagnosis, and autopsy diagnosis is necessary.

Frontotemporal Dementia with Parkinsonism Linked to Chromosome 17

FTD with parkinsonism linked to chromosome 17 (FTDP-17) is an autosomal dominant, early-onset, familial FTD that has been genetically linked to chromosome 17 (77). Parkinsonism and occasionally motor neuron disease are also present. In most kindreds, the genetic defect involves either intronic or exonic portions of the gene for tau, the cytoskeletal protein in the neurofibrillary tangle found in neurons in several neurodegenerative diseases. These mutations alter the binding of microtubules to tau or favor the production of one tau isoform over another (78,79), but their predilection to damage frontal and anterior temporal lobes is not yet understood.

Pick Disease

Considerable confusion has accompanied this disorder in recent years because of the lack of consensus about the pathologic findings. Some pathologists have defined Pick disease based only on focal atrophy, but, more recently, otherwise unique intracellular inclusions called Pick bodies are usually required. This disorder typically causes a clinical syndrome indistinguishable from FTD. It is rare, and the incidence appears to peak in the 50's and 60's. It is not hereditary. Previously reported familial cases do not have Pick bodies, are owing to a tau mutation, and thus are now classified as FTDP-17 (80,81).

Dementias with Parkinsonism

Dementia is often accompanied by slowing of gait, rigidity, impaired balance, and bradykinesia similar to symptoms of Parkinson disease. These symptoms, particularly

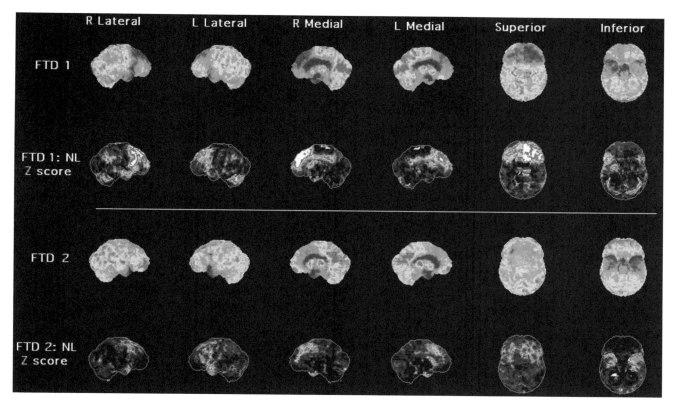

FIGURE 38.5. Cerebral glucose metabolism in two patients with autopsy-confirmed frontotemporal dementia (FTD). The first patient (rows 1 and 2) had Pick bodies in addition to neuronal loss and mild astrocytosis predominantly in the frontal cortex. The second patient (rows 3 and 4) had neuronal loss and gliosis without any inclusions most prominently in the medial temporal lobe, frontal cortex, and anterior cingulate gyrus. Stereotactic surface projection maps are shown of relative regional glucose metabolism (rows 1 and 3) and statistical maps of the differences between each patient and normal elderly controls (NL) in pixel-by-pixel z-scores (rows 2 and 4). These two cases demonstrate the metabolic heterogeneity of FTD. Although the frontal cortex is prominently affected in both, patient 1 has hypometabolism primarily in the frontal cortex with lesser changes limited to the right anterior temporal lobe, whereas patient 2 demonstrates the most extensive hypometabolism in both anterior temporal lobes, best seen in the inferior view of the brain. Analysis methods, abbreviations, and color scales are the same as those used in Figure 38.1. (See Color Figure 38.5 following page 526.)

when they precede or are concurrent with the onset of dementia, suggest that the dementia is caused by one of the disorders that are characterized by parkinsonism (Table 38.9). It should be noted that several disorders are listed both in this category and as Alzheimer mimics. Although these disorders are pathologically distinct, they may be difficult to distinguish from AD. Extrapyramidal symptoms also are frequently seen in AD. More than one-half of patients with moderately severe AD develop new parkinsonian signs over 2 years (82). Resting tremor is common with dementia only in idiopathic Parkinson disease. Although at first this might appear to be helpful, the pathology underlying the dementia in such patients remains in dispute. Some have typical pathologic changes of AD, and others do not. Some have Lewy bodies in the cerebral cortex and should be considered to have DLB. Further research is needed to better understand these patients. For practical purposes, they can be managed like other patients with DLB.

As indicated above, motor symptoms may develop only after dementia has appeared in the disorders listed in this category. This provides another reason that they may be misdiagnosed as AD. FTD may cause parkinsonism, especially FTDP-17. Parkinsonism may be much more evident than FTD in some families with this disorder (83).

Dementia with Lewy Bodies

DLB now appears to be the second most common cause of dementia in the United States, accounting for approximately 20% of all dementia cases in recent autopsy series (84). This condition remains very controversial and is the focus of intensive investigation. It is always accompanied by pathologic changes in the substation nigra characteristic of Parkinson disease, and its relationship to that disease is uncertain. Some would just call this a generalized form of Parkinson disease. Conversely, the changes of DLB are accompanied by pathologic changes of AD in approximately two-thirds of the cases.

TABLE 38.12. CLINICAL FEATURES OF FRONTOTEMPORAL DEMENTIA

Behavioral symptoms
　　Early loss of personal awareness (neglect of personal hygiene
　　　and grooming)
　　Early loss of social awareness (lack of social tact,
　　　misdemeanors such as shoplifting)
　　Disinhibition (such as unrestrained sexuality, violent behavior,
　　　inappropriate jocularity, restless pacing)
　　Mental rigidity and inflexibility
Hyperorality (oral/dietary changes, overeating, food fads,
　　excessive smoking and alcohol consumption, oral
　　exploration of objects)
　　Stereotyped and perseverative behavior (wandering,
　　　mannerisms such as clapping, singing, dancing, ritualistic
　　　preoccupation such as hoarding, toileting, dressing)
　　Utilization behavior (unrestrained exploration of objects in
　　　the environment)
　　Distractibility, impulsivity, and impersistence
Affective symptoms
　　Depression, anxiety, excessive sentimentality, suicidal and
　　　fixed ideation, delusion (early and evanescent)
　　Hypochondriasis, bizarre somatic preoccupation (early and
　　　evanescent)
　　Emotional unconcern (emotional indifference and
　　　remoteness, lack of empathy and sympathy, apathy)
　　Amimia (inertia, aspontaneity)
Speech disorder
　　Progressive reduction of speech (aspontaneity and economy
　　　of utterance)
　　Stereotypy of speech (repetition of limited repertoire of
　　　words, phrases, or themes)
　　Echolalia and perseveration
　　Late mutism

Thus, some have called it a variant of AD. Prominent hallucinations while dementia severity is still mild or moderate in addition to parkinsonism are the best clinical clues of DLB. Clinical criteria for DLB have been proposed and continue to evolve, but they still have significant limitations (85,86).

The disease has important therapeutic implications. These patients are very sensitive to the use of neuroleptic drugs (sometimes even leading to death) and have a cholinergic deficit, suggesting that they might respond to the cholinergic treatments used for AD.

DLB demonstrates primarily temporoparietal hypometabolism similar to AD but, unlike AD, also shows hypometabolism of the occipital cortex (Fig. 38.6). Occipital hypometabolism is seen both with and without coexisting AD pathology (87,88). It can be recognized effectively at postmortem only by using antibodies to ubiquitin and α-synuclein, proteins that accumulate inside the nerve cell to form the unique Lewy bodies.

Progressive Supranuclear Palsy

Patients with PSP usually first develop gait instability and falls. Subsequently, they begin to have dysarthria, dysphagia, and visual complaints. On examination, patients have pseudobulbar palsy, axial and limb rigidity, and slowed saccadic eye movements (89). A peculiar inability to voluntarily move the eyes (supranuclear gaze palsy) is the key to diagnosis. On occasion, however, these patients' dementia is prominent, and gaze palsy develops only later. In these situations, patients can be misdiagnosed as having AD. Diagnostic criteria for PSP have been proposed but have limited sensitivity and specificity in distinguishing PSP from other dementing disorders with parkinsonism (90,91). The dementia associated with PSP primarily reflects frontal dysfunction, as confirmed by functional brain imaging (92).

Corticobasal Degeneration

Corticobasal degeneration (CBD) causes gradual, progressive, asymmetric cognitive and motor impairment. Either cognitive or motor symptoms can develop first. Patients most often develop an asymmetric limb apraxia, which can progress to the "alien limb phenomenon" in which patients feel they have no control over the movements of the limb. Despite good comprehension and being able to describe the task, they are incapable of performing even simple tasks, such as putting on a coat or holding a pencil. The same side of the body becomes rigid and bradykinetic. Frequently, focal limb dystonia accompanies these symptoms, and a variety of other symptoms may emerge, including action tremor, reflex myoclonus, oral apraxia, and dysarthria. Although often initially mistaken for Parkinson disease, motor symptoms do not benefit from levodopa. As the disease progresses, dementia becomes more evident and motor impairments become bilateral, although asymmetry usually remains. Clinical diagnostic criteria have been developed but have limited accuracy, particularly early in the illness (93,94). When patients are severely impaired, their illness may be indistinguishable from AD with extrapyramidal signs. Functional imaging demonstrates asymmetry of both the cerebral cortex and subcortical nuclei (95). Unlike AD, the primary motor cortex may not be spared and the striatum is prominently affected.

Dementia with Focal Findings or Gait Disturbance

Focal motor and sensory abnormalities and nonparkinsonian gait disturbance early in the illness are not features of AD and suggest several other disorders (Table 38.9). Stroke is the most common cause of dementia with focal weakness and asymmetric reflexes. Stroke owing to cerebral infarction and intracerebral hemorrhage can cause this presentation, although cerebral infarction is more common. Cerebral hemorrhage is generally catastrophic, and dementia develops only if the initial insult is survived. Thus, it is of little concern in dementia evaluations because it is usually clinically evident. Closed head injury also can cause dementia with focal findings or gait disturbance. Traumatic brain injury sufficient

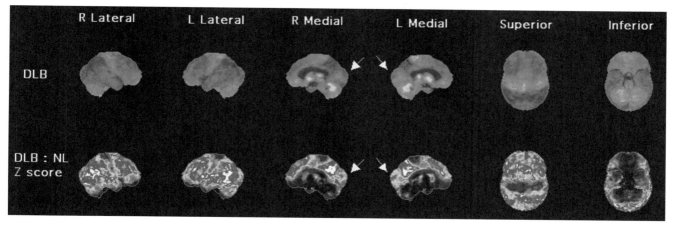

FIGURE 38.6. Cerebral glucose metabolism in a patient with autopsy-confirmed pure dementia with Lewy bodies (DLB). Stereotactic surface projection maps are shown of relative regional glucose metabolism (row 1) and statistical maps of differences between the patient and normal elderly controls in pixel-by-pixel z-scores (row 2). There is diffuse hypometabolism but particularly in the association cortex and posterior cingulate gyrus. There is relative preservation of the primary motor and sensory strips of cortex. The pattern is similar to that seen in Alzheimer disease, except in DLB, hypometabolism is also present in the visual cortex *(arrows in medial views)*. The visual cortex is usually among the most metabolically active areas of the brain, but in this patient, it has values no greater than other areas of the cortex, representing statistically significant hypometabolism. Analysis methods, abbreviations, and color scales are the same as those used in Figure 38.1. (See Color Figure 38.6 following page 526.)

to cause dementia generally follows an extended loss of consciousness or coma and is evident on brain imaging. Subdural hematoma can develop in the elderly after lesser head injury and is associated with focal weakness and reflex changes. It is also evident on CT or MRI. Although on occasion subdural hematoma does not exhibit focal findings or gait impairment, it causes persistent headache and alters alertness so that it usually is not confused with AD.

Dementia and gait disturbance without parkinsonism are early features of multiinfarct dementia (MID) and normal pressure hydrocephalus (NPH). Both MID and NPH cause incontinence and can be suspected on CT and MRI. Gait disturbance can be more evident than parkinsonian features in PSP and CBD, particularly early in the illness.

CBD and CJD can initially present as either a gait disorder or with focal deficits. Asymmetric motor and cognitive impairments are key diagnostic features of CBD. These focal abnormalities or a gait disturbance that initially does not appear parkinsonian can cause diagnostic confusion before other typical features develop. In sporadic CJD, focal cognitive symptoms or a gait disorder develop first. These focal signs may be reflected in focal MRI abnormalities, but they are easily distinguished from changes caused by stroke (34,96). Other clinical features and the rapid progression of symptoms make sporadic CJD easy to distinguish from the other causes of dementia (see section on assessment of rapidly progressive dementia above).

Multiinfarct Dementia

Dementia that results from cerebral infarction is known by a variety of names. The use of the term MID emphasizes the importance of more than one stroke for the development of dementia. Although the term vascular dementia literally applies to dementia associated with all kinds of cerebrovascular disease, in ordinary usage, it usually also refers to dementia as a result of cerebral infarction. A consideration of MID is particularly relevant to any dementia evaluation because it is common and may be difficult to distinguish from other causes of dementia.

Although there are not adequate data, presumably the incidence and prevalence of MID are decreasing as the incidence and prevalence of stroke are decreasing. Nevertheless, MID is the second or third most common cause of dementia in the developed world. Although MID generally causes focal neurologic signs and symptoms, this is not always true at the time of evaluation for dementia. The diagnosis also may not be obvious from the medical history. Consequently, it may mimic AD. Focal signs may be absent because they resolved or because the strokes affected areas of the brain associated with cognition without damaging motor or sensory pathways. Historical information may also be difficult to relate to the patient's dementia. In the face of significant motor deficits, cognitive changes may have been overlooked initially. In other cases, individual strokes may have been small and not recognized. Furthermore, subtle focal deficits may be difficult to recognize if the patient is evaluated only when severely demented. On occasion, patients with MID may present with dementia and parkinsonism. Stroke does not often cause parkinsonism, but in the context of multiple strokes, extrapyramidal pathways are more often affected and may complicate diagnosis.

Strokes can be recognized by structural imaging studies and greatly assist in diagnosis. Both subcortical and cortical strokes can cause dementia. Dementia can occur through a combination of deafferentation of the cerebral cortex and direct cortical injury. Multiple strokes are needed because a single stroke causes a restricted cognitive deficit such as aphasia rather than dementia. Although it is relatively unusual for hippocampal injury to occur with stroke, the effect of bilateral hemispheric injury may lead to memory loss indistinguishable from AD. Conversely, stroke is more likely when there are disproportionate deficits localizable to a single hemisphere. Subcortical strokes may or may not cause dementia. Evidence of impairment of the cerebral cortex on functional imaging studies, rather than the number or location of subcortical infarcts, appears to be the best predictor of dementia (46). The effects of subcortical strokes can be mild, and the most prominent symptoms and metabolic abnormalities often reflect frontal lobe dysfunction. Cognitive changes frequently are obscured by more apparent physical deficits or may be dismissed or overlooked when motor deficits are so much more obvious. Conversely, strokes can be located in strategic areas that cause more cognitive impairment than motor deficits, and, thus, strokes may not be suspected as a cause. Their effects appear to be more than additive and may be most noticeable only several months after the acute stroke. The diagnosis is based on a combination of the history, physical examination, and brain imaging. Beware if there is overreliance on the imaging study because the significance of the lesions to areas responsible for cognition must be assessed and cannot be assumed to have clinical implications. The Hachinski score is based on the typical symptoms and findings on neurologic examination (Table 38.13). This is the best aid in differentiating MID from AD because it has been pathologically validated and appears more reliable than other proposed clinical criteria (97,98).

Normal Pressure Hydrocephalus

NPH causes the triad of mild dementia, urinary incontinence, and a slow, shuffling, unsteady gait, often described as magnetic or apraxic (99,100). In addition, there is a dispro-

TABLE 38.13. CLINICAL FEATURES OF MULTIINFARCT DEMENTIA USED IN THE MODIFIED HACHINSKI ISCHEMIA SCORE (97)

Feature	Point value
Abrupt onset	2
Stepwise deterioration	1
Somatic complaints	1
Emotional incontinence	1
History or presence of hypertension	1
History of strokes	2
Focal neurologic symptoms	2
Focal neurologic signs	2

0–2, suggests Alzheimer' disease; 2–4, indeterminate; >4, suggests multiinfarct dementia or mixed dementia.

portionate ventricular enlargement on CT or MRI. Despite the hydrocephalus, there is normal CSF pressure on lumbar puncture and no obstruction to CSF flow, which can be shown by cisternography or the demonstration of a flow void in the cerebral aqueduct on MRI. The pathology underlying NPH is poorly characterized. Sometimes periventricular gliosis can be seen on MRI as a smooth periventricular rim of increased signal. Other causes of dementia are sometimes found at biopsy or autopsy. In other cases, there is a history of an illness that could explain impaired CSF absorption, such as meningitis, previous head injury, or subarachnoid hemorrhage.

NPH originally generated a great deal of interest because symptoms appeared to improve after ventriculoperitoneal shunting. However, benefits of surgery were poorly documented by current standards, and subsequent experience was often disappointing. A multicenter trial using objective measures found improvement in 36% of patients after shunting, with marked improvement in only 15% (101). Shunt-related complications also are frequent. In this trial, 28% had moderate to severe shunt-related complications. Perhaps surprisingly, the short-term response to ventricular shunting seems to be little influenced by the presence or absence of AD on biopsy (102). Despite many efforts, no clinical measure has been shown to reliably predict which individuals with this syndrome will improve with surgery, although those with a recognizable cause of impaired CSF circulation appear more likely to benefit. A cautious approach to diagnosis and treatment is warranted because NPH appears to be rare and overdiagnosed.

PRINCIPLES OF DEMENTIA TREATMENT

Goals of Treatment

The ultimate goals of dementia treatment are to maximize patients' functional abilities, support caregivers, and protect patients (Table 38.14). To achieve these goals, several different strategies are needed, including drugs, education, social interventions, and nonpharmacologic therapies. The overall approach to dementia treatment draws on the principles of geriatric medical care. Treatment should focus on functional outcomes and use a multidimensional approach recognizing that many problems often coexist.

Taking steps that maximize function means that a patient will require less assistance and supervision and have the best possible quality of life. The essential first step for all therapies is to identify the specific cause of a patient's dementia. Once a specific diagnosis has been reached, it is possible to design appropriate interventions based on what is known about the biology of the disease. It also is important to recognize potential causes of excess disability. Patients with dementia are particularly susceptible to cognitive adverse effects of drugs; therefore, it is best to avoid medications active in the central nervous system whenever possible. The identification and

TABLE 38.14. GOALS OF DEMENTIA TREATMENT

Maximize patient function
  Treat the cause of dementia
  Avoid unnecessary central nervous system–active medications
  Treat coexisting medical illnesses
  Treat behavior disturbances
  Keep patient physically, socially, and mentally active
Support caregivers
  Recognize and treat caregiver concerns
  Encourage use of day care and other respite care
  Identify and involve multiple caregivers early in the illness
  Educate caregivers about the disease, its management, and
    available resources
  Compliment caregiver efforts
  Monitor caregiver health
Protect the patient
  Simplify medications
  Address legal protections (health advocate, durable power of
    attorney, guardianship)
  Evaluate road and home safety
  Help caregivers plan for the unexpected
  Assure the patient has sufficient supervision
  Monitor quality of care and nutrition

treatment of coexisting medical illnesses are an important part of the initial dementia evaluation, but regular medical visits also are helpful in recognizing and treating medical problems that might be contributing to a patient's functional and cognitive difficulties. Behavioral symptoms also contribute to patient disability. Untreated depression, anxiety, and psychosis interfere with everyday activities. There also is increasing evidence that patients are healthier and have fewer behavioral symptoms if they are mentally, physically, and socially active. The quality of sleep, for example, is significantly improved by exposure to sunlight and regular physical activity (103). It is helpful to inquire about exactly what activities patients are involved in and how often they leave their residence. It is reasonable to recommend that patients maintain aerobic exercise equivalent to walking at least 1 mile daily. Patients with memory impairments naturally tend to withdraw from activities, so encouraging socialization is particularly important.

Caregivers play an important role in dementia treatment. All patients with dementia are receiving long-term care. Too often we think of long-term care only as institutionalized care, but actually family members provide most long-term care. There can be a trade-off between family caregiving and institutional care. Inadequate support of caregivers at home will mean earlier institutional care. Most patients strongly prefer care by family members, and, in general, families also prefer to be involved in care. Neither the personal commitment nor the intensity of care that families provide can be duplicated in an institutional setting, particularly when there are financial constraints. It is, therefore, important to recognize and treat caregiver concerns. To focus discussion, it often is helpful to ask the caregivers directly what they

find to be most difficult. Behavioral symptoms are especially important, for they often cause institutionalization (104). However, the physician cannot simply respond to caregiver requests for medication but also must consider environmental and caregiver factors that can be contributing to problems before turning to drugs. Drug treatment will inevitably fail when these other factors are not also addressed.

It is important to set realistic goals for caregivers when extended periods of effort are required. Many individual caregiver factors influence the likelihood of institutionalization, including attitudes toward home care and perceived burden of care (105). It is difficult to generalize about the effects of ethnicity. For example, African-Americans are underrepresented in nursing homes nationally but not in some regions where the average income is higher (106,107). The structure of the caregiver network also differs among different ethnic groups (108). Unlike providing care for acute illnesses, caregivers have to pace themselves and avoid fatigue and exhaustion. Caregivers should be encouraged to use senior citizens' activities, adult day care, and other forms of respite care when they can make adjustments and before they are overwhelmed. Caregivers should never feel strained by their caregiving efforts, or caregiving will not be sustainable over the long run.

Recognizing that increasing care will be needed for progressive dementias, everyone who can help with home care should be identified and utilized. Even when they are unable to provide direct care, all family members should be expected to contribute in some way, either in time or with financial support. It is much easier to involve family members when relatively little is being asked, rather than having a single individual bear the total burden of care alone. The physician can help family members identify and coordinate their contributions by arranging a family meeting. Directly involving as many family members as possible can draw families together by encouraging them to work toward a common goal. They also will have more realistic expectations of the patient's condition and what can be expected. Without early involvement, they may not recognize the changes that have occurred or develop a realistic knowledge of the patient's needs and what can be accomplished. The resulting misunderstandings and disagreements among family members ultimately can be very destructive and extremely time-consuming for physicians.

Education of caregivers about dementia, the role of caregiving, and community resources is important. Caregivers find knowledge empowering, and it increases satisfaction and feelings of competence. Health professionals should support caregivers by verbally recognizing their contributions. Too often caregivers become discouraged when they receive too little guidance or only their faults and failures are identified. Many studies have shown that caregivers themselves suffer adverse health consequences in their role (109–112). Attention to caregivers' health is needed so that they will be there to support and aid the patient.

Physicians sometimes must advocate for patients and protect them from consequences of their impaired decision-making ability. Drug regimens should be simplified to improve compliance and minimize medication errors. Formal legal protections should be recommended to patients early. Families should be involved in these discussions because their assistance may be needed. A health advocate or a durable power of attorney for health decisions should be designated. This can often be easily done in the physician's office with standard forms. It is also important for patients to name a durable power of attorney who can review and oversee financial and legal decision making. These steps will permit patients to participate in designating those of their choice and help to avoid situations in which others can take advantage of their impairments. Guardianship is needed when the patient is incompetent. Addressing these issues early will save the physician later effort and uncertainty. Driving and home safety should be assessed at each visit to avoid injury to the patients and others living in the same building and using the same highways. Patients and caregivers should be asked to consider plans for dealing with emergencies. For example, it is important to identify what would happen if the primary caregiver suddenly became ill or died. Continuing assessment of a patient's degree of supervision and support is needed. A simple way to evaluate care is to observe regularly for physical signs of abuse and to monitor weight and general hygiene. Inadequacies in care may occur when the care system fails or when the needs of an increasingly impaired patient can no longer be met.

Preventive Interventions for Dementia

A passive, laissez-faire approach to dementia problems may seem easiest and most inexpensive in the short term, but failing to intervene eventually costs much more and requires much greater physician effort. Indeed, the best approach is to identify potential problems before they develop fully and address them early. When dementia symptoms are still mild, there is often a "honeymoon period" when there are few care problems and complications have not yet developed. Although there is a tendency to not look for problems, this is an ideal time to address issues that will likely appear in the future.

Specific illnesses have predictable complications that can be considered on follow-up examinations. For example, patients with AD have more frequent behavioral symptoms as the dementia becomes more severe. Consequently, hostile conduct is likely to progress, and treatment should be initiated before a crisis develops. Urinary incontinence is common in AD only when a Mini-Mental State Examination score is 10 or lower. The abrupt onset of incontinence in less impaired patients suggests a urinary tract infection. Many patients with AD develop bradykinesia and rigidity as their disease progresses. Physical activity and dopaminergic drugs may be helpful. Patients with AD lose weight

when they lose the ability to feed themselves and are not adequately fed.

Some complications of progressive dementia can be prevented or at least forestalled. Progressive dementias predictably lead to incompetence, so provisions should be made to have the patient appoint a legal advocate to aid in financial and health decisions as soon as possible. This will avoid the later need for legal proceedings to appoint a guardian and ensure that the patient's wishes will be carried out in the future. When discussing advanced directives, it is helpful to know that tube feeding is generally futile (113). Progressive dementia also leads predictably to the inability to be independent with transportation. Driving competence needs to be repeatedly assessed, and alternate transportation arrangements need to be made. Many communities have transportation services for seniors that can be utilized to maintain mobility within the community. Family members and friends are often happy to provide transportation as a tangible way to provide assistance.

It is important for patients to keep mentally, socially, and physically active. Everyone can benefit from such activities, but that can be particularly advocated for patients with dementia and needs to be recommended because there is a natural tendency to withdraw from friends and all activities. Patients with dementia tend to become increasingly isolated as they find social relationships more threatening to their self-esteem and others become less comfortable with normal social intercourse. Social activities can be promoted by encouraging participation in clubs, senior citizen activities, or day care that can provide activities appropriate for the level of impairment.

Patients may prematurely abandon activities because they are frustrating. Yet patients can often succeed when they are approached differently and caregivers help them by breaking down tasks into discrete steps. Because of relative preservation of reading skills in many dementias, patients may benefit from written instructions. Continued self-sufficiency achieved in this way has great rewards for both the patient and caregiver. It can be particularly beneficial to the self-esteem of a patient who finds him/herself progressively limited. Physical activity promotes conditioning and general physical health. Loss of mobility increases susceptibility to infection, falls, and injury. Traditional classification of patients with dementia as frail reflects the common deconditioning of such individuals because of this withdrawal.

In addition to settling legal arrangements, several other steps can prevent crises. Reinforcing the patient's care system is important. Progressive dementia requires increasing supervision and support. Caregivers risk exhaustion and having their own health problems. An early preventive goal is to identify other individuals, often other family members, who can provide help. Involving others early means that initially less is demanded and so needs are easily met. When the patient is less impaired, it is less difficult to adjust to multiple caregivers. Likewise, others find it natural to gradually adjust their contributions as demands increase. Those who do not serve as primary caregivers are often uncertain about how to help. They need to be directed to participate in care by identifying and then regularly providing enriching activities appropriate for the patient's disability, preferably activities that they also enjoy.

Patient and caregiver education is critical in preventing crises. Information should be provided about available services and options for assistance within their financial means. Sudden, unplanned increases in the burden of caregiving cannot be easily met or sustained. It is helpful to review what will be done if unexpected problems develop, such as the incapacity of the primary caregiver, the patient is suddenly missing, or there is a fire in the home. Plans also should be made for foreseeable situations that would substantially increase caregiver duties, such as the development of incontinence, sleep disturbance, behavioral disorders, or reduced mobility. These plans should involve other family members, friends, and neighbors whenever possible. Caregivers need to be instructed about when and how to contact physicians and social services. Providing continued guidance and education gives caregivers a sense of mastery and permits them to understand what is needed to adequately cope with problems if they arise. This avoids physical and mental abuse by caregivers and eliminates the unfortunate phenomenon of "granny dumping," which occurs when caregivers become overburdened and frustrated.

NONPHARMACOLOGIC TREATMENTS

Dementia increases dependency on others. The environment and the decisions others make increasingly determine how dementia affects the patient as symptoms progress. With inadequate support and encouragement, patients with memory problems typically withdraw from social activities and become increasingly isolated. This is counterproductive because it discourages the involvement of others, decreases the quality of life of the patient, and fails to take advantage of the patient's strength. Inactivity also contributes to physical deconditioning and debility, hastening dependency and increasing the susceptibility to infections. There is increasing evidence that involvement in social activities and physical exercise is beneficial. Initial clinical drug trials for AD, for example, were inadequately powered because the progression of dementia was overestimated by observational studies conducted in institutions with few social or physical activities. Their involvement of regular neuropsychological testing and regular medical visits seemed to have a beneficial effect (114).

While it is easy to recommend simple, apparently logical interventions, randomized studies are needed to ensure that the additional expense and effort are effective and therefore justified. Fortunately, over the past few years, various interventions have been shown convincingly to be effective in controlled trials (Table 38.15). Graded assistance with dress-

TABLE 38.15. EFFECTIVE NONPHARMACOLOGIC TREATMENTS IN DEMENTIA

Problem	Technique
Trouble dressing	Step-by-step retraining (165)
Disruptive vocal outbursts	Scheduled one-on-one interactions (118)
Repetitious questions and nonsensical speech	Use of a memory book (166)
Incontinence	Prompted voiding (116,117)
Inability to provide care at home	Support groups, round-the-clock social work availability, patient and caregiver education (167)
Depressed mood	Problem-solving with caregiver, positive activities, and supervision (168)

ing and other simple tasks can enhance self-sufficiency and self-respect. Graded assistance supplemented by practice and positive reinforcement improves performance of daily activities (115). Prompted voiding reduces urinary incontinence, even among those with severe dementia (116,117). Increasing one-on-one interactions can decrease nonsensical speech and verbal outbursts (118). All these approaches should be used to reduce the need for medication and improve care whenever possible.

PRINCIPLES OF DRUG TREATMENT IN DEMENTIA

Drug regimens should be simplified as much as possible to maximize compliance. A corollary of this principle is that only drugs that are clearly beneficial should be used. It is reasonable to discontinue drugs in severely demented patients that have been prescribed purely to decrease disease risk, such as cholesterol-lowering agents. The use of drugs with potential adverse cognitive effects should be minimized. The importance of medical illnesses in dementia should not be underestimated. In some cases, medical illnesses that can cause dementia are identified. Effective treatment of such illnesses may not reverse the dementia. There are specific treatments for many medical diseases that cause dementia such as vitamin B_{12} deficiency, hypothyroidism, and neurosyphilis. Many nervous system disorders, such as herpes encephalitis and multiple sclerosis, also have specific treatments. This section, however, focuses only on the most common brain diseases in the elderly that are likely to be seen by the neuropsychiatrist.

DRUG TREATMENTS SPECIFIC FOR ALZHEIMER DISEASE

Our understanding of pathophysiology provides the rational basis for the development of drugs to treat AD. *In vitro* studies suggest which drugs might prevent or compensate for biochemical changes caused by disease, and animal research initially evaluates safety and indicates whether expected drug effects are observed. However, there is no substitute for randomized, placebo-controlled clinical trials to show that a treatment is effective. Only when such trials demonstrate safety and efficacy is the use of drugs justified.

Patients and their caregivers understandably are anxious for an effective treatment and may seek any drugs or diet supplements that seem promising, particularly if they seem to have few side effects. However, it is important to await convincing evidence of effectiveness. The importance of waiting for results of randomized trials recently was demonstrated by trials of estrogen replacement therapy. When substantial cellular and epidemiologic evidence accumulated that estrogens might be beneficial in AD and small and open-label trials were encouraging, there began to be widespread use of estrogen replacement therapy for AD (119). However, subsequently several randomized trials showed no improvement of symptoms with treatment and even a trend suggesting worsening of symptoms (120–122). Furthermore, patients receiving estrogen replacement therapy experienced significantly more adverse effects than those receiving placebo. Patients with dementia can be easily exploited, and the cost of drugs and supplements are especially important to consider in view of the great financial burdens of dementia. Using the standard of randomized clinical trials, AD patients clearly benefit from cholinesterase inhibitors and antioxidant drugs.

Cholinesterase Inhibitors

Loss of cholinergic markers is the greatest and most consistent neurotransmitter deficit in AD. Furthermore, cholinergic pathways had long been known to be critical to memory function because anticholinergic drugs can induce amnesia. After inconsistent results for many years, the first convincing evidence of benefit with cholinesterase inhibitors was shown with tacrine (Table 38.16). In retrospect, earlier results had been inconsistent because trials had been too short and too small. Not only were the benefits of treatment less than hoped, but also the rate of progression of symptoms was lower than predicted from observational studies (patients seemed to benefit from study participation). Moreover, to the surprise of investigators at the time, placebo-treated patients had transient improvements (presumably owing to practice effects), requiring longer periods of observation to distinguish benefits from drugs.

These trials were an important landmark for dementia research. They validated methods for evaluating the effect of drugs on memory and cognition. They also demonstrated that cognitive effects were relevant when they could be perceived by physicians and by caregivers blinded to treatment assignment. Furthermore, these trials demonstrated that AD symptoms could be improved, even with drugs that did not reverse neurodegeneration. Previously, many had categorized AD as an untreatable dementia. The success of trials with

TABLE 38.16. CHOLINESTERASES EFFECTIVE IN TREATING ALZHEIMER DISEASE

Drug	Comment
Tacrine[a]	Liver toxicity, four times daily dosing (169,170)
Donepezil[a]	Once daily dosing, long half-life (171,172)
Rivastigmine[a]	Twice daily dosing, also inhibits butyrylcholinesterase (173,174)
Galanthamine[a]	Twice daily dosing, also partial nicotinic agonist (175,176)
Metrifonate	Once a day dosing, very long half-life, neuromuscular blockade (177,178)
Controlled-release physostigmine	(179,180)
Velnacrine	(181,182)

[a]Currently approved for use in the United States.

tacrine opened the possibility that other interventions also could help.

Subsequent studies have demonstrated that several cholinesterase inhibitors can improve patients with AD (Table 38.16). All cholinesterase inhibitors act in the same way, by enhancing the effect of released acetylcholine by reducing its degradation in the synapse. This provides a targeted effect to the physiologically phasic discharges of cholinergic neurons. If neurons are not releasing acetylcholine, they are not affected. The lack of similar targeting probably accounts for the failure of trials with direct cholinergic agonists. It also explains why benefits have not been demonstrated in AD patients with more severe dementia as indicated by a Mini-Mental State Examination score below 10. When there are few functioning cholinergic neurons, the drugs will have little effect. Cholinergic precursors such as lecithin do not have benefit nor add to the effectiveness of cholinesterase inhibitors (123,124).

Cholinesterase inhibitors have only modest benefits. They partially compensate for only a single class of neurons when a diverse population of neurons is damaged. Nevertheless, they have an established role in AD treatment and appear to provide benefits over at least a year and probably longer (125). It is unclear whether more recently developed cholinesterase inhibitors will have advantages. It is now possible to demonstrate the effects of these drugs in living patients with PET and a ligand for cholinesterase, N-^{11}C-methylpiperidin-4-yl propionate (Fig. 38.7). These studies suggest that the effectiveness of these drugs can be improved further (126).

Drugs That Slow Progression

Ideally, we would like to identify drugs that could reverse the symptoms of AD. However, all studies have shown that significant pathologic changes, including substantial neuronal loss, are already present as soon as symptoms are evident. Indeed, there is substantial evidence that significant pathologic changes precede the clinical onset of dementia (127). As a consequence, our most realistic goal is to identify and

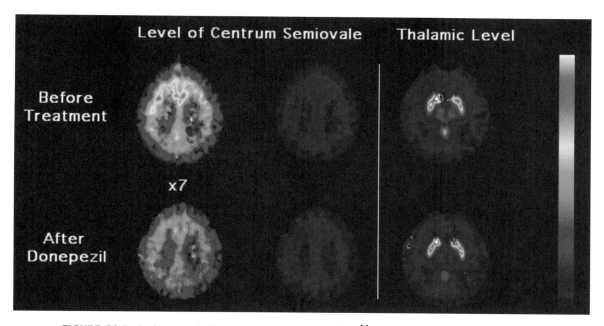

FIGURE 38.7. Positron emission tomography scan with *N*-^{11}C-methylpiperidin-4-yl propionate ([^{11}C]-PMP), a ligand for acetylcholinesterase, in a patient with autopsy-confirmed Alzheimer disease before **(top)** and after **(bottom)** 6 weeks of treatment with 5 mg daily of the cholinesterase inhibitor donepezil. These images reflect enzyme activity and are calculated from a sequence of scans collected over 60 minutes. Two different brain levels are shown. The four images on the left side of the figure are horizontal slices of the brain parallel and approximately 70 mm above the canthomeatal line at the level of the centrum semiovale. The two images on the right side of the figure are horizontal slices approximately 40 mm above the canthomeatal line at a level through the thalamus. The front of the head is shown at the top of each image. Values correspond to the adjacent color bar with white colors representing the highest values. In the two images furthest to the left, values have been multiplied by 7 to better demonstrate binding in the cerebral cortex. The other four images are shown in true scale. Acetylcholinesterase is much more abundant in some subcortical structures, accounting for the much higher values in the striatal images. There is an approximately 20% decrease in the hydrolysis of [^{11}C]-PMP throughout the cerebral cortex with donepezil treatment because of its inhibition of acetylcholinesterase. Although changes with treatment are difficult to discern at this threshold in the striatum, which has much higher baseline values (red and white regions in the far right images), declines are apparent after treatment in the thalamus (blue regions immediately medial and posterior to the striatum) and cerebellar vermis (central green region in the before treatment image). The characteristics of [^{11}C]-PMP are optimal for measuring the lower enzyme levels seen in the cerebral cortex; other radiotracers are better suited to measuring acetylcholinesterase activity in the striatum. (See Color Figure 38.7 following page 526.)

treat AD as early as possible and use a drug that can prevent progression of symptoms. We have not yet been able to reach this goal, but it is possible to slow progression of AD. A multicenter, placebo-controlled clinical trial in moderately severe AD (mean Mini-Mental State Examination score in study groups was 11 to 13) demonstrated that antioxidants vitamin E (2,000 IU daily) and selegiline (10 mg daily) were both superior to placebo (82). These drugs delayed the development of clinically relevant outcomes, including death, institutionalization, loss of basic activities of daily living, and worsening of the clinical dementia rating score. Although vitamin E has not been shown to alter disease progression in less severe dementia, because of its good tolerability and low cost, it is now routinely used in the treatment of AD. Studies are currently underway to determine whether it has a role in preventing or delaying the development of dementia in high-risk populations.

New Drug Treatment Strategies

Our better understanding of the pathophysiology of AD provides many additional strategies for drug development. The development of animal models of AD is also accelerating the evaluation of new drugs. A variety of new strategies are being investigated. Memantine, a NMDA glutamate receptor antagonist has shown benefits in patients with moderate to severe AD (MMSE 5–15). Much current effort targets approaches that decrease Aβ protein. The effects of proposed treatments can be assessed first in transgenic animals engineered to overexpress Aβ (128,129). The enzymes β- and α-secretase are necessary to make Aβ from APP (Fig. 38.3). Inhibitors of these enzymes decrease production of Aβ and may prove valuable. β-Secretase knockout animals appear normal, suggesting that inhibition of this enzyme might be well tolerated clinically. α-Secretase also participates in the

TABLE 38.17. POTENTIAL FUTURE STRATEGIES FOR TREATING ALZHEIMER DISEASE

Alter processing of amuloid precursor protein to decrease Aβ production
Increase clearance of Aβ protein
Inhibit aggregation of Aβ protein
Prevent hyperphosphorylation of tau protein
Inhibit tau aggregation
Enhance neuronal survival and promote maintenance and regrowth of synaptic connections using nerve growth factors
Replace lost neurons and synapses with stem cell implants

processing of other cell membrane proteins including Notch, which is critical for normal development and blood cell production. As a consequence, inhibitors of this enzyme may have significant adverse effects, but further work is warranted because only partial inhibition may be necessary for clinical benefit.

Another approach under intense investigation that would reduce Aβ is methods that would increase clearance of Aβ. Using a strategy similar to that applied in the treatment of systemic amyloidoses, removal of free Aβ reduces its concentration and decreases the formation of aggregation into plaques. In transgenic animals, systemic immunization forming antibodies to Aβ protein prevents the development of plaques and dissolution of plaques that are already present (130). A similar immune response elicited by nasal administration of Aβ has a similar result (131). The precise mechanism by which this can be achieved is not fully understood, but, surprisingly, at least some antibodies appear to cross the blood–brain barrier (132). Using exciting new techniques, the dissolution of the plaques by antibodies can be directly observed in transgenic animals (133). A variety of other approaches can be envisioned that would either remove or increase free Aβ, but we still do not know the normal function of APP or Aβ or whether free or aggregated Aβ is more important in the pathogenesis of AD.

Another approach that is under investigation is the inhibition of Aβ aggregation. This can be easily assessed by *in vitro* experimentation, and many antiaggregants have been identified. Identifying those that will effectively reach the brain and are well tolerated will require significant effort.

Although most effort is currently directed at modifiers of Aβ, there are many other approaches that offer hope of efficacy (Table 38.17). Each of these approaches has its own hurdles to overcome for development, but, unlike in the past, there is a surfeit of ideas that need to be tested.

SPECIFIC DRUG TREATMENTS FOR OTHER DEMENTIAS

It should not be assumed that effective treatments for AD are effective in treating other dementias, irrespective of their clinical similarity. Each dementing disorder has a specific

pathophysiology that determines the applicability of treatment with a specific drug. The major commonality of dementia is neuronal death, but none of our current treatments directly addresses this issue. It is true that clinical drug trials for AD, by necessity using entry criteria based on clinical diagnosis, always include an undetermined number of individuals with AD mimics. However, the overwhelming relative prevalence of AD means that the outcomes are more likely to be diluted in a minor way rather than that the results can be generalized to the treatment of AD mimics.

Frontotemporal Dementia

The behavioral symptoms of FTD can be difficult to manage satisfactorily with drugs. Personality changes typical of FTD, such as unrestrained, inappropriate behavior, poor judgment, and mental inflexibility and impulsivity, can be extremely disruptive and upsetting and respond poorly to drugs. Hypochondriasis can also lead to complications if not recognized as part of the illness. A combination of environmental restrictions and sedative or neuroleptic medications is often needed. Education and supportive involvement of the family are critical.

Society finds such behavior intolerable, and early institutionalization or criminal incarceration is frequent. The neuropsychiatrist may be need to be an advocate for the patient's health needs and testify to the medical rather than criminal nature of this behavior. Although these behaviors can be disabling, they are inconsistent and change as the disease progresses. No single treatment will remain effective, and symptoms and treatment need to be reassessed at least every few months.

Some symptoms of FTD are more amenable to drug therapy. Depression, obsessions, and compulsions respond to usual symptomatic treatments as described below. Unfortunately, no treatments have been identified that alter the underlying causes or course of FTD. It is easy to conceive strategies that might be helpful, particularly for forms of FTD that involve disrupted regulation of tau protein. Transgenic mouse models of FTDP-17 have been developed that might be used for testing drugs (134,135). However, the pathologic diversity of FTD, the current difficulties in accurate diagnosis, and the relatively low prevalence of the disorder are all obstacles for testing treatments in appropriate clinical trials. There is no evidence that they benefit or are likely to benefit from cholinesterase inhibitors or antioxidants.

Dementia with Lewy Bodies

Treating motor symptoms in DLB often proves to be a challenge. The motor symptoms of DLB can be treated with the typical dopaminergic agonist used to treat Parkinson disease. However, adverse effects often limit the doses that can be used and thus the benefits that can be achieved. Exacerbation of hallucinations is particularly a problem. A limited number of controlled clinical trials and studies of cholinergic markers in postmortem brains suggests that cholinesterase inhibitors are likely to be particularly effective in DLB and even may prevent some of the fluctuations in cognition characteristic of this disorder (136,137). The frequent coexistence of AD with DLB provides further justification for the use of AD-specific treatments in this disorder.

Treatment of behavioral symptoms in DLB is particularly challenging. Patients with DLB are especially sensitive to the effects of neuroleptic drugs, including atypical antipsychotics, and they should be avoided (138,139). In particularly difficult situations in which environmental modifications and other medications have failed, quetiapine and clozapine might be considered because they tend to exhibit fewer extrapyramidal side effects.

Multiinfarct Dementia

The most effective treatment of MID is prevention. Dementia owing to stroke can be prevented if stroke, particularly recurrent stroke, is prevented. Several interventions can decrease the risk of stroke, including endarterectomy, antiplatelet drugs, anticoagulants, statins, and effective control of hypertension (140,141). A review of the appropriate use of these drugs is beyond the scope of this chapter, but it is important for neuropsychiatrists to work closely with colleagues who treat stroke. They should help to identify patients who develop even mild dementia after a cerebral infarct and encourage the prevention and stabilization of dementia as explicit goals of stroke therapy.

Although prevention is a worthy goal, the neuropsychiatrist will more often encounter patients with stroke who already have dementia. In this situation, it is reasonable to take steps to prevent the progression of symptoms by aggressively evaluating the cause of stroke and instituting appropriate therapy as soon as possible. The role of cholinesterase inhibitors in MID is under investigation but has not been established. They are appropriate for use in patients with suspected mixed dementia who have a low Hachinski score, a slowly progressive course, or evidence of progression unlinked to further stroke, suggesting that AD is contributing to the patient's symptoms.

DRUG TREATMENT OF BEHAVIORAL SYMPTOMS

Behavioral symptoms become increasingly likely as dementia severity worsens, but they are not inevitable. Some patients never have psychiatric symptoms and others respond to environmental modifications. Drug treatments should be used only when nonpharmacologic approaches have failed to improve symptoms. When nonpharmacologic approaches are not improving symptoms, drug treatment should be started

TABLE 38.18. DRUG TREATMENT OF BEHAVIOR DISTURBANCES IN DEMENTIA

Symptom	Agents of Choice
Depression	Selective serotonin reuptake inhibitor antidepressants
Agitation	Sedating antidepressants such as trazodone, anticonvulsants, β-blockers, typical and atypical antipsychotics
Delusions, paranoia, hallucinations	Antidepressants if depressed, antipsychotics if severe unless extrapyramidal symptoms are present
Sleep disturbance	Sedating antidepressants such as trazodone, sedating antipsychotics, melatonin, intermittent, brief courses of benzodiazepines

early, when symptoms are more responsive to treatment and lower doses may be sufficient. In this situation, drug doses should start low and gradually increase. In general, behavioral symptoms in demented patients can be treated with the same but lower doses of drugs that are used when they are caused by psychiatric disease (Table 38.18). In a crisis, higher doses of drugs are justified, but there should be a planned dose taper, particularly for antipsychotics and drugs with a long half-life. Age and medical problems often decrease drug metabolism and make adverse effects more likely.

Once present, specific behavioral symptoms in demented patients frequently recur and often remain consistent. Studies examining the biologic origin of behavioral symptoms in dementia are only in the early stage of development. However, there is already growing evidence that behavioral symptoms are owing to brain pathology rather than a psychological response to illness or purely owing to environmental factors. For example, profound disturbances in behavior and personality are characteristic of FTD, including agitation. Likewise, agitation has been found to correlate with orbitofrontal and anterior cingulate pathology and left frontal hypometabolism in AD (142,143). Apathy in dementia is associated with frontal and anterior temporal hypoperfusion (144). Anosognosia, the unawareness of deficit, has many behavioral implications, and two independent groups have found it associated with functional impairment of the right frontal cortex (145,146). It is, therefore, understandable that modifications of the environment and caregiver responses are helpful but often insufficient.

Although behavioral symptoms frequently recur, they are not constant. Once symptoms are controlled, psychotropic medications should be tapered gradually every 6 months to see whether they are still needed. If they can be discontinued, side effects can be avoided, and effective medications can be reinitiated if symptoms recur. If they cannot be discontinued without worsening of symptoms, a lower dose might be found effective or the patient always can return to the previously effective dose.

Several types of symptoms can occur simultaneously. This reflects the multifocal nature of most dementias, but it complicates treatment significantly. Each symptom requires its own treatment and possible contributing environmental factors must be assessed, including the responses of the caregiver. One of the most common problems of treating behavioral symptoms is the tendency for caregivers to focus only on the problem at hand. Complaints from caregivers may continue, despite successful treatment of one symptom. For example, lack of motivation may become more apparent when agitation is controlled. It is important to establish with caregivers the target symptom for treatment and judge the response before going on to treat another symptom. Sometimes adverse effects must be tolerated to adequately control a particularly distressing symptom.

Not every behavioral symptom needs to be treated with drugs. Dangerous and careless behaviors are best managed by providing greater supervision and modifying the environment to disable or remove dangerous items such as guns, power tools, and burners. Providing a supervised, open environment can accommodate pacing and wandering. Altering locks and installing alarms at doors are also very helpful. Altered sleep/wake cycles can be managed by day and night supervision. Repetitive behaviors can be distressing to caregivers but do not impair the care or health of the patient. It is helpful to focus on desired activities rather than trying only to extinguish unwanted behaviors. Caregiver attention should be engaged on positive activities rather than just response to behavior disturbances. In general, behavioral problems should be treated only if they interfere with care or are likely to worsen.

Depression

Depression is common, particularly at the onset of dementia. Providing adequate supportive care and positive activities can be very helpful. Depression should be treated with antidepressants lacking anticholinergic properties. Thus, selective serotonin reuptake inhibitors are preferred. Depression should not be overlooked when accompanied by anxiety and agitation. Combinations of antidepressants with antipsychotics or anxiolytics may be needed in these situations. Caregivers may also report depressive symptoms in the patient when they themselves are depressed. Physicians should carefully confirm caregiver reports by considering patient self-reports and observed affective symptoms.

Agitation

Agitation becomes increasingly common as dementia progresses. Increasing degrees of agitation can develop, beginning with threatening and hostile language, physical threats, grabbing, and frank physical assault. Drug treatment should be gauged to the severity of symptoms. Sedating antidepressants, such as trazodone, often are sufficient for mild agi-

tation. Anticonvulsants such as carbamazepine and valproic acid have also been effective. β-Blockers have also been reported to be of benefit in small studies. When agitation is more severe, antipsychotic drugs are needed. Both typical and atypical antipsychotics can reduce the amount of agitation (147,148). Choice of agents is based on side effect profile, ease of administration, and cost. Typical antipsychotics are much less expensive, an important consideration in a chronic disease like most dementias. Many antipsychotics have significant anticholinergic properties that are helpful for sedation but ultimately are counterproductive in dementia, particularly in disorders with cholinergic deficits such as AD and DLB. Haloperidol is favored because of its lack of anticholinergic and hypotensive side effects, but it is very potent. It is particularly helpful in acute behavioral crises because it also can be given by intramuscular injection. Maintenance doses are lower than doses that may be necessary acutely. Because of its long half-life, dose reductions should be planned within a month when larger doses are given to control behavior in a crisis.

Atypical antipsychotics are not free of significant extrapyramidal side effects (149). Clozapine has the lowest incidence of extrapyramidal symptoms but requires laboratory monitoring and is thus difficult to use. Quetiapine has little anticholinergic effect and appears to have the next least likelihood of extrapyramidal side effects. Risperidone is available in a liquid form that is especially convenient for severely impaired patients and permits easy titration of low doses. Severe behavioral disturbances can usually be controlled with sufficiently high doses of drugs. Unfortunately, the use of drugs to control agitation comes at a price. When the symptoms are already well established, the doses of drugs needed to control behavior almost always have some adverse effects. In patients with moderately severe behavioral disturbance, doses of trazodone and haloperidol adjusted to minimize adverse effects are no better than placebo (150). Even if obvious adverse side effects are avoided, these drugs themselves can impair cognition. They should only be used when agitation is interfering with care or leads to assaultive behavior. Frequent reassessment is needed to be certain that the benefits outweigh the potential adverse effects.

Delusions, Paranoia, and Hallucinations

Delusions and paranoia can occur alone or as a symptom of psychotic depression. If sufficiently disruptive, antipsychotic drugs can be used. Delusions that caregivers are impostors or that others have taken objects can often be managed by diversion to other issues. Another common delusion that a stranger rather than the patient themselves is reflected in the mirror is distressing but can be handled easily by removing and hiding mirrors.

Hallucinations should not be treated unless they are threatening and cause the patient great fear. Often hallucinations are pleasant or benign to the patient while being understandably upsetting to the caregiver. The caregiver should be advised to ignore and reassure rather than react to patient reports of hallucinations. Disturbing hallucinations usually respond to antipsychotics, but antipsychotics must be avoided in patients who already have extrapyramidal symptoms. Visual hallucinations are a typical feature of DLB, and, in this disorder particularly, there is danger of sudden worsening of symptoms and even death (139).

All these symptoms are apparent only when the patient is able to communicate them. Either for this reason or because thoughts become less well formed, these symptoms usually resolve as dementia progresses.

Sleep Disturbance

Sleep disturbance is a common cause of institutionalization in dementia. In AD, disrupted sleep may be owing in part to neuronal loss in the suprachiasmatic nucleus, which helps to regulate the sleep/wake cycle (151). The best approach to managing sleep disturbances in dementia is to encourage patients and caregivers to follow good sleep hygiene (Table 38.19). It is particularly helpful to establish daily times for arising and going to bed, reflecting true needs rather than simply caregiver convenience. It is important to provide activities throughout waking hours and periods of physical exertion, particularly outside and with exposure to sunlight. All this may require additional assistance for care. When sleep/wake cycles are sufficiently disrupted, sufficient environmental modification may be needed to permit nighttime activity without causing disruption.

Hypnotic drugs do not provide long-term solutions for this chronic problem. They often also worsen cognition themselves. This is particularly true of drugs with anticholinergic properties that are often found in antihistamines and over-the-counter products. Sedating antidepressants, such as trazodone, may be helpful. Antipsychotic drugs can be helpful, particularly when agitation accompanies sleep disturbance. However, the extrapyramidal side effects of these drugs sometimes also can worsen sleep. Melatonin has been helpful in individual subjects. Treating extrapyramidal symptoms may be very helpful in improving sleep for patients with dementia and parkinsonism. If other approaches fail, short-acting benzodiazepines can be used intermittently for only a few days at a time. Their effectiveness wanes with longer administration, and they can worsen behavioral symptoms by aggravating cognitive impairment.

MEDICAL PRACTICE AND DEMENTIA CARE

Caring for patients with dementia makes demands that are not typical of most medical practice. Evaluating and treating dementia are intellectually demanding and can be enormously gratifying if physicians are prepared for the special challenges of dementia care and adopt practice strategies that increase their success and satisfaction. Particular attention is needed in conducting the medical visit, forming alliances to assist in management, talking about dementia with patients and their families, maintaining patient autonomy, and considering reimbursement issues.

The Medical Visit

Having both patient and caregivers present during medical visits facilitates evaluation and treatment and also prepares for patients who do not adopt the "sick role." The patient should be asked to invite one or more family member to join for the entirety of all medical visits. The patient's wishes in this regard should be honored, but, with few exceptions, patients are more comfortable with others present to help relate the story and recall what was said. For their part, family members usually prefer this approach and appreciate being

---

TABLE 38.19. SUGGESTIONS TO IMPROVE SLEEP HYGIENE AND TREAT SLEEP DISTURBANCE IN DEMENTIA

Treat underlying depression
Restrict time in bed to sleeping, not reading, watching television, or eating
Spend only as much time sleeping as needed to feel refreshed during the next day
Keep a regular schedule. Get up and go to bed at the same time each day
Schedule regular vigorous physical activity every day before evening, preferably outside
Avoid unplanned napping and sitting or reclining for extended periods during the day
Avoid caffeine-containing beverages and alcohol, especially in the evening
A light snack at bedtime may be helpful, but avoid excessive liquids
Insulate the bedroom against sounds that disturb sleep
A night-light can decrease disorientation, and soft music may be calming
On awakening at night, provide reorientation. If trouble falling asleep again, encourage quiet activity outside the bedroom, returning to bed only when sleepy. Get up at the regular time and participate in usual activities the next day, no matter how little sleep
To avoid caregiver concern about wandering or elopement, make sure environment is safe and doors have locks or alarms. Keep a light on in the bathroom or stairs
Patients and caregivers may have different sleep requirements. Obtain assistance for supervision at nighttime, if needed. Caregivers cannot tolerate sustained loss of sleep

included. Arranging visits that accommodate everyone requires planning for sufficient time and space and may necessitate changing office routines for a patient with a memory complaint. It is usually easier to arrange for brief separate discussions during the visit, if they are necessary, than to arrange on short notice a joint meeting with the physician, patient, and caregivers.

Communication strategies used in the first visit are particularly important because they often continue to affect later visits, reinforcing feelings of comfort and sharing or feelings of suspicion and conflict. The patient should be addressed first to establish that he/she is the primary focus of attention. After putting the patient at ease with the recognizable routine of a clinic visit, the patient can be calmly asked about his/her own concerns while the physician notes the reaction of family members. Subsequently, family members can be asked about their concerns, reminding the patient, if necessary, that everyone gets their turn to speak. The evaluation of dementia requires that the physician obtain information about behavior and activities from a collateral source (31). Interviewing family members with the patient present is not easy. The fundamental nature of the interactions are different from the situation in pediatrics and requires the physician to direct questions to caregivers while also including the patient in conversations (152). The physician must establish rapport with the demented patient and each of the caregivers. It takes time and finesse to take a history from several people simultaneously while not neglecting anyone. While the interview proceeds, the physician must remain aware of each person's emotional responses and body language. Either the patient or caregiver can become upset or distraught. The physician must control the situation, not take sides, and remain objective rather than be buffeted by the emotions that may develop. It is also important to observe the quality of the relationship between the patient and caregiver. Each individual will have different questions and concerns that must be addressed. Unlike other medical situations when informants are required, in dementia care, the family may be seeking help for themselves. Although it may at times be difficult, the patient's interests should be foremost and remain the focus of the visit.

Patients who deny or minimize their problems make both diagnosis and management difficult. Meeting with a patient and caregivers together provides reliable information more quickly and helps to enlist families in providing care. Even if the patient has good insight, meeting with the patient and family together has several other advantages to recommend it. Involving family members recognizes their important role and recruits them to become part of the therapeutic team. It quickly reveals the quality of the interactions between the patient and caregivers and resolves conflicting information. It can clarify the role of individual family members and offers the opportunity to mediate disagreements that often arise between caregivers. It also provides a venue where caregiver concerns can be identified and addressed.

Treatment Alliances

A medical practice that works closely with community agencies and other health professionals can provide the comprehensive care needed to effectively treat dementia and at the same time decrease the physician's burden. When dementia progresses, the triad of physician, patient, and caregiver becomes inadequate. Caring for a patient becomes more difficult and additional services are needed. Adult day care, personal care services, and assistance with meal preparation and other activities of daily living can help support families and keep the patient at home, but they are underutilized because there is a lack of knowledge about their availability (153,154). Eventually placement in an assisted living facility, foster care, or nursing home care may be required. At each step along the way, the physician will be called on to intervene and provide advice unless he/she has previously involved other health professionals and community services.

The solution to the daunting challenge of providing care is knowledge of community-based services and how to access them. The physician can help patients and families plan for future care needs by forming an alliance with nurses, social workers, and community agencies early in the patient's illness. Families must arrange and pay for services and coordinate them with their own efforts. Planning ahead is important to prepare for changes in circumstances, which can develop suddenly. These health professionals can provide advice and counseling to families confronting a bewildering array of care options. Developing such broad alliances to provide care is novel for most physicians, but it can also be rewarding. The multidisciplinary team lessens the physician's burdens of providing dementia care and helps with patient education. Rather than feeling inadequate, physicians become leaders of care teams. They can coordinate management of other members of the team and help by providing their specialized knowledge of disease, prognosis, and medication. When caregivers are given adequate support and guidance and prepare for possible contingencies, they can avoid most crises. This reduces emergency department visits, eliminates "granny dumping" by exasperated caregivers, and reduces physician stress and frustration. Dementia care becomes manageable.

Talking About Dementia

Another unusual demand of dementia care is the need to talk about dementia and relate "bad news" to several people at once who are unequally able to understand its implications. Relating information about a serious illness is particularly difficult when the diagnosis is determined without the aid of a definitive diagnostic test, as is usually the situation with dementia. Physicians must learn to accept their own diagnostic uncertainty while providing patients and families with a clear, unambiguous diagnosis and recommendations. Relating a diagnosis and discussing prognosis have great value to a patient and his/her family. These are among the major desires of those with dementia seeking medical advice (Table 38.6). Nevertheless, many physicians communicate poorly about dementia. Caregivers report that a correct diagnosis of AD was provided in only 38% of initial physician consultations, causing them to seek care from others (33). Often a family is not satisfied until they understand what has caused the serious illness that they observe in their loved one and what they can do about it. Instead of leading to despair, even a serious diagnosis can bring relief, a plan of action, and hope through research. The unknown is more difficult to deal with and cannot provide the comfort that everything that should be done is being done (155,156). It is important for the physician to understand his/her own emotional response and adequately prepare to discuss a diagnosis. It is best to tell a patient a diagnosis in a calm setting and in the presence of family members who can be supportive. Family members may be told the diagnosis first, providing an opportunity to gauge their response and deal with their concerns separately. In other cases, the patient may find a meeting with the family separately first threatening or suspicious. In any case, if there is concern about the patient's response or family members have reservations about telling the patient, it is best to inquire directly about the patient's own wishes by asking "Do you want to know the results of the tests and what I think is wrong?" A patient usually wants to hear the doctor's opinion, and this helps both the physician and family understand the patient's desires and ability to comprehend what is happening. There are many strategies that help relate bad news (157,158). It is often useful to "fire a warning shot" by mentioning early in a dementia evaluation that the illness is serious and that AD must be considered. A review of the findings of the examination and test results before revealing a diagnosis can also help by providing its rationale. It is also important to avoid "truth dumping." Adequate time is needed for the patient and family to receive and comprehend bad news and ask questions.

Patient Autonomy

It is important, but difficult, to maintain patient autonomy when the patient has an impaired capacity to make decisions and others must assist in providing a history and providing care. Patient autonomy is best respected when legal protections are established early and patients are involved in discussions as much as possible. Even when legal consent is not possible, patient assent and cooperation should be sought. In general, caregivers should be given wide latitude for making care decisions because they are usually in the best position to judge their situation and the patient's interests. However, a physician should not forget that his/her primary responsibility is the patient's best interest and that he/she has a legal responsibility to report abuse and advise a patient and the family of driving and personal safety risks associated with dementia.

Reimbursement Issues

Dementia care sometimes requires enormous time and effort that are inadequately reimbursed. This can be particularly frustrating because health insurance plans and colleagues often do not recognize what is required for adequate evaluation and management of dementia. Nevertheless, there are steps that can be taken to minimize these concerns and still provide excellent care. Fortunately, practice strategies that address the unique challenges of dementia care are often also advantageous for reimbursement purposes. Usually it is less time-consuming to meet with the patient and caregiver together than to depend on separate patient and family visits. This approach also avoids reimbursement issues when insurance plans only pay for time the physician spent with the patient. Because of their distress and their lack of knowledge about dementia and how to handle behavior disturbances, caregivers may request that their concerns be discussed at separate visits that are difficult to bill to insurance (159). Addressing these issues during regularly scheduled patient visits can meet both caregiver and reimbursement requirements. It is likely that the unique requirements of good dementia care will be recognized and adequately remunerated in the future as the problem of dementia becomes more evident in our society. Until then, doing what is best for the patient will always provide the best guidance for successful medical practice.

ACKNOWLEDGMENTS

Supported in part by the Michigan Alzheimer's Disease Research Center (NIH Grant P50-AG08671), NIH Grant RO1-NS24896, and the Donald W. Reynolds Foundation. I thank Robert Koeppe, Ph.D., Virginia Rogers, and the University of Michigan PET Center for their assistance in preparing Figures 38.2, 38.5, 38.6, and 38.7. The PET images in these figures are provided through the courtesy of David E. Kuhl, M.D., and Satoshi Minoshima, M.D., Ph.D.

REFERENCES

1. Katzman R, Bick K. *Alzheimer disease: the changing view.* San Diego: Academic Press, 2000.
2. Pippenger M, Holloway RG, Vickrey BG. Neurologists' use of ICD-9CM codes for dementia. *Neurology* 2001;56:1206–1209.
3. Barnes RF, Raskind MA. DSM-III criteria and the clinical diagnosis of dementia: a nursing home study. *J Gerontol* 1981;36:20–27.
4. The Michigan Task Force on Alzheimer's Disease and Related Conditions. *Alzheimer's disease and related conditions: reducing uncertainty, volume 1.* Lansing, MI: Michigan Department of Public Health, 1987.
5. Douglass RL, Bartus MF, Waldron-Wilson R. Long-term utilization by black elderly in Detroit: racial representation and equity of care. In: Altman HJ, Altman BN, eds. *Alzheimer's and*

Parkinson's diseases: recent advances in research and clinical management. New York: Plenum Press, 1989.

6. Raiford K, Anton-Johnson S, Haycox Z, et al. CERAD part VII: accuracy of reporting dementia on death certificates of patients with Alzheimer's disease. *Neurology* 1994;44:2208–2209.

7. Ganguli M, Rodriguez EG. Reporting of dementia on death certificates: a community study. *J Am Geriatr Soc* 1999;47:842–849.

8. Heidebrink JL, Foster NL. Alzheimer's disease is particularly underreported in women and minorities. *Gerontologist* 1997;37:22–23.

9. Erkinjuntti T, Ostbye T, Steenhuis R, et al. The effect of different diagnostic criteria on the prevalence of dementia. *N Engl J Med* 1997;337:1667–1674.

10. Gao S, Hendrie HC, Hall KS, et al. The relationships between age, sex, and the incidence of dementia and Alzheimer disease: a meta-analysis. *Arch Gen Psychiatry* 1998;55:809–815.

11. Jorm AF, Korten AE, Henderson AS. The prevalence of dementia: a quantitative integration of the literature. *Acta Psychiatr Scand* 1987;76:465–479.

12. O'Connor DW, Pollitt PA, Hyde JB, et al. Do general practitioners miss dementia in elderly patients? *BMJ* 1988;297:1107–1110.

13. Brookmeyer R, Gray S, Kawas C. Projections of Alzheimer's disease in the United States and the public health impact of delaying disease onset. *Am J Public Health* 1998;88:1337–1342.

14. Khachaturian Z. The five-five, ten-ten plan for Alzheimer's disease. *Neurobiol Aging* 1992;13:197–199.

15. Bishop CE. Where are the missing elders? The decline in nursing home use, 1985 and 1995; despite an aging population, the nation's nursing homes have not seen an expected surge in residents. *Health Aff Millwood* 1999;18:146–155.

16. American Psychiatric Association. *Practice guideline for the treatment of patients with delirium.* Washington, DC: American Psychiatric Association, 1999.

17. Lipowski ZJ. Delirium in the elderly patient. *N Engl J Med* 1989;320:578–582.

18. Lerner AJ, Hedera P, Koss E, et al. Delirium in Alzheimer disease. *Alzheimer Dis Assoc Disord* 1997;11:16–20.

19. Inouye SK, Bogardus ST Jr, Charpentier PA, et al. A multicomponent intervention to prevent delirium in hospitalized older patients. *N Engl J Med* 1999;340:669–676.

20. Petersen RC, Smith GE, Waring SC, et al. Mild cognitive impairment: clinical characterization and outcome. *Arch Neurol* 1999;56:303–308.

21. Berent S, Giordani B, Foster N, et al. Neuropsychological function and cerebral glucose utilization in isolated memory impairment and Alzheimer's disease. *J Psychiatr Res* 1999;33:7–16.

22. Rentz DM, Michalska K, Faust RR, et al. Predicting mild cognitive impairment in high functioning elders. *Neurology* 2001;56[Suppl 3]:A71–A72.

23. Petersen RC, Stevens JC, Ganguli M, et al. Practice parameter: early detection of dementia: mild cognitive impairment (an evidence-based review): report of the Quality Standards Subcommittee of the American Academy of Neurology. *Neurology* 2001;56:1133–1142.

24. Geerlings MI, Jonker C, Bouter LM, et al. Association between memory complaints and incident Alzheimer's disease in elderly people with normal baseline. *Am J Psychiatry* 1999;156:531–537.

25. Graham JE, Rockwood K, Beattie BL, et al. Prevalence and severity of cognitive impairment with and without dementia in an elderly population. *Lancet* 1997;349:1793–1796.

26. Grundman M, Petersen RC, Ernesto C, et al. Rate of dementia

of the Alzheimer type (DAT) in subjects with mild cognitive impairment. *Neurology* 1996;46:A403.

27. Tierney MC, Szalai JP, Snow WG, et al. Prediction of probable Alzheimer's disease in memory-impaired patients: a prospective longitudinal study. *Neurology* 1996;46:661–665.

28. Killiany RJ, Gomez-Isla T, Moss M, et al. Use of structural magnetic resonance imaging to predict who will get Alzheimer's disease. *Ann Neurol* 2000;47:430–439.

29. Bozoki A, Giordani B, Heidebrink JL, et al. Mild cognitive impairments predict dementia in non-demented elderly with memory loss. *Arch Neurol* 2001;58:411–416.

30. Tierney MC, Szalai JP, Snow WG, et al. A prospective study of the clinical utility of ApoE genotype in the prediction of outcome in patients with memory impairment. *Neurology* 1996;46:149–154.

31. Costa PT Jr, Williams TF, Somerfield M, et al. *Recognition and initial assessment of Alzheimer's disease and related dementias. Clinical practice guideline no. 19.* Rockville, MD: U.S. Department of Health and Human Services, Public Health Service, Agency for Health Care Policy and Research; 1996; AHCPR publication 97-0702.

32. O'Connor DW, Pollitt PA, Roth M, et al. Memory complaints and impairment in normal, depressed, and demented elderly persons identified in a community survey. *Arch Gen Psychiatry* 1990;47:224–227.

33. Knopman DS, DeKosky ST, Cummings JL, et al. Practice parameter: diagnosis of dementia (an evidence-based review). Report of the Quality Standards Subcommittee of the American Academy of Neurology. *Neurology* 2001;56:1143–1153.

34. Bahn MM, Parchi P. Abnormal diffusion-weighted magnetic resonance images in Creutzfeldt-Jakob disease. *Arch Neurol* 1999;56:577–583.

35. Johnson RT, Gibbs CJ Jr. Creutzfeldt-Jakob disease and related transmissible spongiform encephalopathies. *N Engl J Med* 1998;339:1994–2004.

36. Parchi P, Giese A, Capellari S, et al. Classification of sporadic Creutzfeldt-Jakob disease based on molecular and phenotypic analysis of 300 subjects. *Ann Neurol* 1999;46:224–233.

37. Brandel JP, Delasnerie-Laupretre N, Laplanche JL, et al. Diagnosis of Creutzfeldt-Jakob disease: effect of clinical criteria on incidence estimates. *Neurology* 2000;54:1095–1099.

38. Satoh J, Kurohara K, Yukitake M, et al. The 14-3-3 protein detectable in the cerebrospinal fluid of patients with prion-unrelated neurological diseases is expressed constitutively in neurons and glial cells in culture. *Eur Neurol* 1999;41:216–225.

39. Lemstra AW, van Meegen MT, Vreyling JP, et al. 14-3-3 testing in diagnosing Creutzfeldt-Jakob disease: a prospective study in 112 patients. *Neurology* 2000;55:514–516.

40. Will RG, Zeidler M, Stewart GE, et al. Diagnosis of new variant Creutzfeldt-Jakob disease. *Ann Neurol* 2000;47:575–582.

41. Krause C. *How healthy is your family tree?: a complete guide to tracing your family's medical and behavioral history.* New York: Simon and Schuster, 1995.

42. Hedera P. Ethical principles and pitfalls of genetic testing for dementia. *J Geriatr Psychiatry Neurol* 2001;14:213–221.

43. Ashizawa T, Wong LJ, Richards CS, et al. CAG repeat size and clinical presentation in Huntington's disease. *Neurology* 1994;44:1137–1143.

44. Pillon B, Dubois B, Ploska A, et al. Severity and specificity of cognitive impairment in Alzheimer's, Huntington's, and Parkinson's diseases and progressive supranuclear palsy. *Neurology* 1991;41:634–643.

45. Bateman D, Boughey AM, Scaravilli F, et al. A follow-up study of isolated cases of suspected Huntington's disease. *Ann Neurol* 1992;31:293–298.

46. Kwan LT, Reed BR, Eberling JL, et al. Effects of subcortical cerebral infarction on cortical glucose metabolism and cognitive function. *Arch Neurol* 1999;56:809–814.
47. Blinkenberg M, Rune K, Jensen CV, et al. Cortical cerebral metabolism correlates with MRI lesion load and cognitive dysfunction in MS. *Neurology* 2000;54:558–564.
48. Whitehouse PJ, Price DL, Clark AW, et al. Alzheimer disease: evidence for selective loss of cholinergic neurons in the nucleus basalis. *Ann Neurol* 1981;10:122–126.
49. Kuhl DE, Minoshima S, Fessler JA, et al. In vivo mapping of cholinergic terminals in normal aging, Alzheimer's disease and Parkinson's disease. *Ann Neurol* 1996;40:399–410.
50. Hof PR, Delacourte A, Bouras C. Distribution of cortical neurofibrillary tangles in progressive supranuclear palsy: a quantitative analysis of six cases. *Acta Neuropathol (Berl)* 1992;84:45–51.
51. Trojanowski JQ, Lee VM. "Fatal attractions" of proteins. A comprehensive hypothetical mechanism underlying Alzheimer's disease and other neurodegenerative disorders. *Ann N Y Acad Sci* 2000;924:62–67.
52. Kuhl DE, Koeppe RA, Minoshima S, et al. In vivo mapping of cerebral acetylcholinesterase activity in aging and Alzheimer's disease. *Neurology* 1999;52:691–699.
53. Lyon G, Adams RA, Kolodny EH. *Neurology of hereditary metabolic diseases of children,* 2nd ed. New York: McGraw-Hill, 1996.
54. Coker SB. The diagnosis of childhood neurodegenerative disorders presenting as dementia in adults. *Neurology* 1991;41:794–798.
55. Cummings JL, Benson DF. *Dementia: a clinical approach,* 2nd ed. Boston: Butterworth-Heineman, 1992.
56. Tierney MC, Fisher RH, Lewis AJ, et al. The NINCDS-ADRDA Work Group criteria for the clinical diagnosis of probable Alzheimer's disease: a clinicopathologic study of 57 cases. *Neurology* 1988;38:359–364.
57. Becker JT, Boller F, Lopez OL, et al. The natural history of Alzheimer's disease. Description of study cohort and accuracy of diagnosis. *Arch Neurol* 1994;51:585–594.
58. DeKosky ST, Scheff SW. Synapse loss in frontal cortex biopsies in Alzheimer's disease: correlation with cognitive severity. *Ann Neurol* 1990;27:457–464.
59. Vogt BA, Vogt LJ, Vrana KE, et al. Multivariate analysis of laminar patterns of neurodegeneration in posterior cingulate cortex in Alzheimer's disease. *Exp Neurol* 1998;153:8–22.
60. Terry RD, Peck A, DeTeresa R, et al. Some morphometric aspects of the brain in senile dementia of the Alzheimer type. *Ann Neurol* 1981;10:184–192.
61. Davies P, Maloney AJR. Selective loss of central cholinergic neurons in Alzheimer's disease. *Lancet* 1976;2:1403.
62. DeKosky ST, Harbaugh RE, Schmitt FA, et al. Cortical biopsy in Alzheimer's disease: diagnostic accuracy and neurochemical, neuropathological, and cognitive correlations. Intraventricular Bethanecol Study Group. *Ann Neurol* 1992;32:625–632.
63. Geula C, Mesulam M. Special properties of cholinesterases in the cerebral cortex of Alzheimer's disease. *Brain Res* 1989;498:185–189.
64. Mesulam MM, Geula C. Butyrylcholinesterase reactivity differentiates the amyloid plaques of aging from those of dementia. *Ann Neurol* 1994;36:722–727.
65. Davis KL, Mohs RC, Marin D, et al. Cholinergic markers in elderly patients with early signs of Alzheimer disease. *JAMA* 1999;281:1401–1406.
66. Perry RH, Candy JM, Perry EK, et al. Extensive loss of choline acetyltransferase activity is not reflected by neuronal loss in the nucleus of Meynert in Alzheimer's disease. *Neurosci Lett* 1982;33:311–315.
67. The Ronald and Nancy Reagan Research Institute of the Alzheimer's Association and the National Institute on Aging Working Group. Consensus report of the Working Group on Molecular and Biochemical Markers of Alzheimer's Disease. *Neurobiol Aging* 1998;19:109–116.
68. Farrer LA, Cupples LA, Haines JL, et al. Effects of age, sex, and ethnicity on the association between apolipoprotein E genotype and Alzheimer disease. A meta-analysis. APOE and Alzheimer Disease Meta Analysis Consortium. *JAMA* 1997;278:1349–1356.
69. American College of Medical Genetics/American Society of Human Genetics Working Group on ApoE and Alzheimer disease. Statement on use of apolipoprotein E testing for Alzheimer disease. *JAMA* 1995;274:1627–1629.
70. Blessed G, Tomlinson BE, Roth M. The association between quantitative measures of dementia and of senile change in the cerebral grey matter of elderly subjects. *Br J Psychiatry* 1968;114:797–811.
71. Constantinidis J. Is Alzheimer's disease a major form of senile dementia? Clinical, anatomical, and genetic data. In: Katzman R, Terry RD, Bick KL, eds. *Alzheimer's disease: senile dementia and related disorders.* New York: Raven Press, 1978:15–25.
72. Cochran EJ, Bennett DA, Cervenáková L, et al. Familial Creutzfeldt-Jakob disease with a five-repeat octapeptide insert mutation. *Neurology* 1996;47:727–733.
73. Nuwer MR. Progressive supranuclear palsy despite normal eye movements. *Arch Neurol* 1981;38:784.
74. Lopez OL, Swihart AA, Becker JT, et al. Reliability of NINCDS-ADRDA clinical criteria for the diagnosis of Alzheimer's disease. *Neurology* 1990;40:1517–1522.
75. Neary D, Snowden JS, Gustafson L, et al. Frontotemporal lobar degeneration: a consensus on clinical diagnostic criteria. *Neurology* 1998;51:1546–1554.
76. The Lund and Manchester Groups. Clinical and neuropathological criteria for frontotemporal dementia. *J Neurol Neurosurg Psychiatry* 1994;57:416–418.
77. Foster NL, Wilhelmsen K, Sima AAF, et al. Frontotemporal dementia and parkinsonism linked to chromosome 17: a consensus conference. *Ann Neurol* 1997;41:706–715.
78. Wilhelmsen KC, Clark LN, Miller BL, et al. Tau mutations in frontotemporal dementia. *Dement Geriatr Cogn Disord* 1999;10:88–92.
79. Forman MS, Lee VMY, Trojanowski JQ. New insights into genetic and molecular mechanisms of brain degeneration in tauopathies. *J Chem Neuroanat* 2000;20:225–244.
80. Heutink P, Stevens M, Rizzu P, et al. Hereditary frontotemporal dementia is linked to chromosome 17q21-q22: a genetic and clinicopathological study of three Dutch families. *Ann Neurol* 1997;41:150–159.
81. Hutton M, Lendon CL, Rizzu P, et al. Association of missense and 5′-splice-site mutations in tau with the inherited dementia FTDP-17. *Nature* 1998;393:702–705.
82. Sano M, Ernesto C, Thomas R, et al. A controlled trial of selegiline, alpha-tocopherol, or both as treatment for Alzheimer's disease. *N Engl J Med* 1997;338:1216–1222.
83. Wszolek ZK, Pfeiffer RF, Bhatt MH, et al. Rapidly progressive autosomal dominant parkinsonism and dementia with pallido-ponto-nigral degeneration. *Ann Neurol* 1992;32:312–320.
84. Verghese J, Crystal HA, Dickson DW, et al. Validity of clinical criteria for the diagnosis of dementia with Lewy bodies. *Neurology* 1999;53:1974–1982.
85. McKeith IG, Perry EK, Perry RH. for the Consortium on Dementia with Lewy bodies. Report of the second dementia with

Lewy body international workshop: diagnosis and treatment. *Neurology* 1999;53:902–905.

86. Papka M, Rubio A, Schiffer RB, et al. Lewy body disease: can we diagnose it? *J Neuropsychiatry Clin Neurosci* 1998;10: 405–412.

87. Albin RL, Minoshima S, D'Amato CJ, et al. Fluoro-deoxyglucose positron emission tomography in diffuse Lewy body disease. *Neurology* 1996;47:462–466.

88. Minoshima S, Foster NL, Sima AAF, et al. Alzheimer's disease versus dementia with Lewy bodies: cerebral metabolic distinction with autopsy confirmation. *Ann Neurol* 2001;50:358–365.

89. Santacruz P, Uttl B, Litvan I, et al. Progressive supranuclear palsy: a survey of the disease course. *Neurology* 1998;50:1637–1647.

90. Litvan I, Agid Y, Calne D, et al. Clinical research criteria for the diagnosis of progressive supranuclear palsy (Steele-Richardson-Olszewski syndrome): report of the NINDS-SPSP international workshop. *Neurology* 1996;47:1–9.

91. Litvan I, Agid Y, Jankovic J, et al. Accuracy of clinical criteria for the diagnosis of progressive supranuclear palsy (Steele-Richardson-Olszewski syndrome). *Neurology* 1996;46:922–930.

92. Foster NL, Gilman S, Berent S, et al. Cerebral hypometabolism in progressive supranuclear palsy studied with positron emission tomography. *Ann Neurol* 1988;24:399–406.

93. Riley DE, Lang AE. Clinical diagnostic criteria. *Adv Neurol* 2000;82:29–34.

94. Litvan I, Agid Y, Goetz C, et al. Accuracy of the clinical diagnosis of corticobasal degeneration: a clinicopathologic study. *Neurology* 1997;48:119–125.

95. Brooks DJ. Functional imaging studies in corticobasal degeneration. *Adv Neurol* 2000;82:209–215.

96. Yee AS, Simon JH, Anderson CA, et al. Diffusion-weighted MRI of right-hemisphere dysfunction in Creutzfeldt-Jakob disease. *Neurology* 1999;52:1514–1515.

97. Rosen WG, Terry RD, Fuld PA, et al. Pathological verification of ischemic score in differentiation of dementia. *Ann Neurol* 1980;7:486–488.

98. Chui HC, Mack W, Jackson JE, et al. Clinical criteria for the diagnosis of vascular dementia: a multicenter study of comparability and interrater reliability. *Arch Neurol* 2000;57:191–196.

99. Vanneste JA. Three decades of normal pressure hydrocephalus: are we wiser now? *J Neurol Neurosurg Psychiatry* 1994;57:1021–1025.

100. Stolze H, Kuhtz-Buschbeck JP, Drucke H, et al. Comparative analysis of the gait disorder of normal pressure hydrocephalus and Parkinson's disease. *J Neurol Neurosurg Psychiatry* 2001;70: 289–297.

101. Vanneste J, Augustijn P, Dirven C, et al. Shunting normal-pressure hydrocephalus: do the benefits outweigh the risks? A multicenter study and literature review. *Neurology* 1992;42:54–59.

102. Golomb J, Wisoff J, Miller DC, et al. Alzheimer's disease comorbidity in normal pressure hydrocephalus: prevalence and shunt response. *J Neurol Neurosurg Psychiatry* 2000;68:778–781.

103. King AC, Oman RF, Brassington GS, et al. Moderate-intensity exercise and self-rated quality of sleep in older adults. A randomized controlled trial. *JAMA* 1997;277:32–37.

104. O'Donnell BF, Drachman DA, Barnes HJ, et al. Incontinence and troublesome behaviors predict institutionalization in dementia. *J Geriatr Psychiatry Neurol* 1992;5:45–52.

105. Young RF, Koslowski K, Montgomery RJ. Psychosocial factors in institutionalization of Alzheimer's patients. *J Clin Geropsychol* 1998;4:241–251.

106. Belgrave LL, Wykle ML, Choi JM. Health, double jeopardy, and culture: the use of institutionalization by African-Americans. *Gerontologist* 1993;33:379–385.

107. Douglass RL, Espino E, Meyers MA, et al. Representation of the black elderly in Detroit metropolitan nursing homes. *J Natl Med Assoc* 1988;80:283–288.

108. Connell CM, Gibson GD. Racial, ethnic, and cultural differences in dementia caregiving: review and analysis. *Gerontologist* 1997;37:355–364.

109. Connell CM, Janevic MR, Gallant MP. The costs of caring: the impact of dementia on family caregivers. *J Geriatr Psychiatry Neurol* 2001;14:179–187.

110. Mittelman MS, Ferris SH, Shulman E, et al. A comprehensive support program: effect on depression in spouse-caregivers of AD patients. *Gerontologist* 1995;35:792–802.

111. Biegel DE, Bass DM, Schulz R, et al. Predictors of in-home and out-of-home service use by family caregivers of Alzheimer's disease patients. *J Aging Health* 1993;5:419–438.

112. Schulz R, O'Brien AT, Bookwala J, et al. Psychiatric and physical morbidity effects of dementia caregiving: prevalence, correlates, and causes. *Gerontologist* 1995;35:771–791.

113. Finucane TE, Christmas C, Travis K. Tube feeding in patients with advanced dementia: a review of the evidence. *JAMA* 1999;282:1365–1370.

114. Albert SM, Sano M, Marder K, et al. Participation in clinical trials and long-term outcomes in Alzheimer's disease. *Neurology* 1997;49:38–43.

115. Doody RS, Stevens JC, Beck C, et al. Practice parameter: management of dementia (an evidence-based review). Report of the Quality Standards Subcommittee of the American Academy of Neurology. *Neurology* 2001;56:1154–1166.

116. Burgio LD, McCormick KA, Scheve AS, et al. The effects of changing prompted voiding schedules in the treatment of incontinence in nursing home residents. *J Am Geriatr Soc* 1994;42:315–320.

117. Ouslander JG, Schnelle JF, Uman G, et al. Predictors of successful prompted voiding among incontinent nursing home residents. *JAMA* 1995;273:1366–1370.

118. Cohen-Mansfield J, Werner P. Management of verbally disruptive behaviors in nursing home residents. *J Gerontol A Biol Sci Med Sci* 1997;52:M369–M377.

119. Monk D, Brodaty H. Use of estrogens for the prevention and treatment of Alzheimer's disease. *Dement Geriatr Cogn Disord* 2000;11:1–10.

120. Henderson VW, Paganini-Hill A, Miller BL, et al. Estrogen for Alzheimer's disease in women: randomized, double-blind, placebo-controlled trial. *Neurology* 2000;54:295–301.

121. Mulnard RA, Cotman CW, Kawas C, et al. Estrogen replacement therapy for treatment of mild to moderate Alzheimer's disease. *JAMA* 2000;283:1007–1015.

122. Wang PN, Liao SQ, Liu RS, et al. Effects of estrogen on cognition, mood, and cerebral blood flow in AD: a controlled study. *Neurology* 2000;54:2061–2066.

123. Heyman A, Schmechel D, Wilkinson W, et al. Failure of long term high-dose lecithin to retard progression of early-onset Alzheimer's disease. *J Neural Transm Suppl* 1987;24:279–286.

124. Foster NL, Petersen RC, Gracon SI, et al. An enriched-population, double-blind, placebo-controlled, crossover study of tacrine and lecithin in Alzheimer's disease. *Dementia* 1996;7:260–266.

125. Winblad B, Engedal K, Soininen H, et al. Donepezil enhances global function, cognition and activities of daily living compared with placebo in a 1-year, double-blind trial in patients with mild to moderate Alzheimer's disease. *Int Psychogeriatr* 1999;11[Suppl 1]:138.

126. Kuhl DE, Minoshima S, Frey KA, et al. Limited donepezil

inhibition of acetylcholinesterase measured with positron emission tomography in living Alzheimer cerebral cortex. *Ann Neurol* 2000;48:391–395.

127. Price JL, Morris JC. Tangles and plaques in nondemented aging and "preclinical" Alzheimer's disease. *Ann Neurol* 1999;45:358–368.

128. Games D, Adams D, Alessandrini R, et al. Alzheimer-type neuropathology in transgenic mice overexpressing V717F beta-amyloid precursor protein. *Nature* 1995;373:523–527.

129. Hsiao KK, Borchelt DR, Olson K, et al. Age-related CNS disorder and early death in transgenic FVB/N mice overexpressing Alzheimer amyloid precursor proteins. *Neuron* 1995;15:1203–1218.

130. Schenk D, Barbour R, Dunn W, et al. Immunization with amyloid-beta attenuates Alzheimer-disease-like pathology in the PDAPP mouse. *Nature* 1999;400:173–177.

131. Weiner HL, Lemere CA, Maron R, et al. Nasal administration of amyloid-beta peptide decreases cerebral amyloid burden in a mouse model of Alzheimer's disease. *Ann Neurol* 2000;48:567–579.

132. Bard F, Cannon C, Barbour R, et al. Peripherally administered antibodies against amyloid beta-peptide enter the central nervous system and reduce pathology in a mouse model of Alzheimer disease. *Nat Med* 2000;6:916–919.

133. Bacskai BJ, Kajdasz ST, Christie RH, et al. Imaging of amyloid-beta deposits in brains of living mice permits direct observation of clearance of plaques with immunotherapy. *Nat Med* 2001;7:369–372.

134. Probst A, Gotz J, Wiederhold KH, et al. Axonopathy and amyotrophy in mice transgenic for human four-repeat tau protein. *Acta Neuropathol* 2000;99:469–481.

135. Lewis J, McGowan E, Rockwood J, et al. Neurofibrillary tangles, amyotrophy and progressive motor disturbance in mice expressing mutant (P301L) tau protein. *Nat Genet* 2000;25:402–405.

136. Kaufer DI, Catt KE, Lopez OL, et al. Dementia with Lewy bodies: response of delirium-like features to donepezil. *Neurology* 1998;51:1512.

137. McKeith I, Ser TD, Anand R, et al. Rivastigmine provides symptomatic benefit in dementia with Lewy bodies: findings from a placebo-controlled international multicenter study. *Neurology* 2000;54[Suppl 3]:A450.

138. McKeith I, Fairbairn A, Perry R, et al. Neuroleptic sensitivity in patients with senile dementia of Lewy body type. *BMJ* 1992;305:673–678.

139. McKeith IG, Ballard CG, Harrison RW. Neuroleptic sensitivity to risperidone in Lewy body dementia. *Lancet* 1995;346:699

140. Gubitz G, Sandercock P. Prevention of ischaemic stroke. *BMJ* 2000;321:1455–1459.

141. Hess DC, Demchuk AM, Brass LM, et al. HMG-CoA reductase inhibitors (statins): a promising approach to stroke prevention. *Neurology* 2000;54:790–796.

142. Tekin S, Mega MS, Masterman DM, et al. Orbitofrontal and anterior cingulate cortex neurofibrillary tangle burden is associated with agitation in Alzheimer disease. *Ann Neurol* 2001;49:355–361.

143. Hirono N, Mega MS, Dinov ID, et al. Left frontotemporal hypoperfusion is associated with aggression in patients with dementia. *Arch Neurol* 2000;57:861–866.

144. Craig AH, Cummings JL, Fairbanks L, et al. Cerebral blood flow correlates of apathy in Alzheimer disease. *Arch Neurol* 1996;53:1116–1120.

145. Reed BR, Jagust WJ, Coulter L. Anosognosia in Alzheimer's disease: relationships to depression, cognitive function, and cerebral perfusion. *J Clin Exp Neuropsychol* 1993;15:231–244.

146. Starkstein SE, Vazquez S, Migliorelli R, et al. A single-photon emission computed tomographic study of anosognosia in Alzheimer's disease. *Arch Neurol* 1995;52:415–420.

147. Schneider LS, Pollock VE, Lyness SA. A metaanalysis of controlled trials of neuroleptic treatment in dementia. *J Am Geriatr Soc* 1990;38:553–563.

148. Katz IR, Jeste DV, Mintzer JE, et al. Comparison of risperidone and placebo for psychosis and behavioral disturbances associated with dementia: a randomized, double-blind trial. Risperidone Study Group. *J Clin Psychiatry* 1999;60:107–115.

149. Rosebush PI, Mazurek MF. Neurologic side effects in neuroleptic-naive patients treated with haloperidol or risperidone. *Neurology* 1999;52:782–785.

150. Teri L, Logsdon RG, Peskind E, et al. Treatment of agitation in Alzheimer's disease patients: a randomized placebo controlled clinical trial. *Neurology* 2000;55:1271–1278.

151. Goudsmit E, Hofman MA, Fliers E, et al. The supraoptic and paraventricular nuclei of the human hypothalamus in relation to sex, age and Alzheimer's disease. *Neurobiol Aging* 1990;11:529–536.

152. Adelman RD, Greene MG, Charon R. The physician-elderly patient-companion triad in the medical encounter: the development of a conceptual framework and research agenda. *Gerontologist* 1987;27:729–734.

153. Jette AM, Tennstedt S, Crawford S. How does formal and informal community care affect nursing home use? *J Gerontol B Psychol Sci Soc Sci* 1995;50:S4–S12.

154. Vetter P, Steiner O, Kraus S, et al. Factors affecting the utilization of homecare supports by caregiving relatives of Alzheimer patients. *Dement Geriatr Cogn Disord* 1998;9:111–116.

155. Wiggins S, Whyte P, Huggins M, et al. The psychological consequences of predictive testing for Huntington's disease. Canadian Collaborative Study of Predictive Testing. *N Engl J Med* 1992;327:1401–1405.

156. O'Connor P, Detsky AS, Tansey C, et al. Effect of diagnostic testing for multiple sclerosis on patient health perceptions. Rochester-Toronto MRI Study Group. *Arch Neurol* 1994;51:46–51.

157. Quill TE, Townsend P. Bad news: delivery, dialogue, and dilemmas. *Arch Intern Med* 1991;151:463–468.

158. Buckman R. *How to break bad news: a guide for health care professionals.* Baltimore: Johns Hopkins University Press, 1992.

159. Coleman WH. Importance of behavioral and psychological symptoms of dementia in primary care. *Int Psychogeriatr* 2000;12[Suppl 1]:67–72.

160. Katzman R. Education and the prevalence of dementia and Alzheimer's disease. *Neurology* 1993;43:13–20.

161. Roberts GW, Gentleman SM, Lynch A, et al. Beta amyloid protein deposition in the brain after severe head injury: implications for the pathogenesis of Alzheimer's disease. *J Neurol Neurosurg Psychiatry* 1994;57:419–425.

162. Plassman BL, Havlik RJ, Steffens DC, et al. Documented head injury in early adulthood and risk of Alzheimer's disease and other dementias. *Neurology* 2000;55:1158–1166.

163. Conrad C, Andreadis A, Trojanowski JQ, et al. Genetic evidence for the involvement of tau in progressive supranuclear palsy. *Ann Neurol* 1997;41:277–281.

164. Morris HR, Janssen JC, Bandmann O, et al. The tau gene A0 polymorphism in progressive supranuclear palsy and related neurodegenerative diseases. *J Neurol Neurosurg Psychiatry* 1999;66:665–667.

165. Beck C, Heacock P, Mercer SO, et al. Improving dressing behavior in cognitively impaired nursing home residents. *Nurs Res* 1997;46:126–132.

166. Bourgeois MS, Burgio LD, Schulz R, et al. Modifying repetitive

verbalizations of community-dwelling patients with AD. *Gerontologist* 1997;37:30–39.

167. Mittelman MS, Ferris SH, Shulman E, et al. A family intervention to delay nursing home placement of patients with Alzheimer's disease: a randomized controlled trial. *JAMA* 1996;276:1725–1731.

168. Teri L, Logsdon RG, Uomoto J, et al. Behavioral treatment of depression in dementia patients: a controlled clinical trial. *J Gerontol B Psychol Sci Soc Sci* 1997;52:P159–P166.

169. Farlow M, Gracon SI, Hershey LA, et al. A controlled trial of tacrine in Alzheimer's disease. *JAMA* 1992;268:2523–2529.

170. Knapp MJ, Knopman DS, Solomon PR, et al. A 30-week randomized controlled trial of high-dose tacrine in patients with Alzheimer's disease. *JAMA* 1994;271:985–991.

171. Rogers SL, Friedhoff LT. The efficacy and safety of donepezil in patients with Alzheimer's disease: results of a US multicentre, randomized, double-blind, placebo-controlled trial. The Donepezil Study Group. *Dementia* 1996;7:293–303.

172. Rogers SL, Farlow MR, Doody RS, et al. A 24-week, double-blind, placebo-controlled trial of donepezil in patients with Alzheimer's disease. Donepezil Study Group. *Neurology* 1998;50:136–145.

173. Corey-Bloom J, Anand R, Veach J, et al. A randomized trial evaluating the efficacy and safety of ENA 713 (rivastigmine tartrate), a new acetylcholinesterase inhibitor, in patients with mild to moderately severe Alzheimer's disease. *Int J Geriatr Psychopharmacol* 1998;1:55–65.

174. Rösler M, Anand R, Cicin-Sain A, et al. Efficacy and safety of rivastigmine in patients with Alzheimer's disease: international randomised controlled trial. *BMJ* 1999;318:633–638.

175. Raskind MA, Peskind ER, Wessel T, et al. Galantamine in AD: A 6-month randomized, placebo-controlled trial with a 6-month extension. The Galantamine USA-1 Study Group. *Neurology* 2000;54:2261–2268.

176. Tariot PN, Solomon PR, Morris JC, et al. A 5-month, randomized, placebo-controlled trial of galantamine in AD. The Galantamine USA-10 Study Group. *Neurology* 2000;54:2269–2276.

177. Cummings JL, Cyrus PA, Bieber F, et al. Metrifonate treatment of the cognitive deficits of Alzheimer's disease. Metrifonate Study Group. *Neurology* 1998;50:1214–1221.

178. Morris JC, Cyrus PA, Orazem J, et al. Metrifonate benefits cognitive, behavioral, and global function in patients with Alzheimer's disease. *Neurology* 1998;50:1222–1230.

179. Thal LJ, Ferguson JM, Mintzer J, et al. A 24-week randomized trial of controlled-release physostigmine in patients with Alzheimer's disease. *Neurology* 1999;52:1146–1152.

180. van Dyke CH, Newhouse P, Falk WE, et al. Extended-release physostigmine in Alzheimer disease: a multicenter, double-blind, 12-week study with dose enrichment. Physostigmine Study Group. *Arch Gen Psychiatry* 2000;57:157–164.

181. Antuono PG. Effectiveness and safety of velnacrine for the treatment of Alzheimer's disease. A double-blind, placebo-controlled study. Mentane Study Group. *Arch Intern Med* 1995;155:1766–1772.

182. Zemlan FP, Keys M, Richter RW, et al. Double-blind placebo-controlled study of velnacrine in Alzheimer's disease. *Life Sci* 1996;58:1823–1832.

183. Minoshima S, Frey KA, Koeppe RA, et al. A diagnostic approach in Alzheimer's disease using 3-dimensional stereotactic surface projections of fluorine-18-FDG PET. *J Nucl Med* 1995;36:1238–1248.

184. Chambers CB, Lee JM, Troncoso JC, et al. Overexpression of four-repeat tau mRNA isoforms in progressive supranuclear palsy but not in Alzheimer's disease. *Ann Neurol* 1999;46:325–332.

185. Van Swieten JC, Stevens M, Rosso SM, et al. Phenotypic variation in hereditary frontotemporal dementia with tau mutations. *Ann Neurol* 1999;46:617–626.

39

NEUROPSYCHIATRIC ASPECTS OF EPILEPSY

MICHELLE V. LAMBERT
E. BETTINA SCHMITZ
HOWARD A. RING
MICHAEL TRIMBLE

CLASSIFICATION OF EPILEPSY

Classifying epileptic seizures (ES) and epileptic syndromes has occupied epileptologists for centuries. The problems of terminology became obvious with increasing communication between international epileptologists in the middle of the twentieth century. The first outlines of international classifications of ES (203) and epileptic syndromes (399) were published in 1970. The proposed classifications by the International League Against Epilepsy from 1981 and 1989 are based on agreements among international epileptologists and compromises between various viewpoints (117). They must not be regarded as definitive; a revision is currently being developed (362).

The changing role of psychiatric symptoms in relation to epilepsy in the past century is a good example of how much classification systems are determined by different approaches. In the middle of the nineteenth century, the epilepsy literature was dominated by psychiatrists who tried to categorize epilepsy based on their experiences with chronic patients in asylums. They spent more time on the psychopathologic classification of psychiatric symptoms in epilepsy than on describing ES (179,225,402). Specific psychiatric syndromes, especially short-lasting episodic psychoses and mood disorders, were considered diagnostically the same category as convulsions. In his classification of epilepsy from 1876, Samt (489,490) distinguished 12 categories of epilepsy. Only three were characterized by ES; the other nine categories were attributed to specific epileptic forms of insanity.

With the introduction of electroencephalography (EEG) in the 1930s, many episodic psychiatric states could be identified as nonepileptic in origin. ES, however, showed very different ictal EEG patterns. Since then, epileptologists have concentrated on the electroclinical differentiation of ES. Interestingly, the 1989 classification does not mention psychiatric criteria in the clinical definitions of epileptic syndromes.

This omission is remarkable because it is widely accepted that epilepsy is linked with particular psychiatric syndromes, which may be crucial in determining prognosis and quality of life (QOL). There are also recognized syndrome-related personality traits in juvenile myoclonic epilepsy (596) and temporal lobe epilepsy (TLE) (210).

Classification of Seizures

Most authors in the nineteenth century simply differentiated seizures according to severity (petit mal, grand mal). Hughlings Jackson was the first to recognize the necessity of anatomic description, physiologic delineation of disturbance of function, and pathologic confirmation (568). In the twentieth century, clinical events could be linked to ictal EEG findings. More detailed analyses of seizures became possible with simultaneous EEG-video monitoring and ictal neuropsychological testing. In the former, patients are monitored by video cameras and continuous EEG recordings for prolonged periods. The data are simultaneously displayed on a split-screen television monitor.

The International Classification of Epileptic Seizures (ICES) from 1981 is based on these improved intensive monitoring capabilities that have permitted accurate recognition of seizure symptoms and their longitudinal evolution. At present, the classification of seizures is weighted clinically and gives no clear definitions in terms of seizure origin. The ICES does not, however, reflect the most recent knowledge of the localizing significance of specific seizure symptoms. This approach has grown significantly since 1981 because of increased data from intensive monitoring and epilepsy surgery. Some parts of the ICES are thus outdated and in need of revision. Complex partial seizure (CPS) types, for example, are not yet distinguished in the ICES according to a probable origin in the frontal or temporal lobe.

The principal feature of the ICES is the distinction between seizures that are generalized from the beginning [primary generalized epilepsy (PGE)] and those that are partial or focal at onset and may or may not evolve to secondary generalized seizures (Table 39.1). In generalized seizures, there is initial involvement of both hemispheres. Consciousness may be impaired, and this impairment may be the initial manifestation. Motor manifestations are bilateral. The ictal EEG patterns initially are bilateral. Spikes, spike-wave complexes, and polyspike-wave complexes are all typical.

Partial seizures are those in which, in general, the first clinical and EEG changes indicate initial activation of a system of neurons limited to part of one cerebral hemisphere. The other important feature of the ICES is the separation of simple seizures and CPS, depending on whether there is preservation or impairment of consciousness.

Classification of Syndromes

Epilepsy is the name for "occasional sudden, excessive, rapid and local discharges of the grey matter" (568). This simple definition was formulated by Jackson in 1866 long before the introduction of EEG. It has not lost its application today. Recurrent ES are pathognomonic for all types of epilepsies (Table 39.2). The clinical spectrum of epilepsy, however, is much more complex, and an epileptic syndrome is characterized by a cluster of signs and syndromes customarily occurring together. These include such items as type of seizure, etiology, anatomy, precipitating factors, age at onset, severity, chronicity, diurnal and circadian cycling, and sometimes prognosis (116).

In contrast to a syndrome, a disease is characterized by a specific etiology and prognosis. Some recognized entities in the ICES are diseases; others are syndromes. Some of them may turn out to be diseases—a common etiology may still be discovered.

The ICES distinguishes generalized and localization-related (focal, local, partial) epilepsies. Generalized epilepsies are syndromes characterized by generalized seizures in which there is an involvement of both hemispheres from the beginning of the seizure. Seizures in localization-related epilepsies start in a circumscribed region of the brain. They may be simple or complex partial and may progress to secondary generalized tonic-clonic seizures.

The other important classification criterion refers to etiology. The ICES distinguishes idiopathic, symptomatic, and cryptogenic epilepsies. Idiopathic means that a disease is not preceded or occasioned by another. Symptomatic epilepsies and syndromes are considered the consequence of a known or suspected disorder of the central nervous system (CNS). In cryptogenic diseases, a cause is suspected but remains obscure, often owing to limited sensitivity of diagnostic techniques.

Localization-Related Epilepsies and Syndromes

Seizure semiology and ictal EEG findings are the most important criteria for anatomic classification of localization-

related epilepsies. As mentioned, the 1981 version of the ICES, with its emphasis on formal structures of seizures, is of limited benefit because it lacks the detailed localization incorporated into the ICES from 1989. The classification has been criticized because reliable localization often requires invasive EEG techniques. It is hoped that with more data from taxonomic studies, clinical symptoms or symptom clusters will be identified that eventually will allow classification on clinical grounds in most cases.

TLE are divided into those with lateral temporal seizures with auditory hallucinations, language disorders (in cases of dominant hemisphere focus), or visual illusions and those with amygdalohippocampal (mesiobasal limbic or rhinencephalic) seizures. The latter are characterized by simple seizure symptoms such as rising epigastric discomfort, nausea, marked autonomic signs, and other symptoms including borborygmi, belching, pallor, fullness of the face, flushing of the face, arrest of respiration, pupillary dilation, fear, panic, and olfactory hallucinations. Complex focal seizures often begin with a motor arrest, typically followed by oroalimentary automatisms. The duration is typically more than 1 minute, and consciousness recovers gradually.

Frontal lobe epilepsies are characterized by seizures of short duration, minimal or no postictal confusion, rapid secondary generalization, prominent motor manifestations that are tonic or postural, complex gestural automatisms, and frequent falling. Ictal scalp EEG may show bilateral or multilobar discharges. Accurate localization of frontal lobe epilepsies may be difficult.

Generalized Epilepsies and Syndromes

Idiopathic generalized epilepsies are characterized by an age-related onset (472). In general, patients are normal between seizures. Radiologic investigations are negative. Frequently, there is an overlap of idiopathic generalized epilepsies, especially of those manifesting in later childhood and adolescence. Response to antiepileptic drug (AED) treatment and the psychosocial prognosis are good.

Symptomatic generalized epilepsies and syndromes usually start in infancy or early childhood. In most children, several seizure types occur. EEG discharges are less rhythmic and less synchronous than in idiopathic generalized epilepsies. There are neurologic, neuropsychological, and radiologic signs of diffuse encephalopathy. The only difference between cryptogenic and symptomatic syndromes is that in cryptogenic syndromes, the presumed cause cannot be identified.

Epilepsies and Syndromes Undetermined as to Whether They Are Focal or Generalized

Two groups of patients cannot be classified as focal or generalized. The first group consists of patients with both generalized and focal seizures (e.g., patients with both focal seizures and absence seizures). The second group consists of patients without unequivocal generalized or focal features (e.g., patients with nocturnal grand mal).

Specific Syndromes

These are situation-related syndromes, such as febrile seizures, isolated seizures, and seizures occurring only when there is an acute metabolic or toxic event resulting from the effects of alcohol, drugs, eclampsia, and nonketotic hyperglycinemia.

EPIDEMIOLOGY OF EPILEPSY

Prevalence and Incidence

Epidemiologic studies of epilepsy are often difficult to compare because of different study designs and definitions of epilepsy. Calculated incidence rates of epilepsy range between 20 and 70 per 100,000 per year. The incidence is age dependent, with the highest rates in early childhood and the lowest rates in early adulthood (522). Incidence figures rise again in older age groups, probably because of the higher prevalence of cerebrovascular disease (494). The overall risk of epilepsy is slightly higher in males than in females. The point prevalence of active epilepsy is approximately 35 per 1,000 (645). The cumulative lifetime prevalence has been estimated to be 3.5% (241).

There is thus a ten-fold difference between the point prevalence and the lifetime prevalence, suggesting that the disease remains active only in a small proportion of cases. The prevalence of epilepsy is higher in Third World countries, with prevalence rates as high as 37 per 1,000 in Africa (125). This is probably related to a higher frequency of infectious diseases of the nervous system, such as neurocysticercosis, and of perinatal complications.

Relative Frequency of Epileptic Syndromes

Research on the distribution of epileptic syndromes and ES population studies have revealed conflicting results. In population-based studies, epilepsies with complex focal and focal seizures secondarily evolving to tonic-clonic seizures are the most frequent, occurring in 69% of all patients. This is followed by primary generalized seizures in 30% and absence or myoclonic seizures in less than 5% (241,291). In hospital-based studies, the numbers for subtle generalized seizures (absences and myoclonic seizures) are generally higher, related to the increased accuracy and sensitivity of diagnosis.

PROGNOSIS OF EPILEPSY

Natural Course of Epilepsy

The natural course of untreated epilepsy is unknown. For ethical reasons, no one has ever conducted a controlled study, and there are no systematic data from the days before AED were introduced. Nineteenth-century epileptologists emphasized the poor prognosis in epilepsy (175), but their experience was limited to institutionalized patients. Gowers (224) believed that seizures may beget seizures: "The tendency of the disease is to self-perpetuation, each attack facilitates the occurrence of another, by increasing the instability of nerve elements." Gowers studied the recurrence of seizures in 160 cases. A second seizure followed the first within 1 month in one-third of patients and within 1 year in two-thirds of patients.

ETIOLOGY OF EPILEPSY

In most patients with epilepsy, the etiology remains unclear. According to an epidemiologic survey in Rochester, UK, 65% of cases are idiopathic or cryptogenic. The most frequent cause of epilepsy is vascular disorder (10%) followed by congenital complications (8%), brain trauma (6%), tumor (4%), degenerative disorders (4%), and infectious diseases (3%). The etiology of epilepsy is largely dependent on the age at onset, with congenital disorders dominating in childhood epilepsies, trauma and tumors being most common in early adulthood, and cerebrovascular and degenerative disorders becoming more frequent with older age (240). More recent studies using advanced imaging techniques have revealed a large number of patients with subtle cortical dysgenesis or hippocampal sclerosis, cases that were formerly classified as cryptogenic.

Recurrence Risk After a First Seizure

It is methodologically extremely difficult to investigate the recurrence risk after an initial ES (170). Figures in the literature range between 27% and 71%, depending on inclusion criteria, duration of follow-up, and whether patients are treated after a first seizure. The most important source of error has been the exclusion of patients with early recurrences of seizures before presentation. The longer the interval between the first seizure and inclusion in the study is, the lower the recurrence rate is (170). The recurrence risk within 24 months after a first seizure calculated from a metaanalysis of 14 studies was 42% (44), with a higher risk in the group of patients with symptomatic seizures and pathologic EEG findings (65%) compared with patients with idiopathic seizures and normal EEG (24%).

Prognostic Studies

Recent prospective and population-based studies have challenged the older views that epilepsy is likely to be a chronic disease in as many as 80% of cases (468). In a population-based survey in Rochester, UK, 20 years after the initial diagnosis of epilepsy, 70% of patients were in 5-year remission and 50% of patients had successfully discontinued

TABLE 39.1. INTERNATIONAL CLASSIFICATION OF EPILEPTIC SEIZURES

Clinical Seizure Type	Ictal EEG
I. Focal (partial, local) seizures	Local contralateral discharge starting over corresponding area of cortical representation (not also recorded on the scalp)
A. Simple partial seizures	
1. With motor symptoms	
a. Focal motor without march	
b. Focal motor with march (jacksonian)	
c. Versive	
d. Postural	
e. Phonatory (vocalization or arrest of speech)	
2. With somatosensory or special sensory symptoms (simple hallucinations, e.g., tingling, light flashes, buzzing)	
a. Somatosensory	
b. Visual	
c. Auditory	
d. Olfactory	
e. Gustatory	
f. Vertiginous	
3. With autonomic symptoms or signs (including epigastric sensation, pallor, sweating, flushing, piloerection, pupillary dilation)	
4. With psychic symptoms (disturbances of higher cortical function); these symptoms rarely occur without impairment of consciousness and are much more commonly experienced as CPS	
a. Dysphasic	
b. Dysmnesic (e.g., déjà vu)	
c. Cognitive (e.g., dreamy states, disturbances of time sense)	
d. Affective (e.g., fear, anger)	
e. Illusions (e.g., macropsia)	
f. Structured hallucinations (e.g., music, scenes)	
B. Complex focal seizures (CPS) (with impairment of consciousness; may sometimes begin with simple symptomatology)	Unilateral or frequently bilateral discharge, diffuse or focal in temporal or frontotemporal regions
1. Simple partial onset followed by impairment of consciousness	
a. With simple partial features (as in A, 1–4) followed by impaired consciousness	
b. With automatisms	
2. With impairment of consciousness at onset	
a. With impairment of consciousness only	
b. With automatisms	
C. Focal seizures evolving to secondarily generalized seizures (this may be generalized tonic-clonic, tonic, or clonic)	Above discharges become secondarily and rapidly generalized
1. SPS (A) evolving to generalized seizures	
2. CPS (B) evolving to generalized seizures	
3. SPS evolving to CPS evolving to generalized seizures	
II. Generalized Seizures	
A.1. Absence seizures	Usually regular and symmetric 3 Hz but may be 2–4 Hz spike-and-slow wave complexes and may have multiple spike-and-slow wave complexes; abnormalities are bilateral
a. Impairment of consciousness only	
b. With mild clonic components	
c. With atonic components	
d. With tonic components	
e. With automatisms	
f. With autonomic components	
A.2. Atypical absence; may have	EEG more heterogeneous; may include irregular spike-and-slow wave complexes, fast activity, or other paroxysmal activity; abnormalities are bilateral but often irregular and asymmetric
a. Changes in tone that are more pronounced than A.1	
b. Onset and/or cessation that is not abrupt	

(*continued*)

TABLE 39.1. (*continued*)

Clinical Seizure Type	Ictal EEG
B. Myoclonic seizures Myoclonic jerks (single or multiple)	Polyspike and wave or sometimes spike and wave or sharp and slow waves.
C. Clonic seizures	Fast activity (10 c/s or more) and slow waves; occasional spike-and-wave patterns
D. Tonic seizures	Low-voltage, fast activity or a fast rhythm of $\geq$9–10 c/s decreasing in frequency and increasing in amplitude
E. Tonic-clonic seizures	Rhythm at $\geq$10 c/s, decreasing in frequency and increasing in amplitude during tonic phase, interrupted by slow waves during the clonic phase
F. Atonic seizures	Polyspikes and wave or flattening or low-voltage fast activity

EEG, electroencephalogram; CPS, complex partial seizures; SPS, simple partial seizures.
From Commission on Classification and Terminology of the International League Against Epilepsy. Proposal for revised clinical and electroencephalographic classification of epileptic seizures. *Epilepsia* 1981;22:489–501, with permission.

medication (16). Goodridge and Shorvon (223) concluded from their study in southeast England that 15 years after diagnosis, 81% of patients were seizure free for at least 1 year.

From 104 patients who were followed after onset of treatment by Elwes et al. (171), 60% were in 1-year remission after a 24-month follow-up period. By 8 years of follow-up, 92% had achieved a 1-year remission. It is recommended that AED be withdrawn after a minimal seizure-free period of 2 years. Relapses occur in 12% to 72% after a 2-year remission and in 11% to 53% after a 3-year remission (168).

Approximately 5% to 10% of all epilepsies eventually include intractable seizures, despite optimal medication (239), most of them occurring in patients with CPS (646). Despite the overall favorable prognosis of epilepsy and the good response to treatment, the mortality of epilepsy is 2.3-fold higher than in the general population, being 3.8-fold higher in the first years of the illness (242). The incidence of sudden unexpected death in epilepsy has been estimated to be approximately one per 525 (342).

The prognosis of epilepsy largely depends on the syndromic diagnosis. Idiopathic localization-related epilepsies, such as rolandic epilepsy, have an excellent prognosis in all respects. The prognosis in terms of seizure remission, social adjustment, and life expectancy, conversely, is extremely poor in symptomatic generalized epilepsies, such as West syndrome, and in progressive myoclonus epilepsies.

Several studies have examined prognostic factors independent of the syndromic diagnosis (Table 39.3). Most have consistently shown that diffuse encephalopathies and neurologic and cognitive deficits are associated with a poor outcome (238,468).

There has been less agreement on the significance of other possible risk factors for poor prognosis, such as EEG features and a positive family history of epilepsy (171). Whether early treatment and medical prevention of seizures really improve the long-term prognosis, as suggested by Gowers and indi-

cated by experimental data from animal epilepsy models, is still unclear and controversial.

BASIC MECHANISMS

Epileptic syndromes are characterized by a tendency for paroxysmal regional or generalized hyperexcitability of the cerebral cortex. Because of the phenomenologic diversity and the etiologic heterogeneity of epilepsies, it is likely that there are multiple underlying cellular and molecular mechanisms.

The mechanisms responsible for the occurrence of seizures (ictogenesis) and the development of epilepsy (epileptogenesis) have been studied in animal models and *in vitro* studies of surgical human brain tissue. The exact mechanisms remain to be clarified. They represent complex changes of normal brain function on multiple levels, involving anatomy, physiology, and pharmacology. There are categorical differences in the pathophysiology of idiopathic generalized and symptomatic focal seizures. The latter are caused by a regional cortical hyperexcitation owing to local disturbances in neuronal connectivity. Synaptic reorganization may be caused by any acquired injury or congenital abnormality such as in the many subforms of cortical dysplasia. Some areas of the brain seem more susceptible than others, the most vulnerable being mesial temporal lobe structures. The pathophysiology of hippocampal sclerosis, the most common etiology of TLE, has been a topic of much controversy. It is most likely that the hippocampal structures are damaged by an early trauma, typically prolonged febrile seizures. The process of hippocampal sclerosis involves a synaptic reorganization with excitotoxic neuronal loss, loss of interneurons, and γ-aminobutyric acid (GABA) deficit.

Epileptic neurons in an epileptogenic focus may produce bursts of action potentials that represent isolated spikes on EEG as long as they remain locally restricted. Depending on the failure of local inhibitory mechanisms, these bursts may lead to ongoing and repetitive discharges. If larger neuronal

TABLE 39.2. INTERNATIONAL CLASSIFICATION OF EPILEPSIES AND EPILEPTIC SYNDROMES

1. Localization-related (focal, local, partial) epilepsies and syndromes
   1.1 Idiopathic (with age-related onset)
   At present, the following syndromes are established, but more may be identified in future
   Benign childhood epilepsy with centrotemporal spikes
   Childhood epilepsy with occipital paroxysms
   Primary reading epilepsy
   1.2 Symptomatic
   Chronic progressive epilepsia partialis continua of childhood (Kozhevnikov syndrome)
   Syndromes characterized by seizures with specific modes of precipitation
   Temporal lobe epilepsy
   With amygdala-hippocampus seizures
   With lateral temporal seizures
   Frontal lobe epilepsy
   With supplementary motor seizures
   With cingulate seizures
   With seizures of the anterior frontopolar region
   With orbitofrontal seizures
   With dorsolateral seizures
   With opercular seizures
   With seizures of the motor cortex
   Parietal lobe epilepsies
   Occipital lobe epilepsies
2. Generalized epilepsies and syndromes
   2.1 Idiopathic, with age-related onset, listed in order of age
   Benign neonatal familial convulsions
   Benign neonatal convulsions
   Benign myoclonic epilepsy in infancy
   Childhood absence epilepsy (pyknolepsy)
   Juvenile absence epilepsy
   Juvenile myoclonic epilepsy (impulsive petit mal)
   Epilepsy with grand mal seizures on awakening
   Other generalized idiopathic epilepsies not defined above
   Epilepsies precipitated by specific modes of action
   2.2 Cryptogenic or symptomatic (in order of age)
   West syndrome (infantile spasms, Blitz-Nick-Salaam Krämpfe)
   Lennox-Gastaut syndrome
   Epilepsy with myoclonic-astatic seizures
   Epilepsy with myoclonic absences
   2.3 Symptomatic
      2.3.1 Nonspecific etiology
      Early myoclonic encephalopathy
      Early infantile epileptic encephalopathy with suppression burst
      Other symptomatic generalized epilepsies not defined above
      2.3.2 Specific syndromes
      Epileptic seizures complicating disease states (including diseases in which seizures are a presenting or predominant feature)
3. Epilepsies and syndromes undetermined as to whether they are focal or generalized
   3.1 With both generalized and focal seizures
   Neonatal seizures
   Severe myoclonic epilepsy in infancy
   Epilepsy with continuous spike waves during slow-wave sleep
   Acquired epileptic asphasia (Landau-Kleffner syndrome)
   Other undetermined epilepsies not defined above
   3.2 Without unequivocal generalized or focal seizures

From Commission on Classification and Terminology of the International League Against Epilepsy. Proposal for revised classification of epilepsies and epileptic syndromes. *Epilepsia* 1989;30:389–399, with permission.

TABLE 39.3. FACTORS AFFECTING THE PROGNOSIS OF EPILEPSY

Negative prognostic factors
  Diffuse encephalopathy
  Psychiatric problems
  Mental retardation
  Early onset of epilepsy
  Multiple seizure types
  Simple and complex partial seizures
  (Progressive) neurologic symptoms
  Long duration of active epilepsy
  High number of seizures
  High seizure frequency before treatment
  Positive family history for epilepsy
  Abnormal background electroencephalographic activity

networks are recruited in this hypersynchronous activity, this leads to an ES, which is either focal or secondary generalized, depending on the extent of seizure spread. The local seizure threshold is regulated by influences on excitatory and inhibitory postsynaptic potentials. The membrane excitability is regulated by ion channels that are modulated by excitatory transmitters, particularly glutamate and aspartate, and the inhibitory transmitter GABA. These three levels are the major targets for AED (Table 39.4).

Primary generalized seizures are accompanied by bilateral synchronous epileptic activity. Therefore, they cannot be explained by cortical dysfunction alone. They are caused by an imbalance of pathways between the thalamus and cerebral cortex. These thalamocortical circuits are modulated by reticular nuclei. Excessive GABA-ergic inhibition and dysfunction of thalamic calcium channels also play a role in the

pathogenesis, at least in the pathogenesis of absence seizures. These cortical subcortical circuits are also responsible for circadian rhythms, which may explain the increased seizure risk after awakening and after sleep withdrawal in idiopathic generalized epilepsies. They largely underlie genetic influences and seem to be vulnerable to brain maturation processes because primary generalized seizures are strongly age related.

BIOCHEMISTRY OF EPILEPSY

The cellular mechanisms responsible for epileptogenesis remain to be clarified. Much recent research on the biochemistry of epilepsy has focused on investigating the effects of inhibitory and excitatory amino acid neurotransmitters on the excitability of neurons (388) together with the role of calcium in this process (433).

The major inhibitory neurotransmitter in the brain is GABA (388), a monocarboxylic amino acid. It is distributed throughout the CNS and is a potent inhibitor of neuronal discharge. It may be used by as many as 40% of neurons (229). Two classes of GABA receptors have been characterized (78,531), termed $GABA_A$ and $GABA_B$ and distinguished based on their pharmacologic characteristics (78). The $GABA_A$ receptor is thought to be responsible for mediating CNS neuronal inhibition (266). In experimental epileptic foci in monkeys, it has been demonstrated that there are decreases in both the number of GABA receptors and the levels of glutamate decarboxylase, a GABA synthetic enzyme. It has further been shown in monkeys that after cortical injection of alumina gel to create a seizure focus, a selective loss

TABLE 39.4. MODE OF ACTION OF OLD AND NEW ANTIEPILEPTIC DRUGS

Drug	Sodium Channel Blockade of Voltage-Gated Sodium Channels	GABA Channel Potentiation of GABA-ergic Mechanisms	Glutamate Channel Blockade of Glutamatergic Mechanisms	Calcium Channel Blockade of Thalamic Calcium Channels
Old AED				
Benzodiazepines	+	++	0	0
Carbamazepine	++	+/−	+/−	?
Ethosuximide	0	0	0	+
Phenobarbitone	+	+	+	0
Phenytoin	++	+/−	+/−	0
Valproate	++	+	+/−	0
New AED				
Felbamate	+	+	++	?
Gabapentin	+/−	+	+/−	0
Lamotrigine	++	0	+/−	0
Levetiracetam	0	0	0	0
Oxcarbazepine	++	0	?	?
Tiagabine	?	++	?	?
Topiramate	+	+	+	?
Vigabatrin	0	++	?	?
Zonisamide	+	+/−	?	?

GABA, γ-aminobutyric acid; AED, antiepileptic drug.

of glutamate decarboxylase–positive neuronal somata occurs before the animal starts to experience seizures (455).

The role of GABA-ergic mechanisms in epilepsy has been the focus of much pharmacologic research on the treatment of the condition. In animal models of epilepsy, enhancing GABA-ergic neurotransmission has been shown to protect against seizures of various origins. However, this is not the case for GABA neurons in general but only for GABA systems, in particular, brain regions. Animal studies have demonstrated that increasing GABA-ergic activity in the substantia nigra controls both tonic and clonic seizures produced in the hindbrain as well as clonic seizures originating in the forebrain (200). However, injections of the GABA agonist vigabatrin into other brain regions of rats, including the caudate, thalamus, and superior colliculus, does not protect against experimental seizures (273).

Just as a deficit in inhibitory neurotransmitter activity has been implicated in generating seizure activity, so has an excess of excitatory activity. The excitatory amino acids glutamate and aspartate mediate the majority of excitatory transmission in the vertebrate nervous system. In a study using receptor autoradiography to investigate excitatory amino acid receptors in surgically excised temporal lobes from humans with epilepsy, it was observed that there was an increase in N-methyl-D-aspartate (NMDA)–sensitive glutamate and in kainic acid binding in the parahippocampal gyrus (208). The authors stated that it was not clear whether the observed changes resulted from or contributed to seizure activity. Nevertheless, their results suggested that vulnerability of the hippocampus in TLE may result, at least in part, from the presence of aberrant excitatory circuits in the parahippocampal gyrus.

Modulation of excitatory neurotransmission provides another avenue for pharmacologic manipulation of seizure activity. Indeed, a new anticonvulsant, lamotrigine (LTG), most likely acts by inhibiting the release of excitatory amino acids (457).

Experimental models of epilepsy have demonstrated that calcium has an important role both in the induction of epilepsy and the generation of seizures. The NMDA subtype of glutamate receptors constitutes an important route for influx of extracellular Ca^{2+} and functions as a receptor-operated channel (433). Studies of experimental models of epilepsy have demonstrated that there is a decrease in extracellular Ca^{2+} at the time of seizure onset, probably because of increases in postsynaptic calcium uptake (444), mediated through NMDA-type glutamate receptors and voltage-operated calcium channels.

Calcium is central to neuronal function, and the increased excitation or diminished inhibition that underlies seizure activity is expressed through a variety of calcium systems (433). This central role of calcium has led to trials of calcium antagonists in the management of epilepsy. At the time of writing, however, this approach has not yet led to the development of any new AED.

GENETICS OF EPILEPSY

Genetic factors play a major role in the etiology of idiopathic epilepsies. Progress in molecular genetics has revealed several susceptibility loci and causative gene mutations in a number of rare human epilepsies with monogenic inheritance, such as the progressive myoclonus epilepsies. Mutations in the gene encoding the α_4 subunit of the neuronal nicotinic acetylcholine receptor gene have been identified to predispose to autosomal dominant nocturnal frontal lobe epilepsy. Mutations in two genes encoding potassium channels cause benign neonatal familial convulsions. A gene encoding the β subunit of the voltage-gated sodium channel has been identified in a rare syndrome called generalized epilepsy with febrile seizures plus. These findings suggest that at least some epilepsies are related to ion channel dysfunction (553).

In the common idiopathic generalized epilepsies, positional cloning of susceptibility genes was less successful owing to the underlying complex genetic disposition. Several chromosomal regions (on chromosomes 3, 6, 8, and 15) are thought to harbor susceptibility genes for generalized seizures (491).

In the genetic counseling of patients with epilepsies with complex modes of inheritance, empirical risk estimates are used. The recurrence risk for the offspring of probands with idiopathic generalized epilepsies is approximately 5% to 10% (13).

DIAGNOSIS OF EPILEPSY
Clinical Diagnosis and Differential Diagnosis

Epilepsy is a clinical diagnosis, defined by recurrent ES. Most important for accurate syndromic classification and optimal application of diagnostic techniques is the clinical interview. This should cover seizure-related information, such as subjective and objective ictal symptomatology, precipitation and frequency of seizures, history of seizures in first-degree relatives, and data relevant for etiology (complications during pregnancy and birth, early psychomotor development, history of brain injuries, and other disorders of the CNS). Other important information that should be obtained includes doses, side effects, efficacy of previous medical or nonmedical treatment, evidence of psychiatric complications in the past, and psychosocial parameters including educational and professional status, social independence, and psychosexual history.

The neurologic examination may reveal signs of localized or diffuse brain damage. One should also look for skin abnormalities and minor stigmata suggestive of genetic diseases and dysontogenetic malformations. There may be signs of injuries caused by ES, such as scars from recurrent falls, burns, and tongue biting. Hirsutism, gingival hyperplasia, or

TABLE 39.5. DIFFERENTIAL DIAGNOSIS OF EPILEPSY

Neurologic disorders
  Transient ischemic attacks
  Migraine
  Paroxysmal dysfunction in multiple sclerosis
  Transient global amnesia
  Movement disorders (hyperexplexia, tics, myoclonus, dystonia, paroxysmal choreoathetosis)
  Drop attacks owing to impaired cerebrospinal fluid dynamics
Sleep disorders
  Physiologic myoclonus
  Pavor nocturnus
  Somnambulism
  Enuresis
  Periodic movements in sleep
  Sleep talking
  Bruxism
  Nightmares
  Sleep apnea
  Narcolepsy (cataplexy, automatic behavior, sleep attacks, hallucinations)
Psychiatric disorders
  Nonepileptic seizures
  Anxiety attacks
  Hyperventilation syndrome
  Dissociative states, fugues
  Episodic dyscontrol, rage attacks
Medical disorders
  Cardiac arrhythmias
  Syncope (cardiac, orthostatic, reflex)
  Metabolic disorders (e.g., hypoglycemia)
  Hypertensive crisis
  Endocrine disorders (e.g., pheochromocytoma)

TABLE 39.6. LOCALIZATION OF PATHOLOGIC ZONES IN FOCAL EPILEPSIES

Epileptogenic zone	Ictal EEG with special electrode placements
	Functional imaging (SPECT, MRI)
Irritative zone	Interictal EEG, MEG
Pacemaker zone	Ictal EEG with special electrode placements
Ictal symptomatic zone	Ictal EEG with special electrode placements
Epileptogenic lesion	CT, MRI
Functional deficit zone	Functional imaging (PET, SPECT, fMRI)
	Neurologic examination, neuropsychology
	Nonepileptiform interictal EEG abnormalities

EEG, electroencephalography; SPECT, single photon emission computed tomography; MRI, magnetic resonance imaging; MEG, magnetoencephalography; CT, computed tomography; fMRI, functional magnetic resonance imaging.
Adapted from Lüders HO, Awad I. Conceptual considerations. In: Lüders H, ed. *Epilepsy surgery.* New York: Raven Press, 1991:51–62, with permission.

5. The functional deficit zone, defined as the region of the cortex that in the interictal period is functioning abnormally.

The diagnostic techniques applied in epilepsy are characterized by a selective specificity for these pathologic regions (Table 39.6).

Electroencephalography

Interictal surface EEG is still the most important method in the diagnosis and assessment of all types of epilepsy. Routine EEG is performed over 30 minutes during a relaxed condition, including photic stimulation procedures and 5 minutes of hyperventilation. Paroxysmal discharges strongly suggestive of epilepsy are spikes, spike waves, and sharp waves. These epileptiform patterns are, however, not epilepsy specific. They may be observed in patients with nonepileptic neurologic diseases and even in a small proportion of normal subjects.

The sensitivity of routine EEG is limited by restrictions of spatial and temporal sampling. Approximately 50% of patients with epilepsy do not show paroxysmal epileptiform discharges on a single EEG recording (59). Their detection depends on the epileptic syndrome and the therapeutic status of the patient. In untreated childhood absence epilepsy, the EEG usually shows generalized spike-wave complexes either occurring spontaneously or provoked by hyperventilation. In mild cryptogenic focal epilepsies, conversely, the interictal EEG is often negative. The temporal sensitivity can be increased by repeating the EEG or by carrying out long-term recordings with mobile EEG. Furthermore, paroxysmal discharges may be brought out by performing EEG after sleep deprivation, while the subject is asleep (Table 39.7).

acne vulgaris are indicative of side effects of long-term AED medication.

In most cases, a clinical interview and a neurologic examination are sufficient to distinguish between epilepsy and its wide spectrum of differential diagnoses (Table 39.5). There is, however, a substantial problem with pseudoseizures (see discussion that follows).

For correct interpretation of functional and structural diagnostic techniques, it is important to understand that in focal epilepsies, different concepts of pathologic cerebral regions must be distinguished (361). The epileptogenic zone is defined as the region of the brain from which the patient's habitual seizures arise. Closely related but not necessarily anatomically identical are

1. The irritative zone, defined as the region of the cortex that generates interictal epileptiform discharges on EEG;

2. The pacemaker zone, defined as the region of the cortex from which the clinical seizures originate;

3. The epileptogenic lesion, defined as the structural lesion that is usually related to epilepsy;

4. The ictal symptomatic zone, defined as the region of the cortex that generates the ictal seizure symptomatology;

**TABLE 39.7. ELECTROENCEPHALOGRAPHIC METH-
ODS IN THE DIAGNOSIS OF EPILEPSY**

Interictal EEG
  Routine surface EEG
  EEG after sleep withdrawal, during sleep
  Mobile long-term EEG
Ictal EEG
  Long-term video EEG
Special electrode placement (in the order of invasiveness)
  Nasopharyngeal electrodes
  Sphenoidal electrodes
  Foramen-ovale electrodes
  Epidural electrodes (strips, grids)
  Subdural electrodes (strips, grids)
  Depth electrodes

EEG, electroencephalography.

Simultaneous video-EEG recordings of seizures are useful for differentiating between different types of ES (Table 39.1) and non-ES (NES). Ictal EEG is also required for exact localization of the epileptogenic focus when epilepsy surgery is considered.

Surface EEG records only a portion of the underlying brain activity. Discharges that are restricted to deep structures or to small cortical regions may not be detected. The spatial resolution of EEG can be improved by special electrode placements, such as pharyngeal and sphenoidal electrodes.

Invasive EEG methods with chronic intracranial electrode placement are necessary for complex analysis in cases with discordant or multifocal results of the ictal surface EEG and imaging techniques. These include foramen ovale electrodes positioned in the subdural space along the amygdalo-hippocampal formation (619), epidural and subdural strip electrodes, and grids to study larger brain areas. Stereotactic depth electrodes provide excellent sensitivity to detect small areas of potentially epileptogenic tissue. The definition of exact location and boundaries of the epileptogenic region, however, is limited by the location and number of electrodes placed. Because of the limited coverage of implanted electrodes, it may be difficult to distinguish whether a seizure discharge originates from a pacemaker zone or represents spread from a distant focus. Complications, the most severe being intracerebral hemorrhages, occur in 4%.

Magnetoencephalography

Multichannel magnetoencephalography has recently been introduced in the presurgical assessment as a supplementary method to EEG. The electric activity, which can be measured by EEG, produces a magnetic field perpendicular to the electric flow. This magnetic signal can be measured by magnetoencephalography. In contrast to EEG, magnetoencephalography is not influenced by intervening tissues with the advantage of noninvasive localization of deep electric sources. Disadvantages include high costs and the suscepti-

bility to movement artifacts, making it extremely difficult to do ictal studies (456,550).

Structural Imaging

Imaging studies should always be performed when a symptomatic etiology is suspected. Cranial computed tomography is a quick, easy, and inexpensive technique. The sensitivity can be improved by scanning in the axis of the temporal lobe (in cases of TLE) and by using intravenous contrast enhancement. Except for a few pathologies, such as calcifications, magnetic resonance imaging (MRI) is superior to cranial computed tomography in terms of sensitivity and specificity in detecting epilepsy-related lesions such as malformations, gliosis, and tumors. With an optimized MRI technique (277), including T2-weighted images, inverse recovery sequences, coronal images perpendicular to the hippocampus, and thin sections, the sensitivity of detecting mesial temporal sclerosis reaches 90% (276). Diagnosis of hippocampal pathology can be further improved by quantitative MR techniques such as T2 relaxometry (278) and volumetric studies (119).

MR spectroscopy (MRS) is a noninvasive method of measuring chemicals in the body. MRS does not produce images but instead generates numerical values for chemicals. With phosphate spectroscopy, it is possible to study energy metabolism in relation to seizure activity (642). Proton MRS measures neuronal density, which has been found to be significantly decreased in the mesial temporal lobe of patients with mesial temporal sclerosis (118).

Functional Imaging

Single photon emission computed tomography (SPECT) in epilepsy has mainly been confined to the imaging of regional cerebral blood flow in focal epilepsy. The tracer most widely used at present is technetium-99 hexamethyl-propyleneamine oxime (^{99m}Tc-HMPAO). Interictally, there is localized hypoperfusion in an area extending beyond the epileptogenic region. Initially, there was considerable skepticism about the clinical value of the technique because the early studies were of low resolution and the correlations with EEG findings were imprecise. Figures on the sensitivity of interictal focus detection of SPECT in the literature range from 40% to 80% (56,478). In recent years, there have been major technical developments in instrumentation. Using brain-dedicated, multiheaded camera systems, the sensitivity of SPECT is comparable with that of [^{18}F]fluoro-deoxy-D-glucose positron emission tomography (PET) (12). ^{99m}Tc-HMPAO is distributed within a few minutes after injection in the brain, where it remains fixed for approximately 2 hours. If the radioisotope is injected during or shortly after an ES, scanning can be carried out postictally without problems from involuntary movements. Postictal and ictal SPECT are more sensitive than interictal SPECT and typically show hyperperfusion ipsilateral to the epileptogenic focus (479). Another ligand used for SPECT is [^{123}I] iomazenil, which is used to demonstrate benzodiazepine receptor binding, which is decreased in the epileptogenic region.

Compared with SPECT, PET is superior with respect to spatial and contrast resolution. PET, however, is expensive and requires an on-site medical cyclotron. Ictal studies are difficult with PET because of the short half-life of positron-emitting radioisotopes. PET has mainly been used to study interictal blood flow with ^{13}N-labeled ammonia and oxygen-15 and glucose metabolism with [^{18}F]fluoro-deoxy-D-glucose in focal epilepsy. Localized hypoperfusion and hypometabolism in the epileptogenic area as shown by PET are seen as a reliable confirmatory finding in the presurgical assessment of TLE (173). PET has also been used for imaging of benzodiazepine receptor binding (501) and opiate receptor binding (198) in epilepsy.

The major application of functional MRI in epilepsy is the noninvasive mapping of eloquent cortex in the presurgical evaluation of patients based on activation studies. The clear advantage over PET is the superior spatial resolution, which is approximately 3 mm. The coregistration of a high-resolution MRI allows an excellent localization of activated regions.

Neuropsychology

Identification of neuropsychological deficits is important for optimizing education, professional training, and rehabilitation in patients with epilepsy. Another aim of neuropsychological testing is to establish the cognitive effects of AED, seizure frequency, and subclinical EEG activity.

The presurgical assessment neuropsychological evaluation is used to identify localizable deficits that can be related to the epileptogenic lesion. Crucial for lateralizing temporal lobe epilepsies is the function of verbal and nonverbal memory (287). Another neuropsychological task in the presurgical assessment is forecasting postsurgical cognitive outcome, which sometimes requires the intracarotid sodium amobarbital procedure (Wada test).

Recently, magnetic cerebral stimulation has been used to lateralize the epileptogenic region in focal epilepsies. The technique of repetitive stimulation is also being evaluated for therapeutic antiepileptic properties.

TREATMENT OF EPILEPSY

After a diagnosis of epilepsy has been made, followed by the decision to treat, several options are available. Pharmacologic, behavioral, and surgical approaches may be taken. Most patients with epilepsy are treated with anticonvulsants. Recently, there has also been renewed interest in attempts at behavioral and nonpharmacologic approaches to the management of seizures. A minority of patients with drug-resistant epilepsy may proceed to have surgery.

These approaches should not be viewed as being mutually exclusive. Hence, although in the discussion that follows these treatments are described separately, in practice, they may be combined. Indeed, if surgery is pursued, it is very likely that patients will be receiving antiepileptic medication both before and after the operation.

Antiepileptic Agents

Ideally, patients should be managed on a single drug, which leads to complete seizure freedom, without causing any side effects. Approximately 75% of patients with epilepsy can be fully controlled on monotherapy (50% with initial monotherapy, 25% with second monotherapy), the choice of agent being determined by the epilepsy syndrome and seizure type (Table 39.8). Carbamazepine (CBZ), valproic acid (VPA), and LTG are the recommended substances for simple, complex partial, and secondary generalized tonic-clonic seizures. For most generalized seizures, VPA or LTG are the most useful treatments. These drugs are generally thought of as the first-line treatments. Of the remaining patients not controlled on monotherapy, addition of another first-line drug will gain control in a further 15%. However, some patients will go on to develop chronic seizures

TABLE 39.8. SELECTION OF ANTIEPILEPTIC DRUGS ACCORDING TO THE EPILEPTIC SYNDROME

Type of Syndrome	First-Line Drugs	Second-Line Drugs
Focal epilepsies All seizure types	Carbamazepine, valproate, lamotrigine	Gabapentin, topiramate, levetiracetam, oxcarbazepine, phenytoin, clobazam, primidone, acetazolamide, vigabatrin, felbamate, zonisamide
Generalized epilepsies: idiopathic		
Absences	Ethosuximide, valproic acid	Lamotrigine
Myoclonic	Valproic acid	Lamotrigine, primidone, levetiracetam
Awakening grand mal	Valproic acid	Lamotrigine, primidone, topiramate
Generalized epilepsies: symptomatic		
West syndrome	Vigabatrin, ACTH, valproate	Clobazam
Lennox-Gastaut syndrome	Valproate, lamotrigine	Clobazam, topiramate, felbamate

ACTH, adrenocorticotropic hormone.

unrelieved by these treatments. In such circumstances, alternative monotherapy or adjunctive therapies will be considered. Several of the more recently introduced anticonvulsants, often considered as second-line treatments, may be introduced, either alone or in combination with a first-line agent.

In Europe and the United States, several novel anticonvulsants have been introduced in the past decade: vigabatrin (VGB), felbamate, tiagabine (TGB), gabapentin (GBP), oxcarbazepine, topiramate (TPM), levetiracetam (LEV), and zonisamide. Because of serious side effects, two of these drugs are only used in a highly selected group of patients. Felbamate has caused fatal hepatotoxic and hematotoxic adverse events and VGB leads to irreversible visual field constriction in 30% of patients.

An important feature in the clinical use of the first-line treatments is therapeutic drug monitoring (TDM). This should not be done as a routine procedure but may be useful in some circumstances, e.g., to detect noncompliance and to explore changes in pharmacokinetics. There are significant interindividual variations in the relationship between drug levels and clinical side effects. Therefore, the measurement of serum concentrations should not be used for the optimization of treatment (with few exceptions); decisions to alter doses should be based on efficacy and tolerability of AED. For appropriate agents, indications for TDM may include drug initiation or dose change, investigation of absence of or change in therapeutic response, investigation of suspected toxicity, and situations in which pharmacokinetics may change such as during pregnancy and related to renal or hepatic disease. Sampling should occur once steady state has been achieved, four to six half-lives after treatment has been modified or introduced. TDM is particularly important for phenytoin (PHT) because its hepatic metabolism is saturable and thus a small increase in dose can result in a disproportionate and unpredictable increase in serum concentration.

Behavioral Treatment

Pharmacologic treatments of epilepsy are not uniformly successful in all patients. Even if good seizure control is obtained, many patients experience troublesome side effects of treatments that must be continued, often for many years. The need to take medication on a long-term basis has obvious implications for women wishing to become pregnant. In addition, many individuals describe feelings of oppression and an increased fear of being labeled as ill because of their ongoing need for regular drug taking. Surgical treatment is an option only for a minority of patients and is not without physical and psychological sequelae. Not surprisingly, then, alternative treatment approaches should be sought.

It has been suggested that many patients with epilepsy have a mental mechanism they use to attempt to inhibit their seizures. In one study of 70 patients, 36% of them claimed that they could sometimes stop their seizures (187). A be-

havioral approach to the treatment of epilepsy is based on observations that epilepsy can be manipulated in a systematic way through environmental, psychological, and physical changes. The initial stage in this approach is a behavioral analysis of the ways in which environmental and behavioral factors interact with seizure occurrence (122).

Significant reductions in seizure frequency may be achieved by teaching patients a specific contingent relaxation technique that they must be able to employ rapidly when they identify a situation in which they are at high risk of having a seizure (122). In a subsequent study of three children with intractable seizures, contingent relaxation alone did not significantly reduce seizures, but such a reduction was obtained after the addition of specific countermeasures aimed at changing the arousal level as it related to early seizure cues. For instance, the patient was instructed to suddenly jerk the head to the right when it would habitually move to the left with a feeling of drowsiness at the onset of a seizure (123). Other specific approaches to seizure inhibition have been reviewed by several authors (187,406).

Some patients have reflex seizures in that their seizures are precipitated by external stimuli. Some patients can identify specific environmental or affective triggers and may be able to develop specific strategies to abort or delay a seizure. These methods may involve motor or sensory activity or may be purely mental. However, in one study, it was found that 50% of patients who inhibited their seizures at times had to "pay the price in subsequent discomfort" (17).

Primary seizure inhibition describes the direct inhibition of seizures by an act of will. For example, a man whose seizures were precipitated by a feeling of unsteadiness tackled this problem by keeping his gaze fixed on a point when walking down an incline (187). The nature of the successful act varies from person to person, and this treatment approach will not be effective unless it is individually tailored, based on an analysis of each patient's seizures and any actions that he/she may already have noticed that modify the seizures.

The term secondary inhibition is employed by Fenwick (187) to describe behavioral techniques that effect change in cortical activity in the partially damaged group 2 neurons around the focus, thereby reducing the risk both of a partial seizure discharge and a generalized seizure discharge. The latter may otherwise follow recruitment of surrounding normal brain by group 2 neurons firing abnormally (356). An example of this is the act of maintaining alertness by a patient whose seizures appear in a state of drowsiness. Treatment in this case starts with trying to identify situations in which the subject reliably tends to have seizures or, alternatively, to be free of them.

In addition to these seizure-related approaches, more general psychological strategies have been investigated. Several anecdotal reports have been published demonstrating benefit from a reward system that aims to reward seizure-free periods.

Based on the observation that some patients with olfactory auras can prevent progression of the seizure by applying a sudden, usually unpleasant olfactory counterstimulus, Betts et al. (53) explored the use of aromatherapy techniques to control epilepsy. So far, the relative differential contributions of specific olfactory stimuli as opposed to the general relaxation that is part of any treatment are not clear.

Specific biofeedback techniques have also been explored. Measurement of scalp electrical activity has demonstrated that there is an increase in surface negative slow cortical potentials in the seconds before a seizure occurs. These slow cortical potentials represent the extent to which apical dendrites of cortical pyramidal cells are depolarized and hence indicate neuronal excitability (467). Studies using visual feedback of this effect have demonstrated that some patients are able to modulate cortical electrical activity with an associated decrease in seizure frequency (61). However, it appears that patients with epilepsy are less able than normal controls to regulate their cortical excitability. This impairment can be minimized by extending the amount of training received by those with epilepsy. In a study that gave 28 1-hour training sessions to 18 patients followed for at least 1 year, six became seizure free (467). However, not every patient who achieved reliable slow cortical potential control experienced a reduction in seizure frequency.

Nonspecific EEG biofeedback has been used to modulate cerebral electrical rhythms, for instance, to increase fast low-voltage activity and to suppress slow-wave activity surrounding the epileptic focus (625).

Ultimately, the teaching of any of these methods of self-control of seizures may increase morale, not only by reducing seizures, but also by providing patients with a sense of control over their epilepsy. An important aspect of many nonmedical treatments of epilepsy is that although they still have very limited proven benefit, they aim to consider seizures in the wider setting of the patient's life. In mainstream clinical management, it is sometimes easier to focus purely on seizure response to the latest change in anticonvulsant therapy.

Surgical Treatment

In all patients with persistent epilepsy not relieved by AED treatment, it becomes appropriate to consider surgical intervention. The aim of epilepsy surgery is the removal of the epileptogenic brain tissue to achieve seizure control without causing additional iatrogenic deficits. Only in rare cases, such as in those patients with diffuse pathology and catastrophic seizures, palliative surgical procedures are performed. Approximately one-third of patients with refractory epilepsy is suitable for epilepsy surgery. Prerequisites of surgery are frequent ES and a minimum of 3 to 5 years of unsuccessful AED treatment, including at least two first-line AED in monotherapy and combination therapy.

These are minimal criteria that only apply to patients who are ideal candidates for epilepsy surgery. These are patients with unilateral mesial TLE plus hippocampal sclerosis, extratemporal epilepsies with localized structural lesions, and some types of catastrophic epilepsies in childhood. The

poorer the individual prognosis of surgery is, the longer are the requested presurgical attempts to achieve seizure control with AED and complementary nonsurgical treatments.

Presurgical Assessment

Treatment centers that engage in routine surgical management of epilepsy generally have a standardized assessment process for patients being considered for such treatment. Although the program may vary a little from place to place, the general procedure is similar. A first hypothesis with respect to the suspected epileptogenic region is based on clinical history, neurologic examination, and interictal EEG. These procedures focus particularly on looking for etiologic factors, evidence of localizing signs and symptoms, and a witnessed description of the seizures. In addition, psychosocial information must be gathered relating to education, employment, social support, and past and present mental state findings. In a second step of presurgical assessment, all patients undergo more specific investigations. In general, this includes several days of continuous videotelemetry using surface electrodes. The aim of telemetry is to obtain multiple ictal recordings, which give more valuable localizing information than interictal records. Often patients will reduce their AED before telemetry to facilitate the occurrence of seizures. Recent advances in structural and functional neuroimaging have made invasive EEG recording less necessary. MRI using optimized techniques has increased the sensitivity to detect subtle structural lesions not seen on cranial computed tomography and standard MRI and is now used routinely. Functional imaging techniques using SPECT and PET are used to further delineate the epileptogenic region. If results from all investigations are concordant, patients will go on to have surgery. If results are discordant, invasive EEG recordings using subdural grid electrodes or intracerebral depth electrodes may be applied to identify the critical epileptogenic region.

All patients being considered for epilepsy surgery should have a neuropsychological assessment. This is important both to detect focal brain dysfunction and to predict the results of surgery, especially temporal lobe surgery. The IQ may be measured using the Wechsler Adult Intelligence Scale. In some centers, a score of less than 75 has been taken as evidence of diffuse neurologic disorder and, hence, as a relative contraindication to surgery. It is important before proceeding to surgery to investigate the hemisphere localization of language and memory. This is generally performed using the Wada test. Sodium amobarbital is injected into one internal carotid artery, and when that hemisphere is briefly suppressed, language and memory tests are performed. The procedure is then repeated for the other hemisphere.

Surgical Procedures

Several surgical procedures have been developed. Most patients who undergo surgery have TLE. Most of these patients undergo one of two standard procedures, two-thirds anterior resection or amygdalohippocampectomy. In extratemporal epilepsies, lobectomies, lesionectomies, or individually tailored topectomies are performed. The latter techniques may involve preoperative stimulation to identify eloquent brain tissue, which should be preserved from resection. A palliative method, which can be applied in functionally crucial cortical regions, is multiple subpial resections that interrupt intracortical connections without destroying the neuronal columns that are necessary for normal cerebral function. Another palliative method, which is performed mainly in patients with epileptic drop attacks, is anterior or total callosotomy.

Prognosis and Complications

The outcome of epilepsy surgery is classified according to four categories: class 1, no disabling seizures; class 2, almost seizure free; class 3, clinical improvement; and class 4, no significant improvement (173). According to a metaanalysis of 30 surgical series with a total of 1,651 patients, seizure outcome was as follows: class 1, 59%; class 2, 14%; class 3, 15%; and class 4, 12%. In this study, predictors of a good surgical outcome were febrile seizures, CPS, low preoperative seizure frequency, lateralized interictal EEG findings, unilateral hippocampal pathology on MRI, and neuropathologic diagnosis of hippocampal sclerosis. Predictors for a poor outcome were generalized seizures, diffuse pathology on MRI, normal histology of removed tissue, and early postoperative seizures (583).

Relative contraindications for temporal lobe resections are extensive or multiple lesions, bilateral hippocampal sclerosis or dual pathology within one temporal lobe, significant cognitive deficits, multiple seizure types, extratemporal foci in the interictal EEG, and normal MRI. The ideal patient has mesial TLE owing to unilateral hippocampal sclerosis of the nondominant hemisphere. A negative MRI decreases the chances of a surgery-free outcome, and a focus in the dominant hemisphere increases the risks of postoperative neuropsychological deficits.

The nature of potential perioperative complications and postsurgical neurologic, cognitive, and psychiatric sequelae depends in part on the site of surgery. Operative complications occur in fewer than 5% of patients. In two-thirds of patients who undergo temporal resection of the dominant hemisphere, verbal memory is impaired. Another possible complication of epilepsy surgery is psychiatric disorder (discussed later in this chapter).

Vagal Nerve Stimulation

A quite different surgical approach is that of vagal nerve stimulation using an implanted stimulator. This approach has the drawback that although the nerve is being stimulated, usually for 30 seconds every 5 to 10 minutes, the voice changes. More intense stimulation may be associated with throat pain or coughing. Nevertheless, in one series of 130 patients, the mean seizure frequency decreased by 30% after 3 months and by 50% after 1 year of therapy. Altogether 60% to 70% of the patients showed some response, but seizure freedom has been achieved very rarely (38). Therefore, vagal nerve stimulation is only indicated in patients who are pharmacoresistant and who are not suitable for resective epilepsy surgery.

STATUS EPILEPTICUS

Status epilepticus (SE) is defined as a condition in which a patient has a prolonged seizure or has recurrent seizures without fully recovering in the interval. With respect to the most severe seizure type (primary or generalized tonic-clonic), a duration of 5 minutes is sufficient for diagnosis of SE. SE may occur in a person with chronic epilepsy (the most frequent cause being noncompliance) or in a person with an acute systemic or brain disease (such as hypoglycemia or stroke). The classification of SE follows the ICES. Clinically, however, often only two major types are distinguished: convulsive and nonconvulsive SE. SE, in particular convulsive SE, is a life-threatening condition that requires emergency treatment. Prolonged seizure activity causes systemic complications and brain damage (523).

The incidence of SE is 50 per 100,000. In all studies, the convulsive type is most common (80%). However, nonconvulsive SE is difficult to recognize and may therefore be underrepresented in epidemiologic studies. The mortality rate of SE is approximately 20%. Prognosis and response to treatment largely depend on the etiology of SE. SE occurring in the context of chronic epilepsy has a much better outcome than SE complicating acute processes (such as hypoxemia). SE should always be managed in a hospital emergency department. Drug treatment of SE includes the immediate intravenous administration of a short-acting benzodiazepine such as diazepam plus phenytoin or the long-acting benzodiazepine lorazepam in monotherapy. If SE persists, phenobarbitone is an alternative, the next step being a general anesthesia using pentobarbital or propofol.

ROLE OF THE LIMBIC SYSTEM

Although definitions of the limbic system vary, most agree that the central components include the amygdala, hippocampus, and their outflow pathways. The latter include, importantly, the limbic forebrain structures, such as the dopamine-rich nucleus accumbens and surrounding nuclei of the ventral striatum, and pathways to the hypothalamus, thalamus, and their connections down to the midbrain tegmentum. A recent description of some of these areas uses the concept of the extended amygdala, noting the close anatomic and neurochemical links between medial temporal structures and limbic forebrain structures (5).

To understand the role of the limbic system structures in epilepsy and in the behavior disorders of epilepsy, keep in mind that the limbic system modulates both behavior and emotion. In the 1930's, Papez (424) was first to suggest that a limbic circuit, which comprises the hippocampus, the mamillary body of the thalamus, the anterior thalamus, the cingulate gyrus, and a loop back to the hippocampus, was a neural substrate for the emotions.

The concept of the limbic system was developed further by MacLean (373). Basing his ideas on comparative anatomy, he introduced the *triune* brain. Essentially, he pointed out how human brains contain a neocortex, a paleocortex, and a reptilian brain. The essential concept was that the limbic system (the paleocortex) was only developed in a very rudimentary way before the rise of mammals and is very poorly represented in reptiles. As MacLean reflected: "The history of the evolution of the limbic system is the history of the evolution of the mammals, while the history of the evolution of the mammals is the history of the development of a family way of life."

In other words, with the development of infant maternal bonding and the evolution of complex social behaviors, there has been a parallel development of the neuroanatomic and arrangements of limbic structures.

The neuropathologic changes in epilepsy are varied, but it has been known for a number of years that a specific form of pathology is found in many patients, particularly those with localization-related TLE. Mesial temporal sclerosis is a specific pathologic lesion affecting largely hippocampal structures, particularly subfields within the hippocampus, such as CA1. It is known to occur secondary to early anoxic brain damage. A key history in many patients with this pathology is an early febrile convulsion, particularly one that has been prolonged or complicated. Patients may then be seizure free for several years, developing simple seizures or CPS in their adolescence.

Other pathologies found in the temporal lobes that affect limbic system structures often reflect developmental abnormalities. These include the hamartomas, a heterogeneous group of pathologies that are benign but arise in the brain during fetal development. Again, patients with such pathologies often remain seizure free for many years, the first seizures erupting in late childhood or early adulthood.

It is common to loosely speak of TLE. However, neuroanatomically, neurochemically, and neurophysiologically, there are clearly a number of subdivisions of the temporal lobes, as there are, for example, of the frontal lobes. One such subdivision is the limbic portion, the medial temporal structures, in particular, the amygdala, hippocampus, and parahippocampal gyrus. Thus, some authors have now more clearly distinguished a medial temporal syndrome, which may be referred to as limbic epilepsy. This syndrome's characteristics are essentially difficult-to-control seizures arising from limbic structures and cognitive, especially memory, complaints. Behavioral disorders appear to be an integral part of the medial temporal syndrome, confirming the central role of limbic system structures in the development of emotions and behaviors.

In clinical practice, patients with TLE are often the most refractory to treat, often find themselves on several anticonvulsant drugs, and probably attend a chronic epilepsy clinic. Limbic epilepsy may therefore be defined as a form of seizure disorder with pathology in limbic system structures, usually arising during an early developmental period, in which patients have little relief from seizures and often present with behavioral disorders. Knowing that the limbic system modulates emotional expression, we should not be surprised that a lesion there would have a profound effect on behavior during crucial periods of psychosocial development.

It is important to recognize the relationship of the anatomic structures of the limbic system and the outflow pathways from the hippocampus and amygdala. Thus, there are significant monosynaptic tracts linking the medial temporal structures to the dopamine-rich limbic forebrain. The latter are thought to be related to the development of psychosis in such conditions as schizophrenia. This finding must be relevant to the development of behavioral disorders in patients with TLE.

As noted previously, the clinical phenomenology of the psychoses associated with TLE are often indistinguishable from schizophrenia in the absence of epilepsy. This may reflect the underlying anatomic homologies.

Finally, numerous neuropathologic studies of schizophrenia reveal that where abnormalities of the hippocampus or parahippocampal gyrus are sought, they are invariably found (see Chapter 30). The pathologic picture differs from that of classic mesial temporal sclerosis, in particular by the absence of a gliosis. Nevertheless, the neuronal disarray that is so frequently reported, which probably reflects abnormal neuronal migration during fetal development, is thought to be linked with the development of the schizophrenia syndrome. The relevance of the limbic system, therefore, is that this neuronal circuitry underlying the development of emotion and behavior is disrupted in patients who develop schizophrenia as well as in patients with one form of epilepsy, namely TLE or limbic system epilepsy. These patients are most likely to develop psychopathology resembling schizophrenia, hence the term SLPE adopted by Slater and Beard (533).

EPIDEMIOLOGY OF PSYCHIATRY AND EPILEPSY

The incidence and prevalence of psychiatric disorders in people with epilepsy are not known. Most studies have been criticized for being retrospective, using nonrepresentative hospital samples and inadequate or absent controls, and not clearly identifying the seizure and syndrome type. Interpretation and comparison are further complicated by differing methods of determining psychiatric cases, ranging from self-report questionnaires to structured clinical interviews using operational criteria.

There are few large prevalence studies. Most have been performed in patients with medically intractable partial epilepsy being evaluated for epilepsy surgery. The findings of recent major studies are summarized here, and additional data are given in subsequent sections.

Cockerell et al. (112) performed a prospective study of severe "acute psychological disorder" in epilepsy of sufficient severity to warrant specialist attention, during a 1-year period, using the British Neurological Surveillance Unit reporting scheme (111). Sixty-four cases were reported, giving a minimal estimate of the annual rate of new cases of severe psychiatric disorder of 0.1%. The main reported diagnoses were delirium (27%), schizophreniform (31%), affective (30%), and delusional (5%) disorder. The majority (81%) occurred in patients with localization-related epilepsy (112).

Jalava and Sillanpää (280) assessed the psychiatric morbidity in 94 patients with childhood-onset epilepsy, after a mean follow-up period of 35 years. Nineteen percent had psychiatric comorbidity (compared with 5% of controls) and 65% a psychosomatic comorbidity (compared with 57% of controls), regardless of whether they were still receiving anticonvulsant medication. Patients received psychotropic drugs 8.6 times more frequently than controls. Thus, it would appear from this follow-up study that epilepsy *per se* rather than medication was associated with increased occurrence of psychiatric comorbidity.

There have been few studies assessing psychiatric disorder in patients with both learning disability and epilepsy. Furthermore, some studies assessing psychiatric comorbidity in people with epilepsy have excluded patients with learning disability (176,446), further limiting the information available in this group. Lund (363) found that among those with learning disability, the addition of epilepsy increased the prevalence of psychiatric illness (including behavioral disorder and autism) and doubled the rate if seizures had occurred in the previous year. Surprisingly, the two other major studies in this group (128,428) did not find an increase of psychiatric disorder in patients with or without epilepsy. This may reflect the difficulty in diagnosing psychiatric comorbidity in patients with learning disability, especially if severely affected.

In summary, approximately 30% to 50% of patients with epilepsy appear to suffer at some time from a psychiatric disorder, usually anxiety or depression. Early studies comparing patients cared for solely by their general practitioner with those managed by hospital specialists revealed that psychological disturbance was a more important factor in determining referral to hospital than the severity of epilepsy (437). More recent work, however, has reported that patients attending hospital clinics tend to have an earlier age at onset of epilepsy and less well-controlled CPS (167). Generally, the rate of psychiatric disorder increases with seizure frequency (279) and is higher in those attending hospital clinics (167). Interestingly, a study assessing the impact of the newer anticonvulsant drugs VGB, LTG, GBP, and TPM found a worse response in terms of seizure freedom in patients with psychiatric disorder (574).

DELIBERATE SELF-HARM AND SUICIDE IN PATIENTS WITH EPILEPSY

The importance of recognizing psychiatric disorder in epilepsy is not only to enable treatment to be considered but also to identify patients at risk of suicide.

The reported incidence of suicide and deliberate self-harm (DSH) is thought to be at least four to five times that of the general population (26,237); although those with TLE have an increased risk by a factor of as much as 25 (26,237). Robertson (462) reviewed 17 studies detailing cause of death in epilepsy and found the average incidence of suicide to be nine to ten times that of the general population (13.2% compared with 1.4%). Suicide has been reported in patients with severe epilepsy and epilepsy with other handicaps and in those seen in specialist clinics (462) but may also follow an improvement in seizure control (74). Two reports have also shown a 5% to 7% increase in self-poisoning in patients with epilepsy (244,372). Kanemoto et al. (295) found that suicide attempts were more frequent during postictal psychosis (PIP) (7%) than during either acute interictal psychosis (AIP) (2%) or postictal confusion. Risk factors for completed suicide have been found to include history of DSH, family history of suicide, stressful life events, poor morale and stigma, and psychiatric disorders, especially alcohol and drug abuse, depression, psychosis, and personality disorder (237,462).

PSYCHIATRIC DISORDERS IN PATIENTS WITH EPILEPSY

Psychiatric disorders in epilepsy can occur peri- or interictally, the latter unrelated in time to seizure occurrence. However, in patients with frequent seizures, differentiation may be difficult.

Affective Disorder

Depression

Preictal Depression
Prodromal moods of depression or irritability occur hours to days before a seizure and are often relieved by the convulsion (140). Blanchet and Frommer (65) found that patients tended to report more depression on the days immediately preceding their seizures than on interictal days along with improvement of mood after convulsions. A more recent study

(270) evaluated 148 adult patients with epilepsy, of which 128 had partial seizures. One-third of the patients with partial seizures reported premonitory symptoms (usually preceding secondarily generalized seizures) compared with none of the patients with PGE. Half of the symptoms were emotional in nature, usually irritability, depression, fear, elation, or anger and lasted between 10 minutes and 3 days.

The mechanism for the development of prodromal mood changes is not known. The same biologic processes may cause both the change in affect, and the seizure or the dysphoric mood itself may precipitate the seizure.

Ictal Depression

Williams (627) observed depression occurring as part of an aura—simple partial seizure (SPS)—in approximately 1% of a series of 2,000 people with epilepsy, although fear occurred more frequently. Ictal depression appears to be more common in patients with TLE (627) in whom rates of more than 10% have been reported (143,617). No association with laterality of seizure focus has been found (143,617,627).

Ictal depression, classically, is of sudden onset and occurs out of context, not related to environmental stimuli. The severity ranges from mild feelings of sadness to profound hopelessness and despair (143). Lim et al. (353) described a patient with nonconvulsive SE presenting as a case of psychotic depression, and both DSH and suicide have been reported during ictal depression (49,392). Complex (formed) hallucinations may accompany the depressive feelings (627), and several workers have found ictal depression to be of prolonged duration, often persisting into the postictal period, which may represent underlying subclinical seizure activity (617,627).

Postictal Depression

Blumer (69) described three patients who developed depression lasting hours to days immediately after seizures. However, these patients also experienced episodes of interictal depression, and Blumer commented that it was rare to find a patient who suffered postictal depression alone. Devinsky et al. (144) also noted depressed affect after CPS originating in right temporal structures in patients with medically intractable epilepsy, and more prominent changes were seen with bilateral limbic dysfunction. However, these findings do not appear to be consistent with those of an earlier single case report (272) of a patient who developed postictal depression after left-sided CPS and hypomania after right-sided seizures.

Treatment of Pre- and Periictal Depression

Although prodromal and periictal depression does not usually require treatment *per se,* improvement in seizure frequency should reduce the occurrence of these forms of depression.

Some patients may regard depressive pre- and periictal feelings as an "early warning system." Patients may have time to alert people that they require assistance, enabling the patient to be moved to a place of safety. Medication such as fast-acting benzodiazepines may abort or prevent the development of the epileptic attack, as may behavioral methods (53,213,222).

Interictal Depression
Epidemiology. Interictal depression is thought to be much more common than periictal; however, the exact prevalence is not known. Moreover, some studies have reported rates of depressive *symptomatology,* whereas others reported depressive *illness.* Mendez et al. (393) found that four times more patients with epilepsy had been hospitalized for depression compared with a control group with similar socioeconomic and disability levels. Studies have found a history of depression or depressive symptomatology in as many as two-thirds of patients with medically intractable epilepsy (480,549,610), whereas community studies have only demonstrated affective disorder in one-fourth (165). Depression tends to occur approximately 10 years after the onset of epilepsy (7).

Etiology. Various causative factors have been proposed for the development of depression in epilepsy. The etiology is most likely multifactorial with several different factors playing a part in an individual patient.

Gender. Few studies have addressed this variable. However, Altshuler et al. (7) found the highest Beck Depression Inventory (BDI) scores in male patients with TLE, and Strauss et al. (558) found that men with left-sided foci were more vulnerable to depression. Other authors have also found men to be overrepresented (313,393,518).

Conversely, Standage and Fenton (549) found a higher (but nonsignificant) number of women to have psychiatric morbidity as measured using the Present State Examination, and more women with epilepsy and depression have been reported by Robertson et al. (465), Hermann and Whitman (249), and Kohler et al. (314).

Others, however, have not found gender to be a significant etiologic factor (466,509,610).

Although the results of the various studies are not congruent, most studies that have addressed gender have found men to be most at risk. This is of particular significance because studies of people with depression without epilepsy have demonstrated that there is an increased prevalence in women.

Sinistrality. Again, this factor has not been intensively explored. Some studies have observed that more left-handed patients with epilepsy tended to have psychiatric morbidity, especially depression (7). However, other studies have found no such association (274). The expression of left-handedness may indicate early brain injury.

Neurologic. Depression may be associated with a neurologic condition that is also responsible for the epilepsy, e.g., multiple sclerosis, cerebrovascular disorder, dementia, and head injury (355,463), and may be more common in patients with a structural lesion (312,564). However, some studies have not found a structural brain lesion to be associated with depression in patients with epilepsy (251,393).

Genetic/Environmental. A family history of depression has been reported in some studies. Robertson et al. (466) found that more than half of 66 patients with both epilepsy and depression had a family history of psychiatric illness, usually depression. Wiegartz et al. (626) found that more patients with epilepsy and depression had first-degree relatives who had been treated for depression than patients with epilepsy but no depression. Others, however, have found no association (274,393,466,509). Only one study has compared the incidence of psychiatric disorders in first-degree relatives of patients with acquired epilepsy (26 probands) with relatives of patients with genetic epilepsy (juvenile myoclonic epilepsy) (411). Depression was the most common psychiatric disorder in both probands and relatives, and almost twice as many patients with juvenile myoclonic epilepsy had first-degree relatives with mental illness. The findings of this small study may be explained by either genetic or environmental hypotheses, and further studies are needed.

Epilepsy Factors
Age at Onset/Duration of Epilepsy. Although some studies have found an association between depression and early-onset (445) or late-onset (400) epilepsy, most have found no relationship with either age at onset (7,8,251,313,393,394,465,466,480,509,549,610,626) or duration of epilepsy (7,8,251,274,466,480,549,626).

Seizure Type. Many studies have found depression to be more common in patients with CPS (121,149,233,274,393,465,480). Others, however, demonstrated that the number of seizure types was correlated with greater risk of psychiatric disorder (154,193).

Several studies have found depression to be more common in patients with TLE (8,90,121,159,212,233,312,432,470,510). Quiske et al. (445) found patients with MRI-detected mesial temporal sclerosis to have higher BDI scores than those with circumscribed neocortical temporal lesions.

Tebartz van Elst et al. (569) compared the volumes of the amygdala (assessed using quantitative MRI) in 48 patients with TLE, 11 of whom had dysthymia, with healthy controls. They found a highly significant enlargement of both amygdalae in the patients with TLE and dysthymia. In addition, none of the patients with dysthymia had amygdala atrophy. Furthermore, they found a positive correlation between left amygdala volumes and depression measured using the BDI. They suggested that increased processing of negative emotional information might increase amygdala blood

flow and result in amygdala enlargement. However, they also speculated that enlargement of the amygdala may predispose to the development of depression or that amygdala atrophy might be protective for the development of dysthymia. The numbers of patients with dysthymia were small, and further studies are needed to clarify these issues.

Others, however, have not confirmed the excess nonpsychotic psychiatric morbidity in patients with TLE (537,549,554). Edeh and Toone (166) did not find higher psychiatric morbidity in patients with TLE compared with nontemporal lobe focal epilepsy but found that focal epilepsy, regardless of the site of focus, had more interictal psychopathology than those with PGE. A study by Hermann et al. (257) found increased perseverative responding on the Wisconsin Card Sorting Test and dysphoric mood state in patients with left TLE, suggesting that frontal lobe dysfunction may be an additional factor in the genesis of depression in patients with TLE. Bromfield et al. (89) confirmed this hypothesis by reporting a bilateral reduction in inferior frontal glucose metabolism in depressed patients with CPS. This link to frontal dysfunction has also been confirmed by Schmitz et al. (see below).

One reason for the excess of TLE in those with psychiatric disorders may merely represent the fact that TLE is the most common type of epilepsy in adults and is more difficult to treat. Furthermore, other studies have found either no association with seizure type (313,376,549,597) or focus (465). Manchanda et al. (378) assessed 30 patients with TLE, 25 with nontemporal lobe focal epilepsy and 19 with multifocal onset/generalized epilepsy referred to a teaching hospital's epilepsy investigation unit. More than half had a *Diagnostic and Statistical Manual of Mental Disorders, Revised Third Edition (DSM-III-R)* psychiatric disorder, but there were no significant differences in psychiatric morbidity observed between the different foci. Only 4.25% of the entire group were diagnosed as having a "mood" disorder (6% with TLE, 4% with nontemporal focal epilepsy, and none with generalized epilepsy).

Lateralization of Seizure Focus. Flor-Henry (194) postulated that depression was associated with nondominant (right) temporal lobe lesions. However, this conclusion was based on the evaluation of nine patients, of whom four had a right, two a left, and three a bilateral focus. Since then, many studies have addressed the issue of laterality.

Some studies have found that right-sided foci are associated with depression (393) and nonsignificant association (314), and several have found that lateralization was not a significant etiologic factor (8,251,274,376,413,445,469,626). Most studies, however, have found that although psychiatric disorder generally, especially hypomania (272,486), tends to be associated with right-sided foci, depression appears to be more common in those with left-sided foci (8,89,158,393,394,398,418,431,465,513,518,558,610,611). Victoroff (611) found that left-sided ictal onset and hypometabolism were associated with a history of and current depression in patients with CPS. However, patients with right-sided hypometabolism also had a history of major depression, suggesting that left-sided ictal onset and the degree of hypometabolism were independent risk factors for the development of depression.

Schmitz et al. (508) noted that in patients with left-sided epilepsy, higher scores on the BDI were associated with contralateral temporal and bilateral frontal hypoperfusion (measured using SPECT). The authors commented that the hypoperfusion may represent the widespread perfusion changes involving the limbic frontal regions observed in people with TLE, which may be related to functional deafferentation, interictal inhibitory activity, or postictal depletion of substrates (89,508). Thus, depression may be more common in patients with left-sided TLE and evidence (radiologic or neuropsychological) of frontal lobe dysfunction.

Seizure Frequency/Severity and Control: The Role of Forced Normalization.

Several studies have reported a decrease in seizure frequency before the onset of the depressive illness (47,159,194,549). Mendez et al. (394) found patients with depression had significantly fewer generalized seizures than those without depression and postulated that nonreactive depression may occur when anticonvulsant therapy prevents generalization from an epileptic focus. Others, however, have not found an association (274,393,465), whereas some have even documented an association with an increase in seizure frequency (152,477).

Recent QOL studies have found frequent seizures to be associated with increased anxiety, depression, and stigma (20,279), as has perceived seizure severity (21). Trostle et al. (604) reported fewer psychosocial problems (assessed using the Washington Psychosocial Seizure Inventory) in patients who were not taking AED and had been seizure free during the preceding year. Those who were seizure free on AED had intermediate scores. Patients continuing to experience seizures despite AED had the most severe psychosocial problems. Severity of seizures was associated with the severity of self-reported psychosocial problems.

Iatrogenic

Anticonvulsant Medication. Polypharmacy in epilepsy has been shown to be associated with depression (193,394), and Shorvon and Reynolds (526) reported an improvement in alertness, concentration, drive, mood, and sociability on reducing polytherapy to monotherapy.

Some anticonvulsants have been found to be psychotropic, whereas others have been associated with behavioral changes and depression. The correlation between fatigue and depression shown by Ettinger et al. (176) may partly explain this association. This is discussed in more detail at the end of the chapter, but in summary, the anticonvulsants most associated with depression would appear to be the barbiturates [phenobarbital (PB) and primidone

(PRM)] along with PHT and VGB. The AED associated with least depression are CBZ, VPA, GBP, LEV, and LTG. The drugs most associated with depressive symptoms are, interestingly, all GABA agonists, which may have relevance for the etiology of the syndrome.

Folate Deficiency. AED therapy has been found to result in a decrease in serum, red blood cell, and cerebrospinal fluid folate levels in 11% to 15% of patients compared with normal controls (165,197). Although reduced folate has been documented with most AED, it is more marked in patients on polytherapy (165,197,454), and it would appear to be more common with the liver-enzyme inducing AED, especially barbiturates and PHT (197). Most studies assessing VPA have found little effect unless the patient was also taking enzyme inducers (197,221), and the one study assessing LTG found no reduction in folate (492). Folate deficiency has been associated with psychiatric morbidity (predominantly depression) in patients both with (165,197,454) and without epilepsy (104,105,220).

Depression After Epilepsy Surgery

Several studies, many several decades old, assessed postoperative depression; they are summarized below.

Most authors have found that complete freedom from seizures is necessary for a postoperative reduction in depression (76,251,256,258,302,450,514,555). Similarly, Altshuler et al. (7) found that overall, 47% of patients with preoperative episodes of depression experienced no recurrence after temporal lobectomy; however, most of these patients had also experienced complete seizure control. Blumer et al. (76) suggested the hypothesis that the psychiatric disorders of epilepsy may be owing to a seizure-suppressing inhibitory mechanism. Thus, the postoperative lessening of excitatory activity and relative predominance of inhibition may predispose to the emergence of dysphoric, affective, and psychotic disorders. If there is a recurrence of seizures, renewed seizure-suppressing activity may also produce psychiatric morbidity. Patients not rendered seizure free (even with more than 75% improvement of seizures) do not appear to have any change in levels of depression or behavioral or emotional adjustment (64,251,256,263,284,514). Patients experiencing only SPS postoperatively also continue to experience depression (609), perhaps because the continued premonitory symptoms may act as a reminder of their seizures and causes alarm that the aura may herald a generalized seizure.

Many investigators have found the side of temporal lobe surgery to be predictive of postoperative depression; however, the various studies have produced contradictory findings. Fenwick (188) assessed 96 patients after temporal lobectomy. One-fourth had postoperative depressive symptoms, of which 72% had undergone a right-sided and 28% a left-sided operation. Other authors have also reported more mood changes after right-sided surgery (414).

Depression has been reported occurring *de novo* after temporal lobectomy in approximately 10% of cases (7,76,93,178,318,564) and after amygdalohippocampectomy (414). Parashos et al. (425) reported the development of bipolar affective disorder and depression/dysthymia after anterior temporal lobectomy in two patients. Postoperative MRI scans revealed ipsilateral degeneration of the thalamus and putamen in both patients. Although Altshuler et al. (7) reported *de novo* depression in patients after left- and right-sided lobectomy, 75% of those developing depression within a year of surgery had left temporal lobe surgery complicated by hemiparesis and aphasia.

Postoperative depression tends to be more common in the first 2 months after surgery (76,110,319) and is often transient (110). Ring et al. (460) observed that at 6 weeks after surgery, 27% had developed *de novo* symptoms of anxiety and 23% of depression and 45% had increased emotional lability. At 3 months after surgery, emotional lability and anxiety symptoms had diminished, but depression tended to persist. Anhoury et al. (15) noted that the tendency to develop emotional lability postoperatively is correlated with the amount of tissue removed at the operation.

Psychosocial. Psychosocial factors are thought to play a major role in the development of depression in patients with epilepsy. Hermann (247) proposed that the exposure to unpredictable, uncontrollable, aversive events (seizures) might produce depression. This has been developed into the concept of locus of control, and preoperatively depression has been correlated with an external locus of control (a perception of events not being attributable to the patient's own efforts but rather to the effects of fate) (251). More recently, a pessimistic attributional style (attributing global difficulties to epilepsy) has been associated with the development of depression in patients with epilepsy (259). Hermann and Whitman (249) showed that increased stressful life events, poor adjustment to seizures, and financial stress were predictive of increased depression. However, other researchers have found no relationship between depression and psychosocial factors such as socioeconomic status (274), education (8,274), and employment status (7).

Diagnosis of Interictal Depression. An essential prerequisite to treating depression in epilepsy is making the correct diagnosis.

Betts (48) differentiated between a depressive *reaction* (or *reactive* depression) and a depressive *illness* (or *endogenous* depression). Depressive feelings usually respond to circumstance and tend to be understandable as a reaction to being given the diagnosis of epilepsy.

The main features of depressive illness are low mood, loss of interest and enjoyment, and reduced energy. In addition, reduced concentration, attention, and self-esteem/self-confidence along with ideas of guilt and unworthiness, pessimistic views of the future, ideas or acts of self-harm, disturbed sleep, and diminished appetite may be present (624). Diagnosis can be difficult, however, and Betts (48) stresses that drug intoxication (especially with PHT) can resemble depression.

Phenomenology of Interictal Depression. Patients with epilepsy and depression tend to have significantly fewer neurotic traits, such as anxiety, guilt, rumination, hopelessness, low self-esteem, and somatization, and more psychotic symptoms, such as paranoia, delusions, and persecutory auditory hallucinations. Between episodes of major depression, they tend to be dysthymic with irritability and humorlessness (393). Betts (48) also reported that depression in epilepsy tends to be endogenous in nature with sudden onset, fluctuating markedly until it suddenly ends.

Interictal Dysphoric Disorder

Blumer (68) noted that during the interictal phase, some patients experienced various affective symptoms (Table 39.9). Although eight symptoms were described, a minimum of three was needed for a diagnosis of interictal dysphoric disorder (IDD) (75). Most patients experienced five (range, three to eight), the most common being depressive mood, anergia, and irritability, each of which occurred in approximately 75% of those with affective symptoms. The depressive episodes tended to last as long as 12 hours, occurring on average five times per month, along with periods of euphoria lasting as long as 4 hours, occurring on average three times per month. The outbursts of rage tended to occur in predominantly "good-natured, very ethical, or even deeply religious" individuals and were characteristically followed by remorse. Blumer et al. (76) reported that 57% of 75 patients

TABLE 39.9. CLASSIFICATION OF INTERICTAL DYS-PHORIC DISORDER

Symptom[a]	Details
Depression	Intermittent, intense to the point of suicidal despair, lasting hours to days (average: 12 h), average frequency: 5/mo, DMV not present, may be depressive baseline
Anergia	May occur without depression
Pain	Atypical, often headache
Insomnia	Difficulty initiating and maintaining sleep, often prolonged sleeping next day
Fear	Episodic agoraphobia
Anxiety	Episodic generalized anxiety
Irritability	Intermittent and paroxysmal; gradual penting up of anger and fury resulting in explosive anger, rage, and verbal outbursts
Euphoria	Sudden, endogenous sense of blissful euphoria; brief, lasting a few hours (average: 4), average frequency: 3/mo; not associated with hyperactivity

[a]For a diagnosis of interictal dysphoric disorder, a person with epilepsy should experience at least three symptoms, each to a troublesome degree.

undergoing neurodiagnostic monitoring for epilepsy surgery had IDD. Blumer (70) noted that IDD tends to worsen before menstruation, and in these women, treatment of the IDD may result in dysphoria occurring solely premenstrually. Blumer stressed that patients with a large number of the above symptoms may be at increased risk of sudden, unexpected suicide attempts and the development of an interictal psychosis. In fact, he believes that a continuum exists between IDD and a more severe dysphoric disorder with more prolonged and prominent psychosis and that interictal psychosis is invariably preceded by a marked dysphoric disorder (72).

Treatment of Interictal Dysphoric Disorder. Blumer (70) noted that all the symptoms, not just the depressive ones, tend to respond rapidly to low-dose antidepressants. Furthermore, he views the psychiatric disorders of epilepsy, in particular IDD, as resulting from the inhibitory activity that develops in reaction to the excessive excitatory activity of the chronic seizure disorder, and the therapeutic effects of the proconvulsant antidepressants are the result of them acting as antagonists to this inhibition. He advocates starting with a low dose of a tricyclic antidepressant (usually 100 mg imipramine at night). If improvement is not observed within a few days, he suggests adding a selective serotonin reuptake inhibitor (70). Severe IDD may require the addition of a low-dose neuroleptic such as risperidone (77).

Psychological. Depressive reactions should be treated with supportive therapy provided by trained therapists, social workers, or epilepsy nurse specialists. Betts (48) stresses that these reactions should not be treated medically unless prolonged and therefore atypical because such episodes may become protracted if treated with antidepressants or tranquilizers. More severe reactions may require specialized psychotherapy such as cognitive/behavioral therapy (127). Psychotherapy can also be used to improve coping skills, and this has been shown to improve mild depressive illness and anxiety as well as reduce seizure frequency (213).

Patient support groups introduce patients to fellow sufferers who can provide emotional support, which has been shown to modify depression and dysthymia in people with epilepsy (36).

Adjustment of Antiepileptic Drugs. In view of the association between polytherapy and some AED and affective disorders, considerable attention should be given to patients' prescriptions. Where possible, polytherapy should be reviewed, and treatment with monotherapy attempted. Some AED are more likely to be associated with depression and are best avoided (see above).

Antidepressant Treatments: General Considerations. Approximately 60% to 70% of acute major depressive episodes will respond to antidepressant treatment (310), and early treatment intervention has been shown to reduce the duration of the episode by almost 50% (326). After complete remission of symptoms, antidepressant therapy should be continued for at least 6 months to reduce the chance of relapse.

Choice of Antidepressant. Choice depends on the efficacy, interactions with concomitant medication, and the side effect profile, in particular, the epileptogenic potential. These factors were recently reviewed (330).

Efficacy. It is widely believed that the newer antidepressants do not significantly differ with respect to efficacy compared with the tricyclic antidepressants (300,519,560). There has only been one double-blind, placebo-controlled study of antidepressants in epilepsy (464) that showed that after 12 weeks, nomifensine was superior to amitriptyline.

Interactions. Most antidepressants are either metabolized by or inhibit to varying degrees one or more of the cytochrome P-450 isoenzymes located in the liver, which has been reviewed by Nemeroff et al. (416). Clinically relevant interactions mainly involve the concomitant use of antidepressants with PHT and CBZ. Most published studies consist of single case reports, and in clinical practice, fluoxetine or fluvoxamine can induce toxic anticonvulsant levels, whereas sertraline, paroxetine, and citalopram should have little effect (199,286,416).

The anticonvulsants PB, PRM, PHT, and CBZ are potent liver enzyme inducers (87), which can result in reduced plasma levels (thus reduced efficacy) of antidepressants metabolized by the same isoenzymes. Clinically significant interactions have been reported with tricyclic antidepressants (545) and paroxetine (83).

Safety: Epileptogenic Potential. The incidence of seizures occurring with therapeutic doses of antidepressants varies from 0.1% to 4% (476), which is not much higher than the annual incidence of first seizures in the general population. Some authors have suggested that various risk factors may predispose to the precipitation of antidepressant-induced seizures and may be either patient or drug related (Table 39.10) (440,476,532,637). Seizures have been found more likely to occur during the first week of antidepressant treatment or after an increase in dose, especially after rapid dose escalation (476).

There have been reports of seizures with the selective serotonin reuptake inhibitors. However, these have mainly been single case studies, and the seizures often occur when the selective serotonin reuptake inhibitor is administered with or shortly after another drug known to lower the seizure threshold or in patients at risk of seizures (for a review, see reference 330).

In summary, the older antidepressants, especially the tricyclic antidepressants, appear to lower seizure threshold, especially in vulnerable individuals. There is less evidence of

TABLE 39.10. RISK FACTORS FOR ANTIDEPRESSANT-INDUCED SEIZURES

Patient related
  History of seizures
  Family history of seizure disorder
  Abnormal pretreatment EEG
  Brain damage
  Dementia
  Learning disability
  Previous electroconvulsive therapy
  Alcohol/substance abuse/withdrawal
  Reduced renal/hepatic drug elimination capacity
Drug related
  High dose/plasma level of antidepressant or metabolites
  Overdose of antidepressant
  Rapid dose escalation
  Concurrent use of drugs that lower seizure threshold
  Concurrent use of drugs that inhibit metabolism of
    antidepressant

From Lambert MV, David AS. Psychiatric comorbidity. In: Oxbury J, Polkey C, Duchowny M, eds. *Intractable focal epilepsy.* London: Harcourt, 2000:445–471, with permission.

such an effect with the newer antidepressants, especially moclobemide and citalopram, and there are no reliable data on the newer agents, such as mirtazapine and venlafaxine. Antidepressants should be introduced at a low dose, gradually increased to therapeutic levels, and continued for at least 6 months after complete clinical recovery. Other medication known to reduce the seizure threshold should be avoided, as should benzodiazepine withdrawal. Patients and anticonvulsant levels should be carefully monitored, and if seizures develop, the patient should be changed to an antidepressant with lower risk. Patients with severe, uncontrolled epilepsy or those who develop an exacerbation of their seizures may be best managed as inpatients.

Electroconvulsive Therapy, Transcranial Magnetic Stimulation, and Vagal Nerve Stimulation. Electroconvulsive therapy (ECT) is not contraindicated in patients with epilepsy and may be life saving in patients with severe or psychotic depression not responding to antidepressants (48). Although there have been reports of spontaneous seizures occurring in patients after ECT (138,226), major studies have found the incidence of spontaneous seizures after ECT to be lower than that of epilepsy in the general population (63,66). The efficacy of ECT may be reduced if AED decrease the intensity of the induced seizures (324); hence, Weiner and Coffey (618) recommend that, with the exception of patients at high risk of SE or with recent generalized tonic-clonic seizures, AED should be omitted the morning before each ECT treatment.

There are no data on the treatment of depression in epilepsy with transcranial magnetic stimulation. Although in the early days of therapy, the possible proconvulsant effect was highlighted, in practice, especially with low-dose stimulation, seizure provocation is not reported (452).

Vagal nerve stimulation is now widely used in the management of patients with refractory epilepsy. The left vagal nerve is stimulated, the afferent signals being thought to affect limbic and other subcortical structures. Trials in epilepsy suggest that it has a beneficial effect on mood, and preliminary data suggest that it may be of value in the treatment of depression in patients without epilepsy (481).

Hypomania and Bipolar Affective Disorder

Maniacal episodes were described in 4.8% of Dongier's (159) series of patients with epilepsy, and Williams (627) reported periictal elation in three of 2,000 patients. Since then, reports of mania and hypomania have mainly consisted of single cases or studies involving small numbers of patients (24,49,97,218,264,271,288,365,403,495,530,566, 634); however, several have noted an association with nondominant TLE (24,97,218,271,495). This may be explained by cases of mania being classified as schizoaffective psychosis and thus included in the epilepsy and psychosis literature. Alternatively, the use of AED, such as CBZ and VPA, which are known to be effective in the treatment of bipolar affective disorder (23,438), may reduce the occurrence of hypomania in these patients.

Hypomania has been described as occurring *de novo* after temporal lobe surgery (293) with a right-sided emphasis.

Treatment of Hypomania/Mania in Epilepsy

The traditional treatment of bipolar affective disorder is lithium. However, despite small open studies of people with epilepsy treated without adverse effect (365,530), controversy remains as to its proconvulsant properties. Several studies have reported seizures in patients treated with lithium at both toxic (510,547,621) and therapeutic serum levels (22,132,289,290,401,621). EEG changes have also been reported in patients (289,475,547) and normal volunteers (573). A recent paper by Bell et al. (37) included preexisting EEG abnormalities as a risk factor for the development of neurotoxicity at therapeutic lithium levels.

CBZ has been shown to be efficacious as an antidepressant in patients without epilepsy and prophylactic in the control of manic-depressive illness (438), as have VPA (23), LTG (100,614), GBP (483,503), and TPM (109,380). Thus, these AED should be used for the treatment of epilepsy in patients with coexisting bipolar disorder, and additional therapy with lithium may be avoided (264).

Anxiety Disorders

Ictal fear and anxiety have been described (627) and can be mistaken for anxiety or panic attacks, especially when there is a personal or family history of mood disorder (331,339,641). Conversely, anxiety attacks may be misdiagnosed as seizures, especially in people with temporal lobe epileptiform EEG activity or a history of epilepsy (209,265). Guidelines have been determined to help to differentiate these conditions (331). Ictal anxiety or fear is usually stereotyped, with rapid onset and shorter duration than panic attacks. Anxiety occurring as part of an SPS may evolve into a typical CPS with disturbance of consciousness during which aphasia may occur and may be followed by fatigue, confusion, and memory disturbance. The most commonly encountered anxiety disorders experienced interictally are generalized anxiety, phobic, and panic disorders.

Anxiety can also coexist with a depressive disorder and is particularly common in patients with late-onset epilepsy (465). Treatment of the underlying depression should also alleviate the anxiety.

Anxiety can be a reaction to acquiring the diagnosis of epilepsy and the accompanying social and family problems (48). A self-reinforcing situation may occur in which the patient is fearful of leaving the house in case he/she has a seizure, becomes anxious, and hyperventilates, thus increasing the likelihood of convulsing. This, in turn, increases anxiety levels, and a phobic anxiety state may ensue (48).

Treatment of anxiety in epilepsy consists of relaxation training, biofeedback, counseling, behavioral/cognitive therapy, and, if necessary, formal psychotherapy. Minor tranquilizers should be avoided because most are also anticonvulsants and thus are particularly difficult to withdraw.

Obsessive-Compulsive Disorder

There have been several single case reports and small studies demonstrating an association between epilepsy and obsessive-compulsive disorder (OCD) in both adults (99,281,316,323) and children (103,307,349). Some studies have also found patients with OCD to have abnormal EEG, predominantly temporal sharp-wave activity (281). However, a review of 50 patients with OCD (360) revealed only one with concomitant epilepsy. One case report described a patient who appeared to have an inverse relationship between seizures and OCD in that he suffered more severely from compulsive cutting behavior when his seizures were well controlled, having no such behavior on the days during which he experienced seizures (99).

Schmitz et al. (508) did not find any association between obsessionality (measured using the Leyton Obsessionality Inventory) and laterality of seizure focus. However, in patients with right-sided foci, higher obsessionality scores were associated with hyperperfusion (measured with HMPAO SPECT) in ipsilateral temporal, thalamic, and basal ganglia regions along with the bilateral frontal regions. The rather pedantic, circumstantial viscous thinking of some patients with long-standing, difficult-to-control epilepsy has obsessional features but must be distinguished from typical OCD.

Treatment of Obsessive-Compulsive Disorder in Epilepsy

Behavioral and pharmacologic therapies (434) and psychosurgery (349) have been found to be effective in the treatment of OCD. Serotonergic drugs such as fluoxetine (283) are most commonly prescribed; however, several case studies have demonstrated the proconvulsant properties of fluoxetine (330), and CBZ may be an effective, safer alternative for these patients (103,282,316,323).

Nonepileptic Attack Disorder

NES have been defined as "paroxysmal events that alter or appear to alter neurologic function to produce motor signs or sensory, autonomic, or psychic symptoms that at least superficially resemble those occurring during epileptic seizures" (612).

The terminology of NES is contentious, with many different pseudonyms including non–ES-like events, pseudoseizures, psychogenic seizures, and nonepileptic attack disorder (NEAD).

NES can be of psychological (emotional or psychiatric) or physiologic (medical) origin. Psychiatric causes of NES include panic, somatization, conversion, and dissociative disorder along with malingering and factitious disorder. Medical causes include migraine and disorders of sleep and the cardiovascular and cerebrovascular systems. The discussion of NEAD in this chapter focuses on the etiology, clinical features, diagnosis, and management of nonepileptic conversion/dissociative disorders.

The clinical significance of diagnosing NEAD includes providing appropriate treatment rather than ever-increasing polypharmacy and avoiding such medical complications as respiratory arrest, septicemia, and pneumonia when emergency interventions are instigated for "pseudostatus" (348). In addition, some patients with apparently medically unresponsive seizures and temporal spikes on the interictal EEG may be needlessly investigated for epilepsy surgery until EEG monitoring reveals only medically intractable NES (246).

Epidemiology

In patients undergoing video-EEG monitoring for medically refractory seizures, NES have been reported in 22% to 24% (75,612). NES may coexist with ES in 4% to 58% (207,267,446,612), further complicating diagnosis and treatment.

Diagnosis of Nonepileptic Attack Disorder

The differentiation between ES and NES may be difficult, even for experienced observers (309), and seizures with atypical features and/or of frontal lobe origin may easily be misdiagnosed as NES (427). Furthermore, as many as 8% of diagnoses of NEAD are later changed to epilepsy (51). Diagnosis entails careful history taking from both the patient and observers, and some special investigations, especially videotelemetry.

Demographic Characteristics

NEAD has been reported in patients of all ages; however, the average age at onset is 14 to 30 years (328). It appears to be much more common in women in whom three-fourths of the cases are reported (348). There is usually a model on which the NES is based, often a witnessed attack of epilepsy; hence, NES often occurs in health care workers. Similarly, a family history of epilepsy was found in 37.6% of patients with NEAD in one study (332).

Phenomenology of Nonepileptic Seizures

The clinical features of NES may help to differentiate them from ES. The most diagnostically helpful of these are shown in Table 39.11.

Etiology

NES usually serve a function. They may be an expression of emotional needs or help the patient cope with emotional difficulties. The attacks may provide "secondary gain" such as increased attention, avoidance of activities, and decreased pressure to succeed academically or socially.

The details and circumstances of the first NES may reveal clues as to the underlying etiology. NES have also been reported to occur *de novo* after epilepsy surgery (319,417,426), especially in patients with low IQ, preoperative psychological problem, and postoperative complications (417).

Some studies have found that patients with NEAD have a history of a neurologic insult [23.7% in one study (332)], usually a head injury with loss of consciousness. It may be that the patients suffered seizures after the head injury or witnessed seizures on the neurologic/neurosurgical ward, again providing a model.

Previous childhood sexual and/or physical abuse has been associated with the development of NES by many researchers (4,6,18,81,82,235,520). Betts and Bowden (52) describe the "swoon" type of NES to be an automatic reaction to intrusion into consciousness of unpleasant memories/flashbacks of childhood sexual abuse (dissociation) and "abreactive" attacks as a symbolic replay of the abuse experience (conversion reaction).

Psychiatric Comorbidity

Psychiatric disorders have been reported in 43% to 100% of patients with NEAD (18,82), and one study reported a mean of 4.4 current and 6.0 lifetime *DSM* axis I psychiatric diag-

noses (82). The most common conditions were anxiety disorders (especially panic disorder with or without agoraphobia, and posttraumatic stress disorder) (52,82,542), mood disorders (mainly major depression) (82,509,552), somatization disorder (4,82,169,542), and personality disorder (4,82,169,423).

Aldenkamp and Mulder (4) developed a model for the development of NEAD. The patients tend to have an underlying personality disorder along with a tendency to somatize when stressed. The development of seizures rather than other somatic symptoms, according to their model, reflects a prior knowledge of epilepsy.

Investigations

Prolactin

Trimble (590) demonstrated an increase in serum prolactin (PRL) within 20 minutes of a spontaneous generalized seizure, contrasting with no change after an NES. Others have also found that most patients with generalized seizures (1,10,447) and 43% to 100% with CPS (442,447,544,601,638) have increases in serum PRL. However, elevations may not be seen after frontal lobe seizures (328), many cases of SPS (601, 638), and SE. Trimble originally suggested that an increase to more than 1,000 IU/L from a normal baseline was indicative of an ES. A serum prolactin threshold of 700 μU/mL has been reported to differentiate between NES and 80% of tonic-clonic and 39% of CPS (22). Tharyan et al. (572) found that an increase in PRL to three times baseline levels should differentiate ES from NES. The baseline level should be taken several hours after the episode because the PRL level may remain elevated for as long as 1 hour (1,10). PRL levels may, however, be misleading in patients with coexistent psychiatric disorders because most neuroleptics, as well as stress, elevate serum PRL levels (328).

Electroencephalography

EEG studies need to be interpreted with caution because 50% of single waking interictal EEG in people with epilepsy are normal. Sleep recordings help to increase the diagnostic yield, but even repeated wake and sleep recordings will fail to reveal epileptiform discharges interictally in 8% (59). EEG recordings during convulsive epileptic attacks should show epileptiform activity; however, they may be obscured by muscle artifacts, but the postictal record usually reveals slowing (40). Movement artifact during NES may even resemble epileptiform activity (95). The EEG may be normal, however, during SPS, especially with visceral or psychic symptomatology (60). The EEG of patients with frontal lobe CPS may also be normal, and ictal frontal slowing may be mistaken for artifact. However, frontal lobe seizures are often nocturnal, and if the seizure onset precedes EEG evidence of waking, the attack can be assumed to be epileptic (33,60). It should also be noted that generalized epileptiform discharges

TABLE 39.11. CLINICAL FEATURES OF NONEPILEPTIC SEIZURES

Characteristic	NES	Comment
Precipitant	Emotion/stress or environmental factors	Similar factors may also precipitate ES
Onset	Often gradual over minutes, rarely occurs when alone, may arise from pseudosleep	ES may be preceded by prodrome/SPS, ES may arise from sleep
Symptoms at onset	Vocalizations, palpitations, malaise, choking, numbness, peripheral sensory disturbances, pain, and olfactory, gustatory, or visual hallucinations	Similar symptoms may occur during prodrome or SPS
Duration	May be prolonged (>5 min), mean: 134 s, range: 20–805 s	GTCS usually lasts <4 min, mean: 70 s, range: 50–92 s; GTCS lasting >5 min usually accompanied by cyanosis and decreased oxygen saturation
Clinical features of seizure	Ranges from unresponsiveness with no motor manifestation to movements including generalized trembling, thrashing, flailing, struggling, side-to-side rocking of head, back arching with forward pelvic thrusting. If subject is placed on one side, the eyes will be directed toward the floor. If patient is turned, the eye deviation is reversed (geotropic eye movement). Avoidance reactions to noxious/painful stimuli (e.g., hand being dropped onto face). Clinical features may fluctuate from one seizure to next. Furthermore, new features may be incorporated over time, especially if inquired about by doctors.	Bizarre movements including hand clapping, pedaling, and pelvic thrusting may occur during frontal lobe seizures. However, frontal lobe seizures are usually brief (<1 min) and nocturnal. Patient tends to assume prone position and tonic abduction of upper limbs may occur. Avoidance reactions may occur during SPS. ES are usually stereotyped. In patients with ES and NES, both events may be stereotyped and distinguishable.
Motor activity	Intermittent, arrhythmic out-of-phase jerking or movements oriented in multiple directions.	In GTCS, jerking is rhythmic and in phase, with alternating brisk contraction and relaxation, usually slows before stopping.
Consciousness	May be preserved with ability to talk during bilateral motor activity. Amnesia is common during NES, but subsequent recall may occur during hypnosis.	Speech may occur during CPS. ES from supplementary motor area may produce bilateral motor activity with preserved consciousness and speech.
Reflexes	Usually normal, may resist attempt to open eyes.	Dilated pupils, reduced corneal reflexes and extensor plantars may occur during NES.
Injury	Occurs in as many as 40%; most injuries consist of bruising. Fractures and other serious injuries may occur in NES.	Biting of hands, lips, or tip of tongue may occur in NES. In ES, usually get biting of side of tongue. Carpet burns on elbows and face are common in NES. Heat-induced burns (e.g., from hot water, falling into fires) more common in ES.
Incontinence	Urinary incontinence occurs in as many as 44%. Bowel incontinence is rare in NES.	Combination of incontinence, injury, and tongue biting during NES, associated with DSH/suicide.
Postictal features	Lack of headache, drowsiness, or confusion, which may be aborted by suggestion; crying common.	Recovery may be rapid after frontal lobe seizures.
Response to medication	Frequent seizures despite therapeutic AED levels.	AED toxicity may exacerbate seizures or cause behavioral symptoms suggestive of NES.

NES, nonepileptic seizure; ES, epileptic seizure; SPS, simple partial seizure; CPS, complex partial seizure; GTCS, generalized tonic-clonic seizure; DSH, deliberate self-harm; AED, antiepileptic drug.
Adapted from Lambert MV, David AS. Psychiatric comorbidity. In: Oxbury J, Polkey C, Duchowny M, eds. *Intractable focal epilepsy.* London: Harcourt, 2000:445–471, with permission.

could occur during AED (especially barbiturate) withdrawal, regardless of whether the patient has epilepsy.

A normal EEG with preserved α rhythm during a typical attack manifested by loss of consciousness is virtually diagnostic of NEAD (328,348).

EEG monitoring along with videorecording, therefore, is often performed to help to confirm the diagnosis. However, it is essential that the recorded seizure is the same as the patient's habitual seizure; therefore, a close family member should either be an observer during the monitoring or, failing that, review the recorded seizure.

Provocative Tests

Despite activation procedures being routinely used in most epilepsy centers for the provocation of ES (hyperventilation, sleep deprivation, and medication withdrawal), the provocation of NES remains the most controversial area of the management of patients with NEAD (137,141).

Generally, suggestion is used along with physical placebo procedures such as intravenous saline (29,32,55, 113,160,536,613,639) or a stressful psychiatric interview (114). Various studies have reported that NES can be provoked in 77% to 91% of patients thought to have NEAD (334). Walczak et al. (613), Slater et al. (536), and many others have found them to be a useful screening method that may reduce the length of video-EEG monitoring, an important factor in units with limited resources (55).

Criticism of the use of provocative tests to diagnose NEAD includes the concern expressed by several authors (32,160) that the provoked seizures may not be typical of the patient's habitual seizures or that NES or even ES may be mimicked by or provoked, resulting in a false-positive diagnosis of NEAD. Walczak et al. (613) provoked typical seizures in two patients with epilepsy (10%), one of whom had reflex epilepsy, and nonepileptic events in three patients (15%). They also found the provocation procedure to be stressful for many patients, resulting in increased heart rate and blood pressure, which may have contributed to the development of ES. However, the most widely expressed criticism of these methods relates to ethical issues. Some workers prefer not to use provocative or suggestive techniques because they believe that they tend to interfere with the formation of a trusting relationship and may alienate the patient from the treatment care team and thus could compromise therapy (190,205).

Neuroimaging

Typically, MRI scans are normal in patients with NES. However, caution must be exercised in diagnosis, as one study has reported a series of four patients with typical NES (verified by a normal "ictal" EEG during habitual episodes) yet who had MRI evidence of mesial temporal sclerosis (41).

The presence of a normal interictal and ictal SPECT scan does however help support a diagnosis of NEAD. Varma

et al. (608) found that HMPAO SPECT scans were normal in seven of ten NEAD patients yet showed the typical hypoperfusion in eight of ten patients with focal epilepsy. One patient with NEAD, however, showed hypoperfusion indistinguishable from that seen in epilepsy. De León et al. (129) performed ictal SPECT scans in patients with suspected NEAD and reported that hyperperfusion did not occur during the NES.

Neuropsychological Tests   Pakalnis et al. (423) found Minnesota Multiphasic Personality Inventory (MMPI) scale elevations indicative of psychopathology in 15 of 16 patients with a history of nonepileptic status. Using the MMPI-Revised, Derry and McLachlan (134) found a classification accuracy of 92% for NES and 94% for epilepsy and demonstrated that it could help to identify patients with epilepsy who also had NES.

Treatment of Nonepileptic Attack Disorder

Most researchers believe that the most important aspect of management after a correct diagnosis of NEAD is the nonjudgmental presentation of the diagnosis (2,94,137,348,520). One study found that 13 of 45 patients with NES stopped having the episodes after being told of their diagnosis, and five of these had received no psychotherapeutic intervention (299).

A multidisciplinary team should usually provide treatment. Fenwick and Aspinal (190) describe giving a patient the "good news" that he/she does not have epilepsy but has "emotional attacks" at a time when he/she has begun to accept that there may be emotional precipitants of their attacks. They stress the importance of ensuring that the patient does not feel rejected by the team and emphasize the possibility of treatment and recovery. McDade and Brown (366) and Betts and Bowden (51) advocate the use of operant conditioning—attempting to prevent "rewarding" of seizure activity by ignoring it, whereas nonseizure activity should be positively reinforced by verbal praise and encouragement. Fenwick and Aspinal (190) describe a behavioral program that addresses the antecedents, the behavior during the attack, and the consequences along with the use of relaxation as a countermeasure. Finally, both the patient and his/her family should be encouraged to readjust to a life without seizures (190). Other therapeutic interventions include anxiety management, abreaction, hypnosis, counseling, formal psychotherapy, and pharmacotherapy (usually with major tranquilizers) (2,51,137,348). Patients with a history of childhood sexual abuse will need specific therapy directed at disclosure and ventilation of the abuse experience (52). Coexisting psychiatric disorders should be treated appropriately.

Prognosis of Nonepileptic Attack Disorder

Buchanan and Snars (94) reported a follow-up of a series of patients with NEAD (some with coexisting epilepsy).

Three-fourths of the patients derived "significant benefit" from management, consisting of confrontation with diagnosis in half, formal psychotherapy in almost a third, with the remainder having supportive therapy or ongoing support. AED were stopped in 32% and reduced in 14%. The patient, family, and treating physician considered that 48% made a complete recovery and 28% improved. Only 16% were unaffected by management; however, in 6% although the NES stopped or improved, the patients developed other conversion symptoms.

Betts and Bowden (51) reported that 63% of their patients no longer had NES on discharge from their unit; however, on follow-up, only 31% were attack free. Many had had their anticonvulsant medication restarted, often after pressure from the family.

Good prognostic factors include recent onset of seizures (345), a short interval between any stressor and diagnosis of NEAD (51,345), normal psychiatric evaluation (345), female gender (387), living an independent life (387), and lack of coexisting epilepsy (387). Poor prognostic factors include low IQ (366), history of violent behavior (366), prolonged history of NEAD (94,642), and personality disorder (423). Patients with both NEAD and epilepsy may be particularly difficult to treat because when the ES are under control, the NES may increase, thus maintaining the same number of events. These patients need to be helped to recognize the different types of attacks so that both can be monitored and treated appropriately.

Psychoses

When the psychiatric aspects of epilepsy were "rediscovered" in the 1950's and 1960's (202,335,571), American and British authors reported an excess of schizophrenialike psychoses in epilepsy (SLPE) patients, especially in those with TLE (212,436,533).

Slater et al. published a detailed analysis of 69 patients from two London hospitals who suffered from epilepsy and interictal psychoses. Based on this case series, the authors challenged the antagonism theory and postulated a positive link between epilepsy and schizophrenia. Although Slater et al. were criticized for drawing conclusions based on insufficient statistics (554), the temporal lobe hypothesis soon became broadly accepted and stimulated extensive research into the role of temporal lobe pathology in schizophrenia. The use of epileptic psychoses as a biologic model or "mock up" of schizophrenia (461) is largely based on the work of Gibbs and Slater (212) work.

The possible impact of research on epileptic psychosis on the understanding of the pathophysiology of endogenous psychoses explains the bias in the literature toward the study of interictal schizophrenialike psychoses. The spectrum of psychotic syndromes in epilepsy is, however, much more complex, and psychotic complications are not restricted to patients with TLE.

Epidemiology

There are several population-based studies of the frequency of mixed psychoses in epilepsy. Krohn (322), in a population-based survey in Norway, found a 2% prevalence of psychoses with epilepsy. Zielinski (644), in a field study of the Warsaw population, found prevalence rates for psychoses in epilepsy of 2% to 3%. In a field study of 2,635 registered epilepsy patients in a district in Poland, only 0.5% were diagnosed as having schizophrenia, but 19.5% suffered from postictal twilight states (57). Gudmundsson (228), in a study of the frequency of mixed psychoses in epilepsy in the population of Iceland, found prevalence rates for males of 6% and for females of 9%. These figures can be compared with findings of an earlier study by Helgason (245), who looked at the risk of psychosis in the general population of Iceland using the same diagnostic criteria as Gudmundsson. Helgason found prevalence rates of 7% for males and 5% for females. The comparison of Helgason's and Gudmundsson's results suggests a similar risk of psychoses in people with and without epilepsy, being only slightly higher in females but slightly lower in males with epilepsy.

Most figures in the literature on the frequency of psychosis in epilepsy derive from clinical case series and are therefore likely to be biased by unknown selection mechanisms. They cannot be regarded as representative of epilepsy in general.

Recently, Bredjkaer et al. (84) carried out a linkage study between the sample of people with epilepsy from a National Patient Register and the Danish Psychiatric Register. They reported an increased incidence of schizophrenia spectrum psychoses in both men and women with epilepsy. In another study, Stefansson et al. (551) compared patients with epilepsy and controls with other somatic diseases drawn from a disability register in Iceland. They found that schizophrenia was significantly overrepresented in males with epilepsy. In Table 39.12, the prevalence rates are presented as a function of the source of patients. The table suggests that psychoses are highly overrepresented in specialized centers.

There is now evidence from epidemiologic studies of an excess of psychoses in people with epilepsy, and clinical case series clearly indicate that psychosis is a significant problem in patients attending specialized centers. This suggests that there are risk factors for the development of psychosis related to complicated epilepsy and/or chronic illness.

Classification

There is no internationally accepted syndromic classification of psychoses in epilepsy. Psychiatric aspects are not considered in the ICES, and the use of operational diagnostic systems for psychiatric disorders such as the *DSM-III-R* (9) is limited because, if applied strictly, a diagnosis of functional psychosis is not allowed in the context of epilepsy.

Most of the previously proposed classification systems for psychosis in epilepsy (92,181,315,593) are based on a

TABLE 39.12. FREQUENCY OF MIXED PSYCHOSES IN EPILEPSY: EUROPE, JAPAN, AND THE UNITED STATES

Source of Patients	Author	Year	Prevalence (%)
Field study	Gudmundsson	1966	7
	Zielinski	1974	2
General practices	Edeh and Toone	1987	5
Neurology	Roger et al.	1956	9
departments	Standage and Fenton	1975	8
Epileptology	Bruens	1973	6
departments	Sengoku et al.	1983	6
	Schmitz	1988	4
Psychiatric	Roger et al.	1956	41
departments	Betts	1974	21
	Bash and Mahnig	1984	60

combination of psychopathologic, etiologic, longitudinal, and EEG parameters. Unfortunately, owing to a lack of taxonomic studies, our knowledge about regular syndromic associations is still limited. It seems, however, that in epileptic psychoses, diagnostic criteria are not strictly intercorrelated. Dongier (159) concluded from her detailed analysis of 536 psychotic episodes in 516 patients that it was not possible to deduce from the type of psychosis the type of epilepsy or vice versa. Atypical syndromes are not unusual, such as ictal and postictal psychoses in clear consciousness (633). Variations of phenomenology and precipitation can also be seen in individual patients who experience recurrent psychotic episodes (27,521). In addition, there is also some evidence of an etiologic overlap between syndromes (635). Simplified classifications on doubtful hypothetical grounds seem therefore not to be justified.

The International League Against Epilepsy has set up a commission to study the classification of the psychiatric disorders of epilepsy, which at present has not reported final results. However, it must be recognized that the psychoses of epilepsy represent a separate psychopathologic state that is subtly different from schizophrenia. Any classification scheme should note the relationship between the onset of psychosis and seizure activity, antiepileptic therapy, and any changes of EEG findings.

For pragmatic reasons in this discussion, psychoses of epilepsy are grouped according to their temporal relationship to seizures (Table 39.13). It should be acknowledged that such a classification does not necessarily imply fundamental differences in terms of pathophysiology.

Periictal Psychoses

Psychotic symptoms may occur as part of the seizure and can be prolonged in cases of nonconvulsive SE in which concurrent EEG studies may be required to make the diagnosis. Usually EEG studies performed during generalized (absence) status reveal generalized bilateral synchronous spike-and-wake activity between 1 and 4 Hz. During complex partial status (epileptic twilight state), the EEG may show focal or bilateral epileptiform patterns with slowed background activity. A wide range of experiential phenomena may occur, including affective, behavioral, and perceptual experiences, often accompanied by automatisms. Consciousness is usually impaired (but not in cases of simple partial status), insight tends to be maintained, but amnesia will often follow.

Absence status typically occurs in patients with a known history of generalized epilepsy, but atypical absence status may occur as a first manifestation of epilepsy, especially in later life (341).

Two types of complex focal status (synonyms: status psychomotoricus, epileptic twilight state) have been distinguished: a continuous form and a discontinuous or cyclic form. The latter consists of frequently recurring CPS. In

TABLE 39.13. CLINICAL CHARACTERISTICS OF PSYCHOSES IN RELATION TO SEIZURE ACTIVITY

	Ictal Psychosis	Postictal Psychosis	Perictal Psychosis	Alternative Psychosis	Interictal Psychosis
Consciousness	Impaired	Impaired or normal	Impaired	Normal	Normal
Duration	Hours to days	Days to weeks	Days to weeks	Days/weeks	Months
EEG	Status epilepticus	Increased epileptic and slow activity	Increased epileptic and slow activity	Normalized	Unchanged
Treatment	AED (I.V.)	Spontaneous recovery in many cases	Improve seizure control	Reduction of AED	Neuroleptic drug

EEG, electroencephalography; AED, antiepileptic drug; I.V., intravenous.

between seizures, patients may or may not experience simple focal seizure symptoms, and consciousness may recover to nearly normal states.

Noncyclic forms of complex partial status consist of prolonged confusional episodes or psychotic behavior. The EEG during complex partial status shows focal or bilateral epileptiform patterns and slowed background activity. Subtle rudiments of motor seizure symptoms such as lid fluttering and bursts of myoclonic jerks in absence status or mild oral activity automatisms in continuous complex partial status may point to the underlying epileptic activity. Mutism and paucity of speech or even speech arrest occur in both absence and complex partial status.

In discontinuous complex partial status, seizure phenomenology may help to localize the status origin in the mesial or lateral temporal lobe or extratemporally. Continuous complex partial status is more often of frontal or extratemporal origin than cyclic status (131).

Simple focal status or aura continua may cause complex hallucinations, thought disorders, and affective symptoms. The continuous epileptic activity is restricted and may escape scalp EEG recordings. Insight usually is maintained, and true psychoses emerging from such a state have not been described.

There is debate as to the mechanism underlying ictal psychosis, i.e., whether the seizure discharge activates a behavioral mechanism originating from part of the limbic system (positive effect) or whether behaviors are released by the inactivation of structures that normally suppress them (negative effect) (484). A recent case study (340) reported periodic left temporal sharp waves, during an episode of psychosis in an alert patient oriented only in person. The psychosis was characterized by agitation, religious delusions, and visual hallucinations and resolved over a period of 24 hours after parenteral diazepam, midazolam, and haloperidol. Although the episode of psychosis occurred after a cluster of six generalized convulsive seizures, Lee (340) believed that the episode should be regarded as ictal rather than postictal because the behavior was different from the patient's usual postictal state of confusion and somnolence and the EEG was topographically different from the interictal (bilaterally synchronous frontal spike-and-wave and sharp wave complexes). Whether this episode was indeed ictal or interictal could be debated, but the appearance of temporal sharp waves only during psychosis in a patient with CPS of frontal lobe origin is interesting, and Lee (340) postulated that the cluster of generalized seizures had potentiated dopaminergic neurotransmission in the temporal lobe.

Nonconvulsive SE requires immediate treatment with intravenous AED. It may closely mimic schizophrenia, although atypical features such as eyelid fluttering or myoclonic jerks in absence status or oral automatisms in partial status may clarify the diagnosis. However, any suspicion of status epilepticus merits an EEG; if unavailable,

an intravenous injection of a benzodiazepine may help to resolve the diagnosis.

Postictal Psychosis

The most common and well-investigated periictal psychosis is that which occurs postictally and is thought to account for one-fourth of the cases of psychosis in epilepsy (159). The incidence and prevalence of PIP are not known; however, rates of as high as 18% have been reported in patients with medically intractable focal epilepsy (298,606).

In 1988, Logsdail and Toone (357) determined operational criteria for the diagnosis of PIP that have been widely accepted (Table 39.14).

The authors described a series of 14 patients who fulfilled these criteria with a mean age at onset of epilepsy of 16.7 years and of psychosis, 32.2 years. The gap between age at onset of epilepsy and of psychosis was 3 to 33 years (mean, 15.5). Most patients had CPS with secondary generalization, but three had primary generalized epilepsy. One-half of the group had seizures at least weekly, and most developed psychosis after an exacerbation of seizure activity (usually a cluster of two to three). Most also had a lucid interval (usually of 1 to 2 days) between the restoration of an apparent normal mental state and the beginning of the psychosis; however, nine were confused at the onset of the psychosis. EEG were obtained in ten patients while psychotic; in eight, there were increased abnormalities (spikes, sharp waves, and/or slow waves), one showed reduced abnormalities, and another did not change. The mean length of the psychosis was 14.3 days (range, 1–90); eight needed major tranquilizers and one needed lithium. Six patients had further generalized seizures while psychotic, five of whom experienced deterioration in psychotic symptomatology. Follow-up ranged from 3 months to 8 years, during which time it was observed that PIP tended to recur; two patients developed a chronic psychosis after 8 years, another patient after 6 months, and four

TABLE 39.14. OPERATIONAL CRITERIA FOR THE DIAGNOSIS OF POSTICTAL PSYCHOSIS

Operational criteria for postictal psychosis
1. Onset of confusion or psychosis within 1 week of the return of apparently normal mental function
2. Duration of 1 day to 3 months
3. Mental state characterized by
   a. Clouding of consciousness, disorientation, or delirium
   b. Delusions, hallucinations in clear consciousness
   c. A mixture of a and b
4. No evidence of factors that may have contributed to the abnormal mental state
   a. Anticonvulsant toxicity
   b. A history of interictal psychosis
   c. Electroencephalographic (EEG) evidence of status epilepticus
   d. Recent history of head injury or alcohol/drug intoxication

From Logsdail SJ, Toone BK. Postictal psychosis. A clinical and phenomenological description. *Br J Psychiatry* 1988;152:246–252, with permission.

died (357). Since then, there have been several case reports and studies of patients with episodes of PIP.

Risk Factors for Postictal Psychosis.

The following are risk factors for the development of PIP:

1. Bilateral cerebral dysfunction (146,357,489,543,606);
2. Ictal fear (297,498);
3. Clusters of seizures (146,297,298,333,357,391,498,543, 606);
4. Absence of a history of febrile convulsions (606) or mesial temporal sclerosis (461);
5. Less hippocampal sclerosis than patients without PIP (607), anterior preservation of the hippocampus (86);
6. Preexisting personality disorder including an interictal psychotic profile on the MMPI (357,498);
7. Family history of psychiatric disorder. [However, although one of the patients in the series of Logsdail and Toone (357) series had a family history of excess alcohol consumption; other studies have not found a positive family psychiatric history in patients with PIP (391,498,544).]

Clinical Features and Phenomenology of Postictal Psychosis.

In most studies, PIP develops in patients with CPS (often with secondary generalization); however, it has also been described in patients with PGE (146,357). There tends to be a delay between the onset of habitual seizures and the development of PIP ranging from 1 month to 56 years (146), with a mean ranging from 13.1 (298) to 21.7 years (146).

Most reports document a lucid interval between the restoration of apparent normal mental state after the seizure and the beginning of the psychosis, which could last as long as 72 hours (146,297,298,333,357,391,498,543). PIP is often preceded by a period of confusion (357,391).

The duration of PIP tends to range from 1 to 90 days (357), with a mean duration varying from approximately 3 (146,298) to 14.3 days (357).

The phenomenology of PIP appears to vary widely both within and between series. Logsdail and Toone (357) found that only one patient had primary delusions and thought disorder; nine had an abnormal mood, and six had paranoid delusions. Visual, auditory, or somatic hallucinations were reported in six, six, and two cases, respectively. Savard et al. (498) reported that seven of nine patients developed a paranoid delusional syndrome with prominent persecutory delusions. Six patients had further generalized seizures while psychotic, five of whom experienced deterioration in psychotic symptomatology. Lancman et al. (333) described paranoid delusions, mysticism, and religious preoccupations along with auditory and visual hallucinations. They noted that in most cases, the patients could recall what had happened during the psychotic episodes. Devinsky et al. (146) documented fluctuating combinations of delirium, persecu-

tory delusions, hallucinations, and affective changes. Kanner et al. (298) reported that most patients exhibited an abnormal affect: depressed in 90% alternating with hypomania in 70%. Seventy percent were irritable, and 20% had suicidal ideation. Delusions were experienced by 90% (paranoid, grandiose, somatic, and religious) and hallucinations were experienced in 40% (mainly auditory). All patients were oriented in time, place, and person. Kanemoto et al. (297) described sexual indiscretions and sudden, unprovoked aggressive behavior along with religious and grandiose delusions in patients with PIP. In a later study, Kanner et al. (299) compared 30 patients with PIP with 33 with AIP. Well-directed violent attacks were observed in 13 of the 57 episodes of psychosis witnessed in the patients with PIP (23%) compared with only three of the 62 witnessed episodes of acute interictal psychosis (5%).

Investigations.

In some studies, EEG was performed when the patient was psychotic, and in some cases, there was an exacerbation of abnormalities (357,391,543). In others, there was an improvement (357) or only minimal diffuse background slowing (333).

One study has performed SPECT on two patients who experienced episodes of PIP while being assessed for epilepsy surgery (195). In both cases, an increased regional blood flow (compared with interictal studies) was observed over the right temporal neocortex and the contralateral basal ganglion. In both cases, ictal EEG confirmed a right temporal focus. Briellmann et al. (86) compared the volume of the hippocampus (assessed using quantitative MRI) of six patients with PIP (five of whom had a left-sided foci) and 45 controls with medically intractable TLE. Although the overall volumes and hippocampal T2 relaxometry were similar, the patients with PIP had a relative sparing of the anterior hippocampus (whereas the controls had diffuse hippocampal loss). In addition, histopathologic examination of the resected temporal lobes revealed that mesial dysplasia was more frequent in the patients with PIP.

Tebartz van Elst et al. (570) reported quantitative MRI studies of the amygdala and hippocampus in PIP, noting increases in the right amygdala volumes compared with patients with interictal psychosis and patients with epilepsy with no psychosis.

Management and Treatment of Postictal Psychosis.

Logsdail and Toone (357) reported that more than one-half of their patients needed to be treated with major tranquilizers and one with lithium. Lancman et al. (333), however, did not demonstrate any benefits from either neuroleptics or psychotherapy and noted that most patients returned to their premorbid state within 1 week regardless of intervention. Some patients responded well to mild sedation (with benzodiazepines or chloral hydrate) given in a supportive environment. Clobazam can be used to abort a cluster of seizures or used after a cluster with the first warning of any developing psychopathology, such as irritability, mood lability, and sleeplessness.

The clinical significance of PIP in patients with intractable focal epilepsy is that it tends to follow a cluster of seizures often provoked by a reduction of medication while being monitored as part of a presurgical assessment. Savard et al. (498) suggested that patients with risk factors for PIP including an interictal psychotic profile on the MMPI should have a more cautious and gradual reduction of AED. Kanner et al. (298) treated one patient with a history of PIP prophylactically with neuroleptics before depth electrode recording, and no psychosis occurred, despite a cluster of seizures.

Psychosis is considered by some centers to be an exclusion factor for epilepsy surgery (594). However, the first-line management of patients who only have episodes of psychosis after seizures should be attaining seizure control. Few studies have addressed the outcome of surgery in patients with preoperative psychosis. Kanemoto et al. (294) reported postoperative mood disorders (depression, hypomania, and mania) in eight of 38 cases. Risk factors for the psychopathology were preoperative episodes of PIP, left-sided operations, and ictal fear.

There have been few long-term follow-up studies; however, PIP would appear to be recurrent, often in a stereotyped way (333,357,498), and some patients (as many as 25%) go on to develop a chronic interictal psychosis (298,357).

Interictal Psychosis

Forced Normalization. At the beginning of the past century, it was observed that there appeared to be an antagonism between psychosis and epilepsy. This influenced von Meduna (386) to introduce convulsive therapy for the treatment of schizophrenia. Since then, brief psychotic episodes, characterized by paranoid delusions and auditory hallucinations, have been observed in epilepsy. They tend to last days to weeks and are more common when seizures are infrequent or fully controlled. Such episodes have been reported to alternate with periods of increased seizure activity and may be terminated by a seizure. Premonitory symptoms such as anxiety and insomnia have been reported, and Wolf (636) suggested that anxiolytic treatment may prevent the development of the psychosis. Landolt (336) noted that the EEG normalized during such episodes of psychosis, generating the term "forced" or "paradoxical normalization" accompanying the "alternative psychosis" (571). It is now acknowledged that the EEG does not have to fully normalize; however, the interictal disturbances tend to decrease. The EEG abnormalities may then return when the psychosis remits.

These episodes of psychosis are considered to be rare, only occurring in three of 697 patients in one series (507). They have, however, been reported as occurring more commonly with some AED; the original observations were occurrences with ethosuximide (636). With the introduction of several newer agents in the past 10 years, this clinical picture has been reported frequently, when patients with habitual seizures have them abruptly terminated. Drugs implicated include VGB (495), TPM, TGB (602), and zonisamide (384), and acute psychotic symptoms have recently been reported with TPM with rapid resolution after withdrawing the AED (308).

Kanemoto et al. (296) suggested a link between forced normalization and the concept of secondary generalization. The "running down" phenomenon refers to the gradual remission of seizures over months to years after temporal lobectomy. It is thought to be related to the ability of secondary epileptogenic areas to generate seizure activity for a limited time after the primary epileptogenic zone has been removed (488). Kanemoto et al. (296) found that patients who exhibited this phenomenon were more likely to develop both pre- and postoperative episodes of AIP. Thus, they concluded that the anterior temporal lobe itself could not generate AIP without involving areas of secondary epileptogenesis.

Chronic Interictal Psychosis: Schizophrenialike Psychoses of Epilepsy

The relationship between epilepsy and chronic psychosis has been well described since the 1950's (262,435), developing in an estimated 7% to 9% of patients with epilepsy (369,395).

In 1963, Slater et al. (535) described the SLPE. They studied 69 patients with epilepsy and schizophrenialike states. Eleven had a chronic psychosis that had been preceded by recurrent confusional states of short duration, 46 had a psychosis suggestive of paranoid schizophrenia, and 12 had hebephrenic schizophrenia. There was no family history of psychiatric illness, and the premorbid personality was normal. The mean age at onset of the psychosis was 29.8 years, and it occurred after the epilepsy had been present for a mean of 14.1 years. In some cases, the psychotic symptoms appeared when the fit frequency was falling. Most patients had delusions without any change in level of consciousness, and mystical delusions were common. The authors found all the features of schizophrenia when they looked at their patients as a whole but found catatonic phenomena to be unusual. They noted that affective responses remained warm, but periodic mood swings were seen in most patients. Air encephalography was carried out in 56 cases; there were abnormalities in 39, usually atrophy; but in 19, there was dilation of one or both temporal horns. They found that TLE was the most common form of epilepsy.

Flor-Henry (194) performed a retrospective study comparing 50 patients with TLE who had experienced an episode of psychosis with nonpsychotic controls, excluding patients with dementia. He found that patients with psychosis tended to have less frequent seizures (especially psychomotor attacks) and had either a dominant hemisphere lesion, if the disorder was unilateral, or bilateral foci.

Taylor (565), in 1975, studied patients who had undergone unilateral temporal lobectomy for epilepsy. Psychosis occurred more frequently in those who had "alien tissue" rather than mesial temporal sclerosis on neuropathologic examination. He believed that this dysfunctional tissue was more likely to produce psychosis than a nonfunctional lobe secondary to medial temporal sclerosis (565).

Perez and Trimble (430) examined 23 patients with active psychosis and epilepsy prospectively with the Present State Examination. They found a schizophreniform psychosis, psychopathologically identical to that found in a nonepileptic group with schizophrenia, to be almost exclusively associated with TLE. There was a trend for the schizophreniform picture to be associated with left-sided lesions.

The mechanism for the development of chronic psychosis in epilepsy is not known; however, theories include the psychosis may be a direct consequence of the epileptic discharge either directly or through the development of neurophysiologic or neurochemical changes, such as kindling. The other major theory is that both the psychosis and the epilepsy have a shared etiology (484). The role of anticonvulsant medication in the etiology of SLPE is not known. A recent study (325) reported persistent symptoms of auditory hallucinations and delusions of persecution and/or reference in 12 patients (two of whom were in remission on antipsychotic medication) an average of 30.8 months (range, 2–145) after withdrawal of AED. These patients developed psychotic symptoms on average 181.7 months (15.1 years) after the onset of their epilepsy, which is consistent with the rest of the published literature and suggests that discontinuation of AED medication does not prevent the development of SLPE. One-half of these patients developed psychotic symptoms within a year (one-third within 4 months) of the withdrawal of AED, and it may be that this acted as a precipitant for the onset of psychosis.

Risk Factors for the Development of Schizophrenialike Psychoses of Epilepsy.
Risk factors for the development of SLPE are shown in Table 39.15.

The pathogenesis of psychotic episodes in epilepsy is likely to be heterogeneous. In most patients, a multitude of chronic and acute factors can be identified that are potentially responsible for the development of a psychiatric disorder. These factors are difficult to investigate in retrospect, and the interpretation as either causally related or simply intercorrelated is arguable.

The literature on risk factors is highly controversial; studies are difficult to compare because of varying definitions of the epilepsy, the psychiatric disorder, and the investigated risk factors. Most studies are restricted to interictal psychoses.

Genetic Predisposition. With few exceptions (284), most authors do not find any evidence of an increased rate of psychiatric disorders in relatives of epilepsy patients with psychoses (194,430,533).

Gender Distribution. There has been a bias toward the female gender in several case series (563), which has not been confirmed in controlled studies (27,320,321).

Duration of Epilepsy. The interval between age at onset of epilepsy and age at first manifestation of psychosis has been remarkably homogeneous in many series, being in the region of 11 to 15 years (593). This interval has been used to postulate the etiologic significance of the seizure disorder and a kindlinglike mechanism. Some authors (92,554) have argued that the supposedly specific interval represents an artifact. They noted a wide range, being significantly shorter in patients with a later onset of epilepsy. They also pointed out that any patients whose psychosis did not succeed their epilepsy were excluded in most series and that there is a tendency in the general population for the age at onset of epilepsy to peak at an earlier age than that of schizophrenia.

Type of Epilepsy. There is a clear excess of TLE in almost all case series of patients with epilepsy and psychosis. Summarizing the data of ten studies, 217 (76%) of 287 patients had TLE (593). The preponderance of this type of epilepsy

TABLE 39.15. RISK FACTORS FOR THE DEVELOPMENT OF CHRONIC INTERICTAL PSYCHOSIS (SCHIZOPHRENIALIKE PSYCHOSIS OF EPILEPSY)

Age of onset	Before/around puberty (10–14 years before onset of psychosis)
Seizures	Lack of history of febrile convulsions, multiple types, CPS > PGE, history of status epilepticus, medically intractable
Epilepsy syndrome	Localization related (TLE)
Seizure frequency	Diminished (especially after temporal lobectomy)
Gender bias	Female > male
Neurology	Sinistrality
Premorbid personality	Normal
Family psychiatric history	None
Electroencephalography	Mesiobasal focus, left > right or bilateral
Functional neuroimaging (SPECT)	Left temporal hypoperfusion, ? independent of seizure focus
Pathology	Ganglioglioma/hamartoma

CPS, complex partial seizure; PGE, primary generalized epilepsy; TLE, temporal lobe epilepsy; SPECT, single photon emission computed tomography.
From Lambert MV, David AS. Psychiatric comorbidity. In: Oxbury J, Polkey C, Duchowny M, eds. *Intractable focal epilepsy.* London: Harcourt, 2000:445–471, with permission.

is, however, not a uniform finding; in Gudmundsson's (228) epidemiologic study, for example, only 7% had "psychomotor" epilepsy.

The nature of a possible link of psychoses to TLE is not entirely clear (505), partly owing to ambiguities in the definition of TLE in the literature, either based on seizure symptomatology (psychomotor epilepsy), involvement of specific functional systems (limbic epilepsy), or anatomic localization as detected by depth EEG or neuroimaging (amygdalohippocampal epilepsy). Unfortunately, most authors have not sufficiently differentiated frontal lobe epilepsy and TLE.

The temporal lobe hypothesis, although widely accepted, has been criticized for being based on uncontrolled case series, such as the studies by Gibbs (212) and Slater and Beard (533). It was argued that TLE is the most frequent type of epilepsy in the general population and that there is an over-representation of this type of epilepsy in patients attending specialized centers. There is a consensus that psychoses are very rare in patients with neocortical extratemporal epilepsies (92,159,212,421,517,504,554).

The findings are less unequivocal regarding TLE and generalized epilepsies. In fact, with only three exceptions (233,517,529), the majority of controlled studies failed to establish significant differences in the frequency of psychoses in generalized epilepsy compared with TLE (91,202,400,504,537,554,549). However, many patients with generalized epilepsy show pathology of temporal structures, making classification difficult.

Several studies show that psychoses in generalized epilepsies differ from psychoses in TLE (593). The former are more likely to be short lasting and confusional (91,159,554). Alternative psychoses, which are especially common in generalized epilepsy, are usually relatively mild and often remit before development of paranoid-hallucinatory symptoms. Schneiderian first-rank symptoms and chronicity are more frequent in patients with TLE (504,598). This has considerable significance for psychiatrists attempting to unravel the underlying "neurology" of schizophrenia.

Type of Seizures.

There is evidence from several studies that focal seizure symptoms, which indicate ictal medial temporal or limbic involvement, are overrepresented in patients with psychosis. Hermann and Chabria (248) noted a relationship between ictal fear and high scores on paranoia and schizophrenia scales of the MMPI. Kristensen and Sindrup (320,321) found an excess of dysmnesic and epigastric auras in their psychotic group. They also reported a higher rate of ictal amnesia. In another controlled study, ictal impairment of consciousness was related to psychosis, but simple seizure symptoms indicating limbic involvement were not (504).

No seizure type is specifically related to psychosis in generalized epilepsies. Most patients with psychosis and generalized epilepsies have absence seizures (504).

Severity of Epilepsy.

The strongest risk factors for psychosis in epilepsy are those that indicate the severity of epilepsy.

These are long duration of active epilepsy (533), multiple-seizure types (92,253,354,422,470,504,517), history of SE (504), and poor response to drug treatment (354). Seizure frequency, however, is reported by most authors to be lower in psychotic epilepsy patients than in nonpsychotic patients (194,517,534,548). It has not been clarified whether seizure frequency was low before or during the psychotic episode. This may represent a variant of forced normalization (see below).

Laterality.

Left lateralization of temporal lobe dysfunction or temporal lobe pathology as a risk factor for schizophreniform psychosis was originally suggested by Flor-Henry (194). Studies supporting the laterality hypothesis have been made using surface EEG (354), depth electrode recordings (521), computed tomography (585), neuropathology (564), neuropsychology (430), and PET (591). The literature has been summarized by Trimble (593). In a synopsis of 14 studies with 341 patients, 43% had left, 23% right, and 34% bilateral abnormalities. This is a striking bias toward left lateralization. However, lateralization of epileptogenic foci was not confirmed in all controlled studies (159,320,321,528). Again, it may be that some symptoms, e.g., some first-rank psychotic symptoms, are associated with a specific side of focus.

Structural Lesions.

The literature on brain damage and epileptic psychosis is very controversial. Some authors have suggested a higher rate of pathologic neurologic examinations, diffuse slowing on the EEG, and mental retardation (320,321), whereas others could not find an association with psychosis (194,284). Neuropathologic studies from resected temporal lobes of patients with TLE have suggested a link between psychosis and the presence of cerebral malformations such as hamartomas and gangliogliomas compared with mesial temporal sclerosis (563). These findings have been seen as consistent with recent findings of structural abnormalities found in the brains of schizophrenic patients without epilepsy, which arise during fetal development.

Information from neuroimaging studies has helped to understand the etiology. Marshall et al. (382) compared patients with SLPE and matched controls with epilepsy using SPECT. They found that the group with SLPE showed a significant reduction in the index of regional cerebral blood flow in the left medial temporal region. Few studies have compared brain volumes in patients with epilepsy with patients with schizophrenia and normal controls. Barr et al. (25) found significantly larger ventricular volumes in both patients with first-episode schizophrenia (n = 32) and patients with TLE (n = 39) compared with normal controls (n = 42). Both patient groups had smaller left hippocampi compared with the healthy controls. However, they did not comment on whether any of the patients with epilepsy developed a psychosis.

A recent study (374) compared the volumes of the amygdalohippocampal complex in 24 patients with epilepsy (12 with psychosis), 26 patients with schizophrenia, and 38 normal controls. In addition, they compared the results from spectroscopy of the anterior hippocampal metabolite concentrations of *N*-acetyl aspartate between these subject groups. Bilateral hippocampal volume reductions (anterior to the fornix) were found in those with epilepsy and psychosis compared with the controls. Spectroscopic abnormalities were also found in the epilepsy group. In these patients, *N*-acetyl aspartate was reduced bilaterally, greater in those with psychosis, especially on the left side (although this latter finding did not reach significance), providing further support for the role of the left (dominant) hemisphere in the development of psychosis, both in epilepsy and schizophrenia.

Clinical Features and Phenomenology of Schizophrenia-like Psychoses of Epilepsy. The clinical picture is usually that of a paranoid psychosis, and most studies have found it difficult to differentiate between SLPE and schizophrenia (317,430). Others, however, have noted the prominence of religious delusions, preservation of affect, and relative lack of negative symptoms (533). A recent study compared the phenomenology of interictal psychosis in epilepsy with temporal and frontal lobe foci (3). Patients with frontal lobe epilepsy tended to show more marked hebephrenic characteristics (emotional withdrawal and blunted affect) compared with patients with TLE (3). However, the number of patients with frontal lobe epilepsy and SLPE was small (8), and the study was performed retrospectively by reviewing case notes. Mellers et al. (390) compared patients with SLPE, schizophrenia, and TLE using standardized neuropsychological tests. Surprisingly, the patients with schizophrenia and those with SLPE showed similar deficits in attention, episodic memory (verbal > visual), and executive function compared with patients with epilepsy but no psychosis. This study supports the belief that patients with SLPE have left-sided temporal lobe dysfunction but also suggests that they have a more global cognitive impairment.

It may be argued that peri- and interictal psychoses are not separate entities but represent different modes of presentation of psychosis in epilepsy or reflect differing levels of severity. Kanemoto et al. (297) evaluated 808 adult patients with CPS and found 13% had experienced psychotic episodes. There appeared to be many shared features between the presentations of psychosis but also some factors that differentiated them. Thirty patients (3.71%) were classified as having PIP, which occurred in close association with clusters of CPS or secondarily generalized tonic-clonic seizures. Thirty-three patients (4.08%) had episodes of AIP that alternated with seizures and were only diagnosed if patients had more than 80% reduction in seizures in the month preceding the psychotic episode. Twenty-five patients (3.09%) had an episode of psychosis lasting more than 3 months and were diagnosed as having chronic psychosis. Comparing the groups, patients with PIP had the longest latent pe-

riod between the onset of epilepsy and the onset of psychosis (20 ± 11.9 years) and the chronic psychosis group had the shortest (14.2 ± 10.6 years); psychic auras (e.g., ictal fear/dysmnesia) were reported more than twice as often in the PIP group. Febrile convulsive status occurred in similar numbers in each group [four of 30 (13.3%) with PIP and four of 25 (16%) with chronic psychosis]; however, the authors did not comment on the numbers of patients having focal febrile convulsions or repeated episodes. Patients with PIP were more likely to have extratemporal EEG foci, although no mention is made of the presence of bilateral foci. Although most patients had EEG evidence of TLE, temporal lobe pathology was only identified neuroradiographically in five of 33 (15.2%) patients with interictal psychosis, ten of 30 (33.3%) with PIP, and seven of 25 (28%) with chronic interictal psychosis. However, some patients only had computed tomography scans, and it is possible that more pathology may have been identified if a high-resolution MRI scan had been performed. All patients with chronic or interictal psychosis and more than one-half with PIP had no evidence of impaired consciousness during the episode. At least one-half of the PIP group developed a lucid interval consisting of normal behavior and clear consciousness before the onset of the psychosis. Secondarily generalized seizures occurred more commonly in the PIP group. They found phenomenologic differences between the groups, with first-rank symptoms of Schneider (delusional perception and voices commenting), auditory hallucinations, delusions of persecution and reference being less common, and sexual indiscretions, sudden, unprovoked aggressive behavior, excessive emotional responses to trivial external stimuli, illusion of familiarity, and religious and grandiose delusions being more common in patients with PIP. No patient in the chronic group responded adequately to antipsychotic medication. Patients with PIP tended to respond more rapidly to medication than the interictal group. The authors did not comment on recurrence of PIP or development into chronic psychosis.

Treatment of Interictal Psychosis. Generally, management of acute psychosis necessitates treatment with neuroleptic drugs in a calm environment, preferably in a ward with staff experienced in managing such neuropsychiatric patients.

All neuroleptic medication reduces the seizure threshold to some degree, with rates of seizures ranging from 0.5% to 1.2% (623). Patients most likely to experience an exacerbation of their epilepsy are those with cerebral damage, and the increase in seizures tends to occur early in treatment or after a dose increment, especially if the escalation was rapid. The choice of neuroleptic depends on many factors; generally, clozapine, chlorpromazine, and loxapine should be avoided because these drugs would appear to be the most epileptogenic. Of the conventional neuroleptics, haloperidol appears to be relatively safe. If long-term treatment is required, sulpiride, quetiapine, olanzapine, and risperidone have little effect on seizure threshold and have fewer extrapyramidal side effects. Although depot neuroleptics improve compliance, they are more difficult to titrate slowly, and if adverse effects do occur, they will last longer. Thus, oral drugs are preferable. When possible, only one neuroleptic drug should be given, starting at the lowest possible dose and gradually increased, while monitoring seizure frequency.

Trimble (595) suggested that in patients in whom their psychosis is clearly related to a fall in seizure frequency (such as cases of "alternative" psychosis), the neuroleptics of choice may be those that lower the seizure threshold (such as clozapine).

PSYCHOSIS AND EPILEPSY SURGERY

De Novo Postoperative Psychosis

Postictal Postoperative Psychosis

Manchanda et al. (377) reported four patients in a series of 298 (1.3%) who developed a PIP after surgery for medically intractable TLE. In three, this occurred within 2 years of surgery, but in one, there was an interval of 13 years. All were male, with normal intelligence and left-hemisphere dominance, and had undergone a right-sided temporal lobectomy. None had a psychiatric history. The psychosis lasted 1 to 8 days and tended to occur after an exacerbation of seizures (at least two in 24 hours), whether CPS or secondarily generalized seizures. The phenomenology of the delusions varied but was predominantly depressive and/or paranoid. Three of these patients had bilateral EEG abnormalities before and/or after surgery. Two of the patients had experienced clusters of seizures preoperatively but did not develop an episode of PIP, and two experienced an exacerbation of seizures subsequently without developing psychosis. Thus, the authors suggested that there may be a period of increased susceptibility to PIP shortly after surgery.

Interictal Postoperative Psychosis

Mace and Trimble (371) reported six cases of *de novo* postoperative psychosis arising over a 10-year period. All patients were male and had undergone right-sided surgery. The psychoses developed between 1 month and 9 years postoperatively, and all but one developed within 5 years. In four patients, a schizophreniform illness developed with paranoid features, and two patients developed a depressive psychosis, one of which also had features suggestive of the Capgras syndrome. Five of the patients had marked passivity phenomena. Most patients were left-hemisphere dominant for speech. Pathologic findings included medial temporal sclerosis (two patients) and oligodendroglioma (one patient); two had drainage of an abscess or extirpation of a cyst with no lobectomy. Four of the patients developed their psychosis after a marked improvement in seizure control, whereas an-

other developed it when seizures recurred after an interval of 2 years. All five of these patients responded well to antipsychotic medication. One patient, however, did not experience a postoperative improvement in seizures. He had bilateral representation of speech and postictal dysphasia. No abnormal pathology was found in the resected specimen, suggesting that the patient may have had bilateral epileptogenic foci. In this patient, the episodes of psychosis alternated with periods of increased seizure frequency.

Leinonen et al. (343) reported that three of 57 patients (5.3%) developed psychosis *de novo* within 6 months of surgery for TLE, which only resulted in seizure freedom in one. Two patients were female, and two had right-sided surgery. Bilateral preoperative EEG abnormalities were recorded in one patient. The pathologic findings were nonspecific in two patients and characteristic of an oligodendroglioma in the other. All had suffered from anxiety symptoms preoperatively. Two had paranoid symptomatology, and all responded to neuroleptics. Callender and Fenton (101) reported a man who developed *de novo* psychosis in clear consciousness 4 years after a right-sided temporal lobectomy. Initially, he experienced a 75% reduction in seizures, and the psychosis developed after his CPS became more frequent, although the psychotic symptoms occurred independently of the seizures. EEG recordings revealed spike-and-sharp wave discharges arising from the right mid- and posterior temporal regions along with occasional independent spikes from the anterior medial structures of the contralateral temporal lobe. The symptoms did not respond to treatment with neuroleptic medication; however, both the seizures and perceptual abnormalities improved with GBP.

Andermann et al. (11) reported a series of six patients from four centers who developed *de novo* psychosis after removal of a foreign-tissue glial lesion (ganglioglioma, hamartoma, or dysembryoplastic neural tumor). The operations were left sided in four patients, three of whom were left-hemisphere dominant for language with the other being left-handed with no hemisphere dominance documented. None had evidence of bilateral pathology. Preoperatively, none was psychotic, although one was dysthymic, another had NES, and another had behavioral problems. Five of the patients were rendered seizure free, and one had a greater than 90% improvement. The psychoses developed within a year of surgery, in one case within a week, and predominantly consisted of paranoid delusions with depressive features. In five cases, the psychosis fully responded to antipsychotic medication (usually haloperidol), but in one case, the psychosis and depressive ideation recurred, and the patient committed suicide.

In summary, psychosis (both post- and interictal) may arise *de novo* after epilepsy surgery. Although there may be an interval of many years, the majority develops within 5 years. Risk factors appear to include foreign-tissue glial lesions and right-sided operations. Bilateral pathology may also be a risk factor, as assessed by bilateral EEG foci (377).

PERSONALITY CHANGES IN EPILEPSY

Interictal behavioral changes have been documented for centuries. It is generally considered that most patients with epilepsy have a normal personality structure, but many authors agree that some undergo personality alteration. This in effect represents a subtle organic personality change.

Waxman and Geschwind (616) described the interictal behavior syndrome of TLE, consisting of religiosity, hypergraphia, viscosity/stickiness, circumstantiality, meticulousness, and attention to detail (the Geschwind syndrome). Bear and Fedio (34) documented 18 traits that were associated with TLE (Table 39.16). In addition, they reported that patients with TLE appeared to differ from normal controls and patients with other neurologic diseases on particular personality traits, recorded using their inventory [the Bear-Fedio Inventory (BFI)]. The authors also noted that patients with right TLE exhibited more emotional traits and tended to deny their negative behavior, whereas those with a left-sided focus displayed more ideational traits and tended to be more self-critical (34). The study has been criticized, however, for assessing small numbers of patients: 15 with right temporal, 12 with left temporal foci, and no patients with other forms of epilepsy. They did not control for the presence or absence of psychiatric disorder nor did they specify anticonvulsant medication. Since then, there has been controversy regarding the replication of these findings. Mungas (408) found that the BFI did not differentiate patients with TLE from other neurologic groups but merely reflected the presence or absence of concomitant psychiatric illness. Dodrill and Batzel (154) reviewed the literature in 1986 and reported that patients with epilepsy show more behavioral problems than normal controls and differ from patients with nonneu-

rologic medical problems in that they show more psychotic-like symptoms. Rodin and Schmaltz (469) concluded that the BFI is a general measure of emotional maladjustment but provides no support for a specific TLE syndrome. It would appear that the number of seizure types is more predictive of behavioral abnormalities than the presence or absence of TLE (151,253,471), as is the presence of cephalic auras (dizziness, pressure, heaviness), especially when not followed by secondary generalization and hence amnesia (396).

The rating scale most commonly used to assess personality is the MMPI. Hermann et al. (254) found that onset of TLE during adolescence was associated with more maladjustment in the MMPI. In a later study (254), they noted that patients with an aura of ictal fear (often associated with epilepsy originating from the periamygdaloid region of the limbic system) scored higher on the MMPI profiles of paranoia, psychasthenia, schizophrenia, and psychopathic deviation. Naugle and Rogers (412) reported differences between male patients with intractable seizures and medical, psychiatric, and normal controls using the MMPI and the California Psychological Inventory but concluded that these differences merely represented logical, adaptive responses to their epilepsy. Benson and Hermann (43) discussed the limitations of using the MMPI to identify behavioral changes in epilepsy because it does not directly assess the core features of Geschwind syndrome.

The existence of the TLE personality syndrome has recently been debated (71,142). Although many studies using the BFI revealed increased behavioral traits in patients with epilepsy compared with normal controls or other patient groups, the BFI has not reliably differentiated between patients with epilepsy (focal or generalized) and psychiatric controls (142). The BFI has also revealed increased traits including humorlessness to occur in patients with frontal lobe epilepsy diagnosed using depth electrodes (620). Personality changes have also been observed in patients with generalized epilepsy. Wirrell et al. (630) found that patients with absence epilepsy had more behavioral problems than patients with juvenile rheumatoid arthritis. Furthermore, many of the behavioral changes observed in epilepsy may be accounted for by variables such as concomitant head injury, social stigmatization, and the behavioral/cognitive effects of AED. The research on personality disorders in epilepsy is complicated by the lack of standardized assessment procedures. A recent study (358) used the *DSM-III-R* structured clinical interview for personality disorders (546) to assess 52 patients with medically refractory epilepsy. This instrument, however, does not detect the behavioral characteristics associated with TLE. Twenty-one percent met *DSM-III-R* criteria for personality disorders, of which the most common were the dependent and avoidant types. The authors commented that these could be the result of the psychosocial consequences of living with uncontrolled epilepsy. Further research needs to be conducted in this area using standardized operational criteria such as the epilepsy-specific version of the structured

TABLE 39.16. PERSONALITY TRAITS ASSOCIATED WITH TEMPORAL LOBE EPILEPSY

Aggression
Anger
Circumstantiality
Dependence/passivity
Depression/sadness
Elation
Emotionality
Guilt
Humorlessness
Hypergraphia
Hypermoralism
Hyperreligiosity
Obsessionalism
Paranoia
Increased sense of personal destiny
Increased philosophical interest
Altered sexuality (hyposexuality)
Viscosity

From Bear DM, Fedio P. Quantitative analysis of interictal behavior in temporal lobe epilepsy. *Arch Neurol* 1977;34:454–467, with permission.

clinical interview (610), along with clear documentation of epilepsy-related variables such as seizure type, lateralization, and frequency in combination with investigative procedures using advanced EEG and neuroimaging investigative techniques.

Recently, there has been a growing interest in the personality correlates of juvenile myoclonic epilepsy (596). The condition often arises in patients with an erratic lifestyle who develop sleep deprivation. The personality is described in terms of impressionability, unreliability, instability, sentimentality, and immaturity. Because frontal lobe changes have been identified in this syndrome, this personality profile contrasts with the personality traits thought to be related to temporal lobe problems.

Some of these personality traits [increased concern with philosophical, moral, or religious issues; hypergraphia (often associated with circumstantiality and viscosity); hyposexuality; and irritability] have been assessed in more detail, and some authors refer to these traits as Geschwind syndrome (42,210). These are discussed below.

Religiosity

The heightened religiosity of epilepsy has been recognized since the nineteenth century [by Esquirol (175) and Morel (402), cited in Saver and Rabin (500)]. Despite this, little attention has been paid to this relationship. Bear et al. (34,35) found that religiosity trait scores could distinguish between patients with TLE and normal controls and patients with other psychiatric or neurologic conditions including extratemporal focal epilepsy and generalized epilepsy. Ecstatic ictal experiences, in some cases accompanied by visions of a religious nature, have been reported in patients with EEG evidence of temporal lobe discharges (500). Patients may explain seizures, especially involving symptoms of depersonalization, derealization, and autoscopy, as religious experiences (attribution theory) (500). Carrazana et al. (108) reported patients who attributed seizures to voodoo possession, thus delaying seeking medical treatment.

Sudden religious conversions have been reported in the postictal period, closely related to the first seizure or an increase (or more rarely a decrease) in seizure frequency (148). Other studies, however, have failed to demonstrate a relationship between religiosity and seizure type or lateralization of TLE (605).

Hypergraphia

Waxman and Geschwind (616) were the first to document the occurrences of hypergraphia (a tendency to excessive and compulsive writing) in patients with TLE. They reported meticulous and detailed writing often concerned with moral, ethical, or religious issues. Since then, it has been observed in 8% of patients with epilepsy (255) and has been associated with previous psychiatric episodes, emotional mal-

adjustment, computed tomography scan abnormalities, and focal epilepsy (especially TLE) (420), along with affective disturbance (especially hypomania) and nondominant foci (495).

Viscosity

This refers to the tendency to talk repetitively and circumstantially about a restricted range of topics and appears to be more common in patients with TLE, especially with left or bilateral seizure foci, and is also correlated with duration of epilepsy and left-handedness (448). It is widely believed that viscosity may result from impaired linguistic skills (385,448).

Sexual Disorders

Despite sexual dysfunction being described in epilepsy since the nineteenth century (67), there is still debate as to the etiology of the dysfunction and even the extent to which epilepsy increases the risk of sexual disorder (285).

Self-mutilation (562), transvestism (126), sadomasochism (562), exhibitionism (268,562), and fetishism (263) have been reported, especially with TLE, and may resolve with cessation of attacks with medical or surgical treatment (584). The most common sexual dysfunction experienced is the interictal disorder of hyposexuality, which Toone (584) defined as "a global reduction in sexual interest, awareness and activity." The prevalence varies from study to study, with rates ranging from 22% (410) to 67% (562); however, some researchers have not found this degree of dysfunction (163,285).

Toone et al. (587) compared the sexual functioning of adult males with epilepsy being treated by their general practitioners with controls. They found that 56% of patients with epilepsy had experienced previous sexual intercourse compared with 98% of controls, with only 43% having sexual activity in the previous month compared with 91% of controls.

Fewer studies have focused on sexual dysfunction in women. Ndegwa et al. (415) found that women with epilepsy (all of whom had a regular heterosexual partner) had less frequent sexual activity, more vaginismus, and generally less interest in sex than controls. Bergen et al. (45) found 34% of 50 female outpatients attending a tertiary referral center were hyposexual.

The sexual dysfunction in both sexes would appear to have a neurophysiologic component (404). Men with epilepsy have an increased rate of erectile dysfunction, with rates of 57% in one study (587), compared with 9% in the general population (196). This can occur in isolation (261,496) or in conjunction with hyposexuality (73,204,529). Nocturnal penile tumescence testing has shown that the erectile dysfunction appears to be physiologic in origin (230). Endocrine changes (increased sex hormone binding globulin and reduced free testosterone) have been reported in epilepsy, especially when treated with liver enzyme–inducing AED (275,586). However, the relationship between these hormonal changes and hyposexuality is not known.

There are many theories regarding the etiology of sexual dysfunction, in particular, hyposexuality in epilepsy. It is likely multifactorial in origin involving neurologic, endocrine, psychiatric, cognitive, and psychosocial factors. Epilepsy-related factors include the age at onset/duration of epilepsy along with seizure type and focus. In addition, seizure frequency might be relevant because successful epilepsy surgery can result in an improvement in sexual functioning despite remaining on anticonvulsant medication (30,562,615). This was recently reviewed (329), and TLE, especially with a prepubertal onset, treated with liver enzyme–inducing AED, appear to be particularly associated with the development of hyposexuality. Many workers have found that patients do not complain of their hyposexuality, especially if their seizures began prepubertally (73,529,562). Thus, physicians and other health care workers should discuss sexuality when interviewing patients because sexual dysfunction, whether physiologically or psychologically caused, may be treatable (511).

DISORDERS OF IMPULSE CONTROL

Irritability and Aggression

Various studies have investigated whether epilepsy is more common in violent and aggressive people. Riley and Niedermeyer (458) reviewed the EEG of more than 200 patients who had been referred for problems with aggression or bursts of anger. Less than 7% had abnormal EEG, and in none were the abnormalities epileptiform. The prevalence of epilepsy among prisoners has been found to be as many as four times that of the general population (231,622). Several reasons for this have been suggested: There may have been differential sentencing for epilepsy, people with disordered brain function are more liable to be imprisoned, prisoners are often from socioeconomic groups III and IV in whom there is also an increased prevalence of epilepsy, antisocial behavior may result in posttraumatic epilepsy, and, finally, there may be a true increase in criminal activity in epilepsy secondary to reduced self-esteem and social stigma (184,231). An increased rate of epilepsy has not been observed in those convicted of violent crimes compared with those committing nonviolent offenses (232).

Periictal Violence

Treiman (589) classified ictal violence; however, its frequency is unclear and remains controversial, and misconceptions about such aggressive outbursts further stigmatize epilepsy. Ictal aggression usually consists of spontaneous, nondirected, stereotyped behavior; however, there are a few reports of cases

in which violent crimes and arson have been thought to be committed during an ictal or postictal confusional state or automatism (79,98,106,183). An automatism has been defined as "an involuntary piece of behaviour over which an individual has no control. The behaviour itself is usually inappropriate to the circumstances, and may be out of character for the individual. It can be complex, coordinated, and apparently purposeful and directed, though lacking in judgement. Afterwards, the individual may have no recollection, or only partial and confused memory, for his actions" (186). A recent study (295), however, found that well-directed violent behavior was more common during postictal episodes (23%) than during acute interictal (5%) and postictal confusion (1%).

Ictal violence is generally considered to be uncommon. Treiman (588) reviewed violent crimes in which epilepsy was used as a defense. In 26 of 60 cases, the crime was premeditated, in 36 of 57 cases, the episode was provoked by anger, and in only 12 of 50 cases, was there any evidence of amnesia. An international panel of epileptologists reviewed video-EEG studies of suspected aggressive behavior during seizures and concluded that directed ictal aggression was extremely rare (130). However, it is now widely acknowledged that the form an ES takes depends not only on the spread of discharge through the brain but also on the thought content and environment of the patient at the time of the attack. Thus, patients having a seizure in the clinical setting of a recording laboratory would not be expected to show ictal aggression (184).

Interictal Violence

Paroxysmal interictal irritability has been reported in 30% of patients with epilepsy compared with 2% of neurologic or normal controls (143). There have also been reports of aggressive interictal behavior (139), although the most likely explanations for this association are the shared risk factors for epilepsy and aggression, including exposure to violence as a child, male gender, low socioeconomic status, cognitive impairment, and neurologic lesions (260,397,556), or the violence being associated with other psychopathology, such as psychosis (397).

Episodic Dyscontrol

This term has been used to describe patients who manifest paroxysmal outbursts of violent behavior with little or no provocation, although there remains doubt as to whether it should be regarded as a separate entity or categorized as a form of personality disorder such as intermittent explosive disorder (IED). Lucas (359) suggested that such behavior may be considered as one extreme of a continuum including normality.

Bach-y-Rita et al. (19) described 130 patients with explosive violent behavior, 25 of whom had a history of childhood febrile convulsions or adult seizures (seven with TLE) and another 30 with seizurelike episodes consisting of loss of contact with the environment. "Soft" neurologic signs such as general awkwardness in gesture and gait, mixed cerebral dominance, left–right disorientation, clumsiness, and dyslexia, suggestive of neurodevelopmental disorders, are often reported in such patients (375).

Some researchers have questioned whether episodic dyscontrol is a form of a complex seizure originating from the temporal lobes because there are several shared characteristics including violent family backgrounds, low socioeconomic status, and underlying brain damage. Prodromal altered mood states, anticipatory fear, hyperacusis, and derealization along with inaccessibility during the episode, subsequent amnesia, drowsiness, and remorse have also been reported (182,559). However, the violent outbursts are often provoked and EEG abnormalities have not been consistently observed. Drake et al. (161) found nonspecific diffuse or focal EEG slowing in only seven of 23 patients, the remainder having normal records. Fenton (182) observed that the slowing of the EEG tended to occur over the posterior temporal lobes, was attenuated by sleep, and was augmented by overbreathing. In contrast, the EEG changes in TLE tended to be maximal over the anterior temporal lobes, were augmented by sleep, and were not affected by overbreathing.

Further evidence of the association between dyscontrol and epilepsy has been the improvement in outbursts after treatment with anticonvulsant medication noted by many authors (201,211,351,375,408,482,559). Treiman (588) concluded that although he believed that the syndrome existed, it was not epileptic in nature but arose from damage to frontal lobe structures, resulting in impairment of normal inhibitory mechanisms that usually govern social and interactive behavior.

Recent neuroimaging studies have helped to advance knowledge of the pathogenesis of aggression in epilepsy. Tebartz van Elst et al. (570) compared 50 patients with TLE and 25 with and 25 without a history of IED. They found that IED was associated with sinistrality, a history of encephalitis, left-sided or bilateral EEG changes, low IQ (especially verbal scores), and high scores on depression and anxiety inventories. Hippocampal sclerosis was less common in the patients with IED. Although there was no overall difference in amygdala volumes between the groups, 20% of the patients with a history of IED had significant amygdala atrophy (>3 standard deviations below the mean). Using amygdala T2 relaxometry as an indicator of amygdala pathology, there were no group differences; however, 28% of the patients with a history of aggression had diverse pathologies affecting the left temporal lobe such as gliomas and dysembryoplastic neural tumors of the amygdala. The same group (631) performed statistic parametric mapping to the T1-weighted MRI scans of these subjects, after automated segmentation of the cerebral gray matter. They found that patients with TLE and IED had an increase in gray matter

in the left extrahippocampal temporal neocortex. In addition, voxel-by-voxel comparisons revealed a decrease in gray matter in the left frontal lobe in these subjects compared with the other groups. Thus, it may be that the circuitry of aggression involves the amygdala and frontal cortices.

The issue of whether episodic dyscontrol does exist as an entity and its relationship with epilepsy remains unresolved. This has important legal implications, however, because if such violent behavior were considered epileptic, it could be used as an example of an insane automatism in legal defenses.

Drug and Alcohol Abuse

Alcohol use can complicate the management of epilepsy. Alcohol abuse can produce poor seizure control by several mechanisms. These include a stimulant effect of the drug, a withdrawal effect associated with decreasing blood alcohol levels, enhancement of AED metabolism through hepatic enzyme induction, alteration in absorption of AED, and noncompliance with medication (243).

Although there have been no recent studies, there appears to be no evidence that people with epilepsy drink alcohol to excess. Lennox (346) showed that more people with epilepsy had never used alcohol and fewer drank a large amount compared with controls.

A double-blind study assessed the effects of vodka over a 16-week period (269). There was no significant change in seizure frequency, epileptiform activity on EEG, or AED levels, and the authors concluded that social drinking had little adverse effect in epilepsy. Most epileptologists agree that moderate alcohol intake is acceptable (49).

Illicit drug abuse has also been found to be a risk factor for seizure exacerbation, especially in women (164,557). High-dose cocaine appears to pose the highest risk, followed by opioids, and, to a lesser extent, amphetamines, phencyclidine, lysergic acid diethylamide, ecstasy (MNDA), and solvent abuse (164,540,557). The arousal and sleep deprivation associated with the use of these drugs exacerbates the proconvulsant properties (540).

Alcohol and drug abuse, however, is associated with increased rates of depression, DSH, and suicide and, therefore, should be considered when assessing patients with epilepsy and psychiatric comorbidity.

IMPAIRMENT OF COGNITIVE FUNCTION AND MEMORY IN EPILEPSY

Impairment of cognitive abilities and a decline of intellectual function have been commented on as a complication of epilepsy for many years. As with the behavioral problems discussed above, the literature started to accumulate in the mid to late nineteenth century. In recent years, some of the factors underlying neuropsychological deficits, as well as the type of psychological deficits associated with various seizure types, have been further clarified.

Several investigators have noted that patients with symptomatic epilepsy are more likely to have impaired intellect than those with epilepsy of an unknown cause (80). Preexisting brain damage, although an important variable, does not, however, completely explain the neuropsychological deficit. There appears to be some "epilepsy factor" that contributes to cognitive change in addition to the structural lesion (311).

Of the seizure variables, age at onset, duration of epilepsy, seizure type, and seizure frequency have been examined. Most investigators report an early age at onset to have a poorer prognosis with regard to intellectual abilities. Patients with generalized seizures tend to show more deficits of attention and concentration compared with patients with focal seizures. The latter, particularly with seizures arising from the temporal lobes, are more likely to show memory impairments. Patients with generalized absence seizures show impaired cognitive function for the duration of the EEG spike-wave abnormality, although generally patients with this form of seizure show minimal interictal impairment. If absences are frequent, impairment of performance in the classroom setting may lead to educational underachievement. Binnie (58) pointed out how subclinical focal discharges may also have clinical significance, showing transient cognitive impairment on selective tasks correlating with brief interictal EEG focal discharges.

Seidenberg et al. (512) compared a group of patients longitudinally whose seizure frequency decreased during a test–retest interval on psychological tests with another group in which it remained unchanged or was increased. Those with a decreasing seizure frequency showed improvements in intellectual quotients. Dodrill (152) also examined the long-term effect of seizures on psychological tasks and noted that a life-long history of more than 100 individual convulsions is associated with decreased functioning in a variety of areas.

In some patients, it is reasonable to talk of a dementia of epilepsy. These patients show a cognitive deterioration, which is thus an acquired intellectual deficit, although this is not progressive in the way that parenchymatous degenerative dementias are. Patients halt or arrest in the progression of the intellectual decline. Identifying these patients prospectively is at present impossible, but retrospective studies (592,601) have suggested that generalized tonic-clonic seizures with recurrent head injury and the prescribing of some anticonvulsant drugs, notably PHT and PRM, are somehow associated.

The role of PHT in provoking an encephalopathy has been discussed for sometime, since the early descriptions of Dilantin dementia by Rosen (474). This affects only a minority of patients, and the possibility that it may be linked with metabolic disturbances, e.g., folate deficiency, has been suggested. It seems that some patients, particularly those

with learning disability, severe intractable epilepsy, and recurrent head injuries, may be most susceptible to this kind of encephalopathy, which may be partially reversible on withdrawal of either polytherapy or the PHT.

The effects of anticonvulsant drugs on cognitive function have been an area of intense investigation in recent years. Extensive reviews are available (600), and only a brief summary is given here. In patients on polytherapy, rationalization with a diminished burden of anticonvulsant prescriptions improves cognitive function over a wide range of cognitive abilities (579).

There has been a debate as to whether differences between individual anticonvulsant drugs exist. Generally, data favor CBZ, VPA, and many of the newer agents and emphasize more cognitive impairments with PHT and PB (538,580,582). A controversy that has recently emerged is whether the cognitive impairments associated with anticonvulsant drugs are a reflection of purely motor impairments rather than of some effect on higher cognitive function. Data bearing on this issue are available in the literature (581) and support the view that most anticonvulsants may bring about some motor slowing, which is reflected in cognitive tasks. However, differences can be discerned, particularly between CBZ and PHT in terms of higher cognitive tasks, the influences favoring CBZ (162). Furthermore, two studies have shown no impairment of cognitive function after 4 to 12 months with oxcarbazepine (OXC), the 10-keto analogue of CBZ (525).

Information on the cognitive effects of the new drugs is rather limited. The only drug to have been investigated in any detail is VGB, and several studies suggest that it appears to have no significant influence on cognitive abilities. If a trend can be detected, it is toward a psychotropic effect with some improvements (155,214,368,443). Furthermore, a small study comparing VGB and placebo in healthy volunteers failed to find any differences, using tests of attention and concentration (575).

Studies of GBP, LTG, and TGB have not shown much in the way of impairments, whereas TPM seems to be associated with some memory and cognitive problems, in part dose related. Several authors have described a speech problem with TPM, but its characteristics have not been well defined.

Patients with epilepsy frequently complain of memory difficulties. Although often this is linked with problems of concentration and attention and indeed may therefore not be a memory defect *per se,* it is clear that memory may be affected, particularly in patients with temporal lobe abnormalities. Most of the work in this area has been undertaken in patients undergoing temporal lobectomy, in whom careful testing of memory function before surgery is mandatory, and deficits may occur after removal of the offending temporal lobe. In other patients, however, memory function is usually improved by the reduction in seizures. The role of anticonvulsant drugs in exacerbating memory problems has yet to be clearly defined because the impairments associated with anticonvulsants tend to reflect more on concentration, attention, and psychomotor abilities than on memory function *per se.* Recent studies found that subjective complaints of memory decline after epilepsy surgery are infrequent and are associated with depression (497) and neuroticism (102).

PSYCHIATRIC COMORBIDITY ASSOCIATED WITH THE TREATMENT OF EPILEPSY

Some of these issues have been discussed in previous sections. We now give a more detailed review of the more salient issues.

Psychiatric Disorders and Antiepileptic Treatments

The behavioral toxicity profile of most drugs is not simple. A particular drug may influence one aspect of behavior positively and another negatively. The physician needs to be aware of these distinctions and also aware that suppressing seizures at the expense of a deterioration in the patient's behavior is not necessarily appropriate clinically and can provoke a great deal of havoc for the patient both in his/her family and as well as vocationally.

Why anticonvulsants should have adverse or beneficial effects on mood and behavior is unclear, but they do differ from each other structurally and therefore can be postulated to have markedly differing neurophysiologic and neurochemical effects. CBZ is related to the tricyclic drugs, and both CBZ and VPA powerfully suppress limbic kindled seizures (439). These data suggest that they have a different effect at an anatomic level to, for example, PHT or PB.

Several anticonvulsants, particularly those that appear to be associated with adverse effects, such as PB and PHT, lower serum folate levels. LTG has not been found to alter serum and red blood cell folate concentrations (492). Folic acid is a crucial CNS factor in various enzymes that link, among other things, with the metabolism of monoamines. Certainly, patients with more severe psychopathology have the lowest folate levels when this is measured.

Another possibility is the differential effect of CBZ and PB on serum tryptophan. This is a precursor of serotonin, and serum tryptophan levels are associated with the regulation of mood. Patients with epilepsy on CBZ monotherapy showed increased tryptophan levels, whereas patients on polytherapy or taking PB show decreased levels.

Older Anticonvulsants

Several of the older anticonvulsants such as the barbiturates and benzodiazepines have predominantly GABA-ergic effects and are therefore sedating (306).

Barbiturates

PB has been associated with depression in adults (465,539). Dodrill (153) reviewed 90 studies in which the behavioral effect of PHT, barbiturates, CBZ, and VPA was assessed in patients and normal volunteers. Barbiturates were most clearly associated with negative behavioral changes including depression. Adverse effects of anticonvulsant drugs are often seen in children. Essentially these include the provocation of conduct disorder or the development of hyperactivity syndrome, similar phenomenologically to attention deficit/hyperactivity disorder. Brent et al. (85) found a higher prevalence of both depression (40% vs. 4%) and suicidal ideation (47% vs. 4%) in adolescents and children taking PB compared with CBZ. Although the drug most frequently implicated is PB, similar responses have been shown after some benzodiazepines (e.g., clonazepam) and VGB. It is of interest that all these drugs have some action on the GABA-benzodiazepine receptor. Some underlying biologic mechanism may be suggested, which requires further investigation.

Cognition. PB has been associated with dose-dependent adverse cognitive effects, especially in children. One study found that children who had been treated for 2 years with PB after febrile convulsions had a seven-point lower IQ than children given placebo. Furthermore, a five-point difference persisted 6 months after discontinuation of the PB (180).

Benzodiazepines

Benzodiazepines have been associated with CNS-related side effects such as somnolence but are also reported to cause depression, hyperactivity, irritability, and aggression (306). One report found clonazepam but not clobazam to cause behavioral problems in children, especially those with structural brain damage (115).

Phenytoin. PHT acts by affecting use-dependent sodium channels, and it has little GABA-ergic or antiglutaminergic action. However, it is associated with many neuropsychiatric adverse events such as somnolence, psychomotor slowing, hyperactivity, irritability, and psychosis (306).

Carbamazepine. Positive behavioral changes have been associated with CBZ (decreased anxiety and depression). Rodin and Schmaltz (469) used the BFI to examine 148 patients with epilepsy, 36 of whom had temporal lobe EEG abnormalities. They examined the relationships between serum anticonvulsant levels and scores on the rating scale and found no relationships for any anticonvulsants except CBZ. A significant inverse correlation was noted with total sum, elation, philosophical interests, sense of destiny, altered sexuality, and hypergraphia. These data suggest that some positive effect on interictal personality syndrome may occur when patients on alternative medications are prescribed CBZ, although in the absence of further studies, it is difficult to be dogmatic. A double-blind, prospective study compared the efficacy and toxicity of PHT, PRM, PB, and CBZ; at 1 year, behavioral toxicity scores were highest for those on PHT and lowest for those on CBZ (648). Dalby (124) reported a psychotropic effect in approximately one-half of the patients treated with CBZ, which has also been shown to be associated with less depression than PRM (471) and PHT (14). Although psychosis has been reported in patients with epilepsy taking CBZ, other studies have shown an improvement in psychotic symptoms, even in patients without epilepsy (306).

CBZ has also been shown to be efficacious as an antidepressant in patients without epilepsy and prophylactic in the control of manic-depressive illness (172,189,438).

Cognition. CBZ is associated with fewer dose-dependent adverse cognitive effects than barbiturates (306).

Valproic Acid

The main adverse effects associated with VPA are CNS (somnolence and tremor), gastrointestinal (nausea), and weight gain (306). Rarely, encephalopathy has been described (381,643). Positive mood and behavioral changes have been associated with VPA (441), and it does not appear to be associated with psychosis (306).

VPA has also been shown to be efficacious as an antidepressant in patients without epilepsy and prophylactic in the control of manic-depressive illness (23,172,189,367).

Vigabatrin

VGB is an irreversible inhibitor of GABA transaminase and thus increases levels of the inhibitory neurotransmitter GABA. In large studies, 2% to 4% discontinued the AED because of severe behavioral disturbances such as depression, anxiety, and psychosis (227). This drug appears to be particularly associated with depression, developing in as many as 12% of patients (350,493). The depression typically occurs within a few weeks of the drug being introduced or after dose increments. A history of psychiatric disturbance has been found to be a major risk factor (459,576). Mood changes, however, were not found in a small placebo-controlled study in normal volunteers (575).

Psychosis has also been reported, with incidence rates varying from 1% to 12% (506). However, a metaanalysis of ten double-blind, placebo-controlled studies (assessing 717 patients) found an incidence rate for psychosis of 2.5% for VGB compared with 0.3% for placebo (350). The psychotic symptoms respond well to discontinuation of VGB or to neuroleptic treatment (350). Psychosis tends to occur after suppression of seizures (forced normalization) or after withdrawal of the AED (576). In addition, PIP may result from relapse of seizures after initial control, which may be related to tolerance (576). Risk factors for the development of psychosis identified in a retrospective study include a history of psychosis (which may have been related to the use of another AED), right-sided or bilateral EEG focus, and severe, often tonic-clonic seizures (576).

Lamotrigine

LTG inhibits presynaptic voltage-dependent sodium channels, resulting in decreased release of the excitatory amino acids glutamate and aspartate. Efficacy data (453) have shown LTG to be at least as effective in producing seizure freedom as CBZ while being better tolerated (fewer adverse effects and withdrawals from the study). In a large metaanalysis of studies, LTG had side effects similar to those of placebo (383). Several studies have shown improvements in well-being with LTG add-on therapy (50,62,234,407,632). LTG has also been found to produce increased happiness and mastery (perceived internal control) compared with placebo, which was not dependent on change in seizure frequency or severity (541). Monotherapy studies have shown that LTG produced improvements in all the Side Effect and Life Satisfaction Inventory scores (cognition, dysphoria, temper, tiredness, and worry) compared with CBZ (215–217) and improvement in the Side Effect and Life Satisfaction Inventory score for dysphoria compared with PHT (215,552). Psychosis is rarely associated with LTG (306). In addition, LTG has few adverse cognitive effects (306).

There have been few reports of adverse psychological reactions to LTG, although it has been associated with insomnia in some (6.4% of 109) patients (487), and in a study assessing patients with learning disability (177), both improvement and deterioration of behavior were described, although the latter may have been dose related. Recently, LTG has been found to be efficacious in bipolar affective disorder (192), especially in patients with EEG showing temporal slow waves (192).

Gabapentin

GBP has been found to have minimal side effects, not significantly different from those with placebo (383) and moreover has been associated with improvements in well-being in double-blind, placebo-controlled studies (150). Similarly, it has been found to improve mood assessed using the Cornell Dysthymia Rating Scale (236). QOL did not alter in one study when GBP was added to baseline AED (405). Psychosis has rarely been reported (306). The use of GBP in patients with affective disorders (anxiety, panic, bipolar, and schizoaffective disorder) was recently reviewed (192), and it appears to have a role in rapid cycling mood disorders. However, behavioral side effects such as aggression have been observed in children with learning disability treated with GBP (338,561).

Cognition

In add-on studies, GBP has been found to have no effect (337) or to improve cognitive function (the word-reading condition of the Stroop test) (405).

Topiramate

The most common side effects reported in placebo-controlled trials of TPM have been CNS related such as dizziness, somnolence, and fatigue (352,383). The incidence and severity of these adverse events may, however, reflect the rapid titration schedules, which were used when TPM was first introduced.

However, TPM has also been associated with cognitive side effects such as mental slowing, confusion, impaired concentration, and speech and language problems such as word-finding difficulties (352,383). One open study found that more patients reported cognitive, mood, and behavioral problems with TPM than with LTG (515).

More recently, severe psychiatric side effects such as depression (54) and psychosis (54,308,524) have been reported; however, these symptoms tend to resolve quickly with discontinuation of the AED. In several cases, this is reflection of alternative psychosis and forced normalization. Furthermore, there have been case reports of TPM being efficacious in bipolar disorder (109,380).

Tiagabine

TGB acts by blocking the uptake of GABA into neurons and glia. Concern has been expressed, however, that TGB may be associated with the development of depression because, like VGB, it increases the cerebral level of GABA. Intravenous GABA has been found to produce dysphoria and anxiety in both normal volunteers and patients with bipolar affective disorder (419). Although some early reports have documented asthenia, nervousness, and depression as adverse events (88,347), other large studies have not confirmed this (156). In a metaanalysis of double-blind, placebo-controlled studies, TGB had similar rates of somnolence and fatigue as placebo (383). In one study, TGB was added to standard AED therapy with the intention of withdrawing the baseline AED (157). Patients who achieved monotherapy with low-dose (6 mg daily) TGB experienced improvements in mood and psychosocial adjustment. However, patients failing to achieve monotherapy reported adverse mood changes with high-dose TGB (36 mg daily), which was believed to be related to the rapid-titration rate used in the study (157). In the large, open trials assessing 2,185 patients, 1% withdrew because of depression and 1% for nonspecific nervousness, although these symptoms were usually transient, occurring during the titration period, or were reversible with reduction/discontinuation of the drug (88). Psychosis was reported in 2% of 675 patients taking TGB, which was not statistically different from that reported with placebo (1% of 363 patients) (88). Furthermore, in three double-blind add-on studies, psychosis occurred at similar rates with both TGB (0.8%) and placebo (0.4%), despite more than one-third of the patients having a history of psychiatric problems (485). Moreover, adjunctive TGB has been shown to relieve

depression in patients with bipolar disorder (301). However, in most of the studies, patients with ongoing or unstable psychiatric disorders were excluded (156); thus, it is too early to make recommendations about TGB. Furthermore, it has been associated with the development of nonconvulsive SE, which itself may lead to psychosis, and at present caution is advised when prescribing it for patients with a history of psychiatric disorder.

Cognition

A large, double-blind, placebo-controlled add-on study did not show any cognitive decline after 12-week treatment with 16, 32, or 56 mg TGB daily (156). Furthermore, patients achieving monotherapy with high-dose (36 mg daily) TGB experienced improvements in tests of ability (158).

Oxcarbazepine

OXC is the 10-keto analogue of CBZ, and its primary mode of action is, like CBZ, blockade of voltage-sensitive sodium channels. However, it is not metabolized to the epoxide derivative, which is thought to be responsible for the toxic effects of CBZ. The monotherapy studies (in which a total of 462 patients taking OXC were compared with those taking CBZ, VPA, and PHT) were recently reviewed (525). The main adverse events were those of the CNS (fatigue, headache, dizziness, ataxia); however, the mean number of side effects was lower with OXC than with CBZ. Although the drug has not been specifically linked with psychiatric adverse events, during the monotherapy comparative trials (n = 462), there was one report each of psychosis, psychic lability, and suicide attempt (525). Like CBZ, OXC has been shown to have antimanic effects (172).

Levetiracetam

The most recently licensed AED is LEV. Although a derivative of piracetam, its mode of action is unknown. In multicenter, double-blind, placebo-controlled trials (39,527) the major side effects were CNS related, predominantly somnolence and asthenia. Depression was reported in 1.9% of patients taking 1,000 mg and 5.7% taking 2,000 daily of LEV compared with 2.7% of the placebo group (527). In a major placebo-controlled add-on study (120), overall QOL and cognitive functioning improved in patients taking LEV, especially in patients with 50% reduction in seizures.

Withdrawal Effects of Anticonvulsants

Several of the older AED have been associated with psychiatric symptoms when withdrawn (304,306). Withdrawal reactions from barbiturates and benzodiazepines include such symptoms as anxiety, irritability, psychosis, and delirium (306). Furthermore, almost 40% of patients withdrawing from PHT, CBZ, and VPA therapy develop moderate to severe psychiatric symptoms, predominantly anxiety and depression (304). These symptoms tend to start during the final week of tapering the AED and thus are not believed to be a result of withdrawal seizures. The symptoms resolve within 2 weeks of restarting the original AED (304). A prospective study (303) monitoring mood and controlling for life events revealed that patients who withdrew from CBZ therapy (but not PHT or VPA) became more anxious, and this finding was independent of seizures. This was thought to reflect the positive effect that CBZ has on mood. Psychiatric symptoms have also been reported after withdrawal of the newer AED, most commonly VGB (647).

In summary, all the currently available AED are associated with some psychiatric adverse events. Ketter et al. (306) put forth a theory that AED with a predominantly GABA-ergic inhibitory action (barbiturates, benzodiazepines, VPA, VGB, GBP, and TGB) tend to have sedating profiles (somnolence, fatigue, cognitive slowing, and possibly anxiolytic or antimanic effects). These AED may be best suited to patients with baseline symptoms of insomnia, agitation, and anxiety. In contrast, other AED such as LTG (which attenuates glutaminergic excitatory neurotransmission) may have more activating properties (yielding alertness and possibly anxiolytic and antidepressant effects). Thus, LTG may be best suited for patients with baseline symptoms of fatigue, apathy, and depression. The authors speculated that TPM had a mixed profile because of its various modes of action.

PSYCHIATRIC DISORDERS ASSOCIATED WITH EPILEPSY SURGERY

Temporal lobectomy is often considered for patients with medically intractable epilepsy. There have been many reports of psychiatric disturbance associated with the surgery, and large series have been reported from the Maudsley Hospital in London (93,565). The main psychiatric disorders developing after surgery are depression and psychosis, and these were discussed in previous sections. The major psychiatric sequelae are summarized here.

The prevalence of psychiatric disorders postoperatively varies from series to series but is clearly a major cause of postoperative morbidity and mortality. Most studies have found that one-third to one-half of patients develop *de novo* psychiatric symptoms postoperatively (15,219,460). Wilson et al. (630) found that the most common reason for postoperative readmission was anxiety, depression, and/or PIP, and Anhoury et al. (15) found that 10% of their sample needed hospitalization for psychiatric problems. Moreover, high rates of suicide are reported in the first few months after surgery, even if the operation was successful. Suicide was reported in 2.4% of the Maudsley series (567), which accounted for 22% of the postoperative deaths. Harris and Barraclough (237) performed a metaanalysis of follow-up

studies to assess suicide rates in psychiatric and neuropsychiatric conditions. They found the highest suicide rate in patients with surgically treated TLE who had an increased risk by a factor of 80 with a standardized mortality rate of 8,750. In patients who had no presurgical psychiatric problems, postoperative IDD has been reported in 42% (76). Furthermore, 75% of the patients who continued to have seizures developed psychiatric problems, usually 5 to 8 months after surgery. Conversely, only 18% of those rendered seizure free developed a *de novo* psychiatric disorder, and these tended to occur in the immediate postoperative period. Twenty percent of those with presurgical symptoms of IDD were able to discontinue their antidepressant medication 6 to 18 months after surgery; however, 80% of these patients had been rendered seizure free. Blumer et al. (76) also found that preexisting IDD may worsen in the immediate postoperative period, even in those who were seizure free. Ring et al. (460) found in a prospective study that one-half of the patients who had no preoperative psychopathology developed symptoms of anxiety and depression 6 weeks after temporal lobe surgery. At 3 months postoperatively, the anxiety tended to have diminished; however, the depression tended to persist.

One of the major problems in determining the prevalence of postoperative psychiatric problems is the lack of an appropriate control group. Many of the published studies used patients who were found to be unsuitable for surgery as controls; however, because the most common reasons for not operating are difficulty in localizing a focus and extratemporal/bilateral foci, these are not appropriate controls. In short-term studies, "waiting list" patients can be used, but in the majority of prospective studies, patients have to act as their own controls.

Other than depression and psychosis, the main psychiatric disorders that can develop *de novo* after epilepsy surgery are anxiety (64,460), emotional lability (increased susceptibility to sudden outbursts of laughter or tears) (460), and NES (319,426).

Prognostic Factors for Psychiatric Outcome

Predictors of a good postoperative outcome with respect to psychiatric and psychosocial functioning have been determined (Table 39.17) (15,135,136,449,564). The most important of these are discussed below.

Seizure Outcome

Most studies have found that seizure freedom is the main factor determining good psychiatric and psychosocial outcome (76,136,258,284,302,370,450,451,516,628). Furthermore, many studies have shown a deterioration in the psychiatric and social status of patients after temporal lobectomy, especially if there is only partial improvement in their seizures (256,263,284,514,555). Despite this,

TABLE 39.17. PROGNOSTIC FACTORS FOR PSYCHIATRIC OUTCOME OF EPILEPSY SURGERY

Good Outcome	Bad Outcome
Complete seizure control or 75%–90% reduction	Severe preoperative psychiatric disorder, especially psychosis or nonepileptic seizure
No or mild preoperative psychopathology	Unrealistic expectations of surgery
Good family/social support	Size of resection (emotional lability)
Good preoperative relationships (work and socially)	Bilateral independent spikes at preoperative telemetry
Age at surgery <30 yr	Substance abuse (exacerbating seizures)
Good schooling and higher preoperative IQ	Left-sided operation (depression and anxiety)
Learned resourcefulness	Right-sided operation (psychosis)
Good self-perceived quality of life	Preoperative neuroticism
	Multiple seizure types
	Preoperative neurologic deficits
	Family history of psychiatric illness or seizures

little research has looked at this surprising and important aspect of epilepsy surgery. Even postoperative freedom from seizures can be associated with depression and behavioral problems (64,425,555), a possible reflection of forced normalization. However, Ferguson and Rayport (191) discussed this in 1965, suggesting that it was the abrupt (surgical) removal of a psychiatrically significant experience, *the seizure,* rather than the removal of temporal lobe tissue that was of major psychological importance. They postulated that the postoperative seizure-free state permits assessment of the extent to which the seizure disorder has been incorporated into the adaptive patterns of the patient. They proposed that epilepsy might afford some patients psychological benefits: The illness may shield them from any hostile feelings that friends and relatives may have toward the patient. Moreover, seizure freedom has in some cases resulted in divorce (64), especially in women (107). This may be because the marriage was based on the roles of patient and caregiver, or, conversely, a spouse may feel less guilty about ending a relationship if the partner is no longer ill. In addition, the patient may use epilepsy as an excuse for failures in personal, social, and occupational areas. It may even allow inappropriate behavior to be accepted. After the operation, the family may expect a radical change in personality and may become critical of the patient's behavior and weaknesses. This may result in the patient becoming depressed and even wishing that surgery had never been performed. Somatization and the development of NEAD may occur. Recent work has, however, focused on assessing patients' expectations of surgery and the ability to adjust to the demands of postoperative life, in particular discussing the "burden of normality" (628). Several recent studies compared the presurgical expectations of surgery with postoperative outcome and QOL (31,628). Not surprisingly, patients who perceived their operation as successful tended to be seizure free, have no worsening of preoperative anxiety or depression, and have made realistic postoperative plans. In addition, the ability to discard the sick role was important, as was the perception of the mother to be supportive rather than overmanagerial (628). Preoperative counseling and postoperative rehabilitation may help patients and their families adjust to life without seizures (577,607) and should be an integral part of the assessment and management program.

Some authors, however, have not demonstrated any change (improvement or deterioration) in psychiatric status postoperatively, regardless of outcome with respect to seizure control (312).

Preoperative Psychiatric Status

Preoperative emotional distress including depression and "neuroticism" (anxiety and hostility) have been shown to be associated with poor postoperative psychological status (135).

Kanemoto et al. (294) determined that preoperative episodes of PIP, especially in patients with an aura of fear, were associated with serious postoperative psychiatric morbidity. An additional risk factor in their work appeared to be having a left-sided operation. Depression, mania, or both tended to start within 1 to 2 months of temporal lobectomy. In all cases, they resolved; the manic episodes resolved after approximately 4 months, although the depressed episodes could last as long as 15 months and some required treatment with antidepressants. None recurred.

In addition, Anhoury et al. (15) found that patients having seizures postoperatively tended to have a psychiatric history, especially of anxiety disorders, which may be explained by bilateral pathology increasing the likelihood of both psychiatric disorders and continued seizures.

Psychosocial Factors

Good preoperative psychosocial adjustment is associated with a favorable outcome of surgery from the psychiatric point of view (258,473). Factors associated with being in employment after surgery includes having full-time work in the year before surgery (451).

Having good social support, learned resourcefulness (taking active responsibility for one's health), and a history of successful adjustment to similar circumstances have been found to be predictive of good psychological outcome (135). The same researchers (136) also showed that postoperative depression was more likely to occur in patients with a poor preoperative emotional adjustment, especially those who are older, have a preoperative neurologic deficit, have multiple seizure types, and have a family history of psychiatric disorder or seizures.

Side of Surgery

Various studies have implicated the side of surgery as being related to postoperative psychiatric complications. Left-sided surgery has been associated with anxiety (64,460) and depression (93,96,294), whereas right-sided operations have been linked with psychosis, especially postictal (371).

Psychiatric Contraindications to Surgery

These vary from center to center and have been widely debated (185,499). Although some centers exclude patients with psychosis (219), others have not thought that this should be a reason to exclude patients. The main contraindication is any psychiatric disorder of such severity that the patient could not tolerate the preoperative investigations or be safely managed on a surgical ward. The coexistence of NES with epilepsy is of concern because several studies have found that the NES persist postoperatively, and this leads to considerable clinical confusion and often to a failure of the patients to progress psychosocially. Risk factors for postoperative NES include longer duration of epilepsy, lower IQ, inadequate coping skills, and a tendency to somatize (319).

Patients who are investigated for epilepsy surgery but are refused the operation because of inability to localize the focus or inoperable/bilateral foci are often used as a control group in investigating the effect of surgery. However, it could be expected that these patients will suffer from a sense of failure or loss that they were not suitable for surgery and thus could be expected to develop high levels of morbidity, especially short term. Surprisingly, these patients have received little or no attention. Interestingly, Kellett et al. (302) found no difference between these patients and those who had more than ten seizures per year postoperatively. Both groups would have had high expectations during the time of presurgical assessment that were not realized.

PRESURGICAL ASSESSMENT

From the above, it can be seen that many psychiatric disorders may occur in epilepsy, and most appear to be more common in those with intractable focal epilepsy. Because this is the group that may be considered for epilepsy surgery, it is important to perform a careful evaluation of the preoperative mental state so that the full impact of surgery on the patient's life can be assessed and, if possible, predicted. Several studies have shown that the preoperative psychosocial status (such as neuroticism measured using the MMPI-2) can predict postoperative psychosocial adjustment, regardless of seizure outcome (133,258,473). In addition, a history of anxiety disorders may predict poor outcome with respect

to seizure control (15). Multicenter studies researching the psychiatric sequelae of surgery may enable patients at particular risk of adverse events to be identified so that treatment, support, and counseling may be planned. All patients being assessed for epilepsy surgery should have a full neuropsychiatric assessment, preferably by a psychiatrist, using a standardized interview that would generate internationally accepted diagnoses such as the World Health Organization's *International Classification of Diseases,* 10th Revision, *DSM-IV (DSM, Fourth Edition),* or follow the International League Against Epilepsy commission guidelines. A psychosocial history including personal and family psychiatric history should also be documented along with some form of personality assessment and record of alcohol and illicit drug use. This interview could be supplemented by additional self-report questionnaires that may be administered by other members of the assessment team. Fenwick (185) also suggested that a detailed assessment of sexual function should be performed including hormonal profiles. Patients may have unrealistic expectations of surgical outcome, in terms of both seizure control and effect on psychosocial functioning (31). This should also be assessed preoperatively because psychological input may help to unify patient and physician expectations of surgery and thus enable the patient to make a fully informed consent to the operation and indicate the degree of postoperative psychological support and counseling that may be required for successful adjustment.

Because postoperative psychopathology is common, especially in the first 2 months after surgery, patients and their families should be warned of the risk, and facilities for psychiatric assessment should be routinely available so that the emergent psychiatric disorders can be identified and treated accordingly. Some studies have found that psychiatric disorders may not appear until 2 to 4 years after surgery (101), and, conversely, it may take as long as 18 months for preexisting problems to resolve (76). Therefore, a long follow-up period would produce a more realistic picture of the psychiatric sequelae associated with surgery (135).

OPTIMIZATION OF PSYCHOSOCIAL POTENTIAL

The concept of QOL has gained popularity in recent years, although its definition is problematic. It is important to recognize that epilepsy involves more than seizures. Improving psychosocial potential requires attention to the many aspects of the patients' lives; their own internal concept of their QOL has thus become an important management goal.

The World Health Organization (624) proposed three levels at which disease can affect an individual: impairment, disability, and handicap. Impairment relates more to abnormalities of body structure and organ systems, and disability is seen as the impact of the illness on a person's functional abilities and activity levels. It is the definition of handicap that most closely relates to QOL, defining disadvantage owing to disease or treatment that individuals experience in relation to their peers and others.

A review of the literature on QOL reveals several life domains or areas thought to be essential to determining QOL in epilepsy (304), covering physical, cognitive, affective, social, and economic aspirations of patients. The word aspiration is used here to emphasize an important concept in QOL research. Thus, future expectations are a major component of perceived QOL, actual abilities being less important than a discrepancy between the patient's position as it is now and his/her expected situation.

The diagnosis of epilepsy brings with it many psychosocial problems, including stigmatization, social isolation, psychological problems, and education and employment difficulties (578). Societal attitudes play a major role in determining QOL of patients with epilepsy. Discrimination and nonacceptance of patients, unfortunately, are still quite common. There are obvious ways to enhance the psychosocial adjustment of these patients. At the outset, when patients are diagnosed, they should explore the concept of epilepsy and discuss their fears and myths about the condition. It is important to provide constant support from known individuals who will look after them as they learn to live with epilepsy.

Measuring QOL and the scientific study of improvements of QOL with various treatments (e.g., monotherapy vs. polytherapy or surgery or even the introduction of new drugs) is in its infancy. There is a lack of appropriate measures for patients with epilepsy, with existing QOL scales being developed for people with other specific physical illnesses, such as cancer.

Using a recently developed QOL schedule based on repertory grid techniques, Kendrick and Trimble (304) explored discrepancies between the way patients view their world now and how they would like to live. This is a measure of the discrepancy between their current situation and their aspirations, which is central to our concept of QOL. They found using such techniques that apparently, although seizures are undeniably significant to some patients in determining QOL, this is not true for many others. Cognitive and emotional factors rate highly, as do social isolation and lack of friendships. These data highlight the need for a broader management of epilepsy, again recognizing that the condition involves much more than having seizures.

REFERENCES

1. Abbott RJ, Browning M, Davidson DLW. Serum prolactin and cortisol concentrations after grand-mal seizures. *J Neurol Neurosurg Psychiatry* 1980;43:163–167.
2. Aboukasm A, Mahr G, Gahry BR, et al. Retrospective analysis of the effects of psychotherapeutic interventions on outcomes of psychogenic nonepileptic seizures. *Epilepsia* 1998;39:470–473.
3. Adachi N, Onuma T, Nishiwaki S, et al. Inter-ictal and post-ictal psychoses in frontal lobe epilepsy: a retrospective comparison with psychoses in temporal lobe. *Seizure* 2000;9:328–335.
4. Aldenkamp AP, Mulder OG. Behavioural mechanisms involved in pseudoepileptic seizures: a comparison between patients with epileptic seizures and patients with pseudoepileptic seizures. *Seizure* 1997;6:275–282.
5. Alheid SF, Heimer L. New perspectives in basal forebrain organisation of special relevance for neuropsychiatric disorders. *Neuroscience* 1988;27:1–39.
6. Alper K, Devinsky O, Perrine K, et al. Nonepileptic seizures and childhood sexual and physical abuse. *Neurology* 1993;43:1950–1953.
7. Altshuler LL, Devinsky O, Post RM, et al. Depression, anxiety, and temporal lobe epilepsy. Laterality of focus and symptoms. *Arch Neurol* 1990;47:284–288.
8. Altshuler L, Rausch R, Delrahim S, et al. Temporal lobe epilepsy, temporal lobectomy, and major depression. *J Neuropsychiatry Clin Neurosci* 1999;11:436–443.
9. American Psychiatric Association. *Diagnostic and statistical manual of mental disorders, revised third edition.* Washington DC: American Psychiatric Association, 1987.
10. Aminoff MJ, Simon RP, Wiedemann E. The hormonal responses to generalised tonic-clonic seizures. *Brain* 1984;107:569–578.
11. Andermann LF, Savard G, Meencke HJ, et al. Psychosis after resection of ganglioglioma or DNET: evidence for an association. *Epilepsia* 1999;40:83–87.
12. Andersen AR. Single photon computerized tomography in temporal lobe epilepsy. In: Dam M, Gram L, eds. *Comprehensive epileptology.* New York: Raven Press, 1990:375–383.
13. Anderson VE, Andermann E, Hauser WA. Genetic counselling. In: Engel J, Pedley TA, eds. *Epilepsy: a comprehensive textbook.* New York: Lippincott–Raven, 1997:225–232.
14. Andrewes DG, Bullen JG, Tomlinson L, et al. A comparative study of cognitive effects of phenytoin and carbamazepine in new referrals with epilepsy. *Epilepsia* 1986;26:128–134.
15. Anhoury S, Brown RJ, Krishnamoorthy ES, et al. Psychiatric outcome after temporal lobectomy: a predictive study. *Epilepsia* 2000;41:1608–1615.
16. Annegers JF, Hauser WA, Elveback LR. Remission of seizures and relapse in patients with epilepsy. *Epilepsia* 1979;20:729–737.
17. Antebi D, Bird J. The facilitation and evocation of seizures. *Br J Psychiatry* 1992;160:154–164.
18. Arnold LM, Privatera MD. Psychopathology and trauma in epileptic and psychogenic seizure patients. *Psychosomatics* 1996;37:438–443.
19. Bach-y-Rita G, Lion JR, Climent CE, et al. Episodic dyscontrol: a study of 130 violent patients. *Am J Psychiatry* 1971;127:1473–1478.
20. Baker GA, Nashef L, van Hout BA. Current issues in the management of epilepsy: the impact of frequent seizures on cost of illness, quality of life and mortality. *Epilepsia* 1995;38[Suppl 1]:S1–S8.
21. Baker GA, Jacoby A, Chadwick DW. The associations of psychopathology in epilepsy: a community study. *Epilepsy Res* 1996;25:29–39.
22. Baldessarini RJ, Stephens JH. Lithium carbonate for affective disorders. *Arch Gen Psychiatry* 1970;22:72–77.
23. Balfour JA, Bryson HM. Valproic acid. A review of its pharmacology and therapeutic potential in indications other than epilepsy. *CNS Drugs* 1994;2:144–173.
24. Barczak P, Edmunds E, Bets T. Hypomania following complex partial seizures: a report of three cases. *Br J Psychiatry* 1988;152:137–139.
25. Barr WB, Ashtari M, Bilder RM, et al. Brain morphometric comparison of first-episode schizophrenia and temporal lobe epilepsy. *Br J Psychiatry* 1997;170:515–519.

26. Barraclough B. Suicide and epilepsy. In: Reynolds FH, Trimble MR, eds. *Epilepsy and psychiatry.* Edinburgh: Churchill Livingstone, 1981:72–76.

27. Bash KW, Mahnig P. Epileptiker in der psychiatrischen Klinik. Von der Daemmerattacke zur Psychose. *Eur Arch Psychiatr Neurol Sci* 1984;234:237–249.

28. Bauer J, Stefan H, Schrell U, et al. Serum prolactin concentrations and epilepsy. A study which compares healthy subjects with a group of patients in presurgical evaluation and circadian variations with those related to seizures. *Eur Arch Psychiatry Clin Neurosci* 1992;241:365–371.

29. Bauer J, Elger CE, Hefner G, et al. Psychogenic seizures provoked by suggestion: an analysis of ictal characteristics in 100 patients documented by video-telemetry. *Epilepsia* 1993;34[Suppl 2]:147.

30. Bauer J, Stoffel-Wagner B, Flügel D, et al. Serum androgens return to normal after temporal lobe epilepsy surgery in men. *Neurology* 2000;55:820–824.

31. Baxendale SA, Thompson PJ. "If I didn't have epilepsy...": patient expectations of epilepsy surgery. *J Neurol* 1996;9:274–281.

32. Bazil CW, Kothari M, Luciano D, et al. Provocation of non-epileptic seizures by suggestion in the general population. *Epilepsia* 1994;35:768–770.

33. Bazil CW, Walczak TS. Effects of sleep and sleep stage on epileptic and nonepileptic seizures. *Epilepsia* 1997;38:56–62.

34. Bear DM, Fedio P. Quantitative analysis of interictal behavior in temporal lobe epilepsy. *Arch Neurol* 1977;34:454–467.

35. Bear D, Levin K, Blumer D, et al. Interictal behaviour in hospitalised temporal lobe epileptics: relationship to idiopathic psychiatric syndromes. *J Neurol Neurosurg Psychiatry* 1982;45:481–488.

36. Becú M, Becú N, Manzur G, et al. Self-help epilepsy groups: an evaluation of effect on depression and schizophrenia. *Epilepsia* 1993;34:841–845.

37. Bell AJ, Cole A, Eccleston D, et al. Lithium neurotoxicity at normal therapeutic levels. *Br J Psychiatry* 1993;162:689–692.

38. Ben-Menachem E. Vagal stimulation for treatment of refractory partial epilepsy: an overview of clinical results. *Seizure* 1992;1[Suppl A]:S21/4(abst).

39. Ben-Menachem E, Falter U, and the European Levetiracetam Study Group. Efficacy and tolerability of levetiracetam 3000 mg/d in patients with refractory partial seizures: a multicenter double-blind, responder-selected study evaluating monotherapy. *Epilepsia* 2000;41:1276–1283.

40. Benbadis SR, Lancman ME, King LM, et al. Preictal pseudosleep: a new finding in psychogenic seizures. *Neurology* 1996;47:63–67.

41. Benbadis SR, Tatum WO IV, Murtagh FR, et al. MRI evidence of mesial temporal sclerosis in patients with psychogenic nonepileptic seizures. *Neurology* 2000;55:1061–1062.

42. Benson DF. The Geschwind syndrome. In: Smith DB, Treiman DM, Trimble MR, eds. *Neurobehavioral problems in epilepsy.* New York: Raven Press, 1991:411–421.

43. Benson DF, Hermann BP. Personality disorders. In: Engle J Jr, Pedley TA, eds. *Epilepsy: a comprehensive textbook.* New York: Lippincott–Raven, 1997:2065–2070.

44. Berg AT, Shinnar S. The risk of seizure recurrence following a first unprovoked seizure: a quantitative review. *Neurology* 1991;41:965–972.

45. Bergen D, Daugherty S, Eckenfels E. Reduction of sexual activities in females taking antiepileptic drugs. *Psychopathology* 1992;25:1–4.

46. Bergen D, Ristanovic R. Weeping as a common element of pseudoseizures. *Arch Neurol* 1993;50:1059–1060.

47. Betts TA. A follow-up study of a cohort of patients with epilepsy admitted to psychiatric care in an English city. In: Harris P, Mawdsley C, eds. *Epilepsy: proceedings of the Hans Berger Centenary Symposium.* 1974:326–338.

48. Betts TA. Depression, anxiety and epilepsy. In: Reynolds EH, Trimble MR, eds. *Epilepsy and psychiatry.* Edinburgh: Churchill Livingstone, 1981.

49. Betts TA. Neuropsychiatry. In: Laidlaw J, Richens A, Chadwick D, eds. *A textbook of epilepsy.* Edinburgh: Churchill Livingstone, 1993:397–457.

50. Betts T. Lamotrigine in the context of non-pharmacological therapies for epilepsy. *Rev Contemp Pharmacother* 1994;5:141–146.

51. Betts T, Bowden S. Diagnosis, management and prognosis of a group of 128 patients with non-epileptic attack disorder. Part I. *Seizure* 1992;1:19–26.

52. Betts T, Bowden S. Diagnosis, management and prognosis of a group of 128 patients with non-epileptic attack disorder. Part II. Previous childhood sexual abuse in the aetiology of these disorders. *Seizure* 1992;127–32.

53. Betts T, Fox C, MacCallum R. Using olfactory stimuli to abort or prevent seizures: Countermeasures or cue-controlled arousal manipulation? Is there something special about smell? *Epilepsia* 1995;36[Suppl 3]:S25.

54. Betts T, Smith K, Khan G. Severe psychiatric reactions to topiramate. *Epilepsia* 1997;38[Suppl 3]:S64.

55. Bhatia M, Sinha PK, Jain S, et al. Usefulness of short-term video EEG recording with saline induction in pseudoseizures. *Acta Neurol Scand* 1997;95:363–366.

56. Biersack HJ, Reichmann K, Winkler C, et al. ^{99}TC-labelled hexamethylpropyleneamine oxime photon emission scans in epilepsy. *Lancet* 1985;2:1436–1437.

57. Bilikiewicz A, Matkowski K, Przybysz K, et al. Untersuchungen zur Epidemiologie und Psychopathologie der zwischen 1976 und 1980 in der Woiwoidschaft Bydgoscz registrierten Epileptiker. *Psychiatr Neurol Med Psychol (Leipzig)* 1988;40:9–15.

58. Binnie CD. Monitoring seizures. In: Trimble MR, Reynolds EH, eds. *What is epilepsy?* Edinburgh: Churchill Livingstone, 1986:82–87.

59. Binnie CD. Electroencephalography. In: Laidlaw J, Richens A, Chadwick D, eds. *A textbook of epilepsy.* Edinburgh: Churchill Livingstone, 1993:277–348.

60. Binnie CD. Non-epileptic attack disorder. *Postgrad Med J* 1994;70:1–4.

61. Birbaumer N. Application of learnt cortical control to seizure behavior. *Seizure* 1992;1[Suppl A]:25/4(abst).

62. Bisgaard C, Dalby M, Mai J. Lamictal as add-on antiepileptic drug in 210 patients with resistant epilepsy. *Epilepsia* 1994;35[Suppl 7]:S98.

63. Blackwood DHR, Cull RE, Freeman CPL, et al. A study of the incidence of epilepsy following ECT. *J Neurol Neurosurg Psychiatry* 1980;43:1098–1102.

64. Bladin PF. Psychosocial difficulties and outcome after temporal lobectomy. *Epilepsia* 1992;33:898–907.

65. Blanchet P, Frommer GP. Mood change preceding epileptic seizures. *J Nerv Ment Dis* 1986;174:471–476.

66. Blumenthal IJ. Spontaneous seizures and related electroencephalographic findings following shock therapy. *J Nerv Ment Disord* 1955;122:581–588.

67. Blumer D. The psychiatric dimension of epilepsy: historical perspective and current significance. In: Blumer D, ed. *Psychiatric aspects of epilepsy.* Washington, DC: American Psychiatric Press, 1985:1–65.

68. Blumer D. Epilepsy and disorders of mood. In: Smith DB,

Treiman DM, Trimble MR, eds. *Neurobehavioral problems in epilepsy.* New York: Raven Press, 1991:185–195.

69. Blumer D. Postictal depression: significance for the neurobehavioral disorder of epilepsy. *J Epilepsy* 1992;5:214–219.

70. Blumer D. Antidepressant and double antidepressant treatment for the affective disorder of epilepsy. *J Clin Psychiatry* 1997;58: 3–11.

71. Blumer D. Evidence supporting the existence of a temporal lobe epilepsy personality syndrome. *Neurology* 1999;53[Suppl 2]:S9–S12.

72. Blumer D. Dysphoric disorders and paroxysmal affects: recognition and treatment of epilepsy-related psychiatric disorders. *Harv Rev Psychiatry* 2000;8:17.

73. Blumer D, Walker A. Sexual behavior in temporal lobe epilepsy. *Arch Neurol* 1967;16:31–43.

74. Blumer D, Benson DF. Psychiatric manifestations of epilepsy. In: Blumer D, Benson DF, eds. *Psychiatric aspects of neurological disease, volume II.* New York: Grune & Stratton, 1982.

75. Blumer D, Montouris G, Hermann B. Psychiatric morbidity in seizure patients on a neurodiagnostic monitoring unit. *J Neuropsychiatry Clin Neurosci* 1995;7:445–456.

76. Blumer D, Wakhlu S, Davies K, et al. Psychiatric outcome of temporal lobectomy for epilepsy: Incidence and treatment of psychiatric complications. *Epilepsia* 1998;39:478–486.

77. Blumer D, Wakhlu S, Montouris G, et al. Treatment of the interictal psychoses. *J Clin Psychiatry* 2000;61:110–122.

78. Bormann J. Electrophysiology of GABA-A and GABA-B receptor subtypes. *Trends Neurosci* 1988;11:112–116.

79. Borum R, Appelbaum KL. Epilepsy, aggression and criminal responsibility. *Psychiatr Serv* 1996;47:762–763.

80. Bourgeois BFD, Presnky AL, Palkes HS, et al. Intelligence in epilepsy: a prospective study in children. *Ann Neurol* 1983;14:438–444.

81. Bowman ES. Etiology and clinical course of pseudoseizures: relationship to trauma, depression and dissociation. *Psychosomatics* 1993;34:333–342.

82. Bowman ES, Markand ON. Psychodynamic and psychiatric diagnoses of pseudoseizure subjects. *Am J Psychiatry* 1996;153:57–63.

83. Boyer WF, Blumhardt CL. The safety profile of paroxetine. *J Clin Psychiatry* 1992;53[Suppl 2]:61–66.

84. Bredkjaer SR, Mortensen PB, Parnas J. Epilepsy and non-organic non-affective psychosis: National Epidemiologic Study. *Br J Psychiatry* 1998;172:235–238.

85. Brent DA, Crumrine PK, Varma RR, et al. Phenobarbital treatment and major depressive disorder in children with epilepsy. *Paediatrics* 1987;80:909–917.

86. Briellmann RS, Kalnins RM, Hopwood MJ, et al. *Neurology* 2000;55:1027–1030.

87. Brodie MJ. Drug interactions in epilepsy. *Epilepsia* 1992;33[Suppl1]:S13–S22.

88. Brodie MJ. Tiagabine in the management of epilepsy. *Epilepsia* 1997;38[Suppl 2]:S23–S27.

89. Bromfield EB, Altshuler L, Leiderman DB, et al. Cerebral metabolism and depression in patients with complex partial seizures. *Arch Neurol* 1992;49:617–623.

90. Brown SW, McGowan MEL, Reynolds EH. The influence of seizure type and medication on psychiatric symptoms in epileptic patients. *Br J Psychiatry* 1986;148:300–304.

91. Bruens JH. Psychoses in epilepsy. *Psychiatria Neurol Neurochirurg* 1971;74:174–192.

92. Bruens JH. Psychoses in epilepsy. In: Vinken PJ, Bruyn GW, eds. *Handbook of clinical neurology, volume 15.* Amsterdam: North Holland, 1974:593–610.

93. Bruton CJ. *The neuropathology of temporal lobe epilepsy* (Maudsley Monographs). Oxford: Oxford University Press, 1988.

94. Buchanan N, Snars J. Pseudoseizures (non epileptic attack disorder)—clinical management and outcome in 50 patients. *Seizure* 1993;2:141–146.

95. Burnstine TH, Lesser RP, Cole AJ. Pseudoepileptiform EEG patterns during pseudoseizures. *J Epilepsy* 1991;4:165–171.

96. Burton LA, Labar D. Emotional status after right vs. left temporal lobectomy. *Seizure* 1999;8:116–119.

97. Byrne A. Hypomania following increased epileptic activity [Letter]. *Br J Psychiatry* 1988;153:573–574.

98. Byrne A, Walsh JB. The epileptic arsonist [Letter]. *Br J Psychiatry* 1989;154:268.

99. Bystritsky A, Strausser BP. Treatment of obsessive-compulsive disorder with naltrexone [Letter]. *J Clin Psychiatry* 1996;57: 423–424.

100. Calabrese JR, Fatemi SH, Woyshville MJ. Antidepressant effects of lamotrigine in rapid cycling bipolar disorder. *Am J Psychiatry* 1996;153:1236.

101. Callender JS, Fenton GW. Psychosis de novo following temporal lobectomy. *Seizure* 1997;6:409–411.

102. Cañizares S, Torres X, Boget T, et al. Does neuroticism influence cognitive self-assessment after epilepsy surgery? *Epilepsia* 2000;41:1303–1309.

103. Caplan R, Comair Y, Shewmon DA, et al. Intractable seizures, compulsions and coprolalia: a pediatric case study. *J Neuropsychiatry Clin Neurosci* 1992;4:315–319.

104. Carney MWP. Serum folate values in 423 psychiatric patients. *BMJ* 1967;4:512–516.

105. Carney MWP, Chary TKN, Laundy M, et al. Red cell folate concentrations in psychiatric patients. *J Affect Disord* 1990;19: 207–213.

106. Carpenter PK, King AL. Epilepsy and arson. *Br J Psychiatry* 1989;154:554–556.

107. Carran MA, Kohler CG, O'Connor MJ, et al. Marital status after epilepsy surgery. *Epilepsia* 1999;40:1755–1760.

108. Carrazana E, DeToledo J, Tatum W, et al. Epilepsy and religious experiences: voodoo possession. *Epilepsia* 1999;40:239–241.

109. Chengappa R, Rathore D, Levine J, et al. Topiramate as add-on treatment for patients with bipolar mania. *Bipolar Disord* 1999;1:42–53.

110. Chovaz CJ, McLachlan RS, Derry PA, et al. Psychosocial function following temporal lobectomy: influence of seizure control and learned helplessness. *Seizure* 1994;3:171–176.

111. Cockerell OC, Catchpole M, Sander JWAS, et al. The British Neurological Surveillance Unit: a nation-wide scheme for the ascertainment of rare neurological disorders. *Neuroepidemiology* 1995;14:182–187.

112. Cockerell OC, Moriarty J, Trimble M, et al. Acute psychological disorders in patients with epilepsy: a nation-wide study. *Epilepsy Res* 1996;25:119–131.

113. Cohen RJ, Suter C. Hysterical seizures: suggestion as a provocative EEG test. *Ann Neurol* 1982;11:391–395.

114. Cohen LM, Howard GF, Bongar B. Provocation of pseudoseizures by psychiatric interview during EEG and video monitoring. *Int J Psychiatry Med* 1992;22:131–140.

115. Commander M, Green SH, Prendergast M. Behavioural disturbances in children treated with clonazepam [Letter]. *Dev Med Child Neurol* 1991;33:362–364.

116. Commission on Classification and Terminology of the International League Against Epilepsy. Proposal for revised clinical and electroencephalographic classification of epileptic seizures. *Epilepsia* 1981;22:489–501.

117. Commission on Classification and Terminology of the International League Against Epilepsy. Proposal for revised classification

of epilepsies and epileptic syndromes. *Epilepsia* 1989;30:389–399.

118. Connelly A, Gadia DG, Jackson GD, et al. 1H MRS in the investigation of intractable temporal lobe epilepsy. *Neurology* 1994;44:850(abst).

119. Cook MJ, Fish DR, Shorvon SD, et al. Hippocampal volumetric and morphometric studies in frontal and temporal lobe epilepsy. *Brain* 1990;115:1001–1015.

120. Cramer JA, Arrigo C, Van Hammée G, et al. for the N132 Study Group. Effect of levetiracetam on epilepsy-related quality of life. *Epilepsia* 2000;41:868–874.

121. Currie S, Heathfield W, Henson R, et al. Clinical course and prognosis of temporal lobe epilepsy: a survey of 666 patients. *Brain* 1971;94:173–190.

122. Dahl J, Melin L, Lund L. Effects of a contingent relaxation program on adults with refractory epileptic seizures. *Epilepsia* 1987;28:125–132.

123. Dahl A, Melin L, Leissner P. Effects of a behavioral intervention on epileptic seizure behavior and paroxysmal activity: a systematic replication of three cases of children with intractable epilepsy. *Epilepsia* 1988;29:172–183.

124. Dalby MA. Behavioural effects of carbamazepine. In: Penry JK, Daly DD, eds. *Complex partial seizures and their treatment.* New York: Raven Press, 1975:331–344.

125. Danesi MA. African aspects. In: Dam M, Gram L, eds. *Comprehensive epileptology.* New York: Raven Press, 1990:795–805.

126. Davies B, Morgenstern FS. A case of cysticercosis, temporal lobe epilepsy and transvestism. *J Neurol Neurosurg Psychiatry* 1960;23:247–249.

127. Davis GR, Armstrong HE, Donovan DM, et al. Cognitive-behavioural treatment of depressed affect amongst epileptics: preliminary findings. *J Clin Psychol* 1984;4:930–935.

128. Deb S, Hunter D. Psychopathology of people with mental handicap and epilepsy. II: psychiatric illness. *Br J Psychiatry* 1991;159:826–830.

129. De León OA, Blend MY, Jobe TH, et al. Application of ictal SPECT for differentiating epileptic from nonepileptic seizures. *J Neuropsychiatry Clin Neurosci* 1997;9:99–101.

130. Delgado-Escueta A, Mattson R, King L. The nature of aggression during epileptic seizures. *N Engl J Med* 1981;305:711–716.

131. Delgado-Escueta AV. Status epilepticus. In: Dam M, Gram L, eds. *Comprehensive epileptology.* New York: Raven Press, 1990:375–383.

132. Demers R, Lukesh R, Prichard J. Convulsion during lithium therapy [Letter]. *Lancet* 1970;2:315–316.

133. Derry PA, McLachlan RS. Causal attributions for seizures: relation to preoperative psychological adjustment and postoperative psychosocial function temporal lobe epilepsy. *J Epilepsy* 1995;8:74–82.

134. Derry PA, McLachlan RS. The MMPI-2 as an adjunct to the diagnosis of pseudoseizures. *Seizure* 1996;5:35–40.

135. Derry PA, Wiebe S. Psychological adjustment to success and to failure following epilepsy surgery. *Can J Neurol Sci* 2000;27[Suppl 1]:S116–S120.

136. Derry PA, Rose KJ, McLachlan RS. Moderators of the effect of preoperative emotional adjustment on postoperative depression after surgery for temporal lobe epilepsy. *Epilepsia* 2000;41:177–185.

137. Devinsky O. Nonepileptic psychogenic seizures: quagmires of pathophysiology, diagnosis, and treatment [Editorial Commentary]. *Epilepsia* 1998;39:458–462.

138. Devinsky O, Duchowny MS. Seizures after convulsive therapy: a retrospective case survey. *Neurology* 1983;33:921–925.

139. Devinsky O, Bear D. Varieties of aggressive behaviour in temporal lobe epilepsy. *Am J Psychiatry* 1984;141:651–656.

140. Devinsky O, Bear DM. Varieties of depression in epilepsy. *Neuropsychiatry Neuropsychol Behav Neurol* 1991;4:49–61.

141. Devinsky O, Fisher R. Ethical use of placebos and provocative testing in diagnosing nonepileptic seizures. *Neurology* 1996;47:866–870.

142. Devinsky O, Najjar S. Evidence against the existence of a temporal lobe epilepsy personality syndrome. *Neurology* 1999;53[Suppl 2]:S13–S25.

143. Devinsky O, Feldmann E, Bromfield E, et al. Structured interview for simple partial seizures: clinical phenomenology and diagnosis. *J Epilepsy* 1991;4:107–116.

144. Devinsky O, Kelley K, Yacubian EMT, et al. Postictal behavior. A clinical and subdural electroencephalographic study. *Arch Neurol* 1994;51:254–259.

145. Devinsky O, Thacker K. Nonepileptic seizures. *Neurol Clin* 1995;13:299–319.

146. Devinsky O, Abramson H, Alper K, et al. Postictal psychosis: a case control series of 20 patients and 150 controls. *Epilepsy Res* 1995;20:247–253.

147. Devinsky O, Sanchez-Villasenor F, Vazquez B, et al. Clinical profile of patients with epileptic and nonepileptic seizures. *Neurology* 1996;46:1530–1533.

148. Dewhurst K, Beard AW. Sudden religious conversions in temporal lobe epilepsy. *Br J Psychiatry* 1970;117:497–507.

149. Dikmen S, Hermann BP, Witensky AJ, et al. Validity of the Minnesota Multiphasic Personality (MMPI) to psychopathology in patients with epilepsy. *J Nerv Ment Dis* 1983;171:114–122.

150. Dimond KR, Pande AC, Lamoreaux L, et al. Effect of gabapentin (Neurontin) on mood and well-being in patients with epilepsy. *Prog Neuropsychopharmacol Bioll Psychiatry* 1996;20:407–417.

151. Dodrill CB. Number of seizure types in relation to emotional and psychosocial adjustment in epilepsy. In: Porter RJ, Ward AA Jr, Mattson RH, et al., eds. *Advances in epileptology: XVth Epilepsy International Symposium.* New York: Raven Press, 1984: 541–544.

152. Dodrill C. Correlates of generalised tonic-clonic seizures with intellectual, neuropsychological, emotional and social function in patients with epilepsy. *Epilepsia* 1986;27:191–197.

153. Dodrill CB. Behavioral effects of antiepileptic drugs. In: Smith DB, Treiman DM, Trimble MR, eds. *Neurobehavioral problems in epilepsy (Advances in neurology, volume 55).* New York: Raven Press, 1991:213–224.

154. Dodrill CB, Batzel LW. Inter-ictal behavioural features of patients with epilepsy. *Epilepsia* 1986;27[Suppl 2]:S64–S76.

155. Dodrill CB, Arnett JL, Sommerville KW, et al. Effects of differing dosage of vigabatrin on partial seizures and cognitive functions. *Epilepsia* 1995;36:164–173.

156. Dodrill CB, Arnett JL, Sommerville KW, et al. Cognitive and quality of life effects of differing dosages of tiagabine in epilepsy. *Neurology* 1997;48:1025–1031.

157. Dodrill CB, Arnett JL, Shu V, et al. Effects of tiagabine monotherapy on abilities, adjustment, and mood. *Epilepsia* 1998;39:33–42.

158. Dominian J, Serafetinides GA, Dewhurst M. A follow-up study of late-onset epilepsy: II. psychiatric and social findings. *BMJ* 1963;1:431–435.

159. Dongier S. Statistical study of clinical and electroencephalographic manifestations of 536 psychotic episodes occurring in 516 epileptic between clinical seizures. *Epilepsia* 1959/60;1:117–142.

160. Drake ME. Saline activation of pseudoepileptic seizures: clinical EEG and neuropsychiatric observations. *Clin Electroencephalogr* 1985;16:171–176.

161. Drake ME, Hietter SA, Pakalnis A. EEG and evoked

potentials in episodic-dyscontrol syndrome. *Neuropsychobiology* 1992;26:125–128.

162. Duncan JS, Shorvon SD, Trimble MR. Effects of removal of phenytoin, carbamazepine and valproate on cognitive function. *Epilepsia* 1990;31:584–591.

163. Duncan S, Blacklaw J, Beastall GH, et al. Sexual function in women with epilepsy. *Epilepsia* 1997;38:1974–1081.

164. Earnest MP. Neurologic complications of drug and alcohol abuse. Seizures. *Neurol Clin* 1993;1:563–575.

165. Edeh J, Toone BK. Antiepileptic therapy, folate deficiency and psychiatric morbidity: a general practice survey. *Epilepsia* 1985;26:434–440.

166. Edeh J, Toone B. Relationship between interictal psychopathology and the type of epilepsy. Results of a survey in general practice. *Br J Psychiatry* 1987;151:95–101.

167. Edeh J, Toone BK, Corney RH. Epilepsy, psychiatric morbidity, and social dysfunction in general practice. Comparison between hospital clinic patients and clinic nonattenders. *Neuropsychiatry Neuropsychol Behav Neurol* 1990;3:180–192.

168. Ehrhardt P, Forsythe WI. Prognosis after grand mal seizures: a study of 187 children with three year remissions. *Dev Med Child Neurol* 1989;31:633–639.

169. Eisendrath SJ, Valan MN. Psychiatric predictors of pseudoepileptic seizures in patients with refractory seizures. *J Neuropsychiatry Clin Neurosci* 1994;6:257–260.

170. Elwes RDC, Reynolds EH. First seizure in adult life. *Lancet* 1988;2:36.

171. Elwes RDC, Johnson AL, Shorvon SD, et al. The prognosis for seizure control in newly diagnosed epilepsy. *N Engl J Med* 1984;311:944–947.

172. Emrich HM, Dose M, von Zerssen D. The use of sodium valproate, carbamazepine and oxcarbazepine in patients with affective disorders. *J Affect Disord* 1985;8:243–250.

173. Engel J, Kuhl DE, Phelps ME, et al. Interictal cerebral glucose metabolism in partial epilepsy and its relation to EEG changes. *Ann Neurol* 1982;12:529–537.

174. Engel J Jr, Van Ness P, Rasmussen T, et al. Outcome with respect to epileptic seizures. In: Engel J, ed. *Surgical treatment of the epilepsies*, 2nd ed. New York: Raven Press, 1993:609–621.

175. Esquirol J. *Mental maladies: a treatise on insanity*. Philadelphia: Lea and Blanchard, 1845. Hunt E, translator.

176. Ettinger AB, Weisbrot DM, Krupp LB, et al. Fatigue and depression in epilepsy. *J Epilepsy* 1998;11:105–109.

177. Ettinger AB, Weisbrot DM, Saracco J, et al. Positive and negative psychotropic effects of lamotrigine in patients with epilepsy and mental retardation. *Epilepsia* 1998;39:874–877.

178. Falconer MA, Serafetinides EA. A follow-up study of surgery in temporal lobe epilepsy. *J Neurol Neurosurg Psychiatry* 1963;26:154–163.

179. Falret J. De l'etat mental des epileptiques. *Arch Gen Med* 1860;16:661–679.

180. Farwell JR, Lee YL, Hirtz DG, et al. Phenobarbital for febrile seizures—effects on intelligence and seizure recurrence. *N Engl J Med* 1990;322:364–369.

181. Fenton GJ. Psychiatric disorders of epilepsy: classification and phenomenology. In: Reynolds EH, Trimble MR, eds. *Epilepsy and psychiatry*. Edinburgh: Churchill Livingstone, 1981:12–26.

182. Fenton GW. The EEG, epilepsy and psychiatry. In: Trimble MR, Reynolds EH, eds. *What is epilepsy?* Edinburgh: Churchill Livingstone, 1986.

183. Fenwick P. Epilepsy and the law. *BMJ* 1984;288:1938–1939.

184. Fenwick P. Aggression and epilepsy. In: Trimble MR, Bolwig TG, eds. *Aspects of epilepsy and psychiatry*. New York: John Wiley & Sons, 1986:31–60.

185. Fenwick PBC. Postscript: what should be included in a standard psychiatric assessment? In: Engel J, ed. *Surgical treatment of the epilepsies*. New York: Raven Press, 1987:505–510.

186. Fenwick P. Automatism, medicine and the law. *Psychol Med* 1990[Suppl]:17.

187. Fenwick P. Evocation and inhibition of seizures. Behavioral treatment. In: Smith DB, Treiman DM, Trimble MR, eds. *Neurobehavioral problems in epilepsy (Advances in neurology, volume 55)*. New York: Raven Press, 1991:163–183.

188. Fenwick P. Long-term psychiatric outcome after epilepsy surgery. In: Lüders H, ed. *Epilepsy surgery*. New York: Raven Press, 1991:647–652.

189. Fenwick PBC. Antiepileptic drugs and their psychotropic effects. *Epilepsia* 1992;33[Suppl 6]:S33–S36.

190. Fenwick P, Aspinal A. Non-epileptic seizures: treatment. In: Gram L, Johannesson SI, Ostermann PO, eds. *Pseudo-epileptic seizures*. Petersfield, UK: Wrightson Biomedical Publishing, 1993:123–132.

191. Ferguson SM, Rayport M. The adjustment to living without epilepsy. *J Nerv Ment Dis* 1965;140:26–37.

192. Ferrier IN. Lamotrigine and gabapentin. Alternatives in the treatment of bipolar disorder. *Neuropsychobiology* 1998;38: 192–197.

193. Fiordelli E, Beghi E, Bogliun G, et al. Epilepsy and psychiatric disturbance. A cross-sectional study. *Br J Psychiatry* 1993;163:446–450.

194. Flor-Henry P. Psychosis and temporal lobe epilepsy: a controlled investigation. *Epilepsia* 1969;10:363–395.

195. Fong GCY, Fong KY, Mak W, et al. Postictal psychosis related regional cerebral hyperperfusion [Letter]. *J Neurol Neurosurg Psychiatry* 2000;68:100–101.

196. Frank E, Anderson C, Rubinstein D. Frequency of sexual dysfunction in normal couples. *N Engl J Med* 1978;299:111–115.

197. Fröscher W, Maier V, Laage M, et al. Folate deficiency, anticonvulsant drugs, and psychiatric morbidity. *Clin Neuropharmacol* 1995;18:165–182.

198. Frost JJ, Mayberg HS, Fisher RS, et al. μ-Opiate receptors measured by positron emission tomography and its relation to EEG changes. *Ann Neurol* 1988;23:231–237.

199. Gailer JL, Edwards SM. SSRIs and anticonvulsants. *Aust J Hosp Pharm* 1996;26:587–588.

200. Gale K. GABA in epilepsy: the pharmacologic basis. *Epilepsia* 1989;30[Suppl 3]:S1–S11.

201. Gardner DL, Cowdry RW. Positive effects of carbamazepine on behavioral dyscontrol in borderline personality disorder. *Am J Psychiatry* 1986;143:519–522.

202. Gastaut H. Colloque de Marseille. 15–19 Octobre 1956. Compte rendu du colloque sur l'etude electroclinique des episodes psychotiques qui surviennent chez les epileptiques en dehors des crises cliniques. *Rev Neurol* 1953;95:587–616.

203. Gastaut H. Clinical and electroencephalographic classification of epileptic seizures. *Epilepsia* 1970;11:102–113.

204. Gastaut H, Collomb H. Etude du comportement sexuel chez les épileptiques psychomotors. *Ann Med Psychol* 1954;112:657–696.

205. Gates JR. Diagnosis and treatment of nonepileptic seizures. In: McConnell HW, Snyder PJ, eds. *Psychiatric comorbidity in epilepsy. Basic mechanisms, diagnosis, and treatment*. Washington, DC: American Psychiatric Press, 1998:187–204.

206. Gates JR, Ramani V, Whalen S, et al. Ictal characteristics of pseudoseizures. *Arch Neurol* 1985;42:1183–1187.

207. Gates JR, Erdahl P. Classification of non-epileptic events. In: Rowan AJ, Gates JR, eds. *Non-epileptic seizures*. Boston: Butterworth-Heineman, 1993:21–30.

208. Geddes JW, Cahan LD, Cooper SM, et al. Altered distribution

of excitatory amino acid receptors in temporal lobe epilepsy. *Exp Neurol* 1990;108:214–220.

209. Genton P, Bartolomei F, Guerrini R. Panic attacks mistaken for relapse of epilepsy. *Epilepsia* 1995;36:48–51.
210. Geschwind N. Behavioural changes in temporal lobe epilepsy. *Psychol Med* 1979;9:217–219.
211. Giakas WJ, Seibyl JP, Mazure CM. Valproate in the treatment of temper outbursts [Letter]. *J Clin Psychiatry* 1990;51:525.
212. Gibbs FA. Ictal and non-ictal psychiatric disorders in temporal lobe epilepsy. *J Nerv Ment Dis* 1951;113:522–528.
213. Gillham RA. Refractory epilepsy: an evaluation of psychological methods in outpatient management. *Epilepsia* 1990;31:427–432.
214. Gillham RA, Blacklaw J, McKee PJW, et al. Effect of vigabatrin on sedation and cognitive function in patients with refractory epilepsy. *J Neurol Neuorsurg Psychiatry* 1993;56:1271–1275.
215. Gillham R, Baker G, Thompson P, et al. Standardisation of a self-report questionnaire for use in evaluating cognitive, affective and behavioural side-effects of anti-epileptic drug treatments. *Epilepsy Res* 1996;24:47–55.
216. Gillham RA, Kane K, Bryant-Comstock L. Relationship between side effect and life satisfaction (SEALS) score, seizure occurrence and side effects to treatment in newly diagnosed patients with epilepsy. *Epilepsia* 1997;38[Suppl 3]:S163.
217. Gillham R, Kane K, Bryant-Comstock L, et al. A double-blind comparison of lamotrigine and carbamazepine in newly diagnosed epilepsy with health-related quality of life as an outcome measure. *Seizure* 2000;9:375–379.
218. Gillig P, Sackellares J, Greenberg H. Right hemisphere partial complex seizures: mania, hallucinations, and speech disturbances during ictal events. *Epilepsia* 1988;29:26–29.
219. Glosser G, Zwil AS, Glosser DS, et al. Psychiatric aspects of temporal lobe epilepsy before and after anterior temporal lobectomy. *J Neurol Neurosurg Psychiatry* 2000;68:53–58.
220. Godfrey PSA, Toone BK, Carney MWP, et al. Enhancement of recovery from psychiatric illness by methylfolate. *Lancet* 1990;336:392–395.
221. Goggin T, Gough H, Bissessar A, et al. A comparative study of the relative effects of anticonvulsant drugs and dietary folate on the red cell folate status of patients with epilepsy. *Q J Med* 1987;65:911–919.
222. Goldstein LH. Effectiveness of psychological interventions for people with poorly controlled epilepsy. *J Neurol Neurosurg Psychiatry* 1997;63:137–143.
223. Goodridge DMG, Shorvon SD. Epileptic seizures in a population of 6000. I: demography, diagnosis and classification, and role of the hospital services. *BMJ* 1983;287:641–644.
224. Gowers WR. *Epilepsy and other chronic convulsive diseases.* New York: William Wood, 1885.
225. Griesinger W. Über einige epileptoide Zustände. *Arch Psychiatrie Nervenkrankheiten* 1868;1:320–333.
226. Grogan R, Wagner DR, Sullivan T, et al. Generalised nonconvulsive status epilepticus after electroconvulsive therapy. *Convulsive Ther* 1995;11:51–56.
227. Guberman A, Bruni J, Desforges C, et al. Vigabatrin in uncontrolled complex partial seizures in adult patients with epilepsy: a double-blind, placebo-controlled, dose-ranging study. *Can J Neurol Sci* 1994;21[Suppl 2]:S17–S18.
228. Gudmundsson G. Epilepsy in Iceland: a clinical and epidemiological investigation. *Acta Neurol Scand* 1966;43[Suppl 25]:64–90.
229. Guidotti I, Corda MG, Wise BC. GABAergic synapses: supramolecular organization and biochemical regulation. *Neuropharmacology* 1983;22:1471–1479.
230. Guldner GT, Morrell MJ. Nocturnal penile tumescence and

rigidity evaluation in men with epilepsy. *Epilepsia* 1996;37:1211–1214.
231. Gunn J. *Epileptics in prison.* London: Academic Press, 1977.
232. Gunn J, Bonn J. Criminality and violence in epileptic prisoners. *Br J Psychiatry* 1971;118:337–343.
233. Gureje O. Interictal psychopathology in epilepsy. Prevalence and pattern in a Nigerian clinic. *Br J Psychiatry* 1991;158:700–705.
234. Gustev EI, Burd GS. Lamictal in treatment of patients with drug-resistant epilepsy. *Epilepsia* 1994;36[Suppl 3]:S165–166.
235. Harden CL. Pseudoseizures and dissociative disorders: a common mechanism involving traumatic experiences. *Seizure* 1997;6:151–155.
236. Harden CL, Lazar LM, Pick LH, et al. A beneficial effect on mood in partial epilepsy patients treated with gabapentin. *Epilepsia* 1999;40:1129–1134.
237. Harris EC, Barraclough B. Suicide as an outcome for mental disorders. A meta-analysis. *Br J Psychiatry* 1997;170:205–228.
238. Hart YM, Sander JWAS, Johnson AL, et al. National general practice study of epilepsy: recurrence after a first seizure. *Lancet* 1992;336:1271–1274.
239. Hauser WA. The natural history of drug-resistant epilepsy: epidemiologic considerations. NIH Consensus Development Conference on Surgery for Epilepsy, 1990:33–35.
240. Hauser WA. Incidence and prevalence. In: Engel J, Pedley TA, eds. *Epilepsy: a comprehensive textbook.* New York: Lippincott–Raven, 1997:47–57.
241. Hauser WA, Kurland LT. The epidemiology of epilepsy in Rochester, Minnesota, 1935 through 1967. *Epilepsia* 1975;16:1–66.
242. Hauser WA, Annegers JF, Elveback LR. Mortality in patients with epilepsy. *Epilepsia* 1980;21:399–412.
243. Hauser WA, Ng SKC, Brust JCM. Alcohol, seizures, and epilepsy. *Epilepsia* 1988;29[Suppl 2]:S66–S78.
244. Hawton K, Fagg J, Marsack P. Association between epilepsy and attempted suicide. *J Neurol Neurosurg Psychiatry* 1980;43:168–170.
245. Helgason T. Epidemiology of mental disorders in Iceland: psychoses. *Acta Psychiatr Scand* 1964;40[Suppl 173]:67–95.
246. Henry TR, Drury I. Non-epileptic seizures in temporal lobectomy candidates with medically refractory seizures. *Neurology* 1997;48:1374–1382.
247. Hermann BP. Psychopathology in epilepsy and learned helplessness. *Med Hypotheses* 1979;5:723–729.
248. Hermann BP, Chabria S. Interictal psychopathology in patients with ictal fear. *Arch Neurol* 1980;37:667–668.
249. Hermann BP, Whitman S. Psychosocial predictors of interictal depression. *J Epilepsy* 1989;2:231–237.
250. Hermann B, Whitman S. Psychopathology in epilepsy. The role of psychology in altering paradigms of research, treatment, and prevention. *Am Psychol* 1992;47:1134–1138.
251. Hermann BP, Wyler AR. Depression, locus of control, and the effects of epilepsy surgery. *Epilepsia* 1989;30:332–338.
252. Hermann BP, Schwartz MS, Karnes WE, et al. Psychopathology in epilepsy: relationship of seizure type to age of onset. *Epilepsia* 1980;21:15–23.
253. Hermann BP, Dikmen S, Wilensky AJ. Increased psychopathology associated with multiple seizure types: fact or artifact? *Epilepsia* 1982;23:587–596.
254. Hermann BP, Dikmen S, Schwartz MS, et al. Psychopathology in patients with ictal fear: a quantitative investigation. *Neurology* 1982;32:7–11.
255. Hermann BP, Whitman S, Wyler AR, et al. The neurological, psychosocial and demographic correlates of hypergraphia in patients with epilepsy. *J Neurol Neurosurg Psychiatry* 1988;51:203–208.

256. Hermann BP, Wyler AR, Ackermann B, et al. Short-term psychological outcome of anterior temporal lobectomy. *J Neurosurg* 1989;71:327–334.

257. Hermann BP, Seidenberg M, Haltiner A, et al. Mood state in unilateral temporal lobe epilepsy. *Biol Psychiatry* 1991;30:1205–1218.

258. Hermann BP, Wyler AR, Somes G. Preoperative psychological adjustment and surgical outcome are determinants of psychosocial status after temporal lobectomy. *J Neurol Neurosurg Psychiatry* 1992;55:491–496.

259. Hermann BP, Trenerry MR, Colligan RC. Learned helplessness, attributional style and depression in epilepsy. *Epilepsia* 1996;37:680–686.

260. Herzberg JL, Fenwick PBC. The aetiology of aggression in temporal-lobe epilepsy. *Br J Psychiatry* 1988;153:50–55.

261. Hierons R, Saunders. M. Impotence in patients with temporal lobe lesions. *Lancet* 1966;2:761–764.

262. Hill D. Psychiatric disorders of epilepsy. *Med Press* 1953;229:473–475.

263. Hill D, Pond DA, Mitchell W, et al. Personality changes following temporal lobectomy for epilepsy. *J Ment Sci* 1957;103:18–27.

264. Hilty DM, Rodriguez GD, Hales RE. Treatment of comorbid bipolar disorder and epilepsy with valproate [Letter]. *J Neuropsychiatry Clin Neurosci* 2000;12:283–285.

265. Hirsch E, Peretti S, Boulay C, et al. Panic attacks misdiagnosed as partial epileptic seizures. *Epilepsia* 1990;31:636.

266. Holland KD, McKeon AC, Canney DJ, et al. Relative anticonvulsant effects of GABAmimetic and GABA modulatory agents. *Epilepsia* 1992;33:981–986.

267. Holmes MD, Wilkus RJ, Dodrill CB, et al. Coexistence of epilepsy in patients with nonepileptic seizures. *Epilepsia* 1993;34[Suppl 2]:13.

268. Hooshmand H, Brawley BW. Temporal lobe seizures and exhibitionism. *Neurology* 1969;19:1119–1124.

269. Höppener RJ, Juger A, van der Ligt PJM. Epilepsy and alcohol: the influence of social alcohol intake on seizures and treatment in epilepsy. *Epilepsia* 1983;26:459.

270. Hughes J, Devinsky O, Feldmann E, et al. Premonitory symptoms in epilepsy. *Seizure* 1993;2:201–203.

271. Humphries SR, Dickinson PS. Hypomania following complex partial seizures [Letter]. *Br J Psychiatry* 1988;152:571–572.

272. Hurwitz TA, Wada JA, Kosaka BD, et al. Cerebral organization of affect suggested by temporal lobe seizures. *Neurology* 1985;35:1335–1337.

273. Iadarola MJ, Gale K. Substantia nigra: site of anticonvulsant activity mediated by gamma-aminobutyric acid. *Science* 1982;218:1237–1240.

274. Indaco A, Carrieri PB, Nappi C, et al. Interictal depression in epilepsy. *Epilepsy Res* 1992;12:45–50.

275. Isojärvi JIT, Repo M, Pakarinen AJ, et al. Carbamazepine, phenytoin, sex hormones, and sexual function in men with epilepsy. *Epilepsia* 1995;36:366–370.

276. Jackson GD, Bercovic SF, Tress BM, et al. Hippocampal sclerosis can be reliably detected by magnetic resonance imaging. *Neurology* 1990;40:1869–1875.

277. Jackson GD, Bercovic SF, Duncan JS, et al. Optimising the diagnosis of hippocampal sclerosis using magnetic resonance imaging. *Am J Neurol* 1993;14:753–762.

278. Jackson GD, Connelly A, Duncan JS, et al. Detection of hippocampal pathology in intractable partial epilepsy: increased sensitivity with quantitative magnetic resonance T2 relaxometry. *Neurology* 1993;43:1793–1799.

279. Jacoby A, Baker GA, Steen N, et al. The clinical course of epilepsy and its psychosocial correlates: findings from a UK community study. *Epilepsia* 1996;37:148–161.

280. Jalava M, Sillanpää M. Concurrent illnesses in adults with childhood-onset epilepsy: a population-based 35-year follow-up study. *Epilepsia* 1996;37:1155–1163.

281. Jenike MA. Obsessive compulsive disorder: a question of a neurologic lesion. *Compr Psychiatry* 1984;25:298–304.

282. Jenike MA, Brotman AW. The EEG in obsessive-compulsive disorder. *J Clin Psychiatry* 1984;45:122–124.

283. Jenike MA, Buttolph L, Baer L, et al. Open trial of fluoxetine in obsessive-compulsive disorder. *Am J Psychiatry* 1989;146:909–911.

284. Jensen I, Larsen JK. Mental aspects of temporal lobe epilepsy. *J Neurol Neurosurg Psychiatry* 1979;42:256–265.

285. Jensen P, Jensen SB, Sorensen PS, et al. Sexual dysfunction in male and female patients with epilepsy: a study of 86 outpatients. *Arch Sex Behav* 1990;19:1–14.

286. Jeppesen U, Gram LF, Vistisen K, et al. Dose-dependent inhibition of CYP1A2, CYP2C19 and CYP2D6 by citalopram, fluoxetine, fluvoxamine and paroxetine. *Eur J Clin Pharmacol* 1996;51:73–78.

287. Jones-Gotman M. Presurgical neuropsychological evaluation for localization and lateralization of seizure focus. In: Lüders H, ed. *Epilepsy surgery.* New York: Raven Press, 1991:469–476.

288. Joseph AB. A hypergraphic syndrome of automatic writing, affective disorder and temporal lobe epilepsy in two patients. *J Clin Psychiatry* 1986;47:255–257.

289. Julius SC, Brenner RP. Myoclonic seizures with lithium. *Biol Psychiatry* 1987;22:1184–1190.

290. Jus A, Villeneuve A, Gauthier J, et al. Influence of lithium carbonate on patients with temporal lobe epilepsy. *Can J Psychiatry Assoc J* 1973;18:77–78.

291. Juul-Jensen P, Foldsprang A. Natural history of epileptic seizures. *Epilepsia* 1983;24:297–312.

292. Kalinowski LB. Organic psychotic syndromes occurring during electric convulsive therapy. *Arch Neurol Psychiatry* 1945;53:269–273.

293. Kanemoto K. Hypomania after temporal lobectomy: a sequela to the increased excitability of the residual temporal lobe? [Letter]. *J Neurol Neurosurg Psychiatry* 1995;59:448–449.

294. Kanemoto K, Kawasaki J, Mori E. Postictal psychosis as a risk factor for mood disorders after temporal lobe surgery. *J Neurol Neurosurg Psychiatry* 1998;65:587–589.

295. Kanemoto K, Kawasaki J, Mori E. Violence and epilepsy: a close relation between violence and postictal psychosis. *Epilepsia* 1999;40:107–109.

296. Kanemoto K, Miyamoto T, Kawasaki J. Running down phenomenon and acute interictal psychosis in medial temporal lobe epilepsy. *Epilepsy Res* 2000;39:33–36.

297. Kanemoto K, Kawasaki J, Kawai I. Postictal psychosis: a comparison with acute interictal and chronic psychoses. *Epilepsia* 1996;37:551–556.

298. Kanner AM, Stagno S, Kotagal P, et al. Postictal psychiatric events during prolonged video-electroencephalographic monitoring studies. *Arch Neurol* 1996;53:258–263.

299. Kanner AM, Parra J, Frey M, et al. Psychiatric and neurologic predictors of psychogenic pseudoseizures outcome. *Neurology* 1999;53:933–938.

300. Kasper S, Fuger J, Möller HJ. Comparative efficacy of antidepressants. *Drugs* 1992;43:11–22.

301. Kaufman KR. Adjunctive tiagabine treatment of psychiatric disorders: three cases. *Ann Clin Psychiatry* 1998;10:181–184.

302. Kellett MW, Smith DF, Baker GA, et al. Quality of life after epilepsy surgery. *J Neurol Neurosurg Psychiatry* 1997;63:52–58.

303. Kendrick AM, Duncan JS, Trimble MR. Effects of discontinuation of individual antiepileptic drugs on mood. *Hum Psychopharmacol* 1993;8:263–270.

304. Kendrick AM, Trimble MR. Repertory grid in the assessment of the quality of life in patients with epilepsy: the quality of life assessment schedule. In: Trimble MR, Dodson WE, eds. *Epilepsy and the quality of life.* New York: Raven Press, 1994:151–164.

305. Ketter TA, Malow BA, Flamini R, et al. Anticonvulsant withdrawal emergent psychopathology. *Neurology* 1994;44:55–61.

306. Ketter TA, Post RM, Theodore WH. Positive and negative psychiatric effects of antiepileptic drugs in patients with seizure disorders. *Neurology* 1999;53[Suppl 2]:S53–S67.

307. Kettl PA, Marks IM. Neurological factors in obsessive compulsive disorder. Two cases and a review of the literature. *Br J Psychiatry* 1986;149:315–319.

308. Khan A, Faught E, Gilliam F, et al. Acute psychotic symptoms induced by topiramate. *Seizure* 1999;8:235–237.

309. King DW, Gallagher BB, Murvin AJ, et al. Pseudoseizures: diagnostic evaluation. *Neurology* 1982;32:18–32.

310. Klerman GL. Treatment of recurrent unipolar major depressive disorder: commentary on the Pittsburgh study. *Arch Gen Psychiatry* 1990;47:1158–1161.

311. Kløve H, Matthews CG. Psychometric and adaptive abilities in epilepsy with different etiology. *Epilepsia* 1966;7:330–338.

312. Koch-Weser M, Garron DC, Gilley DW, et al. Prevalence of psychologic disorders after surgical treatment of seizures. *Arch Neurol* 1988;45:1308–1311.

313. Kogeorgos J, Fonagy P, Scott DF. Psychiatric symptom patterns of chronic epileptics attending a neurological clinic: a controlled investigation. *Br J Psychiatry* 1982;140:236–243.

314. Kohler C, Norstrand JA, Baltuch G, et al. Depression in temporal lobe epilepsy before epilepsy surgery. *Epilepsia* 1999;40:336–340.

315. Köhler GK. Zur Einteilung der Psychosen bei Epilepsia. Zum Begriff "Psychosen bei Epilepsie" bzw. "epileptische Psychosen". In: Wolf P, Köhler GK, eds. *Psychopathologische und pathogenetische Probleme psychotischer Syndrome bei Epilepsie.* Bern: Huber, 11–18.

316. Koopowitz LF, Berk M. Response of obsessive compulsive disorder to carbamazepine in two patients with co-morbid epilepsy. *Ann Clin Psychiatry* 1997;9:171–173.

317. Kraft AM, Price TRP, Peltier D. Complex partial seizures and schizophrenia. *Compr Psychiatry* 1984;25:113–124.

318. Krahn LE, Rummans TA, Peterson GC, et al. Electroconvulsive therapy for depression after temporal lobectomy for epilepsy. Case report. *Convulsive Ther* 1993;9:217–219.

319. Krahn LE, Rummans TA, Sharborough FW, et al. Pseudoseizures after epilepsy surgery. *Psychosomatics* 1995;36:487–493.

320. Kristensen O, Sindrup HH. Psychomotor epilepsy and psychosis. I. Physical aspects. *Acta Neurol Scand* 1978;7:361–369.

321. Kristensen O, Sindrup HH. Psychomotor epilepsy and psychosis. II. Electroencephalographic findings. *Acta Neurol Scand* 1978;57:370–379.

322. Krohn W. A study of epilepsy in northern Norway, its frequency and character. *Acta Psychiatr Scand* 1961;150[Suppl]:215–225.

323. Kroll L, Drummond LM. Temporal lobe epilepsy and obsessive-compulsive symptoms. *J Nerv Ment Disord* 1993;181:457–478.

324. Krystal AD, Weiner RD. ECT seizure adequacy. *Convulsive Ther* 1993;10:153–164.

325. Kubagawa T, Furusho J, Maruyama H. Study of psychiatric symptoms after discontinuation of antiepileptic drugs. *Epilepsia* 1997;38[Suppl 6]:26–31.

326. Kupfer DJ, Frank E, Perel JM. The advantage of early treatment intervention in recurrent depression. *Arch Gen Psychiatry* 1989;46:771–775.

327. Kuyk J, Dunki Jacobs L, Spinhoven P, et al. Diagnosis of pseudoepileptic and epileptic seizures. *Epilepsia* 1995;36:S173.

328. Kuyk J, Leijten F, Meinardi H, et al. The diagnosis of psychogenic non-epileptic seizures: a review. *Seizure* 1997;6:243–253.

329. Lambert MV. Seizures, hormones and sexuality. *Seizure* 2001 *(in press).*

330. Lambert MV, Robertson MM. Depression in epilepsy: etiology, phenomenology, and treatment. *Epilepsia* 1999;40[Suppl 10]:S21–S47.

331. Lambert MV, Binnie CD, Polkey CE, et al. Ictal panic following anterior-temporal lobectomy: the role of the insula. Review and case report *(in press).*

332. Lancman ME, Brotherton TA, Asconape JJ, et al. Psychogenic seizures in adults: a longitudinal study. *Seizure* 1993;2:281–286.

333. Lancman ME, Craven WJ, Asconape JJ, et al. Clinical management of recurrent postictal psychosis. *J Epilepsy* 1994;7:47–51.

334. Lancman ME, Asconape JJ, Craven WJ, et al. Predictive value of induction of psychogenic seizures by suggestion. *Ann Neurol* 1994;35:359–361.

335. Landolt H. Some clinical EEG correlations in epileptic psychoses (twilight states). *Electroencephalogr Clin Neurophysiol* 1953;5:121.

336. Landolt H. Serial electroencephalographic investigations during psychotic episodes in epileptic patients and during schizophrenic attacks. In: deHass L, ed. *Lectures on epilepsy.* London: Elsevier, 1958.

337. Leach JP, Girvan J, Paul A, et al. Gabapentin and cognition: a double blind, dose ranging, placebo controlled study in refractory epilepsy. *J Neurol Neurosurg Psychiatry* 1997;62:372–376.

338. Lee DO, Steingard RJ, Cesena M, et al. Behavioral side effects of gabapentin in children. *Epilepsia* 1996;37:87–90.

339. Lee DO, Helmers SL, Steingard RJ, et al. Case study: seizure disorder presenting as panic disorder with agoraphobia. *J Am Acad Child Adolesc Psychiatry* 1997;36:1295–1298.

340. Lee E-K. Periodic left temporal sharp waves during acute psychosis. *J Epilepsy* 1998;11:79–83.

341. Lee SI. Nonconvulsive status epilepticus. *Arch Neurol* 1985;42:778–781.

342. Leestma JE, Walcak T, Hughes JR, et al. A prospective study on sudden unexpected death in epilepsy. *Ann Neurol* 1989;26:195–203.

343. Leinonen E, Tuunainen A, Lepola U. Postoperative psychoses in epileptic patients after temporal lobectomy. *Acta Neurol Scand* 1994;90:394–399.

344. Leis AA, Ross MA, Summers AK. Psychogenic seizures: ictal characteristics and diagnostic pitfalls. *Neurology* 1992;42:95–99.

345. Lempert T, Schmidt D. Natural history and outcome of psychogenic seizures: a clinical study in 50 patients. *J Neurol* 1991;237:35–38.

346. Lennox WG. Alcohol and epilepsy. *Q J Stud Alcohol* 1941;2:1–11.

347. Leppik IE. Tiagabine: the safety landscape. *Epilepsia* 1995;6[Suppl 6]:S10–S13.

348. Lesser RP. Psychogenic seizures. *Neurology* 1996;46:1499–1507.

349. Levin B, Duchowny M. Childhood obsessive-compulsive disorder and cingulate epilepsy. *Biol Psychiatry* 1991;30:1049–1055.

350. Levinson DF, Devinsky O. Psychiatric adverse events during vigabatrin therapy. *Neurology* 1999;53:1503–1511.

351. Lewin J, Sumners D. Successful treatment of episodic dyscontrol with carbamazepine. *Br J Psychiatry* 1992;161:261.

352. Lhatoo SD, Walker MC. The safety and adverse event profile of topiramate. *Rev Contemp Pharmacother* 1999;10:185–191.

353. Lim J, Yagnik P, Schraeder P, et al. Ictal catatonia as a manifestation of nonconvulsive status epilepticus. *J Neurol Neurosurg Psychiatry* 1986;49:833–836.

354. Lindsay J, Ounsted C, Richards P. Long-term outcome in children with temporal lobe seizures. II. Psychiatric aspects in childhood and adult life. *Dev Med Child Neurol* 1979;21:630–636.

355. Lishman WA. *Organic psychiatry. The psychological consequences of cerebral disorder,* 3rd ed. Oxford: Blackwell Science, 1998.

356. Lockard JS. A primate model of clinical epilepsy: mechanisms of action through quantification of therapeutic effects. In: Lochard JS, Ward AA, eds. *Epilepsy: a window to brain mechanisms.* New York: Raven Press, 1980:11–49.

357. Logsdail SJ, Toone BK. Post-ictal psychosis. A clinical and phenomenological description. *Br J Psychiatry* 1988;152:246–252.

358. Lopez-Rodriguez F, Altshuler L, Kay J, et al. Personality disorders among medically refractory epileptic patients. *J Neuropsychiatry Clin Neurosci* 1999;11:464–469.

359. Lucas P. Episodic dyscontrol: a look back at anger. *J Forensic Psychiatry* 1994;5:371–407.

360. Lucey JV, Butcher G, Clare AW, et al. The clinical characteristics of patients with obsessive compulsive disorder: a descriptive study of an Irish sample. *Irish J Psychol Med* 1994;11:11–14.

361. Lüders HO, Awad I. Conceptual considerations. In: Lüders H, ed. *Epilepsy surgery.* New York: Raven Press, 1991:51–62.

362. Lüders H, Acharya J, Baumgartner C, et al. Expanding the international classification of seizures to provide localization information. *Epilepsia* 1998;39:1006–1013.

363. Lund J. Epilepsy and psychiatric disorder in the mentally retarded adult. *Acta Psychiatr Scand* 1985;72:557–562.

364. Luther JS, McNamara JO, Carwile S, et al. Pseudoepileptic seizures: methods and video analysis to aid diagnosis. *Ann Neurol* 1982;12:458–462.

365. Lyketsos CG, Stoline AM, Longstreet P, et al. Mania in temporal lobe epilepsy. *Neuropsychiatry Neuropsychol Behav Neurol* 1993;6:19–25.

366. McDade G, Brown SW. Non-epileptic seizures: management and predictive factors of outcome. *Seizure* 1992;1:7–10.

367. McElroy SL, Keck PE, Pope HG. Sodium valproate: Its use in primary psychiatric disorders. *J Clin Psychopharmacol* 1987;7:16–24.

368. McGuire A, Duncan JS, Trimble MR. Effects of vigabatrin on cognitive function and mood when used as add-on therapy in patients with intractable epilepsy. *Epilepsia* 1992;33:128–134.

369. McKenna PJ, Kane JM, Parrish K. Psychotic syndromes in epilepsy. *Am J Psychiatry* 1985;142:895–904.

370. McLachlan RS, Rose KJ, Derry PA, et al. Health-related quality of life and seizure control in temporal lobe epilepsy. *Ann Neurol* 1997;41:482–489.

371. Mace CJ, Trimble MR. Psychosis following temporal lobe surgery: a report of six cases. *J Neurol Neurosurg Psychiatry* 1991;54:639–644.

372. Mackay A. Self-poisoning: a complication of epilepsy. *Br J Psychiatry* 1979;34:277–282.

373. MacLean PD. *The triume brain.* New York: Plenum, 1990.

374. Maier M, Mellers J, Toone B, et al. Schizophrenia, temporal lobe epilepsy and psychosis: an *in vivo* magnetic resonance spectroscopy and imaging study of the hippocampus/amygdala complex. *Psychol Med* 2000;30:571–581.

375. Maletsky BM. The episodic dyscontrol syndrome. *Dis Nervous Syst* 1973;34:178–185.

376. Manchanda R, Schaefer B, McLachlan RS, et al. Interictal psychiatric morbidity and focus of epilepsy in treatment-refractory patients admitted to an epilepsy unit. *Am J Psychiatry* 1992;149:1096–1098.

377. Manchanda R, Miller H, McLachlan RS. Post-ictal psychosis after right temporal lobectomy. *J Neurol Neurosurg Psychiatry* 1993;56:277–279.

378. Manchanda R, Schaefer B, McLachlan RS, et al. Relationship of site of seizure focus to psychiatric morbidity. *J Epilepsy* 1995;8:23–28.

379. Manchanda R, Schaefer B, McLachlan RS, et al. Psychiatric disorders in candidates for surgery for epilepsy. *J Neurol Neurosurg Psychiatry* 1996;61:82–89.

380. Marcotte D. Use of topiramate, a new anti-epileptic as a mood stabilizer. *J Affect Disord* 1998;50:245–251.

381. Marescaux C, Warter JM, Micheletti G, et al. Stuporous episodes during treatment with sodium valproate: report of seven cases. *Epilepsia* 1982;23:297–305.

382. Marshall EJ, Syed GMS, Fenwick PBC, et al. A pilot study of schizophrenia-like psychosis in epilepsy using single-photon emission computerized tomography. *Br J Psychiatry* 1993;163:32–36.

383. Marson AG, Kadir ZA, Chadwick DW. New antiepileptic drugs: a systematic review of their efficacy and tolerability. *BMJ* 1996;313:1169–1174.

384. Matsuura M, Trimble MR. Zonisamide and psychosis. *J Epilepsy* 1997;10:52–54.

385. Mayeux R, Brandt J, Rosen J, et al. Interictal memory and language impairment in temporal lobe epilepsy. *Neurology* 1980;30:120–125.

386. Meduna L von. Versuche ueber die bioogische Beeinglussung des Ablaufes der Schizophenie. I. Campher- und cadiazolkraemfe. *Z Gesamte Neurol Psychiatr* 1935;152:235–262.

387. Meierkord H, Will B, Fish D, et al. The clinical features and prognosis of pseudoseizures diagnosed using video-EEG telemetry. *Neurology* 1991;41:1643–1646.

388. Meldrum BS. Anatomy, physiology, and pathology of epilepsy. *Lancet* 1990;336:231–234.

389. Mellers JDC, Adachi N, Takei N, et al. SPET study of verbal fluency in schizophrenia and epilepsy. *Br J Psychiatry* 1998;173:69–74.

390. Mellers JDC, Toone BK, Lishman WA. A neuropsychological comparison of schizophrenia and schizophrenia-like psychosis of epilepsy. *Psychol Med* 2000;30:325–335.

391. Mendez MF, Grau R. The postictal psychosis of epilepsy: investigation in two patients. *Int J Psychiatry Med* 1991;21:85–92.

392. Mendez MF, Doss RC. Ictal and psychiatric aspects of suicide in epileptic patients. *Int J Psychiatry Med* 1992;22:231–237.

393. Mendez MF, Cummings JL, Benson DF. Depression in epilepsy. Significance and phenomenology. *Arch Neurol* 1986;43:766–770.

394. Mendez MF, Doss RC, Taylor JL, et al. Depression in epilepsy. Relationship to seizures and anticonvulsant therapy. *J Nerv Ment Disord* 1993;181:444–447.

395. Mendez MF, Grau R, Doss RC, et al. Schizophrenia in epilepsy: seizure and psychosis variables. *Neurology* 1993;43:1073–1077.

396. Mendez MF, Doss RC, Taylor JL, et al. Relationship of seizure variables to personality disorders in epilepsy. *J Neuropsychiatry Clin Neurosci* 1993;5:283–286.

397. Mendez MF, Doss RC, Taylor JL. Interictal violence in epilepsy. Relationship to behaviour and seizure variables. *J Nerv Ment Disord* 1993;181:566–569.

398. Mendez MF, Taylor JL, Doss RC, et al. Depression in secondary epilepsy: relation to lesion laterality. *J Neurol Neurosurg Psychiatry* 1994;57:232–233.

399. Merlis JK. Proposal for an international classification of the epilepsies. *Epilepsia* 1970;11:114–119.
400. Mignone RJ, Donnelly EF, Sadowsky D. Psychological and neurological comparisons of psychomotor and non-psychomotor epileptic patients. *Epilepsia* 1970;11:345–359.
401. Moore DP. A case of petit mal epilepsy aggravated by lithium. *Am J Psychiatry* 1981;138:690–691.
402. Morel B. D'une forme de delire, suite d'une surexcitation nerveuse se rattachent a une variete non encore d'ecrite d'epilepsie. *Gaz Hebd Med Chir* 1860;7:773–775.
403. Morphew JA. Hypomania following complex partial seizures [Letter]. *Br J Psychiatry* 1988;152:572.
404. Morrell MJ, Sperling MR, Stecker M, et al. Sexual dysfunction in partial epilepsy: a deficit in physiologic sexual arousal. *Neurology* 1994;44:243–247.
405. Mortimore C, Trimble M, Elmers E. Effects of gabapentin on cognition and quality of life in patients with epilepsy. *Seizure* 1998;7:359–364.
406. Mostofsky D, Balaschak BA. Psychobiological control of seizures. *Psychol Bull* 1977;84:723–750.
407. Mouzichouck L, Maryek G, Obukhova H. Lamotrigine in patients with treatment-resistant epilepsy. *Epilepsia* 1995;36[Suppl 3]:S114.
408. Mungas D. Interictal behavior abnormality in temporal lobe epilepsy. *Arch Gen Psychiatry* 1982;39:108–111.
409. Munroe RR. Dyscontrol syndrome: long term follow-up. *Compr Psychiatry* 1989;30:489–497.
410. Murialdo G, Galimberti CA, Fonzi S, et al. Sex hormones and pituitary function in male epileptic patients with altered or normal sexuality. *Epilepsia* 1995;36:360–365.
411. Murray RE, Abou-Khalil B, Griner L. Evidence for familial association of psychiatric disorders and epilepsy. *Biol Psychiatry* 1994;36:428–429.
412. Naugle RI, Rogers DA. Personality inventory responses of males with medically intractable seizures. *J Pers Assess* 1992;59:500–514.
413. Naugle RI, Rodgers DA, Stagno SJ, et al. Unilateral temporal lobe epilepsy: an examination of psychopathology and psychosocial behaviour. *J Epilepsy* 1991;4:157–164.
414. Naylor AS, Rogvi-Hansen B, Kessing L, et al. Psychiatric morbidity after surgery for epilepsy: short-term follow up of patients undergoing amygdalohippocampectomy. *J Neurol Neurosurg Psychiatry* 1994;57:1375–1381.
415. Ndgegwa D, Rust J, Golombok S, et al. Sexual problems in epileptic women. *Sex Marital Ther* 1986;1:175–177.
416. Nemeroff CB, De Vane CL, Pollock BG. Newer antidepressants and the cytochrome P450 system. *Am J Psychiatry* 1996;153:311–320.
417. Ney GC, Barr WB, Napolitano C, et al. New-onset seizures after surgery for epilepsy. *Arch Neurol* 1998;55:726–730.
418. Nielsen H, Kristensen O. Personality correlates of sphenoidal EEG-foci in temporal lobe epilepsy. *Acta Neurol Scand* 1981;64:289–300.
419. Nurnberger JI Jr, Berrettini WH, Simmons-Alling S, et al. Intravenous GABA administration is anxiogenic in man. *Psychiatry Res* 1986;19:113–117.
420. Okamura T, Fukai M, Yamadori A, et al. A clinical study of hypergraphia in epilepsy. *J Neurol Neurosurg Psychiatry* 1993;56:556–559.
421. Onuma T. Limbic lobe epilepsy with paranoid symptoms: analysis of clinical features and psychological tests. *Folia Psychiatr Neurol Jpn* 1983;37:253–258.
422. Ounsted C. Aggression and epilepsy. Rage in children with temporal lobe epilepsy. *J Psychosom Res* 1969;13:237–242.
423. Pakalnis A, Drake ME, Phillips B. Neuropsychiatric aspects of psychogenic status epilepticus. *Neurology* 1991;41:1104–1106.
424. Papez JW. A proposed mechanism of emotion. *Arch Neurol* 1937;38:725–733.
425. Parashos IA, Oxley SL, Boyko OB, et al. *In vivo* quantitation of basal ganglia and thalamic degenerative changes in two temporal lobectomy patients with affective disorder. *J Neuropsychiatry Clin Neurosci* 1993;5:337–341.
426. Parra J, Iriarte J, Kanner AM, et al. *De novo* psychogenic nonepileptic seizures after epilepsy surgery. *Epilepsia* 1998;39:474–477.
427. Parra J, Iriarte J, Kanner AM. Are we overusing the diagnosis of psychogenic non-epileptic events? *Seizure* 1999;8:223–227.
428. Pary R. Mental retardation, mental illness, and seizure diagnosis. *Am J Ment Retard* 1993;98:S58–S62.
429. Peguero E, Abou-Khalil B, Fakhoury T, et al. Self-injury and incontinence in psychogenic seizures. *Epilepsia* 1995;36:586–591.
430. Perez MM, Trimble MR. Epileptic psychosis—psychopathological comparison with process schizophrenia. *Br J Psychiatry* 1980;137:245–249.
431. Perini G, Mendius R. Depression and anxiety in complex partial seizures. *J Nerv Ment Disord* 1984;172:287–290.
432. Perini GI, Tosin C, Carraro C, et al. Interictal mood and personality disorders in temporal lobe epilepsy. *J Neurol Neurosurg Psychiatry* 1996;61:601–605.
433. Perlin JB, DeLorenzo RJ. In: Pedley TA, Meldrum BS, eds. *Recent advances in epilepsy.* Edinburgh: Churchill Livingstone, 1992:15–36.
434. Perse T. Obsessive-compulsive disorder: a treatment review. *J Clin Psychiatry* 1988;49:48–55.
435. Pond DA. Psychiatric aspects of epilepsy. *J Indian Med Prof* 1957;3:1441–1451.
436. Pond DA. Discussion remark. *Proc R Soc Med* 1962;55:316.
437. Pond DA, Bidwell BH. A survey of epilepsy in fourteen general practices. 2: social and psychological aspects. *Epilepsia* 1959/60;1:285–299.
438. Post RM, Uhde TW, Joffe RT, et al. Anticonvulsant drugs in psychiatric illness: new treatment alternatives and theoretical implications. In: Trimble MR, ed. *The psychopharmacology of epilepsy.* Chichester: Wiley, 1985:95–105.
439. Post RM, Weiss SR, Chuang DM. Mechanisms of action of anticonvulsants in affective disorders: comparisons with lithium. *J Clin Psychopharmacol* 1992;12[Suppl 1]:S23–S35.
440. Preskorn SH, Fast GA. Tricyclic antidepressant-induced seizures and plasma drug concentration. *J Clin Psychiatry* 1992;53:160–162.
441. Prevey ML, Mattson RH, Cramer JA. Improvement in cognitive functioning and mood state after conversion to valproate monotherapy. *Neurology* 1989;39:1640–1641.
442. Pritchard PB III, Wannamaker BB, Sagel J, et al. Endocrine function following complex partial seizures. *Ann Neurol* 1983;14:27–32.
443. Provinciali L, Bartolini M, Mari F, et al. Influence of vigabatrin on cognitive performances and behaviour in patients with drug-resistant epilepsy. *Acta Neurol Scand* 1996;94:12–18.
444. Pumain R, Heinemann U. Stimulus- and amino acid-induced calcium and potassium changes in rat cortex. *J Neurophysiol* 1985;53:1–16.
445. Quiske A, Helmstaedter C, Lux S, et al. Depression in patients with temporal lobe epilepsy is related to mesial temporal sclerosis. *Epilepsy Res* 2000;39:121–125.
446. Ramsay RE, Cohen A, Brown MC. Coexisting epilepsy and nonepileptic seizures. In: Rowan AJ, Gates JR, eds. *Non-epileptic seizures.* Boston: Butterworth-Heineman, 1993:47–54.

447. Rao ML, Stefan H, Bauer J. Epileptic but not psychogenic seizures are accompanied by simultaneous elevation of serum pituitary hormones and cortisol levels. *Neuroendocrinology* 1989;49:33–39.

448. Rao SM, Devinsky O, Grafman J, et al. Viscosity and social cohesion in temporal lobe epilepsy. *J Neurol Neurosurg Psychiatry* 1992;55:149–152.

449. Rausch R. Effects of temporal lobe surgery on behavior. In: Smith DB, Treiman DM, Trimble MR, eds. *Neurobehavioral problems in epilepsy (Advances in neurology, volume 55).* New York: Raven Press, 1991:279–292.

450. Rausch R, Crandall PH. Psychological status related to surgical control of temporal lobe seizures. *Epilepsia* 1982;23:191–202.

451. Reeves AL, So EL, Evans RW, et al. Factors associated with work outcome after anterior temporal lobectomy for intractable epilepsy. *Epilepsia* 1997;38:689–695.

452. Reid PD, Shajahan PM, Glabus MF, et al. Transcranial magnetic stimulation in depression. *Br J Psychiatry* 1998;173:449–452.

453. Reunanen M, Dam M, Yuen A. A randomised open multi-centre comparative trial of lamotrigine and carbamazepine as monotherapy in patients with newly diagnosed or recurrent epilepsy. *Epilepsy Res* 1996;23:149–155.

454. Reynolds EH, Chanarin I, Milner G, et al. Anticonvulsant therapy, folic acid and vitamin B_{12} metabolism, and mental symptoms. *Epilepsia* 1966;7:261–270.

455. Ribak CE, Joubran C, Kesslak JP, et al. A selective decrease in the number of GABAergic somata occurs in pre-seizing monkeys with alumina gel granuloma. *Epilepsy Res* 1989;4:126–138.

456. Ricci GB. Magnetencephalography. In: Dam M, Gram L, eds. *Comprehensive epileptology.* New York: Raven Press, 1990:405–421.

457. Richens A. In: Pedley TA, Meldrum BS, eds. *Recent advances in epilepsy.* Edinburgh: Churchill Livingstone, 1992:197–210.

458. Riley T, Niedermeyer E. Rage attacks and episodic violent behaviour. Electroencephalographic findings and general consideration. *J Clin Electroencephalogr* 1978;9:113–139.

459. Ring HA, Crellin R, Kirker S, et al. Vigabatrin and depression. *J Neurol Neurosurg Psychiatry* 1993;56:925–928.

460. Ring HA, Moriarty J, Trimble MR. A prospective study of the early postsurgical psychiatric associations of epilepsy surgery. *J Neurol Neurosurg Psychiatry* 1998;64:601–604.

461. Roberts GW, Done DJ, Crow TJ. A "mock-up" of schizophrenia: temporal lobe epilepsy and schizophrenia-like psychosis. *Biol Psychiatry* 1990;28:127–143.

462. Robertson MM. Depression in neurological disorders. In: Robertson MM, Katona CLE, eds. *Depression and physical illness.* Chichester: John Wiley & Sons, 1997:305–340.

463. Robertson MM. Suicide, parasuicide, and epilepsy. In: Engel J Jr, Pedley TA, eds. *Epilepsy: a comprehensive textbook.* New York: Lippincott–Raven, 1997.

464. Robertson MM, Trimble MR. The treatment of depression in patients with epilepsy. *J Affect Disord* 1985;9:127–136.

465. Robertson MM, Trimble MR, Townsend HRA. Phenomenology of depression in epilepsy. *Epilepsia* 1987;28:364–372.

466. Robertson MM, Channon S, Baker J. Depressive symptomatology in a general hospital sample of outpatients with temporal lobe epilepsy: a controlled study. *Epilepsia* 1994;35:771–777.

467. Rockstroh B, Elbert T, Birbaumer N, et al. Cortical self-regulation in patients with epilepsies. *Epilepsy Res* 1993;14:63–72.

468. Rodin E. *The prognosis of patients with epilepsy.* Springfield, IL: Charles C Thomas, 1968.

469. Rodin E, Schmaltz S. The Bear-Fedio personality inventory and temporal lobe epilepsy. *Neurology* 1984;34:591–596.

470. Rodin EA, Katz M, Lennox K. Differences between patients with temporal lobe seizures and those with other forms of epileptic attacks. *Epilepsia* 1976;17:313–320.

471. Rodin EA, Rim CS, Kitano H, et al. A comparison of the effectiveness of primidone versus carbamazepine in epileptic outpatients. *J Nerv Ment Disord* 1976;163:41–46.

472. Roger J, Bureau M, Dravet C, et al. *Epileptic syndromes in infancy, childhood and adolescence.* London: John Libbey, 1992.

473. Rose KJ, Derry PA, McLachlan RS. Neuroticism in temporal lobe epilepsy: assessment and implications for pre- and post-operative psychosocial adjustment and health related quality of life. *Epilepsia* 1996;37:484–491.

474. Rosen JA. Dilantin dementia. *Trans Am Neurol Assoc* 1966;93:273.

475. Rosen RB, Stevens R. Action myoclonus in lithium toxicity. *Ann Neurol* 1983;13:221–222.

476. Rosenstein DL, Nelson JC, Jacobs SC. Seizures associated with antidepressants: a review. *J Clin Psychiatry* 1993;54:289–299.

477. Roth DL, Goode KT, Williams VL, et al. Physical exercise, stressful life experience, and depression in adults with epilepsy. *Epilepsia* 1994;35:1248–1255.

478. Rowe CC, Bercovic SF, Austin M, et al. Visual and quantitative analysis of interictal SPECT with 99Tc-HMPAOm-HMPAO in temporal lobe epilepsy. *J Nucl Med* 1991;32:1688–1694.

479. Rowe CC, Bercovic SF, Austin M, et al. Patterns of postictal blood flow in temporal lobe epilepsy: qualitative and quantitative analysis. *Neurology* 1991;41:1096–1103.

480. Roy A. Some determinants of affective symptoms in epileptics. *Can J Psychiatry* 1979;24:554–556.

481. Rush AJ, George MS, Sackheim HA, et al. Vagus nerve stimulation (VNS) for treatment-resistant depressions: a multicenter study. *Biol Psychiatry* 2000;47:276–286.

482. Ryback R, Ryback L. Gabapentin for behavioral dyscontrol [Letter]. *Am J Psychiatry* 1995;152:1399.

483. Ryback RS, Brodsky L, Munasilfi F. Gabapentin in bipolar disorder. *J Neuropsychiatry Clin Neurosci* 1997;9:301.

484. Sachdev P. Schizophrenia-like psychosis and epilepsy: the status of the association. *Am J Psychiatry* 1998;155:325–336.

485. Sackellares JC, Deaton R, Sommerville KW. Incidence of psychosis in controlled tiagabine (Gabitril™) trials. *Epilepsia* 1997;38[Suppl 3]:S67(abst).

486. Sackheim H, Greenberg MS, Weiman AL, et al. Hemispheric asymmetry in the expression of positive and negative emotion. *Arch Neurol* 1982;39:210–218.

487. Sadler M. Lamotrigine associated with insomnia. *Epilepsia* 1999;40:322–325.

488. Salanova V, Andermann F, Rasmussen T, et al. The running down phenomenon in temporal lobe epilepsy. *Brain* 1996;119:989–996.

489. Samt P. Epileptische Irreseinsformen. *Arch Psychiatrie* 1875;5:393–444.

490. Samt P. Epileptische Irreseinsformen. *Arch Psychiatrie Nervenkrankheiten* 1876;6:110–216.

491. Sander T. The genetics of idiopathic generalized epilepsy: implications for the understanding of its etiology. *Mol Med Today* 1996;2:173–180.

492. Sander JWAS, Patsalos PN. An assessment of serum and red blood cell folate concentrations in patients with epilepsy on lamotrigine therapy. *Epilepsy Res* 1992;13:89–92.

493. Sander JWAS, Duncan JS. Vigabatrin. In: Shorvan S, Dreifuss F, Fish D, et al., eds. *The treatment of epilepsy.* Oxford: Blackwell Science, 1996:491–499.

494. Sander JWAS, Hart YM, Johnson AL, et al. National General Practice Study of Epilepsy: newly diagnosed epileptic seizures in a general population. *Lancet* 1992;336:1267–1271.

495. Sanders RD, Mathews TA. Hypergraphia and secondary mania

in temporal lobe epilepsy. Case reports and literature review. *Neuropsychiatry Neuropsychol Behav Neurol* 1994;7:114–117.

496. Saunders M, Rawson M. Sexuality in male epileptics. *J Neurol* 1970;10:577–583.

497. Sawrie SM, Martin RC, Kuzniecky R, et al. Subjective versus objective memory change after temporal lobe epilepsy surgery. *Neurology* 1999;53:1511–1517.

498. Savard G, Andermann F, Olivier A, et al. Postictal psychosis after partial complex seizures: a multiple case study. *Epilepsia* 1991;32:225–231

499. Savard G, Manchanda R. Psychiatric assessment of candidates for epilepsy surgery. *Can J Neurol Sci* 2000;27[Suppl 1]:S44–S49.

500. Saver JL, Rabin J. The neural substrates of religious experience. *J Neuropsychiatry Clin Neurosci* 1997;9:498–510.

501. Savic I, Persson A, Roland P, et al. In vivo demonstration of reduced benzodiazepine-receptor binding in human epileptic foci. *Lancet* 1988;8616:863–866.

502. Saygi S, Katz A, Marks DA, et al. Frontal lobe partial seizures and psychogenic seizures: comparison of clinical and ictal characteristics. *Neurology* 1992;42:1274–1277.

503. Schaffer CB, Schaffer LC. Gabapentin in the treatment of bipolar disorder. *Am J Psychiatry* 1997;154:291–292.

504. Schmitz B. *Psychosen bei Epilepsie. Eine epidemiologische Untersuchung* [thesis]. FU Berlin; 1988.

505. Schmitz B. Psychosis and epilepsy: the link to the temporal lobe. In: Trimble MR, Bolwig TG, eds. *The temporal lobes and the limbic system.* Petersfield, UK: Wrightson Biomedical Publishing, 1992:149–167.

506. Schmitz B. Psychiatric syndromes related to antiepileptic drugs. *Epilepsia* 1999;40[Suppl 10]:S65–S70.

507. Schmitz B, Wolf P. Psychosis in epilepsy: frequency and risk factors. *J Epilepsy* 1995;8:295–305.

508. Schmitz EB, Moriarty J, Costa DC, et al. Psychiatric profiles and patterns of cerebral blood flow in focal epilepsy: interactions between depression, obsessionality, and perfusion related to the laterality of the epilepsy. *J Neurol Neurosurg Psychiatry* 1997;62:458–463.

509. Schmitz EB, Roberston MM, Trimble MR. Depression and schizophrenia in epilepsy: social and biological risk factors. *Epilepsy Res* 1999;35:59–68.

510. Schou M, Amdisen A, Trap-Jensen J. Lithium poisoning. *Am J Psychiatry* 1968;125:520–527.

511. Schover LR, Jensen SB. Physiological factors and sexual problems in chronic illness. In: *Sexuality and chronic illness: a comprehensive approach.* New York: Guilford Press. 1988.

512. Seidenberg M, O'Leary DS, Berent S, et al. Changes in seizure frequency and test re-test scores on the WAIS. *Epilepsia* 1981;22:75–83.

513. Seidenberg M, Hermann B, Noe A, et al. Depression in temporal lobe epilepsy. Interaction between laterality of lesion and Wisconsin Card Sort performance. *Neuropsychiatry Neuropsychol Behav Neurol* 1995;8:81–87.

514. Seidman-Ripley JG, Bound VK, Andermann F, et al. Psychosocial consequences of postoperative seizure relief. *Epilepsia* 1993;34:248–254.

515. Selai CE, Smith K, Trimble MR. Adjunctive therapy in epilepsy: a cost-effectiveness comparison of two AEDs. *Seizure* 1999;8: 8–13.

516. Selai CE, Elstner K, Trimble MR. Quality of life pre and post epilepsy surgery. *Epilepsy Res* 2000;38:67–74.

517. Sengoku A, Yagi K, Seino M, et al. Risks of occurrence of psychoses in relation to the types of epilepsies and epileptic seizures. *Folia Psychiatr Neurol Jpn* 1983;37:221–226.

518. Septien L, Grass P, Giroud M, et al. Depression and temporal lobe epilepsy: the possible role of laterality of the epileptic foci and gender. *Neurophysiol Clin* 1993;23:326–327.

519. Series HG. Drug treatment of depression in medically ill patients. *J Psychosom Res* 1992;36:1–16.

520. Shen W, Bowman ES, Markand ON. Presenting the diagnosis of pseudoseizures. *Neurology* 1990;40:756–759.

521. Sherwin I. Differential psychiatric features in epilepsy; relationship to lesion laterality. *Acta Psychiatr Scand* 1984;69[Suppl 313]:92–103.

522. Shorvon SD. Epidemiology, classification, natural course and genetics of epilepsy. *Lancet* 1990;336:93–96.

523. Shorvon S. *Status epilepticus.* Cambridge: Cambridge University Press, 1994.

524. Shorvon SD. Safety of topiramate: adverse events and relationships to dosing. *Epilepsia* 1996;37[Suppl 2]:S18–S22.

525. Shorvon S. Oxcarbazepine; a review. *Seizure* 2000;9:75–79.

526. Shorvon S, Reynolds EH. Reduction in polypharmacy for epilepsy. *BMJ* 1979;2:1023–1025.

527. Shorvon SD, Löwethal A, Janz D, et al. for the European Levetiracetam Study Group. Multicenter double-blind randomised placebo-controlled trial of levetiracetam as add-on therapy in patients with refractory partial seizures. *Epilepsia* 2000;41: 1179–1186.

528. Shukla GD, Katiyar BC. Psychiatric disorders in temporal lobe epilepsy: the laterality effect. *Br J Psychiatr* 1980;137:181–182.

529. Shukla GD, Sirvastava OM, Katiyar BC. Sexual disturbances in temporal lobe epilepsy: a controlled study. *Br J Psychiatry* 1979;134:288–292.

530. Shukla S, Mukherjee S, Decina P. Lithium in the treatment of bipolar disorders associated with epilepsy: an open study. *J Clin Psychopharmacol* 1988;8:201–204.

531. Sivilotti L, Nistri A. GABA receptor mechanisms in the central nervous system. *Prog Neurobiol* 1991;6:35–92.

532. Skowron DM, Stimmel GL. Antidepressants and the risk of seizures. *Pharmacotherapy* 1992;12:18–22.

533. Slater E, Beard AW. The schizophrenia-like psychoses of epilepsy: discussion and conclusions. *Br J Psychiatry* 1963;109:143–150.

534. Slater E, Moran PAP. The schizophrenia-like psychoses of epilepsy: relation between ages of onset. *Br J Psychiatry* 1969;115:599–600.

535. Slater E, Beard AW, Glithero E. The schizophrenia-like psychoses of epilepsy. *Br J Psychiatry* 1963;109:95–150.

536. Slater JD, Brown MC, Jacobs W, et al. Induction of pseudoseizures with intravenous saline placebo. *Epilepsia* 1995;36: 580–585.

537. Small JG, Milstein V, Stevens JR. Are psychomotor epileptics different? A controlled study. *Arch Neurol* 1962;7:187–194.

538. Smith DB. Cognitive effects of antiepileptic drugs. In: Smith DB, Treiman DM, Trimble MR, eds. *Neurobehavioral problems in epilepsy (Advances in neurology, volume 55).* New York: Raven Press, 1991:197–224.

539. Smith DB, Collins JB. Behavioral effects of carbamazepine, phenobarbital, phenytoin and primidone. *Epilepsia* 1987;28:598.

540. Smith PEM, McBride A. Illicit drugs and seizures. *Seizure* 1999;8:441–443.

541. Smith D, Baker G, Davies G, et al. Outcomes of add-on treatment with lamotrigine in partial epilepsy. *Epilepsia* 1993;34:312–322.

542. Snyder SL, Rosenbaum DH, Rowan DH, et al. SCID diagnosis of panic disorder in psychogenic seizure patients. *J Neuropsychiatry Clin Neurosci* 1994;6:261–266.

543. So NK, Savard G, Andermann F, et al. Acute postictal psychosis: a stereo EEG study. *Epilepsia* 1990;31:188–193.

544. Sperling MR, Pritchard PB III, Engel J Jr, et al. Prolactin in partial epilepsy: an indicator of limbic seizures. *Ann Neurol* 1986;20:716–722.

545. Spina E, Pisani F, Perucca E. Clinically significant pharmacokinetic drug interactions with carbamazepine. An update. *Clin Pharmacokinet* 1996;31:198–214.

546. Spitzer RL, Williams JBW. *Structured clinical interview for DSM-III-R Personality Disorders* (SCID-II, 5/1/86). New York: New York State Psychiatric Institute, Biometric Research Department, 1987.

547. Spring GK. EEG observations in confirming neurotoxicity [Letter]. *Am J Psychiatry* 1979136:1099–1100.

548. Standage KF. Schizophreniform psychosis among epileptics in a mental hospital. *Br J Psychiatry* 1973;123:231–232.

549. Standage KF, Fenton GW. Psychiatric symptom profiles of patients with epilepsy: a controlled investigation. *Psychol Med* 1975;5:152–160.

550. Stefan H. Multichannel magnetencephalography: recordings of epileptiform discharges. In: Lüders H, ed. *Epilepsy surgery.* New York: Raven Press, 1991:423–428.

551. Stefansson SB, Olafsson E, Hauser AW. Psychiatric morbidity in epilepsy: a case controlled study of adults receiving disability benefits. *J Neurol Neurosurg Psychiatry* 1998;64:238–241.

552. Steiner T. Comparison of lamotrigine and phenytoin monotherapy in newly diagnosed epilepsy. *Epilepsia* 1994;35[Suppl 8]:S31.

553. Steinlein OK, Noebels JL. Ion channels and epilepsy in man and mouse. *Curr Opin Genet Dev* 2000;10:286–291.

554. Stevens JR. Psychiatric implications of psychomotor epilepsy. *Arch Gen Psychiatry* 1966;14:461–471.

555. Stevens JR. Psychiatric consequences of temporal lobectomy for intractable seizures: a 20–30 year follow-up of 14 cases. *Psychol Med* 1990;20:529–545.

556. Stevens JR, Hermann BP. Temporal lobe epilepsy, psychopathology, and violence: the state of the evidence. *Neurology* 1981;31:1127–1132.

557. Stimmel GL, Dopheide JA. Psychotropic drug-induced reductions in seizure threshold. Incidences and consequences. *CNS Drugs* 1996;5:37–50.

558. Strauss E, Wada J, Moll A. Depression in male and female subjects with complex partial seizures. *Arch Neurol* 1992;49:391–392.

559. Sugarman P. Carbamazepine and episodic dyscontrol [Letter]. *Br J Psychiatry* 1992;161:721.

560. Swinkels JA, de Jonghe F. Safety of antidepressants. *Int Clin Psychopharmacol* 1995;9[Suppl 4]:19–25.

561. Tallian KB, Nahata MC, Lo W, et al. Gabapentin associated with aggressive behavior in paediatric patients with seizures. *Epilepsia* 1996;37:501–502.

562. Taylor DC. Sexual behaviour and temporal lobe epilepsy. *Arch Neurol* 1969;21:510–516.

563. Taylor DC. Ontogenesis of chronic epileptic psychoses. A reanalysis. *Psychol Med* 1971;1:247–253.

564. Taylor DC. Mental state and temporal lobe epilepsy, a correlative account of 100 patients treated surgically. *Epilepsia* 1972;13:727–765.

565. Taylor DC. Factors influencing the occurrence of schizophrenia-like psychosis in patients with temporal lobe epilepsy. *Psychol Med* 1975;5:249–254.

566. Taylor DC. Epilepsy and organic mania. *Biol Psychiatry* 1991;2:202–203.

567. Taylor DC, Marsh SM. Implications of long-term follow-up studies in epilepsy: with a note on the cause of death. In: Penry JK, ed. *Epilepsy, the Eighth International Symposium.* New York: Raven Press, 1977:27–34.

568. Taylor J. *Selected writings of John Hughlings Jackson,* vol. 2. London: Staples Press, 1958.

569. Tebartz van Elst L, Woermann FG, Lemieux L, et al. Amygdala enlargement in dysthymia—a volumetric study of patients with temporal lobe epilepsy. *Biol Psychiatry* 1999;46:1614–1623.

570. Tebartz van Elst L, Woermann FG, Lemieux L, et al. Affective aggression in patients with temporal lobe epilepsy. A quantitative MRI study of the amygdala. *Brain* 2000;123:234–243.

571. Tellenbach H. Epilepsie als Anfallsleiden und als Psychose. Ueber alternative Psychosen paranoider praegung bei "forcierter Normalisierung" (Landolt) des Elektroencephalogramms Epileptischer. *Nervenarzt* 1965;36:190–202.

572. Tharyan P, Kuruvilla K, Prabhakar S. Serum prolactin changes in epilepsy and hysteria. *Indian J Psychiatry* 1988;30:145–152.

573. Thau K, Rappelsberger P, Lovrek A, et al. Effect of lithium on the EEG of healthy males and females. *Neuropsychobiology* 1988;20:158–163.

574. Thijs RD, Kerr MP. The outcome of prescribing novel anticonvulsants in an outpatient setting: factors affecting response to medication. *Seizure* 1998;7:379–383.

575. Thomas L, Trimble M. The effects of vigabatrin on attention, concentration and mood: an investigation in healthy volunteers. *Seizure* 1996;5:205–208.

576. Thomas L, Trimble M, Schmitz B, et al. Vigabatrin and behaviour disorders: a retrospective survey. *Epilepsy Res* 1996;25:21–27.

577. Thompson PJ. The rehabilitation of people with epilepsy. In: Hopkins A, Shorvon S, Cascino G, eds. *Epilepsy,* 2nd ed. London: Chapman and Hall, 1995:573–579.

578. Thompson PJ, Oxley J. Socioeconomic accompaniments of severe epilepsy. *Epilepsia* 1988;29[Suppl 1]:S9–S18.

579. Thompson PJ, Trimble MR. Anticonvulsant drugs and cognitive functions. *Epilepsia* 1982;23:531–544.

580. Thompson PJ, Trimble MR. Comparative effects of anticonvulsant drugs on cognitive functioning. *Br J Clin Pract* 1982;18[Suppl]:154–156.

581. Thompson PJ, Trimble MR.1993 (in press).

582. Thompson PJ, Huppert F, Trimble MR. Phenytoin and cognitive functions; effects on normal volunteers and implications for epilepsy. *Br J Clin Psychol* 1981;20:151–162.

583. Tonini C, Beghi E, Vitezic D. Epilepsy surgery—methodological aspects and prognosis. A critical review of the literature. *Neurosurgery (in press).*

584. Toone BK. Sexual disorders in epilepsy. In: Pedley TA, Meldrum BS, eds. *Recent advances in epilepsy, 3.* Edinburgh: Churchill Livingstone, 1986:233–259.

585. Toone B, Dawson J, Driver MV. Psychoses of epilepsy. A radiological evaluation. *Br J Psychiatry* 1982;140:244–248.

586. Toone BK, Wheeler M, Nanjee M, et al. Sex hormones, sexual drive and plasma anticonvulsant levels in male epileptics. *J Neurol Neurosurg Psychiatry* 1983;46:824–826.

587. Toone BK, Edeh J, Nanjee MN, et al. Hyposexuality and epilepsy: a community survey of hormonal and behavioural changes in male epileptics. *Psychol Med* 1989;19:937–943.

588. Treiman DM. Epilepsy and violence: medical and legal issues. *Epilepsia* 1986;27[Suppl 2]:S77–S104.

589. Treiman D. Psychobiology of ictal aggression. In: Smith DB, Treiman DM, Trimble MR, eds. *Neurobehavioral problems in epilepsy (Advances in neurology, volume 55).* New York: Raven Press, 1991:55:341–356.

590. Trimble MR. Serum prolactin in epilepsy and hysteria. *BMJ* 1978;2:1682.

591. Trimble MR. PET-scanning in epilepsy. In: Trimble MR, Bolwig TG, eds. *Aspects of epilepsy and psychiatry.* Chichester: John Wiley & Sons, 1986:147–162.

592. Trimble MR. Cognitive hazards of seizure disorders. In: Trimble MR, ed. *Chronic epilepsy—its prognosis and management.* Chichester: John Wiley & Sons, 1989:103–111.

593. Trimble M. *The psychoses of epilepsy.* New York: Raven Press, 1991.

594. Trimble MR. Editorial. Behaviour changes following temporal lobectomy, with special reference to psychosis. *J Neurol Neurosurg Psychiatry* 1992;55:89–91.

595. Trimble MR. Long-term treatment of dysfunctional behaviour in epilepsy. In: Ancill RJ, Lader MH, eds. *Baillière's clinical psychiatry. Pharmacological management of chronic psychiatric disorders, volume 1.* London: Baillière Tindall, 1995:667–682.

596. Trimble M. Cognitive and personality profiles in patients with juvenile myoclonic epilepsy. In: Schmitz B, Sander T, eds. *Juvenile myoclonic epilepsy—the Janz syndrome.* Petersfield, UK: Wrightson Biomedical Publishing, 2000:101–109.

597. Trimble M, Perez M. The phenomenology of the chronic psychoses of epilepsy. *Adv Biol Psychiatry* 1980;8:98–105.

598. Trimble MR, Perez MM. The phenomenology of the chronic psychoses of epilepsy. In: Koella WP, Trimble MR, eds. *Temporal lobe epilepsy, mania and schizophrenia and the limbic system.* Basel: Karger, 1982:98–105.

599. Trimble MR, Schmitz B. The psychoses of epilepsy. A neurobiological perspective. In: McConnell HW, Snyder PJ, eds. *Psychiatric comorbidity in epilepsy. Basic mechanisms, diagnosis, and treatment.* Washington, DC: American Psychiatric Press, 1998:169–186.

600. Trimble MR, Thompson PJ. Neuropsychological aspects of epilepsy. In: Grant I, Adams KM, eds. *Neuropsychological assessment of neuropsychiatric disorders.* New York: Oxford University Press, 1986:321–346.

601. Trimble MR, Corbett JA, Donaldson D. Folic acid and mental symptoms in children with epilepsy. *J Neurol Neurosurg Psychiatry* 1980;43:1030–1034.

602. Trimble MR, Dana-Haeri J, Oxley J, et al. Some neuroendocrine consequences of seizures. In: Porter RJ, Ward AA Jr, Mattson RH, et al., eds. *Advances in epileptology: XVth Epilepsy International Symposium.* New York: Raven Press, 1984:201–208.

603. Trimble MR, Rüsch N, Betts T, et al. Psychiatric symptoms after therapy with new antiepileptic drugs: psychopathological and seizure related variables. *Seizure* 2000;9:249–254.

604. Trostle JA, Hauser A, Sharbrough FW. Psychologic and social adjustment to epilepsy in Rochester, Minnesota. *Neurology* 1989;39:633–637.

605. Tucker DM, Novelly RA, Walker PJ. Hyperreligiosity in temporal lobe epilepsy: redefining the relationship. *J Nerv Ment Disord* 1987;175:181–184.

606. Umbricht D, Degreef G, Barr WB, et al. Postictal and chronic psychoses in patients with temporal lobe epilepsy. *Am J Psychiatry* 1995;152:224–231.

607. Usiskin S. Counselling in epilepsy. In: Hopkins A, Shorvon S, Cascino, G, eds. *Epilepsy,* 2nd ed. London: Chapman and Hall, 1995:565–571.

608. Varma AR, Moriaty J, Costa DC, et al. HMPAO SPECT in non-epileptic seizures: preliminary results. *Acta Neurol Scand* 1996;94:88–92.

609. Vickrey BG. A procedure for developing a quality-of-life measure for epilepsy surgery. *Epilepsia* 1993;34[Suppl 4]:S22–S27.

610. Victoroff JI, Benson DF, Engel J Jr, et al. Interictal depression in patients with medically intractable complex partial seizures: electroencephalography and cerebral metabolic correlates. *Ann Neurol* 1990;28:221.

611. Victoroff J. DSMIIIR psychiatric diagnoses in candidates for epilepsy surgery: lifetime prevalence. *Neuropsychiatry Neuropsychol Behav Neurol* 1994;7:87–97.

612. Vossler DG. Nonepileptic seizures of physiologic origin. *J Epilepsy* 1995;8:1–10.

613. Walczak TS, Williams DT, Berten W. Utility and reliability of placebo infusion in the evaluation of patients with seizures. *Neurology* 1994;44:394–399.

614. Walden J, Hesslinger B, van-Calker D, et al. Addition of lamotrigine to valproate may enhance efficacy in the treatment of bipolar affective disorder. *Pharmacopsychiatry* 1996;29:193–195.

615. Walker AE, Blumer D. Long term effects of temporal lobe lesions on sexual behavior and aggressivity. In: Fields WS, Sweet WH, eds. *Neural bases of violence and aggression.* St. Louis: W.H. Green, 1975:392–400.

616. Waxman SG, Geschwind N. The interictal behaviour syndrome of temporal lobe epilepsy. *Arch Gen Psychiatry* 1975;32:1580–1586.

617. Weil AA. Ictal emotions occurring in temporal lobe dysfunction. *Arch Neurol* 1959;1:87–97.

618. Weiner RD, Coffey CE. Electroconvulsive therapy in the medical and neurologic patient. In: Stoudemire A, Fogel BS, eds. *Psychiatric care of the medical patient.* New York: Oxford University Press, 1993:207–224.

619. Wieser HG, Elger CE, Stodieck SRG. The "foramen ovale electrode": a new recording method for the preoperative evaluation of patients suffering from mesio-basal temporal lobe epilepsy. *Electroencephalogr Clin Neurophysiol* 1985;661:314–322.

620. Wieser HG. Selective amygdalohippocampectomy: indications, investigative techniques and results. *Adv Tech Stand Neurosurg* 1986;13:39–133.

621. Wharton RN. Grand mal seizures with lithium treatment. *Am J Psychiatry* 1969;125:152.

622. Whitman S, Coleman TE, Patmon C, et al. Epilepsy in prison: elevated prevalence and no relationship to violence. *Neurology* 1984;141:651–656.

623. Whitworth AB, Fleischhacker WW. Adverse effects of neuroleptic drugs. *Int Clin Psychopharmacol* 1995;9[Suppl 5]:21–27.

624. WHO. *The ICD-10 classification of mental and behavioural disorders: clinical descriptions and diagnostic guidelines.* Geneva: World Health Organization, 1992.

625. Whyler AR, Lockard JS, Ward AA, et al. Condition EEG desynchronisation and seizure occurrence in patients. *Electroencephalogr Clin Neurophysiol* 1976;41:501–512.

626. Wiegartz P, Seidenberg M, Woodard A, et al. Co-morbid psychiatric disorder in chronic epilepsy: recognition and etiology of depression. *Neurology* 1999;53[Suppl 2]:S3–S8.

627. Williams D. The structure of emotions reflected in epileptic experiences. *Brain* 1956;79:29–67.

628. Wilson SJ, Saling MM, Lawrence J, et al. Outcome of temporal lobectomy: expectations and the prediction of perceived success. *Epilepsy Res* 1999;36:1–14.

629. Wilson SJ, Kincade P, Saling MM, et al. Patient readmission and support utilization following anterior temporal lobectomy. *Seizure* 1999;8:20–25.

630. Wirrell EC, Camfield CS, Camfield PR, et al. Long-term psychosocial outcome in typical absence epilepsy. Sometimes a wolf in sheep's clothing. *Arch Pediatr Adolesc Med* 1997;151:152–158.

631. Woermann FG, Tebartz van Elst L, Koepp MJ, et al. Reduction of frontal neocortical grey matter associated with affective aggression in patients with temporal lobe epilepsy: an objective voxel by voxel analysis of automatically segmented MRI. *J Neurol Neurosurg Psychiatry* 2000;68:162–169.

632. Wohlfarth R, Saar J, Reinshagen G. Behavioural effects of lamotrigine. *Epilepsia* 1994;35[Suppl 7]:S72.

633. Wolf P. Psychosen bei Epilepsie. Ihre Bedingungen und

Wechselbeziehungen zu Anfaellen. Habilitationsschrift, Freie Universitaet Berlin, 1976.

634. Wolf P. Manic episodes in epilepsy. In: Akimoto H, Kazamatsuri H, Seino M, et al., eds. *Advances in epileptology: XIIth Epilepsy International Symposium.* New York: Raven Press, 1982: 237–240.

635. Wolf P. Classification of syndromes. In: Dam M, Gram L, eds. *Comprehensive epileptology.* New York: Raven Press, 1991: 87–97.

636. Wolf P. Acute behavioral symptomatology at disappearance of epileptiform EEG abnormality: paradoxical or "forced" normalization. In: Smith DB, Treiman DM, Trimble MR, eds. *Neurobehavioural problems in epilepsy (Advances in neurology, volume 55).* New York: Raven Press, 1991:127–142.

637. Wroblewski BA, McColgan K, Smith K, et al. The incidence of seizures during tricyclic antidepressant drug treatment in a brain-injured population. *J Clin Psychopharmacol* 1990;10:124–128.

638. Wyllie E, Lüders H, MacMillan JP, et al. Serum prolactin levels after epileptic seizures. *Neurology* 1984;34:1601–1604.

639. Wyllie E, Friedman D, Rothner AD, et al. Psychogenic seizures in children and adolescents: outcome after diagnosis by ictal video and electroencephalographic recording. *Pediatrics* 1990;85:480–485.

640. Wyllie E, Friedman D, Luders H, et al. Outcome of psychogenic seizures in children and adolescents compared with adults. *Neurology* 1991;41:742–744.

641. Young GB, Chandarana PC, Blume WT, et al. Mesial temporal lobe seizures presenting as anxiety disorders. *J Neuropsychiatry Clin Neurosci* 1995;7:352–357.

642. Younkin DP, Delivoria-Papadopoulos M, Maris J, et al. Cerebral metabolic effects of neonatal seizures measured with in vivo P-31 NMR spectroscopy. *Ann Neurol* 1986;20:513–519.

643. Zaret BS, Cohen RA. Reversible valproic acid-induced dementia: a case report. *Epilepsia* 1986;27:234–240.

644. Zielinski JJ. Epidemiology and medical-social problems of epilepsy in Warsaw (final report on research program no. 19-P-58325-F-01 DHEW Social and Rehabilitation Services). Washington, DC: U.S. Government Printing Office, 1974.

645. Zielinski JJ. Epidemiology. In: Laidlaw J, Richens A, eds. *A textbook of epilepsy.* Edinburgh: Churchill Livingstone, 1974: 16–33.

646. Juul-Jensen, 1986.

647. Sander et al., 1991.

648. Smith DB, Treiman DM, Trimble MR, eds. *Neurobehavioral problems in epilepsy (Advances in neurology, volume 55).* New York: Raven Press, 1991.

40

MIGRAINE

HUA CHIANG SIOW
STEPHEN D. SILBERSTEIN

Migraine headache and psychiatric disorders have similar classification and diagnostic systems and overlap with regard to their comorbidity, mechanisms, and treatment. The classification system for headache disorders developed by the International Headache Society (IHS) (28) was modeled on the Diagnostic and Statistical Manual of Mental Disorders, Third Edition (DSM-III) classification system for psychiatric disorders (4). Diagnosing migraine is similar to diagnosing psychiatric disorder: both rely on patients' retrospective reporting of their experiences. Results of physical examination and laboratory tests serve primarily to exclude other, more ominous, causes of headache. Like many psychiatric disorders, headache classification uses symptom-based criteria in the absence of a truly objective diagnostic "gold standard."

The IHS system provides clear operational rules for headache diagnosis (28). The headache disorders differ in symptom profiles, severity, and response to treatment. The IHS system separates migraine from tension-type headache, but this distinction is not universally accepted (43). In psychiatry, there are similar arguments about the relationships between major depression and anxiety disorders.

In this chapter, we discuss migraine, which is a disabling headache disorder that is highly comorbid with depression, anxiety disorders, and a number of other conditions. We discuss the mechanisms responsible for the comorbidity of migraine and psychiatric disease and then consider some of the elements of treatment.

MIGRAINE PREVALENCE

Migraine is a very common condition, with a lifetime prevalence of 16%. Migraine prevalence is age and gender dependent. Before puberty, migraine is slightly more common in boys, with the highest incidence between ages 6 and 10 years. In women, the incidence is highest between ages 14 and 19 years. In general, women are affected more commonly than men (lifetime prevalence 12%–17% vs. 4%–6%, respectively) (38,65). In the American Migraine Study, the 1-year prevalence of migraine increased with age among women and men, reached the maximum at ages 35 to 45 years, and declined thereafter (40). Migraine prevalence is inversely proportional to income, with the low-income group having the highest prevalence. Race and geographic region also influence migraine prevalence. It is highest in North America and Western Europe, and more prevalent among Caucasians than African-Americans and Asian-Americans (39). Migraine is influenced by environmental and genetic factors. Migraine with aura has a stronger genetic influence than migraine without aura and is more influenced by environmental factors. Behavioral, emotional, and climatologic changes may trigger migraine, modify the vulnerability to migraine, and impact on its prevalence.

Evidence suggests that migraine incidence may be increasing. In a population-based survey of migraine in Olmsted County, Stang and Osterhaus (61) found that from 1979 to 1981 there was a striking increase in the age-adjusted incidence of migraine in patients younger than 45 years. Migraine incidence increased 34% for women and 100% for men. In this study, the overall age-adjusted incidence was 137 per 100,000 person-years for men and 294 per 100,00 person-years for women. This was confirmed in a study conducted among school children in Finland; the prevalence of migraine increased from 1.9% to 5.7% over an 18-year period. Neither study provided evidence for the cause of this increase.

In contrast, the American Migraine II Study follow-up to the original American Migraine Study showed that the prevalence of migraine in the United States is 18.2% for women and 6.5% for men (40). This is essentially unchanged from the original study (prevalence 17.6% and 5.7%, respectively). The distribution of disease by sociodemographic factors has remained stable over the last decade, and migraine continues to be more prevalent in Caucasians than in other racial groups and in lower-income than higher-income groups.

QUALITY OF LIFE

Migraineurs often are profoundly affected by their condition, and this impacts not only on their lives but their families' lives as well. When children and adolescents have migraines, their education and social development suffer. About 10% of migrainous children missed at least 1 day of school over a 2-week period due to migraine; nearly 1% missed 4 days. Migraineurs were bedridden for about three million days per month and had an estimated 74.2 million days per year of restricted activity due to migraine. In the United States, of the estimated 6,196,378 migraineurs who worked outside the home, the estimates of lost labor time due to migraine ranged from 1.570 to 4.271 days per year per person. Although the cost to society from lost work productivity is substantial, these figures may provide an underestimate of migraine-related economic loss because the costs from lost productivity are determined not only by missed work days but also by reduced work efficiency. Approximately 80% of employed migraineurs report that their job performance has been negatively impacted by migraine. Estimates of the level of efficiency during the attack vary from 56% to 73% of full capacity. The projected amount of lost work resulting from migraine costs employers between $5.6 and $17.2 billion annually. It is difficult to calculate similar estimates for costs of lost productivity in housewives; however, housewives experienced an estimated 38 million days per year of restricted activity. Eighty-five percent of women and 77% of men visited a physician at some point for their migraine. Migraine is a relatively common disease with significant social and financial impact.

Appropriate and timely migraine treatment is a cost-effective strategy. Today there are more pharmacologic, educational, and behavioral approaches to headache treatment than ever before. The ultimate success of new headache therapies depends not only on efficacy but also on the outcome of cost-effectiveness analyses of these agents.

Migraine has profound effects on daily functioning as ascertained by health-related quality-of-life and functional status questionnaires. Functional status questionnaires measure how headache affects physical and emotional functioning; quality-of-life questionnaires evaluate the subjective effects of the condition.

The original American Migraine Study estimated that 23 million US residents had severe migraine headaches. The American Migraine Study II raised this estimate to 28 million (40). Twenty-five percent of migraineurs have more than four severe attacks per month. Interpersonal relationships are impaired in approximately 70% of adult migraine patients. Regular activities are limited during 78% of migraine attacks, and 50% of migraineurs cancel normal activities. About 50% of subjects believed that their headaches had an effect on their families. Headache subjects modify their behavior to avoid precipitating attacks. More than 75% avoided headache-triggering factors, such as smoke, noise, emotional stress, and certain physical activities, even if job responsibilities were affected. Despite the high disability headaches impose on individuals, only 66% of migraineurs have sought medical attention.

ANATOMY OF MIGRAINES

A typical migraine attack may encompass all or part of the following four phases.

Prodrome

A prodrome occurs hours or days before headache onset in up to 60% of migraineurs (6). It occurs with equal frequency in migraine with and migraine without aura. It may consist of an alteration in mental state (depression, hyperactivity, euphoria, irritability, restlessness); an alteration in neurologic state (photo/phonophobia, yawning, hyperosmia); or an alteration in general (food cravings, anorexia, diarrhea, constipation, stiff neck) (5). An electronic diary study of premonitory symptoms was performed in 97 migraineurs who believed that they could predict their headaches (26). Premonitory symptoms correctly predicted headache 72% of the time. The most common premonitory features were lethargy, concentration problems, and stiff neck.

Aura

The migraine aura consists of focal neurologic symptoms, which usually occur up to 60 minutes before the onset of a headache, develop over 5 to 20 minutes, and last no more than 60 minutes. Auras may be visual, sensory, or motor, and they can occur in succession. Auras frequently occur before a headache, but they may occur with the headache or the headache may not follow at all. Visual symptoms range from minor visual disturbances (phosphenes, scotomas) to complex phenomena such as teichopsia, micropsia, macropsia, and palinopsia (32,37,52,59). Sensory auras usually consist of cheiroaural paresthesias with numbness migrating from the hand up the arm to the face. They usually occur with visual auras. Motor auras consist of unilateral weakness and usually occur with sensory auras.

Headache

The features of migraine without aura are outlined in Table 40.1. Not all of the features are required, and the headache can be bilateral and nonthrobbing. Headaches occur more frequently upon arising in the morning and last between 4 and 72 hours. Symptoms associated with the pain include concentration/memory problems, stiff neck, irritability, anorexia, dizziness, diarrhea, polyuria, nausea, vomiting, and sensitivity to light and sound. Some of these symptoms (nausea, photophobia, phonophobia) are obligatory (Table 40.1).

Resolution

In the resolution phase, after the pain has ended, the patient is drained of energy and irritable, and has scalp tenderness or

TABLE 40.1. MIGRAINE WITHOUT AURA

Previously used terms: common migraine, hemicrania simplex
Diagnostic criteria
   A. At least five attacks fulfilling criteria B–D
   B. Headache lasting 4–72 hours (untreated or unsuccessfully treated)
   C. Headache has at least two of the following characteristics:
      1. Unilateral location
      2. Pulsating quality
      3. Moderate or severe intensity (inhibits or prohibits daily activities)
      4. Aggravation by walking stairs or similar routine physical activity
   D. During headache at least one of the following:
      1. Nausea and/or vomiting
      2. Photophobia and phonophobia
   E. Not attributable to another disorder

mood changes. Euphoria or depression can accompany this phase.

CLASSIFICATION

Most prior migraine descriptions stressed three features: (a) the unilateral distribution of the headache, (b) the presence of a warning (often visual), and (c) nausea or vomiting. The Ad Hoc Committee on Classification of Headache described vascular headache of migraine type as "recurrent attacks of headache, widely varied in intensity, frequency, and duration. The attacks are commonly unilateral in onset; are usually associated with anorexia and sometimes with nausea and vomiting; in some are preceded by, or associated with, conspicuous sensory, motor, and mood disturbances; and are often familial" (25).

The IHS proposed and published their classification of headache disorders in an attempt to increase precision. The IHS criteria attempt to diagnose headache types rather than headache syndromes, yet they require a number of attacks over a period of time before a diagnosis can be made. The criteria currently are undergoing revision.

Migraine headaches were formerly divided into two varieties: classic migraine and common migraine. The IHS now calls common migraine "migraine without aura" and classic migraine "migraine with aura," with the aura being the complex of focal neurologic symptoms that precedes or accompanies an attack. At most, only 30% of migraine headaches are "classic." The same patient may have migraine headache without aura, migraine headache with aura, and migraine aura without headache.

To establish a diagnosis of migraine under the IHS classification, certain clinical features must be present and organic disease must be excluded. To diagnose migraine without aura, five attacks are needed, each lasting 4 to 72 hours and having two of the following four characteristics: (a) unilateral location, (b) pulsating quality, (c) moderate-to-severe

intensity, and (d) aggravation by routine physical activity. In addition, the attacks must have at least one of the following symptoms: nausea (and/or vomiting), photophobia, and phonophobia (Table 40.1). Using these criteria, no single associated feature is mandatory for diagnosing migraine, although recurrent episodic attacks must be documented. A patient who has pulsatile pain aggravated by routine activity, photophobia, and phonophobia meets the criteria, as does the more typical patient with unilateral throbbing pain and nausea. The headache is called *migrainous* if one of the criteria is missing.

A migraine attack usually lasts less than 1 day; when it persists for more than 3 days, the term *status migrainosus* is applied. Although migraine often begins in the morning and sometimes awakens the patient from sleep at dawn, it can begin at any time of the day or night. The frequency of attacks is extremely variable, ranging from a few in a lifetime to several a week. The median attack frequency is 1.5 attacks per month; 10% of migraineurs have one attack per week.

To diagnose IHS migraine with aura (Table 40.2), at least two attacks with any three of four features are required: (a) one or more fully reversible aura symptoms; (b) the aura develops over more than 4 minutes but (c) lasts less than 60 minutes; and (d) the headache follows the aura with a free interval of less than 60 minutes. Migraine with aura is subdivided into migraine with typical aura (homonymous visual disturbance, unilateral numbness or weakness, aphasia); migraine with prolonged aura (aura lasting more than 60 minutes); familial hemiplegic migraine; basilar migraine; migraine aura without headache; and migraine with acute-onset aura. Other varieties of migraine include ophthalmoplegic migraine, retinal migraine, and childhood periodic syndromes. The IHS classification system does not have specific criteria for the headache of migraine with aura, but it is reasonable to expect that it would be similar to that of migraine without aura. Most migraine-with-aura sufferers also have attacks without aura. Neurologic symptoms usually develop over 5 to 20 minutes and last less than 60 minutes. The most common auras are visual, but an aura may

TABLE 40.2. MIGRAINE WITH AURA (CLASSIC MIGRAINE)

Diagnostic criteria
   A. At least two attacks fulfilling criteria B
   B. At least three of the following four characteristics:
      1. One or more fully reversible aura symptoms indicating brain dysfunction
      2. At least one aura symptom develops gradually over more than 4 minutes or two or more symptoms occur in succession
      3. No single aura symptom lasts more than 60 minutes
      4. Headache follows aura with a free interval of less than 60 minutes (it may begin before or simultaneously with the aura)
   C. Not attributable to another disorder

consist of essentially any neurologic symptom. These symptoms include changes in visual perception, auditory and olfactory hallucinations (rare), somatosensory auras (common), mild hemiparesis (uncommon), vertigo (common), aphasia (uncommon), and confusion (mainly in children). In contrast to a transient ischemic attack, the aura of migraine evolves gradually and consists of both positive (e.g., scintillations, tingling) and negative (e.g., scotoma, numbness) features (3). If the aura is stereotypical, the diagnosis of migraine with aura is warranted, even if the subsequent headache does not have typical migrainous features. Almost any symptom or sign of brain dysfunction may be a feature of the aura, but the most common aura is a visual phenomenon.

Focal symptoms and signs of the aura may persist beyond the headache phase. Formerly termed *complicated migraine,* the IHS classification has introduced two more specific labels. If the aura lasts for more than 1 hour but less than 1 week, the term *migraine with prolonged aura* is applied. If the signs persist for more than 1 week or a neuroimaging procedure demonstrates a stroke, a *migrainous infarction* has occurred. Particularly in middle or late life, the aura may not be followed by the headache (migraine equivalent or late-life migraine accompaniment)

Familial Hemiplegic Migraine

Familial hemiplegic migraine is an autosomal dominant inherited migraine variant. A gene defect localized to the short arm of chromosome 19p13 results in abnormal coding for the α1A subunit of the P/Q-type calcium channel gene *CACNA1A* (48). Patients with this form of migraine experience hemiparesis with their migraine attacks, which also may be associated with sensory and visual auras.

Basilar Migraine

Basilar migraine, also known as *Bickerstaff migraine* (11), is characterized by aura symptoms characteristic of a brainstem disturbance (e.g., diplopia, tinnitus, ataxia, dysarthria, bilateral paresthesias or weakness, decreased hearing, vertigo, decreased level of consciousness, bilateral occipital lobe dysfunction). The aura often lasts less than 1 hour and is followed by headache. This syndrome usually affects children, adolescents, and young women, although it can occur in all age groups and in both sexes.

Chronic Migraine

We use the term *chronic migraine* to describe the condition previously known as *transformed migraine.* This term includes all migrainous and tension-type headaches that do not meet the diagnostic criteria for hemicrania continua or new daily persistent headache. The history consists of rapidly increasing episodic migraine frequency with lessening of migrainous features over at least 3 months. These headaches become very frequent, occurring more than 15 days per month and lasting for at least 4 hours per day.

Many patients with chronic migraine have a history of analgesic, triptan, or ergot overuse leading to their increased headache frequency. This chronic headache is called *rebound headache* and leads to a further increase in analgesic consumption and a "catch-22" situation. Withdrawing the offending medication leads to increased headache severity. These patients probably benefit from detoxification therapy, which breaks the cycle with either inpatient or outpatient therapy. Until patients complete their detoxification, no preventive medication will help their migraines. The process may take from 6 to 8 weeks to complete. Chronic migraine may develop from episodic migraine without acute treatment overuse.

PATHOPHYSIOLOGY

There are two main competing theories for migraine pathogenesis: the vasogenic theory and the neurogenic theory. At first these two theories appear to be at odds, but novel ways of looking at brain function indicate that they may complement each other.

The premise of the vasogenic theory is that focal ischemia is the root cause of the migraine aura. Auras are due to hypoperfusion secondary to vasoconstriction of the blood vessel that supplies the cortical lobe that corresponds to the aura symptom, be it visual, sensory, or motor. Reactive vasodilation would explain the genesis of pain through stimulation of the perivascular pain sensitive fibers. This theory would explain the throbbing quality of pain, its varied location, and the relief of pain caused by vasoconstrictive agents such as ergotamine. Patients undergoing the aura phase of their attacks were studied using xenon-133 blood flow techniques. These studies revealed from 17% to 35% reduction in cerebral blood flow in the posterior regions of the brain. More recent positron emission tomographic scan studies during the aura phase of migraine confirmed these results and revealed slowly spreading hypoperfusion. Neither group of studies showed evidence of ischemia.

The neurogenic theory has as its basis the cortical spreading depression of Leão (34). This theory predicts that alterations in blood flow develop as a consequence of neuronal events. During a migraine attack, there is an initial brief hyperperfusion phase followed by a relatively sustained phase of hypoperfusion that corresponds to cortical spreading depression (22). This probably reflects a wave of neuronal and glial depolarization followed by longer-lasting suppression of neural activity. Credence has been lent to this theory by magnetic resonance imaging studies using blood oxygenation level-dependent and perfusion-weighted imaging techniques during migraine with aura attacks; these studies revealed signal changes characteristic of cortical spreading depression within the human brain (53,54). These studies revealed an initial vasodilation occurring with the onset of visual aura that progressed contiguously over the occipital cortex at a rate of 3.5 ± 1.1 mm per minute. This initial vasodilation was followed by hypoperfusion. The findings that the blood oxygenation level-dependent signal changes during aura abort at major sulci, that the light-evoked visual responses were suppressed during migraine with aura attacks and took 15 minutes to return to 80% of baseline, and that the areas first affected were the first to recover provide strong evidence that an electrophysiologic event, such as cortical spreading depression, generates the migraine aura in human visual cortex.

The brain is an insensate organ that lacks innervation by pain fibers. However, the dura mater and meningeal blood vessels are richly innervated by sensory nerve fibers that originate from the ophthalmic division of the trigeminal nerve. Together with the trigeminal nucleus, they constitute the trigeminovascular system. During a migraine attack, these sensory fibers release vasodilating and permeability-promoting peptides from perivascular nerve endings, e.g., substance P, calcitonin gene-related peptide, and neurokinin A. These peptides promote a sterile inflammatory response within the dura matter and cause the sensitization of sensory nerve fibers to previously innocuous stimuli (e.g., blood vessel pulsations or venous pressure changes) that manifests itself as increased intracranial mechanosensitivity and hyperalgesia that are worsened by coughing or sudden head movement. Calcitonin gene-related peptide levels have been found to be elevated in jugular venous blood during migraine attacks, but the levels return to normal after administration of sumatriptan and amelioration of the headache. This is consistent with neuropeptide release from activated sensory nerves during the migraine attack and blockade of peptide release by sumatriptan mediated via 5-HT$_{1B/1D}$ prejunctional receptors on sensory terminals. A single photon emission computed tomographic study provided the first direct evidence for the presence of plasma protein extravasation localized to extraparenchymal regions ipsilateral to the side of pain during a spontaneous migraine attack.

Trigeminovascular activation occurs secondary to an initiation factor for migraine attacks. What initiates this is still unclear. Brainstem, cortical structures, or neurochemical dysfunction may play an important role in either the genesis or modulation of migraines, or both. It now appears that the migraine brain is inherently more hyperexcitable due to either genetic factors (point mutations in genes encoding calcium channels, mitochondrial energy impairment, magnesium deficiency) or external factors (stress and hormonal changes).

Brainstem nuclei, including the periaqueductal gray matter, locus caeruleus, and dorsal raphe nuclei, have been shown to either generate or suppress pain symptoms that resemble migraine in animals and humans. Noradrenergic and serotoninergic nuclei participate in stress responses, anxiety, and depressive states. Migraineurs may exhibit central hypersensitivity to dopaminergic stimulation, which has been linked to behaviors observed during migraine, such as yawning, irritability, hyperactivity, gastroparesis, nausea, and vomiting. Molecular genetic studies have provided further evidence for the involvement of the dopaminergic system. Migraine is associated with the dopaminergic hypersensitivity phenotype and genes encoding the DRD2 receptor.

The possibility of a "migraine generator" in the rostral brainstem was raised by a positron emission tomographic blood flow study performed during spontaneous unilateral headache in nine patients without aura. Increased regional cerebral blood flow was found in medial brainstem predominantly contralateral to the headache and persisted after relief of migraine pain with sumatriptan. Whether these brainstem nuclei serve as migraine generators, participate in modifying the threshold for neuronal activation, or are part of the neuronal system that terminates an attack remains to be clarified.

COMORBIDITY

Comorbidity refers to the coexistence of one disorder with another that occurs more commonly than by chance. Epilepsy, stroke, depression, mania, anxiety, and panic disorders are comorbid with migraine. This has important implications for diagnosis and treatment, because depression, anxiety disorders, and epilepsy all can cause headaches, and headaches can alter mood and behavior.

Often in medicine, we try to ascribe just one diagnosis to a patient's complaints. This may not be justified in migraineurs because they often have a concomitant condition. It also is very easy to attribute migrainous symptoms to depression in patients with clinically undiagnosed migraine. This differentiation is important, because often there are opportunities to treat both disorders with one drug and thus avoid polypharmacy. For example, antidepressants can be used to treat both depression and migraine. Bipolar disorder or epilepsy, co-occurring with migraine, can be treated with anticonvulsants. However, the converse also is true. It may be impossible to use certain drugs (e.g., β-blockers) to treat migraineurs who also have depression because these drugs can exacerbate their depression.

In addition to the diagnostic and therapeutic implications, the presence of comorbidity may provide clues to the pathophysiology of migraine. When two conditions occur in the same person, there are a number of alternative causal explanations. First, the apparent associations may arise by coincidence or because of the method of subject ascertainment (10). Second, one condition may cause the other. Third, shared environmental or genetic risk factors might account for the co-occurrence of two disorders. For example, head injury is a risk factor for both migraine and epilepsy and may account for part of the relationship between the disorders. Shared genetic risk factors may account for the association between comorbid disorders. Finally, independent genetic or environmental risk factors may produce a brain state that gives rise to both migraine and a comorbid condition. For example, migraine and depression both have been attributed to a brain state characterized by serotonergic dysregulation.

Migraine and Epilepsy

Migraine shares many similar features with epilepsy. Both are episodic neurologic disorders and may involve hallucinations, mood changes, alteration of consciousness, and focal neurologic deficits. Headaches are not uncommon during the ictal and postictal phases of seizures, especially with occipital lobe seizures (7,35,49). Migraine auras have been shown to trigger seizures, and the incidence of migraine in epileptics is twice that in nonepileptics (7,35,44). This is especially true when epilepsy is caused by head trauma. However, migraine occurred in every subgroup of epilepsy defined by seizure type, age at onset, etiology, and family history of epilepsy (49).

Differentiating migraine from epilepsy can be difficult (7,23,44). The most helpful clinical features are the duration of the aura and the aura symptom profile. An aura that lasts more than 5 minutes suggests a migraine aura; an aura that lasts less than 5 minutes suggests an epileptic aura. Alteration of consciousness, automatism, and positive motor features are suggestive of an epileptic aura, whereas a migraine aura normally consists of a mix of positive and negative features such as a scintillating scotoma.

The comorbidity of migraine and epilepsy presents unique therapeutic opportunities. In the area of preventive therapy for migraines, there are drugs such as divalproate sodium, gabapentin, and topiramate, which are efficacious in treating both migraines and epilepsy. Conversely, neuroleptics and antidepressants, which often are used in preventive and acute therapy of migraines, may be contraindicated in epilepsy patients because they are known to lower the seizure threshold.

Migraine and Psychiatric Disorders

Migraine is comorbid with depression, mania, anxiety, and panic disorder. Major depression is more prevalent in migraineurs; conversely, migraine is more common in patients suffering from major depression.

Merikangas et al. (45) reported on the association of migraine with specific psychiatric disorders in a random sample of 457 adults aged 27 to 28 years in Zurich, Switzerland. Persons with migraine (n = 61) were found to have increased 1-year rates of affective and anxiety disorders. Specifically, the odds ratio (OR) for major depression (OR = 2.2; 95% confidence interval [CI] = 1.1–4.8), bipolar spectrum disorders (OR = 2.9; 95% CI = 1.1–8.6), generalized anxiety disorder (OR = 2.7; 95% CI = 1.5–5.1), panic disorder (OR = 3.3; 95% CI = 0.8–13.8), simple phobia (OR = 2.4; 95% CI = 1.1–5.1), and social phobia (OR = 3.4;

95% CI = 1.1–10.9) were significantly higher in persons with migraine than in persons without migraine (46,47).

Migraine with major depression frequently is complicated by an anxiety disorder. It has been suggested that in persons with all three disorders, the onset of anxiety generally precedes the onset of migraine, whereas the onset of major depression most often follows the onset of migraine.

Stewart et al. (63) studied the relationship of migraine to panic disorder and panic attacks in a population-based telephone interview survey of 10,000 residents aged 12–29 years in Washington County, Maryland. Men and women with a history of panic disorder reported the highest rates of migraine headaches occurring in the week preceding the study. The relative risk of migraine headache occurring during the previous week and associated with a history of panic disorder was 6.96 in men and 3.70 in women. In a follow-up analysis of the same sample, 14.2% of women and 5.8% of men who had experienced headache in the previous 12 months had consulted a physician for the problem. An unexpectedly high proportion of those who had consulted a physician for headache had a history of panic disorder. Of those who had recently seen a physician, 15% of women and 12.8% of men between the ages of 24 and 29 years had a panic disorder. This suggests that comorbid psychiatric disease is associated with seeking care for headache disorders.

Breslau and Davis (18) studied the association of IHS-defined migraine with specific psychiatric disorders in a sample of 1,007 young adults aged 21 to 30 years in southeast Michigan. Persons with a history of migraine (n = 128) had significantly higher lifetime rates of affective disorder, anxiety disorder, illicit drug use disorder, and nicotine dependence. The sex-adjusted ORs were 4.5 for major depression (95% CI = 3.0–6.9), 6.0 for manic episode (95% CI = 2.0–18.0), 3.2 for any anxiety disorder (95% CI = 2.2–4.6), and 6.6 for panic disorder (95% CI = 3.2–13.9). The psychiatric comorbidity odds associated with migraine with aura were generally higher than those associated with migraine without aura. Migraine with aura was associated with an increased lifetime prevalence of both suicidal ideation and suicide attempts, after the factors of sex, major depression, and other co-occurring psychiatric disorders were controlled for (15,19).

Using follow-up data gathered 3.5 years after baseline, Breslau et al. (20) reported on the prospective relationship between migraine and major depression in a cohort of young adults. The relative risk for the first onset of major depression during the follow-up period in persons with prior migraine versus no prior migraine was 4.1 (95% CI = 2.2–7.4). The relative risk for the first onset of migraine during the follow-up period in persons with prior major depression versus no history of major depression was 3.3 (95% CI = 1.6–6.6) (Table 40.3).

In a later study, Swartz et al. (66) conducted a prospective survey on psychiatric disorders and migraine headaches in the Baltimore area in 1981 and again between 1993 and

TABLE 40.3. RELATIVE RISK ESTIMATES FROM TWO COX PROPORTIONAL HAZARDS MODELS (DETROIT AREA)

	RR	95% CI	*p* Value
Risk of migraine in major depression			
Migraine	3.2	2.3–4.6	0.0001
Sex (female)	1.7	1.2–2.5	0.0014
Education (<college)	1.2	0.9–1.7	0.183
Risk of major depression in migraine			
Major depression	3.1	2.0–5.0	0.0001
Sex (female)	2.7	1.8–4.1	0.0001
Education (<college)	1.8	1.3–2.6	0.0014

CI, confidence interval; RR, relative risk.

1996. In the at-risk population of 1,343, there were 118 incident cases of migraine headaches. The age- and sex-specific incident rates of migraine headaches followed the expected patterns, with younger age and female sex identified as risk factors. In cross-sectional analyses, major depression (OR = 3.14; 95% CI = 2.03–4.84) and panic disorder (OR = 5.09; 95% CI = 2.65–9.79) had the strongest associations, and alcohol and other substance abuse were not associated. In logistic regression models including age, sex, and psychiatric illness in 1981, only phobia was predictive of incident migraines (OR = 1.70; 95% CI = 1.11–2.58). The conclusion was that there was strong cross-sectional association between affective disorders and incident migraine headaches in this cohort. The relation between antecedent affective disorders and incident migraine headaches is less tenuous.

In summary, epidemiologic studies support the association between migraine and major depression previously reported in clinic-based studies. The prospective data indicate that the observed cross-sectional or lifetime association between migraine and major depression could result from a bidirectional influence, from migraine to subsequent onset of major depression, and from major depression to first migraine attack. Furthermore, these epidemiologic studies indicate that persons with migraine have increased prevalence of bipolar disorder, panic disorder, and one or more anxiety disorders (17). It has been proposed that major depression in persons with migraine might represent a psychologic reaction to repeated, disabling migraine attacks. Migraine has an earlier mean age at onset than major depression, both in the general population and in persons with comorbid disease. Nonetheless, the bidirectional influence of each condition on the risk for the onset of the other is incompatible with the simple causal model. Furthermore, Breslau et al. (20) reported that the increased risk for a first episode of major depression (and/or panic disorder) did not vary according to the proximity of migraine attacks. These findings dampen the plausibility that the association between migraine and depression results from the demoralizing experience of recurrent and disabling headaches and suggests instead that their association might reflect shared etiologies.

Depression has been linked to other pain conditions. The evidence of the association between depression and pain calls into question the specificity of the connection between migraine and major depression. A report of a higher prevalence of major depression in persons with migraine but not episodic tension-type headache did not address the specificity issue because the difference in the rates of depression between the two headache conditions might have been explained by differences in severity.

Breslau et al. (21) examined migraine–depression comorbidity in a large-scale epidemiologic study, the Detroit Area Study of Headache. The study consisted of three groups: persons with migraine (n = 536); persons with other severe headaches of comparable pain severity and disability (n = 162); and matched controls with no history of severe headache (n = 586). These three representative samples of the population were identified by a random-digit dialing telephone survey of 4,765 persons aged 25 to 55 years. The lifetime prevalence of major depression in persons with migraine was 40.7%, in those with other severe headaches 35.8%, and in controls 16.0%. Sex-adjusted ORs in the two headache groups, relative to controls, were approximately of the same magnitude, 3.5 and 3.2. However, examination of the bidirectional relationship between major depression and each headache type yielded different results. With respect to migraine, a bidirectional relationship was observed: migraine signaled an increased risk for the first onset of major depression, and major depression signaled an increased risk for the first time occurrence of migraine. Sex-adjusted hazard ratios were 2.4 and 2.8, respectively (both statistically significant). In contrast, severe nonmigraine headache signaled an increased risk for major depression, but there was no evidence of a significant influence in the reverse direction, i.e., from major depression to severe headache. Sex-adjusted hazard ratios were 3.6 and 1.6, respectively (only the first statistically significant).

The pattern of the results suggests the possibility that different causal pathways might account for the comorbidity of major depression in these two headache categories. The results for migraine suggest shared causes, whereas those for other headaches of comparable severity suggest a causal effect of headache on depression.

Another line of evidence comes from a study of a biologic marker of depression in migraineurs. Jarman et al. (31) administered the tyramine test to 40 migraine patients, 16 of whom had a lifetime history of major depression. Low tyramine conjugation, a trait marker for endogenous depression, was strongly associated with a lifetime history of major depression in subjects with comorbid disease, regardless of their current psychiatric status. The authors argue that the association of the trait marker with major depression in migraineurs rules out the possibility that the depression is a psychologic reaction to migraine attacks.

Besides depression, bipolar disorders also are found more commonly in migraineurs. Mahmood et al. (42) undertook a study to estimate the prevalence of migraine in people suffering from bipolar affective disorder. Of a total of 81 patients with a diagnosis of bipolar disorder who completed the questionnaire, 21 (25.9%) reported migraine headaches. Of this population of patients, 25% of bipolar men and 27% of bipolar women suffered from migraine. These rates are higher than those reported in the general population, with the rate for bipolar men being almost five times higher than expected. An increased risk of suffering from migraine was particularly noted in bipolar patients with an early onset of the disorder. This may represent a more severe form of bipolar affective disorder and again calls into contention the question of a bidirectional influence of migraine and psychiatric disorders in general and, in this case, of bipolar disorder. It is highly possible that migraine and the different psychiatric disorders share a common etiology. The biggest clue for this may come with genetic studies. It has been shown that the *DRD2* gene is associated with migraineurs, and specifically in migraine with aura (50). It is interesting to note that the *DRD2* NcoI C allele frequency has been found to be significantly higher in individuals with migraine with aura, anxiety disorders, and/or major depression than in individuals who do not have these disorders.

Psychopathology of Migraine and Personality Characteristics

The relationship between migraine and psychopathology has been discussed far more often than it has been systematically studied. Over the years, many studies have focused on particular personality traits of migraineurs. The basic assumptions are as follows: (a) migraineurs share common personality traits; (b) these traits are enduring and measurable; and (c) these traits differentiate migraineurs from control subjects (58). The notion of a "migraine personality" first grew out of clinical observations of the highly selected patients seen in subspecialty clinics (60).

Touraine and Draper (67) reported that migraineurs were deliberate, hesitant, insecure, detailed, perfectionistic, sensitive to criticism, and deeply frustrated emotionally. They were said to lack warmth and to have difficulty making social contacts. Wolff (70) found migraineurs to be rigid, compulsive, perfectionistic, ambitious, competitive, chronically resentful, and unable to delegate responsibility.

Most investigations have used psychometric instruments such as the Minnesota Multiphasic Personality Inventory (MMPI) (27) or the Eysenck Personality Questionnaire (EPQ) (24). The EPQ is a well-standardized measure that includes four scales: (a) psychoticism (P); (b) extroversion (E); (c) neuroticism (N); and (d) lie (L) (Table 40.4).

Brandt et al. (14) used the Washington County Migraine Prevalence Study to conduct the first population-based case control study of personality in patients with migraine (Table 40.5). More than 10,000 12- to 29-year-olds who were selected through random-digit dialing underwent a diagnostic telephone interview. A sample of subjects who met the criteria for migraine with or without aura [n = 162] was compared with a sample of subjects without migraine. Each subject received the EPQ, the 28-item version of the

TABLE 40.4. MIGRAINE AND PERSONALITY: EPIDEMIOLOGIC STUDIES (COMMUNITY)

Author, Year (Reference No.)	Sample Subjects (N. DX.)	Age (yr)	Source	Controls	Instrument	Findings
Brandt et al., 1990 (14)	162 Mig RR = 94.2%	12–29	Community telephone survey of 10,169	162 (matched for sex and age) nonheadache	EPQ	EPQ-N: Both female and male Mig higher than controls EPQ-P: Female Mig higher than controls
Rasmussen, 1992 (51)	77 Mig 167 TTH RR = 76%	25–64	Random sample of 1,000 selected from the National Central Person Registry	496 nonheadache	EPQ	EPQ-N: Higher for TTH and Mig vs. rest; no difference between "pure" Mig vs. rest
Merikangas et al., 1993 (46,47)	11 Cls 91 Com 63 TTH 56 Sxs only	28–29	Community survey in Zurich (longitudinal)	158 nonheadache	Freiburg Personality Inventory	Cls sign higher than other groups on nervousness, depressiveness, inhibition, and decreased levels of resilience

Cls, classic migraine; Com, common migraine; EPQ, Eysenck Personality Questionnaire; EPQ-N, EPQ Neuroticism; Mig, migraine; TTH, tension-type headache.

TABLE 40.5. MIGRAINE AND PERSONALITY: MMPI (NEUROTIC TRIAD[a]) STUDIES

Author, Year (Reference No.)	Sample Subjects (N. DX)	Age (yr)	Source	Controls	Findings
Sternbach, 1980 (62)	83 Vascular 41 MC 58 Mixed	20–70	Patients seen for HA at pain treatment center	Data on MMPI from 50,000 patients, compared on age and sex	Vascular less depressed and less neurotic than MC and Mixed All groups combined vs. controls significantly higher on scales Hs, D, and Hy Males with MC higher on paranoid than females Females in all HA groups higher on Hy and Hs than males
Andrasik et al., 1982 (9)	26 Mig 39 MC 12 Cluster 22 Mig+MC	18–68	Physician referral or self-referral	30 friends or relatives of patients matched on age, sex, ses, and marital status	Control subjects no meaningful elevations on any scale None of HA types showed significant scale elevations Significant χ^2 tests for t-score elevations $\geq$70 vs. <70 by the five HA groups for scales Hs and Hy
Weeks, et al., 1983 (69)	50 Mig 50 Combo	Not specified	Outpatients evaluated for treatment at HA center		Combination headache significantly higher than Mig on Hy, D, and Hs
Levor et al., 1996 (36)	20 Com 9 Cls 5 Mixed	23–63	Patients recruited from physicians specializing in treatment of HA	Data from days prior to a migraine day were matched with the same person's data prior to headache-free days	Mild, subclinical elevations on Hs, D, and Hs for all HA groups vs. normals
Ellertsen and Klove, 1987 (71)	12 TTH 33 Mig	16–70	Neurology patients	34 HA-free, muscle pain in neck/shoulder patients recruited by ads	Muscle pain group: males scored higher on Hs than females TTH: Males and females had elevated scores on Hs and D (male vs. female not significantly different) Female Mig and TTH groups had similar profiles
Dieter and Swerdlow, 1988 (72)	82 Mig 48 SMCH 58 PT 61 Mixed 53 Cluster	Not specified	Random selection from general practice population	68 HA-free people solicited from the staff of the hospital	SMCH, Mig/Cluster, Mixed/PT each significantly different from control group on scales Hs, D, and Hy Clinically elevated scores from Mixed and PT on Hs (both males and females) Clinically elevated scores for Mixed and PT on D (males only) Clinically elevated scores for Mixed and PT on Hy (both males and females)
Invernizzi et al., 1989 (30)	148 Mig 183 TTH 87 Mixed	16–70	Patients referred to HA center		Mixed vs. Mig and Mixed vs. TTH significantly different for Hs, D, and Hy
Inan et al., 1994 (29)	44 Mig 36 TTH	Not specified	Patients referred to clinical psychologist	36 healthy controls	Female TTH higher on Hs, D, and Hy than controls Female Mig higher scores on Hy than female controls Differences between TTH and Mig not significantly different

[a]MMPI Neurotic Triad includes D, depression; Hs, hypochondriasis; Hy, hysteria.
Com, common migraine; Combo, combination; HA, headache; MC, muscle contraction; Mig, migraine; Mixed, mixed headache type; PT, post trauma; SMCH, scalp muscle contraction; TTH, tension-type headache.

General Health Questionnaire, and a question about headache laterality.

Subjects with migraine scored significantly higher than control subjects on the neuroticism scale of the EPQ scale, indicating that they were more tense, anxious, and depressed than the control group. In addition, women with migraine scored significantly higher than control subjects on the psychoticism scale of the EPQ, indicating that they were more hostile, less interpersonally sensitive, and out of step with their peers. Rasmussen (51) screened a population-based sample to identify patients with migraine and those with tension-type headache. Tension-type headache occurring alone was associated with high neuroticism scores on the EPQ. Persons with pure migraine (i.e., without tension-type headache) did not score above the norms on the neuroticism scale, although persons with migraine, with and without tension-type headache, showed a tendency to score above the norms on the neuroticism scale.

Merikangas et al. (45) investigated the cross-sectional association between personality, symptoms, and headache subtypes as part of a prospective longitudinal study of 19- and 20-year-olds in Zurich, Switzerland. Subjects with migraine scored higher on indicators of neuroticism than did subjects without migraine.

In summary, studies that used the EPQ or similar personality measures and compared persons with migraine with control subjects without migraine have generally reported an association between migraine and neuroticism.

Many investigators have used the MMPI to investigate the personalities of subjects with migraine (30,33,62,69). These studies have been limited by several factors (64). MMPI studies usually were clinic based, which limits their generalizability and creates opportunities for selection bias. Most of the studies did not use control groups and relied instead on historical norms. Many did not use explicit diagnostic criteria for migraine. Despite these limitations, most studies showed elevation of the neurotic triad, although this is not statistically significant (Table 40.4).

Studies of migraine and personality have generally not controlled for drug use, headache frequency, and headache-related disability. Furthermore, they have not controlled for major psychiatric disorders (such as major depression or panic disorder), which occur more commonly in migraineurs. The association between major psychiatric disorders and personality disorders may confound the assessment of the relationships between these disorders and migraine. Neuroticism, in particular, is associated with depression and anxiety, which occur with increased prevalence in migraineurs. Differences in neuroticism across studies might reflect variations in the role of comorbid psychiatric disease. The available data suggest that subjects with migraine may be more neurotic than those without migraine. The stereotypical rigid, obsessional migraine personality might reflect the selection bias of a distinct subtype of migraine that is more likely to be seen in the clinic.

Breslau and Andreski (16) examined the association between migraine and personality, taking into account a history of co-occurring psychiatric disorders. Data came from their epidemiologic study of young adults in the Detroit, Michigan, metropolitan area. Migraine was associated with neuroticism but not with extroversion or psychoticism, as measured by the EPQ. The association remained significant when the authors controlled for sex and history of major depression and anxiety disorders. More than 25% of persons with migraine alone who were uncomplicated by psychiatric comorbidity scored in the highest quartile of neuroticism. The results suggest that subjects with migraine might be more likely to have psychopathology and to adjust poorly to their medical condition. The findings also suggest that the association between migraine and neuroticism is not attributable to comorbid depression or anxiety disorders.

In a later report, Breslau et al. (17) presented findings from prospective data on the association between migraine and neuroticism from their epidemiologic study of young adults. In women, neuroticism measured at baseline predicted the first incidence of migraine during the 5-year follow-up. Specifically, controlling for major depression and anxiety disorders at baseline, women scoring in the highest quartile of the neuroticism scale were nearly three times more likely to develop migraine than those scoring in the lowest quartile. In men, neuroticism did not predict migraine, although the small number of incidence cases in men precluded reliable estimates of the risk for migraine associated with neuroticism.

TREATMENT OF MIGRAINES

The goals of treatment are to relieve or prevent the pain and associated symptoms of migraine and to optimize the patient's ability to function normally. To achieve these goals, patients must learn to identify and avoid headache triggers. Acute treatment almost always is indicated, but preventive treatment should be used more selectively and should be limited to patients with frequent or severe headaches. The choice of treatment should be based on the presence of comorbid conditions. A concurrent illness should be treated with a single agent when possible; agents that might aggravate a comorbid illness should be avoided. Biofeedback, relaxation techniques, and other behavioral interventions can be used as adjunctive therapy.

Pharmacologic treatment of migraine may be acute (abortive, symptomatic) or preventive (prophylactic). Patients who experience frequent severe headaches often require both approaches. Symptomatic treatment attempts to abort (i.e., stop the progression of) or reverse a headache once it has started. Preventive therapy is given on a daily basis, even in the absence of a headache, to reduce the frequency and severity of anticipated attacks. Symptomatic treatment is appropriate for most acute attacks and should be used no

TABLE 40.6. ABORTIVE MEDICATIONS: EFFICACY, SIDE EFFECTS, AND RELATIVE CONTRAINDICATIONS AND INDICATIONS

Drug	Efficacy[a]	Side Effects[a]	Comorbid Condition	
			Relative Contraindication[b]	Relative Indication
Aspirin	1+	1+	Kidney disease, ulcer disease, PVD, gastritis (age <15 yr)	CAD, TIA
Acetaminophen	1+	1+	Liver disease	Pregnancy
Caffeine	2+	1+	Frequent headache	
Adjuvant butalbital, caffeine, and analgesics	2+	2+	Use of other sedative; history of medication overuse	
Isometheptene	2+	1+	Uncontrolled HTN, CAD, PVD	
NSAIDs	2+	1+	Kidney disease, ulcer disease, gastritis	
Narcotics	3+	3+	Drug or substance abuse	Pregnancy; rescue medication
Ergotamine				
Tablet	2+	2+	Prominent nausea or vomiting, CAD, PVD, uncontrolled HTN	
Suppository	3+	3+		
Dihydroergotamine				
Injection	4+	2+	CAD, PVD, uncontrolled HTN	Orthostatic hypotension
Intranasal	3+	1+		
Triptans				
Sumatriptan				
SC injection	4+	1+	CAD, PVD, HTN	Nausea or vomiting
Nasal spray	3+	1+	CAD, PVD, HTN	
Tablet	3+	1+	CAD, PVD, HTN	Nausea or vomiting
Zolmitriptan	3+	1+	CAD, PVD, HTN	Nausea
Rizatriptan	3+	1+	CAD, PVD, HTN	Nausea
Naratriptan	2+	+/−	CAD, PVD, HTN	Nausea

[a]Ratings are on a scale from 1+ (lowest) to 4+ (highest).
[b]Caution is required in patients with frequent headaches.
CAD, coronary artery disease; HTN, hypertension; NSAIDs, nonsteroidal antiinflammatory drugs; PVD, peripheral vascular disease; SC, subcutaneous; TIA, transient ischemic attack.

more than 2 to 3 days per week. If attacks occur more frequently, treatment strategies should focus on decreasing the frequency of attacks.

Medications used for the treatment of acute headache include analgesics, antiemetics, anxiolytics, nonsteroidal antiinflammatory drugs (NSAIDs), ergots, steroids, major tranquilizers, narcotics, and, more recently, selective 5-HT$_{1B/D}$ (serotonin) agonists (triptans) (Table 40.6). One or more of these medications can be used for headaches of differing severities. Preventive treatments include a broad range of medications, most notably β-blockers, calcium channel blockers, antidepressants, serotonin antagonists, and anticonvulsants (Table 40.7).

Acute Therapy

Many acute treatments are available for migraine. The choice depends on the severity and frequency of headaches, the pattern of associated symptoms, the presence or absence of comorbid illnesses, and the patient's treatment response profile (Table 40.6). Both prescription and nonprescription medications are available. As a general rule, we start with nonspecific oral medications, such as analgesics, NSAIDs, or a caffeine adjuvant compound, for patients who have mild-to-moderate headaches. If these medications fail or if the

headache is moderate to severe, we prescribe triptans or dihydroergotamine (DHE). If prominent nausea or vomiting is present, we recommend an antiemetic and nonoral treatment. If oral medication is ineffective or cannot be used because of gastrointestinal symptoms, we recommend suppositories, nasal sprays, or injections, depending on the patient's preferred route of administration. Suppositories include ergotamine, indomethacin, and prochlorperazine. Nasal sprays include transnasal butorphanol, sumatriptan, and DHE. Injections include subcutaneous sumatriptan and intramuscular DHE. We often prescribe more than one acute treatment at the time of the initial visit. For example, we may advise patients to use naproxen sodium for mild-to-moderate headaches and a triptan for more severe headaches.

Simple and Combination Analgesics and Nonsteroidal Antiinflammatory Drugs

We often begin with simple analgesics for patients with mild-to-moderate headaches. Many individuals find headache relief with a simple analgesic such as aspirin or acetaminophen, either alone or in combination with caffeine. Butalbital is often tried, although butalbital use is controversial. We also use Midrin, which is a combination of acetaminophen, isometheptene (a sympathomimetic), and

TABLE 40.7. PREVENTIVE MEDICATIONS: EFFICACY, SIDE EFFECTS, AND RELATIVE CONTRAINDICATIONS AND INDICATIONS

Drug	Efficacy[a]	Side Effects[a]	Comorbid Condition	
			Relative Contraindication	Relative Indication
β-blockers	4+	2+	Asthma, depression, CHF, Raynaud disease, diabetes	HTN, angina
Calcium channel blockers				
Verapamil	2+	1+	Constipation, hypotension	Migraine with aura, HTN, angina, asthma
Antidepressants				
Amitriptyline	4+	2+	Urinary retention, heart block	Other pain disorders, depression, anxiety disorders, insomnia
SSRIs	2+	1+	Mania	Depression, OCD
Anticonvulsants				
Divalproex	4+	3+	Liver disease, bleeding disorders	Mania, epilepsy, anxiety disorders
Gabapentin	2+	2+		Epilepsy, mania?
Topiramate	3+	2+		
NSAIDs				
Naproxen	2+	2+	Ulcer disease, gastritis	Arthritis, other pain disorders

[a]Ratings are on a scale from 1+ (lowest) to 4+ (highest).
CHF, congestive heart failure; HTN, hypertension; NSAIDs, nonsteroidal antiinflammatory drugs; OCD, obsessive-compulsive disorder; SSRIs, selective serotonic reuptake inhibitors.

dichloralphenazone (a chloral hydrate derivative). For patients who are nauseated, we use the antiemetic metoclopramide or prochlorperazine. We often try naproxen sodium first, but we use all the NSAIDs, often in combination with metoclopramide. Indomethacin, available as a 50-mg rectal suppository, and intramuscular ketorolac are useful for patients with severe nausea and vomiting. With the recent advent of the cyclooxygenase-2 inhibitors rofecoxib and celecoxib, we now have the ability to treat migraineurs with a history of gastrointestinal ulcers, bleeds, or perforations who would not otherwise tolerate the NSAIDs.

Opioids

More potent opioid analgesics, such as propoxyphene, meperidine, morphine, hydromorphone, and oxycodone, are available alone and in combination with simple analgesics. Because medication overuse and rebound headache pose a threat, these agents are most appropriate for patients who experience infrequent headaches. There is less potential for abuse of agonist-antagonist opioids than of receptor agonists. Transnasal butorphanol tartrate, given in a dose of 1 mg followed by 1 mg 1 hour later, has been shown to be effective for the acute treatment of migraine headache.

Ergotamine and Dihydroergotamine

Ergotamine and its derivative DHE are effective treatments for moderate-to-severe migraine if simple analgesics do not provide satisfactory headache relief or if they produce significant side effects. DHE is a weaker arterial vasoconstrictor but is almost as potent a venoconstrictor as ergotamine. The oral absorption of ergotamine is erratic; rectal absorption is more

reliable. Patients who cannot tolerate ergotamine because of nausea are pretreated with metoclopramide, prochlorperazine, promethazine, or a mixture of a barbiturate and a belladonna alkaloid. Metoclopramide may enhance the absorption of oral ergotamine. DHE has fewer side effects than ergotamine and can be administered intramuscularly, intranasally, subcutaneously, or intravenously.

DHE is given in intramuscular or intravenous doses of up to 1 mg per treatment, with a maximum of 3 mg per day. We typically limit monthly use to 18 ampules or 12 events. DHE is a mainstay of treatment because it is effective in most patients. It is associated with a low (<20%) headache recurrence rate, and it is less likely than ergotamine to exacerbate nausea or produce rebound headache.

We avoid ergotamine and DHE in women who are attempting to become pregnant and in patients with uncontrolled hypertension, sepsis, renal or hepatic failure, or coronary, cerebral, or peripheral vascular disease. Although nausea is a common side effect with ergotamine, it is less common with DHE (unless it is given intravenously). Other side effects include dizziness, paresthesias, abdominal cramps, and chest tightness. Rare idiosyncratic arterial and coronary vasospasm can occur. We recommend electrocardiographic recording in all patients before their first dose of DHE, particularly if they have any cardiac risk factors (including age >40 years).

Serotonin Receptor Agonists (Triptans)

The first selective 5-HT$_{1B/1D}$ agonist to be developed and tested was sumatriptan, followed by zolmitriptan, naratriptan, and rizatriptan. We often prescribe a triptan at the initial consultation as a first-line drug for severe attacks and as an escape medication for less severe attacks that do not adequately respond to simple or combination analgesics. We prefer subcutaneous injection for patients who need rapid relief or have severe nausea or vomiting. None of the triptans should be used for patients who have clinical ischemic heart disease, Prinzmetal angina, uncontrolled hypertension, vertebrobasilar migraine, hemiplegic migraine, or migraines with prolonged auras.

Sumatriptan is available both as an injection (6 mg subcutaneously) and as tablets (25, 50, and 100 mg). Sumatriptan has a more rapid onset of action when it is given subcutaneously than when it is given orally. Sumatriptan relieves headache pain, nausea, photophobia, and phonophobia, and it restores the patient's ability to function normally. Although 80% of patients experience pain relief from an initial dose of subcutaneous sumatriptan, headache recurs in about 40%. Recurrences are most likely in patients with long-duration headaches and respond well to a second dose of sumatriptan or to simple and combination analgesics. Common side effects of sumatriptan include pain at the injection site, tingling, flushing, burning, and warm or hot sensations. Dizziness, heaviness, neck pain, and dysphoria can occur. These side effects generally abate within 45 minutes. Sumatriptan causes noncardiac chest pressure in approximately 4% of patients. Before using any of the triptans, we obtain an electrocardiogram from patients older than 40 years and from those who have risk factors for heart disease. These side effects generally abate within 45 minutes.

Oral naratriptan differs from sumatriptan primarily in its longer half-life, longer T_{max}, higher oral bioavailability (70%), and lipophilicity. Studies of more than 4,000 patients indicate that the drug has a well-defined dose–response relationship for headache relief, with a mean response of 48% at 2 hours after administration, but therapeutic gains are comparatively modest (21%) (55). Rates of relief with naratriptan were lower than with the other oral 5-HT$_{1B/1D}$ agonists. In a direct comparative crossover study with sumatriptan in patients prone to headache recurrence, about one-third fewer patients experienced recurrence when they used naratriptan compared with sumatriptan. However, relief rates were similar in 24-hour comparisons of naratriptan and sumatriptan. The studies showed excellent tolerability for the 2.5-mg dose of naratriptan, with an adverse event rate close to that of placebo.

Rizatriptan is rapidly absorbed and has high oral bioavailability (45%). Four trials reported that rizatriptan was significantly better than placebo for headache relief and complete relief at 2 hours. Doses tested ranged from 5 to 40 mg, with higher rates of relief reported with the higher doses (doses currently available in United States are 5 and 10 mg). Rizatriptan has high consistency from attack to attack in formal blinded consistency studies and is available as a wafer (melt) formulation that many patients, particularly those with nausea as a prominent feature, find convenient because it dis-

solves on the tongue and requires no water. Rizatriptan has a significant interaction with propranolol, which requires that the dose be halved to 5 mg, and it is contraindicated with monoamine oxidase inhibitors because of its route of metabolism.

Zolmitriptan was the second selective 5-HT$_{1B/1D}$ agent marketed in the United States. It has high oral bioavailability (40%) and T_{max} of about 2.5 hours. It undergoes metabolism by the cytochrome P-450 system to an active metabolite that is degraded by monoamine oxidase-A. Therefore, patients who take monoamine oxidase inhibitors are limited to a total zolmitriptan dose of 5 mg per day. Zolmitriptan (2.5 or 5 mg) was shown in three trials to be significantly more effective than placebo for headache relief and complete relief at 2 and 4 hours (56,57). In studies, zolmitriptan demonstrated a headache response of 64% with a therapeutic gain of 34% for the 2.5-mg dose and a headache response of 65% and a 37% therapeutic gain for the 5-mg dose. The recommended starting dose of 2.5 mg provides the best balance of benefit and side effects, although some patients may benefit from the higher 5-mg dose.

Adjunctive Treatment

The symptoms associated with migraine, such as nausea and vomiting, can be as disabling as the actual headache pain. The gastric stasis and delayed gastric emptying associated with migraine can decrease the effectiveness of oral medication. As a result, antiemetics such as metoclopramide, which is available in tablet, syrup, and injectable forms (in doses of 10–20 mg), are extremely useful for the treatment of migraine. In addition, metoclopramide decreases gastric atony and enhances the absorption of coadministered medications (2). Promethazine, available in tablet, liquid, suppository, and injectable forms (in doses of 25–50 mg), is useful to control nausea and vomiting, but, unlike metoclopramide, it does not enhance gastric emptying. Ondansetron, a selective 5-HT$_3$ receptor antagonist approved for chemotherapy-related emesis, can be used as an antiemetic and administered as an intravenous infusion (0.15 mg/kg diluted in 50 mL of 5% dextrose or normal saline) or as an 8-mg tablet (2,13,68).

Preventive Therapy

The following principles guide our decision on the need to start preventive therapy: (a) recurring migraine that, in the patient's opinion, significantly interferes with his/her daily routine despite acute treatment (e.g., two or more attacks per month producing disability lasting 3 or more days or headache attacks that occur infrequently but produce profound disability); (b) failure of, contraindication to, or troublesome side effects from acute medications; (c) overuse of acute medications (required more than twice a week); (d) unusual migraine subtypes such as hemiplegic migraine, basilar migraines, or attacks with a risk of permanent neurologic injury such as those associated with prolonged auras; (e) headache attacks that occur infrequently but produce profound disruption to patients' lives; (f) very frequent headaches (more than two per week) with the risk of rebound headache; (g) patient preference, that is, the patient's desire to minimize the number of acute attacks.

Greater caution should be exercised when deciding on preventive therapy for pregnant patients because of the potential for teratogenicity in fetal development with most of the current preventive therapies. Exceptions are made if the patient suffers from recurrent severe disabling attacks accompanied by nausea, vomiting, and dehydration.

If preventive medication is indicated, the agent should be chosen from one of the major categories based on side effect profiles and comorbid conditions (Table 40.7). The drug should be started at a low dose and increased slowly until therapeutic effects or side effects develop or the ceiling dose for the agent in question is reached. A full therapeutic trial may take 2 to 6 months. To obtain the maximal benefit from preventive medications, the patient should not overuse analgesics or ergot derivatives. Migraine headache may improve with time independent of treatment. If the headaches are well controlled, a drug holiday can be undertaken following a slow taper program.

β-Blockers

These are probably the most widely used class of migraine preventive medications. The mechanism of action of β-blockers is not certain, but it appears that their antimigraine effect is due to inhibition of β_1-mediated mechanisms (1). β-Blockade results in inhibition of norepinephrine release by blocking prejunctional β-receptors. In addition, it results in a delayed reduction in tyrosine hydroxylase activity (the rate-limiting step in norepinephrine synthesis) in the superior cervical ganglia. In the rat brainstem, a delayed reduction of the firing rate of locus caeruleus neurons has been demonstrated after propranolol administration. This could explain the delay in the prophylactic effect of the β-blockers.

All β-blockers can produce behavioral side effects, such as drowsiness, fatigue, lethargy, sleep disorders, nightmares, depression, memory disturbance, and hallucinations, indicating that they affect the central nervous system (8). They are relatively contraindicated in patients with comorbid depression and in certain migraine subtypes such as hemiplegic migraine, basilar migraine, or migraine with prolonged aura.

β-Blockers that are clinically useful in the treatment of migraine consist of both the nonselective blocking agents (propranolol, nadolol, and timolol) and the selective β_1-blockers (metoprolol and atenolol).

Antidepressants

The antidepressants that are most commonly used as migraine preventives include the tricyclic antidepressants and the atypical antidepressants, such as selective serotonin reuptake inhibitors. These antidepressants are especially useful in patients with comorbid depression and anxiety disorders.

Tricyclic antidepressants up-regulate the γ-aminobutyric acid (GABA)-B receptor, down-regulate the histamine receptor, and enhance the neuronal sensitivity to substance P. Some tricyclic antidepressants are 5-HT$_2$ receptor antagonists. Side effects are common with the use of tricyclic antidepressants. Their adverse effects are due to their interaction with multiple neurotransmitters and their receptors. Antimuscarinic side effects include dry mouth, a metallic taste, epigastric distress, constipation, dizziness, mental confusion, tachycardia, palpitations, blurred vision, and urinary retention. Antihistaminic activity may be responsible for carbohydrate cravings, which contribute to weight gain. Adrenergic activity is responsible for the orthostatic hypotension, reflex tachycardia, and palpitations experienced by patients. The tricyclic antidepressants most commonly used for migraine prophylaxis include amitriptyline, nortriptyline, doxepin, and protriptyline. Imipramine and desipramine have been used at times.

Fluoxetine, fluvoxamine, paroxetine, sertraline, and citalopram are specific serotonergic uptake inhibitors with minimal antihistaminic and antimuscarinic activity. In general, they have fewer side effects than the tricyclic antidepressants. They have fewer cardiovascular side effects. They produce less weight gain and even, in some cases, weight loss. The most common side effects include anxiety, nervousness, insomnia, drowsiness, fatigue, tremor, sweating, anorexia, nausea, vomiting, and dizziness or light-headedness.

Antiepileptic Drugs

The antiepileptic drugs have been shown to be effective in migraine prevention in numerous trials. In addition to their effects on headache, valproic acid and topiramate are especially useful in treating comorbid mood disorders.

Valproic acid is the most useful anticonvulsant used for migraine prophylaxis. The most commonly used form of valproic acid is divalproex. The results of several open studies support its role in the treatment of migraine and chronic migraine. Migraine prevention may be correlated with plasma level, with suggested target plasma levels of 70 and 90 units. Divalproex is especially useful when migraine occurs in patients with epilepsy, anxiety disorders, or manic-depressive illness (12). Because of its safety, divalproex can be administered to patients with depression, Raynaud disease, asthma, and diabetes, thus circumventing the contraindications to β-blockers. Valproate at high concentrations increases GABA levels in synaptosomes, perhaps by inhibiting its degradation. It enhances the postsynaptic response to GABA and, at lower concentrations, increases potassium conductance, thus producing neuronal hyperpolarization. Valproate turns off the firing of the 5-HT neurons of the dorsal raphe, which are implicated in controlling head pain. Nausea, vomiting, and gastrointestinal distress are the most common side effects of valproate therapy. These are generally self-limited and are slightly less common with divalproex sodium than with sodium valproate. When the therapy is continued, the incidence of gastrointestinal symptoms decreases, particularly after 6 months. Valproate has little effect on cognitive functions and rarely causes sedation. On rare occasions, valproate administration is associated with severe adverse reactions, such as hepatitis or pancreatitis. The frequency varies with the number of concomitant medications used, the patient's age, the presence of genetic and metabolic disorders, and the patient's general state of health. These idiosyncratic reactions are unpredictable. Valproate is potentially teratogenic and should not be used by pregnant women or women considering becoming pregnant.

Topiramate is a derivative of the naturally occurring monosaccharide D-fructose and contains a sulfamate functionality. Topiramate has been associated with weight loss, not weight gain (a common reason to discontinue preventive medication), with chronic use (41). Topiramate can influence the activity of some types of voltage-activated Na^+ and Ca^{2+} channels, $GABA_A$ receptors, and the α-amino-3-hydroxy-5-methylisoxazole-4-proprionic acid/kainate subtype of glutamate receptors. The effective dosing of topiramate that has been found effective in treating frequent migraines ranges from 25 to 100 mg per day. Other side effects of topiramate include cognitive dysfunction, which may be reduced if the titration dosing is gradual, and minor side effects, such as loss of taste for certain foods and asthenia mostly in the extremities, probably related to the carbonic anhydrase inhibitory activity of topiramate.

CONCLUSION

Migraine and affective disorders share many similarities. Although the causes of both are still uncertain, with the explosion of research in the neurosciences we hope that they will soon be revealed. Such clarification probably will result from discoveries in neuroendocrinology, neurochemistry, neuroradiology, and neurogenetics. Diagnosis and classification of both disorders have been simplified by the development of a clear set of internationally accepted guidelines that share many similarities.

Both families of disorders produce changes in mood, thought, and behavior that are attributed to poorly specified alterations of brain function. Overlapping neural abnormalities are probable keys to their relationship. Many of the drugs used in the treatment of migraine initially were developed for use in affective disorders. These include neuroleptics and antidepressants, which are particularly effective in treating migraineurs with comorbid disease. Other drugs such as the anticonvulsants, which were developed to treat epilepsy, a condition that is incidentally also comorbid with migraine, are effective in treating both migraine and mania.

Clinicians in neurology and psychiatry need to have a heightened index of suspicion for clinical conditions linked to each other's discipline. Psychiatric disease is commonly found in patients presenting with migraine, and vice versa. As the fundamental links in biochemistry and pharmacology between migraine and psychiatric disease increase, there exists a greater potential to effectively treat both conditions with increasing precision.

REFERENCES

1. Ablad B, Dahlof C. Migraine and β-blockade: modulation of sympathetic neurotransmission. *Cephalalgia* 1986;6:7–13.
2. Albibi R, McCallum RW. Metoclopramide: pharmacology and clinical application. *Ann Intern Med* 1983;98:86–95.
3. Alvarez WC. The migrainous scotoma as studied in 618 persons. *Am J Ophthalmol* 1960;49:489–504.
4. American Psychiatric Association. *Diagnostic and statistical manual of mental disorders, fourth edition.* Washington: American Psychiatric Association, 1994.
5. Amery WK, Waelkens J, Caers I. Dopaminergic mechanisms in premonitory phenomena. In: Amery WK, Wauquier A, eds. *The prelude to the migraine attack.* London: Bailliere Tindall, 1986:64–77.
6. Amery WK, Waelkens J, Van den Bergh V. Migraine warnings. *Headache* 1986;26:60–66.
7. Andermann E, Andermann FA. Migraine-epilepsy relationships: epidemiological and genetic aspects. In: Andermann FA, Lugaresi E, eds. *Migraine and epilepsy.* Boston: Butterworths, 1987:281–291.
8. Andersson K, Vinge E. Beta-adrenoceptor blockers and calcium antagonists in the prophylaxis and treatment of migraine. *Drugs* 1990;39:355–373.
9. Andrasik F, Blanchard EB, Arena JG, et al. Psychological functioning in headache sufferers. *Psychosom Med* 1982;44:171–182.
10. Berkson J. Limitations of the application of four-fold table analysis to hospital data. *Biometrics Bull* 1949;2:47–53.
11. Bickerstaff ER. Migraine variants and complications. In: Blau JN, ed. *Migraine: clinical and research aspects.* Baltimore: Johns Hopkins University Press, 1987:55–75.
12. Bowden CL, Brugger AM, Swann AC. Efficacy of divalproex vs lithium and placebo in the treatment of mania. *JAMA* 1994; 271:918–924.
13. Boyle R, Behan PO, Sutton JA. A correlation between severity of migraine and delayed emptying measured by an epigastric impedance method. *Br J Clin Pharmacol* 1990;30:405–409.
14. Brandt J, Celentano D, Stewart WF. Personality and emotional disorder in a community sample of migraine headache sufferers. *Am J Psychiatry* 1990;147:303–308.
15. Breslau N. Migraine, suicidal ideation, and suicide attempts. *Neurology* 1992;42:392–395.
16. Breslau N, Andreski P. Migraine, personality, and psychiatric comorbidity. *Headache* 1995;35:382–386.
17. Breslau N, Davis GC. Migraine, major depression and panic disorder: a prospective epidemiologic study of young adults. *Cephalalgia* 1992;12:85–89.
18. Breslau N, Davis GC. Migraine, physical health and psychiatric disorders: a prospective epidemiologic study of young adults. *J Psychiatr Res* 1993;27:211–221.
19. Breslau N, Davis GC, Andreski P. Migraine, psychiatric disorders and suicide attempts: an epidemiological study of young adults. *Psychiatry Res* 1991;37:11–23.

20. Breslau N, Davis GC, Schultz LR, et al. Migraine and major depression: a longitudinal study. *Headache* 1994;7:387–393.
21. Breslau N, Schultz LR, Stewart WF, et al. Headache and major depression: is the association specific to migraine? *Neurology* 2000;54:308–313.
22. Cutrer FM, Sorenson AG, Weisskoff RM, et al. Perfusion-weighted imaging defects during spontaneous migrainous aura. *Ann Neurol* 1998;43:25–31.
23. Ehrenberg BL. Unusual clinical manifestions of migraine, and "the borderland of epilepsy" re-explored. *Semin Neurol* 1991;11:118–127.
24. Eysenck HJ, Eysenck SB. *Manual of the Eysenck Personality Questionnaire.* San Diego, CA: Educational and Industrial Testing Service, 1975.
25. Friedman AP, Finley KH, Graham JR. Classification of headache. Ad Hoc Committee on Classification of Headache. *Arch Neurol* 1962;6:13–16.
26. Griffin N, Lipton RB, Silberstein SD, et al. Electronic diary study of premonitory symptoms in migraine. Headache World 2000, London, 2000.
27. Hathaway SR, McKinley JC. Minnesota Multiphasic Personality Inventory. Minneapolis, University of Minnesota, 1943.
28. Headache Classification Committee of the International Headache Society. Classification and diagnostic criteria for headache disorders, cranial neuralgia, and facial pain. *Cephalalgia* 1988;8:1–96.
29. Inan L, Soykan C, Tulunay FC. MMPI profiles of Turkish headache sufferers. *Headache* 1994;34:152–154.
30. Invernizzi G, Gala C, Buono M. Neurotic traits and disease duration in headache patients. *Cephalalgia* 1989;9:173–178.
31. Jarman J, Fernandez M, Davies PT, et al. High incidence of endogenous depression in migraine: confirmation by tyramine test. *J Neurol Neurosurg Psychiatry* 1990;53:573–575.
32. Klee A, Willanger R. Disturbances of visual perception in migraine. *Acta Neurol Scand* 1966;42:400–414.
33. Kudrow L, Sutkus GJ. MMPI pattern specificity in primary headache disorders. *Headache* 1979;19:18–24.
34. Leão AAP. Spreading depression of activity in cerebral cortex. *J Neurophysiol* 1944;7:359–390.
35. Lennox WG, Lenox MA. *Epilepsy and related disorders.* Boston: Little, Brown and Company, 1960:451.
36. Levor RM, Cohen MJ, Naliboff BD, et al. Psychosocial precursors and correlates of migraine headache. *J Consult Clin Psychol* 1986;54:347–353.
37. Lippman CV. Certain hallucinations peculiar to migraine. *J Nerv Ment Dis* 1952;116:346.
38. Lipton RB, Silberstein SD, Stewart WF. An update on the epidemiology of migraine. *Headache* 1994;34:319–328.
39. Lipton RB, Stewart WF. Migraine in the United States: epidemiology and healthcare utilization. *Neurology* 1993;43:6–10.
40. Lipton RB, Stewart WF, Diamond S, et al. Prevalence and burden of migraine in the United States: data from the American Migraine Study II. *Headache* 2001;41:646–657.
41. MacDonald RL, McLean MJ. Anticonvulsant drugs: mechanisms of action. In: Escueta AD, Ward AA, Woodbury DM, et al., eds. *Advances in neurology.* New York: Raven Press, 1986: 713–736.
42. Mahmood T, Romans S, Silversteon T. Prevalence of migraine in bipolar disorder. *J Affect Disord* 1995;52:163–169.
43. Marcus DA. Migraine and tension-type headaches: the questionable validity of current classification systems. *Clin J Pain* 1992;8:28–36.
44. Marks DA, Ehrenberg BL. Migraine related seizures in adults with epilepsy, with EEG correlation. *Neurology* 1993;43:2476–2483.
45. Merikangas KR, Angst J, Isler H. Migraine and psychopathology: results of the Zurich cohort study of young adults. *Arch Gen Psychiatry* 1990;47:849–853.
46. Merikangas KR, Merikangas JR, Angst J. Headache syndromes and psychiatric disorders: associations and familial transmission. *J Psychiatr Res* 1993;27:197–210.
47. Merikangas KR, Stevens DE, Angst J. Headache and personality: results of a community sample of young adults. *J Psychiatr Res* 1993;27:187–196.
48. Ophoff RA, Terwindt GM, Vergouwe MN, et al. Involvement of a Ca^{2+} channel gene in familial hemiplegic migraine and migraine with and without aura. Wolff Award 1997. Dutch Migraine Genetics Research Group. *Headache* 1997;37:479–485.
49. Ottman R, Lipton RB. Comorbidity of migraine and epilepsy. *Neurology* 1994;44:2105–2110.
50. Peroutka SJ, Price SC, Wilhoit TL, et al. Comorbid migraine with aura, anxiety, and depression is associated with dopamine D2 receptor (DRD2) NcoI alleles. *Mol Med* 1998;4:14–21.
51. Rasmussen BK. Migraine and tension-type headache in a general population: psychosocial factors. *Int J Epidemiol* 1992;21:1138–1143.
52. Sacks O. *Migraine: understanding a common disorder.* Berkeley, CA: University of California Press, 1985.
53. Sanchez-delRio M, Bakker D, Hadjikhani N, et al. Neurovascular cortical spreading phenomenon during spontaneous visual aura. *Cephalalgia* 1999;19:310.
54. Sanchez-delRio M, Bakker D, Wu O, et al. Perfusion weighted imaging during migraine: spontaneous visual aura and headache. *Cephalalgia* 1999;19:701–707.
55. Saxena PR. Selective vasoconstriction in carotid vascular bed by methysergide: possible relevance to its antimigraine effect. *Eur J Pharmacol* 1974;27:99–105.
56. Saxena PR, Duncker DJ, Bom AH, et al. Effects of MDL72222 and methiothepin on carotid vascular responses to 5-hydroxytryptamine in the pig: evidence for the presence of vascular 5-hydroxytryptamine-like receptors. *Arch Pharmacol* 1986;333:198–204.
57. Saxena PR, VanHouwelingen P, Bonta IL. The effect of mianserin hydrochloride on the vascular responses to 5-hydroxytryptamine and related substances. *Eur J Pharmacol* 1971; 13:295–305.
58. Schmidt FN, Carney P, Fitzsimmons G. An empirical assessment of the migraine personality type. *J Psychosom Res* 1986;30:189–197.
59. Selby G, Lance JW. Observation on 500 cases of migraine and allied vascular headaches. *J Neurol Neurosurg Psychiatry* 1960;23:23–32.
60. Silberstein SD. Migraine symptoms: results of a survey of self-reported migraineurs. *Headache* 1995;35:387–396.
61. Stang PE, Osterhaus JT. Impact of migraine in the United States: data from the National Health Interview Survey. *Headache* 1993;33:29–35.
62. Sternbach RA. MMPI pattern in common headache disorders. *Headache* 1980;20:311–315.
63. Stewart WF, Linet MS, Celentano DD. Migraine headaches and panic attacks. *Psychosom Med* 1989;51:559–569.
64. Stewart WF, Linet MS, Celentano DD, et al. Age and sex-specific incidence rates of migraine with and without visual aura. *Am J Epidemiol* 1991;34:1111–1120.
65. Stewart WF, Lipton RB, Celentano DD, et al. Prevalence of migraine in the United States. Relation to age, income, race, and other sociodemographic factors. *JAMA* 1992;267:64–69.
66. Swartz KL, Pratt LA, Armenian HK, et al. Mental disorders

and the incidence of migraine headaches in a community sample: results from the Baltimore Epidemiologic Catchment area follow-up study. *Arch Gen Psychiatry* 2000;57:945–950.

67. Touraine GA, Draper G. The migrainous patient: a constitutional study. *J Nerv Ment Dis* 1934;80:183–204.

68. Volans GN. Research review: migraine and drug absorption. *Pharmacokinetics* 1978;3:313–318.

69. Weeks R, Baskin S, Sheftell F. A comparison of MMPI personality data and frontalis electromyographic readings in migraine and combination headache patients. *Headache* 1983;23:75–82.

70. Wolff HG. Personality features and reactions of subjects with migraine. *Arch Neurol Psychiatry* 1937;37:895–921.

71. Ellertsen B, Klove H. MMPI patterns in chronic muscle pain, tension headache and migraine. *Cephalagia* 1987;7:65–71.

72. Dieter JN, Swerdlow B. A replicative investigation of the reliability of the MMPI in the classification of chronic headaches. *Headache* 1988;28:212–222.

41

TRAUMATIC BRAIN INJURY

THOMAS A. HAMMEKE
THOMAS A. GENNARELLI

Traumatic brain injury (TBI) is a major public health problem. It is the leading cause of death and lifelong disability in children and young adults. The Center for Disease Control has estimated that 1.5 million Americans sustain TBI annually, of whom 50,000 die, 230,000 are hospitalized and survive, and 80,000 to 90,000 experience long-term disability (1). To translate, roughly every 10 minutes an American will die from TBI and two others will suffer an injury that is permanently disabling. It is estimated that in the United States there are 5.3 million men, women, and children (2% of the population) currently living with permanent TBI-related disabilities. Because the incidence of TBI is disproportionately high in adolescents and young adults, who are individuals with long lives ahead of them and continuing needs for health care and social and vocational services, the burden of TBI on society is great. A study in 1985 estimated the annual cost of TBI to approach $40 billion, more than half of which was related to injury-related work loss and disability (2). Despite these rates of mortality, morbidity, and economic burden, and likely because the disabilities from TBI often are not conspicuous, many health professionals and public policy makers and much of the general public fail to appreciate the magnitude of TBI as a public health problem. Thus, terms such as the *silent* or *invisible epidemic* have been applied to TBI.

Epidemiologic studies indicate that TBI is 3.5 times more common in males than females. Both incidence and death rates from TBI are highest in people older than 75 years, largely due to falls, and in adolescents and young adults, largely due to firearm and motor vehicle accidents. Approximately half of all TBI cases are due to motor vehicle accidents and another fourth due to falls. Fewer than 10% of TBI cases are due to firearms, yet firearms have been the leading cause of TBI-related deaths since 1990, with African-Americans being disproportionately represented. People injured in motor vehicle accidents, falls, or assaults without firearms are more likely to survive than people injured in shootings. Closed head injuries (CHI), in which the dura remains intact and the brain is not penetrated, comprise more than 90% of TBI. In addition to being a male between ages 16 and 24 years or being older than 75 years, other risk factors for TBI include alcohol use and abuse (3,4), having attention-deficit/hyperactivity disorder (5,6) or other psychiatric disorder (3,7), being an amateur or professional athlete (equestrians, boxers, football players, cyclists, and downhill skiers rank particularly high) (8–10), or having a previous head injury (11).

Research initiatives on the pathophysiologic mechanisms, neurobehavioral effects, and treatment of TBI have increased dramatically during the past three decades and have provided a wealth of information. Still, much is not known about fundamental mechanisms, prevention of secondary effects, range of neurobehavioral sequelae, and strategies to optimize neurorestoration, rehabilitation, and community reintegration. Sports-related concussions now are recognized as a major public health problem, and little is known about when it is safe to allow professional athletes or our children to return to competition after concussion. In this chapter, we summarize the current understanding of pathophysiologic mechanisms and describe the neurocognitive and neurobehavioral features of the acute, postacute, and chronic stages of recovery after TBI. Issues associated with mild TBI and the postconcussion syndrome are reviewed. Management of TBI and prospects for future treatments and technologies to aid our understanding and treatment of TBI also are reviewed.

PATHOBIOLOGY AND PATHOPHYSIOLOGY

Historically TBI was construed as an event, a set of injuries to blood vessels and neurons due to shearing force. It now is known that TBI is a process, not simply an event. It is brought about by a dynamic cascade of metabolic processes set into motion by the initial forces of injury. In addition to the brain injury itself, the victim of TBI often is confronted with a host of systemic sequelae, both concomitant events associated with injury and those caused by the brain injury itself, that produce deleterious effects on recovery of function. These secondary factors most commonly include hypoventilation, intracranial hypertension, hypotension, coagulopathy, and cerebral ischemia.

The mechanisms of TBI can be categorized into two broad types: those that produce focal brain injuries, and those that produce diffuse injuries (12,13). Each type gives rise to corresponding brain injury syndromes. Primary and secondary mechanisms of injury are associated with each type.

Focal Brain Injury

Focal brain injuries typically are due to the mechanical forces that produce localized damage to the brain. In the case of penetrating head wounds, the injury is by definition focal, although focal lacerations may be superimposed on distributed injuries. In the case of CHI, injurious forces cause contusions, lacerations, and hemorrhage in the epidural, subdural, subarachnoid, or intracerebral compartments of the cranium. These are commonly referred to as *primary* mechanisms of injury (14). In rapid acceleration/deceleration forces, brain contusions most commonly occur in the frontal and temporal poles and the inferior surface of the frontal and temporal lobes, where bony protuberances at the skull base serve as a less than friendly cushion for the brain moving within the cranium. Contusions occur in these regions with sufficient regularity, providing the substrate for a significant portion of the "typical" neurobehavioral picture after moderate-to-severe CHI. In blunt trauma, e.g., when an object strikes the head but does not penetrate the skull, contusions commonly occur beneath the site of impact. The contusions consist of a combination of hemorrhage and damaged gray matter that forms necrotic cavities over time. Intracranial hemorrhages are relatively infrequent occurrences in sports-related concussions and are more common in assaults and falls. Hemorrhage due to tears in large arteries, often associated with skull fractures, can produce epidural and intracerebral hematomas that rapidly escalate in size due to the high arterial pressure. Such rapid increases in intracranial pressure (ICP) often are life threatening and present a neurosurgical crisis. Subdural hematomas are more commonly due to tears in bridging veins and typically evolve more slowly because of the lower pressure of the venous system. Tears in capillary beds produce petechial hemorrhages that can be concentrated in areas of contusion and, under other circumstances, are distributed throughout the cerebrum. All of these types of hemorrhage require time to evolve, thus creating a mechanism to account for victims of TBI who are able to converse immediately after injury, only to decompensate neurologically a few hours later—the "talk and die" TBI group.

Diffuse Brain Injury

Diffuse brain injuries usually are due to rapid movement of the head, either from acceleration or deceleration or from rotational forces. These forces cause axonal stretch (a primary mechanism) that sets in motion a complex cascade of metabolic events (secondary mechanisms) that evolve over the course of several days and weeks (14–21). The metabolic cascade involves molecular and cellular events that have multiple consequences. These include breakdown of the blood–brain barrier and cerebral autoregulation, edema formation, impairment of energy metabolism, changes in cerebral perfusion, disruption of ionic homeostasis, activation of autodestructive neurochemicals, generation of free radicals, and genomic changes, which all are considered secondary mechanisms of neural dysfunction and cell death. The neuronal injury associated with these processes occurs throughout the neural axis (20); however, one group of researchers has proposed that the distribution of neuronal pathology depends on the severity of injury. Ommaya and Gennarelli (22) proposed a centripetal injury model from their animal research. Their model posits that injuries due to shear strain occur at progressively deeper levels of the brain as a function of severity of the injurious forces. Thus, in mild traumatic injuries, damage will largely be confined to the surface of the brain; in more severe injury, damage will extend to deeper structures, including the corpus callosum and brainstem. In general, support for this model has been found in radiologic studies of adults (23) and children (24). The effects of metabolic cascade are the principal mechanisms involved in motor vehicle accidents and sports-related concussion (12). These mechanisms are thought to explain the loss of consciousness (LOC) after head injury, evolution of secondary brain damage, and increased vulnerability of the brain to additional insults for an interval of time after initial injury. Our understanding of the magnitude of effect of these secondary processes has steadily increased as the technology to measure them has developed. The neurochemical processes of the metabolic cascade, which overlap with those present in cerebral hypoxia and ischemia, have been the source of intense research during the past decade because the interruption of each individual process holds potential promise for meaningful intervention. It is important to recognize that this research is based almost exclusively on animals and fluid percussion models of injury. Although the full range and interrelationship of the processes in the metabolic cascade are not yet understood, several have been targeted for investigation: (a) excess release of excitatory amino acids; (b) generation of free radicals and reactive oxygen species; (c) release of inflammatory cytokines; (d) apoptosis; (e) calpain proteolysis; and (f) axonal stretch (25).

The metabolic cascade is thought to begin with a massive flux of ions across the neural membrane immediately after trauma. There is an influx of sodium, calcium, and chloride and an efflux of potassium. The ionic flux is thought to be largely due to the release of excitatory amino acids, principally glutamate, which opens ligand gated channels in the neural membrane (15–17). Also, stretch of the axon at the time of injury produces transient defects in the cell membrane around the nodes of Ranvier, a phenomenon known as mechanoporation, which contributes to ionic flux (14). These processes produce a dramatic increase in extracellular potassium concentrations immediately after concussion that may last for 3 to 5 minutes (16). When the level of extracellular potassium reaches a critical threshold beyond normal limits (approximately 4–5 to 20–50 mmol/L), action potentials are inhibited and LOC may occur. Elevation of extracellular potassium to this critical level may take several seconds, thus accounting for the athlete who walks off the field after concussion only to collapse on the sideline. Elevated extracellular potassium depolarizes nerve terminals and can trigger further release of glutamate, thereby augmenting the calcium-mediated neurotoxicity described later (14,26).

Central to the proposed mechanisms of neurotoxicity is a perturbation of neuronal calcium homeostasis. The neuroexcitatory storm that is triggered at the moment of injury produces an influx of Ca^{2+} through the neural membrane that activates the calcium-activated cysteine proteases (calpains—calcium-activated neutral proteases) and caspase-3–like proteases. Activation of the proteases is linked to mitochondrial damage that is associated with both apoptotic and necrotic modes of neuronal cell death (27).

Within minutes to hours after injury, the body shifts into a state of increased metabolism that is associated with increased glucose utilization (14,16,18,28,29). The hypermetabolic state is assumed to be due to the brain's attempt to restore ionic homeostasis that was disrupted by the neuroexcitatory storm (20). At the same time cerebral blood flow decreases, producing an accumulation of lactate and intracellular acidosis. This initial phase is followed by a period of cerebral hypometabolism in which protein synthesis is decreased and oxidative capacity is reduced, making cells particularly vulnerable to additional insults. It is thought that this window of vulnerability underlies much of the secondary damage caused by hypoxia and hypotension. In animal research, this later phase of vulnerability may last up to 10 days (26,29). Hovda et al. (29) offer a hypothesis of mismatch between the energy needs of the cell and the decreased availability of glucose to account for the interval of cell vulnerability. It is not surprising that a major focus of TBI treatment in recent years has been on interrupting the various elements of the metabolic cascade before cell death is inevitable (25).

Exactly how far down the path of this metabolic cascade an individual will go hinges in part on how much mechanical deformation of the brain occurs at the time of injury. Significant deformation has been associated with enlarged astrocytes, leaky neural membranes, activation of calpain, and both neuronal and apoptotic cell death (30). Axonal injury, which is a direct effect of the brain deformation, typically is scattered diffusely throughout subcortical white matter, the major commissures, and brainstem (20). Gennarelli (14) proposes four stages of increasing axonal stretch injury.

Stage I is associated with injuries that induce axon stretch of 5% or less of resting length. Stretches of this magnitude are capable of creating pores in the axolemma (mechanoporation) at the nodes of Ranvier sufficient to cause ionic permeability. The ionic imbalances are rapidly restored, and neuronal dysfunction is brief. Stage II axonal injury occurs with stretch in the 5% to 10% range, causing ionic flux in addition to fluid flux to maintain osmotic balance. Swelling of the axon may occur and is associated with mild impairment in axoplasmic transport. This does not result in cell death, but it extends the period of restoration of structure and function to days or weeks. Mild traumatic brain injury (MTBI; see later discussion on grading severity) and most sports-related concussions likely involve stage I and II axonal injuries. Stage III axonal injury is associated with stretch in the range from 15% to 20%. This degree of stretch is not sufficient to tear the axon, but it will interrupt axonal metabolism and axoplasmic flow sufficiently to cause cytoskeletal compromise and mitochondrial swelling that leads to secondary or delayed axotomy. Stage IV axonal injury involving stretch greater than 20% causes immediate structural disruption of the axon at the time of injury, a primary axotomy.

Other mechanisms of posttraumatic morbidity with generalized effects include dysfunction due to *deafferentation* of brain sites as a result of axonal injury and *vascular changes*. The widespread axotomy seen in moderate-to-severe TBI creates significant effects downstream from the dying axon. Disconnected terminal sites can undergo degenerative changes and impair neuronal function in multiple brain regions that are influenced by the injured axons (31).

In addition to the hemorrhage of torn blood vessels due to trauma, changes in the function of blood vessels after trauma are recognized and are thought to underlie secondary hypotensive and hypoxic crises. Both impaired autoregulation and vascular reactivity to normal physiologic challenges have been identified after trauma (32). The specific mechanisms for the vascular changes have not been fully elucidated, although both mechanical (i.e., stretch) and metabolic (i.e., incidental to the metabolic cascade) factors are likely culprits. Although signs of vascular changes are seen most commonly in patients with severe TBI, a vascular mechanism is proposed to account for the catastrophic neurologic injury that can occur from repeated concussions that occur closely in time. The second impact syndrome (SIS) (33) results in severe, diffuse brain swelling after a second concussion while an individual is still symptomatic from an earlier injury (34,35). Although the second injury can be relatively mild, its occurrence in the context of cerebral vulnerability from the first injury sets the stage for catastrophic decompensation. The pathophysiology of SIS is thought to be linked to a loss of cerebral autoregulation that produces diffuse vascular engorgement, markedly increased ICP, and eventual brain herniation and brainstem compromise (34,35).

MECHANISMS OF RECOVERY

Although our understanding of the pathobiology of TBI is incomplete, our understanding of the mechanisms of recovery is even less complete. Again, what is known is largely drawn from animal models of injury. In very mild injuries where the axolemma remains intact, recovery probably is linked to restoration of ionic homeostasis (14). It is generally thought that functional tissue recovery in areas of hemorrhagic contusions does not occur, which implies that the neurobehavioral recovery that occurs in this context is due to functional reorganization with recruitment of contralateral and ipsilateral brain structures. Animal models that have shown increased fiber ingrowth and synaptic reorganization support this interpretation (31). When axonal injury is widespread and diffuse deafferentation occurs, animal studies have shown that the brain attempts to "rewire" itself (36). When the TBI is less severe and axotomy limited, neighboring fibers related to injured axons sprout connections with terminal sites, thereby restoring function to the natural neurotransmitter populations of deafferented neurons. In animals this neuroplastic process requires several months, a time frame that is not unlike that seen in the full recovery of many patients with mild-to-moderate TBI. With more severe injuries where axotomy is widespread and deafferentation extensive, the neuronal milieu may not support this adaptive regeneration. Instead, natural processes of rewiring may be maladaptive, with connections being formed with functionally unrelated fibers in patterns that are dissimilar to the preinjury circuits (31,36). Formation of these maladaptive connections is posited as one explanation for the sustained morbidity of the severely injured.

NEUROBEHAVIORAL SYNDROME

Focal brain injuries due to penetrating head wounds and blunt trauma have neurobehavioral features that correspond to their locus of injury. In this respect, the focal brain syndromes due to trauma are similar to those seen in stroke; each syndrome is unique to the region of ischemia or damage. In contrast, the neurobehavioral features of individuals with CHI, particularly those involving rapid acceleration/deceleration and rotational forces, are amazingly similar, varying largely on a dimension of severity (e.g., speed of recovery) rather than qualitative character. The similarity is likely due to two reasons: (a) the secondary pathophysiologic processes are distributed throughout the neural axis; and (b) contusions, when they occur, more often are located in the frontal and temporal lobes. Moreover, the common sequence of evolving secondary processes gives rise to a somewhat consistent sequence of evolving neurobehavioral features in CHI. The consistency of the sequence of neurobehavioral features is identified most clearly in cases of severe CHI but can be discerned, albeit less distinctly, in mild CHI. It is important to note that it is common to have signs of focal brain injury (e.g., unilateral neglect, hemiparesis) superimposed on the more generalized neurobehavioral syndrome. Both conditions will require attention in rehabilitation.

Because the evolution of neurobehavioral symptoms is quite similar among severe CHI patients, the description of the evolving clinical picture has been used as a means of staging recovery (37,38). Alexander (37), who did not intend to define a scale, described a series of seven neurobehavioral stages of recovery. The Ranchos Los Amigos Scales (38) has eight stages of recovery (Alexander's and Ranchos Los Amigos sequences are listed in Table 41.1). In both schemas, the first three stages characterize varying comatose states. The Ranchos Scale provides more differentiation of the interval of posttraumatic amnesia (PTA; see later). Less severely injured patients move quickly through the stages of recovery, often within hours or days, whereas more severely injured patients have a protracted course that is frequently incomplete. The principal features of the acute stage are described later, in addition to the more common features of the severely injured patients who show permanent residuals.

TABLE 41.1. STAGES OF RECOVERY FROM SEVERE TRAUMATIC BRAIN INJURY

Alexander's Stages[a]	Ranchos Los Amigos Scale[b]
1. Coma	I. Generalized
2. Unresponsive vigilance	II. Generalized response
3. Mute responsiveness	III. Localized response
4. Confusional state (PTA)	IV. Confused, agitated
	V. Confused, inappropriate, non-agitated
	VI. Confused, appropriate
5. Independent self-care	VII. Automatic appropriate
6. Intellectual independence	
7. Complete social recovery	VIII. Purposeful and appropriate

[a]From Alexander MP. Traumatic brain injury. In: Benson BW, Blumer D, eds. *Psychiatric aspects of neurologic disease, volume II.* New York: Grune & Stratton, 1982:219–249.
[b]From Hagen C. Language-cognitive disorganization following closed head injury: a conceptualization. In: Trexler L, ed. *Cognitive rehabilitation: conceptualization and intervention.* New York: Plenum, 1982, adapted with permission.

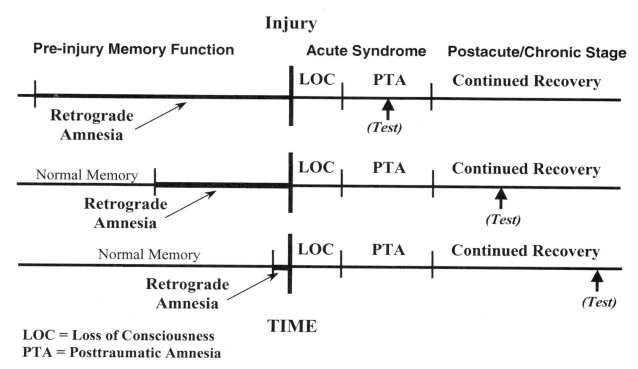

FIGURE 41.1. Retrograde and anterograde amnesia in traumatic brain injury.

Acute Syndrome

The acute period is defined as that time from the moment of injury to the point when the patient's cognitive capacities have recovered sufficiently well that he/she is consistently able to retain new information and remain fully oriented. This includes stages of early unresponsiveness and the time following recovery from coma that commonly is referred to as the interval of PTA (Fig. 41.1). Again, the acute period may range from being so brief so as not to permit prospective study to lasting weeks and months, depending on the severity of injury.

Early in the acute phase of severe injury, the patient is unresponsive to external stimuli, and application of a painful stimulus may elicit only decerebrate or decorticate posturing. Gradually gross signs of wakefulness can be observed with an absence of purposeful movements. Among the most severely injured patients, neurobehavioral recovery may be arrested at this stage. The patient may show features of a persistent vegetative state, i.e., independence in respiration, relatively normal sleep/wake cycles, roving eye movements, mutism, and preservation of startle and orienting reflexes without the capacity to process and respond to command. Fortunately, most patients who survive to this stage begin to show purposeful responding that is more than reflexive in character, e.g., turning the head away from an annoying cloth or spontaneously repositioning the limbs. Recovery of speech typically occurs after the more primitive functions; however, this may be difficult to appreciate, because many patients require intubation to optimize respiration early in their course (see section on Management of Traumatic Brain Injury). In general, patients with left hemisphere lesions tend to have more prolonged periods of coma as assessed by time to follow commands; however, this may be due to the combination of a dominant hemisphere lesion and a verbal method of assessment (39).

Once the patient begins to follow commands consistently, a stage of recovery commonly referred to as the interval of PTA begins [note, however, that early investigators (40) and many clinicians combine the intervals of PTA and coma into a single interval of disturbed consciousness]. Although an anterograde amnesia is a hallmark of this interval, an older characterization of the interval as the "posttraumatic confusional state" (37) is more accurate. Stuss et al. (41) demonstrated that there is significant impairment in both simple and complex attentional capacities during this interval of time, even though attentional functions improve substantively prior to the improvement in memory capacities. Early in this stage of recovery, patients typically show the full range of features of nontraumatic confusional states, including marked impairment in mental tracking, rapid mental fatigue, confabulation, severe perseveration, restlessness, and altered sleep/wake cycles. Late in the comatose stages and early in PTA, both hypokinetic and hyperkinetic states of motor function are common, with agitation and combativeness in the latter frequently of a magnitude that necessitates physical and/or pharmacologic restraint. It is postulated that the confusional features and combativeness may be due to brain swelling and altered levels of neurotransmitters (42). However, agitation and restlessness early in PTA is correlated with frontal temporal contusions (43), and effects from many of the components of the metabolic cascade described earlier may contribute. The more salient features of a confusional state subside dramatically as this stage of recovery approaches its end.

As is characteristic of most acute onset organic amnesias, there are both anterograde and retrograde features during the acute phase of recovery. Retrograde amnesia shows a gradient of memory loss, with the most remote autobiographical memories being the most readily retrieved and memories that were formed shortly before injury being most vulnerable (42). The retrograde loss may extend back years preceding the injury, and this occasionally creates difficult circumstances for relatives and acquaintances who are startled to find that their loved one does not remember getting married, having children, or getting a divorce. Interestingly, the interval of retrograde amnesia decreases dramatically as the features of posttraumatic confusion and anterograde amnesia subside (44). Figure 41.1 shows the characteristic course of receding retrograde amnesia as a function of the time of its assessment after injury. The retrograde interval continues to shrink through the interval of PTA and often for several months thereafter. It ultimately resolves to a gap of permanent memory loss that ranges from a few seconds to a few minutes immediately preceding the injury in most patients (45,46). It is helpful to educate accident investigators, who may be perplexed by the patient's varying accounts of events leading up to an accident given at sequential interviews after injury. The earlier accounts often are confabulated, and later accounts represent varying degrees of recovery of retrograde memories.

It is useful to highlight a few features of retrograde memory loss, because they may assist the clinician in determining whether a patient's report of retrograde memory loss can realistically be attributed to trauma. First, concordant with Ribot's law of regression (47), which posits that the most recently formed memories are the most susceptible to disruption from head trauma, the gradient of shrinkage after head injury always occurs in the direction of the most remotely formed to the most recently formed memories. Thus, if a patient is able to accurately describe events that transpired a few hours prior to trauma but has no recollection for meaningful events 1 month earlier, an etiology other than trauma should be considered to account for the retrograde memory loss, e.g., temporal lobe epilepsy or psychogenic mechanisms. Similarly, report of an extended interval of dense retrograde memory loss, well after resolution of PTA, should raise some suspicion of a nontraumatic mechanism. Permanent memory loss beyond a few days preceding moderate-to-severe injuries (48) and a few seconds preceding mild injuries (49) is rare. A potential exception is a relatively isolated retrograde amnesia for autobiographical information that has been reported infrequently in patients with injury to the anterior

temporal lobes or ventral frontal cortex and underlying white matter, including the uncinate fasciculus (50,51). Finally, the occurrence of an interval of retrograde memory loss, however brief, is essentially universal in moderate-to-severe forms of CHI and all but the most trivial of mild head injuries. Thus, if a patient is able to describe the entire sequence of how he/she sustained the head injury, then one is forced to conclude that the head injury was extremely mild at worst and is unlikely to explain persistent sequelae putatively attributed to brain injury.

The anterograde amnesia apparent after recovery from coma also shows significant resolution during the interval of PTA. Prospective assessment typically is done by sampling personal and temporal orientation (the latter in particular because staying oriented in time requires the capacity to track the passage of time), use of three-word recall and recognition tasks, or the Galveston Orientation and Amnesia Test (GOAT) (52), a scale that combines assessment of preinjury and postinjury autobiographical recall with temporal orientation. Recovery of orientation typically occurs first to person, then place, and finally to time (44). Simple attention processes, such as counting forward and backward, precede three-word recognition and the attainment of GOAT scores of 75 or above, the level generally considered to signal the end of PTA. Three-word recall generally occurs after both of these (41). Recognition and recall of pictures tends to occur an average of 1 day earlier than recognition and recall words (53). This sequence appears to hold for all levels of severity of CHI, but it is difficult to distinguish in the rapid resolution of PTA in mild TBI. Because there is some fluctuation in memory encoding ability near the end of PTA, clinicians generally require consistent memory formation across at least two assessment occasions before they conclude that PTA has resolved (54). Unlike autobiographical memories that were consolidated prior to injury and ultimately become accessible during the recovery process, autobiographical events that occur during the interval of PTA are permanently lost, presumably due to rapid forgetting or failure to adequately consolidate information (55). Thus, personal events that occurred during PTA are unavailable for later retrieval. Exceptions to this are a few isolated memories for events that can occur as the injured individual approaches the end of PTA. These isolated memories are referred to as "islands of memory" and tend to involve events that are emotion laden for the patient, e.g., seeing blood on one's clothes or bedding, going into the bore of a magnetic resonance imaging (MRI) scanner, or having a birthday party in the hospital. Such "islands" do not signal an imminent end to PTA but certainly indicate progress toward the capacity for more consistent memory consolidation.

Anterograde and retrograde memory disturbances of the PTA interval appear most closely associated with the mechanisms of diffuse brain injury in CHI, that is, the axonal injury, and not the mechanisms of focal brain injury. For example, it is not uncommon for a patient with a gunshot wound to the head to later be able to report the entire sequence of events leading up to and following injury until systemic factors interfere with consciousness. Also, while formation of explicit memories is disturbed during PTA, formation of implicit memories including procedural learning and priming effects are generally preserved (56).

Postacute and Chronic Syndrome

After PTA has resolved, most CHI patients are capable of undergoing comprehensive neuropsychological evaluations that reveal a range of neurocognitive deficits, including memory, sustained attention, reaction time, mental processing speed and flexibility, and other signs of executive dysfunction. The neurocognitive deficits show considerable improvement during the first 6 months after injury and relatively subtle additional gain from 6 to 12 months after injury (57,58). Patients with extensive subarachnoid and intraventricular hemorrhage at the time of injury are at increased risk for developing hydrocephalus a few months later. Typically this is recognized by a decline in neurologic function or premature plateau of progress after a period of steady improvement. Except in the most severely injured cases and those that include focal cortical contusions or intracerebral hemorrhage, elementary language skills, spatial judgment, fund of knowledge, verbal abstraction, and spatial reasoning deficits resolve in most patients. However, patients with moderate-to-severe CHI often show varying degrees of permanent cognitive residuals in select functions and manifest affective changes that may persist for years and complicate social and vocational adjustment for the patient and family (59–61).

Deficits in learning and memory functions are among the most salient residuals after severe CHI (57,62,63) and often occur in the presence of average or better scores on standardized intellectual testing (64). These deficits are associated with signs that the severely injured CHI does not spontaneously organize information into semantic categories at the time of learning, which hampers later recall, but shows benefit when information is presented in a semantically organized fashion (65). Memory deficit is thought to be associated with damage to the hippocampal formation caused by reduced cerebral perfusion and focal temporal lesions (57). Signs of hypoxia are a significant predictor of failure to show improvement in memory at 1 year after injury (66). In children, deficits in learning and memory are correlated with cerebral volume loss in the prefrontal lobes (67).

Long-term impairment after severe CHI has been documented in other cognitive capacities as well. Equal in salience to deficits in learning and memory are impairments in speed of information processing and psychomotor response (68–71), a deficit that predicts social and vocational adjustment difficulties (66,72). Although simple attentional abilities as measured by digit span typically show good recovery, measures of sustained (vigilance) and focused (freedom from distraction) attention often show chronic impairment (43,58,68,73).

Executive and working memory functions involved with planning, organization of problem-solving strategies, initiation and inhibition of responses, and flexibility in shifting response strategies are vulnerable to sustained impairment (71,74–78). Although classic signs of aphasia are not common after CHI, mild deficits in naming have been demonstrated (79). Also, analysis of stories told by individuals with CHI have shown a lack of organization and failure to include essential information (80–82).

Chronic disturbances in behavior pose equal if not greater problems than cognitive deficits to long-term social and vocational maladjustment. Common sequelae include irritability or a reduced threshold of tolerance for annoyance, social disinhibition, and other forms of immature social behaviors; diminished sensitivity to social cues; decreased motivation; depression; and poor insight into the magnitude of cognitive and behavioral deficits (83,84). In a group of patients with predominantly moderate injury who were monitored closely, Jorge et al. (85) found that 42% met the criteria for major depression at some point during the first year after injury. Among the more severely injured patients, postinjury depression correlates with the degree of chronic disability and impairment in social function (86–88) but may not correlate with cognitive performance (85,88). Studies of structural lesions and affective and behavior changes have provided additional understanding of brain–behavior correlates. In studies of penetrating wounds in Vietnam veterans, features of agitated depression have been associated with right orbitofrontal lesions, whereas anger and hostility were more characteristic of left orbitofrontal lesions (89). These findings are consistent with civilian studies of traumatic injuries showing increased frequency of major depression with left anterior, typically dorsolateral, lesions (90,91), whereas irritability and anxious depression was more common in right hemisphere lesions (91). Behavioral problems associated with disinhibition, including mania, have been associated with damage to right orbitofrontal and basotemporal regions (92,93). A propensity for violence has been linked to frontal ventromedial lesions in general (94). Persisting affective and characterologic changes are common in children with severe TBI as well (95–97). Personality change identified 2 years after injury was reported in 40% of severely injured children in one large series (61).

MEASURING SEVERITY OF INJURY

Measuring the severity of injury is best done with markers gathered shortly after injury rather than using late-stage morbidity as the gauge of severity, which often encompasses more than injury mechanisms. There are no commonly accepted yardsticks for grading penetrating head injuries, whereas the uniform features and course of CHI afford convenient markers. The most commonly used markers are related to depth and duration of coma and duration of PTA, all of which have repeatedly been validated by their relationship to outcome.

A now widely used yardstick for gauging severity and monitoring neurologic status in CHI is the Glasgow Coma Scale (GCS) (98). The GCS assigns a behavioral rating to each of three dimensions of function (eye opening, motor, and verbal) based on the best response elicited and sums the ratings for a total composite score. The scale ranges from 3 at the worst end of the spectrum, which characterizes an individual who fails to show any response in any of the three dimensions of function (even painful stimulus), to 15, which is assigned to an individual who is alert and fully oriented, follows commands, and speaks normally. GCS scores of 3 to 8 on admission to the emergency department are characterized as severe, 9 to 12 as moderate, and 13 to 15 as mild. Because the patient may decompensate after admission, the lowest GCS score after resuscitation frequently is used as the initial rating. Use of GCS can be complicated by orbital edema or need for intubation, which precludes a complete assessment. To circumvent this, many use the postinjury time that elapses before the first evidence of following a command as the chief index of duration of coma and yardstick of severity. Both of these markers would be affected by use of paralyzing drugs and sedation, which might be necessary in early management.

Duration of PTA is a convenient marker in moderate-to-severe CHI and can be assessed later in the course of recovery when fewer pharmacologic agents are present to complicate its interpretation. It can be prospectively assessed with the GOAT (52) or, in children, the Children's Orientation and Amnesia Test (COAT) (99) using a threshold score of 75/100, which corresponds to the eighth percentile of individuals who recovered from mild TBI (52). Others argue that perfect recall on a three-word memory test is a more reliable and specific index of patients emerging from PTA (53,54). Typically, a threshold performance is required on 2 successive days on the GOAT or three-word test before the first of the two assessments is considered to mark the end of PTA. The end of PTA also can be estimated from medical progress notes on temporal orientation or full recovery on GCS (i.e., a score of 15); however, it should be noted that full temporal orientation typically occurs prior to threshold performances on prospective measures of PTA with the GOAT and three-word test (54). PTA can be estimated in mild TBI when witnesses are not present. Use in this context requires a retrospective analysis of the patient's memory for events after the accident. It hinges on the adequacy of the patient's report and the examiner's inquiry, and likely on how much time has passed since the injury. An interval of PTA less than 24 hours is graded as mild TBI (100). There is no commonly accepted cutoff in PTA for separating moderate and severe TBI, but injuries with PTA that exceed 1 week often are considered severe.

These markers of severity are highly correlated in CHI and similarly share considerable variance in predicting outcome when diffuse brain injury mechanisms are evident; however, the markers are not correlated when signs of only focal brain injury are present (101). Also, individuals older than 40 years tend to show longer intervals of PTA relative to the length of coma compared to individuals 20 to 40 years old (101). The value of GCS and PTA in predicting outcome is enhanced when they are combined with additional information, e.g., findings from radiologic studies, presence of pupillary reaction after injury, and highest elevation of ICP.

As more is learned about the metabolic cascade following head trauma, biologic markers of neural and glial cell injury have been identified that presumably correlate with phases of the cascade and enable a biochemical means of assessing the severity of injury (102). Neuron-specific enolase (NSE), an isoform of the glycolytic enzyme enolase, is normally found in neurons and neuroendocrine cells. After trauma, NSE can be found in serum and cerebrospinal fluid in proportion to clinical and radiologic indices of severity of injury and neurologic outcome (103–105). Protein S-100 is predominantly located in astroglial cells. After trauma, serum concentrations of S-100 correlate with clinical indicators of trauma severity (103,104). In humans, serum S-100B has been shown to peak within the first 24 hours after injury and to be abnormal for up to 3 days after MTBI (103). Although these biochemical markers offer the promise of gauging severity of injury, their value in predicting outcome remains to be demonstrated.

PREDICTING OUTCOME

When predicting outcome after TBI, it is useful to consider preinjury, injury, and postinjury predictor variables, because each may account for unique outcome variance (106). Outcome variables of interest include gross disability, resumption of preinjury lifestyle and activities, return to school or work, and neurocognitive and psychiatric morbidity. A common scale that represents five levels of gross disability is the Glasgow Outcome Scale (GOS) (107). The scale ranges from "death" to "good recovery," the latter essentially reflecting recovery sufficient for resumption of a "normal" lifestyle, including leisure activities and family relationships, even though minor neurologic and psychologic deficits may persist. Return to work is not used as a criterion of "good recovery" because the exigencies of local economic opportunity and various social incentives and disincentives might prevail; however, the "capacity" to sustain gainful employment is generally assumed. Intermediate ranges include persistent vegetative state, severe disability (dependent for daily support because of mental or physical disability), and moderate disability (independence in activities of daily living, capable of using public transportation, capable only of sheltered or restricted employment). The GOS has been criticized for poor definition in the intermediate classifications and a considerable breadth of deficit present in the good recovery category, because it encompasses cognitive and behavioral deficit that is not vocationally disabling. The validity of the scale has generally been established (108), however, and it is a relatively reliable and convenient outcome index for multicenter studies.

Preinjury Parameters

Generally speaking, the better a person was functioning prior to a significant injury, the better he/she will function after the injury. Thus, those who have higher intelligence quotients and socioeconomic status, have completed more education, have achieved higher degrees of vocational success, were consistently employed, and were physically and psychologically healthy prior to injury achieve better outcomes than those less fortunate on these dimensions. In contrast, those with learning disabilities, vocational maladjustment, inconsistent employment, significant developmental or chronic health problems, or psychiatric problems including depression, attention-deficit/hyperactivity disorder, and substance abuse recover less well (97,106,109). In neuropsychological studies of children with moderate-to-severe injury, Yeates et al. (110) found that the preinjury family environment accounted for more of the outcome variance 1 year after injury than did injury severity (25% vs. 20%).

One of the more powerful predictors of outcome is age. Individuals older than 40 years show higher morbidity rates (101) and individuals older than 55 years show higher mortality rates—80% in severe TBI (111). Regression formulas that predict outcome routinely have age as one of the variables accounting for a large portion of outcome variance (112). The relationship of age and outcome may interact with severity of injury such that poorer outcomes occur in severe injuries of individuals younger than 8 years and older than 40 years (101,113,114). Age has consistently proven significant in predicting a patient's return to work after injury (66,115,116).

Having one previous head injury increases the risk threefold for having another head injury, and two previous head injuries increase the risk eight-fold (11). Having a previous head injury may prolong recovery and disability from a second head injury (117). The hypothesis is that head injuries deplete the cognitive reserve capacity that is necessary to sustain vigilance for hazards and to bounce back from recurrent injury.

There is increasing evidence for a genetic vulnerability of the brain to injury. Preliminary data have suggested a link between apolipoprotein E (ApoE) and the brain's response to injury (118,119). The presence of ApoE ϵ4 has been linked to Alzheimer disease, poor outcome after stroke, dementia pugilistica with features combining Alzheimer disease and head trauma, and lower cognitive performances in older professional football players (120,121). It is possible that the gene prevents processes associated with neural repair and thereby increases postinjury morbidity.

Injury Parameters

Studies on long-term outcome from CHI, where mechanisms of diffuse brain injury are primary, have repeatedly demonstrated a strong dose–response relationship, namely, the more severe the acute index of injury, the poorer the outcome. For example, the combined data from four reports show that 45% of patients admitted to the emergency department with GCS of 3 to 8 (severe) did not survive their injuries and another 19% were left in a vegetative or severely disabled state (122–125). In contrast, no patients with GCS of 13 to 15 (mild) died, and only 3% of patients were left in a vegetative or severely disabled state.

Using data from the NIH Trauma Coma Data Bank, Levin et al. (62) found that only 23% of patients whose postresuscitation GCS dipped to 3 to 5 were able to complete neuropsychological testing 1 year after injury, whereas 83% of patients with GCS in the 9 to 15 range were able to complete testing. GSC and time to follow commands correlate well with composite performance on a neuropsychological battery at 1 year (126). GCS more accurately predicts long-term neurobehavioral deficit when it is combined with nonreactive pupils (57,62) and brain lesions on computed tomographic (CT) scans (127). Similar findings have been reported in children showing strong correlation between GCS and neuropsychological, academic, and behavioral adjustment indices at 3 years after injury (128–130).

In persons where diffuse axonal injury is primary, PTA is one of the strongest predictors of outcome (99,101,112,130). Katz and Alexander (101) found that 71% of patients with PTA less than 2 weeks achieved a GOS "good" recovery at 1 year after injury, whereas only 22% of those with PTA greater than 2 weeks achieved a good recovery. Conversely, the incidence of GOS severe disability was four-fold greater in persons with PTA greater than 2 weeks. Dikmen et al. (131) found that only patients with prolonged coma (>1 day), admission GCS less than 8, or PTA greater than 2 weeks showed persistent memory deficits 1 year after injury. Similar findings correlating injury variables and outcome have been reported by other investigators (112,132,133). GCS, length of coma, and PTA share considerable variance in predicting outcome in CHI with diffuse axonal injury in both children (130) and adults (101,112).

Although the predictive power of the acute injury parameters is impressive, it is important to note that the prediction is not perfect. A small proportion of patients with MTBI ultimately have poor outcomes (often older individuals) and a small proportion of patients with very severe TBI show amazingly good recovery (often young healthy individuals). Thus, offering prognoses using these variables should be done with caution and only after careful consideration of whether the basis of the predictive relationship between the acute injury variables and outcome is preserved. The adequacy of GCS, length of coma, and PTA in predicting long-term outcome hinges on whether these measures directly reflect the underlying mechanisms of diffuse brain injury and not other incidental factors. For example, the predictive value of GCS and PTA in TBI is significantly diminished in the setting of acute alcohol intoxication, paralyzing and sedating medication, large doses of analgesics, and systemic illnesses such as infections and respiratory ailments. Each of these "confounding" conditions makes the patient appear more severely injured than the mechanisms of brain injury would in isolation and thus would lead to an erroneously pessimistic prognosis. In keeping with this, prophylactic use of the anticonvulsant phenytoin has been shown to prolong PTA (112) and to have an adverse effect on cognitive outcome 1 month after severe TBI (134). Although GCS in nontraumatic coma has been shown to bear some relationship to 2-week mortality and vegetative state rates (135), long-term predictions from GCS in nontraumatic coma are considerably different than in TBI, reflecting the different underlying mechanisms.

Postinjury Parameters

The longer an individual is in a vegetative state, the less likely he/she will recover from it (136–138). Recovery from a vegetative state after 3 months is rare. Neuropsychological tests have been shown to predict general disability status as measured by GOS (108), with one study showing 80% of the variance in GOS accounted for by a simple measure of fine motor dexterity (77). Using data from the Trauma Coma Data Bank, Ruff et al. (66) found that preservation of language skills and speed of information processing, along with age, predicted who returned to work or school after severe TBI. In a 5-year followup study, persistent deficits in executive functions and psychomotor speed were major factors associated with loss of social autonomy and inability to return to work (72). Deficits in behavioral self-regulation and cognitive speed are significant predictors of persistent problems in interpersonal relationships and overall psychosocial adjustment (139). Using a large series of patients, Dikmen et al. (116) provided probability tables that combine preinjury demographic, injury, and early postinjury neuropsychological data for predicting who will return to work within the first 2 years after injury.

Little is known about the long-term outcome of civilian survivors of penetrating head injuries. Investigators at the National Institutes of Health (NIH) have completed extensive studies on the outcome of patients with penetrating head injuries sustained in Vietnam (140–142). Disabilities that complicated return to work were posttraumatic epilepsy, hemiparesis, visual field loss, verbal memory loss, visual memory loss, psychological problems, and violent behavior. Each of these disabilities contributed equally to predicting unemployment in their regression model (140,143).

Preinjury intelligence, total brain volume loss, and postinjury education also were significant predictors.

Neuroimaging studies show that about 75% of survivors of moderate-to-severe CHI show enlargement of the lateral ventricles relative to noninjured persons, a finding that correlated with cognitive impairment (144). Imaging with MRI late in the course of recovery provides incremental prognostic information relative to early MRI studies (145). The volume of lesions in the prefrontal lobes identified on MRI during the postacute phase is correlated with residual problems in executive functions, including mental flexibility, verbal fluency, semantic clustering in verbal memory, and reaction time on a "go/no go" task (146).

MILD TRAUMATIC BRAIN INJURY

Probably no topic associated with TBI generates more controversy than issues concerning permanent disabilities from MTBI (often referred to as "minor" head injury or concussion). Approximately 90% of head injuries are mild, and it has been estimated that up to 15% of those with MTBI have disabling symptoms (147). The controversial questions are as follows: (a) How severe must a head injury be before symptoms due to brain injury are expected? (b) How severe must a head injury be before permanent changes in brain function occur, e.g., reduction of neural reserve capacity? (c) When symptomatic complaints persist beyond expectation, what mechanisms are responsible and are they amenable to treatment? Much of the controversy is fueled by confusion related to terms, variable definitions, and studies methodologically flawed by sampling bias or inadequate controls that have served to confound issues related to base rate of symptoms. In the end, the controversy often causes the victim of MTBI to suffer twice, first from the head injury itself and then from mismanagement by treating professionals—either failure to recognize and treat brain dysfunction when it exists or failure to recognize treatable psychosocial adjustment difficulties that underlie symptoms and instead forever doom the patient to a diagnosis of "permanently brain damaged."

MTBI is defined by a GCS of 13 to 15 at the time of admission, which connotes an individual who at worst is disoriented and perhaps confused in speech. The American Congress of Rehabilitation Medicine has endorsed a more elaborate set of defining criteria that sets an upper limit of severity with LOC of 30 minutes and PTA less than 24 hours (Table 41.2). Although each of the definitions clearly demarcates mild from more severe forms of TBI, neither provides clear separation of a clinically trivial injury from one that affects brain function. It is recognized that a brain concussion can occur without the head coming into contact with an object. Brain concussion also can occur without LOC or an interval of PTA; however, most experts would agree that, if neither of these occurred, the

TABLE 41.2. MILD TRAUMATIC BRAIN INJURY AS DEFINED BY AMERICAN CONGRESS OF REHABILITATION MEDICINE[a]

A traumatically induced physiologic disruption of brain function, as manifest by at least one of the following:
1. Any period of loss of consciousness,
2. Any loss of memory for events immediately before or after the accident,
3. Any alternation in mental state at the time of the accident (e.g., feeling dazed, disoriented, confused), and
4. Focal neurologic deficit(s) that may or may not be transient

But where the severity of the injury does not exceed
1. Loss of consciousness of 30 minutes
2. After 30 minutes, an initial Glasgow Coma Scale of 13–15, and
3. Posttraumatic amnesia not greater than 24 hours

From Anonymous. Report of the MTBI Committee of the Head Injury Interdisciplinary Special Interest Group of the American Congress of Rehabilitation Medicine (1993). Definition of mild traumatic brain injury. *J Head Trauma Rehabil* 1993;8:86–87, with permission.

degree of brain dysfunction caused by trauma is likely to be quite mild and transient, due to rapid recovery of cellular ionic imbalances. It should be noted that the range of severity of injury captured by these definitions is considerable. The deficits expected after an injury that produced LOC for 25 minutes and an interval of PTA of 23 hours are substantially greater than those from an injury that resulted in the patient feeling dazed without LOC or PTA, the former often requiring months to recover and the latter typically resolving in minutes to occasionally a few days.

Without question, MTBI is associated with some disruption of brain function for an interval of time. Common symptoms of acute MTBI include a disturbance in attention and concentration, headache, dizziness and vertigo, and nausea within a few hours of injury. In subsequent days, report of continuing problems with concentration and headache is typical, along with complaints of memory impairment, fatigue, sleep disturbance, photophobia, blurred vision, irritability and other mood changes, apathy, and anhedonia. A combination of these symptoms occurring shortly after head injury is referred to as the postconcussive syndrome (PCS). In the vast majority of uncomplicated MTBI cases, these symptoms gradually subside within a few weeks, although not all at the same rate (148). For example, disturbance in sleep/wake cycles and vertigo often resolve quickly while concentration problems last longer. When collections of the symptoms persist for months or worsen over time, the diagnostic term "persistent postconcussive syndrome" (PPCS) has been proposed (149,150) as the etiology for persisting symptoms and may be different than that of PCS. (We concur with the restricted use of the PCS terminology only for the acute postconcussive interval, e.g., first 6 months after injury.)

It is useful to note that many of the symptoms of PCS during the acute postinjury interval often are due not to brain concussion but rather to trauma to other organs and tissue

(151). For example, complaints of headache can be due to strain of neck muscle and ligaments, and dizziness and nausea can be due to labyrinthine concussion. Thus, the acute symptoms of PCS usually represent a combination of factors, including brain concussion, peripheral injury, and psychological correlates. It also is important to note that many of the symptoms of PCS (particularly impairment in concentration and memory, sleep disturbance, fatigue, irritability, mood disturbance) are not specific to trauma, or even neurologic disorders in general, but rather are common complaints in psychiatric states (e.g., affective and anxiety disorders, and chronic pain) (152) and the general population (153). Moreover, if naïve adults are asked to pretend they have been in a car accident, symptoms of PCS are among the first to be reported, indicating that expectations may represent an important factor in persisting symptoms (154).

Findings from neuropsychological studies of MTBI vary with the quality of the methodology (155,156). In general, prospective studies using carefully selected control groups have consistently shown resolution of cognitive deficits within 1 to 3 months after a single uncomplicated MTBI. This has been demonstrated in children and adolescents (155), adults (157–160), and elderly subjects (161). In sports-related concussions where the head injuries typically are on the very mild end of the spectrum and preseason baseline testing is obtained to optimize assessment sensitivity, recovery generally occurs over a span from 7 to 10 days (10,162).

A number of investigators believe that each occurrence of MTBI takes a toll, with effects accruing over time. This can be manifested as impaired neuropsychological performance only under conditions of physiologic stress (163) or as slowed or incomplete recovery from subsequent mild head injuries (117,164). It is known that when two concussions occur in close temporal proximity, significant complications in recovery occur in animals (29) and a catastrophic outcome in humans is possible (33) (SIS; see section on Pathobiology and Pathophysiology). The incidence of depression is increased after MTBI for reasons that have not been clearly delineated (165). Also, in animal models of injury, evidence of diffuse axonal injury has been demonstrated with injuries that seemingly are the human equivalent of MTBI (29).

Explanations for the high incidence of persisting problems in MTBI are likely to be multifactorial (166). On the organic side of the spectrum, it must be recognized that the standard yardsticks for MTBI in LOC, GCS, and PTA are indirect measures of pathology and that not all significant deficits may be predicted by these indices. When MTBI is complicated by contusions, prolonged recovery courses and potentially permanent deficits may occur in a fashion not unlike cases of moderate TBI as defined by GCS (127,161). In some cases, the prolonged symptoms may be due to an accumulation of injury effects or a genetic defect in recuperation from injury. On the psychological side, there is no question that sustaining a "disabling" injury serves a psychological need for some individuals in the form of escape from occupational or domestic responsibility. For some, the prospect of a healthy financial settlement seems adequate to sustain a repertoire of unhealthy behaviors. One model holds that persistent deficits evolve from an individual's expectation of what types of symptoms typically occur after head injury combined with tendencies to overestimate the actual postinjury occurrence of the symptoms and to underestimate the premorbid prevalence of these symptoms (154,167). Another model holds that an individual's "vulnerability" to persistent disability from MTBI is a complex interaction of neurologic, personality, family, and vocational systems (166).

In our experience, a significant number of cases with PPCS occur as a psychological reaction that evolves from the early postconcussive symptoms that are anchored in physiologic dysfunction. Patients often are not adequately educated about the nature of these symptoms and their natural course sufficiently early so that they can develop realistic expectations of their course and plan necessary adjustments to lifestyle accordingly. Instead they are told that they "are fine" and capable of resuming all routine activities. When they attempt to resume their activities, adjustment problems arise from mental inefficiencies, limited stamina, and decreased tolerance or frustration. The PCS symptoms are aggravated and conditions are ripe for developing depression. The longer the emotional reaction is left untreated, the more difficult it is to treat. Thus, not unlike the biochemical cascade triggered by trauma, there appears to be a cascade of psychological reactions that potentially can be interrupted early in their course.

We recommend early education, beginning in the emergency room, about the effects of trauma and the typical course. Patients' complaints, mental status, and affect should be closely monitored during the first few weeks after injury. Any symptom that does not resolve should be considered for further investigation. Patients with sleep disturbance for more than 2 weeks should be treated aggressively, particularly those patients with nightmares. Similarly, patients who show signs of depression or anxiety should be treated aggressively with medication and possibly psychotherapy. In our experience, patients who suffer disfiguring injuries from facial bone fractures are at high risk for developing reactive depression and posttraumatic stress disorder (PTSD). The risk of PTSD also is high in patients who had minimal PTA and incurred life-threatening conditions immediately after their accident, e.g., threat of fire from leaking gasoline or asphyxiation from accident debris. Aggressive treatment of such reactions early in their course is hypothesized to prevent the more stable and persistent neurotic habits that can occur later. Return to work should be postponed until symptoms subside substantially, then gradual reintroduction over the course of 1 or 2 weeks is recommended to facilitate adjustment and stamina development.

MANAGEMENT OF TRAUMATIC BRAIN INJURY
Prevention

Although often not stated, preventive measures cannot be emphasized enough. There has been a significant decline in morbidity and mortality from motor vehicle accidents since introduction of safety features within vehicles and on the nation's highways during the past 3 decades (168). Seat belts, when used properly, have been shown to effectively prevent injuries (169). Front and side airbags offer additional protection. Use of helmets on motorcycles and bicycles significantly reduces injury. Motorcycle deaths declined by 27% in California after helmet laws were introduced (25). Actuarial data on teenage motor vehicle accidents imply that extending education and training in driving and use of restricted licenses by adolescents may decrease substantially the risk of accidents. Requiring education and training before ownership and use of firearms may provide similar benefits. Careful consideration about safety and injury prevention should be given to the design of industrial and playground equipment. Impact-absorbing materials on the edge of highways have distinct advantages over unforgiving concrete supports and trees. Preventative actions that target groups and activities that are associated with high risk likely will provide the greatest yield.

Acute Care

The primary focus of managing acute TBI is prevention of secondary injury. Close monitoring and documentation of the patient's neurologic status are critical for acute care and are beneficial to later treatment and disposition planning. Documenting neurologic status ideally should begin in the field because it provides information useful for later management. In the first few hours after injury, the principal concerns are from hypoxia/ischemia and increased ICP from expanding mass lesions. GCS often is combined with measures of blood pressure and respiration to provide a means of tracking the injured patient through the acute interval and to signal deterioration requiring intervention. Use of the GCS and various trauma scales can be complicated by the fact that sedating or paralyzing agents are needed to manage the patients through the early course. Patients in coma (GCS < 8) typically should be intubated and ventilated to ensure adequate oxygenation. Systolic blood pressures less than 90 mm Hg should trigger corrective intervention. Installation of an ICP monitor is recommended in severe TBI with CT abnormalities and in patients older than 40 years with unilateral or bilateral posturing or hypotension. Hyperventilation and mannitol, which are treatments for management of increased ICP, are best reserved for use in patients who show clear signs of intracranial hypertension, e.g., brain herniation or neurologic decompensation, because such treatments have the potential to exacerbate intracranial ischemia (25). High-dose barbiturate therapy is considered when other medical treatment for intracranial hypertension fails.

CT has proven to be invaluable in management of mass lesions in TBI. It should be used in all patients with GCS of 12 or less or when a mild head injury is accompanied by focal neurologic signs. Optimally, scanning should be done as soon as the patient is resuscitated and medically stable, because delays of more than 4 hours in evacuation of hematomas are associated with poorer outcomes. Use of MRI early in the acute setting often is not practical because it is difficult to control movements that degrade image quality and to manage the other needs of the comatose patient in the scanner. Most acute neurosurgical issues are adequately addressed with CT.

Other factors need to be considered. Signs of shock often suggest hemorrhage elsewhere in the body. Injuries that breach the dura have increased risk of central nervous system infection. Nutritional support will need to be implemented by the end of the first week because the starved head-injured patient will lose 15% body weight per week. A 30% loss of body weight increases risk of mortality (25).

Prophylactic treatment of seizures is generally not recommended because risk of epilepsy in CHI is relatively small (2%–5% overall), and at least one antiepileptic drug (phenytoin) has been shown to hamper speed of recovery (170). Higher risks of seizure are associated with severe CHI (11%) (171), depressed skull fractures (15%), intracerebral hematoma (31%), and up to 50% in penetrating head injuries (172,173). The longer the seizure-free interval after TBI, the lower the risk of developing epilepsy. Even in patients with penetrating head injuries in whom the risk is high, 95% are likely to remain seizure free 3 years after injury. In any case, most studies indicate that prophylactic use of antiepileptic drugs does not prevent posttraumatic seizures (25).

Neuropsychological screening during the interval of posttraumatic confusion assists in tracking recovery of anterograde memory during the acute phase and identifying signs of focal brain injury, such as unilateral neglect, aphasia, and prominent executive dysfunction, which will require attention during rehabilitation. Such screening typically is done at bedside using brief procedures because the patient will be unable to sustain attentional focus long enough to enable meaningful results on more lengthy tests. Assessment during this interval should include a survey of preinjury variables that will aid in treatment and disposition planning. Generally, it is not safe to discharge a patient from the hospital while he/she is still in PTA because awareness of deficit is limited and judgment is poor. Discharge should only be done when adequate supervision and safeguards have been arranged and the agent providing supervision is physically capable of intervening if necessary.

Postacute Care

Once patients have recovered from the posttraumatic confusional state, more detailed assessment of neuropsychological status can be undertaken and the results integrated with assessment of fine and gross motor skills, emotional functioning, and daily living skills. The aims of these assessments early in the postacute phase are to stage recovery and guide rehabilitation planning. It is useful to assess the patient's report of retrograde and anterograde amnesia and compare them with hospital records to further substantiate the markers of severity of injury. It is important to monitor progress and intervene when progress is compromised by psychological (e.g., depression) or physical (e.g., hydrocephalus, seizures, infection) factors.

Whereas mild TBI can simply be monitored for adverse reactions during recovery, moderate and severe TBI typically require an interval of more formal rehabilitation procedures, often beginning on an inpatient basis. Treatment will need to be multifaceted and address issues raised by each of the domains of deficit resulting from TBI and the effects the deficits likely will have on posttreatment vocational and psychosocial adjustment. Regression model studies have identified deficits that predict poor social adjustment, community reintegration difficulties, unemployment, and dependence on society. Rehabilitation will need to focus on these contributors if progress in the treatment of brain injury is to be made.

On the whole, the rehabilitation industry has not done a good job of documenting its contribution above and beyond what spontaneously occurs in minimal care, supportive models of treatment. In 1998, the NIH convened a conference that issued a consensus statement acknowledging the limited empirical evidence for the use of cognitive and behavioral treatment techniques in TBI (174). Still, the conference attendees concluded that supportive evidence existed for select cognitive and behavioral procedures and issued a call for increased funding for rehabilitation experts to evaluate treatment procedures. Cicerone et al. (175) systematically reviewed the literature for evidence-based support for cognitive rehabilitation in TBI and stroke. Their review of the literature found 29 Class I studies, of which 20 provided clear evidence supporting the effectiveness of select treatments for TBI and stroke. Several studies showed advantages of cognitive rehabilitation procedures over conventional treatments. Negative and equivocal results occurred in studies where a cognitive procedure was compared with another procedure, frequently cognitive in nature. Only two of 64 Class I and II studies failed to show improvement in patients following cognitive rehabilitation. The domains of recognized remediation were visual perceptual and language function after stroke, and attention, memory, functional communication, and executive function after TBI. Based on their review, the authors offered recommendations for practice standards, guidelines, and options.

Still other reviewers have not been convinced of the durability and clinical relevance of "improvements" found in most rehabilitation studies and have challenged the industry to use standard definitions, control groups, and relevant outcome measures reflecting health and function (176). Consistent with these concerns were the findings from a meta-analysis of rehabilitation procedures for attentional problems from TBI (177). The authors found evidence of significant improvement in pre- and post-treatment paradigms when no control procedures were included, but not when controls were present. Also, the gains in attentional function were not shown to significantly affect outcome.

RECENT DEVELOPMENTS AND FUTURE DIRECTIONS

The more that is understood about the molecular neurobiology of head trauma, the more treatment avenues can be identified for interrupting the metabolic cascade and protecting neurons and support tissues from a fatal course. For example, it is known that mitochondrial swelling has been associated with Ca^{2+}-induced alterations in mitochondrial permeability. In this process there is an uncoupling of the respiratory chain, along with hyperproduction of reactive oxygen radicals and release of apoptogenic proteins and cytochrome C. This activates caspase-3–like protease, which plays a critical role in axonal damage. In animals, cyclosporin A has been shown to block mitochondrial permeability and prevent the subsequent cascade of events. This finding is the basis for clinical trials in humans (20). Several therapeutic studies have been conducted on interruption of the glutamate-based neuroexcitatory mechanism using agents designed to block the *N*-methyl-D-aspartate (NMDA) receptor. Unfortunately, to date these agents have failed in human trials (25). In laboratory animals, use of glutamate-NMDA antagonists seems to interfere with maladaptive neuroplastic changes that occur late in recovery, but this is unproven in humans. Understanding the role that cytokines play in both damaging and inhibiting damage of cells may offer another avenue of treatment.

In addition to treatments that focus on interruption of the metabolic cascade, another important avenue of treatments is aimed at neurorestoration and regeneration. This includes strategies to facilitate dendritic sprouting, axon regeneration, and synaptogenesis and to optimize the function of residual neurons. Genetic manipulations to augment nerve growth factor and other neurotrophic polypeptides are promising areas of investigation (178). Brain tissue transplantation is yet another area of investigation but one that likely is distant on the horizon of treatments for TBI.

Finally, the tools for identification of injury parameters and their effects on function are steadily improving and offer promise for expediting research. Biochemical markers of the metabolic cascade in humans are slowly being identified, e.g., NSE and protein S-100, and offer the hope of bet-

ter understanding and earlier intervention carefully targeted to the metabolic abnormalities of the moment. Magnetic resonance spectroscopy is a promising technology for providing insight into brain cellular metabolism with a medium degree of three-dimensional resolution. Preliminary studies indicate sensitivity to changes in the most basic biochemical elements of cell function following trauma that seem to correlate with neuropsychological function and presumed stage of posttraumatic cellular breakdown and recovery (179,180). Functional imaging techniques, including positron emission tomography and functional magnetic resonance imaging, offer a wide array of techniques and opportunities for investigating the physiologic correlates of trauma and cognitive effects and provide a means of understanding mechanisms and tracking markers of recovery (181–183).

CONCLUSION

The last 3 decades have witnessed a substantial unveiling of the invisible epidemic of TBI, but much remains to be accomplished. TBI is better recognized as one of the major public health concerns; however, this recognition still needs improved translation into preventive action. In TBI, an ounce of prevention is undoubtedly worth a pound of cure. Molecular biology has enabled a better understanding of the basic pathophysiologic mechanisms of TBI and spawned new concepts for preventing the permanent neurologic injury caused by secondary processes and facilitating neurorestoration. Still, we are early in the game of developing these agents. To date, there are no compounds that effectively treat the brain damage of TBI. Nonetheless, there are reasons to be hopeful. Advances in molecular biology, increased understanding of genetic expression, and progress in treatment of related illnesses, such as stroke, subarachnoid hemorrhage, and Alzheimer disease, unquestionably will expedite development of treatments for TBI. New imaging technologies and identification of biochemical markers of injury offer the hope of better tracking and timing of interventions. The human toll of TBI is far too great not to pursue these avenues. We also must work to better assist victims and their families to cope and compensate for losses that so tragically affect their daily lives.

REFERENCES

1. Anonymous. Center for Disease Control. Traumatic brain injury in the United States: a report to Congress. 1-16-2001. Available at: *www.cdc.gov/ncipc/pub-res/tbicongress.htm.*
2. Max W, MacKenzie WJ, Richard MT. Head injuries: costs and consequences. *J Head Trauma Rehabil* 1991;6:76–91.
3. Kerr TA, Kay DW, Lassman LP. Characteristics of patients, type of accident, and mortality in a consecutive series of head injuries admitted to a neurosurgical unit. *Br J Prevent Soc Med* 1971;25:179–185.
4. Dikmen S, Machamer J, Temkin N. Psychosocial outcome in

patients with moderate to severe head injury: 2-year follow-up. *Brain Inj* 1993;7:113–124.

5. Hartsough CS, Lambert NM. Medical factors in hyperactive and normal children: prenatal, developmental, and health history findings. *Am J Orthopsychiatry* 1985;55:190–201.

6. Weiss G, Hechtman L. *Hyperactive children grown up,* 2nd ed. New York: Guilford Press, 1993.

7. Jennett B. Head injuries in children. *Dev Med Child Neurol* 1972;14:137–147.

8. Clarke KS. Epidemiology of athletic neck injury. *Clin Sports Med* 1998;17:83–97.

9. Thurman DJ, Branche CM, Sniezek JE. The epidemiology of sports-related traumatic brain injuries in the United States: recent developments. *J Head Trauma Rehabil* 1998;13:1–8.

10. Barth JT, Alves WM, Ryan TV, et al. Mild head injury in sports: neuropsychological sequelae and recovery of function. In: Levin HS, Eisenberg HM, Benton AL, eds. *Mild head injury.* New York: Oxford University Press, 1989:257–275.

11. Annegers JF, Grabow JD, Kurland LT, et al. The incidence, causes, and secular trends of head trauma in Olmsted County, Minnesota, 1935–1974. *Neurology* 1980;30:912–919.

12. Gennarelli TA. Mechanisms of brain injury. *J Emerg Med* 1993;11[Suppl 1]:5–11.

13. Bailes JE, Cantu RC. Head injury in athletes. *Neurosurgery* 2001;48:26–46.

14. Gennarelli TA. The spectrum of traumatic axonal injury. *Neuropathol Appl Neurobiol* 1996;22:509–513.

15. Bullock R, Zauner A, Myseros JS, et al. Evidence for prolonged release of excitatory amino acids in severe human head trauma. Relationship to clinical events. *Ann NY Acad Sci* 1995;765:290–297.

16. Hovda DA, Lee SM, Smith ML, et al. The neurochemical and metabolic cascade following brain injury: moving from animal models to man. *J Neurotrauma* 1995;12:903–906.

17. Yakovlev AG, Faden AI. Molecular biology of CNS injury. *J Neurotrauma* 1995;12:767–777.

18. Bergsneider M, Hovda DA, Shalmon E, et al. Cerebral hyperglycolysis following severe traumatic brain injury in humans: a positron emission tomography study. *J Neurosurg* 1997;86:241–251.

19. Junger EC, Newell DW, Grant GA, et al. Cerebral autoregulation following minor head injury. *J Neurosurg* 1997;86:425–432.

20. Povlishock JT. Pathophysiology of neural injury: therapeutic opportunities and challenges. *Clin Neurosurg* 2000;46:113–126.

21. Povlishock JT, Jenkins LW. Are the pathobiological changes evoked by traumatic brain injury immediate and irreversible? *Brain Pathol* 1995;5:415–426.

22. Ommaya AK, Gennarelli TA. Cerebral concussion and traumatic unconsciousness. Correlation of experimental and clinical observations of blunt head injuries. *Brain* 1974;97:633–654.

23. Levin HS, Williams D, Crofford MJ, et al. Relationship of depth of brain lesions to consciousness and outcome after closed head injury. *J Neurosurg* 1988;69:861–866.

24. Levin HS, Mendelsohn D, Lilly MA, et al. Magnetic resonance imaging in relation to functional outcome of pediatric closed head injury: a test of the Ommaya-Gennarelli model. *Neurosurgery* 1997;40:432–440.

25. Marshall LF. Head injury: recent past, present, and future. *Neurosurgery* 2000;47:546–561.

26. Maroon JC, Lovell MR, Norwig J, et al. Cerebral concussion in athletes: evaluation and neuropsychological testing. *Neurosurgery* 2000;47:659–669.

27. Buki A, Siman R, Trojanowski JQ, et al. The role of calpain-mediated spectrin proteolysis in traumatically induced axonal injury. *J Neuropathol Exp Neurol* 1999;58:365–375.

28. Hovda DA. Metabolic dysfunction. In: Narayan RK, Wilberger JE, Povlishock JT, eds. *Neurotrauma.* New York: McGraw-Hill, 1996:1459–1478.

29. Hovda D, Prins M, Becker D, et al. Neurobiology of concussion. In: Bailes JE, Lovell MR, Maroon JC, eds. *Sports-related concussion.* St. Louis, MO: Quality Medical Publishers, 1999:12–51.

30. Pike BR, Zhao X, Newcomb JK, et al. Stretch injury causes calpain and caspase-3 activation and necrotic and apoptotic cell death in septo-hippocampal cell cultures. *J Neurotrauma* 2000;17:283–298.

31. Povlishock JT, Christman CW. The pathobiology of traumatically induced axonal injury in animals and humans: a review of current thoughts. *J Neurotrauma* 1995;12:555–564.

32. Lewis SB, Wong ML, Bannan PE, et al. Transcranial Doppler identification of changing autoregulatory thresholds after autoregulatory impairment. *Neurosurgery* 2001;48:369–375.

33. Saunders RL, Harbaugh RE. The second impact in catastrophic contact-sports head trauma. *JAMA* 1984;252:538–539.

34. Cantu RC. Second-impact syndrome. *Clin Sports Med* 1998;17:37–44.

35. McCrory PR, Berkovic SF. Second impact syndrome. *Neurology* 1998;50:677–683.

36. Christman CW, Salvant JB Jr, Walker SA, et al. Characterization of a prolonged regenerative attempt by diffusely injured axons following traumatic brain injury in adult cat: a light and electron microscopic immunocytochemical study. *Acta Neuropathol* 1997;94:329–337.

37. Alexander MP. Traumatic brain injury. In: Benson BW, Blumer D, eds. *Psychiatric aspects of neurologic disease, vol. II.* New York: Grune & Stratton, 1982:219–249.

38. Hagen C. Language-cognitive disorganization following closed head injury: a conceptualization. In: Trexler L, ed. *Cognitive rehabilitation: conceptualization and intervention.* New York: Plenum, 1982.

39. Levin HS, Gary HE, Eisenberg HM. Duration of impaired consciousness in relation to side of lesion after severe head injury. NIH Traumatic Coma Data Bank Research Group. *Lancet* 1989;1:1001–1003.

40. Russell WR. *The traumatic amnesias.* New York: Oxford University Press, 1971.

41. Stuss DT, Binns MA, Carruth FG, et al. The acute period of recovery from traumatic brain injury: posttraumatic amnesia or posttraumatic confusional state? *J Neurosurg* 1999;90:635–643.

42. Levin HS. Prediction of recovery from traumatic brain injury. *J Neurotrauma* 1995;12:913–922.

43. van Zomeren AH, Deelman BG. Differential effects of simple and choice reaction after closed head injury. *Clin Neurol Neurosurg* 1976;79:81–90.

44. High WM Jr, Levin HS, Gary HE Jr. Recovery of orientation following closed-head injury. *J Clin Exp Neuropsychol* 1990;12:703–714.

45. Benson DF, Geschwind N. Shrinking retrograde amnesia. *J Neurol Neurosurg Psychiatry* 1967;30:539–544.

46. Russell WR, Nathan PW. Traumatic amnesia. *Brain* 1946;69:183–187.

47. Ribot T. *Diseases of memory: an essay in the positive psychology.* New York: Appleton, 1882.

48. Russell WR. Amnesias following head injuries. *Lancet* 1935;2:762–763.

49. Paniak C, MacDonald J, Toller-Lobe G, et al. A preliminary normative profile of mild traumatic brain injury diagnostic criteria. *J Clin Exp Neuropsychol* 1998;20:852–855.

50. Levine B, Black SE, Cabeza R, et al. Episodic memory and the

self in a case of isolated retrograde amnesia. *Brain* 1998;121: 1951–1973.

51. Kapur N. Syndromes of retrograde amnesia: a conceptual and empirical synthesis. *Psychol Bull* 1999;125:800–825.

52. Levin HS, O'Donnell VM, Grossman RG. The Galveston Orientation and Amnesia Test. A practical scale to assess cognition after head injury. *J Nerv Ment Dis* 1979;167:675–684.

53. Schwartz ML, Carruth F, Binns MA, et al. The course of post-traumatic amnesia: three little words. *Can J Neurol Sci* 1998;25:108–116.

54. Stuss DT, Binns MA, Carruth FG, et al. Prediction of recovery of continuous memory after traumatic brain injury. *Neurology* 2000;54:1337–1344.

55. Levin HS, High WM Jr, Eisenberg HM. Learning and forgetting during posttraumatic amnesia in head injured patients. *J Neurol Neurosurg Psychiatry* 1988;51:14–20.

56. Ewert J, Levin HS, Watson MG, et al. Procedural memory during posttraumatic amnesia in survivors of severe closed head injury. Implications for rehabilitation. *Arch Neurol* 1989;46: 911–916.

57. Levin HS, Eisenberg HM. Management of head injury. Neurobehavioral outcome. *Neurosurg Clin North Am* 1991;2:457–472.

58. van Zomeren AH, Deelman BG. Long-term recovery of visual reaction time after closed head injury. *J Neurol Neurosurg Psychiatry* 1978;41:452–457.

59. Lezak MD. Living with the characterologically altered brain injured patient. *J Clin Psychiatry* 1978;39:592–598.

60. Lezak MD. Brain damage is a family affair. *J Clin Exp Neuropsychol* 1988;10:111–123.

61. Max JE, Koele SL, Castillo CC, et al. Personality change disorder in children and adolescents following traumatic brain injury. *J Int Neuropsychol Soc* 2000;6:279–289.

62. Levin HS, Gary HE Jr, Eisenberg HM, et al. Neurobehavioral outcome 1 year after severe head injury. Experience of the Traumatic Coma Data Bank. *J Neurosurg* 1990;73:699–709.

63. Tate RL, Fenelon B, Manning ML, et al. Patterns of neuropsychological impairment after severe blunt head injury. *J Nerv Ment Dis* 1991;179:117–126.

64. Levin HS, Goldstein FC, High WM Jr, et al. Disproportionately severe memory deficit in relation to normal intellectual functioning after closed head injury. *J Neurol Neurosurg Psychiatry* 1988;51:1294–1301.

65. Goldstein FC, Gary HE Jr, Levin HS. Assessment of the accuracy of regression equations proposed for estimating premorbid intellectual functioning on the Wechsler Adult Intelligence Scale. *J Clin Exp Neuropsychol* 1986;8:405–412.

66. Ruff RM, Marshall LF, Crouch J, et al. Predictors of outcome following severe head trauma: follow-up data from the Traumatic Coma Data Bank. *Brain Inj* 1993;7:101–111.

67. Di Stefano G, Bachevalier J, Levin HS, et al. Volume of focal brain lesions and hippocampal formation in relation to memory function after closed head injury in children. *J Neurol Neurosurg Psychiatry* 2000;69:210–216.

68. Mattson AJ, Levin HS, Breitmeyer BG. Visual information processing after severe closed head injury: effects of forward and backward masking. *J Neurol Neurosurg Psychiatry* 1994;57:818–824.

69. Ponsford J, Kinsella G. Attentional deficits following closed-head injury [published erratum appears in *J Clin Exp Neuropsychol* 1995;17:640]. *J Clin Exp Neuropsychol* 1992;14:822–838.

70. Spikman JM, van Zomeren AH, Deelman BG. Deficits of attention after closed-head injury: slowness only? *J Clin Exp Neuropsychol* 1996;18:755–767.

71. Spikman JM, Deelman BG, van Zomeren AH. Executive func-

tioning, attention and frontal lesions in patients with chronic CHI. *J Clin Exp Neuropsychol* 2000;22:325–338.

72. Mazaux JM, Masson F, Levin HS, et al. Long-term neuropsychological outcome and loss of social autonomy after traumatic brain injury. *Arch Phys Med Rehabil* 1997;78:1316–1320.

73. Kaufmann PM, Fletcher JM, Levin HS, et al. Attentional disturbance after pediatric closed head injury. *J Child Neurol* 1993;8:348–353.

74. Baddeley AD. *Essentials of human memory.* Hove, UK: Psychology Press, 1999.

75. Stuss DT, Ely P, Hugenholtz H, et al. Subtle neuropsychological deficits in patients with good recovery after closed head injury. *Neurosurgery* 1985;17:41–47.

76. Goldstein FC, Levin HS. Question-asking strategies after severe closed head injury. *Brain Cogn* 1991;17:23–30.

77. Clifton GL, Kreutzer JS, Choi SC, et al. Relationship between Glasgow Outcome Scale and neuropsychological measures after brain injury. *Neurosurgery* 1993;33:34–38.

78. Levine B, Dawson D, Boutet I, et al. Assessment of strategic self-regulation in traumatic brain injury: its relationship to injury severity and psychosocial outcome. *Neuropsychology* 2000;14:491–500.

79. Levin HS, Grossman RG, Kelly PJ. Short-term recognition memory in relation to severity of head injury. *Cortex* 1976;12: 175–182.

80. Chapman SB, Culhane KA, Levin HS, et al. Narrative discourse after closed head injury in children and adolescents. *Brain Lang* 1992;43:42–65.

81. Brookshire BL, Chapman SB, Song J, et al. Cognitive and linguistic correlates of children's discourse after closed head injury: a three-year follow-up. *J Int Neuropsychol Soc* 2000;6:741–751.

82. Chapman SB, Levin HS, Wanek A, et al. Discourse after closed head injury in young children. *Brain Lang* 1998;61:420–449.

83. Brooks DN, Hosie J, Bond MR, et al. Cognitive sequelae of severe head injury in relation to the Glasgow Outcome Scale. *J Neurol Neurosurg Psychiatry* 1986;49:549–553.

84. Lezak MD. Subtle sequelae of brain damage. Perplexity, distractibility, and fatigue. *Am J Phys Med* 1978;57:9–15.

85. Jorge RE, Robinson RG, Arndt SV, et al. Depression following traumatic brain injury: a 1 year longitudinal study. *J Affect Disord* 1993;27:233–243.

86. Gomez-Hernandez R, Max JE, Kosier T, et al. Social impairment and depression after traumatic brain injury. *Arch Phys Med Rehabil* 1997;78:1321–1326.

87. Jorge RE, Robinson RG, Starkstein SE, et al. Influence of major depression on 1-year outcome in patients with traumatic brain injury. *J Neurosurg* 1994;81:726–733.

88. Satz P, Forney DL, Zaucha K, et al. Depression, cognition, and functional correlates of recovery outcome after traumatic brain injury. *Brain Inj* 1998;12:537–553.

89. Grafman J, Vance SC, Weingartner H, et al. The effects of lateralized frontal lesions on mood regulation. *Brain* 1986;109: 1127–1148.

90. Fedoroff JP, Starkstein SE, Forrester AW, et al. Depression in patients with acute traumatic brain injury. *Am J Psychiatry* 1992;149:918–923.

91. Jorge RE, Robinson RG, Starkstein SE, et al. Depression and anxiety following traumatic brain injury. *J Neuropsychiatry Clin Neurosci* 1993;5:369–374.

92. Starkstein SE, Robinson RG. Mechanism of disinhibition after brain lesions. *J Nerv Ment Dis* 1997;185:108–114.

93. Jorge RE, Robinson RG, Starkstein SE, et al. Secondary mania following traumatic brain injury. *Am J Psychiatry* 1993;150: 916–921.

94. Grafman J, Schwab K, Warden D, et al. Frontal lobe injuries,

violence, and aggression: a report of the Vietnam Head Injury Study. *Neurology* 1996;46:1231–1238.

95. Max JE, Roberts MA, Koele SL, et al. Cognitive outcome in children and adolescents following severe traumatic brain injury: influence of psychosocial, psychiatric, and injury-related variables. *J Int Neuropsychol Soc* 1999;5:58–68.

96. Max JE, Koele SL, Smith WL Jr, et al. Psychiatric disorders in children and adolescents after severe traumatic brain injury: a controlled study. *J Am Acad Child Adolesc Psychiatry* 1998;37:832–840.

97. Taylor HG, Yeates KO, Wade SL, et al. Influences on first-year recovery from traumatic brain injury in children. *Neuropsychology* 1999;13:76–89.

98. Teasdale G, Jennett B. Assessment of coma and impaired consciousness. A practical scale. *Lancet* 1974;2:81–84.

99. Ewing-Cobbs L, Levin HS, Fletcher JM, et al. The Children's Orientation and Amnesia Test: relationship to severity of acute head injury and to recovery of memory. *Neurosurgery* 1990;27:683–691.

100. Anonymous. Report of MTBI Committee of the Head Injury Interdisciplinary Special Interest Group of the Am. Congress of Rehabilitation Medicine (1993). Definition of mild traumatic brain injury. *J Head Trauma Rehabil* 1993;8:86–87.

101. Katz DI, Alexander MP. Traumatic brain injury. Predicting course of recovery and outcome for patients admitted to rehabilitation. *Arch Neurol* 1994;51:661–670.

102. Li R, Fujitani N, Jia JT, et al. Immunohistochemical indicators of early brain injury: an experimental study using the fluid-percussion model in cats. *Am J Forens Med Pathol* 1998;19:129–136.

103. Herrmann M, Jost S, Kutz S, et al. Temporal profile of release of neurobiochemical markers of brain damage after traumatic brain injury is associated with intracranial pathology as demonstrated in cranial computerized tomography. *J Neurotrauma* 2000;17:113–122.

104. McKeating EG, Andrews PJD, Mascia L. Relationship of neuron specific enolase and protein S-100 concentrations in systemic and jugular venous serum to injury severity and outcome after traumatic brain injury. *Acta Neurochirurg* 1998;71[Suppl]:117–119.

105. Ogata M, Tsuganezawa O. Neuron-specific enolase as an effective immunohistochemical marker of injured axons after fatal brain injury. *J Legal Med* 1999;113:19–25.

106. Zasler ND. Prognostic indicators in medical rehabilitation of traumatic brain injury: a commentary and review. *Arch Phys Med Rehabil* 1997;78:S12–S16.

107. Jennett B, Bond M. Assessment of outcome after severe brain damage. *Lancet* 1975;1:480–484.

108. Satz P, Zaucha K, Forney DL, et al. Neuropsychological, psychosocial and vocational correlates of the Glasgow Outcome Scale at 6 months post-injury: a study of moderate to severe traumatic brain injury patients. *Brain Inj* 1998;12:555–567.

109. Williams JM, Gomes F, Drudge OW, et al. Predicting outcome from closed head injury by early assessment of trauma severity. *J Neurosurg* 1984;61:581–585.

110. Yeates KO, Taylor HG. Predicting premorbid neuropsychological functioning following pediatric traumatic brain injury. *J Clin Exp Neuropsychol* 1997;19:825–837.

111. Vollmer DG, Torner JC, Eisenberg HM, et al. Age and outcome following traumatic coma: why do older patients fare worse? *J Neurosurg* 1991;75:537–549.

112. Ellenberg JH, Levin HS, Saydjari C. Posttraumatic amnesia as a predictor of outcome after severe closed head injury. Prospective assessment. *Arch Neurol* 1996;53:782–791.

113. Thompson NM, Francis DJ, Stuebing KK, et al. Motor, visual-spatial, and somatosensory skills after closed head injury in children and adolescents: a study of change. *Neuropsychology* 1994;8:333–342.

114. Levin HS, Aldrich EF, Saydjari C, et al. Severe head injury in children: experience of the Traumatic Coma Data Bank. *Neurosurgery* 1992;31:435–443.

115. Godfrey HP, Bishara SN, Partridge FM, et al. Neuropsychological impairment and return to work following severe closed head injury: implications for clinical management. *N Z Med J* 1993;106:301–303.

116. Haaland KY, Temkin N, Randahl G, et al. Recovery of simple motor skills after head injury. *J Clin Exp Neuropsychol* 1994;16:448–456.

117. Gronwall D, Wrightson P. Cumulative effects of concussion. *Lancet* 1975;2:995–997.

118. Graham DI, Horsburgh K, Nicoll JA, et al. Apolipoprotein E and the response of the brain to injury. *Acta Neurochirurg Suppl* 1999;73:89–92.

119. Teasdale GM, Nicoll JA, Murray G, et al. Association of apolipoprotein E polymorphism with outcome after head injury. *Lancet* 1997;350:1069–1071.

120. Erlanger DM, Kutner KC, Barth JT, et al. Neuropsychology of sports-related head injury: dementia pugilistica to post concussion syndrome. *Clin Neuropsychol* 1999;13:193–209.

121. Kutner KC, Erlanger DM, Tsai J, et al. Lower cognitive performance of older football players possessing apolipoprotein E epsilon 4. *Neurosurgery* 2000;47:651–657.

122. Jennett B. Predictors of recovery in evaluation of patients in coma. *Adv Neurol* 1979;22:129–135.

123. Miller JD, Butterworth JF, Gudeman SK, et al. Further experience in the management of severe head injury. *J Neurosurg* 1981;54:289–299.

124. Rimel RW, Giordani B, Barth JT, et al. Disability caused by minor head injury. *Neurosurgery* 1981;9:221–228.

125. Rimel RW, Giordani B, Barth JT, et al. Moderate head injury: completing the clinical spectrum of brain trauma. *Neurosurgery* 1982;11:344–351.

126. Dikmen SS, Machamer JE, Winn HR, et al. Neuropsychological outcome at 1-year post head injury. *Neuropsychology* 1995;9:80–90.

127. Williams DH, Levin HS, Eisenberg HM. Mild head injury classification. *Neurosurgery* 1990;27:422–428.

128. Fay GC, Jaffe KM, Polissar NL, et al. Outcome of pediatric traumatic brain injury at three years: a cohort study. *Arch Phys Med Rehabil* 1994;75:733–741.

129. Kinsella GJ, Prior M, Sawyer M, et al. Predictors and indicators of academic outcome in children 2 years following traumatic brain injury. *J Int Neuropsychol Soc* 1997;3:608–616.

130. McDonald CM, Jaffe KM, Fay GC, et al. Comparison of indices of traumatic brain injury severity as predictors of neurobehavioral outcome in children. *Arch Phys Med Rehabil* 1994;75:328–337.

131. Dikmen S, Temkin N, McLean A, et al. Memory and head injury severity. *J Neurol Neurosurg Psychiatry* 1987;50:1613–1618.

132. Teasdale G, Jennett B. Assessment and prognosis of coma after head injury. *Acta Neurochirurg* 1976;34:45–55.

133. Cifu DX, Keyser-Marcus L, Lopez E, et al. Acute predictors of successful return to work 1 year after traumatic brain injury: a multicenter analysis. *Arch Phys Med Rehabil* 1997;78:125–131.

134. Dikmen S, McLean A Jr, Temkin NR, et al. Neuropsychologic outcome at one-month postinjury. *Arch Phys Med Rehabil* 1986;67:507–513.

135. Sacco RL, VanGool R, Mohr JP, et al. Nontraumatic coma. Glasgow coma score and coma etiology as predictors of 2-week outcome. *Arch Neurol* 1990;47:1181–1184.

136. Adams JH, Jennett B, McLellan DR, et al. The neuropathology of the vegetative state after head injury. *J Clin Pathol* 1999;52:804–806.

137. Adams JH, Graham DI, Jennett B. The neuropathology of the vegetative state after an acute brain insult. *Brain* 2000;123:1327–1338.

138. Jennett B, Adams JH, Murray LS, et al. Neuropathology in vegetative and severely disabled patients after head injury. *Neurology* 2001;56:486–490.

139. Tate RL, Broe GA. Psychosocial adjustment after traumatic brain injury: what are the important variables? *Psychol Med* 1999;29:713–725.

140. Schwab K, Grafman J, Salazar AM, et al. Residual impairments and work status 15 years after penetrating head injury: report from the Vietnam Head Injury Study. *Neurology* 1993;43:95–103.

141. Grafman J, Salazar A, Weingartner H, et al. The relationship of brain-tissue loss volume and lesion location to cognitive deficit. *J Neurosci* 1986;6:301–307.

142. Grafman J, Jonas BS, Martin A, et al. Intellectual function following penetrating head injury in Vietnam veterans. *Brain* 1988;111:169–184.

143. Kraft JF, Schwab KA, Salazar AM, et al. Occupational and educational achievements of head injured Vietnam veterans at 15-year follow-up. *Arch Phys Med Rehabil* 1993;74:596–601.

144. Levin HS, Grossman RG, Sarwar M, et al. Linguistic recovery after closed head injury. *Brain Lang* 1981;12:360–374.

145. Wilson JT, Wiedmann KD, Hadley DM, et al. Early and late magnetic resonance imaging and neuropsychological outcome after head injury. *J Neurol Neurosurg Psychiatry* 1988;51:391–396.

146. Levin HS. Head trauma. *Curr Opin Neurol* 1993;6:841–846.

147. Rutherford WH, Merrett JD, McDonald JR. Symptoms at one year following concussion from minor head injuries. *Injury* 1979;10:225–230.

148. Alexander MP. Mild traumatic brain injury: pathophysiology, natural history, and clinical management. *Neurology* 1995;45:1253–1260.

149. Satz PS, Alfano MS, Light RF, et al. Persistent post-concussive syndrome: a proposed methodology and literature review to determine the effects, if any, of mild head and other bodily injury. *J Clin Exp Neuropsychol* 1999;21:620–628.

150. Alexander MP. Neuropsychiatric correlates of persistent post-concussive syndrome. *J Head Trauma Rehabil* 1992;7:60–69.

151. Rizzo M, Tranel D. *Head injury and postconcussive syndrome*. New York: Churchill Livingstone, 1996.

152. Fox DD, Earnest K, Dolezal-Wood S. Post-concussive symptoms: base rates and etiology in psychiatric patients. *Clin Neuropsychol* 1995;9:89–92.

153. Fox DD, Lees-Haley PR, Earnest K, et al. Base rates of postconcussive symptoms in health maintenance organization patients and controls. *Neuropsychology* 1995;9:606–611.

154. Mittenberg W, DiGiulio DV, Perrin S, et al. Symptoms following mild head injury: expectation as aetiology. *J Neurol Neurosurg Psychiatry* 1992;55:200–204.

155. Satz P, Zaucha K, McCleary C, et al. Mild head injury in children and adolescents: a review of studies (1970–1995). *Psychol Bull* 1997;122:107–131.

156. Binder LM, Rohling ML, Larrabee J. A review of mild head trauma. Part I: meta-analytic review of neuropsychological studies. *J Clin Exp Neuropsychol* 1997;19:421–431.

157. Binder LM. A review of mild head trauma. Part II: clinical implications. *J Clin Exp Neuropsychol* 1997;19:432–457.

158. Bixby-Hammett D, Brooks WH. Common injuries in horseback riding. A review. *Sports Med* 1990;9:36–47.

159. Levin HS, Mattis S, Ruff RM, et al. Neurobehavioral outcome following minor head injury: a three-center study. *J Neurosurg* 1987;66:234–243.

160. Dikmen S, McLean A, Temkin N. Neuropsychological and psychosocial consequences of minor head injury. *J Neurol Neurosurg Psychiatry* 1986;49:1227–1232.

161. Goldstein FC, Levin HS, Goldman WP, et al. Cognitive and neurobehavioral functioning after mild versus moderate traumatic brain injury in older adults. *J Int Neuropsychol Soc* 2001;7:373–383.

162. Macciocchi SN, Barth JT, Alves W, et al. Neuropsychological functioning and recovery after mild head injury in collegiate athletes. *Neurosurgery* 1996;39:510–514.

163. Ewing R, McCarthy D, Gronwall D. Persisting effects of minor head injury observable during hypoxic stress. *J Clin Neuropsychol* 1980;2:147–155.

164. Gronwall D. Cumulative and persisting effects of concussion on attention and cognition. In: Levin H, Eisenberg HM, Benton AL, eds. *Mild head injury*. New York: Oxford University Press, 1989:153–162.

165. Mathias JL, Coats JL. Emotional and cognitive sequelae to mild traumatic brain injury. *J Clin Exp Neuropsychol* 1999;21:200–215.

166. Kay T, Newman B, Cavallo M, et al. Toward a neuropsychological model of functional disability after mild traumatic brain injury. *Neuropsychology* 1992;6:371–384.

167. Ferguson RJ, Mittenberg W, Barone DF, et al. Postconcussion syndrome following sports-related head injury: expectation as etiology. *Neuropsychology* 1999;13:582–589.

168. Sosin DM. Trends in death associated with traumatic brain injury, 1979 through 1992: success and failure. *JAMA* 1995;273:1778–1780.

169. Rivara FP, Koepsell TD, Grossman DC, et al. Effectiveness of automatic shoulder belt systems in motor vehicle crashes. *JAMA* 2000;283:2826–2828.

170. Temkin NR, Dikmen SS, Wilensky AJ, et al. A randomized, double-blind study of phenytoin for the prevention of post-traumatic seizures. *N Engl J Med* 1990;323:497–502.

171. Annegers JF, Grabow JD, Groover RV, et al. Seizures after head trauma: a population study. *Neurology* 1980;30:683–689.

172. Salazar AM, Jabbari B, Vance SC, et al. Epilepsy after penetrating head injury. I. Clinical correlates: a report of the Vietnam Head Injury Study. *Neurology* 1985;35:1406–1414.

173. Weiss GH, Salazar AM, Vance SC, et al. Predicting posttraumatic epilepsy in penetrating head injury. *Arch Neurol* 1986;43:771–773.

174. Rehabilitation of persons with traumatic brain injury. NIH Consensus Statement. 1998;16:1–41.

175. Cicerone KD, Dahlberg C, Kalmar K, et al. Evidence-based cognitive rehabilitation: recommendations for clinical practice. *Arch Phys Med Rehabil* 2000;81:1596–1615.

176. Carney N, Chesnut RM, Maynard H, et al. Effect of cognitive rehabilitation on outcomes for persons with traumatic brain injury: a systematic review. *J Head Trauma Rehabil* 1999;14:277–307.

177. Park NW, Ingles J. Effectiveness of attention rehabilitation after an acquired brain injury. *Neuropsychology* 2001;15:199–210.

178. Mocchetti I, Wrathall JR. Neurotrophic factors in central nervous system trauma. *J Neurotrauma* 1995;12:853–870.

179. Brooks WM, Stidley CA, Petropoulos H, et al. Metabolic and

cognitive response to human traumatic brain injury: a quantitative proton magnetic resonance study. *J Neurotrauma* 2000; 17:629–640.

180. Friedman SD, Brooks WM, Jung RE, et al. Quantitative proton MRS predicts outcome after traumatic brain injury. *Neurology* 1999;52:1384–1391.

181. Ricker JH, Zafonte RD. Functional neuroimaging and quantitative electroencephalography in adult traumatic head injury:

clinical applications and interpretive cautions. *J Head Trauma Rehabil* 2000;15:859–868.

182. Bigler ED. Neuroimaging in pediatric traumatic head injury: diagnostic considerations and relationships to neurobehavioral outcome. *J Head Trauma Rehabil* 1999;14:406–423.

183. Hammeke TA. Functional MRI in neurology. In: Moonen C, Bandettini PA, eds. *Functional MRI.* New York: Springer-Verlag, 1999:475–486.

42

BEHAVIORAL SYNDROMES IN NEUROTOXICOLOGY

ROBERT G. FELDMAN
MARCIA H. RATNER

NEUROTOXICANT EXPOSURES

Exposure to various synthetic and natural chemicals may result in disturbances of the functioning and structural integrity of the nervous system. Synthetic chemicals, known as *neurotoxicants,* are defined as substances that attack nerve cells and are directly and/or indirectly capable of (a) altering nerve cell membranes, thereby affecting excitability, neurotransmitter release, and synaptic activity of neurons; (b) disturbing the flow of axoplasm, thereby interfering with the transport of neurotransmitters and nutrient substances along the axon to and from the cell body; (c) disrupting cellular respiration processes; (d) disrupting protein synthesis; (e) affecting neuronal functions indirectly by damaging Schwann cells and peripheral myelin, oligodendrocytes and central myelin, and astrocytes and microglia; (f) disrupting neurotransmission by binding with enzymes or affecting ion channel receptors; and/or (g) affecting extracellular fluid volume and flow by damaging capillary endothelium, thereby disrupting the blood–brain barrier or blood–nerve barrier (1).

It is not uncommon for a person to be exposed to more than one neurotoxicant resulting in additive or synergistic effects. The clinical outcome of an exposure may be further complicated by (a) reaction of the chemical(s) with other chemicals present in the environment; (b) environmental breakdown of the chemical(s) into constituent elements or other compounds; (c) metabolic activation or biotransformation of the chemical into constituents or other neurotoxic compounds; and (d) individual differences in enzyme activity due to genetics. These factors collectively contribute to the clinical effects of exposure, which may be different for the same chemical(s) under different exposure circumstances and in different individuals.

Neurotoxic effects occur when protective mechanisms (e.g., detoxifying enzymes) fail to prevent or minimize adverse reactions between a neurotoxicant and target cellular macromolecules and/or do not promote the metabolism and elimination of potentially hazardous substances before a critical threshold level is reached. Above this critical threshold, cellular processes (e.g., neurotransmission, aerobic respiration, axonal transport) become impaired, and reversible and irreversible alterations occur.

The severity of behavioral manifestations associated with neurotoxicant exposure depends upon the potency and the amount of chemical absorbed (dose). Acute exposures to high concentrations of neurotoxicants require shorter durations to reach the critical thresholds than do more prolonged exposures at lower levels. The reversibility of neurobehavioral effects depends upon the type, dose, and duration of exposure to the neurotoxicant.

A person exposed to neurotoxicants may be completely unaware of changes in his/her behavior. This unrecognized impairment carries risk for work-related accidents and injuries. Co-workers and/or family members often are the first to recognize changes in the patient's attention, memory function, mood, and affect. If the source of chemical exposure is not immediately suspected or if the possible neurotoxic effects of a particular chemical are not well known, the patient's behavioral changes may be attributed to other possible neurologic conditions. Risk of further exposure before removal from the source(s) of exposure is thus increased. The possibility that a neurotoxic illness underlies the patient's presenting complaints should be fully investigated using the patient's medical history, laboratory findings, and occupational and environmental exposure histories before the diagnosis is made.

A person may report neurologic symptoms immediately after a specific acute exposure episode (e.g., chemical spill accident), or the patient may develop symptoms insidiously in association with repeated or chronic exposures. Early effects of acute and chronic exposures may cause overt abnormal functioning, or they may be subclinical and detectable only by specific sensitive diagnostic tests (e.g., electroencephalography; evoked potential and/or peripheral nerve conduction studies). Damage to neural structures and systems may be reversible or may permanently impair performance of activities of daily living.

NEUROBEHAVIORAL MANIFESTATIONS

Acute neurotoxicant exposure may manifest itself as a reversible *acute intoxication,* such as that is seen after acute ingestion of ethanol, or as an *acute toxic encephalopathy,* as evidenced by overt clinical signs such as seizures or coma (e.g., acute lead encephalopathy seen in children) and possible irreversible sequelae. Chronic exposures to neurotoxicants are associated with the insidious development of *chronic toxic encephalopathy.* Whereas the effects of acute intoxication are reversible, both the acute and chronic toxic encephalopathies are associated with persistent cognitive deficits and behavioral changes that can be measured by formal neuropsychological tests and other measures of brain function and integrity (2).

Neurotoxicants affect brain structures that mediate motor, sensory, and/or cognitive functioning. Many neurotoxic chemicals affect overall neurologic and cognitive performance and emotional behavior to some degree, irrespective of their predilection for producing focal effects. Thus, diverse patterns of symptoms, signs, and neuropsychological performance deficits are not uncommon. Patients exposed to neurotoxicants often complain of symptoms, which include headache, dizziness, changes in mood and affect, irritability, inability to concentrate, and attention and memory problems. Neurotoxicant exposure-induced motor system dysfunction may include spasticity, paralysis, bradykinesia, dyskinesia, dystonia, tremor, and incoordination. Sensory symptoms include numbness and paresthesias.

The observations made by the clinician during a neurologic examination can be used to infer the probable anatomic site(s) of nervous system dysfunction and to describe the patient's functional status. Abnormal neurologic symptoms and signs are expressions of impaired function or damage to particular neural structures, regardless of the specific etiology of the lesion. Thus, neurologic findings arising from the effects of exposures to neurotoxicants may resemble those found in primary or nonneurotoxic neurologic illness. The diagnostic process integrates the clinician's observations of the patient and the results of tests on physiologic, anatomic, and behavioral functions; his/her acumen and judgment accumulated from experience with similar cases; and reference to a background of information contained in previously published literature.

DIFFERENTIAL DIAGNOSIS

Many neurotoxicants produce symptoms and signs that resemble idiopathic neurodegenerative disease processes, which must be included in the differential diagnosis of these conditions. Familiarity with the clinical pictures of idiopathic neurodegenerative diseases, metabolic encephalopathies, and psychiatric illnesses is essential for correct diagnosis. Differentiating neurotoxicant exposure-induced

neurobehavioral syndromes from idiopathic neurologic disease rests on the interpretation of the patient's clinical manifestations in relation to the chronology of events surrounding the onset of symptoms, awareness of chemicals that are known to cause neurologic effects, and documentation of the presence of such neurotoxicants in the patient's environment (Table 42.1) (3–5).

Differential diagnosis of behavioral syndromes associated with toxic exposures can be difficult because chemicals affect the same cerebral structures as do other neuropathologic processes, and they produce similar expressions of cognitive and behavioral impairment clinically and on formal neuropsychological assessment. For example, headache can be due to simple muscle strain, migraine, tumor, or unruptured aneurysm, but it must be differentiated from the symptomatic headache associated with exposure to chemicals [e.g., carbon monoxide (CO), lead, zinc, nitrates, nickel, organic solvents]. Cognitive deficits associated with Alzheimer disease (AD), arteriosclerosis, or pseudodementia of depression must be differentiated from the neurobehavioral effects of exposure to organic solvents, heavy metals, or insecticides. Disturbances of awareness and epileptic seizures must be differentiated from similarly appearing disturbances of consciousness or convulsions due to exposures to neurotoxicants (e.g., trimethyltin). Parkinson disease is characterized by rigidity, slowness of movement, and tremor. However, a similar motor disorder occurs in association with exposure to 1-methyl-4-phenyl-1,2-3,6-tetrahydropyridine (MPTP), carbon disulfide, CO, and manganese (Mn). Individuals with multiple sclerosis (MS) have protean features due to disseminated lesions of brain and spinal cord myelin. Individuals who develop central nervous system demyelination after exposure to tri-*ortho*-cresyl phosphate, toluene, or mercury may present with clinical features similar to those of MS and thus may be misdiagnosed. Peripheral neuropathy occurs in patients with diabetes but also may develop following exposures to methyl-*n*-butyl ketone, *n*-hexane, ethylene oxide (EtO), carbon disulfide, lead or arsenic. Motor neuronopathy and/or anterior horn cell disease have been associated with exposures to lead and mercury. Sensory neuronopathy may occur after exposure to vitamin B_6.

The patterns of impaired neurobehavioral performance associated with exposures to neurotoxicants can be similar to those seen in primary neurologic disorders, but progressive cognitive decline is more common in the nontoxic diagnoses. There are neurobehavioral findings that sometimes are associated with multiple white matter lesions associated with cerebrovascular disease (e.g., severe remote memory deficit, impaired language skills in the context of intact visuospatial ability) that would be usual in neurotoxicant-induced disorders. Similarly, patients with MS or cerebrovascular disease sometimes show impaired language due to a preponderance of left hemisphere lesions, which would not be expected following exposure to neurotoxicants. Generally the differential

TABLE 42.1. NEUROTOXIC SYNDROMES ASSOCIATED WITH EXPOSURES TO SELECTED NEUROTOXICANTS

Neurotoxicant	Sources of Exposure	Clinical Manifestations
Metals		
Lead	Solder, bullets and lead shot, illicit whiskey, autobody fillers, batteries, foundries, smelters, lead-based paints, lead pipes	*Acute:* Encephalopathy *Chronic:* Encephalopathy and peripheral neuropathy
Arsenic	Pesticides, pigments, antifouling paints, electroplating industry, wood preservatives, seafood (organic arsenic), smelters, semiconductors	*Acute:* Encephalopathy *Chronic:* Peripheral neuropathy and encephalopathy
Manganese	Iron, steel industry, welding operations, metal-finishing operations, fertilizers, pesticides (e.g., maneb), fireworks, matches, dry cell batteries	*Acute:* Encephalopathy *Chronic:* Parkinsonism
Mercury	Scientific instruments (e.g., thermometers), electrical equipment (e.g., switches), dental amalgams, electroplating processes, felt industry, photography	*Acute:* Headache, nausea, onset of tremor *Chronic:* Ataxia, encephalopathy, tremor, peripheral neuropathy
Tin	Solders, polyvinyl plastics, canning industry, paints, fungicides, electronic equipment	*Acute:* Memory defects, seizures, disorientation *Chronic:* Encephalomyelopathy
Solvents		
Carbon disulfide	Viscose rayon manufacturing and rubber industries, preservatives, electroplating industry	*Acute:* Encephalopathy *Chronic:* Peripheral neuropathy, parkinsonism
n-Hexane and methyl *n*-butyl ketone	Paints, lacquers, varnishes, inks, degreasing solvents, adhesives, paint removers	*Acute:* Narcosis *Chronic:* Peripheral neuropathy
Trichloroethylene	Degreasing and dry cleaning solvents, varnishes, paints, decaffeination processes	*Acute:* Narcosis *Chronic:* Encephalopathy, cranial neuropathy
Perchloroethylene	Dry cleaning solvents, degreasing solvents, paint removers, extraction agents	*Acute:* Narcosis *Chronic:* Encephalopathy, peripheral neuropathy
Toluene	Rubber solvents, paints, lacquers, paint thinners, glues, gasoline, aviation fuels	*Acute:* Narcosis *Chronic:* Ataxia, encephalopathy, tremor
Pesticides		
Organophosphates	Manufacturing and application processes	*Acute:* Cholinergic crisis *Chronic:* Ataxia, paralysis, peripheral neuropathy, encephalopathy, parkinsonism
Carbamates	Agricultural industry: manufacturing and application	*Acute:* Cholinergic crisis *Chronic:* Tremor, peripheral neuropathy
Gases		
Carbon monoxide	Automobile exhaust, burning of fossil fuels, fires, methylene chloride	*Acute:* Confusion, dizziness, fatigue, loss of consciousness, seizures, death *Chronic:* Parkinsonism, encephalopathy
Hydrogen sulfide	Petroleum industry, sewer gases	*Acute:* Confusion, dizziness, loss of consciousness, seizures, death *Chronic:* Encephalopathy

From Feldman RG. *Occupational and environmental neurotoxicology.* Philadelphia: Lippincott-Raven Publishers, 1999, with permission.

diagnosis can be sorted out by neurologic and laboratory findings and neuropsychological testing.

Cerebellar disorders may be associated with exposure to neurotoxicants (e.g., toluene), as well as nutritional and paraneoplastic etiologies. Patients with cerebellar disorders of any etiology (including tumor) show dysarthria and motor deficits (e.g., tremulousness, gait ataxia, dysmetria); however, they also frequently have deficits in attention and executive function and visuospatial abilities.

The differential diagnosis of idiopathic epilepsy versus a neurotoxicant exposure-induced seizure disorder can be important. The diagnosis rests on the time relationship of the exposure and the seizure. If the first seizure occurs soon after an exposure, there is strong support for a causal relation.

When chemical exposure affects the hippocampus, memory deficits may emerge on neuropsychological testing. Similar memory deficits are seen in patients with mesial temporal lesions. Some patients with a history of exposure may show evidence of dementia, which must be differentiated from dementia of AD based on testing of mood, language, and retrograde memory. In the early stages of AD, the neuropsychological assessment profile can be quite similar to toxic encephalopathy (anterograde memory and visuospatial impairments) but follow-up testing within 1 year or less typically provides the clinical information to make the diagnosis. Whether exposure to neurotoxicants can exacerbate or precipitate the expression of AD in genetically susceptible individuals, or whether chemical exposure can produce a neuropathologic process with the same or similar features to those seen in AD, has not been fully elucidated (6,7).

Preexisting neurologic and psychiatric disorders may be exacerbated by exposures to neurotoxicants. This is especially true of seizure disorders and cerebrovascular disease. Likewise, persons with underlying psychiatric disorders (e.g., major affective disorder or anxiety disorders, paranoid tendencies) may experience an exacerbation of symptom episodes even in the absence of objectively measurable exposure-related brain damage.

DETECTING CLINICAL NEUROTOXIC EFFECTS

The Interview and Questionnaire

The clinical neurologic examination should begin with a face-to-face interview to obtain details about possible occupational and/or environmental exposures, the patient's personal and family medical histories, and comprehensive symptom review (Fig. 42.1). A carefully designed questionnaire (3) provides a standardized format for gathering the data necessary for establishing a chronologic history (time line) detailing the emergence of medical complaints in relation to the circumstances of suspected exposures to neurotoxicants. The questionnaire asks about all chemicals that are currently or were previously encountered in the patient's work and home environments. Because people may be unaware of the risks associated with common occupational or environmental chemical exposures or may not be familiar with the names of the chemicals that they come into contact with at work or in their hobbies or crafts, the questionnaire should include a list of common neurotoxic chemicals. Queries are made about current and past conditions in the patient's workplace, home, and general environment to ensure that the most possible sources of exposure are recognized and documented. The questionnaire seeks information about the patient's current and past use of respirators, protective clothing, and other personal protection equipment and habits. Drinking, smoking, and recreational drug use histories are obtained. In addition, the questionnaire should determine the past medical history of the patient and his/her immediate family members.

Time-Exposure-Symptom Line

A time-exposure-symptom (T-E-S) line constructed from the information obtained from the questionnaire and interview should include the patient's complete medical history. Begin with relevant perinatal, childhood, adolescent, and adult issues through the time of the examination to chrono-

logically identify any possible injuries, illnesses, use or abuse of drugs, and information about suspected or documented exposure to neurotoxicants. Relevant exposure data, including workplace ambient air and other environmental levels, as well as biologic exposure indices for all chemicals suspected to have been present in the patient's environment, should be placed along the T-E-S line. In the absence of any other reasonable explanation to account for the clinical findings and complaints of the patient that appear along the T-E-S line, a probable diagnosis of neurotoxic illness can be made based on the coincidence of documented exposure episodes and the onset and development of symptoms.

Substantiation of the relationships between the clinical manifestations and the chemical exposures can be made by reviewing the relevant scientific literature and referencing reports of the effects of exposure to the same chemicals. If there have been no previous scientific studies or clinical reports of adverse effects associated with the particular chemicals, then exposure data verification and the record of chronologic events along the T-E-S line are the rationale relied upon to make a probable and causative clinical diagnosis (Boston University Environmental Neurology Assessment; Table 42.2) (3).

GENERAL MEDICAL AND NEUROLOGIC EXAMINATIONS

Examination of cardiovascular, gastrointestinal, and pulmonary function is necessary to ascertain the general health of the patient. The clinical neurologic examination of the sensorimotor system evaluates the functional integrity of neurons of the cerebral cortex and their connections with subcortical (e.g., thalamic), brainstem, and cerebellar and spinal cord neurons, as well as with the effector muscles and sensory receptors that permit the perception of stimuli and the initiation and execution of appropriate physical responses and spontaneous actions. Evidence of dysfunction of neurons in the motor cortex (*upper motor neurons*) includes weakness and spasticity of the limb on the opposite side of the body as the lesion (i.e., contralateral limb). Dysfunction of neurons in the basal ganglia alters muscle tone (e.g., dystonia) and speed of response, causing slowness of movement (i.e., bradykinesia). Lesions to, or dysfunction of, neurons in midbrain and brainstem structures result in disturbances of cranial nerve functioning, such as conjugate eye movement and difficulty swallowing. Impaired functioning of neurons in the cerebellum may result in gait ataxia and tremulousness of the trunk, head, and outstretched extremities. The tremor associated with cerebellar and vestibular system dysfunction is present during action and increases upon intention, whereas the tremor associated with Parkinson disease appears during rest and typically disappears during action.

Clinical neurologic assessment of sensory function is readily accomplished with a wisp of cotton, a new safety pin, a

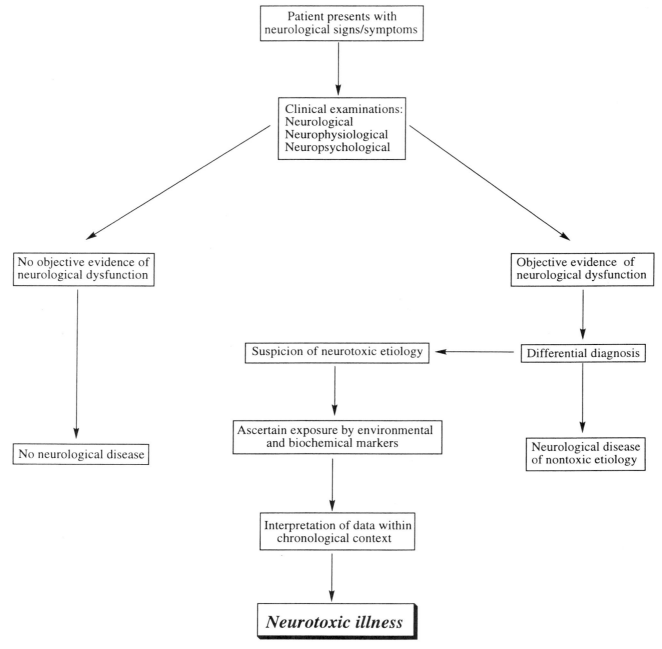

FIGURE 42.1. Algorithm.

tuning fork, and a reflex hammer. Dysfunction in the structures of the brain responsible for processing sensory information (e.g., occipital and parietal cortices, thalamic nuclei) can result in syndromes of contralateral neglect, cortical blindness, and/or visuospatial disturbances. Profound contralateral neglect is a common manifestation of stroke but is an uncommon manifestation of neurotoxicant exposures, which typically are associated with more subtle sensory or visuospatial disturbances. The sensory deficits associated with exposures to neurotoxicants may be detected using brain-

stem auditory and visual evoked response studies and formal neuropsychological tests.

NEUROBEHAVIORAL ASSESSMENT

Mood, Affect, and Cognitive Functioning

The patient's spontaneous remarks and movements, as well as those made in response to the examiner's instructions, are evaluated for any deviations from expected behaviors.

TABLE 42.2. BOSTON UNIVERSITY ENVIRONMENTAL NEUROLOGY ASSESSMENT (BUENA)

I. *Are the data sufficient to identify any or all complaints as being caused by a neurotoxin?*
  A. *List complaints* and relate them on a time line, identifying all possible chemical exposure episodes and their sources (occupational, environmental, hobbies).
    1. Identify symptoms and functional changes expressed, experienced, or observed by others. List examples of mood, anxiety, sleep disturbances, and effect on quality of life.
    2. Cite time of onset, duration, and intensity of all complaints. Characterize symptoms in terms of worsening or remitting in relation to exposure and away from exposure sources (e.g., work week, weekend, time of shift, on vacation).
    3. Evaluate subject's family/genetic health, special sensitivities, and possible congenital factors.
  B. *List all substances* and how they are used (at workplace, home, hobbies).
    1. Obtain chemical names (not trade label names), material safety data sheets, and other identifying data concerning each chemical substance.
    2. Review workplace information available (e.g., OSHA mandated material safety data sheets and employer training program materials; employer's medical records and exposure records, which, if kept by the employer, must be made available under OSHA rules. Review, if available, the following: employer's TSCA 8c and 8e reports to EPA, employer's community right to know reports to local officials regarding hazardous materials made, used, or sorted on site).
  C. *Obtain environmental and industrial hygiene air or drinking water sample measures* to prove the presence of suspected or other chemicals in the alleged sources. *Current levels* are important, and levels taken in relationship to the occurrence of complaints is essential.
  D. *Obtain urine and/or blood samples from the suspected exposed and/or affected* individuals and from known unexposed control patients of similar age and occupation, especially at time of complaints, to *establish body burden of chemical.*
  E. For suspect chemicals, *develop information on dose–response relationships, animal studies, and toxicologic and epidemiologic studies.*
II. *Are the complaints substantiated by clinical neurologic physical examination; standardized neuropsychologic and neurophysiologic tests; and appropropriate blood and urine analyses?* Also, are the complaints corroborated by epidemiologic, toxicologic, or animal studies; by NIOSH, OSHA, or EPA studies of the workforce or community; by employer studies and reports to EPA or OSHA (e.g., TSCA 8c and 8e reports)?
III. *Are the findings due to a primary neurologic disease or other medical condition?*
IV. *Are the findings on examination explained by any other causal factors* in past medical history, or previous and/or current unrelated exposures to substances from sources other than the one under consideration due to a primary neurologic disease or other medical condition?
  A. Time line of past jobs and residences
  B. Time line of past medical history
V. *Analyze individual cases for confirmatory studies;* group data for cluster analysis and/or population statistical study.
VI. *Identify and critically review previously published and/or reported case reports, case control studies, population studies, and animal studies on the alleged neurotoxicants and relate documentation to case data.*
VII. *Estimate the damage consequences for the subject:* disease, anxiety, loss of consort, functional impairments, need for special education or counseling or medical surveillance, need for medical therapeutic measures, job disability, loss of earning.
VIII. *Reevaluate after reasonable absence from all neurotoxic exposure* to assess course of progression, recovery, or persistent impairment and/or disability.

From Feldman RG, White RF, Eriator II, et al. Neurotoxic effects of trichloroethylene in drinking water: Approach to diagnosis. In: Isaacson, Jensen, eds. *The vulnerable brain and environmental risks, volume 3, special hazards from air and water.* 1994, with permission.

The patient's orientation to person, place and time, and circumstances should be noted. Unusual responses that emerge during conversations and which indicate disturbances in mood and affect are noted. Engaging the patient in a conversation about topics familiar to him/her may reveal deficits in the domains of attention and concentration as manifested by an inability to follow the casual verbal exchange. Language comprehension is assessed by challenging the patient to repeat phrases, answer simple questions, and follow simple commands. Expressive speech is assessed by listening to the patient's free speech for dysarthria, paraphasias, and neologisms. Attention and mental control are assessed by asking the patient to spell a word backwards or to recite lists of words or numbers backwards. Remote memory and short-term memory are assessed by the details a patient is able to recall about his/her medical history and by a brief as-

sessment of short-term memory skills (e.g., retention of a telephone number presented by the examiner), respectively. Visual spatial and constructional ability can be assessed by having the patient draw the face of a clock with the hands set at a particular time. These brief tests may reveal the gross cognitive deficits associated with severe acute intoxication, acute toxic encephalopathy, and/or severe chronic toxic encephalopathy. Patients with less severe or "mild chronic toxic encephalopathy" may perform well on these tests even though they may have cognitive deficits. Formal neuropsychological assessment should be performed in all patients exposed to neurotoxicants who present with complaints of central nervous dysfunction. Such assessment may reveal more subtle deficits associated with toxic encephalopathy and provide a more objective measure of the cognitive deficits noted during the clinical examination.

Formal Neuropsychological Testing

Formal neuropsychological assessment allows for documentation of certain cognitive and behavioral complaints. Neuropsychological assessment involves tests of the principal cognitive domains, including language, attention and executive function, learning and memory, motor skills, and visuospatial information processing. Formal measures of personality, mood, and affect [e.g., Minnesota Multiphasic Personality Inventory (MMPI)] also are used (2).

Cognitive Domains

Language/Verbal Function

Language and verbal functioning typically are preserved in adults exposed to neurotoxicants. This aspect of cognitive function is relatively resistant to the effects of neurotoxicant exposure-induced brain damage compared with dynamic cognitive processes such as encoding of new memories. This is in contrast with the effects of stroke and other focal lesions that may have profound impacts on language function. However, patients exposed to certain neurotoxicants (e.g., CO) may show deficits on tests that require the application of verbal and language skills. Motor aspects of writing may be affected in patients with movement disorders (e.g., tremor) resulting from neurotoxic exposures, whereas the grammatical aspects of writing remain intact. Exposure to neurotoxicants is more likely to produce language deficits in children than in adults because the disruption of encoding processes occurs during development and thus can lead to problems with language acquisition. The severity of the deficits depends upon the age of the child at the time of exposure; younger children are more vulnerable.

Attention

Attention capacity is fundamentally dependent on the individual's level of arousal, information processing speed, and his/her ability to select and focus on an object of interest. These functions are mediated in part by the reticular formation or reticular activating system and thalamus, which are involved in maintaining the individual's basal level of arousal and responding to and processing information of interest, respectively. Deficits in attention are seen in patients with reticular activating system and thalamic lesions and in some patients exposed to neurotoxicants. Attention deficits directly impair the patient's ability to selectively focus on relevant stimuli and may have indirect effects on memory function. Exposed patients frequently report that attention deficits disrupt their ability to follow a conversation or to participate in activities of daily living requiring good concentration.

The various measurable components of attention include (a) *immediate or simple attention:* how much information the individual can grasp at once; (b) *divided attention:* the ability of the individual to simultaneously focus on two or more tasks; and (c) *sustained attention or vigilance:* the ability of the individual to remain focused on a single task for an extended duration of time.

All neuropsychological tests have an attention component; therefore, attention deficits can affect performance on most tests. In the severely encephalopathic patient, attention may be so impaired that the individual is essentially stuporous and is unable to perform even the most simplistic tasks. Tests of attention include forward digit spans, which measure simple attention; the Trail-Making Test (Trails A), which measures concentration; and the Continuous Performance Test (CPT), which measures vigilance.

Executive Function

Executive function is a higher-order behavioral process that involves the ability of the subject to appreciate and respond to complex changes in neurobehavioral task demands, including recognizing, maintaining, and shifting set as necessary to carry out such tasks. *Cognitive tracking* and *cognitive flexibility* are two aspects of executive function that may be impaired in patients exposed to neurotoxicants. The Trail-Making Test (Trails A and B) is a cognitive tracking task that has been shown to be sensitive to the effects of exposures to neurotoxicants. Trails A depends on an intact ability to focus attention and maintain set; Trails B depends not only on an intact ability to focus attention and maintain set but also on an intact ability to shift set (i.e., cognitive flexibility). Difficulties in cognitive flexibility also have been revealed by the Wisconsin Card Sorting Test. Although some patients show deficits on tests of both cognitive tracking and cognitive flexibility, other patients exhibit deficits on one test but not the other. It is impossible to predict which type of deficit will be seen in a patient with toxic encephalopathy, and it does not appear to be related to the type or severity of exposure. Patients with severe deficits in executive function may have problems with activities of daily living.

Learning and Memory Function

Learning can be defined as the process of acquiring new information, whereas memory is defined as the persistence of learned information in a capacity that permits its retrieval at a later time. Memory function can be divided into short-term and long-term or remote memory, and semantic and procedural memory. Memory depends on adequate encoding (i.e., storage) and retrieval of information. Tests of free recall are impaired in patients with retrieval deficits. These patients perform disproportionately better on recognition memory tests. Many patients exposed to neurotoxicants show performance deficits on tests of learning and short-term memory for verbal and visual material but have relatively preserved functioning on tests of retrograde or long-term memory function. This dichotomy may reflect the relative sensitivity of the complex and dynamic processes involved in the acquisition of new information (e.g., encoding) to the effects of neurotoxicants. Tests of learning or short-term memory include the Verbal Paired Associates, the California Verbal Learning Test (CVLT), and the Benton Visual Retention Test

(BVRT). Tests of remote memory include Albert's Famous Faces. Tests of memory are sensitive to a patient's level of motivation and thus can be useful in detecting malingering.

Motor Skills

Motor function depends on intact attention and sensory functions (e.g., visuomotor), as well as the integrity of the premotor and motor cortices, basal ganglia, thalamus, and cerebellum. Exposure to neurotoxicants may affect one or all of these structures. The premotor cortex or secondary motor association area is responsible for the integration of basic motor functions and learned and planned actions. Dysfunction in this region of the brain disrupts the normal fluid execution of planned and intended movements but does not result in a loss of the ability to move. Conversely, lesions in the motor cortex can severely disrupt movement and result in paralysis of the innervated limb. Basal ganglia lesions may result in bradykinesia, dystonia, and other movement disorders. Cerebellar lesions result in problems with fine motor control and coordination.

Patients exposed to neurotoxicants may show deficits on tests of motor function such as the finger tapping test and the Santa Ana Form Board. Profound disturbances of motor function (e.g., bradykinesia) can interfere with the patient's performance on other neurobehavioral tasks (e.g., Trail Making Test) and therefore must be considered when interpreting the patient's performance scores, particularly on timed tasks such as the digit symbol test. Patients with basal ganglia dysfunction may perform at normal levels on motor tasks when they are permitted to exceed time limits.

Visuospatial Functions

Visuospatial function is characterized by the ability to discern the structural aspects and relationships among geometric figures and shapes. The posterior association cortex or parieto-occipitotemporal region is involved in this process, which depends on the integration of attention and sensory and motor functions for performing visuospatial and visuomotor tasks (8). Deficits may be seen following exposure to certain neurotoxicants (e.g., mercury). Construction tests, such as block design test, Rey-Osterrieth Complex Figure Test (copy trial), and digit symbol test (symbol copy trial), are sensitive to deficits in this domain. Freehand drawing of a clock can also reveal deficits in this domain. Patients with basal ganglia dysfunction may perform below expectation on timed visuospatial tasks such as the block design test despite having intact visuospatial skills.

Personality, Mood, and Affect

Tests of personality and tests of mood and affect provide information that is useful for interpreting performances on tests of cognitive and neurobehavioral functioning. The Minnesota Multiphasic Personality Inventory-Revised (MMPI-2) is a test used to determine the patient's personality profile. This test was designed to determine the presence of depression, hysteria, hypochondriasis, psycho-pathic deviate, male/female, paranoia, psychasthenia, mania, schizophrenia, and/or social introversion. Another test is the Profile of Mood States (POMS), which is a measure of mood and affect designed to recognize depression, tension, anger, confusion, fatigue, and vigor.

Patients exposed to neurotoxicants may develop depression and other changes in personality and affect that may be secondary to the exposure event itself or to the adverse effects of the exposure. Some patients may develop a reactive depression in response to exposure-related injuries that initially resulted in a loss of previously higher levels of functioning and subsequently affected the patient's ability to work and/or enjoy or participate normally in activities of daily living.

Stress associated with exposure to chemicals can be evaluated with scales and interview-type measures of acute stress disorder and posttraumatic stress disorder (PTSD), both of which may be experienced by some patients exposed to neurotoxicants. Symptoms of PTSD associated with past exposures to chemicals may include avoidance of chemicals, intrusive memories, nightmares or dreams about the exposure situation, and the reexperiencing of symptoms associated with the initial exposure episode, particularly upon encountering relevant stimuli such as low levels of aromatic chemicals that may trigger unwanted emotional or physical reactions. A similar constellation of symptoms has been reported among some patients who develop an apparent sensitivity to chemicals following an initial exposure event. However, the pathologic basis for this later syndrome, commonly referred to as Multiple Chemical Sensitivities (MCS), has not been established (9–15).

NEUROPHYSIOLOGIC TESTING

Neurophysiologic tests of central nervous system functioning include electroencephalography and evoked potentials. These tests provide objective evidence of an organic basis to the neurobehavioral findings. These sensitive tests can detect the effects of primary diseases and disorders attributable to neurotoxicant exposures and provide evidence upon which the differential diagnosis is made.

The *electroencephalogram* (EEG) detects both normal brain electrical activity and disturbances of brain electrical activity that may be associated with pathologic processes. Disruption of the normal firing of neurons produces changes in the symmetry, amplitude, and/or frequencies of the EEG pattern. Marked asymmetry of the EEG pattern reflects the presence of a lateralized neuropathologic process, whereas focal unilateral or bilateral waveform alterations or spikes may reflect the focus of epileptic seizure activity. In contrast, acute neurotoxic and metabolic encephalopathies typically are associated with generalized diffuse bilateral slowing of the background rhythm and an increase in theta activity. The EEG typically normalizes after cessation of exposure, but abnormalities may persist indefinitely following exposure to certain chemicals (e.g., trimethyltin).

Evoked potentials, which include visual evoked potentials, brainstem auditory evoked potentials, and somatosensory evoked potentials, are used to assess the integrity and functional viability of sensory pathways. Evoked potentials are recorded after stimulation of peripheral sensory nerve fibers in the extremities or by direct recordings made over the dorsal columns of the spinal cord.

Visual evoked potentials, which include flash visual evoked potentials and pattern shift visual evoked potentials, have been used to study the effects of exposure to neurotoxicants. Visual evoked potentials are used to assess the integrity of the pathways of the optic system, including the optic nerves, chiasm, and tracts projecting to the calcarine cortex. Visual evoked potentials have been used to document the effects of mercury exposure (3).

Brainstem auditory evoked potentials arise following stimulation of the eighth cranial nerve. They are used to assess the integrity of the auditory pathway and are useful for documenting the effects of ototoxic substances.

Somatosensory evoked potentials (SEPs) are recorded from electrodes placed over the sensory cortex after stimulation of a peripheral sensory or mixed sensorimotor nerve. SEPs are recorded from both upper and lower extremities. Because many neurotoxicants affect peripheral nerves, it is not uncommon for the slowing of sensory conduction associated with neurotoxic damage to the distal-most aspects of the nerve to affect the SEPs in some exposed patients. Therefore, the clinical value of SEPs compared with standard nerve conduction velocities is controversial. Nevertheless, SEPs are of value for determining conduction through the spinal cord posterior columns and in instances where proximal nerve blocks or asymmetric problems are being considered in the differential diagnosis (3).

NEUROIMAGING STUDIES

Computerized radiologic equipment has emerged as an excellent tool for documenting brain damage and dysfunction due to various insults. *Magnetic resonance imaging* (MRI) and *computed tomography* (CT) scans can document structural abnormalities due to strokes and neoplasms; thus, they are extremely useful in the differential diagnosis of primary and neurotoxic diseases (3). Although the sensitivity of structural neuroimaging techniques for documenting the effects of neurotoxicants is limited by spatial resolution (MRI studies of patients with less severe chronic toxic encephalopathies may appear normal despite clinical neurologic and neuropsychological evidence of behavioral abnormalities and performance deficits), there have been reports of clinical correlations between the images obtained with MRI and CT scans of patients exposed to neurotoxicants. MRI and CT scans can reveal the structural changes associated with the cerebral edema of acute lead encephalopathy. The MRIs of patients exposed to CO and manganese may reveal unilateral or

bilateral signal intensity changes in the globus pallidus (3). Cerebral and/or cerebellar atrophy and changes in cerebral white matter may be seen on the brain MRIs of patients with behavioral changes and tremor associated with chronic or repeated exposures to toluene (16,17). MRI changes also have been associated with exposure to lead, tin, and mercury compounds (3).

Functional imaging techniques, including functional MRI, single photon emission computed tomography, and positron emission tomography, may reveal findings consistent with neurotoxicant-induced brain dysfunction occurring both during and after exposure to various neurotoxicants (18–20).

Magnetic resonance spectroscopy (MRS) is emerging as a sensitive measure of the chemical changes associated with brain dysfunction. MRS produces a graph of chemical spectra with peaks reflecting the tissue concentrations of specific chemicals, including *N*-acetyl-L-aspartate (NAA), creatine/phosphocreatine (Cr), choline-containing compounds (Cho), and lactate. NAA is present only in living neurons. A reduction in NAA concentration reflects a loss of neurons as demonstrated by findings in patients diagnosed with neurodegenerative processes such AD (21). The Cr peak seen on MRS reflects levels of creatine and phosphocreatine, which serve as a reserve for high-energy phosphate metabolism in the cytosol of neurons. The Cho peak represents choline-containing compounds. Choline is a precursor for the neurotransmitter acetylcholine and for the membrane constituent phosphatidylcholine; levels of choline can reflect membrane synthesis and degradation. Lactate is an end product of anaerobic respiration and may be elevated following exposure to hypoxia-inducing neurotoxicants such as CO and hydrogen sulfide. Additional chemicals of interest detectable with MRS include glutamate, glutamine, myo-inositol, and γ-aminobutyric acid (GABA) (22). The value of MRS in the diagnosis of exposure to neurotoxicants has not been fully elucidated, but published reports indicate that this technology holds considerable promise as a diagnostic tool (23,24).

DOCUMENTING LEVELS OF EXPOSURE

The concentration of a specific neurotoxicant present in the environment that is potentially absorbable by a person in the area is the *exposure level.* Both the concentration and the duration of exposure to a chemical determine a *cumulative exposure level,* which should be distinguished from a *"dose,"* which is the amount of a chemical that actually has been absorbed by an exposed individual. The dose is determined not only by the exposure levels but also by the personal hygiene habits used on the premises (e.g., washing, protective clothing, respirators). The individual's biologic characteristics, including genetic factors that control the activity and availability of enzymes involved in the metabolism and excretion of neurotoxicants from the body, influence the

TABLE 42.3. ADDRESSES OF INTERNET WEB SITES THAT PROVIDE INFORMATION ON NEUROTOXICANTS

Occupational Safety and Health Administration (OSHA)	*www.osha.gov*
National Institute for Occupational Safety and Health (NIOSH)	*www.cdc.gov/niosh/homepage.html*
Neurotoxic.com	*www.neurotoxic.com*
Industrialhygiene.com	*www.industrialhygiene.com*

biologically active dose and subsequently the effects of exposure. Daily activities (e.g., time spent at the site of exposure and proximity to the source of exposure) that may influence an individual's total exposure are reviewed elsewhere (25,26).

Documentation of chemical exposures can be made by *on-site evaluations* of current exposure levels and circumstances at the workplace or residence. Ambient air, water and soil samples can be analyzed for concentrations of relevant chemicals. Precisely collected quantities are analyzed for concentrations of the constituents, including possible neurotoxicants. On-site monitoring of pollution levels can provide indicators of relative risk to persons who find themselves in the contaminated area, either by residence, employment or accident. Safe levels of exposure, so-called *permissible exposure levels,* have been established by OSHA for various contaminants (27). The American Conference of Governmental Industrial Hygienists (ACGIH) has recommended *biologic exposure indices* for many neurotoxicants, which can be used in conjunction with the permissible exposure levels to more closely monitor at-risk persons (28). Additional information on exposure limits and monitoring protocols is available on the World Wide Web (Table 42.3).

When the patient's presenting clinical symptoms cannot be directly linked to the effects of a neurotoxicant because exposure levels or circumstances have changed, an estimate of the personal exposure level can be made using mathematical modeling methods and previous exposure data. Remote exposures can be estimated by modeling methods using current and previously obtained exposure data, such as ambient air and water samples. Mathematical modeling using "historical" exposure records may be the only way to estimate a patient's cumulative exposure dose. When past monitoring data are not available, ascertaining total exposure is limited to obtaining current exposure levels and circumstances and making assumptions to estimate the person's past exposure.

Personal (breathing zone) sampling is used to determine the concentration of neurotoxicants present in an individual's immediate area and provides the most information about a specific individual's possible exposure to airborne neurotoxicants. Personal breathing zone sampling can be accomplished by following an individual around with a portable air-sampling meter while he/she is in the potentially contaminated area and then determining personal exposure over time. Calculations of drinking and bathing water exposure levels are based on water concentrations of the neurotoxi-

cant; estimated personal intake levels (e.g., glasses of water consumed per day) and usage (e.g., duration and frequency of showers or baths); and length of residence or duration of employment in the at-risk areas. Environmental exposure to other hazardous substances, such as soil or air, are determined by area sampling (29–31).

Biologic markers, unlike environmental concentrations, are direct indicators of total uptake or dose. They reflect intake from multiple routes and fluctuating or cumulative exposures over time (32). In order for biologic markers to be useful in the clinical setting, certain practical issues regarding collection of the appropriate sample should be considered: (a) availability of a marker from an appropriate specimen; (b) appropriate timing of sample collection; and (c) use of the correct device for sample collection.

Neurotoxicant exposure-induced neurologic dysfunction is substantiated by documenting elevated levels of the chemical and/or its metabolites in samples of urine, blood, hair, and nails. These data confirm an increased body burden among individuals suspected to be at risk based on environmental monitoring (e.g., workplace ambient air samples). The ACGIH has recommended biologic exposure indices for many neurotoxicants (28). Remote exposures can be estimated by modeling methods using current and previously obtained data, such as current blood levels and ambient air and water samples. Neurophysiologic documentation of cellular dysfunction and nerve biopsy studies demonstrating neuropathology consistent with known effects of specific neurotoxicants (e.g., abnormal nerve conduction studies and a nerve biopsy showing paranodal axonal swelling in an individual with a suspected exposure to *n*-hexane) can be used to support a causal relationship and may be particularly valuable in the absence of adequate biologic and environmental exposure data. Likewise, laboratory data demonstrating abnormalities in other organ systems, such as liver function tests, can be used to corroborate and support a causal diagnosis.

The clinical relevance of biologic marker data should be interpreted cautiously. It is important to recognize that different chemicals may have similar detoxification pathways and may result in common detectable metabolic byproducts. For example, exposure to both trichloroethylene (TCE) and perchloroethylene results in urinary excretion of the same metabolites (trichloroethanol and trichloroacetic acid). The *sensitivity* and *specificity* of the biologic marker must therefore be considered for it to be used reliably in the clinical setting as an indication of neurotoxicant exposure. Sensitivity is the ability of a marker to identify correctly subjects who, because of exposure to a suspected neurotoxicant, possess symptoms and/or signs of concern, which are detectable by measuring this particular biologic marker. Specificity is the ability to identify correctly and separate out those who have been exposed from those who have not been exposed because they do not possess detectable symptoms and/or signs of disease or effects of exposure to neurotoxicants.

The predictive value of a marker to accurately define an exposure and an effect is determined not only by factors that determine validity of the test itself (i.e., sensitivity and specificity), but also by the characteristics of the population to which the test is applied, particularly the prevalence of the exposure status or the symptoms and signs of disease. Because exposure to neurotoxic substances can result in a range of biologic effects, there is a need to clearly define the critical effect(s) that is(are) being considered a previously observed outcome of exposure to the suspected neurotoxicant. The positive predictive value of a biologic marker is based on its presence (or an increased amount of the marker) among persons who have been exposed and/or who have illness (32,35,36). The higher the prevalence of the exposure or disease, the more likely a positive test (i.e., presence of the marker) is predictive of exposure or disease. Thus, the validation study of markers indicative of exposure or effect requires information on specificity, sensitivity, and prior knowledge of the disease or exposure rate (37).

Recent advances in genetics permit the documentation of previously elusive factors that contribute to individual differences in susceptibility to the effects of exposure to specific classes of neurotoxicants. More specifically, the recognition and documentation of genetic polymorphisms that allow for the synthesis of various forms of a particular enzyme with different levels of activity in different individuals have contributed to our understanding of individual susceptibility to the toxic effects. For example, polymorphisms in the enzyme butyrylcholinesterase contribute to individual differences in susceptibility to the cholinergic effects of the organophosphorus compounds (OPCs) (38). Polymorphisms in the enzyme glutathione-*S*-transferase contribute to individual differences in susceptibility to EtO and other electrophilic compounds that are inactivated by conjugation with glutathione (39).

Three types of biologic markers are used by clinicians to make a diagnosis of neurotoxic illness: (a) *biologic markers of exposure;* (b) *biologic markers of effect;* and (c) *biologic markers of susceptibility* (34,40). Regulatory standards have not yet been established for many of the biologic markers of effect, such as neurophysiologic or neuropsychological test results.

Biologic markers of exposure reflect the individual's body burden (or internal dose) and the biologically effective dose (i.e., the product of an interaction between a substance and some target molecules or other nervous system receptor).

Individual variations in absorption rate, tissue retention and distribution, metabolism, and excretion all affect the internal dose and the biologically effective dose (34).

Biologic markers of effect are measurable biochemical or physiologic alterations within the nervous system that can be associated with an established or potential health impairment or disease. These include markers of altered function and structure related to the development of disease. They can be any qualitative or quantitative changes from baseline or expected levels of appearance or performance that would indicate impairment or damage resulting from exposure to a neurotoxicant (34). In the clinical setting, the results of neurophysiologic studies, neuropsychological tests, and neuroimaging studies serve as important biologic markers of the effects of neurotoxicants on the nervous system (40).

Biomarkers of susceptibility are indicators of increased or decreased risk for the adverse effects exposure. These variations can affect the interpretation of biologic markers of exposure and effect along the continuum from exposure to overt disease. Some individuals have a lower threshold for exposure effects than do others; a preexisting disease state or concurrent condition may affect the metabolism of certain neurotoxicants, thus increasing the internal dose or the biologically effective dose. Biologic markers of susceptibility include genetically determined variations in absorption, metabolism, and excretion; immunologic responsiveness; and low organ reserve capacity due to primary disease or other identifiable factors that can influence the effects of a neurotoxic substance (34). For example, genetic polymorphisms in chemical metabolizing enzymes such as glutathione-*S*-transferase can affect the metabolism and excretion of neurotoxicants and thus increase or decrease an individual's risk for adverse neurologic effects following exposure to a neurotoxicant.

NEUROPATHOGENESIS

The pathogenesis of clinical dysfunction is dependent on the action of the particular chemical to which the patient was exposed. There are several common pathogenic mechanisms that underlie the effects of most neurotoxicants. These include the generation of free radicals (e.g., MPTP); covalent binding of electrophilic compounds to nucleophilic centers of cellular macromolecules (e.g., EtO); molecular mimicry (e.g., OPCs); and disruption of the cellular aerobic respiration process (e.g., hydrogen sulfide).

Factors that contribute to the severity of the effects of exposure include durations and concentrations, and induction or inhibition of chemical metabolizing enzymes such as cytochrome P-450 and glutathione-*S*-transferase (39,41,42). Depletion of critical peptides such as glutathione can also contribute to the development of adverse neurologic effects (43).

Free Radicals

Manganese

The oxidation state of Mn determines its biochemical activity. Mn in the +2 oxidation state is a scavenger of free radicals, whereas Mn in the +3 oxidation state promotes the oxidation of critical substrates such as catecholamines, glutathione, and fatty acids. This generates toxic free radicals that bind to deoxyribonucleic acid (DNA) and ribonucleic acid (RNA), disrupt neuronal membrane integrity, and promote apoptosis (44–46). The capacity of Mn to induce selective lesions in the globus pallidus, substantia nigra, caudate nucleus, and putamen appears to be related to the chemical environment of these regions, in which Mn^{2+} is oxidized to Mn^{3+} (47,48). Accumulation of Mn^{3+} in the dopamine-rich nigrostriatal pathway promotes the autooxidation of dopamine in that system (45,46,49).

The adverse effects of Mn^{3+} are potentiated when protective scavenger enzymes such as superoxide dismutase and DT diaphorase (also known as NAD(P)H:menadione oxidoreductase; NAD(P)H:(quinone acceptor) oxidoreductase; quinone reductase, azo-dye reductase) are depleted or otherwise unable to alter the oxidation potential of critical amounts of reactive oxygen species. Symptoms may be more severe in individuals who are simultaneously exposed to several neurotoxic chemicals or who have polymorphisms in the genes encoding for these enzymes (42,50–53).

Postmortem studies of the brains of Mn-exposed workers reveal reduced neuromelanin content in the substantia nigra pars reticulata. This finding is in stark contrast with those seen in idiopathic Parkinson disease, in which the pars compacta is the most severely affected region of the substantia nigra (54). Neuropathologic studies in humans and primates reveal that Mn also damages basal ganglia neurons that are downstream from the substantia nigra, particularly those in the globus pallidus internus. The putamen and caudate nucleus are affected to a lesser degree (18,55–57).

Paraquat

Paraquat is structurally similar to MPTP, which has been associated with an increased risk for PD (58–60). Exposure to paraquat is associated with nervous system dysfunction (61,62). Paraquat has been shown to cross the blood–brain barrier and to induce a loss of dopaminergic neurons in the substantia nigra of rats (63). Paraquat is metabolized by NADPH-dependent reduction to a free radical that reacts with molecular oxygen to yield a superoxide anion that is converted to hydrogen peroxide by superoxide dismutase. Both the superoxide anion and hydrogen peroxide are capable of inducing lipid peroxidation and thereby disrupting cellular membrane integrity and cellular functioning (64). The potential toxic effects of paraquat are attenuated by the conjugation of its free radical metabolites with glutathione (65).

Electrophiles

Ethylene Oxide

EtO is a simple epoxide that does not require metabolic activation to produce neurotoxic effects. It forms hydroxyethyl adducts with cellular macromolecules such as DNA (66–72). EtO metabolism is catalyzed by epoxide hydrolase and glutathione-S-transferase. A small fraction is exhaled unchanged (73). Studies in workers occupationally exposed to EtO indicate that individuals who are polymorphic for less active forms of the glutathione-S-transferase GSTT1 are at increased risk for toxic effects (39).

The formation of adducts (alkylated amino acids) suggests that alkylation of the amino acids in cellular proteins and DNA occurs to some extent in all tissues, including the brain (72,74). These adducts are used as a biologic marker of the effects of EtO. Damage to the DNA of cells, including neurons, can induce cell death by apoptosis (50,75). Damage to DNA and RNA can disrupt the synthesis of proteins involved in cell structure and neurotransmission. Neuropathologic effects of EtO have been documented in both the central and the peripheral nervous systems (76–81). Neuropathologic studies in the peripheral nervous system demonstrate that EtO exposure induces a distal dying-back axonopathy (77,80).

n-Hexane

Exposure to normal hexane (*n*-hexane) has been associated with neuropathologic changes in the central and peripheral nervous systems. Neurologic damage occurs when its metabolite, 2,5-hexanedione, a γ-diketone, chemically cross-links axonal neurofilaments and interrupts axonal transport mechanisms (82). The formation of these neurofilamentous cross-links is inhibited by glutathione, which reacts with 2,5-hexanedione to detoxify the molecule (83). In 1976, Spencer and Schaumburg (88) coined the term "central-peripheral distal dying back neuropathy" to described the morphologic changes seen in a patient exposed to *n*-hexane. Light microscopic studies of sural nerve biopsies of workers occupationally exposed to *n*-hexane reveal paranodal swelling with associated retraction of the myelin sheath (84–87). Electron microscopic studies of the morphologic changes seen in these peripheral nerves reveal accumulation of neurofilaments within the proximal paranodal region of the axon, suggesting that this process underlies the paranodal swelling (85,87,88). Workers exposed to *n*-hexane may also be at increased risk for developing Parkinson disease (89).

n-Hexane is metabolized by cytochrome P-450 2E1 (CYP2E1) and subsequently conjugated with glutathione for excretion in urine (83). This suggests that workers with genetic polymorphisms for CYP2E1 or who are chronically exposed to ethanol or other chemicals that induce the activity of CYP2E1 may be at increased risk for neurotoxic effects, including the development of PD (41,83,90).

Molecular Mimicry

Inorganic Lead

Many of the acute and chronic neurologic effects of inorganic lead can be attributed to its structural similarity to the calcium ion that allows it to displace this biologically important chemical (91–95). For example, the calcium-dependent regulatory enzyme, protein kinase C, is activated by inorganic lead (96). Lead also interferes with calcium ion currents by acting on voltage-gated calcium channels (95). Lead has an inhibitory effect on Ca^{2+} ATPase activity, resulting in an increase in intracellular calcium (93). Although elevated levels of intracellular calcium normally trigger the exocytosis responsible for release of neurotransmitters, there also is evidence suggesting that excessive levels of intracellular calcium can induce cell death (50).

Lead-induced disruptions of normal intracellular calcium concentrations have been associated with disturbances of learning and neurodevelopment. The ability of lead to bind to the calcium binding sites on *N*-methyl-D-aspartate (NMDA) receptors and thereby possibly disrupt the activity of the glutaminergic system, which is involved in long-term potentiation, may account for its effects on learning (97).

Organophosphorus Compounds

OPCs are structurally similar to acetylcholine. OPCs form bonds with acetylcholinesterase (AChE), the enzyme that normally degrades acetylcholine (98). Inhibition of AChE results in accumulation of acetylcholine in the synaptic cleft (e.g., neuromuscular junction) and symptoms of cholinergic overstimulation, such as lacrimation, salivation, and muscle weakness, that typically are associated with exposure to OPCs.

Many OPCs can induce a central-peripheral neuropathy that develops several weeks after the acute clinical signs and symptoms of cholinergic overstimulation have subsided. Neuropathologic studies of "organophosphate-induced delayed neuropathy" (OPIDN) reveal that the distal-most segments of the longest and largest myelinated nerve fibers degenerate first, after which the axon dies back to the perikaryon. This process of distal-to-proximal axonal degeneration has been referred to as a "distal dying-back neuropathy" (99–102).

The mechanism of OPIDN has not been fully elucidated, but it is known that OPCs phosphorylates the enzyme neuropathy target esterase (103,104). The degree of inhibition of neuropathy target esterase has been found to correlate with the severity of OPIDN (105). Impairment of axonal transport due to abnormal phosphorylation of cytoskeletal proteins have been noted in OPIDN (104). Hyperphosphorylated cytoskeletal proteins are not easily transported by axoplasmic flow; thus, they may accumulate and induce axonal swelling with secondary demyelination (104).

Disruption of Cellular Aerobic Respiration

Carbon Monoxide

CO is an odorless gas that binds to hemoglobin and displaces oxygen. The effects of CO are due in part to its structural similarity to oxygen. The binding of CO to hemoglobin also influences the oxygen dissociation curve, resulting in less oxygen released in the appropriate tissues. The decrease in available oxygen results in a shift of the cell to anaerobic respiration. Because the oxygen demands of the brain are very high, neuronal damage results from the associated tissue hypoxia (3,4).

Hydrogen Sulfide

Hydrogen sulfide is a colorless gas with the characteristic odor of rotten eggs. It disrupts cellular respiration by inhibiting the activity of cytochrome oxidase a3, the terminal enzyme in the electron transport chain. The result is a disruption of the mitochondrial aerobic respiration process and a switch to anaerobic respiration with the associated development of metabolic acidosis. Neurologic damage and cell death result from tissue hypoxia. Hydrogen sulfide also has been shown to alter neurotransmitter release (4).

NEUROBEHAVIORAL SYNDROMES ASSOCIATED WITH SELECTED NEUROTOXICANTS

Metals

Inorganic Lead

Exposure to lead is associated with peripheral neuropathy and overt signs of encephalopathy, such as restlessness, irritability, attention and memory deficits, drowsiness, tremor, and stupor. Left untreated, the syndrome progresses to seizures, coma, and death in the most severe cases of lead poisoning. These overt clinical manifestations typically are associated with blood lead levels greater than 70 mg/dL, but subclinical manifestations can be seen at blood lead levels less than 70 mg/dL (3).

Children are more susceptible to lead encephalopathy than are adults, possibly because of differences in the permeability of the blood–brain barrier. Children exposed to sufficient levels of lead may develop a reversible brain edema associated with severe acute encephalopathy. If they survive, they may have persistent cognitive deficits and behavioral problems (106). The persistence of structural and functional changes after acute childhood lead encephalopathy was documented in a 10-year-old boy who had documented blood lead levels of 51 mg/dL when he was 3 years old (24). Formal neuropsychological assessment of the child revealed deficits in attention and mental control. He also had difficulties with reading, writing, and linguistic performance. MRS studies of this child revealed a reduction in the NAA/creatine ratio suggesting a loss of neurons. Conventional MRI of the brain performed at the same time was normal. In contrast, the neuropsychological and MRS findings in his cousin, a 9-year-old boy, who had no history of exposure to lead, were within normal limits.

Peripheral neuropathy is seen in both adults and children exposed to lead, but it may be less commonly diagnosed in children because the rapidity and severity with which the encephalopathy ensues distracts from the neuropathy that develops with chronic exposures. Electrophysiologic findings in lead-exposed children without overt encephalopathy indicate that slowed nerve conduction velocities occur before overt encephalopathy in this group and that this parameter may be useful in the diagnosis of at-risk populations (107). Lead-exposed children show a somewhat different pattern of deficits than is seen in adults. Deficits in psychomotor speed, verbal concept formation, visuospatial abilities, and attention are seen in children as well as adults, but lead-exposed children also show deficits on tests of language function that lead-exposed adults do not show (108).

Adults exposed to lead typically develop more subtle signs of encephalopathy, such as irritability and attention deficits. Cognitive performance deficits are seen on tests of attention and cognitive tracking, short-term memory, verbal reasoning, visual spatial ability, and concept formation among the lead-exposed workers we have seen in our clinic. Attention deficits interfere with performance on demanding tasks such as counting by threes and completing oral arithmetic problems. Attention deficits also are revealed on digit span test and CPT. Short-term memory deficits are seen on tasks that require recall of verbal and visual material, including paired associate learning, recall of verbal narrative material, drawing of visual designs, and recognition of visual stimuli. Deficits typically are more pronounced on delayed recall, which suggests that the deficits are not entirely due to problems with attention but may reflect disturbances of encoding processes. Performance deficits are revealed on tests of complex verbal reasoning that require abstraction, such as the abstract definitions of vocabulary words. Tasks sensitive to the application of principles of social judgment and tasks involving proverb interpretation and the ability to deduce similarities also are impaired in lead-exposed individuals, suggesting frontal lobe involvement. The digit symbol task reveals evidence of psychomotor slowing, as does the Santa Ana Form Board. Mood disorders are common, with subjective complaints of apathy, irritability, and diminished ability to control anger. The POMS scales of lead-exposed workers reveal increased fatigue. The performance deficits associated with lead exposure seen among adults may persist indefinitely after cessation of exposure (109,110).

Manganese

Mn is an essential trace mineral necessary for normal development (111–113). Mn is widely used for its alloying properties with both ferrous and nonferrous metals. Mn oxide is used in the manufacture of paints, disinfectants, and fertilizers. The toxicity of Mn is related to its oxidation state; divalent Mn (Mn^{2+}) is a powerful antioxidant, whereas the trivalent form (Mn^{3+}) promotes the formation of free radicals (44,45,114–116). Occupational exposure to inorganic Mn occurs among miners and welders (117–120). Organic Mn compounds include manganese ethylene-bis-dithiocarbamate (maneb), which is used as a fungicide (121), and methylcyclopentadienyl manganese tricarbonyl (MMT), which is used as an antiknock additive in gasoline (122,123).

The similarities between the clinical manifestations associated with Mn intoxication and those seen in Parkinson disease suggest that occupational and environmental exposures to this and other neurotoxicants that promote the formation of free radicals may be involved in the pathogenesis of certain neurodegenerative diseases (3,40,50,123,124). Mn has been associated with signs of basal ganglia dysfunction, including rigidity, gait abnormalities, dysarthria, hypomimia, and bradykinesia in patients with chronic liver failure (123–132). Patients who are genetically predisposed to developing idiopathic Parkinson disease may be especially sensitive to the effects of Mn on the basal ganglia and therefore may develop symptoms sooner than expected at a given level of exposure (41,124,133).

The manifestations of Mn poisoning include a constellation of behavioral changes that emerge early in the clinical course and are collectively referred to as *manganese psychosis*. The clinical symptoms of Mn psychosis include mood changes, emotional lability, uncontrolled laughter, and hallucinations (127,134). Performance on formal neuropsychological tests may be impaired at this time. Motor disturbances, characterized by tremor, dysarthria, dystonia, slowness and clumsiness of movement, gait disturbance (awkward high-stepping gait), and postural instability, emerge with continued exposure to Mn (112,134–137). The gait disturbance associated with Mn poisoning is easily distinguishable from the shuffling gait of Parkinson disease. In some cases, the extrapyramidal symptoms of Mn poisoning may progress after cessation of exposure (137).

Formal neuropsychological testing can be used to document the behavioral effects of Mn exposure. Performance deficits may be seen on tests of attention, memory, and psychomotor functioning. Tests of psychomotor functioning, such as the Santa Ana Form Board, digit symbol, and finger tapping test, are sensitive to the neurotoxic effects of Mn (138,139). Language and verbal functioning typically is spared in Mn-exposed adults (140,141), but children exposed to Mn may develop persistent impairments of language function (142,143).

Memory deficits have been reported in Mn-exposed welders (144). Neurobehavioral assessments of 141 workers exposed to a mean ambient air concentration of 1 mg/m^3 for a mean duration of 7.1 years performed below expectation on tests of eye–hand coordination, reaction time, and short-term memory (139). Foundry workers exposed to Mn for a mean duration of 9.9 years at a mean ambient air concentration of only 0.25 mg/m³ performed below expectation on finger tapping, simple reaction time, and digit span tests (140,145). Reaction-time impairments were noted in a group of welders exposed to Mn concentrations ranging from 1.0 to 4.0 mg/m³ for a mean duration of 16 years (146).

Lucchini et al. (147) conducted neurobehavioral testing on 58 workers from a ferromanganese alloy plant. Workers who had been exposed for durations of 1 to 28 years (mean 13 years) were divided into three groups based on the Mn exposure histories. The *low exposure* group consisted of foreman, clerks, and laboratory technicians who were exposed to Mn concentrations of only 0.009 to 0.15 mg/m³ (mean blood Mn: 6.0 fg/L; mean urine Mn: 1.7 fg/L); the *medium exposure* group consisted of maintenance workers with exposure levels ranging from 0.072 to 0.76 mg/m³ (mean blood Mn: 8.6 fg/L; mean urine Mn: 2.3 fg/L); and the *high exposure* group consisted of workers with the highest levels of exposures of up to 2.6 mg/m³ (mean blood Mn: 11.9 fg/L; mean urine Mn: 2.8 fg/L). Cumulative exposure indices were determined for each subject. Cognitive domains assessed included attention (simple and complex reaction times), executive function (arithmetic), psychomotor function (digit symbol and finger tapping), short-term memory (digit span), and verbal understanding (vocabulary). Impairments were found on tests of psychomotor function (finger tapping and digit symbol) and short-term memory (digit span). The results of these tests revealed a correlation between environmental exposure levels as represented by individual cumulative exposure indices.

Solvents

Trichloroethylene

TCE is a colorless, highly volatile, chlorinated hydrocarbon used as degreasing solvent. Neurotoxic effects are associated with exposures to TCE and with its environmental degradation product dichloroacetylene (DCA) (3). Individuals who work with TCE or encounter it in nonoccupational settings are at risk for the neurotoxic effects of DCA. The predilection of TCE and DCA for the trigeminal nerve has been noted by researchers and clinicians (3,148,149). DCA is believed to be the offending agent in those cases of trigeminal nerve damage associated with exposures to TCE. However, in patients who present with trigeminal neuropathy plus peripheral nerve dysfunction and encephalopathy, the offending agent more likely is TCE (3).

Acute exposure to TCE is associated with reversible symptoms of headache, dizziness, confusion, nausea, vomiting, stupor, and loss of consciousness (150–160). Persistent neurologic effects also are associated with severe acute exposures to TCE and may include facial anesthesia, reduced perception of taste, dysarthria, flattening of the nasolabial folds, ptosis, reduced pupillary responses, constricted visual fields, sensorimotor neuropathy, and encephalopathy (154,156,159–161).

Symptoms of chronic exposure to TCE develop insidiously and may include headache, dizziness, fatigue, irritability, insomnia, problems with memory and concentration, and paresthesias (162). Occupationally exposed workers report experiencing forgetfulness, dizziness, headache, sleep disturbances, fatigue, irritability, anorexia, trigeminal nerve symptoms, sexual problems, and peripheral neuropathy (163–168). Speech and hearing disorders have been reported among children exposed to TCE environmentally through contaminated drinking and bathing water (169).

Exposure to TCE is associated with deficits on neuropsychological tests of attention and executive functioning, short-term memory, and visuospatial ability. Children exposed to TCE may develop persistent impairment of verbal functioning. Language and verbal skills typically are spared in adults (161,170).

Serial neurologic examinations and neuropsychological assessments of a man who developed chronic toxic encephalopathy, trigeminal neuropathy, and peripheral neuropathy after he experienced a single severe acute exposure to TCE at age 26 years documented the clinical course and recovery. The initial neuropsychological assessment of this man revealed difficulties with sequential problem solving and short-term memory tasks. His performance improved somewhat over the course of the first several years after the incident, but he continued to experience chronic apathy and problems with attention and short-term memory. These complaints and deficits were documented on follow-up neuropsychological assessments. At follow-up examination 16 years after the exposure incident, the patient's performance IQ (PIQ) was still 19 points lower than his verbal IQ (VIQ) (161). Significant visuospatial deficits accounted for a large part of the discrepancy between the patient's PIQ and VIQ. Tests of short-term verbal and visual memory (i.e., Benton Visual Retention Test, visual reproduction, logical memories, and paired associates) showed deficits on immediate and delayed recall. In addition, the MMPI indicated that the patient was depressed. A similar report of TCE-induced chronic toxic encephalopathy was reported by Steinberg (171), who had examined a 62-year-old machinist who experienced a brief acute exposure to TCE. Follow-up neuropsychological assessment of this patient was performed 5.5 years after the incident and revealed performance deficits on the digit symbol and object assembly sections of the Wechsler Adult Intelligence Scale. Furthermore, the patient exhibited overt impairments in executive function, lack of insight, and apathy, all of which were considered to be permanent residual effects of TCE intoxication.

Grandjean et al. (138) reported on the neurologic symptoms and neurobehavioral findings in 50 mechanical engineers who were chronically exposed to TCE. The workers reported experiencing fatigue, dizziness, and an inability to tolerate alcohol. Neuropsychological tests revealed deficits on tests of general intellectual functioning, executive function, memory, mood, and affect. Neuropsychological performance was within expectation among workers who had been exposed to TCE for less than 1 year.

Neuropsychological testing revealed mild-to-moderate encephalopathy in 24 of 28 individuals who were chronically exposed to TCE through consumption of contaminated drinking and bathing water (162). Tests of attention and executive function, short-term memory, visuospatial ability, and psychomotor functioning revealed deficits. Of the 28 individuals, attention and executive function deficits were seen in 19; significant memory impairments in 24; and visuospatial and manual motor function deficits in 17. Performance on tests of language and verbal functioning was within expectation for the exposed adults, but their children showed a decrease in performance on the Boston Naming Test. These findings suggest that children exposed to TCE may be at risk for exposure-related learning disabilities.

Carbon Disulfide

Carbon disulfide is used as a solvent in the rubber and viscose rayon industries. It also is found in various insecticides used in the agricultural industry. Exposure to carbon disulfide is associated with neuropathologic effects on the basal ganglia and peripheral nervous system. Perhaps the best-known manifestations of carbon disulfide poisoning are polyneuropathy and parkinsonism. Vigliani (172) identified 100 cases of carbon disulfide poisoning among workers in the viscose rayon manufacturing industry. In addition to peripheral neuropathy, these workers experienced headaches, dizziness, psychosis, and symptoms of extrapyramidal system involvement. EEG studies of viscose rayon workers revealed 21 of 54 with abnormal findings compared with six among 50 control subjects (173). The EEG abnormalities have been attributed to vascular changes and to direct toxic effects of the parent chemical or a metabolite on cortical and subcortical structures.

Carbon disulfide exposure has been associated with damage to subcortical structures of the basal ganglia and with impaired psychomotor function, especially manual dexterity and motor speed. Tests sensitive to these parameters include the Santa Ana Dexterity Test and reaction times. Cortical functions, especially visuospatial abilities and concentration, may be assessed with the block design test, digit symbol test, and CPT. Subtle symptoms of carbon disulfide exposure, such as irritability and increased systemic complaints, may precede overt neurologic illness and can be assessed by the POMS and a structured clinical interview.

Hänninen (174) assessed neuropsychological functioning among workers exposed to carbon disulfide and found deficits in attention, vigilance, psychomotor skills, and visuospatial function. Cassito et al. (175) performed serial neuropsychological assessments on a group of 493 viscose rayon workers exposed to concentrations ranging from 20 to 40 ppm in 1974 to less than 10 ppm in 1990. Behavioral domains assessed included perceptual ability, visuospatial ability, visuomotor skills, attention, and memory. The results of this study revealed a marked improvement in neurobehavioral performance following reductions of ambient air concentrations of carbon disulfide. The neurobehavioral test performances among the workers who had been exposed to higher levels of carbon disulfide before improvements in industrial hygiene were implemented were markedly worse than those of workers who had been employed only since exposures had been reduced. These findings suggest that performance deficits are related to exposure levels and that they persist after reductions of carbon disulfide exposure.

Neuropsychological assessment was performed on a 28-year-old male electrician who experienced repeated acute exposures to carbon disulfide vapors over the course of several months. The vapors were emanating from puddles of liquid carbon disulfide near where he was working. He also experienced a single severe acute dermal exposure episode in which he was doused with carbon disulfide that was leaking from some pipes near where he was installing a carbon disulfide monitor. His family began to notice behavioral changes, which included increased irritability, apathy, and memory problems, approximately 2 months after he began working around the carbon disulfide. He developed a tremor that made it difficult for him to hold his tools steady while he was working. The tremor and memory problems eventually became so severe that it was necessary for him to stop working. Formal neuropsychological testing performed after his cessation of exposure to carbon disulfide revealed deficits on tests of attention and executive function, vigilance, short-term memory, motor skills, and visuospatial function. Follow-up examination 5 years later revealed improvement in his affect, but cognitive deficits in the domains of attention and executive function, visuospatial ability, and short-term memory persisted, as did the tremor (3).

Insecticides

Oganophosphorus Compounds

The OPCs are used primarily as pesticides, but some (e.g., Sarin and VX) also are used as chemical warfare agents. The OPCs are readily absorbed following inhalation, oral, and dermal exposures. Once absorbed, these compounds readily cross the blood–brain barrier. The OPCs inhibit the activity of AChE, an enzyme involved in the degradation of the neurotransmitter acetylcholine. Inhibition of AChE results in the persistence of acetylcholine and its continued stimulation of receptors in the central and peripheral nervous systems. Farmers, nursery and greenhouse workers, and workers involved in the manufacture of OPCs are at risk for exposure (3).

Acute exposure to the OPCs is associated with *cholinergic crisis* due to the inhibition of AChE. Clinical manifestations may include weakness, muscle fasciculations, tachycardia, miosis, lacrimation, excessive salivation, seizures, and coma. The symptoms may persist for several days after cessation of exposure. In some cases, the cholinergic crisis is followed by the *intermediate syndrome,* which includes proximal muscle weakness and reduced tendon reflexes. The intermediate syndrome, which reflects damage to muscle tissue, develops approximately 72 hours after removal from exposure and can persist for up to 6 weeks (176,177). The respiratory muscles also may be involved, and weakness of the diaphragm may necessitate intubation and respiratory support. Symptoms and signs of the *OPIDN* emerge about 1 to 5 weeks after exposure and may include flaccid paralysis of the distal muscles of the lower and upper extremities. This clinical picture is in contrast with that of the intermediate syndrome, which involves the more proximal limb muscles. Deep tendon reflexes of patients with OPIDN typically are reduced and may be absent in severe cases. Numbness and paresthesias may be present. Signs and symptoms indicative of central nervous system involvement have been documented among patients with OPIDN (178,179).

Mood changes and cognitive deficits (e.g., forgetfulness) are associated with the acute cholinergic crisis. These neurobehavioral manifestations may persist indefinitely after cessation of exposure. Formal neuropsychological testing of these patients reveals deficits in attention and executive functioning, psychomotor skills, and short-term memory (180–185). Fiedler et al. (185) reported significantly slower reaction times among workers who were chronically exposed to low levels of OPCs, indicating that the occurrence of an acute cholinergic crisis may not be required to induce persistent changes in central nervous system functioning.

Neuropsychological assessment of a 44-year-old Environmental Protection Agency field inspector who had been acutely exposed to OPCs revealed persistent findings. The patient had been doused with phosmet, which was released from a crop duster airplane flying overhead (3). His acute symptoms persisted for several days and included headache, nausea, excessive salivation, blurred vision, anxiety, irritability, weakness, photophobia, and insomnia. He recovered from most of these symptoms of an acute cholinergic crisis within the week, but the blurred vision, dizziness, and attention deficits persisted. Results of neurologic examination at that time were interpreted as normal, as was the EEG and MRI. Detailed neuropsychological assessment 2 years after the exposure incident revealed VIQ of 126 and PIQ of 106. Performance deficits were noted on tests of attention and executive function (Trail Making) and psychomotor speed (digit symbol). His immediate recall of verbal and visual material was within expectation, but his performance on delayed recall was impaired. He missed the overall concept of the task, and his answers tended to be concrete on tests of complex verbal reasoning. POMS revealed depression,

anger, and fatigue. At follow-up assessment 16 months later (40 months after the OPC exposure incident), he continued to show deficits on tests of attention and had difficulty with recall of verbal paired associates, but overall improvement in his functioning was noted. His major problem continued to be difficulty with complex verbal reasoning tasks. The POMS revealed only irritability.

Gases

Carbon Monoxide

CO is an odorless gas that forms as a by-product of the incomplete combustion of organic materials, including petroleum fuels. CO binds to hemoglobin to form carboxyhemoglobin (COHb) and interferes with the delivery of oxygen to various tissues of the body, including the brain, which has a very high oxygen demand. The severity of CO poisoning is determined by the COHb levels reached during the exposure. Subjective awareness of the effects of CO typically occurs at blood COHb levels of 15%. The first symptoms of CO poisoning to emerge include headache, dizziness, fatigue, weakness, disorientation, and hyperventilation. If exposure continues, seizures, coma, and/or death may occur (3).

Neuropathologic studies of cases of CO poisoning reveal necrosis in the hippocampus, globus pallidus, putamen, internal capsule, and centrum semiovale of the white matter of the cerebral hemispheres; the subcortical U fibers, hypothalamus, substantia nigra, and brainstem are typically spared (186–193). The neuropathologic changes seen following exposures to CO have been attributed in part to the release of lactate that occurs during hypoxia and which results in intracellular acidosis in tissues of the brain (194,195). Exposure to CO also increases the extracellular levels of catecholamines, such as dopamine, and production of free radicals, which have been associated with cell damage (50,196,197). Stimulation of the NMDA receptor–ion channel complex has been implicated in the mechanism of neuronal degeneration associated with CO exposure. This hypothesis is based on the observation that administration of NMDA antagonists decreases the glutamate-mediated release of dopamine and provides protection against the neurodegenerative effects of CO (195,198).

Nonfatal acute exposures to high levels of CO may result in permanent damage to the central nervous system. The permanent effects of CO poisoning emerge after a brief period of recovery (pseudo-recovery). Persistent symptoms of CO poisoning include parkinsonism, disorientation, memory impairment, aphasia, agnosia, cortical blindness, ataxia, and apraxia, and increased irritability, aggression, and moodiness (3,189,199,200). Neuropsychological test batteries should be selected on the basis of previously demonstrated sensitivities to the effects of CO exposure (2,201). CO poisoning typically is associated with deficits on tests of memory and psychomotor function (202–204). For example neuropsychological assessments of a 51-year-old man and his 52-year-old sister who were exposed to CO (blood COHb levels were 23.8% in the man and 20.1% in his sister) while they were driving from Maine to Connecticut in an automobile having a faulty exhaust system were performed. The assessments revealed deficits on tests of memory, executive functioning, and psychomotor functioning (3,204).

Hydrogen Sulfide

Hydrogen sulfide, also known as sewer gas, is a colorless gas with the characteristic odor of rotten eggs. It is produced during the bacterial decomposition of proteins and the decay of sulfur-containing products. Industrial exposures to hydrogen sulfide occur in workers in the leather tanning and oil refining industries and in sourgas field workers involved in natural gas production. Hydrogen sulfide disrupts the cellular respiration process and results in tissue hypoxia. Exposure to hydrogen sulfide has been associated with overt "knockdowns" in which the exposed individual loses consciousness due to hypoxia of brain tissues. Severe cases result in death of the exposed individual due to disruption of respiration. Lower levels of exposure to hydrogen sulfide may produce subjective symptoms of irritation of eyes and mucous membranes, nausea, headache, dizziness, confusion, feeling of shortness of breath, bradycardia, and chest pain (4). Acute exposures to hydrogen sulfide have been associated with permanent damage to the nervous system. Delayed onset of neurologic symptoms after a period of pseudo-recovery has been reported (205). Motor symptoms, including ataxia and tremor, have been reported among exposed individuals, which suggests that exposure to hydrogen sulfide damages structures of the basal ganglia (205,206).

Neuropsychological assessment of persons who experienced a knockdown may reveal that deficits in memory and motor function are common among patients exposed to hydrogen sulfide. Tvedt et al. (205) reported on a 30-year-old man who was exposed to hydrogen sulfide while he was working in a tannery. He lost consciousness for approximately 15 minutes while he was working near a waste tank. His EEG 3 months after the exposure episode showed diffuse slow wave activity. Follow-up EEG 8 years later showed slight generalized dysrhythmia. At follow-up 10 years later, he continued to complain of headache and memory problems. Neuropsychological testing at that time revealed deficits on tests of attention and executive function, verbal and visual memory, and visuospatial ability. The patient also reported an increased sensitivity to odors.

Schnieder et al. (206) reported on the persistent cognitive and motor deficits found in a 27-year-old man who had been exposed to hydrogen sulfide. He was admitted to the hospital with a Glasgow Coma Scale (GCS) score of 3. He was treated with hyperbaric oxygen, and on day 7 after exposure he had a GCS score of 15. MRI and CT scans were unremarkable. Neurologic examination 3 years after the exposure episode revealed signs of extrapyramidal dysfunction. Positron emission tomographic scan using [^{18}F]-deoxyglucose performed at that time revealed abnormal (heterogeneously distributed) bilateral uptake in the striatum. Metabolism was decreased bilaterally in the temporal and parietal lobes and in the left thalamus. Single photon emission computed tomographic study perform 6 months later (3.5 years after exposure) revealed decreased flow in the putamen but no cortical abnormalities. Neuropsychological assessment revealed deficits on tests of executive function, memory, and psychomotor ability.

REFERENCES

1. Spencer PS, Schaumburg HH. An expanded classification of neurotoxic responses based on cellular targets of chemical agents. *Acta Neurol Scand* 1984;70[Suppl 100]:9–19.
2. White RF, Feldman RG, Proctor SP. Neurobehavioral effects of toxic exposure. In: White RF, ed. *Clinical syndromes in adult neuropsychology: the practitioner's handbook.* Amsterdam: Elsevier Science Publishers, 1992:1–51.
3. Feldman RG. *Occupational and environmental neurotoxicology.* Philadelphia: Lippincott-Raven, 1999.
4. Goldfrank LR, Flomenbaum NE, Lewin NA, et al., eds. *Goldfrank's toxicologic emergencies,* 6th ed. Connecticut: Appleton and Lange, 1998.
5. Spencer PS, Schaumburg HH, eds. *Experimental and clinical neurotoxicology,* 2nd ed. New York: Oxford University Press, 2000.
6. Chong JP, Turpie I, Haines T, et al. Concordance of occupational and environmental exposure information elicited from patients with Alzheimer's disease and surrogate respondents. *Am J Ind Med* 1989;15:73–89.
7. Hugon J, Esclaire F, Lesort M, et al. Toxic neuronal apoptosis and modifications of tau and APP gene and protein expressions. *Drug Metab Rev* 1999;3:635–47.
8. Lezak MD. *Neuropsychological assessment,* 3rd ed. New York: Oxford University Press, 1995.
9. Bell IR, Peterson JM, Schwartz GE. Medical histories and psychological profiles of middle-aged women with and without self-reported illness from environmental chemicals. *J Clin Psychiatry* 1995;56:151–160.
10. Bell IR, Schwartz GE, Baldwin CM, et al. Individual differences in neural sensitization and the role of cortex in illness from low-level environmental chemical exposures. *Environ Health Perspect* 1997;105[Suppl 2]:457–466.
11. Ashford NA, Miller CS. *Chemical exposures: low levels and high stakes.* New York: Van Nostrand Reinhold, 1991.
12. Schottenfeld RS, Cullen MR. Occupation-induced posttraumatic stress disorders. *Am J Psychiatry* 1985;142:198–202.
13. Sorg BA, Prasad BM. Potential role of stress and sensitization in the development and expression of multiple chemical sensitivity. *Environ Health Perspect* 1997;105[Suppl 2]:467–471.
14. Ziem G, McTamney J. Profile of patients with chemical injury and sensitivity. *Environ Health Perspect* 1997;105[Suppl 2]:417–436.
15. Wiess B. Neurobehavioral properties of chemical sensitivity syndromes. *Neurotoxicology* 1998;19:259–268.
16. Kamran S, Bakshi R. MRI in chronic toluene abuse: low signal in the cerebral cortex on T2-weighted images. *Neuroradiology* 1998;40:519–521.

17. Filley CM, Heaton RK, Rosenberg NL. White matter dementia in chronic toluene abuse. *Neurology* 1990;40:532–534.

18. Shinotoh H, Snow BJ, Hewitt KA, et al. MRI and PET studies of manganese intoxicated monkeys. *Neurology* 1995;45:1199–1204.

19. Heuser G, Mena I. Neurospect in neurotoxic chemicals exposure: demonstration of long-term functional abnormalities. *Toxicol Ind Health* 1998;14:813–827.

20. Levin JM, Ross MH, Mendelson JH, et al. Reduction of BOLD fMRI response to primary visual stimulation following alcohol ingestion. *Psychiatry Res* 1998;82:135–146.

21. Pfefferbaum A, Adalsteinsson E, Spielman D, et al. In vivo brain concentrations of N-acetyl compounds, creatine, and choline in Alzheimer disease. *Arch Gen Psychiatry* 1999;56:185–192.

22. Jenkins BG, Kraft E. Magnetic resonance spectroscopy in toxic encephalopathy and neurodegeneration. *Curr Opin Neurol* 1999;12:753–760.

23. Sakamoto K, Murata T, Omori M, et al. Clinical studies on three cases of the interval form of carbon monoxide poisoning: serial proton magnetic resonance spectroscopy as a prognostic predictor. *Psychiatry Res* 1998;83:179–192.

24. Trope I, Lopez-Villegas D, Lenkinski RE. Magnetic resonance imaging and spectroscopy of regional brain structure in a 10-year-old boy with elevated blood lead levels. *Pediatrics* 1998;101:E7.

25. World Health Organization (WHO). *Early detection of occupational diseases.* Geneva, Switzerland: WHO, 1986:243–251.

26. Cohen BS. Industrial hygiene measurement and control. In: Rom W, ed. *Environmental and occupational medicine,* 2nd ed. Boston: Little Brown and Company, 1992:1389–1404.

27. Occupational Safety and Health Administration (OSHA). Code of Federal Regulations, 29, 1910.1000/.1047. Washington, DC: Office of the Federal Register, National Archives and Records Administration, 1995:411–431.

28. American Conference of Governmental Industrial Hygienists (ACGIH). Threshold limit values (TLVs) for chemical substances and physical agents and biological exposure indices (BEIs). Cincinnati, OH: ACGIH, 1995.

29. Agency for Toxic Substances and Disease Registry (ATSDR). Determining contaminants of concern. In: *Public health assessment guidance manual.* Atlanta, GA: ATSDR, 1992.

30. Agency for Toxic Substances and Disease Registry (ATSDR). Appendix D: estimation of exposure dose. In: *Public health assessment guidance manual.* Atlanta, GA: ATSDR, 1992.

31. Johnson BL. A precision exposure assessment. *J Environ Health* 1992;55:6–9.

32. Wilcosky TC, Griffith JD. Applications of biological markers. In: Hulka BS, Wilcosky TC, Griffith JD, eds. *Biological markers in epidemiology.* New York: Oxford University Press, 1990:16–27.

33. Hennekens CH, Buring JE. Screening. In: Mayrent SL, ed. *Epidemiology in medicine.* Boston: Little, Brown and Company, 1987:327–345.

34. National Research Council (NRC). *Environmental neurotoxicology.* Washington, DC: National Academy Press, 1992.

35. Schulte P. Contribution of biological markers to occupational health. *Am J Ind Med* 1991;20:435–446.

36. National Research Council (NRC). Dimensions of the problem: exposure assessment. In: *Environmental epidemiology, vol. 1, public health and hazardous waste.* Washington, DC: National Academy Press, 1991.

37. Needleman HL. Introduction: biomarkers in neurodevelopmental toxicology. *Environ Health Perspect* 1987;74:149–151.

38. Loewenstein-Lichtenstein Y, Schwarz M, Glick D, et al. Genetic predisposition to adverse consequences of anti-cholinesterases in "atypical" BCHE carriers. *Nat Med* 1995;1:1082–1085.

39. Thier R, Lewalter J, Kempkes M, et al. Haemoglobin adducts of acrylonitrile and ethylene oxide in acrylonitrile workers dependent on polymorphisms of the glutathione transferases GSTT1 and GSTM1. *Arch Toxicol* 1999;73:197–202.

40. Chern C-M, Proctor SP, Feldman RG. Exposure assessment in clinical neurotoxicology: environmental monitoring and biologic markers. In: Chang L, Slokker W Jr, eds. *Neurotoxicology: approaches and methods.* San Diego, CA: Academic Press, 1995:695–671.

41. Feldman RG, Ratner MH. The pathogenesis of neurodegenerative disease: neurotoxic mechanisms of action and genetics. *Curr Opin Neurol* 1999;12:725–731.

42. Nakajima T, Aoyama T. Polymorphism of drug-metabolizing enzymes in relation to individual susceptibility to industrial chemicals. *Ind Health* 2000;38:143–152.

43. Srivastava SP, Das M, Seth PK. Enhancement of lipid peroxidation in rat liver on acute exposure to styrene and acrylamide a consequence of glutathione depletion. *Chem Biol Interact* 1983;45:373–380.

44. Kono Y, Takahashi M, Asada K. Oxidation of manganous pyrophosphate by superoxide radicals and illuminated spinach chloroplasts. *Arch Biochem Biophys* 1976;174:454–461.

45. Donaldson J, LaBella FS, Gesser D. Enhanced autoxidation of dopamine as a possible basis of manganese neurotoxicity. *Neurotoxicology* 1980;2:53–64.

46. Archibald FS, Tyree C. Manganese poisoning and the attack of trivalent manganese upon catecholamines. *Arch Biochem Biophys* 1987;256:638–650.

47. Ambani LM, Van Woert MH, Murphy S. Brain peroxidase and catalase in Parkinson diseases. *Arch Neurol* 1975;32:114–118.

48. Swartz HM, Sarna T, Zecca L. Modulation by neuromelanin of the availability and reactivity of metal ions. *Ann Neurol* 1992;32:S69–S75.

49. Graham DG. Catecholamine toxicity: a proposal for the molecular pathogenesis of manganese neurotoxicity and Parkinson's disease. *Neurotoxicology* 1984;5:83–96.

50. Donaldson J, Barbeau A. Manganese neurotoxicity: possible clues to the etiology of human brain disorders. In: Gabay SJ, Harris J, Ho BT, eds. *Metal ions in neurology and psychiatry.* New York: Alan R. Liss, 1985:259–285.

51. Halliwell B. Oxidants and the central nervous system: some fundamental questions. Is oxidant damage relevant to Parkinson's disease, Alzheimer's disease, traumatic injury and stroke? *Acta Neurol Scand* 1989;126:23–33.

52. Segura-Aguila J, Lind C. On the mechanism of the Mn^{3+}-induced neurotoxicity of dopamine: prevention of quinone-derived oxygen toxicity by DT diaphorase and superoxide dismutase. *Chem Biol Interact* 1989;72:309–324.

53. Hori H, Ohmori O, Shinkai T, et al. Manganese superoxide dismutase gene polymorphism and schizophrenia. Relation to tardive dyskinesia. *Neuropsychopharmacology* 2000;23:170–177.

54. Hutchinson M, Raff U. Structural changes of the substantia nigra in Parkinson's disease as revealed by MR imaging. *AJNR Am J Neuroradiol* 2000;21:697–701.

55. Barbeau A. The role of manganese in dystonia. *Adv Neurol* 1976;14:339–352.

56. Yamada M, Ohno S, Okayasu I, et al. Chronic manganese poisoning: a neuropathological study with determination of manganese distribution in the brain. *Acta Neuropathol (Berl)* 1986;70:273–278.

57. Olanow CW, Good PF, Shinotoh H, et al. Manganese intoxication in the rhesus monkey: a clinical, imaging, pathologic, and biochemical study. *Neurology* 1996;46:492–498.

58. Rajput AH, Uitti RJ, Stern W, et al. Geography, drinking water

chemistry, pesticides and herbicides and the etiology of Parkinson's disease. *Can J Neurol Sci* 1987;14[Suppl 3]:414–418.

59. Barbeau A, Roy M, Bernier G, et al. Ecogenetics of Parkinson's disease: prevalence and environmental aspects in rural areas. *Can J Neurol Sci* 1987;14:36–41.

60. Liou HH, Tsai MC, Chen CJ, et al. Environmental risk factors and Parkinson's disease: a case control study in Taiwan. *Neurology* 1997;48:1583–1588.

61. Mukada T, Sasano N, Sato K. Autopsy findings in a case of acute paraquat poisoning with extensive cerebral purpura. *Tohoku J Exp Med* 1978;125:253–263.

62. Vale TJ, Meredith TJ, Buckley BM. Acute pesticide poisoning in England and Wales. *Health Trends* 1987;19:5–7.

63. Brooks AL, Chadwick CA, Gelbard HA, et al. Paraquat elicited neurobehavioral syndrome caused by dopaminergic neuron loss. *Brain Res* 1999;823:1–10.

64. Ecobichon D. Pesticides. In: Klaassen CD, ed. *Casarett and Doull's toxicology: the basic science of poisons,* 5th ed. New York: McGraw-Hill, 1996:671–675.

65. Di Llio C, Sacchetta P, Iannarelli V, et al. Binding of pesticides to alpha, mu and pi class glutathione transferase. *Toxicol Lett* 1995;76:173–177.

66. Ehrenberg L, Osterman-Golkar S. Alkylation of macromolecules for detecting mutagenic agents. *Teratogen Carcinogen Mutagen* 1980;1:105–127.

67. Ehrenberg L, Hussain S. Genetic toxicity of some important epoxides. *Mutat Res* 1981;86:1–113.

68. Högstedt B, Gullberg B, Hedner K, et al. Chromosomal aberration and micronuclei in bone marrow cells and peripheral blood lymphocytes in human exposed to ethylene oxide. *Hereditas* 1983;98:105–113.

69. Segerbäck D. Alkylation of DNA and hemoglobin in the mouse following exposure to ethene and ethene oxide. *Chem Biol Interact* 1983;45:139–151.

70. Katoh M, Cachiero NLA, Cornett CV, et al. Fetal anomalies produced subsequent to treatment of zygotes with ethylene oxide or ethylene methane sulfonate are not likely due to the usual genetic causes. *Mutat Res* 1989;210:337–344.

71. Segerbäck D. Reaction products in hemoglobin and DNA after in vitro treatment with ethylene oxide and N-(2-hydroxyethyl)-N-nitrosourea. *Carcinogen* 1990;11:307–312.

72. Walker VE, Fennell TR, Boucheron JA, et al. Macromolecular adducts of ethylene oxide: a literature review and time-course study on the formation of 7-(2-hydroxyethyl) guanine following exposure of rats by inhalation. *Mutat Res* 1990;233:151–164.

73. Csanady GA, Denk B, Putz C, et al. A physiological toxicokinetic model for exogenous and endogenous ethylene and ethylene oxide in rat, mouse, and human: formation of 2-hydroxyethyl adducts with hemoglobin and DNA. *Toxicol Appl Pharmacol* 2000;165:1–26.

74. Ehrenberg L, Hiesche KD, Osterman-Golkar S, et al. Evaluation of genetic risks of alkylating agents: tissue doses in mouse from air contaminated with ethylene oxide. *Mutat Res* 1974;24:83–103.

75. Corcoran GB, Fix L, Jones DP, et al. Contemporary issues in toxicology. Apoptosis: molecular control point in toxicity. *Toxicol Appl Pharmacol* 1994;128:169–181.

76. Kuzuhara S, Kanazawa I, Nakanishi T, et al. Ethylene oxide polyneuropathy. *Neurology* 1983;33:377–380.

77. Schröder JM, Hoheneck M, Weis J, et al. Ethylene oxide polyneuropathy: clinical follow up study with morphometric and electron microscopic findings in a sural nerve biopsy. *J Neurol* 1985;232:83–90.

78. Ohnishi A, Inoue N, Yamamoto T, et al. Ethylene oxide induces

79. Ohnishi A, Inoue N, Yamamoto T, et al. Ethylene oxide neuropathy in rats: exposure to 250 ppm. *J Neurol Sci* 1986;74:215–221.

80. Ohnishi A, Murai Y. Polyneuropathy due to ethylene oxide, propylene oxide, and butylene oxide. *Environ Res* 1993;60:242–247.

81. Matsuoka M, Igisu H, Inoue N, et al. Inhibition of creatine kinase activity by ethylene oxide. *Br J Ind Med* 1990;47:44–47.

82. Graham DG. Neurotoxicants and the cytoskeleton. *Curr Opin Neurol* 1999;12:733–737.

83. Zhu M, Spink DC, Yan B, et al. Inhibition of 2,5-hexanedione-induced protein cross-linking by biological thiols: chemical mechanisms and toxicological implications. *Chem Res Toxicol* 1995;8:764–771.

84. Yokoyama K, Feldman RG, Sax DS, et al. Relation of distribution of conduction velocities to nerve biopsy findings in *n*-hexane poisoning. *Muscle Nerve* 1990;13:314–320.

85. Herskowitz A, Ishii N, Schaumburg H. *n*-Hexane neuropathy: a syndrome occurring as a result of industrial exposure. *N Engl J Med* 1971;285:82–85.

86. Shirabe T, Tsuda T, Terao A, et al. Toxic polyneuropathy due to glue-sniffing. Report of two cases with a light and electron microscopic study of the peripheral nerves and muscles. *J Neurol Sci* 1974;21:101–113.

87. Scelsi R, Poggi P, Fera L, et al. Toxic polyneuropathy due to *n*-hexane. A light- and electron-microscopic study of the peripheral nerve and muscle from three cases. *J Neurol Sci* 1980;47:7–19.

88. Spencer PS, Schaumburg HH. Feline nervous system response to chronic intoxication with commercial grades of methyl *n*-butyl ketone, methyl *iso*-butyl ketone and methyl ethyl ketone. *Toxicol Appl Pharmacol* 1976;37:301–311.

89. Pezzoli G, Antonini A, Barbieri S, et al. *n*-Hexane-induced parkinsonism: pathogenetic hypotheses. *Mov Disord* 1995;10:279–282.

90. Jenner P. Oxidative mechanisms in nigral cell death in Parkinson's disease. *Mov Disord* 1998;13[Suppl 1]:24–34.

91. Silbergeld EK, Fales JT, Goldberg AM. Lead: evidence for a prejunctional effect on neuromuscular function. *Nature* 1974;247:49–50.

92. Simons TJ, Pocock G. Lead enters bovine adrenal medullary cells through calcium channels. *J Neurochem* 1987;48:383–389.

93. Sandhir R, Gill KD. Alterations in calcium homeostasis on lead exposure in rat synaptosomes. *Mol Cell Biochem* 1994;131:25–33.

94. Albano E, Bellomo G, Benedetti A, et al. Alterations in hepatocyte Ca^{2+} homeostasis by triethylated lead (Et_3Pb^+): are they correlated with cytotoxicity? *Chem Biol Interact* 1994;90:59–72.

95. Hegg CC, Miletic V. Acute exposure to inorganic lead modifies high-threshold voltage-gated calcium currents in rat PC12 cells. *Brain Res* 1996;738:333–336.

96. Markovac J, Goldstein GW. Lead activates protein kinase C in immature rat brain microvessels. *Toxicol Appl Pharmacol* 1988;96:14–23.

97. Marchioro M, Swanson KL, Aracava Y, et al. Glycine and calcium-dependent effects of lead on N-methyl-D-aspartate receptor function in rat hippocampal neurons. *J Pharmacol Exp Ther* 1996;279:143–153.

98. Ordentlich A, Barak D, Kronman C. The architecture of human acetylcholine esterase active center probed by interactions with selected organophosphate inhibitors. *J Biol Chem* 1996;271:11953–11962.

99. Cavanagh JB. The toxic effects of tri-ortho-cresyl phosphate on

central peripheral distal axonal degeneration of the lumbar primary neurones in rats. *Br J Ind Med* 1985;42:373–379.

the nervous system. *J Neurol Neurosurg Psychiatry* 1954;17:163–172.

100. Prineas J. The pathogenesis of dying-back polyneuropathies. I. An ultrastructural study of experimental tri-ortho-cresyl phosphate intoxication in the cat. *J Neuropathol Exp Neurol* 1969;28:571–597.

101. Bouldin TW, Cavanagh JB. Organophosphorus neuropathy. I. A teased-fiber study of the spatio-temporal spread of axonal degeneration. *Am J Pathol* 1979;94:241–252.

102. Bouldin TW, Cavanagh JB: Organophosphorus neuropathy. II. A fine-structural study of the early stages of axonal degeneration. *Am J Pathol* 1979;94:253–270.

103. Aldridge WN, Barnes JM. Studies on delayed neurotoxicity produced by some organophosphorus compounds. *Ann NY Acad Sci* 1969;160:314–222.

104. Abou-Donia M. The cytoskeleton as a target for organophosphorus ester-induced delayed neurotoxicity (OPIDN). *Chem Biol Interact* 1993;87:383–393.

105. Olajos EJ, Rosenblum I. Measurement of neurotoxic esterase activity in various subcellular fractions of hen brain and sciatic nerve homogenates and the effect of diisopropyl fluorophosphate (DFP) administration. *Ecotoxicol Environ Safety* 1979;3:18–28.

106. Needleman HL, Gunnoe C, Leviton A, et al. Deficits in psychological and classroom performance of children with elevated dentine lead levels. *N Engl J Med* 1979;300:689–695.

107. Feldman RG, Hayes MK, Younes R, et al. Lead neuropathy in adults and children. *Arch Neurol* 1977;34:481–488.

108. Feldman RG, White RF. Lead neurotoxicity and disorders of learning and attention. *J Child Neurol* 1992;7:354–359.

109. Baker EL, Feldman RG, White RF, et al. Occupational lead neurotoxicity: a behavioral and electrophysiological evaluation. Study design and year one results. *Br J Ind Med* 1984;41:352–361.

110. Baker EL, White RF, Pothier LJ, et al. Occupational lead neurotoxicity: improvement in behavioral effects after reduction of exposure. *Br J Ind Med* 1985;42:507–516.

111. National Research Council (NRC). Manganese. In: *Publication of the Panel on Manganese Committee on Medical and Biological Effects of Environmental Pollutants.* Washington, DC: National Academy of Science, 1973.

112. Hurley LS. Teratogenic aspects of manganese, zinc, and copper nutrition. *Physiol Rev* 1981;61:249–295.

113. Mena I. Manganese. In: Waldron H, ed. *Metals in the environment.* New York: Academic Press, 1980:119–220.

114. Graham DG. Oxidative pathways for catecholamines in the genesis of neuromelanin and cytotoxic quinones. *Mol Pharmacol* 1978;14:633–643.

115. Graham DG, Tiffany SM, Bell WR, et al. Autoxidation versus covalent binding of quinones as the mechanism of toxicity of dopamine, 6-hydroxydopamine, and related compounds towards C1300 neuroblastoma cells *in vitro. Mol Pharmacol* 1978;14:644–653.

116. Ali SF, Duhart HM, Newport GD, et al. Manganese-induced reactive oxygen species: comparison between Mn^{+2} and Mn^{+3}. *Neurodegeneration* 1995;4:329–334.

117. Flinn RH, Neal PA, Reinhart WH, et al. *Chronic manganese poisoning in an ore crushing mill.* Public Health Bulletin No. 247. Washington, DC: US Government Printing Office, 1940:1–77.

118. Huang CP, Quist GC. The dissolution of manganese ore in dilute aqueous solution. *Environ Int* 1983;9:379–389.

119. Saric M. Manganese. In: Friberg L, Nordberg GF, Vouk V, eds. *Handbook on the toxicology of metals,* 2nd ed. Amsterdam: Elsevier, 1986:354–386.

120. Sjogren B, Ingren A, Frech W, et al. Effects on the nervous system among welders exposed to aluminum and manganese. *Occup Environ Med* 1996;53:32–40.

121. Ferraz HB, Bertolucci PHF, Pereira JS, et al. Chronic exposure to the fungicide maneb may produce symptoms and signs of CNS manganese intoxication. *Neurology* 1988;38:550–553.

122. Tanaka S. Manganese and its compounds. In: Zenz C, ed. *Occupational medicine: principles and practical applications,* 2nd ed. St. Louis, MO: Yearbook Medical Publishers, 1988:583–589.

123. Mergler D. Manganese: the controversial metal. At what levels can deleterious effects occur. *Can J Neurol Sci* 1996;23:93–94.

124. Tanner CM, Langston JW. Do environmental toxins cause Parkinson's disease? A critical review. *Neurology* 1990;40:17–30.

125. Blake CI, Spitz E, Leehy M, et al. Platelet mitochondrial respiratory chain function in Parkinson's disease. *Mov Disord* 1997;12:3–8.

126. Sherlock S, Summerskill WHJ, White LP, et al. Portal-systemic encephalopathy: neurological complications of liver disease. *Lancet* 1954;267:453–457.

127. Mena I, Marin O, Fuenzalida S, et al. Chronic manganese poisoning: clinical picture and manganese turnover. *Neurology* 1967;17:128–136.

128. Read AE, Sherlock S, Laidlaw J, et al. The neuropsychiatric syndromes associated with chronic liver disease and an extensive portal-systemic collateral circulation. *Q J Med* 1967;36:135–150.

129. Cook DG, Fahn S, Brait KA. Chronic manganese intoxication. *Arch Neurol* 1974;30:59–64.

130. Huang C, Chu N, Lu C, et al. Chronic manganese intoxication. *Arch Neurol* 1989;46:1104–1106.

131. Hauser RA, Zesiewicz TA, Rosemurgy AS, et al. Manganese intoxication and chronic liver failure. *Ann Neurol* 1994;36:871–875.

132. Butterworth RF. Complications of cirrhosis: III. Hepatic encephalopathy. *J Hepatol* 2000;32:171–180.

133. Feldman RG. Manganese as possible ecoetiologic factor in Parkinson's disease. *Ann NY Acad Sci* 1992;648:266–267.

134. Rodier J. Manganese poisoning in Moroccan miners. *Br J Ind Med* 1955;12:21–35.

135. Penalver R. Diagnosis and treatment of manganese intoxication. *Arch Ind Health* 1957;16:64–66.

136. Chandra SV. Neurological consequences of manganese imbalance. In: Dreosti IE, Smith RM, eds. *Neurobiology of the trace elements, vol. 2: neurotoxicology and neuropharmacology.* Clifton, NJ: Humana Press, 1983:167–196.

137. Huang C, Lu C, Chu N, Hochberg F, et al. Progression after chronic manganese exposure. *Neurology* 1993;43:1479–1482.

138. Grandjean E, Munchinger R, Turrian V, et al. Investigations into the effects of exposure to trichloroethylene in mechanical engineering. *Br J Ind Med* 1955;12:131–142.

139. Roels HM, Sarhan J, Hanotiau I, et al. Preclinical toxic effects of manganese in workers from a Mn salts and oxides producing plant. *Sci Total Environ* 1985;42:201–206.

140. Iregren A. Psychological test performance in foundry workers exposed to low levels of manganese. *Neurotoxicol Teratol* 1990;12:673–675.

141. Hua MS, Huang C. Chronic occupational exposure to manganese and neurobehavioral function. *J Clin Exp Neuropsychol* 1991;13:495–507.

142. Bronstein AC, Kadushin FS, Riddle MW, et al. Oral manganese ingestion and atypical organic brain syndrome and autistic behavior. *Vet Hum Toxicol* 1988;30:346.

143. Zhang G, Liu D, He P. Effects of manganese on learning abilities in school children. *Chin J Prevent Med* 1995;29:156–158.

144. Grandjean P. Behavioral toxicology of heavy metals. In: Zhinben G, Cuomo V, Racagni G, et al., eds. *Applications of behavioral pharmacology in toxicology.* New York: Raven Press, 1983:331–339.

145. Wennberg A, Iregren A, Struwe G, et al. Manganese exposure in steel smelters: a health hazard to the nervous system. *Scand J Work Environ Health* 1991;17:255–262.

146. Siegl P, Bergert KD. Eine Frudiagnostische Uberwachungsmethode bei. Manganexposition. *Z Gesamte Hygeine* 1982;28:524–526.

147. Lucchini R, Sells L, Folli D, et al. Neurobehavioral effects of manganese in workers from a ferroalloy plant after temporary cessation of exposure. *Scand J Work Environ Health* 1995;21:143–149.

148. Barret L, Torch S, Usson Y, et al. A morphometric evaluation of the effects of trichloroethylene and dichloracetylene on the rat mental nerve. Preliminary results. *Neurosci Lett* 1991;131:141–144.

149. Barret L, Torch S, Leray C, et al. Morphometric and biochemical studies in trigeminal nerve of rat after trichloroethylene on dichloracetylene oral administration. *Neurotoxicology* 1992;13:601–614.

150. Stephens JA. Poisoning by accidental drinking of trichloroethylene. *Br Med J* 1945;2:218–219.

151. Cohen HP, Cohen MM, Lin S, et al. Tissue levels of trichloroethylene after acute and chronic exposure. *Anesthesiology* 1958;19:188–196.

152. Buxton PH, Hayward M. Polyneuritis cranialis associated with industrial trichloroethylene poisoning. *J Neurol Neurosurg Psychiatry* 1967;30:511–518.

153. Feldman RG, Lessell S. Neuro-ophthalmological aspects of trichloroethylene intoxication. In: Burnett J, Bardeau, eds. *Progress in neuro-ophthalmology, vol. 2.* Amsterdam: Excerpta Medica, 1969:281–286.

154. Feldman RG, Mayer R, Taub A. Evidence for peripheral neurotoxic effect of trichloroethylene. *Neurology* 1970;20:599–606.

155. Sagawa K, Nishitani H, Kawai H, et al. Transverse lesion of the spinal cord after accidental exposure to trichloroethylene. *Int Arch Arbeitsmed* 1973;31:257–264.

156. Lawrence WH, Partyka EK. Chronic dysphagia and trigeminal anesthesia after trichloroethylene exposure. *Ann Intern Med* 1981;95:710.

157. Martinelli P, Gulli MR, Gabellini AS. Acute intoxication of trichloroethylene with complete recovery: a case report. *Ital J Neurol Sci* 1984;5:469.

158. Perbellini L, Olivato D, Zedde A, et al. Acute trichloroethylene poisoning by ingestion: clinical and pharmacokinetic aspects. *Int Care Med* 1991;17:234–235.

159. Leandri M, Schizzi R, Scielzo C, et al. Electrophysiological evidence of trigeminal root damage after trichloroethylene exposure. *Muscle Nerve* 1995;18:467–468.

160. Szlatenyi CS, Wang RY. Encephalopathy and cranial nerve palsies caused by intentional trichloroethylene inhalation. *Am J Emerg Med* 1996;14:464–467.

161. Feldman RG, White RF, Currie JN, et al. Long term follow up after single toxic exposure to trichloroethylene. *Am J Ind Med* 1985;8:119–126.

162. Feldman RG, White RF, Eriator II, et al. Neurotoxic effects of trichloroethylene in drinking water: approach to diagnosis. In: Isaacson RL, Jensen KE, eds. *The vulnerable brain and environmental risks, vol. 3, special hazards from air and water.* New York: Plenum, 1994.

163. Lilis R, Stanescu D, Muica N, et al. Chronic effects of trichloroethylene exposure. *Clin Occup Dis* 1969;60:595–601.

164. Takeuchi Y, Iwata M, Hisanaga N, et al. Polyneuropathy caused by chronic exposure to trichloroethylene. *Ind Health* 1986;24:243–247.

165. Barret L, Garrel S, Daniel V, et al. Chronic trichloroethylene intoxication: a new approach by trigeminal evoked potentials? *Arch Environ Health* 1987;42:297–302.

166. Liu YT, Jin C, Chen Z, et al. Increased subjected symptom prevalence among workers exposed to trichloroethylene at sub-OEL levels. *Tohuku J Exp Med* 1988;155:183–195.

167. Noseworthy JH, Rice GP. Trichloroethylene poisoning mimicking multiple sclerosis. *Can J Neurol Sci* 1988;15:87–88.

168. Rasmussen K, Arlien-Soberg P, Sabroe S. Clinical neurological findings among metal degreasers exposed to chlorinated solvents. *Acta Neurol Scand* 1993;87:200–204.

169. Burg JR, Gist GL, Alldred SL, et al. The National Exposure Registry: morbidity analysis of noncancer outcomes from the trichloroethylene subregistry baseline data. *Int J Occup Med Toxicol* 1995;4:237–257.

170. White RF. Clinical neuropsychological investigation of solvent neurotoxicity. In: Chang LW, Dyer RS, eds. *Handbook of neurotoxicology.* New York: Marcel Dekker, 1995:355–376.

171. Steinberg W. Residual neuropsychological effects following exposure to trichloroethylene (TCE): a case study. *Clin Neuropsychol* 1981;3:1–4.

172. Vigliani EC. Clinical observations on carbon disulfide intoxication in Italy. *Ind Med Surg* 1950;19:240–242.

173. Seppäläinen AM, Tolonen M. Neurotoxicity of long-term exposure to carbon disulfide in the viscose rayon industry. A neurophysiological study. *Work Environ Health* 1974;11:145–153.

174. Hänninen H. Psychological picture of manifest and latent carbon disulphide poisoning. *Br J Ind Med* 1971;28:374–381.

175. Cassitto MG, Camerino D, Imbriani M, et al. Carbon disulfide and the central nervous system: a 15-year neurobehavioral surveillance of an exposed population. *Environ Res* 1993;63:252–263.

176. Wadia RS, Chitra S, Amin RB, et al. Electrophysiological studies in acute organophosphate poisoning. *J Neurol Neurosurg Psychiatry* 1987;50:1442–1448.

177. De Bleecker J. The intermediate syndrome in organophosphate poisoning: an overview of experimental and clinical observations. *Clin Toxicol* 1995;33:683–686.

178. Aring CD. The systemic nervous affinity of triorthocresyl phosphate (Jamaica ginger palsy). *Brain* 1942:65:34–47.

179. Morgan JP, Penovich P. Jamaica ginger paralysis. Forty-seven-year follow-up. *Arch Neurol* 1978:35;530–532.

180. Gershon S, Shaw FH. Psychiatric sequelae of chronic exposure to organophosphorus insecticides. *Lancet* 1961;1:1371–1374.

181. Metcalf DR, Holmes JH. EEG, psychological, and neurological alterations in humans with organophosphorus exposure. *Ann NY Acad Sci* 1969;160:357–365.

182. Savage EP, Keefe TJ, Mounce LM, et al. Chronic neurological sequelae of acute organophosphate pesticide poisoning. *Arch Environ Health* 1988;43:38–45.

183. Steenland K, Jenkins B, Ames RG, et al. Chronic neurological sequelae to organophosphate pesticide poisoning. *Am J Public Health* 1994;84:731–736.

184. Cole DC, Crapio F, Julian J, et al. Neurobehavioral outcomes among farm and nonfarm rural Ecuadorians. *Neurotoxicol Teratol* 1997;19:277–286.

185. Fiedler N, Kipen H, Kelly-McNeil K, et al. Long-term use of organophosphates and neuropsychological performance. *Am J Ind Med* 1997;32:487–496.

186. Grinker RR. Parkinsonism following carbon monoxide poisoning. *J Nerv Ment Dis* 1926;64:18–28.

187. Richardson JC, Chambers RA, Heywood PM. Encephalopathies of anoxia and hypoglycemia. *AMA Arch Neurol* 1959;1:178–190.

188. LaPresle J, Fardeau M. The central nervous system and carbon monoxide poisoning. II. Anatomical study of brain lesions following intoxication with carbon monoxide (22 cases). *Prog Brain Res* 1967;24:31–74.

189. Geschwind N, Quadfasel FA, Segarra JM. Isolation of the speech area. *Neuropsychologia* 1968;6:327–340.

190. Ginsberg MD, Myers RE, McDonagh BF. Experimental carbon monoxide encephalopathy in the primate. II. Clinical aspects, neuropathology, and physiologic correlation. *Arch Neurol* 1974;30:209–216.

191. Okeda R, Funata N, Takano T, et al. The pathogenesis of carbon monoxide encephalopathy in acute phase—physiological and morphological correlation. *Acta Neuropathol* 1981;54:1–10.

192. Kobayashi K, Isaki K, Fukutani Y, et al. CT findings of the interval form of carbon monoxide poisoning compared with neuropathological findings. *Eur Neurol* 1984;23:34–43.

193. Jones JS, Lagasse J, Zimmerman G. Computed tomographic findings after acute carbon monoxide poisoning. *Am J Emerg Med* 1994;12:448–451.

194. Wallace AG, Waugh RA. The pathophysiology of cardiovascular disease. In: Smith LH, Thier SO, eds. *Pathophysiology: the biological principles of disease,* 2nd ed. Philadelphia: WB Saunders, 1985:855–1003.

195. Murata T, Itoh S, Koshino Y, et al. Serial proton magnetic resonance spectroscopy in a patient with the interval form of carbon monoxide poisoning. *J Neurol Neurosurg Psychiatry* 1995;58:100–103.

196. Hiramatsu M, Yokoyama S, Nebeshima J, et al. Changes in concentrations of dopamine, serotonin, and their metabolites induced by carbon monoxide (CO) in rat striatum as determined by in vivo microdialysis. *Pharmacol Biochem Behav* 1994;48:9–15.

197. Nathanson JA, Scavone C, Scanlon C, et al. The cellular Na^+ pump as a site of action for carbon monoxide and glutamate: a mechanism for long-term modulation of cellular activity. *Neuron* 1995;14:781–794.

198. Ishimaru H, Katoh A, Suzuki H, et al. Effects of N-methyl-D-aspartate receptor antagonists on carbon monoxide-induced brain damage in mice. *J Pharmacol Exp Ther* 1992;261:349–352.

199. Ringel SP, Klawans HL. Carbon monoxide-induced parkinsonism. *J Neurol Sci* 1972;16:245–251.

200. Smith JS, Brandon S. Morbidity from acute carbon monoxide poisoning at three-year follow-up. *Br Med J* 1973;1:318–321.

201. Laties VG, Merigan WH. Behavioral effects of carbon monoxide on animals and man. *Annu Rev Pharmacol Toxicol* 1979;19:357–392.

202. Beard RR, Grandstaff N. Carbon monoxide exposure and cerebral function. *Ann NY Acad Sci* 1970;174:385–395.

203. Mihevic PM, Gliner JA, Horvath SM. Carbon monoxide exposure and information processing during perceptual-motor performance. *Int Arch Occup Environ Health* 1983;51:355–363.

204. Deckel AW. Carbon monoxide poisoning and frontal lobe pathology: two case reports and a discussion of the literature. *Brain Inj* 1994;8:345–356.

205. Tvedt B, Skyberg K, Aaserud O, et al. Brain damage caused by hydrogen sulfide: a follow-up study of six patients. *Am J Ind Med* 1991;20:91–101.

206. Schneider JS, Tobe EH, Mozley PD Jr, et al. Persistent cognitive and motor deficits following acute hydrogen sulphide poisoning. *Occup Med* 1998;48:255–260.

43

NEUROPSYCHIATRY OF NEUROMUSCULAR DISEASE AND CHRONIC FATIGUE SYNDROME

WILLIAM S. MUSSER
RANDOLPH B. SCHIFFER

If one defines the focus of neuropsychiatry as the neurobehavioral and cognitive sequelae of all illnesses rather than the sequelae of only central nervous system (CNS) illnesses such as dementia, stroke, epilepsy, and brain injury, a broader spectrum of neuropsychiatric illness appears. Neuromuscular disease and chronic fatigue syndrome (CFS) are two areas in which neuropsychiatry may develop an important role in both research and treatment.

NEUROPSYCHIATRY OF NEUROMUSCULAR DISEASE

As the state of medical knowledge continues to advance, it becomes clear that many neurologic diseases once "localized" to a single level of the nervous system may, in fact, affect multiple levels. Neurologic disease traditionally has been categorized as either central or peripheral. This localization guides development of differential diagnosis, diagnostic investigations, and treatment options. Few neurologic diseases have been associated with both CNS and peripheral nervous system (PNS) disease. Vitamin B_{12} deficiency is an example of an illness with both CNS and PNS effects. Neuromuscular disease is one area where appreciation of CNS involvement is increasing in disorders primarily associated with the PNS. Muscle disease and motor neuron disease (MND) are two examples of this PNS/CNS overlap. Neuromuscular specialists traditionally have focused more on the motor symptoms of these disorders (weakness of limb, respiratory, or cranial nerve muscles; atrophy; cramps; myalgias and fatigue), although they have been aware of the neurobehavioral sequelae. Neuropsychiatry usually has not had a role in the evaluation or treatment of these patients, probably because the severity (or importance) of the motor symptoms outweighs the behavioral effects. Neuropsychiatry of neuromuscular disease has not received as much clinical attention and research as the motor manifestations of these diseases.

There is a scientific literature describing the CNS manifestations of neuromuscular disorders, and it continues to expand slowly, using the relatively new tools of molecular genetics and functional neuroimaging, among others.

This chapter reviews the neuropsychiatric manifestations of Duchenne muscular dystrophy, myotonic dystrophy (DM), and MND. Although there are a few reports in the literature on cognitive impairment in other neuromuscular disorders such as myasthenia gravis (1,2) and Friedreich ataxia (3), more work remains to be done before any conclusions can be drawn.

Fibromyalgia is not discussed in this chapter. It is not a disease of muscle or peripheral nerve and is rarely confused in the clinic with neuromuscular disease. Fibromyalgia is a constellation of vague symptoms, including widespread tenderness, trigger points, and generalized pain. Although its etiology is unclear at present, fibromyalgia appears to be a pain syndrome, specifically a disorder of abnormal sensitivity to pain.

DUCHENNE DYSTROPHY

Definition and Clinical Characteristics

Duchenne and Becker muscular dystrophies are produced by different mutations in the dystrophin gene on the X-chromosome. Dystrophin is one component of a glycoprotein complex located in the sarcolemmal membrane of muscle fibers. In Duchenne dystrophy, there is a complete absence of dystrophin; in Becker dystrophy, there is a partial deficiency of dystrophin. Males are predominantly affected; female carriers typically are asymptomatic. Duchenne dystrophy affects 1 in 3,500 live male births. Proximal muscle weakness and calf muscle hypertrophy develop in boys at age 3 to 4 years. The weakness is progressive and becomes more diffuse, with loss of ambulation by age 12 years and death by age 20 years due to respiratory complications (pneumonia or respiratory failure). In Becker dystrophy, the clinical syndrome is relatively less severe, with symptomatic weakness beginning between ages 5 and 15 years. Loss of ambulation typically occurs after age 15 years, and death usually occurs in the third or fourth decade. Becker dystrophy may present in middle age with myalgias and muscle cramping upon exertion. Cardiomyopathy, symptomatic or asymptomatic, may be present in either disorder. Up to one third of Duchenne dystrophy cases are sporadic.

Cognitive Impairment

Approximately 30% of Duchenne patients have cognitive impairment when tested (4,5). Results of a prospective study of 57 Duchenne patients demonstrated verbal and performance intelligence testing scores that were 1 SD below those of the normal mean; impairment was more pronounced on verbal than performance scales. Reading scores were 1 to 5 SD below the normal mean (6). Prior studies reported an approximate intelligence quotient (IQ) of 85 in Duchenne patients (4,7,8). All reports suggest that the cognitive defect is nonprogressive. A later study that compared the results of cognitive testing in 24 children with Duchenne dystrophy and 17 children with spinal muscular atrophy, an inherited neuropathy, found a statistically significant impairment in verbal IQ score in the Duchenne group (9). In contrast, a study of 16 patients with limb-girdle dystrophy, a heterogeneous group of disorders due to abnormalities in genes encoding for other protein components of the dystrophin–glycoprotein complex, found no impairment in cognitive functioning (10). Mild cognitive impairment is an integral manifestation of Duchenne dystrophy, predominantly affecting verbal performance. The presence of cognitive impairment in Becker dystrophy currently is unclear. If such impairment were present and correlated with the level of dystrophin loss, it could be quite mild. However, the longer duration of illness in Becker dystrophy may allow the cognitive impairment to have a greater effect on the patient as he grows to adulthood. This remains to be studied.

The mechanism of cognitive impairment in Duchenne dystrophy is unknown at present. The dystrophin gene product is expressed in brain as well as muscle (11), but its function in the CNS is unknown. Deletions in the distal end of the dystrophin gene were associated with a lower IQ in a study of 74 patients (12). A subsequent study further localized the possible deletion site to the dystrophin isoform Dp140 (13).

Treatment of cognitive symptoms in Duchenne dystrophy currently is limited to special education programs and may not be the primary focus of those treating an affected child as the disease progresses. No pharmacologic treatment is available. Prednisone has been shown to slow the progression of weakness in Duchenne dystrophy (14), but its effect on cognitive impairment is unknown. Prednisone may produce symptoms of mood lability, irritability, aggression, or attention difficulties in treated children, and these symptoms may exacerbate the preexisting cognitive impairment.

MYOTONIC DYSTROPHY

Definition and Clinical Characteristics

DM is the most common muscular dystrophy; it develops in 15 of 100,000 live births. Males and females are equally affected. Inheritance is autosomal dominant with incomplete penetrance. The genetic basis for the majority of DM cases is an expanded trinucleotide repeat in a protein kinase gene on chromosome 19. *Anticipation,* the development of more severe symptoms in successive generations due to expansion of the length of trinucleotide repeat between generations, may occur. Distal muscle weakness develops more frequently than proximal muscle weakness and arises in adolescence to early adulthood, along with neck, facial, and pharyngeal muscle weakness. Cardiac conduction abnormalities, cataracts, diabetes, myotonia, temporal baldness, and testicular atrophy in males also are present (15). Facial weakness and hair loss occur in both males and females and contribute to a characteristic facial appearance in DM. Respiratory muscle weakness may develop later in the illness.

Proximal myotonic myopathy (PROMM) is a disorder of predominantly proximal muscle weakness and stiffness, compared with the predominantly distal involvement in DM. The systemic manifestations in PROMM are similar to those of DM. The genetics of PROMM have yet to be elucidated fully.

Neuropsychiatric abnormalities in DM have long been commented upon. Mild, generalized cognitive impairment; profound hypersomnia; and a peculiar, indifferent affective state have been noted. A multidisciplinary, scientific literature detailing these abnormalities exists and is reviewed in the following. The presence of neuropsychiatric abnormalities in PROMM is unknown at present.

Cognitive Impairment in Myotonic Dystrophy

Severe intellectual impairment was reported in a study of 20 patients with DM (16). Abnormalities in visuospatial and constructional abilities were the most common and were seen even in patients without clinical evidence of intellectual impairment. Half of the patients had a "significant global intellectual deficit." The severity of cognitive impairment did not correlate with patient age, severity of muscle disease, or maternal versus paternal inheritance. In a study of 55 subjects with DM, the degree of impairment seen on IQ scores was inversely proportional to the length of the trinucleotide repeat expansion and directly proportional to the degree of muscle weakness (17). However, the number of trinucleotide repeats was not predictive of severity of intellectual impairment. A subsequent study using positron emission tomography (PET) demonstrated that the degree of cerebral glucose metabolism in DM is directly proportional to trinucleotide repeat expansion length (18). A lack of progression in cognitive impairment was demonstrated in a study of 16 DM patients with a mean testing interval of 12 years (19). Some reports have described the pattern of cognitive impairment in DM as a "subcortical dementia" (16,20).

Cognitive impairment appears to be maternally inherited. In one study of DM patients with maternal inheritance and paternal inheritance, both groups had slowing of information processing speed compared with normal controls, but the maternally inherited group demonstrated abnormal performance on tests of visual construction, verbal fluency, and divided attention and had a lower overall score on intelligence testing than the paternally inherited and control groups (21). These results confirmed the results of a prior study (22). Cerebral blood flow abnormalities also have been found in DM patients with maternally inherited disease but not in those with paternally inherited disease (23).

Not all studies have found evidence of cognitive impairment in DM (24), but this may be due to selection bias and the clinical heterogeneity of DM. Subjects with DM who receive clinical and research attention usually have significant impairments. Those with less symptomatic or asymptomatic involvement are less apt to seek clinical treatment and therefore are not included in research studies. DM clearly has a wide spectrum of severity and distribution of clinical symptoms, and research studies to date have focused on individuals with the more severe form of the disease. We are left to speculate as to the true distribution of cognitive impairment in DM.

Neuroimaging studies, magnetic resonance imaging (MRI) in particular, have revealed the presence of white matter lesions in DM (Fig. 43.1). These lesions may be temporal or extratemporal and abut the cortex. Their confluent distribution and lack of marked periventricular involvement differentiate them from lesions found in ischemic small-vessel white matter disease and demyelinating illness such as multiple sclerosis. In a prospective study of 14 patients with DM and 14 age-matched controls with headaches, 10 of the 14 DM patients had "moderate or marked ventriculomegaly" compared with none of the control group (25). Ventriculomegaly was present in all four of the DM patients who had symptom onset early in life. In addition, almost 43% of the DM group versus none of the control group had abnormalities described as a "lumpy and/or thick pattern of increased periventricular signal" changes on T2-weighted MRI. The authors concluded that the presence of these two abnormalities was not due to stroke risk factors and that the white matter lesions may lead to the development of ventriculomegaly. These white matter lesions are better visualized using the fluid-attenuated inversion recovery (FLAIR) magnetic MR sequence (26). MR spectroscopy may demonstrate the presence of white matter abnormalities in DM earlier than routine MR (27).

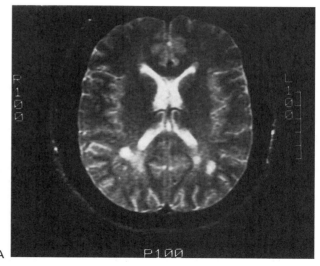

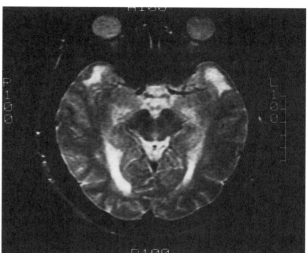

FIGURE 43.1. Magnetic resonance lesions in myotonic dystrophy. Axial T_2-weighted images at the level of the internal capsule **(A)** and midbrain **(B)** demonstrate discrete areas of hyperintensity in the subcortical white matter posteriorly and in the subcortical white matter of the anterior temporal lobes bilaterally. (Images courtesy of John T. Kissel, M.D.)

Neuroimaging studies have sought to establish a link between the presence of cognitive impairment and white matter lesions. A prospective study of 41 DM patients using neuropsychologic testing and MRI found that patients with a severe intellectual deficit were more likely to have neuroimaging abnormalities (i.e., temporal lobe white matter lesions and increased skull thickness) (28). A subcortical distribution of white matter disease was more likely associated with cognitive impairment than a periventricular distribution in an MR study of DM patients. The authors speculated that subcortical white matter disease might lead to cognitive impairment by interrupting subcortical association fibers (29). Another prospective study of 25 DM patients found that, although white matter abnormalities were present on MR in

84% of DM patients, there was no association with cognitive impairment (30).

Neurophysiologic investigations in DM include both electroencephalographic (EEG) and evoked potential studies. EEG findings in DM are consistent with a diffuse encephalopathy. Excessive delta and theta activity is present on EEG in approximately half of DM patients (31,32). The dominant posterior rhythm is of lower voltage, is slower, and is less reactive than normal (33). Whether these EEG changes remain static or progress with the other symptoms of the illness is unknown. In a study of 27 DM patients and 20 control subjects, the P3 peak of the brainstem auditory evoked response was either absent or of decreased amplitude in all of the DM patients (34). A later study of brainstem auditory evoked response results in 11 patients with DM also reported either decreased or absent amplitude in the P3 and N2 peaks of all patients tested (35). Taken together, these two studies suggest an abnormality in central sensory pathways in DM.

Hypersomnia in Myotonic Dystrophy

Patients with DM often report daily hypersomnia lasting 12 or more hours. In a study of 36 DM patients and 13 patients with Charcot-Marie-Tooth disease (CMT), an inherited neuropathy, 39% of the DM patients had hypersomnia on the basis of sleep questionnaire results compared with none of the CMT group. Although the presence of hypersomnia correlated significantly with functional disability, there was no correlation between patient age, duration of illness, or length of trinucleotide repeat sequence (36). An earlier study demonstrated excessive daytime somnolence in 10 DM patients with multiple sleep latency testing, although only five of the group reported having a sleep disturbance (37). The same study suggested an overrepresentation of the human leukocyte antigen haplotype DRW6-DQW1, although this did not reach statistical significance.

The differential diagnosis of hypersomnia in DM includes both central and peripheral etiologies. Several neuropathologic studies have demonstrated serotonergic neuronal cell loss in two brainstem nuclei (dorsal raphe nucleus and superior central nucleus) in postmortem tissue of DM patients with a history of hypersomnia (38,39). Decreased nocturnal secretion of growth hormone also has been suggested as an etiology of hypersomnia in DM (40). With the onset and progression of respiratory and pharyngeal muscle involvement, chronic hypercapnia and sleep apnea may appear in DM patients (41). Both of these conditions may cause or worsen preexisting excessive daytime somnolence. Determination of the precise relationship between these diverse entities has been difficult, and more than one etiology may be present in a particular DM patient (42,43). One study found that although 17 of 22 DM patients reported excessive daytime somnolence, only three of these patients had evidence of sleep apnea on polysomnography. The authors postulated

that "idiopathic hypersomnolence" exists in DM and is due to CNS dysfunction (44).

The presence or development of hypersomnia in a patient with DM requires an evaluation for peripheral etiologies such as chronic hypercapnia due to respiratory muscle weakness, sleep apnea due to pharyngeal weakness, or both. Workup should include pulmonary function testing and polysomnography. If either or both of these etiologies are present, treatment may include respiratory support using continuous positive airway pressure.

Limited studies of drug treatment for central causes of hypersomnia have been reported. A double-blinded crossover study of selegiline (20 mg/day) failed to produce any improvement in DM patients with hypersomnia (45). A trial of higher-dosage selegiline therapy has not been reported. Clinical experience and several reports in the literature suggest a role for stimulants as treatment for hypersomnia in DM. An open trial of methylphenidate (20–40 mg/day) in 11 DM patients with hypersomnia demonstrated a response in 63% (44), and a single case report described a sustained response to methylphenidate over a 3-year span (46). Modafinil (200–400 mg/day) improved excessive daytime somnolence in nine DM patients in an open label trial (47).

Affect in Myotonic Dystrophy

Clinicians frequently comment on the affect of their DM patients. *Affect* is the external representation of an internal emotional state. It may be described as "odd," "indifferent," "unconcerned," or "depressed." The source of this affect is unclear. Is it due to an adjustment to the illness, a DM-induced personality disorder, or apathy? It does not appear that an affective disorder such as depression is responsible.

Is there a personality disorder of DM? The answer is unclear. Several reports have described a high prevalence of abnormal results on structured personality inventories, but they have differed in their results and conclusions (21,48–50). Two early studies reported the presence of heterogeneous personality abnormalities in one third of patients with DM; the latter study ascribed the personality changes to chronic severe depression (48,49). A later study of a group of 17 DM patients reported the presence of dependent personality traits in greater than 50% but concluded that these traits were a reaction to having a progressive illness rather than a direct manifestation of the disease process (21). A study comparing personality inventory scores of 15 adult DM patients with minimal neuromuscular impairment, 14 age-matched healthy controls, and 12 patients with a mild form of facioscapulohumeral dystrophy found that avoidant, passive-aggressive, obsessive-compulsive, and schizotypal traits were present in the DM group. These results were statistically significant. Four of the 15 DM patients fulfilled criteria for a personality disorder with avoidant traits (50). Because these were patients with minimal-to-no neuromuscular symptoms, the author concluded that the findings were not due to a psychological reaction to having DM but rather to a primary manifestation of the disease.

The DM patient's lack of concern to his/her disease is, at times, a striking abnormality. The disease may be associated with a significant impairment in functioning, an inability to live independently, chronic respiratory failure leading to death, and the 50% chance of transmitting the disorder to one's offspring, and yet the DM patient often reacts to these prospects with indifference. This reaction is not denial. The patients usually are able to articulate having an illness, their functional limitations and the risk of passing along the disease. Some authors have theorized that a depressive illness is responsible (28,49). An alternative explanation is that this lack of concern is due to apathy.

Apathy is a lack of motivation that is not due to impairment of consciousness, intellect, or emotion (51). Although the symptom of apathy may appear secondarily as part of dementia, affective illness, psychotic illness, or a frontal lobe process, it also may arise as a primary entity. This duality may lead to diagnostic confusion, and the failure to recognize an apathy syndrome may lead to misdiagnosis of an affective or other psychiatric illness. Could the concept of apathy in DM help explain some of the study results? One study has addressed directly the question of apathy in DM. In a study of two patient groups (one with DM and the other with CMT), apathy ratings, both clinician-rated and self-rated, were significantly higher in the DM group. These results remained significant even after two patients with high apathy ratings were removed from statistical analysis because of a current diagnosis of depression. Severity of apathy did not correlate with illness severity, illness duration, or length of trinucleotide repeat (36).

A compelling reason to consider the presence of an apathy syndrome in DM is that apathy is treatable (52). Although reports to date have not addressed the treatment of apathy in DM, the literature includes reports of successful treatment for apathy in other neuropsychiatric disorders, including dementia, stroke, traumatic brain injury, and alcohol dependence. Successful pharmacologic agents have included methylphenidate, amphetamine, bromocriptine, selegiline, bupropion, and amantadine.

MOTOR NEURON DISEASE

Definition and Clinical Characteristics

In amyotrophic lateral sclerosis (ALS), anterior horn cells in the spinal cord and brainstem and pyramidal cells of the corticospinal tract degenerate. Weakness and atrophy, due to anterior horn cell degeneration, are the disease hallmark and lead to bulbar dysfunction, flaccid quadriparesis, and respiratory failure. Signs of upper motor neuron involvement, for example, spasticity, also develop. The disease usually is sporadic, and death usually occurs within 4 years of diagno-

sis. Primary lateral sclerosis, unlike ALS, is a disorder with upper motor neuron involvement only. It is a rare disorder characterized by a spastic quadriparesis, pseudobulbar affect, and spastic dysarthria (53).

Cognitive Impairment in Motor Neuron Disease

Dementia occurs in approximately 3% of ALS patients (54), but a spectrum of less severe cognitive impairment also may develop. In a study of 22 ALS patients, 18 medical controls, and 17 healthy controls, mean score on the Mini-Mental State Examination (MMSE) (55) was 27.8, 29.38, and 29.29, respectively. These results did not achieve statistical significance (56). A significant negative correlation existed between upper motor neuron symptoms and scores on the MMSE. A later comparison of neuropsychological testing results of 146 ALS patients with normative data revealed that approximately 36% of ALS patients performed at or below the fifth percentile on two or more of eight neuropsychologic measures. This impairment was present in almost 50% of ALS patients with dysarthria, patients with more severe motor impairment, and patients of lower educational status (57). These two studies suggest that cognitive impairment in ALS is a direct effect of the neurodegenerative process. In a study of nine patients with primary lateral sclerosis, mild impairments in executive functions and memory were present, but dementia was absent (58).

Neuropathologic and neuroimaging studies are consistent with a frontotemporal distribution of neurodegenerative changes in ALS. Neuropathologic examination of the brains of five ALS patients and three nondemented controls demonstrated gliosis and neuronal loss in the superficial layers of the neocortex, particularly in the dorsomedial frontal lobe, and in regions of the hippocampus and parahippocampus in the brains of ALS patients. The authors proposed that involvement of these portions of the limbic system was responsible for the cognitive difficulties seen in some ALS patients (59). A greater number of ubiquitin-immunoreactive intraneuronal inclusions, superficial linear spongiosis, and a greater number of dystrophic neurons were seen in the brains of four ALS patients with cognitive impairment compared with four brains of ALS patients without cognitive impairment and four normal control brains. The authors speculated that cognitive impairment in ALS was neuropathologically similar to frontotemporal dementia (60). While performing tests of frontal lobe function in a PET scanner, 18 nondemented ALS patients had significantly reduced glucose metabolism in the entire neocortex, particularly the frontal cortex, compared with a control group (61). Subsequent PET study of 12 ALS patients, half of whom had mild cognitive impairment clinically, and a control group demonstrated a statistically significant decrease in activation of the dorsolateral prefrontal cortex, lateral premotor cortex, medial prefrontal and premotor cortices, insular cortex, and anterior thalamic nuclei in the cognitively impaired ALS group (62). Another PET study demonstrated a statistically significant decrease in regional cerebral blood flow to the anterior cingulate cortex in a group of nondemented ALS patients (63). The authors of these studies postulated impairment in a neural pathway involving the limbic system, thalamus, and cortex as the source of cognitive impairment in ALS.

Currently there are no treatment options for the symptoms of cognitive impairment in MND. The rate of progression of neuromuscular impairment usually is greater than that of cognitive impairment, overshadowing it in the care of the MND patient.

Comorbid Motor Neuron Disease and Frontotemporal Dementia

Evidence of a frontotemporal degenerative process underlying cognitive impairment in ALS has been discussed. A somewhat different and relatively uncommon clinical situation is the development of ALS symptoms in the context of a frontotemporal dementia. Frontotemporal dementias are a heterogeneous group of disorders that may overlap clinically with MND (64). Whether frontal lobe dementia with MND is a separate disease entity or a hybrid of these two disease entities is unclear (64). Cases of these overlapping syndromes have been described (65,66). The clearest descriptions of comorbid MND and frontotemporal symptoms are provided in a case series of seven patients who presented with a rapidly progressive, nonfluent aphasia at the onset of symptoms of MND or shortly before. Over half of the patients were dead within 2 years, a rate of progression that is more rapid than that seen in "typical" ALS. Neuropathologic examination of several of these cases demonstrated superficial cortical neuronal loss and gliosis, as well as loss and gliosis of anterior horn cell nuclei in the spinal cord (67).

Pseudobulbar Affect in Motor Neuron Disease

With degeneration of descending neocortical motor fibers to the brainstem, ALS patients may exhibit pseudobulbar affect and have profound emotional responses to trivial stimuli. In a large database of approximately 1,800 ALS patients, pseudobulbar affect was present in almost 14% of patients at the time of neurologic evaluation (68). Motor symptoms typically precede development of a pseudobulbar affect.

Although the presence of a pseudobulbar affect is not a life-threatening complication of ALS, it can significantly impair a patient's ability to communicate effectively. Consideration often is given to pharmacologic treatment of the symptom. Successful treatment of pseudobulbar affect in an ALS patient has been reported with amitriptyline (69). Successful treatment strategies in other neurologic disorders

have included amitriptyline (70), fluoxetine (71,72), and levodopa (73).

CHRONIC FATIGUE SYNDROME
Definition and Clinical Characteristics

The syndrome of chronic debilitating fatigue has been an important clinical subject for neurology since the post-Civil War era in the nineteenth century. At that time it was known as neurasthenia, or neurocirculatory asthenia. The disorder may have decreased in frequency during the middle years of the twentieth century, but in the mid-1980's reports began to reappear in the United States of patients presenting with pathologic fatigue (74,75). The initial pathophysiologic attribution for this syndrome involved chronic Epstein-Barr virus infection. Although subsequent serologic and virologic investigations failed to confirm the chronic infection hypothesis, the new disease failed to disappear. There has been an explosion of interest in what has been variously termed the *chronic fatigue syndrome* (CFS), the *chronic fatigue and immune deficiency syndrome, myalgic encephalomyelitis,* and *neurasthenia* (76,77).

It is difficult to be sure of the epidemiology of the present CFS epidemic, because the conceptual boundaries of the syndrome remain hazy. Although a 6-month duration of fatigue symptoms is generally required to justify the appellation "chronic," there has been no objectification of the symptom of *fatigue* (78). The fatigue appears not to be due to neuromuscular weakness, although patients commonly are referred to neurology services for that complaint. Depression in its classic sense does not account for the full range of symptoms in most of these patients in whom the fatigue becomes a powerful and personal subjective experience.

The clinical features of CFS often extend beyond fatigue itself. Subjective complaints of memory failure, as well as difficulty with attention and concentration, are common. Gait ataxia and myalgias are frequent complaints. Sleep disturbance, sore throat, and headaches are commonly described.

Comorbid psychiatric symptoms are common in people who complain of chronic fatigue, as is a positive clinical history for lifetime psychiatric disorders (79), but a majority of patients do not meet the criteria for current psychiatric disorders. In a cohort of CFS patients, 35% had a treatable intercurrent psychiatric disorder, most commonly depression (80). In several cases described by the authors, the patients seemed to have experienced a depressive event during the time of the initiation of the fatigue symptoms, which remitted as the CFS syndrome crystallized. There also is considerable overlap between the syndrome of CFS and other emerging psychosomatic syndromes such as fibromyalgia (81). These authors report that as many as 58% of women diagnosed with fibromyalgia also meet criteria for CFS.

TABLE 43.1. WORKING DEFINITION OF CHRONIC FATIGUE SYNDROME

Duration of symptoms	6 months
Functional impairment	Disability
Cognition affected	Mental fatigue required
Time course	New onset in adult life required
Medical evaluation	Exclusion of known physical courses of chronic fatigue
Psychiatric evaluation	Exclusion of major psychiatric diseases

A working definition of the CFS from a British consensus conference is given in Table 43.1 (82).

Prevalence

Simon Wessely and his colleagues at the Kings College School of Medicine in London completed an epidemiologic survey of CFS in a 30,000-person cohort registered with a set of general practices in the National Health Services (83,84). The reported rates of unexplained fatigue are surprisingly high in these United Kingdom populations, approaching 9% (85). One of the interesting findings of this study was that when a within-cohort case control study was done of patients who had recently experienced a viral syndrome, no evidence was found for increased rates of chronic fatigue complaints in the postinfection cohort. Point prevalence rates for chronic fatigue symptoms have been reported as high as 21% in the US primary care patient population (86). This does not mean that all of these patients would meet the full criteria for CFS listed earlier. When the more stringent criteria listed in Table 43.1 are applied to general medical clinic attendees, point prevalence rates of approximately 1% are found for the full CFS (87). There is some evidence that the clinical characteristics of CFS patients in a primary care setting may differ from patients with this diagnosis in an academic medical center (88). CFS patients in university clinics may have more severe fatigue, may have more strongly held beliefs about fatigue attribution to physical illness, and usually are from higher socioeconomic groups.

Clinical Evaluation

Because approximately 50% of CFS patients have significant comorbid depression or anxiety, it is important that these patients be screened for such affective disorders. The Hospital Anxiety and Depression Scale has been reported to be useful as a screening tool for chronic fatigue patients with concurrent depression or anxiety (89), although there is little reason to believe such instruments are superior to a discerning clinical interview.

Neurobiology of Chronic Fatigue

Even as neurasthenia was observed frequently to follow an infectious illness, so CFS experienced its resurgence in the United States and United Kingdom as "postviral fatigue" (90). A variety of agents have been implicated as the cause, including *Brucella,* Epstein-Barr virus, coxsackievirus, enterovirus, and herpes virus. Serologic studies have produced variable and inconstant results.

There are reports of abnormalities noted on muscle biopsies in single cases or small numbers of patients with CFS (91). There are no reports of impaired neuromuscular *function* in CFS, and the pattern of diffuse severe fatigability without weakness fails to correspond with known diseases of the PNS.

A variety of immune system abnormalities have been reported in some patients with CFS. Such abnormalities have included reduction of CD8 cell counts (92), decreased natural killer cell counts (93), and increased circulating immune complexes. Markers of immune activation and inflammation, such as interleukin 2, interleukin-6, C-reactive protein, β_2-microglobulin, and neopterin are elevated over controls in patients with CFS (94). There is limited diagnostic utility for such tests, however, unless individual patients meet clinical or laboratory criteria for actual autoimmune diseases.

Laboratory Studies and Imaging

Cerebral MRI demonstrates subcortical, punctate, high-signal lesions in a significant minority of patients who meet criteria for CFS. Schwartz et al. (95) reported these abnormalities in 50% of a series of 45 patients with CFS. Natelson et al. (96) compared brain MR scans of 52 patients who had CFS with the scans of an equal number of age- and sex-matched controls. They reported that 27% of the CFS patients had abnormal scans versus 2% of controls (96). The most common finding was scattered subcortical hyperintensities on T2 sequences. Single photon emission computed tomographic scans also have been reported to be abnormal in an uncontrolled series of CFS patients, with evidence of decreased perfusion in parietal areas being the most common finding (97).

There are some clinical similarities between CFS patients and patients with documented history of Lyme disease (98). Serum antibodies to *Borrelia burgdorferi* should be checked in patients undergoing evaluation for unexplained fatigue.

Treatment

At the present time we do not have systematic knowledge about the results of treatment interventions in patients suffering from CFS. We are not even sure of the natural history of the disorder. There is some evidence that the rigidly held belief that CFS symptoms are related solely to physical factors is a poor prognostic sign (99). It seems reasonable at the present time for clinicians to attempt treatment of selected patients with CFS and to move these patients toward a greater psychosocial mindset about their symptoms during the course of the therapy.

There is no generally recognized pharmacologic approach to treatment of CSF. Treatment with dialyzable leukocyte extract has not been effective in improving the symptoms of CFS (100). Antidepressants are frequently prescribed, but in a controlled trial fluoxetine improved only depressive symptoms and not exercise tolerance in a group of patients with CFS (101).

Corticosteroids have been prescribed for patients with CFS. In a placebo-controlled study, investigators in London reported that 5 to 10 mg/day of hydrocortisone reduced fatigue and disability over a 1-month period (102). Long-term benefits and risks of such treatment are not known.

Alternative medicines are widely used by CFS sufferers, in view of many reports of limited scientific value that nutritional deficiencies occur in such patients (103). Commonly used supplements include the B vitamins, vitamin C, magnesium, sodium, zinc, L-tryptophan, L-carnitine, and coenzyme Q10. Systematic data on the usefulness of such treatments are lacking.

Deale and Wessely (104) have outlined an initial treatment strategy for nonpharmacologic therapy of CFS that emphasizes psychological engagement of the patient and avoids the debate over biologic causation of the syndrome (Table 43.2). This strategy emphasizes availability on the part of the therapist and opportunity to expand the psychological mindset on the part of the patient. The authors suggest that neurobiologic diagnostic tests and pharmacologic interventions are to be avoided. The same authors have made this approach more specific through a formal randomized trial of cognitive behavior therapy compared with a control therapy of relaxation exercises (105). In the later study, the active treatment had a behavioral, as opposed to an insight, orientation. The therapists used education and negotiation. The aim of the therapy was to increase activity on the part of the patient without exacerbating related symptoms. During

the first three sessions, the patients were engaged and offered a treatment rationale. At the fourth session, a schedule of graded activity was established. Once a structured schedule was established, small increases in activity were negotiated each week. At the eighth session, individualized cognitive alteration strategies were introduced. In the closing sessions (about 12), strategies for dealing with setbacks were rehearsed. The results indicated much greater increases in activity in the cognitive-behavior treated group than in the relaxation group.

REFERENCES

1. Tucker DM, Roeltgen DP, Wann PD, et al. Memory dysfunction in myasthenia gravis: evidence for central cholinergic effects. *Neurology* 1988;38:1173–1177.
2. Iwasaki Y, Kinoshita M, Ikeda K, et al. Neuropsychological function before and after plasma exchange in myasthenia gravis. *J Neurol Sci* 1993;114:223–226.
3. Hart RP, Henry GK, Kwentus JA, et al. Information processing speed of children with Friedreich's ataxia. *Dev Med Child Neurol* 1986;28:310–313.
4. Dubovitz V. Intellectual impairment in muscular dystrophy. *Arch Dis Child* 1965;40:296–301.
5. Worden DK, Vignos PJ. Intellectual function in childhood progressive muscular dystrophy. *Pediatrics* 1962;29:968–977.
6. Leibowitz D, Dubowitz V. Intellect and behavior in Duchenne muscular dystrophy. *Dev Med Child Neurol* 1981;23:577–590.
7. Zellweger H, Niedermeyer E. Central nervous system manifestations in childhood muscular dystrophy. 1. Psychometric and electroencephalographic findings. *Ann Pediatr* 1965;205:25–42.
8. Allen JE, Rodgin DW. Mental retardation in association with progressive muscular dystrophy. *Am J Dis Child* 1960;100:208–211.
9. Billard C, Gillet P, Signoret JL, et al. Cognitive functions in Duchenne muscular dystrophy: a reappraisal and comparison with spinal muscular atrophy. *Neuromuscul Disord* 1992;2:371–378.
10. Miladi N, Bourguignon JP, Hentati F. Cognitive and psychological profile of a Tunisian population of limb girdle muscular dystrophy. *Neuromuscul Disord* 1999;9:352–354.
11. Lidov HGW. Dystrophin in the nervous system. *Brain Pathol* 1996;6:63–77.
12. Bushby KMD, Appleton R, Anderson LVB, et al. Deletion status and intellectual impairment in Duchenne muscular dystrophy. *Dev Med Child Neurol* 1995;37:260–269.
13. Felisari G, Martinelli Boneschi F, Bardoni A, et al. Loss of DP140 dystrophin isoform and intellectual impairment in Duchenne dystrophy. *Neurology* 2000;55:559–564.
14. Mendell JR, Moxley RT, Griggs RC, et al. Randomized, double-blind six month trial of prednisone in Duchenne's muscular dystrophy. *N Engl J Med* 1989;320:1592–1597.
15. Lotz BP, Van der Meyden CH. Myotonic dystrophy, part II: a clinical study of 96 patients. *S Afr Med J* 1985;67:815–817.
16. Censori B, Danni M, Del Pesce M, et al. Neuropsychological profile in myotonic dystrophy. *J Neurol* 1990;237:251–256.
17. Turnpenny P, Clark C, Kelly K. Intelligence quotient profile in myotonic dystrophy, intergenerational deficit, and correlation with CTG amplification. *J Med Genet* 1994;31:300–305.
18. Annane D, Fiorelli M, Mazoyer B, et al. Impaired cerebral glucose metabolism in myotonic dystrophy: a triplet-size dependent phenomenon. *Neuromuscul Disord* 1998;8:39–45.

TABLE 43.2. OUTLINE FOR THERAPY OF CHRONIC FATIGUE SYNDROME

1. Engagement: Build an alliance with the patient; listen to the patient; develop some empathic understanding of his/her distress
2. Develop a therapeutic rationale: Can a shared therapeutic plan be developed? Can the dichotomous debate between physical causation and psychological causation be avoided?
3. Evolution of a treatment plan: Evolve a therapeutic plan that is defined by objective performance targets and time frames
4. Avoid psychopharmacology
5. Avoid invasive and/or expensive medical testing
6. Seek opportunities to clarify the importance of psychological factors as the therapy proceeds

19. Tuikka RA, Laaksonen RK, Somer HVK. Cognitive function in myotonic dystrophy: a follow-up study. *Eur Neurol* 1993; 33:436–441.

20. Kissel JT, Huber SJ, Clapp L, et al. Neuropsychological assessment and magnetic resonance imaging correlates of dementia in myotonic dystrophy. *Ann Neurol* 1988;24:126.

21. Palmer BW, Boone KB, Chang L, et al. Cognitive deficits and personality patterns in maternally versus paternally inherited myotonic dystrophy. *J Clin Exp Neuropsychol* 1994;16;5:785–795.

22. Portwood MM, Wicks JJ, Lieberman JS et al. Intellectual and cognitive function in adults with myotonic muscular dystrophy. *Arch Phys Med Rehabil* 1986;67:299–303.

23. Chang L, Anderson T, Migneco OA, et al. Cerebral abnormalities in myotonic dystrophy: cerebral blood flow, magnetic resonance imaging, and neuropsychological tests. *Arch Neurol* 1993;50:917–923.

24. van Spaendonck KPM, Ter Bruggen JP, Weyn Banningh EWA, et al. Cognitive function in early adult and adult onset myotonic dystrophy. *Acta Neurol Scand* 1995;91:456–461.

25. Glantz RH, Wright RB, Huckkman MS, et al. Central nervous system magnetic resonance imaging findings in myotonic dystrophy. *Arch Neurol* 1988;45:36–37.

26. Abe K, Fujimura H, Soga F. The fluid-attenuated inversion-recovery pulse sequence in assessment of central nervous system involvement in myotonic dystrophy. *Neuroradiology* 1998;40:32–35.

27. Akiguchi I, Nakano S, Shiino A. Brain proton magnetic resonance spectroscopy and brain atrophy in myotonic dystrophy. *Arch Neurol* 1999;56:325–330.

28. Huber SJ, Kissel JT, Shuttleworth EC, et al. Magnetic resonance imaging and clinical correlates of intellectual impairment in myotonic dystrophy. *Arch Neurol* 1989;46:536–540.

29. Damian MS, Schilling G, Bachman G, et al. White matter lesions and cognitive deficits: relevance of lesion pattern? *Acta Neurol Scand* 1994;90:430–436.

30. Censori B, Provinciali L, Chiaramoni L, et al. Brain involvement in myotonic dystrophy: MRI features and their relationship to clinical and cognitive conditions. *Acta Neurol Scand* 1994;90:211–217.

31. Lundervold A, Refsum S, Jacobsen W. The EEG in dystrophica myotonica. *Eur Neurol* 1969;2:279–284.

32. Murri L, Massetani R, Rossi B, et al. Electrophysiological abnormalities in myotonic dystrophy. *Electroencephalogr Clin Neurophysiol* 1990;75:S101.

33. Beijersbergen RSHM, Kemp A, Storm van Leeuwen W. EEG observations in dystrophia myotonica. *Electroencephalogr Clin Neurophysiol* 1980;49:143–151.

34. Perini GI, Colombo G, Armani M, et al. Intellectual impairment and cognitive evoked potentials in myotonic dystrophy. *J Nerv Ment Dis* 1989;177;12:750–754.

35. Ragazzoni A, Pinto F, Taiuti R, et al. Myotonic dystrophy: an electrophysiological study of cognitive deficits. *Can J Neurol Sci* 1991;18:300–306.

36. Rubinsztein JS, Rubinsztein DC, Goodburn S, et al. Apathy and hypersomnia are common features of myotonic dystrophy. *J Neurol Neurosurg Psychiatry* 1998;64:510–515.

37. Manni R, Zucca C, Martinetti M, et al. Hypersomnia in dystrophia myogenica: a neurophysiological and immunogenetic study. *Acta Neurol Scand* 1991;84:498–502.

38. Ono S, Kanda F, Takahashi K, et al. Neuronal cell loss in the dorsal raphe nucleus and the superior central nucleus in myotonic dystrophy: a clinicopathological correlation. *Acta Neuropathol* 1995;89:122–125.

39. Ono S, Takahashi K, Jinnai K, et al. Loss of serotonin-containing neurons in the raphe of patients with myotonic dystrophy: a quantitative immunohistochemical study and relation to hypersomnia. *Neurology* 1998;50:535–538.

40. Culebras A, Podolsky S, Leopold NA. Absence of sleep-related growth hormone elevations in myotonic dystrophy. *Neurology* 1977;27:165–167.

41. Begin P, Mathieu J, Almirall J, et al. Relationship between chronic hypercapnia and inspiratory-muscle weakness in myotonic dystrophy. *Am J Respir Crit Care Med* 1997;156:133–139.

42. Coccagna G, Mantovani M, Parchi C, et al. Alveolar hypoventilation and hypersomnia in myotonic dystrophy. *J Neurol Neurosurg Psychiatry* 1975;38:977–984.

43. Hansotia P, Frens D. Hypersomnia associated with alveolar hypoventilation in myotonic dystrophy. *Neurology* 1981;31:1336–1337.

44. van der Meche RGA, Bogaard JM, van der Sluys JCM, et al. Daytime sleep in myotonic dystrophy is not caused by sleep apnea. *J Neurol Neurosurg Psychiatry* 1994;57:626–628.

45. Antonini G, Morino S, Fiorelli M, et al. Selegiline in the treatment of hypersomnolence in myotonic dystrophy: a pilot study. *J Neurol Sci* 1997;147:167–169.

46. van der Meche FGA, Boogaard JM, van den Berg B. Treatment of hypersomnolence in myotonic dystrophy with a CNS stimulant. *Muscle Nerve* 1986;9:341–344.

47. Damian MS, Gerlach BS, Schmidt F, et al. Modafinil for excessive daytime sleepiness in myotonic dystrophy. *Neurology* 2001;56:794–796.

48. Bird TD, Follett C, Griep E. Cognitive and personality function in myotonic muscular dystrophy. *J Neurol Neurosurg Psychiatry* 1983;46:971–980.

49. Brumback RA, Wilson H. Cognitive and personality function in myotonic muscular dystrophy. *J Neurol Neurosurg Psychiatry* 1984;47:888–890.

50. Delaporte C. Personality patterns in patients with myotonic dystrophy. *Arch Neurol* 1998;55:635–640.

51. Marin RS. Differential diagnosis and classification of apathy. *Am J Psychiatry* 1990;147:22–30.

52. Marin RS, Fogel BS, Hawkins J, et al. Apathy: a treatable syndrome. *J Neuropsychiatry Clin Neurosci* 1995;7:23–30.

53. Pringle CE, Hudson AJ, Munoz DG, et al. Primary lateral sclerosis: clinical features, neuropathology and diagnostic criteria. *Brain* 1992;115:495–520.

54. Kew JJM, Leigh PN. Unusual forms of dementia. In: Burns A, Levy R, ed. *Dementia*. London: Chapman and Hall, 1994:789–811.

55. Folstein MF, Folstein SE, McHugh PR. Mini-mental state: a practical method for grading the mental state of patients for clinician. *J Psychiatr Res* 1975;12:189–198.

56. Iwasaki Y, Kinoshita M, Ikeda K, et al. Neuropsychological dysfunctions in amyotrophic lateral sclerosis: relation to motor disabilities. *Int J Neurosci* 1990;54:191–195.

57. Massman PJ, Sims J, Cooke N, et al. Prevalence and correlates of neuropsychological deficits in amyotrophic lateral sclerosis. *J Neurol Neurosurg Psychiatry* 1996;61:450–455.

58. Caselli RJ, Smith BE, Osborne D. Primary lateral sclerosis: a neuropsychological study. *Neurology* 1995;45:2005–2009.

59. Kato S, Oda M, Hayashi H, et al. Participation of the limbic system and its associated areas in the dementia of amyotrophic lateral sclerosis. *J Neurol Sci* 1994;126:62–69.

60. Wilson CM, Grace GM, Munoz DG, et al. Cognitive impairment in sporadic ALS: a pathologic continuum underlying a multisystem disorder. *Neurology* 2001;57:651–657.

61. Ludolph AC, Langen KJ, Regard M, et al. Frontal lobe function

in amyotrophic lateral sclerosis: a neuropsychologic and positron emission tomography study. *Acta Neurol Scand* 1992;85:81–89.

62. Abrams S, Goldstein LH, Kew JJM, et al. Frontal lobe dysfunction in amyotrophic lateral sclerosis: a PET study. *Brain* 1996;119:2105–2120.

63. Kew JJM, Goldstein LH, Leigh PN, et al. The relationship between abnormalities of cognitive function and cerebral activation in amyotrophic lateral sclerosis: a neuropsychological and positron emission tomography study. *Brain* 1993;116:1399–1423.

64. Talbot PR. Frontal lobe dementia and motor neuron disease. *J Neural Transm* 1996;47[Suppl]:125–132.

65. Sam J, Gutmann L, Schochet S, et al. Pick's disease: a case clinically resembling amyotrophic lateral sclerosis. *Neurology* 1991;41:1831–1833.

66. Hooten WM, Lyketson CG. Frontotemporal dementia: a clinicopathological review of four postmortem studies. *J Neuropsychiatry Clin Neurosci* 1996;8:10–19.

67. Caselli RJ, Windebank AJ, Peterson RC, et al. Rapidly progressive aphasic dementia and motor neuron disease. *Ann Neurol* 1993;33:200–207.

68. Miller RG, Anderson FA, Bradley WG, et al. The ALS patient care database. *Neurology* 2000;54:53–57.

69. Jankovic J. Amitriptyline in amyotrophic lateral sclerosis. *N Engl J Med* 1985;313:1478–1479.

70. Schiffer RB, Herndon RM, Rudick RA. Treatment of pathologic laughing and weeping with amitriptyline. *N Engl J Med* 1985;312:1480–1482.

71. Seliger GM, Hornstein A. Serotonin, fluoxetine, and pseudobulbar affect [Letter]. *Neurology* 1989;39;10:1400.

72. Lauterbach EC, Schweri MM. Amelioration of pseudobulbar affect by fluoxetine: possible alteration of dopamine-related pathophysiology by a selective serotonin reuptake inhibitor [Letter]. *J Clin Psychopharmacol* 1991;11;6:392–393.

73. Udaka T, Yamao S, Nagata H, et al. Pathological laughing and crying treated with levodopa. *Arch Neurol* 1984;41:1095–1096.

74. Holmes G, Kaplan J, Stewart J, et al. A cluster of patients with a chronic mononucleosis-like syndrome: is Epstein-Barr virus the cause? *JAMA* 1987;257:2297–2303.

75. Jones J, Straus S. Chronic Epstein-Barr virus infection. *Annu Rev Med* 1987;38:195–209.

76. Klonoff D. Chronic fatigue syndrome. *Clin Infect Dis* 1992;15:812–823.

77. Leitch A. The chronic fatigue syndrome reviewed. *Proc R Coll Phys Edinb* 1994;24:480–508.

78. Wessely S, Chalder T, Hirsch S, et al. Postinfectious fatigue: prospective cohort study in primary care. *Lancet* 1995;345:1333–1338.

79. Wessely S, Chalder T, Hirsch S, et al. Psychological symptoms, somatic symptoms, and psychiatric disorder in chronic fatigue and chronic fatigue syndrome: a prospective study in the primary care setting. *Am J Psychiatry* 1996;153:1050–1059.

80. Deale A, Wessely S. Diagnosis of psychiatric disorder in clinical evaluation of chronic fatigue syndrome. *J R Soc Med* 2000;93:310–312.

81. White KP, Speechley M, Harth M, et al. Co-existence of chronic fatigue syndrome with fibromyalgia syndrome in the general population. A controlled study. *Scand J Rheumatol* 2000;29:44–51.

82. Sharpe MC, Archard LC, Banatvala JE, et al. A report-chronic fatigue syndrome: guidelines for research. *J R Soc Med* 1991;84:118–121.

83. Pawlikowska T, Chalder T, Hirsch SR, et al. Population based study of fatigue and psychological distress. *BMJ* 1994;308:763–766.

84. Wessely S. The epidemiology of chronic fatigue syndrome. *Epidemiol Rev* 1995;17:139–151.

85. Skapinakis P, Lewis G, Meltzer H. Clarifying the relationship between unexplained chronic fatigue and psychiatric morbidity: results from a community survey in Great Britain. *Am J Psychiatry* 2000;157:1492–1498.

86. Buchwald D, Sullivan JL, Komaroff AL. Frequency of chronic active Epstein-Barr virus infection in a general medical practice. *JAMA* 1987;257:2303–2307.

87. Bates DW, Schmitt W, Buchwald D, et al. Prevalence of fatigue and chronic fatigue syndrome in a primary care practice. *Arch Intern Med* 1993;153:2759–2765.

88. Euga R, Chalder T, Deale A, et al. A comparison of the characteristics of chronic fatigue syndrome in primary and tertiary care. *Br J Psychiatry* 1996;168;1:121–126.

89. Morriss RK, Wearden AJ. Screening instruments for psychiatric morbidity in chronic fatigue syndrome. *J R Soc Med* 1998;91:365–368.

90. Bearn J, Wessely S. Neurobiological aspects of the chronic fatigue syndrome. *Eur J Clin Invest* 1994;24:79–90.

91. Behan W, More I, Behan P. Mitochondrial abnormalities in the post viral fatigue syndrome. *Acta Neuropathol* 1991;83:61–65.

92. Landay A, Jessop C, Lennette E, et al. Chronic fatigue syndrome: clinical condition associated with immune activation. *Lancet* 1991;338:707–712.

93. Buchwald D, Komaroff AL. Review of laboratory findings for patients with chronic fatigue syndrome. *Rev Infect Dis* 1991;13[Suppl 1]:S12–S18.

94. Buchwald D, Wener MH, Pearlman T, et al. Markers of inflammation and immune activation in chronic fatigue and chronic fatigue syndrome. *J Rheumatol* 1997;24:372–376.

95. Schwartz RB, Komaroff AL, Garada BM, et al. SPECT imaging of the brain: comparison of findings in patients with chronic fatigue syndrome, AIDS dementia complex, and major unipolar depression. *AJR Am J Roentgenol* 1994;162:943–951.

96. Natelson BH, Cohen JM, Brasslof I, et al. A controlled study of brain magnetic resonance imaging in patients with the chronic fatigue syndrome. *J Neurol Sci* 1993;120:213–217.

97. Troughton AH, Blacker R, Vivian G. 99m TC HMPAO SPECT in the chronic fatigue syndrome. *Clin Radiol* 1992;45:59.

98. Gaudino EA, Coyle PK, Krupp LB. Post-Lyme syndrome and chronic fatigue syndrome: neuropsychiatric similarities and differences. *Arch Neurol* 197;54:1372–1376.

99. Wilson A, Hickie I, Lloyd A, et al. Longitudinal study of outcome of chronic fatigue syndrome. *BMJ* 1994;308:756–759.

100. Lloyd AR, Hickie I, Brockman A, et al. Immunologic and psychologic therapy for patients with chronic fatigue syndrome: a double-blind, placebo-controlled trial. *Am J Med* 1993;94:197–203.

101. Wearden AJ, Morriss RK, Mullis R, et al. Randomized, double-blind, placebo-controlled treatment trial of fluoxetine and graded exercise for chronic fatigue syndrome. *Br J Psychiatry* 1998;172:485–490.

102. Cleare AJ, Heap E, Malhi GS, et al. Low-dose hydrocortisone in chronic fatigue syndrome: a randomized crossover trial. *Lancet* 1999;353:455–458.

103. Werbach MR. Nutritional strategies for treating chronic fatigue syndrome. *Altern Med Rev* 2000;5:93–108.

104. Deale A, Wessely S. A cognitive-behavioral approach to chronic fatigue syndrome. *Therapist* 1994;11–14.

105. Deale A, Chalder T, Marks I, et al. Cognitive behavior therapy for chronic fatigue syndrome: a randomized controlled trial. *Am J Psychiatry* 1997;154:408–414.

SUBJECT INDEX

SUBJECT INDEX

Page numbers followed by f indicate figures and page numbers followed by t indicate tables.

indications for and side effects of,
143–144
for posttraumatic brain injury depression,
741
for self-injurious behavior, 559
Butalbital
with caffeine and analgesics
for migraine, 1142, 1142t
Butorphanol tartrate
for migraine, 1143
Butyrylcholinesterase
in Alzheimer disease, 1047

C
Calculation, 11, 641
CADSIL. *See* Cerebral autosomal dominant
arteriopathy with subcortical
infarcts and leukoencephalopathy
(CADSIL)
Caffeine, adenosine antagonism and,
207
Caffeinism
generalized anxiety disorder *versus,* 765
CAGE, for alcoholism, 895
Calcarine fissure, 296f
Calcitonin gene related peptide (CGRP)
in immune processes, 258
Calcium
channel blockers
cocaine, heroin and, 855
HIV and, 1022
for migraine prophylaxis, 1143t
in epilepsy, 1078
homeostasis
in brain injury, 1151
inorganic lead and, 1180
intracellular and lead, 1180
Calculation, 11
in learning disorders, 554, 641
California Psychological Inventory (CPI)
in temporal lobe epilepsy, 1106
California Verbal Learning Test (CVLT), 9,
1020t
in attention deficit hyperactivity disorder,
610
neurotoxicant exposure and, 1175
California Verbal Learning Test-II
(CVLT-II), 24, 34
Colostomy, 991, 1084
Cambridge Neuropsychological Test
Automated Battery (CANTAB),
610
Cancer
delirium in, 1024
psychoneuroimmunology and, 261
suicide and, 688
Cancer pain, 829
antidepressants for, 816t
psychostimulants in, 827t, 829
Cannabinoids
activity of, 909
for aggression, 178
receptor, 841t, 847–848
Capsaicin, 397
for neuropathic pain, 402

receptor, 847
substance P and, 210
Carbamates
neurobehavioral syndromes and, 1170t
Carbamazepine
aggression and, 178
for bipolar affective disorder, 1092
choice and initiation of, 148t
in dementia, 1062
dose titration, 151
drug interactions with, 138
for epilepsy, 1077t, 1081, 1081t,
1111–1112
for fragile X syndrome, 564
for GTS, 973
versus lithium, 146
for mania, 744
in HIV infection, 1026
for neuropathic pain, 399
neurotoxicity of, 149
for nonneuropathic pain, 829t
for obsessive-compulsive disorder
in epilepsy, 1093
for panic attacks, 920
for REM sleep behavior disorder, 382
reproductive endocrine effects of,
232–233
for restless legs syndrome, 382
testosterone and, 345
Carbon disulfide
neurobehavioral syndromes and, 1170t
Carbon monoxide, 212
cellular aerobic respiration and, 1180
neurobehavioral syndromes and, 1170t
Cardiac nerves, 273
Cardiac plexus, 273, 274, 275
Cardiovascular system
central autonomic structures and, 277
in fragile X syndrome, 564
function of, 279–280
psychostimulant abuse and, 864
in Williams syndrome, 568
Caregivers, 1055, 1055t, 1057, 1064
Carotid arteries, 274
Carotid sinus, 279
Carpal tunnel syndrome, 400
Carrlson, Arvid, dopamine and, 201
Caspase inhibition
in acute brain injury, 166
Cataplexy, 379, 380, 381
Catastrophic reaction, 725, 727, 728
Catatonia, 138, 143
Catecholamine(s)
alterations with aging, 255
in immune regulation, 255–256
in innervation of lymphoid organs,
254–255
intracellular, 254
opiate withdrawal and, 860
peripheral and panic attack, 757
Category fluency test, 1020t
Category retrieval task, 11, 12t
Category Test, 31
Catecholamines
immune regulation by

in vitro, 255
in vivo, 255–256
Cat studies
aggression and, 665
of locus caeruleus and fear, 752
Caudate
in attention deficit/hyperactivity disorder,
617
autism and, 523
functional magnetic resonance imaging of
in attention deficit/hyperactivity
disorder, 615
head of
lesions in bipolar disorder, 744–745
secondary mania and, 742
Caudate nucleus
acetylcholine neurons in, 200
in attention deficit/hyperactivity disorder,
612
cholecystokinin in, 211
fetal alcohol syndrome and, 54, 55f
in GTS, 956–957
Huntington disease and, 685
in schizophrenia, 56
vasoregulatory mechanisms and, 289
Causalgia, 397–398
Cavernous nerves, 282
Cavum septi pellucidum
fetal alcohol syndrome and, 54
Celecoxib
for migraine, 1143
Celiac ganglion(a), 275, 276f
Celiac plexus, 273
Cell assemblies, 169
Cell death, 166
Central nervous system (CNS)
drug abuse and, 856–870
in HIV/AIDS, 1018–1019
HIV infection and, 924–925
immune system communication with,
259–260
immune system interaction with,
245–248
infection in, 911
dementia in, 1036
drug abuse and, 898t
Central sleep apnea (CSA), 384, 385
Cerebellar vermis
in attention deficit/hyperactivity disorder,
613–614
in fetal alcohol syndrome, 567
fetal alcohol syndrome and, 54, 55f
Joubert syndrome and, 519
Cerebellum
attention deficit/hyperactivity disorder
and, 613–614
autism and, 523, 526
autonomic function of, 277
language processing and, 457
neurotoxic effects on, 1170
schizophrenia and, 457
tumors of, 290
Cerebral artery
poststroke depression and, 730, 734
Cerebral atrophy. *See* Atrophy